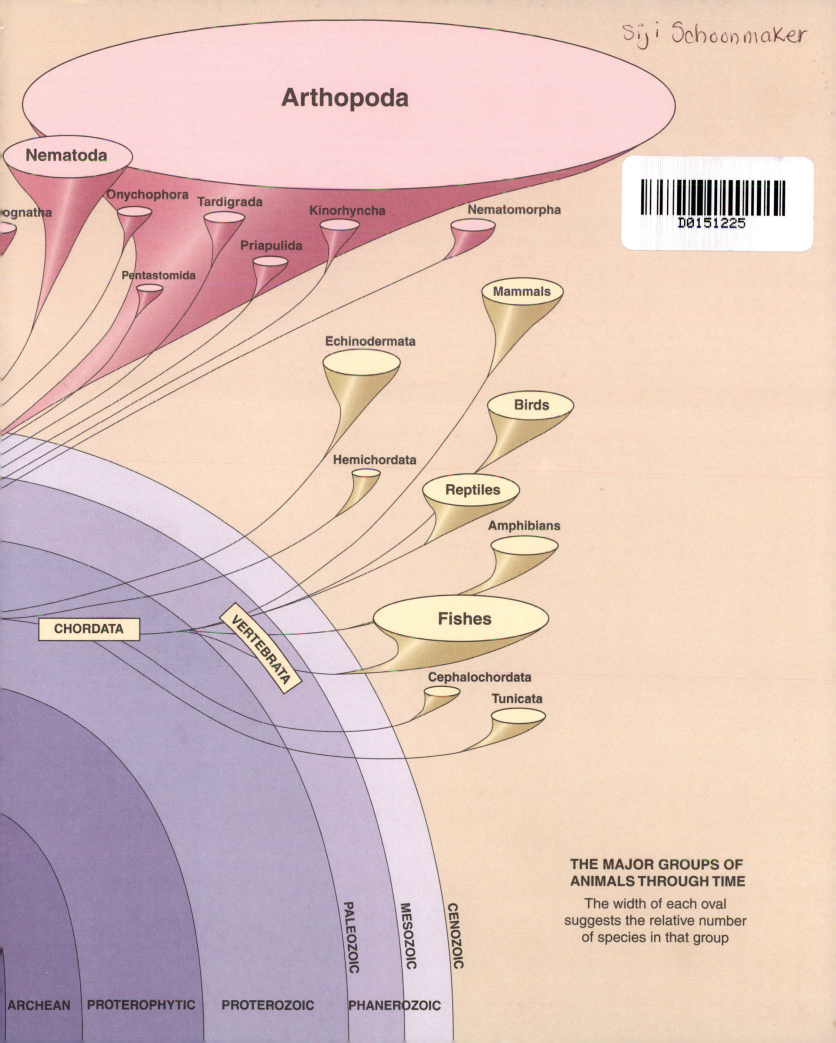

Siji Schoonmaker

Arthopoda

Nematoda

Onychophora

ognatha

Tardigrada

Kinorhyncha

Nematomorpha

Pentastomida

Priapulida

Mammals

Echinodermata

Birds

Hemichordata

Reptiles

Amphibians

CHORDATA

VERTEBRATA

Fishes

Cephalochordata

Tunicata

THE MAJOR GROUPS OF ANIMALS THROUGH TIME

The width of each oval suggests the relative number of species in that group

PALEOZOIC

MESOZOIC

CENOZOIC

ARCHEAN PROTEROPHYTIC PROTEROZOIC PHANEROZOIC

IMPORTANT:

HERE IS YOUR REGISTRATION CODE TO ACCESS
YOUR PREMIUM McGRAW-HILL ONLINE RESOURCES.

For key premium online resources you need THIS CODE to gain access. Once the code is entered, you will be able to use the Web resources for the length of your course.

If your course is using **WebCT** or **Blackboard**, you'll be able to use this code to access the McGraw-Hill content within your instructor's online course.

Access is provided if you have purchased a new book. If the registration code is missing from this book, the registration screen on our Website, and within your WebCT or Blackboard course, will tell you how to obtain your new code.

Registering for McGraw-Hill Online Resources

TO gain access to your McGraw-Hill web resources simply follow the steps below:

1 USE YOUR WEB BROWSER TO GO TO: www.mhhe.com/hickman12

2 CLICK ON **FIRST TIME USER**.

3 ENTER THE REGISTRATION CODE* PRINTED ON THE TEAR-OFF BOOKMARK ON THE RIGHT.

4 AFTER YOU HAVE ENTERED YOUR REGISTRATION CODE, CLICK **REGISTER**.

5 FOLLOW THE INSTRUCTIONS TO SET-UP YOUR PERSONAL UserID AND PASSWORD.

6 WRITE YOUR UserID AND PASSWORD DOWN FOR FUTURE REFERENCE.
   KEEP IT IN A SAFE PLACE.

TO GAIN ACCESS to the McGraw-Hill content in your instructor's **WebCT** or **Blackboard** course simply log in to the course with the UserID and Password provided by your instructor. Enter the registration code exactly as it appears in the box to the right when prompted by the system. You will only need to use the code the first time you click on McGraw-Hill content.

Thank you, and welcome to your McGraw-Hill online Resources!

0-07-292281-8 T/A HICKMAN: INTEGRATED PRINCIPLES OF ZOOLOGY, 12/E

MCGRAW-HILL
ONLINE RESOURCES

REGISTRATION CODE

GOV9-1RDW-RAMZ-0IDV-VE2Q

INTEGRATED PRINCIPLES OF
ZOOLOGY

INTEGRATED PRINCIPLES OF
ZOOLOGY

TWELFTH EDITION

Cleveland P. Hickman, Jr.
WASHINGTON AND LEE UNIVERSITY

Larry S. Roberts
FLORIDA INTERNATIONAL UNIVERSITY

Allan Larson
WASHINGTON UNIVERSITY

Helen I'Anson
WASHINGTON AND LEE UNIVERSITY

Original Artwork by
WILLIAM C. OBER, M.D. and CLAIRE W. GARRISON, R.N.

Boston   Burr Ridge, IL   Dubuque, IA   Madison, WI   New York   San Francisco   St. Louis
Bangkok   Bogotá   Caracas   Kuala Lumpur   Lisbon   London   Madrid   Mexico City
Milan   Montreal   New Delhi   Santiago   Seoul   Singapore   Sydney   Taipei   Toronto

Higher Education

INTEGRATED PRINCIPLES OF ZOOLOGY, TWELFTH EDITION

Published by McGraw-Hill, a business unit of The McGraw-Hill Companies, Inc., 1221 Avenue of the Americas, New York, NY 10020. Copyright © 2004, 2001, 1997 by The McGraw-Hill Companies, Inc. All rights reserved. No part of this publication may be reproduced or distributed in any form or by any means, or stored in a database or retrieval system, without the prior written consent of The McGraw-Hill Companies, Inc., including, but not limited to, in any network or other electronic storage or transmission, or broadcast for distance learning.

Some ancillaries, including electronic and print components, may not be available to customers outside the United States.

 This book is printed on recycled, acid-free paper containing 10% postconsumer waste.

Domestic   3 4 5 6 7 8 9 0   QPD/QPD0 9 8 7 6 5 4 3

ISBN 0–07–243940–8

Publisher: *Margaret J. Kemp*
Senior developmental editor: *Donna Nemmers*
Marketing manager: *Heather K. Wagner*
Senior project manager: *Mary E. Powers*
Senior production supervisor: *Laura Fuller*
Lead media project manager: *Judi David*
Media technology producer: *Renee Russian*
Designer: *David W. Hash*
Cover/interior designer: *Kaye Farmer*
Cover image: *© The Image Bank/Keren Su; Panda cub (Ailuropoda melanoieca)*
Senior photo research coordinator: *John C. Leland*
Photo research: *Roberta Spieckerman*
Supplement producer: *Brenda A. Ernzen*
Compositor: *Precision Graphics*
Typeface: *10/12 Garamond*
Printer: *Quebecor World Dubuque*

The credits section for this book begins on page 845 and is considered an extension of the copyright page.

Library of Congress Cataloging-in-Publication Data

Hickman, Cleveland P.
    Integrated principles of zoology / Cleveland P. Hickman, Jr., Larry S. Roberts, Allan Larson, Helen I'Anson;
original artwork by William C. Ober and Claire W. Garrison; contributions by David Eisenhour . . . [et al.]. — 12th ed.
        p.   cm.
Includes bibliographical references and index.
ISBN 0–07–243940–8 (hard : alk. paper) — ISBN 0–07–121470–4 (international ed.)
1. Zoology. I. Roberts, Larry S., 1935–. II. Larson, Allan. III. Title.

QL47.2.H527   2004
590—dc21
                                                                            2003006561
                                                                            CIP

CONTENTS IN BRIEF

About the Authors *xiii*
Preface *xv*

PART ONE
Introduction to Living Animals

1   Life: Biological Principles and the Science of Zoology 2
2   The Origin and Chemistry of Life 21
3   Cells as Units of Life 36
4   Cellular Metabolism 56

PART TWO
Continuity and Evolution of Animal Life

5   Genetics: A Review 74
6   Organic Evolution 102
7   The Reproductive Process 131
8   Principles of Development 151

PART THREE
Diversity of Animal Life

9   Architectural Pattern of an Animal 176
10 Classification and Phylogeny of Animals 190
11 Protozoan Groups 207
12 Mesozoa and Parazoa 233
13 Radiate Animals 247
14 Acoelomate Bilateral Animals 274
15 Pseudocoelomate Animals 295
16 Molluscs 316
17 Segmented Worms 345
18 Arthropods 364

19 Aquatic Mandibulates 377
20 Terrestrial Mandibulates 397
21 Smaller Protostome Phyla 426
22 Echinoderms and Hemichordates 442
23 Chordates 467
24 Fishes 486
25 Early Tetrapods and Modern Amphibians 514
26 Amniote Origins and Reptilian Groups 534
27 Birds 555
28 Mammals 582

PART FOUR
Activity of Life

29 Support, Protection, and Movement 612
30 Homeostasis 634
31 Internal Fluids and Respiration 655
32 Digestion and Nutrition 677
33 Nervous Coordination 695
34 Chemical Coordination 721
35 Immunity 738
36 Animal Behavior 752

PART FIVE
Animals and Their Environment

37 The Biosphere and Animal Distribution 772
38 Animal Ecology 789

Glossary *807*
Credits *845*
Index *851*

TABLE OF CONTENTS

About the Authors   *xiii*
Preface   *xv*

PART ONE

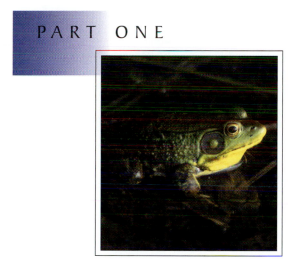

Introduction to Living Animals

CHAPTER 1

Life: Biological Principles and the Science of Zoology   2

Fundamental Properties of Life   3
Zoology as a Part of Biology   10
Principles of Science   10
Theories of Evolution and Heredity   14
Summary   *19*

CHAPTER 2

The Origin and Chemistry of Life   21

Organic Molecular Structure of Living Systems   22
Chemical Evolution   26
Origin of Living Systems   30
Precambrian Life   32
Summary   *34*

CHAPTER 3

Cells as Units of Life   36

Cell Concept   37
Organization of Cells   39
Mitosis and Cell Division   48
Summary   *54*

CHAPTER 4

Cellular Metabolism   56

Energy and the Laws of Thermodynamics   57
The Role of Enzymes   58
Chemical Energy Transfer by ATP   60
Cellular Respiration   61
Metabolism of Lipids   68
Metabolism of Proteins   69
Management of Metabolism   70
Summary   *71*

PART TWO

Continuity and Evolution of Animal Life

CHAPTER 5

Genetics: A Review   74

Mendel's Investigations   75
Chromosomal Basis of Inheritance   75
Mendelian Laws of Inheritance   79
Gene Theory   88
Storage and Transfer of Genetic Information   88
Sources of Phenotypic Variation   97
Molecular Genetics of Cancer   98
Summary   *99*

CHAPTER 6
Organic Evolution   102

Origins of Darwinian Evolutionary Theory   103
Darwinian Evolutionary Theory: The Evidence   106
Revisions of Darwin's Theory   120
Microevolution: Genetic Variation and Change within Species   120
Macroevolution: Major Evolutionary Events   126
Summary   128

CHAPTER 7
The Reproductive Process   131

Nature of the Reproductive Process   132
The Origin and Maturation of Germ Cells   136
Reproductive Patterns   140
Plan of Reproductive Systems   141
Endocrine Events That Orchestrate Reproduction   143
Summary   148

CHAPTER 8
Principles of Development   151

Early Concepts: Preformation Versus Epigenesis   152
Fertilization   153
Cleavage and Early Development   155
Gastrulation and Formation of Germ Layers   158
Mechanisms of Development   160
Gene Expression During Development   162
Vertebrate Development   165
Development of Systems and Organs   168
Summary   171

CHAPTER 9
Architectural Pattern of an Animal   176

Hierarchical Organization of Animal Complexity   177
Extracellular Components of Metazoan Bodies   178
Types of Tissues   179
Animal Body Plans   182
Summary   189

CHAPTER 10
Classification and Phylogeny of Animals   190

Linnaeus and the Development of Classification   191
Species   193
Taxonomic Characters and Phylogenetic Reconstruction   195
Theories of Taxonomy   197
Major Divisions of Life   202
Major Subdivisions of the Animal Kingdom   203
Summary   204

CHAPTER 11
Protozoan Groups   207

Form and Function   209
Phyla Retortamonada and Axostylata   216
Phylum Chlorophyta   216
Phylum Euglenozoa   218
Superphylum Alveolata   219
Phylum Apicomplexa   219
Phylum Ciliophora   222
Phylum Dinoflagellata   226
Amebas   226
Phylogeny and Adaptive Radiation   229
Summary   231

CHAPTER 12
Mesozoa and Parazoa   233

Origin of Metazoa   234
Phylum Mesozoa   234
Phylum Placozoa   236
Phylum Porifera: Sponges   236
Summary   245

CHAPTER 13
Radiate Animals   247

Phylum Cnidaria   248
Phylum Ctenophora   267
Phylogeny and Adaptive Radiation   270
Summary   272

PART THREE

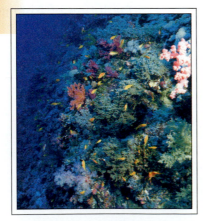

Diversity of Animal Life

CHAPTER 14
Acoelomate Bilateral Animals   274

Phylum Platyhelminthes   275
Phylum Nemertea (Rhynchocoela)   289
Phylum Gnathostomulida   291
Phylogeny and Adaptive Radiation   292
Summary   293

CHAPTER 15
Pseudocoelomate Animals   295

Pseudocoelomates   296
Ecdysozoan Phyla   296
Lophotrochozoan Phyla   306
Phylogeny and Adaptive Radiation   312
Summary   314

CHAPTER 16
Molluscs   316

Molluscs   317
Form and Function   318
Classes of Molluscs   321
Phylogeny and Adaptive Radiation   340
Summary   343

CHAPTER 17
Segmented Worms   345

Body Plan   346
Class Polychaeta   347
Class Oligochaeta   353
Class Hirudinea: Leeches   357
Evolutionary Significance of Metamerism   360
Phylogeny and Adaptive Radiation   360
Summary   362

CHAPTER 18
Arthropods   364

Phylum Arthropoda   365
Subphylum Trilobita   367
Subphylum Chelicerata   367
Phylogeny and Adaptive Radiation   374
Summary   375

CHAPTER 19
Aquatic Mandibulates   377

Subphylum Crustacea   378
A Brief Survey of Crustaceans   386
Phylogeny and Adaptive Radiation   393
Summary   396

CHAPTER 20
Terrestrial Mandibulates   397

Class Chilopoda   398
Class Diplopoda   398
Class Pauropoda   399
Class Symphyla   399
Class Insecta   400
Insects and Human Welfare   416
Phylogeny and Adaptive Radiation   420
Summary   424

CHAPTER 21
Smaller Protostome Phyla   426

Lophotrochozoan Phyla   427
Ecdysozoan Phyla   434
Phylogeny   438
Summary   440

CHAPTER 22
Echinoderms and Hemichordates   442

Phylum Echinodermata   443
Phylogeny and Adaptive Radiation   458
Phylum Hemichordata   460
Phylogeny and Adaptive Radiation   464
Summary   464

CHAPTER 23
Chordates   467

The Chordates   468
Five Chordate Hallmarks   469
Ancestry and Evolution   472
Subphylum Urochordata (Tunicata)   473
Subphylum Cephalochordata   475
Subphylum Vertebrata (Craniata)   477
Summary   483

CHAPTER 24
Fishes 486

Ancestry and Relationships of Major Groups of Fishes 487
Living Jawless Fishes 489
Class Chondrichthyes: Cartilaginous Fishes 493
Osteichthyes: Bony Fishes 497
Structural and Functional Adaptations of Fishes 502
Summary 512

CHAPTER 25
Early Tetrapods and Modern Amphibians 514

Movement onto Land 515
Early Evolution of Terrestrial Vertebrates 515
Modern Amphibians 519
Summary 532

CHAPTER 26
Amniote Origins and Reptilian Groups 534

Origin and Adaptive Radiation of Reptilian Groups 535
Characteristics of Reptiles That Distinguish Them
from Amphibians 538
Characteristics and Natural History of Reptilian Orders 540
Summary 552

CHAPTER 27
Birds 555

Origin and Relationships 556
Form and Function 560
Migration and Navigation 570
Social Behavior and Reproduction 573
Bird Populations 576
Summary 580

CHAPTER 28
Mammals 582

Origin and Evolution of Mammals 583
Structural and Functional Adaptations of Mammals 587
Humans and Mammals 601
Human Evolution 601
Summary 608

PART FOUR

Activity of Life

CHAPTER 29
Support, Protection, and Movement 612

Integument among Various Groups of Animals 613
Skeletal Systems 616
Animal Movement 622
Summary 631

CHAPTER 30
Homeostasis: Osmotic Regulation, Excretion, and Temperature Regulation 634

Water and Osmotic Regulation 635
Invertebrate Excretory Structures 639
Vertebrate Kidney 641
Temperature Regulation 646
Summary 652

CHAPTER 31
Internal Fluids and Respiration 655

Internal Fluid Environment 656
Composition of Blood 657
Circulation 659
Respiration 666
Summary 674

CHAPTER 32
Digestion and Nutrition   677

Feeding Mechanisms   678
Digestion   682
Organization and Regional Function of Alimentary Canals   683
Regulation of Food Intake   689
Nutritional Requirements   690
Summary   693

CHAPTER 33
Nervous Coordination:
Nervous System and Sense Organs   695

Neurons: Functional Units of Nervous Systems   696
Synapses: Junctions Between Nerves   699
Evolution of Nervous Systems   701
Sense Organs   709
Summary   719

CHAPTER 34
Chemical Coordination: Endocrine System   721

Mechanisms of Hormone Action   722
Invertebrate Hormones   724
Vertebrate Endocrine Glands and Hormones   724
Summary   736

CHAPTER 35
Immunity   738

Susceptibility and Resistance   739
Innate Defense Mechanisms   739
Acquired Immune Response in Vertebrates   741
Blood Group Antigens   747
Immunity in Invertebrates   748
Summary   749

CHAPTER 36
Animal Behavior   752

Describing Behavior: Principles of Classical Ethology   754
Control of Behavior   755
Social Behavior   759
Summary   769

PART FIVE

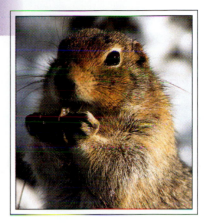

Animals and Their Environment

CHAPTER 37
The Biosphere and Animal Distribution   772

Distribution of Life on Earth   773
Animal Distribution (Zoogeography)   780
Summary   787

CHAPTER 38
Animal Ecology   789

The Hierarchy of Ecology   790
Summary   804

Glossary   807
Credits   845
Index   849

CLEVELAND P. HICKMAN

Cleveland P. Hickman, Jr., Professor Emeritus of Biology at Washington and Lee University in Lexington, Virginia, has taught zoology and animal physiology for more than 30 years. He received his Ph.D. in comparative physiology from the University of British Columbia, Vancouver, B.C., in 1958 and taught animal physiology at the University of Alberta before moving to Washington and Lee University in 1967. He has published numerous articles and research papers in fish physiology, in addition to co-authoring these highly successful texts: *Integrated Principles of Zoology, Biology of Animals, Animal Diversity, Laboratory Studies in Animal Diversity,* and *Laboratory Studies in Integrated Principles of Zoology.*

Over the years Dr. Hickman has led many field trips to the Galápagos Islands. His current research is on intertidal zonation and marine invertebrate systematics in the Galápagos. He has published three field guides in the Galápagos Marine Life Series for the identification of echinoderms, marine molluscs, and marine crustaceans. (To read more about these field guides, visit http://www.galapagosmarine.com.)

His interests include scuba diving, woodworking, and participating in chamber music ensembles.

Dr. Hickman can be contacted at: hickman.c@wlu.edu.

LARRY S. ROBERTS

Larry S. Roberts, Professor Emeritus of Biology at Texas Tech University and an adjunct professor at Florida International University, has extensive experience teaching invertebrate zoology, marine biology, parasitology, and developmental biology. He received his Sc.D. in parasitology at the Johns Hopkins University and is the lead author of Schmidt and Roberts' *Foundations of Parasitology,* sixth edition. Dr. Roberts is also co-author of *Integrated Principles of Zoology, Biology of Animals,* and *Animal Diversity,* and is author of *The Underwater World of Sport Diving.*

Dr. Roberts has published many research articles and reviews. He has served as President of the American Society of Parasitologists, Southwestern Association of Parasitologists, and Southeastern Society of Parasitologists, and is a member of numerous other professional societies. Dr. Roberts also serves on the Editorial Board of the journal, *Parasitology Research.* His hobbies include scuba diving, underwater photography, and tropical horticulture.

Dr. Roberts can be contacted at: RobertsLS@cs.com.

ALLAN LARSON

Allan Larson is a professor at Washington University, St. Louis, MO. He received his Ph.D. in genetics at the University of California, Berkeley. His fields of specialization include evolutionary biology, molecular population genetics and systematics, and amphibian systematics. He teaches courses in macroevolution, molecular evolution, and the history of evolutionary theory, and has organized and taught a special course in evolutionary biology for high-school teachers.

Dr. Larson has an active research laboratory that uses DNA sequences to examine evolutionary relationships among vertebrate species, especially in salamanders and lizards. The students in Dr. Larson's laboratory have participated in zoological field studies around the world, including projects in Africa, Asia, Australia, Madagascar, North America, South America, and the Caribbean Islands. Dr. Larson has authored numerous scientific publications, and has edited for the journals *The American Naturalist, Evolution, Journal of Experimental Biology, Molecular Phylogenetics and Evolution,* and *Systematic Biology.* Dr. Larson serves as an academic advisor to undergraduate students and supervises the undergraduate biology curriculum at Washington University.

Dr. Larson can be contacted at: larson@wustlb.wustl.edu.

HELEN I'ANSON

Helen I'Anson, a native of England, is associate professor of biology at Washington and Lee University in Lexington, Virginia. She received her Ph.D. in physiology at the University of Kentucky, Lexington, KY, and postdoctoral training at the University of Michigan, Ann Arbor, MI. She teaches courses in animal physiology, microanatomy, neuroendocrinology and reproductive physiology. She has an active research program that focuses on the neural regulation of reproductive development. In particular, she is interested in how energy is partitioned in the developing animal, how signals from food and food storage depots are monitored by the brain, and how such signals are transduced to regulate reproductive activity at the onset of puberty in mammals.

Her interests include gardening, hiking, fishing, aromatherapy, music, and participating in choral ensembles.

Dr. I'Anson can be contacted at: iansonh@wlu.edu.

Integrated Principles of Zoology is a college text adaptable to any introductory course in zoology. This twelfth edition, as with previous editions, describes the diversity of animal life and the fascinating adaptations that enable animals to inhabit nearly all conceivable ecological niches. We retain in this revision the basic organization of the eleventh edition and its distinctive features, especially emphasis on the principles of evolution and zoological science. Also retained are several pedagogical features that have made previous editions easily accessible to students: opening chapter dialogues drawn from the chapter's theme; chapter summaries and review questions to aid student comprehension and study; accurate and visually appealing illustrations; in-text derivations of generic names; chapter notes and essays that enhance the text by offering interesting sidelights to the narrative; and an extensive glossary providing pronunciation, derivation, and definition of terms used in the text.

NEW TO THE TWELFTH EDITION

Many changes in this edition were guided by suggestions of the 49 zoology instructors who read and commented on sections of the eleventh edition. For the eleventh edition, the vertebrate chapters of Part Three, and several chapters on functional systems (Part Four) were revised by invited contributors, all experienced zoologists who were solicited for their interest and expertise in the subject matter of specific chapters. This twelfth edition continues to benefit from their participation. In general, we revised all chapters to keep the text current while eliminating excessive detail, and placed emphasis on experimentation and comparative studies in zoology.

Chapter Organization

- Order of chapters in Parts One and Two has been maintained as in the eleventh edition. The presentation is streamlined by discussing basic chemistry in the context of the origin of life.
- Several chapters in Part Three have been reorganized. Because evidence is strong that lophophorate phyla and Chaetognatha are protostomes, separate chapters for each of these are not justified; therefore, material on these phyla has been moved to a single chapter covering all smaller phyla of coelomate protostomes (Chapter 21). Removal of chaetognaths from the chapter covering hemichordates made that chapter quite short, and accordingly, hemichordates are now covered in the same chapter as echinoderms, their closest relatives.

New Pedagogy

- Throughout the text we updated references, revised or replaced many illustrations, and rewrote many of the Review Questions to provoke thought and reduce emphasis on rote memorization.
- Suggested Internet topics are added at the end of each chapter; hyperlinks are available on this text's Online Learning Center website at www.mhhe.com/zoology.
- The art inside the front cover has been completely revised to depict phylogenetic relationships consistent with the Lophotrochozoa/Ecdysozoa hypothesis and other phylogenetic relationships based on sequence analysis of DNA.

The principal revisions in each chapter are explained here.

Part One: Introduction to Living Animals

- Chapter 2, The Origin and Chemistry of Life, now introduces the important concept of symbiogenesis in the context of the origin of eukaryotic cells. A review of basic chemistry (atoms, elements, and molecules; bonding theory; acids, bases, salts, and buffers) is available for reference; it will be found at our Online Learning Center website: www.mhhe.com/zoology.

Part Two: Continuity and Evolution of Animal Life

- Chapter 5, Genetics: A Review, features new art, including a revised meiosis figure that includes homologous chromosomes in synapsis during prophase of meiosis I and clarifications of 5′ and 3′ ends of RNA and DNA, indicating direction of synthesis. We have added new material on publication of the draft sequence of the human genome and its implications, as well as an explanation of proteomics.
- In Chapter 6, Organic Evolution, we have added information on new fossil discoveries from China. We relate the issue of gradualism to new work in developmental genetics.
- Chapter 7, The Reproductive Process, was revised extensively in the eleventh edition. Notes on contraception were updated to reflect new trends.

- Among many revisions and updatings to Chapter 8, Principles of Development, the section on gene expression during development was completely rewritten to describe the three stages in epigenesis: pattern formation, determination of position in the body and involvement of the homeotic genes, and induction of limbs and organs appropriate for that position.

Part Three: Diversity of Animal Life

- Chapter 10, Classification and Phylogeny of Animals, has been reorganized. We present species concepts and diagnosis and then proceed to higher-level taxonomy, rather than vice versa, as in the eleventh edition. We revised coverage of subspecies and added the process of species recognition. A boxed essay has been added to describe how phylogenies are constructed from DNA sequences, and a new taxonomic system called PhyloCode is explained.

- Chapter 11, Protozoan Groups, was completely reorganized in an effort to make it consistent with current classification of unicellular eukaryotes. Traditional taxa such as Sarcomastigophora, Sarcodina, Mastigophora, Zoomastigophora, and Phytomastigophora are all discarded. Flagellated forms that we discuss are distributed among separate phyla: Retortamonada, Axostylata, Chlorophyta, Euglenozoa, and Dinoflagellata. Protozoologists now disperse ameboid forms into numerous lineages, and it is impractical for us to place them all in formal taxa; therefore, we simply use an informal heading "Amebas" to introduce some of the most important ones. The section on phylogeny was rewritten to reflect current evidence, including the likelihood of secondary symbiogenesis in origins of several groups.

- In discussing the phylogeny of Mesozoa in Chapter 12, we note that molecular evidence supports a relationship with flatworms. The description of Hexactinellida was completely rewritten, and a new figure depicting their structure is included.

- Chapter 13, Radiate Animals, underwent several important revisions. Most noticeable to users of earlier editions is a preference for use of Octocorallia over Alcyonaria for this anthozoan subclass. Because members of the subclass always have eight tentacles, we believe that Octocorallia is easier for students to remember. In addition, the discussion of cnidarian phylogeny was rewritten to include evidence from *Hox* genes and sequence data that Anthozoa is basal for the phylum. A new cladogram reflects this relationship.

- In Chapter 14, Acoelomate Animals, the description of both form and function in Nemertea and phylogeny of flatworms were rewritten.

- Chapter 15, Pseudocoelomate Animals, was completely reorganized. We hope that placing all animals with a common body plan in the same chapter will be helpful for students, rather than separating pseudocoelomate lophotrochozoans from pseudocoelomate ecdysozoans in separate chapters. Thus, phyla in each superphylum were grouped together under appropriate headings (see also Chapter 21). Furthermore, because Nematoda is the largest and most important of all pseudocoelomate phyla, their coverage was moved to the beginning of the chapter. Recognizing molecular evidence of a rotifer-acanthocephalan affinity, coverage of Acanthocephala was moved to follow immediately after Rotifera.

- In Chapter 16, Molluscs, a note on the evolution of eyes in animals was added, with the observation that all animals with eyes apparently share two conserved genes: *Pax 6,* the "master control gene for eye morphogenesis," and the gene encoding rhodopsin, a visual pigment. The eyes of cephalopod molluscs are amazingly similar to those of vertebrates. Although this has long been considered an astonishing example of convergence, both groups must have inherited the genes from a common ancestor, and thus their eyes would be homologous.

- Chapter 18, Arthropods, is an introduction to Arthropoda and a more detailed coverage of Chelicerata. In this chapter, many sections on spiders, scorpions, and harvestmen were rewritten and revised.

- Chapter 19, Aquatic Mandibulates, benefits from much new art. The phylogeny section in this chapter was rewritten and includes a discussion of how biramous limbs in Crustacea probably arose by a mutation resulting in a modulation of *Distal-less* gene expression.

- Chapter 20, Terrestrial Mandibulates, is a large chapter because insects are a large group. There were many small rewrites and additions, including the discussion and illustration of sucking mouthparts. The phylogeny discussion was heavily rewritten, including roles of *Hox* genes and *Distal-less* gene and the evolution of flight.

- Chapter 21, Smaller Protostome Phyla, was extensively reorganized, as in Chapter 15, grouping lophotrochozoan and ecdysozoan phyla under their appropriate headings. Lophophorate phyla and Chaetognatha were brought into this chapter because present evidence lends little justification to covering them in separate chapters. Significant changes in the phylogeny discussion were required.

- Chapter 22, Echinoderms and Hemichordates, now includes all deuterostomes other than chordates. After removal of chaetognaths from the chapter covering hemichordates, a very short chapter resulted; therefore, hemichordates were moved to join Echinodermata.

- Chapter 23, Chordates, underwent many changes throughout, incorporating new ideas about chordate origins and character evolution, primarily from new fossil finds in China, molecular phylogenetics, and studies of developmental biology. Specifically, we modified and added figures, revised characteristics of chordates and vertebrates, revised discussion of hypotheses of origins of chordates and vertebrates, and added a discussion on *Hox* genes. Discussions of conodonts, ostracoderms, origin of jaws, and origin of paired appendages were revised to better emphasize critical concepts and include more recent information.

- Coverage of the ecology of amphibian declines in the world is increased in Chapter 25, Early Tetrapods and Modern Amphibians.

- The title of Chapter 26, Amniote Origins and Reptilian Groups, was changed to better reflect expanded coverage of derived features of amniotes, including the amniote egg. Relationships of major groups were updated, especially showing that snakes and amphisbaenians evolved from a lizardlike ancestor. Discussions of dinosaur behavior and extinction, snake feeding methods, snakebite statistics, and ectothermy were revised. A new illustration of a snake skull was added to show the extreme mobility of their head bones.

- Chapter 27, Birds, incorporates material on new fossil dinosaur-bird finds and their significance to flight. The section on mating systems was also modified.

- In Chapter 28, Mammals, coverage on synapsid evolution has been expanded and supporting artwork is modified. We added a section on whale evolution, revised the section on the fermentation system, and updated coverage of human evolution, incorporating information from recent fossil finds and analyses. Coverage of evolution of reproductive systems was changed extensively, including evidence that the chorioallantoic placenta of placental mammals evolved in a common ancestor of marsupials and placentals, and was secondarily lost in marsupials.

Part Four: Activity of Life

- In Chapter 29, Support, Protection, and Movement, we clarified the role of exercise in differential stimulation of fast and slow fiber types, and we updated the microanatomy and physiology of skeletal muscle contraction.

- We elaborated on the concept of homeostatic setpoint in Chapter 30, Homeostasis, and sections on the role of urea in osmoregulation of elasmobranchs and in urine concentration by mammals were revised. Also revised were coverage of aldosterone secretion, glomerular tubular secretion, ADH mechanism of action, and countercurrent multiplication.

- Chapter 31, Internal Fluids and Respiration, includes revisions on excitation and control of heart function, structure and function of the arterial system, respiration in birds, regulation of respiration, and transport of respiratory gases in mammals.

- In Chapter 32, Digestion and Nutrition, we added a consideration of the brainstem as a hunger center and of obesity as a lifestyle problem in developed countries. New information on leptin resistance and obesity, and on the role of inflammation in development of atherosclerosis was added.

- Chapter 33, Nervous Coordination, includes new information on the vomeronasal organ as a pheromonal sense organ in terrestrial vertebrates and the possibility of pheromonal communication in humans. Coverage of taste discrimination was updated, including discovery of a new taste, umami.

- Chapter 34, Chemical Coordination, includes new information on steroid hormone membrane receptors, steroid abuse by young people, insulin action in the brain, and the importance of leptin during periods of decreased food intake.

- Within Chapter 35, Immunity, we significantly updated the section on innate immunity by addition of material on the exciting discoveries of antimicrobial peptides produced by many invertebrates and vertebrates, responding on first exposure to broad categories of invaders such as gram-positive bacteria, gram-negative bacteria, and fungi.

- Chapter 36, Animal Behavior, covers advances in application of phylogenetic approaches to study of animal behavior, which highlights an important recent advancement in the field. Material was added on the important concepts of reciprocal altruism and evolutionarily stable strategies.

Part Five: Animals and Their Environment

- Chapter 37, The Biosphere and Animal Distribution, features new information on Wallace's Line and its illustration of the importance of plate tectonics in biogeography. Connections between material in this chapter and that in Chapters 6 and 10 are strengthened.

- In Chapter 38, Animal Ecology, we elaborate the niche concept and have made it more consistent with current leading ecology texts. The discussion of interactions among species in a community (for example, commensalism and mimicry) has been extended.

TEACHING AND LEARNING AIDS

To help students in **vocabulary development,** key words are boldfaced, and derivations of technical and zoological terms are provided, along with generic names of animals where they first appear in the text. In this way students gradually become familiar with the more common roots that comprise many technical terms. An extensive glossary of almost 1,100 terms provides pronunciation, derivation, and definition of each term. Many new terms were added to the glossary or rewritten for this edition.

A distinctive feature of this text is a **prologue** for each chapter that draws out some theme or fact relating to the subject of the chapter. Some prologues present biological, particularly evolutionary, principles; others (especially those in the survey sections) illuminate distinguishing characteristics of the group treated in the chapter. Each is intended to present an important concept drawn from the chapter in an interesting manner that will facilitate learning by students, as well as engage their interest and pique their curiosity.

Chapter notes, which appear throughout the book, augment the text material and offer interesting sidelights without interrupting the narrative. We prepared many new notes for this edition and revised several existing notes.

To assist students in chapter review, each chapter ends with a **concise summary,** a list of **review questions,** and **annotated selected references.** The review questions enable a student to self-test retention and understanding of the more important chapter material.

The **historical appendix,** unique to this textbook, lists key discoveries in zoology, and separately describes books and publications that have greatly influenced the development of zoology. Many readers have found this appendix an invaluable reference to be consulted long after their formal training in zoology. The historical appendix will be found on this textbook's Online Learning Center website at www.mhhe.com/zoology.

Again, William C. Ober and Claire W. Garrison have enhanced the **art program** for this text with many new full-color paintings that replace older art, or that illustrate new material. Bill's artistic skills, knowledge of biology, and experience gained from an earlier career as a practicing physician have enriched this text through eight of its editions. Claire practiced pediatric and obstetric nursing before turning to scientific illustration as a full-time career. Texts illustrated by Bill and Claire have received national recognition and won awards from the Association of Medical Illustrators, American Institute of Graphic Arts, Chicago Book Clinic, Printing Industries of America, and Bookbuilders West. They are also recipients of the Art Directors Award.

SUPPLEMENTS

The **Instructor's Testing and Resource CD-ROM** provides both an Instructor's Manual and a test bank to support the textbook. The Instructor's Manual includes annotated chapter outlines, chapter-specific changes for this edition, lecture enrichment suggestions, commentaries and lesson plans, questions for advanced classes, and a listing of resource references for each chapter. The Instructor's Manual also includes a listing of transparencies and slides available with the book. A comprehensive **test bank** offers 35 to 50 objective questions per chapter within a computerized test generation system for IBM and Macintosh computers. Using this system, instructors can create tests or quizzes quickly and easily. Questions can be sorted by type or level of difficulty, and instructors also can add their own material to the bank of questions provided. We trust this will be of particular value to first-time users of the text, although experienced teachers may also find much of value.

The **Laboratory Manual** by Cleveland P. Hickman, Jr., and Lee B. Kats, *Laboratory Studies in Integrated Principles of Zoology,* now in its twelfth edition, includes exercises on molecular techniques. This manual can be customized conveniently for two semester, one semester, or term courses by judicious selection of exercises.

A set of 150 full-color **transparency acetates** of important textual illustrations are available with this edition of *Integrated Principles of Zoology.* Labeling is clear, dark, and bold for easy reading.

A set of 148 **animal diversity slides,** photographed by the authors and Bill Ober on their various excursions, are offered in this unique textbook supplement. Both invertebrates and vertebrates are represented. Descriptions, including specific names of each animal and brief overview of the animal's ecology and/or behavior, accompany the slides.

The **Digital Content Manager CD-ROM** is a multimedia collection of visual resources that allows instructors to create powerful presentations for the classroom. The Digital Content Manager CD-ROM contains many photos and all of the illustrations from this textbook, along with 200 additional animal diversity images, 26 animations of key biological concepts, and a collection of underwater video clips displaying a variety of ocean life, habitats, and behaviors. See page xx for additional details.

The **Zoology Essential Study Partner CD-ROM** is an interactive student study tool that features activities, animations, drag-and-drop exercises, self-testing tools, full-color art and graphs, and more, and can be packaged to go with any McGraw-Hill zoology title. For further details, see page xxii, or contact your McGraw-Hill sales representative.

An **Online Learning Center website** is available with this edition and contains additional readings, animations, quizzing, key term flashcards, cladogram exercises, Animal Diversity PowerPoint slides, Zoology Essential Study Partner, and much more (see page xx for specific information). Check it out at www.mhhe.com/zoology.

Digital Zoology

Digital Zoology Version 2.0 Interactive CD-ROM and Student Workbook by Jon Houseman is an interactive guide to the specimens and materials covered in zoology laboratory and lecture sessions. Laboratory modules contain illustrations, photographs, annotations of the major structures of organisms, interactive quizzes, and video clips. Interactive cladograms within lab modules provide links to interactive synapomorphies of the various animal groups. Key terms throughout the program link to an interactive glossary. This CD-ROM is the perfect student study tool to promote learning both in and outside of the zoology laboratory, and also comes with an accompanying student workbook and website to provide additional study tips, exercises, and phyla characteristics. (For further information, see page xxi.)

ACKNOWLEDGMENTS

We wish to thank David Eisenhour, who was engaged by McGraw-Hill to contribute directly to the revision of Chapter 23, Chordates, Chapter 24, Fishes, Chapter 26, Amniote Origins and Reptilian Groups, Chapter 27, Birds, and Chapter 28, Mammals. Leon Harris revised Chapter 8, Principles of Development, and we thank him for his efforts. McGraw-Hill also wishes to thank

Matthew M. Douglas, Science Editor, for his contributions in preparing the Instructor's Manual and Test Bank for the 12th edition.

The authors extend their warmest thanks to reviewers who suggested numerous improvements and whose collective wisdom was of the greatest assistance to us as we approached this edition. Their experience with students of varying backgrounds, and their interest in and knowledge of the subject, helped to shape the text into its final form.

Barbara J. Abraham, *Hampton University*

Amir M. Assadi-Rad, *Delta College*

Linda Barham, *Meridian Community College*

Rose Bast, *Mount Mary College*

Lisa L. Bellows, *North Central Texas College*

Barbara A. Bernardi, *Springfield College in Illinois*

Annalisa Berta, *San Diego State University*

Sandra B. Bobick, *Community College of Allegheny County*

Ty W. Bryan, *Bossier Parish Community College*

John R. Capeheart, *University of Houston–Downtown*

Vincent Cobb, *Northeastern State University*

Marjorie M. Collier, *Louisiana State University–Alexandria*

D. Charles Dailey, *Sierra College*

Paula Dedmon, *Gaston College*

Randall Deere, *Western Kentucky University*

Elizabeth Drumm, *Oakland Community College*

Phillip Eichman, *University of Rio Grande*

Jim Gibson, *Chadron State College*

Jim R. Goetze, *Laredo Community College*

Glenn Gorelick, *Citrus College*

Joseph Goy, *Harding University*

Kenneth D. Hoover, *Jacksonville University*

Dan F. Ippolito, *Anderson University*

Ritin Khan, *Armstrong Atlantic State University*

Gopal Krishna, *Moberly Area Community College*

David Kurjiaka, *Ohio University–Athens*

Zhiming Liu, *Eastern New Mexico University*

William J. Mathena, *Kaskaskia College*

Gregory J. McConnell, *Emory and Henry College*

Tim R. Mullican, *Dakota Wesleyan University*

Mark S. Peterson, *University of Southern Mississippi*

Polly Phillips-Sanborn, *Miami-Dade Community College*

Robert Powell, *Avila College*

Susan K. Reimer, *St. Francis College*

Donald C. Rizzo, *Marygrove College*

Greig Rose, *West Valley College*

Allen F. Sanborn, *Barry University*

Neil B. Schanker, *College of The Siskiyous*

Fred Schindler, *Indian Hills Community College*

Larry Sellers, *Louisiana Tech University*

W. David Sissom, *West Texas A&M University*

F. Christopher Sowers, *Wilkes Community College*

Ed Story, *Maysville Community College*

Danny Wann, *Carl Albert Junior College*

Clayton M. White, *Brigham Young University–Provo*

Mary K. Wicksten, *Texas A&M University–College Station*

Brian Wisenden, *Moorhead State University*

Brenda Zink, *Northeastern Junior College*

The authors express their appreciation to the editors and support staff at McGraw-Hill Higher Education who made this project possible. Special thanks are due Marge Kemp, Publisher, and Donna Nemmers, Senior Developmental Editor, who were the driving forces in piloting this text throughout its development. Mary Powers, Project Manager, somehow kept authors, text, art, and production programs on schedule. John Leland oversaw the extensive photographic program, and David Hash managed the book's interior and cover design. We are indebted to them for their talents and dedication.

Although we make every effort to bring to you an error-free text, errors of many kinds inevitably find their way into a textbook of this scope and complexity. We will be grateful to readers who have comments or suggestions concerning content to send their remarks to Fran Schreiber, Developmental Editor, 2460 Kerper Boulevard, Dubuque, IA 52001. Fran may also be contacted by e-mail: fran_schreiber@mcgraw-hill.com, or through this textbook's website: www.mhhe.com/zoology.

Cleveland P. Hickman, Jr.
Larry S. Roberts
Allan Larson
Helen I'Anson

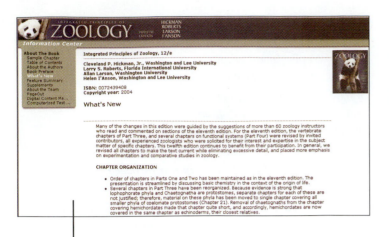

THE ONLINE LEARNING CENTER
Your Password to Success
www.mhhe.com/zoology
(click on cover)

This text-specific website allows students and instructors from all over the world to communicate. Instructors can create a more interactive course with the integration of this site, and students will find tools such as practice quizzing, key term flashcards, and animations that will help them improve their grades and learn that zoology can be fun.

Student Resources

Quizzing with immediate feedback
Hyperlinks to chapter-related websites
Key Term Flashcards
Animations
Interactive Cladogram Exercise
"Development of Zoology" timeline
"Basic Structure of Matter" appendix
Zoology Essential Study Partner

Instructor Resources

Instructor's Manual

- Annotated chapter outlines
- Twelfth edition changes
- Lecture enrichment suggestions
- Commentary/lesson plan
- Advanced class questions
- Source material listings

Links to related websites to expand on particular topics
List of Digital Content Manager images
List of slides
List of transparency acetates
Teaching Animal Molecular Phylogenetics tutorial

DIGITAL CONTENT MANAGER
CD-ROM (ISBN 0-07-254700-6)

This powerful and easy-to-use tool is designed to help instructors easily incorporate text-specific illustrations, photos, and tables, plus animations and other related assets, into lecture presentations and printed classroom materials. Organized by chapter, this cross-platform CD-ROM contains a collection of visual and multimedia resources that can be imported and reproduced in multiple formats to create customized lectures, visually based tests and quizzes, dynamic course website content, or attractive printed support materials. The *Integrated Principles of Zoology* Digital Content Manager features:

- An **art library** of JPEG files featuring EVERY ILLUSTRATION from the twelfth edition (720 images, organized by chapter).

- A **textbook photo library**—218 JPEG files featuring photos from the twelfth edition (organized by chapter).

- A **table library** of JPEG files featuring EVERY TABLE from the text (organized by chapter).

- An **additional photo library of 200 animal diversity images**—comprised of additional photographs not found in the text (labeled by topic, with search function).

- **PowerPoint art slides**—art, photos, and tables from each chapter that have been preinserted into PowerPoint slides.

- **Animations library**—26 full-color animations illustrating key biological concepts. Harness the visual impact of processes in motion by importing these files into classroom presentations or online course materials.

- **Video clips library**—74 underwater video clips showing the ocean's animal diversity, habitats, and behaviors.

DIGITAL ZOOLOGY CD-ROM AND STUDENT WORKBOOK

Digital Zoology is an interactive guide that will help reinforce what students are learning in their zoology lecture and laboratory sessions. Here is what you will find on this easy-to-use CD-ROM:

- Laboratory modules containing illustrations, photographs, annotations of the major structures of organisms, interactive quizzes, and over 70 video clips of animal diversity, habitats, and behaviors

- Interactive cladograms and dendograms within lab modules, along with links to interactive synapomorphies of the various animal groups

- Key terms with links to an interactive glossary of over 750 definitions

- Taxon "Read Abouts" that include information on over 100 taxa, the number of living and extinct species, habitats, and how they function

- Updated taxonomy of the animal phyla that allow for easy comparison of the differences between traditional morphological phylogenies and those created with molecular data

- Image galleries available for many of the major phyla, providing photos that detail diversity within the phyla

- An accompanying student workbook with self-tests, crossword puzzles, and a website to provide additional study tips, exercises, and phyla characteristics

INSTRUCTORS: CUT CLASS PREP TIME IN HALF!

Organizing your class materials has never been so easy! Visit www.digitalzoology.com to take advantage of all the tools available to you on the *Digital Zoology* website. Here is what you will find:

- PowerPoint slides with cladograms and dendrograms illustrating the relationships between the different major taxa, as well as many additional images to show the diversity of each group

- Test bank, sorted by taxonomic group

- *Teaching Animal Molecular Phylogenetics* website

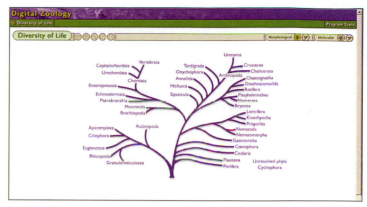

Main dendrogram (tree) for *Digital Zoology.*

Taxon "Read About."

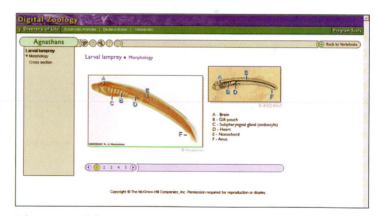

Laboratory module.

ZOOLOGY ESSENTIAL STUDY PARTNER

This study partner engages, investigates, and reinforces what you are learning in your zoology course. You'll find the **Zoology Essential Study Partner** to be a complete, interactive student study tool packed with hundreds of learning activities and animations. From quizzes to interactive diagrams, you'll find that there has never been a better study partner to ensure the mastery of core concepts. This innovative study tool can be accessed through the Online Learning Center for this textbook, or it is also available on CD-ROM, and can be packaged with your zoology textbook. Contact your McGraw-Hill sales representative for further details.

- The **unit pop-up menu** is accessible at any time within the program. Clicking on the current unit will bring up a menu of other units available in the program.

- The **topic menu** contains an interactive list of the available topics. Clicking on any of the listings within this menu will open your selection and will show the specific concepts presented within this topic.

- Various navigational aids appear along the bottom of the screen. At the left are **arrows** that allow you to page forward and backward through text screens or interactive exercise screens. You can also use the LEFT and RIGHT arrows on your keyboard to perform the same function.

- To the right of the arrows is a row of icons that represent the number of screens in a concept. There are three different icons, each representing different functions that a screen in that section will serve. The screen that is currently displayed will highlight in yellow and visited ones will be checked.

- The **film icon** represents an animation screen.

- The **page icon** represents a page of informational text.

- The **activity icon** represents an interactive learning activity.

PAGEOUT

Need a course website? PageOut is your solution. More than 35,000 professors have chosen PageOut to create a course website. Their feedback has led to a powerful new Third Release (3.0) of PageOut. It contains new features based on customer feedback:

- The PageOut Library offers instant access to fully loaded course websites with no work required on your part.
- Courses can now be password protected.
- You can now upload, store, and manage up to 10 MB of data.
- You can copy your course and share it with colleagues or use it as a foundation for next semester.

Short on time? Let us do the work. Our McGraw-Hill service team is ready to build your PageOut website—and provide content and any necessary training. Learn more about PageOut and other McGraw-Hill digital solutions at www.mhhe.com/solutions.

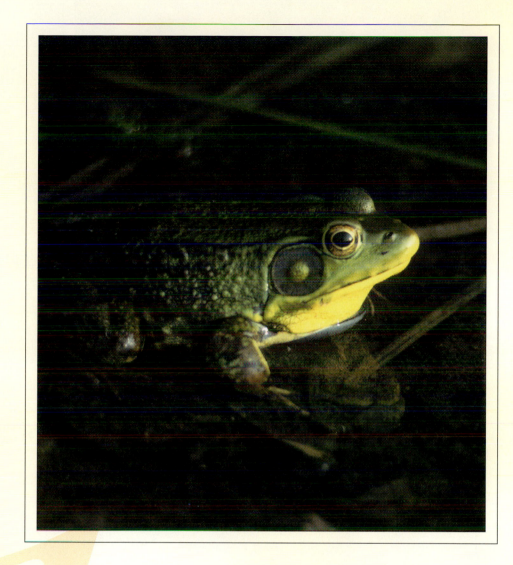

PART ONE

A green frog,
Rana clamitans,
in a Michigan pond.

Introduction to
Living Animals

1   Life: Biological Principles and the Science of Zoology
2   The Origin and Chemistry of Life
3   Cells as Units of Life
4   Physiology of Cells

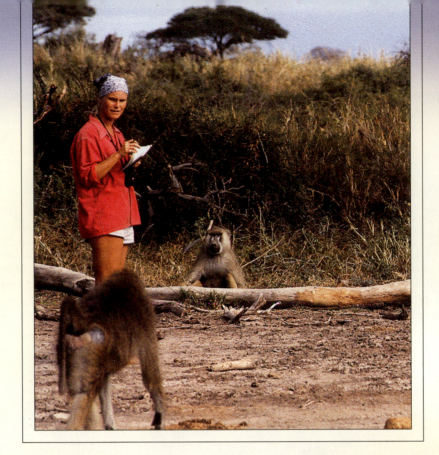

Zoologist studying the behavior of yellow baboons (*Papio cynocephalus*) in the Amboseli Reserve, Kenya.

Life: Biological Principles and the Science of Zoology

The Uses of Principles

We gain knowledge of the animal world by actively applying important guiding principles to our investigations. Just as the exploration of outer space is both guided and limited by available technologies, exploration of the animal world depends critically on our questions, methods, and principles. The body of knowledge that we call zoology makes sense only when the principles that we use to construct it are clear.

The principles of modern zoology have a long history and many sources. Some principles derive from laws of physics and chemistry, which all living systems obey. Others derive from the scientific method, which tells us that our hypotheses regarding the animal world are useless unless they guide us to gather data that potentially can refute them. Many important principles derive from previous studies of the living world, of which animals are one part. Principles of heredity, variation, and organic evolution guide the study of life from the simplest unicellular forms to the most complex animals, fungi, and plants. Because life shares a common evolutionary origin, principles learned from the study of one group often pertain to other groups as well. By tracing the origins of our operating principles, we see that zoologists are not an island unto themselves but part of a larger scientific community.

We begin our study of zoology not by focusing narrowly within the animal world, but by searching broadly for our most basic principles and their diverse sources. These principles simultaneously guide our studies of animals and integrate those studies into the broader context of human knowledge.

Zoology, the scientific study of animal life, builds on centuries of human inquiry into the animal world. Mythologies of nearly every human culture document attempts to solve the mysteries of animal life and its origin. Zoologists now confront these same mysteries with the most advanced methods and technologies developed throughout all branches of science. We start by documenting the diversity of animal life and organizing it in a systematic way. This complex and exciting process builds on the contributions of thousands of zoologists working in all dimensions of the biosphere (Figure 1-1). We strive through this work to understand how animal diversity originated and how animals perform the basic processes of life that permit them to thrive in many diverse environments.

This chapter introduces the fundamental properties of animal life, the methodological principles on which their study is based, and two important theories that guide our research: (1) the theory of evolution, which is the central organizing principle of biology, and (2) the chromosomal theory of inheritance, which guides our study of heredity and variation in animals. These theories unify our knowledge of the animal world.

FUNDAMENTAL PROPERTIES OF LIFE

Does Life Have Defining Properties?

We begin with the difficult question, What is life? Although many attempts have been made to define life, simple definitions are doomed to failure. When we try to give life a simple definition, we look for fixed properties maintained throughout life's history. However, the properties that life exhibits today (pp. 4–9) are very different from those present at its origin. The history of life shows perpetual change, which we call *evolution*. As the genealogy of life progressed and branched from the earliest living form to the millions of species living today, new properties evolved and were passed from parents to their offspring. Through this process, living systems have generated many rare and spectacular features that have no counterparts in the nonliving world. Unexpected properties emerge on many different lineages in life's evolutionary history, producing the great organismal diversity observed today.

A

B

C

D

Figure 1-1

A few of the many dimensions of zoological research: **A,** Observing moray eels in Maui, Hawaii; **B,** Working with tranquilized polar bears; **C,** Banding mallard ducks; **D,** observing *Daphnia pulex* (×150) microscopically.

We might try to define life on the basis of universal properties evident at its origin. Replication of molecules, for example, can be traced to life's origin and represents one of life's universal properties. Defining life based on properties present at its origin faces the major problem that these are the properties most likely to be shared by some nonliving forms. To study the origin of life, we must ask how organic molecules acquired the ability for precise replication. But where do we draw the line between those replicative processes that characterize life and those that are merely general chemical features of the matter from which life arose? Replication of complex crystalline structures in nonliving chemical assemblages might be confused, for example, with the replicative molecular properties associated with life. If we define life using only the most advanced properties that characterize the highly evolved living systems observed today, the nonliving world would not intrude on our definition, but we would eliminate the early forms of life from which all others descended and which give life its historical unity.

Ultimately our definition of life must be based on the common history of life on earth. Life's history of descent with modification gives it an identity and continuity that separates it from the nonliving world. We can trace this common history backward through time from the diverse forms observed today and in the fossil record to their common ancestor that arose in the atmosphere of the primitive earth (see Chapter 2). All organisms forming part of this long history of hereditary descent from life's common ancestor are included in our concept of life.

We do not force life into a simple definition, but we can readily identify the living world through its history of common evolutionary descent and separate it from the nonliving. Many remarkable properties have arisen during life's history and are observed in various combinations among living forms. These properties, discussed in the next section, clearly identify their possessors as part of the unified historical entity called life. All such features occur in the most highly evolved forms of life, such as those that compose the animal kingdom. Because they are so important for maintenance and functioning of living forms that possess them, these properties should persist through life's future evolutionary history.

General Properties of Living Systems

The most outstanding general features that have arisen during life's history include chemical uniqueness; complexity and hierarchical organization; reproduction (heredity and variation); possession of a genetic program; metabolism; development; and environmental interaction.

1. **Chemical uniqueness.** *Living systems demonstrate a unique and complex molecular organization.* Living systems assemble large molecules, known as macromolecules, that are far more complex than the small molecules that constitute nonliving matter. These macromolecules are composed of the same kinds of atoms and chemical bonds that occur in nonliving matter and they obey all fundamental laws of chemistry; it is only the complex organizational structure of these macromolecules that makes them unique.

We recognize four major categories of biological macromolecules: nucleic acids, proteins, carbohydrates, and lipids (see Chapter 2). These categories differ in the structures of their component parts, the kinds of chemical bonds that link their subunits together, and their functions in living systems.

The general structures of these macromolecules evolved and stabilized early in the history of life. With some modifications, these same general structures are found in every form of life today. Proteins, for example, contain about 20 specific kinds of amino acid subunits linked together by peptide bonds in a linear sequence (Figure 1-2). Additional bonds occurring between amino acids that are not adjacent to each other in the protein chain give the protein a complex, three-dimensional structure (see Figures 1-2 and 2-11). A typical protein contains several hundred amino acid subunits. Despite the stability of this basic protein structure, the ordering of the different amino acids in the protein molecule is subject to enormous variation. This variation underlies much of the diversity that we observe among different kinds of living forms. The nucleic acids, carbohydrates, and lipids likewise contain characteristic bonds that link variable subunits (Chapter 2). This organization gives living systems both a biochemical unity and a great potential for diversity.

2. **Complexity and hierarchical organization.** *Living systems demonstrate a unique and complex hierarchical organization.* Nonliving matter is organized at least into atoms and molecules and often has a higher degree of organization as well. However, atoms and molecules are combined into patterns in the living world that do not exist in the nonliving world. In living systems, we find a hierarchy of levels that includes, in ascending order of complexity, macromolecules, cells, organisms, populations, and species (Figure 1-3). Each level builds on the level below it and has its own internal structure, which is also often hierarchical. Within the cell, for example, macromolecules are compounded into structures such as ribosomes, chromosomes, and membranes, and these are likewise combined in various ways to form even more complex subcellular structures called organelles, such as mitochondria (see Chapters 3 and 4). The organismal level also has a hierarchical substructure; cells are combined into tissues, which are combined into organs, which likewise are combined into organ systems (see Chapter 9).

Cells (Figure 1-4) are the smallest units of the biological hierarchy that are semiautonomous in their ability to conduct basic functions, including reproduction. Replication of molecules and subcellular components occurs only within a cellular context, not independently. Cells are therefore viewed as the basic units of living systems (Chapter 3). We can isolate cells from an organism and cause them to grow and multiply under laboratory conditions in the presence of nutrients alone. This semiautonomous replication is not possible for any individual molecules or subcellular components, which require additional cellular constituents for their reproduction. Each successively higher level of the biological hierarchy is composed of units of the preceding lower level in the hierarchy. An important characteristic of this

A

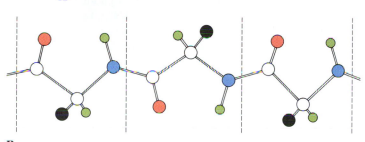

B

Figure 1-2

A computer simulation of the three-dimensional structure of the lysozyme protein **(A)**, which is used by animals to destroy bacteria. The protein is a linear string of molecular subunits called amino acids, connected as shown in **B**, which fold in a three-dimensional pattern to form the active protein. The white balls correspond to carbon atoms, the red balls to oxygen, the blue balls to nitrogen, the yellow balls to sulfur, the green balls to hydrogen, and the black balls **(B)** to molecular groups formed by various combinations of carbon, oxygen, nitrogen, hydrogen, and sulfur atoms that differ among amino acids. Hydrogen atoms are not shown in **A**. The purple molecule in **A** is a structure from the bacterial cell wall that is broken by lysozyme.

Figure 1-3

Volvox globator (see pp. 216–218) is a multicellular chlorophytan that illustrates three different levels of the biological hierarchy: cellular, organismal, and populational. Each individual spheroid (organism) contains cells embedded in a gelatinous matrix. The larger cells function in reproduction, and the smaller ones perform the general metabolic functions of the organism. The individual spheroids together form a population.

Figure 1-4

Electron micrograph of ciliated epithelial cells and mucus-secreting cells (see pp. 180–182). Cells are the basic building blocks of living organisms.

hierarchy is that the properties of any given level cannot be obtained from even the most complete knowledge of the properties of its component parts. A physiological feature, such as blood pressure, is a property of the organismal level; it is impossible to predict someone's blood pressure simply by knowing the physical characteristics of individual cells of the body. Likewise, systems of social interaction, as observed in bees, occur at the populational level; it would not be possible to infer properties of this social system by knowing only properties of individual bees.

The appearance of new characteristics at a given level of organization is called **emergence,** and these characteristics are known as **emergent properties.** These properties arise from interactions among the component parts of a system. For this reason, we must study all levels directly, and subdivisions of the field of biology (molecular biology; cell biology; organismal anatomy, physiology and genetics; population biology) reflect this fact (Table 1-1). We find that

emergent properties expressed at a particular level of the biological hierarchy are certainly influenced and restricted by properties of the lower-level components. For example, it would be impossible for a population of organisms that lack hearing to develop a spoken language. Nonetheless, properties of parts of a living system do not rigidly deter-

TABLE 1.1

Different Hierarchical Levels of Biological Complexity that Display Reproduction, Variation, and Heredity

Level	Timescale of Reproduction	Fields of Study	Methods of Study	Some Emergent Properties
Cell	Hours (mammalian cell = ~16 hours)	Cell biology	Microscopy (light, electron), biochemistry	Chromosomal replication (meiosis, mitosis), synthesis of macromolecules (DNA, RNA, proteins, lipids, polysaccharides)
Organism	Hours to days (unicellular); days to years (multicellular)	Organismal anatomy, physiology, genetics	Dissection, genetic crosses, clinical studies, physiological experimentation	Structure, functions and coordination of tissues, organs and organ systems (blood pressure, body temperature, sensory perception, feeding)
Population	Up to thousands of years	Population biology, population genetics, ecology	Statistical analysis of variation, abundance, geographical distribution	Social structures, systems of mating, age distribution of organisms, levels of variation, action of natural selection
Species	Thousands to millions of years	Systematics and evolutionary biology, community ecology	Study of reproductive barriers, phylogeny, paleontology, ecological interactions	Method of reproduction, reproductive barriers

mine the properties of the whole. Many different spoken languages have emerged in human culture from the same basic anatomical structures that permit hearing and speech. The freedom of the parts to interact in different ways makes possible a great diversity of potential emergent properties at each level of the biological hierarchy.

Different levels of the biological hierarchy and their particular emergent properties are products of evolution. Before multicellular organisms evolved, there was no distinction between the organismal and cellular levels, and it is still absent from single-celled organisms (Chapter 11). The diversity of emergent properties that we see at all levels of the biological hierarchy contributes to the difficulty of giving life a simple definition or description.

3. **Reproduction.** *Living systems can reproduce themselves.* Life does not arise spontaneously but comes only from prior life, through a process of reproduction. Although life certainly originated from nonliving matter at least once (Chapter 2), this required enormously long periods of time and conditions very different from those of the current biosphere. At each level of the biological hierarchy, living forms reproduce to generate others like themselves (Figure 1-5). Genes are replicated to produce new genes. Cells divide to produce new cells. Organisms reproduce, sexually or asexually, to produce new organisms (Chapter 5). Populations can become fragmented to give rise to new populations, and species can give rise to new species through a process known as speciation. Reproduction at any level of the hierarchy usually features an increase in numbers. Individual genes, cells, organisms, populations, or species may fail to reproduce themselves, but reproduction is nonetheless an expected property of these individuals.

Reproduction at each of these levels features the complementary, and yet apparently contradictory, phenomena of **heredity** and **variation.** Heredity is the faithful transmission of traits from parents to offspring, usually (but not necessarily) observed at the organismal level. Variation is the production of *differences* among the traits of different individuals. In the reproductive process, the properties of descendants resemble those of their parents to varying degrees but usually are not identical to them. Replication of deoxyribonucleic acid (DNA) occurs with high fidelity, but errors occur at repeatable rates. Cell division is an exceptionally precise process, especially with regard to the nuclear material, but chromosomal changes occur nonetheless at measurable rates. Organismal reproduction likewise demonstrates both heredity and variation, the latter being obvious especially in sexually reproducing forms. The production of new populations and species also demonstrates conservation of some properties and changes of others. Two closely related frog species may have similar mating calls but differ in the rhythm of repeated sounds.

We will see later in this book that the interaction of heredity and variation in the reproductive process is the basis for organic evolution (Chapter 6). If heredity were perfect, living systems would never change; if variation were uncontrolled by heredity, biological systems would lack the stability that allows them to persist through time.

4. **Possession of a genetic program.** *A genetic program provides fidelity of inheritance* (Figure 1-6). The structures of the protein molecules needed for organismal development and functioning are encoded in **nucleic acids** (Chapter 5). For animals and most other organisms, genetic information is contained in **DNA.** DNA is a very long, linear chain of

A

B

C

D

Figure 1-5

Reproductive processes observed at four different levels of biological complexity: **A,** Molecular level—electron micrograph of a replicating DNA molecule; **B,** Cellular level—micrograph of cell division at mitotic telophase; **C,** Organismal level—a king snake hatching; **D,** Species level—formation of new species in the sea urchin (*Eucidaris*) after geographic separation of Caribbean (*E. tribuloides*) and Pacific (*E. thouarsi*) populations by the formation of a land bridge.

subunits called nucleotides, each of which contains a sugar phosphate (deoxyribose phosphate) and one of four nitrogenous bases (adenine, cytosine, guanine, or thymine, abbreviated A, C, G, and T, respectively). The sequence of nucleotide bases represents a code for the order of amino acids in the protein specified by the DNA molecule. The correspondence between the sequence of bases in DNA and the sequence of amino acids in a protein is known as the **genetic code.**

The genetic code was established early in the evolutionary history of life, and the same code is present in bacteria and in the nuclear genomes of almost all animals and plants. The near constancy of this code among living forms provides strong evidence for a single origin of life. The genetic code has undergone very little evolutionary change since its origin because an alteration would disrupt the structure of nearly every protein, which would in turn severely disrupt cellular functions that require very specific protein structures. Only in the rare instance in which the altered protein structures are still compatible with their cellular functions would such a change have a chance to survive and be reproduced. Evolu-

tionary change in the genetic code has occurred in the DNA contained in animal mitochondria, the organelles that regulate cellular energy. The genetic code in animal mitochondrial DNA therefore is slightly different from the standard code of nuclear and bacterial DNA. Because mitochondrial DNA specifies far fewer proteins than nuclear DNA, the likelihood of getting a change in the code that does not disrupt cellular functions is greater there than in the nucleus.

5. **Metabolism.** *Living organisms maintain themselves by obtaining nutrients from their environments* (Figure 1-7). The nutrients are broken down to obtain chemical energy and molecular components for use in building and maintaining the living system (Chapter 4). We call these essential chemical processes **metabolism.** They include digestion, production of energy (respiration), and synthesis of molecules and structures. Metabolism is often viewed as an interaction of destructive (catabolic) and constructive (anabolic) reactions. The most fundamental anabolic and catabolic chemical processes used by living systems arose early in the evolutionary history of life and are shared by all living

A

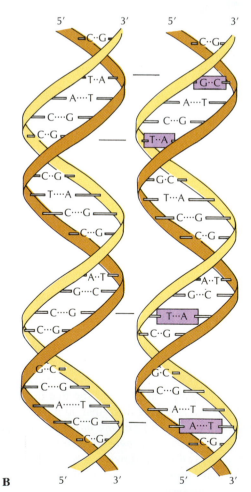

B

Figure 1-6

James Watson and Francis Crick with a model of the DNA double helix **(A)**. Genetic information is coded in the nucleotide base sequence inside the DNA molecule. Genetic variation is shown **(B)** in DNA molecules that are similar in base sequence but differ from each other at four positions. Such differences can encode alternative traits, such as different eye colors.

A

B

Figure 1-7

Feeding processes illustrated by **(A)** an ameba surrounding food and **(B)** a chameleon capturing insect prey with its projectile tongue.

forms. These include synthesis of carbohydrates, lipids, nucleic acids, and proteins and their constituent parts and the cleavage of chemical bonds to recover energy stored in them. In animals, many fundamental metabolic reactions occur at the cellular level, often in specific organelles found throughout the animal kingdom. Cellular respiration occurs, for example, in mitochondria. Cellular and nuclear membranes regulate metabolism by controlling the movement of molecules across the cellular and nuclear boundaries, respectively. The study of the performance of complex metabolic functions is known as **physiology.** We devote a large portion of this book to describing and comparing the diverse tissues, organs, and organ systems that different groups of animals have evolved to perform the basic physiological functions of life (Chapters 11 through 36).

6. **Development.** *All organisms pass through a characteristic life cycle.* Development describes the characteristic changes that an organism undergoes from its origin (usually the fertil-

A   **B**

Figure 1-8

A, Adult monarch butterfly emerging from its pupal case; **B,** Fully formed adult monarch butterfly.

ization of an egg by sperm) to its final adult form (Chapter 8). Development usually features changes in size and shape, and differentiation of structures within an organism. Even the simplest one-celled organisms grow in size and replicate their component parts until they divide into two or more cells. Multicellular organisms undergo more dramatic changes during their lives. Different developmental stages of some multicellular forms are so dissimilar that they are hardly recognizable as belonging to the same species. Embryos are distinctly different from juvenile and adult forms into which they will develop. Even postembryonic development of some organisms includes stages that are dramatically different from each other. The transformation that occurs from one stage to another is called **metamorphosis.** There is little resemblance, for example, among the egg, larval, pupal, and adult

stages of metamorphic insects (Figure 1-8). Among animals, the early stages of development are often more similar among organisms of related species than are later developmental stages. In our survey of animal diversity, we will describe all stages of observed life histories, but we will concentrate on adult stages in which diversity both within and between different animal groups tends to be greatest.

7. **Environmental interaction.** *All animals interact with their environments.* The study of organismal interaction with the environment is known as **ecology.** Of special interest are the factors that influence geographic distribution and abundance of animals (Chapters 37 and 38). The science of ecology permits us to understand how an organism can perceive environmental stimuli and respond in appropriate ways by adjusting its metabolism and physiology (Figure 1-9). All organisms respond to stimuli in their environment, a property called **irritability.** The stimulus and response may be simple, such as a unicellular organism moving from or toward a light source or away from a noxious substance, or it may be quite complex, such as a bird responding to a complicated series of signals in a mating ritual (see Chapter 36). Life and the environment are inseparable. We cannot isolate the evolutionary history of a lineage of organisms from the environments in which it occurred.

Life Obeys Physical Laws

To untrained observers, these seven properties of life may appear to violate basic laws of physics. Vitalism, the idea that life is endowed with a mystical vital force that violates physical and chemical laws, was once widely advocated. Biological research has consistently rejected vitalism, showing instead that all living systems obey basic laws of physics and chemistry. Laws governing energy and its transformations (thermodynamics) are particularly important for understanding life (Chapter 4). The **first law of thermodynamics** is the law of conservation of energy. Energy is neither created nor destroyed, but it can be

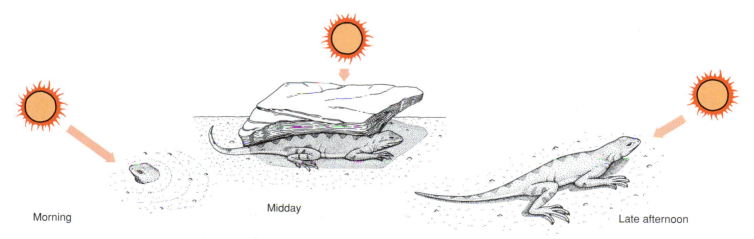

Morning   Midday   Late afternoon

Figure 1-9

A lizard regulates its body temperature by choosing different locations (microhabitats) at different times of day.

transformed from one form to another. All aspects of life require energy and its transformation. The energy to support life on earth flows from the fusion reactions in our sun and reaches the earth as light and heat. Sunlight is captured by green plants and cyanobacteria and transformed by photosynthesis into chemical bonds. Energy in chemical bonds is a form of potential energy that can be released when the bond is broken; the energy is used to perform numerous cellular tasks. Energy transformed and stored in plants is then used by the animals that eat the plants, and these animals may in turn provide energy for other animals that eat them.

The **second law of thermodynamics** states that physical systems tend to proceed toward a state of greater disorder, or **entropy.** Energy obtained and stored by plants is subsequently released by a variety of mechanisms and finally dissipated as heat. The high degree of molecular organization found in living cells is attained and maintained only as long as energy fuels the organization. The ultimate fate of materials in the cells is degradation and dissipation of their chemical bond energy as heat. The process of evolution whereby organismal complexity can increase over time may appear at first to violate the second law of thermodynamics, but it does not. Organismal complexity is achieved and maintained only by the constant use and dissipation of energy flowing into the biosphere from the sun. Survival, growth, and reproduction of animals require energy that comes from breaking complex food molecules into simple organic waste products. The processes by which animals acquire energy through nutrition and respiration command the attention of the many physiological sciences.

ZOOLOGY AS A PART OF BIOLOGY

Animals form a distinct branch on the evolutionary tree of life. It is a large and old branch that originated in the Precambrian seas over 600 million years ago. Animals form part of an even larger limb known as **eukaryotes,** organisms whose cells contain membrane-enclosed nuclei. This larger limb includes plants, fungi and a host of unicellular forms. Perhaps the most distinctive characteristic of the animals as a group is their means of nutrition, which consists of eating other organisms. This basic way of life has led to the evolution of many diverse systems for locomotion and for capturing and processing a wide array of food items.

Animals can be distinguished also by the absence of properties that have evolved in other eukaryotes. Plants, for example, have evolved the ability to use light energy to produce organic compounds (photosynthesis), and they have evolved rigid cell walls that surround their cell membranes; photosynthesis and cell walls are absent from animals. Fungi have evolved the ability to acquire nutrition by absorption of small organic molecules from their environment, and they have a body plan consisting of tubular filaments called *hyphae;* structures of this kind are absent from the animal kingdom.

Some organisms combine the properties of animals and plants. For example, *Euglena* (Figure 1-10) is a motile, single-celled organism that resembles plants in being photosynthetic, but it resembles animals in its ability to eat food particles.

Figure 1-10
Some organisms, such as the single-celled *Euglena* (shown here) and *Volvox* (see Figure 1-3), combine properties that are normally associated with both animals (motility) and plants (photosynthetic ability).

Euglena is part of a separate eukaryotic lineage that diverged from those of plants and animals early in the evolutionary history of eukaryotes. *Euglena* and other unicellular eukaryotes are sometimes grouped into the kingdom Protista, although this kingdom is an arbitrary grouping of unrelated lineages that violates taxonomic principles (see Chapter 10).

The fundamental structural and developmental features evolved by the animal kingdom are presented in detail in Chapters 8 and 9.

PRINCIPLES OF SCIENCE

Nature of Science

We stated in the first sentence of this chapter that zoology is the scientific study of animals. A basic understanding of zoology therefore requires an understanding of what science is, what it is not, and how knowledge is gained using the scientific method.

Science is a way of asking questions about the natural world and sometimes obtaining precise answers to them. Although science, in the modern sense, has arisen recently in human history (within the last 200 years or so), the tradition of asking questions about the natural world is an ancient one. In this section we examine the methodology that zoology shares with science as a whole. These features distinguish sciences from activities that we exclude from the realm of science, such as art and religion.

Despite the enormous impact of science on our lives, many people have only a minimal understanding of the real nature of science. For example, on March 19, 1981, the governor of Arkansas signed into law the Balanced Treatment for Creation-Science and Evolution-Science Act (Act 590 of 1981). This act falsely presented "creation-science" as a valid scientific endeavor. "Creation-science" is actually a religious position advocated by a minority of the American religious community, and it does not qualify as science. The enactment of this law led to a historic lawsuit tried in December 1981 in the court of Judge William R. Overton, U.S. District Court, Eastern District of Arkansas. The suit was brought by the American Civil Liberties Union on behalf of 23 plaintiffs, including a number of religious leaders and groups representing several denominations, individual parents, and educational associations. The plaintiffs contended that the law was a violation of the First Amendment to the U.S. Constitution, which prohibits "establishment of religion" by the government. This prohibition includes passing a law that would aid one religion or prefer one religion over another. On January 5, 1982, Judge Overton permanently enjoined the State of Arkansas from enforcing Act 590.

Considerable testimony during the trial dealt with the nature of science. Some witnesses defined science simply, if not very informatively, as "what is accepted by the scientific community" and "what scientists do." However, on the basis of other testimony by scientists, Judge Overton was able to state explicitly these essential characteristics of science:

1. It is guided by natural law.
2. It has to be explanatory by reference to natural law.
3. It is testable against the observable world.
4. Its conclusions are tentative and therefore not necessarily the final word.
5. It is falsifiable.

The pursuit of scientific knowledge must be guided by the physical and chemical laws that govern the state of existence. Scientific knowledge must explain what is observed by reference to natural law without requiring the intervention of a supernatural being or force. We must be able to observe events in the real world, directly or indirectly, to test hypotheses about nature. If we draw a conclusion relative to some event, we must be ready always to discard or to modify our conclusion if further observations contradict it. As Judge Overton stated, "While anybody is free to approach a scientific inquiry in any fashion they choose, they cannot properly describe the methodology used as scientific if they start with a conclusion and refuse to change it regardless of the evidence developed during the course of the investigation." Science is separate from religion, and the results of science do not favor one religious position over another.

Scientific Method

These essential criteria of science form the basis for the **hypothetico-deductive method.** The first step of this method is the generation of hypotheses or potential answers to the question being asked. These hypotheses are usually based on prior observations of nature or derived from theories based on such observations. Scientific hypotheses often constitute general statements about nature that may explain a large number of diverse observations. Darwin's hypothesis of natural selection, for example, explains the observations that many different species have properties that adapt them to their environments. On the basis of the hypothesis, a scientist must make a prediction about future observations. The scientist must say, "If my hypothesis is a valid explanation of past observations, then future observations ought to have certain characteristics." The best hypotheses are those that make many predictions which, if found erroneous, will lead to rejection, or falsification, of the hypothesis.

The hypothesis of natural selection was invoked to explain variation observed in British moth populations (Figure 1-11). In industrial areas of England having heavy air pollution, many populations of moths contain primarily darkly pigmented (melanic) individuals, whereas moth populations inhabiting clean forests

A

B

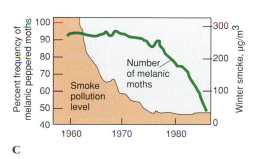

C

Figure 1-11

Light and melanic forms of the peppered moth, *Biston betularia* on **A,** a lichen-covered tree in unpolluted countryside and **B,** a soot-covered tree near industrial Birmingham, England. These color variants have a simple genetic basis. **C,** Recent decline in the frequency of the melanic form of the peppered moth with falling air pollution in industrial areas of England. The frequency of the melanic form still exceeded 90% in 1960, when smoke and sulfur dioxide emissions were still high. Later, as emissions fell and light-colored lichens began to grow again on the tree trunks, the melanic form became more conspicuous to predators. By 1986, only 50% of the moths were still of the melanic form, the rest having been replaced by the light form.

The Animal Rights Controversy

In recent years, the debate surrounding the use of animals to serve human needs has intensified. Most controversial of all is the issue of animal use in biomedical and behavioral research and in testing commercial products.

A few years ago, Congress passed a series of amendments to the Federal Animal Welfare Act, a body of laws covering animal care in laboratories and other facilities. These amendments have become known as the three R's: **Reduction** in the number of animals needed for research; **Refinement** of techniques that might cause stress or suffering; **Replacement** of live animals with simulations or cell cultures whenever possible. As a result, the total number of animals used each year in research and in testing of commercial products has declined. Developments in cellular and molecular biology also have contributed to a decreased use of animals for research and testing. The animal rights movement, composed largely of vocal antivivisectionists, has created an awareness of the needs of animals used in research and has stimulated researchers to discover cheaper, more efficient, and more humane alternatives.

However, computers and culturing of cells can simulate the effects on organismal systems of, for instance, drugs, only when the basic principles involved are well known. When the principles themselves are being scrutinized and tested, computer modeling is not sufficient. A recent report by the National Research Council concedes that although the search for alternatives to the use of animals in research and testing will continue, "the chance that alternatives will completely replace animals in the foreseeable future is nil." Realistic immediate goals, however, are reduction in number of animals used, replacement of mammals with other vertebrates, and refinement of experimental procedures to reduce discomfort of the animals being tested.

Medical and veterinary progress depends on research using animals. Every drug and every vaccine developed to improve the human condition has been tested first on animals. Research using animals has enabled medical science to eliminate smallpox and polio, and to immunize against diseases previously common and

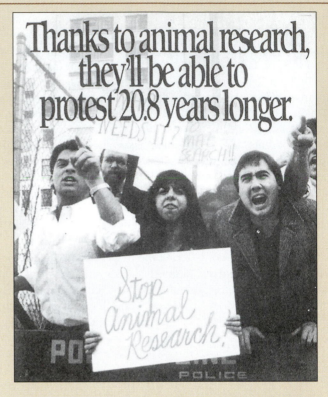

According to the U.S. Department of Health and Human Services, animal research has helped extend our life expectancy by 20.8 years.

often deadly, including diphtheria, mumps, and rubella. It also has helped to create treatments for cancer, diabetes, heart disease, and manic-depressive psychoses, and to develop surgical procedures including heart surgery, blood transfusions, and cataract removal. AIDS research is wholly dependent on studies using animals. The

---

show a much higher frequency of lightly pigmented individuals. The hypothesis suggests that moths can survive most effectively by matching their surroundings, thereby remaining invisible to birds that seek to eat them. Experimental studies have shown that, consistent with this hypothesis, birds are able to locate and then to eat moths that do not match their surroundings. Birds in the same area frequently fail to find moths that match their surroundings, leaving them to reproduce and to increase their numbers relative to conspicuous moths. Another testable prediction of the hypothesis of natural selection is that when polluted areas are cleaned, the moth populations should demonstrate an increase in frequency of lightly pigmented individuals. Observations of such populations confirmed the result predicted by natural selection.

If a hypothesis is very powerful in explaining a wide variety of related phenomena, it attains the status of a theory. Natural selection is a good example. Our example of the use of natural selection to explain observed pigmentation patterns in moth populations is only one of many phenomena to which natural

selection applies. Natural selection provides a potential explanation for the occurrence of many different traits distributed among virtually all animal species. Each of these instances constitutes a specific hypothesis generated from the theory of natural selection. Note, however, that falsification of a specific hypothesis does not necessarily lead to rejection of the theory as a whole. Natural selection may fail to explain origins of human behavior, for example, but it provides an excellent explanation for many structural modifications of the pentadactyl (five-fingered) vertebrate limb for diverse functions. Scientists test many subsidiary hypotheses of their major theories to ask whether their theories are generally applicable. Most useful are theories that explain the largest array of different natural phenomena.

We emphasize that the meaning of the word "theory," when used by scientists, is not "speculation" as it is in ordinary English usage. Failure to make this distinction has been prominent in creationist challenges to evolution. The creationists have spoken of evolution as "only a theory," as if it were little better than a guess.

similarity of simian AIDS, identified in rhesus monkeys, to human AIDS has permitted the disease in monkeys to serve as a model for the human disease. Recent work indicates that cats, too, may be useful models for development of an AIDS vaccine. Skin grafting experiments, first done with cattle and later with other animals, opened a new era in immunological research with vast contributions to treatment of disease in humans and other animals.

Research using animals also has benefited *other animals* through the development of veterinary cures. The vaccines for feline leukemia and canine parvovirus were first introduced to other cats and dogs. Many other vaccinations for serious diseases of animals were developed through research on animals—for example, rabies, distemper, anthrax, hepatitis, and tetanus. No endangered species is used in general research (except to protect that species from total extinction). Thus, research using animals has provided enormous benefits to humans and other animals. Still, much remains to be learned about treatment of diseases such as cancer, AIDS, diabetes, and heart disease, and research with animals will be required for this purpose.

Despite the remarkable benefits produced by research on animals, advocates of animal rights often present an inaccurate and emotionally distorted picture of this research. The ultimate goal of most animal rights activists, who have focused specifically on the use of animals in science rather than on treatment of animals in all contexts, remains the total abolition of all forms of research using animals. The scientific community is deeply concerned about the impact of these attacks on the ability of scientists to conduct important experiments that will benefit people and animals. They argue that if we are justified to use animals for food and fiber and as pets, we are justified in experimentation to benefit human welfare when these studies are conducted humanely and ethically.

The Association for Assessment and Accreditation of Laboratory Animal Care International supports the use of animals to advance medicine and science when nonanimal alternatives are not available and when animals are treated in an ethical and humane way. Accreditation by this organization allows research institutions to demonstrate excellence in their standards of animal care. Nearly all major institutions receiving funding from the National Institutes of Health have sought and received this accreditation. See the website at *http://www.aaalac.org* for more information on accreditation of laboratory animal care.

References on Animal Rights Controversy

Commission on Life Sciences, National Research Council. 1988. Use of laboratory animals in biomedical and behavioral research. Washington, D.C., National Academy Press. *Statement of national policy on guidelines for the use of animals in biomedical research. Includes a chapter on benefits derived from use of animals.*

Goldberg, A. M., and J. M. Frazier. 1989. Alternatives to animals in toxicity testing. Sci. Am. **261:**24–30 (Aug.). *Describes alternatives that are being developed for the costly and time-consuming use of animals in the testing of thousands of chemicals that each year must be evaluated for potential toxicity to humans.*

Pringle, L. 1989. The animal rights controversy. San Diego, California, Harcourt Brace Jovanovich, Publishers. *Although no one writing about the animal rights movement can honestly claim to be totally objective and impartial on such an emotionally charged issue, this book comes as close as any to presenting a balanced treatment.*

Rowan, A. N. and F. M. Loew. 1995. The animal rights controversy: protest, process and public policy: an analysis of strategic issues. North Grafton, Massachusetts, Center for Animals and Public Policy, Tufts University School of Veterinary Medicine. *A good review of the issues.*

Sperling, S. 1988. Animal liberators: research and morality. Berkeley, University of California Press. *Thoughtful and carefully researched study of the animal rights movement, its ideological roots, and the passionate idealism of animal rights activists.*

In fact, the theory of evolution is supported by such massive evidence that biologists view repudiation of evolution as tantamount to repudiation of reason. Nonetheless, evolution, along with all other theories in science, is not proven in a mathematical sense, but it is testable, tentative, and falsifiable. Powerful theories that guide extensive research are called **paradigms.** The history of science has shown that even major paradigms are subject to refutation and replacement when they fail to account for our observations of the natural world. They are then replaced by new paradigms in a process known as a **scientific revolution.** For example, prior to the 1800s, animal species were studied as if they were specially created entities whose essential properties remained unchanged through time. Darwin's theories led to a scientific revolution that replaced these views with the evolutionary paradigm. The evolutionary paradigm has guided biological research for more than 130 years, and to date there is no scientific evidence that falsifies it; it continues to guide active inquiry into the natural world, and it is generally accepted as the cornerstone of biology.

Experimental versus Evolutionary Sciences

The many questions asked about the animal world since the time of Aristotle can be grouped into two major categories.* The first category seeks to understand the **proximate** or **immediate causes** that underlie the functioning of biological systems at a particular time and place. These include the problems of explaining how animals perform their metabolic, physiological, and behavioral functions at the molecular, cellular, organismal, and even populational levels. For example, how is genetic information expressed to guide the synthesis of proteins? What causes cells to divide to produce new cells? How does population density affect the physiology and behavior of organisms?

---

*Mayr, E. 1982. *The Growth of Biological Thought.* Cambridge, Harvard University Press, pp. 67–71.

The biological sciences that address proximate causes are known as **experimental sciences,** and they proceed using the experimental method. This method consists of three steps: (1) predicting how a system being studied will respond to a disturbance, (2) making the disturbance, and then (3) comparing the observed results with the predicted ones. Experimental conditions are repeated to eliminate chance occurrences that might produce erroneous conclusions. **Controls**—repetitions of the experimental procedure that lack the disturbance—are established to eliminate unperceived factors that may bias the outcome of the experiment. The processes by which animals maintain a body temperature under different environmental conditions, digest their food, migrate to new habitats, or store energy are some additional examples of physiological phenomena studied by experiment (Chapters 29 through 36). Subfields of biology that constitute experimental sciences include molecular biology, cell biology, endocrinology, developmental biology, and community ecology.

In contrast to questions concerning the proximate causes of biological systems are questions of the **ultimate causes** that have produced these systems and their distinctive characteristics through evolutionary time. For example, what are the evolutionary factors that caused some birds to acquire complex patterns of seasonal migration between temperate and tropical areas? Why do different species of animals have different numbers of chromosomes in their cells? Why do some animal species maintain complex social systems, whereas individuals of other species are largely solitary?

The biological sciences that address questions of ultimate cause are known as **evolutionary sciences,** and they proceed largely using the **comparative method** rather than experimentation. Characteristics of molecular biology, cell biology, organismal structure, development, and ecology are compared among related species to identify their patterns of variation. The patterns of similarity and dissimilarity are then used to test hypotheses of relatedness, and thereby to reconstruct the evolutionary tree that relates the species being studied. The evolutionary tree is then used to examine hypotheses of the evolutionary origins of the diverse molecular, cellular, organismal, and populational properties observed in the animal world. Clearly, evolutionary sciences rely on results of experimental sciences as a starting point. Evolutionary sciences include comparative biochemistry, molecular evolution, comparative cell biology, comparative anatomy, comparative physiology, and phylogenetic systematics.

THEORIES OF EVOLUTION AND HEREDITY

We turn now to a specific consideration of the two major paradigms that guide zoological research today: Darwin's theory of evolution and the chromosomal theory of inheritance.

Darwin's Theory of Evolution

Darwin's theory of evolution is now over 140 years old (Chapter 6). Darwin articulated the complete theory when he published his famous book *On the Origin of Species by Means of Natural Selection* in England in 1859 (Figure 1-12). Biologists today are

Figure 1-12

Modern evolutionary theory is strongly identified with Charles Robert Darwin, who, with Alfred Russel Wallace, provided the first credible explanation of evolution. This photograph of Darwin was taken in 1854 when he was 45 years old. His most famous book, *On the Origin of Species,* appeared five years later.

frequently asked, "What is Darwinism?" and "Do biologists still accept Darwin's theory of evolution?" These questions cannot be given simple answers, because Darwinism encompasses several different, although mutually compatible, theories. Professor Ernst Mayr of Harvard University has argued that Darwinism should be viewed as five major theories.* These five theories have somewhat different origins and different fates and cannot be discussed accurately as if they were only a single statement. The theories are (1) perpetual change, (2) common descent, (3) multiplication of species, (4) gradualism, and (5) natural selection. The first three theories are generally accepted as having universal application throughout the living world. The theories of gradualism and natural selection are controversial among evolutionists, although both are strongly advocated by a large portion of the evolutionary community and are important components of the Darwinian evolutionary paradigm. Gradualism and natural selection are clearly part of the evolutionary process, but their explanatory power might not be as widespread as Darwin intended. Legitimate controversies regarding gradualism and natural selection often are misrepresented by creationists as challenges to the first three theories presented above, although the validity of those first three theories is strongly supported by all relevant observations.

1. **Perpetual change.** This is the basic theory of evolution on which the others are based. It states that the living world is neither constant nor perpetually cycling, but is always changing. The properties of organisms undergo transformation

---

*Mayr, E. 1985. Chapter 25 in D. Kohn, ed. *The Darwinian Heritage.* Princeton, Princeton University Press.

across generations throughout time. This theory originated in antiquity but did not gain widespread acceptance until Darwin advocated it in the context of his other four theories. "Perpetual change" is documented by the fossil record, which clearly refutes creationists' claims for a recent origin of all living forms. Because it has withstood repeated testing and is supported by an overwhelming number of observations, we now regard "perpetual change" as a scientific fact.

2. **Common descent.** The second Darwinian theory, "common descent," states that all forms of life descended from a common ancestor through a branching of lineages (Figure 1-13). The opposing argument, that the different forms of life arose independently and descended to the present in linear, unbranched genealogies, has been refuted by comparative studies of organismal form, cell structure, and macromolecular structures (including those of the genetic material, DNA). All of these studies confirm the theory that life's history has the structure of a branching evolutionary tree, known as a **phylogeny.** Species that share relatively recent common ancestry have more similar features at all levels than do

species that have only an ancient common ancestry. Much current research is guided by Darwin's theory of common descent toward reconstructing life's phylogeny using the patterns of similarity and dissimilarity observed among species. The resulting phylogeny serves as the basis for our taxonomic classification of animals (Chapter 10).

3. **Multiplication of species.** Darwin's third theory states that the evolutionary process produces new species by splitting and transforming older ones. Species are now generally viewed as reproductively distinct populations of organisms that usually but not always differ from each other in organismal form. Once species are fully formed, interbreeding among members of different species does not occur. Evolutionists generally agree that the splitting and transformation of lineages produces new species, although there is still much controversy concerning details of this process (Chapter 6) and the precise meaning of the term "species" (Chapter 10). The study of the historical processes that generate new species guides much active scientific research.

4. **Gradualism.** Gradualism states that the large differences in anatomical traits that characterize diverse species originate through the accumulation of many small incremental changes over very long periods of time. This theory is important because genetic changes having very large effects on organismal form are usually harmful to an organism. It is possible, however, that some genetic variants that have large effects on an organism are nonetheless sufficiently beneficial to be favored by natural selection. Therefore, although gradual evolution is known to occur, it may not explain the origins of all structural differences that we observe among species (Figure 1-14). Scientists are still actively studying this question.

5. **Natural selection.** Natural selection, Darwin's most famous theory, rests on three propositions. First, there is variation among organisms (within populations) for anatomical, behavioral, and physiological traits. Second, the variation is at least partly heritable so that offspring tend to resemble their parents. Third, organisms with different variant forms are expected to leave different numbers of offspring to future generations. Variants that permit their possessors most effectively to exploit their environments will preferentially survive to be transmitted to future generations. Over many generations, favorable new traits will spread throughout a population. Accumulation of such changes leads, over long periods of time, to production of new organismal features and new species. Natural selection is therefore a creative process that generates novel features from the small individual variations that occur among organisms within a population.

Figure 1-13

An early tree of life drawn in 1874 by the German biologist, Ernst Haeckel, who was strongly influenced by Darwin's theory of common descent. Many of the phylogenetic hypotheses shown in this tree, including the unilateral progression of evolution toward humans (= Menschen, *top*), have since been refuted.

Natural selection explains why organisms are constructed to meet the demands of their environments, a phenomenon called **adaptation** (Figure 1-15). Adaptation is the expected result of a process that accumulates the most favorable variants occurring in a population throughout long periods of evolutionary time. Adaptation was viewed previously as strong evidence against evolution, and Darwin's theory of natural selection was therefore important for convincing people that a natural process,

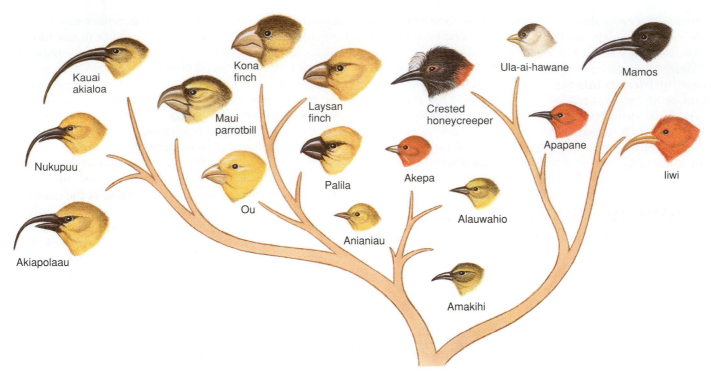

Figure 1-14

Gradualism provides a plausible explanation for the origin of different bill shapes in the Hawaiian honeycreepers shown here. This theory has been challenged, however, as an explanation of the evolution of such structures as vertebrate scales, feathers, and hair from a common ancestral structure. The geneticist Richard Goldschmidt viewed the latter forms as unbridgeable by any gradual transformation series.

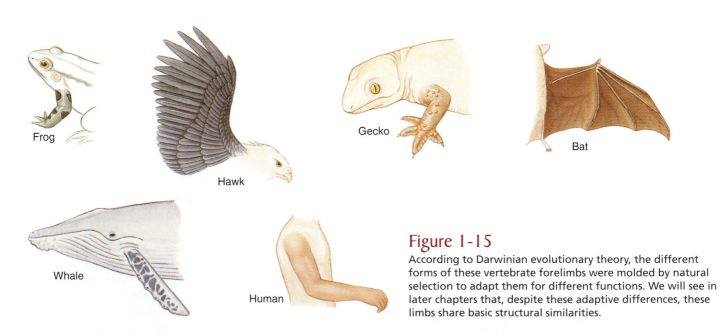

Figure 1-15

According to Darwinian evolutionary theory, the different forms of these vertebrate forelimbs were molded by natural selection to adapt them for different functions. We will see in later chapters that, despite these adaptive differences, these limbs share basic structural similarities.

capable of being studied scientifically, could produce new species. The demonstration that natural processes could produce adaptation was important to the eventual acceptance of all five Darwinian theories.

Darwin's theory of natural selection faced a major obstacle when it was first proposed: it lacked a theory of heredity. People assumed incorrectly that heredity was a blending process, and that any favorable new variant appearing in a population

therefore would be lost. The new variant arises initially in a single organism, and that organism therefore must mate with one lacking the favorable new trait. Under blending inheritance, the organism's offspring would then have only a diluted form of the favorable trait. These offspring likewise would mate with others that lack the favorable trait. With its effects diluted by half each generation, the trait eventually would cease to exist. Natural selection would be completely ineffective in this situation.

Darwin was never able to counter this criticism successfully. It did not occur to Darwin that hereditary factors could be discrete and nonblending and that a new genetic variant therefore could persist unaltered from one generation to the next. This principle is known as **particulate inheritance.** It was established after 1900 with the discovery of Gregor Mendel's genetic experiments, and it was eventually incorporated into what we now call the **chromosomal theory of inheritance.** We use the term **neo-Darwinism** to describe Darwin's theories as modified by incorporating this theory of inheritance.

Mendelian Heredity and the Chromosomal Theory of Inheritance

The chromosomal theory of inheritance is the foundation for current studies of genetics and evolution in animals (Chapters 5 and 6). This theory comes from the consolidation of research done in the fields of genetics, which was founded by the experimental work of Gregor Mendel (Figure 1-16), and cell biology.

Genetic Approach

The genetic approach consists of mating or "crossing" populations of organisms that are true-breeding for contrasting traits, and then following hereditary transmission of those traits through subsequent generations. "True-breeding" means that a population maintains across generations only one of the contrasting states of a particular feature when propagated in isolation from other populations.

Gregor Mendel studied the transmission of seven variable features in garden peas, crossing populations that were true-breeding for alternative traits (for example, tall versus short plants). In the first generation (called the F_1 generation, for "filial"), only one of the alternative parental traits was observed; there was no indication of blending of the parental traits. In the example, the offspring (called F_1 hybrids) formed by crossing the tall and short plants were tall, regardless of whether the tall trait was inherited from the male or the female parent. These F_1 hybrids were allowed to self-pollinate, and both parental traits were found among their offspring (called the F_2 generation), although the trait observed in the F_1 hybrids (tall plants in this example) was three times more common than the other trait. Again, there was no indication of blending of the parental traits (Figure 1-17).

Mendel's experiments showed that the effects of a genetic factor can be masked in a hybrid individual, but that these factors are not physically altered during the transmission process. He postulated that variable traits are specified by paired hereditary factors, which we now call "genes." When **gametes** (eggs or sperm) are produced, the two genes controlling a particular feature are segregated from each other and each gamete receives only one of them. Fertilization restores the paired condition. If an organism possesses different forms of the paired genes for a feature, only one of them is expressed in its appearance, but both genes nonetheless are transmitted unaltered in equal numbers to the gametes produced. Transmission of these genes is particulate, not blending. Mendel observed that inheritance of one pair of traits is independent of inheritance of other paired traits. We now know, however, that not all pairs of traits are inherited independently of each other. Numerous studies, particularly of the fruit fly, *Drosophila melanogaster,* have shown that principles of inheritance discovered initially in plants apply also to animals.

A

B

Figure 1-16

A, Gregor Johann Mendel. **B,** The monastery in Brno, Czech Republic, now a museum, where Mendel performed his experiments with garden peas.

PARTICULATE INHERITANCE (observed)

BLENDING INHERITANCE (not observed)

Figure 1-17

Different predictions of particulate versus blending inheritance regarding the outcome of Mendel's crosses of tall and short plants. The prediction of particulate inheritance is upheld and the prediction of blending inheritance is falsified by the results of the experiments. The reciprocal experiments (crossing short female parents with tall male parents) produced similar results. (P_1 = parental generation; F_1 = first filial generation; F_2 = second filial generation.)

Figure 1-18

An early nineteenth-century micrographic drawing of sperm from (1) guinea pig, (2) white mouse, (3) hedgehog, (4) horse, (5) cat, (6) ram, and (7) dog (Prévost and Dumas, 1821). Some biologists initially interpreted these as parasitic worms in the semen, but on further examination found them to be male gametes.

Figure 1-19

Paired chromosomes being separated before nuclear division in the process of forming gametes.

Contributions of Cell Biology

Improvements in microscopes during the 1800s permitted cytologists to study the production of gametes by direct observation of reproductive tissues. Interpreting the observations was initially difficult, however. Some prominent biologists hypothesized, for example, that sperm were parasitic worms in the semen (Figure 1-18). This hypothesis was soon falsified, and the true nature of gametes was clarified. As the precursors of gametes prepare to divide early in gamete production, the nuclear material condenses to reveal discrete, elongate structures called chromosomes. Chromosomes occur in pairs that are usually similar but not identical in appearance and informational content. The number of chromosomal pairs varies among species. One member of each pair is derived from the female parent and the other from the male parent. Paired chromosomes are physically associated and then segregated into different daughter cells during cell division prior to gamete formation (Figure 1-19). Each resulting gamete receives one chromosome from each pair. Different pairs of chromosomes are sorted into gametes independently of each other. Because the behavior of chromosomal material during gamete formation parallels that postulated for Mendel's genes, Sutton and Boveri in 1903 through 1904 hypothesized that chromosomes were the physical bearers of the genetic material. This hypothesis met with extreme skepticism when first proposed. A long series of tests designed to falsify it nonetheless showed that its predictions were upheld. The chromosomal theory of inheritance is now well established.

SUMMARY

Zoology is the scientific study of animals, and it is part of biology, the scientific study of life. Animals and life in general can be identified by attributes that they have acquired over their long evolutionary histories. The most outstanding attributes of life include chemical uniqueness, complexity and hierarchical organization, reproduction, possession of a genetic program, metabolism, development, and interaction with the environment. Biological systems comprise a hierarchy of integrative levels (molecular, cellular, organismal, populational, and species levels), each of which demonstrates a number of specific emergent properties.

Science is characterized by the acquisition of knowledge by constructing and then testing hypotheses through observations of the natural world. Science is guided by natural law, and its hypotheses are testable, tentative, and falsifiable. Zoological sciences can be subdivided into two categories, experimental sciences and evolutionary sciences. Experimental sciences use the experimental method to ask how animals perform their basic metabolic, developmental, behavioral, and reproductive functions, including investigations of their molecular, cellular, and populational systems. Evolutionary sciences use the comparative method to reconstruct the history of life, and then use that history to understand how diverse species and their molecular, cellular, organismal, and populational properties arose through evolutionary time. Hypotheses that withstand repeated testing and therefore explain many diverse phenomena gain the status of a theory. Powerful theories that guide extensive research are called "paradigms." The major paradigms that guide the study of zoology are Darwin's theory of evolution and the chromosomal theory of inheritance.

The principles given in this chapter illustrate the unity of biological science. All components of biological systems are guided by natural laws and are constrained by those laws. Living organisms can come only from other living organisms, just as new cells can be produced only from preexisting cells. Reproductive processes occur at all levels of the biological hierarchy and demonstrate both heredity and variation. Interaction of heredity and variation at all levels of the biological hierarchy produces evolutionary change and has generated the great diversity of animal life documented throughout this book.

REVIEW QUESTIONS

1. Why is life difficult to define?
2. What are the basic chemical differences that distinguish living from nonliving systems?
3. Describe the hierarchical organization of life. How does this organization lead to the emergence of new properties at different levels of biological complexity?
4. What is the relationship between heredity and variation in reproducing biological systems?
5. Describe how evolution of complex organisms is compatible with the second law of thermodynamics.
6. What are the essential characteristics of science? Describe how evolutionary studies fit these characteristics whereas "scientific creationism" does not.
7. Use studies of natural selection in British moth populations to illustrate the hypothetico-deductive method of science.
8. How do we distinguish the terms hypothesis, theory, paradigm, and scientific fact?
9. How do biologists distinguish experimental and evolutionary sciences?
10. What are Darwin's five theories of evolution (as identified by Ernst Mayr)? Which are accepted as fact and which continue to stir controversy among biologists?
11. What major obstacle confronted Darwin's theory of natural selection when it was first proposed? How was this obstacle overcome?
12. How does neo-Darwinism differ from Darwinism?
13. Describe the respective contributions of the genetic approach and cell biology to formulating the chromosomal theory of inheritance.

SELECTED REFERENCES

Futuyma, D. J. 1995. Science on trial: the case for evolution. Sunderland, Massachusetts, Sinauer Associates, Inc. *A defense of evolutionary biology as the exclusive scientific approach to the study of life's diversity.*

Kitcher, P. 1982. Abusing science: the case against creationism. Cambridge, Massachusetts, MIT Press. *A treatise on how knowledge is gained in science and why creationism does not qualify as science.*

Kuhn, T. S. 1970. The structure of scientific revolutions. ed. 2, enlarged. Chicago, University of Chicago Press. *An influential and controversial commentary on the process of science.*

Mayr, E. 1982. The growth of biological thought: diversity, evolution and inheritance. Cambridge, Massachusetts, The Belknap Press of Harvard University Press. *An interpretive history of biology with special reference to genetics and evolution.*

Medawar, P. B. 1989. Induction and intuition in scientific thought. London, Methuen & Company. *A commentary on the basic philosophy and methodology of science.*

Moore, J. A. 1993. Science as a way of knowing: the foundations of modern biology. Cambridge, Massachusetts, Harvard University Press. *A lively, wide-ranging account of the history of biological thought and the workings of life.*

Perutz, M. F. 1989. Is science necessary? Essays on science and scientists. New York, E. P. Dutton. *A general discussion of the utility of science.*

Rennie, J. 2002. 15 answers to creationist nonsense. Sci. Amer. 287:78–85 (July). *A guide to the most common arguments used by creationists against evolutionary biology, with concise explanations of the scientific flaws of creationists' claims.*

ZOOLOGY LINKS TO THE INTERNET

Visit the textbook's Online Learning Center at www.mhhe.com/zoology to find live Internet links for each of the topics listed here.

Introductory Materials and Governmental Sites

Introductory Sites

The Science of Biology

Writing Papers and Study Tips

Glossaries and Dictionaries

Utility and Organizational Sites

Individual Contributions to Environmental Issues

Scientific Method

Mendelian Genetics

Darwinian Evolutionary Theory

Earth's abundant supply of water was critical for the origin of life.

The Origin and Chemistry of Life

Spontaneous Generation of Life?

From ancient times, people commonly believed that life arose repeatedly by spontaneous generation from nonliving material in addition to parental reproduction. For example, frogs appeared to arise from damp earth, mice from putrefied matter, insects from dew, and maggots from decaying meat. Warmth, moisture, sunlight, and even starlight often were mentioned as factors that encouraged spontaneous generation of living organisms.

Among the accounts of early efforts to synthesize organisms in the laboratory is a recipe for making mice, given by the Belgian plant nutritionist Jean Baptiste van Helmont (1648). "If you press a piece of underwear soiled with sweat together with some wheat in an open jar, after about 21 days the odor changes and the ferment . . . changes the wheat into mice. But what is more remarkable is that the mice which came out of the wheat and underwear were not small mice, not even miniature adults or aborted mice, but adult mice emerge!"

In 1861, the great French scientist Louis Pasteur convinced scientists that living organisms cannot arise spontaneously from nonliving matter. In his famous experiments, Pasteur introduced fermentable material into a flask with a long S-shaped neck that was open to air. The flask and its contents were then boiled for a long time to kill any microorganisms that might be present. Afterward the flask was cooled and left undisturbed. No fermentation occurred because all organisms that entered the open end were deposited in the neck and did not reach the fermentable material. When the neck of the flask was removed, microorganisms in the air promptly entered the fermentable material and proliferated. Pasteur concluded that life could not originate in the absence of previously existing organisms and their reproductive elements, such as eggs and spores. Announcing his results to the French Academy, Pasteur proclaimed, "Never will the doctrine of spontaneous generation arise from this mortal blow."

All living organisms share a common ancestor, most likely a population of colonial microorganisms that lived almost 4 billion years ago. This common ancestor was itself the product of a long period of prebiotic assembly of nonliving matter, including organic molecules and water, to form self-replicating units. All living organisms retain a fundamental chemical composition inherited from their ancient common ancestor.

According to the big-bang model, the universe originated from a primeval fireball and has been expanding and cooling since its inception 10 to 20 billion years ago. The sun and planets formed approximately 4.6 billion years ago from a spherical cloud of cosmic dust and gases. The cloud collapsed under the influence of its own gravity into a rotating disc. As material in the central part of the disc condensed to form the sun, gravitational energy was released as radiation. The pressure of this outwardly directed radiation prevented the collapse of the nebula into the sun. The material left behind cooled and eventually produced the planets, including earth (Figure 2-1).

In the 1920s, Russian biochemist Alexander I. Oparin and British biologist J. B. S. Haldane independently proposed that life originated on earth after an inconceivably long period of "abiogenic molecular evolution." Rather than arguing that the first living organisms miraculously originated all at once, a notion that formerly discouraged scientific inquiry, Oparin and Haldane argued that the simplest form of life arose gradually by the progressive assembly of small molecules into more complex organic molecules. Molecules capable of self-replication eventually would be produced, ultimately leading to assembly of living microorganisms.

ORGANIC MOLECULAR STRUCTURE OF LIVING SYSTEMS

Chemical evolution in the prebiotic environment produced simple organic compounds that ultimately formed the building blocks of living cells. The term "organic" refers broadly to compounds that contain carbon. Many also contain hydrogen, oxygen, nitrogen, sulfur, phosphorus, salts, and other elements. Carbon has a great ability to bond with other carbon atoms in chains of varying lengths and configurations. Carbon-to-carbon combinations introduce the possibility of enormous complexity and variety into molecular structure. More than a million organic compounds have been identified.

We review the kinds of organic molecules found in living systems, followed by further discussion of their origins in earth's primitive reducing atmosphere.

Carbohydrates: Nature's Most Abundant Organic Substance

Carbohydrates are compounds of carbon, hydrogen, and oxygen. These elements are usually present in the ratio of 1 C: 2 H: 1 O and are grouped as H—C—OH. Carbohydrates function in protoplasm mainly as structural elements and as a source of chemical energy. Glucose is the most important of these energy-storing carbohydrates. Familiar examples of carbohydrates include sugars, starches, and cellulose (the woody structure of plants). Cellulose occurs on earth in greater quantities than all other organic materials combined. Carbohydrates are synthesized by green plants from water and carbon dioxide, with the aid of solar energy. This process, called **photosynthesis,** is a reaction upon which all life depends, for it is the starting point in the formation of food.

Carbohydrates are usually categorized into the following three classes: (1) **monosaccharides,** or simple sugars; (2) **disaccharides,** or double sugars; and (3) **polysaccharides,** or complex sugars. Simple sugars are composed of carbon chains containing 4 carbons (tetroses), 5 carbons (pentoses), or 6 carbons (hexoses). Other simple sugars may have up to

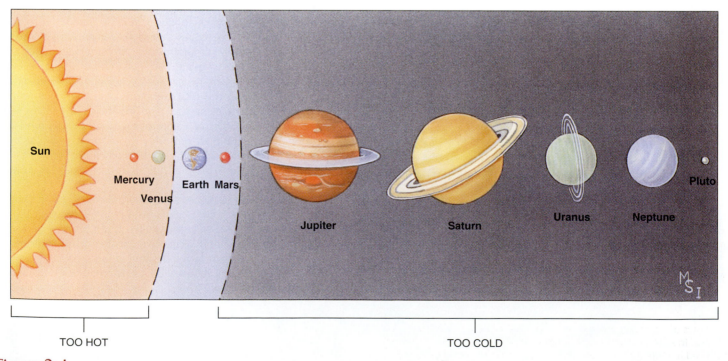

Figure 2-1
Solar system showing narrow range of conditions suitable for life.

10 carbons, but these sugars are not biologically important. Simple sugars, such as glucose, galactose, and fructose, all contain a free sugar group,

in which the double-bonded O may be attached to the terminal or nonterminal carbons of a chain. The hexose **glucose** (also called dextrose) is particularly important to the living world. Glucose is often shown as a straight chain (Figure 2-2A), but in water it tends to form a cyclic compound (Figure 2-2B). The "chair" diagram (Figure 2-3) of glucose best represents its true configuration, but all forms of glucose, however represented, are the same molecule. Other hexoses of biological significance include galactose and fructose, which are compared with glucose in Figure 2-4.

Disaccharides are double sugars formed by bonding two simple sugars. An example is maltose (malt sugar), composed of two glucose molecules. As shown in Figure 2-5, the two glucose molecules are condensed together by the removal of a molecule of water. This condensation reaction, with the sharing of an oxygen atom by the two sugars, characterizes the formation of all disaccharides. Two other common disaccharides are sucrose (ordinary cane, or table, sugar), formed by the linkage of glucose and fructose, and lactose (milk sugar), composed of glucose and galactose.

Polysaccharides are composed of many molecules of simple sugars (usually glucose) linked together in long chains called polymers. Their empirical formula is usually written $(C_6H_{10}O_5)_n$, where n designates the number of simple sugar subunits contained in the polymer. Starch is the common form in which sugar is stored in most plants and is an important food for animals. **Glycogen** is an important form for storing sugar in animals. It is

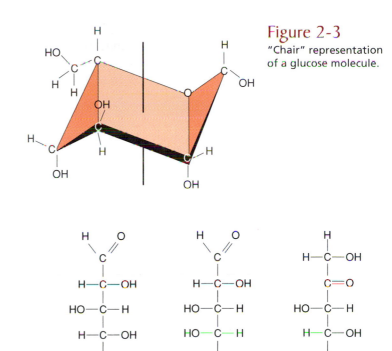

Figure 2-3
"Chair" representation of a glucose molecule.

Glucose   **Galactose**   **Fructose**

Figure 2-4
These three hexoses are the most common monosaccharides.

Figure 2-2
Two ways of depicting the structural formula of the simple sugar glucose. In **A,** the carbon atoms are shown in open-chain form. When dissolved in water, glucose tends to assume a ring form as in **B.** In this ring model the carbon atoms located at each turn in the ring are usually not shown.

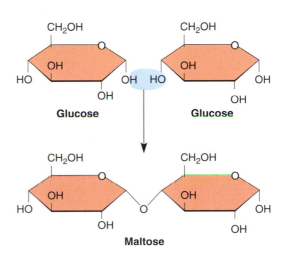

Glucose   **Glucose**

Maltose

Figure 2-5
Formation of a double sugar (disaccharide maltose) from two glucose molecules with the removal of one molecule of water.

found mainly in liver and muscle cells in vertebrates. When needed, glycogen is converted to glucose and delivered by blood to the tissues. Another polymer is **cellulose,** the principal structural carbohydrate of plants.

Lipids: Fuel Storage and Building Material

Lipids are fats and fatlike substances. They are composed of molecules of low polarity; consequently, they are virtually insoluble in water but are soluble in organic solvents, such as acetone and ether. The three principal groups of lipids are neutral fats, phospholipids, and steroids.

Neutral Fats

The neutral or "true" fats are major fuels of animals. Stored fat may be derived directly from dietary fat or indirectly from dietary carbohydrates that are converted to fat for storage. Fats are oxidized and released into the bloodstream as needed to meet tissue demands, especially those of active muscle.

Neutral fats include triglycerides, which are molecules consisting of glycerol and three molecules of fatty acids. Neutral fats are therefore esters, a combination of an alcohol (glycerol) and an acid. Fatty acids in triglycerides are simply long-chain monocarboxylic acids; they vary in size but are commonly 14 to 24 carbons long. Production of a typical fat by the union of glycerol and stearic acid is shown in Figure 2-6A. In this reaction, three fatty-acid molecules can be seen to have united with OH groups of the glycerol to form stearin (a neutral fat) plus three molecules of water.

Most triglycerides contain two or three different fatty acids attached to glycerol, and bear ponderous names such as myristoyl stearoyl glycerol (Figure 2-6B). The fatty acids in this triglyceride are **saturated;** every carbon within the chain holds two hydrogen atoms. Saturated fats, more common in animals than in plants, are usually solid at room temperature. **Unsaturated** fatty acids, typical of plant oils, have two or more carbon atoms joined by double bonds; the carbons are not "saturated" with hydrogen atoms and are able to form bonds with other atoms. Two common unsaturated fatty acids are oleic acid and linoleic acid (Figure 2-7). Plant fats such as peanut oil and corn oil tend to be liquid at room temperature.

Phospholipids

Unlike the fats that are fuels and serve no structural roles in the cell, phospholipids are important components of the molecular organization of tissues, especially membranes. They resemble triglycerides in structure, except that one of the three fatty acids is replaced by phosphoric acid and an organic base. An example is lecithin, an important phospholipid of nerve membranes

Figure 2-6

Neutral fats. **A,** Formation of a neutral fat from three molecules of stearic acid (a fatty acid) and glycerol. **B,** A neutral fat bearing three different fatty acids.

Figure 2-7

Unsaturated fatty acids: oleic acid having one double bond and linoleic acid having two double bonds. The remainder of the hydrocarbon chains of both acids is saturated.

(Figure 2-8). Because the phosphate group on phospholipids is charged and polar and therefore soluble in water, and the remainder of the molecule is nonpolar, phospholipids can bridge two environments and bind water-soluble molecules such as proteins to water-insoluble materials.

Steroids

Steroids are complex alcohols. Although they are structurally unlike fats, they have fatlike properties. The steroids are a large group of biologically important molecules, including cholesterol (Figure 2-9), vitamin D, many adrenocortical hormones, and the sex hormones.

Amino Acids and Proteins

Proteins are large, complex molecules composed of 20 commonly occurring amino acids (Figure 2-10). The amino acids are linked together by **peptide bonds** to form long, chainlike polymers. In the formation of a peptide bond, the carboxyl group of one amino acid is linked by a covalent bond to the amino group of another, with elimination of water, as shown here:

Figure 2-8

Lecithin (phosphatidyl choline), an important phospholipid of nerve membranes.

Figure 2-9

Cholesterol, a steroid. All steroids have a basic skeleton of four rings (three 6-carbon rings and one 5-carbon ring) with various side groups attached.

Figure 2-10

Five of the twenty naturally occurring amino acids.

The combination of two amino acids by a peptide bond forms a dipeptide, and, as is evident, there is still a free amino group on one end and a free carboxyl group on the other; therefore, additional amino acids can be joined until a long chain is produced. The 20 different kinds of amino acids can be arranged in an enormous variety of sequences of up to several hundred amino acid units, accounting for the large diversity of proteins found among living organisms.

A protein is not just a long string of amino acids; it is a highly organized molecule. For convenience, biochemists recognize four levels of protein organization called primary, secondary, tertiary, and quaternary structures.

The **primary structure** of a protein constitutes the sequence of amino acids composing the polypeptide chain. Because the bonds between the amino acids in the chain can form only a limited number of stable angles, certain recurrent structural patterns are assumed by the chain. These bond angles give rise to the **secondary structure,** such as the **alpha-helix,** which makes helical turns in a clockwise direction like a screw (Figure 2-11). The spirals of the chains are stabilized by hydrogen bonds, usually between a hydrogen atom of one amino acid and the peptide-bond oxygen of another amino acid from an adjacent turn of the helix. In addition, the helical and other configurations formed by the polypeptide chain themselves bend and fold, giving the protein its complex, yet stable, three-dimensional **tertiary structure** (Figure 2-11). The folded chains are stabilized by the interactions between side groups of amino acids. One of these interactions is the **disulfide bond,** a covalent bond between the sulfur atoms in two cysteine amino acids that are brought together by folds in the polypeptide chain. Also stabilizing the tertiary structure of proteins are hydrogen bonds, ionic bonds, and hydrophobic bonds.

The term **quaternary structure** describes proteins that contain more than one polypeptide chain. For example, hemoglobin (the oxygen-carrying substance in blood) of higher vertebrates is composed of four polypeptide subunits held together in a single protein molecule (Figure 2-11).

Proteins perform many functions in living organisms. They serve as the structural framework of protoplasm and form many cellular components. Proteins also may function as **enzymes,** the biological catalysts required for almost every reaction in the body. Enzymes lower the activation energy required for specific reactions and enable life processes to proceed at moderate temperatures. They control the reactions by which food is digested, absorbed, and metabolized. They promote the synthesis of structural materials for growth and to replace those lost by wear. They

Figure 2-11

Structure of proteins. The amino acid sequence of a protein (*primary structure*) encourages the formation of hydrogen bonds between nearby amino acids, producing coils and foldbacks (the *secondary structure*). Bends and helices cause the chain to fold back on itself in a complex manner (*tertiary structure*). Individual polypeptide chains of some proteins aggregate together to form the functional molecule composed of several subunits (*quaternary structure*).

determine the release of energy used in respiration, growth, muscle contraction, physical and mental activities, and many other activities. Enzyme action is described in Chapter 4 (p. 58).

Nucleic Acids

Nucleic acids are complex substances of high molecular weight that are a fundamental part of life. The sequence of nitrogenous bases in these polymeric molecules encodes the genetic information necessary for biological inheritance. They store directions for the synthesis of enzymes and other proteins, and are the only molecules that can (with the help of the right enzymes) replicate themselves. The two kinds of nucleic acids in cells are **deoxyribonucleic acid (DNA)** and **ribonucleic acid (RNA).** They are polymers of repeated units called **nucleotides,** each containing a sugar, a nitrogenous base, and a phosphate group. Because the structure of nucleic acids is crucial to the mechanism of inheritance and protein synthesis, detailed information on nucleic acids is presented in Chapter 5 (p. 88).

CHEMICAL EVOLUTION

Both Haldane and Oparin proposed that earth's primitive atmosphere consisted of simple compounds such as water, carbon dioxide, molecular hydrogen, methane, and ammonia, but lacked oxygen. The nature of the primeval atmosphere is critical for understanding life's origin. The organic compounds that compose living organisms are neither synthesized outside cells nor stable in the presence of molecular oxygen, which is abundant in the atmosphere today. The best evidence indicates, however, that the primitive atmosphere contained not more than a trace of molecular oxygen, most of which had reacted with hydrogen to form the water present on the earth's surface. The primeval atmosphere therefore was a reducing one, consisting primarily of molecules in which hydrogen exceeds oxygen; methane (CH_4) and ammonia (NH_3), for example, constitute fully reduced compounds. During this time, the earth was bombarded by large (100 km diameter) comets and meteorites, generating heat that repeatedly vaporized the oceans.

This reducing atmosphere was conducive to the prebiotic synthesis that led to life's beginnings, although totally unsuited for the organisms that exist today. Haldane and Oparin proposed that when such a gas mixture was exposed to ultraviolet radiation, many organic substances such as sugars and amino acids could be formed. Haldane believed that the early organic molecules accumulated in the primitive oceans to form a "hot dilute soup." In this primordial broth, carbohydrates, fats, proteins, and nucleic acids could have assembled to form the earliest structures capable of guiding their own replication.

If the simple gaseous compounds present in the early atmosphere are mixed with methane and ammonia in a closed glass system and kept at room temperature, they never react chemically with each other. To produce a chemical reaction, a continuous source of **free energy** sufficient to overcome reaction-activation barriers must be supplied. Ultraviolet light from the sun must have been intense on the primitive earth before the accumulation of atmospheric oxygen; ozone, a three-atom form of oxygen located high in the atmosphere, now blocks much of the ultraviolet radiation from reaching the earth's surface. Electrical discharges could have provided further energy for chemical evolution. Although the total amount of electrical energy released by lightning is small compared with solar energy, nearly all of the energy of lightning is effective in synthesizing organic compounds in a reducing atmosphere. A single flash of lightning through a reducing atmosphere generates a large amount of organic matter. Thunderstorms may have been one of the most important sources of energy for organic synthesis.

Widespread volcanic activity on the primitive earth is another possible source of energy. One hypothesis maintains, for example, that life did not originate on the surface of the earth, but deep beneath the sea in or around **hydrothermal vents** (p. 801). Hydrothermal vents are submarine hot springs, in which seawater seeps through cracks in the bottom until the water comes close to hot magma. The water is superheated and expelled forcibly, carrying a variety of dissolved molecules from the superheated rocks. These molecules include hydrogen sulfide, methane, iron ions, and sulfide ions. Hydrothermal vents have been discovered in several locations beneath the deep sea, and they would have been much more widely prevalent on the early earth. Interestingly, many heat- and sulfur-loving bacteria grow in hot springs today.

Prebiotic Synthesis of Small Organic Molecules

The Oparin-Haldane hypothesis stimulated experimental work to test the hypothesis that organic compounds characteristic of life could be formed from the simpler molecules present in the prebiotic environment. In 1953, Stanley Miller and Harold Urey in Chicago successfully simulated the conditions thought to prevail on the primitive earth. Miller built an apparatus designed to circulate a mixture of methane, hydrogen, ammonia, and water past an electric spark (Figure 2-12). Water in the flask was boiled to produce steam that helped to circulate the gases. The products formed in the electrical discharge (representing lightning) were condensed in the condenser and collected in the U-tube and small flask (representing an ocean).

After a week of continuous sparking, approximately 15% of the carbon that was originally in the reducing "atmosphere" had been converted into organic compounds that collected in the "ocean." The most striking finding was that many compounds related to life were synthesized. These compounds included four of the amino acids commonly found in proteins, urea, and several simple fatty acids. We can appreciate the astonishing nature of this synthesis when we consider that there are thousands of known organic compounds with structures no more complex than those of the amino acids formed. Yet in Miller's synthesis, most of the relatively few substances formed were compounds found in living organisms. This result was surely no coincidence, and it suggests that prebiotic synthesis on the primitive earth may have occurred under conditions not greatly different from those that Miller chose to simulate.

Figure 2-12

Dr. S. L. Miller with a replica of the apparatus used in his 1953 experiment on the synthesis of amino acids with an electric spark in a strongly reducing atmosphere.

Miller's experiments have been criticized in light of current opinion that the early atmosphere on earth was quite different from Miller's strongly reducing simulated atmosphere. Nevertheless, Miller's work stimulated many other investigators to repeat and extend his experiment. Amino acids were found to be synthesized in many different kinds of gas mixtures that were heated (volcanic heat), irradiated with ultraviolet light (solar radiation), or subjected to electrical discharge (lightning). The only conditions required to produce amino acids were that the gas mixture be reducing and that it be subjected violently to a source of energy. In other experiments, electrical discharges were passed through mixtures of carbon monoxide, nitrogen, and water, yielding amino acids and nitrogenous bases. Although reaction rates were much slower than in atmospheres containing methane and ammonia, and yields were poor in comparison, these experiments support the hypothesis that the chemical beginnings of life can occur in atmospheres that are only mildly reducing. The need for methane and ammonia, however, led to proposals that these substances might have been introduced by comets or meteorites, or that they were synthesized near the hydrothermal vents.

Water and Life

The origin and maintenance of life on earth depend critically upon water. Water is the most abundant of all compounds in cells, comprising 60% to 90% of most living organisms. Water has several extraordinary properties that explain its essential role in living systems and their origin. These properties result largely from hydrogen bonds that form between its molecules.

Water has a **high specific heat capacity:** 1 calorie* is required to elevate the temperature of 1 g of water 1° C, a higher thermal capacity than any other liquid except ammonia. Much of this heat energy is used to rupture some hydrogen bonds in addition to increasing the kinetic energy (molecular movement), and thus the temperature, of the water. Water's high thermal capacity greatly moderates environmental temperature changes, thereby protecting living organisms from extreme thermal fluctuation. Water also has a **high heat of vaporization,** requiring more than 500 calories to convert 1 g of liquid water to water vapor. All hydrogen bonds between a water molecule and its neighbors must be ruptured before that water molecule can escape the surface and enter the air. For terrestrial animals (and plants), cooling produced by evaporation of water is important for expelling excess heat.

Another property of water important for life is its **unique density behavior** during changes of temperature. Most liquids become denser with decreasing temperature. Water, however, reaches its maximum density at 4° C *while still a liquid,* then becomes less dense with further cooling. Therefore, ice *floats* rather than forming on the bottoms of lakes and ponds. If ice were denser than liquid water, bodies of water would freeze solid from the bottom upward in winter and would not necessarily melt completely in summer. Such conditions would severely limit aquatic life. In ice, water molecules form an extensive, open, crystal-like network supported by hydrogen bonds that connect all molecules. The molecules in this lattice are farther apart, and thus less dense, than in liquid water at 4° C.

Water has **high surface tension,** exceeding that of any other liquid but mercury. Hydrogen bonding among water molecules produces a cohesiveness that is important for maintaining protoplasmic form and movement. The resulting surface tension creates an ecological niche (see p. 790) for insects, such as water striders and whirligig beetles, that skate on the surfaces of ponds. Despite its high surface

---

*A calorie is defined as the amount of heat required to heat 1 g of water from 14.5° C to 15.5° C. Although the calorie is the traditional unit of heat widely used in publications and tables, it is not part of the International System of Units (the SI system) which uses the joule (J) as the energy unit (1 cal = 4.184 J).

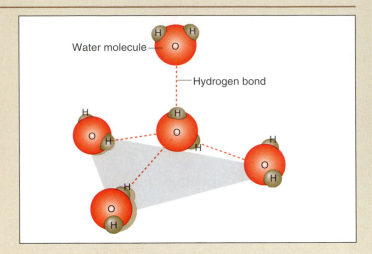

Geometry of water molecules. Each water molecule is linked by hydrogen bonds (*dashed lines*) to four other molecules. If imaginary lines are used to connect the divergent oxygen atoms, a tetrahedron is obtained.

Because of hydrogen bonds between water molecules at the water-air interface, the water molecules cling together and create a high surface tension. Thus some insects, such as this water strider, can literally walk on water.

Thus the experiments of many scientists have shown that highly reactive intermediate molecules such as hydrogen cyanide, formaldehyde, and cyanoacetylene are formed when a reducing mixture of gases is subjected to a violent energy source. These molecules react with water and ammonia or nitrogen to form more complex organic molecules, including amino acids, fatty acids, urea, aldehydes, sugars, and nitrogenous bases (purines and pyrimidines), all of the building blocks required for the synthesis of the most complex organic compounds of living matter.

Formation of Polymers

The next stage in chemical evolution involved the condensation of amino acids, nitrogenous bases, and sugars to yield larger molecules, such as proteins and nucleic acids. Such condensations do not occur easily in dilute solutions, because the presence of excess water tends to drive reactions toward decomposition (hydrolysis). Although the primitive ocean might have been called a "primordial soup," it was probably a rather dilute one containing organic material that was approximately one-tenth to one-third as concentrated as chicken bouillon.

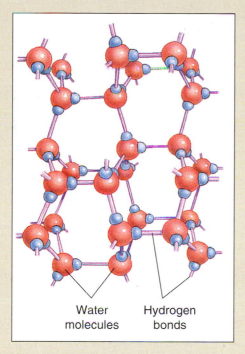

When water freezes at 0° C, the four partial charges of each atom in the molecule interact with the opposite charges of atoms in other water molecules. The hydrogen bonds between all the molecules form a crystal-like lattice structure, and the molecules are farther apart (and thus less dense) than when some of the molecules have not formed hydrogen bonds at 4° C.

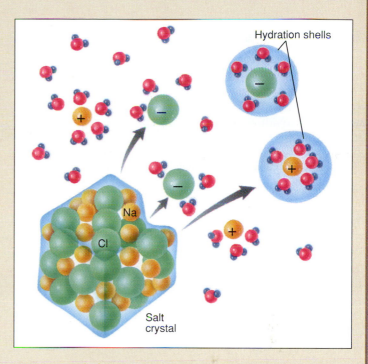

When a crystal of sodium chloride dissolves in water, the negative ends of the dipolar molecules of water surround the Na^+ ions, while the positive ends of water molecules face the Cl^- ions. The ions are thus separated and do not reenter the salt lattice.

tension, water has **low viscosity,** permitting movement of blood through minute capillaries and of cytoplasm inside cellular boundaries.

Water is an excellent **solvent.** Salts dissolve more extensively in water than in any other solvent. This property results from the dipolar nature of water, which causes it to orient around charged particles dissolved in it. When, for example, crystalline NaCl dissolves in water, the Na^+ and Cl^- ions separate. The negative zones of the water dipoles attract the Na^+ ions while the positive zones attract the Cl^- ions. This orientation keeps the ions separated, promoting their dissociation. Solvents lacking this dipolar character are less effective at keeping the ions separated. Binding of water to dissolved protein molecules is essential to the proper functioning of many proteins.

Water also participates in many chemical reactions in living organisms. Many compounds are split into smaller pieces by the addition of a molecule of water, a process called **hydrolysis.** Likewise, larger compounds may be synthesized from smaller components by the reverse of hydrolysis, called **condensation** reactions.

$$R-R + H_2O \xrightarrow[\text{Hydrolysis}]{} R-OH + H-R$$

$$R-OH + H-R \xrightarrow[\text{Condensation}]{} R-R + H_2O$$

Need for Concentration

Prebiotic synthesis must have occurred in restricted regions where concentrations of the reactants were high. Violent weather on the primitive earth would have created enormous dust storms; impacts of meteorites would have lofted great amounts of dust into the atmosphere. The dust particles could have become foci of water droplets. Salt concentration in the particles could have been high and provided a concentrated medium for chemical reactions. Alternatively, perhaps the surface of the earth was too warm to have oceans but not too hot for a damp surface. This condition would have resulted from constant rain and rapid evaporation. Thus, the earth's surface could have become coated with organic molecules, an "incredible scum." Prebiotic molecules might have been concentrated by adsorption on the surface of clay and other minerals. Clay has the capacity to concentrate and condense large amounts of organic molecules. The surface of iron pyrite (FeS_2) also has been suggested as a site for the evolution of biochemical pathways. The positively charged surface of pyrite would attract a variety of negative ions, which would become bound to its surface. Furthermore, pyrite is abundant around hydrothermal vents, compatible with the hydrothermal-vent hypothesis.

Thermal Condensations

Most biological polymerizations are condensation (dehydration) reactions, in which monomers are linked together by the removal of water (p. 29). In living systems, condensation reactions always occur in an aqueous (cellular) environment containing appropriate enzymes. Without enzymes and energy supplied by ATP, macromolecules (proteins and nucleic acids) of living systems soon decompose into their constituent monomers.

One way in which dehydration reactions could have occurred without enzymes in primitive earth conditions is by thermal condensation. The simplest dehydration is accomplished by driving water from solids by direct heating. For example, heating a mixture of all 20 amino acids to 180° C produces a good yield of polypeptides.

The thermal synthesis of polypeptides to form "proteinoids" has been studied extensively by the American scientist Sidney Fox. He showed that heating dry mixtures of amino acids and then mixing the resulting polymers with water forms small spherical bodies. These proteinoid microspheres (Figure 2-13) possess certain characteristics of living systems. Each is not more than 2 μm in diameter and is comparable in size and shape to spherical bacteria. The outer walls of the microspheres appear to have a double layer, and they show osmotic properties and selective diffusion. They may grow by accretion or proliferate by budding like bacteria. We do not know whether proteinoids may have been the ancestors of the first cells, or whether they are

Figure 2-13

Electron micrograph of proteinoid microspheres. These proteinlike bodies can be produced in the laboratory from polyamino acids and may represent precellular forms. They have definite internal ultrastructure. (×1700)

just interesting creations of a chemist's laboratory. Their formation requires conditions likely to have occurred only in volcanoes. Organic polymers might have condensed on or in volcanoes and then, wetted by rain or dew, reacted further in solution to form polypeptides or polynucleotides.

ORIGIN OF LIVING SYSTEMS

The fossil record reveals that life existed by 3.8 billion years ago; therefore, the origin of the earliest life form can be estimated at approximately 4 billion years BP. The first living organisms were protocells, autonomous membrane-bound units with a complex functional organization that permitted the essential activity of self-reproduction. The primitive chemical systems that we have described lack this essential property. The principal problem in understanding the origin of life is explaining how primitive chemical systems could have become organized into living, autonomous, self-reproducing cells.

As we have seen, a lengthy chemical evolution on the primitive earth produced several molecular components of living forms. In a later stage of evolution, nucleic acids (DNA and RNA) began to behave as simple genetic systems that directed the synthesis of proteins, especially enzymes. However, this conclusion has led to a troublesome chicken-egg paradox: (1) How could nucleic acids have appeared without enzymes to synthesize them? (2) How could enzymes have evolved without nucleic acids to direct their synthesis? These questions are based on a long-accepted notion that only proteins could act as enzymes. Startling evidence presented in the 1980s indicates that RNA in some instances has catalytic activity.

Catalytic RNA (ribozymes) can mediate processing of messenger RNA (removal of introns, p. 92), and can catalyze formation of peptide bonds. Strong evidence suggests that translation of mRNA by ribosomes (p. 93) is catalyzed by their RNA, not protein, content.

Therefore the earliest enzymes could have been RNA, and the earliest self-replicating molecules could have been RNA. Investigators are now calling this stage the "RNA world." Nonetheless, proteins have several important advantages over RNA as catalysts, and DNA is a more stable carrier of genetic information than RNA. The first protocells containing protein enzymes and DNA should have had a selective advantage over those with only RNA.

Once this stage of organization was reached, natural selection (pp. 118–120) would have acted on these primitive self-replicating systems. This point was critical. Before this stage, biogenesis was shaped by the favorable environmental conditions on the primitive earth and by the nature of the reacting elements themselves. When self-replicating systems became responsive to natural selection, they began to evolve. The more rapidly replicating and more successful systems were favored, and they replicated even faster. In short, the most efficient forms survived. Evolution of the genetic code and fully directed protein synthesis followed. The system now meets the requirements for being the common ancestor of all living organisms.

Origin of Metabolism

Living cells today are organized systems with complex and highly ordered sequences of enzyme-mediated reactions. How did such vastly complex metabolic schemes develop? The exact history of this phase of life's evolution is unknown. We present here a model of the simplest sequence of events that could explain the origin of observed metabolic properties of living systems.

The traditional view is that the first organisms were primary heterotrophs. Carl Woese finds it easier to visualize membrane-associated molecular aggregates that absorbed visible light and converted it with some efficiency into chemical energy. Thus the first organisms would have been autotrophs. Woese also suggests that the earliest "metabolism" may have consisted of numerous chemical reactions catalyzed by nonprotein cofactors (substances necessary for the function of many of the protein enzymes in living cells). These cofactors would have been associated with membranes.

Organisms that can synthesize their food from inorganic sources using light or another source of energy are called **autotrophs** (Gr. *autos,* self, + *trophos,* feeder) (Figure 2-14). Organisms lacking this ability must obtain their food supplies directly from the environment and are known as **heterotrophs** (Gr. *heteros,* another, + *trophos,* feeder). The earliest microorganisms are sometimes called **primary heterotrophs** because they relied on environmental sources for their food and existed prior to the evolution of any autotrophs. They were probably anaerobic organisms similar to bacteria of the genus *Clostridium.* Because chemical evolution had supplied generous stores of nutrients in the prebiotic soup, the earliest organisms would not have been required to synthesize their own food.

Protocells able to convert inorganic precursors to a required nutrient would have had a tremendous selective advantage over the primary heterotrophs in areas where nutrients became depleted from the environment. Evolution of autotrophic organisms most likely required acquisition of enzymatic activities to catalyze conversion of inorganic molecules to more complex ones, such as carbohydrates. The numerous enzymes of cellular metabolism appeared when cells became able to utilize proteins for catalytic functions.

Appearance of Photosynthesis and Oxidative Metabolism

Autotrophy evolved in the form of photosynthesis. In photosynthesis, hydrogen atoms obtained from water react with carbon dioxide obtained from the atmosphere to generate sugars and molecular oxygen. The sugars provide nutrition to the organism and molecular oxygen is released into the atmosphere.

$$6CO_2 + 6H_2O \xrightarrow{\text{light}} C_6H_{12}O_6 + 6O_2$$

This equation summarizes the many reactions now known to occur in the process of photosynthesis. Undoubtedly these reactions did not appear all at once, and other reduced compounds, such as hydrogen sulfide (H_2S), probably were the early sources of hydrogen.

Gradually, oxygen produced by photosynthesis accumulated in the atmosphere. When atmospheric oxygen reached approximately 1% of its current level, ozone began to accumulate and to absorb ultraviolet radiation, thereby greatly restricting the amount of ultraviolet light that reached the earth. Land and surface waters then were occupied by photosynthetic organisms, thereby increasing oxygen production.

Accumulation of atmospheric oxygen would interfere with anaerobic cellular metabolism that had evolved in the primitive reducing atmosphere. As the atmosphere slowly changed from a somewhat reducing to a highly oxidizing one, a new and highly efficient kind of metabolism appeared: **oxidative (aerobic) metabolism.** By using available oxygen as a terminal electron acceptor (p. 67) and completely oxidizing glucose to carbon dioxide and water, much of the bond energy stored by photosynthesis could be recovered. Most living forms became completely dependent upon oxidative metabolism.

Our atmosphere today is strongly oxidizing. It contains 78% molecular nitrogen, approximately 21% free oxygen, 1% argon, and 0.03% carbon dioxide. Although the time course for production of atmospheric oxygen is much debated, the most important source of oxygen is photosynthesis. Almost all oxygen currently produced comes from cyanobacteria (blue-green algae),

Figure 2-14

Koala, a heterotroph, feeding on a eucalyptus tree, an autotroph. All heterotrophs depend for their nutrients directly or indirectly on autotrophs that capture the sun's energy to synthesize their own nutrients.

eukaryotic algae, and plants. Each day these organisms combine approximately 400 million tons of carbon dioxide with 70 million tons of hydrogen to produce 1.1 billion tons of oxygen. Oceans are a major source of oxygen. Almost all oxygen produced today is consumed by organisms for respiration; otherwise, the amount of oxygen in the atmosphere would double in approximately 3000 years. Because Precambrian fossil cyanobacteria resemble modern cyanobacteria, it is reasonable to suppose that oxygen entering the early atmosphere came from their photosynthesis.

PRECAMBRIAN LIFE

As depicted on the inside back cover of this book, the Precambrian period covers the geological time before the beginning of the Cambrian period some 570 to 600 million years BP. Most major animal phyla appear in the fossil record within a few million years at the beginning of the Cambrian period. This appearance has been called the "Cambrian explosion" because before this time, fossil deposits are mostly devoid of any organisms more complex than single-celled bacteria. Comparative molecular studies (p. 197) now suggest that the rarity of Precambrian fossils may represent poor fossilization rather than absence of animal diversity from the Precambrian period. Nonetheless, animals make a relatively late appearance in the history of life on earth. What were the early forms of life that generated both the oxidizing atmosphere critical for animal evolution and the evolutionary lineage from which animals would arise?

Prokaryotes and the Age of Cyanobacteria (Blue-Green Algae)

The earliest bacterium-like organisms proliferated, giving rise to a great variety of forms, some of which were capable of photosynthesis. From these arose the oxygen-producing **cyanobacteria** approximately 3 billion years ago.

Bacteria are called **prokaryotes,** meaning literally "before the nucleus." They contain a single, large molecule of DNA not located in a membrane-bound nucleus, but found in a nuclear region, or **nucleoid.** The DNA is not complexed with histone proteins, and prokaryotes lack membranous organelles such as mitochondria, plastids, Golgi apparatus, and endoplasmic reticulum (Chapter 3). During cell division, the nucleoid divides and

replicates of the cell's DNA are distributed to the daughter cells. Prokaryotes lack the chromosomal organization and chromosomal (mitotic) division seen in animals, fungi, and plants.

Bacteria and especially cyanobacteria ruled the earth's oceans unchallenged for 1 to 2 billion years. The cyanobacteria reached the zenith of their success approximately 1 billion years BP, when filamentous forms produced great floating mats on the oceans' surface. This long period of cyanobacterial dominance, encompassing approximately two-thirds of the history of life, has been called with justification the "age of blue-green algae." Bacteria and cyanobacteria are so completely different from forms of life that evolved later that they were placed in a separate kingdom, Monera.

Carl Woese and his colleagues at the University of Illinois have discovered that the prokaryotes actually comprise at least two distinct lines of descent: the Eubacteria ("true" bacteria) and the Archaebacteria also called Archaea (p. 203). Although these two groups of bacteria look very much alike when viewed with the electron microscope, they are biochemically distinct. Archaebacteria differ fundamentally from bacteria in cellular metabolism, and their cell walls lack muramic acid, which is found in the cell walls of all Eubacteria. The most compelling evidence for differentiating these two groups comes from the use of one of the newest and most powerful tools at the disposal of the evolutionist, sequencing of nucleic acids (see note). Woese found that archaebacteria differ fundamentally from other bacteria in the sequence of bases in ribosomal RNA (p. 93). Woese considers the archaebacteria so distinctly different from the true bacteria that they should be considered a separate kingdom, Archaea. The Monera then comprise only the true bacteria.

Appearance of the Eukaryotes

The **eukaryotes** ("true nucleus"; Figure 2-15) have cells with membrane-bound nuclei containing **chromosomes** composed of **chromatin.** Constituents of eukaryotic chromatin include proteins called **histones** and RNA, in addition to the DNA. Some nonhistone proteins are found associated with both prokaryotic

---

The name "algae" is misleading because it suggests a relationship to eukaryotic algae, and many scientists prefer the alternative name "cyanobacteria" rather than "blue-green algae." These organisms were responsible for producing oxygen initially released into the atmosphere. Studies of biochemical reactions in extant cyanobacteria suggest that they evolved in a time of fluctuating oxygen concentration. For example, although they can tolerate atmospheric concentrations of oxygen (21%), the optimum concentration for many of their metabolic reactions is only 10%.

Molecular sequencing has emerged as a very successful approach to unraveling the genealogies of ancient forms of life. The sequences of nucleotides in the DNA of an organism's genes are a record of evolutionary relationship, because every gene that exists today is an evolved copy of a gene that existed millions, even billions, of years ago. Genes become altered by mutations through the course of time, but vestiges of the original gene usually persist. With modern techniques, one can determine the sequence of nucleotides in an entire molecule of DNA or in short segments of the molecule. When corresponding genes are compared between two different organisms, the extent to which the genes differ can be correlated with the time elapsed since the two organisms diverged from a common ancestor. Similar comparisons can be made with RNA and proteins.

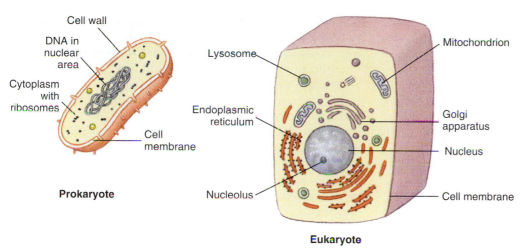

Figure 2-15
Comparison of prokaryotic and eukaryotic cells. Prokaryotic cells are about one-tenth the size of eukaryotic cells.

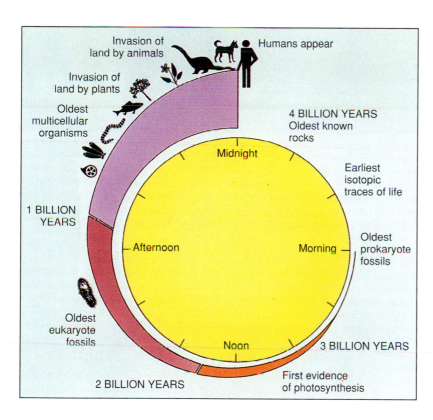

Figure 2-16
The clock of biological time. A billion seconds ago it was 1961, and most students using this text had not yet been born. A billion minutes ago the Roman empire was at its zenith. A billion hours ago Neanderthals were alive. A billion days ago the first bipedal hominids walked the earth. A billion months ago the dinosaurs were at the climax of their radiation. A billion years ago no creature had ever walked on the surface of the earth.

DNA and eukaryotic chromosomes. Eukaryotes are generally larger than prokaryotes and contain much more DNA. Cellular division usually is by some form of mitosis. Within their cells are numerous membranous organelles, including mitochondria, in which the enzymes for oxidative metabolism are packaged. Eukaryotes include animals, fungi, plants, and numerous single-celled forms formerly known as "protozoans" or "protists." Fossil evidence suggests that single-celled eukaryotes arose at least 1.5 billion years ago (Figure 2-16).

Because the organizational complexity of the eukaryotes is much greater than that of the prokaryotes, it is difficult to visualize how a eukaryote could have arisen from any known prokaryote. The American biologist Lynn Margulis and others have proposed that eukaryotes did not arise from any single prokaryote but were derived from a **symbiosis** ("life together")

of two or more types of bacteria. Mitochondria and plastids, for example, each contain their own complement of DNA (apart from the nucleus of the cell), which has some prokaryotic characteristics.

Nuclei, plastids, and mitochondria each contain genes encoding ribosomal RNA. Comparisons of the sequence of bases of these genes show that the nuclear, plastid, and mitochondrial DNAs represent distinct evolutionary lineages. Plastid and mitochondrial DNAs are closer in their evolutionary history to bacterial DNAs than to the eukaryotic nuclear DNA. Plastids are closest evolutionarily to cyanobacteria, and mitochondria are closest to another group of bacteria (purple bacteria), consistent with the symbiotic hypothesis of eukaryotic origins. Mitochondria contain the enzymes of oxidative metabolism, and plastids (a plastid with chlorophyll is a chloroplast) conduct photosynthesis. It is easy to see how a host cell that was able to accommodate such guests in its cytoplasm would have had enormous evolutionary success.

In addition to maintaining that mitochondria and plastids originated as bacterial symbionts, Lynn Margulis argues that eukaryote flagella, cilia (locomotory structures), and even the spindle of mitosis came from a kind of bacterium like a spirochete. Indeed, she suggests that this association (the spirochete with its new host cell) made the evolution of mitosis possible. Margulis's evidence that the organelles are former partners of the ancestral cell is now accepted by most biologists. Such merging of disparate organisms to produce evolutionarily novel forms is called symbiogenesis.

Eukaryotes may have originated more than once. The first eukaryotes were undoubtedly unicellular, and many were photosynthetic autotrophs. Some of these forms lost their photosynthetic ability and became heterotrophs, feeding on the autotrophs and the prokaryotes. As cyanobacteria were cropped, their dense filamentous mats began to thin, providing space for other organisms. Carnivores appeared and fed on herbivores. Soon a balanced ecosystem of carnivores, herbivores, and primary producers appeared. By freeing space, cropping herbivores encouraged a greater diversity of producers, which in turn promoted the evolution of new and more specialized croppers. An ecological pyramid developed with carnivores at the top of the food chain (p. 802).

The burst of evolutionary activity that followed at the end of the Precambrian period and beginning of the Cambrian period was unprecedented. Some investigators hypothesize that the explanation for the "Cambrian explosion" lies in the accumulation of oxygen in the atmosphere to a critical threshold level. Larger, multicellular animals required the increased efficiency of oxidative metabolism; these pathways could not be supported under conditions of limiting oxygen concentration.

SUMMARY

Living organisms show a remarkable uniformity in their chemical constituents and metabolism, reflecting their common descent from an ancient ancestor.

Experiments by Louis Pasteur in the 1860s convinced scientists that organisms do not arise repeatedly from inorganic matter. About 60 years later, A. I. Oparin and J. B. S. Haldane provided an explanation for how a common ancestor of all living forms could have arisen from nonliving matter almost 4 billion years ago. The origin of life followed a long period of "abiogenic molecular evolution" on earth in which organic molecules slowly accumulated in a "primordial soup." The atmosphere of the primitive earth was reducing, with little or no free oxygen present. Ultraviolet radiation, electrical discharges of lightning, or energy from hydrothermal vents could have provided energy for the early formation of organic molecules. Stanley Miller and Harold Urey demonstrated the plausibility of the Oparin-Haldane hypothesis by simple but ingenious experiments. The concentration of reactants necessary for early synthesis of organic molecules might have been provided by damp surfaces, clay particles, iron pyrite, or other conditions. RNA may have been the primordial biomolecule, performing the functions of both genetic coding of information and catalysis. When self-replicating systems became established, evolution by natural selection could have increased their diversity and complexity.

Life on earth could not have appeared without water, the primary component of living cells. The unique structure of water and its ability to form hydrogen bonds between adjacent water molecules are responsible for its special properties: solvency, high heat capacity, boiling point, surface tension, and lower density as a solid than as a liquid.

Life also depends critically on the chemistry of carbon. Carbon is especially versatile in bonding with itself and with other atoms, and it is the only element capable of forming the large molecules found in living organisms. Carbohydrates are composed primarily of carbon, hydrogen, and oxygen grouped as H—C—OH. The simplest carbohydrates are sugars, which serve as immediate sources of energy in living systems. Monosaccharides, or simple sugars, may bond together to form disaccharides or polysaccharides, which serve as storage forms of sugar or perform structural roles. Lipids constitute another class of large molecules featuring chains of carbon compounds; fats exist principally as neutral fats, phospholipids, and steroids. Proteins are large molecules composed of amino acids linked together by peptide bonds. Many proteins function as enzymes that catalyze biological reactions. Each kind of protein has a characteristic primary, secondary, tertiary, and often, quaternary structure critical for its functioning. Nucleic acids are polymers of nucleotide units, each composed of a sugar, a nitrogenous base, and a phosphate group. They contain the material of inheritance and function in protein synthesis.

The first organisms are hypothesized to have been primary heterotrophs, living on the energy stored in molecules dissolved in the primordial soup. Later evolution produced autotrophic organisms, which can synthesize their own organic nutrients (carbohydrates) from inorganic materials. Autotrophs are better protected than heterotrophs from depletion of organic compounds from their environments. Molecular oxygen began to accumulate in the atmosphere as an end product of photosynthesis, an autotrophic process that produces sugars and oxygen by reacting water and carbon dioxide. Cyanobacteria appear to be primarily responsible for generation of atmospheric oxygen early in life's history.

All bacteria are prokaryotes, organisms that lack a membrane-bound nucleus and other organelles in their cytoplasm. The prokaryotes consist of two genetically distinct groups, Archaebacteria and Monera.

The eukaryotes apparently arose from symbiotic unions of two or more types of prokaryotes. The genetic material (DNA) of eukaryotes is borne in a membrane-bound nucleus, and also in mitochondria and sometimes plastids. Mitochondria and plastids have resemblances to bacteria, and their DNA is more closely allied to that of certain bacteria than to eukaryotic nuclear genomes.

REVIEW QUESTIONS

1. Explain each of these properties of water, and tell how each is conferred by the dipolar nature of the water molecule: high specific heat capacity; high heat of vaporization; unique density behavior; high surface tension; good solvent for ions of salts.
2. What was the composition of the earth's atmosphere at the time of the origin of life, and how did it differ from the atmosphere of today?
3. Regarding the experiments of Miller and Urey described in this chapter, explain what constituted the following in each case: observations, hypothesis, deduction, prediction, data, control. (The scientific method is described on p. 11.)
4. Explain the significance of the Miller-Urey experiments.
5. Name three different sources of energy that could have powered reactions on early earth to form organic compounds.
6. By what mechanism might organic molecules have been concentrated in the prebiotic world so that further reactions could occur?
7. Name two simple carbohydrates, two storage carbohydrates, and a structural carbohydrate.
8. What are characteristic differences in molecular structure between lipids and carbohydrates?
9. Explain the difference between the primary, secondary, tertiary, and quaternary structures of a protein.
10. What are the important nucleic acids in a cell, and of what units are they constructed?
11. Distinguish among the following: primary heterotroph, autotroph, secondary heterotroph.
12. What is the origin of the oxygen in the present-day atmosphere, and what is its metabolic significance to most organisms living today?
13. Distinguish between prokaryotes and eukaryotes.
14. Describe Margulis' view on the origin of eukaryotes from prokaryotes.
15. What was the "Cambrian explosion" and how might you explain it?

SELECTED REFERENCES

Berg, J. M., T. L. Tymoczko, and L. Stryer. 2002. Biochemistry. New York, W. H. Freeman. *A current and thorough textbook of biochemistry.*

Conway, Morris, S. 1993. The fossil record and the early evolution of the Metazoa. Nature **361:**219–225. *An important summary correlating fossil and molecular evidence.*

Gesteland, R. F., and J. F. Atkins, editors. 1993. The RNA world. Cold Spring Harbor, New York, Cold Spring Harbor Laboratory Press. *Evidence that there was a period when RNA served in both catalysis and transmission of genetic information.*

Kasting, J. F. 1993. Earth's early atmosphere. Science **259:**920–926. *Most investigators agree that there was little or no oxygen in the atmosphere of early earth and that there was a significant increase about 2 billion years ago.*

Knoll, A. H. 1991. End of the Proterozoic Eon. Sci. Am. **265:**64–73 (Oct.). *Multicellular animals probably originated only after oxygen in the atmosphere accumulated to a critical level.*

Lehninger, A. L., D. L. Nelson, and M. M. Cox. 2000. Principles of biochemistry, ed. 3. New York, Worth Publishers, Inc. *Clearly presented advanced textbook in biochemistry.*

Lodish, H., A. Berk, S. L. Zipursky, P. Matsudira, D. Baltimore, and J. Darnell. 2000. Molecular cell biology, ed. 4. New York, W. H. Freeman & Co. *Thorough treatment; begins with fundamentals such as energy, chemical reactions, bonds, pH, and biomolecules, then proceeds to advanced molecular biology.*

Margulis, L. 1993. Symbiosis in cell evolution, ed. 2. New York, W. H. Freeman. *An important updating of the author's 1981 book on this topic.*

Orgel, L. E. 1994. The origin of life on the earth. Sci. Am. **271:**77–83 (Oct.). *There is growing evidence for an RNA world, but difficult questions are unanswered.*

Rand, R. P. 1992. Raising water to new heights. Science **256:**618. *Cites some ways that water can affect function of protein molecules.*

Wainright, P. O., G. Hinkle, M. L. Sogin, and S. L. Stickel. 1993. Monophyletic origins of the Metazoa: an evolutionary link with fungi. Science **260:**940–942. *Molecular evidence that multicellular animals share a closer common ancestor with Fungi than with plants or other eukaryotes.*

Waldrop, M. M. 1992. Finding RNA makes proteins gives "RNA world" a big boost. Science **256:**1396–1397. *Reports on the significance of Noller et al. (1992, Science 256:1416), who found that ribosomal RNA can catalyze the formation of peptide bonds and hence proteins.*

Woese, C. R. 1984. The origin of life. Carolina Biology Readers, no. 13. Burlington, North Carolina, Carolina Biological Supply Company. *Thought-provoking account of the author's views, critique of traditional concepts, focus on problems.*

ZOOLOGY LINKS TO THE INTERNET

Visit the textbook's Online Learning Center at www.mhhe.com/zoology to find live Internet links for each of the topics listed here.

Inorganic Chemistry
Atoms
The Nature of Molecules
Bonds

Carbon
Water
Organic Chemistry
Carbohydrates
Lipids
Proteins

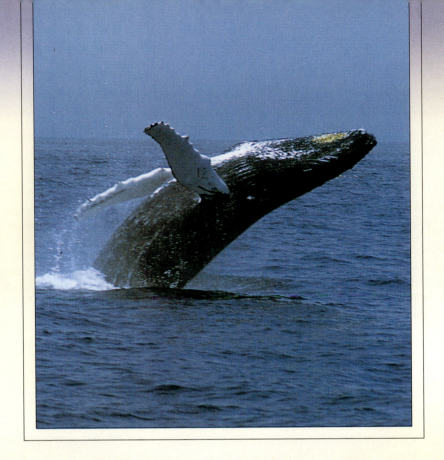

A humpback whale, *Megaptera novaeangliae,* leaps from the water.

Cells as Units of Life

The Fabric of Life

It is a remarkable fact that living forms, from amebas and unicellular algae to whales and giant redwood trees, are formed from a single type of building unit: cells. All animals and plants are composed of cells and cell products. Thus the cell theory is another great unifying concept of biology.

New cells come from division of preexisting cells, and the activity of a multicellular organism as a whole is the sum of activities and interactions of its constituent cells. Energy to support virtually all of life's activities flows from sunlight that is captured by green plants and algae and transformed by photosynthesis into chemical bond energy. Chemical bond energy is a form of potential energy that can be released when the bond is broken; the energy is used to perform electrical, mechanical, and osmotic tasks in the cell. Ultimately, all energy is dissipated, little by little, into heat. This is in accord with the second law of thermodynamics, which states that there is a tendency in nature to proceed toward a state of greater molecular disorder, or entropy. Thus the high degree of molecular organization in living cells is attained and maintained only as long as energy fuels the organization.

CELL CONCEPT

More than 300 years ago the English scientist and inventor Robert Hooke, using a primitive compound microscope, observed boxlike cavities in slices of cork and leaves. He called these compartments "little boxes or cells." In the years that followed Hooke's first demonstration of the remarkable powers of the microscope before the Royal Society of London in 1663, biologists gradually began to realize that cells were far more than simple containers filled with "juices."

Cells are the fabric of life. Even the most primitive cells are enormously complex structures that form the basic units of all living organisms. All tissues and organs are composed of cells. In a human an estimated 60 trillion cells interact, each performing its specialized role in an organized partnership. In single-celled organisms all functions of life are performed within the confines of one microscopic package. There is no life without cells. The idea that a cell represents the basic structural and functional unit of life is an important unifying concept of biology.

With the exception of some eggs, which are the largest cells (in volume) known, cells are small and mostly invisible to the unaided eye. Consequently, our understanding of cells paralleled technical advances in the resolving power of microscopes. The Dutch microscopist A. van Leeuwenhoek sent letters to the Royal Society of London containing detailed descriptions of the numerous organisms he had observed using high-quality single lenses that he had made (1673 to 1723). In the early nineteenth century, improved design of microscopes permitted biologists to see separate objects only 1 μm apart. This advance was quickly followed by new discoveries that laid the groundwork for the **cell theory**—a theory stating that all living organisms are composed of cells.

In 1838 Matthias Schleiden, a German botanist, announced that all plant tissue was composed of cells. A year later one of his countrymen, Theodor Schwann, described animal cells as being similar to plant cells, an understanding that had been long delayed because animal cells are bounded only by a nearly invisible plasma membrane rather than a distinct cell wall characteristic of plant cells. Schleiden and Schwann are thus credited with the unifying cell theory that ushered in a new era of productive exploration in cell biology. Another German, Rudolf Virchow, recognized that all cells came from preexisting cells (1858).

In 1840 J. Purkinje introduced the term **protoplasm** to describe cell contents. Protoplasm was at first thought to be a granular, gel-like mixture with special and elusive life properties of its own; cells were viewed as bags of thick soup containing a nucleus. Later the interior of cells became increasingly visible as microscopes were improved and better tissue-sectioning and staining techniques were introduced. Rather than being a uniform granular soup, a cell's interior is composed of numerous **cellular organelles,** each performing a specific function in the life of a cell. Today we realize that the components of a cell are so highly organized, structurally and functionally, that describing its contents as "protoplasm" is like describing the contents of an automobile engine as "autoplasm."

How Cells Are Studied

Light microscopes, with all their variations and modifications, have contributed more to biological investigation than any other instrument developed by humans. They have been powerful exploratory tools for 300 years, and they continue to be so more than 50 years after invention of the electron microscope. However, electron microscopy has vastly enhanced our appreciation of the delicate internal organization of cells, and modern biochemical, immunological, physical, and molecular techniques have contributed enormously to our understanding of cell structure and function.

Electron microscopes employ high voltages to direct a beam of electrons through objects examined. The wavelength of the electron beam is approximately 0.00001 that of ordinary white light, thus permitting far greater magnification and resolution (compare A and B of Figure 3-1). In preparation for viewing, specimens are cut into extremely thin sections (10 nm to 100 nm thick) and treated with "electron stains" (ions of elements such as osmium, lead, and uranium) to increase contrast between

Figure 3-1

Liver cells. **A,** Magnified approximately 400 times through light microscopy. Note the prominently stained nucleus in each polyhedral cell. **B,** Portion of single liver cell, magnified approximately 5000 times by electron microscopy. A single large nucleus dominates the field; mitochondria **(M),** rough endoplasmic reticulum **(RER),** and glycogen granules **(G),** are also seen.

A   B

different structures. Images are seen on a fluorescent screen and photographed (Figure 3-2). Because electrons pass through a specimen to a photographic plate, the instrument is called a transmission electron microscope.

In contrast, specimens prepared for scanning electron microscopy are not sectioned, and electrons do not pass through them. The whole specimen is bombarded with electrons, causing secondary electrons to be emitted. An apparent three-dimensional image is recorded in the photograph. Although the magnification capability of scanning instruments is not as great as transmission microscopes, much has been learned about surface features of organisms and cells, as well as internal, membrane-bound structures. Examples of scanning electron micrographs are shown on pp. 138, 154, and 655.

A still greater level of resolution can be achieved with X-ray crystallography and nuclear magnetic resonance (NMR) spectroscopy. These techniques reveal a great deal about the shape of biomolecules and the relationship of atoms within them to each other. Both techniques are laborious, but NMR spectroscopy does not require purification and crystallization of a substance, and molecules can be observed in solution.

Advances in techniques of cell study (cytology) are not limited to improvements in microscopes but include new methods of tissue preparation, staining for microscopic study, and

the great contributions of modern biochemistry and molecular biology. For example, the various organelles of cells have differing, characteristic densities. Cells can be broken up with most of the organelles remaining intact, then centrifuged in a density gradient (Figure 3-3), and relatively pure preparations of each organelle may be recovered. Thus the biochemical functions of various organelles may be studied separately. DNA and various types of RNA can be extracted and studied. Many enzymes can be purified and their characteristics determined. We use radioactive isotopes to study many metabolic reactions and pathways in cells. Modern chromatographic techniques can separate chemically similar intermediates and products. A particular protein in cells can be extracted and purified, and specific antibodies (see p. 741) against the protein can be prepared. When the antibody is complexed with a fluorescent

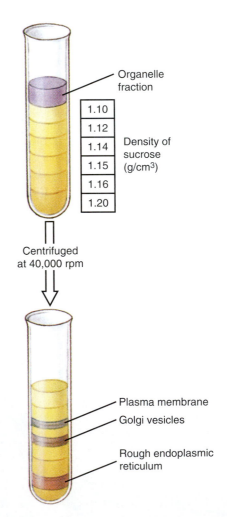

Figure 3-3

Separation of cell organelles in a density gradient by ultracentrifugation. The gradient is formed by layering sucrose solutions in a centrifuge tube, then carefully placing a preparation of mixed organelles on top. The tube is centrifuged at about 40,000 revolutions per minute for several hours, and organelles become separated down the tube according to their density.

Figure 3-2

Comparison of optical paths of light and electron microscopes. To facilitate comparison, the scheme of the light microscope has been inverted from its usual orientation with light source below and image above. In an electron microscope the lenses are magnets to focus the beam of electrons.

substance and the complex is used to stain cells, the complex binds to the protein of interest, and its precise location in cells can be determined. We could cite many more examples that have contributed enormously to our present understanding of cell structure and function.

ORGANIZATION OF CELLS

If we were to restrict our study of cells to fixed and sectioned tissues, we would be left with the erroneous impression that cells are static, quiescent, rigid structures. In fact, a cell's interior is in a constant state of upheaval. Most cells are continually changing shape, pulsing, and heaving; their organelles twist and regroup in a cytoplasm teeming with starch granules, fat globules, and vesicles of various sorts. This description is derived from studies of living cell cultures with time-lapse photography and video. If we could see the swift shuttling of molecular traffic through gates in the cell membrane and the metabolic energy transformations within cell organelles, we would have an even stronger impression of internal turmoil. However, cells are anything but bundles of disorganized activity. There is order and harmony in cell functioning. Studying this dynamic phenomenon through a microscope, we realize that, as we gradually comprehend more and more about these units of life, we are gaining a greater understanding of the nature of life itself.

Prokaryotic and Eukaryotic Cells

We already described the radically different cell plan of prokaryotes and eukaryotes (p. 32). A fundamental distinction, expressed in their names, is that prokaryotes lack the membrane-bound nucleus present in all eukaryotic cells. Among other differences, eukaryotic cells have many membranous organelles

(specialized structures that perform particular functions within cells) (Table 3-1).

Despite these differences, which are of paramount importance in cell studies, prokaryotes and eukaryotes have much in common. Both have DNA, use the same genetic code, and synthesize proteins. Many specific molecules such as ATP perform similar roles in both. These fundamental similarities imply common ancestry. The upcoming discussion is restricted to eukaryotic cells, of which all animals are composed.

Components of Eukaryotic Cells and Their Functions

Typically, eukaryotic cells are enclosed within a thin, selectively permeable **cell membrane** (Figure 3-4). The most prominent organelle is a spherical or ovoid **nucleus,** enclosed within *two* membranes to form a double-layered **nuclear envelope** (Figure 3-4). Cellular material located between the cell membrane and nuclear envelope is collectively called **cytoplasm.** Within the cytoplasm are many organelles, such as mitochondria, Golgi complexes, centrioles, and endoplasmic reticulum. Plant cells typically contain **plastids,** some of which are photosynthetic organelles, and plant cells bear a cell wall containing cellulose outside the cell membrane.

The **fluid-mosaic model** is the currently accepted concept of cell membranes. By electron microscopy, a cell membrane appears as two dark lines, each approximately 3 nm thick, at each side of a light zone (Figure 3-5). The entire membrane is 8 to 10 nm thick. This image is the result of a phospholipid bilayer, two layers of phospholipid molecules, all oriented with their water-soluble ends toward the outside and their fat-soluble portions toward the inside of the membrane (Figure 3-6). An important characteristic of the phospholipid bilayer is that it is liquid,

TABLE 3.1

Comparison of Prokaryotic and Eukaryotic Cells

Characteristic	Prokaryotic Cell	Eukaryotic Cell
Cell size	Mostly small (1–10 μm)	Mostly large (10–100 μm)
Genetic system	DNA with some nonhistone protein; simple, circular DNA molecule in nucleoid; nucleoid is not membrane bound	DNA complexed with histone and nonhistone proteins in complex chromosomes within nucleus with membranous envelope
Cell division	Direct by binary fission or budding; no mitosis	Some form of mitosis; centrioles in many; mitotic spindle present
Sexual system	Absent in most; highly modified if present	Present in most; male and female partners; gametes that fuse to form zygote
Nutrition	Absorption by most; photosynthesis by some	Absorption, ingestion, photosynthesis by some
Energy metabolism	No mitochondria; oxidative enzymes bound to cell membrane, not packaged separately; great variation in metabolic pattern	Mitochondria present; oxidative enzymes packaged therein; more unified pattern of oxidative metabolism
Intracellular movement	None	Cytoplasmic streaming, phagocytosis, pinocytosis
Flagella/cilia	If present, not with "9 + 2" microtubular pattern	With "9 + 2" microtubular pattern
Cell wall	Contains disaccharide chains cross-linked with peptides	If present, not with disaccharide polymers linked with peptides

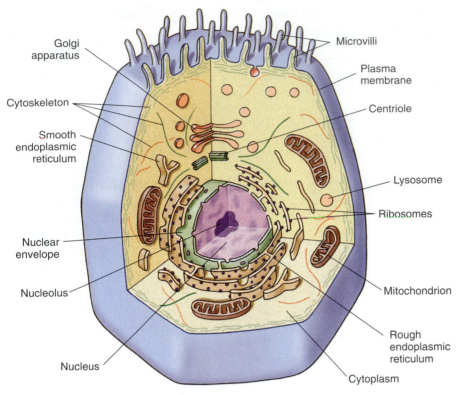

Golgi
apparatus

Microvilli

Plasma
membrane

Cytoskeleton

Centriole

Smooth
endoplasmic
reticulum

Lysosome

Ribosomes

Nuclear
envelope

Nucleolus

Mitochondrion

Nucleus

Rough
endoplasmic
reticulum

Cytoplasm

Figure 3-4
Generalized cell with principal organelles, as might be seen with the electron microscope.

Figure 3-5
Plasma membranes of two adjacent cells. Each membrane (*between arrows*) shows a typical dark-light-dark staining pattern. (×325,000)

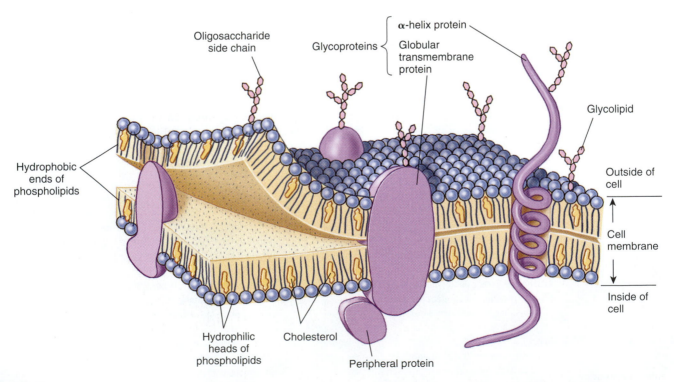

Oligosaccharide
side chain

Glycoproteins

α-helix protein

Globular
transmembrane
protein

Glycolipid

Hydrophobic
ends of
phospholipids

Outside of
cell

Cell
membrane

Inside of
cell

Hydrophilic
heads of
phospholipids

Cholesterol

Peripheral protein

Figure 3-6
Diagram illustrating fluid-mosaic model of a cell membrane.

giving the membrane flexibility and allowing the phospholipid molecules to move sideways freely within their own monolayer. Molecules of cholesterol are interspersed in the lipid portion of the bilayer (Figure 3-6). They make the membrane even less permeable and decrease its flexibility.

Glycoproteins (proteins with carbohydrates attached) are essential components of cell membranes. Some of these proteins catalyze the transport of substances such as charged ions across the membrane. Others act as specific receptors for various molecules or as highly specific markers. For example, the self/nonself recognition that enables the immune system to react to invaders (Chapter 35) is based on proteins of this type. Some aggregations of protein molecules form pores through which small polar molecules may enter. Like the phospholipid molecules, most glycoproteins can move laterally in the membrane, although more slowly.

Nuclear envelopes contain less cholesterol than cell membranes, and pores in the envelope (Figure 3-7) allow molecules to move between nucleus and cytoplasm. Nuclei contain **chromatin,** a complex of DNA, basic proteins called **histones,** and nonhistone protein. Chromatin carries the genetic information, the code that results in most of the components characteristic of a cell after transcription and translation (see Chapter 5). **Nucleoli** are specialized parts of certain chromosomes that stain in a characteristically dark manner. They carry multiple copies of the DNA information to synthesize ribosomal RNA. After transcription from DNA, ribosomal RNA combines with protein to form **ribosomal subunits,** which detach from the nucleolus and pass to the cytoplasm through pores in the nuclear envelope.

The outer membrane of the nuclear envelope is continuous with extensive membranous elements in the cytoplasm called **endoplasmic reticulum (ER)** (Figures 3-7 and 3-8). The space between the membranes of the nuclear envelope communicates with channels **(cisternae)** in the ER. The ER is a complex of membranes that separates products of a cell from the synthetic machinery that produces them, apparently functioning as routes for transport of proteins within a cell. Membranes of the ER may be covered on their outer surfaces with ribosomes and are thus designated **rough ER,** or they may lack ribosomal covering and be called **smooth ER.** Ribosomes on rough ER synthesize peptides or proteins that enter the ER cisternae or membrane and are destined for incorporation into the cell membrane or for export from the cell, or they are bound for the lysosomes (after passing through the Golgi complex, see next paragraph). Smooth ER functions in synthesis of lipids and phospholipids.

The **Golgi complex** (Figures 3-9 and 3-10) is composed of a stack of membranous vesicles that function in storage, modification, and packaging of protein products, especially secretory products. These vesicles do not synthesize protein but may add complex carbohydrates to the molecules. Small vesicles of ER containing protein pinch off and then fuse with sacs on the "forming face" of a Golgi complex. After modification, the proteins bud off vesicles on the "maturing face" of the complex (Figure 3-10). The contents of some of these vesicles may be expelled to the outside of the cell, as secretory products destined to be exported from a glandular cell. Some may bear integral proteins for the cell membrane, such as receptors or carrier proteins. Others may contain digestive enzymes that remain in the same cell that produces them. Such vesicles are called **lysosomes** (literally "loosening body," a body capable of causing lysis, or disintegration). Enzymes that they contain are involved in the breakdown of foreign material, including bacteria engulfed by a cell. Lysosomes also are capable of breaking down injured or diseased cells and worn-out cellular components. Their enzymes are so powerful that they kill the cell that formed them if the lysosome membrane ruptures. In normal cells the enzymes remain safely enclosed within the protective membrane. Lysosomal vesicles may pour their enzymes into a larger membrane-bound body containing an ingested food particle, a **food vacuole** or **phagosome.** Other vacuoles, such as **contractile vacuoles** of some single-celled organisms (p. 213), may contain only fluid and function to regulate ions and water.

Mitochondria (sing., **mitochondrion**) (Figure 3-11) are conspicuous organelles present in nearly all eukaryotic cells. They are diverse in size, number, and shape; some are rodlike, and others are more or less spherical. They may be scattered uniformly throughout the cytoplasm, or they may be localized near cell surfaces and other regions where there is high metabolic activity. A mitochondrion is composed of a double membrane. The outer membrane is smooth, whereas the inner membrane is folded into numerous platelike or fingerlike projections called **cristae** (Figure 3-11),

Nucleolus   Nucleus   Mitochondrion

Endoplasmic reticulum   Glycogen particles

Figure 3-7

Electron micrograph of part of hepatic cell of rat showing portion of nucleus (*left*) and surrounding cytoplasm. Endoplasmic reticulum and mitochondria are visible in cytoplasm, and pores (*arrows*) can be seen in nuclear envelope. (×14,000)

Figure 3-8

Endoplasmic reticulum. **A,** Endoplasmic reticulum is continuous with the nuclear envelope. It may have associated ribosomes (rough endoplasmic reticulum) or not (smooth endoplasmic reticulum). **B,** Electron micrograph showing rough endoplasmic reticulum. (×28,000)

Ribosomes

Hosmer

A

B

Figure 3-9

Golgi complex (= Golgi body, Golgi apparatus). **A,** The smooth cisternae of the Golgi complex have enzymes that modify proteins synthesized by the rough endoplasmic reticulum. **B,** Electron micrograph of a Golgi complex. (×46,000)

Secretory vesicles

A

B

which increases internal surface area where chemical reactions take place. These characteristic features make mitochondria easy to identify among organelles. Mitochondria are often called "powerhouses of cells," because enzymes located on the cristae carry out the energy-yielding steps of aerobic metabolism. ATP (adenosine triphosphate), the most important energy-transfer molecule of all cells, is produced in this organelle. Mitochondria are self-replicating. They have a tiny, circular genome, much like genomes of prokaryotes except that it is much smaller. It contains DNA that specifies some, but not all, proteins of a mitochondrion.

Eukaryotic cells characteristically have a system of tubules and filaments that form a **cytoskeleton** (Figures 3-12 and 3-13). These provide support and maintain the form of cells, and in many cells, they provide a means of locomotion and translocation of organelles within the cell. **Microfilaments** are thin, linear structures, first observed distinctly in muscle cells, where they are responsible for the ability of a cell to contract. They are made of a protein called **actin.** Several dozen other proteins are known that bind with actin and determine its configuration and behavior in particular cells. One of these is **myosin,** whose interaction with actin causes contraction in muscle and other

cells (p. 626). Actin microfilaments also provide a means for moving messenger RNA (p. 91) from the nucleus to particular positions within the cytoplasm. **Microtubules,** somewhat larger than microfilaments, are tubular structures composed of a protein called **tubulin** (Figure 3-13). They play a vital role in moving chromosomes toward the daughter cells during cell division (p. 51), and they are important in intracellular architecture, organization, and transport. In addition, microtubules form essential parts of the structures of cilia and flagella. Microtubules radiate out from a microtubule organizing center, the **centrosome,** near the nucleus. Centrosomes are not membrane bound. Within centrosomes are found a pair of **centrioles** (Figures 3-4 and 3-14), which are themselves composed of microtubules. Microtubules radiating from the centrioles form the **aster.** Each centriole of a pair lies at right angles to the other and is a short cylinder of nine triplets of microtubules. They replicate before cell division. Although cells of higher plants do not have centrioles, a microtubule organizing center is present. **Intermediate filaments** are larger than microfilaments but smaller than microtubules. There are five biochemically distinct types of intermediate filaments, and their composition and arrangement depend on the cell type in which they are found.

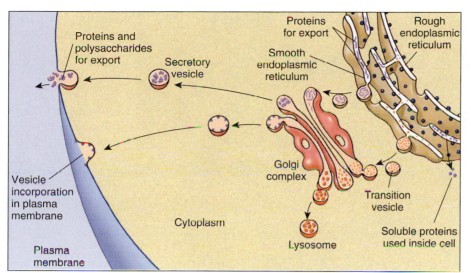

Proteins and polysaccharides for export

Secretory vesicle

Proteins for export

Smooth endoplasmic reticulum

Rough endoplasmic reticulum

Vesicle incorporation in plasma membrane

Golgi complex

Transition vesicle

Cytoplasm

Lysosome

Soluble proteins used inside cell

Plasma membrane

Figure 3-10

System for assembling, isolating, and secreting proteins for export in a eukaryotic cell, for lysosomes, or for incorporation into plasma membrane.

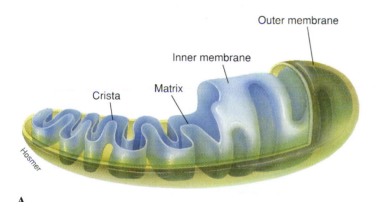

Outer membrane

Inner membrane

Crista

Matrix

Hosmer

A

B

Figure 3-11

Mitochondria. **A,** Structure of a typical mitochondrion. **B,** Electron micrograph of mitochondria in cross and longitudinal section. (×30,000)

Surfaces of Cells and Their Specializations

The free surface of epithelial cells (cells that cover the surface of a structure or line a tube or cavity) sometimes bears either **cilia** or **flagella** (sing., **cilium, flagellum**). These are motile extensions of the cell surface that sweep materials past the cell. In many single-celled organisms and some small multicellular forms, they propel the entire organism through a liquid medium (see pp. 224, 268). Flagella provide the means of locomotion for male reproductive cells (sperm) of most animals and many plants.

Cilia and flagella have different beating patterns (see p. 624), but their internal structure is the same. With few exceptions, the internal structures of locomotory cilia and flagella are composed of a long cylinder of nine pairs of microtubules enclosing a central pair (see Figure 11-3). At the base of each cilium or flagellum is a **basal body (kinetosome),** which is identical in structure to a centriole.

Indeed, cilia and flagella are so alike in details of their structure that it seems highly likely that they had a common evolutionary origin. Whether their origin was the symbiosis of a spirochete-like bacterium and host cell, as suggested by Margulis (see p. 33), is more conjectural. Margulis and others prefer the term *undulipodia* to include both cilia and flagella, and it is less awkward to use one word for organelles that are alike in structure and origin. However, the terms cilia and flagella are so widely used that every student should be familiar with them.

Many cells move neither by cilia nor flagella but by **ameboid movement** using **pseudopodia.** Some groups of protozoa (p. 211), migrating cells in embryos of multicellular animals, and some cells of adult multicellular animals, such as white blood cells, show ameboid movement. Cytoplasmic streaming through the action of actin microfilaments extends a lobe (pseudopodium) outward from the surface of the cell. Continued streaming in the

Intermediate filaments   Microtubules

Microfilaments

Figure 3-12

Cytoskeleton of a cell, showing its complex nature. Three visible cytoskeletal elements, in order of increasing diameter, are microfilaments, intermediate filaments, and microtubules. (×66,600)

Figure 3-13

The microtubules in kidney cells of a baby hamster have been rendered visible by treatment with a preparation of fluorescent proteins that specifically bind to tubulin.

Microtubule triplet

A

B

Figure 3-14

Centrioles. **A,** Each centriole is composed of nine triplets of micro-tubules arranged as a cylinder. **B,** Electron micrograph of a pair of centrioles, one in longitudinal (*right*) and one in cross section (*left*). The normal orientation of centrioles is at right angles to each other.

direction of a pseudopodium brings cytoplasmic organelles into the lobe and accomplishes movement of the entire cell. Some specialized pseudopodia have cores of microtubules (p. 212), and movement is effected by assembly and disassembly of the tubular rods.

Cells covering the surface of a structure (epithelial cells) or cells packed together in a tissue may have specialized junctional complexes between them. Nearest the free surface, the membranes of two cells next to each other appear to fuse, forming a **tight junction** (Figure 3-15). Tight junctions function as seals to prevent the passage of molecules between cells from one side of a layer of cells to another, because there is usually a space of about 20 nm between the cell membranes of adjacent cells. Tight junctions between intestinal cells, for example, force molecules absorbed from the intestinal contents to pass through epithelial cells, rather than between them. At various points beneath tight junctions, small ellipsoid discs occur, just within the cell membrane in each cell. These appear to act as "spot-welds" and are called **desmosomes.** From each desmosome a tuft of intermediate filaments extends into the cytoplasm, and linker proteins extend through the cell membrane into the intercellular space to bind the discs together. Desmosomes are not seals but seem to increase the strength of the tissue. **Gap junctions,** rather than serving as points of attachment, provide a means of intercellular communication. They form tiny canals between cells, so that their cytoplasm becomes continuous, and small molecules and ions can pass from one cell to the other. Gap junctions may occur between cells of epithelial, nervous, and muscle tissues.

Figure 3-15

Two apposing plasma membranes forming the boundary between two epithelial cells. Various kinds of junctional complexes are found. A tight junction is a firm, adhesive band completely encircling the cell. Desmosomes are isolated "spot-welds" between cells. Gap junctions serve as sites of intercellular communication. Intercellular space may be greatly expanded in cells of some tissues.

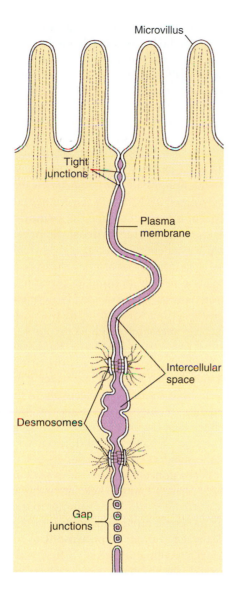

Figure 3-16

Electron micrograph of microvilli. (×59,000)

(also called the plasmalemma) form dynamic structures having remarkable activity and selectivity. They are a permeability barrier that separates the interior from the external environment. They regulate the vital flow of molecular traffic into and out of the cell, and provide many of the unique functional properties of specialized cells.

Membranes inside a cell surround a variety of organelles. Indeed, a cell is a system of membranes that divide it into numerous compartments. It has been estimated that if all membranes present in one gram of liver tissue were spread out flat, they would cover 30 square meters! Internal membranes share many structural features of plasma membranes and are the site for many, perhaps most, of a cell's enzymatic reactions.

A plasma membrane acts as a selective gatekeeper for entrance and exit of many substances involved in cell metabolism. Some substances can pass through with ease, others enter slowly and with difficulty, and still others cannot enter at all. Because conditions outside a cell are different from and more variable than conditions within a cell, it is necessary that passage of substances across the membrane be rigorously controlled.

We recognize three principal ways that a substance may enter across a cell membrane: (1) by **diffusion** along a concentration gradient; (2) by a **mediated transport system,** in which the substance binds to a specific site that assists it across the membrane; and (3) by **endocytosis,** in which the substance is enclosed within a vesicle that forms on the membrane surface and detaches inside the cell.

Another specialization of cell surfaces is the lacing together of adjacent cell surfaces where cell membranes of the cells infold and interdigitate very much like a zipper. These infoldings are especially common in epithelial cells of kidney tubules. The distal or apical boundaries of some epithelial cells, as seen by electron microscopy, show regularly arranged **microvilli** (sing., **microvillus**). They are small, fingerlike projections consisting of tubelike evaginations of the cell membrane with a core of cytoplasm containing bundles of microfilaments (Figure 3-16). They are seen clearly in the lining of the intestine where they greatly increase the absorptive and digestive surface. Such specializations appear as brush borders by light microscopy.

Membrane Function

The incredibly thin, yet sturdy, plasma membrane that encloses every cell is vitally important in maintaining cellular integrity. Once believed to be a rather static entity that defined cell boundaries and kept cell contents from spilling out, plasma membranes

Diffusion and Osmosis

Diffusion is a movement of particles from an area of higher concentration to an area of lower concentration of the particles or molecules, thus tending to equalize the concentration throughout the area of diffusion. If a living cell surrounded by a membrane is immersed in a solution having a higher concentration of solute molecules than the fluid inside the cell, a **concentration gradient** instantly exists between the two fluids. Assuming that the membrane is **permeable** to the solute, there is a net movement of solute toward the inside, the side having the lower concentration. The solute diffuses "downhill" across the membrane until its concentrations on each side are equal.

Most cell membranes are **selectively permeable,** that is, permeable to water but variably permeable or impermeable to solutes. In free diffusion it is this selectiveness that regulates molecular traffic. As a rule, gases (such as oxygen and carbon dioxide), urea, and lipid-soluble solutes (such as hydrocarbons and alcohol) are the only solutes that can diffuse through biological membranes with any degree of freedom. Because many water-soluble molecules readily pass through membranes, such movements cannot be explained by simple diffusion. Sugars, as well as many electrolytes and macromolecules, are moved across membranes by carrier-mediated processes, which are described in the next section.

If we place a membrane between two unequal concentrations of solutes to which the membrane is impermeable, water flows through the membrane from the more dilute to the more concentrated solution. The water molecules move across the membrane down a concentration gradient from an area where the *water* molecules are more concentrated to an area on the other side of the membrane where they are less concentrated. This is **osmosis.**

We can demonstrate osmosis by a simple experiment in which we tie a selectively permeable membrane tightly over the end of a funnel. We fill the funnel with a salt solution and place it in a beaker of pure water so that the water levels inside and outside the funnel are equal. In a short time the water level in the glass tube of the funnel rises, indicating a net movement of water through the membrane into the salt solution (Figure 3-17).

Inside the funnel are salt molecules, as well as water molecules. In the beaker outside the funnel are only water molecules. Thus the concentration of water is less on the inside because some of the available space is occupied by the nondiffusible salt ions. A concentration gradient exists for water molecules in the system. Water diffuses from the region of greater concentration of water (pure water outside) to the region of lesser concentration (salt solution inside).

As water enters the salt solution, the fluid level in the funnel rises. Eventually the pressure produced by the increasing weight of solution in the funnel pushes water molecules out as fast as they enter. The level in the funnel becomes stationary and the system is in equilibrium. The **osmotic pressure** of the solution is equivalent to the **hydrostatic pressure** necessary to prevent further net entry of water.

The concept of osmotic pressure is not without problems. A solution reveals an osmotic "pressure" only when it is separated from solvent by a selectively permeable membrane. It can be disconcerting to think of an isolated bottle of salt solution as having "pressure" much as compressed gas in a bottle (*hydrostatic* pressure) would have. Furthermore, the osmotic pressure is really the hydrostatic pressure that must be applied to a solution to keep it from gaining water *if* the solution were separated from pure water by a selectively permeable membrane. Consequently, biologists frequently use the term **osmotic potential** rather than osmotic pressure. However, since the term "osmotic pressure" is so firmly fixed in our vocabulary, it is necessary to understand the usage despite its potential confusion.

The concept of osmosis is very important in understanding how animals control their internal fluid and solute environment (see Chapter 30). For example, marine bony fishes maintain a solute concentration in their blood about one-third of that in sea-water; they are **hypoosmotic** to seawater. If a fish swims into a river mouth and then up a freshwater stream, as salmon do, it would pass through a region where its blood solutes were equal in concentration to those in its environment **(isosmotic),** then enter fresh water, where its blood solutes were **hyperosmotic** to those in its environment. It must have physiological mechanisms to avoid net loss of water in the sea and gain of water in the river.

Figure 3-17

Simple membrane osmometer. **A,** The end of a tube containing a salt solution is closed at one end by a selectively permeable membrane. The membrane is permeable to water but not to salt. **B,** When the tube is immersed in pure water, water molecules diffuse through the membrane into the tube. Water molecules are in higher concentration in the beaker because they are diluted inside the tube by salt ions. Because the salt cannot diffuse out through the membrane, the volume of fluid inside the tube increases, and the level rises. **C,** When the weight of the column of water inside the tube exerts a downward force (hydrostatic pressure) causing water molecules to leave through the membrane in equal number to those that enter, the volume of fluid inside the tube stops rising. At this point the hydrostatic pressure is equivalent to the osmotic pressure.

Mediated Transport

We have seen that a cell membrane is an effective barrier to free diffusion of most molecules of biological significance. Yet it is essential that such materials enter and leave a cell. Nutrients such as sugars and materials for growth such as amino acids must enter a cell, and wastes of metabolism must leave. Such molecules are moved across a membrane by special proteins called **transporters** or **permeases.** Transporters enable solute molecules to cross the phospholipid bilayer (Figure 3-18A). Transporters are usually quite specific, recognizing and transporting only a limited group of chemical substances or perhaps even a single substance.

At high concentrations of solute, mediated transport systems show a saturation effect. This means simply that the rate of influx reaches a plateau beyond which increasing the solute concentration has no further effect on influx rate (Figure 3-18B). This is evidence that the number of transporters available in a membrane is limited. When all transporters become occupied by solutes, the rate of transport is at a maximum and it cannot be increased. Simple diffusion shows no such limitation; the greater the difference in solute concentrations on the two sides of the membrane, the faster the influx.

Two distinctly different kinds of mediated transport mechanisms are recognized: (1) **facilitated diffusion,** in which a transporter assists a molecule to diffuse through the membrane

A

B

Figure 3-18

Facilitated transport. **A,** A permease molecule binds with a molecule to be transported (substrate) on one side of a plasma membrane, changes shape, and releases the molecule on the other side. Facilitated transport takes place in the direction of a concentration gradient. **B,** Rate of transport increases with increasing substrate concentration until all permease molecules are occupied.

that it cannot otherwise penetrate, and (2) **active transport,** in which energy is supplied to the transporter system to transport molecules in the direction opposite a concentration gradient (Figure 3-19). Facilitated diffusion therefore differs from active transport in that it sponsors movement in a downhill direction (in the direction of a concentration gradient) only and requires no metabolic energy to drive the transport system.

In many animals facilitated diffusion aids in transport of glucose (blood sugar) into body cells that oxidize it as a principal energy source for the synthesis of ATP. The concentration of glucose is greater in blood than in the cells that consume it, favoring inward diffusion, but glucose is a water-soluble molecule that does not, by itself, penetrate cell membranes rapidly enough to support the metabolism of many cells; the carrier system increases the inward flow of glucose.

In active transport, molecules are moved uphill against the forces of passive diffusion. Active transport always involves an expenditure of energy (from ATP) because materials are pumped against a concentration gradient. Among the most important active transport systems in all animals are those that maintain sodium and potassium ion gradients between cells and the surrounding extracellular fluid or external environment. Most animal cells require a high internal concentration of potassium ions for protein synthesis at the ribosome and for certain enzymatic functions. The potassium ion concentration may be 20 to 50 times greater inside a cell than outside. Sodium ions, on the other hand, may be 10 times more concentrated outside a cell than inside. Both of these ionic gradients are maintained by active transport of potassium ions into and sodium ions out of the cell. In many cells outward pumping of sodium is linked to inward pumping of potassium; the same transporter molecule does both. As much as 10% to 40% of all energy produced by cells is consumed by the **sodium-potassium exchange pump** (Figure 3-19).

Endocytosis

Endocytosis, the ingestion of material by cells, is a collective term that describes three similar processes, phagocytosis, potocytosis, and receptor-mediated endocytosis (Figure 3-20). They are pathways for specifically internalizing solid particles, small molecules and ions, and macromolecules, respectively. All require energy and thus may be considered forms of active transport.

Phagocytosis, which literally means "cell eating," is a common method of feeding among protozoa and lower metazoa. It is also the way in which white blood cells (leukocytes) engulf cellular debris and uninvited microbes in the blood. By phagocytosis, an area of the cell membrane, coated internally with actin and actin-binding proteins, forms a pocket that engulfs the solid material. The membrane-enclosed vesicle then detaches from the cell surface and moves into the cytoplasm where it fuses with lysosomes, and its contents are digested by intracellular enzymes.

Potocytosis is similar to phagocytosis except that small areas of the surface membrane are invaginated into cells to form tiny vesicles. The invaginated pits and vesicles are called **caveolae** (ka-vee'o-lee). Specific binding receptors for the molecule or ion to be internalized are concentrated on the cell surface of caveolae. Potocytosis apparently functions for intake of at least some

Figure 3-19

Sodium-potassium pump, powered by bond energy of ATP, maintains the normal gradients of these ions across the cell membrane. The pump works by a series of conformational changes in the permease: *Step 1.* Three ions of Na^+ bind to the interior end of the permease, producing a conformational (shape) change in the protein complex. *Step 2.* The complex binds a molecule of ATP and cleaves it. *Step 3.* The binding of the phosphate group to the complex induces a second conformational change, passing the three Na^+ ions across the membrane, where they are now positioned facing the exterior. This new conformation has a very low affinity for the Na^+ ions, which dissociate and diffuse away, but it has a high affinity for K^+ ions and binds two of them as soon as it is free of the Na^+ ions. *Step 4.* Binding of the K^+ ions leads to another conformational change in the complex, this time leading to dissociation of the bound phosphate. Freed of the phosphate, the complex reverts to its original conformation, with the two K^+ ions exposed on the interior side of the membrane. This conformation has a low affinity for K^+ ions so that they are now released, and the complex has the conformation it started with, having a high affinity for Na^+ ions.

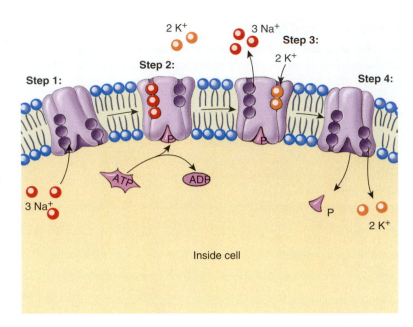

vitamins, and similar mechanisms may be important in translocating substances from one side of a cell to the other (see "exocytosis," next heading) and internalizing signal molecules, such as some hormones or growth factors.

> In phagocytosis, potocytosis, and receptor-mediated endocytosis some amount of extracellular fluid is necessarily trapped in the vesicle and nonspecifically brought within the cell. We describe this as *bulk-phase endocytosis,* and because it is nonspecific, the process corresponds roughly to what we have called traditionally *pinocytosis,* or "cell-drinking." Actually, potocytosis also means "cell-drinking" but was coined to distinguish internalization of specific small molecules or ions.

Receptor-mediated endocytosis is a specific mechanism for bringing large molecules within the cell. Proteins of the plasma membrane specifically bind particular molecules (termed **ligands** in this process), which may be present in the extracellular fluid in very low concentrations. The invaginations of the cell surface that bear the receptors are coated within the cell with a protein called **clathrin;** hence, they are described as **clathrin-coated pits.** As a clathrin-coated pit with its receptor-bound ligand invaginates and is brought within the cell, it is uncoated, the receptor and the ligand are dissociated, and the receptor and membrane material are recycled back to the surface membrane. Some important proteins, peptide hormones, and cholesterol are brought into cells in this manner.

Exocytosis

Just as materials can be brought into a cell by invagination and formation of a vesicle, the membrane of a vesicle can fuse with the plasma membrane and extrude its contents to the surrounding medium. This is the process of **exocytosis.** This process occurs in various cells to remove undigestible residues of substances brought in by endocytosis, to secrete substances such as hormones (Figure 3-10), and to transport a substance completely across a cellular barrier (**transcytosis**), as we just mentioned. For example, a substance may be picked up on one side of the wall of a blood vessel by potocytosis, moved across the cell, and released by exocytosis.

MITOSIS AND CELL DIVISION

All cells arise from the division of preexisting cells. All cells found in most multicellular organisms originated from the division of a single cell, a **zygote,** which is the product of union (fertilization) of an **egg** and a **sperm (gametes).** Cell division provides the basis for one form of growth, for both sexual and asexual reproduction, and for transmission of hereditary qualities from one cell generation to another cell generation.

In the formation of **body cells (somatic cells)** the process of nuclear division is **mitosis.** By mitosis each "daughter cell" is ensured of receiving a complete set of genetic instructions. Mitosis is a delivery system for distributing the chromosomes and the DNA they contain to continuing cell generations. As an animal grows, its somatic cells differentiate and assume different functions and appearances because of differential gene action. Even though most of the genes in specialized cells remain silent and unexpressed throughout the lives of those cells, every cell possesses a complete genetic complement. Mitosis ensures equality of genetic potential; later, other processes direct the orderly expression of genes during embryonic development by selecting from the genetic instructions that each cell contains. (These fundamental properties of cells of multicellular organisms are discussed further in Chapter 8.)

In animals that reproduce **asexually,** mitosis is the only mechanism for the transfer of genetic information from parent to progeny. In animals that reproduce **sexually,** the parents must

Phagocytosis	**Potocytosis**	**Receptor-mediated Endocytosis**

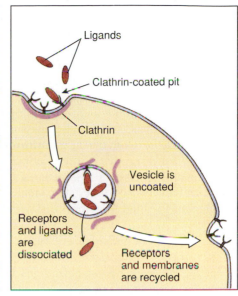

Figure 3-20

Three types of endocytosis. In phagocytosis the cell membrane binds to a large particle and extends to engulf it. In potocytosis small areas of cell membrane, bearing specific receptors for a small molecule or ion, invaginate to form caveolae. Receptor-mediated endocytosis is a mechanism for selective uptake of large molecules in clathrin-coated pits. Binding of the ligand to the receptor on the surface membrane stimulates invagination of pits.

produce **sex cells** (gametes or germ cells) that contain only half the usual number of chromosomes, so that offspring formed by the union of gametes will not contain double the number of parental chromosomes. This requires a special type of *reductional* division called **meiosis,** described in Chapter 5 (p. 75).

Structure of Chromosomes

As mentioned on page 41, DNA in eukaryotic cells occurs in chromatin, a complex of DNA with histone and nonhistone protein. Chromatin is organized into a number of discrete bodies called **chromosomes** (color bodies), so named because they stain deeply with certain biological dyes. In cells that are not dividing, chromatin is loosely organized and dispersed, so that individual chromosomes cannot be distinguished by light microscopy (Chapter 5, p. 56). Before division the chromatin becomes more compact, and chromosomes can be recognized and their individual morphological characteristics determined. They are of varied lengths and shapes, some bent and some rodlike. Their number is constant for a species, and every body cell (but not the germ cells) has the same number of chromosomes regardless of a cell's function. A human, for example, has 46 chromosomes in each somatic cell.

During mitosis (nuclear division) chromosomes shorten further and become increasingly condensed and distinct, and each assumes a shape partly characterized by the position of a constriction, the **centromere** (Figure 3-21). The centromere is the location of the **kinetochore,** a disc of proteins that binds with microtubules of the spindle fibers during mitosis.

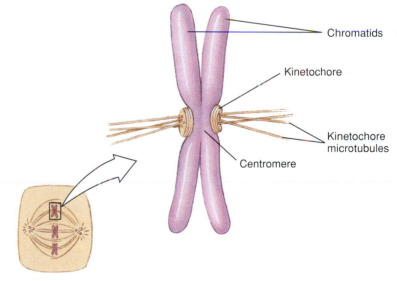

Figure 3-21

Structure of a metaphase chromosome. Sister chromatids are still attached at their centromere. Each chromatid has a kinetochore, to which the kinetochore fibers are attached. Kinetochore microtubules from each chromatid run to one of the centrosomes, which are located at opposite poles.

The problem of packaging a cell's DNA so that the genetic instructions are accessible during the transcription process, and compacting them to a form that makes mitosis possible, is formidable. Transcription is the formation of RNA from DNA (Chapter 5, p. 91).

Phases in Mitosis

There are two distinct stages of cell division: division of nuclear chromosomes **(mitosis)** and division of cytoplasm **(cytokinesis).** Mitosis (that is, chromosomal segregation) is certainly the most obvious and complex part of cell division and that of greatest interest to cytologists. Cytokinesis normally immediately follows mitosis, although occasionally the nucleus may divide a number of times without a corresponding division of the cytoplasm. In such a case the resulting mass of protoplasm containing many nuclei is referred to as a **multinucleate cell.** An example is the giant resorptive cell type of bone (osteoclast), which may contain 15 to 20 nuclei. Sometimes a multinucleate mass is formed by cell fusion rather than nuclear proliferation. This arrangement is called a **syncytium.** An example is vertebrate skeletal muscle, which is composed of multinucleate fibers formed by fusion of numerous embryonic cells.

The process of mitosis is divided into four successive stages or phases, although one stage merges into the next without sharp lines of transition. These phases are prophase, metaphase, anaphase, and telophase (Figure 3-22). When cells are not actively dividing, they are in interphase, during which DNA replicates and genes are transcribed.

Prophase

At the beginning of prophase, centrosomes (along with their centrioles) replicate, the nuclear envelope disintegrates, and the two centrosomes migrate to opposite poles of the cell (Figure 3-22). At the same time, microtubules appear between the two centrosomes to form a football-shaped **spindle,** so named because of its resemblance to nineteenth-century wooden spindles, used to twist thread together in spinning. Other microtubules radiate outward from each centrosome to form **asters.**

At this time the diffuse nuclear chromatin condenses to form visible chromosomes. These actually consist of two identical sister **chromatids** formed during interphase. Sister chromatids are joined together at their centromere. Dynamic spindle fibers repeatedly extend and retract from the centrosome. When a fiber encounters a kinetochore, it binds to the kinetochore, ceases extending and retracting, and is now called a **kinetochore fiber.** Centrosomes send out "feelers" to find chromosomes.

Microtubules are long, hollow, inelastic cylinders composed of the protein tubulin (Figure 3-23). Each tubulin molecule is actually a doublet composed of two globular proteins. The molecules are attached head-to-tail to form a strand, and 13 strands aggregate to form a microtubule. Because the tubulin subunits in a microtubule are always attached head-to-tail, the ends of the microtubule differ chemically and functionally. One end (called the plus end) both adds and deletes tubulin subunits more rapidly than the other end (the minus end). In a mitotic spindle, the plus ends of the kinetochore and polar fibers are away from the centrosome, and the minus ends are at the centrosome. The microtubule grows when the rate of adding subunits exceeds that of removing them, and it becomes shorter when the rate of removal exceeds that of addition.

Figure 3-22

Stages of mitosis, showing division of a cell with two pairs of chromosomes. One chromosome of each pair is shown in red.

Nucleus

Centrosome with pair of centrioles

Microtubules   Aster

Prophase

Interphase

Chromosomes

Prometaphase

Astral fiber

Metaphase plate

Polar fiber

Kinetochore fiber

Spindle

Metaphase

Anaphase

Daughter nuclei

Telophase

(+) end

Tubulin dimer

(–) end

Figure 3-23

A microtubule is composed of 13 strands of tubulin molecules, and each molecule is a dimer. Tubulin dimers are added to and removed from the (+) end of the microtubule more rapidly than at the (–) end.

Metaphase

Each centromere has two kinetochores, and each of the kinetochores is attached to one of the centrosomes by a kinetochore fiber. By a kind of tug-of-war during metaphase, the condensed sister chromatids are moved to the middle of the nuclear region to form a **metaphase plate** (Figure 3-24). The centromeres line up precisely on the plate with the arms of the chromatids trailing off randomly in various directions.

Anaphase

The single centromere that has held two chromatids together now splits so that two independent chromosomes, each with its own centromere, are formed. The chromosomes move toward their respective poles, pulled by their kinetochore fibers. This phase is often called **anaphase A.** The arms of each chromosome trail along behind as the microtubules shorten to drag a complete set of chromosomes toward each pole of the cell. Present evidence indicates that the force moving the chromosomes is disassembly of the tubulin subunits at the kinetochore end of the microtubules (see the boxed note, p. 50).

As chromosomes approach their respective centrosomes, the spindle lengthens, and the centrosomes move farther apart. This is **anaphase B.** The mechanism of this movement appears to involve the interdigitating free ends of the polar fibers. Tubu

lin in these microtubules has other protein molecules associated with it that serve as "motor molecules." These motor molecules interact with the adjacent fiber (or motor molecules on the adjacent fiber) and push the two halves of the spindle away from each other.

Telophase

When daughter chromosomes reach their respective poles, telophase has begun. Daughter chromosomes are crowded together and stain intensely with histological stains. Spindle fibers disappear and chromosomes lose their identity, reverting to a diffuse chromatin network characteristic of an interphase nucleus. Finally, nuclear membranes reappear around the two daughter nuclei.

Cytokinesis: Cytoplasmic Division

During the final stages of nuclear division a **cleavage furrow** appears on the surface of a dividing cell and encircles it at the midline of the spindle. The cleavage furrow deepens and pinches the plasma membrane as though it were being tightened by an invisible rubber band. Microfilaments of actin are present just beneath the surface in the furrow between the cells. Interaction with myosin and other actin-binding proteins, similar to that which occurs when muscle cells contract (p. 626), draws the furrow inward. Finally, the infolding edges of the plasma membrane meet and fuse, completing cell division. As with other aspects of the cytoskeleton, such as the spindle, the centrosomes are responsible for locating and contracting microfilaments equidistant between them and at right angles to the spindle.

Cell Cycle

Cycles are conspicuous attributes of life. The descent of a species through time is in a very real sense a sequence of life cycles. Similarly, cells undergo cycles of growth and replication as they repeatedly divide. A cell cycle is the interval between one cell division and the next (Figure 3-25).

Actual nuclear division occupies only about 5% to 10% of the cell cycle; the rest of the cell's time is spent in **interphase,** the stage between nuclear divisions. For many years it was thought that interphase was a period of rest, because nuclei appeared inactive when observed by ordinary light microscopy. In the early 1950s new techniques for revealing DNA replication in nuclei were introduced at the same time that biologists came to appreciate fully the significance of DNA as the genetic material. It was then discovered that DNA replication occurred during the interphase stage. Further studies revealed that many other protein and nucleic acid components essential to normal cell function, growth, and division were synthesized during the seemingly quiescent interphase period.

Replication of DNA occurs during a phase called the S period (period of synthesis). In mammalian cells in tissue culture, S period lasts about six of the 18 to 24 hours required to complete one cell cycle. In this period both strands of DNA must

Figure 3-24
Stages of mitosis in whitefish.

replicate; new complementary partners are synthesized for both strands so that two identical molecules are produced from the original strand (see Chapter 5).

The S period is preceded and succeeded by G_1 and G_2 periods, respectively (G stands for "gap"), during which no DNA synthesis is occurring. For most cells, G_1 is an important preparatory stage for the replication of DNA that follows. During G_1, transfer RNA, ribosomes, messenger RNA, and several enzymes are synthesized. During G_2, spindle and aster proteins are synthesized in preparation for chromosome separation during mitosis. G_1 is typically of longer duration than G_2, although there is much variation in different cell types. Embryonic cells divide very rapidly because there is no cell growth between divisions, only subdivision of mass. DNA synthesis may proceed a hundred times more rapidly in embryonic cells than in adult cells,

and the G_1 period is very shortened. As an organism develops, the cycle of most of its cells lengthens, and many cells may be arrested for long periods in G_1 and enter a nonproliferative or quiescent phase called G_0. Neurons, for example, divide no further and are essentially in a permanent G_0.

Recent results have yielded much information on the exquisite regulation of events in cell cycles. Transitions during cell cycles are mediated by **cyclin-dependent kinases (cdk's)** and activating subunits of cdk's called **cyclins.** Kinases are enzymes that add phosphate groups to other proteins to activate or inactivate them, and kinases themselves may require activation. Cdk's become active only when they are bound with the appropriate cyclin, and cyclins are synthesized and degraded during cell cycle (Figure 3-26). Mechanisms involved in cdk regulation of cell cycles are mostly not known.

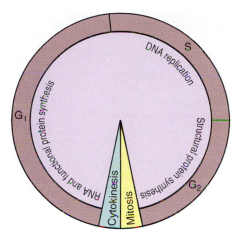

Figure 3-25

Cell cycle, showing relative duration of recognized periods. S, G_1, and G_2 are periods within interphase; S, synthesis of DNA; G_1, presynthetic period; G_2, postsynthetic period. Actual duration of the cycle and the different periods varies considerably in different cell types. After mitosis and cytokinesis the cell may go into an arrested, quiescent stage known as G_0.

Flux of Cells

Cell division is important for growth, for replacement of cells lost to natural attrition and wear and tear, and for wound healing. Cell division is especially rapid during early development of an organism. At birth a human infant has about 2 trillion cells from repeated division of a single fertilized egg. This immense number could be attained by just 42 cell divisions, with each generation dividing once every six to seven days. With only five more cell divisions, the cell number would increase to approximately 60 trillion, the number of cells in a human adult weighing 75 kg. But of course no organism develops in this machinelike manner. Cell division is rapid during early embryonic development, then slows with age. Furthermore, different cell populations divide at widely different rates. In some the average period between divisions is measured in hours, whereas in others it is measured in days, months, or even years. Cells in the central nervous system stop dividing altogether after the early months of fetal development and persist without further division for the life of an individual. Muscle cells also stop dividing during the third month of fetal development, and future growth depends on enlargement of fibers already present.

In other tissues that are subject to wear and tear, lost cells must be constantly replaced. It has been estimated that in humans about 1% to 2% of all body cells—a total of 100 billion—are shed daily. Mechanical rubbing wears away the outer cells of the skin, and food in the alimentary canal rubs off lining cells. The restricted life cycle of blood corpuscles involves enormous numbers of replacements, and during active sex life of males many millions of sperm are produced each day. Such lost cells are replaced by mitosis.

Normal development, however, does entail cell death in which cells are not replaced. They may become senescent, accumulating damage from destructive oxidizing agents and eventually dying. Other cells undergo a programmed cell death, or **apoptosis** (a-puh-TOE-sis) (Gr. *apo-*, from, away from; + *ptosis*, a falling), which is in many cases necessary for continued health and development of an organism. For example, during embryonic development of vertebrates, excess immune cells that would attack the body's own tissues "commit suicide" in this manner, and nerve cells die to create cerebral convolutions. Apoptosis consists of a well-coordinated and predictable series of events: The cells shrink, fragment, and their remains are taken up by surrounding cells.

Apoptosis currently is receiving a great deal of attention from researchers. One of the most valuable laboratory models is a tiny free-living nematode, *Caenorhabditis elegans* (see p. 297). The effects of apoptosis are not always beneficial to the organism. For example, an important disease mechanism in AIDS (acquired immune deficiency syndrome) seems to be an inappropriate triggering of programmed cell death among important cells of the immune system.

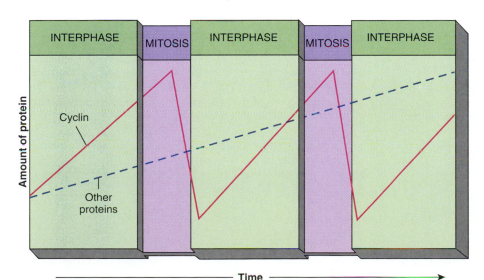

Figure 3-26

Variations in the level of cyclin in dividing cells of early sea urchin embryos. Cyclin binds with its cyclin-dependent kinase to activate the enzyme.

SUMMARY

Cells are the basic structural and functional units of all living organisms. Eukaryotic cells differ from the prokaryotic cells of bacteria and archaebacteria in several respects, the most distinctive of which is presence of a membrane-bound nucleus containing chromosomes that carry the hereditary material.

Cells are surrounded by a plasma membrane that regulates the flow of molecular traffic between the cell and its surroundings. The nucleus, enclosed by a double membrane, contains chromatin and one or more nucleoli. Outside the nuclear envelope is cell cytoplasm, subdivided by a membranous network, the endoplasmic reticulum. Among the organelles within cells are the Golgi complex, mitochondria, lysosomes, and other membrane-bound vesicles. The cytoskeleton is composed of microfilaments (actin), microtubules (tubulin), and intermediate filaments (several types). Cilia and flagella are hairlike, motile appendages that contain microtubules. Ameboid movement by pseudopodia operates by means of actin microfilaments. Tight junctions, desmosomes, and gap junctions are structurally and functionally distinct connections between cells.

Membranes in a cell are composed of a phospholipid bilayer and other materials including cholesterol and proteins. Hydrophilic ends of the phospholipid molecules are on the outer and inner surfaces of membranes, and the fatty acid portions are directed inward, toward each other, to form a hydrophobic core.

Substances can enter cells by diffusion, mediated transport, and endocytosis. Osmosis is diffusion of water through a selectively permeable membrane as a result of osmotic pressure. Solutes to which the membrane is impermeable require a transporter or permease molecule to traverse the membrane. Permease-mediated systems include facilitated diffusion (in the direction of a concentration gradient) and active transport (against a concentration gradient, which requires energy). Endocytosis includes bringing droplets (pinocytosis, potocytosis) or particles (phagocytosis) into a cell. In exocytosis the process of endocytosis is reversed.

Cell division in eukaryotes includes mitosis, or division of the nuclear chromosomes, and cytokinesis, the division of the cytoplasm. Mitosis itself is only a small part of the total cell cycle. In interphase, G_1, S, and G_2 periods are recognized, and the S period is the time when DNA is synthesized (the chromosomes are replicated).

Replicated chromosomes are each held together by a centromere. In prophase, replicated chromosomes condense into recognizable bodies. A spindle forms between the centrosomes as they separate to opposite poles of the cell. At the end of prophase the nuclear envelope disintegrates, and the kinetochores of each chromosome become attached to both centrosomes by microtubules (kinetochore fibers). At metaphase the sister chromatids are moved to the center of the cell. At anaphase the centromeres divide, and a complete set of chromosomes is pulled toward each centrosome by the attached kinetochore fibers. At telophase the chromosomes gather in the position of the nucleus in each cell and revert to a diffuse chromatin network. A nuclear membrane reappears, and cytokinesis occurs.

Cells divide rapidly during embryonic development, then more slowly with age. Some cells continue to divide throughout the life of an animal to replace cells lost by attrition and wear, whereas others, such as nerve and muscle cells, complete their division during early development and never divide again. Some cells undergo a programmed cell death, or apoptosis.

REVIEW QUESTIONS

1. Explain the difference (in principle) between a light microscope and an electron microscope.
2. Briefly describe the structure and function of each of the following: plasma membrane, chromatin, nucleus, nucleolus, rough endoplasmic reticulum (rough ER), Golgi complex, lysosomes, mitochondria, microfilaments, microtubules, intermediate filaments, centrioles, basal body (kinetosome), tight junction, gap junction, desmosome, glycoprotein, microvilli.
3. Name two functions each for actin and for tubulin.
4. Distinguish between cilia, flagella, and pseudopodia.
5. What are the functions of each of the main constituents of the plasma membrane?
6. Our current concept of the plasma membrane is known as the fluid-mosaic model. Why?
7. You place some red blood cells in a solution and observe that they swell and burst. You place some cells in another solution, and they shrink and become wrinkled. Explain what has happened in each case.
8. Explain why a beaker containing a salt solution, placed on a table in your classroom, can have a high osmotic pressure, yet be subjected to a hydrostatic pressure of only one atmosphere.
9. The cell membrane is an effective barrier to molecular movement across it, yet many substances do enter and leave the cell. Explain the mechanisms through which this is accomplished and comment on the energy requirements of these mechanisms.
10. Distinguish between phagocytosis, potocytosis, receptor-mediated endocytosis, and exocytosis.
11. Define the following: chromosome, centromere, centrosome, kinetochore, mitosis, cytokinesis, syncytium.
12. Explain phases of the cell cycle, and comment on important cellular processes that take place during each phase. What is G_0?
13. Name the stages of mitosis in order, and describe the behavior and structure of the chromosomes at each stage.
14. Briefly describe ways that cells may die during the normal life of a multicellular organism.

SELECTED REFERENCES

Anderson, R. G. W., B. A. Kamen, K. G. Rothberg, and S. W. Lacey. 1992. Potocytosis: sequestration and transport of small molecules by caveolae. Science **255**:410–413. *Describes the mechanism of cell internalization of small molecules.*

Barinaga, M. 1996. Forging a path to cell death. Science **273**:735–737. *Researchers are discovering the signal pathways that regulate apoptosis.*

Bayley, H. 1997. Building doors into cells. Sci. Am. **277**:62–67 (Sept.). *Artificial pores in cell membranes can be constructed; they can be a route of drug delivery or act as biosensors to detect toxic chemicals.*

Bretscher, M. S. 1985. The molecules of the cell membrane. Sci. Am. **253**:100–108 (Oct.). *Good presentation of molecular structure of cell membranes, junctions, and mechanism of receptor-mediated endocytosis.*

Bretscher, M. S., and S. Munro, 1993. Cholesterol and the Golgi apparatus. Science **261**:1280–1281. *Good description of behavior of cholesterol in the cell and how it is concentrated in the plasma membrane by the Golgi.*

Dautry-Varsat, A., and H. F. Lodish. 1984. How receptors bring proteins and particles into cells. Sci. Am. **250**:52–58 (May). *Good coverage of receptor-mediated endocytosis.*

Glover, D. M., C. Gonzalez, and J. W. Raff. 1993. The centrosome. Sci. Am. **268**:62–68 (June). *The centrosome of animal cells serves as an organizing center for the cytoskeleton.*

Hartwell, L. H., and M. B. Kastan. 1994. Cell cycle control and cancer. Science **266**:1821–1828. *Genetic changes in the coordination of cyclin-dependent kinases, checkpoint controls, and repair pathways can lead to uncontrolled cell division.*

Lodish, H., D. Baltimore, A. Berk, S. L. Zipursky, P, Matsudira, and J. Darnell. 1995. Molecular biology, ed. 2. New York, Scientific American Books, W. H. Freeman & Company. *Up-to-date, thorough, and readable. Includes both cell biology and molecular biology. Advanced, but highly recommended.*

McIntosh, J. R., and K. L. McDonald. 1989. The mitotic spindle. Sci. Am. **261**:48–56 (Oct.). *Current knowledge and hypotheses on the function of the microtubules of mitosis.*

Miller, L. J., J. Marx. 1998. Apoptosis. Science. **281**:1301. *Introduction to series of articles on apoptosis.*

Murray, A., and T. Hunt. 1993. The cell cycle. An introduction. New York, Oxford University Press. *A good review of our present understanding of the cell cycle.*

Murray, A. W., and M. W. Kirschner. 1991. What controls the cell cycle. Sci. Am. **264**:56–63 (Mar.). *Presents the evidence for the fascinating role of cdc2 kinase and cyclin in the cell cycle.*

ZOOLOGY LINKS TO THE INTERNET

Visit the textbook's Online Learning Center at www.mhhe.com/zoology to find live Internet links for each of the topics listed here.

Cell Structure and Function

DNA Structure and Function

Mitochondria

Cilia, Flagella, and Microvilli

Mitosis in Animal Cells

Chromosomes and Their Structure

The Cell Cycle

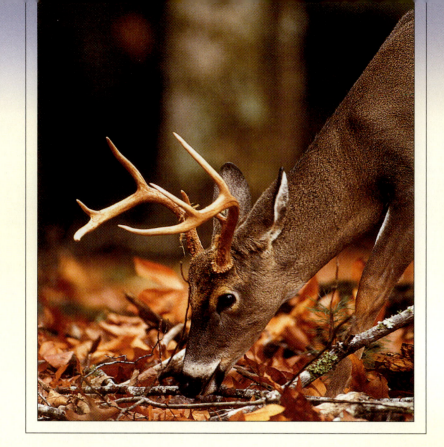

White-tailed deer
(*Odocoileus virginianus*)
foraging for acorns.

Cellular Metabolism

Deferring the Second Law

Living systems appear to contradict the second law of thermodynamics, which states that energy in the universe has direction and that it has been, and always will be, running down. In effect all forms of energy inevitably will be degraded to heat. This increase in disorder, or randomness, in any closed system is termed entropy. Living systems, however, *decrease* their entropy by *increasing* the molecular orderliness of their structure. Certainly an organism becomes vastly more complex during its development from fertilized egg to adult. The second law of thermodynamics, however, applies to closed systems, and living organisms are not closed systems. Animals grow and maintain themselves by borrowing free energy from the environment. When a deer feasts on the acorns and beechnuts of summer, it transfers potential energy, stored as chemical bond energy in the nuts' tissues, to its own body. Then, in step-by-step sequences called biochemical pathways, this energy is gradually released to fuel the deer's many activities. In effect, the deer decreases its own internal entropy by increasing the entropy of its food. The orderly structure of the deer is not permanent, however, but will be dissipated when it dies.

The ultimate source of this energy for the deer—and for almost all life on earth—is the sun (Figure 4-1). Sunlight is captured by green plants, which fortunately accumulate enough chemical bond energy to sustain both themselves and animals that feed on them. Thus the second law is not violated; it is simply held at bay by life on earth, which uses the continuous flow of solar energy to maintain a biosphere of high internal order, at least for the period of time that life exists on earth.

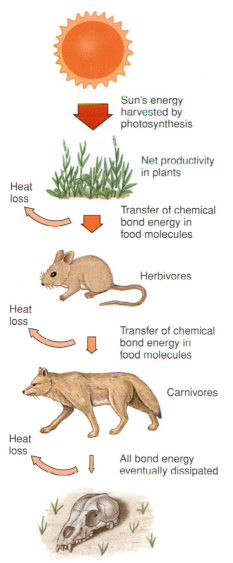

Sun's energy
harvested by
photosynthesis

Net productivity
in plants

Heat
loss

Transfer of chemical
bond energy in
food molecules

Herbivores

Heat
loss

Transfer of chemical
bond energy in
food molecules

Carnivores

Heat
loss

All bond energy
eventually dissipated

Figure 4-1

Solar energy sustains virtually all life on earth. With each energy transfer, however, about 90% of the energy is lost as heat.

A ll cells must obtain energy, synthesize their own internal structure, control much of their own activity, and guard their boundaries. **Cellular metabolism** refers to the collective total of chemical processes that occur within living cells to accomplish these activities. Although the enormous number of reactions in their aggregate are extremely complex, the central metabolic routes through which matter and energy are channeled are not difficult to understand.

ENERGY AND THE LAWS OF THERMODYNAMICS

The concept of energy is fundamental to all life processes. We usually express energy as the capacity to do work, that is, to bring about change. Yet energy is a somewhat abstract quantity that is difficult to define and elusive to measure. Energy cannot be seen; it can be defined and described only by how it affects matter.

Energy can exist in either of two states: kinetic or potential. **Kinetic energy** is the energy of motion. **Potential energy** is stored energy, energy that is not doing work but has the capacity to do so. Energy can be transformed from one state to another. Especially important for living organisms is chemical energy, a form of potential energy that is stored in chemical bonds of molecules. Chemical energy can be tapped when bonds are rearranged to release kinetic energy. Much of the work done by living organisms involves the conversion of potential energy to kinetic energy.

Conversion of one form of energy to another is governed by the two laws of thermodynamics. The **first law of thermodynamics** states that energy cannot be created or destroyed. It can change from one form to another, but the total amount of energy in a system remains the same. In short, energy is conserved. If we burn gasoline in an engine, we do not create new energy but merely convert the chemical energy in gasoline to another form, in this example, mechanical energy and heat. The **second law of thermodynamics,** introduced in the prologue to this chapter, concerns the transformation of energy. This fundamental law states that a closed system moves toward increasing disorder, or entropy, as energy is dissipated from the system (Figure 4-2). Living systems, however, are open systems that not only maintain their organization but also increase it, as during the development of an animal from egg to adult.

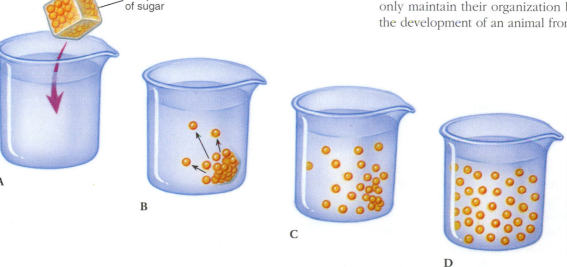

Lump
of sugar

A

B

C

D

Figure 4-2

Diffusion of a solute through a solution, an example of entropy. When the solute (sugar molecules) is first introduced into a solution, the system is ordered and unstable **(B)**. Without energy to maintain this order, the solute particles become distributed into solution, reaching a state of disorder (equilibrium) **(D)**. Entropy has increased from left diagram to right diagram.

Free Energy

To describe the energy changes that take place in chemical reactions, biochemists use the concept of **free energy.** Free energy is simply the energy in a system available for doing work. In a molecule, free energy equals the energy present in chemical bonds minus the energy that cannot be used. The majority of reactions in cells release free energy and are said to be **exergonic** (Gr. *ex,* out, + *ergon,* work). Such reactions are spontaneous and always proceed "downhill" since free energy is lost from the system. Thus:

Loss of some energy

AB
Reactant

A   B

Products, containing less
free energy than reactant

Exergonic reaction

However, many important reactions in cells require the addition of free energy and are said to be **endergonic** (Gr. *endon,* within, + *ergon,* work). Such reactions have to be "pushed uphill" because they end up with more energy than they started with:

Products, containing more
free energy than reactants

Reactants

C   D

CD

Energy input

Endergonic reaction

As we will see on page 60, ATP is the ubiquitous, energy-rich intermediate used by organisms to power important uphill reactions such as those required for active transport of molecules across membranes and cellular synthesis.

THE ROLE OF ENZYMES

Enzymes and Activation Energy

For any reaction to occur, even exergonic ones that tend to proceed spontaneously, chemical bonds first must be destabilized. For example, if a reaction involves splitting a covalent bond, the atoms forming the bond must first be stretched apart to make them less stable. Some energy, termed the **activation energy,** must be supplied before the bond will be stressed enough to break. Only then will an overall loss of free energy and formation of reaction products occur. This requirement can be likened to the energy needed to push a ball over the crest of a hill before it will roll spontaneously down the other side, the ball liberating its potential energy as it descends.

One way to activate chemical reactants is to raise the temperature. By increasing the rate of molecular collisions and pushing chemical bonds apart, heat can impart the necessary activation energy to make a reaction proceed. However metabolic reactions must occur at biologically tolerable temperatures, which are usually too low to allow reactions to proceed beyond imperceptible rates. Instead, living systems have evolved a different strategy: they employ **catalysts.**

Catalysts are chemical substances that accelerate reaction rates without affecting products of the reaction and without being altered or destroyed as a result of the reaction. A catalyst cannot make an energetically impossible reaction happen; it simply accelerates a reaction that would have proceeded at a very slow rate otherwise.

Enzymes are catalysts of the living world. The special catalytic talent of an enzyme is its power to reduce the amount of activation energy required for a reaction. In effect, an enzyme steers the reaction through one or more intermediate steps, each of which requires much less activation energy than that required for a single-step reaction (Figure 4-3). Note that enzymes do not supply the activation energy. Instead they lower the activation energy barrier, making a reaction more likely to proceed. Enzymes affect only the reaction rate. They do not in any way alter the free energy change of a reaction, nor do they change the proportions of reactants and products in a reaction.

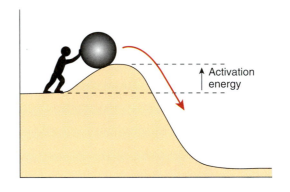

Figure 4-3

Energy changes during enzyme catalysis of a substrate. The overall reaction proceeds with a net release of energy (exergonic). In the absence of an enzyme, substrate is stable because of the large amount of activation energy needed to disrupt strong chemical bonds. The enzyme reduces the energy barrier by forming a chemical intermediate with a much lower internal energy state.

Nature of Enzymes

Enzymes are complex molecules that vary in size from small, simple proteins with a molecular weight of 10,000 to highly complex molecules with molecular weights up to 1 million. Many enzymes are pure proteins—delicately folded and interlinked chains of amino acids. Other enzymes require participation of small nonprotein groups called **cofactors** to perform their enzymatic function. In some cases these cofactors are metallic ions (such as ions of iron, copper, zinc, magnesium, potassium, and calcium) that form a functional part of the enzyme. Examples are carbonic anhydrase, which contains zinc; the cytochromes, which contain iron; and troponin (a muscle contraction enzyme), which contains calcium. Another class of cofactors, called **coenzymes,** is organic. All coenzymes contain groups derived from vitamins, most of which must be supplied in the diet. All of the B-complex vitamins are coenzymatic compounds. Since animals have lost the ability to synthesize the vitamin components of coenzymes, it is obvious that a vitamin deficiency can be serious. However, unlike dietary fuels and nutrients that must be replaced after they are burned or assembled into structural materials, vitamins are recovered in their original form and are used repeatedly. Examples of coenzymes that contain vitamins are nicotinamide adenine dinucleotide (NAD), which contains the vitamin nicotinic acid (niacin); coenzyme A, which contains the vitamin pantothenic acid; and flavin adenine dinucleotide (FAD), which contains riboflavin (vitamin B_2).

Action of Enzymes

An enzyme functions by associating in a highly specific way with its **substrate,** the molecule whose reaction it catalyzes. Enzymes bear an active site that is located within a cleft or pocket and that contains a unique molecular configuration. The active site has a flexible surface that enfolds and conforms to the substrate (Figure 4-4). The binding of enzyme to substrate forms an **enzyme-substrate complex (ES complex),** in which the substrate is secured by covalent bonds to one or more points in the active site of the enzyme. The ES complex is not strong and will quickly dissociate, but during this fleeting moment the enzyme provides a unique chemical environment that stresses certain chemical bonds in the substrate so that much less energy is required to complete the reaction.

If the formation of an enzyme-substrate complex is so rapidly followed by dissociation, how can biochemists be certain that an ES complex exists? The original evidence offered by Leonor Michaelis in 1913 is that, when the substrate concentration is increased while the enzyme concentration is held constant, the reaction rate reaches a maximum velocity. This *saturation effect* is interpreted to mean that all catalytic sites become filled at high substrate concentration. A saturation effect is not seen in uncatalyzed reactions. Other evidence includes the observation that the ES complex displays unique spectroscopic characteristics not displayed by either the enzyme or the substrate alone. Furthermore, some ES complexes can be isolated in pure form, and at least one kind (nucleic acids and their polymerase enzymes) has been directly visualized by electron microscopy.

Enzymes that engage in important main-line sequences—such as the crucial energy-providing reactions of cells that proceed constantly—usually operate in sets rather than in isolation. For example, conversion of glucose to carbon dioxide and water proceeds through 19 reactions, each requiring a specific enzyme. Main-line enzymes are found in relatively high concentrations in cells, and they may implement quite complex and highly

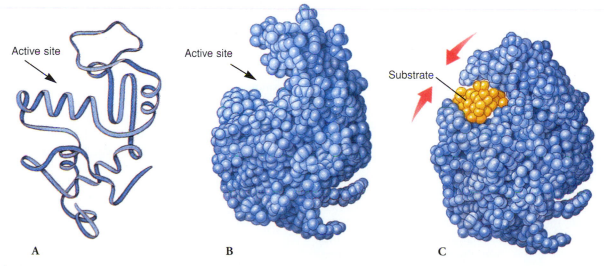

Active site   Active site   Substrate

A   B   C

Figure 4-4

How an enzyme works. This space-filling model shows that the enzyme lysozyme bears a pocket containing the active site. When a chain of sugars (substrate) enters the pocket, the protein enzyme changes shape slightly so that the pocket enfolds the substrate and conforms to its shape. This positions the active site (an amino acid in the protein) next to a bond between adjacent sugars in the chain, causing the sugar chain to break.

integrated enzymatic sequences. One enzyme carries out the first step, then passes the product to another enzyme that catalyzes another step, this process continuing until the end of the enzymatic pathway is reached. The reactions may be said to be coupled. Coupled reactions will be explained in a section on chemical energy transfer by ATP (p. 60).

Specificity of Enzymes

One of the most distinctive attributes of enzymes is their high specificity. Specificity is a consequence of the exact molecular fit that is required between enzyme and substrate. Furthermore, an enzyme catalyzes only one reaction. Unlike reactions carried out in an organic chemist's laboratory, no side reactions or by-products result. Specificity of both substrate and reaction is obviously essential to prevent a cell from being swamped with useless by-products.

However, there is some variation in degree of specificity. Some enzymes catalyze the oxidation (dehydrogenation) of only one substrate. For example, succinic dehydrogenase catalyzes the oxidation of succinic acid only. Others, such as proteases (for example, pepsin and trypsin), will act on almost any protein, although each protease has its particular point of attack in the protein (Figure 4-5). Usually an enzyme will take on one substrate molecule at a time, catalyze its chemical change, release the product, and then repeat the process with another substrate molecule. The enzyme may repeat this process billions of times until it is finally worn out (after a few hours to several years) and is broken down by scavenger enzymes in the cell. Some enzymes undergo successive catalytic cycles at speeds of up to a million cycles per minute, but most operate at slower rates. Many enzymes are repeatedly activated and inactivated; several mechanisms for regulating enzyme activity are well known (p. 69).

Enzyme-Catalyzed Reactions

Enzyme-catalyzed reactions are reversible, which is signified by double arrows between substrate and products. For example:

Fumaric acid + H₂O ⇌ Malic acid

However, for various reasons reactions catalyzed by most enzymes tend to go predominantly in one direction. For example, the proteolytic enzyme pepsin degrades proteins into amino acids (a **catabolic** reaction), but it does not accelerate the rebuilding of amino acids into any significant amount of protein (an **anabolic** reaction). The same is true of most enzymes that catalyze the cleavage of large molecules such as nucleic acids, polysaccharides, lipids, and proteins. There is usually one set of reactions and enzymes that break them down (catabolism; Gr. *kata,* down, + *bole,* throw), but they must be resynthesized by a different set of reactions that are catalyzed by different enzymes (anabolism; Gr. *ana,* up, + *bole,* throw). This apparent irreversibility exists because the chemical equilibrium usually favors formation of the smaller degradation products.

The net **direction** of any chemical reaction depends on the relative energy contents of the substances involved. If there is

Figure 4-5
High specificity of trypsin. It splits only peptide bonds adjacent to lysine or arginine.

little change in chemical bond energy of substrate and products, the reaction is more easily reversible. However, if large quantities of energy are released as the reaction proceeds in one direction, more energy must be provided in some way to drive the reaction in the reverse direction. For this reason many if not most enzyme-catalyzed reactions are in practice irreversible unless the reaction is coupled to another that makes energy available. In cells both reversible and irreversible reactions are combined in complex ways to make possible both synthesis and degradation.

> Hydrolysis literally means "breaking with water." In hydrolysis reactions, a molecule is broken down by the addition of water. A hydrogen is attached to one subunit and a hydroxyl (—OH) unit is attached to another. This breaks the covalent bond between subunits. Hydrolysis is the opposite of condensation (water-losing) reactions in which subunits of molecules are linked together by removal of water. Macromolecules are built by condensation reactions.

CHEMICAL ENERGY TRANSFER BY ATP

We have seen that endergonic reactions are those that will not proceed spontaneously by themselves because their products require an input of free energy. However, an endergonic reaction may be driven by coupling the energy-requiring reaction with an energy-yielding reaction. ATP is the most common intermediate in **coupled reactions,** and because it can drive such energetically unfavorable reactions, it is of central importance in metabolic processes.

An ATP molecule consists of adenosine (the purine adenine and the 5-carbon sugar ribose) and a triphosphate group (Figures 4-6 and 4-7). Most free energy in ATP resides in the triphosphate group, especially in two **phosphoanhydride bonds** between the three phosphate groups. These two bonds are called **"high-energy bonds"** because a great deal of free energy in the bonds is liberated when ATP is hydrolyzed to adenosine diphosphate (ADP) and inorganic phosphate.

ATP + H₂O → ADP + Pᵢ

Figure 4-6

A, Structure of ATP. **B,** ATP formation from ADP.

Figure 4-7

Space-filling model of ATP. In this model, carbon is shown in black; nitrogen in blue; oxygen in red; and phosphorus in yellow.

Figure 4-8

A coupled reaction. The endergonic conversion of substrate A to product A will not occur spontaneously but requires an input of energy from another reaction involving a large release of energy. ATP is the intermediate through which the energy is shuttled.

where P_i represents inorganic phosphate (i = inorganic). The high-energy groups in ATP are designated by the "tilde" symbol ~. A high-energy phosphate bond is shown as ~P and a low-energy bond (such as the bond linking the triphosphate group to adenosine) as —P. ATP may be symbolized as A—P~P~P and ADP as A—P~P.

The way that ATP can act to drive a coupled reaction is shown in Figure 4-8. A coupled reaction is really a system involving two reactions linked by an energy shuttle (ATP). The conversion of substrate A to product A is endergonic because the product contains more free energy than the substrate. Therefore energy must be supplied by coupling the reaction to one that is exergonic, the conversion of substrate B to product B. Substrate B in this reaction is commonly called a **fuel** (for example, glucose or a lipid). Bond energy that is released in reaction B is transferred to ADP, which in turn is converted to ATP. ATP now contributes its phosphate-bond energy to reaction A, and ADP is produced again.

The high-energy bonds of ATP are actually rather weak, unstable bonds. Because they are unstable, the energy of ATP is readily released when ATP is hydrolyzed in cellular reactions. Note that ATP is an **energy-coupling agent** and *not* a fuel. It is not a storehouse of energy set aside for some future need. Rather it is produced by one set of reactions and is almost immediately consumed by another. ATP is formed as it is needed, primarily by oxidative processes in the mitochondria. Oxygen is

not consumed unless ADP and phosphate molecules are available, and these do not become available until ATP is hydrolyzed by some energy-consuming process. *Metabolism is therefore mostly self-regulating.*

CELLULAR RESPIRATION

How Electron Transport Is Used to Trap Chemical Bond Energy

Having seen that ATP is the one common energy denominator by which all cellular machines are powered, we are in a position to ask how this energy is captured from fuel substrates. This

Oxidation   Reduction

Figure 4-9

A redox pair. The molecule at left is oxidized by the loss of an electron. The molecule at right is reduced by gaining an electron.

question directs us to an important generalization: *all cells obtain their chemical energy requirements from oxidation-reduction reactions.* This means that in the degradation of fuel molecules, hydrogen atoms (electrons and protons) are passed from electron donors to electron acceptors with a release of energy. A portion of this energy is trapped and used to form the high-energy bonds in ATP.

Because they are so important, let us review what we mean by oxidation-reduction ("redox") reactions. In these reactions there is a transfer of electrons from an electron donor (the reducing agent) to an electron acceptor (the oxidizing agent). As soon as the electron donor loses its electrons, it becomes oxidized. As soon as the electron acceptor accepts electrons, it becomes reduced (Figure 4-9). In other words, a reducing agent becomes oxidized when it reduces another compound, and an oxidizing agent becomes reduced when it oxidizes another compound. Thus for every oxidation there must be a corresponding reduction.

In an oxidation-reduction reaction the electron donor and electron acceptor form a redox pair:

$$\text{Electron donor} \rightleftharpoons e^- + \text{Electron acceptor}$$
$$+ \text{Energy}$$

(reducing agent;	(oxidizing agent;
becomes oxidized)	becomes reduced)

When electrons are accepted by the oxidizing agent, energy is liberated because the electrons move to a more stable position. In a cell, electrons flow through a series of carriers. Each carrier is reduced by accepting electrons and then is reoxidized by passing electrons to the next carrier in the series. By transferring electrons stepwise in this manner, energy is gradually released, and a maximum yield of ATP is realized.

Aerobic versus Anaerobic Metabolism

Ultimately, the electrons are transferred to a **final electron acceptor.** The nature of this final acceptor is the key that determines the overall efficiency of cellular metabolism. Heterotrophs (organisms that cannot synthesize their own food but must obtain nutrients from the environment, including animals, fungi, and many single-celled organisms) can be divided into two groups: **aerobes,** those that use molecular oxygen as the final electron acceptor, and **anaerobes,** those that employ some other molecule as the final electron acceptor.

As discussed in Chapter 2, life originated in the absence of oxygen, and the abundance of atmospheric oxygen was produced after photosynthetic organisms (autotrophs) evolved. Some strictly anaerobic organisms still exist and indeed play some important roles in specialized habitats. However, evolution has favored aerobic metabolism, not only because oxygen became available, but also because aerobic metabolism is vastly more efficient than anaerobic metabolism. In the absence of oxygen, only a very small fraction of the bond energy present in foodstuffs can be released. For example, when an anaerobic microorganism degrades glucose, the final electron acceptor (such as pyruvic acid) still contains most of the energy of the original glucose molecule. An aerobic organism on the other hand, using oxygen as the final electron acceptor, can completely oxidize glucose to carbon dioxide and water. Almost 20 times as much energy is released when glucose is completely oxidized as when it is degraded only to the stage of pyruvic acid. An obvious advantage of aerobic metabolism is that a much smaller quantity of foodstuffs is required to maintain a given rate of metabolism.

Overview of Respiration

Aerobic metabolism is more familiarly known as true **cellular respiration,** defined as the oxidation of fuel molecules with molecular oxygen as the final electron acceptor. As mentioned on this page, oxidation of fuel molecules describes the *removal of electrons* and *not* the direct combination of molecular oxygen with fuel molecules. Let us look at this process in general before considering it in more detail.

Hans Krebs, the British biochemist who contributed so much to our understanding of respiration, described three stages in the complete oxidation of fuel molecules to carbon dioxide and water (Figure 4-10). In stage I, foodstuffs passing through the intestinal tract are digested into small molecules that can be absorbed into the circulation. There is no useful energy yield during digestion, which is discussed in Chapter 32. In stage II, most of the degraded foodstuffs are converted into two 3-carbon units (pyruvic acid) in the cell cytoplasm. The pyruvic acid molecules then enter mitochondria, where in another reaction they join with a coenzyme (coenzyme A) to form acetyl coenzyme A. Some ATP is generated in stage II, but the yield is small compared with that obtained in stage III of respiration. In stage III the final oxidation of fuel molecules occurs, with a large yield of ATP. This stage takes place entirely in mitochondria. Acetyl coenzyme A is channeled into the Krebs cycle where the acetyl group is completely oxidized to carbon dioxide. Electrons released from acetyl groups are transferred to special carriers that pass them to electron acceptor compounds in the electron transport chain. At the end of the chain the electrons (and the protons accompanying them) are accepted by molecular oxygen to form water.

Glycolysis

We begin our journey through the stages of respiration with glycolysis, a nearly universal pathway in living organisms that converts glucose into pyruvic acid. In a series of reactions, glucose

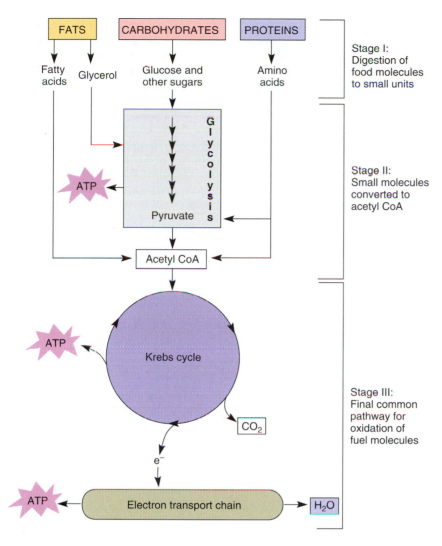

Figure 4-10

Overview of cellular respiration, showing the three stages in the complete oxidation of food molecules to carbon dioxide and water.

and other 6-carbon monosaccharides are split into 3-carbon fragments, **pyruvic acid** (Figure 4-11). A single oxidation occurs during glycolysis, and each molecule of glucose yields two molecules of ATP. In this pathway the carbohydrate molecule is phosphorylated twice by ATP, first to glucose-6-phosphate (not shown in Figure 4-11) and then to fructose-1,6-diphosphate. The fuel has now been "primed" with phosphate groups in this uphill portion of glycolysis and is sufficiently reactive to enable subsequent reactions to proceed. This is a kind of deficit financing that is required for an ultimate energy return many times greater than the original energy investment.

In the downhill portion of glycolysis, fructose-1,6-diphosphate is cleaved into two 3-carbon sugars, which undergo an oxidation (electrons are removed), with the electrons and one of the hydrogen ions being accepted by nicotinamide adenine dinucleotide (NAD, a derivative of the vitamin niacin) to produce a reduced form called NADH. NADH serves as a carrier molecule to convey high-energy electrons to the final electron transport chain, where ATP will be produced.

The two 3-carbon sugars next undergo a series of reactions, ending with the formation of two molecules of pyruvic acid (Figure 4-11). In two of these steps, a

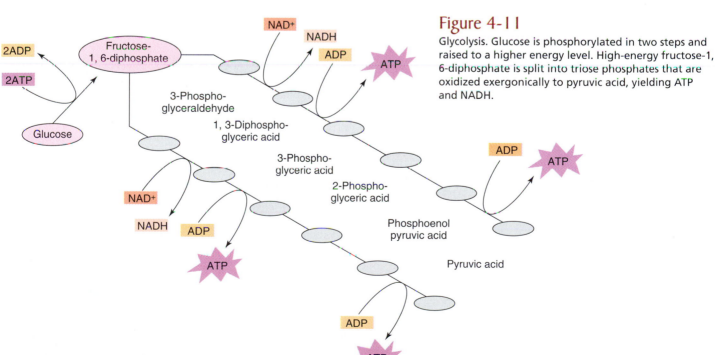

Figure 4-11

Glycolysis. Glucose is phosphorylated in two steps and raised to a higher energy level. High-energy fructose-1, 6-diphosphate is split into triose phosphates that are oxidized exergonically to pyruvic acid, yielding ATP and NADH.

molecule of ATP is produced. In other words, each 3-carbon sugar yields two ATP molecules, and since there are two 3-carbon sugars, four ATP molecules are generated. Recalling that two ATP molecules were used to prime the glucose initially, the net yield up to this point is two ATP molecules. The 10 enzymatically catalyzed reactions in glycolysis can be summarized as:

$$\text{Glucose} + 2\ \text{ADP} + 2\ \text{P}_i + 2\ \text{NAD}^+ \rightarrow 2\ \text{pyruvic acid} + 2\ \text{NADH} + 2\ \text{ATP}$$

Acetyl Coenzyme A: Strategic Intermediate in Respiration

In aerobic metabolism the two molecules of pyruvic acid formed during glycolysis enter a mitochondrion. There, each molecule is oxidized, and one of the carbons is released as carbon dioxide (Figure 4-12). The 2-carbon residue condenses with **coenzyme A** to form **acetyl coenzyme A** (acetyl-CoA).

Pyruvic acid is the undissociated form

$$\underset{\text{O}}{\overset{\text{O}}{\underset{\parallel}{}}}$$

of the acid CH_3—C—COOH. Under physiological conditions pyruvic acid typically dissociates into pyruvate

$$\overset{\text{O}}{\overset{\parallel}{}}$$

(CH_3—C—COO$^-$) and H$^+$. It is correct to use either term in describing this and other organic acids (such as lactic acid, or lactate) in metabolism.

Acetyl coenzyme A is a critically important compound. Its oxidation in the Krebs cycle (following) provides energized electrons to generate ATP, and it is also a crucial intermediate in lipid metabolism (p. 68).

Krebs Cycle: Oxidation of Acetyl Coenzyme A

Degradation (oxidation) of the 2-carbon acetyl group of acetyl coenzyme A occurs in a cyclic sequence called the **Krebs cycle** (also called citric acid cycle and tricarboxylic acid cycle [TCA

cycle]) (Figure 4-13). Acetyl coenzyme A condenses with a 4-carbon acid (oxaloacetic acid), releasing coenzyme A to react again with pyruvic acid. Through a series of reactions the two carbons from the acetyl group are released as carbon dioxide, and oxaloacetic acid is regenerated. Hydrogen ions and electrons in the oxidations transfer to NAD and to FAD (flavine adenine dinucleotide, another electron acceptor), and a pyrophosphate bond is generated in the form of guanosine triphosphate (GTP). This high-energy phosphate readily transfers to ADP to form ATP. The overall products of the Krebs cycle are CO_2, ATP, NADH, and FADH$_2$:

$$\text{Acetyl unit} + 3\ \text{NAD}^+ + \text{FAD} + \text{ADP} + \text{P}_i \rightarrow$$
$$2\ CO_2 + 3\ \text{NADH} + \text{FADH}_2 + \text{ATP}$$

The molecules of NADH and FADH$_2$ formed will yield 11 molecules of ATP when oxidized in the electron transport chain. The other molecules in the cycle behave as intermediate reactants and products which are continuously regenerated as the cycle turns.

Electron Transport Chain

Transfer of hydrogen ions and electrons from reduced NAD and FAD to the final electron acceptor, molecular oxygen, is accomplished in an elaborate electron transport chain embedded in the inner membrane of mitochondria (Figure 4-14, see also p. 42). Each carrier molecule in the chain (labeled I to IV in Figure 4-14) is a large protein-based complex that accepts and releases electrons at lower energy levels than the carrier preceding it in the chain. As electrons pass from one carrier molecule to the next, free energy is released. Some of this energy drives the synthesis of ATP by setting up a H$^+$ gradient across the mitochondrial membrane. This process is called chemiosmotic coupling (Figure 4-14). According to this model, as electrons contributed by NADH and FADH$_2$ are carried down the electron transport chain, they activate proton pumping channels which pump protons (hydrogen ions) outward and into the space between the two mitochondrial membranes. This causes the proton concentration outside to rise, producing a diffusion pressure that drives the protons back into the mitochondrion through special proton channels. These channels are ATP-forming protein complexes (ATP synthase) that use the inward passage of protons to induce the formation of ATP. Exactly how proton movement is coupled to ATP synthesis is not yet understood. By this means, oxidation of one NADH yields three ATP molecules. Reduced FAD from the Krebs cycle enters the electron transport chain at a lower level than NADH and so yields two ATP molecules. This method of energy capture is called **oxidative phosphorylation** because the formation of high-energy phosphate is coupled to oxygen consumption, and these reactions depend on demand for ATP by other metabolic activities within the cell.

Efficiency of Oxidative Phosphorylation

We are now in a position to calculate the ATP yield from the complete oxidation of glucose (Figure 4-15). The overall reaction is:

$$\text{Glucose} + 2\ \text{ATP} + 36\ \text{ADP} + 36\ \text{P} + 6\ O_2 \rightarrow$$
$$6\ CO_2 + 2\ \text{ADP} + 36\ \text{ATP} + 6\ H_2O$$

Figure 4-12

Formation of acetyl coenzyme A from pyruvic acid.

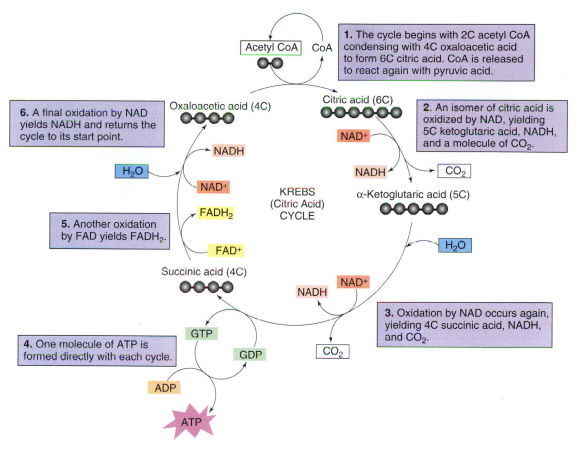

Figure 4-13

The Krebs cycle in outline form, showing production of three molecules of reduced NAD, one molecule of reduced FAD, one molecule of ATP, and two molecules of carbon dioxide. The molecules of NADH and FADH$_2$ will yield 11 molecules of ATP when oxidized in the electron transport system.

ATP has been generated at several points along the way (Table 4-1). The cytoplasmic NADH generated in glycolysis requires a molecule of ATP to fuel transport of each molecule of NADH into a mitochondrion; therefore each NADH from glycolysis results in only two ATP (total of four), compared with the three ATP per NADH (total of six) formed within mitochondria. Accounting for the two ATP used in the priming reactions in glycolysis, the net yield may be as high as 36 molecules of ATP per molecule of glucose. The yield of 36 ATP is a theoretical maximum because some of the H$^+$ gradient produced by electron transport may be used for other functions, such as transporting substances in and out of the mitochondrion. Overall efficiency of aerobic oxidation of glucose is about 38%, comparing very favorably with human-designed energy conversion systems, which seldom exceed 5% to 10% efficiency.

Anaerobic Glycolysis: Generating ATP without Oxygen

Up to this point we have been describing aerobic cellular respiration. We will now consider how animals generate ATP without oxygen, that is, anaerobically.

Under anaerobic conditions, glucose and other 6-carbon sugars are first broken down stepwise to a pair of 3-carbon pyruvic acid molecules, yielding two molecules of ATP and four atoms of hydrogen (four reducing equivalents, represented by 2 NADH + H$^+$). In the absence of molecular oxygen, further oxidation of pyruvic acid cannot occur because the Krebs cycle and electron transport chain cannot operate and cannot, therefore, provide a mechanism for reoxidizing the NADH produced in glycolysis. The problem is neatly solved in most animal cells by reducing pyruvic acid to lactic acid (Figure 4-16). Pyruvic acid becomes the final electron acceptor and lactic acid the end product of anaerobic glycolysis. This frees the hydrogen-bound carrier to recycle and pick up more H$^+$. In **alcoholic fermentation** (as in yeast, for example) the steps are identical to glycolysis down to pyruvic acid. One of its carbons is then released as carbon dioxide, and the resulting 2-carbon compound is reduced to ethanol, thus regenerating the NAD.

Anaerobic glycolysis is only one-eighteenth as efficient as complete oxidation of glucose to carbon dioxide and water, but its key virtue is that it provides *some* high-energy phosphate in situations in which oxygen is absent or in short supply. Many microorganisms live in places where oxygen is severely depleted, such as waterlogged soil, in mud of lake or sea bottom, or within

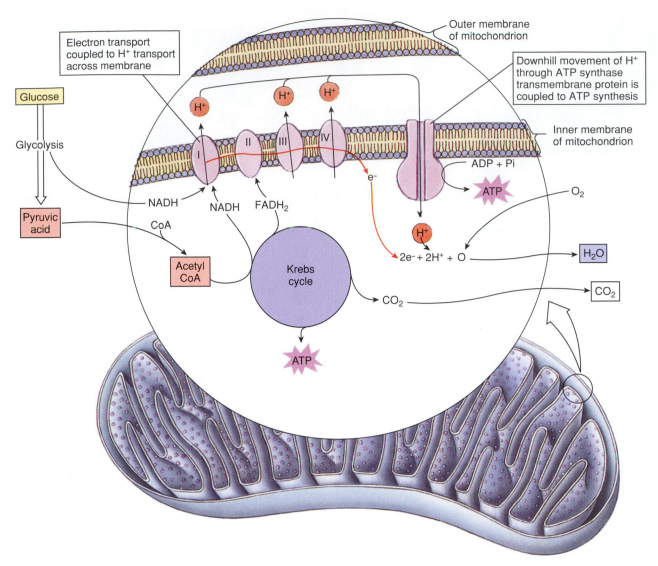

Electron transport coupled to H+ transport across membrane

Outer membrane of mitochondrion

Downhill movement of H+ through ATP synthase transmembrane protein is coupled to ATP synthesis

Inner membrane of mitochondrion

Glucose

Glycolysis

H+   H+   H+

I   II   III   IV

ADP + Pi

ATP

O_2

e^-

NADH

NADH   FADH₂

Pyruvic acid

CoA

H+

$2e^- + 2H^+ + O$

H_2O

Acetyl CoA

Krebs cycle

CO_2

CO_2

ATP

Figure 4-14

Oxidative phosphorylation. Most of the ATP in living organisms is produced in the electron transport chain. Electrons removed from fuel molecules in cellular oxidations (glycolysis and the Krebs cycle) flow through the electron transport chain, the major components of which are four protein complexes (I, II, III, and IV). Electron energy is tapped by the major complexes and used to push H+ outward across the inner mitochondrial membrane. The H+ gradient created drives H+ inward through proton channels that couple H+ movement to ATP synthesis.

a decaying carcass. Vertebrate skeletal muscle may rely heavily on glycolysis during short bursts of activity when contraction is so rapid and powerful that oxygen delivery to tissues is not sufficient to supply energy demands by oxidative phosphorylation alone. At such times an animal has no choice but to supplement oxidative phosphorylation with anaerobic glycolysis. Intense activity is followed by a period of increased oxygen consumption as lactic acid diffuses from muscle to the liver where it is metabolized. Because oxygen consumption increases following heavy activity, the animal is said to have acquired an **oxygen debt** during such activity, which is repaid when activity ceases, and accumulated lactic acid is metabolized.

Some animals rely heavily on anaerobic glycolysis during normal activities. For example, diving birds and mammals fall back on glycolysis almost entirely to give them the energy needed to sustain long dives without breathing (that is, without requiring oxygen). Salmon would never reach their spawning grounds were it not for anaerobic glycolysis providing almost all of the ATP used in the powerful muscular bursts needed to carry them up rapids and falls. Many parasitic animals have dispensed with oxidative phosphorylation entirely at some stages of their life cycles. They secrete relatively reduced end products of their energy metabolism, such as succinic acid, acetic acid, and propionic acid. These compounds are produced in mitochondrial reactions that derive several more molecules of ATP than does the path from glycolysis to lactic acid, although such sequences are still far less efficient than the classical electron transport chain.

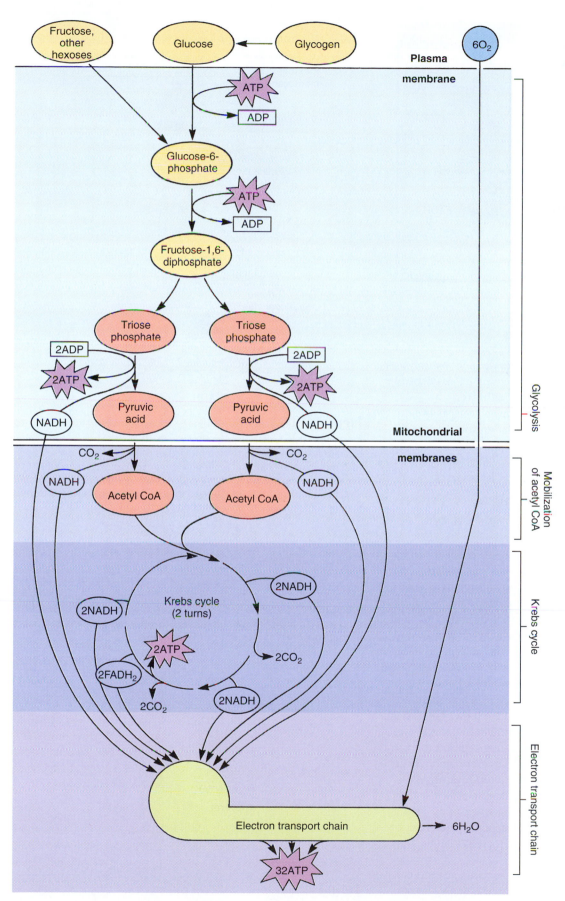

Figure 4-15

Pathway for oxidation of glucose and other carbohydrates. Glucose is degraded to pyruvic acid by cytoplasmic enzymes (glycolytic pathway). Acetyl coenzyme A is formed from pyruvic acid and is fed into the Krebs cycle. An acetyl group (two carbons) is oxidized to two molecules of carbon dioxide with each turn of the cycle. Pairs of electrons are removed from the carbon skeleton of the substrate at several points in the pathway and are carried by oxidizing agents NADH or FADH$_2$ to the electron transport chain where 32 molecules of ATP are generated. Four molecules of ATP are also generated by substrate phosphorylation in the glycolytic pathway, and two molecules of ATP (initially GTP) are formed in the Krebs cycle. This yields a total of 38 molecules of ATP (36 molecules net) per glucose molecule. Molecular oxygen is involved only at the very end of the pathway.

TABLE 4.1

Calculation of Total ATP Molecules Generated in Respiration

ATP Generated	Source
4	Directly in glycolysis
2	As GTP ($\rightarrow$ATP) in Krebs cycle
4	From NADH in glycolysis
6	From NADH produced in pyruvic acid to acetyl coenzyme A reaction
4	From reduced FAD in Krebs cycle
18	From NADH produced in Krebs cycle
38 Total	
−2	Used in priming reactions in glycolysis
36 Net	

METABOLISM OF LIPIDS

The first step in the breakdown of a triglyceride is its hydrolysis to glycerol and three fatty acid molecules (Figure 4-17). Glycerol is phosphorylated and enters the glycolytic pathway.

The remainder of the triglyceride molecule consists of fatty acids. One abundant naturally occurring fatty acid is **stearic acid.**

$$H_3C \quad CH_2 \ CH_2 \ CH_2 \ CH_2 \ CH_2 \ CH_2 \ CH_2 \ CH_2 \ O$$
$$CH_2 \ CH_2 \ CH_2 \ CH_2 \ CH_2 \ CH_2 \ CH_2 \ CH_2 \ C$$
Stearic acid $\qquad OH$

The long hydrocarbon chain of a fatty acid is sliced up by oxidation, two carbons at a time; these are released from the end of the molecule as acetyl coenzyme A. Although two high-energy phosphate bonds are required to prime each 2-carbon fragment, energy is derived both from the reduction of NAD and FAD in the oxidations and from the acetyl group as it is degraded in the Krebs cycle. It can be calculated that the complete oxidation of one molecule of 18-carbon stearic acid will net 146 ATP molecules. By comparison, three molecules of glucose (also totaling 18 carbons) yield 108 ATP molecules. Since there are three fatty acids in each triglyceride molecule, a total of 440 ATP molecules are formed. An additional 22 molecules of ATP are generated in the breakdown of glycerol, giving a grand total of 462 molecules of ATP—little wonder that fat is considered the king of animal fuels! Fats are more concentrated fuels than carbohydrates, because fats are almost pure hydrocarbons; they contain more hydrogen per carbon atom than sugars do, and it is the energized electrons of hydrogen that generate high-energy bonds when they are carried through the mitochondrial electron transport chain.

Fat stores are derived principally from surplus fats and carbohydrates in the diet. Acetyl coenzyme A is the source of carbon atoms used to build fatty acids. Since all major classes of organic molecules (carbohydrates, fats, and proteins) can be

Figure 4-16

Anaerobic glycolysis, a process that proceeds in the absence of oxygen. Glucose is broken down to two molecules of pyruvic acid, generating four molecules of ATP and yielding two, since two molecules of ATP are used to produce fructose-1,6-diphosphate. Pyruvic acid, the final electron acceptor for the hydrogen ions and electrons released during pyruvic acid formation, is converted to lactic acid.

Figure 4-17

Hydrolysis of a triglyceride (neutral fat) by intracellular lipase. The R groups of each fatty acid represent a hydrocarbon chain.

degraded to acetyl coenzyme A, all can be converted into stored fat. The biosynthetic pathway for fatty acids resembles a reversal of the catabolic pathway already described, but it requires an entirely different set of enzymes. From acetyl coenzyme A, the fatty acid chain is assembled two carbons at a time. Because fatty acids release energy when they are oxidized, they obviously require an input of energy for their synthesis. This is provided principally by electron energy from glucose degradation. Thus the total ATP derived from oxidation of a molecule of triglyceride is not as great as calculated, because varying amounts of energy are required for synthesis and storage.

Stored fats are the greatest reserve fuel in the body. Most usable fat resides in adipose tissue that is composed of specialized cells packed with globules of triglycerides. Adipose tissue is widely distributed in the abdominal cavity, in muscles, around deep blood vessels and large organs (for example, heart and kidneys), and especially under the skin. Women average about 30% more fat than men, which is largely responsible for differences in shape between males and females. Humans can only too easily deposit large quantities of fat, generating personal unhappiness and hazards to health.

Physiological and psychological aspects of obesity are now being investigated by many researchers. There is increasing evidence that food intake, and therefore the amount of fat deposition, is regulated by feeding centers located in the brain (lateral and ventral hypothalamus and brain stem). The set point of these regions determines normal food intake and body weight for an individual, which may be maintained above or below what is considered normal for humans. Although evidence is accumulating that there is a genetic component to obesity, the almost epidemic proportions of obesity in the United States are more easily accounted for by lifestyle and feeding habits. Other developed countries show a similar, but less pronounced, trend toward development of an obesity problem.

METABOLISM OF PROTEINS

Since proteins are composed of amino acids, of which 20 kinds commonly occur (p. 25), the central topic of our consideration is amino acid metabolism. Amino acid metabolism is complex. For one thing, each of the 20 amino acids requires a separate pathway for biosynthesis and degradation. For another, amino acids are precursors to tissue proteins, enzymes, nucleic acids, and other nitrogenous constituents that form the fabric of cells. The central purpose of carbohydrate and fat oxidation is to provide energy needed to construct and maintain these vital macromolecules.

Let us begin with the **amino acid pool** in blood and extracellular fluid from which the tissues draw their requirements. When animals eat proteins, most are digested in the gut, releasing their constituent amino acids, which are then absorbed (Figure 4-18). Tissue proteins also are hydrolyzed during normal growth, repair, and tissue restructuring; their amino acids join those derived from protein foodstuffs to enter the amino acid pool. A portion of the amino acid pool is used to rebuild tissue

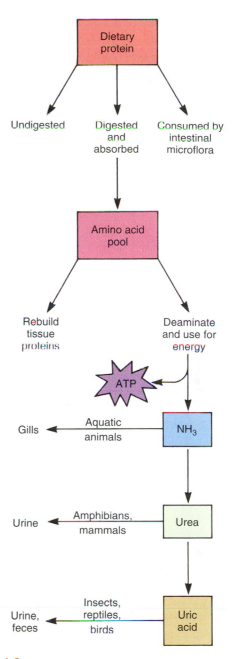

Figure 4-18
Fate of dietary protein.

proteins, but most animals ingest a surplus of protein. Since amino acids are not excreted as such in any significant amounts, they must be disposed of in some other way. In fact, amino acids can be and are metabolized through oxidative pathways to yield high-energy phosphate. In short, excess proteins serve as fuel as do carbohydrates and fats. Their importance as fuel obviously depends on the nature of the diet. In carnivores that ingest a diet of almost pure protein and fat, nearly half of their high-energy phosphate is derived from amino acid oxidation.

Before an amino acid molecule may enter the fuel depot, nitrogen must be removed by deamination (the amino group splits to form ammonia and a keto acid) or by transamination (the amino group is transferred to a keto acid to yield a new amino acid). Thus amino acid degradation yields two main products, carbon skeletons and ammonia, which are handled in different ways. Once nitrogen atoms are removed, the carbon skeletons of amino acids can be completely oxidized, usually by way of pyruvic acid or acetic acid. These residues then enter regular routes used by carbohydrate and fat metabolism.

The other product of amino acid degradation is ammonia. Ammonia is highly toxic because it reacts with α-ketoglutaric acid to form glutamic acid (an amino acid). Any accumulation of ammonia effectively removes α-ketoglutarate from the Krebs cycle (see Figure 4-13) and inhibits respiration. Disposal of ammonia offers little problem to aquatic animals because it is soluble and readily diffuses into the surrounding medium through respiratory surfaces. Terrestrial forms cannot get rid of ammonia so conveniently and must detoxify it by converting it to a relatively nontoxic compound. The two principal compounds formed are **urea** and **uric acid,** although a variety of other detoxified forms of ammonia are excreted by different invertebrate and vertebrate groups. Among vertebrates, amphibians and especially mammals produce urea. Reptiles and birds, as well as many terrestrial invertebrates, produce uric acid (the excretion of uric acid by insects and birds is described on pages 408 and 567, respectively).

The key feature that seems to determine choice of nitrogenous waste is availability of water in the environment. When water is abundant, the chief nitrogenous waste is ammonia. When water is restricted, it is urea. And for animals living in truly arid habitats, it is uric acid. Uric acid is highly insoluble and easily precipitates from solution, allowing its removal in solid form. Embryos of birds and reptiles benefit greatly from excretion of nitrogenous waste as uric acid, because waste cannot be eliminated through their shells. During embryonic development, harmless, solid uric acid is retained in one of the extraembryonic membranes. When a hatchling emerges into its new world, accumulated uric acid, along with the shell and membranes that supported development, is discarded.

MANAGEMENT OF METABOLISM

The complex pattern of enzymatic reactions that constitutes metabolism cannot be explained entirely in terms of physico-chemical laws or chance happenings. Although some enzymes do indeed "flow with the tide," the activity of others is rigidly controlled. In the former case, suppose the function of an enzyme is to convert A to B. If B is removed by conversion into another compound, the enzyme will tend to restore the original ratio of B to A. Since many enzymes act reversibly, either synthesis or degradation may result. For example, an excess of an intermediate in the Krebs cycle would result in its contribution to glycogen synthesis; a depletion of such a metabolite would

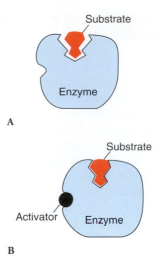

Figure 4-19

Enzyme regulation. **A,** The active site of an enzyme may only loosely fit its substrate in the absence of an activator. **B,** With the regulatory site of the enzyme occupied by an activator, the enzyme binds the substrate, and the site becomes catalytically active.

lead to glycogen breakdown. This automatic compensation (equilibration) is not, however, sufficient to explain all that actually takes place in an organism, as for example, what happens at branch points in a metabolic pathway.

Mechanisms exist for critically regulating enzymes in both *quantity* and *activity.* In bacteria, genes leading to synthesis of an enzyme are switched on or off, depending on the presence or absence of a substrate molecule. In this way the *quantity* of an enzyme is controlled. It is a relatively slow process.

Mechanisms that alter activity of enzymes can quickly and finely adjust metabolic pathways to changing conditions in a cell. The presence or increase in concentration of some molecules can alter the shape (conformation) of particular enzymes, thus activating or inhibiting the enzyme (Figure 4-19). For example, phosphofructokinase, which catalyzes phosphorylation of glucose-6-phosphate to fructose-1, 6-diphosphate (Figure 4-15), is inhibited by high concentrations of ATP or citric acid. Their presence means that a sufficient amount of precursors has reached the Krebs cycle and additional glucose is not needed.

As well as being subject to alteration in physical shape, many enzymes exist in both an active and an inactive form. These forms may be chemically different. Enzymes that degrade glycogen (phosphorylase) and synthesize it (synthase) are examples. Conditions that activate phosphorylase tend to inactivate synthase and vice versa.

Many cases of enzyme regulation are known, but these selected examples must suffice to illustrate the importance of enzyme regulation in integration of metabolism.

SUMMARY

Living systems are subject to the same laws of thermodynamics that govern nonliving systems. The first law states that energy cannot be destroyed, although it may change form. The second law states that the structure of systems proceeds toward total randomness, or increasing entropy, as energy is dissipated. Solar energy trapped by photosynthesis as chemical bond energy is passed through the food chain where it is used for biosynthesis, active transport, and motion, before finally being dissipated as heat. Living organisms are able to decrease their entropy and maintain high internal order because the biosphere is an open system from which energy can be captured and used. Energy available for use in biochemical reactions is termed "free energy."

Enzymes are proteins, often associated with nonprotein cofactors, that vastly accelerate rates of chemical reactions in living systems. An enzyme does this by temporarily binding its reactant (substrate) onto an active site in a highly specific fit. In this configuration, internal activation energy barriers are lowered enough to modify the substrate, and the enzyme is restored to its original form.

Cells use the energy stored in chemical bonds of organic fuels by degrading fuels through a series of enzymatically controlled steps. This bond energy is transferred to ATP and packaged in the form of "high-energy" phosphate bonds. ATP is produced as it is required in cells to power various synthetic, secretory, and mechanical processes.

Glucose is an important source of energy for cells. In aerobic metabolism (respiration), the 6-carbon glucose is split into two 3-carbon molecules of pyruvic acid. Pyruvic acid is decarboxylated to form 2-carbon acetyl coenzyme A, a strategic intermediate that leads to the Krebs cycle. Acetyl coenzyme A can also be derived from breakdown of fat. In the Krebs cycle, acetyl coenzyme A is oxidized in a series of reactions to carbon dioxide, yielding, in the course of the reactions, energized electrons that are passed to electron acceptor molecules (NAD and FAD). In the final stage, the energized electrons are passed along an electron transport chain consisting of a series of electron carriers located in the inner membranes of mitochondria. A hydrogen gradient is produced as electrons are passed from carrier to carrier and finally to oxygen, and ATP is generated as the hydrogen ions flow down their electrochemical gradient through ATP synthase molecules located in the inner mitochondrial membrane. A net total of 36 molecules of ATP may be generated from one molecule of glucose.

In the absence of oxygen (anaerobic glycolysis), glucose is degraded to two 3-carbon molecules of lactic acid, yielding two molecules of ATP. Although anaerobic glycolysis is vastly less efficient than respiration, it provides essential energy for muscle contraction when heavy energy expenditure outstrips the oxygen-delivery system of an animal; it also is the only source of energy generation for microorganisms living in oxygen-free environments.

Triglycerides (neutral fats) are especially rich depots of metabolic energy because the fatty acids of which they are composed are highly reduced and free of water. Fatty acids are degraded by sequential removal of 2-carbon units, which enter the Krebs cycle through acetyl-CoA.

Amino acids in excess of requirements for synthesis of proteins and other biomolecules are used as fuel. They are degraded by deamination or transamination to yield ammonia and carbon skeletons. The latter enter the Krebs cycle to be oxidized. Ammonia is a highly toxic waste product that aquatic animals quickly expel through respiratory surfaces. Terrestrial animals, however, convert ammonia into much less toxic compounds, urea or uric acid, for disposal.

Integration of metabolic pathways is finely regulated by mechanisms that control both amount and activity of enzymes. The quantity of some enzymes is regulated by certain molecules that switch on or off enzyme synthesis. Enzyme activity may be altered by the presence or absence of metabolites that cause conformational changes in enzymes and thus improve or diminish their effectiveness as catalysts.

REVIEW QUESTIONS

1. State the first and second laws of thermodynamics. Living systems may appear to violate the second law of thermodynamics because living things maintain a high degree of organization despite a universal trend toward increasing disorganization. What is the explanation for this apparent paradox?
2. Explain what is meant by "free energy" in a system. Will a reaction that proceeds spontaneously have a positive or negative change in free energy?
3. Many biochemical reactions proceed slowly unless the energy barrier to the reaction is lowered. How is this accomplished in living systems?
4. What happens in the formation of an enzyme-substrate complex that favors the disruption of substrate bonds?
5. What is meant by a "high-energy bond"?
6. Although ATP supplies energy to an endergonic reaction, why is it not considered a fuel?
7. What is an oxidation-reduction reaction and why are such reactions considered so important in cellular metabolism?
8. Give an example of a final electron acceptor found in aerobic and anaerobic organisms. Why is aerobic metabolism more efficient than anaerobic metabolism?
9. Why is it necessary for glucose to be "primed" with a high-energy phosphate bond before it can be degraded in the glycolytic pathway?
10. What happens to the electrons that are removed during the oxidation of triose phosphates during glycolysis?
11. Why is acetyl coenzyme A considered a "strategic intermediate" in respiration?
12. Why are oxygen atoms important in oxidative phosphorylation?
13. Explain how animals can generate ATP *without* oxygen. Given that anaerobic glycolysis is much less efficient than oxidative phosphorylation, why has anaerobic glycolysis not been discarded during animal evolution?

14. Why are animal fats sometimes called "the king of fuels"? What is the significance of acetyl coenzyme A to lipid metabolism?

15. The breakdown of amino acids yields two products: ammonia and carbon skeletons. What happens to these products?

16. Explain the relationship between the amount of water in an animal's environment and the kind of nitrogenous waste it produces.

17. Explain three ways that enzymes may be regulated in cells.

SELECTED REFERENCES

Berg, J., J. Tymoczko, and L. Stryer. 2002. Biochemistry, ed. 5. San Francisco, W. H. Freeman & Company. *One of the best undergraduate biochemistry texts.*

Dickerson, R. E. 1980. Cytochrome c and the evolution of energy metabolism. Sci. Am. **242:**136–153 (Mar.). *How the metabolism of modern organisms evolved.*

Hinkle, P., and R. McCarty. 1978. How cells make ATP. Sci. Am. **238:**104–123 (Mar.). *Good description of the chemiosmotic hypothesis, which explains how ATP is made from electrons removed from foodstuffs.*

Lodish, H., A. Berk, S. L. Zipursky, P. Matsudaira, D. Baltimore, and J. Darnell. 2000. Molecular cell biology, ed. 4. San Francisco, W. H. Freeman & Company. *Chapter 16 is a comprehensive, well-illustrated treatment of energy metabolism.*

Wolfe, S. L. 1995. Introduction to cell and molecular biology. Belmont, California, Wadsworth Publishing Company. *Covers the same topics as Wolfe's big book, but in less detail.*

ZOOLOGY LINKS TO THE INTERNET

Visit the textbook's Online Learning Center at www.mhhe.com/zoology to find live Internet links for each of the topics listed here.

Energy and Laws of Thermodynamics

Enzymes

Energy and Enzymes

Cellular Respiration

Aerobic Respiration

Anaerobic Respiration and Fermentation Pathways

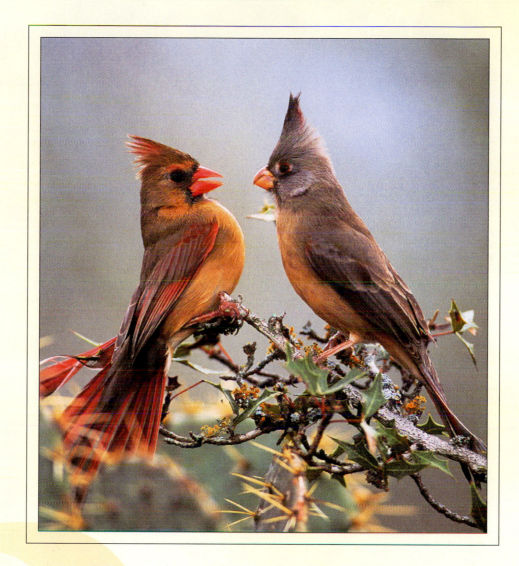

A female *Cardinalis cardinalis* (left) and a female *Cardinalis sinuatus* (right).

PART TWO

Continuity and Evolution of Animal Life

5   Genetics: A Review
6   Organic Evolution
7   The Reproductive Process
8   Principles of Development

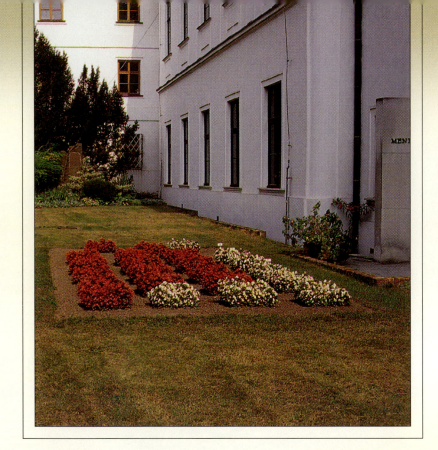

Refectory and site of Gregor Mendel's experimental garden, Brno, Czech Republic.

Genetics: A Review

A Code for All Life

The principle of hereditary transmission is a central tenet of life on earth: all organisms inherit a structural and functional organization from their progenitors. What is inherited by an offspring is not necessarily an exact copy of the parent but a set of coded instructions that gives rise to a certain expressed organization. These instructions are in the form of genes, the fundamental units of inheritance. One of the great triumphs of modern biology was the discovery in 1953 by James Watson and Francis Crick of the nature of the coded instructions in genes. This was followed by the discovery of the way in which the code is translated into the expression of characteristics. The genetic material (deoxyribonucleic acid, DNA) is composed of nitrogenous bases arranged on a backbone of sugar phosphate units. The genetic code lies in the linear order or sequence of bases in the DNA strand.

Because the DNA molecules replicate themselves in their passage from generation to generation, genetic variations can persist once they have happened. Such molecular alterations, called mutations, are the ultimate source of biological variation and the raw material of evolution.

A basic principle of modern evolutionary theory is that organisms attain their diversity through hereditary modifications of preexisting lines of ancestors. All known lineages of plants and animals are related by descent from common ancestral groups.

Heredity establishes the continuity of life forms. Although offspring and parents in a particular generation may look different, there is nonetheless a basic sameness that runs from generation to generation for any species of plant or animal. In other words, "like begets like." Yet children are not precise replicas of their parents. Some of their characteristics show resemblances to one or both parents, but they also demonstrate many traits not seen in either parent. What is actually inherited by an offspring from its parents is a certain type of germinal organization **(genes)** that, under the influence of environmental factors, guides the orderly sequence of differentiation of a fertilized egg into a human being, bearing the unique physical characteristics as we see them. Each generation hands on to the next the instructions required for maintaining continuity of life.

The gene is the unit entity of inheritance, the germinal basis for every characteristic that appears in an organism. The study of what genes are and how they work is the science of genetics. It is a science that deals with the underlying causes of *resemblance,* as seen in the remarkable fidelity of reproduction, and of *variation,* which is the working material for organic evolution. Genetics has shown that all living forms use the same information storage, transfer, and translation system, and thus it has provided an explanation for both the stability of all life and its probable descent from a common ancestral form. This is one of the most important unifying concepts of biology.

MENDEL'S INVESTIGATIONS

The first man to formulate the cardinal principles of heredity was Gregor Johann Mendel (1822–1884) (Figure 5-1 and p. 17), who was an Augustinian monk living in Brünn (Brno), Moravia. At that time Brünn was a part of Austria, but now it is in the eastern part of the Czech Republic. While conducting breeding experiments in a small monastery garden from 1856 to 1864, Mendel examined with great care the progeny of many thousands of plants. He worked out in elegant simplicity the laws governing the transmission of characters from parent to offspring. His discoveries, published in 1866, were of great significance, coming just after Darwin's publication of *On the Origin of Species by Means of Natural Selection.* Yet these discoveries remained unappreciated and forgotten until 1900—some 35 years after the completion of the work and 16 years after Mendel's death.

Mendel's classic observations were based on garden peas because they had been produced in pure strains by gardeners over a long period of time by careful selection. For example, some varieties were definitely dwarf and others were tall. A second reason for selecting peas was that they were self-fertilizing but also capable of cross-fertilization. To simplify his problem Mendel chose single characters that displayed sharply contrasting traits. He carefully avoided mere quantitative and intermediate characteristics. Mendel selected pairs of contrasting traits, such as tall plants versus dwarf plants and smooth seeds versus wrinkled seeds (Figure 5-1).

A giant stride in chromosomal genetics was made when the renowned American geneticist Thomas Hunt Morgan and his colleagues selected a species of fruit fly, *Drosophila melanogaster,* for their studies (1910–1920). Flies were cheaply and easily reared in bottles in the laboratory, fed on a simple medium of bananas and yeast. Most important, they produced a new generation every 10 days, enabling Morgan to proceed at least 25 times more rapidly than with organisms that take longer to mature, such as garden peas. Morgan's work led to the mapping of genes on chromosomes and founded the discipline of cytogenetics.

Mendel crossed plants having one of these traits with others having the contrasting trait. He removed the stamens (male part, containing the pollen) from a flower to prevent self-fertilization and then placed on the stigma (female part of flower) pollen from the flower of the plant that had the contrasting character. Pollination from other sources such as wind and insects was rare and did not affect his results. When the cross-fertilized flower bore seeds, he noted the kind of plants (hybrids) that were produced from the planted seeds. These hybrids then produced offspring by self-pollination.

Mendel knew nothing of the cytological basis of heredity, since chromosomes and genes were unknown to him. Although we can admire Mendel's power of intellect in his discovery of the principles of inheritance without knowledge of chromosomes, these principles are certainly easier to understand if we first review chromosomal behavior, especially in meiosis.

CHROMOSOMAL BASIS OF INHERITANCE

In sexually reproducing organisms, special **sex cells,** or **gametes** (ova and sperm), are responsible for providing genetic information to offspring. A scientific explanation of genetic principles required a study of germ cells and their behavior, which meant working backward from certain visible results of inheritance to the mechanism responsible for such results. Nuclei of sex cells, especially the chromosomes, were early suspected of furnishing the real answer to the mechanism. Chromosomes are apparently the only entities inherited in equal quantities from both parents to offspring.

When Mendel's laws were rediscovered in 1900, their parallelism with the cytological behavior of chromosomes was obvious. Later experiments showed that chromosomes carried the hereditary material.

Meiosis: Reduction Division of Gametes

Every body cell contains *two* chromosomes bearing genes for the same set of characteristics, and the two members of each

Figure 5-1

Seven experiments on which Gregor Mendel based his postulates. These are the results of monohybrid crosses for first and second generations.

Round vs. wrinkled seeds
F_1 = all round
F_2 = 5474 round
    1850 wrinkled
Ratio: 2.96:1

Purple vs. white flowers
F_1 = all purple
F_2 = 705 purple
    224 white
Ratio: 3.15:1

Yellow vs. green seeds
F_1 = all yellow
F_2 = 6022 yellow
    2001 green
Ratio: 3.01:1

Green vs. yellow pods
F_1 = all green
F_2 = 428 green
    152 yellow
Ratio: 2.82:1

Inflated vs. constricted pods
F_1 = all inflated
F_2 = 882 inflated
    299 constricted
Ratio: 2.95:1

Long vs. short stems
F_1 = all long
F_2 = 787 long
    277 short
Ratio: 2.84:1

Axial vs. terminal flowers
F_1 = all axial
F_2 = 651 axial
    207 terminal
Ratio: 3.14:1

pair usually, but not always, have the same size and shape. The members of such a pair are called **homologous** chromosomes; each individual member of a pair is called a **homolog.** One homolog comes from the mother and the other from the father. Meiosis consists of *two* nuclear divisions in which the chromosomes divide only once (Figure 5-2). The result is that mature gametes have only *one* member of each homologous chromosome pair, or a **haploid** (*n*) number of chromosomes. When the gametes unite in any fertilization, a **zygote** is formed. In humans the zygotes and all body cells normally have the **diploid** number (2*n*), or 46 chromosomes; the gametes have the haploid number (*n*), or 23, and meiosis reduces the number of chromosomes from diploid to haploid.

Thus each cell normally has two copies of each gene coding for a given trait, one on each of the homologous chromosomes. Alternative forms of genes for the same trait are **allelic** forms, or **alleles.** Sometimes only one of the alleles has an effect on the organism, although both are present in each cell, and either may be passed to the progeny as a result of meiosis and subsequent fertilization.

Alleles are alternative forms of the same gene that have arisen by mutation of the DNA sequence. Like a baseball team with several pitchers, only one of whom can occupy the pitcher's mound at a time, only one allele can occupy a chromosomal locus (position). Alternative alleles for the locus may be on homologous chromosomes of a single individual, making that individual heterozygous for the gene in question. Numerous allelic forms of a gene may be found among different individuals in the population of the species.

During an individual's growth, all the chromosomes of mitotically dividing cells contain the double set of chromosomes (mitosis is described on p. 48). In the reproductive organs,

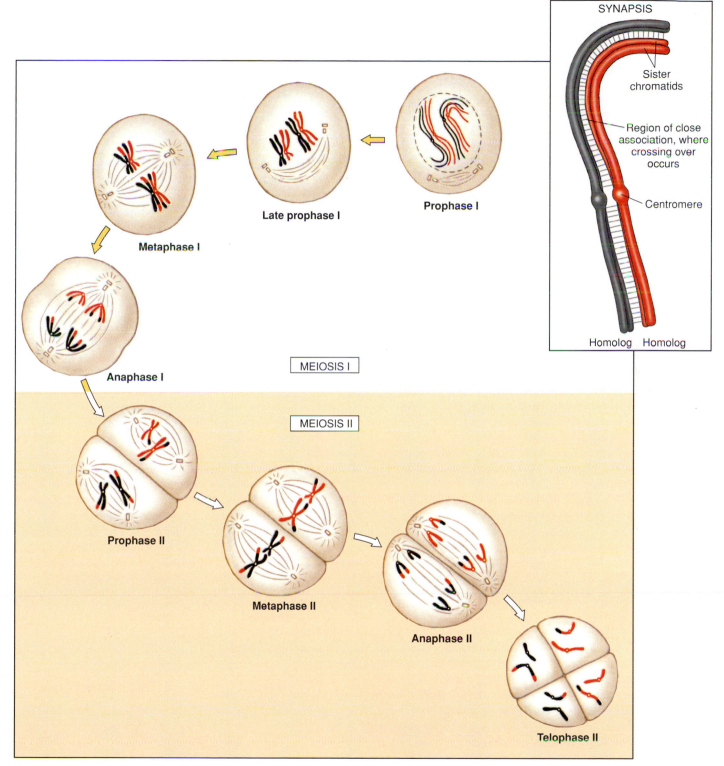

Figure 5-2

A, Meiosis in a sex cell with two pairs of chromosomes. Prophase I, homologous chromosomes come to lie with side-to-side contact, or synapsis, forming bivalents. Late prophase I, bivalents are composed of two pairs of chromatids. Metaphase I, bivalents align at the spindle equator. Anaphase I, dyads of former bivalents pulled toward opposite poles. Prophase II, daughter cells contain one of each homologous chromosome (haploid) but with diploid amount of DNA. Metaphase II, dyads align at the spindle equator. Anaphase II, chromatids of dyads separate. Telophase II, four haploid cells (gametes) formed, each with haploid amount of DNA. **B,** Synapsis occurs in prophase I, in which homologous chromosomes can break and exchange equivalent portions.

gametes (germ cells) are formed after meiosis, which *separates* the homologous pairs of chromosomes. If it were not for this reductional division, the union of ovum (egg) and sperm would produce an individual with twice as many chromosomes as the parents. Continuation of this process in just a few generations could yield astronomical numbers of chromosomes per cell.

Most unique features of meiosis occur during prophase of the first meiotic division (Figure 5-2). The two members of each pair of homologous chromosomes come into side-by-side contact **(synapsis)** to form a **bivalent.** Each chromosome of the bivalent has already replicated to form two chromatids, each of which will become a new chromosome. The two chromatids are joined at one point, the centromere, so that each bivalent is composed of two pairs of chromatids (each pair is a **dyad**), or *four* future chromosomes, and is thus called a **tetrad.** The position or location of any gene on a chromosome is the gene **locus** (pl., **loci**), and in synapsis all gene loci on a chromatid normally lie exactly opposite the corresponding loci on the sister chromatid. Toward the end of prophase, the chromosomes shorten and thicken and are ready to enter into the first meiotic division. In contrast to mitosis, the centromeres holding the chromatids together *do not divide* at the beginning of anaphase. As a result, each of the dyads is pulled toward each pole by microtubules of the division spindle. Therefore at the end of the first meiotic division, the daughter cells contain *one* of *each* of the homologous chromosomes, so the total chromosome number has been reduced to haploid. However, because the chromatids are still joined by their centromeres, each cell contains $2n$ amount of DNA.

The second meiotic division more closely resembles events in mitosis. The dyads are split at the beginning of anaphase by division of their centromeres, and single-stranded chromosomes move toward each pole. Thus by the end of the second meiotic division, the cells have the haploid number of chromosomes and n amount of DNA. Each chromatid of the original tetrad exists in a separate nucleus. Four products are formed, each containing one complete haploid set of chromosomes and only one allele of each gene. Only one of the four products in female gametogenesis will become a functional gamete (p. 139).

Sex Determination

Before the importance of chromosomes in heredity was realized in the early 1900s, how gender was determined was totally unknown. The first really scientific clue to the determination of sex came in 1902 when C. McClung observed that bugs (Hemiptera) produced two kinds of sperm in approximately equal numbers. One kind contained among its regular set of chromosomes a so-called accessory chromosome that was lacking in the other kind of sperm. Since all eggs of these species had the same number of haploid chromosomes, half the sperm would have the same number of chromosomes as the eggs, and half of them would have one chromosome less. When an egg was fertilized by a spermatozoon carrying the accessory (sex) chromosome, the resulting offspring was a female; when fertilized by a spermatozoon without an accessory chromosome, the offspring was a male. Therefore a distinction was made between sex chromosomes, which determine sex (and sex-linked traits);

and **autosomes,** the remaining chromosomes, which do not influence sex. The particular type of sex determination just described is often called the XX-XO type, which indicates that females have two X chromosomes and males only one X chromosome (the O indicates absence of the chromosome). The XX-XO method of sex determination is depicted in Figure 5-3.

Speculation on how sex was determined in animals produced several incredible theories, for example, that the two testicles of males contained different types of semen, one begetting males, the other females. It is not difficult to imagine the abuse and mutilation of domestic animals that occurred when attempts were made to alter sex ratios of herds. Another conjecture asserted that sex of offspring was determined by the more heavily sexed parent. An especially masculine father should produce sons, an effeminate father only daughters. Such mistaken ideas have lingered until recently.

Later, other types of sex determination were discovered. In humans and many other animals each sex contains the same number of chromosomes; however, the sex chromosomes (XX) are alike in females but unlike (XY) in males. Hence a human egg contains 22 autosomes + 1 X chromosome. Sperm are of two kinds; half carry 22 autosomes + 1 X and half bear 22 autosomes + 1 Y. The Y chromosome is much smaller than the X and carries very little genetic information. At fertilization, when the 2 X chromosomes come together, offspring are female; when X and Y come together, offspring are male. The XX-XY kind of determination is shown in Figure 5-4.

A third type of sex determination is found in birds, moths, butterflies, and some fish in which the male has 2 X (or sometimes called ZZ) chromosomes and the female an X and Y (or ZW). Finally, there are both invertebrates (p. 428) and vertebrates (p. 552) in which sex is determined by environmental or behavioral conditions rather than by sex chromosomes, or by

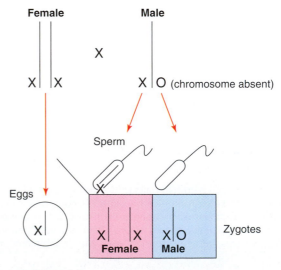

Figure 5-3
XX-XO sex determination.

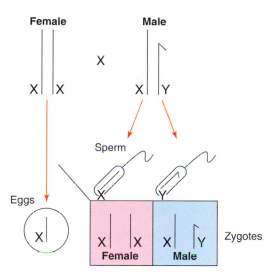

Figure 5-4
XX-XY sex determination.

genetic loci whose variation is not associated with visible difference in chromosomal structure.

In the case of X and Y chromosomes, homologous chromosomes are unlike in size and shape. Therefore, they do not both carry the same genes. Genes of the X chromosome often do not have allelic counterparts on the diminutive Y chromosome. This fact is very important in sex-linked inheritance (p. 84).

MENDELIAN LAWS OF INHERITANCE

Mendel's First Law

Mendel's **law of segregation** states that *in the formation of gametes, paired factors specifying alternative phenotypes (visible traits) segregate independently of one another.* In one of Mendel's original experiments, he pollinated pure-line tall plants with the pollen of pure-line dwarf plants. Thus the visible characteristics, or **phenotypes,** of the parents were tall and dwarf. Mendel found that all progeny in the first generation (F_1) were tall, just as tall as the tall parents of the cross. The reciprocal cross—dwarf plants pollinated with tall plants—gave the same result. The tall phenotype was observed in progeny no matter which way the cross was made. Obviously, this kind of inheritance was not a blending of two traits, since none of the progeny was of intermediate size.

Next Mendel self-fertilized ("selfed") the tall F_1 plants and raised several hundred progeny, the second (F_2) generation. This time, *both* tall and dwarf plants appeared. Again, there was no blending (no plants of intermediate size), but the appearance of dwarf plants from all tall parental plants was surprising. The dwarf trait, present in the grandparents but not in the parents, had reappeared. When he counted the actual number of tall and dwarf plants in the F_2 generation, he discovered that there were almost exactly three times more tall plants than dwarf ones.

Mendel then repeated this experiment for the six other contrasting traits that he had chosen, and in every case he obtained ratios very close to 3:1 (see Figure 5-1). At this point it must have been clear to Mendel that he was dealing with hereditary determinants for the contrasting traits that did not blend when brought together. Even though the dwarf trait disappeared in the F_1 generation, it reappeared fully expressed in the F_2 generation. He realized that the F_1 generation plants carried determinants (which he called "factors") of both tall and dwarf parents, even though only the tall trait was expressed in the F_1 generation.

Mendel called the tall factor **dominant** and the short **recessive.** Similarly, the other pairs of traits that he studied showed dominance and recessiveness. Whenever a dominant factor is present, the recessive one cannot produce its effect. The recessive trait will appear only when both factors are recessive, or in other words, in a pure condition.

In representing his crosses, Mendel used letters as symbols; dominant traits were represented by capital letters, and for recessive traits he used the corresponding lowercase letters. Modern geneticists still follow this custom. Thus the factors for pure tall plants might be represented by *T/T,* the pure recessive by *t/t,* and the mix, or hybrid, of the two plants by *T/t.* The slash mark is to indicate that the alleles are on homologous chromosomes. The zygote bears the complete genetic constitution of the organism. All the gametes produced by *T/T* must necessarily be *T,* whereas those produced by *t/t* must be *t.* Therefore a zygote produced by union of the two must be *T/t,* or a **heterozygote.** On the other hand, the pure tall plants (*T/T*) and pure dwarf plants (*t/t*) are **homozygotes,** meaning that the paired factors (alleles) are alike on the homologous chromosomes. A cross involving only one pair of contrasting traits is called a **monohybrid cross.**

In the cross between tall and dwarf plants there were two phenotypes: tall and dwarf. On the basis of genetic formulas there are three *hereditary* types: *T/T, T/t,* and *t/t.* These are called **genotypes.** A genotype is an allelic combination (*T/T, T/t,* or *t/t*), and the phenotype is the corresponding appearance of the organism (tall or dwarf).

One of Mendel's original crosses (tall plant and dwarf plant) could be represented as follows:

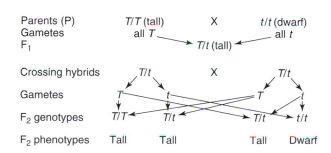

In other words, all possible combinations of F_1 gametes in the F_2 zygotes will yield a 3:1 phenotypic ratio and a 1:2:1 genotypic ratio. It is convenient in such crosses to use the checkerboard method devised by Punnett (Punnett square) for representing the various combinations resulting from a cross. In the F_2 cross this scheme would apply:

Ratio: 3 tall to 1 dwarf

The next step was an important one because it enabled Mendel to test his hypothesis that every plant contained non-blending factors from both parents. He self-fertilized the plants in the F₂ generation; that is, the stigma of a flower was fertilized by the pollen of the same flower. The results showed that self-pollinated F₂ dwarf plants produced only dwarf plants, whereas one-third of the F₂ tall plants produced tall and the other two-thirds produced both tall and dwarf in the ratio of 3:1, just as the F₁ plants had done. Genotypes and phenotypes were as follows:

$$
F_2 \text{ plants: } \text{Tall} \begin{cases} \tfrac{1}{4}\ T/T \xrightarrow{\text{Selfed}} \text{all } T/T \text{ (homozygous tall)} \\ \tfrac{1}{2}\ T/t \xrightarrow{\text{Selfed}} 1\ T/T: 2\ T/t: 1\ t/t\ (3 \text{ tall}: 1 \text{ dwarf}) \end{cases}
$$

$$
\text{Dwarf } \tfrac{1}{4}\ t/t \xrightarrow{\text{Selfed}} \text{all } t/t \text{ (homozygous dwarf)}
$$

This experiment showed that the dwarf plants were pure because they at all times gave rise to short plants when self-pollinated; the tall plants contained both pure tall and hybrid tall. It also demonstrated that, although the dwarf trait disappeared in the F₁ plants, which were all tall, dwarfness appeared in the F₂ plants.

Mendel reasoned that the factors for tallness and dwarfness were units that did not blend when they were together. The F₁ generation contained both of these units or factors, but when these plants formed their germ cells, the factors separated so that each germ cell had only one factor. In a pure plant both factors were alike; in a hybrid they were different. He concluded that individual germ cells were always pure with respect to a pair of contrasting factors, even though the germ cells were formed from hybrid individuals possessing both contrasting factors.

This idea formed the basis for his law of segregation, that is, that whenever two factors are brought together in a hybrid, they segregate into separate gametes that are produced by the hybrid. Either one of the paired factors or alleles of the parent pass with equal frequency to the gametes. We now understand that the factors segregate because there are two alleles for the character, one on each chromosome of a homologous pair, but the gametes receive only one of each in meiosis. Thus in current usage the law of segregation refers to the parting of homologous chromosomes during meiosis.

Mendel's great contribution was his quantitative approach to inheritance. This really marks the birth of genetics, because before Mendel, people believed that traits were blended like mixing together two colors of paint, a notion that unfortunately still lingers in the minds of many and was a problem for Darwin's theory of natural selection when he first proposed it (p. 17). If traits were blended, variability would be lost in hybridiza-

tion between individuals. With particulate inheritance, on the other hand, different variations are retained and can be shuffled about and resorted like blocks.

In not reporting conflicting findings, which must surely have arisen as they do in any original research, Mendel has been accused of "cooking" his results. The chances are, however, that he carefully avoided ambiguous material to strengthen his central message, which we still regard as an exemplary achievement in experimental analysis.

Testcross

When one of the alleles is dominant, heterozygous individuals are identical in phenotype to individuals homozygous for the dominant allele. Therefore you cannot determine the genotypes of these individuals just by observing their phenotypes. For instance, in Mendel's experiment of tall and dwarf traits, it is impossible to determine the genetic constitution of the tall plants of the F₂ generation by mere inspection of the tall plants. Three-fourths of this generation are tall, but which of them are heterozygotes?

As Mendel reasoned, the test is to cross the questionable individuals with pure recessives. If the tall plant is homozygous, all the offspring in such a testcross are tall, thus:

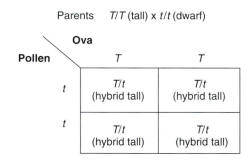

All of the offspring are *T/t* (hybrid tall). If, on the other hand, the tall plant is heterozygous, half of the offspring are tall and half dwarf, thus:

Parents   *T/t* (hybrid tall) x *t/t* (dwarf)

Pollen \ Ova	T	t
t	*T/t* (hybrid tall)	*t/t* (homozygous dwarf)
t	*T/t* (hybrid tall)	*t/t* (homozygous dwarf)

The **testcross** is often used in modern genetics for the analysis of the genetic constitution of the offspring, as well as for a quick way to make desirable homozygous stocks of animals and plants.

Intermediate Inheritance

In some cases neither allele is completely dominant over the other, and the heterozygous phenotype appears either intermediate between or even quite distinct from those of the parents. This is called **intermediate inheritance,** or **incomplete dominance.** In the four-o'clock flower (*Mirabilis*), two allelic variants determine red versus pink or white flowers; homozygotes are red or white flowered, but heterozygotes have pink flowers. In a certain strain of chickens, a cross between those with black and splashed white feathers produces offspring that are not gray but a distinctive color called Andalusian blue (Figure 5-5). In each case, if the F$_1$s are crossed, the F$_2$s have a ratio of 1:2:1 in colors, or 1 red: 2 pink: 1 white in four-o'clock flowers and 1 black: 2 blue: 1 white for Andalusian chickens. This can be illustrated for the chickens as follows:

Parents	*B/B*	(black feathers)	X	*B'/B'*	(white feathers)
Gametes	all *B*			all *B'*	
F$_1$			*B/B'* (all blue)		
Crossing hybrids		*B/B'*	X	*B/B'*	
Gametes		*B,B'*		*B,B'*	
F$_2$ genotypes	*B/B*	*B/B'*		*B/B'*	*B'/B'*
F$_2$ phenotypes	Black	Blue		Blue	White

When neither of the alleles is recessive, it is customary to represent both by capital letters and to distinguish them by the addition of a "prime" sign (*B'*) or by superscript letters, for example, B^b (equals black feathers) and B^w (equals white feathers).

In this kind of a cross, the heterozygous phenotype is indeed a blending of both parental types. It is easy to see how such observations would encourage the notion of the blending concept of inheritance. However, in the cross of black and white chickens or red and white flowers, *only* the hybrid is a phenotypic blend; the homozygous strains breed true to the parental phenotypes.

Mendel's Second Law

According to Mendel's **law of independent assortment,** *genes located on different pairs of homologous chromosomes assort independently during meiosis.* Thus the law deals with genes for two different characters that are borne on two different pairs of chromosomes. Mendel carried out experiments on peas that differed from each other at two or more genes, that is, experiments involving two or more phenotypic characters.

Mendel had already established that tall plants were dominant to dwarf. He also noted that crosses between plants bearing yellow seeds and plants bearing green seeds produced plants with yellow seeds in the F$_1$ generation; therefore yellow was dominant to green. The next step was to make a cross between plants differing in these two characteristics. When a tall

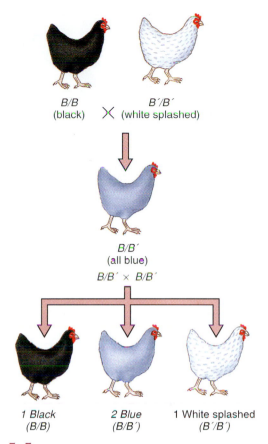

Figure 5-5

Cross between chickens with black and splashed white feathers. Black and white are homozygous; Andalusian blue is heterozygous.

plant with yellow seeds (*T/T Y/Y*) was crossed with a dwarf plant with green seeds (*t/t y/y*), the F$_1$ plants were tall and yellow as expected (*T/t Y/y*).

The F$_1$ hybrids were then crossed with each other, and the F$_2$ results are shown in Figure 5-6.

Parents	*T/T Y/Y*	x	*t/t y/y*
	(tall, yellow)		(dwarf, green)
Gametes	all *TY*		all *ty*
F$_1$		*T/t Y/y*	
		(tall, yellow)	

Mendel already knew that a cross between two plants bearing a single pair of alleles of the genotype *T/t* would yield a 3:1 ratio. Similarly, a cross between two plants with the genotypes *Y/y* would yield the same 3:1 ratio. If we examine *only* the tall and dwarf phenotypes expected in the outcome of the dihybrid experiment, they produce a ratio of 12 tall to 4 dwarf, which reduces to a ratio of 3:1. Likewise, a total of 12 plants have yellow seeds for every 4 plants that have green—again a 3:1 ratio. Thus the monohybrid ratio prevails for both traits when they are considered independently. The 9:3:3:1 ratio is nothing more than a combination of the two 3:1 ratios.

$$3:1 \times 3:1 = 9:3:3:1$$

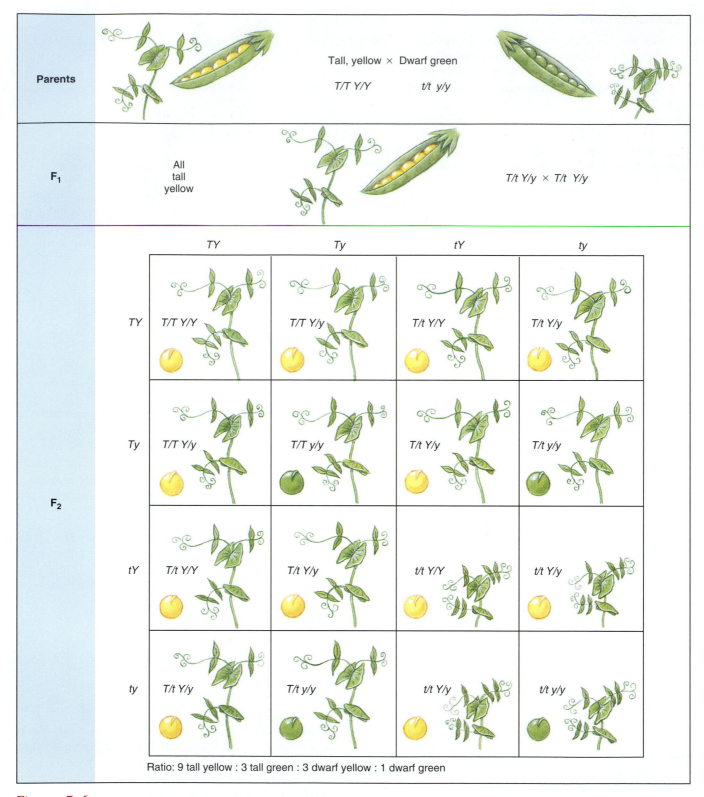

Ratio: 9 tall yellow : 3 tall green : 3 dwarf yellow : 1 dwarf green

Figure 5-6
Punnett square method for determination of genotypes and phenotypes expected in a dihybrid cross for independently assorting genes.

When one of the alleles is unknown, it can be designated by a dash ($T/$—). This designation can also be used when it is immaterial whether the genotype is heterozygous or homozygous, as when we total all of a certain phenotype. The dash could be either T or t.

The F_2 genotypes and phenotypes are as follows:

1	T/T Y/Y		
2	T/t Y/Y		
2	T/T Y/y	9 $T/$—$Y/$—	9 Tall yellow
4	T/t Y/y		
1	T/T y/y	3 $T/$—y/y	3 Tall green
2	T/t y/y		
1	t/t Y/Y	3 t/t—$Y/$—	3 Dwarf yellow
2	t/t Y/y		
1	t/t y/y	1 t/t y/y	1 Dwarf green

The results of this experiment show that the segregation of alleles for plant height is entirely independent of the segregation of alleles for seed color. Neither has any influence on the other. Thus another way to state Mendel's law of independent assortment is that *allelic variants of different genes on different chromosomes segregate independently of one another.* The reason is that during meiosis the member of any pair of homologous chromosomes received by a gamete is independent of which member of any other pair of chromosomes it receives. Of course, if the genes were on the same chromosome, they would assort together (be linked) unless crossing over occurred. Linked genes are discussed on p. 85.

One way to estimate proportions of progeny with a given genotype or phenotype is to construct a Punnett square. With a monohybrid cross, this is easy; with a dihybrid cross, a Punnett square is rather laborious; and with a trihybrid cross, it is very tedious. We can make such estimates much more easily by taking advantage of simple probability calculations. The basic assumption is that all the genotypes of gametes of one sex have an equal chance of uniting with all the genotypes of gametes of the other sex, in proportion to the numbers of each present. This is generally true when the sample size is large enough, and the actual numbers observed come close to those predicted by the laws of probability.

We may define probability as follows:

$$\text{Probability (p)} = \frac{\text{Number of times an event happens}}{\text{Total number of trials or possibilities for the event to happen}}$$

For example, the probability (p) of a coin falling heads when tossed is 1/2, because the coin has two sides. The probability of rolling a three on a die is 1/6, because the die has six sides.

The probability of independent events occurring together (ordered events) involves the **product rule,** which is simply the product of their individual probabilities. When two coins are tossed together, the probability of getting two heads is $1/2 \times 1/2$ = 1/4, or 1 chance in 4. The probability of rolling two threes simultaneously with two dice is as follows:

Probability of two threes = $1/6 \times 1/6 = 1/36$

Note, however, that a small sample size may give a result quite different from that predicted. Thus if we tossed the coin three times and it fell heads each time, we would not be much surprised. But if we tossed the coin 1000 times, and the number of times it fell heads diverged very much from 500, we would strongly suspect that there was something wrong with the coin. However, probability has no "memory." The probability of a coin toss yielding heads remains 1/2, no matter how many times the coin was tossed previously or results of the tosses.

We can use the product rule to predict the ratios of inheritance in monohybrid or dihybrid (or larger) crosses if the genes sort independently in the gametes (as they did in all of Mendel's experiments) (Table 5-1).

Multiple Alleles

On page 76 we defined alleles as alternate forms of a gene. Whereas an individual can have no more than two alleles at a given locus (one each on each chromosome of the homologous pair, p. 76), many more dissimilar alleles can exist in a population. An example is the set of multiple alleles that affects coat color in rabbits. The different alleles are C (normal color), c^{cb} (chinchilla color), c^b (Himalayan color), and c (albino). The four alleles fall into a dominance series with C dominant over everything. The dominant allele is always written to the left and the recessive to the right:

$$C/c^b = \text{Normal color}$$
$$c^{cb}/c^b = \text{Chinchilla color}$$
$$c^b/c = \text{Himalayan color}$$
$$c/c = \text{albino}$$

Multiple alleles arise through mutations at the same gene locus over periods of time. Any gene may mutate (p. 97) if given time and thus can give rise to slightly different alleles at the same locus.

Gene Interaction

The types of crosses previously described are simple in that the character variation involved results from the action of a single gene, but many cases are known in which the variation of a character is a result of two or more genes. Mendel probably did not appreciate the real significance of the genotype, as contrasted with the visible character—the phenotype. We now know that many different genotypes may affect a single phenotype **(polygenic inheritance).**

Also, many genes have more than a single effect on organismal phenotypes, a phenomenon called **pleiotropy.** A gene for eye color, for instance, may be the ultimate cause of eye color, yet at the same time it may be responsible for influencing the development of other characters as well. An allele at one locus may mask or prevent the expression of an allele at another locus acting on the same trait, a phenomenon called **epistasis.** Another case of gene interaction is that in which several sets of alleles may produce a cumulative effect on the same character.

TABLE 5.1

Use of Product Rule for Determination of Genotype and Phenotype Ratios in a Dihybrid Cross for Independently Assorting Genes

Parents' genotypes	T/t Y/y		x	T/t Y/y
Equivalent monohybrid crosses	T/t x T/t		and	Y/y x Y/y
Genotype ratios in F$_1$s of monohybrid crosses		1/4 T/T		1/4 Y/Y
		2/4 T/t		2/4 Y/y
		1/4 t/t		1/4 y/y
Combine two monohybrid ratios to determine dihybrid genotype ratios	1/4 T/T	×	{	1/4 Y/Y = 1/16 T/T Y/Y
				2/4 Y/y = 2/16 T/T Y/y
				1/4 y/y = 1/16 T/T y/y
	2/4 T/t	×	{	1/4 Y/Y = 2/16 T/t Y/Y
				2/4 Y/y = 4/16 T/t Y/y
				1/4 y/y = 2/16 T/t y/y
	1/4 t/t	×	{	1/4 Y/Y = 1/16 t/t Y/Y
				2/4 Y/y = 2/16 t/t Y/y
				1/4 y/y = 1/16 t/t y/y
Phenotype ratios in F$_1$s of monohybrid crosses				3/4 T/— (tall), 1/4 t/t (dwarf)
				3/4 Y/— (yellow), 1/4 y/y (green)
Combine two monohybrid ratios to determine phenotype ratios	3/4 T/—	×	{	3/4 Y/— = 9/16 T/—Y/— (tall, yellow)
				1/4 y/y = 3/16 T/—y/y (tall, green)
	1/4 t/t	×	{	3/4 Y/— = 3/16 t/t Y/— (dwarf, yellow)
				1/4 y/y = 1/16 t/t y/y (dwarf, green)
Therefore phenotype ratios = 9 tall, yellow: 3 tall, green: 3 dwarf, yellow: 1 dwarf, green				

Several characters in humans are polygenic. In such cases the characters, instead of having discrete alternative phenotypes, show continuous variation between two extremes. This is sometimes called **blending,** or **quantitative inheritance.** In this kind of inheritance the children are often more or less intermediate between the two parents.

One illustration of such a type is the degree of pigmentation in matings between the black and white human races. The cumulative genes in such matings have a quantitative expression. Three or four genes are probably involved in skin pigmentation, but we will simplify our explanation by assuming that there are only two pairs of independently assorting genes. Thus a person with very dark pigment has two genes for pigmentation on separate chromosomes (A/A B/B). Each dominant allele contributes one unit of pigment. A person with very light pigment has alleles (a/a b/b) that contribute no color. (Freckles that commonly appear in the skin of very light people represent pigment contributed by entirely separate genes.) The offspring of very dark and very light parents would have an intermediate skin color (A/a B/b).

Children of parents having intermediate skin color show a range of skin color, depending on the number of genes for pigmentation that they inherit. Their skin color ranges from very dark (A/A B/B), to dark (A/A B/b or A/a B/B), intermediate (A/A b/b or A/a B/b or a/a B/B), light (A/a b/b or a/a B/b), to very light (a/a b/b). It is thus possible for parents heterozygous for skin color to produce children with darker or lighter colors than themselves.

Inheritance of eye color in humans is another example of gene interaction. One allele (B) determines whether pigment is present in the front layer of the iris. This allele is dominant over the allele for the absence of pigment (b). The genotypes B/B and B/b pigment generally produce brown eyes, and b/b produces blue eyes. However, these phenotypes are greatly affected by many modifier genes influencing, for example, the amount of pigment present, the tone of the pigment, and its distribution. Thus a person with B/b may even have blue eyes if modifier genes determine a lack of pigment, thus explaining the rare instances of a brown-eyed child of blue-eyed parents.

Sex-Linked Inheritance

It is known that inheritance of some characters depends on the sex of the parent carrying the gene and the sex of the offspring. One of the best-known sex-linked traits of humans is hemophilia (Chapter 31, p. 659). Another example is red-green color blindness in which red and green colors are indistinguishable to varying degrees. Color-blind men greatly outnumber color-blind women. When color blindness does appear in women, their fathers are color blind. Furthermore, if a woman with normal vision who is a carrier of color blindness (a **carrier** is heterozygous for the gene and is phenotypically normal) bears sons, half of them are likely to be color blind, regardless of whether the father had normal or affected vision. How are these observations explained?

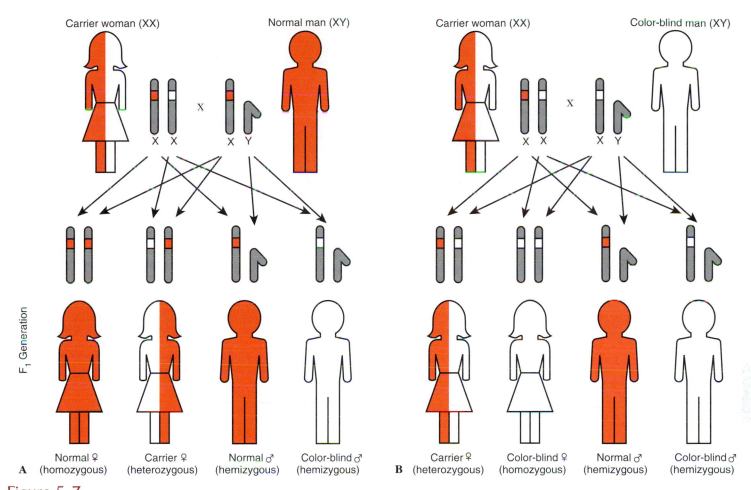

Figure 5-7

Sex-linked inheritance of red-green color blindness in humans. **A,** Carrier mother and normal father produce color blindness in one-half of their sons but in none of their daughters. **B,** Half of both sons and daughters of carrier mother and color-blind father are color blind.

Color blindness and hemophilia defects are recessive traits carried on the X chromosome. They are phenotypically expressed either when both genes are defective in the female or when only one defective gene is present in the male. The inheritance pattern of these defects is shown for color blindness in Figure 5-7. When the mother is a carrier and the father is normal, half of the sons but none of the daughters are color blind. However, if the father is color blind and the mother is a carrier, half of the sons *and* half of the daughters are color blind (on the average and in a large sample). It is easy to understand then why such defects are much more prevalent in males: a single sex-linked recessive gene in the male has a visible effect because he has only one X chromosome. What would be the outcome of a mating between a homozygous normal woman and a color-blind man?

Another example of a sex-linked character was discovered by Thomas Hunt Morgan (1910) in *Drosophila*. Normal eye color of this fly is red, but mutations for white eyes do occur (Figure 5-8). A gene for eye color is carried on the X chromosome. If truebreeding white-eyed males and red-eyed females are crossed, all F_1 offspring have red eyes because this trait is dominant (Figure 5-8). If these F_1 offspring are interbred, all F_2 females have red eyes; half of the males have red eyes and the other half have white eyes. No white-eyed females are found in this generation; only males have the recessive character (white eyes). The allele for white eyes is recessive and should affect eye color only in a homozygous condition. However, since the male has only one X chromosome (the Y does not carry a gene for eye color), white eyes appear whenever the X chromosome carries the gene for this trait. Males are said to be **hemizygous** (only one copy of a genetic locus is present) for traits carried on the X chromosome.

If the reciprocal cross is made in which females are white eyed and males red eyed, all F_1 females are red eyed and all males are white eyed (Figure 5-9). If these F_1 offspring are interbred, the F_2 generation shows equal numbers of red-eyed and white-eyed males and females.

Autosomal Linkage and Crossing Over

Linkage

Since Mendel's laws were rediscovered in 1900, it became clear that, contrary to Mendel's second law, not all factors segregate independently. Indeed, many traits are inherited together. Since

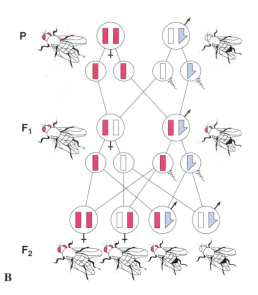

Figure 5-8

Sex-linked inheritance of eye color in fruit fly *Drosophila melanogaster.* **A,** White and red eyes of *D. melanogaster.* **B,** Genes for eye color are carried on X chromosome; Y carries no genes for eye color. Normal red is dominant to white. Homozygous red-eyed female mated with white-eyed male gives all red-eyed in F_1. F_2 ratios from F_1 cross are one homozygous red-eyed female and one heterozygous red-eyed female to one red-eyed male and one white-eyed male.

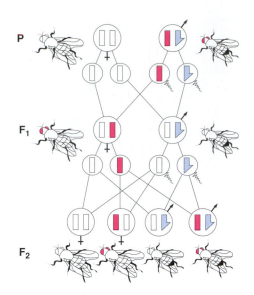

Figure 5-9

Reciprocal cross of Figure 5-8 (homozygous white-eyed female with red-eyed male) gives white-eyed males and red-eyed females in F_1. F_2 shows equal numbers of red-eyed and white-eyed females and red-eyed and white-eyed males.

the number of chromosomes in any organism is relatively small compared with the number of traits, each chromosome must contain many genes. All genes present on a chromosome are said to be **linked.** Linkage simply means that the genes are on the same chromosome, and all genes present on homologous chromosomes belong to the same linkage groups. Therefore there should be as many linkage groups as there are chromosome pairs.

Geneticists commonly use the word "linkage" in two somewhat different senses. Sex linkage refers to inheritance of a trait on the sex chromosomes, and thus its phenotypic expression depends on the sex of the organism and the factors already discussed. Autosomal linkage, or simply, linkage, refers to inheritance of genes on a given autosomal chromosome. Letters used to represent such genes are normally written without a slash mark between them, indicating that they are on the same chromosome. For example, *AB/ab* shows that genes *A* and *B* are on the same chromosome. Interestingly, Mendel studied seven characteristics of garden peas, which assorted independently because they were on seven different chromosomes. If he had studied eight characteristics, he would not have found independent assortment in two of the traits because garden peas have only seven pairs of homologous chromosomes.

In *Drosophila,* in which this principle has been studied most extensively, there are four linkage groups that correspond to the four pairs of chromosomes found in these fruit flies. Usually, small chromosomes have small linkage groups, and large chromosomes have large groups.

Crossing Over

Linkage, however, is usually not complete. If we perform an experiment in which animals such as *Drosophila* are crossed, we find that linked traits separate in some percentage of the offspring. Separation of alleles located on the same chromosome occurs because of **crossing over.**

During the protracted prophase of the first meiotic division, paired homologous chromosomes break and exchange equivalent portions; genes "cross over" from one chromosome to its homolog, and vice versa (Figure 5-10). Each chromosome consists of two sister chromatids held together by means of a proteinaceous structure called a **synaptonemal complex.** Breaks and exchanges occur at corresponding points on nonsister chromatids. (Breaks and exchanges also occur between sister chromatids but usually have no genetic significance because sister chromatids are identical.) Crossing over is a means for exchanging genes between homologous chromosomes and as such greatly increases the amount of genetic recombination. The frequency of crossing over varies depending on the species, but usually at least one and often several crossovers occur each time chromosomes pair.

Because the frequency of recombination is proportional to the distance between loci, the comparative linear position of each locus can be determined. Genes located far apart on very large chromosomes may assort independently because the probability of a crossover occurring between them in each meiosis is close to 100%. Such genes are found to be carried on the same chromosome only because each one is genetically linked to additional genes located physically between them on the chromosome. Laborious genetic experiments over many years have

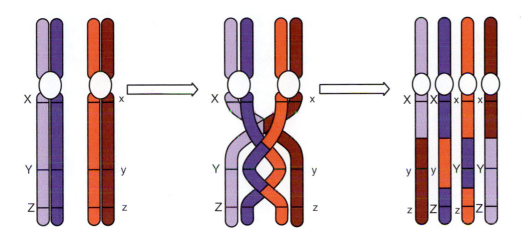

Figure 5-10

Crossing over during meiosis. Nonsister chromatids exchange portions, so that none of the resulting gametes is genetically the same as any other. Gene X is farther from gene Y than Y is from Z; therefore gene X is more frequently separated from Y in crossing over than Y is from Z.

produced gene maps that indicate the positions of more than 500 genes distributed on the four chromosomes of *Drosophila melanogaster.*

Chromosomal Aberrations

Structural and numerical deviations from the norm that affect many genes at once are called chromosomal aberrations. They are sometimes called chromosomal mutations, but most cytogeneticists prefer to use the term "mutation" to refer to qualitative changes within a gene; gene mutations are discussed on page 97.

Despite the incredible precision of meiosis, chromosomal aberrations do occur, and they are more common than one might think. They are responsible for great economic benefit in agriculture. Unfortunately, they are also responsible for many human genetic malformations. It is estimated that five out of every 1000 humans are born with *serious* genetic defects attributable to chromosomal anomalies. An even greater number of embryos with chromosomal defects are aborted spontaneously, far more than ever reach term.

Changes in chromosome numbers are called **euploidy** when there is the addition or deletion of whole sets of chromosomes and **aneuploidy** when a single chromosome is added to or subtracted from a diploid set. A "set of chromosomes contains one member of each homologous pair as would be present in the nucleus of a gamete. The most common kind of euploidy is **polyploidy,** the carrying of one or more additional sets of chromosomes. Such aberrations are much more common in plants than in animals. Animals are much less tolerant of chromosomal aberrations, because sex determination requires a delicate balance between the numbers of sex chromosomes and autosomes. Many domestic plant species are polyploid (cotton, wheat, apples, oats, tobacco, and others), and perhaps 40% of flowering plants may have originated in this manner. Horticulturists favor polyploids and often try to develop them because they have more intensely colored flowers and more vigorous vegetative growth.

Aneuploidy is usually caused by failure of chromosomes to separate during meiosis **(nondisjunction).** If a pair of chromosomes fails to separate during the first or second meiotic divi-

sions, both members go to one pole and none to the other. This results in one gamete having $n - 1$ number of chromosomes and another having $n + 1$ number of chromosomes. If the $n - 1$ gamete is fertilized by a normal n gamete, the result is a **monosomic** animal. Survival is rare because the lack of one chromosome gives an uneven balance of genetic instructions. **Trisomy,** the result of the fusion of a normal n gamete and an $n + 1$ gamete, is much more common, and several kinds of trisomic conditions are known in humans. Perhaps the most familiar is **trisomy 21,** or **Down syndrome.** As the name indicates, it involves an extra chromosome 21 combined with the chromosome pair 21, and it is caused by nondisjunction of that pair during meiosis. It occurs spontaneously, and there is seldom any

A *syndrome* is a group of symptoms associated with a particular disease or abnormality, although every symptom is not necessarily shown by every patient with the condition. An English physician, John Langdon Down, described the syndrome in 1866 that we now know is caused by trisomy 21. Because of Down's belief that the facial features of affected individuals were mongoloid in appearance, the condition has been known as mongolism. The resemblances are superficial, however, and currently accepted names are trisomy 21 and Down syndrome. Among the numerous characteristics of the condition, the most disabling is severe mental retardation. This, as well as other conditions caused by chromosomal aberrations and several other birth defects, can be diagnosed *prenatally* by a procedure involving *amniocentesis.* The physician inserts a hypodermic needle through the abdominal wall of the mother and into the fluids surrounding the fetus (*not into* the fetus) and withdraws some of the fluid, which contains some fetal cells. The cells are grown in culture, their chromosomes are examined, and other tests done. If a severe birth defect is found, the mother has the option of having an abortion performed. As an extra "bonus," the sex of the fetus is learned after amniocentesis. How? Alternatively, determination of concentrations of certain substances in the maternal serum, which is less invasive than amniocentesis, can detect about 60% of Down syndrome fetuses. Ultrasound scanning may be more than 80% accurate.

family history of the abnormality. However, the risk of its appearance rises dramatically with increasing age of the mother; it occurs 40 times as often in women over 40 years old as among women between the ages of 20 and 30. In cases where maternal age is not a factor, 20% to 25% of trisomy 21 is due to nondisjunction during spermatogenesis; it is paternal in origin and is apparently independent of the father's age.

In all diploid species, normal development requires exactly two of each kind of autosome (not sex chromosomes). Nondisjunction can result in trisomies of other chromosomes, but because these lead to imbalance of gene products, they almost always lead to death before or soon after birth. However, each cell requires only one functional X chromosome (the other is inactivated in females). Nondisjunction of sex chromosomes is more well tolerated, but usually results in sterility and abnormalities of sex organs. For example, a human with XXY (Klinefelter syndrome) is a phenotypic male, usually infertile with some female sexual characteristics. Presence of only one X (and no Y) is usually lethal in embryos, but the occasional live birth results in a phenotypic female with a variety of developmental abnormalities (Turner syndrome).

Structural aberrations involve whole sets of genes within a chromosome. A portion of a chromosome may be reversed, placing the linear arrangement of genes in reverse order **(inversion);** nonhomologous chromosomes may exchange sections **(translocation);** entire blocks of genes may be lost **(deletion** usually resulting in serious developmental defects); or an extra section of chromosome may attach to a normal chromosome **(duplication)**. These structural changes often produce phenotypic changes. Duplications, although rare, are important for evolution because they supply additional genetic information that may enable new functions.

GENE THEORY

Gene Concept

The term "gene" (Gr. *genos,* descent) was coined by W. Johannsen in 1909 to refer to the hereditary factors of Mendel. Initially, they were regarded as indivisible units of chromosomes on which they were located. Later studies with multiple mutant alleles demonstrated that alleles are in fact divisible by recombination; that is, *portions* of a gene are separable. Furthermore, parts of many genes in eukaryotes are separated by sections of DNA that do not specify a part of the finished product **(introns).**

As the chief unit of genetic information, genes encode products essential for specifying the basic architecture of every cell, nature and life of the cell, specific protein syntheses, enzyme formation, self-reproduction of the cell, and, directly or indirectly, the entire metabolic function of the cell. Because of their ability to mutate, to be assorted and shuffled in different combinations, genes have become the basis for our modern interpretation of evolution. Genes are units of molecular information that can maintain their identities for many generations, can be self-duplicated in each generation, and can control processes by allowing their specificities to be copied.

One Gene–One Enzyme Hypothesis

Since genes act to produce different phenotypes, we may infer that their action follows the scheme: gene → gene product → phenotypic expression. Furthermore, we may suspect that the gene product is usually a protein, because proteins act as enzymes, antibodies, hormones, and structural elements throughout the body.

The first clear, well-documented study to link genes and enzymes was carried out on the common bread mold *Neurospora* by Beadle and Tatum in the early 1940s. This organism was ideally suited to a study of gene function for several reasons: these molds are much simpler to handle than fruit flies, they grow readily in well-defined chemical media, and they are haploid organisms that are consequently unencumbered with dominance relationships. Furthermore, mutations were readily induced by irradiation with ultraviolet light. Ultraviolet-light-induced mutants, grown and tested in specific nutrient media, had single-gene mutations that were inherited. Each mutant strain was defective in one enzyme, which prevented that strain from synthesizing one or more complex molecules. Putting it another way, the ability to synthesize a particular molecule was controlled by a single gene.

From these experiments Beadle and Tatum set forth an important and exciting formulation: **one gene produces one enzyme.** For this work they were awarded the Nobel Prize in 1958. The new hypothesis was soon validated by the research of others who studied other biosynthetic pathways. Hundreds of inherited disorders, including dozens of human hereditary diseases, are caused by single mutant genes that result in the loss of a specific enzyme. We now know that a particular protein may be made of several chains of amino acids (polypeptides), each of which may be specified by a different gene, and not all proteins specified by genes are enzymes (for example, structural proteins, antibodies, transport proteins, and hormones). Furthermore, genes directing the synthesis of various kinds of RNA were not included in Beadle and Tatum's formulation. Therefore a gene now may be defined more inclusively as **a nucleic acid sequence (usually DNA) that encodes a functional polypeptide or RNA sequence.**

STORAGE AND TRANSFER OF GENETIC INFORMATION

Nucleic Acids: Molecular Basis of Inheritance

Cells contain two kinds of nucleic acids: deoxyribonucleic acid (DNA), which is the genetic material, and ribonucleic acid (RNA), which functions in protein synthesis. Both DNA and RNA are polymers built of repeated units called **nucleotides.** Each nucleotide contains three parts: a **sugar,** a **nitrogenous base,** and a **phosphate group.** The sugar is a pentose (5-carbon) sugar; in DNA it is **deoxyribose** and in RNA it is **ribose** (Figure 5-11).

Nitrogenous bases of nucleotides are also of two types: pyrimidines, which consist of a single, 6-membered ring, and purines, which are composed of two fused rings. Both of these types of compounds contain nitrogen as well as carbon in their rings, which is why they are called "nitrogenous" bases. The purines in both RNA and DNA are adenine and guanine (Table 5-2). The pyrimidines in DNA are thymine and cytosine, and in RNA they are uracil and cytosine. Carbon atoms in the bases are numbered (for identification) according to standard biochemical notation (Figure 5-12). Carbons in ribose and deoxyribose are also numbered, but to distinguish them from the carbons in the bases, numbers for carbons in the sugars are given prime signs (see Figure 5-11).

The sugar, phosphate group, and nitrogenous base are linked as shown in the generalized scheme for a nucleotide:

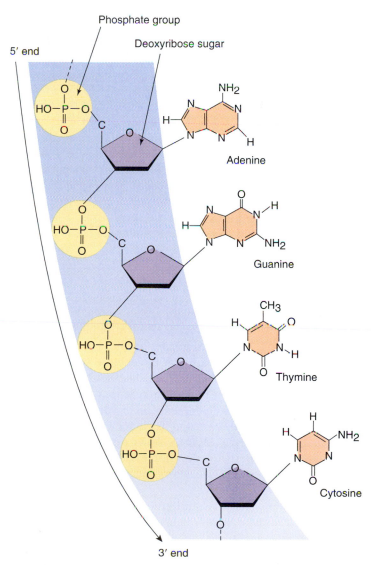

In DNA the backbone of the molecule is built of phosphoric acid and deoxyribose; to this backbone are attached the nitrogenous bases (Figure 5-13). The **5′ end** of the backbone has a free phosphate group on the **5′** carbon of the ribose, and

Figure 5-11
Ribose and deoxyribose, the pentose sugars of nucleic acids. A carbon atom lies in each of the four corners of the pentagon (labeled 1′ to 4′). Ribose has a hydroxyl group (—OH) and a hydrogen on the number 2′ carbon; deoxyribose has two hydrogens at this position.

TABLE 5.2		
Chemical Components of DNA and RNA		
	DNA	**RNA**
Purines	Adenine	Adenine
	Guanine	Guanine
Pyrimidines	Cytosine	Cytosine
	Thymine	Uracil
Sugar	2-Deoxyribose	Ribose
Phosphate	Phosphoric acid	Phosphoric acid

Figure 5-12
Purines and pyrimidines of DNA and RNA.

Figure 5-13
Section of a strand of DNA. Polynucleotide chain is built of a backbone of phosphoric acid and deoxyribose sugar molecules. Each sugar holds a nitrogenous base side arm. Shown from top to bottom are adenine, guanine, thymine, and cytosine.

Figure 5-14

Positions of hydrogen bonds between thymine and adenine and between cytosine and guanine in DNA.

Adenine ━ Thymine

Guanine ━ Cytosine

the **3′ end** has a free hydroxyl group on the **3′** carbon. However, one of the most interesting and important discoveries about the nucleic acids is that DNA is not a single polynucleotide chain; rather it consists of *two* complementary chains that are precisely cross-linked by specific hydrogen bonding between purine and pyrimidine bases. The number of adenines is equal to the number of thymines, and the number of guanines equals the number of cytosines. This fact suggested a pairing of bases: adenine with thymine (AT) and guanine with cytosine (GC) (Figures 1-6 and 5-14).

The result is a ladder structure (Figure 5-15). The upright portions are the sugar phosphate backbones, and the connecting rungs are the paired nitrogenous bases, AT or GC. However, the ladder is twisted into a **double helix** with approximately 10 base pairs for each complete turn of the helix (Figure 5-16). The two DNA strands run in opposite directions, that is they are **antiparallel,** and the 5′ end of one strand is opposite the 3′ end of the other. This is evident from an examination of Figure 5-16. The two strands are also **complementary**—the sequence of bases along one strand specifies the sequence of bases along the other strand.

Determination of the structure of DNA has been widely acclaimed as the single most important biological discovery of this century. It was based on x-ray diffraction studies of Maurice H. F. Wilkins and Rosalind Franklin and on ingenious proposals of Francis H. C. Crick and James D. Watson published in 1953. Watson, Crick, and Wilkins were later awarded the Nobel Prize for Physiology or Medicine for their momentous work. Rosalind Franklin was not included because she died prior to the award.

RNA is similar to DNA in structure except that it consists of a *single* polynucleotide chain (except in some viruses), has ribose instead of deoxyribose, and has uracil instead of thymine. Ribosomal, transfer, and messenger RNAs are the most abundant and well-known types (function described on pp. 91–93), but many structural and regulatory RNAs, such as micro RNAs, have been reported.

Every time a cell divides, the structure of DNA must be precisely copied in the daughter cells. This is called **replication** (Figure 5-17). During replication, the two strands of the double helix unwind, and each separated strand serves as a **template** against which a complementary strand is synthesized. That is, an enzyme (DNA polymerase) assembles a new strand of polynucleotides with a thymine group going next to the adenine group in the template strand, a guanine group next to the cyto-

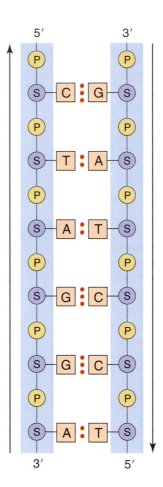

Figure 5-15

DNA, showing how the complementary pairing of bases between the sugar-phosphate "backbones" keeps the double helix at a constant diameter for the entire length of the molecule. Dots represent the three hydrogen bonds between each cytosine and guanine and the two hydrogen bonds between each adenine and thymine.

sine group, and so on. But DNA polymerase only synthesizes new strands in the direction of 5′ to 3′. Because the parent DNA strands are antiparallel, one of which runs 5′ to 3′, the other running 3′ to 5′, synthesis along one of the strands is continuous, and the other must be performed in a series of fragments, each of which begins with a 5′ end running toward a 3′ end (Figure 5.17).

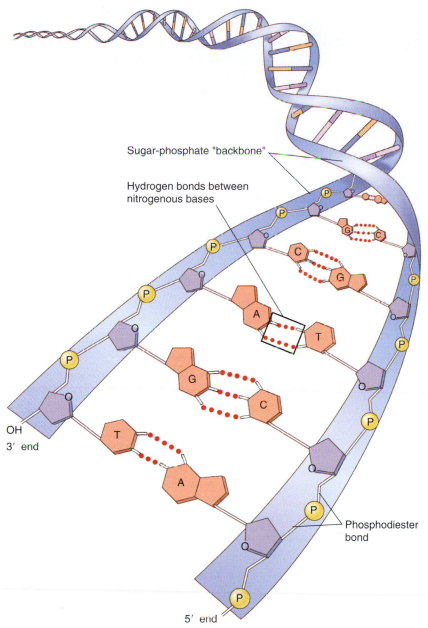

Sugar-phosphate "backbone"

Hydrogen bonds between
nitrogenous bases

OH
3' end

Phosphodiester
bond

5' end

Figure 5-16
DNA molecule.

DNA Coding by Base Sequence

Since DNA is the genetic material and is composed of a linear sequence of base pairs, an obvious extension of the Watson-Crick model is that the sequence of base pairs in DNA codes for, and is colinear with, the sequence of amino acids in a protein. The coding hypothesis had to account for the way a string of four different bases—a four-letter alphabet—could dictate the sequence of 20 different amino acids.

In the coding procedure, obviously there cannot be a 1:1 correlation between four bases and 20 amino acids. If a coding unit (often called a word, or **codon**) consisted of two bases, only 16 words (4^2) could be formed, which could not account

for 20 amino acids. Therefore the codon had to consist of at least three bases or three letters, because 64 possible words (4^3) could be formed by four bases when taken as triplets. This means that there could be a considerable redundancy of triplets (codons), since DNA codes for just 20 amino acids. Later work confirmed that nearly all of the amino acids are specified by more than one triplet code (Table 5-3).

DNA shows a surprising stability, both in prokaryotes and in eukaryotes. Interestingly, it is susceptible to damage by harmful chemicals in the environment and radiation. Such damage is usually not permanent, because cells have an efficient repair system. Various types of damage and repair are known, one of which is called **excision repair.** Ultraviolet irradiation often causes adjacent pyrimidines to link together by covalent bonds (dimerize), preventing transcription and replication. A series of several enzymes "recognizes" the area of the damaged strand and excises the pair of dimerized pyrimidines and several bases following them. DNA polymerase then synthesizes the missing strand along the remaining one, according to the base-pairing rules, and the enzyme **DNA ligase** joins the end of the new strand to the old one.

Transcription and the Role of Messenger RNA

Information is coded in DNA, but DNA does not participate directly in protein synthesis. It is obvious that an intermediary is required. This intermediary is another nucleic acid called **messenger RNA (mRNA).** The triplet codes in DNA are **transcribed** into mRNA, with uracil substituting for thymine (Table 5-3).

Ribosomal, transfer, and messenger RNAs are transcribed directly from DNA, each encoded by different sets of genes. In this process of making a complementary copy of one strand or gene of DNA in the formation of mRNA, an enzyme, **RNA polymerase,** is needed. (In eukaryotes each type of RNA [ribosomal, transfer, and messenger] is transcribed by a specific type of RNA polymerase.) The mRNA contains a sequence of bases that complements the bases in one of the two DNA strands, just as the DNA strands complement each other. Thus A in the coding DNA strand is replaced by U in RNA; C is replaced by G; G is replaced by C; and T is replaced by A. Only one of the two chains is used as the template for RNA synthesis because only one bears the AUG codon that initiates a message (Table 5-3). The reason why only one strand of the double-stranded DNA is a "coding strand" is that mRNA otherwise would always be formed in complementary pairs, and enzymes also would be synthesized in complementary pairs. In other words, two different enzymes would be produced for every DNA coding sequence instead of one. This certainly would lead to metabolic chaos.

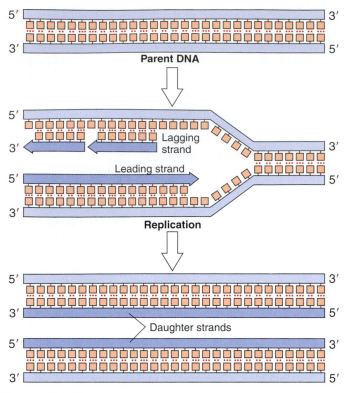

Figure 5-17

Replication of DNA. Parent strands of DNA part, and DNA polymerase synthesizes daughter strands using the base sequence of parent strands as a template. Because synthesis always proceeds in a 5′ to 3′ direction, synthesis of one strand is continuous, and the other strand must be synthesized as a series of fragments.

Only one strand of DNA serves as the coding strand in all DNA except that found in plasmids (see p. 92). Messenger RNA can be transcribed from both DNA strands in one region of plasmid DNA, and this is the only known example of proteins being encoded in overlapping parts of both DNA strands.

A prokaryotic gene is encoded on a continuous stretch of DNA, transcribed into mRNA, and then translated (see the next section). It was assumed that this was also the case for eukaryotic genes until the surprising discovery that some stretches of DNA are transcribed in the nucleus but are not found in the corresponding mRNA in the cytoplasm. In other words, pieces of the nuclear mRNA were removed in the nucleus before the finished mRNA was transported to the cytoplasm (Figure 5-18). It was thus discovered that many genes are split, interrupted by sequences of bases that do not code for the final product, and mRNA transcribed from them must be edited or "matured" before translation in the cytoplasm. The intervening segments of DNA are now known as **introns,** while those that code for part of the mature RNA and are translated into protein are called **exons.** Before mRNA leaves the nucleus, a methylated guanine "cap" is added at the 5′ end, and a tail of adenine nucleotides (poly-*A*) is often added at the 3′ end (Figure 5-18). The cap and the poly-*A* tail are characteristic of mRNA molecules.

In mammals genes coding for histones and for interferons are on continuous stretches of DNA. However, we now know that genes coding for many proteins are split. In lymphocyte dif-

TABLE 5.3

The Genetic Code: Amino Acids Specified by Codons of Messenger RNA

First Letter		Second Letter								Third Letter
		U		**C**		**A**		**G**		
U	UUU UUC	Phenylalanine	UCU UCC	Serine	UAU UAC	Tyrosine	UGU UGC	Cysteine		U C
	UUA UUG	Leucine	UCA UCG		UAA UAG	End chain	UGA UGG	End chain Tryptophane		A G
C	CUU CUC CUA CUG	Leucine	CCU CCC CCA CCG	Proline	CAU CAC	Histidine	CGU CGC CGA CGG	Arginine		U C A G
					CAA CAG	Glutamine				
A	AUU AUC AUA	Isoleucine	ACU ACC ACA	Threonine	AAU AAC	Asparagine	AGU AGC	Serine		U C A
	AUG	Methionine*	ACG		AAA AAG	Lysine	AGA AGG	Arginine		G
G	GUU GUC GUA GUG	Valine	GCU GCC GCA GCG	Alanine	GAU GAC	Aspartic acid	GGU GGC GGA GGG	Glycine		U C A G
					GAA GAG	Glutamic acid				

*Also, begin chain.

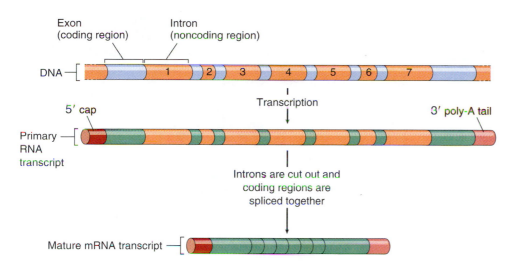

Figure 5-18

Transcription and maturation of ovalbumin gene of chicken. The entire gene of 7700 base pairs is transcribed to form the primary mRNA, then the 5′ cap of methyl guanine and the 3′ polyadenylate tail are added. After the introns are spliced out, the mature mRNA is transferred to the cytoplasm.

ferentiation the parts of the split genes coding for immunoglobulins are actually *rearranged* during development, so that different proteins result from subsequent transcription and translation. This partly accounts for the enormous diversity of antibodies manufactured by descendants of the lymphocytes (p. 741).

Base sequences in some introns are complementary to other base sequences in the intron, suggesting that the intron could fold so that complementary sequences would pair. This may be necessary to control proper alignment of intron boundaries before splicing. Most surprising of all has been the discovery that, at least in some cases, RNA can "self-catalyze" the excision of introns. The ends of the intron join; the intron thus becomes a small circle of RNA, and the exons are spliced together. This process does not fit the classical definition of an enzyme or other catalyst since the molecule itself is changed in the reaction.

Translation: Final Stage in Information Transfer

The **translation** process takes place on **ribosomes,** granular structures composed of protein and **ribosomal RNA (rRNA).** Ribosomal RNA is composed of a large and a small subunit, and the small subunit comes to lie in a depression of a large subunit to form a functional ribosome (Figure 5-19). Messenger RNA molecules attach themselves to ribosomes to form a messenger RNA-ribosome complex. Since only a short section of mRNA molecule is in contact with a single ribosome, the mRNA usually attaches to several ribosomes at once. The entire complex, called a **polyribosome** or **polysome,** allows several molecules of the same kind of polypeptide to be synthesized at once, one on each ribosome of the polysome (Figure 5-19).

Assembly of polypeptides on the mRNA-ribosome complex requires the action of another kind of RNA called **transfer RNA (tRNA).** Transfer RNAs are surprisingly large molecules that are folded in a complicated way in the form of a cloverleaf (Figure 5-20). Molecules of tRNA collect free amino acids from the cytoplasm and deliver them to the polysome, where they are assembled into a polypeptide. There are special tRNA molecules for every amino acid. Furthermore, each tRNA is accompanied by a

specific tRNA synthetase. Transfer RNA synthetases are enzymes necessary to sort and attach the correct amino acid to the terminal adenine on the 3′ end of each tRNA by a process called **charging.**

On the cloverleaf-shaped molecule of tRNA, a special sequence of three bases (the **anticodon**) is exposed in just the right way to form base pairs with complementary bases (the codon) in the mRNA. The codons are read and polypeptides assembled along the mRNA in a 5′ to 3′ direction. The anticodon of each tRNA is the key to the correct sequencing of amino acids in the polypeptide being assembled.

For example, alanine is assembled into a polypeptide when it is signaled by the codon GCG in an mRNA. The translation is accomplished by alanine tRNA in which the anticodon is CGC. Alanine tRNA is first charged with alanine by its tRNA synthetase. The alanine tRNA complex enters the ribosome where it fits precisely into the right place on the mRNA strand. Then the next charged tRNA specified by the mRNA code (glycine tRNA, for example) enters the ribosome and attaches itself beside the alanine tRNA. The two amino acids are united with a peptide bond (with the energy from a molecule of guanosine triphosphate), and the alanine tRNA falls off. The process continues stepwise as the polypeptide chain is built (Figure 5-21). A polypeptide of 500 amino acids can be assembled in less than 30 seconds.

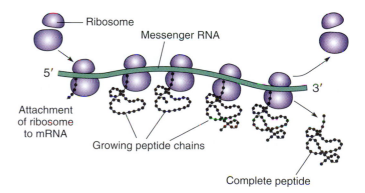

Figure 5-19

How the polypeptide chain is formed. As ribosomes move along messenger RNA in a 5′ to 3′ direction, amino acids are added stepwise to form the polypeptide chain.

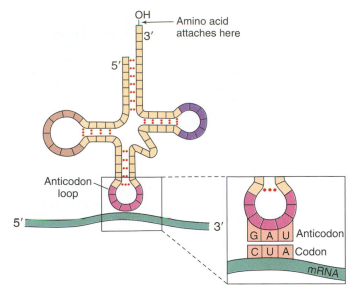

Figure 5-20

Structure of a tRNA molecule. The anticodon loop bears bases complementary to those in the mRNA codon. The other two loops function in binding to the ribosomes in polypeptide synthesis. The amino acid is added to the free single-stranded 3′ end by tRNA synthetase.

Regulation of Gene Expression

In Chapter 8 we will see how the orderly differentiation of an organism from fertilized ovum to adult requires the involvement of genetic material at every stage of development. Developmental biologists have provided convincing evidence that every cell in a developing embryo is genetically equivalent. Thus it is clear that as tissues differentiate (change developmentally), they use only a part of the genetic instruction present in every cell. Certain genes express themselves only at certain times and not at others. Indeed, there is reason to believe that in a particular cell or tissue, most of the genes are inactive at any given moment. The problem in development is to explain how, if every cell has a full gene complement, certain genes are "turned on" and produce proteins that are required for a particular developmental stage while the other genes remain silent.

Actually, although the developmental process brings the question of gene activation clearly into focus, gene regulation is necessary throughout an organism's existence. The cellular enzyme systems that control all functional processes obviously require genetic regulation because enzymes have powerful effects even in minute amounts. Enzyme synthesis must be responsive to the influences of supply and demand.

Gene Regulation in Eukaryotes

There are a number of different phenomena in eukaryotic cells that can serve as control points, and we now discuss a few examples.

Transcriptional Control   This may be the most important mechanism. **Transcription factors** are molecules that may

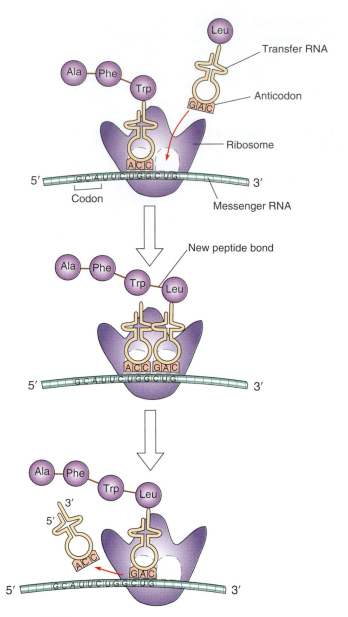

Figure 5-21

Formation of polypeptide chain on messenger RNA. As ribosome moves down messenger RNA molecule, transfer RNA molecules with attached amino acids enter ribosomes *(top)*. Amino acids are joined together into polypeptide chain, and transfer RNA molecules leave ribosome *(bottom)*.

have a positive or a negative effect on transcription of RNA from the DNA of target genes. The factors may act within cells that produce them or they may be transported to different parts of the body prior to action. Examples of transcription factors are steroid receptors when bound to a steroid hormone. Steroid hormones produced by endocrine glands elsewhere in the body enter the cell and bind with a receptor protein in the nucleus. The steroid-receptor complex then binds with DNA near the target gene (p. 723). Progesterone, for example, binds with a nuclear receptor in cells of the chicken oviduct; the hormone-receptor complex then activates transcription of genes encoding egg albumin and other substances.

Translational Control   Genes can be transcribed and the mRNA sequestered in some way so that translation is delayed. This commonly happens in development of eggs of many animals. Oocytes accumulate large quantities of messenger RNA during their development; then, fertilization activates metabolism and initiates translation of maternal mRNA.

Gene Rearrangement   Vertebrates contain cells called lymphocytes that bear genes coding for proteins called antibodies (p. 741). Each type of antibody has the capacity to bind specifically with a particular foreign substance (antigen). Because the number of different antigens is enormous, genetic diversity of antibody genes must be equally great. One source of this diversity is rearrangement of DNA sequences coding for antibodies during development of lymphocytes.

DNA Modification   An important mechanism for turning genes off appears to be methylation of cytosine residues, that is, adding a methyl group (CH$_3$—) to the carbon in the 5 position in the cytosine ring (Figure 5-22A). This usually happens when the cytosine is next to a guanine residue; thus, the bases in the complementary DNA strand would also be a cytosine and a guanine (Figure 5-22B). When the DNA is replicated, an enzyme recognizes the CG sequence and quickly methylates the daughter strand, maintaining the gene in an inactive state.

Molecular Genetics

Progress in our understanding of genetic mechanisms on the molecular level, as discussed in the last few pages, has been almost breathtaking in the last few years. We can expect many more discoveries in the near future. This progress has been due largely to the effectiveness of many biochemical techniques now used in molecular biology. We have space to describe only a few briefly.

Recombinant DNA

One of the most important tools in this technology is a series of enzymes called **restriction endonucleases.** Each of these enzymes, derived from bacteria, cleaves double-stranded DNA at particular sites determined by the particular base sequences at that point. Many of these endonucleases cut the DNA strands so that one has several bases projecting farther than the other strand (Figure 5-23), leaving what are called "sticky ends." When these DNA fragments are mixed with others that have been cleaved by the same endonuclease, they tend to anneal (join) by the rules of complementary base pairing. They are sealed into their new position by the enzyme **DNA ligase.**

> Besides their chromosomes, most prokaryote and at least some eukaryote cells have small circles of double-stranded DNA called *plasmids.* Though comprising only 1% to 3% of the bacterial genome, they may carry important genetic information, for example, resistance to an antibiotic. Plastids in plant cells (for example, chloroplasts) and mitochondria, found in most eukaryotic cells, are self-replicating and have their own complement of DNA in the form of small circles reminiscent of plasmids. The DNA of mitochondria and plastids codes for some of their proteins, and some of their proteins are specified by nuclear genes.

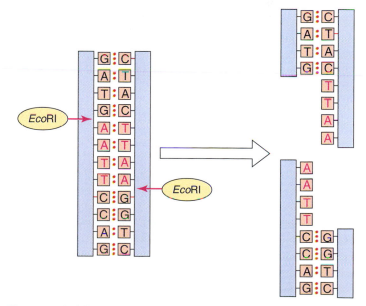

Figure 5-22
Some genes in eukaryotes are turned off by methylation of some cytosine residues in the chain. **A,** Structure of 5-methyl cytosine. **B,** Cytosine residues next to guanine are those that are methylated in a strand, thus allowing both strands to be symmetrically methylated.

Figure 5-23
Action of restriction endonuclease, *Eco*RI. Such enzymes recognize specific base sequences that are palindromic (a palindrome is a word spelled the same backward and forward). *Eco*RI leaves "sticky ends," which anneal to other DNA fragments cleaved by the same enzyme. The strands are joined by DNA ligase.

If the DNA annealed after cleavage by the endonuclease is from two different sources, for example, a plasmid (see note above) and a mammal, the product is **recombinant DNA.** To make use of the recombinant DNA, the modified plasmid must be cloned in bacteria. The bacteria are treated with dilute calcium chloride to make them more susceptible to taking up the recombinant DNA, but plasmids do not enter most cells present. Bacterial cells that have taken up the recombinant DNA can be identified if the plasmid has a marker, for example, resistance to an antibiotic. Then, only bacteria that can grow in the presence of the antibiotic are those that have absorbed the recombinant DNA. Some bacteriophages (bacterial viruses) have also been used as carriers for recombinant DNA. Plasmids and bacteriophages that carry recombinant DNA are called **vectors.** The vectors retain the ability to replicate in the bacterial cells; therefore the recombinant insert is amplified.

A clone is a collection of individuals or cells all derived by asexual reproduction from a single individual. When we speak of cloning a gene or plasmid in bacteria, we mean that we isolate a colony or group of bacteria derived from a single ancestor into which the gene or plasmid was inserted.

Polymerase Chain Reaction

Recent advances have made it a simple task to clone a specific gene enzymatically from any organism as long as part of the sequence of that gene is known. The technique is called the **polymerase chain reaction (PCR).** Two short chains of nucleotides called primers are synthesized; primers are complementary to different DNA strands in the known sequence. A large excess of each primer is added to a sample of DNA from the organism, and the mixture is heated to separate the double helix into single strands. When the mixture is cooled, there is a much greater probability that each strand of the gene of interest will anneal to a primer than to the other strand of the gene—because there is so much more primer present. A heat-stable DNA polymerase is added along with the four deoxyribonucleotide triphosphates, and DNA synthesis proceeds from the 3′ end of each primer, extending the primer in the 3′ to 5′ direction. If the primers are chosen so that each anneals toward the 3′ end of each of the complementary strands, entire new complementary strands will be synthesized, and the number of copies of the gene has doubled (Figure 5-24). The reaction mixture is then reheated and cooled again to allow more primers to bind original and new copies of each strand. With each cycle of DNA synthesis, the number of copies of the gene doubles. Since each cycle can take less than five minutes, the number of copies of a gene can increase from one to over one million in less than two hours! The PCR allows cloning a known gene from an individual patient, identification of a drop of dried blood at a crime scene, or cloning DNA of a 40,000-year-old woolly mammoth.

Recombinant DNA technology and the PCR are currently being used in many areas with great positive potential and many practical uses.

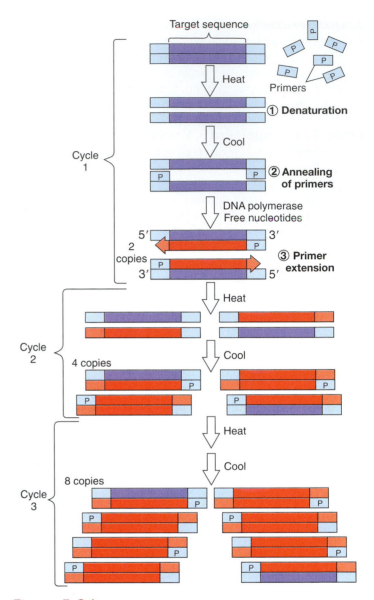

Figure 5-24
Steps in the polymerase chain reaction (PCR).

The techniques of molecular biology have allowed scientists to accomplish feats of which few could dream only a decade or so ago. These accomplishments will bring enormous benefits for humanity in the form of enhanced food production and treatment of disease. Progress with crop plants has been so rapid that genetically engineered soybean, cotton, rice, corn, sugarbeet, tomato, and alfalfa have already reached the market in the United States. There is resistance to sale of genetically altered produce in Europe, apparently because of widespread fears that such vegetables can somehow harm consumers.

Development of transgenic animals of potential use has not progressed as far as development of such plants. Gene therapy for inherited diseases presents many difficulties, but research in this area is vigorous, and clinical trials for certain conditions are under way.

Genomics and Proteomics

The scientific field of mapping, sequencing and analyzing genomes is now known as **genomics,** a term that has come into wide use only in recent years. Some researchers divide genomic analysis into "structural genomics" (mapping and sequencing) and "functional genomics" (development of genome-wide or system-wide experimental approaches to understand gene function).

In the 1970s Allan Maxam and Walter Gilbert in the United States and Frederick Sanger in England reported practical techniques for determination of the sequence of bases in DNA. By 1984 and 1985 scientists proposed to sequence and map the entire human genome, an effort that came to be known as the Human Genome Project. It was a most ambitious undertaking: the genome was estimated at 50,000 to 100,000 genes and regulatory subunits encoded in a linear sequence of about 3 to 6 billion pairs of bases. Using the techniques available in 1988, it would have taken until the year 2700 to sequence the genome completely, but biologists then expected that technical improvements would make it possible to finish in the twenty-second century. In fact, development and improvement of automated sequencers, as well as competition between the publicly supported Human Genome Sequencing Consortium and a large group of privately supported scientists (Celera Genomics and collaborators) led to publication of draft sequences in 2001!

Whether determination of the draft sequence was "the greatest scientific discovery of our time," as claimed by Davies's book (in Selected References), is debatable. Nevertheless, it was very exciting and yielded many surprises. For example, the human genome is much smaller than thought previously, now estimated at between 30,000 and 40,000 protein-encoding genes. About 740 genes code for RNAs. About 90% of sequences in **euchromatin** (gene-rich portions of chromosomes), contrasted with **heterochromatin** (areas where there are few genes), have been determined completely. Only 5% of the 28% that is actually transcribed into RNA encodes protein. More than half the DNA present is repeated sequences of several types, including 45% in parasitic DNA elements. Parasitic DNA (also called "selfish" and "junk" DNA) is DNA that seems to serve no function save its own propagation, but it may have utility in ways that are as yet unclear.

A thousand human diseases, such as cystic fibrosis and Huntington's chorea, result from defects in single genes. Almost 300 disease-associated genes have already been identified. Information developed from knowledge of gene sequences can lead to new diagnostic tests, treatments, possible preventive strategies, and advances in the molecular understanding of genetic diseases. However, to realize such benefits it is not sufficient simply to know the sequence of amino acids encoded by a nucleotide sequence in a gene. The human genome of some 40,000 genes is responsible for hundreds of thousands of different proteins **(proteome)** in a typical cell. Clearly, a single gene can, by some means, give rise to many differing proteins. Many scientists are now engaged in the difficult field of **proteomics:** to identify all the proteins in a cell, tissue, or organism; to determine how the proteins interact to accomplish their functions; and to outline the folding structure of the proteins.

SOURCES OF PHENOTYPIC VARIATION

The creative force of evolution is natural selection acting on biological variation. Without variability among individuals, there could be no continued adaptation to a changing environment and no evolution (Chapter 6).

There are actually several sources of variability, some of which we have already described. The independent assortment of chromosomes during meiosis is a random process that creates new chromosomal recombinations in the gametes. In addition, chromosomal crossing over during meiosis allows recombination of linked genes between homologous chromosomes, further increasing variability. The random fusion of gametes from both parents produces still another source of variation.

There is a story that George Bernard Shaw once received a letter from a famous actress who suggested that they conceive a perfect child who would combine her beauty and his brains. He declined the offer, pointing out that the child could just as well inherit her brains and his beauty. Shaw was correct; the fusion of parental gametes is random and thus unpredictable.

Thus sexual reproduction multiplies variation and provides the diversity and plasticity necessary for a species to survive environmental change. Sexual reproduction with its sequence of gene segregation and recombination, generation after generation, is, as the geneticist T. Dobzhansky has said, the "master adaptation which makes all other evolutionary adaptations more readily accessible."

Although sexual reproduction reshuffles and amplifies whatever genetic diversity exists in the population, there must be ways to generate *new* genetic variation. This happens through gene mutations and, sometimes, through chromosomal aberrations and possibly by participation of parasitic DNA.

Gene Mutations

Gene mutations are chemicophysical changes in genes resulting in an alteration of the sequence of bases in the DNA. These mutations can be studied directly by determining the DNA sequence and indirectly through their effects on organismal phenotype, if such effects are present. A mutation may result in a codon substitution as, for example, in the condition in humans known as **sickle cell anemia.** Homozygotes with sickle cell trait often die before the age of 30 because the ability of their red blood cells to carry oxygen is greatly impaired, a result of the substitution of only a single amino acid in the amino acid sequence of their hemoglobin. Other mutations may involve the deletion of one or more bases or the insertion of additional bases into the DNA chain. Translation of mRNA will thus be shifted, leading to codons that specify incorrect amino acids and to a nonfunctional or dysfunctional protein product.

Once a gene is mutated, it faithfully reproduces its new self just as it did before it was mutated. Many mutations are harmful, many are neither helpful nor harmful, and sometimes mutations are advantageous. Helpful mutations are of great significance to evolution because they furnish new possibilities on which natural selection works to build adaptations. Natural selection determines which new alleles merit survival; the environment imposes a screening process that passes the beneficial and eliminates the harmful.

When an allele of a gene is mutated to a new allele, it tends to be recessive and its effects are normally masked by its partner allele. Only in the homozygous condition can such mutant alleles influence phenotype. Thus a population carries a reservoir of mutant recessive alleles, some of which are homozygous lethals but which are rarely present in the homozygous condition. Inbreeding encourages the formation of homozygotes and increases the probability of recessive mutants being expressed in the phenotype.

Most mutations are destined for a brief existence. There are cases, however, in which mutations may be harmful or neutral under one set of environmental conditions and helpful under a different set. Should the environment change, there could be a new adaptation beneficial to the species. The earth's changing environment has provided numerous opportunities for new gene combinations and mutations, as evidenced by the great diversity of animal life today.

Frequency of Mutations

Although mutation occurs randomly, different mutation rates prevail at different loci. Some *kinds* of mutations are more likely to occur than others, and individual genes differ considerably in length. A long gene (more base pairs) is more likely to have a mutation than a short gene. Nevertheless, it is possible to estimate average spontaneous rates for different organisms and traits.

Relatively speaking, genes are extremely stable. In the well-studied fruit fly *Drosophila* there is approximately one detectable mutation per 10,000 loci (rate of 0.01% per locus per generation). The rate for humans is one per 10,000 to one per 100,000 loci per generation. If we accept the latter, more conservative figure, then a single normal allele is expected to go through 100,000 generations before it is mutated. However, since human chromosomes contain 30,000 to 40,000 loci, about every third person carries approximately one new mutation. Similarly, each ovum or spermatozoon produced contains, on the average, one mutant allele.

Since most mutations are deleterious, these statistics are anything but cheerful. Fortunately, most mutant genes are recessive and are not expressed in heterozygotes. Only a few will by chance increase enough in frequency for homozygotes to be produced.

MOLECULAR GENETICS OF CANCER

The crucial defect in cancer cells is that they proliferate in an unrestrained manner **(neoplastic growth).** The mechanism that controls the rate of division of normal cells has somehow broken down, and cancer cells multiply much more rapidly, invading other tissues in the body. Cancer cells originate from normal cells that lose their constraint on division and become dedifferentiated (less specialized) to some degree. Thus there are many kinds of cancer, depending on the original founder cells of the tumor. In recent years mounting evidence has indicated that the change in many cancerous cells, perhaps all, has a genetic basis, and investigation of the genetic damage that causes cancer is now a major thrust of cancer research.

Oncogenes and Tumor Suppressor Genes

We now recognize that cancer is a result of a series of specific genetic changes that take place in a particular clone of cells. These include alterations in two types of genes: **oncogenes** and **tumor suppressor genes,** and there are numerous specific genes of each type now known.

Oncogenes (Gr. *onkos,* bulk, mass; + *genos,* descent) are genes whose activity has been associated for some time with the production of cancer. They are genes that are normally found in cells, and in their normal form they are called **proto-oncogenes.** One of these codes for a protein known as **Ras.** Ras protein is a guanosine triphosphatase (GTPase) that is located just beneath the cell membrane. When a receptor on the cell surface binds a growth factor, Ras is activated and initiates a cascade of reactions, ultimately leading to cell division. The oncogene form codes for a protein that initiates the cell-division cascade even when the growth factor has not bound to the surface receptor; that is, the growth factor is absent.

Of the many ways that cellular DNA can sustain damage, the three most important are ionizing radiation, ultraviolet radiation, and chemical mutagens. The high energy of ionizing radiation (x rays and gamma rays) causes electrons to be ejected from the atoms it encounters, resulting in ionized atoms with unpaired electrons (free radicals). The free radicals (principally from water) are highly reactive chemically, and they react with molecules in the cell, including DNA. Some damaged DNA is repaired, but if the repair is inaccurate, a mutation results. Ultraviolet radiation is of much lower energy than ionizing radiation and does not produce free radicals. It is absorbed by pyrimidines in DNA and causes formation of a double covalent bond between the adjacent pyrimidines. UV repair mechanisms can also be inaccurate. Chemical mutagens react with the DNA bases and cause mispairing during replication.

Gene products of tumor suppressor genes act as a constraint on cell proliferation. One such product is called **p53** (for "53-kilodalton protein," a reference to its molecular weight). Mutations in the gene coding for p53 are present in about half of the 6.5 million cases of human cancer diagnosed each year. Normal p53 has a number of crucial functions, depending on the circumstances of the cell. It can trigger apoptosis (p. 53), act as a transcription activator or repressor (turning genes on or off), control progression from G_1 to S phase in the cell cycle, and promote repair of damaged DNA. Many of the mutations known in p53 interfere with its binding to DNA and thus its function.

SUMMARY

In sexual animals genetic material is distributed to offspring via gametes (ova and sperm), produced by meiosis. Each somatic cell in an organism has two chromosomes of each kind (homologous chromosomes) and is thus diploid. Meiosis separates homologous chromosomes, so that each gamete has half the somatic chromosome number (haploid). In the first meiotic division, centromeres do not divide, and each daughter cell receives one of each pair of replicated homologous chromosomes with sister chromatids still attached to the centromere. At the beginning of the first meiotic division, replicated homologous chromosomes come to lie alongside each other (synapsis), forming a bivalent. The gene loci on one set of chromatids lie opposite the corresponding loci on the homologous chromatids. Portions of adjacent chromatids can exchange with the nonsister chromatids (crossing over) to produce new genetic combinations. At the second meiotic division, the centromeres divide, completing the reduction in chromosome number and amount of DNA. The diploid number is restored when male and female gametes fuse to form a zygote. Gender is determined in most animals by the sex chromosomes; in humans, fruit flies, and many other animals, females have two X chromosomes, and males have an X and a Y.

Genes are the unit entities that influence all characteristics of an organism and are inherited by offspring from their parents. Allelic variants of genes may be dominant, recessive, or intermediate; a recessive allele in the heterozygous genotype will not be expressed in the phenotype but requires the homozygous condition for overt expression. In a monohybrid cross involving a dominant allele and its recessive allele (both parents homozygous), the F_1 generation will be all heterozygous, whereas F_2 genotypes will occur in a 1:2:1 ratio, and phenotypes in a 3:1 ratio. This demonstrates Mendel's law of segregation. Heterozygotes in intermediate inheritance show phenotypes intermediate between the homozygous phenotypes, or sometimes they show a different phenotype altogether, with corresponding alterations in phenotypic ratios. Dihybrid crosses (in which genes for two different characteristics are carried on separate pairs of homologous chromosomes) demonstrate Mendel's law of independent assortment, and phenotypic ratios will be 9:3:3:1 with dominant and recessive characters. Ratios for monohybrid and dihybrid crosses can be determined by construction of a Punnett square, but the laws of probability allow calculation of the ratios in crosses of two or more characters much more easily.

Genes can have more than two alleles in a population, and different combinations of alleles can produce different phenotypic effects. Alleles of different genes can interact in producing a phenotype, as in polygenic inheritance, in which one gene affects the expression of another gene.

A gene on the X chromosome shows sex-linked inheritance and will produce an effect in males, even if a recessive allele is present, because the Y chromosome does not carry a corresponding allele. All genes on a given autosomal chromosome are linked, and their variants do not assort independently unless they are very far apart on the chromosome, so that crossing over occurs between them in nearly every meiosis. Crossing over increases the amount of genetic recombination in a population.

Occasionally, a pair of homologous chromosomes may fail to separate in meiosis and one of the gametes gets one chromosome too many and the other gets $n - 1$ chromosomes. Resulting zygotes usually do not survive; humans with $2n + 1$ chromosomes may live, but they are born with serious abnormalities, such as Down syndrome.

One gene most commonly controls the production of one polypeptide (one gene–one polypeptide hypothesis), but ribosomal and transfer RNAs are also encoded by genes.

Nucleic acids in the cell are DNA and RNA, which are large polymers of nucleotides composed of a nitrogenous base, pentose sugar, and phosphate group. The nitrogenous bases in DNA are adenine (A), guanine (G), thymine (T), and cytosine (C), and those in RNA are the same except that uracil (U) is substituted for thymine. DNA is a double-stranded, helical molecule in which the bases extend toward each other from the sugar-phosphate backbone: A always pairs with T and G with C. Thus the strands are antiparallel and complementary, being held in place by hydrogen bonds between the paired bases. In DNA replication the strands part, and the enzyme DNA polymerase synthesizes a new strand along each parental strand, using the parental strand as a template.

The sequence of bases in DNA is a code for the amino acid sequence in the ultimate product protein. Each triplet of three bases specifies a particular amino acid.

Proteins are synthesized by transcription of DNA into the base sequence of a molecule of messenger RNA (mRNA), which functions in concert with ribosomes (containing ribosomal RNA [rRNA] and protein) and transfer RNA (tRNA). Ribosomes attach to the strand of mRNA and move along it, assembling the amino acid sequence of the protein. Each amino acid is brought into position for assembly by a molecule of tRNA, which itself bears a base sequence (anticodon) complementary to the respective codons of the mRNA. In eukaryotic nuclear DNA the sequences of bases in DNA coding for amino acids in a protein (exons) are interrupted by intervening sequences (introns). The introns are removed from the primary mRNA before it leaves the nucleus, and the protein is synthesized in the cytoplasm.

Genes, and the synthesis of the products for which they are responsible, must be regulated: turned on or off in response to varying environmental conditions or cell differentiation. Gene regulation in eukaryotes is complex, with control of transcription being a particularly important point of regulation.

Modern methods in molecular genetics have made spectacular advances possible. Restriction endonucleases cleave DNA at specific base sequences, and such cleaved DNA from different sources can be rejoined to form recombinant DNA. Combining mammalian with plasmid or viral DNA, a mammalian gene can be introduced into bacterial cells, which then multiply and express the mammalian gene. The polymerase chain reaction (PCR) makes it relatively simple to clone specific genes if only a small sequence of the gene is known. Efforts of a publicly supported consortium and a privately supported group of scientists led to publication of draft sequences of the human genome in 2001. Among many exciting results of these efforts was a revision of estimated protein-encoding genes to 30,000 to 40,000, down from previous estimates of 100,000. These genes are somehow responsible for hundreds of thousands of proteins in a typical cell.

A mutation is a physicochemical alteration in the bases of DNA that may change the phenotypic effect of a gene. Although rare and usually detrimental to survival and reproduction of an organism, mutations are occasionally beneficial and provide new genetic material on which natural selection can work.

Cancer (neoplastic growth) is associated with a series of genetic changes in a clone of cells that allow unrestrained proliferation of those cells. Oncogenes (such as the gene coding for Ras protein) and inactivation of tumor suppressor genes (such as that coding for p53 protein) have been implicated in many cancers.

REVIEW QUESTIONS

1. What is the relationship between homologous chromosomes and alleles?
2. Describe or diagram the sequence of events in meiosis (both divisions).
3. What are the designations of the sex chromosomes in males of bugs, humans, and butterflies?
4. How do the chromosomal mechanisms of determining sex differ in the three taxa in question 3?
5. Diagram by Punnett square a cross between individuals with the following genotypes: $A/a \times A/a$; $A/a\ B/b \times A/a\ B/b$.
6. Concisely state Mendel's law of segregation and his law of independent assortment.
7. Assuming brown eyes (B) are dominant over blue eyes (b), determine the genotypes of all the following individuals. The blue-eyed son of two brown-eyed parents marries a brown-eyed woman whose mother was brown eyed and whose father was blue eyed. Their child is blue eyed.
8. Recall that red color (R) in four-o'clock flowers is incompletely dominant over white (R'). In the following crosses, give the genotypes of the gametes produced by each parent and the flower color of the offspring: $R/R' \times R/R'$; $R'/R' \times R/R'$; $R/R \times R/R'$; $R/R \times R'/R'$.
9. A brown male mouse is mated with two female black mice. When each female has produced several litters of young, the first female has had 48 black and the second female has had 14 black and 11 brown young. Can you deduce the pattern of inheritance of coat color and the genotypes of the parents?
10. Rough coat (R) is dominant over smooth coat (r) in guinea pigs, and black coat (B) is dominant over white (b). If a homozygous rough black is mated with a homozygous smooth white, give the appearance of each of the following: F_1; F_2; offspring of F_1 mated with smooth, white parent; offspring of F_1 mated with rough, black parent.
11. Assume right-handedness (R) is genetically dominant over left-handedness (r) in humans, and that brown eyes (B) are genetically dominant over blue (b). A right-handed, blue-eyed man marries a right-handed, brown-eyed woman. Their two children are right handed, blue eyed and left handed, brown eyed. The man marries again, and this time the woman is right handed and brown eyed. They have 10 children, all right handed and brown eyed. What are the probable genotypes of the man and his two wives?
12. In *Drosophila*, red eyes are dominant to white and the variation for this characteristic is on the X chromosome. Vestigial wings (v) are recessive to normal (V) for an autosomal gene. What will be the appearance of the following crosses: $X^W/X^w\ V/v \times X^w/Y\ v/v$, $X^w/X^w\ V/v \times X^W/Y\ V/v$.
13. Assume that color blindness is a recessive character on the X chromosome. A man and woman with normal vision have the following offspring: daughter with normal vision who has one color-blind son and one normal son; daughter with normal vision who has six normal sons; and a color-blind son who has a daughter with normal vision. What are the probable genotypes of all the individuals?
14. Distinguish the following: euploidy, aneuploidy, and polyploidy; monosomy and trisomy.
15. Name the purines and pyrimidines in DNA and tell which pair with each other in the double helix. What are the purines and pyrimidines in RNA and to what are they complementary in DNA?
16. Explain how DNA is replicated.
17. Why is it not possible for a codon to consist of only two bases?
18. Explain the transcription and processing of mRNA in the nucleus.
19. Explain the role of mRNA, tRNA, and rRNA in polypeptide synthesis.
20. What are four ways that genes can be regulated in eukaryotes?
21. In modern molecular genetics, what is recombinant DNA, and how is it prepared?
22. Name three sources of phenotypic variation.
23. Distinguish between proto-oncogene and oncogene. What are two mechanisms whereby cancer can be caused by genetic changes?
24. What are Ras protein and p53? How can mutations in the genes for these proteins contribute to cancer?
25. Outline the essential steps in the procedure for the polymerase chain reaction.
26. Draft sequences of the human genome have been published. What are some interesting observations based on the publications? What are some potential benefits that may be derived? What is the proteome?

SELECTED REFERENCES

Cavenee, W. K., and R. L. White. 1995. The genetic basis of cancer. Sci. Am. **272**:72–79 (Mar.). *Describes mutations in cells of colorectal cancer and brain tumors.*

Conery, J. S., and M. Lynch. 2000. The evolutionary fate and consequences of duplicate genes. Science **290**:1151–1155. *Gene duplications are an important source of genetic variation.*

Davies, K. 2001. Cracking the genome: inside the race to unlock human DNA. Craig Venter, Francis Collins, James Watson, and the story of the greatest scientific discovery of our time. New York, The Free Press. *Fascinating story of competition between the Human Genome Sequencing Project and Craig Venter's Celera Genomics. Of course, the genome is not "cracked" until its meaning is deciphered, and we are a long way from that.*

Dhand, R. 2000. Functional genomics. Nature **405**:819. *Introduction to a number of articles on functional genomics.*

Ezzell, C. 2000. Beyond the human genome. Sci. Am. **283**:64–69 (July). *The transcriptome is all the mRNAs being produced by a cell at a given time, and a proteome is all the proteins being made according to the instructions in the transcriptome.*

Ezzell, C. 2002. Proteins rule. Sci. Am. **286:**40–47 (April). *An excellent explanation of the current status and problems of proteomics.*

Futreal, P.A., A. Kasprzyk, E. Birney, J. C. Mullikin, R. Wooster, and M. R. Stratton. 2001. Cancer and genomics. Nature **409:**850–852. *Good list of cancer genes.*

Griffiths, A. J., W. M. Gelbart, R. C. Lewontin, and J. H. Miller. 2002. Modern genetic analysis, ed. 2. New York, W. H. Freeman & Company. *A good and recent general genetics text.*

Hall, A. 1994. A biochemical function for Ras—at last. Science **264:**1413–1414. *Ras protein is an enzyme in a signal transduction cascade stimulating a cell to divide.*

Jimenez-Sanchez, G., B. Childs, and D. Valle. 2001. Human disease genes. Nature **409:**853–855. *They found "striking correlation between the function of the gene product and features of disease. . . ."*

Lewin, B. 2000. Genes VII. New York, Oxford University Press. *Thorough, up-to-date coverage on molecular biology of genes.*

Mange, E. J., and A. P. Mange. 1999. Basic human genetics, ed. 2. Sunderland, Massachusetts, Sinauer Associates. *A readable, introductory text concentrating on the genetics of the animal species of greatest concern to most of us.*

Mullis, K. B. 1990. The unusual origin of the polymerase chain reaction. Sci. Am. **262:**56–65 (Apr.). *How the author had the idea for the simple production of unlimited copies of DNA while driving through the mountains of California.*

Pennisi, E. 2000. Genomics comes of age. Science **290:**2220–2221. *Determination of genomes of a number of organisms is "Breakthrough of the Year."*

Roberts, L. 2001. A history of the Human Genome Project. Science **291:**chart. *Each major discovery from double helix to full genome (1953–2001). Has glossary.*

The International Human Genome Mapping Consortium. 2001. A physical map of the human genome. Nature **409:**934–941. *The draft sequence published by the publicly supported consortium.*

Venter, J. C., M. D. Adams, E. W. Myers, P. W. Li, R. J. Mural, and 265 others. 2001. The sequence of the human genome. Science **291:**1304–1351. *The draft sequence published by Celera Genomics and its collaborators.*

Verma, I. M. 1990. Gene therapy. Sci. Am. **263:**68–84 (Nov.). *A review of prospects for treating and preventing genetic diseases by putting healthy genes into the body.*

ZOOLOGY LINKS TO THE INTERNET

Visit the textbook's Online Learning Center at
www.mhhe.com/zoology to find live Internet links
for each of the topics listed here.

Genetics
Mendelian Genetics
Meiosis and Sexual Reproduction
Molecular Basis of Genetics
Protein Synthesis
Gene Control and Expression
Recombinant Technology
Cloning
Plant Tissue Culture and Transgenic Crops
Human Genome Project

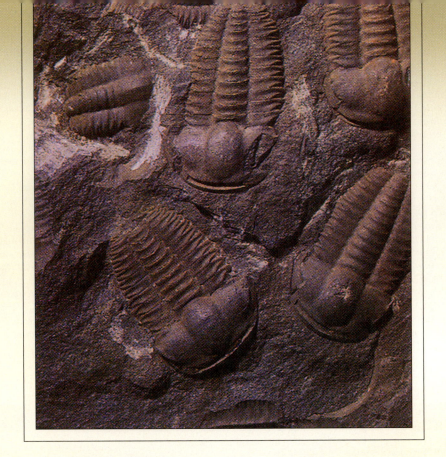

Trilobites fossilized
in Paleozoic rock.

Organic Evolution

A Legacy of Change

Life's history is a legacy of perpetual change. Despite the apparent permanence of the natural world, change characterizes all things on earth and in the universe. Earth's rock strata record the irreversible, historical change that we call organic evolution. Countless kinds of animals and plants have flourished and disappeared, leaving behind a sparse fossil record of their existence. Many, but not all, have left living descendants that bear some resemblance to them.

Life's changes are observed and measured in many ways. On a short evolutionary timescale, we see changes in the frequencies of different genetic traits within populations. Evolutionary changes in the relative frequencies of light- and dark-colored moths were observed within a single human lifetime in the polluted towns of industrial England. The formation of new species and dramatic changes in the appearances of organisms, as seen in the evolutionary diversification of Hawaiian birds, requires longer timescales covering 100,000 to 1 million years. Major evolutionary trends and periodic mass extinctions occur on even larger timescales, covering tens of millions of years. The fossil record of horses through the past 50 million years shows a series of different species replacing older ones through time and ending with the horses alive today. The fossil record of marine invertebrates shows us a series of mass extinctions separated by intervals of approximately 26 million years.

Because every feature of life as we know it today is a product of the evolutionary process, biologists consider organic evolution the keystone of all biological knowledge.

A   B

Figure 6-1

Founders of the theory of natural selection. **A,** Charles Robert Darwin (1809 to 1882), as he appeared in 1881, the year before his death. **B,** Alfred Russel Wallace (1823 to 1913) in 1895. Darwin and Wallace independently developed the same theory. A letter and essay from Wallace written to Darwin in 1858 spurred Darwin into writing *On The Origin of Species,* published in 1859.

In Chapter 1, we introduced Darwinian evolutionary theory as the dominant paradigm of biology. Charles Robert Darwin and Alfred Russel Wallace (Figure 6-1) first established evolution as a powerful scientific theory. Today the reality of organic evolution can be denied only by abandoning reason. As the noted English biologist Sir Julian Huxley wrote, "Charles Darwin effected the greatest of all revolutions in human thought, greater than Einstein's or Freud's or even Newton's, by simultaneously establishing the fact and discovering the mechanism of organic evolution." Darwinian theory helps us to understand both the genetics of populations and long-term trends in the fossil record. Darwin and Wallace were not the first, however, to consider the basic idea of organic evolution, which has an ancient history. We review the history of evolutionary thinking as it led to Darwin's theory, evidence supporting it, and changes to the theory that have produced our modern synthetic theory of evolution.

ORIGINS OF DARWINIAN EVOLUTIONARY THEORY

Pre-Darwinian Evolutionary Ideas

Before the eighteenth century, speculation on the origin of species rested on mythology and superstition, not on anything resembling a testable scientific theory. Creation myths viewed the world remaining constant after its creation. Nevertheless, some people approached the idea that nature has a long history of perpetual and irreversible change.

Early Greek philosophers, notably Xenophanes, Empedocles, and Aristotle, developed an early idea of evolutionary change. They recognized fossils as evidence for former life that they believed had been destroyed by natural catastrophe. Despite their intellectual inquiry, the Greeks failed to establish an evolutionary concept, and the issue declined well before the rise of Christianity. The opportunity for evolutionary thinking became even more restricted as the biblical account of the earth's creation became accepted as a tenet of faith. The year 4004 B.C. was fixed by Archbishop James Ussher (mid-seventeenth century) as the date of life's creation. Evolutionary views were considered rebellious and heretical. Still, some speculation continued. The French naturalist Georges Louis Buffon (1707 to 1788) stressed the influence of environment on the modifications of animal form. He also extended the age of the earth to 70,000 years.

Lamarckism: The First Scientific Explanation of Evolution

French biologist Jean Baptiste de Lamarck (1744 to 1829; Figure 6-2) authored the first complete explanation of evolution in 1809, the year of Darwin's birth. He made a convincing case that fossils were remains of extinct animals. Lamarck's proposed evolutionary mechanism, **inheritance of acquired characteristics,** was engagingly simple: organisms, by striving to meet the demands of their environments, acquire adaptations and pass them by heredity to their offspring. According to Lamarck, the giraffe evolved its long neck because its ancestors lengthened their necks by stretching to obtain food and then passed the lengthened neck to their offspring. Over many generations, these changes accumulated to produce the long necks of modern giraffes.

We call Lamarck's concept of evolution **transformational,** because it claims that individual organisms transform their characteristics to produce evolution. We now reject transformational theories because genetic studies show that traits acquired by an organism during its lifetime, such as strengthened muscles, are not inherited by offspring. Darwin's evolutionary theory differs from Lamarck's in being a **variational** theory, based on the distribution of genetic variation in populations. Evolutionary change is caused by differential survival and reproduction among organisms that differ in hereditary traits, not by inheritance of acquired characteristics.

Figure 6-2

Jean Baptiste de Lamarck (1744 to 1829), French naturalist who offered the first scientific explanation of evolution. Lamarck's hypothesis that evolution proceeds by inheritance of acquired characteristics has been disproven.

Figure 6-3

Sir Charles Lyell (1797 to 1875), English geologist and friend of Darwin. His book *Principles of Geology* greatly influenced Darwin during Darwin's formative period. This photograph was made about 1856.

Charles Lyell and Uniformitarianism

The geologist Sir Charles Lyell (1797 to 1875; Figure 6-3) established in his *Principles of Geology* (1830 to 1833) the principle of uniformitarianism. Uniformitarianism encompasses two important principles that guide the scientific study of the history of nature: (1) that the laws of physics and chemistry remain the same throughout the history of the earth, and (2) that past geological events occurred by natural processes similar to those observed today. Lyell showed that natural forces, acting over long periods of time, could explain the formation of fossil-bearing rocks. Lyell's geological studies led him to conclude that the earth's age must be measured in millions of years. These principles were important for discrediting miraculous and supernatural explanations of the history of nature and replacing them with scientific explanations. Lyell also stressed the gradual nature of geological changes that occur through time, and he argued further that such changes have no inherent tendency to occur in any particular direction. Both of these claims left important marks on Darwin's evolutionary theory.

Darwin's Great Voyage of Discovery

"After having been twice driven back by heavy southwestern gales, Her Majesty's ship *Beagle,* a ten-gun brig, under the command of Captain Robert FitzRoy, R.N., sailed from Devonport on the 27th of December, 1831." Thus began Charles Darwin's account of the historic five-year voyage of the *Beagle* around the world (Figure 6-4). Darwin, not quite 23 years old, had been asked to accompany Captain FitzRoy on the *Beagle,* a small vessel only 90 feet in length, which was about to depart on an extensive surveying voyage to South America and the Pacific (Figure 6-5). It was the beginning of one of the most important voyages of the nineteenth century.

During the voyage (1831 to 1836), Darwin endured seasickness and the erratic companionship of the authoritarian Captain FitzRoy, but Darwin's youthful physical strength and early training as a naturalist equipped him for his work. The *Beagle* made many stops along the harbors and coasts of South America and adjacent regions. Darwin made extensive collections and observations on the fauna and flora of these regions. He unearthed numerous fossils of animals long extinct and noted the resemblance between fossils of the South American pampas and the known fossils of North America. In the Andes he encountered seashells embedded in rocks at 13,000 feet. He experienced a severe earthquake and watched mountain torrents that relentlessly wore away the earth. These observations strengthened his conviction that natural forces were responsible for the geological features of the earth.

Figure 6-4

Five-year voyage of H.M.S. *Beagle.*

A   **B**

Figure 6-5

Charles Darwin and H.M.S. *Beagle*. **A,** Darwin in 1840, four years after the *Beagle* returned to England, and a year after his marriage to his cousin, Emma Wedgwood. **B,** The H.M.S. *Beagle* sails in Beagle Channel, Tierra del Fuego, on the southern tip of South America in 1833. The watercolor was painted by Conrad Martens, one of two official artists during the voyage of the *Beagle*.

In mid-September of 1835, the *Beagle* arrived at the Galápagos Islands, a volcanic archipelago straddling the equator 600 miles west of Ecuador (Figure 6-6). The fame of the islands stems from their infinite strangeness. They are unlike any other islands on earth. Some visitors are struck with awe and wonder, others with a sense of depression and dejection. Circled by capricious currents, surrounded by shores of twisted lava, bearing skeletal brushwood baked by the equatorial sun, almost devoid of vegetation, inhabited by strange reptiles and by convicts stranded by the Ecuadorian government, the islands had few admirers among mariners. By the middle of the seventeenth century, the islands were already known to Spaniards as "Las Islas Galápagos"—the tortoise islands. The giant tortoises, used for food first by buccaneers and later by American and British whalers, sealers, and ships of war, were the islands' principal attraction. At the time of Darwin's visit, the tortoises already were heavily exploited.

During the *Beagle*'s five-week visit to the Galápagos, Darwin began to develop his views of the evolution of life on earth. His original observations of the giant tortoises, marine iguanas, mockingbirds, and ground finches, all contributed to the turning point in Darwin's thinking.

Darwin was struck by the fact that, although the Galápagos Islands and the Cape Verde Islands (visited earlier in this voyage of the *Beagle*) were similar in climate and topography, their fauna and flora were altogether different. He recognized that Galápagos plants and animals were related to those of the South American mainland, yet differed from them in ecology and adaptive traits. Each island often contained a unique species related to forms on other islands. In short, Galápagos life must have originated in continental South America and then undergone modification in the various environmental conditions of the different islands. He concluded that living forms were neither divinely created nor immutable; they were, in fact, products of

Figure 6-6

The Galápagos Islands viewed from the rim of a volcano.

evolution. Although Darwin devoted only a few pages to Galápagos animals and plants in his monumental *On the Origin of Species,* published more than two decades later, his observations on the unique character of the animals and plants were, in his own words, the "origin of all my views."

On October 2, 1836, the *Beagle* returned to England, where Darwin conducted the remainder of his scientific work (Figure 6-7). Most of Darwin's extensive collections had preceded him there, as had most of his notebooks and diaries kept during the cruise. Darwin's journal was published three years after the *Beagle*'s return to England. It was an instant success and required two additional printings within the first year. In later versions, Darwin made extensive changes and titled his book *The Voyage of the Beagle*. The fascinating account of his observations written in a simple, appealing style has made the book one of the most lasting and popular travel books.

Figure 6-7
Darwin's study at Down House in Kent, England, is preserved today much as it was when Darwin wrote *On The Origin of Species.*

Curiously, the main product of Darwin's voyage, his theory of evolution, did not appear in print for more than 20 years after the *Beagle*'s return. In 1838, he "happened to read for amusement" an essay on populations by T. R. Malthus (1766 to 1834), who stated that animal and plant populations, including human populations, tend to increase beyond the capacity of the environment to support them. Darwin already had been gathering information on artificial selection of animals under domestication by humans. After reading Malthus's article, Darwin realized that a process of selection in nature, a "struggle for existence" because of overpopulation, could be a powerful force for evolution of wild species.

He allowed the idea to develop in his own mind until it was presented in 1844 in a still-unpublished essay. Finally in 1856, he began to assemble his voluminous data into a work on the origin of species. He expected to write four volumes, a very big book, "as perfect as I can make it." However, his plans were to take an unexpected turn.

In 1858, he received a manuscript from Alfred Russel Wallace (1823 to 1913), an English naturalist in Malaya with whom he was corresponding. Darwin was stunned to find that in a few pages, Wallace summarized the main points of the natural selection theory on which Darwin had been working for two decades. Rather than withhold his own work in favor of Wallace as he was inclined to do, Darwin was persuaded by two close friends, the geologist Lyell and the botanist Hooker, to publish his views in a brief statement that would appear together with Wallace's paper in the *Journal of the Linnean Society.* Portions of both papers were read before an unimpressed audience on July 1, 1858.

For the next year, Darwin worked urgently to prepare an "abstract" of the planned four-volume work. This book was published in November 1859, with the title *On the Origin of Species by Means of Natural Selection, or the Preservation of Favoured Races in the Struggle for Life.* The 1250 copies of the first printing sold the first day! The book instantly generated a storm that has

never completely abated. Darwin's views were to have extraordinary consequences on scientific and religious beliefs and remain among the greatest intellectual achievements of all time.

> "Whenever I have found that I have blundered, or that my work has been imperfect, and when I have been contemptuously criticized, and even when I have been overpraised, so that I have felt mortified, it has been my greatest comfort to say hundreds of times to myself that 'I have worked as hard and as well as I could, and no man can do more than this.'" *Charles Darwin, in his autobiography, 1876.*

Once Darwin's caution had been swept away by the publication of *On the Origin of Species,* he entered an incredibly productive period of evolutionary thinking for the next 23 years, producing book after book. He died on April 19, 1882, and was buried in Westminster Abbey. The little *Beagle* had already disappeared, having been retired in 1870 and presumably dismantled for scrap.

DARWINIAN EVOLUTIONARY THEORY: THE EVIDENCE

Perpetual Change

The main premise underlying Darwinian evolution is that the living world is neither constant nor perpetually cycling, but always changing. Perpetual change in the form and diversity of animal life throughout its 600- to 700-million-year history is seen most directly in the fossil record. A **fossil** is a remnant of past life uncovered from the crust of the earth (Figure 6-8). Some fossils constitute complete remains (insects in amber and mammoths), actual hard parts (teeth and bones), and petrified skeletal parts that are infiltrated with silica or other minerals (ostracoderms and molluscs). Other fossils include molds, casts, impressions, and fossil excrement (coprolites). In addition to documenting organismal evolution, fossils reveal profound changes in the earth's environment, including major changes in the distributions of lands and seas. Because many organisms left no fossils, a complete record of the past is always beyond our reach; nonetheless, discovery of new fossils and reinterpretation of existing ones expand our knowledge of how the form and diversity of animals changed through geological time.

> Fossil remains may on rare occasions include soft tissues preserved so well that recognizable cellular organelles can be viewed by electron microscopy! Insects are frequently found entombed in amber, the fossilized resin of trees. One study of a fly entombed in 40-million-year-old amber revealed structures corresponding to muscle fibers, nuclei, ribosomes, lipid droplets, endoplasmic reticulum, and mitochondria (Figure 6-8D). This extreme case of mummification probably occurred because chemicals in the plant sap diffused into the embalmed insect's tissues.

A   B

C   D

Figure 6-8

Four examples of fossil material. **A,** Fish fossil from rocks of the Green River Formation, Wyoming. Such fish swam here during the Eocene epoch of the Tertiary period, approximately 55 million years ago. **B,** Stalked crinoids (class Crinoidea, p. 456) from 85-million-year-old Cretaceous rocks. The fossil record of these echinoderms shows that they reached their peak millions of years earlier and began a slow decline to the present. **C,** An insect fossil that got stuck in the resin of a tree 40 million years ago and that has since hardened into amber. **D,** Electron micrograph of tissue from a fly fossilized as shown in **C**; the nucleus of a cell is marked in red.

Interpreting the Fossil Record

The fossil record is biased because preservation is selective. Vertebrate skeletal parts and invertebrates with shells and other hard structures left the best records (Figure 6-8). Soft-bodied animals, including jellyfishes and most worms, are fossilized only under very unusual circumstances such as those that formed the Burgess Shale of British Columbia (Figure 6-9). Exceptionally favorable conditions for fossilization produced the Precambrian fossil bed of South Australia, the tar pits of Rancho La Brea (Hancock Park, Los Angeles), the great dinosaur beds (Alberta, Canada, and Jensen, Utah; Figure 6-10), the Olduvai Gorge of Tanzania, and the Yunnan and Lianoning provinces of China.

Fossils are deposited in stratified layers with new deposits forming on top of older ones. If left undisturbed, which is rare, a sequence is preserved with the ages of fossils being directly proportional to their depth in the stratified layers. Characteristic fossils often serve to identify particular layers. Certain widespread marine invertebrate fossils, including various foraminiferans (p. 228) and echinoderms (p. 443), are such good indicators of specific geological periods that they are called "index," or "guide," fossils. Unfortunately, the layers are usually tilted or show faults (cracks). Old deposits exposed by erosion may be covered with new deposits in a different plane. When exposed to tremendous pressures or heat, stratified sedimentary rock metamorphoses into crystalline quartzite, slate, or marble, which destroys fossils.

Geological Time

Long before the earth's age was known, geologists divided its history into a table of succeeding events based on the ordered layers of sedimentary rock. The "law of stratigraphy" produced a relative dating with the oldest layers at the bottom and the youngest at the top of the sequence. Time was divided into

eons, eras, periods, and epochs as shown on the endpaper inside the back cover of this book. Time during the last eon (Phanerozoic) is expressed in eras (for example, Cenozoic), periods (for example, Tertiary), epochs (for example, Paleocene), and sometimes smaller divisions of an epoch.

In the late 1940s, radiometric dating methods were developed for determining the absolute age in years of rock formations. Several independent methods are now used, all based on the radioactive decay of naturally occurring elements into other elements. These "radioactive clocks" are independent of pressure and temperature changes and therefore are not affected by often violent earth-building activities.

One method, potassium-argon dating, depends on the decay of potassium-40 (^{40}K) to argon-40 (^{40}Ar) (12%) and calcium-40 (^{40}Ca) (88%). The half-life of potassium-40 is 1.3 billion years; half of the original atoms will decay in 1.3 billion years, and half of the remaining atoms will be gone at the end of the next 1.3 billion years. This decay continues until all radioactive potassium-40 atoms are gone. To measure the age of the rock, one calculates the ratio of remaining potassium-40 atoms to the amount of potassium-40 originally there (the remaining potassium-40 atoms plus the argon-40 and calcium-40 into which other potassium-40 atoms have decayed). Several such isotopes exist for dating purposes, some for dating the age of the earth itself. One of the most useful radioactive clocks depends on the decay of uranium into lead. With this method, rocks over 2 billion years old can be dated with a probable error of less than 1%.

The fossil record of macroscopic organisms begins near the start of the Cambrian period of the Paleozoic era, approximately 600 million years BP. Geological time before the Cambrian is called the Precambrian era or Proterozoic eon. Although the Precambrian era occupies 85% of all geological time, it has received much less attention than later eras, partly because oil, which

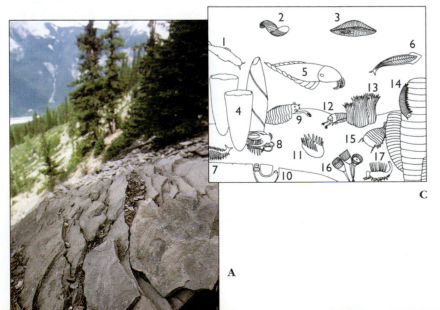

Figure 6-9

A, Fossil trilobites visible at the Burgess Shale Quarry, British Columbia. **B,** Animals of the Cambrian period, approximately 580 million years ago, as reconstructed from fossils preserved in the Burgess Shale of British Columbia, Canada. The main new body plans that appeared rather abruptly at this time established the body plans of animals familiar to us today. **C,** Key to Burgess Shale drawing. *Amiskwia* (1), from an extinct phylum; *Odontogriphus* (2), from an extinct phylum; *Eldonia* (3), a possible echinoderm; *Halichondrites* (4), a sponge; *Anomalocaris canadensis* (5), from an extinct phylum; *Pikaia* (6), an early chordate; *Canadia* (7), a polychaete; *Marrella splendens* (8), a unique arthropod; *Opabinia* (9), from an extinct phylum; *Ottoia* (10), a priapulid; *Wiwaxia* (11), from an extinct phylum, *Yohoia* (12), a unique arthropod; *Xianguangia* (13), an anemone-like animal; *Aysheaia* (14), an onychophoran or extinct phylum; *Sidneyia* (15), a unique arthropod; *Dinomischus* (16), from an extinct phylum; *Hallucigenia* (17), from an extinct phylum.

Figure 6-10

A dinosaur skeleton partially excavated from rock at Dinosaur Provincial Park, Alberta.

provides the commercial incentive for much geological work, seldom exists in Precambrian formations. The Precambrian era contains well-preserved fossils of bacteria and algae, and casts of jellyfishes, sponge spicules, soft corals, segmented flatworms, and worm trails. Most, but not all, are microscopic fossils.

Evolutionary Trends

The fossil record allows us to view evolutionary change across the broadest scale of time. Species arise and go extinct repeatedly throughout the fossil record. Animal species typically survive approximately 1 million to 10 million years, although their duration is highly variable. When we study patterns of species or taxon replacement through time, we observe trends. Trends are directional changes in the characteristic features or patterns of diversity in a group of organisms. Fossil trends clearly demonstrate Darwin's principle of perpetual change.

A well-studied fossil trend is the evolution of horses from the Eocene epoch to the present. Looking back at the Eocene epoch, we see many different genera and species of horses that were replaced by others through time (Figure 6-11). George Gaylord Simpson (p. 200) showed that this trend is compatible with Darwinian evolutionary theory. The three characteristics that show the clearest trends in horse evolution are body size, foot structure, and tooth structure. Compared to modern horses, those of extinct genera were small, their teeth had a relatively small grinding surface, and their feet had a relatively large number of toes (four). Throughout the subsequent Oligocene, Miocene, Pliocene, and Pleistocene epochs, there were continuing patterns of new genera arising and old ones going extinct. In each case, a net increase in body size, expansion of the grinding surface of the teeth, and reduction in the number of toes occurred. As the number of toes was reduced, the central digit became increasingly more prominent in the foot, and eventually only this central digit remained.

The fossil record shows a net change not only in the characteristics of horses but also variation in the numbers of different horse genera (and numbers of species) that have existed through time. The many horse genera of past epochs have been lost to extinction, leaving only a single survivor, *Equus.* Evolutionary trends in diversity are observed in fossils of many different groups of animals (Figure 6-12).

Trends in fossil diversity through time are produced by different rates of species formation versus extinction through time. Why do some lineages generate large numbers of new species whereas others generate relatively few? Why do different lineages undergo higher or lower rates of extinction (of species, genera, or taxonomic families) throughout evolutionary time? To answer these questions, we must turn to Darwin's other four theories of evolution. Regardless of how we answer these questions, however, the observed trends in animal diversity clearly illustrate Darwin's principle of perpetual change. Because the remaining four theories of Darwinism rely on the theory of perpetual change, evidence supporting these theories strengthens Darwin's theory of perpetual change.

Common Descent

Darwin proposed that all plants and animals have descended from an ancestral form into which life was first breathed. Life's history is depicted as a branching tree, called a **phylogeny.** Pre-Darwinian evolutionists, including Lamarck, advocated multiple independent origins of life, each of which gave rise to lineages that changed through time without extensive branching. Like all good scientific theories, common descent makes several important predictions that can be tested and potentially used to reject it. According to this theory, we should be able to trace the genealogies of all modern species backward until they converge on ancestral lineages shared with other species, both living and extinct.

We should be able to continue this process, moving farther backward through evolutionary time, until we reach the primordial ancestor of all life on earth. All forms of life, including many extinct forms that represent dead branches, will connect to this tree somewhere. Although reconstructing the history of life in this manner may seem almost impossible, phylogenetic research has been extraordinarily successful. How has this difficult task been accomplished?

Homology and Phylogenetic Reconstruction

Darwin recognized the major source of evidence for common descent in the concept of **homology.** Darwin's contemporary, Richard Owen (1804 to 1892), used this term to denote "the same organ in different organisms under every variety of form and function." A classic example of homology is the limb skeleton of vertebrates. Bones of vertebrate limbs maintain characteristic structures and patterns of connection despite diverse modifications for different functions (Figure 6-13). According to Darwin's theory of common descent, the structures that we call homologies represent characteristics inherited with some modification from a corresponding feature in a common ancestor.

Darwin devoted an entire book, *The Descent of Man and Selection in Relation to Sex,* largely to the idea that humans share common descent with apes and other animals. This idea was repugnant to the Victorian world, which responded with

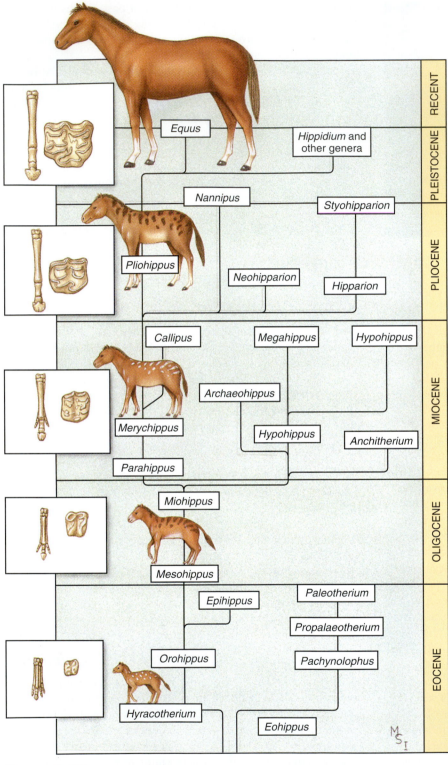

Figure 6-11

A reconstruction of genera of horses from Eocene to present. Evolutionary trends toward increased size, elaboration of molars, and loss of toes are shown together with a hypothetical genealogy of extant and fossil genera.

predictable outrage (Figure 6-14). Darwin built his case mostly on anatomical comparisons revealing homology between humans and apes. To Darwin, the close resemblances between apes and humans could be explained only by common descent.

Throughout the history of all forms of life, evolutionary processes generate new characteristics that are then inherited by subsequent generations. Every time a new feature arises on an evolving lineage, we see the origin of a new homology. That homology gets transmitted to all descendant lineages unless it is subsequently lost. The pattern formed by the sharing of homologies among species provides evidence for common descent and allows us to reconstruct the branching evolutionary history of life. We can illustrate such evidence using a phylogenetic tree for a group of large, ground-dwelling birds (Figure 6-15). A new skeletal homology arises on each of the lineages shown (descriptions of specific homologies are not included because they are highly technical). The different groups of species located at the tips of the branches contain different combinations of these homologies, which reflect ancestry. For example, ostriches show homologies 1 through 5 and 8, whereas kiwis show homologies 1, 2, 13, and 15. Branches of the tree combine these species into a **nested hierarchy** of groups within groups (see Chapter 10). Smaller groups (species grouped near terminal branches) are contained within larger ones (species grouped by basal branches, including the trunk of the tree). If we erase the tree structure but retain patterns of homology observed in the living species, we are able to reconstruct the branching structure of the entire tree. Evolutionists test the theory of common descent by observing patterns of homology present within all groups of organisms. The pattern formed by all homologies taken together should specify a single branching tree that represents the evolutionary genealogy of all living organisms.

The nested hierarchical structure of homology is so pervasive in the living world that it forms the basis for our systematic classification of all forms of life (genera grouped into families, families grouped into orders, and other categories). Hierarchical classification even preceded Darwin's theory because this pattern is so evident, but it was not explained adequately before Darwin. Once the idea of common descent was accepted, biologists began investigating the structural, molecular, and chromosomal homologies of animal groups. Taken together, the nested hierarchical patterns uncovered by these studies have permitted us to reconstruct evolutionary trees of many groups and to continue investigating others. Use of Darwin's theory of common descent to

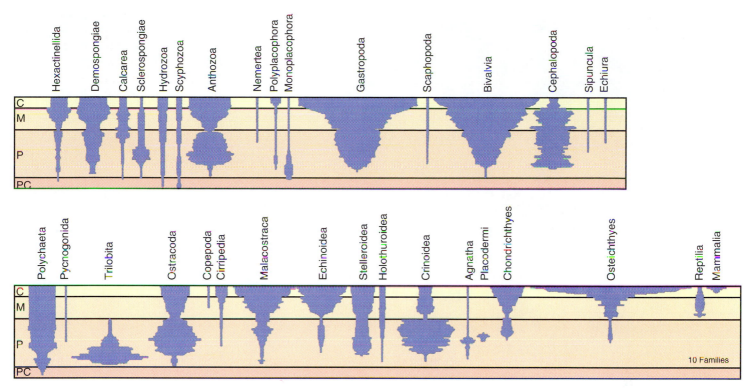

Figure 6-12

Diversity profiles of taxonomic families from different animal groups in the fossil record. The scale marks the Precambrian (*PC*), Paleozoic (*P*), Mesozoic (*M*), and Cenozoic (*C*) eras. The relative number of families is indicated from the width of the profile.

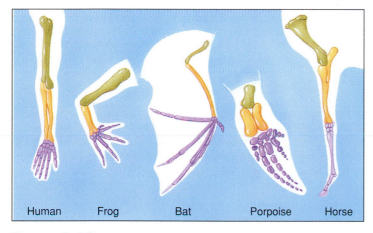

Human   Frog   Bat   Porpoise   Horse

Figure 6-13

Forelimbs of five vertebrates show skeletal homologies: *green,* humerus; *yellow,* radius and ulna; *purple,* "hand" (carpals, metacarpals, and phalanges). Clear homologies of bones and patterns of connection are evident despite evolutionary modification for various particular functions.

reconstruct the evolutionary history of life and to classify animals is the subject of Chapter 10.

Note that the earlier evolutionary hypothesis that life arose many times, forming unbranched lineages, predicts linear sequences of evolutionary change with no nested hierarchy of homologies among species. Because we do observe nested hierarchies of homologies, that hypothesis is rejected. Note also

that because the creationist argument is not a scientific hypothesis, it can make no testable predictions about any pattern of homology.

Ontogeny, Phylogeny, and Recapitulation

Ontogeny is the history of the development of an organism through its entire life. Early developmental and embryological features contribute greatly to our knowledge of homology and common descent. Comparative studies of ontogeny show how the evolutionary alteration of developmental timing generates new characteristics, thereby producing evolutionary divergence among lineages.

The German zoologist Ernst Haeckel, a contemporary of Darwin, believed that each successive stage in the development of an individual represented one of the adult forms that appeared in its evolutionary history. The human embryo with gill depressions in the neck was believed, for example, to resemble the adult appearance of a fishlike ancestor. On this basis Haeckel gave his generalization: *ontogeny (individual development) recapitulates (repeats) phylogeny (evolutionary descent).* This notion later became known simply as **recapitulation** or the **biogenetic law.** Haeckel based his biogenetic law on the flawed premise that evolutionary change occurs by successively adding new features onto the end of an unaltered ancestral ontogeny while condensing the ancestral ontogeny into earlier developmental stages. This notion was based on Lamarck's concept of the inheritance of acquired characteristics (p. 103).

Figure 6-14

This 1873 advertisement for Merchant's Gargling Oil ridicules Darwin's theory of the common descent of humans and apes, which received only limited acceptance by the general public during Darwin's lifetime.

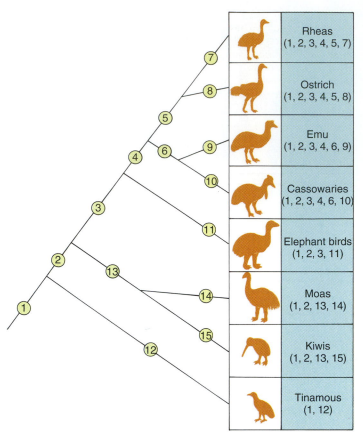

Figure 6-15

The phylogenetic pattern specified by 15 homologous structures in the skeletons of a group of flightless birds. Homologous features are numbered 1 through 15 and are marked both on the branches of the tree on which they arose and on the birds that have them. If you were to erase the tree structure, you would be able to reconstruct it without error from the distributions of homologous features shown for the birds at the terminal branches.

The nineteenth-century embryologist, K. E. von Baer, gave a more satisfactory explanation of the relationship between ontogeny and phylogeny. He argued that early developmental features were simply more widely shared among different animal groups than later ones. Figure 6-16 shows, for example, the early embryological similarities of organisms whose adult forms are very different (see Figure 8-19, p. 166). The adults of animals with relatively short and simple ontogenies often resemble pre-adult stages of other animals whose ontogeny is more elaborate, but embryos of descendants do not necessarily resemble the adults of their ancestors. Even early development undergoes evolutionary divergence among lineages, however, and it is not as stable as von Baer believed.

We now know that there are many parallels between ontogeny and phylogeny, but features of an ancestral ontogeny can be shifted either to earlier or later stages in descendant ontogenies. Evolutionary change in timing of development is called **heterochrony,** a term initially used by Haeckel to denote exceptions to recapitulation. If a descendant's ontogeny extends beyond its ancestral one, new characteristics can be added late in development, beyond the point at which development would have terminated in the evolutionary ancestor. Features observed in the

ancestor often are moved to earlier stages of development in this process, and ontogeny therefore does recapitulate phylogeny to some degree. Ontogeny also can be shortened during evolution, however. Terminal stages of the ancestor's ontogeny may be deleted, causing adults of descendants to resemble pre-adult stages of their ancestors (Figure 6-17). This outcome reverses the parallel between ontogeny and phylogeny (reverse recapitulation) producing **paedomorphosis** (the retention of ancestral juvenile characters by descendant adults). Because lengthening or shortening of ontogeny can change different parts of the body independently, we often see a mosaic of different kinds of developmental evolutionary change in a single lineage. Therefore, cases in which an entire ontogeny recapitulates phylogeny are rare.

Multiplication of Species

Multiplication of species through time is a logical corollary to Darwin's theory of common descent. A branch point on the evolutionary tree means that an ancestral species has split into two different ones. Darwin's theory postulates that genetic variation

Figure 6-16

Comparison of gill arches of different embryos. All are shown separated from the yolk sac. Note the remarkable similarity of the four embryos at this early stage in development.

Fish   Reptile

Bird   Human

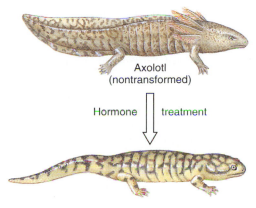

Axolotl
(nontransformed)

Hormone   treatment

Transformation to adult form

Figure 6-17

Aquatic and terrestrial forms of axolotls. Axolotls retain the juvenile, aquatic morphology (*top*) throughout their lives unless forced to metamorphose (*bottom*) by hormone treatment. Axolotls evolved from metamorphosing ancestors, an example of paedomorphosis.

present within a species, especially variation that occurs between geographically separated populations, provides the material from which new species are produced. Because evolution is a branching process, the total number of species produced by evolution increases through time, although most of these species eventually become extinct. A major challenge for evolutionists is to discover the process by which an ancestral species "branches" to form two or more descendant species.

Before we explore multiplication of species, we must decide what we mean by "species." As explained in Chapter 10, no consensus exists regarding definition of species. Most biologists agree, however, that important criteria for recognizing species include (1) descent of all members from a common ancestral population, (2) reproductive compatibility (ability to interbreed) within and reproductive incompatibility between species, and (3) maintenance within species of genotypic and phenotypic cohesion (lack of abrupt differences among populations in allelic frequencies and organismal characteristics). The criterion of reproductive compatibility has received the greatest attention in studies of species formation, also called **speciation.**

Biological features that prevent different species from interbreeding are called **reproductive barriers.** The primary problem of speciation is to discover how two initially compatible populations evolve reproductive barriers that cause them to become distinct, separately evolving lineages. How do populations diverge from each other in their reproductive properties while maintaining complete reproductive compatibility within each population?

Reproductive barriers between populations usually evolve gradually. Evolution of reproductive barriers requires that diverging populations must be kept physically separate for long periods of time. If diverging populations reunite before reproductive barriers are completely formed, interbreeding occurs between the populations and they merge. Speciation by gradual divergence in animals may require extraordinarily long periods of time, perhaps 10,000 to 100,000 years or more. Geographical isolation followed by gradual divergence is the most effective way for reproductive barriers to evolve, and many evolutionists consider geographical separation a prerequisite for branching speciation.

Allopatric Speciation

Allopatric ("in another land") populations of a species are those that occupy separate geographical areas. Because of their geographical separation, they cannot interbreed, but would be expected to do so if the geographic barriers between them were removed. Speciation that results from evolution of reproductive barriers between geographically separated populations is called **allopatric speciation** or geographic speciation. The separated populations evolve independently and adapt to their different environments, generating reproductive barriers between them as a result of their separate evolutionary paths. Ernst Mayr (Figure 6-18) has contributed greatly to our knowledge of allopatric speciation through his studies of speciation in birds.

Allopatric speciation begins when a species splits into two or more geographically separated populations. This splitting can happen in either of two ways: by **vicariant speciation** or by a **founder event.** Vicariant speciation is initiated when climatic or

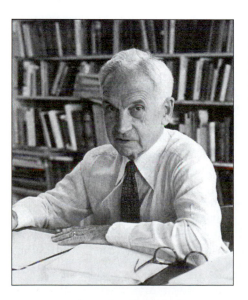

Figure 6-18

Professor Ernst Mayr, a major contributor to our knowledge of speciation and of evolution in general.

geological changes fragment a species' habitat, producing impenetrable barriers that separate different populations. For example, a mammalian species inhabiting a lowland forest could be divided by uplifting of a mountain barrier, sinking and flooding of a geological fault, or climatic changes that cause prairie or desert conditions to encroach on the forest.

Vicariant speciation has two important consequences. Although the ancestral population is fragmented, individual fragments are usually left fairly intact. The vicariant process itself does not induce genetic change by reducing populations to a small size or by transporting them to unfamiliar environments. Another important consequence is that the same vicariant events may fragment several different species simultaneously. For example, fragmentation of the lowland forest described above most likely would disrupt numerous and diverse species, including salamanders, frogs, snails, and many other forest dwellers. Indeed, the same geographic patterns are observed among closely related species in different groups of organisms whose habitats are similar. Such patterns provide strong evidence for vicariant speciation.

The alternative means of initiating allopatric speciation is for a small number of individuals to disperse to a distant place where no other members of their species are present. The dispersing individuals may establish a new population in what is called a founder event. Allopatric speciation caused by founder events has been observed, for example, in the native fruit flies of Hawaii. Hawaii contains numerous patches of forest separated by volcanic lava flows. On rare occasions, strong winds can transport a few flies from one forest to another, geographically isolated forest where the flies are able to start a new population. Sometimes, a single fertilized female may found a new population. Unlike what happens in vicariant speciation, the new population initially has a very small size, which can cause its genetic structure to change dramatically from that of its ancestral population (see p. 122). When this event happens, phenotypic characteristics that were stable in the ancestral population often reveal unprecedented variation in the new population. As the newly expressed variation is sorted by natural selection, large changes in phenotype and reproductive properties occur, hastening the evolution of reproductive barriers between the ancestral and newly founded populations.

Surprisingly, we often learn most about the genetics of allopatric speciation from cases in which formerly separated populations regain geographic contact following evolution of incipient reproductive barriers that are not absolute. The occurrence of mating between divergent populations is called **hybridization** and offspring of these matings are called **hybrids** (Figure 6-19). By studying the genetics of hybrid populations, we can identify the genetic bases of reproductive barriers.

Biologists often distinguish between reproductive barriers that impair fertilization (premating barriers) and those that impair growth and development, survival, or reproduction of hybrid individuals (postmating barriers). Premating barriers may cause members of divergent populations either not to recognize each other as potential mates or not to complete the mating ritual successfully. In some cases, female and male genitalia of the different populations will be incompatible. In others, premating barriers may be strictly behavioral, with members of different

A   **B**   **C**

Figure 6-19

Pure and hybrid salamanders. Hybrids are intermediate in appearance between parental populations. **A,** Pure white-spotted *Plethodon teyahalee;* **B,** a hybrid between white-spotted *P. teyahalee* and red-legged *P. shermani,* intermediate in appearance for both spotting and leg color; **C,** pure red-legged *P. shermani.*

species being otherwise nearly identical in phenotype. Different species that are indistinguishable in organismal appearance are called **sibling species.** Sibling species arise when allopatric populations diverge in the seasonal timing of reproduction or in auditory, behavioral, or chemical signals required for mating. Evolutionary divergence in these features can produce effective premating barriers without obvious changes in organismal appearance. Sibling species occur in groups as diverse as ciliates, flies, and salamanders.

Nonallopatric Speciation

Can speciation ever occur without prior geographic separation of populations? Allopatric speciation may seem an unlikely explanation for situations where many closely related species occur together in restricted areas that have no traces of physical barriers to animal dispersal. For example, several large lakes around the world contain very large numbers of closely related species of fish. The great lakes of Africa (Lake Malawi, Lake Tanganyika, and Lake Victoria) each contain many species of cichlid fishes that are found nowhere else. Likewise, Lake Baikal in Siberia contains many different species of sculpins that occur nowhere else in the world (Figure 6-20). It is difficult to conclude that these species arose anywhere other than in the lakes they inhabit, and yet those lakes are young on an evolutionary timescale and have no obvious environmental barriers that would fragment fish populations.

To explain speciation of fish in freshwater lakes and other examples like these, **sympatric** ("same land") **speciation** has been hypothesized. According to this hypothesis, different individuals within a species become specialized for occupying different components of the environment. By seeking out and using very specific habitats in a single geographic area, different populations achieve sufficient physical and adaptive separation to

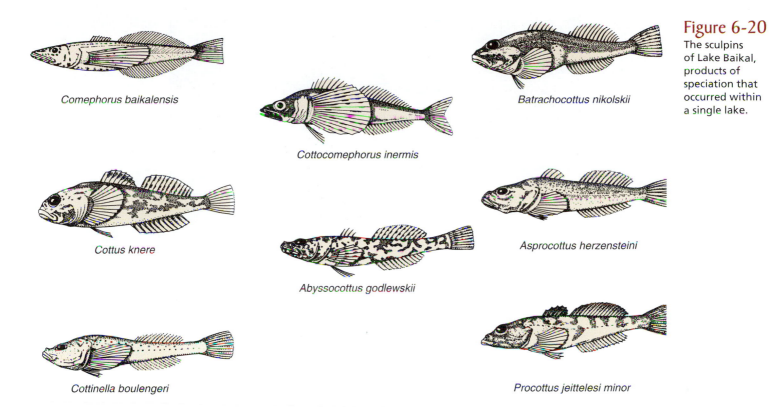

Figure 6-20
The sculpins of Lake Baikal, products of speciation that occurred within a single lake.

Comephorus baikalensis

Cottocomephorus inermis

Batrachocottus nikolskii

Cottus knere

Abyssocottus godlewskii

Asprocottus herzensteini

Cottinella boulengeri

Procottus jeittelesi minor

evolve reproductive barriers. For example, cichlid species of African lakes are very different from each other in their feeding specializations. In many parasitic organisms, particularly parasitic insects, different populations may use different host species, thereby providing the physical separation necessary for reproductive barriers to evolve. Supposed cases of sympatric speciation have been criticized, however, because the reproductive distinctness of the different populations often is not well demonstrated, so that we may not be observing formation of distinct evolutionary lineages that will become different species.

The occurrence of sudden sympatric speciation is perhaps most likely among higher plants. Between one-third and one-half of flowering plant species may have evolved by polyploidy (doubling of chromosome numbers), without prior geographic isolation of populations. In animals, however, speciation through polyploidy is an exceptional event.

Adaptive Radiation

The production of ecologically diverse species from a common ancestral stock is called **adaptive radiation.** Some of our best examples of adaptive radiation are associated with lakes and young islands, which are sources of new evolutionary opportunities for aquatic and terrestrial organisms, respectively. Oceanic islands formed by volcanoes are initially devoid of life. They are gradually colonized by plants and animals from a continent or from other islands in separate founder events. The founders encounter ideal situations for evolutionary diversification, because environmental resources that were heavily exploited by other species on the mainland are free for colonization on the sparsely populated island. Archipelagoes, such as the Galápagos Islands, greatly increase opportunities for both founder events and eco-

logical diversification. The entire archipelago is isolated from the continent and each island is geographically isolated from the others by the sea; moreover, each island is different from every other one in its physical, climatic, and biotic characteristics.

Galápagos finches clearly illustrate adaptive radiation on an oceanic archipelago (Figures 6-21 and 6-22). Galápagos finches (the name "Darwin's finches" was popularized in the 1940s by the British ornithologist David Lack) are closely related to each other, but each species differs from others in size and shape of the beak and in feeding habits. If the finches were specially created, it would require the strangest kind of coincidence for 13 similar kinds of finches to be created on the Galápagos Islands and nowhere else. Darwin's finches descended from a single ancestral population that arrived from the mainland and subsequently colonized the different islands of the Galápagos archipelago. The finches underwent adaptive radiation, occupying habitats that on the mainland would have been denied to them by the presence of other species that are better able to exploit those habitats. Galápagos finches thus assumed characteristics of mainland families as diverse and unfinchlike as warblers and woodpeckers. A fourteenth Darwin's finch, found on isolated Cocos Island far north of the Galápagos archipelago, is similar in appearance to the Galápagos finches and almost certainly descended from the same ancestral stock.

Gradualism

Darwin's theory of gradualism opposes arguments for the sudden origin of species. Small differences, resembling those that we observe among organisms within populations today, are the raw material from which the different major forms of life evolved. This theory shares with Lyell's uniformitarianism the

Figure 6-21

Tentative model for evolution of the 13 Darwin's finches on the Galápagos Islands. The model postulates three steps: (1) Immigrant finches from the South American mainland reach the Galápagos and colonize an island; (2) once the population becomes established, finches disperse to other islands where they adapt to new conditions and change genetically; and (3) after a period of isolation, secondary contact is established between different populations. The two populations are then recognized as separate species if they cannot interbreed successfully.

Figure 6-22

A, Adaptive radiation in ten species of Darwin's finches from Santa Cruz, one of the Galápagos Islands. Differences in bills and feeding habits are shown. All apparently descended from a single common ancestral finch from the South American continent. **B,** Woodpecker finch, one of the 13 species of Galápagos Islands finches, using a slender twig as a tool for feeding. This finch worked for about 15 minutes before spearing and removing a wood roach from a break in the tree.

notion that we must not explain past changes by invoking unusual catastrophic events that are not observed today. If new species originated in single, catastrophic events, we should be able to see such events happening today and we do not. Instead, what we observe in natural populations are usually small, continuous changes in phenotypes. Such continuous changes can produce major differences among species only by accumulating over many thousands to millions of years. A simple statement of Darwin's theory of gradualism is that accumulation of quantitative changes leads to qualitative change.

Mayr (see Figure 6-18) makes an important distinction between populational gradualism and phenotypic gradualism. **Populational gradualism** states that new traits become established in a population by increasing their frequency initially from a small fraction of the population to a majority of the population. Populational gradualism is well established and is not controversial. **Phenotypic gradualism** states that new traits, even those that are strikingly different from ancestral ones, are produced in a series of small, incremental steps.

Phenotypic Gradualism

Phenotypic gradualism was controversial when Darwin first proposed it, and it is still controversial. Not all phenotypic changes are small, incremental ones. Some mutations that appear during artificial breeding change the phenotype substantially in a single mutational step. Such mutations traditionally are called "sports." Sports that produce dwarfing are observed in many species, including humans, dogs, and sheep, and have been used by animal breeders to achieve desired results; for example, a sport that deforms the limbs was used to produce ancon sheep, which cannot jump hedges and are therefore easily contained (Figure 6-23). Many colleagues of Darwin who accepted his other theories considered phenotypic gradualism too extreme. If sporting mutations can be used in animal breeding, why must we exclude them from our evolutionary theory? In favor of gradualism, Darwin and

others have replied that sporting mutations always have negative side-effects that would cause selection to eliminate them from natural populations. Indeed, it is questionable whether ancon sheep, despite their attractiveness to farmers, would propagate successfully in the presence of their long-legged relatives without human intervention. Recent work in evolutionary developmental genetics (p. 165) illustrates the continuing controversy surrounding phenotypic gradualism.

Punctuated Equilibrium

When we view Darwinian gradualism on a geological timescale, we may expect to find in the fossil record a long series of intermediate forms connecting the phenotypes of ancestral and descendant populations (Figure 6-24). This predicted pattern is called **phyletic gradualism.** Darwin recognized that phyletic gradualism is not often revealed by the fossil record. Studies conducted since Darwin's time generally have not revealed the continuous series of fossils predicted by phyletic gradualism. Is the theory of gradualism therefore refuted by the fossil record? Darwin and others claim that it is not, because the fossil record is too imperfect to preserve transitional series. Although evolution is a slow process by our standards, it is rapid relative to the rate at which good fossil deposits accumulate. Others have argued, however, that abrupt origins and extinctions of species in the fossil record force us to conclude that phyletic gradualism is rare.

Niles Eldredge and Stephen Jay Gould proposed **punctuated equilibrium** to explain the discontinuous evolutionary changes observed throughout geological time. Punctuated equilibrium states that phenotypic evolution is concentrated in relatively brief events of branching speciation, followed by much longer intervals of evolutionary stasis (Figure 6-25). Speciation is an episodic event, having a duration of approximately 10,000 to 100,000 years. Because species may survive for 5 million to 10 million years, the speciation event is a "geological instant," representing 1% or less of a species' life span. Ten thousand years is plenty of time, however, for Darwinian evolution to accomplish dramatic changes. A small fraction of the evolutionary history of a group therefore contributes most of the morphological evolutionary change that we observe.

Figure 6-23
The ancon breed of sheep arose from a "sporting mutation" that caused dwarfing of legs. Many of his contemporaries criticized Darwin for his claim that such mutations are not important in the process of evolution by natural selection.

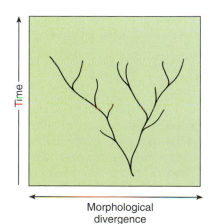

Figure 6-24
The gradualist model of evolutionary change in morphology, viewed as proceeding more or less steadily through geological time (millions of years). Bifurcations followed by gradual divergence led to speciation.

Figure 6-25

The punctuated equilibrium model sees evolutionary change being concentrated in relatively rapid bursts of branching speciation (lateral lines) followed by prolonged periods of no change throughout geological time (millions of years).

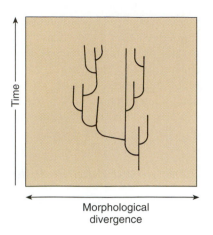

The process of allopatric speciation by founder events provides a possible explanation for punctuated equilibria. Remember that founder-induced speciation requires the breaking of genetic equilibrium in a small, geographically isolated population. Such small populations have very little chance of being preserved in the fossil record. After a new genetic equilibrium forms and stabilizes, the new population may increase in size, thereby increasing the likelihood that some of its members will be preserved as fossils. Founder-induced speciation cannot be the exclusive cause of punctuated equilibrium, however, because punctuated equilibrium may be observed in groups where speciation by founder events is unlikely.

Evolutionists who lamented the imperfect state of the fossil record were treated in 1981 to the opening of an uncensored page of fossil history in Africa. Peter Williamson, a British paleontologist working in fossil beds 400 m deep near Lake Turkana, documented a remarkably clear record of speciation in freshwater snails. The geology of the Lake Turkana basin reveals a history of instability. Earthquakes, volcanic eruptions, and climatic changes caused the waters to rise and fall periodically, sometimes by hundreds of feet. Thirteen lineages of snails show long periods of stability interrupted by relatively brief periods of rapid change in shell shape when snail populations were fragmented by receding waters. These populations diverged to produce new species that then remained unchanged through thick deposits before becoming extinct and being replaced by descendant species. The transitions occurred within 5000 to 50,000 years. In the few meters of sediment where speciation occurred, transitional forms were visible. Williamson's study conforms well to the punctuated equilibrium model of Eldredge and Gould.

Natural Selection

Natural selection is the centerpiece of Darwin's theory of evolution. It gives us a natural explanation for the origins of **adaptation,** including all developmental, behavioral, anatomical, and physiological attributes that enhance the organism's ability to use environmental resources to survive and to reproduce. Darwin developed his theory of natural selection as a series of five observations and three inferences drawn from them:

Observation 1—Organisms have great potential fertility. All populations produce large numbers of gametes and potentially large numbers of offspring each generation. Population size would increase exponentially at an enormous rate if all individuals that were produced each generation survived and reproduced. Darwin calculated that, even in slow-breeding animals such as elephants, a single pair breeding from age 30 to 90 and having only six young could produce 19 million descendants in 750 years.

Observation 2—Natural populations normally remain constant in size, except for minor fluctuations. Natural populations fluctuate in size across generations and sometimes go extinct, but no natural populations show the continued exponential growth that their reproductive biology theoretically could sustain.

Observation 3—Natural resources are limited. Exponential growth of a natural population would require unlimited natural resources to provide food and habitat for the expanding population, but natural resources are finite.

Inference 1—A continuing *struggle for existence* exists among members of a population. Survivors represent only a part, often a very small part, of the individuals produced each generation. Darwin wrote in *On The Origin of Species* that "it is the doctrine of Malthus applied with manifold force to the whole animal and vegetable kingdoms." The struggle for food, shelter, and space becomes increasingly severe as overpopulation develops.

Observation 4—Populations show *variation* among organisms. No two individuals are exactly alike. They differ in size, color, physiology, behavior, and many other ways.

Observation 5—Some variation is heritable. Darwin noted that offspring tend to resemble their parents, although he did not understand how. The hereditary mechanism discovered by Gregor Mendel would be applied to Darwin's theory many years later.

Inference 2—Varying organisms show *differential survival and reproduction* favoring advantageous traits. Survival in the struggle for existence is not random with respect to hereditary variation present in the population. Some traits give their possessors an advantage in using the environment for effective survival and reproduction.

Inference 3—Over many generations, differential survival and reproduction generate new adaptations and new species. The differential reproduction of varying organisms gradually transforms species and results in the long-term "improvement" of types. Darwin knew that people often use hereditary variation to produce useful new breeds of livestock and plants. *Natural* selection acting over millions of years should

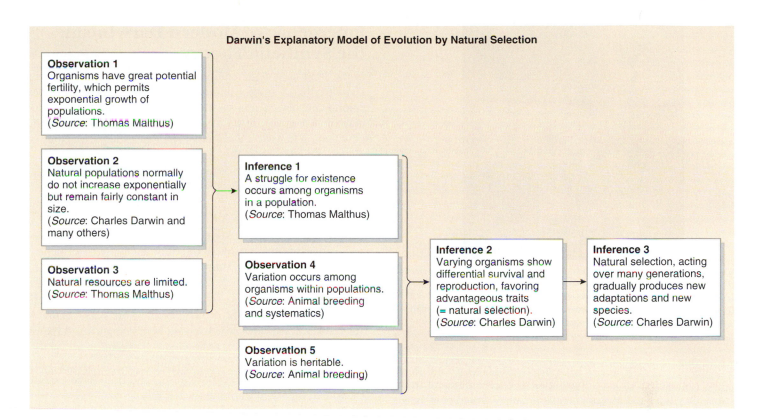

Darwin's Explanatory Model of Evolution by Natural Selection

Observation 1
Organisms have great potential fertility, which permits exponential growth of populations.
(*Source*: Thomas Malthus)

Observation 2
Natural populations normally do not increase exponentially but remain fairly constant in size.
(*Source*: Charles Darwin and many others)

Observation 3
Natural resources are limited.
(*Source*: Thomas Malthus)

Inference 1
A struggle for existence occurs among organisms in a population.
(*Source*: Thomas Malthus)

Observation 4
Variation occurs among organisms within populations.
(*Source*: Animal breeding and systematics)

Observation 5
Variation is heritable.
(*Source*: Animal breeding)

Inference 2
Varying organisms show differential survival and reproduction, favoring advantageous traits (= natural selection).
(*Source*: Charles Darwin)

Inference 3
Natural selection, acting over many generations, gradually produces new adaptations and new species.
(*Source*: Charles Darwin)

be even more effective in producing new types than the *artificial* selection imposed during a human lifetime. Natural selection acting independently on geographically separated populations would cause them to diverge from each other, thereby generating reproductive barriers that lead to speciation.

The popular phrase "survival of the fittest" was not originated by Darwin but was coined a few years earlier by the British philosopher Herbert Spencer, who anticipated some of Darwin's principles of evolution. Unfortunately the phrase later came to be coupled with unbridled aggression and violence in a bloody, competitive world. In fact, natural selection operates through many other characteristics of living organisms. The fittest animal may be one that enhances the living conditions of its population. Fighting prowess is only one of several means toward successful reproductive advantage.

Natural selection can be viewed as a two-step process with a random component and a nonrandom component. Production of variation among organisms is the random component. The mutational process does not preferentially generate traits that are favorable to the organism; new variants are probably more likely to be unfavorable. The nonrandom component is the survival of different traits. This differential survival is determined by the effectiveness of different traits in permitting their possessors to use environmental resources to survive and to reproduce. The phenomenon of differential survival and reproduction among varying organisms is now called **sorting** and should not be equated with natural selection. We now know that even random

processes (genetic drift, p. 122) can produce sorting among varying organisms. Selection states that sorting occurs *because certain traits give their possessors advantages in survival and reproduction* relative to others that lack those traits. Selection is therefore a specific cause of sorting.

Darwin's theory of natural selection has been challenged repeatedly. One challenge claims that directed (nonrandom) variation governs evolutionary change. In the decades around 1900, diverse evolutionary hypotheses collectively called **orthogenesis** proposed that variation has momentum that forces a lineage to evolve in a particular direction that is not always adaptive. The extinct Irish elk was a popular example of orthogenesis. Newly produced variation was considered biased toward enlarging their antlers, thereby generating an evolutionary momentum for producing larger antlers. Natural selection was considered ineffective at stopping the antlers eventually from becoming so large and cumbersome that they forced the Irish elk into extinction (Figure 6-26). Orthogenesis explained apparently nonadaptive evolutionary trends that supposedly forced species into decline. Because extinction is the expected evolutionary fate of most species, disappearance of the Irish elk is not extraordinary and probably not related to large antlers. Subsequent genetic research on the nature of variation clearly has rejected the genetic predictions of orthogenesis.

Another recurring criticism of natural selection is that it cannot generate new structures or species but can only modify old ones. Most structures in their early evolutionary stages could not have performed the biological roles that the fully formed structures perform, and it is therefore unclear how natural selection could have favored them. What use is half a wing or the rudiment

Figure 6-26
The Irish elk, a fossil species that once was used to support the orthogenetic idea that momentum in variation caused the antlers to become so large that the species was forced into extinction.

of a feather for a flying bird? To answer this criticism, we propose that many structures evolved initially for purposes different from the ones they have today. Rudimentary feathers would have been useful in thermoregulation, for example. The feathers later became useful for flying after they incidentally acquired some aerodynamic properties. Natural selection then could act to improve the usefulness of feathers for flying. Because structural changes that separate members of different species are similar in kind to variation that we observe within species, it is reasonable to propose that selection can produce new species.

REVISIONS OF DARWIN'S THEORY

Neo-Darwinism

The most serious weakness in Darwin's theory was his failure to identify correctly the mechanism of inheritance. Darwin saw heredity as a blending phenomenon in which the characteristics of parents melded together in their offspring. Darwin also invoked the Lamarckian hypothesis that an organism could alter its heredity through use and disuse of body parts and through the direct influence of the environment. August Weismann rejected Lamarckian inheritance by showing experimentally that modifications of an organism during its lifetime do not change its heredity (see Chapter 5), and he revised Darwin's theory accordingly. We now use the term **neo-Darwinism** to denote Darwin's theory as revised by Weismann.

Mendelian genetics eventually clarified the particulate inheritance that Darwin's theory of natural selection required (p. 79). Ironically, when Mendel's work was rediscovered in 1900, it was viewed as antagonistic to Darwin's theory of natural selection. When mutations were discovered in the early 1900s, most geneticists thought that they produced new species in single large steps. These geneticists relegated natural selection to the role of executioner, a negative force that merely eliminated the obviously unfit.

Emergence of Modern Darwinism: the Synthetic Theory

In the 1930s a new generation of geneticists began to reevaluate Darwin's theory from a different perspective. These were population geneticists, scientists who studied variation in natural populations using statistics and mathematical models. Gradually, a new comprehensive theory emerged that brought together population genetics, paleontology, biogeography, embryology, systematics, and animal behavior in a Darwinian framework.

Population geneticists study evolution as a change in the genetic composition of populations. With the establishment of population genetics, evolutionary biology became divided into two different subfields. **Microevolution** pertains to evolutionary changes in frequencies of different allelic forms of genes (p. 76) within populations. **Macroevolution** refers to evolution on a grand scale, encompassing the origins of new organismal structures and designs, evolutionary trends, adaptive radiation, phylogenetic relationships of species, and mass extinction. Macroevolutionary research is based in systematics and the comparative method (p. 196). Following the evolutionary synthesis, both macroevolution and microevolution have operated firmly within the tradition of neo-Darwinism, and both have expanded Darwinian theory in important ways.

MICROEVOLUTION: GENETIC VARIATION AND CHANGE WITHIN SPECIES

Microevolution is the study of genetic change occurring within natural populations. The observation of different allelic forms of a gene in a population is called **polymorphism.** All alleles of all genes possessed by members of a population collectively form the **gene pool** of that population. The amount of polymorphism present in large populations is potentially enormous, because at observed mutation rates, many different alleles are expected for all genes.

Population geneticists study polymorphism by identifying the different allelic forms of a gene that are present in a population and then measuring the relative frequencies of the different alleles in the population. The relative frequency of a particular allelic form of a gene in a population is known as its **allelic frequency.** For example, in the human population, there are three different allelic forms of the gene encoding the ABO blood types (p. 747). Using the symbol I to denote the gene encoding the ABO blood types, I^A and I^B denote genetically codominant alleles encoding blood types A and B, respectively. Allele i is a recessive allele encoding blood group O. Therefore genotypes $I^A I^A$ and $I^A i$ produce type A blood, genotypes $I^B I^B$ and $I^B i$ produce type B blood, genotype $I^A I^B$ produces type AB blood, and genotype ii produces type O blood. Because each individual contains two copies of this gene, the total number of copies present in the population is twice the number of individuals. What fraction of this total is represented by each of the three different allelic forms? In France, we find the following

Figure 6-27

Frequencies of the blood-type B allele among humans in Europe. The allele is more common in the east and rarer in the west. The allele may have arisen in the east and gradually diffused westward through the genetic continuity of human populations. This allele has no known selective advantage; its changing frequency probably represents the effects of random genetic drift.

allelic frequencies: I^A = .46, I^B = .14, and i = .40. In Russia, the corresponding allelic frequencies differ (I^A = .38, I^B = .28, and i = .34), demonstrating microevolutionary divergence between these populations (see Figure 6-27). Although alleles I^A and I^B are dominant to i, i is nearly as frequent as I^A and exceeds the frequency of I^B in both populations. Dominance describes the *phenotypic effect* of an allele in heterozygous individuals, not its relative abundance in a population of individuals. We will demonstrate that Mendelian inheritance and dominance do not alter allelic frequencies directly or produce evolutionary change in a population.

Genetic Equilibrium

In many human populations, genetically recessive traits, including the O blood type, blond hair, and blue eyes, are very common. Why have not the genetically dominant alternatives gradually supplanted these recessive traits? It is a common misconception that a characteristic associated with a dominant allele increases in frequency because of its genetic dominance. This misconception is refuted by a principle called **Hardy-Weinberg equilibrium** (see box), which forms the foundation for population genetics. According to this theorem, the hereditary process alone does not produce evolutionary change. In large biparental populations, allelic frequencies and genotypic ratios attain an equilibrium in one generation and *remain constant* thereafter *unless* disturbed by recurring mutations, natural selection, migration, nonrandom mating, or genetic drift (random sorting). Such disturbances are the sources of microevolutionary change.

A rare allele, according to this principle, does not disappear from a large population merely because it is rare. Certain rare traits, such as albinism and cystic fibrosis, persist for endless generations. For example, albinism in humans is caused by a rare recessive allele a. Only one person in 20,000 is an albino, and this individual must be homozygous (a/a) for the recessive allele. Obviously the population contains many carriers, people with normal pigmentation who are heterozygous (A/a) for albinism. What is their frequency? A convenient way to calculate the frequencies of genotypes in a population is with the binomial expansion of $(p + q)^2$ (see box). We will let p represent the allelic frequency of A and q the allelic frequency of a.

Assuming that mating is random (a questionable assumption, but one that we will accept for our example), the distribution of genotypic frequencies is p^2 = A/A, $2pq$ = A/a, and q^2 = a/a. Only the frequency of genotype a/a is known with certainty, 1/20,000; therefore:

$$q^2 = 1/20,000$$
$$q = (1/20,000)^{1/2} = 1/141$$
$$p = 1 - q = 140/141$$

The frequency of carriers is:

$$A/a = 2pq = 2 \times 140/141 \times 1/141 = 1/70$$

Hardy-Weinberg Equilibrium: Why the Hereditary Process Does Not Change Allelic Frequencies

The Hardy-Weinberg law is a logical consequence of Mendel's first law of segregation and expresses the tendency toward equilibrium inherent in Mendelian heredity.

Let us select for our example a population having a single locus bearing just two alleles T and t. The phenotypic expression of this gene might be, for example, the ability to taste a chemical compound called phenylthiocarbamide. Individuals in the population will be of three genotypes for this locus, T/T, T/t (both tasters), and t/t (nontasters). In a sample of 100 individuals, let us suppose we have determined that there are 20 of T/T genotype, 40 of T/t genotype, and 40 of t/t genotype. We could then make a table showing the allelic frequencies (remember that every individual has two copies of the gene):

Genotype	Number of Individuals	Copies of the T Allele	Copies of the t Allele
T/T	20	40	
T/t	40	40	40
t/t	40		80
Total	100	80	120

Of the 200 copies, the proportion of the T allele is 80/200 = 0.4 (40%); and the proportion of the t allele is 120/200 = 0.6 (60%). It is customary in presenting this equilibrium to use "p" and "q" to represent the two allelic frequencies. The genetically dominant allele is represented by p, and the genetically recessive by q. Thus:

$$p = \text{frequency of } T = 0.4$$
$$q = \text{frequency of } t = 0.6$$
$$\text{Therefore } p + q = 1$$

Having calculated allelic frequencies in the sample, let us determine whether these frequencies will change spontaneously in a new generation of the population. Assuming that mating is random (and this is important; all mating combinations of genotypes must be equally probable), each individual will contribute an equal number of gametes to the "common pool" from which the next generation is formed. Frequencies of gametes in the "pool" then will be

proportional to allelic frequencies in the sample: 40% of the gametes will be T, and 60% will be t (ratio of 0.4:0.6). Both ova and sperm will, of course, show the same frequencies. The next generation is formed:

Sperm	Ova	
	$T = 0.4$	$t = 0.6$
$T = 0.4$	$T/T = 0.16$	$T/t = 0.24$
$t = 0.6$	$T/t = 0.24$	$t/t = 0.36$

Collecting genotypes, we have:

$$\text{frequency of } T/T = 0.16$$
$$\text{frequency of } T/t = 0.48$$
$$\text{frequency of } t/t = 0.36$$

Next, we determine the values of p and q from the randomly mated populations. From the table above, we see that the frequency of T will be the sum of genotypes T/T, which is 0.16, and one-half of the genotype T/t, which is 0.24:

$$T(p) = 0.16 + .5(0.48) = 0.4$$

Similarly, the frequency of t will be the sum of genotypes t/t, which is 0.36, and one-half the genotype T/t, which is 0.24:

$$t(p) = 0.36 + .5(0.48) = 0.6$$

The new generation bears exactly the same allelic frequencies as the parent population! Note that there has been no increase in the frequency of the genetically dominant allele T. Thus *in a freely interbreeding, sexually reproducing population, the frequency of each allele would remain constant generation after generation in the absence of natural selection, migration, recurring mutation and genetic drift* (see text). The more mathematically minded reader will recognize that the genotype frequencies T/T, T/t, and t/t are actually a binomial expansion of $(p + q)^2$:

$$(p + q)^2 = p^2 + 2pq + q^2 = 1$$

One person in every 70 is a carrier! Although a recessive trait may be rare, it is amazing how common a recessive allele may be in a population. There is a message here for anyone proposing to eliminate a "bad" recessive allele from a population by controlling reproduction. It is practically impossible. Because only the homozygous recessive individuals reveal the phenotype against which artificial selection could act (by sterilization, for example), the allele would persist through heterozygous carriers. For a recessive allele present in 2 of every 100 persons (but homozygous in only 1 in 10,000 persons), it would require 50 generations of complete selection against the homozygotes just to reduce its frequency to one in 100 persons.

How Genetic Equilibrium Is Upset

Genetic equilibrium is disturbed in natural populations by (1) random genetic drift, (2) nonrandom mating, (3) recurring mutation, (4) migration, (5) natural selection, and interactions among

these factors. Recurring mutation is the ultimate source of variability in all populations, but it usually requires interaction with one or more of the other factors to upset genetic equilibrium. We will look at these other factors individually.

Genetic Drift

Some species, such as cheetahs (Figure 6-28), contain very little genetic variation, probably because their ancestral lineages passed through periods when the total number of individuals in the population was very small. A small population clearly cannot contain large amounts of genetic variation. Each individual organism has at most two different allelic forms of each gene, and a single breeding pair contains at most four different allelic forms of each gene. Suppose that we have such a breeding pair. We know from Mendelian genetics (Chapter 5) that chance decides which of the different allelic forms of a gene gets passed to offspring. It is therefore possible by chance alone that one or two of the parental alleles in this example will not be passed to

Figure 6-28

The cheetah, a species whose genetic variability has been depleted to very low levels because of small population size in the past.

any offspring. It is highly unlikely that the different alleles present in a small ancestral population are all passed to descendants without any change of allelic frequency. This chance fluctuation in allelic frequency from one generation to the next, including loss of alleles from the population, is called **genetic drift.**

Genetic drift occurs to some degree in all populations of finite size. Perfect constancy of allelic frequencies, as predicted by Hardy-Weinberg equilibrium, occurs only in infinitely large populations, and such populations occur only in mathematical models. All populations of animals are finite and therefore experience some effect of genetic drift, which becomes greater, on average, as population size declines. Genetic drift erodes genetic variability of a population. If population size remains small for many generations in a row, genetic variation can be greatly depleted. This loss is harmful to a species' evolutionary success because it restricts potential genetic responses to environmental change. Indeed, biologists are concerned that cheetah populations may have insufficient variation for continued survival.

Nonrandom Mating

If mating is nonrandom, genotypic frequencies will deviate from the Hardy-Weinberg expectations. For example, if two different alleles of a gene are equally frequent ($p = q = .5$), we expect half of the genotypes to be heterozygous ($2pq = 2 [.5] [.5] = .5$) and one-quarter to be homozygous for each of the respective alleles ($p^2 = q^2 = [.5]^2 = .25$). If we have **positive assortative mating,** individuals mate preferentially with others of the same genotype, such as albinos mating with other albinos. Matings among individuals homozygous for the same allele generate offspring that are homozygous like themselves. Matings among individuals heterozygous for the same pair of alleles produce on average 50% heterozygous offspring and 50% homozygous offspring (25% of each alternative type) each generation. Positive assortative mating increases the frequency of homozygous genotypes

and decreases the frequency of heterozygous genotypes in the population but does not change allelic frequencies.

Preferential mating among close relatives also increases homozygosity and is called **inbreeding.** Whereas positive assortative mating usually affects one or a few traits, inbreeding simultaneously affects all variable traits. Strong inbreeding greatly increases chances that rare recessive alleles will become homozygous and be expressed.

Because inbreeding and genetic drift are both promoted by small population size, they are often confused with each other. Their effects are very different, however. Inbreeding alone cannot change allelic frequencies in the population, only the ways that alleles are combined into genotypes. Genetic drift changes allelic frequencies and consequently also changes genotypic frequencies. Even very large populations have the potential for being highly inbred if there is a behavioral preference for mating with close relatives, although this situation rarely occurs in nature. Genetic drift, however, will be relatively weak in very large populations.

Migration

Migration prevents different populations of a species from diverging. If a large species is divided into many small populations, genetic drift and selection acting separately in the different populations can produce evolutionary divergence among them. A small amount of migration each generation keeps the different populations from becoming too distinct genetically. For example, the French and Russian populations whose ABO allele frequencies were discussed previously show some genetic divergence, but continuing migration between them prevents them from becoming completely distinct.

Natural Selection

Natural selection can change both allelic frequencies and genotypic frequencies in a population. Although the effects of selection are often reported for particular polymorphic genes, we must stress that natural selection acts on the whole animal, not on isolated traits. An organism that possesses a superior combination of traits will be favored. An animal may have traits that confer no advantage or even a disadvantage, but it is successful overall if its combination of traits is favorable. When we claim that a genotype at a particular gene has a higher **relative fitness** than others, we state that on average that genotype confers an advantage in survival and reproduction in the population. If alternative genotypes have unequal probabilities of survival and reproduction, the Hardy-Weinberg equilibrium will be upset.

Some traits and combinations of traits are advantageous for certain aspects of an organism's survival or reproduction and disadvantageous for others. Darwin used the term **sexual selection** to denote the selection of traits that are advantageous for obtaining mates but may be harmful for survival. Bright colors and elaborate feathers may enhance a male bird's competitive ability in obtaining mates while simultaneously increasing his vulnerability to predators (Figure 6-29). Changes in the environment can alter the selective value of different traits. The action of selection on character variation is therefore very complex.

Figure 6-29

A pair of wood ducks. Brightly-colored feathers of male birds probably confer no survival advantage and might even be harmful by alerting predators. Such colors nonetheless confer advantage in attracting mates, which overcomes, on average, the negative consequences of these colors for survival. Darwin used the term "sexual selection" to denote traits that give an individual an advantage in attracting mates, even if the traits are neutral or harmful for survival.

Interactions of Selection, Drift, and Migration

Subdivision of a species into small populations that exchange migrants is an optimal situation for promoting rapid adaptive evolution of the species as a whole. Interaction of genetic drift and selection in different populations permits many different genetic combinations of many polymorphic genes to be tested against natural selection. Migration among populations permits particularly favorable new genetic combinations to spread throughout the species as a whole. Interaction of selection, genetic drift, and migration in this example produces evolution-

ary change that is qualitatively different from what would result if any of these three factors acted alone. Natural selection, genetic drift, mutation, nonrandom mating, and migration interact in natural populations to create an enormous opportunity for evolutionary change; perpetual stability, as predicted by Hardy-Weinberg equilibrium, almost never occurs across any significant amount of evolutionary time.

Measuring Genetic Variation within Populations

How do we measure the genetic variation that occurs in natural populations? Genetic dominance, interactions between alleles of different genes, and environmental effects on a phenotype make it difficult to quantify genetic variation indirectly by observing organismal phenotypes. Variability can be quantified, however, at the molecular level.

Protein Polymorphism

Different allelic forms of genes encode proteins that may differ slightly in their amino acid sequence. This phenomenon is called **protein polymorphism.** If these differences affect the protein's net electric charge, the different allelic forms can be separated using protein electrophoresis (Figure 6-30). We can identify the genotypes of particular individuals for protein-coding genes and measure allelic frequencies in the population.

Over the last 35 years, geneticists using this approach have discovered far more variation than was previously expected. Despite the high levels of polymorphism discovered using protein electrophoresis (Table 6-1), these studies underestimate both protein polymorphism and the total genetic variation present in a population. For example, protein polymorphism that does not involve charge differences is not detected. Further-

Figure 6-30

The study of genetic variation in proteins using gel electrophoresis. **A,** An electrophoretic apparatus separates allelic variants of proteins that differ in charge because of differences in their sequence of amino acids. **B,** Genetic variation in the protein leucine aminopeptidase for nine brown snails, *Helix aspersa.* Two different sets of allelic variants are revealed. The top set contains two alleles [denoted fast (F) and slow (S) according to their relative movement in the electric field]. Individuals homozygous for the fast allele show only a single fast band on the gel (FF), those homozygous for the slow allele show only a single slow band (SS), and heterozygous individuals have both bands (FS). The lower set contains three different alleles denoted fast (F), medium (M), and slow (S). Note that no individuals shown are homozygous for the medium (M) allele.

more, because the genetic code is degenerate (more than one codon for most amino acids, p. 92), protein polymorphism does not reveal all of the genetic variation present in protein-coding genes. Genetic changes that do not alter protein structure may alter patterns of protein synthesis during development and can be very important to an organism. When all kinds of variation are considered, it is evident that most species have an enormous potential for further evolutionary change.

Quantitative Variation

Quantitative traits are those that show continuous variation with no obvious pattern of Mendelian segregation in their inheritance. The values of the trait in offspring often are intermediate between the values in the parents. Such traits are influenced by variation at many genes, each of which follows Mendelian inheritance and contributes a small, incremental amount to the total phenotype. Examples of traits that show quantitative variation include tail length in mice, length of a leg segment in grasshoppers, number of gill rakers in sunfishes, number of peas in pods, and height of adult males of the human species. When the values are graphed with respect to frequency distribution, they often approximate a normal, or bell-shaped, probability curve (Figure 6-31A). Most individuals fall near the average; fewer fall somewhat above or below the average, and extremes form the "tails" of the frequency curve with increasing rarity. Usually, the

larger the population sample, the more closely the frequency distribution resembles a normal curve.

Selection can act on quantitative traits to produce three different kinds of evolutionary response (see Figure 6-31B, C, and D). One outcome is to favor average values of the trait and to disfavor extreme ones; this outcome is called **stabilizing selection** (Figure 6-31B). **Directional selection** favors an extreme value of the phenotype and causes the population average to shift toward it over time (Figure 6-31C). When we think about natural selection producing evolutionary change, it is usually directional selection that we have in mind, although we must remember that this is not the only possibility. A third alternative is **disruptive selection** in which two different extreme phenotypes are simultaneously favored, but the average is disfavored (Figure 6-31D). The population will become bimodal, meaning that two very different phenotypes will predominate.

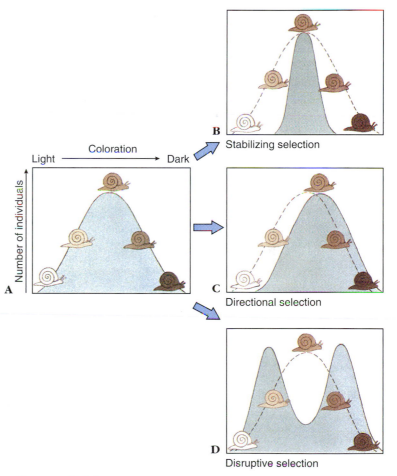

Figure 6-31

Responses to selection on a continuous (polygenic) character, coloration in a snail. **A,** The frequency distribution of coloration before selection. **B,** Stabilizing selection culls extreme variants from the population, in this case eliminating individuals that are unusually light or dark, thereby stabilizing the mean. **C,** Directional selection shifts the population mean, in this case by favoring darkly colored variants. **D,** Disruptive selection favors both extremes but not the mean; the mean is unchanged but the population no longer has a bell-shaped distribution of phenotypes.

(a) Species	Number of Proteins	P*	H*
Humans	71	0.28	0.067
Northern elephant seal	24	0.0	0.0
Horseshoe crab	25	0.25	0.057
Elephant	32	0.29	0.089
Drosophila pseudoobscura	24	0.42	0.12
Barley	28	0.30	0.003
Tree frog	27	0.41	0.074
(b) Taxa	Number of Species	P*	H*
Plants	—	0.31	0.10
Insects (excluding *Drosophila*)	23	0.33	0.074
Drosophila	43	0.43	0.14
Amphibians	13	0.27	0.079
Reptiles	17	0.22	0.047
Birds	7	0.15	0.047
Mammals	46	0.15	0.036
Average		0.27	0.078

TABLE 6.1

Values of Polymorphism (P) and Heterozygosity (H) for Various Animals and Plants as Measured Using Protein Electrophoresis

Source: Data from P.W. Hedrick, *Population biology.* Jones and Bartlett, Boston, 1984.
*P, the average number of alleles per gene per species; H, the proportion of heterozygous genes per individual.

MACROEVOLUTION: MAJOR EVOLUTIONARY EVENTS

Macroevolution describes large-scale events in organic evolution. Speciation links macroevolution and microevolution. Major trends in the fossil record (see Figures 6-11 and 6-12) are clearly within the realm of macroevolution. Patterns and processes of macroevolutionary change emerge from those of microevolution, but they acquire some degree of autonomy in doing so. The emergence of new adaptations and species, and the varying rates of speciation and extinction observed in the fossil record go beyond the fluctuations of allelic frequencies within populations.

Stephen Jay Gould recognized three different "tiers" of time at which we observe distinct evolutionary processes. The first tier constitutes the timescale of population genetic processes, from tens to thousands of years. The second tier covers millions of years, the scale on which rates of speciation and extinction can be measured and compared among different groups of organisms. The third tier covers tens to hundreds of millions of years, and is marked by occurrence of periodic mass extinctions. In the fossil record of marine organisms, mass extinctions recur at intervals of approximately 26 million years. Five of these mass extinctions have been particularly disastrous (Figure 6-32). The study of long-term changes in animal diversity focuses on the third-tier timescale (see Figures 6-12 and 6-32).

Speciation and Extinction through Geological Time

Evolutionary change at the second tier provides a new perspective on Darwin's theory of natural selection. A species has two possible evolutionary fates: it may give rise to new species or become extinct without leaving descendants. Rates of speciation and extinction vary among lineages, and lineages that have the highest speciation rates and lowest extinction rates produce the greatest diversity of living forms. The characteristics of a species may make it more or less likely than others to undergo speciation or extinction events. Because many characteristics are passed from ancestral to descendant species (analogous to heredity at the organismal level), lineages whose characteristics increase the probability of speciation and confer resistance to extinction should come to dominate the living world. This species-level process that produces differential rates of speciation and extinction among lineages is analogous in many ways to natural selection. It represents an expansion of Darwin's theory of natural selection.

Species selection encompasses the differential survival and multiplication of species through geological time based on variation among lineages, especially in emergent, species-level properties. These species-level properties include mating rituals, social structuring, migration patterns, geographic distribution, and all other properties that emerge at the species level (see p. 6). Descendant species usually resemble their ancestors in these properties. For example, a "harem" system of mating in which a single male and several females compose a breeding unit characterizes some mammalian lineages but not others. We expect speciation rates to be enhanced by social systems that promote founding of new populations by small numbers of individuals. Certain social systems may increase the likelihood that a species will survive environmental challenges through cooperative action. Such properties would be favored by species selection over geological time.

Differential speciation and extinction among lineages also may be caused by variation in organismal-level properties (such as specialized versus generalized feeding) rather than species-level properties (see p. 6). Organisms that specialize in eating a restricted range of foods, for example, may be subjected more readily than generalized feeders to geographic isolation among populations, because areas where their preferred food is scarce or absent will function as geographic barriers to dispersal. Such geographic isolation could generate more frequent opportunities

Figure 6-32

Changes in numbers of families of marine animals through time from the Cambrian period to the present. Sharp drops represent five major extinctions of skeletonized marine animals. Note that despite the extinctions, the overall number of marine families has increased to the present.

for speciation to occur throughout geological time. The fossil records of two major groups of African antelopes demonstrate this result (Figure 6-33). A lineage of specialized grazers that contains blesboks, hartebeests, and wildebeests shows high speciation and extinction rates. Since the late Miocene, 33 extinct and 7 living species are found, representing at least 18 events of branching speciation and 12 terminal extinctions. In contrast, a lineage of generalist grazers and browsers that contains impalas shows neither branching speciation nor terminal extinction during this same interval of time. Interestingly, although these two lineages differ greatly in speciation rates, extinction rates, and species diversity, they do not differ significantly in total number of individual animals alive today.

Paleontologist Elisabeth Vrba, whose research produced the results in Figure 6-33, uses the term **effect macroevolution** to describe differential speciation and extinction rates among lineages caused by organismal-level properties. She reserves the term **species selection** for cases where species-level emergent properties are of primary importance. Some other evolutionary paleontologists consider effect macroevolution a subset of species selection because fitness differences occur among different species lineages rather than among varying organisms within species. Our coverage uses species selection in this more inclusive sense.

Mass Extinctions

When we study evolutionary change on an even larger timescale, we observe periodic events in which large numbers of taxa go extinct simultaneously. These events are called **mass extinctions** (see Figure 6-32). The most cataclysmic of these extinction episodes happened about 225 million years ago, when at least half of the families of shallow-water marine invertebrates, and fully 90% of marine invertebrate species disappeared within a few million years. This event was the **Permian extinction.** The **Cretaceous extinction,** which occurred about 65 million years ago, marked the end of the dinosaurs, as well as numerous marine invertebrates and many small reptilian taxa.

The causes of mass extinctions and their occurrence at intervals of approximately 26 million years are difficult to explain. Some people have proposed biological explanations for these periodic mass extinctions and others consider them artifacts of our statistical and taxonomic analyses. Walter Alvarez proposed that the earth was periodically bombarded by asteroids, causing these mass extinctions (Figure 6-34). The drastic effects of such bombardment of a planet were observed in July 1994 when fragments of Comet Shoemaker-Levy 9 bombarded Jupiter. The first fragment to hit Jupiter was estimated to have the force of 10 million hydrogen bombs. Twenty additional fragments hit Jupiter within the following week, one of which was 25 times more powerful than the first fragment. This bombardment was the most violent event in

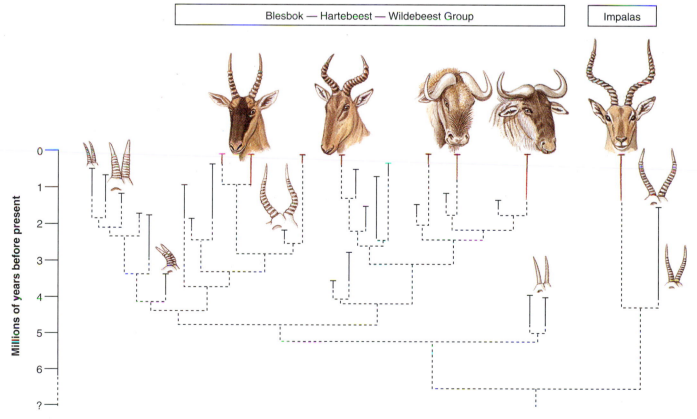

Figure 6-33
Contrasting diversity between two major groups of African antelopes. Higher speciation and extinction rates in the group containing the blesboks, hartebeests, and wildebeests are attributed to greater specialization in feeding relative to the impalas, an example of effect macroevolution, a kind of species selection.

the recorded history of the solar system. A similar bombardment on earth would send debris into the atmosphere, blocking sunlight and causing drastic changes of climate. Temperature changes would challenge ecological tolerances of many species. Alvarez's hypothesis is being tested in several ways, including a search for impact craters left by asteroids and for altered mineral content of rock strata where mass extinctions occurred. Atypical concentrations of the rare-earth element iridium in strata at the Cretaceous-Tertiary boundary imply that this element entered the earth's atmosphere through asteroid bombardment.

Sometimes, lineages favored by species selection are unusually susceptible to mass extinction. Climatic changes produced by the hypothesized asteroid bombardments could produce selective challenges very different from those encountered at other times in the earth's history. Selective discrimination of particular biological traits by events of mass extinction is termed **catastrophic species selection.** For example, mammals survived the end-Cretaceous mass extinction that destroyed the dinosaurs and other prominent vertebrate and invertebrate groups. Following this event, mammals were able to use environmental resources that previously had been denied them, leading to their adaptive radiation.

Natural selection, species selection, effect macroevolution, and catastrophic species selection interact to produce the macroevolutionary trends seen in the fossil record. Studies of these interacting causal processes have made modern evolutionary paleontology an active and exciting field.

Figure 6-34

Twin craters of Clearwater Lakes in Canada show that multiple impacts on the earth are not as unlikely as they might seem. Evidence suggests that at least two impacts within a short time were responsible for the Cretaceous mass extinction.

SUMMARY

Organic evolution explains the diversity of living organisms as the historical outcome of gradual change from previously existing forms. Evolutionary theory is strongly identified with Charles Robert Darwin, who presented the first credible explanation for evolutionary change. Darwin derived much of the material used to construct his theory from his experiences on a five-year voyage around the world aboard the H.M.S. *Beagle.*

Darwin's evolutionary theory has five major components. Its most basic proposition is *perpetual change,* the theory that the world is neither constant nor perpetually cycling but is steadily undergoing irreversible change. The fossil record amply demonstrates perpetual change in the continuing fluctuation of animal form and diversity following the Cambrian explosion 600 million years ago. Darwin's theory of *common descent* states that all organisms descend from a common ancestor through a branching of genealogical lineages. This theory explains morphological homologies among organisms as characteristics inherited with modification from a corresponding feature in their common evolutionary ancestor. Patterns of homology formed by common descent with modification permit us to classify organisms according to their evolutionary relationships.

A corollary of common descent is the *multiplication of species* through evolutionary time. Allopatric speciation describes the evolution of reproductive barriers between geographically separated populations to generate new species. In some animals, especially parasitic insects that specialize on different host species, speciation may occur without geographical isolation, which is known as sympatric speciation. Adaptive radiation is the proliferation of many adaptively diverse species from a single ancestral lineage. Oceanic archipelagoes, such as the Galápagos Islands, are particularly conducive to adaptive radiation of terrestrial organisms.

Darwin's theory of *gradualism* states that large phenotypic differences between species are produced by accumulation through evolutionary time of many individually small changes. Gradualism is still controversial. Mutations that have large effects on an organism have been useful in animal breeding, leading some to dispute Darwin's claim that such mutations are not important in evolution. On a macroevolutionary perspective, punctuated equilibrium states that most evolutionary change occurs in relatively brief events of branching speciation, separated by long intervals in which little phenotypic change accumulates.

Darwin's fifth major statement is that *natural selection* is the guiding force of evolution. This principle is founded on observations that all species overproduce their kind, causing a struggle for the limited resources that support existence. Because no two organisms are exactly alike, and because variable traits are at least partially heritable, those whose hereditary endowment enhances their use of resources for survival and reproduction contribute disproportionately to the next generation. Over many generations, the sorting of variation by selection produces new species and new adaptations.

Mutations are the ultimate source of all new variation on which selection acts. Darwin's theory emphasizes that variation is produced at random with respect to an organism's needs and that differential survival and reproduction provide the direction for evolutionary change. Darwin's theory of natural selection was modified in this century by correction of his genetic errors. This modified theory became known as neo-Darwinism.

Population geneticists discovered the principles by which genetic properties of populations change through time. A particularly important discovery, known as Hardy-Weinberg equilibrium, showed that the hereditary process itself does not change the genetic composition of populations. Important sources of evolutionary change include mutation, genetic drift, nonrandom mating, migration, natural selection, and their interactions.

Neo-Darwinism, as elaborated by population genetics, formed the basis for the Evolutionary Synthesis of the 1930s and 1940s. Genetics, natural history, paleobiology, and systematics were unified by the common goal of expanding our knowledge of Darwinian evolution. Microevolution comprises studies of genetic change within contemporary populations. These studies show that most natural populations contain enormous amounts of variation. Macroevolution comprises studies of evolutionary change on a geological timescale. Macroevolutionary studies measure rates of speciation, extinction, and changes of diversity through time. These studies have expanded Darwinian evolutionary theory to include higher-level processes that regulate rates of speciation and extinction among lineages, including species selection and catastrophic species selection.

REVIEW QUESTIONS

1. Briefly summarize Lamarck's concept of the evolutionary process. What is wrong with this concept?
2. What is "uniformitarianism"? How did it influence Darwin's evolutionary theory?
3. Why was the *Beagle*'s journey so important to Darwin's thinking?
4. What was the key idea contained in Malthus's essay on populations that was to help Darwin formulate his theory of natural selection?
5. Explain how each of the following contributes to Darwin's evolutionary theory: fossils; geographic distributions of closely related animals; homology; animal classification.
6. How do modern evolutionists view the relationship between ontogeny and phylogeny? Explain how the observation of paedomorphosis conflicts with Haeckel's "biogenetic law."
7. What are the important differences between the vicariant and founder-effect modes of allopatric speciation?
8. What are reproductive barriers? How do premating and postmating barriers differ?
9. Under what conditions is sympatric speciation proposed?
10. What is the main evolutionary lesson provided by Darwin's finches on the Galápagos Islands?
11. How is the observation of "sporting mutations" in animal breeding used to challenge Darwin's theory of gradualism? Why did Darwin reject such mutations as having little evolutionary importance?
12. What does the theory of punctuated equilibrium state about the occurrence of speciation throughout geological time? What observation led to this theory?
13. Describe the observations and inferences that compose Darwin's theory of natural selection.
14. Identify the random and nonrandom components of Darwin's theory of natural selection.
15. Describe some recurring criticisms of Darwin's theory of natural selection. How can these criticisms be refuted?
16. It is a common but mistaken belief that because some alleles are dominant and others are recessive, the dominants will eventually replace all the recessives in a population. How does the Hardy-Weinberg equilibrium refute this notion?
17. Assume that you are sampling a trait in animal populations; the trait is controlled by a single allelic pair A and a, and you can distinguish all three phenotypes AA, Aa, and aa (intermediate inheritance). Your sample includes:

Population	AA	Aa	aa	TOTAL
I	300	500	200	1000
II	400	400	200	1000

Calculate the distribution of phenotypes in each population as expected under Hardy-Weinberg equilibrium. Is population I in equilibrium? Is population II in equilibrium?
18. If after studying a population for a trait determined by a single pair of alleles you find that the population is not in equilibrium, what possible reasons might explain the lack of equilibrium?
19. Explain why genetic drift is more powerful in small populations.
20. Describe how the effects of genetic drift and natural selection can interact in a subdivided species.
21. Is it easier for selection to remove a deleterious recessive allele from a randomly mating population or a highly inbred population? Why?
22. Distinguish between microevolution and macroevolution.

SELECTED REFERENCES

Avise, J. C. 2000. Phylogeography: the history and formation of species. Cambridge, Massachusetts, Harvard University Press. *An exciting and readable account of using molecular studies to help us understand speciation.*

Darwin, C. 1859. On the origin of species by means of natural selection, or the preservation of favoured races in the struggle for life. London, John Murray. *There were five subsequent editions by the author.*

Desmond, A., and J. Moore. 1991. Darwin. New York, Warner Books. *An interpretive biography of Charles Darwin.*

Freeman, S., and J. C. Herron. 2001. Evolutionary analysis, ed. 2. Upper Saddle River, New Jersey, Prentice-Hall. *An introductory textbook on evolutionary biology designed for undergraduate biology majors.*

Futuyma, D. J. 1998. Evolutionary biology, ed. 3. Sunderland, Massachusetts, Sinauer Associates. *A very thorough introductory textbook on evolution.*

Glen, W. 1994. The mass extinction debates: how science works in a crisis. Stanford, Stanford University Press. *A discussion of mass extinction presented in the form of a debate and panel discussion among concerned scientists.*

Gould, S. J. 2002. The structure of evolutionary theory. Cambridge, Massachusetts, Belknap Press of Harvard University Press. *An insightful discussion of what fossils tell us about the nature of life's evolutionary history.*

Grauer, D., and W. H. Li. 2000. Fundamentals of molecular evolution. Sunderland, Massachusetts, Sinauer Associates. *A current textbook on molecular evolution.*

Hall, B. K. 1998. Evolutionary developmental biology. New York, Chapman and Hall. *An excellent textbook on the emerging field of developmental evolutionary biology.*

Hartl, D. L., and A. G. Clark. 1997. Principles of population genetics. Sunderland, Massachusetts, Sinauer Associates. *A current textbook on population genetics.*

Levinton, J. S. 2001. Genetics, paleontology and macroevolution, ed. 2. Cambridge, U.K., Cambridge University Press. *A provocative discussion on the Darwinian basis of macroevolutionary theory.*

Magurran, A. E., and R. M. May. 1999. Evolution of biological diversity. Oxford, U.K., Oxford University Press. *An edited volume covering recent issues in the study of speciation, with contributions from many active evolutionary biologists.*

Mayr, E. 2001. What evolution is. New York, Basic Books. *A general survey of evolution by a leading evolutionary biologist.*

Ruse, M. 1998. Philosophy of biology. Amherst, New York, Prometheus Books. *A collection of essays on evolutionary biology, including information on the Arkansas Balanced Treatment for Creation-Science and Evolution-Science Act.*

Stokstad, E. 2001. Exquisite Chinese fossils add new pages to book of life. Science **291**: 232–236. *Exciting new fossil discoveries help to complete our understanding of life's evolutionary history. Some related articles immediately follow this one.*

ZOOLOGY LINKS TO THE INTERNET

Visit the textbook's Online Learning Center at www.mhhe.com/zoology to find live Internet links for each of the topics listed here.

Evolution

Darwinian Evolutionary Theory

Neo-Darwinism

Evidence for Evolution

Natural Selection

Speciation

Punctuated vs Gradualist Evolution

Macroevolution

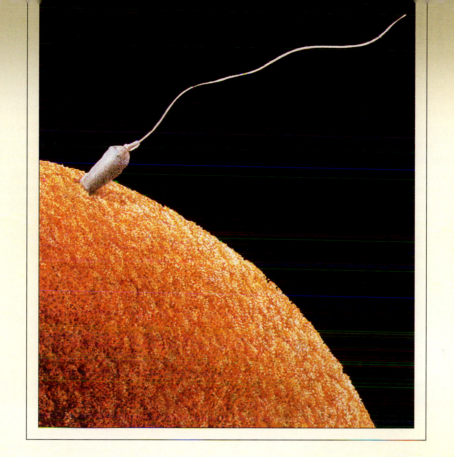

Human egg and sperm
at the moment
of fertilization.

The Reproductive Process

"Omne vivum ex ovo"

In 1651, late in a long life, William Harvey, the English physiologist who earlier had ushered in experimental physiology by explaining the circuit of blood, published a treatise on reproduction. He asserted that all life developed from the egg—*omne vivum ex ovo*. This was curiously insightful, since Harvey had no means for visualizing eggs of many animals, in particular the microscopic mammalian egg, which is no larger than a speck of dust to the unaided eye. Further, argued Harvey, eggs are launched into their developmental course by some influence from semen, a conclusion that was either remarkably perceptive or a lucky guess, since sperm also were invisible to Harvey. Such ideas differed sharply from existing notions of biogenesis, which saw life springing from many sources of which eggs were but one. Harvey was describing characteristics of sexual reproduction in which two parents, male and female, must come together physically to ensure fusion of gametes from each.

Despite the importance of Harvey's aphorism that all life arises from eggs, it was too sweeping to be wholly correct. Life springs from reproduction of preexisting life, and reproduction may not be restricted to eggs and sperm. Nonsexual reproduction, the creation of new, genetically identical individuals by budding or fragmentation or fission from a single parent, is common, indeed characteristic, among some phyla. Most animals have found sex to be the winning strategy, probably because sexual reproduction promotes diversity, enhancing long-term survival of the lineage in a world of perpetual change.

Reproduction is one of the ubiquitous miracles of life. Evolution is inextricably linked to reproduction, because the ceaseless replacement of aging predecessors with new life gives animals the means to respond and evolve in a changing environment as the earth itself has changed over the ages. In this chapter we distinguish asexual and sexual reproduction and explore the reasons why, for multicellular animals at least, sexual reproduction appears to offer important advantages over asexual. We then consider, in turn, the origin and maturation of germ cells; plan of reproductive systems; reproductive patterns in animals; and, finally, the endocrine events that orchestrate reproduction.

NATURE OF THE REPRODUCTIVE PROCESS

Two modes of reproduction are recognized: asexual and sexual. In **asexual** reproduction (Figure 7-1A and B) there is only one parent and there are no special reproductive organs or cells. Each organism is capable of producing genetically identical copies of itself as soon as it becomes an adult. The production of copies is marvelously simple, direct, and typically rapid. **Sexual** reproduction (Figure 7-1C and D) as a rule involves two parents, each of which contributes special **germ cells** (**gametes** or **sex cells**) that in union (fertilization) develop into a new individual. The **zygote** formed from this union receives genetic material from both parents, and the combination of genes produces a genetically unique individual, still bearing characteristics of the species but also bearing traits that make it different from its parents. Sexual reproduction, by recombining parental characters, tends to multiply variations and makes possible a richer and more diversified evolution.

Mechanisms for interchange of genes between individuals are more limited in organisms with only asexual reproduction. Of course, in asexual organisms that are haploid (bear only one set of genes, p. 76), mutations are immediately expressed and evolution can proceed quickly. In sexual animals, on the other hand, a gene mutation is often not expressed immediately, since it may be masked by its normal partner on the homologous chromosome. (Homologous chromosomes, discussed on p. 76, are those that pair during meiosis and carry genes encoding the same characteristics.) There is only a remote chance that both members of a gene pair will mutate in the same way at the same moment.

A   Binary fission in *Euglena*

Barnacles mating

Hydra budding

B

C

Frogs in amplexus

D

Figure 7-1

Examples of asexual and sexual reproduction in animals. **A,** Binary fission in *Euglena*, a single-celled eukaryote, results in two individuals. **B,** Budding, a simple form of asexual reproduction as shown in a hydra, a radiate animal. The buds eventually detach themselves and grow into fully formed individuals. **C,** Barnacles reproduce sexually, but are hermaphroditic, with each individual bearing both male and female organs. Each barnacle possesses a pair of enormously elongated penises—an obvious advantage to a sessile animal—that can be extended many times the length of the body to inseminate another barnacle some distance away. The partner may reciprocate with its own penises. **D,** Frogs, here in mating position (amplexus), represent bisexual reproduction, the most common form of sexual reproduction involving separate male and female individuals.

Asexual Reproduction: Reproduction without Gametes

Asexual reproduction (see Figure 7-1A and B) is the production of individuals without gametes (eggs or sperm). It includes a number of distinct processes, all without involving sex or a second parent. Offspring produced by asexual reproduction all have the same genotype (unless mutations occur) and are called **clones.**

Asexual reproduction appears in bacteria and unicellular eukaryotes and in many invertebrate phyla, such as cnidarians, bryozoans, annelids, echinoderms, and hemichordates. In animal phyla in which asexual reproduction occurs, most members also employ sexual reproduction. In these groups, asexual reproduction ensures rapid increase in numbers when differentiation of the organism has not advanced to the point of forming gametes. Asexual reproduction is absent among vertebrates (although some forms of parthenogenesis have been interpreted as asexual by some authors; see p. 135).

It would be a mistake to conclude that asexual reproduction is in any way a "defective" form of reproduction relegated to the minute forms of life that have not yet discovered the joys of sex. Given the facts of their abundance, that they have persisted on earth for 3.5 billion years, and that they form the base of the food chain on which all higher forms depend, single-celled asexual organisms are both resoundingly abundant and supremely important. For these forms the advantages of asexual reproduction are its rapidity (many bacteria divide every half hour) and simplicity (no germ cells to produce and no time and energy expended in finding a mate).

The basic forms of asexual reproduction are fission (binary and multiple), budding, gemmulation, and fragmentation.

Binary fission is common among bacteria and protozoa (Figure 7-1A). In binary fission the body of the parent divides by mitosis (p. 48) into two approximately equal parts, each of which grows into an individual similar to the parent. Binary fission may be lengthwise, as in flagellate protozoa, or transverse, as in ciliate protozoa. In **multiple fission** the nucleus divides repeatedly before division of the cytoplasm, producing many daughter cells simultaneously. Spore formation, called sporogony, is a form of multiple fission common among some parasitic protozoa, for example, malarial parasites.

Budding is an unequal division of an organism. A new individual arises as an outgrowth (bud) from its parent, develops organs like those of the parent, and then detaches itself. Budding occurs in several animal phyla and is especially prominent in cnidarians (Figure 7-1B).

Gemmulation is the formation of a new individual from an aggregation of cells surrounded by a resistant capsule, called a gemmule. In many freshwater sponges, gemmules develop in the fall and survive the winter in the dried or frozen body of the parent. In spring, the enclosed cells become active, emerge from the capsule, and grow into a new sponge.

In **fragmentation** a multicellular animal breaks into two or more parts, with each fragment capable of becoming a complete individual. Many invertebrates can reproduce asexually by simply breaking into two parts and then regenerating the missing parts of the fragments.

Sexual Reproduction: Reproduction with Gametes

Sexual reproduction is the production of individuals from gametes. It includes **bisexual** (or **biparental**) reproduction as the most common form, involving two separate individuals. **Hermaphroditism** and **parthenogenesis** are less common forms of sexual reproduction that also will be discussed.

Bisexual Reproduction

Bisexual reproduction is the *production of offspring formed by the union of gametes from two genetically different parents* (Figures 7-1C and D, and 7-2). Offspring will thus have a new genotype different from either parent. Individuals sharing parenthood are characteristically of different **sexes,** male and female (there are exceptions among sexually reproducing organisms, such as bacteria and some protozoa in which sexes are lacking). Each has its own reproductive system and produces only one kind of germ cell, spermatozoon or ovum, but never both. Nearly all vertebrates and many invertebrates have separate sexes, and such a condition is called **dioecious** (Gr. *di-*, two, + *oikos,* house). An exception to this is found in individual animals that have both male and female reproductive organs, a condition which is called **monoecious** (Gr. *monos,* single, + *oikos,* house). These animals are called **hermaphrodites** (from a combination of the names of the Greek god Hermes and goddess Aphrodite) and this form of reproduction will be described on p. 134.

Distinctions between male and female are based, not on any differences in parental size or appearance, but on the size and mobility of the gametes they produce. The **ovum** (egg) is produced by the female. Ova are large (because of stored yolk to sustain early development), nonmotile, and produced in relatively small numbers. The **spermatozoon** (sperm) is produced by the male. Sperm are small, motile, and produced in enormous numbers. Each is a stripped-down package of highly condensed genetic material designed for the single purpose of reaching and fertilizing an egg.

There is another crucial event that distinguishes sexual from asexual reproduction: **meiosis,** a distinctive type of gamete-producing nuclear division (described in detail on p. 75). Meiosis differs from ordinary cell division (mitosis) in being a double division. Chromosomes split once, but the cell divides *twice,* producing four cells, each with half the original number of chromosomes (the **haploid** number). Meiosis is followed by **fertilization** in which two haploid gametes are combined to restore the normal (**diploid**) chromosomal number of the species.

The new cell (zygote), which now begins to divide by mitosis, has equal numbers of chromosomes from each parent and accordingly is different from each. It is a unique individual bearing a recombination of parental characteristics. Genetic recombination is the great strength of sexual reproduction that keeps feeding new genetic combinations into the population.

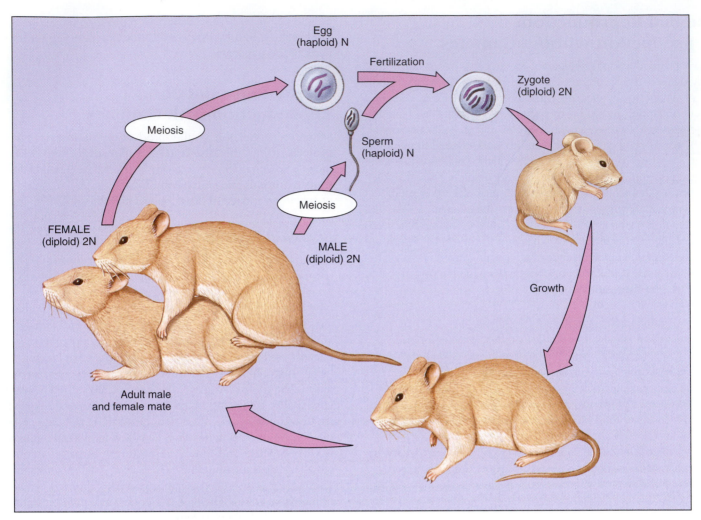

Figure 7-2

A sexual life cycle. The life cycle begins with haploid germ cells, formed by meiosis, combining to form a diploid zygote, which grows by mitosis to an adult. Most of the life cycle is spent as a diploid organism.

Many unicellular organisms reproduce both sexually and asexually. When sexual reproduction does occur, it may or may not involve male and female gametes. Sometimes two mature sexual parents merely join together to exchange nuclear material or merge cytoplasm (**conjugation,** p. 215 in Chapter 11). Distinct sexes do not exist in these cases.

The male-female distinction is more clearly evident in most animals. Organs that produce germ cells are called **gonads.** The gonad that produces sperm is a **testis** (see Figure 7-12) and that which forms eggs is an **ovary** (see Figure 7-13). Gonads represent the **primary sex organs,** the only sex organs found in certain groups of animals. Most metazoa, however, have various **accessory sex organs** (such as penis, vagina, uterine tubes, and uterus) that transfer and receive germ cells. In the primary sex organs germ cells undergo many complicated changes during their development, the details of which are described on p. 138.

Hermaphroditism

Animals that have both male and female organs in the same individual are called **hermaphrodites,** and the condition is called **hermaphroditism.** In contrast to the dioecious state of separate sexes, hermaphrodites are **monoecious,** meaning that the same organism bears both male and female organs. Many sessile, burrowing, or endoparasitic invertebrate animals (for example, most flatworms, some hydroids and annelids, and all barnacles and pulmonate snails) and a few vertebrates (some fishes), are hermaphroditic. Some hermaphrodites fertilize themselves, but most avoid self-fertilization by exchanging germ cells with another member of the same species (Figure 7-3). An advantage is that with every individual producing eggs, a hermaphroditic species could potentially produce twice as many offspring as could a dioecious species in which half the individuals are nonproductive males. Some fishes are **sequential hermaphrodites,** in which there is a genetically programmed sex change during their life. In many species of reef fishes, for example, wrasses, an animal begins life as either a female or a male (depending on the species) but later becomes the opposite sex.

Parthenogenesis

Parthenogenesis ("virgin origin") is the development of an embryo from an unfertilized egg or one in which the male and female nuclei fail to unite following fertilization. There are many

Figure 7-3

Hermaphroditic snails mating. Pulmonate snails are "simultaneous" hermaphrodites; during mating each partner inserts its penis into the female opening of the other.

patterns of parthenogenesis. In one type, called **ameiotic parthenogenesis,** no meiosis occurs, and the egg is formed by mitotic cell division. This "asexual" form of parthenogenesis is known to occur in some species of flatworms, rotifers, crustaceans, insects, and probably others. In these cases, the offspring are clones of the parent because, without meiosis, the parent's chromosomal complement is passed intact to offspring.

In **meiotic parthenogenesis** a haploid ovum is formed by meiosis, and it may or may not be activated by the influence of a male. For example, in some species of fishes, a female may be inseminated by a male of the same or related species, but the sperm serves only to activate the egg; the male's genome is rejected before it can penetrate the egg. In several species of flatworms, rotifers, annelids, mites, and insects, the haploid egg begins development spontaneously; no males are required to stimulate activation of an ovum. The diploid condition may be restored by chromosomal duplication or by autogany (rejoining of meiotic nuclei). A variant of this type of parthenogenesis occurs in many bees, wasps, and ants. In honey bees, for example, the queen bee can either fertilize eggs as she lays them or allow them to pass unfertilized. Fertilized eggs become diploid females (queens or workers), and unfertilized eggs develop parthenogenetically to become haploid males (drones); this type of sex determination is known as **haplodiploidy.** In some animals meiosis may be so severely modified that the offspring are clones of the parent. This happens in certain populations of whiptail lizards of the American southwest, which are clones consisting solely of females (Cole, 1984).

Parthenogenesis is surprisingly widespread in animals. It is an abbreviation of the usual steps required of bisexual reproduction. It may have evolved to avoid the problem—which may be great in some animals—of bringing together males and females at the right moment for successful fertilization. The disadvantage of parthenogenesis is that if the environment should suddenly change, as it often does, parthenogenetic species have limited capacity to shift gene combinations to adapt to any new

conditions. Bisexual species, by recombining parental characteristics, have a better chance of producing variant offspring that can utilize new environments.

From time to time claims arise that spontaneous parthenogenetic development to term has occurred in humans. A British investigation of about 100 cases in which the mother denied having had intercourse revealed that in nearly every case the child possessed characteristics not present in the mother, and consequently must have had a father. Nevertheless, mammalian eggs very rarely will spontaneously start developing into embryos without fertilization. In certain strains of mice, such embryos will develop into fetuses and then die. The most remarkable instance of parthenogenetic development among the vertebrates has been found in turkeys in which ova of certain strains, selected for their ability to develop without sperm, grow to reproducing adults.

Why Do So Many Animals Reproduce Sexually Rather Than Asexually?

Because sexual reproduction is so nearly universal among animals, it might be inferred to be highly advantageous. Yet it is easier to list disadvantages to sex than advantages. Sexual reproduction is complicated, requires more time, and uses much more energy than asexual reproduction. Mating partners must come together and coordinate their activities to produce young. Many biologists believe that an even more troublesome problem is the "cost of meiosis." A female that reproduces asexually passes all of her genes to her offspring. But when she reproduces sexually the genome is divided during meiosis and only half her genes flow to the next generation. Another cost is wastage in production of males, many of which fail to reproduce and thus consume resources that could be applied to production of females. Whiptail lizards of the American southwest offer a fascinating example of the potential advantage of parthenogenesis. When unisexual and bisexual species of the same genus are reared under similar conditions in the laboratory, the population of the unisexual species grows more quickly because all unisexual lizards (all females) deposit eggs, whereas only 50% of the bisexual lizards do so (Figure 7-4).

Clearly, the costs of sexual reproduction are substantial. How are they offset? Biologists have disputed this question for years without producing an answer that satisfies everyone. Many biologists believe that sexual reproduction, with its breakup and recombination of genomes, keeps producing novel genotypes that *in times of environmental change* may survive and reproduce, whereas most others die. Variability, advocates of this viewpoint argue, is sexual reproduction's trump card.

Is variability worth the biological costs of sexual reproduction? The underlying problem keeps coming back: asexual organisms, because they can have more offspring in a given time, appear to be more fit in Darwinian terms. And yet most metazoan animals are determinedly committed to sexuality. Considerable evidence suggests that asexual reproduction is most successful in colonizing new environments. When habitats are

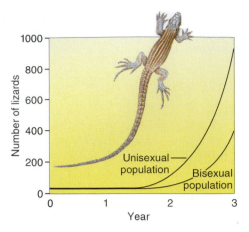

Figure 7-4

Comparison of the growth of a population of unisexual whiptail lizards with a population of bisexual lizards. Because all individuals of the unisexual population are females, all produce eggs, whereas only half the bisexual population are egg-producing females. By the end of the third year the unisexual lizards are more than twice as numerous as the bisexual ones.

empty what matters most is rapid reproduction; variability matters little. But as habitats become more crowded, competition between species for resources increases. Selection becomes more intense, and genetic variability—new genotypes produced by recombination in sexual reproduction—furnishes the diversity that permits a population to resist extinction. Therefore, on a geological timescale, asexual lineages, because they lack genetic flexibility, may be more prone to extinction than sexual lineages. Sexual reproduction is therefore favored by species selection (species selection is described on p. 126). There are many invertebrates that use both sexual and asexual reproduction, thus enjoying the advantages each has to offer.

Variety may make sexual reproduction a winning strategy for the unstable environment, but some biologists believe that for many vertebrates sexual reproduction is unnecessary and may even be maladaptive. In animals in which most of the young survive to reproductive age (humans, for example), there is no demand for novel recombinations to cope with changing habitats. One offspring appears as successful as the next in each habitat. Significantly, parthenogenesis has evolved in several species of fish and in a few amphibians and reptiles. Such species are exclusively parthenogenetic, suggesting that where it has been possible to overcome the numerous constraints to making the transition, bisexual reproduction loses out.

THE ORIGIN AND MATURATION OF GERM CELLS

Most sexually reproducing organisms are composed of nonreproductive **somatic cells,** which are differentiated for specialized functions and die with the individual, and **germ cells,** which form the gametes: eggs and sperm. Germ cells provide continuity of life between generations and ensure the species' survival. Germ cells, or their precursors, the **primordial germ cells,** are set aside at the beginning of embryonic development, usually in the endoderm, and migrate to the gonads. Here they develop into eggs and sperm—nothing else. The continuity of germ cells from one generation to the next is called the **germ cell line.** The other cells of the gonads are somatic cells. They cannot form eggs or sperm, but they are necessary for the support, protection, and nourishment of the germ cells during their development **(gametogenesis).**

A traceable germ cell line, as present in vertebrates, is distinguishable in some invertebrates, such as nematodes and arthropods. In many invertebrates, however, germ cells develop directly from somatic cells at some period in the life of an individual.

Migration of Germ Cells

In vertebrates, the actual tissue from which gonads arise appears in early development as a pair of **genital ridges,** growing into the coelom from the dorsal coelomic lining on each side of the hind-gut near the anterior end of the kidney (mesonephros).

Surprisingly perhaps, primordial germ cells do not arise in the developing gonad, but in the yolk-sac endoderm (p. 165). From studies with frogs and toads, it has been possible to trace the germ cell line back to the fertilized egg, in which a localized area of germinal cytoplasm (called **germ plasm**) can be identified in the vegetal pole of the uncleaved egg mass. This material can be followed through subsequent cell divisions of the embryo until it becomes situated in primordial germ cells in gut endoderm. From here the cells migrate by ameboid movement to the genital ridges. A similar migration of primordial germ cells occurs in mammals (Figure 7-5). Primordial germ cells are the future stock of gametes for an animal. Once in the genital ridges and during subsequent gonadal development, germ cells begin to divide by mitosis, increasing their numbers from a few dozen to several thousand.

Sex Determination

At first gonads are sexually indifferent. In normal human males, a "male-determining gene" on the Y chromosome called **SRY** (**sex-determining region Y**) organizes the developing gonad into a testis instead of an ovary. Once formed, the testis secretes the steroid **testosterone.** This hormone, and its metabolite, **dihydrotestosterone (DHT),** masculinizes the fetus, causing the differentiation of penis, scrotum, and the male ducts and glands. It also destroys the incipient breast primordia, but leaves behind the nipples as a reminder of the indifferent ground plan from which both sexes develop. Testosterone is also responsible for the masculinization of the brain, but it does so indirectly. Surprisingly, testosterone is enzymatically converted to estrogen in the brain, and it is **estrogen** that determines the organization of the brain for male-typical behavior.

Biologists have often stated that in mammals the indifferent gonad has an inherent tendency to become an ovary. Classic

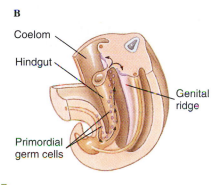

Figure 7-5

Migration of mammalian primordial germ cells. **A,** From the yolk sac the primordial germ cells migrate through the region of the hindgut into the genital ridges **(B).** In human embryos, migration is complete by the end of the fifth week of gestation.

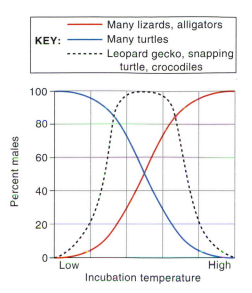

Figure 7-6

Temperature-dependent sex determination. In many reptiles that lack sex chromosomes incubation temperature of the nest determines gender. The graph shows that embryos of many turtles develop into males at low temperature, whereas embryos of many lizards and alligators become males at high temperatures. Embryos of crocodiles, leopard geckos, and snapping turtles become males at intermediate temperatures, and become females at higher or lower temperatures.

Source: Data from David Crews, "Animal Sexuality," Scientific American 270(1):108–114, January 1994.

experiments performed in rabbits provide support for the idea that the female is the default sex during development. Removal of the fetal gonads before they have differentiated will invariably produce a female with uterine tubes, uterus, and vagina, even if the rabbit is a genetic male. Localization in 1994 of a region on the X chromosome named **DDS (dosage-sensitive sex reversal)** or **SRVX (sex-reversing X),** which promotes ovary formation, has challenged this view. In addition, the presence of such a region may help to explain feminization in some XY males. It is clear, however, that absence of testosterone in a genetic female embryo promotes development of female sexual organs: vagina, clitoris, and uterus. The developing female brain does require special protection from the effects of estrogen because, as mentioned above, estrogen causes masculinization of the brain. In rats, a blood protein (alpha-fetoprotein) binds to estrogen and keeps the hormone from reaching the brain. This does not appear to be the case in humans, however, and even though circulating fetal estrogen levels can be quite high, the developing female brain does not become masculinized. One possible explanation is that the level of brain estrogen receptors in the developing female brain is low, and therefore, high levels of circulating estrogen would have no effect.

The genetics of sex determination are treated in Chapter 5 (p. 78). Sex determination is strictly chromosomal in mammals, birds, amphibians, many reptiles, and probably most fishes. However, many fishes and reptiles lack sex chromosomes altogether; in these groups, gender is determined by nongenetic factors such as temperature or behavior. In crocodilians, many tur-

tles, and some lizards the incubation temperature of the nest determines the sex ratio by an unknown sex-determining mechanism. Alligator eggs, for example, incubated at low temperature all become females; those incubated at higher temperature all become males (Figure 7-6). Sex determination of many fishes is behavior dependent. Most of these species are hermaphroditic, possessing both male and female gonads. Sensory stimuli from the animal's social environment determine whether it will be male or female.

> For every structure in the reproductive system of males or females, there is a homologous structure in the other. This happens because during early development male and female characteristics begin to differentiate from the embryonic genital ridge and two duct systems, which at first are identical in both sexes, develop. Under the influence of sex hormones, the genital ridge develops into the testes of males and the ovaries of females. One duct system (mesonephric or Wolffian) becomes ducts of the testes in males and a vestigial structure adjacent to the ovaries in females. The other duct (paramesonephric or Müllerian) develops into the uterine tubes, uterus, and vagina of females and into the small, vestigial appendix of the testes in males. Similarly, the clitoris and labia of females are homologous to the penis and scrotum of males, since they develop from the same embryonic structures.

Gametogenesis

The series of transformations that results in the formation of mature gametes is called gametogenesis. Although the same essential processes are involved in the maturation of both sperm and eggs in vertebrates, there are some important differences. Gametogenesis in testes is called **spermatogenesis,** and in ovaries, **oogenesis.**

Spermatogenesis

The walls of the seminiferous tubules contain differentiating germ cells arranged in a stratified layer five to eight cells deep (Figure 7-7). Germ cells develop in close contact with large **sustentacular (Sertoli) cells,** which extend from the periphery of the seminiferous tubules to the lumen and provide nourishment during germ-cell development and differentiation (Figure 7-8). The outermost layers contain **spermatogonia,** diploid cells that have increased in number by ordinary mitosis. Each spermatogonium increases in size and becomes a **primary spermatocyte.** Each primary spermatocyte then undergoes the first meiotic division, as described in Chapter 5 (p. 78), to become two **secondary spermatocytes.**

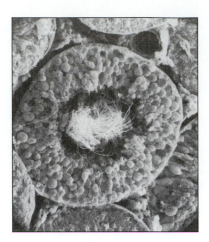

Figure 7-7

Section of a seminiferous tubule containing male germ cells. More than 200 long, highly coiled seminiferous tubules are packed in each human testis. This scanning electron micrograph reveals, in the tubule's central cavity, numerous tails of mature spermatozoa that have differentiated from germ cells in the periphery of the tubule. (×525)

From R. G. Kessel and R. H. Kardon, Tissues and Organs: A Text-Atlas of Scanning Electron Microscopy, *1979, W. H. Freeman and Co.*

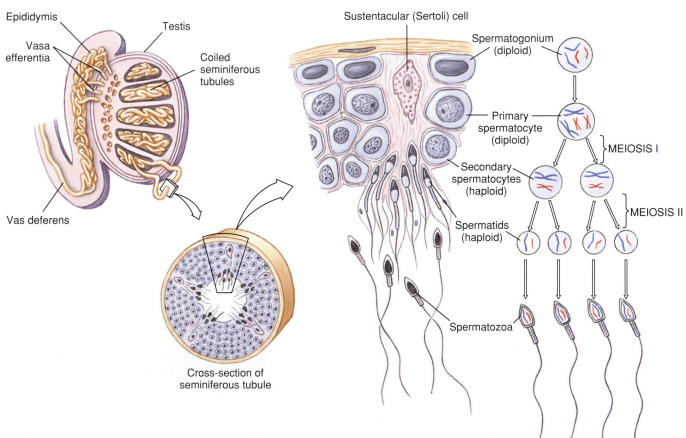

Figure 7-8

Spermatogenesis. Section of seminiferous tubule showing spermatogenesis. Germ cells develop within the recesses of large sustentacular (or Sertoli) cells, that extend from the periphery of seminiferous tubules to their lumen, and that provide nourishment to the germ cells. Stem germ cells from which sperm differentiate are the spermatogonia, diploid cells located peripherally in the tubule. These divide by mitosis to produce either more spermatogonia or primary spermatocytes. Meiosis begins when primary spermatocytes divide to produce haploid secondary spermatocytes with double-stranded chromosomes. The second meiotic division forms four haploid spermatids with single-stranded chromosomes. As sperm develop they are gradually pushed toward the lumen of the seminiferous tubule.

Each secondary spermatocyte enters the second meiotic division without intervention of a resting period. In the two steps of meiosis each primary spermatocyte gives rise to four **spermatids,** each containing the haploid number (23 in humans) of chromosomes. A spermatid may contain all chromosomes that the male inherited from his mother, those he inherited from his father, or most likely, a combination of his parents' chromosomes. Without further divisions the spermatids are transformed into mature **spermatozoa** or **(sperm)** (Figure 7-8). Modifications include great reduction of cytoplasm, condensation of the nucleus into a head, formation of a middle piece containing mitochondria, and a whiplike, flagellar tail for locomotion (Figure 7-8 and 7-9). The head consists of a nucleus containing the chromosomes for heredity and an **acrosome,** a distinctive feature of nearly all the metazoa (exceptions are teleost fishes and certain invertebrates). In many species, both invertebrate and vertebrate, the acrosome contains lysins that serve to clear an entrance through the layers that surround an egg. In mammals at least, one of the lysins is the enzyme hyaluronidase, which allows a sperm to penetrate the follicular cells surrounding an egg. A striking feature of many invertebrate spermatozoa is the acrosome filament, an extension of varying length in different species that projects suddenly from the sperm head when the latter first contacts the surface of an egg. Fusion of the egg and sperm plasma membranes is the initial event of fertilization (See Contact and Recognition between Egg and Sperm, p. 153).

The total length of a human sperm is 50 to 70 μm. Some toads have sperm that exceed 2 mm (2000 μm) in length (Figure 7-9) and are easily visible to the unaided eye. Most sperm, however, are microscopic in size (see p. 152 for an early seventeenth-century drawing of mammalian sperm, interpreted by biologists of the time as parasitic worms in the semen). In all sexually reproducing animals the number of sperm in males is far greater than the number of eggs in corresponding females. The number of eggs produced is correlated to the chances of young to hatch and reach maturity.

Oogenesis

Early germ cells in the ovary, called **oogonia,** increase in number by ordinary mitosis. Each oogonium contains the diploid number of chromosomes. After the oogonia cease to increase in number, they grow in size and become **primary oocytes** (Figure 7-10). Before the first meiotic division, the chromosomes in each primary oocyte meet in pairs, paternal and maternal homologues, just as in spermatogenesis. When the first maturation (reduction) division occurs, the cytoplasm is divided unequally. One of the two daughter cells, the **secondary oocyte,** is large and receives most of the cytoplasm; the other is very small and is called the **first polar body** (Figure 7-10). Each of these daughter cells, however, has received half of the chromosomes.

In the second meiotic division, the secondary oocyte divides into a large **ootid** and a small polar body. If the first polar body also divides in this division, which sometimes happens, there are three polar bodies and one ootid (Figure 7-10). The ootid develops into a functional **ovum.** Polar bodies are nonfunctional, and they disintegrate. Formation of nonfunctional polar

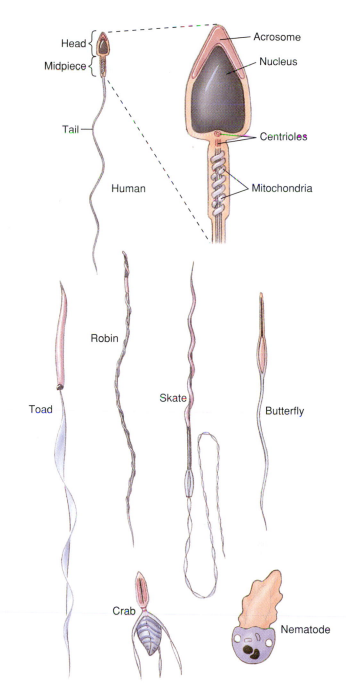

Figure 7-9
Types of vertebrate and invertebrate sperm.

bodies is necessary to enable an egg to dispose of excess chromosomes, and the unequal cytoplasmic division makes possible a large cell with the cytoplasm containing sufficient yolk for the development of young. Thus a mature ovum has N (haploid) number of chromosomes, the same as a sperm. However, each primary oocyte gives rise to only *one* functional gamete instead of four as in spermatogenesis.

In most vertebrates and many invertebrates the egg does not actually complete all the meiotic divisions before fertilization occurs. The general rule is that development is arrested during prophase I of the first meiotic division. Meiosis resumes and is

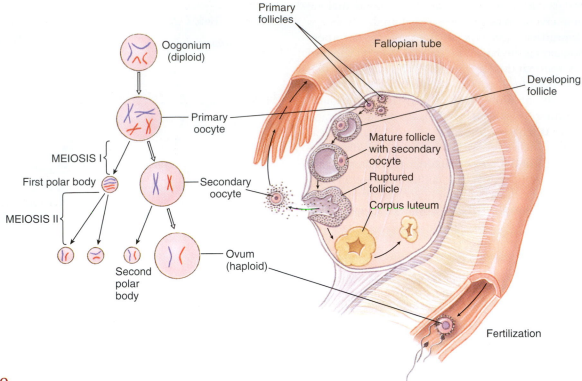

Figure 7-10

Oogenesis in humans. Early germ cells (oogonia) increase by mitosis during embryonic development to form diploid primary oocytes. After puberty, each menstrual month a diploid primary oocyte divides in the first meiotic division into a haploid secondary oocyte and a haploid polar body. If the secondary oocyte is fertilized, it enters the second meiotic division. The double-stranded chromosomes separate into a large ootid and small second polar body. Both ootid and second polar body now contain the N amount of DNA. Fusion of the haploid egg nucleus with a haploid sperm nucleus produces a diploid (2N) zygote.

completed either at the time of ovulation (birds and most mammals) or shortly after fertilization (many invertebrates, teleost fishes, amphibians, and reptiles). In humans, the ova begin the first meiotic division at about the thirteenth week of fetal development. Then their development arrests in prophase I as the primary oocyte until puberty, at which time one of these primary oocytes typically develops into a functional egg each menstrual month. Thus, in humans meiosis II is completed only when the ovum is penetrated by a spermatozoon.

The most obvious feature of egg maturation is deposition of yolk. Yolk, usually stored as granules or more organized platelets, is not a definite chemical substance but may be lipid or protein or both. In insects and vertebrates, all having more or less yolky eggs, yolk may be synthesized within an egg from raw materials supplied by surrounding follicle cells, or preformed lipid or protein yolk may be transferred by pinocytosis from follicle cells to the oocyte.

Enormous accumulation of yolk granules and other nutrients (glycogen and lipid droplets) causes an egg to grow well beyond the normal limits that force ordinary body (somatic) cells to divide. A young frog oocyte 50 μm in diameter, for example, grows to 1500 μm in diameter when mature after 3 years of growth in the ovary, and its volume has increased by a factor of 27,000. Bird eggs attain even greater absolute size; a hen egg will increase 200 times in volume in only the last 6 to 14 days of rapid growth preceding ovulation.

Thus eggs are remarkable exceptions to the otherwise universal rule that organisms are composed of relatively minute cellular units. An egg's large size creates a problematic surface area-to-cell volume ratio, since everything that enters and leaves the ovum (nutrients, respiratory gases, wastes, and so on) must pass through the cell membrane. As the egg becomes larger, the available surface per unit of cytoplasmic volume (mass) becomes smaller. As we would anticipate, the metabolic rate of an egg gradually diminishes until, when mature, an ovum is in suspended animation awaiting fertilization.

REPRODUCTIVE PATTERNS

The great majority of invertebrates, as well as many vertebrates, lay their eggs in the environment for development; these animals are called **oviparous** ("egg-birth"). Fertilization may be either internal (eggs are fertilized inside the body of a female before she lays them) or external (eggs are fertilized by a male after a female lays them). While many oviparous animals simply abandon their eggs rather indiscriminately, others display extreme care in finding places that will provide immediate and suitable sources of food for the young when they hatch.

Some animals retain eggs in their body (usually the oviduct) while they develop, with embryos deriving all their nourishment from yolk stored within the egg. These animals are called **ovo-**

viviparous ("egg-live-birth"). Ovoviviparity occurs in several invertebrate groups (for example, various annelids, brachiopods, insects, and gastropod molluscs) and is common among certain fishes and reptiles.

In the third pattern, **viviparous** ("live-birth"), eggs develop in the oviduct or uterus with embryos deriving their nourishment directly from the mother. Usually some kind of intimate anatomical relationship is established between developing embryos and their mother. In both ovoviviparity and viviparity, fertilization must be internal (within the body of the female) and the mother gives birth to young in an advanced stage of development. Viviparity is confined mostly to mammals and elasmobranch fishes, although viviparous invertebrates (scorpions, for example), amphibians, and reptiles are known. Development of embryos within a mother's body, whether ovoviviparous or viviparous, obviously affords more protection to the offspring than egg-laying.

PLAN OF REPRODUCTIVE SYSTEMS

The basic components of reproductive systems are similar in sexual animals, although differences in reproductive habits and methods of fertilization have produced many variations. Sexual systems consist of two components: (1) **primary organs,** which are the gonads that produce sperm and eggs and sex hormones; and (2) **accessory organs,** which assist the gonads in formation and delivery of gametes, and may also serve to support the embryo. They are of great variety, and include gonoducts (sperm ducts and oviducts), accessory organs for transferring spermatozoa into the female, storage organs for spermatozoa or yolk, packaging systems for eggs, and nutritional organs such as yolk glands and placenta.

Invertebrate Reproductive Systems

Invertebrates that transfer sperm from male to female for internal fertilization require organs and plumbing to facilitate this function that may be as complex as those of any vertebrate. In contrast, reproductive systems of invertebrates that simply release their gametes into the water for external fertilization may be little more than centers for gametogenesis. Polychaete annelids, for example, have no permanent reproductive organs. Gametes arise by proliferation of cells lining the body cavity. When mature the gametes are released through coelomic or nephridial ducts or, in some species, may spill out through ruptures in the body wall.

Insects have separate sexes (dioecious), practice internal fertilization by copulation and insemination, and consequently have complex reproductive systems (Figure 7-11). Sperm from the testes pass through sperm ducts to seminal vesicles (where the sperm are stored) and then through a single ejaculatory duct to a penis. Seminal fluid from one or more accessory glands is added to the semen in the ejaculatory duct. Females have a pair of ovaries formed from a series of egg tubes (ovarioles). Mature ova pass through oviducts to a common genital chamber and

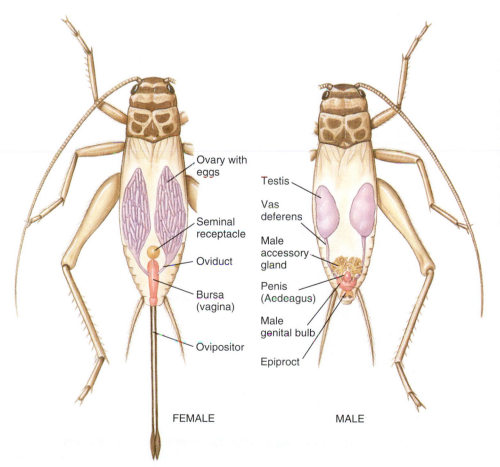

Ovary with eggs

Seminal receptacle

Oviduct

Bursa (vagina)

Ovipositor

FEMALE

Testis

Vas deferens

Male accessory gland

Penis (Aedeagus)

Male genital bulb

Epiproct

MALE

Figure 7-11

Reproductive system of crickets. Sperm from the paired testes of males pass through sperm tubes (vas deferens) to an ejaculatory duct housed in the penis. In females, eggs from the ovaries pass through oviducts to the genital bursa. At mating sperm enclosed in a membranous sac (spermatophore) formed by the secretions of the accessory gland are deposited in the genital bursa of the female, then migrate to her seminal receptacle where they are stored. The female controls the release of a few sperm to fertilize her eggs at the moment they are laid, using the needlelike ovipositor to deposit the eggs in the soil.

then to a short copulatory bursa (vagina). In most insects, the male transfers sperm by inserting the penis directly into the female system where sperm are stored in a seminal receptacle. Often a single mating provides sufficient sperm to last the reproductive life of a female.

Vertebrate Reproductive Systems

In vertebrates the reproductive and excretory systems are together called a **urogenital system** because of their close anatomical connection, especially in males. This association is very striking during embryonic development. In male fishes and amphibians the duct that drains the kidney **(mesonephric duct or Wolffian duct)** also serves as a sperm duct. In male reptiles, birds, and mammals in which the kidney develops its own independent duct **(ureter)** to carry away waste, the old **mesonephric duct** becomes exclusively a sperm duct or **vas deferens.** In all these forms, with the exception of most mammals, the ducts open into a **cloaca** (derived, appropriately, from the Latin meaning "sewer"), a common chamber into which intestinal, reproductive, and excretory canals empty. Almost all placental mammals have no cloaca; instead the urogenital system has its own opening separate from the anal opening. The **uterine duct** or **oviduct** of the female is an independent duct that does, however, open into the cloaca in forms having a cloaca.

Male Reproductive System

The male reproductive system of vertebrates, such as that of human males (Figure 7-12) includes testes, vasa efferentia, vas deferens, accessory glands, and (in some birds and reptiles, and all mammals) a penis.

Paired **testes** are the sites of sperm production. Each testis is composed of numerous **seminiferous tubules,** in which the sperm develop (Figure 7-8). The sperm are surrounded by **sustentacular cells** (or **Sertoli cells**), which nourish the developing sperm. Between the tubules are **interstitial cells,** which produce the male sex hormone **(testosterone).** In most mammals the two testes are housed permanently in a sac-like scrotum suspended outside the abdominal cavity, or the testes descend into the scrotum during the breeding season. This odd and seemingly insecure arrangement provides an environment of slightly lower temperature, since in most mammals (including humans) sperm apparently do not form at temperatures maintained within the body. In marine mammals and all other vertebrates the testes are positioned permanently within the abdomen.

The sperm travel from the seminiferous tubules to the **vasa efferentia,** small tubes passing to a coiled **epididymis** (one for each testis), where final sperm maturation takes place, and then to a **vas deferens,** the ejaculatory duct (Figure 7-8). In mammals the vas deferens joins the **urethra,** a duct that serves to carry both sperm and urinary products through the **penis,** or external intromittent organ.

Most aquatic vertebrates have no need for a penis, since sperm and eggs are liberated into the water in close proximity to each other. However, in terrestrial (and some aquatic) vertebrates that bear their young alive or enclose the egg within a shell, sperm must be transferred to the female. Few birds have a true penis (examples of exceptions are the ostrich and the Argentine lake duck) and the mating process simply involves presenting cloaca to cloaca. Reptiles and mammals have a true penis. In mammals the normally flaccid organ becomes erect when engorged with blood. Many other mammals possess a bone in the penis (baculum), which presumably helps with rigidity.

In most mammals three sets of accessory glands open into the reproductive channels: a pair of **seminal vesicles,** a single **prostate gland,** and the pair of **bulbourethral glands** (Figure 7-12). Fluid secreted by these glands furnishes food to the sperm, lubricates the passageways for sperm, and counteracts the acidity of the vagina so that the sperm are not harmed.

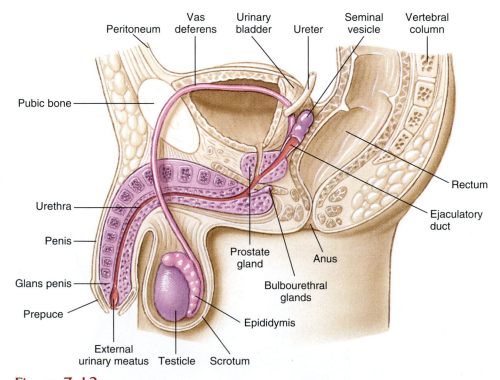

Figure 7-12

Human male reproductive system showing the reproductive structures in sagittal view.

Female Reproductive System

The ovaries of female vertebrates produce both ova and female sex hormones (estrogens and progesterone). In all jawed vertebrates, mature ova from each ovary enter the funnel-like opening of a **uterine tube** or **oviduct,** which typically has a fringed margin (fimbriae) that envelops the ovary at the time of ovulation. The terminal end of the uterine tube is unspecialized in most fishes and amphibians, but in cartilaginous fishes, reptiles, and birds that produce a large, shelled egg, special regions have developed for production of albumin and shell. In amniotes (reptiles, birds, and mammals; see Amniotes and the Amniotic Egg, p. 165) the terminal portion of the uterine tube is expanded into a muscular **uterus** in which shelled eggs are held before laying or in which embryos complete their development. In placental mammals, the walls of the uterus establish a close vascular association with the embryonic membranes through a **placenta** (see p. 167).

The paired ovaries of the human female (Figure 7-13), slightly smaller than the male testes, contain many thousands of oocytes. Each oocyte develops within a **follicle** that enlarges and finally ruptures to release a secondary oocyte (Figure 7-10). During a woman's fertile years, except following fertilization, approximately 13 oocytes mature each year, and usually the ovaries alternate in releasing oocytes. Because a woman is fertile for only about 30 years, of the approximately 400,000 primary oocytes in her ovaries at birth, only 300 to 400 have a chance to reach maturity; the others degenerate and are resorbed.

The **uterine tubes,** or **oviducts,** are lined with cilia for propelling the egg in its course. The two ducts open into the upper corners of the **uterus,** or womb, which is specialized for housing the embryo during the 9 months of its intrauterine existence. It is provided with thick muscular walls, many blood vessels, and a specialized lining: the **endometrium.** The uterus varies among different mammals, and in many it is designed to hold more than one developing embryo. Ancestrally it was paired but is fused in many eutherian mammals.

The **vagina** is a muscular tube adapted for receiving the male's penis and for serving as birth canal during expulsion of a fetus from the uterus. Where vagina and uterus meet, the uterus projects down into the vagina to form a **cervix.**

The external genitalia of human females, or **vulva,** include folds of skin, the **labia majora** and **labia minora,** and a small erectile organ, the **clitoris** (the female homolog of the glans penis of males). The opening into the vagina is often reduced in size in the virgin state by a membrane, the **hymen,** although in today's more physically active females, this membrane may be much reduced in extent.

ENDOCRINE EVENTS THAT ORCHESTRATE REPRODUCTION

Hormonal Control of Timing of Reproductive Cycles

From fish to mammals, reproduction in vertebrates is usually a seasonal or cyclic activity. Timing is crucial, because offspring should appear when food is available and other environmental conditions are optimal for survival. The sexual reproductive process is controlled by hormones, which are regulated by environmental cues, such as food intake, and seasonal changes in photoperiod, rainfall, or temperature, and by social cues. Neurosecretory centers of the brain (hypothalamus) regulate the release of anterior pituitary gland hormones, which in turn stimulate tissues of the gonads (neurosecretion and the pituitary gland are described in Chapter 34). This delicately balanced hormonal system controls development of the gonads, accessory sex structures, and secondary sexual characteristics (see p. 144), as well as timing of reproduction.

The cyclic reproductive patterns of mammals are of two types: **estrous cycle,** characteristic of most mammals, and **menstrual cycle,** characteristic only of the anthropoid primates (monkeys, apes, and humans). These two cycles differ in two important ways. First, in estrous cycles, females are receptive to males only during brief periods of **estrus,** or "heat," whereas in the menstrual cycle receptivity may

Figure 7-13

Human female reproductive system showing the pelvis in sagittal section.

occur throughout the cycle. Second, a menstrual cycle, but not an estrous cycle, ends with breakdown and discharge of the inner portion of the endometrium. In an estrous animal, each cycle ends with the endometrium simply reverting to its original state, without the discharge characteristic of the menstrual cycle.

Gonadal Steroids and Their Control

The ovaries of vertebrates produce two kinds of steroid sex hormones (GPH)—**estrogens** and **progesterone** (Figure 7-14). There are three kinds of estrogens: estradiol, estrone and estriol, of which estradiol is secreted in the highest amounts during reproductive cycles. Estrogens are responsible for development of female accessory sex structures (oviducts, uterus, and vagina) and for stimulating female reproductive activity. Secondary sex characters, those characteristics that are not primarily involved in formation and delivery of ova (or sperm in males), but that are essential for behavioral and functional success of reproduction, are also controlled or maintained by estrogens. These include characteristics such as distinctive skin or feather coloration, bone development, body size and, in mammals, initial development of the mammary glands. In mammals, both estrogen and progesterone are responsible for preparing the uterus to receive a developing embryo. These hormones are controlled by **pituitary gonadotropins: follicle-stimulating hormone (FSH),** and **luteinizing hormone (LH)** (Figure 7-15). The two gonadotropins are in turn governed by **gonadotropin-releasing hormone (GnRH)** produced by neurosecretory cells in the **hypothalamus** (see p. 725 and Table 34.1). Through this control system environmental factors such as light, nutrition, and stress may influence reproductive cycles.

The male sex steroid, **testosterone** (Figure 7-14), is manufactured by the **interstitial cells** of the testes. Testosterone, and its metabolite, **dihydrotestosterone (DHT),** are necessary for the growth and development of the male accessory sex structures (penis, sperm ducts, and glands), development of secondary male sex characters (such as bone and muscle growth, male plumage or pelage coloration, antlers in deer, and, in humans, voice quality), and male sexual behavior. Development of the testes and secretion of testosterone is controlled by FSH and LH, the same pituitary hormones that regulate the female reproductive cycle, and ultimately by GnRH from the hypothalamus. Testosterone and DHT feedback to the hypothalamus and anterior pituitary to keep secretion of GnRH and FSH and LH in check (see Chapter 34, for a discussion of negative feedback of hormones).

Both the ovary and testes secrete a peptide hormone, **inhibin,** which is secreted by the developing follicles in the female (see p. 725) and by the **sustentacular cells** (or Sertoli cells) in the male. This hormone is an additional regulator of the secretion of FSH from the anterior pituitary in a negative feedback manner.

The Menstrual Cycle

The human menstrual cycle (L. *mensis,* month) consists of two distinct phases within the ovary: follicular phase and luteal phase, and three distinct phases within the uterus: menstrual phase, proliferative phase and secretory phase (Figure 7-15). Menstruation (the "period") signals the **menstrual phase,** when part of the lining of the uterus (endometrium) degenerates and sloughs off, producing the menstrual discharge. Meanwhile, the **follicular phase** within the ovary is occurring, and by day 3 of the cycle blood levels of FSH and LH begin to rise slowly, prompting some of the ovarian follicles to begin growing and to secrete estrogen. As estrogen levels in the blood increase, the uterine endometrium heals and begins to thicken, and uterine glands within the endometrium enlarge **(proliferative phase).** By day 10 most of the ovarian follicles that began to develop at day 3 now degenerate (become **atretic**), leaving only one (sometimes two or three) to continue ripening until it appears like a blister on the surface of the ovary. This is a mature follicle or **graafian follicle.** During the latter part of the follicular phase, the graafian follicle secretes more estrogen, and also inhibin. As the levels of inhibin rise, the levels of FSH fall.

At day 13 or 14 in the cycle, the now high levels of estrogen from the graafian follicle stimulate a surge of GnRH from the hypothalamus, which induces a surge of LH (and to a lesser extent, FSH) from the anterior pituitary. The LH surge causes the largest follicle to rupture **(ovulation),** releasing an oocyte from the ovary. Now follows a critical period, for unless the mature oocyte is fertilized, it will die. The oocyte remains viable for approximately 12 hours. During the ovarian **luteal phase,** a **corpus luteum** ("yellow body" for its appearance in cow ovaries) forms from the remains of the ruptured follicle that released the oocyte at ovulation (Figures 7-10 and 7-15). The corpus luteum, responding to continued stimulation of LH, becomes a transitory endocrine gland that secretes progesterone (and estrogen in primates). Progesterone ("before carrying [gestation]"), as its name implies, stimulates the uterus to undergo final maturational changes that prepare it for gestation **(secretory phase).** The uterus is now fully ready

Figure 7-14

Sex hormones. These three sex hormones show the basic four-ring steroid structure. The main female sex hormone, estradiol (an estrogen) is a C_{18} (18-carbon) steroid with an aromatic A ring (first ring to left). The main male sex hormone testosterone (an androgen) is a C_{19} steroid with a carbonyl group (C=O) on the A ring. The female sex hormone progesterone is a C_{21} steroid, also bearing a carbonyl group on the A ring.

Testosterone   **Progesterone**   **Estradiol-17β**

to house and nourish an embryo. If fertilization has *not* occurred, the corpus luteum degenerates, and its hormones are no longer secreted. Since the uterine lining (endometrium) depends on progesterone and estrogen for its maintenance, their declining levels cause the uterine lining to deteriorate, leading to menstrual discharge of the next cycle.

> Oral contraceptives (the "Pill") usually are combined preparations of estrogen and progesterone that act to decrease the output of pituitary gonadotropins FSH and LH. This prevents the ovarian follicles from ripening and ovulation from occurring. Oral contraceptives are highly effective, with a failure rate of less than 1% if the treatment procedure is followed properly. Progesterone-only contraceptives ("mini-pill," Depo-Provera, Norplant) may not block follicular development or ovulation. Rather, they act on the reproductive tract as a whole, making it inhospitable for sperm and any fertilized oocyte.

GnRH from the hypothalamus, and LH and FSH from the anterior pituitary, are controlled by **negative feedback** of ovarian steroids (and inhibin). This negative feedback occurs throughout the menstrual cycle, except for a few days before ovulation. As just mentioned, ovulation is due to the *high levels of estrogen* causing a surge of GnRH, LH (and FSH). Such **positive feedback** mechanisms are rare in the body, since they move events away from stable set points. (Feedback mechanisms are described in Chapter 34, p. 724). This event is terminated by ovulation when estrogen levels fall as an oocyte is released from the follicle.

> While women in more than 90 other countries benefit from safe, recently developed, easier-to-use contraceptives, American couples have until recently been limited to the standby contraceptives developed more than 40 years ago: the Pill, condom, IUD, diaphragm, and surgical sterilization. Progesterone-only methods of contraception have more recently been made available in this country, including the "mini-pill," Depo-Provera and Norplant. Contraception for men (other than condoms) is still unavailable. The new contraceptive additions have significantly reduced the risk of unwanted pregnancies, but the cost of contraception is often prohibitive and not made available to younger, sexually active individuals. An unfortunate consequence is that lack of contraception, together with contraceptive failures, account for some 2 million unwanted pregnancies each year in the United States and for about half the 1.5 million abortions, one of the highest abortion rates in the industrialized world. Without a change in the present adverse policies, there is little hope of reducing unwanted pregnancies.

Figure 7-15

Human menstrual cycle, showing changes in blood-hormone levels and uterine endometrium during the 28-day cycle. FSH promotes maturation of ovarian egg follicles, which secrete estrogen. Estrogen prepares the uterine endometrium and causes a surge in LH, which in turn stimulates the corpus luteum to secrete progesterone and estrogen. Progesterone and estrogen production will persist only if an egg is fertilized; without pregnancy progesterone and estrogen levels decline and menstruation follows.

Hormones of Human Pregnancy and Birth

If fertilization occurs, it normally does so in the first third of the uterine tube **(ampulla).** The **zygote** travels from here to the uterus, dividing by mitosis to form a **blastocyst** (see Chapter 8, p. 167) by the time it reaches the uterus. The developing blastocyst will contact the uterine surface after about 6 days and bury itself in the endometrium. This process is called **implantation.** Growth of the embryo continues, producing a spherically shaped **trophoblast.** This embryonic stage contains three distinct tissue layers, the amnion, chorion, and embryo proper, the inner cell mass (Figure 8-23, p. 168). The **chorion** becomes the source of **human chorionic gonadotropin (hCG),** which appears in the bloodstream soon after implantation. hCG stimulates the corpus luteum to continue to synthesize and release both estrogen and progesterone (Figure 7-16).

The point of attachment between trophoblast and uterus becomes the **placenta** (evolution and development of the placenta is described in Chapter 8, p. 167). Besides serving as a medium for the transfer of materials between maternal and fetal bloodstreams, the placenta also serves as an endocrine gland. The placenta continues to secrete hCG and also produces estrogen (mainly estriol) and progesterone. After about the third month of pregnancy, the corpus luteum degenerates, but by then the placenta itself is the main source of both progesterone and estrogen (Figure 7-17).

Preparation of the mammary glands for secretion of milk requires two additional hormones, **prolactin (PRL)** and **human placental lactogen (hPL)** (or **human chorionic somatomammotropin**). PRL is produced by the anterior pituitary, but in nonpregnant women its secretion is inhibited. During pregnancy, elevated levels of progesterone and estrogen depress the inhibitory signal, and PRL begins to appear in the blood. PRL, in combination with hPL, prepare the mammary glands for secretion. hPL, together with maternal growth hormone, also stimulates an increase in available nutrients in the mother, so that more are provided to the developing embryo. Later the placenta begins to synthesize a peptide hormone called **relaxin;** this hormone allows some expansion of the pelvis by increasing the flexibility of the pubic symphysis, and also dilates the cervix in preparation for delivery.

Birth, or **parturition,** begins with a series of strong, rhythmic contractions of the uterine musculature, called **labor.** The exact signal that triggers birth is not fully understood in humans, but several important factors have been identified in other mammals. Just before birth, secretion of estrogen, which stimulates uterine contractions, rises sharply, while the level of progesterone, which inhibits uterine contractions, declines (Figure 7-17). This removes the "progesterone block" that keeps the uterus quiescent throughout pregnancy. **Prostaglandins,** a large group of hormones (long-chain fatty acid derivatives), also increase at this time, making the uterus more "irritable" (see Chapter 34, p. 729, for more on prostaglandins). Finally, stretching of the uterus sets in motion neural reflexes that stimulate secretion of **oxytocin** from the posterior pituitary. Oxytocin also stimulates uterine smooth muscle, leading to stronger and more frequent labor contractions. Secretion of oxytocin during childbirth is another example of **positive feedback.** This time the event is terminated by birth of the baby.

Figure 7-16

The multiple roles of progesterone and estrogen in normal human pregnancy. After implantation of an embryo in the uterus, the trophoblast (the future embryo and placenta) secretes human chorionic gonadotropin (hCG) which maintains the corpus luteum until the placenta, at about the seventh week of pregnancy, begins producing the sex hormones progesterone and estrogen.

Given the intricacy of pregnancy it may seem remarkable that healthy babies are ever born! In fact we are the lucky survivors of pregnancy, for miscarriages are quite common and serve as a mechanism to reject prenatal abnormalities such as chromosomal damage and other genetic errors, exposure to drugs or toxins, immune irregularities, or improper hormonal priming of the uterus. Modern hormonal tests show that about 30 percent of fertile zygotes are spontaneously aborted before or right after implantation; such miscarriages are unknown to the mother or are expressed as a slightly late menstrual period. Another 20 percent of established pregnancies end in miscarriage (those known to the mother), giving a spontaneous abortion rate of about 50 percent.

Childbirth occurs in three stages. In the first stage the neck (cervix), or opening of the uterus into the vagina, is enlarged by pressure from the baby in its bag of amniotic fluid, which may be ruptured at this time (Figure 7-18B). In the second stage, the baby is forced out of the uterus and through the vagina to the outside (Figure 7-18C). In the third stage, the placenta, or **afterbirth,** is expelled from the mother's body, usually within 10 minutes after the baby is born (Figure 7-18D).

After birth, secretion of milk is triggered when the infant sucks on its mother's nipple. This leads to a reflex release of oxytocin from the pituitary; when oxytocin reaches the mammary

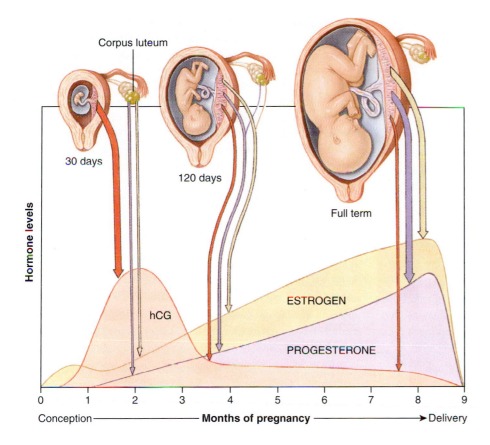

Figure 7-17
Hormone levels released from the corpus luteum and placenta during pregnancy. The width of the arrows suggests the relative amounts of hormone released; hCG (human chorionic gonadotropin) is produced solely by the placenta. Synthesis of progesterone and estrogen shifts during pregnancy from the corpus luteum to the placenta.

Corpus luteum

30 days

120 days

Full term

Hormone levels

hCG

ESTROGEN

PROGESTERONE

0 1 2 3 4 5 6 7 8 9

Conception —————— **Months of pregnancy** ——————→ Delivery

Umbilical cord

Wall of uterus

Placenta

Vagina

Intestine

A Human fetus just before birth

B First stage of labor: dilation

Figure 7-18
Birth, or parturition, in humans.

C Second stage of labor: expulsion

Placenta

D Third stage of labor: placental delivery

glands it causes contraction of smooth muscles lining ducts and sinuses of the mammary glands and ejection of milk. Suckling also stimulates release of prolactin from the anterior pituitary gland, which stimulates continued production of milk by the mammary glands.

Multiple Births

Many mammals give birth to more than one offspring at a time or to a litter (**multiparous**), each member of which has come from a separate egg. There are some mammals, however, that have only one offspring at a time (**uniparous**), although occasionally they may have more than one young. The armadillo (*Dasypus*) is almost unique among mammals in giving birth to four young at one time—all of the same sex, either male or female, and all derived from the same zygote.

Human twins may come from one zygote (**identical,** or **monozygotic** twins; Figure 7-19A) or two zygotes (**nonidentical, dizygotic,** or **fraternal** twins; Figure 7-19B). Fraternal twins do not resemble each other any more than other children born separately in the same family, but identical twins are, of course, strikingly alike and always of the same sex. Triplets, quadruplets, and quintuplets may include a pair of identical twins. The other babies in such multiple births usually come from separate zygotes. About 33% of identical twins have separate placentas, indicating that the blastomeres separated at an early, possibly the two-cell, stage (Figure 7-19A, *top*). All other identical twins share a common placenta, indicating that splitting occurred after formation of the inner cell mass (see Figure 8-23 on p. 168). If splitting were to happen after placenta formation, but before the amnion forms, the twins would have individual amniotic sacs (Figure 7-19A, *middle*), as observed in the great majority of identical twins. Finally, a very small percentage of identical twins share one amniotic sac and a single placenta (Figure 7-19A, *bottom*), indicating that separation occurred after day 9 of pregnancy, by which time the amnion has formed. In these cases, the twins are at risk of becoming conjoined, a condition known as Siamese twinning. Embryologically, each member of fraternal twins has its own placenta and amnion (Figure 7-19B).

The frequency of twin births in comparison to single births is approximately 1 in 86, that of triplets 1 in 86^2, and that of quadruplets approximately 1 in 86^3. Frequency of identical twin births to all births is about the same the world over, whereas frequency of fraternal births varies with race and country. In the United States, three-fourths of all twin births are dizygotic (fraternal), whereas in Japan only a little more than one-fourth are dizygotic. The tendency for fraternal twinning (but apparently not identical twinning) seems to run in family lines; fraternal twinning (but not identical twinning) also increases in frequency as mothers get older.

SUMMARY

Reproduction is the production of new life and provides an opportunity for evolution to occur. Asexual reproduction is a rapid and direct process by which a single organism produces genetically identical copies of itself. It may occur by fission, budding, gemmulation, or fragmentation. Sexual reproduction involves production of germ cells (sex cells or gametes), usually by two parents (bisexual reproduction), which combine by fertilization to form a zygote that develops into a new individual. Germ cells are formed by meiosis, reducing the number of chromosomes to haploid, and the diploid chromosome number is restored at fertilization. Sexual reproduction recombines parental characters and thus reshuffles and amplifies genetic diversity. Genetic recombination is important for evolution. Two alternatives to typical bisexual reproduction are hermaphroditism, the presence of both male and female organs in the same individual, and parthenogenesis, the development of an unfertilized egg.

Sexual reproduction exacts heavy costs in time and energy, requires cooperative investments in mating, and results in a 50% loss of genetic representation of each parent in the offspring. The classical view of why sex is needed is that it maintains variable offspring within the population, which may help the population to survive environmental change.

In vertebrates the primordial germ cells arise in the yolk-sac endoderm, then migrate to the gonad. In mammals, a gonad becomes a testis in response to masculinizing signals encoded on the Y chromosome of the male, and the reproductive tract masculinizes in response to circulating male sex steroids. Female reproductive structures (ovary, uterine tubes, uterus, and vagina) develop in the absence of signals encoded on the Y chromosome in females, although recent data suggests a female-determining region on the X chromosome may play an important role in differentiation of female reproductive organs.

Germ cells mature in the gonads by a process called gametogenesis (spermato-genesis in males and oogenesis in females), involving both mitosis and meiosis. In spermatogenesis, each primary spermatocyte gives rise by meiosis and growth to four motile sperm, each bearing the haploid number of chromosomes. In oogenesis, each primary oocyte gives rise to only one mature, nonmotile, haploid ovum. The remaining nuclear material is discarded in polar bodies. During oogenesis an egg accumulates large food reserves within its cytoplasm.

Sexual reproductive systems vary enormously in complexity, ranging from some invertebrates, such as polychaete worms that lack any permanent reproductive structures to the complex systems of vertebrates and many invertebrates consisting of permanent gonads and various accessory structures for transferring, packaging, and nourishing gametes and embryos.

The male reproductive system of humans includes testes, composed of seminiferous tubules in which millions of sperm develop, and a duct system (vasa efferentia and vas deferens) that joins the urethra, glands (seminal vesicles, prostate, bulbourethral), and penis. The human female system includes ovaries, containing thousands of eggs within follicles; egg-carrying uterine tubes; uterus; and vagina.

The seasonal or cyclic nature of reproduction in vertebrates has required evolution of precise hormonal mechanisms that control production of germ cells, signal readiness for mating, and prepare ducts and glands for successful fertilization of eggs. Neurosecretory centers of the brain secrete gonadotropin-releasing hormone (GnRH), which stimulates endocrine cells of the anterior pituitary to

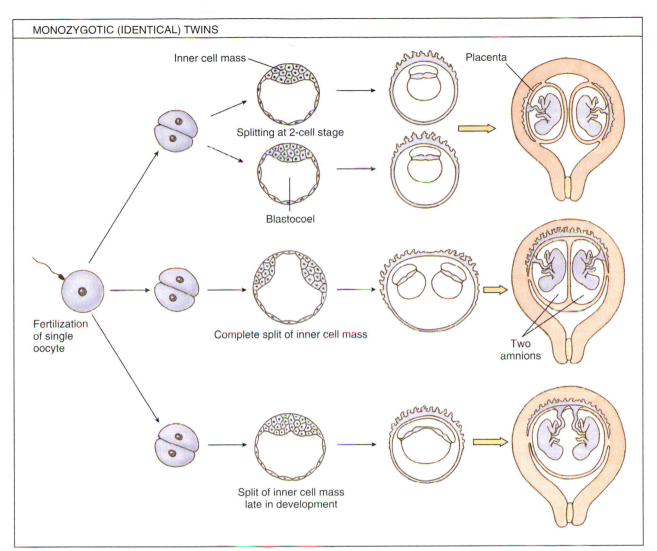

Figure 7-19

Formation of
human twins.
A, Monozygotic
(identical) twin
formation.
B, Dizygotic
(fraternal) twin
formation.
See text for
explanation.

MONOZYGOTIC (IDENTICAL) TWINS

Inner cell mass

Placenta

Splitting at 2-cell stage

Blastocoel

Fertilization
of single
oocyte

Complete split of inner cell mass

Two
amnions

Split of inner cell mass
late in development

A

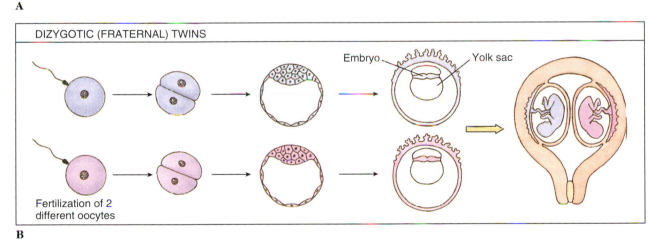

DIZYGOTIC (FRATERNAL) TWINS

Embryo

Yolk sac

Fertilization of 2
different oocytes

B

release follicle-stimulating hormone (FSH) and luteinizing hormone (LH), which in turn stimulate the gonads. Estrogens and progesterone in females, and testosterone and dihydrotestosterone (DHT) in males, control the growth of accessory sex structures and secondary sex characteristics.

In the human menstrual cycle, estrogen induces the initial proliferation of uterine endometrium. A surge in GnRH and LH midway in the cycle induces ovulation and causes the corpus luteum to secrete progesterone (and estrogen in humans), which completes preparation of the uterus for implantation. If an egg is fertilized, pregnancy is maintained by hormones produced by the placenta and mother. Human chorionic gonadotropin (hCG) maintains secretion of progesterone and estrogen from the corpus luteum, while the placenta grows and eventually secretes estrogen, progesterone, hGC, and human placental lactogen (hPL). Estrogen, progesterone, and hPL, as well as maternal prolactin, induce development of the mammary glands in preparation for lactation. hPL and maternal growth hormone also increase nutrient availability for the developing embryo.

How is it possible that a tiny, spherical fertilized human egg, scarcely visible to the naked eye, can develop into a fully formed, unique person, consisting of thousands of billions of cells, each cell performing a predestined functional or structural role? How is this marvelous unfolding controlled? Clearly all information needed must originate from the nucleus and in the surrounding cytoplasm. But knowing where the control system lies is very different from understanding how it guides the conversion of a fertilized egg into a fully differentiated animal. Despite intense scrutiny by thousands of scientists over many decades, it seemed until very recently that developmental biology, almost alone among the biological sciences, lacked a satisfactory conceptual coherence. This now has changed. During the last two decades the combination of genetics with modern techniques of cellular and molecular biology produced an avalanche of information that solved many questions. Causal relationships between development and evolution have also become the focus of research. We do at last appear to have a conceptual framework to account for development.

EARLY CONCEPTS: PREFORMATION VERSUS EPIGENESIS

Early scientists and laypeople alike speculated at length about the mystery of development long before the process was submitted to modern techniques of biochemistry, molecular biology, tissue culture, and electron microscopy. An early and persistent idea was that young animals were preformed in eggs and that development was simply a matter of unfolding what was already there. Some claimed they could actually see a miniature of the adult in the egg or sperm (Figure 8-1). Even the more cautious argued that all parts of the embryo were in the egg, needing only to unfold, but so small and transparent they could not be seen. The concept of **preformation** was strongly advocated by most seventeenth- and eighteenth-century naturalist-philosophers.

In 1759 German embryologist Kaspar Friedrich Wolff clearly showed that in the earliest developmental stages of the chick, there was no preformed individual, only undifferentiated granular material that became arranged into layers. These layers continued to thicken in some areas, to become thinner in others, to fold, and to segment, until the body of the embryo appeared. Wolff called this process **epigenesis** ("origin upon or after"), an idea that a fertilized egg contains building material only, somehow assembled by an unknown directing force. Current ideas of development are essentially epigenetic in concept, although we know far more about what directs growth and differentiation.

Development describes the progressive changes in an individual from its beginning to maturity (Figure 8-2). In sexual multicellular organisms, development usually begins with a fertilized egg that divides mitotically to produce a many-celled embryo. These cells then undergo extensive rearrangements and interact with one another to generate an animal's body plan and all of the many kinds of specialized cells in its body. This generation of cellular diversity is not defined all at once but is formed as the result of a **hierarchy of developmental decisions.** The many familiar cell types that make up the body do not simply "unfold" at some point, but arise from conditions created in preceding stages. At each stage of development new structures arise from the interaction of less committed rudiments. Each interaction is increasingly restrictive, and the decision made at each stage in the hierarchy further limits developmental fate. Once cells embark on a course of differentiation, they become irrevocably committed to that course. They no longer depend on the stage that preceded them, nor do they have the option of becoming something different. Once a structure becomes committed it is said to be **determined.** Thus the hierarchy of commitment is progressive and it is usually irreversible. The two

Figure 8-1

Preformed human infant in sperm as imagined by seventeenth-century Dutch histologist Niklass Hartsoeker, one of the first to observe sperm, using a microscope of his own construction. Other remarkable pictures published during this period depicted the figure sometimes wearing a nightcap!

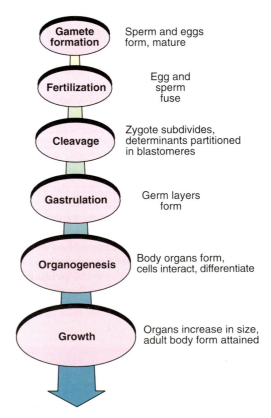

Gamete formation	Sperm and eggs form, mature
Fertilization	Egg and sperm fuse
Cleavage	Zygote subdivides, determinants partitioned in blastomeres
Gastrulation	Germ layers form
Organogenesis	Body organs form, cells interact, differentiate
Growth	Organs increase in size, adult body form attained

Figure 8-2

Key events in animal development.

basic processes that are responsible for this progressive subdivision are **cytoplasmic localization** and **induction.** We will discuss both processes as we proceed through this chapter.

FERTILIZATION

The initial event in development in sexual reproduction is **fertilization,** the union of male and female gametes to form a **zygote.** Fertilization accomplishes two things: it provides for recombination of paternal and maternal genes, thus restoring the original diploid number of chromosomes characteristic of a species, and it activates the egg to begin development. However, sperm are not always required for development. Eggs of some species can be artificially induced to initiate development without sperm fertilization (artificial parthenogenesis), but in the great majority of cases an embryo will not be able to progress very far down the developmental path before lethal developmental abnormalities arise. However, some species have natural parthenogenesis (p. 135). Of these, some have eggs that develop normally in the absence of sperm. In other species (some fishes and salamanders), sperm is required for egg activation, but the sperm contributes no genetic material. Thus neither sperm contact nor the parental genome is always essential for egg activation.

Oocyte Maturation

During oogenesis, described in the preceding chapter, an egg prepares itself for fertilization and for the beginning of development. Whereas a sperm eliminates all its cytoplasm and condenses its nucleus to the smallest possible dimensions, an egg grows in size by accumulating yolk reserves to support future growth. An egg's cytoplasm also contains vast amounts of messenger RNA, ribosomes, transfer RNA, and other elements that will be required for protein synthesis. In addition, eggs of most species contain **morphogenetic determinants** that direct activation and repression of specific genes later in postfertilization development. The nucleus also grows rapidly in size during egg maturation, becoming bloated with RNA and so changed in appearance that it is given a special name, the **germinal vesicle.**

Most of this intense preparation occurs during an arrested stage of meiosis. In mammals, for example, it occurs during the prolonged prophase of the first meiotic division. The oocyte is now poised to resume meiotic divisions that are essential to produce a haploid female pronucleus that will join a male haploid pronucleus at fertilization. After resumption of meiosis, the egg rids itself of excess chromosomal material in the form of polar bodies (described in Chapter 7, p. 139). A vast amount of synthetic activity has preceded this stage. The oocyte is now a highly structured system, provided with a dowry which, after fertilization, will support nutritional requirements of the embryo and direct its development through cleavage.

Fertilization and Activation

Our current understanding of fertilization and activation derives in large part from more than a century of research on marine invertebrates, especially sea urchins. Sea urchins produce large numbers of eggs and sperm, which can be combined in the laboratory for study. Fertilization also has been studied in many vertebrates and, more recently, in mammals, using sperm and eggs of mice, hamsters, and rabbits.

Contact and Recognition between Egg and Sperm

Most marine invertebrates and many marine fishes simply release their gametes into the ocean. Although an egg is a large target for a sperm, the enormous dispersing effect of the ocean and limited swimming range of a spermatozoon conspire against an egg and a sperm coming together by chance encounter. To improve likelihood of contact, eggs of numerous marine species release a chemotactic factor that attracts sperm to eggs. The chemotactic molecule is species-specific, attracting to eggs only sperm of the same species.

In sea urchin eggs, sperm first penetrate a jelly layer surrounding the egg, then contact an egg's vitelline envelope, a thin membrane lying just above the egg plasma membrane (Figure 8-3). At this point, egg-recognition proteins on the acrosomal process of the sperm (Figure 8-4) bind to species-specific sperm receptors on the vitelline envelope. This mechanism ensures that an egg will recognize only sperm of the same species; all others are screened out. This is important in the marine environment where many closely related species may be spawning at the same time. Similar recognition proteins have been found on sperm of vertebrate species (including mammals) and presumably are a universal property of animals.

Prevention of Polyspermy

At the point of sperm contact with the egg vitelline envelope a **fertilization cone** appears into which the sperm head is later drawn (see Figure 8-4). This event is followed immediately by important changes in the egg surface that block entrance of additional sperm, which, in marine eggs especially, may quickly surround the egg in swarming numbers (Figure 8-5). Entrance of

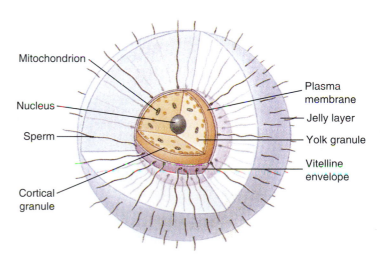

Figure 8-3

Structure of sea urchin egg during fertilization.

Figure 8-4
Sequence of events during sperm contact and penetration of a sea urchin egg.

Figure 8-5
Binding of sperm to the surface of a sea urchin egg. Only one sperm penetrates the egg surface, the others being blocked from entrance by rapid changes in the egg membranes. Unsuccessful sperm are soon lifted away from the egg surface by a newly formed fertilization membrane.

more than one sperm, called **polyspermy,** must be prevented because union of more than two haploid nuclei would be ruinous for normal development. In a sea urchin egg, contact of the first sperm with the egg membrane is instantly followed by an electrical potential change in the egg membrane that prevents additional sperm from fusing with the membrane. This event, called the **fast block,** is followed immediately by the **cortical reaction,** in which thousands of enzyme-rich cortical granules, located just beneath the egg membrane, fuse with the membrane and release their contents into the space between the egg membrane and the overlying vitelline envelope (see Figure 8-4). The cortical reaction creates an osmotic gradient, causing water to rush into this space, elevating the envelope and lifting away all sperm bound to it, except the one sperm that has successfully fused with the egg membrane. One of the cortical granule enzymes causes the vitelline envelope to harden, and it is now called a **fertilization membrane.** The block to polyspermy is complete. Timing sequence of these early events is summarized in Figure 8-6. Mammals have a similar security system that is erected within seconds after the first sperm fuses with the egg membrane.

Fusion of Pronuclei and Egg Activation

Once sperm and egg membranes have fused, the sperm loses its flagellum, which disintegrates. Its nuclear envelope then breaks apart, allowing the sperm chromatin to expand from its extremely condensed state. The enlarged sperm nucleus, now called a **pronucleus,** migrates inward to contact the female pronucleus. Their fusion forms the diploid **zygote nucleus.** Nuclear fusion takes only about 12 minutes in sea urchin eggs (Figure 8-6), but requires about 12 hours in mammals.

Fertilization sets in motion several important changes in the cytoplasm of an egg—now properly called a zygote—that prepare for cleavage. It serves to remove one or more inhibitors that had blocked metabolism and kept the egg in its quiescent, suspended-animation state. Fertilization is immediately followed by a burst of DNA and protein synthesis, the latter utilizing the abundant sup-

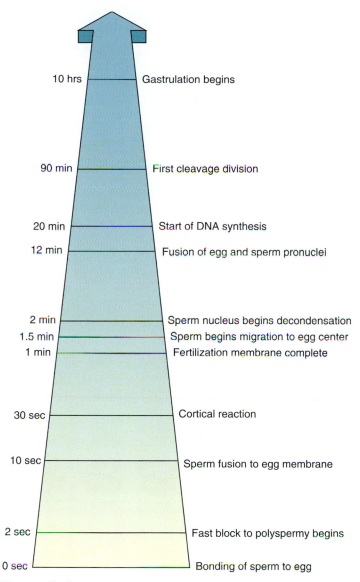

Figure 8-6
Timing of events during fertilization and early development in a sea urchin.

Timeline labels (from top to bottom):

Time	Event
10 hrs	Gastrulation begins
90 min	First cleavage division
20 min	Start of DNA synthesis
12 min	Fusion of egg and sperm pronuclei
2 min	Sperm nucleus begins decondensation
1.5 min	Sperm begins migration to egg center
1 min	Fertilization membrane complete
30 sec	Cortical reaction
10 sec	Sperm fusion to egg membrane
2 sec	Fast block to polyspermy begins
0 sec	Bonding of sperm to egg

ply of messenger RNA previously stored in the egg cytoplasm. Fertilization also initiates an almost complete reorganization of the cytoplasm within which are morphogenetic determinants that will activate or repress specific genes as development proceeds. Movement of cytoplasm repositions the determinants into new and correct spatial arrangements that are essential for proper development. The zygote now enters cleavage.

CLEAVAGE AND EARLY DEVELOPMENT

During cleavage the embryo divides repeatedly to convert the large, unwieldy cytoplasmic mass into a large cluster of small, maneuverable cells (called **blastomeres**). There is no growth during this period, only subdivision of mass, which continues until normal somatic cell size is attained. At the end of cleavage

the zygote has been divided into many hundreds or thousands of cells (about 1000 in polychaete worms, 9000 in amphioxus, and 700,000 in frogs). **Polarity** is present in the egg in the form of a polar axis, which establishes the direction of cleavage and subsequent differentiation of the embryo.

Patterns of Cleavage

Although cleavage is usually very regular within a species, there is considerable variation between species with regard to cleavage pattern. The pattern of cleavage is greatly affected by (1) quantity and distribution of yolk present and (2) genes controlling the symmetry of cleavage. Four principal types of cleavage are shown in Figure 8-7.

How Amount and Distribution of Yolk Affect Cleavage

Eggs with very little yolk that is evenly distributed in the egg are called **isolecithal** (Gr. *isos,* equal, + *lekithos,* yolk). In such eggs, cleavage is **holoblastic** (Gr. *holo,* whole, + *blastos,* germ), meaning that the cleavage furrow extends completely through the egg (see Figure 8-7A, C, and E). Isolecithal eggs are found in a great diversity of animals, including echinoderms, tunicates, cephalochordates, nemerteans, most molluscs, as well as marsupial and placental mammals (including humans).

Amphibian eggs (Figure 8-7B) are called **mesolecithal** (Gr. *mesos,* middle, + *lekithos,* yolk) because they have a moderate amount of yolk concentrated in the **vegetal pole.** The opposite **animal pole** contains mostly cytoplasm and very little yolk. Mesolecithal eggs also cleave holoblastically, with each cleavage furrow beginning at the animal pole and extending toward the vegetal pole. However, cleavage is slowed in the yolk-rich vegetal pole. Thus, as cleavage progresses, the animal region becomes packed with numerous small cells, while the vegetal region contains relatively few, large, yolk-filled cells.

Eggs of birds, reptiles, most fishes, a few amphibians, cephalopod molluscs, and monotreme mammals are called **telolecithal** (Gr. *telos,* end, + *lekithos,* yolk) because they contain an abundance of yolk that is densely concentrated at the vegetal pole of the egg (refer to chick development in Figure 8-7D). Actively dividing cytoplasm is confined to a narrow disc-shaped mass lying on top of the yolk. Cleavage is partial, or **meroblastic** (Gr. *meros,* part, + *blastos,* germ), because the cleavage furrows cannot cut through the heavy yolk concentration, but instead stop at the border between the cytoplasm and yolk below.

Centrolecithal eggs, typical of insects and many other arthropods, also exhibit meroblastic cleavage (see Figure 8-8). These eggs have a large mass of centrally located yolk, and cytoplasmic cleavage is limited to a surface layer of yolk-free cytoplasm while the yolk-rich inner cytoplasm remains uncleaved.

The amount of yolk is related not only to cleavage pattern, but also to whether a larval stage occurs during development. Animals in which the zygote is telolecithal generally have **direct development** into a small version of the adult. Species with isolecithal or mesolecithal zygotes generally have **indirect development** into larvae, which, by definition, differ from adults. Larvae compensate for their lack of yolk by having adaptations for feeding themselves as they continue developing (see Figure 8-21, p. 167). A larva will then undergo a **metamorphosis** into the adult body form. A major exception to this generalization occurs in most mammals, as well as in other animals in which mothers provide nourishment to embryos by means of a placenta. In those animals there is little yolk, but embryos are able to develop directly without requiring a larval stage capable of feeding itself.

How Cleavage Is Affected by Different Inherited Patterns

Another important influence on a species' pattern of cleavage is its inherited pattern of cell division. This effect is most apparent in isolecithal eggs with holoblastic cleavage, in which four major patterns of cleavage can be observed: radial, spiral, bilateral, and rotational. In mesolecithal and telolecithal eggs with meroblastic cleavage, there are two other major patterns: discoidal and superficial. These different cleavage patterns are characteristic of different phylogenetic groups of animals.

In **radial cleavage** (so called because the embryonic cells are arranged in radial symmetry around the animal-vegetal axis), each cleavage furrow is oriented either parallel or perpendicular to the animal-vegetal axis of the egg. In sea stars (Figure 8-7A), the first cleavage plane passes right through the animal-vegetal axis, yielding two identical **blastomeres.** For the second cleavage division, furrows form simultaneously in both blastomeres, and these also are oriented parallel to the animal-vegetal axis (but perpendicular to the first cleavage furrow). Cleavage furrows next form simultaneously in the four daughter blastomeres, this time oriented perpendicular to the animal-vegetal axis, yielding two tiers of four cells each. Subsequent cleavages yield an embryo composed of several tiers of cells.

Spiral cleavage (represented by nemertean worm development in Figure 8-7C) is different from radial in two important ways. Rather than dividing parallel or perpendicular to the animal-vegetal axis, blastomeres cleave oblique to this axis and typically produce quartets of cells that come to lie, not on top of each other, but in the furrows between the cells. In addition, spirally cleaving blastomeres pack themselves tightly together much like a group of soap bubbles, rather than just lightly contacting each other as do many radially cleaving blastomeres.

The importance of these two cleavage patterns extends well beyond the differences we have described. They are signals of a fundamental dichotomy, the early evolutionary divergence of bilateral metazoan animals into two separate lineages. Spiral cleavage is found in annelids, nemerteans, turbellarian flatworms, all molluscs except cephalopods, some brachiopods, and echiurans. These and several other invertebrate phyla are included in the **Protostomia** division of the animal kingdom. These phyla with spiral cleavage are further restricted to one of the two clades within Protostomia, the **Lophotrochozoa** (see p. 204). Radial cleavage is characteristic of the **Deuterostomia** division of the animal kingdom, a grouping that traditionally includes echinoderms (sea stars and their kin), hemichordates, and chordates. Other distinguishing developmental hallmarks of these two divisions are summarized in Figure 8-9.

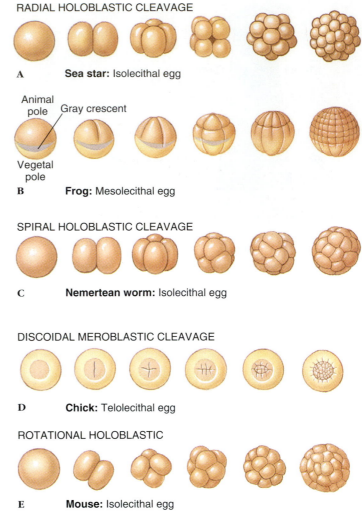

RADIAL HOLOBLASTIC CLEAVAGE

A   Sea star: Isolecithal egg

Animal pole   Gray crescent

Vegetal pole

B   Frog: Mesolecithal egg

SPIRAL HOLOBLASTIC CLEAVAGE

C   Nemertean worm: Isolecithal egg

DISCOIDAL MEROBLASTIC CLEAVAGE

D   Chick: Telolecithal egg

ROTATIONAL HOLOBLASTIC

E   Mouse: Isolecithal egg

Figure 8-7
Cleavage stages in sea star, frog, nemertean worm, chick, and mouse.

Cephalopod molluscs and ascidian chordates (also called tunicates) exhibit **bilateral cleavage.** In ascidian eggs, the antero-posterior axis is defined prior to fertilization by asymmetrical distribution of several cytoplasmic components (Figure 8-10). The first cleavage furrow passes through the animal-vegetal axis, dividing the asymmetrically distributed cytoplasm equally between the first two blastomeres. Thus, this first cleavage division separates the embryo into its future right and left sides, establishing its bilateral symmetry (hence the name bilateral holoblastic cleavage). Each successive division orients itself to this plane of symmetry, and the half-embryo formed on one side of the first cleavage is the mirror image of the half embryo on the other side.

Most mammals possess isolecithal eggs and a unique cleavage pattern called **rotational cleavage,** so called because of the orientation of blastomeres with respect to each other during the second cleavage division (see mouse development in Figure 8-7E). Cleavage in mammals is slower than in any other animal group. In humans, the first division is completed about 36 hours after fertilization (compared with about a minute and a half in sea urchins), and the next divisions follow at 12- to 24-hour intervals. As in most other animals, the first cleavage plane runs

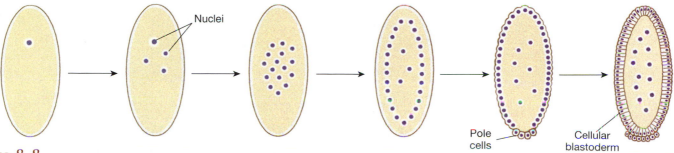

Figure 8-8

Superficial cleavage in a *Drosophila* embryo. The zygote nucleus at first divides repeatedly in the yolk-rich endoplasm by mitosis without cytokinesis. After several rounds of mitosis, most nuclei migrate to the surface where they are separated by cytokinesis into separate cells. Some nuclei migrate to the posterior pole to form the primordial germ cells, called pole cells. Several nuclei remain in the endoplasm where they will regulate breakdown of yolk products. The cellular blastoderm stage corresponds to the blastula stage of other embryos.

through the animal-vegetal axis to yield a two-cell embryo. However, during the second cleavage one of these blastomeres divides meridionally (through the animal-vegetal axis) while the other divides equatorially (perpendicular to the animal-vegetal axis). Thus, the cleavage plane in one blastomere is rotated 90 degrees with respect to the cleavage plane of the other blastomere (hence the name rotational cleavage). Furthermore, early divisions are asynchronous; all blastomeres do not divide at the same time. Thus, mammalian embryos may not increase regularly from two to four to eight blastomeres, but often contain odd numbers of cells. After the third division, the cells suddenly close into a tightly packed configuration, which is stabilized by tight junctions that form between outermost cells of the embryo. These outer cells form the **trophoblast.** The trophoblast is not part of the embryo proper but will form the embryonic portion of the placenta when the embryo implants in the uterine wall. Cells that actually give rise to the embryo proper form from the inner cells, called the **inner cell mass** (see blastula stage in Figure 8-11E).

Among animals that exhibit meroblastic cleavage, there are two major inherited patterns of cleavage. Telolecithal eggs of reptiles, birds, and most fish divide by **discoidal cleavage.** Because of the great mass of yolk in these eggs, cleavage is confined to a small disc of cytoplasm lying atop a mound of yolk (see chick development in Figure 8-7D). Early cleavage furrows carve this cytoplasmic disc to yield a single layer of cells called the blastoderm. Further cleavages divide the blastoderm into five to six layers of cells. By contrast, centrolecithal eggs of insects undergo **superficial cleavage** (Figure 8-8). The centrally located mass of yolk restricts cleavage to the cytoplasmic rim of the egg. This pattern is highly unusual because cytoplasmic cleavage (cytokinesis)

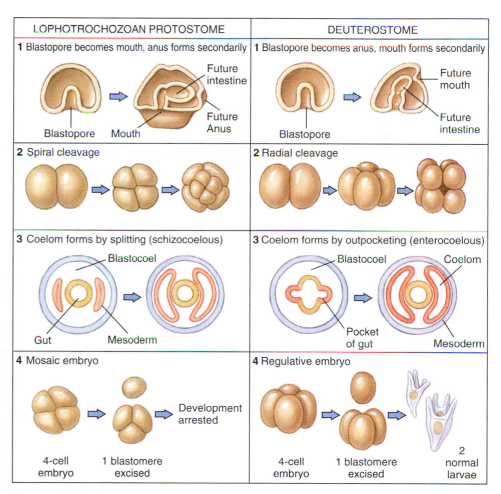

Figure 8-9

Developmental tendencies of lophotrochozoan protostomes (flatworms, annelids, molluscs, etc.) and deuterostomes. These tendencies are much modified in some groups, for example vertebrates. Cleavage in mammals is rotational rather than radial; in reptiles, birds, and many fishes cleavage is discoidal. Vertebrates have also evolved a derived form of coelom formation that is basically schizocoelous.

does not occur until after many rounds of nuclear division. After roughly eight rounds of mitosis in the absence of cytoplasmic division (yielding 256 nuclei), the nuclei migrate to the yolk-free periphery of the egg. A few nuclei at the posterior end of the egg become surrounded by cytoplasm to form pole cells, which will

Figure 8-10

Bilateral cleavage in tunicate embryos. The first cleavage division divides the asymmetrically distributed cytoplasm evenly between the first two blastomeres, establishing the future right and left sides of the adult animal. Bilateral symmetry of the embryo is maintained through subsequent cleavage divisions.

give rise to germ cells of the adult. Next, the entire egg cell membrane folds inward, partitioning each nucleus into a single cell, and yielding a layer of cells at the periphery surrounding the mass of yolk (Figure 8-8). Thus, different groups of animals have evolved different mechanisms for dealing with large volumes of yolk. Because yolk is an impediment to cleavage, both these patterns avoid cleaving the yolk and instead confine cytoplasmic division to small regions of yolk-free cytoplasm.

Blastulation

Cleavage, however modified by different cleavage patterns and by the presence of varying amounts of yolk, results in a cluster of cells called a **blastula** (commonly called a blastocyst in mammals) (Figure 8-11). In many animals the cells arrange themselves around a central fluid-filled cavity called a **blastocoel** (blast-oh-seal). At this point, an embryo consists of a few hundred to several thousand cells poised for further development. There has been a great increase in total DNA content, since each of the many daughter cell nuclei, by chromosomal replication at mitosis, contains as much DNA as the original zygote nucleus. The whole embryo, however, has not increased in size above the zygote; it has been subdivided into smaller and smaller cells.

GASTRULATION AND FORMATION OF GERM LAYERS

Gastrulation involves extensive and highly integrated cell and tissue movements, resulting in dramatic rearrangement of cells of the blastula. In most animals, gastrulation converts the spherical blastula into a more complex configuration of three germ layers. At the end of gastrulation, ectoderm covers the embryo, and mesoderm and endoderm have been brought inside. As a result, cells have new positions and new neighbors, and interaction of these cells and tissues will generate the embryonic body plan. Patterns of gastrulation vary enormously between different groups of animals, and these differences depend very much on amount and distribution of yolk. As with cleavage, yolk impedes gastrulation. Thus, gastrulation is relatively simple in most non-yolky embryos, but it is more complex in embryos developing from yolk-laden eggs.

In sea stars, gastrulation begins when the entire vegetal area of the blastula flattens to form a **vegetal plate.** This event is followed by a process called **invagination,** in which the vegetal plate (a sheet of epithelial tissue) bends inward and extends about one-third of the way into the blastocoel, forming a new internal cavity, the archenteron (Figure 8-11A). The archenteron is the primitive gut and its opening to the outside is called a blastopore. In sea stars and other members of the Deuterostomia ("mouth second"), the blastopore becomes the anus, while the mouth forms secondarily (see Figure 8-9). The archenteron continues to elongate toward the animal pole and its anterior end expands into two pouchlike **coelomic vesicles,** which pinch off to form left and right coelomic compartments (Figure 8-11A).

The gastrula is now an embryo of three **germ layers.** The outer layer is **ectoderm;** it will give rise to the epithelium of the body surface and to the nervous system. The inner layer that forms the archenteron is **endoderm;** it will give rise to the epithelial lining of the digestive tube. The outpocketing of the archenteron is the origin of **mesoderm.** This third germ layer will form the muscular system, reproductive system, peritoneum (lining of the coelomic compartments), and the calcareous plates of the sea star's endoskeleton.

Gastrulation in nemertean worms (see Figure 8-11C) resembles gastrulation in sea stars, in that the archenteron is formed by invagination. However, in nemerteans and other members of the Protostomia ("mouth first"), the blastopore becomes the mouth and the anus forms secondarily (see Figure 8-9). In addition, mesoderm forms differently in protostomes and deuterostomes. In protostomes, cells destined to become mesoderm arise ventrally at the lip of the blastopore and proliferate between the walls of the archenteron (endoderm) and outer body wall (ectoderm). Meticulous cell lineage studies by early embryologists established that in many protostomes (for example, flatworms, annelids, and molluscs) these mesodermal precursors arise from a single large blastomere at the 29- to 64-cell stage embryo called the 4d cell (see Figure 10-13, p. 205). In most nemerteans, the precise origin of mesoderm is not yet known; in some it is probably the 4d cell, but in others it apparently derives from an earlier blastomere.

In frogs, deuterostomes with radial cleavage (see Figure 8-7B), morphogenetic movements of gastrulation are greatly influenced by the mass of inert yolk in the vegetal half of the embryo. Cleavage divisions are slowed in this half so that the resulting blastula consists of many small cells in the animal half and a few large cells in the vegetal half (see Figure 8-11B). Gastrulation in amphibians begins when cells located at the future dorsal side of the embryo invaginate to form a slitlike blastopore. Thus, as in sea stars, invagination initiates archenteron formation, but amphibian gastrulation begins in the marginal zone of the blastula, where animal and vegetal hemispheres come together, and where there is less yolk than in the vegetal region. Gastrulation progresses as sheets of cells in the marginal zone turn inward over the blastopore lip and move inside the gastrula to form mesoderm and endoderm (see opening figure of this chapter, p. 151). The three germ layers now formed are the primary structural layers that play crucial roles in further differentiation of the embryo.

In bird and reptile embryos (see Figure 8-11D), gastrulation begins with a thickening of the blastoderm at the caudal end of the embryo that migrates forward to form a **primitive streak** (Figure 8-12). The primitive streak becomes the anteroposterior axis of the embryo and the center of early growth. The primitive streak is homologous to the blastopore of frog embryos, but in chicks it does not open into the gut cavity because of the obstructing mass of yolk. The blastoderm consists of two layers (epiblast and hypoblast) with a blastocoel between them. Cells of the epiblast move as a sheet toward the primitive streak, then roll over the edge and migrate as individual cells into the blastocoel. These migrating cells separate into two streams. One stream of cells moves deeper (displacing the hypoblast along the midline)

and forms endoderm. The other stream moves between the epiblast and hypoblast to form mesoderm. Cells on the surface of the embryo compose the ectoderm. The embryo now has three germ layers, at this point arranged as sheetlike layers with ectoderm on top and endoderm at the bottom. This arrangement changes, however, when all three germ layers lift from the underlying yolk (Figure 8-12), then fold under to form a three-layered embryo that is pinched off from the yolk except for a stalk attachment to the yolk at midbody.

Gastrulation in mammals is remarkably similar to gastrulation in reptiles and birds (see Figure 8-11E). Gastrulation movements in the inner cell mass produce a primitive streak. Epiblast cells move medially through the primitive streak into the blastocoel, and individual cells then migrate laterally through the blastocoel to form mesoderm and endoderm. Endoderm cells (derived from the hypoblast) form a yolk sac devoid of yolk (since mammalian embryos derive nutrients directly from the mother via the placenta).

Amphibians, reptiles, and birds, which have moderate to large amounts of yolk concentrated in the vegetal region of the

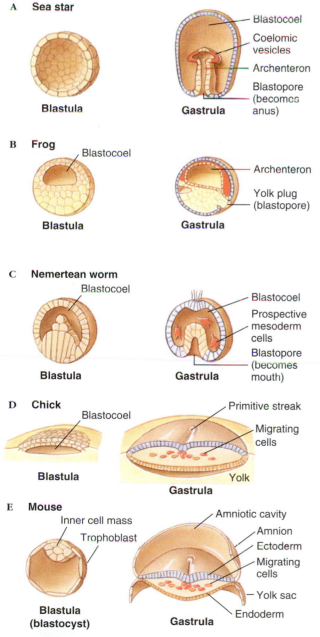

Figure 8-11
Blastula and gastrula stages in embryos of sea star, frog, nemertean worm, chick, and mouse.

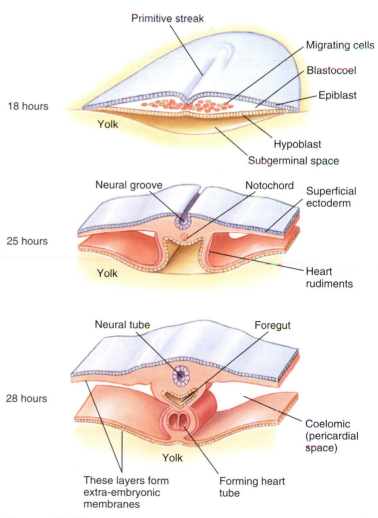

Figure 8-12
Gastrulation in a chick. Transverse sections through the heart-forming region of the chick show development at 18, 25, and 28 hours of incubation.

egg, have evolved derived gastrulation patterns in which the yolk does not participate in gastrulation. Yolk is an impediment to gastrulation and consequently the gastrulation process occurs around (amphibians) or on top (reptiles and birds) of the vegetal yolk. Mammalian eggs are isolecithal, and thus one might expect them to have a gastrulation pattern similar to that of sea stars. Instead they have a pattern more suited to telolecithal eggs. The best explanation for this feature of mammalian egg development is common ancestry with birds and reptiles. Reptiles, birds, and mammals share a common ancestor whose eggs were telolecithal. Thus, all three groups inherited their gastrulation patterns from this common ancestor, and mammals subsequently evolved isolecithal eggs but retained the telolecithal pattern.

In Cnidaria and Ctenophora, only two germ layers are formed, endoderm and ectoderm. These animals are **diploblastic.** In all other metazoa except sponges, the mesoderm also appears, either from pouches of the archenteron or from other cells associated with endoderm formation. This three-layered condition is called **triploblastic.**

Formation of the Coelom

A coelom, or true body cavity that contains the viscera, may be formed by one of two methods (see Figure 8-9)—**schizocoely** (Gr. *schizein,* to split, + *koilos,* hollow or cavity) or **enterocoely** (Gr. *enteron,* gut)—or by modification of these methods. In schizocoelous formation, the coelom arises, as the word implies, from splitting of mesodermal bands that originate from the blastopore region and grow between ectoderm and endoderm; in enterocoelous formation, the coelom comes from pouches of the archenteron, or primitive gut.

These two quite different origins for a coelom are another expression of the deuterostome-protostome dichotomy of bilateral animals. The coelom of protostomes develops by the schizocoelous method. Deuterostomes primitively follow the enterocoelous plan. Vertebrates, however, are exceptions to this distinction because their coelom is formed by mesodermal splitting (schizocoelous). This is a derived condition that evolved in early vertebrates to accommodate large stores of yolk during development.

MECHANISMS OF DEVELOPMENT

Nuclear Equivalence

How does a developing embryo generate the multitude of many cell types of a complete multicellular organism from the starting point of a single diploid nucleus of a zygote? To many nineteenth- century embryologists there seemed only one acceptable answer: as cell division ensued, hereditary material had to be parceled unequally to daughter cells. In this view, the genome gradually became broken into smaller and smaller units until finally only the information required to impart the characteristics of a single cell type remained. This became known as the Roux-Weismann hypothesis, after the two German embryologists who developed the concept.

However, in 1892 Hans Driesch discovered that if he mechanically shook apart a two-celled sea urchin into separate cells, both half-embryos developed into normal larvae. Driesch concluded that both cells contained all the genetic information of the original zygote. Still, this experiment did not settle the argument, because many embryologists believed that even if all cells contained complete genomes, the nuclei might become progressively modified in some way to dispense with the information they do not use in forming differentiated cells.

The efforts of Hans Driesch to disrupt egg development are poetically described by Peattie: "Behold Driesch grinding the eggs of Loeb's favorite sea urchin up between plates of glass, pounding and breaking and deforming them in every way. And when he ceased from thus abusing them, they proceeded with their orderly and normal development. Is any machine conceivable, Driesch asks, which could thus be torn down . . . have its parts all disarranged and transposed, and still have them act normally? One cannot imagine it. But of the living egg, fertilized or not, we can say that there lie latent within it all the potentialities presumed by Aristotle, and all of the sculptor's dream of form, yes, and the very power in the sculptor's arm." From Peattie, D. C. 1935. *An Almanac for Moderns.* New York, G. P. Putnam's Sons.

Around the turn of the century Hans Spemann introduced a new approach to testing the Roux-Weismann hypothesis. Spemann placed minute ligatures of human hair around salamander zygotes just as they were about to divide, constricting them until they were almost, but not quite, separated into two halves. The nucleus lay in one half of the partially divided zygote; the other side was anucleate, containing only cytoplasm. The zygote then completed its first cleavage division on the side containing the nucleus; the anucleate side remained undivided. Eventually, when the nucleated side had divided into about 16 cells, one of the cleavage nuclei would wander across the narrow cytoplasmic bridge to the anucleate side. Immediately this side began to divide and developed normally.

Sometimes, however, Spemann observed that the nucleated half of the embryo developed only into an abnormal ball of "belly" tissue. The explanation, Spemann discovered, depended on the presence of the gray crescent, a pigment-free area shown in Figure 8-7B. The gray crescent is required for normal development because it is the precursor of the Spemann organizer discussed in the opening essay on p. 151.

Spemann's experiment demonstrated that every blastomere contains sufficient genetic information for the development of a complete animal. In 1938 he suggested another experiment that would demonstrate that even somatic cells of an adult contain a complete genome. The experiment, which Spemann characterized as being "somewhat fantastical" at that time, would be to remove the nucleus of an egg cell and replace it with the nucleus from a somatic cell from a different individual. If all cells contained the same genetic information as a zygote, then the embryo should develop into an individual that is genetically identical to the animal from which the nucleus was obtained. It took several decades to solve the technical difficulties, but the experiment was successfully performed on amphibians, and today it is done in a variety of mammals. The procedure is now familiarly known as **cloning.** One of the most famous cloned mammals, Dolly the sheep, got the genetic material in her nuclei from the mammary glands of a six-year-old ewe.

If all nuclei are equivalent, what causes some cells to develop into neurons while others develop into skeletal muscle? In most animals (excluding insects), there are two major ways by which cells become committed to particular developmental fates: (1) cytoplasmic segregation of determinative molecules during cleavage and (2) interaction with neighboring cells (inductive interactions). All animals use both of these mechanisms to some extent to specify different cell types. However, in some animals cytoplasmic specification is dominant, whereas others rely predominantly on inductive interactions.

Cytoplasmic Specification

A fertilized egg contains cytoplasmic components that are unequally distributed within the egg. These different cytoplasmic components are thought to contain morphogenetic determinants that control commitment of the cell to a particular cell type. These morphogenetic determinants are partitioned among different blastomeres as a result of cleavage, and the developmental fate of each cell becomes specified by the type of cytoplasm it acquires during development.

This process is especially striking (and easily visualized) in some tunicate species in which the fertilized egg contains as many as five differently colored types of cytoplasm (Figure 8-10). These differently pigmented cytoplasms are segregated into different blastomeres which then proceed to form distinct tissues or organs. For example, yellow cytoplasm gives rise to muscle cells while gray equatorial cytoplasm produces notochord and neural tube. Clear cytoplasm produces larval epidermis and gray vegetal cytoplasm gives rise to the gut.

Another characteristic of this type of specification is that cell fate is determined without reference to neighboring cells. When a particular blastomere is isolated from the rest of the embryo, it still forms its characteristic structure (Figure 8-13B). In the absence of a particular blastomere, the animal lacks just those structures normally formed by that blastomere. This pattern is called **mosaic development,** since the embryo seems to be a mosaic of self-differentiating parts. Mosaic development is characteristic of most protostomes (see Figure 8-9).

In many animals, the fate of a cell depends on its interactions with neighboring cells, rather than on what piece of cytoplasm it acquired during cleavage. In these embryos, at least early in development, each cell is able to produce an entire embryo if separated from other cells (see Figure 8-13A). In other words, an early blastomere originally has the ability to follow more than one path of differentiation, but its interaction with other cells restricts its fate. If a blastomere is removed from an early embryo, the remaining blastomeres can alter their normal fates to compensate for the missing blastomere and produce a complete

organism. This adaptability is termed **regulative development.** Regulative development occurs in most deuterostomes (excluding tunicates) (see Figure 8-9).

Embryonic Induction

Induction, the capacity of some cells to evoke a specific developmental response in others, is a widespread phenomenon in development. The classic experiments, cited in the opening essay on p. 151, were reported by Hans Spemann and Hilde Mangold in 1924. When a piece of dorsal blastopore lip from a salamander gastrula was transplanted into a ventral or lateral position of another salamander gastrula, it invaginated and developed a notochord and somites. It also induced the *host* ectoderm to form a neural tube. Eventually a whole system of organs developed where the graft was placed, and then grew into a nearly complete secondary embryo (Figure 8-14). This creature was composed partly of grafted tissue and partly of induced host tissue.

It was soon found that *only* grafts from the dorsal lip of the blastopore were capable of inducing the formation of a complete or nearly complete secondary embryo. This area corresponds to the presumptive areas of notochord, somites, and prechordal plate. It was also found that only ectoderm of the host would develop a nervous system in the graft and that the reactive ability was greatest at the early gastrula stage and declined as the recipient embryo got older.

Spemann designated the dorsal lip area the **primary organizer** because it was the only tissue capable of inducing the

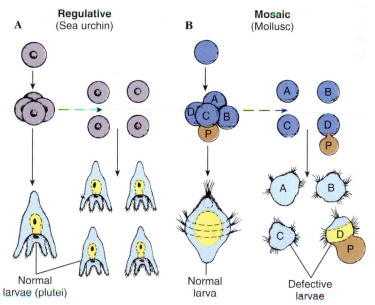

Figure 8-13

Regulative and mosaic cleavage. **A,** Regulative cleavage. Each of the early blastomeres (such as that of a sea urchin) when separated from the others develops into a small pluteus larva. **B,** Mosaic cleavage. In a mollusc, when blastomeres are separated, each gives rise to only a part of an embryo. The larger size of one defective larva is the result of the formation of a polar lobe (P) composed of clear cytoplasm of the vegetal pole, which this blastomere alone receives.

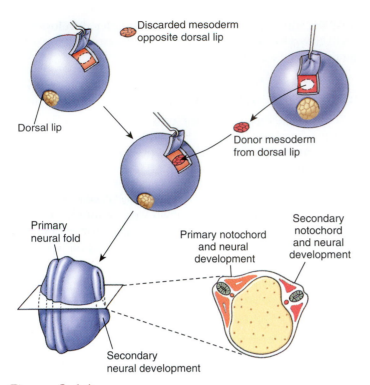

Figure 8-14
The Spemann-Mangold primary organizer experiment.

development of a secondary embryo in the host. It is now often called the Spemann organizer. Spemann also termed this inductive event **primary induction** because he believed it to be the first inductive event in development. Subsequent studies showed that many other cell types originate by inductive interactions, a process called **secondary induction.**

Usually cells that have differentiated act as inductors for adjacent undifferentiated cells. Timing is important. Once a primary inductor sets in motion a specific developmental pattern in some cells, numerous secondary inductions follow. What emerges is a sequential pattern of development involving not only inductions but cell movement, changes in adhesive properties of cells, and cell proliferation. There is no "hard-wired" master control panel directing development, but rather a sequence of local patterns in which one step in development is a subunit of another. In showing that each step in the developmental hierarchy is a necessary preliminary for the next, Hans Spemann's induction experiments were among the most significant events in experimental embryology.

GENE EXPRESSION DURING DEVELOPMENT

Since every cell with few exceptions receives the same genetic material, cytoplasmic specification and induction must involve the activation of different combinations of genes in different cells. Understanding development is therefore ultimately a problem of understanding the genetics involved. It is not surprising that developmental genetics was first studied in the geneticists'

favorite model organism, the fruit fly *Drosophila*. These studies have been repeated in several other model animals, such as the nematode worm *Caenorhabditis elegans,* zebra fish *Danio rerio,* frog *Xenopus laevis,* chick *Gallus gallus,* and mouse *Mus musculus.* This research suggests that epigenesis proceeds in three general stages: pattern formation, determination of position in the body, and induction of limbs and organs appropriate for that position. Each stage is guided by gradients of gene products that function as **morphogens.**

Pattern Formation

The first step in organizing development of an embryo is pattern formation: determination of the front-to-rear (anteroposterior), left-to-right, and back-to-front (dorsoventral) axes. As Spemann demonstrated in salamanders, the anteroposterior axis of the embryo is determined by the Spemann organizer, located in the gray crescent of a zygote. In *Drosophila* the anteroposterior axis is determined even before an egg is fertilized. Christiane Nüsslein-Volhard and her colleagues in Germany found that this determination is due to a gradient of mRNA that is secreted into the egg by nurse cells in the mother. The end of the egg that receives the highest level of this mRNA is fated to become the anterior of the embryo and eventually of the adult. The mRNA is transcribed from a gene called *bicoid* (pronounced BICK-oyd) in the nurse cells. After an egg is fertilized, *bicoid* mRNA is translated into a protein morphogen called bicoid (not italicized) that binds to certain other genes. The products of these genes in turn activate others in a cascade that ultimately causes the production of structures appropriate to the anterior. *Bicoid* is one of about 30 maternal genes that control pattern formation in an embryo. Some of these determine the dorsoventral axis. The gene *short gastrulation* leads to development of ventral structures, such as the nerve cord.

One of the most exciting discoveries in developmental genetics has been that the developmental genes of vertebrates and many other animals are similar to those of *Drosophila;* they are conserved over a wide range of animals. A gene similar to *bicoid* is also important in pattern formation in vertebrates. In vertebrates, however, the gene, called *Pitx2,* determines positioning of certain internal organs to either the left or right side of the body. Mutations in *Pitx2* in frogs, chicks, and mice can result in the heart and stomach being on the right instead of the left side. Such mutations may also be responsible for a reversal of organ position that sometimes occurs in humans. *Pitx2* is in turn activated by a protein produced by the gene *sonic hedgehog (Shh),* which is similar to a *Drosophila* gene called *hedgehog.* (The name *hedgehog* refers to the bristly appearance of fruit flies lacking the gene. The *"sonic"* comes from the video-game character Sonic the Hedgehog.) In vertebrates, *sonic hedgehog* is active in the left side only at the anterior end of the primitive streak (Figure 8-11). *Short gastrulation* also has a counterpart in vertebrates—the gene *chordin,* which produces one of the proteins from the Spemann organizer.

In *Drosophila,* as well as other arthropods, annelid worms, chordates, and a few other groups, one important aspect of pattern formation along the anteroposterior axis is **segmentation,** also called **metamerism.** Segmentation is a division of the body into discrete segments or metameres (Figure 9.14, p. 188). The segments are identical early in development, but later activation of different combinations of genes causes each segment to form different structures. For example, the anterior segment of insect embryos will form antennae, eyes, and mouthparts, while segments farther back will form legs. Segments are obvious in insects, but in chordates segmentation is apparent only in somites that produce such structures as vertebrae and repeated muscle bands (myomeres) of fishes (Figure 24-24, p. 502). In *Drosophila* the number and orientation of segments is controlled by **segmentation genes.** There are three classes of segmentation genes: gap, pair-rule, and segment polarity. **Gap genes** are activated first and divide an embryo into regions such as head, thorax, and abdomen. **Pair-rule genes** divide these regions into segments. Finally, **segment-polarity genes,** such as *hedgehog,* organize the anterior-to-posterior structures within each segment.

Homeotic and *Hox* Genes

Segmentation genes apparently regulate expression of other genes, ensuring that they are active only in appropriate segments. Such segment-specific genes are called homeotic genes. Mutations in homeotic genes, called **homeotic mutations,** result in formation of appendages or other structures in the wrong part of the body. For example, in *Drosophila* the homeotic gene *Antennapedia,* which helps trigger development of legs, is normally active only in the thorax. If the *Antennapedia* gene is activated by a homeotic mutation in the head of a maggot, the adult will have legs in place of antennae (Figure 8-15). *Antennapedia* and some other homeotic genes, as well as many other genes involved in development, include a sequence of 180 DNA base pairs, called the **homeobox.** The homeobox produces the part of a protein that attaches to the DNA of other genes, activating or blocking their expression.

Several other homeotic and nonhomeotic genes that are clustered close to *Antennapedia* on the same chromosome in *Drosophila* also include a homeobox. Genes in this cluster are called HOM genes. HOM genes do not encode specific limbs and organs. Instead, they function by specifying the location in the body along the anteroposterior axis. Intriguingly, the order of the HOM genes within the cluster on the chromosome is the same as the order in which they are expressed along the length of the body (Figure 8-16). One of the most exciting discoveries of the late twentieth century was that genes similar to HOM genes of *Drosophila* occur in other insects, as well as in chordates and unsegmented animals such as hydra and nematode worms. They also occur in plants and yeasts, and perhaps in all eukaryotes. These genes in organisms other than *Drosophila* are usually called *Hox* genes. Like HOM genes of *Drosophila,* most *Hox* genes occur in a cluster on one chromosome. Mammals have four clusters, each on a different chromosome, with from 9 to 11 *Hox* genes each. As in *Drosophia,* the sequence of *Hox* genes within a cluster is the same as the front-to-rear order in which they are expressed in the body.

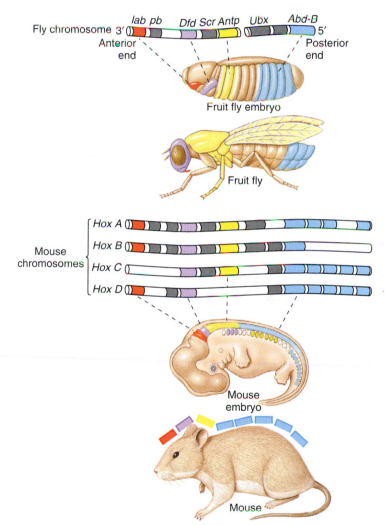

Figure 8-16

Homology of *Hox* genes in insects and mammals. These genes in both insects (fruit fly) and mammals (mouse) control the subdivision of the embryo into regions of different developmental fates along the anterior-posterior axis. The homeobox-containing genes lie on a single chromosome of the fruit fly and on four separate chromosomes in the mouse. Clearly defined homologies between the two, and the parts of the body in which they are expressed, are shown in color. The open boxes denote areas where it is difficult to identify specific homologies between the two. The *Hox* genes shown here are only a small subset of all the homeobox genes.

Figure 8-15

Head of a fruit fly with a pair of legs growing out of head sockets where antennae normally grow. The *Antennapedia* homeotic gene normally specifies the second thoracic segment (with legs), but the dominant mutation of this gene leads to this bizarre phenotype.

Morphogenesis of Limbs and Organs

Hox and other homeobox genes also play a role in shaping individual organs and limbs. As shown in Figure 8-16 and 8-17, for example, regions of the brain and identity of somites are specified by particular *Hox* and homeobox genes. Many other developmental genes that are also involved in pattern formation for the entire body also help shape individual limbs and organs by producing gradients of morphogens. One example, which has been studied by Cheryll Tickle and her coworkers at University College in London, is formation and development of limb buds in chicks. They have found that a new limb bud can be induced to grow from the side of a chick by implanting a bead soaked in fibroblast growth factor (FGF). This result implies that limbs are normally induced to develop by activation of the gene for FGF in appropriate parts of the body. Whether the limb bud develops into a wing or a leg depends on whether the FGF is applied toward the front or the rear of the chick.

FGF also plays a role in shaping the limb. It is secreted by cells in an **apical ectodermal ridge** at the end of the limb bud. FGF acts as a morphogen that forms a gradient from the apical ectodermal ridge to the base of a limb bud. This gradient helps establish a proximodistal axis—one of three axes that guide development of a limb (Figure 8-18). Fingers or toes develop at the end of the proximodistal axis with the highest level of FGF. An anteroposterior axis is established by a gradient of sonic hedgehog and ensures that fingers or toes develop in the appropriate order. Finally, Wnt7a, a protein produced by a gene that is similar to the segment-polarity gene *wingless* in *Drosophila*, helps determine the dorsoventral axis. Wnt7a makes the dorsal side of the wing or foot different from the ventral side.

Evolutionary Developmental Biology

Zoologists have always looked to embryology for clues to the evolutionary history, or phylogeny, of animals. Developmental features such as the number of germ layers and the fate of the blastopore do suggest evolutionary relationships among different phyla. Advances in development genetics, such as those just described in the previous section, have made the relationship between development and evolution even closer and have given rise to an exciting new field called evolutionary developmental biology. Evolutionary developmental biology, often nicknamed evo-devo, is based on a realization that evolution is essentially a process in which organisms become different as a result of changes in the genetic control of development. The fact that the genes that control development are similar in animals as different as fruit flies and mice offers hope that we can reconstruct the evolutionary history of animals by understanding how functioning of those genes came to differ. Evolutionary developmental biology has already contributed several exciting concepts to our thinking about evolution, but the field is so new that it would be premature to accept these concepts as established. It is best to state them as questions for further study.

Are the body plans of all bilaterally symmetric animals fundamentally similar? As noted on page 151, *chordin*, one of the genes responsible for development of the nervous system in

Control tadpole

Tadpole injected with antibodies
to *X1Hbox 1* protein

Figure 8-17

How the inhibition of a homeodomain regulatory protein alters normal development of the central nervous system of a frog tadpole. When the protein (encoded by a homeobox DNA sequence known as *X1Hbox 1*) was inactivated by antibodies directed against it, the area that should have become anterior spinal cord transformed into hindbrain instead.

the dorsal part of a frog, is similar to *short gastrulation,* which is necessary for development of the ventral nerve cord in *Drosophila.* In addition, the gene *decapentaplegic* promotes dorsal development in *Drosophila,* and the similar gene *bone morphogenetic protein-4* promotes ventral development in frogs. In other words, insects and amphibians, whose body plans look so different, actually share a similar control of dorsoventral patterning, except that one is upside down compared with the other. This finding has prompted a reappraisal of an idea first proposed by the French naturalist Etienne Geoffroy St. Hilaire in 1822 after he noticed that in a dissected lobster on its back the nerve cord was above the gut, and the heart was below it, as in a vertebrate in its normal position. The idea that a vertebrate is like an inverted invertebrate was quickly rejected, but now biologists are once more considering whether the body plans of protostomes and deuterostomes are simply inverted with respect to each other.

Can the anatomy of extinct ancestral species be inferred from the developmental genes shared by their descendants? The fact that dorsoventral patterning is similar in protostomes and deuterostomes suggests that the most recent common ancestor of these two branches had a similar dorsoventral patterning with a heart and nervous system separated by the gut. One can also infer from the similarity in HOM/*Hox* clusters in insects and chordates that the most recent common ancestor of protostomes and deuterostomes may have been segmented and that its segments differentiated by similar genes. It may also have had at least rudimentary eyes, judging from the fact that similar genes, *eyeless/Pax-6,* are involved in eye formation in a wide range of both protostomes and deuterostomes.

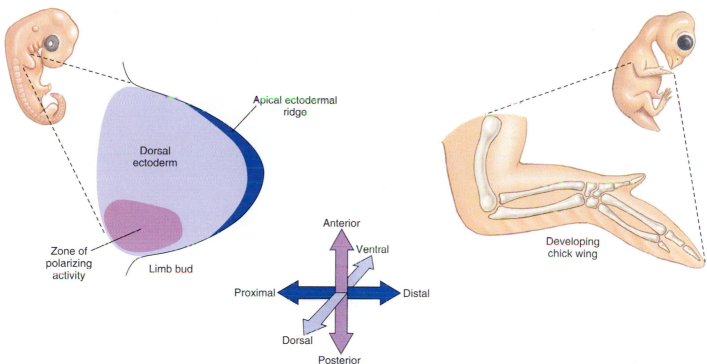

Figure 8-18

Morphogenesis in a vertebrate limb bud. The skeleton of a mature chicken limb is shown for orientation. Three axes are established in the limb bud: a proximal-distal axis by fibroblast growth factor (FGF) from the apical ectodermal ridge; an anterior-posterior axis by sonic hedgehog protein from the zone of polarizing activity; and a dorsal-ventral axis by Wnt7a protein from dorsal ectoderm.

Instead of evolution proceeding by the gradual accumulation of numerous small mutations, could it proceed by relatively few mutations in a few developmental genes? The fact that formation of legs or eyes can be induced by a mutation in one gene suggests that these and other organs develop as modules. If so, then entire limbs and organs could have been lost or acquired during evolution as a result of one or a few mutations, which would challenge Darwin's theory of gradualism (p. 115). If this is correct, then the apparently rapid evolution of numerous groups of animals during the few million years of the Cambrian Explosion and at other times is more easily explained. Instead of requiring mutations in numerous genes, each with a small effect, evolution of different groups could be a result of changes in timing, number, or expression of relatively few developmental genes.

VERTEBRATE DEVELOPMENT

The Common Vertebrate Heritage

A prominent outcome of shared ancestry of vertebrates is their common pattern of development. This common pattern is best seen in the remarkable similarity of postgastrula vertebrate embryos (Figure 8-19). The likeness occurs at a brief moment in the development of vertebrates when shared chordate hallmarks of dorsal neural tube, notochord, pharyngeal gill pouches with aortic arches, ventral heart, and postanal tail are present at about the same stage of development. Their moment of similarity—when the embryos seem almost interchangeable—is all the more extraordinary considering the great variety of eggs and widely

different types of early development that have converged toward a common design. Then, as development continues, the embryos diverge in pace and direction, becoming recognizable as members of their class, then their order, then family, and finally their species. The important contribution of early vertebrate development to our understanding of homology and evolutionary common descent is described in Chapter 6 in the section on Ontogeny, Phylogeny, and Recapitula, p. 111.

Amniotes and the Amniotic Egg

Reptiles, birds, and mammals form a monophyletic grouping of vertebrates called **amniotes,** so named because their embryos develop within a membranous sac, the **amnion.** The amnion is one of four **extraembryonic membranes** that compose a sophisticated support system within the **amniotic egg** (Figure 8-20), which evolved when the first amniotes appeared in the late Paleozoic era.

The **amnion** is a fluid-filled sac that encloses the embryo and provides an aqueous environment in which the embryo floats, protected from mechanical shock and adhesions. The shelled, amniotic egg could be buried in nests on land, thus freeing early amniotes from the aquatic environment and making possible unfettered conquest of land by vertebrates.

Evolution of the second extraembryonic membrane, the **yolk sac,** actually predates appearance of amniotes many millions of years. The yolk sac with its enclosed yolk is a conspicuous feature of all fish embryos. After hatching, a growing fish larva depends on the remaining yolk provisions to sustain it until

FISH   SALAMANDER   TORTOISE   CHICK   HUMAN

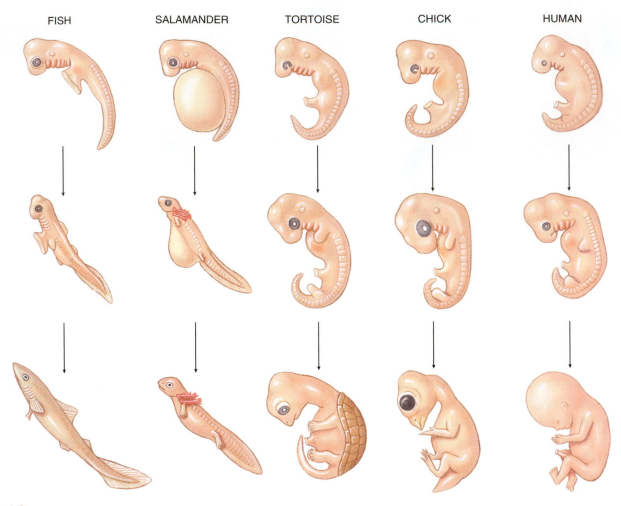

Figure 8-19

Early vertebrate embryos drawn from photographs. Embryos as diverse as fish, salamander, tortoise, bird, and human show remarkable similarity following gastrulation. At this stage (top row) they reveal features common to the entire subphylum Vertebrata. As development proceeds they diverge, each becoming increasingly recognizable as belonging to a specific class, order, family, and finally, species.

it can begin to feed itself (Figure 8-21). The mass of yolk is an extraembryonic structure because it is not a part of the embryo proper, and the yolk sac is an extraembryonic membrane because it is an accessory structure that develops outside the embryo and is discarded after the yolk is consumed.

The **allantois** is a sac that grows out of the hindgut and serves as a repository for metabolic wastes during development. It also functions as a respiratory surface for exchange of oxygen and carbon dioxide.

The **chorion** lies just beneath the eggshell and completely encloses the rest of the embryonic system. As the embryo grows and its need for oxygen increases, the allantois and chorion fuse to form the **chorioallantoic membrane.** This double membrane is provided with a rich vascular network connected to the embryonic circulation. Lying just beneath the porous shell, the vascular chorioallantois serves as a provisional "lung" across which oxygen and carbon dioxide can freely exchange. Thus an amniotic egg provides a complete life-support system for the embryo, enclosed by a tough outer shell. The amniotic egg is one of the most important adaptations to have evolved in vertebrate ancestry.

The evolution of a shelled amniotic egg made internal fertilization a reproductive requirement. A male must introduce sperm directly into the female reproductive tract, since sperm must reach and fertilize the egg before the egg shell is wrapped around it.

The Mammalian Placenta and Early Mammalian Development

Rather than developing within an egg shell like other amniotes, most mammalian embryos evolved the propitious strategy of developing within the mother's body. We have already seen that mammalian gastrulation closely parallels that of egg-laying amniotes. The earliest mammals were egg layers, and even today some mammals retain this primitive character; **monotremes** (duck-billed platypus and spiny anteater) lay large yolky eggs that closely resemble bird eggs. In **marsupials** (pouched mammals such as opossums and kangaroos), embryos develop for a time within the mother's uterus. But an embryo does not "take root" in the uterine wall, and consequently it receives little nourishment from the mother before birth. The young of marsupials

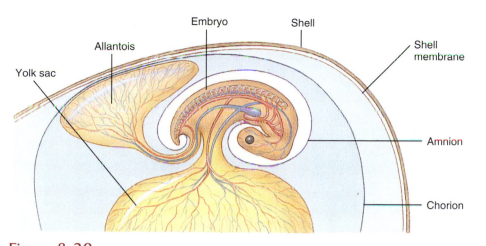

Figure 8-20

Amniotic egg at an early stage of development showing a chick embryo and its extraembryonic membranes.

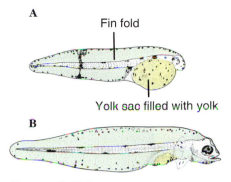

Figure 8-21

Fish larvae showing yolk sac. **A,** The one-day-old larva of a marine flounder has a large yolk sac. **B,** After 10 days of growth the larva has developed mouth, sensory organs, and a primitive digestive tract. With its yolk supply now exhausted, it must capture food to grow and survive.

are therefore born immature and are sheltered in a pouch in the mother's abdominal wall and nourished with milk (reproduction in marsupials is described on p. 597).

All other mammals, composing 94% of class Mammalia, are **placental mammals.** These mammals have evolved a **placenta,** a remarkable fetal structure through which an embryo is nourished. Evolution of this fetal organ required substantial restructuring, not only of extraembryonic membranes to form the placenta but also of the maternal oviduct, part of which had to expand into long-term housing for embryos, the **uterus.** Despite these modifications, development of extra-embryonic membranes in placental mammals is remarkably similar to their development in egg-laying amniotes (compare Figures 8-20 and 8-22).

Early stages of mammalian cleavage, shown in Figure 8-7E, occur while a **blastocyst** is traveling down the oviduct toward the uterus, propelled by ciliary action and muscular peristalsis. When a human blastocyst is about six days old and composed of about 100 cells, it contacts the uterine endometrium (uterine lining) (Figure 8-23). On contact, the trophoblast cells proliferate rapidly and produce enzymes that break down the epithelium of the uterine endometrium. These changes allow the blastocyst to

implant in the endometrium. By the eleventh or twelfth day the blastocyst is completely buried and surrounded by a pool of maternal blood. The trophoblast thickens, sending out thousands of tiny, fingerlike projections, the **chorionic villi.** These projections sink like roots into the uterine endometrium after the embryo implants. As development proceeds and embryonic demands for nutrients and gas exchange increase, the great proliferation of chorionic villi vastly increases the total surface area of the placenta. Although a human placenta at term measures only 18 cm (7 inches) across, its total absorbing surface is approximately 13 square meters—50 times the surface area of the skin of the newborn infant.

Since a mammalian embryo is protected and nourished through the placenta rather than with stored yolk, what becomes

One of the most intriguing questions the placenta presents is, why is it not immunologically rejected by the mother? Both placenta and embryo are genetically alien to the mother because they contain proteins (called major histocompatibility proteins, p. 741) that differ from those of the mother. We would expect uterine tissues to reject the embryo just as the mother would reject an organ transplanted from her own child. The placenta is a uniquely successful foreign transplant, or **allograft,** because it has evolved measures for suppressing the immune response that normally would be mounted against it and the fetus by the mother. Experiments suggest that the chorion produces proteins and lymphocytes that block the normal immune response by suppressing formation of specific antibodies by the mother.

Figure 8-22

Generalized diagram of extraembryonic membranes of a mammal, showing how their development parallels that of a chick (compare with Figure 8-20). Most extraembryonic membranes of mammals have been redirected to new functions.

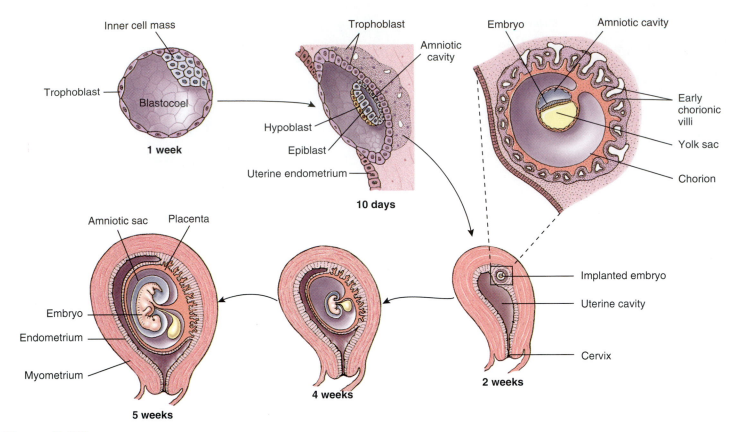

Figure 8-23

Early development of the human embryo and its extraembryonic membranes.

of the four extraembryonic membranes it has inherited from early amniotes? The amnion remains unchanged, a protective water jacket in which the embryo floats. A fluid-filled yolk sac is also retained, although it contains no yolk. It has acquired a new function: during early development it is the source of stem cells that give rise to blood and lymphoid cells. These stem cells later migrate into the developing embryo. The two remaining extraembryonic membranes, allantois and the chorion, are recommitted to new functions. The allantois is no longer needed for storage of metabolic wastes. Instead it contributes to the **umbilical cord,** which links the embryo physically and functionally with the placenta. The chorion, the outermost membrane, forms most of the placenta itself. The rest of the placenta is formed by the adjacent uterine endometrium.

The embryo grows rapidly, and in humans all major organs of the body have begun their formation by the end of the fourth week of development. The embryo is now about 5 mm in length and weighs approximately 0.02 g. During the first two weeks of development (**germinal period**) the embryo is quite resistant to outside influences. However, during the next eight weeks, when all major organs are being established and body shape is forming (**embryonic period**), an embryo is more sensitive to disturbances that might cause malformations (such as exposure to alcohol or drugs taken by the mother) than at any other time in its development. The embryo becomes a **fetus** at approximately two months after fertilization. This ushers in the **fetal period,** which is primarily a growth phase, although some organ systems (especially the nervous and endocrine systems) will continue to develop. The fetus grows from approximately 28 mm and 2.7 g at 60 days to approximately 350 mm and 3000 g at term (nine months).

DEVELOPMENT OF SYSTEMS AND ORGANS

During vertebrate gastrulation the three germ layers are formed. These differentiate, as we have seen, first into primordial cell masses and then into specific organs and tissues. During this process, cells become increasingly committed to specific directions of differentiation. Derivatives of the three germ layers are diagrammed in Figure 8-24.

Assignment of early embryonic layers to specific "germ layers" (not to be confused with "germ cells," which are the eggs and sperm) is for the convenience of embryologists and is of no concern to the embryo. Whereas the three germ layers normally differentiate into the tissue and organs described here, it is not the germ layer itself that determines differentiation, but rather the precise position of an embryonic cell with relation to other cells.

Figure 8-24
Derivatives of the primary germ layers in mammals.

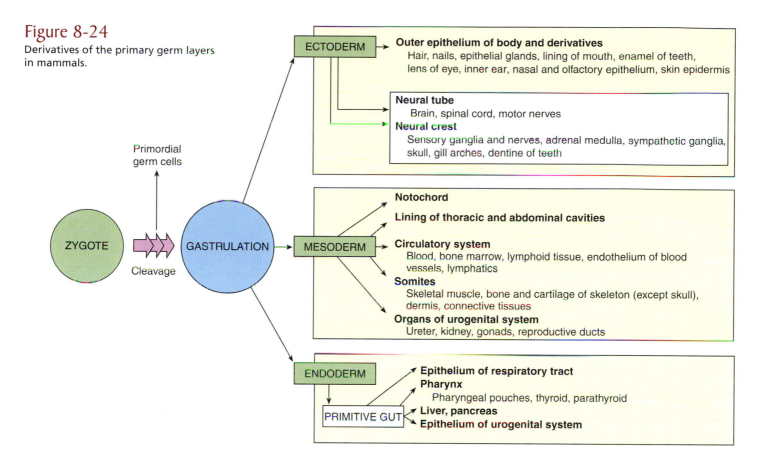

Derivatives of Ectoderm:
Nervous System and Nerve Growth

The brain, spinal cord, and nearly all outer epithelial structures of the body develop from primitive ectoderm. They are among the earliest organs to appear. Just above the notochord, the ectoderm thickens to form a **neural plate.** The edges of this plate rise up, fold, and join together at the top to create an elongated, hollow **neural tube.** The neural tube gives rise to most of the nervous system: anteriorly it enlarges and differentiates into the brain and cranial nerves; posteriorly it forms the spinal cord and spinal motor nerves. Much of the rest of the peripheral nervous system is derived from **neural crest cells,** which pinch off from the neural tube before it closes (Figure 8-25). Among the multitude of different cell types and structures that originate with the neural crest are portions of the cranial nerves, pigment cells, cartilage and bone of most of the skull (including jaws), ganglia of the autonomic nervous system, medulla of the adrenal gland, and contributions to several other endocrine glands. Neural crest tissue is unique to vertebrates and was probably of prime importance in evolution of the vertebrate head and jaws.

How are the billions of nerve axons in the body formed? What directs their growth? Biologists were intrigued with these questions, which seemed to have no easy solutions. Since a single nerve axon may be more than a meter in length (for example, motor nerves running from the spinal cord to the toes), it seemed impossible that a single cell could reach out so far. The answer had to await the development of one of the most powerful tools available to biologists, the cell culture technique.

In 1907 embryologist Ross G. Harrison discovered that he could culture living neuroblasts (embryonic nerve cells) for weeks outside the body by placing them in a drop of frog lymph hung from the underside of a cover slip. Watching nerves grow for periods of days, he saw that each axon was an outgrowth of a single cell. As the axon extended outward, materials for growth flowed down the axon center to the growing tip (growth cone) where they were incorporated into new protoplasm (Figure 8-26).

The second question—what directs nerve growth—has taken longer to unravel. An idea held well into the 1940s was that nerve growth is a random, diffuse process. A major hypothesis proposed that the nervous system developed as an equipotential network, or blank slate, that later would be shaped by usage into a functional system. The nervous system just seemed too incredibly complex for us to imagine that nerve fibers could find their way selectively to so many predetermined destinations. Yet it appears that this is exactly what they do! Research with invertebrate nervous systems indicated that each of the billions of nerve cell axons acquires a distinct identity that in some way directs it along a specific pathway to its destination. Many years ago Harrison observed that a growing nerve axon terminated in a growth cone, from which extend numerous tiny threadlike pseudopodial processes (filopodia) (Figure 8-26). Recent research has shown that the growth cone is steered by an array of guidance molecules secreted along the pathway and by the axon's target. This chemical guidance system, which must, of course, be genetically directed, is just one example of the amazing flexibility that characterizes the entire process of differentiation.

The tissue culture technique developed by Ross G. Harrison is now used extensively by scientists in all fields of active biomedical research, not just by developmental biologists. The great impact of the technique has been felt only in recent years. Harrison was twice considered for the Nobel Prize (1917 and 1933), but he failed ever to receive the award because, ironically, the tissue culture method was then believed to be "of rather limited value."

Derivatives of Endoderm: Digestive Tube and Survival of Gill Arches

In frog embryos the primitive gut makes its appearance during gastrulation with the formation of the **archenteron.** From this simple endodermal cavity develop the lining of the digestive tract, lining of the pharynx and lungs, most of the liver and pancreas, the thyroid and parathyroid glands, and the thymus (Figure 8-24).

In other vertebrates the **alimentary canal** develops from the primitive gut and is folded off from the yolk sac by growth and folding of the body wall (Figure 8-27). The ends of the tube open to the exterior and are lined with ectoderm, whereas the rest of the tube is lined with endoderm. **Lungs, liver,** and **pancreas** arise from the foregut.

Among the most intriguing derivatives of the digestive tract are the pharyngeal pouches, which make their appearance in the early embryonic stages of all vertebrates (see Figure 8-19). During development the endodermally-lined pharyngeal pouches interact with overlying ectoderm to form gill arches. In fishes, gill arches develop into gills and supportive structures and serve as respiratory organs. When early vertebrates moved onto land, gills were unsuitable for aerial respiration and were replaced by lungs.

Why then do gill arches persist in embryos of terrestrial vertebrates? Certainly not for the convenience of biologists who use these and other embryonic structures to reconstruct lines of vertebrate descent. Although gill arches serve no respiratory function in either embryos or adults of terrestrial vertebrates, they remain as necessary primordia for a variety of other structures. For example, the first arch and its endoderm-lined pouch (the space between adjacent arches) form the upper and lower jaws and inner ear of vertebrates. The second, third, and fourth gill pouches contribute to the tonsils, parathyroid gland, and thymus. We can understand then why gill arches and other fishlike structures appear in early mammalian embryos. Their original function has been abandoned, but the structures are retained for new uses. The great conservatism of early embryonic development has conveniently provided us with a telescoped evolutionary history.

Derivatives of Mesoderm: Support, Movement, and Beating Heart

The mesoderm forms most skeletal and muscular tissues, the circulatory system, and urinary and reproductive organs (Figure 8-24). As vertebrates have increased in size and complexity, mesodermally derived supportive, movement, and transport structures have become an even greater proportion of the body.

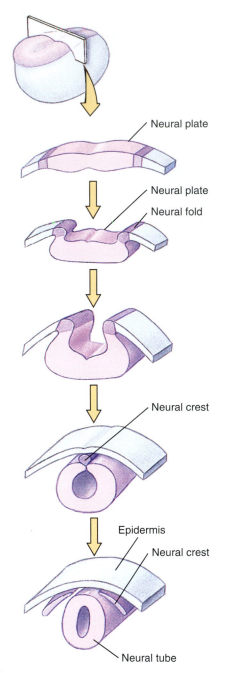

Figure 8-25

Development of neural tube and neural crest cells from neural plate ectoderm.

Most **muscles** arise from the mesoderm along each side of the neural tube (Figure 8-28). This mesoderm divides into a linear series of blocklike somites (38 in humans), which by splitting, fusion, and migration become the axial skeleton, dermis of the dorsal skin, and muscles of the back, body wall, and limbs.

Mesoderm gives rise to the first functional organ, the embryonic heart. Guided by the underlying endoderm, two clusters of precardiac mesodermal cells move ameba-like into position on either side of the developing gut. These clusters differentiate into a pair of double-walled tubes, which later fuse to form a single, thin tube (see Figure 8-12, p. 159).

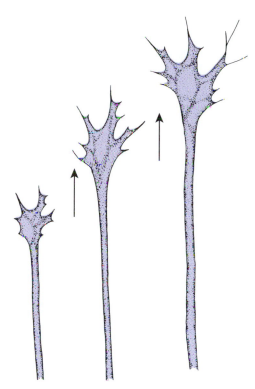

Figure 8-26
Growth cone at the growing tip of a nerve axon. Materials for growth flow down the axon to the growth cone from which numerous threadlike filopodia extend. These serve as a pioneering guidance system for the developing axon. Direction of growth is shown by arrows.

Even while the cells group together, the first twitchings are evident. In a chick embryo, a favorite animal for experimental embryology studies, the primitive heart begins to beat on the second day of the 21-day incubation period; it begins beating before any true blood vessels have formed and before there is any blood to pump. As the ventricle primordium develops, the spontaneous cellular twitchings become coordinated into a feeble but rhythmical beat. New heart chambers, each with a beat faster than its predecessor, then develop.

Finally a specialized area of heart muscle called the **sinoatrial (SA) node** develops and takes command of the entire heartbeat (the role of the SA node in the excitation of the heart is described on p. 663). The SA node becomes the heart's primary **pacemaker.** As the heart builds up a strong and efficient beat, vascular channels open within the embryo and across the yolk. Within the vessels are the first primitive blood cells suspended in plasma.

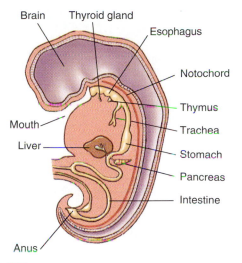

Figure 8-27
Derivatives of the alimentary canal of a human embryo.

Figure 8-28
Human embryo showing somites, which differentiate into skeletal muscles and axial skeleton.

Early development of the heart and circulation is crucial to continued embryonic development, because without a circulation an embryo could not obtain materials for growth. Food is absorbed from the yolk and carried to the embryonic body, oxygen is delivered to all tissues, and carbon dioxide and other wastes are carried away. An embryo is totally dependent on these extraembryonic support systems, and the circulation is the vital link between them.

SUMMARY

Developmental biology is concerned with the emergence of order and complexity during the development of a new individual from a fertilized egg, and with control of this process. The early preformation concept of development gave way in the eighteenth century to the theory of epigenesis, which holds that development is the progressive appearance of new structures that arise as the products of antecedent development. Fertilization of an egg by a sperm restores the diploid number of chromosomes and activates the egg for development. Both sperm and egg have evolved devices to promote efficient fertilization. The sperm is a highly condensed haploid nucleus provided with a locomotory flagellum. Many eggs release chemical sperm attractants, most have surface receptors that recognize and bind only with sperm of their own species, and all have developed devices to prevent polyspermy.

During cleavage an embryo divides rapidly and usually synchronously, producing a multicellular blastula. Cleavage is greatly influenced by quantity and distribution of yolk in the egg. Eggs with little yolk, such as those of many marine invertebrates, divide completely (holoblastic) and usually have indirect development with a larval stage interposed between the embryo and adult. Eggs having an abundance of yolk, such as those of birds, reptiles, and most arthropods divide only partially (meroblastic) and birds and reptiles have no larval stage.

Based on several developmental characteristics, bilateral metazoan animals are divided into two major groups. The Protostomia are characterized by mosaic cleavage and the mouth forming at or near the embryonic blastopore. The Deuterostomia are characterized by regulative cleavage and the mouth forming secondarily and not from the blastopore.

At gastrulation, cells on an embryo's surface move inward to form germ layers (endoderm, ectoderm, mesoderm) and the embryonic body plan. Like cleavage, gastrulation is much influenced by the quantity of yolk.

Despite the different developmental fates of embryonic cells, every cell contains a complete genome and thus the same nuclear information. Early development is governed by products of the maternal genome because the cortex of eggs contain cytoplasmic determinants, deposited during oogenesis, that guide development through cleavage. With the approach of gastrulation, control gradually shifts from maternal to embryonic as an embryo's own nuclear genes begin transcribing mRNA.

Harmonious differentiation of tissues proceeds in three general stages: pattern formation, determination of position in the body, and induction of limbs and organs appropriate for each position. Each stage is guided by morphogens. Pattern formation refers to determination of the anteroposterior, dorsoventral, and left-to-right body axes. In amphibians the anteroposterior axis is established by morphogens such as chordin from the Spemann organizer in the gray crescent of the zygote. In *Drosophila* that axis is determined by the morphogen bicoid, which is transcribed from maternal mRNA deposited at the anterior of the egg. In these and other segmented animals, such morphogens activate genes that divide the body into head, thorax and abdomen, and then into correctly oriented segments. The structures appropriate to each segment are then induced by homeotic genes, which are characterized by a particular sequence of DNA bases called the homeobox. Mutations in homeotic genes result in the development of inappropriate structures on a segment: legs on the head, for example.

The anteroposterior position of an embryo is determined by homeotic and other homeobox-containing genes that are found in one or more clusters on particular chromosomes. These genes, called *Hox* genes, occur not only in *Drosophila* and amphibians, but apparently in all animals. Each *Hox* gene is active in a particular region of the body, depending on its position within the cluster. Thus a *Hox* gene at one end of a cluster will be active only at the anterior of the embryo, producing morphogens that induce development of structures appropriate for the head. Dorsoventral and left-right axes are similarly determined by morphogens that are produced only in the appropriate regions of the embryo. Similarly, morphogens guide the development of limbs along three body axes. Morphogens have been found to be remarkably similar in animals as different as *Drosophila* and amphibians. This realization has given rise to the field of evolutionary developmental biology, which is based on the idea that the evolution of the enormous variety of animals has been the result of changes in the position and timing of relatively few genes that control development.

The postgastrula stage of vertebrate development represents a remarkable conservation of morphology when jawed vertebrates from fish to humans exhibit features common to all. As development proceeds, species-specific characteristics are formed.

Amniotes are terrestrial vertebrates that develop extraembryonic membranes during embryonic life. The four membranes are amnion, allantois, chorion, and yolk sac, each serving a specific life-support function for the embryo that develops within a self-contained egg (as in birds and reptiles) or within the maternal uterus (mammals).

Mammalian embryos are nourished by a placenta, a complex fetal-maternal structure that develops in the uterine wall. During pregnancy the placenta becomes an independent nutritive, endocrine, and regulatory organ for the embryo.

Germ layers formed at gastrulation differentiate into tissues and organs. The ectoderm gives rise to skin and nervous system; endoderm gives rise to alimentary canal, pharynx, lungs, and certain glands; and mesoderm forms muscular, skeletal, circulatory, reproductive, and excretory organs.

REVIEW QUESTIONS

1. What is meant by epigenesis? How did Kaspar Friedrich Wolff's concept of epigenesis differ from the early notion of preformation?
2. How is an egg (oocyte) prepared during oogenesis for fertilization? Why is preparation essential to development?
3. Describe events that follow contact of a spermatozoon with an egg. What is polyspermy and how is it prevented?
4. What is meant by the term "activation" in embryology?
5. How does amount of yolk affect cleavage? Compare cleavage in a sea star with that in a bird.
6. What is the difference between radial and spiral cleavage?
7. What other developmental hallmarks are often associated with spiral or radial cleavage?
8. What is indirect development?
9. Using sea star embryos as an example, describe gastrulation. Explain how the mass of inert yolk affects gastrulation in frog and bird embryos.
10. What is the difference between schizocoelous and enterocoelous origins of a coelom?
11. Describe two different experimental approaches that serve as evidence for nuclear equivalence in animal embryos.
12. What is meant by "induction" in embryology? Describe the famous organizer experiment of Spemann and Mangold and explain its significance.
13. What are homeotic genes and what is the "homeobox" contained in such genes? What is the function of the homeobox? What are *Hox* genes? What is the significance of their apparently universal occurrence in animals?

14. What is the embryological evidence that vertebrates form a monophyletic group?
15. What are the four extraembryonic membranes of amniotic eggs of birds and reptiles and what is the function of each membrane?
16. What is the fate of the four extraembryonic membranes in embryos of placental mammals?

17. Explain what the "growth cone" that Ross Harrison observed at the ends of growing nerve fibers does to influence direction of nerve growth.
18. Name two organ system derivatives of each of the three germ layers.

SELECTED REFERENCES

Cibelli, J. B., R. P. Lanza, and M. D. West. 2002. The first human cloned embryo. Sci. Am. **286**:44–51 (Jan.). *Describes the first cloning of human embryos—but only to the 6-cell stage. Many scientists remain skeptical.*

De Robertis, E. M., O. Guillermo, and C. V. E. Wright. 1990. Homeobox genes and the vertebrate body plan. Sci. Am. **263**:46–52 (July). *How a family of regulatory genes, first discovered in fruit flies, determines the shape of the vertebrate body.*

Gilbert, S. F. 2000. Developmental biology, ed. 6. Sunderland, Massachusetts, Sinauer Associates. *Combines descriptive and mechanistic aspects; good selection of examples from many animal groups.*

Gilbert, S. F., and A. M. Raunio (eds.). 1997. Embryology: constructing the organism. Sunderland, Massachusetts, Sinauer Associates. *The embryology of numerous animal groups.*

Goodman, C. S., and M. J. Bastiani. 1984. How embryonic nerve cells recognize one another. Sci. Am. **251**:58–66 (Dec.). *Research with insect larvae shows that developing neurons follow pathways having specific molecular labels.*

McGinnis, W., and M. Kuziora. 1994. The molecular architects of body design. Sci. Am. **270**:58–66 (Feb.). *Describes the nearly identical molecular mechanisms that define the body shapes in all animals.*

Nüsslein-Volhard, C. 1996. Gradients that organize embryo development. Sci. Am. **275**:54–61 (Aug.). *An account of the author's Nobel Prize–winning research.*

Riddle, R. D., and C. J. Tabin. 1999. How limbs develop. Sci. Am. **280**:74–79 (Feb.). *The morphogens that determine the orientation of limbs.*

Rosenberg, K. R., and W. R. Trevathan. 2001. The evolution of human birth. Sci. Am. **285**:72–77 (Nov.). *Examines reasons why humans are the only primates to seek assistance during childbirth.*

Wolpert, L. 1991. The triumph of the embryo. Oxford, Oxford University Press. *Written for the nonspecialist, this engaging book is rich in detail and insight for all biologists interested in the development of life.*

ZOOLOGY LINKS TO THE INTERNET

Visit the textbook's Online Learning Center at www.mhhe.com/zoology to find live Internet links for each of the topics listed here.
Fertilization and Development of the Embryo
Hormonal Control of Events During Gestation
Principles of Development/Embryology in Invertebrates
Principles of Development/Embryology in Vertebrates
Vertebrate Laboratory Exercises

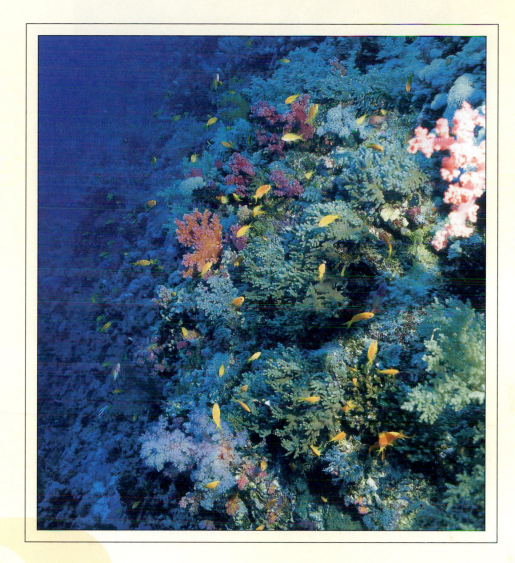

PART THREE

A view of coral reef biodiversity.

Diversity of Animal Life

9   Architectural Pattern of an Animal

10 Classification and Phylogeny of Animals

11 Protozoan Groups

12 Mesozoa and Parazoa

13 Radiate Animals

14 Acoelomate Bilateral Animals

15 Pseudocoelomate Animals

16 Molluscs

17 Segmented Worms

18 Arthropods

19 Aquatic Mandibulates

20 Terrestrial Mandibulates

21 Smaller Protostome Phyla

22 Echinoderms and Hemichordates

23 Chordates

24 Fishes

25 Early Tetrapods and Modern Amphibians

26 Amniote Origins and Reptilian Groups

27 Birds

28 Mammals

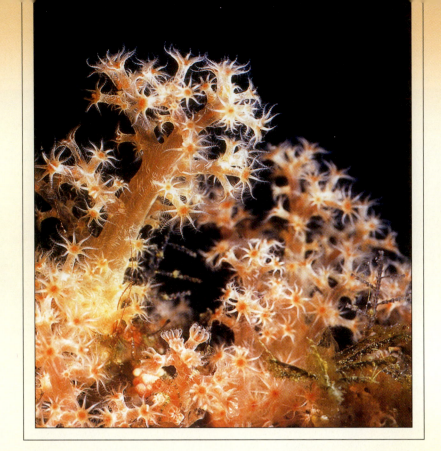

Cnidarian polyps have radial symmetry and cell-tissue grade of organization, (*Dendronephthya* sp.).

Architectural Pattern of an Animal

New Designs for Living

Zoologists today recognize 32 phyla of multicellular animals, each phylum characterized by a distinctive body plan and biological properties that set it apart from all other phyla. All are survivors of perhaps 100 phyla that appeared 600 million years ago during the Cambrian explosion, the most important evolutionary event in the geological history of animal life. Within the space of a few million years virtually all major body plans that we see today, together with many other novel plans that we know only from the fossil record, were established. Entering a world sparse in species and mostly free of competition, these new life forms began widespread experimentation, producing new themes in animal architecture. Nothing since has equaled the Cambrian explosion. Later bursts of speciation that followed major extinction events produced only variations on established themes.

Once forged, a major body plan becomes a limiting determinant of body form for descendants of that ancestral line. Molluscs beget only molluscs and birds beget birds, nothing else. Despite the appearance of structural and functional adaptations for distinctive ways of life, the evolution of new forms always develops within the architectural constraints of the phylum's ancestral pattern. This is why we shall never see molluscs that fly or birds confined within a protective shell.

The English satirist Samuel Butler proclaimed that the human body was merely "a pair of pincers set over a bellows and a stewpan and the whole thing fixed upon stilts." While human attitudes toward the human body are distinctly ambivalent, most people less cynical than Butler would agree that the body is a triumph of intricate, living architecture. Less obvious, perhaps, is that the architecture of humans and most other animals conforms to the same well-defined plan. The basic uniformity of biological organization derives from the common ancestry of animals and from their basic cellular construction. Despite vast differences of structural complexity of organisms ranging from unicellular forms to humans, all share an intrinsic material design and fundamental functional plan. In this introduction to the diversity chapters (Chapter 11–28), we consider the limited number of body plans that underlie the apparent diversity of animal form and examine some of the common architectural themes that animals share.

HIERARCHICAL ORGANIZATION OF ANIMAL COMPLEXITY

Among the different unicellular and metazoan groups, we recognize five major grades of organization (Table 9-1). Each grade is more complex than the one preceding, and builds on it in a hierarchical manner.

The unicellular protozoan groups are the simplest eukaryotic organisms. They are nonetheless complete organisms that perform all of the basic functions of life as seen in the more complex animals. Within the confines of their cell, they show remarkable organization and division of labor, possessing distinct supportive structures, locomotor devices, fibrils, and simple sensory structures. The diversity observed among unicellular organisms is achieved by varying the architectural patterns of subcellular structures, organelles, and the cell as a whole (Chapter 11).

The **metazoa,** or multicellular animals, evolved greater structural complexity by combining cells into larger units. A metazoan cell is a specialized part of the whole organism and, unlike a protozoan cell, it is not capable of independent existence. Cells of a multicellular organism are specialized for performing the various tasks accomplished by subcellular elements in unicellular forms. The simplest metazoans show the *cellular* grade of organization in which cells demonstrate division of labor but are not strongly associated to perform a specific collective function (Table 9-1). In the more complex **tissue** grade, similar cells are grouped together and perform their common functions as a highly coordinated unit. In animals of the tissue-organ grade of organization, tissues are assembled into still larger functional units called **organs.** Usually one type of tissue carries the burden of an organ's chief function, as muscle tissue does in the heart; other tissues—epithelial, connective, and nervous—perform supportive roles. The chief functional cells of an organ are called **parenchyma** (pa-ren´ka-ma; Gr. *para,* beside, + *enchyma,* infusion). The supportive tissues are its **stroma** (Gr. bedding). For instance, in the vertebrate pancreas the secreting cells are the parenchyma; capsule and connective tissue framework represent stroma.

Most metazoa (nemerteans and all more structurally complex phyla) have an additional level of complexity in which different organs operate together as **organ systems.** Eleven different kinds of organ systems are observed in metazoans: skeletal, muscular, integumentary, digestive, respiratory, circulatory, excretory, nervous, endocrine, immune, and reproductive. The great evolutionary diversity of these organ systems is covered in Chapters 14 through 28.

Complexity and Body Size

The opening essay (p. 176) suggests that size is a major consideration in the design of animals. The most complex grades of metazoan organization permit and to some extent even promote evolution of large body size (Figure 9-1). Large size confers several important physical and ecological consequences for an organism. As animals become larger, the body surface increases much more slowly than body volume because surface area increases as the square of body length (length^2), whereas volume (and therefore mass) increases as the cube of body length (length^3). In other words, a large animal will have less surface area relative to its volume than will a small animal of the same shape. The surface area of a large animal may be inadequate for respiration and nutrition by cells located deep within its body. There are two possible solutions to this problem. One solution is to fold or invaginate the body surface to increase the surface area or, as exploited by flatworms, flatten the body into a ribbon or disc so that no internal space is far from the surface. This solution allows a body to become large without internal complexity. However, most large animals adopted a second solution; they developed internal transport systems to shuttle nutrients, gases, and waste products between cells and the external environment.

Larger size buffers an animal against environmental fluctuations; it provides greater protection against predation and enhances offensive tactics; and it permits a more efficient use of metabolic energy. A large mammal uses more oxygen than a small mammal, but the cost of maintaining its body temperature is less per gram of weight for a large mammal than for a small one. Large animals also can move at less energy cost than can small animals. For example, a large mammal uses more oxygen in running than a small mammal, but the energy cost of moving 1 g of its body over a given distance is much less for a large mammal than for a

The tendency for maximum body size to increase within lines of descent is known as "Cope's law of phyletic increase," named after nineteenth-century American paleontologist and naturalist Edward Drinker Cope. Cope noted that lineages begin with small organisms that give rise to larger and ultimately to giant forms. Large forms frequently become extinct, providing opportunities for new lineages, which in turn evolve larger forms. Cope's rule holds well for nonflying vertebrates and many invertebrate groups, even though Cope's Lamarckian explanation for the trend—that organisms evolved from an inner urge to attain a higher state of being (and larger size)—was preposterous. Exceptions to Cope's rule are few (but the insects are a particularly large one).

TABLE 9.1

Levels of Organization in Organismal Complexity

1. *Protoplasmic grade of organization.* Protoplasmic organization is found in unicellular organisms. All life functions are confined within the boundaries of a single cell, the fundamental unit of life. Within a cell, protoplasm is differentiated into organelles capable of performing specialized functions.

2. *Cellular grade of organization.* Cellular organization is an aggregation of cells that are functionally differentiated. A division of labor is evident, so that some cells are concerned with, for example, reproduction, others with nutrition. Such cells have little tendency to become organized into tissues (a tissue is a group of similar cells organized to perform a common function). Some flagellates, such as *Volvox*, that have distinct somatic and reproductive cells might be placed at the cellular level of organization. Many authorities also place sponges at this level.

3. *Cell-tissue grade of organization.* A step beyond the preceding is an aggregation of similar cells into definite patterns or layers, thus becoming a **tissue.** Sponges are considered by some authorities to belong to this grade, although jellyfishes and their relatives (Cnidaria) more clearly demonstrate the tissue plan. Both groups are still largely of the cellular grade of organization because most cells are scattered and not organized into tissues. An excellent example of a tissue in cnidarians is the **nerve net,** in which nerve cells and their processes form a definite tissue structure, with the function of coordination.

4. *Tissue-organ grade of organization.* An aggregation of tissues into organs is a further step in complexity. Organs are usually composed of more than one kind of tissue and have a more specialized function than tissues. This is the organizational level of flatworms (Platyhelminthes), in which there are well-defined organs such as eyespots, proboscis, and reproductive organs. In fact, the reproductive organs are well organized into a reproductive system.

5. *Organ-system grade of organization.* When organs work together to perform some function, we have the highest level of organization—an organ system. Systems are associated with basic body functions such as circulation, respiration, and digestion. The simplest animals having this type of organization are nemertean worms, which have a complete digestive system distinct from the circulatory system. Most animal phyla demonstrate this type of organization.

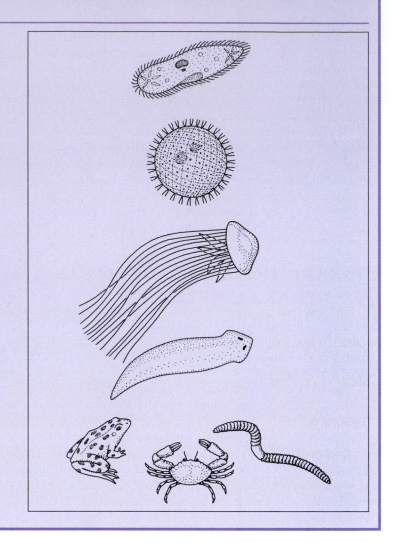

small one (Figure 9-2). For all of these reasons, ecological opportunities of larger animals are very different from those of small ones. In subsequent chapters we will describe the extensive adaptive radiations observed in taxa of large animals.

EXTRACELLULAR COMPONENTS OF METAZOAN BODIES

In addition to hierarchically arranged cellular structures discussed in the preceding section, metazoan animals contain two important noncellular components: body fluids and extracellular structural elements. In all eumetazoans, body fluids are subdivided into two fluid "compartments": those that occupy **intracellular space,** within the body's cells, and those that occupy **extracellular space,** outside the cells. In animals with closed vascular systems (such as segmented worms and vertebrates), the extracellular fluids are subdivided further into **blood plasma** (the fluid portion of blood outside the blood cells) and **interstitial fluid.**

Interstitial fluid, also called tissue fluid, occupies the space surrounding cells. Many invertebrates have open blood systems, however, with no true separation of blood plasma from interstitial fluid. We explore these relationships further in Chapter 31.

The term "intercellular," meaning "between cells," should not be confused with the term "intracellular," meaning "within cells."

If we were to remove all specialized cells and body fluids from the interior of the body, we would be left with the third element of the animal body: extracellular structural elements. This is the supportive material of the organism, including loose connective tissue (especially well developed in vertebrates but present in all metazoa), cartilage (molluscs and chordates), bone (vertebrates), and cuticle (arthropods, nematodes, annelids, and others). These elements provide mechanical stability and protection (Chapter 29). In some instances, they act also as a depot of materials for exchange, and serve as a medium for extracel-

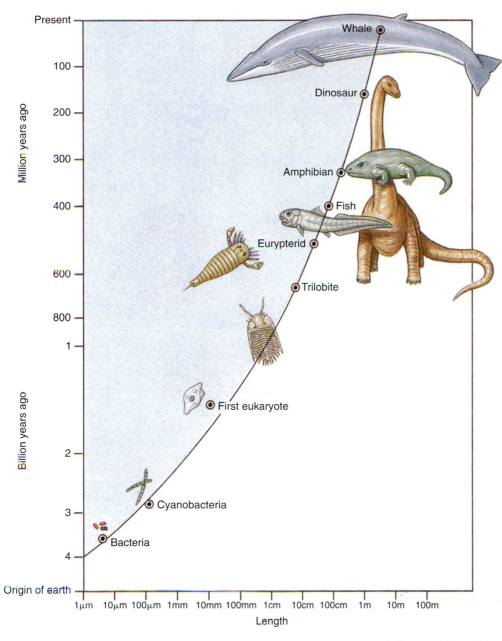

Figure 9-1
Graph showing the evolution of size (length) increase in organisms at different periods of life on earth. Note that both scales are logarithmic.

Figure 9-2
Net cost of running for mammals of various sizes. Each point represents the cost (measured in rate of oxygen consumption) of moving 1 g of body over 1 km. Cost decreases with increasing body size.

lular reactions. We describe diversity of extracellular skeletal elements characteristic of different groups of animals in Chapters 15 through 28.

TYPES OF TISSUES

A **tissue** is a group of similar cells (together with associated cell products) specialized for performance of a common function. The study of tissues is called **histology** (Gr. *histos,* tissue, + *logos,* discourse). All cells in metazoan animals form tissues. Sometimes cells of a tissue may be of several kinds, and some tissues have many intercellular materials.

During embryonic development, the germ layers become differentiated into four kinds of tissues. These are epithelial, connective, muscular, and nervous tissues (Figure 9-3). This is a surprisingly short list of only four basic tissue types that are able to meet the diverse requirements of animal life.

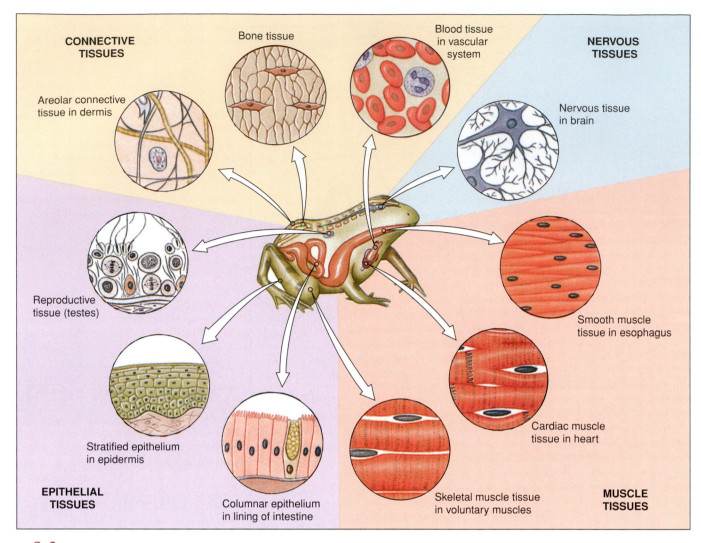

CONNECTIVE TISSUES

Areolar connective tissue in dermis

Bone tissue

Blood tissue in vascular system

NERVOUS TISSUES

Nervous tissue in brain

Reproductive tissue (testes)

Smooth muscle tissue in esophagus

Stratified epithelium in epidermis

Cardiac muscle tissue in heart

EPITHELIAL TISSUES

Columnar epithelium in lining of intestine

Skeletal muscle tissue in voluntary muscles

MUSCLE TISSUES

Figure 9-3
Types of tissues in a vertebrate, showing examples of where different tissues are located in a frog.

Epithelial Tissue

An **epithelium** (pl., epithelia) is a sheet of cells that covers an external or internal surface. Outside the body, epithelium forms a protective covering. Inside, epithelium lines all organs of the body cavity, as well as ducts and passageways through which various materials and secretions move. Thus, ions and molecules must pass through epithelial cells as they move to and from all other cells of the body. Consequently a large variety of transport molecules are located on epithelial cell membranes (see Chapter 3). On many surfaces epithelial cells are modified into glands that produce lubricating mucus or specialized products such as hormones or enzymes.

Epithelia are classified by cell form and number of cell layers. Simple epithelia (a single layer of cells; Figure 9-4) are found in all metazoan animals, while stratified epithelia (many cell layers; Figure 9-5) are mostly restricted to vertebrates. All types of epithelia are supported by an underlying basement membrane, which is a condensed region of ground substance of connective tissue, but is secreted by both epithelial and connective tissue

cells. Blood vessels never penetrate into epithelial tissues, which depend on diffusion of oxygen and nutrients from underlying tissues.

Connective Tissue

Connective tissues are a diverse group of tissues that serve various binding and supportive functions. They are so widespread in the body that removal of other tissues would still leave the complete form of the body clearly apparent. Connective tissue is composed of relatively few cells, a great many extracellular fibers, and a **ground substance** (also called **matrix**), in which the fibers are embedded. We recognize several different types of connective tissue. Two kinds of **connective tissue proper** occur in vertebrates. **Loose connective tissue** is composed of fibers and both fixed and wandering cells suspended in a viscous fluid ground substance. **Dense connective tissue,** such as tendons and ligaments, is composed largely of densely packed fibers (Figure 9-6). Much of the fibrous tissue of connective tissue is composed of **collagen** (Gr. *kolla,* glue, + *genos,* descent),

Simple squamous epithelial cell   Basement membrane   Nucleus   Free surface

Simple squamous epithelium

Simple squamous epithelium, composed of flattened cells that form a continuous delicate lining of blood capillaries, lungs, and other surfaces where it permits the passive diffusion of gases and tissue fluids into and out of cavities.

Simple cuboidal epithelial cell   Basement membrane   Lumen (free space)

Simple cuboidal epithelium

Simple cuboidal epithelium is composed of short, boxlike cells. Cuboidal epithelium usually lines small ducts and tubules, such as those of the kidney and salivary glands, and may have active secretory or absorptive functions.

Basement membrane   Epithelial cells   Microvilli on cell surface   Nuclei

Simple columnar epithelium

Figure 9-4
Types of simple epithelium.

Simple columnar epithelium resembles cuboidal epithelium, but the cells are taller and usually have elongate nuclei. This type of epithelium is found on highly absorptive surfaces such as the intestinal tract of most animals. The cells often bear minute, fingerlike projections called microvilli that greatly increase the absorptive surface. In some organs, such as female reproductive tract, the cells may be ciliated.

a protein material of great tensile strength. Collagen is the most abundant protein in the animal kingdom, found in animal bodies wherever both flexibility and resistance to stretching are required. Connective tissue of invertebrates, as in vertebrates, consists of cells, fibers, and ground substance, but it is not as elaborately developed.

Other types of specialized connective tissue include **blood, lymph, tissue fluid** (collectively considered vascular tissue), **adipose** (fat) tissue, **cartilage** and **bone.** Vascular tissue is com-posed of distinctive cells in a fluid ground substance, the plasma. Vascular tissue lacks fibers under normal conditions. Blood composition is discussed in Chapter 31.

Cartilage is a semi-rigid form of connective tissue with closely packed fibers embedded in a gel-like ground substance (matrix). **Bone** is a calcified connective tissue containing calcium salts organized around collagen fibers (see Figure 9-6). Structure of cartilage and bone is covered in the section on skeletons in Chapter 29.

Free surface

Stratified squamous
epithelial cell

Nuclei

Basement
membrane

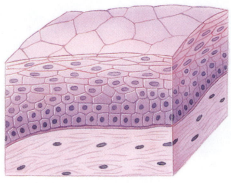

Stratified squamous epithelium
consists of two to many layers of cells
adapted to withstand mild mechanical
abrasion and distortion. The basal layer
of cells undergoes continuous mitotic
divisions, producing cells that are pushed
toward the surface where they are
sloughed off and replaced by new cells
from beneath. This type of epithelium
lines the oral cavity, esophagus, and anal
canal of many vertebrates, and the vagina
of mammals.

Stratified squamous epithelium

Free surface   Basement
membrane

Connective   Nucleus   Transitional
tissue   epithelial cell

Transitional epithelium—unstretched

Transitional epithelium is a type of
stratified epithelium specialized to
accommodate great stretching. This type
of epithelium is found in the urinary
tract and bladder of vertebrates. In the
relaxed state it appears to be four or
five cell layers thick, but when stretched
it appears to have only two or three
layers of extremely flattened cells.

Transitional epithelium—Stretched

Figure 9-5
Types of stratified epithelium.

Muscular Tissue

Muscle is the most abundant tissue in the body of most animals. It originates (with few exceptions) from mesoderm, and its unit is the cell or **muscle fiber,** specialized for contraction. When viewed with a light microscope, **striated muscle** appears transversely striped (striated), with alternating dark and light bands (Figure 9-7). In vertebrates we recognize two types of striated muscle: **skeletal** and **cardiac muscle.** A third kind of muscle is **smooth** (or visceral) **muscle,** which lacks the characteristic alternating bands of the striated type (Figure 9-7). Unspecialized cytoplasm of muscles is called **sarcoplasm,** and contractile elements within the fiber are **myofibrils.** Muscular movement is covered in Chapter 29.

Nervous Tissue

Nervous tissue is specialized for reception of stimuli and conduction of impulses from one region to another. Two basic types of cells in nervous tissue are **neurons** (Gr. nerve), the basic functional unit of nervous systems, and **neuroglia** (nu-rog′le-a;

Gr. nerve, + *glia,* glue), a variety of nonnervous cells that insulate neuron membranes and serve various supportive functions. Figure 9-8 shows functional anatomy of a typical nerve cell. The functional roles of nervous tissue are treated in Chapter 33.

ANIMAL BODY PLANS

As mentioned in the prologue to this chapter, the diversity of animal body form is constrained by evolutionary history, habitat, and way of life. Although a worm that adopts a parasitic life in the intestine of a vertebrate looks and functions very differently from a free-living member of the same group, both share distinguishing hallmarks of their phylum. We consider here the limited number of basic body plans and architectural themes that underlie diversity of animal form.

Major evolutionary innovations in animal form include multicellularity, bilateral symmetry, "tube-within-a-tube" plan, and eucoelomate (true coelom) body plan. The evolutionary distributions of these body plans are shown in Figure 9-9.

Nucleus   Collagen fiber   Elastic fiber

Nucleus   Fibers

Loose connective tissue, also called areolar connective tissue, is the "packing material" of the body that anchors blood vessels, nerves, and body organs. It contains fibroblasts that synthesize the fibers and ground substance of connective tissue and wandering macrophages that phagocytize pathogens or damaged cells. The different fiber types include strong collagen fibers (thick and red in micrograph) and thin elastic fibers (black and branching in micrograph) formed of the protein elastin.

Dense connective tissue forms tendons, ligaments, and fasciae (fa'sha), the latter arranged as sheets or bands of tissue surrounding skeletal muscle. In a tendon (shown here) the collagenous fibers are extremely long and tightly packed together.

Chondrocyte   Lacuna   Matrix

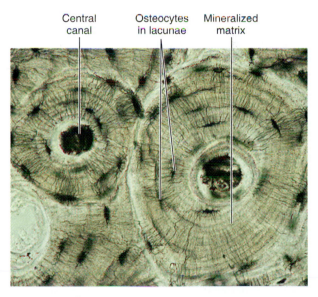

Central canal   Osteocytes in lacunae   Mineralized matrix

Cartilage is a vertebrate connective tissue composed of a firm gel ground substance (matrix) containing cells (chondrocytes) living in small pockets called lacunae, and collagen or elastic fibers (depending on type of cartilage). In hyaline cartilage shown here, both collagen fibers and matrix are stained uniformly purple and cannot be distinguished one from the other. Because cartilage lacks a blood supply, all nutrients and waste materials must diffuse through the ground substance from surrounding tissues.

Bone, strongest of vertebrate connective tissues, contains mineralized collagen fibers. Small pockets (lacunae) within the matrix contain bone cells, called osteocytes. The osteocytes communicate with each other by means of a tiny network of channels called canaliculi. Blood vessels, extensive in bone, are located in larger channels, including central canals. Unlike cartilage, bone undergoes remodeling during an animal's life, and can repair itself following even extensive damage.

Figure 9-6
Types of connective tissue.

Animal Symmetry

Symmetry refers to balanced proportions, or correspondence in size and shape of parts on opposite sides of a median plane.

Spherical symmetry means that any plane passing through the center divides a body into equivalent, or mirrored, halves (Figure 9-10, *top left*). This type of symmetry is found chiefly among some unicellular forms and is rare in animals. Spherical forms are best suited for floating and rolling.

Radial symmetry (Figure 9-10, *top right*) applies to forms that can be divided into similar halves by more than two planes passing through the longitudinal axis. These are tubular, vase, or bowl shapes found in some sponges and in hydras, jellyfish, sea

Nuclei of smooth muscle cells

Smooth muscle is nonstriated muscle found in both invertebrates and vertebrates. Smooth muscle cells are long, tapering strands, each containing a single nucleus. Smooth muscle is the most common type of muscle in invertebrates in which it serves as body wall musculature and lines ducts and sphincters. In vertebrates, smooth muscle lines the walls of blood vessels and surrounds internal organs such as intestine and uterus. It is called involuntary muscle in vertebrates because its contraction is usually not consciously controlled.

Skeletal muscle fiber   Nucleus   Striations

Skeletal muscle is a type of striated muscle found in both invertebrates and vertebrates. It is composed of extremely long, cylindrical fibers, which are multinucleate cells that may reach from one end of the muscle to the other. Viewed through the light microscope, the cells appear to have a series of stripes, called striations, running across them. Skeletal muscle is called voluntary muscle (in vertebrates) because it contracts when stimulated by nerves under conscious central nervous system control.

Note striations   Nucleus of cardiac   Intercalated discs
                   muscle cell   (special junctions
                                         between cells)

Cardiac muscle is another type of striated muscle found only in the vertebrate heart. The cells are much shorter than those of skeletal muscle and have only one nucleus per cell (uninucleate). Cardiac muscle tissue is a branching network of fibers with individual cells interconnected by junctional complexes called intercalated discs. Cardiac muscle is considered involuntary muscle because it does not require nerve activity to stimulate contraction. Instead, heart rate is controlled by specialized pacemaker cells located in the heart itself. However, autonomic nerves from the brain may alter pacemaker activity.

Figure 9-7
Types of muscle tissue.

urchins, and related groups, in which one end of the longitudinal axis is usually the mouth. A variant form is **biradial symmetry** in which, because of some part that is single or paired rather than radial, only two planes passing through the longitudinal axis produce mirrored halves. Sea walnuts (phylum Ctenophora, p. 267), which are globular but have a pair of tentacles, are an example. Radial and biradial animals are usually sessile, freely floating, or weakly swimming. Radial animals, with no front or back end, can interact with their environment in all directions—an advantage to sessile or free-floating forms with feeding structures arranged to snare prey approaching from any direction.

The two phyla that are primarily radial, Cnidaria and Ctenophora, are called the **Radiata.** Echinoderms (sea stars and their kin) are primarily bilateral animals (their larvae are bilateral) that have become secondarily radial as adults.

Bilateral symmetry applies to animals that can be divided along a sagittal plane into two mirrored portions—right and left halves (Figure 9-10, *bottom*). The appearance of bilateral sym-

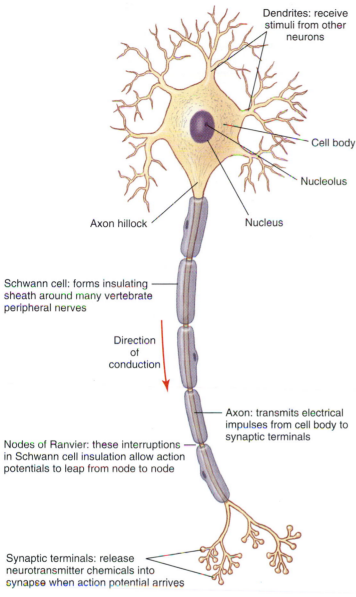

Figure 9-8

Functional anatomy of a neuron. From the nucleated cell body, or **soma,** extend one or more **dendrites** (Gr. *dendron,* tree), which receive electrical impulses from receptors or other nerve cells, and a single **axon** that carries impulses away from the cell body to other nerve cells or to an effector organ. The axon is often called a **nerve fiber.** Nerves are separated from other nerves or from effector organs by specialized junctions called synapses.

metry in animal evolution was a major innovation, because bilateral animals are much better fitted for directional (forward) movement than are radially symmetrical animals. Bilateral animals form a monophyletic group of phyla called the **Bilateria.** Bilateral symmetry is strongly associated with **cephalization,** discussed on p. 188.

Some convenient terms used for locating regions of bilaterally symmetrical animals (Figure 9-11) are **anterior,** used to designate the head end; **posterior,** the opposite or tail end; **dorsal,** the back side; and **ventral,** the front or belly side. **Medial** refers

to the midline of the body; **lateral,** to the sides. **Distal** parts are farther from the middle of the body; **proximal** parts are nearer. A **frontal plane** (sometimes called coronal plane) divides a bilateral body into dorsal and ventral halves by running through the anteroposterior axis and the right-left axis at right angles to the **sagittal plane,** the plane dividing an animal into right and left halves. A **transverse plane** (also called a cross section) would cut through a dorsoventral and a right-left axis at right angles to both the sagittal and frontal planes and would result in anterior and posterior portions (Figure 9-11). In vertebrates **pectoral** refers to the chest region or area supported by the forelegs, and **pelvic** refers to the hip region or area supported by the hind legs.

Body Cavities

A major evolutionary innovation appearing within bilateral animals is a fluid-filled space that surrounds the gut. Such a space, either a pseudocoel or a coelom, provides a tube-within-a-tube arrangement (Figure 9-12) that allows much greater flexibility of the body cavity. A pseudocoel or coelom also provides space for visceral organs and permits greater size and complexity by exposing more cells to surface exchange. These fluid-filled spaces additionally serve as hydrostatic skeletons in some forms, especially many worms, aiding in such activities as movement and burrowing.

As shown in Figure 9-9, the presence or absence of a coelom or pseudocoel is a key factor in evolution of bilateral metazoan body plans.

Acoelomate Bilateria

Many bilateral animals do not have a true coelom. In fact, flatworms and a few others have *no body cavity* surrounding the gut (Figure 9-12, *top*); they are "acoelomate" (Gr. *a,* without, + *koiloma,* cavity). The region between the ectodermal epidermis and the endodermal digestive tract is completely filled with mesoderm in the form of a spongy mass of space-filling cells, the **parenchyma.** Parenchyma is derived from an inwandering of ectodermal cells from the general surface of the early embryo. In at least some acoelomates, the parenchymal cells are cell bodies of muscle cells (see p. 276).

Pseudocoelomate Bilateria

Nematodes and several other phyla have a body cavity called a **pseudocoel** surrounding the gut. These animals have a tube-within-a-tube arrangement as do animals with a true coelom. However, unlike a true coelom, the pseudocoel is derived from the embryonic blastocoel and is in fact a persistent blastocoel. It also lacks a **peritoneum,** a thin cellular membrane derived from mesoderm that, in animals with a true coelom, lines the body cavity (Figure 9-12, *center*).

Eucoelomate Bilateria

The remaining bilateral animals possess a **true coelom** lined with mesodermal peritoneum (Figure 9-12, *bottom*). The true coelom arises within the mesoderm itself and may be formed by one of two methods, **schizocoelous** or **enterocoelous**

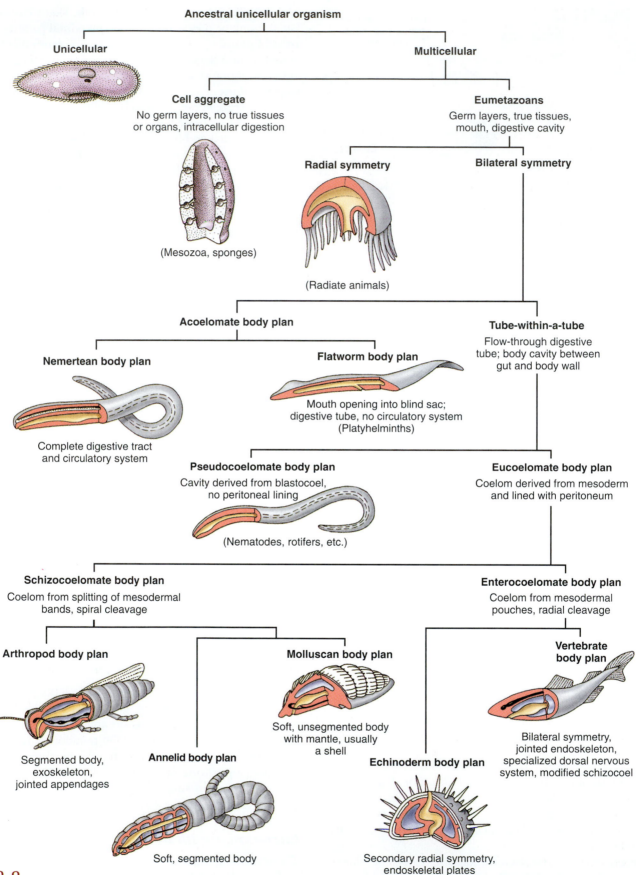

Figure 9-9

Architectural patterns of animals. These basic body plans have been variously modified during evolutionary descent to fit animals to a great variety of habitats. Ectoderm is shown in gray, mesoderm in red, and endoderm in yellow.

Spherical symmetry   Radial symmetry   Bilateral symmetry

Figure 9-10

Animal symmetry. Illustrated are animals showing spherical, radial, and bilateral symmetry.

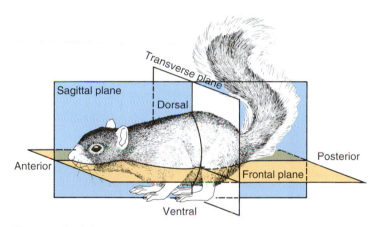

Figure 9-11

The planes of symmetry as illustrated by a bilaterally symmetrical animal.

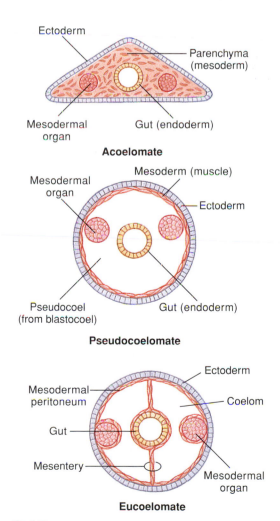

Acoelomate

Pseudocoelomate

Eucoelomate

Figure 9-12

Acoelomate, pseudocoelomate, and eucoelomate body plans.

development (Figure 9-13), or by modifications of these methods. The two terms are descriptive, for *schizo* comes from the Greek *schizein,* to split; *entero* is derived from the Greek *enteron,* meaning gut; and *coelous* comes from the Greek *koilos,* meaning hollow or cavity. In schizocoelous formation a coelom arises, as the word implies, from splitting of mesodermal bands that originate from cells in the blastopore region. (Mesoderm is one of three primary germ layers that appear very early in the development of all bilateral animals, lying between the innermost endoderm and outermost ectoderm; see Figure 8-24, p. 169). In enterocoelous formation the coelom comes from pouches of the archenteron, or primitive gut.

Once development is complete, the results of schizocoelous and enterocoelous formations are indistinguishable. Both give rise to a true coelom lined with mesodermal peritoneum (Gr. *peritonaios,* stretched around) and having mesenteries in which visceral organs are suspended.

Figure 9-13

Types of mesoderm and coelom formation. In schizocoelous formation, the mesoderm originates from the wall of the archenteron near the blastopore and proliferates into a band of tissue that splits to form the coelom. In enterocoelous formation, most mesoderm originates as a series of pouches from the archenteron; these pinch off and enlarge to form the coelom. In both formations, the coeloms expand to obliterate the blastocoel.

Metamerism (Segmentation)

Metamerism is a serial repetition of similar body segments along the longitudinal axis of the body. Each segment is called a **metamere,** or **somite.** In forms such as earthworms and other annelids, in which metamerism is most clearly represented, the segmental arrangement includes both external and internal structures of several systems. There is repetition of muscles, blood vessels, nerves, and setae of locomotion. Some other organs, such as those of sex, may be repeated in only a few somites. Evolutionary changes have obscured much of the segmentation in many animals, including humans.

The appearance of metamerism in body plans was a highly significant evolutionary event. Metamerism permits greater body mobility and complexity of structure and function. Its potential is amply displayed in phylum Arthropoda, the largest assemblage of animals on earth. Metamerism is found in phyla Annelida and Chordata in addition to Arthropoda (Figure 9-14), although superficial segmentation of ectoderm and body wall may appear among diverse groups of animals. The importance and potential of metamerism is discussed in Chapters 17 and 18.

Cephalization

Differentiation of a head is called **cephalization** and is found in bilaterally symmetrical animals. The concentration of nervous tissue and sense organs in a head bestows obvious advantages to an animal moving through its environment head first. This is the most efficient positioning of organs for sensing the environment and responding to it. Usually the mouth of an animal is located on the head as well, since so much of an animal's activity is concerned with procuring food. Cephalization is always accompanied by differentiation along an anteroposterior axis **(polarity).** Polarity usually involves gradients of activities between limits, such as between anterior and posterior ends.

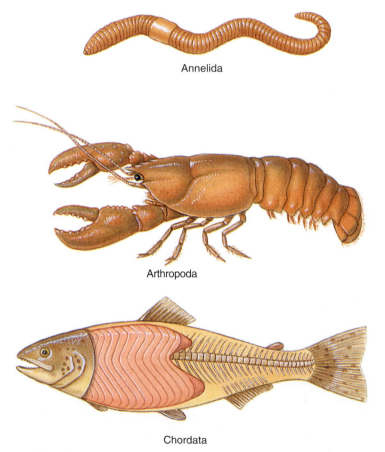

Figure 9-14

Segmented phyla. These three phyla have all made use of an important principle in nature: metamerism, or repetition of structural units. Segmentation in annelids and arthropods is homologous, but chordates apparently derived their segmentation independently. Segmentation brings more varied specialization because segments, especially in arthropods, have become modified for different functions.

SUMMARY

From the relatively simple organisms that mark the beginnings of life on earth, animal evolution has progressed through a history of more intricately organized forms. Organelles are integrated into cells, cells into tissues, tissues into organs, and organs into systems. Whereas a unicellular organism performs all life functions within the confines of a single cell, a multicellular animal is an organization of subordinate units united at successive levels.

One correlate of increased anatomical complexity is an increase in body size, which offers certain advantages such as more effective predation, reduced energy cost of locomotion, and improved homeostasis.

A metazoan body consists of cells, most of which are functionally specialized; body fluids, divided into intracellular and extracel-lular fluid compartments; and extracellular structural elements, which are fibrous or formless elements that serve various structural functions in the extracellular space. The cells of metazoa develop into various tissues composed of similar cells performing common functions. Basic tissue types are epithelial, connective, muscular, and nervous. Tissues are organized into larger functional units called organs, and organs are associated to form systems.

Every organism has an inherited body plan that may be described in terms of broadly inclusive characteristics, such as symmetry, presence or absence of body cavities, partitioning of body fluids, presence or absence of segmentation, degree of cephalization, and type of nervous system.

REVIEW QUESTIONS

1. Name the five levels of organization in organismal complexity and explain how each successive level is more complex than the one preceding it.
2. Can you suggest why, during the evolutionary history of animals, there has been a tendency for maximum body size to increase? Do you think it inevitable that complexity should increase along with body size? Why or why not?
3. What is the meaning of the terms "parenchyma" and "stroma" as they relate to body organs?
4. Body fluids of eumetazoan animals are separated into fluid "compartments." Name these compartments and explain how compartmentalization may differ in animals with open and closed circulatory systems.
5. What are the four major types of tissues in metazoans?
6. How would you distinguish between simple and stratified epithelium? What characteristic of stratified epithelium might explain why it, rather than simple epithelium, is found lining the oral cavity, esophagus, and vagina?
7. What three elements are present in all connective tissue? Give some examples of different types of connective tissue.

8. What are three different kinds of muscle found among animals? Explain how each is specialized for particular functions.
9. Describe the principal structural and functional features of a neuron.
10. Match the animal group with its body plan:
    - ____ Unicellular
    - ____ Cell aggregate
    - ____ Blind sac, acoelomate
    - ____ Tube-within-a-tube, pseudocoelomate
    - ____ Tube-within-a-tube, eucoelomate

    a. Nematode
    b. Vertebrate
    c. Protozoan
    d. Flatworm
    e. Sponge
    f. Arthropod
    g. Nemertean
11. Distinguish among spherical, radial, biradial, and bilateral symmetry.
12. Use the following terms to identify regions on your body and on the body of a frog: anterior, posterior, dorsal, ventral, lateral, distal, proximal.
13. How would frontal, sagittal, and transverse planes divide your body?
14. What is meant by metamerism? Name three phyla showing metamerism.

SELECTED REFERENCES

Arthur, W. 1997. *The origin of animal body plans.* Cambridge, U.K., Cambridge University Press. *Explores genetic, developmental, and population-level processes involved in the evolution of the 35 or so body plans that arose in the geological past.*

Bonner, J. T. 1988. *The evolution of complexity by means of natural selection.* Princeton, New Jersey, Princeton University Press. *Levels of complexity in organisms and how size affects complexity.*

Grene, M. 1987. Hierarchies in biology. Am. Sci. **75:**504–510. *The term "hierarchy" is used in many different senses in biology. Current evolutionary theory carries the hierarchical concept beyond the Darwinian restriction to the two levels of gene and organism.*

Kessel, R. G. 1998. *Basic medical histology: the biology of cells, tissues and organs.* New York, Oxford University Press. *A current textbook of animal histology.*

McGowan, C. 1999. *A practical guide to vertebrate mechanics.* New York, Cambridge University Press. *Using many examples from his earlier book,* Diatoms to dinosaurs, *the author describes principles of biomechanics that underlie functional anatomy. Includes practical experiments and laboratory exercises.*

McMahon, T. A., and J. T. Bonner. 1983. *On size and life.* New York, Scientific American Books, Inc. *A well-illustrated book about size and scale in the living world; clear examples and explanations.*

Welsch, U., and V. Storch. 1976. *Comparative animal cytology and histology.* London, Sidgwick & Jackson. *Comparative histology with good treatment of invertebrates.*

Willmer, P. 1990. *Invertebrate relationships: patterns in animal evolution.* Cambridge, U.K., Cambridge University Press. *Chapter 2 is an excellent discussion of animal symmetry, developmental patterns, origin of body cavities, and segmentation.*

ZOOLOGY LINKS TO THE INTERNET

Visit the textbook's Online Learning Center at www.mhhe.com/zoology to find live Internet links for each of the topics listed here.
Architectural Pattern and Diversity of Animals
Cells and Tissues: Histology

Basic Tissue Types
Vertebrate Laboratory Exercises
Human Organization
Animal Systems

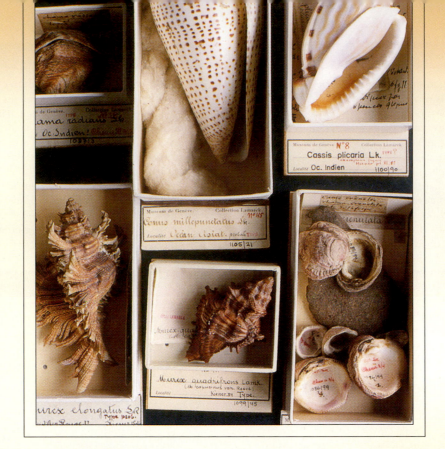

Molluscan shells from the collection of Jean Baptiste de Lamarck (1744 to 1829).

Classification and Phylogeny of Animals

Order in Diversity

Zoologists have named more than 1.5 million species of animals, and thousands more are described each year. Some zoologists estimate that the species named so far constitute less than 20% of all living animals and less than 1% of all those that have existed in the past.

Despite its magnitude, the diversity of animals is not without limits. There are many conceivable forms that do not exist in nature as our myths of minotaurs and winged horses demonstrate. Characteristic features of humans and cattle never occur together in nature as they do in the mythical minotaur. Nor do characteristic wings of birds and bodies of horses occur naturally as they do in the mythical horse, Pegasus. Humans, cattle, birds, and horses are distinct groups of animals, yet they do share some important features, including vertebrae and homeothermy, that separate them from even more dissimilar forms such as insects and flatworms.

All human cultures classify their familiar animals according to various patterns in animal diversity. These classifications have many purposes. Animals may be classified in some societies according to their usefulness or destructiveness to human endeavors. Others may group animals according to their roles in mythology. Biologists group animals according to their evolutionary relationships as revealed by ordered patterns in their sharing of homologous features. This classification is called a "natural system" because it reflects relationships that exist among animals in nature, outside the context of human activity. Systematic zoologists have three major goals: to discover all species of animals, to reconstruct their evolutionary relationships, and then to classify them accordingly.

Darwin's theory of common descent (Chapters 1 and 6) is the underlying principle that guides our search for order in the diversity of animal life. Our science of taxonomy ("arrangement law") produces a formal system for naming and classifying species that reflects this order. Animals that have very recent common ancestry share many features in common and are grouped most closely in our taxonomic classification. Taxonomy is part of the broader science of systematics, or comparative biology, in which everything that is known about animals is used to understand their evolutionary relationships. The study of taxonomy predates evolutionary biology, however, and many taxonomic practices are remnants of a pre-evolutionary world view. Adjusting our taxonomic system to accommodate evolution has produced many problems and controversies. Taxonomy has reached an unusually active and controversial point in its development in which several alternative taxonomic systems are competing for use. To understand this controversy, it is necessary first to review the history of animal taxonomy.

Figure 10-1
Carolus Linnaeus (1707 to 1778). This portrait was made of Linnaeus at age 68, three years before his death.

LINNAEUS AND THE DEVELOPMENT OF CLASSIFICATION

The Greek philosopher and biologist Aristotle was the first to classify organisms on the basis of their structural similarities. The flowering of systematics in the eighteenth century culminated in the work of Carolus Linnaeus (Figure 10-1), who gave us our current scheme of classification.

Linnaeus was a Swedish botanist at the University of Uppsala. He had a great talent for collecting and classifying objects, especially flowers. Linnaeus produced an extensive system of classification for both plants and animals. This scheme, published in his great work, *Systema Naturae,* used morphology (the comparative study of organismal form) for arranging specimens in collections. He divided the animal kingdom into species and gave each one a distinctive name. He grouped species into genera, genera into orders, and orders into classes. Because his knowledge of animals was limited, his lower categories, such as genera, often were very broad and included animals that are only distantly related. Much of his classification has been drastically altered, but the basic principle of his scheme is still followed.

Linnaeus's scheme of arranging organisms into an ascending series of groups of ever-increasing inclusiveness is a **hierarchical system** of classification. Major **taxa** (sing., **taxon**), into which organisms are grouped were given one of several standard **taxonomic ranks** to indicate the general degree of inclusiveness of the group. The hierarchy of taxonomic ranks has been expanded considerably since Linnaeus's time (Table 10-1). It now includes seven mandatory ranks for the animal kingdom, in descending series: kingdom, phylum, class, order, family, genus, and species. All organisms being classified must be placed into at least seven

TABLE 10.1

Examples of Taxonomic Categories to Which Representative Animals Belong

	Human	Gorilla	Southern Leopard Frog	Katydid
Kingdom	Animalia	Animalia	Animalia	Animalia
Phylum	Chordata	Chordata	Chordata	Arthropoda
Subphylum	Vertebrata	Vertebrata	Vertebrata	Uniramia
Class	Mammalia	Mammalia	Amphibia	Insecta
Subclass	Eutheria	Eutheria	—	Pterygota
Order	Primates	Primates	Anura	Orthoptera
Suborder	Anthropoidea	Anthropoidea	—	Ensifera
Family	Hominidae	Hominidae	Ranidae	Tettigoniidae
Subfamily	—	—	Raninae	Phaneropterinae
Genus	*Homo*	*Gorilla*	*Rana*	*Scudderia*
Species	*Homo sapiens*	*Gorilla gorilla*	*Rana sphenocephala*	*Scudderia furcata*
Subspecies	—	—	—	*Scudderia furcata furcata*

The hierarchical system of classification applied to four species (human, gorilla, Southern leopard frog, and katydid). Higher taxa generally are more inclusive than lower-level taxa, although taxa at two different levels may be equivalent in content. Closely-related species are united at a lower point in the hierarchy than are distantly related species. For example, humans and gorillas are united at the level of the family (Hominidae) and above; they are united with the Southern leopard frog at the subphylum level (Vertebrata) and with the katydid at the level of the kingdom (Animalia).

taxa, one at each of the mandatory ranks. Taxonomists have the option of subdividing these seven ranks further to recognize more than seven taxa (superclass, subclass, infraclass, superorder, suborder, etc.) for any particular group of organisms. In all, more than 30 taxonomic ranks are recognized. For very large and complex groups, such as fishes and insects, these additional ranks are needed to express different degrees of evolutionary divergence. Unfortunately, they also contribute complexity to the system.

Linnaeus's system for naming species is known as **binomial nomenclature.** Each species has a latinized name composed of two words (hence binomial) printed in italics (or underlined if handwritten or typed). The first word is the name of the **genus,** which is capitalized; the second word is the **species epithet,** which is peculiar to the species within the genus and is written in lowercase (see Table 10-1). The genus name is always a noun, and the species epithet is usually an adjective that must agree in gender with the genus. For instance, the scientific name of the common robin is *Turdus migratorius* (L. *turdus,* thrush; *migratorius,* of migratory habit). The species epithet never stands alone; the complete binomial must be used to name a species. Names of genera must refer only to single groups of organisms; the same name cannot be given to two different genera of animals. The same species epithet may be used in different genera, however, to denote different and unrelated species. For example, the scientific name of the white-breasted nuthatch is *Sitta carolinensis.* The species epithet *"carolinensis"* is used in other genera for the species *Parus carolinensis* (Carolina chickadee) and *Anolis carolinensis* (green anole, a lizard) to mean "of Carolina." All ranks above the species are designated using uninomial nouns, written with a capital initial letter.

Sometimes a species is divided into subspecies using a trinomial nomenclature (see katydid example, Table 10-1, and salamander example, Figure 10-2); such species are called polytypic. The generic, specific, and subspecific names are printed in italics (underlined if handwritten or typed). A polytypic species contains one subspecies whose subspecific name is a repetition of the species epithet and one or more additional subspecies whose names differ. Thus, to distinguish geographic variants of *Ensatina eschscholtzii,* one subspecies is named *Ensatina eschscholtzii eschscholtzii,* and different subspecies names are used for each of six other subspecies (Figure 10-2). Both the genus name and species epithet may be abbreviated as shown in Figure 10-2. Formal recognition of subspecies has lost popularity among taxonomists because the boundaries between subspecies are rarely distinct. Recognition of subspecies may be based on one or a few superficial characters and does not necessarily denote an evolutionarily distinct unit. Subspecies designations, therefore, should be viewed as tentative statements indicating that the species status of the populations needs further investigation.

Figure 10-2

Geographic variation of color patterns in the salamander genus *Ensatina.* The species status of these populations has puzzled taxonomists for generations and continues to do so. Current taxonomy recognizes only a single species (*Ensatina eschscholtzii*) divided into subspecies as shown. Hybridization is evident between most adjacent populations, but studies of variation in proteins and DNA show large amounts of genetic divergence among populations. Furthermore, populations of the subspecies *E. e. eschscholtzii* and *E. e. klauberi* can overlap geographically without interbreeding.

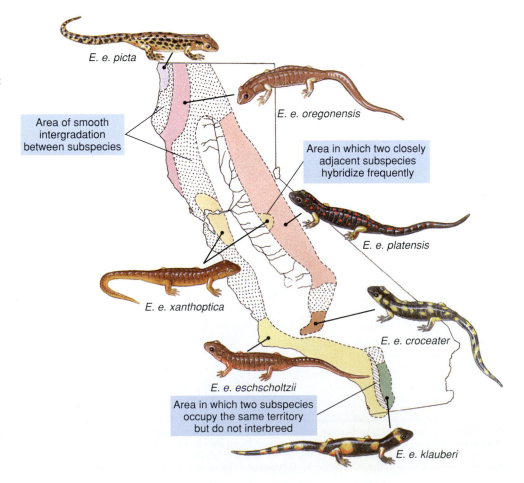

E. e. picta

E. e. oregonensis

Area of smooth intergradation between subspecies

Area in which two closely adjacent subspecies hybridize frequently

E. e. platensis

E. e. xanthoptica

E. e. croceater

E. e. eschscholtzii

Area in which two subspecies occupy the same territory but do not interbreed

E. e. klauberi

SPECIES

While discussing Darwin's book, *On the Origin of Species,* in 1859, Thomas Henry Huxley asked, "In the first place, what is a species? The question is a simple one, but the right answer to it is hard to find, even if we appeal to those who should know most about it." We have used the term "species" so far as if it had a simple and unambiguous meaning. Actually, Huxley's commentary is as valid today as it was over 140 years ago. Our concepts of species have become more sophisticated, but the diversity of different concepts and disagreements surrounding their use are as evident now as they were in Darwin's time.

Despite widespread disagreement about the nature of species, biologists have repeatedly designated certain criteria as being important to their identification of species. First, **common descent** is central to nearly all modern concepts of species. Members of a species must trace their ancestry to a common ancestral population although not necessarily to a single pair of parents. Species are thus historical entities. A second criterion is that species must be the **smallest distinct groupings** of organisms sharing patterns of ancestry and descent; otherwise, it would be difficult to separate species from higher taxa whose members also share common descent. Morphological characters traditionally have been important in identifying such groupings, but chromosomal and molecular characters increasingly are being used for this purpose. A third important criterion is that of **reproductive community.** Members of a species must form a reproductive community that excludes members of other species. For sexually reproducing populations, interbreeding is critical for maintaining a reproductive community. For organisms whose reproduction is strictly asexual, reproductive community entails occupation of a particular ecological habitat in a particular place so that a reproducing population responds as a unit to evolutionary forces such as natural selection and genetic drift.

Any species has a distribution through space, known as its **geographic range,** and a distribution through time, known as its **evolutionary duration.** Species differ greatly from each other in both of these dimensions. Species having very large geographic ranges or worldwide distributions are called **cosmopolitan,** whereas those with very restricted geographic distributions are called **endemic.** If a species were restricted to a single point in space and time, we would have little difficulty recognizing it, and nearly every species concept would lead us to the same decision. We have little difficulty distinguishing from each other the different species of animals that we can find living in our local park or woods. However, when we compare a local population of a species to similar but not identical populations located hundreds of miles away, it may be hard to determine whether these populations represent parts of a single species or different species (Figure 10-2).

Throughout the evolutionary duration of a species, its geographic range may change many times. A geographic range may be either continuous or disjunct, the latter having breaks within it where the species is absent. Suppose that we find two similar but not identical populations living 300 miles apart with no related populations between them. Are we observing a single species with a disjunct distribution or two different but closely related species? Suppose that these populations have been separated historically for 50,000 years. Is this enough time for them to have evolved separate reproductive communities, or can we still view them as being part of the same reproductive community? The answers to such questions are very hard to find. Much of the disagreement among different species concepts relates to solving these problems.

Typological Species Concept

Before Darwin, a species was considered a distinct and immutable entity. Species were defined by fixed, essential features (usually morphological) that were interpreted as a divinely created pattern or archetype. This practice constitutes the **typological (or morphological) species concept.** Scientists recognized species formally by designating a **type specimen** that was labeled and deposited in a museum to represent the ideal form or morphology for the species (Figure 10-3). When scientists obtained additional specimens and wanted to assign them to a species, the type specimens of described species were consulted. The new specimens were assigned to a previously described species if they possessed the essential features of its type specimen. Small differences from the type specimen were considered accidental imperfections. Large differences from existing type specimens would lead the scientist to describe a new species with the designation of its own type specimen. In this manner, the living world was categorized into species.

Evolutionists discarded the typological species concept, but some of its traditions remain. Scientists still name species by describing type specimens deposited in museums, and the type specimen formally bears the name of the species. Organismal morphology is likewise still important in recognizing species; however, species are no longer viewed as classes defined by possession of certain morphological features. The basis of the evolutionary world view is that species are historical entities whose properties are subject always to change. Variation that we observe among organisms within a species is not the imperfect manifestation of an eternal "type"; the type itself is only an

Figure 10-3

Specimens of birds from the Smithsonian Institution (Washington D.C.), including birds originally collected by John J. Audubon, Theodore Roosevelt, John Gould, and Charles Darwin.

abstraction taken from the very real and important variation present within the species. The type is at best an average form that will change as organismal variation is sorted through time by natural selection. The type specimen serves only as a guide to the general kinds of morphological features that we may expect to find in the species as we observe it today.

The person who first describes a type specimen and publishes the name of a species is called the authority. This person's name and date of publication are often written after the species name. Thus, *Didelphis marsupialis* Linnaeus, 1758, tells us that Linnaeus was the first person to publish the species name of the opossum. Sometimes, the generic status of a species is revised following its initial description. In this case, the name of the authority is presented in parentheses. The Nile monitor lizard is denoted *Varanus niloticus* (Linnaeus, 1766) because the species originally was named by Linnaeus as *Lacerta nilotica,* and subsequently placed into a different genus.

Biological Species Concept

The most influential concept of species inspired by Darwinian evolutionary theory is the **biological species concept** formulated by Theodosius Dobzhansky and Ernst Mayr. This concept solidified during the evolutionary synthesis of the 1930s and 1940s from earlier ideas, and it has been refined and reworked several times since then. In 1982, Mayr stated the biological species concept as follows: *"A species is a reproductive community of populations (reproductively isolated from others) that occupies a specific niche in nature."* Note that the species is identified here according to reproductive properties of populations, not according to possession of any specific organismal characteristics. The species is an **interbreeding population** of individuals having common descent and sharing intergrading characteristics. The study of populational variation in organismal morphology, chromosomal structure, and molecular genetic features will be very useful for evaluating the geographical boundaries of interbreeding populations in nature. The criterion of the "niche" (Chapter 38) recognizes that members of a reproductive community are expected also to have common ecological properties.

Because a reproductive community should maintain genetic cohesiveness, we expect that organismal variation will be relatively smooth and continuous within species and discontinuous between them. Although the biological species is based on reproductive properties of populations rather than organismal morphology, morphology nonetheless can help us to diagnose biological species. Sometimes species status can be evaluated directly by conducting breeding experiments. Controlled breeding is practical only in a minority of cases, however, and our decisions regarding species membership are usually made by studying character variation. Variation in molecular characters is very useful for identifying geographical boundaries of reproductive communities. Molecular studies have revealed the presence of cryptic or **sibling species** (p. 114) that are too similar in morphology to be diagnosed as separate species by morphological characters alone.

The biological species concept has received strong criticism because of several perceived problems. First, the concept lacks an explicit temporal dimension. It provides a means for diagnosing species status of contemporary populations but gives little guidance regarding the species status of ancestral populations relative to their evolutionary descendants. Proponents of the biological species concept often disagree on the degree of reproductive isolation necessary for considering two populations separate species, thereby revealing some ambiguity in the concept. Another problem is that because the biological species concept emphasizes interbreeding as the criterion of reproductive community, it denies the existence of species in groups of organisms that reproduce only asexually. It is common systematic practice, however, to describe species in all groups of organisms, regardless of whether reproduction is sexual or asexual.

Evolutionary Species Concept

The time dimension creates obvious problems for the biological species concept. How do we assign fossil specimens to biological species that are recognized today? If we trace a lineage backward through time, how far must we go before we have crossed a species boundary? If we could follow the unbroken genealogical chain of populations backward through time to the point where two sister species converge on their common ancestor, we would need to cross at least one species boundary somewhere. It would be very hard to decide, however, where to draw a sharp line between the two species.

To address this problem, the **evolutionary species concept** was proposed by Simpson in the 1940s to add an evolutionary time dimension to the biological species concept. This concept persists in a modified form today. A current definition of the evolutionary species is *a single lineage of ancestor-descendant populations that maintains its identity from other such lineages and that has its own evolutionary tendencies and historical fate.* Note that the criterion of common descent is retained here in the need for a lineage to have a distinct historical identity. Reproductive cohesion is the means by which a species maintains its identity from other such lineages and keeps its evolutionary fate separate from other species. The same kinds of diagnostic features discussed for the biological species concept are relevant for identifying evolutionary species, although in most cases only morphological features will be available from fossils. Unlike the biological species concept, the evolutionary species concept applies both to sexually and asexually reproducing forms. As long as continuity of diagnostic features is maintained by the evolving lineage, it will be recognized as a species. Abrupt changes in diagnostic features will mark the boundaries of different species in evolutionary time.

Phylogenetic Species Concept

The last concept that we present is the **phylogenetic species concept.** The phylogenetic species concept is defined as an *irreducible (basal) grouping of organisms diagnosably distinct from other such groupings and within which there is a parental pattern of ancestry and descent.* This concept emphasizes most strongly the criterion of common descent. Both asexual and sexual groups are covered.

A phylogenetic species is a single population lineage with no detectable branching. The main difference in practice between the evolutionary and phylogenetic species concepts is that the latter emphasizes recognizing as separate species the smallest groupings of organisms that have undergone independent evolutionary change. The evolutionary species concept would group into a single species geographically disjunct populations that demonstrate some phylogenetic divergence but are judged similar in their "evolutionary tendencies," whereas the phylogenetic species concept would treat them as separate species. In general, a greater number of species would be described using the phylogenetic species concept than any other species concept, and many taxonomists consider it impractical for this reason. For strict adherence to cladistic systematics (p. 201), the phylogenetic species concept is ideal because only this concept guarantees strictly monophyletic units at the species level.

The phylogenetic species concept intentionally disregards details of evolutionary process and gives us a criterion that allows us to describe species without first needing to conduct detailed studies on evolutionary processes. Advocates of the phylogenetic species concept do not necessarily disregard the importance of studying evolutionary process. They argue, however, that the first step in studying evolutionary process is to have a clear picture of life's history. To accomplish this task, the pattern of common descent must be reconstructed in the greatest detail possible by starting with the smallest taxonomic units that have a history of common descent.

Dynamism of Species Concepts

Current disagreements concerning concepts of species should not be considered discouraging. Whenever a field of scientific investigation enters a phase of dynamic growth, old concepts will be reevaluated and either refined or replaced with newer, more progressive ones. The active debate occurring within systematics shows that this field has acquired unprecedented activity and importance in biology. Just as Thomas Henry Huxley's time was one of enormous advances in biology, so is the present time. Both times are marked by fundamental reconsiderations of the meaning of species. We cannot predict which concept of species will be dominant 10 years from now. The conflicts between the current concepts, however, will lead us into the future. Understanding the conflicting perspectives, rather than learning a single species concept, is therefore of greatest importance for people now entering the study of zoology.

TAXONOMIC CHARACTERS AND PHYLOGENETIC RECONSTRUCTION

A major goal of systematics is to reconstruct the evolutionary tree or **phylogeny** that relates all extant and extinct species. This task is accomplished by identifying organismal features, for-

mally called **characters,** that vary among species. A character is any feature that the taxonomist uses to study variation within and among species. We find useful taxonomic characters in morphological, chromosomal, and molecular features (see p. 196). Taxonomists find characters by observing patterns of similarity among organisms and less frequently in behavioral and ecological ones. Phylogenetic analysis depends upon finding among organisms shared features that are inherited from a common ancestor. Character similarity that results from common ancestry is called **homology** (see Chapter 6). Similarity does not always reflect common ancestry, however. Independent evolutionary origin of similar features on different lineages produces patterns of similarity among organisms that do not reflect common descent; this occurrence complicates the work of taxonomists. Character similarity that misrepresents common descent is called nonhomologous similarity or **homoplasy.**

Using Character Variation to Reconstruct Phylogeny

To reconstruct the phylogeny of a group using characters that vary among its members, the first step is to determine which variant form of each character was present in the common ancestor of the entire group. This character state is called **ancestral** for the group as a whole. We presume that all other variant forms of the character arose later within the group, and these are called evolutionarily **derived character states.** Determining the **polarity** of a character refers to identifying which one of its contrasting states is ancestral and which one(s) derived. For example, if we consider as a character the dentition of amniotic vertebrates (reptiles, birds, and mammals), presence versus absence of teeth in the jaws constitute two different character states. Teeth are absent from modern birds but present in the other amniotes. To evaluate the polarity of this character, we must determine which character state, presence or absence of teeth, characterized the most recent common ancestor of amniotes and which state was derived subsequently within the amniotes.

The method that we use to examine the polarity of a variable character is called **outgroup comparison.** We consult an additional group of organisms, called an **outgroup,** that is phylogenetically close but not within the group being studied. We infer that any character state found both within the group being studied and in the outgroup is ancestral for the study group. Amphibians and different groups of bony fishes constitute appropriate outgroups to the amniotes for polarizing variation in the dentition of amniotes. Teeth are usually present in amphibians and bony fishes; therefore, we infer that presence of teeth is ancestral for amniotes and absence of teeth is derived. The polarity of this character indicates that teeth were lost in the ancestral lineage of all modern birds. Polarity of characters is evaluated most effectively when several different outgroups are used. All character states found in the study group that are absent from appropriate outgroups are considered derived.

Organisms or species that share derived character states form subsets within the study group called **clades** (Gr. *klados,* branch). A derived character shared by the members of a clade is formally called a **synapomorphy** (Gr. *synapsis,* joining together, + *morphē,* form) of that clade. Taxonomists use synapomorphies as evidence of homology to infer that a particular group of organisms forms a clade. Among extant amniotes, absence of teeth and presence of feathers are synapomorphies that identify the birds as a clade. A clade corresponds to a unit of evolutionary common descent; it includes all descendants of a particular ancestral lineage. The pattern formed by the derived states of all characters within our study group will take the form of a **nested hierarchy** of clades within clades. The goal is to identify all of the different clades nested within the study group, which would give a complete account of the patterns of common descent among the species in the group.

Character states ancestral for a taxon are often called **plesiomorphic** for that taxon and the sharing of ancestral states among organisms is termed **symplesiomorphy.** Unlike synapomorphies, however, symplesiomorphies do not provide useful information on nesting of clades within clades. In the example just given, we found that presence of teeth in jaws was plesiomorphic for amniotes. If we grouped together mammalian and reptilian groups, which possess teeth, to the exclusion of modern birds, we would not obtain a valid clade. Birds also descend from all common ancestors of reptiles and mammals and must be included within any clade that includes all reptiles and mammals. Errors in determining polarity of characters therefore clearly can produce errors in inference of phylogeny. It is important to note, however, that character states that are plesiomorphic at one taxonomic level can be synapomorphies at a more inclusive level. For example, the presence of jaws bearing teeth is a synapomorphy of gnathostome vertebrates (p. 482), a group that includes the amniotes plus amphibians, bony fishes, and cartilaginous fishes, although teeth have been lost in birds and some other gnathostomes. The goal of phylogenetic analysis therefore can be restated as one of finding the appropriate taxonomic level at which any given character state is a synapomorphy. The character state is then used at that level to identify a clade.

The nested hierarchy of clades is presented as a branching diagram called a **cladogram** (Figure 10-4; see also Figure 6-15). Taxonomists often make a technical distinction between a cladogram and a **phylogenetic tree.** The branches of a cladogram are only a formal device for indicating the nested hierarchy of clades within clades. The cladogram is not strictly equivalent to a phylogenetic tree, whose branches represent real lineages that occurred in the evolutionary past. To obtain a phylogenetic tree, we must add to the cladogram important additional information concerning ancestors, the durations of evolutionary lineages, or the amounts of evolutionary change that occurred on the lineages. A cladogram is often used, however, as a first approximation of the branching structure of the corresponding phylogenetic tree.

Sources of Phylogenetic Information

We find characters used to construct cladograms in comparative morphology (including embryology), comparative cytology, and comparative biochemistry. **Comparative morphology** examines the varying shapes and sizes of organismal structures, including their developmental origins. Both macroscopic and

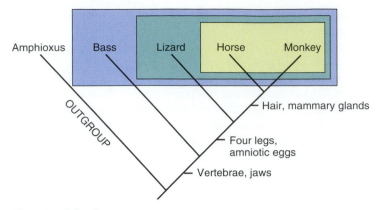

Figure 10-4

A cladogram as a nested hierarchy of taxa. Amphioxus is the outgroup, and the study group comprises four vertebrates (bass, lizard, horse, and monkey). Four characters that vary among vertebrates are used to generate a simple cladogram: presence versus absence of four legs, amniotic eggs, hair, and mammary glands. For all four characters, absence is the ancestral state in vertebrates because this is the condition found in the outgroup, Amphioxus; for each character, presence is the derived state in vertebrates. Because they share presence of four legs and amniotic eggs as synapomorphies, the lizard, horse, and monkey form a clade relative to the bass. This clade is subdivided further by two synapomorphies (presence of hair and mammary glands) that unite the horse and monkey relative to the lizard. We know from comparisons involving even more distantly related animals that presence of vertebrae and jaws constitute synapomorphies of vertebrates and that Amphioxus, which lacks these features, falls outside the vertebrate clade.

microscopic characters are used, including details of cell structure revealed by histology. As we will see in chapters 23–28, the variable structures of skull bones, limb bones, and integument (scales, hair, feathers) are particularly important for reconstructing the phylogeny of vertebrates. Comparative morphology uses specimens obtained from both living organisms and fossilized remains. **Comparative biochemistry** uses sequences of amino acids in proteins and the sequences of nucleotides in nucleic acids (see Chapter 5) to identify variable characters for constructing a cladogram (Figure 10-5). Direct sequencing of DNA is regularly applied to phylogenetic studies; however, comparisons of protein sequences are usually indirect, involving immunological or allozymic (see Figure 6-30) methods, or inference from DNA sequences of protein-coding genes. Recent studies have shown that comparative biochemistry can be applied to some fossils in addition to living organisms. **Comparative cytology** uses variation in the numbers, shapes, and sizes of chromosomes and their parts (Chapter 3) to obtain variable characters for constructing cladograms. Comparative cytology is used almost exclusively on living rather than fossilized organisms.

To add an evolutionary timescale necessary for producing a phylogenetic tree, we must consult the fossil record. We can look for the earliest appearance in fossils of derived morphological characters to estimate the ages of clades distinguished by those characters. The age of a fossil showing the derived characters of a particular clade is determined by radioactive dating (p. 107). An example of a phylogenetic tree constructed using these methods is Figure 14-28 p. 293.

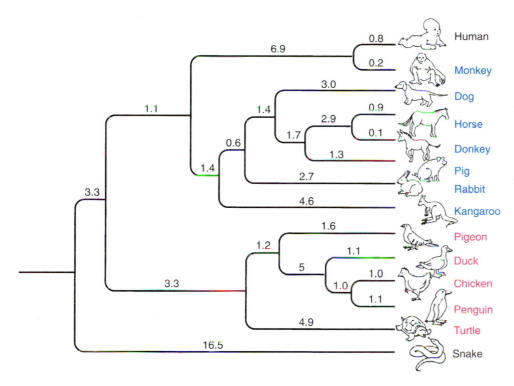

Figure 10-5
A phylogenetic tree of representative amniotes based on inferred base substitutions in the gene that encodes the respiratory protein, cytochrome *c*. Numbers on the branches are the expected numbers of mutational changes that occurred in this gene along the different evolutionary lineages.

We can use comparative biochemical data to estimate the ages of different lineages on a phylogenetic tree. Some protein and DNA sequences undergo approximately linear rates of divergence through evolutionary time. The age of the most recent common ancestor of two species is therefore proportional to the differences measured between their proteins and DNA sequences. We calibrate evolution of proteins and DNA sequences by measuring their divergence between species whose most recent common ancestor has been dated using fossils. We then use the molecular evolutionary calibration to estimate ages of other branches on the phylogenetic tree.

THEORIES OF TAXONOMY

A theory of taxonomy establishes the principles that we use to recognize and to rank taxonomic groups. There are two currently popular theories of taxonomy: (1) traditional evolutionary taxonomy and (2) phylogenetic systematics (cladistics). Both are based on evolutionary principles. We will see, however, that these two theories differ on how evolutionary principles are used. These differences have important implications for how we use a taxonomy to study evolutionary processes.

The relationship between a taxonomic group and a phylogenetic tree or cladogram is important for both of these theories. This relationship can take one of three forms: **monophyly, paraphyly,** or **polyphyly** (Figure 10-6). A taxon is monophyletic if it includes the most recent common ancestor of the group and all descendants of that ancestor (see Figure 10-6A). A taxon is paraphyletic if it includes the most recent common ancestor of all members of a group and some but not all of the descendants of that ancestor (see Figure 10-6B). A taxon is polyphyletic if it does not include the most recent common ancestor of all members of a group; this condition requires that the group has had at least two

separate evolutionary origins, usually requiring independent evolutionary acquisition of similar features (see Figure 10-6C). Both evolutionary and cladistic taxonomy accept monophyletic groups and reject polyphyletic groups in their classifications. They differ on the acceptance of paraphyletic groups, however, and this difference has important evolutionary implications.

Traditional Evolutionary Taxonomy

Traditional **evolutionary taxonomy** incorporates two different evolutionary principles for recognizing and ranking higher taxa: (1) common descent and (2) amount of adaptive evolutionary change, as shown on a phylogenetic tree. Evolutionary taxa must have a single evolutionary origin, and must show unique adaptive features.

The mammalian paleontologist George Gaylord Simpson (Figure 10-7) and Ernst Mayr (Figure 6-18) were highly influential in developing and formalizing the procedures of evolutionary taxonomy. According to Simpson and Mayr, a particular branch on the evolutionary tree is given the status of a higher taxon if it represents a distinct **adaptive zone.** Simpson describes an adaptive zone as "a characteristic reaction and mutual relationship between environment and organism, a way of life and not a place where life is led." By entering a new adaptive zone through a fundamental change in organismal structure and behavior, an evolving population can use environmental resources in a completely new way.

A taxon that comprises a distinct adaptive zone is termed a **grade.** Simpson gives the example of penguins as a distinct adaptive zone within birds. The lineage immediately ancestral to all penguins underwent fundamental changes in the form of the body and wings to permit a switch from aerial to aquatic locomotion (Figure 10-8). Aquatic birds that can fly both in air and

Phylogenies from DNA Sequences

A simple example illustrates cladistic analysis of DNA sequence data to examine phylogenetic relationships among species. The study group in this example contains three species of chameleons, two from the island of Madagascar (*Brookesia theili* and *B. brygooi*) and one from Equatorial Guinea (*Chamaeleo feae*). The outgroup is a lizard of genus *Uromastyx*, which is a distant relative of chameleons. Do the molecular data in this example confirm or reject the prior taxonomic hypothe-

sis that the two Madagascaran chameleons are more closely related to each other than either one is to the Equatorial Guinean species?

The molecular information in this example comes from a piece of the mitochondrial DNA sequence (57 bases) for each species. Each sequence encodes amino acids 221–239 of a protein called "NADH dehydrogenase subunit 2" in the species from which it was obtained. These DNA base sequences are aligned and numbered as:

Each column in the aligned sequences constitutes a character that takes one of four states: A, C, G or T (a fifth possible state, deletion of the base, is not observed in this example). Only characters that vary among the three chameleon species potentially contain information

on which pair of species is most closely related. Twenty-three of the 57 aligned bases show variation *among chameleons,* as shown here in bold letters:

To be useful for constructing a cladogram, a character must demonstrate sharing of derived characters (= synapomorphy). Which of these 22 characters demonstrate synapomorphies for chameleons? For each of the 22 variable characters, we must ask whether one of the states

observed in chameleons is shared with the outgroup, *Uromastyx*. If so, this state will be judged ancestral for chameleons and the alternative state(s) derived. Derived characters are identified for 21 of the 23 characters just identified; derived states are underlined here:

Note that polarity is ambiguous for two variable characters (at positions 23 and 54) whose alternative states in chameleons are not observed in the outgroup.

Of the characters showing derived states, 10 of them show synapomorphies among chameleons. These characters are marked here with numbers 1, 2 or 3 in the appropriate column.

underwater are somewhat intermediate in habitat, morphology, and behavior between aerial and aquatic adaptive zones. Nonetheless, the obvious modifications of the wings and body of penguins for swimming represent a new grade of organization. Penguins are therefore recognized as a distinct taxon within the birds, the family Spheniscidae. The broader the adaptive

zone when fully occupied by a group of organisms, the higher the rank given to the corresponding taxon.

Evolutionary taxa may be either monophyletic or paraphyletic. Recognition of paraphyletic taxa requires, however, that our taxonomies distort patterns of common descent. An evolutionary taxonomy of the anthropoid primates provides a

The eight characters marked 1 show synapomorphies grouping the two Madagascaran species (*Brookesia theili* and *B. brygooi*) to the exclusion of the Equatorial Guinean species, *Chamaeleo feae*. We can represent these relationships as a cladogram:

We can explain evolution of all characters favoring this cladogram by placing a single mutational change on the branch ancestral to the two *Brookesia* species. This is the simplest explanation for evolutionary change of these characters.

Characters marked 2 and 3 disagree with our cladogram and favor alternative relationships as shown here:

To explain evolutionary changes in characters favoring cladograms 2 or 3 using cladogram 1, we need at least two changes per character. Likewise, if we try to explain evolution of characters favoring cladogram 1 on cladograms 2 or 3, we need at least two changes for each of these characters. These two diagrams show the minimum numbers of changes required for character 5 (which favors cladogram 1) and character 41 (which favors cladogram 1) on cladogram 1; the ancestral state of each character is shown at the root of the tree and the states observed in each species at the tips of the branches:

character 5 (1 change)

character 41 (2 changes)

Systematics often use a principle called **parsimony** to resolve conflicts among taxonomic characters, as seen here. We choose as our best working hypothesis the cladogram that requires the smallest total amount of character change. In our example, cladogram 1 is favored by parsimony. For all 10 phylogenetically informative characters, cladogram 1 requires a total of 12 changes of character state (one for each of the 8 characters favoring it and two for each of the other 2 characters). Cladograms 2 and 3 each require at least 19 character-state changes, 7 steps longer than cladogram 1. By choosing cladogram 1, we claim that characters favoring cladograms 2 and 3 show homoplasy in their evolution.

The molecular sequences shown in this example therefore confirm predictions of the prior hypothesis, based on appearance and geography of these chameleons, that the *Brookesia* species shared a common ancestor with each other more recently than either one did with *Chamaeleo feae*.

As a further exercise, you should convince yourself that the 12 characters that vary among chameleons but which do not demonstrate unambiguous sharing of derived states are equally compatible with each of the three possible cladograms. For each character, find the minimum total number of changes that must occur to explain its evolution on each cladogram. You will see, if you do this exercise correctly, that the three cladograms do not differ in minimum numbers of changes required for each of these characters. For this reason, the characters are phylogenetically uninformative by the parsimony criterion.

Data from Townsend, T., and A. Larson. 2002. Molecular phylogenetics and mitochondrial genomic evolution in the Chamaeleonidae (Reptilia, Squamata). Molecular Phylogenetics and Evolution 23:22–36.

good example (Figure 10-9). This taxonomy places humans (genus *Homo*) and their immediate fossil ancestors in the family Hominidae, and it places the chimpanzees (genus *Pan*), gorillas (genus *Gorilla*), and orangutans (genus *Pongo*) in the family Pongidae. However, the pongid genera *Pan* and *Gorilla* share more recent common ancestry with the Hominidae than they do with the remaining pongid genus, *Pongo*. This arrangement makes the family Pongidae paraphyletic because it does not include humans, who also descend from the most recent common ancestor of all pongids (see Figure 10-9). Evolutionary taxonomists nonetheless recognize the pongid genera as a single, family-level grade of arboreal, herbivorous primates having

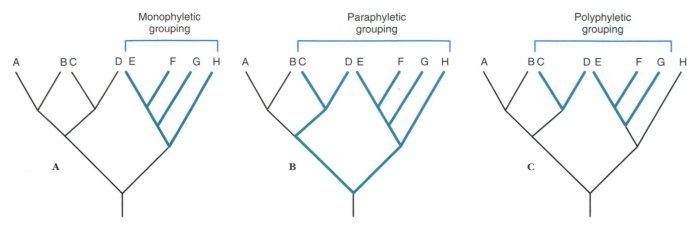

Figure 10-6

Relationships between phylogeny and taxonomic groups illustrated for a hypothetical phylogeny of eight species (A through H).
A, *Monophyly*—a monophyletic group contains the most recent common ancestor of all members of the group and all of its descendants.
B, *Paraphyly*—a paraphyletic group contains the most recent common ancestor of all members of the group and some but not all of its descendants. **C,** *Polyphyly*—a polyphyletic group does not contain the most recent common ancestor of all members of the group, thereby requiring that the group have at least two separate phylogenetic origins.

Figure 10-7

George Gaylord Simpson (1902 to 1984) formulated the principles of evolutionary taxonomy.

A

B

Figure 10-8

A, Penguin. **B,** Diving petrel. Penguins (avian family Spheniscidae) were recognized by George G. Simpson as a distinct adaptive zone within birds because of their adaptations for submarine flight. Simpson believed that the adaptive zone ancestral to penguins resembled that of diving petrels, which display adaptations for combined aerial and aquatic flight. Adaptive zones of penguins and diving petrels are distinct enough to be recognized taxonomically as different families within a common order (Ciconiiformes).

limited mental capacity; in other words, they show the same family-level adaptive zone. Humans are terrestrial, omnivorous primates who have greatly expanded mental and cultural attributes, thereby comprising a distinct adaptive zone at the taxonomic level of the family. Unfortunately, if we want our taxa to constitute adaptive zones, we compromise our ability to present common descent in the most straightforward taxonomic manner.

Traditional evolutionary taxonomy has been challenged from two opposite directions. One challenge states that because phylogenetic trees can be very difficult to obtain, it is impractical to base our taxonomic system on common descent and adaptive evolution. We are told that our taxonomy should represent a more easily measured feature, the overall similarity of organisms evaluated without regard to phylogeny. This principle is known as **phenetic taxonomy.** Phenetic taxonomy did not have a strong impact on animal classification, and scientific interest in this approach is in decline. Despite the difficulties of reconstructing phylogeny, zoologists still consider this endeavor a central goal of their systematic work, and they are unwilling to compromise this goal for purposes of methodological simplicity.

Phylogenetic Systematics/Cladistics

A second and stronger challenge to evolutionary taxonomy is one known as **phylogenetic systematics** or **cladistics.** As the first name implies, this approach emphasizes the criterion of common descent and, as the second name implies, it is based on the cladogram of the group being classified. This approach to taxonomy was first proposed in 1950 by the German entomologist, Willi Hennig (Figure 10-10) and therefore is sometimes called "Hennigian systematics." All taxa recognized by Hennig's cladistic system must be monophyletic. We saw on Figure 10-9 how evolutionary taxonomists' recognition of the primate families Hominidae and Pongidae distorts genealogical relationships to emphasize adaptive uniqueness of the Hominidae. Because the most recent common ancestor of the paraphyletic family Pongidae is also an ancestor of the Hominidae, recognition of the Pongidae is incompatible with cladistic taxonomy. To avoid paraphyly, cladistic taxonomists have discontinued use of the traditional family Pongidae, placing chimpanzees, gorillas, and orangutans with humans in the family Hominidae. We adopt the cladistic classification in this book.

The disagreement on the validity of paraphyletic groups may seem trivial at first, but its important consequences become clear when we discuss evolution. For example, claims that amphibians evolved from bony fish, that birds evolved from reptiles, or that humans evolved from apes may be made by an evolutionary taxonomist but are meaningless to a cladist. We imply by these statements that a descendant group (amphibians, birds, or humans) evolved from part of an ancestral group (bony fish, reptiles, and apes, respectively) to which the descendant does not belong. This usage automatically makes the ancestral group paraphyletic, and indeed bony fish, reptiles, and apes as traditionally recognized are paraphyletic groups. How are such paraphyletic groups recognized? Do they share distinguishing features that are not shared by the descendant group?

Paraphyletic groups are usually defined in a negative manner. They are distinguished only by lacking features found in a particular descendant group, because any traits that they share from their common ancestry are symplesiomorphies present also in the excluded descendants (unless secondarily lost). For example, apes are those "higher" primates that are not humans. Likewise, fish are those vertebrates that lack the distinguishing characteristics of tetrapods (amphibians and amniotes). What does it mean then to say that humans evolved from apes? To the evolutionary taxonomist, apes and humans are different adaptive zones or grades of organization; to say that humans evolved from

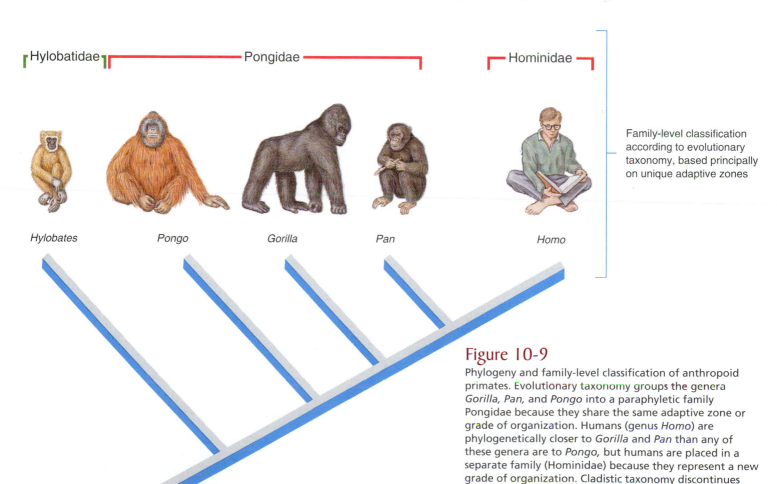

Family-level classification according to evolutionary taxonomy, based principally on unique adaptive zones

Figure 10-9

Phylogeny and family-level classification of anthropoid primates. Evolutionary taxonomy groups the genera *Gorilla, Pan,* and *Pongo* into a paraphyletic family Pongidae because they share the same adaptive zone or grade of organization. Humans (genus *Homo*) are phylogenetically closer to *Gorilla* and *Pan* than any of these genera are to *Pongo,* but humans are placed in a separate family (Hominidae) because they represent a new grade of organization. Cladistic taxonomy discontinues recognition of paraphyletic family Pongidae, consolidating *Pongo, Gorilla, Pan,* and *Homo* in family Hominidae.

Figure 10-10

Willi Hennig (1913 to 1976), German entomologist who formulated the principles of phylogenetic systematics/cladistics.

apes states that bipedal, tailless organisms of large brain capacity evolved from arboreal, tailed organisms of smaller brain capacity. To the cladist, however, the statement that humans evolved from apes says essentially that humans evolved from something that they are not, a trivial statement that lacks useful information. To the cladist, any statement that a particular monophyletic group descends from a paraphyletic one is nothing more than a claim that the descendant group evolved from something that it is not. Extinct ancestral groups are always paraphyletic because they exclude a descendant that shares their most recent common ancestor. Although many such groups have been recognized by evolutionary taxonomists, none are recognized by cladists.

Zoologists often construct paraphyletic groups because they are interested in a terminal, monophyletic group (such as humans), and they want to ask questions about its ancestry. It is often convenient to lump together organisms whose features are considered approximately equally distant from the group of interest and to ignore their own unique features. It is significant in this regard that humans have never been placed in a paraphyletic group, whereas most other organisms have been. Apes, reptiles, fishes, and invertebrates are all terms that traditionally designate paraphyletic groups formed by combining various "side branches" found when human ancestry is traced backward through the tree of life. Such a taxonomy can give the erroneous impression that all of evolution is a progressive march toward humanity or, within other groups, a progressive march toward whatever species humans designate as being most "advanced." Such thinking is a relic of pre-Darwinian views that there is a linear scale of nature having "primitive" creatures at the bottom and humans near the top just below angels. Darwin's theory of common descent states, however, that evolution is a branching process with no linear scale of increasing perfection along a single branch. Nearly every branch will contain its own combination of ancestral and derived features. In cladistics, this perspective is emphasized by recognizing taxa only by their own unique properties and not grouping organisms only because they lack the unique properties found in related groups.

Fortunately, there is a convenient way to express the common descent of groups without constructing paraphyletic taxa. It is done by finding what is called the **sister group** of the taxon of

interest to us. Two different monophyletic taxa are each other's sister group if they share common ancestry more recently than either one does with any other taxa. The sister group of humans appears to be chimpanzees, with gorillas forming the sister group to humans and chimpanzees combined. Orangutans are the sister group of a clade that includes humans, chimpanzees, and gorillas; gibbons form the sister group of the clade that includes orangutans, chimpanzees, gorillas, and humans (see Figure 10-9).

Current State of Animal Taxonomy

The formal taxonomy of animals that we use today was established using the principles of evolutionary systematics and has been revised recently in part using the principles of cladistics. Introduction of cladistic principles initially has the effect of replacing paraphyletic groups with monophyletic subgroups while leaving the remaining taxonomy mostly unchanged. A thorough revision of taxonomy along cladistic principles, however, will require profound changes, one of which almost certainly will be abandonment of Linnaean ranks. A new taxonomic system called PhyloCode is being developed as an alternative to Linnean taxonomy. In our coverage of animal taxonomy, we try as much as possible to use taxa that are monophyletic and therefore consistent with criteria of both evolutionary and cladistic taxonomy. We continue, however, to use Linnaean ranks. In some cases in which commonly recognized taxa are clearly paraphyletic grades, we note this fact and suggest alternative taxonomic schemes that contain only monophyletic taxa.

In discussing patterns of descent, we avoid statements such as "mammals evolved from reptiles" that imply paraphyly and instead specify appropriate sister-group relationships. We avoid referring to groups of organisms as being primitive, advanced, specialized, or generalized because all groups of animals contain combinations of primitive, advanced, specialized, and generalized features; these terms are best restricted to describing specific characteristics and not an entire group.

Revision of taxonomy according to cladistic principles can cause confusion. In addition to new taxonomic names, we see old ones used in unfamiliar ways. For example, cladistic use of "bony fishes" includes amphibians and amniotes (including reptilian groups, birds, and mammals) in addition to finned, aquatic animals that we normally term "fish." Cladistic use of "reptiles" includes birds in addition to snakes, lizards, turtles, and crocodilians; however, it excludes some fossil forms, such as synapsids, that were traditionally placed in the Reptilia (see Chapters 26 through 28). Taxonomists must be very careful to specify when using these seemingly familiar terms whether the traditional evolutionary taxa or newer cladistic taxa are being discussed.

MAJOR DIVISIONS OF LIFE

From Aristotle's time to the late 1800s it was traditional to assign every living organism to one of two kingdoms: plant or animal. However, the two-kingdom system had serious problems. Although it was easy to place rooted, photosynthetic organisms such as trees and herbs among the plants and to place food-ingesting, motile

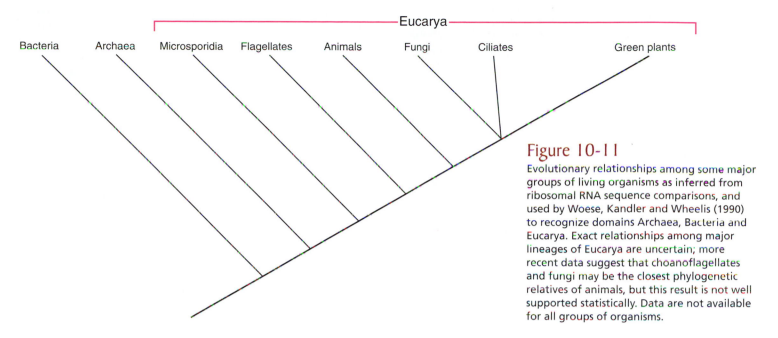

Figure 10-11
Evolutionary relationships among some major groups of living organisms as inferred from ribosomal RNA sequence comparisons, and used by Woese, Kandler and Wheelis (1990) to recognize domains Archaea, Bacteria and Eucarya. Exact relationships among major lineages of Eucarya are uncertain; more recent data suggest that choanoflagellates and fungi may be the closest phylogenetic relatives of animals, but this result is not well supported statistically. Data are not available for all groups of organisms.

forms such as insects, fishes, and mammals among the animals, unicellular organisms presented difficulties (Chapter 11). Some forms were claimed both for the plant kingdom by botanists and for the animal kingdom by zoologists. An example is *Euglena* (p. 218), which is motile, like animals, but has chlorophyll and photosynthesis, like plants. Other groups, such as bacteria, were rather arbitrarily assigned to the plant kingdom.

Several alternative systems have been proposed to solve the problem of classifying unicellular forms. In 1866 Haeckel proposed the new kingdom Protista to include all single-celled organisms. At first the bacteria and cyanobacteria (blue-green algae), forms that lack nuclei bounded by a membrane, were included with nucleated unicellular organisms. Finally, the important differences between the anucleate bacteria and cyanobacteria (prokaryotes) and all other organisms that have membrane-bound nuclei (eukaryotes) were recognized. In 1969 R. H. Whittaker proposed a five-kingdom system that incorporated the basic prokaryote-eukaryote distinction. The kingdom Monera contained the prokaryotes. The kingdom Protista contained the unicellular eukaryotic organisms (protozoa and unicellular eukaryotic algae). Multicellular organisms were split into three kingdoms on the basis of mode of nutrition and other fundamental differences in organization. The kingdom Plantae included multicellular photosynthesizing organisms, higher plants, and multicellular algae. Kingdom Fungi contained molds, yeasts, and fungi that obtain their food by absorption. Invertebrates (except the protozoa) and vertebrates compose the kingdom Animalia. Most of these forms ingest their food and digest it internally, although some parasitic forms are absorptive.

All these different systems were proposed without regard to the phylogenetic relationships needed to construct evolutionary or cladistic taxonomies. The oldest phylogenetic events in the history of life have been obscure, because the different forms of life share very few characters that can be compared among them to reconstruct phylogeny. Recently, however, a cladistic classification of all life-forms has been proposed based on phylogenetic information

obtained from molecular data (the nucleotide base sequence of DNA encoding ribosomal RNA, Figure 10-11). According to this tree, Woese, Kandler, and Wheelis (1990) recognized three monophyletic domains above the kingdom level: Eucarya (all eukaryotes), Bacteria (the true bacteria), and Archaea (prokaryotes differing from bacteria in membrane structure and ribosomal RNA sequences). They did not divide the Eucarya into kingdoms, although if we retain Whittaker's kingdoms Plantae, Animalia, and Fungi, Protista becomes a paraphyletic group (Figure 10-12). To maintain a cladistic classification, Protista must be discontinued by recognizing as separate kingdoms Ciliata, Flagellata, and Microsporidia as shown in Figure 10-11, and phylogenetic information must be gathered for additional protistan groups, including amebas. This taxonomic revision has not been made; however, if the phylogenetic tree in Figure 10-11 is supported by further evidence, revision of the taxonomic kingdoms will be necessary.

Until a few years ago, animal-like protistans were traditionally studied in zoology courses as animal phylum Protozoa. Given current knowledge and the principles of phylogenetic systematics, this taxonomy commits two errors; "protozoa" are neither animals nor are they a valid monophyletic taxon at any level of the Linnaean hierarchy. Kingdom Protista is likewise invalid because it is not monophyletic. Animal-like protistans, now divided into seven or more phyla, are nonetheless of interest to students of zoology because of their animal-like properties.

MAJOR SUBDIVISIONS OF THE ANIMAL KINGDOM

The phylum is the largest formal taxonomic category in the Linnaean classification of the animal kingdom. Animal phyla are often grouped together to produce additional, informal taxa intermediate between the phylum and the animal kingdom. These taxa are based on embryological and anatomical characters that reveal phylogenetic affinities of different animal phyla. Zoologists in the past have recognized subkingdom Protozoa, which contains the primarily unicellular phyla, and the subkingdom Metazoa, which contains multicellular phyla. As just noted, however, Protozoa is not a valid taxonomic group and does not belong within the animal kingdom, which is synonymous with Metazoa. The traditional higher-level groupings of true animal phyla are:

Branch A (Mesozoa): phylum Mesozoa, the mesozoa

Branch B (Parazoa): phylum Porifera, the sponges, and phylum Placozoa

Branch C (Eumetazoa): all other phyla
   Grade I (Radiata): phyla Cnidaria, Ctenophora
   Grade II (Bilateria): all other phyla
      Division A (Protostomia): characteristics in Figure 10-13
         Acoelomates: phyla Platyhelminthes, Gnathostomulida, Nemertea
         Pseudocoelomates: phyla Rotifera, Gastrotricha, Kinorhyncha, Nematoda, Nematomorpha, Acanthocephala, Entoprocta, Priapulida, Loricifera
         Eucoelomates: phyla Mollusca, Annelida, Arthropoda, Echiurida, Sipunculida, Tardigrada, Pentastomida, Onychophora, Pogonophora
      Division B (Deuterostomia): characteristics in Figure 10-13 phyla Phoronida, Ectoprocta, Chaetognatha, Brachiopoda, Echinodermata, Hemichordata, Chordata

As in the outline, bilateral animals are customarily divided into protostomes and deuterostomes by their embryological development (Figure 10-13). However, some of the phyla are difficult to place into one of these two categories because they possess some characteristics of each group (Chapter 21).

   Recent molecular phylogenetic studies have challenged traditional classification of the Bilateria. Molecular phylogenetic results place four phyla classified above as deuterostomes (Brachiopoda, Chaetognatha, Ectoprocta, and Phoronida) in the Protostomia. Furthermore, the traditional major groupings of protostome phyla (acoelomates, pseudocoelomates, and eucoelomates) appear not to be monophyletic. Instead, protostomes are divided into two major monophyletic groups called the Lophotrochozoa and Ecdysozoa. Reclassification of the Bilateria is summarized:

Grade II: Bilateria
   Division A (Protostomia):
      Lophotrochozoa: phyla Platyhelminthes, Nemertea, Rotifera, Gastrotricha, Acanthocephala, Mollusca, Annelida, Echiurida, Sipunculida, Pogonophora, Phoronida, Ectoprocta, Entoprocta, Gnathostomulida, Chaetognatha, Brachiopoda
      Ecdysozoa: phyla Kinorhyncha, Nematoda, Nematomorpha, Priapulida, Arthropoda, Tardigrada, Onychophora, Loricifera, Pentastomida
   Division B (Deuterostomia): phyla Chordata, Hemichordata, Echinodermata

Further study is needed to confirm these new groupings. We organize our survey of animal diversity using the traditional classification, but discuss implications of this new one.

Figure 10-12

Whittaker's five-kingdom classification superimposed on a phylogenetic tree showing living representatives of these kingdoms. Note that the kingdoms Monera and Protista constitute paraphyletic groups and are therefore unacceptable to cladistic systematics.

SUMMARY

Animal systematics has three major goals: (1) to identify all species of animals, (2) to evaluate evolutionary relationships among animal species, and (3) to group animal species in a hierarchy of taxonomic groups (taxa) that conveys evolutionary relationships. Taxa are ranked to denote increasing inclusiveness as follows: species, genus, family, order, class, phylum, and kingdom. All of these ranks can be subdivided to signify taxa that are intermediate between them. Names of species are binomial, with the first name designating the genus to which the species belongs (capitalized) followed by a

species epithet (lowercase), both written in italics. Taxa at all other ranks are given single nonitalicized names.

   The biological species concept has guided the recognition of most animal species. A biological species is defined as a reproductive community of populations (reproductively isolated from others) that occupies a specific niche in nature. It is not immutable through time but changes during the course of evolution. Because the biological species concept may be difficult to apply in spatial and temporal dimensions, and because it excludes asexually repro-

PROTOSTOMES		DEUTEROSTOMES	
Spiral cleavage	Cleavage mostly spiral	Cleavage mostly radial	Radial cleavage
Cell from which mesoderm will derive	Endomesoderm usually from a particular blastomere designated 4d	Endomesoderm from enterocoelous pouching (except chordates)	Endomesoderm from pouches from primitive gut
Primitive gut / Mesoderm / Coelom / Blastopore	In coelomate protostomes the coelom forms as a split in mesodermal bands (schizocoelous)	All coelomate, coelom from fusion of enterocoelous pouches (except chordates, which are schizocoelous)	Coelom / Mesoderm / Primitive gut / Blastopore
Anus	Mouth forms from or near blastopore; anus a new formation	Anus forms from or near blastopore, mouth a new formation	Mouth
Annelid (earthworm)	Embryology mostly determinate (mosaic)	Embryology usually indeterminate (regulative)	
Mouth	Includes phyla Platyhelminthes, Nemertea, Annelida, Mollusca, Arthropoda, Chaetognatha, Phoronida, Ectoprocta, Brachiopoda, minor phyla	Includes phyla Echinodermata, Hemichordata, Chordata	Anus

Figure 10-13

Basis for the distinction between divisions of bilateral animals.

ducing forms, alternative concepts have been proposed. These alternatives include the evolutionary species concept and the phylogenetic species concept. No single concept of species is universally accepted by all zoologists.

Two major schools of taxonomy are currently active. Traditional evolutionary taxonomy groups species into higher taxa according to the joint criteria of common descent and adaptive evolution; such taxa have a single evolutionary origin and occupy a distinctive adaptive zone. A second approach, known as phylogenetic systematics or cladistics, emphasizes common descent exclusively in grouping species into higher taxa. Only monophyletic taxa (those having a single evolutionary origin and containing all descendants of the group's most recent common ancestor) are used in cladistics. In addition to monophyletic taxa, evolutionary taxonomy recognizes some taxa that are paraphyletic (having a single evolutionary origin but excluding some descendants of the most recent common ancestor of the group). Both schools of taxonomy exclude polyphyletic taxa (those having more than one evolutionary origin).

Both evolutionary taxonomy and cladistics require that patterns of common descent among species be assessed before higher

taxa are recognized. Comparative morphology (including development), cytology, and biochemistry are used to reconstruct nested hierarchical relationships among taxa that reflect the branching of evolutionary lineages through time. The fossil record provides estimates of the ages of evolutionary lineages. Comparative studies and the fossil record jointly permit us to reconstruct a phylogenetic tree representing the evolutionary history of the animal kingdom.

Traditionally, all living forms were placed into two kingdoms (animal and plant) but more recently, a five-kingdom system (animals, plants, fungi, protistans, and monerans) has been followed. Neither of these systems conforms to the principles of evolutionary or cladistic taxonomy because they place single-celled organisms into either paraphyletic or polyphyletic groups. Based on our current knowledge of the phylogenetic tree of life, "protozoa" do not form a monophyletic group and they do not belong within the animal kingdom.

Phylogenetic relationships among animal phyla have been clarified by molecular phylogenetic studies, although many of these higher-level groupings remain tentative. Particularly controversial is the grouping of bilaterally symmetrical animals into clades Deuterostomia, Protostomia, Ecdysozoa, and Lophotrochozoa.

REVIEW QUESTIONS

1. List in order, from most inclusive to least inclusive, the principal categories (taxa) in Linnean classification as currently applied to animals.
2. Explain why the system for naming species that originated with Linnaeus is "binomial."
3. How does the biological species concept differ from earlier typological concepts of a species? Why do evolutionary biologists prefer it to typological species concepts?
4. How do monophyletic, paraphyletic, and polyphyletic taxa differ? How do these differences affect the validity of such taxa for both evolutionary and cladistic taxonomies?
5. How are taxonomic characters recognized? How are such characters used to construct a cladogram?
6. What is the difference between a cladogram and a phylogenetic tree? Given a cladogram for a group of

species, what additional information is needed to obtain a phylogenetic tree?
7. How would cladists and evolutionary taxonomists differ in their interpretations of the statement that humans evolved from apes, which evolved from monkeys?
8. What taxonomic practices based on the typological species concept are retained in systematics today? How has their interpretation changed?
9. What problems have been identified with the biological species concept? How do other species concepts attempt to overcome these problems?
10. What are the five kingdoms distinguished by Whittaker? How does their recognition conflict with the principles of cladistic taxonomy?

SELECTED REFERENCES

Aguinaldo, A. M. A., J. M. Turbeville, L. S. Linford, M. C. Rivera, J. R. Garey, R. A. Raff, and J. A. Lake. 1997. Evidence for a clade of nematodes, arthropods and other moulting animals. Nature **387:**489–493. *This molecular phylogenetic study challenges traditional classification of the Bilateria.*

Ereshefsky, M. (ed.). 1992. The units of evolution. Cambridge, Massachusetts, MIT Press. *A thorough coverage of concepts of species, including reprints of important papers on the subject.*

Ereshefsky, M. 2001. The poverty of the Linnean hierarchy. Cambridge, U.K., Cambridge University Press. *A philosophical critique of Linnean taxonomy illustrating its problems with cladistic taxonomy.*

Felsenstein, J. 2002. Inferring phylogenies. Sunderland, Massachusetts, Sinauer Associates. *A thorough coverage of phylogenetic methods.*

Hall, B. K. 1994. Homology: the hierarchical basis of comparative biology. San Diego, Academic Press. *A collection of papers discussing the many dimensions of homology, the central concept of comparative biology and systematics.*

Hillis, D. M., C. Moritz and B. K. Mable (eds.). 1996. Molecular systematics, ed. 2. Sunderland, Massachusetts, Sinauer Associates, Inc. *A detailed coverage of the biochemical and analytical procedures of comparative biochemistry.*

Hull, D. L. 1988. Science as a process. Chicago, University of Chicago Press. *A study of the working methods and interactions of systematists, containing a thorough review of the principles of evolutionary, phenetic, and cladistic taxonomy.*

Maddison, W. P., and D. R. Maddison. 2001. MacClade version 4.03. Sunderland, Massachusetts, Sinauer Associates, Inc. *A computer program for the MacIntosh that conducts phylogenetic analyses of systematic characters. The instruction manual stands alone as an excellent introduction to phylogenetic procedures. The computer program is user-friendly and excellent for instruction in addition to serving as a tool for analyzing real data.*

Mayr, E., and P. D. Ashlock. 1991. Principles of systematic zoology. New York, McGraw-Hill. *A detailed survey of systematic principles as applied to animals.*

Panchen, A. L. 1992. Classification, evolution, and the nature of biology. New York, Cambridge University Press. *Excellent explanations of the methods and philosophical foundations of biological classification.*

Swofford, D. 2002. Phylogenetic analysis using parsimony (and other methods) PAUP* version 4. Sunderland, Massachusetts, Sinauer Associates. *A powerful computer package for constructing phylogenetic trees from data.*

Wagner, G. P. (ed.). 2001. The character concept in evolutionary biology. San Diego, Academic Press. *A thorough coverage of evolutionary character concepts.*

Woese, C. R., O. Kandler, and M. L. Wheelis. 1990. Towards a natural system of organisms: proposal for the domains Archaea, Bacteria, and Eucarya. Proceedings of the National Academy of Sciences, USA, **87:**4576–4579. *Proposed cladistic classification for the major taxonomic divisions of life.*

ZOOLOGY LINKS TO THE INTERNET

Visit the textbook's Online Learning Center at www.mhhe.com/zoology to find live Internet links for each of the topics listed here.

Classification and Phylogeny of Animals
Classification of Living Things
Methods of Classification
Trees of Life

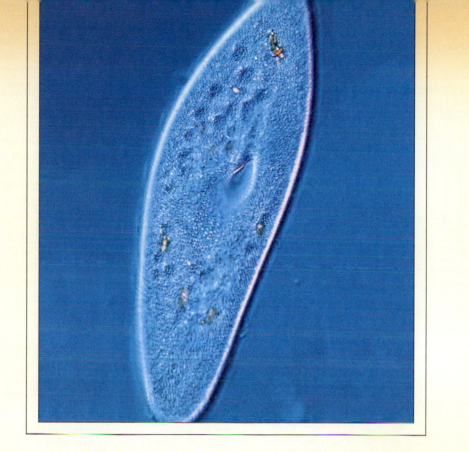

A paramecium.

Protozoan Groups

Emergence of Eukaryotes and a New Life Pattern

The first reasonable evidence for life on earth dates from approximately 3.5 billion years ago. These first cells were prokaryotic, bacteria-like organisms. After an enormous time span of evolutionary diversification at the prokaryotic level, unicellular eukaryotic organisms appeared. Although the origin of single-celled eukaryotes may never be known with certainty, it clearly involved a process of symbiogenesis. Certain aerobic bacteria may have been engulfed by other bacteria that were unable to cope with the increasing concentrations of oxygen in the atmosphere. The aerobic bacteria had the enzymes necessary for deriving energy in the presence of oxygen, and they would have become the ancestors of mitochondria. Most, but not all, genes of the mitochondria would come to reside in the host-cell nucleus. Almost all present-day eukaryotes have mitochondria and are aerobic.

Some ancestral unicellular eukaryotic cells engulfed photosynthetic bacteria, which evolved to become chloroplasts, and those eukaryotes thereby were able to manufacture their own food molecules using energy from sunlight. The descendants of one line, green algae, eventually gave rise to multicellular plants. Other groups apparently originated by *secondary* endosymbiosis, in which one eukaryotic cell engulfed another eukaryotic cell, and the latter became evolutionarily transformed into an organelle.

Some eukaryotes that did not become residences for chloroplasts, and even some that did, evolved animal-like characteristics and gave rise to a variety of phyla that are collectively called protozoa. Protozoa are a diverse assemblage of unicellular organisms with puzzling affinities. They are distinctly animal-like in several respects: they lack a cell wall, have at least one motile stage in the life cycle, and most ingest their food. Throughout their long history, protozoa have radiated to generate a bewildering array of morphological forms within the constraints of a single cell.

Position Relative to the Animal Kingdom

A protozoan is a complete organism in which all life activities occur within the limits of a single plasma membrane. Because their protoplasmic mass is not subdivided into cells, protozoa sometimes have been termed "acellular," but most people prefer "unicellular" to emphasize the many structural similarities to the cells of multicellular animals.

Evidence from electron microscopy, life cycle studies, genetics, biochemistry, and molecular biology has shown that the former phylum Protozoa encompassed numerous phyla of varying evolutionary relationships. Combining all animal-like unicellular eukaryotes with the unicellular algae into a kingdom **Protista** (or **Protoctista**) simply created another, more massive, paraphyletic taxon. Thus we will use the terms *protozoa* and *protozoan* informally, covering these organisms in a single chapter as a convenience and not implying that they form a monophyletic group.

Biological Contributions

1. **Intracellular specialization** (division of labor within a cell) involves organization of functional organelles in the cell.
2. The simplest example of **division of labor between cells** is seen in certain colonial protozoa that have both somatic and reproductive zooids (individuals) in the colony.
3. **Asexual reproduction** by mitotic division appears in unicellular eukaryotes.
4. **True sexual reproduction** with zygote formation is found in some protozoa.
5. The responses (taxes) of protozoa to stimuli represent the **simplest reflexes and instincts** as we know them in metazoans.
6. The simplest animal-like organisms with **exoskeletons** are certain shelled protozoa.
7. **All types of nutrition** are developed in protozoa; autotrophic, saprozoic, and holozoic. **Basic enzyme systems** to accomplish these types of nutrition are developed.
8. Means of **locomotion** in aqueous media are developed.

A protozoan is a complete organism in which all life activities are carried on within the limits of a single plasma membrane. For many years, "Protozoa" was considered a single phylum, but phylogenetic studies show that protozoa do not form a monophyletic group. The various phyla of protozoa now recognized are included in this single chapter only for convenience.

The traditional phylum Protozoa contained four classes: flagellates, amebas, sporozoans (an important parasitic group including malarial organisms), and ciliates. An enormous amount of information on protozoan structure, life histories, and physiology accumulated, and the Society of Protozoologists published a new classification of protozoa in 1980, recognizing seven separate phyla, the most important of which were Sarcomastigophora (flagellates and amebas), Apicomplexa (sporozoans and related

organisms), and Ciliophora (ciliates). However, analyses of sequences of bases in genes, primarily those encoding the small subunit of ribosomal RNA, but also those encoding several proteins, have revolutionized our concepts of phyletic affinities and relationships, not only of protozoan groups, but all eukaryotes. According to some scientists, if we retain classical kingdoms such as animals (Metazoa), fungi, and plants, we must recognize no fewer than *twelve additional* kingdoms.* In some cases, however, DNA sequence data have been collected for very few (or only one) species in a group; this seems slender evidence upon which to establish a kingdom of living organisms. Thus we will generally follow the system of phyla in comprehensive recent monographs, such as Hausmann and Hülsmann (1996). Ameboid organisms (former Sarcodina) fall into numerous lineages, but for the sake of simplicity, we will discuss these together under an informal heading "Amebas."

Protozoan phyla do demonstrate a basic body plan or grade—a single eukaryotic cell—and they amply demonstrate the enormous adaptive potential of that grade. Over 64,000 species have been named, and over half of these are fossils. Although they are unicellular, protozoa are functionally complete organisms with many complicated, microanatomical structures. Their various organelles tend to be more specialized than those of the average cell in a multicellular organism. Particular organelles may perform as skeletons, sensory structures, conducting mechanisms, and other functions.

Protozoa are found wherever life exists. They are highly adaptable and easily distributed from place to place. They require moisture, whether they live in marine or freshwater habitats, soil, decaying organic matter, or plants and animals. They may be sessile or free swimming, and they form a large part of the floating plankton. The same species are often found widely separated in time as well as in space. Some species may have spanned geological eras exceeding 100 million years.

Despite their wide distribution, many protozoa can live successfully only within narrow environmental ranges. Species adaptations vary greatly, and successions of species frequently occur as environmental conditions change. These changes may be caused by physical factors, such as drying of a pond or seasonal changes in temperature, or by biological changes, such as predator pressure.

Protozoa play an enormous role in the economy of nature. Their fantastic numbers are attested by the gigantic ocean soil deposits formed over millions of years by their skeletons. About 10,000 species of protozoa are symbiotic in or on animals or plants, sometimes even other protozoa. The relationship may be **mutualistic** (both partners benefit), **commensalistic** (one partner benefits, no effect on the other), or **parasitic** (one partner benefits at the expense of the other), depending on the species involved. Parasitic protozoa cause some of the most important diseases of humans and domestic animals.

---

*Baldauf, S. L., A. J. Roger, I. Wenk-Siefert, and W. F. Doolittle. 2000. A kingdom-level phylogeny of eukaryotes based on combined protein data. Science **290**:972–997.

FORM AND FUNCTION

Structures and physiology of protozoan cells are largely the same as those of cells of multicellular organisms. However, because they must conduct all functions of life as individual organisms, and because they show such enormous diversity in form, habitat, and feeding, various protozoan cells have unique features.

Nucleus and Cytoplasm

As in other eukaryotes, the nucleus is a membrane-bound structure whose interior communicates with the cytoplasm by small pores. Within the nucleus the genetic material (DNA) is borne on chromosomes. Except during cell division, chromosomes are not usually condensed in a form that can be distinguished, although during fixation of the cells for light microscopy, chromosomal material (chromatin) often clumps together irregularly, leaving some areas within the nucleus relatively clear. The appearance is described as **vesicular** and is characteristic of many protozoan nuclei (Figure 11-1). Condensations of chromatin may be distributed around the periphery of the nucleus or internally in distinct patterns. In most dinoflagellates (p. 226) chromosomes are visible through interphase as they would appear during prophase of mitosis.

Also within the nucleus, one or more **nucleoli** are often present (see Figures 11-1, 11-11, and 11-15).

Characteristics of Protozoan Phyla

1. **Unicellular;** some colonial, and some with multicellular stages in their life cycles
2. **Mostly microscopic,** although some are large enough to be seen with the unaided eye
3. All symmetries represented in the group; shape variable or constant (oval, spherical, or other)
4. **No germ layer present**
5. No organs or tissues, but **specialized organelles** are found; nucleus single or multiple
6. Free-living, mutualism, commensalism, parasitism all represented in the groups
7. Locomotion by **pseudopodia, flagella, cilia,** and direct cell movements; some sessile
8. Some provided with a **simple endoskeleton** or **exoskeleton,** but most are naked
9. **Nutrition of all types:** autotrophic (manufacturing own nutrients by photosynthesis), heterotrophic (depending on other plants or animals for food), saprozoic (using nutrients dissolved in the surrounding medium)
10. Aquatic or terrestrial habitat, free-living or symbiotic mode of life
11. Reproduction **asexually** by fission, budding, and cysts and **sexually** by conjugation or by syngamy (union of male and female gametes to form a zygote)

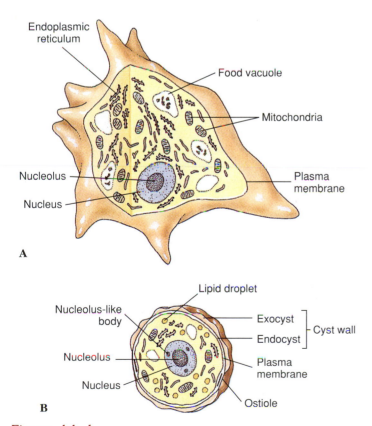

Figure 11-1
Structure of *Acanthamoeba palestinensis.* **A,** Active, feeding form. **B,** Cyst.

Macronuclei of ciliates are described as **compact** or **condensed** because the chromatin material is more finely dispersed and clear areas cannot be observed with the light microscope (see Figure 11-20).

Cellular organelles like those in cells of multicellular animals can be distinguished in the cytoplasm of many protozoa. These organelles include mitochondria, endoplasmic reticulum, Golgi apparatus, and various vesicles. Chloroplasts, the membrane-bound organelles in which photosynthesis takes place, are found in many groups (see Figure 11-14).

Sometimes peripheral and central areas of cytoplasm can be distinguished as **ectoplasm** and **endoplasm** (see Figure 11-4). Endoplasm appears more granular and contains the nucleus and cytoplasmic organelles. Ectoplasm appears more transparent (hyaline) by light microscopy, and it bears the bases of the cilia or flagella. Ectoplasm is often more rigid and is in the gel state of a colloid, whereas the more fluid endoplasm is in the sol state.

Colloidal systems are permanent suspensions of finely divided particles that do not precipitate, such as milk, blood, starch, soap, ink, and gelatin. Colloids in living systems are commonly proteins, lipids, and polysaccharides suspended in the watery fluid of cells (cytoplasm). Such systems may undergo sol-gel transformations, depending on whether the fluid or particulate components become continuous. In the sol state of cytoplasm, solids are suspended in a liquid, and in the semisolid gel state, liquid is suspended in a solid.

Locomotor Organelles

Protozoa move chiefly by cilia and flagella and by pseudopodial movement. These mechanisms are extremely important in the biology of higher animals as well.

Cilia and Flagella

Many small metazoans use cilia not only for locomotion but also to create water currents for their feeding and respiration. Ciliary movement is vital to many species in such functions as handling food, reproduction, excretion, and osmoregulation (as in flame cells, p. 278).

No real morphological distinction exists between cilia and flagella (Figure 11-2), and some investigators have preferred to call them both undulipodia (L. dim. of *unda,* a wave, 1 Gr. *podos,* a foot). However, a cilium propels water parallel to the surface to which the cilium is attached, whereas a flagellum propels water parallel to the main axis of the flagellum. Each flagellum or cilium contains nine pairs of longitudinal microtubules arranged in a circle around a central pair (Figure 11-3), and this is true for all motile flagella and cilia in the animal kingdom, with a few notable exceptions. This "9 + 2" tube of microtubules in a flagellum or cilium is its **axoneme;** an axoneme is covered by a membrane continuous with the cell membrane covering the rest of the organism. At about the point where an axoneme enters the cell proper, the central pair of microtubules ends at a small plate within the circle of nine pairs (Figure 11-3A). Also at about that point, another microtubule joins each of the nine pairs, so that these form a short tube extending from the base of the flagellum into the cell. The tube consists of nine *triplets* of microtubules and is known as a **kinetosome (**or **basal body).** Kinetosomes are exactly the same in structure as **centrioles** that organize mitotic spindles during cell division (see p. 42) and Figure 3-22, p. 50). Centrioles of some flagellates may give rise to kinetosomes, or kinetosomes may function as centrioles. All typical flagella and cilia have a kinetosome at their base, regardless of whether they are borne by a protozoan or metazoan cell.

Description of the axoneme as "9 + 2" is traditional, but it is also misleading because there is only a single pair of microtubules in the center. If we were consistent, we would have to describe the axoneme as "9 + 1."

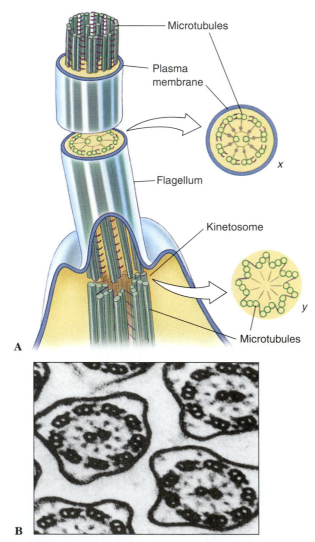

A, A flagellum illustrating the central axoneme, which is composed of nine pairs of microtubules plus a central pair. The axoneme is enclosed within the cell membrane. The central pair of microtubules ends near the level of the cell surface in a basal plate (axosome). The peripheral microtubules continue inward for a short distance to compose two of each of the triplets in the kinetosome (basal body) (at level *y* in **A**). **B,** Electron micrograph of a section through several cilia, corresponding to section *x* in **A.** (×133,000)

Figure 11-3

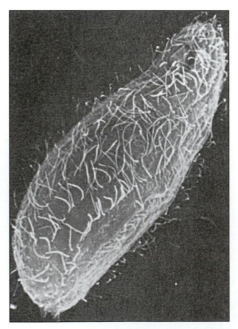

Figure 11-2
Scanning electron micrograph of a free-living ciliate *Tetrahymena thermophila* showing rows of cilia (×2000). Beating of flagella either pushes or pulls the organism through its medium, while cilia propel the organism by a "rowing" mechanism. Their structure is similar, whether viewed by scanning or transmission electron microscopy.

The current explanation for ciliary and flagellar movement is the **sliding microtubule hypothesis.** The movement is powered by a release of chemical bond energy in ATP (p. 60). Two little arms composed of the protein, dynein, are visible in electron micrographs on each of the pairs of peripheral tubules in the axoneme (level *X* in Figure 11-3), and these bear the enzyme adenosine triphosphatase (ATPase), which cleaves the ATP. When bond energy in ATP is released, the arms "walk along" one of the filaments in the adjacent pair, causing it to slide relative to the other filament in the pair. Shear resistance, causing the axoneme to bend when the filaments slide past each other, is provided by "spokes" from each doublet to the central pair of fibrils. These spokes are visible in electron micrographs. Direct evidence for the sliding microtubule hypothesis was obtained by attaching tiny gold beads to axonemal microtubules and observing their movement microscopically.

Pseudopodia

Although pseudopodia are the chief means of locomotion in amebas, they can be formed by a variety of flagellate protozoa, as well as by ameboid cells of many invertebrates. In fact, much defense against disease in the human body depends on ameboid white blood cells, and ameboid cells in many other animals, vertebrate and invertebrate, play similar roles.

In protozoa, pseudopodia exist in several forms. The most familiar are **lobopodia** (Figures 11-4 and 11-5), which are rather large, blunt extensions of the cell body containing both endoplasm and ectoplasm. Some amebas characteristically do not extend individual pseudopodia, but move the whole body with pseudopodial motion; this movement is known as the **limax** form (for a genus of slugs, *Limax*). **Filopodia** are thin extensions, usually branching, and containing only ectoplasm. They are found in some amebas, such as *Euglypha* (see Figure 11-10). **Reticulopodia** (see Figure 11-26) are distinguished from filopodia in that reticulopodia repeatedly rejoin to form a netlike mesh, although some protozoologists believe that the distinction between filopodia and reticulopodia is artificial. Members of superclass Actinopoda have **axopodia** (see Figure 11-29), which are long, thin pseudopodia supported by axial rods of microtubules (Figure 11-6). The microtubules are arranged in a definite spiral or geometrical array, depending on the species, and

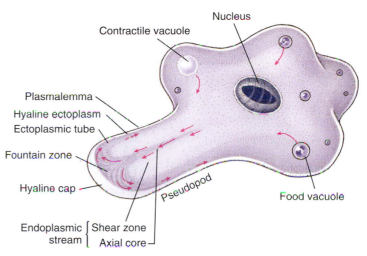

Figure 11-4

Ameba in active locomotion. Arrows indicate the direction of streaming endoplasm. The first sign of a new pseudopodium is thickening of the ectoplasm to form a clear hyaline cap, into which the fluid endoplasm flows. As the endoplasm reaches the forward tip, it fountains out and is converted into ectoplasm, forming a stiff outer tube that lengthens as the forward flow continues. Posteriorly the ectoplasm is converted into fluid endoplasm, replenishing the flow. Substratum is necessary for ameboid movement.

constitute the axoneme of the axopod. Axopodia can be extended or retracted, apparently by addition or removal of microtubular material. Since the tips can adhere to the substrate, the organism can progress by a rolling motion, shortening the axonemes in front and extending those in the rear. Cytoplasm can flow along the axonemes, toward the body on one side and in the reverse direction on the other.

How pseudopodia work has long attracted the interest of zoologists, but we have only recently gained some insight into the phenomenon. When a typical lobopodium begins to form, an extension of ectoplasm called a **hyaline cap** appears, and endoplasm begins to flow toward and into the hyaline cap (Figures 11-4 and 11-7). The flowing endoplasm contains actin subunits attached to regulatory, actin-binding proteins that prevent actin from polymerizing. As endoplasm flows into the hyaline cap, it fountains out to the periphery. Interaction with phospholipids in the cell membrane releases the actin subunits from their

Figure 11-5

Ameboid movement. At left and center, the ameba extends a pseudopodium toward a *Pandorina* colony. At right, the ameba surrounds the *Pandorina* before engulfing it by phagocytosis.

A

Axopodium

Actinosphaerium

B

Figure 11-6

A, Electron micrograph of axopodium (from *Actinosphaerium nucleofilum*) in cross section. **B,** Diagram of axopodium to show orientation of **A.** The axoneme of an axopodium is composed of an array of microtubules, which may vary from three to many in number depending on the species. Some species can extend or retract their axopodia quite rapidly. (×99,000)

Figure 11-7

Proposed mechanism of pseudopodial movement. In endoplasm, actin subunits are bound to regulatory proteins that keep them from assembling **(A).** Upon stimulation, hydrostatic force carries the subunits through a weakened gel to the hyaline cap. The actin subunits are freed from the regulatory proteins by lipids in the cell membrane **(B).** Subunits quickly assemble into filaments and, upon interaction with actin-binding protein (ABP), form gel-like ectoplasm **(C).** At the trailing edge, calcium ions activate actin-severing proteins, loosening the network enough that myosin molecules can pull on it **(D).** Subunits pass up through the tube of ectoplasm to be reused.

regulatory proteins and allows them to polymerize into actin microfilaments. The microfilaments become cross-linked to each other by actin-binding protein (ABP) to form a semisolid gel, transforming the ectoplasm into a tube through which the fluid endoplasm flows as the pseudopodium extends. Near the trailing edge of the gel, calcium ions activate an actin-severing protein, releasing microfilaments from the gel and permitting myosin to associate with and pull on these microfilaments. Thus contraction at the trailing edge results in a pressure that forces the fluid endoplasm, along with its now-dissociated actin subunits, back toward the hyaline cap.

Excretion and Osmoregulation

Vacuoles can be seen by light microscopy in the cytoplasm of many protozoa. Some of these vacuoles periodically fill with a fluid substance that is then expelled. Evidence is strong that these **contractile vacuoles** (see Figures 11-4, 11-14, and 11-20) function principally in osmoregulation. They are more prevalent and fill and empty more frequently in freshwater protozoa than in marine and endosymbiotic species, where their surrounding medium would be more nearly isosmotic (having the same osmotic pressure) to their cytoplasm. Smaller species, which have a greater surface-to-volume ratio, generally have more rapid filling and expulsion rates in their contractile vacuoles. Excretion of metabolic wastes, on the other hand, is almost entirely by diffusion. The main end product of nitrogen metabolism is ammonia, which readily diffuses out of the small bodies of protozoa.

Although it seems clear that contractile vacuoles function to remove excess water that has entered cytoplasm by osmosis, a reasonable explanation for such removal has been elusive. A recent hypothesis suggests that proton pumps (p. 64) on the vacuolar surface and on tubules radiating from it actively transport H^+ and cotransport bicarbonate (HCO_3^-) (Figure 11-8), which are osmotically active particles. As these particles accumulate within a vacuole, water would be drawn into the vacuole. Fluid within the vacuole would remain isosmotic to the cytoplasm. Then as the vacuole finally joins its membrane to the surface membrane and empties its contents to the outside, it would expel water, H^+, and HCO_3^-. These ions can be replaced readily by action of carbonic anhydrase on CO_2 and H_2O. Carbonic anhydrase is present in the cytoplasm of amebas.

Some ciliates, such as *Blepharisma*, have contractile vacuoles with structure and filling mechanisms apparently similar to those described for amebas. Others, such as *Paramecium*, have more complex contractile vacuoles. Such vacuoles are located in a specific position beneath the cell membrane, with an "excretory" pore leading to the outside, and surrounded by ampullae of about six feeder canals (see Figure 11-20). Feeder canals, in turn, are surrounded by fine tubules about 20 nm in diameter, which connect with the canals during filling of ampullae and at their lower ends connect with the tubular system of endoplasmic reticulum. Ampullae and contractile vacuoles are surrounded by bundles of fibrils, which may function in contraction of these structures. Contraction of ampullae fills the vacuole. When the vacuole contracts to discharge its contents to the outside, the ampullae become disconnected from the vacuole, so that backflow is prevented. Tubules, ampullae, or vacuoles may be supplied with proton pumps to draw water into their lumens by the mechanism already described.

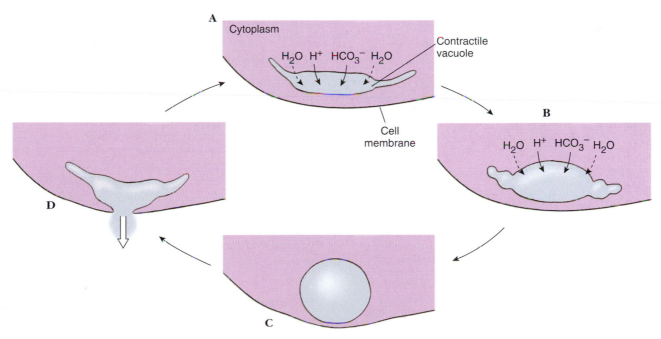

Figure 11-8

Proposed mechanism for operation of contractile vacuoles. **A, B,** Vacuoles are composed of a system of cisternae and tubules. Proton pumps in their membranes transport H^+ and cotransport HCO_3^- into the vacuoles. Water diffuses in passively to maintain an osmotic pressure equal to that in the cytoplasm. When the vacuole fills **C,** its membrane fuses with the cell's surface membrane, expelling water, H^+, and HCO_3^-. **D,** Protons and bicarbonate ions are replaced readily by action of carbonic anhydrase on carbon dioxide and water.

Nutrition

Protozoa can be categorized broadly into **autotrophs** (which synthesize their own organic constituents from inorganic substrates) and **heterotrophs** (which must obtain organic molecules synthesized by other organisms (p. 31)). Another kind of classification, usually applied to heterotrophs, involves those that ingest visible particles of food (**phagotrophs,** or **holozoic** feeders) as contrasted with those ingesting food in a soluble form (**osmotrophs,** or **saprozoic** feeders). However, reality is not so simple, even among one-celled organisms. Autotrophic protozoa (phototrophs) use light energy to synthesize their organic molecules, but they often practice phagotrophy and osmotrophy as well. Even among heterotrophs, few are exclusively either phagotrophic or osmotrophic. A single class Euglenoidea (phylum Euglenozoa) contains some forms that are mainly phototrophs, some that are mainly osmotrophs, and some that are mainly phagotrophs. Species of *Euglena* show considerable variety in nutritional capability. Some species require certain preformed organic molecules, even though they are autotrophs, and some lose their chloroplasts if maintained in darkness, thus becoming permanent osmotrophs.

Holozoic nutrition implies phagocytosis (Figure 11-9), in which an infolding or invagination of the cell membrane surrounds a food particle. As the invagination extends farther into the cell, it is pinched off at the surface (p. 47). The food particle

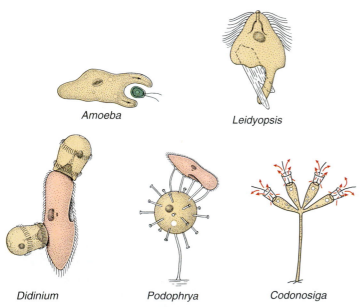

Figure 11-9

Some feeding methods among protozoa. *Amoeba* surrounds a small flagellate with pseudopodia. *Leidyopsis,* a flagellate living in the intestine of termites, forms pseudopodia and ingests wood chips. *Didinium,* a ciliate, feeds only on *Paramecium,* which it swallows through a temporary cytostome in its anterior end. Sometimes more than one *Didinium* feed on the same *Paramecium. Podophrya* is a suctorian ciliophoran. Its tentacles attach to its prey and suck prey cytoplasm into the body of the *Podophrya,* where it is pinched off to form food vacuoles. *Codonosiga,* a sessile flagellate with a collar of microvilli, feeds on particles suspended in the water drawn through its collar by the beat of its flagellum. Technically, all of these methods are types of phagocytosis.

thus is contained in an intracellular, membrane-bound vesicle, a **food vacuole** or **phagosome.** Lysosomes, small vesicles containing digestive enzymes, fuse with the phagosome and pour their contents into it, where digestion begins. As digested products are absorbed across the vacuolar membrane, the phagosome becomes smaller. Any undigestible material may be released to the outside by exocytosis, the vacuole again fusing with the cell surface membrane. In most ciliates, many flagellates, and many apicomplexans, the site of phagocytosis is a definite mouth structure, the **cytostome** (Figure 11-20). In amebas, phagocytosis can occur at almost any point by envelopment of a particle with pseudopodia. Particles must be ingested through the opening of the test, or shell, in amebas that have tests. Flagellates may form a temporary cytostome, usually in a characteristic position, or they may have a permanent cytostome with specialized structure. Many ciliates have a characteristic structure for expulsion of waste matter, the **cytopyge** or **cytoproct,** found in a characteristic location. In some, the cytopyge also serves as the site for expulsion of the contents of the contractile vacuole.

Saprozoic feeding may be by pinocytosis or by transport of solutes directly across the outer cell membrane. Pinocytosis and transport across a cell membrane are discussed on p. 48. Direct transport across a membrane may be by diffusion, facilitated transport, or active transport. Diffusion is probably of little or no importance in nutrition of protozoa, except possibly in some endosymbiotic species. Some important food molecules, such as glucose and amino acids, may be brought into a cell by facilitated diffusion and active transport.

It has been shown that a stimulatory substance, or "inducer," must be present in the surrounding medium for many protozoa to initiate pinocytosis. Several proteins act as inducers, as can some salts and other substances; it appears that the inducer must be a positively charged molecule. Pinocytosis takes place at the inner end of the cytopharynx in protozoa possessing that structure.

Reproduction

Sexual phenomena occur widely among protozoa, and sexual processes may precede certain phases of asexual reproduction, but embryonic development does not occur; protozoa do not have embryos. The essential features of sexual processes include a reduction division of the chromosome number to half (diploid number to haploid number), the development of sex cells (gametes) or at least gamete nuclei, and usually a fusion of gamete nuclei (p. 225).

Fission

The cell multiplication process that produces more individuals in protozoa is called fission. The most common type of fission is **binary,** in which two essentially identical individuals result (Figure 11-10). When a progeny cell is considerably smaller than the parent and then grows to adult size, the process is called **budding.** Budding occurs in some ciliates. In **multiple fission,** divi-

A *Arcella*

B *Euglypha*

C *Trypanosoma*   **D** *Euglena*

Figure 11-10

Binary fission in some sarcodines and flagellates. **A,** The two nuclei of *Arcella* divide as some of its cytoplasm is extruded and begins to secrete a new test for the daughter cell. **B,** The test of another sarcodine, *Euglypha,* is constructed of secreted platelets. Secretion of platelets for the daughter cell is begun before cytoplasm begins to move out of the aperture. As these are used to construct the test of the daughter cell, the nucleus divides. **C,** *Trypanosoma* has a kinetoplast (part of the mitochondrion) near the kinetosome of its flagellum close to its posterior end in the stage shown. All of these parts must be replicated before the cell divides. **D,** Division of *Euglena.* Compare **C** and **D** with Figure 11-22, fission in a ciliophoran.

sion of the cytoplasm (cytokinesis) is preceded by several nuclear divisions, so that a number of individuals are produced almost simultaneously (see Figure 11-17). Multiple fission, or **schizogony,** is common among the Apicomplexa and some amebas. If the multiple fission is preceded by or associated with union of gametes, it is referred to as **sporogony.**

The foregoing types of division are accompanied by some form of mitosis (p. 48). However, this mitosis is often somewhat unlike that found in metazoans. For example, the nuclear membrane often persists through mitosis, and the microtubular spindle may be formed within the nuclear membrane. Centrioles have not been observed in nuclear division of ciliates; the nuclear membrane persists in micronuclear mitosis, with the spindle within the nucleus. The macronucleus of ciliates seems simply to elongate, constrict, and divide without any recognizable mitotic phenomena **(amitosis).**

Sexual Processes

Although all protozoa reproduce asexually, and some are apparently exclusively asexual, the widespread occurrence of sex among protozoa testifies to its importance as a means of genetic recombination. Gamete nuclei, or pronuclei, which fuse in fertilization to restore the diploid number of chromosomes, are usually borne in special gametic cells. When gametes all look alike,

they are called **isogametes,** but most species have two dissimilar types, or **anisogametes.**

In animals meiosis usually occurs during or just before gamete formation (called gametic meiosis, p. 133). Such is indeed the case in Ciliophora and some flagellated and amebic groups. However, in other flagellated groups and in Apicomplexa, the first divisions *after* fertilization are meiotic **(zygotic meiosis),** and all individuals produced asexually (mitotically) in the life cycle up to the next zygote are haploid. Most protozoa that do not reproduce sexually probably are haploid, although demonstration of ploidy is difficult in the absence of meiosis. In some amebas (foraminiferans) haploid and diploid generations alternate **(intermediary meiosis),** a phenomenon widespread among plants.

Fertilization of an individual gamete by another is **syngamy,** but some sexual phenomena in protozoa do not involve syngamy. Examples are **autogamy,** in which gametic nuclei arise by meiosis and fuse to form a zygote within the same organism that produced them, and **conjugation,** in which an exchange of gametic nuclei occurs between paired organisms (conjugants). We will describe conjugation further in the discussion of *Paramecium.*

Encystment and Excystment

Separated as they are from their external environment only by their delicate external cell membrane, it seems astonishing that protozoa could be so successful in habitats frequently subjected to extremely harsh conditions. Survival under harsh conditions surely is related to the ability to form **cysts,** dormant forms marked by possession of resistant external coverings and a more or less complete shutdown of metabolic machinery. Cyst formation is also important to many parasitic forms that must survive a harsh environment between hosts (Figure 11-1). However, some parasites do not form cysts, apparently depending on direct transfer from one host to another. Reproductive phases such as fission, budding, and syngamy may occur in cysts of some species. Encystment has not been found in *Paramecium,* and it is rare or absent in marine forms.

Cysts of some soil-inhabiting and fresh-water protozoa have amazing durability. Cysts of the soil ciliate *Colpoda* can survive 12 days in liquid nitrogen and 3 hours at 100°C. Survival of *Colpoda* cysts in dried soil has been shown for up to 38 years, and those of a certain small flagellate *(Podo)* can survive up to 49 years! Not all cysts are so sturdy, however. Those of *Entamoeba histolytica* will tolerate gastric acidity but not desiccation, temperature above 50°C, or sunlight.

The conditions stimulating encystment are incompletely understood, although in some cases cyst formation is cyclic, occurring at a certain stage in the life cycle. In most free-living forms, adverse environmental change favors encystment. Such conditions may include food deficiency, desiccation, increased environmental osmotic pressure, decreased oxygen concentration, or change in pH or temperature.

During encystment a number of organelles, such as cilia or flagella, are resorbed, and the Golgi apparatus secretes cyst wall material, which is carried to the surface in vesicles and extruded.

Although the exact stimulus for excystation (escape from cysts) is usually unknown, a return of favorable conditions initiates excystment in those protozoa in which the cysts are a resistant stage. In parasitic forms excystment stimulus may be more specific, requiring conditions similar to those found in the host.

PHYLA RETORTAMONADA AND AXOSTYLATA

These groups are rather small in numbers of species, but some of each are of importance to humans. They may represent groups derived very early from an ancestral eukaryote before mitochondria and plastids were acquired by symbiogenesis (p. 207). However, typically mitochondrial enzymes have been reported in both groups; therefore, they might have acquired the symbionts and then lost them. Members of both phyla are parasitic.

Retortamonads lack both mitochondria and Golgi bodies. *Giardia* (Figure 11-11) is a common intestinal parasite of humans and other animals. It is often asymptomatic but may cause a rather discomfiting, but not fatal, diarrhea. Cysts are passed in the feces, and new hosts are infected by ingestion of cysts, often in contaminated water.

Members of Axostylata have a stiffening rod composed of microtubules, the **axostyle,** that extends along the longitudinal axis of their body. Members of class Parabasalea have a **parabasal body,** which is a Golgi apparatus connected by a fiber to one of the kinetosomes. Although mitochondria are absent, members of order Trichomonadida possess **hydrogenosomes,** which are organelles analogous to mitochondria but which produce molecular hydrogen when oxygen is absent.

Some trichomonads (Figure 11-11) are of medical or veterinary importance. *Trichomonas vaginalis* infects the urogenital tract of humans and is sexually transmitted. It produces no symptoms in males but is the most common cause of vaginitis in females. **Pentatrichomonas hominis** lives in the cecum and colon of humans and *Trichomonas tenax* lives in the mouth; they apparently cause no disease. Other species of Trichomonadida are widely distributed through vertebrates of all classes and many invertebrates.

Giardia lamblia is commonly transmitted through water supplies contaminated with sewage. The same species, however, lives in a variety of mammals other than humans. Beavers seem to be an important source of infection in mountains of the western United States. When one has hiked for miles in the wild on a hot day, it can be very tempting to fill a canteen and drink from a crystal-clear beaver pond. Many cases of infection have been acquired that way.

PHYLUM CHLOROPHYTA

"Plantlike" protozoa were formerly included by zoologists as a class (Phytomastigophorea) under the phylum Sarcomastigophora. Molecular evidence clearly indicates that neither of these taxa is acceptable, and both must be abandoned. Single-celled (and multicellular) green algae (Chlorophyta) form a clade with "higher" plants (bryophytes and vascular plants). Flagellated, single-celled algae (such as *Chlamydomonas,* Figure 11-12) and colonial forms (such as *Gonium,* Figure 11-12, and *Volvox,* Figure 11-13) belong with this group. They are autotrophic and contain one or more characteristic **chloroplasts,** colored bodies that contain chlorophylls.

Volvox globator

Volvox (Figure 11-13) is a multicellular chlorophytan that contains separate somatic and reproductive cells (see p. 5). It is often studied in introductory courses because its mode of development is somewhat similar to embryonic development of some metazoa. The order to which *Volvox* belongs (Volvocida) includes many freshwater flagellates, mostly green, with a cellulose cell wall through which two short flagella project. Many are colonial forms (Figure 11-12, *Pandorina, Eudorina, Gonium*), in which a single organism contains more than one cell but separate somatic and reproductive types do not exist.

Volvox (Figure 11-13) is a green, hollow sphere that may reach a diameter of 0.5 to 1 mm. A single organism contains many

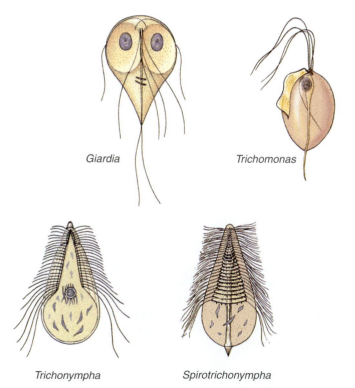

Giardia   *Trichomonas*

Trichonympha   *Spirotrichonympha*

Figure 11-11

Examples of phyla Retortamonada (*Giardia*) and Axostylata (*Trichomonas, Trichonympha,* and *Spirotrichonympha*). *Giardia lamblia* often causes diarrhea in humans. *Trichomonas vaginalis* is transmitted sexually and is a frequent cause of vaginitis in humans. *Trichonympha* and *Spirotrichonympha* are mutualistic symbionts in termites.

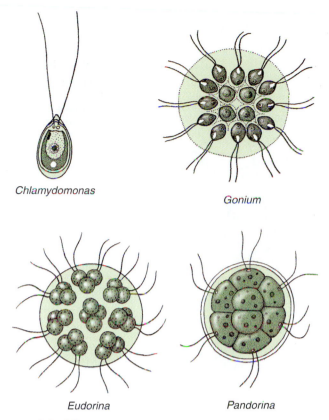

Chlamydomonas

Gonium

Eudorina

Pandorina

Figure 11-12
Examples of phylum Chlorophyta. They are all photoautotrophs.

thousands of zooids (up to 50,000) embedded in the gelatinous surface of a jelly ball. Each cell is much like a euglenid (p. 218), with a nucleus, a pair of flagella, a large chloroplast, and a red stigma. Adjacent cells are connected with each other by cytoplasmic strands. At one pole (usually in front as the colony moves), the stigmata are a little larger. Coordinated action of the flagella causes the colony to move by rolling over and over.

In *Volvox* we have a division of labor to the extent that most of the zooids are somatic cells concerned with nutrition and locomotion, and a few germ cells located in the posterior half are responsible for reproduction. Reproduction is asexual or sexual. In either case only certain zooids located around the equator or in the posterior half take part.

The original polarity of zooids in *Volvox* is such that their flagella are protruding into the interior cavity of the developing organism. To move the flagella on the outside so that locomotion is possible, the entire spheroid must turn itself inside out. This process, called inversion, is *very unusual*. Of all other living organisms, only the sponges (phylum Porifera) have a comparable developmental process.

Asexual reproduction in *Volvox* occurs by repeated mitotic division of one of the germ cells to form a hollow sphere

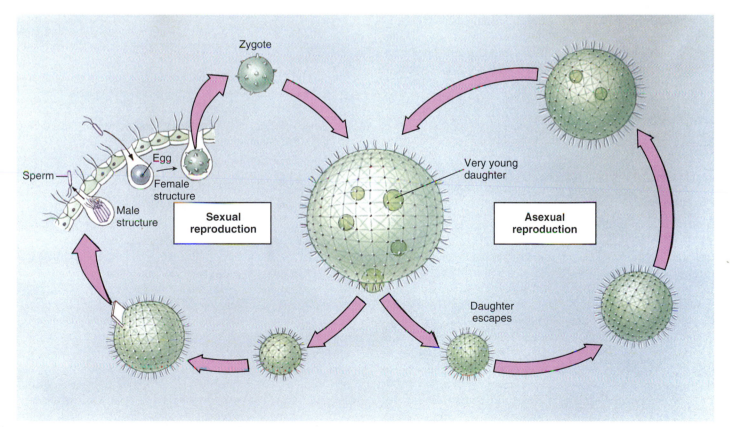

Figure 11-13
Life cycle of *Volvox*. Asexual reproduction occurs in spring and summer when specialized diploid reproductive cells divide to form young organisms that remain in the mother organism until large enough to escape. Sexual reproduction occurs largely in autumn when haploid sex cells develop. The fertilized ova may encyst and so survive the winter, developing into a mature asexual organism in the spring. In some species the organisms have separate sexes; in others both eggs and sperm are produced in the same organism.

of cells, with the flagellated ends of the cells inside. The sphere then turns itself inside out to form a daughter colony similar to the parent colony. Several daughter colonies are formed inside the parent colony before they escape by rupture of the parent.

In **sexual reproduction** some of the zooids differentiate into **macrogametes** or **microgametes** (Figure 11-13). Macrogametes are fewer and larger and are loaded with food for nourishment of the young organism. Microgametes, by repeated division, form bundles or balls of small flagellated sperm that leave the mother organism when they mature and swim about to find a mature ovum. After fertilization, the zygote secretes a hard, spiny, protective shell around itself. When released by the rupture of a parent, a zygote remains quiescent during the winter. Within its shell the zygote undergoes repeated division, producing a small organism that breaks out in the spring. A number of asexual generations may follow, during the summer, before sexual reproduction occurs again.

PHYLUM EUGLENOZOA

Members of this phylum have a series of longitudinal microtubules just beneath the cell membrane that help to stiffen the membrane into a **pellicle.** *Euglena,* which zoologists formerly considered a phytomastigophorean, is an autotrophic member of this phylum. It has chloroplasts containing chlorophylls and a light-sensitive **stigma,** or eyespot, which is a shallow pigment cup that allows light from only one direction to strike a light-sensitive receptor (Figure 11-14). Euglenozoa also contains Kinetoplasta, which are all parasitic in plants and animals. They have a unique organelle known as a **kinetoplast,** which is part of their mitochondrion and is composed of a large disc of DNA. Various species of *Trypanosoma* (Gr. *trypanon,* auger, + *soma,* body) (Figure 11-15) cause serious diseases of humans and other animals.

Subphylum Euglenida
Euglena viridis

Euglena viridis (Figure 11-14) is a flagellate commonly studied in introductory zoology courses. Its natural habitat is fresh-water streams and ponds where there is considerable vegetation. The organisms are spindle shaped and about 60 μm long, but some species of *Euglena* are smaller and some larger (*E. oxyuris* is 500 μm long). Just beneath the outer membrane of *Euglena* are proteinaceous strips and microtubules that form a pellicle. In *Euglena* the pellicle is flexible enough to permit bending, but in other euglenids it may be more rigid. A **flagellum** extends from a flask-shaped **reservoir** at the anterior end, and another, short flagellum ends within the reservoir. A **kinetosome** is found at the base of each flagellum, and a **contractile vacuole** empties into the reservoir. A red eyespot, or **stigma,** apparently functions in orientation to light. Within the cytoplasm are oval **chloroplasts** that bear chlorophyll and give the organism its greenish color. **Paramylon granules** of various shapes are masses of a starchlike food storage material.

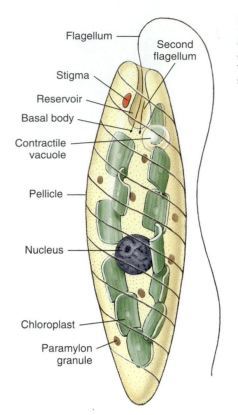

Figure 11-14
Euglena. Features shown are a combination of those visible in living and stained preparations.

Labels: Flagellum, Second flagellum, Stigma, Reservoir, Basal body, Contractile vacuole, Pellicle, Nucleus, Chloroplast, Paramylon granule

Nutrition of *Euglena* is normally autotrophic (holophytic), but if kept in the dark the organism makes use of saprozoic nutrition, absorbing nutrients through its body surface. Mutants of *Euglena* can be produced that have permanently lost their photosynthetic ability. Although *Euglena* does not ingest solid food, some euglenids are phagotrophic. *Peranema* has a cytostome that opens alongside its flagellar reservoir.

Euglena reproduces by binary fission and can encyst to survive adverse environmental conditions.

Subphylum Kinetoplasta

Some of the most important protozoan parasites are kinetoplastans. Many of them belong to the genus *Trypanosoma* (Figure 11-15) and live in the blood of fish, amphibians, reptiles, birds, and mammals. Some are nonpathogenic, but others produce severe diseases in humans and domestic animals. *Trypanosoma brucei gambiense* and *T. brucei rhodesiense* cause African sleeping sickness in humans, and *T. brucei brucei* causes a related disease in domestic animals. Trypanosomes are transmitted by tsetse flies (*Glossina* spp.). *Trypanosoma b. rhodesiense,* the more virulent of the sleeping sickness trypanosomes, and *T. b. brucei* have natural reservoirs (antelope and other wild mammals) that are apparently not harmed by the parasites. Some 10,000 new cases of human sleeping sickness are diagnosed each year, of which about half are fatal, and many of the remainder sustain permanent brain damage.

Trypanosoma cruzi causes Chagas disease in humans in Central America and South America. It is transmitted by "kissing bugs" (Triatominae), a name arising from the bug's habit of biting its sleeping victim on the face. Acute Chagas disease is most

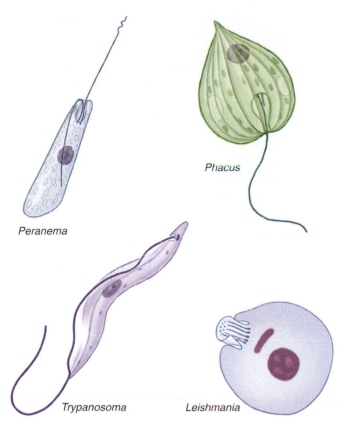

Figure 11-15

Examples of phylum Euglenozoa. *Peranema* is a colorless, free-living phagotroph, and *Phacus* is a green, free-living photoautotroph. *Trypanosoma* and *Leishmania* are parasitic, and some species cause serious diseases of humans and domestic animals. *Leishmania* is shown as its intracellular form, without an external flagellum.

common and severe among children less than five years old, while the chronic disease is seen most often in adults. Symptoms are primarily a result of central and peripheral nervous dysfunction. Two to three million people in South and Central America show chronic Chagas disease, and 45,000 of these die each year.

Several species of *Leishmania* (Figure 11-15) cause disease in humans. Infection with some species may result in a serious visceral disease affecting especially the liver and spleen; others can cause disfiguring lesions in the mucous membranes of the nose and throat, and the least serious result is a skin ulcer. *Leishmania* spp. are transmitted by sand flies. Visceral leishmaniasis and cutaneous leishmaniasis are common in parts of Africa and Asia, and the mucocutaneous form is found in Central America and South America.

SUPERPHYLUM ALVEOLATA

Based on molecular and mophological evidence, the next three phyla (Apicomplexa, Ciliophora, Dinoflagellata) are often grouped together in a superphylum Alveolata. Alveolates possess **alveoli** or related structures, which are membrane-bound sacs that lie beneath the cell membrane and serve structural functions or produce pellicles (ciliates) or thecal plates (dinoflagellates).

PHYLUM APICOMPLEXA

Class Coccidea

All apicomplexans are endoparasites, and their hosts are found in many animal phyla. The presence of a certain combination of organelles, the **apical complex,** distinguishes this phylum (Figure 11-16A). The apical complex is usually present only in certain developmental stages of the organisms; for example, **merozoites** and **sporozoites** (Figure 11-17). Some structures, especially the **rhoptries** and **micronemes,** apparently aid in penetrating the host's cells or tissues.

Locomotor organelles are less obvious in this group than in other protozoa. Pseudopodia occur in some intracellular stages, and gametes of some species are flagellated. Tiny contractile fibrils can form waves of contraction across the body surfaces to propel the organism through a liquid medium.

The life cycle usually includes both asexual and sexual reproduction, and sometimes an invertebrate intermediate host. At some point in the life cycle, the organisms develop a **spore (oocyst),** which is infective for the next host and is often protected by a resistant coat.

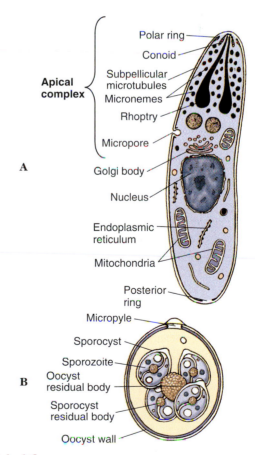

Figure 11-16

A, Diagram of an apicomplexan sporozoite or merozoite at the electron-microscope level, illustrating the apical complex. The polar ring, conoid, micronemes, rhoptries, subpellicular microtubules, and micropore (cytostome) are all considered components of the apical complex. **B,** Infective oocyst of *Eimeria*. The oocyst is the resistant stage and has undergone multiple fission after zygote formation (sporogony).

Coccidia are intracellular parasites in invertebrates and vertebrates, and the group includes species of very great medical and veterinary importance.

Eimeria species

The name "coccidiosis" is generally applied only to infections with *Eimeria* or *Isospora*. Humans can be infected with species of *Isospora,* but there is usually little disease. However, *Isospora* infections can be very serious in AIDS patients. Some species of *Eimeria* may cause serious disease in some domestic animals. Symptoms usually include severe diarrhea or dysentery.

Eimeria tenella is often fatal to young fowl, producing severe pathogenesis in the intestine. The organisms undergo schizogony (p. 215) in the intestinal cells, finally producing gametes. After fertilization the zygote forms an oocyst that passes out of its host in the feces (Figure 11-16B). Sporogony occurs within the oocyst outside the host, producing eight sporozoites in each oocyst. Infection occurs when a new host accidentally ingests a sporulated oocyst and the sporozoites are released by digestive enzymes.

Toxoplasma gondii

A similar life cycle occurs in *Toxoplasma gondii,* a parasite of cats, but this species produces extraintestinal stages as well. When rodents, cattle, sheep, humans, many other mammals, or even birds, ingest sporozoites, the sporozoites cross from the intestine and begin rapid, asexual reproduction in a variety of tissues. As the host mounts an immune response (see Chapter 35), reproduction of the zoites slows, and they become enclosed in tough **tissue cysts.** The zoites, now called **bradyzoites,** accumulate in large numbers in each tissue cyst. Bradyzoites are infective for other hosts, including cats, where they can initiate the intestinal cycle in a cat that eats infected prey. Bradyzoites can remain viable and infective for months or years, and it is estimated that one-third of the world's human population carries tissue cysts containing bradyzoites in their body. The normal

Some 20% or more of adults in the United States are infected with *Toxoplasma gondii;* we have no symptoms because the parasite is held in check by our immune systems. However, *T. gondii* is one of the most important opportunistic infections in AIDS patients. The latent infection is activated in between 5% and 15% of AIDS patients, often in the brain, with serious consequences. Another coccidian, *Cryptosporidium parvum,* first was reported in humans in 1976. We now recognize it as a major cause of diarrheal disease worldwide, especially in children in tropical countries. Waterborne outbreaks have occurred in the United States, and the diarrhea can be life-threatening in immunocompromised patients (such as those with AIDS). The latest coccidian pathogen to emerge has been *Cyclospora cayetanensis*. About 850 cases of diarrhea due to *Cyclospora* in the United States and Canada were reported in May and June, 1996. We do not yet know how the parasite is transmitted or even the normal host.

route of infection for humans is apparently consumption of infected meat that is insufficiently cooked.

In humans *Toxoplasma* causes little or no ill effects except in AIDS patients or in women infected during pregnancy, particularly in the first trimester. Such infection greatly increases the chances of a birth defect in the baby; it is now believed that 2% of all mental retardation in the United States is a result of congenital toxoplasmosis. Toxoplasmosis can also be a serious disease in persons who are immunosuppressed, either with drugs or by AIDS. In such patients rupture of a tissue cyst, which would be contained easily in a person with a normal immune system, becomes a source of life-threatening infection.

Plasmodium: *The Malarial Organism*

The best known coccidians are *Plasmodium* spp., causative organisms of the most important infectious disease of humans: **malaria.** Malaria is a very serious disease, difficult to control and widespread, particularly in tropical and subtropical countries. Four species of *Plasmodium* infect humans: *P. falciparum, P. vivax, P. malariae,* and *P. ovale*. Although each species produces its own peculiar clinical picture, all four have similar cycles of development in their hosts (Figure 11-17).

The parasite is carried by mosquitoes (*Anopheles*), and sporozoites are injected into a human with the insect's saliva during its bite. Sporozoites penetrate liver cells and initiate schizogony. The products of this division then enter other liver cells to repeat the schizogonous cycle, or in *P. falciparum* they penetrate red blood cells after only one cycle in the liver. The period when the parasites are in the liver is the **incubation period,** and it lasts from 6 to 15 days, depending on the species of *Plasmodium*.

Merozoites released as a result of liver schizogony enter red blood cells, where they begin a series of schizogonous cycles. When they enter red blood cells, they become ameboid **trophozoites,** feeding on hemoglobin. The end product of the parasite's digestion of hemoglobin is a dark, insoluble pigment: **hemozoin.** Hemozoin accumulates in the host cell, is released when the next generation of merozoites is produced, and eventually accumulates in the liver, spleen, or other organs. A trophozoite within a red blood cell grows and undergoes schizogony, producing 6 to 36 merozoites, which, depending on the species, burst forth to infect new red cells. When a red blood cell containing merozoites bursts, it releases the parasite's metabolic products, which have accumulated there. Release of these foreign substances into the patient's circulation results in the chills and fever characteristic of malaria.

Since the populations of schizonts maturing in red blood cells are synchronized to some degree, the episodes of chills and fever have a periodicity characteristic of the particular species of *Plasmodium*. In *P. vivax* (benign tertian) malaria and *P. ovale* malaria, episodes occur every 48 hours; in *P. malariae* (quartan) malaria, every 72 hours; and in *P. falciparum* (malignant tertian) malaria, about every 48 hours, although synchrony is less well defined in this species. People usually recover from infections with the first three species, but mortality may be high in untreated cases of *P. falciparum* infection. Sometimes grave

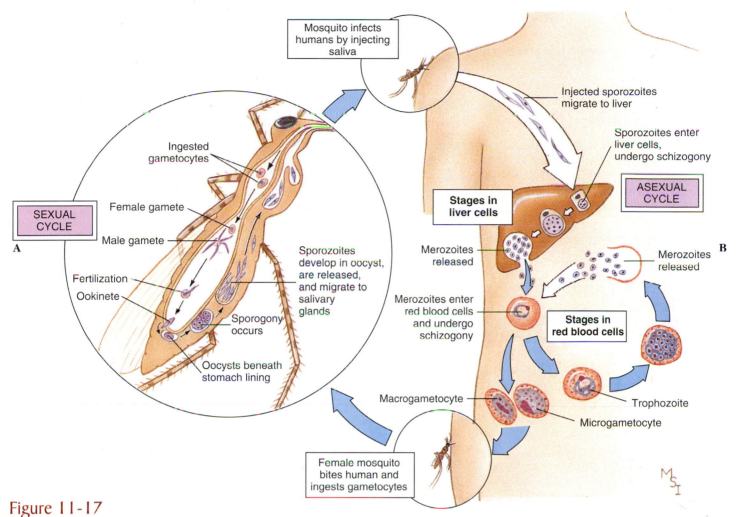

Figure 11-17

Life cycle of *Plasmodium vivax,* one of the protozoa (class Coccidia) that causes malaria in humans. **A,** Sexual cycle produces sporozoites in body of mosquito. Meiosis occurs just after zygote formation (zygotic meiosis). **B,** Sporozoites infect a human and reproduce asexually, first in liver cells and then in red blood cells. Malaria is spread by *Anopheles* mosquito, which ingests gametocytes along with human blood, then, when biting another victim, leaves sporozoites in new wound.

complications, such as **cerebral malaria,** occur. Unfortunately, *P. falciparum* is the most common species, accounting for 50% of all malaria in the world. Certain genes, for example the gene for sickle cell hemoglobin (p. 97 and p. 659), confer some resistance to malaria on people that carry them.

After some cycles of schizogony in red blood cells, infection of new cells by some of the merozoites results in production of **microgametocytes** and **macrogametocytes** rather than another generation of merozoites. When gametocytes are ingested by a mosquito feeding on a patient's blood, they mature into **gametes,** and fertilization occurs. The zygote becomes a motile **ookinete,** which penetrates the stomach wall of the mosquito and becomes an **oocyst.** Within the oocyst, sporogony occurs, and thousands of **sporozoites** are produced. The oocyst ruptures, and the sporozoites migrate to the salivary glands, from which they are transferred to a human by a bite of the mosquito. Development in a mosquito requires 7 to 18 days but may be longer in cool weather.

Elimination of mosquitoes and their breeding places by insecticides, drainage, and other methods has been effective in controlling malaria in some areas. However, difficulties in carrying out such activities in remote areas and areas suffering civil unrest, and acquisition of resistance to insecticides by mosquitoes and to antimalarial drugs by *Plasmodium* (especially *P. falciparum*),

A *disease* is any illness or disorder that can be recognized by a given set of signs and symptoms. *Epidemiology* is the study of all factors that influence transmission, geographic distribution, incidence, and prevalence of a disease. Epidemiology of parasitic diseases often involves poor sanitation and contamination of water or food with infectious stages. That is not the case with arthropod-borne diseases, such as malaria. Transmission and distribution of malaria depend on presence of a suitable *Anopheles* species, as well as its breeding, feeding, and resting habits. The climate (whether the mosquito can breed and feed throughout the year) is important, as are the prevalence of infected humans (especially asymptomatic individuals). It has nothing to do with improper waste disposal or poverty.

mean that malaria will be a serious disease of humans for a long time to come.

Other species of *Plasmodium* parasitize birds, reptiles, and mammals. Those of birds are transmitted chiefly by *Culex* mosquitoes.

PHYLUM CILIOPHORA

Ciliates are a large and interesting group, with a great variety of forms living in all types of freshwater and marine habitats. They are the most structurally complex and diversely specialized of all protozoa. The majority are free living, but some are commensal or parasitic. They are usually solitary and motile, but some are sessile and some colonial. There is great diversity of shape and size. In general, they are larger than most other protozoa, but they range from very small (10 to 12 μm) up to 3 mm long. All have cilia that beat in a coordinated rhythmical manner, although the arrangement of cilia varies and some lack cilia as adults.

Ciliates are always multinucleate, possessing at least one **macronucleus** and one **micronucleus,** but varying from one to many of either type. The macronuclei are apparently responsible for metabolic and developmental functions and for maintaining all the visible traits, such as the pellicular apparatus. Macronuclei vary in shape among the different species (Figures 11-18 and 11-20). Micronuclei participate in sexual reproduction and give rise to macronuclei after exchange of micronuclear material between individuals. Micronuclei divide mitotically, and macronuclei divide amitotically (see p. 215).

The pellicle of ciliates may consist only of a cell membrane or in some species may form a thickened armor. Cilia are short and usually arranged in longitudinal or diagonal rows. Cilia may cover the surface of the organism or may be restricted to the oral region or to certain bands. In some forms cilia are fused into a sheet called an **undulating membrane** or into smaller **membranelles,** both used to propel food into the **cytopharynx** (gullet). In other forms there may be fused cilia forming stiffened tufts called **cirri,** often used in locomotion by the creeping ciliates (Figure 11-18).

An apparently structural system of fibers, in addition to the kinetosomes, makes up the **infraciliature,** just beneath the pellicle (Figure 11-19). Each cilium terminates beneath the pellicle in its kinetosome, and from each kinetosome a fibril arises and passes along beneath the row of cilia, joining with the other fibrils of that row. The cilia, kinetosomes, and other fibrils of that ciliary row make up what is called a **kinety** (Figure 11-19). All ciliates seem to have kinety systems, even those that lack cilia at some stage. The infraciliature apparently does not coordinate ciliary beat, as formerly thought. Coordination of ciliary movement seems to be by waves of depolarization of the cell membrane moving down the organism, similar to the phenomenon in a nerve impulse.

Most ciliates are holozoic. Most of them possess a cytostome (mouth) that in some forms is a simple opening and in others is connected to a gullet or ciliated groove. The mouth in some is strengthened with stiff, rodlike trichites for swallowing larger prey; in others, such as paramecia, ciliary water currents carry microscopic food particles toward the mouth. *Didinium* has a proboscis for engulfing paramecia on which it feeds (Figure 11-

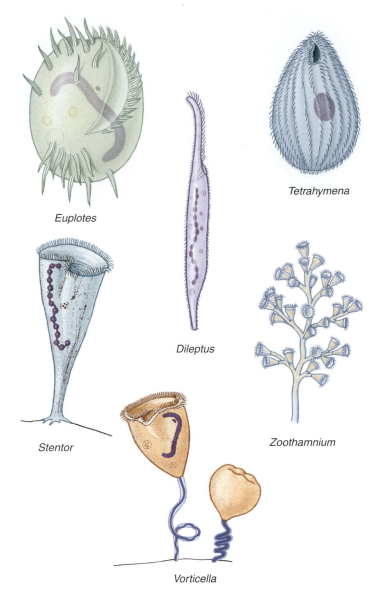

Euplotes

Tetrahymena

Dileptus

Stentor

Zoothamnium

Vorticella

Figure 11-18

Some representative ciliates. *Euplotes* have stiff cirri used for crawling about. Contractile fibrils in ectoplasm of *Stentor* and in stalks of *Vorticella* allow great expansion and contraction. Note the macronuclei, long and curved in *Euplotes* and *Vorticella,* shaped like a string of beads in *Stentor.*

9). Suctorians paralyze their prey and then ingest the contents through tubelike tentacles by a complex feeding mechanism that apparently combines phagocytosis with a sliding filament action of microtubules in the tentacles (Figure 11-9).

Some ciliates have curious small bodies in their ectoplasm between the bases of the cilia. Examples are **trichocysts** (Figure 11-19 and 11-20) and **toxicysts.** Upon mechanical or chemical stimulation, these bodies explosively expel a long, threadlike structure. The mechanism of expulsion is unknown. The function of trichocysts is thought to be defensive. When attacked by a *Didinium,* a paramecium expels its trichocysts but to no avail. Toxicysts, however, release a poison that paralyzes the prey of carnivorous ciliates. Toxicysts are structurally quite distinct from trichocysts. Many dinoflagellates have trichocysts.

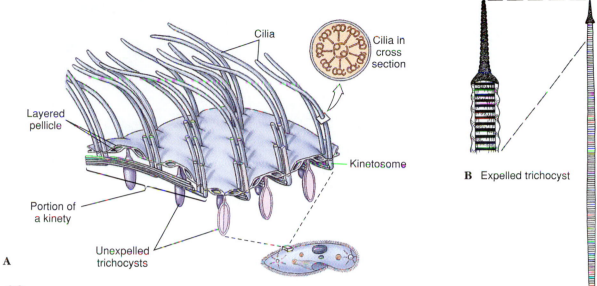

Figure 11-19

Infraciliature and associated structures in ciliates. **A,** Structure of the pellicle and its relation to the infraciliature system. **B,** Expelled trichocyst.

Among the more striking and familiar of ciliates are *Stentor* (Gr. herald with a loud voice), trumpet shaped and solitary, with a beadshaped macronucleus (Figure 11-18); *Vorticella* (L. dim. of *vortex,* a whirlpool), bell shaped and attached by a contractile stalk (Figure 11-18); and *Euplotes* (Gr. *eu,* true, good, + *ploter,* swimmer) with a flattened body and groups of fused cilia (cirri) that function as legs.

Paramecium: A Representative Ciliate

Paramecia are usually abundant in ponds or sluggish streams containing aquatic plants and decaying organic matter.

Form and Function

Paramecia are often described as slipper shaped. *Paramecium caudatum* is 150 to 300 µm in length and is blunt anteriorly and somewhat pointed posteriorly (Figure 11-20). The organism has an asymmetrical appearance because of the **oral groove,** a depression that runs obliquely backward on the ventral side.

The **pellicle** is a clear, elastic membrane that may be ornamented by ridges or papillalike projections (Figure 11-19), and its entire surface is covered with cilia arranged in lengthwise rows. Just below the pellicle is the thin clear **ectoplasm** that surrounds the larger mass of granular **endoplasm** (Figure 11-20). Embedded in ectoplasm just below the surface are spindle-shaped **trichocysts,** which alternate with the bases of cilia. The infraciliature can be seen only with special fixing and staining methods.

A **cytostome** at the end of the oral groove leads into a tubular **cytopharynx,** or **gullet.** Along the gullet an undulating membrane of modified cilia keeps food moving. Fecal material is discharged through a **cytoproct** posterior to the oral groove (Figure 11-20). Within the endoplasm are food vacuoles contain-ing food in various stages of digestion. There are two **contractile vacuoles,** each consisting of a central space surrounded by several **radiating canals** (Figure 11-20) that collect fluid and empty it into the central vacuole. We described excretion and osmoregulation on p. 213.

Paramecium caudatum has two nuclei: a large kidney-shaped **macronucleus** and a smaller **micronucleus** fitted into the depression of the former. These can usually be seen only in stained specimens. The number of micronuclei varies in different species; for example, *P. multimicronucleatum* may have as many as seven.

Paramecia are holozoic, living on bacteria, algae, and other small organisms. Cilia in the oral groove sweep food particles in the water into the cytostome, from which point they are carried into the cytopharynx by the undulating membrane. From the cytopharynx food is collected into a food vacuole that is constricted into the endoplasm. Food vacuoles circulate in a definite course through the cytoplasm while the food is being digested by enzymes from the endoplasm. Indigestible parts of the food are ejected through the cytoproct.

The body is elastic, allowing it to bend and squeeze its way through narrow places. Its cilia can beat either forward or backward, so that the organism can swim in either direction. The cilia beat obliquely, causing the organism to rotate on its long axis. In the oral groove the cilia are longer and beat more vigorously than the others so that the anterior end swerves aborally. As a

Figure 11-20

Left, enlarged section of a contractile vacuole (water expulsion vesicle) of *Paramecium.* Water is apparently collected by endoplasmic reticulum, emptied into feeder canals and then into the vesicle. The vesicle contracts to empty its contents to the outside, thus serving as an osmoregulatory organelle. *Right, Paramecium,* showing cytopharynx, food vacuoles, and nuclei.

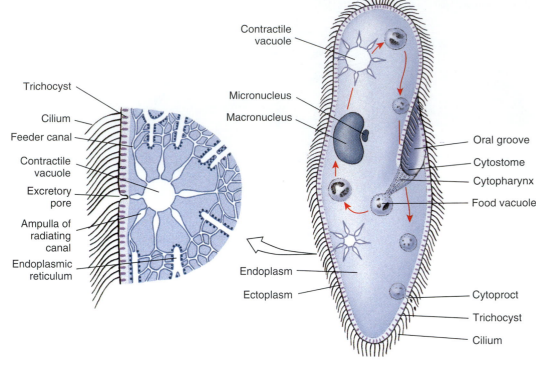

Figure 11-21

A, Spiral path of swimming *Paramecium.*
B, Avoidance reaction of *Paramecium.*

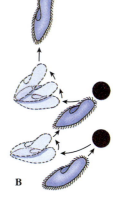

result of these factors, the organism moves forward in a spiral path (Figure 11-21A).

When a ciliate, such as a paramecium, comes in contact with a barrier or a disturbing chemical stimulus, it reverses its cilia, backs up a short distance, and swerves the anterior end as it pivots on its posterior end. This behavior is called an **avoiding reaction** (Figure 11-21B). A paramecium may continue to change its direction to keep itself away from a noxious stimulus, and it may react in a similar fashion to keep itself within the zone of an attractant. A paramecium may also change its swimming speed. How does a paramecium "know" when to change directions or swimming speed? Interestingly, reactions of the organism depend on effects of the stimulus on the electrical

potential difference across its cell membrane. Paramecia slightly hyperpolarize in attractants and depolarize in repellents that produce the avoiding reaction. Hyperpolarization increases the rate of the forward ciliary beat, and depolarization results in ciliary reversal and backward swimming.

Locomotor responses, by which an organism more or less continuously orients itself with respect to a stimulus, are called *taxes* (sing., *taxis*). Movement toward the stimulus is a positive taxis; movement away is a negative taxis. Some examples are thermotaxis, response to heat; phototaxis, response to light; thigmotaxis, response to contact; chemotaxis, response to chemical substances; rheotaxis, response to currents of air or water; galvanotaxis, response to constant electric current; and geotaxis, response to gravity. Some stimuli do not cause an orienting response but simply a change in movement: more rapid movement, more frequent random turning, or slowing or cessation of movement. Such responses are known as kineses. Is the avoiding reaction of a paramecium a taxis or a kinesis?

Reproduction

Paramecia reproduce only by binary fission across kineties (ciliary rows) but have certain forms of sexual phenomena called conjugation and autogamy.

In **binary fission** the micronucleus divides mitotically into two daughter micronuclei, which move to opposite ends of the cell (Figure 11-22). The macronucleus elongates and divides amitotically.

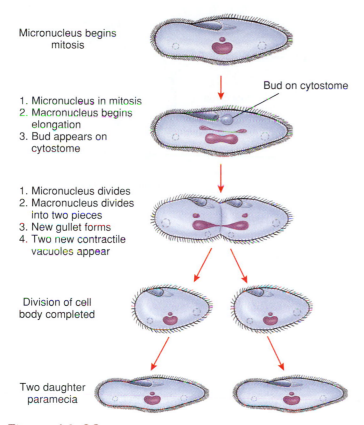

Micronucleus begins
mitosis

1. Micronucleus in mitosis
2. Macronucleus begins
   elongation
3. Bud appears on
   cytostome

Bud on cytostome

1. Micronucleus divides
2. Macronucleus divides
   into two pieces
3. New gullet forms
4. Two new contractile
   vacuoles appear

Division of cell
body completed

Two daughter
paramecia

Figure 11-22

Binary fission in a ciliophoran *(Paramecium)*. Division is across rows
of cilia.

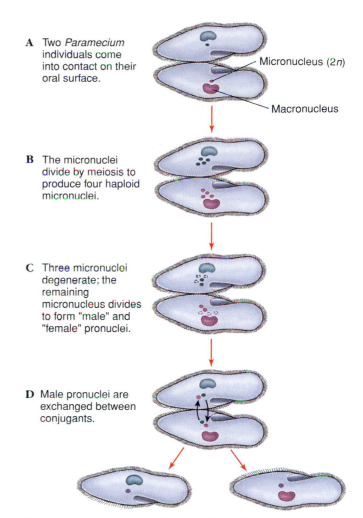

A Two *Paramecium*
individuals come
into contact on their
oral surface.

Micronucleus (2*n*)

Macronucleus

B The micronuclei
divide by meiosis to
produce four haploid
micronuclei.

C Three micronuclei
degenerate; the
remaining
micronucleus divides
to form "male" and
"female" pronuclei.

D Male pronuclei are
exchanged between
conjugants.

E Male and female pronuclei fuse, and individuals separate.
Subsequently old macronuclei are absorbed and replaced
by new macronuclei.

Figure 11-23

Scheme of conjugation in *Paramecium*.

Conjugation occurs at intervals in ciliates. Conjugation is
the temporary union of two individuals to exchange chromosomal material (Figure 11-23). During the union the macronucleus
disintegrates and the micronucleus of each individual undergoes
meiosis, giving rise to four haploid micronuclei, three of which
degenerate (Figure 11-23A to C). The remaining micronucleus
then divides into two haploid pronuclei, one of which is
exchanged with the other conjugant. The pronuclei fuse to
restore the diploid number of chromosomes, followed by several more nuclear events detailed in Figure 11-23. Following this
complicated process, the organisms may continue to reproduce
by binary fission without the necessity of conjugation.

The result of conjugation is similar to that of zygote formation, for each exconjugant contains hereditary material from two
individuals. The advantage of sexual reproduction is that it permits gene recombinations, thus increasing genetic variation in
the population. Although ciliates in clone cultures can apparently reproduce repeatedly and indefinitely without conjugation,
the stock seems eventually to lose vigor. Conjugation restores
vitality to a stock. Seasonal changes or a deteriorating environment will usually stimulate sexual reproduction.

Autogamy is a process of self-fertilization that is similar to
conjugation except that there is no exchange of nuclei. After the
disintegration of the macronucleus and the meiotic divisions of
the micronucleus, two haploid pronuclei fuse to form a
synkaryon that is completely homozygous (Chapter 5, p. 79).

Symbiotic Ciliates

Many symbiotic ciliates live as commensals, but some can be
harmful to their hosts. *Balantidium coli* lives in the large intestine of humans, pigs, rats, and many other mammals (Figure 11-
24). There seem to be host-specific strains, and the organism is
not easily transmitted from one species to another. Transmission
is by fecal contamination of food or water. Usually the organisms are not pathogenic, but in humans they sometimes invade
the intestinal lining and cause a dysentery similar to that caused
by *Entamoeba histolytica* (p. 227). The disease can be serious
and even fatal. Infections are common in parts of Europe, Asia,
and Africa but are rare in the United States.

Other species of ciliates live in other hosts. *Entodinium* (Figure 11-24) belongs to a group that has very complex structure
and lives in the digestive tract of ruminants, where they may be
very abundant. *Nyctotherus* live in the colon of frogs and toads.
In aquarium and wild freshwater fishes, *Ichthyophthirius* causes
a disease known to many fish culturists as "ick." Untreated, it can
cause much loss of exotic fishes.

Figure 11-24

Some symbiotic ciliates. *Balantidium coli* is a parasite of humans and other mammals. *Ichthyophthirius* causes a common disease in aquarium and wild freshwater fishes. *Entodinium* is found in the rumen of cows and sheep.

Ichthyophthirius

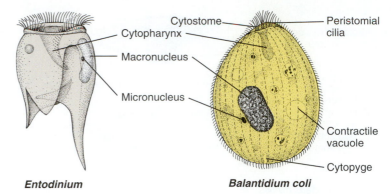

Entodinium   **Balantidium coli**

Suctorians

Suctorians are ciliates in which the young possess cilia and are free swimming, and the adults grow a stalk for attachment, become sessile, and lose their cilia. They have no cytostome but feed by long, slender, tubelike tentacles. The suctorian captures living prey, usually a ciliate, by the tip of one or more tentacles and paralyzes it. The cytoplasm of the prey then flows through the attached tentacles, forming food vacuoles in the feeding suctorian (Figure 11-9).

One of the best places to find freshwater suctorians is in algae that grow on the carapace of turtles. Common genera of suctorians found there are *Anarma* (without stalk or test) and *Squalorophrya* (with stalk and test). Other freshwater representatives are *Podophrya* (Figure 11-9) and *Dendrosoma*. *Acinetopsis* and *Ephelota* are saltwater forms.

Suctorian parasites include *Trichophrya*, the species of which are found on a variety of invertebrates and freshwater fish; *Allantosoma*, which live in the intestine of certain mammals; and *Sphaerophrya*, which are found in *Stentor*.

PHYLUM DINOFLAGELLATA

Dinoflagellates are another group that was formerly included by zoologists among Phytomastigophorea, and about half are photoautotrophic with chromatophores bearing chlorophyll. The rest are colorless and heterotrophic. It is believed that ancestral dinoflagellates were heterotrophic, and some acquired chloroplasts by endosymbiosis from a variety of algal sources. Ecologically, some species are among the most important primary producers in marine environments. They commonly have two flagella, one equatorial and one longitudinal, each borne at least partially in grooves on the body (Figure 11-25). The body may be naked or covered by cellulose plates or valves. Many species can ingest prey through a mouth region between the plates near the posterior area of the body. *Ceratium* (Figure 11-25), for example, has a thick covering with long spines, into which the body extends, but it can catch food with posterior pseudopodia and ingest it between the flexible plates in the posterior groove. *Noctiluca* (Figure 11-25), a colorless dinoflagellate, is a voracious predator and has a long, motile tentacle, near the base of which its single, short flagellum emerges. *Noctiluca* is one of many marine organisms that can produce light (bioluminescence).

Several groups of autotrophic flagellates are planktonic primary producers (p. 800) in freshwater and marine environments;

however, dinoflagellates are the most important, particularly in the sea. Zooxanthellae are dinoflagellates that live in mutualistic association in tissues of certain invertebrates, including other protozoa, sea anemones, horny and stony corals, and clams. The association with stony corals is of ecological and economic importance because only corals with symbiotic zooxanthellae can form coral reefs (Chapter 13).

Dinoflagellates can damage other organisms, such as when they produce a "red tide." Although this name originally was applied to situations in which the organisms reproduced in such profusion (producing a "bloom") that the water turned red from their color, any instance of a bloom producing detectable levels of toxic substances is now called a red tide. The water may be red, brown, yellow, or not remarkably colored at all. The toxic substances are apparently not harmful to the organisms that produce them, but they may be highly poisonous to fish and other marine life. Several different types of dinoflagellates and one species of cyanobacterium have been responsible for red tides. Red tides have resulted in considerable economic losses to the shellfish industry. Another flagellate produces a toxin that is concentrated in the food chain, especially in large, coral reef fishes. The illness produced in humans after eating such fish is known as ciguatera.

AMEBAS

Members of the former Sarcodina do not form a monophyletic group, but for sake of simplicity, we will discuss them together under this informal heading. Ameboid organisms characteristically move and feed by means of pseudopodia (Figures 11-4 and 11-26). Classification of amebas was based in part on the characteristics of their pseudopodia and on characteristics of their protective **tests** (skeletons), if any. Amebas are found in both fresh and salt water and in moist soils. Some are planktonic; some prefer a substratum. A few are parasitic.

Nutrition in amebas is holozoic; that is, they ingest and digest liquid or solid foods. Most amebas are omnivorous, living on algae, bacteria, protozoa, rotifers, and other microscopic organisms. An ameba may take in food at any part of its body surface merely by producing a pseudopodium to enclose the food (**phagocytosis**). The enclosed food particle, along with some environmental water, becomes a food vacuole, which is

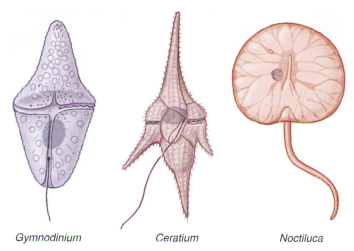

Gymnodinium   *Ceratium*   *Noctiluca*

Figure 11-25

Examples of phylum Dinoflagellata. *Gymnodinium* bears no cellulose plates. Some members of its family are autotrophic and some phagotrophic. *Ceratium* bears plates and is both autotrophic and phagotrophic. *Noctiluca* is entirely phagotrophic, can be very large (more than 1 mm wide), and has a large tentacle involved in feeding.

carried about by the streaming movements of endoplasm. As digestion occurs within the vacuole by enzymatic action, water and digested materials pass into the cytoplasm. Undigested particles are eliminated through the cell membrane.

Most amebas reproduce by binary fission. Sporulation and budding occur in some.

Amoeba proteus: A Rhizopodan

The most commonly studied species of ameba is *Amoeba proteus*. These amebas live in slow streams and ponds of clear water, often in shallow water on aquatic vegetation or on sides of ledges. They are rarely found free in water, for they require a substratum on which to crawl. They have an irregular shape because lobopodia may be formed at any point on their bodies. They are colorless and about 250 to 600 µm in greatest diameter. Unlike *Euglena,* their **pellicle** consists only of a cell membrane. **Ectoplasm** and **endoplasm** are prominent. Organelles such as **nucleus, contractile vacuole, food vacuoles,** and small **vesicles** can be observed easily with a light microscope. Amebas live on algae, protozoa, rotifers, and even other amebas, upon which they feed by phagocytosis. An ameba can live for many days without food but decreases in volume during starvation. The time necessary for digestion by a food vacuole varies with the kind of food but is usually around 15 to 30 hours. When an ameba reaches full size, it divides by binary fission with typical mitosis.

Other Rhizopodans

There are many species of amebas; for example, *A. verrucosa* has short pseudopodia; *Chaos carolinense (Pelomyxa carolinensis)* is several times as large as *A. proteus;* and *A. radiosa* has many slender pseudopodia.

Amoeba   *Arcella*

Difflugia

Globigerina   *Chlamydophrys*

Figure 11-26

Examples of amebas. *Amoeba, Difflugia,* and *Arcella* have lobopodia; *Chlamydophrys* has filopodia; and the foraminiferan *Globigerina* bears reticulopodia.

There are many entozoic amebas, most of which live in intestines of humans or other animals. Two common genera are *Endamoeba* and *Entamoeba. Endamoeba blattae* is an endocommensal in the intestine of cockroaches, and related species are found in termites. *Entamoeba histolytica* is the most important rhizopodan parasite of humans. It lives in the large intestine and on occasion can invade the intestinal wall by secreting enzymes that attack the intestinal lining. If this occurs, a serious and sometimes fatal amebic dysentery may result. The organisms may be carried by the blood to the liver and other organs and cause abscesses there. Many infected persons show few or no symptoms but are carriers, passing cysts in their feces. Diagnosis is complicated by the existence of a nonpathogenic species, *E. dispar,* which is morphologically identical to *E. histolytica.* Infection is spread by contaminated water or food containing cysts. *Entamoeba histolytica* is found throughout the world, but clinical amebiasis is most prevalent in tropical and subtropical areas.

Other species of *Entamoeba* found in humans are *E. coli* in the intestine and *E. gingivalis* in the mouth. Neither of these species is known to cause disease.

Not all rhizopodans are "naked" as are amebas. Some, such as *Arcella* and *Difflugia,* have their delicate plasma membrane covered with a protective **test** or shell (Figure 11-26). They have a test of secreted siliceous or chitinoid material that may be reinforced with grains of sand. They move by means of pseudopodia that project from openings in the shell.

Foraminiferans

Foraminiferans (granuloreticulosans) are an ancient group of shelled amebas found in all oceans, with a few in fresh and brackish water. Most foraminiferans live on the ocean floor in incredible numbers, having perhaps the largest biomass of any animal group on earth. Their tests are of numerous types (Figure 11-26 and 11-27). Most tests are many chambered and are made of calcium carbonate, although they sometimes use silica, silt, and other foreign materials. Slender pseudopodia extend through openings in the test, then branch and run together to form a protoplasmic net **(reticulopodia)** in which they ensnare their prey. Here captured prey is digested, and digested products are carried into the interior by flowing protoplasm. Life cycles of foraminiferans are complex, for they have multiple fission and alternation of haploid and diploid generations (intermediary meiosis).

Actinopodans

Here we consider organisms that move by axopodia, including the freshwater heliozoans and several marine groups known as radiolarians. Except for some heliozoans, they have tests (Figure 11-28). These protozoa are beautiful little organisms.

Biological characteristics of freshwater heliozoans are somewhat better known than those of other actinopods. Examples are *Actinosphaerium,* which is about 1 mm in diameter and can be seen with the unassisted eye, and *Actinophrys* (Figure 11-29), only 50 μm in diameter; neither has a test. *Clathrulina* (Figure 11-29) secretes a latticed test.

The oldest known protozoa are found among marine actinopodans known as radiolarians. Radiolarians are nearly all pelagic (live in open water). Most of them are planktonic in shallow water, although some live in deep water. Their highly specialized skeletons are intricate in form and of great beauty (Figure 11-28). The body is divided by a central capsule that separates inner and outer zones of cytoplasm. The central capsule, which may be spherical, ovoid, or branched, is perforated to allow cytoplasmic continuity. The skeleton is made of silica, strontium sulfate, or a combination of silica and organic matter and usually has a radial arrangement of spines that extend through the capsule from the center of the body. At the surface a shell may be fused with the spines. Around the capsule is a frothy mass of cytoplasm from which axopodia arise (p. 212). These are sticky to catch prey, which are carried by the streaming protoplasm to the central capsule to be digested. The ectoplasm on one side of the axial rod moves outward, or toward the tip, while on the other side it moves inward, or toward the test.

Radiolarians may have one or many nuclei. Their life history is not completely known, but binary fission, budding, and sporulation have been observed in them.

Role of Amebas in Building Earth Deposits

Foraminiferans and radiolarians have existed since Precambrian times and have left excellent fossil records. In many instances their hard shells have been preserved unaltered. Many extinct

A   B

Figure 11-27

A, Living foraminiferan, showing thin pseudopodia extending from test. **B,** Test of foraminiferan, *Vertebralima striata.* Foraminiferans are ameboid marine protozoa that secrete a calcareous, many-chambered test in which to live and then extrude protoplasm through pores to form a layer over the outside. The animal begins with one chamber, and as it grows, it secretes a succession of new and larger chambers, continuing this process throughout life. Many foraminiferans are planktonic, and when they die, their shells are added to the ooze on the ocean's bottom.

Figure 11-28

Types of radiolarian tests. In his study of these beautiful forms collected on the famous *Challenger* expedition of 1872 to 1876, Haeckel proposed our present concepts of symmetry.

species closely resemble those of the present day. They were especially abundant during the Cretaceous and Tertiary periods. Some were among the largest protozoa that have ever existed, measuring up to 100 mm (about 4 in) or more in diameter.

For untold millions of years tests of dead foraminiferans have been sinking to the bottom of the ocean, building up a characteristic ooze rich in lime and silica. About one-third of the sea bottom is covered with ooze that is made up of shells of the genus *Globigerina*. This ooze is especially abundant in the Atlantic Ocean.

Radiolarians (Figure 11-28), with their less soluble siliceous shells, are usually found at greater depths (4600 to 6100 meters), mainly in the Pacific and Indian oceans. Radiolarian ooze probably covers about 5 to 8 million square kilometers to a thickness of 700 to 4000 m. Under certain conditions, radiolarian ooze forms rocks (chert). Many fossil radiolarians are found in Tertiary rocks of California.

Of equal interest and of greater practical importance are the limestone and chalk deposits that were laid down by the accumulation of foraminiferans when sea covered what is now land. Later, through a rise in the ocean floor and other geological changes, this sedimentary rock emerged as dry land. The chalk deposits of many areas of England, including the White Cliffs of Dover, were formed in this way. The great pyramids of Egypt were made from stone quarried from limestone beds that were formed by a very large foraminiferan population that flourished during the early Tertiary period.

Since fossil foraminiferans and radiolarians can be found in well drillings, their identification is often important to oil geologists for correlation of rock strata.

PHYLOGENY AND ADAPTIVE RADIATION

Phylogeny

Traditionally, amebas (former Sarcodina) and flagellates (former Mastigophora) were considered separate classes in phylum Protozoa. Observations that some flagellates could form pseudopodia, that some species of amebas had flagellated stages, and that a supposed ameba was really a flagellate without a flagellum, all seemed to support the concept of a phylum Sarcomastigophora. However, analyses of sequences of bases in genes, particularly the gene encoding the small subunit of ribosomal RNA (p. 93), have provided strong evidence that neither Sarcodina nor Mastigophora are monophyletic groups. For example, unicellular green algae (members of class Phytomastigophorea, subphylum Mastigophora in the traditional classification) and higher plants are more closely related to animals than they are to other protozoan groups. Ameboid forms arose independently many times, and there is no justification for placing them all in a subphylum Sarcodina.

In fact, molecular evidence has almost completely revised our concepts of protozoan phylogeny. Such evidence has strengthened the hypothesis of a symbiotic origin of eukaryotes and that origins of nuclei, mitochondria, and plastids represent

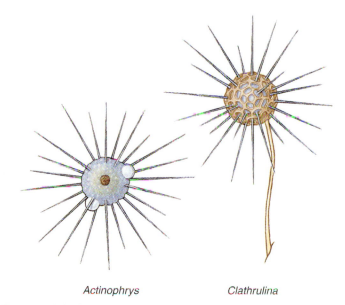

Actinophrys   Clathrulina

Figure 11-29
Actinophrys and *Clathrulina* are amebas with axopodia.

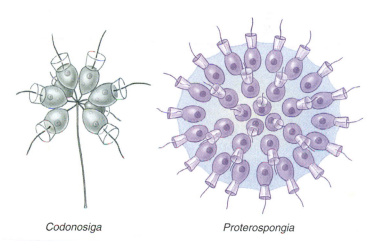

Codonosiga   Proterospongia

Figure 11-30
Codonosiga is a choanoflagellate, whose cells so resemble choanocytes of sponges (phylum Porifera, Chapter 12) that a phylogentic relationship has long been suspected. Present evidence suggests that choanoflagellates and sponges are sister groups.

separate endosymbiogenetic events. Lack of mitochondria in Diplomonadida and Parabasalea suggests that these groups diverged before the common ancestor of other eukaryotes merged with its mitochondrial partner. This point is controversial: discovery of mitochondrial enzymes in some trichomonads and diplomonads suggests that mitochondria were lost during evolution of these groups, and some scientists contend that there are no primitively amitochondriate protozoa.

Soon after mitochondria were added to eukaryote cells, lines leading to Fungi and Metazoa split from those leading to the remaining phyla (or kingdoms). Some protozoa, such as choanoflagellates (Figure 11-30), are phylogenetically related to fungi and animals and may share a near common ancestor with

Classification of Protozoan Phyla

This classification primarily follows Hausmann and Hülsmann (1996) and is abridged from Roberts and Janovy (2000). It represents a major departure from that in the last edition of *Integrated Principles of Zoology*. With few exceptions, we are including only taxa of examples discussed in this chapter.

Much strong evidence indicates that phylum Sarcomastigophora and its subphyla are no longer tenable. Newer monographs consider amebas as belonging to several taxa with various affinities, not all yet determined. Organisms previously assigned to phylum Sarcomastigophora, subphylum Sarcodina should be placed in at least two phyla, if not more. Nevertheless, amebas fall into a number of fairly recognizable morphological groups, which we will use for the convenience of readers and not assign such groups to specific taxonomic levels.

Phylum Chlorophyta (klor-of'i-ta) (Gr. *chlōros,* green, + *phyton,* plant). Unicellular and multicellular algae; photosynthetic pigments of chlorophyll *a* and *c,* reserve food is starch (characters in common with "higher" plants: bryophytes and vascular plants); all with biflagellated stages; flagella of equal length and smooth; mostly free-living photoautotrophs. Examples: *Chlamydomonas, Volvox.*

Phylum Retortamonada (re-tor'ta-mo'nad-a) (L. *retorqueo,* to twist back, + *monas,* single, unit). Mitochondria and Golgi bodies lacking; three anterior and one recurrent (running toward posterior) flagellum lying in a groove; intestinal parasites or free living in anoxic environments.

    Class Diplomonadea (di'plo-mon-a'de-a) (Gr. *diploos,* double + L. *monas,* unit). One or two karyomastigonts (group of kinetosomes with a nucleus); individual mastigonts with one to four flagella; mitotic spindle within nucleus; cysts present; free living or parasitic.

        Order Diplomonadida (di'plo-mon-a'di-da). Two karyomastigonts, each with four flagella, one recurrent; with variety of microtubular bands. Example: *Giardia.*

Phylum Axostylata (ak-so-sty-la'ta) (Gr. *axōn,* axle + *stylos,* style, stake). With an axostyle made of microtubules.

    Class Parabasalea (par'a-bas-al'e-a) (Gr. *para,* beside + *basis,* base). With very large Golgi bodies associated with karymastigont; up to thousands of flagella.

        Order Trichomonadida (tri'ko-mon-a'di-da) (Gr. *trichos,* hair + *monas,* unit). Typically at least some kinetosomes associated with rootlet filaments characteristic of trichomonads; parabasal body present; division spindle extranuclear; hydrogenosomes present; no sexual reproduction; true cysts rare; all parasitic. Examples: *Dientamoeba, Trichomonas.*

Phylum Euglenozoa (yu-glen-a-zo'a) (Gr. *eu-,* good, true, + *glēnē,* cavity, socket, + *zöon,* animal). With cortical microtubules; flagella often with paraxial rod (rodlike structure accompanying axoneme in flagellum); mitochondria with discoid cristae; nucleoli persist during mitosis.

    Subphylum Euglenida (yu-glen'i-da) With pellicular microtubules that stiffen pellicle.

        Class Euglenoidea (yu-glen-oyd'e-a) (Gr. *eu-,* good, true, + *glēnē,* cavity, socket, + *-ōideos,* form of, type of). Two heterokont flagella (flagella with different structures) arising from apical reservoir; some species with light-sensitive stigma and chloroplasts. Example: *Euglena.*

    Subphylum Kinetoplasta (ky-neet'o-plas'ta) (Gr. *kinētos,* to move + *plastos,* molded, formed). With a unique mitochondrion containing a large disc of DNA; paraxial rod.

        Class Trypanosomatidea (try-pan'o-som-a-tid'e-a) (Gr. *trypanon,* a borer, + *sōma,* the body). One or two flagella arising from pocket; flagella typically with paraxial rod that parallels axoneme; single mitochondrion (nonfunctional in some forms) extending length of body as tube, hoop, or network of branching tubes, usually with single conspicuous DNA-containing kinetoplast located near flagellar kinetosomes;

---

sponges (Chapter 12). Another important branch led to a major group called **straminopiles,** which contains brown algae, yellow algae, diatoms, and many others.

The common ancestor of green plants and green algae acquired its chlorophyll-bearing plastids by symbiogenesis with a cyanobacterium. These plastids have two membranes, one apparently derived from the ancestral cyanobacterium, and the other from the host vacuole containing it. Some investigators believe that the last common ancestor of the Alveolata and the Euglenoidea (among the Euglenozoa) both added plastids by secondary symbiogenesis from the same group of unicellular, eukaryotic algae. However, the presence of different chlorophylls in dinoflagellates and euglenoids would argue against a common origin of their plastids. In Euglenozoa acquisition of their plastids apparently occurred after divergence from ancestors of Kinetoplasta. Among alveolates, many dinoflagellates are photoautotrophs, maintaining their chloroplasts, but some either lost them or never acquired them. Apicomplexans have a small circular, plastidlike DNA, localized to a discrete organelle surrounded by four membranes, and evidently a relict plastid, possibly inherited from their common ancestor with dinoflagellates. Ancestors of ciliophorans either lost their plastid symbionts or diverged from their common ancestor with dinoflagellates before the secondary symbiogenesis occurred.

Adaptive Radiation

We have described some of the wide range of adaptations of protozoan groups in this chapter. Amebas range from bottom-dwelling, naked species to planktonic forms such as foraminiferans and radiolarians with beautiful, intricate tests. There are many symbiotic species of amebas. Flagellated forms likewise show adaptations for a similarly wide range of habitats, with the added variation of photosynthetic ability in many groups.

Within a single-cell body plan, the division of labor and specialization of organelles are carried furthest by ciliates. These have become the most complex of all protozoa. Specializations for intracellular parasitism have been adopted by Apicomplexa.

Golgi body typically in region of flagellar pocket, not connected to kinetosomes and flagella; all parasitic. Examples: *Leishmania, Trypanosoma.*

Phylum Apicomplexa (ap'i-compleks'a) (L. *apex,* tip or summit, + *complex,* twisted around). Characteristic set of organelles (apical complex) associated with anterior end present in some developmental stages; cilia and flagella absent except for flagellated microgametes in some groups; cysts often present; all parasitic.

Class Gregarinea (gre-ga-ryn'e-a) (L. *gregarius,* belonging to a herd or flock). Mature gamonts (individuals that produce gametes) large, extracellular; gametes usually alike in shape and size; zygotes forming oocysts within gametocysts; parasites of digestive tract or body cavity of invertebrates, life cycle usually with one host. Examples: *Monocystis, Gregarina.*

Class Coccidea (kok-sid'e-a) (Gr. *kokkos,* kernel, grain). Mature gamonts small, typically intracellular; life cycle typically with merogony, gametogony, and sporogony; most species live inside vertebrates. Examples: *Cryptosporidium, Cyclospora, Eimeria, Toxoplasma, Plasmodium, Babesia.*

Phylum Ciliophora (sil-i-of'-or-a) (L. *cilium,* eyelash, + Gr. *phora,* bearing). Cilia or ciliary organelles in at least one stage of life cycle; two types of nuclei, with rare exception; binary fission across rows of cilia, budding and multiple fission also occur; sexuality involving conjugation, autogamy, and cytogamy; nutrition heterotrophic; contractile vacuole typically present; most species free living, but many commensal, some parasitic. (This is a very large group, now divided by the Society of Protozoologists classification into three classes and numerous orders and suborders. The classes are separated on the basis of technical characteristics of the ciliary patterns, especially around the cytostome, the development of the cytostome, and other characteristics.) Examples: *Paramecium, Colpoda, Tetrahymena, Balantidium, Stentor, Blepharisma, Epidinium, Euplotes, Vorticella, Carchesium, Trichodina, Podophrya, Ephelota.*

Phylum Dinoflagellata (dy'no-fla-jel-at'a) (Gr. *dinos,* whirling + *flagellum,* little whip). Typically with two flagella, one transverse and one trailing; body usually grooved transversely and longitudinally, each groove containing a flagellum; chromoplasts usually yellow or dark brown, occasionally green or blue-green, bearing chlorophylls *a* and *c;* nucleus unique among eukaryotes in having chromosomes that lack or have low levels of histones; mitosis intranuclear; body form sometimes of spherical unicells, colonies, or simple filaments; sexual reproduction present; members free living, planktonic, parasitic, or mutualistic. Examples: *Zooxanthella, Ceratium, Noctiluca, Ptychodiscus.*

Amebas Although members of the former Sarcodina do not form a monophyletic group, we are considering them under this informal heading for the sake of simplicity. Amebas move by pseudopodia or locomotive protoplasmic flow without discrete pseudopodia; flagella, when present, usually restricted to developmental or other temporary stages; body naked or with external or internal test or skeleton; asexual reproduction by fission; sexuality, if present, associated with flagellated or, more rarely, ameboid gametes; most free living.

Rhizopodans Locomotion by lobopodia, filopodia (thin pseudopodia that often branch but do not rejoin), or by protoplasmic flow without production of discrete pseudopodia. Examples: *Amoeba, Entamoeba, Difflugia, Arcella, Chlamydophrys.*

Granuloreticulosans Locomotion by reticulopodia (thin pseudopodia that branch and often rejoin [anastomose]); includes foraminiferans. Examples: *Globigerina, Vertebralima.*

Actinopodans Locomotion by axopodia; includes radiolarians. Examples: *Actinophrys, Clathrulina.*

SUMMARY

"Animal-like," single-celled organisms were formerly assigned to the phylum Protozoa. It is now recognized that the "phylum" was composed of numerous phyla of varying phylogenetic relationships. We use the terms *protozoa* and *protozoan* informally to refer to all these highly diverse organisms. They demonstrate the great adaptive potential of the basic body plan: a single eukaryotic cell. They occupy a vast array of niches and habitats. Many species have complex and specialized organelles.

All protozoa have one or more nuclei, and these often appear vesicular with light microscopy. Macronuclei of ciliates are compact. Nucleoli are often evident in the nuclei. Many protozoa have organelles similar to those found in metazoan cells.

Pseudopodial or ameboid movement is a locomotory and food-gathering mechanism in protozoa and plays a vital role as a defense mechanism in metazoa. It is accomplished by assembly of actin subunits into microfilaments and interaction of microfilaments with actin binding protein and myosin, and it requires expenditure of energy from ATP. Ciliary movement is likewise important in both protozoa and metazoa. Currently, the most widely accepted mechanism to account for ciliary movement is the sliding-microtubule hypothesis.

Various protozoa feed by holophytic, holozoic, or saprozoic means. The excess water that enters their bodies is expelled by contractile vacuoles (water-expulsion vesicles). Respiration and waste elimination are through the body surface. Protozoa can reproduce asexually by binary fission, multiple fission, and budding; sexual processes are common. Cyst formation to withstand adverse environmental conditions is an important adaptation in many protozoa.

Members of several phyla have photoautotrophic species, including Chlorophyta, Euglenozoa, and Dinoflagellata. Some of these are very important planktonic organisms. Euglenozoa includes many nonphotosynthetic species, and some of these cause serious diseases of humans, such as African sleeping sickness and Chagas' disease. Apicomplexa are all parasitic, including *Plasmodium,* which causes malaria. Ciliophora move by means of cilia or ciliary organelles. They are a large and diverse group, and many are complex in structure. Amebas move by pseudopodia and are now assigned to a number of phyla.

REVIEW QUESTIONS

1. Explain why a protozoan may be very complex, even though it is composed of only one cell.
2. Distinguish among the following protozoan groups: Euglenozoa, Apicomplexa, Ciliophora, Dinoflagellata.
3. Distinguish vesicular and compact nuclei.
4. Explain the transitions of endoplasm and ectoplasm in ameboid movement. What is a current hypothesis regarding the role of actin in ameboid movement?
5. Distinguish lobopodia, filipodia, reticulopodia, and axopodia.
6. Contrast the structure of an axoneme of a cilium with that of a kinetosome.
7. What is the sliding-microtubule hypothesis?
8. Explain how protozoa eat, digest their food, osmoregulate, and respire.
9. Distinguish the following: binary fission, budding, multiple fission, and sexual and asexual reproduction.
10. What is the survival value of encystment?
11. Contrast and give an example of autotrophic and heterotrophic protozoa.
12. Name three kinds of amebas, and tell where they are found (their habitats).
13. Outline the general life cycle of malaria organisms. How do you account for the resurgence of malaria in recent years?
14. What is the public-health importance of *Toxoplasma,* and how do humans become infected with it? What is the public health importance of *Cryptosporidium* and *Cyclospora?*
15. Define the following with reference to ciliates: macronucleus, micronucleus, pellicle, undulating membrane, cirri, infraciliature, trichocysts, conjugation.
16. Outline the steps in conjugation of ciliates.
17. What are indications that apicomplexons descended from a photoautotrophic ancestor?
18. Distinguish primary endosymbiogenesis from secondary endosymbiogenesis.

SELECTED REFERENCES

Allen, R. D. 1987. The microtubule as an intracellular engine. Sci. Am. **256:**42–49 (Feb.). *The action of microtubules accounts for the movement of chromosomes in mitosis and pseudopodial movement of filopodia and reticulopodia.*

Baldauf, S. L., A. J. Roger, I. Senk-Siefert, and W. F. Doolittle. 2000. A kingdom-level phylogeny of eukaryotes based on combined protein data. Science **290:**972–977. *They contend that combining sequence data for genes encoding several proteins indicate there are 15 kingdoms of organisms.*

Cavalier-Smith, T. 1999. Principles of protein and lipid targeting in secondary symbiogenesis: euglenoid, dinoflagellate, and sporozoan plastid origins and the eukaryote family tree. J. Euk. Microbiol. **46:**347–366. *Many organisms are the products of secondary symbiogenesis (a eukaryote is consumed by another eukaryote, both products of primary symbiogenesis, and symbiont becomes an organelle), but tertiary symbiogenesis also has occurred (product of secondary symbiogenesis itself becomes a symbiont . . . and organelle).*

Harrison, G. 1978. Mosquitoes, malaria and man: a history of the hostilities since 1880. New York, E. P. Dutton. *A fascinating story, well told.*

Hausmann, K., and N. Hülsmann. 1996. Protozoology. New York, Thieme Medical Publishers, Inc. *This was the most up-to-date, comprehensive treatment available before the release of Lee et al. (2000).*

Lee, J. J., G. F. Leedale, and P. Bradbury (eds). 2000. An illustrated guide to the protozoa, ed. 2, 1432 pp., 2 vols. Lawrence Kansas, Society of Protozoologists. *This long-awaited guide appeared in 2002. It is an essential reference for students of protozoa.*

Roberts, L. S., and J. J. Janovy, Jr. 2000. Foundations of parasitology, ed. 6. Dubuque, Iowa, McGraw-Hill Higher Education. *Up-to-date and readable information on parasitic protozoa.*

Sleigh, M. A. 1989. Protozoa and other protists. London, Edward Arnold. *Extensively updated version of the author's* The biology of protozoa.

Sogin, M. L., and J. D. Silberman. 1998. Evolution of the protists and protistan parasites from the perspective of molecular systematics. Int. J. Parasitol. **28:**11–20. *Studies in molecular systematics are leading to drastic revision of our former concepts of protozoan relationships.*

Stossel, T. P. 1994. The machinery of cell crawling. Sci. Am. **271:**54–63 (Sept.). *Ameboid movement—how cells crawl—is important throughout the animal kingdom, as well as in protozoa. We now understand quite a bit about its mechanism.*

Wainright, P. O., G. Hinkle, M. L. Sogin, and S. K. Stickel. 1993. Monophyletic origins of the Metazoa: an evolutionary link with fungi. Science **260:**340–342. *Their data suggest that green algae and plants are closer to animals and fungi than green algae are to other unicellular groups.*

ZOOLOGY LINKS TO THE INTERNET

Visit the textbook's Online Learning Center www.mhhe.com/zoology to find live Internet links for each of the topics listed here.

Protistan Groups
Phylum Sarcomastigophora (new taxonomic nomenclature: Phylum Euglenozoa, several phyla of amebas)
Phylum Apicomplexa
Phylum Ciliophora
Other Protozoan Phyla
Parasitic Protists
Free-living Protists
Dinoflagellates
Red Tides

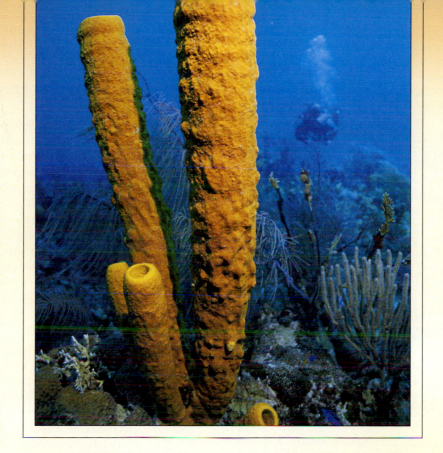

A Caribbean demosponge, *Aplysina fistularis*.

Mesozoa and Parazoa

Phylum Mesozoa
Phylum Placozoa
Phylum Porifera: Sponges

The Advent of Multicellularity

Sponges are the simplest multicellular animals. Because cells are the elementary units of life, organisms larger than unicellular organisms arose as aggregates of such building units. Nature experimented with producing larger organisms without cellular differentiation— certain large, single-celled marine algae, for example—but such examples are rarities. There are many advantages to multicellularity as opposed to simply increasing the mass of a single cell. Since it is at cell surfaces that exchange takes place, dividing a mass into smaller units greatly increases the surface area available for metabolic activities. It is impossible to maintain a workable surface-to-mass ratio by simply increasing the size of a single-celled organism. Thus multicellularity is a highly adaptive path toward increasing body size.

Strangely, while sponges are multicellular, their organization is quite distinct from other metazoans. A sponge body is an assemblage of cells embedded in a gelatinous matrix and supported by a skeleton of minute needlelike spicules and protein. Because sponges neither look nor behave like other animals, it is understandable that they were not completely accepted as animals by zoologists until well into the nineteenth century. Nonetheless, molecular evidence suggests that sponges share a common ancestor with other metazoa.

Position in Animal Kingdom

Multicellular organisms (metazoa) are typically divided into three grades: (1) Mesozoa (a single phylum), (2) Parazoa (phylum Porifera, sponges; and phylum Placozoa), and (3) Eumetazoa (all other phyla).

Although Mesozoa and Parazoa are multicellular, their plan of organization is distinct from that in eumetazoan phyla. Such cellular layers as they possess are not homologous to the germ layers of Eumetazoa, and neither group has developmental patterns in line with other metazoa. The name Parazoa means the "beside-animals."

Biological Contributions

1. Although the simplest in organization of all metazoa, these groups do compose a higher level of morphological and physiological integration than that found in protozoan colonies. Mesozoa and Parazoa may be said to belong to a **cellular level of organization.**
2. Mesozoans, although composed simply of an outer layer of somatic cells and an inner layer of reproductive cells, nevertheless have a very complex reproductive cycle somewhat suggestive of that of trematodes (flukes). Mesozoans are entirely parasitic.
3. Placozoans are essentially composed of two epithelia with fluid and some fibrous cells between them.
4. Sponges (poriferans) are more complex, with several types of cells differentiated for various functions, some of which are organized into **incipient tissues** of a low level of integration.
5. Developmental patterns of these three phyla are different from those of other phyla, and their embryonic layers are not homologous to the germ layers of Eumetazoa.
6. Sponges have developed a unique system of **water currents** on which they depend for food and oxygen.

ORIGIN OF METAZOA

Unraveling the origin of multicellular animals (metazoans) has presented many problems for zoologists. Three prominent hypotheses for the origin of metazoans from unicellular ancestors are that (1) metazoans arose from a syncytial (multinucleate) ciliated form in which cell boundaries later evolved, (2) they arose from a colonial flagellated form in which cells gradually became more specialized and interdependent, and (3) metazoans are polyphyletic, having been derived independently from more than one group of unicellular organisms.

Proponents of the **syncytial ciliate hypothesis** believe that metazoans arose from an ancestor shared with the single-celled ciliates. The common ancestor of metazoans acquired multiple nuclei within a single cell membrane and later became compartmentalized into the multicellular condition. It is assumed that the body form of the ancestor resembled that of modern ciliates and thus tended toward bilateral symmetry. Therefore the earliest metazoans would have been bilateral and similar to some present flatworms. There are several objections to this hypothesis. It ignores embryology of the flatworms in which nothing similar to cellularization occurs; it does not explain the presence of flagellated sperm in metazoans; and, perhaps more important, it implies that the radial symmetry of cnidarians is derived from a primary bilateral symmetry, for which there is no evidence.

The **colonial flagellate hypothesis**—first proposed by Haeckel in 1874—is the classical scheme, which, with various revisions, still has many followers. According to this hypothesis, metazoans descended from ancestors characterized by a hollow, spherical, colony of flagellated cells. Individual cells within the colony became differentiated for specific functional roles (reproductive cells, nerve cells, somatic cells, and so on), thus subordinating cellular independence to welfare of the colony as a whole. The colonial ancestral form was at first radially symmetrical, similar perhaps to the free-swimming planula larvae of the cnidarians (jellyfishes and others, p. 247). This larva is radially symmetrical and has no mouth. Cnidarians, with their radial symmetry, could have evolved from this form.

Bilateral symmetry could have evolved later when some of these planula-like ancestors became adapted for a creeping form of locomotion on the ocean floor. Dorsal and ventral surfaces would have differentiated, a ventral mouth would have appeared, and a start would have been made toward cephalization (a concentration of neurons and sensory structures at the anterior). These adaptations for creeping locomotion would have led to primitive bilateral symmetry, resembling that of flatworms.

Some zoologists prefer the idea that metazoans had a **polyphyletic origin** and suggest that the sponges, cnidarians, ctenophores, and remaining eumetazoans evolved independently. Thus no single scheme might account for them all.

We now have phylogenetic evidence based on small-subunit ribosomal RNA sequences and on similarities in complex biochemical pathways.[*] This evidence generally supports the colonial flagellate hypothesis that metazoans represent a monophyletic assemblage including choanoflagellates ("collared" flagellates such as *Codonosiga* and *Proterospongia,* see p. 229). The sister group of metazoans appears to be fungi. The molecular evidence excludes the syncytial ciliate hypothesis because metazoa are apparently closer to eukaryotic algae and higher plants than they are to ciliates.

PHYLUM MESOZOA

The name Mesozoa (mes-o-zo′a) (Gr. *mesos,* in the middle, + *zōon,* animal) was coined by an early investigator (van Beneden, 1876) who believed that the group was a "missing link" between protozoa and metazoa. These minute, ciliated, wormlike animals represent an extremely simple level of organization. All mesozoans live as parasites in marine invertebrates, and the majority of them are only 0.5 to 7 mm in length. Most are composed of only 20 to 30 cells arranged basically in two layers. The layers are not homologous to the germ layers of higher metazoans.

The two classes of mesozoans, Rhombozoa and Orthonectida, differ so much from each other that some authorities place them in separate phyla.

---

*Wainwright, P. O., et al. 1993. Science **260:**340–342.

Rhombozoans (Gr. *rhombos,* a spinning top, + *zōon,* animal) live in kidneys of benthic cephalopods (bottom-dwelling octopuses, cuttlefishes, and squids). Adults, called **vermiforms** (or nematogens), are long and slender (Figure 12-1). Their inner, reproductive cells give rise to vermiform larvae that grow and then reproduce. When a population becomes crowded, reproductive cells of some adults develop into gonadlike structures producing male and female gametes. Zygotes grow into minute (0.04 mm) ciliated infusoriform larvae (Figure 12-1B), quite unlike the parent. These larvae are shed with host urine into the seawater. The next part of the life cycle is unknown because infusoriform larvae are not immediately infective to a new host.

Orthonectids (Gr. *orthos,* straight, + *nektos,* swimming) (Figure 12-2) parasitize a variety of invertebrates, such as brittle stars, bivalve molluscs, polychaetes, and nemerteans. Their life cycles involve sexual and asexual phases, and the asexual stage is quite different from that of rhombozoans. It consists of a multinucleated mass called a **plasmodium,** which by division ultimately gives rise to males and females.

Phylogeny of Mesozoans

There is still much to learn about these mysterious little parasites, but probably one of the most intriguing questions is the place of mesozoans in the evolutionary picture. Some investigators believe they represent primitive or degenerate flatworms and even place them in phylum Platyhelminthes. Present molecular evidence supports a phylogenetic relationship of mesozoans and flatworms and inclusion of Mesozoa in superphylum Lophotrochozoa.

Figure 12-1

Two methods of reproduction by mesozoans. **A,** Asexual development of vermiform larvae from reproductive cells in the axial cell of an adult. **B,** Under crowded conditions in the host kidney, reproductive cells develop into gonads with gametes that produce infusoriform dispersal larvae that emerge in the host urine.

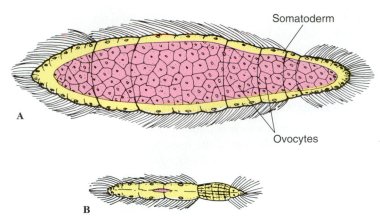

Figure 12-2

A, Female and, **B,** male orthonectid (*Rhopalura*). This mesozoan parasitizes such forms as flatworms, molluscs, annelids, and brittle stars. The structure is a single layer of ciliated epithelial cells surrounding an inner mass of sex cells.

PHYLUM PLACOZOA

The phylum Placozoa (Gr. *plax, plakos,* tablet, plate, + *zōon,* animal) was proposed in 1971 by K. G. Grell to contain a single species, *Trichoplax adhaerens* (Figure 12-3A), a tiny (2 to 3 mm) marine form that had been considered either a mesozoan or a cnidarian larva by various workers. The body is platelike and has no symmetry, no organs, and no muscular or nervous system. It is composed of a dorsal epithelium of cover cells and shiny spheres, a thick ventral epithelium containing monociliated cells (cylinder cells) and nonciliated gland cells, and a space between the epithelia containing fluid and fibrous cells (Figure 12-3B). The organisms glide over their food, secrete digestive enzymes on it, and then absorb the products. Grell considers *Trichoplax* diploblastic (see p. 160), with dorsal epithelium representing ectoderm and ventral epithelium representing endoderm because of its nutritive function. The phylogenetic position of placozoans is uncertain, although recent molecular evidence places them as a sister group to the phylum Cnidaria (p. 247).

PHYLUM PORIFERA: SPONGES

Sponges belong to phylum Porifera (po-rif'-er-a) (L. *porus,* pore, + *fera,* bearing). Sponges bear myriads of tiny pores and canals that constitute a filter-feeding system adequate for their inactive life habit. They are sessile animals that depend on water currents carried through their unique canal systems to bring them food and oxygen and to carry away their body wastes. Their bodies are little more than masses of cells embedded in a gelatinous matrix and stiffened by a skeleton of minute **spicules** of calcium carbonate or silica and collagen (p. 180). They have no organs or true tissues, and even their cells show a certain degree of independence. As sessile animals with only negligible body movement, they have not evolved a nervous system or sense organs and have only the simplest of contractile elements.

So, although they are multicellular, sponges share few of the characteristics of other metazoan phyla. They have been considered outside the clade containing choanoflagellates and metazoa. For this reason they are often called Parazoa (Gr. *para,* beside or alongside of, + *zōon,* animal).

Sponges vary in size from a few millimeters to the great loggerhead sponges, which may reach 2 m or more across. Many sponge species are brightly colored because of pigments in their dermal cells. Red, yellow, orange, green, and purple sponges are not uncommon. However, color fades quickly when sponges are removed from water. Some sponges, including the simplest, are radially symmetrical, but many are quite irregular in shape. Some stand erect, some are branched or lobed, and others are low, even encrusting, in form (Figure 12-4). Some bore holes into shells or rocks.

Most of the 5000 or more sponge species are marine, although some 150 species live in fresh water. Marine sponges are abundant in all seas and at all depths, and a few even exist in brackish water. Although their embryos are free swimming, adults are always attached, usually to rocks, shells, corals, or other submerged objects. Some bottom-dwelling forms even grow on sand or mud. Their growth patterns often depend on shape of the substratum, direction and speed of water currents, and availability of space, so that the same species may differ markedly in appearance under different environmental condi-

Figure 12-3

A, *Trichoplax adhaerens* is a marine, platelike animal only 2 to 3 mm in diameter. The only member of phylum Placozoa, it has the most primitive features of any known metazoan. **B,** Section through *Trichoplax adhaerens,* showing histological structure.

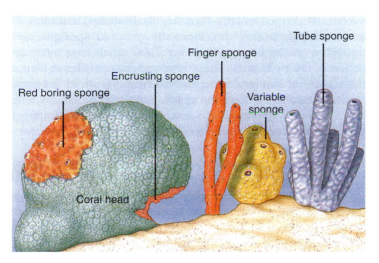

Figure 12-4
Some growth habits and forms of sponges.

tions. Sponges in calm waters may grow taller and straighter than those in rapidly moving waters.

Many animals (crabs, nudibranchs, mites, bryozoans, and fish) live as commensals or parasites in or on sponges. Larger sponges tend to harbor a large variety of invertebrate commensals. On the other hand, sponges grow on many other living animals, such as molluscs, barnacles, brachiopods, corals, or hydroids. Some crabs attach pieces of sponge to their carapace for camouflage and for protection, since most predators seem to find sponges distasteful. Some reef fishes, however, graze on shallow-water sponges.

Sponges are an ancient group, with an abundant fossil record extending back to the early Cambrian period and even, according to some claims, the Precambrian. Living poriferans traditionally have been assigned to three classes: Calcarea (with calcareous spicules), Hexactinellida (six-rayed siliceous spicules), and Demospongiae (with a skeleton of siliceous spicules or **spongin** [a specialized collagen] or both). A fourth class (Sclerospongiae) was erected to contain sponges with a massive calcareous skeleton and siliceous spicules. Some zoologists maintain that known species of sclerosponges can be placed in the traditional classes of sponges (Calcarea and Demospongiae); thus we do not need a new class.

Certainly one reason for the success of sponges as a group is that they have few enemies. Because of a sponge's elaborate skeletal framework and often noxious odor, most potential predators find sampling a sponge about as pleasant as eating a mouthful of glass splinters embedded in evil-smelling gristle.

Form and Function

The only body openings of these unusual animals are pores, usually many tiny ones called **ostia** for incoming water, and a few large ones called **oscula** (sing., **osculum**) for water outlet.

Characteristics of Phylum Porifera

1. Multicellular; body a loose aggregation of cells of mesenchymal origin
2. Body with pores (ostia), canals, and chambers that serve for passage of water
3. Mostly marine; all aquatic
4. Radial symmetry or none
5. Epidermis of flat pinacocytes; most interior surfaces lined with flagellated collar cells (choanocytes) that create water currents; a gelatinous protein matrix called mesohyl (mesoglea) contains amebocytes of various types and skeletal elements
6. Skeletal structure of fibrillar collagen (a protein) and calcareous or siliceous crystalline spicules, often combined with variously modified collagen (spongin)
7. No organs or true tissues; digestion intracellular; excretion and respiration by diffusion
8. Reactions to stimuli apparently local and independent; nervous system probably absent
9. All adults sessile and attached to substratum
10. Asexual reproduction by buds or gemmules and sexual reproduction by eggs and sperm; free-swimming ciliated larvae

These openings are connected by a system of canals, some of which are lined with peculiar flagellated collar cells called **choanocytes,** whose flagella maintain a current of environmental water through the canals. Water enters the canals through a multitude of tiny incurrent pores **(dermal ostia)** and leaves by way of one or more large oscula. Choanocytes not only keep the water moving but also trap and phagocytize food particles that are carried in the water. Cells lining the passageways are very loosely organized. Collapse of the canals is prevented by the skeleton, which, depending on the species, may be composed of needlelike calcareous or siliceous spicules, a meshwork of organic spongin fibers, or a combination of the two.

Sessile animals make few movements and therefore need little in the way of nervous, sensory, or locomotor parts. Sponges apparently have been sessile from their earliest appearance and have never acquired specialized nervous or sensory structures, and they have only the very simplest of contractile systems.

Types of Canal Systems

Most sponges have one of three types of canal systems: asconoid, syconoid, or leuconoid (Figure 12-5).

Asconoids: Flagellated Spongocoels　Asconoid sponges have the simplest organization. They are small and tube shaped. Water enters through microscopic dermal pores into a large cavity called a **spongocoel,** which is lined with choanocytes. Choanocyte flagella pull water through the pores and expel it through a single large osculum (see Figure 12-5). *Leucosolenia* (Gr. *leukos,* white, + *solen,* pipe) is an asconoid type of sponge. Its slender, tubular individuals grow in groups attached by a common stolon,

Figure 12-9

Small section through sponge wall, showing four types of sponge cells. Pinacocytes are protective and contractile; choanocytes create water currents and engulf food particles; archaeocytes have a variety of functions; collencytes secrete collagen.

through chambers slows to 0.001 cm/second. Such a flow rate allows ample opportunity for food capture by collar cells. All water is expelled through a single osculum at a velocity of 8.5 cm/second: a jet force capable of carrying waste products some distance away from the sponge. Some large sponges can filter 1500 liters of water a day. At least some sponges can crawl (move laterally over their supporting substratum) at speeds of up to 4 mm per day. This ability may give them an advantage over more sessile encrusting organisms in competition for space.

Sponges feed primarily on particles suspended in the water pumped through their canal systems. Detritus particles, planktonic organisms, and bacteria are consumed nonselectively in the size range from 50 μm (average diameter of ostia) to 0.1 μm (width of spaces between microvilli of choanocyte collar). Pinacocytes may phagocytize particles at the surface, but most larger particles are consumed in the canals by archaeocytes that move close to the lining of the canals. The smallest particles, accounting for about 80% of the particulate organic carbon, are phagocytized by choanocytes. Sponges also absorb dissolved nutrients from the water passing through the system. Protein molecules are taken into choanocytes by pinocytosis (p. 48).

Digestion is entirely **intracellular** (occurs within cells), and present evidence indicates that archaeocytes perform this chore. Choanocytes pass particles of food to archaeocytes for digestion.

There are no respiratory or excretory organs; both functions apparently occur by diffusion in individual cells. Contractile vacuoles are found in archaeocytes and choanocytes of freshwater sponges.

The only visible activities and responses in sponges, other than propulsion of water, are slight alterations in shape and closing and opening of incurrent and excurrent pores, and these movements are very slow. The most common response is closure of the oscula. Apparently excitation spreads from cell to cell, although some zoologists point to the possibility of coordination by means of substances carried in the water currents, and some zoologists have tried, not very successfully, to demonstrate presence of nerve cells.

functions. They can phagocytize particles at the pinacoderm and receive particles for digestion from choanocytes. Archaeocytes apparently can differentiate into any of the other types of more specialized cells in the sponge. Some, called **sclerocytes,** secrete spicules. Others, called **spongocytes,** secrete the spongin fibers of the skeleton, and **collencytes** secrete fibrillar collagen (p. 180). **Lophocytes** secrete large quantities of collagen but are distinguishable morphologically from collencytes.

Types of Skeletons

Its skeleton gives support to a sponge, preventing collapse of canals and chambers. The major structural protein in the animal kingdom is collagen, and fibrils of collagen are found throughout the intercellular matrix of all sponges. In addition, various Demospongiae secrete a form of collagen traditionally known as spongin. Several types of spongin, differing in chemical composition and form (fibers, spicules, filaments, spongin surrounding spicules, and so on) are found in various demosponges. Demospongiae also secrete siliceous spicules. Calcareous sponges secrete spicules composed mostly of crystalline calcium carbonate and have one, three, or four rays (Figure 12-11). Glass sponges have siliceous spicules with six rays arranged in three planes at right angles to each other. There are many variations in the shape of spicules, and these structural variations are of taxonomic importance.

Sponge Physiology

All activities of a sponge depend on the current of water flowing through its body. A sponge pumps a remarkable amount of water. *Leuconia* (Gr. *leukos,* white), for example, is a small leuconoid sponge about 10 cm tall and 1 cm in diameter. It is estimated that water enters through some 81,000 incurrent canals at a velocity of 0.1 cm/second. However, because *Leuconia* has more than 2 million flagellated chambers whose combined diameter is much greater than that of the canals, water flow

Reproduction

Sponges reproduce both asexually and sexually. **Asexual reproduction** occurs by means of bud formation and by regeneration following fragmentation. **External buds,** after reaching a certain size, may become detached from the parent and float away to form new sponges, or they may remain to form colonies. **Internal buds,** or **gemmules** (Figure 12-12), are formed in freshwater sponges and some marine sponges. Here, archaeocytes collect in the mesohyl and become surrounded by a tough spongin coat incorporating siliceous spicules. When the parent animal dies, the gemmules survive and remain dormant, preserving the species during periods of freezing or severe drought. Later, cells in the gemmules escape through a special opening, the **micropyle,** and develop into new sponges. Gemmulation in freshwater sponges (Spongillidae) is thus an adaptation to changing seasons. Gemmules are also a means of colonizing new habitats, since they can spread by streams or animal carriers. What

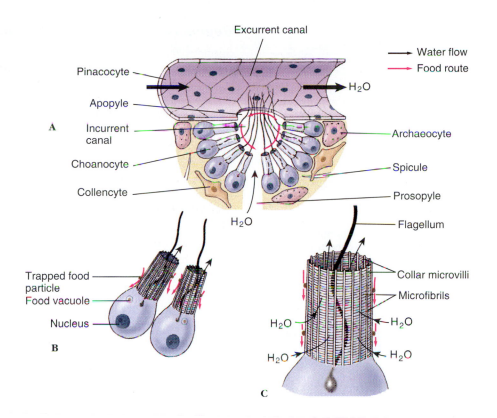

Figure 12-10

Food trapping by sponge cells. **A,** Cutaway section of canals showing cellular structure and direction of water flow. **B,** Two choanocytes and **C,** structure of the collar. Small red arrows indicate movement of food particles.

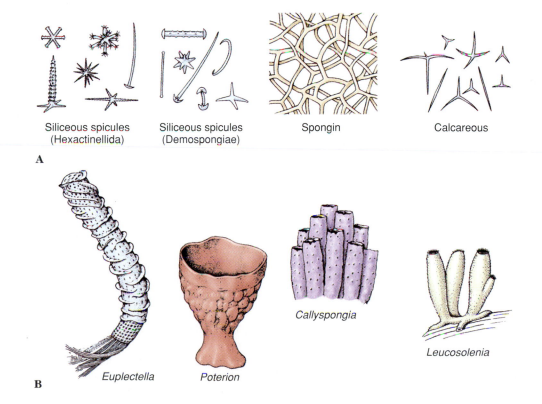

Figure 12-11

A, Types of spicules found in sponges. An amazing diversity, complexity, and beauty of form occurs among the many types of spicules. **B,** Some sponge body forms. *Euplectella* is in Hexactinellida, *Poterion* and *Callyspongia* are members of Demospongiae, and *Leucosolenia* is in Calcarea.

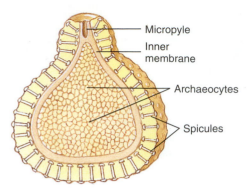

Micropyle

Inner
membrane

Archaeocytes

Spicules

Figure 12-12

Section through a gemmule of a freshwater sponge (Spongillidae). Gemmules are a mechanism for survival of the harsh conditions of winter. On return of favorable conditions, the archaeocytes exit through the micropyle to form a new sponge. The archaeocytes of the gemmule give rise to all cell types of the new sponge structure.

prevents gemmules from hatching during the season of formation rather than remaining dormant? Some species secrete a substance that inhibits early germination of gemmules, and gemmules do not germinate as long as they are held in the body of the parent. Other species undergo maturation at low temperatures (as in winter) before they germinate. Gemmules in marine sponges also seem to be an adaptation to pass the cold of winter; they are the only form in which *Haliclona loosanoffi* exists during the colder parts of the year in the northern part of its range.

In **sexual reproduction** most sponges are **monoecious** (have both male and female sex cells in one individual). Sperm arise from transformation of choanocytes. In Calcarea and at least some Demospongiae, oocytes also develop from choanocytes; in other demosponges oocytes apparently are derived from archaeocytes. Most sponges are viviparous; after fertilization the zygote is retained in and derives nourishment from its parent, and a ciliated larva is released. In such sponges, sperm are released into the water by one individual and taken into the canal system of another. There choanocytes phagocytize the sperm, then the choanocytes transform into carrier cells, which carry the sperm through the mesohyl to oocytes. Other sponges are oviparous, and both oocytes and sperm are expelled into the water. The free-swimming larva of most sponges is a solid-bodied **parenchymula** (Figure 12-13A). The outwardly directed, flagellated cells migrate to the interior after the larva settles and become choanocytes in the flagellated chambers.

Calcarea and a few Demospongiae have a very strange developmental pattern. A hollow blastula, called an **amphiblastula** (Figure 12-13B), develops, with flagellated cells toward the interior. The blastula then turns *inside out* **(inversion)**, the flagellated ends of the cells becoming directed to the outside! Flagellated cells **(micromeres)** of the larva are at one end, and larger, nonflagellated cells **(macromeres)** are at the other. In contrast to other metazoan embryos, the micromeres invaginate into and are overgrown by the macromeres. The flagellated micromeres become choanocytes, archeocytes, and collencytes of the new sponge, and the nonflagellated cells give rise to pinacoderm and sclerocytes.

Regeneration and Somatic Embryogenesis

Sponges have a tremendous ability to repair injuries and to restore lost parts, a process called **regeneration.** Regeneration does not imply a reorganization of the entire animal, but only of the wounded portion.

On the other hand, if a sponge is cut into small fragments, or if the cells of a sponge are entirely dissociated and are allowed to fall into small groups, or aggregates, entire new sponges can develop from these fragments or aggregates of cells. This process has been termed **somatic embryogenesis.** Somatic embryogenesis involves a complete reorganization of the structure and functions of participating cells or bits of tissue. Isolated from influence of adjoining cells, they can realize their own potential to change in shape or function as they develop into a new organism.

A great deal of experimental work has been done in this field. The process of reorganization appears to differ in sponges of differing complexity. There is still some controversy concerning just what mechanisms cause adhesion of the cells and the share that each type of cell plays in the formative process.

Class Calcarea (Calcispongiae)

Calcarea (also called Calcispongiae) are calcareous sponges, so called because their spicules are composed of calcium carbonate. Spicules are straight (monaxons) or have three or four rays. These sponges tend to be small—10 cm or less in height—and tubular or vase shaped. They may be asconoid, syconoid, or leuconoid in structure. Though many are drab in color, some are bright yellow, red, green, or lavender. *Leucosolenia* and *Sycon* (often called *Scypha* or *Grantia* by biological supply companies) are marine shallow-water forms commonly studied in the laboratory. *Leucosolenia* is a small asconoid sponge that grows in branching colonies, usually arising from a network of horizontal, stolonlike tubes. *Clathrina* is small with intertwined tubes (Figure 12-6). *Sycon* is a solitary sponge that may live singly or form clusters by budding. The vase-shaped, typically syconoid animal is 1 to 3 cm long, with a fringe of straight spicules around the osculum to discourage small animals from entering.

Class Hexactinellida (Hyalospongiae): Glass Sponges

Glass sponges make up class Hexactinellida (or Hyalospongiae). Nearly all are deep-sea forms that are collected by dredging. Most are radially symmetrical, with vase- or funnel-shaped bodies usually attached by stalks of root spicules to a substratum (Figure 12-11, *Euplectella*) (N. L. from Gr. *euplektos,* well plaited). They range from 7.5 cm to more than 1.3 m in length. Their distinguishing features include a skeleton of six-rayed siliceous spicules that are commonly bound together into a network forming a glasslike structure.

Their tissue structure differs so dramatically from other sponges that some scientists advocate placing hexactinellids in a subphylum separate from other sponges. The body of hexactinellids is composed of a single, continuous syncytial tissue (a tissue not divided into separate cells) called a **trabecular retic-**

A, Developing demosponge

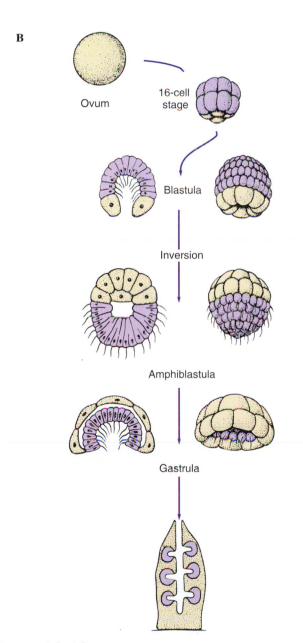

Figure 12-13
A, Development of demosponges. **B,** Development of the syconoid sponge *Sycon.*

ulum. The trabecular reticulum is the largest, continuous syncytial tissue known in Metazoa. It is bilayered and encloses a thin, collagenous mesohyl between the two layers, as well as cellular elements such as archaeocytes, sclerocytes, and **choanoblasts.** Choanoblasts are associated with flagellated chambers, where the layers of the trabecular reticulum separate into a **primary reticulum** (incurrent side) and a **secondary reticulum** (excurrent, or atrial side) (Figure 12-14). The spherical choanoblasts are borne by the primary reticulum, and each choanoblast has one or more processes extending to **collar bodies,** the bases of which are also supported by the primary reticulum. Each collar body with its flagellum extends into the flagellated chamber through an opening in the secondary reticulum. Water is drawn into the space between primary and secondary reticula through prosopyles in the primary reticulum, then through the collars into the lumen of the flagellated chamber. Collar bodies do not participate in phagocytosis; rather that process is accomplished by the primary and secondary reticula.

The latticelike network of spicules found in many glass sponges is of exquisite beauty, such as that of *Euplectella,* or Venus' flower basket (Figure 12-11), a classic example of Hexactinellida.

Class Demospongiae

Class Demospongiae contains over 90% of living sponge species, including most larger sponges. Spicules are siliceous but are not six rayed, and they may be bound together by spongin or may be absent altogether. All members of the class are leuconoid, and all are marine except one family, the Spongillidae, or freshwater sponges.

Freshwater sponges are widely distributed in well-oxygenated ponds and streams, where they encrust plant stems and old pieces of submerged wood. They may resemble a bit of wrinkled scum, be pitted with pores, and be brownish or greenish in color. Common genera are *Spongilla* (L. *spongia,* from Gr. *spongos,* sponge) and *Myenia.* Freshwater sponges are most common in midsummer, although some are more easily found in the fall. They die and disintegrate in late autumn, leaving gemmules to produce the next year's population. They also reproduce sexually.

Marine Demospongiae are quite varied and may be quite striking in color and shape (Figure 12-15). Some are encrusting; some are tall and fingerlike; some are low and spreading; some bore into shells; and some are shaped like fans, vases, cushions, or balls (Figure 12-15). Loggerhead sponges may grow several meters in diameter.

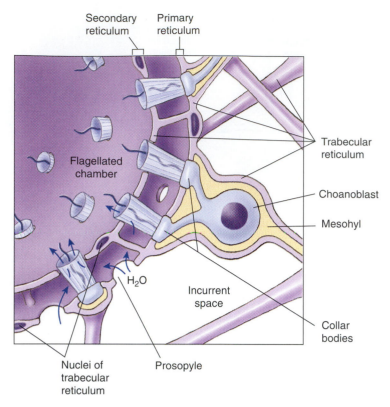

Secondary reticulum
Primary reticulum

Flagellated chamber

Trabecular reticulum

Choanoblast

Mesohyl

H₂O

Incurrent space

Collar bodies

Nuclei of trabecular reticulum

Prosopyle

Figure 12-14

Diagram of part of a flagellated chamber of hexactinellids. The primary and secondary reticula are branches of the trabecular reticulum, which is syncytial. Cell bodies of the choanoblasts and their processes are borne by the primary reticulum and are embedded in a thin, collagenous mesohyl. Processes of the choanoblasts end in collar bodies, whose collars extend up through the secondary reticulum. Flagellar action propels water (*arrows*) to be filtered through the mesh of collar microvilli (see Figure 12-10).

So-called bath sponges (*Spongia, Hippospongia*) belong to the group called horny sponges, which have spongin skeletons and lack siliceous spicules entirely.

Phylogeny and Adaptive Radiation

Phylogeny

Sponges originated before the Cambrian period. Two groups of calcareous spongelike organisms occupied early Paleozoic reefs. The Devonian period saw rapid development of many glass sponges. The possibility that sponges arose from choanoflagellates (protozoa that bear collars and flagella, p. 229) earned support for a time. However, many zoologists opposed that hypothesis because sponges do not acquire collars until later in their embryological development. The outer cells of the larvae are flagellated but not collared, and they do not become collar cells until they become internal. Also, collar cells are found in certain corals and echinoderms, so they are not unique to sponges.

However, these objections are countered by evidence derived from base sequences of ribosomal RNA genes. This evidence supports the hypothesis that choanoflagellates and metazoans are sister groups. It suggests also that sponges and Eumetazoa are sister groups, with Porifera having separated before the origin of radiates and placozoans, but sharing a common ancestor.

Adaptive Radiation

Porifera are a highly successful group that includes several thousand species and a variety of marine and freshwater habitats. Their diversification centers largely on their unique water-current system and its various degrees of complexity. The leuconoid body plan yields a great increase in the ratio of surface to volume in a sponge colony. Such an increase made possible greater efficiency in gaseous exchange, proliferation of flagellated chambers for feeding, and overall increase in body size.

A

B

C

Figure 12-15

Marine Demospongiae on Caribbean coral reefs. **A,** *Pseudoceratina crassa* is a colorful sponge growing at moderate depths. **B,** *Ectyoplasia ferox* is irregular in shape and its oscula form small, volcano-like cones. It is toxic and may cause skin irritation if touched. **C,** *Monanchora unguifera* with commensal brittle star, *Ophiothrix suensoni* (phylum Echinodermata, class Ophiuroidea).

Classification of Phylum Porifera

Class Calcarea (cal-ca're-a) (L. *calcis,* lime) **(Calcispongiae).** Have spicules of calcium carbonate that often form a fringe around the osculum (main water outlet); spicules needle shaped or three or four rayed; all three types of canal systems (asconoid, syconoid, leuconoid) represented; all marine. Examples: *Sycon, Leucosolenia, Clathrina.*

Class Hexactinellida (hex-ak-tin-el'i-da) (Gr. *hex, six,* + aktis, ray, + L. *-ellus,* dim. suffix) **(Hyalospongiae).** Have six-rayed, siliceous spicules extending at right angles from a central point; spicules often united to form network; body often cylindrical or funnel shaped; flagellated chambers in simple syconoid or leuconoid arrangement; habitat mostly deep water; all marine. Examples: Venus' flower basket *(Euplectella), Hyalonema.*

Class Demospongiae (de-mo-spun'je-e) (Gr. *demos,* people, + *spongos,* sponge). Have siliceous spicules that are not six rayed, or spongin, or both; leuconoid-type canal systems; one family found in fresh water; all others marine. Examples: *Thenea, Cliona, Spongilla, Myenia, Poterion, Callyspongia,* and all bath sponges.

SUMMARY

Members of phylum Mesozoa are very simply organized animals that are parasitic in kidneys of cephalopod molluscs (class Rhombozoa) and in several other invertebrate groups (class Orthonectida). They have only two cell layers, but these are not homologous to the germ layers of higher metazoans. They have a complicated life history that is still incompletely known. Their simple organization may have been derived from a more complex platyhelminth-like ancestor.

Phylum Placozoa has only one member, a small platelike marine organism. It too has only two cell layers, but some workers believe that these layers are homologous to ectoderm and endoderm of more complex metazoans. The closet relatives of placozoans seem to be cnidarians.

Sponges (phylum Porifera) are an abundant marine group with some freshwater representatives. They have various specialized cells, but these are not organized into tissues or organs. They depend on the flagellar beat of their choanocytes to circulate water through their bodies for food gathering and respiratory gas exchange. They are supported by secreted skeletons of fibrillar collagen, collagen in the form of large fibers or filaments (spongin), calcareous or siliceous spicules, or a combination of spicules and spongin in most species.

Sponges reproduce asexually by budding, fragmentation, and gemmules (internal buds). Most sponges are monoecious but produce sperm and oocytes at different times. Embryogenesis is unusual, with a migration of flagellated cells at the surface to the interior (parenchymella) or the production of an amphiblastula with inversion and growth of macromeres over micromeres. Sponges have great regenerative abilities.

Sponges are an ancient group, seemingly remote phylogenetically from other metazoa, but molecular evidence suggests that they are a sister group to Eumetazoa. Their adaptive radiation is centered on elaboration of the water circulation and filter-feeding system.

REVIEW QUESTIONS

1. Briefly describe and contrast the syncytial ciliate hypothesis, the colonial flagellate hypothesis, and the polyphyletic origin of the metazoa. Which hypothesis seems most compatible with available data?
2. Describe the body plan of Mesozoa and Placozoa.
3. Give eight characteristics of sponges.
4. Briefly describe asconoid, syconoid, and leuconoid body types in sponges.
5. What sponge body type is most efficient and makes possible the largest body size?
6. Define the following: ostia, osculum, spongocoel, apopyles, prosopyles.
7. Define the following: pinacocytes, choanocytes, archaeocytes, sclerocytes, spongocytes, collencytes.
8. What material is found in the skeleton of all sponges?
9. Describe the skeletons of each class of sponges.
10. Describe how sponges feed, respire, and excrete.
11. What is a gemmule?
12. Describe how gametes are produced and the process of fertilization in most sponges.
13. Contrast embryogenesis in most Demospongiae with that in the Calcarea.
14. What is the largest class of sponges, and what is its body type?
15. Describe possible ancestors to sponges. Justify your answer.

SELECTED REFERENCES

Bergquist, P. R. 1978. Sponges. Berkeley, University of California Press. *Excellent monograph on sponge structure, classification, evolution, and general biology.*

Bond, C. 1997. Keeping up with the sponges. Nat. Hist. **106:** 22–25. *Sponges are not fixed in permanent position; at least some can crawl on their substrate.* Haliclona loosanoffi *can move over 4 mm/day.*

Grell, K. G. 1982. Placozoa. In S. P. Parker (ed). Synopsis and classification of living organisms, vol. 1. New York, McGraw-Hill Book Company. *Synopsis of placozoan characteristics.*

Kobayashi, M., H. Furuya, and P. W. H. Holland. 1999. Dicyemids are higher animals. Nature. **401:**762. *Sequence analysis of the gene for Hox protein is evidence that mesozoans are members of superphylum Lophotrochozoa and that they are derived from a more complex ancestor that has undergone simplification during its parasitic evolution. They ". . . are not basal and primitive animals and should not be excluded from Metazoa."*

Leys, S. P. 1999. The choanosome of hexactinellid sponges. Invert. Biol. **118:**221–235. *Choanosomes are flagellated chambers and associated tissues. This author supports the position that Hexactinellida should constitute a separate subphylum. She includes an excellent description of trabecular reticulum.*

Reiswig, H. M., and T. L. Miller. 1998. Freshwater sponge gemmules survive months of anoxia. Invert. Biol. **117:**1–8. *Hatchability of gemmules kept in the absence of oxygen was equal to controls, but they would not hatch unless oxygen was present.*

Wainright, P. O., G. Hinkle, M. L. Sogin, S. K. Stickel. 1993. Monophyletic origins of the Metazoa: an evolutionary link with Fungi. Science **260:**340–342. *Reports molecular evidence that the sister group of metazoans is Fungi and that multicellular animals, including sponges, are monophyletic.*

Winnepenninckx, B. M. H., Y. Van de Peer, and T. Backeljau. 1998. Metazoan relationships on the basis of 18S rRNA sequences: A few years later . . . Amer. Zool. **38:**888–906. *Some evidence is provided for a common ancestry of Platyhelminthes and Mesozoa and for monophyly of Mesozoa.*

Wood, R. 1990. Reef-building sponges. Am. Sci. **78:**224–235. *The author presents evidence that known sclerosponges belong to either the Calcarea or the Demospongiae and that a separate class Sclerospongiae is not needed.*

Wyeth, R. C. 1999. Video and electron microscopy of particle feeding in sandwich cultures of the hexactinellid sponge, *Rhabdocalyptus dawsoni.* Invert. Biol. **118:**236–242. *Phagocytosis is not by choanoblasts but by trabecular reticulum, especially primary reticulum. He places Hexactinellida in subphylum Symplasma and the rest of Porifera in subphylum Cellularia.*

ZOOLOGY LINKS TO THE INTERNET

Visit the textbook's Online Learning Center at www.mhhe.com/zoology to find live Internet links for each of the topics listed here.
Mesozoa and Parazoa
Phylum Placozoa
Phylum Porifera

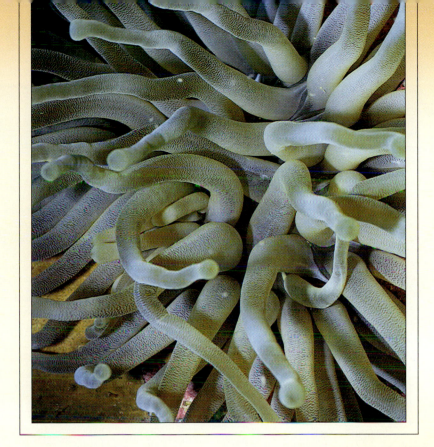

Tentacles of a Caribbean sea anemone, *Condylactis gigantea*.

Radiate Animals

Phylum Cnidaria
Phylum Ctenophora

A Fearsome Tiny Weapon

Although members of phylum Cnidaria are more highly organized than sponges, they are still relatively simple animals. Most are sessile; those that are unattached, such as jellyfish, can swim only feebly. None can chase their prey. Indeed, we might easily get the false impression that cnidarians were placed on earth to provide easy meals for other animals. The truth is, however, many cnidarians are very effective predators that are able to kill and eat prey that are much more highly organized, swift, and intelligent. They manage these feats because they possess tentacles that bristle with tiny, remarkably sophisticated weapons called nematocysts.

As it is secreted within the cell that contains it, the nematocyst is endowed with potential energy to power its discharge. It is as though a factory manufactured a gun, cocked and ready with a bullet in its chamber, as it rolls off the assembly line. Like a cocked gun, a completed nematocyst requires only a small stimulus to make it fire. Rather than a bullet, a tiny thread bursts from a nematocyst. Achieving a velocity of 2 meters/sec and an acceleration of 40,000 × gravity, it instantly penetrates its prey and injects a paralyzing toxin. A small animal unlucky enough to brush against one of the tentacles is suddenly speared with hundreds or even thousands of nematocysts and quickly immobilized. Some nematocyst threads can penetrate human skin, resulting in sensations ranging from minor irritation to great pain, even death, depending on the species. A fearsome, but wondrous, tiny weapon.

Position in Animal Kingdom

The phyla Cnidaria and Ctenophora are characterized by **primary radial** or **biradial symmetry,** which we believe is ancestral for eumetazoans. Radial symmetry, in which the body parts are arranged concentrically around the oral-aboral axis, is particularly suitable for sessile or sedentary animals and for free-floating animals because they approach their environment (or it approaches them) from all sides equally. Biradial symmetry is basically a type of radial symmetry in which only two planes through the oral-aboral axis divide the animal into mirror images because some part is paired. All other eumetazoans have a primary bilateral symmetry; they are bilateral or were derived from an ancestor that was bilateral.

Neither phylum has advanced beyond the **tissue level of organization,** although a few organs occur. In general, ctenophores are structurally more complex than cnidarians.

Biological Contributions

1. Both phyla have developed two well-defined **germ layers,** ectoderm and endoderm; a third, or mesodermal, layer, which is derived embryologically from the ectoderm, is present in some. The body plan is saclike, and the body wall is composed of two distinct layers, epidermis and gastrodermis, derived from the ectoderm and endoderm, respectively. A gelatinous matrix, mesoglea, between these layers may be structureless, may contain a few cells and fibers, or may be composed largely of mesodermal connective tissue and muscle fibers.
2. An internal body cavity, the **gastrovascular cavity,** is lined by gastrodermis and has a single opening, the mouth, which also serves as an anus.

3. **Extracellular digestion** occurs in the gastrovascular cavity, and intracellular digestion takes place in the gastrodermal cells. Extracellular digestion allows ingestion of larger food particles.
4. Most radiates have **tentacles,** or extensible projections around the oral end, that aid in food capture.
5. Radiates are the simplest animals to possess true **nerve cells** (protoneurons), but the nerves are arranged as a nerve net, with no central nervous system.
6. Radiates are the simplest animals to possess sense organs, which include well-developed statocysts (organs of equilibrium) and ocelli (photosensitive organs).
7. Locomotion in free-moving forms is achieved either by **muscular contractions** (cnidarians) or **ciliary comb plates** (ctenophores). However, both groups are still better adapted to floating or being carried by currents than to strong swimming.
8. **Polymorphism*** in cnidarians has widened their ecological possibilities. In many species the presence of both a polyp (sessile and attached) stage and a medusa (free-swimming) stage permits occupation of a benthic (bottom) and a pelagic (open-water) habitat by the same species. Polymorphism also widens the possibilities of structural complexity.
9. Some unique features are found in these phyla, such as **nematocysts** (stinging organelles) in cnidarians and **colloblasts** (adhesive organelles) and **ciliary comb plates** in ctenophores.

---

*Note that polymorphism here refers to more than one structural form of individual within a species, as contrasted with the use of the word in genetics (p. 120), in which it refers to different allelic forms of a gene in a population.

PHYLUM CNIDARIA

Phylum Cnidaria (ny-dar'e-a) (Gr. *knide,* nettle, + L. *aria* [pl. suffix], like or connected with) is an interesting group of more than 9000 species. It takes its name from cells called **cnidocytes,** which contain the stinging organelles **(nematocysts)** characteristic of the phylum. Nematocysts are formed only by cnidarians. Another name for the phylum, Coelenterata (se-len'te-ra'ta) (Gr. *koilos,* hollow, + *enteron,* gut, + L. *ata* [pl. suffix], characterized by), is used less than formerly, and it sometimes now refers to both radiate phyla, since its meaning is equally applicable to both.

Cnidarians are generally regarded as originating close to the basal stock of the metazoan line. They are an ancient group with the longest fossil history of any metazoan, reaching back more than 700 million years. Although their organization has a structural and functional simplicity not found in other metazoans, they form a significant proportion of the biomass in some locations. They are widespread in marine habitats, and there are a few in fresh water. Although they are mostly sessile, or at best, fairly slow moving or slow swimming, they are quite efficient predators of organisms that are much swifter and more complex. The phylum includes some of nature's strangest and loveliest creatures: branching, plantlike hydroids; flowerlike sea anemones;

jellyfishes; and those architects of the ocean floor, horny corals (sea whips, sea fans, and others) and stony corals whose thousands of years of calcareous house-building have produced great reefs and coral islands (p. 264).

Cnidarians are found most abundantly in shallow marine habitats, especially in warm temperatures and tropical regions. There are no terrestrial species. Colonial hydroids are usually found attached to mollusc shells, rocks, wharves, and other animals in shallow coastal water, but some species are found at great depths. Floating and free-swimming medusae are found in open seas and lakes, often far from shore. Floating colonies such as the Portuguese man-of-war and *Velella* (L. *velum,* veil; + *ellus,* dim. suffix) have floats or sails by which the wind carries them.

Some ctenophores, molluscs, and flatworms eat hydroids bearing nematocysts and use these stinging structures for their own defense. Some other animals feed on cnidarians, but cnidarians rarely serve as food for humans.

Cnidarians sometimes live symbiotically with other animals, often as commensals on the shell or other surface of their host. Certain hydroids (Figure 13-1) and sea anemones commonly live on snail shells inhabited by hermit crabs, providing the crabs some protection from predators. Algal cells frequently live as mutuals in the tissues of cnidarians, notably in some fresh-

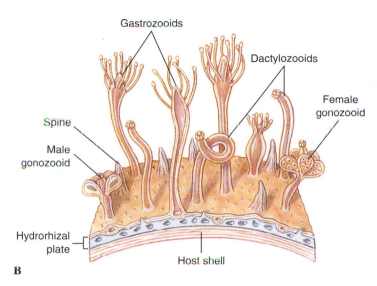

Figure 13-1

A, A hermit crab with its cnidarian mutuals. The shell is blanketed with polyps of the hydrozoan *Hydractinia milleri.* The crab gets some protection from predation by the cnidarians, and the cnidarians get a free ride and bits of food from their host's meals. **B,** Portion of a colony of *Hydractinia,* showing the types of zooids and the stolon (hydrorhiza) from which they grow.

water hydras and in reef-building corals. The presence of the algae in reef-building corals limits the occurrence of coral reefs to relatively shallow, clear water where there is sufficient light for the photosynthetic requirements of the algae. These kinds of corals are an essential component of coral reefs, and reefs are extremely important habitats in tropical waters. Coral reefs are discussed further on p. 264.

Although many cnidarians have little economic importance, reef-building corals are an important exception. Fish and other animals associated with reefs provide substantial amounts of food for humans, and reefs are of economic value as tourist attractions. Precious coral is used for jewelry and ornaments, and coral rock serves for building purposes.

Planktonic medusae may be of some importance as food for fish that are of commercial value; the reverse is also true—young fish fall prey to cnidarians.

Four classes of Cnidaria are commonly recognized: Hydrozoa (most variable class, including hydroids, fire corals, Portuguese man-of-war, and others), Scyphozoa ("true" jellyfishes), Cubozoa (cube jellyfishes), and Anthozoa (largest class, including sea anemones, stony corals, soft corals, and others).

Form and Function

Dimorphism and Polymorphism in Cnidarians

One of the most interesting—and sometimes puzzling—aspects of this phylum is the dimorphism and often polymorphism displayed by many of its members. All cnidarian forms fit into one of two morphological types (dimorphism): a **polyp,** or hydroid form, which is adapted to a sedentary or sessile life, and a **medusa,** or jellyfish form, which is adapted for a floating or free-swimming existence (Figure 13-2).

Most polyps have tubular bodies with a mouth at one end surrounded by tentacles. The aboral end is usually attached to a

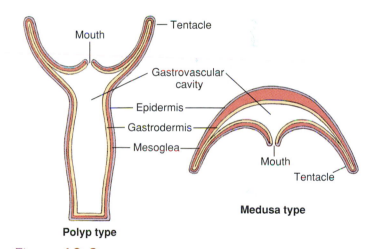

Figure 13-2

Comparison between polyp and medusa types of individuals.

substratum by a pedal disc or other device. Polyps may live singly or in colonies. Colonies of some species include morphologically differing individuals (polymorphism) each specialized for a certain function, such as feeding, reproduction, or defense (Figure 13-1).

Medusae are usually free swimming and have bell-shaped or umbrella-shaped bodies and tetramerous symmetry (body parts arranged in fours). The mouth is usually centered on the concave side, and tentacles extend from the rim of the umbrella.

Sea anemones and corals (class Anthozoa) are all polyps: hence, they are not dimorphic. True jellyfishes (class Scyphozoa) have a conspicuous medusoid form, but many have a polypoid larval stage. Colonial hydroids of class Hydrozoa, however, sometimes have life histories that feature both a polyp stage and a free-swimming medusa stage—rather like a Jekyll-and-Hyde existence. A species that has both an attached polyp and a floating

Characteristics of Phylum Cnidaria

1. Entirely aquatic, some in fresh water but mostly marine
2. **Radial symmetry** or biradial symmetry around a longitudinal axis with **oral** and **aboral** ends; no definite head
3. Two basic types of individuals: **polyps** and **medusae**
4. Exoskeleton or endoskeleton of chitinous, calcareous, or protein components in some
5. Body with two layers, epidermis and gastrodermis, with mesoglea **(diploblastic);** mesoglea with cells and connective tissue (ectomesoderm) in some
6. **Gastrovascular cavity** (often branched or divided with septa) with a single opening that serves as both mouth and anus; extensible tentacles usually encircling the mouth or oral region
7. Special stinging cell organelles called **nematocysts** in either epidermis or gastrodermis or in both; nematocysts abundant on tentacles, where they may form batteries or rings
8. **Nerve net** with symmetrical and asymmetrical synapses; with some sensory organs; diffuse conduction
9. Muscular system (epitheliomuscular type) of an outer layer of longitudinal fibers at base of epidermis and an inner one of circular fibers at base of gastrodermis; modifications of this plan in more complex cnidarians, such as separate bundles of independent fibers in the mesoglea
10. Asexual reproduction by budding (in polyps) or sexual reproduction by gametes (in all medusae and some polyps); sexual forms monoecious or dioecious; **planula larva;** holoblastic indeterminate cleavage
11. No excretory or respiratory system
12. No coelomic cavity

medusa within its life history can take advantage of feeding and distribution possibilities of both pelagic (open-water) and benthic (bottom) environments. Many hydrozoans are also polymorphic, with several distinct types of polyps in a colony.

Superficially the polyp and medusa seem very different. But actually each has retained the saclike body plan basic to the phylum (Figure 13-2). A medusa is essentially an unattached polyp with the tubular portion widened and flattened into a bell shape.

Both polyp and medusa possess the three body-wall layers typical of cnidarians, but the jellylike layer of mesoglea is much thicker in a medusa, constituting the bulk of the animal and making it more buoyant. It is because of this mass of mesoglea "jelly" that medusae are commonly called jellyfishes.

Nematocysts: Stinging Organelles

Stinging organelles called **nematocysts** (Figure 13-3) are the most characteristic structures in the entire cnidarian group. Over 20 different types of nematocysts (Figure 13-3) have been described in cnidarians; they are important in taxonomic determinations. Nematocysts are tiny capsules composed of material similar to chitin and containing a coiled tubular "thread" or filament, which is a continuation of the narrowed end of the capsule. This end of the capsule is covered by a little lid, or **operculum.** The inside of the undischarged thread may bear tiny barbs, or spines.

A nematocyst is enclosed in the cell that has produced it, the **cnidocyte** (during its development, a cnidocyte is properly called a **cnidoblast**). Except in Anthozoa, cnidocytes are equipped with a triggerlike **cnidocil,** which is a modified cilium. Anthozoan cnidocytes have a somewhat different ciliary mechanoreceptor. In some sea anemones, and perhaps other

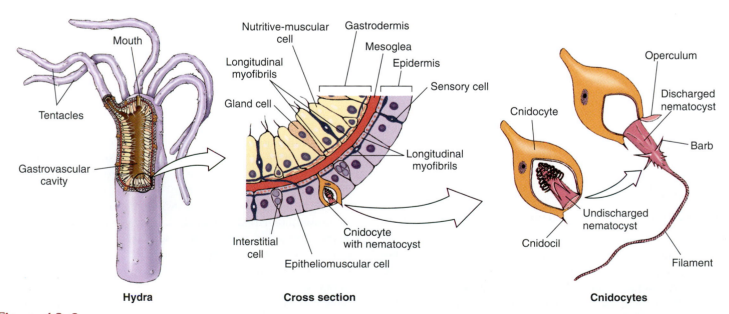

Figure 13-3

At right, structure of a stinging cell. Center, portion of the body wall of a hydra. Cnidocytes, which contain the nematocysts, arise in the epidermis from interstitial cells.

cnidarians, small organic molecules from the prey "tune" the mechanoreceptors, sensitizing them to the frequency of vibration caused by prey swimming. Tactile stimulation causes the nematocyst to discharge. Cnidocytes are borne in invaginations of ectodermal cells and, in some forms, in gastrodermal cells, and they are especially abundant on the tentacles. When a nematocyst is discharged, its cnidocyte is absorbed and a new one replaces it. Not all nematocysts have barbs or inject poison. Some, for example, do not penetrate prey but rapidly recoil like a spring after discharge, grasping and holding any part of the prey caught in the coil (Figure 13-4). Adhesive nematocysts usually do not discharge in food capture.

The mechanism of nematocyst discharge is remarkable. Evidence indicates that discharge is due to a combination of tensional forces generated during nematocyst formation and to an astonishingly high osmotic pressure within the nematocyst: 140 atmospheres. When stimulated to discharge, the high internal osmotic pressure causes water to rush into the capsule. The operculum opens, and the rapidly increasing *hydrostatic pressure* within the capsule forces the thread out with great force, turning inside out as it goes. At the everting end of the thread, the barbs flick to the outside like tiny switchblades. This minute but awesome weapon then injects poison when it penetrates prey.

> Note again the distinction between osmotic and hydrostatic pressure (p. 46). The nematocyst is never required actually to contain 140 atmospheres of hydrostatic pressure within itself; such a hydrostatic pressure would doubtless cause it to explode. As the water rushes in during discharge, the osmotic pressure falls rapidly, while the hydrostatic pressure rapidly increases.

Nematocysts of most cnidarians are not harmful to humans and are a nuisance at worst. However, the stings of a Portuguese man-of-war (see Figure 13-14) and certain jellyfishes are quite painful and sometimes dangerous (see note, p. 260).

Nerve Net

The nerve net of cnidarians is one of the best examples of a diffuse nervous system in the animal kingdom. This plexus of nerve cells is found both at the base of the epidermis and at the base of the gastrodermis, forming two interconnected nerve nets. Nerve processes (axons) end on other nerve cells at synapses or at junctions with sensory cells or effector organs (nematocysts or epitheliomuscular cells). Nerve impulses are transmitted from one cell to another by release of a neurotransmitter from small vesicles on one side of the synapse or junction (p. 699). One-way transmission between nerve cells in higher animals is ensured because the vesicles are located on only one side of the synapse. However, cnidarian nerve nets are peculiar in that many of the synapses have vesicles of neurotransmitters on both sides, allowing transmission across the synapse in either direction. Another peculiarity of cnidarian nerves is the absence of any sheathing material (myelin) on the axons.

Figure 13-4

Several types of nematocysts shown after discharge. At bottom are two views of a type that does not impale prey; it recoils like a spring, catching any small part of the prey in the path of the recoiling thread.

There is no concentrated grouping of nerve cells to suggest a "central nervous system." Nerves are grouped, however, in the "ring nerves" of hydrozoan medusae and in marginal sense organs of scyphozoan medusae. In some cnidarians the nerve nets form two or more systems: in Scyphozoa there is a fast conducting system to coordinate swimming movements and a slower one to coordinate movements of tentacles.

> Note that there is little adaptive value for a radially symmetrical animal to have a central nervous system with a brain. The environment approaches from all sides equally, and there is no control over the direction of approach to a prey organism.

Nerve cells of the net have synapses with slender sensory cells that receive external stimuli, and the nerve cells have junctions with epitheliomuscular cells and nematocysts. Together with the contractile fibers of the epitheliomuscular cells, the sensory-nerve cell net combination is often termed a **neuromuscular system,** an important landmark in the evolution of nervous systems. The nerve net arose early in metazoan evolution, and it has never been completely lost phylogenetically. Annelids have it in their digestive systems. In the human digestive system it is represented by nerve plexuses in the musculature. Rhythmical peristaltic movements of the stomach and intestine are coordinated by this counterpart of the cnidarian nerve net.

Class Hydrozoa

The majority of Hydrozoa are marine and colonial in form, and a typical life cycle includes both an asexual polyp and a sexual medusa stage. Some, however, such as freshwater hydras, have no medusa stage. Some marine hydroids do not have free medusae (Figure 13-5), whereas some hydrozoans occur only as medusae and have no polyp.

Hydras, although not typical hydrozoans, are widely used in zoology laboratories as an introduction to Cnidaria because of their small size and ready availability. Combining study of a hydra with that of a representative colonial marine hydroid such as *Obelia* (Gr. *obelias,* round cake) gives an excellent overview of class Hydrozoa.

Hydra: A Freshwater Hydrozoan

Common freshwater hydras (Figure 13-6) are solitary polyps and some of the few cnidarians found in fresh water. Their normal habitat is the underside of aquatic leaves and lily pads in cool, clean fresh water of pools and streams. The hydra family is found throughout the world, with 16 species occurring in North America.

Body Plan   The body of a hydra can extend to a length of 25 to 30 mm or can contract to a tiny, gelatinous mass. It is a cylindrical tube with the aboral end drawn out into a slender stalk, ending in a **basal** (or pedal) **disc** for attachment. This basal disc is provided with gland cells to enable a hydra to

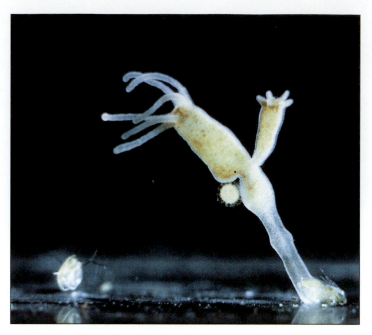

Figure 13-6
Hydra with developing bud and ovary.

adhere to a substratum and also to secrete a gas bubble for floating. In the center of the disc there may be an excretory pore. The mouth, located on a conical elevation called the **hypostome,** is encircled by 6 to 10 hollow tentacles that, like the body, can greatly extend when the animal is hungry.

The mouth opens into the **gastrovascular cavity** (also called a coelenteron) which communicates with cavities in the tentacles. In some individuals **buds** may project from the sides, each with a mouth and tentacles like the parent. Testes or ovaries, when present, appear as rounded projections on the surface of the body (Figure 13-6).

Body Wall   The body wall surrounding the gastrovascular cavity consists of an outer **epidermis** (ectodermal) and an inner **gastrodermis** (endodermal) with **mesoglea** between them (Figure 13-3).

Epidermis.   The epidermal layer contains epitheliomuscular, interstitial, gland, cnidocyte, and sensory and nerve cells.

Epitheliomuscular cells make up most of the epidermis and serve both for covering and for muscular contraction (Figure 13-7). The bases of most of these cells are extended parallel to the tentacle or body axis and contain myofibrils, thus forming a layer of longitudinal muscle next to the mesoglea. Contraction of these fibrils shortens the body or tentacles.

Interstitial cells are undifferentiated stem cells found among the bases of the epitheliomuscular cells. Differentiation of interstitial cells produces cnidoblasts, sex cells, buds, nerve cells, and others, but generally not epitheliomuscular cells (which reproduce themselves).

Gland cells are tall cells located around the basal disc and mouth, that secrete an adhesive substance for attachment and sometimes a gas bubble for floating.

Reduced medusae (gonophores)

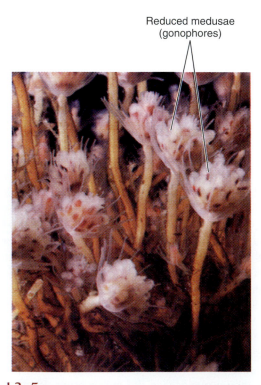

Figure 13-5
In some hydroids, such as this *Tubularia crocea,* medusae are reduced to gonadal tissue and do not detach. These reduced medusae are known as gonophores.

Over 230 years ago, Abraham Trembley was astonished to discover that isolated sections of the stalk of hydra could regenerate and each become a complete animal. Since then, over 2000 investigations of hydra have been published, and the organism has become a classic model for morphological differentiation. The mechanisms governing morphogenesis have great practical importance, and the simplicity of hydra lends itself to these investigations. Substances controlling development (morphogens), such as those determining which end of a cut stalk will develop a mouth and tentacles, have been discovered, and they may be present in the cells in extremely low concentrations (10^{-10} M).

Cnidocytes containing nematocysts occur throughout the epidermis. Hydras have three functional types of nematocysts: those that penetrate prey and inject poison (penetrants, Figure 13-3), those that recoil and entangle prey (volvents), and those that secrete an adhesive substance used in locomotion and attachment (glutinants).

Sensory cells are scattered among the other epidermal cells, especially near the mouth and tentacles and on the basal disc. The free end of each sensory cell bears a flagellum, which is the sensory receptor for chemical and tactile stimuli. The other end branches into fine processes that synapse with nerve cells.

Nerve cells of the epidermis are generally multipolar (have many processes), although in more highly organized cnidarians the cells may be bipolar (with two processes). Their processes (axons) form synapses with sensory cells and other nerve cells and junctions with epitheliomuscular cells and cnidocytes. There are both one-way (morphologically asymmetrical) and two-way synapses with other nerve cells.

Gastrodermis.   The gastrodermis, a layer of cells lining the gastrovascular cavity, contains chiefly large, ciliated, columnar epithelial cells with irregular flat bases. The cells of the gastrodermis include nutritive-muscular, interstitial, and gland cells.

Nutritive-muscular cells are usually tall columnar cells and have laterally extended bases containing myofibrils. The myofibrils run at right angles to the body or tentacle axis and so form a circular muscle layer. However, this muscle layer in hydras is

very weak, and longitudinal extension of the body and tentacles is brought about mostly by increasing the volume of water in the gastrovascular cavity. Water is brought in through the mouth by beating of cilia on the nutritive muscular cells. Thus, water in the gastrovascular cavity serves as a **hydrostatic skeleton.** The two cilia on the free end of each cell also serve to circulate food and fluids in the digestive cavity. The cells often contain large numbers of food vacuoles. Gastrodermal cells in green hydras (*Chlorohydra*) (Gr. *chloros*, green, + *hydra*, a mythical nine-headed monster slain by Hercules) bear green algae (zoochlorellae, phylum Chlorophyta), which give the hydras their color. This existence is probably a case of symbiotic mutualism, since the algae use the respiratory carbon dioxide from the hydra to form organic compounds useful to the host. Algae receive shelter and probably other physiological requirements in return.

Interstitial cells are scattered among the bases of the nutritive cells. They transform into other types of cells when the need arises.

Gland cells in the hypostome and in the column secrete digestive enzymes. Mucous glands surrounding the mouth aid in ingestion.

Nematocysts are not found in the gastrodermis because cnidocytes are lacking in this layer.

Mesoglea.   The mesoglea lies between the epidermis and gastrodermis and is attached to both layers. It is gelatinous, or jellylike, and both epidermal and gastrodermal cells send processes into it. It is a continuous layer that extends over both body and tentacles, thickest in the stalk portion and thinnest in the tentacles. This arrangement allows the pedal region to withstand great mechanical strain and gives the tentacles more flexibility. The mesoglea helps to support the body and acts as a type of elastic skeleton.

Locomotion   Unlike colonial polyps, which are permanently attached, hydras can move about freely by gliding on a basal disc, aided by mucous secretions. Using an "inchworm" movement, they can bend over and attach tentacles to the substratum. They may even turn end over end or detach themselves and, by forming a gas bubble on the basal disc, float to the surface.

Feeding and Digestion   Hydras feed on a variety of small crustaceans, insect larvae, and annelid worms. A hydra awaits its prey with tentacles extended (Figure 13-8). The food organism that brushes against its tentacles may find itself harpooned by scores of nematocysts that render it helpless, even though it may be larger than the hydra. The tentacles move toward the mouth, which slowly widens. Well moistened with mucous secretions, the mouth glides over and around the prey, totally engulfing it.

The activator that actually causes the mouth to open is the reduced form of glutathione, which is found to some extent in all living cells. Glutathione escapes from the prey through wounds made by nematocysts, but only animals releasing enough of the chemical to activate a feeding response are eaten by a hydra. This mechanism explains how a hydra distinguishes between *Daphnia*, which it relishes, and some other forms that it refuses. If we place glutathione in water containing hydras,

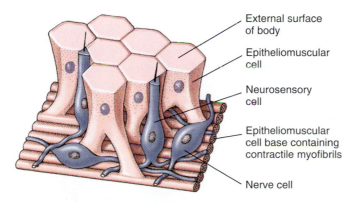

External surface
of body

Epitheliomuscular
cell

Neurosensory
cell

Epitheliomuscular
cell base containing
contractile myofibrils

Nerve cell

Figure 13-7
Epitheliomuscular and nerve cells in hydra.

Figure 13-8

Hydra catches an unwary water flea with the nematocysts of its tentacles. This hydra already contains one water flea eaten previously.

each hydra will go through the motions of feeding even though no prey is present.

Inside the gastrovascular cavity, gland cells discharge enzymes on the food. Digestion starts in the gastrovascular cavity (extracellular digestion), but many food particles are drawn by pseudopodia into nutritive-muscular cells of the gastrodermis, where intracellular digestion occurs. Ameboid cells may carry undigested particles to the gastrovascular cavity, where they are eventually expelled with other indigestible matter.

Reproduction   Hydras reproduce sexually and asexually. In asexual reproduction, buds appear as outpocketings of the body wall and develop into young hydras that eventually detach from the parent. Most species are dioecious. Temporary gonads (Figure 13-6) usually appear in autumn, stimulated by lower temperatures and perhaps also by reduced aeration of stagnant waters. Eggs in the ovary usually mature one at a time and are fertilized by sperm shed into the water.

Zygotes undergo holoblastic cleavage to form a hollow blastula. The inner part of the blastula delaminates to form the endoderm (gastrodermis), and the mesoglea is laid down between ectoderm and endoderm. A cyst forms around the embryo before it breaks loose from the parent, enabling it to survive winter. Young hydras hatch in spring when weather is favorable.

Hydroid Colonies

Far more representative of class Hydrozoa than hydras are hydroids that have a medusa stage in their life cycle. We often use *Obelia* in laboratory exercises to illustrate the hydroid type (Figure 13-9).

A typical hydroid has a base, a stalk, and one or more terminal zooids. The base by which colonial hydroids attach to the substratum is a rootlike stolon, or **hydrorhiza,** which gives rise to one or more stalks called **hydrocauli.** The living cellular part of the hydrocaulus is a tubular **coenosarc,** composed of the three typical cnidarian layers surrounding the coelenteron (gastrovascular cavity). The protective covering of the hydrocaulus is

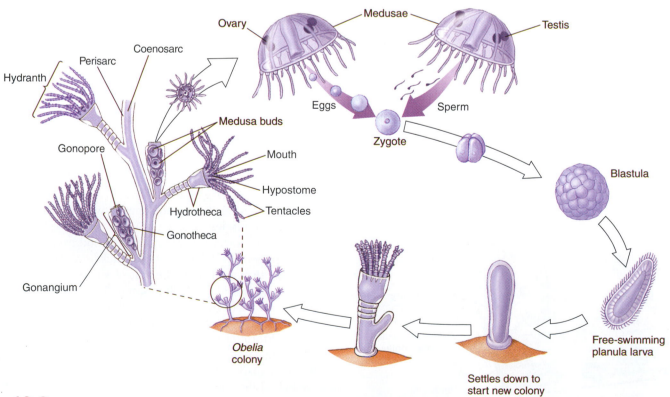

Figure 13-9

Life cycle of *Obelia,* showing alternation of polyp (asexual) and medusa (sexual) stages. *Obelia* is a thecate hydroid, its polyps as well as its stems being protected by continuations of the nonliving covering.

a nonliving chitinous sheath, or **perisarc.** Attached to the hydro-caulus are individual polyp animals, or zooids. Most zooids are feeding polyps called **hydranths,** or **gastrozooids.** They may be tubular, bottle shaped, or vaselike, but all have a terminal mouth and a circle of tentacles. In **thecate** forms, such as *Obelia,* the perisarc continues as a protective cup around the polyp into which it can withdraw for protection (Figure 13-9). In others the polyp is **(athecate)** naked (Figure 13-10). In some forms the perisarc is an inconspicuous, thin film.

Hydranths, much like hydras, capture and ingest prey, such as tiny crustaceans, worms, and larvae, thus providing nutrition for the entire colony. After partial extracellular digestion in a hydranth, the digestive broth passes along the common gastrovascular cavity where it is taken up by gastrodermal cells, and intracellular digestion occurs.

Circulation within the gastrovascular cavity is a function of the ciliated gastrodermis but is also aided by rhythmical contractions and pulsations of the body, which occur in hydroids.

Just as hydras reproduce asexually by budding, colonial hydroids bud off new individuals, thus increasing the size of the colony. New feeding polyps arise by budding, and medusa buds also arise on the colony. In *Obelia* these medusae bud from a reproductive polyp called a **gonangium.** Young medusae leave the colony as free-swimming individuals that mature and produce gametes (eggs and sperm) (Figure 13-9). In some species medusae remain attached to the colony and shed their gametes there. In other species medusae never develop and gametes are shed by male and female gonophores (Figure 13-10). Embryonation of the zygote results in a ciliated planula larva that swims about for a time. Then it settles down to a substratum to develop into a minute polyp that gives rise, by asexual budding, to the hydroid colony, thus completing the life cycle.

Hydroid medusae are usually smaller than scyphozoan medusae, ranging from 2 to 3 mm to several centimeters in diameter (Figure 13-11). The margin of the bell projects inward as a shelflike **velum,** which partly closes the open side of the bell and is used in swimming (Figure 13-12). Muscular pulsations that alternately fill and empty the bell propel the animal forward, aboral side first, with a weak "jet propulsion." Tentacles attached to the bell margin are rich in nematocysts.

The mouth opening at the end of a suspended **manubrium** leads to a stomach and four radial canals that connect with a ring canal around the margin. This ring canal connects with the hollow tentacles. Thus the gastrovascular cavity is continuous from mouth to tentacles, and gastrodermis lines the entire system. Nutrition is similar to that of hydranths.

The nerve net is usually concentrated into two nerve rings at the base of the velum. The bell margin has a liberal supply of sensory cells. It usually also bears two kinds of specialized sense organs: **statocysts,** which are small organs of equilibrium (Figure 13-12B), and **ocelli,** which are light-sensitive organs.

A

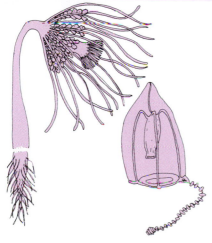

B

Figure 13-10

Athecate hydroids. **A,** *Ralpharia gorgoniae,* a large (up to 2.5 cm) solitary polyp with naked hydranths and gonophores. There are no free medusae. **B,** *Corymorpha* is a solitary hydroid that produces free-swimming medusae, each with a single trailing tentacle.

Figure 13-11

Bell medusa, *Polyorchis penicillatus,* medusa stage of an unknown attached polyp.

Figure 13-12

Structure of *Gonionemus*.
A, Medusa with typical tetramerous arrangement.
B, Cutaway view showing morphology. **C,** Portion of a tentacle with its adhesive pad and ridges of nematocysts. **D,** Tiny polyp, or hydroid stage, that develops from the planula larva. It can produce more polyps by budding (frustules) or produce medusa buds.

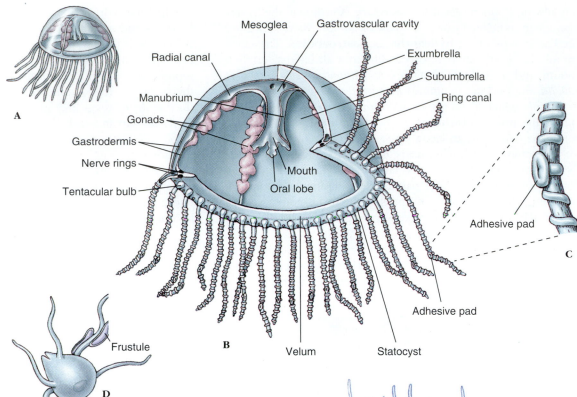

Freshwater Medusae

The freshwater medusa *Craspedacusta sowberii* (Figure 13-13) (order Hydroida) may have evolved from marine ancestors in the Yangtze River of China. Probably introduced with shipments of aquatic plants, this interesting form has now been found in many parts of Europe, throughout the United States, and in parts of Canada. Medusae may attain a diameter of 20 mm.

The polyp phase of this animal is tiny (2 mm) and has a very simple form with no perisarc and no tentacles. It occurs in colonies of a few polyps. For a long time its relation to the medusa was not recognized, and thus the polyp was given a name of its own, *Microhydra ryderi*. On the basis of its relationship to the jellyfish and the law of priority, both polyp and medusa should be called *Craspedacusta* (N.L. *craspedon,* velum, + Gr. *kystis,* bladder).

The polyp has three methods of asexual reproduction, as shown in Figure 13-13.

Other Hydrozoans

Members of orders Siphonophora and Chondrophora are among the most specialized Hydrozoa. They form polymorphic swimming or floating colonies containing several types of modified medusae and polyps.

Physalia (Gr. *physallis,* bladder), the Portuguese man-of-war (Figure 13-14), is one such colony with a rainbow-hued float of blues and pinks that carries it along the surface waters of tropical seas. Many are blown to shore on the eastern coast of the United States. The long, graceful tentacles, actually zooids, are laden with nematocysts and are capable of inflicting painful

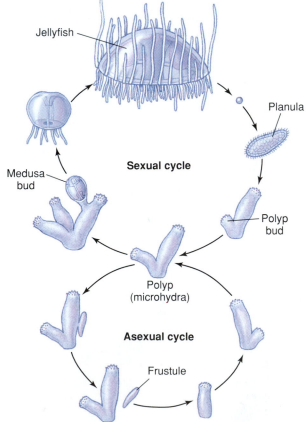

Figure 13-13

Life cycle of *Craspedacusta,* a freshwater hydrozoan. The polyp has three methods of asexual reproduction: by budding off new individuals, which may remain attached to the parent (colony formation); by constricting off nonciliated planula-like larvae (frustules), which can move around and give rise to new polyps; and by producing medusa buds, which develop into sexual jellyfish.

Figure 13-14

A Portuguese man-of-war colony, *Physalia physalis* (order Siphonophora, class Hydrozoa). Colonies often drift onto southern ocean beaches, where a hazard to bathers. Each colony of medusa and polyp types is integrated to act as one individual. As many as 1000 zooids may be found in one colony. The nematocysts secrete a powerful neurotoxin.

A

B

Figure 13-15

These hydrozoans form calcareous skeletons that resemble true coral. **A,** *Stylaster roseus* (order Stylasterina) occurs commonly in caves and crevices in coral reefs. These fragile colonies branch in only a single plane and may be white, pink, purple, red, or red with white tips. **B,** Species of *Millepora* (order Milleporina) form branching or platelike colonies and often grow over the horny skeleton of gorgonians (see p. 266), as is shown here. They have a generous supply of powerful nematocysts that produce a burning sensation on human skin, justly earning the common name fire coral.

stings. The float, called a **pneumatophore,** is believed to have expanded from the original larval polyp. It contains a sac arising from the body wall and is filled with a gas similar to air. The float acts as a type of nurse-carrier for future generations of individuals that bud from it and hang suspended in the water. Some siphonophores, such as *Stephalia* and *Nectalia,* possess swimming bells as well as a float.

> An interesting mutualistic relationship exists between *Physalia* and a small fish called *Nomeus* (Gr. herdsman) that swims among the tentacles with perfect safety. Why the fish is not stung to death by its host's nematocysts is unclear, but like the anemone fish to be discussed later, *Nomeus* is probably protected by a skin mucus that does not stimulate nematocyst discharge.

There are several types of polyp individuals. Gastrozooids are feeding polyps with a single long tentacle arising from the base of each. Some of these long, stinging tentacles become separated from the feeding polyp and are called **dactylozooids,** or fishing tentacles. These tentacles sting prey and lift it to the lips of feeding polyps. Among the modified medusoid individuals are the **gonophores,** which are little more than sacs containing either ovaries or testes.

Other hydrozoans secrete massive calcareous skeletons that resemble true corals (Figure 13-15). They are sometimes called **hydrocorals.**

Class Scyphozoa

Class Scyphozoa (si-fo-zo'a) (Gr. *skyphos,* cup) includes most of the larger jellyfishes, or "cup animals." A few, such as *Cyanea* (Gr. *kyanos,* dark-blue substance), may attain a bell diameter

exceeding 2 m and tentacles 60 to 70 m long (Figure 13-16). Most scyphozoans, however, range from 2 to 40 cm in diameter. Most are found floating in the open sea, some even at depths of 3000 m, but one unusual order is sessile and attaches by a stalk to seaweeds and other objects on the sea bottom (Figure 13-17). Their coloring may range from colorless to striking orange and pink hues.

Bells of different species vary in depth from a shallow saucer shape to a deep helmet or goblet shape. The jelly (mesoglea) layer is unusually thick, giving the bell a fairly firm

Figure 13-16

Giant jellyfish, *Cyanea capillata* (order Semaeostomeae, class Scyphozoa). A North Atlantic species of *Cyanea* reaches a bell diameter exceeding 2 m. It is known as the "sea blubber" by fishermen.

Figure 13-17

Thaumatoscyphus hexaradiatus (order Stauromedusae, class Scyphozoa). Members of this order are unusual scyphozoans in that the medusae are sessile and attached to seaweed or other objects.

consistency. Despite the bell's firmness the jelly is 95% to 96% water. Unlike hydromedusae, this layer in scyphomedusae also contains ameboid cells and fibers. Movement is by rhythmical pulsations of the umbrella. There is no velum as in hydromedusae. Tentacles may be numerous or few, and they may be short as in *Aurelia* (L. *aurum,* gold) or long as in *Cyanea. Aurelia aurita* (Figures 13-18 and 13-19) is a familiar species 7 to 10 cm in diameter, commonly found in waters off both the east and west coasts of the United States, and widely used for study.

The margin of the umbrella is scalloped, usually with each indentation bearing a pair of **lappets,** and between them is a sense organ called a **rhopalium** (tentaculocyst). *Aurelia* has

eight such notches. Some scyphozoans have 4, others 16. Each rhopalium is club shaped and contains a hollow statocyst for equilibrium and one or two pits lined with sensory epithelium. In some species the rhopalia also bear ocelli.

The mouth is centered on the subumbrella side. The manubrium usually forms four frilly **oral arms** that are used in capturing and ingesting prey.

Tentacles, manubrium, and often the entire body surface are well supplied with nematocysts that can deliver painful stings. However, the primary function of scyphozoan nematocysts is not to attack humans but to paralyze prey animals, which are conveyed to the mouth lobes with the help of other tentacles or by bending of the umbrella margin.

Aurelia, which has comparatively short tentacles, feeds on small planktonic animals. These are caught in mucus of the umbrella surface, are carried to "food pockets" on the umbrella margin by cilia, and are picked up from the pockets by the oral lobes whose cilia carry the food to the gastrovascular cavity. Cilia in the gastrodermis layer keep a current of water moving to bring food and oxygen into the stomach and to expel wastes.

Cassiopeia (L. mythical queen of Ethiopia), the "upside-down jellyfish" common to Florida waters, and *Rhizostoma* (Gr. *rhiza,* root, + *stoma,* mouth), which can be found in colder temperatures, belong to a group differing from that of *Aurelia* both in their lack of tentacles on the umbrella margin and in the structure of the oral arms. During development, edges of the oral lobes fold over and fuse, forming canals **(arms** or **brachial canals)** that become highly branched. These canals open to the surface at frequent intervals by pores called "mouths"; the original mouth is obliterated in the fusion of the oral lobes. Planktonic organisms caught in the mucus of the frilly oral arms are transported by cilia to the mouths and then up the brachial canals to the gastric cavity. In contrast to the usual swimming habit of medusae, *Cassiopeia* is usually found lying on its "back" in shallow lagoons. Its umbrella margin contracts about 20 times a minute, creating water currents to bring plankton into contact with the mucus and nematocysts of its oral lobes. Its tissues are abundantly supplied with symbiotic dinoflagellates (p. 226) **(zooxanthellae).** As they lie sunning themselves in shallow water, *Cassiopeia* are thus reminiscent of large flowers in more ways than one.

Internally, extending out from the stomach of scyphozoans are four **gastric pouches** in which gastrodermis extends down in little tentacle-like projections called **gastric filaments.** These filaments are covered with nematocysts to quiet further any prey that may still be struggling. Gastric filaments are lacking in hydromedusae. A complex system of **radial canals** branches out from the pouches to a **ring canal** in the margin and forms a part of the gastrovascular cavity.

The **nervous system** in scyphozoans is a nerve net, with a subumbrella net that controls bell pulsations and another, more diffuse net that controls local reactions such as feeding.

Sexes are separate, with gonads located in the gastric pouches. Fertilization is internal, with sperm being carried by ciliary currents into the gastric pouch of females. Zygotes may develop in seawater or may be brooded in folds of the oral arms. The ciliated planula larva becomes attached and develops into a **scyphistoma,** a hydralike form (Figure 13-18). By a process of

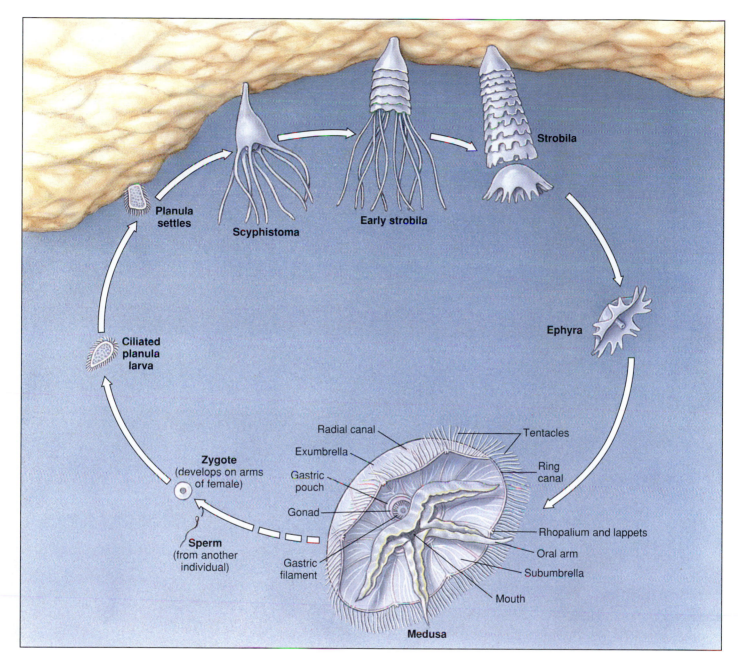

Figure 13-18

Life cycle of *Aurelia,* a marine scyphozoan medusa.

strobilation the scyphistoma of *Aurelia* forms a series of saucer-like buds, **ephyrae,** and is now called a **strobila** (Figure 13-18). When ephyrae break loose, they grow into mature jellyfish.

Class Cubozoa

The Cubozoa until recently were considered an order (Cubomedusae) of Scyphozoa. The medusoid is the predominant form (Figure 13-20); the polypoid is inconspicuous and in most cases unknown. Some cubozoan medusae may range up to 25 cm tall, but most are about 2 to 3 cm. In transverse section the bells are almost square. A tentacle or group of tentacles is found at each corner of the square at the umbrella margin. The base of each

tentacle is differentiated into a flattened, tough blade called a **pedalium** (Figure 13-20). Rhopalia are present. The umbrella margin is not scalloped, and the subumbrella edge turns inward to form a **velarium.** The velarium functions as a velum does in hydrozoan medusae, increasing swimming efficiency, but it differs structurally. Cubomedusae are strong swimmers and voracious predators, feeding mostly on fish. Stings of some species can be fatal to humans.

The complete life cycle is known for only one species, *Tripedalia cystophora* (L. *tri,* three, + Gr. *pedalion,* rudder). The polyp is tiny (1 mm tall), solitary, and sessile. New polyps bud laterally, detach and creep away. Polyps do not produce ephyrae but metamorphose directly into medusae.

Figure 13-19
Moon jellyfish *Aurelia aurita* (class Scyphozoa) is cosmopolitan in distribution. It feeds on planktonic organisms caught in mucus on its umbrella.

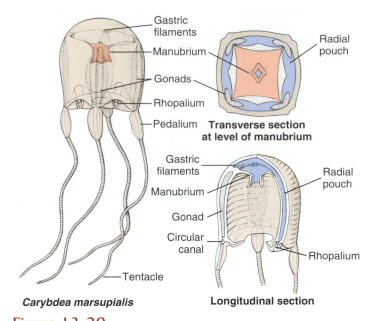

Carybdea marsupialis

Figure 13-20
Carybdea, a cubozoan medusa.

Chironex fleckeri (Gr. *cheir,* hand, + *nexis,* swimming) is a large cubomedusa known as the sea wasp. Its stings are quite dangerous and sometimes fatal. Most fatal stings have been reported from tropical Australian waters, usually following quite massive stings. Witnesses have described victims as being covered with "yards and yards of sticky wet string." Stings are very painful, and death, if it is to occur, ensues within a matter of minutes. If death does not occur within 20 minutes after stinging, complete recovery is likely.

Class Anthozoa

Anthozoans, or "flower animals," are polyps with a flowerlike appearance (Figure 13-21). There is no medusa stage. Anthozoa are all marine and are found in both deep and shallow water and in polar seas as well as tropical seas. They vary greatly in size and may be solitary or colonial. Many forms are supported by skeletons.

The class has three subclasses: **Hexacorallia** (or **Zoantharia**), containing sea anemones, hard corals, and others; **Ceriantipatharia,** containing only tube anemones and thorny corals; and **Octocorallia** (or **Alcyonaria**), containing soft and horny corals, such as sea fans, sea pens, sea pansies, and others. Zoantharians and ceriantipatharians have a **hexamerous** plan (of six or multiples of six) or polymerous symmetry and have simple tubular tentacles arranged in one or more circlets on the oral disc. Octocorallians are **octomerous** (built on a plan of eight) and always have eight pinnate (featherlike) tentacles arranged around the margin of the oral disc (Figure 13-22).

The gastrovascular cavity is large and partitioned by septa, or mesenteries, that are inward extensions of the body wall. Where one septum extends into the gastrovascular cavity from

Figure 13-21
Sea anemones are the familiar and colorful "flower animals" of tide pools, rocks, and pilings of the intertidal zone. Most, however, are subtidal, their beauty seldom revealed to human eyes. These are rose anemones. *Tealia piscivora* (subclass Hexacorallia, class Anthozoa).

the body wall, another extends from the diametrically opposite side; thus, they are said to be **coupled.** In Hexacorallia, the septa are not only coupled, they are also **paired** (Figure 13-23). The muscular arrangement varies among different groups, but usually features circular muscles in the body wall and longitudinal and transverse muscles in the septa.

The mesoglea is a mesenchyme containing ameboid cells. A general tendency toward biradial symmetry in the septal arrangement occurs in the shape of the mouth and pharynx. There are no special organs for respiration or excretion.

Figure 13-22

A, Orange sea pen *Ptilosarcus gurneyi* (order Pennatulacea, class Anthozoa). Sea pens are colonial forms that inhabit soft bottoms. The base of the fleshy body of the primary polyp is buried in the bottom. It gives rise to numerous secondary, branching polyps. **B,** Close-up of a gorgonian. The pinnate tentacles characteristic of subclass Octocorallia are apparent.

Sea Anemones

Sea anemone (order Actiniaria) polyps are larger and heavier than hydrozoan polyps (Figure 13-21). Most range from 5 mm or less to 100 mm in diameter, and from 5 mm to 200 mm long, but some grow much larger. Some sea anemones are quite colorful. Anemones are found in coastal areas all over the world, especially in warmer waters. They attach by means of their pedal

discs to shells, rocks, timber, or whatever submerged substrata they can find. Some burrow in mud or sand.

Sea anemones are cylindrical in form with a crown of tentacles arranged in one or more circles around the mouth of the flat **oral disc** (Figure 13-23). The slit-shaped mouth leads into a **pharynx.** At one or both ends of the mouth is a ciliated groove called a **siphonoglyph,** which extends into the pharynx. The siphonoglyph creates a water current directed into the pharynx. Cilia elsewhere on the pharynx direct water outward. Currents thus created carry in oxygen and remove wastes. They also help maintain an internal fluid pressure, providing a hydrostatic skeleton that serves in lieu of a true skeleton as a support for opposing muscles.

The pharynx leads into a large **gastrovascular cavity** that is divided into six radial chambers by means of six pairs of **primary (complete) septa,** or **mesenteries,** extending vertically from the body wall to the pharynx (Figure 13-23). Openings between chambers (septal perforations) in the upper part of the pharyngeal region help in water circulation. Smaller **(incomplete)** septa partially subdivide the large chambers and provide a means of increasing the surface area of the gastrovascular cavity. The free edge of each incomplete septum forms a type of sinuous cord called a **septal filament,** which contains nematocysts and gland cells for digestion. In some anemones (such as *Metridium*) the lower ends of the septal filaments are prolonged into **acontia threads,** also provided with nematocysts and gland cells, that can be protruded through the mouth or through pores in the body wall to help overcome prey or provide defense. The pores also aid in rapid discharge of water from the body when the animal is endangered and contracts to a small size.

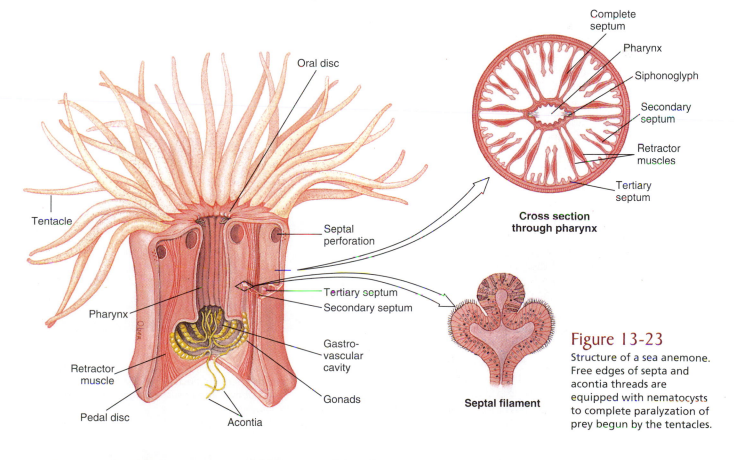

Figure 13-23

Structure of a sea anemone. Free edges of septa and acontia threads are equipped with nematocysts to complete paralyzation of prey begun by the tentacles.

A   B

Figure 13-24

A sea anemone that swims. When attacked by a predatory sea star *Dermasterias*, the anemone *Stomphia didemon* (subclass Hexacorallia, class Anthozoa) detaches from the bottom and rolls or swims spasmodically to a safer location.

Sea anemones are carnivorous, feeding on fish or almost any live (and sometimes dead) animals of suitable size. Some species live on minute forms caught by ciliary currents.

Feeding behavior in many zoantharians is under chemical control. Some respond to reduced glutathione. In certain others two compounds are involved: asparagine, the feeding activator, causes a bending of tentacles toward the mouth; then reduced glutathione induces swallowing of food.

Muscles are well developed in sea anemones, but the arrangement is quite different from that in Hydrozoa. Longitudinal fibers of the epidermis occur only in the tentacles and oral disc of most species. The strong longitudinal muscles of the column are gastrodermal and are located in the septa (Figure 13-23). Gastrodermal circular muscles in the column are well developed.

Most anemones can glide slowly along the substrate on their pedal discs. They can expand and stretch their tentacles in search of small vertebrates and invertebrates, which they overpower with tentacles and nematocysts and carry to their mouth. When disturbed, sea anemones contract and withdraw their tentacles and oral discs. Some anemones are able to swim to a limited extent by rhythmical bending movements, which may be a mechanism for escape from enemies such as sea stars and nudibranchs. *Stomphia*, for example, at the touch of a predatory sea star, will detach its pedal disc and make creeping or swimming movements to escape (Figure 13-24). This escape reaction is elicited not only by the touch of the star but also by exposure to drippings exuded by the star or to crude extracts made from its tissues. The sea drippings contain steroid saponins that are toxic and irritating to most invertebrates. Extracts from nudibranchs also can provoke this reaction in some sea anemones.

Anemones form some interesting mutualistic relationships with other organisms. Many species harbor symbiotic dinoflagellates (zooxanthellae) within their tissues, similar to the hard coral–zooxanthellae association (p. 265), and the anemones profit from the product of algal photosynthesis. Some anemones

habitually attach to the shells occupied by certain hermit crabs. The hermit encourages the relationship and, finding its favorite species, which it recognizes by touch, it massages the anemone until it detaches. The hermit crab holds the anemone against its own shell until the anemone is firmly attached. The crab derives some protection against predators by the anemone. The anemone gets free transportation and particles of food dropped by the hermit crab.

Certain damselfishes (anemone fishes) (family Pomacentridae) form associations with large anemones, especially in tropical Indo-Pacific waters (Figure 13-25). An unknown property of the skin mucus of the fish causes the anemone's nematocysts not to discharge, but if some other fish is so unfortunate as to brush the anemone's tentacles, it is likely to become a meal. The anemone obviously provides shelter for the anemone fish, and the fish may help ventilate the anemone by its movements, keep

Figure 13-25

Orangefin anemone fish (*Amphiprion chrysopterus*) nestles in the tentacles of its sea anemone host. Anemone fishes do not elicit stings from their hosts but may lure unsuspecting other fish to become meals for the anemone.

A   **B**   **C**

Figure 13-26

A, Cup coral *Tubastrea* sp. Its polyps form clumps resembling groups of sea anemones. Although often found on coral reefs, *Tubastrea* is not a reef-building coral (ahermatypic) and has no symbiotic zooxanthellae in its tissues. **B,** Polyps of *Montastrea cavernosa* are tightly withdrawn during daytime but open to feed at night, as in **C** (subclass Hexacorallia).

the anemone free of sediment, and even lure an unwary victim to seek the same shelter.

Sexes are separate in some sea anemones, and some are hermaphroditic. Monoecious species are **protandrous** (produce sperm first, then eggs). Gonads are arranged on the margins of the septa, and fertilization takes place externally or in the gastrovascular cavity. The zygote develops into a ciliated larva. Asexual reproduction commonly occurs by **pedal laceration** or by longitudinal fission, occasionally by transverse fission or by budding. In pedal laceration, small pieces of the pedal disc break off as the animal moves, and each of these regenerates a small anemone.

Hexacorallian Corals

Hexacorallian corals belong to the order Scleractinia, sometimes known as true or stony corals. Stony corals might be described as miniature sea anemones that live in calcareous cups they themselves have secreted (Figures 13-26 and 13-27). Like that of anemones, a coral polyp's gastrovascular cavity is subdivided by septa arranged in multiples of six (hexamerous) and its hollow tentacles surround the mouth, but there is no siphonoglyph.

Instead of a pedal disc, the epidermis at the base of the column secretes a limy skeletal cup, including sclerosepta, which project up into the polyp between its true septa (Figure 13-27). Living polyps can retract into the safety of their cup when not feeding. Since the skeleton is secreted below the living tissue rather than within it, the calcareous material is an exoskeleton. In many colonial corals, the skeleton may become massive, building up over many years, with the living coral forming a sheet of tissue over the surface (Figure 13-28). The gastrovascular cavities of the polyps are all connected through this sheet of tissue.

Three other small orders of Zoantharia are recognized.

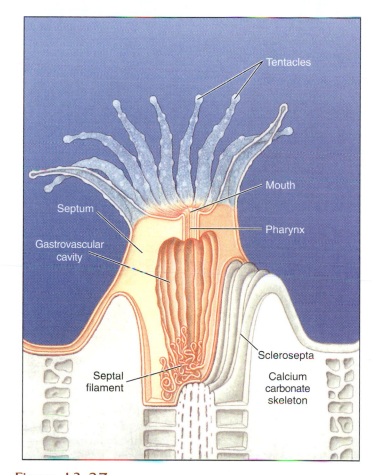

Figure 13-27

Polyp of a hexacorallian coral (order Scleractinia) showing calcareous cup (exoskeleton), gastrovascular cavity, sclerosepta, septa, and septal filaments.

Figure 13-28
Boulder star coral, *Montastrea annularis,* (subclass Hexacorallia, class Anthoza). Colonies can grow up to 10 feet (3 m) high.

Tube Anemones and Thorny Corals

Members of subclass Ceriantipatharia have unpaired septa. Tube anemones (order Ceriantharia) (Figure 13-29) are solitary and live buried to the level of the oral disc in soft sediments. They occupy tubes constructed of secreted mucus and threads of nematocyst-like organelles, into which they can withdraw. Thorny or black corals (order Antipatharia) (Figure 13-30) are colonial and attached to a firm substratum. Their skeleton is of a horny material and has thorns. Both of these orders are small in numbers of species and are limited to warmer waters of the sea.

Octocorallian Corals

Octocorallia (Alcyonaria) have strict octomerous symmetry, with eight pinnate tentacles and eight unpaired, complete septa (Figure 13-22). They are all colonial, and gastrovascular cavities of the polyps communicate through a system of gastrodermal tubes called **solenia** (Figure 13-31). The solenia run through an exten-

Figure 13-29
A tube anemone (subclass Ceriantipatharia, order Ceriantharia) extends from its tube at night. Its oral disc bears long tentacles around the margin and short tentacles around the mouth.

sive mesoglea **(coenenchyme)** in most octocorallians, and the surface of the colony is covered by epidermis. The skeleton is secreted in the coenenchyme and consists of limy spicules, fused spicules, or a horny protein, often in combination. Thus the skeletal support of most alcyonarians is an endoskeleton. The variation in pattern among the species of octocorallians lends great variety to the form of the colonies: from soft corals such as *Dendronephthya* (Figure 13-32), with their spicules scattered through the coenenchyme, to the tough, axial supports of sea fans and other gorgonian corals (Figure 13-33), to the fused spicules of organ-pipe coral. *Renilla* (L. *ren,* kidney, + *illa,* suffix), the sea pansy, is a colony reminiscent of a pansy flower. Its polyps are embedded in the fleshy upper side, and a short stalk that supports the colony is embedded in the sea floor. *Ptilosarcus* (Gr. *ptilon,* feather, + *sarkos,* flesh), a sea pen, is a member of the same order and may reach a length of 50 cm (Figure 13-22).

The graceful beauty of octocorallians—in hues of yellow, red, orange, and purple—helps create the "submarine gardens" of the coral reefs.

Coral Reefs

Most students will have seen photographs or movies giving a glimpse of the vibrant color and life found on coral reefs, and some may have been fortunate enough to visit a reef. Coral reefs are among the most productive of all ecosystems, and they have a diversity of life forms rivaled only by tropical rain forests. They are large formations of calcium carbonate (limestone) in shallow tropical seas laid down by living organisms over thousands of years; living plants and animals are confined to the top layer of reefs where they add calcium carbonate to that deposited by their predecessors. The most important organisms that precipitate calcium carbonate from seawater to form reefs are the scleractinian, **hermatypic** (reef-building) **corals** (Figure 13-28) and **coralline algae.** Not only do coralline algae contribute to the total mass of calcium carbonate, but their precipitation of the substance helps to hold the reef together. Some octocorallians and hydrozoa (especially *Millepora* [L. *mille,* thousand, + *porus,*

Figure 13-30
A, Colony of *Antipathes,* a black or thorny coral (order Antipatharia, subclass Ceriantipatharia class Anthozoa). Most abundant in deep waters in the tropics, black corals secrete a tough, proteinaceous skeleton that can be worked into jewelry. **B,** The polyps of Antipatharia have six simple, nonretractile tentacles. The spiny processes in the skeleton are the origin of the common name thorny corals.

Tentacle   Spiny skeleton

Enlargement of single polyp

B

A

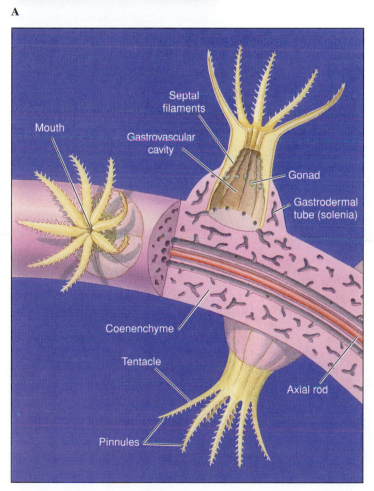

Figure 13-31
Polyps of an octocorallian coral. Note the eight pinnate tentacles, coenenchyme, and solenia. They have an endoskeleton of limy spicules often with a horny protein, which may be in the form of an axial rod.

Mouth
Septal filaments
Gastrovascular cavity
Gonad
Gastrodermal tube (solenia)
Coenenchyme
Tentacle
Axial rod
Pinnules

pore] spp., "fire coral," Figure 13-15B) contribute in some measure to the calcareous material, and an enormous variety of other organisms contributes small amounts. However, hermatypic (Gr. *herma,* support, mound, + *typos,* type) corals seem essential to formation of large reefs, since such reefs do not occur where these corals cannot live.

Hermatypic corals require warmth, light, and the salinity of undiluted seawater. These requirements limit coral reefs to shallow waters between 30 degrees north and 30 degrees south latitude and excludes them from areas with upwelling of cold water or areas near major river outflows with attendant low salinity and high turbidity. These corals require light because they have mutualistic dinoflagellates (zooxanthellae) living in their tissues. The microscopic zooxanthellae are very important to the corals; their photosynthesis and fixation of carbon dioxide furnish food molecules for their hosts, they recycle phosphorus and nitrogenous waste compounds that otherwise would be lost, and they enhance the ability of the coral to deposit calcium carbonate.

Because zooxanthellae are vital to hermatypic corals, and water absorbs light, hermatypic corals rarely live at depths greater than 100 feet (30 m). Interestingly, some deposits of coral reef limestone, particularly around Pacific islands and atolls, reach great thickness—even thousands of feet. Clearly, corals and other organisms could not have grown from the bottom in the abyssal blackness of deep sea and reached shallow water where light could penetrate. Charles Darwin was the first to realize that such reefs began their growth in *shallow* water around volcanic islands; then as the islands slowly sank beneath the sea, the growth of the reefs kept pace with the rate of sinking, thus accounting for the depth of the deposits.

Several types of reefs are commonly recognized. **Fringing reefs** are close to a landmass with either no lagoon or a narrow lagoon between reef and shore. A **barrier reef** (Figure 13-34) runs roughly parallel to shore and has a wider and deeper lagoon than does a fringing reef. **Atolls** are reefs that encircle a lagoon but not an island. These types of reefs typically slope rather steeply into deep water at their seaward edge. **Patch** or

Figure 13-32
A soft coral, *Dendronephthya* sp. (order Alcyonacea, subclass Octocorallia, class Anthozoa), on a Pacific coral reef. The showy hues of this soft coral vary from pink and yellow to bright red and contribute much color to Indo-Pacific reefs.

bank reefs occur some distance back from the steep, seaward slope in lagoons of barrier reefs or atolls. The so-called Great Barrier Reef, extending 2027 km long and up to 145 km from shore off the northeast coast of Australia, is actually a complex of reef types.

Fringing, barrier, and atoll reefs all have distinguishable zones that are characterized by different groups of corals and other animals. The side of the reef facing the sea is the **reef front** or **fore reef slope** (Figure 13-34). The reef front is more or less parallel to the shore and perpendicular to the predominant direction of wave travel. It slopes downward into deeper water, sometimes gently at first, then precipitously. Characteristic assemblages of scleractinian corals grow deep on the slope, high near the crest, and in intermediate zones. In shallow water or slightly emergent at the top of the reef front is a **reef crest.** The upper front and the crest bear the greatest force of waves and must absorb great energy during storms. Pieces of coral and other organisms are broken off at such times and thrown shoreward onto the **reef flat,** which slopes down into the lagoon. The reef flat thus receives a supply of calcareous material that is eventually broken down into coral sand. The sand is stabilized by growth of plants such as turtle grass and coralline algae and ultimately becomes cemented into the mass of the reef by precipitation of carbonates. A reef is not an unbroken wall facing the sea but is highly irregular, with grooves, caves, crevices, channels through from the flat to the front, and deep, cup-shaped holes ("blue holes"). Octocorallians tend to grow in these areas that are more protected from the full force of waves, as well as on the flat and the deeper areas of the fore reef slope. Many other kinds of organisms inhabit cryptic locations such as caves and crevices.

Enormous numbers of species and individuals of invertebrate groups and fishes populate the reef ecosystem. For example, there are 300 *common* species of fishes on Caribbean reefs and more than 1200 on the Great Barrier Reef complex of Australia. It is marvelous that such diversity and productivity can be maintained, since reefs are washed by nutrient-poor waves of the open ocean. Although relatively little nutrient enters the ecosystem, little is lost because the interacting organisms are so efficient in recycling. The corals even feed on feces of fish swimming over them!

A   B   C

Figure 13-33
Colonial gorgonian, or horny, corals (order Gorgonacea, subclass Octocorallia, class Anthozoa) are conspicuous components of reef faunas. These examples are from the western Pacific. **A,** Red gorgonian *Melithaea* sp. **B,** A sea fan, *Subergorgia mollis.* **C,** Red whip coral, *Ellisella* sp.

Classification of Phylum Cnidaria

Strong molecular and morphological evidence now indicates that members of the former phylum Myxozoa, commonly occurring fish parasites, are in fact highly derived cnidarians.* At this time, we cannot place them with confidence in the following classification; it is possible they are hydrozoans or a separate class.

Class Hydrozoa (hi-dro-zo′a) (Gr. *hydra*, water serpent, + *zōon*, animal). Solitary or colonial; asexual polyps and sexual medusae, although one type may be suppressed; hydranths with no mesenteries; medusae (when present) with a velum; both freshwater and marine. Examples: *Hydra, Obelia, Physalia, Tubularia*.

Class Scyphozoa (si-fo-zo′a) (Gr. *skyphos*, cup, + *zōon*, animal). Solitary; polyp stage reduced or absent; bellshaped medusae without velum; gelatinous mesoglea much enlarged; margin of bell or umbrella typically with eight notches that are provided with sense organs; all marine. Examples: *Aurelia, Cassiopeia, Rhizostoma*.

Class Cubozoa (ku′bo-zo′a) (Gr. *kybos*, a cube + *zōon*, animal). Solitary; polyp stage reduced; bell-shaped medusae square in cross section, with tentacle or group of tentacles hanging from a bladelike pedalium at each corner of the umbrella; margin of umbrella entire, without velum but with velarium; all marine. Examples: *Tripedalia, Carybdea, Chironex, Chiropsalmus*.

Class Anthozoa (an-tho-zo′a) (Gr. *anthos*, flower, + *zōon*, animal). All polyps; no medusae; solitary or colonial; gastrovascular cavity subdivided by at least eight mesenteries or septa bearing nematocysts; gonads endodermal; all marine.

    Subclass Hexacorallia (heks′a-ko-ral′e-a) (Gr. *hex*, six, + *korallion*, coral) **(Zoantharia)**. With simple unbranched tentacles; mesenteries in pairs; sea anemones, hard corals, and others. Examples: *Metridium, Anthopleura, Tealia, Astrangia, Acropora*.

    Subclass Ceriantipatharia (se-re-an-tip′a-tha′re-a) (N. L. combination of Ceriantharia and Antipatharia). With simple unbranched tentacles; mesenteries unpaired; tube anemones and black or thorny corals. Examples: *Cerianthus, Antipathes, Stichopathes*.

    Subclass Octocorallia (ok′to-ko-ral′e-a) (L. *octo*, + Gr. *korallion*, coral) **(Alcyonaria)**. With eight pinnate tentacles; eight complete, unpaired mesenteries; soft and horny corals. Examples: *Tubipora, Alcyonium, Gorgonia, Plexaura, Renilla*.

*Siddall, M. E., et al. 1995. J. Parasitol. 81:961–967.

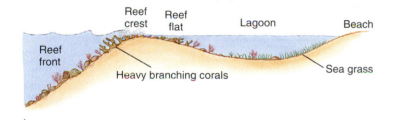

A

Figure 13-34

A, Profile of a barrier reef. **B,** Portion of an atoll from the air. Reef slope plunges into deep water at left (dark blue), lagoon at right.

B

Despite their great intrinsic and economic value, coral reefs in many areas are threatened by a variety of factors, mostly of human origin. These include enrichment with nutrients (from sewage and runoff of agricultural fertilizer from nearby land) and overfishing of herbivorous fishes, both of which contribute to overgrowth of multicellular algae. Agricultural pesticides, sediment from tilled fields and dredging, and oil spills contribute to reef degradation. When such environmental stresses do not kill corals directly, they may make the organisms more susceptible to the numerous coral diseases that have been observed in recent years. Coral reefs are apparently suffering from effects of global warming. When their surrounding water becomes too warm, corals expel their zooxanthellae (coral "bleaching") for reasons that are not yet clear. Instances of coral bleaching are becoming increasingly common around the world. Furthermore, higher atmospheric concentrations of carbon dioxide (from burning hydrocarbon fuels) tends to acidify ocean water, which makes precipitation of $CaCO_3$ by corals more difficult metabolically.

PHYLUM CTENOPHORA

Ctenophora (te-nof′o-ra) (Gr. *kteis, ktenos*, comb, + *phora*, pl. of bearing) is composed of fewer than 100 species. All are marine forms occurring in all seas but especially in warm waters. They take their name from the eight rows of comblike plates they bear for locomotion. Common names for ctenophores are "sea walnuts" and "comb jellies." Ctenophores, along with cnidarians, represent the only two phyla having primary radial symmetry, in contrast to other metazoans, which have primary bilateral symmetry.

    Ctenophores do not have nematocysts, except in one species (*Haeckelia rubra*, after Ernst Haeckel, nineteenth-century German zoologist) that carries nematocysts on certain regions of its tentacles but lacks colloblasts. These nematocysts are apparently appropriated from cnidarians on which it feeds.

Characteristics of Phylum Ctenophora

1. Symmetry **biradial;** arrangement of internal canals and position of the paired tentacles change the radial symmetry into a combination of the two (radial + bilateral)
2. Usually ellipsoidal or spherical in shape, with **radially arranged rows of comb plates for swimming**
3. Ectoderm, endoderm, and a mesoglea (ectomesoderm) with scattered cells and muscle fibers; may be considered **triploblastic**
4. Nematocysts absent but **adhesive cells (colloblasts)** present
5. Digestive system consisting of mouth, pharynx, stomach, a series of canals, and anal pores
6. Nervous system consisting of a subepidermal plexus concentrated around the mouth and beneath the comb plate rows; an **aboral sense organ** (statocyst)
7. No polymorphism or dimorphism
8. Reproduction monoecious; gonads (endodermal origin) on the walls of the digestive canals, which are under the rows of comb plates; mosaic cleavage; cydippid larva
9. Luminescence common

Comparison with Cnidaria

Ctenophores resemble cnidarians in these ways:
1. Form of radial symmetry
2. Aboral-oral axis around which the parts are arranged
3. Well-developed gelatinous ectomesoderm (collenchyme)
4. No coelomic cavity
5. Diffuse nerve plexus
6. Lack of organ systems

They differ from the cnidarians in these ways:
1. They do not form nematocysts
2. Development of distinct muscle cells from mesenchyme
3. Presence of comb plates and colloblasts
4. Mosaic, or determinate type of development
5. Presence of pharynx generally
6. No polymorphism or dimorphism
7. Never colonial
8. Presence of anal openings

Like cnidarians, ctenophores have not advanced beyond the tissue grade of organization. There are no definite organ systems in the strict meaning of the term.

Except for a few creeping and sessile forms, ctenophores are free-swimming. Although they are feeble swimmers and are more common in surface waters, ctenophores are sometimes found at considerable depths. They are often at the mercy of tides and strong currents, but they avoid storms by swimming into deeper water. In calm water they may rest vertically with little movement, but when moving they use their ciliated comb plates to propel themselves mouth-end forward. Highly modified forms such as *Cestum* (L. *cestus,* girdle) use sinuous body movements as well as their comb plates in locomotion.

The fragile, transparent bodies of ctenophores are easily seen at night when they emit light (luminesce).

Class Tentaculata

Representative Type: Pleurobrachia

Pleurobrachia (Gr. *pleuron,* side, + L. *brachia,* arms) is a representative of this group. Its transparent body is about 1.5 to 2 cm in diameter (Figure 13-35A). The oral pole bears the mouth opening, and the aboral pole has a sensory organ, the **statocyst.**

Comb Plates   On the surface are eight equally spaced bands called **comb rows,** which extend as meridians from the aboral pole and end before reaching the oral pole (Figure 13-36). Each band consists of transverse plates of long fused cilia called **comb plates** (Figure 13-36A). Ctenophores are propelled by beating of cilia on the comb plates. The beat in each row starts at the aboral end and proceeds successively along the combs to the oral end. All eight rows normally beat in unison. The animal is thus driven forward with the mouth in advance. The animal can swim backward by reversing the direction of the wave.

Tentacles   The two **tentacles** are long, solid and very extensible, and they can be retracted into a pair of **tentacle sheaths.** When completely extended, they may measure 15 cm in length. The surface of the tentacles bears **colloblasts,** or glue cells (Figure 13-36C), which secrete a sticky substance that is used for catching and holding small animals.

Body Wall   The cellular layers of ctenophores are generally similar to those of cnidarians. Between the epidermis and gastrodermis is a gelatinous **collenchyme** that fills most of the interior of the body and contains muscle fibers and ameboid cells. Although they are derived from ectodermal cells, muscle cells

A *Pleurobrachia*   B *Mnemiopsis*

Figure 13-35

A, Comb jelly *Pleurobrachia* sp. (order Cydippida, class Tentaculata). Its fragile beauty is especially evident at night when it luminesces from its comb rows. **B,** *Mnemiopsis* sp. (order Lobata, class Tentaculata).

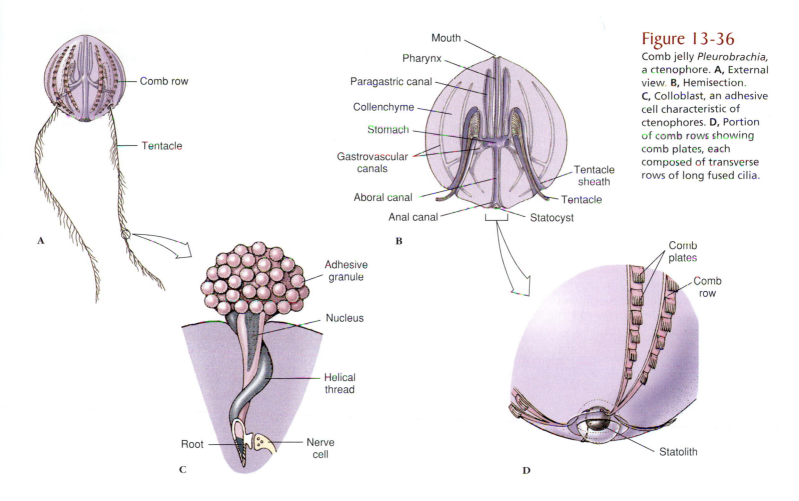

Figure 13-36
Comb jelly *Pleurobrachia,* a ctenophore. **A,** External view. **B,** Hemisection. **C,** Colloblast, an adhesive cell characteristic of ctenophores. **D,** Portion of comb rows showing comb plates, each composed of transverse rows of long fused cilia.

are distinct and are not contractile portions of epitheliomuscular cells (in contrast to Cnidaria).

Digestive System and Feeding   The **gastrovascular system** consists of a mouth, a pharynx, a stomach, and a system of gastrovascular canals that branch through the jelly to extend to the comb plates, tentacular sheaths, and elsewhere (Figure 13-36). There are two blind canals that terminate near the mouth, and an aboral canal that passes near the statocyst and then divides into two small **anal canals** through which undigested material is expelled.

Ctenophores prey on small planktonic organisms such as copepods. Glue cells on the tentacles stick to small prey and enable the tentacles to carry prey to the ctenophore's mouth. Digestion is both extracellular and intracellular.

Respiration and Excretion   Respiration and excretion occur through the body surface.

Nervous and Sensory Systems   Ctenophores have a nervous system similar to that of cnidarians. It features a subepidermal plexus, which is concentrated under each comb plate, but no central control as is found in more complex animals.

The sense organ at the aboral pole is a statocyst. Tufts of cilia support a calcareous statolith, with the whole being enclosed in a bell-like container. Alterations in the position of the animal change the pressure of the statolith on the tufts of cilia. The sense organ is also concerned in coordinating the beating of the comb rows but does not trigger their beat.

The epidermis of ctenophores bears abundant sensory cells, so the animals are sensitive to chemical and other stimuli. When a ctenophore contacts an unfavorable stimulus, it often reverses the beat of its comb plates and moves backward. Comb plates are very sensitive to touch, which often causes them to be withdrawn into the jelly.

Reproduction and Development   *Pleurobrachia,* like other ctenophores, is monoecious. Gonads are located on the lining of the gastrovascular canals under the comb plates. Fertilized eggs are discharged through the epidermis into the water.

Cleavage in ctenophores is determinate (mosaic), since the various parts of the animal that will be formed by each blastomere are determined early in embryogenesis. If one of the blastomeres is removed in the early stages, the resulting embryo will be deficient. This type of development differs from that of cnidarians, which is regulative (indeterminate). The free-swimming **cydippid larva** is superficially similar to the adult ctenophore and develops directly into an adult.

Some biologists have regarded the ctenophores and some more complex cnidarians (for example, some anthozoans) as triploblastic because the highly cellular nature of the mesoglea would constitute a mesoderm. However, others define mesoderm strictly as a layer derived from endoderm; thus both cnidarians and ctenophores would be diploblastic.

Other Ctenophores

Ctenophores are fragile and beautiful creatures. Their transparent bodies glisten like fine glass, brilliantly iridescent during the day and luminescent at night.

One of the most striking ctenophores is *Beroe* (L. a nymph), which may be more than 100 mm in length and 50 mm in breadth (Figure 13-37A). It is conical or thimble shaped and is flattened in the tentacular plane. The tentacular plane in *Beroe* is defined as where the tentacles would have been, because it has a large mouth but no tentacles. The animal is pink or rusty brown. Its body wall is covered with an extensive network of canals formed by union of the paragastric and meridional canals. Venus' girdle (*Cestum,* Figure 13-37B) is highly compressed in the tentacular plane. Bandlike, it may be more than 1 m long and presents a graceful appearance as it swims in the oral direction. The highly modified *Ctenoplana* (Gr. *ktenos,* comb, + L. *planus,* flat) and *Coeloplana* (Gr. *koilos,* hollow, + L. *planus,* flat) (Figure 13-37C) are rare but are interesting because they have disc-shaped bodies flattened in the oral-aboral axis

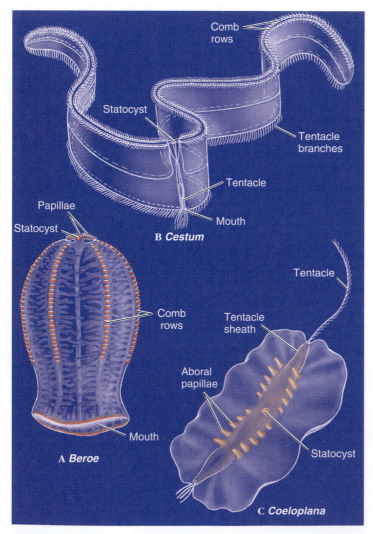

Figure 13-37

Diversity among phylum Ctenophora. **A,** *Beroe* sp. (order Beroida, class Nuda). **B,** *Cestum* sp. (order Cestida, class Tentaculata). **C,** *Coeloplana* sp. (order Platyctenea, class Tentaculata).

Classification of Phylum Ctenophora

Class Tentaculata (ten-tak'yu-la'ta) (L. *tentaculum,* feeler, + *ata,* group suffix). With tentacles; tentacles may have sheaths into which they retract; some types flattened in oral-aboral axis for creeping; others compressed in tentacular plane to a bandlike form; in some the comb plates may be confined to the larva. Examples: *Pleurobrachia, Cestum.*

Class Nuda (nu-da) (L. *nudus,* naked). Without tentacles, but flattened in tentacular plane; wide mouth and pharynx; gastrovascular canals much branched. Example: *Beroe.*

---

and are adapted for creeping rather than swimming. A common ctenophore along the Atlantic and Gulf coasts is *Mnemiopsis* (Gr. *mneme,* memory, + *opsis,* appearance) (Figure 13-35B), which has a laterally compressed body with two large oral lobes and unsheathed tentacles.

Nearly all ctenophores emit flashes of luminescence at night, especially such forms as *Mnemiopsis* (Figure 13-35B). The vivid flashes of light seen at night in southern seas are often caused by members of this phylum.

Since the 1980s population explosions of *Mnemiopsis leidyi* in the Black and Azov Seas have led to catastrophic declines in fisheries there. Inadvertently introduced from the coast of the Americas with ballast water of ships, the ctenophores feed on zooplankton, including small crustaceans and eggs and larvae of fish. The normally inoffensive *M. leidyi* is kept in check in the Atlantic by certain specialized predators, but introduction of such predators into the Black Sea carries its own dangers.

PHYLOGENY AND ADAPTIVE RADIATION

Phylogeny

Although the origin of cnidarians and ctenophores is obscure, the most widely supported hypothesis is that radiate phyla arose from a radially symmetrical, planula-like ancestor. Such an ancestor could have been common to radiates and to bilateral metazoans, the latter having been derived from a branch whose members habitually crept about on the sea bottom. Such a habit would select for bilateral symmetry. Others became sessile or free floating, conditions for which radial symmetry is a selective advantage. A planula larva in which an invagination formed to become the gastrovascular cavity would correspond roughly to a cnidarian with an ectoderm and an endoderm.

Increasing complexity in homeobox (Hox) genes (p. 163) may have been part of the genetic basis for such transitions. Hox genes are highly conserved throughout almost all Metazoa and control expression of other genes determining body axis and morphogenesis along the body axis. Sponges, which essentially have no body axis, also appear to lack the Hox cluster com-

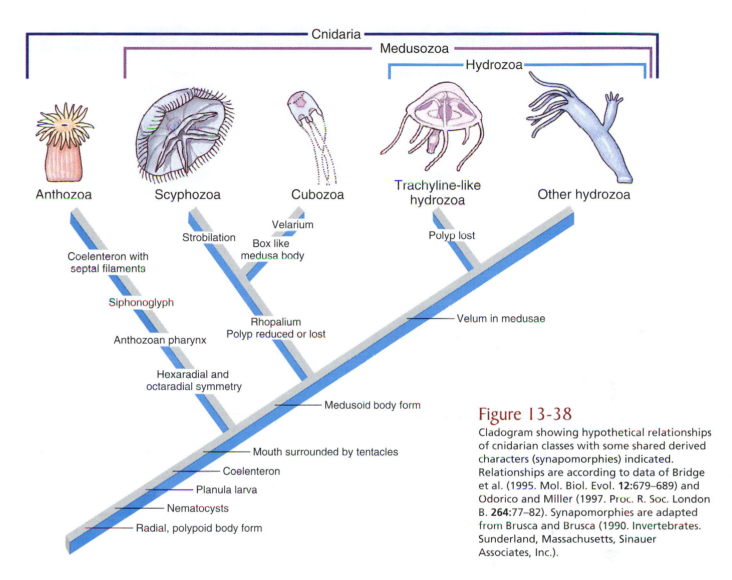

Figure 13-38

Cladogram showing hypothetical relationships of cnidarian classes with some shared derived characters (synapomorphies) indicated. Relationships are according to data of Bridge et al. (1995. Mol. Biol. Evol. **12**:679–689) and Odorico and Miller (1997. Proc. R. Soc. London B. **264**:77–82). Synapomorphies are adapted from Brusca and Brusca (1990. Invertebrates. Sunderland, Massachusetts, Sinauer Associates, Inc.).

pletely. Cnidarians have a definite body axis (oral-aboral) and possess anterior and posterior Hox genes; however, they lack intermediate (group 3 or central) Hox genes that are present in bilaterians.

Potential antecedents of those hallmark organelles of cnidarians, nematocysts, can be found among some single-celled groups, for example, trichocysts and toxicysts in ciliates (p. 222) and trichocysts in dinoflagellates (p. 226). In fact, some dinoflagellates have organelles that are strikingly similar in structure to nematocysts.

For some years, prevalent opinion was that trachyline medusae (order Trachylina, class Hydrozoa) resemble the ancestral cnidarian in their direct development from planula and actinula larvae to medusae. However, molecular evidence suggests that Anthozoa is basal in phylum Cnidaria (Figure 13-38). Development of medusae would then become a synapomorphy of the other classes, with a subsequent loss of a polyp stage in ancestors of Trachylina.

In the past it was assumed that ctenophores arose from a medusoid cnidarian, but this assumption has been challenged. Similarities between the groups are mostly of a general nature and do not seem to indicate a close relationship. Some molecular evidence suggests that ctenophores branched off the metazoan line after sponges but before cnidarians. Ctenophores have a more derived and stereotypical cleavage pattern than cnidarians. Other analyses indicate a nearest common ancestor of ctenophores and cnidarians, which together would form the sister taxon to bilateral metazoans.

Adaptive Radiation

Neither phylum has deviated far from its basic structural plan. In Cnidaria, both polyp and medusa are constructed on the same scheme. Likewise, ctenophores have adhered to the arrangement of the comb plates and their biradial symmetry.

Nonetheless, cnidarians have achieved large numbers of individuals and species, demonstrating a surprising degree of diversity considering the simplicity of their basic body plan. They are efficient predators, many feeding on prey quite large in relation to themselves. Some are adapted for feeding on small particles. The colonial form of life is well explored, with some colonies growing to great size among corals, and others, such as siphonophores, showing astonishing polymorphism and specialization of individuals within a colony.

SUMMARY

Phyla Cnidaria and Ctenophora have a primary radial symmetry; radial symmetry is an advantage for sessile or free-floating organisms because environmental stimuli come from all directions equally. Cnidaria are surprisingly efficient predators because they possess stinging organelles called nematocysts. Both phyla are essentially diploblastic (some triploblastic, depending on the definition of mesoderm), with a body wall composed of epidermis and gastrodermis separated by a mesoglea. The digestive-respiratory (gastrovascular) cavity has a mouth and no anus. Cnidarians are at the tissue level of organization. They have two basic body types (polypoid and medusoid), and in many hydrozoans and scyphozoans the life cycle involves both an asexually reproducing polyp and a sexually reproducing medusa.

That unique organelle, the nematocyst, is produced by a cnidoblast (which becomes the cnidocyte) and is coiled within a capsule. When discharged, some types of nematocysts penetrate prey and inject poison. Discharge is effected by a change in permeability of the capsule and an increase in internal hydrostatic pressure because of the high osmotic pressure within the capsule.

Most hydrozoans are colonial and marine, but the freshwater hydras are commonly demonstrated in class laboratories. They have a typical polypoid form but are not colonial and have no medusoid stage. Most marine hydrozoans are in the form of a branching colony of many polyps (hydranths). Hydrozoan medusae may be free-swimming or remain attached to their colony.

Scyphozoans are typical jellyfishes, in which the medusa is the dominant body form, and many have an inconspicuous polypoid stage. Cubozoans are predominantly medusoid. They include the dangerous sea wasps.

Anthozoans are all marine and are polypoid; there is no medusoid stage. The most important subclasses are Hexacorallia (with hexamerous or polymerous symmetry) and Octocorallia (with octomerous symmetry). The largest hexacorallian orders contain sea anemones, which are solitary and do not have a skeleton, and stony corals, which are mostly colonial and secrete a calcareous exoskeleton. Stony corals are a critical component in coral reefs, which are habitats of great beauty, productivity, and ecological and economic value. Octocorallia contain the soft and horny corals, many of which are important and beautiful components of coral reefs.

Ctenophora are biradial and swim by means of eight comb rows. Colloblasts, with which they capture small prey, are characteristic of the phylum.

Cnidaria and Ctenophora are probably derived from an ancestor that resembled the planula larva of cnidarians. Despite their relatively simple level of organization, cnidarians are an important phylum economically, environmentally, and biologically.

REVIEW QUESTIONS

1. Explain the selective advantage of radial symmetry for sessile and free-floating animals.
2. What characteristics of phylum Cnidaria are most important in distinguishing it from other phyla?
3. Name and distinguish the taxonomic classes in phylum Cnidaria.
4. Distinguish between polyp and medusa forms.
5. Explain the mechanism of nematocyst discharge. How can a hydrostatic pressure of one atmosphere be maintained within the nematocyst until it receives an expulsion stimulus?
6. What is an unusual feature of the nervous system of cnidarians?
7. Diagram a hydra and label the main body parts.
8. Name and give functions of the main cell types in the epidermis and in the gastrodermis of hydra.
9. What stimulates feeding behavior in hydras?
10. Define the following with regard to hydroids: hydrorhiza, hydrocaulus, coensosarc, perisarc, hydranth, gonangium, manubrium, statocyst, ocellus.
11. Give an example of a highly polymorphic, floating, colonial hydrozoan.
12. Distinguish the following from each other: statocyst and rhopalium; scyphomedusae and hydromedusae; scyphistoma, strobila, and ephyrae; velum, velarium, and pedalium; Hexacorallia and Octocorallia.
13. Define the following with regard to sea anemones: siphonoglyph; primary septa or mesenteries; incomplete septa; septal filaments; acontia threads; pedal laceration.
14. Describe three specific interactions of anemones with nonprey organisms.
15. Contrast the skeletons of hexacorallian and alcyonarian corals.
16. Coral reefs generally are limited in geographic distribution to shallow marine waters. How do you account for this?
17. Specifically, what kinds of organisms are most important in deposition of calcium carbonate on coral reefs?
18. How do zooxanthellae contribute to the welfare of hermatypic corals?
19. Distinguish each of the following from each other: fringing reefs; barrier reefs; atolls; patch or bank reefs.
20. What characteristics of Ctenophora are most important in distinguishing it from other phyla?
21. How do ctenophores swim, and how do they obtain food?
22. Compare cnidarians and ctenophores, giving five ways in which they resemble each other and five ways in which they differ.
23. What is a widely held hypothesis on the origin of radiate phyla?

SELECTED REFERENCES

Buddemeier, R. W., and S. V. Smith. 1999. Coral adaptation and acclimatization: a most ingenious paradox. Amer. Zool. **39:**1–9. *First of a series of papers in this issue dealing with effects of climatic and temperature changes on coral reefs.*

Crossland, C. J., B. G. Hatcher, and S. V. Smith. 1991. Role of coral reefs in global ocean production. Coral Reefs **10:**55–64. *Because of extensive recycling of nutrients within reefs, their net energy production for export is relatively minor. However, they play a major role in inorganic carbon precipitation by biologically-mediated processes.*

Finnerty, J. R. 2001. Cnidarians reveal intermediate stages in the evolution of Hox clusters and axial complexity. Amer. Zool. **41:**608–620. *Sponges appear to lack Hox cluster completely. Cnidarians have anterior and posterior Hox genes, but lack intermediate (group 3 or central) Hox genes that are present in bilaterians.*

Kenchington, R., and G. Kelleher. 1992. Crown-of-thorns starfish management conundrums. Coral Reefs **11:**53–56. *The first article of an entire issue on the starfish:* Acanthaster planci, *a predator of corals. Another entire issue was devoted to this predator in 1990 (p. 447).*

Lesser, M. P. 1997. Oxidative stress causes coral bleaching during exposure to elevated temperatures. Coral Reefs **16:**187–192. *Evidence that reactive types of oxygen molecules, perhaps produced by the zooxanthellae, cause cell damage and expulsion of zooxanthellae. Stress on the zooxanthellae may be induced by increased temperature or UV irradiation.*

Odorico, D. M., and D. J. Miller. 1997. Internal and external relationships of the Cnidaria: implications of primary and predicted secondary structure of the 5'-end of the 23S-like rDNA. Proc. R. Soc. London B. **264:**77–82. *Results support a sister group relationship between Scyphozoa and Cubozoa and a basal position of Anthozoa among Cnidaria. A close relationship of Placozoa and Cnidaria is not supported.*

Pennisi, E. 1998. New threat seen from carbon dioxide. Science **279:**989. *Increase in atmospheric CO_2 is acidifying ocean water, making it more difficult for corals to deposit $CaCO_3$. If CO_2 doubles in the next 70 years, as expected, reef formation will decline by 40%, and by 75% if CO_2 doubles again.*

Rosenberg, E., and Y. Loya, 1999. Vibrio shiloi is the etiological (causative) agent of Oculina patagonica bleaching: general implications. Reef Encounter **25:**8–10. *These investigators believe all coral bleaching is due to bacteria, not just that of O. patagonica. The bacterium that they report (Vibrio shiloi) needs high temperatures.*

Winnepenninckx, B. M. H., Y. Van de Peer, and T. Backeljau. 1998. Metazoan relationships on the basis of 18S rRNA sequences: a few years later . . . Amer. Zool. **38:**888–906. *Some analyses suggest that sponges and ctenophores form a clade that branched off from their common ancestor with cnidarians, all of them forming the sister group of metazoans.*

ZOOLOGY LINKS TO THE INTERNET

Visit the textbook's Online Learning Center at www.mhhe.com/zoology to find live Internet links for each of the topics listed here.
Radiate Animals
Phylum Cnidaria
Class Hydrozoa
Class Scyphozoa
Class Anthozoa
Coral Reefs
Coral Bleaching
Class Cubozoa
Phylum Ctenophora

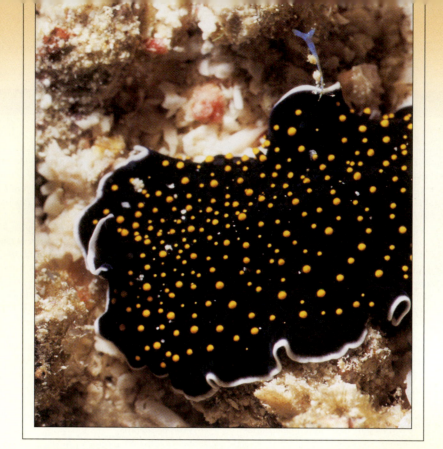

Thysanozoon nigropapillosum, a marine turbellarian (order Polycladida).

Acoelomate Bilateral Animals

Phylum Platyhelminthes
Phylum Nemertea
Phylum Gnathostomulida

Getting Ahead

For animals that spend their lives sitting and waiting, as do most members of the two radiate phyla we considered in the preceding chapter, radial symmetry is ideal. One side of the animal is just as important as any other for snaring prey coming from any direction. But if an animal is active in seeking food, shelter, home sites, and reproductive mates, it requires a different set of strategies and a new body organization. Active, directed movement requires an elongated body form with head (anterior) and tail (posterior) ends. In addition, one side of the body is kept up (dorsal) and the other side, specialized for locomotion, is kept down (ventral). What results is a bilaterally symmetrical animal in which the body can be divided along only one plane of symmetry to yield two halves that are mirror images of each other. Furthermore, since it is better to determine where one is going than where one has been, sense organs and centers for nervous control have come to be located on the head. This process is called cephalization. Thus cephalization and primary bilateral symmetry evolved together.

The three acoelomate phyla considered in this chapter are not greatly more complex in organization than radiates except in symmetry. The evolutionary consequence of that development alone was enormous, however, for it is the type of symmetry assumed by all more complex animals.

The three phyla considered in this chapter have the simplest organization within the Bilateria (unless Mesozoa are regarded as bilateral, see Chapter 13), a grouping of phyla that includes all the rest of the animal kingdom. These three are Platyhelminthes (Gr. *platys*, flat, + *belmins*, worm), or flatworms; Nemertea (Gr. *Nemertes*, one of the nereids, unerring one), or ribbon worms; and Gnathostomulida (Gr. *gnathos*, jaw, + *stoma*, mouth, + L. *ulus*, dim.), or jaw worms. They have only one internal space, a digestive cavity, with the region between the ectoderm and endoderm filled with mesoderm in the form of muscle fibers and mesenchyme (parenchyma). Since they lack a coelom or a pseudocoel, they are termed **acoelomate bilateral** animals (Figure 14-1), and because they have three well-defined germ layers, they are **triploblastic.** Acoelomates show more specialization and division of labor among their organs than do radiate animals because the mesoderm makes more elaborate organs possible; thus, acoelomates are said to have reached the organ-system level of organization.

These phyla belong to superphylum Lophotrochozoa of the Protostomia and typically have spiral cleavage. They have some centralization of the nervous system, with a concentration of nerves anteriorly and a ladder-type arrangement of trunks and connectives down the body. They have an excretory (or osmoregulatory) system, and the nemerteans also have a circulatory system and a one-way digestive system, with an anus as well as a mouth.

PHYLUM PLATYHELMINTHES

The word "worm" is loosely applied to elongated, bilateral invertebrate animals without appendages. At one time zoologists considered worms (Vermes) a group in their own right. Such a group included a highly diverse assortment of forms. This unnatural assemblage was reclassified into various phyla. By tradition, however, zoologists still refer to the various groups of these animals as flatworms, ribbon worms, roundworms, and segmented worms.

Platyhelminthes were derived from an ancestor that probably had many cnidarian-like characteristics, including a gelatinous mesoglea. Nonetheless, replacement of the gelatinous mesoglea with a cellular, mesodermal **parenchyma** laid the basis for a more complex organization. Parenchyma is a form of "packing" tissue containing more cells and fibers than the mesoglea of cnidarians. In at least some platyhelminths, the parenchyma is made up of noncontractile cell bodies of muscle cells; the cell body containing the nucleus and other organelles is connected to an elongated contractile portion in somewhat the same manner as epitheliomuscular cells of cnidarians (see Figure 13-7).

Flatworms range in size from a millimeter or less to some tapeworms that are many meters in length. Their flattened bodies may be slender, broadly leaflike, or long and ribbonlike.

Flatworms include both free-living and parasitic forms, but free-living members are found exclusively in class Turbellaria. A few turbellarians are symbiotic or parasitic, but the majority are adapted as bottom dwellers in marine or fresh water or live in moist places on land. Many, especially of the larger species, are found on the underside of stones and other hard objects in freshwater streams or in tidal and subtidal zones (p. 779) of the ocean.

Position in Animal Kingdom

1. Platyhelminthes, or flatworms, Nemertea, or ribbon worms, and Gnathostomulida, or jaw worms, are the simplest animals to have **primary bilateral symmetry.**
2. These phyla have only one internal space, a digestive cavity, with the region between the ectoderm and endoderm filled with mesoderm in the form of muscle fibers and mesenchyme (parenchyma). Since they lack a coelom or a pseudocoelom, they are termed **acoelomate** animals, and because they have three well-defined germ layers, they are termed **triploblastic.**
3. Acoelomates show more specialization and division of labor among their organs than do radiate animals because the mesoderm makes more elaborate organs possible. Thus acoelomates are said to have reached the **organ-system level of organization.**
4. They belong to the protostome division of the Bilateria and have spiral cleavage, and at least platyhelminths and nemerteans have mosaic (determinate) cleavage.
5. They are members of superphylum Lophotrochozoa.

Biological Contributions

1. Acoelomates developed the basic **bilateral** plan of organization that has been widely exploited in the animal kingdom.
2. **Mesoderm** developed into a well-defined embryonic germ layer **(triploblastic),** making available a great source of tissues, organs, and systems.
3. Along with bilateral symmetry, **cephalization** was established. Some centralization of the nervous system evident in the **ladder type of system** found in flatworms.
4. Along with the subepidermal musculature, there is also a mesenchymal system of muscle fibers.
5. They are the simplest animals with an **excretory system.**
6. Nemerteans are the simplest animals to have a **circulatory system** with blood and a **one-way alimentary canal.** Although not stressed by zoologists, the rhynchocoel cavity in ribbon worms is technically a true coelom, but because it is merely a part of the proboscis mechanism, it is probably not homologous to the coelom of eucoelomate animals.
7. Unique and specialized structures occur in all three phyla. The parasitic habit of many flatworms has led to many specialized adaptations, such as organs of adhesion.

Figure 14-1
Diagram of acoelomate body plan (cross section).

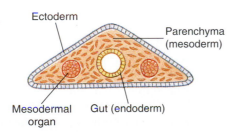

Ectoderm

Parenchyma (mesoderm)

Mesodermal organ

Gut (endoderm)

Figure 14-2
Stained planarian.

Most species of turbellarians are marine, but there are many freshwater species. Planarians (Figure 14-2) and some others frequent streams and spring pools; others prefer flowing water of mountain streams. Some species occur in fairly hot springs. Terrestrial turbellarians are found in rather moist places under stones and logs. There are about six species of terrestrial turbellarians in the United States.

All members of classes Monogenea and Trematoda (flukes) and Cestoda (tapeworms) are parasitic. Most Monogenea are ectoparasites, but all trematodes and cestodes are endoparasitic. Many species have indirect life cycles with more than one host; the first host is often an invertebrate, and the final host is usually a vertebrate. Humans serve as hosts for a number of species. Certain larval stages may be free living.

Many animals covered in this chapter and Chapters 11, 15, 18, 19, and 20 are parasites. Humans have suffered greatly through the centuries from their parasites and those of their domestic animals. Fleas and bacteria conspired to destroy a third of the European population in the seventeenth century, and malaria, schistosomiasis, and African sleeping sickness have sent untold millions to their graves. Even today, after successful campaigns against yellow fever, malaria, and hookworm infections in many parts of the world, parasitic diseases in association with nutritional deficiencies are the primary killers of humans. Civil wars and environmental changes wrought by humans have led to resurgences in malaria, trypanosomiasis, and leishmaniasis, and global prevalences of intestinal roundworms are unchanged in the last 50 years.

Form and Function

Tegument, Muscles

Most turbellarians have a cellular, ciliated epidermis. Freshwater planarians, such as *Dugesia,* belong to order Tricladida and are used extensively in introductory laboratory courses. Their ciliated epidermis rests on a basement membrane. It contains rod-

---

Characteristics of Phylum Platyhelminthes

1. Three germ layers **(triploblastic)**
2. **Bilateral symmetry;** definite polarity of anterior and posterior ends
3. **Body flattened dorsoventrally;** oral and genital apertures mostly on ventral surface
4. Epidermis may be cellular or syncytial (ciliated in some); **rhabdites** in epidermis of most Turbellaria; epidermis a syncytial **tegument** in Monogenea, Trematoda, Cestoda, and some Turbellaria
5. Muscular system primarily of a sheath form and of mesodermal origin; layers of circular, longitudinal, and sometimes oblique fibers beneath the epidermis
6. No internal body space other than digestive tube (acoelomate); spaces between organs filled with parenchyma
7. Digestive system incomplete (gastrovascular type); absent in some
8. Nervous system consisting of a **pair of anterior ganglia** with **longitudinal nerve cords** connected by transverse nerves and located in the mesenchyme in most forms; similar to cnidarians in primitive forms
9. Simple sense organs; eyespots in some
10. Excretory system of two lateral canals with branches bearing **flame cells (protonephridia);** lacking in some forms
11. Respiratory, circulatory, and skeletal systems lacking; lymph channels with free cells in some trematodes
12. Most forms monoecious; reproductive system complex, usually with well-developed gonads, ducts, and accessory organs; internal fertilization; development direct in free-swimming forms and those with single hosts; complicated life cycle often involving several hosts in many internal parasites
13. Class Turbellaria mostly free living; classes Monogenea, Trematoda, and Cestoda entirely parasitic

---

shaped **rhabdites,** which swell and form a protective mucous sheath around the body when discharged with water. Single-cell mucous glands open on the surface of the epidermis (Figure 14-3). Most orders of turbellarians have **dual-gland** adhesive organs in the epidermis. These organs consist of three cell types: viscid and releasing gland cells and anchor cells (Figure 14-4). Secretions of the viscid gland cells apparently fasten microvilli of the anchor cells to the substrate, and secretions of the releasing gland cells provide a quick, chemical detaching mechanism.

In the body wall below the basement membrane of flatworms are layers of **muscle fibers** that run circularly, longitudinally, and diagonally. A meshwork of **parenchyma** cells, developed from mesoderm, fills the spaces between muscles and visceral organs. Parenchyma cells in some, perhaps all, flatworms are not a separate cell type but are the noncontractile portions of muscle cells.

A few turbellarians have a **syncytial** epidermis (nuclei are not separated from each other by intervening cell membranes).

All members of Trematoda, Monogenea, and Cestoda are parasitic, and their body covering, as adults, is unusual among animals. Furthermore, they lack cilia. Rather than "epidermis,"

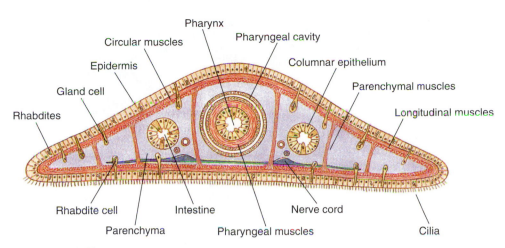

Figure 14-3
Cross section of planarian through pharyngeal region, showing relationships of body structures.

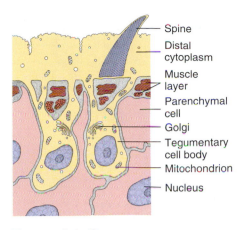

Figure 14-5
Diagrammatic drawing of the structure of the tegument of a trematode *Fasciola hepatica*.

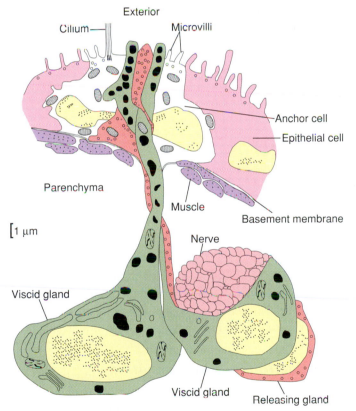

Figure 14-4
Reconstruction of dual-gland adhesive organ of the turbellarian *Haplopharynx* sp. There are two viscid glands and one releasing gland, which lie beneath the body wall. The anchor cell lies within the epidermis, and one of the viscid glands and the releasing gland are in contact with a nerve.

their body covering is designated by a more noncommital term **tegument** (Figure 14-5). This distinctive tegumental plan is the basis for uniting trematodes, monogeneans, and cestodes in a taxon known as **Neodermata.** It is a peculiar epidermal arrangement and may be related to adaptations for parasitism in ways that are still unclear.

Nutrition and Digestion

Other than in cestodes, which have no digestive system, platyhelminth digestive systems include a mouth, a pharynx, and an intestine (Figure 14-6). In planarians the pharynx is enclosed in a **pharyngeal sheath** (Figure 14-6) and opens posteriorly just inside the mouth, through which it can extend. The intestine has three many-branched trunks, one anterior and two posterior. The whole forms a **gastrovascular cavity** lined with columnar epithelium (Figure 14-6). The mouth of trematodes and monogeneans usually opens at or near the anterior end of their body into a muscular, nonextensible pharynx (see Figures 14-7 and 14-16). Posteriorly, their esophagus opens into a blindly ending intestine, which is commonly Y-shaped but may be highly branched or unbranched, depending on the species.

Planarians are mainly carnivorous, feeding largely on small crustaceans, nematodes, rotifers, and insects. They can detect food from some distance by means of chemoreceptors. They entangle prey in mucous secretions from the mucous glands and rhabdites. A planarian grips prey with its anterior end, wraps its body around prey, extends its proboscis, and sucks up food in small amounts. Monogeneans and trematodes graze on host cells, feeding on cellular debris and body fluids.

Intestinal secretions contain proteolytic enzymes for some **extracellular digestion.** Bits of food are sucked into the intestine, where phagocytic cells of the gastrodermis complete digestion **(intracellular).** Undigested food is egested through the pharynx. Because cestodes have no digestive tract, they must depend on host digestion, and absorption is confined to small molecules.

Excretion and Osmoregulation

Except in some turbellarians, the osmoregulatory system of flatworms consists of **protonephridia** (excretory or osmoregulatory organs closed at the inner end) with **flame cells** (Figure 14-6A). A flame cell is cup shaped with a tuft of flagella extending from the inner face of the cup. In some turbellarians and in all Neodermata, the protonephridia form a **weir** (Old English *wer*,

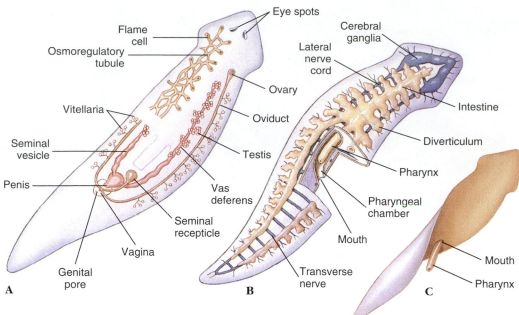

Figure 14-6

Structure of a planarian. **A,** Reproductive and osmoregulatory systems, shown in part. Inset at left is enlargement of flame cell. **B,** Digestive tract and ladder-type nervous system. Pharynx is shown in resting position. **C,** Pharynx extended through ventral mouth.

a fence placed in a stream to catch fish); the rim of the cup is elongated into fingerlike projections that extend between similar projections of a tubule cell. The space (lumen) enclosed by the tubule cell continues into collecting ducts that finally open to the outside by pores. Beating flagella (resembling a flickering flame) drive fluid down the collecting ducts and provide a negative pressure to draw fluid through the delicate interlacing projections of the weir. The wall of the duct beyond the flame cell commonly bears folds or microvilli that probably function in resorption of certain ions or molecules.

In planarians collecting ducts join and rejoin into a network along each side of the animal (Figure 14-6) and may empty through many nephridiopores. This system is mainly osmoregulatory because it is reduced or absent in marine turbellarians, which do not have to expel excess water. Monogeneans usually have two excretory pores opening laterally, near the anterior. Collecting ducts of trematodes empty into an excretory bladder that opens to the exterior by a terminal pore (Figure 14-7). In cestodes there are two main excretory canals on each side that are continuous through the entire length of the worm (see Figure 14-20). They join in the last segment (proglottid, see p. 285) to form an excretory bladder that opens by a terminal pore. When the terminal proglottid is shed, the two canals open separately.

Metabolic wastes are removed largely by diffusion through the body wall.

Nervous System

The most primitive flatworm nervous system, found in some turbellarians, is a **subepidermal nerve plexus** resembling the nerve net of cnidarians. Other flatworms have, in addition to a nerve plexus, one to five pairs of **longitudinal nerve cords** lying under the muscle layer. More derived flatworms tend to have the lesser number of nerve cords. Freshwater planarians have one ventral pair (Figure 14-6B). Connecting nerves form a "ladder-type" pattern. Their brain is a bilobed mass of ganglion cells aris-

ing anteriorly from the ventral nerve cords. Except in acoel turbellarians, which have a diffuse system, neurons are organized into sensory, motor, and association types—an important development in evolution of nervous systems.

Sense Organs

Active locomotion in flatworms has favored not only cephalization in the nervous system but also further evolution of sense organs. **Ocelli,** or light-sensitive eyespots, are common in turbellarians (Figure 14-6A), monogeneans, and larval trematodes.

Tactile cells and chemoreceptive cells are abundant over the body, and in planarians they form definite organs on the auricles (the earlike lobes on the sides of the head). Some species also have statocysts for equilibrium and rheoreceptors for sensing direction of the water current. Sensory endings are abundant around the oral sucker of trematodes and holdfast organ (scolex, p. 286) of cestodes and around genital pores in both groups.

Reproduction and Regeneration

Many turbellarians reproduce both asexually (by fission) and sexually. Asexually, freshwater planarians merely constrict behind the pharynx and separate into two animals, each of which regenerates the missing parts—a quick means of population increase. Evidence suggests that a reduced population density results in an increase in the rate of fissioning. In some fissioning forms individuals may remain temporarily attached, forming chains of zooids (Figure 14-8).

Trematodes undergo asexual reproduction in their intermediate hosts, snails. Details of their astonishing life cycles are described on p. 281. Some juvenile cestodes show asexual reproduction, budding off hundreds, or in some cases, even millions, of offspring (p. 289).

Virtually all flatworms are monoecious (hermaphroditic) but practice cross-fertilization. In some turbellarians the yolk for

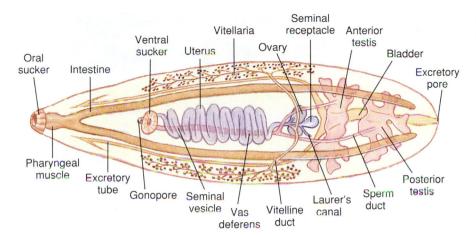

Figure 14-7
Structure of human liver fluke *Clonorchis sinenesis*.

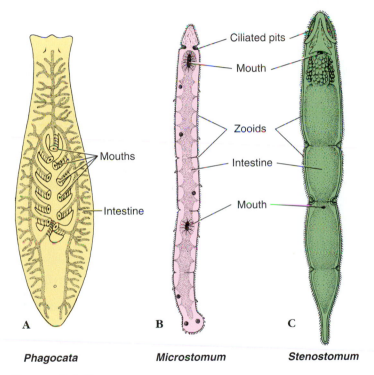

Phagocata   **Microstomum**   **Stenostomum**

Figure 14-8
Some small freshwater turbellarians. **A,** *Phagocata* has numerous pharynges. **B** and **C,** Incomplete fission results for a time in a series of attached zooids.

nutrition of a developing embryo is contained within the egg cell itself **(endolecithal),** and embryogenesis shows spiral determinate cleavage typical of protostomes (p. 205). Possession of endolecithal eggs is considered ancestral for flatworms. Other turbellarians plus all trematodes, monogeneans, and cestodes share a derived condition in which female gametes contain little or no yolk, and yolk is contributed by cells released from separate organs called **vitellaria.** Yolk cells are conducted toward a juncture with the **oviduct** by **vitelline ducts** (Figures 14-6 and 14-7). Usually a number of yolk cells surrounds the zygote within the eggshell; thus development is **ectolecithal.** Cleavage is affected in such a way that a spiral pattern cannot be distinguished. The entire package consisting of yolk cells and zygote,

surrounded by the eggshell, moves into the **uterus** and finally is released through a common genital pore or a separate uterine pore (see Figures 14-7).

Male reproductive organs include one, two, or more **testes** connected to **vasa efferentia** that join to become a single **vas deferens.** The vas deferens commonly leads into a **seminal vesicle** and hence to a papilla-like **penis** or an extensible copulatory organ called a **cirrus.**

During breeding season turbellarians develop both male and female organs, which usually open through a common genital pore (Figure 14-6A). After copulation one or more fertilized eggs and some yolk cells become enclosed in a small cocoon. The cocoons are attached by little stalks to the underside of stones or plants. Embryos emerge as juveniles that resemble mature adults. In some marine forms embryos develop into ciliated free-swimming larvae.

Monogeneans hatch as free-swimming larvae that attach to the next host and develop into juveniles. Larval trematodes emerge from the eggshell as ciliated larvae that penetrate a snail intermediate host, or they may hatch only after being eaten by a snail. Most cestodes hatch only after being consumed by an intermediate host, which may be any of many different animals, depending on species of tapeworm.

Class Turbellaria

Turbellarians are mostly free-living worms that range in length from 5 mm or less to 50 cm. They typically are creeping forms that combine muscular with ciliary movements to achieve locomotion. Their mouth is on the ventral side. Unlike trematodes and cestodes, they have simple life cycles.

Very small planaria swim by means of their cilia. Others move by gliding, head slightly raised, over a slime track secreted by the marginal adhesive glands. Beating of epidermal cilia in the slime track moves the animal forward, while rhythmical muscular waves can be seen passing backward from the head. Large polyclads and terrestrial turbellarians crawl by muscular undulations, much in the manner of a snail.

As traditionally recognized, turbellarians form a paraphyletic group. Several synapomorphies, such as "insunk" epidermis and ectolecithal development, show that some turbellarians are phylogenetically closer to Trematoda, Monogenea, and Cestoda than they are to other turbellarians. Ectolecithal turbellarians therefore appear to form a clade with trematodes, monogeneans, and cestodes to the exclusion of endolecithal turbellarians. Endolecithal turbellarians also are paraphyletic; presence of a dual-gland adhesive system in some endolecithal turbellarians indicates a clade with ectolecithal flatworms to the exclusion of other endolecithal turbellarian lineages. Turbellaria therefore describes an artificial group and the term is used here only because it is prevalent in zoological literature.

Characteristics used to distinguish orders of turbellarians are form of the gut (present or absent; simple or branched; pattern of branching) and pharynx (simple; folded; bulbous). Except for order Polycladida (Gr. *poly*, many, + *klados*, branch), turbellarians with endolecithal eggs have a simple gut or no gut and a simple pharynx. In a few turbellarians there is no recognizable pharynx. Polyclads have a folded pharynx and a gut with many branches (Figure 14-9). Polyclads include many marine forms of moderate to large size (3 to more than 40 mm) (Figure 14-10), and a highly branched intestine is correlated with larger size in turbellarians. Members of order Tricladida (Gr. *treis*, three, + *klados*, branch), which are ectolecithal and include freshwater planaria, have a three-branched intestine (Figure 14-9).

Members of order Acoela (Gr. *a*, without, + *koilos*, hollow) (Figure 14-11) have been regarded as having changed least from the ancestral form. In fact, molecular evidence suggests that acoels should not be placed in phylum Platyhelminthes and that they represent the earliest divergent Bilateria (p. 292). In body form they are small and have a mouth but no gastrovascular cavity or excretory system. Food is merely passed through the mouth into temporary spaces that are surrounded by mesenchyme, where gastrodermal phagocytic cells digest food intracellularly.

The considerable powers of regeneration in planarians have provided an interesting system for experimental studies of development. For example, a piece excised from the middle of a planarian can regenerate both a new head and a new tail. However,

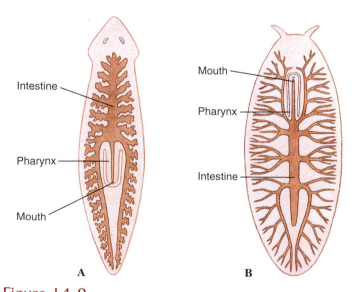

Figure 14-9
Intestinal pattern of two orders of turbellarians. **A,** Tricladida. **B,** Polycladida.

Labels in figure A: Intestine, Pharynx, Mouth. Labels in figure B: Mouth, Pharynx, Intestine.

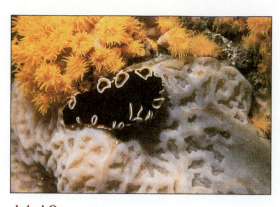

Figure 14-10
Pseudoceros hancockanum, a marine polyclad turbellarian. Marine polyclads are often large and beautifully colored. The orange polyps of *Tubastrea aurea,* an ahermatypic coral, and *Aplidium cratiferum,* a colonial tunicate (Chapter 23) that looks something like cartilage, are also in the photograph.

Figure 14-11
An acoel, *Amphiscolops* sp., swimming.

the piece retains its original polarity: a head grows at the anterior end and a tail at the posterior end. An extract of heads added to a culture medium containing headless worms will prevent regeneration of new heads, suggesting that substances in one region will suppress regeneration of the same region at another level of the body. Many other experiments could be cited.

Class Trematoda

Trematodes are all parasitic flukes, and as adults they are almost all found as endoparasites of vertebrates. They are chiefly leaflike in form and are structurally similar in many respects to ectolecithal turbellarians. A major difference is in the tegument, which lacks cilia in the adult.

Other structural adaptations for parasitism are apparent: various penetration glands or glands to produce cyst material, organs for adhesion such as suckers and hooks, and increased reproductive capacity. Otherwise, trematodes share several characteristics with turbellarians, such as a well-developed alimentary canal (but with the mouth at the anterior, or cephalic, end) and similar reproductive, excretory, and nervous systems, as well as a musculature and parenchyma that are only slightly modified from those of turbellarians. Sense organs are poorly developed.

Of the subclasses of Trematoda, Aspidogastrea and Didymozoidea are small and poorly known groups, but Digenea (Gr. *dis,* double, + *genos,* race) is a large group with many species of medical and economic importance.

Subclass Digenea

With rare exceptions, digeneans have a complex life cycle, the first **(intermediate)** host being a mollusc and the **definitive** host (the host in which sexual reproduction occurs, sometimes called the **final** host) being a vertebrate. In some species a second, and sometimes even a third, intermediate host intervenes. The group

has radiated greatly, and its members parasitize almost all kinds of vertebrate hosts. Digeneans inhabit, according to species, a wide variety of sites in their hosts: all parts of the digestive tract, respiratory tract, circulatory system, urinary tract, and reproductive tract.

Among the world's most amazing biological phenomena are digenean life cycles. Although cycles of different species vary widely in detail, a typical example would include an adult, egg (shelled embryo), miracidium, sporocyst, redia, cercaria, and metacercaria stages (Figure 14-12). The shelled embryo or larva usually passes from the definitive host in excreta and must reach water to develop further. There, it hatches to a free-swimming, ciliated larva, the **miracidium.** The miracidium penetrates the tissues of a snail, where it transforms into a **sporocyst.** Sporocysts reproduce asexually to yield either more sporocysts or a number of **rediae.** Rediae, in turn, reproduce asexually to produce more rediae or to produce **cercariae.** In this way a single egg can give rise to an enormous number of progeny. Cercariae

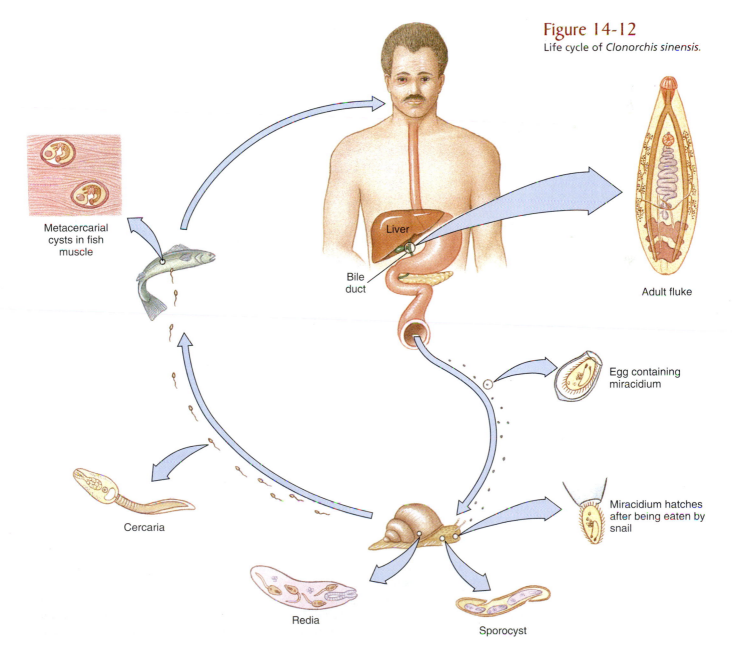

Figure 14-12
Life cycle of *Clonorchis sinensis.*

Metacercarial cysts in fish muscle

Liver

Bile duct

Adult fluke

Egg containing miracidium

Miracidium hatches after being eaten by snail

Cercaria

Redia

Sporocyst

TABLE 14.1

Examples of Flukes Infecting Humans

Common and Scientific Names	Means of Infection; Distribution and Prevalence in Humans
Blood flukes (*Schistosoma* spp.); three widely prevalent species, others reported	Cercariae in water penetrate skin; 200 million people infected with one or more species
S. mansoni	Africa, South and Central America
S. haematobium	Africa
S. japonicum	Eastern Asia
Chinese liver flukes (*Clonorchis sinensis*)	Eating metacercariae in raw fish; about 30 million cases in eastern Asia
Lung flukes (*Paragonimus* spp.), seven species, most prevalent is *P. westermani*	Eating metacercariae in raw freshwater crabs, crayfish; Asia and Oceania, sub-Saharan Africa, South and Central America; several million cases in Asia
Intestinal fluke (*Fasciolopsis buski*)	Eating metacercariae on aquatic vegetation; 10 million cases in eastern Asia
Sheep liver fluke (*Fasciola hepatica*)	Eating metacercariae on aquatic vegetation; widely prevalent in sheep and cattle, occasional in humans

emerge from the snail and penetrate a second intermediate host or encyst on vegetation or other objects to become **metacercariae,** which are juvenile flukes. Adults grow from metacercariae when that stage is eaten by a definitive host.

Some of the most serious parasites of humans and domestic animals belong to Digenea (Table 14-1). The first digenean life cycle to be worked out was that of *Fasciola hepatica* (L. *fasciola,* a small bundle, band), which causes "liver rot" in sheep and other ruminants. Adult flukes live in the bile passage of the liver, and eggs are passed in feces. After hatching, a miracidium penetrates a snail to become a sporocyst. There are two generations of rediae, and the cercaria encysts on vegetation. When infested vegetation is eaten by a sheep or other ruminant (or sometimes humans), the metacercariae excyst and grow into young flukes.

Clonorchis sinensis: *Liver Fluke in Humans*

Clonorchis (Gr. *clon,* branch, + *orchis,* testis) is the most important liver fluke of humans and is common in many regions of eastern Asia, especially in China, Southeast Asia, and Japan. Cats, dogs, and pigs are also often infected.

Structure   The worms vary from 10 to 20 mm in length (Figure 14-7). Their structure is typical of many trematodes in most respects. They have an **oral sucker** and a **ventral sucker.** The **digestive system** consists of a pharynx, a muscular esophagus, and two long, unbranched intestinal ceca. The **excretory system** consists of two protonephridial tubules, with branches provided with flame cells. The two tubules unite to form a single median bladder that opens to the outside. The nervous system, like that of other flatworms, is made up of two cerebral ganglia connected to longitudinal cords that have transverse connectives.

The **reproductive system** is hermaphroditic and complex. They have two branched **testes** and two **vasa efferentia** that unite to form a single **vas deferens,** which widens into a **seminal vesicle.** The seminal vesicle leads into an **ejaculatory duct,** which terminates at the genital opening. Unlike most trematodes, *Clonorchis* does *not* have a protrusible copulatory organ, the cirrus. The female system contains a branched **ovary** with a short **oviduct,** which is joined by ducts from the **seminal receptacle** and **vitellaria** at an **ootype.** The ootype is surrounded by a glandular mass, **Mehlis' gland,** of uncertain function. From Mehlis' gland the much-convoluted **uterus** runs to the genital pore. Cross-fertilization between individuals is usual, and sperm are stored in the seminal receptacle. When an oocyte is released from the ovary, it is joined by a sperm and a group of vitelline cells and is fertilized. The vitelline cells release a proteinaceous shell material, which is stabilized by a chemical reaction; the Mehlis' gland secretions are added, and the egg passes into the uterus.

Life Cycle   The normal habitat of the adults is in the bile passageways of humans and other fish-eating mammals (Figure 14-12). Eggs, each containing a complete miracidium, are shed into water with the feces but do not hatch until they are ingested by the snail *Parafossarulus* or related genera. Eggs, however, may live for some weeks in water. In a snail a miracidium enters the tissues and transforms into a sporocyst (a baglike structure with embryonic germ cells), which produces one generation of rediae. A redia is elongated, with an alimentary canal, a nervous system, an excretory system, and many germ cells in the process of development. Rediae pass into the liver of the snail where the germ cells continue embryonation and give rise to tadpolelike cercariae.

Cercariae escape into the water, swim about until they encounter a fish of family Cyprinidae, and then bore under scales into the fish's muscles. Here cercariae lose their tails and encyst as metacercariae. If a mammal eats raw infected fish, the metacercarial cyst dissolves in the intestine, and young flukes apparently migrate up the bile duct, where they become adults. There the flukes may live for 15 to 30 years.

The effect of these flukes on humans depends mainly on the extent of the infection but includes abdominal pain and other abdominal symptoms. A heavy infection can cause a pronounced cirrhosis of the liver and result in death. Cases are diagnosed through fecal examinations. To avoid infection, all fish used as food should be thoroughly cooked. Destruction of snails that carry larval stages is a method of control.

Schistosoma: *Blood Flukes*

Schistosomiasis, infection with blood flukes of genus *Schistosoma* (Gr. *schistos,* divided, + *soma,* body), ranks as one of the major infectious diseases in the world, with 200 million people infected. The disease is widely prevalent over much of Africa and parts of South America, West Indies, Middle East, and Far East. The old generic name for the worms was *Bilharzia* (from Theodor Bilharz, German parasitologist who discovered *Schistosoma haematobium*), and the infection was called bilharziasis, a name still used in many areas.

Unfortunately, some projects intended to raise the standard of living in some tropical countries, such as the Aswan High Dam in Egypt, have increased the prevalence of schistosomiasis by creating more habitats for the snail intermediate hosts. Before the dam was constructed, the 500 miles of the Nile River between Aswan and Cairo was subjected to annual floods; alternate flooding and drying killed many snails. Four years after dam completion, prevalence of schistosomiasis had increased sevenfold along that segment of the river. Prevalence in fishermen around the lake above the dam increased from a very low level to 76%.

Blood flukes differ from most other flukes in being dioecious and having the two branches of the digestive tube united into a single tube in the posterior part of the body. Males are broader and heavier and have a large, ventral groove, the **gynecophoric canal,** posterior to the ventral sucker. The gynecophoric canal embraces the long, slender female (Figure 14-13).

Three species account for most schistosomiasis in humans: *S. mansoni,* which lives primarily in venules draining the large intestine; *S. japonicum,* which is found mostly in venules of the small intestine; and *S. haematobium,* which lives in venules of the urinary bladder. *Schistosoma mansoni* is common in parts of Africa, Brazil, northern South America, and the West Indies; species of *Biomphalaria* are the principal snail intermediate hosts. *Schistosoma haematobium* is widely prevalent in Africa, using snails of the genera *Bulinus* and *Physopsis* as the main intermediate hosts. *Schistosoma japonicum* is confined to the Far East, and its hosts are several species of *Oncomelania*.

The life cycle of blood flukes is similar in all species. Eggs are discharged in human feces or urine; if they get into water, they hatch as ciliated miracidia, which must contact the required kind of snail within a few hours to survive. In the snail, they transform into sporocysts, which produce another generation of sporocysts. Daughter sporocysts give rise to cercariae directly, without formation of rediae. Cercariae escape from the snail and swim about until they contact bare skin of a human. They penetrate the skin, shedding their tails in the process, and reach a blood vessel where they enter the circulatory system. There is no metacercarial stage. The young schistosomes make their way to the hepatic portal system of blood vessels and undergo a period of development in the liver before migrating to their characteristic sites. As eggs are released by adult females, they are somehow extruded through the wall of venules and through the gut or bladder lining, to be voided with feces or urine, accord-

A

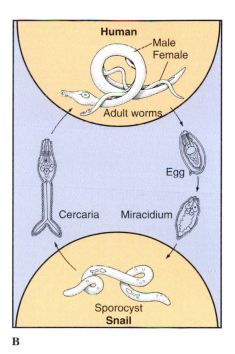

B

Figure 14-13

A, Adult male and female *Schistosoma japonicum* in copulation. The male has a long gynecophoric canal that holds the female (the darkly stained individual). Humans are usually hosts of adult parasites, found mainly in Africa but also in South America and elsewhere. Humans become infected by wading or bathing in cercaria-infested waters. **B,** Life cycle of *Schistosoma mansoni*.

ing to species. Many eggs do not make this difficult transit and are swept by blood flow back to the liver or other areas, where they become centers of inflammation and tissue reaction (see Figure 35-7).

The main ill effects of schistosomiasis result from the eggs. With *S. mansoni* and *S. japonicum,* eggs in the intestinal wall cause ulceration, abscesses, and bloody diarrhea with abdominal pain. Similarly, *S. haematobium* causes ulceration of the bladder wall with bloody urine and pain on urination. Eggs swept to the liver or other sites cause symptoms associated with the organs where they lodge. When they are caught in the capillary bed of the liver, they impede circulation and cause cirrhosis, a fibrotic reaction that interferes with liver function (Figure 14-14). Of the three species, *S. haematobium* is considered least

Figure 14-14

Cut surface of a liver showing severe fibrosis. The patient was a 27-year-old man who died from hematemesis (vomiting blood) associated with spleen and liver enlargement. Over 180 pairs of adult *Schistosoma mansoni* were counted at autopsy.

Courtesy A. W. Cheever/From H. Zaiman, A Pictorial Presentation of Parasites.

serious and *S. japonicum* most severe. The prognosis is poor in heavy infections of *S. japonicum* without early treatment.

Control is best achieved by educating people to dispose of body wastes hygienically and avoid exposure to contaminated water, difficult problems for poor people living under primitive conditions.

Although proper disposal of body wastes is the best control for schistosomiasis, other strategies are being pursued with varying success: chemotherapy, vector control, and vaccination. Development of a vaccine is the subject of much research, but an effective vaccine is not yet available. Vector control by environmental management and by biological means appears promising. Biological controls include introduction of species of snails, crayfish, and fish that prey on the snail vectors.

Schistosome Dermatitis (Swimmer's Itch)   Various species of schistosomes in several genera cause a rash or dermatitis when their cercariae penetrate hosts that are unsuitable for further development. Cercariae of several genera whose normal hosts are North American birds cause dermatitis in bathers in northern lakes. Severity of the rash increases with an increasing number of contacts with the organisms, or sensitization. After penetration, cercariae are attacked and killed by the host's immune mechanisms, and they release allergenic substances, causing itching. The condition is more an annoyance than a serious threat to health, but there may be economic losses to persons depending on vacation trade around infested lakes.

Paragonimus: *Lung Flukes*

Several species of *Paragonimus* (Gr. *para,* beside, + *gonimos,* generative), a fluke that lives in the lungs of its host, are known

from a variety of mammals. *Paragonimus westermani* (Figure 14-15), found in East Asia and the southwest Pacific, parasitizes a number of wild carnivores, humans, pigs, and rodents. Its eggs are coughed up in the sputum, swallowed, then eliminated with feces. Metacercariae develop in freshwater crabs, and the infection is acquired by eating uncooked crab meat. The infection causes respiratory symptoms, with breathing difficulties and chronic cough. Fatal cases are common. A closely related species, *P. kellicotti,* occurs in mink and similar animals in North America, but only one human case has been recorded. Its metacercariae are in crayfish.

Some Other Trematodes

Fasciolopsis buski (L. *fasciola,* small bundle, + Gr. *opsis,* appearance) parasitizes the intestine of humans and pigs in India and China. Larval stages occur in several species of planorbid snails, and cercariae encyst on water chestnuts, an aquatic vegetation eaten raw by humans and pigs.

Leucochloridium is noted for its remarkable sporocysts. Snails (*Succinea*) eat vegetation infected with eggs from bird droppings. Sporocysts become much enlarged and branched, and cercariae encyst within the sporocyst. Sporocysts enter the snail's head and tentacles, become brightly striped with orange and green bands, and pulsate at frequent intervals. Birds are attracted by the enlarged and pulsating tentacles, eat the snails, and so complete the life cycle.

Class Monogenea

Monogenetic flukes traditionally were placed as an order of Trematoda, but they are sufficiently different to deserve a separate class. Cladistic analysis places them closer to the Cestoda. Monogeneans are all parasites, primarily of gills and external surfaces of fish. A few are found in the urinary bladder of frogs and turtles, and one parasitizes the eye of a hippopotamus. Although widespread and common, monogeneans seem to cause little damage to their hosts under natural conditions. However, like numerous other fish pathogens, they become a serious threat when their hosts are crowded together, as, for example, in fish farming.

Life cycles of monogeneans are direct, with a single host. The egg hatches to produce a ciliated larva, called an **oncomiracidium,** that attaches to its host or swims around awhile before attachment. The oncomiracidium bears hooks on its posterior, which in many species become the hooks on the large posterior attachment organ (**opisthaptor**) of the adult. Because monogeneans must cling to the host and withstand the force of water flow over the gills or skin, adaptive radiation has produced a wide array of opisthaptors in different species. Opisthaptors may bear large and small hooks, suckers, and clamps, often in combination with each other.

Common genera are *Gyrodactylus* (L. *gyro,* a circle, + Gr. *daktylos,* toe, finger) (Figure 14-16) and *Dactylogyrus* (Gr. *daktylos,* toe, finger, + L. *gyro,* a circle), both of economic importance to fish culturists, and *Polystoma* (Gr. *polys,* many, + *stoma,* mouth), found in the urinary bladder of frogs.

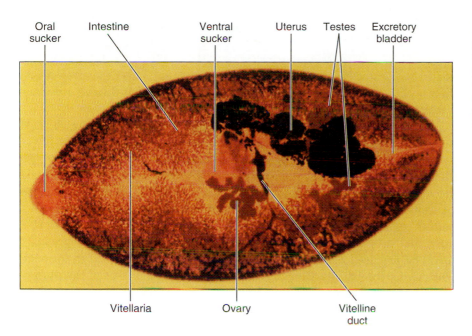

Oral sucker | Intestine | Ventral sucker | Uterus | Testes | Excretory bladder

Vitellaria | Ovary | Vitelline duct

Figure 14-15
Lung fluke *Paragonimus westermani.* Adults are up to 2 cm long. Eggs discharged in sputum or feces hatch into free-swimming miracidia that enter snails. Cercariae from snails enter freshwater crabs and encyst in soft tissues. Humans are infected by eating poorly cooked crabs or by drinking water containing larvae freed from dead crabs.

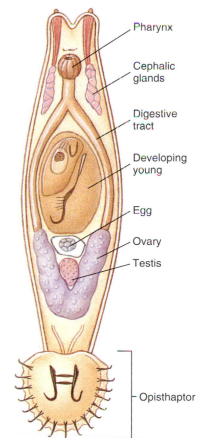

Pharynx

Cephalic glands

Digestive tract

Developing young

Egg

Ovary

Testis

Opisthaptor

Figure 14-16
A monogenetic fluke *Gyrodactylus cylindriformis,* ventral view.

Class Cestoda

Cestoda, or tapeworms, differ in many respects from the preceding classes. They usually have long flat bodies in which there is a linear series of sets of reproductive organs. Each set is called a **proglottid** and usually has at its anterior and posterior ends zones of muscle weakness, marked externally by grooves. No digestive system is present. As in Monogenea and Trematoda, no external, motile cilia occur in adults, and the tegument is of a distal cytoplasm with sunken cell bodies beneath the superficial muscle layer (Figure 14-17). In contrast to monogeneans and trematodes, however, the entire surface of cestodes is covered with minute projections similar to microvilli of the vertebrate small intestine (p. 45). These **microtriches** (sing. **microthrix**) greatly enlarge the surface area of the tegument, which is a vital adaptation for a tapeworm since it must absorb all its nutrients across its tegument.

Tapeworms are nearly all monoecious. They have well-developed muscles, and their excretory system and nervous system are somewhat similar to those of other flatworms. They have no special sense organs but do have sensory endings in the tegument that are modified cilia (Figure 14-17). One of their most specialized structures is their **scolex,** or holdfast, which is their organ of attachment. It is usually provided with suckers or suckerlike organs and often with hooks or spiny tentacles (Figure 14-18).

With rare exceptions, all cestodes require at least two hosts, and adults are parasites in the digestive tract of vertebrates. Often their intermediate host is an invertebrate.

Subclass Eucestoda contains the great majority of species in the class. With the exception of two small orders, members of this subclass have the body divided into a series of proglottids and are thus termed **polyzoic.** Larval forms of all eucestodes have six hooks. The main body of the worms, the chain of proglottids, is called a **strobila** (Figure 14-18). Typically, there is a **germinative zone** just behind the scolex where new proglottids are formed. As younger proglottids are differentiated in front

of it, each individual proglottid moves posteriorly in the strobila, and its gonads mature. A proglottid is usually fertilized by another proglottid in the same or a different strobila. The shelled embryos form in the uterus of the proglottid, and they are expelled through a uterine pore or the entire proglottid is shed from the worm as it reaches the posterior end.

Some zoologists have maintained that the proglottid formation of cestodes represents "true" segmentation (metamerism), but we do not support this view. Segmentation of tapeworms is best considered a replication of sex organs to increase reproductive capacity and is not related to the metamerism found in Annelida, Arthropoda, and Chordata (see pp. 188 and 360).

More than 1000 species of tapeworms are known to parasitologists. Almost all vertebrate species are infected. Normally, adult tapeworms do little harm to their hosts. The most common tapeworms found in humans are given in Table 14-2.

Figure 14-17

Schematic drawing of a longitudinal section through a sensory ending in the tegument of *Echinococcus granulosus*.

GUTLESS WONDER

Though lacking skeletal strengths
Which we associate with most
Large forms, tapeworms go to great
Lengths to take the measure of a host.

Monotonous body sections
In a limp mass-production line
Have nervous and excretory connections

And the means to sexually combine
And to coddle countless progeny
But no longer have the guts
To digest for themselves or live free
Or know a meal from soup to nuts.

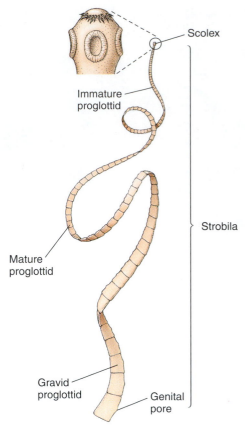

Figure 14-18

A tapeworm, showing strobila and scolex. The scolex is the organ of attachment.

Taenia saginata: *Beef Tapeworm*

Structure   *Taenia saginata* (Gr. *tainia*, band, ribbon) is called the beef tapeworm, but it lives as an adult in the intestine of humans. Juvenile forms are found primarily in intermuscular tissue of cattle. A mature adult may reach a length of 10 m or more. Its scolex has four suckers for attachment to the intestinal wall, but no hooks. A short neck connects the scolex to the strobila, which may have as many as 2000 proglottids. Gravid proglottids bear shelled, infective larvae (Figure 14-19) and become detached and pass in feces.

Tapeworms show some unity in their organization. **Excretory canals** in the scolex are also connected to the canals, two on each side, in the proglottids. Two longitudinal **nerve cords** from a **nerve ring** in the scolex run back into the proglottids (Figure 14-20). Attached to the excretory ducts are flame cells. Each mature proglottid also contains muscles and parenchyma as well as a complete set of male and female organs similar to those of a trematode.

In the order to which this species belongs, however, vitellaria are typically a single, compact **vitelline gland** located just posterior to the ovaries. When gravid proglottids break off and pass out with the feces, they usually crawl out of the fecal mass and onto vegetation nearby. There they may be picked up by grazing cattle. A proglottid ruptures as it dries, further scattering

TABLE 14.2

Common Cestodes of Humans

Common and Scientific Name	Means of Infection; Prevalence in Humans
Beef tapeworm (*Taenia saginata*)	Eating rare beef; most common of all tapeworms in humans
Pork tapeworm (*Taenia solium*)	Eating rare pork; less common than *T. saginata*
Fish tapeworm (*Diphyllobothrium latum*)	Eating rare or poorly cooked fish; fairly common in Great Lakes region of United States, and other areas of world where raw fish is eaten
Dog tapeworm (*Dipylidium caninum*)	Unhygienic habits of children (juveniles in flea and louse); moderate frequency
Dwarf tapeworm (*Hymenolepis nana*)	Juveniles in flour beetles; common
Unilocular hydatid (*Echinococcus granulosus*)	Cysts of juveniles in humans; infection by contact with dogs; common wherever humans are in close relationship with dogs and ruminants
Multilocular hydatid (*Echinococcus multilocularis*)	Cysts of juveniles in humans; infection by contact with foxes; less common than unilocular hydatid

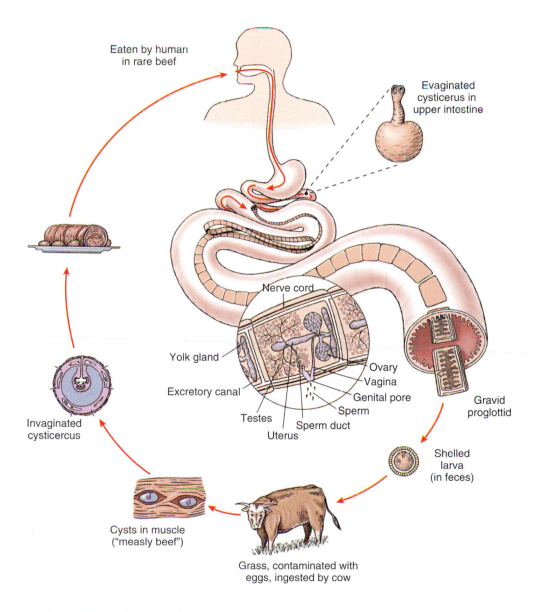

Figure 14-19

Life cycle of beef tapeworm, *Taenia saginata*. Ripe proglottids detach in the human intestine, leave the body in feces, crawl onto grass, and are ingested by cattle. Eggs hatch in the cow's intestine, freeing oncospheres, which penetrate into muscles and encyst, developing into "bladder worms." A human eats infected rare beef, and cysticercus is freed in intestine where it attaches to the intestinal wall, forms a strobila, and matures.

Excretory
canal

Sperm
duct

Testes

Genital
pore

Uterus

Vagina

Ovary

Mehlis'
gland

Vitelline
gland

Figure 14-20

Mature proglottid of *Taenia pisiformis,* a dog tapeworm. Portions of two other proglottids also shown.

the embryos on soil and grass. Embryos may remain viable on grass for as long as 5 months.

Life Cycle   Shelled larvae **(oncospheres)** swallowed by cattle hatch and use their hooks to burrow through the intestinal wall into blood or lymph vessels and finally reach voluntary muscle, where they encyst to become **bladder worms** (juveniles called **cysticerci**). There the juveniles develop an invaginated scolex but remain quiescent. When infected "measly" meat is eaten by a suitable host, the cyst wall dissolves, the scolex evaginates and attaches to the intestinal mucosa, and new proglottids begin to develop. It takes 2 to 3 weeks for a mature worm to form. When a person is infected with one of these tapeworms, numerous gravid proglottids are expelled daily, sometimes crawling out the anus. Humans become infected by eating rare roast beef, steaks, and barbecues. Considering that about 1% of American cattle are infected, that 20% of all cattle slaughtered are not federally inspected, and that even in inspected meat one-fourth of infections are missed, it is not surprising that tapeworm infection is fairly common. Infection is precluded when meat is thoroughly cooked.

Some Other Tapeworms

Taenia solium: *Pork Tapeworm*   Adult *Taenia solium* (Gr. *tainia,* band, ribbon) live in the small intestine of humans, whereas juveniles occur in the muscles of pigs. The scolex has both suckers and hooks arranged on its tip (Figure 14-18), the **rostellum.** The life history of this worm is similar to that of the beef tapeworm, except that humans become infected by eating insufficiently cooked pork.

Taenia solium is much more dangerous than *T. saginata* because cysticerci, as well as adults, can develop in humans. If eggs or proglottids are accidentally ingested by a human, the liberated embryos migrate to any of several organs and form cysticerci (Figure 14-21). The condition is called **cysticercosis.** Common sites are eyes and brain, and infection in such locations can result in blindness, serious neurological symptoms, or death.

Diphyllobothrium latum: *Fish Tapeworm*   Adult *Diphyllobothrium* (Gr. *dis,* double, + *phyllon,* leaf, + *bothrion,* hole, trench) are found in the intestine of humans, dogs, cats, and other mammals; immature stages are in crustaceans and fish. With a length of up to 20 m, it is the largest cestode that infects humans. Fish tapeworm infections can occur anywhere in the world where people commonly eat raw fish; in the United States infections are most common in the Great Lakes region. In Finland, but apparently not other areas, the worm can cause a serious anemia.

Figure 14-21
Section through the brain of a person who died of cerebral cysticercosis, an infection with cysticerci of *Taenia solium*.

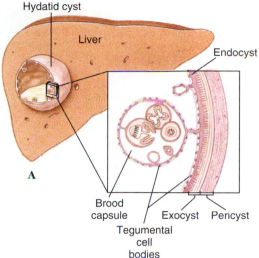

Echinococcus granulosus: *Unilocular Hydatid*

Echinococcus granulosus (Gr. *echinos,* hedgehog, + *kokkos,* kernel) (Figure 14-22B), a dog tapeworm, causes hydatidosis in humans, a very serious disease in many parts of the world. Adult worms develop in canines, and juveniles grow in more than 40 species of mammals, including humans, monkeys, sheep, reindeer, and cattle. Thus humans may serve as an intermediate host in the case of this tapeworm. The juvenile stage is a special kind of cysticercus called a **hydatid cyst** (Gr. *hydatis,* watery vesicle). It grows slowly, but it can grow for a long time—up to 20 years—reaching the size of a basketball in an unrestricted site such as the liver. If a hydatid grows in a critical location, such as heart or central nervous system, serious symptoms may appear in a much shorter time. The main cyst maintains a single (or unilocular) chamber, but within the main cyst, daughter cysts bud off, and each contains thousands of scolices. Each scolex will produce a worm when eaten by a canine. The only treatment is surgical removal of the hydatid.

PHYLUM NEMERTEA (RHYNCHOCOELA)

Nemerteans (nem-er'te-ans) are often called ribbon worms. Their name (Gr. *Nemertes,* one of the Nereids, unerring one) refers to the unerring aim of the proboscis, a long muscular tube (Figures 14-23 and 14-24) that can be thrust out swiftly to grasp their prey. The phylum is also called Rhynchocoela (ring'ko-se'la) (Gr. *rhynchos,* beak, + *koilos,* hollow), which also refers to the proboscis. They are thread-shaped or ribbon-shaped worms; nearly all are marine. Some live in secreted gelatinous tubes. There are about 650 species in the group.

Nemertean worms are usually less than 20 cm long, although a few are several meters in length (Figure 14-25). *Lineus longissimus* (L. *linea,* line) is said to reach 30 m. Their

Figure 14-22
Echinococcus granulosus, a dog tapeworm, which may be dangerous to humans. **A,** Early hydatid cyst or bladder-worm stage found in cattle, sheep, hogs, and sometimes humans produces hydatid disease. Humans acquire disease by unsanitary habits in association with dogs. When eggs are ingested, liberated larvae encyst in the liver, lungs, or other organs. Brood capsules containing scolices are formed from the inner layer of each cyst. The cyst enlarges, developing other cysts with brood pouches. It may grow for years to the size of a basketball, necessitating surgery. **B,** The adult tapeworm lives in intestine of a dog or other carnivore.

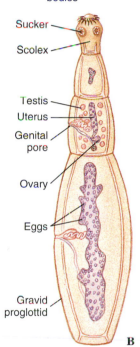

colors are often bright, although most are dull or pallid. In the odd genus *Gorgonorhynchus* (Gr. *Gorgo,* name of a female monster of terrible aspect, + *rhynchos,* beak, snout) the proboscis is divided into many proboscides, which appear as a mass of wormlike structures when everted.

With few exceptions, the general body plan of nemerteans is similar to that of turbellarians. Like the latter, their epidermis is ciliated and has many gland cells. Another striking similarity is the presence of flame cells in the excretory system. Rhabdites have been found in several nemerteans, including *Lineus.* However, nemerteans differ from flatworms in their reproductive system. They are mostly dioecious. In marine forms there is a ciliated larva that has some resemblance to trochophore larvae found in annelids and molluscs. Other flatworm characteristics are bilateral symmetry, and a mesoderm and no coelom. Present evidence indicates that nemerteans share a close common ancestor with Platyhelminthes.

Figure 14-23

Ribbon worm *Amphiporus bimaculatus* (phylum Nemertea) is 6 to 10 cm long, but other species range up to several meters. The proboscis of this specimen is partially extended at the top; the head is marked by two brown spots.

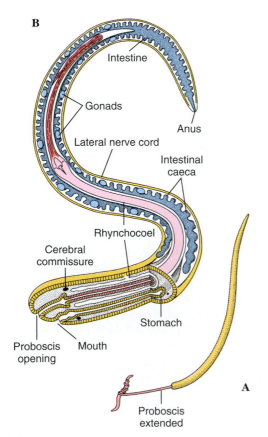

Figure 14-24

A, *Amphiporus,* with proboscis extended to catch prey. **B,** Structure of female nemertean worm *Amphiporus* (diagrammatic). Dorsal view to show proboscis.

Characteristics of Phylum Nemertea

1. Bilateral symmetry; highly contractile body that is cylindrical anteriorly and flattened posteriorly
2. Three germ layers
3. Epidermis with cilia and gland cells; rhabdites in some
4. Body spaces with parenchyma, which is partly gelatinous
5. An **eversible proboscis,** which lies free in a cavity (rhynchocoel) above the alimentary canal
6. **Complete digestive system** (mouth to anus)
7. Body-wall musculature of outer circular and inner longitudinal layers with diagonal fibers between the two; sometimes another circular layer inside the longitudinal layer
8. **Blood-vascular system with two or three longitudinal trunks**
9. Acoelomate, although the rhynchocoel technically may be considered a true coelom
10. Nervous system usually a four-lobed brain connected to paired longitudinal nerve trunks or, in some, middorsal and midventral trunks
11. Excretory system of two coiled canals, which are branched with **flame cells**
12. Sexes separate with simple gonads; asexual reproduction by fragmentation; few hermaphrodites; **pilidium larvae** in some
13. No respiratory system
14. Sensory **ciliated pits** or **head slits** on each side of head, which communicate between the outside and the brain; tactile organs and ocelli (in some)
15. In contrast to Platyhelminthes, there are few parasitic nemerteans

Nemerteans show some derived features absent from flatworms. One of these is the eversible **proboscis** and its sheath, for which there are no counterparts among Platyhelminthes. Another difference is the presence of an **anus** in adults, producing a **complete digestive system.** A digestive system with an anus is more efficient because ejection of waste materials back through the mouth is not necessary. Nemerteans are also the simplest animals to have a **blood-vascular system.**

A few nemerteans occur in moist soil and fresh water, but by far the larger number are marine. At low tide they are often coiled under stones. It seems probable that they are active at high tide and quiescent at low tide. Some nemerteans such as *Cerebratulus* (L. *cerebrum,* brain, + *ulus,* dim. suffix) often live in empty mollusc shells. The small species live among seaweed, or they may be found swimming near the surface of the water. Nemerteans are often secured by dredging at depths of 5 to 8 m or deeper. A few are commensals or parasites. *Prostoma rubrum* (Gr. *pro,* before, in front of, + *stoma,* month) which is 20 mm or less in length, is a well-known freshwater species.

Form and Function

Many nemerteans are difficult to examine because they are so long and fragile. *Amphiporus* (Gr. *amphi,* on both sides, + *poros,* pore), a genus of smaller forms that ranges from 2 to 10 cm in length, is fairly typical of nemertean structure (see Figure 14-24).

Figure 14-25
Baseodiscus is a genus of nemerteans whose members typically measure several meters in length. This *B. mexicanus* is from the Galápagos Islands.

Its body wall consists of ciliated epidermis and layers of circular and longitudinal muscles (Figure 14-26). Locomotion consists largely of gliding over a slime track, although larger species move by muscular contractions.

The mouth is anterior and ventral, and the digestive tract is complete, extending the full length of the body and ending at an anus. Development of an anus marks a significant advancement over the gastrovascular systems of flatworms and radiates. Regurgitation of wastes is no longer necessary; ingestion and egestion can occur simultaneously. Cilia move food through the intestine. Digestion is largely extracellular.

Nemerteans are carnivorous, feeding primarily on annelids and other small invertebrates. They seize their prey with a proboscis that lies in an interior cavity of its own, the **rhynchocoel,** above the digestive tract (but not connected with it). The proboscis itself is a long, blind muscular tube that opens at the anterior end at a proboscis pore above the mouth. (In a few nemerteans the esophagus opens through the proboscis pore rather than through a separate

Classification of Phylum Nemertea

Class Enopla (en'o-pla) (Gr. *enoplos,* armed). Proboscis usually armed with stylets; mouth opens in front of brain. Examples: *Amphiporus, Prostoma.*
Class Anopla (an'o-pla) (Gr. *anoplos,* unarmed). Proboscis lacks stylets; mouth opens below or posterior to brain. Examples: *Cerebratulus, Tubulanus, Lineus.*

mouth.) Muscular pressure on fluid in the rhynchocoel causes the long tubular proboscis to be everted rapidly through the proboscis pore. Eversion of the proboscis exposes a sharp barb, called a stylet (absent in some nemerteans). The sticky, slime-covered proboscis coils around the prey and stabs it repeatedly with the stylet, while pouring a toxic secretion on the prey. Then, retracting its proboscis, a nemertean draws the prey near its mouth (Figure 14-24), through which the esophagus is thrust to engulf the prey.

Unlike other acoelomates, nemerteans have a true circulatory system, and an irregular flow is maintained by the contractile walls of the vessels. Many flame-bulb protonephridia are closely associated with the circulatory system, so that their function appears to be truly excretory (for disposal of metabolic wastes), in contrast to their apparently osmoregulatory role in Platyhelminthes.

Nemerteans have a pair of nerve ganglia, and one or more pairs of longitudinal nerve cords are connected by transverse nerves.

Some species reproduce asexually by fragmentation and regeneration. In contrast to flatworms, most nemerteans are dioecious.

PHYLUM GNATHOSTOMULIDA

The first known species of Gnathostomulida (nath'o-sto-myu'lid-a) (Gr. *gnathos,* jaw, + *stoma,* mouth, + L. *ulus,* dim. suffix) was observed in 1928 in the Baltic, but its description was not published until 1956. Since then jaw worms have been found in many

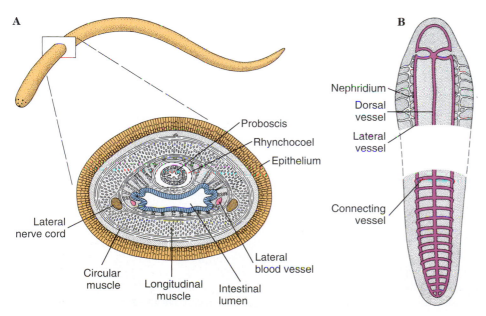

Figure 14-26
A, Diagrammatic cross section of female nemertean worm. **B,** Excretory and circulatory systems of nemertean worm. Flame bulbs along nephridial canal are closely associated with lateral blood vessels.

Figure 14-27

Gnathostomula jenneri (phylum Gnathostomulida) is a tiny member of the interstitial fauna between grains of sand or mud. Species in this family are among the most commonly encountered jaw worms, found in shallow water and down to depths of several hundred meters.

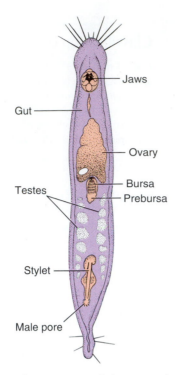

parts of the world, including the Atlantic coast of the United States, and over 80 species in 18 genera have been described.

Gnathostomulids are delicate wormlike animals and are 0.5 to 1 mm long (Figure 14-27). They live in interstitial spaces of very fine sandy coastal sediments and silt and can endure conditions of very low oxygen. They often occur in large numbers and frequently in association with gastrotrichs, nematodes, ciliates, tardigrades, and other small forms.

Lacking a pseudocoel, a circulatory system, and an anus, gnathostomulids show some similarities to turbellarians and were at first included in that group. However, their parenchyma is poorly developed, and their pharynx is reminiscent of a rotifer mastax. Their pharynx is armed with a pair of lateral jaws used to scrape fungi and bacteria off the substratum. And, although the epidermis is ciliated, each epidermal cell has but one cilium, a condition rarely found in the less derived bilateral animals except in some gastrotrichs (p. 310).

Gnathostomulids can glide, swim in loops and spirals, and bend the head from side to side. Sexual stages may include males, females, and hermaphrodites. Fertilization is internal.

PHYLOGENY AND ADAPTIVE RADIATION

Phylogeny

There can be little doubt that bilaterally symmetrical animals were derived from a radial ancestor, perhaps one very similar to the planula larva of cnidarians. Some investigators believe that this **planuloid ancestor** may have given rise to one branch of descendants that were sessile or free floating and radial, which became the Cnidaria, and another branch that acquired a creeping habit and bilateral symmetry. Bilateral symmetry is a selective advantage for creeping or swimming animals because sensory

structures are concentrated on the anterior end (cephalization), which is the end that first encounters environmental stimuli.

A recent report* cites sequence data from small-subunit rDNA, embryonic cleavage patterns and mesodermal origins, and nervous system structure as evidence that acoels are not, in fact, members of the phylum Platyhelminthes. These investigators conclude that acoels are the sister group to all other Bilateria. If this interpretation is sound, then Platyhelminthes as currently constituted is polyphyletic. Other scientists, however, interpreting sequence data from large-subunit ribosomal DNA, concluded that another order of Turbellaria (not Acoela) was basal for Platyhelminthes.** Neither report disputed paraphyly of Turbellaria.

Relationships of Nemertea and Gnathostomulida are not clear, but they apparently belong in Lophotrochozoa (see p. 204). Ultrastructural similarities of gnathostomulid jaws and trophi of rotifers (p. 306) suggest that these two phyla are sister groups. Although Turbellaria is clearly paraphyletic, we are retaining the taxon for the present because presentation based on thorough cladistic analysis would require introduction of many more taxa and characteristics that are beyond the scope of this book and not yet common in zoological literature. For example, ectolecithal turbellarians should be allied with trematodes, monogeneans, and cestodes to the sister group to endolecithal turbellarians. Some ectolecithal turbellarians share a number of other derived characters with trematodes and cestodes and have been placed by Brooks (1989) in a group designated Cercomeria (Gr. *kerkos,* tail, + *meros,* part) (Figure 14-28). Several synapomorphies, including the unique architecture of the tegument, indicates that neodermatans (trematodes, monogeneans, and cestodes) form a monophyletic group, and monophyly of Neodermata is supported by sequence data.**

Adaptive Radiation

The flatworm body plan, with its creeping adaptation, placed a selective advantage on bilateral symmetry and further development of cephalization, ventral and dorsal regions, and caudal differentiation. Because of their body shape and metabolic requirements, early flatworms must have been predisposed toward parasitism and gave rise to symbiotic descendants in the Neodermata. These descendants radiated abundantly as parasites, and many flatworms became very highly specialized for that mode of existence.

Ribbon worms have stressed the proboscis apparatus in their evolutionary diversity. Its use in capturing prey may have been secondarily evolved from its original function as a highly sensitive organ for exploring the environment. Although ribbon worms have evolved beyond flatworms in their complexity of organization, they have been dramatically less abundant as a group. Perhaps the proboscis was so efficient as a predator tool that there was little selective pressure to explore parasitism, or perhaps some critical preconditions were simply not present.

Likewise, jaw worms have neither radiated nor become nearly as abundant or diverse as flatworms. However, they have exploited the marine interstitial environment, particularly zones of very low oxygen concentration.

*Ruiz-Trillo et al. 1999. Science **283:**1919–1923.
Livaitis and Rohde. 1999. Invert. Biol. **118:42–56.

Jaws

Gut

Ovary

Bursa
Preburse

Testes

Stylet

Male pore

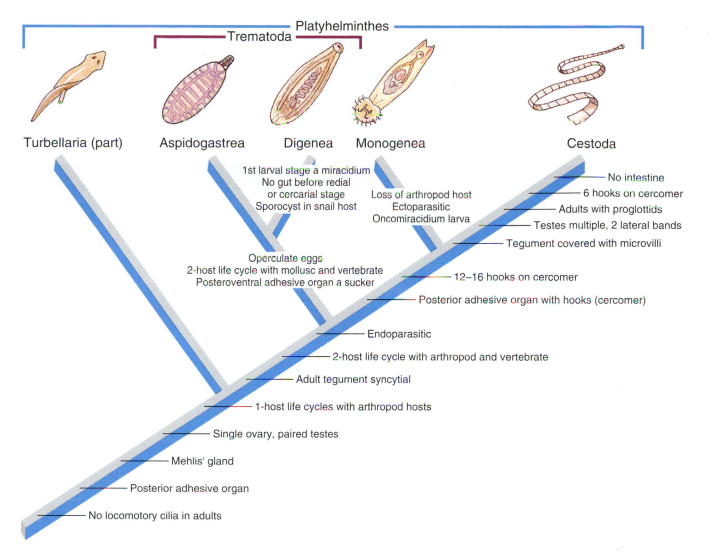

Figure 14-28

Hypothetical relationships among parasitic Platyhelminthes. The traditionally accepted class Turbellaria is paraphyletic. Some turbellarians have ectolecithal development and, together with the Trematoda, Monogenea, and Cestoda, form a clade and a sister group of the endolecithal turbellarians. For the sake of simplicity, the synapomorphies of those turbellarians and of the Aspidogastrea, as well as many others given by Brooks (1989) are omitted. All of these organisms form a clade (called Cercomeria) with a posterior adhesive organ.

Source: Modified from D. R. Brooks. The phylogeny of the Cercomeria (Platyhelminthes: Rhabdocoela) and general evolutionary principles. Journal of Parasitology 75:606–616, 1989.

SUMMARY

With the possible exception of Mesozoa, Platyhelminthes, Nemertea, and Gnathostomulida are the simplest forms that are bilaterally symmetrical, a condition of adaptive value for actively crawling or swimming animals. They have neither a coelom nor a pseudocoel and are thus acoelomate. They are triploblastic and at the organ-system level of organization.

The body surface of turbellarians is usually a cellular epithelium, at least in part ciliated, containing mucous cells and rod-shaped rhabdites that function together in locomotion. Members of all other classes of flatworms are covered by a nonciliated, syncytial tegument with a vesicular distal cytoplasm and cell bodies beneath superficial muscle layers. Digestion is extracellular and intracellular in most; cestodes must absorb predigested nutrients across their tegument because they have no digestive tract. Osmoregulation is by flame-cell protonephridia, and removal of metabolic wastes and

respiration occur across the body wall. Except for Acoela, flatworms have a ladder-type nervous system with motor, sensory, and association neurons. Most flatworms are hermaphroditic, and asexual reproduction occurs in some groups.

Class Turbellaria is a paraphyletic group with mostly free-living and carnivorous members. Digenetic trematodes have a mollusc intermediate host and almost always a vertebrate definitive host. The great amount of asexual reproduction that occurs in their intermediate host helps to increase the chances that some of their offspring will reach a definitive host. Aside from the tegument, digeneans share many basic structural characteristics with turbellarians. Digenea includes a number of important parasites of humans and domestic animals. Digeneans contrast with Monogenea, which are important ectoparasites of fishes and have a direct life cycle (without intermediate hosts).

Cestodes (tapeworms) generally have a scolex at their anterior end, followed by a long chain of proglottids, each of which contains a complete set of reproductive organs of both sexes. Cestodes live as adults in the digestive tract of vertebrates. They have microvillus-like microtriches on their tegument, which increase its surface area for absorption. Shelled larvae are passed in the feces, and juveniles develop in a vertebrate or invertebrate intermediate host.

Members of Nemertea have a complete digestive system with an anus and a true circulatory system. They are free living, mostly marine, and they capture prey by ensnaring it with their long, eversible proboscis.

Gnathostomulida are a curious phylum of little wormlike marine animals living among sand grains and silt. They have no anus, and they share certain characteristics with such widely diverse groups as turbellarians and rotifers.

Flatworms and the cnidarians both probably evolved from a common ancestor (planuloid), some of whose descendants became sessile or free floating and radial (cnidarians), while others became creeping and bilateral (flatworms).

Sequence analysis of rDNA, as well as some developmental and morphological criteria, suggest that Acoela, heretofore considered an order of turbellarians, diverged from a common ancestor shared with other Bilateria and are the sister group of all other bilateral phyla.

REVIEW QUESTIONS

1. Why is bilateral symmetry of adaptive value for actively motile animals?
2. Match the terms in the right column with the classes in the left column:

    ___ Turbellaria   a. Endoparasitic
    ___ Monogenea   b. Free living and commensal
    ___ Trematoda   c. Ectoparasitic
    ___ Cestoda

3. Give several characteristics that distinguish Platyhelminthes.
4. Distinguish two mechanisms by which flatworms supply yolk for their embryos. Which system is evolutionarily ancestral for flatworms and which one is derived?
5. Briefly describe the body plan of most turbellarians.
6. What do planarians eat, and how do they digest it?
7. Briefly describe the osmoregulatory system, the nervous system, and the sense organs of turbellarians, trematodes, and cestodes.
8. Contrast asexual reproduction in Turbellaria, Trematoda, and Cestoda.
9. Contrast a typical life cycle of a monogenean with that of a digenetic trematode.
10. Describe and contrast the tegument of most turbellarians and the other classes of platyhelminths. Does the tegument provide evidence that trematodes, monogeneans, and cestodes form a clade within Platyhelminthes? Why?
11. Answer the following questions with respect to both *Clonorchis* and *Schistosoma:* (a) how do humans become infected? (b) what is the general geographical distribution? (c) what are the main disease conditions produced?
12. Why is *Taenia solium* a more dangerous infection than *Taenia saginata?*
13. What are two cestodes for which humans can serve as intermediate hosts?
14. Define each of the following with reference to cestodes: scolex, microtriches, proglottids, strobila.
15. Give three differences between nemerteans and platyhelminths.
16. Where do gnathostomulids live?
17. Explain how a planuloid ancestor could have given rise to both Cnidaria and Bilateria.
18. Recent evidence suggests that acoels are not members of Platyhelminthes but constitute a sister group for all other Bilateria. If so, what is a consequence to phylogenetic integrity of Platyhelminthes if Acoela remains within that phylum? What evidence indicates that the traditional class Turbellaria is paraphyletic?

SELECTED REFERENCES

Brooks, D. R. 1989. The phylogeny of the Cercomeria (Platyhelminthes: Rhabdocoela) and general evolutionary principles. J. Parasitol. **75:**606–616. *Cladistic analysis of parasitic flatworms.*

Desowitz, R. S. 1981. New Guinea tapeworms and Jewish grandmothers. New York, W. W. Norton & Company. *Accounts of parasites and parasitic diseases of humans. Entertaining and instructive. Recommended for all students.*

Ehlers, U. 1985. Phylogenetic relationships within the Platyhelminthes. In Conway Morris, S., J. D. George, R. Gibson, and H. M. Platt (eds). The origins and relationships of lower invertebrates. Oxford, Clarendon Press. *Presents relationships of groups normally assigned to the paraphyletic Turbellaria.*

Livaitis, M. K., and K. Rohde. 1999. A molecular test of platyhelminth phylogeny: inferences from partial 28S rDNA sequences. Invert. Biol. **118:**42–56. *This report does not support a basal position for Acoela and presents evidence that Monogenea is paraphyletic.*

Rieger, R. M., and S. Tyler. 1995. Sister-group relationship of Gnathostomulida and Rotifera-Acanthocephala. Invert. Biol. **114:**186–188. *Evidence that gnathostomulids are the sister group of a clade containing rotifers and acanthocephalans.*

Roberts, L. S., and J. Janovy, Jr. 2000. Foundations of parasitology, ed. 6. Dubuque, Iowa, McGraw-Hill Higher Education. *Up-to-date and readable accounts of parasitic flatworms.*

Ruiz-Trillo, I., M. Riutort, D. Timoth, J. Littlewood, E. A. Herniou, and J. Baguñà. 1999. Acoel flatworms: earliest extant bilaterian metazoans, not members of Platyhelminthes. Science **283:**1919–1923. *The case for Acoela serving as a sister group for the remaining Bilateria.*

Strickland, G. T. 2000. Hunter's tropical medicine and emerging infectious diseases, ed. 8. Philadelphia, W. B. Saunders Company. *A valuable source of information on parasites of medical importance.*

ZOOLOGY LINKS TO THE INTERNET

Visit the textbook's Online Learning Center at www.mhhe.com/zoology to find live Internet links for each of the topics listed here.
Phylum Platyhelminthes
Class Turbellaria
Class Trematoda
Class Cestoda
Class Monogenea
Phylum Nemertea
Phylum Gnathostomulida

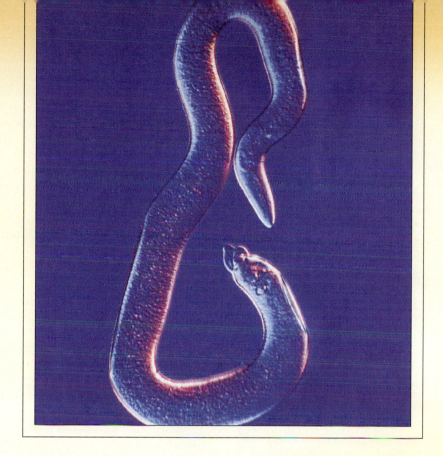

CHAPTER 15

Male *Trichinella spiralis*, a nematode.

Pseudocoelomate Animals

ECDYSOZOAN PHYLA
Nematoda
Nematomorpha
Kinorhyncha
Loricifera
Priapulida

LOPHOTROCHOZOAN PHYLA
Rotifera
Acanthocephala
Gastrotricha
Entoprocta

A World of Nematodes

Without any doubt, nematodes are the most important pseudo-coelomate animals, in terms of both numbers and their impact on humans. Nematodes are abundant over most of the world, yet most people are only occasionally aware of them as parasites of humans or of their pets. We are not aware of the millions of these worms in the soil, in ocean and freshwater habitats, in plants, and in all kinds of other animals. Their dramatic abundance moved N. A. Cobb* to write in 1914:

> If all the matter in the universe except the nematodes were swept away, our world would still be dimly recognizable, and if, as disembodied spirits, we could then investigate it, we should find its mountains, hills, vales, rivers, lakes and oceans represented by a thin film of nematodes. The location of towns would be decipherable, since for every massing of human beings there would be a corresponding massing of certain nematodes. Trees would still stand in ghostly rows representing our streets and highways. The location of the various plants and animals would still be decipherable, and, had we sufficient knowledge, in many cases even their species could be determined by an examination of their erstwhile nematode parasites.

*From N. A. Cobb. 1914. Yearbook of the United States Department of Agriculture, p. 472.

Position in Animal Kingdom

In the nine phyla covered in this chapter, the original blastocoel of the embryo persists as a space, or body cavity, between the enteron and body wall. Because this cavity lacks the peritoneal lining found in true coelomates, it is called a **pseudocoel**, and the animals possessing it are called **pseudocoelomates.** Pseudocoelomates belong to the Protostomia division of bilateral animals, but some pseudocoelomate phyla are assigned to superphylum Ecdysozoa and some to Lophotrochozoa.

Biological Contributions

1. The pseudocoel is a distinct gradation in body plan compared with the solid body structure of acoelomates. The pseudocoel may be filled with fluid or may contain a gelatinous substance with some mesenchyme cells. In common with a true coelom, it presents certain adaptive potentials, although these are by no means realized in all members: (1) greater freedom of movement; (2) space for development and differentiation of digestive, excretory, and reproductive systems; (3) a simple means of circulation or distribution of materials throughout the body; (4) a storage place for waste products to be discharged to the outside by excretory ducts; and (5) a hydrostatic organ. Since most pseudocoelomates are quite small, the most important functions of the pseudocoel are probably in circulation and as a means to maintain a high internal hydrostatic pressure.
2. A complete, mouth-to-anus digestive tract is found in these phyla and in all more complex phyla.

PSEUDOCOELOMATES

Vertebrates and more complex invertebrates have a true **coelom**, or peritoneal cavity, which is formed in the mesoderm during embryonic development and is therefore lined with a layer of mesodermal epithelium, the **peritoneum** (Figure 15-1; see p. 312). Pseudocoelomate phyla have a pseudocoel rather than a true coelom. It is derived from the embryonic blastocoel rather than from a secondary cavity within the mesoderm. It is a space between the gut and the mesodermal and ectodermal components of the body wall, and it is not lined with peritoneum.

Phylogenetic analysis based on base sequences in the small-subunit ribosomal RNA genes group Nematoda, Nematomorpha, Kinorhyncha, and Priapulida in superphylum Ecdysozoa, and Rotifera, Acanthocephala, Gastrotricha, and Entoprocta in Lophotrochozoa. Analysis based on morphological criteria suggests affinity of Loricifera to other pseudocoelomate phyla in Ecdysozoa. Entoprocta have sometimes been grouped with Ectoprocta, together called Bryozoa (moss animals). However, because ectoprocts have a true coelom, they are usually considered a separate phylum, and the term "bryozoans" is currently taken to exclude entoprocts.

Pseudocoelomates are a heterogeneous group of animals, considered in this chapter only because they have a common body plan. Such a diversified group of phyla, however, does share some characteristics. All have a body wall of epidermis (often syncytial), a dermis, and muscles surrounding the pseudocoel. The digestive tract is complete (except in Acanthocephala), and it, along with gonads and excretory organs, is within the pseudocoel and bathed in perivisceral fluid. The epidermis in many secretes a nonliving cuticle with some specializations such as bristles or spines.

A constant number of cells or nuclei in individuals of a species, or in parts of their bodies, is known as **eutely,** which is common to several groups. In most of them there is an emphasis on the longitudinal muscle layer.

ECDYSOZOAN PHYLA

Phylum Nematoda: Roundworms

Approximately 12,000 species of Nematoda (nem-a-to´da) (Gr., *nematos,* thread) have been named, but it has been estimated that if all species were known, the number would be nearer 500,000. They live in the sea, in fresh water, and in soil, from polar regions to the tropics, and from mountaintops to the depths of the sea. Good topsoil may contain billions of nematodes per acre. Nematodes also parasitize virtually every type of animal and many plants. Effects of nematode infestation on crops, domestic animals, and humans make this phylum one of the most important of all parasitic animal groups.

Free-living nematodes feed on bacteria, yeasts, fungal hyphae, and algae. They may be saprozoic or coprozoic (live in fecal material). Predatory species may eat rotifers, tardigrades,

Figure 15-1
Acoelomate, pseudocoelomate, and eucoelomate body plans.

Acoelomate

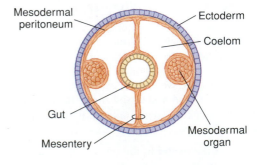

Pseudocoelomate

Eucoelomate

small annelids, and other nematodes. Many species feed on plant juices from higher plants, which they penetrate, sometimes causing agricultural damage of great proportions. Nematodes themselves may be prey for mites, insect larvae, and even nematode-capturing fungi. *Caenorhabditis elegans,* a free-living nematode, is easy to culture in the laboratory and has become an invaluable model for studies of basic developmental biology.

In 1963 Sydney Brenner started studying a free-living nematode, *Caenorhabditis elegans,* the beginning of some extremely fruitful research. Now this small worm has become one of the most important experimental models in biology. The origin and lineage of all the cells in its body (959) have been traced from zygote to adult, and the complete "wiring diagram" of its nervous system is known—all neurons and all connections between them. The genome has been completely mapped and sequenced. Many basic discoveries have been made and will be made using *C. elegans.*

Virtually every species of vertebrate and many invertebrates serve as hosts for one or more types of parasitic nematodes. Nematode parasites in humans cause much discomfort, disease, and death, and in domestic animals they are a source of great economic loss.

Form and Function

Distinguishing characteristics of this large group of animals are their cylindrical shape; their flexible, nonliving cuticle; their lack of motile cilia or flagella (except in one species); the muscles of their body wall, which have several unusual features, such as running in a longitudinal direction only, and eutely. Correlated with their lack of cilia, nematodes do not have protonephridia; their excretory system consists of one or more large gland cells opening by an excretory pore, or a canal system without gland cells, or both cells and canals together. Their pharynx is characteristically muscular with a triradiate lumen and resembles the pharynx of gastrotrichs and of kinorhynchs. Use of the pseudocoel as a hydrostatic organ is highly developed in nematodes, and much functional morphology of nematodes can be best understood in the context of the high **hydrostatic pressure** (turgor) in the pseudocoel.

Most nematode worms are less than 5 cm long, and many are microscopic, but some parasitic nematodes are more than 1 m in length.

Their outer body covering is a relatively thick, noncellular **cuticle** secreted by the underlying epidermis **(hypodermis).** The hypodermis is syncytial, and its nuclei are located in four **hypodermal cords** that project inward (Figure 15-2). Dorsal and ventral hypodermal cords bear longitudinal dorsal and ventral nerves, and the lateral cords bear excretory canals. The cuticle is of great functional importance to the worm, serving to contain the high hydrostatic pressure exerted by fluid in the pseudocoel. The several layers of the cuticle are primarily of **collagen,** a structural protein also abundant in vertebrate connective tissue. Three of the layers are composed of crisscrossing fibers, which confer some longitudinal elasticity on the worm but severely limit its capacity for lateral expansion.

Characteristics of Pseudocoelomate Phyla

1. Symmetry bilateral; unsegmented; triploblastic (three germ layers)
2. Body cavity a **pseudocoel**
3. Size mostly small; some microscopic; a few a meter or more in length
4. Body vermiform; body wall a **syncytial** or cellular epidermis with thickened cuticle, sometimes molted; muscular layers mostly of **longitudinal fibers;** cilia absent in several phyla
5. Digestive system (lacking in acanthocephalans) complete with mouth, enteron, and anus; pharynx muscular and well developed: **tube-within-a-tube arrangement;** digestive tract usually only an epithelial tube with **no definite muscle layer**
6. Circulatory and respiratory organs lacking
7. Excretory system of canals and protonephridia in some; cloaca that receives excretory, reproductive, and digestive products may be present
8. Nervous system of cerebral ganglia or of a circumenteric nerve ring connected to anterior and posterior nerves; sense organs of ciliated pits, papillae, bristles, and some eyespots
9. Reproductive system of gonads and ducts that may be single or double; sexes nearly always separate, with males usually smaller than females; eggs microscopic with shell often containing chitin
10. Development may be direct or within a complicated life history; cleavage mostly mosaic; **constancy in number of cells or nuclei common**

Body wall muscles of nematodes are very unusual. They lie beneath the hypodermis and contract longitudinally only. There are no circular muscles in the body wall. The muscles are arranged in four bands, or quadrants, marked off by the four hypodermal cords (Figure 15-2). Each muscle cell has a contractile **fibrillar** portion (or **spindle**) and a noncontractile **sarcoplasmic** portion (cell body). The spindle is distal and abuts the hypodermis, and the cell body projects into the pseudocoel. The spindle is striated with bands of actin and myosin, reminiscent of vertebrate skeletal muscle (see Figure 9-7, p. 184, and p. 626). The cell bodies contain the nuclei and are a major depot for glycogen storage in the worm. From each cell body a process or **muscle arm** extends either to the ventral or the dorsal nerve. Though not unique to nematodes, this arrangement is very curious; in most animals nerve processes (axons, p. 696) extend to the muscle, rather than the other way around.

Their fluid-filled pseudocoel, in which the internal organs lie, constitutes a hydrostatic skeleton. Hydrostatic skeletons, found in many invertebrates, lend support by transmitting the force of muscle contraction to the enclosed, noncompressible fluid. Normally, muscles are arranged antagonistically, so that movement is effected in one direction by contraction of one group of muscles, and movement back in the opposite direction is effected by the antagonistic set of muscles. However, nematodes do not have circular body wall muscles to antagonize the longitudinal muscles; therefore the cuticle must serve that function. As muscles on one side of the body contract, they compress the cuticle on that side, and the force of the contraction is

Figure 15-2

A, Structure of a nematode as illustrated by *Ascaris* female. *Ascaris* has two ovaries and uteri, which open to the outside by a common genital pore. **B,** Cross section. **C,** Single muscle cell; spindle abuts hypodermis, muscle arm extends to dorsal or ventral nerve.

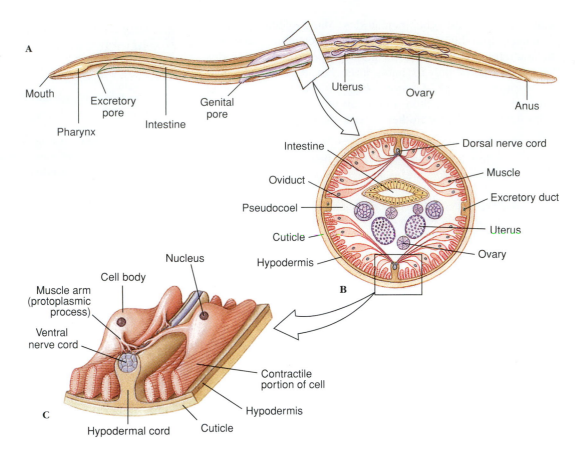

transmitted (by the fluid in the pseudocoel) to the other side of the nematode, stretching the cuticle on that side. This compression and stretching of the cuticle serve to antagonize the muscle and are the forces that return the body to resting position when the muscles relax; this action produces the characteristic thrashing motion seen in nematode movement. An increase in efficiency of this system can be achieved only by an increase in hydrostatic pressure. Consequently, hydrostatic pressure in the nematode pseudocoel is much higher than is usually found in other kinds of animals that have hydrostatic skeletons but that also have antagonistic muscle groups.

The alimentary canal of nematodes consists of a mouth (Figure 15-2), a muscular pharynx, a long nonmuscular intestine, a short rectum, and a terminal anus. Food is sucked into the pharynx when the muscles in its anterior portion contract rapidly and open the lumen. Relaxation of the muscles anterior to the food mass closes the lumen of the pharynx, forcing the food posteriorly toward the intestine. The intestine is one cell-layer thick. Food matter moves posteriorly by body movements and by additional food being passed into the intestine from the pharynx. Defecation is accomplished by muscles that simply pull the anus open, and expulsive force is provided by the high pseudocoelomic pressure that surrounds the gut.

Adults of many parasitic nematodes have an anaerobic energy metabolism; thus, a Krebs cycle and cytochrome system characteristic of aerobic metabolism are absent. They derive energy through glycolysis and probably through some incompletely known electron-transport sequences. Interestingly, some free-living nematodes and free-living stages of parasitic nema-

todes are obligate aerobes and have a Krebs cycle and cytochrome system.

A **ring of nerve tissue and ganglia** around the pharynx gives rise to small nerves to the anterior end and to two **nerve cords,** one dorsal and one ventral. **Sensory papillae** are concentrated around the head and tail. The **amphids** (Figure 15-3) are a pair of somewhat more complex sensory organs that open on each side of the head at about the same level as the cephalic circle of papillae. The amphidial opening leads into a deep cuticular pit with sensory endings of modified cilia. Amphids are usually reduced in nematode parasites of animals, but most parasitic nematodes bear a bilateral pair of **phasmids** near the posterior end. They are rather similar in structure to amphids.

Most nematodes are dioecious. Males are smaller than females, and their posterior end usually bears a pair of **copulatory spicules** (Figure 15-4). Fertilization is internal, and eggs are usually stored in the uterus until deposition. After embryonation a juvenile worm hatches. The four juvenile stages are each separated by a molt, or shedding, of the cuticle. Many parasitic nematodes have free-living juvenile stages. Others require an intermediate host to complete their life cycles.

Some Nematode Parasites

As mentioned on p. 297, nearly all vertebrates and many invertebrates are parasitized by nematodes. A number of these are very important pathogens of humans and domestic animals. A few nematodes are common in humans in North America (Table 15-1), but they and many others usually abound in tropical countries. Space permits mention of only a few in this discussion.

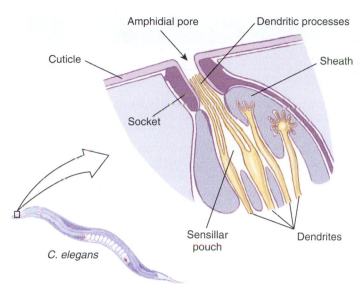

Figure 15-3

Diagram of an amphid in *Caenorhabditis elegans*.

Redrawn from Wright, K. A. 1980. Nematode sense organs. In B. M. Zuckerman (ed.), Nematodes as biological models, Vol. 2, Aging and other model systems. Copyright © Academic Press, New York.

Copulatory spicules of male nematodes are not true intromittent organs, since they do not conduct sperm, but are another adaptation to cope with high internal hydrostatic pressure. Spicules must hold the vulva of a female open while the ejaculatory muscles overcome the hydrostatic pressure in the female and rapidly inject sperm into her reproductive tract. Furthermore, nematode spermatozoa are unique among those studied in the animal kingdom in that they lack a flagellum and acrosome. Within a female's reproductive tract, sperm become ameboid and move by pseudopods. Could this be another adaptation to the high hydrostatic pressure in the pseudocoel?

Ascaris lumbricoides: *The Large Roundworm of Humans*

Because of its size and availability, *Ascaris* (Gr. *askaris,* intestinal worm) is usually selected as a type for study in zoology, as well as in experimental work. Thus it is probable that parasitologists know more about structure, physiology, and biochemistry of *Ascaris* than of any other nematode. This genus includes several species. One of the most common, *A. megalocephala,* is found in the intestine of horses. *Ascaris lumbricoides* (Figure 15-5) is one of the most common parasites found in humans; recent surveys have shown a prevalence of up to 64% in some areas of the southeastern United States, and more than 1.2 *billion* people are infected worldwide. The large roundworm of pigs, *A. suum,* is morphologically close to *A. lumbricoides,* and they were long considered the same species.

A female *Ascaris* may lay 200,000 eggs a day, passing out in the host's feces. Given suitable soil conditions, embryonation is complete within 2 weeks. Direct sunlight and high temperatures are rapidly lethal, but the eggs have an amazing tolerance to other adverse conditions, such as desiccation or lack of oxygen. Shelled juveniles can remain viable for many months or even years in soil. Infection usually occurs when eggs are ingested with uncooked vegetables or when children put soiled fingers or toys in their mouths. Unsanitary defecation habits "seed" the soil, and viable eggs remain long after all signs of the fecal matter have disappeared.

When a host swallows embryonated eggs, the tiny juveniles hatch. They burrow through the intestinal wall into veins or lymph vessels and are carried through the heart to the lungs. There they break out into alveoli and are carried up to the trachea. If the infection is large, they may cause a serious pneumonia at this stage. On reaching the pharynx, juveniles are swallowed, passed through the stomach, and finally mature about 2 months after the eggs were ingested. In the intestine, where they feed on intestinal contents, worms cause abdominal symptoms and allergic reactions, and in large numbers they may cause intestinal blockage. Perforation of the intestine with resultant peritonitis is not uncommon, and wandering worms may occasionally emerge from the anus or throat or may enter the trachea or eustachian tubes and middle ears.

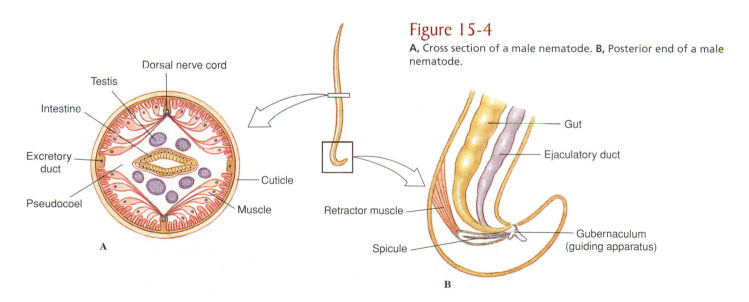

Figure 15-4

A, Cross section of a male nematode. **B,** Posterior end of a male nematode.

TABLE 15.1

Common Parasitic Nematodes of Humans in North America

Common and Scientific Names	Mode of Infection; Prevalence
Hookworm (*Ancylostoma duodenale* and *Necator americanus*)	Contact in soil with juveniles that burrow into skin; common in southern states
Pinworm (*Enterobius vermicularis*)	Inhalation of dust with ova and by contamination with fingers; most common worm parasite in United States
Intestinal roundworm (*Ascaris lumbricoides*)	Ingestion of embryonated ova in contaminated food; common in rural areas of Appalachia and southeastern states
Trichina worm (*Trichinella* spp.)	Ingestion of infected muscle; occasional in humans throughout North America
Whipworm (*Trichuris trichiura*)	Ingestion of contaminated food or by unhygienic habits; usually common wherever *Ascaris* is found

A

B

Figure 15-5

A, Intestinal roundworm *Ascaris lumbricoides,* male and female. Male, *top,* is smaller and has characteristic sharp kink in the end of the tail. Females of this large nematode may be over 30 cm long. **B,** Intestine of a pig, nearly completely blocked by *Ascaris suum.* Such heavy infections are also fairly common with *A. lumbricoides* in humans.

Other ascarids are common in wild and domestic animals. Species of *Toxocara,* for example, are found in dogs and cats. Their life cycle is generally similar to that of *Ascaris,* but juveniles often do not complete their tissue migration in adult dogs, remaining in the host's body in a stage of arrested development. Pregnancy in a female dog, however, stimulates juvenile worms to wander, and they infect the embryos in the uterus. Puppies are then born with worms. These ascarids also survive in humans but do not complete their development, leading to an occasionally serious condition in children known as *visceral larva migrans.* This is a good argument for pet owners to practice hygienic disposal of canine wastes!

Hookworms   Hookworms are so named because the anterior end curves dorsally, suggesting a hook. The most common species is *Necator americanus* (L. *necator,* killer), whose females are up to 11 mm long. Males can reach 9 mm in length. Large plates in their mouths (Figure 15-6) cut into the intestinal

mucosa of the host where they suck blood and pump it through their intestine, partially digesting it and absorbing the nutrients. They suck much more blood than they need for food, and heavy infections cause anemia in patients. Hookworm disease in children may result in retarded mental and physical growth and a general loss of energy.

Eggs pass in the feces, and juveniles hatch in the soil, where they live on bacteria (Figure 15-7). When human skin comes in contact with infected soil, infective juveniles burrow through the skin to the blood, and reach the lungs and finally the intestine in a manner similar to that described for *Ascaris.*

Trichina Worm   *Trichinella spiralis* (Gr. *trichinos,* of hair, + *-ella,* diminutive) is one of several species of tiny nematodes responsible for the potentially lethal disease trichinosis. Adult worms burrow in the mucosa of the small intestine where females produce living young. Juveniles penetrate blood vessels and are carried throughout the body, where they may be found in almost any tissue or body space. Eventually, they penetrate skeletal muscle cells, becoming one of the largest known intra-

Figure 15-6

Section through anterior end of hookworm attached to dog intestine. Note cutting plates of mouth pinching off mucosa from which the thick muscular pharynx sucks blood. Esophageal glands secrete anticoagulant to prevent blood clotting.

Plates

cellular parasites. Juveniles cause astonishing redirection of gene expression in their host cell, which loses its striations and becomes a **nurse cell** that nourishes the worm (Figure 15-8). When meat containing live juveniles is swallowed, the worms are liberated into the intestine where they mature.

Trichinella spp. can infect a wide variety of mammals in addition to humans, including hogs, rats, cats, and dogs. Hogs become infected by eating garbage containing pork scraps with juveniles or by eating infected rats. In addition to *T. spiralis*, we now know there are four other sibling species in the genus. They differ in geographic distribution, infectivity to different host species, and freezing resistance.

Heavy infections may cause death, but lighter infections are much more common—about 2.4% of the population of the United States is infected.

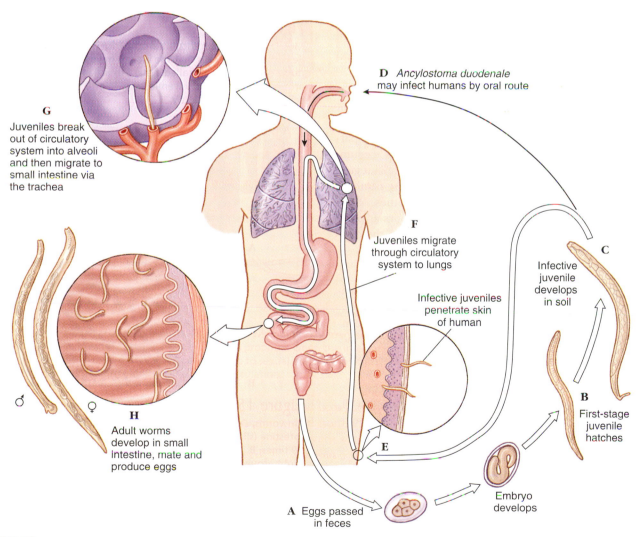

G
Juveniles break out of circulatory system into alveoli and then migrate to small intestine via the trachea

D *Ancylostoma duodenale* may infect humans by oral route

F
Juveniles migrate through circulatory system to lungs

C
Infective juvenile develops in soil

Infective juveniles penetrate skin of human

♂   ♀

H
Adult worms develop in small intestine, mate and produce eggs

E

B
First-stage juvenile hatches

A Eggs passed in feces

Embryo develops

Figure 15-7

The life cycle of hookworms. **A,** Shelled embryo passed in feces. **B,** First-stage juvenile hatches. **C,** Two molts ensue and then third-stage juvenile (infective) enters developmental arrest until it reaches a new host. **D,** *Ancylostoma duodenale* may infect humans by oral route. **E,** Infective juveniles penetrate skin of humans. **F,** Juveniles migrate through circulatory system to lungs. **G,** Juveniles break out of circulatory system into alveoli and then migrate to small intestine via the trachea. **H,** Adult worms develop in the small intestine, mate, and produce eggs.

Drawing by William Ober and Claire Garrison.

Figure 15-11

Elephantiasis of leg caused by adult filarial worms of *Wuchereria bancrofti,* which live in lymph passages and block the flow of lymph. Tiny juveniles, called microfilariae, are ingested with blood meal of mosquitos, where they develop to infective stage and are transmitted to a new host.

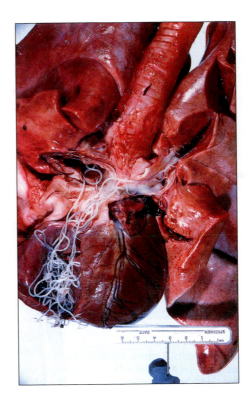

Figure 15-12

Dirofilaria immitis in right ventricle, extending up into right and left pulmonary arteries of an eight-year-old Irish setter.

nematodes, with which they share the structure of the cuticle, presence of epidermal cords, longitudinal muscles only, and pattern of nervous system. However, since the early larval form of some species has a striking resemblance to Priapulida, it is impossible to say to what group the nematomorphs are most closely related.

About 250 species of horsehair worms have been named. Worldwide in distribution, they are free living as adults and parasitic in arthropods as juveniles. Adults do not feed but will live almost anywhere in wet to moist surroundings if oxygen is adequate. Some juveniles, such as *Gordius* (named for an ancient king who tied an intricate knot), a cosmopolitan genus, are believed to encyst on vegetation that may later serve as food for a grasshopper or other arthropod. In the marine form *Nectonema* (Gr. *nektos,* swimming, + *nema,* thread), juveniles occur in hermit crabs and other crabs.

Form and Function

Horsehair worms are extremely long and slender, with a cylindrical body. Their length ranges from 10 to 70 cm, but their diameter is only 0.3 to 2.5 mm. Their anterior end is usually rounded, and their posterior end is rounded or has two or three caudal lobes (Figure 15-13).

Their body wall is much like that of nematodes: a secreted cuticle, a hypodermis, and musculature of **longitudinal muscles** only. Ventral, or dorsal and ventral, but not lateral, hypodermal cords are present. In most nematomorphs the ventral nerve cord is connected to the ventral hypodermal cord by a **nervous lamella.**

Their digestive system is vestigial. The pharynx is a solid cord of cells, and the intestine does not open to the cloaca. Lar-

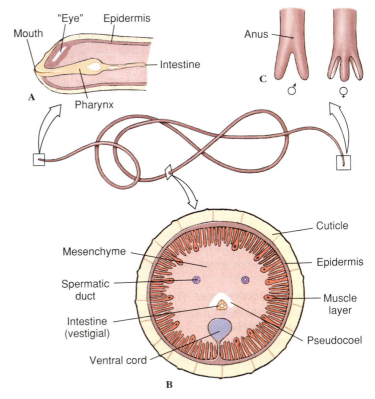

Figure 15-13

Structure of *Paragordius,* a nematomorph. **A,** Longitudinal section through the anterior end. **B,** Transverse section. **C,** Posterior end of male and female worms. Nematomorphs, or "horsehair worms," are very long and very thin. Their pharynx is usually a solid cord of cells and is nonfunctional. *Paragordius,* whose pharynx opens through to the intestine, is unusual in this respect and also in the possession of a photosensory organ ("eye").

val forms absorb food from their arthropod hosts through their body wall, and adults apparently live on stored nutrients.

Circulatory, respiratory, and excretory systems are lacking. There are a nerve ring around the pharynx and a midventral nerve cord.

Juveniles do not emerge from their arthropod host unless water is nearby. Adults are often seen wriggling slowly about in ponds or streams, with males being more active than females. Each sex has a pair of gonads and a pair of gonoducts that empty into a cloaca. Females discharge eggs into water in long strings. Juveniles hatch from eggs and somehow gain entry into an arthropod host. After several months in the hemocoel of the host, the matured worm emerges into the water. Curiously, if the host is a terrestrial insect, the parasite stimulates its host by an unknown mechanism to seek water.

Phylum Kinorhyncha

Kinorhyncha (kin'o-ring'ka) (Gr. *kinein*, to move, + *rhynchos*, beak) are marine worms a little larger than rotifers and gastrotrichs but usually not more than 1 mm long. This phylum has also been called Echinodera, meaning spiny necked. About 75 species have been described.

Kinorhynchs are cosmopolitan, living from pole to pole, from intertidal areas to 6000 m in depth. Most live in mud or sandy mud, but some have been found in algal holdfasts, sponges, or other invertebrates. They feed mainly on diatoms. About 100 species have been reported. Among the best-known genera of Kinorhyncha are *Echinoderes*, *Pycnophyes*, and *Kinorhynchus*.

Form and Function

The body of kinorhynchs is divided into 13 segments, which bear spines but have no cilia (Figure 15-14). The retractile head has a circlet of spines with a small retractile proboscis. Their body is flat ventrally and arched dorsally. Their body wall is composed of a cuticle, a syncytial epidermis, and longitudinal epidermal cords, much like those of nematodes. The arrangement of muscles is correlated with the segments, and circular, longitudinal, and diagonal muscle bands are all represented.

A kinorhynch cannot swim. In silt and mud where it commonly lives, it burrows by extending the head into the mud and anchoring it with spines. It then draws its body forward until its head is retracted into its body. When disturbed, a kinorhynch draws in its head and protects it with a closing apparatus of cuticular plates (Figure 15-14).

Their digestive system is complete, with a mouth at the tip of a proboscis, a pharynx, an esophagus, a stomach-intestine, and an anus. Kinorhynchs feed on diatoms or on organic material in the mud where they burrow.

Their pseudocoel is filled with amebocytes containing fluid. A multinucleated solenocyte protonephridium on each side of the tenth and eleventh segments serve as their excretory system. Each solenocyte has one long and one short flagellum.

The nervous system is in contact with the epidermis, with a multilobed brain encircling their pharynx, and with a ventral ganglionated nerve cord extending throughout the body. Sense organs are represented by eyespots in some and by sensory bristles.

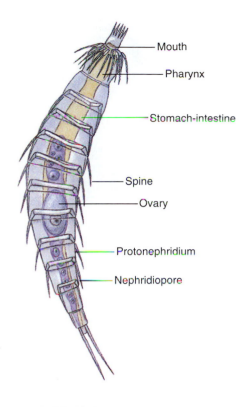

Figure 15-14

Echinoderes, a kinorhynch, is a minute marine worm. Segmentation is superficial. The head, with its circle of spines, is retractile.

Labels: Mouth, Pharynx, Stomach-intestine, Spine, Ovary, Protonephridium, Nephridiopore

Sexes are separate, with paired gonads and gonoducts. There is a series of about six juvenile stages and a definitive, nonmolting adult.

Phylum Loricifera

Loricifera (L., *lorica*, corselet, + Gr., *phora*, bearing) are a very recently described phylum of animals (1983). The tiny animals (0.25 mm long) have a protective external case (lorica) and live in spaces between grains of marine gravel, to which they cling tightly. Though they were described from specimens collected off the coast of France, they are apparently widely distributed in the world.

Form and Function

Loriciferans have oral spines rather similar to those of kinorhynchs, and the entire forepart of their body can be retracted into the circular lorica (Figure 15-15). Their diet is unknown. Their brain fills most of the head, and oral spines are innervated by nerves from the brain and other ganglia. Sexes are separate, but details of reproduction are unknown. Juveniles resemble adults in several respects but have a pair of tapering toes believed to function in locomotion.

Phylum Priapulida

Priapulida (pri'a-pyu'li-da) (Gr. *priapos*, phallus, + *ida*, pl. suffix) are a small group (only 18 species) of marine worms found chiefly in colder waters of both hemispheres. They have been reported along the Atlantic coast from Massachusetts to Greenland and along the Pacific coast from California to Alaska. They live in mud and sand of the sea floor and range from intertidal zones to depths of several thousand meters. *Tubiluchus* (L. *tubulus*, dim.

Figure 15-15

Dorsal view of adult loriciferan, *Nanoloricus mysticus*.

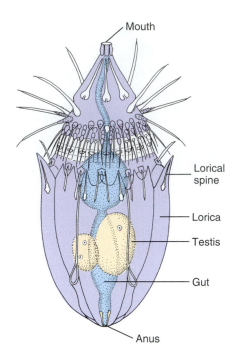

Mouth

Lorical spine

Lorica

Testis

Gut

Anus

Figure 15-16

Major internal structures of *Priapulus.*

Mouth

Proboscis

Pharynx

Retractor muscles

Intestine

Urogenital organ

Urogenital duct

Rectum

Caudal appendage

of *tubus,* waterpipe) is a minute detritus feeder adapted to interstitial life in warm coralline sediments. *Maccabeus* (named for a Judean patriot who died in 160 B.C.) is a tiny tube-dweller discovered in muddy Mediterranean bottoms.

Form and Function

Priapulids have cylindrical bodies, usually less than 12 to 15 cm long, but *Halicryptus higginsi* is up to 39 cm in length. Most are burrowing predaceous animals that usually orient themselves upright in mud with their mouth at the surface. They are adapted for burrowing by body contractions.

The body includes a proboscis, trunk, and usually one or two caudal appendages (Figure 15-16). Their eversible proboscis is ornamented with papillae and ends with rows of curved spines that surround their mouth. The proboscis is used in sampling surroundings as well as for capturing small, soft-bodied prey. *Maccabeus* has a crown of brachial tentacles around its mouth.

Their trunk is not metameric but is superficially divided into 30 to 100 rings and is covered with tubercles and spines. The tubercles are probably sensory in function. The anus and urogenital pores are located at the posterior end of the trunk. Caudal appendages are hollow stems believed to be respiratory and probably chemoreceptive in function. A chitinous cuticle, molted periodically throughout life, covers their body.

Their digestive system contains a muscular pharynx and a straight intestine and rectum (Figure 15-16). There is a nerve ring around the pharynx and a midventral nerve cord. Amebocytes occur in their pseudocoel and, at least in *Priapulus caudatus,* corpuscles containing a respiratory pigment called hemerythrin.

Sexes are separate. Paired urogenital organs are each made up of a gonad and clusters of solenocytes, both connected to a protonephridial tubule that carries both gametes and excretory products outside the body. Embryology is poorly known. In some the zygote undergoes radial cleavage and develops into a stereogastrula. Larvae of *Priapulus* dig into mud and become detritus feeders.

Long considered pseudocoelomate, priapulids were erroneously judged coelomate when nuclei were found in membranes lining their body cavity, the membranes thus interpreted as a peritoneum. However, electron microscopy showed that nuclei of their muscle cells were peripheral, and the muscles secreted a extracellular membrane. The muscle nuclei and extracellular membrane gave the false appearance of an epithelial lining.

LOPHOTROCHOZOAN PHYLA

Phylum Rotifera

Rotifera (ro-tif′e-ra) (L. *rota,* wheel, + *fera,* those that bear) derive their name from the characteristic ciliated crown, or **corona,** that, when beating, often gives the impression of rotating wheels. Rotifers range from 40 μm to 3 mm in length, but most are between 100 and 500 μm long. Some have beautiful colors, although most are transparent, and some have bizarre shapes (Figure 15-17). Their shapes are often correlated with their mode of life. Floaters are usually globular and saclike; creepers and swimmers are somewhat elongated and wormlike; and sessile types are commonly vaselike, with a thickened outer epidermis (lorica). Some are colonial. One of the best-known genera is *Philodina* (Gr. *philos,* fond of, + *dinos,* whirling) (Figure 15-18), which is often used for study.

Rotifers are a cosmopolitan group of about 1800 species, some of which are found throughout the world. Most species are freshwater inhabitants, a few are marine, some are terrestrial, and some are epizoic (live on the body of another animal) or parasitic.

Rotifers are adapted to many ecological conditions. Most species are benthic, living on the bottom or in vegetation of

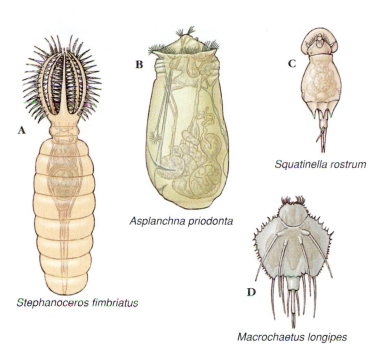

Figure 15-17
Variety of form in rotifers, **A,** *Stephanoceros* has five long, fingerlike coronal lobes with whorls of short bristles. It catches its prey by closing its funnel when food organisms swim into it, and the bristly lobes prevent prey from escaping. **B,** *Asplanchna* is a pelagic, predatory genus with no foot. **C,** *Squatinella* has a semicircular nonretractable, transparent hoodlike extension covering the head. **D,** *Machrochaetus* is dorsoventrally flattened.

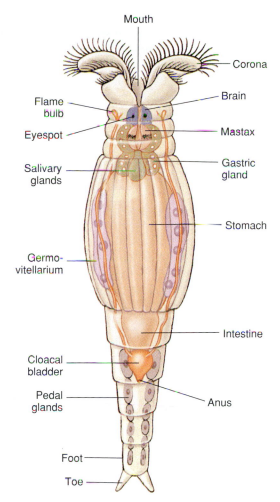

Figure 15-18
Structure of *Philodina* rotifer.

ponds or along the shores of freshwater lakes where they swim or creep about on the vegetation. A large proportion of species that live in the water film between sand grains of beaches (meiofauna) are rotifers. Pelagic forms (Figure 15-17B) are common in surface waters of freshwater lakes and ponds, and they may exhibit cyclomorphosis, variations in body form resulting from seasonal or nutritional changes.

Many species of rotifers can endure long periods of desiccation, during which they resemble grains of sand. While in a desiccated condition, rotifers are very tolerant of temperature variations, especially those rotifers that dwell in mosses. True encystment occurs in only a few rotifers. On addition of water, desiccated rotifers resume their activity.

Strictly marine species are rather few in number. Some littoral (intertidal) species of the sea may be freshwater ones that are able to adapt to seawater.

Form and Function

External Features   The body of a rotifer is composed of a head bearing a ciliated corona, a trunk, and a posterior tail, or foot. It is covered with a cuticle and is nonciliated except for the corona.

Their ciliated corona, or crown, surrounds a nonciliated central area of their head, which may bear sensory bristles or papillae. The appearance of the head end depends on which of several types of corona it has—usually a circlet of some sort, or a pair of trochal (coronal) discs (the term *trochal* comes from a Greek word meaning wheel). Cilia on the corona beat in succes-

sion, giving the appearance of a revolving wheel or pair of wheels. Their **mouth** is located in the corona on the midventral side. Coronal cilia are used in both locomotion and feeding.

The trunk may be elongated, as in *Philodina* (Figure 15-18), or saccular in shape (see Figure 15-17). It contains visceral organs and often bears sensory antennae. The body wall of many species is superficially ringed to simulate segmentation. Though some rotifers have a true, secreted cuticle, all have a fibrous layer within their epidermis. The fibrous layer in some is quite thick and forms a caselike **lorica,** which is often arranged in plates or rings.

Their **foot** is narrower and usually bears one to four **toes.** Its cuticle may be ringed so that it is telescopically retractile. It is tapered gradually in some forms (Figure 15-18) and sharply set off in others (see Figure 15-17). The foot is an attachment organ and contains **pedal glands** that secrete an adhesive material used by both sessile and creeping forms. In swimming pelagic forms, the foot is usually reduced. Rotifers move by creeping with leechlike movements aided by the foot, or by swimming with the coronal cilia, or both.

Internal Features   Underneath the cuticle is a **syncytial epidermis,** which secretes the cuticle, and bands of **subepidermal muscles,** some circular, some longitudinal, and some running through the pseudocoel to visceral organs. The **pseudocoel** is large, occupying the space between body wall and viscera. It is filled with fluid, some of the muscle bands, and a network of mesenchymal ameboid cells.

The digestive system is complete. Some rotifers feed by sweeping minute organic particles or algae toward the mouth by the beating of the coronal cilia. Cilia are able to dispose of larger unsuitable particles. Their pharynx **(mastax)** is fitted with a muscular portion that is equipped with hard jaws **(trophi)** for sucking in and grinding up food particles. The constantly chewing mastax is often a distinguishing feature of these tiny animals. Carnivorous species feed on protozoa and small metazoans, which they capture by trapping or grasping. Trappers have a funnel-shaped area around the mouth. When small prey swim into the funnel, the lobes fold inward to capture and hold them until they are drawn into the mouth and pharynx. Hunters have trophi that can be projected and used like forceps to seize prey, bring it back into the pharynx, and then pierce it or break it up so that edible parts can be sucked out and the rest discarded. **Salivary** and **gastric glands** are believed to secrete enzymes for extracellular digestion. Absorption occurs in the stomach.

Their excretory system typically consists of a pair of **protonephridial tubules,** each with several **flame cells,** that empty into a common bladder. The bladder, by pulsating, empties into a **cloaca**—into which the intestine and oviducts also empty. The fairly rapid pulsation of the protonephridia—one to four times per minute—would indicate that protonephridia are important osmoregulatory organs. Water apparently enters through the mouth rather than across the epidermis; even marine species empty their bladder at frequent intervals.

The nervous system consists of a bilobed **brain,** dorsal to the mastax, that sends paired nerves to the sense organs, mastax, muscles, and viscera. Sensory organs include paired **eyespots** (in some species such as *Philodina*), sensory bristles and papillae, and ciliated pits and dorsal antennae.

Reproduction   Rotifers are dioecious, and males are usually smaller than females. In the class Bdelloidea males are entirely unknown, and in Monogononta they seem to occur only for a few weeks of the year.

The female reproductive system in the Bdelloidea and Monogononta consists of combined ovaries and yolk glands **(germovitellaria)** and oviducts that open into the cloaca. Yolk is supplied to developing ova by way of flow-through cytoplasmic bridges, rather than as separate yolk cells as in ectolecithal Platythelminthes.

Mictic (Gr., *miktos,* mixed, blended) refers to the capacity of haploid eggs to be fertilized (that is, "mixed") with the male's sperm nucleus to form a diploid embryo. Amictic ("without mixing") eggs are already diploid and can develop only parthenogenetically.

In Bdelloidea (*Philodina,* for example), all females are parthenogenetic and produce diploid eggs that hatch into diploid females. These females reach maturity in a few days. In class Seisonidea females produce haploid eggs that must be fertilized

and that develop into either males or females. In Monogononta, however, females produce two kinds of eggs (Figure 15-19). During most of the year diploid females produce thin-shelled, **diploid amictic eggs.** Such eggs, called **amictic,** develop parthenogenetically into diploid (amictic) females. However, such rotifers often live in temporary ponds or streams and are cyclic in their reproductive patterns. Any one of several environmental factors—for example, crowding, diet, or photo-period (according to species)—may induce amictic eggs to develop into diploid mictic females

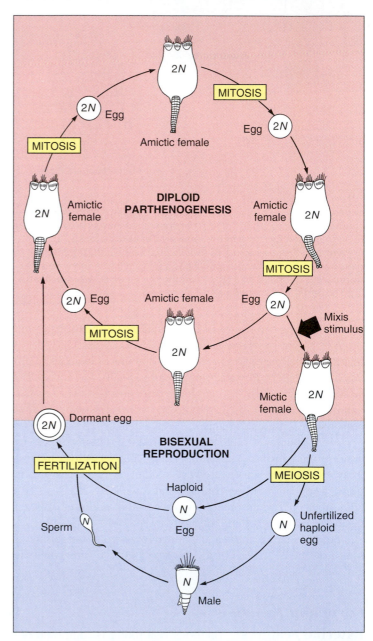

Figure 15-19

Reproduction of some rotifers (class Monogononta) is parthenogenetic during the part of the year when environmental conditions are suitable. In response to certain stimuli, females begin to produce haploid (*N*) eggs. If haploid eggs are not fertilized, they hatch into haploid males. Males provide sperm to fertilize other haploid eggs, which then develop into diploid (2*N*), dormant eggs that can resist the rigors of winter. When suitable conditions return, dormant eggs resume development, and a female hatches.

Classification of Phylum Rotifera

Class Seisonidea (sy'son-id'e-a) (Gr. *seison*, earthen vessel, + *eidos*, form). Marine; elongated form; corona vestigial; sexes similar in size and form; females with pair of ovaries and no vitellaria; single genus *(Seison)* with two species; epizoic on gills of a crustacean *(Nebalia)*.

Class Bdelloidea (del-oyd'e-a) (Gr. *bdella*, leech, + *eidos*, form). Swimming or creeping forms; anterior end retractile; corona usually with pair of trochal discs; males unknown; parthenogenetic; two germovitellaria. Examples: *Philodina* (Figure 15-18), *Rotaria*.

Class Monogononta (mon'o-go-non'ta) (Gr. *monos*, one, + *gonos*, primary sex gland). Swimming or sessile forms; single germovitellarium; males reduced in size; eggs of three types (amictic, mictic, dormant). Examples: *Asplanchna* (Figure 15-17B), *Epiphanes*.

that will produce thin-shelled **haploid eggs.** If these eggs are not fertilized, they develop into haploid males. But if fertilized, the eggs, called **mictic** eggs, develop a thick, resistant shell and become dormant. They survive over winter ("winter eggs") or until environmental conditions are again suitable, at which time they hatch into diploid females. Dormant eggs are often dispersed by winds or birds, which may account for the peculiar distribution patterns of rotifers.

The male reproductive system includes a single testis and a ciliated sperm duct that runs to a genital pore (males usually lack a cloaca). The end of the sperm duct is specialized as a copulatory organ. Copulation is usually by hypodermic impregnation; the penis can penetrate any part of a female's body wall and inject sperm directly into her pseudocoel.

Females hatch with adult features, needing only a few days' growth to reach maturity. Males often do not grow and are sexually mature at hatching.

Nuclear Constancy   Most structures in rotifers are syncytial, but nuclei in various organs show a remarkable constancy in numbers in any given species (eutely). For example, E. Martini (1912) reported that in one species of rotifer he always found 183 nuclei in the brain, 39 in the stomach, 172 in the corona epithelium, and so on. Organisms with eutely show a highly precise genetic control of nuclear division and differentiation. Nuclei are programmed to differentiate and divide an exact number of times, then halt when the appointed number is reached.

Phylum Acanthocephala

Members of phylum Acanthocephala (a-kan'tho-sef'a-la) (Gr. *akantha*, spine or thorn, + *kephalē*, head) are commonly known as "spiny-headed worms." The phylum derives its name from one of its most distinctive features, a cylindrical, invaginable proboscis bearing rows of recurved spines, by which it attaches itself to the intestine of its host. All acanthocephalans are endoparasitic, living as adults in the intestine of vertebrates.

Various species range in size from less than 2 mm to more than 1 m in length, with females of a species usually larger than males. Their body is usually bilaterally flattened, with numerous transverse wrinkles. Worms are typically cream color but may absorb yellow or brown pigments from the intestinal contents.

Acanthocephalans inflict traumatic damage by penetrating their host's intestinal wall with their spiny proboscis. In many cases there is remarkably little inflammation, but in some species the inflammatory response of the host is intense. Infection with these worms can cause great pain, particularly if the gut wall is completely perforated.

The phylum is cosmopolitan, and more than 500 species are known, most of which parasitize fish, birds, and mammals. However, no species is normally a parasite of humans, although species that usually occur in other hosts occasionally infect humans. *Macracanthorhynchus hirudinaceus* (Gr. *makros*, long, large, + *akantha*, spine, thorn, + *rhynchos*, beak) occurs throughout the world in the small intestine of pigs and sometimes in other mammals.

Larvae of spiny-headed worms develop in arthropods, either crustaceans or insects, depending on the species.

Form and Function

In life the body is somewhat flattened, although it is usual for specimens to be treated with tap water before fixation so that fixed specimens are turgid and cylindrical (Figure 15-20C).

The body wall is syncytial, and its surface is punctured 4 to 6 µm deep, by minute crypts which greatly increase the surface area of the tegument. About 80% of the thickness of the tegument is the radial fiber zone, which contains a **lacunar system** of ramifying fluid-filled canals (Figure 15-20A and B). Curiously, body-wall muscles are tubelike and filled with fluid. Tubes in the muscles are continuous with the lacunar system; therefore, circulation of lacunar fluid may well bring nutrients to and remove wastes from the muscles. There is no heart or other circulatory system, and contraction of the muscles would serve to move lacunar fluid through the canals and muscles. Both longitudinal and circular body-wall muscles are present.

Their proboscis, which bears rows of recurved hooks, is attached to the neck region (Figure 15-20) and can be inverted into a **proboscis receptacle** by retractor muscles. Attached to the neck region (but not within the proboscis) are two elongated **lemnisci** (extensions of the tegument and lacunar system) that may serve as reservoirs of lacunar fluid from the proboscis when that organ is invaginated.

There is no respiratory system. When present, the excretory system consists of a pair of **protonephridia** with flame cells. These unite to form a common tube opening into the sperm duct or uterus.

Their nervous system has a central ganglion within the proboscis receptacle and nerves to the proboscis and body. There are sensory endings on the proboscis and genital bursa.

Acanthocephalans have no digestive tract, and they must absorb all nutrients through their tegument. They can absorb various molecules by specific membrane transport mechanisms, and other substances can cross the tegumental membrane by

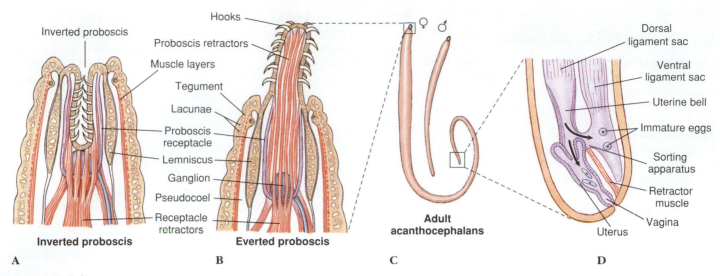

Figure 15-20

Structure of a spiny-headed worm (phylum Acanthocephala). **A** and **B,** Eversible spiny proboscis by which the parasite attaches to the intestine of its host, often doing great damage. Because they lack a digestive tract, food is absorbed through the tegument. **C,** Male is typically smaller than female. **D,** Scheme of the genital selective apparatus of a female acanthocephalan. It is a unique device for separating immature from mature fertilized eggs. Eggs containing larvae enter the uterine bell and pass on to the uterus and exterior. Immature eggs are shunted into the ventral ligament sac or into the pseudocoel to undergo further development.

pinocytosis (probably potocytosis). Their tegument bears some enzymes, such as peptidases, which can cleave several dipeptides, and the amino acids are then absorbed by the worm. Like cestodes, acanthocephalans require host dietary carbohydrate, but their mechanism for absorption of glucose is different. As glucose is absorbed, it is rapidly phosphorylated and compartmentalized, so that a metabolic "sink" is created into which glucose from the surrounding medium can flow. Glucose diffuses down the concentration gradient into the worm because it is constantly removed as soon as it enters.

Acanthocephalans are dioecious. A pair of tubular **genital ligaments,** or **ligament sacs,** extends posteriorly from the end of the proboscis receptacle. Males have a pair of testes, each with a vas deferens, and a common ejaculatory duct that ends in a small penis. During copulation sperm are ejected into the vagina, travel up the genital duct, and escape into the pseudocoel.

In females the ovarian tissue in the ligament sac breaks up into **ovarian balls** that rupture the ligament sacs and float free in the pseudocoel. One of the ligament sacs leads to a funnel-shaped **uterine bell** that receives the developing shelled embryos and passes them on to the uterus (Figure 15-20). An interesting and unique selective apparatus operates here. Fully developed embryos are slightly longer than immature ones, and they are passed on into the uterus, while immature eggs are retained for further maturation.

The shelled embryos, which are discharged in feces of their vertebrate host, do not hatch until eaten by an intermediate host. For *M. hirudinaceus* this is any of several species of soil-inhabiting beetle larvae, especially scarabeids. Grubs of the June beetle (*Phyllophaga*) are frequent hosts. Here the larva (**acanthor**) burrows through the intestine and develops into a juvenile (**cystacanth**) in the insect's hemocoel. Pigs become infected by eating the grubs. Multiple infections may do considerable damage to a pig's intestine, and perforations can occur.

Phylum Gastrotricha

Gastrotricha (gas-tro-tri′ka) (N. L. fr. Gr., *gaster, gastros,* stomach or belly, + *thrix, trichos,* hair) includes small, ventrally flattened animals about 65 to 500 µm long, somewhat like rotifers but lacking a corona and mastax and having a characteristically bristly or scaly body. They are usually found gliding on the substrate, or the surface of an aquatic plant or animal, by means of their ventral cilia, or they compose part of the meiofauna in interstitial spaces between substrate particles.

Gastrotrichs are found in both fresh and salt water. The 400 or so species are about equally divided between the two media. Many species are cosmopolitan, but only a few occur in both fresh water and the sea. Much is yet to be learned about their distribution.

Form and Function

A gastrotrich (Figures 15-21 and 15-22) is usually elongated, with a convex dorsal surface bearing a pattern of bristles, spines, or scales, and a flattened ciliated ventral surface. Cells on the ventral surface may be monociliated or multiciliated. The head is often lobed and ciliated, and the tail end may be forked.

A syncytial epidermis is found beneath the cuticle. Longitudinal muscles are better developed than are circular ones, and in most cases they are unstriated. Adhesive tubes secrete a substance for attachment. A dual-gland system for attachment and release is present, similar to that described for Turbellaria (p. 276). The pseudocoel is somewhat reduced and contains no amebocytes.

Their digestive system is complete and is made up of a mouth, a muscular pharynx, a stomach-intestine, and an anus (Figure 15-21B). Food is largely algae, protozoa, and detritus, which are directed to their mouth by their head cilia. Digestion appears to be extracellular. Protonephridia are equipped with **solenocytes** rather than flame cells. Solenocytes have a single flagellum enclosed in a cylinder of cytoplasmic rods.

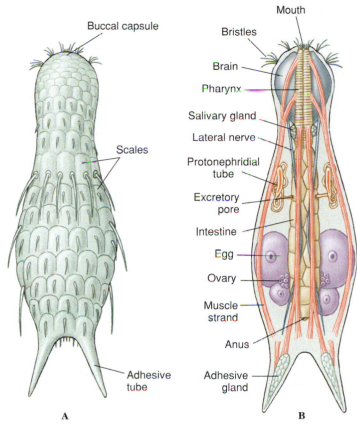

Figure 15-21

Chaetonotus, a gastrotrich. **A,** Dorsal surface. **B,** Internal structure, ventral view.

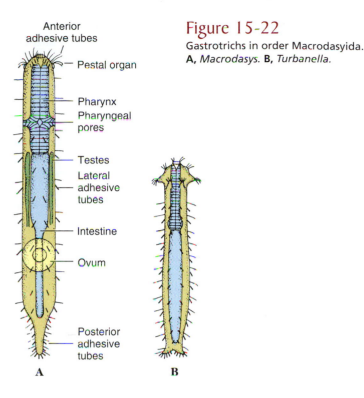

Figure 15-22

Gastrotrichs in order Macrodasyida. **A,** *Macrodasys.* **B,** *Turbanella.*

Figure 15-23

A, *Urnatella,* a freshwater entoproct, forms small colonies of two or three stalks from a basal plate. **B,** *Loxosomella,* a solitary entoproct. Both solitary and colonial entoprocts can reproduce asexually by budding, as well as sexually.

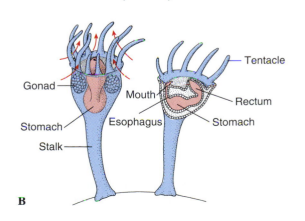

Their nervous system includes a brain near the pharynx and a pair of lateral nerve trunks. Sensory structures are similar to those in rotifers, except that eyespots are generally lacking. Sensory bristles, often concentrated on the head, are modified from cilia.

Gastrotrichs are hermaphroditic, although the male system of some is so rudimentary that they are functionally parthenogenetic females. Like rotifers, some gastrotrichs produce thin-walled, rapidly developing eggs and thick-shelled, dormant eggs. The thick-shelled eggs can withstand harsh environments and may survive dormancy for some years. Development is direct, and juveniles have the same form as adults.

Phylum Entoprocta

Entoprocta (en′to-prok′ta) (Gr. *entos,* within, + *proktos,* anus) is a small phylum of about 150 species of tiny, sessile animals that superficially resemble hydroid cnidarians but have ciliated tentacles that tend to roll inward (Figure 15-23). Most entoprocts are microscopic, and none is more than 5 mm long. They are all stalked and sessile forms; some are colonial, and some are solitary. All are ciliary feeders.

With the exception of *Urnatella* (L. *urna,* urn, + *ellus,* dim. suffix), all entoprocts are marine forms that have a wide distribution from polar regions to tropics. Most marine species are restricted to coastal and brackish waters and often grow on shells and algae. Some are commensals on marine annelid worms. Freshwater entoprocts occur on the underside of rocks in running water. *U. gracilis* is the only common freshwater species in North America (Figure 15-23A).

Form and Function

The body, or **calyx,** of an entoproct is cup shaped, bears a crown, or circle, of ciliated tentacles, and may be attached to a substratum by a single stalk and an attachment disc with adhesive glands, as in the solitary *Loxosoma* and *Loxosomella* (Gr. *loxos,* crooked, + *soma,* body) (Figure 15-23B), or by two or more stalks in colonial forms. Both tentacles and stalk are continuations of the body wall. The 8 to 30 tentacles making up the crown are ciliated on their lateral and inner surfaces, and each can move individually. Tentacles can roll inward to cover and protect the mouth and anus but cannot be retracted into the calyx.

Movement is usually restricted in entoprocts, but *Loxosoma,* which lives in the tubes of marine annelids, is quite active, moving over the annelid and its tube freely.

The gut is U-shaped and ciliated, and both the mouth and the anus open within the circle of tentacles. Entoprocts are **ciliary filter feeders.** Long cilia on the sides of the tentacles keep a current of water containing protozoa, diatoms, and particles of detritus moving in between the tentacles. Short cilia on the inner surfaces of the tentacles capture food and direct it downward toward the mouth.

Their body wall consists of a cuticle, cellular epidermis, and longitudinal muscles. The pseudocoel is largely filled with a gelatinous parenchyma in which is embedded a pair of protonephridia and their ducts, which unite and empty near the mouth. There is a well-developed **nerve ganglion** on the ventral side of the stomach, and the body surface bears sensory bristles and pits. Circulatory and respiratory organs are absent. Exchange of gases occurs through the body surface, probably much of it through the tentacles.

Some species are monoecious, some dioecious, and some appear to be protandrous, the gonad at first producing sperm and later eggs. The gonoducts open within the circle of tentacles.

Fertilized eggs develop in a depression, or brood pouch, between the gonopore and the anus. Entoprocts have a modified spiral cleavage pattern with mosaic blastomeres. The embryo gastrulates by invagination. The trochophore-like larva (see p. 321) is ciliated and free swimming. It has an apical tuft of cilia at the anterior end and a ciliated girdle around the ventral margin of the body. Eventually the larva settles to a substratum and inverts to form an adult.

PHYLOGENY AND ADAPTIVE RADIATION

Phylogeny

Evidence from a sequence analysis of the small-subunit ribosomal gene suggests that some time after ancestral deuterostomes diverged from ancestral protostomes in the Precambrian, proto-

stomes split again into two large groups (or superphyla): Ecdysozoa, containing phyla that go through a series of molts during development, and Lophotrochozoa, including lophophorate phyla (see Chapter 21) and phyla many of whose larvae are trochophore-like (p. 321).* Some pseudocolomates (Nematoda, Nematomorpha, Kinorhyncha, and Priapulida) fall into Ecdysozoa; sequences of others place them in Lophotrochozoa (Rotifera, Acanthocephala, Gastroticha, and Entoprocta). We will discuss the Ecdysozoa/Lophotrochozoa hypothesis further in Chapters 16, 17, and 18.

Loriciferans bear some similarity to kinorhynchs, larval Priapulida, larval nematomorphs, rotifers, and tardigrades (p. 436). Although loriciferans are poorly known, cladistic analysis suggests that they form a sister group to kinorhynchs and that these two phyla together are a sister group of priapulids. If so, we would expect sequence analysis to place Loricifera in Ecdysozoa.

Acanthocephalans are highly specialized parasites with a unique morphology and have doubtless been so for millions of years. Any ancestral or other related group that would shed a clue to phylogenetic relationships of Acanthocephala is probably long since extinct. However, genetic sequence analysis can provide hypothetical phylogenetic relationships when morphological or developmental similarities are virtually or completely absent. Such analysis has led to the startling conclusion that acanthocephalans are highly derived rotifers (Figure 15-24).**

Entoprocts were once included with phylum Ectoprocta in a phylum called Bryozoa, but ectoprocts are true coelomate animals, and many zoologists prefer to place them in a separate group. Ectoprocts are still often called bryozoans. Sequence analysis places both entoprocts and ectoprocts among lophotrochozoan phyla.

In December 1995, P. Funch and R. M. Kristensen reported that they had found some very strange little creatures clinging to the mouthparts of the Norway lobster (*Nephrops norvegicus*), so strange that they did not fit into any known phylum (Nature **378:**711–714). Funch and Kristensen concluded that the organisms, only 0.35 mm long, represented a new phylum, which was named **Cycliophora.** The name refers to a crown of compound cilia, reminiscent of rotifers, with which the organisms feed. They are described as "acoelomate," although whether they might have a pseudocoel is unclear, and they have a cuticle. Their life cycle seems bizarre. The sessile feeding stages on the lobster's mouthparts undergo internal budding to produce motile stages: (1) larvae containing new feeding stages; (2) dwarf males, which become attached to feeding stages that contain developing females; and (3) females, which also attach to the lobster's mouthparts, then produce dispersive larvae and degenerate.

*Aguinaldo, A. M. A., et al. 1997. Nature **387:**489–493; Balavoine, G., and A. Adoutte. 1998. Science **280:**397–398.
Welch, W. D. B. 2000. Invert. Biol. **199:17–26.

Figure 15-24

Hypothetical relationships among pseudocoelomate phyla according to Nielsen (1995). Nielsen placed all phyla with a radial pharynx and nerve ring around the pharynx in a monophyletic group he called Cycloneuralia, and he moved Entroprocta, along with Ectoprocta (p. 431), to form a protostome group (Bryozoa). The synapomorphy for nematodes "6 + 6 + 4 sensilla" refers to the rings of sensory papillae (sensilla) on the anterior end. The outgroup for this cladogram is Spiralia (protostomes with spiral cleavage, trophophore larva, and mesoderm from the 4d cell). If interpretation of sequence data (cited in text) is supported by further research, relationships shown here must be substantially revised because some phyla have been assigned to Lophotrochozoa (Rotifera, Gastrotricha) and others to Ecdysozoa (Nematoda, Nematomorpha, Priapulida, Kinorhyncha).

Whether the proposed new phylum will withstand the scrutiny of further research is unknown, and its relationships to other phyla are quite unclear. Funch and Kristensen think that the organisms are protostomes and see affinities with Entoprocta and Ectoprocta. Little short of astonishing, however, is their abundance on the mouthparts of a host as well known as Norway lobsters. How could biologists have failed to notice them before? At a time when habitat destruction drives many species to extinction every year, we wonder if there are phyla suffering the same fate. S. Conway Morris ponders the possibility of further undiscovered phyla (Nature **378:**661–662), suggesting you may need a couple of zoology textbooks and a decent microscope when you next dine at your favorite seafood restaurant: "Who knows what might be found lurking under the lettuce?"

Adaptive Radiation

Certainly the most impressive adaptive radiation in this group of phyla is shown by nematodes. They are by far the most numerous in terms of both individuals and species, and they have been able to adapt to almost every habitat available to animal life. Their basic pseudocoelomate body plan, with the cuticle, hydrostatic skeleton, and longitudinal muscles, has proved generalized and plastic enough to adapt to an enormous variety of physical conditions. Free-living lines gave rise to parasitic forms on at least several occasions, and virtually all potential hosts have been exploited. All types of life cycle occur: from simple and direct to complex, with intermediate hosts; from normal dioecious reproduction to parthenogenesis, hermaphroditism, and alternation of free-living and parasitic generations. A major factor contributing to evolutionary opportunism of nematodes has been their extraordinary capacity to survive suboptimal conditions, for example, developmental arrests in many free-living and animal parasitic species and ability to undergo cryptobiosis (survival in harsh conditions by assuming a very low metabolic rate) in many free-living and plant parasitic species.

SUMMARY

Phyla covered in this chapter possess a body cavity called a pseudocoel, which is derived from the embryonic blastocoel, rather than a secondary cavity in the mesoderm (coelom). Several groups exhibit eutely, a constant number of cells or nuclei in adult individuals of a given species.

Analysis of base sequences in the gene for small-subunit rRNA provides evidence that some pseudocoelomate phyla belong to superphylum Ecdysozoa (Nematoda, Nematomorpha, Kinorhyncha, and Priapulida) and some to Lophotrochozoa (Rotifera, Acanthocephala, Gastrotricha, and Entoprocta). Loricifera is probably an ecdysozoan based on cladistic analysis.

Nematoda is the largest and most important of these phyla, and may contain as many as 500,000 species. They are more or less cylindrical, tapering at the ends, and covered with a tough, secreted cuticle. Their body-wall muscles are longitudinal only, and to function well in locomotion, such an arrangement must enclose a volume of fluid in the pseudocoel at high hydrostatic pressure. This fact of nematode life has a profound effect on most of their other physiological functions, for example, ingestion of food, egestion of feces, excretion, copulation, and others. Most nematodes are dioecious, and there are four juvenile stages, each separated by a molt of the cuticle. Almost all invertebrate and vertebrate animals and many plants have nematode parasites, and many other nematodes are free living in soil and aquatic habitats. Some parasitic nematodes have part of their life cycle free living, some undergo a tissue migration in their host, and some have an intermediate host in their life cycle. Some parasitic nematodes cause severe diseases in humans and other animals.

Nematomorpha or horsehair worms are related to nematodes and have parasitic juvenile stages in arthropods, followed by a free-living, aquatic, nonfeeding adult stage.

Kinorhyncha and Loricifera are small phyla of tiny, aquatic pseudocoelomates. Kinorhynchs anchor and then pull themselves by spines on their head. Loriciferans can withdraw their bodies into their lorica. Priapulids are marine burrowing worms of moderate size.

Phylum Rotifera is composed of small, mostly freshwater organisms with a ciliated corona, which creates currents of water to draw planktonic food toward the mouth. Their mouth opens into a muscular pharynx, or mastax, which is equipped with jaws.

Acanthocephalans are all parasitic in the intestine of vertebrates as adults, and their juvenile stages develop in arthropods. They have an anterior, invaginable proboscis armed with spines, which they embed in the intestinal wall of their host. They do not have a digestive tract and so must absorb all nutrients across their tegument. Molecular evidence suggests a phylogenetic affinity of acanthocephalans and rotifers.

Gastrotrichs are also tiny, aquatic pseudocoelomates. They move by cilia or adhesive glands. Entoprocta are small, sessile, aquatic animals with a crown of ciliated tentacles encircling both the mouth and anus.

Of all these phyla, nematodes have undergone greater adaptive radiation than other pseudocoelomate phyla.

REVIEW QUESTIONS

1. What are some adaptive advantages of a pseudocoel compared to the acoelomate condition?
2. Explain the difference between a true coelom and a pseudocoel.
3. What is a hydrostatic skeleton?
4. Distinguish a solenocyte from a flame cell protonephridium.
5. Explain two peculiar features of the body-wall muscles in nematodes.
6. What feature of body-wall muscles in nematodes requires a high hydrostatic pressure in the pseudocoelomic fluid for efficient function?
7. Explain the interaction of cuticle, body-wall muscles, and pseudocoelomic fluid in locomotion of nematodes.
8. Explain how the high pseudocoelomic pressure affects feeding and defecation in nematodes.
9. Outline the life cycle of each of the following: *Ascaris lumbricoides*, hookworm, *Enterobius vermicularis*, *Trichinella spiralis*, *Wuchereria bancrofti*.
10. Where in the human body are adults of each species in question 9 found?
11. Outline the life cycle of a typical nematomorph.
12. How are nematodes and nematomorphs alike, and how are they different?
13. What is the normal size of a rotifer; where is it found; and what are its major features?
14. Explain the difference between mictic and amictic eggs of rotifers. What is the adaptive value of each?
15. What is eutely?
16. Describe the major features of the acanthocephalan body.
17. How do acanthocephalans get food?
18. The evolutionary ancestry of acanthocephalans is particularly obscure. Describe some characters of acanthocephalans that make it surprising that they could be highly derived rotifers.
19. About how big are loriciferans, priapulids, gastrotrichs, and kinorhynchs? Where are each of them found?
20. What are distinguishing characteristics of entoprocts?
21. What phylum covered in this chapter has radiated into the most diversity? How do the members of this phylum impact humans?

SELECTED REFERENCES

Aguinaldo, A. M. A., J. M. Turbevill, L. S. Linford, M. C. Rivera, J. J. F. R. Garey, R. A. Raff, and J. A. Lake. 1997. Evidence for a clade of nematodes, arthropods and other moulting animals. Nature **387**:489–493. *Sequence analysis to support a superphylum Ecdysozoa.*

Balavoine, G., and A. Adoutte. 1998. One or three Cambrian radiations? Science **280**:397–398. *Discusses radiation into superphyla Ecdysozoa, Lophotrochozoa, and Deuterostomia.*

Bird, A. F., and J. Bird. 1991. The structure of nematodes, ed. 2. New York, Academic Press. *The most authoritative reference available on nematode morphology. Highly recommended.*

Chan, M.-S. 1997. The global burden of intestinal nematode infections—fifty years on. Parasitol. Today **13**:438–443. *According to this author, most recent estimates are 1.273 billion infections (24% prevalence) with Ascaris, 0.902 billion (17% prevalence) with Trichuris, and 1.277 billion (24% prevalence) with hookworms. Worldwide prevalence of these nematodes has remained essentially unchanged in 50 years!*

Despommier, D. D. 1990. *Trichinella spiralis*: the worm that would be virus. Parasitol. Today **6**:193–196. *Juveniles of* Trichinella *are among the largest of all intracellular parasites.*

Duke, B. O. L. 1990. Onchocerciasis (river blindness)—can it be eradicated? Parasitol. Today **6**:82–84. *Despite the introduction of a very effective drug, the author predicts that this parasite will not be eradicated in the foreseeable future.*

Ogilvie, B. M., M. E. Selkirk, and R. M. Maizels. 1990. The molecular revolution and nematode parasitology: yesterday, today, and tomorrow. J. Parasitol. **76**:607–618. *Modern molecular biology has wrought enormous changes in investigations on nematodes.*

Poinar, G. O., Jr. 1983. The natural history of nematodes. Englewood Cliffs, New Jersey, Prentice-Hall, Inc. *Contains a great deal of information about these fascinating worms.*

Welch, M. D. B. 2000. Evidence from a protein-coding gene that acanthocephalans are rotifers. Invert. Biol. **119**:17–26. *Sequence analysis of a gene coding for a heat-shock protein supports a position of acanthocephalans within Rotifera. Other molecular and morphological evidence is cited that supports this position.*

Taylor, M. J., and A. Hoerauf. 1999. *Wolbachia* bacteria of filarial nematodes. Parasitol. Today **15**:437–442. *All filarial parasites of humans have endosymbiotic* Wolbachia, *and most filarial nematodes of all kinds are infected. Nematodes can be "cured" by treatment with the antibiotic tetracycline. If cured, they cannot reproduce. Bacteria are apparently passed vertically from females to offspring.*

ZOOLOGY LINKS TO THE INTERNET

Visit the textbook's Online Learning Center at www.mhhe.com/zoology to find live Internet links for each of the topics listed here.

Pseudocoelomate Animals
Phylum Rotifera
Phylum Kinorhyncha
Phylum Loricifera

Phylum Priapulida
Phylum Nematoda
Human Diseases Caused by Nematodes
Caenorhabditis elegans
Phylum Nematomorpha
Phylum Acanthocephala

Fluted giant clam,
Tridacna squamosa.

Molluscs

Phylum Mollusca

A Significant Space

Long ago in the Precambrian era, the most complex animals populating the seas were acoelomate. They must have been inefficient burrowers, unable to exploit the rich subsurface ooze. Any that developed fluid-filled spaces within the body would have had a substantial selective advantage because these spaces could serve as a hydrostatic skeleton and improve burrowing efficiency.

The simplest, and probably the first, mode of achieving a fluid-filled space within the body was retention of the embryonic blastocoel, as in pseudocoelomates. This was not the best evolutionary solution because, for example, the organs lay loose in the body cavity.

Some descendants of Precambrian acoelomate organisms evolved a more elegant arrangement: a fluid-filled space *within* the mesoderm, the *coelom*. This meant that the space was lined with mesoderm and the organs were suspended by mesodermal membranes, the *mesenteries*. Not only could the coelom serve as an efficient hydrostatic skeleton, with circular and longitudinal body-wall muscles acting as antagonists, but a more stable arrangement of organs with less crowding resulted. Mesenteries provided an ideal location for networks of blood vessels, and the alimentary canal could become more muscular, more highly specialized, and more diversified without interfering with other organs.

Development of a coelom was a major step in the evolution of larger and more complex forms. All major groups in chapters to follow are coelomates.

Position in Animal Kingdom

1. Molluscs are one of the major groups of true **coelomate** animals.
2. They belong to the **protostome,** lophotrochozoan branch of schizocoelous coelomates, and have spiral cleavage and determinate (mosaic) development.
3. Many molluscs have a **trochophore larva** similar to trochophore larvae of marine annelids and other marine protostomes. Developmental evidence thus indicates that molluscs and annelids share a more recent common ancestry than either phylum does with arthropods or deuterostomes.
4. Because molluscs are not metameric, they must have diverged from their common ancestor with annelids before the advent of metamerism.
5. All **organ systems** are present and well developed.

Biological Contributions

1. In molluscs gaseous exchange occurs not only through the body surface as in phyla discussed previously, but also in specialized **respiratory organs** in the form of **gills** or **lungs.**
2. Most classes have an **open circulatory system** with pumping **heart,** vessels, and blood sinuses. In most cephalopods the circulatory system is closed.
3. The efficiency of the respiratory and circulatory systems in the cephalopods has made greater body size possible. Invertebrates reach their largest size in some cephalopods.
4. They have a fleshy **mantle** that in most cases secretes a shell and is variously modified for a number of functions.
5. Features unique to the phylum are the **radula** and muscular **foot.**
6. The highly developed direct **eye** (photosensitive cells in retina face light source) of cephalopods is similar to the indirect eye (photosensitive cells face away from light source) of vertebrates but arises as a skin derivative in contrast to the brain eye of vertebrates.

MOLLUSCS

Mollusca (mol-lus′ka) (L. *molluscus,* soft) is one of the largest animal phyla after Arthropoda. There are nearly 50,000 living species and some 35,000 fossil species. The name Mollusca indicates one of their distinctive characteristics, a soft body. This very diverse group (Figure 16-1) includes chitons, tooth shells, snails, slugs, nudibranchs, sea butterflies, clams, mussels, oysters, squids, octopuses, and nautiluses. The group ranges from fairly simple organisms to some of the most complex of invertebrates, and in size from almost microscopic to the giant squid *Architeuthis.* These huge molluscs may grow to 18 m long, including their tentacles. They may weigh 450 kg (1,000 pounds). The shells of some giant clams, *Tridacna gigas,* which inhabit Indo-Pacific coral reefs, reach 1.5 m in length and weigh more than 225 kg. These are extremes, however, for probably 80% of all molluscs are less than

5 cm in maximum shell size. The phylum includes some of the most sluggish and some of the swiftest and most active invertebrates. It includes herbivorous grazers, predaceous carnivores, filter feeders, detritus feeders, and parasites.

Molluscs are found in a great range of habitats, from the tropics to polar seas, at altitudes exceeding 7000 m, in ponds, lakes, and streams, on mud flats, in pounding surf, and in open ocean from the surface to abyssal depths. Most of them live in the sea, and they represent a variety of lifestyles, including bottom feeders, burrowers, borers, and pelagic forms.

According to fossil evidence, molluscs originated in the sea, and most of them have remained there. Much of their evolution occurred along the shores, where food was abundant and habitats were varied. Only bivalves and gastropods moved into brackish and freshwater habitats. As filter feeders, bivalves were unable to leave aquatic surroundings. Only snails (gastropods) actually invaded the land. Terrestrial snails are limited in their range by their need for humidity, shelter, and presence of calcium in the soil.

Many kinds of molluscs are used as food. Pearl buttons are obtained from shells of bivalves. The Mississippi and Missouri river basins have furnished material for most of this industry in the United States; however, supplies are becoming so depleted that attempts are being made to propagate bivalves artificially. Pearls, both natural and cultured, are produced in the shells of clams and oysters, most of them in a marine oyster, *Meleagrina,* found around eastern Asia.

Some molluscs are destructive. Burrowing shipworms, which are bivalves of several species (see Figure 16-27), do great damage to wooden ships and wharves. To prevent the ravages of shipworms, wharves must be either creosoted or built of concrete (unfortunately, some ignore creosote, and some bivalves bore into concrete). Snails and slugs frequently damage garden and other vegetation. In addition, snails often serve as intermediate hosts for serious parasites of humans and domestic animals. Boring snails of genus *Urosalpinx* rival sea stars in destroying oysters.

In this chapter we explore the various major groups of molluscs, including those that apparently met with little evolutionary success (classes Caudofoveata, Solenogastres, Monoplacophora, and Scaphopoda). Members of class Polyplacophora (chitons) are common to abundant marine animals, especially in the intertidal zone. Bivalves (class Bivalvia) have evolved many species, both marine and freshwater. Largest and most intelligent of all invertebrates are in class Cephalopoda (squids, octopuses, and others). Most abundant and widespread of molluscs, however, are snails and their relatives (class Gastropoda). Although enormously diverse, molluscs have in common a basic body plan (p. 318–321). It seems peculiar, though, that molluscs have failed to exploit their coelom. The coelom in molluscs is limited to a space around the heart, and perhaps around the gonads and part of the kidneys. Although it develops embryonically in a manner similar to the coelom of annelids, the functional consequences of this space are quite different. Some zoologists believe that molluscs arose from a flatworm-type ancestor separately from annelids and that their coeloms are not homologous.

Class Gastropoda

Among molluscs class Gastropoda is by far the largest and most diverse, containing about 40,000 living and 15,000 fossil species. It contains so much diversity that there is no single general term in our language that can apply to them as a whole. They include snails, limpets, slugs, whelks, conchs, periwinkles, sea slugs, sea hares, and sea butterflies. They range from marine molluscs with many primitive characters to highly evolved terrestrial, air-breathing snails and slugs. These animals are basically bilaterally symmetrical, but because of **torsion,** a twisting process that occurs in the veliger stage, their visceral mass has become asymmetrical.

The shell, when present, is always of one piece **(univalve)** and may be coiled or uncoiled. Starting at the **apex,** which contains the oldest and smallest **whorl,** the whorls become successively larger and spiral around the central axis, or **columella** (Figure 16-12). The shell may be right handed **(dextral)** or left handed **(sinistral),** depending on the direction of coiling. Dextral shells are far more common. Direction of coiling is genetically controlled.

Gastropods range from microscopic forms to giant marine forms such as *Pleuroploca gigantea,* a snail with a shell up to 60 cm long, and sea hares, *Aplysia* (see Figure 16-21), some species of which reach 1 m in length. Most of them, however, are between 1 and 8 cm in length. Some fossil gastropods are as much as 2 m long.

The range of gastropod habitats is large. In the sea gastropods are common both in littoral zones and at great depths, and some are even pelagic. Some are adapted to brackish water and others to fresh water. On land they are restricted by such factors as mineral content of the soil and extremes of temperature, dryness, and acidity. Even so, they are widespread, and some have been found at great altitudes and some even in polar regions. Snails occupy all kinds of habitats: in small pools or large bodies of water, in woodlands, in pastures, under rocks, in mosses, on cliffs, in trees, underground, and on the bodies of other animals. They have successfully undertaken every mode of life except aerial locomotion.

Gastropods are usually sluggish, sedentary animals because most of them have heavy shells and slow locomotion. Some are specialized for climbing, swimming, or burrowing. Shells are their chief defense, although they are also protected by coloration and by secretive habits. Many snails have an **operculum,** a horny plate that covers the shell **aperture** when the body is withdrawn into the shell. Others lack shells altogether. Some are distasteful to other animals, and a few such as *Strombus* can deal an active blow with their foot, which bears a sharp operculum. Nevertheless, they are eaten by birds, beetles, small mammals, fish, and other predators. Serving as intermediate hosts for many kinds of parasites, especially trematodes (p. 280), snails are often harmed by larval stages of parasites.

Torsion

Of all molluscs, only gastropods undergo torsion. Torsion is a peculiar phenomenon that moves the mantle cavity, which was originally (primitively) posterior, to the front of the body, thus twisting the visceral organs as well through a 90- to 180-degree rotation. It occurs during the veliger stage, and in some species the first part may take only a few minutes. The second 90 degrees typically takes a longer period. Before torsion occurs, the embryo's mouth is anterior and the anus and mantle cavity are posterior (Figure 16-13). The change is brought about by an uneven growth of the right and left muscles that attach the shell to the head-foot.

After torsion, the anus and mantle cavity become anterior and open above the mouth and head. The left gill, kidney, and heart auricle are now on the right side, whereas the original right gill, kidney, and heart auricle are now on the left, and the nerve cords have been twisted into a figure eight. Because of the space available in the mantle cavity, the animal's sensitive head end can now be withdrawn into the protection of the shell, with the tougher foot forming a barrier to the outside.

Varying degrees of **detorsion** are seen in opisthobranchs and pulmonates, and the anus opens to the right side or even to the posterior. However, both of these groups were derived from torted ancestors.

The curious arrangement that results from torsion poses a serious sanitation problem by creating the possibility of wastes being washed back over the gills **(fouling)** and causes us to wonder what strong evolutionary pressures selected for such a strange realignment of body structures. Several explanations have been proposed, none entirely satisfying. For example, sense

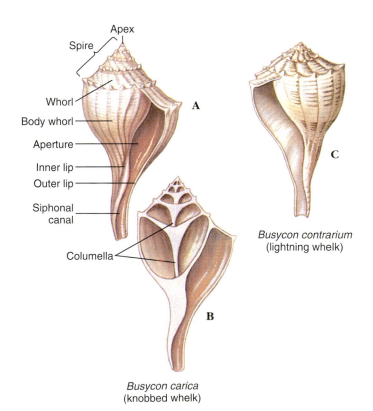

Apex
Spire
Whorl
Body whorl
Aperture
Inner lip
Outer lip
Siphonal canal
Columella

A

C

B

Busycon contrarium
(lightning whelk)

Busycon carica
(knobbed whelk)

Figure 16-12

Shell of the whelk *Busycon.* **A** and **B,** *Busycon carica,* a dextral, or right-handed, shell. A dextral shell has the aperture on the right side when the shell is held with the apex up and the aperture facing the observer. **C,** *B. contrarium,* a sinistral, or left-handed, shell.

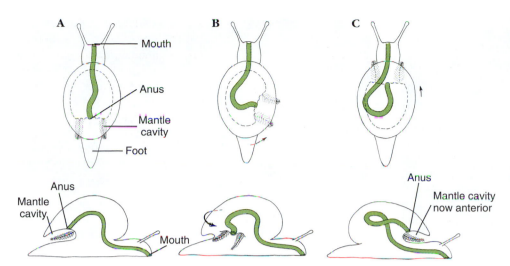

Figure 16-13

Torsion in gastropods. **A,** Ancestral condition before torsion. **B,** Hypothetical intermediate condition. **C,** Early gastropod, torsion complete; direction of crawling now tends to carry waste products back into mantle cavity, resulting in fouling.

organs of the mantle cavity (osphradia) would better sample water when turned in the direction of travel. Certainly the consequences of torsion and the resulting need to avoid fouling have been very important in the subsequent evolution of gastropods. These consequences cannot be explored, however, until we describe another unusual feature of gastropods—coiling.

Coiling

Coiling, or spiral winding, of the shell and visceral mass is not the same as torsion. Coiling may occur in the larval stage at the same time as torsion, but the fossil record shows that coiling was a separate evolutionary event and originated in gastropods earlier than torsion did. Nevertheless, all living gastropods have descended from coiled, torted ancestors, whether or not they now show these characteristics.

Early gastropods had a bilaterally symmetrical **planospiral** shell, in which all whorls lay in a single plane (Figure 16-14A). Such a shell was not very compact, since each whorl had to lie completely outside the preceding one. Curiously, a few modern species have secondarily returned to a planospiral form. The compactness problem of a planospiral shell was solved by the **conispiral** shape, in which each succeeding whorl is at the side of the preceding one (Figure 16-14B). However, this shape was clearly unbalanced, hanging as it was with much weight over to one side. Better weight distribution was achieved by shifting the shell upward and posteriorly, with the shell axis oblique to the longitudinal axis of the foot (Figure 16-14C). The weight and bulk of the main body whorl, the largest whorl of the shell, pressed on the right side of the mantle cavity, however, and apparently interfered with the organs on that side. Accordingly, the gill, auricle, and kidney of the right side have been lost in most living gastropods, leading to a condition of *bilateral asymmetry*.

Although loss of the right gill was probably an adaptation to the mechanics of carrying a coiled shell, that condition displayed in most modern prosobranchs made possible a way to avoid the problem caused by torsion—fouling. Water is brought into the left side of the mantle cavity and out the right side, carrying with it wastes from the anus and nephridiopore, which lie near the right side. Ways in which fouling is avoided in other gastropods are mentioned on p. 327–328.

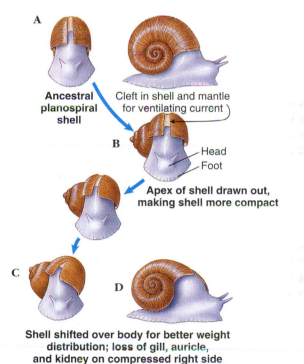

Apex of shell drawn out, making shell more compact

Shell shifted over body for better weight distribution; loss of gill, auricle, and kidney on compressed right side

Figure 16-14

Evolution of shell in gastropods. **A,** Earliest coiled shells were planospiral, each whorl lying completely outside the preceding whorl. **B,** Better compactness was achieved by snails in which each whorl lay partially to the side of the preceding whorl. **C** and **D,** Better weight distribution resulted when shell was moved upward and posteriorly.

Feeding Habits

Feeding habits of gastropods are as varied as their shapes and habitats, but all include use of some adaptation of the radula. Most gastropods are herbivorous, rasping off particles of algae from hard surfaces. Some herbivores are grazers, some are browsers, and some are planktonic feeders. *Haliotis,* the abalone (Figure 16-15A), holds seaweed with its foot and breaks off pieces with its radula. Land snails forage at night for green vegetation.

Some snails, such as *Bullia* and *Buccinum,* are scavengers living on dead and decaying flesh; others are carnivores that tear

A

B

Figure 16-15

A, Red abalone, *Haliotis rufescens.* This huge, limpetlike snail is prized as food and extensively marketed. Abalones are strict vegetarians, feeding especially on sea lettuce and kelp. **B,** Moon snail, *Polinices lewisii.* A common inhabitant of West Coast sand flats, the moon snail is a predator of clams and mussels. It uses its radula to drill neat holes through its victim's shell, through which the proboscis is then extended to eat the bivalve's fleshy body.

their prey with radular teeth. *Melongena* feeds on clams, especially *Tagelus,* the razor clam, thrusting its proboscis between the gaping shell valves. *Fasciolaria* and *Polinices* (Figure 16-15B) feed on a variety of molluscs, preferably bivalves. *Urosalpinx cinerea,* oyster borers, drill holes through the shell of oysters. Their radula, bearing three longitudinal rows of teeth, is used first to begin the drilling action, then the snails glide forward, evert an accessory boring organ through a pore in the anterior sole of their foot, and hold it against the oyster's shell, using a chemical agent to soften the shell. Short periods of rasping alternate with long periods of chemical activity until a neat round hole is completed. With its proboscis inserted through the hole, a snail may feed continuously for hours or days, using its radula to tear away the soft flesh. *Urosalpinx* is attracted to its prey at some distance by sensing some chemical, probably one released in metabolic wastes of the prey.

Cyphoma gibbosum (see Figure 16-20B) and related species live and feed on gorgonians (phylum Cnidaria, Chapter 13) in shallow, tropical coral reefs. These snails are commonly known as flamingo tongues. During normal activity their brightly colored mantle entirely envelops the shell, but it can be quickly withdrawn into the shell aperture when the animal is disturbed.

Members of genus *Conus* (Figure 16-16) feed on fish, worms, and molluscs. Their radula is highly modified for prey capture. A gland charges the radular teeth with a highly toxic venom. When *Conus* senses the presence of its prey, a single radular tooth slides into position at the tip of the proboscis. Upon striking the prey, the proboscis expels a tooth like a harpoon, and the poison quiets the prey at once. This is an effective adaptation for a slowly moving predator to prevent escape of a swiftly moving prey. Some species of *Conus* can deliver very painful stings, and in several species the sting is lethal to humans. The venom consists of a series of toxic peptides, and each *Conus* species carries peptides **(conotoxins)** that are specific for the neuroreceptors of its preferred prey. Conotoxins have become valuable tools in research on the various receptors and ion channels of nerve cells.

Some gastropods feed on organic deposits on the sand or mud. Others collect the same sort of organic debris but can digest only microorganisms contained in it. Some sessile gastropods, such as some limpets, are ciliary feeders that use gill cilia to draw in particulate matter, roll it into a mucous ball, and carry it to their mouth. Some sea butterflies secrete a mucous net to catch small planktonic forms; then they draw the web into their mouth.

After maceration by the radula or by some grinding device, such as a gizzard in the sea hare *Aplysia,* digestion is usually extracellular in the lumen of the stomach or digestive glands. In ciliary feeders the stomachs are sorting regions, and most digestion is intracellular in digestive glands.

Internal Form and Function

Respiration in most gastropods is carried out by a **ctenidium** (two ctenidia is the primitive condition, found in some prosobranchs) located in the mantle cavity, though some aquatic forms, lacking gills, depend on the mantle and skin. After the more derived prosobranchs lost one gill, most of them lost half of the remaining one, and the central axis became attached to the wall of the mantle cavity (Figure 16-17). Thus they attained the most efficient gill arrangement for the way the water circulated through the mantle cavity (in one side and out the other).

Pulmonates have a highly vascular area in their mantle that serves as a **lung** (Figure 16-18). Most of the mantle margin seals to the back of the animal, and the lung opens to the outside by a small opening called a **pneumostome.** Many aquatic pulmonates must surface to expel a bubble of gas from their lung. To inhale, they curl the edge of the mantle around the pneumostome to form a siphon.

Most gastropods have a single nephridium (kidney). The circulatory and nervous systems are well developed (Figure 16-18). The latter incorporates three pairs of ganglia connected by nerves. Sense organs include eyes or simple photoreceptors, statocysts, tactile organs, and chemoreceptors. The simplest type of

A

B

Figure 16-16

Conus extends its long, wormlike proboscis **(A)**. When a fish attempts to consume this tasty morsel, the *Conus* stings it in the mouth and kills it. The snail engulfs the fish with its distensible stomach **(B)**, then regurgitates the scales and bones some hours later.

Figure 16-17

Evolution of ctenidia in gastropods. **A,** Primitive condition with two ctenidia and excurrent water leaving the mantle cavity by a dorsal slit or hole. **B,** Condition after one ctenidium had been lost. **C,** Derived condition found in most marine gastropods, in which filaments on one side of remaining gill are lost, and axis is attached to mantle wall.

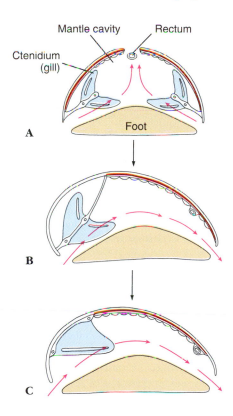

ulation each partner deposits its eggs in shallow burrows in the ground. Gastropods with the most primitive reproductive characteristics discharge ova and sperm into seawater where fertilization occurs, and embryos soon hatch as free-swimming trochophore larvae. In most gastropods fertilization is internal.

Fertilized eggs encased in transparent shells may be emitted singly to float among the plankton or may be laid in gelatinous layers attached to a substratum. Some marine forms enclose their eggs, either in small groups or in large numbers, in tough egg capsules, or in a wide variety of egg cases (Figure 16-19). Offspring generally emerge as veliger larvae (Figure 16-7), or they may spend the veliger stage in the case or capsule and emerge as young snails. Some species, including many freshwater snails, are ovoviviparous, brooding their eggs and young in their oviduct.

Major Groups of Gastropods

Traditional classification of class Gastropoda recognizes three subclasses: Prosobranchia, much the largest subclass, almost all of which are marine; Opisthobranchia, an assemblage including sea slugs, sea hares, nudibranchs, and canoe shells, all marine; and Pulmonata, containing most freshwater and terrestrial species. Currently, gastropod taxonomy is in flux. Evidence suggests that Prosobranchia is paraphyletic. Opisthobranchia may or may not be paraphyletic, but Opisthobranchia and Pulmonata together apparently form a monophyletic grouping. For convenience and organization, we will continue to use the words "prosobranchs" and "opisthobranchs," recognizing that they may not represent valid taxa.

Prosobranchs   This group contains most marine snails and some freshwater and terrestrial gastropods. The mantle cavity is anterior as a result of torsion, with the gill or gills lying in front of the heart. Water enters the left side and exits from the right side, and the edge of the mantle often extends into a long siphon to separate incurrent from excurrent flow. In prosobranchs with two gills (for example, the abalone *Haliotis* and keyhole limpets *Diodora*, Figures 16-15A and 16-20A), fouling is

gastropod eye is simply a cuplike indentation in the skin lined with pigmented photoreceptor cells. In many gastropods the eyecup contains a lens covered with a cornea. A sensory area called an **osphradium,** located at the base of the incurrent siphon of most gastropods, is chemosensory in some forms, although its function may be mechanoreceptive in some and remains unknown in others.

There are both dioecious and monoecious gastropods. Many gastropods perform courtship ceremonies. During copulation in monoecious species there is an exchange of spermatozoa or spermatophores (bundles of sperm). Many terrestrial pulmonates eject a dart from a dart sac (Figure 16-18) into the partner's body to heighten excitement before copulation. After cop-

Figure 16-18
Anatomy of a pulmonate snail.

A

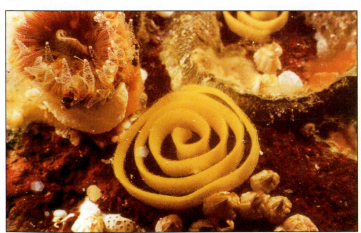

B

Figure 16-19
Eggs of marine gastropods. **A,** The wrinkled whelk, *Nucella lamellosa*, lays egg cases resembling grains of wheat; each contains hundreds of eggs. **B,** Egg ribbon of a dorid nudibranch.

avoided by having the excurrent water go up and out through one or more holes in the shell above the mantle cavity.

Prosobranchs have one pair of tentacles. Sexes are usually separate. An operculum is often present.

They range in size from periwinkles and small limpets (*Patella* and *Diodora*) (Figure 16-20A) to horse conchs (*Pleuroploca*), the largest gastropods in the Atlantic Ocean. Familiar examples of prosobranchs are abalones (*Haliotis*), which have an ear-shaped shell; whelks (*Busycon*), which lay their eggs in double-edged, disc-shaped capsules attached to a cord a meter long; common periwinkles (*Littorina*); moon snails (*Polinices*, Figure 16-15B); oyster borers (*Urosalpinx*), which bore into oysters and suck out their juices; rock shells (*Murex*), a European species that was used to make the royal purple of the ancient Romans; and some freshwater forms (*Goniobasis* and *Viviparus*).

Opisthobranchs   Opisthobranchs are an odd assemblage of molluscs that include sea slugs, sea hares, sea butterflies, and canoe shells. They are nearly all marine; most are shallow-water forms, hiding under stones and seaweed; a few are pelagic. Currently nine or more orders of opisthobranchs are recognized. Opisthobranchs show partial or complete detorsion; thus the anus and gill (if present) are displaced to the right side or rear of the body. Clearly, the fouling problem is obviated if the anus is moved away from the head toward the posterior. Two pairs of tentacles are usually found, and the second pair is often further modified (**rhinophores,** Figure 16-21), with platelike folds that apparently increase the area for chemoreception. Their shell is typically reduced or absent. All are monoecious.

Sea hares (*Aplysia*, Figure 16-21), have large, earlike anterior tentacles and vestigial shells. In pteropods or sea butterflies (*Cavolina* and *Clione*) the foot is modified into fins for swimming; thus, they are pelagic and form a part of the plankton fauna.

Nudibranchs are carnivorous and often brightly colored (Figure 16-22). Plumed sea slugs (Aeolidae) which live on sea anemones and hydroids, have elongate papillae (**cerata**) covering their back. They ingest their prey's nematocysts and transport the nematocysts undischarged to the tips of their cerata. There the nematocysts are placed in cnidosacs that open to the outside, and the aeolid can use its highjacked nematocysts for its own defense. *Hermissenda* is one of the more common West Coast nudibranchs.

Pulmonates   Pulmonates show some detorsion and include land and most freshwater snails and slugs (and a few brackish and saltwater forms). They have lost their ancestral ctenidia, but their vascularized mantle wall has become a lung, which fills with air by contraction of the mantle floor (some aquatic species have developed secondary gills in the mantle cavity). The anus and nephridiopore open near the pneumostome, and waste is expelled forcibly with air or water from the lung. They are

A   B

Figure 16-20

A, *Diodora aspera,* a gastropod with a hole in its apex through which water leaves the mantle cavity. **B,** Flamingo tongues, *Cyphoma gibbosum,* are showy inhabitants of Caribbean coral reefs, where they are associated with gorgonians. These snails have a smooth, creamy, orange to pink shell that is normally covered by the brightly marked mantle.

Rhinophore   Oral tentacle

A   B

Figure 16-21

A, The sea hare, *Aplysia dactylomela,* crawls and swims across a tropical seagrass bed, assisted by large, winglike parapodia, here curled above the body. **B,** When attacked, sea hares squirt a copious protective secretion from their "purple gland" in the mantle cavity.

Figure 16-22

Phyllidia ocellata, a nudibranch. Like other *Phyllidia* spp., it has a hard body with dense calcareous spicules and bears its gills along the sides, between its mantle and foot.

monoecious. Aquatic species have one pair of nonretractile tentacles, at the base of which are eyes; land forms have two pairs of tentacles, with the posterior pair bearing eyes (Figure 16-23). Among the thousands of land species, some of the most familiar American forms are *Helix, Polygyra, Succinea, Anguispira, Zonitoides, Limax,* and *Agriolimax.* Aquatic forms are represented by *Helisoma, Lymnaea,* and *Physa. Physa* is a left-handed (sinistral) snail.

Class Bivalvia (Pelecypoda)

Bivalvia are also known as Pelecypoda (pel-e-sip'o-da), or "hatchet-footed" animals, as their name implies (Gr. *pelekys,* hatchet, + *pous, podos,* foot). They are bivalved molluscs that include mussels, clams, scallops, oysters, and shipworms (Figures 16-24 to 16-27) and they range in size from tiny seed shells 1 to 2 mm in length to giant South Pacific clams, *Tridacna,* which may reach more than 1 m in length and as much as 225 kg (500

Figure 16-23

A, Pulmonate land snail. Note two pairs of tentacles; the second, larger pair bears the eyes. **B,** Banana slug, *Ariolimax columbianus.* Note pneumostome.

Pneumostome

A

B

A

B

Figure 16-24

Bivalve molluscs. **A,** Mussels, *Mytilus edulis,* occur in northern oceans around the world; they form dense beds in the intertidal zone. A host of marine creatures live protected beneath attached mussels. **B,** Scallops *(Chlamys opercularis)* swim to escape attack by starfish *(Asterias rubens).* When alarmed, these most agile of bivalves swim by clapping the two shell valves together.

pounds) in weight (see Figure 16-32). Most bivalves are sedentary **filter feeders** that depend on ciliary currents produced by cilia on their gills to bring in food materials. Unlike gastropods, they have no head, no radula, and very little cephalization.

Most bivalves are marine, but many live in brackish water and in streams, ponds, and lakes.

Freshwater clams were once abundant and diverse in streams throughout the eastern United States, but they are now easily the most jeopardized group of animals in the country. Of more than 300 species once present, 12 are extinct, 42 are listed as threatened or endangered, and as many as 88 more may be listed soon. A combination of causes is responsible, of which a decline in water quality is among the most important. Pollution and sedimentation from mining, industry, and agriculture are among the culprits. Poaching to supply the Japanese cultured pearl industry is partially to blame. And in addition to everything else, the prolific zebra mussels (see note, p. 333) attach in great numbers to the native clams, exhausting food supplies (phytoplankton) in the surrounding water.

Form and Function

Shell Bivalves are laterally compressed, and their two shells **(valves)** are held together dorsally by a hinge ligament that causes the valves to gape ventrally. The valves are drawn together by adductor muscles that work in opposition to the hinge ligament (Figure 16-26C and D). The **umbo** is the oldest part of the shell, and growth occurs in concentric lines around it (Figure 16-26A).

Pearl production is a by-product of a protective device used by the animals when a foreign object (grain of sand, parasite, or

Figure 16-25

Representing a group that has evolved from burrowing ancestors, the surface-dwelling bivalve *Aequipecten irradians* has developed sensory organs along its mantle edges (tentacles and a series of blue eyes).

other) becomes lodged between the shell and mantle. The mantle secretes many layers of nacre around the irritating object (Figure 16-5). Pearls are cultured by inserting particles of nacre, usually taken from the shells of freshwater clams, between the shell and mantle of a certain species of oyster and by keeping the oysters in enclosures for several years. *Meleagrina* is an oyster used extensively by the Japanese for pearl culture.

Body and Mantle   The **visceral mass** is suspended from the dorsal midline, and the muscular foot is attached to the visceral mass anteroventrally. The ctenidia hang down on each

side, each covered by a fold of the mantle. The posterior edges of the mantle folds are modified to form dorsal excurrent and ventral incurrent openings (Figure 16-28A). In some marine bivalves the mantle is drawn out into long muscular siphons that allow the clam to burrow into the mud or sand and extend the siphons to the water above (Figure 16-28B to D).

Locomotion   Bivalves initiate movement by extending a slender muscular foot between the valves (Figure 16-28D). They pump blood into their foot, causing it to swell and to act as an anchor in the mud or sand, then longitudinal muscles contract to shorten the foot and pull the animal forward.

Scallops and file shells are able to swim with a jerky motion by clapping their valves together to create a sort of jet propulsion. The mantle edges can direct the stream of expelled water, so that the animals can swim in virtually any direction (Figure 16-24).

Gills   Gaseous exchange occurs through both mantle and gills. Gills of most bivalves are highly modified for filter feeding; they are derived from primitive ctenidia by a great lengthening of filaments on each side of the central axis (Figure 16-29). As ends of long filaments became folded back toward the central axis, ctenidial filaments took the shape of a long, slender **W**. Filaments lying beside each other became joined by ciliary junctions or tissue fusions, forming platelike **lamellae** with many vertical water tubes inside. Thus water enters the incurrent siphon, propelled by ciliary action, then enters the water tubes through pores between the filaments in the lamellae, proceeds dorsally into a common **suprabranchial chamber** (Figure 16-30), and then out the excurrent aperture.

Figure 16-26

Tagelus plebius, stubby razor clam (class Bivalvia). **A,** External view of left valve. **B,** Inside of right shell showing scars where muscles were attached. The mantle was attached at its insertion area. **C** and **D,** Sections showing function of adductor muscles and hinge ligament. In **C,** the adductor muscle is relaxed, allowing the hinge ligament to pull the valves apart. In **D,** the adductor muscle is contracted, pulling the valves together.

A

Umbo
Excurrent siphon
Anterior end
Foot
Incurrent siphon

B
Insertion area of anterior adductor muscle
Insertion area of anterior retractor
Insertion area of posterior adductor muscle
Insertion area of mantle
Location of siphons

Hinge ligament
Adductor muscle relaxed
C

Adductor muscle contracted
D

A   **B**

Figure 16-27

A, Shipworms are bivalves that burrow in wood, causing great damage to unprotected wooden hulls and piers. They are nicknamed "termites of the sea." **B,** The two small, anterior valves, seen at left, are used as rasping organs to extend the burrow.

A

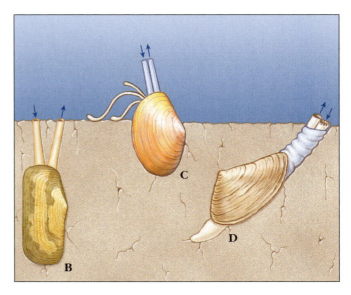

Figure 16-28

Adaptations of siphons in bivalves. **A,** In northwest ugly clams *Entodesma saxicola,* incurrent and excurrent siphons are clearly visible. **B to D,** In many marine forms the mantle is drawn out into long siphons. In **A, B,** and **D,** the incurrent siphon brings in both food and oxygen. In **C,** *Yoldia,* the siphons are respiratory; long ciliated palps feel about over the mud surface and convey food to the mouth.

Feeding   Most bivalves are filter feeders. Respiratory currents bring both oxygen and organic materials to the gills where ciliary tracts direct them to the tiny pores of the gills. Gland cells on the gills and labial palps secrete copious amounts of mucus, which entangles particles suspended in water going through gill pores. These mucous masses slide down the outside of the gills toward food grooves at the lower edge of the gills (Figure 16-31). Heavier particles of sediment drop off the gills as a result of gravitational pull, but smaller particles travel along the food grooves toward the labial palps. The palps, being also grooved and ciliated, direct the mucous mass into the mouth.

Some bivalves, such as *Nucula* and *Yoldia,* are deposit feeders and have long proboscides attached to the labial palps (Figure 16-28C). These can be protruded onto sand or mud to collect food particles, in addition to particles attracted by gill currents.

Shipworms (Figure 16-27) burrow in wood and feed on particles they excavate. Symbiotic bacteria live in a special organ in the bivalve and produce cellulase to digest wood. Other bivalves such as giant clams gain much of their nutrition from the photosynthetic products of symbiotic dinoflagellates living in their mantle tissue (Figure 16-32).

Septibranchs, another group of bivalves, draw small crustaceans or bits of organic debris into the mantle cavity by sudden inflow of water created by the pumping action of a muscular septum in the mantle cavity.

Internal Structure and Function   The floor of the stomach of filter-feeding bivalves is folded into ciliary tracts for sorting a continuous stream of particles. In most bivalves a cylindrical **style sac** opening into the stomach secretes a gelatinous

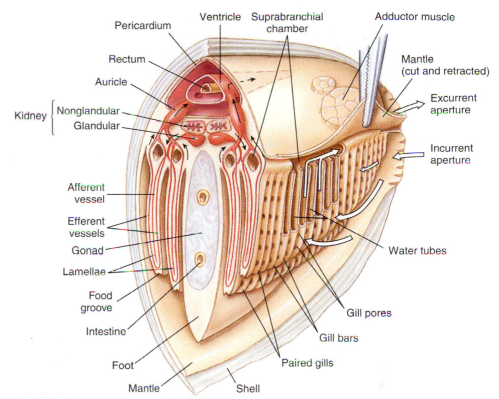

Figure 16-29

Evolution of bivalve ctenidia. By a great lengthening of individual filaments, ctenidia became adapted for filter feeding and separated the incurrent chamber from the excurrent, suprabranchial chamber.

Figure 16-30

Section through heart region of a freshwater clam to show relation of circulatory and respiratory systems. Respiratory water currents: water is drawn in by cilia, enters gill pores, and then passes up water tubes to suprabranchial chambers and out excurrent aperture. Blood in gills exchanges carbon dioxide for oxygen. Blood circulation: ventricle pumps blood forward to sinuses of foot and viscera, and posteriorly to mantle sinuses. Blood returns from mantle to auricles; it returns from viscera to the kidney, and then goes to the gills, and finally to the auricles.

rod called a **crystalline style,** which projects into the stomach and is kept whirling by means of cilia in the style sac (Figure 16-33). Rotation of the style helps to dissolve its surface layers, freeing digestive enzymes (especially amylase) that it contains, and to roll the mucous food mass. Dislodged particles are sorted, and suitable ones are directed to the digestive gland or engulfed by amebocytes. Further digestion is intracellular.

The three-chambered heart, which lies in the pericardial cavity (Figure 16-31), has two auricles and a ventricle and beats slowly, ranging from 0.2 to 30 times per minute. Part of the blood is oxygenated in the mantle and returns to the ventricle through the auricles; the rest circulates through sinuses and passes in a vein to the kidneys, from there to the gills for oxygenation, and back to the auricles.

A pair of U-shaped kidneys (nephridial tubules) lies just ventral and posterior to the heart (Figure 16-31B). The glandular portion of each tubule opens into the pericardium; the bladder portion empties into the suprabranchial chamber.

Zebra mussels, *Dreissena polymorpha,* are a recent and disastrous biological introduction into North America. They were apparently picked up as veligers with ballast water by one or more ships in freshwater ports in northern Europe and then expelled between Lake Huron and Lake Erie in 1986. This 4 cm bivalve spread throughout the Great Lakes by 1990, and by 1994 it was as far south on the Mississippi as New Orleans, as far north as Duluth, Minnesota, and as far east as the Hudson River in New York. It attaches to any firm surface and filter feeds on phytoplankton. Populations rapidly increase in size. They foul water intake pipes of municipal and industrial plants, impede intake of water for municipal supplies, and have far-reaching effects on the ecosystem (see note, p. 330). Zebra mussels will cost billions of dollars to control.

Another freshwater clam, *Corbicula fluminea,* was introduced into the United States from Asia more than 50 years ago by unknown means. It is now a pest in at least 35 states, infesting water systems and clogging pipes. Efforts to control *Corbicula* cost over a billion dollars per year.

The nervous system consists of three pairs of widely separated ganglia connected by commissures and a system of nerves. Sense organs are poorly developed. They include a pair of statocysts in

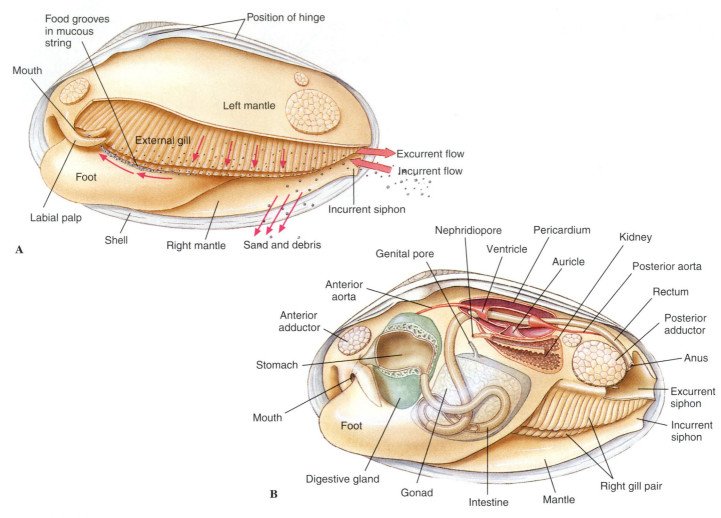

Figure 16-31

A, Feeding mechanism of freshwater clam. Left valve and mantle are removed. Water enters the mantle cavity posteriorly and is drawn forward by ciliary action to the gills and palps. As water enters the tiny openings of the gills, food particles are sieved out and caught in strings of mucus that are carried by cilia to the palps and directed to the mouth. Sand and debris drop into the mantle cavity and are removed by cilia. **B,** Clam anatomy.

the foot, a pair of osphradia of uncertain function in the mantle cavity, tactile cells, and sometimes simple pigment cells on the mantle. Scallops (*Aequipecten, Chlamys*) have a row of small blue eyes along each mantle edge (Figure 16-25). Each eye has a cornea, lens, retina, and pigmented layer. Tentacles on the margin of the mantle of *Aequipecten* (Figure 16-25) and *Lima* have tactile and chemoreceptor cells.

Reproduction and Development

Sexes are usually separate. Gametes are discharged into the suprabranchial chamber to be carried out with the excurrent flow. An oyster may produce 50 million eggs in a single season. In most bivalves fertilization is external. The embryo develops into trochophore, veliger, and spat stages (Figure 16-34).

In most freshwater clams fertilization is internal. Eggs drop into the water tubes of the gills where they are fertilized by sperm entering with the incurrent flow. They develop there into a bivalved **glochidium larva** stage, which is a specialized veliger (Figure 16-35). When discharged, glochidia are carried

by water currents, and if they come in contact with a passing fish, they attach to its gills or skin and live as parasites for several weeks. Then they sink to the bottom to begin independent lives. Larval "hitchhiking" helps distribute a form whose locomotion is very limited.

Boring   Many bivalves can burrow into mud or sand, but some have evolved a mechanism for burrowing into much harder substances, such as wood or stone.

Teredo, Bankia, and some other genera are called shipworms. They can be very destructive to wooden ships and wharves. These strange little clams have a long, wormlike appearance, with a pair of slender posterior siphons that keep water flowing over the gills, and a pair of small globular valves on the anterior end with which they burrow (Figure 16-27). The valves have microscopic teeth that function as very effective wood rasps. The animals extend their burrows with an unceasing rasping motion of the valves. This motion sends a continuous flow of fine wood particles into the digestive tract where

Some clams bore into rock. The piddock *(Pholas)* bores into limestone, shale, sandstone, and sometimes wood or peat. It has strong valves that bear spines, which it uses to cut away the rock gradually while anchoring itself with its foot. *Pholas* may grow to 15 cm long and make rock burrows up to 30 cm long.

Class Cephalopoda

Cephalopoda (Gr. *kephalē,* head, + *pous, podos,* foot) include squids, octopuses, nautiluses, devilfish, and cuttlefish. All are marine, and all are active predators.

Their modified foot is concentrated in the head region. It takes the form of a funnel for expelling water from the mantle cavity, and the anterior margin is drawn out into a circle or crown of arms or tentacles.

Cephalopods range upward in size from 2 or 3 cm. The common squid of markets, *Loligo,* is about 30 cm long. Giant squids, *Architeuthis,* are the largest invertebrates known.

Fossil records of cephalopods go back to Cambrian times. The earliest shells were straight cones; others were curved or coiled, culminating in the coiled shell similar to that of the modern *Nautilus,* the only remaining member of the once flourishing nautiloids (Figure 16-36). Cephalopods without shells or with internal shells (such as octopuses and squids) apparently evolved from some early straight-shelled ancestor. Many ammonoids, which are extinct, had quite elaborate shells (Figure 16-36C).

The natural history of some cephalopods is fairly well known. They are marine animals and appear sensitive to the degree of salinity. Few are found in the Baltic Sea, where the water has a low salt content. Cephalopods are found at various depths. Octopuses are often seen in the intertidal zone, lurking among rocks and crevices, but occasionally they are found at great depths. The more active squids are rarely found in very shallow water, and some have been taken at depths of 5000 m. *Nautilus* is usually found near the bottom in water 50 to 560 m deep, near islands in the southwestern Pacific.

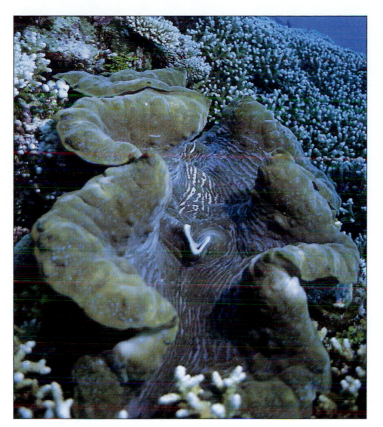

Figure 16-32
Clam *(Tridacna gigas)* lies buried in coral rock with greatly enlarged siphonal area visible. These tissues are richly colored and bear enormous numbers of symbiotic single-celled algae (zooxanthellae) that provide much of the clam's nutriment.

they are attacked by cellulase produced by symbiotic bacteria. Interestingly, these bacteria also fix nitrogen, an important property for their hosts, which live on a diet (wood) high in carbon but deficient in nitrogen.

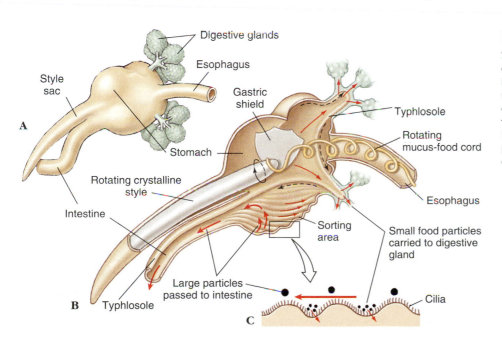

Figure 16-33
Stomach and crystalline style of ciliary-feeding clam. **A,** External view of stomach and style sac. **B,** Transverse section showing direction of food movements. Food particles in incoming water are caught in a cord of mucus that is kept rotating by the crystalline style. Ridged sorting areas direct large particles to the intestine and small food particles to digestive glands. **C,** Sorting action of cilia.

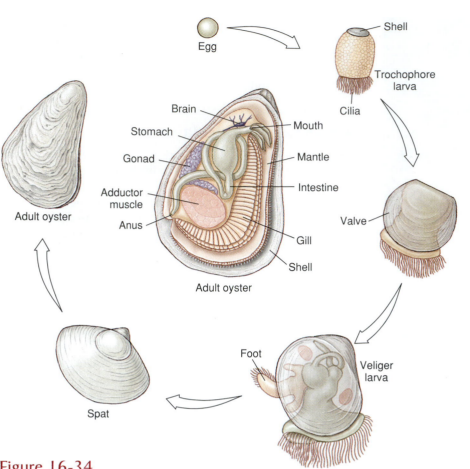

Figure 16-34

Life cycle of oysters. Oyster larvae swim about for approximately 2 weeks before settling down for attachment to become spats. Oysters take about 4 years to grow to commercial size.

Figure 16-35

A, Glochidium, or larval form, for some freshwater clams. When larvae are released from the mother's brood pouch, they may become attached to a fish's gill by clamping their valves closed. They remain as parasites on the fish for several weeks. Their size is approximately 0.3 mm. **B,** Some clams have adaptations that help their glochidia find a host. The mantle edge of this female pocketbook mussel *(Lampsilis ovata)* mimics a small minnow, complete with eye. When a smallmouth bass comes to dine, it gets doused with glochidia.

The enormous giant squid, *Architeuthis,* is very poorly known because no one has ever been able to study a living specimen. The anatomy has been studied from stranded animals, from those captured in nets of fishermen, and from specimens found in the stomach of sperm whales. The mantle length is 5 to 6 m, and the head is up to one meter. They have the largest eyes in the animal kingdom: up to 25 cm (10 inches) in diameter. They apparently eat fish and other squids, and they are an important food item for sperm whales. They are thought to live on or near the sea bottom at a depth of 1000 m, but some have been observed swimming at the surface.

Form and Function

Shell   Although early nautiloid and ammonoid shells were heavy, they were made buoyant by a series of **gas chambers,** as is that of *Nautilus* (Figure 16-36B), enabling the animal to swim while carrying its shell. The shell of *Nautilus,* although coiled, is quite different from that of a gastropod. The shell is divided by transverse septa into internal chambers (Figure 16-36B), only the last inhabited by the living animal. As it grows, it moves forward, secreting behind its body a new septum. The chambers are connected by a cord of living tissue called a **siphuncle,** which extends from the visceral mass. Cuttlefishes (Figure 16-37) also have a small, curved shell, but it is entirely enclosed by the mantle. In squids most of the shell has disappeared, leaving only a thin, horny strip called a pen, which is enclosed by the mantle. In *Octopus* (Gr. *oktos,* eight, + *pous, podos,* foot) the shell has disappeared entirely.

Locomotion   Cephalopods swim by forcefully expelling water from the mantle cavity through a ventral **funnel** (or **siphon**)—a sort of jet propulsion. The funnel is mobile and can be pointed forward or backward to control direction; the force of water expulsion controls speed.

Squids and cuttlefishes are excellent swimmers. The squid body is streamlined and built for speed (Figure 16-38). Cuttlefishes swim more slowly. The lateral fins of squids and cuttlefishes serve as stabilizers, but they are held close to the body for rapid swimming.

Nautilus is active at night; its gas-filled chambers keep the shell upright. Although not as fast as squids, it moves surprisingly well.

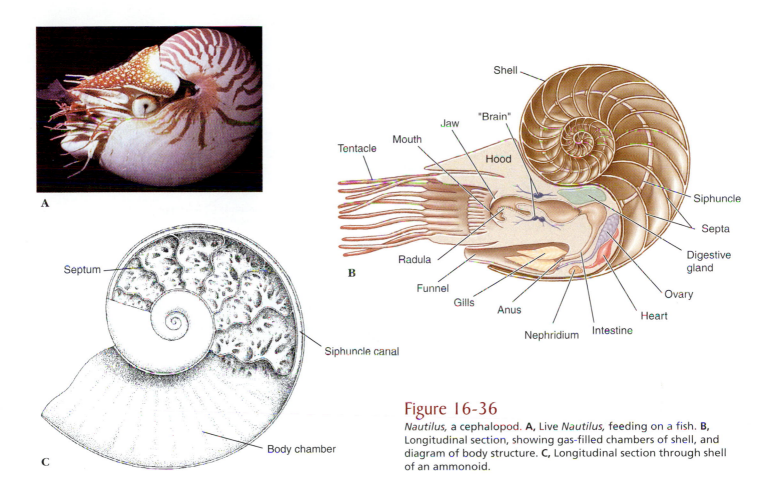

Figure 16-36
Nautilus, a cephalopod. **A,** Live *Nautilus,* feeding on a fish. **B,** Longitudinal section, showing gas-filled chambers of shell, and diagram of body structure. **C,** Longitudinal section through shell of an ammonoid.

Octopus has a rather globular body and no fins (Figure 16-1E). An octopus can swim backward by spurting jets of water from its funnel, but it is better adapted to crawling about over rocks and coral, using suction discs on its arms to pull or to anchor itself. Some deep-water octopods have the arms webbed like an umbrella and swim in a medusa-like fashion.

Internal Features   The active habits of cephalopods are reflected in their internal anatomy, particularly their respiratory, circulatory, and nervous systems.

Respiration and Circulation.   Except for nautiloids, cephalopods have one pair of gills. Because ciliary propulsion would not circulate enough water for their high oxygen requirements, there are no cilia on the gills. Instead, radial muscles in the mantle wall compress the wall and enlarge the mantle cavity, drawing water inside. Strong circular muscles contract and expel water forcibly through the funnel. A system of one-way valves prevents water from being taken in through the funnel and expelled around the mantle margin.

Likewise, the open circulatory system of ancestral molluscs would be inadequate for cephalopods. Their circulatory system consists of a closed network of vessels, and capillaries conduct blood through the gill filaments. Furthermore, the molluscan plan of circulation places the entire systemic circulation before the blood reaches the gills (in contrast to vertebrates, in which the

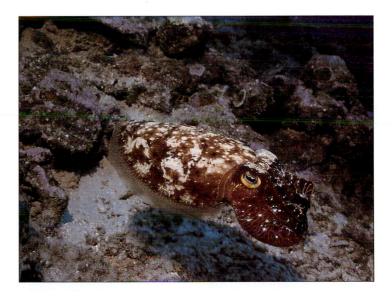

Figure 16-37
Cuttlefish, *Sepia latimanus,* has an internal shell familiar to keepers of caged birds as "cuttlebone."

blood leaves the heart and goes directly to the gills or lungs). This functional problem was solved by the development of **accessory** or **branchial hearts** (Figure 16-38B) at the base of each gill to increase the pressure of blood going through the capillaries there.

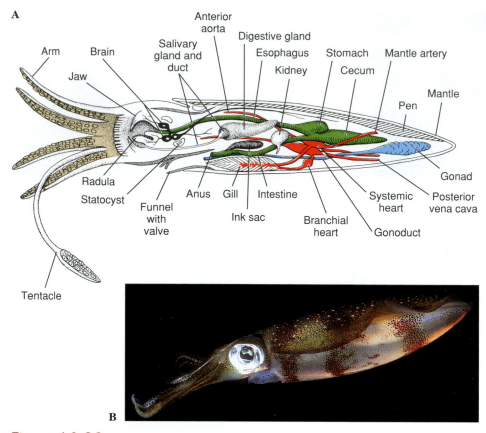

A

Arm
Brain
Jaw
Anterior aorta
Salivary gland and duct
Digestive gland
Esophagus
Kidney
Stomach
Cecum
Mantle artery
Pen
Mantle

Radula
Statocyst
Funnel with valve
Anus
Gill
Intestine
Ink sac
Branchial heart
Systemic heart
Gonoduct
Gonad
Posterior vena cava

Tentacle

B

Figure 16-38
A, Lateral view of squid anatomy, with the left half of the mantle removed. **B**, Reef squid *Sepioteuthis lessoniana.*

After a member of genus *Nautilus* secretes a new septum, the new chamber is filled with fluid similar in ionic composition to that of *Nautilus'* blood (and of seawater). Fluid removal involves active secretion of ions into tiny intercellular spaces in the siphuncular epithelium, so that a very high local osmotic pressure is produced, and water is drawn out of the chamber by osmosis. The gas in the chamber is just the respiratory gas from the siphuncle tissue that diffuses into the chamber as fluid is removed. Thus gas pressure in the chamber is 1 atmosphere or less because it is in equilibrium with gases dissolved in the seawater surrounding the *Nautilus,* which are in turn in equilibrium with air at the surface of the sea, despite the fact that the *Nautilus* may be swimming at 400 m beneath the surface. That the shell can withstand implosion by the surrounding 41 atmospheres (about 600 pounds per square inch), and that the siphuncle can remove water against this pressure are marvelous feats of natural engineering!

Nervous and Sensory Systems.   Nervous and sensory systems are more elaborate in cephalopods than in other molluscs. The brain, the largest in any invertebrate, consists of several lobes with millions of nerve cells. Squids have giant nerve fibers (among the largest known in the animal kingdom), which are activated when the animal is alarmed and that initiate maximal contractions of the mantle muscles for a speedy escape.

Squid nerves played an important role in early biophysical studies. Our current understanding of transmission of action potentials along and between nerve fibers (see Chapter 33) is based primarily on work performed using the giant nerve fibers of squids, *Loligo* spp. A. Hodgkin and A. Huxley received the Nobel Prize in Physiology or Medicine, 1963, for their achievements in this field.

Sense organs are well developed. Except for *Nautilus,* which has relatively simple eyes, cephalopods have highly complex eyes with cornea, lens, chambers, and retina (Figure 16-39). Orientation of the eyes is controlled by the statocysts, which are larger and more complex than in other molluscs. The eyes are held in a constant relation to gravity, so that the slit-shaped pupils are always in a horizontal position.

Octopods are apparently colorblind but can be taught to discriminate between shapes—for example, a square and a rectangle—and to remember such a discrimination for a considerable time. Experimenters find it easy to modify their behavior patterns by devices of reward and punishment. Octopods are capable of observational learning; when one octopus observes another being rewarded by making a correct choice, the observer learns which choice is rewarded and consistently makes the same selection when given the opportunity.

When similar structures that are not inherited from a common ancestor evolve in unrelated animals, we call it **convergence,** or **convergent evolution.** For many years cephalopod eyes and vertebrate eyes have been cited as a marvelous example of convergent evolution. Cephalopod and vertebrate eyes are similar in many details of structure but differ in development. Compound eyes of arthropods (p. 383), differing in both structure and development, were viewed as examples of other, independently derived eyes in animals. Now we recognize that all triploblastic animals with eyes, even those with the most simple eyespots, such as platyhelminths, share at least two conserved genes: that for rhodopsin, a visual pigment, and *Pax 6,* now sometimes called the "master control gene for eye morphogenesis." Once these two genes originated, natural selection eventually produced the specialized organs of vertebrates, molluscs, and arthropods. Thus, eyes of Bilateria may have been derived from a common ancestral type of photoreceptive cell.

Octopods use their arms for tactile exploration and can discriminate between textures by feel but apparently not between shapes. Their arms are well supplied with both tactile and chemoreceptor cells. Cephalopods seem to lack a sense of hearing.

Communication   Little is known of social behavior of nautiloids or deep-water cephalopods, but inshore and littoral forms such as *Sepia, Sepioteuthis, Loligo,* and *Octopus* have been studied extensively. Although their tactile sense is well developed and they have some chemical sensitivity, visual signals are the predominant means of communication. These signals consist of a host of movements of the arms, fins, and body, as well as many color changes. Movements may range from minor body motions to exaggerated spreading, curling, raising, or lowering of some or all of the arms. Color changes are effected by **chromatophores,** cells in the skin that contain pigment granules (see Chapter 29, p. 616). Tiny muscle cells surround each elastic chromatophore, whose contractions pull the cell boundary of the chromatophore outward, causing it to expand greatly. As the cell expands, the pigment becomes dispersed, changing the color pattern of the animal. When the muscles relax, chromatophores return to their original size and pigment becomes concentrated again. By means of the chromatophores, which are under nervous and probably hormonal control, an elaborate system of changes in color and pattern is possible, including general darkening or lightening; flushes of pink, yellow, or lavender; and formation of bars, stripes, spots, or irregular blotches. These colors may be used variously as danger signals, as protective coloring, in courtship rituals, and probably in other ways.

By assuming different color patterns of different parts of the body, a squid can transmit three or four different messages *simultaneously* to different individuals and in different directions, and it can instantaneously change any or all of the messages. Probably no other system of communication in invertebrates can convey so much information so rapidly.

Deep-water cephalopods may have to depend more on chemical or tactile senses than their littoral or surface cousins,

Figure 16-39
Eye of a cuttlefish *(Sepia).* The structure of cephalopod eyes is very similar to that of eyes of vertebrates (see note, this page).

but they also produce their own type of visual signals, for they have evolved many elaborate luminescent organs.

Most cephalopods other than nautiloids have another protective device. An ink sac that empties into the rectum contains an **ink gland** that secretes **sepia,** a dark fluid containing the pigment melanin, into the sac. When the animal is alarmed, it releases a cloud of ink, which may hang in the water as a blob or be contorted by water currents. The animal quickly departs from the scene, leaving the ink as a decoy to the predator.

Reproduction   Sexes are separate in cephalopods. Spermatozoa are encased in spermatophores and stored in a sac that opens into the mantle cavity. One arm of adult males is modified as an intromittent organ, called a **hectocotylus,** used to pluck a spermatophore from his own mantle cavity and insert it into the mantle cavity of a female near the oviduct opening (Figure 16-40). Before copulation males often undergo color displays, apparently directed against rival males. Eggs are fertilized as they leave the oviduct and are then usually attached to stones or other objects. Some octopods tend their eggs. Females of *Argonauta,* the paper nautilus, secrete a fluted "shell," or capsule, in which eggs develop.

The large yolky eggs undergo meroblastic cleavage. During embryonic development, the head and foot become indistinguishable. The ring around the mouth, which bears the arms, or tentacles, may be derived from the anterior part of the foot. Juveniles hatch from eggs; no free-swimming larva exists in cephalopods.

Figure 16-40
Copulation in cephalopods. **A,** Mating cuttlefishes. **B,** Male octopus uses modified arm to deposit spermatophores in female mantle cavity to fertilize her eggs. Octopuses often tend their eggs during development.

Major Groups of Cephalopods

There are three subclasses of cephalopods: Nautiloidea, which have two pairs of gills; the entirely extinct Ammonoidea; and Coleoidea, which have one pair of gills. Nautiloidea populated Paleozoic and Mesozoic seas, but there survives only one genus, *Nautilus* (see Figure 16-36), of which there are five or six species. *Nautilus'* head, with its 60 to 90 or more tentacles, can be extended from the opening of the body compartment of its shell. Its tentacles have no suckers but are made adhesive by secretions. They are used in searching for, sensing, and grasping food. Beneath its head is the funnel. Mantle, mantle cavity, and visceral mass are sheltered by the shell.

Ammonoids were widely prevalent in the Mesozoic era but became extinct by the end of the Cretaceous period. They had chambered shells analogous to nautiloids, but the septa were more complex, and the septal sutures (where septa contact the inside of the shell) were frilled (compare shells in Figure 16-36B and C). The reasons for their extinction remain a mystery. Present evidence suggests that they were gone before the asteroid bombardment at the end of the Cretaceous period (p. 127), whereas some nautiloids, which some ammonoids closely resembled, survive to the present.

Subclass Coleoidea includes all living cephalopods except *Nautilus.* There are four orders of coleoids. Members of order Sepioidea (cuttlefishes and their relatives) have a rounded or compressed, bulky body bearing fins (Figure 16-37). They have eight arms and two tentacles. Both arms and tentacles have suckers, but tentacles bear suckers only at their ends (Figure 16-37). Members of order Teuthoidea (squids, Figure 16-38) have a more cylindrical body but also have eight arms and two tentacles. Order Vampyromorpha (vampire squid) contains only a single, deep-water species. Members of order Octopoda have eight arms and no tentacles (see Figure 16-1E). Their bodies are short and saclike, with no fins. The suckers in squids are stalked (pedunculated), with horny rims bearing teeth; in octopuses the suckers are sessile and have no horny rims.

PHYLOGENY AND ADAPTIVE RADIATION

The first molluscs probably arose during Precambrian times because fossils attributed to Mollusca have been found in geological strata as old as the early Cambrian period. On the basis of such shared features as spiral cleavage, mesoderm from the 4d blastomere, and trochophore larva, most zoologists have accepted Mollusca as protostomes, allied with the annelids and arthropods. Opinions differ, however, as to whether molluscs were derived from a flatwormlike ancestor independent of annelids, share an ancestor with annelids after the advent of the coelom, or share a metameric common ancestor with annelids. This last hypothesis is strengthened if *Neopilina* (class Monoplacophora) can be considered metameric, as some scientists have contended. However, it is unlikely that such a successful adaptation as metamerism would have been lost in all later molluscs, and there is no trace of metamerism in development of any known molluscan larva. The most reasonable hypothesis is that molluscs branched off from the annelid line after the coelom arose but before the advent of metamerism. Some analyses suggest that molluscs and annelids are more closely related to each other than either is to the arthropods. This contention is strengthened by molecular evidence that places annelids and molluscs in Lophotrochozoa and arthropods in Ecdysozoa (p. 204). The Lophotrochozoa/Ecdysozoa hypothesis, however, requires that metamerism arose among protostomes at least twice independently.

Fossils are remains of past life uncovered from the crust of the earth (Chapter 6). They can be actual parts or products of animals (teeth, bones, shells, and so on), petrified skeletal parts, molds, casts, impressions, footprints, and others. Soft and fleshy parts rarely leave recognizable fossils. Therefore we have no record of molluscs before they had shells, and there can be some doubt that certain early fossil shells are really remains of molluscs, particularly if the group they represent is now extinct. The issue of how to define a mollusc from hard parts alone was emphasized by Yochelson (1978, *Malacologia* **17**:165), who said, "If scaphopods were extinct and soft parts were unknown, would they be called mollusks? I think not."

A "hypothetical ancestral mollusc" (see Figure 16-2) was long viewed as representing the original mollusc ancestor, but neither a solid shell nor a broad, crawling foot are now considered universal characters for Mollusca. The primitive ancestral mollusc was probably a more or less wormlike organism with a ventral gliding surface and a dorsal mantle with a chitinous cuticle and calcareous scales (Figure 16-41). It had a posterior mantle cavity with two gills, a radula, a ladderlike nervous system, and an open circulatory system with a heart. Among living molluscs the primitive condition is most nearly approached by caudofoveates, although the foot is reduced to an oral shield in members of this class. Solenogasters have lost the gills, and the

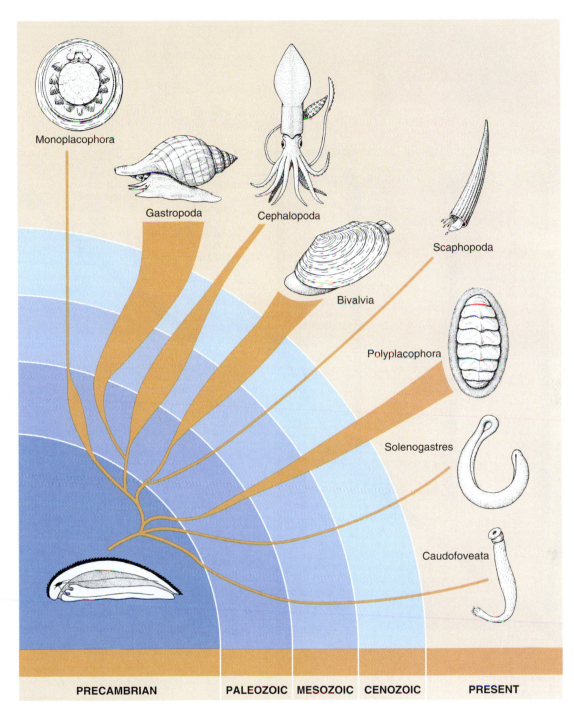

Figure 16-41
Classes of Mollusca, showing their derivations and relative abundance.

foot is represented by their ventral groove. Both these classes probably branched from primitive ancestors before the development of a solid shell, a distinct head with sensory organs, and a ventral muscularized foot. Polyplacophorans probably also branched early from the main lines of molluscan evolution before the veliger was established as the larva. Some workers believe that shells of polyplacophorans are not homologous to shells of other molluscs because they differ structurally and developmentally. Polyplacophora and the remaining classes are sister groups (Figure 16-42).

Cladistic analysis suggests that Gastropoda and Cephalopoda form the sister group to Monoplacophora (see Figure 16-42). Both gastropods and cephalopods have a greatly expanded visceral mass. The mantle cavity was brought toward the head by torsion in gastropods, but in cephalopods the mantle cavity was extended ventrally. Evolution of a chambered shell in cephalopods was a very important contribution to their freedom from the substratum and their ability to swim. Elaboration of their respiratory, circulatory, and nervous systems is correlated with their predatory and swimming habits.

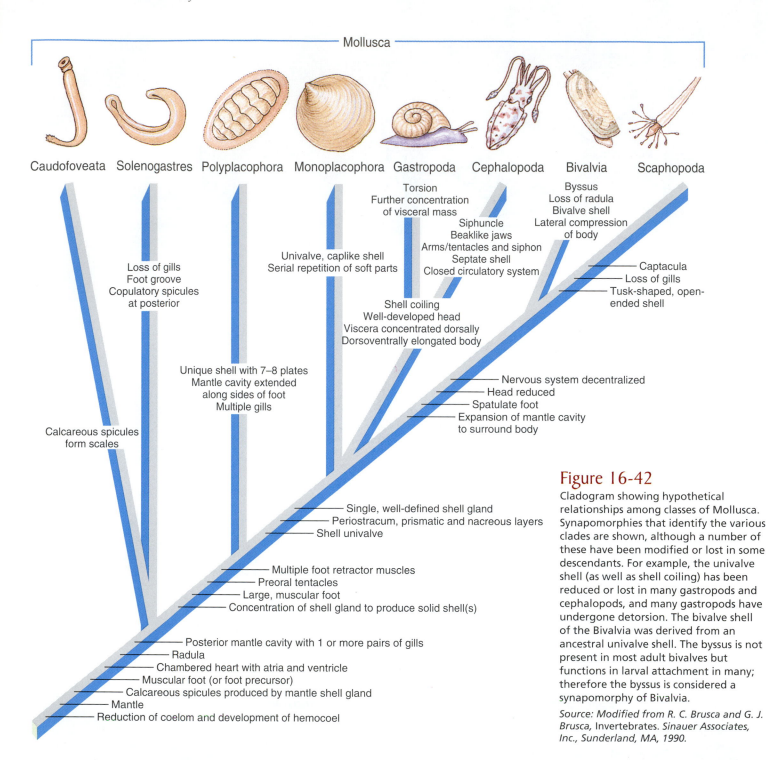

Mollusca

Caudofoveata Solenogastres Polyplacophora Monoplacophora Gastropoda Cephalopoda Bivalvia Scaphopoda

Torsion
Further concentration
of visceral mass

Byssus
Loss of radula
Bivalve shell
Lateral compression
of body

Siphuncle
Beaklike jaws
Arms/tentacles and siphon
Septate shell
Closed circulatory system

Loss of gills
Foot groove
Copulatory spicules
at posterior

Univalve, caplike shell
Serial repetition of soft parts

Captacula
Loss of gills
Tusk-shaped, open-
ended shell

Shell coiling
Well-developed head
Viscera concentrated dorsally
Dorsoventrally elongated body

Unique shell with 7–8 plates
Mantle cavity extended
along sides of foot
Multiple gills

Nervous system decentralized
Head reduced
Spatulate foot
Expansion of mantle cavity
to surround body

Calcareous spicules
form scales

Single, well-defined shell gland
Periostracum, prismatic and nacreous layers
Shell univalve

Multiple foot retractor muscles
Preoral tentacles
Large, muscular foot
Concentration of shell gland to produce solid shell(s)

Posterior mantle cavity with 1 or more pairs of gills
Radula
Chambered heart with atria and ventricle
Muscular foot (or foot precursor)
Calcareous spicules produced by mantle shell gland
Mantle
Reduction of coelom and development of hemocoel

Figure 16-42

Cladogram showing hypothetical relationships among classes of Mollusca. Synapomorphies that identify the various clades are shown, although a number of these have been modified or lost in some descendants. For example, the univalve shell (as well as shell coiling) has been reduced or lost in many gastropods and cephalopods, and many gastropods have undergone detorsion. The bivalve shell of the Bivalvia was derived from an ancestral univalve shell. The byssus is not present in most adult bivalves but functions in larval attachment in many; therefore the byssus is considered a synapomorphy of Bivalvia.

Source: Modified from R. C. Brusca and G. J. Brusca, Invertebrates. *Sinauer Associates, Inc., Sunderland, MA, 1990.*

Scaphopods and bivalves have an expanded mantle cavity that essentially envelops the body. Adaptations for burrowing characterize this clade: the spatulate foot and reduction of the head and sense organs.

Most diversity among molluscs is related to their adaptation to different habitats and modes of life and to a wide variety of feeding methods, ranging from sedentary filter feeding to active predation. There are many adaptations for food gathering within the phylum and an enormous variety in radular structure and function, particularly among gastropods.

The versatile glandular mantle has probably shown more plastic adaptive capacity than any other molluscan structure. Besides secreting the shell and forming the mantle cavity, it is variously modified into gills, lungs, siphons, and apertures, and it sometimes functions in locomotion, in the feeding processes, or in a sensory capacity. The shell, too, has undergone a variety of evolutionary adaptations.

Classification of Phylum Mollusca

Class Caudofoveata (kaw′do-fo-ve-at′ a) (L. *cauda*, tail, + *fovea*, small pit). Wormlike; shell, head, and excretory organs absent; radula usually present; mantle with chitinous cuticle and calcareous scales; oral pedal shield near anterior mouth; mantle cavity at posterior end with pair of gills; sexes separate; formerly united with solenogasters in class Aplacophora. Examples: *Chaetoderma, Limifossor.*

Class Solenogastres (so-len′o-gas′ trez) (Gr. *solen*, pipe, + *gaster*, stomach): **solenogasters.** Wormlike; shell, head, and excretory organs absent; radula present or absent; mantle usually covered with scales or spicules; mantle cavity posterior, without true gills, but sometimes with secondary respiratory structures; foot represented by long, narrow, ventral pedal groove; hermaphroditic. Example: *Neomenia.*

Class Monoplacophora (mon′o-pla-kof′o-ra) (Gr. *monos*, one, + *plax*, plate, + *phora*, bearing). Body bilaterally symmetrical with a broad flat foot; a single limpetlike shell; mantle cavity with five or six pairs of gills; large coelomic cavities; radula present; six pairs of nephridia, two of which are gonoducts; separate sexes. Example: *Neopilina* (see Figure 16-8).

Class Polyplacophora (pol′y-pla-kof′o-ra) (Gr. *polys*, many, several, + *plax*, plate, + *phora*, bearing): **chitons.** Elongated, dorsoventrally flattened body with reduced head; bilaterally symmetrical; radula present; shell of eight dorsal plates; foot broad and flat; gills multiple along sides of body between foot and mantle edge; sexes usually separate, with a trochophore but no veliger larva. Examples: *Mopalia* (see Figure 16-10), *Tonicella* (see Figure 16-1A).

Class Scaphopoda (ska-fop′o-da) (Gr. *skaphē*, trough, boat, + *pous, podos,* foot): **tusk shells.** Body enclosed in a one-piece tubular shell open at both ends; conical foot; mouth with radula and tentacles; head absent; mantle for respiration; sexes separate, trochophore larva. Example: *Dentalium* (see Figure 16-11).

Class Gastropoda (gas-trop′o-da) (Gr. *gaster*, stomach, + *pous, podos,* foot): **snails and slugs.** Body asymmetrical and shows effects of torsion; body usually in a coiled shell (shell uncoiled or absent in some); head well developed, with radula; foot large and flat; one or two gills, or with mantle modified into secondary gills or a lung; most with single auricle and single nephridium; nervous system with cerebral, pleural, pedal, and visceral ganglia; dioecious or monoecious, some with trochophore, typically with veliger, some without pelagic larva. Examples: *Busycon, Polinices* (see Figure 16-15B), *Physa, Helix, Aplysia* (see Figure 16-21).

Class Bivalvia (bi-val′ve-a) (L. *bi*, two, + *valva*, folding door, valve) **(Pelecypoda): bivalves.** Body enclosed in a two-lobed mantle; shell of two lateral valves of variable size and form, with dorsal hinge; head greatly reduced, but mouth with labial palps; no radula; no cephalic eyes, a few with eyes on mantle margin; foot usually wedge shaped; gills platelike; sexes usually separate, typically with trochophore and veliger larvae. Examples: *Anodonta, Venus, Tagelus* (see Figure 16-26), *Teredo* (see Figure 16-27).

Class Cephalopoda (sef′a-lop′o-da) (Gr. *kephalē*, head, + *pous, podos,* foot): **squids and octopuses.** Shell often reduced or absent; head well developed with eyes and a radula; head with arms or tentacles; foot modified into siphon; nervous system of well-developed ganglia, centralized to form a brain; sexes separate, with direct development. Examples: *Sepioteathis* (see Figure 16-38), *Octopus* (see Figure 16-1E), *Sepia* (see Figure 16-37).

SUMMARY

Mollusca is the largest lophotrochozoan phylum and one of the largest and most diverse of all phyla, its members ranging in size from very small organisms to the largest of invertebrates. Their basic body divisions are head-foot, and visceral mass, which is usually covered by a shell. The majority are marine, but some are freshwater, and a few are terrestrial. They occupy a wide variety of niches. A number are economically important, and a few are medically important as hosts of parasites.

Molluscs are coelomate (have a coelom), although their coelom is limited to the area around the heart and gonads. Evolutionary development of a coelom was important because it enabled better organization of visceral organs and, in many of the animals that have it, an efficient hydrostatic skeleton.

The mantle and mantle cavity are important characteristics of molluscs. The mantle secretes the shell and overlies a part of the visceral mass to form a cavity housing the gills. The mantle cavity has been modified into a lung in some molluscs. The foot is usually a ventral, solelike, locomotory organ, but it may be variously modified, as in cephalopods, where it has become arms and a funnel. The radula is found in all molluscs except bivalves and solenogasters and is a protrusible, tonguelike organ with teeth used in feeding. Except in cephalopods, which have a closed circulatory system, the circulatory system of molluscs is open, with a heart and blood sinuses. Molluscs usually have a pair of nephridia connecting with the coelom and a complex nervous system with a variety of sense organs. The primitive larva of molluscs is the trochophore, and most marine molluscs have a more advanced larva, the veliger.

Classes Caudofoveata and Solenogastres are small groups of wormlike molluscs with no shell. Scaphopoda is a slightly larger class with a tubular shell, open at both ends, and the mantle wrapped around the body.

Class Monoplacophora is a tiny, univalve marine group showing pseudometamerism. Polyplacophora are more common, marine organisms with shells in the form of a series of eight plates. They are rather sedentary animals with a row of gills along each side of their foot.

Gastropoda are the most successful and largest class of molluscs. Their interesting evolutionary history includes torsion, or twisting of the posterior end to the anterior, so that anus and head are at the same end, and coiling, an elongation and spiraling of the visceral mass. Torsion has led to the problem of fouling, which is the release of excreta over the head and in front of the gills, and this has been solved in various ways among different gastropods. Among the solutions to fouling are bringing water into one side of the mantle cavity and out the other (many gastropods), some degree of detorsion (opisthobranchs), and conversion of the mantle cavity into a lung (pulmonates).

Class Bivalvia are marine and freshwater, and they have their shell divided into two valves joined by a dorsal ligament and held together by an adductor muscle. Most of them are filter feeders, drawing water through their gills by ciliary action.

Members of class Cephalopoda are the most advanced molluscs; they are all predators and many can swim rapidly. Their tentacles capture prey by adhesive secretions or by suckers. They swim by forcefully expelling water from their mantle cavity through a funnel, which was derived from the foot.

There is strong embryological and molecular evidence that molluscs share a common ancestor with annelids more recently than either of these phyla do with arthropods or deuterostome phyla, although molluscs are not metameric.

REVIEW QUESTIONS

1. Members of such a large and diverse phylum as Mollusca impact humans in many ways. Discuss this statement.
2. How does a molluscan coelom develop embryologically? Why was the evolutionary development of a coelom important?
3. What are characteristics of Mollusca that distinguish it from other phyla?
4. Briefly describe characteristics of the ancestral mollusc, and tell how each class of molluscs (Caudofoveata, Solenogastres, Monoplacophora, Polyplacophora, Scaphopoda, Gastropoda, Bivalvia, Cephalopoda) differs from the ancestral condition with respect to each of the following: shell, radula, foot, mantle cavity and gills, circulatory system, and head.
5. Define the following: ctenidia, odontophore, periostracum, prismatic layer, nacreous layer, metanephridia, nephrostome, trochophore, veliger, glochidium, osphradium.
6. Briefly describe the habitat and habits of a typical chiton.
7. Define the following with respect to gastropods: operculum, columella, torsion, fouling, bilateral asymmetry, rhinophore, pneumostome.

8. What survival problem did torsion create? How have gastropods evolved to avoid this problem?
9. Gastropods have radiated enormously. Illustrate this statement by describing variations in feeding habits found in gastropods.
10. Distinguish between opisthobranchs and pulmonates.
11. Briefly describe how a typical bivalve feeds and how it burrows.
12. How is the ctenidium modified from the ancestral form in a typical bivalve?
13. What is the function of the siphuncle of cephalopods?
14. Describe how cephalopods swim and how they eat.
15. Describe adaptations in the circulatory and neurosensory systems of cephalopods that are particularly valuable for actively swimming, predaceous animals.
16. Distinguish between ammonoids and nautiloids.
17. Which other invertebrate groups are likely to be the closest relatives of molluscs? Give some evidence for these relationships.

SELECTED REFERENCES

Abbott, R. T., and P. A. Morris. 2001. R. T. Peterson (ed.). A field guide to shells: Atlantic coasts and the West Indies, ed. 5. Boston, Houghton Mifflin Company. *An excellent revision of a popular handbook.*

Barinaga, M. 1990. Science digests the secrets of voracious killer snails. Science **249**:250–251. *Describes current research on the toxins produced by cone snails.*

Fleischman, J. 1997. Mass extinctions come to Ohio. Discover **18**(5):84–90. *Of the 300 species of freshwater bivalves in the Mississippi River basin, 161 are extinct or endangered.*

Gehring, W. J., and I. Kazuho. 1999. *Pax 6:* mastering eye morphogenesis and eye evolution. Trends Genet. **15**:371–377. *The authors discuss morphogenetic pathways by which various animal eyes could have evolved from a common ancestral type of photoreceptive cell.*

Gosline, J. M., and M. D. DeMont. 1985. Jet-propelled swimming in squids. Sci. Am. **252**:96–103 (Jan.). *Mechanics of swimming in squid are analyzed; elasticity of collagen in mantle increases efficiency.*

Gould, S. J. 1994. Common pathways of illumination. Nat. Hist. **103**:10–20. Pax6 *gene controls eye morphogenesis in* Drosophila *and vertebrates.*

Hanlon, R. T., and J. B. Messenger. 1996. Cephalopod behaviour. Cambridge, U.K., Cambridge University Press. *Intended for nonspecialists and specialists.*

Holloway, M. 2000. Cuttlefish say it with skin. Nat. Hist. **109**(3):70–76. *Cuttlefish and other cephalopods can change texture and color of their skin with astonishing speed. Fifty-four components of cuttlefish "vocabulary" have been described, including color display, skin texture, and a variety of arm and fin signals.*

Roper, C. R. E., and K. J. Boss. 1982. The giant squid. Sci. Am. **246**:96–105 (April). *Many mysteries remain about the deep-sea squid,* Architeuthis,

because it has never been studied alive. It can reach a weight of 1000 pounds and a length of 18 m, and its eyes are as large as automobile headlights.

Ross, J. 1994. An aquatic invader is running amok in U.S. waterways. Smithsonian **24**(11):40–50 (Feb.). *A small bivalve, the zebra mussel apparently introduced into the Great Lakes with ballast water from ships, is clogging up intake pipes and municipal water supplies. It will take billions of dollars to control.*

Ward, P., L. Greenwald, and O. E. Greenwald. 1980. The buoyancy of the chambered nautilus. Sci. Am. **243**:190–203 (Oct.). *Reviews discoveries on how the nautilus removes water from a chamber after secreting a new septum.*

Ward, P. D. 1998. Coils of time. Discover **19**(3):100–106. *Present* Nautilus *has apparently existed essentially unchanged for 100 million years, and all other species known were derived from it, including king nautilus (*Allonautilus*), a recent derivation.*

Woodruff, D. S., and M. Mulvey. 1997. Neotropical schistosomiasis: African affinities of the host snail *Biomphalaria glabrata* (Gastropoda: Planorbidae). Biol. J. Linn. Soc. **60**:505–516. *The pulmonate snail* Biomphalaria glabrata *is the intermediate host in the New World for* Schistosoma mansoni, *an important trematode of humans (p. 283). Allozyme analysis shows that B. glabrata clusters with African species rather than the neotropical ones. Thus, when S. mansoni was brought to New World in African slaves, it found a compatible host.*

Zorpette, G. 1996. Mussel mayhem, continued. Sci. Am. **275**:22–23 (Aug.). *Some benefits, though dubious, of the zebra mussel invasion have been described, but these are outweighed by the problems created.*

ZOOLOGY LINKS TO THE INTERNET

Visit the textbook's Online Learning Center at www.mhhe.com/zoology to find live Internet links for each of the topics listed here.
Phylum Mollusca

Primitive Classes of Molluscs
Class Bivalvia
Class Gastropoda
Class Cephalopoda

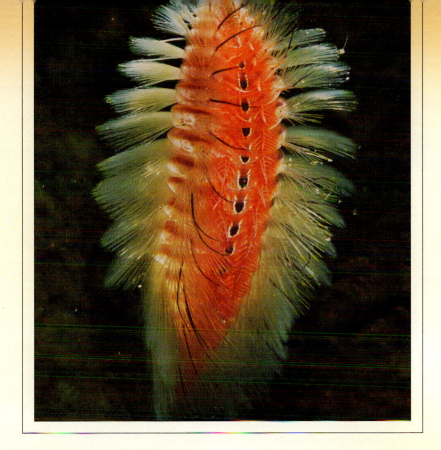

Chloeia sp., a polychaete.

Segmented Worms

Phylum Annelida

Dividing the Body

Although a spacious, fluid-filled coelom provided an efficient hydrostatic skeleton for burrowing, precise control of body movements was not possible in the earliest coelomates. The force of muscle contraction in one area was carried throughout the body by the fluid in the undivided coelom. This limitation was remedied when a series of partitions (septa) evolved in an ancestor of annelids and arthropods. When septa divided the coelom into a series of compartments, components of most other body systems, such as circulatory, nervous, and excretory, were repeated in each segment. This body plan is known as *metamerism*.

The evolutionary advent of metamerism was highly significant because it made possible development of much greater complexity in structure and function. Metamerism not only increased efficiency of burrowing but also made possible independent movements by separate segments. A need for fine control of movements led, in turn, to evolution of a more sophisticated nervous system. Moreover, repetition of body parts gave the organisms a built-in redundancy, as in some human-made systems. Redundancy provided a safety factor: if one segment should fail, others could still function. Thus an injury to one part would not necessarily be fatal.

The evolutionary potential of the metameric body plan is amply demonstrated by the large and diverse phyla Annelida, Arthropoda and Chordata, which apparently represent three separate evolutionary origins of metamerism.

Phylum Annelida (an-nel'i-da) (L. *annelus,* little ring, + *ida,* pl. suffix) consists of the segmented worms. It is a large phylum, numbering approximately 15,000 species, the most familiar of which are earthworms and freshwater worms (class Oligochaeta) and leeches (class Hirudinea). However, approximately two-thirds of the phylum is composed of marine worms (class Polychaeta), which are less familiar to most people. Among the latter are many curious members; some are strange, even grotesque, whereas others are graceful and beautiful. They include clamworms, plumed worms, parchment worms, scaleworms, lugworms, and many others. Annelids are true coelomates and belong to superphylum Lophotrochozoa, with spiral cleavage and mosaic development. They are a highly developed group in which the nervous system is more centralized and the circulatory system more complex than those of phyla we have studied thus far.

Annelida are worms whose bodies are divided into similar rings, or **segments,** arranged in linear series and externally marked by circular grooves called **annuli;** the name of the phylum refers to this characteristic. Body segmentation, or **metamerism,** is a division of the body into a serial succession of segments, each of which contains similar representatives of all major organ systems. In annelids the segments (also called **metameres** or **somites**) are delimited by septa. Metamerism is not limited to annelids, but is shared by arthropods (insects, crustaceans, and others) and vertebrates, in which it evolved independently.

Annelids are sometimes called "bristle worms" because, with the exception of leeches, most annelids bear tiny chitinous bristles called **setae** (L. *seta,* hair or bristle). Short needlelike setae help anchor somites during locomotion to prevent backward slipping; long, hairlike setae aid aquatic forms in swimming. Since many annelids either are burrowers or live in secreted tubes, stiff setae also aid in preventing the worm from being pulled out or washed out of its home. Robins know from experience how effective earthworms' setae are.

Annelids have a worldwide distribution, and a few species are cosmopolitan. Polychaetes are chiefly marine forms. Most are benthic, but some live pelagic in the sea. Oligochaetes and leeches occur predominantly in fresh water or terrestrial soils. Some freshwater species burrow in mud and sand and others among submerged vegetation. Many leeches are predators, and many are specialized for piercing their prey and feeding on blood or soft tissues. A few leeches are marine, but most live in fresh water or in damp regions. Suckers are typically found at both ends of the body for attachment to the substratum or to their prey.

BODY PLAN

The annelid body typically has an anterior **prostomium,** a segmented body, and a terminal portion bearing the anus **(pygidium).** The prostomium and pygidium are not considered metameres, but anterior segments often fuse with the prostomium to form a head. New metameres differentiate during development just in front of the pygidium; thus the oldest

Position in Animal Kingdom

1. Annelids belong to the lophotrochozoan **protostome** branch of the animal kingdom and have **spiral cleavage** and **mosaic (determinate) development,** characters shared with and indicating relationship to molluscs.
2. Annelids as a group show a primitive metamerism with comparatively few differences between the different somites.
3. Characters shared with arthropods include an outer secreted cuticle and a similar nervous system.

Biological Contributions

1. **Metamerism** represents the greatest innovation seen in this phylum. A more highly specialized metamerism is seen in arthropods.
2. A true coelomic cavity reaches a high stage of development in this group.
3. Specialization of the head region into differentiated organs, such as tentacles, palps, and eyespots of polychaetes, is carried further in some annelids than in other invertebrates so far considered.
4. There are modifications of the **nervous system,** with cerebral ganglia (brain), two closely fused ventral nerve cords with giant fibers running the length of the body, and various ganglia with their lateral branches.
5. The circulatory system is much more complex than any we have so far considered. It is a closed system with muscular blood vessels and aortic arches ("hearts") for propelling the blood.
6. The appearance of the fleshy **parapodia,** with their respiratory and locomotor functions, introduces a suggestion of the paired appendages and specialized gills found in arthropods.
7. Well-developed **nephridia** in most of the somites have reached a differentiation that involves removal of waste from the blood as well as from the coelom.
8. Annelids are the most highly organized animals capable of complete regeneration. However, this ability varies greatly within the group.

segments are at the anterior end and the youngest segments are at the posterior.

The body wall has strong circular and longitudinal muscles adapted for swimming, crawling, and burrowing and is covered with epidermis and a thin, outer layer of nonchitinous cuticle (Figure 17-1).

In most annelids the coelom develops embryonically as a split in the mesoderm on each side of the gut **(schizocoel),** forming a pair of coelomic compartments in each segment. **Peritoneum** (a layer of mesodermal epithelium) lines the body wall of each compartment, forms dorsal and ventral **mesenteries,** and covers all organs (Figure 17-1). Peritonea of adjacent segments meet to form **septa.** Septa are perforated by the gut and longitudinal blood vessels. Not only is the coelom metamerically arranged, but practically every body system is affected in some way by this segmental arrangement.

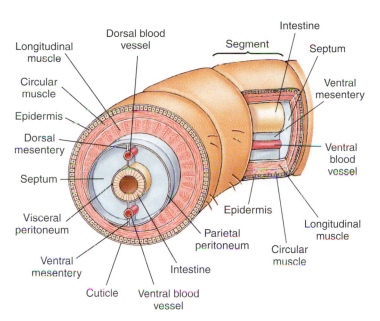

Figure 17-1
Annelid body plan.

A

B

Figure 17-2
Tube-dwelling sedentary polychaetes. **A,** One of the featherduster worms (called a Christmas-tree worm), *Spirobranchus giganteus,* has a double crown of radioles and lives in a calcareous tube. **B,** Sabellid polychaetes, *Bispira brunnea,* live in leathery tubes.

Except in leeches, the coelom is filled with fluid and serves as a **hydrostatic skeleton.** Because the volume of fluid is essentially constant, contraction of the longitudinal body-wall muscles causes the body to shorten and become larger in diameter, whereas contraction of the circular muscles causes it to lengthen and become thinner. Separation of the hydrostatic skeleton into a metameric series of coelomic cavities increases its efficiency greatly, because the force of local muscle contraction is not transferred throughout the length of the worm. Widening and elongation can occur in restricted areas. Crawling motions are effected by alternating waves of contraction of longitudinal and circular muscles (peristaltic contraction) passing down the body. Segments in which longitudinal muscles are contracted widen and anchor themselves against burrow walls or other substratum while other segments, in which circular muscles are contracted, elongate and stretch forward. Forces powerful enough for burrowing as well as locomotion can thus be generated. Swimming forms use undulatory rather than peristaltic movements in locomotion.

CLASS POLYCHAETA

Polychaetes form the largest class of annelids with more than 10,000 species, most of them marine. Although most are 5 to 10 cm long, some are less than 1 mm, and others may be as long as 3 m. Some are brightly colored in reds and greens; others are dull or iridescent. Some are picturesque, such as the "featherduster" worms (Figure 17-2).

Polychaetes differ from other annelids in having a well-differentiated head with specialized sense organs; paired appendages, called **parapodia,** on most segments; and no clitellum (see p. 357) (Figure 17-3). As their name implies, they have many setae, usually arranged in bundles on the parapodia. They

show a pronounced differentiation of some body somites and a specialization of sensory organs practically unknown among clitellates (see p. 361).

Many polychaetes are euryhaline (can tolerate a wide range of environmental salinity) and occur in brackish water. The freshwater polychaete fauna is more diversified in warmer regions than in temperate zones.

Polychaetes live under rocks, in coral crevices, or in abandoned shells, or they burrow into mud or sand; some build their own tubes on submerged objects or in bottom sediment; some adopt the tubes or homes of other animals; some are planktonic. They are extremely abundant in some areas; for example, a square meter of mudflat may contain thousands of polychaetes. They play a significant part in marine food chains because they are eaten by fish, crustaceans, hydroids, and many others.

They are often divided for convenience into two groups (formerly the basis of subclasses): sedentary polychaetes, and

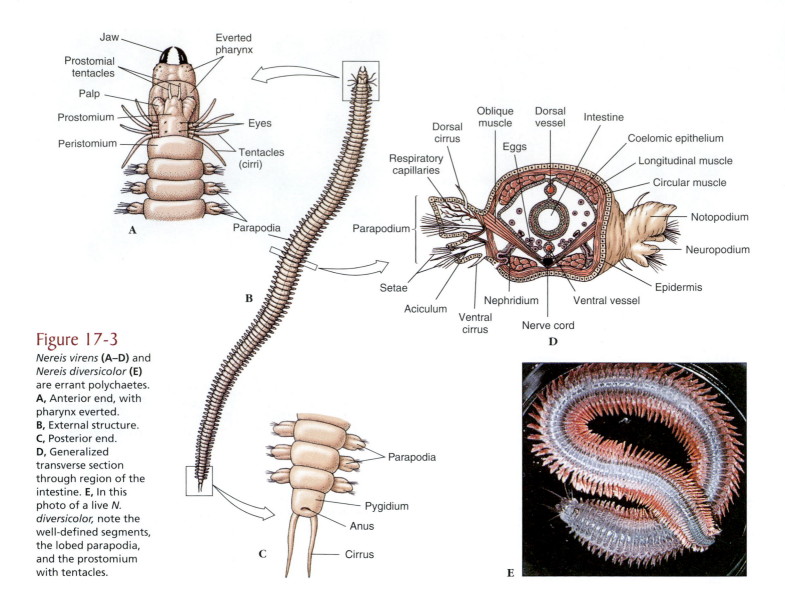

Figure 17-3

Nereis virens (**A–D**) and *Nereis diversicolor* (**E**) are errant polychaetes. **A,** Anterior end, with pharynx everted. **B,** External structure. **C,** Posterior end. **D,** Generalized transverse section through region of the intestine. **E,** In this photo of a live *N. diversicolor,* note the well-defined segments, the lobed parapodia, and the prostomium with tentacles.

errant or free-moving polychaetes. Sedentary polychaetes are mainly tubicolous, spending all or much of their time in tubes or permanent burrows. Many of them, especially those that live in tubes, have elaborate devices for feeding and respiration. Errant polychaetes (L. *errare,* to wander), include free-moving pelagic forms, active burrowers, crawlers, and tube worms that leave their tubes for feeding or breeding. Most of these, like clam worms *Nereis* (Gr. name of a sea nymph) (Figure 17-3), are predatory and equipped with jaws or teeth. They have a muscular eversible pharynx armed with teeth that can be thrust out with surprising speed and dexterity for capturing prey.

Form and Function

A polychaete typically has a head, or **prostomium,** which may or may not be retractile and which often bears eyes, tentacles, and sensory palps (Figure 17-3). The first segment **(peristomium)** surrounds the mouth and may bear setae, palps, or, in predatory forms, chitinous jaws. Ciliary feeders may bear a tentacular crown that can be opened like a fan or withdrawn into the tube.

The trunk is segmented, and most segments bear parapodia, which may have lobes, cirri, setae, and other parts on them (see Figure 17-3). Parapodia are used in crawling, swimming, or anchoring in tubes. They usually serve as the chief respiratory organs, although some polychaetes also have gills. *Amphitrite* (Gr. a mythical sea nymph), for example, has three pairs of branched gills and long extensible tentacles (Figure 17-4). *Arenicola* (L. *arena,* sand + *colo,* inhabit), the lugworm (Figure 17-5), which burrows through sand leaving characteristic castings at the entrance to its burrow, has paired gills on certain somites.

Nutrition

A polychaete's digestive system consists of a foregut, midgut, and hindgut. The foregut includes a stomodeum, pharynx, and anterior esophagus. It is lined with cuticle, and the jaws, where present, are constructed of cuticular protein. The midgut is derived from endoderm. More anterior portions secrete digestive enzymes, and absorption takes place toward the posterior. The short, ectodermally-derived hindgut connects the midgut to the exterior via the anus, which is on the pygidium.

Characteristics of Phylum Annelida

1. Body **metameric**; symmetry bilateral
2. Body wall with outer circular and inner longitudinal muscle layers; outer transparent moist cuticle secreted by epithelium
3. **Chitinous setae** often present; setae absent in leeches
4. Coelom (schizocoel) well developed and divided by septa, except in leeches; coelomic fluid supplies turgidity and functions as hydrostatic skeleton
5. **Circulatory system closed** and segmentally arranged; respiratory pigments (hemoglobin, hemerythrin, or chlorocruorin) often present; amebocytes in blood plasma
6. Digestive system complete and not metamerically arranged
7. Respiratory gas exchange through skin, **gills**, or **parapodia**
8. Excretory system typically a **pair of nephridia for each metamere**
9. Nervous system with a double ventral nerve cord and a pair of ganglia with lateral nerves in each metamere; brain a pair of dorsal cerebral ganglia with connectives to cord
10. Sensory system of tactile organs, taste buds, statocysts (in some), photoreceptor cells, and eyes with lenses (in some)
11. Hermaphroditic or separate sexes; larvae, if present, are trochophore type; asexual reproduction by budding in some; spiral cleavage and mosaic development

Errant polychaetes are mostly predators and scavengers. Sedentary polychaetes feed on suspended particles, or they are deposit feeders, consuming particles on or in the sediment.

Circulation and Respiration

Polychaetes show considerable diversity in both circulatory and respiratory structure and function. As mentioned before, parapodia and gills serve for gaseous exchange in various species. In some polychaetes there are no special organs for respiration, and gaseous exchange takes place across the body surface.

The circulatory pattern varies greatly. In *Nereis* (Gr. name of a sea nymph) a dorsal longitudinal vessel carries blood anteriorly, and a ventral longitudinal vessel conducts it posteriorly (Figure 17-3D). Blood flows between these two vessels via segmental networks in the parapodia, septa, and around the intestine. In *Glycera* (Gr. *Glykera*, a feminine proper name) the circulatory system is reduced and joins with the coelom. Septa are incomplete, and thus the coelomic fluid assumes the function of circulation.

Many polychaetes have respiratory pigments such as hemoglobin, chlorocruorin, or hemerythrin (p. 673).

Excretion

Although there is some variety in excretory organs, including possession of protonephridia and mixed proto- and metanephridia in some, most polychaetes have metanephridia (Figure 17-3). There is one pair per metamere, with the inner end of each (**nephrostome**) opening into a coelomic compartment. Coelomic fluid passes into the nephrostome, and selective resorption occurs along the nephridial duct, as in oligochaetes (see Figure 17-14).

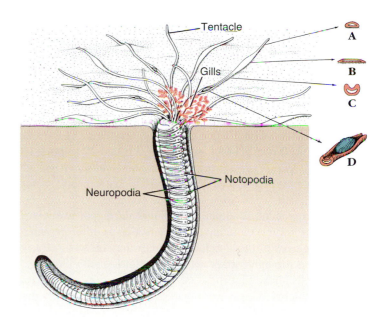

Figure 17-4

Amphitrite, which builds its tubes in mud or sand, extends long grooved tentacles out over the mud to pick up bits of organic matter. The smallest particles are moved along food grooves by cilia, larger particles by peristaltic movement. Its plumelike gills are blood red. **A,** Section through exploratory end of tentacle. **B,** Section through tentacle in area adhering to substratum. **C,** Section showing ciliary groove. **D,** Particle being carried toward mouth.

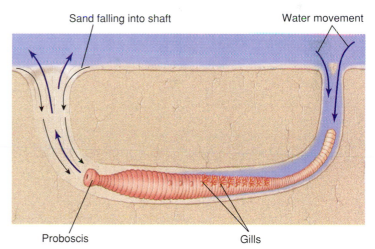

Figure 17-5

Arenicola, the lugworm, lives in an L-shaped burrow in intertidal mudflats. It burrows by successive eversions and retractions of its proboscis. By peristaltic movements it keeps water filtering through the sand. The worm then ingests the food-laden sand.

Nervous System and Sense Organs

Organization of the central nervous system in polychaetes follows the basic annelid plan (see Figure 17-15). Dorsal cerebral ganglia connect with a subpharyngeal ganglion via a circumpharyngeal commissure. A double ventral nerve cord courses the length of the worm, with metamerically arranged ganglia.

Sense organs are more highly developed in polychaetes than in oligochaetes and include eyes, nuchal organs, and statocysts. Eyes, when present, may range from simple eyespots to well-developed organs. Eyes are most conspicuous in errant worms. Usually the eyes are retinal cups, with rodlike photoreceptor cells lining the cup wall and directed toward the lumen of the cup. The highest degree of development occurs in the family Alciopidae, which has large, image-resolving eyes similar in structure to those of some cephalopod molluscs (Figure 16-39, p. 339), with cornea, lens, retina, and retinal pigment. Alciopid eyes also have accessory retinas, a characteristic shared by deep-sea fishes and some deep-sea cephalopods. Different wavelengths of light penetrate to different depths in water, and the accessory retinas of alciopids are sensitive to different wavelengths. The eyes of these pelagic animals may be well adapted to function as the light varies with depth. Studies with electroencephalograms show that they are sensitive to dim light of the deep sea. Nuchal organs are ciliated sensory pits or slits that appear to be chemoreceptive, an important factor in food gathering. Some burrowing and tube-building polychaetes have statocysts that function in body orientation.

Reproduction and Development

In contrast to clitellates, polychaetes have no permanent sex organs, and they usually have separate sexes. Reproductive systems are simple. Gonads appear as temporary swellings of the peritoneum and shed their gametes into the coelom. They are carried outside through gonoducts, through the metanephridia, or by rupture of the body wall. Fertilization is external, and the early larva is a trochophore (see Figure 16-6).

Some polychaetes live most of the year as sexually immature animals called atokes, but during the breeding season a portion of the body becomes sexually mature and swollen with gametes (Figure 17-6). An example is the palolo worm, which lives in burrows among coral reefs. During the swarming period, the sexually mature portions, now called epitokes, break off and swim to the surface. Just before sunrise, the sea is literally covered with them, and at sunrise they burst, freeing eggs and sperm for fertilization. Anterior portions of the worms regenerate new posterior sections. Swarming is of great adaptive value because the synchronous maturation of all the epitokes ensures the maximum number of fertilized eggs. However, this reproductive strategy is very hazardous; many types of predators have a feast on the swarming worms. In the meantime, the atoke remains safely in its burrow to produce another epitoke at the next cycle. In some polychaetes, epitokes arise from atokes by asexual budding (Figure 17-7) and become complete worms.

Clam Worms: *Nereis*

Clam worms (Figure 17-3), or sand worms as they are sometimes called, are errant polychaetes that live in mucus-lined burrows in or near low tide. Sometimes they are found in temporary hiding places, such as under stones, where they stay with their bodies covered and their heads protruding. They are most active at night, when they wiggle out of their hiding places and swim about or crawl over the sand in search of food.

Figure 17-6
Eunice viridis, the Samoan palolo worm. The posterior segments make up the epitokal region, consisting of segments packed with gametes. Each segment has an eyespot on the ventral side. Once a year the worms swarm, and the epitokes detach, rise to the surface, and discharge their ripe gametes, leaving the water milky. By the next breeding season, the epitokes are regenerated.

Atoke

Epitoke

The body, containing about 200 somites, may grow to 30 or 40 cm in length. The head is made up of a prostomium and a peristomium. The prostomium bears a pair of stubby palps, sensitive to touch and taste; a pair of short sensory tentacles; and two pairs of small dorsal eyes that are light sensitive. The peristomium bears the ventral mouth, a pair of chitinous jaws, and four pairs of sensory tentacles (Figure 17-3A).

Each parapodium has two lobes: a dorsal **notopodium** and a ventral **neuropodium** (Figure 17-3D). One or more chitinous spines **(acicula)** supports each lobe. Parapodia bear setae and are abundantly supplied with blood vessels. Parapodia are used for both creeping and swimming and are manipulated by oblique muscles that run from the midventral line to the parapodia in each somite. The worm swims by lateral undulatory wriggling of the body—unlike the peristaltic movement of earthworms. It can dart through the water with considerable speed. These undulatory movements can also be used to suck water into or pump it out of the burrow. The worm will usually adapt some kind of burrow if it can find one. When a worm is placed near a glass tube, it will wriggle in without hesitation.

Clam worms feed on small animals, other worms, and larval forms. They seize food with their chitinous jaws, which they protrude through the mouth when they evert their pharynx. As they withdraw the pharynx, they swallow the food. Movement of food through the alimentary canal is by peristalsis.

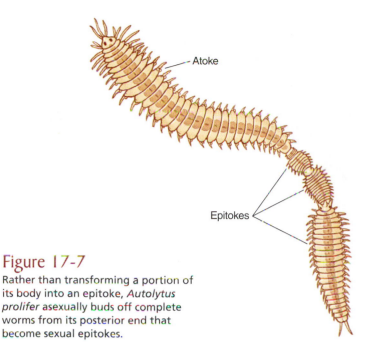

Figure 17-7
Rather than transforming a portion of its body into an epitoke, *Autolytus prolifer* asexually buds off complete worms from its posterior end that become sexual epitokes.

Other Interesting Polychaetes

Scale worms (Figure 17-8) are members of the family Polynoidae (Gr. *Polynoë*, daughter of Nereus and Doris, a sea god and goddess), one of the most abundant and widespread of polychaete families. Their rather flattened bodies are covered with broad scales, modified from dorsal parts of parapodia. Most are of modest size, but some are enormous (up to 190 mm long and 100 mm wide). They are carnivorous and feed on a wide variety of animals. Many are commensal, living in burrows of other polychaetes or in association with cnidarians, molluscs, or echinoderms.

Hermodice carunculata (Gr. *herma*, reef, + *dex*, a worm found in wood) (Figure 17-9) and related species are called fireworms. Their setae are hollow, brittle, and contain a poisonous secretion. When touched, the setae break off in the wound and cause skin irritation. They feed on corals, gorgonians, and other cnidarians.

Tube dwellers secrete many types of tubes. Some are parchmentlike or leathery (Figure 17-2B); some are firm, calcareous tubes attached to rocks or other surfaces (Figure 17-2A); and some are simply grains of sand or bits of shell or seaweed cemented together with mucous secretions. Many burrowers in sand and mud flats simply line their burrows with mucus (Figure 17-5).

Most sedentary tube and burrow dwellers are particle feeders, using ciliary or mucoid methods of obtaining food. The principal food source is plankton and detritus. Some, like *Amphitrite* (Figure 17-4), with head peeping out of the mud, send out long extensible tentacles over the surface to deposit feed. Cilia and mucus on the tentacles entrap particles found on the sea bottom and move them toward the mouth. Lugworms, *Arenicola*, employ an interesting combination of suspension and deposit feeding. They live in an L-shaped burrow through which, by peristaltic movements, they cause water to flow. Food particles are filtered out by the sand at the front of the burrow, and they ingest the food-laden sand (Figure 17-5).

Figure 17-8
A scale worm, *Hesperonoe adventor*, normally lives as a commensal in the tubes of *Urechis* (phylum Echiura, p. 428).

Figure 17-9
A fireworm, *Hermodice carunculata*, feeds on gorgonians and stony corals. Its setae are like tiny glass fibers and serve to ward off predators.

Fanworms, or "featherduster" worms, are beautiful tubeworms, fascinating to watch as they emerge from their secreted tubes and unfurl their lovely tentacular crowns to feed (Figure 17-2). A slight disturbance, sometimes even a passing shadow, causes them to duck quickly into the safety of the homes they have built. Food attracted to the feathery arms, or **radioles,** by ciliary action is trapped in mucus and is carried down ciliated food grooves to the mouth (Figure 17-10). Particles too large for the food grooves pass along the margins and drop off. Further sorting may occur near the mouth where only small particles of food enter the mouth, and sand grains are stored in a sac to be used later in enlarging the tube.

The parchment worm *Chaetopterus* (Gr. *chaitē*, long hair, + *pteron*, wing) feeds on suspended particles by an entirely different mechanism (Figure 17-11). It lives in a U-shaped, parchmentlike tube buried, except for the tapered ends, in sand or mud along the shore. The worm attaches to the side of the tube by ventral suckers. Fans (modified parapodia) on segments 14 to 16 pump water through the tube by rhythmical movements. A pair of

Figure 17-10

Sabella, a polychaete ciliary feeder, extends its crown of feeding radioles from its leathery secreted tube, reinforced with sand and debris. **A,** Anterior view of the crown. Cilia direct small food particles along grooved radioles to mouth and discard larger particles. Sand grains are directed to storage sacs and later are used in tube building. **B,** Distal portion of radiole showing ciliary tracts of pinnules and food grooves.

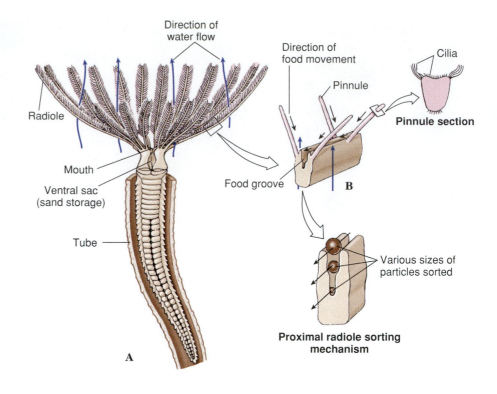

Figure 17-11

Chaetopterus, a sedentary polychaete, lives in a U-shaped tube in the sea bottom. It pumps water through the parchmentlike tube (of which one-half has been cut away here) with its three pistonlike fans. The fans beat 60 times per minute to keep water currents moving. The winglike notopodia of the twelfth segment continuously secrete a mucous net that strains out food particles. As the net fills with food, the food cup rolls it into a ball, and when the ball is large enough (about 3 mm), the food cup bends forward and deposits the ball in a ciliated groove to be carried to the mouth and swallowed.

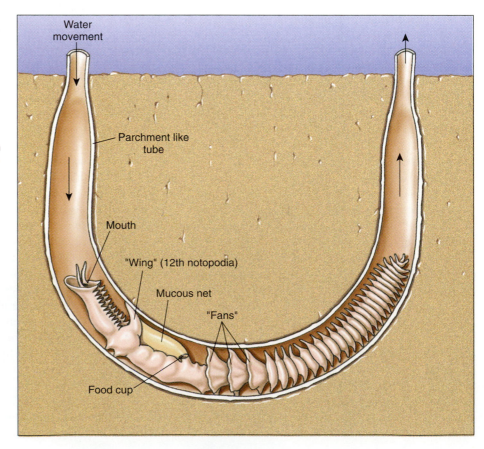

enlarged parapodia in the twelfth segment secretes a long mucous bag that reaches back to a small food cup just in front of the fans. All water passing through the tube is filtered through this mucous bag, the end of which is rolled into a ball by cilia in the cup. When the ball is about the size of a BB shot (about 3 mm diameter), the fans stop beating and the ball of food and mucus is rolled forward by ciliary action to the mouth and swallowed.

CLASS OLIGOCHAETA

More than 3000 species of oligochaetes are found in a great variety of sizes and habitats. They include the familiar earthworms and many species that live in fresh water. Most are terrestrial or freshwater forms, but some are parasitic, and a few live in marine or brackish water.

With few exceptions, oligochaetes bear setae, which may be long or short, straight or curved, blunt or needlelike, or arranged singly or in bundles. Whatever the type, they are less numerous in oligochaetes than in polychaetes, as is implied by the class name, which means "few long hairs." Aquatic forms usually have longer setae than do earthworms.

Earthworms

The most familiar of oligochaetes are earthworms ("night crawlers"), which burrow in moist, rich soil, emerging at night to explore their surroundings. In damp, rainy weather they stay near the surface, often with mouth or anus protruding from the burrow. In very dry weather they may burrow several feet underground, coil up in a slime chamber, and become dormant. *Lumbricus terrestris* (L. *lumbricum,* earthworm), the form commonly studied in school laboratories, is 12 to 30 cm long (Figure 17-12). Giant tropical earthworms may have from 150 to 250 or more segments and may grow to as much as 4 m in length. They usually live in branched and interconnected tunnels.

Aristotle called earthworms the "intestines of the soil." Some 22 centuries later Charles Darwin published his observations in his classic *The Formation of Vegetable Mould Through the Action of Worms.* He showed how worms enrich soil by bringing subsoil to the surface and mixing it with topsoil. An earthworm can ingest its own weight in soil every 24 hours, and Darwin estimated that from 10 to 18 tons of dry earth per acre pass through their intestine annually, thus bringing potassium and phosphorus from the subsoil and also adding to the soil nitrogenous products from their own metabolism. They expose the mold to air and sift it into small particles. They also drag leaves, twigs, and organic substances into their burrows closer to the roots of plants. Their activities are important in aerating soil. Darwin's views were at odds with his contemporaries, who thought earthworms were harmful to plants. But recent research has amply confirmed Darwin's findings, and earthworm management is now practiced in many countries.

Form and Function

In earthworms the mouth is overhung by a fleshy prostomium at the anterior end, and the anus is on the posterior end (Figure 17-12B). In most earthworms each segment bears four pairs of chitinous setae (Figure 17-12C), although in some oligochaetes each segment may have up to 100 or more. Each seta is a bristle-like rod set in a sac within the body wall and moved by tiny muscles (Figure 17-13). The setae project through small pores in the cuticle to the outside. In locomotion and burrowing, setae anchor parts of the body to prevent slipping. Earthworms move by peristaltic movement. Contractions of circular muscles in the anterior end lengthen the body, pushing the anterior end forward where it is anchored by setae; contractions of longitudinal muscles then shorten the body, pulling the posterior end forward. As these waves of contraction pass along the entire body, it gradually moves forward.

Nutrition   Most oligochaetes are scavengers. Earthworms feed mainly on decaying organic matter, bits of leaves and vegetation, refuse, and animal matter. After being moistened by secretions from the mouth, food is drawn inward by a sucking action of their muscular pharynx. The liplike prostomium aids in manipulating food into position. Calcium from soil swallowed with food tends to produce a high blood calcium level. **Calciferous glands** along the esophagus secrete calcium ions into the gut and so reduce the calcium ion concentration of their blood. Calciferous glands are really ionoregulatory, rather than digestive, organs. They also function in regulating acid-base balance of body fluids, maintaining the pH at a fairly stable value.

Leaving the esophagus, food is stored temporarily in the thin-walled **crop** before being passed on into the **gizzard,** which grinds food into small pieces. Digestion and absorption occur in the **intestine.** The wall of the intestine is infolded dorsally to form a **typhlosole,** which greatly increases the absorptive and digestive surface (Figure 17-12C). The digestive system secretes various enzymes to break down their food.

Surrounding the intestine and dorsal vessel and filling much of the typhlosole is a layer of yellowish **chloragogen tissue** (Gr. *chlōros,* green, + *agōgē,* a carrying away) derived from the peritoneum. This tissue serves as a center for synthesis of glycogen and fat, a function roughly equivalent to that of liver cells. Chloragogen cells when ripe (full of fat) are released into the coelom where they float free as cells called **eleocytes** (Gr. *elaio,* oil, + *kytos,* hollow vessel [cell]), which transport materials to the body tissues. They apparently can pass from segment to segment and have been found to accumulate around wounds and regenerating areas, where they break down and release their contents into the coelom. Chloragogen cells also function in excretion.

Circulation and Respiration   Annelids have a double transport system: coelomic fluid and circulatory system. Food, wastes, and respiratory gases are carried by both coelomic fluid and blood in varying degrees. Blood circulates in a closed system of vessels, including capillary systems in the tissues. Five main blood trunks run lengthwise through the body.

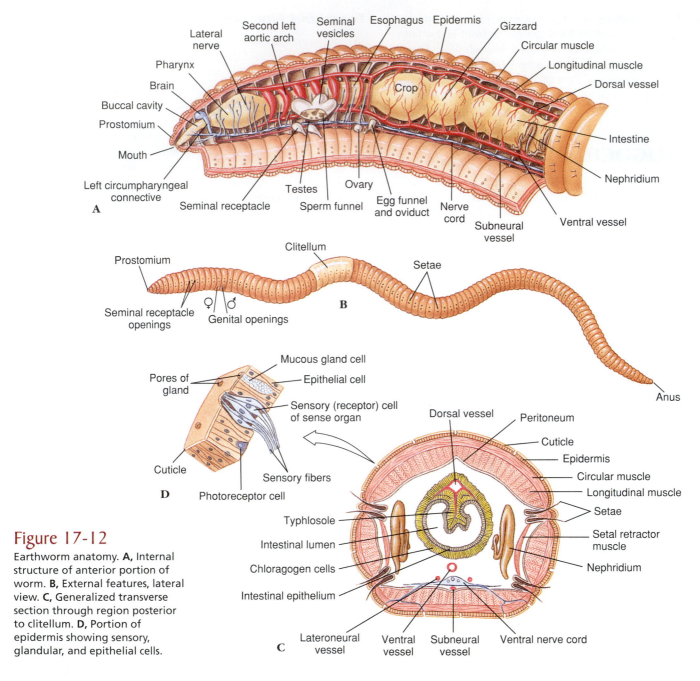

Figure 17-12

Earthworm anatomy. **A,** Internal structure of anterior portion of worm. **B,** External features, lateral view. **C,** Generalized transverse section through region posterior to clitellum. **D,** Portion of epidermis showing sensory, glandular, and epithelial cells.

A single **dorsal vessel** runs above the alimentary canal from the pharynx to the anus. It is a pumping organ, provided with valves, and it functions as a true heart. This vessel receives blood from vessels of the body wall and digestive tract and pumps it anteriorly into five pairs of **aortic arches.** The function of aortic arches is to maintain a steady pressure of blood into the ventral vessel.

A single **ventral vessel** serves as an aorta. It receives blood from the aortic arches and delivers it to the brain and rest of the body, providing segmental vessels to the walls, nephridia, and digestive tract.

Their blood contains colorless ameboid cells and a dissolved respiratory pigment, hemoglobin (p. 673). The blood of some annelids may have respiratory pigments other than hemoglobin, as noted on p. 349.

Earthworms have no special respiratory organs, but gaseous exchange takes place across their moist skin.

Excretion   Each somite except the first three and the last one bears a pair of **metanephridia.** Each nephridium occupies parts of two successive somites (Figure 17-14). A ciliated funnel, the **nephrostome,** lies just anterior to an intersegmental septum and leads by a small ciliated tubule through the septum into the somite behind, where it connects with the main part of the nephridium. Several complex loops of increasing size compose the nephridial duct, which terminates in a bladderlike structure leading to an aperture, the **nephridiopore.** The nephridiopore opens to the outside near the ventral row of setae. By means of cilia, wastes from the coelom are drawn into the nephrostome and tubule, where they are joined by salts and organic wastes transported from blood capillaries in the glandular part of the nephridium. Waste is discharged to the outside through the nephridiopore. Resorption of important materials into the capillary network occurs before waste accumulation in the bladder.

Aquatic oligochaetes excrete ammonia; terrestrial oligochaetes excrete the much less toxic urea. *Lumbricus* produces both, the level of urea depending somewhat on environmental conditions. Both urea and ammonia are produced by chloragogen cells, which may break off and enter nephridia directly, or their products may be carried by the blood. Some nitrogenous waste is eliminated through the body surface.

Oligochaetes are largely freshwater animals, and even such terrestrial forms as earthworms must exist in a moist environment. Osmoregulation is a function of the body surface and the nephridia, as well as the gut and dorsal pores. *Lumbricus* will

gain weight when placed in tap water and lose it when returned to soil. Salts as well as water can pass across the integument, salts apparently being carried by active transport.

Nervous System and Sense Organs   The nervous system in earthworms (Figure 17-15) consists of a central system and peripheral nerves. The central system reflects the typical annelid pattern: a pair of **cerebral ganglia** (the brain) above the pharynx, a pair of **connectives** passing around the pharynx connecting the brain with the first pair of ganglia in the nerve cord; a **ventral nerve cord,** really double, running along the floor of the coelom to the last somite; and a pair of fused ganglia on the nerve cord in each somite. Each pair of fused ganglia provides nerves to the body structures, which contain both sensory and motor fibers.

Neurosecretory cells have been found in the brain and ganglia of annelids, both oligochaetes and polychaetes. They are endocrine in function and secrete neurohormones concerned with regulation of reproduction, secondary sex characteristics, and regeneration.

For rapid escape movements most annelids have from one to several very large axons commonly called **giant axons** (Figure 17-16), or giant fibers, located in the ventral nerve cord. Their large diameter increases rate of conduction (see p. 699) and makes possible simultaneous contractions of muscles in many segments.

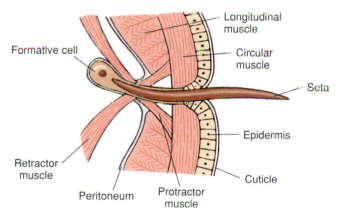

Figure 17-13

Seta with its muscle attachments showing relation to adjacent structures. Setae lost by wear are replaced by new ones, which develop from formative cells.

In the dorsal median giant fiber of *Lumbricus,* which is 90 to 160 μm in diameter, speed of conduction has been estimated at 20 to 45 m/second, several times faster than in ordinary neurons of this species. This is also much faster than in polychaete giant fibers, probably because in earthworms the giant fibers are enclosed in myelinated sheaths. The speed of conduction may be altered by changes in temperature.

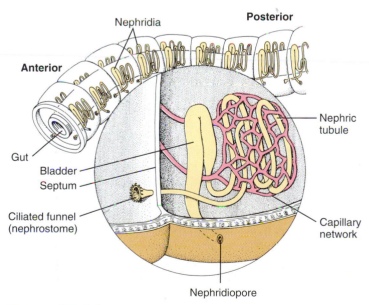

Figure 17-14

Nephridium of earthworm. Wastes are drawn into the ciliated nephrostome in one segment, then passed through the loops of the nephridium, and expelled through the nephridiopore of the next segment.

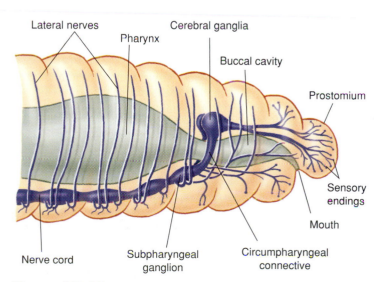

Figure 17-15

Anterior portion of earthworm and its nervous system. Note concentration of sensory endings in this region.

Figure 17-16

Portion of nerve cord of earthworm showing arrangement of simple reflex arc *(in foreground)* and the three dorsal giant fibers that are adapted for rapid reflexes and escape movements. Ordinary crawling involves a succession of reflex acts, the stretching of one somite stimulating the next to stretch. Impulses are transmitted much faster in giant fibers than in regular nerves so that all segments can contract simultaneously when quick withdrawal into a burrow is necessary.

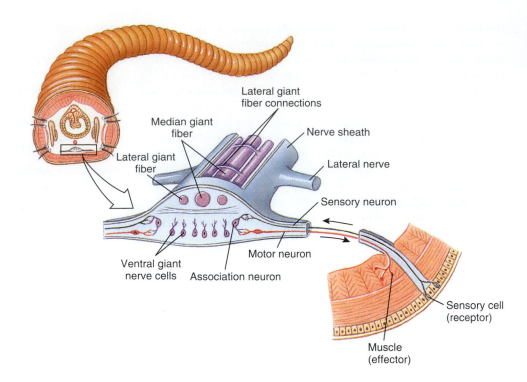

Simple sense organs are distributed all over the body. Earthworms have no eyes but do have many lens-shaped photoreceptors in their epidermis. Most oligochaetes are negatively phototactic to strong light but positively phototactic to weak light. Many single-celled sense organs are widely distributed in the epidermis. What are presumably chemoreceptors are most numerous on the prostomium. There are many free nerve endings in the integument, which are probably tactile.

General Behavior   Earthworms are among the most defenseless of creatures, yet their abundance and wide distribution indicate their ability to survive. Although they have no specialized sense organs, they are sensitive to many stimuli. They react positively to mechanical stimuli when such stimuli are moderate and negatively to a strong vibration (such as footfall near them), which causes them to retire quickly into their burrows. They react to light, which they avoid unless it is very weak. Chemical responses aid them in the choice of food.

Chemical as well as tactile responses are very important to earthworms. They not only must sample the organic content of the soil to find food, but also must sense its texture, acidity, and calcium content.

Experiments show that earthworms have some learning ability. They can be taught to avoid an electric shock, and thus an association reflex can be produced in them. Darwin credited earthworms with a great deal of intelligence in pulling leaves into their burrows: he observed that they seized the leaves by the narrow end, the easiest way for drawing a leaf-shaped object into a small hole. Darwin assumed that seizure of leaves by worms did not result from random handling or from chance but was purposeful in its mechanism. However, investigations since Darwin's time have shown that the process is mainly one of trial and error, for earthworms often seize a leaf several times before getting it right.

Reproduction and Development   Earthworms are monoecious (hermaphroditic); both male and female organs are found in the same animal (Figure 17-12A). In *Lumbricus* reproductive systems are found in somites 9 to 15. Two pairs of small testes and two pairs of sperm funnels are surrounded by three pairs of large seminal vesicles. Immature sperm from the testes mature in seminal vesicles, then pass into sperm funnels and down sperm ducts to the male genital pores in somite 15, where they are expelled during copulation. Eggs are discharged by a pair of small ovaries into the coelomic cavity, where ciliated funnels of the oviducts carry them outside through female genital pores on somite 14. Two pairs of seminal receptacles in somites 9 and 10 receive and store sperm from the mate during copulation.

Reproduction in earthworms may occur at any season, but they usually copulate at night during warm, moist weather (Figure 17-17). When mating, worms extend their anterior ends from their burrows and bring their ventral surfaces together (Figure 17-17). They are held together by mucus secreted by their **clitellum** (L. *clitellae,* packsaddle) and by special ventral setae, which penetrate each other's bodies in the regions of contact. After discharge, sperm travel to seminal receptacles of the other worm in its seminal grooves. After copulation each worm secretes first a mucous tube and then a tough, chitinlike band that forms a **cocoon** around its clitellum. As the cocoon passes forward, eggs from the oviducts, albumin from skin glands, and sperm from the mate (stored in the seminal receptacles) pour into it. Fertilization of eggs then takes place within the cocoon. When the cocoon slips past the anterior end of the worm, its ends close, producing a sealed, lemon-shaped body. Embryogenesis occurs within the cocoon, and the form that hatches from the egg is a young worm similar to the adult. Thus development is direct with no metamorphosis. Juveniles do not develop a clitellum until they are sexually mature.

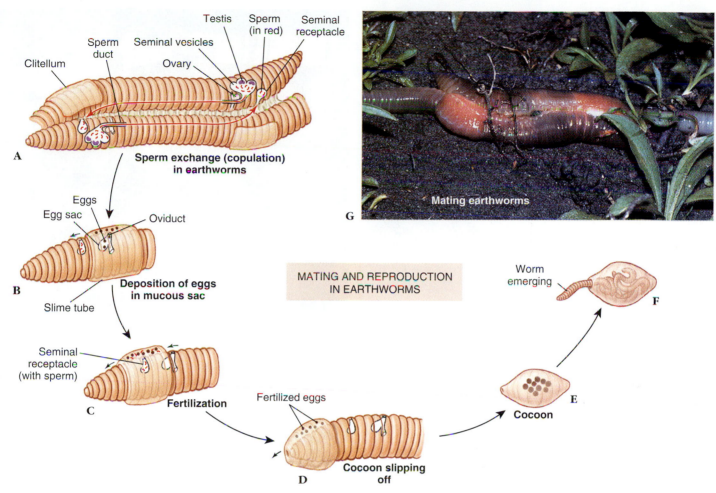

Figure 17-17

Earthworm copulation and formation of egg cocoons. **A,** Mutual insemination; sperm from genital pore (somite 15) pass along seminal grooves to seminal receptacles (somites 9 and 10) of each mate. **B** and **C,** After worms separate, a slime tube formed over the clitellum passes forward to receive eggs from oviducts and sperm from seminal receptacles. **D,** As cocoon slips off over anterior end, its ends close and seal. **E,** Cocoon is deposited near burrow entrance. **F,** Young worms emerge in 2 to 3 weeks. **G,** Two earthworms in copulation. Their anterior ends point in opposite directions as their ventral surfaces are held together by mucous bands secreted by the clitella.

Freshwater Oligochaetes

Freshwater oligochaetes usually are smaller and have more conspicuous setae than do earthworms. They are more mobile than earthworms and tend to have better-developed sense organs. They are generally benthic forms that creep about on the substrate or burrow in soft mud. Aquatic oligochaetes are an important food source for fishes. A few are ectoparasitic.

Some of the more common freshwater oligochaetes are the 1 mm long *Aeolosoma* (Gr. *aiolos,* quick-moving, + *soma,* body) (Figure 17-18B), which contains red or green pigments, has bundles of setae, and is often found in hay cultures; the 2 to 4 mm long *Nais* (L. *nais,* water nymph), which is brownish and has two bundles of setae on anterior segments and four bundles of setae on each posterior segment; the 10 to 25 mm long *Stylaria* (Gr. *stylos,* pillar) (Figure 17-18A), with setae arranged like those of *Nais,* a prostomium extended into a long process, and black eyespots; the 5 to 10 mm long *Dero* (Gr. *dere,* neck or throat), which is reddish, lives in tubes, and usually has 3 to 4 pairs of tail gills (Figure 17-18D); the 30 to 40 mm long *Tubifex* (L. *tubus,*

tube, + *faciens,* to make or do) (Figure 17-18C), which is reddish and lives with its head in mud at the bottom of ponds and its tail waving in the water; the 10 to 15 mm long *Chaetogaster* (N.L. *chaeta,* bristle, + *gastrula,* belly), which has only ventral bundles of setae; and *Enchytraeus* (Gr. *enchytraeus,* living in an earthen pot), small whitish worms that live both in moist soil and in water. Some oligochaetes, such as *Aeolosoma,* may form chains of zooids asexually by transverse fission (Figure 17-18B). *Tubifex* is an alternate host necessary in the life cycle of *Myxobolus cerebralis,* a parasite that causes a very serious condition in rainbow trout called whirling disease in North America.

CLASS HIRUDINEA: LEECHES

Leeches occur predominantly in freshwater habitats, but a few are marine, and some have even adapted to terrestrial life in warm, moist places. They are more abundant in tropical countries than in temperate zones. Some leeches attack human beings and are a nuisance.

Figure 17-18

Some freshwater oligochaetes.
A, *Stylaria* has the prostomium drawn out into a long snout.
B, *Aeolosoma* uses cilia around the mouth to sweep in food particles, and it buds off new individuals asexually.
C, *Tubifex* lives head down in long tubes.
D, *Dero* has ciliated anal gills.

Stylaria

Aeolosoma

Tubifex

Dero

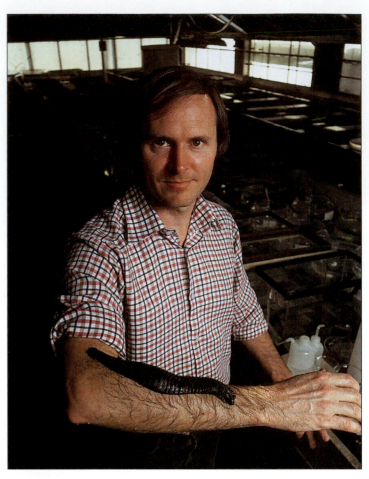

Figure 17-19

The world's largest leech, *Haementeria ghilianii,* on the arm of Dr. Roy K. Sawyer, who found it in French Guiana, South America.

Most leeches are between 2 and 6 cm in length, but some are smaller; some, including "medicinal" leeches, reach 20 cm, but the giant of all is the Amazonian *Haementeria* (Gr. *haimateros,* bloody) (Figure 17-19), which reaches 30 cm.

Leeches occur in a variety of patterns and colors: black, brown, red or olive green. They are usually flattened dorsoventrally. Some are adapted for forcing their pharynx or proboscis into soft tissues such as the gills of fish. The most specialized leeches, however, have sawlike chitinous jaws with which they can cut through tough skin. Many leeches live as carnivores on small invertebrates; some are temporary parasites; and some are permanent parasites, never leaving their host.

Like oligochaetes, leeches are hermaphroditic and have a clitellum, which appears only during breeding season. The clitellum secretes a cocoon for reception of eggs. Leeches are more highly specialized than oligochaetes. As fluid feeders and bloodsuckers, they have lost the setae used by oligochaetes in locomotion and have developed suckers for attachment while sucking blood; their gut is specialized for storage of large quantities of blood.

Form and Function

Unlike other annelids, leeches have a fixed number of somites (usually 34; 15 or 30 in some groups), but they appear to have many more because each somite is marked by transverse grooves to form from two to 16 superficial rings **(annuli)** (Figure 17-20).

The coelom represents another difference between leeches and other annelids; leeches lack distinct coelomic compartments.

In all but one species the septa have disappeared, and the coelomic cavity is filled with connective tissue and a system of spaces called **lacunae.** The coelomic lacunae form a regular system of channels filled with coelomic fluid, which in some leeches serves as an auxiliary circulatory system.

Most leeches creep with looping movements of the body, by attaching first one sucker and then the other and pulling the body along the surface. Aquatic leeches swim with a graceful undulatory movement.

Nutrition

Leeches are popularly considered parasitic, but many are predaceous. Even the true bloodsuckers rarely remain on the host for a long period of time. Most freshwater leeches are active predators or scavengers equipped with a proboscis that can be extended to ingest small invertebrates or to take blood from cold-blooded vertebrates. Some freshwater leeches are true bloodsuckers, preying on cattle, horses, humans, and others. Some terrestrial leeches feed on insect larvae, earthworms, and slugs, which they hold by an oral sucker while using a strong sucking pharynx to ingest food. Other terrestrial forms climb bushes or trees to reach warm-blooded vertebrates such as birds or mammals.

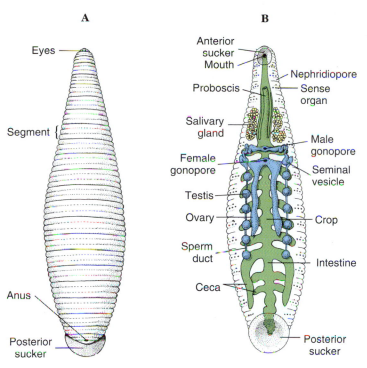

A

- Eyes
- Segment
- Anus
- Posterior sucker

B

- Anterior sucker
- Mouth
- Proboscis
- Salivary gland
- Female gonopore
- Testis
- Ovary
- Sperm duct
- Ceca
- Nephridiopore
- Sense organ
- Male gonopore
- Seminal vesicle
- Crop
- Intestine
- Posterior sucker

Figure 17-20
Structure of a leech, *Placobdella*. **A,** External appearance, dorsal view. **B,** Internal structure, ventral view.

Figure 17-21
Hirudo medicinalis feeding on blood from human arm.

Most leeches are fluid feeders. Many prefer to feed on tissue fluids and blood pumped from wounds already open. True bloodsuckers, which include the so-called medicinal leech, *Hirudo medicinalis* (L. *hirudo*, a leech) (Figure 17-21), have cutting plates, or "jaws," for cutting tissues. Some parasitic leeches leave their hosts only during breeding season, and certain fish parasites are permanently parasitic, depositing their cocoons on their host fish.

For centuries "medicinal leeches" *(Hirudo medicinalis)* were used for bloodletting because of the mistaken idea that a host of bodily disorders and fevers were caused by an excess of blood. A 10- to 12-cm-long leech can extend to a much greater length when distended with blood, and the amount of blood it can suck is considerable. Leech collecting and leech culture in ponds were practiced in Europe on a commercial scale during the nineteenth century. Wordsworth's poem "The Leech-Gatherer" was based on this use of leeches.

Leeches are once again being used medically. When fingers, toes, or ears are severed, microsurgeons can reconnect arteries but not all the more delicate veins. Leeches are used to relieve congestion until the veins can grow back into the healing digit.

Respiration and Excretion

Gas exchange occurs only through the skin except in some fish leeches, which have gills. There are 10 to 17 pairs of nephridia, in addition to coelomocytes and certain other specialized cells that also may be involved in excretory functions.

Nervous and Sensory Systems

Leeches have two "brains"; one is anterior and is composed of six pairs of fused ganglia forming a ring around the pharynx, and one is posterior and is composed of seven pairs of fused ganglia. An additional 21 pairs of segmental ganglia occur along the double nerve cord. In addition to free sensory nerve endings and photoreceptor cells in the epidermis, there is a row of sense organs, called **sensillac,** in the central annulus of each segment; pigment-cup ocelli also are present.

Leeches are highly sensitive to stimuli associated with the presence of a prey or host. They are attracted by and will attempt to attach to an object smeared with appropriate host substances, such as fish scales, oil secretions, or sweat. Those that feed on the blood of mammals are attracted by warmth; terrestrial haemadipsids of the tropics will converge on a person standing in one place.

Reproduction

Leeches are hermaphroditic but practice cross-fertilization during copulation. Sperm are transferred by a penis or by hypodermic impregnation (a spermatophore is expelled from one worm and penetrates the integument of the other). After copulation their clitellum secretes a cocoon that receives eggs and sperm. They bury their cocoons in mud, attach them to submerged objects, or, in terrestrial species, place them in damp soil. Development is similar to that of oligochaetes.

Circulation

In leeches the coelom has been reduced by the invasion of connective tissue and, in some, by a proliferation of chloragogen tissue, to a system of coelomic sinuses and channels. Some orders of leeches retain a typical oligochaete circulatory system, and in these the coelomic sinuses act as an auxiliary blood-vascular system. In other orders the traditional blood vessels are lacking and

Classification of Phylum Annelida

Higher classification of annelids is based primarily on the presence or absence of parapodia, setae, and other morphological features. Because both oligochaetes and hirudineans (leeches) bear a clitellum, these two groups are often placed under the heading Clitellata (cli-tel-la'ta) and members are called clitellates. Alternatively, because both Oligochaeta and Polychaeta possess setae, some authorities place them together in a group called Chaetopoda (ke-top'o-da) (N.L. *chaeta*, bristle, from Gr. *chaitē*, long hair, + *pous, podos,* foot).

Class Polychaeta (pol'e-ke'ta) (Gr. *polys,* many, + *chaitē,* long hair). Mostly marine; head distinct and bearing eyes and tentacles; most segments with parapodia (lateral appendages) bearing tufts of many setae; clitellum absent; sexes usually separate; gonads transitory; asexual budding in some; trochophore larva usually present; mostly marine. Examples: *Nereis, Aphrodita, Glycera, Arenicola, Chaetopterus, Amphitrite.*

Class Oligochaeta (ol'i-go-ke'ta) (Gr. *oligos,* few, + *chaitē,* long hair). Body with conspicuous segmentation; number of segments variable; setae few per metamere; no parapodia; head absent; coelom spacious and usually divided by intersegmental septa; hermaphroditic; development direct, no larva; chiefly terrestrial and freshwater. Examples: *Lumbricus, Stylaria, Aeolosoma, Tubifex.*

Class Hirudinea (hir'u-din'e-a) (L. *hirudo,* leech, + *ea,* characterized by): **leeches.** Body with fixed number of segments (normally 34; 15 or 30 in some groups) with many annuli; oral and posterior suckers usually present; clitellum present; no parapodia; setae absent (except in *Acanthobdella*); coelom closely packed with connective tissue and muscle; development direct; hermaphroditic; terrestrial, freshwater, and marine. Examples: *Hirudo, Placobdella, Macrobdella.*

Branchiobdellida, a group of small annelids that are parasitic or commensal on crayfish and show similarities to both oligochaetes and leeches, are here placed with oligochaetes, but they are considered a separate class by some authorities. They have 15 segments and bear a head sucker.

One genus of leech, *Acanthobdella,* has some characteristics of leeches and some of oligochaetes; it is sometimes separated from other leeches into a special class, Acanthobdellida, that characteristically has 30 somites, setae on the first five segments, and no anterior sucker.

---

the system of coelomic sinuses forms the only blood-vascular system. In those orders contractions of certain longitudinal channels provide propulsion for the blood (the equivalent of coelomic fluid).

EVOLUTIONARY SIGNIFICANCE OF METAMERISM

No truly satisfactory explanation has yet been given for the origins of metamerism and the coelom, although the subject has stimulated much speculation and debate. All classical explanations of the origin of metamerism and the coelom have had important arguments leveled against them, and more than one may be correct, or none, as suggested by R. B. Clark.* The coelom and metamerism may have evolved independently in more than one group of animals, as, for example, in chordates and probably twice in protostomes. Clark stressed the functional and evolutionary significance of these features to the earliest animals that possessed them. He argued forcefully that the adaptive value of a coelom, at least in protostomes, was as a **hydrostatic skeleton** in a burrowing animal. Thus contraction of muscles in one part of the animal could act antagonistically on muscles in another part by transmission of the force of contraction through the enclosed constant volume of fluid in the coelom.

Although the original function of the coelom may have been burrowing in the substrate, certain other advantages accrued to its possessors. Some of these were mentioned in the prologue to

Chapter 16. In addition, coelomic fluid would have acted as a circulatory fluid for nutrients and wastes, making large numbers of flame cells distributed throughout the tissues unnecessary. Gametes could be stored in the spacious coelom for release simultaneously with other individuals in the population, thus enhancing chances of fertilization, and synchronous release of gametes would have selected for greater nervous and endocrine control. Finally, separation of the coelom into a series of compartments by septa (metamerism) would have increased burrowing efficiency and made possible independent and separate movements by separate metameres, as mentioned in the prologue to this chapter. Independent movements of metameres in different parts of the body would have placed selective value on a more sophisticated nervous system for control of movements, thereby leading to elaboration of the central nervous system.

PHYLOGENY AND ADAPTIVE RADIATION

Phylogeny

There are so many similarities in early development of molluscs, annelids, and primitive arthropods that few biologists have doubted their close relationship. These three phyla were considered the sister group of flatworms. Many marine annelids and molluscs have an early embryogenesis typical of protostomes, in common with some marine flatworms, and that developmental pattern is probably a shared ancestral trait (p. 196). Annelids share with arthropods a similar body plan and nervous system, as well as similarities in development. The most important resemblance probably lies in the metameric plans of annelid and arthropod

---

*Clark, R. B. 1964. Dynamics in metazoan evolution. The origin of the coelom and segments. Oxford, U.K., Clarendon Press.

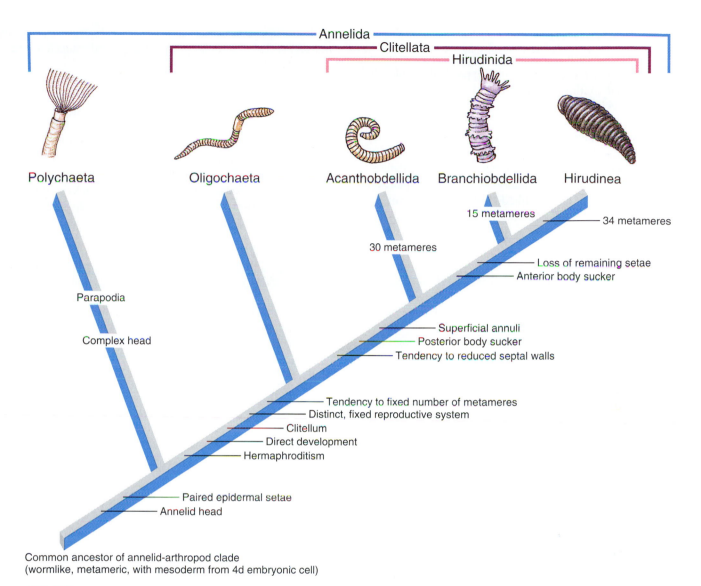

Figure 17-22

Cladogram of annelids, showing the appearance of shared derived characters that specify the five monophyletic groups (based on Brusca and Brusca, 1990). The Acanthobdellida and the Branchiobdellida are two small groups discussed briefly in the note on p. 360. Brusca and Brusca place both groups, together with the Hirudinea ("true" leeches), within a single taxon, the Hirudinida. This clade has several synapomorphies: tendency toward reduction of septal walls, the appearance of a posterior sucker, and the subdivision of body segments by superficial annuli. Note also that, according to this scheme, the Oligochaeta have no defining synapomorphies; that is, they are defined solely by retention of plesiomorphies (retained primitive characters, p. 196), and thus might be paraphyletic.

Source: Modified from R. C. Brusca and G. J. Brusca, Invertebrates. *Sinauer Associates, Inc., Sinderland, MA, 1990.*

body structures. These long-accepted evolutionary relationships are not supported, however, by a recent hypothesis based on analysis of the base sequence in the gene encoding small-subunit ribosomal RNA (p. 198), which places annelids and molluscs in a superphylum Lophotrochozoa and arthropods in another protostome superphylum, Ecdysozoa.

Regardless of its relationship to other phyla, Annelida remains a well-accepted monophyletic group. What can we infer about the common ancestor of annelids? Most hypotheses of annelid origin have assumed that metamerism arose in connection with development of lateral appendages (parapodia) resembling those of polychaetes. However, the oligochaete body is adapted to vagrant burrowing in a substratum with a peristaltic movement that is highly benefited by a metameric coelom. On the other hand, polychaetes with well-developed parapodia are generally adapted to swimming and crawling in a medium too fluid for effective peristaltic locomotion. While parapodia do not prevent such locomotion, they do little to further it, and they seem likely to have evolved as an adaptation for swimming.

Although polychaetes have the most primitive reproductive system, some authorities argue that ancestral annelids were more similar to oligochaetes in overall body plan and that those of polychaetes and leeches are more evolutionarily derived. Leeches are closely related to oligochaetes but have diverged from them by having a swimming existence and no burrowing. This relationship is shown by the cladogram in Figure 17-22.

Adaptive Radiation

Annelids are an ancient group that has undergone extensive adaptive radiation. The basic body structure, particularly of polychaetes, lends itself to almost endless modification. As marine worms, polychaetes have a wide range of habitats in an environment that is not physically or physiologically demanding. Unlike earthworms, whose environment imposes strict physical and physiological demands, polychaetes have been free to experiment and thus have achieved a wide range of adaptive features.

A basic adaptive feature in evolution of annelids is their septal arrangement, resulting in fluid-filled coelomic compartments. Fluid pressure in these compartments is used as a hydrostatic skeleton in precise movements such as burrowing and swimming. Powerful circular and longitudinal muscles can flex, shorten, and lengthen the body.

Feeding adaptations show great variation, from the sucking pharynx of oligochaetes and the chitinous jaws of carnivorous polychaetes to the specialized tentacles and radioles of particle feeders.

In polychaetes the parapodia have been adapted in many ways and for a variety of functions, chiefly locomotion and respiration.

In leeches many adaptations, such as suckers, cutting jaws, pumping pharynx, distensible gut, and production of hirudin, relate to their predatory and bloodsucking habits.

SUMMARY

Phylum Annelida is a large, cosmopolitan group containing marine polychaetes, earthworms and freshwater oligochaetes, and leeches. Certainly the most important structural innovation underlying diversification of this group is metamerism, a division of the body into a series of similar segments, each of which contains a repeated arrangement of many organs and systems. The coelom also is highly developed in annelids, and this, together with the septal arrangement of fluid-filled compartments and a well-developed body-wall musculature, is an effective hydrostatic skeleton for precise burrowing and swimming movements. Further metameric specialization occurs in arthropods, the subjects of Chapters 18, 19, and 20.

Polychaetes, the largest class of annelids, are mostly marine. On each somite they have many setae, which are borne on paired parapodia. Parapodia show a wide variety of adaptations among polychaetes, including specialization for swimming, respiration, crawling, maintaining position in a burrow, pumping water through a burrow, and accessory feeding. Some polychaetes are mostly predaceous and have an eversible pharynx with jaws. Other polychaetes rarely leave the burrows or tubes in which they live. Several styles of deposit and filter feeding are shown among members of this group. Polychaetes are dioecious, have a primitive reproductive system, no clitellum, external fertilization, and a trochophore larva.

Class Oligochaeta contains earthworms and many freshwater forms; they have a small number of setae per segment (compared to Polychaeta) and no parapodia. They have a closed circulatory system, and the dorsal blood vessel is the main pumping organ. Paired nephridia occur in most somites. Earthworms contain the typical annelid nervous system: dorsal cerebral ganglia connected to a double, ventral nerve cord with segmental ganglia running the length of the worm. Oligochaetes are hermaphroditic and practice cross-fertilization. The clitellum plays an important role in reproduction, including secretion of mucus to surround the worms during copulation and secretion of a cocoon to receive eggs and sperm and in which embryonation occurs. A small, juvenile worm hatches from the cocoon.

Leeches (class Hirudinea) are mostly freshwater, although a few are marine and a few are terrestrial. They feed mostly on fluids; many are predators, some are temporary parasites, and a few are permanent parasites. The hermaphroditic leeches reproduce in a fashion similar to oligochaetes, with cross-fertilization and cocoon formation by the clitellum.

Embryological evidence places annelids with molluscs and arthropods in the Protostomia. Recent molecular evidence suggests that annelids and molluscs are more closely related to each other (in superphylum Lophotrochozoa) than either phylum is to arthropods (in superphylum Ecdysozoa).

REVIEW QUESTIONS

1. What characteristics of phylum Annelida distinguish it from other phyla?
2. Distinguish among the classes of phylum Annelida.
3. Describe the annelid body plan, including body wall, segments, coelom and its compartments, and coelomic lining.
4. Explain how the hydrostatic skeleton of annelids helps them to burrow. How is the efficiency for burrowing increased by metamerism?
5. Describe three ways that various polychaetes obtain food.
6. Define each of the following: prostomium, peristomium, pygidium, radioles, parapodium, neuropodium, notopodium.
7. Explain functions of each of the following in earthworms: pharynx, calciferous glands, crop, gizzard, typhlosole, chloragogen tissue.
8. Compare the main features of each of the following in each class of annelids: circulatory system, nervous system, excretory system.
9. Describe the function of the clitellum and cocoon.
10. How are freshwater oligochaetes generally different from earthworms?
11. Describe the ways in which leeches obtain food.
12. What are the main differences in reproduction and development among the three classes of annelids?
13. What was the evolutionary significance of metamerism and the coelom to its earliest possessors?
14. What are the phylogenetic relationships between the molluscs, annelids, and arthropods? Give evidence for these relationships.

SELECTED REFERENCES

Fischer, A., and U. Fischer. 1995. On the life-style and life-cycle of the luminescent polychaete *Odontosyllis enopla* (Annelida: Polychaeta). Invert. Biol. **114**:236–247. *If epitokes of this species survive their spawning swarm, they can return to a benthic existence.*

Lent, C. M., and M. H. Dickinson. 1988. The neurobiology of feeding in leeches. Sci. Am. **258**:98–103 (June). *Feeding behavior in leeches is controlled by a single neurotransmitter (serotonin).*

McClintock, J. 2001. Blood suckers. Discover **22**:56–61 (Dec.). *Describes modern medical uses for leeches.*

Mirsky, S. 2000. When good hippos go bad. Sci. Am. **282**:28 (Jan.). *Placobdelloides jaegerskioeldi is a parasitic leech that breeds only in the rectum of hippopotomuses.*

Pernet, B. 2000. A scaleworm's setal snorkel. Invert. Biol. **119**:147–151. *Sthenelais berkeleyi is an apparently rare but large (20 cm) polychaete that buries its body in sediment and communicates with water above just by its anterior end. Ciliary movement on parapodia pumps water into the burrow for ventilation. The worm remains immobile for long periods, except when prey comes near; it then rapidly everts its pharynx to capture prey.*

Rouse, G. W., and K. Fauchald. 1998. Recent views on the status, delineation and classification of the Annelida. Am. Zool. **38**:953–964. *A discussion of analyses that make Polychaeta paraphyletic, as well as other groups that should/could go into phylum Annelida. These authors conclude that sequence analysis is "clearly no panacea."*

Winnepenninckx, B. M. H., Y. Van de Peer, and T. Backeljau. 1998. Metazoan relationships on the basis of 18S rRNA sequences: A few years later . . . Am. Zool. **38**:888–906. *Their calculations and analysis support monophyly of Clitellata but cast doubt on monophyly of Polychaeta.*

ZOOLOGY LINKS TO THE INTERNET

Visit the textbook's Online Learning Center at www.mhhe.com/zoology to find live Internet links for each of the topics listed here.
Phylum Annelida
Class Polychaeta
Class Oligochaeta
Class Hirudinea

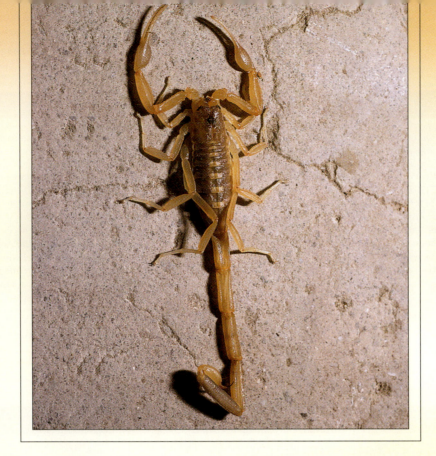

A scorpion.

Arthropods

Phylum Arthropoda
Subphylum Trilobita
Subphylum Chelicerata

A Suit of Armor

Sometime, somewhere in the Precambrian era, a major milestone in the evolution of life on earth was passed. The soft cuticle in a metameric ancestor of animals we now call arthropods was stiffened by deposition of additional protein and an inert polysaccharide called chitin. The cuticular exoskeleton was some protection against predators and other environmental hazards, and it conferred on its possessors a formidable array of other selective advantages. Of course, a suit of armor could not be uniformly stiff; the animal would be as unable to move as the rusted tin woodsman in the *Wizard of Oz*. Stiff sections of cuticle were separated from each other by thin, flexible sections, which formed joints. The cuticular exoskeleton had enormous evolutionary potential. Jointed extensions on each metamere became appendages.

Once the stiffened cuticle evolved, or perhaps concurrently with it, many other changes were necessary in the bodies of protoarthropods. Growth required a sequence of cuticular molts controlled by hormones. Coelomic compartments lost their hydrostatic skeletal function, causing a regression of the coelom and its replacement with an open system of sinuses (hemocoel). Motile cilia were lost. These changes and others are called "arthropodization." Some zoologists argue that all changes in arthropodization follow from development of a cuticular exoskeleton. If several different ancestors had independently evolved a cuticular exoskeleton, then they independently would have evolved the identical suite of characters we associate with arthropodization. The huge phylum we call Arthropoda would be in reality polyphyletic. However, we agree with other zoologists who feel that the weight of evidence still supports single-phylum status.

PHYLUM ARTHROPODA

Phylum Arthropoda (ar-throp'o-da) (Gr. *arthron,* joint, + *pous, podos,* foot) is the most extensive phylum in the animal kingdom, composed of more than three-fourths of all known species. Approximately 900,000 species of arthropods have been recorded, and probably at least as many more remain to be classified. However, based on surveys of insect fauna in the canopy of rain forests, many estimates of yet undescribed species are much higher. Arthropods include spiders, scorpions, ticks, mites, crustaceans, millipedes, centipedes, insects, and some others. In addition, there is a rich fossil record extending to the very late Precambrian period.

Arthropods are eucoelomate protostomes with well-developed organ systems, and they share with annelids the property of conspicuous metamerism.

Arthropods have an exoskeleton containing chitin, and their primitive pattern is that of a linear series of similar somites, each with a pair of jointed appendages. However, the pattern of somites and appendages varies greatly in the phylum. There is a tendency for somites to be combined or fused into functional groups, called **tagmata** (sing., **tagma**), for specialized purposes; appendages are frequently differentiated and specialized for pronounced division of labor.

Few arthropods exceed 60 cm in length, and most are far below this size. The largest, a Japanese crab *Macrocheira* (Gr. *makros,* large, + *cheir,* hand), spans approximately 4 m; the smallest is a parasitic mite *Demodex* (Gr. *dēmos,* body, frame, + *dex,* a wood worm), which is less than 0.1 mm long.

Arthropods are usually active, energetic animals. They utilize all modes of feeding—carnivory, herbivory, and omnivory—although most are herbivorous. Most aquatic arthropods depend on algae for their nourishment, and the majority of land forms live chiefly on plants. In diversity of ecological distribution, arthropods have no rivals.

Although arthropods compete with humans for food and spread serious diseases, they are essential in pollination of many food plants, and they also serve as food, yield drugs and dyes, and generate products such as silk, honey, and beeswax.

Arthropods are more widely and more densely distributed throughout all regions of the earth than are members of any other phylum. They are found in all types of environment from low ocean depths to very high altitudes, and from the tropics far into both north and south polar regions. Different species are adapted for life in the air; on land; in fresh, brackish, and marine waters; and in or on the bodies of plants and other animals. Some species live in places where no other animal could survive.

We cover subphyla Trilobita (all extinct) and Chelicerata in this chapter, and Chapters 19 and 20 are devoted to subphyla Crustacea and Uniramia (classification of Arthropoda on p. 374).

Why Have Arthropods Achieved Such Great Diversity and Abundance?

Arthropods have achieved a great diversity, number of species, wide distribution, variety of habitats and feeding habits, and power of adaptation to changing conditions. In our discussion we briefly summarize some structural and physiological patterns that have been helpful to them.

1. A **versatile exoskeleton.** Arthropods possess an exoskeleton that is highly protective without sacrificing mobility. This skeleton is the **cuticle,** an outer covering secreted by the underlying epidermis. The cuticle is made up of an inner and thicker **procuticle** and an outer, relatively thin **epicuticle.** The procuticle is divided into an **exocuticle,** which is secreted before a molt, and **endocuticle,** which is secreted after molting. Both layers of the procuticle contain **chitin** bound with protein. Chitin is a tough, resistant, nitrogenous polysaccharide that is insoluble in water, alkalis, and weak acids. Thus the procuticle not only is flexible and lightweight but also affords protection, particularly against dehydration. In some crustaceans the chitin may form 60% to 80% of the

Characteristics of Phylum Arthropoda

1. Bilateral symmetry; **metameric body** divided into **tagmata** consisting of head and trunk; head, thorax, and abdomen; or cephalothorax and abdomen
2. **Jointed appendages;** primitively, one pair to each somite, but number often reduced; appendages often modified for specialized functions
3. **Exoskeleton of cuticle** containing protein, lipid, chitin, and often calcium carbonate secreted by underlying epidermis and shed (molted) at intervals
4. **Complex muscular system,** with exoskeleton for attachment, **striated muscles** for rapid actions, smooth muscles for visceral organs; no cilia
5. **Reduced coelom** in adult; most of body cavity consisting of hemocoel (sinuses, or spaces, in the tissues) filled with blood
6. **Complete digestive system;** mouthparts modified from appendages and adapted for different methods of feeding
7. **Open circulatory system,** with dorsal **contractile heart,** arteries, and hemocoel (blood sinuses)
8. Respiration by **body surface, gills, tracheae** (air tubes), or **book lungs**
9. Paired excretory glands called **coxal, antennal,** or **maxillary glands** present in some, homologous to metameric nephridial system of annelids; some with other excretory organs, called **Malpighian tubules**
10. **Nervous system** of **annelid plan,** with dorsal brain connected by a ring around the gullet to a double nerve chain of ventral ganglia; fusion of ganglia in some species; well-developed sensory organs
11. **Sexes usually separate,** with paired reproductive organs and ducts; usually internal fertilization; oviparous, ovoviviparous, or viviparous; often with **metamorphosis;** parthenogenesis in some

Comparison of Arthropoda with Annelida

Similarities between Arthropoda and Annelida are as listed here.

1. External segmentation marked
2. Segmental arrangement of muscles
3. Ventral nerve cord with metamerically arranged ganglia and dorsal cerebral ganglia
4. Spiral cleavage (found in some arthropods)
5. Some parapodia resemble arthropod limbs

Arthropods differ from annelids in having the following:

1. Usually lack intersegmental septa
2. Pronounced tagmatization (compared with limited tagmatization in annelids)
3. Coelomic cavity reduced; main body cavity a hemocoel
4. Open (lacunar) circulatory system
5. Exoskeleton containing chitin
6. Jointed appendages
7. Compound eyes (also present in a few annelids) and other well-developed sense organs
8. Absence of cilia

procuticle, but in insects it is probably not more than 50% (the remainder being protein). In most crustaceans the procuticle is also impregnated with **calcium salts,** which reduce its flexibility. In the hard shells of lobsters and crabs, for instance, this calcification is extreme. The outer epicuticle is composed of protein and lipid. The protein is stabilized and hardened by chemical cross-linking **(sclerotized),** adding further protection. Both the procuticle and epicuticle are laminated, that is, composed of several layers each (see Figure 29-1, p. 614). In many insects the outer layer of epicuticle is composed of waxes, which retards water loss.

The cuticle may be soft and permeable or may form a veritable coat of armor. Between body segments and between the segments of appendages it is thin and flexible, creating movable joints and permitting free movements. In crustaceans and insects the cuticle forms ingrowths **(apodemes)** that serve for muscle attachment. It may also line foregut and hindgut, line and support the trachea, and be adapted for biting mouthparts, sensory organs, copulatory organs, and ornamental purposes. It is indeed a versatile material.

The nonexpansible cuticular exoskeleton does, however, impose important restrictions on growth. To grow, an arthropod must shed its outer covering at intervals and grow a larger one—a process called **ecdysis,** or **molting.** Arthropods molt two to ten times before reaching adulthood, and some continue to molt after that. An exoskeleton is also relatively heavy and becomes proportionately heavier with increasing size, thereby limiting ultimate body size.

2. **Segmentation and appendages for more efficient locomotion.** Typically each somite bears a pair of jointed appendages, but this arrangement is often modified, with both segments and appendages specialized for adaptive functions. Limb segments are essentially hollow levers moved by internal muscles, most of which are striated for rapid action. The appendages have sensory hairs (as well as bristles and spines) and may be modified and adapted for sensory functions, food handling, swift and efficient walking, and swimming.

3. **Air piped directly to cells.** Most terrestrial arthropods have a highly efficient tracheal system of air tubes, which delivers oxygen directly to the tissues and cells and makes a high metabolic rate possible during their periods of activity. This system also tends to limit body size. Aquatic arthropods breathe mainly by some form of gill that is quite efficient.

4. **Highly developed sensory organs.** Sensory organs are found in great variety, from the compound (mosaic) eye to those accomplishing touch, smell, hearing, balancing, and chemical reception. Arthropods are keenly alert to what happens in their environment.

5. **Complex behavior patterns.** Arthropods exceed most other invertebrates in complexity and organization of their activities. Innate (unlearned) behavior unquestionably controls much of what they do, but learning also plays an important part in the lives of many.

6. **Limiting intraspecific competition through metamorphosis.** Many arthropods pass through metamorphic changes, including a larval form quite different from the adult in structure. Larval forms often are adapted for eating food different from that of adults and occupy a different space, resulting in less competition within a species.

SUBPHYLUM TRILOBITA

Trilobites probably had their beginnings before the Cambrian period, in which they flourished. They have been extinct for 200 million years, but were abundant during the Cambrian and Ordovician periods. Their name refers to the trilobed shape of the body in cross section, caused by a pair of longitudinal grooves. They were bottom dwellers and probably scavengers (Figure 18-1A). Most of them could roll up like pill bugs, and they ranged from 2 to 67 cm in length.

Their exoskeleton contained chitin, strengthened in some areas by calcium carbonate. There were three tagmata in the body: head (also called cephalon), trunk, and pygidium. Their cephalon was one piece but showed signs of former segmentation; their trunk had a variable number of somites; and somites of the pygidium, at the posterior end, were fused into a plate. Their cephalon bore a pair of antennae, compound eyes, a mouth, and four pairs of leglike appendages. There were no mouthparts. Each body somite except the last also bore a pair of biramous (two-branched) appendages. One of the branches had a fringe of filaments that may have served as gills.

SUBPHYLUM CHELICERATA

Chelicerate arthropods are an ancient group that includes eurypterids (extinct), horseshoe crabs, spiders, ticks and mites, scorpions, and sea spiders. They are characterized by six pairs of cephalothoracic appendages that include a pair of chelicerae (mouthparts), a pair of pedipalps, and four pairs of walking legs (a pair of chelicerae and five pairs of legs in horseshoe crabs). They have no antennae. Most chelicerates suck liquid food from their prey.

Class Merostomata

Class Merostomata is represented by eurypterids, all now extinct, and xiphosurids, or horseshoe crabs, an ancient group sometimes called "living fossils."

Subclass Eurypterida

The eurypterids, or giant water scorpions (Figure 18-1B) were the largest of all fossil arthropods, some reaching a length of 3 m. Their fossils occur in rocks from the Ordovician to the Permian periods. They had many resemblances to marine horseshoe crabs (Figure 18-2) and also to scorpions. Their head had

A   B

Figure 18-1

Fossils of early arthropods. **A,** Trilobite fossils, dorsal view. These animals were abundant in mid-Cambrian period. **B,** Eurypterid fossil. Eurypterids flourished in Europe and North America from Ordovician to Permian periods.

six fused segments and bore both simple and compound eyes and chelicerae, pedipalps, and four pairs of walking legs. Their abdomen had 12 segments and a spikelike telson.

Eurypterids were the dominant predators of their time. It is possible that development of dermal armor in early fishes (p. 481) resulted from selection pressure of eurypterid predation.

Subclass Xiphosurida: Horseshoe Crabs

Xiphosurids are an ancient marine group that dates from the Cambrian period. Our common horseshoe crab *Limulus* (L. *limus,* sidelong, askew) (Figure 18-2) goes back practically unchanged to the Triassic period. Only three genera (five species) survive today: *Limulus,* which lives in shallow water along the North American Atlantic coast; *Carcinoscorpius* (Gr. *karkinos,* crab, + *skorpiōn,* scorpion), along the southern shore of Japan; and *Tachypleus* (Gr. *tachys,* swift, + *pleutēs,* sailor), in the East Indies and along the coast of southern Asia. They usually live in shallow water.

Xiphosurids have an unsegmented, horseshoe-shaped **carapace** (hard dorsal shield) and a broad abdomen, which has a long **telson,** or tailpiece. Their cephalothorax bears a pair of chelicerae and five pairs of walking legs, whereas their abdomen has six pairs of broad, thin appendages that are fused in the median line (Figure 18-2). On some abdominal appendages, **book gills** (flat, leaflike gills) are exposed. There are two compound and two simple eyes on the carapace. The horseshoe crab swims by means of its abdominal plates and can walk with its walking legs. It feeds at night on worms and small molluscs, which it seizes with its chelicerae.

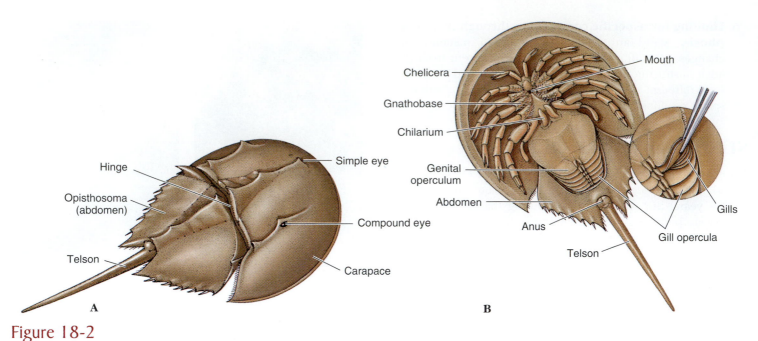

Figure 18-2

A, Dorsal view of horseshoe crab *Limulus* (class Merostomata). They grow to 0.5 m in length. **B,** Ventral view.

During the mating season horseshoe crabs come to shore at high tide to mate. A female burrows into sand where she lays eggs, with one or more smaller males following closely to add sperm to the nest before the female covers it with sand. American *Limulus* mate and lay eggs during high tides of full and new moons in spring and summer. Eggs are warmed by the sun and protected from waves until young larvae hatch and return to the sea by another high tide. Larvae are segmented and are often called "trilobite larvae" because they resemble trilobites, to which xiphosurids may be related.

Class Pycnogonida: Sea Spiders

Some sea spiders are only a few millimeters long, but others are much larger. They have small, thin bodies and usually four pairs of long, thin walking legs. In addition, they have a feature unique among arthropods: somites are duplicated in some groups, so that they possess five or six pairs of legs instead of the four pairs normally characteristic of arachnids. Males of many species bear a subsidiary pair of legs (**ovigers**) (Figure 18-3) on which they carry developing eggs, and ovigers are often absent in females. Many species also are equipped with chelicerae and palps.

Their mouth is at the tip of a long **proboscis,** which sucks juices from cnidarians and soft-bodied animals. Most pycnogonids have four simple eyes. Their circulatory system is limited to a simple dorsal heart, and excretory and respiratory systems are absent. The long, thin body and legs provide a large surface, in proportion to volume, that is evidently sufficient for diffusion of gases and wastes. Because of the small size of the body, the digestive system sends branches into the legs, as do the gonads.

Sea spiders are found in all oceans, but they are most abundant in polar waters. *Pycnogonum* (Figure 18-3B) is a common intertidal genus found on both Atlantic and Pacific coasts of the United States; it has relatively short, heavy legs. *Nymphon* (Fig-

ure 18-3A) is the largest genus of pycnogonids, with over 200 species. It occurs from subtidal depths to 6800 m in all seas except the Black and Baltic seas.

Some authorities have suggested that pycnogonids were more closely related to crustaceans than to other arthropods, but morphological and molecular evidence strongly supports placement of pycnogonids in Chelicerata.

Class Arachnida

Arachnids (Gr. *arachnē,* spider) show wide anatomical variation. In addition to spiders, the group includes scorpions, pseudoscorpions, whip scorpions, ticks, mites, daddy longlegs (harvestmen), and others. There are many differences among these with respect to form and appendages. They are mostly free living and are far more common in warm, dry regions than elsewhere.

Arachnid tagmata are a cephalothorax and abdomen, and the cephalothorax usually bears a pair of chelicerae, a pair of pedipalps, and four pairs of walking legs (Figure 18-4). Most arachnids are predaceous and have claws, fangs (claws and fangs are modified pedipalps and chelicerae), poison glands, or stingers. They usually have a strong sucking pharynx with which they ingest the fluids and soft tissues from the bodies of their prey. Among their interesting adaptations are spinning glands of spiders.

Arachnids have become extremely diverse. More than 80,000 species have been described so far. They were among the first arthropods to move into terrestrial habitats. Scorpions are among Silurian fossils, and by the end of the Paleozoic period mites and spiders had appeared.

Most arachnids are harmless to humans and actually do much good by destroying injurious insects. A few, such as black widow and brown recluse spiders, can give painful or even dangerous bites. Stings of scorpions may be quite painful, and those of a few species can be fatal. Some ticks and mites are carriers

A

Proboscis

Figure 18-3

A, Pycnogonid, *Nymphon* sp. In this genus all anterior appendages (chelicerae, palps, and ovigers) are present in both sexes, although ovigers are often not present in females of other genera.
B, *Pycnogonum hancockii,* a pycnogonid with relatively short legs. Females of this genus have neither chelicerae nor ovigers and males have ovigers.

of diseases as well as causes of annoyance and painful irritations. Certain mites damage a number of important food and ornamental plants by sucking their juices.

Several smaller orders are not included in our discussion.

Order Araneae: Spiders

Spiders are a large group of 32,000 species, distributed throughout the world. The spider body is compact: a **cephalothorax (prosoma)** and **abdomen (opisthosoma),** both unsegmented and joined by a slender pedicel. A few spiders have a segmented abdomen, a primitive character.

Anterior appendages are a pair of **chelicerae** (Figure 18-4), which have terminal **fangs** through which run ducts from poison glands, and a pair of **pedipalps** having basal parts with which they chew (Figure 18-4). Four pairs of **walking legs** ter-

Figure 18-4

External anatomy of a jumping spider, with anterior view of head (at *right*).

minate in claws. Pedipalps of males are modified, sometimes elaborately for sperm transfer.

All spiders are predaceous, feeding largely on insects. They effectively dispatch prey with their fangs and poison. Some spiders chase prey, others ambush them, and many trap them in a net of silk. After a spider seizes prey with its chelicerae and injects venom, it liquefies the tissues with a digestive fluid and sucks the resulting broth into its stomach. Spiders with teeth at the bases of chelicerae crush or chew prey, aiding digestion by enzymes from their mouth.

Spiders breathe by means of book lungs or tracheae or both. Book lungs consist of many parallel air pockets extending into a blood-filled chamber (Figure 18-5). Air enters the chamber by a slit in the body wall. Tracheae form a system of air tubes that carry air directly to the blood from an opening called a spiracle. The tracheae are similar to those in insects (p. 407) but are much less extensive.

Spiders and insects have a unique **excretory system of Malpighian tubules** (Figure 18-5), which work in conjunction with specialized resorptive cells in the intestinal epithelium. Potassium and other solutes and waste materials are secreted into the tubules, which drain the fluid, or "urine," into the intestine. Resorptive cells recapture most potassium and water, leaving behind such wastes as uric acid. By this cycling of water and potassium, species living in dry environments may conserve body fluids, producing a nearly dry mixture of urine and feces. Many spiders also have **coxal glands,** which are modified nephridia that open at the coxa, or base, of the first and third walking legs.

Spiders usually have eight **simple eyes,** each with a lens, optic rods, and a retina (Figure 18-4). They are used chiefly for perception of moving objects, but some, such as those of hunting and jumping spiders, may form images. Since a spider's vision is usually poor, its awareness of its environment depends largely on cuticular mechanoreceptors, such as **sensory setae** (sensilla), **trichobothria,** and **slit sense organs** (slit sensilla). Many setae (as well as bristles and spines) are attached at their bases to one to three neurons. Trichobothria are extremely fine setae in special sockets. Slit sensilla are areas of quite thin, deformable cuticle 1 to 2 μm wide by 8 to 200 μm long, ennervated by a single dendrite, especially numerous on a spider's legs. These receptors can detect vibrations in the web, struggling prey, or even air movements.

Figure 18-5
Spider, internal anatomy.

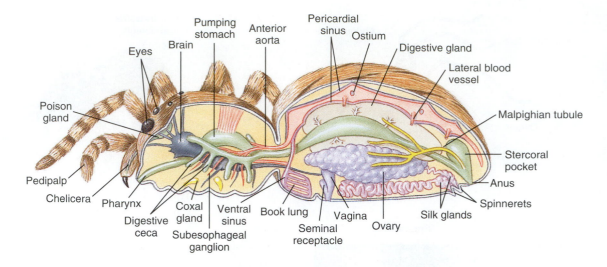

Web-Spinning Habits   The ability to spin silk is central to a spider's life, as it is in some other arachnids. Two or three pairs of spinnerets containing hundreds of microscopic tubes run to special abdominal **silk glands** (Figure 18-5). A scleroprotein secretion emitted as a liquid apparently hardens as a result of being pulled from the spinnerets and forms a silk thread. Spiders' silk threads are stronger than steel threads of the same diameter and are considered second in strength only to fused quartz fibers. The threads will stretch one-fifth of their length before breaking.

A web used for trapping insects is the use of silk familiar to most people. The kind of net varies among species. Some are simple and consist merely of a few strands of silk radiating out from a spider's burrow or place of retreat. Others spin beautiful, geometrical orb webs. However, spiders use silk threads for many purposes besides web making. They use them to line their nests; form sperm webs or egg sacs; build draglines; make bridge lines, warning threads, molting threads, attachment discs, or nursery webs; or to wrap their prey securely (Figure 18-6). Not all spiders spin webs for traps. Some, such as wolf spiders, jumping spiders (Figure 18-4), and fisher spiders (Figure 18-7), simply chase and catch their prey.

Reproduction   Before mating, a male spins a small web, deposits a drop of sperm on it, and then picks up the sperm to be stored in special cavities of his pedipalps. When he mates, he inserts his pedipalps into the female genital opening to store the sperm in his mate's seminal receptacles. A courtship ritual usually precedes mating. A female lays her eggs in a silken net, which she may carry about or attach to a web or plant. A cocoon may contain hundreds of eggs, which hatch in approximately two weeks. Young usually remain in the egg sac for a few weeks and molt once before leaving it. Between five and ten molts occur before adulthood.

Are Spiders Really Dangerous?   It is truly amazing that such small and innocuous creatures as spiders have generated so much unreasoning fear in human minds. Spiders are timid creatures that, rather than being dangerous enemies to humans, are actually allies in the continuing battle with insects. Venom

Figure 18-6
Grasshopper, snared and helpless in the web of a golden garden spider (*Argiope aurantia*), is wrapped in silk while still alive. If the spider is not hungry, the prize will be saved for a later meal.

Figure 18-7
Fisher spider, *Dolomedes triton*, feeds on a minnow. This handsome spider feeds mostly on aquatic and terrestrial insects but occasionally captures small fishes and tadpoles. It pulls its paralyzed victim from the water, pumps in digestive enzymes, then sucks out the predigested contents.

produced to kill prey is usually harmless to humans. Even the most poisonous spiders bite only when threatened or when defending their eggs or young. American tarantulas (Figure 18-8), despite their fearsome size, are *not* dangerous. They rarely bite, and their bite is not serious.

There are, however, two genera in the United States that can give severe or even fatal bites: *Latrodectus* (L. *latro*, robber, + *dectes*, biter; **black widow,** five species) and *Loxosceles* (Gr. *loxos*, crooked, + *skelos*, leg; **brown recluse,** 13 species). Black widows are moderate to small in size and shiny black, usually with a bright

Figure 18-8

A tarantula, *Brachypelma vagans*.

A

B

Figure 18-9

A, Black widow spider, *Latrodectus mactans,* suspended on her web. Note the red "hourglass" on the ventral side of her abdomen. **B,** Brown recluse spider, *Loxosceles reclusa,* is a small venomous spider. Note the small violin-shaped marking on its cephalothorax. The venom is hemolytic and dangerous.

orange or red spot on the underside of their abdomen. The spot is commonly in the shape of an hourglass (Figure 18-9A). Their venom is neurotoxic, acting on the nervous system. About four or five of each 1000 bites reported have proved fatal.

Brown recluse spiders are brown and bear a violin-shaped dorsal stripe on their cephalothorax (Figure 18-9B). Their venom is hemolytic rather than neurotoxic, producing death of tissues and skin surrounding the bite. Their bite can be mild to serious and occasionally fatal.

Some spiders in other parts of the world are dangerous, for example, funnelweb spiders *Atrax* spp. in Australia. Most dangerous of all are certain spiders in South and Central America, *Phoneutria* spp. They are large (10 to 12 cm leg span), and, in contrast to most spiders, they are quite aggressive. Their venom is among the most pharmacologically active of spider venoms, and their bites cause intense pain, neurotoxic effects, sweating, priapism, respiratory paralysis and spasm, and histamine-like effects.

W. S. Bristowe (1971) estimated that at certain seasons a field in Sussex, England, that had been undisturbed for several years had a population of 2 million spiders to the acre. He concluded that so many could not successfully compete except for the many specialized adaptations they had evolved. These include adaptations to cold and heat, wet and dry conditions, and light and darkness.

Some spiders capture large insects, some only small ones; web-builders snare mostly flying insects, whereas hunters seek those that live on the ground. Some lay eggs in the spring, others in the late summer. Some feed by day, others by night, and some have developed flavors that are distasteful to birds or to certain predatory insects. As it is with spiders, so has it been with other arthropods; their adaptations are many and diverse and contribute in no small way to their long success.

Order Scorpiones: Scorpions

Although scorpions are more common in tropical and subtropical regions, some occur in temperate zones. Scorpions are generally secretive, hiding in burrows or under objects by day and feeding at night. They feed largely on insects and spiders, which they seize with their pedipalps and tear up with their chelicerae.

Sand-dwelling scorpions locate prey by sensing surface waves generated by movements of insects on or in the sand. These waves are picked up by compound slit sensilla located on the basitarsal segments of the legs. A scorpion can locate a burrowing cockroach 50 cm away and reach it in three or four quick orientation movements.

Scorpion tagmata are a rather short **cephalothorax,** which bears chelicerae, pedipalps, legs, a pair of large median eyes, and usually two to five pairs of small lateral eyes; a **preabdomen (or mesosoma)** of seven segments; and a long slender **postabdomen (or metasoma)** of five segments, which ends in a stinging apparatus (Figure 18-10A). Their chelicerae are small and three jointed; their pedipalps are large, chelate (pincerlike), and six jointed; and the four pairs of walking legs are eight jointed.

On the ventral side of the abdomen are curious comblike **pectines,** which are tactile organs used for exploring the ground and for sex recognition. The stinger on the last segment consists of a bulbous base and a curved barb that injects venom. Venom of most species is not harmful to humans but may produce a painful swelling. However, the sting of certain species of *Androctonus* in Africa and *Centruroides* (Gr. *kenteō,* to prick, + *oura,* tail, + *oides,* form) in Mexico can be fatal unless antivenin is administered.

Scorpions perform a complex mating dance, the male holding the female's chelae and stepping back and forth. He kneads her chelicerae with his own and, in some species, stings her on her pedipalp or on the edge of her cephalothorax. The stinging action is slow and deliberate, and the stinger remains in the female's body for some minutes. Both individuals are motionless during that time. Finally, the male deposits a spermatophore and pulls the female over it until the sperm mass is taken into the female orifice. Scorpions are truly viviparous; females brood their young within their reproductive tract. After several months to a year of development anywhere from 1 to over 100 young are produced, depending on the species. The young, only a few millimeters long, crawl onto their mother's back until after their first molt (Figure 18-10A). They mature in from 1 to 7 or 8 years and may live for as long as 15 years.

Order Opiliones: Harvestmen

Harvestmen (Figure 18-10B), often known as "daddy longlegs," are common in the United States and other parts of the world. These curious creatures are easily distinguished from spiders: their abdomen and cephalothorax are broadly joined, without constriction of a pedicel; their abdomen shows external segmentation; and they have only two eyes, mounted on a tubercle on their cephalothorax. They have four pairs of long, spindly legs, and they can cast off one or more of these without apparent ill effect if they are grasped by a predator (or human hand). The ends of their chelicerae are pincerlike, and, while carnivorous, they feed much more as scavengers than do spiders.

Harvestmen are not venomous and are harmless to humans. Odoriferous glands that open on the cephalothorax deter some predators. Other than some mites, opilionids are unique among arachnids in having a penis for sperm transfer; all are oviparous.

Traditionally allied with Acari, more recent studies indicate that Opiliones forms a clade with scorpions and two smaller orders. They are the sister group of scorpions.

Order Acari: Ticks and Mites

Members of order Acari are without doubt the most medically and economically important group of arachnids. They far exceed other orders in numbers of individuals and species. Although about 40,000 species have been described, some authorities estimate that from 500,000 to 1 million species exist. Hundreds of individuals of several species of mites may be found in a small portion of leaf mold in forests. They occur throughout the world in both terrestrial and aquatic habitats, even extending into such inhospitable regions as deserts, polar areas, and hot springs. Many acarines are parasitic during one or more stages of their life cycle.

Most mites are 1 mm or less in length. Ticks, which are only one suborder of Acari, range from a few millimeters to occasionally 3 cm. A tick may become enormously distended with blood after feeding on its host.

Acarines differ from all other arachnids in having complete fusion of the cephalothorax and abdomen, with no sign of external division or segmentation (Figure 18-11). They carry their mouthparts on a little anterior projection, the **capitulum.** The capitulum mainly consists of the feeding appendages surrounding the mouth. On each side of their mouth is a chelicera, which functions in piercing, tearing, or gripping food. The form of the chelicerae varies greatly in different families. Lateral to the che-

Figure 18-10

A, An emperor scorpion (order Scorpiones), *Paninus imperator,* with young, which stay with the mother until their first molt. B, Harvestmen, *Mitopus* sp. (order Opiliones). Harvestmen run rapidly on their stiltlike legs. They are especially noticeable during the harvesting season, hence the common name.

Figure 18-11

A, Wood tick, *Dermacentor variabilis* (order Acari). Larvae, nymphs, and adults are all parasitic but drop off their hosts to molt to the next stage. B, Red velvet (harvest) mite, *Trombidium* sp. As with chiggers (*Trombicula*), only larvae of *Trombidium* are parasitic. Nymphs and adults are free living and feed on insect eggs and small invertebrates.

licerae is a pair of segmented pedipalps, which also vary greatly in form and function related to feeding. Ventrally the bases of the pedipalps fuse to form a **hypostome,** whereas a **rostrum,** or **tectum,** extends dorsally over their mouth. Adult mites and ticks usually have four pairs of legs, although there may be only one to three in some specialized forms.

Most acarines transfer sperm directly, but many species use a spermatophore. A larva with six legs hatches from the egg, and one or more eight-legged nymphal stages follow before the adult stage is reached.

Many species of mites are entirely free living. *Dermatophagoides farinae* (Gr. *dermatos,* skin, + *phagō,* to eat, + *eidos,* likeness of form) (Figure 18-12) and related species are denizens of house dust all over the world, sometimes causing allergies and dermatoses. There are some marine mites, but most aquatic species are found in fresh water. They have long, hairlike setae on their legs for swimming, and their larvae may be parasitic on aquatic invertebrates. Such abundant organisms must be important ecologically, but many acarines have more direct effects on our food supply and health. Spider mites (family Tetranychidae) are serious agricultural pests on fruit trees, cotton, clover, and many other plants. They suck out the contents of plant cells, causing a mottled appearance to the leaves (Figure 18-13), and construct a protective web from silk glands opening near the base of the chelicerae. Larvae of genus *Trombicula* are called chiggers or redbugs. They feed on the dermal tissues of terrestrial vertebrates, including humans, and may cause an irritating dermatitis; some species of chiggers transmit a disease called Asiatic scrub typhus. Hair follicle mites, *Demodex* (Figure 18-14), are apparently nonpathogenic in humans; they infect most of us although we are unaware of them. Other species of *Demodex* and other genera of mites cause mange in domestic animals. Human itch mites, *Sarcoptes scabiei* (Figure 18-15), cause intense itching as they burrow beneath the skin.

The inflamed welt and intense itching that follows a chigger bite is not the result of the chigger burrowing into the skin, as is popularly believed. Rather a chigger bites through the skin with its chelicerae and injects a salivary secretion containing powerful enzymes that liquefy skin cells. Human skin responds defensively by forming a hardened tube that the larva uses as a drinking straw and through which it gorges itself with host cells and fluid. Scratching usually removes the chigger but leaves the tube, which is a source of irritation for several days.

In addition to disease conditions that they themselves cause, ticks are among the world's premier disease vectors, ranking second only to mosquitos. They surpass other arthropods in carrying a great variety of infectious agents including apicomplexans, rickettsial, viral, bacterial, and fungal organisms. Species of *Ixodes* carry the most common arthropod-borne infection in the United States, Lyme disease (see note). Species of *Dermacentor* (Figure 18-11A) and other ticks transmit Rocky Mountain spotted fever, a poorly named disease because most cases occur in the eastern United States. *Dermacentor* also transmits tularemia and

Figure 18-12
Scanning electron micrograph of house dust mite, *Dermatophagoides farinae.*

Figure 18-13
Damage to *Chamaedorea* sp. palm caused by mites of the family Tetranychidae (order Acari). Over 130 species of this family occur in North America, and some are serious agricultural pests. Mites pierce plant cells and suck out contents, giving leaves the mottled appearance shown here.

Figure 18-14
Demodex folliculorum, human follicle mite.

Figure 18-15
Sarcoptes scabiei, itch mite.

Figure 18-16

Boophilus annulatus, a tick that carries Texas cattle fever.

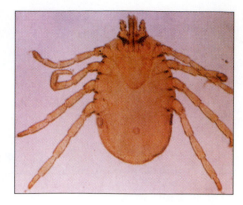

agents of several other diseases. Texas cattle fever, also called red-water fever, is caused by a protozoan parasite transmitted by cattle ticks, *Boophilus annulatus* (Figure 18-16). Many more examples could be cited.

An epidemic of arthritis occurred in the 1970s in the town of Lyme, Connecticut. Subsequently known as Lyme disease, it is caused by a bacterium and carried by ticks of the genus *Ixodes*. There are now thousands of cases a year in Europe and North America, and other cases have been reported from Japan, Australia, and South Africa. Many people bitten by infected ticks recover spontaneously or do not get the disease. Others, if not treated at an early stage, develop a chronic, disabling disease.

PHYLOGENY AND ADAPTIVE RADIATION

Phylogeny

Shared derived characters between annelids and arthropods gave strong support to the hypothesis that both phyla originated from a line of coelomate segmented protostomes, which in time diverged to form a protoannelid line with laterally located parapodia and one or more protoarthropod lines with more ventrally located appendages.

The molecular evidence supporting alignment of annelids and arthropods with separate superphyla is at dramatic variance with our long-held belief that the two phyla are closely related. Separation into separate superphyla implies that metamerism in the two groups arose independently and is a convergent character. However, the analyses supporting an Ecdysozoa-Lophotrochozoa hypothesis (p. 204) suffer an important defect: failure to support monophyly of Annelida and Mollusca.

Whether phylum Arthropoda itself is monophyletic has also been controversial. Some scientists have contended that Arthropoda is polyphyletic and that some or all the present subphyla are derived from different annelid-like ancestors that underwent "arthropodization." The crucial development is the hardening of the cuticle to form an arthropod exoskeleton, and most of the features that distinguish arthropods from annelids (p. 366) result from the stiffened exoskeleton (see the prologue for this chapter). For example, once the vital role of the coelomic compartments as

---

Classification of Phylum Arthropoda

Subphylum Trilobita (tri′lo-bi′ta) (Gr. *tri*, three, + *lobos*, lobe): **trilobites.** All extinct forms; Cambrian to Carboniferous; body divided by two longitudinal furrows into three lobes; distinct head, trunk, and abdomen, biramous (two-branched) appendages.

Subphylum Chelicerata (ke-lis′e-ra′ta) (Gr. *chēlē*, claw, + *keras*, horn, + *ata*, group suffix): **eurypterids, horseshoe crabs, spiders, ticks.** First pair of appendages modified to form chelicerae; pair of pedipalps and four pairs of legs; no antennae, no mandibles; cephalothorax and abdomen usually unsegmented.

Subphylum Crustacea (crus-ta′she-a) (L. *crusta*, shell, + *acea*, group suffix): **crustaceans.** Mostly aquatic, with gills; cephalothorax usually with dorsal carapace; biramous appendages, modified for various functions. Head appendages consisting of two pairs of antennae, one pair of mandibles, and two pairs of maxillae. Development primitively with nauplius stage (see classification of crustaceans, p. 392).

Subphylum Uniramia (yu-ni-ra′me-a) (L. *unus*, one, + *ramus*, a branch): **insects and myriapods.** All appendages uniramous; head appendages consisting of one pair of antennae, one pair of mandibles, and one or two pairs of maxillae (see classification of uniramians, p. 418).

Subphylum Chelicerata

Class Merostomata (mer′o-sto′ma-ta) (Gr. *mēros*, thigh, + *stoma*, mouth, + *ata*, group suffix): **aquatic chelicerates.** Cephalothorax and abdomen; compound lateral eyes; appendages with gills; sharp telson; subclasses Eurypterida (all extinct) and Xiphosurida, horseshoe crabs. Example: *Limulus.*

Class Pycnogonida (pik′no-gon′i-da) (Gr. *pyknos*, compact, + *gony*, knee, angle): **sea spiders.** Small (3 to 4 mm), but some reach 500 mm; body chiefly cephalothorax; tiny abdomen; usually four pairs of long walking legs (some with five or six pairs); mouth on long proboscis; four simple eyes; no respiratory or excretory system. Example: *Pycnogonum.*

Class Arachnida (ar-ack′ni-da) (Gr. *arachnē*, spider): **scorpions, spiders, mites, ticks, harvestmen.** Four pairs of legs; segmented or unsegmented abdomen with or without appendages and generally distinct from cephalothorax; respiration by gills, tracheae, or book lungs; excretion by Malpighian tubules and/or coxal glands; dorsal bilobed brain connected to ventral ganglionic mass with nerves, simple eyes; chiefly oviparous; no true metamorphosis. Examples: *Argiope, Centruroides.*

---

a hydrostatic skeleton was gone, intersegmental septa were unnecessary, as was a closed circulatory system. Jointed appendages, of course, are necessary if the external surface is hard, and body-wall muscles of annelids could be converted and inserted on the considerable inner surfaces of the cuticle for efficient movement of body parts. Compared with annelids, there was a great restriction in permeable surfaces for respiration and excretion. Thus arthropods *could* have evolved more than once. However, other zoologists argue strongly that the derived similarities of the arthropod subphyla strongly support monophyly of the phylum. The phylum Tardigrada may be the sister taxon to arthropods, with phylum Onychophora being the sister taxon to the combined Arthropoda

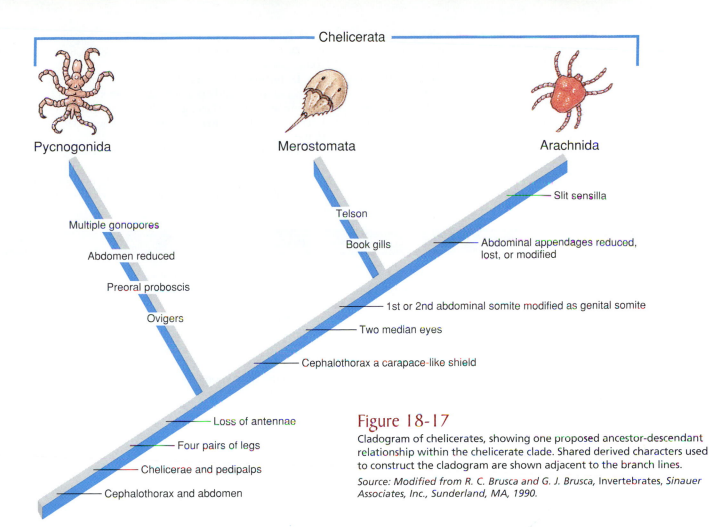

Pycnogonida

Merostomata

Arachnida

Chelicerata

Slit sensilla

Telson

Book gills

Multiple gonopores

Abdominal appendages reduced, lost, or modified

Abdomen reduced

Preoral proboscis

Ovigers

1st or 2nd abdominal somite modified as genital somite

Two median eyes

Cephalothorax a carapace-like shield

Loss of antennae

Four pairs of legs

Chelicerae and pedipalps

Cephalothorax and abdomen

Figure 18-17

Cladogram of chelicerates, showing one proposed ancestor-descendant relationship within the chelicerate clade. Shared derived characters used to construct the cladogram are shown adjacent to the branch lines.

Source: Modified from R. C. Brusca and G. J. Brusca, Invertebrates, Sinauer *Associates, Inc., Sunderland, MA, 1990.*

and Tardigrada (Chapter 21). A cladogram depicting possible relationships is presented in Chapter 21 (p. 439).

Some evidence based on ribosomal RNA sequences supports monophyly of Arthropoda and the inclusion of Onychophora in the phylum.* These data also suggest that myriapods (millipedes and centipedes) are a sister group to all other arthropods and that crustaceans and insects form a monophyletic group! If these conclusions are supported by further investigations, our concepts of arthropod phylogeny and classification are subject to major revision.

Controversy on phylogeny within Chelicerata also exists, especially on the position of Pycnogonida (Figure 18-17). Some workers place pycnogonids as the sister group to chelicerates in a larger grouping called Cheliceriformes.

*Ballard, J. W. O., et al. 1992. Science **258**:1345–1348.

Adaptive Radiation

Annelids show limited tagmatization and little differentiation of appendages. However, in arthropods an adaptive trend has been toward pronounced tagmatization by differentiation or fusion of somites, giving rise in more derived groups to such tagmata as head and trunk; head, thorax, and abdomen; or cephalothorax (fused head and thorax) and abdomen. The primitive arthropod condition is to have similar appendages on each somite, with each somite bearing a pair of appendages. More derived forms have appendages specialized for specific functions, or some somites lack appendages entirely.

Much of the amazing diversity in arthropods seems to have developed because of modification and specialization of their cuticular exoskeleton and their jointed appendages, resulting in a wide variety of locomotor and feeding adaptations.

SUMMARY

Arthropoda is the largest, most abundant and diverse phylum in the world. Arthropods are metameric, coelomate, ecdysozoan protostomes with well-developed organ systems. Most show marked tagmatization. They are extremely diverse and occur in all habitats capable of supporting life. Perhaps more than any other single factor, prevalence of arthropods is explained by adaptations made pos-

sible by their cuticular exoskeleton. Other important elements are jointed appendages, tracheal respiration, efficient sensory organs, complex behavior, and metamorphosis.

Trilobites were a dominant Paleozoic subphylum, now extinct. Members of subphylum Chelicerata have no antennae, and their main feeding appendages are chelicerae. In addition, they have a pair of pedipalps (which may be similar to the walking legs) and four pairs of walking legs. Class Merostomata includes the extinct eurypterids and the ancient, although still extant, horseshoe crabs. Class Pycnogonida contains the sea spiders, which are odd little animals with a large suctorial proboscis and vestigial abdomen. The great majority of living chelicerates are in class Arachnida: spiders (order Araneae), scorpions (order Scorpiones), harvestmen (order Opiliones), ticks and mites (order Acari), and others.

Tagmata of most spiders (cephalothorax and abdomen) show no external segmentation and are joined by a waistlike pedicel. Spiders are predaceous, and their chelicerae are provided with poison glands for paralyzing or killing prey. They breathe by book lungs, tracheae, or both. Spiders can spin silk, which they use for a variety of purposes, including webs for trapping prey in some cases.

Distinctive characters of scorpions are their large, clawlike pedipalps and their clearly segmented abdomen, which bears a terminal stinging apparatus. Harvestmen have small, ovoid bodies with very long, slender legs. Their abdomen is segmented and broadly joined to their cephalothorax.

The cephalothorax and abdomen of ticks and mites are completely fused and mouthparts are borne on an anterior capitulum. They are the most numerous of any arachnids; some are important carriers of disease, and others are serious plant pests.

REVIEW QUESTIONS

1. What are important distinguishing features of arthropods?
2. Name the subphyla of arthropods, and give a few examples of each.
3. How do arthropods differ from annelids, and how are they alike?
4. Briefly discuss the contribution of the cuticle to the success of arthropods, and name some other factors that have contributed to their success.
5. What is a trilobite?
6. What appendages are characteristic of chelicerates?
7. Briefly describe the appearance of each of the following: eurypterids, horseshoe crabs, pycnogonids.
8. What are the tagmata of arachnids, and which tagmata bear appendages?
9. Describe the mechanism of each of the following with respect to spiders: feeding, excretion, sensory reception, web-spinning, reproduction.
10. What are the most important spiders in the United States that are dangerous to humans?
11. Distinguish each of the following orders from each other: Araneae, Scorpiones, Opiliones, Acari.
12. Discuss the importance of members of order Acari to human well-being.
13. Some biologists suggest that the Arthropoda is polyphyletic. Explain why this could be so despite the characteristics shared by all arthropods.

SELECTED REFERENCES

Bowman, A. S., J. W. Dillwith, J. R. Sauer. 1996. Tick salivary prostaglandins: presence, origin and significance. Parasitol. Today **12:**388–396. *Tick prostaglandins act as immunosuppressants, anticoagulants, and analgesics. They allow the tick to feed over an extended time without the blood clotting, an inflammatory reaction occurring, or the host dislodging them.*

Foelix, R. F. 1996. Biology of spiders. New York, Oxford University Press. *Attractive, comprehensive book with extensive references; of interest to both amateurs and professionals.*

Hubbell, S. 1997. Trouble with honeybees. Nat. Hist. **106:**32–43. *Parasitic mites (*Varroa jacobsoni *on bee larvae and* Acarapis woodi *in the trachea of adults) cause serious losses among honey bees.*

Kaston, B. J. 1978. How to know the spiders, ed. 3. Dubuque, Iowa, William C. Brown Publishers. *Spiral-bound identification manual.*

Lane, R. P., and R. W. Crosskey (eds). 1993. Medical insects and arachnids. London, Chapman and Hall. *This is the best book currently available on medical entomology.*

Luoma, J. R. 2001. The removable feast. Audubon **103**(3):48–54. *During May and June large numbers of horseshoe crabs ascend the shores of U.S. Atlantic states to breed and lay eggs. Since the 1980s they have been heavily harvested to be chopped up and used for bait. This practice has led to serious declines in* Limulus *populations, with accompanying declines in populations of migrating shore birds that feed on* Limulus *eggs.*

McDaniel, B. 1979. How to know the ticks and mites. Dubuque, Iowa, William C. Brown Publishers. *Useful, well-illustrated keys to genera and higher categories of ticks and mites in the United States.*

Ostfeld, R. S. 1997. The ecology of Lyme-disease risk. Am. Sci. **85:**338–346. *Lyme disease, caused by a bacterium transmitted by ticks, has been reported in 48 of the 50 United States and seems to be increasing in frequency and geographic range.*

Polis, G. A. (ed). 1990. The biology of scorpions. Stanford, California, Stanford University Press. *The editor brings together a readable summary of what is known about scorpions.*

Schultz, J. W. 1990. Evolutionary morphology and phylogeny of Arachnida. Cladistics **6:**1–38. *A cladistic analysis of arachnid orders based on morphological data; this study disrupted the traditional views that scorpions are the sister group of other arachnids or were the sister group of eurypterids.*

Shear, W. A. 1994. Untangling the evolution of the web. Am. Sci. **82:**256–266. *Fossil spider webs are nonexistent. Evolution of the web must be studied by comparing modern spider webs to each other and correlating studies of spider anatomy.*

Suter, R. B. 1999. Walking on water. Am. Sci. **87:**154–159. *Fishing spiders (*Dolomedes) *depend on surface tension to walk on water.*

Weaver, D. C. 1999. Mysterious fevers. Discover **20:**37–40. *Ehrlichiosis is caused by a bacterial parasite of white blood cells transmitted by ticks.*

Wheeler, W. C., and C. Y. Hayashi. 1998. The phylogeny of the extant chelicerate orders. Cladistics **14:**173–192. *A cladistic analysis of Chelicerata based on morphological and molecular characters.*

ZOOLOGY LINKS TO THE INTERNET

Visit the textbook's Online Learning Center at www.mhhe.com/zoology to find live Internet links for each of the topics listed here.

Phylum Arthropoda
Subphylum Chelicerata
Class Arachnida

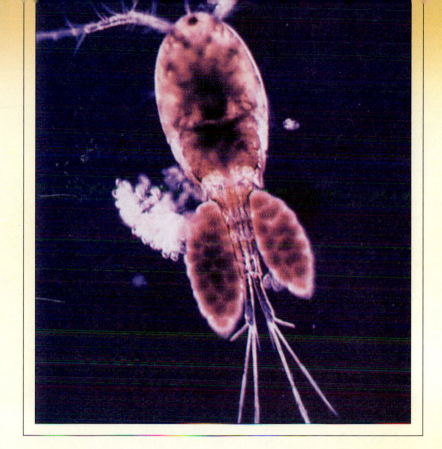

CHAPTER 19

A female copepod bearing eggsacs.

Aquatic Mandibulates

Phylum Arthropoda
Subphylum Crustacea

"Insects of the Sea"

Subphylum Crustacea (L. *crusta,* shell) gets its name from the hard shell that most crustaceans bear. Over 30,000 species have been described, and several times that number probably exist. Most familiar to people are edible ones, for example, lobsters, crayfishes, shrimps, and crabs. In addition to these "crusty" crustaceans, there is an astonishing array of less familiar forms such as copepods, ostracods, water fleas, whale lice, tadpole shrimp, and krill. They fill a wide variety of ecological roles and show enormous variation in morphological characteristics, making a satisfactory description of the group singularly hard to frame.

We live in the age of arthropods, notwithstanding our anthropocentric attachment to our tradition of calling the current era the age of mammals. Together, insects and crustaceans compose more than 80% of all named animal species. Just as insects pervade terrestrial habitats (more than a million named species and countless billions of individuals), crustaceans abound in oceans, lakes, and rivers. Some walk or creep on the bottom, some burrow, and some (such as barnacles) are sessile. Some swim upright, others swim upside down, and many are delicate microscopic forms that drift as plankton in oceans or in lakes. Indeed, it is probable that the *most abundant animals in the world* are members of the copepod genus *Calanus*. In recognition of their dominance of marine habitats, it is understandable that crustaceans have been called "insects of the sea."

Arthropods that possess mandibles (jawlike appendages) are known as mandibulates and traditionally have been united in subphylum Mandibulata. In addition, many investigators believe that there are sufficient differences between crustaceans and uniramians (insects, millipedes, centipedes, pauropods, and symphylans) to justify separation to subphylum level. Both Crustacea and Uniramia have, at least, a pair of **antennae,** a pair of **mandibles,** and a pair of **maxillae** on the head. These appendages perform sensory, masticatory, and food-handling functions, respectively. The body may consist of a head and trunk, but in the more derived forms, a high degree of tagmatization (p. 365) has occurred so that there is a well-defined head, thorax, and abdomen. In most Crustacea one or more thoracic segments are fused with the head to form a **cephalothorax.** Thoracic and abdominal appendages are mainly for walking or swimming, but in some groups they are highly specialized in function. Crustacea are mainly marine; however, there are many freshwater and a few terrestrial species, whereas uniramians are mainly terrestrial. There are numerous species of insects in freshwater habitats, but only a few in marine.

SUBPHYLUM CRUSTACEA

General Nature of a Crustacean

Crustaceans differ from other arthropods in a variety of ways, but the only truly distinguishing characteristic is that crustaceans are the only arthropods with **two pairs of antennae.** In addition to two pairs of antennae and a pair of mandibles, crustaceans have two pairs of maxillae on the head, followed by a pair of appendages on each body segment or somite. In some crustaceans not all somites bear appendages. All appendages, except perhaps the first antennae, are primitively **biramous** (two main branches), and at least some appendages of present-day adults show that condition. Organs specialized for respiration, if present, function as **gills.**

Most crustaceans have between 16 and 20 somites, but some forms have 60 somites or more. A larger number of somites is a primitive feature. The more derived condition is to have fewer segments and increased tagmatization (see p. 365). Major tagmata are head, thorax, and abdomen, but these are not homologous throughout the class (or even within some subclasses) because of varying degrees of fusion of somites, for example, as in the cephalothorax.

By far the largest group of crustaceans is class Malacostraca, which includes lobsters, crabs, shrimps, beach hoppers, sow bugs, and many others. These show a surprisingly constant arrangement of body segments and tagmata, which is considered the ancestral plan of this class (Figure 19-1). This typical body plan has a head of five (six embryonically) fused somites, a thorax of eight somites, and an abdomen of six somites (seven in a few species). At the anterior end is a nonsegmented **rostrum** and at the posterior end is a nonsegmented **telson,** which with the last abdominal somite and its **uropods** forms a tail fan in many forms.

In many crustaceans the dorsal cuticle of the head may extend posteriorly and around the sides of the animal to cover

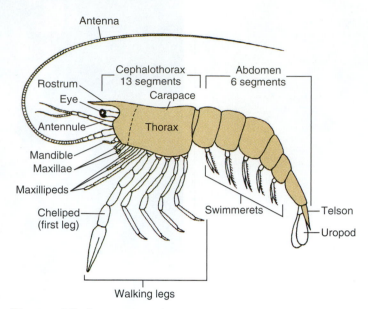

Figure 19-1

Archetypical plan of Malacostraca. The two maxillae and three maxillipeds have been separated diagrammatically to illustrate the general plan.

or be fused with some or all of the thoracic and abdominal somites. This covering is called a **carapace.** In some groups the carapace forms clamshell-like valves that cover most or all of the body. In decapods (including lobsters, shrimp, crabs, and others), the carapace covers the entire cephalothorax but not the abdomen.

Form and Function

Because of their size and easy availability, large crustaceans such as crayfishes have been studied more than other groups. They are also commonly studied in introductory laboratory courses. Therefore many of the comments that are included here apply specifically to crayfishes and their relatives.

External Features

Bodies of crustaceans are covered with a secreted cuticle composed of chitin, protein, and calcareous material. The harder, heavy plates of larger crustaceans are particularly high in calcareous deposits. The hard protective covering is soft and thin at the joints between somites, allowing flexibility of movement. The carapace, if present, covers much or all of the cephalothorax; in decapods such as crayfishes, all head and thoracic segments are enclosed dorsally by the carapace. Each somite not enclosed by the carapace is covered by a dorsal cuticular plate, or **tergum** (Figure 19-2A), and a ventral transverse bar, or **sternum,** lies between the segmental appendages (Figure 19-2B). The abdomen terminates in a telson, which is not considered a somite and bears the anus. (The telson may be homologous to the annelid pygidium.)

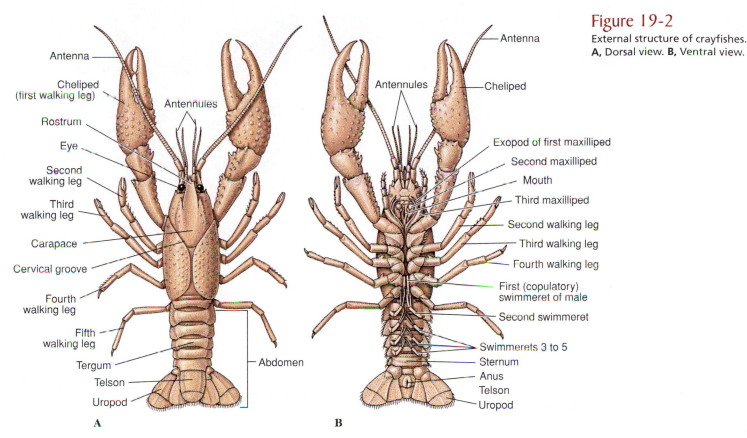

Figure 19-2
External structure of crayfishes.
A, Dorsal view. **B,** Ventral view.

Figure 19-3
Parts of a biramous crustacean appendage (third maxilliped of a crayfish).

Position of the **gonopores** varies according to sex and group of crustaceans. They may be on or at the base of a pair of appendages, at the terminal end of the body, or on somites without legs. In crayfishes openings of the vasa deferentia are on the median side at the base of the fifth pair of walking legs, and those of the oviducts are at the base of the third pair. In females an opening to the seminal receptacle is usually located in the midventral line between the fourth and fifth pairs of walking legs.

Appendages Members of classes Malacostraca (including crayfishes) and Remipedia typically have a pair of jointed appendages on each somite (Figure 19-3), although abdominal somites in the other classes do not bear appendages. Considerable specialization is evident in appendages of derived crustaceans such as crayfishes. However, all are variations of the basic, biramous plan, illustrated by a crayfish appendage such as a maxilliped (a thoracic limb modified to become a head appendage) (Figures 19-3 and 19-4). The basal portion, or **protopod,** bears a lateral **exopod** and a medial **endopod.** The protopod is made up of one or two joints (**basis** and **coxa),** whereas the exopod and endopod have from one to several joints each. Some appendages, such as walking legs of crayfishes, have become secondarily uniramous. Medial or lateral processes, called **endites** and **exites,** respectively, sometimes occur on crustacean limbs, and an exite on the protopod is called an **epipod.** Epipods are often modified as gills. Table 19-1 shows how various appendages have become modified from the biramous plan to fit specific functions.

Terminology applied by various workers to crustacean appendages has not been blessed with uniformity. At least two systems are in wide use. Alternative terms to those we have used, for example, are protopodite, exopodite, endopodite, basipodite, coxopodite, and epipodite. The first and second pairs of antennae may be termed antennules and antennae, and first and second maxillae are often called maxillules and maxillae. A rose by any other name. . .

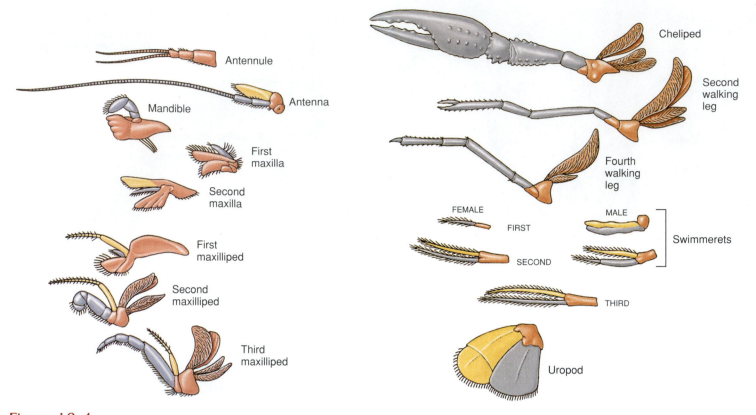

Figure 19-4

Appendages of a crayfish showing how they have become modified from the basic biramous plan, as found in a swimmeret. Protopod, brown; endopod, blue; exopod, yellow.

Structures that have a similar basic plan and have descended from a common form are said to be homologous, whether they have the same function or not. Since specialized walking legs, mouthparts, chelipeds, and swimmerets have all developed from a common biramous type but have become modified to perform different functions, they are all homologous to each other, a condition known as **serial homology.** Primitively limbs were all very similar, but during evolution of structural modifications, some branches have been reduced, some lost, some greatly altered, and some new parts added. Crayfishes and their allies possess the most elaborate serial homology in the animal kingdom, having 17 distinct but serially homologous types of appendages (Table 19-1).

Internal Features

The muscular and nervous systems and segmentation in thorax and abdomen clearly show metamerism, but there are marked modifications in other systems. Most changes involve concentration of parts in a particular region or else reduction or complete loss of parts, such as the intersepta.

Hemocoel   The major body space in arthropods is not a coelom but a blood-filled **hemocoel.** During embryonic development of most arthropods, vestigial coelomic cavities open within the mesoderm of at least some somites. These are soon obliterated or become continuous with space between develop-ing mesodermal and ectodermal structures and yolk. This space becomes the hemocoel and is thus not lined by a mesodermal peritoneum. In crustaceans the only coelomic compartments remaining are end sacs of excretory organs and space around the gonads.

Muscular System   Striated muscles make up a consider-able part of the body of most Crustacea. Muscles are usually arranged in antagonistic groups: **flexors,** which draw a part toward the body, and **extensors,** which extend it outward. The abdomen of a crayfish has powerful flexors (Figure 19-5), which are used when the animal swims backward with a burst of speed that is its best means of escape. Strong muscles on either side of the stomach control its mandibles.

Respiratory System   Respiratory gas exchange in smaller crustaceans occurs over thinner areas of cuticle (for example, in the legs) or the entire body, and specialized structures may be absent. Larger crustaceans have gills, which are delicate, feather-like projections with very thin cuticle. In decapods the sides of the carapace enclose the gill cavity, which is open anteriorly and ventrally (Figure 19-6). Gills may project from the pleural wall into the gill cavity, from the articulation of thoracic legs with the body, or from thoracic coxae. The latter two types are typical of crayfishes. The "bailer," a part of the second maxilla, draws water over the gill filaments, into the gill cavity at the bases of the legs, and out of the gill cavity at the anterior.

TABLE 19.1

Crayfish Appendages

Appendage	Protopod	Endopod	Exopod	Function
First antenna (antennule)	3 segments, statocyst in base	Many-jointed feeler	Many-jointed feeler	Touch, taste, equilibrium
Second antenna (antenna)	2 segments, excretory pore in base	Long, many-jointed feeler	Thin, pointed blade	Touch, taste
Mandible	2 segments, heavy jaw and base of palp	2 distal segments of palp	Absent	Crushing food
First maxilla (maxillule)	2 segments with 2 thin endites	Small unjointed lamella	Absent	Food handling
Second maxilla (maxilla)	2 segments, with 2 endites and 1 scaphognathite (epipod)	1 small pointed segment	Part of scaphognathite (bailer)	Drawing currents of water into gills
First maxilliped	2 medial plates and epipod	2 small segments	1 basal segment, plus many-jointed filament	Touch, taste, food handling
Second maxilliped	2 segments plus gill (epipod)	5 short segments	2 slender segments	Touch, taste, food handling
Third maxilliped	2 segments plus gill (epipod)	5 larger segments	2 slender segments	Touch, taste, food handling
First walking leg (cheliped)	2 segments plus gill (epipod)	5 segments with heavy pincer	Absent	Offense and defense
Second walking leg	2 segments plus gill (epipod)	5 segments plus small pincer	Absent	Walking and prehension
Third walking leg	2 segments plus gill (epipod); genital pore in female	5 segments plus small pincer	Absent	Walking and prehension
Fourth walking leg	2 segments plus gill (epipod)	5 segments, no pincer	Absent	Walking
Fifth walking leg	2 segments; genital pore in male; no gill	5 segments, no pincer	Absent	Walking
First swimmeret	In female reduced or absent; in male fused with endopod to form tube			In male, transferring sperm to female
Second swimmeret				
Male	Structure modified for transfer of sperm to female	Structure modified for transfer of sperm to female		
Female	2 segments	Jointed filament	Jointed filament	Creating water currents; carrying eggs and young
Third, fourth, and fifth swimmerets	2 short segments	Jointed filament	Jointed filament	Creating water currents; in female carrying eggs and young
Uropod	1 short, broad segment	Flat, oval plate	Flat, oval plate; divided into 2 parts with hinge	Swimming; egg protection in female

Circulatory System   Crustaceans and other arthropods have an "open" or lacunar type of circulatory system. This means that there are no veins and no separation of blood from interstitial fluid, as there is in animals with closed systems (see p. 660). Hemolymph (blood) leaves the heart by way of arteries, circulates through the hemocoel, and returns to venous sinuses, or spaces, instead of veins before it reenters the heart. Thus, crustaceans contrast with annelids, which have a closed system, as do vertebrates.

A dorsal heart is the chief propulsive organ. It is a single-chambered sac of striated muscle. Hemolymph enters the heart from the surrounding **pericardial sinus** through paired ostia, with valves that prevent backflow into the sinus (Figure 19-6). From the heart hemolymph enters one or more arteries. Valves

Figure 19-5

Internal structure of a male crayfish.

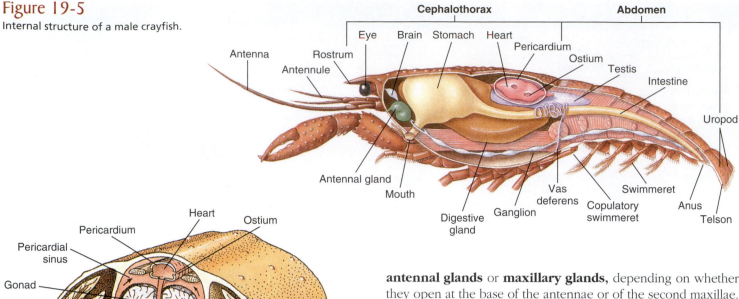

Figure 19-6

Diagrammatical cross section through heart region of a crayfish showing direction of blood flow in this "open" blood system. Heart pumps blood to body tissues through arteries, which empty into tissue sinuses. Returning blood enters sternal sinus, then goes through gills for gas exchange, and finally back to pericardial sinus by efferent channels. Note absence of veins.

in the arteries prevent a backflow of hemolymph. Small arteries empty into tissue sinuses, which in turn often discharge into a large **sternal sinus** (Figure 19-6).

From there, afferent sinus channels carry hemolymph to the gills, if present, for oxygen and carbon dioxide exchange. Hemolymph then returns to the pericardial sinus by efferent channels (Figure 19-6).

Hemolymph in arthropods is largely colorless. It includes ameboid cells of at least two types. Hemocyanin, a copper-containing respiratory pigment, or hemoglobin, an iron-containing pigment, may be carried in solution. Hemolymph has the property of clotting, which prevents its loss in minor injuries. Some ameboid cells release a thrombinlike coagulant that precipitates clotting.

Excretory System   Excretory organs of adult crustaceans are a pair of tubular structures located in the ventral part of their head anterior to the esophagus (Figure 19-5). They are called

antennal glands or **maxillary glands,** depending on whether they open at the base of the antennae or of the second maxillae. A few adult crustaceans have both. Excretory organs of decapods are antennal glands, also called **green glands** in this group. Crustaceans do not have Malpighian tubules, the excretory organs of spiders and insects.

The **end sac** of the antennal gland, which is derived from an embryonic coelomic compartment, consists of a small vesicle **(saccule)** and a spongy mass called a **labyrinth.** The labyrinth connects by an **excretory tubule** to a dorsal **bladder,** which opens to the exterior by a pore on the ventral surface of the basal antennal segment (Figure 19-7). Hydrostatic pressure within the hemocoel provides force for filtration of fluid into the end sac. As filtrate passes through the excretory tubule and bladder, it is modified by resorption of salts, amino acids, glucose, and some water and is finally excreted as urine.

Excretion of nitrogenous wastes (mostly ammonia) takes place by diffusion across thin areas of cuticle, especially gills, and the so-called excretory organs function principally to regulate ionic and osmotic composition of body fluids. Freshwater crustaceans, such as crayfishes, are constantly threatened with overdilution of their blood by water, which diffuses across the gills and other water-permeable surfaces. Green glands, by forming a dilute, low-salt urine, act as an effective "flood-control" device. Some Na^+ and Cl^- are lost in the urine, but this loss is compensated by active absorption of dissolved salt by the gills. In marine crustaceans, such as lobsters and crabs, the kidney functions to adjust salt composition of hemolymph by selective modification of salt content of tubular urine. In these forms urine remains isosmotic to the blood.

Nervous and Sensory Systems   Nervous systems of crustaceans and annelids have much in common, although those of crustaceans have more fusion of ganglia (Figure 19-5). The brain is a pair of **supraesophageal ganglia** that supplies nerves to the eyes and two pairs of antennae. It is joined by connectives to the **subesophageal ganglion,** a fusion of at least five pairs of ganglia that supply nerves to mouth, appendages, esophagus, and antennal glands. The double ventral nerve cord has a pair of ganglia for each somite and nerves serving the

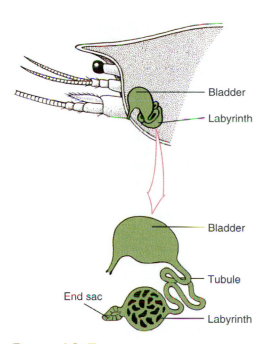

Figure 19-7
Scheme of antennal gland (green gland) of crayfishes. (In natural position organ is much folded.) Some crustaceans lack a labyrinth, and the excretory tubule (nephridial canal) is a much-coiled tube.

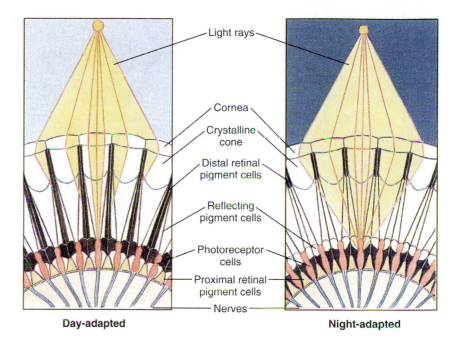

Figure 19-8
Portion of compound eye of an arthropod showing migration of pigment in ommatidia for day and night vision. Five ommatidia are represented in each diagram. In daytime each ommatidium is surrounded by a dark pigment collar so that each ommatidium is stimulated only by light rays that enter its own cornea (mosaic vision); in nighttime, pigment forms incomplete collars and light rays can spread to adjacent ommatidia (continuous, or superposition, image).

appendages, muscles, and other parts. In addition to this central system, there may be a sympathetic nervous system associated with the digestive tract.

Crustaceans have better-developed sense organs than do annelids. The largest sense organs of crayfishes are eyes and statocysts. Widely distributed over the body are **tactile hairs,** delicate projections of cuticle that are especially abundant on chelae, mouthparts, and telson. Chemical senses of taste and smell are found in receptors on antennae, mouthparts, and other places.

A saclike **statocyst,** opening to the surface by a dorsal pore, is found on the basal segment of each first antenna of crayfishes. Statocysts contain a ridge that bears sensory setae formed from the chitinous lining and grains of sand that serve as **statoliths.** Whenever the animal changes its position, corresponding changes in the position of the grains on the sensory setae are relayed as stimuli to the brain, and the animal can adjust itself accordingly. Each molt (ecdysis) of cuticle results in loss of the cuticular lining of statocysts and sand grains they contain. New grains are picked up through the dorsal pore after ecdysis.

Eyes in many crustaceans are compound, composed of many photoreceptor units called **ommatidia** (Figure 19-8). Covering the rounded surface of each eye is a transparent area of cuticle, the **cornea,** which is divided into many small squares or hexagons known as facets. These facets are the outer ends of the ommatidia. Each ommatidium behaves like a tiny eye and contains several kinds of cells arranged in a columnar fashion (Figure 19-8). Black pigment cells are found between adjacent ommatidia.

Movement of pigment in an arthropod compound eye permits it to adjust for different amounts of light. There are three sets of pigment cells in each ommatidium: distal retinal, proximal retinal, and reflecting; these are so arranged that they can form a more or less complete collar or sleeve around each ommatidium. For strong light or day adaptation the distal retinal pigment moves inward and meets the outward-moving proximal retinal pigment so that a complete pigment sleeve forms around the ommatidium (Figure 19-8). In this condition only rays that strike the cornea directly will reach the photoreceptor (retinular) cells, for each ommatidium is shielded from others. Thus each ommatidium will see only a limited area of the field of vision (a mosaic, or apposition, image). In dim light distal and proximal pigments separate so that light rays, with the aid of reflecting pigment cells, have a chance to spread to adjacent ommatidia and to form a continuous, or superposition, image. This second type of vision is less precise but takes maximum advantage of the limited amount of light received.

Reproduction, Life Cycles, and Endocrine Function

Most crustaceans have separate sexes, and there are various specializations for copulation among different groups. Barnacles are monoecious but generally practice cross-fertilization. In some ostracods males are scarce, and reproduction is usually parthenogenetic. Most crustaceans brood their eggs in some manner: branchiopods and barnacles have special brood chambers, copepods have brood sacs attached to the sides of their abdomen (see Figure 19-19), and many malacostracans carry eggs and young attached to their abdominal appendages.

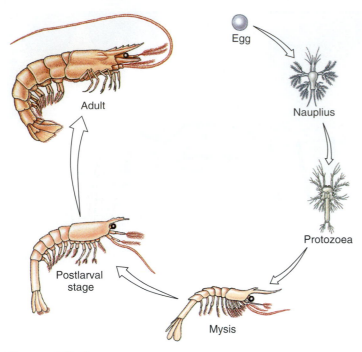

Figure 19-9

Life cycle of a Gulf shrimp, *Penaeus.* Penaeids spawn at depths of 40 to 90 m. Young larval forms are planktonic and move inshore to water of lower salinity to develop as juveniles. Older shrimp return to deeper water offshore.

Crayfishes have direct development: there is no larval form. A tiny juvenile with the same form as the adult and a complete set of appendages and somites hatches from the egg. However, development is indirect in most crustaceans, and larvae quite unlike adults in structure and appearance hatch from eggs. Change from larva ultimately to an adult is **metamorphosis.** The primitive and most widely occurring larva in Crustacea is the **nauplius** (Figure 19-9 and 19-20). Nauplii bear only three pairs of appendages: uniramous first antennules, biramous antennae, and biramous mandibles. All function as swimming appendages at this stage. Subsequent development may involve a gradual change to adult body form, and appendages and somites are added through a series of molts, or assumption of the adult form may involve more abrupt changes. For example, metamorphosis of a barnacle proceeds from a free-swimming nauplius to a larva with a bivalve carapace called a cyprid and finally to a sessile adult with calcareous plates.

Ecdysis   Ecdysis (ek'duh-sis) (Gr. *ekdyein,* to strip off), or molting, is necessary for the body to increase in size because the exoskeleton is nonliving and does not grow as the animal grows. Much of a crustacean's functioning, including its reproduction, behavior, and many metabolic processes, is directly affected by the physiology of the molting cycle.

Cuticle, which is secreted by underlying epidermis, has several layers (Figure 19-10). The outermost is **epicuticle,** a very thin layer of lipid-impregnated

protein. The bulk of cuticle is the several layers of **procuticle:** (1) **exocuticle,** which lies just beneath the epicuticle and contains protein, calcium salts, and chitin; (2) **endocuticle,** which itself is composed of (3) a **principal layer,** which contains more chitin and less protein and is heavily calcified, and (4) an uncalcified **membranous layer,** a relatively thin layer of chitin and protein.

Some time before actual ecdysis, epidermal cells enlarge considerably. They separate from the membranous layer, secrete a new epicuticle, and begin secreting a new exocuticle (Figure 19-11). Enzymes are released into the area above the new epicuticle. These enzymes begin to dissolve old endocuticle, and soluble products are resorbed and stored within the body of the crustacean. Some calcium salts are stored as **gastroliths** (mineral accretions) in the walls of the stomach. Finally, only exocuticle and epicuticle of the old cuticle remain, underlain by new epicuticle and new exocuticle. The animal swallows water, which it absorbs through its gut, and its blood volume increases greatly. Internal pressure causes the cuticle to split, and the animal pulls itself out of its old exoskeleton (Figure 19-12). Then follow a stretching of the still soft new cuticle, deposition of new endocuticle, redeposition of salvaged inorganic salts and other constituents, and hardening of the new cuticle. During the period of molting, the animal is defenseless and remains hidden away.

When a crustacean is young, ecdysis must occur frequently to allow growth, and the molting cycle is relatively short. As the animal approaches maturity, intermolt periods become progressively longer, and in some species molting ceases entirely. During intermolt periods, increase in tissue mass occurs as living tissue replaces water.

Hormonal Control of the Ecdysis Cycle   Although ecdysis is hormonally controlled, the cycle is often initiated by an environmental stimulus perceived by the central nervous system. Such stimuli may include temperature, day length, and humidity (in the case of land crabs). The signal from the central nervous system decreases production of a **molt-inhibiting hormone** by the **X-organ.** The X-organ is a group of neurosecretory cells in the medulla terminalis of the brain. In crayfishes and

Figure 19-10

Structure of crustacean cuticle.

Figure 19-11
Cuticle secretion and resorption in ecdysis.

Intermolt condition

In precedysis, old procuticle separates from epidermis, which secretes new epicuticle

As new exocuticle is secreted, molting fluid dissolves old endocuticle, and solution products are reabsorbed

At ecdysis, the old epicuticle and exocuticle are discarded

In postecdysis, new cuticle is stretched and unfolded, and endocuticle is secreted

Figure 19-12
Molting sequence in a lobster, *Homarus americanus.* **A,** Membrane between carapace and abdomen ruptures, and carapace begins slow elevation. This step may take up to 2 hours. **B** and **C,** Head, thorax, and finally abdomen withdraw. This process usually takes no more than 15 minutes. Immediately after ecdysis, chelipeds are desiccated and body is very soft. Lobster continues rapid absorption of water so that within 12 hours the body increases about 20% in length and 50% in weight. Tissue water will be replaced by protein in succeeding weeks.

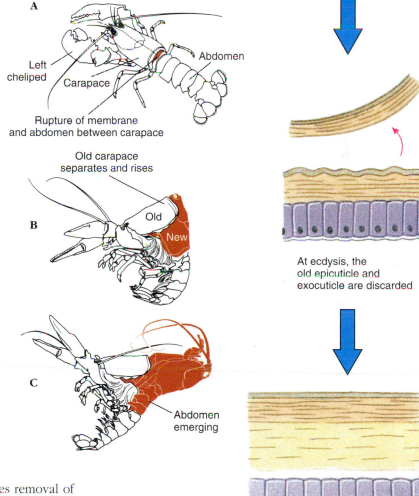

other decapods, the medulla terminalis is found in the eyestalk. The hormone is carried in axons of the X-organ to the **sinus gland** (which itself is probably not glandular in function), also in the eyestalk, where it is released into the hemolymph.

A drop in level of molt-inhibiting hormone promotes release of a **molting hormone** from the **Y-organs.** Y-organs lie beneath the epidermis near the adductor muscles of the mandibles, and they are homologus to prothoracic glands of insects, which produce the hormone ecdysone. Action of molting hormone is to initiate processes leading to ecdysis (proecdysis). Once initiated, the cycle proceeds automatically without further action of hormones from either X- or Y-organs.

Other Endocrine Functions   Not only does removal of eyestalks accelerate molting, but crustaceans whose eyestalks have been removed can no longer adjust body coloration to background conditions. Over 50 years ago it was discovered that the defect was caused not by loss of vision but by loss of hormones in the eyestalks. Body color of crustaceans is largely a result of pigments in special branched cells (chromatophores) in the epidermis. Concentration of pigment granules in the center of the cells causes lightening, and dispersal of pigment throughout the cells causes darkening. Pigment behavior is controlled by hormones from neurosecretory cells in the eyestalk, as is migration of retinal pigment for light and dark adaptation in the eyes (Figure 19-8).

Neurosecretory cells are nerve cells that are modified for secretion of hormones. They are widespread in invertebrates and also occur in vertebrates. Cells in the vertebrate hypothalamus and posterior pituitary are good examples (see p. 728).

Release of neurosecretory material from the pericardial organs in the wall of the pericardium causes an increase in the rate and amplitude of the heartbeat.

Androgenic glands, first found in an amphipod (*Orchestia,* a common beach hopper), occur in male malacostracans. Unlike most other endocrine organs in crustaceans, these are not neurosecretory organs. Their secretion stimulates expression of male sexual characteristics. Young malacostracans have rudimentary androgenic glands, but in females these glands fail to develop. If they are artificially implanted in a female, her ovaries transform to testes and begin to produce sperm, and her appendages begin to acquire male characteristics at the next molt. In isopods the androgenic glands are found in testes; in all other malacostracans they are between muscles of the coxopods of the last thoracic legs and partly attached near ends of the vasa deferentia. Although females do not possess organs similar to androgenic glands, their ovaries produce one or two hormones that influence secondary sexual characteristics.

Hormones that influence other body processes in Crustacea may be present, and evidence suggests that a neurosecretory substance produced in the eyestalk regulates the level of blood sugar.

Feeding Habits

Feeding habits and adaptations for feeding vary greatly among crustaceans. Many forms can shift from one type of feeding to another depending on environment and food availability, but all use the same fundamental set of mouthparts. Mandibles and maxillae function to ingest food; maxillipeds hold and crush food. In predators walking legs, particularly chelipeds, serve in food capture.

Many crustaceans, both large and small, are predatory, and some have interesting adaptations for killing prey. Mantis shrimps, malacostracan order Stomatopoda, have on one of their walking legs a specialized digit that can be drawn into a groove and released suddenly to pierce passing prey. Pistol shrimps (*Alpheus* spp.) have an enormously enlarged chela that can be cocked like the hammer of a gun and snapped shut at great speed, forming a cavitation bubble that implodes with a force sufficient to stun their prey.

Food of **suspension feeders** ranges from plankton and detritus to bacteria. **Predators** consume larvae, worms, crustaceans, snails, and fishes. **Scavengers** eat dead animal and plant matter. Suspension feeders, such as fairy shrimps, water fleas, and barnacles, use their legs, which bear a thick fringe of setae, to create water currents that sweep food particles through the setae. Mud shrimps (*Upogebia* spp.) use long setae on their first two pairs of thoracic appendages to strain food from water circulated through their burrow by movements of their swimmerets.

Crayfishes have a two-part stomach (Figure 19-13). The first part contains a **gastric mill** in which food, already torn up by their mandibles, can be further ground up by three calcareous teeth into particles fine enough to pass through a filter of setae in the second part; food particles then pass into the intestine for chemical digestion.

A BRIEF SURVEY OF CRUSTACEANS

Crustaceans are an extensive group with many subdivisions. They have many structures, habitats, and modes of living. Some are much larger than crayfishes; others are smaller, even microscopic. Some are highly developed and specialized; others have simpler organization.

You should realize that the following summary of crustaceans and the classification on p. 392 are misleadingly brief. Although we mention all classes, a complete presentation of taxa in the hierarchy below class level would require coverage beyond the scope of this textbook.

Class Remipedia

Remipedia (Figure 19-14A) is a very small, recently described class of Crustacea. The 10 species described so far have come from caves with connections to the sea. Remipedes have some very primitive features. There are 25 to 38 trunk segments (thorax and abdomen), all bearing paired, biramous, swimming appendages that are essentially alike. Antennules are biramous. Both pairs of maxillae and a pair of maxillipeds, however, are prehensile and apparently adapted for feeding. The shape of the swimming appendages is similar to that found in Copepoda, but unlike copepods and cephalocarids, swimming legs are directed laterally rather than ventrally.

Figure 19-13

Malacostracan stomach showing gastric "mill" and directions of food movements. Mill has chitinous ridges, or teeth, for mastication, and setae for straining food before it passes into the pyloric stomach.

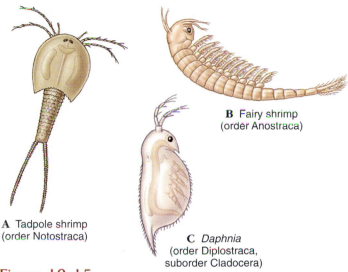

Figure 19-14

A, A remipede crustacean of class Remipedia.
B, A cephalocarid crustacean of class Cephalocarida.

Figure 19-15

Examples of class Branchiopoda.

Class Cephalocarida

Cephalocarida (Figure 19-14B) is also a small group, with only nine species known. Cephalocarids occur along both coasts of the United States, in the West Indies, and in Japan. They are 2 to 3 mm long and have been found in bottom sediments from the intertidal zone to a depth of 300 m. Some features are quite primitive. Thoracic limbs are very similar to each other, and second maxillae are similar to thoracic limbs. Second maxillae and the first seven thoracic legs have a large epipod on their protopod, and the protopod is a single joint. Cephalocarids have no eyes, carapace, or abdominal appendages. True hermaphrodites, they are unique among Arthropoda in discharging both eggs and sperm through a common duct.

Class Branchiopoda

Branchiopoda also represents a crustacean type with some primitive characters. Three orders are recognized: **Anostraca** (fairy shrimp and brine shrimp, Figure 19-15B), with no carapace; **Notostraca** (tadpole shrimp, Figure 19-15A), whose carapace forms a large dorsal shield; and **Diplostraca** (water fleas, Figure 19-15C), typically with a carapace that encloses the body but not the head, or the carapace may enclose the entire body (clam shrimps). Branchiopods have reduced first antennae and second maxillae. Their legs are flattened and leaflike **(phyllopodia)** and are the chief respiratory organs (hence the name branchiopods). Most branchiopods also use their legs for suspension feeding, and in groups other than cladocerans, they use their legs for locomotion as well.

Most branchiopods are freshwater forms. Most important and abundant are water fleas (cladocerans), which often form a large segment of freshwater zooplankton. Their reproduction is very interesting and is reminiscent of that occurring in some rotifers (Chapter 15). During summer cladocerans often produce only females, by parthenogenesis, rapidly increasing the population. With onset of unfavorable conditions, some males are produced, and eggs that must be fertilized are produced by normal meiosis. Fertilized eggs are highly resistant to cold and desiccation, and they are very important for survival of the population over winter and for passive transfer to new habitats. Most cladocerans have direct development, whereas other branchiopods have gradual metamorphosis.

Class Ostracoda

Members of Ostracoda are, like diplostracans, enclosed in a bivalve carapace and resemble tiny clams, 0.25 to 10 mm long (Figure 19-16A). Ostracods show considerable fusion of trunk somites, and numbers of thoracic appendages are reduced to two or none. Feeding and locomotion are principally by use of the head appendages. Most ostracods live on the bottom or climb on plants, but some are planktonic or burrowing, and a few are parasitic. Feeding habits are diverse; there are particle, plant, and carrion feeders and predators. They are widespread in both marine and fresh-water habitats. Development is gradual metamorphosis.

Class Maxillopoda

Class Maxillopoda includes a number of crustacean groups traditionally considered classes themselves. Specialists have recognized evidence that these groups descended from a common ancestor and thus form a monophyletic group within Crustacea. They basically have five cephalic, six thoracic, and usually four abdominal somites plus a telson, but reductions are common.

There are no typical appendages on the abdomen. The eye of the nauplius (when present) has a unique structure and is referred to as a **maxillopodan eye.**

Subclass Mystacocarida

Mystacocarida is a class of tiny crustaceans (less than 0.5 mm long) that live in interstitial water between sand grains of marine beaches (psammolittoral habitat) (Figure 19-16B). Only 10 species have been described, but mystacocarids are widely distributed through many parts of the world.

Subclass Copepoda

This group is second only to Malacostraca in numbers of species. Copepods are small (usually a few millimeters or less in length) and rather elongate, tapering toward the posterior. They lack a carapace and retain a simple, median, nauplius (maxillopodan) eye in adults (Figure 19-16C). They have a single pair of uniramous maxillipeds and four pairs of rather flattened, biramous, thoracic swimming appendages. The fifth pair of legs is reduced. The posterior part of the body is usually separated from the anterior, appendage-bearing portion by a major articulation. Antennules are often longer than other appendages. Copepoda have become very diverse and evolutionarily enterprising, with large numbers of symbiotic as well as free-living species. Many parasites are highly modified, and adults may be

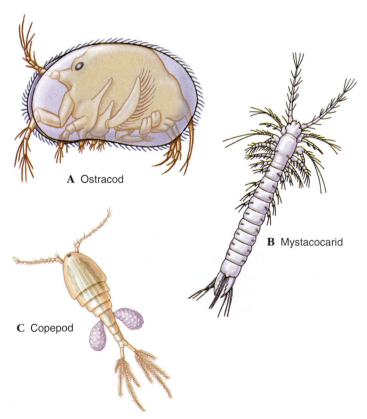

A Ostracod

B Mystacocarid

C Copepod

Figure 19-16

A, An ostracod of class Ostracoda. **B,** A mystacocarid crustacean of subclass Mystacocarida, class Maxillopoda. **C,** A copepod with attached ovisacs; subclass Copepoda, class Maxillopoda.

so highly modified (and may depart so far from the description just given) that they can hardly be recognized as arthropods, let alone crustaceans.

Ecologically, free-living copepods are of extreme importance, often dominating the primary consumer level (p. 800) in aquatic communities. In many marine localities the copepod *Calanus* is the most abundant organism in zooplankton and has the greatest proportion of total biomass (p. 800). In other localities it may be surpassed in biomass only by euphausids (p. 391). *Calanus* forms a major portion of the diet of such economically and ecologically important fish as herring, menhaden, sardines, and larvae of larger fish and (along with euphausids) is an important food item for some whales and sharks. Other genera commonly occur in marine zooplankton, and some forms such as *Cyclops* and *Diaptomus* may form an important segment of freshwater plankton. Many species of copepods are parasites of a wide variety of other marine invertebrates and marine and freshwater fish, and some of the latter are of economic importance. Some species of free-living copepods serve as intermediate hosts of parasites of humans, such as *Diphyllobothrium* (a tapeworm) and *Dracunculus* (a nematode), and of other animals.

Development in copepods is indirect, and some highly modified parasites show striking metamorphoses.

Subclass Tantulocarida

Tantulocarida (Figure 19-17) is the most recently described class (here considered a subclass) of crustaceans (1983). Only about 12 species are known so far. They are tiny (0.15 to 0.2 mm) copepod-like ectoparasites of other deep-sea benthic crustaceans. They have no recognizable head appendages except one pair of antennae on sexual females. Their life cycle is not known with certainty, but present evidence suggests that there is a parthenogenetic cycle and a bisexual cycle with fertilization. **Tantulus** larvae penetrate the cuticle of their hosts by a mouth tube. Then their abdomen and all thoracic limbs are lost during metamorphosis to an adult. Alone among maxillopodans, juveniles bear six to seven abdominal somites, but other evidence supports inclusion in this class.

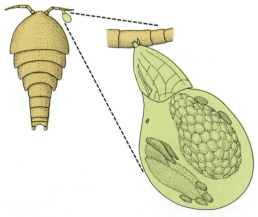

Figure 19-17

A tantulocarid. This curious little parasite is shown attached to the first antenna of its copepod host at left; subclass Tantulocarida, class Maxillopoda.

Figure 19-18

Fish louse; subclass Branchiura, class Maxillopoda.

Subclass Branchiura

Branchiurans are a small group of primarily fish parasites, which, despite their name, have no gills (Figure 19-18). Members of this group are usually between 5 and 10 mm long and may be found on marine or freshwater fish. They typically have a broad, shieldlike carapace, compound eyes, four biramous thoracic appendages for swimming, and a short, unsegmented abdomen. Second maxillae have become modified as suction cups, enabling the parasites to move about on their fish host or even from fish to fish. Development is almost direct: there is no nauplius, and young resemble adults except in size and degree of development of appendages.

Subclass Cirripedia

Cirripedia includes barnacles (order Thoracica), which are usually enclosed in a shell of calcareous plates, as well as three smaller orders of burrowing or parasitic forms. Barnacles are sessile as adults and may be attached to the substrate by a stalk (gooseneck barnacles) (Figure 19-19B) or directly (acorn barnacles) (Figure 19-19A). Typically their carapace (mantle) surrounds their body and secretes a shell of calcareous plates. The head is reduced, they have no abdomen, and the thoracic legs are long, many-jointed cirri with hairlike setae. The cirri are extended through an opening between the calcareous plates to filter from the water small particles on which the animal feeds (Figure 19-19). Although all barnacles are marine, they are often found in the intertidal zone and are therefore exposed to drying and sometimes fresh water for some periods of time. During these periods the aperture between the plates closes to a very narrow slit.

Barnacles frequently foul ship bottoms by settling and growing there. So great may be their number that the speed of the ship may be reduced 30% to 40%, necessitating drydocking the ship to clean them off.

Barnacles are hermaphroditic and undergo a striking metamorphosis during development. Most hatch as nauplii, which soon become cyprid larvae, so called because of their resemblance to an ostracod genus *Cypris*. They have a bivalve carapace and compound eyes. Cyprids attach to the substrate by means of their first antennae, which have adhesive glands, and begin their metamorphosis. This involves several dramatic changes, including secretion of calcareous plates, loss of eyes, and transformation of swimming appendages to cirri.

A

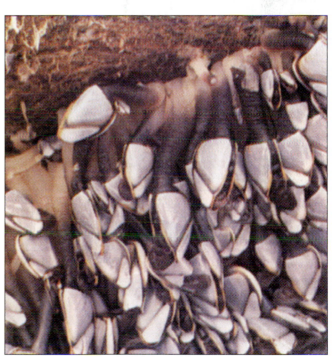

B

Figure 19-19

Barnacles; order Thoracica, subclass Cirripedia, class Maxillopoda. **A,** Acorn barnacles, *Balanus balanoides,* on an intertidal rock await the return of the tide. **B,** Common gooseneck barnacles, *Lepas anatifera.* Note the feeding legs, or cirri, on *Lepas.* Barnacles attach themselves to a variety of firm substrates, including rocks, pilings, and boat bottoms.

In this chapter we introduce animals commonly placed in subphylum Uniramia (insects and myriapods). Some scientists now question the validity of a taxon "Uniramia," but we tentatively retain the grouping. We discuss the controversy further under Phylogeny (p. 420). Uniramians are primarily terrestrial arthropods. Only a few have returned to aquatic life, usually in fresh water.

The term "myriapod," meaning "many footed," is often used for a group of four classes of uniramians that have evolved a pattern of two tagmata—head and trunk—with paired appendages on most or all trunk somites. Myriapods include Chilopoda (centipedes), Diplopoda (millipedes), Pauropoda (pauropods), and Symphyla (symphylans).

Insects have evolved a pattern of three tagmata—head, thorax, and abdomen—with appendages on the head and thorax but greatly reduced on or absent from the abdomen. The common ancestor of insects may have resembled a myriapod in general body form.

Uniramians have only one pair of antennae, and their appendages are always uniramous, never biramous like those of crustaceans. Although some insect young are aquatic and have gills, their gills are not homologous with those of crustaceans.

Insects and myriapods use tracheae to carry respiratory gases directly to and from all body cells in a manner similar to onychophorans and some arachnids.

Excretion is usually by Malpighian tubules.

CLASS CHILOPODA

Chilopoda (ki-lop'o-da) (Gr. *cheilos,* margin, lip, + *pous, podos,* foot), or centipedes, are land forms with somewhat flattened bodies that may contain from a few to 177 somites (Figure 20-1). Each somite, except the one behind the head and the last two in the body, bears a pair of jointed legs. Appendages of the first body segment are modified to form poison claws. The last pair of legs is longer than the others and serve a sensory function.

The head appendages are similar to those of an insect (Figure 20-1B). There are a pair of antennae, a pair of mandibles, and one or two pairs of maxillae. A pair of eyes on the dorsal side of the head consists of groups of ocelli.

The digestive system is a straight tube into which salivary glands empty at the anterior end. Two pairs of Malpighian tubules empty into the hind part of the intestine. There is an elongated heart with a pair of arteries to each somite. The heart has a series of ostia to provide for return of blood to the heart from the hemocoel. Respiration is by means of a tracheal system of branched air tubes that come from a pair of spiracles in each somite. The nervous system is typically arthropod, and there is also a visceral nervous system.

Sexes are separate, with unpaired gonads and paired ducts. Some centipedes lay eggs and others are viviparous. The young are similar in form to adults.

Centipedes prefer moist places such as under logs, bark, and stones. They are very agile and are carnivorous in their eating habits, living on cockroaches and other insects, and earthworms. They kill their prey with their poison claws and then chew it with their mandibles. Common house centipedes *Scutig-*

A

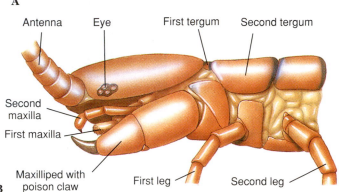

B

Figure 20-1

A, A centipede, *Scolopendra* (class Chilopoda) from the Amazon Basin, Peru. Most segments have one pair of appendages each. First segment bears a pair of poison claws, which in some species can inflict serious wounds. Centipedes are carnivorous. **B,** Head of centipede.

era (L. *scutum,* shield, + *gera,* bearing), which have 15 pairs of legs, are often seen scurrying around bathrooms and damp cellars, where they catch insects. Most species of centipedes are harmless to humans, although many tropical centipedes, some of which may reach a length of 30 cm, are dangerous.

CLASS DIPLOPODA

Diplopoda (Gr. *diploo,* double, two + *pous, podos,* foot) are commonly called millipedes, which literally means "thousand feet" (Figure 20-2). Even though they do not have that many legs, they do have a large number of appendages, since each abdominal somite has two pairs, a condition that may have arisen from fusion of pairs of somites. However, this "diplodous condition" may arise from modulation in *Distal-less* gene expression, as in biramous appendages of Crustacea (p. 394). Their cylindrical bodies are made up of 25 to 100 somites. Their short thorax consists of four somites, each bearing one pair of legs.

Their head bears two clumps of simple eyes and a pair each of antennae, mandibles, and maxillae (Figure 20-2B). The general body structures are similar to those of centipedes, with a few variations here and there. Two pairs of spiracles on each abdominal somite open into air chambers that connect to tracheal air tubes. There are two genital apertures toward the anterior end.

Figure 20-3
A, Pauropod. Pauropods are minute, whitish myriapods with three-branched antennae and nine pairs of legs. They live in leaf litter and under stones. They are eyeless but have sense organs that resemble eyes. B, *Scutigerella*, a symphylan, is a minute whitish myriapod that is sometimes a greenhouse pest.

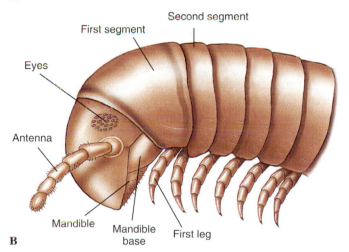

Figure 20-2
A, A tropical millipede with warning coloration. Note the typical doubling of appendages on most segments, hence diplosegments. B, Head of millipede.

In most millipedes the appendages of the seventh somite are specialized as copulatory organs. After millipedes copulate, females lay eggs in a nest and guard them carefully. Larval forms have only one pair of legs to each somite.

Millipedes are not as active as centipedes. They walk with a slow, graceful motion, not wriggling as centipedes do. They prefer dark, moist places under logs or stones. Most are herbivorous, feeding on decayed plant matter, although sometimes they eat living plants. Because they are slow-moving animals, many millipedes roll into a coil when disturbed. Many millipedes also protect themselves from predation by secreting toxic or repellent fluids from special glands (**repugnatorial glands**) positioned along the sides of the body. Common examples of this class are *Spirobolus* and *Julus*, both of which have wide distribution.

CLASS PAUROPODA

Pauropoda (Gr. *pauros,* small, + *pous, podos,* foot) are a group of minute (2 mm or less), soft-bodied myriapods, numbering almost 500 species. They have a small head with branched

antennae and no eyes, but they have a pair of sense organs that resemble eyes (Figure 20-3A). Their 12 trunk segments usually bear nine pairs of legs (none on the first or the last two segments). They have only one tergal plate covering each two segments.

Tracheae, spiracles, and circulatory system are lacking. Pauropods are probably most closely related to diplopods.

Although widely distributed, pauropods are the least well known myriapods. They live in moist soil, leaf litter, or decaying vegetation and under bark and debris. Representative genera are *Pauropus* and *Allopauropus.*

CLASS SYMPHYLA

Symphyla (Gr. *sym,* together, + *phylon,* tribe) are small (2 to 10 mm) and have centipede-like bodies (Figure 20-3B). They live in humus, leaf mold, and debris. *Scutigerella* (L. dim. of *Scutigera*) are often pests on vegetables and flowers, particularly in greenhouses. They are soft bodied, with 14 segments, 12 of which bear legs and one a pair of spinnerets. The antennae are long and unbranched. Only 160 species have been described.

Mating behavior of *Scutigerella* is unusual. Males place a spermatophore at the end of a stalk. When a female finds it, she takes it into her mouth, storing sperm in special buccal pouches. Then she removes eggs from her gonopore with her mouth and attaches them to moss or lichen, or to walls of crevices, smearing them during handling with some of the semen and so fertilizing them. Young at first have only six or seven pairs of legs.

Symphylans are eyeless but have sensory pits at the bases of the antennae. Their tracheal system is limited to a pair of spiracles on their head and tracheal tubes to anterior segments only.

CLASS INSECTA

Insecta (L. *insectus,* cut into) are the most diverse and abundant of all groups of arthropods. There are more species of insects than species of all other classes of animals combined. The number of insect species has been estimated at up to 30 million. There is also striking evidence of continuing evolution among insects at the present time.

It is difficult to appreciate fully the significance of this extensive group and its role in the biological pattern of animal life. The study of insects (**entomology**) occupies the time and resources of skilled men and women all over the world. The struggle between humans and their insect competitors seems to be endless, yet paradoxically insects have so interwoven themselves into the economy of nature in so many useful roles that we would have a difficult time without them.

Insects differ from other arthropods in having **three pairs of legs** and usually **two pairs of wings** on the thoracic region of the body, although some have one pair of wings or none. Insects range in size from less than 1 mm to 20 cm in length, the majority being less than 2.5 cm long. Generally, the largest insects live in tropical areas.

Distribution

Insects are among the most abundant and widespread of all land animals. They have spread into practically all habitats that will support life except most of the sea. Relatively few are marine. Marine water striders (*Halobates*), which live on the surface of the ocean, are the only marine invertebrates that live on the sea-air interface. Insects are common in brackish water, in salt marshes, and on sandy beaches. They are abundant in fresh water, in soil, in forests (especially the tropical forest canopy), and in plants, and they are found even in deserts and wastelands, on mountaintops, and as parasites in and on plants and animals.

Their wide distribution is made possible by their powers of flight and their highly adaptable nature. In most cases they can easily surmount barriers that are virtually impassable to many other animals. Their small size allows them to be carried by currents of both wind and water to far regions. Their well-protected eggs can withstand rigorous conditions and can be carried long distances by birds and other animals. Their agility and ecological aggressiveness enable them to fight for every possible niche in a habitat. No single pattern of biological adaptation can be applied to them.

Adaptability

Insects, during their evolution, have shown an amazing adaptability, as evidenced by their wide distribution and enormous diversity of species. Most of their structural modifications have taken place in wings, legs, antennae, mouthparts, and alimentary canal. Such wide diversity enables this vigorous group to take advantage of all available food and shelter. Some are parasitic, some suck the sap of plants, some chew the foliage of plants, some are predaceous, and some live on the blood of various animals. Within these different groups, specialization occurs, so that a particular kind of insect will eat, for instance, leaves of only one kind of plant. This specificity of eating habits lessens competition with other species and to a great extent accounts for their biological diversity.

Insects are well adapted to dry and desert regions. Their hard and protective exoskeleton helps prevent evaporation. Some insects also extract most of the water from food, fecal material, and by-products of cell metabolism.

As in other arthropods, the exoskeleton is made up of a complex system of plates known as **sclerites,** connected by concealed, flexible hinge joints. Muscles between sclerites enable insects to make precise movements. Rigidity of their exoskeleton is attributable to unique scleroproteins and not to its chitin component. Its lightness makes flying possible. By contrast, the cuticle of crustaceans is stiffened mostly by minerals.

External Form and Function

Insects show a remarkable variety of morphological characteristics, but they are much more homogeneous in tagmatization than are Crustacea. Some insects are fairly generalized in body structure; some are highly specialized. Grasshoppers, or locusts, are a generalized type often used in laboratories to demonstrate general features of insects (Figure 20-4).

Insect tagmata are head, thorax, and abdomen. The cuticle of each body somite typically is composed of four plates (sclerites), a dorsal notum (tergum), a ventral sternum, and a pair of lateral pleura. Pleura of abdominal segments are membranous rather than sclerotized.

The head usually bears a pair of relatively large compound eyes, a pair of antennae, and usually three ocelli (see Figure 20-19A). Antennae, which vary greatly in size and form (Figure 20-5), act as tactile organs, olfactory organs, and in some cases as auditory organs. Mouthparts, formed from specially hardened cuticle, typically consist of a labrum, a pair each of mandibles and maxillae, a labium, and a tonguelike hypopharynx. The type of mouthparts an insect possesses determines how it feeds. We discuss some of these modifications later.

The thorax is composed of three somites: prothorax, mesothorax, and metathorax, each bearing a pair of legs (Figure 20-4). In most insects the mesothorax and metathorax each bear a pair of wings. Wings are cuticular extensions formed by the epidermis. They consist of a double membrane containing veins of thicker cuticle that serve to strengthen the wing. Although these veins vary in their patterns among different taxa, they are constant within a family, genus, or species and serve as one means of classification and identification.

Legs of insects often are modified for special purposes. Terrestrial forms have walking legs with terminal pads and claws as in beetles. These pads may be sticky for walking upside down, as in house flies. Hindlegs of grasshoppers and crickets are adapted for jumping (Figure 20-6). Mole crickets have the first pair of legs modified for burrowing in the ground. Water bugs and many beetles have paddle-shaped appendages for swimming. For grasping prey, forelegs of a praying mantis are long and strong (Figure 20-7). Legs of honey bees show complex adaptations for collecting pollen (Figure 20-8).

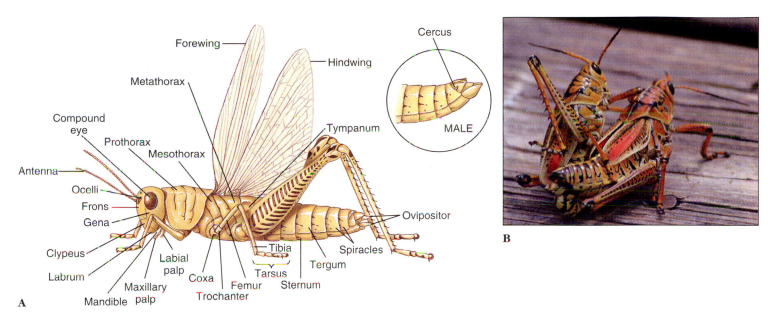

Figure 20-4

A, External features of a female grasshopper. The terminal segment of a male with external genitalia is shown in inset. **B,** A pair of lubber grasshoppers, *Romalea guttata* (order Orthoptera), copulating. The African desert locust mentioned in the chapter prologue (p. 397) is *Schistocerca gregaria*.

Figure 20-5

A few types of insect antennae.

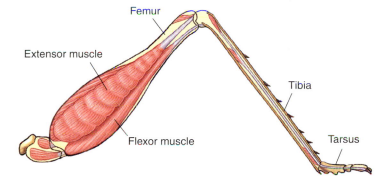

Figure 20-6

Hindleg of grasshopper. Muscles that operate the leg are found within a hollow cylinder of exoskeleton. Here they are attached to the internal wall, from which they manipulate segments of limb on the principle of a lever. Note pivot joint and attachment of tendons of extensor and flexor muscles, which act reciprocally to extend and flex the limb.

There are innumerable variations in body form among insects. Beetles are usually thick and plump (Figure 20-9A); damselflies, ant lions, and walking sticks are long and slender (Figure 20-9B); many aquatic beetles are streamlined; and cockroaches are flat, adapted to living in crevices. The ovipositor of female ichneumon wasps is extremely long (Figure 20-10). Their cerci form horny forceps in earwigs and are long and many jointed in stoneflies and mayflies. Antennae are long in cockroaches and katydids, short in dragonflies and most beetles, knobbed in butterflies, and plumed in most moths. Other variations exist (Figure 20-5).

Locomotion

Walking   When walking, most insects use a triangle of legs involving the first and last leg of one side together with the middle

The abdomen of insects is composed of 9 to 11 segments; the eleventh, when present, is reduced to a pair of **cerci** (appendages at the posterior end). Larval or nymphal forms have a variety of abdominal appendages, but these appendages are lacking in adults. The end of the abdomen bears external genitalia (Figure 20-4A).

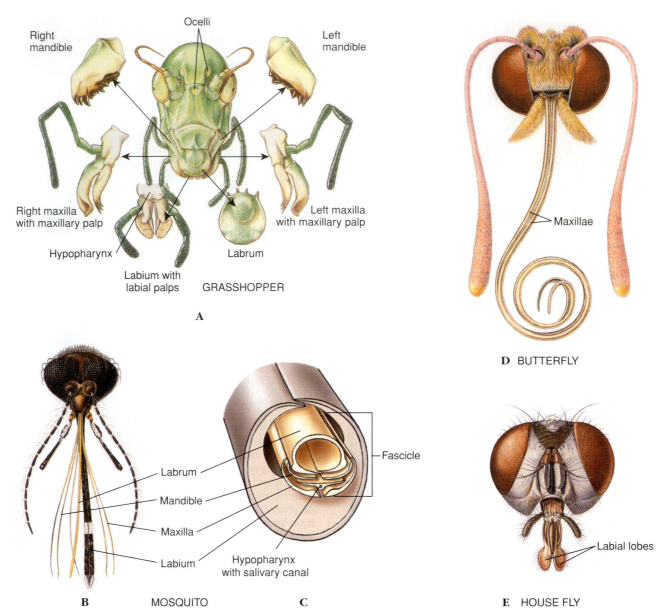

Figure 20-19
Four types of insect mouthparts. **A,** Chewing mouthparts of a grasshopper. **B** and **C,** Sucking mouthparts of a mosquito. Parts of piercing fascicle are shown in cross section (**C**). **D,** Sucking mouthparts of a butterfly. Mandibles are absent, and maxillae form a long proboscis. **E,** Sponging mouthparts of a house fly. A pair of large lobes with grooves on their lower surface are on the end of the labium.

sponge up surface liquids but bite into skin with slender, tapering mandibles and then sponge up blood.

Chewing mouthparts such as those of grasshoppers and many other herbivorous insects are adapted for seizing and crushing food (Figure 20-19A); those of most carnivorous insects are sharp and pointed for piercing their prey. Mandibles of chewing insects are strong, toothed plates whose edges can bite or tear while the maxillae hold the food and pass it toward the mouth. Enzymes secreted by the salivary glands add chemical action to the chewing process.

Circulation

A tubular heart in the pericardial cavity (Figure 20-14) moves hemolymph (blood) forward through the only blood vessel, a dorsal aorta. The heartbeat is a peristaltic wave. Accessory pulsatory organs help move hemolymph into the wings and legs, and flow is also facilitated by various body movements. Hemolymph consists of plasma and amebocytes and apparently has little to do with oxygen transport in most insects, but hemoglobin is present in hemolymph of some and functions in oxygen transport.

Gas Exchange

Terrestrial animals require efficient respiratory systems that permit rapid oxygen and carbon dioxide exchange but at the same time restrict water loss. In insects this is the function of the **tracheal system,** an extensive network of thin-walled tubes that branch into every part of the body (Figure 20-20). The tracheal

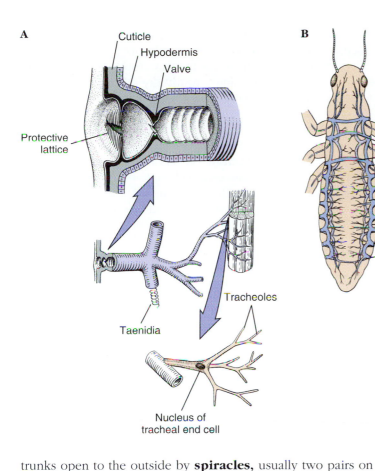

A

Cuticle
Hypodermis
Valve
Protective lattice
Tracheoles
Taenidia
Nucleus of tracheal end cell

B

Commissural tracheae
Ventral trachea
Dorsal trachea
Spiracles

Figure 20-20

A, Relationship of spiracle, tracheae, taenidia (chitinous bands that strengthen the tracheae), and tracheoles (diagrammatic). **B,** Generalized arrangement of insect tracheal system (diagrammatic). Air sacs and tracheoles not shown.

trunks open to the outside by **spiracles,** usually two pairs on the thorax and seven or eight pairs on the abdomen. A spiracle may be merely a hole in the integument, as in primary wingless insects, but there is usually a valve or some sort of closing mechanism that reduces water loss. Evolution of such a device must have been very important in enabling insects to move into drier habitats. Spiracles may also possess a filtering device such as a sieve plate or a set of interlocking bristles that may prevent entrance of water, parasites, or dust into the tracheae.

Tracheae are composed of a single layer of cells and are lined with cuticle that is shed during molts along with the outer cuticle. Spiral thickenings of cuticle (called **taenidia**) support the tracheae and prevent their collapse. Tracheae branch out into smaller tubes, ending in very fine, fluid-filled tubules called **tracheoles** (lined with cuticle, but not shed at ecdysis), which branch into a fine network over the cells. In large insects the largest tracheae may be several millimeters in diameter but taper down to 1 to 2 µm. Tracheoles then taper to 0.5 to 0.1 µm in diameter. In one stage of silkworm larvae it is estimated that there are 1.5 million tracheoles! Scarcely any living cell is more than a few micrometers away from a tracheole. In fact, the ends of some tracheoles actually indent the membranes of cells they supply, so that they terminate close to mitochondria. The tracheal system affords efficient transport usually without use of oxygen-carrying pigments in hemolymph, although hemoglobin is present in some.

The tracheal system may also include **air sacs,** which are apparently dilated tracheae without taenidia (Figure 20-20A). They are thin walled and flexible and are mostly in the body cavity but also in appendages. In many insects the air sacs

increase the volume of air inspired and expired. Muscular movements in the abdomen draw air into the tracheae and expand the sacs, which collapse on expiration. In some insects—locusts, for example—additional pumping is provided by telescoping the abdomen, pumping with the prothorax, or thrusting the head forward and backward. In some insects, air sacs have functions other than respiratory. For example, they may allow internal organs to change in volume during growth without changing the shape of the insect, and they reduce the weight of large insects.

In some very small insects, gas transport occurs entirely by diffusion along a concentration gradient. Consumption of oxygen causes a reduced pressure in their tracheae that sucks air in through the spiracles.

The tracheal system is an adaptation for air breathing, but many insects (nymphs, larvae, and adults) live in water. In small, soft-bodied aquatic nymphs, gaseous exchange may occur by diffusion through the body wall, usually into and out of a tracheal network just under the integument. Aquatic nymphs of stoneflies, mayflies, and damselflies have **tracheal gills,** which are thin extensions of the body wall containing a rich tracheal supply. Gills of dragonfly nymphs are ridges in the rectum (rectal gills) where gas exchange occurs as water enters and leaves.

Although diving beetles, *Dytiscus* (Gr. *dytikos,* able to swim), can fly, they spend most of their life in water as excellent swimmers. They use an "artificial gill" in the form of a bubble of air held under the first pair of wings. The bubble is kept stable by a layer of hairs on top of the abdomen and is in contact with spiracles on the abdomen. Oxygen from the bubble diffuses into their tracheae and is replaced by diffusion of oxygen from the surrounding water. However, nitrogen from the bubble diffuses into the water, slowly decreasing the size of the bubble; therefore, diving beetles must surface every few hours to replace the air. Mosquito larvae are not good swimmers but live just below the surface, putting out short breathing tubes like snorkels to the surface for air (see Figure 20-25B). Spreading oil on the water, a favorite method of mosquito control, clogs their tracheae with oil and so suffocates the larvae. "Rat-tailed maggots" of certain syrphid flies have an extensible tail that can stretch as much as 15 cm to the water surface.

Excretion and Water Balance

Insects and spiders have a unique excretory system consisting of **Malpighian tubules** that operate in conjunction with specialized glands in the wall of the rectum. Malpighian tubules, variable in number, are thin, elastic, blind tubules attached to the juncture between the midgut and hindgut (Figures 20-14 and 20-21A). Free ends of the tubules lie in the hemocoel and are bathed in hemolymph.

The mechanism of urine formation in Malpighian tubules of herbivorous insects appears to depend on active secretion of potassium into the tubules (Figure 20-21B). This primary secretion of ions pulls water along with it by osmosis to produce a potassium-rich fluid. Other solutes and waste materials also are secreted or diffuse into the tubule. The predominant waste product of nitrogen metabolism in most insects is uric acid, which is virtually insoluble in water (see p. 641). Uric acid enters the upper end of tubules, where the pH is slightly alkaline, as relatively soluble potassium and urate (abbreviated KHUr in Figure 20-21). As formative urine passes into the lower end of tubules, potassium combines with carbon dioxide, is reabsorbed as potassium bicarbonate ($KHCO_3$), the pH changes to acidic (pH 6.6), and insoluble uric acid (HUr) precipitates out. As urine drains into the intestine and passes through the hindgut, specialized rectal glands absorb chloride, sodium (and in some cases potassium), and water.

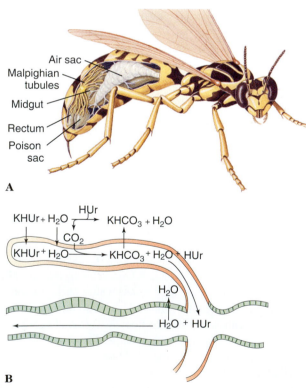

A

B

Figure 20-21

Malpighian tubules of insect. **A,** Malpighian tubules are located at the juncture of the midgut and hindgut (rectum) as shown in the cutaway view of a wasp. **B,** Function of Malpighian tubules. Solutes, especially potassium, are actively secreted into upper tubules. Water and potassium acid urate (KHUr) follow. Potassium is resorbed in the lower tubules, and water and other solutes are resorbed in the rectum.

Labels in figure A: Air sac; Malpighian tubules; Midgut; Rectum; Poison sac

Labels in figure B: $KHUr + H_2O$; HUr; $KHCO_3 + H_2O$; CO_2; $KHUr + H_2O$; $KHCO_3 + H_2O + HUr$; H_2O; $H_2O + HUr$

Since water requirements vary among different types of insects, this ability to cycle water and salts is very important. Insects living in dry environments may resorb nearly all water from the rectum, producing a nearly dry mixture of urine and feces. Freshwater larvae need to excrete water and conserve salts. Insects that feed on dry grains need to conserve water and excrete salt. Leaf-feeding insects ingest and excrete quantities of fluid. For example, aphids and other homopterans (p. 420) pass the excess fluid in the form of a sweetish material called honeydew, which is relished by other insects, especially ants (see Figure 20-35A). Honeydew promotes growth of sooty mold (fungus) on leaves of infested plants and "rains" on cars parked beneath infested trees.

Nervous System

The nervous system in general resembles that of larger crustaceans, with a similar tendency toward fusion of ganglia (Figure 20-14). A number of insects have a giant fiber system. There is also a stomodeal nervous system that corresponds in function to the autonomic nervous system of vertebrates. Neurosecretory cells located in various parts of the brain have an endocrine function, but, except for their role in molting and metamorphosis, little is known of their activity.

Sense Organs

Along with neuromuscular coordination, insects have unusually keen sensory perception. Their sense organs are mostly microscopic and are located chiefly in the body wall. Each type usually responds to a specific stimulus. Various organs are receptive to mechanical, auditory, chemical, visual, and other stimuli.

Mechanoreception   Mechanical stimuli, those involving touch, pressure, and vibration, are picked up by **sensilla.** A sensillum may be simply a seta, or hairlike process, connected with a nerve cell, a nerve ending just under the cuticle and lacking a seta, or a more complex organ (scolopophorous organ) consisting of sensory cells with their endings attached to the body wall. Such organs are widely distributed over the antennae, legs, and body.

Auditory Reception   Very sensitive setae (hair sensilla) or tympanal organs may detect airborne sounds. In tympanal organs a number of sensory cells (ranging from a few to hundreds) extend to a very thin tympanic membrane that encloses an air space in which vibrations can be detected. Tympanal organs occur in certain Orthoptera (Figure 20-4), Homoptera, and Lepidoptera. Some insects are fairly insensitive to airborne sounds but can detect vibrations reaching them through the substrate. Organs on the legs usually detect vibrations of the substrate. Some nocturnal moths can detect ultrasonic pulses emitted by bats for echolocation (p. 596).

Chemoreception   Chemoreceptors (for taste or smell) are usually bundles of sensory cell processes often located in sensory pits. These are often on mouthparts, but in ants, bees, and wasps they are also on the antennae, and butterflies, moths, and flies also have them on the legs. Chemical sense is generally keen, and some insects can detect certain odors for several kilometers. Many patterns of insect behavior such as feeding, mating, habitat selection, and host-parasite relations are mediated through chemical senses. These senses play a crucial role in responses of insects to artificial repellents and attractants. For example, an increase in carbon dioxide concentration, such as would be caused by a potential host nearby, causes a resting mosquito to begin flying, then it follows gradients of warmth and moisture to find its host. Diethyl toluamide, a repellent, blocks the mosquito's ability to sense water vapor, thus preventing host location.

Visual Reception   Insect eyes are of two types, simple and compound. Simple eyes are found in some nymphs and larvae and in many adults. Most insects have three ocelli on their head. Honey bees probably use ocelli to monitor light intensity but not to form images.

Most adult insects have compound eyes, which may cover much of the head. They consist of thousands of ommatidia—6300 in the eye of a honey bee, for example. The structure of the compound eye is similar to that of crustaceans (Figure 20-22). An insect such as a honey bee can see simultaneously in almost all directions around its body, but it is more myopic than humans, and images, even of nearby objects, are fuzzy. However, most flying insects rate much higher than humans in flicker-fusion tests. Flickers of light become fused in human eyes at a frequency of 45 to 55 per second, but bees and blow flies can distinguish as many as 200 to 300 separate flashes of light per second. This is probably an advantage in analyzing a fast-changing landscape during flight.

A bee can distinguish colors, but its sensitivity begins in the ultraviolet range, which human eyes cannot see. Although uniform in color to our perception, bee-pollinated flowers often have petals with zones differing in ultraviolet (UV) light absorption and reflection. The zone of UV absorption acts as a "nectar guide," leading bees to nectar in the flower. Many insects have vision sensitive to red wavelengths, but honey bees are red-blind.

Other Senses   Insects also have well-developed senses for temperature, especially on the antennae and legs, and for humidity, proprioception (sensation of muscle stretch and body position), gravity, and other physical properties.

Neuromuscular Coordination

Insects are active creatures with excellent neuromuscular coordination. Arthropod muscles are typically cross-striated, just as vertebrate skeletal muscles are. A flea can leap a distance of 100 times its own length, and an ant can carry in its jaws a load greater than its own weight. This sounds as though insect muscle were stronger than that of other animals. Actually, however, the force a particular muscle can exert is related directly to its cross-sectional area, not its length. Based on maximum load moved per square centimeter of cross section, the strength of insect muscle is relatively the same as that of vertebrate muscle. The illusion of great strength of insects (and other small animals) is simply a consequence of small body size.

In terms of proportionate body length, a flea's jump would be the equivalent of a 6-foot human executing a standing high jump of 600 feet! Actually, a flea's muscles are not entirely responsible for its jump; they cannot contract rapidly enough to reach the required acceleration. Fleas depend on pads of *resilin*, a protein with unusual elastic properties, which is also found in wing-hinge ligaments of many other insects. Resilin releases 97% of its stored energy on returning from a stretched position, compared with only 85% in most commercial rubber. When a flea prepares to jump, it rotates its hind femurs and compresses the resilin pads, then engages a "catch" mechanism. In effect, it has cocked itself. To take off, the flea needs to exert a relatively small muscular action to unhook the catches, allowing the resilin to expand.

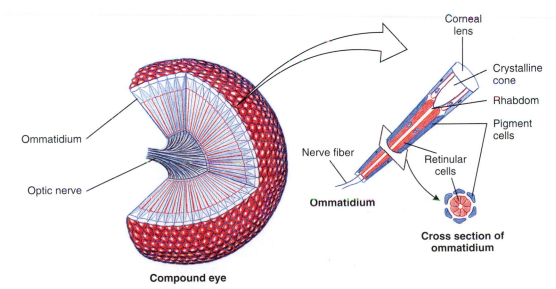

Compound eye

Figure 20-22

Compound eye of an insect. A single ommatidium is shown enlarged to the right.

Reproduction

Sexes are separate in insects, and fertilization is usually internal. Parthenogenesis occurs prominently in Homoptera and Hymenoptera (see insect orders, p. 419). Insects have various means of attracting mates. A female moth releases a powerful pheromone that males can detect for a great distance. Fireflies use flashes of light; some insects find each other by sounds or color signals and by various kinds of courtship behavior.

Males usually deposit sperm in a female's vagina during copulation (Figures 20-14 and 20-23). In some orders sperm are encased in spermatophores that may be transferred at copulation or deposited on the substratum to be picked up by a female. A male silverfish deposits a spermatophore on the ground, then spins signal threads to guide a female to it. During evolutionary transition of ancestral insects from aquatic to terrestrial life, spermatophores were used widely and copulation evolved much later.

Usually sperm are stored in the seminal receptacle of a female in numbers sufficient to fertilize more than one batch of eggs. Many insects mate only once during their lifetime, but male damselflies copulate several times per day.

Insects usually lay a great many eggs. A queen honey bee, for example, may lay more than 1 million eggs during her lifetime. On the other hand, some flies are viviparous and bring forth only a single offspring at a time. Insects that make no provision for care of their young may lay many more eggs than do insects that provide for their young or those that have a very short life cycle.

Most species lay their eggs in a particular habitat to which visual, chemical, or other cues guide them. Butterflies and moths lay their eggs on the specific kind of plant on which the caterpillar must feed. A tiger moth may look for a pigweed, a sphinx moth for a tomato or tobacco plant, and a monarch butterfly for a milkweed plant (Figure 20-24). Insects whose immature stages are aquatic characteristically lay their eggs in water (Figure 20-25). A tiny braconid wasp lays her eggs on the caterpillar of the sphinx moth where they will feed and pupate in tiny white cocoons (Figure 20-18). An ichneumon wasp, with unerring accuracy, seeks out a certain kind of larva in which her young will live as parasitoids. Her long ovipositor may have to penetrate 1 to 2 cm of wood to find a larva of a wood wasp or a wood-boring beetle in which she will deposit her eggs (Figure 20-10).

Metamorphosis and Growth

Early development occurs within eggs, and hatching young escape from eggs in various ways. During postembryonic development most insects change in form, undergoing **metamorphosis** (Figure 20-24). During this period they must undergo a series of molts to grow, and each stage between molts is called an **instar.**

Although metamorphosis occurs in many animals, insects illustrate it more dramatically than any other group. The transformation, for instance, of a hickory horned devil caterpillar into a beautiful royal walnut moth represents an astonishing morphological change. In insects metamorphosis is associated with evolution of wings, which are restricted to the reproductive stage where they are most beneficial.

Figure 20-23

Copulation in insects (see also Figure 20-4B and 20-9B). **A,** *Omura congrua* (order Orthoptera) are a kind of grasshopper found in Brazil. **B,** Bluet damselflies, *Enallagma* sp. (order Odonata), are common throughout North America. Here, a male still grasps a female after copulation as the female (white abdomen) lays eggs.

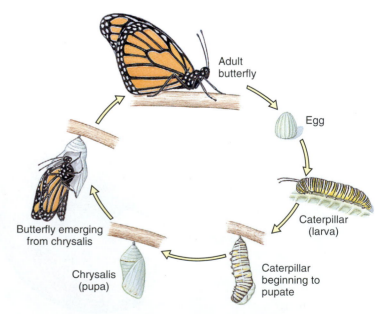

Figure 20-24

Complete (holometabolous) metamorphosis in a butterfly, *Danaus plexippus.* Eggs hatch to produce first of several larval instars. Last larval instar molts to become a pupa. Adult emerges at pupal molt.

Figure 20-25

A, *Mosquito Culex* (order Diptera) lays her eggs in small packets or rafts on the surface of standing or slowly moving water. **B,** Mosquito larvae are the familiar wrigglers of ponds and ditches. To breathe, they hang head down, with respiratory tubes projecting through the surface film of water. Motion of vibratile tufts of fine hairs on the head brings a constant supply of food.

Figure 20-26

A, Ecdysis in a cicada, *Tibicen davisi* (order Homoptera). The old cuticle splits along a dorsal midline as a result of increased blood pressure and of air forced into the thorax by muscle contraction. The emerging insect is pale, and its new cuticle is soft. The wings will be expanded by blood pumped into veins, and the insect enlarges by taking in air. **B,** An adult *Tibicen davisi*.

Holometabolous Metamorphosis

Approximately 88% of insects undergo a **holometabolous** (Gr. *holo*, complete, + *metabolē*, change) metamorphosis, which separates physiological processes of growth (larva) from those of differentiation (pupa) and reproduction (adult) (Figure 20-24). Each stage functions efficiently without competition with the other stages, for larvae often live in entirely different surroundings and eat different foods from adults. The wormlike larvae, which usually have chewing mouthparts, are known as caterpillars, maggots, bagworms, fuzzy worms, or grubs. After a series of instars, a larva molts to enter a transitional stage called a pupa. Pupae are usually inactive and enveloped by a case, which may take several different forms. Pupae are nonfeeding and are the stage in which many insects pass the winter. When the final molt occurs over winter, a full-grown adult emerges, pale and with wings wrinkled. In a short time its wings expand and harden, and the insect is on its way. The stages, then, are egg, larva (several instars), pupa, and adult (Figure 20-24). Adults undergo no further molting.

Hemimetabolous Metamorphosis

Some insects undergo a **hemimetabolous** (Gr. *hemi,* half, + *metabolē*, change), or gradual (incomplete), metamorphosis. These include grasshoppers, cicadas, mantids, and terrestrial bugs, which have terrestrial young, and mayflies, stoneflies, dragonflies, and aquatic bugs that lay their eggs in water and whose young are aquatic. The young are called **nymphs,** and their wings develop externally as budlike outgrowths in early instars and increase in size as the animal grows by successive molts and becomes a winged adult (Figures 20-26 and 20-27). Aquatic nymphs in some orders have tracheal gills or other modifications for aquatic life (Figure 20-28). Stages are egg, nymph (several instars), and adult (Figure 20-27).

The *biological* meaning of the word "bug" is a great deal more restrictive than in common English usage. People often refer to all insects as "bugs," even extending the word to include such non-animals as bacteria, viruses, and glitches in computer programs. Strictly speaking, however, a bug is a member of order Hemiptera and nothing else.

Direct Development

A few insects, such as silverfish and springtails, undergo direct development. The young, or juveniles, are similar to adults except in size and sexual maturation. Stages are egg, juvenile, and adult. Such insects include the primitively wingless insects.

Physiology of Metamorphosis

Hormones regulate metamorphosis in insects. Major endocrine organs that are involved in development are the **brain, prothoracic (ecdysial) glands, corpora cardiaca,** and **corpora allata** (Figure 34-4, p. 725).

Figure 20-34

A, Termite workers, *Reticulitermes flavipes* (order Isoptera), eating yellow pine. Workers are wingless sterile adults and nymphs that tend the nest. B, Termite queen becomes a distended egg-laying machine. The queen and several workers and soldiers are shown here.

A

B

serve for defense of the colony. As in bees and ants, extrinsic factors cause caste differentiation. Reproductive individuals and soldiers secrete inhibiting pheromones that pass throughout the colony to nymphs through a mutual feeding process, called **trophallaxis,** so that they become sterile workers. Workers also produce pheromones, and if the level of "worker substance" or "soldier substance" falls, as might happen after an attack by marauding predators, for example, the next generation produces compensating proportions of the appropriate caste.

Ants also have highly organized societies. Superficially, they resemble termites, but they are quite different (belong to a different order) and can be distinguished easily. In contrast to termites, ants are usually dark in color, are hard bodied, and have a constriction posterior to their first abdominal somite. Antennae of ants are elbowed, while those of termites are threadlike or resemble a string of beads (moniliform).

In ant colonies males die soon after mating and the queen either starts her own new colony or joins some established colony and does the egg laying. Sterile females are wingless workers and soldiers that do work of the colony: gather food, care for young, and protect the colony. In many larger colonies there may be two or three types of individuals within each caste.

Ants have evolved some striking patterns of "economic" behavior, such as making slaves, farming fungi, herding "ant cows" (aphids or other homopterans) (Figure 20-35A), sewing their nests together with silk (Figure 20-35B), and using tools.

INSECTS AND HUMAN WELFARE

Beneficial Insects

Although most of us think of insects primarily as pests, humanity would have great difficulty in surviving if all insects were suddenly to disappear. Some produce useful materials: honey and beeswax from bees, silk from silkworms, and shellac from a wax secreted by lac insects. More importantly, however, insects are necessary for cross-fertilization of many crops. Bees pollinate almost $10 billion worth of food crops per year in the United States alone, and this figure does not include pollination of forage crops for livestock or pollination by other insects.

Very early in their evolution, insects and flowering plants formed a relationship of mutual adaptations that have been to each other's advantage. Insects exploit flowers for food, and flowers exploit insects for pollination. Each floral development

A

B

Figure 20-35

A, An ant (order Hymenoptera) tending a group of aphids (order Homoptera). The aphids feed copiously on plant juices and excrete the excess as a clear liquid rich in carbohydrates ("honeydew"), which is cherished as a food by ants. B, A weaver ant nest in Australia.

A   B   C

Figure 20-36

Some beneficial insects. **A,** A predaceous stink bug (order Hemiptera) feeds on a caterpillar. Note the sucking proboscis of the bug. **B,** A ladybird beetle ("ladybug," order Coleoptera). Adults (and larvae of most species) feed voraciously on plant pests such as mites, aphids, scale insects, and thrips. **C,** A parasitic wasp (*Larra bicolor*) attacking a mole cricket. The wasp drives the cricket from its burrow, then stings and paralyzes it. After the wasp deposits her eggs, the mole cricket recovers and resumes an active life–until it is killed by developing wasp larvae.

of petal and sepal arrangement is correlated with the sensory adjustment of certain pollinating insects. Among these mutual adaptations are amazing devices of allurements, traps, specialized structures, nectar guides (p. 409), and precise timing.

Many predaceous insects, such as tiger beetles, aphid lions, ant lions, praying mantids, and lady bird beetles, destroy harmful insects (Figure 20-36). Parasitoid insects are very important in controlling populations of many harmful insects. Dead animals are quickly consumed by maggots hatched from eggs laid in carcasses.

Insects serve as an important source of food for many birds, fishes and other animals.

Harmful Insects

Harmful insects include those that eat and destroy plants and fruits, such as grasshoppers, chinch bugs, corn borers, boll weevils, grain weevils, San Jose scale, and scores of others (Figure 20-37). Practically every cultivated crop has several insect pests. Humans expend enormous resources in all agricultural activities, in forestry, and in the food industry to counter insects and the damage they engender. Outbreaks of bark beetles or defoliators such as spruce budworms and gypsy moths have generated tremendous economic losses and have become a major element in determining the composition of forests in the United States. Gypsy moths, introduced into the United States in 1869 in an ill-advised attempt to breed a better silkworm, have spread throughout the northeast as far south as Virginia. They defoliate oak forests in years when there are outbreaks. In 1981, they defoliated 13 million acres in 17 northeastern states.

Ten percent of all arthropod species are parasitic insects. Lice, bloodsucking flies, warble flies, bot flies, and many others attack humans or domestic animals or both. Malaria, carried by *Anopheles* mosquitos (Figure 20-38), is still one of the world's major diseases; mosquitos also transmit yellow fever and lymphatic filariasis. Fleas carry plague, which at times in history has wiped out significant portions of human populations. House flies are vectors of typhoid, as are lice for typhus fever; tsetse flies carry African sleeping sickness; and bloodsucking bugs, *Rhodnius* and related genera, transmit Chagas' disease.

A

B

C

Figure 20-37

Some insect pests. **A,** Japanese beetles, *Popillia japonica* (order Coleoptera), are serious pests of fruit trees and ornamental shrubs. They were introduced into the United States from Japan in 1917. **B,** Longtailed mealybug, *Pseudococcus longispinus* (order Homoptera). Many mealybugs are pests of commercially valuable plants. **C,** Corn ear worms, *Heliothis zea* (order Lepidoptera). An even more serious pest of corn is the infamous corn borer, an import from Europe in 1908 or 1909.

Figure 20-38

A mosquito, *Anopheles quadrimaculatus* (order Diptera). *Anopheles* spp. are vectors of malaria.

There is tremendous destruction of food, clothing, and property by weevils, cockroaches, ants, clothes moths, termites, and carpet beetles. Not the least of insect pests are bed bugs, *Cimex,* bloodsucking hemipterous insects that humans contracted from bats that shared their caves early in human evolution.

Control of Insects

Because all insects are an integral part of the ecological communities to which they belong, their total destruction would probably do more harm than good. Food chains would be disturbed, some of our favorite birds would disappear, and the biological cycles by which dead animal and plant matter disintegrates and returns to enrich the soil would be seriously impeded. The beneficial role of insects in our environment is often overlooked, and in our zeal to control the pests we spray the landscape indiscriminately with extremely effective "broad-spectrum" insecticides that eradicate good, as well as harmful, insects. We have also found, to our dismay, that many chemical insecticides persist in the environment and accumulate as residues in the bodies of animals higher in the food chain. Furthermore, many insects have developed a resistance to insecticides in common use. Honey bees are especially susceptible to insecticides, and resistance mainly is developed by harmful insects.

In recent years, methods of control other than chemical insecticides have been under intense investigation, experimentation, and development. Economics, concern for the environment, and consumer demand are causing thousands of farmers across the United States to use alternatives to strict dependence on chemicals.

Several types of biological controls have been developed and are under investigation. All of these areas present problems but show great possibilities. One is the use of bacterial, viral, and fungal pathogens. A bacterium, *Bacillus thuringiensis,* is quite effective in control of lepidopteran pests (cabbage looper, imported cabbage worm, tomato worm, gypsy moth). Other strains of *B. thuringiensis* attack insects in other orders, and the species diversity of target insects is being widened by techniques of genetic engineering. Genes coding for the toxin produced by *B. thuringiensis* also have been introduced into other bacteria and even into the plants themselves, which makes the plants resistant to insect attack.

A number of viruses and fungi that have potential as insecticides have been isolated. Difficulties and expense in rearing

Classification of Subphylum Uniramia

Class Diplopoda (di-plop'o-da) (Gr. *diploos,* double, + *pous, podos,* foot): **millipedes.** Body almost cylindrical; head with short antennae and simple eyes; body with variable number of somites; short legs, usually two pairs of legs to a somite; oviparous. Examples: *Julus, Spirobolus.*
Class Chilopoda (ki-lop'o-da) (Gr. *cheilos,* lip, + *pous, podos,* foot): **centipedes.** Dorsoventrally flattened body; variable number of somites, each with one pair of legs; one pair of long antennae; oviparous. Examples: *Cermatia, Lithobius, Geophilus.*
Class Pauropoda (pau-ro'po-da) (Gr. *pauros,* small, + *pous, podos,* foot): **pauropods.** Minute (1 to 1.5 mm); cylindrical body consisting of double segments and bearing 9 or 10 pairs of legs; no eyes. Example: *Pauropus.*
Class Symphyla (sym'fy-la) (Gr. *syn,* together, + *phylē,* tribe): **garden centipedes.** Slender (1 to 8 mm) with long, filiform antennae; body consisting of 15 to 22 segments with 10 to 12 pairs of legs; no eyes. Example: *Scutigerella.*
Class Insecta (in-sek'ta) (L. *insectus,* cut into): **insects.** Body with distinct head, thorax, and abdomen; pair of antennae; mouthparts modified for different food habits; head of six fused somites; thorax of three somites; abdomen with variable number, usually 11 somites; thorax with two pairs of wings (sometimes one pair or none) and three pairs of jointed legs; usually oviparous; gradual or abrupt metamorphosis.

these agents are being overcome in certain cases, and some are nearing commercial production.

Introduction of natural predators or parasitoids of the insect pests has met with some success. In the United States vedalia beetles from Australia help control cottony-cushion scale on citrus plants, and numerous instances of control by use of insect parasitoids have been recorded.

Another approach to biological control is to interfere with reproduction or behavior of insect pests with sterile males or with naturally occurring organic compounds that act as hormones or pheromones. Such research, although very promising, is slow because of our limited understanding of insect behavior and the problems of isolating and identifying complex compounds that are produced in such minute amounts. Nevertheless, pheromones will probably play an important role in biological pest control in the future.

A systems approach referred to as **integrated pest management** is practiced with many crops. This approach involves integrated utilization of all possible, practical techniques to contain pest infestations at a tolerable level, for example, cultural techniques (resistant plant varieties, crop rotation, tillage techniques, timing of sowing, planting or harvesting, and others), use of biological controls, and sparing use of insecticides.

The sterile-male approach has been used effectively in eradicating screwworm flies, a livestock pest. Large numbers of male insects, sterilized by irradiation, are introduced into the natural population; females that mate with the sterile flies lay infertile eggs.

Classification of Class Insecta

Entomologists do not all agree on names of orders or on the limits of each order. Some choose to combine and others to divide the groups. However, this synopsis of orders is one that is rather widely accepted.

Order Protura (pro-tu′ra) (Gr. *protos,* first, + *oura,* tail). Minute (1 to 1.5 mm); no eyes or antennae; appendages on abdomen as well as thorax; live in soil and dark, humid places; direct development.

Order Diplura (dip-lu′ra) (Gr. *diploos,* double, + *oura,* tail): **japygids** and **campodeids.** Usually less than 10 mm; pale, eyeless; a pair of long terminal filaments or pair of caudal forceps; live in damp humus or rotting logs; development direct.

Order Collembola (col-lem′bo-la) (Gr. *kolla,* glue, + *embolon,* peg, wedge): **springtails, jumping bristletails,** and **snow fleas.** Small (5 mm or less); compound eyes lacking, eye patches of 1 to several lateral ocelli; respiration by trachea or body surface; a springing organ folded under the abdomen for leaping; abundant in soil; sometimes swarm on pond surface film or on snowbanks in spring; development direct.

Order Thysanura (thy-sa-nu′ra) (Gr. *thysanos,* tassel, + *oura,* tail): **silverfish** (Figure 20-39) and **bristletails.** Small to medium size; large eyes; long antennae; three long terminal cerci; live under stones and leaves and around human habitations; development direct.

Order Ephemeroptera (e-fem-er-op′ter-a) (Gr. *ephēmeros,* lasting but a day, + *pteron,* wing): **mayflies** (Figure 20-40). Wings membranous; forewings larger than hindwings; adult mouthparts vestigial; nymphs aquatic, with lateral tracheal gills.

Order Odonata (o-do-na′ta) (Gr. *odontos,* tooth, + *ata,* characterized by): **dragonflies, damselflies** (Figure 20-23B, and 20-28B). Large; membranous wings are long, narrow, net veined, and similar in size; long and slender body; aquatic nymphs with gills and prehensile labium for capture of prey.

Order Orthoptera (or-thop′ter-a) (Gr. *orthos,* straight, + *pteron,* wing): **grasshoppers** (Figure 20-4), **locusts, crickets, cockroaches, walking sticks** (Figure 20-9B), **praying mantids** (Figure 20-7). Wings, when present, with forewings thickened and hindwings folded like a fan under forewings; chewing mouthparts. Many entomologists divide Orthoptera as given here into additional orders, such as Orthoptera (limited to grasshoppers, crickets, and related forms), Blattaria (cockroaches), Mantodea (praying mantids), Phasmida (walking sticks), and Grylloblattaria (rockcrawlers).

Order Dermaptera (der-map′ter-a) (Gr. *derma,* skin, + *pteron,* wing): **earwigs.** Very short forewings; large and membranous hindwings folded under forewings when at rest; chewing mouthparts; forcepslike cerci.

Order Plecoptera (ple-kop′ter-a) (Gr. *plekein,* to twist, + *pteron,* wing): **stoneflies** (Figure 20-28A). Membranous wings; larger and fanlike hindwings; aquatic nymph with tufts of tracheal gills.

Order Isoptera (i-sop′ter-a) (Gr. *isos,* equal, + *pteron,* wing): **termites** (Figure 20-34). Small; membranous, narrow wings similar in size with few veins; wings shed at maturity; erroneously called "white ants"; distinguishable from true ants by broad union of thorax and abdomen; complex social organization.

Order Embioptera (em-bi-op′ter-a) (Gr. *embios,* lively, + *pteron,* wing): **webspinners.** Small; male wings membranous, narrow, and similar in size; wingless females; chewing mouthparts; colonial; make silk-lined channels in tropical soil.

Order Psocoptera (so-cop′ter-a) (Gr. *psoco,* rub away, + *pteron,* wing) **(Corrodentia): psocids, book lice, bark lice.** Body usually small, may be as large as 10 mm; membranous, narrow wings with few veins, usually held rooflike over abdomen when at rest; some wingless species; found in books, bark, bird nests, on foliage.

Order Zoraptera (zo-rap′ter-a) (Gr. *zōros,* pure, + *apterygos,* wingless): **zorapterans.** As large as 2.5 mm; membranous, narrow wings usually shed at maturity; colonial and termitelike.

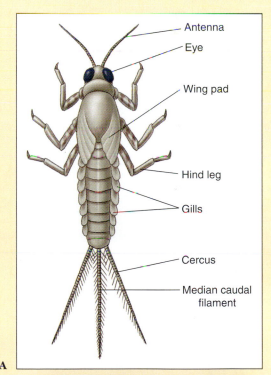

Antenna

Eye

Wing pad

Hind leg

Gills

Cercus

Median caudal filament

A

B

Figure 20-40

Mayfly (order Ephemeroptera). **A,** Nymph. **B,** Adult.

Figure 20-39

Silverfish *Lepisma* (order Thysanura) is often found in homes.

Classification of Class Insecta

Order Mallophaga (mal-lof'a-ga) (Gr. *mallos,* wool, + *phagein,* to eat): **chewing lice** (Figure 20-16). As large as 6 mm; wingless; chewing mouthparts; legs adapted for clinging to host; live on birds and mammals.

Order Anoplura (an-o-plu'ra) (Gr. *anoplos,* unarmed, + *oura,* tail): **sucking lice** (Figure 20-17). Depressed body; as large as 6 mm; wingless; mouthparts for piercing and sucking; adapted for clinging to warm-blooded host; includes head lice, body lice, crab lice, others.

Order Thysanoptera (thy-sa-nop'ter-a) (Gr. *thysanos,* tassel, + *pteron,* wing): **thrips.** Length 0.5 to 5 mm (a few longer); wings, if present, long, very narrow, with few veins, and fringed with long hairs; sucking mouthparts; destructive plant-eaters, but some feed on insects.

Order Hemiptera (he-mip'ter-a) (Gr. *hemi,* half, + *pteron,* wing) **(Heteroptera): true bugs.** Size 2 to 100 mm; wings present or absent; forewings with basal portion leathery, apical portion membranous; hindwings membranous; at rest, wings held flat over abdomen; piercing-sucking mouthparts; many with odorous scent glands; includes water scorpions, water striders (Figure 20-11), bed bugs, squash bugs, assassin bugs, chinch bugs, stink bugs, plant bugs, lace bugs, others.

Order Homoptera (ho-mop'ter-a) (Gr. *homos,* same, + *pteron,* wing): **cicadas** (Figure 20-26), **aphids** (Figure 20-35A), **scale insects, leafhoppers, treehoppers** (Figure 20-30 and 20-41). (Often included as suborder under Hemiptera.) If winged, either

Figure 20-41
Oak treehoppers, *Platycotis vittata* (order Homoptera).

Figure 20-42
Adult ant lion (order Neuroptera).

membranous or thickened front wings and membranous hindwings; wings held rooflike over body; piercing-sucking mouthparts; all plant-eaters; some destructive; a few serving as source of shellac, dyes, and so on; some with complex life histories.

Order Neuroptera (neu-rop'ter-a) (Gr. *neuron,* nerve, + *pteron,* wing): **dobsonflies, ant lions** (Figure 20-42), **lacewings.** Medium to large size; similar, membranous wings with many cross veins; chewing mouthparts; dobsonflies with greatly enlarged mandibles in males, and with aquatic larvae; ant lion larvae (doodlebugs) make craters in sand to trap ants.

Order Coleoptera (ko-le-op'ter-a) (Gr. *koleos,* sheath, + *pteron,* wing): **beetles** (Figure 20-9A, 20-31, 20-37A), including **fireflies** (Figure 20-32), **weevils** (Figure 20-43). The largest order of animals in the world; front wings (elytra) thick, hard, opaque; membranous hindwings folded under front wings at rest; mouthparts for biting and chewing; includes ground beetles, carrion beetles, whirligig beetles, darkling beetles, stag beetles, dung beetles, diving beetles, boll weevils, others.

Order Strepsiptera (strep-sip'ter-a) (Gr. *strepsis,* a turning, + *pteron,* wing): **stylops** or **twisted wing parasites.** Females with no wings, eyes, or antennae; males with vestigial forewings and fan-shaped hindwings; females and larvae parasitic in bees, wasps, and other insects.

Order Mecoptera (me-kop'ter-a) (Gr. *mekos,* length, + *pteron,* wing): **scorpionflies.** Small to medium size; wings long, slender,

PHYLOGENY AND ADAPTIVE RADIATION

Insect fossils, although not abundant, have been found in numbers sufficient to give a general idea of the evolutionary history of insects. Although several groups of marine arthropods, such as trilobites, crustaceans, and xiphosurans, were present in the Cambrian period, the first terrestrial arthropods—scorpions and millipedes—did not appear until the Silurian period. The first insects, which were wingless, date from the Devonian period. By the Carboniferous period, several orders of winged insects, most of which are now extinct, had appeared.

Opinions of arthropodologists on relationships among the animals composing Uniramia are diverse. Many scientists are dis-

inclined to use the name at all because it was originally proposed to include Onychophora (p. 435). We are using the term as a convenience, but we suggest it should be accepted with suitable caution. Figure 20-44 is only one hypothesis. Some scientists believe that Myriapoda is paraphyletic and that the Diplopoda/Pauropoda clade is the sister group of insects. A report based on mitochondrial genome sequences supports a sister group relationship between Myriapoda and Chelicerata, with the myriapod/chelicerate clade forming the sister group of Mandibulata (insects and crustaceans).* On the other hand,

---

*Hwang, U. W., M. Friedrich, D. Tautz, C. J. Park, and W. Kim. 2001. Mitochondrial protein phylogeny joins myriapods with chelicerates. Nature 413:154–157.

Classification of Class Insecta

with many veins; at rest, wings held rooflike over back; scorpion-like male clasping organ at end of abdomen; carnivorous; live in most woodlands.

Order Lepidoptera (lep-i-dop'ter-a) (Gr. *lepidos,* scale, + *pteron,* wing): **butterflies and moths.** Membranous wings covered with overlapping scales, wings coupled at base; mouthparts a sucking tube, coiled when not in use; larvae (caterpillars) with chewing mandibles for plant eating, stubby prolegs on abdomen, and silk glands for spinning cocoons; antennae knobbed in butterflies and usually plumed in moths (Figure 20-43).

Order Diptera (dip'ter-a) (Gr. *dis,* two, + *pteron,* wing): **true flies** (Figure 20-12). Single pair of wings, membranous and narrow; hindwings reduced to inconspicuous balancers (halteres); sucking mouthparts or adapted for sponging, lapping, or piercing; legless larvae (maggots); includes crane flies, mosquitos, moth flies, midges, fruit flies, flesh flies, house flies, horse flies, bot flies, blow flies, and many others.

Order Trichoptera (tri-kop'ter-a) (Gr. *trichos,* hair, + *pteron,* wing): **caddisflies.** Small, soft bodies; wings well veined and hairy, folded rooflike over hairy body; chewing mouthparts, mandibles much reduced; aquatic larvae of many species construct cases of leaves, sand, gravel, bits of shell, or plant matter, bound together with secreted silk or cement; some make silk feeding nets attached to rocks in stream.

Order Siphonaptera (si-fon-ap'ter-a) (Gr. *siphon,* a siphon, + *apteros,* wingless); **fleas** (Figure 20-15). Small; wingless; bodies laterally compressed; legs adapted for leaping; ectoparasitic on birds and mammals; larvae legless and scavengers.

Order Hymenoptera (hi-men-op'ter-a) (Gr. *hymen,* membrane, + *pteron,* wing): **ants, bees, wasps** (Figure 20-43). Very small to large; membranous, narrow wings coupled distally; subordinate hindwings; mouthparts for chewing and lapping up liquids; ovipositor sometimes modified into stinger, piercer, or saw (Figure 20-10); both social and solitary species, most larvae legless, blind, and maggotlike.

A

C

B

D

Figure 20-43

A, *Papilio krishna* (order Lepidoptera) is a beautiful swallowtail butterfly from India. Members of the Papilionidae grace many areas of the world, both tropical and temperate, including North America. Compare knobbed antennae with plumed antennae in **B,** *Rothschildia jacobaea,* a saturniid moth from Brazil. *Hyalophora cecropia* is a common saturniid in North America. **C,** Paper wasps (order Hymenoptera) attending their pupae and larvae. **D,** *Curculio proboscideus,* the chestnut weevil, is a member of the largest family (Curculionidae) of the largest insect order (Coleoptera). This family includes many serious agricultural pests.

another analysis, based on a comprehensive data set, including sequences of nuclear and mitochondrial genes, and morphological, developmental, ultrastructural, and gene-order characters, supports a clade consisting of mandibulates (myriapods, insects, and crustaceans) with chelicerates as their sister group.[†] This report further suggests that pycnogonids are the sister group of all other arthropods.

Whether or not some or all myriapods are closely related to insects, it is likely that ancestral insects had a head and trunk of many similar somites, most or all of which bore limbs. Early fossil insects had small abdominal appendages (and some multiramous appendages, Figure 20-45), and some modern, apterygote (wingless) orders have abdominal styli that are considered vestigial legs. We now understand that absence of abdominal legs in most insects results from modulation of expression of certain *Hox* genes that prevents expression of the *Distal-less* gene in the abdomen of insects but not crustaceans or onychophorans (p. 394).[‡§]

Apterygotes have traditionally been regarded as having the most primitive characteristics, but subclass Apterygota is apparently paraphyletic (Figure 20-46). Three apterygote orders

[†]Giribet, G., G. D. Edgecombe, and W. C. Wheeler. 2001. Arthropod phylogeny based on eight molecular loci and morphology. Nature **413:**157–161.

[‡]Galant, R., and S. B. Carroll. 2002. Evolution of a transcriptional repression domain in an insect Hox protein. Nature **415:**910–913.
[§]Ronshaugen, M., N. McGinnis, and W. McGinnis. 2002. Hox protein mutation and macroevolution of the insect body plan. Nature **415:**914–917.

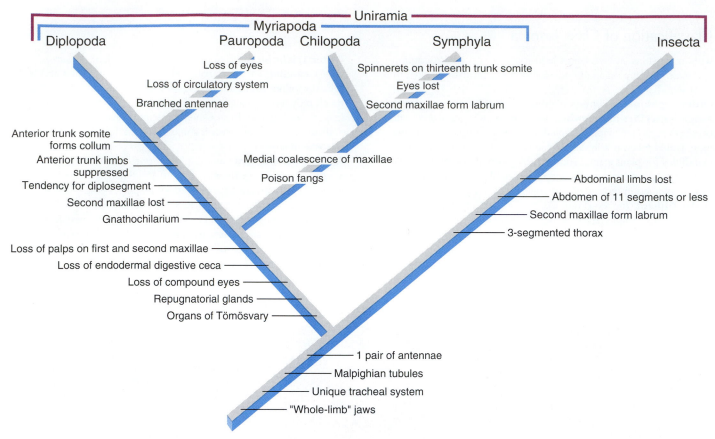

Figure 20-44

Cladogram showing hypothetical relationships of uniramians. Here myriapods and insects are sister groups; therefore Diplopoda, Pauropoda, Chilopoda, and Symphyla would become subclasses under class Myriapoda. Another hypothesis would make Diplopoda and Pauropoda the sister group of insects. Organs of Tömösvary are unique sensory organs opening at the bases of the antennae, and repugnatorial glands, located on certain somites or legs, secrete an obnoxious substance for defense. The gnathochilarium is formed in diplopods and pauropods by fusion of the first maxillae, and the collum is the collarlike tergite of the first trunk segment. Formation of a labrum from the second maxillae has been sometimes considered evidence of sister-group relationship of symphylans and insects; it is viewed here as convergence. Outgroups for this cladogram would be non-uniramian arthropod lineages.

Source: Modified from R. C. Brusca and G. J. Brusca, Invertebrates, *Sinauer Associates, Inc., Sunderland, MA, 1990.*

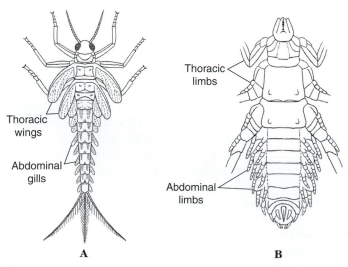

Figure 20-45

A, A Paleozoic mayfly nymph with thoracic and abdominal gills. The thoracic gills could have been precursors of wings. **B,** A Paleozic bug with multiramous thoracic legs and vestigal, multiramous abdominal limbs.

(Diplura, Collembola, Protura) have their mandibles and first maxillae located deeply in pouches in the head, a condition known as **endognathy.** They share other primitive and derived characters, and there are many similarities between endognathous insects and myriapods. All other insects are **ectognathous,** including the wingless order Thysanura (Figure 20-39). Endognathous and ectognathous insects form sister groups.

The evolutionary origin of insect wings has long been a puzzle. The adaptive value of wings for flight is clear, but such structures did not spring into existence fully developed. One hypothesis was that wings developed from lateral thoracic expansions that were useful in gliding. However, this hypothesis did not explain an origin or function of articulations and neuromusculature in the protowings that would provide raw material for selection and eventual evolution of flapping wings to support flight. An alternative hypothesis is that ancestral flying insects were derived from aquatic insects or insects with aquatic juveniles that bore external gills on their thorax from which wings could have been derived. Thoracic and abdominal gills on Paleozoic insects were apparently articulated and movable,

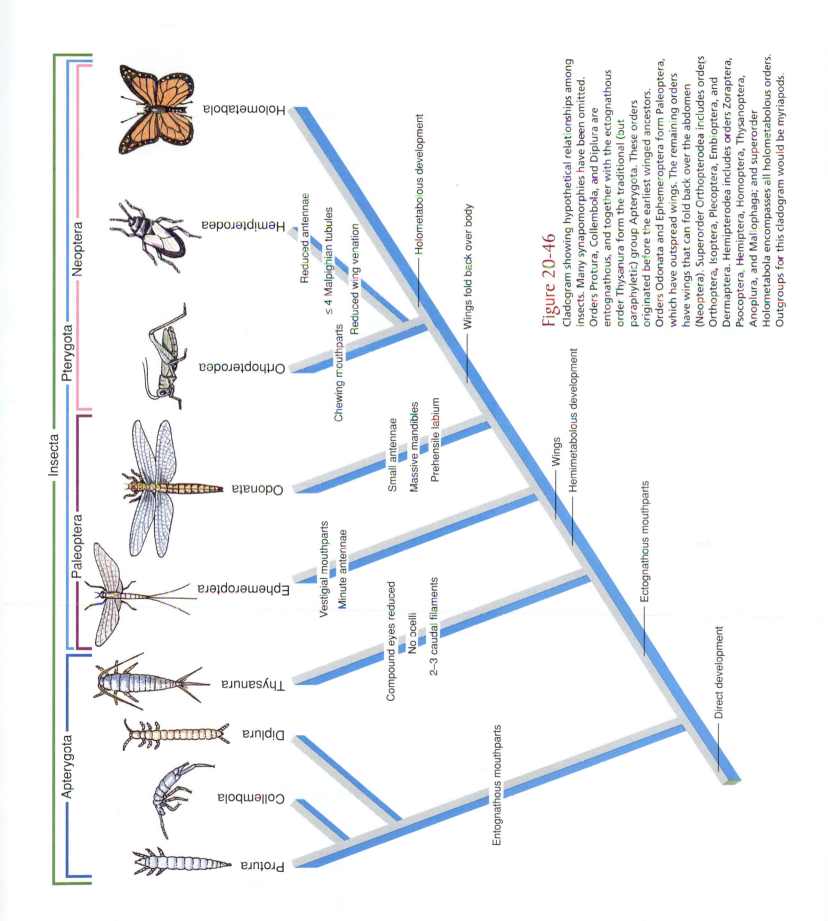

Figure 20-46

Cladogram showing hypothetical relationships among insects. Many synapomorphies have been omitted. Orders Protura, Collembola, and Diplura are entognathous, and together with the ectognathous order Thysanura form the traditional (but paraphyletic) group Apterygota. These orders originated before the earliest winged ancestors. Orders Odonata and Ephemeroptera form Paleoptera, which have outspread wings. The remaining orders have wings that can fold back over the abdomen (Neoptera). Superorder Orthopterodea includes orders Orthoptera, Isoptera, Plecoptera, Embioptera, and Dermaptera. Hemipterodea includes orders Zoraptera, Psocoptera, Hemiptera, Homoptera, Thysanoptera, Anoplura, and Mallophaga; and superorder Holometabola encompasses all holometabolous orders. Outgroups for this cladogram would be myriapods.

capable of ventilation and swimming movements. They would have been excellent precursor structures for protowings. A reasonable intermediate stage is illustrated by stoneflies (Figure 20-28). Stonefly nymphs are aquatic, and adults are at best poor flyers. Upon emergence, legs of adults are supported by surface tension of the water, and they skim rapidly across the surface propelled by aerodynamic forces of their flapping wings. Even with small protowings, ancestral insects could skim faster than they could swim, and natural selection would have resulted in longer wings, stronger muscles, and better neural coordination.

The ancestral winged insect gave rise to three lines, which differed in their ability to flex their wings. Two of these (Odonata and Ephemeroptera) have outspread wings or hold their wings vertically over the abdomen. The other line has wings that can fold back horizontally over the abdomen. It branched into three groups, all of which were present by the Permian period. One group with hemimetabolous metamorphosis, chewing mouthparts, and cerci includes Orthoptera, Dermaptera, Isoptera, and Embioptera; another group with hemimetabolous metamorphosis and a tendency toward sucking mouthparts includes Thysanoptera, Hemiptera, and Homoptera and perhaps also Psocoptera, Zoraptera, Mallophaga, and Anoplura, although there is some disagreement among authorities about the last group. Insects with holometabolous metamorphosis have the most specialized life history, and these apparently form a clade.

Adaptive properties of insects have been stressed throughout this chapter. The directions and ranges of their adaptive radiation, both structurally and physiologically, have been amazingly varied. Whether it be in the area of habitat, feeding adaptations, means of locomotion, reproduction, or general mode of living, the adaptive achievements of insects are truly remarkable.

SUMMARY

Members of subphylum Uniramia have uniramous appendages and bear one pair of antennae, a pair of mandibles, and two pairs of maxillae (one pair of maxillae in millipedes) on the head. Tagmata are head and trunk in myriapods and head, thorax, and abdomen in insects.

Insecta is the largest class of the world's largest phylum. Insects are easily recognized by the combination of their tagmata and possession of three pairs of thoracic legs.

The evolutionary success of insects is largely explained by several features allowing them to exploit terrestrial habitats, such as waterproofing their cuticle and other mechanisms to minimize water loss and the ability to become dormant during adverse conditions.

Most insects bear two pairs of wings on their thorax, although some have one pair and some are wingless. Wing movements in some insects are controlled by direct flight muscles, which insert directly on the base of the wings in the thorax, whereas others have indirect flight muscles, which move the wings by changing the shape of the thorax. Each contraction of synchronous flight muscles requires a nerve impulse, while asynchronous flight muscles contract many times for each nerve impulse.

Feeding habits vary greatly among insects, and there is an enormous variety of specialization of mouthparts reflecting the particular feeding habits of a given insect. They breathe by means of a tracheal system, which is a system of tubes that opens by spiracles on the thorax and abdomen. Excretory organs are Malpighian tubules.

Sexes are separate in insects, and fertilization is usually internal. Almost all insects undergo metamorphosis during development. In hemimetabolous (gradual) metamorphosis, larval instars are called nymphs, and adults emerge at the last nymphal molt. In holometabolous (complete) metamorphosis, the last larval molt gives rise to a nonfeeding stage (pupa). A winged adult emerges at the final, pupal, molt. Both types of metamorphosis are hormonally controlled.

Insects are important to human welfare, particularly because they pollinate food and forage crop plants, control populations of other, harmful insects by predation and parasitism, and serve as food for other animals. Many insects are harmful to human interests because they inflict great damage on crops, food, forests, clothing, and property, and many are carriers of important diseases affecting humans and domestic animals.

Although some authorities believe that myriapods form a clade with chelicerates, the bulk of evidence supports a mandibulate clade containing myriapods, insects and crustaceans. Legs have been lost from the abdomen of insects, being now confined to each of the three thoracic somites. Endognathous and ectognathous insects are sister groups. Wings of ancestral winged insects may have been derived from external gills of aquatic nymphs or adults, and there may have been a stage in which wings were used for skimming across the surface of water, rather than for flight.

Adaptive diversity and the numbers of both species and individuals in Insecta are enormous.

REVIEW QUESTIONS

1. Distinguish the following from each other: Diplopoda, Chilopoda, Insecta.
2. What characteristics of insects distinguish them from *all* other arthropods?
3. Explain why indirect flight muscles can beat much more rapidly than direct flight muscles.
4. How do insects walk?
5. What are parts of an insect's gut, and what are functions of each?
6. Describe three different types of mouthparts found in insects, and tell how they are adapted for feeding on different foods.
7. Describe the tracheal system of a typical insect and explain why it is able to function efficiently without oxygen-carrying pigments in the hemolymph. Why would a tracheal system not be suitable for humans?

8. Describe the unique excretory system of insects. How is uric acid formed?

9. Describe sensory receptors that insects have to various stimuli.

10. Explain the difference between holometabolous and hemi-metabolous metamorphosis in insects, including stages of each.

11. Describe hormonal control of metamorphosis in insects, including the action of each hormone and where each is produced.

12. What is diapause, and what is its adaptive value?

13. Briefly describe three features that insects have evolved to avoid predation.

14. Describe and give an example of each of four ways insects can communicate with each other.

15. What are castes found in honey bees and in termites, and what is the function of each?

16. What are mechanisms of caste determination in honey bees and termites?

17. What is trophallaxis? What function(s) does it serve in termites?

18. Name several ways in which insects are beneficial to humans and several ways they are detrimental.

19. What are ways in which detrimental insects can be controlled? What is integrated pest management?

20. What are the most probable characteristics of the most recent common ancestor of insects? What major lineages descended from it?

21. What is a plausible scenario for evolution of wings and flying insects?

SELECTED REFERENCES

Beckage, N. E. 1997. The parasitic wasp's secret weapon. Sci. Am. **277**:82–87 (Nov.). *This parasitoid wasp carries a virus that invades the host insect when the wasp lays its eggs, then paralyzes the host.*

Bennet-Clark, H. C. 1998. How cicadas make their noise. Sci. Am. **278**:58–61 (May). *Male cicadas are the loudest known insects.*

Berenbaum, M. R. 1995. Bugs in the system. Reading, Massachusetts, Addison-Wesley Publishing Company. *How insects impact human affairs. Well written for a wide audience, highly recommended.*

Blum, M. S. (ed). 1985. Fundamentals of insect physiology. New York, John Wiley & Sons. *Good, multi-authored text on insect physiology. Recommended.*

Downs, A. M. R., K. A. Stafford, and G. C. Coles. 1999. Head lice: Prevalence in schoolchildren and insecticide resistance. Parasitol. Today **15**:1–4. *This report is mainly concerned with England, but head lice are one of the most common parasites of American schoolchildren.*

Hayashi, A. M. 1999. Attack of the fire ants. Sci. Am. **280**:26, 28 (Feb.). *Fire ants have invaded Galápagos, Melanesia, and West Africa, where they may be blinding elephants and otherwise disrupting the ecosystem.*

Heinrich, B., and H. Esch. 1994. Thermoregulation in bees. Am. Sci. **82**:164–170. *Fascinating behavioral and physiological adaptations for increasing and decreasing body temperature allow bees to function in a surprisingly wide range of environmental temperatures.*

Hölldobler, B. H., and E. O. Wilson. 1990. The ants. Cambridge, Massachusetts, Harvard University Press. *The fascinating story of social organization in ants.*

Hubbell, S. 1997. Trouble with honey bees. Nat. Hist. **106(4)**:32–43. *Infection of honey bees with* Varroa *mites is a big problem for beekeepers.*

Levine, M. 2002. How insects lose their limbs. Nature **415**:848–849. *A Hox gene product in the abdomen of insects inhibits action of another gene product that is necessary for limb formation.*

Marden, J. H. 1995. Flying lessons from a flightless insect. Nat. Hist. **104(2)**:4–8. *The earliest winged insects may have used their wings for skimming on the surface of water, rather than flying, just as do modern stoneflies.*

McMasters, J. H. 1989. The flight of the bumble bee and related myths of entomological engineering. Am. Sci. **77**:164–169. *There is a popular myth about an aerodynamicist who "proved" that a bumble bee cannot fly—but his assumptions were wildly wrong.*

O'Brochta, D. A., and P. W. Atkinson. 1998. Building the better bug. Sci. Am. **279**:90–95 (Dec.). *Inserting new genes into certain insect species could render them incapable of being vectors for disease, help agriculture, and have other applications.*

Raff, R. A. 1996. The shape of life: genes, development, and the evolution of animal form. Chicago, University of Chicago Press. *Includes a good account of how insect wings may have evolved.*

Romoser, W. S., and J. G. Stoffolano, Jr. 1998. The science of entomology, ed. 4. Dubuque, Iowa, McGraw-Hill. *A good entomology text.*

Tallamy, D. W. 1999. Child care among the insects. Sci. Am. **280**:72–77 (Jan.). *Most insects render no assistance to eggs or young, but females, and in a few cases, males, of a variety of species, care for and protect eggs and young.*

Topoff, H. 1990. Slave-making ants. Am. Sci. **78**:520–528. *An amazing type of social parasitism in which certain species of ants raid the colonies of related species, abduct their pupae, then exploit them to do all the work in the host colony.*

ZOOLOGY LINKS TO THE INTERNET

Visit the textbook's Online Learning Center at www.mhhe.com/zoology to find live Internet links for each of the topics listed here.

Phylum Arthropoda
Subphylum Uniramia
Minor Classes of Uniramids
Class Insecta

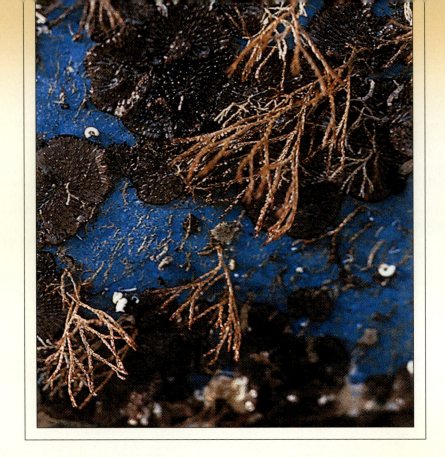

Ectoprocts and other animals fouling a boat bottom.

Smaller Protostome Phyla

LOPHOTROCHOZOAN PHYLA
Sipuncula
Echiura
Pogonophora
Phoronida
Ectoprocta
Brachiopoda

ECDYSOZOAN PHYLA
Pentastomida
Onychophora
Tardigrada
Chaetognatha

Some Evolutionary Experiments

During the Cambrian period, about 535 to 530 million years ago, a most fertile time occurred in evolutionary history. For over 3 billion years before this time, evolution had forged little more than prokaryotes and unicellular eukaryotes. Then, within the space of a few million years, all of the major phyla of macroscopic invertebrates, and probably all of the smaller phyla, became established. This was the Cambrian explosion, the greatest evolutionary "bang" the world has known. In fact, the fossil record suggests that more phyla existed in the Paleozoic Era than exist now, but some disappeared during major extinction events that punctuated the evolution of life on earth. Greatest of these disruptions was the Permian extinction about 230 million years ago. Thus evolution has led to many "experimental models." Some of these models failed because they were unable to survive in changing conditions. Others gave rise to abundant and dominant species and individuals that inhabit the world today. Still others produced a small number of species, some of which persist, while others were formerly more abundant but are now in decline.

The great evolutionary flow that began with the appearance of a coelom and led to the three huge phyla of molluscs, annelids, and arthropods produced other lines as well. Most of those that have survived are small and lack great economic and ecological importance. Because they are all coelomate protostomes, we will consider them in a single chapter.

Position in the Animal Kingdom

The phyla discussed in this chapter are all coelomate protostomes, although some also have some deuterostome characteristics in their embryological development. Their relationship to each other and to major protostome phyla is controversial. We mention hypothetical phylogenetic relationships later in this chapter.

Figure 21-1
Themiste,
a sipunculan.

This chapter includes a brief discussion of ten coelomate phyla whose position in the phylogeny of the animal kingdom has long been problematic. Sipuncula, Echiura, and Pogonophora have some annelid-like characters, and molecular evidence now supports their placement with annelids in superphylum Lophotrochozoa. Ectoprocta, Phoronida, and Brachiopoda, grouped together by virtue of having a structure known as a lophophore (p. 431), likewise are positioned in Lophotrochozoa, despite their having been considered deuterostomes based on developmental and morphological criteria. Pentastomida, Onychophora, and Tardigrada show some arthropod-like characters; they often have been grouped together and called Pararthropoda because they have unjointed limbs with claws (at some stage) and a cuticle that undergoes molting. Molecular and other evidence now supports placement of these phyla near Arthropoda in Ecdysozoa. Chaetognatha is another phylum formerly positioned in Deuterostomia, but DNA sequence analysis indicates placement of chaetognaths in Ecdysozoa near nematodes.

LOPHOTROCHOZOAN PHYLA

Phylum Sipuncula

Phylum Sipuncula (sigh-pun′kyu-la) (L. *sipunculus,* little siphon) consists of benthic marine worms, predominantly littoral or sublittoral. They live sedentary lives in burrows in mud or sand, occupy borrowed snail shells, or live in coral crevices or among vegetation. Some species construct their own rock burrows by chemical and perhaps mechanical means. More than half the species are restricted to tropical zones. Some are tiny, slender worms, but the majority range from 15 to 30 cm in length. Some are commonly known as "peanut worms" because, when disturbed, they can contract to a peanut shape (Figure 21-1).

Sipunculans have no segmentation or setae. They are most easily recognized by a slender retractile **introvert,** or **proboscis,** which is continually and rapidly being run in and out of the anterior end. Walls of the **trunk** are muscular. When the introvert is everted, the mouth can be seen at its tip surrounded by a crown of ciliated tentacles. Undisturbed sipunculans usually extend the anterior end from their burrow or hiding place and stretch out their tentacles to explore and feed. They are largely deposit feeders living on organic matter collected in mucus on the tentacles and moved to the mouth by ciliary action. The introvert is extended by hydrostatic pressure produced by contraction of body-wall muscles against the coelomic fluid. The lumen of the hollow tentacles is not connected to the coelom but rather to one or two blind, tubular compensation sacs that lie along their esophagus (Figure 21-2). These sacs receive fluid from the tenta-

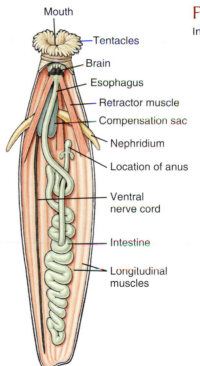

Figure 21-2
Internal structure of *Sipunculus.*

cles when the introvert is retracted. Retraction is effected by special retractor muscles. The surface of the introvert is often rough because of surface spines, hooks, or papillae.

There is a large, fluid-filled coelom traversed by muscle and connective tissue fibers. Their digestive tract is a long tube that doubles back on itself to end in an anus near the base of the introvert (Figure 21-2). A pair of large nephridia opens to the outside to expel waste-filled coelomic amebocytes; the nephridia also serve as gonoducts. Circulatory and respiratory systems are lacking, but coelomic fluid contains red corpuscles that have a respiratory pigment, hemerythrin, used in transportation of oxygen. Their nervous system has a bilobed cerebral ganglion just behind the tentacles and a ventral nerve cord extending the length of the body. Sexes are separate. Permanent gonads are lacking, and ovaries or testes develop seasonally in the connective tissue covering the origins of one or more of the retractor muscles. Sex cells are released through the nephridia. The larval form is usually a trochophore. Asexual reproduction also occurs by transverse fission, the posterior one-fifth of the parent constricting off to become a new individual.

There are approximately 330 species and 16 genera, which are placed by some authorities into four families. The best-known genera are probably *Sipunculus, Phascolosoma* (Gr. *phaskōlos,* leather bag, pouch, + *sōma,* body), *Aspidosiphon* (Gr. *aspidos,* shield, + *siphōn,* siphon), and *Golfingia* (named by E. R. Lankester in remembrance of an afternoon of golfing at St. Andrews, Scotland).

Phylum Echiura

Phylum Echiura (ek-ee-yur'a) (Gr. *echis,* viper, serpent, + *oura* tail, + *ida,* pl. suffix) consists of marine worms that burrow into mud or sand, live in empty snail shells or sand-dollar tests, or rocky crevices. They are found in all oceans—most commonly in littoral zones of warm waters—but some are found in polar waters or dredged from depths of 2000 m. They vary in length from a few millimeters to 40 or 50 cm.

Echiurans have only about one-third as many species (140) as sipunculans. There are two classes: Echiurida and Sactosomatida. Echiurida is much larger and includes two orders and five families.

Echiurans are cylindrical and somewhat sausage shaped (Figure 21-3). Anterior to the mouth is a flattened, extensible proboscis which, unlike that of sipunculids, cannot be retracted into the trunk. Echiurids are often called "spoonworms" because of the shape of the contracted proboscis in some species. Their proboscis, which contains the brain, is actually a cephalic lobe,

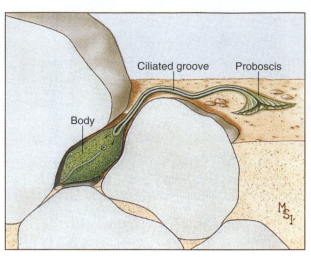

Figure 21-4

Bonellia (phylum Echiura) is a detritus feeder. Lying in its burrow, it explores the surface with its long proboscis, which picks up organic particles and carries them along a ciliated groove to the mouth.

probably homologous to an annelid's prostomium. The proboscis has a ciliated groove leading to the mouth. While they lie buried, the proboscis can extend out over the mud for exploration and deposit feeding (Figure 21-4). *Bonellia viridis* picks up very small particles and moves them along its proboscis by cilia; larger particles are moved by a combination of cilia and muscular action or by muscular action alone. Unwanted particles can be rejected along the route to the mouth. The proboscis is short in some forms and long in others. *Bonellia,* which is only 8 cm long, can extend its proboscis up to 2 m.

In some species sexual dimorphism is pronounced, with the female being much the larger of the two. *Bonellia* has an extreme sexual dimorphism, and tiny males live on the body of the female or in her nephridia. Determination of sex in *Bonellia* is very interesting. Free-swimming larvae are sexually undifferentiated. Those that settle on the proboscis of a female become males (1 to 3 mm long). About 20 males are usually found in a single female. Larvae that do not contact a female proboscis metamorphose into females. The stimulus for development into males is apparently a hormone produced by a female's proboscis.

One common form, *Urechis* (Gr. *oura,* tail, + *echis,* viper, serpent), lives in a U-shaped burrow in which it secretes a funnel-shaped mucous net. It pumps water through the net, capturing bacteria and fine particulate material in it. *Urechis* periodically swallows the food-laden net. *Lissomyema* (Gr. *lissos,* smooth, + *mys,* muscle) lives in empty gastropod shells in which it constructs galleries irrigated by rhythmical pumping of water and feeds on sand and mud drawn in by this process.

Cuticle and epithelium, which may be smooth or ornamented with papillae (Figure 21-3), cover the muscular body wall. There may be a pair of anterior setae or a row of bristles around the posterior end. The coelom is large. The digestive tract is long and

Figure 21-3

Echiurus, an echiurian common on both Atlantic and Pacific coasts of North America. The shape of their proboscis lends them the common name of "spoon worms."

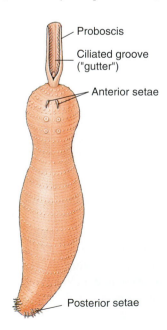

Proboscis

Ciliated groove ("gutter")

Anterior setae

Posterior setae

coiled and terminates at the posterior end (Figure 21-5). A pair of anal sacs may have an excretory and osmoregulatory function. Most echiurans have a closed circulatory system with colorless blood but contain hemoglobin in coelomic corpuscles and certain body cells. Two to many nephridia serve mainly as gonoducts. A nerve ring runs around the pharynx and forward into the proboscis, and there is a ventral nerve cord. There are no specialized sense organs.

Sexes are separate, with a single gonad in each sex. Mature sex cells break loose from gonads and leave the body cavity by way of the nephridia, and fertilization is usually external.

Early cleavage and trochophore stages are very similar to those of annelids and sipunculans. The trochophore stage, which may last from a few days to 3 months, according to species, is followed by gradual metamorphosis to a wormlike adult.

Phylum Pogonophora

Phylum Pogonophora (po'go-nof'e-ra) (Gr. *pōgōn,* beard, + *phora,* bearing), or beardworms, was entirely unknown before the twentieth century. The first specimens to be described were collected from deep-sea dredgings off the coast of Indonesia in 1900. They have since been discovered in several seas, including the western Atlantic off the U.S. eastern coast. Some 145 species have been described so far. We recognize two classes, Perviata and Vestimentifera, but some authorities consider Vestimentifera a separate phylum. The usual length of perviatans is from 5 to 85 cm, with a diameter usually of less than a millimeter. Vestimentiferans, however, live around deepwater hydrothermal vents and grow much larger: up to 1.5 m long and 5 cm in diameter (Figure 21-6).

> Among the most amazing animals found in deep-water, Pacific rift communities (Chapter 38, p. 801) are giant pogonophorans, *Riftia pachyptila,* much larger than any other pogonophores reported. The trophosome of other pogonophores is confined to the posterior part of the trunk, which is buried in sulfide-rich sediments, but the trophosome of *Riftia* occupies most of its large trunk. It has a much larger supply of hydrogen sulfide, enough to nourish its large body, in the effluent of the hydrothermal vents.

These elongated tube-dwelling forms have left no known fossil record. Their closest affinity seems to be to annelids.

Most pogonophores live in ooze on the ocean floor, always below the intertidal zone and usually at depths of more than 200 m. This location accounts for their delayed discovery, for they are obtained only by dredging. They are sessile animals that secrete very long chitinous tubes in which they live, probably extending the anterior end of their body only for absorbing nutrients. The tubes are generally oriented upright in the bottom ooze. A tube is usually about the same length as the animal, which can move up or down inside its tube but cannot turn around.

Beardworms have a long, cylindrical body covered with cuticle. The body is divided into a short anterior **forepart;** a long, very slender **trunk;** and a small, segmented **opisthosoma** (Fig-

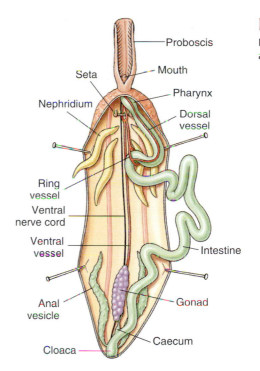

Figure 21-5
Internal anatomy of an echiuran.

Proboscis
Seta
Mouth
Nephridium
Pharynx
Dorsal vessel
Ring vessel
Ventral nerve cord
Ventral vessel
Intestine
Anal vesicle
Gonad
Cloaca
Caecum

Figure 21-6
A colony of giant beardworms (phylum Pogonophora) at great depth near a hydrothermal vent along the Galápagos Trench, eastern Pacific Ocean.

ure 21-7). At its anterior, a cephalic lobe bears from 1 to 260 long tentacles (the "beard" that gives this phylum its name), depending on species. Tentacles are hollow extensions of the coelom and bear minute pinnules. For a part or all of their length, tentacles lie parallel with each other, enclosing a cylindrical intertentacular space into which the pinnules project (Figure 21-8).

The long trunk bears papillae and, about midway back, two rings of short-toothed setae called **girdles,** which grip the tube wall, allowing the two halves of the body to contract or extend independently in the tube. Posterior to the girdles, the trunk is very thin and easily broken when animals are collected. In fact, the segmented tail end, or opisthosoma, was not found and described until after 1963! It is thicker than the trunk and has 5 to 23 short segments that bear setae.

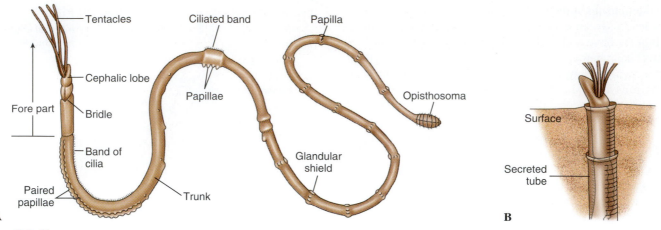

A   **B**

Figure 21-7

Diagram of a typical pogonophoran. **A,** External features. In life, the body is much more elongated than shown in this diagram. **B,** Position in tube.

Figure 21-8

Cross section of tentacular crown of pogonophore *Lamellisabella.* Tentacles arise from ventral side of forepart at base of cephalic lobe. Tentacles (which vary in number in different species) enclose a cylindrical space, with the pinnules forming a kind of food-catching network. Food molecules may be absorbed into the blood supply of tentacles and pinnules.

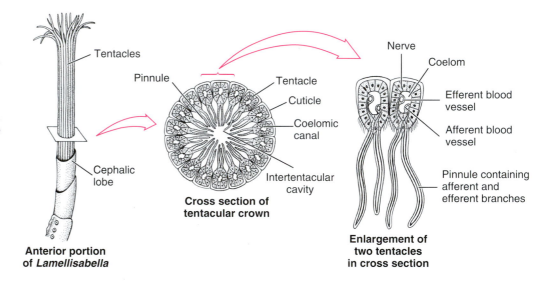

Anterior portion of *Lamellisabella*

Cross section of tentacular crown

Enlargement of two tentacles in cross section

Cuticle, epidermis, and circular and longitudinal muscles compose the body wall. Their cuticle is similar in structure to that of annelids and sipunculans.

Pogonophores are remarkable in having no mouth or digestive tract, making their mode of nutrition a puzzling matter. They absorb some nutrients dissolved in seawater, such as glucose, amino acids, and fatty acids, through the pinnules and microvilli of their tentacles. Most of their energy, however, apparently is derived from a mutualistic association with chemoautotrophic bacteria. These bacteria oxidize hydrogen sulfide to provide energy to produce organic compounds from carbon dioxide. Pogonophores bear the bacteria in an organ called a **trophosome,** which is derived embryonically from the midgut (all traces of foregut and hindgut are absent in adults).

Sexes are separate, with a pair of gonads and a pair of gonoducts in the trunk section. Cleavage is unequal and atypical. It seems to be closer to radial than to spiral. Development of the apparent coelom is schizocoelic, not enterocoelic as was originally described. The worm-shaped embryo is ciliated but a poor swimmer. It is probably swept along by water currents until it settles.

Pogonophore tubes were originally thought to resemble those of hemichordate pterobranchs, but analysis of their amino acid and chitin content shows no relationship to pterobranchs. Pogonophores have photoreceptor cells very similar to those of annelids (oligochaetes and leeches), and structure of the cuticle, makeup of the setae, and segmentation of the opisthosoma all strongly suggest a relationship with the annelids.

Adaptive radiation has not been extensive. The chief areas of diversity are in the structure of the tentacular crown and the tube.

Lophophorates

Phoronida are wormlike marine forms that live in secreted tubes in sand or mud or attached to rocks or shells. Ectoprocta are minute forms, mostly colonial, whose protective cases often form encrusting masses on rocks, shells, or plants. Brachiopoda are bottom-dwelling marine forms that superficially resemble molluscs because of their bivalved shells.

One might wonder why these three apparently different types of animals are lumped together in a group called lophophorates. Actually they have more in common than first appears. They are all coelomate; they have some deuterostome and some protostome characteristics; and none has a distinct head. But other phyla share these characteristics. What really sets this group apart from other phyla is the common possession of a ciliary feeding device called a **lophophore** (Gr. *lophos,* crest or tuft, + *phorein,* to bear).

A lophophore is a unique arrangement of ciliated tentacles borne on a ridge (a fold of the body wall), which surrounds the mouth but not the anus. The lophophore with its crown of tentacles contains within it an extension of the coelom, and the thin, ciliated walls of the tentacles are not only an efficient feeding device but also serve as a respiratory surface for exchange of gases between environmental water and coelomic fluid. A lophophore can usually be extended for feeding or withdrawn for protection.

In addition, all three phyla have a U-shaped alimentary canal, with their anus placed near their mouth but outside the lophophore. Their coelom is primitively divided into three compartments, **protocoel, mesocoel** and **metacoel,** and the mesocoel extends into the hollow tentacles of the lophophore. Their protocoel, where present, forms a cavity in a flap over their mouth, called an **epistome.** The portion of the body containing their mesocoel is known as a **mesosome,** and that containing their metacoel is a **metasome.**

Phylum Phoronida

Phylum Phoronida (fo-ron′i-da) (L. *Phoronis,* in mythology, surname of Io, who was turned into a white heifer) contains approximately 10 species of small, wormlike animals that live on the bottom of shallow coastal waters, especially in temperate seas. They range from a few millimeters to 30 cm in length. Each worm secretes a leathery or chitinous tube in which it lies free, but which it never leaves. The tubes may be anchored singly or in a tangled mass on rocks, shells, or pilings or buried in sand. They thrust out the tentacles on the lophophore for feeding, but a disturbed animal can withdraw completely into its tube.

A lophophore has two parallel ridges curved in a horseshoe shape, the bend located ventrally and the mouth lying between the two ridges (Figure 21-9). Horns of the ridges often coil into twin spirals. Each ridge carries hollow ciliated tentacles, which, like the ridges themselves, are extensions of the body wall.

Cilia on the tentacles direct a water current toward a groove between the two ridges, which leads toward the mouth. Plankton and detritus caught in this current become entangled in mucus and are carried by cilia to the mouth. The anus lies dorsal to the mouth, outside the lophophore, flanked on each side by a nephridiopore (Figure 21-9). Water leaving the lophophore passes over the anus and nephridiopores, carrying away wastes. Cilia in the stomach area of the U-shaped gut aid in food movement.

The body wall consists of cuticle, epidermis, and both longitudinal and circular muscles. The protocoel is present as a small cavity in the epistome; it connects to the mesocoel along the lateral aspects of the epistome. A septum separates the metacoel from the mesocoel. Phoronids have a closed system of contractile blood vessels but no heart; their blood contains hemoglobin within nucleated cells. There is a pair of metanephridia. A nerve

ring sends nerves to tentacles and body wall; a single giant motor fiber lies in the epidermis; and an epidermal nerve plexus supplies the body wall and epidermis.

There are both monoecious (the majority) and dioecious species of Phoronida, and at least one species reproduces asexually. Cleavage is radial, and development is regulative. Coelom formation is by a highly modified enterocoelous route, but the blastopore becomes the mouth. A free-swimming, ciliated larva, called an actinotroch, metamorphoses into an adult, which sinks to the bottom, secretes a tube, and becomes sessile.

Phoronopsis californica is a large, orange form about 30 cm long found along the United States west coast. *Phoronis architecta* is a smaller (approximately 12 cm long) Atlantic coast species that has a very wide distribution.

Phylum Ectoprocta (Bryozoa)

Ectoprocta (ek-to-prok′ta) (Gr. *ektos,* outside, + *proktos,* anus) have long been called bryozoans, or moss animals (Gr. *bryon,* moss, + *zōon,* animal), a term that originally included Entoprocta also. However, because entoprocts are pseudocoelomates and have their anus located within the tentacular crown, they are commonly separated from ectoprocts, which, like other lophophorates, are eucoelomate and have their anus outside the circle of tentacles. Some authors continue to use the name "Bryozoa" but exclude entoprocts from the group.

Of the 4000 or so species of ectoprocts, few are more than 0.5 mm long. All are aquatic, both freshwater and marine, but

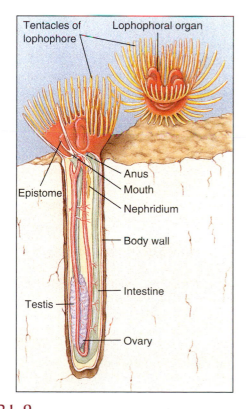

Figure 21-9

Internal structure of *Phoronis* (phylum Phoronida), in diagrammatic vertical section.

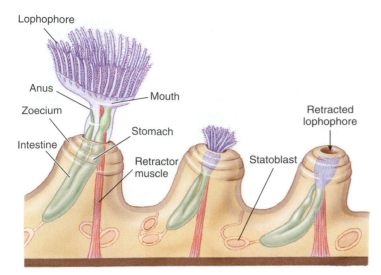

Lophophore

Anus — Mouth

Zoecium

Stomach

Intestine

Retractor muscle

Statoblast

Retracted lophophore

Figure 21-10

Small portion of freshwater colony of *Plumatella* (phylum Ectoprocta), which grows on the underside of rocks. These tiny individuals disappear into their chitinous zoecia when disturbed. Statoblasts are resistant capsules containing germinative cells.

A

B

Figure 21-11

Colonies of marine ectoprocts. **A,** The zooids are extended in this lacy colony of *Triphyllozoon* sp. **B,** *Reteporella graffei* has upright, branching colonies.

are found largely in shallow waters. With very few exceptions they are colony builders. Ectoprocts have become diverse and abundant. They have left a rich fossil record since the Ordovician period. Marine forms today exploit all kinds of firm surfaces, such as shells, rocks, large brown algae, mangrove roots, and ship bottoms.

Each member of a colony lives in a tiny chamber, called a **zoecium,** which is secreted by its epidermis (Figure 21-10). Each individual, or **zooid,** consists of a feeding polypide and a case-forming cystid. A **polypide** includes the lophophore, digestive tract, muscles, and nerve centers. A **cystid** includes the body wall of an animal, together with its secreted exoskeleton. The exoskeleton, or zoecium, may, according to species, be gelatinous, chitinous, or stiffened with calcium and possibly also impregnated with sand. Its shape may be boxlike, vaselike, oval, or tubular.

Some colonies form limy encrustations on seaweed, shells, and rocks; others form fuzzy or shrubby growths on erect, branching colonies that look like seaweed (Figure 21-11). Some ectoprocts might easily be mistaken for hydroids but can be distinguished under a microscope by presence of an anus (Figure 21-10). In some freshwater forms individuals are borne on finely branching stolons that form delicate tracings on the underside of rocks or plants. Other freshwater ectoprocts are embedded in large masses of gelatinous material. Although zooids are minute, the colonies may be several centimeters in diameter, some encrusting colonies may be a meter or more in width (Figure 21-12), and erect forms may reach 30 cm or more in height. Freshwater ectoprocts may form mosslike colonies on stems of plants or on rocks, usually in shallow ponds or pools. They may be able to slide along slowly on the object that supports them.

Polypides live a type of jack-in-the-box existence, popping up to feed and then quickly withdrawing into their little chamber, which often has a tiny trapdoor (operculum) that shuts to conceal its inhabitant. To extend its tentacular crown, certain muscles contract, which increases hydrostatic pressure within the body cavity and pushes the lophophore out by a hydraulic mechanism. Other muscles can contract to withdraw the crown to safety with great speed.

The lophophore ridge tends to be circular in marine ectoprocts (Figure 21-13A) and U-shaped in freshwater species (Figure 21-13B). When feeding, an animal extends its lophophore and spreads its tentacles out into a funnel. Cilia on the tentacles draw water into the funnel and out between tentacles. Food particles caught by cilia in the funnel are drawn into the mouth, both by a pumping action of the muscular pharynx and by action of cilia in the pharynx. Undesirable particles can be rejected by reversing the ciliary action, by drawing the tentacles close together, or by retracting the whole lophophore into the zoecium. Digestion in the ciliated, U-shaped digestive tract appears to be extracellular for protein and starches and intracellular for fats.

Respiratory, vascular, and excretory organs are absent. Gaseous exchange is through the body surface, and since ectoprocts are small, coelomic fluid is adequate for internal transport. Coelomocytes engulf and store waste materials. A ganglionic mass and a nerve ring surround the pharynx, but no sense organs are present. A septum divides the mesocoel in the lophophore from the larger posterior metacoel. A protocoel and

Figure 21-12

Skeletal remains of a colony of *Membranipora*, a marine encrusting form of Ectoprocta. Each little oblong zoecium is the calcareous former home of a tiny ectoproct.

Figure 21-14

Statoblast of a freshwater ectoproct *Cristatella*. This statoblast is about 1 mm in diameter and bears hooked spines.

A

B

Figure 21-13

A, Ciliated lophophore of *Electra pilosa,* a marine ectoproct. **B,** *Plumatella repens,* a freshwater bryozoan (phylum Ectoprocta). It grows on the underside of rocks and vegetation in lakes, ponds, and streams.

Most ectoprocts are hermaphroditic. Some species shed eggs into seawater, but most brood their eggs, some within the coelom and some externally in a special ovicell, which is a modified zoecium in which an embryo develops. Cleavage is radial but apparently mosaic. Little is known of mesoderm derivation. Larvae of nonbrooding species have a functional gut and swim about for a few months before settling; larvae of brooding species do not feed and settle after a brief free-swimming existence. They attach to the substratum by mucopolysaccharide and protein secretions from an **adhesive sac,** then metamorphose to their adult form.

Ectoprocts reproduce asexually by budding and form colonies. Fresh-water ectoprocts have another type of budding that produces **statoblasts** (Figure 21-14), which are hard, resistant capsules containing a mass of germinative cells. Statoblasts are formed during summer and fall. When a colony dies in late autumn, statoblasts remain, and in spring they give rise to new polypides and eventually to new colonies.

Phylum Brachiopoda

Brachiopoda (brak-i-op′o-da) (Gr. *brachiōn,* arm, + *pous, podos,* foot), or lamp shells, are an ancient group. Although about 325 species are now living, some 12,000 fossil species, which once flourished in Paleozoic and Mesozoic seas, have been described. Modern forms have changed little from early ones. Genus *Lingula* (L. tongue) (Figure 21-15A) is probably the most ancient of these "living fossils," having existed virtually unchanged since Ordovician times. Most modern brachiopod shells range between 5 to 80 mm in length, but some fossil forms reached 30 cm.

Brachiopods are attached, bottom-dwelling, marine forms that mostly prefer shallow water. Externally brachiopods resemble bivalved molluscs in having two calcareous shell valves secreted by a mantle. They were, in fact, classified with molluscs until the middle of the nineteenth century, and their name refers to the arms of the **lophophore,** which were thought homologous to the mollusc foot. Brachiopods, however, have dorsal and ventral valves instead of right and left lateral valves as do bivalve molluscs and, unlike bivalves, most of them are attached to a substrate either directly or by means of a fleshy stalk called a **pedicel** (or pedicle). Some, such as *Lingula,* live in vertical burrows in sand or mud. Muscles open and close the valves and provide movement for the stalk and tentacles.

epistome occur only in freshwater ectoprocts. Pores in the walls between adjoining zooids permit exchange of materials by way of the coelomic fluid.

Most colonies contain only feeding individuals, but polymorphism occurs in some species. One type of modified zooid resembles a bird beak that snaps at small invading organisms that might foul a colony. Another type has a long bristle that sweeps away foreign particles.

Figure 21-15

Brachiopods. **A,** *Lingula,* an inarticulate brachiopod that normally occupies a burrow. The contractile pedicel can withdraw the body into the burrow. **B,** An articulate brachiopod, *Terebratella.* The valves have a tooth-and-socket articulation, and a short pedicel projects through the pedicel valve to attach to the substratum.

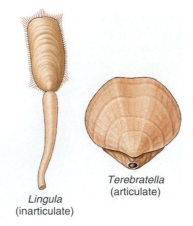

Terebratella
(articulate)

Lingula
(inarticulate)

In most brachiopods the ventral (pedicel) valve is slightly larger than the dorsal (brachial) valve, and one end projects in the form of a short, pointed beak that is perforated where the fleshy stalk passes through the shell to attach to the substratum (Figure 21-15B). In many the pedicel valve is shaped like a classic oil lamp of ancient Greece and Rome, so that brachiopods came to be known as "lamp shells."

There are two classes of brachiopods, based on shell structure. Shell valves of Articulata have a connecting hinge with an interlocking tooth-and-socket arrangement, as in *Terebratella* (L. *terebratus,* a boring, + *ella,* dim. suffix); those of Inarticulata lack the hinge and are held together by muscles only, as in *Lingula* and *Glottidia* (Gr. *glōttidos,* mouth of windpipe).

Their body occupies only the posterior part of the space between the valves (Figure 21-16), and extensions of the body wall form mantle lobes that line and secrete the shell. Their large horseshoe-shaped lophophore in the anterior mantle cavity bears long, ciliated tentacles used in respiration and feeding. Ciliary water currents carry food particles between the gaping valves and over the lophophore. Tentacles catch food particles, and ciliated grooves carry the particles along the arm of the lophophore to their mouth. Rejection tracts carry unwanted particles to the mantle lobe where they are swept out in ciliary cur-

rents. Organic detritus and some algae are apparently primary food sources. A brachiopod's lophophore not only can create food currents, as do other lophophorates, but also seems able to absorb dissolved nutrients directly from environmental seawater.

There is no cavity in the epistome of articulates, but in inarticulates there is a protocoel in the epistome that opens into the mesocoel. As in other lophophorates, the posterior metacoel bears the viscera. One or two pairs of nephridia open into the coelom and empty into the mantle cavity. Coelomocytes, which ingest particulate wastes, are expelled by nephridia. There is an open circulatory system with a contractile heart. Lophophore and mantle are probably the chief sites of gaseous exchange. There is a nerve ring with a small dorsal and a larger ventral ganglion.

Sexes are separate, and paired gonads discharge gametes through the nephridia. Most fertilization is external, but a few species brood their eggs and young.

Cleavage is radial, and coelom and mesoderm formation in at least some brachiopods is enterocoelic. The blastopore closes, but its relationship to the mouth is uncertain. In articulates, metamorphosis of larvae occurs after they have attached by a pedicel. In inarticulates, juveniles resemble a minute brachiopod with a coiled pedicel in the mantle cavity. There is no metamorphosis. As a larva settles, its pedicel attaches to the substratum, and adult existence begins.

ECDYSOZOAN PHYLA

Phylum Pentastomida

Pentastomida (pen-ta-stom'i-da) (Gr. *pente,* five, + *stoma,* mouth), or tongue worms, are a phylum of about 90 species of wormlike parasites of the respiratory system of vertebrates. Adults live mostly in lungs of reptiles, such as snakes, lizards, and crocodiles, but one species, *Reighardia sternae,* lives in air sacs of terns and gulls, and another, *Linguatula serrata* (Gr. *lingua,* tongue), lives in the nasopharynx of canines and felines (and occasionally humans). Although more common in tropical areas, they also occur in North America, Europe, and Australia.

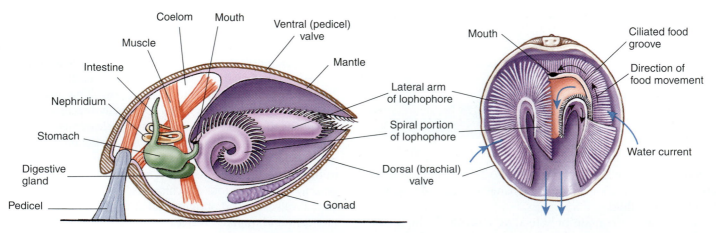

Figure 21-16

Phylum Brachiopoda. **A,** An articulate brachiopod (longitudinal section). **B,** Feeding and respiratory currents. Large arrows show water flow over lophophore; small arrows indicate food movement toward mouth in ciliated food groove. Note that its pedicel emerges from the ventral valve, so that when it is attached to a substrate, an articulate brachiopod is "upside down," with its ventral valve on top and its dorsal valve below.

Adults range from 1 to 13 cm in length. Transverse rings give their bodies a segmented appearance (Figure 21-17). Their body is covered with a chitinous cuticle that is molted periodically during larval stages. The anterior end may bear five short protuberances (hence the name Pentastomida). Four of these bear claws, and the fifth bears the mouth (Figure 21-18). There is a simple straight digestive system, adapted for sucking. The nervous system, similar to that of annelids and arthropods, has paired ganglia along the ventral nerve cord. The only sense organs appear to be papillae. There are no circulatory, excretory, or respiratory organs.

Sexes are separate, and females are usually larger than males. A female may produce several million eggs, which pass up the trachea of the host, are swallowed, and pass out with feces. Larvae hatch as oval, tailed creatures with four stumpy legs. Most pentastomid life cycles require an intermediate vertebrate host such as a fish, a reptile, or, rarely, a mammal, that is eaten by the definitive vertebrate host. After ingestion by an intermediate host, larvae penetrate the intestine, migrate randomly in the body, and finally metamorphose into nymphs. After growth and several molts, a nymph finally becomes encapsulated and dormant. When eaten by a final host, a juvenile finds its way to a lung, feeds on blood and tissue, and matures.

Several species have been found encysted in humans, the most common being *Armillifer armillatus* (L. *armilla,* ring, bracelet, + *fero,* to bear), but usually they cause few symptoms. *Linguatula serrata* is a cause of nasopharyngeal pentastomiasis, or "halzoun," a disease of humans in the Middle East and India.

Phylum Onychophora

Members of phylum Onychophora (on-y-kof'o-ra) (Gr. *onyx,* claw, + *pherein,* to bear) are commonly called "velvet worms," or "walking worms." They compose approximately 70 species of caterpillar-like animals, ranging from 1.4 to 15 cm in length. They live in rain forests and other moist, leafy habitats in tropical and subtropical regions and in some temperate regions of the Southern Hemisphere.

Their fossil record shows that they have changed little in their 500-million-year history. A fossil form, *Aysheaia,* discovered in the Burgess shale deposit of British Columbia and dating back to mid-Cambrian times, is very much like modern onychophorans (Figure 6-9, p. 108). Onychophorans were probably far more common at one time than they are now. Today they are terrestrial and extremely retiring, becoming active only at night or when the air is nearly saturated with moisture.

Form and Function

External Features   Onychophorans are more or less cylindrical and show no external segmentation except for the paired appendages (Figure 21-19). The skin is soft, velvety, and covered with a thin, flexible cuticle that contains protein and chitin. In structure and chemical composition it resembles arthropod cuticle; however, it never hardens like arthropod cuticle, and it is molted in patches rather than all at one time. The body is studded with tiny **tubercles,** some of which bear sensory bristles. The color may be green, blue, orange, dark gray, or black, and minute scales on the tubercles give the body an iridescent and velvety appearance. The head bears a pair of large **antennae,** each with an annelid-like eye at the base (Figure 21-19). The ventral mouth has a pair of clawlike **mandibles** and is flanked by a pair of **oral papillae** which can expel a defensive secretion.

Their **unjointed legs** are short, stubby, and clawed. Onychophorans crawl by passing waves of contraction from anterior to posterior. When a segment extends, the legs lift up and move forward. The legs are more ventrally located than are parapodia of annelids.

Internal Features   The body wall is muscular like that of annelids. The body cavity is a **hemocoel,** imperfectly divided into compartments, or sinuses, much like those of arthropods. **Slime glands** on each side of the body cavity open on the oral papillae. When disturbed by a predator, the animal can eject

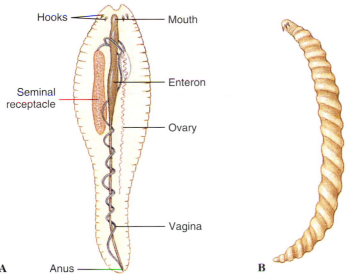

Figure 21-17

Two pentastomids. **A,** *Linguatula,* found in nasal passages of carnivorous mammals. Female is shown with some internal structures. **B,** Female *Armillifer,* a pentastomid with pronounced body rings. In parts of Africa and Asia, humans are parasitized by immature stages; adults (10 cm long or more) live in lungs of snakes. Human infection may occur from eating snakes or from contaminated food or water.

Figure 21-18

Anterior end of a pentastome. Note both the mouth (*arrow*), between the middle hooks, and the apical sensory papillae.

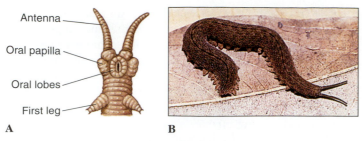

Figure 21-19

Peripatus, a caterpillar-like onychophoran that has characteristics in common with both annelids and arthropods. **A,** Ventral view of head. **B,** In natural habitat.

from the slime glands two streams of a sticky substance that rapidly hardens.

The mouth, surrounded by lobes of skin, contains a dorsal tooth and a pair of lateral mandibles for grasping and cutting prey. There is a muscular pharynx and a straight digestive tract (Figure 21-20). Most velvet worms are predaceous, feeding on caterpillars, insects, snails, and worms. Some onychophorans live in termite nests and feed on termites.

Each segment contains a pair of **nephridia,** each nephridium with a vesicle, ciliated funnel and duct, and nephridiopore opening at the base of a leg. Absorptive cells in the midgut excrete crystalline uric acid, and certain pericardial cells function as nephrocytes, storing excretory products taken from the blood.

For respiration there is a **tracheal system** that ramifies to all parts of the body and communicates with the outside by many openings, or **spiracles,** scattered all over the body. Onychophorans cannot close their spiracles to prevent water loss, so although the tracheae are efficient, these animals are restricted to moist habitats. Their tracheal system is somewhat different from that of arthropods and probably originated independently.

The open circulatory system has, in the pericardial sinus, a dorsal, tubular heart with a pair of ostia in each segment.

There are a pair of cerebral ganglia with connectives and a pair of widely separated nerve cords with connecting commissures. Nerves to antennae and head region extend from the brain, and the nerve cords supply nerves to legs and body wall. Sense organs include pigment cup ocelli, taste spines around the mouth, tactile papillae on the integument, and hygroscopic receptors that orient the animal toward water vapor.

Onychophorans are dioecious, with paired reproductive organs. Males usually deposit their sperm in spermatophores in a female's seminal receptacle. A male deposits the spermatophores on a female's back, which may accumulate a number of them. White blood cells dissolve the skin beneath the spermatophores. Sperm can then enter the body cavity and migrate in the blood to the ovaries to fertilize eggs. Onychophorans may be oviparous, ovoviviparous, or viviparous. Only two Australian genera are oviparous, laying shell-covered eggs in moist places. In all other onychophorans eggs develop in the uterus, and living young are produced. In some species there is a placental attachment between mother and young (viviparous); in others young develop in the uterus without attachment (ovoviviparous).

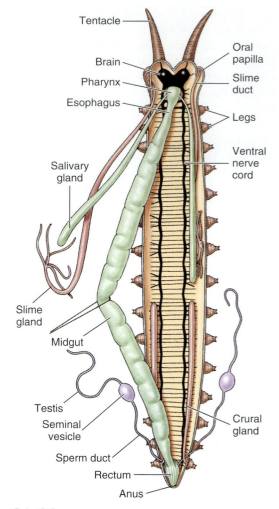

Figure 21-20

Internal anatomy of an onychophoran.

Phylum Tardigrada

Tardigrada (tar-di-gray′da) (L. *tardus,* slow, + *gradus,* step), or "water bears," are minute organisms usually less than a millimeter in length. Most of the 300 to 400 species are terrestrial forms that live in the water film surrounding mosses and lichens. Some live in freshwater algae or mosses or in bottom debris, and a few are marine, inhabiting interstitial spaces between sand grains, in both deep and shallow seawater. They share many characteristics with arthropods.

They have an elongated, cylindrical, or a long oval body that is unsegmented. The head is merely the anterior part of the trunk. The trunk bears four pairs of short, stubby, unjointed legs, each armed with four to eight claws (Figure 21-21). They are covered by a nonchitinous cuticle that is molted along with claws and buccal apparatus four or more times in the life history. Cilia are absent. Common American genera are *Macrobiotus* (Gr. *makros,* large, + *biotos,* life), *Echiniscus* (Gr. *echinos,* hedgehog, + *iskos,* dim. suffix), and *Hypsibius* (Gr. *hypsos,* high, height, + *bios,* life).

The mouth of tardigrades opens into a buccal tube that empties into a muscular pharynx adapted for sucking (Figure 21-22). Two needlelike stylets flanking the buccal tube can be protruded

Figure 21-21
Scanning electron micrograph of an aquatic tardigrade, *Pseudobiotus.*

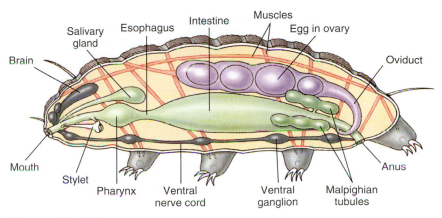

Figure 21-22
Internal anatomy of a tardigrade.

through the mouth. These stylets pierce cellulose walls of plant cells, and the pharynx sucks in the liquid contents. Some tardigrades suck body juices of nematodes, rotifers, and other small animals. Some, such as *Echiniscus,* expel feces when molting, leaving the feces in the discarded cuticle. At the junction of intestine and rectum, three glands, thought to be excretory and often called Malpighian tubules, empty into the digestive system.

Most of the body cavity is a hemocoel, with their true coelom restricted to the gonadal cavity. There are no circulatory or respiratory systems, gaseous exchange occurring through the body surface.

Their muscular system consists of a number of long muscle bands. Circular muscles are absent, but hydrostatic pressure of the body fluid may act as a skeleton. Being unable to swim, water bears creep about with apparent awkwardness, clinging to the substrate with their claws.

Their brain is relatively large and covers most of the dorsal surface of the pharynx. Circumpharyngeal connectives link it to the subpharyngeal ganglion, from which the double ventral nerve cord extends posteriorly as a chain of four ganglia.

Sexes are separate in tardigrades. In some freshwater and moss-dwelling species, males are unknown and parthenogenesis seems to be the rule. In marine species, however, males and females occur with approximately equal frequency. Eggs of some species are highly ornate (Figure 21-23). Egg laying, like defecation, apparently occurs only at molting, when the volume of coelomic fluid is reduced. Females of some species deposit eggs in the molted cuticle (Figure 21-24). Males gather around the old cuticle and shed sperm into it. Fertilization in other species is internal but only at the time of molting.

Cleavage is holoblastic but atypical, and a stereogastrula is formed. Six pairs of coelomic pouches arise from the gut, but all except the last pair disaggregate to form the buccal apparatus, pharynx, and body musculature. The last pair fuses to form the gonad. Thus the gonocoel (which is enterocoelic) is the only true coelom left in adults. Development is direct.

One of the most intriguing features of terrestrial tardigrades is their capacity to enter a state of suspended animation, called cryptobiosis, during which metabolism is virtually imperceptible; such an organism can withstand harsh environmental conditions.

Figure 21-23
Scanning electron micrograph of a highly ornate egg of a tardigrade, *Macrobiotus hufelandii.*

Figure 21-24
Molted cuticle of a tardigrade, containing a number of fertilized eggs.

Under gradual drying conditions, water content of the body decreases from 85% to only 3%, movement ceases, and the body becomes barrel shaped. In a cryptobiotic state tardigrades can resist temperature extremes, ionizing radiation, oxygen deficiency, and other adverse conditions and may survive for years. Activity resumes when moisture is again available. Some nematodes and rotifers can also undergo cryptobiosis.

Phylum Chaetognatha

A common name for chaetognaths is arrowworms. They are all marine animals and are highly specialized for their planktonic existence. Their relationship to other groups is obscure, although embryological characters indicate deuterostome affinities. Therefore, they traditionally have been included among deuterostomes, but this arrangement is not supported by molecular evidence.

The name Chaetognatha (ke-tog'na-tha) (Gr. *chaitē,* long flowing hair, + *gnathos,* jaw) refers to the sickle-shaped bristles on each side of their mouth. This is not a large group, for only some 65 species are known. Their small, straight bodies resemble miniature torpedoes, or darts, ranging from 2.5 to 10 cm in length.

Arrowworms are all adapted for a planktonic existence, except for *Spadella* (Gr. *spadix,* palm frond, + *ella,* dim. suffix), a benthic genus. They usually swim to the surface at night and descend during daytime. Much of the time they drift passively, but they can dart forward in swift spurts, using their caudal fin and longitudinal muscles—a fact that no doubt contributes to their success as planktonic predators. Horizontal fins bordering the trunk are used in flotation rather than in active swimming.

Form and Function

Arrowworms are unsegmented, and their body includes a head, trunk, and postanal tail (Figure 21-25A). Beneath the head is a large vestibule leading to the mouth. The vestibule contains teeth and is flanked on both sides by curved chitinous spines used in seizing prey. A pair of eyes is located dorsally. A peculiar hood formed from a fold of the neck can be drawn forward over the head and spines. When a chaetognath captures prey, it retracts its hood, and its teeth and raptorial spines spread apart and then snap shut with startling speed. Arrowworms are voracious feeders, living on planktonic forms, especially copepods, and even small fish (Figure 21-25B). When they are abundant, as they often are, they may have a substantial ecological impact. They are nearly transparent, a characteristic of adaptive value in their role as planktonic predators.

A thin cuticle covers the body, and their epidermis is single layered except along the sides, where it is stratified in a thick layer. These are the only invertebrates with a many-layered epidermis.

Arrowworms have a complete digestive system, a well-developed coelom, and a nervous system with a nerve ring connecting a cerebral ganglion above the esophagus to a number of lateral ganglia and a large ventral ganglion. Sense organs include eyes, sensory bristles, and a unique U-shaped ciliary loop that extends over their neck from the back of the head. The ciliary loop may detect water currents or may be chemosensory. However, vascular, respiratory, and excretory systems are entirely lacking.

Arrowworms are hermaphroditic with either cross- or self-fertilization. Eggs of *Sagitta* (L. arrow) bear a coat of jelly and are planktonic. Eggs of other arrowworms may be attached to the parent and carried about for a time. Juveniles develop directly without metamorphosis. Chaetognath embryogenesis suggests deuterostome affinities but differs from that of typical deuterostomes because their coelom is formed by a backward extension from the archenteron rather than by pinched-off coelomic sacs.

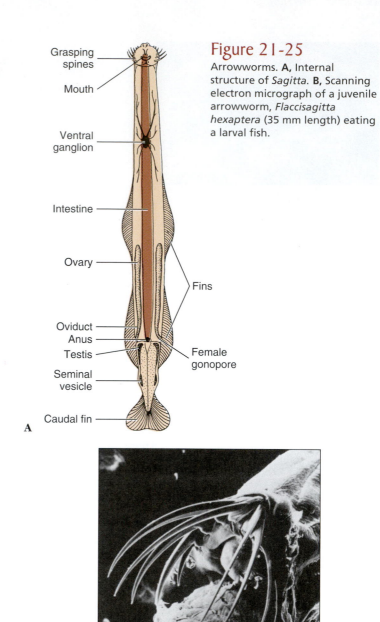

Figure 21-25
Arrowworms. **A,** Internal structure of *Sagitta*. **B,** Scanning electron micrograph of a juvenile arrowworm, *Flaccisagitta hexaptera* (35 mm length) eating a larval fish.

Labels for A:
Grasping spines
Mouth
Ventral ganglion
Intestine
Ovary
Fins
Oviduct
Anus
Testis
Seminal vesicle
Caudal fin
Female gonopore

There is no true peritoneum lining the coelom. Cleavage is radial, complete, and equal.

A common arrowworm is *Sagitta* (Figure 21-25A).

PHYLOGENY

Early embryological development of sipunculans, echiurans, and annelids is almost identical, showing a very close relationship among the three. It is also similar to molluscan development. Some authors group the four phyla into a supraphyletic assem-

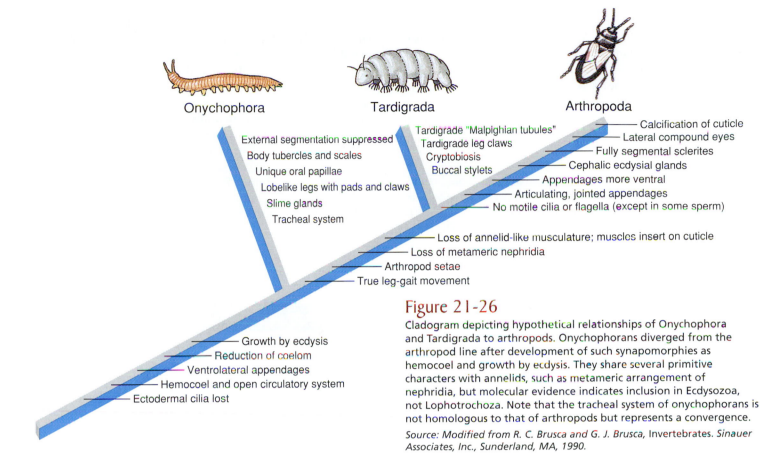

Figure 21-26

Cladogram depicting hypothetical relationships of Onychophora and Tardigrada to arthropods. Onychophorans diverged from the arthropod line after development of such synapomorphies as hemocoel and growth by ecdysis. They share several primitive characters with annelids, such as metameric arrangement of nephridia, but molecular evidence indicates inclusion in Ecdysozoa, not Lophotrochoza. Note that the tracheal system of onychophorans is not homologous to that of arthropods but represents a convergence.

Source: Modified from R. C. Brusca and G. J. Brusca, Invertebrates. *Sinauer Associates, Inc., Sunderland, MA, 1990.*

blage called "Trochozoa" because of common possession of a trochophore larva. Other similarities, too, point to close relationship of sipunculans to echiurans and annelids, such as the nature of the nervous system and body wall. Sipunculans and echiurans are not metameric and thus are more primitive in that characteristic than annelids. They probably represent collateral evolutionary lines that branched from protoannelid stock before the origin of metamerism.

Molecular sequence analysis supports close relationships of Echiura, Sipuncula, and Pogonophora to Annelida and Mollusca in Lophotrochozoa.

The phylogenetic position of lophophorates (Phoronida, Ectapracta, and Brachiopoda) has generated much controversy and debate. Sometimes they have been considered protostomes with some deuterostome characters, and at other times deuterostomes with some protostome characters. Sequence analysis of the gene coding for small-subunit ribosomal RNA provides evidence that they are protostomes. They appear clearly allied with Annelida and Mollusca and very close to Entoprocta within superphylum Lophotrochozoa. Their common possession of a lophophore is a unique synapomorphy. Other features, such as a U-shaped digestive tract, metanephridia, and tendency to secrete outer casings may be homologous within the clade, but they are convergent with many other taxa.

Division of the coelom into three parts (**trimerous, or tripartite**) is a feature shared with deuterostomes, but the character must be convergent if lophophorates are protostomes. Furthermore, some authors question the trimerous nature and homolo-

gies of the coelom in some lophophorates (for example, whether the space in the epistome of inarticulate brachiopods is a protocoel, whether the mesocoel and metacoel in brachiopods are homologous to those spaces in other lophophorates, and whether the body coelom of ectoprocts is homologous to that of brachiopods and phoronids). The blastopore origin of the mouth in phoronids and mosaic development in ectoprocts are typical protostome characters. Their larvae have been called trochophore-type in the past, although the resemblance to trochophores of annelids and molluscs is not close.

Phylogenetic affinities of Pentastomida have been controversial. Their larval appendages and molting cuticle, however, are arthropod characteristics. Their larvae resemble tardigrade larvae. Most modern taxonomists align them with arthropods, however, and evidence is accumulating that they are most closely related to the crustacean subclass Branchiura (p. 389). This evidence includes similarities in morphology of their sperm and in base sequences of ribosomal RNA. If pentastomids really are close to branchiurans, then their status as a phylum should be revoked, and they should be classified as crustacean arthropods.

Onychophorans share a number of characteristics with annelids: metamerically arranged nephridia, muscular body wall, pigment cup ocelli, and ciliated reproductive ducts. Characteristics shared with arthropods include a cuticle, tubular heart and hemocoel with open circulatory system, presence of tracheae (probably not homologous), and large size of their brain. Unique characteristics include oral papillae, slime glands, body tubercles, and suppression of external segmentation.

Some authors have advocated that onychophorans should be included with myriapods and insects in a phylum Uniramia. However, most authors believe that their differences seem to warrant keeping them in a separate phylum (Figure 21-26). Sequence analysis supports placement of Onychophora in Ecdysozoa.

Tardigrades have some similarities to rotifers, particularly in their reproduction and their cryptobiotic tendencies, and some authors have called them pseudocoelomates. Their embryogenesis, however, would seem to put them among coelomates. The enterocoelic origin of the mesoderm is a deuterostome characteristic. Other authors identify several important synapomorphies that suggest grouping them with arthropods (Figure 21-16). DNA sequence analysis supports alignment with arthropods in Ecdysozoa.

Recent discoveries of Cambrian fossil pentastomids and tardigrades and additional fossil onychophorans strongly suggest that these small phyla arose during the Cambrian explosion, just as did the major phyla. Because this period was long before terrestrial vertebrates evolved, identity of hosts for Cambrian pentastomids remains enigmatic; some authors have suggested that they might have been conodonts (see p. 481).

Chaetognatha have a number of deuterostome characters and have long been considered members of that division of the animal kingdom. However, sequence analysis of the gene encoding small-subunit rRNA supports placement of chaetognaths among protostomes, apparently closely related to Nematoda in superphylum Ecdysozoa.

SUMMARY

The ten small phyla covered in this chapter are grouped together here for convenience. They are all coelomate protostomes, but molecular sequence analysis places some in superphylum Lophotrochozoa (Sipuncula, Echiura, Pogonophora, Brachiopoda, Ectoprocta, and Phoronida) and some in superphylum Ecdysozoa (Pentastomida, Onychophora, Tardigrada, and Chaetognatha).

Sipunculans are small, burrowing marine worms with an eversible introvert at their anterior end. The introvert bears tentacles used for deposit-feeding. Sipunculans are not metameric.

Echiurans are more diverse than sipunculans, but there are fewer species. They are also burrowing marine worms, and most are deposit-feeders, with a proboscis anterior to their mouth. They also lack metamerism.

Pogonophorans live in tubes on the deep ocean floor, and they are metameric. They have no mouth or digestive tract but apparently absorb some nutrient by the crown of tentacles at their anterior end. Much of their energy is due to chemoautotrophy of bacteria in their trophosome.

Phoronida, Ectoprocta, and Brachiopoda all bear a lophophore, which is a crown of ciliated tentacles surrounding the mouth but not the anus and containing an extension of the mesocoel. They are also sessile as adults, have a U-shaped digestive tract, and have a free-swimming larva. The lophophore functions as both a respiratory and a feeding structure, its cilia creating water currents from which food particles are filtered.

Phoronida are the least common lophophorates, living in tubes mostly in shallow coastal waters. They thrust the lophophore out of the tube for feeding.

Ectoprocts are abundant in marine habitats, living on a variety of submerged substrata, and a number of species are common in fresh water. Ectoprocts are colonial, and although each individual is quite small, colonies are commonly several centimeters or more in width or height. Each individual lives in a chamber (zoecium), which is a secreted exoskeleton of chitinous, calcium carbonate, or gelatinous material.

Brachiopods were very abundant in the Paleozoic era but have been declining in numbers and species since the early Mesozoic era.

Their bodies and lophophores are covered by a mantle, which secretes a dorsal and a ventral valve (shell). They are usually attached to the substrate directly or by means of a pedicel.

Pentastomida are wormlike parasites in the lungs and nasal passages of carnivorous vertebrates. They are probably most closely related to arthropods.

Onychophora are caterpillar-like animals found in humid, mostly tropical habitats. They are metameric and crawl by means of a series of unjointed, clawed appendages.

Tardigrades are minute animals, mostly terrestrial, living in the water film that surrounds mosses and lichens. They have eight unjointed legs and a nonchitinous cuticle. Their chief body cavity is a hemocoel, as in arthropods. They can undergo cryptobiosis, withstanding adverse conditions for long periods.

Arrowworms (phylum Chaetognatha) are a small group but an important component of marine plankton. They have a well-developed coelom and are effective predators, catching other planktonic organisms with the teeth and chitinous spines around their mouth.

Sipuncula and Echiura apparently diverged from the protostome line before the advent of metamerism, but Pogonphora and Annelida may share a metameric ancestor. Lophophorates (Phoronida, Ectoprocta and Brachiopoda) have coelomic compartments that may be convergent with the three compartments protocoel, mesocoel, and metacoel, found in many deuterostomes. Embryogenesis of lophophorates shows both protostome and deuterostome characteristics, but phylogenetic analysis of DNA sequences places them in superphylum Lophotrochozoa of protostomes.

Pentastomida have certain arthropod-like characteristics, and available evidence indicates an ancestor shared with the crustacean subclass Branchiura. If so, they should lose their status as a phylum and be classified with Crustacea.

Onychophora and Tardigrada are probably phylogenetically close to Arthropoda (Figure 21-26). Chaetognaths have been long considered deuterostomes on developmental and morphological criteria, but molecular evidence indicates that they are ecdysozoan protostomes.

REVIEW QUESTIONS

1. Distinguish the following phyla from each other, and describe each one's habitat: Sipuncula, Echiura, Pogonophora, Ectoprocta, Phoronida, Brachiopoda, Pentastomida, Onychophora, Tardigrada, and Chaetognatha.
2. What do members of each of the aforementioned groups eat?
3. What evidence suggests that Sipuncula and Echiura diverged from the protostome line before the origin of annelids? Why are these phyla considered closely related?
4. What is the largest pogonophoran known? Where is it found, and how is it nourished?
5. What characters do lophophorate phyla have in common? What characters distinguish them from each other?
6. Define each of the following: lophophore, zoecium, zooid, polypide, cystid, statoblasts.
7. What are some protostome characters found among lophophorates? What are their deuterostome characters?
8. What are the coelomic compartments found in lophophorates?
9. What is the difference in orientation of the valves of brachiopods compared with bivalve molluscs?
10. How is the lophophore of ectoprocts extended?
11. Briefly describe the life cycle of a typical pentastomid.
12. What is the survival value of cryptobiosis in tardigrades?
13. Onychophora and Tardigrada seem to be related to Arthropoda, but some workers regard Pentastomida as arthropods. Why?
14. What is evidence that Chaetognatha are deuterostomes? What is evidence that conflicts with this hypothesis?
15. What is the ecological importance of arrowworms?

SELECTED REFERENCES

Childress, J. J., H. Felbeck, and G. N. Somero. 1987. Symbiosis in the deep sea. Sci. Am. **256:**114–120 (May). *The amazing story of how the animals around deep-sea vents, including* Riftia pachyptila, *absorb hydrogen sulfide and transport it to their mutualistic bacteria. For most animals, hydrogen sulfide is highly toxic.*

Conway Morris, S., B. L. Cohen, A. B. Gawthrop, T. Cavalier-Smith. 1996. Lophophorate phylogeny. Science **272:**282–283. *These authors urged caution in acceptance of the taxon Lophotrochozoa proposed by Halanych et al. (1995).*

Cutler, E. B. 1995. The Sipuncula. Their systematics, biology, and evolution. Ithaca, New York, Cornell University Press. *The author tried to "bring together everything known about" sipunculans.*

Garey, J. R., M. Krotec, D. R. Nelson, and J. Brooks. 1996. Molecular analysis supports a tardigrade-arthropod association. Invert. Biol. **115:**79–88. *Relationship of tardigrades and arthropods based on morphological characters is supported by sequence analysis of the gene encoding small-subunit rRNA.*

Gould, S. J. 1995. Of tongue worms, velvet worms, and water bears. Natural History **104**(1):6–15. *Intriguing essay on affinities of Pentastomida, Onychophora, and Tardigrada and how they, along with larger phyla, were products of the Cambrian explosion.*

Halanych, K. M. 1996. Testing hypothesis of chaetognath origins: long branches revealed by 18S ribosomal DNA. Syst. Biol. **45:**223–246. *Analysis suggests chaetognaths are most closely related to nematodes.*

Halanych, K. M., J. D. Bacheller, A. M. A. Aguinaldo, S. M. Liva, D. M. Hillis, and J. A. Lake. 1995. Evidence from 18S ribosomal DNA that lophophorates are protostome animals. Science **267:**1641–1643. *Despite much morphological and developmental evidence that lophophorates are deuterostomes, they clustered with annelids and molluscs in this analysis. The authors proposed Lophotrochozoa, defined as the last common ancestor of lophophorate taxa, annelids, and molluscs, and all descendants of that ancestor.*

Haugerud, R. E. 1989. Evolution in the pentastomids. Parasitol. Today **5:**126–132. *Much remains to be learned of this puzzling group, but there is strong evidence of its crustacean affinities.*

Hoffman, P. F., and D. P. Schrag. 2000. Snowball Earth. Sci. Am. **282:**68–75 (Jan.). *It appears that an extreme ice age prevailed on earth 600 million years ago, followed by an extremely warm period fueled by a brutal greenhouse effect. Did these events precipitate the Cambrian explosion?*

Menon, J., and A. J. Arp. 1998. Ultrastructural evidence of detoxification in the alimentary canal of Urechis caupo. Invert. Biol. **117:**307–317. *This curious echiuran has detoxification bodies in its gut cells and epithelial cells that allow it to live in a highly toxic sulfide environment.*

Telford, M. J., and P. W. H. Holland. 1993. The phylogenetic affinities of the chaetognaths: a molecular analysis. Mol. Biol. Evol. **10:**660–676. *This article and Wada and Satoh (1994) present evidence that chaetognaths are protosomes.*

Wada, H., and N. Satoh. 1994. Details of the evolutionary history from invertebrates to vertebrates, as deduced from the sequence of 18S rDNA. Proc. Nat. Acad. Sci. **91:**1801–1804.

ZOOLOGY LINKS TO THE INTERNET

Visit the textbook's Online Learning Center at www.mhhe.com/zoology to find live Internet links for each of the topics listed here.

Phylum Sipuncula
Phylum Echiura
Phylum Pogonophora
The Marine Deep Sea Zone
Phylum Onychophora
Phylum Tardigrada
Phylum Ectoprocta
Phylum Phoronida
Phylum Brachiopoda
Phylum Chaetognatha

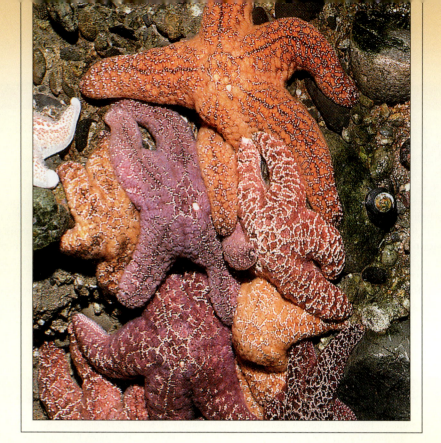

A mass of sea stars (*Pisaster ochraceus*) above the waterline at low tide.

Echinoderms and Hemichordates

Phylum Echinodermata
Phylum Hemichordata

A Design to Puzzle the Zoologist

Libbie Hyman, a distinguished American zoologist, once described echinoderms as a "noble group especially designed to puzzle the zoologist." With a combination of characteristics that should delight the most avid reader of science fiction, echinoderms would seem to confirm Lord Byron's observation that

Tis strange—but true;
for truth is always strange;
Stranger than fiction.

Despite the adaptive value of bilaterality for free-moving animals, and the merits of radial symmetry for sessile animals, echinoderms confounded the rules by becoming free moving but radial. That they evolved from a bilateral ancestor there can be no doubt, for their larvae are bilateral. They undergo a bizarre metamorphosis to a radial adult in which there is a 90° reorientation in body axis.

A compartment of the coelom has been transformed in echinoderms into a unique water-vascular system that uses hydraulic power to operate a multitude of tiny tube feet used in food gathering and locomotion. Dermal ossicles may fuse together to invest an echinoderm in armor or may be reduced to microscopic bodies. Many echinoderms have miniature jawlike pincers (pedicellariae) scattered on their body surface, often stalked and sometimes equipped with poison glands.

This constellation of characteristics is unique in the animal kingdom. It has both defined and limited evolutionary potential of echinoderms. Despite the vast amount of research that has been devoted to them, we are still far from understanding many aspects of echinoderm biology.

PHYLUM ECHINODERMATA

Echinoderms are marine forms and include sea stars (also called starfishes), brittle stars, sea urchins, sea cucumbers, and sea lilies. They represent a bizarre group sharply distinguished from all other animals. Their name is derived from their external spines or protuberances. A calcareous endoskeleton is found in all members of the phylum, either in the form of plates or represented by scattered tiny ossicles.

The most noticeable characteristics of echinoderms are (1) spiny endoskeleton of plates, (2) water-vascular system, (3) pedicellariae, (4) dermal branchiae, and (5) radial or biradial symmetry. No other group with such complex organ systems has radial symmetry.

Echinoderms are an ancient group of animals extending back to the Cambrian period. Despite an excellent fossil record, the origin and early evolution of echinoderms are still obscure. It seems clear that they descend from bilateral ancestors because their larvae are bilateral but become radially symmetrical later in development. Many zoologists believe that early echinoderms were sessile and evolved radiality as an adaptation to sessile existence. Bilaterality is of adaptive value to animals that travel through their environment, while radiality is of value to animals whose environment meets them on all sides equally. Hence, the body plan of present-day echinoderms seems to have been derived from one that was attached to the bottom by a stalk, had radial symmetry and radiating grooves (ambulacra) for food gathering, and had an upward-facing oral side. Attached forms were once plentiful, but only about 80 species, all in class Crinoidea, still survive. Oddly, conditions have favored survival of their free-moving descendants, although they are still quite radial, and among them are some of the most abundant marine animals. Nevertheless, in the exception that proves the rule (that bilaterality is adaptive for free-moving animals), at least three groups of echinoderms (sea cucumbers and two groups of sea urchins) have evolved back toward bilaterality.

Echinoderms have no ability to osmoregulate and thus rarely venture into brackish waters. They occur in all oceans of the world and at all depths, from intertidal to abyssal regions. Often the most common animals in deep ocean are echinoderms. The most abundant species found in the Philippine Trench (10,540 m) was a sea cucumber. Echinoderms are virtually all bottom dwellers, although there are a few pelagic species.

No parasitic echinoderms are known, but a few are commensals. On the other hand, a wide variety of other animals make their homes in or on echinoderms, including parasitic or commensal algae, protozoa, ctenophores, turbellarians, cirripedes, copepods, decapods, snails, clams, polychaetes, fish, and other echinoderms.

Asteroids, or sea stars (Figure 22-1), are commonly found on hard, rocky surfaces, but numerous species are at home on sandy or soft bottoms. Some species are particle feeders, but many are predators, feeding particularly on sedentary or sessile prey, since sea stars themselves are relatively slow moving.

Ophiuroids—brittle stars, or serpent stars (see Figure 22-10)—are by far the most active echinoderms, moving by their arms rather than by tube feet. A few species are reported to have swimming ability, and some burrow. They may be scavengers,

Position in Animal Kingdom

1. Phylum Echinodermata (e-ki´no-der´ma-ta) (L. *echinatus*, prickly, + Gr. *derma*, skin, + *ata*, characterized by) belongs to the **Deuterostomia** branch of the animal kingdom, the members of which are enterocoelous coelomates. The other phyla assigned to this group are Hemichordata and Chordata.
2. Primitively, deuterostomes have the following embryological features in common: anus developing from or near the blastopore, and mouth developing elsewhere; coelom budded off from the archenteron (enterocoel); radial and regulative (indeterminate) cleavage; and endomesoderm (mesoderm derived from or with endoderm) from enterocoelic pouches.
3. Thus echinoderms, chordates, and hemichordates are presumably derived from a common ancestor. Nevertheless, their evolutionary history has taken echinoderms to the point where they are very much unlike any other animal group.

Biological Contributions

1. There is one word that best describes echinoderms: strange. They have a unique constellation of characteristics found in no other phylum. Among the more striking features shown by echinoderms are these.
   a. A system of channels composing a **water-vascular system,** derived from a coelomic compartment.
   b. A **dermal endoskeleton** composed of calcareous ossicles.
   c. A **hemal system,** whose function remains mysterious, also enclosed in a coelomic compartment.
   d. Their **metamorphosis,** which changes a bilateral larva to a radial adult.

browsers, or deposit or filter feeders. Some are commensal in large sponges, in whose water canals they may live in great numbers.

Holothurians, or sea cucumbers (see Figure 22-20), are widely prevalent in all seas. Many are found on sandy or mucky bottoms, where they lie concealed. Compared with other echinoderms, holothurians are greatly extended in the oral-aboral axis. They are oriented with that axis more or less parallel to the substrate and lying on one side. Most are suspension or deposit feeders.

Echinoids, or sea urchins (see Figure 22-15), are adapted for living on the ocean bottom and always keep their oral surface in contact with the substratum. "Regular" sea urchins prefer hard bottoms, but sand dollars and heart urchins ("irregular" urchins) are usually found on sand. Regular urchins, which are radially symmetrical, feed chiefly on algae or detritus, while irregulars, which are secondarily bilateral, feed on small particles.

Crinoids (see Figure 22-25) stretch their arms out and up like a flower's petals and feed on plankton and suspended particles. Most living species become detached from their stems as adults, but they nevertheless spend most of their time on the substrate, holding on by means of aboral appendages called cirri.

A zoologist who admires the fascinating structure and function of echinoderms can share with a layperson an admiration of the beauty of their symmetry, often enhanced by bright colors. Many species are rather drab, but others may be orange, red, purple, blue, and often multicolored.

Figure 22-1

Some sea stars (class Asteroidea) from the Pacific. **A,** Pincushion star, *Culcita navaeguineae,* preys on coral polyps and also eats other small organisms and detritus. **B,** *Choriaster granulatus* scavenges dead animals on shallow Pacific reefs. **C,** *Tosia queenslandensis* from the Great Barrier Reef System browses encrusting organisms. **D,** Crown-of-thorns star, *Acanthaster planci,* is a major coral predator (see note, p. 447).

A

B

C

D

Because of the spiny nature of their structure, echinoderms are not often prey of other animals—except other echinoderms (sea stars). Some fishes have strong teeth and other adaptations that enable them to feed on echinoderms. A few mammals, such as sea otters, feed on sea urchins. In scattered parts of the world, humans relish sea urchin gonads, either raw or roasted on the half shell. Trepang, the cooked, protein-rich body wall of certain large sea cucumbers, is a delicacy in many east Asian countries. Unfortunately, the intense, often illegal, fishery for sea cucumbers has severely depleted their populations in many areas of the tropical world.

Sea stars feed on a variety of molluscs, crustaceans, and other invertebrates. In some areas they may perform an important ecological role as a top carnivore in a habitat. Their chief economic impact is on clams and oysters. A single starfish may eat as many as a dozen oysters or clams in a day. To rid shellfish beds of these pests, quicklime is sometimes spread over areas where they abound. Quicklime damages the delicate epidermal membrane, destroying the dermal branchiae and ultimately the animal itself. Unfortunately, other soft-bodied invertebrates are also damaged. However, oysters remain with their shells tightly closed until the quicklime is degraded.

Echinoderms have been widely used in developmental studies, for their gametes are usually abundant and easy to collect and handle in a laboratory. Investigators can follow embryonic developmental stages with great accuracy. We know more about molecular biology of sea urchin development than that of almost any other embryonic system. Artificial parthenogenesis was first discovered in sea urchin eggs, when it was found that, by treating eggs with hypertonic seawater or subjecting them to a variety of other stimuli, development would proceed without sperm.

Class Asteroidea

Sea stars, often called starfishes, demonstrate basic features of echinoderm structure and function very well, and they are easily obtainable. Thus we will consider them first, then comment on major differences shown by other groups.

Characteristics of Phylum Echinodermata

1. Body unsegmented (nonmetameric) with **radial, pentamerous symmetry;** body rounded, cylindrical, or star shaped, with five or more radiating areas, or **ambulacra,** alternating with interambulacral areas
2. **No head or brain;** few specialized sensory organs; sensory system of tactile and chemoreceptors, podia, terminal tentacles, photoreceptors, and statocysts
3. Nervous system with circumoral ring and radial nerves; usually two or three systems of networks located at different levels in the body, varying in degree of development according to group
4. **Endoskeleton** of **dermal calcareous ossicles** with **spines** or of calcareous **spicules** in dermis; covered by an epidermis (ciliated in most); **pedicellariae** (in some)
5. A unique **water-vascular system** of coelomic origin that extends from the body surface as a series of tentacle-like projections **(podia,** or **tube feet)** that are protracted by increase of fluid pressure within them; an opening to the exterior **(madreporite** or **hydropore)** usually present
6. Locomotion by tube feet, which project from ambulacral areas, by movement of spines, or by movement of arms, which project from central disc of body
7. Digestive system usually complete; axial or coiled; anus absent in ophiuroids
8. Coelom extensive, forming the perivisceral cavity and the cavity of the water-vascular system; coelom of enterocoelous type; coelomic fluid with amebocytes
9. Blood-vascular system **(hemal system)** much reduced, playing little if any role in circulation, and surrounded by extensions of coelom **(perihemal sinuses);** main circulation of body fluids (coelomic fluids) by peritoneal cilia
10. Respiration by **dermal branchiae, tube feet, respiratory tree** (holothuroids), and **bursae** (ophiuroids)
11. **Excretory organs absent**
12. Sexes separate (except a few hermaphroditic) with large gonads, single in holothuroids but multiple in most; simple ducts, with no elaborate copulatory apparatus or secondary sexual structures; fertilization usually external; eggs brooded in some
13. Development through **free-swimming, bilateral, larval stages** (some with direct development); metamorphosis to radial adult or subadult form; radial cleavage and regulative development
14. Autotomy and regeneration of lost parts conspicuous

Sea stars are familiar along the shoreline where large numbers may aggregate on rocks. Sometimes they cling so tenaciously that they are difficult to dislodge without tearing off some tube feet. They also live on muddy or sandy bottoms and among coral reefs. They are often brightly colored and range in size from a centimeter in greatest diameter to about a meter across. *Asterias* (Gr. *asteros,* a star) is a common genus of the east coast of the United States and is commonly studied in zoology laboratories. *Pisaster* (Gr. *pisos,* a pea, + *asteros,* a star, p. 442) is common on the west coast of the United States, as is *Dermasterias* (Gr. *dermatos,* skin, leather, + *asteros,* a star), the leather star.

Form and Function

External Features

Sea stars are composed of a central disc that merges gradually with the tapering arms (rays). The body is somewhat flattened, flexible, and covered with a ciliated, pigmented epidermis. Their mouth is centered on the under, or oral, side, surrounded by a soft peristomial membrane. An **ambulacrum** (pl., **ambulacra,** L. *ambulacrum,* a covered way, an alley, a walk planted with trees) or **ambulacral area,** runs from the mouth on the oral side of each arm to the tip of the arm. Sea stars typically have five arms, but they may have more (Figure 22-1D), and there are as many ambulacral areas as there are arms. An **ambulacral groove** is found along the middle of each ambulacral area, and the groove is bordered by rows of **tube feet (podia)** (Figure 22-2). These in turn are usually protected by movable **spines.** A large **radial nerve** can be seen in the center of each ambulacral groove (Figure 22-3C), between the rows of tube feet. The nerve is very superficially located, covered only by thin epidermis. Under the nerve is an extension of the coelom and the radial canal of the water-vascular system, all of which are external to the underlying ossicles (Figure 22-3C). In all other classes of living echinoderms except crinoids, these structures are covered by ossicles or other dermal tissue; thus ambulacral grooves in asteroids and crinoids are said to be *open,* and those of the other groups are *closed.*

The aboral surface is usually rough and spiny, although spines of many species are flattened, so that the surface appears smooth (Figure 22-1C). Around the bases of spines are groups of minute, pincerlike **pedicellariae,** bearing tiny jaws manipulated by muscles (Figure 22-4). These jaws help keep the body surface free of debris, protect papulae, and sometimes aid in food capture. **Papulae (dermal branchiae** or **skin gills)** are soft delicate projections of the coelomic cavity, covered only with epidermis and lined internally with peritoneum; they extend out through spaces between ossicles and are concerned with respiration (Figures 22-3C, and 22-4F). Also on the aboral side are an inconspicuous anus and a circular **madreporite** (Figure 22-2A), a calcareous sieve leading to the water-vascular system.

Endoskeleton

Beneath the epidermis of sea stars is a mesodermal endoskeleton of small calcareous plates, or **ossicles,** bound together with connective tissue. From these ossicles project spines and tubercles that make up the spiny surface. Ossicles are penetrated by a meshwork of spaces, usually filled with fibers and dermal cells. This internal meshwork structure is described as **stereom** and is unique to echinoderms.

Muscles in the body wall move the rays and can partially close the ambulacral grooves by drawing their margins together.

Coelom, Excretion, and Respiration

Coelomic compartments of larval echinoderms give rise to several structures in adults, one of which is a spacious body coelom filled with fluid. This fluid contains amebocytes (coelomocytes), bathes internal organs, and projects into papulae. The ciliated peritoneal lining of the coelom circulates the fluid around the body cavity and into the papulae. Exchange of respiratory gases and excretion of nitrogenous waste, principally ammonia, take place by diffusion through the thin walls of papulae and tube feet. Some wastes

Figure 22-2
External anatomy of asteroid. **A,** Aboral view. **B,** Oral view.

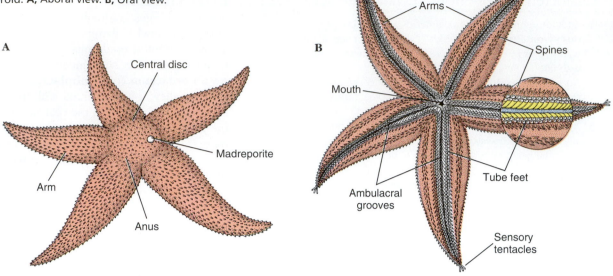

may be picked up by coelomocytes, which migrate through the epithelium of papulae or tube feet to the exterior, or tips of papulae containing waste-laden coelomocytes may pinch off.

Water-Vascular System   The water-vascular system is another coelomic compartment and is unique to echinoderms. Showing exploitation of hydraulic mechanisms to a greater degree than in any other animal group, it is a system of canals and specialized tube feet that, together with their dermal ossicles, has determined the evolutionary potential and limitations of this phylum. In sea stars the primary functions of the water-vascular system are locomotion and food gathering, in addition to respiration and excretion.

Structurally, the water-vascular system opens to the outside through small pores in the madreporite. The madreporite of asteroids is on the aboral surface (Figure 22-2A) and leads into a **stone canal,** which descends toward a **ring canal** around the mouth (Figure 22-3B). **Radial canals** diverge from the ring canal, one into the ambulacral groove of each ray. Also attached to the ring canal are four or five pairs of folded, pouchlike **Tiedemann's bodies** and from one to five **polian vesicles** (polian vesicles are absent in some sea stars, such as *Asterias*). Tiedemann's bodies may produce coelomocytes, and polian vesicles are apparently for fluid storage.

A series of small **lateral canals,** each with a one-way valve, connects the radial canal to the cylindrical podia, or tube feet, along the sides of the ambulacral groove in each ray. Each podium is a hollow, muscular tube, the inner end of which is a muscular sac, or **ampulla,** which lies within the body coelom (Figure 22-3A and C), and the outer end of which usually bears a **sucker.** Some species lack suckers. Podia pass to the outside between ossicles in the ambulacral groove.

The water-vascular system operates hydraulically and is an effective locomotor mechanism. Valves in the lateral canals prevent backflow of fluid into the radial canals. Each tube foot has in its walls connective tissue that maintains the cylinder at a rela-

tively constant diameter. Contraction of muscles in the ampulla forces fluid into the podium, extending it. Conversely, contraction of the longitudinal muscles in the tube foot retracts the podium, forcing fluid back into the ampulla. Contraction of muscles in one side of the podium bends the organ toward that side. Small muscles at the end of the tube foot can raise the middle of the disclike end, creating suction when the end is applied to a firm substrate. It has been estimated that by combining mucous adhesion with suction, a single podium can exert a pull equal to 25 to 30 g. Coordinated action of all or many of the tube feet is sufficient to draw the animal up a vertical surface or over rocks. The ability to move while firmly adhering to the substrate is a clear advantage to an animal living in a sometimes wave-churned environment.

On a soft surface, such as muck or sand, suckers are ineffective (numerous sand-dwelling species have no suckers), so the tube feet are employed as legs. Locomotion becomes mainly a stepping process. Most sea stars can move only a few centimeters per minute, but some very active ones—*Pycnopodia* (Gr. *pyknos,* compact, dense, + *pous, podos,* foot) (Figure 22-5B), for example—can move 75 to 100 cm per minute. When inverted, a sea star bends its rays until some tubes reach the substratum and attach as an anchor; then it slowly rolls over.

Tube feet are innervated by the central nervous system (ectoneural and hyponeural systems, see following text). Nervous coordination enables tube feet to move in a single direction, although not in unison, so that the sea star may progress. If the radial nerve in an arm is cut, podia in that arm lose coordination, although they can still function. If the circumoral nerve ring is cut, podia in all arms become uncoordinated, and movement ceases.

Feeding and Digestive System   The mouth on the oral side leads through a short esophagus to a large stomach in the central disc. The lower (cardiac) part of the stomach can be everted through the mouth during feeding (Figure 22-2B), and excessive eversion is prevented by gastric ligaments. The upper (pyloric) part is smaller and connects by ducts to a pair of large

Figure 22-3

A, Internal anatomy of a sea star. **B,** Water-vascular system. Podia penetrate between ossicles. (Polian vesicles are not present in *Asterias*.) **C,** Cross section of arm at level of gonads, illustrating open ambulacral grooves.

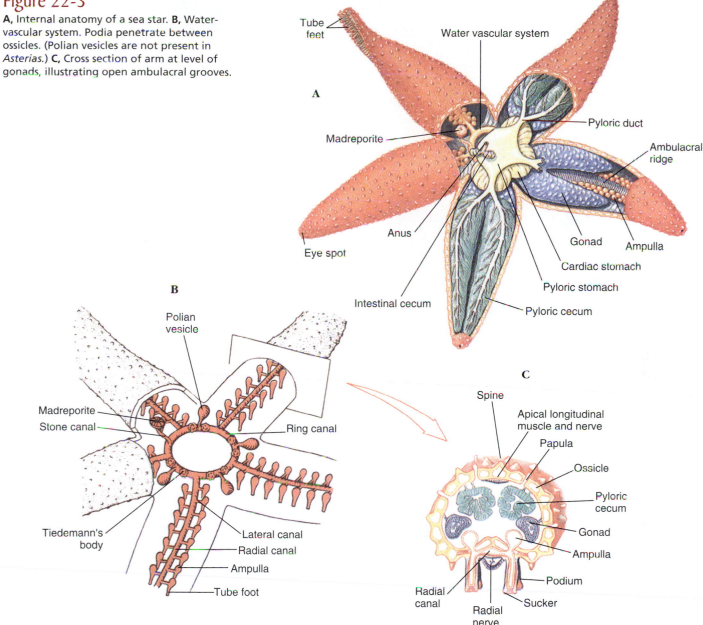

pyloric ceca (digestive glands) in each arm (Figure 22-3A). Digestion is mostly extracellular, although some intracellular digestion may occur in the ceca. A short intestine leads aborally from the pyloric stomach, and there are usually a few small, saclike **intestinal ceca** (Figure 22-3A). The anus is inconspicuous, and some sea stars lack an intestine and anus.

Many sea stars are carnivorous and feed on molluscs, crustaceans, polychaetes, echinoderms, other invertebrates, and sometimes small fish. Sea stars consume a wide range of food items, but many show particular preferences (Figures 22-1D and 22-5). Some select brittle stars, sea urchins, or sand dollars, swallowing them whole and later regurgitating undigestible ossicles and spines (Figure 22-5B). Some attack other sea stars, and if they are small compared with their prey, they may attack and begin eating at the end of one arm.

Since 1963 there have been numerous reports of increasing numbers of the crown-of-thorns starfish (*Acanthaster planci* [Gr. *akantha*, thorn, + *asteros*, star]) (Figure 22-1D) that were damaging large areas of coral reef in the Pacific Ocean. Crown-of-thorns stars feed on coral polyps, and they sometimes occur in large aggregations, or "herds." There is some evidence that outbreaks have occurred in the past, but an increase in frequency during the past 40 years suggests that some human activity may be affecting the starfish. Of reefs surveyed in 2002, 12% had outbreaks, compared with 1988 when 10% had outbreaks, and there was extensive damage. Efforts to control these organisms are very expensive and of questionable effectiveness. The controversy continues, especially in Australia, where it is exacerbated by extensive media coverage.

F

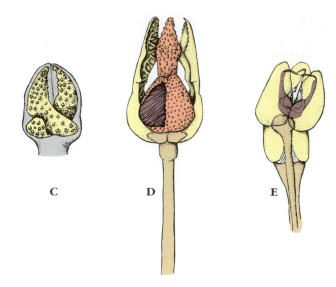

Figure 22-4

Pedicellariae of sea stars and sea urchins. **A,** Forceps-type pedicellaria of *Asterias*. **B,** and **C,** Scissors-type pedicellariae of *Asterias;* size relative to spine is shown in **B. D,** Tridactyl pedicellaria of *Strongylocentrotus*. **E,** Globiferous pedicellaria of *Strongylocentrotus*. **F,** Close-up view of the aboral surface of the sea star *Pycnopodia helianthoides.* Note the large pedicellariae, as well as groups of small pedicellariae around the spines. Many thin-walled papulae can be seen.

Some asteroids feed heavily on molluscs (Figure 22-5A), and *Asterias* is a significant predator on commercially important clams and oysters. When feeding on a bivalve, a sea star will hump over its prey, attaching its podia to the valves, and then exert a steady pull, using its feet in relays. A force of some 1300 g can thus be exerted. In half an hour or so the bivalve's adductor muscles fatigue and relax. With a very small gap available, the star inserts its soft everted stomach into the space between the valves and wraps it around the soft parts of the shellfish. After feeding, the sea star draws its stomach inward by contraction of its stomach muscles and relaxation of body-wall muscles.

Some sea stars feed on small particles, either entirely or in addition to carnivorous feeding. Plankton and other organic particles coming in contact with an animal's surface are carried by epidermal cilia to the ambulacral grooves and then to the mouth.

Hemal System   The so-called hemal system is not very well developed in asteroids, and its function in all echinoderms is unclear. The hemal system has little or nothing to do with circulation of body fluids. It is a system of tissue strands enclosing unlined sinuses and is itself enclosed in another coelomic compartment, or **perihemal channels** (Figure 22-6). The hemal system may be useful in distributing digested products, but its specific functions are not really known.

Nervous System   The nervous system consists of three units at different levels in the disc and arms. Chief of these systems is an **oral (ectoneural)** system composed of a **nerve ring**

around the mouth and a main **radial nerve** into each arm. It appears to coordinate the tube feet. A **deep (hyponeural)** system lies aboral to the oral system, and an **aboral** system consists of a ring around the anus and radial nerves along the roof of each ray. An **epidermal nerve plexus** or nerve net freely connects these systems with the body wall and related structures. The epidermal plexus coordinates responses of the dermal branchiae to tactile stimulation—the only instance known in echinoderms in which coordination occurs through a nerve net.

Sense organs are not well developed. Tactile organs and other sensory cells are scattered over the surface, and an ocellus is at the tip of each arm. Their reactions are mainly to touch, temperature, chemicals, and differences in light intensity. Sea stars are usually more active at night.

Reproductive System, Regeneration, and Autotomy
Most sea stars have separate sexes. A pair of gonads lies in each interradial space (Figure 22-3A). Fertilization is external and occurs in early summer when eggs and sperm are shed into the water. A secretion from neurosecretory cells located on the radial nerves stimulates maturation and shedding of asteroid eggs.

Echinoderms can regenerate lost parts. Sea-star arms can regenerate readily, even if all are lost. Sea stars also have the power of autotomy and can cast off an injured arm near its base. Regeneration of a new arm may take several months.

Some species can regenerate a complete new sea star (Figure 22-7) from a detached arm that contains a part (about one-fifth) of the central disc. In former times fishermen used to dis-

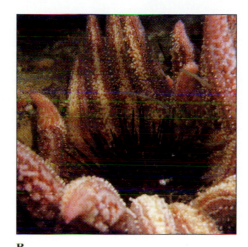

Figure 22-5

A, *Orthasterias koehleri* eating a clam. **B,** This *Pycnopodia helianthoides* has been overturned while eating a large sea urchin, *Strongylocentrotus franciscanus.* This sea star has 20 to 24 arms and can range up to 1 m in diameter (arm tip to arm tip).

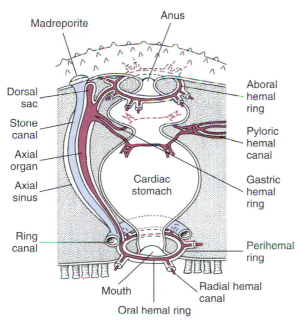

Figure 22-6

Hemal system of asteroids. The main perihemal channel is the thin-walled axial sinus, which encloses both the axial organ and stone canal. Other features of the hemal system are shown.

Figure 22-7

Pacific sea star, *Echinaster luzonicus,* can reproduce itself by splitting across the disc, then regenerating missing arms. The one shown here has evidently regenerated six arms from the longer one at top left.

patch sea stars they collected from their oyster beds by chopping them in half with a hatchet—a worse than futile activity. Some sea stars reproduce asexually under normal conditions by cleaving their central disc, each part regenerating the rest of the disc and missing arms.

Development   In some species liberated eggs are brooded, either under the oral side of the animal or in specialized aboral structures, and development is direct, but in most species embryonating eggs are free in the water and hatch to free-swimming larvae.

Early embryogenesis shows a typical primitive deuterostome pattern (see Figures 8-7A and 8-11A). Gastrulation is by invagination, and the anterior end of the archenteron pinches off to become a coelomic cavity, which expands in a U shape to fill the blastocoel. Each leg of the U, at the posterior, constricts to become a separate vesicle, and these eventually give rise to the main coelomic compartments of the body (metacoels, called **somatocoels** in echinoderms). The anterior portion of the U undergoes subdivision to form protocoels and mesocoels (called **axocoels** and **hydrocoels** in echinoderms) (Figure 22-8). The left hydrocoel will become the water-vascular system, and the left axocoel will give rise to the stone canal and perihemal channels. The right axocoel and hydrocoel will disappear. The free-swimming larva has cilia arranged in bands and is called a **bipinnaria** (Figure 22-9A). These ciliated tracts become extended into larval arms. Soon the larva grows three adhesive arms and a sucker at its anterior end and is then called a **brachiolaria.** At that time it attaches to the substratum, forms a temporary attachment stalk, and undergoes metamorphosis.

Metamorphosis involves a dramatic reorganization of a bilateral larva into a radial juvenile. The anteroposterior axis of the larva is lost, and *what was the left side becomes the oral surface, and the larval right side becomes the aboral surface* (Figure 22-8). Correspondingly, larval mouth and anus disappear, and a new mouth and anus form on what were originally the left and right sides, respectively. The portion of the anterior coelomic compartment from the left side expands to form the ring canal of the water-vascular system around the mouth, and then it grows branches to form radial canals. As the short, stubby arms and the first podia appear, the animal detaches from its stalk and begins

Figure 22-8

Asteroid metamorphosis. The left somatocoel becomes the oral coelom, and the right somatocoel becomes the aboral coelom. The left hydrocoel becomes the water-vascular system and the left axocoel the stone canal and perihemal channels. The right axocoel and hydrocoel are lost.

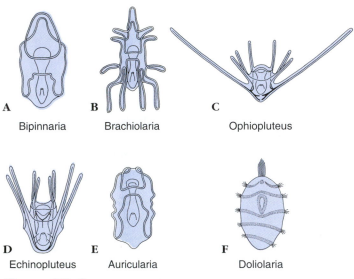

A Bipinnaria   **B** Brachiolaria   **C** Ophiopluteus

D Echinopluteus   **E** Auricularia   **F** Doliolaria

Figure 22-9

Larvae of echinoderms. **A,** Bipinnaria of asteroids. **B,** Brachiolaria of asteroids. **C,** Ophiopluteus of ophiuroids. **D,** Echinopluteus of echinoids. **E,** Auricularia of holothuroids. **F,** Doliolaria of crinoids.

life as a young sea star. A number of regulatory genes found in bilateral animals are conserved in echinoderms and have surprisingly similar functions. For example, *Distal-less* and its homolog in vertebrates regulate outgrowth of limbs in these animals; its homolog in echinoderms is active in development of tube feet.

Class Ophiuroidea

Brittle stars are largest of the major groups of echinoderms in numbers of species, and they are probably the most abundant also. They abound in all types of benthic marine habitats, even carpeting the abyssal sea bottom in many areas.

Form and Function

Apart from a typical possession of five arms, brittle stars are surprisingly different from asteroids. Arms of brittle stars are slender and sharply set off from the central disc (Figure 22-10). They have no pedicellariae or papulae, and their ambulacral grooves

A

B

Figure 22-10

A, Brittle star, *Ophiura lutkeni* (class Ophiuroidea). Brittle stars do not use their tube feet for locomotion but can move rapidly (for an echinoderm) by means of their arms. **B,** Basket star, *Astrophyton muricatum* (class Ophiuroidea). Basket stars extend their many-branched arms to filter feed, usually at night. They show a strongly negative phototropic response.

Figure 22-11
Oral view of spiny brittle star, *Ophiothrix*.

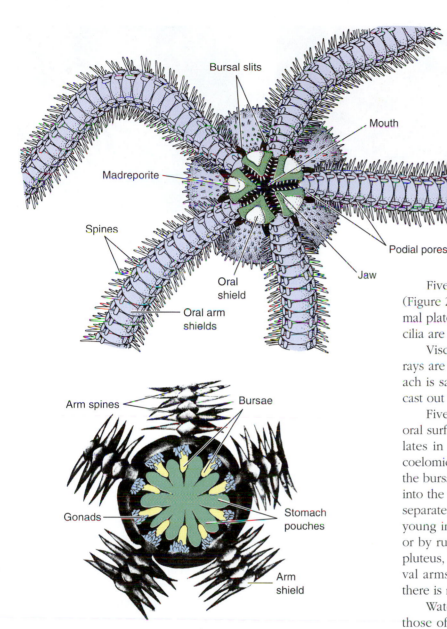

Figure 22-12

Ophiuroid with aboral disc wall cut away to show principal internal structures. Bursae are fluid-filled sacs in which water constantly circulates for respiration. They also serve as brood chambers. Only bases of arms are shown.

Five movable plates that serve as jaws surround the mouth (Figure 22-11). There is no anus. Their skin is leathery, with dermal plates and spines arranged in characteristic patterns. Surface cilia are mostly lacking.

Visceral organs are confined to the central disc, since their rays are too slender to contain them (Figure 22-12). Their stomach is saclike, and there is no intestine. Indigestible material is cast out of the mouth.

Five pairs of invaginations called **bursae** open toward the oral surface by genital slits at the bases of the arms. Water circulates in and out of these sacs for exchange of gases. On the coelomic wall of each bursa are small gonads that discharge into the bursa their ripe sex cells, which pass through the genital slits into the water for fertilization (Figure 22-13A). Sexes are usually separate; a few ophiuroids are hermaphroditic. Some brood their young in the bursae; the young escape through the genital slits or by rupturing the aboral disc. Their larva is called an ophiopluteus, and its ciliated bands extend onto delicate, beautiful larval arms (Figure 22-9C). During metamorphosis to a juvenile, there is no temporarily attached phase, as there is in asteroids.

Water-vascular, nervous, and hemal systems are similar to those of sea stars. Each arm contains a small coelom, a radial nerve, and a radial canal of the water-vascular system.

Biology

Brittle stars tend to be secretive, living on hard bottoms in locations where little or no light penetrates. They are often negatively phototropic and insinuate themselves into small crevices between rocks, becoming more active at night. They are commonly fully exposed on the bottom in the permanent darkness of deep seas. Ossicles in arms of at least some photosensitive ophiuroids show a remarkable adaptation for photoreception. Tiny, rounded structures on their aboral surface serve as microlenses, focusing light on nerve bundles just beneath them. Related species that are indifferent to light have no such structures.

Ophiuroids feed on a variety of small particles, either browsing food from the bottom or suspension feeding. Podia are important in transferring food to the mouth. Some brittle stars extend arms into the water and catch suspended particles in mucous strands between arm spines.

are closed and covered with arm ossicles. Their tube feet are without suckers; they aid in feeding but are of limited use in locomotion. In contrast to asteroids, the madreporite of ophiuroids is located on the oral surface, on one of the oral shield ossicles (Figure 22-11). Ampullae are absent from podia, and force for protrusion of a podium is generated by a proximal muscular portion of the podium.

Each jointed arm consists of a column of articulated ossicles (so-called **vertebrae**), connected by muscles and covered by plates. Locomotion is by arm movement. Arms are moved forward in pairs and are placed against the substratum, while one (any one) is extended forward or trailed behind, and the animal is pulled or pushed along in a jerky fashion.

Figure 22-13

A, This brittle star, *Ophiopholis aculeata,* has its bursae swollen with eggs, which it is ready to expel. The arms have been broken and are regenerating. **B,** Oral view of a basket star, *Gorgonocephalus eucnemis,* showing pentaradial symmetry.

A

B

Regeneration and autotomy are even more pronounced in brittle stars than in sea stars. Many seem very fragile, releasing an arm or even part of the disc at the slightest provocation. Some can reproduce asexually by cleaving the disc; each progeny then regenerates missing parts.

Some common ophiuroids along the coast of the United States are *Amphipholis* (Gr. *amphi,* both sides of, + *pholis,* horny scale) (viviparous and hermaphroditic), *Ophioderma* (Gr. *ophis,* snake, + *dermatos,* skin), *Ophiothrix* (Gr. *ophis,* snake, + *thrix,* hair), and *Ophiura* (Gr. *ophis,* snake, + *oura,* tail) (Figure 22-10). Basket stars *Gorgonocephalus* (Gr. *Gorgo,* name of a female monster of terrible aspect, + *kephalē,* a head) (Figure 22-13B) and *Astrophyton* (Gr. *asteros,* star, + *phyton,* creature, animal) (Figure 22-10B) have arms that branch repeatedly. Most ophiuroids are drab, but some are attractive, with bright color patterns (Figure 22-13A).

Class Echinoidea

Echinoids have a compact body enclosed in an endoskeletal test, or shell. Dermal ossicles, which have become closely fitting plates, make up a test. Echinoids lack arms, but their tests reflect a typical pentamerous plan of echinoderms in their five ambulacral areas. The most notable modification of the ancestral body plan is that the oral surface has expanded around to the aboral side, so that ambulacral areas extend up to an area close to the anus **(periproct).** A majority of living species of sea urchins are "regular"; they have a hemispherical shape, radial symmetry, and medium to long spines (Figures 22-14 and 22-15). Sand dollars (Figure 22-16) and heart urchins (Figure 22-17) are "irregular" because members of their orders have become secondarily bilateral; their spines are usually very short. Regular urchins move by means of their tube feet, with some assistance from their spines, and irregular urchins move chiefly by their spines (Figure 22-16). Some echinoids are quite colorful.

Echinoids have a wide distribution in all seas, from intertidal regions to deep oceans. Regular urchins often prefer rocky or hard bottoms, whereas sand dollars and heart urchins like to burrow into a sandy substrate. Distributed along one or both coasts of North America are common genera of regular urchins (*Arbacia* [Gr. *Arbakēs,* first king of Media], *Strongylocentrotus* [Gr. *strongylos,* round, compact, + *kentron,* point, spine] [Figure 22-14], and *Lytechinus* [Gr. *lytos,* dissolvable, broken, + *echinos,* sea urchin]) and sand dollars (*Dendraster* [Gr. *dendron,* tree, stick, + *asteros,* star] and *Echinarachnius* [Gr. *echinos,* sea urchin, + *arachnē,* spider]). The West Indies-Florida region is rich in echinoderms, including echinoids, of which *Diadema* (Gr. *diadeō,* to bind around), with its long, needle-sharp spines, is a notable example (Figure 22-15D).

Form and Function

An echinoid test is a compact skeleton of 10 double rows of plates that bear movable, stiff spines (Figure 22-18). The plates are sutured firmly. The five pairs of ambulacral rows are homologous to the five arms of sea stars and have pores (Figure 22-18B) through which long tube feet extend. The plates bear small tubercles on which the round ends of spines articulate as ball-and-socket joints. Spines are moved by small muscles around the bases.

There are several kinds of pedicellariae, most common of which are three jawed and are mounted on long stalks (Figure 22-4D and E). Pedicellariae help keep the body clean, especially by preventing marine larvae from settling on the body surface. Pedicellariae of many species bear poison glands, and their toxin paralyzes small prey.

Figure 22-14

Purple sea urchin, *Strongylocentrotus purpuratus,* is common along the Pacific coast of North America where there is heavy wave action.

A

B

C

Figure 22-15

Diversity among regular sea urchins (class Echinoidea). **A,** Pencil urchin, *Eucidaris tribuloides.* Members of this order have many primitive characters and have survived since the Paleozoic era. They may be closest in resemblance to the common ancestor of all other extant echinoids. **B,** Slate-pencil urchin, *Heterocentrotus mammilatus.* The large, triangular spines of this urchin were formerly used for writing on slates. **C,** Aboral spines of the intertidal urchin, *Colobocentrotus atratus,* are flattened and mushroom shaped, while the marginal spines are wedge shaped, giving the animal a streamlined form to withstand pounding surf. **D,** *Diadema antillarum* is a common species in the West Indies and Florida. **E,** *Astropyga magnifica* is one of the most spectacularly colored sea urchins, with bright-blue spots along its interambulacral areas.

D

E

Figure 22-16

Two sand dollar species. **A,** *Encope grandis* as normally found burrowing near the surface on a sandy bottom. **B,** Removed from the sand. The short spines and petaloids on the aboral surface of this *Encope micropora* are easily seen.

A

B

Diadema antillarum is not nearly as prominent as it once was. In January 1983, an epidemic swept through the Caribbean and along the Florida Keys. Its cause has never been determined, but it decimated the *Diadema* population, leaving less than 5% of the original numbers. Other species of sea urchins were unaffected. However, various types of algae, formerly grazed heavily by *Diadema* have increased greatly on the reefs, and *Diadema* populations have not recovered. This abundance of algae has had a disastrous effect on coral reefs around Jamaica. Herbivorous fish around that island had been chronically overharvested, and then, after the *Diadema* epidemic, there was nothing left to control algal overgrowth. Coral reefs around Jamaica have been largely destroyed. As of this writing, *Diadema* populations remain at a small fraction of their former densities.

Five converging teeth surround the mouth of regular urchins. In some sea urchins branched gills (modified podia) encircle the peristome. Anus, genital pores, and madreporite are located aborally in the periproct region (Figure 22-18). Sand dollars also have teeth, and their mouth is located at about the center of the oral side, but their anus has shifted to the posterior margin or even the oral side of the disc, so that an anteroposterior axis and bilateral symmetry can be recognized. Bilateral symmetry is even more accentuated in heart urchins, with their anus near the posterior on the oral side and their mouth moved away from the oral pole toward the anterior (Figure 22-17).

Inside the test (Figure 22-18) are the coiled digestive system and a complex chewing mechanism (in regular urchins and in sand dollars), called **Aristotle's lantern** (Figure 22-19), to which the teeth are attached. A ciliated **siphon** connects the esophagus to the intestine and enables water to bypass the stomach to concentrate food for digestion in the intestine. Sea urchins eat algae and other organic material, which they graze with their

Figure 22-17

An irregular echinoid, *Meoma,* one of the largest heart urchins (test up to 18 cm). *Meoma* occurs in the West Indies and from the Gulf of California to the Galápagos Islands. **A,** Aboral view. Anterior ambulacral area is not modified as a petaloid in heart urchins, although it is in sand dollars. **B,** Oral view. Note curved mouth at anterior end and periproct at posterior end.

teeth. Sand dollars have short club-shaped spines that move sand and its organic contents over the aboral surface and down the sides. Fine food particles drop between the spines, and ciliated tracts on the oral side carry the particles to their mouth.

Hemal and nervous systems are basically similar to those of asteroids. Ambulacral grooves are closed, and radial canals of the water-vascular system run just beneath the test, one in each ambulacral radius (Figure 22-18). Ampullae for the podia are within the test, and each ampulla usually communicates with its podium by *two* canals through pores in the ambulacral plate; consequently, such pores in the plates are in pairs. Peristomial gills, where present, are of little or no importance in respiratory gas exchange, this function being carried out principally by other podia. In irregular urchins respiratory podia are thin walled, flattened, or lobulate and are arranged in ambulacral fields called **petaloids** on the aboral surface. Irregular urchins also have short, suckered, single-pored podia in ambulacral and sometimes interambulacral areas; these podia function in food handling.

Sexes are separate, and both eggs and sperm are shed into the sea for external fertilization. Some, such as certain pencil urchins, brood their young in depressions between the spines. **Echinopluteus larvae** (Figure 22-9D) of nonbrooding echinoids may live a planktonic existence for several months and then metamorphose quickly into young urchins.

Class Holothuroidea

In a phylum characterized by odd animals, class Holothuroidea (sea cucumbers) contains members that both structurally and physiologically are among the strangest. These animals bear a remarkable resemblance to the vegetable after which they are

Figure 22-18

A, Internal structure of a sea urchin; water-vascular system in tan. **B,** Detail of portion of endoskeleton.

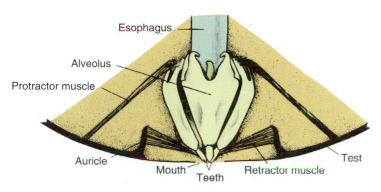

Figure 22-19

Aristotle's lantern, a complex mechanism used by sea urchins for masticating their food. Five pairs of retractor muscles draw the lantern and teeth up into the test; five pairs of protractors push the lantern down and expose the teeth. Other muscles produce a variety of movements. Only major skeletal parts and muscles are shown in this diagram.

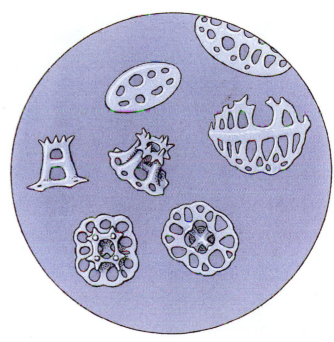

Figure 22-21

Ossicles of sea cucumbers are usually microscopic bodies buried in leathery dermis. They can be extracted from this tissue with commercial bleach and are important taxonomic characteristics. The ossicles shown here, called tables, buttons, and plates, are from *Holothuria difficilis*. They illustrate the meshwork (stereom) structure observed in ossicles of all echinoderms at some stage in their development (×250).

named (Figure 22-20). Compared with other echinoderms, holothurians are greatly elongated in the oral-aboral axis, and ossicles are much reduced in most; consequently, these animals are soft bodied. Some species crawl on the sea floor, others are found beneath rocks, and some are burrowers.

Common species along the east coast of North America are *Cucumaria frondosa* (L. *cucumis*, cucumber), *Sclerodactyla briareus* (Gr. *skleros*, hard, + *daktylos*, finger) (see Figure 22-22), and the translucent, burrowing *Leptosynapta* (Gr. *leptos*, slender, + *synapsis*, joining together). Along the Pacific coast there are several species of *Cucumaria* (Figure 22-20C) and the striking reddish brown *Parastichopus* (Gr. *para*, beside, + *stichos*, line or row, + *pous, podos*, foot) (Figure 22-20A), with very large papillae.

Form and Function

Their body wall is usually leathery, with tiny ossicles embedded in it (Figure 22-21), although a few species have large ossicles forming a dermal armor (Figure 22-20B). Because of the elongate body form of sea cucumbers, they characteristically lie on one side. In some species locomotor tube feet are equally distributed to the five ambulacral areas (Figure 22-20C) or all over the body, but most have well-developed tube feet only in the ambulacra normally applied to the substratum (Figure 22-20A and B). Thus a secondary bilaterality is present, albeit of quite different origin from that of irregular urchins. The side applied to the substratum has three ambulacra and is called a **sole;** tube feet in the dorsal ambulacral areas, if present, are usually without suckers and may be modified as sensory papillae. All tube feet, except oral tentacles, may be absent in burrowing forms.

Figure 22-20

Sea cucumbers (class Holothuroidea). **A,** Common along the Pacific coast of North America, *Parastichopus californicus* grows to 50 cm in length. Its tube feet on the dorsal side are reduced to papillae and warts. **B,** In sharp contrast to most sea cucumbers, the surface ossicles of *Psolus chitonoides* are developed into a platelike armor. The ventral surface is a flat, soft, creeping sole, and the mouth (surrounded by tentacles) and anus are turned dorsally. **C,** Tube feet are found in all ambulacral areas of *Cucumaria miniata* but are better developed on its ventral side, shown here.

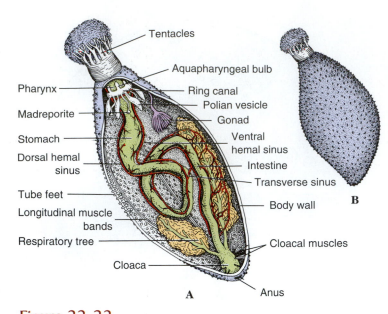

Figure 22-22

Anatomy of a sea cucumber *Sclerodactyla.* **A,** Internal. **B,** External. *Red,* hemal system.

Oral tentacles are 10 to 30 retractile, modified tube feet around the mouth. The body wall contains circular and longitudinal muscles along the ambulacra.

A sea cucumber's coelomic cavity is spacious and fluid filled and has many coelomocytes. Dermal ossicles no longer function as an endoskeleton because of their small size, and the fluid-filled coelom now serves as a hydrostatic skeleton.

The digestive system empties posteriorly into a muscular **cloaca** (Figure 22-22). A **respiratory tree** composed of two long, many-branched tubes also empties into the cloaca, which pumps seawater into it. The respiratory tree serves for both respiration and excretion and is not present in any other group of living echinoderms. Gas exchange also occurs through skin and tube feet.

The hemal system is more well developed in holothurians than in other echinoderms. Their water-vascular system is peculiar in that the madreporite lies free in the coelom.

Sexes are usually separate, but some holothurians are hermaphroditic. Among echinoderms, only sea cucumbers have a single gonad, and this is considered a primitive character. The gonad is usually in the form of one or two clusters of tubules that join at the gonoduct. Fertilization is external, and the free-swimming larva is called an **auricularia** (Figure 22-9E). Some species brood their young either inside their body or somewhere on the body surface.

Biology

Sea cucumbers are sluggish, moving partly by means of their ventral tube feet and partly by waves of contraction in the muscular body wall. More sedentary species trap suspended food particles in mucus of their outstretched oral tentacles or pick up particles from the surrounding bottom. They then stuff their tentacles into the pharynx, one by one, sucking off food material (Figure 22-23A). Others crawl along, grazing the bottom with their tentacles (Figure 22-23B).

Sea cucumbers have a peculiar power of what appears to be self-mutilation but is in reality a defense mechanism. When irritated or subjected to unfavorable conditions, many species can cast out a part of their viscera by a strong muscular contraction that may either rupture the body wall or evert its contents through their anus. Lost parts are soon regenerated. Certain species have organs of Cuvier (cuvierian tubules), which are attached to the posterior part of the respiratory tree and can be expelled in the direction of an enemy (Figure 22-23C). These tubules become long and sticky after expulsion, and some contain toxins.

There is an interesting commensal relationship between some sea cucumbers and a small fish, *Carapus,* that uses the cloaca and respiratory tree of the sea cucumber as shelter.

Class Crinoidea

Crinoids include sea lilies and feather stars. They have several primitive characters. As fossil records reveal, crinoids were once far more numerous than they are now. They differ from other echinoderms by being attached during a substantial part of their lives. Sea lilies have a flower-shaped body that is placed at the tip of an attached stalk (Figure 22-24). Feather stars have long, many-branched arms, and adults are free moving, though they may remain in the same spot for long periods (Figure 22-25). During metamorphosis feather stars become sessile and stalked, but after several months they detach and become free moving. Many crinoids are deep-water forms, but feather stars may inhabit shallow waters, especially in Indo-Pacific and West-Indian–Caribbean regions, where the largest numbers of species are found.

Form and Function

Their body disc, or **calyx,** is covered with a leathery skin **(tegmen)** containing calcareous plates. Epidermis is poorly developed. Five flexible arms branch to form many more arms, each with many lateral **pinnules** arranged like barbs on a feather (Figure 22-24). Calyx and arms together are called the **crown.** Sessile forms have a long, jointed **stalk** attached to the aboral side of the body. This stalk is composed of plates, appears jointed, and may bear **cirri.** Madreporite, spines, and pedicellariae are absent.

Their upper (oral) surface bears a mouth, which opens into a short esophagus, from which the long intestine with diverticula proceeds aborally for a distance and then makes a complete turn to an **anus,** which may be on a raised cone (Figure 22-24B). With the aid of tube feet and mucous nets, crinoids feed on small organisms that are caught in their ambulacral grooves. **Ambulacral grooves** are open and ciliated and serve to carry food to their mouth (Figure 22-24B). Tube feet in the form of tentacles are also found in the grooves.

Their water-vascular system has the basic echinoderm plan. The nervous system has an oral ring and a radial nerve that runs to each arm. The aboral or entoneural system is more highly developed in crinoids than in most other echinoderms. Sense organs are scanty and primitive.

Sexes are separate. Gonads are simply masses of cells in the genital cavity of arms and pinnules. Gametes escape without

A

B

C

Figure 22-23

A, *Eupentacta quinquesemita* extends its tentacles to collect particulate matter in the water, then puts them one by one into its mouth and cleans the food from them. **B,** Moplike tentacles of *Parastichopus californicus* are used for deposit feeding on the bottom. **C,** *Bohadschia argus* expels its cuvierian tubules, modified parts of its respiratory tree, when it is disturbed. These sticky strands, containing a toxin, discourage potential predators.

Pinnules

Arm

A

Calyx

Cirri

Stalk

Skeletal plates

Figure 22-24

Crinoid structure. **A,** Sea lily (stalked crinoid) with portion of stalk. Modern crinoid stalks rarely exceed 60 cm, but fossil forms were as much as 20 m long. **B,** Oral view of calyx of a crinoid, *Antedon,* showing direction of ciliary food currents. Ambulacral grooves with podia extend from mouth along arms and branching pinnules. Food particles touching podia are tossed into ambulacral grooves and carried, tangled in mucus, by strong ciliary currents toward mouth. Particles falling on interambulacral areas are carried by cilia first toward mouth and then outward and finally dropped off the edge, thus keeping the oral disc clean.

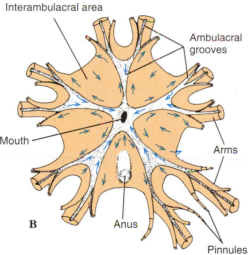

Interambulacral area

Ambulacral grooves

Mouth

B

Arms

Anus

Pinnules

Figure 22-25

Comantheria briareus are crinoids found on Pacific coral reefs. They extend their arms into the water to catch food particles both during the day and at night.

ducts through a rupture in pinnule walls. Brooding occurs in some forms. **Doliolaria** larvae (Figure 22-9F) are free swimming for a time before they become attached and metamorphose. Most living crinoids are from 15 to 30 cm long, but some fossil species had stalks 20 m in length.

Class Concentricycloidea

Strange little (less than 1 cm diameter), disc-shaped animals (Figure 22-26) were discovered in water over 1000 m deep off New Zealand. Sometimes called sea daisies, they are the most recently described (1986) class of echinoderms, and only two species are known so far. They are pentaradial in symmetry but have no arms. Their tube feet are located around the periphery of the disc, rather than along ambulacral areas, as in other echinoderms. Their water-vascular system includes two concentric ring canals; the outer ring may represent radial canals since podia arise from it. A hydropore, homologous to the madreporite, connects the inner ring canal to the aboral surface. One species has no digestive tract; its oral surface is covered by a membranous **velum,** by which it apparently absorbs nutrients. The other species has a shallow, sac-like stomach but no intestine or anus.

Figure 22-26

Xyloplax spp. (class Concentricycloidea) are bizarre little disc-shaped echinoderms. With their podia around the margin, they are the only echinoderms not having podia distributed along ambulacral areas.

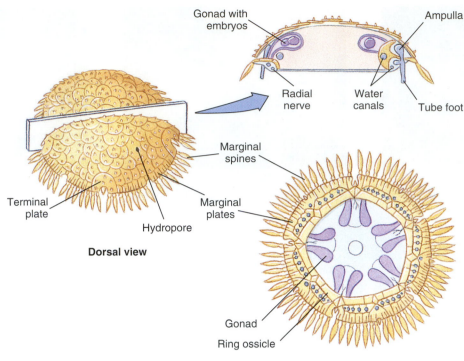

Gonad with embryos

Ampulla

Radial nerve · Water canals · Tube foot

Terminal plate

Marginal spines

Marginal plates

Hydropore

Dorsal view

Gonad

Ring ossicle

Ventral view

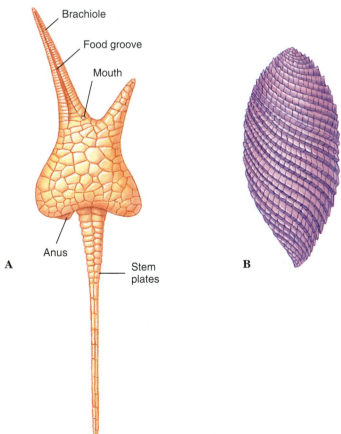

Brachiole

Food groove

Mouth

Anus

Stem plates

A

B

Figure 22-27

A, *Dendrocystites,* a carpoid (subphylum Homalozoa) with one brachiole. Brachioles are so called to distinguish them from the heavier arms of asteroids, ophiuroids and crinoids. This group bore some characters interpreted as chordate in nature. It is called Calcichordata by some investigators (p. 473). **B,** *Helicoplacus,* a helicoplacoid, had three ambulacral areas and apparently a water-vascular system. It is the sister group to modern echinoderms.

PHYLOGENY AND ADAPTIVE RADIATION

Phylogeny

Echinoderms left an extensive fossil record and evolved about 25 anatomically distinct body forms. Most of these became extinct by the end of the Paleozoic, and only six (including Concentricycloidea) survive today. Despite their fossil record, numerous contesting hypotheses on their phylogeny have been proposed. Based on their bilateral larvae, there can be little doubt that their ancestors were bilateral and that their coelom had three pairs of spaces (trimeric). Some investigators have held that radial symmetry arose in a free-moving echinoderm ancestor and that sessile groups were derived several times independently from free-moving ancestors. However, this view does not account for the adaptive significance of radial symmetry as an adaptation for a sessile existence. The more traditional view is that the first echinoderms were sessile, became radial as an adaptation to that existence, and then gave rise to the free-moving groups. Figure 22-28 is consistent with this hypothesis. It views evolution of endoskeletal plates with stereom structure and of external ciliary grooves for feeding as early echinoderm (or pre-echinoderm) developments. Extinct carpoids (Figures 22-27A, 22-28) had stereom ossicles but were not radially symmetrical, and the status of their water-vascular system, if any, is uncertain. Some investigators regard carpoids as a separate subphylum of echinoderms (Homalozoa) and consider them closer to chordates (Calcichordata, p. 473). Fossil helicoplacoids (Figures 22-27B, 22-28) show evidence of three, true ambulacral grooves, and their mouth was on the side of their body.

Figure 22-28

Cladogram showing inferred relationships among echinoderm groups. The extinct Homalozoa (carpoids), which were not radial in symmetry but had stereom endoskeletal plates, represent an early split from echinoderms. An intermediate form is represented by the extinct helicoplacoids, which had three ambulacral grooves that wound around their bodies in a spiral fashion. Helicoplacoids are the sister group of modern echinoderms. Evolution of pentaradial symmetry was an adaptation to sessile existence and is a synapomorphy of modern echinoderms. The scheme depicted here views ophiuroids as having arisen separately from asteroids, after evolution of closed ambulacral grooves, and the possession of five arms would thus have been of separate origin. Alternatively, if Asteroidea and Ophiuroidea form a monophyletic group, with five arms being synapomorphic, then closed ambulacral grooves in the ophiuroids would have evolved separately from that character in echinoids and holothuroids.

†Extinct groups.

Classification of Phylum Echinodermata

There are about 6,000 living and 20,000 extinct or fossil species of Echinodermata. The traditional classification placed all free-moving forms that were oriented with oral side down in subphylum Eleutherozoa, containing most living species. The other subphylum, Pelmatozoa, contained mostly forms with stems and oral side up; most extinct classes and living Crinoidea belong to this group. Although alternative schemes have strong supporters, cladistic analysis provides evidence that the two traditional subphyla are monophyletic. This list includes only groups with living members.

Subphylum Pelmatozoa (pel-ma′to-zo′a) (Gr. *pelmatos;* a stalk, + *zōon,* animal). Body in form of cup or calyx, borne on aboral stalk during part or all of life; oral surface directed upward; open ambulacral grooves; madreporite absent; both mouth and anus on oral surface; several fossil classes plus living Crinoidea.

    Class Crinoidea (krin-oi′de-a) (Gr. *krinon,* lily; + *eidos,* form; + *ea,* characterized by): **sea lilies** and **feather stars.** Five arms branching at base and bearing pinnules; ciliated ambulacral grooves on oral surface with tentacle-like tube feet for food gathering; spines, madreporite, and pedicellariae absent. Examples: *Antedon, Comantheria* (Figure 22-25).

Subphylum Eleutherozoa (e-lu′ther-o-zo′a) (Gr. *eleutheros,* free, not bound, + *zōon,* animal). Body form star-shaped, globular, discoidal, or cucumber-shaped; oral surface directed toward substratum or oral-aboral axis parallel to substratum; body with or without arms; ambulacral grooves open or closed.

    Class Concentricycloidea (kon-sen′tri-sy-kloy′de-a) (L. *cum,* together, + *centrum,* center [having a common center], + Gr. *kyklos,* circle, + *eidos,* form, + *ea,* characterized by): **sea daisies.** Disc-shaped body, with marginal spines but no arms; concentrically arranged skeletal plates; ring of suckerless podia near body margin; hydropore present; gut present or absent, no anus. Example: *Xyloplax* (Figure 22-26).

Class Asteroidea (as′ter-oy′de-a) (Gr. *aster,* star, + *eidos,* form, + *ea,* characterized by): **sea stars (starfish).** Star-shaped, with arms not sharply marked off from central disc; ambulacral grooves open, with tube feet on oral side; tube feet often with suckers; anus and madreporite aboral; pedicellariae present. Examples: *Asterias, Pisaster* (p. 442).

Class Ophiuroidea (o′fe-u-roy′de-a) (Gr. *ophis,* snake, + *oura,* tail, + *eidos,* form, + *ea,* characterized by): **brittle stars** and **basket stars.** Star shaped, with arms sharply marked off from central disc; ambulacral grooves closed, covered by ossicles; tube feet without suckers and not used for locomotion; pedicellariae absent; anus absent. Examples: *Ophiura* (Figure 22-10A), *Gorgonocephalus* (Figure 22-13B).

Class Echinoidea (ek′i-noy′de-a) (Gr. *echinos,* sea urchin, hedgehog, + *eidos,* form, + *ea,* characterized by): **sea urchins, sea biscuits,** and **sand dollars.** More or less globular or disc-shaped, with no arms; compact skeleton or test with closely fitting plates; movable spines; ambulacral grooves closed; tube feet with suckers; pedicellariae present. Examples: *Arbacia, Strongylocentrotus* (Figure 22-14), *Lytechinus, Mellita.*

Class Holothuroidea (hol′o-thu-roy′de-a) (Gr. *holothourion,* sea cucumber, + *eidos,* form, + *ea,* characterized by): **sea cucumbers.** Cucumber-shaped, with no arms; spines absent; microscopic ossicles embedded in thick muscular wall; anus present; ambulacral grooves closed; tube feet with suckers; circumoral tentacles (modified tube feet); pedicellariae absent; madreporite internal. Examples: *Sclerodactyla, Parastichopus, Cucumaria* (Figure 22-20C).

Attachment to a substratum by their aboral surface would have selected for radial symmetry and origin of Pelmatozoa. Both Cystoidea (extinct) and Crinoidea primitively were attached to a substratum by an aboral stalk. An ancestor that became free-moving and applied its oral surface to the substratum would have given rise to Eleutherozoa. Phylogeny within Eleutherozoa is controversial. Most investigators agree that echinoids and holothuroids are related and form a clade, but opinions diverge on the relationship of ophiuroids and asteroids. Figure 22-28 illustrates the view that ophiuroids arose after closure of their ambulacral grooves, but this scheme treats evolution of five ambulacral rays (arms) in ophiuroids and asteroids as independently evolved. Alternatively, if ophiuroids and asteroids are a single clade, then closed ambulacral grooves must have evolved separately in ophiuroids and in the common ancestor of echinoids and holothuroids.

Data on Concentricycloidea are insufficient to place this group on a cladogram, although they are tentatively placed in Eleutherozoa.

Adaptive Radiation

Radiation of echinoderms has been determined by limitations and potentials inherent in their most important characters: radial symmetry, water-vascular system, and dermal endoskeleton. If their ancestors had a brain and specialized sense organs, these were lost in adoption of radial symmetry. Thus it is not surprising that there are large numbers of creeping, benthic forms with filter-feeding, deposit-feeding, scavenging, and herbivorous habits, comparatively few predators, and very rare pelagic forms. In this light the relative success of asteroids as predators is impressive and probably attributable to the extent to which they have exploited the hydraulic mechanism of their tube feet.

The basic body plan of echinoderms has severely limited their evolutionary opportunities to become parasites. Indeed, the most mobile of echinoderms, ophiuroids, which are also the ones most able to insert their bodies into small spaces, are the only group with significant numbers of commensal species.

PHYLUM HEMICHORDATA

Hemichordata (hem′i-kor-da′ta) (Gr. *hemi,* half, + *chorda,* string, cord) are marine animals that were formerly considered a subphylum of chordates, based on their possession of gill slits and a rudimentary notochord. However, the so-called hemichordate notochord is really a buccal diverticulum (called a stomochord, meaning "mouth-cord") and not homologous with a chordate notochord, so hemichordates are ranked as a separate phylum.

Position in Animal Kingdom

1. Hemichordates belong to the deuterostome branch of the animal kingdom and are enterocoelous coelomates with radial cleavage.
2. A chordate plan of structure is suggested by gill slits and a restricted dorsal tubular nerve cord.
3. Similarity to echinoderms is shown in larval characteristics.

Biological Contributions

1. A **tubular dorsal nerve cord** in the collar zone may represent an early stage of the condition in chordates; a diffused net of nerve cells is similar to the uncentralized, subepithelial plexus of echinoderms.
2. **Gill slits** in the pharynx, which are also characteristic of chordates, serve primarily for filter feeding and only secondarily for breathing and are thus comparable to those in protochordates.

Hemichordates are vermiform bottom dwellers, living usually in shallow waters. Some colonial species live in secreted tubes. Most are sedentary or sessile. Their distribution is almost cosmopolitan, but their secretive habits and fragile bodies make collecting them difficult.

Members of class Enteropneusta (Gr. *enteron,* intestine, + *pneustikos,* of, or for, breathing) (acorn worms) range from 20 mm to 2.5 m in length. Members of class Pterobranchia (Gr. *pteron,* wing, + *branchia,* gills) are smaller, usually 1 to 12 mm, not including their stalk. About 70 species of enteropneusts and two small genera of pterobranchs are recognized.

Hemichordates have the typical tricoelomate structure of deuterostomes.

Class Enteropneusta

Enteropneusts, or acorn worms, are sluggish, wormlike animals that live in burrows or under stones, usually in mud or sand flats of intertidal zones. *Balanoglossus* (Gr. *balanos,* acorn, + *glōssa,* tongue) and *Saccoglossus* (Gr. *sakkos,* sac, strainer, + *glōssa,* tongue) (Figure 22-29) are common genera.

Form and Function

Their mucus-covered body is divided into a tonguelike proboscis, a short collar, and a long trunk (protosome, mesosome, and metasome).

Proboscis　Their proboscis is the active part of the animal. It probes about in the mud, examining its surroundings and collecting food in mucous strands on its surface. Cilia carry particles to the groove at the edge of the collar, direct them to the mouth on the underside, and then the particles are swallowed. Large particles can be rejected by covering the mouth with the edge of the collar (Figure 22-30).

Burrow dwellers use their proboscis to excavate, thrusting it into mud or sand and allowing cilia and mucus to move sand backward. They also may ingest sand or mud as they go, extracting its organic contents. They build U-shaped, mucus-lined burrows, usually with two openings 10 to 30 cm apart and with the base of the U 50 to 75 cm below the surface. They can thrust their proboscis out the front opening for feeding. Defecation at the back opening builds characteristic spiral mounds of feces that leave a telltale clue to location of burrows.

In the posterior end of their proboscis is a small coelomic sac (protocoel) into which extends a **buccal diverticulum,** a slender, blindly ending pouch of the gut that reaches forward into the buccal region and was formerly considered a notochord. A slender canal connects the protocoel with a **proboscis pore**

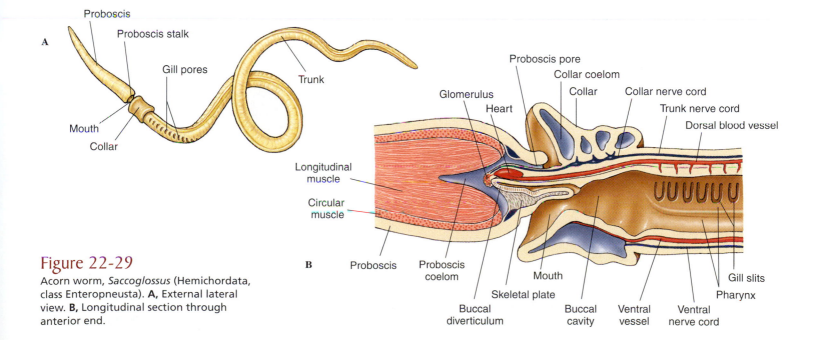

Figure 22-29

Acorn worm, *Saccoglossus* (Hemichordata, class Enteropneusta). **A,** External lateral view. **B,** Longitudinal section through anterior end.

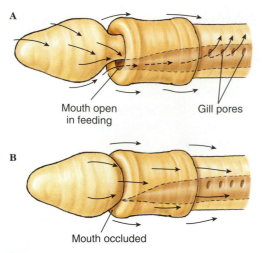

Figure 22-30

Food currents of enteropneust hemichordate. **A,** Side view of acorn worm with mouth open, showing direction of currents created by cilia on proboscis and collar. Food particles are directed toward mouth and digestive tract. Rejected particles move toward outside of collar. Water leaves through gill pores. **B,** When mouth is occluded, all particles are rejected and passed onto the collar. Nonburrowing and some burrowing hemichordates use this feeding method.

to the outside (Figure 22-29B). Paired coelomic cavities in the collar also open by pores. By taking in water through these pores into the coelomic sacs, their proboscis and collar can be stiffened to aid in burrowing. Contraction of body musculature then forces excess water out through the gill slits, reducing the hydrostatic pressure and allowing the animal to move forward.

Branchial System   A row of **gill pores** is located dorsolaterally on each side of the trunk just behind the collar (Figure 22-30A). Pores open from a series of gill chambers that in turn connect with a series of **gill slits** in the sides of the pharynx. There are no gills on the gill slits, but some respiratory gaseous exchange occurs in the vascular branchial epithelium, as well as in the body surface. Ciliary currents keep a fresh supply of water moving from their mouth through the pharynx and out the gill slits and branchial chambers to the outside.

Feeding and the Digestive System   Hemichordates are largely ciliary-mucus feeders. Behind their buccal cavity lies a large pharynx containing in its dorsal part the **U**-shaped gill slits (Figure 22-29B). Since there are no gills, the primary function of the branchial mechanism of the pharynx is presumably food gathering. Having been caught in mucus and brought to the mouth by ciliary action on the proboscis and collar, food particles are strained from the branchial water that leaves through the gill slits. Food then passes to the ventral part of the pharynx and esophagus to the intestine, where digestion and absorption occur (Figure 22-30).

Circulatory and Excretory Systems   A middorsal vessel carries the colorless blood forward above the gut. In the collar the vessel expands into a sinus and a heart vesicle above the buccal diverticulum. Blood then enters a network of blood sinuses called the **glomerulus,** which partially surrounds these structures. The glomerulus is assumed to have an excretory function (Figure 22-29B). Blood travels posteriorly through a ventral vessel below the gut, passing through extensive sinuses to the gut and body wall.

Nervous and Sensory Systems   Their nervous system consists mostly of a subepithelial network, or plexus, of nerve cells and fibers to which processes of epithelial cells are attached. Thickenings of this net form dorsal and ventral nerve cords that are united posterior to the collar by a ring connective. The dorsal cord continues into the collar and furnishes many fibers to the plexus of the proboscis. The collar cord is hollow in some species and contains giant nerve cells with processes running to nerve trunks. This nerve plexus system is quite reminiscent of that of cnidarians and echinoderms.

Sensory receptors include neurosensory cells throughout the epidermis (especially in the proboscis, a preoral ciliary organ that may be chemoreceptive) and photoreceptor cells.

Reproductive System and Development   Sexes are separate in enteropneusts. A dorsolateral row of gonads runs along each side of the anterior trunk. Fertilization is external, and in some species a ciliated **tornaria** larva develops and at

Figure 22-31

Comparison of a hemichordate tornaria **(A)** to an echinoderm bipinnaria **(B).**

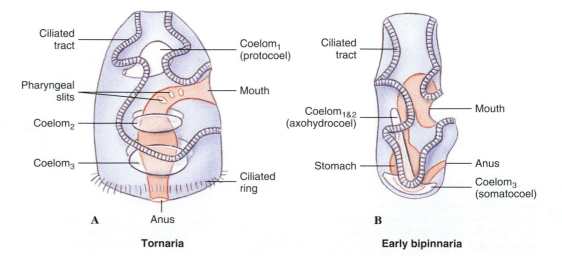

certain stages is so similar to an echinoderm bipinnaria that it was once believed to be an echinoderm larva (Figure 22-31). The familiar *Saccoglossus* of American waters has direct development without a tornaria stage.

Class Pterobranchia

The basic plan of class Pterobranchia is similar to that of Enteropneusta, but certain structural differences are correlated with the sedentary life-style of pterobranchs. The first pterobranch ever reported was obtained by the famed *Challenger* expedition of 1872 to 1876. Although first placed among Polyzoa (Entoprocta and Ectoprocta), its affinities to hemichordates were later recognized. Only two genera (*Cephalodiscus* and *Rhabdopleura*) are known in any detail.

Pterobranchs are small animals, usually within a range of 1 to 7 mm in length, although the stalk may be longer. Many individuals of *Cephalodiscus* (Gr. *kephalē,* head,+ *diskos,* disc) (Figure 22-32) live together in collagenous tubes, which often form an anastomosing system. Zooids are not connected, however,

and live independently in the tubes. Through apertures in these tubes, they extend their crown of tentacles. They are attached to the tube walls by extensible stalks that can jerk the owners back into the tubes when necessary.

The body of *Cephalodiscus* is divided into the three regions—proboscis, collar, and trunk—characteristic of hemichordates. There is only one pair of gill slits, and the alimentary canal is U-shaped, with their anus near their mouth. The proboscis is shield shaped. At the proboscis base are five to nine pairs of branching arms with tentacles containing an extension of the coelomic compartment of the mesosome, as in a lophophore. Ciliated grooves on the tentacles and arms collect food. Some species are dioecious, and others are monoecious. Asexual reproduction by budding may also occur.

In *Rhabdopleura* (Gr. *rhabdos,* rod, + *pleura,* a rib, the side), which is smaller than *Cephalodiscus,* the members remain together to form a colony of zooids connected by a stolon and enclosed in secreted tubes (Figure 22-33). The collar in these forms bears two branching arms. No gill clefts or glomeruli are present. New individuals are produced by budding from a

Cephalodiscus colony

Figure 22-32

Cephalodiscus, a pterobranch hemichordate. These tiny (5 to 7 mm) forms live in tubes in which they can move freely. Ciliated tentacles and arms direct currents of food and water toward mouth.

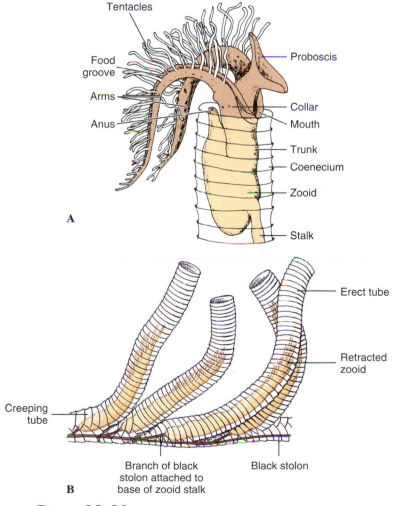

Figure 22-33

A, *Rhabdopleura,* a pterobranch hemichordate in its tube. Individuals live in branching tubes connected by stolons, and protrude ciliated tentacles for feeding. **B,** Portion of a colony.

Characteristics of Phylum Hemichordata

1. Soft bodied; wormlike or short and compact with stalk for attachment
2. Body divided into **proboscis, collar,** and **trunk;** coelomic pouch single in proboscis, but paired in other two; **buccal diverticulum** in posterior part of proboscis
3. Enteropneusta free moving and of burrowing habits; pterobranchs sessile, mostly colonial, living in secreted tubes
4. Circulatory system of dorsal and ventral vessels and dorsal heart
5. Respiratory system of **gill slits** (few or none in pterobranchs) connecting the pharynx with outside as in chordates
6. No nephridia; a single **glomerulus** connected to blood vessels may have excretory function
7. A subepidermal nerve plexus thickened to form dorsal and ventral nerve cords, with a ring connective in the collar; **dorsal nerve cord** of collar hollow in some
8. Sexes separate in Enteropneusta, with gonads projecting into body cavity; in pterobranchs reproduction may be sexual or asexual (in some) by budding; tornaria larva in some Enteropneusta

creeping basal stolon, which branches on a substratum. No pterobranch has a tubular nerve cord in the collar, but otherwise their nervous system is similar to that of Enteropneusta.

Fossil graptolites of the middle Paleozoic era often are placed as an extinct class under Hemichordata. They are important index fossils of Ordovician and Silurian geological strata. Alignment of graptolites with hemichordates has been very controversial, but discovery of an organism that seems to be a living graptolite lends strong support to the hypothesis. It has been described as a new species of pterobranch, called *Cephalodiscus graptolitoides.*

PHYLOGENY AND ADAPTIVE RADIATION

Phylogeny

Hemichordate phylogeny has long been puzzling. Hemichordates share characters with both echinoderms and chordates. With chordates they share gill slits, which serve primarily for filter feeding and secondarily for breathing, as they do in some protochordates. In addition, a short dorsal, somewhat hollow nerve cord in the collar zone may be homologous to the nerve cord of chordates (Figure 22-34). The buccal diverticulum in the hemichordate mouth cavity, long thought homologous to the notochord of chordates, is now considered a synapomorphy of hemichordates themselves. Early embryogenesis of hemichordates is remarkably like that of echinoderms, and early tornaria larvae are almost identical to bipinnaria larvae of asteroids, suggesting that echinoderms form the sister group of hemichordates and chordates (Figure 22-34). Sequence analysis of the gene encoding the small subunit of rRNA supports a deuterostome clade, albeit smaller than traditional Deuterostomia, being shorn of lophophorates and chaetognaths (p. 439).

Adaptive Radiation

Because of their sessile lives and their habitat in secreted tubes in ocean bottoms, where conditions are fairly stable, pterobranchs have undergone little adaptive divergence. They have retained a tentacular type of ciliary feeding. Enteropneusts, on the other hand, although sluggish, are more active than pterobranchs. Having lost their tentaculated arms, they use a proboscis to trap small organisms in mucus, or they eat sand as they burrow and digest organic sediments from the sand. Their evolutionary divergence, although greater than that of pterobranchs, is still modest.

SUMMARY

Phylum Echinodermata shows characteristics of the Deuterostomia division of the animal kingdom. They are an important marine group sharply distinguished from other phyla of animals. They have radial symmetry but were derived from bilateral ancestors.

Sea stars (class Asteroidea) can be used to illustrate echinoderms. Sea stars usually have five arms, which merge gradually with a central disc. Like other echinoderms, they have no head and few specialized sensory organs. Their mouth is directed toward the substratum. They have stereom dermal ossicles, respiratory papulae, and open ambulacral grooves. Many sea stars have pedicellariae. Their water-vascular system is an elaborate hydraulic system derived embryonically from one of their coelomic compartments. Along the ambulacral areas, branches of the water-vascular system (tube feet) are important in locomotion, food gathering, respiration, and excretion. Many sea stars are predators, whereas others feed on small particles. Sexes are separate, and reproductive systems are very simple. The bilateral, free-swimming larva becomes attached, transforms to a radial juvenile, then detaches and becomes a motile sea star.

Arms of brittle stars (class Ophiuroidea) are slender and sharply set off from the central disc. Ophiuroids have no pedicellariae or ampullae and their ambulacral grooves are closed. Their tube feet have no suckers, and their madreporite is on the oral side. They crawl by means of arm movements, and their tube feet function in food gathering.

Dermal ossicles of sea urchins (class Echinoidea) are closely fitting plates, the body is compact, and there are no arms. Ambulacral areas are closed and extend around their body toward the aboral pole. Sea urchins move by means of tube feet or by their spines. Some urchins (sand dollars and heart urchins) have returned to adult bilateral symmetry.

Dermal ossicles in sea cucumbers (class Holothuroidea) are very small; therefore the body wall is soft. Their ambulacral areas also are closed and extend toward the aboral pole. Holothuroids are greatly elongated in the oral-aboral axis and lie on their side. Because certain ambulacral areas are characteristically against the substratum, sea cucumbers have also undergone some return to bilateral symmetry. Tube feet around the mouth are modified into tentacles, with which they feed. They have an internal respiratory tree, and their madreporite hangs free in the coelom.

Sea lilies and feather stars (class Crinoidea) are the only group of living echinoderms, other than asteroids, with open ambulacral grooves. They are mucociliary particle feeders and lie with their oral side up.

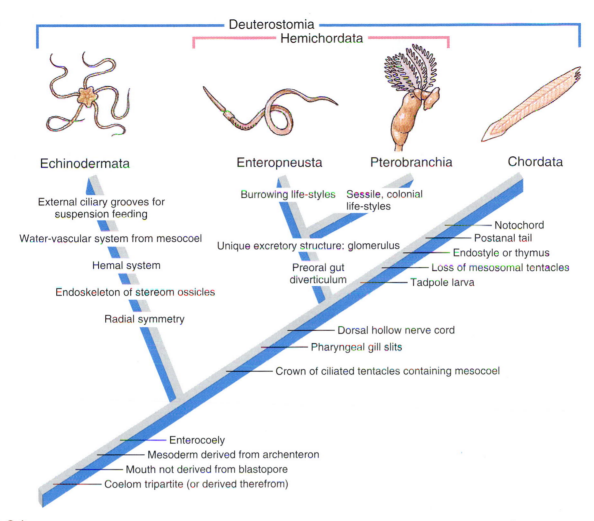

Figure 22-34

Cladogram showing inferred relationships among deuterostome phyla. Brusca and Brusca considered the crown of ciliated tentacles (containing extensions of the mesocoel) a character borne by ancestors of lophophorates, hemichordates and chordates. The tentacular crown would have become the lophophore in lophophorate phyla and retained as a primitive character in pterobranchs. Because molecular evidence indicates that lophophores are protostomes, we removed them from this cladogram; the ciliated tentacular crown in pterobranchs and lophophorates can be considered a convergent character.

Source: Modified from R. C. Brusca and G. S. Brusca, Invertebrates. *Sinauer Associates, Inc., Sunderland, MA, 1990.*

Sea daisies (class Concentricycloidea) are a newly discovered class of very small echinoderms that are circular in shape, have marginal tube feet, and two concentric ring canals in their water-vascular system.

Ancestors of echinoderms were bilaterally symmetrical, and they probably evolved through a sessile stage that became radially symmetrical and then gave rise to free-moving forms.

Members of phylum Hemichordata are marine worms that were formerly considered chordates because their buccal diverticulum was considered a notochord. However, like chordates, some do have gill slits and a hollow, dorsal nerve cord. Divisions of their body (proboscis, collar, trunk) contain typical deuterostome coelomic compartments (protocoel, mesocoel, metacoel). Hemichordate class Enteropneusta contains burrowing worms that feed on particles strained out of water by gill slits. Members of class Pterobranchia are tube dwellers, filter feeding with tentacles. Hemichordates are important phylogenetically because they show affinities with chordates and echinoderms, and they are the likely sister group of chordates.

REVIEW QUESTIONS

1. What constellation of characteristics is possessed by echinoderms and is found in no other phylum?
2. How do we know that echinoderms were derived from an ancestor with bilateral symmetry?
3. Distinguish the following groups of echinoderms from each other: Crinoidea, Asteroidea, Ophiuroidea, Echinoidea, Holothuroidea, Concentricycloidea.

4. What is an ambulacrum, and what is the difference between open and closed ambulacral grooves?

5. Trace or make a rough copy of Figure 22-3B without labels; then from memory label the parts of the water-vascular system of sea stars.

6. Briefly explain the mechanism of action of a sea star's tube foot.

7. What structures are involved in the following functions in sea stars? Briefly describe the action of each: respiration, feeding and digestion, excretion, reproduction.

8. Compare the structures and functions in question 7 as they are found in brittle stars, sea urchins, sea cucumbers, and crinoids.

9. Briefly describe development in sea stars, including metamorphosis.

10. Match groups in the left column with *all* correct answers in the right column.

_____ Crinoidea	a. Closed ambulacral grooves
_____ Asteroidea	b. Oral surface generally upward
_____ Ophiuroidea	c. With arms
_____ Echinoidea	d. Without arms
_____ Holothuroidea	e. Approximately globular or disc-shaped
_____ Concentri-cycloidea	f. Elongated in oral-aboral axis
	g. With pedicellariae
	h. Madreporite internal
	i. Madreporite on oral plate

11. Define the following: pedicellariae, madreporite, respiratory tree, Aristotle's lantern.

12. What evidence suggests that ancestral echinoderms were sessile?

13. Give four examples of how echinoderms are important to humans.

14. What is a major difference in function of the coelom in holothurians compared with other echinoderms?

15. Describe a reason for the hypothesis that the ancestor of eleutherozoan groups was a radial, sessile organism.

16. What characteristics do Hemichordata share with Chordata, and how do the two phyla differ?

17. Distinguish Enteropneusta from Pterobranchia.

18. What evidence suggest that Hemichordata are related to echinoderms?

SELECTED REFERENCES

Aizenberg, J., A. Tkachenkoo, S. Weiner, L. Addadi, and G. Hendler. 2001. Calcitic microlenses as part of the photoreceptor system in brittlestars. Nature **412**:819–822. *Tiny bumps on stereom ossicles in arms serve as microlenses to focus light on nerve photoreceptors.*

Baker, A. N., F. W. E. Rowe, and H. E. S. Clark. 1986. A new class of Echinodermata from New Zealand. Nature **321**:862–864. *Describes the strange class Concentricycloidea.*

Birkeland, C. 1989. The Faustian traits of the crown-of-thorns starfish. Am. Sci. **77**:154–163. *Fast growth in the early years of the life of an* Acanthaster planci *results in loss of body integrity in later life.*

Gilbert, S. F. 2000. Developmental biology, ed. 6. Sunderland, Massachusetts, Sinauer Associates. *Any modern text in developmental biology, such as this one, provides a multitude of examples in which studies on echinoderms have contributed (and continue to contribute) to our knowledge of development.*

Hendler, G., J. E. Miller, D. L. Pawson, and P. M. Kier. 1995. Sea stars, sea urchins, and allies: Echinoderms of Florida and the Caribbean. Washington, Smithsonian Institution Press. *An excellent field guide for echinoderm identification.*

Hickman, C. P., Jr. 1998. A field guide to sea stars and other echinoderms of Galápagos. Lexington, VA, Sugar Spring Press. *Provides descriptions and nice photographs of members of classes Asteroidea, Ophiuroidea, Echinoidea, and Holothuroidea in the Galápagos Islands.*

Hughes, T. P. 1994. Catastrophes, phase shifts and large-scale degradation of a Caribbean coral reef. Science **265**:1547–1551. *Describes the sequence of events, including the die-off of sea urchins, leading to the destruction of the coral reefs around Jamaica.*

Lane, D. J. W. 1996. A crown-of-thorns outbreak in the eastern Indonesian Archipelago, February 1996. Coral Reefs. **15**:209–210. *This is the first report of an outbreak of* Acanthaster planci *in Indonesia. It includes a good photograph of an aggregation of these sea stars.*

Lawrence, J. 1989. A functional biology of echinoderms. Baltimore, The Johns Hopkins Press. *Well-researched book on echinoderm biology with emphasis on feeding, maintenance, and reproduction.*

Mooi, R., and B. David. 1998. Evolution within a bizarre phylum: homologies of the first echinoderms. Am. Zool. **38**:965–974. *These authors contend that "The familiarity of a seastar or a sea urchin belies their overall weirdness." They describe the Etraxial/Axial Theory (EAT) of echinoderm skeletal homologies.*

Moran, P. J. 1990. *Acanthaster planci* (L.): biographical data. Coral Reefs **9**:95–96. *Presents a summary of essential biological data on* A. planci. *This entire issue of Coral Reefs is devoted to* A. planci.

Svitii, K. W. 1993. It's alive, and it's a graptolite. Discover **14**(7):18–19. *Short account of the discovery of a "living fossil."* Cephalodiscus graptolitoides.

Woodley, J. D., P. M. H. Gayle, and N. Judd. 1999. Sea-urchins exert top-down control of macroalgae on Jamaican coral reefs (2). Coral Reefs **18**:193. *In areas where* Tripneustes *(another sea urchin) have invaded fore reefs there is much less macroalgae, and such areas present a better chance that corals can recolonize. Diadema recovery has been slow.*

Wray, G. A., and R. A. Raff. 1998. Body builders of the sea. Nat. Hist. **107**:38–47. *Regulatory genes in bilateral animals have assumed new but analogous roles in radial echinoderms.*

ZOOLOGY LINKS TO THE INTERNET

Visit the textbook's Online Learning Center at www.mhhe.com/zoology to find live Internet links for each of the topics listed here.

Phylum Echinodermata
Class Asteroidea
Phylum Hemichordata

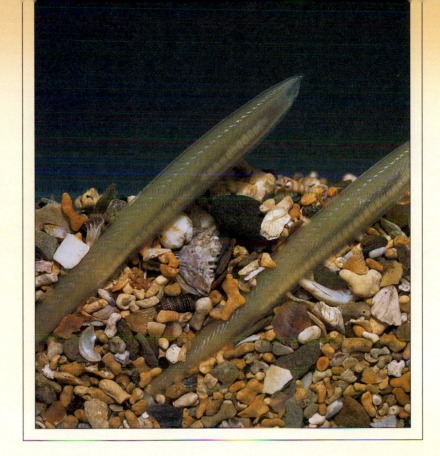

Two amphioxus
in feeding posture.

Chordates

General Characteristics, Protochordates, and Origin of the Early Vertebrates

It's a Long Way From Amphioxus

Along the more southern coasts of North America, half buried in sand on the sea floor, lives a small fishlike translucent animal quietly filtering organic particles from seawater. Inconspicuous, of no commercial value and largely unknown, this creature is nonetheless one of the famous animals of classical zoology. It is amphioxus, an animal that wonderfully exhibits the five distinctive hallmarks of the phylum Chordata—(1) dorsal, tubular nerve cord overlying (2) a supportive notochord, (3) pharyngeal slits and (4) endostyle for filter feeding, and (5) a postanal tail for propulsion—all wrapped up in one creature with textbook simplicity. Amphioxus is an animal that might have been designed by a zoologist for the classroom. During the nineteenth century, with interest in vertebrate ancestry running high, amphioxus was considered by many to resemble closely the direct ancestor of vertebrates. Its exalted position was later acknowledged by Philip Pope in a poem sung to the tune of "Tipperary." It ends with the refrain:

It's a long way from amphioxus
It's a long way to us.
It's a long way from amphioxus
To the meanest human cuss.
Well, it's good-bye to fins and gill slits
And its welcome lungs and hair,
It's a long, long way from amphioxus
But we all came from there.

But amphioxus' place in the sun was not to endure. For one thing, amphioxus lacks one of the most important of vertebrate characteristics, a distinct head with special sense organs and the equipment for shifting to an active predatory mode of life. Absence of a head, together with several specialized features, suggests to zoologists today that amphioxus represents an early departure from vertebrate ancestry the main line of chordate descent. It seems that we are a very long way indeed from amphioxus. Nevertheless, while amphioxus is denied the vertebrate ancestral award, it resembles the chordate condition immediately preceding the origin of vertebrates more closely than any other living animal we know.

THE CHORDATES

The animals most familiar to most people belong to phylum Chordata (kor-da′ta) (L. *chorda*, cord). Humans are members and share with other chordates the characteristic from which the phylum derives its name—the **notochord** (Gr. *nōton,* back, + L. *chorda,* cord) (Figure 23-1). All members of the phylum possess this structure, either restricted to early development or present throughout life. The notochord is a rodlike, semirigid body of cells enclosed by a fibrous sheath, which extends, in most cases, the length of the body just ventral to the central nervous system. Its primary purpose is to support and stiffen the body to provide support for muscles.

The structural plan of chordates shares features of many nonchordate invertebrates, such as bilateral symmetry, anteroposterior axis, coelom, tube-within-a-tube arrangement, metamerism, and cephalization. However, the exact phylogenetic position of chordates within the animal kingdom is unclear.

Two possible lines of descent have been proposed. Earlier speculations that focused on the arthropod-annelid-mollusc group (Protostomia branch) of the invertebrates have fallen from favor. It is now believed that only members of the echinoderm-hemichordate assemblage (Deuterostomia branch) deserve serious consideration as a chordate sister group. Chordates share with other Deuterostomes several important characteristics: radial cleavage (p. 156), anus derived from the first embryonic opening (blastopore) and mouth derived from an opening of secondary origin, and a coelom primitively formed by fusion of enterocoelous pouches (except in vertebrates in which coelom formation is schizocoelus, but independently derived, as an accommodation for their large yolks). These uniquely shared characteristics indicate a natural unity among the Deuterostomia.

As a whole, there is more fundamental unity of plan throughout all the organs and systems of this phylum than there is in many other phyla. Ecologically the chordates are among the most adaptable of organic forms and are able to occupy most kinds of habitat. They illustrate perhaps better than any other animal group the basic evolutionary processes of origin of new structures, adaptive strategies, and adaptive radiation.

Traditional and Cladistic Classification of the Chordates

Traditional Linnaean classification of chordates (p. 482) provides a simple and convenient way to indicate the taxa included in each major group. However, in cladistic usage, some of the traditional taxa, such as Agnatha and Reptilia, are no longer recognized. Such taxa do not satisfy the requirement of cladistics that only **monophyletic** groups are valid taxonomic entities, that is, groups that contain all known descendants of a single common ancestor. The reptiles, for example, are considered a **paraphyletic** grouping because this group does not contain all of the descendants of their most recent common ancestor (p. 538). The common ancestor of reptiles as traditionally recognized is also the ancestor of birds and mammals. As shown in the cladogram (Figure 23-3), reptiles, birds, and mammals compose a monophyletic group called Amniota, so named because all

Position in the Animal Kingdom

Phylum Chordata (kor-da′ta) (L. *chorda*, cord) belongs to the Deuterostomia branch of the animal kingdom, which includes the phyla Echinodermata and Hemichordata. These three phyla share many embryological features and are probably descended from an ancient common ancestor. From humble beginnings, chordates have evolved a vertebrate body plan of enormous adaptability that always remains distinctive, while providing almost unlimited scope for specialization in life habitat, form, and function.

Biological Contributions

1. The **endoskeleton** of vertebrates permits continuous growth without molting and attainment of large body size, and it provides an efficient framework for muscle attachment.
2. The **perforated pharynx** of protochordates that originated as a suspension-feeding device served as the framework for subsequent evolution of true internal gills with pharyngeal muscular pump, and jaws.
3. Adoption of a **predatory habit** by the early vertebrates and accompanying evolution of a **highly differentiated brain** and **paired special sense organs** contributed in large measure to successful adaptive radiation of vertebrates.
4. **Paired appendages** that appeared in aquatic vertebrates were successfully adapted later as jointed limbs for efficient locomotion on land or as wings for flight.

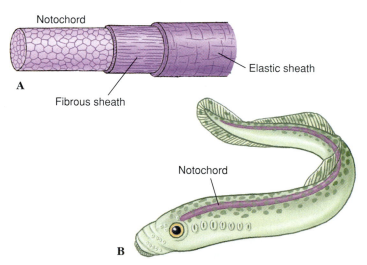

Figure 23-1

A, Structure of the notochord and its surrounding sheaths. Cells of the notochord proper are thick walled, pressed together closely, and filled with semifluid. Stiffness is caused mainly by turgidity of fluid-filled cells and surrounding connective tissue sheaths. This primitive type of endoskeleton is characteristic of all chordates at some stage of the life cycle. The notochord provides longitudinal stiffening of the main body axis, a base for trunk muscles, and an axis around which the vertebral column develops. **B,** In hagfishes and lampreys it persists throughout life, but in other vertebrates it is largely replaced by vertebrae. In mammals slight remnants are found in nuclei pulposi of intervertebral discs. The method of notochord formation is different in the various groups of animals. In amphioxus it originates from endoderm; in birds and mammals it arises as an anterior outgrowth of the embryonic primitive streak.

Characteristics of Phylum Chordata

1. Bilateral symmetry; segmented body; three germ layers; well-developed coelom
2. **Notochord** (a skeletal rod) present at some stage in the life cycle
3. **Single, dorsal, tubular nerve cord;** anterior end of cord usually enlarged to form brain
4. **Pharyngeal pouches** present at some stage in the life cycle; in aquatic chordates these develop into pharyngeal slits
5. **Endostyle** in floor of pharynx or a **thyroid gland** derived from the endostyle
6. **Postanal tail,** usually projecting beyond the anus at some stage but may or may not persist
7. Complete digestive system
8. **Segmentation,** if present, restricted to outer body wall, head, and tail and not extending into coelom

develop from an egg having special extraembryonic membranes, one of which is the amnion. Therefore according to cladistics, the reptiles can only be used as a term of convenience to refer to amniotes that are not birds or mammals; there are no derived characters that unite reptiles to the exclusion of birds and mammals. The reasons why nonmonophyletic groups are not used in cladistic taxonomy are explained in Chapter 10 (p. 200).

The phylogenetic tree of chordates (Figure 23-2) and the cladogram of chordates (Figure 23-3) provide different kinds of information. The cladogram shows a nested hierarchy of taxa grouped by their sharing of derived characters. These characters may be morphological, physiological, embryological, behavioral, chromosomal, or molecular in nature. By contrast, the branches of a phylogenetic tree are intended to represent real lineages that occurred in the evolutionary past. Geological information regarding ages of lineages is added to information from the cladogram to generate a phylogenetic tree for the same taxa.

In our treatment of chordates, we have retained the traditional Linnaean classification (p. 482) because of its conceptual usefulness and because the alternative—thorough revision following cladistic principles—would require extensive change and virtual abandonment of familiar rankings. However, we have tried to use monophyletic taxa as much as possible, because such usage is necessary to reconstruct the evolution of morphological characters in chordates.

Several traditional divisions of phylum Chordata used in Linnaean classifications are shown in Table 23-1. A fundamental separation is Protochordata from Vertebrata. Since the former lack a well-developed head, they are also called Acraniata. All vertebrates have a well-developed skull enclosing the brain and are called Craniata. We should note that some cladistic classifications exclude Myxini (hagfishes) from the group Vertebrata because they lack vertebrae, although retaining them in Craniata since they do have a cranium. The vertebrates (craniates) may be variously subdivided into groups based on shared possession of characteristics. Two such subdivisions shown in Table 23-1 are: (1) Agnatha, vertebrates lacking jaws (hagfishes and lampreys), and Gnathostomata, vertebrates having jaws (all other

vertebrates) and (2) Amniota, vertebrates whose embryos develop within a fluid-filled sac, the amnion (reptiles, birds, and mammals), and Anamniota, vertebrates lacking this adaptation (fishes and amphibians). The Gnathostomata in turn can be subdivided into Pisces, jawed vertebrates with appendages, if any, in the shape of fins; and Tetrapoda (Gr. *tetras,* four, + *podos,* foot), jawed vertebrates with two pairs of limbs. Note that several of these groupings are paraphyletic (Protochordata, Acraniata, Agnatha, Anamniota, Pisces) and consequently are not accepted in cladistic classifications. Accepted monophyletic taxa are shown at the top of the cladogram in Figure 23-3 as a nested hierarchy of increasingly more inclusive groupings.

FIVE CHORDATE HALLMARKS

The five distinctive characteristics that, taken together, set chordates apart from all other phyla are **notochord, dorsal tubular nerve cord, pharyngeal pouches** or slits, **endostyle,** and **postanal tail.** These characteristics are always found at some embryonic stage, although they may be altered or may disappear in later stages of the life cycle. All but pharyngeal pouches or slits are unique to chordates; hemichordates also have pharyngeal slits.

Notochord

Notochord

The notochord is a flexible, rodlike structure, extending the length of the body. It is the first part of the endoskeleton to appear in the embryo. The notochord is an axis for muscle attachment, and because it can bend without shortening, it permits undulatory movements of the body. In most protochordates and in jawless vertebrates, the notochord persists throughout life (Figure 23-1). In all vertebrates except hagfishes a series of cartilaginous or bony vertebrae are formed from mesenchymal cells derived from blocks of mesodermal cells (somites) lateral to the notochord. In most vertebrates, the notochord is displaced by vertebrae, although remains of the notochord may persist between or within the vertebrae.

Brain   Dorsal nerve cord

Dorsal Tubular Nerve Cord

In most invertebrate phyla that have a nerve cord, it is ventral to the alimentary canal and is solid, but in chordates the single cord is dorsal to the alimentary canal and is a tube (although the hollow center may be nearly obliterated during growth). The anterior end becomes enlarged to form the brain in most chordates.

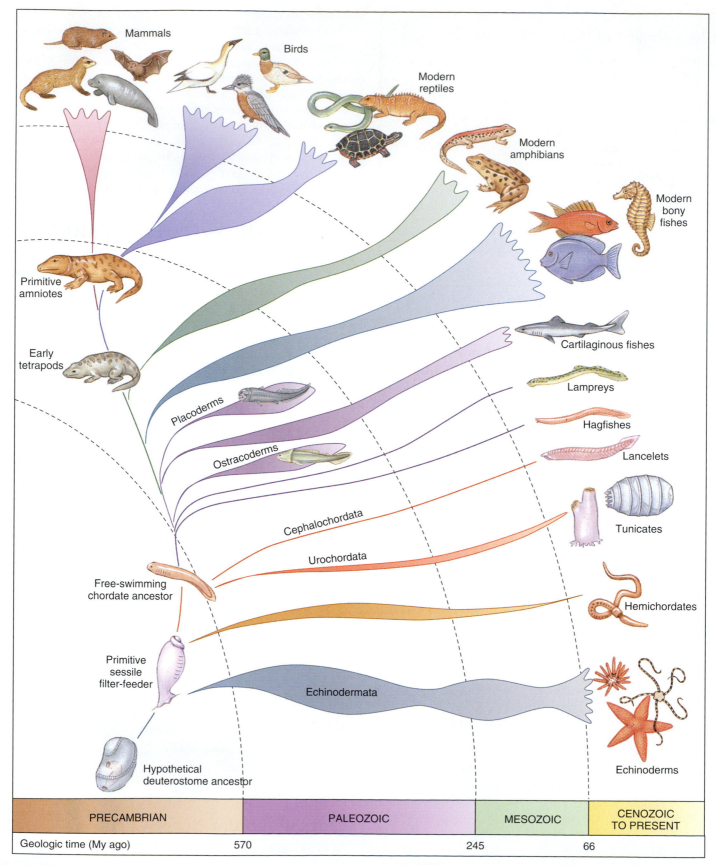

Figure 23-2
Phylogenetic tree of the chordates, suggesting probable origin and relationships. Other schemes have been suggested and are possible. The relative abundance in numbers of species of each group through geological time, as indicated by the fossil record, is suggested by the bulging and thinning of that group's line of descent.

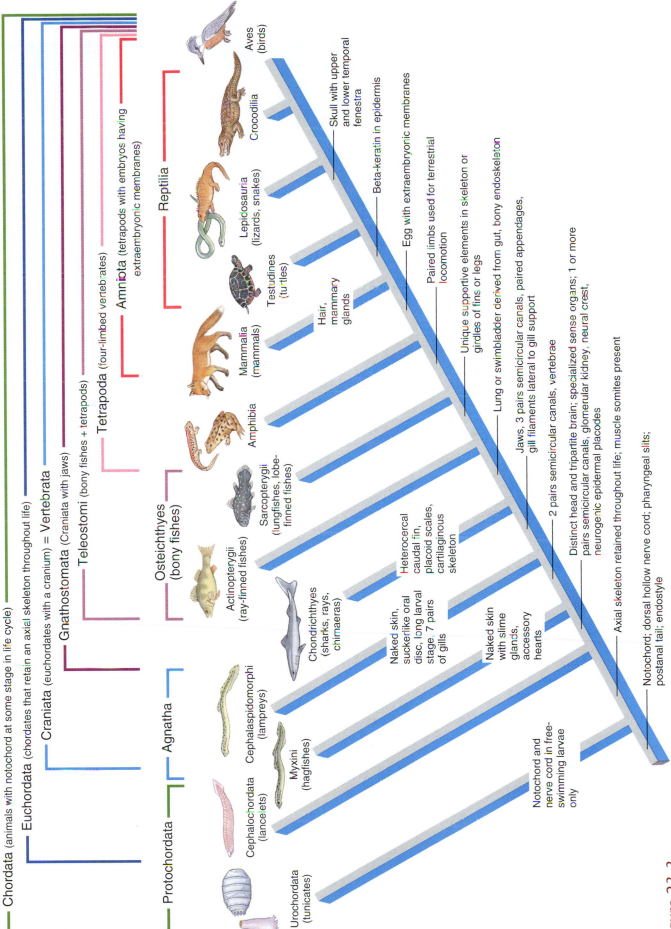

Figure 23-3

Cladogram of living members of phylum Chordata showing probable relationships of monophyletic groups composing the phylum. Each branch in the cladogram represents a monophyletic group. Some derived character states that identify the branchings are shown at right of the branch points. Nesting brackets across the top of the cladogram identify monophyletic groupings within the phylum. The term Craniata, although commonly equated with Vertebrata, is preferred by many authorities because it recognizes that some jawless vertebrates have a cranium but no vertebrae. The lower set of brackets identify the traditional groupings Protochordata, Agnatha, Osteichthyes, and Reptilia. These paraphyletic groups are not recognized in cladistic treatments, but are shown because of widespread use.

TABLE 23.1

Traditional Divisions of the Phylum Chordata

Urochordata (tunicates)	Cephalochordata (lancelets)	Myxini (hagfishes)	Cephalaspidomorphi (lampreys)	Chondrichthyes (sharks)	Osteichthyes (bony fishes)	Amphibia (amphibians)	Reptilia (reptiles)	Aves (birds)	Mammalia (mammals)
◄——————————————————————————————— Chordata ———————————————————————————————►									
◄— Protochordata —►		◄——————————————————————— Vertebrata ———————————————————————►							
◄— Acraniata —►		◄——————————————————————— Craniata ————————————————————————►							
		◄——————— Agnatha ———————►		◄——————————————— Gnathostomata ———————————————►					
				◄——— Pisces ———►		◄————————————— Tetrapoda —————————————►			
		◄————————————————— Anamniota —————————————————►					◄————————— Amniota —————————►		

The hollow cord is produced in the embryo by infolding of ectodermal cells on the dorsal side of the body above the notochord. Among vertebrates, the nerve cord passes through the protective neural arches of the vertebrae, and the anterior brain is surrounded by a bony or cartilaginous cranium.

Pharyngeal slits

Pharyngeal Pouches and Slits

Pharyngeal slits are openings that lead from the pharyngeal cavity to the outside. They are formed by inpocketing of the outside ectoderm (pharyngeal grooves) and evagination, or outpocketing, of the endodermal lining of the pharynx (pharyngeal pouches). In aquatic chordates, the two pockets break through the pharyngeal cavity where they meet to form the pharyngeal slit. In amniotes some pockets may not break through the pharyngeal cavity and only grooves are formed instead of slits. In tetrapod (four-footed) vertebrates the pharyngeal pouches give rise to several different structures, including the Eustachian tube, middle ear cavity, tonsils, and parathyroid glands (see p. 170).

The perforated pharynx evolved as a filter-feeding apparatus and is used as such in protochordates. Water with suspended food particles is drawn by ciliary action through the mouth and flows out through pharyngeal slits where food is trapped in mucus. Later, in vertebrates, ciliary action was replaced by a muscular pump that drives water through the pharynx by expanding and contracting the pharyngeal cavity. Also modified were the aortic arches that carry blood through the pharyngeal bars. In protochordates these are simple vessels surrounded by connective tissue. Early fishes added a capillary network having only thin, gas-permeable walls, thus improving efficiency of gas transfer between blood and the water outside. These adaptations led to the evolution of **internal gills,** completing conversion of the pharynx from a filter-feeding apparatus in protochordates to a respiratory organ in aquatic vertebrates.

Endostyle

Endostyle or Thyroid Gland

Until recently, the endostyle was not recognized as a chordate character. However, it or its derivative, the thyroid gland, is found in all chordates, but in no other animals. The endostyle, in the pharyngeal floor, secretes mucus that traps small food particles brought into the pharyngeal cavity. An endostyle is found in protochordates and lamprey larvae. Some cells in the endostyle secrete iodinated proteins. These cells are homologous with the iodinated-hormone-secreting thyroid gland of adult lampreys and the remainder of vertebrates. In primitive chordates, endostyle and perforated pharynx work together to create an efficient filter-feeding apparatus.

Muscular, postanal tail

Postanal Tail

The postanal tail, together with somatic musculature and the stiffening notochord, provides the motility that larval tunicates and amphioxus need for their free-swimming existence. As a structure added to the body behind the end of the digestive tract, it clearly has evolved specifically for propulsion in water. Its efficiency is later increased in fishes with the addition of fins. The tail is evident in humans only as a vestige (the coccyx, a series of small vertebrae at the end of the spinal column) but most other mammals have a waggable tail as adults.

ANCESTRY AND EVOLUTION

Since the mid-nineteenth century when the theory of organic evolution became the focal point for ferreting out relationships among groups of living organisms, zoologists have debated the question of chordate origins. It has been very difficult to reconstruct lines of descent because the earliest protochordates were probably soft-bodied creatures that stood little chance of being preserved as fossils even under the most ideal conditions. Consequently, such reconstructions largely come from the study of living organisms, especially from an analysis of early developmental stages, which tend to be more evolutionarily conserved than the differentiated adult forms they become.

Most early efforts to identify kinship of chordates to other phyla are now recognized as based on similarities related to analogy rather than homology. Analogous structures are those that perform similar functions but have different origins (such as wings of birds and butterflies). Homologous structures, on the other hand, share a common origin but may look different (at least superficially) and perform quite different functions. For example, all vertebrate forelimbs are homologous because they are derived from a pentadactyl limb of the same ancestor, even though they may be modified as differently as a human's arm and a bird's wing. Homologous structures share a genetic heritage; analogous structures do not. Obviously, only homologous similarities reveal common ancestry.

Zoologists at first speculated that chordates evolved within the protostome lineage (annelids and arthropods) but discarded such ideas when they realized that supposed morphological similarities had no developmental basis. Early in the twentieth century when further theorizing became rooted in developmental patterns of animals, it became apparent that the chordates must have originated within the deuterostome branch of the animal kingdom. As explained in Chapter 8 (p. 156 and Figure 8-9), Deuterostomia, a grouping that includes the echinoderms, hemichordates, and chordates, has several important embryological features that clearly separate it from the Protostomia and establish its monophyly. Thus the deuterostomes are almost certainly a natural grouping of interrelated animals that have their common origin in ancient Precambrian seas. Several lines of anatomical, developmental, and molecular evidence suggest that somewhat later, at the base of the Cambrian period some 570 million years ago, the first distinctive chordates arose from a lineage related to echinoderms and hemichordates (Figure 23-2; see also Figure 22-34, p. 465). Some workers suggest that Hemichordata is the sister group to Chordata, citing pharyngeal slits as a shared derived character. Others suggest that the chordate ancestry lies with an extinct free-swimming echinoderm.

While modern echinoderms look nothing at all like modern chordates, evolutionary affinity between chordates and echinoderms gains support from fossil evidence. One curious group of fossil echinoderms, the Calcichordata (Figure 23-4, see also p. 458) have been suggested as basal chordates. These small, nonsymmetrical forms have a head resembling a long-toed medieval boot and structures that have been interpreted as a series of pharyngeal slits, a postanal tail, notochord, and muscle blocks. However, homologies of these internal structures with those of chordates is highly speculative and the minerals of the hard skeleton also differ, being calcium carbonate in calcichordates

Pharyngeal slits　　Mouth

Anus　Tail

Figure 23-4

Fossil of an early echinoderm, a calcichordate, that lived during the Ordovician period (450 million years BP). It shows affinities with both echinoderms and chordates.

and calcium phosphate in chordates. An additional problem is that the proposed stem calcichordates are younger (Ordovician) than the earliest chordate fossils (early Cambrian). With the difficulties of the calcichordate hypothesis we prefer to treat a different deuterostome group, perhaps hemichordates, as the sister group of chordates (see Figure 22-34, p. 465). Thus, despite the uncertainty of the identity of the long-sought chordate ancestor, we do know two living protochordate groups that descended from it. These we will now consider.

SUBPHYLUM UROCHORDATA (TUNICATA)

Urochordates ("tail-chordates"), more commonly called tunicates, include about 2000 species. They are found in all seas from near shoreline to great depths. Most are sessile as adults, although some are free living. The name "tunicate" is suggested by the usually tough, nonliving **tunic,** or test, that surrounds the animal and contains cellulose (Figure 23-5). As adults, tunicates are highly specialized chordates, for in most species only the larval form, which resembles a microscopic tadpole, bears all the chordate hallmarks. During adult metamorphosis, the notochord (which, in the larva, is restricted to the tail, hence the group name Urochordata) and tail disappear, while the dorsal nerve cord becomes reduced to a single ganglion.

Urochordata is divided into three classes: **Ascidiacea** (Gr. *askiolion,* little bag, + *acea,* suffix), **Larvacea** (L. *larva,* ghost, + *acea,* suffix), and **Thaliacea** (Gr. *thalia,* luxuriance, + *acea,* suffix). Of these members of Ascidiacea are by far the most common, diverse, and best known. They are often called "sea squirts" because some species forcefully discharge a jet of water from the excurrent siphon when irritated. All but a few ascidian species are sessile animals, attached to rocks or other hard substrates such as pilings or bottoms of ships. In many areas, they are among the most abundant of intertidal animals.

Ascidians may be solitary, colonial, or compound. Each of the solitary and colonial forms has its own test, but among compound forms many individuals may share the same test (Figure 23-6). In

Figure 23-5

Structure of a common tunicate, *Ciona* sp.

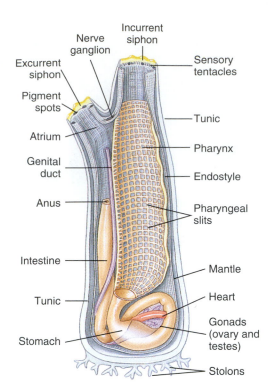

Incurrent siphon — Nerve ganglion — Excurrent siphon — Sensory tentacles — Pigment spots — Tunic — Atrium — Pharynx — Genital duct — Endostyle — Anus — Pharyngeal slits — Intestine — Mantle — Tunic — Heart — Stomach — Gonads (ovary and testes) — Stolons

A

B

Figure 23-6

A, Group of simple tunicates, *Clavelina puertosecensis,* on a Caribbean reef. Note separate incurrent and excurrent siphons for each animal. **B,** Seven colonies of compound tunicates, *Atriolum robustum,* on a Pacific reef. Individuals in a colony share a common tunic (yellow), but each has a separate incurrent (oral) siphon. Each colony has a single, large excurrent (atrial) siphon on top.

some compound ascidians each member has its own incurrent siphon, but the excurrent opening is common to the group.

Solitary ascidians (Figure 23-5) are usually spherical or cylindrical forms. Lining the tunic is an inner membrane, the **mantle.** On the outside are two projections: an **incurrent siphon,** or oral siphon, which corresponds to the anterior end of the body, and an **excurrent siphon,** or atrial siphon, that marks the dorsal side. When a sea squirt is expanded, water enters the incurrent siphon and passes into a capacious ciliated **pharynx** that is minutely subdivided by gill slits to form an elaborate basketwork. Water passes through the gill slits into an **atrial cavity** and out through the excurrent siphon.

Feeding depends on formation of a mucous net that is secreted by a glandular groove, the **endostyle,** located along the midventral side of the pharynx. Cilia on gill bars of the pharynx pull the mucus into a sheet that spreads dorsally across the inner face of the pharynx. Food particles brought in the incurrent opening are trapped on the mucous net, which is then worked into a rope and carried posteriorly by cilia into the esophagus and stomach. Nutrients are absorbed in the midgut and indigestible wastes are discharged from the anus, located near the excurrent siphon. The terminal part of the gut referred to as the intestine actually is not homologous with the intestine of other chordates, but rather with the hepatic cecum of amphioxus.

The circulatory system consists of a ventral heart and two large vessels, one on either side of the heart; these vessels connect to a diffuse system of smaller vessels and spaces serving the pharyngeal basket (where respiratory exchange occurs), the digestive organs, gonads, and other structures. An odd feature found in no other chordate is that the heart drives the blood first in one direction for a few beats, then pauses, reverses its action, and drives the blood in the opposite direction for a few beats. Another remarkable feature is the presence of strikingly high

amounts of rare elements in the blood, such as vanadium and niobium. The vanadium concentration in the sea squirt *Ciona* may reach 2 million times its concentration in seawater. The function of these rare metals in the blood is a mystery.

The nervous system is restricted to a **nerve ganglion** and plexus of nerves that lie on the dorsal side of the pharynx. Beneath the nerve ganglion is located the **subneural gland,** connected by a duct to the pharynx.

Sea squirts are hermaphroditic, with usually a single ovary and a single testis in the same animal. Germ cells are carried by ducts into the atrial cavity, and then into the surrounding water where fertilization occurs.

Of the five chief characteristics of chordates, adult sea squirts have only two: pharyngeal slits and endostyle. However, the larval form reveals the secret of their true relationship. The

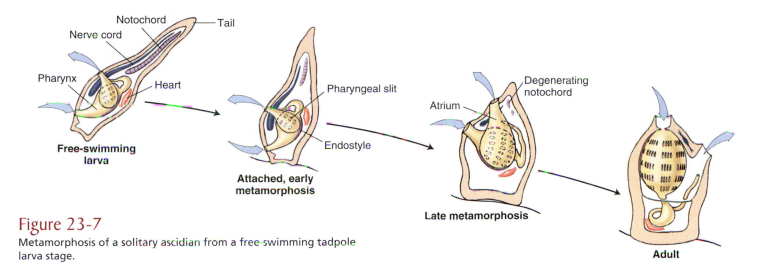

Figure 23-7
Metamorphosis of a solitary ascidian from a free-swimming tadpole larva stage.

tadpole larva (Figure 23-7) is an elongate, transparent form with all five chordate characteristics: notochord, hollow dorsal nerve cord, propulsive postanal tail, and a large pharynx with endostyle and pharyngeal slits. The larva does not feed but swims for some hours before fastening itself vertically by adhesive papillae to a solid object. It then undergoes a dramatic metamorphosis (Figure 23-7) to become a sessile adult, so modified as to become almost unrecognizable as a chordate.

Tunicates of class Thaliacea, known as thaliaceans or salps, are barrel- or lemon-shaped pelagic forms with transparent, gelatinous bodies that, despite the considerable size that some species reach, are nearly invisible in sunlit surface waters. They occur singly or in colonial chains that may reach several meters in length (Figure 23-8). The cylindrical thaliacean body is typically surrounded by bands of circular muscle, with incurrent and excurrent siphons at opposite ends. Water pumped through the body by muscular contraction (rather than by cilia as in ascidians) is used for locomotion by a sort of jet propulsion, for respiration, and as a source of particulate food that is filtered on mucous surfaces. Many are provided with luminous organs, which give a brilliant light at night. Most of the body is hollow, with the viscera forming a compact mass on the ventral side.

The life histories of thaliaceans are often complex and are adapted to respond to sudden increases in their food supply. The appearance of a phytoplankton bloom, for example, is met by an explosive population increase leading to extremely high density of thaliaceans. Common forms include *Doliolum* and *Salpa,* both of which reproduce by an alternation of sexual and asexual generations.

The third tunicate class, the Larvacea (Appendicularia in some classifications) are curious larvalike pelagic creatures shaped like a bent tadpole. In fact their resemblance to the larval stages of other tunicates has given them their class name of Larvacea. They feed by a method unique in the animal world. Each builds a delicate house, a transparent hollow sphere of mucus interlaced with filters and passages through which water enters (Figure 23-9). Particulate food trapped on a feeding filter inside the house is drawn into the animal's mouth through a strawlike tube. When the filters become clogged with waste, which happens about every 4 hours, the larvacean abandons its house and builds a new house, a process that takes only a few minutes. Like thaliaceans, larvaceans can quickly build up dense populations when food is abundant. At such times scuba diving among the houses, which are about the size of walnuts, is likened to swimming through a snowstorm! Larvaceans are paedomorphic; they are sexually mature animals that have retained the larval body form of their evolutionary ancestors (see the boxed note explaining paedomorphosis on p. 480).

SUBPHYLUM CEPHALOCHORDATA

Cephalochordates are lancelets: slender, laterally compressed, translucent animals about 5 to 7 cm in length (Figure 23-10) that inhabit the sandy bottoms of coastal waters around the world. Lancelets originally bore the generic name *Amphioxus* (Gr. *amphi,* both ends, + *oxys,* sharp), later surrendered by priority to *Branchiostoma* (Gr. *branchia,* gills, + *stoma,* mouth). Amphioxus is still used, however, as a convenient common name for all of approximately 25 species in this diminutive subphylum. Four species of amphioxus occur in North American coastal waters.

Amphioxus is especially interesting because it has the five distinctive characteristics of chordates in simple form. Water

Figure 23-8
Colonial thaliacean. The transparent individuals of this delicate, planktonic species are grouped in a chain. Visible within each individual is an orange gonad, an opaque gut, and a long serrated gill bar.

Figure 23-9

Larvacean adult *(left)* and as it appears within its transparent house *(right),* which is about the size of a walnut. When the feeding filters become clogged with food, the tunicate abandons its house and builds a new one.

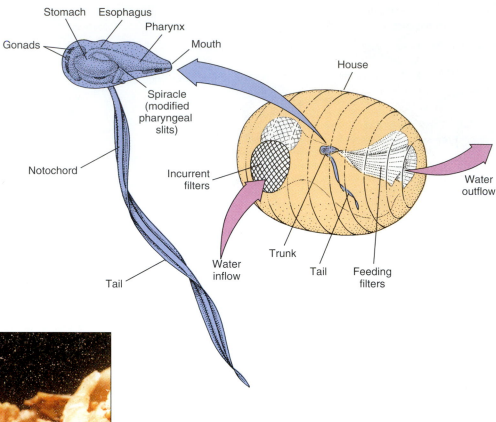

Stomach   Esophagus
Pharynx
Gonads
Mouth
Spiracle (modified pharyngeal slits)
House
Notochord
Incurrent filters
Water outflow
Trunk
Water inflow
Tail   Feeding filters
Tail

A

Figure 23-10

Amphioxus. This interesting bottom-dwelling cephalochordate illustrates the five distinctive chordate characteristics (notochord, dorsal nerve cord, pharyngeal slits, endostyle, and postanal tail). The vertebrate ancestor is thought to have had a similar body plan. **A,** Living amphioxus in typical position for filter feeding. Note the oral hood with tentacles surrounding the mouth. **B,** Internal structure.

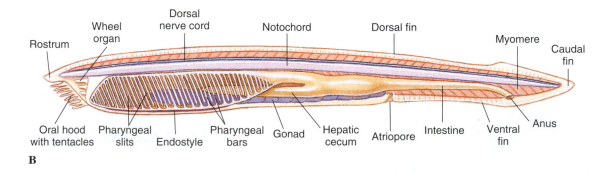

Dorsal nerve cord
Wheel organ
Notochord
Dorsal fin
Myomere
Caudal fin
Rostrum
Oral hood with tentacles
Pharyngeal slits
Endostyle
Pharyngeal bars
Gonad
Hepatic cecum
Atriopore
Intestine
Ventral fin
Anus

B

enters the mouth, driven by cilia in the buccal cavity, then passes through numerous pharyngeal slits where food is trapped in mucus, which is then moved by cilia into the intestine. Here the smallest food particles are separated from the mucus and passed into the **hepatic cecum** where they are phagocytized and digested intracellularly. As in tunicates, filtered water passes first into an **atrium,** then leaves the body by an **atriopore** (equivalent to the excurrent siphon of tunicates).

The closed circulatory system is complex for so simple a chordate. The flow pattern is remarkably similar to that of primitive fishes, although there is no heart. Blood is pumped forward in the **ventral aorta** by peristaltic-like contractions of the vessel wall, then passes upward through branchial arteries (aortic arches) in the pharyngeal bars to paired **dorsal aortas** which join to become a single dorsal aorta. From here blood is distributed to body tissues by microcirculation and then is collected in veins, which return it to the ventral aorta. Lacking both erythrocytes and hemoglobin, their blood is thought to transport nutrients but play little role in gas exchange.

The nervous system is centered around a hollow nerve cord lying above the notochord. Pairs of **spinal nerve roots** emerge at each trunk myomeric (muscle) segment. Sense organs are simple, unpaired bipolar receptors located in various parts of the body. The "brain" is a simple vesicle at the anterior end of the nerve cord.

Sexes are separate. Sex cells are set free in the atrial cavity, then pass out the atriopore to the outside where fertilization occurs. Cleavage is total (holoblastic) and a gastrula is formed by invagination. Larvae hatch soon after egg deposition and gradually assume the shape of adults.

No other chordate shows the basic diagnostic chordate characteristics as clearly as amphioxus. In addition to the five chordate anatomical hallmarks, amphioxus possesses several structural features that suggest the vertebrate plan. Among these are a hepatic cecum, a diverticulum that resembles the vertebrate pancreas in secreting digestive enzymes, **segmented trunk musculature,** and the basic circulatory plan of more advanced chordates. As discussed on p. 480, most zoologists consider amphioxus a living descendant of an ancestor that gave rise to both cephalochordates and vertebrates. Therefore cephalochordates are, in cladistic terms, the sister group of vertebrates (Figure 23-3).

SUBPHYLUM VERTEBRATA (CRANIATA)

The third subphylum of chordates is the large and diverse Vertebrata. This monophyletic group shares the basic chordate characteristics with the other two subphyla, but in addition it demonstrates a number of novel characters that the others do not share. The alternative name of the subphylum, Craniata, more accurately describes the group since all have a cranium (bony or cartilaginous braincase) whereas some jawless fishes lack vertebrae.

Adaptations That Have Guided Early Vertebrate Evolution

The earliest vertebrates were substantially larger and considerably more active than the protochordates. Increased speed and mobility resulted from modifications of the skeleton and muscles. The higher activity level and size of vertebrates also requires structures specialized in the location, capture, and digestion of food and adaptations designed to support a high metabolic rate.

Musculoskeletal Modifications

Most vertebrates possess both an exoskeleton and endoskeleton of cartilage or bone. Growing within the body as it does, the endoskeleton permits almost unlimited body size with much greater economy of building materials. Some vertebrates have become the most massive organisms on earth. The endoskeleton forms an excellent jointed scaffolding for attachment of segmented muscles.

The segmented body muscles (myomeres) changed from the V-shaped muscles of cephalochordates to the W-shaped muscles of vertebrates. This increased complexity of folding in the myomeres provides powerful control over an extended length of the body. Also unique to vertebrates are the presence of fin rays of dermal origin in the fins, aiding in swimming.

The endoskeleton probably was composed initially of cartilage that later gave rise to bone. Cartilage, with its fast growth and flexibility, is ideal for constructing the first skeletal framework of all vertebrate embryos. The endoskeleton of living hagfishes, lampreys, sharks and their kin, and even in some "bony" fishes, such as sturgeons, is mostly composed of cartilage. Bone may have been adaptive in early vertebrates in several ways. The presence of bone in the skin of ostracoderms and other ancient fishes certainly provided protection from predators, although there are some more important benefits of bone. The structural strength of bone is superior to cartilage, making it ideal for muscle attachment in areas of high mechanical stress. One of the most interesting ideas is that the function associated with the origin of bone was for mineral regulation. Phosphorus and calcium are used for many physiological processes and are in particularly high demand in organisms with high metabolic rates. Storage and regulation of calcium and phosphorus ions were likely important functions of bone in the earliest vertebrates.

We should note that most vertebrates possess an extensive exoskeleton, although it is highly modified in advanced forms. Some of the most primitive fishes, including ostracoderms and placoderms, were partly covered in a bony, dermal armor. This armor is modified in later fishes as scales. Many of the bones encasing the brain of advanced vertebrates develop from tissue that originates from the dermis! Most vertebrates are further protected with keratinized structures derived from the epidermis, such as reptilian scales, hair, feathers, claws, and horns.

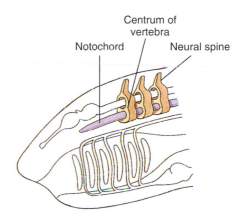

Physiology

Vertebrates have modifications to the digestive, respiratory, circulatory, and excretory systems required to meet an increased metabolic demand. The perforated pharynx evolved as a filter-feeding device in early chordates. Water with suspended food particles was drawn through the pharynx by ciliary action and trapped by mucus secreted by the endostyle. In larger, predatory vertebrates, the pharynx was modified into a muscular apparatus that pumped water through the pharynx. With the origin of highly vascularized gills, the function of the pharynx shifted to primarily gas exchange. Changes in the gut, including a shift from movement of food by ciliary action to muscular action and addition of accessory digestive glands, the liver and pancreas, were necessary to manage the increased amount of food ingested. A ventral three-chambered heart consisting of a sinus venosus, atrium, and ventricle, and erythrocytes with hemoglobin enhanced transportation of nutrients, gases, and other substances. Protochordates have no distinct kidneys, but vertebrates possess paired, glomerular kidneys that remove metabolic waste products and regulate body fluids and ions.

New Head, Brain, and Sensory Systems

When vertebrate ancestors shifted from filter feeding to active predation, new sensory, motor, and integrative controls became essential for location and capture of larger prey. The anterior end of the nerve cord became enlarged as a **tripartite brain** (forebrain, midbrain, and hindbrain) and protected by a cartilaginous or bony cranium. Paired special sense organs designed for distance reception evolved. These included eyes with lenses and inverted retinas; pressure receptors, such as paired inner ears designed for equilibrium and sound reception; chemical receptors including taste and exquisitely sensitive olfactory organs; lateral-line receptors for detecting water vibrations; and electroreceptors for detecting electrical currents that signal prey.

Neural Crest, Neural Placodes, and Hox Genes

Development of the vertebrate head and special sense organs was largely the result of two embryonic innovations present only in vertebrates: the **neural crest** and **epidermal placodes.** The neural crest, a population of ectodermal cells lying along the length of the embryonic neural tube, contributes to the forma-

tion of many different structures, among them most of the cranium, pharyngeal skeleton, teeth dentine, some cranial nerves, some ganglia, Schwann cells, and some endocrine glands. In addition they may regulate the development of adjacent tissue, such as tooth enamel and pharyngeal muscles (branchiomeres). The epidermal placodes (Gr. *placo,* plate) are platelike ectodermal thickenings that appear on either side of the neural tube. These give rise to the olfactory epithelium, lens of the eye, inner ear epithelium, some ganglia, some cranial nerves, lateral-line mechanoreceptors, and electroreceptors. The placodes also induce the formation of taste buds. Thus the vertebrate head with its sensory structures located adjacent to the mouth (later equipped with prey-capturing jaws), stemmed from the creation of completely new cell types.

Recent studies of the distribution of homeobox-containing genes that control the body plan of chordate embryos (homeobox genes are described on p. 163) suggest that the *Hox* genes were duplicated at about the time of the origin of vertebrates. One copy of *Hox* genes is found in amphioxus and other invertebrates whereas living gnathostomes have four copies. Perhaps these additional copies of body-plan-controlling genes provided genetic material free to evolve a more complex kind of animal.

The Search for the Vertebrate Ancestral Stock

Fossil invertebrate chordates are rare and known primarily from two fossil beds—the well-known middle Cambrian Burgess Shales of Canada and the recently discovered early Cambrian fossil beds of Chengjiang and Haikou, China. An ascidian tunicate and *Yunnanozoon,* a probable cephalochordate, are known from Chengjiang. Slightly better known is *Pikaia,* a ribbon-shaped, somewhat fishlike creature about 5 cm in length discovered in the Burgess Shales (Figure 23-11). The presence of V-shaped myomeres and a notochord clearly identifies *Pikaia* as a chordate. The superficial resemblance of *Pikaia* to living amphioxus suggests it may be an early cephalochordate.

Closer to the origin of vertebrates is *Haikouella,* recently discovered in 530-million-year-old sediments near Haikou. *Haikouella* is known from over 300 specimens, including 32 nearly complete fossils. This amazing amount of material provided considerable insight into the anatomy of these animals. It possessed several characters that clearly identify it as a chordate, including notochord, pharynx, and dorsal nerve cord, but also had several characters that are more typical of vertebrates (Figure 23-12). *Haikouella* seems to have had dorsal and ventral aortae, a heart, gill filaments, and a tripartite brain, although there is no evidence

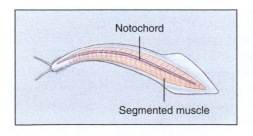

Figure 23-11

Pikaia, an early chordate, from the Burgess Shale of British Columbia, Canada.

Characteristics of Subphylum Vertebrata

1. Chief diagnostic features of chordates—**notochord, dorsal tubular nerve cord, pharyngeal pouches, endostyle** or **thyroid gland,** and **postanal tail**—all present at some stage of the life cycle
2. **Integument** basically of two divisions, an outer epidermis of stratified epithelium from ectoderm and an inner dermis of connective tissue derived from mesoderm; many modifications of skin among the various classes, such as glands, scales, feathers, claws, horns, and hair
3. Distinctive cartilage or bone **endoskeleton** consisting of vertebral column (except in hagfishes, which lack vertebrae) and a head skeleton (cranium and pharyngeal skeleton) derived largely from **neural crest cells**
4. **Muscular pharynx;** in fishes pharyngeal pouches open to the outside as slits and bear gills; in tetrapods pharyngeal pouches are sources of several glands
5. Complex, W-shaped muscle segments or **myomeres** to provide movement
6. Complete, **muscularized digestive tract** ventral to vertebral column with distinct liver and pancreas
7. Circulatory system consisting of a **ventral heart** of multiple chambers; closed blood system of arteries, veins, and capillaries; blood fluid containing **erythrocytes** with **hemoglobin;** paired aortic arches connecting ventral and dorsal aortas and giving off branches to the gills among aquatic vertebrates; in terrestrial forms; aortic arches modified into pulmonary and systemic systems
8. Well-developed **coelom** divided into a pericardial cavity and a pleuroperitoneal cavity
9. Excretory system consisting of **paired, glomerular kidneys** provided with ducts to drain waste to the cloaca
10. Highly differentiated **tripartite brain;** 10 or 12 pairs of **cranial nerves;** a pair of spinal nerves for each primitive myotome; **paired special sense organs** derived from **epidermal placodes**
11. **Endocrine system** of ductless glands scattered throughout the body
12. Nearly always separate sexes; each sex containing gonads with ducts that discharge their products either into the cloaca or into special openings near the anus
13. Most vertebrates with two pairs of appendages supported by limb girdles and appendicular skeleton

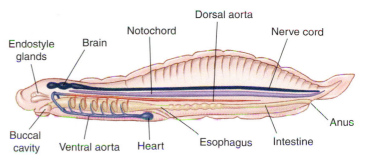

Figure 23-12

Haikouella, a chordate with several vertebrate features from early Cambrian shales in Haikou, China.

Figure 23-13

Garstang's hypothesis of larval evolution. Adult tunicates live on the sea floor but produce a free-swimming tadpole larva. More than 550 million years ago, some larvae began to reproduce in the swimming stage. These evolved into the first vertebrates.

of a cranium. The mix of vertebrate and protochordate characters suggests that evolution of "vertebrate" soft characters may have preceded evolution of an endoskeleton. Despite recent fossil discoveries of early chordates, most speculations regarding vertebrate ancestry have focused on living protochordates, in part because they are much better known than the fossil forms.

Garstang's Hypothesis of Chordate Larval Evolution

The chordates have pursued two paths in their early evolution, one path leading to the sedentary urochordates, the other to active, mobile cephalochordates and vertebrates. At the time of its

discovery in 1869, the tadpole larva of tunicates was considered a descendant of an ancient free-swimming chordate ancestor. An alternate hypothesis, proposed in 1928 by Walter Garstang of England, suggested that the chordate ancestral stock was derived by retaining into adulthood the larval form of sessile tunicate-like animals. The tadpole larva of tunicates does indeed bear all the right attributes to qualify as a possible vertebrate ancestor; notochord, hollow dorsal nerve cord, pharyngeal slits, endostyle, and postanal tail. At some point, Garstang suggested, the tadpole larva failed to metamorphose into an adult tunicate, instead developing gonads and reproducing in the larval stage. With continued evolution, a new group of free-swimming animals appeared, the ancestors of cephalochordates and vertebrates (Figure 23-13).

Garstang called this process **paedomorphosis** (Gr. *pais,* child + *morphē,* form), a term describing the evolutionary retention of juvenile or larval traits in the adult body. Garstang departed from previous thinking by suggesting that evolution may occur in larval stages of animals—and in this case, lead to the vertebrate lineage. Paedomorphosis is a well-known phenomenon in several different animal groups (paedomorphosis in amphibians is described on p. 522). Furthermore, Garstang's hypothesis agrees with the embryological evidence. Recent workers have revived the idea that the chordate ancestor was free swimming, based on phylogenies generated from molecular data. These analyses suggest that the sessile ascidians represent a derived body form, and that free-swimming larvaceans are perhaps most similar in body form to the ancestral chordates.

Position of Amphioxus

Zoologists believe that the cephalochordate amphioxus is the closest living relative of vertebrates. Cephalochordates share several characters with vertebrates that are absent from tunicates, including segmented myomeres, dorsal and ventral aortas, branchial or aortic arches, and podocytes, specialized excretory cells. However, as noted in the prologue to the chapter (p. 467), amphioxus is unlike the most recent common ancestor of vertebrates because it lacks the tripartite brain, chambered heart, special sensory organs, muscular gut and pharynx, and neural crest tissue inferred to have been present in that ancestor. In addition, the larger fins of some extinct cephalochordates suggest they were more free-swimming than modern amphioxus.

Paedomorphosis, the displacement of ancestral larval or juvenile features into a descendant adult, can be produced by three different evolutionary-development processes: neoteny, progenesis, and post-displacement. In neoteny, the growth rate of body form is slowed so that the animal does not attain the ancestral adult form when it reaches maturity. Progenesis is the precocious maturation of gonads in a larval (or juvenile) body that then stops growing and never attains the adult body form. In postdisplacement, the onset of a developmental process is delayed relative to reproductive maturation, so that the ancestral adult form is not attained by the time of reproductive maturation. Neoteny, progenesis and postdisplacement thus describe different ways in which paedomorphosis can happen. Biologists use the inclusive term paedomorphosis to describe results of these evolutionary-developmental processes.

Despite these specializations most zoologists believe that amphioxus has largely retained the body structure of the immediate prevertebrate condition. Thus cephalochordates are probably the sister group of vertebrates (Figure 23-3).

The Ammocoete Larva of Lampreys as a Model of the Primitive Vertebrate Body Plan

Lampreys (jawless fishes of the class Cephalaspidomorphi, discussed in the next chapter) have a freshwater larval stage known as the **ammocoete** (Figure 23-14). In body form, appearance, life habit, and many anatomical details, the ammocoete larva resembles amphioxus. In fact, lamprey larvae were given the genus name *Ammocoetes* (Gr. *ammos*, sand, + *koitē*, bed, referring to the preferred larval habitat) in the nineteenth century when it was erroneously thought to be an adult cephalochordate, closely allied with amphioxus. Ammocoete larvae are so different from adult lampreys that the mistake is understandable; the exact relationship was not explained until metamorphosis into an adult lamprey was observed.

Ammocoete larvae have a long, slender body with an oral hood surrounding the mouth much like amphioxus (Figure 23-14). Ammocoetes are filter feeders, but instead of drawing water by ciliary action into the pharynx as amphioxus does, ammocoetes produce a feeding current by muscular pumping action much like modern fishes. The arrangement of body muscle into myomeres, the presence of a notochord serving as chief skeletal axis, and the plan of the circulatory system all resemble these features in amphioxus.

Ammocoetes do have several characteristics lacking in amphioxus that are homologous to those of vertebrates. These include a chambered heart, a tripartite brain, special sense organs derived from epidermal placodes, and a pituitary gland. The kidney is pronephric (p. 641) and conforms to the basic vertebrate plan. Instead of the numerous pharyngeal slits of amphioxus, there are only seven pairs of pharyngeal pouches and slits in ammocoetes. From pharyngeal bars separating the pharyngeal slits project gill filaments bearing secondary lamellae much like the more extensive gills of modern fishes (see Figure 24-28, p. 505). Ammocoetes also have a true liver replacing the hepatic cecum of amphioxus, a gallbladder, and pancreatic tissue (but no distinct pancreatic gland).

Ammocoete larvae display the most primitive condition for these characteristics of any living vertebrate. They clearly illustrate many shared derived characters of vertebrates that are

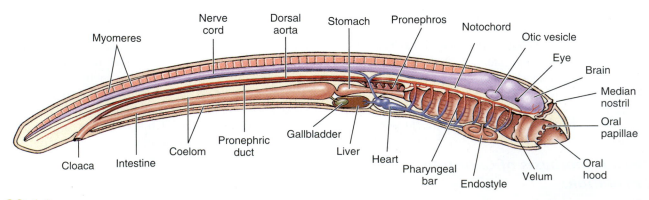

Figure 23-14

Ammocoete larva, freshwater larval stage of a sea lamprey. Although they resemble amphioxus in many ways, ammocoetes have a well-developed brain, paired eyes, pronephric kidney, and other features lacking in amphioxus but representative of the vertebrate body plan.

obscured in the development of other vertebrates. They may approach most closely the supposed body plan of the ancestral vertebrate.

> The term "ostracoderm" does not denote a natural evolutionary assemblage but rather is a term of convenience for describing several groups of heavily armored extinct jawless fishes.

The Earliest Vertebrates

The earliest known vertebrate fossils, until recently, were armored jawless fishes called **ostracoderms** (os-trak'o-derm) (Gr. *ostrakon,* shell, + *derma,* skin) from late Cambrian and Ordovician deposits. In 1999 researchers described two fishlike 530-million-year-old vertebrates, *Myllokunmingia* (Gr. *myllo,* sea fish + Kunming, a city in China) and *Haikouichthys* (Haikou, a city in China + Gr. *ichthy,* fish), from the amazing Chengjiang deposits. These fossils push back the origin of vertebrates to at least the early Cambrian. Although they possess many typically vertebrate characters, such as W-shaped myomeres, a heart, cranium, and fin rays, they lack evidence of mineralized tissues. This lack of mineralized tissue may explain the extreme rarity of vertebrate fossils prior to the late Cambrian.

The earliest ostracoderms were armored with bone in their dermis and lacked paired fins that later fishes found so important for stability (Figure 23-15). The swimming movements of one of the early groups, the **heterostracans** (Gr. *heteros,* different, + *ostrakon,* shell) must have been imprecise, although sufficient to propel them along the ocean bottom where they searched for food. With fixed circular or slitlike mouth openings they may have filtered small food particles from the water or ocean bottom. However, unlike the ciliary filter-feeding protochordates, ostracoderms sucked water into the pharynx by muscular pumping, an important innovation that suggests to some authorities that ostracoderms may have been mobile predators that fed on soft-bodied animals.

During the Devonian period, the heterostracans underwent a major radiation, resulting in the appearance of numerous peculiar-looking forms. Without ever evolving paired fins or jaws, these earliest vertebrates flourished for 150 million years until becoming extinct near the end of the Devonian period.

Coexisting with heterostracans throughout much of the Devonian period were **osteostracans** (Gr. *osteon,* bone + *ostrakon,* shell). Osteostracans had paired pectoral fins, an innovation that functioned to improve swimming efficiency by controlling yaw, pitch, and roll. A typical osteostracan such as *Cephalaspis* (Gr. *kephalē,* head, + *aspis,* shield) (Figure 23-15), was a small animal, seldom exceeding 30 cm in length. It was covered with a heavy, dermal armor of cellular bone, including a single-piece head shield. Examination of internal features of the braincase reveal a sophisticated nervous system and sense organs, similar to those of modern lampreys.

Another group of ostracoderms, the **anaspids,** (Figure 23-15) were more streamlined and more agile than other ostracoderms. These and other ostracoderms enjoyed an impressive radiation in the Silurian and Devonian periods. However, all ostracoderms became extinct by the end of the Devonian period.

For decades geologists have used strange, microscopic toothlike fossils called **conodonts** (Gr. *kōnos,* cone, + *odontos,* tooth) to date Paleozoic marine sediments without having any idea what kind of creature originally possessed these elements. The discovery in the early 1980s of fossils of complete conodont

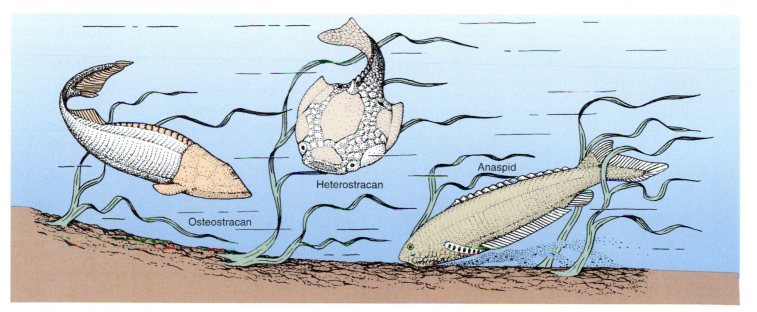

Figure 23-15

Three ostracoderms, jawless fishes of Silurian and Devonian times. They are shown as they might have appeared while searching for food on the floor of a Devonian sea. All were probably filter-feeders, but employed a strong pharyngeal pump to circulate water rather than the much more limiting mode of ciliary feeding used by their protovertebrate ancestors (presumably resembling amphioxus for this feature).

Traditional Linnean Classification of Living Members of Phylum Chordata

Phylum Chordata

Subphylum Urochordata (u'ro-kor-da'ta) (Gr. *oura,* tail, + L. *chorda,* cord, + *ata,* characterized by) **(Tunicata): tunicates.** Notochord and nerve cord in free-swimming larva only; ascidian adults sessile, encased in tunic. About 2000 species.

Subphylum Cephalochordata (sef'a-lo-kor-da'ta) (Gr. *kephalē,* head, + L. *chorda,* cord): **lancelets (amphioxus).** Notochord, nerve cord, and postanal tail persist throughout life; fishlike in form. 22 species.

Subphylum Vertebrata (ver'te-bra'ta) (L. *vertebratus,* backboned). **(Craniata): vertebrates.** Bony or cartilaginous vertebrae surrounding spinal cord; notochord in embryonic stages, persisting in some fishes; also may be divided into two groups (superclasses) according to presence of jaws.

Superclass Agnatha (ag'na-tha) (Gr. *a,* without, + *gnathos,* jaw) **hagfishes, lampreys.** Without true jaws or paired appendages. (Probably a paraphyletic group.)

Class Myxini (mik-sin'y) (Gr. *myxa,* slime): **hagfishes.** Terminal mouth with four pairs of tentacles; buccal funnel absent; nasal sac with duct to pharynx; 5 to 15 pairs of pharyngeal pouches. About 65 species.

Class Cephalaspidomorphi (sef-a-lass'pe-do-morf'e) (Gr. *kephalē,* head, + *aspidos,* shield, *morphē,* form): **lampreys.** Suctorial mouth with keratinized teeth; nasal sac not connected to mouth; seven pairs of pharyngeal pouches. 41 species.

Superclass Gnathostomata (na'tho-sto'ma-ta) (Gr. *gnathos,* jaw, + *stoma,* mouth): **jawed fishes, all tetrapods.** With jaws and (usually) paired appendages.

Class Chondrichthyes (kon-drik'thee-eez) (Gr. *chondros,* cartilage, + *ichthys,* a fish): **sharks, skates, rays, chimaeras.** Cartilaginous skeleton; intestine with spiral valve; claspers present in males; no swim bladder. About 850 species.

Class Actinopterygii (ak'ti-nop-te-rij'ee-i) Gr. *aktis,* ray, + *pteryx,* fin, wing): **ray-finned fishes.** Skeleton ossified; single gill opening covered by operculum; paired fins supported primarily by dermal rays; limb musculature within body; swim bladder mainly a hydrostatic organ, if present; atrium and ventricle not divided. About 23,700 species.

Class Sarcopterygii (sar-cop-te-rij'ee-i) (Gr. *sarkos,* flesh, + *pteryx,* fin, wing): **lobe-finned fishes.** Skeleton ossified; single gill opening covered by operculum; paired fins with sturdy internal skeleton and musculature within limb; diphycercal tail; intestine with spiral valve; usually with lunglike swim bladder; atrium and ventricle at least partly divided. 8 species. Paraphyletic unless tetrapods are included.

Class Amphibia (am-fib'e-a) (Gr. *amphi,* both or double, + *bios,* life): **amphibians.** Ectothermic tetrapods; respiration by lungs, gills, or skin; development through larval stage; skin moist, containing mucous glands, and lacking scales. About 4900 species.

Class Reptilia (rep-til'e-a) (L. *repere,* to creep): **reptiles.** Ectothermic tetrapods possessing lungs; embryo develops within shelled egg; no larval stage; skin dry, lacking mucous glands, and covered by epidermal scales. (A paraphyletic group.) About 7100 species.

Class Aves (ay'veez) (L. pl. of *avis,* bird): **birds.** Endothermic vertebrates with front limbs modified for flight; body covered with feathers; scales on feet. About 9600 species.

Class Mammalia (ma-may'lee-a) (L. *mamma,* breast): **mammals.** Endothermic vertebrates possessing mammary glands; body more or less covered with hair; well-developed neocerebrum; three middle ear bones. About 4600 species.

The Swedish paleozoologist Erik Stensiö was the first to approach fossil anatomy with the same painstaking attention to minute detail that morphologists have long applied to the anatomical study of living fishes. He developed novel and exacting methods for gradually grinding away a fossil, a few micrometers at a time, to reveal internal features. He was able to reconstruct not only bone anatomy, but nerves, blood vessels, and muscles in numerous groups of Paleozoic and early Mesozoic fishes. His innovative methods are widely used today by paleozoologists.

animals has changed this situation. With their phosphatized toothlike elements, W-shaped myomeres, cranium, notochord, and extrinsic eye muscles, conodont animals clearly belong to the vertebrate lineage (Figure 23-16). Although their exact position in the vertebrate lineage is unclear, they are important in understanding the origin of vertebrates.

Early Jawed Vertebrates

All jawed vertebrates, whether extinct or living, are collectively called **gnathostomes** ("jaw mouth") in contrast to the jawless vertebrates, the **agnathans** ("without jaw"). The gnathostomes are a monophyletic group since presence of jaws is a derived character state shared by all jawed fishes and tetrapods. Agnathans, however, are defined principally by the absence of jaws, a character that is not unique to jawless fishes since jaws are lacking in vertebrate ancestors. Agnatha may be paraphyletic.

The origin of jaws was one of the most important events in vertebrate evolution. The utility of jaws is obvious: they allow predation on large and active forms of food not available to jawless vertebrates and permit manipulation of objects. Ample evidence suggests that jaws arose through modifications of the first or second of the serially repeated cartilaginous gill arches. The mandibular arch may have first become enlarged to assist gill ventilation, perhaps to meet the increasing metabolic demands of early vertebrates. Later, the anterior gill arches became hinged

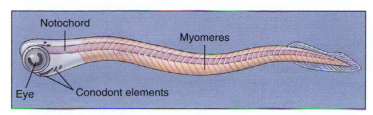

Figure 23-16

Restoration of a living conodont animal. Conodonts superficially resembled amphioxus, but they possessed a much greater degree of encephalization (large, paired eyes, possible auditory capsules) and bonelike mineralized elements—all indicating that conodont animals were vertebrates. Conodont elements are believed to be gill-supporting structures or part of a food-handling apparatus.

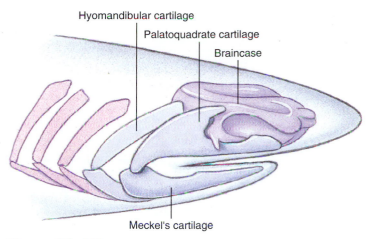

Figure 23-17

How vertebrates got their jaw. The resemblance between jaws and the gill supports of the primitive fishes such as this carboniferous shark suggests that the upper jaw (palatoquadrate) and lower jaw (Meckel's cartilage) evolved from structures that originally functioned as gill supports. The gill supports immediately behind the jaws are hinged like jaws and served to link the jaws to the braincase. Relics of this transformation are seen during the development of modern sharks.

and bent forward into the characteristic position of vertebrate jaws. Evidence for this remarkable transformation is threefold. First, both gill arches and jaws form from upper and lower bars that bend forward and are hinged in the middle (Figure 23-17). Second, both gill arches and jaws are derived from neural crest cells. Third, the jaw musculature is homologous to the original gill support musculature as evidenced by cranial nerve distribution. Nearly as remarkable as this drastic morphological remodeling is the subsequent evolutionary fate of jawbone elements—their transformation into ear ossicles of the mammalian middle ear (see the note on p. 713).

An additional feature characteristic of all gnathostomes is the presence of paired pectoral and pelvic appendages in the form of fins or limbs. These likely originated as stabilizers to check yaw, pitch, and roll generated during active swimming. The fin-fold hypothesis has been proposed to explain the origin of paired fins. According to this hypothesis, paired fins arose from paired continuous, ventrolateral folds or fin-forming zones. The addition of skeletal supports in the fins served to enhance their properties of providing stability during swimming. Evidence for this hypothesis is found in the paired flaps of *Myllokunmingia, Haikouichthys,* and anaspids and in the multiple paired fins of acanthodians, also described in this section. However, pectoral fins appear in the fossil record before pelvic fins, suggesting a more complex evolutionary scenario. In one fish lineage the muscle and skeletal supports in the paired fins became strengthened, allowing them to become adapted for locomotion on land as limbs. The origin of jaws and paired

appendages may be linked to a second *Hox* duplication, near the origin of the gnathostomes. The appearance of both jaws and paired fins were major innovations in vertebrate evolution, among the most important reasons for the subsequent major radiations of vertebrates that produced the modern fishes and all tetrapods, including you, the reader of this book.

Among the first jawed vertebrates were the heavily armored **placoderms** (plak'o-derm) (Gr. *plax,* plate, + *derma,* skin). These first appear in the fossil record in the early Silurian period (Figure 23-18). Placoderms evolved a great variety of forms, some very large (one was 10 m in length!) and grotesque in appearance. They were armored fish covered with diamond-shaped scales or with large plates of bone. All became extinct by the end of the Devonian period and appear to have left no descendants. However, contemporary with placoderms were the **acanthodians** (Figure 23-18), a group of early jawed fishes characterized by fins with large spines, may have given rise to the great radiation of bony fishes that dominates the waters of the world today.

SUMMARY

Phylum Chordata is named for the rodlike notochord that forms a stiffening body axis at some stage in the life cycle of every chordate. All chordates share five distinctive hallmarks that set them apart from all other phyla: notochord, dorsal tubular nerve cord, pharyngeal pouches, endostyle, and postanal tail. Two of the three chordate subphyla are invertebrates and lack a well-developed head. They are Urochordata (tunicates), most of which are sessile as adults but all of which have a free-swimming larval stage, and Cephalochordata (lancelets), fishlike forms that include the famous amphioxus.

The chordates apparently descended from hemichordate or echinoderm-like ancestors, probably in the Precambrian period, but the precise origin of the chordates is not yet known. Taken as a whole, chordates have a greater fundamental unity of organ systems and body plan than have many other phyla.

Subphylum Vertebrata includes the backboned members of the animal kingdom the hagfishes actually lack vertebrae but are included with the Vertebrata by tradition because they share numerous homologies with vertebrates). As a group vertebrates are characterized by

Figure 23-18

Early jawed fishes of the Devonian period, 400 million years ago. Shown are a placoderm *(left)* and a related acanthodian *(right)*. Jaws and the gill supports from which the jaws evolved develop from neural crest cells, a diagnostic character of vertebrates. Most placoderms were bottom dwellers that fed on benthic animals although some were active predators. Acanthodians carried less armor than placoderms and had a bony endoskeleton and prominent spines on paired fins. Most were marine but several species entered fresh water.

having a well-developed head, and by their comparatively large size, high degree of motility, and a distinctive body plan that embodies several distinguishing features that permitted the exceptional adaptive radiation of the group. Most important of these are the living endoskeleton that allows continuous growth and provides a sturdy framework for efficient muscle attachment and action, a muscular pharynx with slits (lost or greatly modified in higher vertebrates) with vastly increased respiratory efficiency, a muscularized gut, chambered heart, and glomerular kidney for meeting higher metabolic demands, and an advanced nervous system with a distinct brain and paired sense organs. Evolution of jaws and paired appendages likely contributed to the incredible success of one group of vertebrates, the gnathostomes.

REVIEW QUESTIONS

1. What characteristics are shared by the three deuterostome phyla that indicate a monophyletic group of interrelated animals?

2. Explain how the use of a cladistic classification for the vertebrates results in important regroupings of the traditional vertebrate taxa (refer to Figure 23-3). Why are certain traditional groupings such as Reptilia and Agnatha not recognized in cladistic usage?

3. Name five hallmarks shared by all chordates, and explain the function of each.

4. In debating the question of chordate origins, zoologists eventually agreed that chordates must have evolved within the deuterostome assemblage rather than from a protostome group as earlier argued. What embryological evidences support this view? What characteristics does the fossil echinoderm group Calcichordata possess that suggest it might closely resemble the ancestor of chordates?

5. Offer a description of an adult tunicate that would identify it as a chordate, yet distinguish it from any other chordate group.

6. Amphioxus long has been of interest to zoologists searching for a vertebrate ancestor. Explain why amphioxus captured such interest and why it no longer is considered to resemble closely the most recent common ancestor of all vertebrates.

7. Both sea squirts (urochordates) and lancelets (cephalochordates) are filter-feeding organisms. Describe the filter-feeding apparatus of a sea squirt and explain in what ways its mode of feeding is similar to, and different from, that of amphioxus.

8. Explain why it is necessary to know the life history of a tunicate to understand why tunicates are chordates.

9. List three groups of adaptations that guided vertebrate evolution, and explain how each has contributed to the success of vertebrates.

10. In 1928 Walter Garstang hypothesized that tunicates resemble the ancestral stock of the vertebrates. Explain this hypothesis.

11. Distinguish between ostracoderms and placoderms. What important evolutionary advances did each contribute to vertebrate evolution? What are conodonts?

12. Explain how zoologists think the vertebrate jaw evolved.

SELECTED REFERENCES

Alldredge, A. 1976. Appendicularians. Sci. Am. **235**:94–102 (July). *Describes the biology of larvaceans, which build delicate houses for trapping food.*

Bone, Q. 1979. The origin of chordates. Oxford Biology Readers, No. 18, New York, Oxford University Press. *Synthesis of hypotheses and range of disagreements bearing on an unsolved riddle.*

Bowler, P. J. 1996. Life's splendid drama: evolutionary biology and the reconstruction of life's ancestry 1860–1940. Chicago, University of Chicago Press. *Thorough and eloquent exploration of scientific debates over reconstruction of history of life on earth; chapter 4 treats theories of chordate and vertebrate origins.*

Carroll, R. L. 1997. Patterns and processes of vertebrate evolution. New York, Cambridge University Press. *A comprehensive analysis of the evolutionary processes that have influenced large-scale changes in vertebrate evolution.*

Forey, P., and P. Janvier. 1994. Evolution of the early vertebrates. Am. Sci. **82**:554–565. *Summarizes the biology and evolution of many groups of ostracoderms and other primitive craniates.*

Gee, H. 1996. Before the backbone: views on the origin of the vertebrates. New York, Chapman & Hall. *Outstanding review of the many vertebrate origin hypotheses. Gee links much of the recent genetic, developmental and molecular evidence in his discussion.*

Gans, C. 1989. Stages in the origin of vertebrates: analysis by means of scenarios. Biol. Rev. **64**:221–268. *Reviews the diagnostic characters of protochordates and ancestral vertebrates and presents a scenario for the protochordate-vertebrate transition.*

Gould, S. J. 1989. Wonderful life: the Burgess Shale and the nature of history. New York, W.W. Norton & Company. *In this book describing the marvelous Cambrian fossils of the Burgess Shale, Gould "saves the best for last" by inserting an epilogue on* Pikaia, *the first known chordate.*

Gould, S. J. (ed.) 1993. The book of life. New York, W. W. Norton & Company. *A sweeping, handsomely illustrated view of (almost entirely) vertebrate life.*

Jeffries, R. P. S. 1986. The ancestry of the vertebrates. Cambridge, Cambridge University Press. *Jeffries argues that the Calcichordata are the direct ancestors of the vertebrates, a view that most zoologists are not willing to accept. Still, this book is an excellent summary of the deuterostome groups and of the various competing hypotheses of vertebrate ancestry.*

Long, J. A. 1995. The rise of fishes: 500 million years of evolution. Baltimore, The Johns Hopkins University Press. *An authoritative, liberally illustrated evolutionary history of fishes.*

Maisey, J. G. 1996. Discovering fossil fishes. New York, Henry Holt & Company. *Handsomely illustrated chronology of fish evolution with cladistic analysis of evolutionary relationships.*

Pough, F. H., C. M. Janis, and J. B. Heiser. Vertebrate life, ed. 6. Upper Saddle River, New Jersey, Prentice Hall. *Vertebrate morphology, physiology, ecology, and behavior cast in a cladistic framework.*

Stokes, M. D., and N. D. Holland. 1998. The lancelet. Am. Sci. **86**(6):552–560. *Describes the historical role of amphioxus in early hypotheses of vertebrate ancestry and summarizes recent molecular data that has rekindled interest in amphioxus.*

ZOOLOGY LINKS TO THE INTERNET

Visit the textbook's Online Learning Center at www.mhhe.com/zoology to find live Internet links for each of the topics listed here.

General Chordate References

Phylum Chordata

Subphylum Urochordata

Subphylum Cephalochordata

Subphylum Vertebrata

Systematics and Characteristics of the Craniates

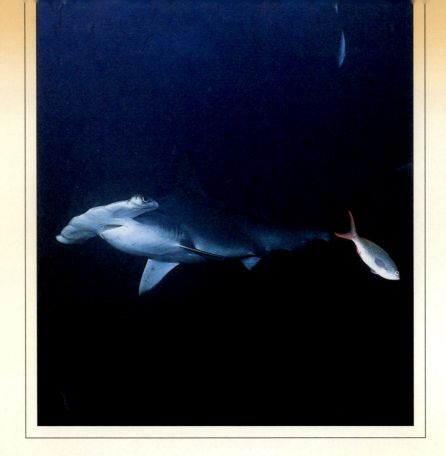
Hammerhead shark near the Galápagos Islands.

Fishes

Phylum Chordata
Class Myxini
Class Cephalaspidomorphi
Class Chondrichthyes
Class Actinopterygii
Class Sarcopterygii

What Is a Fish?

In common (and especially older) usage, the term fish denotes a mixed assortment of water-dwelling animals. We speak of jellyfish, cuttlefish, starfish, crayfish, and shellfish, knowing full well that when we use the word "fish" in such combinations, we are not referring to a true fish. In earlier times, even biologists did not make such a distinction. Sixteenth-century natural historians classified seals, whales, amphibians, crocodiles, even hippopotamuses, as well as a host of aquatic invertebrates, as fish. Later biologists were more discriminating, eliminating first the invertebrates and then the amphibians, reptiles, and mammals from the narrowing concept of a fish. Today we recognize a fish as an aquatic vertebrate with gills, limbs, if present, in the form of fins, and usually with a skin covered in scales of dermal origin. Even this modern concept of the term "fish" is used for convenience, not as a taxonomic unit, because fishes do not compose a monophyletic group. The common ancestor of fishes is also an ancestor to land vertebrates, which we exclude from the term "fish," unless we use the term in an exceedingly nontraditional way. Because fishes live in a habitat that is basically alien to humans, people have rarely appreciated the remarkable diversity of these vertebrates. Nevertheless, whether appreciated by humans or not, the world's fishes have enjoyed an effusive proliferation that has produced an estimated 24,600 living species—more than all other species of vertebrates combined—with adaptations that have fitted them to almost every conceivable aquatic environment. No other animal group threatens their domination of the seas.

The life of a fish is bound to its body form. Their mastery of stream, lake, and ocean is revealed in the many ways that fishes have harmonized their life design to the physical properties of their aquatic surroundings. Suspended in a medium that is 800 times more dense than air, a trout or pike can remain motionless, varying its neutral buoyancy by adding or removing air from its swim bladder. Or it may dart forward or at angles, using its fins as brakes and tilting rudders. With excellent organs for salt and water exchange, fishes can steady and finely tune their body fluid composition in their chosen freshwater or seawater environment. Their gills are the most effective respiratory devices in the animal kingdom for extracting oxygen from a medium that contains less than 1/20 as much oxygen as air. Fishes have excellent olfactory and visual senses and a unique lateral line system, which with its exquisite sensitivity to water currents and vibrations provides a "distance touch" in water. Thus in mastering the physical problems of their element, early fishes evolved a basic body plan and set of physiological strategies that both shaped and constrained the evolution of their descendants.

ANCESTRY AND RELATIONSHIPS OF MAJOR GROUPS OF FISHES

Fishes are of ancient ancestry, having descended from an unknown free-swimming protochordate ancestor (hypotheses of chordate and vertebrate origins are discussed in Chapter 23). The earliest vertebrates were a paraphyletic assemblage of jawless **agnathan** fishes, including the ostracoderms (Figure 23-15, p. 481). One group of ostracoderms gave rise to the jawed **gnathostomes** (Figure 24-1).

> The use of *fishes* as the plural form of *fish* may sound odd to most people accustomed to using *fish* in both the singular and the plural. *Fish* refers to one or more individuals of the same species; *fishes* refers to more than one species.

The jawless agnathans include along with the extinct ostracoderms the living **hagfishes** and **lampreys,** fishes adapted as scavengers or parasites. Although hagfishes have no vertebrae and lampreys have only rudimentary vertebrae, they nevertheless are included with the subphylum Vertebrata because they have a cranium and many other vertebrate homologies. Although hagfishes and more derived lampreys superficially look much alike, they are in fact so different from each other that they have been assigned to separate classes by ichthyologists.

All remaining fishes have paired appendages and jaws and are included, along with tetrapods (land vertebrates) in the monophyletic group of gnathostomes. They appear in the fossil record in the late Silurian period with fully formed jaws, and no forms intermediate between agnathans and gnathostomes are known. By the Devonian period, the Age of Fishes, several distinct groups of jawed fishes were well represented. One of these, the placoderms (p. 483), became extinct in the following Carboniferous period, leaving no direct descendants. A second

Position in the Animal Kingdom

Fishes are a vast array of distantly related gill-breathing aquatic vertebrates with fins. Fishes are the most ancient and most diverse of the monophyletic subphylum Vertebrata within phylum Chordata, constituting five of the nine living vertebrate classes and one-half of the approximately 48,000 recognized vertebrate species. Although they are a heterogeneous assemblage, they exhibit phylogenetic continuity within the group and with the tetrapod vertebrates. Jawless fishes—hagfishes and lampreys—are the living forms that resemble most closely the armored ostracoderms that appeared in the Cambrian period of the Paleozoic. The living jawed fishes, cartilaginous and bony fishes, are related phylogenetically to the acanthodians, a group of jawed fishes that were contemporary with the placoderms of the Silurian and Devonian periods of the Paleozoic. Tetrapod vertebrates—amphibians, reptiles, birds, and mammals—arose from one lineage of bony fishes, the sarcopterygians (lobe-finned fishes). The evolution of fishes paralleled the appearance of numerous advances in vertebrate history.

Biological Contributions

1. The basic vertebrate body plan was established in the common ancestor of all vertebrates. Foremost was the evolution of **cellular bone.** The **vertebral column** replaced the notochord as the main stiffening axis in most adult vertebrates and provided attachment for the skull, many muscles, and the appendages.
2. With the **brain and spinal cord enclosed** and protected within the cranium and vertebral column, early fishes were the first animals to house the central nervous system separate from the rest of the body. **Specialized sense organs** for smell, vision, and hearing evolved with a tripartite brain. Other sensory innovations include an inner ear with semicircular canals, an electrosensory system, and intricate lateral line sensory systems.
3. The development of **jaws with teeth** permitted predation of large and active foods. This gave rise to a predator-prey arms race that became a major shaping element in vertebrate evolution through the ages.
4. The evolution of **paired pectoral and pelvic fins** supported by shoulder and hip girdles provided greatly improved maneuverability and became the precursors of limbs of tetrapod vertebrates.
5. Fishes developed physiological adaptations that enabled them to invade every conceivable type of aquatic habitat. The origin of the homologous **swim bladder** and **lung** resulted in refined buoyancy control and use of atmospheric oxygen, and was a precursor to the invasion of land.

group, **cartilaginous fishes** of class Chondrichthyes (sharks, rays, and chimaeras), lost the heavy dermal armor of early jawed fishes and adopted cartilage rather than bone for the skeleton. Most are active predators with sharklike or raylike body forms that have undergone only minor changes over the ages. As a group, sharks and their kin flourished during the Devonian and Carboniferous periods of the Paleozoic era but declined danger-

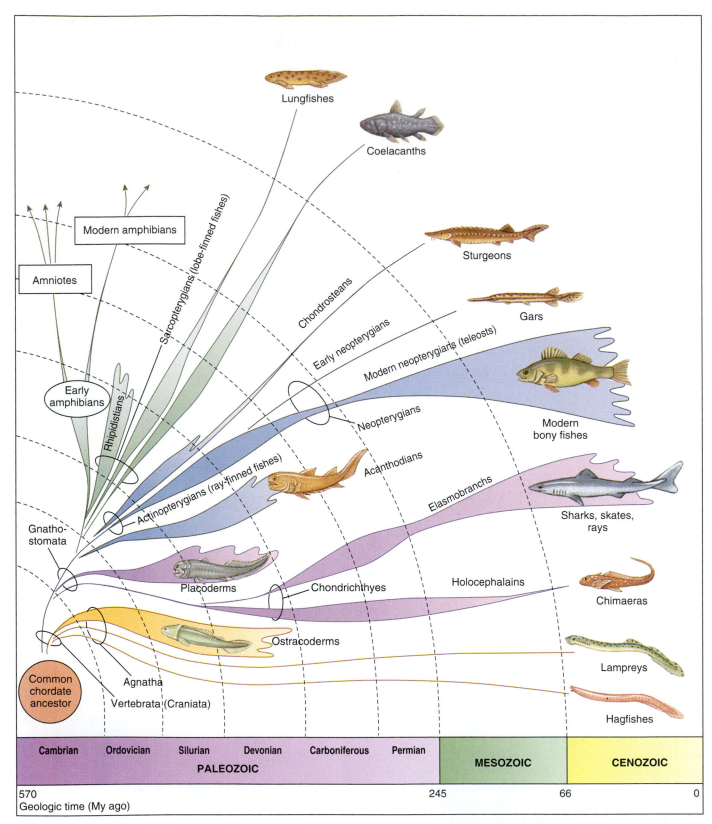

Figure 24-1

Graphic representation of the family tree of fishes, showing evolution of major groups through geological time. Numerous lineages of extinct fishes are not shown. Widened areas in the lines of descent indicate periods of adaptive radiation and relative number of species in each group. The lobe-finned fishes (sarcopterygians), for example, flourished in the Devonian period, but declined and are today represented by only four surviving genera (lungfishes and coelacanths). Homologies shared by sarcopterygians and tetrapods suggest that they form a clade. Sharks and rays radiated during the Carboniferous period, declined in the Permian, then radiated again in the Mesozoic era. Johnny-come-latelies in fish evolution are the spectacularly diverse modern fishes, or teleosts, which make up most living fishes.

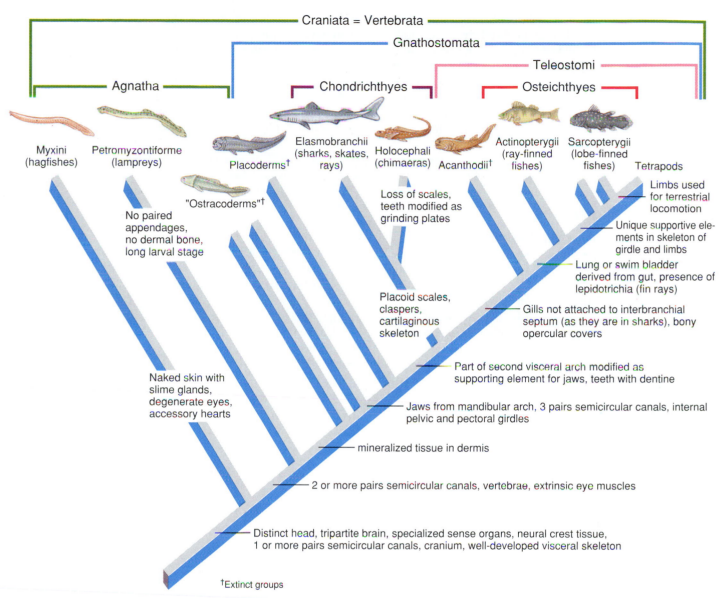

Figure 24-2

Cladogram of the fishes, showing the probable relationships of major monophyletic fish taxa. Several alternative relationships have been proposed. Extinct groups are designated by a dagger (†). Some of the shared derived characters that mark the branchings are shown to the right of branch points. The groups Agnatha and Osteichthyes, although paraphyletic structural grades considered undesirable in cladistic classification, are conveniently recognized in systematics because they share broad structural and functional patterns of organization.

ously close to extinction at the end of the Paleozoic. They staged a recovery in the early Mesozoic and radiated to form the modest but thoroughly successful assemblage of modern sharks and rays (Figure 24-1).

The other two groups of gnathostome fishes, **acanthodians** (p. 483) and **bony fishes,** were well represented in the Devonian period. Acanthodians somewhat resembled bony fishes but were distinguished by having heavy spines on all fins except the caudal fin. They became extinct in the lower Permian period. Although affinities of the acanthodians are much debated, many authors believe that they are the sister group of bony fishes. **Bony fishes** (Osteichthyes, Figure 24-2) are the dominant fishes today. We can recognize two distinct groups of bony fishes. Of these two, by far the most diverse are **ray-finned fishes** (class

Actinopterygii), which radiated to form most modern bony fishes. The other group, the **lobe-finned fishes** (class Sarcopterygii), contains few species today but includes the sister group of the tetrapods. Lobe-finned fishes are represented today by **lungfishes** and **coelacanths**—meager remnants of important stocks that flourished in the Devonian period (Figure 24-1). A classification of the major fish taxa is on p. 510.

LIVING JAWLESS FISHES

Living jawless fishes are represented by approximately 106 species divided between two classes: Myxini (hagfishes) with about 65 species and Cephalaspidomorphi (lampreys) with 41 species (Figures 24-3 and 24-4). Members of both groups lack jaws, internal ossification, scales, and paired fins, and both groups share porelike gill openings and an eel-like body form. In other respects, however, the two groups are morphologically very different. Hagfishes are certainly the least derived of the two, while lampreys bear many derived morphological characters that place them phylogenetically much closer to gnathostomes than to hagfishes. Because of these differences, hagfishes and lampreys have been assigned to separate vertebrate classes, leaving the grouping "agnatha" as a paraphyletic assemblage of jawless fishes.

Class Myxini: Hagfishes

Hagfishes are an entirely marine group that feeds on annelids, molluscs, crustaceans, and dead or dying fishes. Thus they are not parasitic like lampreys but are scavengers and predators. There are about 65 species of hagfishes, of which the best known in North America are the Atlantic hagfish *Myxine glutinosa* (Gr. *myxa*, slime) (Figure 24-3) and the Pacific hagfish *Eptatretus stouti* (N. L. *ept*, Gr. *hepta*, seven + *tretos*, perforated). Although almost completely blind, hagfishes are quickly attracted to food, especially dead or dying fishes, by their keenly developed senses of smell and touch. A hagfish enters a dead or dying animal through an orifice or by digging inside. Using two toothed, keratinized plates on its tongue that fold together in a pincerlike action, the hagfish rasps

away bits of flesh from its prey. For extra leverage, the hagfish often ties a knot in its tail, then passes the knot forward along its body until it is pressed securely against the side of its prey.

While the unique anatomical and physiological features of the strange hagfishes are of interest to biologists, hagfishes have not endeared themselves to either sports or commercial fishermen. In earlier days of commercial fishing mainly by gill nets and set lines, hagfish often bit into the bodies of captured fish and ate out the contents, leaving behind a useless sack of skin and bones. But as large and efficient otter trawls came into use, hagfishes ceased to be an important pest.

Hagfishes are renowned for their ability to generate enormous quantities of slime. If disturbed or roughly handled, a hagfish exudes a milky fluid from special glands positioned along its body. On contact with seawater, the fluid forms a slime so slippery that the animal is almost impossible to grasp.

Unlike any other vertebrate, the body fluids of hagfishes are in osmotic equilibrium with seawater, as in most marine invertebrates. Hagfishes have several other anatomical and physiological peculiarities, including a low-pressure circulatory system served by three accessory hearts in addition to the main heart positioned behind the gills.

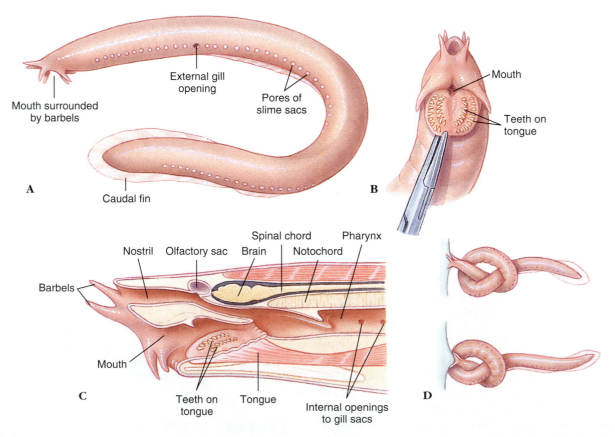

Figure 24-3

Atlantic hagfish *Myxine glutinosa* (class Myxini). **A,** External anatomy; **B,** Ventral view of head, showing horny plates used to grasp food during feeding; **C,** Sagittal section of head region (note retracted position of rasping tongue and internal openings into a row of gill sacs); **D,** Hagfish knotting, showing how it obtains leverage to tear flesh from prey.

Characteristics of Class Myxini

1. Body slender, eel-like, rounded, with **naked skin containing slime glands**
2. **No paired appendages,** no dorsal fin (the caudal fin extends anteriorly along the dorsal surface)
3. **Fibrous and cartilaginous skeleton;** notochord persistent
4. Biting mouth with two rows of eversible teeth, but no jaws
5. Heart with sinus venosus, atrium, and ventricle; **accessory hearts,** aortic arches in gill region
6. Five to 16 pairs of gills with a variable number of gill openings
7. Segmented **mesonephric kidney;** marine, **body fluids isosmotic with seawater**
8. Digestive system **without stomach;** no spiral valve or cilia in intestinal tract
9. Dorsal nerve cord with differentiated brain; **no cerebellum;** 10 pairs of cranial nerves; dorsal and ventral nerve roots united
10. Sense organs of taste, smell, and hearing; **eyes degenerate; one pair semicircular canals**
11. Sexes separate (ovaries and testes in same individual but only one is functional); external fertilization; large yolky eggs, **no larval stage**

Figure 24-4
Sea lamprey, *Petromyzon marinus*, feeding on body fluids of a dying fish.

The reproductive biology of hagfishes remains largely a mystery, despite a still unclaimed prize offered more than 100 years ago by the Copenhagen Academy of Science for information on the animal's breeding habits. It is known that females, which in some species outnumber males 100 to one, produce small numbers of surprisingly large, yolky eggs 2 to 7 cm in diameter depending on the species. There is no larval stage.

Class Cephalaspidomorphi (Petromyzontes): Lampreys

All lampreys of the Northern Hemisphere belong to family Petromyzontidae (Gr. *petros*, stone, + *myzon,* sucking). The group name refers to a lamprey's habit of grasping a stone with its mouth to hold position in a current. The destructive marine lamprey *Petromyzon marinus* is found on both sides of the Atlantic Ocean (in America and Europe) and may attain a length of 1 m (Figure 24-4). *Lampetra* (L. *lambo,* to lick or lap up) also has a wide distribution in North America and Eurasia and ranges from 15 to 60 cm long. There are 22 species of lampreys in North America. About half of these belong to the nonparasitic brook type; the others are parasitic. The genus *Ichthyomyzon* (Gr. *ichthyos,* fish, + *myzon,* sucking), which includes three parasitic and three nonparasitic species, is restricted to eastern North America. On the west coast of North America the chief marine form is *Lampetra tridentatus,* commonly mislabeled as *P. marinus* by biological supply companies.

All lampreys ascend freshwater streams to breed. The marine forms are anadromous (Gr. *anadromos,* running upward); that is, they leave the sea where they spend their adult lives to swim up streams to spawn. In North America all lampreys spawn in winter or spring. Males begin nest building and are joined later by females. Using their oral discs to lift stones and pebbles and vig-

orous body vibrations to sweep away light debris, they form an oval depression (Figure 24-5). At spawning, with the female attached to a rock to maintain her position over the nest, the male attaches to the dorsal side of her head. As eggs are shed into the nest, they are fertilized by the male. The sticky eggs adhere to pebbles in the nest and quickly become covered with sand. Adults die soon after spawning.

Eggs hatch in about 2 weeks, releasing small larvae called **ammocoetes,** which are so unlike their parents that early biologists thought they were a separate species. The larva bears a remarkable resemblance to amphioxus and possesses the basic chordate characteristics in such simplified and easily visualized form that it has been considered a chordate archetype (p. 480).

Characteristics of Class Cephalaspidomorphi

1. Body slender, eel-like, rounded with naked skin
2. One or two **dorsal fins, no paired appendages**
3. **Fibrous and cartilaginous skeleton;** notochord persistent
4. Suckerlike oral disc and tongue with well-developed keratinized teeth
5. Heart with sinus venosus, atrium, and ventricle; aortic arches in gill region
6. Seven pairs of gills each with external gill opening
7. **Opisthonephric kidney;** anadromous and fresh water; **body fluids osmotically and ionically regulated**
8. Dorsal nerve cord with differentiated brain, **small cerebellum present;** 10 pairs cranial nerves; dorsal and ventral nerve roots separated
9. Digestive system without distinct stomach; intestine with **spiral fold**
10. Sense organs of taste, smell, hearing; **eyes well developed** in adult; **two pairs semicircular canals**
11. Sexes separate; single gonad without duct; external fertilization; **long larval stage** (ammocoete)

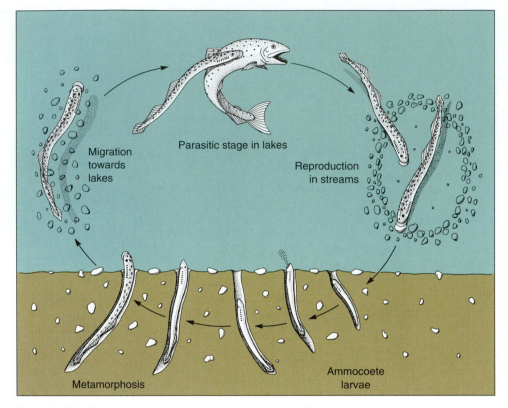

Figure 24-5

Life cycle of the "landlocked" form of the sea lamprey *Petromyzon marinus.*

After absorbing the remainder of their yolk supply, young ammocoetes, now about 7 mm long, leave the nest gravel and drift downstream to burrow in some suitable sandy, low-current area. The larvae take up a suspension-feeding existence while growing slowly for 3 to 7 or more years, then rapidly metamorphose into adults. This change involves the eruption of eyes, replacement of the hood by the oral disc with keratinized teeth, enlargement of fins, maturation of gonads, and modification of the gill openings.

Parasitic lampreys either migrate to the sea, if marine, or remain in fresh water, where they attach themselves by their suckerlike mouth to a fish and, with their sharp keratinized teeth, rasp away the flesh and suck out body fluids (Figure 24-6). To promote the flow of blood, the lamprey injects an anticoagulant into the wound. When gorged, the lamprey releases its hold but leaves the fish with a large, gaping wound that is sometimes fatal. Parasitic freshwater adults live 1 to 2 years before spawning and then die; anadromous forms live 2 to 3 years.

Nonparasitic lampreys do not feed after emerging as adults and their alimentary canal degenerates to a nonfunctional strand of tissue. Within a few months they also spawn and die.

Invasion of the Great Lakes by the landlocked sea lamprey *Petromyzon marinus* in this century has had a devastating effect on fisheries. No lampreys were present in the Great Lakes west of Niagara Falls until the Welland Ship Canal was built in 1829. Even then nearly 100 years elapsed before sea lampreys were first seen in Lake Erie. After that sea lampreys spread rapidly and were causing extraordinary damage in all the Great Lakes by the middle 1940s. No fish species was immune from attack, but lam-

**Attachment to fish with
keratinized teeth and suction**

**Tongue protruded for
rasping flesh**

Figure 24-6

How a lamprey uses its keratinized tongue to feed. After firmly attaching to a fish by its sucker, the protrusible tongue rapidly rasps an opening through the fish's integument. Body fluid, abraded skin, and muscle are eaten.

preys preferred lake trout, and this multimillion-dollar fishing industry was brought to total collapse in the late 1950s. Lampreys then turned to rainbow trout, whitefish, burbot, yellow perch, and lake herring, all important commercial species. These stocks were decimated in turn. Lampreys then began attacking chubs and suckers. Coincident with decline in attacked species, sea lampreys themselves began to decline after reaching a peak abundance in 1951 in Lakes Huron and Michigan and in 1961 in Lake Superior. The fall has been attributed both to depletion of food and to effectiveness of control measures (mainly chemical larvicides in selected spawning streams). Lake trout, aided by a restocking program, are now recovering. Wounding rates are low in Lake Michigan but still high in some lakes. Fishery organizations are now experimenting with release of sterilized male lampreys into spawning streams; when fertile females mate with sterilized males the female's eggs fail to develop.

CLASS CHONDRICHTHYES: CARTILAGINOUS FISHES

There are nearly 850 living species in class Chondrichthyes, an ancient, compact, and highly developed group. Although a much smaller and less diverse assemblage than bony fishes, their impressive combination of well-developed sense organs, powerful jaws and swimming musculature, and predaceous habits ensures them a secure and lasting place in the aquatic community. One of their distinctive features is their cartilaginous skeleton. Although calcification may be extensive in their skeletons, bone is entirely absent throughout the class—a curious evolutionary feature, since Chondrichthyes are derived from ancestors having well-developed bone. Almost all chondrichthyans are marine; only 28 species live primarily in fresh water.

With the exception of whales, sharks include the largest living vertebrates. The larger sharks may reach 12 m in length. Dogfish sharks so widely studied in zoological laboratories rarely exceed 1 m.

Subclass Elasmobranchii: Sharks, Skates, and Rays

The nine living orders of elasmobranchs number about 815 species. Coastal waters are dominated by requiem sharks, order Carcharhiniformes, which consist of typical-looking sharks such as tiger and bull sharks and more bizarre forms, including hammerheads (Figure 24-7). Order Lamniformes contains several large, pelagic sharks dangerous to humans, including great white and mako sharks. Dogfish sharks, familiar to generations of comparative anatomy students, are in order Squaliformes. Skates and several groups of rays (sawfish rays, electric rays, stingrays, eagle rays, manta rays, and devil rays) belong to order Rajiformes.

Much has been written about the propensities of sharks to attack humans, both by those exaggerating their ferocious nature and by those seeking to write them off as harmless. It is true, as the latter group of writers argues, that sharks are by nature timid and cautious. But it also is a fact that certain of them are dangerous to humans. There are numerous authenticated cases of shark

---

Characteristics of Class Chondrichthyes

1. Large (average about 2 m), **body fusiform,** or dorsoventrally depressed, with a **heterocercal** caudal fin (diphycercal in chimaeras) (see Figure 24-16); paired pectoral and pelvic fins, two dorsal median fins; pelvic fins in male modified for "claspers"
2. **Mouth ventral;** two olfactory sacs that do not open into the mouth cavity in elasmobranchs; nostrils open into mouth cavity in chimaeras; jaws modified from pharyngeal arch
3. Skin with **placoid scales** or naked in elasmobranchs (see Figure 24-18); skin naked in chimaeras; teeth of modified placoid scales and serially replaced in elasmobranchs; teeth modified as grinding plates in chimaeras
4. **Endoskeleton entirely cartilaginous;** notochord persistent but reduced; vertebrae complete and separate in elasmobranchs; vertebrae present but centra absent in chimaeras; appendicular, girdle, and visceral skeletons present
5. Digestive system with J-shaped stomach (stomach absent in chimaeras); **intestine with spiral valve;** often with large oil-filled liver for buoyancy
6. Circulatory system of several pairs of aortic arches; dorsal and ventral aorta, hepatic portal and renal portal systems; four-chambered heart with sinus venosus, atrium, ventricle, and conus arteriosus
7. Respiration by means of five to seven pairs of gills leading to exposed gill slits in elasmobranchs; four pairs of gills covered by an operculum in chimaeras
8. No swim bladder or lung
9. Opisthonephric kidney and rectal gland; blood isosmotic or slightly hyperosmotic to sea water; **high concentrations of urea and trimethylamine oxide in blood**
10. Brain of two olfactory lobes, two cerebral hemispheres, two optic lobes, cerebellum, medulla oblongata; 10 pairs of cranial nerves; **three pairs of semicircular canals**
11. Senses of smell, vibration reception (lateral line system), vision, and electroreception well developed; inner ear opens to outside via endolymphatic duct
12. Sexes separate; gonads paired; reproductive ducts open into cloaca (separate urogenital and anal openings in chimaeras); oviparous, ovoviviparous, or viviparous; direct development; **fertilization internal**

The worldwide shark fishery is experiencing unprecedented pressure, driven by the high price of shark fins for shark-fin soup, an Asian delicacy (which commonly sells for $50.00 per bowl). Coastal shark populations in general have declined so rapidly that "finning" has been outlawed in the United States; other countries, too, are setting quotas to protect threatened shark populations. Even in the Marine Resources Reserve of the Galápagos Islands, one of the world's exceptional wild places, tens of thousands of sharks have been killed illegally for the Asian shark-fin market. That illegal fishery continues at this writing. Contributing to the threatened collapse of shark fisheries worldwide is the low fecundity of sharks, and the long time required by most sharks to reach sexual maturity; some species take as long as 35 years.

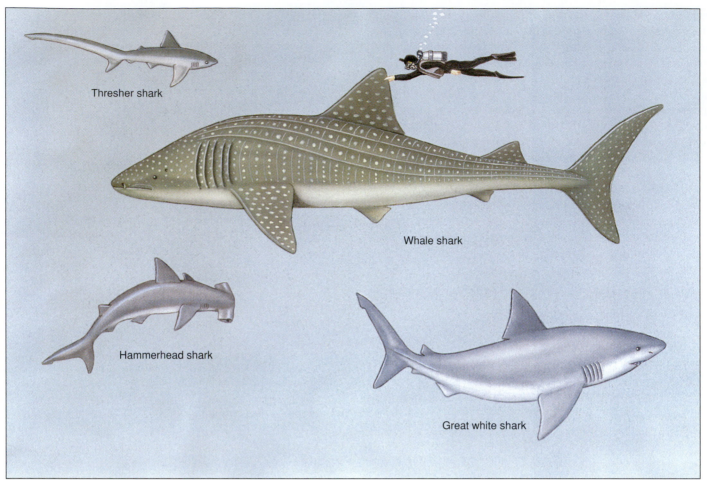

Figure 24-7

Diversity in sharks of the subclass Elasmobranchii. The thresher shark, *Alopias vulpinus,* exceptional because of its long upper tail lobe, may exceed 4 m in length. The great white shark, *Carcharodon carcharias,* largest and most notorious of dangerous sharks, is a heavy-bodied, spindle-shaped shark that may reach 6 m in length. The nine species of hammerheads (genus *Sphyrna*) are distinguished from all other sharks by the flattened head with hammerlike lobes bearing eyes and nostrils on the ends. The whale shark, *Rhincodon typus,* is the world's largest fish, reaching 12 m in length. It is a suspension feeder that feeds on plankton collected on a sievelike mesh over its gills.

attacks by great white sharks, *Carcharodon* (Gr. *karcharos,* sharp, + *odous,* tooth) (reaching 6 m); mako sharks, *Isurus* (Gr. *is,* equal, + *ouros,* tail); tiger sharks, *Galeocerdo* (Gr. *galeos,* shark, + *kerdō,* fox); bull sharks, *Carcharhinus levcas* (Gr. *Karcharos,* sharp, + *rhinos,* nose); and hammerhead sharks, *Sphyrna* (Gr. *sphyra,* hammer). More shark casualties have been reported from tropical and temperate waters of the Australian region than from any other. During World War II there were several reports of mass shark attacks on victims of ship sinkings in tropical waters.

Form and Function

Although to most people sharks have a sinister appearance and fearsome reputation, they are at the same time among the most gracefully streamlined of all fishes. The body of a dogfish shark (Figure 24-8) is fusiform (spindle shaped). In front of the ventral mouth is a pointed **rostrum;** at the posterior end the vertebral column turns up to end in the longer upper lobe of the tail. This type of tail is called **heterocercal.** There are paired **pectoral** and

pelvic fins supported by appendicular skeletons, one or two median **dorsal** fins (each with a spine in *Squalus* [L. a kind of sea fish]), and a median **caudal** fin. A median **anal** fin is present in most sharks, including the smooth dogfish, *Mustelus* (L. *mustela,* weasel). In males, the medial part of the pelvic fin is modified to form a **clasper,** which is used in copulation. Paired **nostrils** (blind pouches) are ventral and anterior to the mouth (Figure 24-9). The lateral eyes are lidless, and behind each eye is a spiracle (remnant of the first gill slit). Five gill slits are found anterior to each pectoral fin. The tough, leathery skin is covered with toothlike, dermal **placoid scales** arranged to reduce the turbulence of water flowing along the body surface during swimming.

Sharks are well equipped for their predatory life. They track their prey using highly sensitive senses in an orderly sequence. Sharks may initially detect prey from a kilometer or more away with their large olfactory organs, capable of detecting chemicals as low as 1 part per 10 billion. The laterally placed nostrils of hammerhead sharks (Figure 24-7) may enhance odor localization by improving stereo-olfaction. Prey also may be located

Figure 24-8

Male spiny dogfish shark, *Squalus acanthias*. Inset: section of lower jaw shows new teeth developing inside the jaw. These move forward to replace lost teeth. Rate of replacement varies in different species.

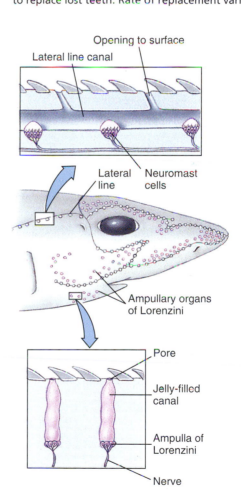

Figure 24-9

Head of sand tiger shark, *Carcharias* sp. Note the series of successional teeth. Also visible below the eye are the ampullae of Lorenzini.

Figure 24-10

Sensory canals and receptors in a shark. The ampullae of Lorenzini respond to weak electric fields, and possibly to temperature, water pressure, and salinity. The lateral line sensors, called neuromasts, are sensitive to disturbances in the water, enabling a shark to detect nearby objects by reflected waves in the water.

from long distances by sensing low-frequency vibrations with mechanoreceptors in the **lateral line system.** This system is composed of special receptor organs **(neuromasts)** in interconnected tubes and pores extending along the sides of the body and over the head (Figure 24-10). At closer range a shark switches to vision as the primary method of tracking prey. Contrary to popular belief, most sharks have excellent vision, even in dimly lit waters. During the final stage of attack, sharks are guided to their prey by the bioelectric fields that surround all animals. Electroreceptors, the **ampullae of Lorenzini** (Figure 24-9), are located primarily on the shark's head. In addition, sharks may use electroreception to find prey buried in the sand.

Both upper and lower jaws of sharks are provided with many sharp, triangular teeth. The front row of functional teeth on the edge of the jaw is backed by rows of developing teeth that replace worn teeth throughout the life of the shark (Figure 24-8 and 24-9). The mouth cavity opens into a large **pharynx,** which contains openings to separate gill slits and spiracles. A short, wide esophagus runs to the J-shaped stomach. A **liver** and **pancreas** open into a short, straight **intestine,** which contains the **spiral valve** that slows passage of food and increases the absorptive surface (Figure 24-11). Attached to a short rectum is a

rectal gland, unique to chondrichthyans, which secretes a colorless fluid containing a high concentration of sodium chloride. The rectal gland assists the **opisthonephric kidney** in regulating salt concentration of the blood. Chambers of the **heart** are arranged in tandem formation, and blood circulates in the same pattern seen in other gill-breathing vertebrates (Figure 24-11).

All chondrichthyans have internal fertilization, but maternal support of embryos is highly variable. Many elasmobranchs lay large, yolky eggs immediately after fertilization; these species are termed **oviparous.** Some oviparous sharks and rays deposit

Figure 24-11

Internal anatomy of a dogfish shark, *Squalus acanthias*.

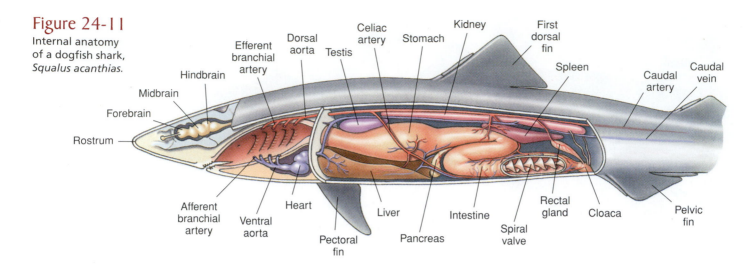

their eggs in a horny capsule called a "mermaid's purse," which often is provided with tendrils that wrap around the first firm object it contacts, much like tendrils of grape vines. Embryos are nourished from the yolk for a prolonged period—6 to 9 months in some, as much as 2 years in one species—before hatching as miniature replicas of adults. Many sharks, however, retain embryos in their reproductive tract for prolonged periods. Some are **ovoviviparous** species, which retain developing young in the uterus while they are nourished by contents of their yolk sac until born. Still other species have true **viviparous** reproduction. In these, embryos receive nourishment from the maternal bloodstream through a **placenta,** or from nutritive secretions, "uterine milk," produced by the mother. Some sharks (sand tigers) exhibit a grisly type of reproduction in which embryos receive additional nutrition by eating eggs and siblings. The evolution of prolonged retention of embryos by many elasmobranchs was an important innovation that contributed to their success. However, regardless of the form of maternal support, once the eggs are laid, or the young born, all parental care ends.

Marine elasmobranchs have developed an interesting solution to the physiological problem of living in a salty medium. To prevent water from being drawn out of the body osmotically, elasmobranchs retain nitrogenous compounds, especially urea and trimethylamine oxide, in their extracellular fluid. These solutes, combined with the blood salts, raise the blood solute concentration to exceed slightly that of seawater, eliminating an osmotic inequality between their bodies and surrounding seawater.

A little more than half of all elasmobranchs are rays, a group that includes skates, electric rays, sawfishes, stingrays, eagle rays, and manta rays. Most are specialized for bottom dwelling, with a dorsoventrally flattened body and greatly enlarged pectoral fins that are fused to the head and used like wings in swimming (Figure 24-12). Gill openings are on the underside of the head, but the large spiracles are on top. Water for breathing is taken in through these spiracles to prevent clogging the gills, for the mouth is often buried in sand. Their teeth are adapted for crushing prey: molluscs, crustaceans, and an occasional small fish.

In the order containing skates and rays (Rajiformes), we commonly refer to members of only one family (Rajidae) as skates. Alone among members of Rajiformes, skates do not bear living young but lay large, yolky eggs enclosed within a horny covering ("mermaid's purse") that often wash up on beaches. Although their tails are slender, skates have a somewhat more muscular tail than most rays, and they usually have two dorsal fins and sometimes a caudal fin.

Stingrays have a slender and whiplike tail armed with one or more saw-edged spines with venom glands at the base. Wounds from the spines are excruciatingly painful, and may heal slowly and with complications. Electric rays are sluggish fish with large electric organs on each side of their head (Figure 24-13). Each organ is composed of numerous vertical stacks of disclike cells

Figure 24-12

Skates and rays are specialized for life on the sea floor. Both the clearnose skates, *Raja eglanteria* **(A)**, and southern stingrays, *Dasyatis americana* **(B)**, are flattened dorsoventrally and move by undulations of winglike pectoral fins. This stingray **(B)** is followed by a pilot fish.

A   B

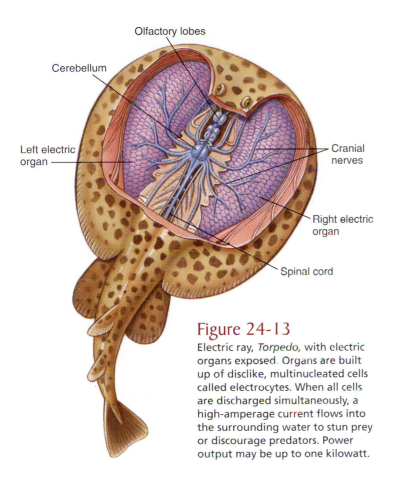

Figure 24-13
Electric ray, *Torpedo,* with electric organs exposed. Organs are built up of dislike, multinucleated cells called electrocytes. When all cells are discharged simultaneously, a high-amperage current flows into the surrounding water to stun prey or discourage predators. Power output may be up to one kilowatt.

connected in parallel so that when all cells discharge simultaneously, a high-amperage current is produced that flows out into the surrounding water. The voltage produced is relatively low (50 volts) but power output may be almost one kilowatt—quite sufficient to stun prey or discourage predators. Electric rays were used by ancient Egyptians for a form of electrotherapy in the treatment of afflictions such as arthritis and gout.

Subclass Holocephali: Chimaeras

Members of the small subclass Holocephali, distinguished by such suggestive names as ratfish (Figure 24-14), rabbitfish, spookfish, and ghostfish, are remnants of a line that diverged from the shark lineage at least 360 million years ago. Fossil chimaeras (ky-meer'-uz) first occurred in the Devonian period, reached their zenith in the Cretaceous and early Tertiary periods (120 million to 50 million years ago), and have declined ever since. Today there are only about 31 species extant.

Anatomically chimaeras have several features linking them to elasmobranchs, but possess a suite of unique characters, too. Instead of a toothed mouth, their jaws bear large flat plates. The upper jaw is completely fused to the cranium, a most unusual feature in fishes. Their food is seaweed, molluscs, echinoderms, crustaceans, and fishes—a surprisingly mixed diet for such a specialized grinding dentition. Chimaeras are not commercial species and are seldom caught. Despite their grotesque shape, they are beautifully colored with a pearly iridescence.

Figure 24-14
Spotted ratfish, *Hydrolagus collei,* of North American west coast. This species is one of the most handsome of chimaeras, which tend toward bizarre appearances.

OSTEICHTHYES: BONY FISHES

Origin, Evolution, and Diversity

In the early to middle Silurian, a lineage of fishes with bony endoskeletons gave rise to a clade of vertebrates that contains 96% of living fishes and all living tetrapods. Fishes of this clade have traditionally been termed "bony fishes" (**Osteichthyes**), because it was originally believed these were the only fishes with bony skeletons. Although it is now recognized that bone occurs in many other early fishes (ostracoderms, placoderms, and acanthodians), bony fishes and tetrapods are united by the presence of **endochondral bone** (bone that replaces cartilage developmentally), presence of lungs or a swim bladder derived from the gut, and several cranial and dental characters. Because traditional usage of Osteichthyes does not describe a monophyletic (natural) group (Figure 24-2), most recent classifications, including the one presented on p. 510, do not recognize this term as a valid taxon. Rather, it is used as a term of convenience to describe vertebrates with endochondral bone that are conventionally termed fishes.

Fossils of the earliest bony fishes show similarities in several craniopharyngeal structures, including a bony operculum and branchiostegal rays, with acanthodians (p. 483 and Figure 23-18), indicating they likely descended from a common ancestor. By the middle of the Devonian bony fishes already had radiated extensively into two major groups, with adaptations that fitted them for every aquatic habitat except the most inhospitable. One of these groups, ray-finned fishes (class Actinopterygii), includes modern bony fishes (Figure 24-15), the most species-rich group of living vertebrates. A second group, lobe-finned fishes (class Sarcopterygii), is represented today by only eight fishlike vertebrates, the lungfishes and coelacanths (see Figures 24-22 and 24-23); however, it includes the sister group of land vertebrates (tetrapods).

Several key adaptations contributed to their radiation. Bony fishes have an **operculum** over the gill composed of bony plates and attached to a series of muscles. This feature increased respiratory efficiency because outward rotation of the operculum created a negative pressure so that water would be drawn

Figure 24-15

Anatomy of a yellow
perch, *Perca flavescens,*
a freshwater teleost fish.

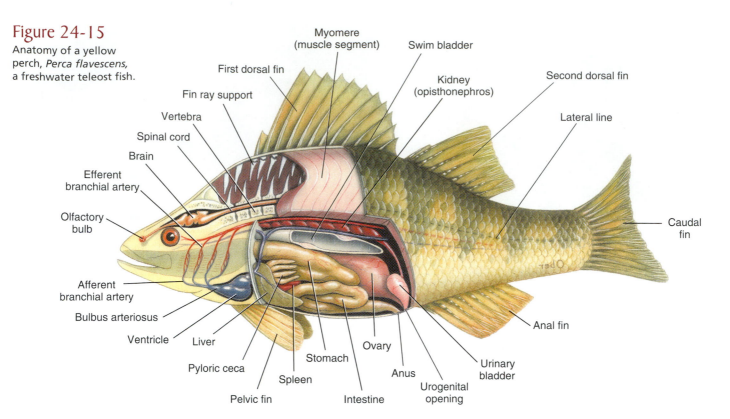

Class Actinopterygii: Ray-Finned Fishes

Ray-finned fishes are an enormous assemblage containing all our familiar bony fishes—more than 23,600 species. Earliest actinopterygians, known as **palaeoniscids** (pay'lee-o-nis'ids), were small fishes, with large eyes, heterocercal tail (Figure 24-16), and thick, interlocking scales with an outer layer of **ganoin** (Figure 24-18). These fishes had a single dorsal fin and numerous bony rays derived from scales stacked end to end, distinctively different in appearance from the lobe-finned fishes with which they shared the Devonian waters. Palaeoniscids are represented by fossil fragments as early as the late Silurian, and flourished throughout the late Paleozoic era, during the same period that ostracoderms, placoderms, and acanthodians disappeared and sarcopterygians declined in abundance (Figure 24-1). This suggests morphological specializations evolving in the actinopterygian lineage gave them ecological superiority over most other fishes.

From these earliest ray-finned fishes, two major groups appeared. Those with the most primitive characteristics are **chondrosteans** (Gr. *chondros,* cartilage, + *osteon,* bone), represented today by freshwater and anadromous sturgeons, paddlefishes, and bichirs (Figure 24-19). Chondrosteans show many characters similar to their paleoniscid ancestors, including a het-

across the gills, as well as pushed across by the mouth pump. A gas-filled derivative of the esophagus provided an additional means of gas exchange in hypoxic waters and an efficient means for achieving neutral buoyancy. Progressive specialization of jaw musculature and skeletal elements involved in feeding is another key feature of bony fish evolution.

erocercal tail and ganoid scutes or scales. The bichir *Polypterus* (Gr. *poly,* many, + *pteros,* winged) of African waters is an interesting relict with lungs and other primitive characters that make it resemble palaeoniscids more than any other living fish.

The second major group of ray-finned fishes to emerge from palaeoniscid stock were the **neopterygians** (Gr. *neos,* new, + *pteryx,* fin). Neopterygians appeared in the late Permian and radiated extensively during the Mesozoic era (Figure 24-1). During the Mesozoic one lineage gave rise to a secondary radiation that led to modern bony fishes, the teleosts. There are two surviving genera of early neopterygians, the bowfin *Amia* (Gr. tunalike fish) of shallow, weedy waters of the Great Lakes and Mississippi Valley, and gars *Lepisosteus* (Gr. *lepidos,* scale, + *osteon,* bone) of eastern and southern North America (Figure 24-20). The seven species of gars are large, ambush predators with

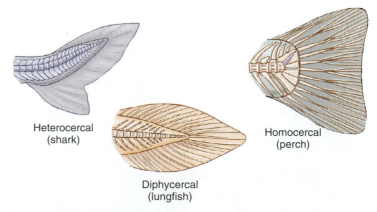

Heterocercal
(shark)

Diphycercal
(lungfish)

Homocercal
(perch)

Figure 24-16

Types of caudal fins among fishes.

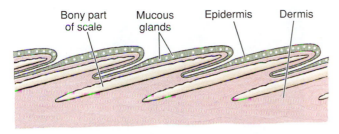

Figure 24-17

Section through the skin of a bony fish, showing the overlapping scales (yellow). The scales lie in the dermis and are covered by epidermis.

elongate bodies and jaws filled with needlelike teeth. Gars and bowfin may surface to gulp air, filling their vascularized swim bladder with air to supplement oxygen obtained in the gills.

The major lineage of neopterygians are **teleosts** (Gr. *teleos*, perfect, + *osteon*, bone), the modern bony fishes (Figure 24-15). Teleost diversity is astounding, with about 23,600 described species, representing about 96% of all living fishes or about half of all vertebrates (Figure 24-21). In addition, it has been estimated there are an additional 5,000 to 10,000 undescribed species. Although most of the 200 or so new species of teleosts described each year are from poorly sampled areas such as South America or deep oceanic waters, several new species are described each year from areas as well known as the fresh waters of North America! Teleosts range in size from 10 mm adult gobies to 17 m oarfish and 900 kg, 4.5 m blue marlin (Figure 24-21). These fishes occupy almost every conceivable habitat, from elevations up to 5200 m in Tibet to 8000 m below the surface of the ocean. Some species live in hot springs at 44°C, while others live under the Antarctic ice at −2° C. They may live in lakes with salt concentrations three times that of seawater, caves of total darkness, swamps devoid of oxygen, or even make extended excursions onto land, as do mudskippers (Figure 24-21).

Several morphological trends in the teleost lineage allowed them to diversify into this truly incredible variety of habitats and

> ## Characteristics of Class Actinopterygii
>
> 1. **Skeleton with bone of endochondral origin;** caudal fin heterocercal in ancestral forms, usually **homocercal** in descendant forms (Figure 24-16); skin with mucous glands and embedded dermal scales (Figure 24-17); scales **ganoid** in ancestral forms, scales **cycloid, ctenoid** or absent in advanced forms (Figure 24-18)
> 2. Paired and median fins present, **supported by long dermal rays (lepidotrichia);** muscles controlling fin movement within body
> 3. Jaws present; teeth usually present with enamaloid covering; olfactory sacs do not open into mouth; spiral valve present in ancestral forms, absent in advanced forms
> 4. Respiration primarily by gills supported by arches and covered with an **operculum**
> 5. **Swim bladder** often present with or without a duct connecting to esophagus, usually functioning in buoyancy
> 6. Circulation consisting of a heart with a sinus venosus, an undivided atrium, and an undivided ventricle; single circulation; typically four aortic arches; nucleated erythrocytes
> 7. Excretory system of paired opisthonephric kidneys; sexes usually separate; fertilization usually external; larval forms may differ greatly from adults
> 8. Nervous system of a brain with small cerebrum, optic lobes, and cerebellum; 10 pairs of cranial nerves; three pairs of semicircular canals

forms. Heavy dermal armor of primitive ray-finned fishes was replaced by light, thin, flexible **cycloid** and **ctenoid** scales (Figure 24-18). Some teleosts, such as most eels and catfishes, completely lack scales. The increased mobility and speed that resulted from loss of heavy armor improved predator avoidance and feeding efficiency. Changes in the fins of teleosts increased maneuverability and speed and allowed fins to serve a variety of other functions. The symmetrical shape of the **homocercal** tail

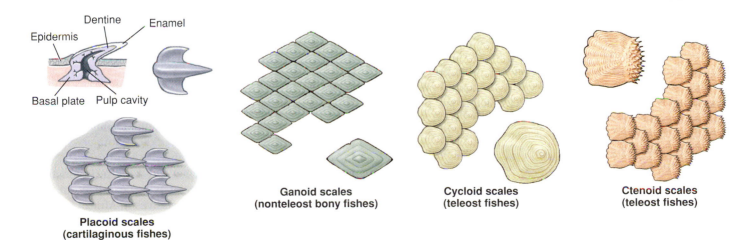

Figure 24-18

Types of fish scales. Placoid scales are small, conical toothlike structures characteristic of Chondrichthyes. Diamond-shaped ganoid scales, present in early bony fishes such as the gar, are composed of layers of silvery enamel (ganoin) on the upper surface and bone on the lower. Teleosts have either cycloid or ctenoid scales. These are thin and flexible and are arranged in overlapping rows.

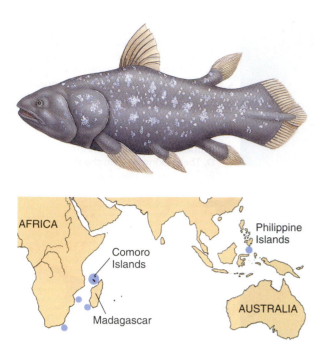

Figure 24-23
The coelacanth genus *Latimeria* is a surviving marine relict of a group of lobe-finned fishes that flourished some 350 million years ago.

therefore a paraphyletic group). **Coelacanths** also arose in the Devonian period, radiated somewhat, and reached their evolutionary peak in the Mesozoic era. At the end of the Mesozoic era they nearly disappeared but left one remarkable surviving genus, *Latimeria* (Figure 24-23). Since the last coelacanths were believed to have become extinct 70 million years ago, the astonishment of the scientific world may be imagined when the remains of a coelacanth were found on a dredge off the coast of South Africa in 1938. An intensive search to locate more specimens was successful off the coast of the Comoro Islands. There fishermen occasionally catch them at great depths with hand lines, providing specimens for research. This was believed to be the only population of *Latimeria* until 1998, when the scientific world was again surprised by the capture of a new species of coelacanth in Sulawesi, Indonesia, 10,000 km from the Comoros!

The "modern" marine coelacanths are descendants of the Devonian freshwater stock. The tail is diphycercal (Figure 24-16) but possesses a small lobe between the upper and lower caudal lobes, producing a three-pronged structure (Figure 24-23).

Coelacanths are a deep metallic blue with irregular white or brassy flecks, providing camouflage against the dark lava-cave reefs they inhabit. Young are born fully formed after hatching internally from eggs 9 cm in diameter—the largest among bony fishes.

STRUCTURAL AND FUNCTIONAL ADAPTATIONS OF FISHES

Locomotion in Water

To the human eye, some fishes appear capable of swimming at extremely high speeds. But our judgment is unconsciously tempered by our own experience that water is a highly resistant medium through which to move. Most fishes, such as a trout or a minnow, can swim maximally about 10 body lengths per second, obviously an impressive performance by human standards. Yet when these speeds are translated into kilometers per hour it means that a 30 cm (1 foot) trout can swim only about 10.4 km (6.5 miles) per hour. As a general rule, the larger the fish the faster it can swim.

Measuring fish cruising speeds accurately is best done in a "fish wheel," a large ring-shaped channel filled with water that is turned at a speed equal and opposite to that of the fish. Much more difficult to measure are the sudden bursts of speed that most fish can make to capture prey or to avoid being captured. A hooked bluefin tuna was once "clocked" at 66 km per hour (41 mph); swordfish and marlin are thought to be capable of incredible bursts of speed approaching, or even exceeding, 110 km per hour (68 mph). Such high speeds can be sustained for no more than 1 to 5 seconds.

The propulsive mechanism of a fish is its trunk and tail musculature. The axial, locomotory musculature is composed of zigzag bands, called **myomeres.** Muscle fibers in each myomere are relatively short and connect the tough connective tissue partitions that separate each myomere from the next. On the surface myomeres take the shape of a **W** lying on its side (Figure 24-24) but internally the bands are complexly folded and nested so that the pull of each myomere extends over several vertebrae. This arrangement produces more power and finer control of movement since many myomeres are involved in bending a given segment of the body.

Figure 24-24
Trunk musculature of a teleost fish, partly dissected to show internal arrangement of the muscle bands (myomeres). The myomeres are folded into a complex, nested grouping, an arrangement that favors stronger and more controlled swimming.

Eel   **Trout**

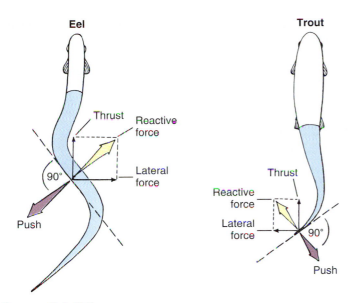

Figure 24-25
Movements of swimming fishes, showing the forces developed by an eel-shaped and spindle-shaped fish.

Source: From Vertebrate Life, by Pough et al., 1996. Reprinted by permission of Prentice-Hall, Inc., Upper Saddle River, NJ.

Figure 24-26
Bluefin tuna, showing adaptations for fast swimming. Powerful trunk muscles pull on the slender tail stalk. Since the body does not bend, all of the thrust comes from beats of the stiff sickle-shaped tail.

Understanding how fishes swim can be approached by studying the motion of a very flexible fish such as an eel (Figure 24-25). The movement is serpentine, not unlike that of a snake, with waves of contraction moving backward along its body by alternate contraction of myomeres on either side. The anterior end of the body bends less than the posterior end, so that each undulation increases in amplitude as it travels along the body. While undulations move backward, bending of the body pushes laterally against the water, producing a **reactive force** that is directed forward, but at an angle. It can be analyzed as having two components: **thrust,** which is used to overcome drag and propels the fish forward, and **lateral force,** which tends to make the fish's head "yaw," or deviate from the course in the same direction as its tail. This side-to-side head movement is very obvious in a swimming eel or shark, but many fishes have a large, rigid head with enough surface resistance to minimize yaw.

The movement of an eel is reasonably efficient at low speed, but its body shape generates too much frictional drag for rapid swimming. Fishes that swim rapidly, such as trout, are less flexible and limit body undulations mostly to the caudal region (Figure 24-25). Muscle force generated in the large anterior muscle mass is transferred through tendons to the relatively nonmuscular caudal peduncle and tail where thrust is generated. This form of swimming reaches its highest development in tunas, whose bodies do not flex at all. Virtually all thrust is derived from powerful beats of the tail fin (Figure 24-26). Many fast oceanic fishes such as marlin, swordfish, amberjacks, and wahoo have swept-back tail fins shaped much like a sickle. Such fins are the aquatic counterpart of the high speed wings of the swiftest birds (p. 000).

Swimming is the most economical form of animal locomotion, largely because aquatic animals are almost perfectly supported by their medium and need expend little energy to overcome the force of gravity. If we compare the energy cost per kilogram of body weight of traveling 1 km by different forms of locomotion, we find swimming costs only 0.39 kcal (salmon) as compared with 1.45 kcal for flying (gull) and 5.43 for walking (ground squirrel). However, part of the unfinished business of biology is understanding how fish and aquatic mammals are able to move through water while creating almost no turbulence. The secret lies in the way aquatic animals bend their bodies and fins (or flukes) to swim and in the friction-reducing properties of the body surface.

Neutral Buoyancy and the Swim Bladder

All fishes are slightly heavier than water because their skeletons and other tissues contain heavy elements present only in trace amounts in natural waters. To keep from sinking, sharks must always keep moving forward in the water. The asymmetrical (heterocercal) tail of a shark provides the necessary tail lift as it sweeps to and fro in the water, and its broad head and flat pectoral fins (Figure 24-8) act as angled planes to provide lift. Sharks also are aided in buoyancy by having very large livers containing a special fatty hydrocarbon called **squalene** with a density of only 0.86 grams per milliliter. The liver thus acts like a large sack of buoyant oil that helps to compensate the shark's heavy body.

By far the most efficient flotation device is a gas-filled space. The **swim bladder** serves this purpose in bony fishes (Figure 24-27). It arose from the paired lungs of primitive Devonian bony fishes. Lungs were probably a ubiquitous feature of Devonian freshwater bony fishes when, as we have seen, warm, swampy habitats would have made such an accessory respiratory structure advantageous. Swim bladders are present in most pelagic bony fishes but are absent in tunas, most abyssal fishes, and most bottom dwellers, such as flounders and sculpins.

Without a swim bladder, bony fishes sink because their tissues are more dense than water. To compensate and achieve neutral buoyancy, they displace additional water by a volume of gas in a swim bladder, thus adjusting their total density to that of the surrounding water. This allows fishes with a swim bladder to remain suspended indefinitely at any depth with no muscular effort. Unlike bone, blood, and other tissues, gas is compressible and changes volume as a fish changes its depth. If a fish swims to a greater depth, the greater pressure exerted by the surrounding water compresses the gas in the swim bladder, so that the fish becomes less buoyant and tends to sink. The volume of gas

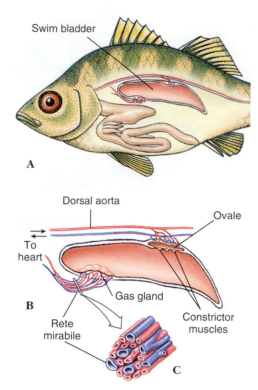

Figure 24-27

A, Swim bladder of a teleost fish. The swim bladder lies in the coelom just beneath the vertebral column. **B,** Gas is secreted into the swim bladder by the gas gland. Gas from the blood is moved into the gas gland by the rete mirabile, a complex array of tightly-packed capillaries that act as a countercurrent multiplier to increase oxygen concentration. The arrangement of venous and arterial capillaries in the rete is shown in **C.** To release gas during ascent, a muscular valve opens, allowing gas to enter the ovale from which the gas is removed by diffusion into the blood.

in the swim bladder must be increased to establish a new equilibrium buoyancy. When a fish swims to a lesser depth, gas in the bladder expands because of the reduced surrounding water pressure, so that the fish becomes more buoyant. Unless gas is removed, the fish will continue to ascend with increasing speed as the swim bladder continues to expand.

Gas may be removed from the swim bladder in one of two ways. The more primitive **physostomous** (Gr., *phys,* bladder, *stoma,* mouth) fishes (trout, for example) have a pneumatic duct that connects the swim bladder to the esophagus. These fishes may simply expel air through the pneumatic duct. More advanced teleosts exhibit the **physoclistous** (Gr., *phys,* bladder, *clist,* closed) condition in which the pneumatic duct is lost in adults. In physoclistous fishes, gas must be absorbed by blood from the **ovale,** a vascularized area of the swim bladder (Figure 24-27). Both types of fishes require gas to be secreted into the swim bladder from the blood, although a few shallow-water-inhabiting physostomes may gulp air to fill their swim bladder.

Physiologists who were at first baffled by the secretion mechanism now understand how it operates. In brief, the gas gland secretes lactic acid, which enters the blood, causing a localized high acidity in the **rete mirabile** that forces hemoglo-

bin to release its load of oxygen. The capillaries in the rete are arranged so that the released oxygen accumulates in the rete, eventually reaching such a high pressure that the oxygen diffuses into the swim bladder. The final gas pressure attained in the swim bladder depends on the length of the rete capillaries; they are relatively short in fishes living near the surface, but are extremely long in deep-sea fishes.

The amazing effectiveness of this device is exemplified by a fish living at a depth of 2400 m (8000 feet). To keep the bladder inflated at that depth, the gas inside (mostly oxygen, but also variable amounts of nitrogen, carbon dioxide, argon, and even some carbon monoxide) must have a pressure exceeding 240 atmospheres, which is much greater than the pressure in a fully charged steel gas cylinder. Yet the oxygen pressure in the fish's blood cannot exceed 0.2 atmosphere—in equilibrium with the oxygen pressure in the atmosphere at the sea surface.

Respiration

Fish gills are composed of thin filaments, each covered with a thin epidermal membrane that is folded repeatedly into platelike **lamellae** (Figure 24-28). These are richly supplied with blood vessels. The gills are located inside the pharyngeal cavity and are covered with a movable flap, the **operculum.** This arrangement provides excellent protection to delicate gill filaments, streamlines the body, and makes possible a pumping system for moving water through the mouth, across the gills, and out the operculum. Instead of opercular flaps as in bony fishes, elasmobranchs have a series of **gill slits** (Figure 24-8) out of which the water flows. In both elasmobranchs and bony fishes the branchial mechanism is arranged to pump water continuously and smoothly over the gills, although to an observer it appears that fish breathing is pulsatile. The flow of water is opposite the direction of blood flow (countercurrent flow), the best arrangement for extracting the greatest possible amount of oxygen from water. Some bony fishes can remove as much as 85% of the dissolved oxygen from water passing over their gills. Very active fishes, such as herring and mackerel, can obtain sufficient water for their high oxygen demands only by swimming forward continuously to force water into their open mouth and across their gills. This process is called ram ventilation. Such fish will be asphyxiated if placed in an aquarium that restricts free swimming movements, even if the water is saturated with oxygen.

A surprising number of fishes can live out of water for varying lengths of time by breathing air. Several devices are employed by different fishes. We already have described the lungs of lungfishes, gars, and the extinct rhipidistians. Freshwater eels often make overland excursions during rainy weather, using the skin as a major respiratory surface. Electric eels, *Electrophorus* (Gr. *ēlektron,* something bright, + *phoros,* to bear), have degenerate gills and must supplement gill respiration by gulping air through a vascular mouth cavity. One of the best air breathers of all is the Indian climbing perch *Anabas* (Gr. *anabainō,* to go up), which spends most of its time on land near the water's edge, breathing air through special air chambers above much-reduced gills.

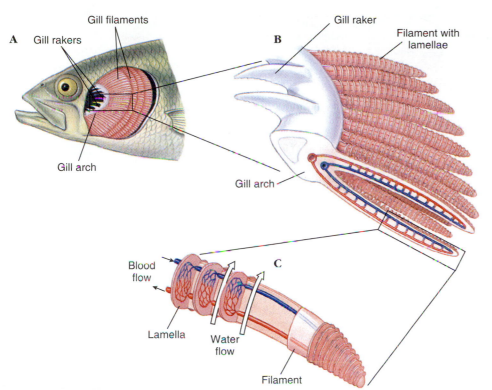

Figure 24-28

Gills of fish. The bony, protective flap covering the gills (operculum) has been removed, **A,** to reveal branchial chamber containing the gills. There are four gill arches on each side, each bearing numerous filaments. A portion of gill arch **(B)** shows gill rakers that project forward to strain out food and debris, and gill filaments that project to the rear. A single gill filament **(C)** is dissected to show the blood capillaries within the platelike lamellae. Direction of water flow *(large arrows)* is opposite the direction of blood flow.

Osmotic Regulation

Fresh water is an extremely dilute medium with a salt concentration (0.001 to 0.005 gram moles per liter [M]) much below that of the blood of freshwater fishes (0.2 to 0.3 M). Water therefore tends to enter their bodies osmotically, and salt is lost by diffusion outward. Although the scaled and mucus-covered body surface is almost totally impermeable to water, water gain and salt loss do occur across thin membranes of the gills. Freshwater fishes are **hyperosmotic regulators** with several defenses against these problems (Figure 24-29). First, excess water is pumped out by the **opisthonephric** kidney (p. 641), which is capable of forming very dilute urine. Second, special **salt-absorbing cells** located in the gill epithelium actively move salt ions, principally sodium and chloride, from water to the blood. This, together with salt present in the fish's food, replaces diffusive salt loss. These mechanisms are so efficient that a freshwater fish devotes only a small part of its total energy expenditure to keeping itself in osmotic balance.

Perhaps 90% of all bony fishes are restricted to either a freshwater or a seawater habitat because they are incapable of osmotic regulation in the "wrong" habitat. Most fresh-water fishes quickly die if placed in seawater, as will marine fishes placed in fresh water. However, some 10% of all teleosts can pass back and forth with ease between both habitats. These **euryhaline fishes** (Gr. *eurys,* broad, + *hals,* salt) are of two types: those such as many flounders, sculpins, and killifish that live in estuaries or certain intertidal areas where the salinity fluctuates throughout the day; and those such as salmon, shad, and eels, that spend part of their life cycle in fresh water and part in seawater.

Marine bony fishes are **hypoosmotic regulators** that encounter a completely different set of problems. Having a much lower blood salt concentration (0.3 to 0.4 M) than the seawater around them (about 1 M), they tend to lose water and gain salt. A marine teleost fish quite literally risks drying out, much like a desert mammal deprived of water. Again, marine bony fishes, like their freshwater counterparts, have evolved an appropriate set of defenses (Figure 24-29). To compensate for water loss, a marine teleost drinks seawater. Although this behavior obviously brings needed water into its body, it is unfortunately accompanied by a great deal of unneeded salt. Unwanted salt is disposed in two ways: (1) major sea salt ions (sodium, chloride, and potassium) are carried by the blood to the gills where they are secreted outward by special **salt-secretory cells;** and (2) the remaining ions, mostly divalent ions (magnesium, sulfate, and calcium), are left in the intestine and voided with feces. However, a small but significant fraction of these residual divalent salts in the intestine, some 10% to 40% of the total, penetrates the intestinal mucosa and enters the bloodstream. These ions are excreted by the kidney. Unlike a freshwater fish's kidney, which forms its urine by the usual filtration-resorption sequence typical of most vertebrate kidneys (pp. 643–645), a marine fish's kidney excretes divalent ions by tubular secretion. Since very little if any filtrate is formed, the glomeruli have lost their importance and disappeared in some marine teleosts. Pipefishes, and the goosefish shown in Figure 24-31, are examples of "aglomerular" marine fishes.

Feeding Behavior

For any fish, feeding is one of its main concerns in day-to-day living. Although many a luckless angler would swear otherwise, the fact is that a fish devotes more time and energy to eating, or

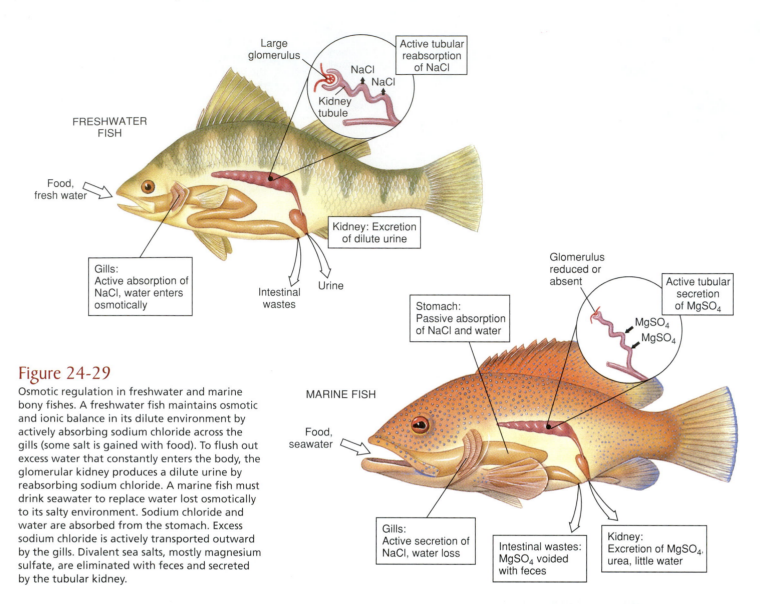

Figure 24-29

Osmotic regulation in freshwater and marine bony fishes. A freshwater fish maintains osmotic and ionic balance in its dilute environment by actively absorbing sodium chloride across the gills (some salt is gained with food). To flush out excess water that constantly enters the body, the glomerular kidney produces a dilute urine by reabsorbing sodium chloride. A marine fish must drink seawater to replace water lost osmotically to its salty environment. Sodium chloride and water are absorbed from the stomach. Excess sodium chloride is actively transported outward by the gills. Divalent sea salts, mostly magnesium sulfate, are eliminated with feces and secreted by the tubular kidney.

searching for food to eat, than to anything else. Throughout the long evolution of fishes, there has been unrelenting selective pressure for those adaptations that enable a fish to win the eat-or-be-eaten contest. Certainly the most far-reaching single event was the evolution of jaws. Jaws freed fishes from a largely passive filter-feeding existence, enabling them to adopt a predatory mode of life. Improved means of capturing larger prey demanded stronger muscles, more agile movement, better balance, and improved special senses. More than any other aspect of its life habit, feeding behavior shapes the fish.

Most fishes are **carnivores** and prey on a myriad of animal foods from zooplankton and insect larvae to large vertebrates. Some deep-sea fishes are capable of eating victims nearly twice their own size—an adaptation for life in a world where meals are infrequent. Most advanced ray-finned fishes cannot masticate their food as we can because doing so would block the current of water across the gills. Some, however, such as wolf eels (Figure 24-30), have molarlike teeth in their jaws for crushing prey, which may include hard-bodied crustaceans. Others that do grind their food use powerful pharyngeal teeth in their throat. Most carnivorous fish swallow their prey whole, using sharp-pointed

Figure 24-30

Wolf eel, *Anarrhichthys ocellatus*, feeding on a sea cucumber it has captured and pulled to the opening of its den.

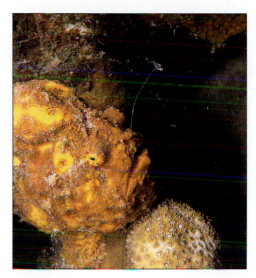

Figure 24-31
Goosefish, *Lophius piscatorius*, awaits its meal. Above its head swings a modified dorsal fin spine ending in a fleshy tentacle that contracts and expands in a convincing wormlike manner. When a fish approaches the alluring bait, the huge mouth opens suddenly, creating a strong inward current that sweeps the prey inside. In a split second all is over.

teeth in their jaws and on the roof of the mouth to seize prey. The incompressibility of water assists many large-mouthed predators in capturing prey. When the mouth is suddenly opened, water rushes in, sweeping the victim inside (Figure 24-31).

A second group of fishes are **herbivores** that eat plants and algae. Although plant eaters are relatively few in number, they are crucial intermediates in the food chain, especially in freshwater rivers, lakes, and ponds that contain very little plankton.

Suspension-feeders that crop the abundant microorganisms of the sea form a third and diverse group of fishes ranging from fish larvae to basking sharks. However, the most characteristic group of plankton feeders are herringlike fishes (menhaden, herring, anchovies, capelin, pilchards, and others), mostly **pelagic** (open-sea dwellers) fishes that travel in large schools. Both phytoplankton and smaller zooplankton are strained from the water with sievelike gill rakers (Figure 24-28). Because plankton feeders are the most abundant of all marine fishes, they are important food for numerous larger but less abundant carnivores. Many freshwater fishes also depend on plankton for food.

A fourth group of fishes contains **omnivores** that feed on both plant and animal food. Finally there are **scavengers** that feed on organic debris (detritus) and **parasites** that suck body fluids of other fishes.

Digestion in most fishes follows the vertebrate plan. Except in several fishes that lack distinct stomachs, food proceeds from stomach to tubular intestine, which tends to be short in carnivores (Figure 24-15) but may be extremely long and coiled in herbivorous forms. In herbivorous grass carps, for example, the intestine may be nine times the body length, an adaptation for the lengthy digestion required for plant carbohydrates. In carnivores, some protein digestion may be initiated in the acid medium of the stomach, but the principal function of the stomach is to store often large and infrequent meals while awaiting their reception by the intestine.

Digestion and absorption proceed simultaneously in the intestine. A curious feature of ray-finned fishes, especially teleosts, is the presence of numerous **pyloric ceca** (Figure 24-15) found in no other vertebrate group. Their primary function appears to be fat absorption, although all classes of digestive enzymes (protein-, carbohydrate-, and fat-splitting) are secreted there.

Migration

Freshwater Eels

For centuries naturalists had been puzzled about the life history of freshwater eels, *Anguilla* (an-gwil′la) (L. eel), a common and commercially important species of coastal streams of the North Atlantic. Eels are **catadromous** (Gr. *kata*, down, + *dromos*, running), meaning that they spend most of their lives in fresh water but migrate to the sea to spawn. Each fall, large numbers of eels were seen swimming down rivers toward the sea, but no adults ever returned. Each spring countless numbers of young eels, called "elvers" (Figure 24-32), each about the size of a wooden matchstick, appeared in coastal rivers and began swimming upstream. Beyond the assumption that eels must spawn somewhere at sea, location of their breeding grounds was completely unknown.

The first clue was provided by two Italian scientists, Grassi and Calandruccio, who in 1896 reported that elvers were not larval eels but rather were relatively advanced juveniles. True larval eels, they discovered, were tiny, leaf-shaped, completely transparent creatures that bore absolutely no resemblance to an eel. They had been called **leptocephali** (Gr. *leptos*, slender, + *kephalē*, head) by early naturalists, who never suspected their true identity. In 1905 Johann Schmidt, supported by the Danish government, began a systematic study of eel biology, which he continued until his death in 1933. With cooperation of captains of commercial vessels plying the Atlantic, thousands of leptocephali were caught in different areas of the Atlantic with plankton nets Schmidt supplied. By noting where larvae in different stages of development were captured, Schmidt and his colleagues eventually reconstructed the spawning migrations.

When adult eels leave the coastal rivers of Europe and North America, they swim steadily and apparently at great depth for 1 to 2 months until they reach the Sargasso Sea, a vast area of warm oceanic water southeast of Bermuda (Figure 24-32). Here, at depths of 300 m or more, the eels spawn and die. Minute larvae then begin an incredible journey back to the coastal rivers of Europe. Drifting with the Gulf Stream and preyed on constantly by numerous predators, they reach the middle of the Atlantic after 2 years. By the end of the third year they arrive in the coastal waters of Europe where the leptocephali metamorphose into elvers, with an unmistakable eel-like body form (Figure 24-32). Here males and females part company; males remain in the brackish waters of coastal rivers and estuaries while females continue up the rivers, often traveling hundreds of miles upstream. After 8 to 15 years of growth, the females, now 1 m or more long, return to the sea to join the smaller males; both return to ancestral breeding grounds thousands of miles away to complete the life cycle.

Figure 24-32

Life histories of the European eel, *Anguilla anguilla,* and American eel, *Anguilla rostrata*. Migration patterns of European species are shown in pink. Migration patterns of American species are shown in black. Boxed numbers refer to stages of development. Note that the American eel completes its larval metamorphosis and sea journey in one year. It requires nearly three years for the European eel to complete its much longer journey.

Recent enzyme electrophoretic analysis of eel larvae confirmed not only the existence of separate European and American species but also Schmidt's belief that the European and American eels spawn in partially overlapping areas of the Sargasso Sea.

Schmidt found that American eels *(Anguilla rostrata)* could be distinguished from European eels *(A. vulgaris)* because they had fewer vertebrae—an average of 107 in American eels as compared with an average 114 in the European species. Since American eels are much closer to the North American coastline, the larvae require only about 8 months to make their journey.

Homing Salmon

The life history of salmon is nearly as remarkable as that of eels and certainly has received far more popular attention. Salmon are **anadromous** (Gr. *anadromos,* running upward); they spend their adult lives at sea but return to fresh water to spawn. Atlantic salmon *(Salmo salar)* (L. *salmo,* salmon, *sal,* salt) and Pacific salmon (six species in the genus *Oncorhynchus* [on-ko-rink′us] [Gr. *onkos,* hook, + *rhynchos,* snout]) have this practice, but there are important differences among the seven species.

Atlantic salmon may make repeated upstream spawning runs. The six Pacific salmon species (king, sockeye, silver, humpback, chum, and Japanese masu) each make a single spawning run (Figure 24-33), after which they die.

The virtually infallible homing instinct of the Pacific species is legendary: after migrating downstream as a smolt, a sockeye salmon ranges many hundreds of miles over the Pacific for nearly 4 years, grows to 2 to 5 kg in weight, and then returns almost unerringly to spawn in the headwaters of its parent stream. Some straying does occur and is an important means of increasing gene flow and populating new streams.

Experiments by A. D. Hasler and others have shown that homing salmon are guided upstream by the characteristic odor of their parent stream. When salmon finally reach the spawning beds of their parents (where they themselves were hatched), they spawn and die. The following spring, newly hatched fry transform into smolts before and during the downstream migration. At this time they are imprinted (p. 758) with the distinctive odor of the stream, which is apparently a mosaic of compounds released by the characteristic vegetation and soil in the watershed of the parent stream. They also seem to imprint on odors of other streams they pass while migrating downriver and use these odors in reverse sequence as a map during the upriver migration as returning adults.

Salmon runs in the Pacific Northwest have been devastated by a lethal combination of spawning stream degradation by logging, pollution and, especially, by more than 50 hydroelectric dams, which obstruct upstream migration of adult salmon and kill downstream migrants as they pass through the dams' power-generating turbines. In addition, the chain of reservoirs behind the dams, which has converted the Columbia and Snake Rivers into a series of lakes, increases mortality of young salmon migrating downstream by slowing their passage to the sea. The result is that the annual run of wild salmon is today only about 3% of the 10 to 16 million fish that ascended the rivers 150 years ago. While recovery plans have been delayed by the power industry, environmental groups argue that in the long run losing the salmon will be more expensive to the regional economy than making changes now that will allow salmon stocks to recover.

How do salmon find their way to the mouth of a coastal river from the trackless miles of open ocean? Salmon move hundreds of miles away from the coast, much too far to be able to detect the odor of their parent stream. Experiments suggest that some migrating fish, like birds, can navigate by orienting to the position of the sun. However, migrant salmon can navigate on cloudy days and at night, indicating that sun navigation, if used at all, cannot be a salmon's only navigational cue. Fish also (again, like birds) appear able to detect and navigate to the earth's magnetic field. Finally, fishery biologists concede that salmon may not require precise navigational abilities at all, but instead may use ocean currents, temperature gradients, and food availability to reach the general coastal area where "their" river is located. From this point, they would navigate by their imprinted odor map, making correct turns at each stream junction until they reach their natal stream.

Figure 24-33
Migrating Pacific sockeye salmon *(Oncorhynchus nerka)*.

Reproduction and Growth

In a group as diverse as fishes, it is no surprise to find extraordinary variations on the basic theme of sexual reproduction. Most fishes favor a simple theme: they are **dioecious,** with **external fertilization** and **external development** of their eggs and embryos (oviparity). However, as tropical fish enthusiasts are well aware, the ever-popular ovoviviparous guppies and mollies of home aquaria bear their young alive after development in the ovarian cavity of the mother (Figure 24-34). As described earlier in this chapter (p. 496), some viviparous sharks develop a kind of placental attachment through which the young are nourished during gestation.

Let us return to the much more common oviparous mode of reproduction. Many marine fishes are extraordinarily profligate egg producers. Males and females come together in great schools and release vast numbers of gametes into the water to drift with currents. Large female cod may release 4 to 6 million eggs at a single spawning. Less than one in a million will survive the numerous perils of the ocean to reach reproductive maturity.

Unlike the minute, buoyant, transparent eggs of pelagic marine teleosts, those of many near-shore bottom-dwelling (benthic) species are larger, typically yolky, nonbuoyant, and adhesive. Some bury their eggs, many attach them to vegetation, some deposit them in nests, and some even incubate them in their mouths (Figure 24-35). Many benthic spawners guard their eggs. Intruders expecting an easy meal of eggs may be met with a vivid and often belligerent display by the guard, which is almost always the male.

Figure 24-34
Rainbow surfperch, *Hypsurus caryi,* giving birth. All of the West Coast surfperches (family Embiotocidae) are ovoviviparous.

Figure 24-35
Male banded jawfish, *Opistognathus macrognathus,* orally brooding its eggs. The male retrieves the female's spawn and incubates the eggs until they hatch. During brief periods when the jawfish is feeding, the eggs are left in the burrow.

Classification of Living Fishes

The following Linnaean classification of major fish taxa follows that of Nelson (1994). The probable relationships of these traditional groupings together with the major extinct groups of fishes are shown in a cladogram in Figure 24-2. Other schemes of classification have been proposed. Because of the difficulty of determining relationships among the numerous living and fossil species, we can appreciate why fish classification has undergone, and will continue to undergo, continuous revision.

Phylum Chordata

Subphylum Vertebrata (Craniata)

Superclass Agnatha (ag'na-tha) (Gr. *a*, not + *gnathos*, jaw). No jaws; cartilaginous skeleton; paired fins absent; one or two semicircular canals; notochord persistent. Not a monophyletic taxon.

Class Myxini (mik-sy'ny) (Gr. *myxa*, slime): **hagfishes.** Four pairs of tentacles around mouth; nasal sac with duct to pharynx; 5 to 16 pairs of gill pouches, accessory hearts and slime glands present; poorly developed eyes. Examples. *Myxine, Bdellostoma;* about 65 species, marine.

Class Cephalaspidomorphi (sef-a-lass'pe-do-morf'e) (Gr. *kephalē*, head, + *aspidos*, shield, + *morphē*, form): **lampreys.** Buccal funnel with keratinized teeth; nasal sac not connected to mouth; seven pairs of gill pouches; well-developed eyes. Examples: *Petromyzon, Ichthyomyzon, Lampetra;* 41 species, freshwater and anadromous.

Superclass Gnathostomata (na'tho-sto'-ma-ta) (Gr. *gnathos*, jaw, + *stoma*, mouth). Jaws present; paired appendages present (secondarily lost in a few forms); three pairs of semicircular canals; notochord partly or completely replaced by centra.

Class Chondrichthyes (kon-drik'thee-eez) (Gr. *chondros*, cartilage + *ichthys*, fish): **cartilaginous fishes.** Cartilaginous skeleton; teeth not fused to jaws and usually replaced; no swim bladder; intestine with spiral valve; claspers present in males.

Subclass Elasmobranchii (e-laz'mo-bran'kee'i) (Gr. *elasmos*, plated, + *branchia*, gills): **sharks, skates, and rays.** Placoid scales or derivatives (scutes and spines) usually present; five to seven gill arches and gill slits in separate clefts along pharynx; upper jaw not fused to cranium. Examples: *Squalus, Raja, Charcarodon, Sphyrna.* About 815 species, mostly marine.

Subclass Holocephali (hol'o-sef'a-li) (Gr. *holos*, entire, + *kephalē*, head): **chimaeras, ratfishes.** Scales absent; four gill slits covered by operculum; jaws with tooth plates; accessory clasping organ (tentaculum) in males; upper jaw fused to cranium. Examples: *Chimaera, Hydrolagus;* 31 species, marine.

Class Actinopterygii (ak'ti-nop-te-rij'ee-i) (Gr. *aktis*, ray, + *pteryx*, fin, wing): **ray-finned fishes.** Skeleton ossified; single gill opening covered by operculum; paired fins supported primarily by dermal rays; limb musculature within body; swim bladder mainly a hydrostatic organ, if present; atrium and ventricle not divided; teeth with enamaloid covering.

Subclass Chrondrostei (kon-dros'tee-i) (Gr. *chondros*, cartilage, + *osteon*, bone): **bichirs, paddlefishes, sturgeons.** Skeleton primarily cartilage; caudal fin heterocercal; scales ganoid, if present; spiral valve present; spiracle usually present; more fin rays than ray supports. Examples: *Polypterus, Polyodon, Acipenser.* 34 species, freshwater and anadromous.

Subclass Neopterygii (nee'op-te-rij'ee-i) (Gr. *neo*, new, + *pteryx*, fin, wing): **gars, bowfin, teleosts.** Skeleton primarily bone; caudal fin usually homocercal; scales cycloid, ctenoid, absent, or rarely, ganoid. Fin ray number equal to their supports in dorsal and anal fins. Examples: *Amia, Lepisosteus, Anguilla, Oncorhynchus, Perca.* About 23,600 species, nearly all aquatic habitats.

Class Sarcopterygii (sar-cop-te-rij'ee-i) (Gr. *sarkos*, flesh, + *pteryx*, fin, wing): **lobe-finned fishes.** Skeleton ossified; single gill opening covered by operculum; paired fins with sturdy internal skeleton and musculature within limb; diphycercal tail; intestine with spiral valve; usually with lunglike swim bladder; atrium and ventricle at least partly divided; teeth with enamel covering. Examples: *Latimeria* (coelacanths); *Neoceratodus, Lepidosiren, Protopterus* (lungfishes). 8 species, marine and freshwater. Not monophyletic unless tetrapods are included.

Freshwater fishes almost invariably produce nonbuoyant eggs. Those, such as perch, that provide no parental care simply scatter their myriads of eggs among weeds or along the bottom. Freshwater fishes that do provide egg care, such as bullhead catfishes and some darters, produce fewer, larger eggs that enjoy a better chance for survival.

Elaborate preliminaries to mating are the rule for freshwater fishes. A female Pacific salmon, for example, performs a ritualized mating "dance" with her breeding partner after arriving at the spawning bed in a fast-flowing, gravel-bottomed stream (Figure 24-36). She then turns on her side and scoops a nest hole with her tail. As eggs are laid by a female, they are fertilized by

a male (Figure 24-36). After a female covers the eggs with gravel, the exhausted fish dies and drifts downstream.

Soon after an egg of an oviparous species is laid and fertilized, it absorbs water and the outer layer hardens. Cleavage follows, and a blastoderm forms, astride a relatively enormous yolk mass. Soon the yolk mass is enclosed by the developing blastoderm, which then begins to assume a fishlike shape. Many fish hatch as larvae, carrying a semitransparent sac of yolk, which provides their food supply until the mouth and digestive tract have developed. The larvae then begin searching for their own food. After a period of growth a larva undergoes a metamorphosis, especially dramatic in many marine species such as

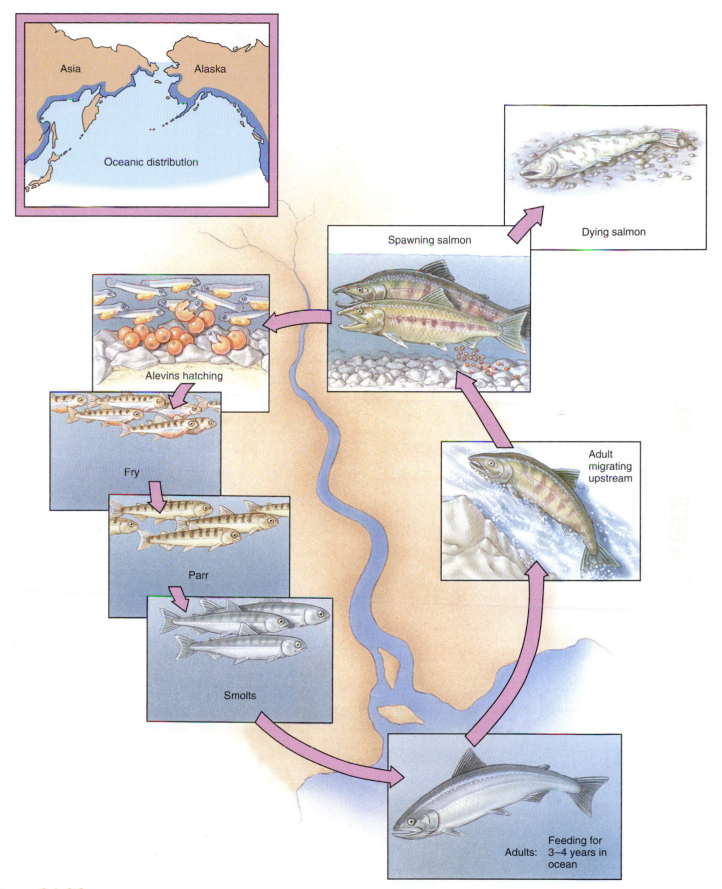

Asia

Alaska

Oceanic distribution

Spawning salmon

Dying salmon

Alevins hatching

Fry

Parr

Smolts

Adult migrating upstream

Adults: Feeding for 3–4 years in ocean

Figure 24-36

Spawning Pacific salmon and development of eggs and young.

freshwater eels (Figure 24-32). Body shape is refashioned, fin and color patterns change, and the animal becomes a juvenile bearing the unmistakable definitive body form of its species.

Growth is temperature dependent. Consequently, fish living in temperate regions grow rapidly in summer when temperatures are high and food is abundant but nearly stop growing in winter. Annual rings in the scales, otoliths, and other bony parts reflect this seasonal growth (Figure 24-37), a distinctive record of convenience to fishery biologists who wish to determine a fish's age. Unlike birds and mammals, which stop growing after reaching adult size, most fishes after attaining reproductive maturity continue to grow for as long as they live. This may be a selective advantage, since the larger the fish, the more gametes it produces and the greater its contribution to future generations.

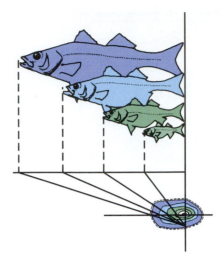

Figure 24-37
Scale growth. Fish scales disclose seasonal changes in growth rate. Growth is interrupted during winter, producing year marks (annuli). Each year's increment in scale growth is a ratio to the annual increase in body length. Otoliths (ear stones) and certain bones can also be used in some species to determine age and growth rate.

SUMMARY

Fishes are poikilothermic, gill-breathing aquatic vertebrates with fins for limbs. They include the oldest vertebrate groups, having originated from an unknown chordate ancestor in the Cambrian period or possibly earlier. Five classes of living fishes are recognized. The jawless hagfishes (class Myxini) and lampreys (class Cephalaspidomorphi), have an eel-like body form without paired fins, a cartilaginous skeleton, a notochord that persists throughout life, and a disclike mouth adapted for sucking or biting. All other vertebrates have jaws, a major development in vertebrate evolution. Members of class Chondrichthyes (sharks, rays, skates, and chimaeras) have a cartilaginous skeleton (a degenerative feature), paired fins, excellent sensory equipment, and an active, characteristically predaceous habit. Bony fishes may be divided into two classes. Lobe-finned fishes of class Sarcopterygii, represented today by lungfishes and coelacanths, form a paraphyletic group if tetrapods are excluded, as done in traditional classification. Terrestrial vertebrates arose from one lineage within this group. The second is ray-finned fishes (class Actinopterygii), a huge and diverse modern assemblage containing nearly all familiar freshwater and marine fishes. Modifications of the skeletal and muscular systems in this group increased locomotion and feeding efficiency.

Modern bony fishes (teleosts) have radiated to form approximately 23,600 species that reveal an enormous diversity of adapta-

tions, body form, behavior, and habitat preference. Fishes swim by undulatory contractions of the body muscles, which generate thrust (propulsive force) and lateral force. Flexible fishes oscillate the whole body, but in more rapid swimmers undulations are limited to the caudal region or tail fin alone.

Most pelagic bony fishes achieve neutral buoyancy in water using a gas-filled swim bladder, the most effective gas-secreting device known in the animal kingdom. Gills of fishes, having efficient countercurrent flow between water and blood, facilitate high rates of oxygen exchange. All fishes show well-developed osmotic and ionic regulation, achieved principally by the kidneys and gills.

With the exception of agnathans, all fishes have jaws that are variously modified for carnivorous, herbivorous, planktivorous, and omnivorous feeding modes.

Many fishes are migratory, and some, such as catadromous freshwater eels and anadromous salmon, make remarkable migrations of great length and precision. Fishes reveal an extraordinary range of sexual reproductive strategies. Most fishes are oviparous, but ovoviviparous and viviparous fishes are not uncommon. Reproductive investment may be in large numbers of eggs with low survival (many marine fishes) or in fewer eggs with greater parental care for better survival (freshwater fishes).

REVIEW QUESTIONS

1. Provide a brief description of the fishes citing characteristics that would distinguish them from all other animals.
2. What characteristics distinguish hagfishes and lampreys from all other fishes?
3. Describe feeding behavior in hagfishes and lampreys. How do they differ?
4. Describe the life cycle of sea lampreys, *Petromyzon marinus,* and the history of their invasion of the Great Lakes.
5. In what ways are sharks well equipped for a predatory life habit?

6. The lateral line system has been described as a "distant touch" system for sharks. What function does the lateral line system serve? Where are receptors located?
7. Explain how bony fishes differ from sharks and rays in the following systems or features: skeleton, scales, buoyancy, respiration, reproduction.
8. Match ray-finned fishes in the right column with the group to which each belongs in the left column:

____ Chondrosteans	a. Perch	e. Paddlefish
____ Nonteleost	b. Sturgeon	f. Bowfin
neopterygians	c. Gar	
____ Teleosts	d. Salmon	

9. Although chondrosteans are today a relic group, they were one of two major groups that emerged from early ray-finned fishes of the Devonian period. Give examples of living chondrosteans. What does the term "Actinopterygii", the class to which the chondrosteans belong, literally mean (refer to the Classification of Living Fishes on p. 510)?

10. List four characteristics of teleosts that contributed to their incredible diversity and success.

11. Only seven species of lobe-finned fishes are alive today, remnants of a group that flourished in the Devonian period of the Paleozoic. What morphological characteristics distinguish lobe-finned fishes? What is the literal meaning of Sarcopterygii, the class to which lobe-finned fishes belong?

12. Give the geographical locations of the three surviving genera of lungfishes and explain how they differ in their ability to survive out of water. Which of the three is the least specialized?

13. Describe discovery of living coelacanths. What is the evolutionary significance of the group to which they belong?

14. Compare the swimming movements of eels with those of trout, and explain why the latter are more efficient for rapid locomotion.

15. Sharks and bony fishes approach or achieve neutral buoyancy in different ways. Describe the methods evolved in each group. Why must a teleost fish adjust the gas volume in its swim bladder when it swims upward or downward? How is gas volume adjusted?

16. What is meant by "countercurrent flow" as it applies to fish gills?

17. Compare the osmotic problem and the mechanism of osmotic regulation in freshwater and marine bony fishes.

18. Two principal groups of fishes, with respect to feeding behavior, are the carnivores and the suspension-feeders. How are these two groups adapted for their feeding behavior?

19. Describe the life cycle of a European eel. How does the life cycle of an American eel differ from that of the European?

20. How do adult Pacific salmon find their way back to their parent stream to spawn?

21. What mode of reproduction in fishes is described by each of the following terms: oviparous, ovoviviparous, viviparous?

22. Reproduction in marine pelagic fishes and in freshwater fishes is distinctively different. How and why do they differ?

SELECTED REFERENCES

Bond, C. E. 1996. Biology of fishes, ed. 2. Fort Worth, Harcourt College Publishers. *A superior treatment of fish biology, anatomy, and genetics.*

Bone, Q., N. B. Marshall, and J. H. S. Blaxter. 1995. Biology of fishes, ed. 2. New York, Chapman & Hall. *Concise, well-written, and well-illustrated primer on the functional process of fishes.*

Conniff, R. 1991. The most disgusting fish in the sea. Audubon **93**(2):100–108 (March). *Recent discoveries shed light on the life history of the enigmatic hagfish that fishermen loathe.*

Helfman, G. J., B. B. Collette, and D. E. Facey. 1997. The diversity of fishes. Malden, Massachusetts, Blackwell Science. *This delightful and information-packed textbook focuses on adaptation and diversity and is particularly strong in evolution, systematics, and history of fishes.*

Horn, M. H., and R. N. Gibson. 1988. Intertidal fishes. Sci. Am. **258**:64–70 (Jan.). *Describes the special adaptations of fishes living in a demanding environment.*

Long, J. A. 1995. The rise of fishes: 500 million years of evolution. Baltimore, The Johns Hopkins University Press. *A lavishly illustrated evolutionary history of fishes.*

Moyle, P. B., and J. J. Cech, Jr. 2000. Fishes: an introduction to ichthyology, ed. 4. Englewood Cliffs, New Jersey, Prentice-Hall, Inc. *Textbook written in a lively style and stressing ecology rather than morphology.*

Nelson, J. S. 1994. Fishes of the world, ed. 3. New York, John Wiley & Sons, Inc. *Authoritative classification of all major groups of fishes.*

Page, L. M., and B. M. Burr. 1991. A field guide to freshwater fishes: North America north of Mexico. Boston, Houghton Mifflin. *Range maps and color illustrations for most species.*

Paxton, J. R., and W. N. Eschmeyer. 1998. Encyclopedia of fishes, ed. 2. San Diego, Academic Press. *Excellent authoritative reference that focuses on diversity and is spectacularly illustrated.*

Stevens, J. D. (ed.) 1987. Sharks. New York, Facts on File Publications. *Evolution, biology, and behavior of sharks, handsomely illustrated.*

Thomson, K. S. 1991. Living fossil. The story of the coelacanth. New York, W.W. Norton.

Webb, P. W. 1984. Form and function in fish swimming. Sci. Am. **251**:72–82 (July). *Specializations of fish for swimming and analysis of thrust generation.*

ZOOLOGY LINKS TO THE INTERNET

Visit the textbook's Online Learning Center at www.mhhe.com/zoology to find live Internet links for each of the topics listed here.
Class Myxini
Class Cephalaspidomorphi
Subclass Elasmobranchii
Dissection Guides for Elasmobranchs

Class Osteichthyes
(new taxonomic nomenclature: Class Actinopterygii)
Primitive Bony Fish
(new taxonomic nomenclature: Class Sarcopterygii)
Teleosts
Dissection Guides for Teleosts
Fisheries and Conservation Issues Concerning Teleosts
Fisheries

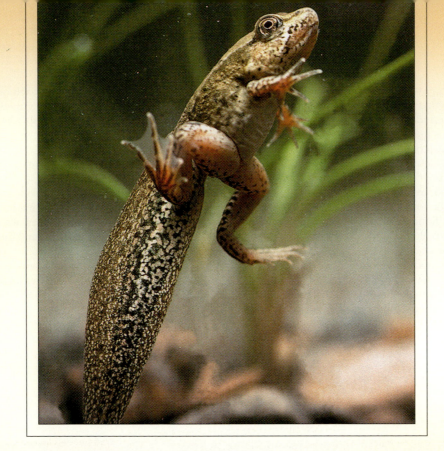

A pickeral frog,
Rana palustris,
during metamorphosis.

Early Tetrapods and Modern Amphibians

Phylum Chordata
Class Amphibia

From Water to Land in Ontogeny and Phylogeny

A chorus of frogs beside a pond on a spring evening heralds one of nature's dramatic events. Mating frogs produce masses of eggs, which will hatch into limbless, gill-breathing, fishlike tadpole larvae that feed and grow. Then, almost imperceptibly, a remarkable transformation occurs. Hindlegs appear and gradually lengthen. The tail shortens. Larval teeth are lost, and gills are replaced by lungs. Eyelids develop. Forelimbs emerge. In a few weeks the aquatic tadpole has completed its metamorphosis to an adult frog.

The evolutionary transition from water to land occurred not in weeks but over millions of years. A lengthy series of alterations cumulatively fitted the vertebrate body plan for life on land. The origin of land vertebrates is no less a remarkable feat for this fact—

a feat that would have a poor chance of succeeding today because well-established terrestrial competitors would exclude poorly adapted transitional forms.

Amphibians are the only living vertebrates that have a transition from water to land in both their ontogeny and phylogeny. Even after 350 million years of evolution, few amphibians are completely land adapted; most are quasiterrestrial, hovering between aquatic and land environments. This double life is expressed in their name. Even the amphibians best adapted for a terrestrial existence cannot stray far from moist conditions. Many, however, have developed ways to keep their eggs out of open water where their larvae would be exposed to enemies.

Position in the Animal Kingdom

Amphibians are ectothermic, primitively quadrupedal vertebrates, with glandular skin and dependence on water for their reproduction. They are one of two major groups of living descendants of early Devonian tetrapods, the first vertebrates to evolve adaptations to breathe, support themselves, move, and detect airborne sounds and odors on land, while minimizing water loss. The other group is amniotes: reptiles, birds, and mammals that completed movement onto land by evolving adaptations that freed them from dependence on water for reproduction.

Biological Contributions

1. **Strong skeletal framework** to support body weight on land, and the **tetrapod limb** with associated shoulder/hip girdle for walking on land.
2. A respiratory system with **lungs** (some modern amphibians are gilled, and some lack both lungs and gills) and paired **internal nostrils** (choanae), which enable breathing through the nose.
3. **Double circulation** with functionally separated pulmonary and systemic circuits and a **three-chambered heart. Pulmonary arteries and veins** supply the lungs and return oxygenated blood to the heart.
4. Ancestral aquatic sensory receptors were modified for life on land. The ear with **tympanic membrane** (eardrum) and **stapes** (columella) for transmitting vibrations to the inner ear is designed to detect airborne sounds. For vision in air, the cornea rather than the lens became the principal refractive surface for bending light; **eyelids** and **lachrymal glands** evolved to protect and wash the eyes. A well-developed **olfactory epithelium** lining the nasal cavity evolved to detect airborne odors.

Adaptation for life on land is a major theme of the remaining vertebrate groups. These animals form a monophyletic unit known as **tetrapods.** Amphibians and amniotes (including reptiles, birds, and mammals) represent the two major extant branches of tetrapod phylogeny. In this chapter, we review what is known about the origins of terrestrial vertebrates and discuss the amphibian branch in detail. We discuss the major amniote groups in Chapters 26 through 28.

MOVEMENT ONTO LAND

The movement from water to land is perhaps the most dramatic event in animal evolution because it involves the invasion of a physically hazardous habitat. Life originated in water. Animals are mostly water in composition, and all cellular activities occur in water. Nevertheless, organisms eventually invaded land, carrying their watery composition with them. Vascular plants, pulmonate snails, and tracheate arthropods made the transition much earlier than vertebrates, and winged insects were diversifying at approximately the same time that the earliest terrestrial vertebrates evolved. Although invasion of land required modifi-

cation of almost every system in the vertebrate body, aquatic and terrestrial vertebrates retain many basic structural and functional similarities. We see a transition between aquatic and terrestrial vertebrates most clearly today in the many living amphibians that make this transition during their own life histories.

Beyond the obvious difference in water content, there are several important physical differences that animals must accommodate when moving from water to land. These include (1) oxygen content, (2) density, (3) temperature regulation, and (4) habitat diversity. Oxygen is at least 20 times more abundant in air, and it diffuses much more rapidly through air than through water. Consequently, terrestrial animals can obtain oxygen far more easily than aquatic ones once they possess appropriate adaptations, such as lungs. Air, however, has approximately 1000 times less buoyant density than water and is approximately 50 times less viscous. It therefore provides relatively little support against gravity, requiring terrestrial animals to develop strong limbs and to remodel their skeleton to achieve adequate structural support. Air fluctuates in temperature more readily than water does, and terrestrial environments therefore experience harsh and unpredictable cycles of freezing, thawing, drying, and flooding. Terrestrial animals require behavioral and physiological strategies to protect themselves from thermal extremes; one such important strategy is homeothermy (regulated constant body temperature) of birds and mammals.

Despite its hazards, the terrestrial environment offers a great variety of habitats including coniferous, temperate, and tropical forests, grasslands, deserts, mountains, oceanic islands, and polar regions. Provision of safe shelter for protection of vulnerable eggs and young may be accomplished much more readily in many of these terrestrial habitats than in aquatic ones.

EARLY EVOLUTION OF TERRESTRIAL VERTEBRATES

Devonian Origin of Tetrapods

The Devonian period, beginning some 400 million years ago, was a time of mild temperatures and alternating droughts and floods. During this period, some primarily aquatic vertebrates evolved two features that would be important for permitting subsequent evolution for life on land: lungs and limbs.

Devonian freshwater environments were unstable. During dry periods, many pools and streams evaporated, water became foul, and dissolved oxygen disappeared. Only those fishes able to acquire atmospheric oxygen survived such conditions. Gills were unsuitable because in air the filaments collapsed, dried, and quickly lost their function. Virtually all freshwater fishes surviving this period, including lobe-finned fishes and lungfishes, had a kind of lung that developed as an outgrowth of the pharynx. The efficiency of the air-filled cavity was enhanced by improving its vascularity with a rich capillary network, and by supplying it with arterial blood from the last (sixth) pair of aortic arches. Oxygenated blood returned directly to the heart by a pulmonary vein to form a complete pulmonary circuit. Thus the **double circulation** characteristic of all tetrapods originated: a systemic circulation serving the body and a pulmonary circulation supplying the lungs.

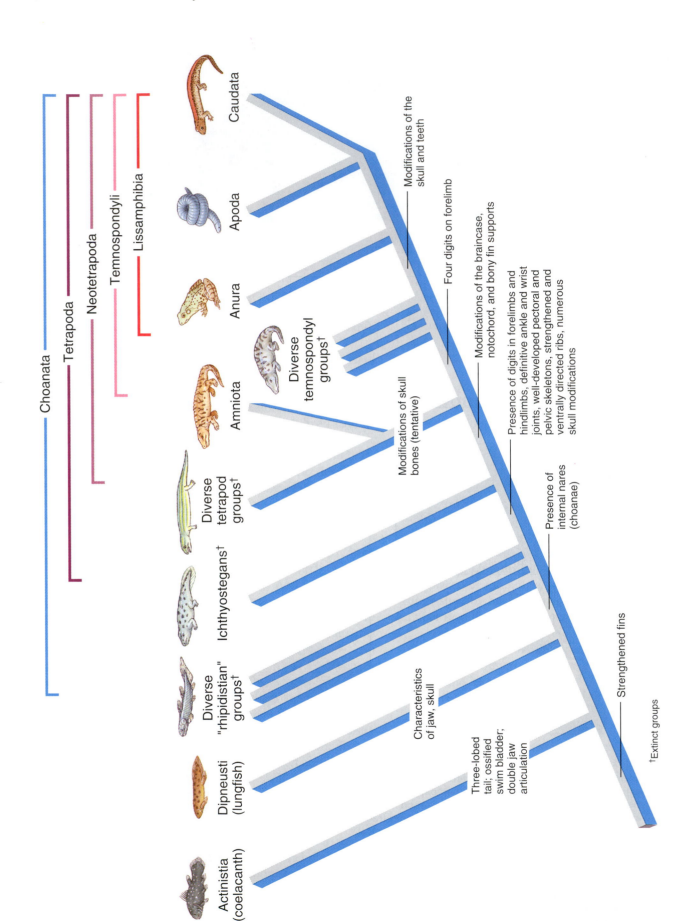

Choanata
Tetrapoda
Neotetrapoda
Temnospondyli
Lissamphibia

Caudata
Apoda
Anura
Amniota
Diverse temnospondyl groups†
Diverse tetrapod groups†
Ichthyostegans†
Diverse "rhipidistian" groups†
Dipneusti (lungfish)
Actinistia (coelacanth)

Modifications of the skull and teeth

Four digits on forelimb

Modifications of the braincase, notochord, and bony fin supports

Presence of digits in forelimbs and hindlimbs, definitive ankle and wrist joints, well-developed pectoral and pelvic skeletons, strengthened and ventrally directed ribs, numerous skull modifications

Modifications of skull bones (tentative)

Presence of internal nares (choanae)

Characteristics of jaw, skull

Strengthened fins

Three-lobed tail; ossified swim bladder; double jaw articulation

†Extinct groups

Figure 25-2

Tentative cladogram of the Tetrapoda with emphasis on descent of amphibians. Especially controversial are relationships of major tetrapod groups (Amniota, Anthracosauria, Lepospondyli, Temnospondyli) and outgroups (Actinistia, Dipneusti, Rhipidistia). All aspects of this cladogram are controversial, however, including monophyly of Lissamphibia. Relationships shown for the three groups of Lissamphibia are based on recent molecular evidence.

Source: Modified from E. W. Gaffney in the Bulletin of the Carnegie Museum of Natural History 13:92–105 (1979).

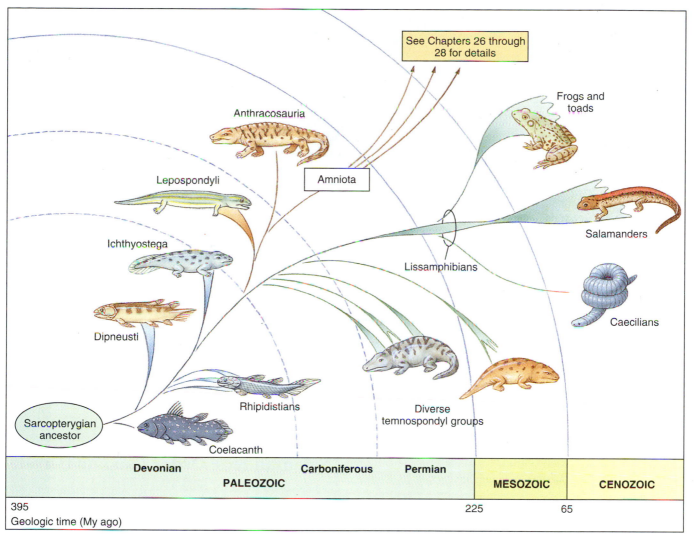

Figure 25-3

Early tetrapod evolution and the descent of amphibians. Tetrapods share most recent common ancestry with Devonian rhipidistians. Amphibians share most recent common ancestry with diverse temnospondyls of the Carboniferous and Permian periods of the Paleozoic, and Triassic period of the Mesozoic.

MODERN AMPHIBIANS

The three living amphibian orders comprise more than 4200 species. Most share general adaptations for life on land, including skeletal strengthening and a shifting of special sense priorities from the ancestral lateral-line system to the senses of smell and hearing. The olfactory epithelium and ear are redesigned to improve sensitivities to airborne odors and sounds, respectively.

Nonetheless, most amphibians meet problems of independent life on land only halfway. In the ancestral life history of amphibians, eggs are aquatic and hatch to produce an aquatic larval form that uses gills for respiration. A metamorphosis follows in which gills are lost and lungs, which are present throughout larval life, are then activated for respiration. Many amphibians retain this general pattern, but important exceptions include some salamanders that lack a complete metamorphosis and retain a permanently aquatic, larval morphology throughout life. Some other salamanders live entirely on land and have no

aquatic larval phase. Both of these are evolutionarily derived conditions. Some frogs also have acquired a strictly terrestrial existence by eliminating an aquatic larval stage. Some frogs, salamanders, and caecilians that undergo a complete metamorphic life cycle nonetheless remain in water as adults rather than moving onto land during their metamorphosis.

Even the most terrestrial amphibians remain dependent on very moist if not aquatic environments. Their skin is thin, and it requires moisture for protection against desiccation in air. An intact frog loses water nearly as rapidly as a skinless frog. Amphibians also require moderately cool environments. Being ectothermic, their body temperature is determined by and varies with the environment, greatly restricting where they can live. Cool and wet environments are especially important for reproduction. Eggs are not well protected from desiccation, and they must be shed directly into water or onto moist terrestrial surfaces. Completely terrestrial amphibians may lay eggs under logs or rocks, in the moist forest floor, in flooded tree holes, in

Figure 25-4

Reproductive strategies of anurans. **A,** Female South American pygmy marsupial frog, *Flectonotus pygmaeus,* carries developing larvae in a dorsal pouch. **B,** Female Surinam frog carries eggs embedded in specialized brooding pouches on the dorsum; froglets emerge and swim away when development is complete. **C,** Male poison arrow frog, *Phyllobates bicolor,* carries tadpoles adhering to its back. **D,** Tadpoles of a male Darwin's frog, *Rhinoderma darwinii,* develop into froglets in its vocal pouch. When ready to emerge, a froglet crawls into the parent's mouth, which the parent opens to allow the froglet's escape.

A   B

C   D

pockets on the mother's back (Figure 25-4), or in folds of the body wall. One species of Australian frog even broods its young in its vocal pouch.

We now highlight special characteristics of the three major groups of amphibians. We expand coverage of general amphibian features when discussing groups in which particular features have been studied most extensively. For most features, this group is frogs.

Caecilians: Order Gymnophiona (Apoda)

The order Gymnophiona (jim'no-fy'o-na) (Gr. *gymnos,* naked, + *opineos,* of a snake) contains approximately 160 species of elongate, limbless, burrowing creatures commonly called **caecilians** (Figure 25-5). They occur in tropical forests of South America (their principal home), Africa, and Southeast Asia. Caecilians possess a long, slender body, small scales in the skin of some, many vertebrae, long ribs, no limbs, and a terminal anus. Eyes are small, and most species are totally blind as adults. Special sensory tentacles occur on the snout. Because they are almost entirely burrowing or aquatic, they seldom are seen by humans. Their food consists mostly of worms and small invertebrates, which they find underground. Fertilization is internal, and males have a protrusible copulatory organ. Eggs usually are deposited in moist ground near water. Larvae may be aquatic, or complete larval development may occur in the egg. In some species eggs are carefully guarded during their development in folds of the body. Viviparity also is common in some caecilians, with embryos obtaining nourishment by eating the wall of the oviduct.

Salamanders: Order Caudata (Urodela)

As its name suggests, order Caudata (L. *caudatus,* having a tail) consists of tailed amphibians, approximately 360 species of salamanders. Salamanders are found in almost all northern temperate regions of the world, and they are abundant and diverse in North America. Salamanders are found also in tropical areas of Central and northern South America. Salamanders are typically small; most of the common North American salamanders are less than 15 cm long. Some aquatic forms are considerably longer, and Japanese giant salamanders may exceed 1.5 m in length.

Most salamanders have limbs set at right angles to the body, with forelimbs and hindlimbs of approximately equal size. In some aquatic and burrowing forms, limbs are rudimentary and some may be absent.

Salamanders are carnivorous both as larvae and adults, preying on worms, small arthropods, and small molluscs. Most

Figure 25-5

Female caecilian coiled around eggs in burrow.

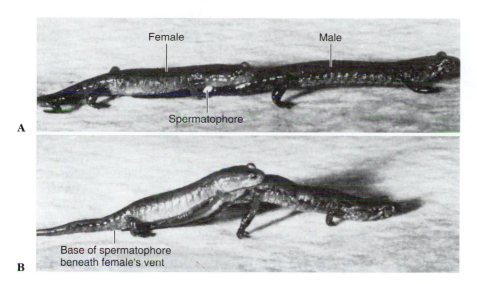

A

B

Female   Male

Spermatophore

Base of spermatophore
beneath female's vent

Figure 25-6

Courtship and sperm transfer in pygmy salamanders, *Desmognathus wrighti*. After judging the female's receptivity by the presence of her chin on his tail base, the male deposits a spermatophore on the ground, then moves forward a few paces. **A,** The white mass of the sperm atop a gelatinous base is visible at the level of the female's forelimb. The male moves ahead, the female following until the spermatophore is at the level of her vent. **B,** The female has recovered the sperm mass in her vent, while the male arches his tail, tilting the female upward and presumably facilitating recovery of the sperm mass.

eat only animals that are moving. Since their food is rich in proteins, they do not store great quantities of fat or glycogen. Like all amphibians, they are ectotherms with a low metabolic rate.

Breeding Behavior

Some salamanders are aquatic throughout their life cycle, but most are metamorphic, having aquatic larvae and terrestrial adults that live in moist places under stones and rotten logs. Eggs of most salamanders are fertilized internally, usually after the female recovers in her vent a packet of sperm **(spermatophore)** that previously has been deposited by the male on a leaf or stick (Figure 25-6). Aquatic species lay their eggs in clusters or stringy masses in water. Their eggs hatch to produce an aquatic larva having external gills and a finlike tail. Completely terrestrial species deposit eggs in small, grapelike clusters under logs or in excavations in soft moist earth, and in many species adults guard their eggs (Figure 25-7). Terrestrial species have **direct development:** they bypass the larval stage and hatch as miniature versions of their parents. The most complex of salamander life cycles is observed in some American newts, whose aquatic larvae metamorphose to form terrestrial juveniles that later metamorphose again to produce secondarily aquatic, breeding adults (Figure 25-8). Many newt populations skip the terrestrial "red eft" stage, however, remaining entirely aquatic.

Respiration

Salamanders demonstrate an unusually diverse array of respiratory mechanisms. They share the general amphibian condition of having in their skin extensive vascular nets that serve respiratory exchange of oxygen and carbon dioxide. At various stages of their life history, salamanders also may have external gills, lungs, both, or neither of these. Salamanders with an aquatic larval stage hatch with gills, but lose them later if a metamorphosis occurs. Several lineages of salamanders have evolved permanently aquatic forms that fail to undergo a complete metamorphosis and retain their gills and finlike tail throughout life. Lungs, the most widespread respiratory organ of terrestrial vertebrates, are present from birth in salamanders that have them, and become active following metamorphosis.

Although we normally associate lungs with terrestrial organisms and gills with aquatic ones, salamander evolution has produced aquatic forms that breathe primarily with lungs and terrestrial forms that lack them completely. Amphiumas of the salamander family Amphiumidae have evolved a completely aquatic life history with a greatly reduced metamorphosis. Amphiumas nonetheless lose their gills before adulthood and then breathe primarily by lungs. They periodically point their nostrils above the surface of the water to get air.

Amphiumas provide a curious contrast to many species of the family Plethodontidae that are entirely terrestrial but have no lungs. This large family contains more than 340 species, including many familiar North American salamanders (see Figures 25-6, 25-7, and 25-9). The efficiency of cutaneous respiration is increased by penetration of a capillary network into the epidermis or by thinning of the epidermis over superficial dermal capillaries. Cutaneous respiration is supplemented by pumping air through the mouth where respiratory gases are exchanged across the vascularized membranes of the buccal (mouth) cavity (buccopharyngeal breathing). Lungless plethodontids probably originated in swift streams where lungs would have been a disadvantage by providing too much buoyancy, and where water is so cool and well oxygenated that cutaneous respiration alone

Figure 25-7

Female dusky salamander (*Desmognathus* sp.) attending eggs. Some salamanders exercise parental care of eggs, which includes rotating eggs and protecting them from fungal infections and predation by various arthropods and other salamanders.

Characteristics of Modern Amphibians

1. Skeleton mostly bony, with varying numbers of vertebrae; ribs present in some, absent or fused to vertebrae in others; notochord does not persist; exoskeleton absent

2. Body forms vary greatly from an elongated trunk with distinct head, neck, and tail to a compact, depressed body with fused head and trunk and no intervening neck

3. **Limbs usually four (tetrapod),** although some are legless; forelimbs of some much smaller than hindlimbs, in others all limbs small and inadequate; webbed feet often present; no true nails or claws; **forelimb usually with four digits** but sometimes five and sometimes fewer

4. **Skin smooth and moist with many glands,** some of which may be poison glands; pigment cells (chromatophores) common, of considerable variety; no scales, except concealed dermal ones in some

5. Mouth usually large with small teeth in upper or both jaws; two nostrils open into anterior part of mouth cavity

6. Respiration by lungs (absent in some salamanders), skin, and gills in some, either separately or in combination; external gills in larval forms and may persist throughout life in some

7. **Circulation with three-chambered heart,** two atria and one ventricle, and a **double circulation through the heart;** skin abundantly supplied with blood vessels

8. Ectothermal

9. Excretory system of paired mesonephric kidneys; urea main nitrogenous waste

10. Ten pairs of cranial nerves

11. Separate sexes; fertilization mostly internal in salamanders and caecilians, mostly external in frogs and toads; predominantly oviparous, some ovoviviparous or viviparous; metamorphosis usually present; **moderately yolky eggs (mesolecithal) with jellylike membrane coverings**

was sufficient for life. Some plethodontids have aquatic larvae whose gills are lost at metamorphosis. Others retain a permanently larval form with gills throughout life. Many others are completely terrestrial and bear the distinction of being the only vertebrates to have neither lungs nor gills at any stage of their life history. It is odd that the most completely terrestrial lineage of salamanders evolved in a group that lacks lungs.

Paedomorphosis

A persistent phylogenetic trend observed in salamander evolution is for descendants to retain into adulthood features that were present only in pre-adult stages of their ancestors. Some characteristics of ancestral adult morphology are consequently eliminated. This condition is called **paedomorphosis** (Gr. "child form"; see Chapter 6, p. 112). The most dramatic form of paedomorphosis occurs in species that become sexually mature while retaining their gills, aquatic life habit, and other larval characteristics. These nonmetamorphic species are said to be **perennibranchiate** ("permanently gilled"). Mud puppies of genus *Necturus* (Figure 25-10), which live on bottoms of ponds and lakes, are an extreme example. These and many other salamanders are obligately perennibranchiate; they have never been observed to metamorphose under any conditions.

Some other species of salamanders reach sexual maturity with larval morphology but, unlike *Necturus*, may metamorphose to terrestrial forms under certain environmental conditions. Good examples are found in *Ambystoma* species from Mexico and the United States. Gilled individuals are called **axolotls** (Figure 25-10). Their typical habitat consists of small ponds that can disappear through evaporation in dry weather. When its pond evaporates, an axolotl metamorphoses to a terrestrial form, losing its gills and developing lungs. It can then travel across land in search of new ponds, to which it must

Figure 25-8

Life history of a red-spotted newt, *Notophthalmus viridescens* of the family Salamandridae. In many habitats the aquatic larva metamorphoses into a brightly colored "red eft" stage, which remains on land from 1 to 3 years before transforming into a secondarily aquatic adult.

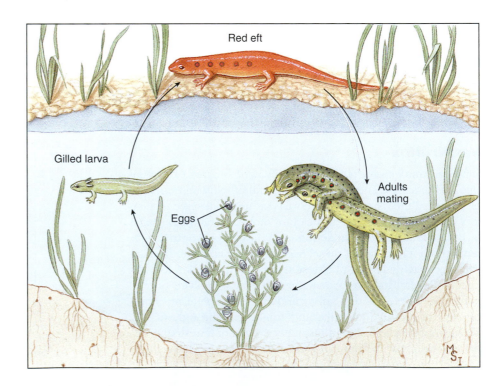

Figure 25-9

Longtail salamander, *Eurycea longicauda*, a common plethodontid salamander.

Common mud puppy
(*Necturus maculosus*)

Axolotl
(*Ambystoma mexicanum*)

Figure 25-10

Paedomorphosis in salamanders. **A,** The mud puppy *Necturus* sp., is a permanently gilled (perennibranchiate) aquatic form. **B,** An axolotl (*Ambystoma mexicanum*) may remain permanently gilled, or, should its pond habitat evaporate, metamorphose to a terrestrial form that loses its gills and breathes by lungs.

return to reproduce. Axolotls are forced to metamorphose artificially when they are treated with the thyroid hormone, thyroxine (T_4). Thyroid hormones (T_3 and T_4) are essential for amphibian metamorphosis. The pituitary gland appears not to become fully active in nonmetamorphosing forms, thereby failing to release the hormone thyrotropin, which is required to stimulate the thyroid gland to produce thyroid hormones.

Paedomorphosis takes many different forms in different salamanders. It may affect the body as a whole or may be restricted to one or a few structures. The amphiumas lose their gills and activate their lungs before maturity, but they retain many general features of larval body form. Paedomorphosis is important even in terrestrial plethodontids, which never have an aquatic larval stage. We can see the effects of paedomorphosis, for example, in the shapes of the hands and feet of the tropical plethodontid genus *Bolitoglossa* (Figure 25-11). Ancestral morphology of *Bolitoglossa* features well-formed digits that grow outward from the pad of the hand or foot during development. Some species have enhanced their ability to climb smooth vegetation, such as banana trees, by halting growth of the digits and

retaining throughout life a padlike foot. This padlike foot can produce adhesion and suction to attach the salamander to smooth vertical surfaces, and thereby serves an important adaptive function.

Frogs and Toads: Order Anura (Salientia)

The more than 3450 species of frogs and toads that compose order Anura (Gr. *an,* without, + *oura,* tail) are for most people the most familiar amphibians. Anura is an old group, known from the Jurassic period, 150 million years ago. Frogs and toads occupy a great variety of habitats. Their aquatic mode of reproduction and water-permeable skin prevent them from wandering too far from sources of water, however, and their ectothermy bars them from polar and subarctic habitats. The name of the order, Anura, refers to an obvious group characteristic, absence of tails in adults. Although all pass through a tailed larval stage during development, only genus *Ascaphus* contains a tail-like structure in adults. Frogs and toads are specialized for jumping, as suggested by the alternative order name, Salientia, which means leaping.

We see in the appearance and life habit of their larvae further distinctions between Anura and Caudata. Eggs of most frogs hatch into a tadpole ("polliwog"), having a long, finned tail, both internal and external gills, no legs, specialized mouthparts for herbivorous feeding (salamander larvae and some tadpoles are carnivorous), and a highly specialized internal anatomy. They look and act entirely different from adult frogs. Metamorphosis of a frog tadpole to an adult frog is thus a striking transformation. The perennibranchiate condition never occurs in frogs and toads as it does in salamanders.

In addition to their importance in biomedical research and education, frogs have long served the epicurean frog-leg market. Mainstay of this market is the bullfrog, which is in heavy demand in Europe (especially France) and the United States—the worldwide harvest is an estimated 200 million bullfrogs (about 10,000 metric tons) annually—that its populations have fallen drastically from excessive exploitation and the draining and pollution of wetlands. Major bullfrog suppliers include Bangladesh, China, Indonesia, and Japan, with about 80 million collected each year from rice fields in Bangladesh alone. With so many insect-eating frogs removed from the ecosystem, rice production is threatened from uncontrolled, flourishing insect populations. In the United States, attempts to raise bullfrogs in farms have not been successful, mainly because bullfrogs are voracious eating machines that normally will only accept living prey, such as insects, crayfish, and other frogs.

Frogs and toads are divided into 21 families. The best-known frog families in North America are Ranidae, which contains most of our familiar frogs (Figure 25-12A), and Hylidae, tree frogs (Figure 25-12B). True toads, belonging to family Bufonidae, have short legs, stout bodies, and thick skins, usually with prominent warts (Figure 25-13). However, the term "toad" is used informally to refer also to some terrestrial members of several other families.

Figure 25-11

Foot structure of representatives of three different species of the tropical plethodontid salamander genus *Bolitoglossa*. These specimens have been treated chemically to clear the skin and muscles and to stain the bone red and cartilage blue. The species having the most fully ossified and distinct digits **(A, C)** live primarily on the forest floor. The species having the padlike foot caused by restricted digital growth **(B)** climbs smooth leaves and stems using the foot surface to produce suction or adhesion for attachment. The padlike foot evolved by paedomorphosis; it was derived evolutionarily by truncating development of the foot to prevent full digital development.

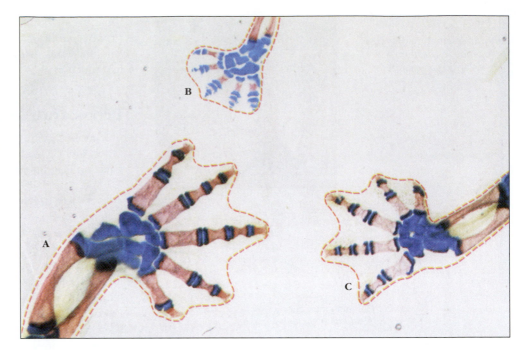

Figure 25-12

Two common North American frogs. **A**, Bullfrog, *Rana catesbeiana*, largest American frog and mainstay of the frog-leg epicurean market (family Ranidae). **B**, Green tree frog, *Hyla cinerea*, a common inhabitant of swamps of the southeastern United States (family Hylidae). Note adhesive pads on the feet.

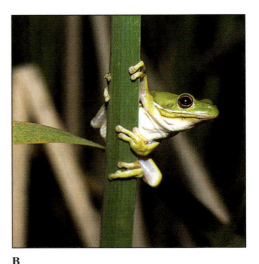

A   B

The largest anuran is the West African *Conraua goliath*, which is more than 30 cm long from tip of nose to anus (Figure 25-14). This giant eats animals as big as rats and ducks. The smallest frogs recorded are *Eleutherodactylus iberia* and *Psyllophryne didactyla*, measuring less than 1 cm in length; they are also the smallest known tetrapods. These tiny frogs, which can be covered by a dime, are found respectively in Cuba and in the Brazilian rain forest. The largest American frog is the bullfrog, *Rana catesbeiana* (see Figure 25-12A), which reaches a head and body length of 20 cm.

Habitats and Distribution

Probably the most abundant frogs are the approximately 260 species of genus *Rana* (Gr. frog), found throughout temperate and tropical regions of the world except in New Zealand, the oceanic islands, and southern South America. They usually are found near water, although some, such as wood frogs, *R. sylvat-* ica, spend most of their time on damp forest floors. Wood frogs probably return to pools only for breeding in early spring. The larger bullfrogs, *R. catesbeiana*, and green frogs, *R. clamitans*, are nearly always found in or near permanent water or swampy regions. Leopard frogs, *R. pipiens* and related species, have a wider variety of habitats and are the most widespread of North American frogs; they are commonly used in biology laboratories and for classical electrophysiological research. They have been found in some form in nearly every state, although sparingly represented along the extreme western part of the Pacific coast. They also extend far into northern Canada and as far south as Panama.

Frog species are often patchy in distribution, being restricted to certain localities (for instance, to specific streams or pools) and absent or scarce in similar habitats elsewhere. Pickerel frogs (*R. palustris*) are especially noteworthy in this respect because they are known to be abundant only in certain localized regions. Recent studies have shown that many populations of frogs

Figure 25-13

American toad, *Bufo americanus* (family Bufonidae). This principally nocturnal yet familiar amphibian feeds on large numbers of insect pests and on snails and earthworms. The warty skin contains numerous glands that produce a surprisingly poisonous milky fluid, providing excellent protection from a variety of potential predators.

Figure 25-14

Conraua (Gigantorana) goliath (family Ranidae) of West Africa, the world's largest frog. This specimen weighed 3.3 kg (approximately 7½ pounds).

worldwide may be suffering declines in numbers and becoming even more patchy than usual in their distributions.

Most larger frogs are solitary in their habits except during breeding season. During breeding periods most of them, especially males, are very noisy. Each male usually takes possession of a particular perch near water, where he may remain for hours or even days, trying to attract a female to that spot. At times frogs are mainly silent, and their presence is not detected until they are disturbed. When they enter the water, they dart swiftly to the bottom of the pool and kick the substrate to conceal themselves in a cloud of muddy water. In swimming, they hold their forelimbs near their body and kick backward with their webbed hindlimbs, propelling themselves forward. When they surface to breathe, only the head and foreparts are exposed and, since they usually take advantage of any protective vegetation, they are difficult to see.

Amphibian populations are falling in various parts of the world, whereas in other areas they are doing well. No single explanation fits all declines. In some populations, changes are simply random fluctuations caused by periodic droughts and other naturally occurring phenomena. Frog and toad eggs exposed on the surface of ponds are especially sensitive to the damaging action of ultraviolet radiation. Climatic changes that reduce water depth at oviposition sites increase ultraviolet exposure of embryos and make them more susceptible to fungal infection. Declines in population survival may be accompanied by an increased incidence of malformed individuals, such as frogs with extra limbs.

During winter months most frogs in temperate climates hibernate in the soft mud underlying pools and streams. Their life processes are at a very low ebb during their hibernation period, and such energy as they need is derived from glycogen and fat stored in their bodies during the spring and summer months. More terrestrial frogs, such as tree frogs, hibernate in humus of the forest floor. They are tolerant of low temperatures, and many actually survive freezing all extracellular fluid, representing 35% of the body water. Such frost-tolerant frogs prepare for winter by accumulating glucose and glycerol in body fluids, thereby protecting tissues from the normally damaging effects of ice-crystal formation.

Declines of some amphibian populations may be caused by other amphibians. While native American amphibians continue to disappear as wetlands are drained, an exotic frog introduced into southern California has found the climate quite to its liking. African clawed frogs, *Xenopus laevis* (Figure 25-15), are voracious, aggressive, primarily aquatic frogs that rapidly displace native frogs and fish from waterways. This species was introduced into North America in the 1940s when it was used extensively in human pregnancy tests. When more efficient tests appeared in the 1960s, some hospitals simply dumped surplus frogs into nearby streams, where these prolific breeders have become almost indestructible pests. Similar results occurred when giant toads, *Bufo marinus* (to 23 cm in length), were introduced to Queensland, Australia, and southern Florida to control agricultural pests. They are rapidly spreading, producing numerous ecological problems, including displacement of native anurans.

Adult frogs have numerous enemies, such as snakes, aquatic birds, turtles, raccoons, and humans; fish prey on tadpoles, and only a few tadpoles survive to maturity. Although usually defenseless, many frogs and toads in the tropics and subtropics are aggressive, jumping and biting at predators. Some defend themselves by feigning death. Most anurans can inflate their lungs so that they are difficult to swallow. When disturbed along the margin of a pond or brook, a frog often remains quite still; when it thinks it is detected, it jumps, not always into the water where

Figure 25-15
African clawed frog, *Xenopus laevis*. The claws, an unusual feature in frogs, are on the hind feet. This frog has been introduced into California, where it is considered a serious pest.

enemies may be lurking, but into grassy cover on the bank. When held in the hand, a frog may cease its struggles for an instant to put its captor off guard and then leap violently, voiding its urine. A frog's best protection is its ability to leap and, in some species, to use poison glands. Bullfrogs in captivity do not hesitate to snap at tormentors and are capable of inflicting painful bites.

Integument and Coloration

The skin of a frog is thin and moist, and it is attached loosely to the frog's body only at certain points. Histologically the skin is composed of two layers: an outer stratified **epidermis** and an inner spongy **dermis** (Figure 25-16). The outer layer of epidermal cells (which is shed periodically when a frog or toad "molts") contains deposits of **keratin,** a tough, fibrous protein that provides protection against abrasion and loss of water from the skin. More terrestrial amphibians such as toads have especially heavy deposits of keratin, although amphibian keratin is soft, unlike the hard keratin that forms scales, claws, feathers, horns, and hair of amniotes.

The inner layer of epidermis gives rise to two types of integumentary glands that grow into the loose dermal tissues below. Small **mucous** glands secrete a protective mucous waterproofing onto the skin surface, and large **serous** glands produce a whitish, watery poison that is highly irritating to would-be predators. All amphibians produce a skin poison, but its effectiveness varies from species to species and with different predators. The extremely toxic poison of three species of *Phyllobates,* a genus of small South American dendrobatid frogs, is used by a western Colombian Indian tribe to poison points of blowgun darts. Most species of the family Dendrobatidae produce toxic skin secretions, some of which are among the most lethal animal secretions known, more dangerous even than venoms of sea snakes or any of the most venomous arachnids.

Skin color in frogs is produced, as in other amphibians, by special pigment cells, **chromatophores,** located mainly in the

Figure 25-16
Section through frog skin.

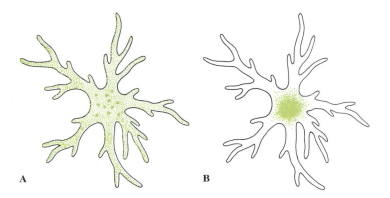

Figure 25-17
Pigment cells (chromatophores). **A,** Pigment dispersed. **B,** Pigment concentrated. The pigment cell does not contract or expand; color effects are produced by streaming of cytoplasm, carrying pigment granules into cell branches for maximum color effect or to the center of the cell for minimum effect. Control over dispersal or concentration of pigment is mostly by light stimuli acting through a pituitary hormone.

dermis. Amphibian chromatophores, like those of many other vertebrates, are branched cells containing pigment that may be concentrated in a small area or dispersed throughout the branching processes to control skin coloration (Figure 25-17; see also p. 616). Most amphibians have three types of chromatophores: uppermost in the dermis are **xanthophores,** containing yellow, orange, or red pigments; beneath these lie **iridophores,** containing a silvery, light-reflecting pigment; and lowermost are **melanophores,** containing black or brown melanin. Iridophores act like tiny mirrors, reflecting light back through the xanthophores to produce the brightly conspicuous colors of many tropical frogs. Surprisingly perhaps, green hues so common in North American frogs are produced not by green pigment but by an interaction of xanthophores containing a yellow pigment and underlying iridophores that, by reflecting and scattering light (Tyndall scattering), produce a blue color. Blue light is filtered by the overlying yellow pigment and thus appears green. Many frogs can adjust their color to blend with their background and thus camouflage themselves (Figure 25-18).

Figure 25-18

Cryptic coloration of the gray tree frog, *Hyla versicolor.* Camouflage is so good that presence of this frog usually is disclosed only at night by its resonant, flutelike call.

Skeletal and Muscular Systems

In amphibians, as in other vertebrates, a well-developed **endoskeleton** of bone and cartilage provides a framework for muscles in movement and protection for viscera and nervous systems. Movement onto land and the necessity of transforming paddlelike fins into tetrapod limbs capable of supporting the body's weight introduced a new set of stress and leverage problems. Metamorphosis is most noticeable in anurans, whose entire musculoskeletal system is specialized for jumping and swimming by simultaneous extensor thrusts of the hindlimbs.

The amphibian vertebral column assumes a new role as a support from which the abdomen is slung and to which limbs are attached. Since amphibians move with limbs instead of swimming with serial contractions of the trunk musculature, the vertebral column has lost much of the original flexibility characteristic of fishes. It has become a rigid frame for transmitting force from the hindlimbs to their body. Anurans are further specialized by an extreme shortening of the body. Typical frogs have only nine trunk vertebrae and a rodlike **urostyle,** which represents several fused caudal vertebrae (coccyx) (Figure 25-19). The limbless caecilians, which obviously have not shared these specializations for tetrapod locomotion, may have as many as 285 vertebrae.

A frog's skull is also vastly altered as compared with its vertebrate ancestors; it is much lighter in weight and more flattened in profile and has fewer bones and less ossification. The front part of the skull, wherein are located the nose, eyes, and brain, is better developed, whereas the back of the skull, which contains the gill apparatus in fishes, is much reduced (see Figure 25-19).

Bones and muscles in limbs are of typical tetrapod pattern, with three main joints in each limb (hip, knee, and ankle; or shoulder, elbow, and wrist). The foot is typically five-rayed (pentadactyl) and the hand is four-rayed with both foot and hand having several joints in each digit (see Figure 25-19). It is a repetitive system that can be derived from one resembling the bone structure of lobe-fins, which are distinctly suggestive of amphibian limbs (see Figure 25-1). It is not difficult to imagine how selective pressures through millions of years remodeled ancestral lobe-fins into limbs.

Muscles of the limbs are presumably homologous to radial muscles that move the fins of fishes, but the muscular arrangement has become so complex in tetrapod limbs that its exact correspondence with fin musculature is unclear. Despite this complexity, we can recognize two major groups of muscles on any limb: an anterior and ventral group that pulls the limb forward and toward the midline (protraction and adduction), and a second set of posterior and dorsal muscles that draws the limb back and away from the body (retraction and abduction).

Trunk musculature, which in fishes is segmentally organized into powerful muscular bands (myomeres, p. 502) for locomotion by lateral flexion, has been much modified during amphibian evolution. Dorsal (epaxial) muscles are arranged to support the head and brace the vertebral column. Ventral (hypaxial) muscles are more developed in amphibians than in fishes, since they must support the viscera in air without the buoying assistance of water.

Respiration and Vocalization

Amphibians use three respiratory surfaces for gas exchange in air: skin (cutaneous breathing), mouth (buccal breathing), and lungs. Frogs and toads are more dependent on lung breathing

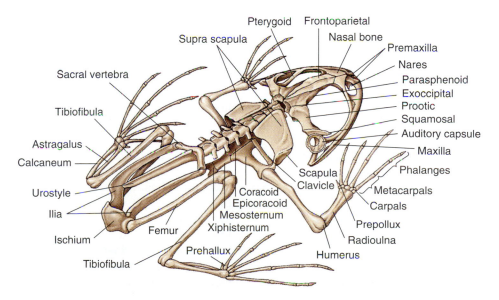

Figure 25-19

Skeleton of a bullfrog, *Rana catesbeiana.*

Pterygoid
Frontoparietal
Nasal bone
Premaxilla
Nares
Parasphenoid
Exoccipital
Prootic
Squamosal
Auditory capsule
Maxilla
Phalanges
Metacarpals
Carpals
Prepollux
Radioulna
Humerus
Scapula
Clavicle
Coracoid
Epicoracoid
Mesosternum
Xiphisternum
Femur
Prehallux
Tibiofibula
Ischium
Ilia
Urostyle
Calcaneum
Astragalus
Tibiofibula
Sacral vertebra
Supra scapula

than are salamanders; nevertheless, skin provides an important supplementary avenue for gas exchange in anurans, especially during hibernation in winter. Even when lung breathing predominates, carbon dioxide is lost primarily across the skin while oxygen is absorbed primarily across the lungs.

Lungs are supplied by pulmonary arteries (derived from the sixth aortic arches) and blood returns directly to the left atrium by pulmonary veins. Frog lungs are ovoid, elastic sacs with their inner surfaces divided into a network of septa that are subdivided into small terminal air chambers called alveoli. Alveoli of frog lungs are much larger than those of amniote vertebrates, and consequently frog lungs have a smaller relative surface available for gas exchange: the respiratory surface of *Rana pipiens* is about 20 cm² per cubic centimeter of air contained, compared with 300 cm² for humans. The problem in lung evolution was not development of a good internal vascular surface, but rather the problem of moving air. A frog is a positive-pressure breather that fills its lungs by forcing air into them; this system contrasts with the negative-pressure system of amniotes. The sequence and explanation of breathing in a frog are shown in Figure 25-20. One can easily follow this sequence in a living frog at rest: rhythmical throat movements of mouth breathing may continue some time before flank movements indicate that the lungs are being emptied and refilled.

Both male and female frogs have **vocal cords,** but those of males are much better developed. They are located in the **larynx,** or voice box. A frog produces sound by passing air back and forth over the vocal cords between the lungs and a large pair of sacs (vocal pouches) in the floor of the mouth. The latter also serve as effective resonators in males, which use their voices to attract mates. Most species identify themselves with characteristic sounds. Many people are familiar with springtime calls of spring peepers, which produce high-pitched sounds surprisingly strident for such tiny frogs. The bass notes of green frogs are banjolike, those of leopard frogs are long and guttural, and bullfrogs produce resonant "jug-o-rum" calls.

Figure 25-20

Breathing in a frog. Frogs are positive-pressure breathers that fill their lungs by forcing air into them. **A,** Floor of mouth is lowered, drawing air in through nostrils. **B,** With nostrils closed and glottis open, the frog forces air into its lungs by elevating floor of mouth. **C,** Mouth cavity rhythmically ventilates for a period. **D,** Lungs are emptied by contraction of body-wall musculature and by elastic recoil of lungs.

Circulation

As in fishes, circulation in amphibians is a closed system of arteries and veins serving a vast peripheral network of capillaries through which blood is forced by a single pressure pump, the heart. The principal changes in circuitry involve the shift from gill to lung breathing. With the elimination of gills, a major obstacle to blood flow was removed from the arterial circuit. But two new problems arose. The first was to provide a blood circuit to the lungs. As we have seen, this problem was solved by converting the sixth aortic arch into pulmonary arteries to serve the lungs and by developing new pulmonary veins for returning oxygenated blood to the heart. The second and evidently more difficult evolutionary problem was to separate pulmonary circulation from the rest of the body's circulation, so that oxygenated blood from the lungs would be sent to the body and deoxygenated venous blood returning from the body would be sent to the lungs. Solving this problem required a double circulation consisting of separate pulmonary and systemic circuits. Tetrapods solved the problem by evolving a partition down the center of the heart, creating a dou-

ble pump, one for each circuit. However, partitioning is incomplete in amphibians and most reptilian taxa. Birds and mammals have the most completely divided hearts containing two atria and two ventricles.

Frog hearts (Figure 25-21) have two separate atria and a single undivided ventricle. Blood from the body (systemic circuit) first enters a large receiving chamber, the sinus venosus, which forces blood into the right atrium. The left atrium receives freshly oxygenated blood from the lungs and skin. The right and left atria contract asynchronously so that although the **ventricle** is undivided, blood remains mostly separated when it enters this chamber. When the ventricle contracts, oxygenated pulmonary blood enters the systemic circuit and deoxygenated systemic blood enters the pulmonary circuit. This separation is aided by a **spiral valve,** which divides the systemic and pulmonary flows in the **conus arteriosus** (Figure 25-21) and by different blood pressure in the pulmonary and systemic blood vessels leaving the conus arteriosus.

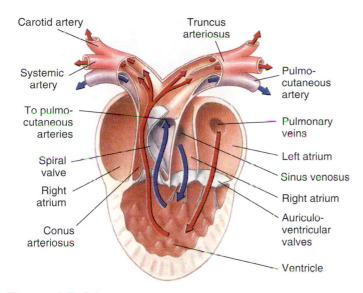

Figure 25-21
Structure of a frog heart. *Red arrows*, oxygenated blood. *Blue arrows*, deoxygenated blood.

Feeding and Digestion

Frogs are carnivorous, as are most other adult amphibians, and they feed on insects, spiders, worms, slugs, snails, millipedes, and nearly anything else that moves and is small enough to swallow whole. They snap at moving prey with their protrusible tongue, which is attached to the front of the mouth and is free behind. The highly glandular free end of the tongue produces a sticky secretion that adheres to prey. When teeth are present on the premaxillae, maxillae, and vomers, they are used to prevent escape of prey, not for biting or chewing. The digestive tract is relatively short in adult amphibians, a characteristic of most carnivores, and it produces a variety of enzymes for digesting proteins, carbohydrates, and fats.

Larval stages of anurans (tadpoles) are usually herbivorous, feeding on pond algae and other vegetable matter; they have a relatively long digestive tract because their bulky food must be submitted to time-consuming fermentation before useful products can be absorbed.

Nervous System and Special Senses

Three fundamental parts of the brain—forebrain (telencephalon), concerned with the sense of smell; midbrain (mesencephalon), concerned with vision; and hindbrain (rhombencephalon), concerned with hearing and balance—have undergone dramatic developmental trends as vertebrates moved onto land and improved their environmental awareness (p. 704). Cephalization increases with emphasis on information processing by the brain and a corresponding loss of independence of the spinal ganglia, which are capable only of stereotyped reflexive behavior. Nonetheless, a headless frog preserves an amazing degree of purposive and highly coordinated behavior. With only the spinal cord intact, it maintains normal body posture and can accurately raise its leg to wipe an irritant from its skin. It will even use the opposite leg if the closer leg is held.

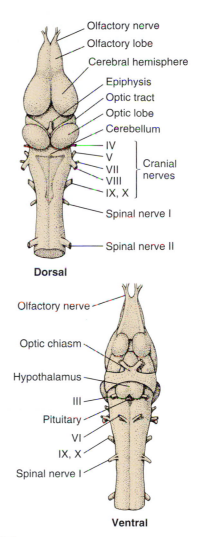

Figure 25-22
Brain of a frog, dorsal and ventral views.

The forebrain (Figure 25-22) contains the olfactory center, which assumes greatly increased importance for detection of dilute airborne odors on land. The sense of smell is in fact one of the dominant special senses in frogs. The remainder of the forebrain, or cerebrum, is of little importance in amphibians. Instead, complex integrative activities of frogs are located in the midbrain optic lobes. The hindbrain is divided into an anterior cerebellum and a posterior medulla. The cerebellum (see Figure 25-22), which is concerned with equilibrium and movement coordination, is not well developed in amphibians, especially in terrestrial species, which stay close to the ground and are not noted for dexterity of movement. The medulla is really the enlarged anterior end of the spinal cord through which pass all sensory neurons except those of vision and smell. Here are located centers for auditory reflexes, respiration, swallowing, and vasomotor control.

Evolution of a semiterrestrial life by amphibians has necessitated a reordering of sensory receptor priorities on land. The pressure-sensitive lateral line (acousticolateral) system of fishes remains only in aquatic larvae of amphibians and in a few strictly aquatic adult amphibian species. This system can serve no useful purpose on land, because it was designed to detect and localize objects in water by reflected pressure waves. Instead the task of detecting airborne sounds falls on the ear.

A frog's ear is by amniote standards a simple structure: a middle ear closed externally by a large **tympanic membrane** (eardrum) and containing a **columella** (stapes) that transmits vibrations to the inner ear (Figure 25-23). The latter contains a **utricle,** from which arise three semicircular canals, and a **saccule** bearing a diverticulum, or **lagena.** The lagena is partly covered with a **tectorial membrane,** which in its fine structure is not unlike that of the much more complex mammalian cochlea. In most frogs this structure is sensitive to low-frequency sound energy not greater than 4000 Hz (cycles per second); in bullfrogs the main frequency response is in the 100 to 200 Hz range, which matches the energy of a male frog's low-pitched call.

Vision is the dominant special sense in many amphibians (the mostly blind caecilians are obvious exceptions). Several modifications of ancestral aquatic eyes were required to adapt them for use in air. Lachrymal glands and eyelids evolved to keep eyes moist, wiped free of dust, and shielded from injury. Since the cornea is exposed to air, it is an important refractive surface, removing much of the burden from the lens of bending light rays and focusing the image on the retina. As in fishes, accommodation (adjusting focus for near and distant objects) is accomplished by moving the lens. Unlike eyes of most fishes, amphibian eyes at rest are adjusted for distant objects and the lens is moved forward to focus on nearby objects.

Keeping a sharp image on the retina for approaching or receding objects requires accommodation, which is accomplished in different ways by different vertebrates. Eyes of bony fishes and lampreys are adjusted for near vision; to focus on distant objects, the lens must be moved backward. In amphibians, sharks, and snakes, the relaxed eye is focused on distant objects and the lens is moved *forward* to focus on nearby objects. In birds, mammals, and all reptiles except snakes, the lens accommodates by changing its *curvature* rather than by being moved forward or backward. The resting eye in these forms is adjusted for distant vision, and to focus on nearby objects the lens curvature is increased by being squeezed (or, in some, allowed to relax) into a rounded shape.

A **retina** contains both **rods and cones,** the latter providing frogs with color vision. The iris contains well-developed circular and radial muscles and can rapidly expand or contract the aperture (pupil) to adjust to changing illumination. The upper lid of the eye is fixed, but the lower one is folded into a transparent **nictitating membrane** capable of moving across the eye surface (Figure 25-24). Frogs and toads generally possess good vision, a property of crucial importance to animals that rely on quick escape to avoid their numerous predators and on accurate movements to capture rapidly moving prey.

Other sensory receptors include tactile and chemical receptors in skin, taste buds on the tongue and palate, and a well-developed olfactory epithelium lining the nasal cavity.

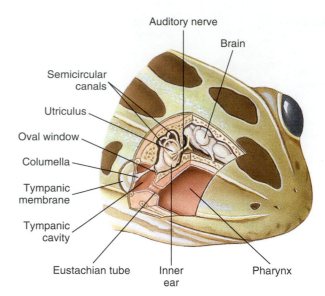

Figure 25-23

Cutaway of frog head showing ear structure. Sound vibrations are transmitted from the tympanic membrane by way of the columella to the inner ear. The eustachian tube allows pressure equilibration between the tympanic cavity and the pharynx.

Figure 25-24

Amphibian eye.

Reproduction

Because frogs and toads are ectothermic, they breed, feed, and grow only during warm seasons of the year. One of the first drives after the dormant period is breeding. In spring males call vociferously to attract females. When their eggs are mature, females enter water and are clasped by males in a process called **amplexus** (Figure 25-25). As a female lays eggs, a male discharges sperm over the eggs to fertilize them. After fertilization, the jelly layers absorb water and swell. Eggs are laid in large masses, usually anchored to vegetation.

A fertilized egg (zygote) begins development almost immediately (Figure 25-26). By repeated division (cleavage) an egg is

Figure 25-25

A male green tree frog, *Hyla cinerea*, clasps a larger female during breeding season in a South Carolina swamp. Clasping (amplexus) is maintained until the female deposits her eggs. Like most tree frogs, these are capable of rapid and marked color changes; the male here, normally green, has darkened during amplexus.

Classification of Class Amphibia

Order Gymnophiona (jim'no-fy'o-na) (Gr. *gymnos*, naked, + *ophioneos*, of a snake) **(Apoda): caecilians.** Body elongate; limbs and limb girdle absent; mesodermal scales present in skin of some; tail short or absent; 95 to 285 vertebrae; pantropical, 6 families, 34 genera, approximately 160 species.

Order Caudata (caw-dot'uh) (L. *caudatus*, having a tail) **(Urodela): salamanders.** Body with head, trunk, and tail; no scales; usually two pairs of equal limbs; 10 to 60 vertebrae; predominantly holarctic; 10 living families, 62 genera, approximately 480 species.

Order Anura (uh-nur'uh) (Gr. *an*, without, + *oura*, tail) **(Salientia): frogs, toads.** Head and trunk fused; no tail; no scales; two pairs of limbs; large mouth; lungs; 6 to 10 vertebrae including urostyle (coccyx); cosmopolitan, predominantly tropical; 21 living families; 301 genera; approximately 3450 species.

Figure 25-26

Life cycle of a leopard frog.

converted into a hollow ball of cells (blastula). The blastula undergoes gastrulation and then continues to differentiate to form an embryo with a tail bud. At 6 to 9 days, depending on temperature, a tadpole hatches from the protective jelly coats that had surrounded the original fertilized egg.

At hatching, a tadpole has a distinct head and body with a compressed tail. Its mouth is located on the ventral side of the head and is provided with keratinized jaws for feeding by scraping vegetation from hard objects. Behind the mouth is a ventral adhesive disc for clinging to objects. In front of the mouth are two deep pits, which later develop into nostrils. Swellings found on each side of the head later become external gills. There are three pairs of external gills, which later transform into internal gills and become covered with a flap of skin (operculum) on each side. On the right side the operculum completely fuses with the body wall, but on the left side a small opening, the spiracle (L. *spiraculum,* air hole) remains. Water flows through the spiracle after entering the mouth and passing the internal gills. Hindlimbs appear first during metamorphosis, while forelimbs remain temporarily hidden by folds of the operculum. The tail is resorbed, the intestine becomes much shorter, the mouth undergoes a transformation into the adult condition, lungs develop, and gills are resorbed (Figure 25-26). Leopard frogs usually complete metamorphosis within 3 months, whereas bullfrogs take 2 or 3 years to complete the process.

Migration of frogs and toads is correlated with their breeding habits. Males usually return to a pond or stream before females, which they then attract by their calls. Some salamanders also have a strong homing instinct, returning each year to reproduce in the same pool, to which they are guided by olfactory cues. The initial stimulus for migration in many cases is attributable to a seasonal cycle in the gonads plus hormonal changes that increase frogs' sensitivity to changes in temperature and humidity.

SUMMARY

Amphibians are ectothermic, primitively quadrupedal vertebrates that have glandular skin and breathe by lungs, gills, or skin. They are survivors of one of two major branches of tetrapod phylogeny, the other one being represented today by amniotes. Modern amphibians consist of three major evolutionary groups. Caecilians (order Gymnophiona) are a small tropical group of limbless, elongate forms. Salamanders (order Caudata) are tailed amphibians that have retained the generalized four-limbed body plan of their Paleozoic ancestors. Frogs and toads (order Anura) are the largest group of modern amphibians, all of which are specialized for a jumping mode of locomotion on land.

Most amphibians have a biphasic life cycle, beginning with an aquatic larva that later metamorphoses to produce a terrestrial adult, which returns to the water to lay eggs. Some frogs, salamanders, and caecilians have evolved direct development that omits the aquatic larval stage, and some caecilians have evolved viviparity. Salamanders are unique among amphibians in having evolved several perennibranchiate species that retain a permanently larval morphology throughout life, eliminating a terrestrial phase completely. The perennibranchiate condition is obligate in some species, but others metamorphose to a terrestrial form if the pond habitat evaporates.

Although amphibians have evolved adaptations to the aquatic phase of their life history, adaptations to their terrestrial existence are particularly noteworthy. Respiratory exchange of gases occurs across porous skin in all amphibians and is supplemented in most amphibians by lungs. Oddly, the most highly terrestrial salamanders lack lungs whereas some aquatic forms use lungs as their major respiratory structure. Life on land also required strengthening and redirection of skeletal elements, especially ribs, pectoral and pelvic girdles, and limbs. Derived features of amphibian auditory and visual systems and associated regions of the brain facilitate sensory perception on land.

Despite their adaptations for terrestrial life, adults and eggs of all amphibians require cool, moist environments if not actual pools or streams. Eggs and adult skin have no effective protection against very cold, hot, or dry conditions, greatly restricting adaptive radiation of amphibians to environments that have moderate temperatures and abundant water.

REVIEW QUESTIONS

1. Compared with aquatic habitats, terrestrial habitats offer both advantages and problems for an animal making a transition from water to land. Summarize how these differences might have influenced the early evolution of tetrapods.
2. Describe the different modes of respiration used by amphibians. What paradox do amphiumas and terrestrial plethodontids present regarding the association of lungs with life on land?
3. Evolution of the tetrapod limb was one of the most important advances in vertebrate history. Describe the supposed sequence in its evolution.
4. Compare the general life history patterns of salamanders with those of frogs. Which group shows a greater variety of evolutionary changes of the ancestral biphasic amphibian life cycle?
5. Give the literal meaning of the name Gymnophiona. What animals are included in this amphibian order, what do they look like, and where do they live?
6. What is the literal meaning of the order names Caudata and Anura? What major features distinguish members of these two orders from each other?

7. Describe the breeding behavior of a typical woodland salamander.
8. How has paedomorphosis been important to evolutionary diversification of salamanders?
9. Describe the integument of a frog. What is responsible for skin color in frogs?
10. Describe amphibian circulation.

11. Explain how the forebrain, midbrain, hindbrain, and the sensory structures with which each brain division is concerned have developed to meet sensory requirements for amphibian life on land.
12. Briefly describe the reproductive behavior of frogs. In what important ways do frogs and salamanders differ in their reproduction?

SELECTED REFERENCES

Conant, R., and J. T. Collins. 1998. A field guide to reptiles and amphibians: Eastern and Central North America. The Peterson field guide series. Boston, Houghton Mifflin Company. *Updated version of a popular field guide; color illustrations and distribution maps for all species.*

del Pino, E. M. 1989. Marsupial frogs. Sci. Am. **260:**110–118 (May). *Several species of tropical frogs incubate their eggs on the female's back, often in a special pouch, and emerge as advanced tadpoles or fully formed froglets.*

Duellman, W. E. 1992. Reproductive strategies of frogs. Sci. Am. **267:**80–87 (July). *Many frogs have evolved improbable reproductive strategies that have permitted colonization of land.*

Duellman, W. E., and L. R. Trueb. 1994. Biology of amphibians. Baltimore, Johns Hopkins University Press. *Important comprehensive sourcebook of information on amphibians, extensively referenced and illustrated.*

Halliday, T. R., and K. Adler (eds). 1986. The encyclopedia of reptiles and amphibians. New York, Facts on File, Inc. *Excellent authoritative reference work with high-quality illustrations.*

Hanken, J. 1989. Development and evolution in amphibians. Am. Sci. **77:**336–343 (July–Aug.). *Explains how diversity in amphibian morphology has been achieved by modifications in development.*

Kiesecker, J. M., A. R. Blaustein, and L. K. Belden. 2001. Complex causes of amphibian declines. Nature **410:**681–683. *A complex interaction of climatic changes, ultraviolet radiation and fungal infections may explain declines of some amphibian populations.*

Lewis, S. 1989. Cane toads: an unnatural history. New York, Dolphin/Doubleday. *Based on an amusing and informative film of the same title, this book describes the introduction of cane toads to Queensland, Australia and the unexpected consequences of their population explosion there. "If Monty Python teamed up with National Geographic, the result would be Cane Toads."*

Moffett, M. W. 1995. Poison-dart frogs: lurid and lethal. National Geographic **187**(5):98–111 (May). *Photographic essay of frogs that can be lethal even to the touch.*

Narins, P. M. 1995. Frog communication. Sci. Am. **273:**78–83 (Aug.). *Frogs employ several strategies to hear and be heard amid the cacophony of chorusing of many frogs.*

Petranka, J. W. 1998. Salamanders of the United States and Canada. Washington, DC, Smithsonian Institution Press. *A comprehensive coverage of life history and ecology of American and Canadian salamanders.*

Pough, F. H., R. M. Andrews, J. E. Cadle, M. L. Crump, A. H. Savitsky, and K. D. Wells. 2001. Herpetology, ed 2. Upper Saddle River, New Jersey, Prentice-Hall. *A current general textbook of herpetology.*

Sever, D. M. (ed). 2003. Reproductive biology and phylogeny of Urodela (Amphibia). Enfield, New Hampshire, Science Publishers, Inc. *A thorough review of reproductive biology and evolutionary relationships among salamanders.*

Stebbins, R. C., and N. W. Cohen. 1995. A natural history of amphibians. Princeton, New Jersey, Princeton University Press. *Worldwide treatment of amphibian biology, emphasizing physiological adaptations, ecology, reproduction, behavior, and a concluding chapter on amphibian declines.*

Zug, G. R., L. J. Vitt, and J. P. Caldwell. 2001. Herpetology: an introduction to the biology of amphibians and reptiles. San Diego, Academic Press. *A current general textbook of herpetology.*

ZOOLOGY LINKS TO THE INTERNET

Visit the textbook's Online Learning Center at www.mhhe.com/zoology to find live Internet links for each of the topics listed here.
Class Amphibia
Order Salienta (new taxonomic nomenclature: Order Anura)
Dissection Guides for Amphibians
Conservation Issues Concerning Amphibians

Hatching Komodo lizard
(*Voronus komodoensis*).

Amniote Origins and Reptilian Groups

Phylum Chordata
Class Reptilia

Enclosing the Pond

The amphibians, with well-developed limbs, redesigned sensory and respiratory systems, and modifications of the postcranial skeleton for supporting the body in air, have made a notable conquest of land. But, with shell-less eggs and often gill-breathing larvae, their development remains hazardously tied to water. The lineage containing reptiles, birds, and mammals developed an egg that could be laid on land. This shelled egg, perhaps more than any other adaptation, unshackled early reptiles from the aquatic environment by freeing the developmental process from dependence on aquatic or very moist terrestrial environments. In fact, the "pond-dwelling" stages were not eliminated but enclosed within a series of extraembryonic membranes that provided complete support for embryonic development. One membrane, the amnion, encloses a fluid-filled cavity, the "pond," within which the developing embryo floats. Another membranous sac, the allantois, serves both as a respiratory surface and as a chamber for the storage of nitrogenous wastes. Enclosing these membranes is a third membrane, the chorion, through which oxygen and carbon dioxide freely pass. Finally, surrounding and protecting everything is a porous, parchmentlike or leathery shell.

With the last ties to aquatic reproduction severed, conquest of land by vertebrates was ensured. Paleozoic tetrapods that developed this reproductive pattern were ancestors of a single, monophyletic assemblage called the Amniota, named after the innermost of the three extraembryonic membranes, the amnion. Before the end of the Paleozoic era amniotes had diverged into multiple lineages that gave rise to all the reptilian groups, the birds, and the mammals.

Members of the paraphyletic class Reptilia (rep-til'e-a) (L. *repto,* to creep) include the first truly terrestrial vertebrates. With nearly 8000 species (approximately 340 species in the United States and Canada) occupying a great variety of aquatic and terrestrial habitats, they are diverse and abundant. Nevertheless, reptiles are perhaps remembered best for what they once were, rather than for what they are now. The Age of Reptiles, which lasted for more than 165 million years, saw the appearance of a great radiation of reptilian lineages into a bewildering array of terrestrial and aquatic forms. Among these were herbivorous and carnivorous dinosaurs, many of huge stature and awesome appearance, that dominated animal life on land. Then, during a mass extinction at the end of the Mesozoic era, many reptilian lineages became extinct. Among the lineages to emerge from the Mesozoic extinction are today's reptiles. One of these, the tuataras (*Sphenodon*) of New Zealand, are the sole survivors of a group that otherwise disappeared 100 million years ago. But others, especially lizards and snakes, have radiated since the Mesozoic extinction into diverse and abundant groups. Understanding the 300-million-year-old history of reptile life on earth has been complicated by widespread convergent and parallel evolution among the many lineages and by large gaps in the fossil record.

ORIGIN AND ADAPTIVE RADIATION OF REPTILIAN GROUPS

As mentioned in the prologue to this chapter, amniotes are a monophyletic group that evolved in the late Paleozoic. Most paleontologists agree that the amniotes arose from a group of amphibian-like tetrapods, the anthracosaurs, during the early Carboniferous period of the Paleozoic. By the late Carboniferous (approximately 300 million years ago), amniotes had separated into three groups. The first group, the **anapsids** (Gr. *an,* without, + *apsis,* arch), is characterized by a skull having no temporal opening behind the orbits, the skull behind the orbits being completely roofed with dermal bone (see Figure 26-2). This group is represented today only by turtles. Their morphology is an odd mix of ancestral and derived characters that has scarcely changed at all since turtles first appeared in the fossil record in the Triassic some 200 million years ago.

The second group, the **diapsids** (Gr. *di,* double, + *apsis,* arch), includes all other reptilian groups and birds (Figure 26-1). The diapsid skull is characterized by two temporal openings: one pair located low on the cheeks, and a second pair positioned above the lower pair and separated from them by a bony arch (Figure 26-2). Four subgroups of diapsids appeared. The **lepidosaurs** include all modern reptiles except turtles and crocodilians. The **archosaurs** comprised dinosaurs and their relatives, and living crocodilians and birds. A third, smaller subgroup, the **sauropterygians** included several extinct aquatic groups, most conspicuous of which were the large, long-necked plesiosaurs. **Ichthyosaurs,** represented by extinct, aquatic dolphinlike forms (Figure 28-1), comprise the fourth subgroup.

Position in the Animal Kingdom

Modern reptilian groups comprise two of three lineages of amniote vertebrates that arose from amphibian-like tetrapods of the late Paleozoic era. Reptiles traditionally comprise anapsid amniotes, represented by turtles, and diapsid amniotes, represented by lizards, snakes, crocodilians, and tuataras. They are survivors of an enormous radiation of Mesozoic amniotes, including the dinosaurs, most of which became extinct at the end of the Mesozoic. As traditionally defined, class Reptilia is a paraphyletic group because it excludes the birds, which are descendants of the common ancestor of the diapsid lineage. A third lineage of amniotes, the synapsids, gave rise to modern mammals.

Biological Contributions

1. The **shelled, amniotic egg** that evolved with the earliest Paleozoic amniotes is supplied with extraembryonic membranes that provide a complete life-support system for the enclosed embryo. This innovation allowed amniotes to lay larger eggs and in drier habitats. In some viviparous reptiles the extraembryonic membranes are restructured into a placenta suggestive of and somewhat paralleling the evolution in the synapsid lineage of the more complex mammalian placenta.
2. A **tough, dry, heavily keratinized skin** that provides protection against desiccation and injury. Scales in reptiles and feathers in birds arise as epidermal elevations overlying a nourishing dermal layer.
3. Larger and stronger jaw muscles permit **powerful jaw closure.** Temporal openings in the diapsid skull provide space for bulging temporal muscles.
4. **Internal fertilization,** with sperm introduced directly into the female reproductive tract with a copulatory organ.
5. Effective **adaptations for water conservation** include a metanephric kidney that excretes nitrogenous wastes as uric acid (diopsids and desert turtles) or urea (most turtles). Such adaptations allowed reptiles (and birds) to occupy many terrestrial habitats.

The third group is the **synapsids** (Gr. *syn,* together, + *apsis,* arch), the mammals and extinct forms traditionally termed mammal-like reptiles. The synapsid skull had a single pair of temporal openings located low on the cheeks and bordered by a bony arch (Figure 26-2). Synapsids were the first amniote group to diversify, giving rise first to pelycosaurs, later to therapsids, and finally to mammals (Figure 26-1).

Changes in Traditional Classification of Reptilian Groups

With increasing use of cladistic methodology in zoology, and its insistence on hierarchical arrangement of monophyletic groups (see p. 197), important changes have been made in the traditional classification of reptiles. Class Reptilia is no longer recognized by cladists as a valid taxon because it is not monophyletic.

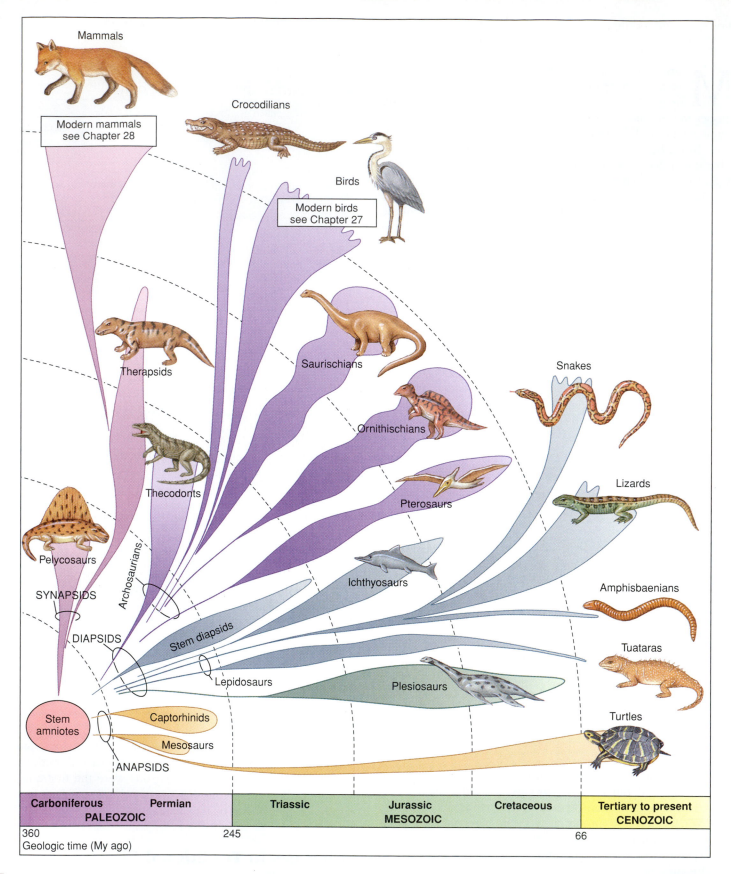

Mammals

Crocodilians

Modern mammals
see Chapter 28

Birds

Modern birds
see Chapter 27

Therapsids

Saurischians

Snakes

Thecodonts

Ornithischians

Lizards

Pelycosaurs

Pterosaurs

SYNAPSIDS

Archosaurians

Amphisbaenians

DIAPSIDS

Stem diapsids

Ichthyosaurs

Lepidosaurs

Tuataras

Plesiosaurs

Stem
amniotes

Captorhinids

Turtles

Mesosaurs

ANAPSIDS

Carboniferous	Permian	Triassic	Jurassic	Cretaceous	Tertiary to present
PALEOZOIC		**MESOZOIC**			**CENOZOIC**

360
Geologic time (My ago)

245

66

Figure 26-1

Evolution of amniotes. The evolutionary origin of amniotes occurred by evolution of an amniotic egg that made reproduction on land possible, although this egg may well have developed before the earliest amniotes had ventured far on land. The amniote assemblage, which includes reptiles, birds, and mammals, evolved from a lineage of small, lizardlike forms that retained the anapsid skull pattern of early tetrapods. First to diverge from the primitive stock was a lineage that evolved a skull pattern termed the synapsid condition. All other amniotes, including birds and all living reptiles except turtles, have a skull pattern known as diapsid. Turtles retain the primitive, anapsid skull pattern. The great Mesozoic radiation of reptiles may have resulted partly from the increased variety of ecological habitats that the amniotes could exploit.

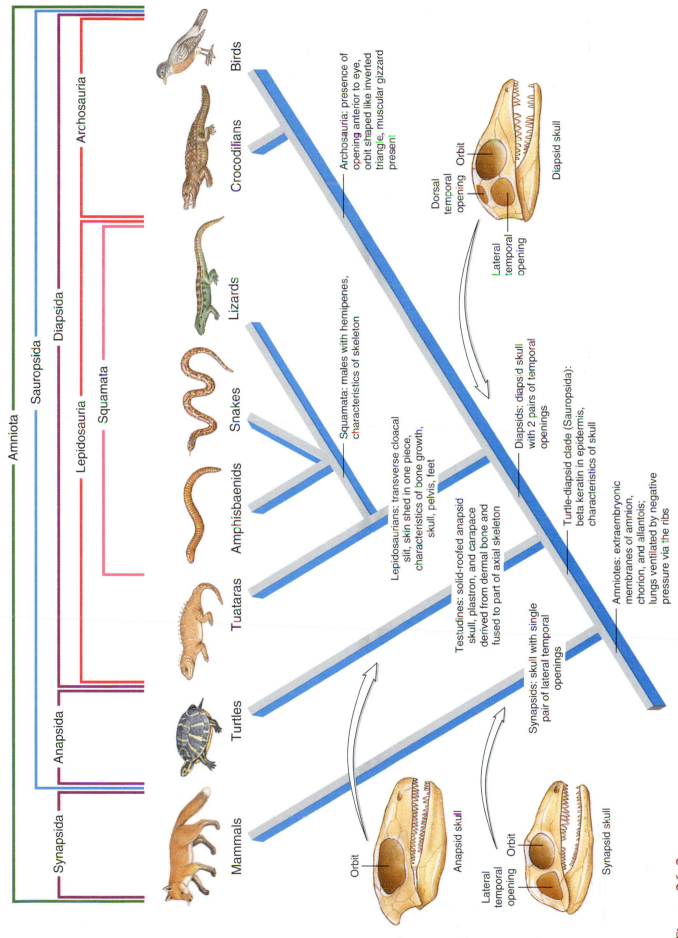

Figure 26-2

Cladogram of living Amniota showing monophyletic groups. Some shared derived characters (synapomorphies) of the groups are given. The skulls represent the ancestral condition of the three groups. Skulls of modern diapsids and synapsids are often highly modified by loss or fusion of skull bones that obscures the ancestral condition. Representative skulls for anapsids are *Nyctiphruetus* of the upper Permian; for diapsids, *Youngina* of the upper Permian; for synapsids, *Aerosaurus*, a pelycosaur of the lower Permian. The relationships expressed in this cladogram are tentative and controversial. Relationships among lizards, snakes, and amphisbaenians are uncertain, but evidence suggests both amphisbaenians and snakes evolved within the lizard clade.

Birds

Crocodilians

Lizards

Snakes

Amphisbaenids

Tuataras

Turtles

Mammals

Amniota

Sauropsida

Diapsida

Archosauria

Anapsida

Lepidosauria

Squamata

Synapsida

Archosauria: presence of opening anterior to eye, orbit shaped like inverted triangle, muscular gizzard present

Squamata: males with hemipenes, characteristics of skeleton

Lepidosaurians: transverse cloacal slit, skin shed in one piece, characteristics of bone growth, skull, pelvis, feet

Diapsids: diapsid skull with 2 pairs of temporal openings

Turtle-diapsid clade (Sauropsida): beta keratin in epidermis, characteristics of skull

Testudines: solid-roofed anapsid skull, plastron, and carapace derived from dermal bone and fused to part of axial skeleton

Amniotes: extraembryonic membranes of amnion, chorion, and allantois; lungs ventilated by negative pressure via the ribs

Synapsids: skull with single pair of lateral temporal openings

---

Dorsal temporal opening   Orbit

Lateral temporal opening

Diapsid skull

Orbit

Anapsid skull

Lateral temporal opening

Orbit

Synapsid skull

As customarily defined, class Reptilia excludes birds, which descended from the most recent common ancestor of the reptiles. Consequently, reptiles are a **paraphyletic** group because they do not include all descendants of their most recent common ancestor. Reptiles and birds share several derived characters, including several skull characteristics and a largely aglandular skin with a special type of harder keratin called beta keratin, which unites them as a monophyletic group (Figure 26-2). Although we recognize that Reptilia, as historically used, is not a monophyletic group, we can use it as a term of convenience to refer to all amniotes that have beta keratin in their epidermis that are not birds. Thus we are using the word "reptiles" to refer to the living turtles, snakes, lizards, amphisbaenians, tuataras, and crocodilians in addition to a number of extinct groups such as plesiosaurs, ichthyosaurs, pterosaurs, and dinosaurs.

Crocodilians and birds are sister groups; they are more recently descended from a common ancestor than either is from any other living reptilian lineage. In other words, birds and crocodilians belong to a monophyletic group apart from other reptiles and, according to the rules of cladistics, should be assigned to a clade that separates them from the remaining reptiles. This clade is in fact recognized; it is Archosauria (Figures 26-1 and 26-2), a grouping that also includes the extinct dinosaurs. Therefore, according to cladistics, birds should be classified as reptiles. Archosaurs plus their sister group, the lepidosaurs (tuataras,

lizards, snakes, and amphisbaenids), comprise a monophyletic group that cladists call Reptilia. The term "Reptilia" is thereby redefined to include birds in contrast to its traditional usage. However, evolutionary taxonomists argue that birds represent a novel adaptive zone and grade of organization whereas crocodilians remain within the traditionally recognized reptilian adaptive zone and grade. In this view, the morphological and ecological novelty of birds has been recognized by maintaining the traditional classification that places crocodilians in class Reptilia and birds in class Aves. Such conflicts of opinion between proponents of the two major competing schools of taxonomy (cladistics and evolutionary taxonomy) have had the healthy effect of forcing zoologists to reevaluate their views of amniote genealogy and how vertebrate classifications should represent genealogy and degrees of divergence. In our treatment, "reptilian group," "reptile," and "reptilian" refer to members of four living monophyletic groups (turtles, crocodilians, squamates, tuataras) that are combined into the paraphyletic class Reptilia.

CHARACTERISTICS OF REPTILES THAT DISTINGUISH THEM FROM AMPHIBIANS

1. **Reptiles have tough, dry, scaly skin offering protection against desiccation and physical injury.** The skin consists of a thin epidermis and a much thicker, well-developed **dermis** (Figure 26-3). The dermis is provided with **chromatophores,** color-bearing cells that give many lizards and snakes their colorful hues. This layer, unfortunately for their bearers, is converted into alligator and snakeskin leather, so esteemed for expensive pocketbooks and shoes. Resistance to desiccation is provided by hydrophobic lipids in the epidermis. The epidermis also contains a hard form of keratin called beta keratin that is unique to reptiles. The characteristic **scales** of reptiles, formed largely of beta keratin, provide protection against wear in terrestrial environments. They are derived mostly from the epidermis and thus are not homologous to fish scales, which are bony, dermal structures (see

Characteristics of Class Reptilia

1. Body varied in shape, compact in some, elongated in others; **body covered with keratinized epidermal scales** with the addition sometimes of bony dermal plates; **integument with few glands**
2. **Two paired limbs, usually with five toes,** and adapted for climbing, running, or paddling; limbs vestigial or absent in snakes and some lizards and amphisbaenians
3. Skeleton well ossified; ribs with sternum (sternum absent in snakes) forming a complete thoracic basket; **skull with one occipital condyle**
4. Respiration by lungs; **no gills;** cloaca used for respiration by some; branchial arches in embryonic life
5. Circulatory system functionally divided into pulmonary and systemic circuits; heart typically consisting of a sinus venosus, an atrium completely divided into two chambers, and a ventricle incompletely divides into three chambers; crocodilians with a sinus venosus, two atria, and a ventricle completely divided into two chambers
6. Ectothermic; many thermoregulate behaviorally
7. **Metanephric kidney (paired); uric acid main nitrogenous waste**
8. Nervous system with the optic lobes on the dorsal side of brain; **12 pairs of cranial nerves** in addition to nervus terminalis; enlarged cerebrum
9. Sexes separate; **fertilization internal**
10. **Eggs covered with calcareous or leathery shells; extraembryonic membranes (amnion, chorion, and allantois)** present during embryonic life; **no aquatic larval stages**

Figure 26-3
Section of the skin of a reptile showing overlapping epidermal scales.

Figure 29-2, p. 615). In some reptiles, such as alligators, the scales remain throughout life, growing gradually to replace wear. In others, such as snakes and lizards, new scales grow beneath the old, which are then shed at intervals. Turtles add new layers of keratin under the old layers of the plate-like scutes, which are modified scales. In snakes the old skin (epidermis and scales) is turned inside out when discarded; lizards split out of the old skin leaving it mostly intact and right side out, or it may slough off in pieces. Crocodiles and many lizards possess bony plates called **osteoderms** located beneath the keratinized scale in the dermis.

2. **The amniotic egg of reptiles permits rapid development of large young in relatively dry environments.** Membranes around the anamniotic eggs of amphibians are insufficient to permit gas exchange necessary for large, rapidly growing embryos. Two membranes of amniotes, the chorion and allantois (Figure 26-4), assist in exchange of oxygen and carbon dioxide with the environment, permitting development of large young. High energy requirements of these young are supplied by the large, extraembryonic yolk sac. The amnion and fibrous or calcareous shell provide support for the growing embryo and reduce the amount of water lost to the external environment. Although reptile eggs must still be maintained at relatively high humidity to avoid desiccation, they can be laid in much drier environments than even the most terrestrial amphibian eggs. For many reptile species, egg development takes place in a female's reproductive tract, providing even greater protection from predators and dehydration, and potential for the mother to manage the embryo's nutritional and other physiological needs.

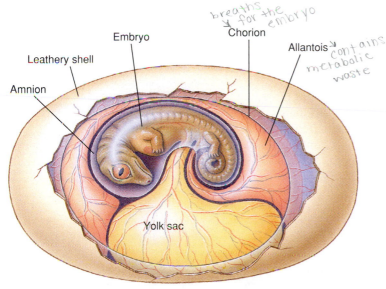

Figure 26-4

Amniotic egg. The embryo develops within the amnion and is cushioned and supported by amniotic fluid. Food is provided by yolk from the yolk sac and metabolic wastes are deposited within the allantois. As development proceeds, the allantois fuses with the chorion, a membrane lying against the inner surface of the shell; both membranes are supplied with blood vessels that assist in the exchange of oxygen and carbon dioxide across the porous shell. Because this kind of egg is an enclosed, self-contained system, it is often called a "cleidoic" egg (Gr. *kleidoun*, to lock in).

3. **Reptilian jaws are efficiently designed for applying crushing or gripping force to prey.** Jaws of fishes and amphibians are designed for quick closure, but once the prey is seized, little static force can be applied. In reptiles, jaw muscles became larger, longer, and arranged for much better mechanical advantage.

4. **Reptiles have some form of copulatory organ, permitting internal fertilization.** Internal fertilization is obviously a requirement for a shelled egg, because sperm must reach an egg before it is enclosed. Sperm from paired testes are carried by the vasa deferentia to a copulatory organ, a penis or hemipenes, which is an evagination of the cloacal wall. The female system consists of paired ovaries and oviducts. Glandular walls of the oviducts secrete albumin (source of amino acids, minerals, and water for the embryo) and shells for large eggs.

5. **Reptiles have an efficient and flexible circulatory system and higher blood pressure than amphibians.** In all reptiles the right atrium, which receives unoxygenated blood from the body, is completely partitioned from the left atrium, which receives oxygenated blood from the lungs. Crocodilians have two completely separated ventricles as well (Figure 26-5); in other reptiles the ventricle is incompletely partitioned into multiple chambers. Even in reptiles with incomplete separation of the ventricles, flow patterns within the heart prevent admixture of pulmonary (oxygenated) and systemic (unoxygenated) blood; all reptiles therefore have two functionally separate circulations. This incomplete separation between the right and left sides of the heart provides an added benefit of permitting blood to bypass the lungs when pulmonary respiration is not occurring (for example, diving or aestivation).

6. **Reptilian lungs are better developed than those of amphibians.** Reptiles depend almost exclusively on lungs for gas exchange, supplemented by respiration through pharyngeal membranes in some aquatic turtles. Unlike amphibians, which *force* air into their lungs with mouth muscles, reptiles *suck* air into lungs by enlarging the thoracic cavity, either by expanding their rib cage (snakes and lizards) or by movement of internal organs (turtles and crocodilians). Reptiles have no muscular diaphragm, a structure found only in mammals. Cutaneous respiration (gas exchange across the skin), so important to amphibians, has been abandoned by reptiles.

7. **Reptiles have evolved efficient strategies for water conservation.** All amniotes have a metanephric kidney, which is drained by its own passageway, the ureter. However, nephrons of the reptilian metanephros lack the specialized intermediate section of the tubule, the loop of Henle (p. 646), which enables a kidney to concentrate solutes in urine. Many reptiles have salt glands located near the nose or eyes (in the tongue of saltwater crocodiles), which secrete a salty fluid that is strongly hyperosmotic to body fluids. Nitrogenous wastes are excreted as uric acid, rather than urea or ammonia. Uric acid has a low solubility and precipitates out of solution readily, allowing water to be conserved; urine of many reptiles is a semisolid suspension.

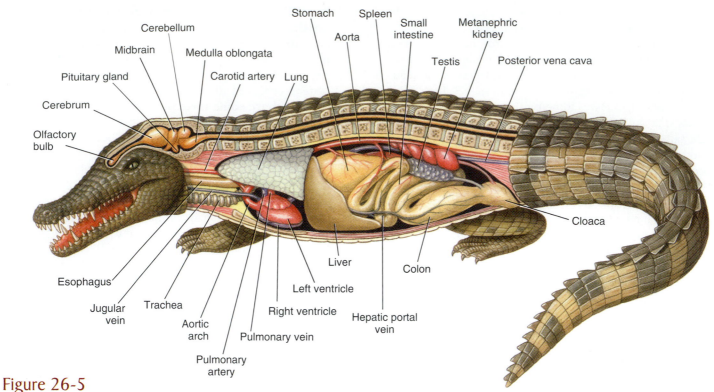

Figure 26-5

Internal structure of a male crocodile.

8. **All reptiles, except limbless members, have better body support than amphibians and more efficiently designed limbs for travel on land.** Nevertheless, most modern reptiles walk with their legs splayed outward and their belly close to the ground. Most dinosaurs, however, (and some modern lizards) walked on upright legs held beneath the body, the best arrangement for rapid movement and for support of body weight. Many dinosaurs walked on powerful hindlimbs alone.

9. **Reptilian nervous systems are considerably more complex than amphibian systems.** Although a reptile's brain is small, the cerebrum is larger relative to the rest of the brain. Connections to the central nervous system are more advanced, permitting complex behaviors unknown in amphibians. With exception of hearing, sense organs in general are well developed. Jacobson's organ, a specialized olfactory chamber present in many tetrapods, is highly developed in lizards and snakes. Odors are carried to Jacobson's organ by the tongue in some reptiles.

CHARACTERISTICS AND NATURAL HISTORY OF REPTILIAN ORDERS

Anapsid Reptiles: Subclass Anapsida

Order Testudines (Chelonia): Turtles

Turtles descended from one of the earliest anapsid lineages, probably a group known as procolophonids of the late Permian, but turtles themselves do not appear in the fossil record until the

Upper Triassic, some 200 million years ago. From the Triassic, turtles plodded on to the present with very little change to their early morphology. They are enclosed in shells consisting of a dorsal **carapace** (Fr., from Sp., *carapacho,* covering) and a ventral **plastron** (Fr., breastplate). The shell is composed of two layers: an outer horny layer of keratin and an inner layer of bone. New layers of keratin are laid down beneath the old as the turtle grows and ages. The bony layer is a fusion of ribs, vertebrae, and many dermally-ossifying elements (Figure 26-6). Unique among vertebrates, turtle limbs and limb girdles are located *inside* the ribs! Lacking teeth, a turtle's jaw is provided with tough, horny (keratinized) plates for gripping food (Figure 26-7).

Clumsy and unlikely as they appear to be within their protective shells, turtles are nonetheless a varied and ecologically diverse group that seems able to adjust to human presence.

The terms "turtle," "tortoise," and "terrapin" are applied variously to different members of the turtle order. In North American usage, they are all correctly called turtles. The term "tortoise" is frequently given to land turtles, especially large forms. British usage of the terms is different: "tortoise" is the inclusive term, whereas "turtle" is applied only to the aquatic members.

One consequence of living in a rigid shell with fused ribs is that a turtle cannot expand its chest to breathe. Turtles solved this problem by employing certain abdominal and pectoral muscles as a "diaphragm." Air is drawn in by increasing abdominal cavity volume by contracting limb flank muscles. Exhalation is also active and is accomplished by drawing the shoulder girdle

Figure 26-6
Skeleton and shell of a turtle, showing fusion of vertebrae and ribs with the carapace. The long and flexible neck allows the turtle to withdraw its head into its shell for protection.

Figure 26-7
Snapping turtle, *Chelydra serpentina*, showing the absence of teeth. Instead, the jaw edges are covered with a keratinized plate.

Figure 26-8
Mating Galápagos tortoises. The male has a concave plastron that fits over the highly convex carapace of the female, helping to provide stability during mating. Males utter a roaring sound during mating, the only time they are known to emit vocalizations.

back into the shell, thus compressing the viscera and forcing air out of the lungs. Breathing is visible as the bellowslike movements of the turtle's "limb pockets": folds of skin between the limbs and the rigid shell. Movements of the limbs during walking also help to ventilate the lungs. Many aquatic turtles gain enough oxygen by just pumping water in and out of a vascularized mouth cavity; this activity enables them to remain submerged for long periods when inactive. When active they must lung-breathe more frequently.

A turtle's brain, like that of other reptiles, is small, never exceeding 1% of body weight. The cerebrum, however, is larger than that of an amphibian, and turtles are able to learn a maze about as quickly as a rat. Turtles have both a middle and an inner ear, but perception of sound is poor. Not unexpectedly, therefore, turtles are virtually mute (the biblical "voice of the turtle" refers to the turtledove, a bird), although many tortoises utter grunting or roaring sounds during mating (Figure 26-8). Compensating for poor hearing are a good sense of smell, acute vision, and color perception evidently as good as that of humans.

Turtles are oviparous. Fertilization is internal and all turtles, even marine forms, bury their shelled, amniotic eggs in the ground. Usually they exercise considerable care in constructing their nest, but once eggs are deposited and covered, the female deserts them. An odd feature of turtle reproduction is that in some turtle families, as in all crocodilians and some lizards, nest temperature determines sex of the hatchlings. In turtles, low temperatures during incubation produce males and high temperatures produce females. All reptiles with temperature-dependent sex determination during embryogenesis lack sex chromosomes.

Marine turtles, buoyed by their aquatic environment, may reach great size. Leatherbacks are largest, attaining a length of 2 m and weight of 725 kg. Green turtles (Figure 26-9), so named because of their greenish body fat, may exceed 360 kg, although most individuals of this economically valuable and heavily exploited species seldom live long enough to reach anything approaching this size. Some land tortoises may weigh several hundred kilograms, such as the giant tortoises of the Galápagos Islands that so intrigued Darwin during his visit there in 1835. Most tortoises are rather slow moving; an hour of determined trudging carries a large Galápagos tortoise approximately 300 m (although they may move much more rapidly for short distances). Their low metabolism probably explains their longevity, for some are believed to live more than 150 years.

The shell, like a medieval coat of armor, offers obvious advantages. The head and appendages can be drawn in for protection. The familiar box tortoise (*Terrapene carolina*) has a

Figure 26-9

Green sea turtle, *Chelonia mydas*. Green turtles are herbivores that subsist on marine grasses and algae. Sea turtles range widely in the oceans, returning to land only to deposit their eggs. Sea turtles are found in all tropical oceans.

Figure 26-10

Alligator snapping turtle *Macroclemys temmincki* of the southeastern United States lies on the bottom, mouth agape, luring fishes and other unwary prey by undulating a pink, wormlike protrusion from its tongue. Any prey attempting to eat the bait is instantly captured in powerful jaws.

plastron that is hinged, forming two movable parts that can be pulled up against the carapace so tightly that one can hardly force a knife blade between the shells. Some turtles, such as the large eastern snapping turtle (*Chelydra serpentina*), have reduced shells, making complete withdrawal for protection quite impossible. Snappers, however, have another formidable defense, as their name implies (Figure 26-7). They are entirely carnivorous, living on fishes, frogs, waterfowl, or almost anything that comes within reach of their powerful jaws. An alligator snapper lures unwary fish into its mouth with a pink, wormlike extension of its tongue that serves as a "bait" (Figure 26-10). Snappers are wholly aquatic and come ashore only to lay their eggs.

Diapsid Reptiles: Subclass Diapsida

Diapsid reptiles, those reptiles having a skull with two pairs of temporal openings (Figure 26-2), are classified into three lineages (superorders; see the Classification of Amniotes and Living Reptiles on p. 551). Superorders with living representatives are Lepidosauria, containing lizards, snakes, worm lizards, and tuataras; and Archosauria, containing the crocodilians.

Order Squamata: Lizards, Snakes, and Worm Lizards

Squamates are the most recent and diverse products of diapsid evolution, comprising approximately 95% of all known living reptiles. Lizards appeared in the fossil record as early as the Jurassic, but they did not begin their radiation until the Cretaceous period of the Mesozoic era when dinosaurs were at the climax of their radiation. Snakes appeared during the late Jurassic period, probably from a group of lizards whose descendants include the Gila monster and monitor lizards. Two specializations in particular characterize snakes: extreme elongation of their body and accompanying displacement and rearrangement

of internal organs; and specializations for eating large prey. Amphisbaenians (worm lizards), which first appear in the fossil record of the early Cenozoic era, have structural specializations associated with a burrowing habit.

Viviparity in living reptiles is limited to squamates, and has evolved at least 100 separate times. Evolution of viviparity is usually associated with cold climates and occurs by increasing the length of time eggs are kept within the oviduct. Developing young respire through extraembryonic membranes and obtain nutrition from yolk sacs (**lecithotrophy**) or via the mother (**placentotrophy**), or some combination of each.

Skulls of squamates are modified from the ancestral diapsid condition by loss of dermal bone ventral and posterior to the lower temporal opening. This modification has allowed evolution in most lizards and snakes of a mobile skull having movable joints. Such a skull is called a **kinetic skull.** The quadrate, which in other reptiles is fused to the skull, has a joint at its dorsal end, as well as its usual articulation with the lower jaw. In addition, joints in the palate and across the roof of the skull allow the snout to be tilted upward (Figure 26-11). Specialized mobility of the skull enables squamates to seize and manipulate their prey; it also increases the effective closing force of the jaw musculature. The skull of snakes is even more kinetic than that of lizards. Such exceptional skull mobility is considered a major factor in diversification of lizards and snakes.

Suborder Sauria: Lizards   Lizards are an extremely diverse group, including terrestrial, burrowing, aquatic, arboreal and aerial members. Among the more familiar groups in this varied suborder are **geckos** (Gekkonidae) (Figure 26-12), small, agile, mostly nocturnal forms with adhesive toe pads that enable them to walk upside down and on vertical surfaces; **iguanids**

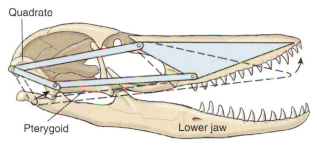

Figure 26-11

Kinetic diapsid skull of a modern lizard (monitor lizard, *Varanus* sp.) showing the joints that allow the snout and upper jaw to move on the rest of the skull. The quadrate can move at its dorsal end and ventrally at both the lower jaw and the pterygoid. The front part of the braincase is also flexible, allowing the snout to be raised. Note that the lower temporal opening is very large with no lower border; this modification of the diapsid condition, common in modern lizards, provides space for expansion of large jaw muscles. The upper temporal opening lies dorsal and medial to the postorbital-squamosal arch and is not visible in this drawing.

Figure 26-12

Tokay, *Gekko gecko*, of Southeast Asia has a true voice and is named after the strident repeated *to-kay, to-kay* call.

Figure 26-13

A large male marine Iguana, *Amblyrhynchus cristatus*, of the Galápagos Islands, feeding underwater on algae. This is the only marine lizard in the world. It has special salt-removing glands in the eye orbits and long claws that enable it to cling to the bottom while feeding on small red and green algae, its principal diet. It may dive to depths exceeding 10 m (33 feet) and remain submerged more than 30 minutes.

(Iguanidae), including most familiar New World lizards, often brightly colored with ornamental crests, frills, and throat fans, and a group that includes the remarkable marine iguana of the Galápagos Islands (Figure 26-13); **skinks** (Scincidae), with elongate bodies and reduced limbs in many species; and **chameleons** (Chamaeleonidae), a group of arboreal lizards, mostly of Africa and Madagascar. Chameleons are entertaining creatures that catch insects with a sticky-tipped tongue that can be flicked accurately and rapidly to a distance greater than the length of their body (Figure 26-14). The great majority of lizards have four limbs and relatively short bodies, but in many the limbs are reduced, and a few such as glass lizards (Figure 26-15) are completely limbless.

Most lizards have movable eyelids, whereas a snake's eyes are permanently covered with a transparent cap. Lizards have keen vision for daylight (retinas rich in both cones and rods; see p. 718 for discussion of color vision), although one group, the nocturnal geckos, has retinas composed entirely of rods. Most lizards have an external ear that snakes lack. The inner ear of lizards is variable in structure, but as with other reptiles, hearing does not play an important role in the lives of most lizards. Geckos are exceptions because males are strongly vocal (to announce territory and discourage approach of other males) and they must, of course, hear their own vocalizations. Other species of lizards vocalize in defensive behavior.

Many lizards live in the world's hot and arid regions. Since their skin lacks glands, water loss by this avenue is much reduced. They produce a semisolid urine with a high content of crystalline uric acid. This excellent mechanism for conserving water is found in other groups living successfully in arid habitats (birds, insects, and pulmonate snails). Some, such as Gila monsters of southwestern United States deserts, store fat in their tails, which they use during drought to provide energy and metabolic water (Figure 26-16). Gila monsters, and their close relatives, beaded lizards, are the only lizards capable of delivering a venomous bite.

Lizards, like nearly all reptiles, are ectotherms, adjusting their body temperature by moving among different microclimates (see Chapter 30, p. 648). Cold climates provide limited opportunities for ectotherms to raise their body temperature to preferred levels. As a result, there are relatively few reptile species in cold climates. However, because ectotherms use considerably less energy than endotherms, reptiles are successful in ecosystems with low productivity and warm climates, such as tropical deserts and grasslands. Thus, ectothermy is not

Figure 26-14

A chameleon snares a dragonfly. After cautiously edging close to its target, the chameleon suddenly lunges forward, anchoring its tail and feet to the branch. A split second later, it launches its sticky-tipped, foot-long tongue to trap the prey. The eyes of this common European chameleon (*Chamaeleo chamaeleon*) are swiveled forward to provide binocular vision and excellent depth perception.

Figure 26-16

Gila monster, *Heloderma suspectum*, of southwestern United States desert regions and the related Mexican beaded lizard are the only venomous lizards known. These brightly colored, clumsy-looking lizards feed principally on birds' eggs, nesting birds, mammals, and insects. Unlike venomous snakes, the Gila monster secretes venom from glands in its lower jaw. The chewing bite is painful to humans but seldom fatal.

Figure 26-15

A glass lizard, *Ophisaurus* sp., of the southeastern United States. This legless lizard feels stiff and brittle to the touch and has an extremely long, fragile tail that readily fractures when the animal is struck or seized. Most specimens, such as this one, have only a partly regenerated tip to replace a much longer tail previously lost. Glass lizards can be readily distinguished from snakes by the deep, flexible groove running along each side of the body. They feed on worms, insects, spiders, birds' eggs, and small reptiles.

an "inferior" characteristic of reptiles, but rather a successful strategy for coping with specific environmental challenges.

Suborder Amphisbaenia: Worm Lizards

The common name "worm lizards" describes a group of highly specialized, burrowing forms that are neither worms nor true lizards but certainly are related to the latter. The name of the suborder literally means "double walk," in reference to their peculiar ability to move backward nearly as effectively as forward. They have elongate, cylindrical bodies of nearly uniform diameter, and most lack any trace of external limbs (Figure 26-17). The skull of amphisbaenians is solidly built and specially shaped to aid in burrowing through the soil. Their soft skin is divided into numerous rings, which combined with absence of visible eyes and ears (both are hidden under skin) make amphisbaenians look like earthworms. The resemblance, although superficial, is the kind of structural convergence that often occurs when two unrelated groups come to occupy similar habitats. Amphisbaenians have an extensive distribution in South America and tropical Africa. In the United States, one species, *Rhineura floridana*, is found in Florida where it is known as the "graveyard snake."

Suborder Serpentes: Snakes

Snakes are limbless and usually lack both pectoral and pelvic girdles (the latter persists as a vestige in pythons, boas, and some other snakes). The numerous vertebrae of snakes, shorter and wider than those of tetrapods, permit quick lateral undulations through grass and over rough terrain. The ribs increase rigidity of the vertebral column, providing more resistance to lateral stresses. Elevation of the neural spine gives the numerous muscles more leverage.

Many lineages of lizards and amphisbaenians exhibit reduction or loss of limbs, but none of these lineages experienced the remarkable radiation of snakes. Because of the lack of or limited kinesis of the amphisbaenian and lizard skulls (Figure 26-11), these squamates are able to consume only relatively small food items. In contrast, the highly kinetic skull and feeding apparatus of snakes, which enable them to eat prey several times their own diameter, are remarkable specializations, which may be responsible for their incredible success. Unlike lizard jaws, the two halves of the lower jaw (mandibles) are joined only by muscles and skin, allowing them to spread widely apart. Many skull bones are so loosely articulated that the entire skull can flex asymmetrically to accommodate oversized prey (Figure 26-18). Since a snake must keep breathing during the slow process of swallowing, its tracheal opening (glottis) is thrust forward between the two mandibles.

The cornea of a snake's eye is permanently protected with a transparent membrane called a spectacle, which, together with

Figure 26-17

A worm lizard of the suborder Amphisbaenia. Worm lizards are burrowing forms with a solidly constructed skull used as a digging tool. The species pictured, *Amphisbaena alba*, is widely distributed in South America.

Labels: Prefrontal, Supratemporal, Quadrate, Pterygoid, Ectopterygoid, Maxilla, Flexible tissues join halves of lower jaw, Lower jaw

A

B

Figure 26-18

A, Lateral view of python skull. Each side of the extremely kinetic skull has several movable joints (labeled) which permit extraordinary movements of the jaw in feeding. The halves of the lower jaw are united by flexible soft tissues, permitting wide separation and independent movement of each side. **B,** The great mobility of the snake jaw and skull elements is evident in this snake swallowing an egg.

reduced eyeball mobility, gives snakes the cold, unblinking stare that many people find unnerving. Most snakes have relatively poor vision, arboreal snakes of the tropical forest being a conspicuous exception (Figure 26-19). Some arboreal snakes possess excellent binocular vision that helps them track prey through branches where scent trails would be impossible to follow.

Snakes have no external ears or tympanic membrane. This condition, together with absence of any obvious response to aerial sounds, led to a widespread opinion that snakes are totally deaf. But snakes do have internal ears, and recent work has shown quite clearly that within a limited range of low frequencies (100 to 700 Hz), hearing in snakes compares favorably with that of most lizards. Snakes are also quite sensitive to vibrations carried in the ground.

Nevertheless, for most snakes it is the chemical senses and not vision and hearing that are employed to hunt their prey. In addition to the usual olfactory areas in the nose, which are not well developed, there are **Jacobson's organs** (vomeronasal organs), a pair of pitlike organs in the roof of the mouth. These are lined with richly innervated chemosensory epithelium. The forked tongue, flicking through the air, picks up scent molecules and conveys them to the mouth; the tongue is then drawn past Jacobson's organs and information is then transmitted to the brain where scents are identified (Figure 26-20).

Snakes have evolved several solutions to the obvious problem of movement without legs. The most typical pattern of movement is **lateral undulation** (Figure 26-21B). Movement follows an S-shaped path, with a snake propelling itself by exerting lateral force against surface irregularities. A snake seems

to "flow," since the moving loops appear stationary with respect to the ground. Lateral undulatory movement is fast and efficient under most but not all circumstances. **Concertina movement** (Figure 26-21A) enables a snake to move in a narrow passage, as when climbing a tree by using irregular channels in the bark. A snake extends forward while bracing S-shaped loops against the sides of the channel. To advance in a straight line as when stalking prey, many heavy-bodied snakes employ **rectilinear movement.** Two or three sections of the body rest on the ground to support the snake's weight. Intervening sections are lifted free of the ground and pulled forward by muscles (shown in red in Figure 26-21C) that originate on ribs and insert on the ventral skin. Rectilinear movement is a slow but effective way of moving inconspicuously toward prey, even when there are no surface irregularities. **Side-winding** is a fourth form of movement that enables desert vipers to move with surprising speed across loose, sandy surfaces with minimum surface contact (Figure 26-21D). The sidewinder rattlesnake moves by throwing its

The Mesozoic World of Dinosaurs

When in 1841 the English anatomist Richard Owen coined the term *dinosaur* ("terrible lizard") to describe fossil Mesozoic reptiles of gigantic size, only three poorly known dinosaur genera were distinguished. But with new and marvelous fossil discoveries quickly following, by 1887 zoologists were able to distinguish two groups of dinosaurs based on differences in structure of the pelvic girdles. The Saurischia ("lizard-hipped") had a simple, three-pronged pelvis with hip bones arranged much as they are in other reptiles. The large bladelike ilium is attached to the backbone by stout ribs. The pubis and ischium extend anteriorly and ventrally respectively, and all three bones meet at the hip socket, a deep opening on the side of the pelvis. The Ornithischia ("bird-hipped") had a somewhat more complex pelvis. The ilium and ischium were arranged similarly in ornithischians and saurischians, but the ornithischian pubis was a narrow, rod-shaped bone with anteriorly and posteriorly directed processes lying alongside the ischium. Oddly, while the ornithischian pelvis, as the name suggests, was similar to that of birds, birds are of the saurischian lineage.

Dinosaurs and their living relatives, the birds, are archosaurs ("ruling lizards"), a group that includes thecodonts (early archosaurs restricted to the Triassic), crocodiles, and pterosaurs (refer to the classification of amniotes on p. 551). As traditionally recognized, dinosaurs are a paraphyletic group because they do not include birds, which are descended from the most recent common ancestor of dinosaurs.

From among the various archosaurian radiations of the Triassic there emerged a thecodont lineage with limbs drawn under the body to provide an upright posture. This lineage gave rise to the earliest dinosaurs of the late Triassic. In *Herrerasaurus,* a bipedal dinosaur from Argentina, we see one of the most distinctive characteristics of dinosaurs: walking upright on pillarlike limbs, rather than on legs splayed outward as with modern amphibians and reptiles. This arrangement allowed the legs to support the great weight of the body while providing an efficient and rapid stride.

Although their ancestry is unclear, two groups of saurischian dinosaurs have been proposed based on differences in feeding habits and locomotion: the carnivorous and bipedal theropods, and the herbivorous and quadrupedal sauropods (sauropodomorphs). *Coelophysis* was an early theropod with a body form typical of all theropods: powerful hindlegs with three-toed feet; long, heavy counterbalancing tail; slender, grasping forelimbs; flexible neck; and a large head with jaws armed with dagger-like teeth. Large predators such as *Allosaurus,* common during the Jurassic, were replaced by even more massively built carnivores of the Cretaceous, such as *Tyrannosaurus,* which reached a length of 14.5 m (47 ft), stood nearly 6 m high, and weighed more than 7200 kg (8 tons). Not all predatory saurischians were massive; several were swift and nimble, such as *Velociraptor* ("speedy predator") of the Upper Cretaceous.

Herbivorous saurischians, the quadrupedal sauropods, appeared in the late Triassic. Although early sauropods were small- and medium-sized dinosaurs, those of the Jurassic and Cretaceous attained gigantic proportions, the largest terrestrial vertebrates ever to have lived. *Brachiosaurus* reached 25 m (82 ft) in length and may have weighed in excess of 30,000 kg (33 tons). Even larger sauropods have been discovered; *Supersaurus* was 43 m (140 ft) long. With long necks and long front legs, sauropods were the first vertebrates adapted to feed on trees. They reached their greatest diversity in the Jurassic and began to decline in overall abundance and diversity during the Cretaceous.

The second group of dinosaurs, the Ornithischia, were all herbivorous. Although more varied, even grotesque, in appearance than saurischians, the ornithischians are united by several derived skeletal features that indicate common ancestry. The huge back-plated *Stegosaurus* of the Jurassic is a well-known example of armored ornithischians, which comprised two of the five major groups of ornithischians. Even more shielded with bony plates than stegosaurs were the heavily built ankylosaurs, "armored tanks" of the dinosaur world. As the Jurassic gave way to the Cretaceous, several groups of unarmored ornithischians appeared, although many bore impressive horns. The steady increase in ornithiscian diversity in the Cretaceous paralleled a concurrent gradual decline in giant sauropods, which had flourished in the Jurassic. *Triceratops* is representative of horned dinosaurs that were common in the Upper Cretaceous. Even more prominent in the Upper Cretaceous were the hadrosaurs, such as *Parasaurolophus,* which are believed to have lived in large herds. Many hadrosaurs had skulls elaborated with crests that probably functioned as vocal resonators to produce species-specific calls.

Dinosaurs likely had considerably more complex parental care than most other reptilian groups. Support for dinosaurs as caregivers can be found by examining the phylogenetic relationships of archosaurs. Two living groups, birds and crocodilians, are members of the clade Archosauria in which dinosaurs are contained (Figure 26-1). Because both crocodilians and birds share well-developed parental care, it is likely that dinosaurs exhibited similar behavior. In addition, fossil nests of dinosaurs are known for several groups. In one case, a fossil adult of the small theropod *Oviraptor* was found apparently guarding a nest of eggs. Originally, it was believed that the adult was a predator on the eggs (*Oviraptor* means "egg seizer"). Later, an embryo in similar eggs was found and identified as *Oviraptor,* indicating that the adult was probably with its own eggs! Examination of baby *Maiasaura* (a hadrosaur) found in a nest revealed considerable wear on their teeth. This suggests that the babies had remained in a nest and were possibly fed by adults during part of their early life.

Sixty-five million years ago, the last Mesozoic dinosaurs became extinct, leaving birds as the only surviving lineage of archosaurs. The demise of dinosaurs coincided with a large asteroid impact on the Yucatan peninsula that would have produced worldwide environmental upheaval. However, this impact hypothesis does not explain why dinosaurs became extinct while most other vertebrate lineages persisted. Many paleontologists suggest the extinctions were caused by changing climates and landforms at the close of the Cretaceous. We continue to be fascinated by the awe-inspiring, often staggeringly large creatures that dominated the Mesozoic era for 165 million years—an incomprehensibly long period of time. Today, inspired by clues from fossils and footprints from a lost world, scientists continue to piece together the puzzle of how various dinosaur groups arose, behaved, and diversified.

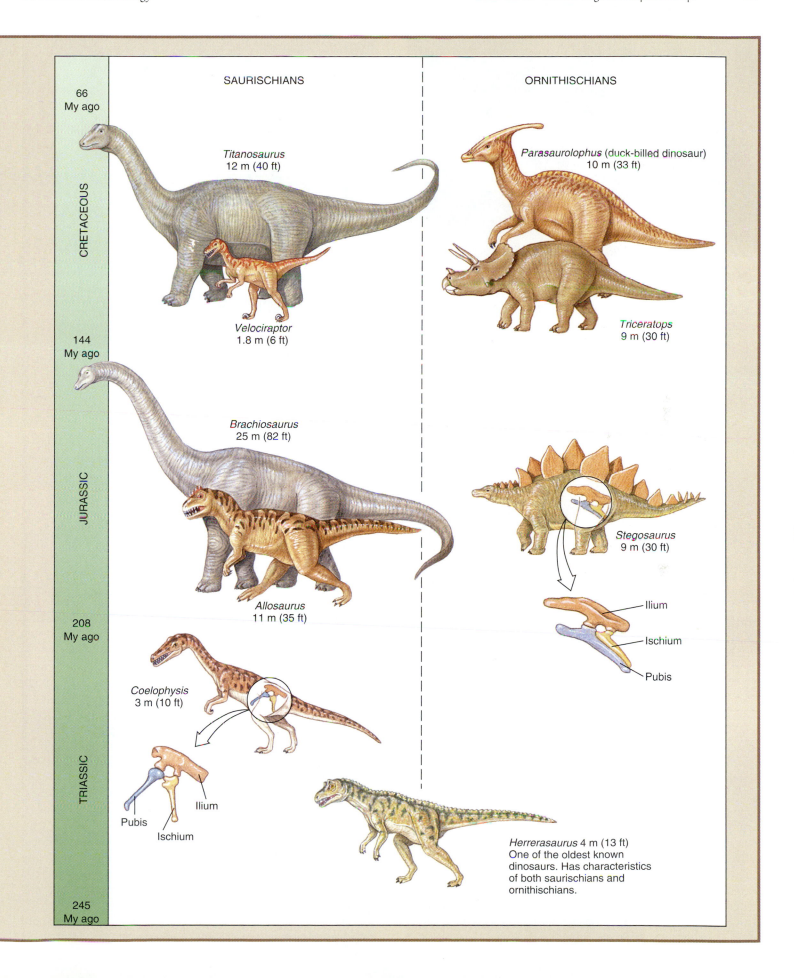

SAURISCHIANS

ORNITHISCHIANS

66
My ago

CRETACEOUS

Titanosaurus
12 m (40 ft)

Parasaurolophus (duck-billed dinosaur)
10 m (33 ft)

Velociraptor
1.8 m (6 ft)

Triceratops
9 m (30 ft)

144
My ago

JURASSIC

Brachiosaurus
25 m (82 ft)

Stegosaurus
9 m (30 ft)

Ilium

Ischium

Pubis

Allosaurus
11 m (35 ft)

208
My ago

TRIASSIC

Coelophysis
3 m (10 ft)

Pubis

Ilium

Ischium

Herrerasaurus 4 m (13 ft)
One of the oldest known
dinosaurs. Has characteristics
of both saurischians and
ornithischians.

245
My ago

Figure 26-19

Parrot snake, *Leptophis ahaetulla.* The slender body of this Central American tree snake is an adaptation for sliding along branches without weighing them down.

Figure 26-20

A blacktail rattlesnake, *Crotalus molossus,* flicks its tongue to smell its surroundings. Scent particles trapped on the tongue's surface are transferred to Jacobson's organs, olfactory organs in the roof of the mouth. Note the heat-sensitive pit organ between the nostril and eye.

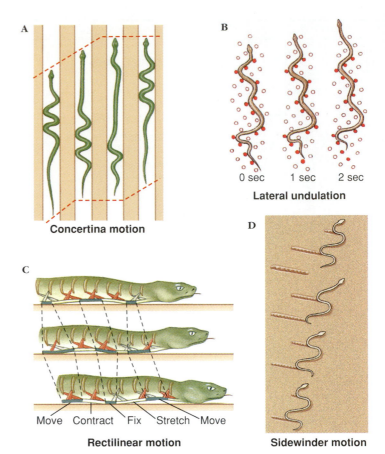

Figure 26-21

Snake locomotion. **A,** Concertina motion. **B,** Lateral undulation. **C,** Rectilinear motion. **D,** Sidewinder motion. Refer to text for explanation.

body forward in loops with its body lying at an angle of about 60 degrees to its direction of travel.

Most snakes capture their prey by grasping it with their mouth and swallowing it while still alive. Because swallowing a struggling, protesting animal may be hazardous, most snakes that swallow prey alive specialize on smaller prey, such as worms, insects, fish, frogs, and, less frequently, small mammals. Many of these snakes, which may be quite fast, locate prey by actively foraging. Snakes that first kill their prey by constriction (Figure 26-22) often specialize on large mammalian prey. The largest constrictors are able to kill and swallow prey as large as deer, leopards, and crocodilians. However, because muscle rearrangements that permit constricting prey also reduce speed of travel, most constrictors tend to ambush prey.

Other snakes kill their prey before swallowing by injecting it with venom. Less than 20 percent of all snakes are venomous, although venomous species outnumber nonvenomous species 4 to 1 in Australia. Venomous snakes are usually divided into five families, based in part on type of fangs. Vipers (family Viperidae) have highly developed, moveable, tubular fangs at the front of their mouth. This family includes American pit vipers and Old World true vipers, which lack facial heat-sensing pits. Among the latter are the common European adder and African puff adder. A second family of venomous snakes (family Elapidae) has short, permanently erect fangs in the front of their mouth. In this group are cobras (Figure 26-23), mambas, coral snakes, and kraits. The highly venomous sea snakes are usually placed in a third family (Hydrophiidae). The fossorial mole vipers (family Atractaspididae) usually have fangs similar to vipers, although their relationship to other venomous snakes is not clear. The very large family Colubridae, which contains most familiar (and nonvenomous) snakes, does include a few snakes that have been responsible for human fatalities. Two examples are the African boomslang and the African twig snake, both rear-fanged snakes that normally use their venom to quiet struggling prey.

Snakes of subfamily Crotalinae within family Viperidae are called **pit vipers** because they possess special heat-sensitive pit

Figure 26-22
Nonvenomous African house snake, *Boaedon fuluginosus*, constricting a mouse before swallowing it.

Figure 26-23
Spectacled, or Indian, cobra, *Naja naja*. Cobras erect the front of the body and flatten the neck as a threat display and before attacking. Although a cobra's strike range is limited, all cobras are dangerous because of the extreme toxicity of their venom.

Opening to pit organ
Outer chamber
Membrane
Inner chamber
Receptor nerve endings packed with mitochondria

Figure 26-24
Pit organ of rattlesnake, a pit viper. Cutaway shows location of a deep membrane that divides the pit into inner and outer chambers. Heat-sensitive nerve endings are concentrated in the membrane.

organs on their heads, located between their nostrils and eyes (Figures 26-20 and 26-24). All best-known North American venomous snakes are pit vipers, such as cottonmouths, and copperheads, and several species of rattlesnakes. The pits are supplied with a dense packing of free nerve endings from the fifth cranial nerve. These respond to radiant energy in the long-wave infrared (5000 to 15,000 nm) and are especially sensitive to heat emitted by warm-bodied birds and mammals that are their food (infrared wavelengths of about 10,000 nm). Some measurements suggest that pit organs can distinguish temperature differences of only 0.003° C from a radiating surface. Pit vipers use pit organs to track warm-blooded prey and to aim strikes with great accuracy, as effectively in total darkness as in daylight. Boa constrictors and pythons also have heat receptors (in their lips), but the anatomy is quite different from that of pit vipers, suggesting that they probably evolved independently.

All vipers have a pair of teeth on the maxillary bones modified as fangs. These lie in a membrane sheath when the mouth is closed. When a viper strikes, a special muscle and bone lever system erects the fangs when its mouth opens (Figure 26-25). The fangs are driven into the prey by the thrust of the strike, and venom is injected into the wound through a channel in the fangs. A viper immediately releases its prey after the bite and waits until it is paralyzed or dead. Then the snake swallows its prey whole. The bite of a pit viper can be dangerous to humans, although in many instances the snake injects very little venom when it bites. Each year in the United States approximately 8000 bites from pit vipers are reported, resulting in only about 12 deaths.

Even the saliva of harmless snakes possesses limited toxic qualities, and it is logical that there was a natural selection for this toxic tendency as snakes evolved. Snake venoms have traditionally been divided into two types. The **neurotoxic** type acts mainly on the nervous system, affecting the optic nerves (causing blindness) or the phrenic nerve of the diaphragm (causing paralysis of respiration). The **hemorrhagin** type breaks down red blood corpuscles and blood vessels and produces extensive hemorrhaging of blood into tissue spaces. In fact, most snake venoms are complex mixtures of various fractions that attack different organs in specific ways; they seldom can be assigned categorically to one or the other of the traditional types. In addition, all venoms possess enzymes that speed up the process of digestion.

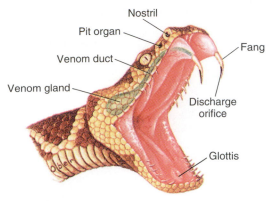

Figure 26-25
Head of rattlesnake showing the venom apparatus. The venom gland, a modified salivary gland, is connected by a duct to the hollow fang.

The toxicity of a venom is measured by the median lethal dose on laboratory animals (LD$_{50}$). By this standard the venoms of the Australian tiger snake and some of the sea snakes appear to be the most deadly of venomous drop for drop. However, several larger snakes are more dangerous. The aggressive king cobra, which may exceed 5.5 m in length, is the largest and perhaps the most dangerous of all venomous snakes. It is estimated worldwide that 50,000 to 60,000 people die from snakebite each year. Most deaths occur in India, Pakistan, Myanmar, and nearby countries where poorly shod people frequently come into contact with venomous snakes and frequently do not get immediate medical attention following snakebite. The snakes primarily responsible for deaths in these areas are Russell's viper, saw-scaled viper, and several species of cobras.

The LD$_{50}$ (median lethal dose) has been the standardized procedure for assaying toxicity of drugs; it was originally developed in the 1920s by pharmacologists. In practice, small samples of laboratory animals, usually mice, are exposed to a graded series of doses of the drug or toxin. The dose that kills 50% of the animals in the test period is recorded as the LD$_{50}$. Expensive and time consuming, this classical procedure is being replaced by alternative methods that greatly reduce the number of animals needed. Among these alternatives are cytotoxicity tests that evaluate the ability of test substances to kill cells, and toxikinetic procedures that measure interaction of a drug or toxin with a living system.

Most snakes are **oviparous** (L. *ovum,* egg, + *parere,* to bring forth) species that lay their shelled, elliptical eggs beneath rotten logs, under rocks, or in holes in the ground. Most of the remainder, including all American pit vipers, except the tropical bushmaster, are ovoviviparous (L. *ovum,* egg, + *vivus,* living, + *parere,* to bring forth), giving birth to well-formed young. Very few snakes are viviparous (L. *vivus,* living, + *parere,* to bring forth); in these snakes a primitive placenta forms, permitting exchange of materials between the embryonic and maternal bloodstreams. Snakes are able to store sperm and can lay several clutches of fertile eggs at long intervals after a single mating.

Figure 26-26
Tuatara, *Sphenodon* sp., a living representative of order Sphenodonta. This "living fossil" reptile has, on top of the head, a well-developed parietal "eye" with retina, lens, and nervous connections to the brain. Although covered with scales, this third eye is sensitive to light. The parietal eye may have been an important sense organ in early reptiles. The tuatara is found today only on certain islands off the coastline of New Zealand.

Order Sphenodonta: *Tuataras*

The order Sphenodonta is represented by two living species of genus *Sphenodon* (Gr. *sphenos,* wedge, + *odontos,* tooth) of New Zealand (Figure 26-26). Tuataras are sole survivors of the sphenodontid lineage that radiated modestly during the early Mesozoic era but declined toward the end of the Mesozoic. Several species of tuataras were once widespread throughout the two main islands of New Zealand but the two living species are now restricted to small islets of Cook Strait and off the northeast coast of North Island. On some of these islands, under protection by the New Zealand government, they are prospering.

Tuataras are lizardlike forms 66 cm long or less that live in burrows often shared with petrels. They are slow-growing animals with long lives; one is recorded to have lived 77 years.

Tuataras have captured the interest of zoologists because of numerous features that are almost identical to those of Mesozoic animals that lived 200 million years ago. These features include a diapsid skull with two temporal openings bounded by complete arches. Tuataras also bear a well-developed median parietal eye complete with elements of cornea, lens, and retina (although since it is buried beneath opaque skin this "third eye" can register only changes in light intensity, and its function, if any, remains unknown). In many other respects *Sphenodon* resembles diapsids of the early Mesozoic. *Sphenodon* represents one of the slowest rates of morphological evolution known among vertebrates.

Order Crocodilia: *Crocodiles and Alligators*

Modern crocodilians are the only surviving reptiles of the archosaurian lineage that gave rise to the great Mesozoic radiation of dinosaurs and their kin and to birds. Although modern crocodiles belong to a lineage that began its radiation in the late

Classification of Amniotes Emphasizing Early Amniotes and Living Reptiles

The following Linnaean classification is adapted primarily from Carroll (1988)* and agrees with the genealogical relationships of living reptiles shown in Figure 26-2. Extinct groups are indicated by a dagger.

Subclass Anapsida (a-nap′se-duh) (Gr. *an*, without, + *apsis*, arch): **anapsids.** Amniotes having some primitive features, such as a skull with no temporal opening.

> **Order Captorhinida**[†] (kap-to-rine′uh-duh) (Gr. *kapto*, to seize, + *rhinos*, nose). Amniotes of Carboniferous and early Permian.
>
> **Order Testudines** (tes-tu′din-eez) (L. *testudo*, tortoise) **(Chelonia): turtles.** Body in a bony case of dorsal carapace and ventral plastron; jaws with horny beaks instead of teeth; vertebrae and ribs fused to overlying carapace; tongue not extensible; neck usually retractable; approximately 300 species.

Subclass Diapsida (di-ap′se-duh) (Gr. *di*, double, + *apsis*, arch): **diapsids.** Amniotes having a skull with two temporal openings.

> **Superorder Lepidosauria** (lep-i-do-sor′ee-uh) (Gr. *lepidos*, scale, + *sauros*, lizard). Diapsid lineage appearing in the Triassic; characterized by sprawling posture; no bipedal specializations; diapsid skull often modified by loss of one or both temporal arches; transverse cloacal slit, skin shed in one piece.
>
> > **Order Squamata** (skwa-ma′ta) (L. *squamatus*, scaly, + *ata*, characterized by): **snakes, lizards, amphisbaenians.** Skin of horny epidermal scales or plates, which is shed; quadrate movable; skull kinetic (except amphisbaenians); vertebrae usually concave in front; paired copulatory organs.
> >
> > > **Suborder Lacertilia** (lay-sur-till′ee-uh) (L. *lacerta*, lizard) **(Sauria): lizards.** Body slender, usually with four limbs; rami of lower jaw fused; eyelids movable; external ear present; this paraphyletic suborder contains approximately 4600 species.
> > >
> > > **Suborder Amphisbaenia** (am′fis-bee′nee-a) (L. *amphis*, double, + *baina*, to walk): worm lizards. Body elongate and of nearly uniform diameter; no legs (except one genus with short front legs); skull bones interlocked for burrowing (not kinetic); limb girdles vestigial; eyes hidden beneath skin; only one lung; approximately 160 species.
> > >
> > > **Suborder Serpentes** (sur-pen′tes) (L. *serpere*, to creep): **snakes.** Body elongate; limbs, ear openings, and middle ear absent; mandibles joined anteriorly by ligaments; eyelids fused into transparent spectacle; tongue forked and protrusible; left lung reduced or absent; approximately 2900 species.
> >
> > **Order Sphenodonta** (sfen′o-don′tuh) (Gr. *sphen*, wedge, + *odontos*, tooth) **(Rhynchocephalia).** Primitive diapsid skull; vertebrae biconcave; quadrate immovable; parietal eye present; two extant species in *Sphenodon*.

*Carroll, R. L. 1988. *Vertebrate paleontology and evolution*. New York, W. H. Freeman and Company.

> **Superorder Ichthyosauria**[†] (ik′thee-o-sor′ee-uh) (Gr. *ichthys*, fish, + *sauros*, lizard). Mesozoic marine dolphin-shaped diapsids with reduced limbs.
>
> **Superorder Sauropterygia**[†] (sor-op-ter-ig′ee-uh) (Gr. *sauros*, lizard, + *pteryginos*, winged). Mesozoic marine reptiles.
>
> > **Order Plesiosauria**[†] (plees′ee-o-sor′ee-uh) (Gr. *plesios*, near, + *sauros*, lizard). Long-necked Mesozoic marine reptiles with paddlelike limbs.
>
> **Superorder Archosauria** (ark′uh-sor′ee-uh) (Gr. *archōn*, ruling, + *sauros*, lizard). Advanced diapsids, mostly terrestrial, but some specialized for flight; gizzard present.
>
> > **Order Thecodontia**[†] (thek′uh-dont′ee-uh) (Gr. *theke*, encased [socket], + *odontos*, tooth). Paraphyletic group of dominant Triassic archosaurs with teeth set in sockets; bipedal tendency.
> >
> > **Order Crocodilia** (crok′uh-dil′ee-uh) (L. *crocodilus*, crocodile): **crocodilians.** Skull elongate and massive; nares terminal; secondary palate present; four-chambered heart; vertebrae usually concave in front; forelimbs usually of five digits; hindlimbs of four digits; quadrate immovable; advanced social behavior; 23 species.
> >
> > **Order Pterosauria**[†] (ter′uh-sor′ee-uh) (Gr. *pteron*, winged, + *sauros*, lizard). Flying, Mesozoic archosaurs with membranous wings; extensive radiation.
> >
> > **Order Saurischia** (sor-ish′ee-uh) (Gr. *sauros*, lizard, + *ischion*, hip). Mesozoic dinosaurs; bipedal carnivores and quadrupedal herbivores; primitive (reptilian) hip structure.
> >
> > > **Suborder Sauropodomorpha**[†] (sor′uh-pod-uh-morf′uh) (Gr. *sauros*, lizard, + *podos*, foot, + *morphē*, form). Herbivorous saurischians including Mesozoic giants such as *Brachiosaurus*, *Apatosaurus* and *Diplodocus*.
> > >
> > > **Suborder Theropoda** (the-ro′po-duh) (Gr. *thēr*, wild beast, + *podos*, foot). Carnivorous saurischians including huge predators such as *Tyrannosaurus* and small, agile predators such as *Deinonychus* and *Velociraptor*. Birds are descended from this lineage.
> >
> > **Order Ornithischia**[†] (orn′uh-thish′ee-uh) (Gr. *ornis*, bird, + *ischion*, hip). Mesozoic dinosaurs; bipedal and quadrupedal herbivores such as *Stegosaurus*, *Triceratops*, and *Parasaurolophus*; advanced (birdlike) hip structure.

Subclass Synapsida (sin-ap′si-duh) (Gr. *syn*, together, + *apsis*, arch). Amniotes having skull with one pair of lateral temporal openings.

> **Order Pelycosauria**[†] (pel′uh-ko-sor′ee-uh) (Gr. *pelyx*, wooden bowl, + *sauros*, lizard). Carboniferous and Permian synapsids with many primitive amniote characteristics; carnivorous and herbivorous.
>
> **Order Therapsida** (ther-ap′-si-duh) (Gr. *thēr*, wild beast, + *apsis*, arch). Permian and Triassic synapsids with many mammal-like characteristics; both carnivores and herbivores. Living mammals are descended from this lineage.

Cretaceous period, they differ little in structural details from primitive crocodilians of the early Mesozoic. Having remained mostly unchanged for nearly 200 million years, crocodilians face an uncertain future in a world dominated by humans. Modern crocodilians are divided into three families: alligators and caimans, mostly a New World group; crocodiles, which are widely distributed and include the saltwater crocodile, one of the largest living reptiles; and gavials, represented by a single species in India and Burma.

All crocodilians have an elongate, robust, well-reinforced skull and massive jaw musculature arranged to provide a wide gape and rapid, powerful closure. Teeth are set in sockets, a type of dentition called **thecodont** that was typical of all archosaurs as well as the earliest birds. Another adaptation, found in no other vertebrate except mammals, is a complete secondary palate. This innovation allows crocodilians to breathe when their mouth is filled with water or food (or both). Crocodilians also share with birds and mammals a four-chambered heart with completely divided atria and ventricles.

The estuarine crocodile (*Crocodylus porosus*), found in southern Asia, and the Nile crocodiles (*C. niloticus;* Figure 26-27A) grow to great size (adults weighing 1000 kg have been reported) and are swift and aggressive. Crocodiles are known to attack animals as large as cattle, deer, and people. Alligators (Figure 26-27B) are usually less aggressive than crocodiles and far less dangerous to humans. In the United States, *Alligator mississippiensis* (Figure 26-27B) is the only species of alligator; *Crocodylus acutus,* restricted to extreme southern Florida, is the only species of crocodile. Large alligators are powerful animals nevertheless, and adults have almost no enemies but humans. The chink in their formidable armor is their developmental stages. Nests left unguarded by the mother are almost certain to be discovered and raided by any of several mammals that relish eggs, and the young hatchlings may be devoured by large fishes.

Alligators are able to make definite vocalizations. Male alligators give loud bellows in the mating season. Alligators and crocodiles are oviparous. Usually 20 to 50 eggs are laid in a mass of dead vegetation and guarded by the mother. The mother hears vocalizations from hatching young and responds by opening the nest to allow the hatchlings to escape. As with many turtles and some lizards, incubation temperature of the eggs determines sex ratio of the offspring. However, unlike turtles (p. 541) low nest temperatures produce only females, whereas high nest temperatures produce only males. This results in highly unbal-

A

B

Figure 26-27

Crocodilians. **A,** Nile crocodile (*Crocodylus niloticus*) basking. The fourth tooth of the lower jaw fits outside the slender upper jaw; alligators lack this feature. **B,** American alligator (*Alligator mississippiensis*), an increasingly noticeable resident of rivers, bayous, and swamps of the southeastern United States.

anced sex ratios in some areas. For example, in one study area in Louisiana, female hatchlings outnumbered males five to one.

Crocodiles and alligators can be distinguished on the basis of head morphology. Crocodiles have a relatively narrow snout, and when their mouths are closed, the fourth lower jaw tooth is visible. Alligators generally have a broader snout, and their fourth lower jaw tooth is hidden by fitting into a notch in the upper jaw (Figure 26-27). Gavials have very narrow snouts, and are largely fish eaters.

SUMMARY

Reptiles diverged phylogenetically from a group of labyrinthodont amphibians during the late Paleozoic era, some 300 million years ago. Their success as terrestrial vertebrates is attributed in large part to evolution of the amniotic egg, which, with its three extraembryonic membranes, provided support for full embryonic development within the protection of a shell. Thus reptiles could lay their eggs on land. Reptiles are also distinguished from amphibians by their dry, scaly skin, which limits water loss; more powerful jaws; internal fertilization; and advanced circulatory, respiratory, excretory, and nervous systems. Like amphibians, reptiles are ectotherms, but most exercise considerable behavioral control over their body temperature.

Before the end of the Paleozoic era, amniotes began a radiation that separated into three groups: anapsids, which gave rise to the turtles; synapsids, a lineage that led to the modern mammals; and diapsids, which led to all other reptiles and to birds. The great burst of reptilian radiation during the Mesozoic era produced a worldwide fauna of great diversity, including the extinct ichthyosaurs, plesiosaurs, pterosaurs, and dinosaurs.

Turtles (order Testudines), with their distinctive shells, have changed little in design since the Triassic period. Turtles are a small group of long-lived terrestrial, semiaquatic, aquatic, and marine species. They lack teeth. All are oviparous and all, including marine forms, bury their eggs.

Lizards, snakes, and worm lizards (order Squamata) make up 95% of all living reptiles. Lizards (suborder Lacertilia) are a diversified and successful group adapted for walking, running, climbing, swimming, and burrowing. They are distinguished from snakes by typically having two pairs of limbs (some species are limbless), united lower jaw halves, movable eyelids, and external ears. Many lizards are well adapted for survival under hot and arid desert conditions.

Worm lizards (suborder Amphisbaenia) are a small tropical group of limbless squamates highly adapted for burrowing.

Snakes (suborder Serpentes), in addition to being entirely limbless, are characterized by their elongate bodies and a highly kinetic skull that permits them to swallow whole prey that may be much larger than the snake's own diameter. Most snakes rely on their chemical senses, especially Jacobson's organs, to hunt prey, rather than on weakly developed visual and auditory senses. Two groups of snakes (pit vipers and boids) have unique infrared-sensing organs for tracking warm-bodied prey. Some snakes are venomous.

The tuataras of New Zealand (order Sphenodonta) are relics and sole survivors of a group that otherwise disappeared 100 million years ago. They bear several features that are almost identical to those of Mesozoic fossil diapsids.

Crocodiles, alligators, and caimans (order Crocodilia) are the only living reptilian representatives of the archosaurian lineage that gave rise to the extinct dinosaurs and the living birds. Crocodilians have several adaptations for a carnivorous, semiaquatic life, including a massive skull with powerful jaws, and a secondary palate. They have the most complex social behavior of any reptile.

REVIEW QUESTIONS

1. What were the three major amniote radiations of the Mesozoic and from which lineage or lineages did the birds and mammals descend? How could you distinguish the skulls characteristic of these different radiations?
2. What changes in egg design allowed reptiles to lay eggs on land? Why is their egg often called an "amniotic" egg? What are "amniotes"?
3. Why are reptiles considered a paraphyletic rather than a monophyletic group? How have cladistic taxonomists revised the content of this taxon to make it monophyletic?
4. Describe ways in which reptiles are more functionally or structurally suited for terrestriality than amphibians.
5. What are the main characteristics of reptile skin and how would you distinguish it from frog skin?
6. Describe the principal structural features of turtles that would distinguish them from any other reptilian order.
7. How might nest temperature affect egg development in turtles? In crocodilians?
8. What is meant by a "kinetic" skull and what benefit does it confer? How are snakes able to eat such large prey?
9. In what ways are the special senses of snakes similar to those of lizards, and in what ways have they evolved for specialized feeding strategies?

10. What are amphisbaenians? What morphological adaptations do they have to aid in burrowing?
11. Distinguish ornithischian and saurischian dinosaurs based on their hip anatomy. Which lineage gave rise to birds?
12. How do snakes and crocodilians breathe when their mouths are full of food?
13. What is the function of Jacobson's organ of snakes?
14. What is the function of the "pit" of pit vipers?
15. What is the difference in structure or location of the fangs of a rattlesnake, a cobra, and an African boomslang?
16. Most snakes are oviparous, but some are ovoviviparous or viviparous. What do these terms mean and what would you have to know to be able to assign a particular snake to one of these reproductive modes?
17. Describe how a snake moves by lateral undulation. Why might this form of locomotion be inefficient on an unstable surface (such as sand) or a surface lacking irregularities? What forms of locomotion would work for a snake under these conditions?
18. Why are tuataras (Sphenodon) of special interest to biologists? Where would you have to go to see one in its natural habitat?
19. From which diapsid lineage have crocodilians descended? What other major fossil and living vertebrate groups belong to this same lineage? In what structural and behavioral ways are crocodilians more advanced than other living reptiles?

SELECTED REFERENCES

Alexander, R. M. 1991. How dinosaurs ran. Sci. Am. **264:**130–136 (April). *By applying the techniques of modern physics and engineering, a zoologist calculates that the large dinosaurs walked slowly but were capable of a quick run; none required the buoyancy of water for support.*

Alvarez, W., and F. Asaro. 1990. An extraterrestrial impact. Sci. Am. **263:**78–84 (Oct.). *This article and an accompanying article by V. E. Courtillot, "A volcanic eruption," present opposing interpretations of the cause of the Cretaceous mass extinction that led to the demise of the dinosaurs.*

Appenzeller, T. 1999. *T. rex* was fierce, yes, but feathered, too. Science **285:**2052–2053. *Evidence from fossils from Chinese beds suggest that several predacious dinosaurs related to bird lineage had feathers at some time in their lives.*

Cogger, H. G., and R. G. Zweifel (eds). 1998. Encyclopedia of reptiles and amphibians. San Diego, Academic Press. *This comprehensive, up-to-date, and lavishly illustrated volume was written by some of the best-known herpetologists in the field.*

Crews, D. 1994. Animal sexuality. Sci. Am. **270:**108–114 (Jan.). *The reproductive strategies of reptiles, including nongenetic sex determination, provide insights into the origins and functions of sexuality.*

Erickson, G. M. 1999. Breathing life into *Tyrannosaurus rex*. Sci. Am. **281:**42–49 (Sept.). *Present evidence suggests that* T. rex *was gregarious, but whether they were predominantly scavengers or predators is still unsettled.*

Greene, H. W. 1997. Snakes: The evolution of mystery in nature. Berkeley, University of California Press. *Beautiful photographs accompany a well-written volume for the scientist or novice.*

Halliday, T. R., and K. Adler (eds). 1986. The encyclopedia of reptiles and amphibians. New York, Facts on File, Inc. *Comprehensive and beautifully illustrated treatment of reptilian groups with helpful introductory sections on origins and characteristics.*

Hutchinson, J. R., and M. Garcia. 2002. *Tyrannosaurus* was not a fast runner. Nature **415:**1018–1021. *Analysis concludes that* Tyrannosaurus *would have had less than half enough leg-muscle mass to run and therefore could only walk.*

King, G. 1996. Reptiles and herbivory. London, Chapman & Hall. *Explains the adaptations reptiles use in obtaining nutrients from a herbivorous diet.*

Lillywhite, H. B. 1988. Snakes, blood circulation and gravity. Sci. Am. **259:**92–98 (Dec.). *Even long snakes are able to maintain blood circulation when the body is extended vertically (head-up posture) through special circulatory reflexes that control blood pressure.*

Lohmann, K. J. 1992. How sea turtles navigate. Sci. Am. **266:**100–106 (Jan.). *Recent evidence suggests that sea turtles use the earth's magnetic field and the direction of ocean waves to navigate back to their natal beaches to nest.*

Mattison, C. 1995. The encyclopedia of snakes. New York, Facts on File, Inc. *Generously illustrated book treating evolution, physiology, behavior, and classification of snakes.*

Norman, D. 1991. Dinosaur! New York, Prentice-Hall. *Highly readable account of the life and evolution of dinosaurs, with fine illustrations.*

Pough, F. H., R. M. Andrews, J. E. Cadle, M. L. Crump, A. H. Savitzky, and K. D. Wells. 2001. Herpetology, ed. 2. Upper Saddle River, New Jersey, Prentice-Hall. *A comprehensive textbook treating diversity, physiology, behavior, ecology, and conservation of reptiles and amphibians.*

Zug, G. R., L. J. Vitt, and J. P. Caldwell. 2001. Herpetology: an introduction to the biology of amphibians and reptiles. San Diego, Academic Press.

ZOOLOGY LINKS TO THE INTERNET

Visit the textbook's Online Learning Center at www.mhhe.com/zoology to find live Internet links for each of the topics listed here.

Class Reptilia
Order Testudines
Marine Turtles
Order Crocodilia
Suborder Serpentes
Superorder Archosauria and Related Mesozoic Reptiles
Dissection Guides for Reptiles
Conservation Issues Concerning Reptiles

Storks during
night migration.

Birds

Phylum Chordata
Class Aves

Long Trip to a Summer Home

Perhaps it was ordained that birds, having mastered flight, would use this power to make the long seasonal migrations that have captured human wonder and curiosity. For the advantages of migration are many. Moving between southern wintering regions and northern summer breeding regions with long summer days and an abundance of insects provides parents with ample food to rear their young. Predators of birds are not so abundant in the far North, and a brief once-a-year appearance of vulnerable young birds does not encourage buildup of predator populations. Migration also vastly increases the amount of space available for breeding and reduces aggressive territorial behavior. Finally, migration favors homeostasis—the balancing of physiological processes that maintains internal stability—by allowing birds to avoid climatic extremes.

Still, the wonder of the migratory pageant remains, and there is much yet to learn about its mechanisms. What times migration, and what determines that each bird shall store sufficient fuel for the journey? How did the sometimes difficult migratory routes originate, and what cues do birds use in navigation? And what was the origin of this instinctive force to follow the retreat of winter northward? For it is instinct that drives the migratory waves in spring and fall, instinctive blind obedience that carries most birds successfully to their northern nests, while countless others fail and die, winnowed by the ever-challenging environment.

Of the vertebrates, birds of class Aves (ay'veez) (L. pl. of *avis,* bird) are the most noticeable, the most melodious, and many think the most beautiful. With more than 9000 species distributed over nearly the entire earth, birds outnumber any other vertebrate group except fishes. Birds are found in forests and deserts, in mountains and prairies, and on all oceans. Four species are known to have visited the North Pole, and one, a skua, was seen at the South Pole. Some birds live in total blackness in caves, finding their way by echolocation, and others dive to depths greater than 45 m to prey on aquatic life. The "bee" hummingbird of Cuba, weighing in at only 1.8 g, is one of the smallest vertebrate endotherms.

The single unique feature that distinguishes birds from other animals is their flight feathers. If an animal has feathers, it is a bird; if it lacks feathers, it is not a bird. No other living vertebrate group bears such an easily recognizable and foolproof identification tag. Feathers also were present in some theropod dinosaurs, although these feathers were not capable of supporting flight.

There is great uniformity of structure among birds. Despite approximately 150 million years of evolution, during which they proliferated and adapted to specialized ways of life, we have little difficulty recognizing a living bird as a bird. In addition to feathers, all birds have forelimbs modified into wings (although not always used for flight); all have hindlimbs adapted for walking, swimming, or perching; all have keratinized beaks; and all lay eggs. The reason for this great structural and functional uniformity is that birds evolved into flying machines. This fact greatly restricts diversity, so much more evident in other vertebrate classes. For example, birds do not approach the diversity seen in their endothermic evolutionary peers, mammals, a group that includes forms as dissimilar as whales, porcupines, bats, and giraffes.

A bird's entire anatomy is designed around flight. An airborne life for a large vertebrate is a highly demanding evolutionary challenge. A bird must, of course, have wings for support and propulsion. Bones must be light yet serve as a rigid airframe. The respiratory system must be highly efficient to meet the intense metabolic demands of flight and serve also as a thermoregulatory device to maintain a constant body temperature. A bird must have a rapid and efficient digestive system to process an energy-rich diet; it must have a high metabolic rate; and it must have a high-pressure circulatory system. Above all, birds must have a finely tuned nervous system and acute senses, especially superb vision, to handle the complex problems of headfirst, high-velocity flight.

ORIGIN AND RELATIONSHIPS

Approximately 147 million years ago, a flying animal drowned and settled to the bottom of a shallow marine lagoon in what is now Bavaria, Germany. It was rapidly covered with a fine silt and eventually fossilized. There it remained until discovered in 1861 by a workman splitting slate in a limestone quarry. The fossil was approximately the size of a crow, with a skull not unlike that of modern birds except that the beaklike jaws bore small bony teeth set in sockets like those of reptiles (Figure 27-1). The skeleton was decidedly reptilian with a long bony tail, clawed

Position in the Animal Kingdom

Birds are a lineage of endothermic, diapsid amniotes that evolved flight in the Jurassic period of the Mesozoic. Phylogenetically, they are most closely related to theropod dinosaurs, a group of bipedal carnivores with birdlike skeletal characteristics. Their closest living relatives are crocodilians (refer to p. 538 for a discussion of the shared ancestry of birds and crocodiles). The morphological characteristics and great uniformity of structure of birds relate almost entirely to the strict demands of flight, and the mobility that flight provides is responsible for many distinctive aspects of their behavior and ecology.

Biological Contributions

1. Feathers distinguish birds from all other animals. The evolution of feathers was the single most important event leading to the capacity for flight in birds.
2. In addition to feathers, several other essential adaptations contribute to the two prime requirements for flight: increase in power and decrease in weight. These adaptations include forelimbs modified as strong wings, hollow bones, keratinized bill (rather than heavy jaws and teeth), endothermy, high metabolic rate (six to ten times as high as reptiles of similar weight and body temperature), large hearts and high-pressure circulation, highly efficient respiratory system, keen vision, and excellent neuromuscular coordination.
3. Birds occupy almost every available habitat on the earth's surface and, within the constraints imposed by requirements for flight, have radiated modestly in body form, especially in bill adaptations.
4. The unparalleled mobility of birds has enabled many to benefit from the advantages of making seasonal and long-distance migrations. Migration enables birds to secure seasonal habitats most beneficial for breeding, finding food, avoiding predators, and reducing interspecific competition.

fingers, and abdominal ribs. It might have been classified as a theropod dinosaur except that it carried an unmistakable imprint of **feathers,** those marvels of biological engineering that only birds possess. *Archaeopteryx lithographica* (ar-kee-op'ter-ix lith-o-graf'e-ca, Gr., meaning "ancient wing inscribed in stone"), as the fossil was named, was an especially fortunate discovery because the fossil record of birds is disappointingly meager. The finding was also dramatic because it demonstrated beyond reasonable doubt the phylogenetic relatedness of birds and theropod dinosaurs.

Zoologists had long recognized the similarity of birds and reptiles. The skulls of birds and reptiles abut against the first neck vertebra by a single occipital condyle (mammals have two condyles). Birds and reptiles have a single middle ear bone, the stapes (mammals have three middle ear bones). Birds and reptiles have a lower jaw composed of five or six bones, whereas the lower jaw of mammals has one bone, the dentary. Birds and reptiles excrete their nitrogenous wastes as uric acid whereas mammals excrete theirs as urea. Birds and most reptiles lay sim-

Figure 27-1
Archaeopteryx, a 147-million-year-old ancestor of modern birds.
A, Cast of the second and most nearly perfect fossil of *Archaeopteryx,*
which was discovered in a Bavarian stone quarry. Seven specimens of
Archaeopteryx have been discovered, the most recent one in 1992.
B, Reconstruction of *Archaeopteryx.*

ilar yolked eggs with the early embryo developing on the surface by shallow cleavage divisions.

The distinguished English zoologist Thomas Henry Huxley was so impressed with these and many other anatomical and physiological affinites that he called birds "glorified reptiles" and classified them with a group of dinosaurs called theropods that displayed several birdlike characteristics (Figures 27-2 and 27-3). Theropod dinosaurs share many derived characters with birds, the most obvious of which is the elongate, mobile, S-shaped neck.

Dromeosaurs, a group of theropods that includes *Velociraptor,* share many additional derived characters with birds, including a furcula (fused clavicles) and lunate wrist bones that permit swiveling motions used in flight (Figure 27-3). Additional evidence linking birds to dromeosaurs comes from recently described fossils from late Jurassic and early Cretaceous deposits in Liaoning Province, China. These spectacular fossils, including *Protarchaeopteryx* and *Caudipteryx,* are dromeosaur-like theropods, but with feathers! It is unlikely these feathered dinosaurs could fly, however, as they had short forelimbs and symmetrical vaned feathers (the flight feathers of modern flying birds are asymmetrical). Clearly these feathers were not used for flight or thermoregulation; perhaps they were used as snares to trap or knock down prey or were colorful and used in social displays. Additional theropod dinosaurs recently unearthed in China, such as *Sinosauropteryx,* are covered with filaments that appear to be homologous with feathers. The filamentous covering of these dinosaurs likely served as insulation and was a precursor to vaned feathers. Other fossils from Spain and Argentina

of birds more derived than *Archaeopteryx* document the development of the keeled sternum and alula (Figure 27-7), loss of teeth, and fusion of bones characteristic of modern birds. Clearly, a phylogenetic approach to classification would include birds with theropod dinosaurs. With this view, dinosaurs are not extinct—they are with us today as birds!

Living birds (Neornithes) are divided into two groups: (1) **Paleognathae** (Gr. *palaios,* ancient, + *gnathos,* jaw), the large flightless ostrichlike birds and the kiwis, often called ratite birds, which have a flat sternum with poorly developed pectoral muscles, and (2) **Neognathae** (Gr. *neos,* new, + *gnathos,* jaw), flying birds that have a keeled sternum on which powerful flight muscles insert. This division assumes flightless ratites form a monophyletic group. However, evidence supporting this grouping is weak, and relationships of ratites to other birds are controversial. Ostrichlike paleognathids clearly have descended from flying ancestors. Furthermore, not all neognathous birds can fly and many of them even lack keels (Figure 27-4). Flightlessness has appeared independently among many groups of birds; the fossil record reveals flightless wrens, pigeons, parrots, cranes, ducks, auks, and even a flightless owl. Penguins are flightless although they use their wings to "fly" through water (p. 200). Flightlessness almost always has evolved on islands where few terrestrial predators are found. Flightless birds living on continents today are the large paleognathids (ostrich, rhea, cassowary, emu), which can run fast enough to escape predators. An ostrich can run 70 km (42 miles) per hour, and claims of speeds of 96 km (60 miles) per hour have been made. The evolution and dispersal of flightless birds are discussed on pp. 110 and 783, respectively.

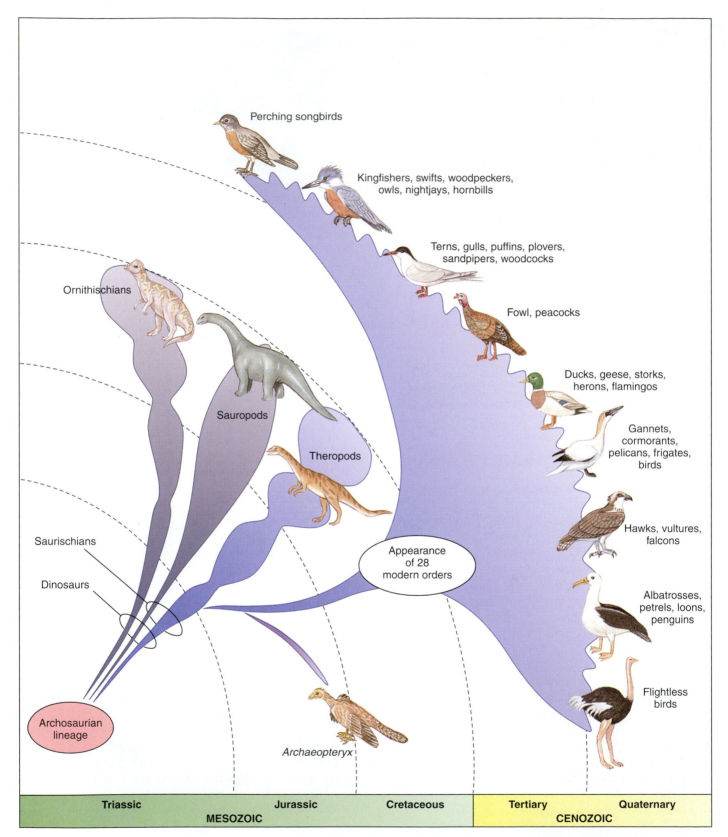

Figure 27-2

Evolution of modern birds. Of 28 living bird orders, 9 of the largest are shown. The earliest known bird, *Archaeopteryx,* lived in the Upper Jurassic, about 147 million years ago. *Archaeopteryx* uniquely shares many specialized aspects of its skeleton with the smaller theropod dinosaurs and is considered to have evolved within the theropod lineage. Evolution of modern bird orders occurred rapidly during the Cretaceous and early Tertiary periods.

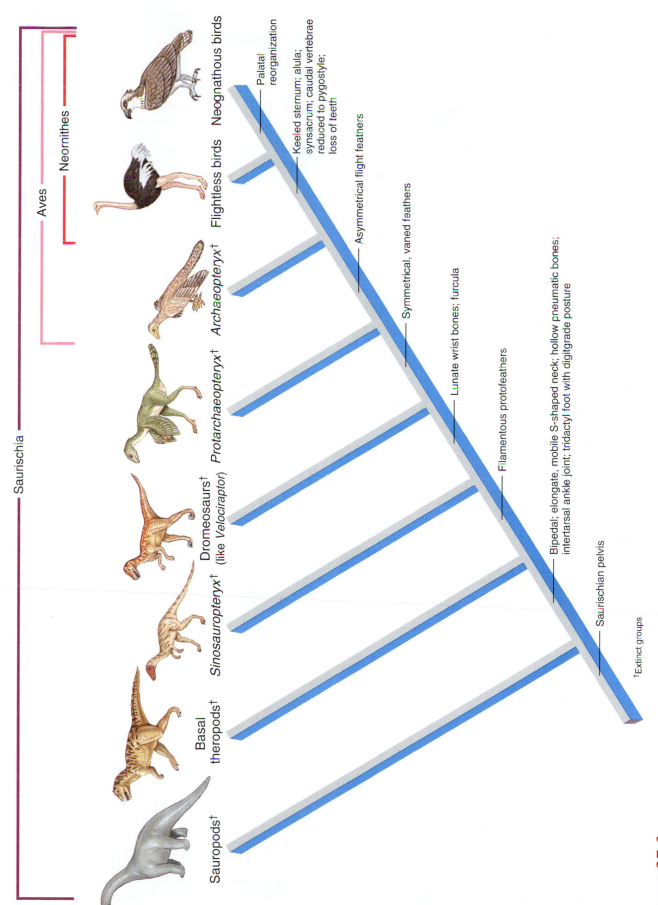

Neognathous birds

Flightless birds

Archaeopteryx†

Protarchaeopteryx†

Dromeosaurs† (like Velociraptor)

Sinosauropteryx†

Basal theropods†

Sauropods†

Palatal reorganization

Keeled sternum; alula; synsacrum; caudal vertebrae reduced to pygostyle; loss of teeth

Asymmetrical flight feathers

Symmetrical, vaned feathers

Lunate wrist bones; furcula

Filamentous protofeathers

Bipedal; elongate, mobile S-shaped neck; hollow pneumatic bones; intertarsal ankle joint; tridactyl foot with digitgrade posture

Saurischian pelvis

†Extinct groups

Saurischia

Aves

Neornithes

Neornithes

Figure 27-3

Cladogram of the Saurischia, showing the relationship of several taxa to modern birds. Shown are a few of the shared derived characters, mostly related to flight, that were used to construct the genealogy. The ornithischians are the sister group to the saurishcians and all are members of the clade Archosauria (see Figures 26-1, p. 536, and 26-2, p. 537).

Figure 27-4

One of the strangest birds in a strange land, a flightless cormorant, *Nannopterum harrisi,* of the Galápagos Islands dries its wings after a fishing forage. It is a superb swimmer, propelling itself through the water with its feet to catch fish and octopuses. The flightless cormorant is an example of a carinate bird (having a keeled sternum) that has lost the keel and the ability to fly.

The bodies of flightless birds are dramatically redesigned to remove all of the restrictions of flight. The keel of the sternum is lost, and heavy flight muscles (as much as 17% of the body weight of flying birds), as well as other specialized flight apparatus, disappear. Since body weight is no longer a constraint, flightless birds tend to become large. Several extinct flightless birds were enormous: giant moas of New Zealand weighed more than 225 kg (500 pounds) and elephant-birds of Madagascar, the largest bird that ever lived, probably weighed nearly 450 kg (about 1000 pounds) and stood nearly 2 m tall.

FORM AND FUNCTION

Just as an airplane must be designed and built according to rigid aerodynamic specifications if it is to fly, so too must birds meet stringent structural requirements if they are to stay airborne. All the special adaptations found in flying birds contribute to two things: more power and less weight. Flight by humans became possible when they developed an internal combustion engine and learned how to reduce the weight-to-power ratio to a critical point. Birds accomplished flight millions of years ago. But birds must do much more than fly. They must feed themselves and convert food into high-energy fuel; they must escape predators; they must be able to repair their own injuries; they must be able to air-condition themselves when overheated and heat themselves when too cool; and, most important of all, they must reproduce themselves.

Feathers

Feathers are very lightweight, yet possess remarkable toughness and tensile strength. Most typical of bird feathers are **contour feathers,** vaned feathers that cover and streamline the bird's body. A contour feather consists of a hollow **quill,** or calamus, emerging from a skin follicle, and a **shaft,** or rachis, which is a continuation of the quill and bears numerous **barbs** (Figure 27-5). The barbs are arranged in a closely parallel fashion and

Characteristics of Class Aves

1. Body usually spindle shaped, with four divisions: head, neck, trunk, and tail; **neck disproportionately long** for balancing and food gathering
2. Limbs paired with the **forelimbs usually modified for flying;** posterior pair variously adapted for perching, walking, or swimming; foot with four toes (2 or 3 toes in some)
3. Epidermal **covering of feathers** and **leg scales;** thin integument of epidermis and dermis; no sweat glands; oil or preen gland at base of tail; **pinna of ear rudimentary**
4. **Fully ossified skeleton with air cavities;** skull bones fused with **one occipital condyle;** skull diapsid with antorbital fenestra; each jaw covered with a keratinized sheath, forming a **beak; no teeth;** ribs with strengthening processes, the uncinate process attaching ribs with one another; **tail not elongate;** sternum well developed with keel or reduced with no keel; **single bone in middle ear**
5. Nervous system well developed, with brain and 12 pairs of cranial nerves
6. Circulatory system of **four-chambered heart** of two atria and two ventricles; completely separate pulmonary and systemic circuits; with the **right aortic arch persisting** as the dorsal aorta; reduced renal portal system; nucleated red blood cells
7. Endothermic
8. Respiration by slightly expansible lungs, with thin **air sacs** among the visceral organs and skeleton; **syrinx (voice box)** near junction of trachea and bronchi
9. Excretory system of metanephric kidney; ureters open into cloaca; **no bladder;** semisolid urine; uric acid main nitrogenous waste
10. Sexes separate; testes paired, with the vas deferens opening into the cloaca; **females with functional left ovary and oviduct only;** copulatory organ (penis) in ducks, geese, paleognathids, and a few others
11. Fertilization internal; **amniotic eggs with much yolk and hard calcareous shells;** embryonic membranes in egg during development; **incubation external;** young active at hatching **(precocial)** or helpless and naked **(altricial);** sex determination by females (females heterogametic)

spread diagonally outward from both sides of the central shaft to form a flat, expansive, webbed surface, the **vane.** There may be several hundred barbs in a vane.

If we examine a feather with a microscope, each barb appears to be a miniature replica of the feathers with numerous parallel filaments called **barbules** set in each side of the barb and spreading laterally from it. There may be 600 barbules on each side of a barb, adding up to more than 1 million barbules for the feather. Barbules of one barb overlap barbules of a neighboring barb in a herringbone pattern and are held together with great tenacity by tiny hooks. Should two adjoining barbs become separated—and considerable force is needed to pull the vane apart—they are instantly zipped together again by drawing the feather through the fingertips. A bird, of course, does this preening with its bill, and much of a bird's time is occupied with preening to keep its feathers in perfect condition.

Figure 27-5

Types of bird feathers and their development. **A** to **E**, Successive stages in development of a vaned, or contour, feather. Growth occurs within a protective sheath, **D**, that splits open when growth is complete, allowing the mature feather to spread flat. **F** to **H**, Other feather varieties, including a pheasant feather with aftershaft, **F**, filoplumes, **G**, and down feathers, **H**.

Types of Feathers

Different types of bird feathers serve different functions. **Contour feathers** (Figure 27-5E) give the bird its outward form and are the type we have already described. Contour feathers that extend beyond the body and are used in flight are called **flight feathers. Down feathers** (Figure 27-5H) are soft tufts hidden beneath contour feathers. They are soft because their barbules lack hooks. They are especially abundant on the breast and abdomen of water birds and on young quail and grouse and function principally to conserve heat. **Filoplume feathers** (Figure 27-5G) are hairlike, degenerate feathers; each is a weak shaft with a tuft of short barbs at the tip. They are the "hairs" of a plucked fowl. They have no known function. Bristles around the mouths of flycatchers and whippoorwills are probably modified filoplumes. A fourth type of highly modified feather, the **powder-down feather,** is found on herons, bitterns, hawks, and parrots. Tips of these disintegrate as they grow, releasing a talclike powder that helps to waterproof the feathers and give them metallic luster.

Origin and Development

Like the reptiles' scale to which it is homologous, a feather develops from an epidermal elevation overlying a nourishing dermal core (Figure 27-5A). However, rather than flattening like a scale, a feather bud rolls into a cylinder and sinks into the follicle from which it is growing. During growth, pigments (lipochromes and melanin) are added to the epidermal cells. As a feather enlarges and nears the end of its growth, the soft rachis and barbs are transformed into hard structures by deposition of keratin. The protective sheath splits apart, allowing the end of a feather to protrude and barbs to unfold.

Molting

When fully grown, a feather, like mammalian hair, is a dead structure. Shedding, or molting, of feathers is a highly orderly process. Except in penguins, which molt all at once, feathers are discarded gradually to avoid appearance of bare spots. Flight and tail feathers are lost in exact pairs, one from each side, so that balance is maintained (Figure 27-6). Replacements emerge before the next pair is lost, and most birds can continue to fly unimpaired during the molting period; however, many water birds (ducks, geese, loons, and others) lose all their primary feathers at once and are grounded during the molt. Many prepare for molting by moving to isolated bodies of water where they can find food and more easily escape enemies. Nearly all birds molt at least once a year, usually in late summer after nesting season.

The vivid color of feathers is of two kinds: pigmentary and structural. Red, orange, and yellow feathers are colored by pigments, called lipochromes, deposited in feather barbules as they are formed. Black, brown, red-brown, and gray colors are from a different pigment, melanin. Blue feathers of blue jays, indigo buntings, and bluebirds depend not on pigment but on scattering of shorter wavelengths of light by particles within the feather; these are structural colors. Blue feathers are usually underlain by melanin, which absorbs certain wavelengths, thus intensifying the blue. Such feathers look the same from any angle of view. Green colors are almost always a combination of yellow pigment and blue feather structure. Another kind of structural color is the beautiful iridescent color of many birds, which ranges from red, orange, copper, and gold to green, blue, and violet. Iridescent color is based on interference that causes light waves to reinforce, weaken, or eliminate each other. Iridescent colors may change with the angle of view; quetzals, for example, look blue from one angle and green from another. Among vertebrates, only tropical reef fishes can vie with birds for intensity and vividness of color.

Skeleton

A major structural requirement for flight is a light, yet sturdy, skeleton. As compared with the earliest known bird, *Archaeopteryx*, bones of modern birds (Figure 27-7A) are phenomenally light, delicate, and laced with air cavities. Such **pneumatized** bones (Figure 27-8) are nevertheless strong. The skeleton of a frigate bird with a 2.1 m (7-foot) wingspan weighs only 114 grams (4 ounces), less than the weight of all its feathers.

As archosaurs, birds evolved from ancestors with diapsid skulls (p. 537). However, skulls of modern birds are so specialized that it is difficult to see any trace of the original diapsid condition. A bird's skull is built lightly and mostly fused into one piece. The braincase and orbits are large to accommodate a

Figure 27-6

Osprey, *Pandion haliaetus* (order Falconiformes) lands while holding a freshly captured fish. Feathers are molted in sequence in exact pairs so that balance is maintained during flight.

bulging brain and large eyes needed for quick motor coordination and superior vision. Yet, a pigeon skull weighs only 0.21% of its body weight; by comparison a rat's skull weighs 1.25% of its body weight. As a whole, however, the skeleton of a bird is not lighter than that of a mammal of similar size. The difference is in distribution of mass: whereas the skull and pneumatized wing bones are especially light, the leg bones are heavier than those of mammals. This helps lower the bird's center of gravity as required for aerodynamic stability.

In *Archaeopteryx*, both jaws contained teeth set in sockets, an archosaurian characteristic. Modern birds are completely toothless, having instead a horny (keratinous) beak molded around the bony jaws. The mandible is a complex of several bones hinged to provide a double-jointed action that permits the mouth to gape widely. Most birds have kinetic skulls (kinetic skulls of lizards are described on p. 542). The attachment of the upper jaw to the skull is flexible; this allows the upper jaw to move slightly, thus increasing the gape. In some birds, such as parrots, the upper jaw is especially flexible because it is hinged to the skull.

The most distinctive feature of the vertebral column is its rigidity. Most vertebrae except the **cervicals** (neck vertebrae) are fused together. Most caudal vertebrae are fused into a **pygostyle** (Figure 27-7A), while many of the remaining vertebrae in the trunk are fused as the **synsacrum.** These fused vertebrae, with the pelvic girdle form a stiff but light framework to support the legs and provide rigidity for flight. To assist in this rigidity, ribs are braced against each other with uncinate processes (Figure 27-7A). Except in flightless birds, the sternum bears a large, thin keel that provides an attachment for powerful flight muscles. Fused clavicles form an elastic **furcula** that apparently stores energy as it flexes during wing beats. Examination of the anatomy of *Archaeopteryx* permits some insight to its flight abilities. The asymmetrical feathers and large furcula of *Archaeopteryx* lend strong support that it was a flying bird. However, as compared to modern birds, it probably was not a strong flier, because its small sternum offered little area for attachment of flight muscles (Figure 27-7B).

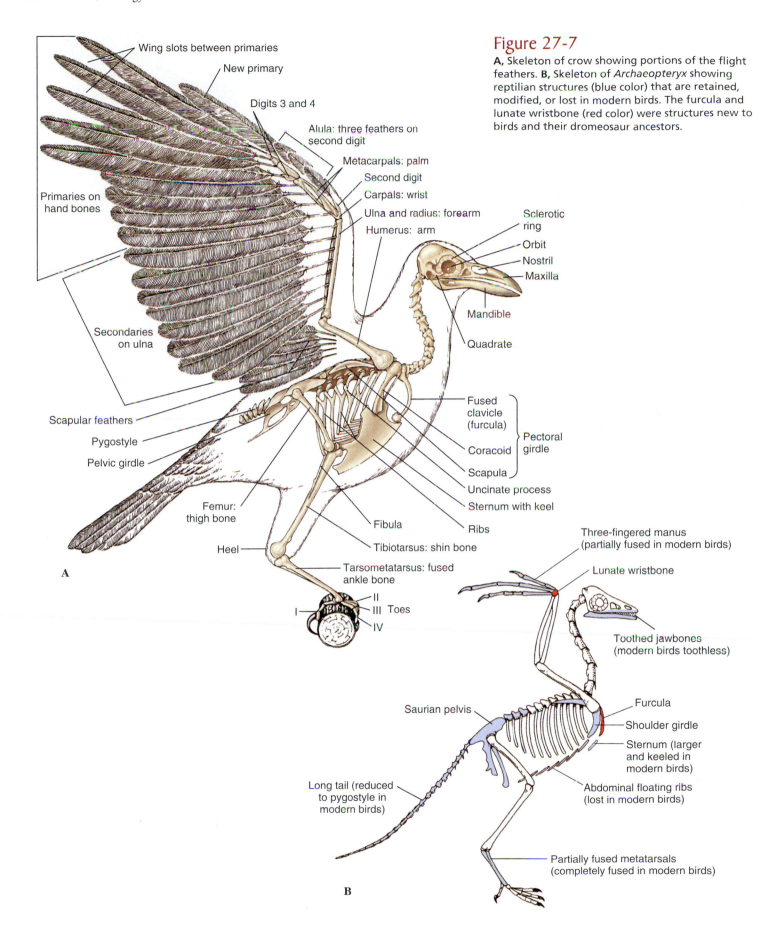

Figure 27-7
A, Skeleton of crow showing portions of the flight feathers. **B,** Skeleton of *Archaeopteryx* showing reptilian structures (blue color) that are retained, modified, or lost in modern birds. The furcula and lunate wristbone (red color) were structures new to birds and their dromeosaur ancestors.

Wing slots between primaries

New primary

Digits 3 and 4

Alula: three feathers on second digit

Metacarpals: palm

Second digit

Carpals: wrist

Ulna and radius: forearm

Humerus: arm

Sclerotic ring

Orbit

Nostril

Maxilla

Mandible

Quadrate

Primaries on hand bones

Secondaries on ulna

Scapular feathers

Pygostyle

Pelvic girdle

Fused clavicle (furcula)

Coracoid

Scapula

Uncinate process

Sternum with keel

Pectoral girdle

Femur: thigh bone

Fibula

Ribs

Tibiotarsus: shin bone

Heel

Tarsometatarsus: fused ankle bone

II

III   Toes

I

IV

A

Three-fingered manus (partially fused in modern birds)

Lunate wristbone

Toothed jawbones (modern birds toothless)

Saurian pelvis

Furcula

Shoulder girdle

Sternum (larger and keeled in modern birds)

Abdominal floating ribs (lost in modern birds)

Long tail (reduced to pygostyle in modern birds)

Partially fused metatarsals (completely fused in modern birds)

B

Figure 27-8
Hollow wing bone of a songbird showing stiffening struts and air spaces that replace bone marrow. Such "pneumatized" bones are remarkably light and strong.

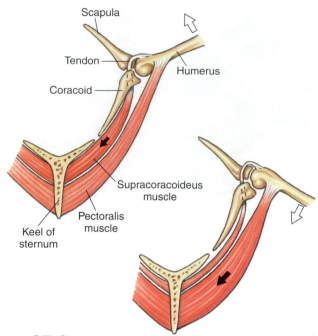

Scapula

Tendon

Coracoid

Humerus

Supracoracoideus muscle

Pectoralis muscle

Keel of sternum

Figure 27-9
The arrangement of flight muscles keeps the center of gravity low in the body. Both major flight muscles are anchored on the sternum keel. Contraction of the pectoralis muscle pulls the wing downward. Then, as the pectoralis relaxes, the supracoracoideus muscle contracts and, acting as a pulley system, pulls the wing upward.

Bones of the forelimbs are highly modified for flight. They are reduced in number, and several are fused together. Despite these alterations, a bird's wing is clearly a rearrangement of the basic vertebrate tetrapod limb from which it arose (p. 517), and all the elements—arm, forearm, wrist, and fingers—are represented in modified form (Figure 27-7).

Muscular System

The locomotor muscles of wings are relatively massive to meet demands of flight. Largest of these is the **pectoralis,** which depresses the wings in flight. Its antagonist is the **supracoracoideus** muscle, which raises the wing (Figure 27-9). Surprisingly, perhaps, this latter muscle is not located on the backbone (anyone who has been served the back of a chicken knows that it offers little meat) but is positioned under the pectoralis on the breast. It is attached by a tendon to the upper side of the humerus of the wing so that it pulls from below by an ingenious "rope-and-pulley" arrangement. Both pectoralis and supracoracoideus are anchored to the keel. Positioning the main muscle mass low in the body improved aerodynamic stability.

From the main leg muscle mass in the thigh, thin but strong tendons extend downward through sleevelike sheaths to the toes. Consequently the feet are nearly devoid of muscles, explaining the thin, delicate appearance of a bird's leg. This arrangement places the main muscle mass near a bird's center of gravity and at the same time allows great agility to the slender,

lightweight feet. Because the feet are composed mostly of bone, tendon, and tough, scaly skin, they are highly resistant to damage from freezing. When a bird perches on a branch, an ingenious toe-locking mechanism (Figure 27-10) is activated, which prevents the bird from falling off its perch when asleep. The same mechanism causes the talons of a hawk or owl automatically to sink deeply into its prey as the legs bend under the impact of a strike. The powerful grip of a bird of prey was described by L. Brown*

> When an eagle grips in earnest, one's hand becomes numb, and it is quite impossible to tear it free, or to loosen the grip of the eagle's toes with the other hand. One just has to wait until the bird relents, and while waiting one has ample time to realize that an animal such as a rabbit would be quickly paralyzed, unable to draw breath, and perhaps pierced through and through by the talons in such a clutch.

Birds have lost the long reptilian tail, still fully evident in *Archaeopteryx,* and have substituted a pincushion-like muscle mound into which the tail feathers are rooted. It contains a perplexing array of tiny muscles, as many as 1000 in some species, which control the crucial tail feathers. The most complex muscular system of all is found in the neck of birds; the thin and stringy muscles, elaborately interwoven and subdivided, provide the bird's neck with the ultimate in vertebrate flexibility.

*From Brown, L. 1970. *Eagles.* New York, Arco Publishing.

Figure 27-10

Perching mechanism of a bird. When a bird settles on a branch, tendons automatically tighten, closing the toes around the perch.

Food, Feeding, and Digestion

In their early evolution, most birds were carnivorous, feeding principally on insects, already well established on the earth's surface in both variety and numbers long before birds made their appearance. With the advantage of flight, birds could hunt insects on the wing and carry their assault to insect refuges mostly inaccessible to their earthbound tetrapod peers. Today, there is a bird to hunt nearly every insect; they probe the soil, search the bark, scrutinize every leaf and twig, and drill into insect galleries hidden in tree trunks.

Other animal foods (worms, molluscs, crustaceans, fish, frogs, reptiles, mammals, as well as other birds) all found their way into the diet of birds. A very large group, nearly one-fifth of all birds, feeds on nectar. Some birds are omnivores (often termed **euryphagous,** or "wide-eating" species) that will eat whatever is seasonally abundant. However, omnivorous birds must compete with numerous other omnivores for the same broad spectrum of food. Others are specialists (called **stenophagous,** or "narrow-eating" species) that have the pantry to themselves—but at a price. Should the food specialty be reduced or destroyed for some reason (disease, adverse climate, and the like), their very survival may be jeopardized.

The beaks of birds are strongly adapted to specialized food habits—from generalized types such as strong, pointed beaks of ravens, to grotesque, highly specialized ones in flamingoes, pelicans, and avocets (Figure 27-11). The beak of a woodpecker is a straight, hard, chisel-like device. Anchored to a tree trunk with its tail serving as a brace, the woodpecker delivers powerful, rapid blows to excavate nest cavities or expose burrows of wood-boring insects. It then uses its long, flexible, barbed tongue to seek out insects in their galleries. A woodpecker's skull is especially thick to absorb shock.

How much do birds eat? By a peculiar twist of reality, the commonplace "to eat like a bird" is supposed to signify a diminutive appetite. Yet birds, because of their intense metabolism, are voracious feeders. Small birds with their high metabolic rate eat more food relative to their body mass than large birds. This happens because oxygen consumption increases only about three-fourths as rapidly as body weight. For example, the resting metabolic rate (oxygen consumed per gram of body weight) of a hummingbird is 12 times that of a pigeon and 25 times that of a chicken. A 3 g hummingbird may eat 100% of its body weight in food each day, an 11 g blue tit about 30%, and a 1880 g domestic chicken, 3.4%. Obviously the weight of food consumed also depends on water content of the food, since water has no nutritive value. A 57 g Bohemian waxwing was estimated to eat 170 g of watery *Cotoneaster* berries in one day—three times its body weight! Seed-eaters of equivalent size might eat only 8 g of dry seeds per day.

Birds process their food rapidly and thoroughly with efficient digestive equipment. A shrike can digest a mouse in 3 hours, and berries will pass completely through the digestive tract of a thrush in just 30 minutes. Because birds lack teeth, foods that require grinding are reduced in the gizzard. The poorly developed salivary glands mainly secrete mucus for lubricating food and the slender, horn-covered tongue. There are few taste buds, although all birds can taste to some extent. Many birds have an enlargement **(crop)** of the esophagus at its lower end that serves as a storage chamber.

In pigeons, doves, and some parrots, the crop not only stores food but also produces a lipid- and protein-rich "milk," composed of sloughed epithelial cells of the crop lining. For a few days after hatching, the helpless young are fed regurgitated crop milk by both parents.

The stomach proper consists of two compartments, a **proventriculus,** which secretes gastric juice, and the muscular **gizzard,** which is lined with keratinized plates that serve as millstones for grinding food. To assist in the grinding process, birds swallow coarse, gritty objects or pebbles, which lodge in the gizzard. Certain birds of prey such as owls form pellets of indigestible materials, for example, bones and fur, in the proventriculus by sloughing the gut lining to enclose this material and ejecting it through the mouth. At the junction of the small intestine with the colon are paired **ceca;** these are well developed in herbivorous birds in which they serve as fermentation chambers.

In young birds the dorsal wall of the cloaca bears the **bursa of Fabricius,** which processes the B lymphocytes that are important in the immune response (p. 740).

Circulatory System

The general plan of bird circulation is not greatly different from that of mammals, although their shared derived characteristics were evolved in parallel. Their four-chambered heart is large, with strong ventricular walls; thus, birds share with mammals a complete separation of respiratory and systemic circulations.

Figure 27-11
Some bills of birds showing variety of adaptations.

Raven
Generalized bill

Cardinal
Seed cracker

American avocet
Worm burrow probe

Pelican
Dip net

Parrot
Nut cracker

Flamingo
Mud sifter

Anhinga
Fish spear

Eagle
Meat tearer

However, the right aortic arch, instead of the left as in the mammals, leads to the dorsal aorta. The two jugular veins in the neck are connected by a cross vein, an adaptation for shunting blood from one jugular to the other as the head rotates. The brachial and pectoral arteries to the wings and breast are unusually large.

The heartbeat is extremely fast, and, as in mammals, there is an inverse relationship between heart rate and body weight. For example, a turkey has a heart rate at rest of about 93 beats per minute, a chicken has a resting rate of 250 beats per minute, and a blackcapped chickadee has 500 beats per minute when asleep, which may increase to a phenomenal 1000 beats per minute during exercise. Blood pressure in birds is roughly equivalent to that in mammals of similar size.

Bird's blood contains **nucleated, biconvex erythrocytes.** (Mammals, the only other endothermic vertebrates, have enucleated, biconcave erythrocytes that are somewhat smaller than those of birds.) The **phagocytes,** or mobile ameboid cells of the blood, are very active and efficient in birds in repairing wounds and destroying microbes.

Respiratory System

The respiratory system of birds differs radically from the lungs of reptiles and mammals and is marvelously adapted for meeting the high metabolic demands of flight. In birds the finest branches of the bronchi, rather than ending in saclike alveoli as in mammals, are developed as tubelike **parabronchi** through which air flows continuously. Also unique is an extensive system of nine interconnecting **air sacs** located in pairs in the thorax and abdomen and even extended by tiny tubes into the centers of the long bones (Figure 27-12). Air sacs connect to the lungs in such a way that most of the inspired air bypasses the lungs and flows directly into the posterior air sacs, which serve as reservoirs for fresh air. On expiration, this oxygenated air is passed through the lung and collected in the anterior air sacs. From there it flows directly to the outside. Thus, it takes two respiratory cycles for a single breath of air to pass through the respiratory system, allowing for continuous one-way flow through the respiratory exchange chambers, the parabronchi. The airflow sequence is shown in Figure 27-12. The advantage of such a system is that an almost continuous stream of oxygenated air is passed through a system of richly vascularized parabronchi. Although many details of a bird's respiratory system are not yet understood, it is clearly the most efficient respiratory system of any terrestrial vertebrate.

The remarkable efficiency of a bird's respiratory system is emphasized by bar-headed geese that routinely migrate over the Himalayan mountains and have been sighted flying over Mt. Everest (8848 meters or 29,141 feet) under conditions that are severely hypoxic to humans. They reach altitudes of 9000 meters in less than a day, without the acclimatization that is absolutely essential for humans even to approach the upper reaches of Mt. Everest.

In addition to performing its principal respiratory function, the air sac system helps cool the bird during vigorous exercise. A pigeon, for example, produces about 27 times more heat when flying than when at rest. Air sacs have numerous diverticula that extend inside the larger pneumatic bones of the pectoral and pelvic girdles, wings, and legs. Because they contain warmed air, they provide considerable buoyancy to the bird.

Excretory System

Urine is formed in the relatively large paired metanephric kidneys by glomerular filtration followed by selective modification of the filtrate in the tubule (the details of this sequence are given on pp. 643 to 645). Urine passes by way of **ureters** to the **cloaca.** There is no urinary bladder.

Birds, like reptiles, excrete their nitrogenous wastes as uric acid, rather than urea. In shelled eggs, all excretory products must remain within the eggshell with the growing embryo. Uric acid crystallizes out of solution and can be stored harmlessly within the eggshell. Because of uric acid's low solubility, a bird can excrete 1 g of uric acid in only 1.5 to 3 ml of water, whereas a mammal may require 60 ml of water to excrete 1 g of urea. The concentration of uric acid occurs almost entirely in the cloaca, where it is combined with fecal material, and the water reabsorbed.

Bird kidneys are much less efficient than mammalian kidneys in removal of solutes, primarily ions of sodium, potassium, and chloride. Most mammals can concentrate solutes to 4 to 8 times that of blood concentration, and some such as desert rodents can concentrate urine to nearly 25 times that of blood. By comparison, most birds concentrate solutes only slightly greater than that of blood (the best that any bird can concentrate is about 6 times that of blood).

To compensate for weak solute-concentrating ability of the kidney, some birds, especially marine birds that must excrete large salt loads from the food they eat and seawater they drink, use extrarenal mechanisms to remove salt from the body. **Salt glands,** one located above each eye of sea birds (Figure 27-13), can excrete highly concentrated solutions of sodium chloride, up to twice the concentration of seawater. The salt solution runs out the internal or external nostrils, giving gulls, petrels, and other sea birds a perpetual runny nose. The size of the salt gland in some birds depends on how much salt the bird takes in its diet. For example, a race of mallard ducks living a semimarine life in Greenland has salt glands 10 times larger than those of ordinary fresh-water mallards.

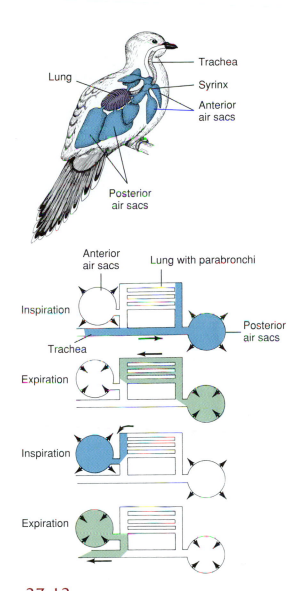

Figure 27-12

Respiratory system of a bird. **A,** Lungs and air sacs. One side of the bilateral air sac system is shown. **B,** Movement of a single volume of air through a bird's respiratory system. Two full respiratory cycles are required to move air through the system.

Nervous and Sensory Systems

The design of a bird's nervous and sensory system reflects the complex problems of flight and a highly visible existence, in which it must gather food, mate, defend territory, incubate and rear young, and correctly distinguish friend from foe. The brain of a bird has well-developed **cerebral hemispheres, cerebellum,** and **optic lobes** (Figure 27-14). The **cerebral cortex—** chief coordinating center of a mammalian brain—is thin, unfissured, and poorly developed in birds. But the core of the cerebrum, the **dorsal ventricular ridge,** has enlarged into the principal integrative center, controlling such activities as eating, singing, flying, and all complex instinctive reproductive activities. Relatively intelligent birds, such as crows and parrots, have larger cerebral hemispheres than do less intelligent birds such as

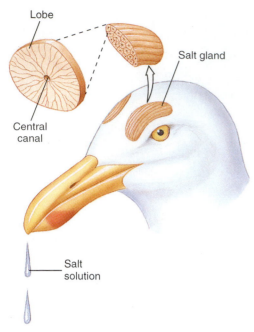

Figure 27-13
Salt glands of a marine bird (gull). One salt gland is located above each eye. Each gland consists of several lobes arranged in parallel. One lobe is shown in cross section, much enlarged. Salt is secreted into many radially arranged tubules, then flows into a central canal that leads into the nose.

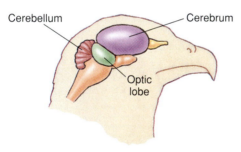

Figure 27-14
Bird brain showing principal divisions.

chickens and pigeons. The **cerebellum** is much larger in birds than in reptiles and serves as a crucial coordinating center where muscle-position sense, equilibrium sense, and visual cues are assembled and used to coordinate movement and balance. The **optic lobes,** laterally bulging structures of the midbrain, form a visual association apparatus comparable to the visual cortex of mammals.

The senses of smell and taste of some birds are poor, but are relatively well developed in many other birds, such as carnivorous birds, flightless birds, oceanic birds, and waterfowl. Birds have good hearing and superb vision, the keenest in the animal kingdom. As in mammals, a bird's ear consists of three regions: (1) **external ear,** a sound-conducting canal extending to the **eardrum,** (2) **middle ear,** containing a rodlike **col-**

Figure 27-15
A hawk eye has all the structural components of a mammalian eye, plus a peculiar pleated structure, the pecten, believed to provide nourishment to the retina. The extraordinarily keen vision of hawks is attributed to the extreme density of cone cells in the foveae: 1.5 million per fovea compared to 0.2 million for humans.

umella that transmits vibrations, and (3) **inner ear,** where the organ of hearing, the **cochlea,** is located. A bird's cochlea is much shorter than the coiled mammalian cochlea, yet birds can hear roughly the same range of sound frequencies as humans. However, birds do not hear high-frequency sounds as well as similar-sized mammals. Actually, a bird's ear far surpasses that of humans in capacity to distinguish differences in intensities and to respond to rapid fluctuations in pitch.

A bird's eye resembles that of other vertebrates in gross structure but is relatively larger; for a given body size, less spherical, and almost immobile; instead of turning their eyes, birds turn their heads with their long and flexible necks to scan the visual field. The light-sensitive **retina** (Figure 27-15) is generously equipped with rods (for dim light vision) and cones (for good acuity and color vision). Cones predominate in diurnal birds, and rods are more numerous in nocturnal birds. A distinctive feature of a bird's eye is the **pecten,** a highly vascularized organ attached to the retina near the optic nerve and jutting into the vitreous humor (Figure 27-15). The pecten is thought to provide nutrients and oxygen to the eye. It may do more, but its function remains largely a mystery. On the anterior side of the eye is a **sclerotic ring** of platelike bones that serve to strengthen and focus the large eye (Figure 27-7).

The position of a bird's eyes in its head is correlated with its life habits. Vegetarians that must avoid predators have eyes placed laterally to give a wide view of the world; predaceous birds such as hawks and owls have eyes directed to the front, allowing more binocular vision for better depth perception. In birds of prey and some others, the **fovea,** or region of keenest vision on the retina, is placed in a deep pit, which makes it necessary for the bird to focus exactly on the source. Many birds, moreover, have two foveae on the retina (Figure 27-15): the central one for sharp monocular views and the posterior one for binocular vision. Woodcocks can probably see binocularly both forward and backward. The visual acuity of a hawk is about 8 times that of a human (enabling it to see clearly a crouching rabbit more than a mile away), and an owl's ability to see in dim light is more than 10 times that of a human. Birds have good color vision, especially toward the red end of the spectrum.

Many birds can see into the ultraviolet, enabling them to view environmental features inaccessible to us but accessible to insects (such as flowers with ultraviolet-reflecting "nectar guides" that attract pollinating insects). Several species of ducks, hummingbirds, kingfishers, and passerines (songbirds) can see in the near ultraviolet (UV) down to 370 nm (the human eye filters out ultraviolet light below 400 nm). For what purpose do birds use their UV-sensitivity? Some, such as hummingbirds, may be attracted to nectar-guiding flowers, like insects. But, for the others, the benefit derived from UV-sensitivity is a matter of conjecture.

Flight

What prompted the evolution of flight in birds, the ability to rise free of earthbound concerns, as almost every human has dreamed of doing? The air was a relatively unexploited habitat stocked with flying insects for food. Flight also offered escape from terrestrial predators and opportunity to travel rapidly and widely to establish new breeding areas and to benefit from year-round favorable climate by migrating north and south with the seasons.

Two competing hypotheses of the origin of bird flight have been offered: birds began to fly either by climbing to a high place and gliding down, or by flapping their wings into the air from the ground. The first hypothesis ("trees down") has been long favored. Proponents of this view envision an arboreal ancestor of *Archaeopteryx* gliding from tree to tree. Modifications permitting lift and powered flight would be highly advantageous for this kind of life. Indeed, there are many arboreal squirrels and lizards that use gliding to move among trees. However, the feathered dromeosaurs were ground-dwelling, not arboreal, casting serious doubt on the "trees down" hypothesis.

Proponents of the "ground up" hypothesis suggest that the feathered wings of bipedal, ground-dwelling ancestors of the first flying birds may have been used as snares to capture insects or to refine aerodynamic control during leaps to capture flying insects. Thus, as the wings became larger, they would have been capable of powered flight. Examination of the foot anatomy of *Archaeopteryx* helps little in this debate as it has been variously interpreted as being either well suited or unsuitable for perching in trees. One thing seems certain: the debate about origin of flight has not been settled. Interestingly enough, feathers were certainly a requirement for bird flight, but were not required for powered flight in two other lineages, bats and extinct pterosaurs, which lack feathers and, notably, convincing flight-origin hypotheses of their own.

Bird Wing as a Lift Device

To fly, birds must generate lift forces greater than their own mass to become airborne and they must provide propulsion in order to move. They use their wings to provide both. In general, the distal part of the wing, the modified hand bones with the attached primaries, acts as a propeller to provide propulsion. Lift is provided by feathers in the more medial part of the wing, the secondaries, associated with the forearm. The wing is streamlined in cross section, with a slightly concave lower surface **(cambered)** and with small, tight-fitting feathers where the leading edge meets the air (Figure 27-16). Air slips smoothly over the wing, creating lift with minimum drag. Some lift is produced by positive pressure against the undersurface of the wing. But on the upper side, where the airstream must travel farther and faster over the convex surface, a negative pressure is created that provides more than two-thirds of the total lift.

The lift-to-drag ratio of an airfoil is determined by the angle of tilt (angle of attack) and the airspeed (Figure 27-16). A wing carrying a given load can pass through the air at high speed and small angle of attack or at low speed and larger angle of attack. As speed decreases, lift can be increased by increasing the angle of attack, but drag forces also increase. Finally a point is reached (usually around 15 degrees) at which the angle of attack becomes too steep; turbulence appears on the upper surface, lift is destroyed, and stalling occurs. Stalling can be delayed or prevented by placing a **wing slot** along the leading edge; this structure directs a layer of rapidly moving air across the upper wing surface. Wing slots were and still are used in aircraft traveling at a low speed. In birds, two kinds of wing slots have developed: (1) the **alula,** or group of small feathers on the thumb (Figures 27-6 and 27-7), which provides a midwing slot, and (2) **slotting between the primary feathers,** which provides a wing-tip slot. In a number of songbirds, these together provide stall-preventing slots for nearly the entire outer (and aerodynamically more important) half of the wing.

Flapping Flight

Two forces are required for flapping flight: a vertical *lifting* force to support the bird's weight, and a horizontal thrusting force to move the bird forward against the resistive forces of friction. Thrust is provided mainly by primary feathers at the wing tips, while secondary feathers of the inner wing, which do not move so far or so fast, act as an airfoil, providing mainly lift. Greatest power is applied on the downstroke. The primary feathers are bent upward and twist to a steep angle of attack, biting into the air like a propeller (Figure 27-17). The entire wing (and the bird's body) is pulled forward. On the upstroke, the primary feathers bend in the opposite direction so that their upper surfaces twist into a positive angle of attack to produce thrust, just as the lower surfaces did on the downstroke. A powered upstroke is essential for hovering flight, as in hummingbirds (Figure 27-18), and is important for fast, steep takeoffs by small birds with elliptical wings.

Basic Forms of Bird Wings

Bird wings vary in size and form because successful exploitation of different habitats has imposed special aerodynamic requirements. Four types of bird wings are easily recognized.*

*Saville, D. B. O. 1957. Adaptive evolution in the avian wing. Evolution **11:**212–224.

Lowest pressure and greatest lift where air flow is fastest

Smaller area of high pressure and lift beneath wing

Angle of attack

Air flow around wing

Lift-destroying turbulence

Stalling at low speed

Wing slot directs fast-moving air over wing surface

Preventing stall with wing slots

Formation of wing tip vortex

Figure 27-16

Air patterns formed by an airfoil, or wing, moving from right to left. At low speed the angle of attack (α) must increase to maintain lift but this increases the threat of stalling. The upper figures show how low-speed stalling can be prevented with wing slots. Wing tip vortex (*bottom*), a turbulence that tends to develop at high speeds, reduces flight efficiency. The effect is reduced in wings that sweep back and taper to a tip.

Elliptical Wings   Birds that must maneuver in forested habitats, such as sparrows, warblers, doves, woodpeckers, and magpies (Figure 27-19A), have elliptical wings. This type has a **low aspect ratio** (ratio of length to average width). The wings of the highly maneuverable British Spitfire fighter plane of World War II fame conformed closely to the outline of a sparrows wing. Elliptical wings are slotted between the primary feathers; this arrangement helps prevent stalling during sharp turns, low-speed flight, and frequent landing and takeoff. Each separated primary feather behaves as a narrow wing with a high angle of attack, providing high lift at low speed. High maneuverability of elliptical wings is exemplified by the tiny chickadee, which, if frightened, can change course within 0.03 second.

High-Speed Wings   Birds that feed during flight, such as swallows, hummingbirds, and swifts, or that make long migrations, such as plovers, sandpipers, terns and gulls, (Figure 27-19B), have wings that sweep back and taper to a slender tip. They are rather flat in section, have a moderately high aspect ratio, and lack wing-tip slotting characteristic of elliptical wings. Sweepback and wide separation of wing tips reduce "tip vortex," a drag-creating turbulence that tends to develop at wing tips at faster speeds (Figure 27-16). This type of wing is aerodynamically efficient for high-speed flight but cannot easily keep a bird airborne at low speeds. The fastest birds, such as sandpipers, clocked at 175 km (109 miles) per hour, belong to this group.

Dynamic Soaring Wings   Oceanic soaring birds have **high-aspect ratio** wings resembling those of sailplanes. This group includes albatrosses, frigate birds, and gannets (Figure 27-19C). Such long, narrow wings lack wing slots and are adapted for high speed, high lift, and dynamic soaring. They have the highest aerodynamic efficiency of all wings but are less maneuverable than the wide, slotted wings of land soarers. Dynamic soarers exploit the highly reliable sea winds, using adjacent air currents of different velocities.

High-Lift Wings   Vultures, hawks, eagles, owls, and ospreys (Figure 27-19D)—predators that carry heavy loads—have wings with slotting, alulas, and pronounced camber, all of which promote high lift at low speed. Many of these birds are land soarers, with broad, slotted wings that provide the sensitive response and maneuverability required for static soaring in capricious air currents over land.

MIGRATION AND NAVIGATION

We described advantages of migration in the prologue to this chapter. Not all birds migrate, of course, but most North American and European species do, and the biannual journeys of some are truly extraordinary undertakings.

Migration Routes

Most migratory birds have well-established routes trending north and south. Since most birds (and other animals) breed in the Northern Hemisphere, where most of the earth's landmass is

Figure 27-17
In normal flapping flight of strong fliers like ducks, the wings sweep downward and forward fully extended. Thrust is provided by the primary feathers at the wing tips. To begin the upbeat, the wing is bent, bringing it upward and backward. The wing then extends, ready for the next downbeat.

Figure 27-18
The secret of a hummingbird's ability to change direction instantly, or hang motionless in the air while sipping nectar from a flower, lies in its wing structure. The wing is nearly rigid, but hinged at the shoulder by a swivel joint and powered by a supracoracoideus muscle that is unusually large for the bird's size. When hovering the wing moves in a sculling motion. The leading edge of the wing moves forward on the forward stroke, then swivels nearly 180 degrees at the shoulder to move backward on the backstroke. The effect is to provide lift without propulsion on *both* forward and backstrokes.

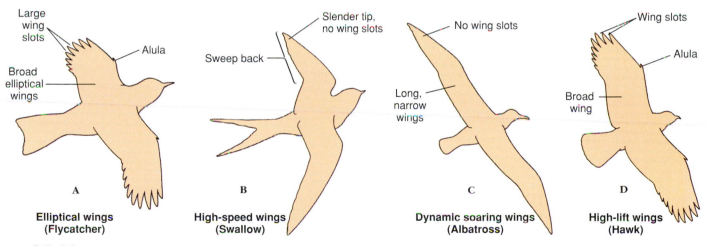

A Elliptical wings (Flycatcher)

B High-speed wings (Swallow)

C Dynamic soaring wings (Albatross)

D High-lift wings (Hawk)

Figure 27-19
Four basic forms of bird wings.

concentrated, most birds migrate south in the northern winter and north to nest in the northern summer. Of the 4000 or more species of migrant birds (a little less than half the total bird species), most breed in the more northern latitudes of the hemisphere. Some use different routes in the fall and spring (Figure 27-20). Some, especially certain aquatic species, complete their migratory routes in a very short time. Others, however, make a leisurely trip, often stopping along the way to feed. Some war-

blers are known to take 50 to 60 days to migrate from their winter quarters in Central America to their summer breeding areas in Canada. Many smaller species migrate at night and feed by day; others migrate chiefly in daytime; and many swimming and wading birds migrate by either day or night.

Many birds are known to follow landmarks, such as rivers and coastlines, but others do not hesitate to fly directly over large bodies of water in their routes. Some birds have very wide

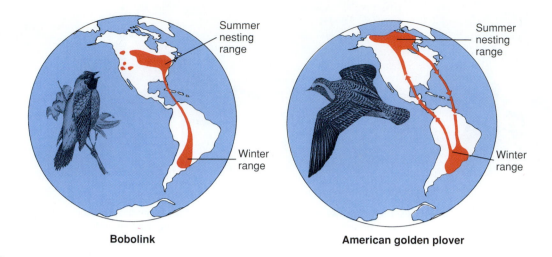

Bobolink   **American golden plover**

Figure 27-20

Migrations of bobolinks, *Dolichonyx oryzivorus,* and American golden plovers, *Pluvialis dominica.* Bobolinks commutes 22,500 km (14,000 miles) each year between nesting sites in North America and their range in Argentina, where they spend the northern winters, a phenomenal feat for such a small bird. Although the breeding range has extended to colonies in western areas, these birds take no shortcuts but adhere to the ancestral seaboard route. American golden plovers fly a loop migration, striking out across the Atlantic in their southward autumnal migration but returning in the spring by way of Central America and the Mississippi Valley because ecological conditions are more favorable at that time.

migration lanes, whereas others, such as certain sandpipers, are restricted to very narrow ones, keeping well to the coastlines because of their food requirements.

Some species are known for their long-distance migrations. Arctic terns, greatest globe spanners of all, breed north of the Arctic Circle during the northern summer, then migrate to Antarctic regions for the northern winter. This species also is known to take a circuitous route in migrations from North America, passing over to the coastlines of Europe and Africa and then to their winter quarters, a trip that may exceed 18,000 km (11,200 miles).

Many small songbirds also make great migratory treks (Figure 27-20). Many birds that nest in Europe or Central Asia spend the northern winter in Africa.

Stimulus for Migration

Humans have known for centuries that onset of reproductive cycles of birds is closely related to season. Only relatively recently, however, has it been demonstrated that the lengthening days of late winter and early spring stimulate development of the gonads and accumulation of fat—both important internal changes that predispose birds to migrate northward. Long day length stimulates the anterior lobe of the pituitary into activity. Release of pituitary gonadotropic hormone in turn sets in motion a complex series of physiological and behavioral changes, resulting in gonadal growth, fat deposition, migration, courtship and mating behavior, and care of the young.

Direction Finding in Migration

Numerous experiments suggest that most birds navigate chiefly by sight. Birds recognize topographical landmarks and follow familiar migratory routes—a behavior assisted by flock migra-

tion, during which navigational resources and experience of older birds can be pooled. But in addition to visual navigation, birds make use of a variety of orientation cues at their disposal. Birds have a highly accurate innate sense of time. Recent work adds credence to an old, much debated hypothesis that birds can detect and navigate by the earth's magnetic field. The navigational abilities of birds are largely inborn, but they may require calibration with existing navigational landmarks. In addition, learning can play a role because a bird's navigational abilities may improve with experience.

In the early 1970s W. T. Keeton showed that flight bearings of homing pigeons were significantly disturbed by magnets attached to the birds' heads, or by minor fluctuations in the geomagnetic field. But until recently the nature and position of a magnetic receptor in pigeons remained a mystery. Deposits of a magnetic substance called magnetite (Fe_3O_4) have been discovered in the neck musculature of pigeons and migratory white-crowned sparrows. If this material were coupled to sensitive muscle receptors, as has been proposed, the structure could serve as a magnetic compass that would enable birds to detect and orient their migrations to the earth's magnetic field.

Experiments by German ornithologists G. Kramer and E. Sauer and American ornithologist S. Emlen demonstrated convincingly that birds can navigate by celestial cues: the sun by day and the stars by night. Using special circular cages, Kramer concluded that birds maintain compass direction by referring to the sun, (Figure 27-21). This is called **sun-azimuth orientation** (*azimuth,* compass bearing of the sun). To use the sun as a compass birds must know the time of day because the sun's position changes throughout the day. By exposing birds to altered light cycles to shift their perception of daybreak, researchers showed that birds do use an internal clock in this fashion. Sauer's and Emlen's ingenious planetarium experiments strongly suggest that some birds, probably many, are able to detect and navigate by the North Star axis around which the constellations appear to rotate.

In an elegant set of experiments designed to determine whether nocturnal migrants have an innate sense of direction or learn direction as nestlings, Stephen Emlen raised indigo buntings under three sets of conditions in a planetarium in which star patterns could be modified. One group of nestlings was allowed to see stars in a normal night sky rotating around the North Star. A second group of nestlings saw a night sky star pattern that was rotating around Betelgeuse, a bright star in the constellation Orion, as if Betelgeuse were the North Star. A third group of nestlings was raised seeing only points of light at night that did not rotate.

When the birds grew to an age for migration, they were placed in cages under a normal night sky that allowed recording of the direction in which they tried to orient or migrate. Birds that had seen only points of light during their development, with no rotation of the sky, showed no ability to detect direction and moved randomly. Birds that had developed seeing the normal sky rotated around the North Star oriented correctly for migration; and the group that developed seeing the sky rotated about Orion showed consistent orientation as if Betelgeuse were the North Star, even though now exposed to a normal night sky with stars rotating around the North Star. Thus, Emlen elegantly showed that these birds do not hatch with an innate sense of direction but must learn direction by seeing the sky rotate around a "pole" star!

Some remarkable feats of bird navigation still defy understanding. Most birds undoubtedly use a combination of environmental and innate cues to migrate. Migration is a rigorous undertaking. The target is often small, and natural selection relentlessly eliminates individuals making errors in migration, leaving only the best navigators to propagate the species.

SOCIAL BEHAVIOR AND REPRODUCTION

The adage says "birds of a feather flock together," and many birds are indeed highly social creatures. Especially during the breeding season, sea birds gather, often in enormous colonies, to nest and rear young (Figure 27-22). Land birds, with some conspicuous exceptions, such as starlings and rooks, tend to be less gregarious than sea birds during breeding and to seek isolation for rearing their brood. But species that covet separation from their kind during breeding may aggregate for migration or feeding. Togetherness offers advantages: mutual protection from enemies, greater ease in finding mates, less opportunity for individual straying during migration, and mass huddling for protection against low night temperatures during migration. Certain species, such as pelicans (Figure 27-23), may use highly organized cooperative behavior to feed. At no time are the

Figure 27-21
Gustav Kramer's experiments with sun-compass navigation in starlings. **A,** In a windowed, circular cage, the bird fluttered to align itself in the direction it would normally follow if it were free. **B,** When the true angle of the sun is deflected with a mirror, the bird maintains the same relative position to the sun. This shows that these birds use the sun as a compass. The bird navigates correctly throughout the day, changing its orientation to the sun as the sun moves across the sky.

highly organized social interactions of birds more evident than during the breeding season, as they stake out territorial claims, select mates, build nests, incubate and hatch their eggs, and rear their young.

Reproductive System

During most of the year the **testes** of males are tiny bean-shaped bodies. But during the breeding season they enlarge greatly, as much as 300 times their nonbreeding size, afterward shrinking again to tiny bodies. Since males of most species lack a penis, copulation is a matter of bringing cloacal surfaces into contact, usually while the male stands on the back of the female (Figure 27-24). Some swifts and hawks copulate in flight.

Figure 27-22

Part of a colony of northern gannets, *Morus bassanus,* showing extremely close spacing between pairs in this highly social bird. Order Pelecaniformes.

Figure 27-24

Copulation in birds. In most bird species males lack a penis. A male copulates by standing on the back of a female, pressing his cloaca against that of the female, and passing sperm to the female.

A

B

Figure 27-23

Cooperative feeding behavior by white pelicans, *Pelecanus onocrotalus.* **A,** Pelicans form a horseshoe to drive fish together. **B,** Then they plunge simultaneously to scoop fish in their huge bills. These photographs were taken 2 seconds apart.

In females of most birds, only the left ovary and oviduct develop; those on the right dwindle to vestigial structures (Figure 27-25). Eggs discharged from the ovary are picked up by the expanded end of the oviduct, the **infundibulum.** The oviduct runs posteriorly to the cloaca. While eggs are passing down the oviduct, **albumin,** or egg white, from special glands is added to them; farther down the oviduct, shell membrane, shell, and shell pigments are also secreted around the egg. Fertilization takes place in the upper oviduct several hours before the layers of albumin, shell membranes, and shell are added. Sperm remain alive in a female's oviduct for many days after a single mating. Hen eggs show good fertility for 5 or 6 days after mating, but then fertility drops rapidly. However, an egg occasionally may be fertile as long as 30 days after separation of hens from a rooster.

Mating Systems

Two types of mating systems in animals are **monogamy,** in which an individual has only one mate and **polygamy,** in which an individual has more than one mate during a breeding period. Monogamy is rare in most animal groups, but is common in birds; more than 90% of birds are monogamous. In a few bird species, such as swans and geese, partners are chosen for life and often remain together throughout the year. Seasonal monogamy is more common as the great majority of migrant birds pair only during the breeding season, living independent lives the rest of the year and perhaps choosing a different mate the next breeding season.

The term *"polygamy"* ("many marriages") is used when the sex of the individual possessing a plurality of mates is not specified. The most common form of polygamy is polygyny ("many females"), in which a male mates with more than one female. Much less common is polyandry ("many males"), in which a female mates with more than one male per breeding season.

Although most birds have a monogamous mating system, either member of a pair may mate with an individual that is not the partner. Recent DNA analyses have shown many passerine species frequently are "unfaithful," engaging in extra-pair copulations. As a result, nests of many of these monogamous species contain a sizeable portion (30% or more) of young with fathers other than the attendant male.

One reason that monogamy is much more common among birds than among mammals is that male and female birds are equally adept at most aspects of parental care. Male mammals do not gestate the young and do not lactate and thus can provide little help in caring for the young. Female and male birds can alternate care of the nest and young, which permits one parent to be at the nest at all times. For some species, a female remains on the nest for months at a time, but is brought food by

Figure 27-25
Reproductive system of a female bird. For most birds, only the left ovary and reproductive tract are functional. Structures on the right dwindle to vestiges.

Figure 27-26
Dominant male sage grouse, *Centrocercus urophasianus,* surrounded by several hens that have been attracted by his "booming" display.

the male. This constant attendance to the nest may be particularly important in species that would experience high loss of eggs or young to predators or rival birds if a nest were left unguarded. For many bird species, the high demands on a male of caring for the young or his mate preclude the establishment of nests with additional females.

The most common form of polygamy in birds, when it occurs, is **polygyny** ("many females"), in which a male has more than one female mate. In many species of grouse, males gather in a collective display ground, the **lek,** which is divided into individual territories, each vigorously defended by a displaying male (Figure 27-26). There is nothing of value in a lek to the female except the male, and all he can offer are his genes, for only females care for the young. Usually there are a dominant male and several subordinate males in a lek. Competition among males for females is intense, but females appear to choose the dominant male for mating because, presumably, social rank correlates with genetic quality.

Polyandry ("many males"), in which a female mates with several males and the male incubates the eggs, is relatively rare in birds. It is practiced by several shorebird species, including spotted sandpipers (*Actitis macularia*). Female spotted sandpipers defend territories and mate with multiple males. Males incubate eggs within the female's territory and provide most

parental care. This unusual reproductive strategy and clustering of individuals may be in response to high predation on spotted sandpiper nests.

Nesting and Care of Young

To produce offspring, all birds lay eggs that must be incubated by one or both parents. Most duties of incubation fall on females, although in many instances both parents share the task, and occasionally only males incubate the eggs.

Most birds build some form of nest in which to rear their young. Some birds simply lay their eggs on bare ground or rocks, making no pretense of nest building. Others build elaborate nests such as pendant nests constructed by orioles, delicate lichen-covered mud nests of hummingbirds (Figure 27-27) and flycatchers, chimney-shaped mud nests of cliff swallows, and floating nests of rednecked grebes. Most birds take considerable pains to conceal their nests from enemies. Woodpeckers, chickadees, bluebirds, and many others place their nests in tree hollows or other cavities; kingfishers excavate tunnels in the banks of streams for their nests; and birds of prey build high in lofty trees or on inaccessible cliffs. Nest parasites such as the brown-headed cowbird and the European cuckoo build no nests at all but simply lay their eggs in the nests of birds smaller than themselves. When the eggs hatch, the foster parents care for the cowbird young, which outcompete the host's own hatchlings.

The developmental state of newly hatched birds varies among species. **Precocial** young, such as quail, fowl, ducks, and most water birds, are covered with down when hatched and can run or swim as soon as their plumage is dry (Figure 27-28). The most precocial birds are brush turkeys or megapodes of Australia, which incubate eggs in sand and vegetation mounds like crocodilians. The young can fly at hatching. However, most precocial young, even those able to leave the nest soon after hatching, are still fed or protected from predators by their parents for some time. **Altricial** hatchlings, on the other hand, are naked and unable to see or walk at birth and remain in the nest

Figure 27-27
Anna's hummingbird, *Calypte anna,* feeding its young in its nest of plant down and spiderwebs and decorated on the outside with lichens. The female builds the nest, incubates two pea-sized eggs, and rears the young with no assistance from a male. Anna's hummingbird is a common resident of California. It is the only hummingbird to overwinter in the United States.

Altricial
One-day-old meadowlark

Precocial
One-day-old ruffed grouse

Figure 27-28
Comparison of 1-day-old altricial and precocial young. The altricial meadowlark (*left*) hatches nearly naked, blind, and helpless. The precocial ruffed grouse (*right*) is covered with down, alert, strong legged, and able to feed itself.

for a week or more. Parents of altricial species must carry food to their young almost constantly, for young birds may eat more than their weight each day.

Although it may seem that precocial young have all the advantages, with their greater ability to find food and escape predation, altricial birds have some advantages of their own. Because altricial birds lay relatively small eggs with minimal yolk supplies, a mother has a relatively small investment in her eggs. Eggs lost to predation or extreme weather conditions are easily replaced. Altricial young also grow faster, perhaps due to the higher growth potential of immature tissue. Many birds are not easily categorized as precocial or altricial because their young are intermediate in development at birth. For example, gulls and terns are born covered with down and with eyes open, but are unable to leave the nest for some time.

BIRD POPULATIONS

Bird populations, like those of other animal groups, vary in size from year to year. Snowy owls, for example, are subject to population cycles that closely follow cycles in their food supply, mainly rodents. Voles, mice, and lemmings in the north have a fairly regular 4-year cycle of abundance; at population peaks, predator populations of foxes, weasels, and buzzards, as well as snowy owls, increase because there is abundant food for rearing their young. After a crash in the rodent population, snowy owls move south, seeking alternative food supplies. They occasionally appear in large numbers in southern Canada and the northern United States, where their total absence of fear of humans makes them easy targets for thoughtless hunters.

Occasionally activities of people may cause spectacular changes in bird distribution. Both starlings (Figure 27-29) and house sparrows have been accidentally or deliberately introduced into numerous countries, to become the two most abundant bird species on earth, with the exception of domestic fowl.

Humans also are responsible for the extinction of many bird species. More than 80 species of birds have, since 1695, followed the last dodo to extinction. Most were victims of changes in their habitat or competition with introduced species. But several have been hunted to extinction, among them passenger pigeons, which only a century ago darkened the skies over North America in incredible numbers estimated in the billions (Figure 27-30).

Today, game bird hunting is a well-managed renewable resource in the United States and Canada, and while hunters kill millions of game birds each year, none of the 74 bird species legally hunted is endangered. Hunting interests, by acquiring large areas of wetlands for migratory bird refuges and sanctuaries, have contributed to the recovery of both game and nongame birds.

Lead poisoning of waterfowl is a side effect of hunting. Before long-delayed federal regulations went into effect in 1991, requiring the use of nonlead shot for all inland and coastal waterfowl hunting, shotguns scattered more than 3000 tons of lead each year in the United States alone. When waterfowl eat the pellets (which they mistake for seeds or grist), the pellets are ground and eroded in their gizzards, facilitating absorption of lead into their blood. Lead poisoning paralyzes or weakens birds, leading to death by starvation. Today, birds are still dying from ingesting lead shot that has accumulated over the years.

Of particular concern is the recent sharp decline of songbirds in the United States and southern Canada. Amateur birdwatchers and ornithologists have recorded that many songbird species that were abundant as recently as 40 years ago are now suddenly scarce. There are several reasons for the decline. Intensification of agriculture, permitted by use of herbicides, pesticides, and fertilizers, has deprived ground-nesting birds of fields that were left fallow before use of these agents. Excessive fragmentation of forests throughout much of the United States has increased exposure of nests of forest-dwelling species to nest

A

B

Figure 27-29

A, Starling, *Sturnus vulgaris*. Starlings are omnivorous, eating mostly insects in spring and summer and shifting to wild fruits in the fall. **B,** Colonization of North America by starlings after the introduction of 120 birds into Central Park in New York City in 1890. There are now perhaps 100 million starlings in the United States alone, testimony to the great reproductive potential of birds.

Figure 27-30

Sport-shooting passenger pigeons in Louisiana during the nineteenth century. Relentless sport and market hunting, before establishment of state and federal hunting regulations, eventually dropped the population too low to sustain colonial breeding. The last passenger pigeon died in captivity in 1914.

predators such as blue jays, raccoons, and opossums, and to nest parasites such as brown-headed cowbirds. House cats also kill millions of small birds every year. From a study of radio-collared farm cats in Wisconsin, researchers estimated that in that state alone, cats may kill 19 million songbirds in a single year.

The rapid loss of tropical forests—approximately 170,000 square kilometers each year, an area about the size of the state of Washington—is depriving some 250 species of songbird migrants of their wintering homes. Recent studies indicate that stressors on the wintering grounds are seriously decreasing the physiological condition of birds, particularly songbirds, prior to

northward migration. Of all long-term threats facing songbird populations, tropical deforestation is the most serious and most intractable to change. If the rate of deforestation accelerates in the next few decades as expected, the world's tropical forests will have disappeared by 2040 (Terborgh, 1992).

Some birds, such as robins, house sparrows, and starlings, can accommodate to these changes, and may even thrive on them. But for most the changes are adverse. Terborgh (1992) warns that unless we take leadership in managing our natural resources wisely we soon could be facing the silent spring that Rachel Carson envisioned in 1962.

Classification of Living Birds of Class Aves*

Class Aves contains more than 9600 species distributed among some 28 orders of living birds and a few fossil orders. Very few birds remain to be discovered. Of the 28 orders, four (or five depending on the classification system) are **ratite,** or flightless birds of superorder Paleognathae that lack a keeled sternum (ostriches, rheas, cassowaries and emus, and kiwis), although flightlessness is not restricted to this superorder. The remaining 23 orders are birds with a keeled sternum.

Class Aves (L. *avis,* bird)

Subclass Archaeornithes (Gr. *archaios,* ancient, + *ornis,* bird). Birds of the late Jurassic and early Cretaceous bearing many primitive characteristics. *Archaeopteryx.*

Subclass Neornithes (Gr. *neos,* new, + *ornis,* bird). Extinct and living birds with well-developed sternum and usually with keel; tail reduced; metacarpals and some carpals fused together. Cretaceous to Recent.

Superorder Paleognathae (Gr. *palaios,* ancient, + *gnathos,* jaw). Modern birds with primitive archosaurian palate. Ratites (with unkeeled sternum) and tinamous (with keeled sternum).

Order Struthioniformes (stroo'thi-on-i-for'meez) (L. *struthio,* ostrich, + *forma,* form): **ostrich.** One species, the flightless ostrich of Africa (*Struthio camelus*) (Figure 27-31) is the largest of living birds, with some specimens being 2.4 m tall and weighing 135 kg. The feet are provided with only two toes of unequal size covered with pads, which enable the birds to travel rapidly through sandy country.

Order Rheiformes (re'i-for'meez) (Gr. mythology, *Rhea,* mother of Zeus, + form): **rheas.** Two species of flightless birds restricted to South America; often called American ostriches.

Order Casuariiformes (kazh'u-ar'ee-i-for'meez) (N.L. *Casuarius,* type genus, + form): **cassowaries, emus.** Four species of flightless birds found in Australia, New Guinea, and a few other islands. Some specimens may reach a height of 1.5 m.

Order Apterygiformes (ap'te-rij'i-for'meez) (Gr. *a,* not, + *pteryx,* wing, + form): **kiwis.** Three species of kiwis, flightless birds about the size of domestic fowl, found only in New Zealand. Only the merest vestige of a wing is present. The egg is extremely large for the size of the bird.

Order Tinamiformes (tin-am'i-for'meez) (N.L. *Tinamus,* type genus, + form): **tinamous.** Ground-dwelling, grouse-like birds of Central and South America. About 60 species.

Superorder Neognathae (Gr. *neos,* new, + *gnathos,* jaw). Modern birds with flexible palate.

Order Sphenisciformes (sfe-nis'i-for'meez) (Gr. *Sphēniskos,* dim. of *sphen,* wedge, from the shortness of the wings, + form): **penguins.** Web-footed marine swimmers of southern seas from Antarctica north to the Galápagos Islands. Although penguins are carinate birds, they use their wings as paddles for swimming rather than for flight. About 17 species.

Order Gaviiformes (gay'vee-i-for'meez) (L. *gavia,* bird, probably sea mew, + form): **loons.** The four species of loons are remarkable swimmers and divers with short legs and heavy bodies. They live exclusively on fish and small aquatic forms. The familiar great northern diver (*Gavia immer*) is found mainly in northern waters of North America and Eurasia.

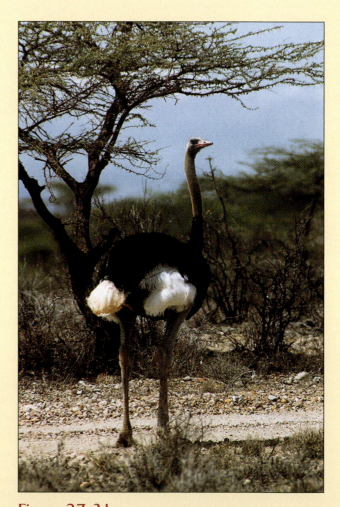

Figure 27-31

Ostrich, *Struthio camelus,* of Africa, the largest of all living birds. Order Struthioniformes.

---

*The traditional bird classification given here, called a morphological taxonomy, is based on careful comparison of shared derived anatomical characters within and between bird groups. A new and still controversial biochemical classification based on degrees of similarity between DNAs of living birds from all over the world is believed by its proponents to represent true phylogenetic relationships much better than the traditional morphological classification. Biochemical taxonomy has produced several astonishing realignments. Most prominent of these is the sweeping revision of order Ciconiiformes, which, as revised, includes penguins, loons, grebes, albatrosses, and birds of prey, all previously placed in separate orders. DNA hybridization studies establish close relatedness of these groups, whose true genetic affinities are masked by divergent evolution. Biochemical taxonomy, now under review by the American Ornithological Union, is certain to produce significant revision of the traditional taxonomy, which has been the standard for more than a century. Proctor and Lynch (1993) compare the biochemical classification reported by Sibley and Ahlquist (1990) with the traditional morphological classification.

Order Podicipediformes (pod′i-si-ped′i-for′meez) (L. *podex,* rump, + *pes, pedis,* foot): **grebes.** These are short-legged divers with lobate-webbed toes. The pied-billed grebe (*Podilymbus podiceps*) is a familiar example of this order. Grebes are most common in old ponds where they build their raftlike floating nests. Eighteen species, worldwide distribution.

Order Procellariiformes (pro-sel-lar′ee-i-for′meez) (L. *procella,* tempest, + form): **albatrosses, petrels, fulmars, shearwaters.** All are marine birds with hooked beak and tubular nostrils. In wingspan (more than 3.6 m in some), albatrosses are the largest of flying birds. About 100 species, worldwide distribution.

Order Pelecaniformes (pel-e-can-i-for′meez) (Gr. *pelekan,* pelican, + form): **pelicans, cormorants, gannets, boobies, and others.** These are colonial fish-eaters with throat pouch and all four toes of each foot included within the web. About 55 species, worldwide distribution, especially in the tropics.

Order Ciconiiformes (si-ko′nee-i-for′meez) (L. *ciconia,* stork, + form): **herons, bitterns, storks, ibises, spoonbills, flamingos** (Figure 27-32) **and vultures.** These are long-necked, long-legged, mostly colonial waders and vultures. A familiar eastern North American representative is the great blue heron (*Ardea herodias*), which frequents marshes and ponds. About 90 species, worldwide distribution.

Order Anseriformes (an′ser-i-for′meez) (L. *anser,* goose, + form): **swans, geese, ducks.** The members of this order have broad bills with filtering ridges at their margins, a foot web restricted to the front toes, and a long breastbone with a low keel. About 150 species, worldwide distribution.

Order Falconiformes (fal′ko-ni-for′meez) (L. *falco,* falcon, + form): **eagles, hawks, falcons, condors, buzzards.** Diurnal birds of prey. All are strong fliers with keen vision. About 270 species, worldwide distribution.

Order Galliformes (gal′li-for′meez) (L. *gallus,* cock, + form): **quail, grouse, pheasants, ptarmigan, turkeys, domestic fowl.** Chickenlike ground-nesting herbivores

Figure 27-32
Greater flamingos, *Phoenicopterus ruber,* on an alkaline lake in East Africa. Order Ciconiiformes.

Figure 27-33
Laughing gulls, *Larus atricilla,* in flight. Order Charadriiformes.

with strong beaks and heavy feet. The bobwhite quail (*Colinus virginianus*) is found all over the eastern half of the United States. The ruffed grouse (*Bonasa umbellus*) is found in about the same region, but in woods instead of the open pastures and grain fields, which the bobwhite frequents. About 250 species, worldwide distribution.

Order Gruiformes (groo′i-for′meez) (L. *grus,* crane, + form): **cranes, rails, coots, gallinules.** Mostly prairie and marsh breeders. About 215 species, worldwide distribution.

Order Charadriiformes (ka-rad′ree-i-for′meez) (N.L. *Charadrius,* genus of plovers, + form): **gulls** (Figure 27-33), **oyster catchers, plovers, sandpipers, terns, woodcocks, turnstones, lapwings, snipe, avocets, phalaropes, skuas, skimmers, auks, puffins.** All are shorebirds. They are strong fliers and are usually colonial. About 330 species, worldwide distribution.

Order Columbiformes (co-lum′bi-for′meez) (L. *columba,* dove, + form): **pigeons, doves.** All have short necks, short legs, and a short, slender bill. About 290 species, worldwide distribution.

Order Psittaciformes (sit′ta-si-for′meez) (L. *psittacus,* parrot, 1 form): *parrots, parakeets.* Birds with hinged and movable upper beak, fleshy tongue. About 320 species, pantropical distribution.

Order Musophagiformes (myu′-so-fa-ji-for′meez): (L. *musa,* banana, + Gr. *phagō,* to eat + form): **turacos.** Medium to large birds of dense forest or forest edge with a conspicuous patch of crimson on the spread wing. Bill brightly colored, wings short and rounded. Six species restricted to Africa.

Order Cuculiformes (ku-koo′li-for′meez) (L. *cuculus,* cuckoo, + form): **cuckoos, roadrunners.** European cuckoos (*Cuculus canorus*) lay their eggs in nests of smaller birds, which rear the young cuckoos. American cuckoos, black billed and yellow billed, usually rear their own young. About 150 species, worldwide distribution.

Order Strigiformes (strij′i-for′meez) (L. *strix,* screech owl, + form): **owls.** Nocturnal predators with large eyes, powerful beaks and feet, and silent flight. About 135 species, worldwide distribution.

Classification of Living Birds of Class Aves

Order Caprimulgiformes (kap'ri-mul'ji-for'meez) (L. *caprimulgus,* goatsucker, + form): **goatsuckers, nighthawks, whippoorwills.** Night and twilight feeders with small, weak legs and wide mouths fringed with bristles. Whippoorwills (*Antrostomus vociferus*) are common in the woods of the eastern states, and nighthawks (*Chordeiles minor*) are often seen and heard in the evening flying around city buildings. About 95 species, worldwide distribution.

Order Apodiformes (up-pod'i-for'meez) (Gr. *apous,* footless, + form): **swifts, hummingbirds.** These are small birds with short legs and rapid wingbeat. The familiar chimney swift (*Chaetura pelagia*) fastens its nest in chimneys by means of saliva. A swift found in China builds a nest of saliva that is used by Chinese people for soup making. Most species of hummingbirds are found in the tropics, but there are 14 species in the United States, of which only one, the ruby-throated hummingbird, is found in the eastern part of the country. About 400 species, worldwide distribution.

Order Coliiformes (ka-ly'i-for'meez) (Gr. *kolios,* green woodpecker, + form): **mousebirds.** Small birds of uncertain relationship. Six species restricted to southern Africa.

Order Trogoniformes (tro-gon'i-for'meez) (Gr. *trōgon,* gnawing, + form): **trogons.** Richly colored, long-tailed birds. About 35 species, pantropical distribution.

Order Coraciiformes (ka-ray'see-i-for'meez or kor'uh-sigh'uh-for'meez) (N.L. *coracii* from Gr. *korakias,* a kind of raven, + form): **kingfishers, hornbills, and others.** Birds with strong, prominent bills that nest in cavities. In the eastern half of the United States, belted kingfishers (*Megaceryle alcyon*) are common along most waterways of any size. About 200 species, worldwide distribution.

Order Piciformes (pis'i-for'meez) (L. *picus,* woodpecker, + form): **woodpeckers, toucans, puffbirds, honeyguides.** Birds with highly specialized bills and having two toes extending forward and two backward. All nest in cavities. There are many species of woodpeckers in North America, most common of which are flickers and downy, hairy, red-bellied, redheaded, and yellow-bellied woodpeckers. Largest is the pileated woodpecker, which is usually found in deep and remote woods. About 380 species, worldwide distribution.

Order Passeriformes (pas'er-i-for'meez) (L. *passer,* sparrow, + form): **perching songbirds** (Figure 27-34). This is the largest order of birds, containing 56 families and 60% of all birds. Most have a highly developed syrinx. Their feet are adapted for perching on thin stems and twigs. The young are altricial. To this order belong many birds with beautiful songs such as thrushes, warblers, mockingbird, meadowlark, and hosts of others. Others of this order, such as swallows, magpie, starling, crows, raven, jays, nuthatch, and creeper, have no songs worthy of the name. More than 5000 species, worldwide distribution.

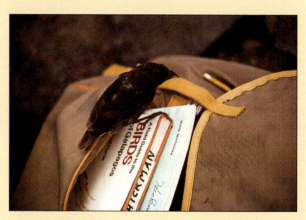

Figure 27-34
Ground finch, *Geospiza fuliginosa,* one of the famous Darwin's finches of the Galápagos Islands. Order Passeriformes.

SUMMARY

The more than 9600 species of living birds are egg-laying, endothermic vertebrates with feathers and having forelimbs modified as wings. Birds are closest phylogenetically to theropods, a group of Mesozoic dinosaurs with several birdlike characteristics. The oldest known fossil bird, *Archaeopteryx* from the Jurassic period of the Mesozoic era, had numerous reptilian characteristics and was almost identical to certain theropod dinosaurs except that it had feathers. It is probably the sister group of modern birds.

Adaptations of birds for flight are of two basic kinds: those reducing body weight and those promoting more power for flight. Feathers, the hallmark of birds, are complex derivatives of reptilian scales and combine lightness with strength, water repellency, and high insulative value. Body weight is further reduced by elimination of some bones, fusion of others (to provide rigidity for flight). The light, horny bill, replacing the heavy jaws and teeth of reptiles,

serves as both hand and mouth for all birds and is variously adapted for different feeding habits.

Adaptations that provide power for flight include high metabolic rate and body temperature coupled with an energy-rich diet; a highly efficient respiratory system consisting of a system of air sacs arranged to pass air through the lungs during both inspiration and expiration; powerful flight and leg muscles arranged to place muscle weight near the bird's center of gravity; and an efficient, high-pressure circulation.

Birds have keen eyesight, good hearing, and superb coordination for flight. The metanephric kidneys produce uric acid as the principal nitrogenous waste.

Birds fly by applying the same aerodynamic principles as an airplane and using similar equipment: wings for lift, support, and propulsion, a tail for steering and landing control, and wing slots for control at low flight speed. Flightlessness in birds is unusual but has evolved independently in several bird orders, usually on islands where terrestrial predators are absent; all are derived from flying ancestors.

Bird migration refers to regular movements between summer nesting places and wintering regions. Spring migration to the north, where more food is available for nestlings, enhances reproductive success. Many cues are used for finding direction during migration, including innate sense of direction and ability to navigate by the sun, stars, or the earth's magnetic field.

The highly developed social behavior of birds is manifested in vivid courtship displays, mate selection, territorial behavior, and incubation of eggs and care of the young.

REVIEW QUESTIONS

1. Explain the significance of the discovery of *Archaeopteryx*. Why did this fossil demonstrate beyond reasonable doubt that birds share an ancestor with some reptilian groups?
2. The special adaptations of birds contribute to two essentials for flight: more power and less weight. Explain how each of the following contributes to one or both of these two essentials: feathers, skeleton, muscle distribution, digestive system, circulatory system, respiratory system, excretory system, reproductive system.
3. How do marine birds rid themselves of excess salt?
4. In what ways are a bird's ears and eyes specialized for the demands of flight?
5. Explain how a bird's wing is designed to provide lift. What design features help to prevent stalling at low flight speeds?
6. Describe four basic forms of bird wings. How does wing shape correlate with bird size and nature of flight (whether powered or soaring)?
7. What are advantages of seasonal migration for birds?
8. Describe different navigational resources birds may use in long-distance migration.
9. What are some advantages of social aggregation among birds?
10. More than 90% of all bird species are monogamous. Explain why monogamy is much more common among birds than among mammals.
11. Briefly describe an example of polygyny and an example of polyandry among birds.
12. Define the terms precocial and altricial as they relate to birds.
13. Offer some examples of how human activities have affected bird populations.

SELECTED REFERENCES

Ackerman, J. 1998. Dinosaurs take wing. Nat. Geog. **194**(1):74–99. *Beautifully illustrated synopsis of dinosaur to bird evolution.*

Brooke, M., and T. Birkhead (eds.). 1991. The Cambridge encyclopedia of ornithology. New York, Cambridge University Press. *Comprehensive, richly illustrated treatment that includes a survey of all modern bird orders.*

Elphick, J. (ed.). 1995. The atlas of bird migration: tracing the great journeys of the world's birds. New York, Random House. *Lavishly illustrated collection of maps of birds' breeding and wintering areas, migration routes, and many facts about each bird's migration journey.*

Emlen, S. T. 1975. The stellar-orientation system of a migratory bird. Sci. Am. **233**:102–111 (Aug.). *Describes fascinating research with indigo buntings, revealing their ability to navigate by the center of celestial rotation at night.*

Feduccia, A. 1996. The origin and evolution of birds. New Haven, Yale University Press. *An updated successor to the author's* The Age of Birds *(1980) but more comprehensive; rich source of information on evolutionary relationships of birds.*

Norbert, U. M. 1990. Vertebrate flight. New York, Springer-Verlag. *Detailed review of the mechanics, physiology, morphology, ecology, and evolution of flight. Covers bats as well as birds.*

Padian, K., and L. M. Chiappe. 1998. The origin of birds and their flight. Sci. Am. **279**:38–47 (Feb.). *The authors argue that birds evolved from small, predatory dinosaurs that lived on the ground.*

Proctor, N. S., and P. J. Lynch. 1993. Manual of ornithology: avian structure and function. New Haven, Connecticut, Yale University Press.

Sibley, C. G., and J. E. Ahlquist. 1990. Phylogeny and classification of birds: a study in molecular evolution. New Haven, Yale University Press. *A comprehensive application of DNA annealing experiments to the problem of resolving avian phylogeny.*

Terborgh, J. 1992. Why American songbirds are vanishing. Sci. Am. **266**:98–104 (May). *The number of songbirds in North America has been dropping sharply. The author suggests reasons why.*

Terres, J. K. 1980. The Audubon Society encyclopedia of North American birds. New York, Alfred A. Knopf, Inc. *Comprehensive, authoritative, and richly illustrated.*

Waldvogel, J. A. 1990. The bird's eye view. Am. Sci. **78**:342–353 (July–Aug.). *Birds possess visual abilities unmatched by humans. So how can we know what they really see?*

Wellnhofer, P. 1990. Archaeopteryx. Sci. Am. **262**:70–77 (May). *Description of perhaps the most important fossil ever discovered.*

Welty, J. C., and L. Baptista. 1988. The life of birds, ed. 4. Philadelphia, Saunders College Publishing. *Among the best of the ornithology texts; lucid style and well illustrated.*

ZOOLOGY LINKS TO THE INTERNET

Visit the textbook's Online Learning Center at www.mhhe.com/zoology to find live Internet links for each of the topics listed here.

Class Aves
Marine Birds
Dissection Guides for Birds
Conservation Issues Concerning Birds
Movement of Populations

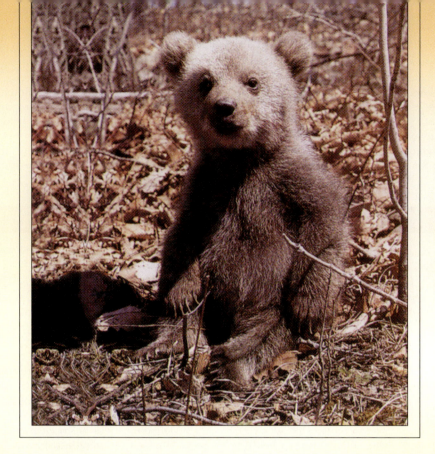

Juvenile grizzly bear.

Mammals

Phylum Chordata
Class Mammalia

The Tell-Tale Hair

If Fuzzy Wuzzy, the bear that had no hair (according to the children's rhyme), was truly hairless, he could not have been a mammal or a bear. For hair is as much an unmistakable characteristic of mammals as feathers are of birds. If an animal has hair it is a mammal; if it lacks hair it must be something else. It is true that many aquatic mammals are nearly hairless (whales, for example) but hair can usually be found (with a bit of searching) at least in vestigial form somewhere on the body of an adult. Unlike feathers, which evolved from converted reptilian scales, mammalian hair is a completely new epidermal structure. Mammals use their hair for protection from harsh environments, for protective coloration and concealment, for waterproofing and buoyancy, and for behavioral signaling; they have turned hairs into sensitive vibrissae on their snouts and into prickly quills. Perhaps most important, mammals use their hair for thermal insulation, which allows them to enjoy the great advantages of homeothermy. Warm-blooded animals in most climates and at sunless times benefit from this natural and controllable protective insulation.

Hair, of course, is only one of several features that together characterize a mammal and help us to understand the mammalian evolutionary achievement. Most mammals have a highly developed placenta for feeding the embryo; mammary glands for nourishing the newborn; and a surpassingly advanced nervous system that far exceeds in performance that of any other animal group. It is doubtful, however, that even with this winning combination of adaptations, mammals could have triumphed as they have without their hair.

Position in the Animal Kingdom

Modern mammals are descendants of the synapsid lineage of amniotes that appeared in the Permian period. The synapsid lineage is characterized in the primitive condition by having a skull with a single temporal opening (Figure 28-1B). Modern mammals are endothermic and homeothermic, have bodies partially or wholly covered with hair, and have mammary glands that secrete milk for the nourishment of the young. These derived characteristics, together with several distinctive skeletal characteristics, a highly developed nervous system, and complex individual and social behavior, distinguish mammals from all other living amniotes. Their genetic plasticity and numerous derived adaptations have enabled mammals to invade almost every environment that supports life on earth.

Biological Contributions

1. Mammals share with birds both **endothermy** and **homeothermy,** which permit a high level of activity at night, and year-round penetration into low temperature habitats denied to ectothermic vertebrates.
2. The **placenta** in placental mammals allows developing young to obtain nourishment and grow in a protected environment during the most vulnerable period of their lives. After birth the young continue to feed by suckling from **mammary glands.** A long period of parental care and education allows the young to acquire skills necessary for survival.
3. **Specialization of mammalian teeth** for different functions permitted evolution of many different feeding specializations in mammals. The **secondary palate,** which separates the air passageway from the food passageway, enables mammals to hold and partially break down food in their mouths without interrupting breathing.
4. Convoluted **turbinate bones** in the nasal cavity provide a high surface area for warming and moistening inspired air and for reducing moisture loss during exhalation.
5. The highly evolved brain, especially the large **neocortex** (p. 706), has bequeathed mammals with a well-developed memory and capacity to learn rapidly and to respond appropriately to problems not previously encountered. Highly elaborated **sense organs** and **special senses,** particularly those of hearing, smell, and touch, contribute an inflow of environmental information that, together with their processing brain centers, provide mammals with a level of environmental awareness and responsiveness unequaled in the animal kingdom.

Mammals, with their highly developed nervous system and numerous ingenious adaptations, occupy almost every environment on earth that supports life. Although not a large group (about 4600 species as compared with more than 9000 species of birds, approximately 24,600 species of fishes, and 800,000 species of insects), class Mammalia (mam-may′lee-a) (L. *mamma,* breast) may be the most biologically differentiated group in the animal kingdom. Mammals are exceedingly diverse in size, shape, form, and function. They range in size from the recently discovered Kitti's hognosed bat in Thailand, weighing only 1.5 g, to blue whales, exceeding 130 metric tons.

Yet, despite their adaptability and in some instances because of it, mammals have been influenced by the presence of humans more than any other group of animals. We have domesticated numerous mammals for food and clothing, as beasts of burden, and as pets. We use millions of mammals each year in biomedical research. We have introduced alien mammals into new habitats, occasionally with benign results but more frequently with unexpected disaster. Although history provides numerous warnings, we continue to overcrop valuable wild stocks of mammals. The whaling industry has threatened itself with total collapse by exterminating its own resource—a classic example of self-destruction in the modern world, in which competing segments of an industry are intent only on reaping all they can today as though tomorrow's supply were of no concern whatever. In some cases destruction of a valuable mammalian resource has been deliberate, such as the officially sanctioned (and tragically successful) policy during the Indian wars of exterminating bison to drive the Plains Indians into starvation. Although commercial hunting has declined, the ever-increasing human population with accompanying destruction of wild habitats has harassed and disfigured mammalian faunas. Approximately 300 species and subspecies of mammals are considered endangered by the International Union for the Conservation of Nature and Natural Resources (IUCN), including most cetaceans, cats (except domestic cats), otters, and primates (except humans).

An international moratorium on all commercial whaling took effect in 1986. However some countries that objected to the moratorium, notably Japan, are still killing hundreds of whales each year under the guise of "scientific" whaling.

We are becoming increasingly aware that our presence on this planet makes us responsible for the character of our natural environment. Since our welfare has been and continues to be closely related to that of the other mammals, it is clearly in our interest to preserve the natural environment of which all mammals, ourselves included, are a part. We need to remember that nature can do without humans but humans cannot exist without nature.

ORIGIN AND EVOLUTION OF MAMMALS

The evolutionary descent of mammals from their earliest amniote ancestors is perhaps the most fully documented transition in vertebrate history. From the fossil record, we can trace the derivation over 150 million years of endothermic, furry mammals from their small, ectothermic, hairless ancestors. Skull structures and especially teeth are the most abundant fossils, and it is largely from these structures that we can identify the evolutionary descent of mammals.

The structure of the skull roof permits us to identify three major groups of amniotes that diverged in the Carboniferous period of the Paleozoic era, the **synapsids, anapsids,** and **diapsids.** The synapsid group which includes the mammals and their ancestors, has a pair of openings in the skull roof for attachment of jaw muscles (Figure 28-1B). Synapsids were the first amniote group to radiate widely into terrestrial habitats. As discussed in Chapter 26, the anapsid group is characterized by solid skulls and includes turtles and their ancestors (Figure 28-1A). The diapsids have two pairs of openings in the skull roof (Figure 28-1C; see also Figure 26-2, p. 537) and this group contains dinosaurs, lizards, snakes, crocodilians, birds, and their ancestors.

The earliest synapsids radiated extensively into diverse herbivorous and carnivorous forms often collectively called **pelycosaurs** (Figures 28-2 and 28-3). These early synapsids were the most common amniotes in the early Permian. Pelycosaurs share a general outward resemblance to lizards, but this resemblance is misleading. Pelycosaurs are not closely related to lizards, which are diapsids, nor are they a monophyletic group. From one group of early carnivorous synapsids arose the **therapsids** (Figure 28-3), the only synapsid group to survive beyond the Paleozoic. With therapsids we see for the first time an efficient erect gait with upright limbs positioned beneath the body. Since stability was reduced by raising the animal from the ground, the cerebellum, muscular coordination center of the brain, assumed an expanded role. Changes in the morphology of the skull and mandibular adductor muscles associated with increased feeding efficiency began with the early therapsids. Therapsids radiated into numerous herbivorous and carnivorous forms; however most early forms disappeared during the great extinction event at the end of the Permian. Previously, pelycosaurs and therapsids have been referred to as "mammal-like reptiles," but this phrase has fallen out of use, since they are not part of the reptile lineage.

One therapsid group to survive into the Mesozoic era was the **cynodonts.** Cynodonts evolved several features that supported a high metabolic rate; increased and specialized jaw musculature, permitting a stronger bite; several skeletal changes, supporting greater agility; **turbinate bones,** in the nasal cavity, aiding reten- tion of body heat (Figure 28-4); and a secondary bony palate (Figure 28-4), enabling an animal to breathe while holding prey or chewing food. The secondary palate would be important to subsequent mammalian evolution by permitting the young to breathe while suckling. Along with improved biomechanical shift to upright posture in cynodonts, the long bones became more slender and developed bony processes at the joints for firmer muscle attachment. Loss of lumbar ribs in cynodonts is correlated with the evolution of a diaphragm and also may have provided greater dorsoventral flexibility of the spinal column. Within the diverse cynodont clade (Figure 28-3), a small carnivorous group called trithelodontids most closely resembles the mammals, sharing with them several derived features of the skull and teeth.

The earliest mammals of the late Triassic period were small mouse- or shrew-sized animals with enlarged crania, redesigned jaws, and a new type of dentition, called **diphyodont,** in which teeth are replaced only once (deciduous and permanent teeth). This contrasts with the primitive amniote pattern of continual tooth replacement throughout life (polyphyodont teeth). With evolution of a new jaw joint between the dentary and squamosal

Figure 28-1

Skulls of early amniotes, showing the pattern of temporal openings that distinguish the three groups.

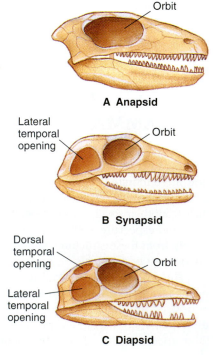

A Anapsid

B Synapsid

C Diapsid

Characteristics of Class Mammalia

1. **Body mostly covered with hair,** but reduced in some
2. **Integument** with **sweat, scent, sebaceous,** and **mammary glands**
3. Skull with **two occipital condyles** and **secondary palate; turbinate bones** in nasal cavity; middle ear with **three ossicles** (malleus, incus, stapes); **seven cervical vertebrae** (except some xenarthrans [edentates] and manatees); **pelvic bones fused**
4. Mouth with **diphyodont teeth** (milk, or deciduous, teeth replaced by a permanent set); teeth heterodont in most (varying in structure and function); lower jaw a **single enlarged bone (dentary)**
5. **Movable eyelids** and **fleshy external ears (pinnae)**
6. Circulatory system of a four-chambered heart (two atria and two ventricles), **persistent left aorta,** and **nonnucleated, biconcave red blood corpuscles**
7. Respiratory system of lungs with alveoli, and larynx; **secondary palate** (anterior bony palate and posterior continuation of soft tissue, the soft palate) separates air and food passages (Figure 28-4); **muscular diaphragm** for air exchange separates thoracic and abdominal cavities
8. Excretory system of metanephric kidneys with ureters that usually open into a bladder
9. Brain highly developed, especially **cerebral cortex;** 12 pairs of cranial nerves
10. Endothermic and homeothermic
11. Cloaca present only in monotremes (present but shallow in marsupials)
12. Separate sexes; reproductive organs of a penis, testes (usually in a scrotum), ovaries, oviducts, and uterus; sex determination by males (heterogametic)
13. Internal fertilization; **embryos develop in a uterus** with **placental attachment** (placenta absent in monotremes); **fetal membranes (amnion, chorion, allantois)**
14. Young nourished by **milk from mammary glands**

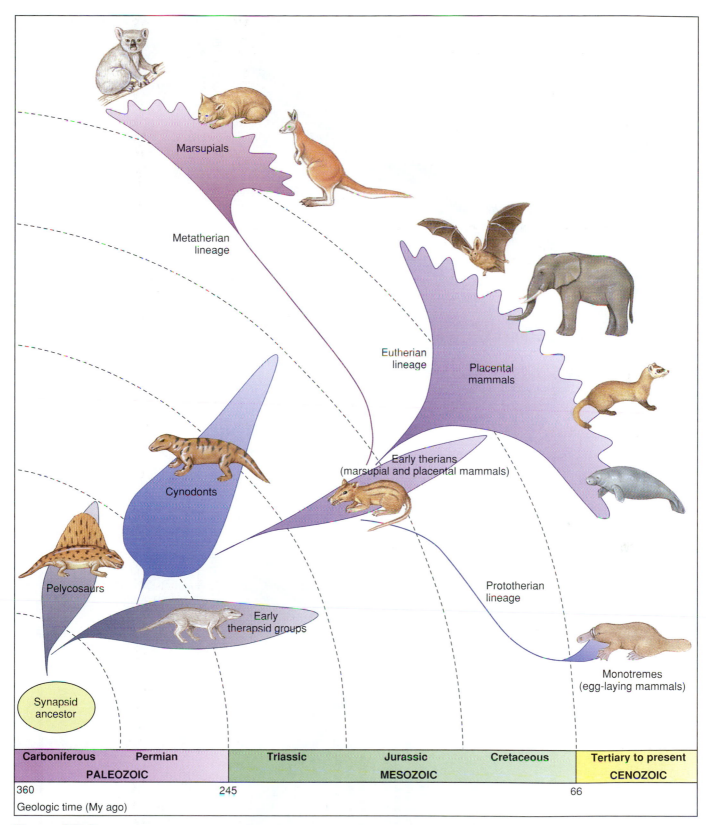

Figure 28-2

Evolution of major groups of synapsids. The synapsid lineage, characterized by a lateral temporal opening, began with the pelycosaurs, dominant amniotes of the early Permian. Pelycosaurs radiated extensively and evolved changes in jaws, teeth, and body form that presaged several mammalian characteristics. These trends continued in their successors, the therapsids, especially in the cynodonts. One lineage of cynodonts gave rise in the Triassic to therians, the *placental* mammals. Fossil evidence, as currently interpreted, indicates that all three groups of living mammals—monotremes, marsupials, and placentals—are derived from the same lineage. The great radiation of modern placental orders occurred during the Cretaceous and Tertiary periods.

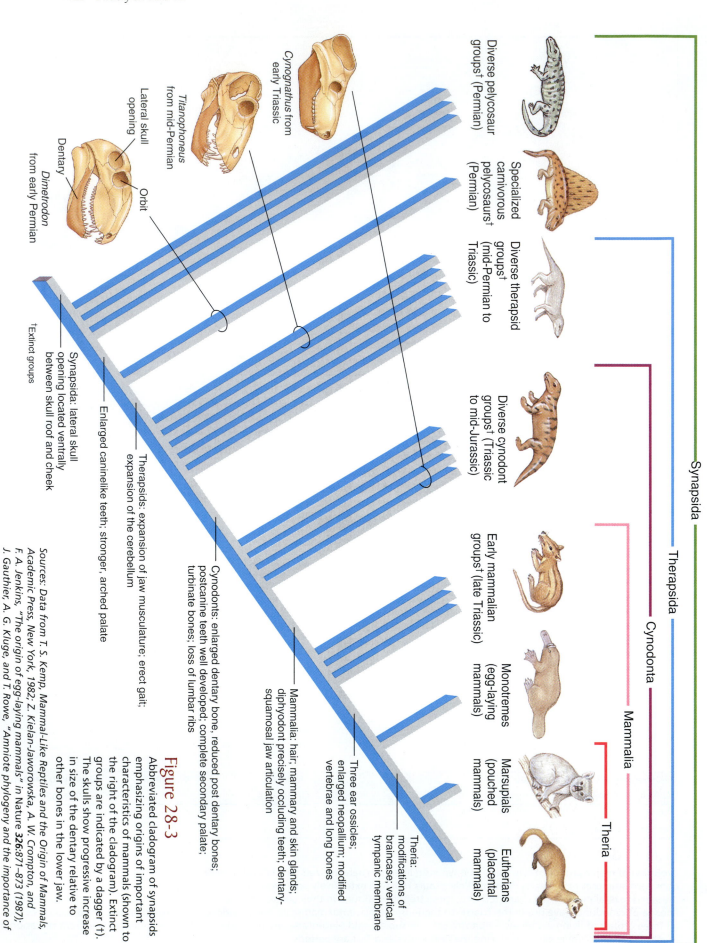

Diverse pelycosaur groups† (Permian)

Specialized carnivorous pelycosaurs† (Permian)

Diverse therapsid groups† (mid-Permian to Triassic)

Diverse cynodont groups† (Triassic to mid-Jurassic)

Early mammalian groups† (late Triassic)

Monotremes (egg-laying mammals)

Marsupials (pouched mammals)

Eutherians (placental mammals)

Cynognathus from early Triassic

Titanophoneus from mid-Permian

Lateral skull opening

Dentary

Orbit

Dimetrodon from early Permian

Synapsida: lateral skull opening located ventrally between skull roof and cheek

†Extinct groups

Enlarged caninelike teeth; stronger, arched palate

Therapsids: expansion of jaw musculature; erect gait; expansion of the cerebellum

Cynodonts: enlarged dentary bone, reduced post dentary bones; postcanine teeth well developed; complete secondary palate; turbinate bones; loss of lumbar ribs

Mammalia: hair; mammary and skin glands; diphyodont precisely occluding teeth; dentary-squamosal jaw articulation

Three ear ossicles; enlarged neopallium; modified vertebrae and long bones

Theria: modifications of braincase; vertical tympanic membrane

Synapsida

Therapsida

Cynodonta

Mammalia

Theria

Figure 28-3

Abbreviated cladogram of synapsids emphasizing origins of important characteristics of mammals (shown to the right of the cladogram). Extinct groups are indicated by a dagger (†). The skulls show progressive increase in size of the dentary relative to other bones in the lower jaw.

Sources: Data from T. S. Kemp, Mammal-Like Reptiles and the Origin of Mammals, Academic Press, New York, 1982; Z. Kielan-Jaworowska, A. W. Crompton, and F. A. Jenkins, "The origin of egg-laying mammals" in Nature 326:871–873 (1987); J. Gauthier, A. G. Kluge, and T. Rowe, "Amniote phylogeny and the importance of fossils" in Cladistics 4:105–209 (1988); R. L. Carroll, Vertebrate Paleontology and Evolution, W. H. Freeman, New York, 1988; and F. H. Pough, C. M. Janis, J. B. Heiser, Vertebrate Life, 6th edition, Prentice Hall, New Jersey, 2002.

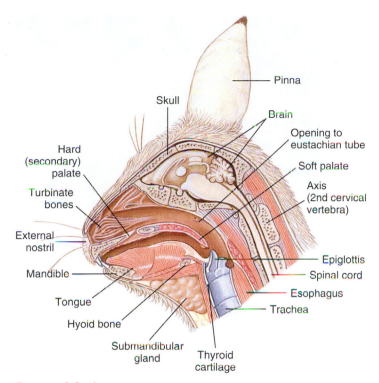

Figure 28-4
Sagittal section of the head of a rabbit.

Class Mammalia includes 21 orders: one order containing monotremes, one order containing marsupials, and 19 orders of placentals. A complete classification is on pp. 606–608.

STRUCTURAL AND FUNCTIONAL ADAPTATIONS OF MAMMALS

Integument and Its Derivatives

Mammalian skin and especially its modifications distinguish mammals as a group. As the interface between an animal and its environment, the skin is strongly molded by an animal's way of life. In general the skin is thicker in mammals than in other classes of vertebrates, although as in all vertebrates it is composed of **epidermis** and **dermis** (see Figure 28-5). The epidermis is thinner where it is well protected by hair, but in places that are subject to much contact and use, such as palms or soles, its outer layers become thick and cornified with keratin.

Hair

Hair is especially characteristic of mammals, although humans are not very hairy creatures and, in whales, hair is reduced to only a few sensory bristles on the snout. A hair grows from a hair follicle that, although an epidermal structure, is sunk into the dermis of the skin (Figure 28-5). Hair grows continuously by rapid proliferation of cells in the follicle. As the hair shaft is pushed upward, new cells are carried away from their source of nourishment and die, filled with the same dense type of fibrous protein, called **keratin,** that constitutes nails, claws, hooves, and feathers. Thus, true hair, found only in mammals, is composed of dead, keratin-packed epidermal cells.

Mammals characteristically have two kinds of hair forming their **pelage** (fur coat): (1) dense and soft **underhair** for insulation and (2) coarse and longer **guard hair** for protection against wear and to provide coloration. Underhair traps a layer of insulating air. In aquatic mammals, such as fur seals, otters, and beavers, it is so dense that it is almost impossible to wet. In water, guard hairs become wet and mat down, forming a protective blanket over the underhair (Figure 28-6).

When a hair reaches a certain length, it stops growing. Normally it remains in its follicle until a new growth starts, whereupon it falls out. In most mammals there are periodic molts of

(temporal) bones in mammals, bones of the previous jaw joint, the articular and quadrate, continued their gradual reduction in size and became relocated in the middle ear, where they became the malleus and incus, respectively. The earliest mammals were almost certainly endothermic, although their body temperature would have been rather lower than modern placental mammals. Hair was essential for insulation, and the presence of hair implies that sebaceous and sweat glands must also have evolved at this time to lubricate the hair and promote heat loss. The fossil record is silent on the appearance of mammary glands, but they must have evolved before the end of the Triassic. The young of early mammals probably hatched from eggs in a very immature condition, totally dependent on maternal milk, warmth, and protection. This mode of reproduction occurs today only in monotremes (echidnas and platypus).

Oddly, early mammals of the mid-Triassic, having developed nearly all novel attributes of modern mammals, had to wait for another 150 million years before they could achieve their great diversity. While dinosaurs became diverse and abundant, all non-mammalian synapsid groups became extinct. But mammals survived, first as shrewlike, probably nocturnal, creatures. Then, in the Cretaceous period, but especially during the Eocene epoch that began about 54 million years ago, modern mammals began to diversify rapidly. The great Cenozoic radiation of mammals is partly attributed to numerous habitats vacated by extinction of many amniote groups at the end of the Cretaceous. Mammalian radiation was almost certainly promoted by the facts that mammals were agile, endothermic, intelligent, adaptable, and gave birth to living young, which they protected and nourished from their own milk supply, thus dispensing with vulnerable eggs laid in nests.

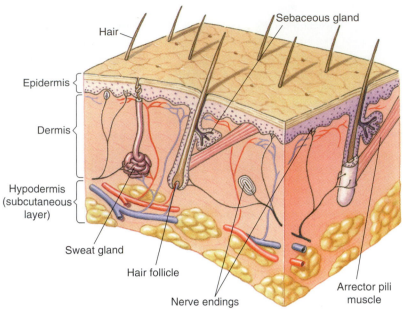

Figure 28-5
Structure of human skin (epidermis and dermis) and hypodermis, showing hair and glands.

Figure 28-6
American beaver, *Castor canadensis,* gnawing on an aspen tree. This second largest rodent (the South American capybara is larger) has a heavy waterproof pelage consisting of long, tough guard hairs overlying the thick, silky underhair so valued in the fur trade. Order Rodentia, family Castoridae.

the entire coat. In humans, hair is shed and replaced throughout life (although balding males confirm that replacement is not assured!).

A hair is more than a strand of keratin. It consists of three layers: the medulla or pith in the center of the hair, the cortex with pigment granules next to the medulla, and the outer cuticle composed of imbricated scales. Hair of different mammals shows a considerable range of structure. It may be deficient in cortex, such as the brittle hair of deer, or it may be deficient in medulla, such as the hollow, air-filled hairs of wolverines. Hairs of rabbits and some others are scaled to interlock when pressed together. Curly hair, such as that of sheep, grows from curved follicles.

In the simplest cases, such as foxes and seals, the coat is shed once every summer. Most mammals have two annual molts, one in the spring and one in the fall. Summer coats are always much thinner than winter coats and in some mammals it may be a different color. Several northern mustelid carnivores, for example, weasels, have white winter coats and brown-colored summer coats. It was once believed that the white inner pelage of arctic animals conserved body heat by reducing radiation loss; in fact, dark and white pelages radiate heat equally well. Winter white pelage of arctic animals is simply camouflage in a land of snow. The varying hare of North America has three annual molts: the white winter coat is replaced by a brownish gray summer coat, and this is replaced in autumn by a grayer coat, which is soon shed to reveal the winter white coat beneath (Figure 28-7). White fur of arctic mammals in winter (leukemism) is not to be confused with albinism, caused by a recessive gene that blocks pigment (melanin) formation. Albinos have red eyes and pinkish

A

B

Figure 28-7
Snowshoe, or varying, hare, *Lepus americanus* in **A,** brown summer coat and, in **B,** white winter coat. In winter, extra hair growth on the hind feet broadens the animal's support in snow. Snowshoe hares are common residents of the taiga and are an important prey for lynxes, foxes, and other carnivores. Order Lagomorpha, family Leporidae.

skin, whereas arctic animals in their winter coats have dark eyes and often dark-colored ear tips, noses, and tail tips.

Outside the Arctic, most mammals wear somber colors that are protective. Often the species is marked with "salt-and-pepper" coloration or a disruptive pattern that helps make it inconspicuous in its natural surroundings. Examples are spots of leopards and fawns and stripes of tigers. Skunks advertise their presence with conspicuous warning coloration.

The hair of mammals has become modified to serve many purposes. Bristles of hogs, spines of porcupines and their kin,

Figure 28-8

Dogs are frequent victims of a porcupine's impressive quills. Unless removed (usually by a veterinarian) quills will continue to work their way deeper in the flesh causing great distress and may lead to the victim's death.

Figure 28-9

Annual growth of buck deer antlers. **A,** Antlers begin growth in late spring, stimulated by pituitary gonadotropins. **B,** Bone grows very rapidly until halted by a rapid rise in testosterone production by the testes. **C,** The skin (velvet) dies and sloughs off. **D,** Testosterone levels peak during the fall breeding season. Antlers are shed in January as testosterone levels subside.

and vibrissae on the snouts of most mammals are examples. **Vibrissae,** commonly called "whiskers," are really sensory hairs that provide a tactile sense to many mammals. The slightest movement of a vibrissa generates impulses in sensory nerve endings that travel to special sensory areas in the brain. Vibrissae are especially long in nocturnal and burrowing animals.

Porcupines, hedgehogs, echidnas, and a few other mammals have developed an effective and dangerous spiny armor. When cornered, the common North American porcupine turns its back toward its attacker and lashes out with its barbed tail. The lightly attached quills break off at their bases when they enter the skin and, aided by backward-pointed hooks on their tips, work deeply into tissues. Dogs are frequent victims (Figure 28-8) but fishers, wolverines, and bobcats are able to flip a porcupine onto its back to expose vulnerable underparts.

Horns and Antlers

Several kinds of horns or hornlike structures are found in mammals. **True horns,** found in members of family Bovidae (for example, sheep and cattle), are hollow sheaths of keratinized epidermis that embrace a core of bone arising from the skull. True horns are not shed, are not branched (although they may be greatly curved), grow continuously, and are found in both sexes.

Antlers of the deer family Cervidae are branched and composed of solid bone when mature. During their annual spring growth, antlers develop beneath a covering of highly vascular soft skin called **velvet** (Figure 28-9). Except for caribou (Figure 28-16A), only males of the species produce antlers. When growth of antlers is complete just before the fall breeding season, blood vessels constrict and a stag removes the velvet by rubbing its antlers against trees. Antlers are shed after the breeding season. New buds appear a few months later to herald the next set of antlers. For several years each new pair of antlers is larger and more elaborate than the previous set. Annual growth of antlers places a strain on mineral metabolism, since during the growing season an older moose or elk must accumulate 50 or more pounds of calcium salts from its vegetable diet.

Horns of the pronghorn antelope (family Antilocapridae) are similar to true horns of bovids except the keratinized portion

is forked and shed annually. Giraffe horns are similar to antlers but are not shed each year. Rhinoceros horn consists of hairlike keratinized filaments that arise from dermal papillae cemented together but are not attached to the skull.

An escalating trade in rhino products—especially rhino horn—during the last three decades, is pushing Asian and African rhinos to the brink of extinction. Rhino horn is valued in China as an agent for reducing fever, and for treating heart, liver, and skin disease, and in North India as an aphrodisiac. Such supposed medicinal values are totally without pharmacological basis. The principal use of rhino horns, however, is to fashion handles for daggers in the Middle East. Because of their phallic shape, rhino horn daggers are traditional gifts at puberty rites. Between 1969 and 1977, horns from 8000 slaughtered rhinos were imported into North Yemen alone.

Glands

Of all vertebrates, mammals have the greatest variety of integumentary glands. Most fall into one of four classes: sweat, scent, sebaceous, and mammary. All are derivatives of epidermis (Figure 28-5).

Sweat glands are tubular, highly coiled glands that occur over much of the body surface in most mammals (Figure 28-5). They are not present in other vertebrates. There are two kinds of sweat glands: eccrine and apocrine. **Eccrine glands** secrete a watery fluid that, if evaporated on the skin's surface, draws heat away from the skin and cools it. Eccrine glands occur in hairless regions, especially foot pads, in most mammals, although in horses and most primates they are scattered over the body. They are either reduced or absent in rodents, rabbits, and whales. **Apocrine glands** are larger than eccrine glands and have longer and more convoluted ducts. Their secretory coil is in the dermis and extends deep into the hypodermis. They always open into a hair follicle or where a hair once was. Apocrine gland development occurs near puberty and is restricted (in humans) to the axillae (armpits), mons pubis, breasts, prepuce, scrotum, and external auditory canals. In contrast to watery secretions of eccrine glands, apocrine secretions are milky fluids, whitish or yellow in color, that dry on the skin to form a film. Apocrine glands are not involved in heat regulation. Their activity is correlated with certain aspects of the reproductive cycle.

Scent glands are present in nearly all mammals. Their location and functions vary greatly. They are used for communication with members of the same species, for marking territorial boundaries, for warning, or for defense. Scent-producing glands are located in orbital, metatarsal, and interdigital regions (deer); behind the eyes and on the cheek (pica and woodchuck); penis (muskrats, beavers, and many canines); base of the tail (wolves and foxes); back of the head (dromedary); and anal region (skunks, minks, and weasels). The latter, the most odoriferous of all glands, open by ducts into the anus; their secretions can be discharged forcefully for 2 to 3 meters. During mating season many mammals give off strong scents for attracting the opposite sex. Humans also are endowed with scent glands. However civilization has taught us to dislike our own scent, a concern that has stimulated a lucrative deodorant industry to produce an endless output of soaps and odor-masking concoctions.

Sebaceous glands (Figure 28-5) are intimately associated with hair follicles, although some are free and open directly onto the surface. The cellular lining of the gland is discharged in the secretory process and must be renewed for further secretion. These gland cells become distended with a fatty accumulation, then die, and are expelled as a greasy mixture called **sebum** into the hair follicle. Called a "polite fat" because it does not turn rancid, it serves as a dressing to keep skin and hair pliable and glossy. Most mammals have sebaceous glands over their entire body; in humans they are most numerous in the scalp and on the face.

Mammary glands, which provide the name for mammals, occur on all female mammals and in a rudimentary form on all male mammals. They develop by thickening of the epidermis to form a milk line along each side of the abdomen in the embryo. On certain parts of these lines the mammae appear while the intervening parts of the ridge disappear. Mammary glands increase in size at maturity, becoming considerably larger during pregnancy and subsequent nursing of young. In human females, adipose tissue begins to accumulate around the mammary glands at puberty to form the breast. In most mammals milk is secreted from mammary glands via nipples or teats, but monotremes lack nipples and simply secrete milk into a depression on the mother's belly where it is lapped by the young.

Food and Feeding

Mammals exploit an enormous variety of food sources; some mammals require highly specialized diets, whereas others are opportunistic feeders that thrive on diversified diets. Food habits and physical structure are thus inextricably linked. A mammal's adaptations for attack and defense and its specializations for finding, capturing, chewing, swallowing, and digesting food all determine a mammal's shape and habits.

Teeth, perhaps more than any other single physical characteristic, reveal the life habit of a mammal (Figure 28-10). With certain exceptions (monotremes, anteaters, certain whales), all mammals have teeth, and their modifications are correlated with what the mammal eats.

As mammals evolved during the Mesozoic, major changes occurred in teeth and jaws. Unlike the uniform **homodont** dentition of the first synapsids, mammalian teeth became differentiated to perform specialized functions such as cutting, seizing, gnawing, tearing, grinding, and chewing. Teeth differentiated in this manner are called **heterodont.** Mammalian dentition is differentiated into four types: **incisors,** with simple crowns and sharp edges, used mainly for snipping or biting; **canines,** with long conical crowns, specialized for piercing; **premolars and molars** with compressed crowns and one or more cusps, suited for shearing, slicing, crushing, or grinding. The primitive tooth formula for most mammals, which expresses the number of each tooth type in one-half of the upper and lower jaw, was I 3/3, C 1/1, PM 4/4, M 3/3 = 44. Members of order Insectivora (shrews), some omnivores, and carnivores come closest to this primitive pattern (Figure 28-10).

Unlike reptiles, mammals do not continuously replace their teeth throughout their lives. Most mammals grow just two sets of teeth: a temporary set, called **deciduous,** or **milk,** teeth, which is replaced by a permanent set when the skull has grown large enough to accommodate a full set. Only incisors, canines, and premolars are deciduous; molars are never replaced and the single permanent set must last a lifetime.

Feeding Specializations

The feeding, or trophic, apparatus of a mammal—teeth and jaws, tongue, and alimentary canal—are adapted to its particular feeding habits. Mammals are customarily divided among four basic trophic categories—insectivores, carnivores, omnivores, and herbivores—but many other feeding specializations have evolved in mammals, as in other living organisms, and the feeding habits of many mammals defy exact classification. The principal feeding specializations of mammals are shown in Figure 28-10.

Insectivorous mammals, such as shrews, moles, anteaters, and most bats, are small. They feed on insects, as well as a variety of small invertebrates, such as worms and grubs. Since insectivores eat little fibrous vegetable matter that requires prolonged fermentation, their intestinal tract tends to be short (Figure 28-11). Insectivorous mammals have teeth with pointed cusps, permitting them to puncture the exoskeleton of their prey. The insectivorous category is not a sharply distinguished one because carnivores and omnivores may include insects in their diets.

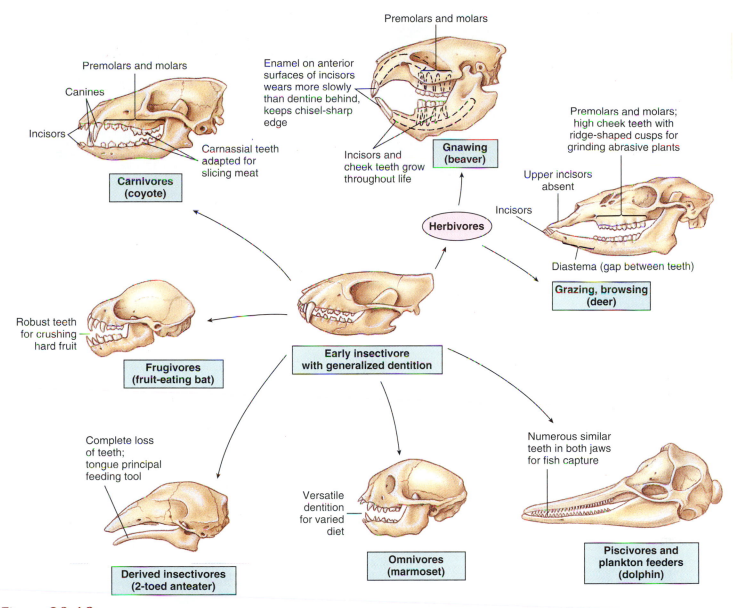

Figure 28-10
Feeding specializations of major trophic groups of eutherian mammals. Early eutherians were insectivores; all other types are descended from them.

Herbivorous mammals that feed on grasses and other vegetation form two main groups: **browsers** and **grazers,** such as ungulates (hooved mammals including horses, deer, antelope, cattle, sheep, and goats), and **gnawers,** such as many rodents, and rabbits and hares. In herbivores, canines are absent or reduced in size, whereas molars, which are adapted for grinding, are broad and usually high-crowned. Rodents (for example, beavers) have chisel-sharp incisors that grow throughout life and must be worn away to keep pace with their continual growth (Figure 28-10).

Herbivorous mammals have a number of interesting adaptations for dealing with their fibrous diet of plant food. **Cellulose,** the structural carbohydrate of plants, is composed of long chains of glucose molecules, and therefore is a potentially nutritious food resource. However, the glucose molecules in cellulose are linked by chemical bonds that few enzymes can attack. No vertebrates synthesize cellulose-splitting enzymes. Instead, herbivorous vertebrates harbor anaerobic bacteria and protozoa in fermentation chambers in their gut. These microorganisms metabolize cellulose, releasing a variety of fatty acids, sugars, and starches that the host animal can absorb and utilize.

Some herbivores, such as horses, zebras, rabbits, elephants, some primates, and many rodents, have a gut with a spacious sidepocket, or diverticulum at the junction of the small and large intestine, called a **cecum,** which serves as a fermentation chamber and absorptive area (Figure 28-11). In these hindgut fermenters, fermentation takes place after the primary absorptive area. Hares, rabbits, and some rodents often eat their fecal pellets **(coprophagy),** giving the food a second pass through the gut to extract additional nutrients.

Figure 28-11

Digestive systems of mammals, showing different morphology with different diets.

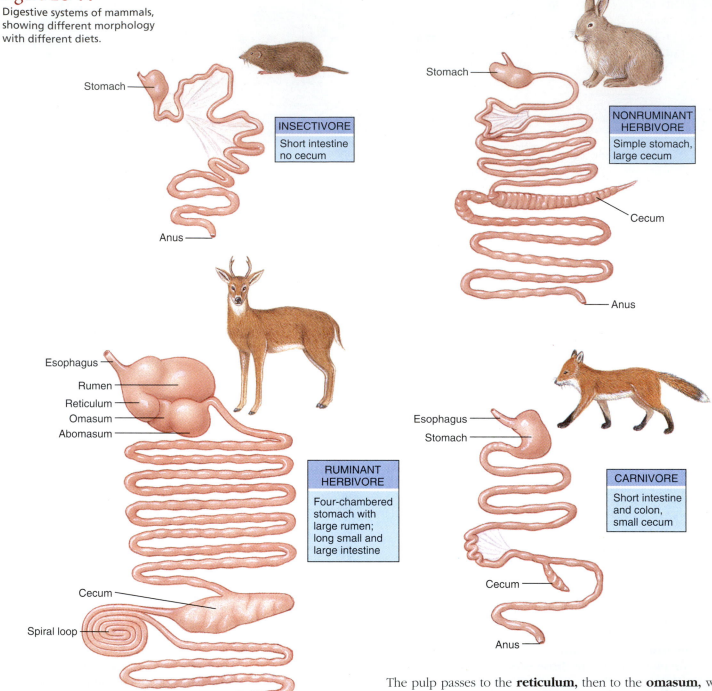

Stomach

INSECTIVORE

Short intestine
no cecum

Anus

Stomach

NONRUMINANT
HERBIVORE

Simple stomach,
large cecum

Cecum

Anus

Esophagus
Rumen
Reticulum
Omasum
Abomasum

RUMINANT
HERBIVORE

Four-chambered
stomach with
large rumen;
long small and
large intestine

Cecum

Spiral loop

Anus

Esophagus
Stomach

CARNIVORE

Short intestine
and colon,
small cecum

Cecum

Anus

Ruminants (cattle, bison, buffalo, goats, antelopes, sheep, deer, giraffes, and okapis) have a huge **four-chambered stomach** (Figure 28-11). As a ruminant feeds, grass passes down the esophagus to the **rumen,** where it is broken down by microorganisms and then formed into small balls of cud. At its leisure, a ruminant returns a cud to its mouth where the cud is deliberately chewed at length to crush the fiber. Swallowed again, food returns to the rumen where it is digested by cellulolytic microorganisms.

The pulp passes to the **reticulum,** then to the **omasum,** where water, soluble food, and microbial products are absorbed. The remainder proceeds to the **abomasum** ("true" acid stomach), where proteolytic enzymes are secreted and normal digestion occurs. Because ruminants are particularly good at extracting nutrients from forage, they are the primary large herbivores in ecosystems with a shortage of food, such as tundras and deserts.

Herbivores generally have large, long digestive tracts and must eat a considerable amount of plant food to survive. An African elephant weighing 6 tons must consume 135 to 150 kg (300 to 400 pounds) of rough fodder each day to obtain sufficient nourishment for life.

Carnivorous mammals feed mainly on herbivores. This group includes foxes, dogs, weasels, wolverines, fishers, cats, lions, and tigers. Carnivores are well-equipped with biting and piercing teeth and powerful clawed limbs for killing their prey. Since their protein diet is more easily digested than the woody food of herbivores, their digestive tract is shorter and the cecum small or absent (Figure 28-11). Carnivores organize their feeding into discrete meals rather than feeding continuously (as do most herbivores) and therefore have much more leisure time.

Note that the terms "insectivores" and "carnivores" have two different uses in mammals: to describe diet and to denote specific taxonomic orders of mammals. For example, not all carnivores belong to order Carnivora (many marsupials and cetaceans are carnivorous) and not all members of order Carnivora are carnivorous. Many are opportunistic feeders and some, such as pandas, are strict vegetarians.

In general, carnivores lead more active—and by human standards more interesting—lives than do herbivores. Since a carnivore must find and catch its prey, there is a premium on intelligence; many carnivores, such as cats, are noted for their stealth and cunning in hunting prey (Figure 28-12). This has led to a selection of herbivores capable either of defending themselves or of detecting and escaping carnivores. Thus for herbivores, there has been a premium on keen senses, speed and agility. Some herbivores, however, survive by virtue of their sheer size (rhinos, elephants) or by defensive group behavior (for example, muskoxen).

Humans have changed the rules in the carnivore-herbivore contest. Carnivores, despite their intelligence, have suffered much from human presence and have been virtually exterminated in some areas. Small herbivores, on the other hand, with their potent reproductive ability, have consistently defeated our most ingenious efforts to banish them from our environment. The problem of rodent pests in agriculture has intensified (Figure 28-29, p. 601); we have removed carnivores, which served as the herbivores' natural population control, but have not been able to devise a suitable substitute.

Figure 28-12
Lionesses, *Panthera leo,* eating a wildebeest. Lacking stamina for a long chase, lions stalk prey and then charge suddenly, surprising their prey. Lions gorge themselves with their kill, then sleep and rest for periods as long as one week before eating again. Order Carnivora, family Felidae.

Figure 28-13
Eastern chipmunk, *Tamias striatus,* with cheek pouches stuffed with seeds to be carried to a hidden cache. It will try to store several liters of food for the winter. It hibernates but awakens periodically to eat some of its cached food. Order Rodentia, family Sciuridae.

Omnivorous mammals—pigs, raccoons, many rodents, bears, and most primates, including humans—use both plants and animals for food. Many carnivorous forms also eat fruits, berries, and grasses when hard pressed. Foxes, which usually feed on mice, small rodents, and birds, eat frozen apples, beechnuts, and corn when their normal food sources are scarce. Other mammals often considered herbivores, such as some rodents, have a mixed diet of insects, seeds, and fruit.

Many mammals cache food stores during periods of plenty. This habit is most pronounced in rodents, such as squirrels, chipmunks, gophers, and certain mice. Tree squirrels collect nuts, conifer seeds, and fungi and store these in caches for winter use. Often each item is hidden in a different place (scatter hoarding) and marked by a scent to assist relocation in the future. Some of the caches of chipmunks and red squirrels can be quite large (Figure 28-13).

Body Weight and Food Consumption

The relationship between body size and metabolic rate was discussed in relation to food consumption of birds (p. 565). The smaller the mammal, the greater is its metabolic rate and the more it must consume relative to its body size (Figure 28-14). This happens because the metabolic rate of a mammal—and therefore the amount of food it must eat to sustain this metabolic rate—varies in rough proportion to the relative surface area rather than to body weight. Surface area is proportional to approximately 0.7 power of body weight, and the amount of food a mammal (or bird) eats also

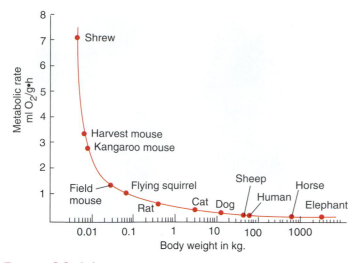

Figure 28-14

Relationship between body weight and metabolic rate for mammals. This relationship, often called the "mouse-to-elephant" curve, shows that metabolic rate is intense for small mammals like shrews and mice, and declines with increasing body weight of the species.

Source: From Eckert Animal Physiology: Mechanisms and Adaptations, 4/e by D. Randall, W. Burggren, K. French. © 1978, 1983, 1988, and 1997 by W. H. Freeman and Company. Used with permission.

Figure 28-15

Shorttail shrew, *Blarina brevicauda,* eating a grasshopper. This tiny but fierce mammal, with a prodigious appetite for insects, mice, snails, and worms, spends most of its time underground and so is seldom seen. Shrews are believed to resemble insectivorous ancestors of placental mammals. Order Insectivora, family Soricidae.

is roughly proportional to a 0.7 power of its body weight. For example, a 3 g mouse will consume *per gram body weight* five times more food than does a 10 kg dog and about 30 times more food than does a 5,000 kg elephant. Thus small mammals (shrews, bats, and mice) must spend much more time hunting and eating food than do large mammals. The smallest shrews weighing only 2 g may eat more than their body weight each day and will starve to death in a few hours if deprived of food (Figure 28-15). In contrast, large carnivores can remain fat and healthy with only one meal every few days. Mountain lions are known to kill an average of one deer a week, although they will kill more frequently when game is abundant.

Migration

Migration is a more difficult undertaking for mammals than for birds. Not surprisingly, few mammals make regular seasonal migrations, preferring instead to center their activities in a defined and limited home range. Nevertheless, there are some striking examples of mammalian migrations. More migrators are found in North America than on any other continent.

An example is barren-ground caribou of Canada and Alaska, which undertakes direct and purposeful mass migrations spanning 160 to 1100 km (100 to 700 miles) twice annually (Figure 28-16). From winter ranges in boreal forests (taiga), they migrate rapidly in late winter and spring to calving ranges on the barren grounds (tundra). Calves are born in mid-June. As summer progresses, caribou are increasingly harassed by warble and nostril flies that bore into their flesh, by mosquitoes that drink their blood (estimated at a liter per caribou each week during the height of the mosquito season), and by wolves that prey on their calves. They move southward in July and August, feeding little along the way. In September they reach the taiga and feed there almost continuously on low ground vegetation. Mating (rut) occurs in October.

Caribou have suffered a drastic decline in numbers since early times when their population reached several million. By 1958 less than 200,000 remained in Canada. The decline has been attributed to several factors, including habitat alteration from exploration and development in the North, but especially to excessive hunting. For example the western Arctic herd in Alaska exceeded 250,000 caribou in 1970. Following five years of heavy unregulated hunting, a 1976 census revealed only about 65,000 animals remaining. After restricting hunting, the herd had increased to 140,000 by 1980 and 160,000 in 1997. However, the proposed scheme to open the Arctic National Wildlife Refuge to petroleum development threatens this recovery.

Plains bison, before their deliberate near extinction by humans, made huge circular migrations to separate summer and winter ranges.

The longest mammalian migrations are made by the oceanic seals and whales. Gray whales, for example, migrate between Alaska in summer and Baja California, Mexico, in winter, an annual migration of over 18,000 km (11,250 miles). One of the most remarkable migrations is that of northern fur seals, which breed on the Pribilof Islands approximately 300 km (185 miles) off the coast of Alaska and north of the Aleutian Islands. From wintering grounds off southern California females journey as much as 2800 km (1740 miles) across open ocean, arriving in the spring at the Pribilofs where they congregate in enormous numbers (Figure 28-17). Young are born within a few hours or days after arrival of the cows. Then the bulls, having already arrived and established territories, collect harems of cows, which they guard with vigilance. After calves have been nursed for approximately three months, cows and juveniles leave for their long migration southward. Bulls do not follow but remain in the Gulf of Alaska during the winter.

Although we might expect bats, the only winged mammals, to use their gift of flight to migrate, few of them do. Most spend winters in hibernation. Four species of American bats that migrate spend their summers in northern or western states and their winters in the southern United States or Mexico.

A

B

Figure 28-16

Barren-ground caribou, *Rangifer tarandus,* of Canada and Alaska. **A,** Adult male caribou in autumn pelage and antlers in velvet. **B,** Summer and winter ranges of some major caribou herds in Canada and Alaska (other herds not shown occur on Baffin Island and in western and central Alaska). Principal spring migration routes are indicated by arrows; routes vary considerably from year to year. The same species is known as reindeer in Europe. Order Artiodactyla, family Cervidae.

Figure 28-17

Annual migrations of northern fur seals, *Callorhinus ursinus,* showing separate wintering grounds of males and females. Both males and females of the larger Pribilof population migrate in early summer to the Pribilof Islands, where females give birth to their pups and then mate. Order Carnivora, family Otariidae.

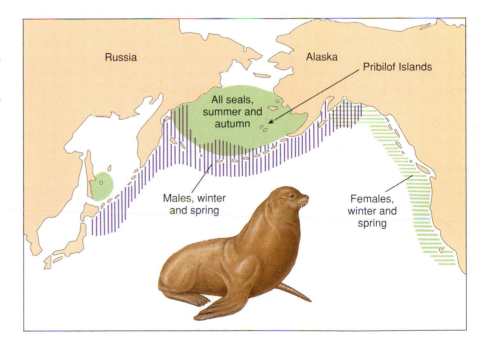

Flight and Echolocation

Many mammals scamper about in trees with amazing agility; some can glide from tree to tree (Figure 28-18) and one group, the bats, is capable of full flight. Gliding and flying evolved independently in several groups of mammals, including marsupials, rodents, flying lemurs, and bats. Flying squirrels (Figure 28-18) actually glide rather than fly, using the gliding skin (patagium) that extends from the sides of the body.

Bats are mostly nocturnal or crepuscular (active at twilight) and thus hold a niche unoccupied by most birds. Their achievement is attributed to two attributes: flight and capacity to navi-

gate by echolocation. Together these adaptations enable bats to fly and avoid obstacles in absolute darkness, to locate and catch insects with precision, and to find their way deep into caves (a habitat largely unexploited by other mammals and birds) where they sleep during the daytime hours.

Research has been concentrated on members of the family Vespertilionidae, to which most common North American bats belong. When in flight, bats emit short pulses 5 to 10 msec in duration in a narrow directed beam from the mouth or nose (Figure 28-19). Each pulse is frequency modulated; that is, it is highest at the beginning, up to 100,000 Hz (hertz, cycles per

Figure 28-18
Northern flying squirrel, *Glaucomys sabrinus,* gliding in for a landing. Area of undersurface is nearly trebled when gliding skin is spread. Glides of 40 to 50 m are possible. Good maneuverability during flight is achieved by adjusting position of the gliding skin with special muscles. Flying squirrels are nocturnal and have superb night vision. Order Rodentia, family Sciuridae.

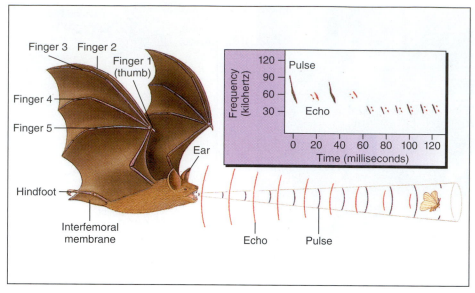

Figure 28-19
Echolocation of an insect by a little brown bat, *Myotis lucifugus.* Frequency modulated pulses are directed in a narrow beam from the bat's mouth. As the bat nears its prey, it emits shorter, lower signals at a faster rate. Order Chiroptera, family Vespertilionidae.

second), and sweeps down to perhaps 30,000 Hz at the end. Sounds of this frequency are ultrasonic to human ears which have an upper limit of about 20,000 Hz. When bats are searching for prey, they produce about 10 pulses per second. If prey is detected, the rate increases rapidly up to 200 pulses per second in the final phase of approach and capture. Pulses are spaced so the echo of each is received before the next pulse is emitted, an adaptation that prevents jamming. Since transmission-to-reception time decreases as a bat approaches an object, it can increase pulse frequency to obtain more information about an object. Pulse length is also shortened as a bat nears an object. It is interesting that some prey of bats, certain nocturnal moths for example, have evolved ultrasonic detectors used to detect and avoid approaching bats (p. 712).

Many insectivores (for example, shrews and tenrecs) use echolocation, but it is crudely developed as compared with bats. Toothed whales, however, have a highly developed capacity to locate objects by echolocation. Totally blind sperm whales in perfect health have been captured with food in their stomachs. Although the mechanism of sound production and reception remains imperfectly understood, it is thought that low- and high-frequency clicks produced in the sinus passages are focused into a narrow beam by a lens-shaped body in the forehead (the "melon"). Returning echos are channeled through oil-filled sinuses in the lower jaw to the inner ear. Toothed whales can apparently determine size, shape, speed, distance, direction, and density of objects in the water and know the position of every whale in their pod.

External ears of bats are large, like hearing trumpets, and shaped variously in different species. Less is known about the inner ear of bats, but it obviously is capable of receiving the ultrasonic sounds emitted. Biologists believe bat navigation is so refined that a bat builds up a mental image of its surroundings from echo scanning that approaches the resolution of a visual image from eyes of diurnal animals.

Not all bats use echolocation to navigate. The approximately 170 species of bats in suborder Megachiroptera lack echolocation abilities. They are primarily nocturnal, although several species are diurnal. These bats feed on fruit and nectar, using large eyes and olfaction to find their food. Flowers of plants that are pollinated by bats open at night, are white or light in color, and emit a musky, batlike odor that nectar-feeding bats find attractive.

The famed tropical vampire bats have razor-sharp incisors used to shave away the epidermis of their prey, exposing underlying capillaries. After infusing an anticoagulant to aid blood flow, they lap up and store their meal in a specially modified stomach.

Reproduction

Reproductive Cycles

Most mammals have definite mating seasons, usually in winter or spring and timed to coincide with the most favorable time of the year for rearing young after birth. Many male mammals are capable of fertile copulation at any time, but female fertility is restricted to a specific time during a periodic cycle, or **estrous cycle.** Females copulate with males only during a relatively brief period in this cycle known as heat or **estrus.** (Figure 28-20).

How often females are in estrus varies greatly among different mammals. Animals that have only a single estrus during their breeding season are called **monestrous;** those having a recurrence of estrus during their breeding season are called **polyestrous.** Dogs, foxes, and bats belong to the first group; field

Figure 28-20

African lions, *Panthera leo*, mating. Lions breed at any season, although predominantly in spring and summer. During the short period a female is receptive, she may mate repeatedly. Three or four cubs are born after gestation of 100 days. Once the mother introduces the cubs into the pride, they are treated with affection by both adult males and females. Cubs go through an 18- to 24-month apprenticeship learning how to hunt and then are frequently driven from the pride to manage themselves. Order Carnivora, family Felidae.

mice and squirrels are all polyestrous as are many mammals living in the more tropical regions of the earth. Old World monkeys and humans have a somewhat different cycle in which the postovulation period is terminated by **menstruation,** during which the, endometrium (lining of the uterus) collapses and is discharged with some blood. This **menstrual cycle** is described in Chapter 7 (pp. 144–145).

> A curious phenomenon that lengthens the gestation period of many mammals is delayed implantation. The blastocyst remains dormant while its implantation in the uterine wall is postponed for periods of a few weeks to several months. For many mammals (for example, bears, seals, weasels, badgers, bats, and many deer) delayed implantation is a device for extending gestation so the young are born at the time of year best for their survival.

Reproductive Patterns

There are three different patterns of reproduction in mammals. One pattern is represented by egg-laying (oviparous) mammals, the **monotremes.** The duck-billed platypus has one breeding season each year. Ovulated eggs, usually two, are fertilized in the oviduct. Embryos continue to develop in the uterus for 10 to 12 days, where they are nourished by yolk supplies deposited prior to ovulation, and secretions from the mother. A thin, leathery shell is secreted around the embryos prior to the eggs being laid. The platypus lays its eggs in a burrow where they hatch at a relatively undeveloped state after about 12 days. Echidnas incubate their eggs in an abdominal pouch. After hatching, the young feed on milk produced by the mother's mammary glands.

Because monotremes have no nipples, the young lap milk secreted onto the belly fur of the mother.

Marsupials are pouched, viviparous mammals that exhibit a second pattern of reproduction. Although only eutherians are called "placental mammals," marsupials do have a primitive type of placenta, called a choriovitelline (or yolk sac) placenta. An embryo (blastocyst) of a marsupial is at first encapsulated by shell membranes and floats free for several days in the uterine fluid. After "hatching" from the shell membranes, embryos of most marsupials do not implant, or "take root" in the uterus as they would in eutherians, but erode shallow depressions in the uterine wall in which they lie and absorb nutrient secretions from the mucosa by way of a vascularized yolk sac. Gestation (the intrauterine period of development) is brief in marsupials, and therefore all marsupials give birth to tiny young that are effectively still embryos, both anatomically and physiologically. However, early birth is followed by a prolonged interval of lactation and parental care (Figure 28-21).

Although it is tempting to view the ephemeral choriovitelline placenta of marsupials as transitional between absence of a placenta in monotremes and the persistent chorioallantoic placenta of placental mammals, cladistic analysis indicates this not to be true. All marsupials and placental mammals have a choriovitelline placenta, and a chorioallantoic placenta is present in the most primitive marsupials. This suggests that a chorioallantoic placenta was present in the common ancestor of marsupials and placental mammals, but later was lost in most marsupials.

In red kangaroos (Figure 28-22) the first pregnancy of the season begins with a 33-day gestation, after which the young (joey) is born, crawls to the pouch without assistance from its mother, and attaches to a nipple. The mother immediately becomes pregnant again, but the presence of a suckling young in the pouch arrests development of the new embryo in the uterus at about the 100-cell stage. This period of arrest, called **embryonic diapause,** lasts approximately 235 days during which time the first joey is growing in the pouch. When the joey leaves the pouch, the uterine embryo resumes development and is born about a month later. The mother again becomes pregnant, but because the second joey is suckling, once again development of the new embryo is arrested. Meanwhile, the first joey returns to the pouch from time to time to suckle. At this point the mother has three young of different ages dependent upon her for nourishment: a joey on foot, a joey in the pouch, and a diapause embryo in the uterus. There are variations on this remarkable sequence—not all marsupials have developmental delays like kangaroos, and some do not even have pouches—but in all, the young are born at an extremely early stage of development and undergo prolonged development while dependent on a teat (Figure 28-23).

The third pattern of reproduction is that of viviparous **placental mammals,** the eutherians. In placentals, reproductive investment is in prolonged gestation, unlike marsupials in which reproductive investment is in prolonged lactation (Figure 28-21). The embryo remains in the uterus, nourished by food supplied initially by a choriovitelline placenta and later by a chorioallantoic type of placenta (described on pp. 167–168), an intimate connection between mother and young. Length of gestation is

Figure 28-21

Comparison of gestation and lactation periods between matched pairs of ecologically similar species of marsupial and placental mammals. The graph shows marsupials have shorter intervals of gestation and much longer intervals of lactation than in similar species of placentals.

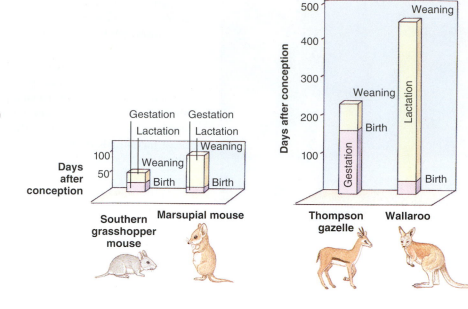

Figure 28-22

Kangaroos have a complicated reproductive pattern in which the mother may have three young in different stages of development dependent on her at once. Order Diprotodontia, family Macropodidae.

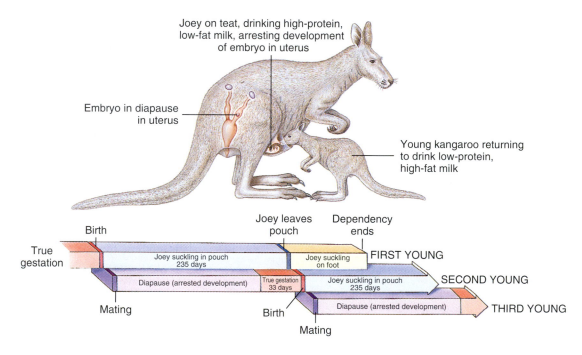

Figure 28-23

Opossums, *Didelphis marsupialis,* 15 days old, fastened to teats in mother's pouch. When born after a gestation period of only 12 days, they are the size of honey bees. They remain attached to the nipples for 50 to 60 days. Order Marsupialia, family Didelphidae.

longer in placentals than marsupials, and in large mammals it is much longer (Figure 28-21). For example, mice have a gestation period of 21 days; rabbits and hares, 30 to 36 days; cats and dogs, 60 days; cattle, 280 days; and elephants, 22 months (the longest). But there are important exceptions (nature seldom offers perfect correlations). Baleen whales, the largest mammals, carry their young for only 12 months, while bats, no larger than mice, have gestation periods of 4 to 5 months. The condition of the young at birth also varies. An antelope bears its young well furred, eyes open, and able to run about. Newborn mice, however, are blind, naked, and helpless. We all know how long it takes a human baby to gain its footing. Human growth is in fact slower than that of any other mammal, and this is one of the distinctive attributes that sets us apart from other mammals.

The number of young produced by mammals in a season depends on mortality rate, which, for some mammals such as mice, may be high at all age levels. Usually, the larger the animal, the smaller the number of young in a litter. Small rodents, which serve as prey for many carnivores, usually produce more than one litter of several young each season. Meadow mice are known to produce as many as 17 litters of four to nine young in a year. Most carnivores have but one litter of three to five young per year. Large mammals, such as elephants and horses, give birth to a single young with each pregnancy. An elephant produces, on average, four calves during her reproductive life of perhaps 50 years.

The renowned fecundity of meadow mice, and the effect of removing natural predators from rodent populations, is felicitously expressed in this excerpt from Thornton Burgess's **Portrait of a Meadow Mouse.**

He's fecund to the nth degree
In fact this really seems to be
His one and only honest claim
To anything approaching fame.
In just twelve months, should all survive,
A million mice would be alive—
His progeny. And this, 'tis clear,
Is quite a record for a year.
Quite unsuspected, night and day
They eat the grass that would be hay.
On any meadow, in a year,
The loss is several tons, I fear,
Yet man, with prejudice for guide,
The checks that nature doth provide
Destroys. The meadow mouse survives
And on stupidity he thrives.

Territory and Home Range

Many mammals have territories—areas from which individuals of the same species are excluded. In fact, many wild mammals, like some humans are basically unfriendly to their own kind, especially so to their own sex during the breeding season. If a mammal dwells in a burrow or den, this area forms the center of its territory. If it has no fixed address, the territory is marked out, usually with the highly developed scent glands (p. 590). Territories vary greatly in size depending on the size of the animal and its feeding habits. Grizzly bears have territories of several square miles, which they guard zealously against all other grizzlies.

Mammals usually use natural features of their surroundings in staking their claims. These are marked with secretions from the scent glands or by urinating or defecating. When an intruder knowingly enters another's marked territory, it is immediately placed at a psychological disadvantage. Should a challenge follow, the intruder almost invariably breaks off the encounter in a submissive display characteristic for the species. Territoriality and aggressive and submissive displays are described in more detail in Chapter 36; pp. 759–763.

A beaver colony is a family unit, and beavers are among several mammalian species in which males and females form

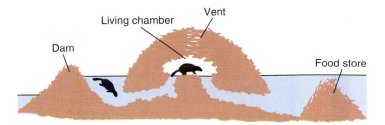

Figure 28-24
Each beaver colony constructs its own lodge in a pond created by damming a stream. Each year the mother bears four or five young; when the third litter arrives, 2-year-olds are driven out of the colony to establish new colonies elsewhere. Order Rodentia, family Castoridae.

Figure 28-25
Immature black-tailed prairie dogs, *Cynomys ludovicianus,* greeting adult. These highly social prairie dwellers are herbivores that serve as an important prey for many animals. They live in elaborate tunnel systems so closely interwoven that they form "towns" of as many as 1000 individuals. Towns are subdivided into family units, each with one or two males, several females, and their litters. Although prairie dogs display ownership of burrows with territorial calls, they are friendly with inhabitants of adjacent burrows. The name "prairie dog" derives from the sharp, doglike bark they make when danger threatens. Order Rodentia, family Sciuridae.

strong monogamous bonds that last a lifetime. Because beavers invest considerable time and energy in constructing a lodge and dam and storing food for winter (Figure 28-24), the family, especially the adult male, vigorously defends its real estate against intruding beavers. Most of the work of building dams and lodges is undertaken by male beavers, but females assist when not occupied with their young.

An interesting exception to the strong territorial nature of most mammals is the prairie dog, which lives in large, friendly communities called prairie dog "towns" (Figure 28-25). When a new litter has been reared, adults relinquish the old home to the young and move to the edge of the community to establish a new home. Such a practice is totally antithetical to the behavior of most mammals, which drive off the young when they are self-sufficient.

A **home range** of a mammal is a much larger foraging area surrounding a defended territory. Home ranges are not defended in the same way as are territories; home ranges may, in fact, overlap, producing a neutral zone used by owners of several territories for seeking food.

Mammalian Populations

A population of animals includes all members of a species that share a particular space and potentially interbreed (Chapter 38). All mammals (like other organisms) live in ecological communities, each composed of numerous populations of different animal and plant species. Each species is affected by the activities of other species and by other changes, especially climatic, that occur. Thus populations are always changing in size. Populations of small mammals are lowest before the breeding season and greatest just after addition of new members. Beyond these expected changes in population size, mammalian populations may fluctuate from other causes.

Irregular fluctuations are commonly produced by variations in climate, such as unusually cold, hot, or dry weather, or by natural catastrophes, such as fires, hailstorms, and hurricanes. These are **density-independent** factors because they affect a population whether it is crowded or dispersed. However, the most spectacular fluctuations are **density dependent;** that is, they correlate with population crowding. These extrinsic limits to growth are discussed in Chapter 38.

Cycles of abundance are common among many rodent species. One of the best-known examples is the mass migrations of Scandinavian and arctic North American lemmings following population peaks. Lemmings (Figure 28-26) breed all year, although more in summer than in winter. The gestation period is only 21 days; young born at the beginning of summer are weaned in 14 days and are capable of reproducing by end of summer. At the peak of their population density, having devastated the vegetation by tunneling and grazing, lemmings begin long, mass migrations to find new undamaged habitats for food and space. They swim across streams and small lakes as they go but cannot distinguish these from large lakes, rivers, and the sea, in which they drown. Since lemmings are the main diet of many carnivorous mammals and birds, any change in lemming population density affects all their predators as well.

In his book *The Arctic* (1974. Montreal, Infacor, Ltd.), Canadian naturalist Fred Bruemmer describes the growth of lemming populations in arctic Canada:

"After a population crash one sees few signs of lemmings; there may be only one to every 10 acres. The next year, they are evidently numerous; their runways snake beneath the tundra vegetation, and frequent piles of rice-sized droppings indicate the lemmings fare well. The third year one sees them everywhere. The fourth year, usually the peak year of their cycle, the populations explode. Now more than 150 lemmings may inhabit each acre of land and they honeycomb it with as many as 4000 burrows. Males meet frequently and fight instantly. Males pursue females and mate after a brief but ardent courtship. Everywhere one hears the squeak and chitter of the excited, irritable, crowded animals. At such times they may spill over the land in manic migrations."

Varying hares (snowshoe rabbits, Figure 28-7) of North America show 10-year cycles in abundance. The well-known fecundity of rabbits enables them to produce litters of three or four young as many as five times per year. Their density may increase to 4000 hares competing for food in each square mile of northern forest. Predators (owls, minks, foxes, and especially lynxes) also increase (Figure 28-27). Then the population crashes precipitously for reasons that have long been a puzzle to scientists. Rabbits die in great numbers, not from lack of food or from an epidemic disease (as was once believed) but evidently from some density-dependent psychogenic cause. As crowding increases, hares become more aggressive, show signs of fear and defense, and stop breeding. The entire population reveals symptoms of pituitary-adrenal gland exhaustion, an endocrine imbalance called "shock disease," which results in death. These dramatic crashes are not well understood. Whatever the causes, population crashes that follow superabundance, although harsh, permit vegetation to recover, providing survivors with a much better chance for successful breeding.

Figure 28-26

Collared lemming, *Dicrostonyx* sp., a small rodent of the far north. Populations of lemmings fluctuate widely. Order Rodentia, family Muridae.

Figure 28-27

Changes in population size of varying hare and lynx in Canada as indicated by pelts received by the Hudson's Bay Company over a 90-year period. Abundance of lynx (predator) follows that of the hare (prey).

HUMANS AND MAMMALS

Some 10,000 years ago, at the time people developed agricultural methods, they also began domestication of mammals. Dogs were certainly among the first to be domesticated, probably entering voluntarily into their human dependence. Dogs are an extremely adaptable and genetically plastic species derived from wolves. Much less genetically variable and certainly less social than dogs are domestic cats, probably derived from an African race of wildcat. Wildcats look like oversized domestic cats and are still widespread in Africa and Eurasia. Domestication of cattle, buffaloes, sheep, and pigs probably came much later. Beasts of burden—horses, camels, oxen, and llamas—probably were subdued by early nomadic peoples. Certain domestic species no longer exist as wild animals, for example, one-humped dromedary camels of North Africa and llamas and alpacas of South America. All truly domestic animals breed in captivity; many have been molded by selective breeding to yield characteristics desirable for human purposes.

Some mammals hold special positions as "domestic" animals. Elephants have never been truly domesticated because they will seldom breed in captivity. In Asia, adults are captured and submit to a life of toil with astonishing docility. Reindeer of northern Scandinavia are domesticated only in the sense that they are "owned" by nomadic peoples who follow them in their seasonal migrations (Figure 28-28). Elands of Africa are undergoing experimental domestication in several places. They are placid, gentle, and immune to native diseases and produce excellent meat.

Activities of mammals can in some instances conflict with human activities. Rodents and rabbits are capable of inflicting staggering damage to growing crops and stored food (Figure 28-29). We have provided an inviting forage for rodents with our agriculture and convenienced them further by removing their natural predators. Rodents also carry various diseases. Bubonic plague and typhus are carried by various rodents, including house rats and prairie dogs. Tularemia (rabbit fever), is transmitted to humans by wood ticks carried by rabbits, woodchucks, muskrats, and other rodents. Rocky Mountain spotted fever is

Figure 28-29

Brown rat, *Rattus norvegicus,* living all too successfully beside human habitations. Brown rats not only cause great damage to food stores but also spread disease, including bubonic plague (a disease, carried by infected fleas, that greatly influenced human history in medieval Europe), typhus, infectious jaundice, *Salmonella* food poisoning, and rabies. Order Rodentia, family Muridae.

carried to humans by ticks from ground squirrels and dogs; Lyme disease is transmitted by ticks from white-tailed deer. Trichina worms and tapeworms are acquired by humans who eat the meat of infected hogs, cattle, and other mammals.

In the introduction to this chapter, we alluded to the discouraging exploitation of whales as one example of our inability to reconcile human needs with preservation of wildlife. Extermination of a species for commercial gain is so totally indefensible that no debate is required. Once a species is extinct, no amount of scientific or technical ingenuity will bring it back. What has taken millions of years to evolve can be destroyed in a decade of thoughtless exploitation. Many people are concerned with the awesome impact we have on wildlife, and there is more determination today to reverse a regrettable trend than ever before. If given a chance, mammals will usually make spectacular recoveries from human depredations, as have sea otters and saiga antelopes, both once in danger of extinction and now numerous.

HUMAN EVOLUTION

Darwin devoted an entire book, *The Descent of Man and Selection in Relation to Sex* (1871), largely to human evolution. The idea that humans shared common descent with apes and other animals was repugnant to the Victorian world, which responded with predictable outrage (Figure 6-14; p. 112). Because at that time virtually no fossil evidence linking humans with apes existed, Darwin built his case mostly on anatomical comparisons between humans and apes. To Darwin, the close resemblances between apes and humans could be explained only by common descent.

The search for fossils, especially for a "missing link" that would provide a connection between apes and humans, began when two skeletons of Neandertals* were collected in the 1880s.

Figure 28-28

Herd of reindeer, *Rangifer tarandus,* during annual roundup by Laplanders in northern Sweden. The same species is known as caribou in North America. Order Artiodactyla, family Cervidae.

*The traditional spelling "Neanderthal" has been replaced by "Neandertal" in many recent publications. In old German, *thal* (valley) is pronounced with a hard *t,* rather than with a soft *th.* Modern spellings of many German words were changed to reflect their pronunciation, resulting in *thal* becoming *tal.*

Then in 1891, Eugene Dubois discovered the famous Java man (*Homo erectus*). The most spectacular discoveries, however, have been made in Africa, especially between 1967 and 1977, which American paleoanthropologist Donald C. Johanson calls the "golden decade." During this same period, comparative biochemical studies demonstrated humans and chimpanzees are as similar genetically as many sibling species. Comparative cytology provided evidence that chromosomes of humans and apes are homologous. We are no longer searching for a mythical "missing link" to establish common descent of humans and apes, our closest living relatives.

Evolutionary Radiation of Primates

Humans are primates, a fact that even pre-evolutionist Linnaeus recognized. All primates share certain significant characteristics: grasping fingers on all four limbs, flat fingernails instead of claws, and forward-pointing eyes with binocular vision and excellent depth perception. Details of primate phylogeny are not entirely clear. The accompanying synopsis will highlight probable relationships of major primate groups.

The earliest primate was probably a small, nocturnal animal similar in appearance to tree shrews. This ancestral primate stock split into two major lineages, one of which gave origin to **prosimians,** including **lemurs, tarsiers** (Figure 28-30), and **lorises;** and the other to **simians,** which include the monkeys (Figure 28-31) and apes (Figure 28-32). Prosimians and many simians are arboreal (tree-dwellers), which is probably the ancestral life-style for both groups. Flexible limbs are essential for active animals moving through trees. Grasping hands and feet, in contrast to clawed feet of squirrels and other rodents, enable primates to grip limbs, hang from branches, seize and manipulate food, and, most significantly, use tools. Primates have highly developed sense organs, especially good binocular vision, and proper coordination of limb and finger muscles to assist their active arboreal life. Of course, sense organs are no better than the brain processing sensory information. Precise timing, judgment of distance, and alertness require a large cerebral cortex.

The earliest simian fossils appeared in Africa in late Eocene deposits, some 40 million years ago. Many of these primates became diurnal rather than nocturnal, making vision the dominant special sense, now enhanced by color vision. We recognize three major simian clades. These are (1) New World monkeys of Central and South America (ceboids; Figure 28-31A), including howler monkeys, spider monkeys, and tamarins, (2) Old World monkeys (cercopithecoids), including baboons (Figure 28-31B), mandrills, and colobus monkeys, and (3) anthropoid apes (Figure 28-32). Old World monkeys and anthropoid apes (including humans) are sister taxa, and together form the sister group of New World monkeys. In addition to their geographic separation, Old World monkeys differ from New World monkeys in lacking a grasping tail, while having close-set nostrils, better opposable, grasping thumbs, and more advanced teeth. Apes differ from Old World monkeys in having a larger cerebrum, a more dorsally placed scapula, and loss of the tail.

Apes first appear in 25-million-year-old fossils. At this time woodland savannas were arising in Africa, Europe, and North America. Perhaps motivated by greater abundance of food on the ground, these apes left the trees and became largely terrestrial.

A

B

Figure 28-31

Monkeys. **A,** Red howler monkeys, *Alovatta seniculus,* order Primates, family Cebidae, an example of New World monkeys. **B,** Olive baboons, *Papio homadryas,* order Primates, family Cercopithecidae, an example of Old World monkeys.

Figure 28-30

A prosimian, the Mindanao tarsier, *Tarsius syrichta carbonarius,* of Mindanao Island in the Philippines. Order Primates, family Tarsiidae.

Figure 28-33
Lucy (*Australopithecus afarensis*), the most nearly complete skeleton of an early hominid ever found. Lucy is dated at 3.0 million years old. A nearly complete skull of *A. afarensis* was discovered in 1994.

Figure 28-32
Gorillas, *Gorilla gorilla*, order Primates, family Hominidae, examples of anthropoid apes.

Figure 28-34
A reconstruction of the appearance of Lucy (*right*) compared with a modern human (*Homo sapiens*) order Primates (*left*).

First Hominids

The gradual replacement of forests with grasslands in eastern Africa provided an impetus for apes to adapt to an open environment, the savannas. Because of the benefits of standing upright (better view of predators, freeing of hands for using tools, defense, caring for young, and gathering food) emerging hominids gradually evolved upright posture. This important transition was an enormous leap because it required extensive redesigning of the skeleton and muscle attachments.

Evidence of the earliest hominids of this period is extremely sparse. Yet in 2001 the desert sands of Chad yielded one of the most astonishing and important discoveries of modern paleontology, a remarkably complete skull of a hominid dated at nearly 7 million years ago. Names *Sahelanthropus tchadensis* ("Sahel hominid of Chad") by its discoverer, French paleontologist Michel Brunet, this chimplike creature is by far the most ancient hominid yet discovered (Figure 28-35). Although its brain is no larger than that of a chimp (between 320 and 380 cm³), its relatively small canine teeth, massive brow ridges on a short face, and mouth and jaw that protrude less than in most apes, confirm that the skull truly is that of a hominid. Until this skull was discovered, the earliest hominid fossil was *Ardipithecus ramidus* from the sands of Ethiopia, originally dated at 4.4 million years (Figure 28-35). *Ardipithecus ramidus* is a mosaic of primitive apelike and derived hominid traits, with indirect (and controversial) evidence that it may have been bipedal. Between 1997 and 2001 additional fossils of *A. ramidus* were discovered that extend its existence back to at least 5.5 million years ago. These fossils have tentatively been assigned to a new subspecies, *Ardipithecus ramidus kadabba*.

Until the discovery of *Sahelanthropus tchadensis,* the most celebrated fossil was a 40% complete skeleton of a female *Australopithecus afarensis* (Figures 28-33 and 28-34). Unearthed in 1974 and named "Lucy" by its discoverer Donald Johanson, *A. afarensis* was short, bipedal hominid with face and brain size resembling those of a chimpanzee. Numerous fossils of this species have now been discovered. The time range for *A. afarensis* is 3.7 to 3.0 million years ago.

In 1995 *Australopithecus anamensis* was discovered in the Rift Valley of Kenya. Many researchers believe that this species, which lived between 4.2 and 3.9 million years ago, is an intermediate between *A. ramidus* and *A. afarensis* (Lucy). The extremely humanlike lower leg bones of *Australopithecus anamensis* provide strong evidence that this species was bipedal.

In the last decade there has been an explosion of australopithecine fossil finds, with eight putative species requiring interpretation. Most of these finds are known as gracile australopithecines because of their relatively light build, especially in skull and teeth (although all were more robust than modern humans). The gracile australopithecines are generally considered direct ancestors of populations of early *Homo* and, by extension, of the lineage leading to modern humans. Following *A. anamensis* and Lucy (*Australopithecus afarensis;* 3.7 to 3 million years ago) there appeared the bipedal *Australopithecus africanus,* which lived between 3 and 2.3 million years ago. This species was similar to *A. afarensis,* but had a more humanlike face, slightly larger body, and a brain size about one-third as large as that of modern humans. In 1998 a partial skull, dated at 2.5 million years ago, was discovered in Ethiopia. Named *Australopithecus garhi,* this species differs from other australopithecines in its combination of features, especially the large size of its molars.

Coexisting with the earliest species of *Homo* was a different line of large and robust australopithecines that existed between 2.5 and 1.2 million years ago. One of these was *Paranthropus robustus* (Figure 28-35), which probably approached the size of a gorilla. The "robust" australopithecines were heavy jawed with skull crests and large back molars, used for chewing coarse roots and tubers. They are a side branch in hominid evolution and not part of our own lineage.

Emergence of *Homo,* True Humans

Although researchers are divided over who the first members of *Homo* were, and indeed how to define the genus *Homo,* the earliest *known* species of the genus was *Homo habilis,* a fully erect hominid. *Homo habilis* ("handy man") is thought to have been about 127 cm (5 ft) tall and weigh about 45 kg (100 lbs), although females were probably smaller. This species was larger brained than the australopithecines and its brain was more humanlike in shape; one brain cast shows a bulge representing Broca's motor speech area, suggesting that *Homo habilis* was capable at least of rudimentary speech.

About 1.8 million years ago *Homo erectus* appeared, a large hominid standing 150 to 170 cm (5 to 5.5 feet) tall, with a low but distinct forehead and strong browridges. Its brain capacity was around 1000 cm^3, intermediate between the brain capacity of *Homo habilis* and modern humans (Figure 28-35). *Homo erectus* was a social species living in tribes of 20 to 50 individuals. *Homo erectus* had a successful and complex culture and became widespread throughout the tropical and temperate Old World.

Homo sapiens: Modern Humans

Recent molecular genetic studies indicate that human populations have formed a single evolutionary lineage for the past 1.7 million years. During this time, populations on different continents have shown some geographic differentiation, but they maintained at least low levels of gene exchange and all have made genetic contributions to modern humans. Several major expansions of populations out of Africa occurred during this time.

Anthropologists have named fossil species to denote spatial and temporal variation in phenotypic characters within this lineage; however, this lineage would constitute a single species according to most biological criteria of species. The earliest human remains originally classified as *Homo sapiens* ("wise man"), from 500,000 to 300,000 years ago, now are identified by anthropologists as *H. heidelbergensis.* One group of well-known humans, the **Neandertals,** arose about 150,000 years ago. These morphologically robust humans were originally classified as a subspecies of *H. sapiens,* but recent evidence pointing to their distinctiveness has led anthropologists to recognize them as *H. neanderthalensis.* With a brain capacity well within range of modern humans, the Neandertals were proficient hunters and tool-users. Neandertals were not homogeneous but varied geographically in response to local conditions and the isolation of populations from one another. They dominated the Old World in the late Pleistocene epoch.

Fossil and mitochondrial DNA (mtDNA) evidence indicates that characteristics of *H. sapiens,* as presently defined arose in Africa about 200,000 years ago (Figure 28-35). About 30,000 years ago Neandertals and remaining *H. erectus* disappeared approximately 10,000 years after the first appearance of *H. sapiens* in Europe and Eastern Asia. Modern humans were tall people with a culture very different from that of Neandertals. Implement crafting developed rapidly, and human culture became enriched with aesthetics, artistry, and sophisticated language.

In closing our discussion of human evolution, it is important to note that recognition of species in *Homo* is based entirely on morphology. Recognition of three or more distinct species of *Homo* does not necessarily imply the occurrence of branching speciation in this lineage; it is possible that we are observing phyletic change within a single species through time, and using the species names only to denote different grades of evolution. Nevertheless, there is clearly only a single species of *Homo* alive today.

The Unique Human Position

Biologically, *Homo sapiens* is a product of the same processes that have directed the evolution of every organism from the time of life's origin. Mutation, isolation, genetic drift, and natural selection have operated for us as they have for other animals. Yet we have what no other animal species has, a nongenetic cultural evolution that provides a constant feedback between past and future experience. Our symbolic languages, capacities for conceptual thought, knowledge of our history, and abilities to manipulate our environment emerge from this nongenetic cultural endowment. Finally, we owe much of our cultural and intellectual achievements to our arboreal ancestry which bequeathed us with binocular vision, superb visuotactile discrimination, and manipulative skills in use of our hands. If the horse (with one toe instead of five fingers) had human mental capacity, could it have accomplished what humans have?

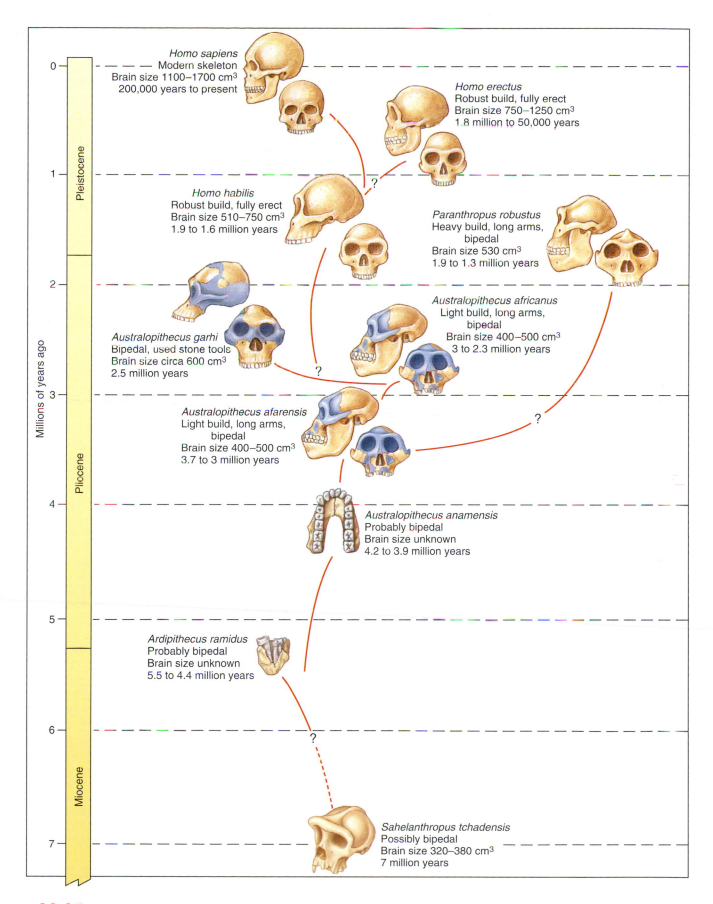

Figure 28-35

Hominid skulls, showing several of the best-known hominid lines preceding modern humans (*Homo sapiens*).

Classification of Living Mammalian Orders*

Class Mammalia

  Subclass Prototheria (pro′to-thir′ee-a) (Gr. *prōtos,* first, + *thēr,* wild animal).

    Infraclass Ornithodelphia (or′ni-tho-del′fee-a) (Gr. *ornis,* bird, + *delphys,* womb). Monotreme mammals.

      Order Monotremata (mon′o-tre′ma-tah) (Gr. *monos,* single, + *trēma,* hole): **egg-laying (oviparous) mammals: duck-billed platypus, echidnas.** Three species in this order are from Australia, Tasmania, and New Guinea. The most noted member of the order is the duck-billed platypus (*Ornithorhynchus anatinus*). Spiny anteaters, or echidnas (*Tachyglossus*), have a long, narrow snout adapted for feeding on ants, their chief food.

  Subclass Theria (thir′ee-a) (Gr. *thēr,* wild animal).

    Infraclass Metatheria (met′a-thir′ee-a) (Gr. *meta,* after, + *thēr,* wild animal). Marsupial mammals.

      Order Marsupialia (mar-su′pe-ay′le-a) (Gr. *marsypion,* little pouch): **viviparous pouched mammals: opossums, kangaroos, koalas, Tasmanian wolves, wombats, bandicoots, numbats, and others.** These mammals are characterized by an abdominal pouch, or **marsupium,** in which they rear their young. Young are nourished in the uterus for a short time by way of a yolk-sac placenta. About 80 species are found in South America, but only one species, the Virginia opossum (*Didelphis virginiana*), is found in North America. Members of this order form the dominant group of mammals in Australia; 260 species.

    Infraclass Eutheria (yu-thir′ee-a) (Gr. *eu,* true, + *thēr,* wild animal). Viviparous placental mammals.

      Order Insectivora (in-sec-tiv′o-ra) (L. *insectum,* an insect, + *vorare,* to devour): **insect-eating mammals: shrews, hedgehogs, tenrecs, moles.** The principal food is insects. Insectivores, widely distributed over the world except Australia and New Zealand, are small, sharp-snouted animals with primitive characters that spend a great part of their lives underground. Shrews are among the smallest mammals known; 419 species.

      Order Macroscelidea (mak-ro-sa-lid′e-a) (Gr. *makros,* large, + *skelos,* leg): **elephant shrews.** These are secretive mammals with long legs, a snoutlike nose adapted for foraging for insects, large eyes. Widespread in Africa; 15 species.

      Order Dermoptera (der-mop′ter-a) (Gr. *derma,* skin, + *pteron,* wing): **flying lemurs.** These are related to true bats and consist of a single genus *Galeopithecus.* They are not lemurs (which are primates) and cannot fly but glide like flying squirrels. They are found in the Malay peninsula in the East Indies; two species.

      Order Chiroptera (ky-rop′ter-a) (Gr. *cheir,* hand, + *pteron,* wing): **bats.** Wings of bats, the only true flying mammals, are modified forelimbs in which the second to fifth digits are elongated to support a thin integumental membrane for flying. The first digit (thumb) is short with a

claw. Common North American forms are little brown bats (*Myotis*), free-tailed bats (*Tadarida*), and big brown bats (*Eptesicus*). In Old World tropics fruit bats, or "flying foxes" (*Pteropus*), are largest of all bats, with a wingspread of 1.2 to 1.5 m; they live chiefly on fruits; 925 species.

      Order Scandentia (skan-dent′e-a) (L. *scandentis,* climbing): **tree shrews.** Tree shrews are small, squirrel-like mammals of the tropical rain forests of southern and southeastern Asia. Despite their name, many are not especially well-adapted for life in trees, and some are almost completely terrestrial; 16 species.

      Order Primates (pry-may′teez) (L. *prima,* first): **prosimians, monkeys, apes, humans.** This order stands first in the animal kingdom in brain development, with especially large cerebral hemispheres. Most species are arboreal, apparently derived from tree-dwelling insectivores. Primates represent the end product of a line that branched off early from other mammals and have retained many primitive characteristics. It is believed that their tree-dwelling habits of agility in capturing food or avoiding enemies were largely responsible for their advances in brain structure. As a group they are generalized with five digits (usually provided with flat nails) on both forelimbs and hindlimbs. All except humans have their bodies covered with hair. Forelimbs are often adapted for grasping, as are the hindlimbs sometimes. The group is singularly lacking in claws, scales, horns, and hoofs. There are two suborders; 223 species.

    Suborder Strepsirhini (strep′suh-ry-nee) (Gr. *strepsō,* to turn, twist, + *rhinos,* nose): **lemurs, aye-aye, lorises, pottos, bush babies.** Seven families of arboreal primates, formerly called prosimians, concentrated on Madagascar, but with species in Africa, southeast Asia, and Malay peninsula. All have a wet, naked region (rhinarium) surrounding comma-shaped nostrils, a long nonprehensile tail, and a second toe provided with a claw. Their food is both plants and animals. 47 species.

    Suborder Haplorhini (hap′lo-ry-nee) (Gr. *haploos,* single, simple + *rhinos,* nose): **tarsiers, marmosets, New and Old World monkeys, gibbons, gorilla, chimpanzees, orangutan, humans.** Six families, four of which were formerly called Anthropoidea. Haplorhine primates have dry, hairy noses, ringed nostrils and differences in uterine anatomy, placental development, and skull morphology that distinguish them from strepsirhine primates. Family **Tarsiidae** contains crepuscular and nocturnal tarsiers (Figure 28-30), with large, forward-facing eyes and reduced snout (five species). New World Monkeys, sometimes called Platyrrhine monkeys because the nostrils are widely separated, are contained in two families: **Callitrichidae** (marmosets and tamarins; 26 species) and **Cebidae** (capuchinlike monkeys; 58 species). Callitrichids, which include the colorful lion tamarins, have prehensile hands and quadrupedal locomotion. Cebid monkeys are much larger than any callitrichid. They include capuchin monkeys (*Cebus*), spider monkeys (*Ateles*), and howler monkeys (*Alouatta*). Some cebids (including spider and howler

*Based on Nowak, R. M. 1991. Walker's Mammals of the World, ed. 5, Baltimore, The Johns Hopkins University Press.

monkeys) have prehensile tails, used like an additional hand for grasping and swinging.

Old World monkeys, termed catarrhine monkeys because their nostrils are set close together and open to the front, are placed in family **Cercopithecidae,** with 81 species. They include mandrills (*Mandrillus*), baboons (*Papio*), macaques (*Macaca*), and langurs (*Presbytis*). The thumb and large toe are opposable. Some have internal cheek pouches; none have prehensile tails. Family **Hylobatidae** contains gibbons and siamang (11 species of genus *Hylobates*), with arms much longer than legs, prehensile hands with fully opposable thumb, and locomotion by true brachiation. Family **Hominidae** contains four living genera and five species: *Gorilla* (one species), *Pan* (two species of chimpanzees), *Pongo* (one species of orangutan), and *Homo* (one species, humans). The first three of these four genera were formerly placed in the paraphyletic family Pongidae; family Hominidae contained only humans. This separation is not recognized by cladistic taxonomy because the most recent common ancestor of the family Pongidae is also the ancestor of humans.

Order Xenarthra (ze-nar'thra) (Gr. *xenos*, intrusive, + *arthron*, joint) (formerly Edentata [L. *edentatus*, toothless]): **anteaters, armadillos, sloths.** Species of this order are either toothless (anteaters) or have simple, peglike teeth (sloths and armadillos). Most live in South and Central America, although nine-banded armadillos (*Dasypus novemcinctus*) are common in the southern United States; 29 species.

Order Pholidota (fol'i-do'ta) (Gr. *pholis*, horny scale): **pangolins.** An odd group of mammals whose bodies are covered with overlapping horny scales that have arisen from fused bundles of hair. Their home is in tropical Asia and Africa; seven species.

Order Lagomorpha (lag'o-mor'fa) (Gr. *lagos*, hare, + *morphē*, form): **rabbits, hares, pikas** (Figure 28-36). Lagomorphs have long, constantly growing incisors, like rodents, but unlike rodents, they have an additional pair of peglike incisors growing behind the first pair. All lagomorphs are herbivores with cosmopolitan distribution; 80 species.

Order Rodentia (ro-den'che-a) (L. *rodere*, to gnaw): **gnawing mammals: squirrels** (Figure 28-37), **rats, woodchucks.** Rodents, comprising nearly 40% of all mammalian species, are characterized by two pairs of razor-sharp incisors used for gnawing through the toughest pods and shells for food. With their impressive reproductive powers, adaptability, and capacity to invade all terrestrial habitats, they are of great ecological significance. Important families of this order are **Sciuridae** (squirrels and woodchucks), **Muridae** (rats and house mice), **Castoridae** (beavers), **Erethizontidae** (porcupines), **Geomyidae** (pocket gophers), and **Cricetidae** (hamsters, deer mice, gerbils, voles, lemmings); 1935 species.

Order Carnivora (car-niv'o-ra) (L. *caro*, flesh, + *vorare*, to devour): **flesheating mammals: dogs, wolves, cats, bears** (Figure 28-38), **weasels seals, sea lions**

Figure 28-37
Eastern gray squirrel, *Sciurus carolinensis*. This common resident of eastern towns and hardwood forests serves as an important reforestation agent by planting numerous nuts and seeds that sprout into trees. Order Rodentia, family Sciuridae.

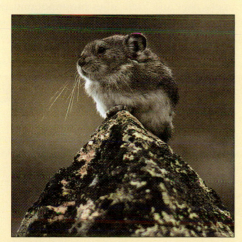

Figure 28-36
A pika, *Ochotona princeps*, atop a rockslide in Alaska. This little rat-sized mammal does not hibernate but prepares for winter by storing dried grasses beneath boulders. Order Lagomorpha, family Ochotonidae.

Figure 28-38
Grizzly bear, *Ursus horribilis*, of Alaska. Grizzlies, once common in the lower 48 states, are now confined largely to wilderness areas. Order Carnivora, family Ursidae.

Figure 28-39

A Galápagos sea lion bull, *Zalophus californianus,* barks to indicate his territorial ownership. Order Carnivora, family Otariidae.

(Figure 28-39), **walruses.** All Carnivora, except the giant panda, have predatory habits, and their teeth are especially adapted for tearing flesh. They are distributed all over the world except in Australian and Antarctic regions where there are no native forms (besides seals). Among more familiar families are **Canidae** (dog family), consisting of dogs, wolves, foxes, and coyotes; **Felidae** (cat family), whose members include domestic cats, tigers, lions, cougars, and lynxes; **Ursidae** (bears); **Procyonidae** (raccoons); and **Mustelidae** (fur-bearing family), containing martens, skunks, weasels, otters, badgers, minks, and wolverines; **Otariidae** (eared seals), containing fur seals and sea lions; 280 species.

Order Tubulidentata (tu'byu-li-den-ta'ta) (L. *tubulus,* tube, *dens,* tooth): **aardvark.** "Aardvark" is Dutch for earth pig, a peculiar animal with a piglike body found in Africa; one species.

Order Proboscidea (pro'ba-sid'e-a) (Gr. *proboskis,* elephant's trunk, from *pro,* before, + *boskein,* to feed): **proboscis mammals: elephants.** These largest of living land animals, have two upper incisors elongated as tusks, and well-developed molar teeth. Asiatic or Indian elephants (*Elephas maximus*) have long been partly domesticated and trained to do heavy tasks. Taming of African elephants (*Loxodonta africana*) is more difficult but was done extensively by ancient Carthaginians and Romans, who employed them in their armies; two species.

Order Hyracoidea (hy'ra-coi'de-a) (Gr. *hyrax,* shrew): **hyraxes** (coneys) Coneys are herbivores restricted to Africa and Syria. They have some resemblance to short-eared rabbits but have teeth like rhinoceroses, with hooves on their toes and pads on their feet. They have four toes on the front feet and three toes on the back; 11 species.

Order Sirenia (sy-re'ne-a) (Gr. *seiren,* sea nymph): **sea cows and manatees.** Sirenians are large, clumsy, aquatic mammals with large head, no hindlimbs, and forelimbs modified into flippers. The sea cow (dugong) of tropical coastlines of East Africa, Asia, and Australia and three species of manatees of the Caribbean area and Florida, Amazon River, and West Africa are the only living species. A fifth species, the large Steller's sea cow, was hunted to extinction by humans in the mid-eighteenth century; four species.

Order Perissodactyla (pe-ris'so-dak'ti-la) (Gr. *perissos,* odd, + *dactylos,* toe): **odd-toed hoofed mammals: horses, asses, zebras, tapirs, rhinoceroses.** Odd-toed hoofed mammals have an odd number of toes (one or three), each with a cornified hoof. Both Perissodactyla and Artiodactyla are often referred to as **ungulates** (L. *ungula,* hoof), or hoofed mammals, with teeth adapted for grinding plants. The horse family (Equidae), which also includes asses and zebras, has only one functional toe. Tapirs have a short proboscis formed from the upper lip and nose. The rhinoceros (*Rhinoceros*) includes several species found in Africa and Southeast Asia. All are herbivorous; 18 species.

Order Artiodactyla (ar'te-o-dak'ti-la) (Gr. *artios,* even, + *daktylos,* toe): **even-toed hoofed mammals: swine, camels, deer and their allies hippopotamuses, antelopes, cattle, sheep, goats.** Most of these ungulates

SUMMARY

Mammals are endothermic and homeothermic vertebrates whose bodies are insulated by hair and who nurse their young with milk. The approximately 4600 species of mammals are descended from the synapsid lineage of amniotes that arose during Carboniferous period of the Paleozoic era. Their evolution can be traced from pelycosaurs of the Permian period to therapsids of the late Permian and Triassic periods of the Mesozoic era. One group of therapsids, the cynodonts, gave rise during the Triassic to mammals. Mammalian evolution was accompanied by the appearance of many important derived characteristics, among these an enlarged brain with greater sensory integration, high metabolic rate, endothermy, and many changes in the skeleton that supported a more active life. Mammals diversified rapidly during the Tertiary period of the Cenozoic era.

Mammals are named for the glandular milk-secreting organs of females (rudimentary in males), a unique adaptation which, combined with prolonged parental care, buffers infants from demands of for-aging for themselves and eases the transition to adulthood. Hair, an integumentary outgrowth that covers most mammals, serves variously for mechanical protection, thermal insulation, protective coloration, and waterproofing. Mammalian skin is rich in glands: sweat glands that function in evaporative cooling, scent glands used in social interactions, and sebaceous glands that secrete lubricating skin oil. All placental mammals have deciduous teeth that are replaced by permanent teeth (diphyodont dentition). Four kinds of teeth—incisors, canines, premolars, and molars—may be highly modified in different mammals for specialized feeding tasks, or they may be absent.

Food habits of mammals strongly influence their body form and physiology. Insectivores feed mainly on insects and other small invertebrates. Herbivorous mammals have specialized teeth for grinding cellulose and silica-rich plants and have specialized regions of the gut for harboring bacteria that can break down cellulose. Carnivorous mammals have adaptations, including special-

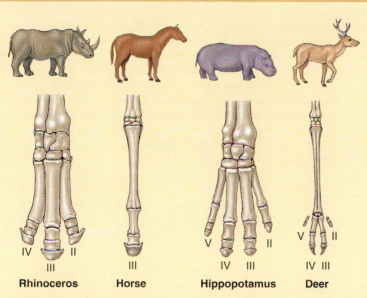

Rhinoceros   **Horse**   **Hippopotamus**   **Deer**

Figure 28-40

Odd-toed and even-toed ungulates. Rhinoceros and horse (order Perissodactyla) are odd-toed; hippopotamus and deer (order Artiodactyla) are even-toed. The lighter, faster mammals run on only one or two toes.

Figure 28-41

Humpback whale, *Megaptera novaeangliae,* breaching. Among the most acrobatic of whales, humpbacks appear to breach to stun fish schools or to communicate information to other pod members. Order Cetacea, family Balaenopteridae.

have two toes, although the hippopotamus and some others have four (Figure 28-40). Each toe is sheathed in a cornified hoof. Many, such as cattle, deer, and sheep have horns or antlers. Many are ruminants. Like Perissodactyla, they are strictly herbivorous. The group is divided into nine living families and many extinct ones and includes some of the most valuable domestic animals. Artiodactyla is commonly divided into three suborders: the **Suina** (pigs, peccaries, and hippopotamuses), the **Tylopoda** (camels), and the **Ruminantia** (deer, giraffes, sheep, cattle); 217 species.

Order Cetacea (see-tay'she-a) (L. *cetus,* whale): **whales** (Figure 28-41), **dolphins, porpoises.** Anterior limbs of cetaceans are modified into broad flippers; posterior limbs are absent. Some have a fleshy dorsal fin and the tail is divided into transverse fleshy flukes. Nostrils are represented by a single or double blowhole on top of the head. They have no hair except for a few on the muzzle, no skin glands except the mammary and those of the eye, no external ear, and small eyes. The order is divided into **toothed whales** (suborder Odontoceti), represented by dolphins, porpoises, and sperm whales; and **baleen whales** (suborder Mysticeti) represented by rorquals, right whales, and gray whales. Baleen whales are generally larger than toothed whales. The blue whale, a rorqual, is the heaviest animal that has ever lived. Rather than teeth, baleen whales have a straining device of whalebone (baleen) attached to the palate, used to filter plankton; 78 species.

ized jaw muscles and teeth, for killing and processing their prey, mainly herbivorous mammals. Omnivores feed on both plant and animal foods.

Some marine, terrestrial, and aerial mammals migrate; some migrations, such as those of fur seals and caribou, are extensive. Migrations are usually made toward favorable climatic and optimal food and calving conditions, or to bring the sexes together for mating.

Mammals with true flight, the bats, are mainly nocturnal and thus avoid direct competition with birds. Most employ ultrasonic echolocation to navigate and feed in darkness.

Living mammals with the most primitive reproductive characters are egg-laying monotremes of the Australian region. After hatching, the young are nourished with their mother's milk. All other mammals are viviparous. Embryos of marsupials have brief gestation periods, are born underdeveloped, and complete their early growth in the mother's pouch, nourished by milk. The remaining mammals are eutherians, mammals that develop an advanced placental attachment between mother and embryos through which embryos are nourished for a prolonged period.

Mammal populations fluctuate from both density-dependent and density-independent causes and some mammals, particularly rodents, may experience extreme cycles of abundance in population density. The unqualified success of mammals as a group cannot be attributed to greater organ system perfection, but rather to their impressive overall adaptability—the capacity to fit more perfectly in total organization to environmental conditions and thus exploit virtually every habitat on earth.

Darwinian evolutionary principles give us great insight into our own origins. Humans are primates, a mammalian group that descended from a shrewlike ancestor. The common ancestor of all modern primates was arboreal and had grasping fingers and forward-facing eyes capable of binocular vision. Primates radiated over the last 80 million years to form two major lines of descent: prosimians (lemurs, lorises, and tarsiers) and simians (monkeys, apes, and hominids). Earliest hominids appeared about 5.5 million years ago; these were the australopithecines. Australopithecines gave rise to, and coexisted with, species *Homo habilis,* the first user of stone tools. *Homo erectus* appeared about 1.8 million years ago and was eventually replaced by modern humans, *Homo sapiens.*

1. Describe the evolution of mammals, tracing their synapsid lineage from early amniote ancestors to true mammals. How would you distinguish pelycosaurs, early therapsids, cynodonts, and mammals?

2. Describe structural and functional adaptations that appeared in early amniotes that foreshadowed the mammalian body plan. Which mammalian attributes do you think were especially important to successful radiation of mammals?

3. Hair is believed to have evolved in therapsids as an adaptation for insulation, but modern mammals have adapted hair for several other purposes. Describe these.

4. What is distinctive about each of the following: horns of bovids, antlers of the deer family, and horns of rhinos? Describe the growth cycle of antlers. What is velvet? Why is it called velvet?

5. Describe location and principal function(s) of each of the following skin glands: sweat glands (eccrine and apocrine), scent glands, sebaceous glands, and mammary glands.

6. Define "diphyodont" and "heterodont" and explain why both terms apply to mammalian dentition.

7. Describe food habits of insectivorous, herbivorous, carnivorous, and omnivorous mammals. Give the common names of some mammals belonging to each group.

8. Most herbivorous mammals depend on cellulose as their main energy source, yet no mammal synthesizes cellulose-splitting enzymes. How are the digestive tracts of mammals specialized for symbiotic digestion of cellulose?

9. How does fermentation differ between horses and cattle?

10. What is the relationship of body mass to metabolic rate of mammals?

11. Describe the annual migrations of barren-ground caribou and fur seals.

12. Explain what is distinctive about the life habit and mode of navigation in bats.

13. Describe and distinguish patterns of reproduction in monotremes, marsupials, and placental mammals. What aspects of mammalian reproduction are present in *all* mammals but in no other vertebrates?

14. Distinguish between territory and home range for mammals.

15. What is the difference between density-dependent and density-independent causes of fluctuations of the size of mammalian populations?

16. Describe the hare-lynx population cycle, considered a classic example of a prey-predator relationship (Figure 28-27). From your examination of the cycle, formulate a hypothesis for the explanation of the oscillations.

17. What do the terms Theria, Metatheria, Eutheria, Monotremata, and Marsupialia mean? List mammals that are grouped under each taxon.

18. What anatomical characteristics set primates apart from other mammals?

19. What role does a fossil named "Lucy" play in the reconstruction of human evolutionary history?

20. In what ways do the genera *Australopithecus* and *Homo*, which coexisted for at least 1 million years, differ?

21. When did the different species of *Homo* appear and how did they differ socially?

22. What major attributes make the human position in animal evolution unique?

SELECTED REFERENCES

Feldhamer, G. A., L. C. Drickamer, S. H. Vessey, and J. F. Merritt. 1999. Mammalogy: adaptation, diversity, and ecology. Dubuque, Iowa, WCB/McGraw-Hill. *Modern well-illustrated textbook.*

Grzimek's encyclopedia of mammals. 1990. vol. 1-5. New York, McGraw-Hill Publishing Company. *Valuable source of information on all mammal orders.*

Johanson, D. C., and M. A. Edey. 1981. Lucy, the beginnings of humankind. New York, Simon & Schuster. *Entertaining account of Johanson's discovery of the famous, nearly complete* Australopithecus afarensis *skeleton.*

Jones, S., R. Martin, and D. Pilbeam. 1992. Cambridge encyclopedia of human evolution. Cambridge, U.K., Cambridge University Press. *Comprehensive and informative encyclopedia written for the nonspecialist. Highly readable and highly recommended.*

Macdonald, D. (ed.). 1984. The encyclopedia of mammals. New York, Facts on File Publications. *Coverage of all mammal orders and families, enhanced with fine photographs and color artwork.*

Nowak, R. M. 1991. Walker's mammals of the world, ed. 5. Baltimore, The Johns Hopkins University Press. *The definitive illustrated reference work on mammals, with descriptions of all extant and recently extinct species.*

Preston-Mafham, R., and K. Preston-Mafham. 1992. Primates of the world. New York, Facts on File Publications. *A small "primer" with high quality photographs and serviceable descriptions.*

Rice, J. A. (ed.). 1994. The marvelous mammalian parade. Nat. Hist. **103**(4):39–91. *A special multi-authored section on mammalian evolution.*

Rismiller, P. D., and R. S. Seymour. 1991. The echidna. Sci. Am. **294**:96–103 (Feb.). *Recent studies of this fascinating monotreme have revealed many secrets of its natural history and reproduction.*

Stringer, C. B. 1990. The emergence of modern humans. Sci. Am. **263**:98–104 (Dec.). *A review of the geographical origins of modern humans.*

Suga, N. 1990. Biosonar and neural computation in bats. Sci. Am. **262**:60–68 (June). *How the bat nervous system processes echolocation signals.*

Tattersall, I. 2001. How we came to be human. Sci. Amer. **285**:56–63 (Dec.). *How acquiring language and capacity for symbolic art sets* Home sapiens *apart from the Neandertals.*

Templeton, A. R. 2002. Out of Africa again and again. Nature **416**:45–51. *A comprehensive analysis of molecular genetic data indicating that humans have evolved as a cohesive lineage for the past 1.7 million years, with multiple migrations of populations out of Africa.*

Wong, K. 2002. The mammals that conquered the seas. Sci. Am. **286**:70–79 (May). *New fossils and DNA evidence help to unravel the evolutionary history of whales.*

Wuethrich, B. 1997. Will fossil from down under upend mammal evolution? Science **278**:1401–1402. *If this fossil jaw is from a placental , then placentals were in Australia 115 mya rather than 5 mya, and they became extinct there until reintroduction.*

ZOOLOGY LINKS TO THE INTERNET

Visit the textbook's Online Learning Center at www.mhhe.com/zoology to find live Internet links for each of the topics listed here.
Class Mammalia
Marine Mammals

Cat Dissections
Rat Dissections
Fetal Pig Dissections
Vertebrate Laboratory Exercises
Conservation Issues Concerning Mammals

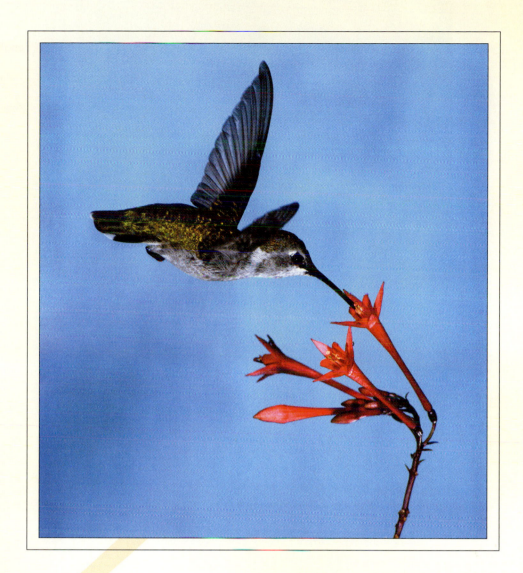

Female broad-billed hummingbird sipping nectar.

Activity of Life

29 Support, Protection, and Movement
30 Homeostasis: Osmotic Regulation, Excretion, and Temperature Regulation
31 Internal Fluids and Respiration
32 Digestion and Nutrition
33 Nervous Coordination: Nervous System and Sense Organs
34 Chemical Coordination: Endocrine System
35 Immunity
36 Animal Behavior

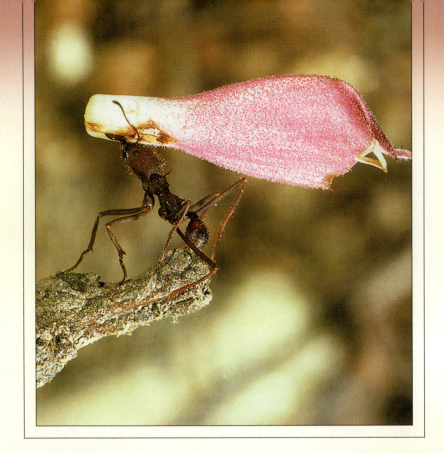

An ant carries with ease a flower petal that is heavier than the ant's body weight.

Support, Protection, and Movement

Of Grasshoppers and Superman

"A dog," remarked Galileo in the seventeenth century, "could probably carry two or three such dogs upon his back; but I believe that a horse could not carry even one of its own size." Galileo was referring to the principle of scaling, a procedure that allows us to understand the consequences of changing body size. A grasshopper can jump to a height of 50 times the length of its body, yet a man in a standing jump cannot clear an obstacle that is no higher than he is tall. Without an understanding of scaling, this comparison could easily lead us to the erroneous conclusion that there is something very special about the musculatures of insects. To the authors of a nineteenth-century entomology text it seemed that "This wonderful strength of insects is doubtless the result of something peculiar in the structure and arrangement of their muscles, and principally their extraordinary power of contraction." But grasshopper muscles are in fact no more powerful than human muscles because *muscles of small and large animals exert the same*

force per cross-sectional area. Grasshoppers leap high in proportion to their size because they are small, not because they possess extraordinary muscles.

The authors of this nineteenth-century text further suggested that it was fortunate that higher animals were withheld the powers of insects, for they would surely have "caused the early desolation of the world." More probably, such powers would have led to their own desolation. For earthly mortals would need more than superhuman muscles were they to leap in the proportions of a grasshopper. They would require superhuman tendons, superhuman ligaments, and superhuman bones to withstand the stresses of mighty contractions, not to mention the crushing strains of landing again on earth at terminal velocity. The feats of Superman would be quite impossible were he built of the structural materials available to earthbound animals, rather than of the wondrous materials available to inhabitants of the mythical planet Krypton.

INTEGUMENT AMONG VARIOUS GROUPS OF ANIMALS

The integument is the outer covering of the body, a protective wrapping that includes the skin and all structures derived from or associated with skin, such as hair, setae, scales, feathers, and horns. In most animals it is tough and pliable, providing mechanical protection against abrasion and puncture and forming an effective barrier against invasion of bacteria. It may provide moisture proofing against fluid loss or gain. Skin helps protect underlying cells against the damaging action of the sun's ultraviolet rays. In addition to being a protective cover, the skin serves a variety of important regulatory functions. For example, in endothermic animals, it is vitally concerned with temperature regulation, since most of the body's heat is lost through the skin; it contains mechanisms that cool the body when it is too hot and slow heat loss when the body is too cold. Skin contains sensory receptors that provide essential information about the immediate environment. It has excretory functions and in some animals respiratory functions as well. Through skin pigmentation some organisms can make themselves more or less conspicuous. Skin secretions can make an animal sexually attractive or repugnant or provide olfactory cues that influence behavioral interactions between individuals.

Invertebrate Integument

Many protozoa have only the delicate cell or plasma membranes for external coverings; others, such as *Paramecium,* have developed a protective pellicle. Most multicellular invertebrates, however, have more complex tissue coverings. The principal covering is a single-layered **epidermis.** Some invertebrates have added a secreted noncellular **cuticle** over the epidermis for additional protection.

Molluscan epidermis is delicate and soft and contains mucous glands, some of which secrete the calcium carbonate of the shell. Cephalopod molluscs (squids and octopuses) have developed a more complex integument, consisting of cuticle, simple epidermis, layer of connective tissue, layer of reflecting cells (iridocytes), and thicker layer of connective tissue.

Arthropods have the most complex of invertebrate integuments, providing not only protection but also skeletal support. Development of a firm exoskeleton and jointed appendages suitable for attachment of muscles has been a key feature in the extraordinary diversity of this phylum, largest of animal groups. Arthropod integument consists of a single-layered **epidermis** (also called more precisely **hypodermis**), which secretes a complex cuticle of two zones (Figure 29-1A). The thicker inner zone, the **procuticle,** is composed of protein and chitin (a polysaccharide) laid down in layers (lamellae) much like veneers of plywood. The outer zone of cuticle, lying on the external surface above the procuticle, is the thin **epicuticle.** The epicuticle is a nonchitinous complex of proteins and lipids that provides a protective moisture-proofing barrier to the integument.

Arthropod cuticle may remain as a tough but soft and flexible layer, as it is in many microcrustaceans and insect larvae. However, it may be hardened by either of two ways. In decapod crustaceans, for example, crabs and lobsters, the cuticle is stiffened by **calcification,** the deposition of calcium carbonate in the outer layers of the procuticle. In insects hardening occurs when protein molecules bond together with stabilizing cross-linkages within and between adjacent lamellae of the procuticle. The result of this process, called **sclerotization,** is formation of a highly resistant and insoluble protein, **sclerotin.** Arthropod cuticle is one of the toughest materials synthesized by animals; it is strongly resistant to pressure and tearing and can withstand boiling in concentrated alkali, yet it is light, having a specific mass of only 1.3 (1.3 times the weight of water).

When arthropods molt, epidermal cells first divide by mitosis. Enzymes secreted by the epidermis digest most of the procuticle. Digested materials are then absorbed and consequently are not lost to the body. Then in the space beneath the old cuticle a new epicuticle and procuticle are formed. After the old cuticle is shed, the new cuticle is thickened and calcified or sclerotized.

Vertebrate Integument and Derivatives

The basic plan of the vertebrate integument, as exemplified by frog and human skin (Figure 29-1B and C), includes a thin, outer stratified epithelial layer, the **epidermis,** derived from ectoderm and an inner, thicker layer, the **dermis,** or true skin, which is of mesodermal origin. (Ectoderm and mesoderm are germ layers, described in Figure 8-24, p. 169.)

Although the epidermis is thin and appears simple in structure, it gives rise to most derivatives of the integument, such as hair, feathers, claws, and hooves. The dermis is a dense connective tissue layer (p. 183) containing blood vessels, collagenous fibers, nerves, pigment cells, fat cells, and connective tissue cells called fibroblasts. These elements support, cushion, and nourish the epidermis, which is devoid of blood vessels. In addition, other cells present in this connective tissue layer (macrophages, mast cells, and lymphocytes, see Chapter 35), provide a first line of defense if the outer epidermal layer is breached.

The epidermis is a stratified squamous epithelium (p. 182) consisting usually of several layers of cells. Cells of the basal part undergo frequent mitosis to renew layers that lie above. As outer layers of cells are displaced upward by new generations of cells beneath, an exceedingly tough, fibrous protein called **keratin** accumulates in the interior of the cells. Gradually, keratin replaces all metabolically active cytoplasm. The cell dies and is eventually shed, lifeless and scalelike. Such is the origin of dandruff as well as a significant fraction of household dust. This process is called **keratinization,** and the cell, thus transformed, is said to be **cornified.** Cornified cells, highly resistant to abrasion and water diffusion, comprise the outermost **stratum corneum.** This epidermal layer becomes especially thick in areas exposed to persistent pressure or wear such as calluses, foot pads of mammals, and the scales of reptiles and birds.

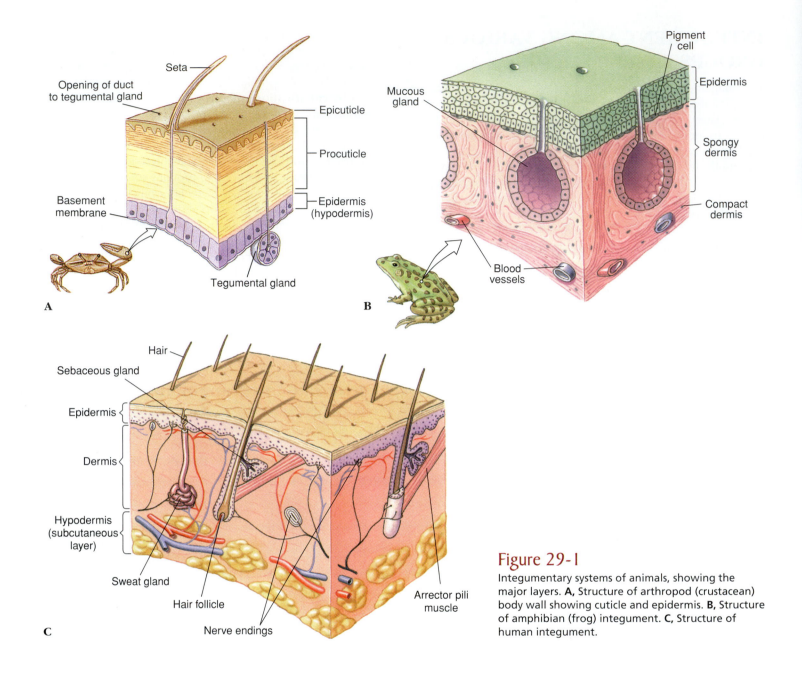

Seta

Opening of duct
to tegumental gland

Epicuticle

Procuticle

Basement
membrane

Epidermis
(hypodermis)

Tegumental gland

A

Pigment
cell

Mucous
gland

Epidermis

Spongy
dermis

Compact
dermis

Blood
vessels

B

Hair

Sebaceous gland

Epidermis

Dermis

Hypodermis
(subcutaneous
layer)

Sweat gland

Hair follicle

Nerve endings

Arrector pili
muscle

C

Figure 29-1

Integumentary systems of animals, showing the
major layers. **A,** Structure of arthropod (crustacean)
body wall showing cuticle and epidermis. **B,** Structure
of amphibian (frog) integument. **C,** Structure of
human integument.

The **dermis,** as already mentioned, mainly serves a supportive role for the epidermis. Nevertheless, true bony structures, where they occur in the integument, are always dermal derivatives. Heavy bony plates were common in ostracoderms and placoderms of the Paleozoic era (Figure 23-18, p. 484) and persist in some living fishes, such as sturgeons (Figure 24-19, p. 500). Scales of contemporary fishes are bony dermal structures that have evolved from the bony armor of the Paleozoic fishes but are much smaller and more flexible. Fish scales are thin bony slivers covered with a mucus-secreting epidermis (Figure 29-2). Most amphibians lack dermal bones in their skin, except for vestiges of dermal scales found in a few tropical caecilian species. In reptiles dermal bones provide the armor of crocodilians, the beaded skin of many lizards, and contribute to the shell of turtles. Dermal bone also gives rise to antlers, as well as the bony core of horns.

Lizards, snakes, turtles, and crocodilians were among the first to exploit the adaptive possibilities of the remarkably tough protein keratin. Reptilian epidermal scale that develops from keratin is a much lighter and more flexible structure than the bony, dermal scale of fishes, yet it provides excellent protection from abrasion and desiccation. Scales may be overlapping structures, as in snakes and some lizards, or develop into plates, as in turtles and crocodilians. In birds, keratin found new uses. Feathers, beaks, and claws, as well as scales, are all epidermal structures composed of dense keratin. Mammals continued to capitalize on keratin's virtues by turning it into hair, hooves, claws, and nails. As a result of its keratin content, hair is by far the strongest material in the body. It has a tensile strength comparable to that of rolled aluminum and is nearly twice as strong, weight for weight, as the strongest bone.

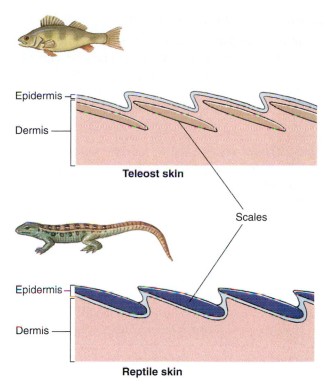

Figure 29-2

Integument of bony fishes and lizards. Bony (teleost) fishes have bony scales from dermis, and lizards have horny scales from epidermis. Thus they are not homologous structures. Dermal scales of fishes are retained throughout life. Since a new growth ring is added to each scale each year, fishery biologists use scales to tell the age of fishes. Epidermal scales of reptiles are shed periodically.

Structures such as claws, beaks, nails, and horns contain combinations of epidermal (keratinized) and dermal components. Their basic structure is the same, with a central bony core covered by a vascularized nutritive layer of the dermis, and an outer epithelial layer. This epithelial layer has a germinative component responsible for the continual growth of horns, hooves, claws, and beaks. The outer epithelial layer is keratinized. Overgrowth of these structures is prevented by constant wear and abrasion (Figure 29-3).

Animal Coloration

The colors of animals may be vivid and dramatic when serving as important recognition marks or as warning coloration, or they may be subdued or cryptic when used for camouflage. Integumentary color is usually produced by pigments, but in many insects and in some vertebrates, especially birds, certain colors are produced by the physical structure of the surface tissue, which reflects certain light wavelengths and eliminates others. Colors produced this way are called **structural color,** and they are responsible for the most beautifully iridescent and metallic hues to be found in the animal kingdom. Many butterflies and beetles and a few fishes thus share with birds the distinction of being the earth's most resplendent animals. Certain structural colors of feathers are caused by minute, air-filled spaces or pores

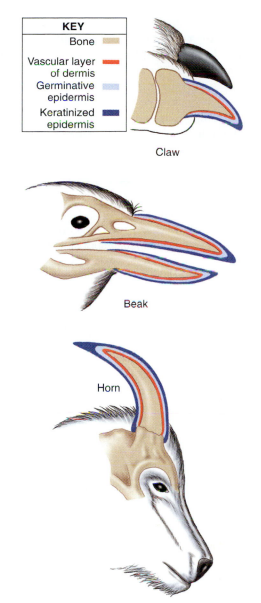

Figure 29-3

Similarity of structure of integumentary derivatives. Claws, beaks, and horns are all built of similar combinations of epidermal (keratinized) and dermal components. A central bony core is covered by a vascularized nutritive layer of dermis. An outer epithelial layer has a basal germinative component which proliferates to allow these structures to grow continually. The thickened surface epithelium is keratinized or cornified. Note that the relative thickness of each component is not drawn to scale.

that reflect white light (white feathers) or some portion of the spectrum (for example, Tyndall blue coloration produced by scattering of light [see note, p. 562]). Iridescent colors that change hue as an animal's angle shifts with respect to an observer are produced when light is reflected from several layers of thin, transparent film. By phase interference, light waves reinforce, weaken, or eliminate each other to produce some of the purest and most brilliant colors we know.

More common than structural colors in animals are **pigments** (biochromes), an extremely varied group of large molecules that reflect light rays. In crustaceans and ectothermic vertebrates these pigments are contained in large cells with branching processes, called **chromatophores** (Figure 29-4A). The pigment may concentrate in the center of the cell in an aggregate too small to be visible, or it may disperse throughout the cell and its processes, providing maximum display. Chromatophores of cephalopod molluscs are entirely different (Figure 29-4B). Each is a small sac-like cell filled with pigment granules and surrounded by muscle cells that, when contracted, stretch the whole cell into a pigmented sheet. When muscles relax, the elastic chromatophore quickly shrinks to a small sphere. With such pigment cells squids and octopuses can alter their color more rapidly than any other animal.

Most widespread of animal pigments are the **melanins,** a group of black or brown polymers responsible for various earth-colored shades that most animals wear. Yellow and red colors are often caused by **carotenoid** pigments, which are frequently contained within special pigment cells called **xanthophores.** Most vertebrates are incapable of synthesizing their own carotenoid pigments but must obtain them directly or indirectly from plants. Two entirely different classes of pigments called ommochromes and pteridines are usually responsible for the yellow pigments of molluscs and arthropods. Green colors are rare; when they occur, they are usually produced by yellow pigment overlying blue structural color. **Iridophores,** a third type of chromatophore, contain crystals of guanine or some other purine, rather than pigment. They produce a silvery or metallic effect by reflecting light.

By vertebrate standards, mammals are a somber-colored group (p. 588). Most mammals are largely color blind, a deficiency that is doubtless connected with lack of bright colors in the group. Exceptions are the brilliantly colored skin patches of some baboons and mandrills. Significantly, primates have color vision and thus can appreciate such eye-catching ornaments. The muted colors of mammals are caused by melanin, which is deposited in growing hair by dermal melanophores.

Figure 29-4

Chromatophores. **A,** The crustacean chromatophore showing the pigment dispersed (*left*) and concentrated (*right*). Vertebrate chromatophores are similar. **B,** The cephalopod chromatophore is an elastic capsule surrounded by muscle fibers that, when contracted (*left*), stretch the capsule to expose the pigment.

Injurious Effects of Sunlight

The familiar vulnerability of human skin to sunburn reminds us of the potentially damaging effects of ultraviolet radiation on protoplasm. Many animals, such as protozoa and flatworms, if exposed to the sun in shallow water are damaged or killed by ultraviolet radiation. Most land animals are protected from such damage by the screening action of special body coverings, for example, the cuticle of arthropods, scales of reptiles, and feathers and fur of birds and mammals. Humans, however, are "naked apes" that lack the furry protection of most other mammals. We must depend on thickening of the epidermis **(stratum corneum)** and on epidermal pigmentation for protection. Most ultraviolet radiation is absorbed in the epidermis, but about 10% penetrates the dermis. Damaged cells in both epidermis and dermis release histamine and other vasodilator substances that cause blood-vessel enlargement in the dermis and the characteristic red coloration of sunburn. Light skins suntan through the formation of the pigment **melanin** in the deeper epidermis and by "pigment darkening," that is, the photooxidative blackening of bleached pigment already present in the epidermis. Unfortunately, tanning does not bestow perfect protection. Sunlight still ages the skin prematurely, and tanning itself causes the skin to become dry and leathery. Sunlight also is responsible for approximately 1 million new cases of skin cancer annually in the United States alone, making skin cancer the most common of malignancies among Caucasians. There is now strong evidence that genetic mutations caused by high doses of sunlight received during the pre-adult years are responsible for skin cancers that appear after middle age.

SKELETAL SYSTEMS

Skeletons are supportive systems that provide rigidity to the body, surfaces for muscle attachment, and protection for vulnerable body organs. The familiar bone of the vertebrate skeleton is only one of several kinds of supportive and connective tissues serving various binding and weight-bearing functions, which are described in this discussion.

Hydrostatic Skeletons

Not all skeletons are rigid; many invertebrate groups use their body fluids as an internal hydrostatic skeleton. Muscles in the body wall of an earthworm, for example, have no firm base for attachment but develop muscular force by contracting against incompressible coelomic fluids, enclosed within a limited space, much like the hydraulic brake system of an automobile.

Alternate contractions of the circular and longitudinal muscles of the body wall enable a worm to thin and thicken, producing backward-moving waves of motion that propel the animal forward (Figure 29-5). Earthworms and other annelids are helped by septa that separate their body into more or less independent compartments (Figure 17-1, p. 347). An obvious advantage is that if a worm is punctured or even cut into pieces, each

Figure 29-5

How an earthworm moves forward. When circular muscles contract, longitudinal muscles are stretched by internal fluid pressure and the worm elongates. Then, by alternate contraction of longitudinal and circular muscles, a wave of contraction passes from anterior to posterior. Bristlelike setae are extended to anchor the animal and prevent slippage as it moves forward.

Figure 29-6

Muscular trunk of an elephant, an example of a muscular hydrostat.

part can still develop pressure and move. Worms that lack internal compartments, for example, the lugworm *Arenicola* (Figure 17-5, p. 349), are rendered helpless if body fluid is lost through a wound.

There are many examples in the animal kingdom of muscles that not only produce movement but also provide a unique form of skeletal support. The elephant's trunk is an excellent example of a structure that lacks any obvious form of skeletal support, yet is capable of bending, twisting, elongating, and lifting heavy weights (Figure 29-6). The elephant's trunk, tongues of mammals and reptiles, and tentacles of cephalopod molluscs are examples of **muscular hydrostats.** Like hydrostatic skeletons of worms, muscular hydrostats work because they are composed of incompressible tissues that remain at constant volume. The remarkably diverse movements of muscular hydrostats depend on muscles arranged in complex patterns.

Rigid Skeletons

Rigid skeletons differ from hydrostatic skeletons in one fundamental way: rigid skeletons consist of rigid elements, usually jointed, to which muscles can attach. Muscles can only contract and relax; they only provide movement in one direction, and are therefore often grouped in an antagonistic set of muscles. Rigid

skeletons provide the anchor points required by opposing sets of muscles, such as flexors and extensors.

Antagonistic muscles are functional opposites that oppose each other's action. For example, the biceps brachii on one side of the upper arm is opposed in its action by the triceps brachii on the opposite side of the arm. By contracting against each other, they balance and smooth rapid movements.

There are two principal types of rigid skeletons: **exoskeleton,** typical of molluscs, arthropods and many other invertebrates; and **endoskeleton,** characteristic of echinoderms, vertebrates, and some cnidarians. An invertebrate's exoskeleton may be mainly protective, but it may also perform a vital role in locomotion. An exoskeleton may take the form of a shell, a spicule, or a calcareous, proteinaceous, or chitinous plate. It may be rigid, as in molluscs, or jointed and movable, as in arthropods. Unlike an endoskeleton, which grows with the animal, an exoskeleton is often a limiting coat of armor that must be periodically molted to make way for an enlarged replacement (molting in crustaceans is described on p. 384). Some invertebrate exoskeletons, such as the shells of snails and bivalves, grow with the animal.

The arthropod-type exoskeleton is perhaps a better arrangement for small animals than a vertebrate-type endoskeleton because a hollow cylindrical tube can support much more weight without collapsing than can a solid cylindrical rod of the same material and weight. Arthropods can thus enjoy both protection and structural support from their exoskeleton. For larger animals the hollow cylinder would be completely impractical. If made thick enough to support the body weight, it would be too heavy to lift; but if kept thin and light, it would be extremely sensitive to buckling or shattering on impact. Finally, can you imagine the sad plight of an animal the size of an elephant when it shed its exoskeleton to molt?

The vertebrate endoskeleton is formed inside the body and is composed of bone and cartilage, which are forms of specialized connective tissue. Bone not only supports and protects but is also the major body reservoir for calcium and phosphorus. In amniote vertebrates red blood cells, platelets, and certain white blood cells are formed in the bone marrow.

Notochord and Cartilage

The **notochord** (see Figure 23-1, p. 468) is a semirigid supportive axial rod of protochordates and all vertebrate larvae and embryos. It is composed of large, vacuolated cells and is surrounded by layers of elastic and fibrous sheaths. It is a stiffening device, preserving body shape during locomotion. Except in jawless vertebrates (lampreys and hagfishes), the notochord is surrounded or replaced by the backbone during embryonic development.

Cartilage is a major skeletal element of some vertebrates. Jawless fishes (for example, lampreys) and elasmobranchs (sharks, skates, and rays) have purely cartilaginous skeletons, which oddly enough is a derived feature, since their Paleozoic ancestors had bony skeletons. Other vertebrates as adults have principally bony skeletons with some cartilage interspersed. Cartilage is a soft, pliable tissue, and unlike most connective tissues, which are quite variable in form, cartilage is basically the same wherever it is found. The basic form, **hyaline cartilage,** has a clear, glassy appearance (see Figure 9-6, p. 183). It is composed of cartilage cells **(chondrocytes)** surrounded by firm complex protein gel interlaced with a meshwork of collagenous fibers. Blood vessels are virtually absent—the reason that sports injuries involving cartilage heal slowly. In addition to forming the cartilaginous skeleton of some vertebrates and that of all vertebrate embryos, hyaline cartilage forms articulating surfaces of many bone joints of most adult vertebrates and supporting tracheal, laryngeal, and bronchial rings of the respiratory system.

Cartilage similar to hyaline cartilage occurs in some invertebrates, for example the radula of gastropod molluscs and lophophore of brachiopods. The cartilage of cephalopod molluscs is of a special type with long, branching processes that resemble cells of vertebrate bone.

Bone

Bone is a living tissue that differs from other connective and supportive tissues by having significant deposits of inorganic calcium salts laid down in an extracellular matrix. Its structural organization is such that bone has nearly the tensile strength of cast iron, yet is only one-third as heavy.

Bone is never formed in vacant space but is always laid down by replacement in areas occupied by some form of connective tissue. Most bone develops from cartilage and is called **endochondral** ("within-cartilage") or **replacement bone.** Embryonic cartilage is gradually eroded leaving it extensively honeycombed; bone-forming cells then invade these areas and begin depositing calcium salts around strandlike remnants of the cartilage. A second type of bone is **intramembranous bone,** which develops directly from sheets of embryonic cells. Dermal bone, mentioned on p. 614, is a type of intramembranous bone. In tetrapod vertebrates intramembranous bone is restricted mainly to bones of the face, cranium and clavicle; the remainder of the skeleton is endochondral bone. Whatever the embryonic origin, once fully formed, endochondral and intramembranous bone look the same.

Fully formed bone, however, may vary in density. **Cancellous** (or spongy) **bone** consists of an open, interlacing framework of bony tissue, oriented to give maximum strength under normal stresses and strains that bone receives. All bone develops first as cancellous bone, but some bones, through further deposition of bone salts, become **compact.** Compact bone is dense, appearing solid to the unaided eye. Both cancellous and compact bone are found in typical long bones of tetrapods (Figure 29-7).

Microscopic Structure of Bone   Compact bone is composed of a calcified bone matrix arranged in concentric rings. The rings contain cavities **(lacunae)** filled with bone cells **(osteocytes),** which are interconnected by many minute passages **(canaliculi).** These passages serve to distribute nutrients throughout the bone. This entire organization of lacunae and canaliculi is arranged into an elongated cylinder called an **osteon** (also called **haversian system**) (Figure 29-7). Bone consists of bundles of osteons cemented together and interconnected with blood vessels and nerves. Because of blood vessels and nerves throughout bone, it is living tissue, although nonliving "ground substance" predominates. As a result of its living state, bone breaks can heal, and bone diseases can be as painful as any other tissue disease.

Following menopause, a woman loses 5% to 6% of her bone mass annually, often leading to the disease osteoporosis and increasing the risk of bone fractures. Dietary supplementation with calcium has been advocated to prevent such losses, but even large doses of calcium alone have little effect in slowing demineralization unless accompanied by therapy with the female sex hormone estrogen (because ovarian production of estrogen drops significantly after menopause). Among animals, only humans, especially females, are troubled with osteoporosis, perhaps a consequence of the long postreproductive life of the human species.

Figure 29-7

Structure of compact bone. **A,** Adult long bone with a cut into the medullary cavity. **B,** Enlarged section showing osteons, the basic histological unit of bone. **C,** Enlarged view of an osteon showing the concentric lamellae and the osteocytes (bone cells) arranged within lacunae. **D,** An osteocyte within a lacuna. Bone cells receive nutrients from the circulatory system via tiny canaliculi that interlace the calcified matrix. Bone cells are known as osteoblasts when they are building bone, but, in mature bone shown here, they become resting osteocytes. Bone is covered with dense connective tissue called periosteum.

Bone

Like muscle, bone is subject to "use and disuse." When we exercise our muscles, our bones respond by producing new bone tissue to give added strength. In fact, the bumps and processes to which muscles attach are produced by bone in response to the action of muscle forces. Conversely, when bones are not subject to stress, as in space flight, the body resorbs the mineral, and the bones become weak. Astronauts who spend many months in space must exercise to a greater degree than on earth to prevent such resorption and bone weakness.

Bone is a dynamic tissue, and bone remodelling and growth are complex restructuring processes, involving both destruction internally by bone-resorbing cells **(osteoclasts)** and deposition externally by bone-building cells **(osteoblasts).** Both processes occur simultaneously so that the marrow cavity inside grows larger by bone resorption while new bone is laid down outside by bone deposition. Bone growth responds to several hormones, in particular **parathyroid hormone** from the parathyroid gland, which stimulates bone resorption, and **calcitonin** from the thyroid gland, which inhibits bone resorption. These two hormones, together with a derivative of vitamin D, **1,25-dihydroxyvitamin D,** are responsible for maintaining a constant level of calcium in the blood. The effect of hormones on bone growth and resorption is described in more detail on p. 732.

Plan of the Vertebrate Skeleton

The vertebrate skeleton is composed of two main divisions: **axial skeleton,** which includes skull, vertebral column, sternum, and ribs, and **appendicular skeleton,** which includes the limbs (or fins or wings) and pectoral and pelvic girdles (Figures 29-8 and 29-9). Not surprisingly, the skeleton has undergone extensive remodeling in the course of vertebrate evolution. The move from water to land forced dramatic changes in body form. With increased cephalization, further concentration of brain, sense organs, and food-gathering apparatus in the head, the skull became the most intricate portion of the skeleton. Some early fishes had as many as 180 skull bones (a source of frustration to paleontologists) but through loss of some bones and

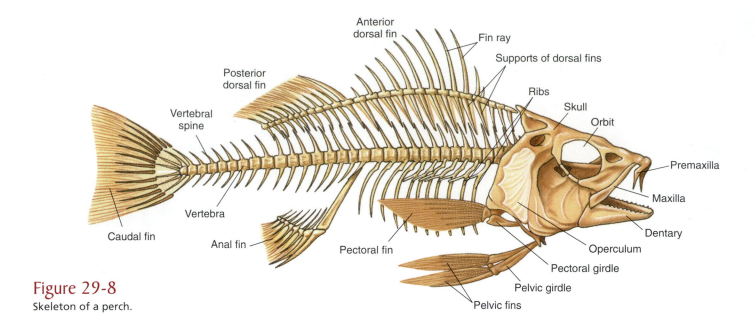

Figure 29-8
Skeleton of a perch.

fusion of others, skull bones became greatly reduced in number during evolution of the tetrapods. Amphibians and lizards have 50 to 95, and mammals, 35 or fewer. Humans have 29.

The vertebral column is the main stiffening axis of the post-cranial skeleton. In fishes it serves much the same function as the notochord from which it is derived; that is, it provides points for muscle attachment and prevents telescoping of the body during muscle contraction. With evolution of amphibious and terrestrial tetrapods, the vertebrate body was no longer buoyed by the aquatic environment. The vertebral column became structurally adapted to withstand new regional stresses transmitted to the column by the two pairs of appendages. In amniote tetrapods (reptiles, birds, and mammals), the vertebrae are differentiated into **cervical** (neck), **thoracic** (chest), **lumbar** (back), **sacral** (pelvic), and **caudal** (tail) vertebrae. In birds and also in humans caudal vertebrae are reduced in number and size, and sacral vertebrae are fused. The number of vertebrae varies among the different vertebrates. Pythons seems to lead the list with more than 400. In humans (Figure 29-9) there are 33 in a young child, but in adults 5 are fused to form the **sacrum** and 4 to form the **coccyx.** Besides the sacrum and coccyx, humans have 7 cervical, 12 thoracic, and 5 lumbar vertebrae. The number of cervical vertebrae (7) is constant in nearly all mammals, whether the neck is short as in dolphins, or long as in giraffes.

The first two cervical vertebrae, **atlas** and **axis,** are modified to support the skull and permit pivotal movements. The atlas bears the globe of the head much as the mythological Atlas bore the earth on his shoulders. The axis, the second vertebra, permits the head to turn from side to side.

Ribs are long or short skeletal structures that articulate medially with vertebrae and extend into the body wall. Fishes have a pair of ribs for every vertebra (Figure 29-8); they serve as stiffening elements in the connective tissue septa that separate the muscle segments and thus improve the effectiveness of muscle contractions. Many fishes have both dorsal and ventral ribs, and

some have numerous riblike intermuscular bones as well—all of which increase the difficulty and reduce the pleasure of eating certain kinds of fish. Other vertebrates have a reduced number of ribs, and some, such as the familiar leopard frog, have no ribs at all. In mammals the ribs together form the thoracic basket, which supports the chest wall and prevents collapse of the lungs. Mammals such as sloths have 24 pairs of ribs, whereas horses possess 18 pairs. Primates other than humans have 13 pairs of ribs; humans have 12 pairs, although approximately 1 person in 20 has a thirteenth pair.

Most vertebrates, fishes included, have paired appendages. All fishes except agnathans have thin pectoral and pelvic fins that are supported by the pectoral and pelvic girdles, respectively (Figure 29-8). Tetrapods (except caecilians, snakes, and limbless lizards) have two pairs of **pentadactyl** (five-toed) limbs, also supported by girdles. The pentadactyl limb is similar in all tetrapods, alive and extinct; even when highly modified for various modes of life, the elements are rather easily homologized (the evolution of the pentadactyl limb is illustrated in Figure 25-1, p. 517).

Modifications of the basic pentadactyl limb for life in different environments involve distal elements much more frequently than proximal, and it is far more common for bones to be lost or fused than for new ones to be added. Horses and their relatives evolved a foot structure for fleetness by elongation of the third toe. In effect, a horse stands on its third fingernail (hoof), much like a ballet dancer standing on the tips of the toes. A bird wing is a good example of distal modification. A bird embryo bears 13 distinct wrist and hand bones (carpals and metacarpals), but most of these and the finger bones (phalanges) regress during development, leaving four bones in three digits in an adult (see p. 563). The proximal bones (humerus, radius, and ulna), however, are only slightly modified in the bird wing.

In nearly all tetrapods the pelvic girdle is firmly attached to the axial skeleton, since the greatest locomotory forces transmitted to the body come from the hindlimbs. The pectoral girdle,

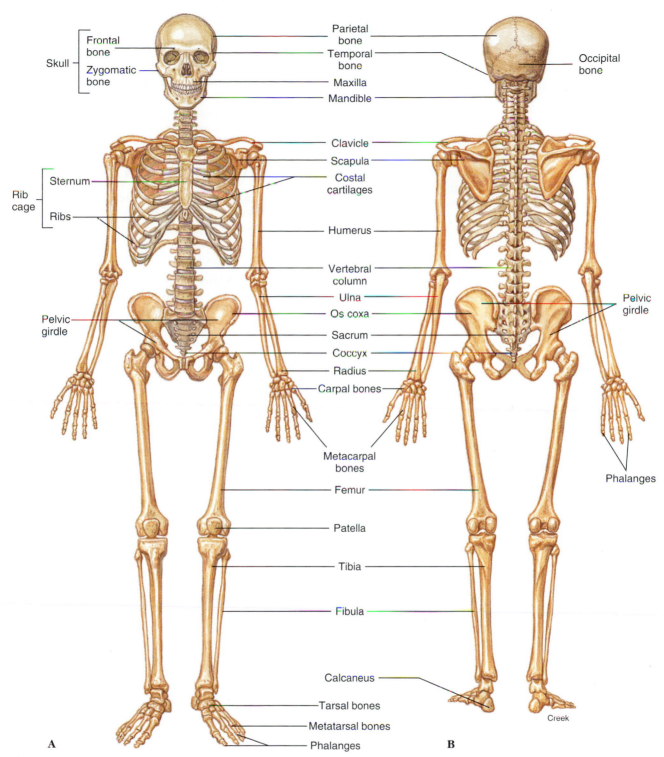

Figure 29-9

Human skeleton. **A,** Ventral view. **B,** Dorsal view. In comparison with other mammals, the human skeleton is a patchwork of primitive and specialized parts. Erect posture, produced by specialized changes in legs and pelvis, enabled the primitive arrangement of arms and hands (arboreal adaptation of human ancestors) to be used for manipulation of tools. Development of the skull and brain followed as a consequence of the premium natural selection put on dexterity and ability to appraise the environment.

however, is much more loosely attached to the axial skeleton, providing the forelimbs with greater freedom for manipulative movements.

Effect of Body Size on Bone Stress

As Galileo realized in 1638, the ability of animals' limbs to support a load decreases as animals increase in size (chapter opening essay, p. 612). Imagine two animals, one twice as long as the other, that are proportionally identical. The larger animal is twice as long, twice as wide, and twice as tall as the smaller. The volume (and the weight) of the larger animal will be eight times the volume of the smaller ($2 \times 2 \times 2 = 8$). However, the strength of the larger animal's legs will be only four times the strength of the smaller, because bone, tendon, and muscle strength are proportional to cross-sectional area. So, as Galileo noted, eight times the weight would have to be carried by only four times the strength. Because the maximum strength of mammalian bone is rather uniform per unit of cross-sectional area, how can animals become larger without placing unbearable stresses on long limb bones? One obvious solution is to make bones stouter and therefore stronger. However, throughout much of their size range, bone shape in different sized mammals does not change much. Instead, mammals have adapted limb posture so that stresses are shifted to align with the long axis of the bones, rather than transversely. Small animals the size of a chipmunk run in a crouched limb posture, whereas a large mammal such as a horse, has adopted an upright posture (Figure 29-10). Bones and muscles are capable of carrying far more weight when aligned more closely with the ground reaction force, as they are in a horse's leg. In this way, peak bone stresses during strenuous activity are no greater for a galloping horse than for a running chipmunk or dog.

For animals larger than horses, further mechanical advantage by changing limb posture is not possible because the limbs are fully upright. Instead, the long bones of an elephant weigh-

ing 2.5 metric tons, and those of the enormous dinosaur *Apatosaurus,* weighing an estimated 34 metric tons, are (were) extremely thick and robust (Figure 29-10), providing the safety factor these massive animals require(d). However, top running speeds of the largest terrestrial mammals decline with increasing size, and recent analysis of one of the most formidable dinosaurs, *Tyrannosaurus,* concluded that it could not run (Hutchinson and Garcia, 2002).

ANIMAL MOVEMENT

Movement is an important characteristic of animals. Animal movement occurs in many forms in animal tissues, ranging from barely discernible streaming of cytoplasm to extensive movements of powerful striated muscles. Most animal movement depends on a single fundamental mechanism: **contractile proteins,** which can change their form to elongate or contract. Contractile machinery is always composed of ultrafine fibrils—fine filaments, striated fibrils, or tubular fibrils (microtubules)—arranged to contract when powered by **ATP.** By far the most important protein contractile system is the **actomyosin system,** composed of two proteins, **actin** and **myosin.** This is an almost universal biomechanical system found from protozoa to vertebrates; it performs a long list of diverse functional roles. Cilia and flagella, however, are composed of different proteins, and thus are exceptions to the rule. In this discussion we examine the three principal kinds of animal movement: ameboid, ciliary and flagellar, and muscular.

Ameboid Movement

Ameboid movement is a form of movement especially characteristic of amebas and other unicellular forms; it is also found in many wandering cells of metazoans, such as white blood cells,

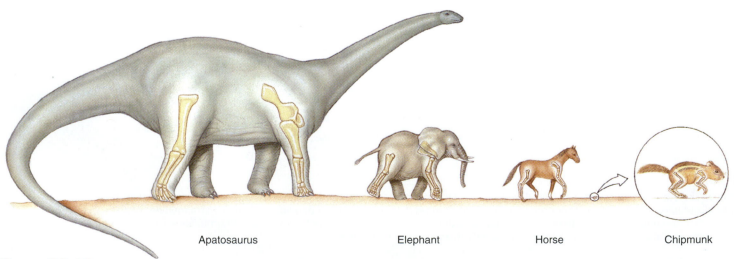

Figure 29-10

Comparison of postures in small and large mammals, showing the effect of scale. Because of its more upright posture, bone stresses in the horse are similar to those in the chipmunk. In mammals larger than horses (above about 300 kg), greatly increased stresses require that bones become exceedingly robust and that the animal lose agility.

Apatosaurus   Elephant   Horse   Chipmunk embryonic mesenchyme, and numerous other mobile cells that move through tissue spaces. Ameboid cells change their shape by extending and withdrawing **pseudopodia** (false feet) from any point on the cell surface. Beneath the plasma membrane lies a nongranular layer, the gel-like **ectoplasm,** which encloses the more liquid **endoplasm** (see Figure 11-4, p. 211).

Research with a variety of ameboid cells, including pathogen-fighting phagocytes present in blood, has produced a consensus model to explain pseudopodial extension and ameboid crawling. Optical studies of an ameba in movement suggest the outer layer of ectoplasm surrounds a rather fluid core of endoplasm. Movement depends on actin and other regulatory proteins. According to one hypothesis (Stossel, 1994), as the pseudopod extends, hydrostatic pressure forces actin subunits in the flowing endoplasm into the pseudopod where they assemble into a network to form the gel-like ectoplasm. At the trailing edge of the gel, where the network disassembles, freed actin interacts with myosin to create a contractile force that pulls the cell forward behind the extending pseudopod. Locomotion is assisted by membrane-adhesion proteins that attach temporarily to the substrate to provide traction, enabling the cell to crawl steadily forward (see Figure 11-7, p. 212).

Ciliary and Flagellar Movement

Cilia are minute, hairlike, motile processes that extend from surfaces of cells of many animals. They are a particularly distinctive feature of ciliate protistans, but except for nematodes in which motile cilia are absent and arthropods in which they are rare, cilia are found in all major groups of animals. Cilia perform many roles either in moving small organisms such as unicellular ciliates, flagellates, and ctenophores (Figure 29-12B) through their aquatic environment or in propelling fluids and materials across epithelial surfaces of larger animals.

Cilia are of remarkably uniform diameter (0.2 to 0.5 μm) wherever they are found. Electron microscopy has shown that each cilium has at its base a **basal body** (kinetosome) that is structurally similar to a centriole (see Figure 3-14, p. 44). Each basal body gives rise to a peripheral circle of nine double microtubules arranged around two single microtubules in the center (Figure 29-11) forming the structural support and machinery for movement within each cilium. (Several exceptions to the 9 + 2 arrangement have been noted; for example, sperm tails of flatworms have but one central microtubule, and sperm tails of a mayfly have no central microtubule.) Each microtubule is composed of a spiral array of protein subunits called **tubulin.** The microtubule doublets around the periphery are connected to each other and to the central pair of microtubules by a complex system of connective elements. Also extending from each doublet is a pair of arms composed of the protein **dynein.** The dynein arms, which act as cross bridges between doublets, operate to produce a sliding force between microtubules.

A **flagellum** is a whiplike structure longer than a cilium and usually present singly or in small numbers at one end of a cell. They are found in many single-celled eukaryotes, in animal spermatozoa, and in sponges. The main difference between a cilium and a flagellum is in their beating pattern rather than in their structure, since both look alike internally. A flagellum beats symmetrically with snakelike undulations so that water is propelled parallel to the long axis of the flagellum. A cilium, in contrast, beats asymmetrically with a fast power stroke in one direction followed by a slow recovery during which the cilium bends as it returns to its original position (Figure 29-12A). Water is propelled parallel to the ciliated surface (Figure 29-12B).

Figure 29-11

A, Longitudinal and cross section of a cilium showing the microtubules and connecting elements of the 9 + 2 arrangement typical of both cilia and flagella. The central pair of microtubules end near the level of the cell surface. The peripheral microtubules continue inward for a short distance to compose two of each of the triplets in the basal body (kinetosome). **B,** Electron micrograph of section through several cilia. (×133,000)

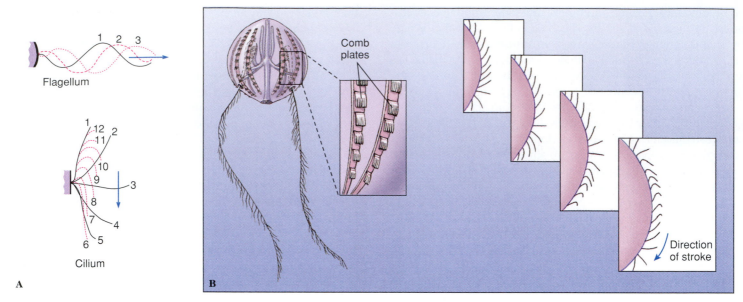

Figure 29-12

A, Flagellum beats in wavelike undulations, propelling water parallel to the main axis of itself. Cilium propels water in direction parallel to the cell surface. **B,** Movement of cilia in comb plates of a ctenophore. Note how the waves of beating comb plates pass down a comb row, opposite the direction of the power stroke of individual cilia. The movement of one comb plate lifts the plate below it and so triggers the next lower plate and so on.

Although the mechanism of ciliary movement is not completely understood, it is known that microtubules behave as "sliding filaments" that move past one another much like the sliding filaments of vertebrate skeletal muscle described in the discussion on the sliding filament hypothesis, p. 626. During ciliary flexion, dynein arms link to adjacent microtubules, then swivel and release in repeated cycles, causing microtubules on the concave side to slide outward past microtubules on the convex side. This process increases curvature of the cilium. During the recovery stroke microtubules on the opposite side slide outward to bring the cilium back to its starting position.

Muscular Movement

Contractile tissue is most highly developed in muscle cells called **fibers.** Although muscle fibers themselves can do work only by contraction and cannot actively lengthen, they can be arranged in so many different configurations and combinations that almost any movement is possible.

Types of Vertebrate Muscle

Vertebrate muscle is broadly classified on the basis of the appearance of muscle cells (fibers) when viewed with a light microscope. **Skeletal muscle** appears transversely striped **(striated),** with alternating dark and light bands (Figure 29-13). **Cardiac muscle** also possesses striations like skeletal muscle but is uninucleate and with branching cells. A third type of vertebrate muscle is **smooth** (or visceral) **muscle,** which lacks the characteristic alternating bands of the striated type.

Skeletal muscle is typically organized into sturdy, compact bundles or bands (Figure 29-13A). It is called skeletal muscle because it is attached to skeletal elements and is responsible for movements of the trunk, appendages, respiratory organs, eyes, mouthparts, and other structures. Skeletal muscle **fibers** are extremely long, cylindrical, multinucleate cells that usually reach from one end of the muscle to the other. They are packed into bundles called **fascicles** (L. *fasciculus,* small bundle), which are enclosed by tough connective tissue. The fascicles are in turn grouped into a discrete **muscle** surrounded by a thick connective tissue layer. Most skeletal muscles taper at their ends, where they connect to bones by tendons. Other muscles, such as the ventral abdominal muscles, are flattened sheets.

In most fishes, amphibians, and to some extent lizards and snakes, muscles are organized into segments that alternate with the vertebrae. Skeletal muscles of other vertebrates, by splitting, fusion, and shifting, have developed into specialized muscles best suited for manipulating jointed appendages that have evolved for locomotion on land. Skeletal muscle contracts powerfully and quickly but fatigues more rapidly than does smooth muscle. Skeletal muscle is sometimes called **voluntary muscle** because it is stimulated by motor fibers and is under conscious control.

Smooth muscle lacks the striations typical of skeletal muscle (Figure 29-13B). The cells are long, tapering strands, each containing a single, central nucleus. The cells interdigitate with each other such that the tapered end of one cell lies close to the central nuclear region of the next. Smooth muscle cells are organized into sheets of muscle circling the walls of the alimentary canal, blood vessels, respiratory passages, and urinary and genital ducts. Smooth muscle is typically slow acting and can main-

Nuclei

A

Nuclei

B

Nuclei

Intercalcated discs

C

Figure 29-13

Photomicrographs of types of vertebrate muscle. **A,** Skeletal muscle (human) showing several striated fibers (cells) lying side by side. Note that nuclei are peripheral. **B,** Smooth muscle (human) showing absence of striations. Note elongate nuclei in the long fibers. **C,** Cardiac muscle (monkey). Note the vertical bars, called intercalated discs, joining separate fibers end to end.

tain prolonged contractions with very little energy expenditure. It is under the control of the autonomic nervous system, hormones and local mechanisms; thus, unlike skeletal muscle, its contractions are involuntary and unconscious. The principal functions of smooth muscles are to push material in a tube, such as the intestine, along its way by active contractions or to regulate the diameter of a tube, such as a blood vessel, by sustained contraction or relaxation.

Cardiac muscle, the seemingly tireless muscle of the vertebrate heart, combines certain characteristics of both skeletal and smooth muscle (Figure 29-13C). It is fast acting and striated like skeletal muscle, but contraction is under involuntary autonomic and hormonal control like smooth muscle. Actually external control mechanisms serve only to speed up or slow down the intrinsic rate of contraction; the heartbeat originates within specialized cardiac muscle, and the heart continues to beat even if removed from the body (heart excitation is described on p. 663). Cardiac muscle is composed of closely opposed, but separate, uninucleate cell fibers.

Types of Invertebrate Muscle

Smooth and striated muscles are also characteristic of invertebrate animals, but there are many variations of both types and even instances in which structural and functional features of vertebrate smooth and striated muscle are combined. Striated muscle appears in invertebrate groups as diverse as cnidarians and arthropods. The thickest muscle fibers known, approximately 3 mm in diameter and 6 cm long, are those of giant barnacles and of Alaska king crabs living along the Pacific coast of North America. Such large muscle cells lend themselves well to physiological studies and are understandably popular with muscle physiologists.

In the limited space available to treat the great diversity of muscle structure and function in the invertebrate assemblage, we have selected for discussion two functional extremes: the specialized adductor muscles of molluscs and the fast flight muscles of insects.

Bivalve molluscan muscles contain fibers of two types. One kind is striated muscle that can contract rapidly, enabling the bivalve to snap shut its valves when disturbed. Scallops use these "fast" muscle fibers to swim in their awkward manner (see Figure 16-24B, p. 330). The second muscle type is smooth muscle, capable of slow, long-lasting contractions. Using these fibers, a bivalve can keep its valves tightly shut for hours or even days. Such adductor muscles use little metabolic energy and receive remarkably few nerve impulses to maintain the activated state. The contracted state has been likened to a "catch mechanism" involving a low rate of cross-bridge cycling (see p. 628) between contractile proteins within the fiber with a low energy expenditure. However, research continues to clarify the mechanism.

Insect flight muscles are virtually the functional antithesis of the slow, holding muscles of bivalves. The wings of some small flies operate at frequencies greater than 1000 beats per second. The so-called **fibrillar muscle,** which contracts at these frequencies—far greater than even the most active of vertebrate muscles—shows unique characteristics. It has very limited extensibility; that is, the wing leverage system is arranged so that the muscles shorten only slightly during each downbeat of the wings. Furthermore, muscles and wings operate as a rapidly oscillating system in an elastic thorax (see Figure 20-13, p. 404). Since the muscles rebound elastically and are activated by stretch during flight, they receive impulses only periodically rather than one impulse per contraction; one reinforcement impulse for every 20 or 30 contractions is enough to keep the system active. Insect flight muscles are described in more detail in Chapter 20 (pp. 402–404).

Structure of Striated Muscle

As mentioned on p. 624, striated muscle is so named because of periodic bands, plainly visible under the light microscope, that pass across the widths of muscle cells. Each cell, or **fiber,** is a multinucleated tube containing numerous **myofibrils,** packed together and invested by a cell membrane, the **sarcolemma** (Figure 29-14). The myofibril contains two types of **filaments,** composed of the protein **myosin,** and the protein **actin.** These are the contractile proteins of the muscle. Actin filaments are held together by a dense structure called the **Z** line. The functional unit of the myofibril, the **sarcomere,** extends between successive **Z** lines. These anatomical relationships are diagrammed in Figure 29-14.

Human muscle tissue develops before birth, and a newborn child's complement of skeletal muscle fibers is all that he or she will ever have. But while an adult male weight lifter and a young boy have a similar number of muscle fibers, the weight lifter may be several times the boy's strength because repeated high-intensity, short-duration exercise has induced the synthesis of additional actin and myosin filaments. Each fiber has hypertrophied, becoming larger and stronger. This type of exercise favors hypertrophy of fast fibers (see p. 630) that fatigue quickly. Endurance exercise such as long-distance running produces a very different response. Slower fiber types are stimulated (see p. 630), but without becoming greatly stronger develop more mitochondria and myoglobin and become adapted for a high rate of oxidative phosphorylation. These changes, together with the development of more capillaries serving the fibers, lead to increased capacity for long-duration activity.

Each myosin filament is composed of many myosin molecules packed together in an elongate bundle (Figure 29-15). Each myosin molecule contains two polypeptide chains, each having a club-shaped head. They are lined up in a bundle to form a myosin filament, the double heads of each myosin molecule face outward from the center of the filament. These heads act as binding sites for high-energy ATP and during muscle contraction they form molecular cross bridges that interact with the actin filaments.

Actin filaments are more complex because they are composed of a backbone of a double strand of the protein actin, twisted into a double helix. In addition, two actin-binding proteins, tropomyosin and troponin, form part of the actin filament complex. They are important in regulating the interaction of actin and myosin during muscle contraction. Two thin strands of **tropomyosin** lie near the grooves between the actin strands. Each tropomyosin strand is itself a double helix as shown in Figure 29-15C. **Troponin,** a complex of three globular proteins, is located at intervals along the actin filament. Troponin is a calcium-dependent switch that acts as the control point in the contraction process.

The actin filament complexes extend outward from both sides of the **Z** line and overlap with myosin bundles toward the certes of each sarcomere (Figure 29-16).

Sliding Filament Hypothesis of Muscle Contraction

In the 1950s the English physiologists A. F. Huxley and H. E. Huxley independently proposed the **sliding filament hypothesis** to explain striated muscle contraction. According to this

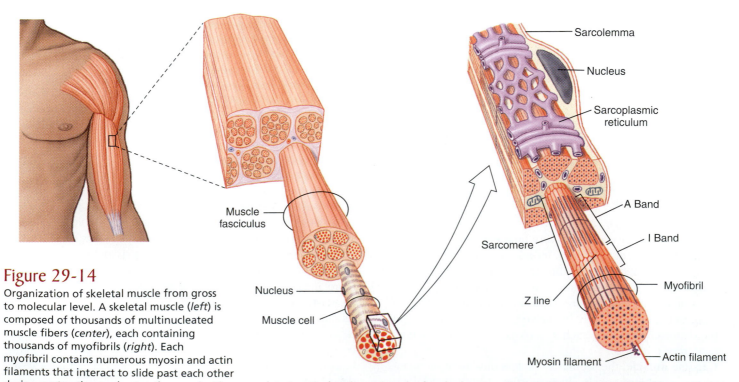

Figure 29-14

Organization of skeletal muscle from gross to molecular level. A skeletal muscle (*left*) is composed of thousands of multinucleated muscle fibers (*center*), each containing thousands of myofibrils (*right*). Each myofibril contains numerous myosin and actin filaments that interact to slide past each other during contraction to shorten the muscle. The sarcoplasmic reticulum is a network of endoplasmic reticulum tubules surrounding the myofibrils and serves as a reservoir for stored calcium that is released during a depolarization, and initiates filament sliding during muscle contraction.

Figure 29-15

Molecular structure of actin and myosin filaments of skeletal muscle. **A,** The myosin molecule is composed of two polypeptides coiled together and expanded at their ends into a globular head. **B,** The myosin filament is composed of a bundle of myosin molecules with the globular heads extended outward toward the actin filaments on both sides. **C,** The actin filament consists of a double strand of actin surrounded by two tropomyosin strands. A globular protein complex, troponin, occurs in pairs at every seventh actin unit. Troponin is a calcium-dependent switch that controls the interaction between actin and myosin.

Figure 29-16

Sliding filament hypothesis, showing how actin and myosin filaments interact during contraction. **A,** Muscle relaxed. **B,** Muscle contracted.

hypothesis, the actin and myosin filaments become linked together by molecular cross bridges, which act as levers to pull the filaments past each other. During contraction, cross bridges on the myosin filaments swing rapidly back and forth, alternately attaching to and releasing from special receptor sites on the actin filaments, and drawing actin filaments past the myosin filaments in a kind of ratchet action. As contraction continues, the Z lines are pulled closer together (Figure 29-16). Thus the sarcomere shortens. Because all sarcomere units shorten together, the muscle contracts. Relaxation is a passive process. When cross bridges between the actin and myosin filaments release, the sarcomeres are free to lengthen. This requires some force, which is usually supplied by antagonistic muscles or the force of gravity.

Control of Contraction

Muscle contracts in response to nerve stimulation. If the nerve supply to a muscle is severed, the muscle **atrophies,** or wastes away. Skeletal muscle fibers are innervated by motor neurons whose cell bodies are located in the spinal cord. Each cell body gives rise to a motor axon that leaves the spinal cord to travel by way of a peripheral nerve trunk to a muscle where it branches repeatedly into many terminal branches. Each terminal branch innervates a single muscle fiber. Depending on the type of muscle, a single motor axon may innervate as few as three or four muscle fibers (where very precise control is needed, such as the muscles that control eye movement) or as many as 2000 muscle fibers (where precise control is not required, such as large leg muscles). A motor neuron and all muscle fibers it innervates is called a **motor unit.** The motor unit is the functional unit of skeletal muscle. When a motor neuron fires, the action potential passes to all fibers of the motor unit and each is stimulated to contract simultaneously. Total force exerted by a muscle depends on the number of motor units activated. Precise control of movement is achieved by varying the number of motor units activated at any one time. A smooth and steady increase in muscle tension is produced by increasing the number of motor units brought into play; this is called motor unit **recruitment.**

The Neuromuscular Junction

The place where a motor axon terminates on a muscle fiber is called the **neuromuscular (or myoneural) junction** (Figure 29-17). At the junction is a tiny gap, or **synaptic cleft,** that thinly separates a nerve fiber and muscle fiber. Close to the junction, the neuron stores a chemical, **acetylcholine,** in minute vesicles known as **synaptic vesicles.** Acetylcholine is released when a nerve impulse reaches a synapse. This substance is a chemical mediator that diffuses across the synaptic cleft and acts on the muscle fiber membrane to generate an electrical depolarization. The depolarization spreads rapidly through the muscle fiber, causing it to contract. Thus the synapse is a special chemical bridge that couples together the electrical activities of nerve and muscle fibers.

Built into vertebrate skeletal muscle is an elaborate conduction system that serves to carry the depolarization from the neuromuscular junction to the densely packed filaments within the fiber. Along the surface of the sarcolemma are numerous invaginations that project into the muscle fiber as a system of tubules, called **T-tubules** (Figure 29-17). The T-tubules are closely associated with the **sarcoplasmic reticulum,** a system of modified endoplasmic reticulum (p. 41) that runs parallel to the actin and myosin filaments. The sarcoplasmic reticulum stores calcium and releases it around the actin and myosin filaments in response to depolarization.

Excitation-Contraction Coupling

How does electrical depolarization activate the contractile machinery? In resting, unstimulated muscle, shortening does not occur because thin tropomyosin strands surrounding the actin myofilaments lie in a position that prevents the myosin heads from attaching to actin. When muscle is stimulated and the electrical depolarization arrives at the sarcoplasmic reticulum surrounding the fibrils, calcium ions are released (Figure 29-17). The calcium binds to the control protein troponin. Troponin immediately undergoes changes in shape that allow tropomyosin to move out of its blocking position, exposing active sites on the actin filaments. Myosin heads then bind to these sites, forming cross bridges between adjacent myosin and actin filaments. This sets in motion an **attach-pull-release cycle** that occurs in a series of steps as shown in Figure 29-18. Release of bond energy from ATP hydrolysis activates the myosin head,

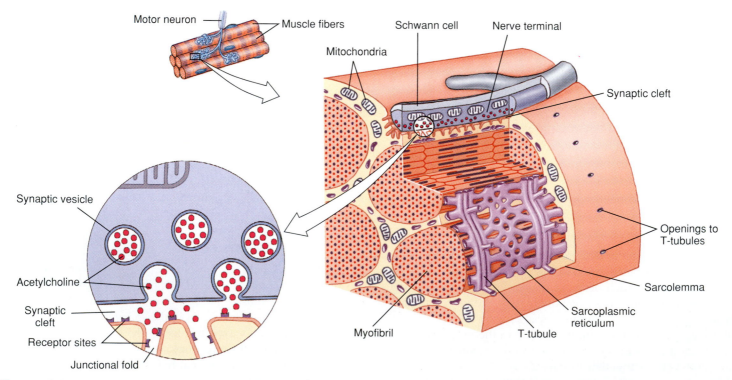

Figure 29-17

Section of vertebrate skeletal muscle showing nerve-muscle synapse (neuromuscular or myoneural junction), sarcoplasmic reticulum, and connecting transverse tubules (T-tubules). Arrival of a nerve impulse at the synapse triggers a release of acetylcholine into the synaptic cleft (*inset at left*). Transmitter molecules binding to receptors generate membrane depolarization. This depolarization spreads across the sarcolemma, into the T-tubules, and to the sarcoplasmic reticulum where the sudden release of calcium sets in motion the contractile machinery of the myofibril.

Figure 29-18

Excitation-contraction coupling in vertebrate skeletal muscle. **Step 1:** An action potential spreads along the sarcolemma and is conducted inward to the sarcoplasmic reticulum by way of T-tubules (T-tubule system). Calcium ions released from the sarcoplasmic reticulum diffuse rapidly into the myofibrils and bind to troponin molecules on the actin filament. Troponin and tropomyosin molecules are moved away from the active sites. **Step 2:** Myosin cross bridges bind to the exposed active sites. **Step 3:** Using the energy stored in ATP, the myosin head swings toward the center of the sarcomere. ADP and a phosphate group are released. **Step 4:** The myosin head binds another ATP molecule; this frees the myosin head from the active site on actin. **Step 5:** The myosin head splits ATP, retaining the energy released as well as the ADP and the phosphate group. The cycle can now be repeated as long as calcium is present to open active sites on the actin molecules.

which swings 45 degrees, at the same time releasing a molecule of ADP. This is the power stroke that pulls the actin filament a distance of about 10 nm, and it comes to an end when phosphate is released and another ATP molecule binds to the myosin head, releasing it from the active site. Thus each cycle requires expenditure of energy in the form of ATP (Figure 29-18).

Shortening will continue as long as nerve impulses arrive at the neuromuscular junction and free calcium remains available around the actin and myosin filaments. The attach-pull-release cycle can repeat again and again, 50 to 100 times per second, pulling actin and myosin filaments past each other. While the distance each sarcomere can shorten is very small, this distance is multiplied by the thousands of sarcomeres lying end to end in a muscle fiber. Consequently, a strongly contracting muscle may shorten by as much as one-third its resting length.

When stimulation stops, calcium is quickly pumped back into the sarcoplasmic reticulum. Troponin resumes its original configuration; tropomyosin moves back into its blocking position on actin, and the muscle relaxes.

Energy for Contraction

Muscle contraction requires large amounts of energy. ATP is the immediate source of energy, and its levels within muscle are held almost constant due to immediate replenishment from an energy reserve, **creatine phosphate.** Creatine phosphate is a high-energy phosphate compound that stores bond energy during periods of rest. As ADP is produced during contraction, creatine phosphate releases its stored bond energy to convert ADP to ATP. This reaction can be summarized as:

$$\text{Creatine phosphate} + \text{ADP} \rightarrow \text{ATP} + \text{Creatine}$$

A contracting muscle also uses stored energy from the breakdown of **glycogen.** Glycogen is a polysaccharide chain of glucose molecules (p. 23) stored in both liver and muscle. Muscle has by far the larger store—some three-fourths of all the glycogen in the body is stored in muscle. As a supply of energy for contraction, glycogen has three important advantages: it is relatively abundant, it can be mobilized quickly, and it can provide energy under anoxic conditions. As soon as the muscle's store of creatine phosphate declines, enzymes break down glycogen, converting it into glucose-6-phosphate molecules, the first stage of glycolysis that leads into mitochondrial respiration and the generation of ATP (p. 60).

If muscular contraction is not too vigorous or too prolonged, the glucose released from glycogen can be completely oxidized to carbon dioxide and water by **aerobic metabolism.** During prolonged or heavy exercise, however, blood flow to the muscles, although greatly increased above resting levels, cannot supply oxygen to the mitochondria rapidly enough to complete oxidation of glucose. The contractile machinery then receives its energy largely by **anaerobic glycolysis,** a process that does not require oxygen (p. 65). The ability to take advantage of this anaerobic pathway, although not nearly as efficient as the aerobic one, is of great importance; without it, all forms of heavy muscular exertion would be impossible.

During anaerobic glycolysis, glucose is degraded to lactic acid with release of energy. This energy is used to resynthesize creatine phosphate, which in turn passes the energy to ADP for the resynthesis of ATP. Lactic acid accumulates in the muscle and diffuses rapidly into the general circulation. If muscular exertion continues, the buildup of lactic acid causes enzyme inhibition and fatigue. Thus the anaerobic pathway is a self-limiting one, since continued heavy exertion leads to exhaustion. The muscles incur an **oxygen debt** because accumulated lactic acid must be oxidized by extra oxygen. After a period of exertion, oxygen consumption remains elevated until all lactic acid has been oxidized or consumed in resynthesis of glycogen.

Muscle Performance

Fast and Slow Fibers

Skeletal muscles of vertebrates consist of more than one type of fiber. **Slow fibers,** which are specialized for slow, sustained contractions without fatigue, are important in maintaining posture in terrestrial vertebrates. Such muscles are often called **red muscles** because they contain an extensive blood supply, a high density of mitochondria for supplying ATP, and abundant stored myoglobin which supplies oxygen reserves, all of which give the muscle a red color.

Two kinds of **fast fibers,** capable of fast, powerful contractions are known. One kind of fast fiber lacks an efficient blood supply and a high density of mitochondria and myoglobin. Muscles composed of these fibers (often referred to as white muscles) are usually pale in color, function anaerobically, and fatigue rapidly. The "white meat" of chicken is a familiar example. The other kind of fast fiber has an extensive blood supply and a high density of mitochondria and myoglobin, and functions largely aerobically. Animals use these muscles for rapid, sustained activities. Most muscles possess a mixture of these different fiber types to provide for a range of activity. Geese, dogs, and ungulates (hoofed mammals), for example, have limb (or wing) muscles with a high percentage of fast aerobic fibers, and are capable of active locomotion for long periods of time. Members of the cat family, however, have running muscles formed almost entirely of fast fibers that operate anaerobically. During a chase, such muscles accumulate a substantial oxygen debt, which is replenished after the chase. For example, a cheetah after a high-speed chase lasting less than a minute, will pant heavily for 30 to 40 minutes before its oxygen debt is paid off.

Importance of Tendons in Energy Storage

When mammals walk or run, much kinetic energy is stored from step to step as elastic strain energy in the tendons. For example, during running the Achilles tendon is stretched by a combination of downward force of the body on the foot and contraction of the calf muscles. The tendon then recoils, extending the foot while the muscle is still contracted, propelling the leg forward

Figure 29-19

Energy storage in the Achilles tendon of human and kangaroo legs. During running, stretching of the Achilles tendon when the foot strikes the ground stores kinetic energy that is released to propel the leg forward.

(Figure 29-19). An extreme example of this bouncing-ball principle is the bounding of a kangaroo, which essentially bounces along on its tendons, utilizing the effect of gravity (Figure 29-19). This type of movement uses far less energy than would be required if every step relied solely on alternate muscle contraction and relaxation.

There are many examples of elastic storage in the animal kingdom. It is used in the ballistic jumps of grasshoppers and fleas, in the wing hinges of flying insects, in the hinge ligaments of bivalve molluscs, and in the highly elastic large dorsal ligament (ligamentum nuchae) that helps support the head of hoofed mammals.

SUMMARY

An animal is wrapped in a protective covering, the integument, which may be as simple as the delicate plasma membrane of an ameba or as complex as the skin of a mammal. The arthropod exoskeleton is the most complex of invertebrate integuments, consisting of a two-layered cuticle secreted by a single-layered epidermis. It may be hardened by calcification or sclerotization and must be molted at intervals to permit body growth. Vertebrate integument consists of two layers: the epidermis, which gives rise to various derivatives such as hair, feathers, and claws; and the dermis, which supports and nourishes the epidermis. It also is the origin of bony derivatives such as fish scales and deer antlers.

Integument color is of two kinds: structural color, produced by refraction or scattering of light by particles in the integument, and pigmentary color, produced by pigments that are usually confined to special pigment cells (chromatophores).

Skeletons are supportive systems that may be hydrostatic or rigid. The hydrostatic skeletons of several soft-walled invertebrate groups depend on body-wall muscles that contract against a noncompressible internal fluid of constant volume. In a similar manner, muscular hydrostats, such as the tongue of mammals and reptiles, and the trunk of elephants, rely on muscle bundles arranged in complex patterns to produce movement without either skeletal support or a liquid-filled cavity. Rigid skeletons have evolved with attached muscles that act with the supportive skeleton to produce movement. Arthropods have an external skeleton, which must be shed periodically to make possible an enlarged replacement. The vertebrates developed an internal skeleton, a framework formed of cartilage or bone, that can grow with the animal, while, in the case of bone, additionally serving as a reservoir of calcium and phosphate.

Animal movement, whether in the form of cytoplasmic streaming, ameboid movement, or the contraction of an organized muscle mass, depends on specialized contractile proteins. The most important of these is the actomyosin system, which is usually organized into elongate actin and myosin filaments that slide past one another during contraction. When a muscle is stimulated, an electrical depolarization is conducted into the muscle fibers through T tubules to the sarcoplasmic reticulum, causing the release of calcium. Calcium binds to a protein-troponin complex associated with the actin filament. This binding causes tropomyosin to shift out of its blocking position and allows the myosin heads to cross-bridge with the actin filament. Powered by ATP, the myosin heads swivel back and forth to pull the actin and myosin filaments past each other. Phosphate bond energy for contraction is supplied by carbohydrate fuels through a storage intermediate, creatine phosphate.

Vertebrate skeletal muscle consists of variable percentages of both slow fibers, used principally for sustained postural contractions, and fast fibers, used in locomotion. Tendons are important in locomotion because the kinetic energy stored in stretched tendons at one stage of a locomotory cycle is released at a subsequent stage.

REVIEW QUESTIONS

1. The arthropod exoskeleton is the most complex of invertebrate integuments. Describe its structure, and explain the difference in the way cuticle is hardened in crustaceans and in insects.

2. Distinguish between epidermis and dermis in vertebrate integument, and describe structural derivatives of these two layers.

3. What is the difference between structural colors and colors based on pigments? How do the chromatophores of vertebrates and cephalopod molluscs differ in structure and function?

4. As "naked apes" humans lack the protective investment of fur that shields other mammals from the damaging effects of sunlight. How does human skin respond to ultraviolet radiation in the short term and with continued exposure?

5. Hydrostatic skeletons have been defined as a mass of fluid enclosed within a muscular wall. How would you modify this definition to make it apply to a muscular hydrostat? Offer examples of both hydrostatic skeleton and muscular hydrostat.

6. One of the special qualities of vertebrate bone is that it is a living tissue that permits continuous remodeling. Explain how the structure of bone allows this remodeling to happen.

7. What is the difference between endochondral and membranous bone? Between spongy and compact bone?

8. Discuss the role of osteoclasts, osteoblasts, parathyroid hormone, and calcitonin in bone growth.

9. The laws of scaling tell us that doubling the length of an animal will increase its weight eightfold while the force its bones can bear increases only fourfold. What solutions to this problem have evolved that allow animals to become large, while maintaining bone stresses within margins of safety?

10. Name the major skeletal components included in the axial and in the appendicular skeleton.

11. An unexpected discovery from studies of ameboid movement is that the same proteins found in the contractile system of metazoan muscle—actin and myosin—are present in ameboid cells. Explain how these and other proteins are believed to interact during ameboid movement.

12. A "9 + 2" arrangement of microtubules is typical of both cilia and flagella. Explain how this system is thought to function to produce a bending motion. What is the difference between a cilium and a flagellum?

13. What functional features of molluscan smooth muscle and insect fibrillar muscle set them apart from any known vertebrate muscle?

14. The sliding filament model of skeletal muscle contraction assumes a sliding or slipping of interdigitating filaments of actin and myosin. Electron micrographs show that during contraction the actin and myosin filaments remain of constant length while the distance between Z lines shortens. Explain how this happens in terms of the molecular structure of the muscle filaments. What is the role of regulatory proteins (troponin and tropomyosin) in contraction?

15. While the sarcoplasmic reticulum of muscle was first described by nineteenth-century microscopists, its true significance was not appreciated until its intricate structure was revealed much later by the electron microscope. What could you tell a nineteenth-century microscopist to enlighten him or her about the structure of the sarcoplasmic reticulum and its role in the coupling of excitation and contraction?

16. The filaments of skeletal muscle are moved by free energy derived from the hydrolysis of ATP. During sustained muscle contraction ATP levels remain fairly constant, while levels of creatine phosphate fall. Explain why this is so. Under what circumstances is an oxygen debt incurred during muscle contraction?

17. During evolution, skeletal muscle became adapted to functional demands ranging from sudden, withdrawal movements of a startled worm, to the sustained contractions required to maintain mammalian posture, to supporting a long, fast chase across the African savanna. What are some of the fiber types in vertebrate muscle that evolved to support these kinds of activities?

SELECTED REFERENCES

Alexander, R. M. 1992. The human machine. New York, Columbia University Press. *Describes all kinds of human movement with the human body viewed as an engineered machine. Well chosen illustrations.*

Anderson, J. L., Schjerling, P., and Saltin, B. 2000. Muscle, genes and athletic performance. Sci. Am. **283:**48–55 (Sept.). *Good discussion of muscle structure and function related to human athletic performance.*

Caplan, A. J. 1984. Cartilage. Sci. Am. **251:**84–94 (Oct.). *Structure, aging and development of vertebrate cartilage.*

Hadley, N. F. 1986. The arthropod cuticle. Sci. Am. **255:**104–112 (July). *Describes properties of this complex covering that account for much of the adaptive success of arthropods.*

Hutchinson, J. R., and M. Garcia. 2002. *Tyrannosaurus* was not a fast runner. Nature **415:**1018–1021. *Analysis concludes that Tyrannosaurus would have had less than half enough leg muscle mass to run and could only walk.*

Leffell, D. J., and D. E. Brash. 1996. Sunlight and skin cancer. Sci. Am. **275:**52–59 (July). *Skin cancer that appears in older people begins with damage received decades earlier. Many cases are caused by a mutation in a single gene.*

Randall, D., W. Burggren, and K. French. 2001. Eckest animal physiology: mechanisms and adaptations. W. H. Freeman and Company, New York. *A comprehensive and comparative treatment of animal physiology.*

Shipman, P., A. Walker, and D. Bichell. 1985. The human skeleton. Cambridge, Massachusetts, Harvard University Press. *Comprehensive view of the human skeleton.*

Stossel, T. P. 1994. The machinery of cell crawling. Sci. Am. **271:**54–63 (Sept.). *Cell crawling depends on the orderly assembly and disassembly of an actin protein scaffold.*

Willmer, P., G. Stone, and I. Johnston. 2000. Environmental physiology of animals. Oxford, U.K., Blackwell Science Ltd. *Well-written information regarding environmental adaptations in both vertebrates and invertebrates.*

ZOOLOGY LINKS TO THE INTERNET

Visit the textbook's Online Learning Center at www.mhhe.com/zoology to find live Internet links for each of the topics listed here.
Integumentary Systems
Vertebrates: Macroscopic Anatomy of the Integument
Human Integumentary System Topics
Skeletal Systems
Vertebrates: Macroscopic Anatomy of the Skeleton
Axial Skeletons
Appendicular Skeleton; Gills, Fins, and Limbs
Ligaments and Joints
Human Skeletal System Topics
Animal Movement and Musculature

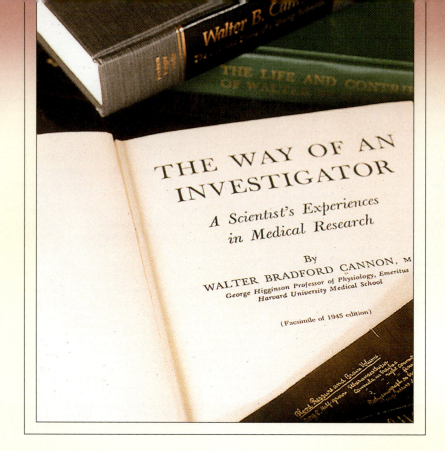

Title page of Walter B. Cannon's autobiography.

Homeostasis

Osmotic Regulation, Excretion, and Temperature Regulation

Homeostasis: Birth of a Concept

The tendency toward internal stabilization of the animal body was first recognized by Claude Bernard, great French physiologist of the nineteenth century who, through his studies of blood glucose and liver glycogen, first discovered internal secretions. From a lifetime of study and experimentation gradually grew the principle for which this retiring and lonely man is best remembered, that of constancy of the internal environment, a principle that in time would pervade physiology and medicine. Years later, at Harvard University, American physiologist Walter B. Cannon (Figure 30-1) reshaped and restated Bernard's idea. From his studies of the nervous system and reactions to stress, he described the ceaseless balancing and rebalancing of physiological processes that maintain stability and restore the normal state when it has been disturbed. He also gave it a name: **homeostasis.** The term soon flooded the medical literature of the 1930s. Physicians spoke of getting their patients back into homeostasis. Even politicians and sociologists saw what they considered deep nonphysiological implications. Cannon enjoyed this broadened application of the concept and later suggested that democracy was the form of government that took a homeostatic middle course. Despite the enduring importance of the homeostasis concept, Cannon never received the Nobel Prize—one of several acknowledged oversights of the Nobel Committee. Late in life, Cannon expressed his ideas about scientific research in his autobiography, *The Way of an Investigator.* This engaging book describes the resourceful career of a homespun man whose life embodied the traits that favor successful research.

Figure 30-1

Walter Bradford Cannon (1871 to 1945), Harvard professor of physiology who coined the term "homeostasis" and developed the concept originated by French physiologist Claude Bernard (Figure 31-2, p. 657).

The concept of homeostasis, described in the chapter opening essay, permeates all physiological thinking and is the theme of this chapter and Chapter 31. Although this concept was first developed from studies with mammals, it applies to single-celled organisms as well as to multicellular organisms. Potential changes in the internal environment arise from two sources. First, metabolic activities require a constant supply of materials, such as oxygen, nutrients, and salts, that cells withdraw from their surroundings and that must be continuously replenished. Cellular activity also produces waste products that must be expelled. Second, the internal environment responds to changes in the organism's external environment. Changes from either source must be stabilized by the physiological mechanisms of homeostasis.

In more complex metazoans, homeostasis is maintained by the coordinated activities of the circulatory, nervous, and endocrine systems, and also by organs that serve as sites of exchange with the external environment. These last include kidneys, lungs or gills, digestive tract, and integument. Through these organs oxygen, nutrients, minerals, and other constituents of body fluids enter, water is exchanged, heat is lost, and metabolic wastes are eliminated.

Thus, systems within an organism function in an integrated way to maintain a constant internal environment around a setpoint. Small deviations from this setpoint in variables such as pH, temperature, osmotic pressure, glucose levels, and oxygen levels activate physiological mechanisms that return the variable to its setpoint.

We look first at the problems of controlling the internal fluid environment of animals living in aquatic habitats. Next we briefly examine how these problems are solved by terrestrial animals and consider the function of the organs that regulate their internal state. Finally we look at strategies that have evolved for living in a world of changing temperatures.

WATER AND OSMOTIC REGULATION

How Marine Invertebrates Meet Problems of Salt and Water Balance

Most marine invertebrates are in osmotic equilibrium with their seawater environment. With body surfaces permeable to salts and water, their body fluid concentration rises or falls in conformity with changes in concentrations of seawater. Because such animals are incapable of regulating osmotic pressure of their body fluid, they are called **osmotic conformers.** Invertebrates living in the open sea are seldom exposed to osmotic fluctuations because the ocean is a highly stable environment. Oceanic invertebrates have, in fact, very limited abilities to withstand osmotic change. Should they be exposed to dilute seawater, they die quickly because their body's cells cannot tolerate dilution and are helpless to prevent it. These animals are restricted to living in a narrow salinity range and are said to be **stenohaline** (Gr. *stenos,* narrow, + *hals,* salt). An example is the marine *spider crab* (*Maia*) (Figure 30-2).

Conditions along coasts and in estuaries and river mouths are much less constant than those of the open ocean. Here animals must cope with large and often abrupt changes in salinity as the tides ebb and flow and mix with fresh water draining from rivers. These animals are termed **euryhaline** (Gr. *eurys,* broad, + *hals,* salt), meaning that they can survive a wide range of salinity changes, mainly by demonstrating varying powers of **osmotic regulation.** For example, a brackish-water *shore crab* (*Eriocheir*) can resist dilution of body fluids by dilute (brackish) seawater (Figure 30-2). Although the concentration of salts in

Figure 30-2

Salt concentration of body fluids of two crabs as affected by variations in seawater concentration. The 45-degree line represents equal concentration between body fluids and seawater. Since the spider crab (*Maia* sp.) cannot regulate the salt concentration of its fluids, it conforms to whatever changes happen in the seawater. The shore crab (*Eriocheir* sp.), however, can regulate osmotic concentration of its fluids to some degree in dilute seawater. For example, when seawater is 200 mM (millimolar), the shore crab's body fluid is approximately 430 mM. Crosses at ends of lines indicate limits of tolerance for each species.

the body fluids falls, it does so less rapidly than the fall in seawater concentration. This crab is a **hyperosmotic regulator,** meaning that it maintains its body fluids more concentrated (hence *hyper-*) than the surrounding water.

By regulating against excessive dilution, thus protecting the cells from extreme changes, brackish-water shore crabs can live successfully in the physically unstable but biologically rich coastal environment. Nevertheless, with limited capacity for osmotic regulation, they will die if exposed to greatly diluted seawater. To understand how a brackish-water shore crab and other coastal invertebrates achieve hyperosmotic regulation, let us examine the problems they face. First, because a crab's body fluids are osmotically more concentrated than the dilute seawater outside, water flows into its body, especially across the thin, permeable membranes of the gills. As with the membrane osmometer containing a salt solution (p. 46), water diffuses inward because of higher solute concentration inside. For the crab, were this inflow of water allowed to continue unchecked, its body fluids would soon become diluted and unbalanced. The problem is solved by the kidneys (antennal glands located in the crab's head), which excrete excess water as dilute urine.

The second problem is salt loss. Again, because the animal is saltier than its environment, it cannot avoid loss of ions by outward diffusion across its gills. Salt is also lost in urine. To compensate for solute loss, special salt-secreting cells in the gills actively remove ions from dilute seawater and move them into the blood, thus maintaining the internal osmotic concentration. This is an **active transport** (p. 47) process that requires energy because ions must be transported against a concentration gradient from a lower salt concentration (in dilute seawater) to an already higher one (in blood).

Invasion of Fresh Water

Some 400 million years ago, during the Silurian and Lower Devonian periods, the major groups of jawed fishes began to penetrate into brackish-water estuaries and then gradually into freshwater rivers. Before them lay a new, unexploited habitat already stocked with food in the form of insects and other invertebrates, which had preceded them into fresh water. However, advantages this new habitat offered were balanced by a tough physiological challenge: the necessity of developing effective osmotic regulation.

Freshwater animals face similar, though more extreme, problems to the shore crab just described. They must keep the salt concentration of their body fluids higher than that of the water in which they live. Water enters their bodies osmotically, and salt is lost by diffusion outward. Unlike the habitat of the brackish-water shore crab, fresh water is much more dilute than are coastal estuaries, and there is no retreat, no salty sanctuary into which a freshwater animal can retire for osmotic relief. It must and has become a permanent and highly efficient hyperosmotic regulator.

The scaled and mucus-covered body surface of a fish is about as waterproof as any flexible surface can be. In addition, freshwater fishes have several defenses against the problems of water gain and salt loss. First, water that inevitably enters the body by osmosis is pumped out by the kidney, which is capable of forming very dilute urine (Figure 30-3). Second, special salt-absorbing cells located in the gills transport salt ions, principally sodium and chloride (present in small quantities even in fresh water), from the water to the blood. This process, together with salt present in the fish's food, replaces diffusive salt loss. These mechanisms are so efficient that a freshwater fish devotes only a small part of its total energy expenditure to maintain osmotic balance.

Crayfishes, aquatic insect larvae, clams, and other freshwater animals are also hyperosmotic regulators and face the same hazards as freshwater fishes; they tend to gain too much water and lose too much salt. Like freshwater fishes, they excrete excess water as urine and replace lost salt by active ion transport across the gills.

Amphibians living in water also must compensate for salt loss by actively absorbing salt from the water (Figure 30-4). They use their skin for this purpose. Physiologists learned some years ago that pieces of frog skin continue to transport sodium and chloride actively for hours when removed and placed in a specially balanced salt solution. Fortunately for biologists, but unfortunately for frogs, these animals are so easily collected and maintained in the laboratory that frog skin became a favorite membrane system for studies of ion-transport phenomena.

Return of Fishes to the Sea

Marine bony fishes maintain the salt concentration of their body fluids at approximately one-third that of seawater (body fluids = 0.3 to 0.4 gram mole per liter [M]; seawater = 1 M). They are **hypoosmotic regulators** because they maintain their body fluids at a lower concentration (hence *hypo-*) than their seawater environment. Bony fishes living in the oceans today are descendants of earlier freshwater bony fishes that moved back into the sea during the Triassic period approximately 200 million years ago. Over millions of years freshwater fishes established an ionic concentration in the body fluid equivalent to approximately one-third that of seawater. Body fluid of terrestrial vertebrates is remarkably similar to that of dilute seawater too, a fact that is undoubtedly related to their ancient marine heritage.

By expressing concentration of salt in seawater or body fluids in molarity, we are saying that the osmotic strength is equivalent to the molar concentration of an ideal solute having the same osmotic strength. In fact, seawater and animal body fluids are not ideal solutions because they contain electrolytes that dissociate in solution. A 1 M solution of sodium chloride (which dissociates in solution) has a much greater osmotic strength than a 1 M solution of glucose, an ideal solute (nonelectrolyte) that does not dissociate in solution. Consequently, biologists usually express osmotic strength of a biological solution in osmolarity rather than molarity. A 1 osmolar solution exerts the same osmotic pressure as a 1 M solution of a nonelectrolyte.

When some freshwater bony fishes of the Triassic period ventured back to the sea, they encountered a new set of problems. Having a much lower internal osmotic concentration than

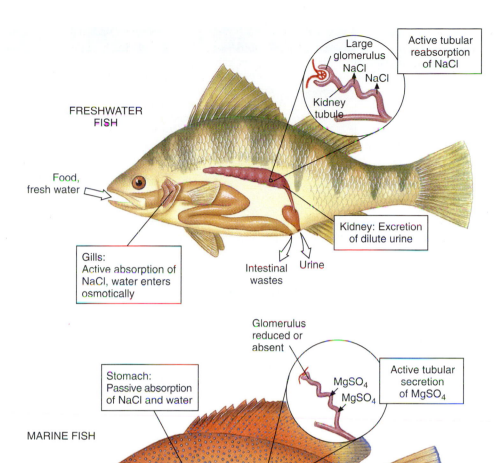

Figure 30-3

Osmotic regulation in freshwater and marine bony fishes. A freshwater fish maintains osmotic and ionic balance in its dilute environment by actively absorbing sodium chloride across the gills (some salt enters with food). To flush out excess water that constantly enters the body, the glomerular kidney produces dilute urine by reabsorbing sodium chloride. A marine fish must drink seawater to replace water lost osmotically to its salty environment. Sodium chloride and water are absorbed from the stomach. Excess sodium chloride is secreted outward by the gills. Divalent sea salts, mostly magnesium sulfate, are eliminated with feces and secreted by the tubular kidney.

the seawater around them, they lost water and gained salt. Indeed a marine bony fish literally risks desiccation, much like a desert mammal deprived of water.

To compensate for water loss a marine fish drinks seawater (Figure 30-3). This seawater is absorbed from the intestine, and the major sea salt, sodium chloride, is carried by the blood to the gills, where specialized salt-secreting cells transport it back into the surrounding sea. Ions remaining in the intestinal residue, especially magnesium, sulfate, and calcium, are voided with feces or excreted by kidneys. In this indirect way, marine fishes rid themselves of excess sea salts they have drunk, and replace water lost by osmosis. Samuel Taylor Coleridge's ancient mariner, surrounded by "water, water, everywhere, nor any drop to drink" undoubtedly would have been tormented even more had he known of the marine fishes' ingenious solution to thirst. A marine fish regulates the amount of seawater it drinks, consuming only enough to replace water loss and no more.

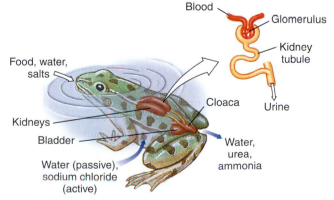

Figure 30-4

Exchange of water and solute in a frog. Water enters the highly permeable skin and is excreted by kidneys. The skin also actively transports ions (sodium chloride) from the environment. Kidneys form dilute urine by reabsorbing sodium chloride. Urine flows into the urinary bladder, where, during temporary storage, most of the remaining sodium chloride is removed and returned to the blood.

Sharks and rays (elasmobranchs) achieve osmotic balance differently. This group is almost totally marine. The salt composition of shark's blood is similar to that of bony fishes, but the blood also carries a large content of organic compounds, especially urea and trimethylamine oxide (TMAO). Urea is a metabolic waste that most animals quickly excrete. The shark kidney, however, conserves urea, allowing it to accumulate in the blood and raising the blood osmolarity to equal or slightly exceed that of seawater. With osmotic difference between blood and seawater eliminated, water balance is not a problem for sharks and their kin; they are in osmotic equilibrium with their environment.

The high concentration of urea in the blood of sharks and rays—more than 100 times as high as in mammals—could not be tolerated by most other vertebrates. In the latter, such high concentrations of urea disrupt peptide bonds of proteins, altering protein configuration. This would also occur in sharks except that TMAO has the opposite effect and serves to stabilize proteins in the presence of high urea levels. So accommodated are elasmobranchs to urea that their tissues cannot function without it, and their heart will stop beating in its absence.

How Terrestrial Animals Maintain Salt and Water Balance

The problems of living in an aquatic environment seem small indeed compared with problems of life on land. Since animal bodies are mostly water, all metabolic activities proceed in water, and life itself was conceived in water, it might seem that animals were meant to stay in water. Yet many animals, like the plants preceding them, moved onto land, carrying their watery composition with them. Once on land, terrestrial animals continued their adaptive radiation, solving the threat of desiccation, until they became abundant even in some of the most arid parts of the earth.

Terrestrial animals lose water by evaporation from respiratory and body surfaces, excretion in urine, and elimination in feces. They replace such losses by water in food, drinking water when available, and retaining **metabolic water** formed in cells by oxidation of foods, such as carbohydrates and fats. Stored fat becomes an important source of metabolic water in diving mammals. Certain arthropods—for example, desert roaches, certain ticks and mites, and mealworms—are able to absorb water vapor directly from atmospheric air. In some desert rodents, metabolic water gain may constitute most of the animals' water intake.

Particularly revealing is a comparison of water balance in human beings, nondesert mammals that drink water, with that of kangaroo rats, desert rodents that may drink no water at all (Table 30-1). Kangaroo rats acquire all their water from their food: 90% is metabolic water derived from oxidation of foods (see Figure 4-14, p. 66, and accompanying discussion of water yield during oxidative phosphorylation) and 10% as free moisture in food. Although we eat foods with a much higher water content than the dry seeds of a kangaroo rat's diet, we still must drink half our total water requirement.

TABLE 30-1

Water Balance in a Human and a Kangaroo Rat, a Desert Rodent

	Human (%)	Kangaroo Rat (%)
Gains		
Drinking	48	0
Free water in food	40	10
Metabolic water	12	90
Losses		
Urine	60	25
Evaporation (lungs and skin)	34	70
Feces	6	5

Source: Some data from K. Schmidt-Nielsen, *How animals work*. Cambridge University Press, 1972.

Given ample water to drink, humans can tolerate extremely high temperatures while preventing a rise in body temperature. Our ability to keep cool by evaporation was impressively demonstrated more than 200 years ago by a British scientist who remained for 45 minutes in a room heated to 260° F (126° C). A steak he carried in with him was thoroughly cooked, but he remained uninjured and his body temperature did not rise. Sweating rates may exceed 3 liters of water per hour under such conditions and cannot be tolerated unless the lost water is replaced by drinking. Without water, a human continues to sweat unabatedly until the water deficit exceeds 10% of the body weight, when collapse occurs. With a water deficit of 12% a human is unable to swallow even if offered water, and death occurs when the water deficit reaches about 15% to 20%. Few people can survive more than a day or two in a desert without water. Thus humans are not physiologically well adapted for desert climates but prosper there nonetheless by virtue of their technological culture.

Excretion of wastes presents a special problem in water conservation. The primary end product of protein breakdown is ammonia, a highly toxic material. Fishes easily excrete ammonia by diffusion across their gills, since there is an abundance of water to wash it away. Terrestrial insects, reptiles, and birds have no convenient way to rid themselves of toxic ammonia; instead, they convert it into uric acid, a nontoxic, almost insoluble compound. This conversion enables them to excrete a semisolid urine with little water loss. Uric acid as an end product has another important benefit. Reptiles and birds lay amniotic eggs enclosing their embryos (Figure 26-4, p. 539), together with their stores of food and water, and wastes that accumulate during development. By converting ammonia to uric acid, a developing embryo's waste can be precipitated into solid crystals, which are stored harmlessly within the egg until hatching.

Marine birds and turtles have evolved an effective solution for excreting large loads of salt eaten with their food. Located above each eye is a special **salt gland** capable of excreting a highly concentrated solution of sodium chloride—up to twice the concentration of seawater. In birds the salt solution runs out the nares (see p. 567 and Figure 27-13). Marine lizards and turtles, like *Alice in Wonderland's* Mock Turtle, shed their salt gland secretion as salty tears. Salt glands are important accessory organs of salt excretion in these animals because their kidneys cannot produce a concentrated urine, as can mammalian kidneys.

INVERTEBRATE EXCRETORY STRUCTURES

Many protozoan groups and some freshwater sponges have special excretory organelles called contractile vacuoles. More complex invertebrates have excretory organs that are basically tubular structures forming urine by first producing an ultrafiltrate or fluid secretion of the blood. This fluid secretion enters the proximal end of the tubule and is modified continuously as it flows down the tubule. The final product is urine.

Contractile Vacuole

The tiny, spherical, intracellular vacuole of protozoa and freshwater sponges is not a true excretory organ, since ammonia and other nitrogenous wastes of metabolism readily diffuse into the surrounding water. The contractile vacuole is an organ of water balance in freshwater protozoa, in which it expels excess water gained by osmosis. As water enters the protozoan, a vacuole grows and finally collapses, emptying its contents through a pore on the surface. The cycle is repeated rhythmically. Although the mechanism for filling the vacuole is not fully understood, recent research suggests that contractile vacuoles are surrounded by a network of membranous channels populated with numerous proton pumps (proton pumps were described in connection with the electron transport chain in Chapter 4, p. 64 and following). Proton pumps apparently create H^+ and HCO^- gradients that draw water into the vacuole, forming an isosmotic solution. These ions are excreted when the vacuole empties (see Figure 11-8, p. 213).

Contractile vacuoles are common in freshwater protozoa, sponges, and radiate animals (such as hydra), but rare or absent in marine forms of these groups, which are isosmotic with seawater and consequently neither lose nor gain too much water.

Nephridium

The most common type of invertebrate excretory organ is the nephridium, a tubular structure designed to maintain appropriate osmotic balance. One of the simplest arrangements is the flame cell system (or **protonephridium**) of acoelomates (flatworms) and some pseudocoelomates.

In planaria and other flatworms the protonephridial system takes the form of two highly branched duct systems distributed throughout the body (Figure 30-5). Fluid enters the system through specialized "flame cells," moves slowly into and down the tubules, and is excreted through pores that open at intervals on the body surface. The rhythmical beat of the flagellar tuft, suggestive of a tiny flickering flame, creates a negative pressure that draws fluid into the tubular portion of the system. In the tubule, water and metabolites valuable to the body are recovered by reabsorption, leaving wastes behind to be expelled. Nitrogenous wastes (mainly ammonia) diffuse across the surface of the body.

The flame-cell system is extensively branched throughout a flatworm's body because these acoelomate animals have no circulatory system to deliver wastes to a centralized excretory system (such as the kidneys of vertebrates and many invertebrates).

The protonephridium just described is a **closed** system. The tubules are closed on the inner end and urine is formed from a fluid that must first enter the tubules by being transported across flame cells. A more advanced type of nephridium is the **open,** or "true," nephridium (**metanephridium**) found in several eucoelomate phyla such as annelids (Figure 30-6), molluscs, and several smaller phyla. A metanephridium is more advanced than a protonephridium in two important ways. First, the tubule is open at *both* ends, allowing fluid to be swept into the tubule

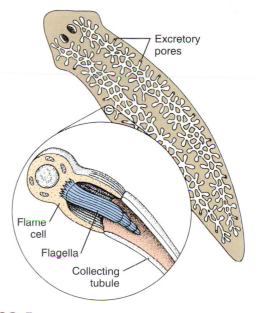

Figure 30-5

Flame cell system of a flatworm. Body fluids collected by flame cells (protonephridia) are passed down a system of ducts to excretory pores on the body surface.

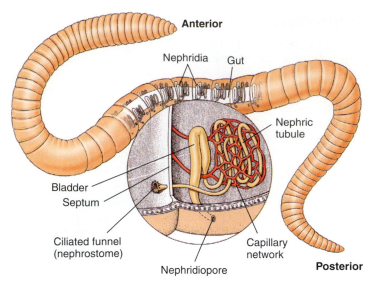

Figure 30-6
Excretory system of an earthworm. Each segment has a pair of large nephridia suspended in a fluid-filled coelom. Each nephridium occupies two segments because the ciliated funnel (nephrostome) drains the segment anterior to the segment containing the rest of the nephridium. The nephric tubule reabsorbs valuable materials from the tubular fluid into the capillary network.

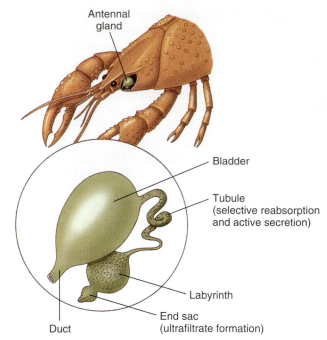

Figure 30-7
Antennal glands of a crayfish. These are filtration kidneys in which a filtrate of the blood is formed in the end sac. The filtrate is converted into urine as it passes down the tubule toward the bladder.

through a ciliated funnel-like opening, the **nephrostome.** Second, a metanephridium is surrounded by a network of blood vessels that assists reclamation of water and valuable materials such as salts, sugars, and amino acids from the tubular fluid.

Despite these differences, the basic process of urine formation is the same in protonephridia and metanephridia: fluid enters and flows continuously through a tubule where the fluid is selectively modified by (1) withdrawing valuable solutes from it and returning these to the body (reabsorption) and (2) adding waste solutes to it (secretion). The sequence ensures removal of wastes from the body without loss of materials valuable to the body. We will see that kidneys of vertebrates operate in basically the same way.

Arthropod Kidneys

The paired **antennal glands** of crustaceans, located in the ventral part of the head (Figure 30-7), are an advanced design of the basic nephridial organ. However, they lack open nephrostomes. Instead, hydrostatic pressure of the blood forms a protein-free filtrate of the blood (ultrafiltrate) in the end sac. In the tubular portion of the gland, selective reabsorption of certain salts and active secretion of others modifies the filtrate as it moves toward the bladder. Thus crustaceans have excretory organs that are basically vertebrate-like in the functional sequence of urine formation.

Insects and spiders have a unique excretory system consisting of **Malpighian tubules** that operate in conjunction with specialized glands in the wall of the rectum (Figure 30-8). These thin, elastic, blind Malpighian tubules are closed and lack an arterial supply. Urine formation is initiated by active secretion of

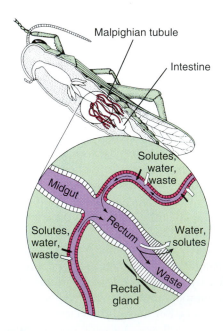

Figure 30-8
Malpighian tubules of insects. Malpighian tubules are located at the juncture of the midgut and hindgut (rectum). Solutes, especially potassium, are actively secreted into the tubules from the surrounding arthropod hemolymph (blood). Water, uric acid, and other wastes follow. This fluid drains into the rectum, where solutes and water are actively reabsorbed, leaving wastes to be excreted.

salts, largely potassium, into the tubules from the hemolymph (blood). This primary secretion of ions creates an osmotic drag that pulls water, solutes, and nitrogenous wastes, especially uric acid, into the tubule. Uric acid enters the upper end of the tubule as soluble potassium urate, which precipitates as insoluble uric acid in the proximal end of the tubule. Once the formative urine drains into the rectum, most of the water and potassium are reabsorbed by specialized rectal glands, leaving behind uric acid and other wastes, which are expelled in feces. The Malpighian tubule excretory system is ideally suited for life in dry environments and has contributed to the adaptive radiation of insects on land.

VERTEBRATE KIDNEY

Ancestry and Embryology

From comparative studies of development, biologists believe that the kidney of the earliest vertebrates extended the length of the coelomic cavity and was composed of segmentally arranged tubules, each resembling an invertebrate nephridium. Each tubule opened at one end into the coelom by a nephrostome and at the other end into a common **archinephric duct.** This ancestral kidney is called an **archinephros** ("ancient kidney"), and we find a segmented kidney very similar to an archinephros in embryos of hagfishes and caecilians (Figure 30-9). Almost from the beginning, the reproductive system, which develops beside the excretory system from the same segmental blocks of trunk mesoderm, used the nephric ducts as a convenient conducting system for reproductive products. Thus even though the two systems have nothing functionally in common, they are closely associated in their use of common ducts (Figure 30-9).

Kidneys of living vertebrates developed from this primitive plan. During embryonic development of amniote vertebrates, there is a succession of three developmental stages of kidneys: **pronephros, mesonephros,** and **metanephros** (Figure 30-9). Some, but not all, of these stages are observed in other vertebrate groups. In all vertebrate embryos, the pronephros is the first kidney to appear. It is located anteriorly in the body and becomes part of the persistent kidney only in adult hagfishes. In all other vertebrates the pronephros degenerates during development and is replaced by a more centrally located mesonephros. The mesonephros is the functional kidney of embryonic amniotes (reptiles, birds, and mammals), and contributes to the adult kidney (called an **opisthonephros**) of fishes and amphibians.

The metanephros, characteristic of adult amniotes, is distinguished in several ways from the pronephros and mesonephros. It is more caudally located and it is a much larger, more compact structure containing a very large number of nephric tubules. It is drained by a new duct, the **ureter,** which developed when the old archinephric duct was relinquished to the reproductive system of the male for sperm transport. Thus three successive kidney types—pronephros, mesonephros, metanephros—succeed each other embryologically, and to some extent phylogenetically, in amniotes.

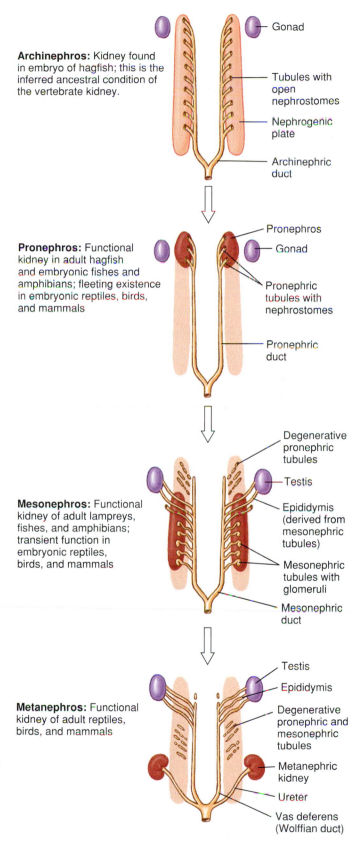

Archinephros: Kidney found in embryo of hagfish; this is the inferred ancestral condition of the vertebrate kidney.
— Gonad
— Tubules with open nephrostomes
— Nephrogenic plate
— Archinephric duct

Pronephros: Functional kidney in adult hagfish and embryonic fishes and amphibians; fleeting existence in embryonic reptiles, birds, and mammals
— Pronephros
— Gonad
— Pronephric tubules with nephrostomes
— Pronephric duct

Mesonephros: Functional kidney of adult lampreys, fishes, and amphibians; transient function in embryonic reptiles, birds, and mammals
— Degenerative pronephric tubules
— Testis
— Epididymis (derived from mesonephric tubules)
— Mesonephric tubules with glomeruli
— Mesonephric duct

Metanephros: Functional kidney of adult reptiles, birds, and mammals
— Testis
— Epididymis
— Degenerative pronephric and mesonephric tubules
— Metanephric kidney
— Ureter
— Vas deferens (Wolffian duct)

Figure 30-9

Comparative development of male vertebrate kidney. *Red,* functional structures. *Light red,* degenerative or undeveloped parts.

Vertebrate Kidney Function

The vertebrate kidney is part of many interlocking mechanisms that maintain homeostasis. The kidney plays a prominent role in this regulatory council because it is the principal organ that regulates volume and composition of the internal fluid environment. While we commonly describe the vertebrate kidney as an organ of excretion, the removal of metabolic wastes is incidental to its regulatory function.

The organization of kidneys differs somewhat in different groups of vertebrates, but in all the basic functional unit is the **nephron,** and urine is formed by three well-defined physiological processes: **filtration, reabsorption,** and **secretion.** This discussion focuses mainly on the mammalian kidney, which is the most completely understood regulatory organ.

The two human kidneys are small organs comprising less than 1% of the body weight. Yet they receive a remarkable 20% to 25% of the total cardiac output, some 2000 liters of blood each day. This vast blood flow is channeled to approximately 2 million nephrons, which form the bulk of the two kidneys. Each nephron begins with an expanded chamber, the **Bowman's capsule,** containing a tuft of capillaries, the **glomerulus** (glo-mer'yoo-lus), which together are called the **renal corpuscle.** Blood pressure in the capillaries forces a protein-free **filtrate** into a Bowman's capsule and along a **renal tubule,** consisting of several segments that perform different functions in the process of urine formation. The filtrate passes first into a **proximal convoluted tubule,** then into a long, thin-walled **loop of Henle,** which drops deep into the inner portion of the kidney (the **medulla**) before returning to the outer portion (the **cortex**) where it joins a **distal convoluted tubule.** From the distal tubule the fluid empties into a **collecting duct** which drains into the **renal pelvis.** Here the urine is collected before being carried by the **ureter** to the **urinary bladder.** These anatomical relationships are shown in Figure 30-10.

Urine that leaves the collecting duct is very different from the filtrate produced in the renal corpuscle. During its travels through the renal tubule and collecting duct, both the composition and concentration of the original filtrate change. Some solutes such as glucose, amino acids, and sodium have been reabsorbed while other materials, such as hydrogen ions and urea, have been concentrated in urine.

The nephron, with its pressure filter and tubule, is intimately associated with blood circulation (Figure 30-11). Blood from the aorta enters each kidney through a large **renal artery,** which divides into a branching system of smaller arteries. The arterial blood reaches the glomerulus through an **afferent arteriole** and leaves by an **efferent arteriole.** From the efferent arteriole the blood travels to an extensive capillary network that surrounds and supplies the proximal and distal convoluted tubules and the loop of Henle (Figure 30-10). This capillary network provides a means for the pickup and delivery of materials that are reabsorbed or secreted by the kidney tubules. From these capillaries blood is collected by veins that unite to form the **renal vein.** This vein returns blood to the vena cava.

Figure 30-10

Urinary system of humans, with enlargements showing detail of the kidney and a single nephron.

Figure 30-11

Scanning electron micrograph of a cast of the microcirculation of the mammalian kidney, showing several glomeruli and associated blood vessels. The capsular epithelium, which normally surrounds each glomerulus, has been digested away in preparing the cast.

From R. G. Kessel and R. H. Kardon, Tissues and Organs: A Text-Atlas of Scanning Electron Microscopy, 1979, W. H. Freeman and Co.

Glomerular Filtration

Let us now return to the glomerulus, where the process of urine formation begins. The glomerulus acts as a specialized mechanical filter in which a protein-free filtrate of the plasma is driven by the blood pressure across capillary walls and into the fluid-filled space of the Bowman's capsule. Solute molecules small enough to pass through the slit pores of the capillary wall are carried through with the water in which they are dissolved. Red blood cells and plasma proteins, however, are withheld because they are too large to pass through these pores (Figure 30-12).

The filtrate continues through the renal tubular system where it undergoes extensive modification before becoming urine. Human kidneys form approximately 180 liters (nearly 50 gallons) of filtrate each day, a volume many times exceeding the total blood volume. If this volume of water and the valuable nutrients and salts it contains were lost, the body would soon be depleted of these compounds. Depletion does not happen because nearly all of the filtrate is reabsorbed. The final urine volume in humans averages 1.2 liters per day.

Conversion of filtrate into urine involves two processes: (1) modification of the composition of the filtrate through tubular reabsorption and secretion, and (2) changes in the total osmotic concentration of urine through regulation of water excretion.

Tubular Reabsorption

Approximately 60% of the filtrate volume and virtually all of the glucose, amino acids, vitamins and other valuable nutrients are

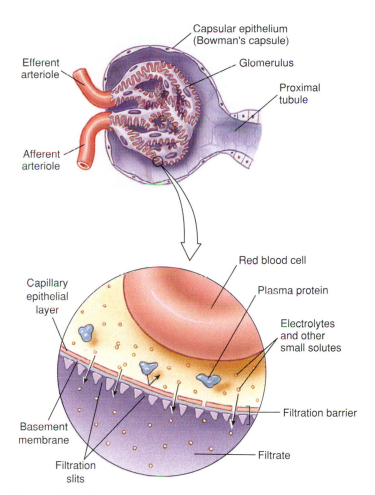

Figure 30-12

Renal corpuscle, showing (*enlargement*) the filtration of fluid through the glomerular capillary membrane. Water, electrolytes, and other small molecules pass the porous filtration barrier, but the plasma proteins are too large to pass the barrier. The filtrate is thus protein free.

reabsorbed in the proximal convoluted tubule. Much of this reabsorption is by **active transport,** in which cellular energy is used to transport materials from tubular fluid to the surrounding capillary network from which they will reenter the blood circulation. Electrolytes such as sodium, potassium, calcium, bicarbonate, and phosphate are reabsorbed by ion pumps, which are carrier proteins driven by the hydrolysis of ATP (ion pumps are described on p. 47). Because an essential function of the kidney is to regulate plasma concentrations of electrolytes, all are individually reabsorbed by ion pumps specific for each electrolyte. Some are strongly reabsorbed and others weakly reabsorbed, depending on the body's need to conserve each mineral. Some materials are passively reabsorbed. Negatively charged chloride ions, for example, passively accompany active reabsorption of positively charged sodium ions in the proximal convoluted tubule. Water, too, is withdrawn passively from the tubule, as it osmotically follows the active reabsorption of solutes.

In the disease diabetes mellitus ("sweet running through"), glucose rises to abnormally high concentrations in the blood plasma (hyperglycemia) because the hormone insulin, which enables body cells to take up glucose, is deficient. As blood glucose rises above a normal level of about 100 mg/100 ml of plasma, the concentration of glucose in the filtrate also rises, and more glucose must be reabsorbed by the proximal tubule. Eventually a point is reached (about 300 mg/100 ml of plasma) at which reabsorptive capacity of the tubular cells is saturated. This point is the transport maximum for glucose. Should plasma glucose continue to rise, glucose spills into the urine. In untreated diabetes mellitus the victim's urine tastes sweet, thirst is unrelenting, and the body wastes away despite a large food intake. In England the disease for centuries was appropriately called the "pissing evil."

For most substances there is an upper limit to the amount of substance that can be reabsorbed. This upper limit is termed the **transport maximum** (renal threshold) for that substance. For example, glucose normally is reabsorbed completely by the kidney because the transport maximum for glucose is poised well above the amount of glucose usually present in the plasma filtrate. Should the plasma glucose concentration exceed this threshold level, as in the disease diabetes mellitus, glucose appears in the urine (Figure 30-13).

Unlike glucose, most electrolytes are excreted in the urine in variable amounts. Reabsorption of sodium, the dominant cation in the plasma, illustrates the flexibility of the reabsorption process. The human kidney filters approximately 600 g of sodium every 24 hours. Nearly all of this sodium is reabsorbed, but the exact amount is matched precisely to sodium intake. With a normal sodium intake of 4 g per day, the kidney excretes 4 g and reabsorbs 596 g each day. A person on a low-salt diet of 0.3 g of sodium per day still maintains salt balance because only 0.3 g escapes reabsorption. But with a very high salt intake, much above 20 g per day, the kidney cannot excrete sodium as fast as it enters. The unexcreted sodium chloride holds additional water in the body fluids, and the person begins to gain weight. (The salt intake of the average North American is about 6 to 18 g per day, approximately 20 times more than the body

needs, and three times more than is considered acceptable for those predisposed to high blood pressure.)

A human kidney can adapt to excrete large quantities of salt (sodium chloride) under conditions of high salt intake. In societies accustomed to widespread use of foods heavily salted for preservation (for example, salted pork and salt herring) daily intakes may approach or even exceed 100 g. Body weight remains normal under such conditions. However, the acute ingestion of 20 to 40 g/day by volunteers unadapted to such large intakes of salt caused swelling of tissues, increase in body weight, and some increase in blood pressure.

Distal convoluted tubules make the final adjustment of filtrate composition. Sodium reabsorbed by the proximal convoluted tubule—some 85% of the total filtered—is obligatory reabsorption; this amount will be reabsorbed independent of sodium intake. In the distal convoluted tubule, however, sodium reabsorption is controlled by **aldosterone,** a steroid hormone from the adrenal gland (p. 733). Aldosterone increases both active reabsorption of sodium and secretion of potassium by the distal tubules; the hormone thus decreases loss of sodium and increases loss of potassium in the urine. The secretion of aldosterone is regulated (1) by the enzyme **renin,** produced by the **juxtaglomerular apparatus,** a complex of cells located in the afferent arteriole at its junction with the glomerulus (Figure 30-10), and (2) by elevated blood potassium levels. Renin is released in response to a low blood sodium level, to low blood pressure (which can occur if the blood volume drops too low), or to low sodium in the glomerular filtrate. Renin then initiates a series of enzymatic events culminating in the production of **angiotensin,** a blood protein that has several related effects. First, it stimulates the release of aldosterone, which acts in turn to increase sodium reabsorption by the distal tubule. Second, it increases the secretion of **antidiuretic hormone** (vasopressin, discussed on p. 645), which promotes water conservation by the kidney. Third, it increases blood pressure. Finally, it stimulates thirst, which also is stimulated by decreased blood volume or increased blood osmolarity. These actions of angiotensin tend to

Figure 30-13

The mechanism for the tubular reabsorption of glucose can be likened to a conveyor belt running at constant speed. **A,** When glucose concentrations in the filtrate are low, all is reabsorbed. **B,** When glucose concentrations in the filtrate have reached the transport maximum, all carrier sites for glucose are occupied. If glucose levels rise further, **C,** as in the disease diabetes mellitus, some glucose escapes the carriers and appears in urine.

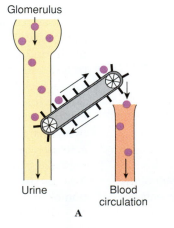

Glomerulus

Urine   Blood circulation

A

Urine   Blood

B

Urine   Blood

C

reverse the circumstances (low blood sodium and low blood pressure and/or blood volume) that triggered the secretion of renin. Sodium and water are conserved, and blood volume and blood pressure are restored to normal.

The flexibility of distal reabsorption of sodium varies considerably in different animals: it is restricted in humans but very broad in many rodents. These differences have appeared because selective pressures during evolution have resulted in rodents adapted for dry environments. They must conserve water and at the same time excrete considerable sodium. Humans, however, were not designed to accommodate the large salt appetites many have. Our closest relatives, the great apes, are vegetarians with an average salt intake of less than 0.5 g per day.

Tubular Secretion

In addition to reabsorbing materials from plasma filtrate, the nephron can secrete materials across the tubular epithelium and *into* the filtrate. In this process, the reverse of tubular reabsorption, carrier proteins in the tubular epithelial cells selectively transport substances from blood in capillaries outside the tubule to the filtrate inside the tubule. Tubular secretion enables a kidney to increase the urine concentrations of materials to be excreted, such as hydrogen and potassium ions, drugs, and various foreign organic materials. The tubular epithelium is able to recognize foreign organic materials, because they are metabolized by the liver to form cationic or anionic proteins. These proteins are transported by the tubular epithelium, which has both cationic and anionic transport proteins in its membrane. The distal convoluted tubule is the site of most tubular secretion.

In the kidneys of bony marine fishes, reptiles, and birds, tubular secretion is a much more highly developed process than it is in mammalian kidneys. Marine bony fishes actively secrete large amounts of magnesium and sulfate, seawater salts that are by-products of their mode of osmotic regulation. Reptiles and birds excrete uric acid instead of urea as their major nitrogenous waste. The material is actively secreted by the tubular epithelium. Since uric acid is nearly insoluble, it forms crystals in the urine and requires little water for excretion. Thus excretion of uric acid is an important adaptation for water conservation.

Water Excretion

Kidneys closely regulate the osmotic pressure of the blood. When fluid intake is high, the kidney excretes dilute urine, saving salts and excreting water. When fluid intake is low, kidneys conserve water by forming concentrated urine. A dehydrated person can concentrate urine to approximately four times blood osmotic concentration. This important ability to concentrate urine enables us to excrete wastes with minimal loss of water.

The capacity of mammalian and some avian kidneys to produce a concentrated urine involves an interaction between the loop of Henle and the collecting ducts. This interplay results in formation of an osmotic gradient in the kidney, as shown in Figure 30-14. In the cortex, interstitial fluid is isosmotic with blood, but deep in the medulla the osmotic concentration is 4 times greater than that of blood (in rodents and desert mammals that can produce highly concentrated urine the osmotic gradient is much greater than in humans). The high osmotic concentrations in the medulla are produced by an exchange of ions in the loop of Henle by **countercurrent multiplication.** "Countercurrent" refers to the opposite directions of fluid movement in the two limbs of the loop of Henle: down in the descending limb and up the ascending limb. "Multiplication" describes the increasing osmotic concentration in the medulla surrounding the loops of Henle and collecting ducts resulting from ion exchange between the two limbs of the loop.

The functional characteristics of this system are as follows. The descending limb of the loop of Henle is permeable to water but impermeable to solutes. The ascending limb is nearly impermeable to water. Sodium chloride passively moves from the lower ascending limb and is actively transported from the thick portion of the ascending limb into the surrounding tissue fluid (Figure 30-14). As the interstitium surrounding the loop becomes more concentrated with solute, water is withdrawn from the descending limb by osmosis. The tubular fluid in the base of the loop, now more concentrated, moves up the ascending limb, where still more sodium chloride diffuses or is pumped out. In this way the effect of ion movement in the ascending limb is multiplied as more water is withdrawn from the descending limb and more concentrated fluid is presented to the ascending limb (Figures 30-14 and 30-15). The blood capillaries surrounding the loops of Henle, the **vasa recta,** are also arranged in a countercurrent fashion. This arrangement of blood vessels is important in maintaining the osmotic concentration gradient of the medulla and cortex.

Final adjustment of urine concentration occurs not in the loops of Henle but in the collecting ducts. Formative urine that enters the distal tubule from the loop of Henle is dilute (because of salt withdrawal) and is diluted still more by active reabsorption of more sodium chloride in the distal tubule. The formative urine, low in solutes but carrying urea, now flows down into the collecting duct. Because of the high concentration of solutes surrounding the collecting duct, water is withdrawn from the urine. As the urine becomes more concentrated, urea also diffuses out of it. The lower regions of the collecting duct are permeable to urea, and thus far, four different urea transporters have been discovered. Since the loop of Henle is less permeable to urea, its concentrations build within the medulla tissue fluid. This buildup of urea contributes significantly to high osmotic concentration of the medulla (Figure 30-15).

The amount of water reabsorbed and the final concentration of the urine depend on permeability of the walls of the distal convoluted tubule and the collecting duct. This is controlled by the **antidiuretic hormone** (ADH, or vasopressin), which is released by the posterior pituitary gland (neurohypophysis). In turn, special receptors in the brain that constantly sense the osmotic pressure of body fluids govern the release of this hormone. When the blood osmotic pressure increases, as during dehydration, the pituitary gland releases more ADH. ADH increases permeability of the collecting duct by increasing the number of water channels in the collecting duct epithelial cells. Then, as the fluid in the collecting duct passes through the hyperosmotic region of the kidney medulla, water diffuses through the channels into the surrounding interstitial fluid and is carried away

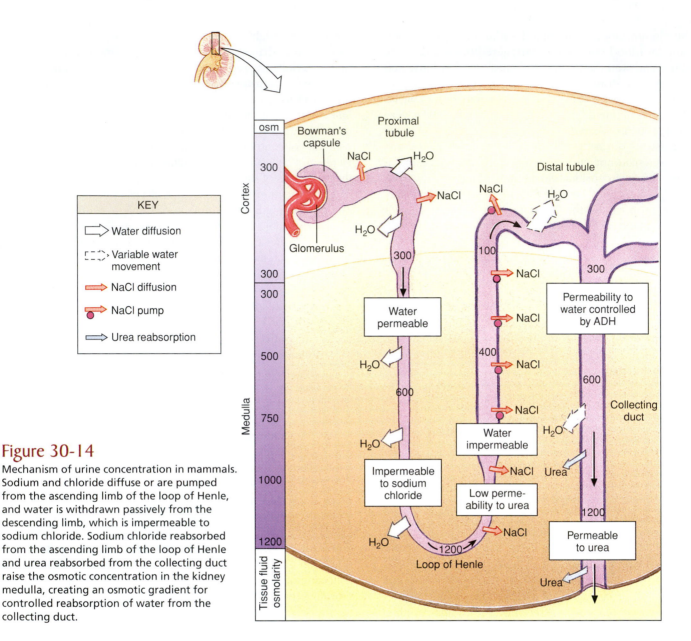

Figure 30-14

Mechanism of urine concentration in mammals. Sodium and chloride diffuse or are pumped from the ascending limb of the loop of Henle, and water is withdrawn passively from the descending limb, which is impermeable to sodium chloride. Sodium chloride reabsorbed from the ascending limb of the loop of Henle and urea reabsorbed from the collecting duct raise the osmotic concentration in the kidney medulla, creating an osmotic gradient for controlled reabsorption of water from the collecting duct.

by the blood circulation. The urine loses water and becomes more concentrated. Given this sequence of events for dehydration, it is not difficult to anticipate how the system responds to overhydration: the pituitary stops releasing ADH, the channels in the collecting duct walls decrease in number, and a large volume of dilute urine is excreted.

The varying ability of different mammals to form a concentrated urine correlates closely with length of the loops of Henle. Beavers, which have no need to conserve water in their aquatic environment, have short loops and can concentrate their urine only to about twice the osmolarity of the blood. Humans, with relatively longer loops, can concentrate urine 4.2 times that of the blood. As we would anticipate, desert mammals have much greater urine-concentrating powers. A camel can produce a urine 8 times the plasma concentration, a gerbil 14 times, and an Australian hopping mouse 22 times. In this creature, the greatest urine concentrator of all, the loops of Henle extend to the tip of a long renal papilla that pushes out into the mouth of the ureter.

TEMPERATURE REGULATION

We have seen that a fundamental problem facing an animal is keeping its internal environment in a state that permits normal cell function. Biochemical activities are sensitive to the chemical environment and our discussion thus far has examined how the chemical environment is stabilized. Biochemical reactions are also extremely sensitive to temperature. All enzymes have an optimum temperature; at temperatures above or below this optimum, enzyme function is impaired. Temperature therefore is a severe constraint for animals, all of which must maintain biochemical stability. When body temperature drops too low, metabolic processes slow, reducing the amount of energy the animal can muster for activity and reproduction. If body temperature rises too high, metabolic reactions become unbalanced and enzymatic activity is hampered or even destroyed. Thus animals can succeed only in a restricted range of temperature, usually between 0° to 40° C. Animals must either find a habitat where

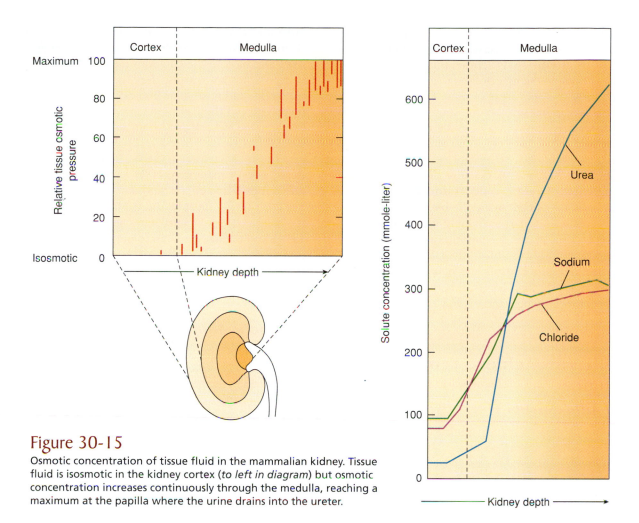

Figure 30-15

Osmotic concentration of tissue fluid in the mammalian kidney. Tissue fluid is isosmotic in the kidney cortex (*to left in diagram*) but osmotic concentration increases continuously through the medulla, reaching a maximum at the papilla where the urine drains into the ureter.

they do not have to contend with temperature extremes, or they must develop means of stabilizing their metabolism independent of temperature extremes.

A temperature difference of 10° C has become a standard used to measure the temperature sensitivity of a biological function. This value, called the Q_{10}, is determined (for temperature intervals of exactly 10° C) simply by dividing the value of a rate function (such as metabolic rate or rate of an enzymatic reaction) at the higher temperature by the value of the rate function at the lower temperature. In general, metabolic reactions have Q_{10} values of about 2.0 to 3.0. Purely physical processes, such as diffusion, have much lower Q_{10} values, usually close to 1.0.

Ectothermy and Endothermy

The terms "cold-blooded" and "warm-blooded" have long been used to divide animals into two groups: invertebrates and vertebrates that feel cold to the touch, and those, such as humans, other mammals, and birds, that do not. It is true that body temperature of mammals and birds is usually (though not always) warmer than the air temperature, but a "cold-blooded" animal is not necessarily cold. Tropical fishes, and insects and reptiles

basking in the sun, may have body temperatures equaling or surpassing those of mammals. Conversely, many "warmblooded" mammals hibernate, allowing their body temperature to approach the freezing point of water. Thus the terms "warmblooded" and "cold-blooded" are hopelessly subjective and nonspecific but are so firmly entrenched in our vocabulary that most biologists find it easier to accept the usage than to try to change people.

The term **poikilothermic** (variable body temperature) and **homeothermic** (constant body temperature) are frequently used by zoologists as alternatives to "cold-blooded" and "warmblooded," respectively. These terms, which refer to variability of body temperature, are more precise and more informative, but still offer difficulties. For example, deep-sea fishes live in an environment having no perceptible temperature change. Even though their body temperature is absolutely stable, day in and day out, to call such fishes homeotherms would distort the intended application of the term. Furthermore, among homeothermic birds and mammals many allow their body temperature to change between day and night, or, as with hibernators, between seasons.

Physiologists prefer yet another way to describe body temperatures, one that reflects the fact that an animal's body temperature is a balance between heat gain and heat loss. All animals produce heat from cellular metabolism, but in most the heat is conducted away as fast as it is produced. In these animals, the **ectotherms**—and the overwhelming majority of animals belong to this group—body temperature is determined solely by the environment. Many ectotherms exploit their environment behaviorally to select areas of more favorable temperature (such as basking in the sun) but the source of energy used to increase body temperature comes from the environment, not from within the body. Alternatively some animals are able to generate and retain enough heat to elevate their own body temperature to a high but stable level. Because the source of their body heat is internal, they are called **endotherms.** These favored few in the animal kingdom are the birds and mammals, as well as a few reptiles and fast-swimming fishes, and certain insects that are at least partially endothermic. Endothermy allows birds and mammals to stabilize their internal temperature, allowing biochemical processes and nervous system functions to proceed at steady high levels of activity. Endotherms can thus remain active in winter and exploit habitats denied to ectotherms.

How Ectotherms Achieve Temperature Independence

Behavioral Adjustments

Although ectotherms cannot control their body temperature physiologically, many are able to regulate their body temperature behaviorally with considerable precision. Ectotherms often have the option of seeking areas in their environment where the temperature is favorable to their activities. Some ectotherms, such as desert lizards, exploit hour-to-hour changes in solar radiation to keep their body temperatures relatively constant (Figure 30-16). In the early morning they emerge from their burrows and bask in the sun with their bodies flattened to absorb heat. As the day warms, they turn to face the sun to reduce exposure, and raise their bodies from the hot substrate. In the hottest part of the day they may retreat to their burrows. Later they emerge to bask as the sun sinks lower and the air temperature drops.

These behavioral patterns help to maintain a relatively steady body temperature of 36° to 39° C while the air temperature varies between 29° and 44° C. Some lizards can tolerate intense midday heat without shelter. The desert iguana of the southwestern United States prefers a body temperature of 42° C when active and can tolerate a rise to 47° C, a temperature that is lethal to all birds and mammals and most other lizards. The term "cold-blooded" clearly does not apply to these animals!

Metabolic Adjustments

Even without the help of the behavioral adjustments just described, most ectotherms can adjust their metabolic rates to the prevailing temperature such that the intensity of metabolism remains mostly unchanged. This is called **temperature compensation** and involves complex biochemical and cellular adjustments. These adjustments enable a fish or a salamander, for example, to benefit from almost the same level of activity in both warm and cold environments. Thus, whereas endotherms achieve metabolic homeostasis by maintaining their body temperature independent of environmental temperature, ectotherms accomplish much the same by directly maintaining their metab-

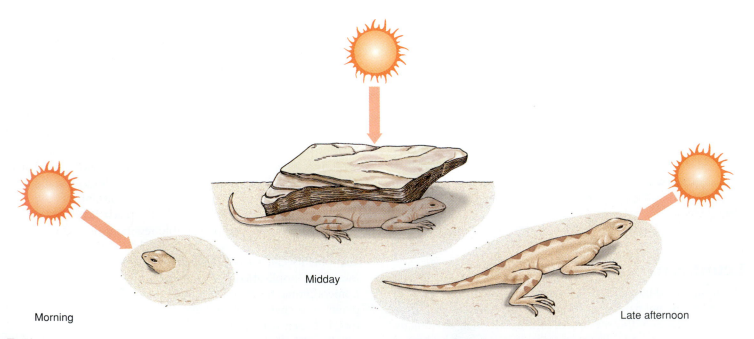

Midday

Morning

Late afternoon

Figure 30-16

How a lizard regulates its body temperature behaviorally. In the morning the lizard absorbs the sun's heat through its head while keeping the rest of its body protected from cool morning air. Later it will emerge to bask. At noon, with its body temperature high, it seeks shade from the hot sun. When the air temperature drops in the late afternoon, it emerges and lies parallel to the sun's rays.

olism independent of body temperature. This metabolic regulation also is a form of homeostasis.

Temperature Regulation in Endotherms

Most mammals have body temperatures between 36° and 38° C, somewhat lower than those of birds, which range between 40° and 42° C. Constant temperature is maintained by a delicate balance between heat production and heat loss—not a simple matter when these animals are alternating between periods of rest and bursts of heat-producing activity.

Heat is produced by the animal's metabolism. This includes oxidation of foods, basal cellular metabolism, and muscular contraction. Because much of an endotherm's daily caloric intake is required to generate heat, especially in cold weather, the endotherm must eat more food than an ectotherm of the same size. Heat is lost by radiation, conduction, and convection (air movement) to a cooler environment and by evaporation of water (Figure 30-17). A bird or mammal can control both processes of heat production and heat loss within rather wide limits. If the animal becomes too cool, it can generate heat by increasing muscular activity (exercise or shivering) and by decreasing heat loss by increasing its insulation. If it becomes too warm, it decreases heat production and increases heat loss. We will examine these processes in the examples presented here.

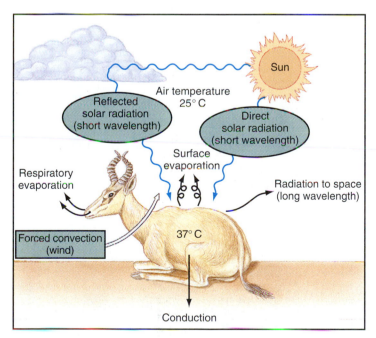

Figure 30-17

Exchange of heat between a large mammal and its environment on a warm day. Blue arrows indicate sources of net heat gain by the animal (all radiation); black arrows are avenues of net heat loss (evaporative cooling, conduction to the ground, longwave radiation into space, and forced convection by the wind). If air and ground temperatures were warmer than the animal, the arrows for forced convection, conduction, and radiation would be reversed. Then the animal could lose heat only by evaporative cooling.

Adaptations for Hot Environments

Despite the harsh conditions of deserts—intense heat during the day, cold at night, and scarcity of water, vegetation, and cover—many kinds of animals live there successfully. The smaller desert mammals are mostly **fossorial** (living mainly in the ground) or **nocturnal** (active at night). The lower temperature and higher humidity of burrows help to reduce water loss by evaporation. As explained on p. 638, desert animals such as the kangaroo rat and the American desert ground squirrels can, if necessary, derive the water they need from their dry food, drinking no water at all. Such animals produce a highly concentrated urine and form almost completely dry feces.

Large desert ungulates (hooved mammals that chew their cud) obviously cannot escape desert heat by living in burrows. Animals such as camels and desert antelopes (gazelle, oryx, and eland) possess a number of adaptations for coping with heat and dehydration. Figure 30-18 shows those of the eland. Mechanisms for controlling water loss and preventing overheating are closely linked. The glossy, pallid color of fur reflects direct sunlight, and fur itself is an excellent insulation that resists heat. Heat is lost by convection and conduction from the underside of elands where the fur is very thin. Fat tissue, an essential food reserve, is concentrated in a single hump on the back, instead of being uniformly distributed under the skin where it would impair loss of heat by radiation. Elands avoid evaporative water loss—the only means an animal has for cooling itself when the environmental temperature is higher than that of the body—by permitting their body temperature to drop during the cool night and then to rise slowly during the day as the body stores heat. Only when the body temperature reaches 41° C must elands prevent further rise through **evaporative cooling** by sweating and panting. They conserve water by producing a concentrated urine and dry feces. Camels have all of these adaptations developed to a similar or even greater degree; they are perhaps the most perfectly adapted of all large desert mammals.

Adaptations for Cold Environments

In cold environments mammals and birds use two major mechanisms to maintain homeothermy: (1) **decreased conductance,** reduction of heat loss by increasing the effectiveness of the insulation, and (2) **increased heat production.**

In all mammals living in the cold regions of the earth, fur thickness increases in winter, sometimes by as much as 50%. Thick underhair is the principal insulating layer, whereas the longer and more visible guard hair serves as protection against wear and for protective coloration. However, unlike the well-insulated trunk of the body, the body extremities (legs, tail, ears, nose) of arctic mammals are thinly insulated and exposed to rapid cooling. To prevent these parts from becoming major avenues of heat loss, they are allowed to cool to low temperatures, often approaching the freezing point. The heat in the warm arterial blood is not lost from the body, however. Instead, a **countercurrent heat exchange** between the outgoing warm blood and the returning cold blood prevents heat loss. Arterial blood in the leg of an arctic mammal or bird passes in close contact with a network of small veins. Because arterial blood flow is

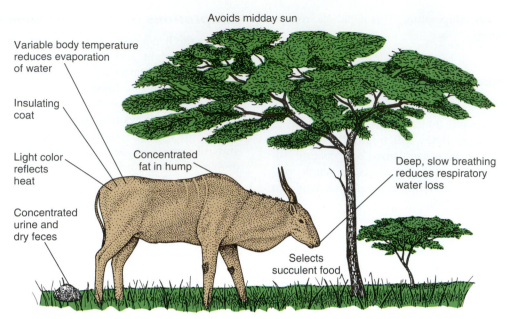

Figure 30-18

Physiological and behavioral adaptations of the common eland for regulating temperature in the hot, arid savanna of central Africa.

opposite to that of returning venous blood, heat is exchanged very efficiently from artery to veins. When arterial blood reaches the foot it has transferred nearly all of its heat to the veins returning blood to the body core (Figure 30-19). Thus little heat is lost to the surrounding cold air from poorly insulated distal regions of the leg. Countercurrent heat exchangers in appendages also are common in aquatic mammals such as seals and whales, which have thinly insulated flippers and flukes that would be avenues of excessive heat loss in the absence of this heat-salvaging arrangement.

A consequence of peripheral heat exchange is that legs and feet of mammals and birds living in cold environments must function at low temperatures. Temperatures of the feet of arctic foxes and barren-ground caribou are just above the freezing point; in fact, the temperature may be below 0° C in footpads and hooves. To keep feet supple and flexible at such low temperatures, fats in the extremities have very low melting points, perhaps 30° C lower than ordinary body fats.

In severely cold conditions all mammals can produce more heat by **augmented muscular activity** through exercise or shivering. We are all familiar with the effectiveness of both activities. A person can increase heat production as much as 18-fold by violent shivering when maximally stressed by cold. Another source of heat is increased oxidation of foods, especially from stores of brown fat (brown fat is described on p. 689). This mechanism is called **nonshivering thermogenesis.**

Small mammals the size of lemmings, voles, and mice meet the challenge of cold environments in a different way. Small mammals are not as well insulated as large mammals because thickness of fur is limited by the need to maintain mobility. Consequently these forms exploit the excellent insulating qualities of snow by living under it in runways on the forest floor, where

Figure 30-19

Countercurrent heat exchange in the leg of an arctic wolf. The diagram to the left shows how the extremities cool when the animal is exposed to low air temperatures. The diagram to the right depicts a portion of the front leg artery and vein, showing how heat is exchanged between arterial and venous blood. Heat is shunted back into the body and conserved.

incidentally, their food also is located. In this **subnivean environment** the temperature seldom drops below −5° C even though the air temperature above may fall to −50° C. Snow insulation decreases thermal conductance from small mammals just as thick pelage does for large mammals. Living beneath the snow is really a type of avoidance response to cold.

Adaptive Hypothermia in Birds and Mammals

Endothermy is energetically expensive. Whereas an ectotherm can survive for weeks in a cold environment without eating, an endotherm must always have energy resources to supply its high metabolic rate. The problem is especially acute for small birds and mammals which, because of their intense metabolism, may require a daily intake of food each day approaching their own body weight to maintain homeothermy (food consumption by birds is related on p. 565, and by mammals on p. 593). It is not surprising then that a few small birds and mammals have evolved ways to abandon homeothermy for periods ranging from a few hours to several months, allowing their body temperature to fall until it approaches or equals the temperature of surrounding air.

Some very small mammals, such as bats, maintain high body temperatures when active but allow their body temperature to drop profoundly when inactive and asleep. This is called **daily torpor,** an adaptive hypothermia that provides enormous saving of energy to small endotherms that are never more than a few hours away from starvation at normal body temperatures. Hummingbirds also may drop their body temperature at night when food supplies are low (Figure 30-20).

Many small and medium-sized mammals in northern temperate regions solve the problem of winter scarcity of food and low temperature by entering a prolonged and controlled state of dormancy: **hibernation.** True hibernators, such as ground squirrels, jumping mice, marmots, and woodchucks (Figure 30-21), prepare for hibernation by storing body fat. Entry into hibernation is gradual. After a series of "test drops" during which

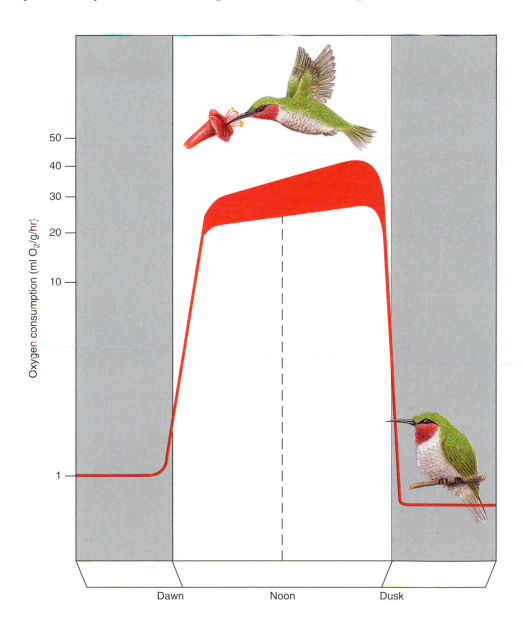

Figure 30-20

Torpor in hummingbirds. Body temperature and oxygen consumption (red line) are high when hummingbirds are active during the day but may drop to one-twentieth these levels during periods of food shortage. Torpor vastly lowers demands on the bird's limited energy reserves.

body temperature decreases a few degrees and then returns to normal, the animal cools to within a degree or less of the ambient temperature. Metabolism decreases to a fraction of normal. In ground squirrels, for example, the respiratory rate decreases from a normal rate of 200 per minute to 4 or 5 per minute, and the heart rate from 150 to 5 beats per minute. During arousal a hibernator both shivers violently and employs nonshivering thermogenesis to produce heat.

Some mammals, such as bears, badgers, raccoons, and opossums, enter a state of prolonged sleep in winter with little or no decrease in body temperature. Prolonged sleep is not true hibernation. Bears of the northern forest sleep for several months. A bear's heart rate may decrease from 40 to 10 beats per minute, but body temperature remains normal and the bear is awakened if sufficiently disturbed. One intrepid but reckless biologist narrowly escaped injury when he crawled into a den and attempted to measure the bear's rectal temperature with a thermometer!

Figure 30-21
Hibernating woodchuck *Marmota monax* (order Rodentia) in den exposed by road-building work sleeps on, unaware of the intrusion. Woodchucks begin hibernating in late September while the weather is still warm and may sleep six months. The animal is rigid and decidedly cold to the touch. Breathing is imperceptible, as slow as one breath every five minutes. Although it appears to be dead, it will awaken if the den temperature drops dangerously low.

SUMMARY

Throughout life, matter and energy pass through the body, potentially disturbing the internal physiological state. Homeostasis, the ability of an organism to maintain internal stability despite such challenges, is a characteristic of all living systems. Homeostasis involves the coordinated activity of several physiological and biochemical mechanisms, and it is possible to relate some major events in animal evolution to increasing internal independence from the consequences of environmental change. In this chapter we have examined two aspects of homeostasis: (1) the varying ability of animals to stabilize the osmotic and chemical composition of the blood, and (2) the capacity of animals to regulate their temperatures in thermally challenging environments.

Most marine invertebrates must either depend on the osmotic stability of the ocean to which they conform, or be able to tolerate wide fluctuations in environmental salinity. Some of the latter show limited powers of osmotic regulation, the capacity to resist internal osmotic change through the evolution of specialized regulatory organs. All animals living in fresh water are hyperosmotic to their environment and have developed mechanisms for recovering salt from the environment and eliminating excess water that enters the body osmotically.

All vertebrate animals, except hagfishes, show excellent osmotic homeostasis. Marine bony fishes maintain their body fluids distinctly hypoosmotic to their environment by drinking seawater and physiologically distilling it. Elasmobranchs (sharks and rays) have adopted a strategy of near-osmotic conformity by retaining urea and trimethyl-amine oxide (TMAO) in the blood.

The kidney is the most important organ for regulating the chemical and osmotic composition of the blood. In all metazoa kidneys are some variation on a basic theme: a tubular structure that forms urine by introducing a fluid secretion or filtrate of the blood or interstitial fluid into a tubule in which it is selectively modified to form urine. Terrestrial vertebrates have especially sophisticated kidneys, since they must be able to regulate closely the water content of the blood by balancing gains and expenditures. The basic excretory unit is the nephron, composed of a glomerulus in which an ultrafiltrate of the blood is formed, and a long nephric tubule in which the formative urine is selectively modified by the tubular epithelium. Water, salts, and other valuable materials pass by reabsorption to the peritubular circulation, and certain wastes pass by secretion from the circulation to the tubular urine. All mammals and some birds can produce urine more concentrated than blood by means of a countercurrent multiplier system localized in the loops of Henle, a specialization not found in other vertebrates.

Temperature has a profound effect on the rate of biochemical reactions and, consequently, on the metabolism and activity of all animals. Animals may be classified according to whether body temperature is variable (poikilothermic) or stable (homeothermic), or by the source of body heat, whether external (ectothermic) or internal (endothermic).

Ectotherms partially free themselves from thermal constraints by seeking out habitats with favorable temperatures, by behavioral thermoregulation, or by adjusting their metabolism to the prevailing temperature through biochemical alterations.

Endothermic birds and mammals differ from ectotherms in having a much higher production of metabolic heat and a much lower conductance of heat from the body. They maintain constant body temperature by balancing heat production with loss.

Small mammals in hot environments for the most part escape intense heat and reduce evaporative water loss by burrowing. Large mammals employ several strategies for dealing with direct exposure to heat, including reflective insulation, heat storage by the body, and evaporative cooling.

Endotherms in cold environments maintain body temperature by decreasing heat loss with thickened pelage or plumage, by peripheral cooling, and by increasing heat production through shivering or nonshivering thermogenesis. Small endotherms may avoid exposure to low temperatures by living under the snow.

Adaptive hypothermia is a strategy used by small mammals and birds to blunt energy demands during periods of inactivity (daily torpor) or periods of prolonged cold and minimal food availability (hibernation).

REVIEW QUESTIONS

1. Define homeostasis. What evolutionary advantages for a species might result from successful maintenance of internal homeostasis?
2. The problems of water balance may have arisen when early metazoan animals began invading estuaries and rivers. Describe the physiological challenges confronting marine invertebrates entering fresh water and, using crustaceans as an example, suggest solutions to these challenges.
3. Distinguish between the following pairs of terms: osmotic conformity and osmotic regulation; stenohaline and euryhaline; hyperosmotic and hypoosmotic.
4. Young downstream salmon migrants moving from their freshwater natal streams into the sea leave an environment nearly free of salt to enter one containing three times as much salt as their body fluids. Describe osmotic challenges of each environment and physiological adjustments salmon must make in moving from fresh water to the sea.
5. Most marine invertebrates are osmotic conformers. How does their body fluid differ from that of sharks and rays, which are also in near osmotic equilibrium with their environment?
6. What strategy does a kangaroo rat use that allows it to exist in the desert without drinking any water?
7. In what animals would you expect to find a salt gland? What is its function?
8. Relate the function of contractile vacuoles to the following experimental observations: to expel an amount of fluid equal in volume to the volume of the animal required 4 to 53 minutes for some freshwater protozoa, and between 2 and 5 hours for some marine species.
9. How does a protonephridium differ structurally and functionally from a true nephridium (metanephridium)? In what ways are they similar?
10. Describe the developmental stages of kidneys in amniotes. How does the developmental sequence for amniotes differ from that of amphibians and fishes?
11. In what ways does the nephridium of an earthworm parallel the human nephron in structure and function?
12. Describe what happens during the following stages in urine formation in the mammalian nephron: filtration, tubular reabsorption, tubular secretion.
13. Explain how the cycling of sodium chloride between the descending and ascending limbs of the loop of Henle in the mammalian kidney, and special permeability of these tubules, produces high osmotic concentrations in interstitial fluids in the kidney medulla. Explain the role of urea in producing high osmotic concentrations in the interstitial fluid of the medulla.
14. Explain how antidiuretic hormone (vasopressin) controls excretion of water in mammalian kidneys.
15. Define the following terms and comment on the limitations (if any) of each in describing the thermal relationships of animals to their environments: poikilothermy, homeothermy, ectothermy, endothermy.
16. Defend the statement: "Both ectotherms and endotherms achieve metabolic homeostasis in unstable thermal environments, but they do so by employing different physiological strategies."
17. Large mammals live successfully in deserts and in the arctic. Describe different adaptations mammals use to maintain homeothermy in each environment.
18. Explain why it is advantageous for certain small birds and mammals to abandon homeothermy during brief or extended periods of their lives.

SELECTED REFERENCES

Beauchamp, G. K. 1987. The human preference for excess salt. Am. Sci. **75:**27–33. *Humans consume much more salt than nutritionally required; such preference for elevated salt level in food is learned from early dietary experience.*

Cossins, A. R., and K. Bowler. 1987. Temperature biology of animals. London, Chapman and Hall. *Comprehensive treatment of both ectotherms and endotherms.*

Dantzler, W. H. 1989. Comparative physiology of the vertebrate kidney. Berlin, Springer-Verlag. *Comprehensive review of vertebrate renal function.*

Heinrich, B. 1996. The thermal warriors: strategies of insect survival. Cambridge, Massachusetts, Harvard University Press. *Describes the many fascinating ways that insects respond to their temperature environment.*

Louw, G. N. 1993. Physiological animal ecology. New York, Longman Scientific & Technical. *Clearly presented survey with emphasis on thermoregulation and water relations in animals.*

Randall, D., W. Burggren, and K. French. 2001. Eckert animal physiology: mechanisms and adaptations. New York, W. H. Freeman and Company. *A comprehensive and comparative treatment of animal physiology.*

Riegel, J. A. 1972. Comparative physiology of renal excretion. New York, Hafner Publishing Company. *Excellent survey of excretory systems both vertebrate and invertebrate.*

Sands, J. M. 1999. Urea transport: it's not just "freely diffusible" anymore. News Physiol. Sci. **14**:46–47. *Summarizes studies that have reported urea transport proteins in the collecting duct of the nephron.*

Schmidt-Nielsen, K. 1981. Countercurrent systems in animals. Sci. Am. **244**:118–128 (May). *Explains how countercurrent systems transfer heat, gases, or ions between fluids moving in opposite directions.*

Schultz, S. G. 1996. Homeostasis, Humpty Dumpty and integrative biology. News Physiol. Sci. **11**:238–246. *Describes the central role that homeostasis plays in the study of physiological systems.*

Smith, H. W. 1953. From fish to philosopher. Boston, Little, Brown & Company. *Classic account of vertebrate kidney evolution.*

Storey, K. B., and J. M. Storey. 1990. Frozen and alive. Sci. Am. **263**:92–97 (Dec.). *Explains how many animals have evolved strategies for surviving complete or almost complete freezing during the winter months.*

Willmes, P., G. Stone, and I. Johnston. 2000. Environmental physiology of animals. Oxford, Blackwell Science Ltd. *Well-written information regarding environmental adaptations in both vertebrates and invertebrates.*

ZOOLOGY LINKS TO THE INTERNET

Visit the textbook's Online Learning Center at www.mhhe.com/zoology to find live Internet links for each of the topics listed here.

Homeostasis

Water and Osmotic Regulation in Terrestrial Organisms

Water and Osmotic Regulation in Aquatic Organisms

Vertebrates: Macroscopic Anatomy of Excretory Organs

Vertebrates: Microscopic Anatomy of Excretory Organs

Temperature Regulation

Thermal Relations

Biological Design

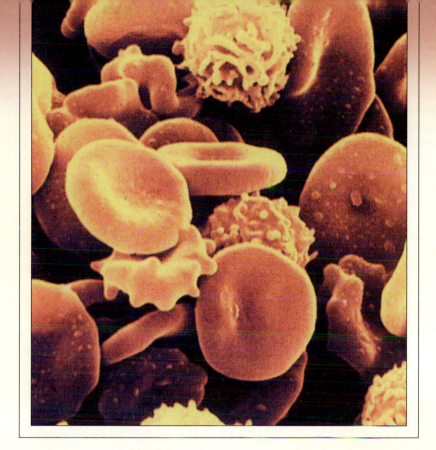

Scanning electron micrograph of blood cells.

Internal Fluids and Respiration

William Harvey's Discovery

Ceaselessly, during a human life, the heart pumps blood through arteries, capillaries, and veins: about 5 liters per minute, until by the end of a normal life the heart has contracted some 2.5 billion times and pumped 300,000 tons of blood. When the heart stops its contractions, life also ends.

The crucial importance of the heart and its contractions for human life has been known since antiquity, probably almost as long as humans have existed. However, the circuit flow of blood, the notion that the heart pumps blood into arteries through the circulation and receives it back in veins became known only a few hundred years ago. The first correct description of blood flow by the English physician William Harvey initially received vigorous opposition when published in 1628. Centuries earlier, the Greek anatomist Galen had taught that air enters the heart from the windpipe and that blood was able to pass from one ventricle to the other through "pores" in the interventricular septum. He also believed that blood first flowed out of the heart into all vessels, then returned—a kind of ebb and flow of blood. Even though there was almost nothing correct about this concept, it was still doggedly trusted at the time of Harvey's publication. Harvey's conclusions were based on sound experimental evidence. He used a variety of animals for his experiments and chided human anatomists, saying that if only they had acquainted themselves with anatomy of lower vertebrates, they would have understood the blood's circuit. By tying ligatures on arteries, he noticed that the region between the heart and ligature swelled. When veins were tied off, the swelling occurred beyond the ligature. When blood vessels were cut, blood flowed in arteries from the cut end nearest the heart; the reverse happened in veins. By means of such experiments, Harvey discovered the correct scheme of blood circulation, even though he could not see the capillaries that connected the arterial and venous flows.

ingle-celled organisms live in direct contact with their environments. They obtain nutrients and oxygen and release wastes directly across their cell surface. These organisms are so small that no special internal system of transport, beyond normal streaming movements of cytoplasm, is required. Even some simple multicellular forms, such as sponges, cnidarians, and flatworms, lack the internal complexity and metabolic demands that would require a circulatory system. Most other multicellular organisms, because of their increased size, activity, and complexity, need a specialized circulatory system to transport nutrients and respiratory gases to and from all tissues of their body. In addition to serving these primary transport needs, circulatory systems have acquired additional functions; hormones are moved from glands that produce them, to target organs where they act in concert with the nervous system to integrate organismal function. Water, electrolytes, and the many other constituents of body fluids are distributed and exchanged between different organs and tissues. An effective response to disease and injury is vastly accelerated by an efficient circulatory system. Homeothermic birds and mammals depend heavily on blood circulation to conserve or dissipate heat as required for maintenance of constant body temperature.

INTERNAL FLUID ENVIRONMENT

The body fluid of a single-celled organism is cellular cytoplasm, a liquid-gel substance in which the various membrane systems and organelles are suspended. In multicellular animals body fluids are divided into two main phases, **intracellular** and **extracellular.** The intracellular phase (also called intra-

cellular fluid) is the collective fluid inside all the body's cells. The extracellular phase (or fluid) is the fluid outside and surrounding the cells (Figure 31-1A). Thus the cells, sites of the body's crucial metabolic activities, are bathed by their own aqueous environment, the extracellular fluid that buffers them from often harsh physical and chemical changes occurring outside the body. The importance of extracellular fluid was first emphasized by the great French physiologist Claude Bernard (Figure 31-2). In animals having closed circulatory systems (vertebrates, annelids, and a few other invertebrate groups; see p. 337) extracellular fluid is further subdivided into **blood plasma** and **interstitial (intercellular) fluid** (Figure 31-1A). Blood vessels contain plasma, whereas interstitial fluid, or tissue fluid as it is sometimes called, occupies spaces surrounding the cells in the body. Nutrients and gases passing between vascular plasma and cells must traverse this narrow fluid separation. Interstitial fluid is constantly formed from plasma by filtration through capillary walls.

Composition of the Body Fluids

All these fluid spaces—plasma, interstitial, and intracellular—differ from each other in solute composition, but all have one feature in common: they are mostly water. Despite their firm appearance, animals are 70% to 90% water. Humans, for example, are approximately 70% water by weight. Of this, 50% is cell water, 15% is interstitial-fluid water, and the remaining 5% is in blood plasma. Plasma serves as the pathway of exchange between the cells of the body and the outside world. This exchange of respiratory gases, nutrients, and wastes is accom-

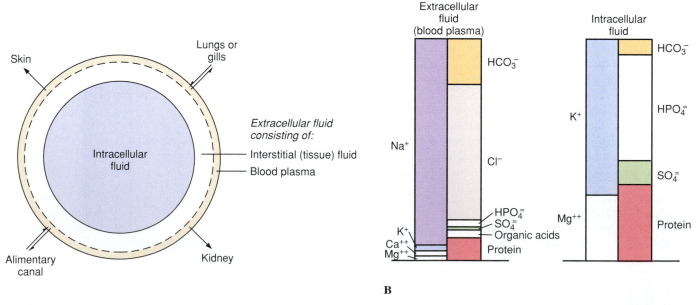

A   **B**

Figure 31-1

Fluid compartments of the body. **A,** All body cells can be represented as belonging to a single large fluid compartment that is completely surrounded and protected by extracellular fluid (*milieu intérieur*). This fluid is further subdivided into plasma and interstitial fluid. All exchanges with the environment occur across the plasma compartment. **B,** Electrolyte composition of extracellular and intracellular fluids. Total equivalent concentration of each major constituent is shown. Equal amounts of anions (negatively charged ions) and cations (positively charged ions) are in each fluid compartment. Note that sodium and chloride, major plasma electrolytes, are virtually absent from intracellular fluid (actually they are present in low concentration). Note the much higher concentration of protein inside cells.

Figure 31-2
French physiologist Claude Bernard (1813 to 1878), one of the most influential of nineteenth-century physiologists. Bernard believed in the constancy of the *milieu intérieur* ("internal environment"), which is the extracellular fluid bathing the cells. He emphasized that it is through the *milieu intérieur* that foods and wastes and gases are exchanged and through which chemical messengers are distributed. He wrote, "The living organism does not really exist in the external environment (the outside air or water) but in the liquid *milieu intérieur* . . . that bathes the tissue elements."

plished by specialized organs (kidney, lung, gill, alimentary canal), as well as by skin (Figure 31-1A).

Body fluids contain many inorganic and organic substances in solution. Principal among these are inorganic electrolytes and proteins. **Sodium, chloride,** and **bicarbonate ions** are the chief extracellular electrolytes, whereas **potassium, magnesium,** and **phosphate ions** and **proteins** are the major intracellular electrolytes (Figure 31-1B). These differences are dramatic; they are always maintained despite continuous flow of materials into and out of cells of the body. The two subdivisions of extracellular fluid—plasma and interstitial fluid—have similar compositions except that plasma has more proteins, which are mostly too large to filter through capillary walls into interstitial fluid.

COMPOSITION OF BLOOD

Among invertebrates that lack a circulatory system (such as flatworms and cnidarians) it is not possible to distinguish a true "blood." These organisms possess a clear, watery tissue fluid containing some phagocytic cells, a little protein, and a mixture of salts similar to seawater. The "blood" of invertebrates with open circulatory systems is more complex and is often called hemolymph (Gr. *haimo,* blood, + L. *lympha,* water). Invertebrates with closed circulatory systems, on the other hand, maintain a clear separation between blood contained within blood vessels and tissue (interstitial) fluid surrounding blood vessels.

In vertebrates, blood is a complex liquid tissue composed of plasma and formed elements, mostly red cells (also called corpuscles), suspended in plasma. If we separate red blood cells and other formed elements from the fluid components by centrifugation, we find that blood is approximately 55% plasma and 45% formed elements.

The composition of mammalian blood is as listed here:

Plasma

1. Water 90%
2. Dissolved solids, consisting of plasma proteins (albumin, globulins, fibrinogen), glucose, amino acids, electrolytes, various enzymes, antibodies, hormones, metabolic wastes, and traces of many other organic and inorganic materials
3. Dissolved gases, especially oxygen, carbon dioxide, and nitrogen

Formed elements (Figure 31-3)

1. Red blood cells (erythrocytes), containing hemoglobin for transport of oxygen and carbon dioxide
2. White blood cells (leukocytes), serving as scavengers and as defensive cells
3. Cell fragments (platelets in mammals) or cells (thrombocytes in other vertebrates) that function in blood coagulation

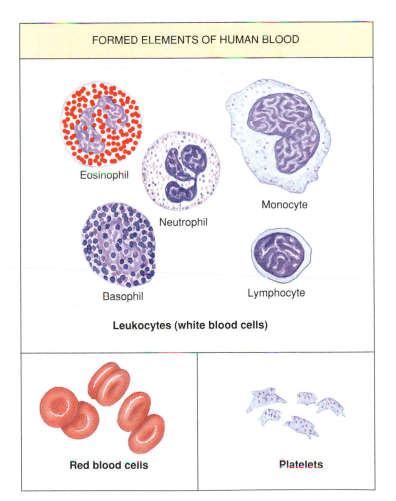

Figure 31-3
Formed elements of human blood. Hemoglobin-containing red blood cells of humans and other mammals lack nuclei, but those of all other vertebrates have nuclei. Various leukocytes provide a wandering system of protection for the body. Platelets participate in the blood's clotting mechanism.

Plasma proteins are a diverse group of large and small proteins that perform numerous functions. The major protein groups are (1) **albumins,** the most abundant group, constituting 60% of the total, which help to keep plasma in osmotic equilibrium with the cells of the body; (2) **globulins,** a diverse group of high-molecular weight proteins (35% of total) immunoglobulins (see Chapter 35, p. 741) and various metal-binding proteins; and (3) **fibrinogen,** a very large protein that functions in blood coagulation. Blood **serum** is plasma minus the proteins involved in clot formation.

Red blood cells, or **erythrocytes,** are present in enormous numbers in blood, approximately 54 million per milliliter of blood in adult men and 48 million in adult women. In mammals and birds, red cells form continuously from large nucleated **erythroblasts** in red bone marrow (in other vertebrates kidneys and spleen are the principal sites of red blood cell production). During erythrocyte formation hemoglobin is synthesized and the precursor cells divide several times. In mammals the nucleus shrinks during development to a small remnant and eventually is lost from the cell by exocytosis. Many other characteristics of a typical cell also are lost: ribosomes, mitochondria, and most enzyme systems. What is left is a biconcave disc consisting of a baglike membrane packed with about 280 million molecules of the blood-transporting pigment hemoglobin. Approximately 33% of an erythrocyte by weight is hemoglobin. The biconcave shape (Figure 31-3) is a mammalian innovation that provides a larger surface for gas diffusion than would a flat or spherical shape. All other vertebrates have nucleated erythrocytes that are usually ellipsoidal in shape (Figure 31-4).

An erythrocyte enters the circulation for an average life span of approximately 4 months. During this time it may journey 11,000 km, squeezing repeatedly through capillaries, which are sometimes so narrow that the erythrocyte must bend to pass through. At last it fragments and is quickly engulfed by large scavenger cells called **macrophages** located in the liver, bone marrow, and spleen. Iron from hemoglobin is salvaged to be used again; the rest of the heme is converted to **bilirubin,** a bile pigment. It is estimated that a human body produces 10 million erythrocytes and destroys another 10 million every second.

White blood cells, or **leukocytes,** form a wandering system of protection for the body. In adults they number only approximately 50,000 to 100,000 per milliliter of blood, a ratio of 1 white cell to 500 to 1000 red cells. There are several kinds of white blood cells: **granulocytes** (subdivided into **neutrophils, basophils,** and **eosinophils**), and **agranulocytes,** the lymphocytes and monocytes (Figure 31-3). We discuss the role of leukocytes in the body's defense mechanisms in Chapter 35.

Hemostasis: Prevention of Blood Loss

It is essential that animals have ways of preventing rapid loss of body fluids after an injury. Since blood is flowing and is under considerable hydrostatic pressure, it is especially vulnerable to hemorrhagic loss.

When a vessel is damaged, smooth muscle in the wall of the vessel contracts, which causes the vessel lumen to narrow, sometimes so strongly that blood flow is completely stopped.

Figure 31-4

Mammalian and amphibian red blood cells. **A,** Erythrocytes of a gerbil are biconcave discs containing hemoglobin and surrounded by a tough stroma. **B,** Frog erythrocytes are convex discs, each containing a nucleus, which is plainly visible in the scanning electron micrograph as a bulge in the center of each cell. (Magnifications: mammalian erythrocytes, ×6300; frog erythrocytes, ×2400.)

This simple but highly effective means of preventing hemorrhage is used by invertebrates and vertebrates alike. Beyond this first defense against blood loss, all vertebrates, as well as some larger, active invertebrates with high blood pressures, have in the blood special cellular elements and proteins that are capable of forming plugs, or clots, at an injury site.

In vertebrates **blood coagulation** is the dominant hemostatic defense. Blood clots form as a tangled network of fibers from one of the plasma proteins, **fibrinogen.** The transformation of fibrinogen into a **fibrin** meshwork (Figure 31-5) that entangles blood cells to form a gel-like clot is catalyzed by the enzyme thrombin. Thrombin is normally present in blood in an inactive form called **prothrombin,** which must be activated for coagulation to occur.

In this process, blood platelets (Figure 31-3) play a vital role. Platelets form in red bone marrow from large, multinucleated

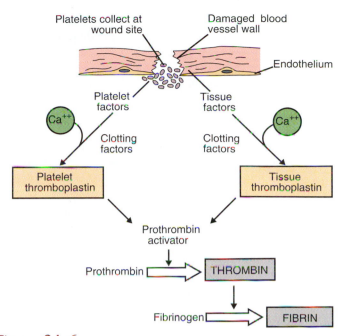

Figure 31-6
Stages in formation of fibrin.

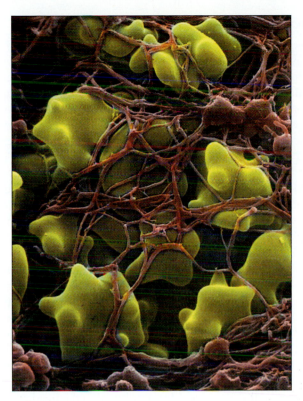

Figure 31-5
Human red blood cells trapped in fibrin clot. Clotting is initiated after tissue damage by disintegration of platelets in blood, resulting in a complex series of intravascular reactions that end with conversion of a plasma protein, fibrinogen, into long, tough, insoluble polymers of fibrin. Fibrin and entangled erythrocytes form the blood clot, which arrests bleeding.

cells that regularly pinch off bits of their cytoplasm; thus they are fragments of cells. There are 150,000 to 300,000 platelets per cubic millimeter of blood. When the normally smooth inner surface of a blood vessel is disrupted, either by a break or by deposits of a cholesterol-lipid material, platelets rapidly adhere to the surface and release **thromboplastin** and other clotting factors. These factors, as well as factors released from damaged tissue and calcium ions, initiate conversion of prothrombin to active thrombin (Figure 31-6).

The catalytic sequence in this scheme is unexpectedly complex, involving a series of plasma protein factors, each normally inactive until activated by a previous factor in the sequence. The sequence behaves like a "cascade" with each reactant in the sequence leading to a large increase in the amount of the next reactant. At least 13 different plasma coagulation factors have been identified. A deficiency of only a single factor can delay or prevent the clotting process. Why has such a complex clotting mechanism evolved? Probably it is necessary to provide a fail-safe system capable of responding to any kind of internal or external hemorrhage that might occur and yet a system that cannot be activated into forming dangerous intravascular clots in the absence of injury.

Several kinds of clotting abnormalities in humans are known. One of these, hemophilia, is a condition characterized by failure of blood to clot, so that even insignificant wounds can cause continuous severe bleeding. It is caused by a rare mutation (the condition occurs in about 1 in 10,000 males) on the X sex chromosome, resulting in an inherited lack of one of the platelet factors in males and in homozygous females. Called the "disease of kings," it once ran through several interrelated royal families of Europe, apparently having originated from a mutation in one of Queen Victoria's parents.

Hemophilia is one of the best known cases of sex-linked inheritance in humans (p. 84). Actually two different loci on the X chromosome are involved. Classic hemophilia (hemophilia A) accounts for about 80% of persons with the condition, and the remainder are caused by Christmas disease (hemophilia B). The allele at each locus results in a deficiency of a different platelet factor.

CIRCULATION

Most animals have evolved mechanisms, in addition to simple diffusion, for transporting materials among various regions of their body. For sponges and radiates the water in which they live provides the medium for transport. Water, propelled by ciliary, flagellar, or body movements, passes through channels or compartments to facilitate the movement of food, respiratory gases, and wastes. True circulatory systems—containing vessels through which blood moves—are essential to animals so large or so active that diffusional processes alone cannot supply their oxygen needs. An animal's shape obviously is important. Flattened and leaflike acoelomate flatworms, even though many are relatively large animals, have no need for a circulatory system because the distance of any body part from the surface is short; respiratory gases and metabolic wastes transfer by simple diffusion.

A circulatory system having a full complement of components—propulsive organ, arterial distribution system, capillaries, and venous reservoir and return system—is fully recognizable in annelid worms. In earthworms (Figure 31-7) there are two main vessels, a dorsal vessel carrying blood toward the head, and a ventral vessel that flows posteriorly, delivering blood throughout the body by way of segmental vessels and a dense capillary network. The dorsal vessel drives the blood forward by peristalsis (see p. 683) and thus serves as a heart. Five aortic arches that on each side connect the dorsal and ventral vessels are also contractile and serve as accessory hearts to maintain a steady flow of blood into the ventral vessel. Many smaller segmental vessels that deliver blood to tissue capillaries are actively contractile as well. We see then that there is no localized pump pushing the blood through a system of passive tubes; instead the power of contraction is widely distributed throughout the vascular system.

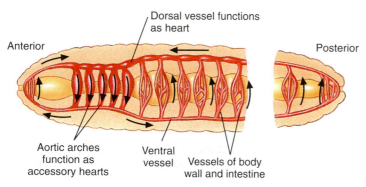

Anterior

Dorsal vessel functions as heart

Posterior

Aortic arches function as accessory hearts

Ventral vessel

Vessels of body wall and intestine

Figure 31-7

Blood flow through the closed vascular system of an earthworm.

Open and Closed Circulations

The system just described is a **closed circulation** because the circulating medium, **blood,** is confined to vessels throughout its journey through the vascular system. Many invertebrates have an **open circulation** in which there are no small blood vessels or capillaries connecting arteries with veins. In insects and other arthropods, in most molluscs, and in many smaller invertebrate groups blood sinuses, collectively called a **hemocoel,** replace capillary beds found in animals with closed systems. During development of the body cavity in these groups, the blastocoel is not completely obliterated by the expanding mesoderm. This space becomes the hemocoel, which is nothing more than the primary body cavity (persistent blastocoel) through which blood (also called **hemolymph**) freely circulates (bottom diagrams in Figure 31-8). Since there is no separation of the extracellular fluid into blood plasma and lymph (as there is in a closed circulation, p. 666), blood volume is large and may constitute 20% to 40% of body volume. By contrast, blood volume in animals with closed circulations (vertebrates, for example) is only about 5% to 10% of body volume.

In arthropods, the heart and all viscera lie in the hemocoel, bathed by blood (Figure 31-8). Blood enters the heart through valved openings, the ostia, and the heart's contractions, which resemble a forward-moving peristaltic wave, propel blood into a limited arterial system. Blood is distributed to the head and other organs, then escapes into the hemocoel. It is routed through the body and appendages by a system of baffles and longitudinal membranes (septa) before returning to the heart. Because blood pressure is very low in open systems, seldom exceeding 4 to 10 mm Hg, many arthropods have auxiliary hearts or contractile vessels to boost blood flow (Figure 31-9).

During embryonic development of animals with closed circulatory systems (most annelids, cephalopod molluscs, and all vertebrates) the coelom increases in size to obliterate the blastocoel

Figure 31-8

Diagrams showing how open and closed circulatory systems develop. The principal body cavity of arthropods is the persistent blastocoel, which becomes a hemocoel; the true coelom remains mostly undeveloped.

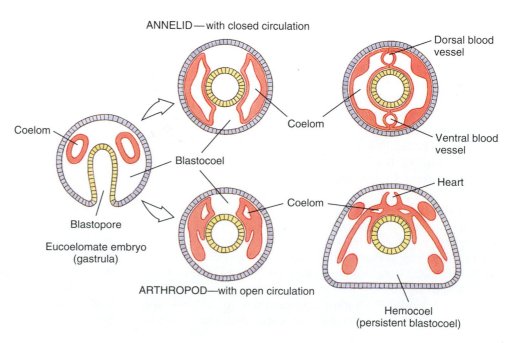

ANNELID—with closed circulation

Coelom

Blastopore

Eucoelomate embryo (gastrula)

Coelom

Blastocoel

Dorsal blood vessel

Ventral blood vessel

Coelom

Heart

ARTHROPOD—with open circulation

Hemocoel (persistent blastocoel)

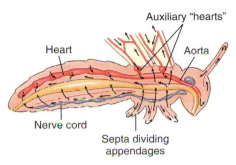

Figure 31-9

Circulatory system of an insect. Although the circulatory system is open, blood is directed through the appendages in channels formed by longitudinal septa. Arrows indicate the course of circulation.

and forms a secondary body cavity (top diagrams in Figure 31-8). A system of continuously connected blood vessels develops within the mesoderm. All closed systems have certain features in common. A **heart** pumps blood into **arteries** that branch and narrow into **arterioles** and then into a vast system of **capillaries.** Blood leaving capillaries enters **venules** and then **veins** that return the blood to the heart. Capillary walls are thin, permitting rapid rates of transfer of materials between blood and tissues. Closed systems are more suitable for large and active animals because blood can be moved rapidly to tissues needing it. In addition, flow to various organs can be readjusted to meet changing needs by varying the diameters of blood vessels.

Because blood pressures are much higher in closed than in open systems, fluid is constantly filtered across capillary walls into the surrounding tissue spaces. Most of this fluid is drawn back into capillaries by osmosis (see p. 665). The remainder is recovered by a **lymphatic system,** which has evolved separately but in conjunction with the high-pressure system of vertebrates.

Plan of Vertebrate Circulatory Systems

In vertebrates the principal differences in the blood-vascular system involve the gradual separation of the heart into two separate pumps as vertebrates evolved from aquatic life with gill breathing to fully terrestrial life with lung breathing. These changes are shown in Figure 31-10, which compares the circulation of fish, amphibians, and mammals.

A fish heart contains two main chambers in series, an **atrium** and a **ventricle.** The atrium is preceded by an enlarged chamber, the **sinus venosus,** which collects blood from the venous system to assure a smooth delivery of blood to the heart. Blood makes a single circuit through a fish's vascular system; it is pumped from the heart to the gills, where it is oxygenated, then flows into the dorsal aorta to be distributed to body organs, and finally returns by veins to the heart. In this circuit the heart must provide sufficient pressure to push the blood through two sequential capillary systems, first that of the gills, and then that of the remainder of the body. The principal disadvantage of the single-circuit system is that gill capillaries offer so much resistance to blood flow that blood pressures to body tissues are greatly reduced.

With evolution of lung breathing and elimination of gills between the heart and aorta, vertebrates developed a high-pressure **double circulation:** a **systemic circuit** that provides oxygenated blood to the capillary beds of the body organs; and a **pulmonary circuit** that serves the lungs. The beginning of this major evolutionary change probably resembled the condition seen in lungfishes and amphibians. In modern amphibians (frogs, toads, salamanders) the atrium is completely separated by a partition into two atria (Figure 31-11). The right atrium receives venous blood from the body while the left atrium receives oxygenated blood from the lungs and skin. The ventricle is undivided, but venous and arterial blood remain mostly separate by the arrangement of vessels leaving the heart and

Figure 31-10

Circulatory systems of fish, amphibian, and mammal, showing evolution of separate systemic and pulmonary circuits in lung-breathing vertebrates.

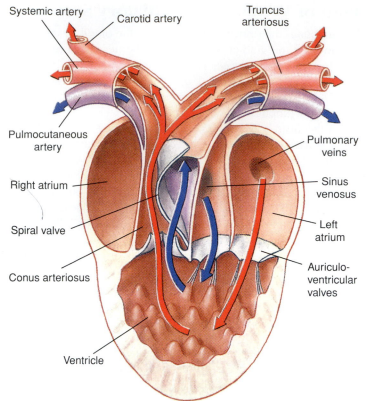

Figure 31-11

Route of blood through a frog heart. Atria are completely separated, and the spiral valve helps to route blood to lungs and systemic circulation.

Labels for Figure 31-11: Systemic artery, Carotid artery, Truncus arteriosus, Pulmocutaneous artery, Right atrium, Spiral valve, Conus arteriosus, Ventricle, Pulmonary veins, Sinus venosus, Left atrium, Auriculo-ventricular valves

Labels for Figure 31-12: (Anterior), Superior vena cava, right pulmonary arteries, Right atrium, Tricuspid Valve, Right atrioventricular (AV) valve, Papillary muscles, (Posterior), Inferior vena cava, (3) right ventricle, Aortic semilunar valve, Aorta (8), Left pulmonary arteries (6), Pulmonary semilunar valve, (5), Pulmonary veins, Left atrium (2), Bicuspid Valve, Left atrioventricular (AV) valve, (4), Left ventricle, Aorta, top of body, down feet, legs

Figure 31-12

Human heart. Deoxygenated blood enters right side of heart and is pumped to the lungs. Oxygenated blood returning from the lungs enters left side of the heart and is pumped to the body. The left ventricular wall is thicker than that of the right ventricle, which needs less muscular force to pump blood into the nearby lungs.

differential blood pressures in these vessels. Separation of the ventricles is nearly complete in some reptiles (crocodilians) and is completely separate in birds and mammals (Figure 31-12). Systemic and pulmonary circuits are now separate circulations, each served by one half of a dual heart (Figure 31-12).

Mammalian Heart

The four-chambered mammalian heart (Figure 31-12) is a muscular organ located in the thorax and covered by a tough, fibrous sac, the **pericardium.** Blood returning from the lungs collects in the **left atrium,** passes into the **left ventricle,** and is pumped into the body (systemic) circulation. Blood returning from the body flows into the **right atrium,** and passes into the **right ventricle,** which pumps it into the lungs. Backflow of blood is prevented by two sets of valves that open and close passively in response to pressure differences between the heart chambers. The **left atrioventricular** (bicuspid) and the **right atrioventricular** (tricuspid) valves separate the cavities of the atrium and ventricle in each half of the heart. Where the great arteries, the **pulmonary** from the right ventricle and the **aorta** from the left ventricle, leave the heart, **semilunar valves** prevent backflow into the ventricles.

Contraction is called **systole** (sis′to-lee), and relaxation, **diastole** (dy-as′to-lee) (Figure 31-13). When the atria contract (atrial systole), the ventricles relax (ventricular diastole), and ventricular systole is accompanied by atrial diastole. Rate of the heartbeat depends on age, sex, and especially exercise. Exercise may increase **cardiac output** (volume of blood forced from either ventricle each minute) more than fivefold. Both heart **rate** and **stroke volume** (volume of blood forced from either ventricle per beat) increase. Heart rates among vertebrates vary with general level of metabolism and body size. Ectothermic codfish have a heart rate of approximately 30 beats per minute; endothermic rabbits of about the same weight have a rate of 200 beats per minute. Small animals have higher heart rates than do large animals. The heart rate in an elephant is 25 beats per minute, in a human 70 per minute, in a cat 125 per minute, in a mouse 400 per minute, and in the tiny 4 g shrew, the smallest mammal, the heart rate approaches a prodigious 800 beats per minute. We must marvel that a shrew's heart can sustain such a frantic pace throughout this animal's life, brief as it is.

**Atrial systole
Ventricular diastole**

**Atrial diastole
Ventricular systole**

Figure 31-13
Human heart in systole and diastole.

Excitation and Control of the Heart

The vertebrate heart is a muscular pump composed of **cardiac muscle.** Cardiac muscle resembles skeletal muscle—both are types of striated muscle—but cardiac cells are branched and joined end-to-end by junctional complexes (utercalated discs) to form a complex branching network (see Figure 9-7, p. 184).

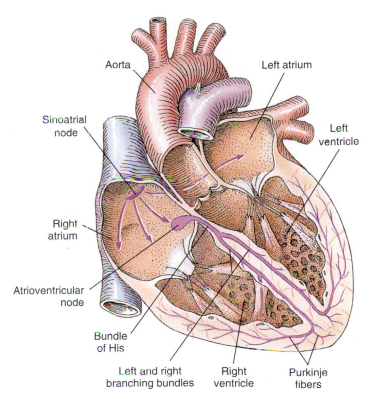

Figure 31-14
Neural mechanisms controlling heartbeat. Arrows indicate spread of excitation from the sinoatrial node, across the atria, to the atrioventricular node. Wave of excitation is then conducted very rapidly to ventricular muscle over the specialized conducting bundles and Purkinje fiber system.

Unlike skeletal muscle, vertebrate cardiac muscle does not depend on nerve activity to initiate a contraction. Instead, regular contractions are established by specialized cardiac muscle cells, called **pacemaker cells.** In a tetrapod heart the pacemaker is in the **sinoatrial (SA) node,** a remnant of the sinus venosus in the fishlike ancestor. Electrical activity initiated in the pacemaker spreads over the muscle of the two atria and then, after a slight delay, to the secondary pacemaker, the atrioventricular (AV) node, at the top of the ventricles. At this point electrical activity is conducted rapidly through the **bundle of His** and left and right branching bundles to the apex of the ventricle and then continues through specialized fibers **(Purkinje fibers)** up the walls of the ventricles (Figure 31-14). This arrangement allows the contraction to begin at the apex or "tip" of the ventricles and spread upward to squeeze out the blood in the most efficient way; it also ensures that both ventricles contract simultaneously. Structural specializations in Purkinje fibers, such as well-developed intercalated discs (see Figure 9-7, p. 184) and numerous gap junctions, facilitate rapid conduction through these fibers.

The **control (cardiac) center** in the brain is located in the medulla and connects to the heart by two sets of nerves. Impulses sent along the parasympathetic **vagus** nerves, apply a braking action to the heart rate, and impulses sent along the sympathetic nerves, speed it up. Both sets of nerves terminate in the SA node, thus guiding the activity of the pacemaker.

The cardiac center in turn receives sensory information about a variety of stimuli. Pressure receptors (sensitive to blood pressure) and chemical receptors (sensitive to carbon dioxide and pH) are located at strategic points in the vascular system. The cardiac center uses this information to increase or reduce heart rate and cardiac output in response to activity or changes in body position. Feedback mechanisms thus control the heart and keep its activity constantly attuned to needs of the body.

Because the heartbeat is initiated in specialized muscle cells, vertebrate hearts, together with the hearts of molluscs and several other invertebrates, are called **myogenic** ("muscle origin") hearts. Although the nervous system does alter pacemaker activity to slow down or speed up heart rate, a myogenic heart will beat spontaneously and involuntarily even if completely removed from the body. An isolated turtle or frog heart beats for hours if placed in a balanced salt solution. Some invertebrates, for example decapod crustaceans, have **neurogenic** ("nerve origin") hearts. In these hearts a cardiac ganglion located on the heart serves as pacemaker. If this ganglion is separated from the heart, the heart stops beating, even though the ganglion itself remains rhythmically active.

Coronary Circulation

It is no surprise that an organ as active as the heart needs a generous blood supply of its own. The heart muscle of frogs and other amphibians is so thoroughly channeled with spaces between muscle fibers that the heart's own pumping action squeezes through sufficient oxygenated blood. In birds and mammals, however, the thickness of the heart muscle and its high rate of metabolism require that the heart have its own vascular supply, the **coronary circulation.** Coronary arteries divide to form an extensive capillary network surrounding the muscle fibers and provide them with oxygen and nutrients. Heart muscle has an extremely high oxygen demand. Even at rest the heart removes 70% of oxygen from the blood, in contrast to most other body tissues, which remove only about 25%. Therefore, an increase in the work of the heart must be met by a massive increase in coronary blood flow—up to nine times the resting level during strenuous exercise. Any reduction in coronary circulation due to partial or complete blockage (coronary artery disease) may lead to a heart attack (myocardial infarction) in which heart cells die from lack of oxygen.

Arteries

All vessels leaving the heart are called arteries whether they carry oxygenated blood (aorta) or deoxygenated blood (pulmonary artery). To withstand high, pounding pressures, the largest arteries closest to the heart are invested with thick layers of elastic fibers, very little smooth muscle **(elastic arteries),** and tough inelastic connective fibers (Figure 31-15). The elasticity of these arteries allows them to yield to the surge of blood leaving the heart during ventricular systole and then to compress the fluid column during ventricular diastole. This elasticity maintains the high blood pressure created by the heartbeat. Thus the normal arterial pressure in humans varies only between 120 mm Hg

Figure 31-15
Artery and vein, showing layers. Note greater thickness of the muscularis layer (tunica media) in the artery.

(systole) and 80 mm Hg (diastole) (usually expressed as 120/80 or 120 over 80), rather than dropping to zero during diastole as we might expect in a fluid system with an intermittent pump. Arteries farther away from the heart possess more smooth muscle and less elastic fibers. These arteries, call **muscular arteries,** can increase or decrease their diameter and dampen the high pressure and flow oscillations that occur with each heartbeat before the blood reaches the body organs.

Thickening and loss of elasticity in arteries is known as *arteriosclerosis.* When arteriosclerosis is characterized by fatty deposits of cholesterol in artery walls, the condition is *atherosclerosis.* Scientists now think that inflammation precedes the accumulation of fat in the arteries. Irregularities in the walls of blood vessels often cause blood to clot around them, forming a *thrombus.* When a bit of the thrombus breaks off and is carried by the blood to lodge elsewhere, it is an *embolus.* If the embolus blocks one of the coronary arteries, the person has a heart attack (a "coronary"). The portion of the heart muscle served by the branch of the coronary artery that is blocked is starved for oxygen. It may be replaced by scar tissue if the person survives.

As arteries branch and narrow into **arterioles,** the walls are composed primarily of only one or two layers of smooth muscle. Contraction of this muscle narrows the arterioles and reduces the flow of blood to body organs, diverting it to where it is most needed. Blood must be pumped with a hydrostatic pressure sufficient to overcome resistance of the narrow passages through which it must flow. Consequently, large animals tend to have higher blood pressure than do small animals.

Blood pressure was first measured in 1733 by Stephan Hales, an English clergyman with unusual inventiveness and curiosity. He tied his mare, which was "to have been killed as unfit for service," on her back and exposed the femoral artery. This he cannulated with a brass tube, connecting it to a tall glass tube with the windpipe of a goose. The use of the windpipe was both imaginative and practical; it gave the apparatus flexibility "to avoid inconveniences that might arise if the mare struggled."

The blood rose 8 feet in the glass tube and bobbed up and down with systolic and diastolic beats of the heart. The weight of the 8-foot column of blood was equal to the blood pressure. We now express blood pressure as the height of a column of mercury (Hg), which is 13.6 times heavier than water. Hales' figures, expressed in millimeters of mercury, indicate that he measured a blood pressure of 180 to 200 mm Hg, about normal for a horse.

Today, we measure blood pressure in humans most commonly and easily with an instrument called a **sphygmomanometer.** We inflate a cuff on the upper arm with air to a pressure sufficient to close the arteries in the arm. Holding a stethoscope over the brachial artery (in the crook of the elbow) and slowly releasing air from the cuff, we can hear the first spurts of blood through the artery as it opens slightly. This is equivalent to systolic pressure. As pressure in the cuff decreases, the sound finally disappears as blood runs smoothly through the artery. The pressure at which the sound disappears is diastolic pressure.

Capillaries

The Italian Marcello Malpighi was the first to describe capillaries in 1661, thus confirming the existence of the minute links between the arterial and venous systems that Harvey knew must exist but could not see. Malpighi studied capillaries of a living frog's lung, which is still one of the simplest and most vivid preparations for demonstrating capillary blood flow.

Capillaries are present in enormous numbers, forming extensive networks in nearly all tissues (Figure 31-16). In muscle there are more than 2000 per square millimeter (1,250,000 per square inch), but not all are open at the same time. Indeed, perhaps less than 1% are open in resting skeletal muscle. But when muscle is active, all capillaries may open to bring oxygen and nutrients to the working muscle fibers and to carry away metabolic wastes.

Capillaries are extremely narrow, averaging in mammals about 8 µm in diameter, which is only slightly wider than the red blood cells that must pass through them. Their walls are formed by a single layer of thin **endothelial** cells, held together by a delicate basement membrane and few connective-tissue fibers.

Capillary Exchange

Capillaries are quite permeable to small ions, nutrients, and water. Blood pressure within a capillary tends to force fluids out through the capillary walls and into the surrounding interstitial space (p. 656). Because larger molecules such as plasma proteins cannot pass through the capillary wall, an almost protein-free filtrate is forced outward. This fluid movement is important in irrigating the interstitial space, in providing tissue cells with oxygen, glucose, amino acids, and other nutrients, and in carrying away metabolic wastes. For capillary exchange to be effective, fluids that leave the capillaries must at some point reenter the circulation. If they did not, fluid would quickly accumulate in tissue spaces, causing edema. The delicate balance of fluid exchange across the capillary wall is achieved by the two opposing forces of hydrostatic (blood) pressure and osmotic pressure (Figure 31-17).

In a capillary, the blood pressure that pushes water molecules and solutes across the capillary wall is greatest at the arteriolar end of the capillary and declines along its length as blood pressure falls (Figure 31-17). Opposing blood hydrostatic pressure is an osmotic pressure created by the proteins that cannot pass across the capillary wall. This **colloid osmotic pressure,**

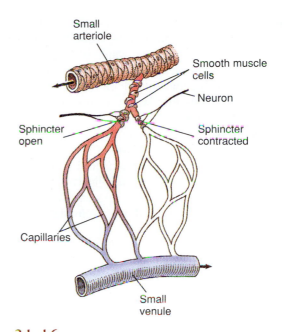

Figure 31-16

Capillary bed. Precapillary sphincters (muscles encircling an opening) control blood flow through capillaries.

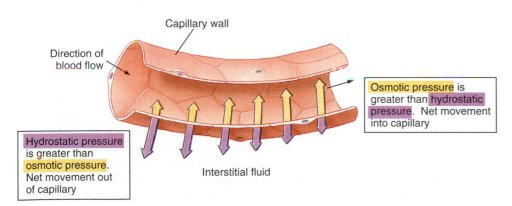

Hydrostatic pressure is greater than osmotic pressure. Net movement out of capillary

Osmotic pressure is greater than hydrostatic pressure. Net movement into capillary

Figure 31-17

Fluid movement across the wall of a capillary. At the arterial end of the capillary, hydrostatic (blood) pressure exceeds colloid osmotic pressure contributed by plasma proteins, and a plasma filtrate is forced outside the capillary. At the venous end, colloid osmotic pressure exceeds the hydrostatic pressure, and fluid is drawn inside the capillary. In this way plasma nutrients are carried into the interstitial space where they can enter cells, and metabolic end products from the cells are drawn into the plasma and carried away.

which is about 25 mm Hg in mammalian plasma, tends to draw water back into the capillary from the tissue fluid. The result of these two opposing forces is that water and solutes tend to be filtered out of the arteriolar end of the capillary where hydrostatic pressure exceeds colloid osmotic pressure, and to be drawn in again at the venous end where colloid osmotic pressure exceeds hydrostatic pressure.

The amount of fluid filtered across the capillary wall fluctuates greatly among different capillaries. Usually outflow exceeds inflow, and some excess fluid remains in the interstitial spaces between tissue cells. This excess is collected and removed by **lymph capillaries** of the lymphatic system and eventually this fluid, called **lymph,** is returned to the circulatory system via larger lymph vessels.

Veins

Venules and veins into which the capillary blood drains for its return journey to the heart are thinner walled, less elastic, and of considerably larger diameter than their corresponding arteries and arterioles (Figure 31-15). Blood pressure in the venous system is low, from approximately 10 mm Hg, where capillaries drain into venules, to approximately zero in the right atrium. Because pressure is so low, venous return gets assistance from valves in the veins, body muscles surrounding the veins, and the rhythmical action of the lungs. Without these mechanisms, blood will pool in the lower extremities of a standing animal—a very real problem for people who must stand for long periods. Veins that lift blood from the extremities to the heart contain valves that divide the long column of blood into segments. When skeletal muscles contract, as in even slight activity, veins are squeezed, and blood within them moves toward the heart because valves within the veins keep blood from slipping back. The well-known risk of fainting while standing at stiff attention in hot weather usually can be prevented by deliberately pumping leg muscles. Negative pressure in the thorax created by inspiratory movements of the lungs also speeds venous return by sucking blood up the large vena cava into the heart.

Lymphatic System

The lymphatic system of vertebrates is an extensive network of thin-walled vessels that arise as blind-ended lymph capillaries in most tissues of the body. These unite to form a treelike structure of increasingly larger lymph vessels, which finally drain into veins in the lower neck (Figure 31-18). A principal function of the lymphatic system is to return to the blood the excess fluid (lymph) filtered across capillary walls into interstitial spaces. This fluid-filtrate, called lymph, is similar to plasma but has a much lower concentration of protein. Large molecules, especially fats absorbed from the gut, also reach the circulatory system by way of the lymphatic system. The rate of lymph flow is very low, a minute fraction of blood flow.

The lymphatic system also plays a central role in the body's defenses. Located at intervals along the lymph vessels are **lymph nodes** (Figure 31-18) that have several defense-related

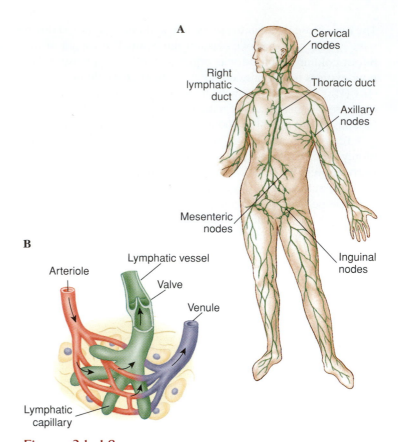

Figure 31-18

Human lymphatic system, showing major vessels, **A,** and a detail of the blood and lymphatic capillaries, **B.**

functions (Chapter 35). Cells in the lymph glands such as macrophages remove foreign particles, especially bacteria, which might otherwise enter the general circulation. They are also centers (together with bone marrow and thymus gland) for production, maintenance, and distribution of lymphocytes—essential components of a body's defense mechanisms.

RESPIRATION

Energy from food is released by oxidative processes, usually with molecular oxygen as the terminal electron acceptor. Oxygen for this purpose is taken into the body across a respiratory surface. Physiologists distinguish two separate but interrelated respiratory processes: **cellular respiration,** the oxidative processes that occur within cells (p. 61), and **external respiration,** the exchange of oxygen and carbon dioxide between the organism and its environment. In this section we describe external respiration and transport of gases from respiratory surfaces to body tissues.

In single-celled organisms, oxygen is acquired and carbon dioxide liberated by direct diffusion across surface membranes. Gas exchange by diffusion alone is possible only for very small organisms less than 1 mm in diameter, where diffusion paths are short and the surface area of the organism is large relative to volume. As animals became larger and evolved a waterproof covering, specialized devices such as lungs and gills evolved to increase the effective surface for gas exchange. In addition, because gases diffuse so slowly through living tissue, a circulatory system was necessary to distribute gases to and from deep tissues of the body. Even these adaptations were inadequate for complex animals with high rates of cellular respiration. The solubility of oxygen in the blood plasma is so low that plasma alone cannot carry enough oxygen to support metabolic demands. With evolution of special oxygen-transporting blood proteins such as hemoglobin, which seems to have evolved in conjunction with the circulatory system, the oxygen-carrying capacity of blood increased greatly. Thus what began as a simple and easily satisfied requirement resulted in evolution of several complex and essential respiratory and circulatory adaptations.

Problems of Aquatic and Aerial Breathing

How an animal respires is determined largely by the nature of its environment. The two great arenas of animal evolution—water and land—are vastly different in their physical characteristics. The most obvious difference is that air contains far more oxygen—at least 20 times more—than does water. For example, water at 5° C (41° F) fully saturated with air contains approximately 9 ml of oxygen per liter (0.9%); by comparison air contains 209 ml of oxygen per 1000 ml (21%). The density and viscosity of water are approximately 800 and 50 times greater, respectively, than that of air. Furthermore, gas molecules diffuse 10,000 times more rapidly in air than in water. These differences mean that aquatic animals must have evolved very efficient ways of removing oxygen from water. Yet even the most advanced fishes with highly efficient gills and pumping mechanisms may use as much as 20% of their energy just extracting oxygen from water. By comparison, the cost for mammals to breathe is only 1% to 2% of their resting metabolism.

Respiratory surfaces must be thin and always kept wet with a fine film of fluid to allow diffusion of gases across an aqueous phase between the environment and the underlying circulation. This is hardly a problem for aquatic animals, immersed as they are in water, but it is a challenge for air breathers. To keep respiratory membranes moist and protected from injury, air breathers have in general developed invaginations of the body surface and then added pumping mechanisms to move air in and out of the body. The lung is the best example of a successful solution to breathing on land. In general **evaginations** of the body surface, such as gills, are most suitable for aquatic respiration; **invaginations,** such as lungs and tracheae, are best for air breathing. We now consider some important examples of respiratory organs employed by animals.

Respiratory Organs

Gas Exchange by Direct Diffusion

Protozoa, sponges, cnidarians, and many worms respire by direct diffusion of gases between organism and environment. We have noted that this kind of **cutaneous respiration** is not adequate when the cellular mass exceeds approximately 1 mm in diameter. However, by greatly increasing the surface of the body relative to its mass, many multicellular animals can supply part or all of their oxygen requirements by direct diffusion. Flatworms are an example of this strategy. Cutaneous respiration frequently supplements gill or lung breathing in larger animals such as amphibians and fishes. For example, an eel can exchange 60% of its oxygen and carbon dioxide through its highly vascular skin. During winter hibernation, frogs and even turtles exchange all their respiratory gases through their skin while submerged in ponds or springs. Lungless salamanders comprise the largest family of salamanders. Some lungless salamanders have larvae with gills, and gills persist in the adults of some, but adults of most species have neither lungs nor gills.

Gas Exchange Through Tubes: Tracheal Systems

Insects and certain other terrestrial arthropods (centipedes, millipedes, and some spiders) have a highly specialized type of respiratory system, in many respects the simplest, most direct, and most efficient respiratory system found in active animals. It is a branching system of tubes **(tracheae)** that extends to all parts of the body (Figure 31-19). The smallest end channels are fluid-filled **tracheoles,** less than 1 μm in diameter, that terminate in close association with the plasma membranes of body cells. Air enters the tracheal system through valvelike openings **(spiracles).** Carbon dioxide diffuses out through spiracles. Some insects can ventilate the tracheal system with body movements; the familiar telescoping movement of the abdomen of bees on hot summer days is an example. Respiratory pigments are present in insect blood, but because the cells have a direct pipeline to the outside, bringing oxygen in and carrying carbon dioxide

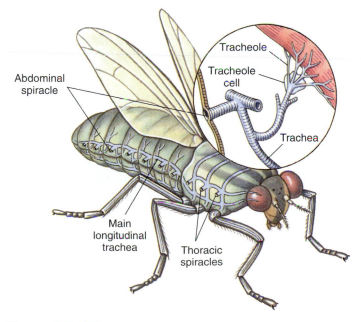

Figure 31-19

Tracheal system of insects. Air enters through spiracles, then travels through tracheae to reach tissues at tracheoles.

out, an insect's respiration is independent of its circulatory system. Consequently, insect blood plays no direct role in oxygen transport.

Efficient Exchange in Water: Gills

Gills of various types are effective respiratory devices for life in water. Gills may be simple **external** extensions of the body surface, such as **dermal papulae** of sea stars (p. 448) or **branchial tufts** of marine worms (p. 348) and aquatic amphibians (p. 520). Most efficient are **internal gills** of fishes (p. 504) and arthropods. Fish gills are thin filamentous structures, richly supplied with blood vessels arranged so that blood flow is opposite to the flow of water across the gills. This arrangement, called **countercurrent flow** (p. 645), provides the greatest possible extraction of oxygen from water. Water flows over the gills in a steady stream, pulled and pushed by an efficient, two-valved, branchial pump (Figure 31-20). Gill ventilation is often assisted by the fish's forward movement through the water (ram ventilation).

Lungs

Gills are unsuitable for life in air because, when removed from the buoying water medium, gill filaments collapse, dry, and stick together; a fish out of water rapidly asphyxiates despite the abundance of oxygen around it. Consequently most air-breathing vertebrates possess lungs, highly vascularized internal cavities. Lungs of a sort are found in certain invertebrates (pulmonate snails, scorpions, some spiders, some small crustaceans), but these structures cannot be very efficiently ventilated.

Lungs that can be ventilated by muscle movements to produce a rhythmic exchange of air are characteristic of terrestrial vertebrates. Most rudimentary of vertebrate lungs are those of lungfishes (Dipneusti), which use them to supplement, or even replace, gill respiration during periods of drought. Although of simple construction, a lungfish lung is supplied with a capillary network in its largely unfurrowed walls, a tubelike connection to

the pharynx, and a primitive ventilating system for moving air in and out of the lung.

Amphibian lungs vary from simple, smooth-walled, baglike lungs of some salamanders to the subdivided lungs of frogs and toads (Figure 31-21). Total surface available for gas exchange is much increased in lungs of reptiles which are subdivided into numerous interconnecting air sacs. Most elaborate of all are mammalian lungs containing millions of small sacs, called **alveoli** (Figure 31-22), each veiled by a rich vascular network. Human lungs have a total surface area of from 50 to 90 m²—50 times the area of the skin surface—and contain 1000 km of capillaries. A large surface area is essential for the high oxygen uptake required to support the elevated metabolic rate of endothermic mammals.

A disadvantage of lungs is that gas is exchanged between blood and air only in the alveoli, located at the ends of a branching tree of air tubes (trachea, bronchi, and bronchioles [Figure 31-22]). Unlike the efficient one-way flow of water across fish gills, air must enter and exit a lung through the same channel. After exhalation, the air tubes are filled with "used" air from the alveoli which, during the following inhalation, is pulled back into the lungs. The volume of air in a lung's passageways is called "dead space." This air shuttles back and forth with each breath, adding to the difficulty of properly ventilating lungs. In fact, lung ventilation in humans is so inefficient that in normal breathing only approximately one-sixth of the air in the lungs is replenished with each inspiration. Even after forced expiration, 20% to 35% of the air remains in the lungs.

In birds, lung efficiency is improved vastly by adding an extensive system of air sacs (Figure 31-21 and p. 566) that serve as air reservoirs during ventilation. On inspiration, about 25% of incoming air passes over the lung **parabronchi** (one-cell-thick air capillaries) where gas exchange takes place. The remaining 75% of incoming air bypasses the lungs to enter the air sacs (gas exchange does not occur here). At expiration some of this fresh air passes directly through the lung passages and eventually into the lung parabronchi. Thus the parabronchi receive nearly fresh air during both inspiration and expiration. The beautifully

Figure 31-20

How a fish ventilates its gills. Through the action of two skeletal muscle pumps, one in the mouth cavity, the other in the opercular cavity, water is drawn into the mouth, passes over the gills, and exits through the gill covers (opercular clefts).

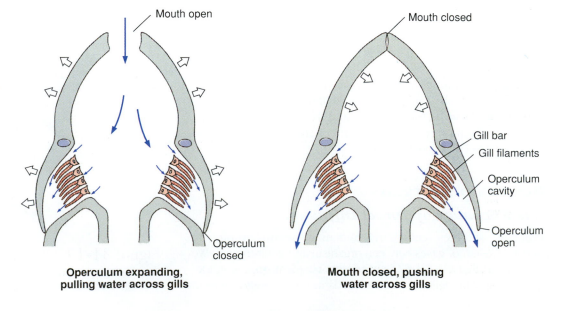

Mouth open

Mouth closed

Gill bar

Gill filaments

Operculum cavity

Operculum open

Operculum closed

Operculum expanding, pulling water across gills

Mouth closed, pushing water across gills

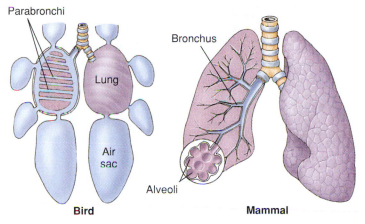

Figure 31-21

Variations among internal structures of lungs among vertebrate groups ranging from simple sacs with little exchange surface between blood and air spaces in amphibians to complex, lobulated structures, each with complex divisions and extensive exchange surfaces in birds and mammals.

designed bird lung is a result of selective pressures during evolution of flight with its high metabolic demands.

Amphibians employ a **positive pressure** action to force air into their lungs, unlike most reptiles, birds, and mammals, which ventilate their lungs by **negative pressure,** in which air is pulled into the lungs by expansion of the thoracic cavity. Frogs ventilate the lungs by first drawing air into the mouth through the **external nares** (nostrils). Then, closing the nares and raising the floor of the mouth, they drive air into the lungs (Figure 31-23). Much of the time, however, frogs rhythmically ventilate only the mouth cavity, a well-vascularized respiratory surface that supplements cutaneous and pulmonary respiration.

Structure and Function of the Mammalian Respiratory System

Air enters a mammalian respiratory system through nostrils (external nares), passes through a **nasal chamber,** lined with mucus-secreting epithelium, and then through **internal nares,** nasal openings connected to the **pharynx.** Here, where pathways of digestion and respiration cross, in-haled air leaves the pharynx by passing into a narrow opening, the **glottis;** food enters the esophagus to pass to the stomach (see Figure 32-10, p. 684). The glottis opens into the **larynx,** or voice box, and then into the **trachea,** or windpipe. The trachea branches into two **bronchi,** one to each lung (Figure 31-22). Within the lungs each bronchus divides and subdivides into small tubes **(bronchioles)** that lead via **alveolar ducts** to the air sacs **(alveoli)** (Figure 31-22). The single-layered endothelial walls of the alveoli and alveolar ducts are thin and moist to facilitate exchange of

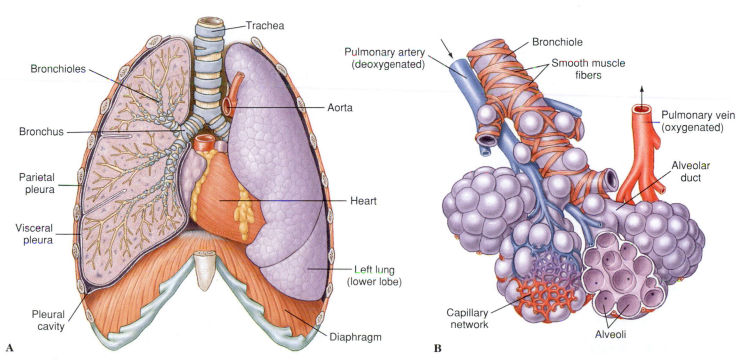

Figure 31-22

A, Lungs of human with right lung shown in section. **B,** Terminal portion of bronchiole showing air sacs with their blood supply. Arrows show direction of blood flow.

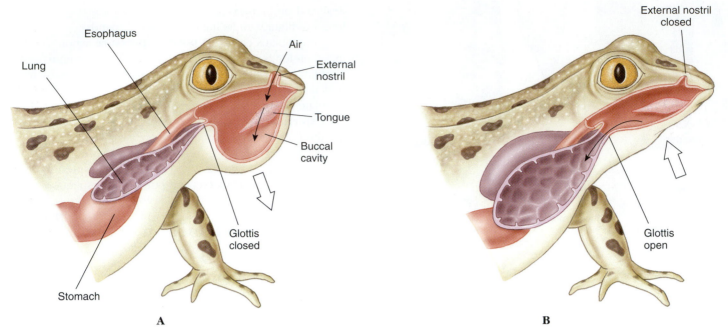

Figure 31-23

Breathing in frogs. Frogs, positive-pressure breathers, fill their lungs by forcing air into them. **A,** The buccal cavity floor is lowered, drawing air in through the nostrils. **B,** With nostrils closed and glottis open, frogs force air into lungs by elevating the buccal cavity floor. The buccal cavity may be ventilated rhythmically for a period before the lungs are emptied by contraction of body-wall musculature and by elastic recoil of lungs.

gases between air and adjacent blood capillaries. Air passageways are lined with both mucus-secreting and ciliated epithelial cells, which play an important role in conditioning the air before it reaches the alveoli. Partial cartilage rings in the walls of the tracheae, bronchi, and even some of the larger bronchioles prevent those structures from collapsing.

In its passage to the air sacs, air undergoes three important changes: (1) it is filtered free from most dust and other foreign substances, (2) it is warmed to body temperature, and (3) it is saturated with moisture.

The lungs consist of a great deal of elastic connective tissue. They are covered by a thin layer of tough epithelium known as the **visceral pleura.** A similar layer, the **parietal pleura,** lines the inner surface of the walls of the chest (Figure 31-22). The two layers of the pleura are in contact and slide over one another as the lungs expand and contract. The "space" between the pleura, called the **pleural cavity,** maintains a partial vacuum, which helps keep the lungs expanded to fill the pleural cavity. Therefore no real pleural space exists; the two pleura rub together, lubricated by tissue fluid. The chest cavity is bounded by the spine, ribs, and breastbone, and floored by the **diaphragm,** a dome-shaped, muscular partition between the chest cavity and abdomen. A muscular diaphragm is found only in mammals.

Ventilating the lungs

The chest cavity is an air-tight chamber. During inspiration the ribs are pulled upward and the diaphragm flattens. The resultant increase in volume of the chest cavity (Figure 31-24) causes air pressure in the lungs to fall below atmospheric pressure: air rushes in through passageways to equalize the pressure. Normal **expiration** is a less active process than inspiration. When the muscles relax, the ribs and diaphragm return to their original position, and the chest cavity decreases in size, the elastic lungs deflate, and air exits (Figure 31-24).

How Breathing Is Coordinated

Breathing is normally involuntary and automatic but can come under voluntary control. Neurons in the medulla of the brain regulate normal, quiet breathing. They spontaneously produce rhythmical bursts that stimulate contraction of the diaphragm and external intercostal muscles. However, respiration must adjust itself to changing requirements of the body for oxygen. Oddly, carbon dioxide rather than oxygen has the greatest effect on respiratory rate because under normal conditions arterial oxygen does not decline enough to stimulate oxygen receptors. Even a small rise in carbon dioxide level in the blood, however, has a powerful effect on respiratory activity. Actually, the stimulatory effects of carbon dioxide are due in part to an increase in hydrogen-ion concentration in cerebrospinal fluid.

$$CO_2 + H_2O \leftrightarrow H_2CO_3 \leftrightarrow H^+ + HCO_3^-$$

This reaction shows that carbon dioxide combines with water to form carbonic acid. Carbonic acid then dissociates to release hydrogen ions, making the cerebrospinal fluid more acidic, and stimulating respiratory receptors in the medulla of the brain.

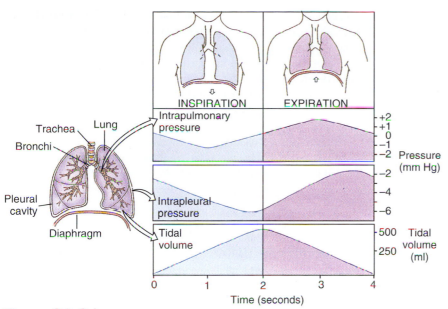

Figure 31-24
Mechanism of breathing in humans.

Gaseous Exchange in Lungs and Body Tissues: Diffusion and Partial Pressure

Air (the atmosphere) is a mixture of gases: about 71% nitrogen, 20.9% oxygen, in addition to fractional percentages of other gases, such as carbon dioxide (0.03%). Gravity attracts the atmospheric gases to the earth. At sea level the atmosphere exerts a hydrostatic pressure due to gravity equal to the weight of a column of mercury (Hg) 760 mm high. Thus we can speak of atmospheric pressure (1 atm) as being equal to 760 mm Hg. But because air is not a single gas but a mixture, *part* of the 760 mm Hg pressure (**partial pressure**) is due to each component gas. For example, the partial pressure of oxygen is $0.209 \times 760 = 159$ mm, and that for carbon dioxide is $0.0003 \times 760 = 0.23$ mm in dry air. (In fact, atmospheric air is never completely dry, and the varying amount of water vapor present exerts a pressure in proportion to its concentration, like other gases.)

As soon as air enters the respiratory tract, its composition changes (Table 31-1, Figure 31-25). Inspired air becomes saturated with water vapor as it travels through the air-filled passageways toward the alveoli. When inspired air reaches the alveoli, it mixes with residual air remaining from the previous respiratory cycle. Partial pressure of oxygen drops and that of carbon dioxide rises. Upon expiration, air from the alveoli mixes with air in the dead space to produce still a different mixture (Table 31-1). Although no significant gas exchange takes place in the dead space, the air it contains is the first air to leave the body when expiration begins.

Because the partial pressure of oxygen in lung alveoli is greater (100 mm Hg) than it is in venous blood of lung capillaries (40 mm Hg), oxygen diffuses into the lung capillaries. In a similar manner carbon dioxide in blood of the lung capillaries has a higher concentration (46 mm Hg) than has this same gas in lung alveoli (40 mm Hg), so carbon dioxide diffuses from the blood into the alveoli.

Both rate and depth of respiration increase. Peripheral chemoreceptors, located close to the heart and in the neck region, monitor changes in blood levels of carbon dioxide and hydrogen ions and send stimulatory signals to the medulla respiratory centers if these levels rise.

It is well known that swimmers can remain submerged much longer if they vigorously hyperventilate first to blow off carbon dioxide from the lungs, thereby delaying the overpowering urge to surface and breathe. The practice is dangerous because blood oxygen is depleted just as rapidly as without prior hyperventilation, and the swimmer may lose consciousness when the oxygen supply to the brain drops below a critical point. Several documented drownings among swimmers attempting long underwater swimming records have been caused by this practice.

TABLE 31.1

Partial Pressures and Gas Concentrations in Air and Body Fluids

	Nitrogen (N_2)	Oxygen (O_2)	Carbon Dioxide (CO_2)	Water Vapor (H_2O)
Inspired air (dry)	600 (79%)	159 (20.9%)	0.2 (0.03%)	—
Alveolar air (saturated)	573 (75.4%)	100 (13.2%)	40 (5.2%)	47 (6.2%)
Expired air (saturated)	569 (74.8%)	116 (15.3%)	28 (3.7%)	47 (6.2%)
Arterial blood	573	100	40	
Peripheral tissues	573	0–30	45–68	
Venous blood	573	40	46	

Note: Values expressed in millimeters of mercury (mm Hg). Percentages indicate proportion of total atmospheric pressure at sea level (760 mm Hg). Inspired air is shown as dry, although atmospheric air always contains variable amounts of water. If, for example, atmospheric air at 20° C were half saturated (relative humidity 50%), the partial pressures and percentages would be N_2 593.5 (78.1%); O_2 157 (20.6%); CO_2 0.2 (0.03%); and H_2O 8.75 (1.1%).

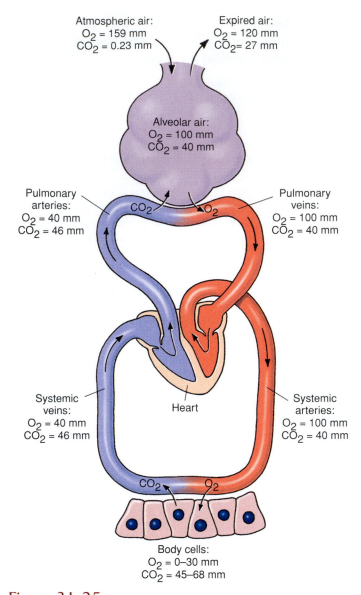

Atmospheric air:
O_2 = 159 mm
CO_2 = 0.23 mm

Expired air:
O_2 = 120 mm
CO_2= 27 mm

Alveolar air:
O_2 = 100 mm
CO_2 = 40 mm

Pulmonary
arteries:
O_2 = 40 mm
CO_2 = 46 mm

CO_2 O_2

Pulmonary
veins:
O_2 = 100 mm
CO_2 = 40 mm

Systemic
veins:
O_2 = 40 mm
CO_2 = 46 mm

Heart

Systemic
arteries:
O_2 = 100 mm
CO_2 = 40 mm

CO_2 O_2

Body cells:
O_2 = 0–30 mm
CO_2 = 45–68 mm

Figure 31-25

Exchange of respiratory gases in lungs and tissue cells. Numbers are partial pressures in millimeters of mercury (mm Hg).

In tissues respiratory gases also move along their concentration gradients (Figure 31-25). Partial pressure of oxygen in the blood (100 mm Hg) is greater than in the tissues (0 to 30 mm Hg), and partial pressure of carbon dioxide in tissues (45 to 68 mm Hg) is greater than that in blood (40 mm Hg). In each case gases diffuse from a location of higher concentration to one of lower concentration.

How Respiratory Gases Are Transported

In some invertebrates respiratory gases are simply carried dissolved in body fluids. However, solubility of oxygen is so low in water that it is adequate only for animals with low rates of metabolism. For example, only approximately 1% of a human's

oxygen requirement can be transported in this way. Consequently in many invertebrates and in virtually all vertebrates, nearly all oxygen and a significant amount of carbon dioxide are transported by special colored proteins, or **respiratory pigments,** in the blood. In all vertebrates these respiratory pigments are packaged into blood cells.

Because of the weight of water, hydrostatic pressure increases the equivalent of 1 atmosphere for every 10 m of depth in seawater, and the pressure of the air supplied to a diver must be increased correspondingly so that it can be drawn into the lungs. Under the increased pressure, additional air dissolves in the blood, the amount depending on depth and time at depth of a dive. If a diver ascends slowly, the gas comes out of solution imperceptibly and is breathed out from the lungs. However, if the ascent is too rapid, the air comes out of solution and forms bubbles in the blood and other tissues, a condition known as *decompression sickness* or *the bends*. The result is painful and, if severe, can cause paralysis or death.

Sickle cell anemia is an up-to-now incurable, inherited condition (p. 97) in which a single amino acid (glutamic acid) in normal hemoglobin (HbA) is replaced by a valine in sickle cell hemoglobin (HbS). The ability of HbS to carry oxygen is severely impaired, and erythrocytes tend to crumple during periods of oxygen stress (for example, during exercise). Capillaries become clogged with misshapen red cells; the affected area is very painful, and the tissue may die. About 1 in 10 African-Americans carry the trait (heterozygous). Heterozygotes do not have sickle cell anemia and live normal lives, but if both parents are heterozygous, each of their offspring has a 25% chance of inheriting the disease.

The most widespread respiratory pigment in the animal kingdom is **hemoglobin,** a red, iron-containing protein present in all vertebrates and many invertebrates. Each molecule of hemoglobin is 5% **heme,** an iron-containing compound giving the red color to blood, and 95% **globin,** a colorless protein. The heme portion of hemoglobin has a great affinity for oxygen; each gram of hemoglobin can carry a maximum of approximately 1.3 ml of oxygen. Because there are approximately 15 g of hemoglobin in each 100 ml of blood, fully oxygenated blood contains approximately 20 ml of oxygen per 100 ml. Of course, for hemoglobin to be of value to the body it must hold oxygen in a loose, reversible chemical combination so that it can be released to tissues. The actual amount of oxygen that combines with hemoglobin depends on the shape or conformation of the hemoglobin molecule, which is affected by several factors, including the concentration of oxygen itself. When the oxygen concentration is high, as it is in the capillaries of the lung alveoli, hemoglobin binds oxygen; in tissues where the prevailing oxygen partial pressure is low, hemoglobin releases its stored oxygen reserves (Figure 31-26).

A   **B**

Figure 31-26

Hemoglobin saturation curves. Curves show how the amount of oxygen that can bind to hemoglobin is related to oxygen partial pressure. **A,** At the higher partial pressure in the lungs, hemoglobin can load with more oxygen. In the tissues the oxygen concentration is less, so hemoglobin can carry less and unloads more oxygen. **B,** Hemoglobin is also sensitive to carbon dioxide partial pressure (Bohr effect). As carbon dioxide enters blood from the tissues, it shifts the curve to the right, decreasing affinity of hemoglobin for oxygen. Thus the hemoglobin unloads more oxygen in the tissues where carbon dioxide concentration is higher.

Although hemoglobin is the only vertebrate respiratory pigment, several other respiratory pigments are known among invertebrates. *Hemocyanin,* a blue, copper-containing protein, occurs in crustaceans and most molluscs. Among other pigments is *chlorocruorin* (klor-a-cru'-o-rin), a green-colored, iron-containing pigment found in four families of polychaete tube worms. Its structure and oxygen-carrying capacity are very similar to those of hemoglobin, but it is carried free in the plasma rather than being enclosed in blood corpuscles. *Hemerythrin* is a red pigment found in some polychaete worms. Although it contains iron, this metal is not present in a heme group (despite the name of the pigment!), and its oxygen-carrying capacity is poor compared to hemoglobin.

We can express the relationship of carrying capacity to surrounding oxygen concentration as **hemoglobin saturation curves** (also called oxygen dissociation curves [Figure 31-26]). As these curves show, the lower the surrounding oxygen tension, the greater the quantity of oxygen released. This important characteristic of hemoglobin allows more oxygen to be released to those tissues which need it most (those having the lowest partial pressure of oxygen).

Another factor that affects conformation of hemoglobin and therefore its release of oxygen to tissues is the sensitivity of oxyhemoglobin (hemoglobin with bound oxygen) to carbon dioxide. Carbon dioxide shifts the hemoglobin saturation curve to the right (Figure 31-26B), a phenomenon called the **Bohr effect** after the Danish scientist who first described it. As carbon dioxide enters the blood from respiring tissues, it causes hemoglobin to unload more oxygen. The opposite event occurs in the lungs; as carbon dioxide diffuses from venous blood into alveolar space, the hemoglobin saturation curve shifts back to the left, allowing more oxygen to be loaded onto hemoglobin.

Unfortunately for humans and many other animals, hemoglobin has an affinity for carbon monoxide that is about 200 times greater than its affinity for oxygen. Consequently, even when carbon monoxide is present in the atmosphere at lower concentrations than oxygen, it tends to displace oxygen from hemoglobin to form a stable compound called carboxyhemoglobin. Air containing only 0.2% carbon monoxide may be fatal. Because of their higher respiratory rate, children and small animals are poisoned more rapidly than adults. Carbon monoxide is becoming an atmospheric contaminant of ever-increasing proportions as the world's population and industrialization continue to increase rapidly.

Figure 31-27

Transport of carbon dioxide in the blood. **A,** Carbon dioxide produced by metabolic oxidation of glucose diffuses from the tissues into plasma and red blood cells. Carbonic anhydrase in red blood cells catalyzes conversion of carbon dioxide into carbonic acid, then bicarbonate and hydrogen ions. The bicarbonate diffuses out of the cells, and diffusion inward of chloride ions maintains electrical balance. Hydrogen ions associate with hemoglobin. **B,** The lower partial pressure of carbon dioxide in the alveoli of the lungs favors reversal of these reactions.

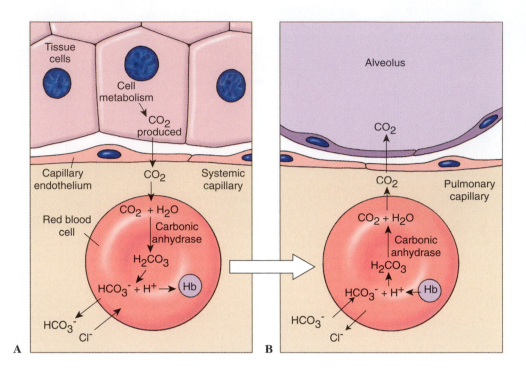

The same blood that transports oxygen to the tissues from the lungs must carry carbon dioxide back to the lungs on its return trip. However, unlike oxygen that is transported almost exclusively in combination with hemoglobin, carbon dioxide is transported in three different forms. A small fraction of the blood-borne carbon dioxide, only about 5%, is carried as gas physically dissolved in the plasma. The remainder diffuses into red blood cells. In red blood cells, most carbon dioxide, approximately 70%, becomes carbonic acid through action of the enzyme carbonic anhydrase. Carbonic acid immediately dissociates into hydrogen ion and bicarbonate ion. We can summarize the entire reaction as follows:

$$\overset{\text{carbonic anhydrase}}{CO_2 + H_2O \rightleftarrows H_2CO_3 \rightleftarrows H^+ + HCO_3^-}$$

The hydrogen ions combine with hemoglobin to form **deoxyhemoglobin,** thus preventing a severe decrease in blood pH, and releasing oxygen simultaneously. Biocarbonate ions are transported out of the red blood cells in exchange for chloride ions **(the chloride shift).** Bicarbonate ions remain in solution in plasma because unlike carbon dioxide, bicarbonate is extremely soluble (Figure 31-27).

Another fraction of the carbon dioxide, approximately 25%, combines reversibly with hemoglobin. Carbon dioxide does not combine with the heme group but with amino groups of several amino acids to form a compound called **carbaminohemoglobin.**

All of these reactions are reversible. When venous blood reaches the lungs, carbon dioxide diffuses from red blood cells into alveolar air.

SUMMARY

Fluid in the body, whether intracellular, plasma, or interstitial, is mostly water, but contains many dissolved substances, including electrolytes and proteins. Vertebrate blood consists of fluid plasma and formed elements, including red and white blood cells and platelets. Plasma has many dissolved solids, as well as dissolved gases. Mammalian red blood cells lose their nucleus during development and contain the oxygen-carrying pigment, hemoglobin. White blood cells are important defensive elements. Platelets are vital in the process of clotting, necessary to prevent excess blood loss when a blood vessel is damaged. Platelets release a series of factors that activate prothrombin to thrombin, an enzyme that causes fibrinogen to change to the gel form, fibrin.

In open circulatory systems, such as those of arthropods and most molluscs, blood escapes from arteries into a hemocoel, which is a primary body cavity derived from the blastocoel. In closed cir-culatory systems, such as those of annelids, vertebrates, and cephalopod molluscs, the heart pumps blood into arteries, then into arterioles of smaller diameter, through a bed of fine capillaries, through venules, and finally through veins, which lead back to the heart. In fishes, which have a two-chambered heart with a single atrium and a single ventricle, blood is pumped to gills and then directly to systemic capillaries throughout the body without first returning to the heart. With evolution of lungs, vertebrates developed a double circulation consisting of a systemic circuit serving the body, and a pulmonary circuit serving the lungs. To be fully efficient, this change required partitioning of both atrium and ventricle to form a double pump; partial partitioning occurs in lung-fishes and amphibians which have two atria but an undivided ventricle, and is complete in birds and mammals, which have four-chambered hearts.

One-way flow of blood during the heart's contraction (ventricular systole) and relaxation (ventricular diastole) is assured by valves between the atria and ventricles and between the ventricles and pulmonary arteries and aorta. Although the heart can beat spontaneously, its rate is controlled by hormones and by nerves from the central nervous system. Heart muscle uses much oxygen and has a well-developed coronary blood circulation. The walls of arteries are thicker than those of veins, and the elastic connective tissue in the walls of larger arteries allows them to expand during ventricular systole and contract during ventricular diastole. Normal arterial blood pressure (hydrostatic) of humans in systole is 120 mm Hg and in diastole, 80 mm Hg. Because capillary walls are permeable to water, a protein-free filtrate crosses capillary walls, its movement determined by a balance between opposing forces of hydrostatic and protein osmotic pressure. Tissue fluid that does not reenter the capillary system is collected by the lymphatic system (now called lymph) and returned to blood by lymph ducts.

Very small animals can depend on diffusion between the external environment and their tissues or cytoplasm for transport of respiratory gases, but larger animals require specialized organs, such as gills, tracheae, or lungs, for this function. Gills and lungs provide an increased surface area for exchange of respiratory gases between blood and environment. Many animals have special respiratory pigments and other mechanisms to help transport oxygen and carbon dioxide in blood. The most widespread respiratory pigment in the animal kingdom, hemoglobin, has a high affinity for oxygen at high oxygen concentrations but releases it at lower concentrations. Vertebrate hemoglobin, which is packaged in red blood cells, combines readily with oxygen in gills or lungs, then releases it in respiring body tissues where the oxygen partial pressure is low. Blood carries carbon dioxide from tissues to lungs as bicarbonate ion and dissolved gas in the plasma, and, in combination with hemoglobin, in red blood cells.

REVIEW QUESTIONS

1. Name the chief intracellular electrolytes and the chief extracellular electrolytes.
2. What is the fate of spent erythrocytes in the body?
3. Outline or briefly describe the sequence of events that leads to blood coagulation.
4. Two distinctly different styles of circulatory systems have evolved among animals: open and closed. What is "open" about an open circulatory system? Closed systems sometimes are cited as adaptive for actively moving animals with (at least at times) high metabolic demand. Can you suggest possible reasons for this assertion?
5. Place the following in correct order to describe the circuit of blood through the vascular system of a fish: ventricle, gill capillaries, sinus venosus, body tissue capillaries, atrium, dorsal aorta.
6. Trace the flow of blood through the heart of a mammal, naming the four chambers, their valves, and explaining where the blood entering each atrium comes from and where blood leaving each ventricle goes. When the ventricles contract, what prevents blood from reentering the atria?
7. Explain the origin and conduction of the excitation that leads to a heart contraction. Why is the vertebrate heart said to be a myogenic heart? If the heart is myogenic, how do you account for alterations in rate of the heartbeat?
8. Define the terms systole and diastole. Distinguish atrial and ventricular systole and diastole.
9. Explain the movement of fluid across the walls of capillaries. How does balance of hydrostatic pressure and colloid osmotic pressure determine direction of net fluid flow?
10. Hydrostatic pressure at the arterial end of capillaries is about 40 mm Hg in humans. If hydrostatic pressure at the venous end is about 15 mm Hg, and colloid osmotic pressure is 25 mm Hg throughout, what is the net effect on fluid movement between capillaries and tissue spaces?
11. Provide a brief description of the lymphatic system. What are its principal functions? Why is movement of lymph through the lymphatic system very slow?

12. What is an advantage of a fish's gills for breathing in water and a disadvantage for breathing on land?
13. Describe the tracheal system of insects. What is the advantage of such a system for a small animal?
14. Trace the route of inspired air in humans from the nostrils to the smallest chamber of the lungs. What is the "dead air space" of a mammalian lung and how does it affect the partial pressure of oxygen reaching the alveoli?
15. The amount of time that scuba divers can spend underwater is limited by several factors, including time required to deplete the air supply in their tanks. To make their air last longer novice divers may be instructed to breathe slowly and exhale as much as possible on each breath. Can you suggest a reason why this behavior would lengthen a diver's air supply?
16. How does a frog ventilate its lungs? Contrast an amphibian's positive-pressure breathing with a mammal's negative-pressure breathing.
17. What is the role of carbon dioxide in the control of rate and depth of mammalian breathing of a mammal?
18. The air pressure supplied to a scuba diver must equal that exerted by the surrounding seawater, and for each 10 m increase in depth, pressure of the surrounding seawater increases one full atmosphere. Assuming the partial pressure of oxygen in air at sea level (one atmosphere) is 0.209×760 mm Hg (= 159 mm Hg), what partial pressure of oxygen would a diver be breathing at a depth of 30 m?
19. Explain how oxygen is carried in blood, including specifically the role of hemoglobin. Answer the same question with regard to carbon dioxide transport.
20. The ability of hemoglobin to bind oxygen decreases with decreasing oxygen concentration and also decreases with increasing carbon dioxide concentration. What effect do these phenomena have on the delivery of oxygen to tissues?

SELECTED REFERENCES

Bartecchi, C. E. 1998. If you don't have a defibrillator. Sci. Am. **278:**91. *Describes "cough" and "thump" techniques that can be used instead of cardiopulmonary resuscitation (CPR).*

Burggren, W. W. 1997. Identifying and evaluating patterns in cardiorespiratory physiology. Am. Zool. **37:**109–115. *One of several papers in this issue which is composed of a symposium on cardiorespiratory physiology.*

Eisenberg, M. S. 1998. Defibrillation: The spark of life. Sci. Am. **278:**86–90. *When the pacemaker loses its rhythm, heart muscle commences uncoordinated contractions. Application of a brief electrical shock from a defibrillator often can "reset" the pacemaker. Defibrillators have saved many lives.*

Feder, M. E., and W. W. Burggren. 1985. Skin breathing in vertebrates. Sci. Am. **253:**126–142 (Nov.). *In many amphibians and reptiles the skin supplements gas exchange and may even replace the work of gills and lungs.*

Hardison, R. 1999. The evolution of hemoglobin. Am. Sci. **87:**126–137. *Comparison of amino-acid sequences in hemoglobins from animals, plants, unicellular eukaryotes, and eubacteria suggests that all share a common ancestor early in organismal evolution.*

Jain, R. K., and P. F. Carmeliet. 2001. Vessels of death or life. Sci. Am. **285:**38–43. *Overview of blood vessel development (argiogenisis) and its role in many serious diseases.*

Kiberstis, P., and J. Marx. 1996. Cardiovascular medicine. Science **272:**663. *Introduction to a series of news and articles on current research on heart development, genetics of blood pressure, genetics of cardiovascular disease, mouse models of atherosclerosis, molecular therapies for vascular diseases, new drugs for stroke.*

Libby, P. 2002. Atherosclerosis: the new view. Sci. Am. **286:**47–55, May. *Describes the most recent data on development of atherosclerosis.*

Lillywhite, H. B. 1988. Snakes, blood circulation and gravity. Sci. Am. **259:**92–98 (Dec.). *How a snake's vascular system is designed to counter the effects of gravity.*

Nucci, M. L., and A. Abuchowski. 1998. The search for blood substitutes. Sci. Am. **278:**73–77. *Shortage of blood supplies and risk of contamination have made the search for substitutes more urgent.*

Randall, D. W. Burggren, and K. French. 2001 Eckert animal physiology: mechanisms and adaptations. New York, W. H. Freeman and Company. *A readable, comparative approach to animal physiology.*

ZOOLOGY LINKS TO THE INTERNET

Visit the textbook's Online Learning Center at www.mhhe.com/zoology to find live Internet links for each of the topics listed here.

Vertebrates: Macroscopic Anatomy of Circulatory Elements
Vertebrates: Microscopic Anatomy of Circulatory Elements
Human Circulatory System Topics
Blood and Vertebrate Circulation
Vertebrates: Macroscopic Anatomy of the Respiratory System
Vertebrates: Microscopic Anatomy of the Respiratory System
Human Respiratory Topics

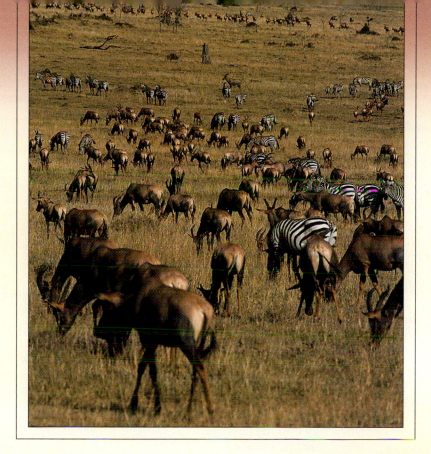

Topi and zebras on the African savannah.

Digestion and Nutrition

A Consuming Cornucopia

Sir Walter Raleigh observed that the difference between a rich man and a poor man is that the former eats when he pleases while the latter eats when he can get it. In today's crowded world, with nearly 80 million people added each year to the world's population of over 6 billion, the separation between the well-fed affluent and the hungry and malnourished poor reminds us that time has not diminished the shrewdness of Sir Walter's remark. Unlike the affluent for whom food acquisition requires only the selection of prepackaged foods at a well-stocked supermarket, the world's poor can appreciate that for them, as for the rest of the animal kingdom, procuring food is fundamental to survival. For most animals, eating is the main business of living.

Potential food is everywhere and little remains unexploited. Animals bite, chew, nibble, crush, graze, browse, shred, rasp, filter, engulf, enmesh, suck, and absorb foods of incredible variety. What an animal eats and how it eats profoundly affect an animal's feeding specialization, its behavior, its physiology, and its internal and external anatomy—in short, both its body form and its role in the web of life. The endless evolutionary jostling between predator and prey has provided compromise adaptations for eating and adaptations for avoiding being eaten. By whatever means food may be secured, there is far less variation among animals in the subsequent digestive simplification of foods. Vertebrates and invertebrates alike use similar digestive enzymes. Even more uniform are the final biochemical pathways for nutrient use and energy transformation. The nourishment of animals is like a cornucopia in which the food flows in rather than out. A great diversity of foods procured by countless feeding adaptations streams into the mouth of the horn, is simplified, and finally applied to the common purpose of survival and reproduction.

All organisms require energy to maintain their highly ordered and complex structure. This energy is chemical bond energy released by transforming complex compounds acquired from an organism's environment into simpler ones.

The ultimate source of energy for life on earth is the sun. Sunlight is captured by chlorophyll molecules in green plants, which transform a portion of this energy into chemical bond energy (food energy). Green plants are **autotrophic** organisms; they require only inorganic compounds absorbed from their surroundings to provide the raw material for synthesis and growth. Most autotrophic organisms are the chlorophyll-bearing **phototrophs,** although some, the chemosynthetic bacteria, are **chemotrophs;** they gain energy from inorganic chemical reactions.

Almost all animals are **heterotrophic organisms** that depend on already synthesized organic compounds of plants and other animals to obtain the materials they will use for growth, maintenance, and reproduction of their kind. Since the food of animals, normally the complex tissues of other organisms, is usually too bulky to be absorbed directly by cells, it must be digested, into soluble molecules small enough to be used.

Animals may be divided into categories based on dietary habits. **Herbivorous** animals feed mainly on plant life. **Carnivorous** animals feed mainly on herbivores and other carnivores. **Omnivorous** forms eat both plants and animals. **Saprophagous** animals feed on decaying organic matter.

Ingestion of foods and their simplification by digestion are only initial steps in nutrition. Foods reduced by digestion to soluble, molecular form are **absorbed** into the circulatory system and **transported** to the body's tissues. There they are **assimilated** into the structures of cells. Oxygen is also transported by blood to tissues, where food products are **oxidized,** or burned to yield energy and heat. Food not immediately used is **stored** for future use. Wastes produced by oxidation must be **excreted.** Food products unsuitable for digestion are **egested** in feces.

We first examine feeding adaptations of animals and then discuss digestion and absorption of food. We close with a consideration of nutritional requirements of animals.

FEEDING MECHANISMS

Few animals can absorb nutrients directly from their external environments. Exceptions are some blood parasites (p. 221), certain intestinal protozoan parasites (p. 227), and tapeworms and acanthocephalans, (p. 285) that nourish themselves on primary organic molecules absorbed directly across their body surfaces. These nutrients have already been digested by their host organism. Most animals, however, must work for their meals. They are active feeders that have evolved numerous specializations for obtaining food. With food procurement as one of the most potent driving forces in animal evolution, natural selection has placed a high priority on adaptations for exploiting new sources of food and the means of food capture and intake. In this brief discussion we consider some of the major food-gathering devices.

Feeding on Particulate Matter

Drifting microscopic particles are found in the upper hundred meters of ocean water. Most of this multitude is **plankton,** organisms too small to do anything but drift with the ocean's currents. The rest is organic debris, disintegrating remains of dead plants and animals. Although this oceanic swarm of plankton forms a rich life domain, it is unevenly distributed. The heaviest plankton growth occurs in estuaries and areas of upwelling, where there is an abundant nutrient supply. It is consumed by numerous larger animals, invertebrates and vertebrates, using a variety of feeding mechanisms.

One of the most important and widely employed methods for feeding is **suspension feeding** (Figure 32-1). The majority of suspension feeders use ciliated surfaces to produce currents that draw drifting food particles into their mouths. Most suspension-feeding invertebrates, such as tube-dwelling polychaete worms, bivalve molluscs, hemichordates, and most protochordates, entrap particulate food on mucous sheets that convey the food into the digestive tract. Others, such as fairy shrimps, water fleas, and barnacles, use sweeping movements of their setae-fringed legs to create water currents and entrap food, which is transferred to their mouth. In freshwater developmental stages of certain insect orders, the organisms use fanlike arrangements of setae or spin silk nets to entrap food.

One form of suspension feeding, often called **filter feeding,** has evolved frequently as a secondary modification among representatives of groups that are primarily selective feeders. Examples include many microcrustaceans, fishes such as herring, menhaden, and basking sharks, certain birds such as flamingos, and the largest of all animals, baleen (whalebone) whales. The vital importance of one component of plankton, diatoms, in supporting a great pyramid of suspension-feeding animals is stressed by N. J. Berrill:[*]

> A humpback whale . . . needs a ton of herring in its stomach to feel comfortably full—as many as five thousand individual fish. Each herring, in turn, may well have 6000 or 7000 small crustaceans in its own stomach, each of which contains as many as 130,000 diatoms. In other words, some 400 billion yellow-green diatoms sustain a single medium-sized whale for a few hours at most.

Another type of particulate feeding exploits deposits of disintegrated organic material (detritus) that accumulates on and in the substratum; this type is called **deposit feeding.** Some deposit feeders, such as many annelids and some hemichordates, simply pass the substrate through their bodies, removing from it whatever provides nourishment. Others, such as scaphopod molluscs, certain bivalve molluscs, and some sedentary and tube-dwelling polychaete worms, use appendages to gather organic deposits some distance from the body and move them toward the mouth (Figure 32-2).

---

*Berrill, N. J. 1958. You and the universe. New York, Dodd, Mead & Co.

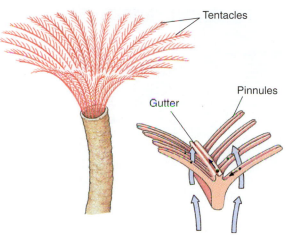

A, Marine fan worms (class Polychaeta, phylum Annelida) have a crown of tentacles. Numerous cilia on the edges of the tentacles draw water (*solid arrows*) between pinnules where food particles are entrapped in mucus; particles are then carried down a "gutter" in the center of the tentacle to the mouth (*broken arrows*).

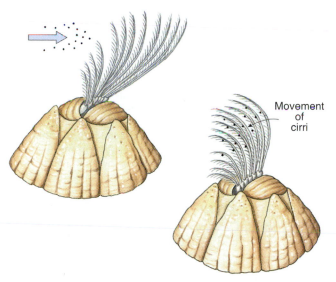

C, Barnacles sweep their thoracic appendages (cirri) through the water to trap plankton and other organic particles on fine bristles that fringe the cirri. Food is transferred to the barnacle's mouth by the first, short cirri. Class Maxillopoda, subphylum Crustacea, phylum Arthropoda.

B, Bivalve molluscs (class Bivalvia, phylum Mollusca) use their gills as feeding devices, as well as for respiration. Water currents created by cilia on the gills carry food particles into the incurrent siphon and between slits in the gills where they are entangled in a mucous sheet covering the gill surface. Ciliated food grooves then transport the particles to the mouth (not shown). Arrows indicate direction of water movement.

D, Herring and other suspension-feeding fishes (class Actinopterygii, phylum Chordata) use gill rakers that project forward from the gill arches into the pharyngeal cavity to strain plankton. Herring swim almost constantly, forcing water and suspended food into their mouth; food is strained by their gill rakers, and water passes through the gill openings.

E, Whalebone whales (class Mammalia, phylum Chordata) filter out plankton, principally large crustaceans called krill, with whalebone, or baleen. Water enters the swimming whale's open mouth by force of the animal's forward motion and is strained out through more than 300 keratinized baleen plates that hang like a curtain from the roof of the mouth. Krill and other plankton caught in the baleen are periodically collected by the huge tongue and swallowed.

Figure 32-1

Some suspension feeders and their feeding mechanisms.

Figure 32-2
The annelid *Amphitrite* is a deposit feeder that lives in a mucus-lined burrow and extends long feeding tentacles in all directions across the surface. Food trapped on mucus is conveyed along the tentacles to the mouth.

Feeding on Food Masses

Among the most interesting animal adaptations are those that have evolved for procuring and manipulating solid food. Such adaptations and animals bearing them are largely shaped by what the animal eats.

Predators must be able to locate, capture, hold, and swallow prey. Most carnivorous animals simply seize food and swallow it intact, although some employ toxins that paralyze or kill prey at time of capture. Although no true teeth appear among invertebrates, many have beaks or toothlike structures for biting and holding. A familiar example is the carnivorous polychaete *Nereis,* which possesses a muscular pharynx armed with chitinous jaws that can be everted with great speed to seize prey (Figure 17-3A, p. 348). Once a capture is made, the pharynx is retracted and prey swallowed. Fish, amphibians, and reptiles use their teeth principally to grip prey and prevent its escape until they can swallow it whole. Snakes and some fishes can swallow enormous meals. Gripping of prey, together with absence of limbs, is associated with some striking feeding adaptations in these groups: recurved teeth for seizing and holding prey and distensible jaws and stomachs to accommodate their large and infrequent meals (Figure 32-3). Birds lack teeth, but their bills are often provided with serrated edges or the upper bill is hooked for seizing and tearing prey (see Figure 27-11, p. 566).

Figure 32-3
This African egg-eating snake, *Dasypeltis,* subsists entirely on hard-shelled birds' eggs, which it swallows whole. Its special adaptations are reduced size and number of teeth, enormously expansible jaw provided with elastic ligaments, and teethlike vertebral spurs that puncture the shell. Shortly after the second photograph was taken, the snake punctured and collapsed the egg, swallowed its contents, and regurgitated the crushed shell.

Many invertebrates are able to reduce food size by shredding devices (such as shredding mouthparts of many crustaceans) or by tearing devices (such as beaklike jaws of cephalopod molluscs). Insects have three pairs of appendages on their heads that serve variously as jaws, chitinous teeth, chisels, tongues, or sucking tubes. Usually the first pair serves as crushing teeth; the second as grasping jaws; and the third, as a probing and tasting tongue.

True mastication, chewing of food as opposed to tearing or crushing, is found only among mammals. Mammals usually have four different types of teeth, each adapted for specific functions. **Incisors** are designed for biting, cutting, and stripping; **canines** are for seizing, piercing and tearing; **premolars** and **molars,** at the back of the jaw, are for grinding and crushing (Figure 32-4). This basic pattern is often greatly modified in animals having specialized food habits (Figure 32-5; see also Figure 28-10, p. 591). Herbivores have suppressed canines but well-developed molars with enamel ridges for grinding. Well-developed, self-sharpening incisors of rodents grow throughout life and must be worn away by gnawing to keep pace with growth. Some teeth have become so highly modified that they are no longer useful for biting or chewing food. An elephant's

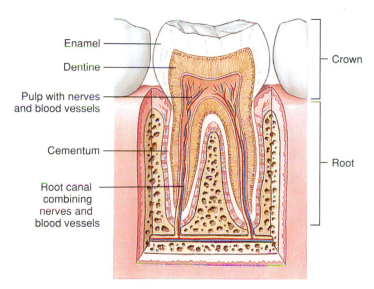

Figure 32-4

Structure of human molar tooth. A tooth is built of three layers of calcified tissue covering: enamel, which is 98% mineral and the hardest material in the body; dentine, which composes the mass of the tooth and is approximately 75% mineral; and cementum, which forms a thin covering over the dentine in the root of the tooth and is very similar to dense bone in composition. The pulp cavity contains loose connective tissue, blood vessels, nerves, and tooth-building cells.

tusk (Figure 32-6) is a modified upper incisor used for defense, attack, and rooting, and a male wild boar has modified canines that are used as weapons. Many feeding specializations of mammals are described on pp. 590–593.

Herbivorous, or plant-eating, animals have evolved special devices for crushing and cutting plant material. Some invertebrates have scraping mouthparts, such as the radula of snails (Figure 16-3, p. 319). Insects such as locusts have grinding and cutting mandibles; herbivorous mammals such as horses and cattle use wide, corrugated molars for grinding. All these mechanisms disrupt the tough cellulose cell wall to accelerate its digestion by intestinal microorganisms, as well as to release cell contents for direct enzymatic breakdown. Thus herbivores are able to digest food that carnivores cannot, and in doing so, convert plant material into protein for consumption by carnivores and omnivores.

Feeding on Fluids

Fluid feeding is especially characteristic of parasites, but it is practiced among many free-living forms as well. Some internal parasites (endoparasites) simply absorb the nutrient surrounding them, unwittingly provided by the host. Others bite and rasp host tissue, suck blood, and feed on contents of the host's intestine. External parasites (ectoparasites) such as leeches, lampreys, parasitic crustaceans, and insects use a variety of efficient piercing and sucking mouthparts to feed on blood or other body fluid. There are numerous arthropods that feed on fluids, for example, fleas, mosquitoes, sucking lice, bedbugs, ticks and mites, to name some of the more troublesome that assault

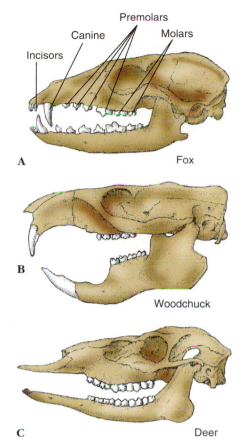

Figure 32-5

Mammalian dentition. **A,** Teeth of gray fox, a carnivore, showing the four types of teeth; **B,** Woodchuck, a rodent, has chisel-like incisors that continue to grow throughout life to replace wear; **C,** White-tailed deer, a browsing ungulate, with flat premolars and molars bearing complex ridges suited for grinding.

Figure 32-6

An African elephant loosening soil from a salt lick with its tusk. Elephants use their powerful modified incisors in many ways to search for food and water: plowing the ground for roots, prying apart branches to reach the edible cambium, and drilling into dry riverbeds for water.

humans as well as other vertebrate hosts. Many are vectors of serious diseases of humankind and thus qualify as far more than pesky annoyances.

Unfortunately for humans and other warm-blooded animals, the ubiquitous mosquito excels in its blood-sucking habit. Alighting gently, the mosquito punctures its prey with an array of six needlelike mouthparts (Figure 20-19B and C, p. 406). One of these is used to inject an anticoagulant saliva (responsible for the irritating itch that follows the "bite" and serving as a vector for microorganisms causing malaria, yellow fever, encephalitis, and other diseases); another mouthpart is a channel through which blood is sucked. It is of little comfort that only females dine on blood to obtain nutrients necessary for formation of their eggs.

DIGESTION

In the process of digestion, which means literally "carrying asunder," organic foods are mechanically and chemically broken into small units for absorption. Although food solids consist principally of carbohydrates, proteins, and fats, the very components that form the body of the consumer, these components must first be reduced to their simplest molecular units and dissolved before they can be assimilated. Each animal reassembles some of these digested and absorbed units into organic compounds of the animal's own unique pattern. Cannibalism confers no special metabolic benefit; victims of an animal's own kind are digested just as thoroughly as food composed of another species.

In protozoa and sponges digestion is entirely **intracellular** (Figure 32-7). A food particle is enclosed within a food vacuole by phagocytosis (see p. 47). Digestive enzymes are added and the products of digestion, the simple sugars, amino acids, and other molecules, are absorbed into the cell cytoplasm where they may be used directly or, in the case of multicellular animals, may be transferred to other cells. Food wastes are simply extruded from the cell by exocytosis (p. 48).

There are important limitations to intracellular digestion. Only particles small enough to be phagocytized can be accepted, and every cell must be capable of secreting all necessary enzymes, and of absorbing the products into their cytoplasm. These limitations were resolved with evolution of an **alimentary system** in which **extracellular** digestion of large food masses could take place. In extracellular digestion certain cells lining the **lumen** (cavity) of alimentary canals specialize in forming various digestive secretions, whereas other cells function largely, or entirely, in absorption. Many simpler metazoans, such as radiates, turbellarian flatworms, and ribbon worms (nemerteans), practice both intracellular and extracellular digestion. With evolution of greater complexity and appearance of complete mouth-to-anus alimentary systems, extracellular digestion became emphasized, together with increasing regional specialization of the digestive tract. For arthropods and vertebrates, digestion is almost entirely extracellular. Ingested food is exposed to various mechanical, chemical, and bacterial treatments, to different acidic and alkaline phases, and to digestive juices that are added at appropriate stages as food passes through the alimentary canal.

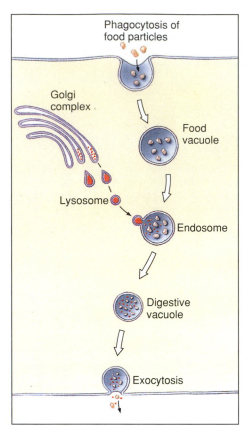

Figure 32-7

Intracellular digestion. Lysosomes containing digestive enzymes (lysozymes) are produced within a cell by the Golgi complex. Lysosomes fuse with food vacuoles and release enzymes that digest the enclosed food. Usable products of digestion are absorbed into the cytoplasm, and indigestible wastes are expelled.

Action of Digestive Enzymes

Mechanical processes of cutting and grinding by teeth and muscular mixing by an intestinal tract are important in digestion. However, reduction of foods to small, absorbable units relies principally on chemical breakdown by **enzymes,** discussed in Chapter 4 (p. 58–60). Digestive enzymes are **hydrolytic** enzymes, or **hydrolases,** so called because food molecules are split by a process of **hydrolysis,** breaking of a chemical bond by adding the components of water across it:

$$R—R + H_2O \xrightarrow[\text{enzyme}]{\text{digestive}} R—OH + H—R$$

In this general enzymatic reaction, R—R represents a food molecule split into two products, R—OH and R—H. Usually these reaction products must in turn be split repeatedly before the original molecule is reduced to its numerous subunits. Proteins, for example, are composed of hundreds, or even thousands, of interlinked amino acids, which must be completely separated before individual amino acids can be absorbed. Similarly, carbohydrates must be reduced to simple sugars. Fats (lipids) are reduced to molecules of glycerol, fatty acids, and monoglycerides, although some fats, unlike proteins and carbohydrates, may be absorbed without first being completely hydrolyzed. There are specific enzymes for each class of organic compounds. These enzymes are located in specific regions of an alimentary canal in an "enzyme chain," in which one enzyme may complete what another has started. Products then move posteriorly for still further hydrolysis.

Motility in Alimentary Canal

Food is moved through digestive tracts by **cilia** or by specialized **musculature,** and often by both. Movement is usually by cilia in acoelomate and pseudocoelomate metazoa that lack mesodermally derived gut musculature of true coelomates. Cilia move intestinal fluids and materials also in some eucoelomates, such as most molluscs, in which the coelom is weakly developed. In animals with well-developed coeloms, the gut is usually lined with two opposing layers of smooth muscle: a longitudinal layer, in which smooth muscle fibers run parallel with the length of the gut, and a circular layer, in which muscle fibers embrace the circumference of the gut. The most characteristic gut movement is **segmentation,** the alternate constriction of rings of smooth muscle of the intestine that constantly divide and squeeze contents back and forth (Figure 32-8A). Walter B. Cannon of homeostasis fame (p. 634), while still a medical student at Harvard in 1900, was the first to use X rays to watch segmentation in experimental animals that had been fed suspensions of barium sulfate. Segmentation serves to mix food but does not move it through the gut. Another kind of muscular action, called **peristalsis,** sweeps food down the gut with waves of contraction of circular muscle (Figure 32-8B).

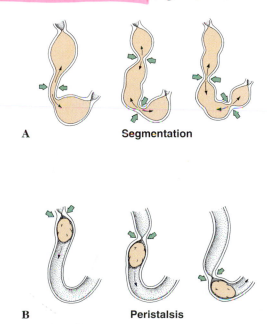

A   **Segmentation**

B   **Peristalsis**

Figure 32-8

Movement of intestinal contents by segmentation and peristalsis. **A,** Segmentational movements of food showing how constrictions squeeze food back and forth, mixing it with enzymes. The sequential mixing movements occur at about 1-second intervals. **B,** Peristaltic movement, showing how food is propelled forward by a traveling wave of contraction.

ORGANIZATION AND REGIONAL FUNCTION OF ALIMENTARY CANALS

Metazoan alimentary canals can be divided into five major regions: (1) reception, (2) conduction and storage, (3) grinding and early digestion, (4) terminal digestion and absorption, and (5) water absorption and concentration of solids. Food progresses from one region to the next, allowing digestion to proceed in sequential stages (Figures 32-9 and 32-10).

Receiving Region

The first region of an alimentary canal consists of devices for feeding and swallowing. These include **mouthparts** (for example, mandibles, jaws, teeth, radula, bills), **buccal cavity** and muscular **pharynx.** Most metazoans other than suspension feeders have **salivary glands** (buccal glands) that produce lubricating secretions containing mucus to assist swallowing (Figure 32-9). Salivary glands often have other specialized functions such as secretion of toxic enzymes for quieting struggling prey and secretion of salivary enzymes to begin digestion. Salivary secretions of leeches, for example, are complex mixtures containing an anesthetic substance (making its bite nearly painless) and several enzymes that prevent blood coagulation and increase blood flow by dilating veins and dissolving tissue cement that binds cells together.

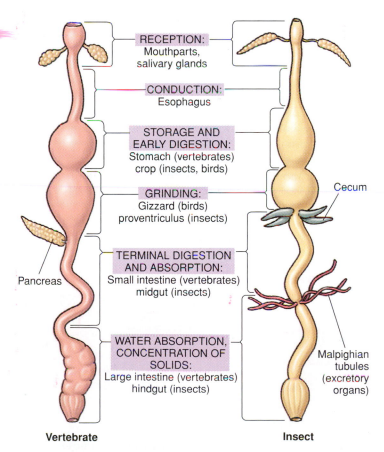

RECEPTION:
Mouthparts, salivary glands

CONDUCTION:
Esophagus

STORAGE AND EARLY DIGESTION:
Stomach (vertebrates)
crop (insects, birds)

GRINDING:
Gizzard (birds)
proventriculus (insects)

Cecum

Pancreas

TERMINAL DIGESTION AND ABSORPTION:
Small intestine (vertebrates)
midgut (insects)

WATER ABSORPTION, CONCENTRATION OF SOLIDS:
Large intestine (vertebrates)
hindgut (insects)

Malpighian tubules (excretory organs)

Vertebrate   **Insect**

Figure 32-9

Generalized digestive tracts of a vertebrate and an insect, showing the major functional regions of metazoan digestive systems.

Figure 32-10

Oral cavity and throat in humans in sagittal section (**A**), and sequence of swallowing (**B** to **D**).

Salivary **amylase** is a carbohydrate-splitting enzyme that begins hydrolysis of plant and animal starches. It is found only in certain herbivorous molluscs, some insects, and in primate mammals, including humans. Starches are long polymers of glucose. Salivary amylase does not completely hydrolyze starch, but breaks it mostly into two-glucose fragments called **maltose**. Some free glucose and longer fragments of starch are also produced. When food mass (bolus) is swallowed, salivary amylase continues to act for some time, digesting perhaps half of the starch before the enzyme is inactivated by the acidic environment of the stomach. Further starch digestion resumes beyond the stomach in the intestine.

A tongue is a vertebrate innovation, usually attached to the floor of the mouth, that assists in food manipulation and swallowing. Tongues are also used as chemosensors and possess taste buds that are used to determine palatability of foods (Chapter 33, p. 710) It may be used for other purposes, however, such as food capture (for example, chameleons, woodpeckers, anteaters) or as an olfactory sensor (many lizards and snakes).

In humans, swallowing begins with the tongue pushing moistened food toward the pharynx. The nasal cavity closes reflexively by raising the soft palate. As food slides into the pharynx, the epiglottis tips down over the trachea, nearly closing it (Figure 32-10). Some particles of food may enter the opening of the trachea but contraction of laryngeal muscles prevents it from going farther. Once food is in the esophagus, peristaltic contraction of esophageal muscles forces it smoothly toward the stomach.

Conduction and Storage Region

The **esophagus** of vertebrates and many invertebrates serves to transfer food to the digestive region. In many invertebrates (annelids, insects, octopods) the esophagus is expanded into a crop (Figure 32-9), used for food storage before digestion. Among vertebrates, only birds have a crop. This crop serves to store and soften food (grain, for example) before it passes to the stomach, or to allow mild fermentation of food before it is regurgitated to feed nestlings.

Region of Grinding and Early Digestion

In most vertebrates, and in some invertebrates, the **stomach** provides initial digestion as well as storage and mixing of food with digestive juices. Mechanical breakdown of food, especially plant food with its tough cellulose cell walls, often continues in herbivorous animals by grinding and crushing devices in the stomach. The muscular **gizzard** of terrestrial oligochaete worms and birds is assisted by stones and grit swallowed along with food or, in arthropods, by hardened linings (for example, chitinous teeth of an insect's proventriculus [Figure 32-9], and calcareous teeth of gastric mills of crustaceans).

Digestive diverticula—blind tubules or pouches arising from the main passage—often supplement stomachs of many invertebrates. They are usually lined with a multipurpose epithelium having cells specialized for secreting mucus or digestive enzymes, or absorption or storage. Examples include ceca of polychaete annelids, digestive glands of bivalve molluscs, hepatopancreas of crustaceans, and pyloric ceca of sea stars.

Herbivorous vertebrates have evolved several strategies for exploiting cellulose-splitting microorganisms to derive maximal nutrition from plant food. Despite its abundance on earth, woody cellulose that encloses plant cells can be digested only by an enzyme, **cellulase,** that has limited distribution in the living world. No metazoan animals can produce intestinal cellulase for direct digestion of cellulose. However many herbivorous metazoans harbor microorganisms (bacteria and protozoa) in their gut that do produce cellulase. These microorganisms ferment cellulose under anaerobic conditions of the gut, producing fatty acids and sugars that herbivores can use. While the ultimate fermentation machine is the multi-chambered stomach of the cud-chewing ruminants described on p. 592, many other animals harbor microorganisms in other parts of the gut, such as their intestine proper or their cecum.

Stomachs of carnivorous and omnivorous vertebrates are typically a U-shaped muscular tube provided with glands that produce proteolytic enzymes and strong acids, the latter an adaptation that probably arose for killing prey and halting bacterial activity. When food arrives at a stomach, a **cardiac sphincter** opens reflexively to allow food to enter, then closes to prevent regurgitation back into the esophagus. In humans, gentle peristaltic waves pass over the filled stomach at a rate of approximately three each minute. Churning is most vigorous at the intestinal end where food is steadily released into the **duodenum,** first region of the small intestine. A **pyloric sphincter** regulates flow of food into the intestine and prevents regurgitation into the stomach. Deep tubular glands in the stomach wall secrete **gastric juice,** in humans approximately 2 liters each day. Two types of cells line these glands: **chief cells,** which secrete **pepsinogen,** and **parietal cells,** which secrete **hydrochloric acid.** Pepsinogen is the precursor of **pepsin,** a **protease** (protein-splitting enzyme) that is produced from pepsinogen only in an acid medium (pH 1.6 to 2.4). This highly specific enzyme splits large proteins by preferentially splitting certain peptide bonds scattered along the peptide chain of a protein molecule. Although pepsin, because of its specificity, cannot completely degrade proteins, it effectively hydrolyzes them into smaller polypeptides. Other proteases that together can split all peptide bonds complete digestion of protein in the intestine. Pepsin is present in stomachs of nearly all vertebrates.

That stomach mucosa is not digested by its own powerful acid secretions results from another gastric secretion, mucin, a highly viscous organic compound that coats and protects the mucosa from both chemical and mechanical injury. We should note that despite a popular misconception that an "acid stomach" is unhealthy, a notion nourished in advertising, stomach acidity is normal and essential. Sometimes, however, the protective mucous coating fails. This failure is often associated with an infection from a bacterium (*Helicobacter pylori*) that secretes toxins causing inflammation of the stomach's lining. This inflammation may lead to a stomach ulcer.

Rennin (not to be confused with renin, an enzyme produced by vertebrate kidneys, p. 644) is a milk-curdling enzyme found in the stomachs of ruminant mammals. It probably occurs in many other mammals. By clotting and precipitating milk proteins, it slows movement of milk through the stomach. Rennin extracted from stomachs of calves is used in making cheese. Human infants, lacking rennin, digest milk proteins with acidic pepsin, just as adults do.

Figure 32-11

Dr. William Beaumont at Fort Mackinac, Michigan Territory, collecting gastric juice from Alexis St. Martin.

Secretion of gastric juices is intermittent. Although a small volume of gastric juice is secreted continuously, even during prolonged periods of starvation, secretion normally increases when stimulated by sight and smell of food, by presence of food in the stomach, and by emotional states such as anxiety and anger.

A most unusual and classic investigation of digestion was made by U.S. Army surgeon William Beaumont during the years 1825 to 1833. His subject was a young, hardliving French-Canadian voyageur named Alexis St. Martin, who in 1822 accidentally shot himself in the abdomen with a musket, the blast "blowing off integuments and muscles of the size of a man's hand, fracturing and carrying away the anterior half of the sixth rib, fracturing the fifth, lacerating the lower portion of the left lobe of the lungs, the diaphragm, and perforating the stomach." Miraculously the wound healed, but a permanent opening, or fistula, formed that permitted Beaumont to see directly into St. Martin's stomach (Figure 32-11). St. Martin became a permanent, although temperamental, patient in Beaumont's care, which included food and housing. Over a period of 8 years, Beaumont was able to observe and record how the lining of the stomach changed under different psychological and physiological conditions, how foods changed during digestion, effects of emotional states on stomach motility, and many other facts about the digestive process of his famous patient.

Region of Terminal Digestion and Absorption: Intestine

The importance of an intestine varies widely among animal groups. In invertebrates that have extensive digestive diverticula in which food is digested and phagocytized, an intestine may serve only as a pathway for conducting wastes out of the body. In other invertebrates with simple stomachs, and in all vertebrates, intestines are equipped for both digestion and absorption.

Devices for increasing the internal surface area of an intestine are highly developed in vertebrates, but are generally absent among invertebrates. Perhaps the most direct way to increase the absorptive surface of a gut is to increase its length. Coiling of their intestine is common among all vertebrate groups and reaches its highest development in mammals, in which the length of the intestine may exceed eight times the length of their body. Although a coiled intestine is rare among invertebrates, other strategies for increasing surface sometimes occur. For example, the **typhlosole** of terrestrial oligochaete worms (see Figure 17-12, p. 354), an inward folding of the dorsal intestinal wall that runs the full length of the intestine, effectively increases internal surface area of a gut in a narrow body lacking space for a coiled intestine.

Lampreys and sharks have longitudinal or spiral folds in their intestine. Other vertebrates have developed elaborate folds (amphibians, reptiles, birds, and mammals) and minute fingerlike projections called **villi** (birds and mammals), which give the inner surface of fresh intestinal tissue the appearance of velvet (Figure 32-12). Electron microscopy reveals that each cell lining the intestinal cavity additionally is bordered by hundreds of short, delicate processes called **microvilli** (Figure 32-13C and D). These processes, together with larger villi and intestinal folds, may increase the internal surface area of an intestine more than a million times as compared to a smooth cylinder of the same diameter. This elaborate surface greatly facilitates absorption of food molecules.

Digestion in the Vertebrate Small Intestine

Food is released into the small intestine through a **pyloric sphincter,** which relaxes at intervals to allow entry of acidic stomach contents into the initial segment of the small intestine, or **duodenum.** Two secretions pour into this region:

Figure 32-12
Scanning electron micrograph of a rat intestine showing numerous fingerlike villi that project into the lumen and vastly increase effective absorptive and secretory surface of the intestine. (×21)

From R. G. Kessel and R. H. Kardon, Tissues and Organs; A Text-Atlas of Scanning Electron Microscopy, 1979, W. H. Freeman and Co.

pancreatic juice and **bile** (Figure 32-14). Both secretions have a high bicarbonate content, especially pancreatic juice, which effectively neutralizes gastric acid, raising the pH of the liquefied food mass, now called **chyme,** from 1.5 to 7 as it enters the duodenum. This change in pH is essential because all intestinal enzymes are effective only in a neutral or slightly alkaline medium.

> Cells of intestinal mucosa, like those of the stomach mucosa, are subjected to considerable wear and are constantly undergoing replacement. Cells deep in the crypt between adjacent villi divide rapidly and migrate up the villus. In mammals the cells reach the tip of the villus in about two days. There they are shed, along with their membrane enzymes, into the lumen at a rate of some 17 billion a day along the length of the human intestine. Before they are shed, however, these cells differentiate into absorptive cells that transport nutrients into the network of blood and lymph vessels, once digestion is complete.

Pancreatic Enzymes   Pancreatic secretion of vertebrates contains several enzymes of major importance in digestion (Figure 32-14). Two powerful proteases, **trypsin** and **chymotrypsin,** continue enzymatic digestion of proteins begun by pepsin, which is now inactivated by the alkalinity of the intestine. Trypsin and chymotrypsin, like pepsin, are highly specific proteases that split peptide bonds deep inside a protein molecule. Hydrolysis of a peptide linkage may be shown as:

Pancreatic juice also contains **carboxypeptidase,** which removes amino acids from carboxyl ends of polypeptides; **pancreatic lipase,** which hydrolyzes fats into fatty acids and glycerol; **pancreatic amylase,** a starch-splitting enzyme identical to salivary amylase in its action; and **nucleases,** which degrade RNA and DNA to nucleotides.

Membrane Enzymes   Cells lining the intestine have digestive enzymes embedded in their surface membrane that continue digestion of carbohydrates, proteins, and phosphate compounds (Figure 32-14). These enzymes of microvillus membranes (Figure 32-13D) include **aminopeptidase** that splits terminal amino acids from the amino end of short peptides, and several **disaccharidases,** enzymes that split 12-carbon sugar molecules into 6-carbon units. The disaccharidases include **maltase,** which splits maltose into two molecules of glucose; **sucrase,** which splits sucrose to fructose and glucose; and **lactase,** which breaks lactose (milk sugar) into glucose and galactose. Also present is **alkaline phosphatase,** an enzyme that attacks a variety of phosphate compounds.

Figure 32-13

Organization of vertebrate digestive tract, showing **A,** the successive layers of mucosa, submucosa, muscle, and enveloping serosa; an enzyme-secreting digestive gland (e.g., pancreas); and the thin mesentery that positions the intestine within the body cavity. **B,** Portion of mucosal lining of intestine, showing fingerlike villi. **C,** Section of a single cell from the mucosal lining. **D,** Microvilli on surface of mucosal cell, rat intestine. (×16,400)

Figure 32-14

Secretions of a mammalian alimentary canal with the principal components and pH of each secretion.

REGION	SECRETION	pH	COMPOSITION
Salivary glands	Saliva	6.5	Amylase Bicarbonate
Stomach	Gastric juice	1.5	Pepsinogen HCl Rennin in ruminant mammals
Liver and gallbladder	Bile	7–8	Bile salts and pigments Cholesterol
Pancreas	Pancreatic juice	7–8	Trypsin, Chymotrypsin, Carboxypeptidase, Lipase, Amylase, Nucleases Bicarbonate
Small intestine	Membrane enzymes	7–8	Aminopeptidase Maltase Lactase Sucrase Alkaline Phosphatase

Although milk is the universal food of newborn mammals and one of the most complete human foods, many adult humans cannot digest milk because they are deficient in lactase, the enzyme that hydrolyzes lactose (milk sugar). Lactose intolerance is genetically determined. It is characterized by abdominal bloating, cramps, flatulence, and watery diarrhea, all appearing within 30 to 90 minutes after ingesting milk or its unfermented by-products. (Fermented dairy products, such as yogurt and cheese, create no intolerance problems.)

Northern Europeans and their descendants, which include a majority of European-Americans, are most tolerant of milk. Many other ethnic groups are generally intolerant to lactose, including the Japanese, Chinese, Inuit (Eskimos), South American Indians, and most Africans. Only about 30% of African-Americans are tolerant; those who are tolerant trace their ancestry to east and central Africa where dairying is traditional and tolerance to lactose is high.

Bile   The liver secretes bile into the **bile duct,** which drains into the upper intestine (duodenum). Between meals bile collects in the **gallbladder,** an expansible storage sac that releases bile when stimulated by the presence of fatty food in the duodenum. Bile contains water, bile salts, and pigments, but no enzymes. **Bile salts** (mainly sodium taurocholate and sodium glycocholate) are essential for digestion of fats. Fats, because of their tendency to remain in large, water-insoluble globules, are especially resistant to enzymatic digestion. Bile salts reduce surface tension of fat globules, allowing the churning action of the intestine to break fats into tiny droplets (emulsification). With total surface exposure of fat particles greatly increased, fat-splitting lipases are able to reach and hydrolyze the triglyceride molecules. The yellow-green color of bile is produced by **bile pigments,** breakdown products of hemoglobin from aged red blood cells. Bile pigments also give feces its characteristic color.

Bile production is only one of the liver's many functions. This highly versatile organ is a storehouse for glycogen, production center for plasma proteins, site of protein synthesis and detoxification of protein wastes, site for destruction of aged red blood cells, and center for metabolism of fat, amino acids, and carbohydrates.

Absorption

Little food is absorbed in the stomach because digestion is still incomplete and because of limited absorptive surface area. However, some materials, such as drugs and alcohol, are absorbed mostly there, which contributes to their rapid action. Most digested food is absorbed from the small intestine where numerous finger-shaped villi provide an enormous surface area through which materials can pass from the intestinal lumen into the circulation.

Carbohydrates are absorbed almost exclusively as simple sugars (monosaccharides, for example, glucose, fructose, and galactose) because the intestine is virtually impermeable to poly-

saccharides. Proteins are absorbed principally as their amino acid subunits, although a limited amount of small proteins or peptide fragments may be absorbed. Both active and passive transport processes transfer simple sugars and amino acids across the intestinal epithelium.

Immediately after a meal these materials are in such high concentration in the gut that they readily diffuse by facilitated transport into the blood, where their concentration is initially lower. However, if absorption were passive only, we would expect transfer to cease as soon as concentrations of a substance became equal on both sides of the intestinal epithelium. Passive transfer alone would permit valuable nutrients to be lost in feces. In fact, very little is lost because passive transfer is supplemented by an **active transport** mechanism located in epithelial cells that transfers food molecules into the blood. Materials thus are moved *against* their concentration gradient, a process requiring energy. Although not all food products are actively transported, those that are, such as glucose, galactose, and most amino acids, are handled by transport mechanisms specific for each kind of molecule.

As mentioned previously, fat droplets are emulsified by bile salts and then digested by pancreatic lipase. Triglycerides are broken into fatty acids and monoglycerides, which complex with bile salts to form minute droplets called **micelles.** When micelles contact the microvilli of the intestinal epithelium, fatty acids and monoglycerides are absorbed by simple diffusion. They then enter the endoplasmic reticulum of absorptive cells, where they are resynthesized into triglycerides before passing into **lacteals** (Figure 32-13B). From the lacteals, fat droplets enter the lymphatic system (Figure 31-18, p. 666) and eventually pass into the blood circulation through the thoracic duct. After a fatty meal, even a peanut butter sandwich, the presence of numerous fat droplets in the blood imparts a milky appearance to blood plasma.

Region of Water Absorption and Concentration of Solids

The large intestine consolidates the indigestible remnants of digestion by reabsorption of water to form solid or semisolid feces for removal from the body by defecation. Reabsorption of water is of special significance in insects, especially those living in dry environments, which must (and do) conserve nearly all water entering the rectum. Specialized **rectal glands** absorb water and ions as needed, leaving behind fecal pellets that are almost completely dry. In reptiles and birds, which also produce nearly dry feces, most water is reabsorbed in the cloaca. A white pastelike feces is formed containing both indigestible food wastes and uric acid.

The colon of humans contains enormous numbers of bacteria, which first enter the sterile colon of a newborn infant with its food. In adults approximately one-third of the dry weight of feces is bacteria; these include harmless bacteria as well as bacteria that can cause serious illness should they escape into the abdomen or bloodstream. Normally the body's defenses prevent invasion of such bacteria. Bacteria degrade organic wastes in feces and provide some nutritional benefit by synthesizing certain vitamins (vitamin K and small quantities of some of B vitamins), which are absorbed by the body.

REGULATION OF FOOD INTAKE

Most animals unconsciously adjust intake of food to balance energy expenditure. If energy expenditure is increased by greater physical activity, more food is consumed. Most vertebrates, from fish to mammals, eat for calories rather than bulk because, if their diet is diluted with fiber, they respond by eating more. Similarly, intake is adjusted downward following a period of several days when caloric intake is too high.

Hunger centers located in the hypothalamus and brainstem of the brain regulate the intake of food. A drop in blood glucose level stimulates a craving for food. While most animals seem able to stabilize their weight at normal levels with ease, many humans cannot. Obesity is rising throughout the industrial world and is a major health problem in many countries today. According to recent surveys some 60% of adults in the United States are either overweight or obese. (Assessment of overweight relies on body mass index [weight in kilograms divided by the square of height in meters], waist circumference, and risk factor for diseases associated with obesity.)

Some obese people do not eat significantly more food than thin people, but rather they have an inherited genetic predisposition to gain weight on a high-fat diet. The increase in fast-food meals, larger portion sizes, and more sedentary life-style is, however, associated with the prevalence of obesity in developed countries. Obese people may also have a reduced capacity to burn excess calories by "nonshivering thermogenesis" (p. 650). Placental mammals are unique in having a dark adipose tissue called **brown fat,** specialized for generation of heat. Newborn mammals, including human infants, have much more brown fat than adults. In human infants brown fat is located in the chest, upper back, and near the kidneys. Abundant mitochondria in brown fat contain a membrane protein called **thermogenin** that acts to uncouple the production of ATP during oxidative phosphorylation (p. 64). In people of average weight, increased caloric intake induces brown fat to dissipate excess energy as heat through the uncoupling action of thermogenin. We call this process "diet-induced thermogenesis." In many people tending toward obesity, this capacity is diminished.

There are other reasons for obesity in addition to the fact that many people simply eat too much and get too little exercise. Fat stores are supervised by the hypothalamus and brain stem, which may be set at a point higher or lower than the norm. A high setting can be lowered somewhat by exercise, but as dieters are painfully aware, the body defends its fat stores with remarkable tenacity. In 1995, a hormone produced by fat cells was found to cure obesity in mutant mice lacking the gene that produces the hormone. The hormone, called **leptin,** appears to operate through a feedback system that tells the hypothalamus and brain stem how much fat the body carries. If levels are high, release of leptin by fat cells leads to diminished appetite and increased thermogenesis. The discovery of leptin has initiated a flurry of research on obesity and a resurgence of commercial interest in producing a weight-loss drug based on leptin. Unfortunately, a majority of obese humans do not respond to infusions of leptin, and actually produce higher than normal amounts themselves. It appears that their brain has become resistant to these high leptin levels and does not respond by decreasing appetite. Current research now focuses on ways to decrease brain leptin resistance.

Regulation of Digestion

The digestive process is coordinated by a family of hormones (see Chapter 34) produced by the body's most diffuse endocrine tissue, the gastrointestinal tract. These hormones are examples of the many substances produced by a vertebrate's body that have hormonal function, yet are not necessarily produced by discrete endocrine glands. Because of their diffuse origins gastrointestinal (GI) hormones have been difficult to isolate and study and only recently have they been researched in depth.

Among the principal GI hormones are gastrin, cholecystokinin (CCK), and secretin (Figure 32-15). **Gastrin** is a small polypeptide hormone produced by endocrine cells in the pyloric portion of the stomach. Gastrin is secreted in response to stimulation by parasympathetic nerve endings (vagus nerve), and when protein food enters the stomach. Its main actions are to stimulate hydrochloric acid secretion and to increase gastric motility. Gastrin is an unusual hormone because it exerts its action on the same organ from which it is secreted. **CCK** is also a polypeptide hormone, and it has a striking structural resemblance to gastrin, suggesting that the two arose by duplication of ancestral genes. CCK is secreted by endocrine cells in the walls of the upper small intestine in response to the presence of fatty acids and amino acids in the duodenum. It has at least three distinct functions. It stimulates gallbladder contraction and thus increases the flow of bile salts into the intestine; it stimulates an enzyme-rich secretion from the pancreas; and it acts on the brain to contribute a feeling of satiety after a meal, particularly one rich in fats. The first hormone to be discovered, **secretin** (see the opening essay for Chapter 34 on p. 721), is produced by endocrine cells in the duodenal wall. It is secreted in response to food and strong acid in the stomach and small intestine, and

The body of many mammals contains two kinds of adipose tissue that perform completely different functions. **White adipose tissue,** the bulk of body fat, is adapted for the storage of fat derived mainly from surplus fats and carbohydrates in the diet. It is distributed throughout the body, particularly in the deep layers of skin. **Brown adipose tissue** is highly specialized for mediating nonshivering thermogenesis rather than for the storage of fat. Brown fat, unique to placental mammals, is especially well developed in hibernating species of bats and rodents, but is present also in many nonhibernating species such as rabbits, artiodactyls, carnivores, and primates (including humans). It is brown because it is packed with mitochondria containing large quantities of iron-bearing cytochrome molecules. In ordinary body cells, ATP is generated by a flow of electrons down the respiratory chain (p. 64). This ATP then powers various cellular processes. In brown fat cells heat is generated instead of ATP. Thermogenesis is activated by the sympathetic nervous system, which responds to signals from the hypothalamus and brain stem.

Figure 32-15

Three hormones of digestion. Shown are principal actions of the hormones gastrin, CCK (cholecystokinin), and secretin.

its principal action is to stimulate release of an alkaline pancreatic fluid that neutralizes stomach acid as it enters the intestine. It also aids fat digestion by inhibiting gastric motility and increasing production of an alkaline bile secretion from the liver.

GI hormones continue to be isolated and their structure determined. So far, all are peptides, and many are present in both the GI tract and in the central nervous system. One of these is CCK, which has been found in high concentrations in the cerebral cortex and hypothalamus of mammals. By providing a feeling of satiety after eating (just mentioned) it may play some role in regulating appetite. Several other GI peptides, for example, vasoactive intestinal peptide (VIP) and gastric inhibitory peptide (GIP) appear to play neurotransmitter roles in the brain. This unexpected versatility has served to broaden our concept of hormones as molecules capable of functioning in several different ways.

NUTRITIONAL REQUIREMENTS

Food of animals must include **carbohydrates, proteins, fats, water, mineral salts,** and **vitamins.** Carbohydrates and fats are required as fuels for energy and for the synthesis of various substances and structures. Proteins (actually the amino acids of which they are composed) are needed for synthesis of specific proteins and other nitrogen-containing compounds. Water is required as the solvent for body chemistry and as a major component of all fluids of the body. Inorganic salts are required as anions and cations of body fluids and tissues and form important structural and physiological components throughout the body. Vitamins are accessory factors from food that are often built into the structure of many enzymes.

A **vitamin** is a relatively simple organic compound that is not a carbohydrate, fat, protein, or mineral and is required in very small amounts in the diet for some specific cellular func-

tion. Vitamins are not sources of energy but are often associated with activity of important enzymes that serve vital metabolic roles. Plants and many microorganisms synthesize all the organic compounds they need; animals, however, have lost certain synthetic abilities during their long evolution and depend ultimately on plants to supply these compounds. Vitamins therefore represent synthetic gaps in the metabolic machinery of animals.

Vitamins are usually classified as fat soluble (soluble in fat solvents such as ether) or water soluble. The **water-soluble vitamins** include the B complex and vitamin C (Table 32-1). Vitamins of the B complex, so grouped because the original B vitamin was subsequently found to consist of several distinct molecules, tend to be found together in nature. Almost all animals, vertebrate and invertebrate, require B vitamins; they are "universal" vitamins. A dietary need for vitamin C and the **fat-soluble vitamins** A, D_3, E, and K is mostly restricted to vertebrates, although some are required by certain invertebrates. Even within groups of close relationship, requirements for vitamins are relative, not absolute. A rabbit does not require vitamin C, but guinea pigs and humans do. Some songbirds require vitamin A, but others do not.

The recognition years ago that many human diseases and those of domesticated animals were caused by or associated with dietary deficiencies led biologists to search for specific nutrients that would prevent such diseases. These studies eventually yielded a list of **essential nutrients** for human beings and other animal species studied. Essential nutrients are those needed for normal growth and maintenance and that *must* be supplied in the diet. In other words, it is "essential" that these nutrients be in the diet because an animal cannot synthesize them from other dietary constituents. Nearly 30 organic compounds (amino acids and vitamins) and 21 elements are essential for humans (Table 32-1). Considering that the body contains

TABLE 32.1
Human Nutrient Requirements

Water-Soluble Vitamins
Thiamine (B$_1$)
Riboflavin (B$_2$)
Niacin (nicotinic acid)
Pyridoxine (B$_6$)
Pantothenic acid
Folacin (folic acid)
Vitamin B$_{12}$ (cobalamin)
Biotin
Ascorbic acid (C)

Fat-Soluble Vitamins
A, D$_3$, E, and K

Minerals

Major	Trace
Calcium	Iron
Phosphorus	Fluorine
Sulfur	Zinc
Potassium	Copper
Chlorine	Silicon
Sodium	Vanadium
Magnesium	Tin
	Nickel
	Selenium
	Manganese
	Iodine
	Molybdenum
	Chromium
	Cobalt

Amino Acids
Phenylalanine
Lysine
Isoleucine
Leucine
Valine
Methionine
Tryptophan
Threonine
Arginine*
Histidine*

Polyunsaturated Fatty Acids
Arachidonic
Linoleic
Linolenic

*Required for normal growth of children.

thousands of different organic compounds, the list in Table 32-1 is remarkably short. Animal cells have marvelous powers of synthesis, enabling them to build compounds of enormous variety and complexity from a small, select group of raw materials.

In the average diet of North Americans approximately 50% of the total calories (energy content) comes from carbohydrates and 40% comes from lipids. Proteins, essential as they are for structural needs, supply only a little more than 10% of total calories of an average diet of North Americans. Carbohydrates are widely consumed because they are more abundant and cheaper than proteins or lipids. Actually humans and many other animals can subsist on diets devoid of carbohydrates, provided sufficient total calories and essential nutrients are present. Inuits, before the decline of their native culture, lived on a diet that was high in fat and protein and very low in carbohydrate.

Lipids are needed principally to provide energy. However, at least three fatty acids are essential for humans because we cannot synthesize them. Much interest and research have been devoted to lipids in our diets because of the association between fatty diets and the disease **atherosclerosis.** The matter is complex, but evidence suggests that atherosclerosis may occur when a diet is high in saturated lipids (lipids with no double bonds in the carbon chains of fatty acids) but low in polyunsaturated lipids (two or more double bonds in the carbon chains).

Atherosclerosis (Gr. *atheroma,* tumor containing gruel-like matter, + *sclerosis,* to harden) is a degenerative disease in which fatty substances are deposited in the lining of arteries, resulting in narrowing of the passage and eventual hardening and loss of elasticity. Current evidence suggests that inflammation of the artery wall precedes deposition of fat. Elevated cholesterol levels can fuel such inflammation.

Proteins are expensive foods and limited in the diet. Proteins, of course, are not themselves the essential nutrients but rather contain essential amino acids. Of the 20 amino acids commonly found in proteins, eight and possibly 10 are essential to humans (Table 32-1). We can synthesize the rest from other amino acids. Generally, animal proteins have more essential amino acids than do proteins of plant origin. All eight essential amino acids must be present simultaneously in the diet for protein synthesis. If one or more is missing, use of the other amino acids will be reduced proportionately; they cannot be stored and are metabolized for energy. Thus heavy reliance on a single plant source as a diet will inevitably lead to protein deficiency. This problem can be corrected if two kinds of plant proteins having complementary strengths in essential amino acids are ingested together. For example, a balanced protein diet can be prepared by mixing wheat flour, which is deficient only in lysine, with a legume (peas or beans), which is a good source of lysine but deficient in methionine and cysteine. Each plant complements the other by having adequate amounts of those amino acids that are deficient in the other.

Because animal proteins are rich in essential amino acids, they are in great demand in all countries. North Americans eat far more animal proteins than do Asians and Africans. In 1989 the annual per capita consumption of red meat was 76 kg in the United States, 27 kg in Japan, 12 kg in Egypt, and 1 kg in India.* Seventy percent of protein in the diet of Americans comes from

*Brown, L. R. 1991. State of the World 1991. New York, Worldwatch Institute/W.W. Norton & Company, p. 159.

animal products and 30% from plants. By comparison, in China only 11% comes from animal sources and 89% from plants. North Americans consume approximately one-quarter of all beef produced in the world. High consumption of meat in North America and Europe carries the price of a high death rate from so-called diseases of affluence: heart disease, stroke, and certain kinds of cancer.

Two different types of severe food deficiency are recognized: marasmus, general undernourishment from a diet low in both calories and protein, and kwashiorkor, protein malnourishment from a diet adequate in calories but deficient in protein. Marasmus (Gr. *marasmos,* to waste away) is common in infants weaned too early and placed on low-calorie–low-protein diets; these children are listless, and their bodies waste away. Kwashiorkor is a West African word describing a disease a child gets when displaced from the breast by a newborn sibling. This disease is characterized by retarded growth, anemia, weak muscles, a bloated body with typical pot belly, acute diarrhea, susceptibility to infection, and high mortality.

Undernourishment and malnourishment rank as two of the world's oldest problems and remain major health problems today, afflicting an eighth of the human population. Growing children and pregnant and lactating women are especially vulnerable to the devastating effects of malnutrition. Cell proliferation and growth in the human brain are most rapid in the terminal months of gestation and the first year after birth. Adequate protein for neuron development is a requirement during this critical time to prevent neurological dysfunction. Brains of children who die of protein malnutrition during the first year of life have 15% to 20% fewer brain cells than those of normal children (Figure 32-16). Malnourished children who survive this period suffer permanent brain damage and cannot be helped by later corrective treatment (Figure 32-17). Recent studies suggest that poverty, with attendant lack of educational and medical resources, and

Figure 32-17
Refugee child suffering severe malnutrition.

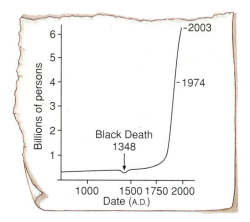

Figure 32-18
Portion of a graph for human population growth since A.D. 800, as it appeared in the 1979 edition of this book when the earth's population had passed 4 billion five years earlier, and updated to show the 2003 figure of 6.3 billion.

lowered expectations, exacerbates effects of malnutrition by delaying intellectual development.[†]

The world's precarious food supply is threatened by rapid population growth. The world population was 2 billion in 1927, reached 4 billion in 1974, passed 6.3 billion in January of 2003 (Figure 32-18), and is expected to pass 8 billion by the year 2030,[‡] several years ahead of earlier estimates. Approximately 78 million people are added each year. An equivalent of the total 1999 United States population of 274 million people is added to the world every 42 months. Yet today, as demand for food increases, world per capita production of grain and world fish catch are in decline.[§] Furthermore, the world each year loses billions of tons of topsoil and trillions of gallons of groundwater needed to grow food crops. In the view of many, the exploding human population is a major force driving the global environmental crisis.

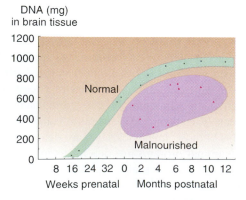

Figure 32-16
Effect of early malnutrition on cell number (measured as total DNA content) in a human brain. This graph shows that malnourished infants (*purple oval*) have far fewer brain cells than do normal infants (*green growth curve*).

†Brown, J. L, and E. Pollitt. 1996. Malnutrition, poverty and intellectual development. Sci. Am. **274**:38–43 (Feb.).
‡See U.S. Bureau of the Census. http://www.census.gov/ipc/www/world pop.html
§According to State of the World 1996, Worldwatch Institute, the world grain harvest has not grown at all since 1990, and 13 of the 15 leading oceanic fisheries are in decline. See also: Safina, C. 1995. The world's imperiled fish. Sci. Am. **273**:46–53 (Nov.).

SUMMARY

Autotrophic organisms (mostly green plants), using inorganic compounds as raw materials, capture the energy of sunlight through photosynthesis and produce complex organic molecules. Heterotrophic organisms (bacteria, fungi, and animals) use organic compounds synthesized by plants, and chemical bond energy stored therein, for their own nutritional and energy needs.

A large group of animals with very different levels of complexity feed by filtering minute organisms and other particulate matter suspended in water. Others feed on organic detritus deposited in the substrate. Selective feeders, on the other hand, have evolved mechanisms for manipulating larger food masses, including various devices for seizing, scraping, boring, tearing, biting, and chewing. Fluid feeding is characteristic of endoparasites, which may absorb food across the general body surface, and of ectoparasites, herbivores, and predators that have developed specialized mouthparts for piercing and sucking.

Digestion is a process of breaking down food mechanically and chemically into molecular subunits for absorption. Digestion is intracellular in protozoan groups and sponges. In more complex metazoans it is supplemented, and finally replaced entirely, by extracellular digestion, which takes place in sequential stages in a tubular cavity, or alimentary canal. The mouth receives food, mixes it with lubricating saliva, then passes it down the esophagus to regions where food may be stored (crop), or ground (gizzard), or acidified and subjected to early digestion (vertebrate stomach). Among vertebrates, most digestion occurs in the small intestine. Enzymes from the pancreas and intestinal mucosa hydrolyze proteins, carbohydrates, fats, nucleic acids, and various phosphate compounds. The liver secretes bile, containing salts that emulsify fats. Once foods are digested, their products are absorbed as molecular subunits (monosaccharides, amino acids, fatty acids and glycerol) into blood or lymph vessels of villi of the small intestine. The large intestine (colon) serves mainly to absorb water and minerals from the food wastes as they pass through it. It also contains symbiotic bacteria that produce certain vitamins.

Most animals balance food intake with energy expenditure. Food intake is regulated primarily by hunger centers located in the hypothalamus and brain stem. In mammals, should caloric intake exceed requirements for energy, excess calories normally are dissipated as heat in specialized brown fat tissue.

Several gastrointestinal hormones coordinate digestive functions. They include gastrin, which stimulates acid secretion by the stomach; CCK, which stimulates gallbladder and pancreatic secretion and leads to satiety; and secretin, which stimulates bicarbonate secretion from the pancreas and inhibits gastric motility.

All animals require a balanced diet containing both fuels (mainly carbohydrates and lipids) and structural and functional components (proteins, minerals, and vitamins). For every multicellular animal, certain amino acids, lipids, vitamins, and minerals are "essential" dietary factors that cannot be produced by an animal's own synthetic machinery. Animal proteins are better-balanced sources of amino acids than are plant proteins, which tend to lack one or more essential amino acids. Undernourishment and protein malnourishment are among the world's major health problems, afflicting millions of people.

REVIEW QUESTIONS

1. Distinguish between the following pairs of terms: autotrophic and heterotrophic; phototrophic and chemotrophic; herbivores and carnivores; omnivores and insectivores.
2. Suspension feeding is one of the most important methods of feeding among animals. Explain the characteristics, advantages, and limitations of suspension feeding, and name three different groups of animals that are suspension feeders.
3. An animal's feeding adaptations are an integral part of an animal's behavior and usually shape the appearance of the animal itself. Discuss contrasting feeding adaptations of carnivores and herbivores.
4. Explain how food is propelled through the digestive tract.
5. Compare intracellular with extracellular digestion and suggest why there has been a phylogenetic trend in some animals from intracellular to extracellular digestion.
6. Which structural modifications vastly increase the internal surface area of the intestine (both invertebrate and vertebrate), and why is this large surface area important?
7. Trace digestion and final absorption of a carbohydrate (starch) in the vertebrate gut, naming the carbohydrate-splitting enzymes, where they are found, the breakdown products of starch digestion, and in what form they are finally absorbed.
8. As in question 7, trace digestion and final absorption of a protein.
9. Explain how fats are emulsified and digested in the vertebrate gut. Explain how bile aids the digestive process even though it contains no enzymes. Provide an explanation for the following observation: fats are broken into fatty acids and monoglycerides in the intestinal lumen, but appear later in the blood as fat droplets.
10. Explain the phrase "diet-induced thermogenesis" and relate it to the problem of obesity in some people. What are other factors that may contribute to human obesity?
11. Name three hormones of the gastrointestinal tract and explain how they assist in the coordination of gastrointestinal function.
12. Name the basic classes of foods that serve mainly as (a) fuels and as (b) structural and functional components.
13. If vitamins are neither biochemically similar compounds nor sources of energy, what characteristics distinguish vitamins as a distinct group of nutrients? What are the water-soluble and the fat-soluble vitamins?
14. Why are some nutrients considered "essential" and others "nonessential" even though both types of nutrients are used in growth and tissue repair?
15. Explain the difference between saturated and unsaturated lipids, and comment on the current interest in these compounds as they relate to human health.
16. What is meant by "protein complementarity" among plant foods?

SELECTED REFERENCES

Blaser, M. J. 1996. The bacteria behind ulcers. Sci. Am. **274**:104–107 (Jan.). *We now know that most cases of stomach ulcers are caused by acid-loving microbes. At least one-third of the human population are infected although most do not become ill.*

Magee, D. F., and A. F. Dalley, II. 1986. Digestion and the structure and function of the gut. Basel, Switzerland, S. Karger AG. *Comprehensive treatment of mammalian (mostly human) digestion.*

Milton, K. 1993. Diet and primate evolution. Sci. Am. **269**:86–93 (Aug.). *Studies with primates suggest that modern human diets often diverge greatly from those to which the human body may be adapted.*

Randall, D., W. Burggren, and K. French. 2001. Eckert animal physiology: mechanisms and adaptations. New York, W. H. Freeman and Company. *A comprehensive and comparative treatment of animal physiology.*

Sanderson, S. L., and R. Wassersug. 1990. Suspension-feeding vertebrates. Sci. Am. **262**:96–101 (Mar.). *A variety of vertebrates, some enormous in size, eat by filtering out small organisms from massive amounts of water passed through a feeding apparatus.*

Stevens, C. E. 1988. Comparative physiology of the vertebrate digestive system. New York, Cambridge University Press. *Lucid and balanced treatment of anatomical characteristics of vertebrate digestive systems and the physiology and biochemistry of food digestion.*

Weindrach, R. 1996. Caloric restriction and aging. Sci. Am. **274**:46–52 (Jan.). *Organisms from single-celled protists to mammals live longer on well-balanced but low-calorie diets. The potential benefits for humans are examined.*

Willmes, P., G. Stone, and I. Johnston. 2000. Environmental physiology of animals. Oxford, Blackwell Science Ltd., U.K. *Well-written information regarding environmental adaptations of both vertebrates and invertebrates.*

ZOOLOGY LINKS TO THE INTERNET

Visit the textbook's Online Learning Center at www.mhhe.com/zoology to find live Internet links for each of the topics listed here.

Vertebrates: Macroscopic Anatomy of the Digestive System

Vertebrates: Microscopic Anatomy of the Digestive System

Nutrition

Digestion and Nutrition

Human Digestive System Topics

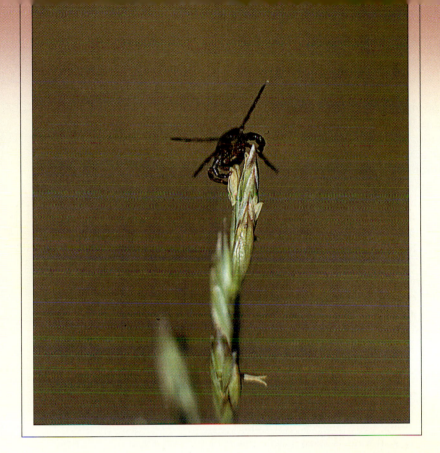

Wood tick on a grass stem awaits its host.

Nervous Coordination

Nervous System and Sense Organs

The Private World of the Senses

By any measure, people enjoy a rich sensory world. We are continually assailed by information from the senses of vision, hearing, taste, olfaction, and touch. These classic five senses are supplemented by sensory inputs of cold, warmth, vibration, and pain, as well as by information from numerous internal sensory receptors that operate silently and automatically to keep our interior domain working smoothly.

The world revealed by our senses is uniquely human even if our individual senses are not. We share this exclusive world with no other animal, nor can we venture into the sensory world of any other animal except as an abstraction through our imagination.

The idea that each animal enjoys an unshared sensory world was first conceived by Jakob von Uexküll, a seldom cited German biologist of the early part of the last century. Von Uexküll asks us to try to enter the world of a tick through our imagination, supplemented by what we know of tick biology. It is a world of temperature, of light and dark, and of the odor of butyric acid, a chemical common to all mammals. Insensible to all other stimuli, the tick climbs up a blade of grass to wait, for years if necessary, for cues that will betray the presence of a potential host. Later, swollen with blood, she drops to the earth, lays her eggs, and dies. The tick's impoverished sensory world, devoid of sensory luxuries and fine-tuned by natural selection for the world she will encounter, has ensured her single goal, reproduction.

A bird and a bat may share for a moment precisely the same environment. The worlds of their perceptions, however, are vastly different, structured by the limitations of the sensory windows each employs and by the brain that garners and processes what it needs for survival. For one it is a world dominated by vision; for the other, echolocation. The world of each is alien to the other, just as their worlds are to us.

The nervous system originated in a fundamental property of life: **irritability,** the ability to respond to environmental stimuli (Chapter 1, p. 9). The response may be simple, such as a protozoan moving to avoid a noxious substance, or quite complex, such as a vertebrate animal responding to elaborate signals of courtship. A protistan receives and responds to a stimulus, all within the confines of a single cell. Evolution of multicellularity and more complex levels of animal organization required increasingly complex mechanisms for communication between cells and organs. Relatively rapid communication is by **neural mechanisms** and involves propagated electrochemical changes in cell membranes. The basic plan of a nervous system is to code information and to transmit and process it for appropriate action. These functions are examined in this chapter. Relatively less rapid or long-term adjustments in animals are governed by **hormonal mechanisms,** subject of Chapter 34.

NEURONS: FUNCTIONAL UNITS OF NERVOUS SYSTEMS

A **neuron,** or nerve cell, may assume many shapes, depending on its function and location; a typical kind is shown diagrammatically in Figure 33-1. From the nucleated cell body extend cytoplasmic processes of two types: one or more **dendrites** in all but the simplest neuron, and a single **axon.** As the name dendrite suggests (Gr. *dendron,* tree), these neurons are often profusely branched. They, and the entire cell body surface, are the nerve cell's receptive apparatus, often receiving information from several different sources at once. Some of these inputs are excitatory, others inhibitory.

The single axon (Gr. *axon,* axle), often a long fiber (meters in length in the largest mammals), is relatively uniform in diameter, and typically carries impulses away from the cell body. In vertebrates and some complex invertebrates, the axon is often covered with an insulating sheath of **myelin.**

Neurons are commonly classified as **afferent,** or sensory; **efferent,** or motor; and **interneurons,** which are neither sensory nor motor but connect neurons with other neurons. Afferent and efferent neurons lie mostly outside the central nervous system (brain and nerve cord) while interneurons, which in humans are 99% of all neurons in the body, lie entirely within the central nervous system. Afferent neurons are connected to **receptors.** Receptors function to convert external and internal environmental stimuli into nerve impulses, which are carried by afferent neurons into the central nervous system. Here impulses may be perceived as conscious sensation. Impulses also move to efferent neurons, which carry them via the peripheral system to **effectors,** such as muscles or glands.

In vertebrates, nerve processes (usually axons) are often bundled together in a well-formed wrapping of connective tissue to form a **nerve** (Figure 33-2). Cell bodies of these nerve processes are located either in the central nervous system or in **ganglia,** which are discrete bundles of nerve cell bodies located outside the central nervous system.

Surrounding neurons are non-nervous **neuroglial cells** (often simply called **"glial"** cells) that have a special relation-

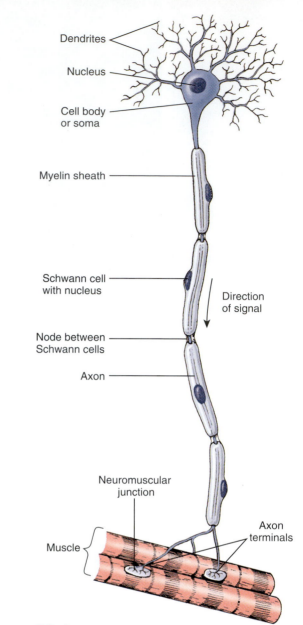

Figure 33-1
Structure of a motor (efferent) neuron.

Figure 33-2
Structure of a nerve showing nerve fibers surrounded by various layers of connective tissue. A nerve may contain thousands of both efferent and afferent fibers.

Figure 33-3

Development of the myelin sheath in a myelinated neuron of the peripheral nervous system. The whole Schwann cell grows around an axon, then rotates around it, enclosing the axon in a tight, multilayered sheath. The myelin sheath insulates a nerve axon and facilitates transmission of nerve impulses.

ship to neurons. Glial cells are extremely numerous in the vertebrate brain, where they outnumber neurons 10 to 1 and may form almost half the volume of the brain. Some glial cells form intimate insulating sheaths of lipid-containing **myelin** around nerve fibers. Vertebrate nerves are often enclosed by concentric rings of myelin, produced by special glial cells called **Schwann cells** (Figure 33-3) in the peripheral nervous system, and **oligodendrocytes** in the central nervous system. Certain glial cells, called **astrocytes** because of their radiating, starlike shape, serve as nutrient and ion reservoirs for neurons, as well as a scaffold during brain development, enabling migrating neurons to find their destinations from points of origin. Astrocytes, and smaller **microglial** cells, are essential for the regenerative process that follows brain injury. Unfortunately, astrocytes also participate in several diseases of the nervous system, including

Parkinsonism and multiple sclerosis. Other functional roles of glial cells are still being determined.

Nature of a Nerve Impulse

A **nerve** impulse is an electrochemical message of neurons, the common functional denominator of all nervous system activity. Despite the incredible complexity of nervous systems of many animals, nerve impulses are basically alike in all neurons and in all animals. An impulse, or **action potential** (see p. 698), is an "all-or-none" phenomenon; either the fiber is conducting an impulse, or it is not. Because all impulses are alike, the only way a nerve fiber can vary its signal is by changing the frequency of impulse conduction. Frequency change is the language of a nerve fiber. A fiber may conduct no impulses at all or very few per second up to a maximum approaching 1000 per second. The higher the frequency (or rate) of conduction, the greater is the level of excitation.

Resting Membrane Potential

Membranes of neurons, like all cellular membranes, have a special permeability that creates ionic imbalances. The interstitial fluid surrounding neurons contains relatively high concentrations of sodium (Na^+) and chloride (Cl^-) ions, but a low concentration of potassium ions (K^+) and large impermeable anions with negative charge, such as proteins. Inside the neuron, the ratio is reversed: K^+ and impermeable anion concentration is high, but Na^+ and Cl^- concentrations are low (Figure 33-4; see also Figure 31-1B, p. 656). These differences are pronounced; there is approximately 10 times more Na^+ outside than in and 25 to 30 times more K^+ inside than out.

When at rest, the membrane of a neuron is selectively permeable to K^+, which can traverse the membrane through special potassium channels. The permeability to Na^+ is nearly zero because Na^+ channels are closed in a resting membrane. Potassium ions tend to diffuse outward through the membrane, following the gradient of potassium concentration. Very quickly the

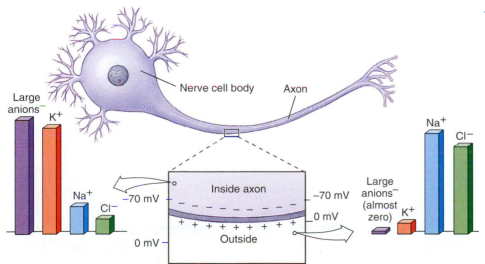

→ ## Figure 33-4

Ionic composition inside and outside a resting nerve cell. An active sodium-potassium exchange pump located in the cell membrane drives sodium to the outside, keeping its concentration low inside. Potassium concentration is high inside. Although the membrane is "leaky" to potassium, this ion is held inside by the repelling positive charge outside the membrane, and its attraction to large negatively charged anions inside the membrane, which cannot leave the cell.

positive charge outside reaches a level that prevents any more K+ from diffusing out of the axon (because like charges repel each other), and because large anions cannot pass through the membrane, positively charged potassium ions are drawn back into the cell. Now the resting membrane is at equilibrium, with an electrical gradient that exactly balances the concentration gradient. This **resting membrane potential** is usually –70 mV (millivolts), with the inside of the membrane negative with respect to the outside.

Action Potential

A nerve impulse is a rapidly moving change in electrical membrane potential called an **action potential** (Figure 33-5). It is a very rapid and brief depolarization of the membrane of the nerve fiber. In most nerve fibers, the action potential does not simply return the membrane potential to zero, but instead overshoots zero. In other words, the membrane potential reverses for an instant so that the outside becomes negative compared with the inside. Then, as the action potential moves ahead, the membrane returns to its normal resting membrane potential, ready to conduct another impulse. The entire event occupies approximately a millisecond. Perhaps the most significant property of the nerve impulse is that it is **self-propagating;** once started the impulse moves ahead automatically and unchanged in intensity, much like the burning of a fuse.

What causes the reversal of polarity in the cell membrane during passage of an action potential? We have seen that the resting membrane potential depends on the high membrane permeability (leakiness) to K+, some 50 to 70 times greater than the permeability to Na+. When the action potential arrives at a given point, Na+ channels suddenly open, permitting a flood of Na+ to diffuse into the axon from the outside, moving down the concentration gradient for Na+. Actually only a very minute amount of Na+ moves across the membrane—less than one-millionth of the Na+ outside—but this sudden rush of positive ions cancels the local resting membrane potential. The membrane is **depolarized,** creating a minute electrical "hole." Potassium ions, finding their electrical barrier gone, begin to move out of the cell. Then,

as the action potential passes, the membrane quickly regains its resting properties. It becomes once again practically impermeable to Na+, the outward movement of K+ is checked, and the membrane again becomes leaky to inward movement of K+.

Thus, the rising phase of an action potential is associated with rapid influx (inward movement) of Na+ (Figure 33-5). When the action potential reaches its peak, Na+ permeability is restored to normal, and K+ permeability briefly increases above the resting level and K+ ions move out. Increased potassium permeability causes the action potential to drop rapidly toward the resting membrane level, during the **repolarization** phase. The membrane is now ready to transmit another nerve impulse.

Sodium Pump

A resting cell membrane has a very low permeability to Na+. Nevertheless due to the high concentration gradient and electrical attraction, some Na+ leaks through it into the cell, even in the resting condition. When the axon is active, during an action potential, Na+ flows inward with each passing impulse. If not removed, the accumulation of Na+ inside the axon would cause the resting membrane potential of the fiber to decay. This decay is prevented by **sodium pumps,** each a complex of protein subunits embedded in the plasma membrane of the axon (see Figure 3-19, p. 48). Each sodium pump uses energy in ATP to transport sodium from the inside to the outside of the membrane. The sodium pump in nerve axons, as in many other cell membranes, also moves K+ into the axon while it is moving Na+ out. Thus, it is a **sodium-potassium exchange pump** that helps to restore the ion gradients of both Na+ and K+. Astrocytes (mentioned on p. 697) help to maintain the correct balance of ions surrounding neurons by storing excess potassium produced during neuronal activity.

High-Speed Conduction

Although ionic and electrical events associated with action potentials are much the same throughout the animal kingdom, conduction velocities vary enormously from nerve to nerve and

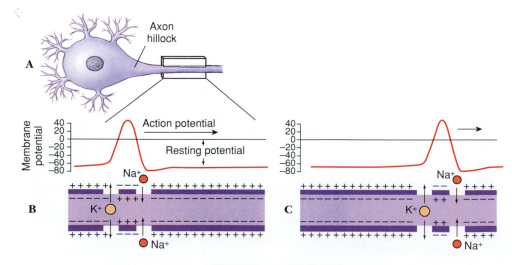

→ # Figure 33-5

Conduction of an action potential or nerve impulse. The action potential originates in the axon hillock of the neuron (**A**) and moves toward the right. **B** and **C** show the electrical event and associated changes in localized membrane permeability to sodium and potassium. The position of the action potential in **C** is shown about 4 milliseconds after **B**. When the impulse arrives at a point, sodium channels open, allowing sodium ions to enter. Sodium inflow reverses the membrane polarity, making the inner surface of the axon positive and the outside negative. Sodium channels then close and potassium channels open. Potassium ions move out and restore the normal resting potential.

from animal to animal—from as slow as 0.1 m/sec in sea anemones to as fast as 120 m/sec in some mammalian motor axons. The speed of conduction is highly correlated with diameter of the axon. Small axons conduct slowly because internal resistance to current flow is high. In most invertebrates, where fast conduction velocities are important for quick response, such as in locomotion to capture prey or to avoid capture, axon diameters are larger. Giant axons of squids are nearly 1 mm in diameter and carry impulses 10 times faster than ordinary fibers in the same animal. A squid's giant axon innervates the animal's mantle musculature and is used for powerful mantle contractions when it swims by jet propulsion. Similar giant axons enable earthworms, which are normally slow-moving animals, to withdraw almost instantaneously into their burrows when startled.

Although vertebrates do not possess giant axons, they achieve high-conduction velocities by a cooperative relationship between axons and the investing layers of myelin laid down by the Schwann cells or oligodendrocytes described on p. 697. Insulating myelin sheaths are interrupted at intervals by nodes (called **nodes of Ranvier**) where the surface of the axon is exposed to fluid surrounding the nerve. In these **myelinated fibers** action potentials depolarize the axon membrane only at nodes because the myelin sheath prevents depolarization elsewhere (Figure 33-6). Ion pumps and channels that move ions across a membrane are concentrated in each node. Once an action potential starts down an axon, depolarization of the first node initiates an electrical current that stretches to the neighboring node, causing it to depolarize and trigger an action potential.

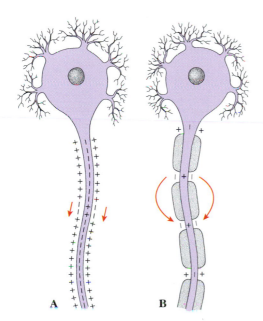

Figure 33-6

Impulse conduction in unmyelinated and myelinated fibers. In unmyelinated fibers (**A**), the action potential spreads continuously along, and must depolarize, the entire length of the axon. In myelinated fibers (**B**), the action potential leaps from node to node, bypassing the insulated portions of the fiber. This is saltatory conduction, which is much faster than continuous conduction.

Thus the action potential leaps from node to node, a kind of conduction called **saltatory** (L. *salto*, to dance, leap). The gain in efficiency as compared with nonmyelinated fibers is impressive. For example, a frog myelinated axon only 12 µm in diameter conducts nerve impulses at the same speed as a squid axon 350 µm in diameter.

> Some invertebrates, including prawns and insects, also have fast fibers invested with multiple layers of a myelin-like substance that is interrupted at intervals much like myelinated fibers of vertebrates. Conduction rates, though not as fast as vertebrate saltatory conduction, are much faster than unmyelinated fibers of the same diameter in other invertebrates.

SYNAPSES: JUNCTIONS BETWEEN NERVES

When an action potential passes down an axon to its terminal, it must cross a small gap, or **synapse** (Gr. *synapsis*, contact, union), separating it from another neuron or an effector organ. Two distinct kinds of synapses are known: electrical and chemical.

Electrical synapses, although much less common than chemical synapses, have been demonstrated in both invertebrate and vertebrate groups. Electrical synapses are points at which ionic currents flow directly across a narrow gap junction (see Figure 3-15, p. 45) from one neuron to another. Electrical synapses show no time lag and consequently are important for escape reactions. They also have been observed in other excitable cell types, and form an important method of communication between cardiac muscle cells of the heart (p. 663) and smooth muscle cells (for example, the uterus, p. 147).

Much more complex than electrical synapses are **chemical synapses,** which contain packets of specialized chemicals called **neurotransmitters.** Neurons bringing impulses toward chemical synapses are called **presynaptic neurons;** those carrying impulses away are **postsynaptic neurons.** At a synapse, membranes are separated by a narrow gap, or **synaptic cleft,** having a width of approximately 20 nm.

The axon of most neurons divides at its end into many branches, each of which bears a synaptic knob that sits on dendrites or cell body of the next neuron (Figure 33-7A). Because a single impulse coming down a nerve axon is transmitted along these many branches and synaptic endings, many impulses may converge on the cell body at one instant or may diverge onto more than one postsynaptic neuron. In addition, axon terminations of many neurons may almost cover a nerve cell body and its dendrites with thousands of synapses.

The 20 nm fluid-filled gap between presynaptic and postsynaptic membranes prevents action potentials from spreading directly to a postsynaptic neuron. Instead synaptic knobs secrete a specific neurotransmitter that communicates chemically with the postsynaptic cell. One of the most common neurotransmitters of the peripheral nervous system is **acetylcholine,** which

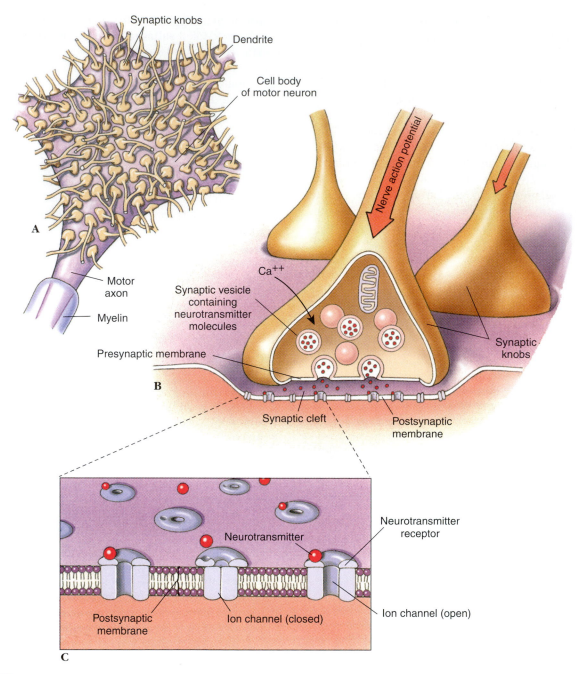

→ Figure 33-7

Transmission of impulses across nerve synapses. **A,** A cell body of a motor nerve is shown with the axon terminals of interneurons. Each terminal ends in a synaptic knob; thousands of synaptic knobs may rest on a single nerve cell body and its dendrites. **B,** A synaptic knob enlarged 60 times more than in **A.** An impulse traveling down the axon causes movement of synaptic vesicles to the presynaptic membrane where exocytosis occurs, releasing neurotransmitter molecules into the cleft. **C,** Diagram of a synaptic cleft at the ultrastructural level. Upon vesicular exocytosis, neurotransmitter molecules move rapidly across the gap to bind briefly with receptor molecules in the postsynaptic membrane. Binding of neurotransmitter to receptor produces a change in the potential of the postsynaptic membrane, in this case, caused by opening of the ion channels.

illustrates typical synaptic transmission. Inside synaptic knobs are numerous tiny **synaptic vesicles,** each containing several thousand molecules of acetylcholine. When an impulse arrives at a terminal knob a sequence of events occurs as portrayed in Figure 33-8. An action potential causes an inward movement of calcium (Ca^+) ions through channels in the synaptic knob membrane and this induces exocytosis of some neurotransmitter-

filled synaptic vesicles. Acetylcholine molecules diffuse across the gap in a fraction of a millisecond and bind briefly to receptor molecules on ion channels in the postsynaptic membrane. This creates a voltage change in the postsynaptic membrane. Whether the voltage change is large enough to trigger a postsynaptic potential depends on how many acetylcholine molecules are released and how many channels are opened. Acetylcholine

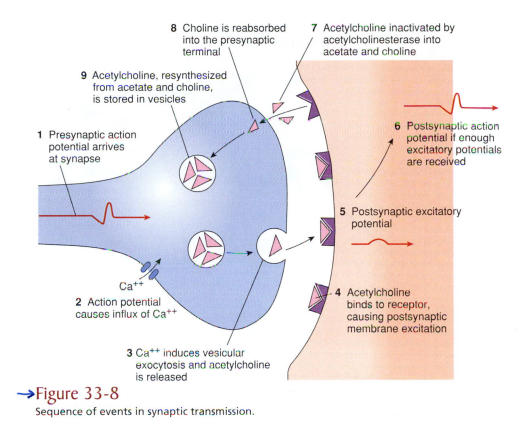

8 Choline is reabsorbed into the presynaptic terminal

9 Acetylcholine, resynthesized from acetate and choline, is stored in vesicles

1 Presynaptic action potential arrives at synapse

Ca⁺⁺

2 Action potential causes influx of Ca^{++}

3 Ca^{++} induces vesicular exocytosis and acetylcholine is released

7 Acetylcholine inactivated by acetylcholinesterase into acetate and choline

6 Postsynaptic action potential if enough excitatory potentials are received

5 Postsynaptic excitatory potential

4 Acetylcholine binds to receptor, causing postsynaptic membrane excitation

→Figure 33-8

Sequence of events in synaptic transmission.

is rapidly destroyed by the enzyme **acetylcholinesterase,** which converts acetylcholine into acetate and choline. If not inactivated in this way, the neurotransmitter would continue to stimulate indefinitely. Organophosphate insecticides (such as malathion) and certain military nerve gases are poisonous for precisely this reason; they block acetylcholinesterase. The final step in the sequence is reabsorption of choline into the presynaptic terminal, resynthesis of acetylcholine and its storage in synaptic vesicles, ready to respond to another impulse.

Many different chemical neurotransmitters have been identified in both vertebrate and invertebrate nervous systems. Some, such as acetylcholine, norepinephrine, and glutamate, depolarize postsynaptic membranes; they are released at **excitatory synapses.** Other neurotransmitters, such as gamma aminobutyric acid (GABA), hyperpolarize postsynaptic membranes, thereby stabilizing them against depolarization. These neurotransmitters are released at **inhibitory synapses.** Neurons in the central nervous system have both excitatory and inhibitory synapses among the hundreds or thousands of synaptic knobs on the dendrites and cell body of each neuron.

The net balance of all excitatory and inhibitory inputs received by a postsynaptic cell determines whether it generates an action potential (Figure 33-8). If many excitatory impulses are received at one time, they can reduce the resting membrane potential enough in the postsynaptic membrane to elicit an action potential. Inhibitory impulses, however, stabilize the postsynaptic membrane, making it less likely that an action potential will be generated. The synapse is a crucial part of the decision-making equipment of the central nervous system, modulating flow of information from one neuron to the next.

EVOLUTION OF NERVOUS SYSTEMS

Invertebrates: Development of Centralized Nervous Systems

Various metazoan phyla reveal a progressive increase in complexity of nervous systems that probably reflects in a general way stages in evolution of nervous systems. The simplest pattern of invertebrate nervous systems is the nerve net of radiate animals, such as sea anemones, jellyfishes, hydras, and comb jellies (Figure 33-9A). A nerve net is a quantum leap in complexity beyond sensory systems of unicellular forms, which lack nerves. A nerve net forms an extensive network in and under the epidermis over all the body. An impulse starting in one part of this net spreads in all directions, since synapses in most radiates do not restrict transmission to one-way movement, as often occurs in animals with more complex nervous systems. There are no differentiated sensory, motor, or connector components in the strict meaning of those terms. Branches of a nerve net connect to receptors in the epidermis and to epithelial cells that have contractile properties, and there is evidence of organization into reflex arcs (p. 703). Although most responses tend to be generalized, many are astonishingly complex for so simple a nervous system. This type of nervous system is found among vertebrates in nerve plexuses located, for example, in the intestinal wall; such nerve plexuses govern generalized intestinal movements such as peristalsis and segmentation (p. 683).

Bilateral nervous systems, the simplest of which occur in flatworms, represent a distinct increase in complexity over the

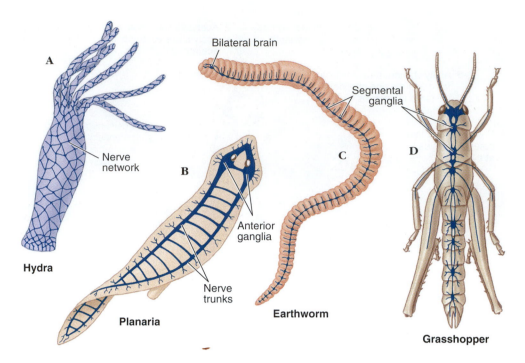

Figure 33-9
Invertebrate nervous systems. **A,** Nerve net of radiates, the simplest neural organization. **B,** Flatworm system, the simplest linear-type nervous system of two nerves connected to a complex neuronal network. **C,** Annelid nervous system, organized into a bilobed brain and ventral cord with segmental ganglia. **D,** Arthropod nervous system, also segmental, with large ganglia and more elaborate sense organs.

nerve net of radiate animals. Flatworms have two anterior ganglia, composed of groups of nerve cell bodies from which two main nerve trunks run posteriorly, with lateral branches extending throughout the body (Figure 33-9B). This is the simplest nervous system showing differentiation into a **peripheral nervous system** (a communication network extending to all parts of the body) and a **central nervous system** (a concentration of nerve cell bodies), which coordinates everything. More complex invertebrates exhibit a more centralized nervous system (brain), with two longitudinal fused nerve cords and many ganglia. Elaborate nervous systems of annelids contain a bilobed brain, a double nerve cord with segmental ganglia, and distinctive **afferent** (sensory) and **efferent** (motor) neurons (Figure 33-9C). Segmental ganglia are relay stations for coordinating regional activity.

The basic plan of molluscan nervous systems is a series of three pairs of well-defined ganglia, but in cephalopods (such as octopus and squid), these ganglia have burgeoned into textured nervous centers of great complexity; those of octopuses contain more than 160 million cells. Sense organs, too, are highly developed. Consequently, cephalopod behavior far outstrips that of any other invertebrate.

The basic plan of arthropod nervous systems (Figure 33-9D) resembles that of annelids, but ganglia are larger and sense organs are much better developed. Social behavior is often elaborate, particularly in hymenopteran insects (bees, wasps, and ants), and most arthropods are capable of considerable manipulation of their environment. Despite the complexity of much insect behavior, insects are nevertheless reflex-bound animals incapable of involved learned behavior principally because of their small size.

Vertebrates: Fruition of Encephalization

The basic plan of the vertebrate nervous system is a hollow, *dorsal* nerve cord terminating anteriorly in a large ganglionic mass, or brain. This pattern contrasts with the nerve cord of bilateral invertebrates, which is solid and ventral to their alimentary canal. By far the most important trend in evolution of vertebrate nervous systems is the great elaboration of size, configuration, and functional capacity of the brain, a process called **encephalization.** Vertebrate encephalization has brought to full fruition several functional capabilities including fast responses, great capacity for storage of information, and enhanced complexity and flexibility of behavior. Another consequence of encephalization is the ability to form associations between past, present, and (at least in humans) future events.

Spinal Cord

The **brain** and **spinal cord** compose the central nervous system. During early embryonic development, the spinal cord and brain begin as an ectodermal neural groove, which by folding and enlarging becomes a long, hollow neural tube (Figure 8-12, p. 159). The cephalic end enlarges to form brain vesicles, and the rest becomes the spinal cord. Unlike any invertebrate nerve cord, segmental nerves of spinal cords of vertebrates (31 pairs in humans) are separated into dorsal sensory roots and ventral motor roots. Sensory nerve cell bodies are gathered together into dorsal root (spinal) ganglia. Both dorsal (sensory) and ventral (motor) roots meet beyond the spinal cord to form a mixed spinal nerve (Figure 33-10).

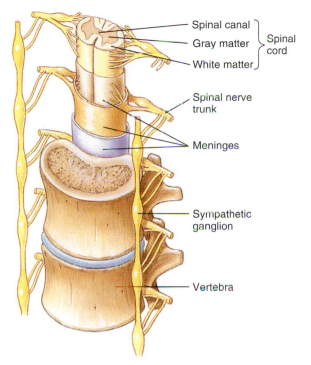

Figure 33-10

Human spinal cord and its protection. Two vertebrae show position of the spinal cord, emerging spinal nerves, and the sympathetic trunk. The cord is wrapped by three layers of membrane (meninges) between two of which lies a protective bath of cerebrospinal fluid.

The spinal cord encloses a central spinal canal and is additionally wrapped in three layers of membranes called **meninges** (men-in′jeez; Gr. *meningos,* membrane). In cross section the cord shows two zones (Figure 33-10). An inner zone of gray matter, resembling in shape the wings of a butterfly, contains the cell bodies of motor neurons and interconnecting interneurons. An outer zone of white matter contains bundles of axons and dendrites linking different levels of the cord with each other and with the brain.

Reflex Arc

Many neurons work in groups called **reflex arcs.** A reflex arc contains at least two neurons, but usually there are more. Parts of a typical reflex arc are (1) a **receptor,** a sense organ in skin, muscle, or another organ; (2) an **afferent,** or sensory, neuron, which carries impulses toward the central nervous system; (3) the **central nervous system,** where synaptic connections are made between sensory neurons and interneurons; (4) an **efferent,** or motor, neuron, which makes a synaptic connection with the interneuron and carries impulses from the central nervous system; and (5) an **effector,** by which an animal responds to environmental changes. Examples of effectors are muscles, glands, ciliated cells, nematocysts of radiate animals, electric organs of fish, and certain pigmented cells called chromatophores.

A reflex arc in its simplest form contains only two neurons— a sensory (afferent) neuron and a motor (efferent) neuron (for example, the "knee-jerk" or stretch reflex, Figure 33-11A). Usually,

Figure 33-11

The reflex arc. **A,** The "knee-jerk" or stretch reflex, a simple reflex arc. Sudden pressure on the patellar ligament stretches muscles in the upper leg. Impulses generated in stretch receptors are conducted along afferent (sensory) neurons to the spinal cord and relayed directly to an efferent (motor) nerve cell body. Impulses pass along efferent neurons to leg muscles (effectors), stimulating them to contract. **B,** Multisynaptic relfex arc. A more common reflex arc includes interneurons between the sensory and motor neuron. Tack puncture is sensed by pain receptors in the skin and the signal is conducted along afferent fibers to the spinal cord where synaptic connections are made with interneurons. Here, an interneuron is shown making connections with motor neurons on both sides of the spinal cord, such that stimulation of muscle fibers in more than one part of the body (both legs for example) allows coordination of muscle responses to the tack puncture.

however, interneurons are interposed between sensory and motor neurons (Figure 33-11B). An interneuron may connect afferent and efferent neurons on the same side of the spinal cord or on opposite sides, or it may connect them on different levels of the spinal cord, either on the same or opposite sides.

A **reflex act** is a response to a stimulus acting over a reflex arc. It is involuntary, meaning that it is often not under the control of the will. For example, many vital processes of the body, such as control of breathing, heartbeat, diameter of blood vessels, and sweat secretion are reflex acts. Some reflex acts are innate; others are acquired through learning.

In almost any reflex act, a number of reflex arcs are involved. For instance, a single afferent neuron may make synaptic connections with many efferent neurons. In a similar way an efferent neuron may receive impulses from many afferent neurons. Afferent neurons also make connections with ascending sensory neurons, which travel in the white matter of the spinal cord, bringing information about peripheral reflexes to the brain. Reflex activity may then be modified by descending motor neurons, which impinge on the final efferent motor neurons before they leave the spinal cord for the periphery.

Brain

Unlike the spinal cord, which has changed little in structure during vertebrate evolution, the brain has changed dramatically. A primitive linear brain, as seen in fishes and amphibians, expanded to form a deeply fissured and enormously intricate brain in the lineage leading to mammals (Figure 33-12). It reaches its greatest complexity in the human brain, which contains some 35 billion nerve cells, each of which may receive information from tens of thousands of synapses at one time. The ratio between weight of the brain and that of the spinal cord affords a fair criterion of an animal's intelligence. In fish and amphibians this ratio is approximately 1:1; in humans the ratio is 55:1—in other words, the brain is 55 times heavier than the spinal cord. Although the human brain is not the largest (the sperm whale's brain is seven times heavier) nor the most convoluted (that of the porpoise is even more folded), it is by all odds the best in overall performance. This "great ravelled knot," as the British physiologist Sir Charles Sherrington called the human brain, in fact may be so complex that it will never be able to understand its own function!

Brains of early vertebrates had three principal divisions: a forebrain, or **prosencephalon;** a midbrain, or **mesencephalon;** and a hindbrain, or **rhombencephalon** (Figure 33-13). Each part was concerned with one or more special senses: the forebrain with smell, the midbrain with vision, and the hindbrain with hearing and balance. These primitive but very fundamental concerns of the brain have been in some instances amplified and in others reduced or overshadowed during continued evolution as sensory priorities were shaped by an animal's habitat and way of life.

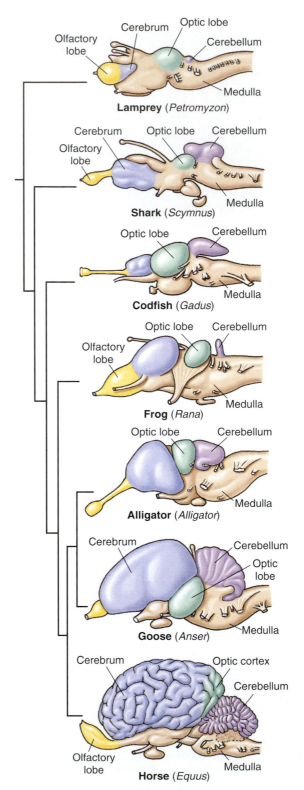

Figure 33-12

Evolution of the vertebrate brain. Note the progressive increase in size of the cerebrum. The cerebellum, concerned with equilibrium and motor coordination, is largest in animals whose balance and precise motor movements are well developed (fishes, birds, and mammals).

Figure 33-13
Divisions of the vertebrate brain.

Embryonic vesicle		Main components in adults	Function
Early embryo	**Late embryo**		
Forebrain (Prosencephalon)	Telencephalon	Cerebrum	Motor area controls voluntary muscle movements; sensory cortex is center of conscious perception of touch, pressure, vibration, pain, temperature, and taste; association areas integrate and process sensory data
	Diencephalon	Thalamus	Part of limbic system; integrates sensory information arriving at thalamus, projects to cerebral frontal lobes
		Hypothalamus	Controls autonomic functions; sets appetitive drives (thirst, hunger, sexual desire) and reproductive behavior; participates in emotional responses; secretes ADH, oxytocin; secretes releasing hormones for anterior pituitary regulation
Midbrain (Mesencephalon)	Mesencephalon	Optic lobes (tectum)	Integrates visual information with other sensory inputs; relays auditory information
		Midbrain nuclei	Involuntary control of muscle tone; processing of incoming sensations and outgoing motor commands
Hindbrain (Rhombencephalon)	Metencephalon	Cerebellum	Involuntary coordination and control of outgoing movements for equilibrium, muscle tone, posture
		Pons	Links cerebellum with other brain centers and with medulla and spinal cord; modifies output of respiratory centers in medulla
	Myelencephalon	Medulla oblongata	Regulates heart rate and force of contraction; vasomotor control; sets rate of respiration; relays information to the cerebellum; integrates feeding and satiety stimuli

Spinal cord

Although the large size of their brain undoubtedly makes humans the wisest of animals, it is apparent that they can do without much of it and still remain wise. Brain scans of persons with hydrocephalus (enlargement of the head as a result of pressure disturbances that cause the brain ventricles to enlarge many times their normal size) show that although many such persons are functionally disabled, others are nearly normal. The cranium of one person with hydrocephalus was nearly filled with cerebrospinal fluid and the only remaining cerebral cortex was a thin layer of tissue, 1 mm thick, pressed against his cranium. Yet this young man, with only 5% of his brain, had achieved first-class honors in mathematics at a British university and was socially normal. This and other similarly dramatic observations suggest that there is enormous redundancy and spare capacity in corticocerebral function. It also suggests that the deep structures of the brain, which are relatively spared in hydrocephalus, may perform functions once believed to be performed solely by the cortex.

Hindbrain The **medulla,** the most posterior division of the brain, is really a conical continuation of the spinal cord. The medulla, together with the more anterior midbrain, constitutes the "brain stem," an area that controls numerous vital and largely subconscious activities such as heartbeat, respiration, vascular tone, gastric secretions, and swallowing. The brain stem also contains centers that appear to integrate information regarding satiety and feeding stimuli. The **pons,** also a part of the hindbrain, contains a thick bundle of fibers that carry impulses from one side of the cerebellum to the other, and also connects both medulla and cerebellum to other brain regions.

The **cerebellum,** lying dorsal to the medulla, controls equilibrium, posture, and movement (Figure 33-14). Its development is directly correlated with an animal's mode of locomotion, agility of limb movement, and balance. It is usually weakly developed in amphibians and reptiles, forms that live close to the ground, and well developed in the more agile bony fishes. It reaches its apogee in birds and mammals in which it is greatly expanded and folded. The cerebellum does not initiate movement but operates as a precision error-control center, or servomechanism, that programs a movement initiated somewhere else, such as the motor cortex of the cerebrum (Figure 33-14). Primates and especially humans, who possess a manual dexterity far surpassing that of other animals, have the most complex cerebellum. Movements of hands and fingers may involve cerebellar coordination of simultaneous contraction and relaxation of hundreds of individual muscles.

Figure 33-14

External view of the human brain, showing lobes of the cerebrum and localization of major function of the cerebrum and cerebellum.

Midbrain   The midbrain consists mainly of the **tectum** (including the optic lobes), which contains nuclei that serve as centers for visual and auditory reflexes. (In neurophysiological usage a nucleus is a small aggregation of nerve-cell bodies within the central nervous system.) The midbrain has undergone little evolutionary change in structure among vertebrates but has changed markedly in function. It mediates the most complex behavior of fishes and amphibians, integrating visual, tactile, and auditory information. Such functions have been gradually assumed by the forebrain in amniotes. In mammals, the midbrain is mainly a relay center for information on its way to higher brain centers.

Forebrain   Just anterior to the midbrain lie the **thalamus** and **hypothalamus,** the most posterior elements of the forebrain. The egg-shaped thalamus is a major relay station that analyzes and passes sensory information to higher brain centers. In the hypothalamus are several "housekeeping" centers that regulate body temperature, water balance, appetite, and thirst—all functions concerned with maintenance of internal constancy (homeostasis). Neurosecretory cells located in the hypothalamus produce several neurohormones (described in Chapter 34). The hypothalamus also contains centers for regulating reproductive function and sexual behavior, and it participates in emotional behaviors.

The anterior portion of the forebrain, or **cerebrum** (Figure 33-14), can be divided into two anatomically distinct areas, the **paleocortex** and **neocortex.** Originally concerned with smell, it became well developed in advanced fishes and early terrestrial vertebrates, which depend on this special sense. In mammals and especially in primates the paleocortex is a deep-lying area called a rhinencephalon ("nose brain"), because many of its functions depend on olfaction. Better known as the **limbic system,** it

mediates several species-specific behaviors that relate to fulfilling needs such as feeding and sex. One region of the limbic system, the **hippocampus,** has been extensively studied as a site of spatial learning and memory. Recently, the hippocampus has gained notoriety since its neurons have been shown to divide in adults, a previously unknown occurrence in neurons.

Although a late arrival in vertebrate evolution, the neocortex completely overshadows the paleocortex and has become so expanded that it envelops much of the forebrain and all of the midbrain (Figure 33-14). Almost all integrative activities primitively assigned to the midbrain are transferred to the neocortex, or **cerebral cortex** as it is usually called.

Functions in the cerebrum have been localized by direct stimulation of exposed brains of people and experimental animals, postmortem examination of persons suffering from various lesions, and surgical removal of specific brain areas in experimental animals. The cortex contains discrete motor and sensory areas (Figures 33-14 and 33-15) as well as large "silent" regions, called **association areas,** concerned with memory, judgment, reasoning, and other integrative functions. These regions are not directly connected to sense organs or muscles.

Thus in mammals, and especially in humans, separate parts of the brain mediate conscious and unconscious functions. The unconscious mind, all of the brain except the cerebral cortex, governs numerous vital functions that are removed from conscious control: respiration, blood pressure, heart rate, hunger, thirst, temperature balance, salt balance, sexual drive, and basic (sometimes irrational) emotions. It is also a complex endocrine gland that regulates and receives feedback from the body's endocrine system (see Chapter 34). The conscious mind, or cerebral cortex, is the site of higher mental activities (for example, planning and reasoning), memory, and integration of sensory information. Memory appears to transcend all parts of the brain rather than being a property of any particular part of the brain as was once believed.

The right and left hemispheres of the cerebral cortex are bridged through the corpus callosum, a neural connection through which the two hemispheres are able to transfer information and coordinate mental activities. In humans, the two hemispheres of the brain are specialized for entirely different functions: the left hemisphere for language development, mathematical and learning capabilities, and sequential thought processes; and the right hemisphere for spatial, musical, artistic, intuitive, and perceptual activities. Each hemisphere also controls the opposite side of the body. It has been known for a long time that even extensive damage to the right hemisphere may cause varying degrees of left-sided paralysis but has little effect on intellect and speech. Conversely, damage to the left hemisphere usually causes loss of speech and may have disastrous effects on intellect. Because these differences in brain symmetry and function exist at birth, they appear to be inborn rather than the result of developmental or environmental effects as once believed.

Hemispheric specialization has long been considered a unique human trait, but was recently discovered in the brains of songbirds in which one side of the brain is specialized for song production.

Peripheral Nervous System

The peripheral nervous system includes all nervous tissue outside the central nervous system. It has two functional divisions: **sensory** or **afferent division,** which brings sensory information to the central nervous system, and **motor** or **efferent division,** which conveys motor commands to muscles and glands. The efferent division has two components: (1) **somatic nervous system,** which innervates skeletal muscle, and (2) **autonomic nervous system,** which innervates smooth muscle, cardiac muscle, and glands.

Autonomic Nervous System   The autonomic system governs involuntary, internal functions of the body that do not ordinarily affect consciousness, such as movements of alimentary canal and heart, contraction of smooth muscle of blood vessels, urinary bladder, iris of the eye, and others, plus secretions of various glands.

Autonomic nerves originate in the brain or spinal cord as do nerves of the somatic nervous system, but unlike the latter, autonomic fibers consist of not one but two motor neurons. They synapse once after leaving the cord and before arriving at the effector organ. These synapses are located outside the spinal cord in ganglia. Fibers passing from the cord to the ganglia are called preganglionic autonomic fibers; those passing from the ganglia to the effector organs are called postganglionic fibers. These relationships are illustrated in Figure 33-16.

Subdivisions of the autonomic system are the **parasympathetic** and **sympathetic** systems. Most organs in the body are innervated by both sympathetic and parasympathetic fibers,

Figure 33-15

Arrangement of sensory and motor cortices. Localizations of sensory terminations from different parts of the body are shown at right; origins of descending motor pathways are shown at left. The motor cortex lies in front of the sensory cortex, so the two are not superimposed. These maps grew out of the 1930s work of Canadian neurosurgeon Wilder Penfield. Recent research shows that the motor cortex is not as orderly as the map suggests; rather correspondence between cortical areas and areas of the body they control is more diffuse.

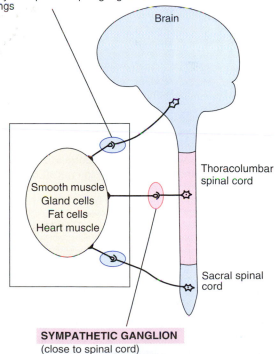

PARASYMPATHETIC GANGLION
(close to effector organ)
Acetylcholine is neurotransmitter released by both pre- and postganglionic fiber endings

Brain

Thoracolumbar spinal cord

Smooth muscle
Gland cells
Fat cells
Heart muscle

Sacral spinal cord

SYMPATHETIC GANGLION
(close to spinal cord)
Acetylcholine released from preganglionic fiber ending; norepinephrine released from postganglionic fiber ending

Figure 33-16

General organization of the autonomic nervous system.

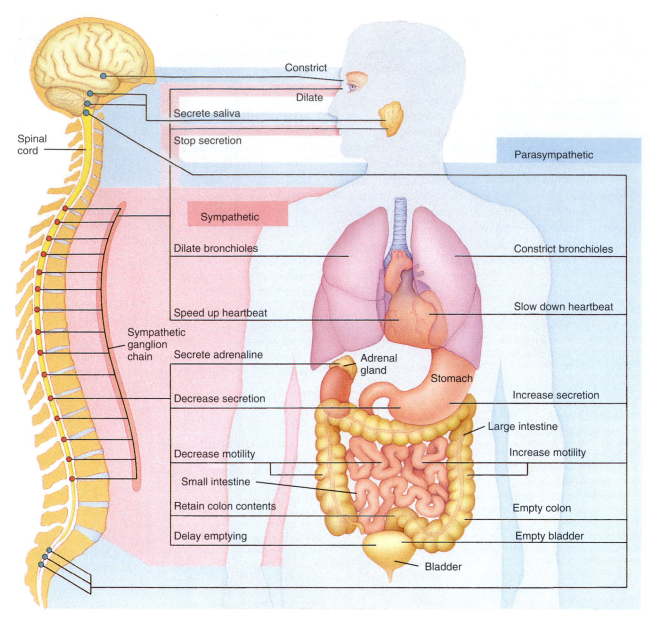

Constrict

Dilate

Secrete saliva

Stop secretion

Spinal cord

Parasympathetic

Sympathetic

Dilate bronchioles

Constrict bronchioles

Speed up heartbeat

Slow down heartbeat

Sympathetic ganglion chain

Secrete adrenaline

Adrenal gland

Stomach

Decrease secretion

Increase secretion

Large intestine

Decrease motility

Increase motility

Small intestine

Retain colon contents

Empty colon

Delay emptying

Empty bladder

Bladder

Figure 33-17

Autonomic nervous system in humans. Outflow of autonomic nerves from the central nervous system is shown at left. Sympathetic (*red*) outflow is from the thoracic and lumbar areas of the spinal cord by way of a chain of sympathetic ganglia. Parasympathetic (*blue*) outflow is from the cranial and sacral regions of the central nervous system; parasympathetic ganglia (not shown) are located in or adjacent to the organs innervated. Most organs are innervated by fibers from both sympathetic and parasympathetic divisions.

whose actions are antagonistic (Figure 33-17). If one fiber stimulates an activity, the other inhibits it. However, neither kind of nerve is exclusively excitatory or inhibitory. For example, parasympathetic fibers inhibit heartbeat but excite peristaltic movements of the intestine; sympathetic fibers increase heartbeat but inhibit intestinal peristaltic movement.

The parasympathetic system consists of motor neurons, some of which emerge from the brain stem by certain cranial nerves and others of which emerge from the sacral (pelvic) region of the spinal cord (Figures 33-16 and 33-17). In the sympathetic division nerve cell bodies of all preganglionic fibers are

located in thoracic and upper lumbar areas of the spinal cord. Their fibers exit through the ventral roots of spinal nerves, separate from these, and go to sympathetic ganglia (Figure 33-17), which are paired and form a chain on each side of the spinal column.

All *preganglionic* fibers, whether sympathetic or parasympathetic, release acetylcholine at their synapse with postganglionic cells. However, parasympathetic *postganglionic* fibers release acetylcholine at their endings, whereas sympathetic *postganglionic* fibers with few exceptions release norepinephrine (also called noradrenaline). This difference is another important characteristic distinguishing the two parts of the autonomic nervous system.

As a general rule the parasympathetic division is associated with nonstressful activities, such as resting, eating, digestion, and urination. The sympathetic division is active under conditions of physical or emotional stress. Under such conditions heart rate increases, blood vessels to the skeletal muscles dilate, blood vessels in the viscera constrict, activity of the intestinal tract decreases, and metabolic rate increases. The importance of these responses in emergency reactions (sometimes called the fright, fight or flight response) are described in the next chapter (p. 734). It should be noted, however, that the sympathetic division is active also during resting conditions in maintaining normal blood pressure and body temperature.

SENSE ORGANS

Animals require a constant inflow of information from the environment to regulate their lives. Sense organs are specialized receptors designed for detecting environmental status and change. An animal's sense organs are its first level of environmental perception; they are channels for bringing information to the brain.

A **stimulus** is some form of energy—electrical, mechanical, chemical, or radiant. A sense organ transforms energy from a stimulus into nerve impulses, the common language of the nervous system. In a very real sense, then, sense organs are biological transducers. A microphone, for example, is a transducer that converts mechanical (sound) energy into electrical energy. Like a microphone, which is sensitive only to sound, sense organs are, as a rule, specific for one kind of stimulus. Thus eyes respond only to light, ears to sound, pressure receptors to pressure, and chemoreceptors to chemicals, converting all forms of energy into nerve impulses.

Since all nerve impulses are qualitatively alike, how do animals perceive and distinguish different sensations of varying stimuli? The answer is that real perception of sensation is done in localized regions of the brain, where each sensory organ has its own hookup. This concept of "labeled lines" of communication to specific brain regions was first described in the 1830s by Johannes Müller who called this the **law of specific nerve energies.** Impulses arriving at a particular sensory area of the brain can be interpreted in only one way. For example, pressure on the eye causes us to see "stars" or other visual patterns; mechanical distortion of the eye initiates impulses in the optic nerve fibers that are perceived as light sensations. Although such an operation probably could never be done, a deliberate surgical switching of optic and auditory nerves would cause the recipient literally to see thunder and hear lightning!

Classification of Receptors

Receptors are traditionally classified by their location. Those near the external surface, called **exteroceptors,** keep an animal informed about its external environment. Internal parts of the body are provided with **interoceptors,** which receive stimuli from internal organs. Muscles, tendons, and joints have **proprioceptors,** which are sensitive to changes in tension of muscles and provide an organism with a sense of body position. Sometimes receptors are classified by the form of energy to which the receptors respond, such as **chemical, mechanical, light,** or **thermal.**

Chemoreception

Chemoreception is the oldest and most universal sense in the animal kingdom. It probably guides behavior of animals more than any other sense. Unicellular forms use **contact chemical receptors** to locate food and adequately oxygenated water and to avoid harmful substances. These receptors elicit an orientation behavior, called **chemotaxis,** toward or away from a chemical source. Most metazoans have specialized **distance chemical receptors,** which are often developed to a remarkable degree of sensitivity. Distance chemoreception, usually called smell or olfaction, guides feeding behavior, location and selection of sexual mates, territorial and trail marking, and alarm reactions of numerous animals.

Social insects and many other animals, including mammals, produce species-specific compounds, called **pheromones,** that constitute a highly developed chemical language. Pheromones are a diverse group of organic compounds that an animal releases to affect the physiology or behavior of another individual of the same species. Ants, for example, are walking batteries of glands (Figure 33-18) that produce numerous chemical signals. These include releaser pheromones, such as alarm and trail pheromones, and primer pheromones, which alter endocrine and reproductive systems of different castes in a colony. Insects bear a variety of chemoreceptors on the surface of their body for sensing specific pheromones, as well as other, nonspecific odors. Recent research has focused on the possibility of pheromonal communication in humans. The best established example is that of menstrual cycle synchronization that occurs in women living together in close association, such as in dormitories.

In all vertebrates and in insects, the senses of **taste** and **smell** are clearly distinguishable. Although there are similarities between taste and smell receptors, in general taste is more restricted in response and is less sensitive than smell. Central nervous system centers for taste and smell are located in different parts of the brain.

Figure 33-18
Pheromone-producing glands of an ant.

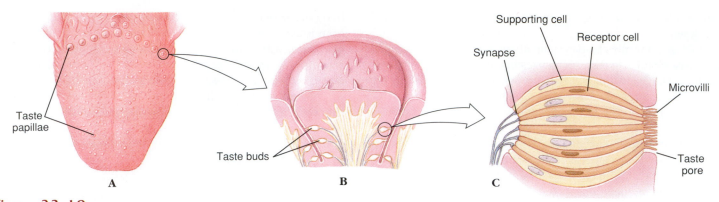

Figure 33-19

Taste receptors. **A,** Surface of human tongue showing location of taste papillae. **B,** Position of taste buds on a taste papilla. **C,** Structure of a taste bud.

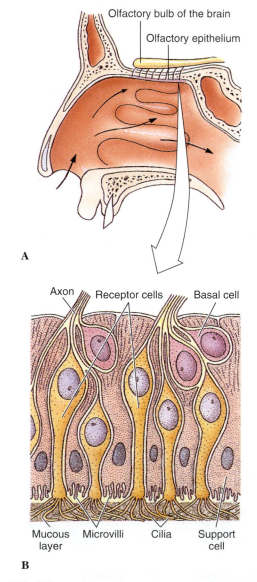

In vertebrates, taste receptors are found in the mouth cavity and especially on the tongue (Figure 33-19), where they provide a means for judging foods before they are swallowed. A **taste bud** consists of a cluster of receptor cells surrounded by supporting cells; it is provided with a small external pore through which slender tips of sensory cells project. Chemicals being tasted apparently combine with specific receptor sites on microvilli of receptor cells. Taste sensations are categorized as sweet, salty, acid, bitter, and possibly umami (Japanese, roughly translated to mean "meaty" or "savory"). Although the mechanisms are different for each basic taste sensation, receptor cells are depolarized by the specific chemical and action potentials are generated. Contrary to what was originally thought, taste receptors can respond to different types of taste categories, although they may respond more strongly to one particular type. These action potentials are transmitted across chemical synapses (p. 699) and travel along sensory neurons to specific brains regions. Taste discrimination depends on assessment by the brain of the relative activity of many different taste receptors. This assessment is similar to color vision in vertebrates, where a whole rainbow of colors can be distinguished by relative excitation of only three types of color photoreceptors (see p. 718). Because receptor cells are subject to wear and tear by abrasive foods, taste buds have a short life (5 to 10 days in mammals) and are continually being replaced.

Although olfactory sense is a primal sense for many animals, used for identification of food, sexual mates, and predators, olfaction is most highly developed in mammals. Even humans, although a species not celebrated for detecting smells, can discriminate perhaps 20,000 different odors. A human nose can detect 1/25 of one-millionth of 1 mg of mercaptan, the odoriferous substance of skunks. Even so, our olfactory abilities compare poorly with those of other mammals that rely on olfaction for survival. A dog explores new surroundings with its nose much as we do with our eyes. A dog's nose is justifiably renowned; with some odorous sources a dog's nose is at least a million times more sensitive than ours. Dogs are assisted in their proficiency by having a nose located close to the ground where odors from passing creatures tend to linger.

Olfactory endings are located in a special epithelium covered by a thin film of mucus, positioned deep in the nasal cavity (Figure 33-20). Within this epithelium lie millions of olfactory

Figure 33-20

Human olfactory epithelium. **A,** The epithelium is a patch of tissue positioned in the roof of the nasal cavity. **B,** It is composed of supporting cells, basal cells, and olfactory receptor cells with cilia protruding from their free ends.

neurons, each with several hairlike cilia protruding from the free end. Odor molecules entering the nose bind to receptor proteins located in the cilia; this binding generates an electrical signal that travels along axons to the olfactory bulb of the brain. From here odor information is sent to the olfactory cortex where odors are analyzed. Odor information is then projected to higher brain centers that influence emotions, thoughts, and behavior.

Recently, using techniques of gene cloning and molecular hybridization (p. 95), researchers discovered a large family of genes that appears to code for odor reception in mammals (including humans). Each of the 500 to 1000 genes discovered encodes a separate type of odor receptor. Since mammals can detect at least 20,000 different odors, each receptor must respond to several odor molecules, and each odor molecule must bind with several types of receptors, each of which responds to a part of the molecule's structure. Brain-mapping techniques have shown that each olfactory neuron projects to a characteristic location on the olfactory bulb, providing a two-dimensional map that identifies which receptors have been activated in the nose. In addition, olfactory neurons expressing the same odor-receptor gene converge to a fixed olfactory bulb region, which might provide an explanation for the extremely high sensitivity of smell. Projected to the brain, odor information is recognized as a unique scent.

Because flavor of food depends on odors reaching the olfactory epithelium through the throat passage, taste and smell are easily confused. All "tastes" other than the five basic ones (sweet, sour, acid, salty, and possibly umami) result from flavor molecules reaching the olfactory epithelium in this manner. Food loses its appeal during a common cold because a stuffy nose blocks odors rising from the mouth.

Many terrestrial vertebrates possess an additional olfactory organ, the **vomeronasal organ (VNO** or **Jacobson's organ).** The VNO is also lined with olfactory epithelium and is located in paired, blind-ending passages that open into the nasal or oral cavity. The VNO olfactory receptor cells respond to chemical signals carried by **pheromones** which are chemically different from the odor molecules that stimulate nasal olfactory receptor cells. Information regarding territory, societal hierarchy, sex and reproductive state are transmitted via this system. Equivocal information currently exists as to the presence of a functional VNO system in humans.

Mechanoreception

Mechanoreceptors are sensitive to quantitative forces such as touch, pressure, stretching, sound, vibration, and gravity—in short, they respond to motion. To interact with their environments, feed themselves, maintain normal postures, and to walk, swim, or fly, animals require a steady flow of information from mechanoreceptors.

Touch

Pacinian corpuscles, relatively large mechanoreceptors that register deep touch and pressure in mammalian skin, illustrate the general properties of mechanoreceptors. These corpuscles are common in deep layers of skin, connective tissue surrounding

muscles and tendons, and abdominal mesenteries. Each corpuscle consists of a nerve terminus surrounded by a capsule of numerous, concentric, onionlike layers of connective tissue (Figure 33-21). Pressure at any point on a capsule distorts its nerve ending, producing a graded **receptor potential,** a local flow of electric current. Progressively stronger stimuli lead to correspondingly stronger receptor potentials until a **threshold current** is produced; this current initiates an action potential in a sensory nerve fiber. A second action potential is initiated at removal of the pressure, but not during the duration of the pressure. This response is called adaptation (not to be confused with the evolutionary meaning of this term [Chapter 6]) and is characteristic of many touch receptors, which are admirably suited to detecting a sudden mechanical change but readily adapt to new conditions. We are aware of new pressures when we put on our shoes in the morning, but we are glad not to be reminded of these pressures all day.

Invertebrates, especially insects, have many kinds of receptors sensitive to touch. Such receptors are well endowed with tactile hairs sensitive to both touch and vibrations. Superficial touch receptors of vertebrates are distributed over the body but tend to be concentrated in areas especially important for exploring and interpreting the environment. In most vertebrates these areas are the face and extremities of limbs. Of the more than half-million separate touch-sensitive spots on the surface of a human body, most are found on the tongue, and fingertips as might be expected based on the large portion of sensory cortex which receives information from these regions (Figure 33-15). The simplest touch receptors are bare nerve-fiber terminals found in skin, but there is an assortment of other kinds of receptors of varying shapes and sizes. Each hair follicle is crowded with receptors sensitive to touch.

Pain

Pain receptors are relatively unspecialized nerve fiber endings that respond to a variety of stimuli signaling possible or real

Figure 33-21

Response of pacinian corpuscle to applied pressure. Progressively stronger pressure produces stronger receptor potentials. When the threshold stimulus is reached, an all-or-none action potential is generated in the afferent nerve fiber.

damage to tissues. These free nerve endings also respond to other stimuli, such as mechanical movement of a tissue and temperature changes. Pain fibers respond to small peptides, such as substance P and bradykinins, which are released by injured cells. This type of response is termed *slow pain*. *Fast pain* responses (for example, a pin prick, cold or hot stimuli) are a more direct response of the nerve endings to mechanical or thermal stimuli.

> Pain is a distress call from the body signaling some noxious stimulus or internal disorder. Although there is no cortical pain center, discrete areas have been located in the brain stem where pain messages from the periphery terminate. These areas contain two kinds of small peptides, endorphins and enkephalins, that have morphinelike or opiumlike activity. When released, they bind with specific opiate receptors in the midbrain. They are the body's own analgesics.

Just as pain is a sign of danger, sensory pleasure is a sign of a stimulus useful to the subject. Pleasure depends on the internal state of an animal and is judged with reference to homeostasis and some physiological set point. Pleasure and pain states may be produced by release of small neuropeptides, called endogenous opioids, within the central nervous system.

Lateral-Line System of Fish and Amphibians

A lateral line is a distant touch reception system for detecting wave vibrations and currents in water. Receptor cells, called **neuromasts,** are located on the body surface in aquatic amphibians and some fishes, but in many fishes they are located within canals running beneath the epidermis; these canals open at intervals to the surface (Figure 33-22). Each neuromast is a collection of **hair cells** with sensory endings, or cilia, embedded in a gelatinous, wedge-shaped mass, the **cupula.** The cupula projects into the center of the lateral-line canal so that it bends in response to any disturbance of water on the body surface. The lateral-line system is one of the principal sensory systems that guide fishes in their movements and in location of predators, prey, and social partners (p. 495).

Hair cells form an important sensory component of several mechanoreceptors found in both invertebrate (statocysts) and vertebrate (vestibular organs) organs of equilibrium, discussed in the next section.

Hearing

An ear is a specialized receptor for detecting sound waves in the surrounding environment. Because sound communication and reception are integral to lives of terrestrial vertebrates, we may be surprised to discover that most invertebrates inhabit a silent world. Only certain arthropod groups—crustaceans, spiders, and insects—have developed true sound-receptor organs. Even among insects, only locusts, cicadas, crickets, grass-hoppers, and most moths possess ears, and these are of simple design: a pair of air pockets, each enclosed by a tympanic membrane that

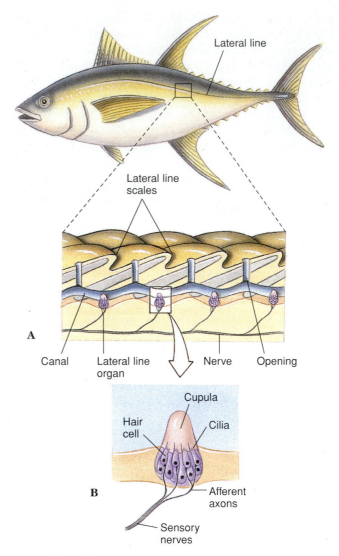

Figure 33-22
Lateral-line system. **A,** Lateral line of a bony fish with both exposed and hidden neuromasts. **B,** Structure of a neuromast (lateral-line organ).

passes sound vibrations to sensory cells. Despite their spartan construction, insect ears are beautifully designed to detect the sound of a potential mate, a rival male, or a predator.

Especially interesting are ultrasonic detectors of certain nocturnal moths. These have evolved specifically to detect approaching bats and thus lessen the moth's chance of becoming a bat's evening meal (echolocation in bats is described on p. 595). Each moth ear possesses just two sensory receptors, known as A_1 and A_2 (Figure 33-23). The A_1 receptor will respond to ultrasonic cries of a bat that is still too far away to detect the moth. As the bat approaches and its cries increase in intensity, the receptor fires more rapidly, informing the moth that the bat is coming nearer. Since the moth has two ears, its nervous system can determine the bat's position by comparing firing rates from the two ears. The moth's strategy is to fly away before the bat detects it. But if the bat continues its approach, the second (A_2) receptor in each ear, which responds only to high-intensity

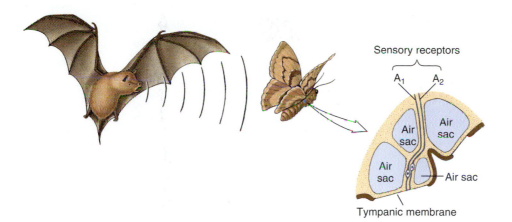

Figure 33-23

Ear of a moth used to detect approaching bats. See text for explanation.

sounds, will fire. The moth responds immediately with an evasive maneuver, usually making a power dive to a bush or the ground where it is safe because the bat cannot distinguish a moth's echo from those of its surroundings.

In their evolution, vertebrate ears originated as a balance organ, or **labyrinth.** In all jawed vertebrates, from fishes to mammals, the labyrinth has a similar structure, consisting of two small chambers called the **saccule** and the **utricle,** and three **semicircular canals** (Figure 33-24). In fish the base of the saccule is extended into a tiny pocket (**lagena**) that, during evolution of vertebrates, developed into the hearing receptor of tetrapods. With continued elaboration and elongation in birds and mammals, the fingerlike lagena was modified to form a **cochlea.**

A human ear (Figure 33-25) is representative of mammalian ears. The outer, or external, ear collects sound waves and funnels them through an **auditory canal** to an eardrum or **tympanic membrane** lying next to the middle ear. The middle ear is an air-filled chamber containing a remarkable chain of three tiny bones, or ossicles, known as the **malleus** (hammer), **incus** (anvil), and **stapes** (stirrup), named because of their fancied resemblance to these objects. These bones conduct sound waves across the middle ear (Figure 33-25B). The bridge of bones is so

arranged that the force of sound waves pushing against the tympanic membrane is amplified as much as 90 times where the stapes contacts the **oval window** of the inner ear. Muscles attached to the middle ear bones contract when the ear receives very loud noises, providing the inner ear some protection from damage. The middle ear connects with the pharynx by means of an **eustachian tube,** which permits pressure equalization on both sides of the tympanic membrane.

The origin of the three tiny bones of the mammalian middle ear—the malleus, incus, and stapes—is one of the most extraordinary and well-documented transitions in vertebrate evolution. Amphibians, reptiles, and birds have a single rodlike ear ossicle, the stapes (also called the columella), which originated as a jaw support (hyomandibular) as seen in fishes (see Figure 23-17, p. 483). With evolution of the earliest tetrapods, the braincase became firmly sutured to the skull, and the hyomandibular, no longer needed to brace the jaw, became converted into the stapes. In a similar way, the two additional ear ossicles of the mammalian middle ear—the malleus and incus—originated from parts of the jaw of early vertebrates. The quadrate bone of reptilian upper jaws became the incus, and the articular bone of the lower jaw became the malleus. Homology of reptilian jaw bones to mammalian ear bones is clearly documented in the fossil record and in embryological development of mammals.

Figure 33-24

Vestibular apparatus of a teleost fish, containing three semicircular canals, which respond to angular acceleration; two balance organs (utricle and saccule), which are static receptors that signal the fish's position in relation to gravity; and a small chamber, the lagena, which is specialized for sound reception.

Within the inner ear is the organ of hearing, or **cochlea** (Gr. *cochlea,* snail's shell), which is coiled in mammals, making two and one half turns in humans (Figure 33-25B). The cochlea is divided longitudinally into three tubular canals running parallel with one another. This relationship is indicated in Figure 33-26. These canals become progressively smaller from the base of the cochlea to the apex. One of these canals is called the **vestibular canal;** its base is closed by the oval window. The **tympanic canal,** which is in communication with the vestibular canal at the tip of the cochlea, has its base closed by the **round window.** Between these two canals is a **cochlear duct,** which contains the **organ of Corti,** the actual sensory apparatus (Figure 33-25C). Within the organ of Corti are fine rows of hair cells that run lengthwise from the base to the tip of the cochlea. At least

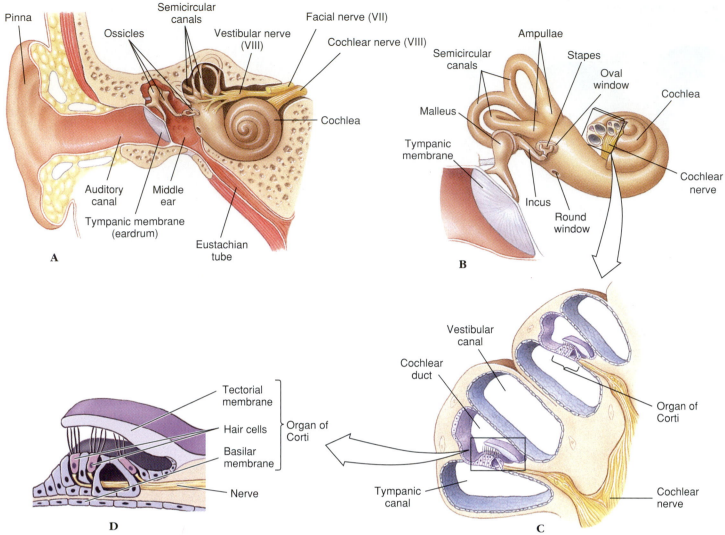

Figure 33-25

Human ear. **A,** Longitudinal section showing external, middle, and inner ear. **B,** Enlargement of middle ear and inner ear. The cochlea of the inner ear has been opened to show the arrangement of canals within. **C,** Enlarged cross section of cochlea showing the organ of Corti. D, Detail of ultrastructure of the organ of Corti.

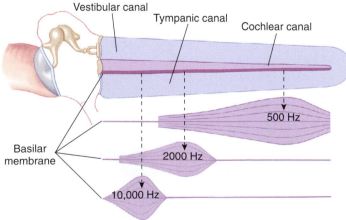

Figure 33-26

Frequency localization in the cochlea of the mammalian ear as it would appear with the cochlea stretched out. Sound waves transmitted to the oval window produce vibration waves that travel down the basilar membrane. High-frequency vibrations cause the membrane to resonate near the oval window. Low-frequency tones travel farther down the basilar membrane.

24,000 hair cells are present in a human ear. The 80 to 100 "hairs" on each cell are actually microvilli and a single large cilium, which project into the endolymph of the cochlear canal. Each cell is connected with neurons of the auditory nerve. Hair cells rest on the **basilar membrane,** which separates the tympanic canal and cochlear duct, and they are covered by the **tectorial membrane,** lying directly above them (Figure 33-25D).

When a sound wave strikes the ear, its energy is transmitted through the ossicles of the middle ear to the oval window, which oscillates back and forth, driving the fluid of the vestibular and tympanic canals before it. Because these fluids are noncompressible, an inward movement of the oval window produces a corresponding outward movement of the round window. The fluid oscillations also cause the basilar membrane with its hair cells to vibrate simultaneously.

According to the **place hypothesis of pitch discrimination** formulated by Georg von Békésy, different areas of the basilar membrane respond to different frequencies; for every sound frequency, there is a specific "place" on the basilar membrane where hair cells respond to that frequency. Initial displacement of the

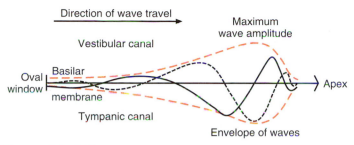

Figure 33-27

Traveling waves along the basilar membrane. The oval window is at left, and the cochlear apex at right. The two wave formations (*solid* and *dashed lines*) occur at separate instants of time. The curves in color represent the extreme displacements of the membrane by traveling waves.

basilar membrane starts a wave traveling down the membrane, much as flipping a rope at one end starts a wave moving down the rope (Figure 33-27). The displacement wave increases in amplitude as it moves from the oval window toward the apex of the cochlea, reaching a maximum at the region of the basilar membrane where the natural frequency of the membrane corresponds to the sound frequency. Here, the membrane vibrates with such ease that the energy of the traveling wave is completely dissipated. Hair cells within the organ of Corti in that region are stimulated and impulses conveyed to the fibers of the auditory nerve. Isolated hair cells have been shown to respond to a particular band of frequencies depending on their location within the cochlea. Thus, impulses that are carried by certain fibers of the auditory nerve are interpreted by the hearing center as particular tones. **Loudness** of a tone depends on the number of hair cells stimulated, whereas **timbre,** or quality, of a tone is produced by the pattern of hair cells stimulated by sympathetic vibration. This latter characteristic of tone enables us to distinguish between different human voices and different musical instruments, although the notes in each case may be of the same pitch and loudness.

Most recent auditory research has focused on a more active role for hair cells within the organ of Corti. Experiments have demonstrated that outer hair cells may respond to sound waves by changing their length and thus mechanically altering the position of the basilar and tectorial membranes. Although a function of such movements is not yet established in vivo, it has been suggested that this active response of these receptor cells in the organ of Corti might increase both sensitivity and selectivity of hearing.

Equilibrium

In invertebrates, specialized sense organs for monitoring gravity and low-frequency vibrations often appear as **statocysts.** Each is a simple sac lined with hair cells and containing a heavy calcareous structure, the **statolith** (Figure 33-28). The delicate, hairlike filaments of sensory cells are activated by the shifting position of the statolith when the animal changes position. Statocysts are found in many invertebrate phyla from radiates to arthropods. All are built on similar principles.

The vertebrate organ of equilibrium is the **labyrinth,** or **vestibular organ.** It consists of two small chambers (**saccule**

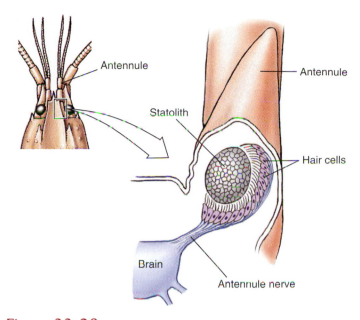

Figure 33-28

Statolith of the crayfish, a static balance organ of invertebrates.

and **utricle)** and three **semicircular canals** (Figure 33-25B). The utricle and saccule are static balance organs that, like invertebrate statocysts, give information about position of the head or body with respect to the force of gravity. As the head is tilted in one direction or another, stony accretions press on different groups of hair cells; these cells send nerve impulses to the brain, which interprets this information with reference to head position.

The semicircular canals of vertebrates are designed to respond to **rotational acceleration** and are relatively insensitive to linear acceleration. The three semicircular canals are at right angles to each other, one for each axis of rotation. They are filled with fluid (endolymph), and within each canal is a bulblike enlargement, the **ampulla,** which contains hair cells. The hair cells are embedded in a gelatinous membrane, the **cupula,** which projects into the fluid. The cupula is similar in structure to the cupula of the lateral-line system of fishes (p. 712). When the head rotates, fluid in the canal at first tends not to move because of inertia. Since the cupula is attached, its free end is pulled in the direction opposite the direction of rotation (Figure 33-29). Bending of the cupula distorts and excites the hair cells embedded in it, and this stimulation increases the discharge rate over afferent nerve fibers leading from the ampulla to the brain. This increased discharge rate produces a sensation of rotation. Since the three canals of each ear are in different planes, acceleration in any direction stimulates at least one ampulla.

Photoreception: Vision

Light-sensitive receptors are called **photoreceptors.** These receptors range from simple light-sensitive cells scattered randomly on the body surface of many invertebrates (dermal light

Figure 33-29

How the semicircular canals respond to angular acceleration. Because of inertia, endolymph in the semicircular canal corresponding to the plane of motion moves past the cupula in a direction opposite to that of angular acceleration. Movement of the cupula stimulates hair cells.

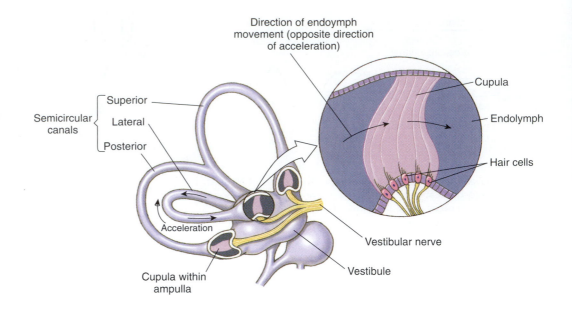

sense) to the exquisitely developed camera-type eye of vertebrates and cephalopods. Eyespots of astonishingly advanced organization appear even in some unicellular forms. That of a dinoflagellate *Nematodinium* bears a lens, a light-gathering chamber, and a photoreceptive pigment cup—all developed within a single-celled organism (Figure 33-30). The dermal light receptors of many invertebrates are of much simpler design. They are far less sensitive than optic receptors, but they are important in locomotory orientation, pigment distribution in chromatophores, photoperiodic adjustment of reproductive cycles, and other behavioral changes.

More highly organized eyes, many capable of excellent image formation, are based on one of two different principles: either a single-lens, camera-type eye such as those of cephalopod molluscs and vertebrates; or a multifaceted (compound) eye as in arthropods. Arthropod **compound eyes** are composed of many independent visual units called **ommatidia** (Figure 33-31).

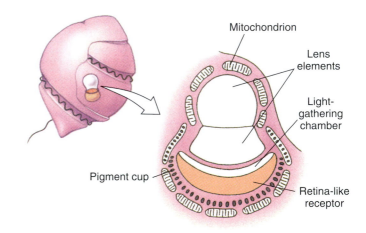

Figure 33-30

Eyespot of the dinoflagellate *Nematodinium.*

Figure 33-31

Compound eye of an insect. A single ommatidium is shown enlarged at right.

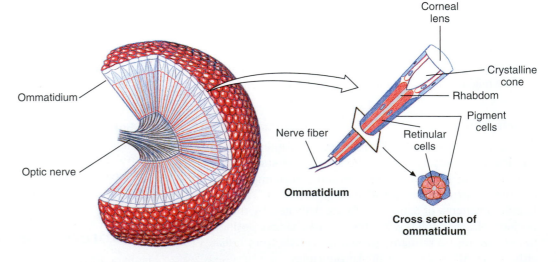

Eyes of bees contain about 15,000 of these units, each of which views a separate narrow sector of the visual field. Such eyes form a mosaic of images of varying brightness from the separate units. Resolution (the ability to see objects sharply) is poor compared with that of a vertebrate eye. A fruit fly, for example, must be closer than 3 cm to see another fruit fly as anything but a single spot. However, a compound eye is especially well suited to detecting motion, as anyone who has tried to swat a fly knows.

Eyes of certain annelids, molluscs, and all vertebrates are built like a camera—or rather we should say that a camera is modeled somewhat after vertebrate eyes. A camera-type eye contains in the front a light-tight chamber and lens system, which focuses an image of the visual field on a light-sensitive surface (the retina) in the back (Figure 33-32).

The spherical eyeball is built of three layers: (1) a tough outer white **sclera** for support and protection, (2) middle **choroid coat,** containing blood vessels for nourishment, and (3) light-sensitive **retina** (Figure 33-32). The **cornea** is a transparent anterior modification of the sclera. A circular, pigmented curtain, or **iris,** regulates the size of the light opening, or **pupil.** Just behind the iris is a **lens,** a transparent, elastic oval disc that, with the aid of **ciliary muscles,** can alter curvature of the lens and bend light rays to focus an image on the retina. In terrestrial vertebrates the cornea actually does most of the bending of light rays, whereas the lens adjusts focus for near and far objects. Between cornea and lens is an **outer chamber** filled with watery **aqueous humor;** between lens and retina is a much larger **inner chamber** filled with viscous **vitreous humor.**

The retina is composed of several cell layers (Figure 33-33). The outermost layer, closest to the sclera, consists of pigment cells. Adjacent to this layer are the photoreceptors, **rods** and **cones.** Approximately 125 million rods and 1 million cones are present in each human eye. Cones are primarily concerned with color vision in ample light and rods with colorless vision in dim light. Next is a network of **intermediate neurons** (bipolar, horizontal, and amacrine cells) that process and relay visual information from the photoreceptors to ganglion cells whose axons form the optic nerve. The network permits much convergence, especially for rods. Information from several hundred rods may converge on a single ganglion cell, an adaptation that greatly increases the effectiveness of rods in dim light. Cones show very little convergence. By coordinating activities between different ganglion cells, and adjusting sensitivities of bipolar cells, horizontal and amacrine cells improve overall contrast and quality of a visual image.

The **fovea centralis,** or **fovea,** the region of keenest vision, is located in the center of the retina (Figure 33-32), in direct line with the center of the lens and cornea. It contains only cones, a vertebrate specialization for diurnal (daytime) vision. Acuity of an animal's eyes depends on the density of cones in the fovea. A human fovea and that of a lion contain approximately 150,000 cones per square millimeter, but many water and field birds have up to 1 million cones per square millimeter. Their eyes are as good as our eyes would be if aided by eight-power binoculars.

Figure 33-32
Structure of the human eye.

Figure 33-33
Structure of a primate retina, showing organization of intermediate neurons that connect photoreceptor cells to ganglion cells of the optic nerve.

One of several marvels of the vertebrate eye is its capacity to compress the enormous range of light intensities presented to it into a narrow range that can be handled by optic nerve fibers. Light intensity between a sunny noon and starlit night differs by more than 10 billion to 1. Rods quickly saturate with high light intensity, but cones do not; they shift their operating range with changing ambient light intensity so that a high-contrast image is perceived over a broad range of light conditions. This shift is made possible by complex interactions among the network of nerve cells that lie between the cones and the ganglion cells that generate retinal output to the brain.

At the peripheral parts of the retina only rods are found. Rods are high-sensitivity receptors for dim light. At night, the cone-filled fovea is unresponsive to low levels of light and we become functionally color blind ("at night all cats are gray"). Under nocturnal conditions, the position of greatest visual acuity is not at the center of the fovea but at its edge. Thus it is easier to see a dim star at night by looking slightly to one side of it.

Chemistry of Vision

Both rods and cones contain light-sensitive pigments known as **rhodopsins.** Each rhodopsin molecule consists of a large protein, **opsin,** which behaves as an enzyme, and a small carotenoid molecule, **retinal,** a derivative of vitamin A. When a quantum of light strikes the photopigment and is absorbed by the rhodopsin molecule, retinal is isomerized, changing the shape of the molecule. This molecular change triggers the enzymatic activity of opsin, which sets in motion a biochemical sequence of several steps. This complex sequence behaves as an excitatory cascade that vastly amplifies the energy of a single photon to generate a nerve impulse in a rod or cone.

The amount of intact rhodopsin in a retina depends on the intensity of light reaching the eye. A dark-adapted eye contains much rhodopsin and is very sensitive to weak light. Conversely in a light-adapted eye, most of the rhodopsin is split into retinal and opsin. It takes approximately half an hour for a light-adapted eye to accommodate to darkness, while the rhodopsin level gradually increases.

Color vision

Cones function to perceive color and require 50 to 100 times more light for stimulation than do rods. Consequently, night vision is almost totally rod vision. Unlike humans, who have both day and night vision, some vertebrates specialize for one or the other. Strictly nocturnal animals, such as bats and owls,

have pure rod retinas. Purely diurnal forms, such as common gray squirrels and some birds, have only cones; they are virtually blind at night.

In 1802 an English physician and physicist Thomas Young speculated that we see color by relative excitation of three kinds of photoreceptors: one each for red, green, and blue. In the 1960s Young's prescient hypothesis was eventually supported through the combined work of several groups of researchers. Humans have three types of cones, each containing a visual pigment that responds to a particular wavelength of light (Figure 33-34). Blue cones absorb the most light at 430 nm, green cones at 540 nm, and red cones at 575 nm. Variation in the structure of opsin produces the different visual pigments found in rods and the three types of cones. Colors are perceived by comparing levels of excitation of the three different kinds of cones. For example, a light having a wavelength of 530 nm would excite green cones 95%, red cones about 70%, and blue cones not at all. This comparison is made both in nerve circuits in the retina and in the visual cortex of the brain, and our brain interprets this combination as green.

Color vision occurs in some members of all vertebrate groups with the possible exception of amphibians. Bony fishes and birds have particularly good color vision. Surprisingly, most mammals are color blind; exceptions are primates and a few other species such as squirrels.

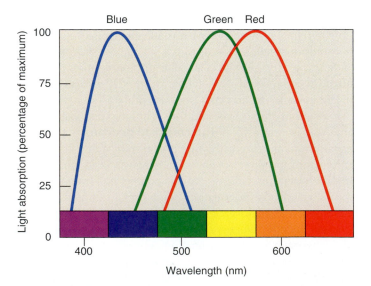

Figure 33-34
The absorption spectrum of human vision. Three types of visual pigments in cones absorb maximally at 430 nm (blue cones), 540 nm (green cones), and 575 nm (red cones).

SUMMARY

The nervous system is a rapid communication system that interacts continuously with the endocrine system to control coordination of body function. The basic unit of nervous integration in all animals is the neuron, a highly specialized cell designed to conduct self-propagating impulses, called action potentials, to other cells. Action potentials are transmitted from one neuron to another across synapses which may be either electrical or chemical. The thin gap between neurons at chemical synapses is bridged by a chemical neurotransmitter molecule, which is released from the synaptic knob, and can be either stimulatory or inhibitory.

The simplest organization of neurons into a system is the nerve net of cnidarians, basically a plexus of nerve cells that, with additions, is the basis of nervous systems of several invertebrate phyla. With the appearance of ganglia (nerve centers) in bilateral flatworms, nervous systems differentiated into central and peripheral divisions. In vertebrates, the central nervous system consists of a brain and spinal cord. Fishes and amphibians have a three-part linear brain, whereas in mammals, the cerebral cortex has become a vastly enlarged multicomponent structure that has assumed the most important integrative activities of the nervous system. It completely overshadows the ancient brain, which is consigned to the role of relay center and to serving numerous unconscious but nonetheless vital functions such as breathing, blood pressure, and heart rate.

In humans the left cerebral hemisphere is usually specialized for language and mathematical skills while the right hemisphere is specialized for visual-spatial and musical skills.

The peripheral nervous system connects the central nervous system to receptors and effector organs. It is divided broadly into an afferent system, which conducts sensory signals to the central nervous system, and an efferent system, which conveys motor impulses to effector organs. This efferent system is subdivided into a somatic nervous system, which innervates skeletal muscle, and an autonomic nervous system, which innervates smooth and cardiac muscle and glands. The autonomic nervous system is subdivided into anatomically distinct sympathetic and parasympathetic systems, each of which sends fibers to most body organs. Generally the sympathetic system governs excitatory activities and the parasympathetic system governs maintenance and restoration of body resources.

Sensory organs are receptors designed especially to respond to internal or environmental change. The most primitive and ubiquitous sense is chemoreception. Chemoreceptors may be contact receptors such as the vertebrate sense of taste, or distance receptors such as olfaction, which detects airborne molecules. In both cases, a specific chemical interacts with a particular receptor and results in impulses transmitted to, and interpreted by, the brain. In spite of the similarity between these two senses, the sense of olfaction is far more sensitive and complex.

Receptors for touch, pain, equilibrium, and hearing are all mechanical force receptors. Touch and pain receptors are characteristically simple structures, but hearing and equilibrium are highly specialized senses based on special hair cells that respond to mechanical deformation. Sound waves received by the ear are mechanically amplified and transmitted to the inner ear where different areas of the cochlea respond to different sound frequencies. Equilibrium receptors, also located in the inner ear, consist of two saclike static balance organs and three semicircular canals that detect rotational acceleration.

Vision receptors (photoreceptors) are associated with special pigment molecules that photochemically decompose in the presence of light and, in doing so, trigger nerve impulses in optic fibers. The advanced compound eye of arthropods is especially well suited to detecting motion in the visual field. Vertebrates have a camera eye with focusing optics. Photoreceptor cells of the retina are of two kinds: rods, designed for high sensitivity with dim light, and cones, designed for color vision in daylight. Cones predominate in the fovea centralis of human eyes, the area of keenest vision. Rods are more abundant in peripheral areas of the retina.

REVIEW QUESTIONS

1. Define the following terms: neuron, axon, dendrite, myelin sheath, afferent neuron, efferent neuron, association neuron.
2. Glial cells far outnumber neurons and contribute roughly half the weight of the mammalian nervous system. Offer examples of functions glial cells perform in the peripheral nervous system and in the central nervous system.
3. The concentration of potassium inside a nerve cell membrane is higher than the concentration of sodium outside the membrane. Yet the inside of the membrane (where the cation concentration is higher) is negative to the outside. Explain this observation in terms of permeability properties of the membrane.
4. What ionic and electrical changes occur during passage of an action potential along a nerve fiber?
5. Explain different ways in which invertebrates and vertebrates have achieved high velocities for conduction of nerve impulses. Can you suggest why the invertebrate solution would not be suitable for the homeothermic birds and mammals?
6. Why is the sodium pump *indirectly* important to the action potential and to maintaining the resting membrane potential?
7. Describe the microstructure of a chemical synapse. Summarize what happens when an action potential arrives at a synapse.
8. Describe the cnidarian (radiate) nervous system. How is a tendency toward centralization of the nervous system manifested in flatworms, annelids, molluscs, and arthropods?
9. How does a vertebrate spinal cord differ morphologically from nerve cords of invertebrates?
10. The knee-jerk reflex is often called a stretch reflex because a sharp tap on the patellar ligament stretches the quadriceps femoris, the extensor muscle of the leg. Describe the components and sequence of events that lead to a "knee jerk." Why is this reflex simpler than most reflex arcs? What is the difference between a reflex arc and a reflex act?
11. Name major functions associated with the following brain structures: medulla, cerebellum, tectum, thalamus, hypothalamus, cerebrum, limbic system.

12. What functional activities are associated with the left and the right hemispheres of the cerebral cortex?

13. What is the autonomic nervous system and what activities does it perform that distinguish it from the central nervous system? Why can the autonomic nervous system be described as a "two-neuron" system?

14. Give the meaning of the statement, "The idea that all sense organs behave as biological transducers is a unifying concept in sensory physiology."

15. Chemoreception in vertebrates and insects is mediated through clearly distinguishable senses of taste and smell. Contrast these two senses in humans in terms of anatomical location and nature of the receptors and sensitivity to chemical molecules.

16. What is the vomeronasal organ and what activity does it perform? Why is its functioning considered distinct from the sense of smell, but a component of the olfactory system of vertebrates?

17. Explain how ultrasonic detectors of certain nocturnal moths are adapted to help them escape an approaching bat.

18. Outline the place theory of pitch discrimination as an explanation of the human ear's ability to distinguish between sounds of different frequencies.

19. Explain how the semicircular canals of the ear are designed to detect rotation of a human head in any directional plane.

20. Explain what happens when light strikes a dark-adapted rod that leads to the generation of a nerve impulse. What is the difference between rods and cones in their sensitivity to light?

21. In 1802 Thomas Young hypothesized that we see color because the retina contains three kinds of receptors. What evidence substantiates Young's hypothesis? How can we perceive any color in the visible spectrum when our retinas contain only three classes of color cones?

SELECTED REFERENCES

Axel, R. 1995. The molecular logic of smell. Sci. Am. **273:**154–159 (Oct.). *Recent research reveals a surprisingly large family of genes that encodes odor-detecting molecules. This and other findings help to illuminate how the nose and brain may perceive scents.*

Changeux, J-P. 1993. Chemical signaling in the brain. Sci. Am. **269:**58–62 (Nov.). *Studies of the electric organs of fishes provide insights into how neurons in the human brain transmit information from one to the next.*

Freeman, W. J. 1991. The physiology of perception. Sci. Am. **264:**78–85 (Feb.). *How the brain transforms sensory messages almost instantly into conscious perceptions.*

Hudspeth, A. J. 1983. The hair cells of the inner ear. Sci. Am. **248:**54–64 (Jan.). *How these biological transducers work.*

Jacobson, M. 1993. Foundations of neuroscience. New York, Plenum Press. *The historical development of neuroscience and its outstanding personages—and the dangers of hero worship of individual neuroscientists.*

Nathan, P. 1997. The nervous system, ed. 4. London, Whurr Publications Ltd. *One of the best of several semipopular accounts of the nervous system.*

Nathans, J. 1989. The genes for color vision. Sci. Am. **260:**42–49 (Feb.). *Isolation of genes that encode color-detecting proteins of the human eye provide clues about the evolution of color vision.*

Nef, P. 1998. How we smell: the molecular and cellular bases of olfaction. News Physiol. Sci. **13:**1–5 (Feb.). *Describes three models for odor perception, each based on experimental data.*

Smith, D. V., and R. F. Margolskee. 2001. Making sense of taste. Sci. Am. **284:**32–39 (March). *Describes the mechanism of taste reception in a very readable way.*

Stebbins, W. C. 1983. The acoustic sense of animals. Cambridge, Massachusetts, Harvard University Press. *Broadly comparative introduction to the physics, physiology, natural history, and evolution of hearing.*

Stryer, L. 1987. The molecules of visual excitation. Sci. Am. **257:**42–50 (July). *Describes the cascade of molecular events following light absorption by a rod cell that leads to a nerve signal.*

Ulfendahl, M., and A. Flock. 1998. Outer hair cells provide active tuning in the organ of Corti. News Physiol. Sci. **13:**107–111 (July). *Describes recent experiments that suggest a more active role for the sensory hair cells within the auditory system of mammals.*

ZOOLOGY LINKS TO THE INTERNET

Visit the textbook's Online Learning Center at www.mhhe.com/zoology to find live Internet links for each of the topics listed here.

Nervous Systems and Sense Organs
Neurons
The Brain and Spinal Cord
Cranial Nerves
Sense Organs; Vision, Hearing and Other Senses
Human Nervous System Topics

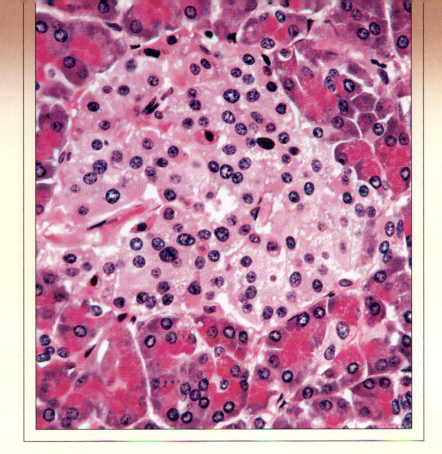

An endocrine islet of Langerhans, site of insulin synthesis, within the human pancreas.

Chemical Coordination

Endocrine System

The Birth of Endocrinology

The birth date of endocrinology as a science is usually given as 1902, the year two English physiologists, W. H. Bayliss and E. H. Starling (Figure 34-1), demonstrated the action of a hormone in a classic experiment that is still considered a model of the scientific method. Bayliss and Starling were interested in determining how the pancreas secreted its digestive juice into the small intestine at the proper time of the digestive process. They tested the hypothesis that acidic food entering the intestine triggered a nervous reflex that released pancreatic juice. To test this hypothesis, Bayliss and Starling cut away all nerves serving a tied-off loop of the small intestine of an anesthetized dog, leaving the isolated loop connected to the body only by its circulation. Injecting acid into the nerveless loop, they saw a pronounced flow of pancreatic juice. Thus, rather than a

nervous reflex, some chemical messenger had circulated from the intestine to the pancreas, causing the pancreas to secrete. Yet acid itself could not be the factor because it had no effect when injected directly into the circulation.

Bayliss and Starling then designed the crucial experiment that was to begin the new science of endocrinology. Suspecting that the chemical messenger originated in the mucosal lining of the intestine, they next prepared an extract of scrapings from the mucosa, injected it into the dog's circulation, and were rewarded with an abundant flow of pancreatic juice. They named the messenger present in the intestinal mucosa *secretin*. Later Starling coined the term **hormone** to describe all such chemical messengers, since he correctly surmised that secretin was only the first of many hormones awaiting discovery.

A   B

Figure 34-1

Founders of endocrinology. **A,** Sir William H. Bayliss (1860 to 1924). **B,** Ernest H. Starling (1866 to 1927).

The endocrine system, the second great integrative system controlling an animal's activities, communicates by chemical messengers called **hormones** (Gr. *hormōn,* to excite). Hormones are chemical compounds released into the blood in small amounts and transported by the circulatory system throughout the body to distant **target cells** where they initiate physiological responses.

Many hormones are secreted by **endocrine glands,** small, well-vascularized ductless glands composed of groups of cells arranged in cords or plates. Since endocrine glands have no ducts, their only connection with the rest of the body is by blood or other body fluid; they receive their raw materials from their extensive blood supply and secrete their finished hormonal products into it. **Exocrine glands,** in contrast, are provided with ducts for discharging their secretions onto a free surface. Examples of exocrine glands are sweat glands and sebaceous glands of skin, salivary glands, and the various enzyme-secreting glands lining the walls of the stomach and intestine (Chapter 32).

The classical definitions of hormones and endocrine glands just given, like so many other generalizations in biology, are changing as new information appears. Some hormones, such as certain neurosecretions, may never enter the general circulation. Furthermore, evidence suggests that many hormones, such as insulin, are synthesized in minute amounts in a variety of nonendocrine tissues (nerve cells, for example), and some, such as cytokines, are secreted by cells of the immune system (p. 743). Such hormones may function as **neurotransmitters** in the brain or as local tissue factors **(parahormones),** which stimulate cell growth or some biochemical process. Most hormones, however, are blood borne and therefore diffuse into every tissue space in the body.

Compared with nervous systems, endocrine systems are slow acting because of the time required for a hormone to reach the appropriate tissue, cross the capillary endothelium, and diffuse through tissue fluid to, and sometimes into, cells. The minimum response time is seconds and may be much longer. Hormonal responses in general are long lasting (minutes to days) whereas those under nervous control are short term (milliseconds to minutes). We expect to find endocrine control where a sustained effect is required, as in many metabolic, growth, and reproductive processes. Despite such differences, nervous and endocrine systems function without sharp separation as a single, interdependent system. Endocrine glands often receive directions from the brain. Conversely, many hormones act on the nervous system and significantly affect a wide array of animal behaviors.

The first formal experiment in endocrinology was performed in 1849 by Professor Arnold Adolph Berthold at the University of Gottingen. He conclusively demonstrated that a blood-borne signal was produced by the testes, and that this chemical was responsible for producing both physical and behavioral characteristics that distinguished an adult male rooster from immature chickens and adult male chickens that had been castrated (capons). Berthold castrated male chicks and divided them into three groups. He left one group of controls to grow normally without their testes, and he reimplanted the testes into the second group. The third group was implanted with testes from different chicks. As the chicks grew, he observed that the castrated group developed into capons, with no interest in hens, lacking rooster plumage and male aggressive behavior. The second and third groups of birds were indistinguishable from each other, with full male plumage, normal aggressive behavior and interest in hens. Berthold then killed the birds and dissected them. He found that the transplanted testes had developed their own blood supply and were functioning normally. From this classic experiment, Berthold concluded that testes must produce a blood-borne signal, since there was no nerve supply to the testes, which produced all characteristics of maleness.

All hormones are low-level signals. Even when an endocrine gland is secreting maximally, its hormone is so greatly diluted by the large volume of blood it enters that its plasma concentration seldom exceeds 10^{-9} M (or one billionth of a 1 M concentration). Some target cells respond to plasma concentrations of hormone as low as 10^{-12} M. Since hormones have far-reaching and often powerful influences on cells, it is evident that their effects are vastly amplified at the cellular level.

MECHANISMS OF HORMONE ACTION

Widespread distribution of hormones in an animal permits certain hormones, such as growth hormone from the vertebrate pituitary gland, to affect most, if not all, cells during specific stages of cellular differentiation. Whether hormones produce widespread or highly specific responses only in certain cells and at certain times depends on the presence of **receptor molecules** on or in target cells. A hormone will engage only those cells that display the receptor that, by virtue of its specific molecular shape, will bind with the hormone molecule. Other cells are insensitive to the hormone's presence because they lack specific receptors. Hormones act through two kinds of receptors: **membrane-bound receptors** and **nuclear receptors.**

Membrane-Bound Receptors and the Second-Messenger Concept

Many hormones, such as most amino acid derivatives, and peptide hormones that are too large, or too polar, to pass through cell membranes, bind to receptor sites present on the surfaces of target-cell membranes. The hormone and receptor form a complex that triggers a cascade of molecular events within a cell. The hormone thus behaves as a **first messenger** that causes activation of a **second-messenger** system in the cytoplasm. At least six different molecules have been identified as second messengers. Each works via a specific **kinase,** which causes activation or inactivation of rate-limiting enzymes that modify the direction and rate of cytoplasmic processes (Figure 34-2). Since many molecules of the second messenger are activated after a single hormone molecule has been bound, the message is amplified, perhaps many thousands of times.

Second-messenger systems known to participate in hormone actions are **cyclic AMP** (cAMP), **cyclic GMP** (cGMP), **Ca^{++}/calmodulin, inositoltrisphosphate** (IP$_3$), and **diacylglycerol** (DAG). Cyclic AMP was the first to be investigated, and has been shown to mediate actions of many peptide hormones, including parathyroid hormone, glucagon, adrenocorticotropic hormone (ACTH), thyrotropic hormone (TSH), melanophore-stimulating hormone (MSH), and vasopressin. It also mediates action of epinephrine (also called adrenaline), an amino acid derivative.

Nuclear Receptors

Unlike peptide hormones and epinephrine, which are much too large to pass through cell membranes, **steroid hormones** (for example, estrogen, testosterone, and aldosterone), are lipid-soluble molecules that readily diffuse through cell membranes. Once inside the cytoplasm of target cells, steroid hormones bind selectively to receptor molecules. While these receptor molecules may be located in either cytoplasm or nucleus, their ultimate site of activity is nuclear. The hormone-receptor complex, now known as a **gene regulatory protein,** then activates or inhibits specific genes. As a result, gene transcription is altered, since messenger RNA molecules are synthesized on specific sequences of DNA. Stimulation or inhibition of mRNA formation modifies production of key enzymes, thus setting in motion a hormone's observed effect (Figure 34-2). Thyroid hormones and the insect-molting hormone, ecdysone, also act through nuclear receptors.

Compared with peptide hormones that act *indirectly* through second-messenger systems, steroids have a *direct* effect on protein synthesis because they bind a nuclear receptor that modifies specific gene activity.

Recent evidence suggests that lipid-soluble hormones, such as estrogen, may also possess membrane-bound receptors that activate second messenger systems in the same way as peptide hormones, providing multiple and complex control of target cells.

Control of Secretion Rates of Hormones

Hormones influence cellular functions by altering rates of many different biochemical processes. Many affect enzymatic activity and thus alter cellular metabolism, some change membrane permeability, some regulate synthesis of cellular proteins, and some stimulate release of hormones from other endocrine glands. Because these are all dynamic processes that must adapt to changing metabolic demands, they must be regulated, not merely activated, by appropriate hormones. This regulation is achieved by precisely controlled release of a hormone into the blood. However, concentration of a hormone in body fluid depends on two factors: its rate of secretion and the rate at which it is inactivated and removed from the circulation. Consequently, if secretion is to be correctly controlled, an endocrine gland requires information about the level of its own hormone(s) in the plasma.

Many hormones are controlled by negative feedback systems that operate between glands secreting hormones and target cells (Figure 34-3). A feedback pattern is one in which output is constantly compared with a set point, like a thermostat. For example, CRH (corticotropin-releasing hormone), secreted by the hypothalamus, stimulates the pituitary (containing target cells) to release ACTH. ACTH stimulates the adrenal gland (containing target cells) to secrete cortisol. As the level of ACTH rises in the plasma, it acts on, or "feeds back" on, the hypothalamus to inhibit release of CRH. Similarly, as cortisol levels rise in the

Figure 34-2

Mechanisms of hormone action. Peptide hormones and epinephrine act through second-messenger systems, as for example, cyclic AMP, shown here. The combination of hormone with a membrane receptor stimulates the enzyme adenylate cyclase to catalyze formation of cyclic AMP (second messenger). Steroid hormones and thyroid hormones penetrate the cell membrane to combine with cytoplasmic or nuclear receptors that alter gene transcription.

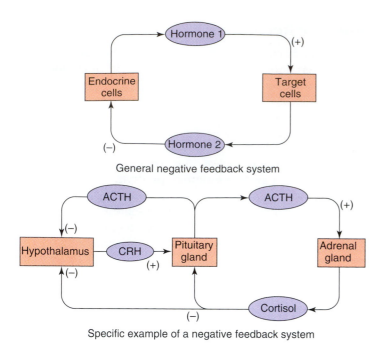

Figure 34-3
Negative feedback systems.

plasma, it "feeds back" on the hypothalamus and pituitary to inhibit release of both CRH and ACTH, respectively. Thus any deviation from the set point (a specific plasma level of each hormone) leads to corrective action in the opposite direction (Figure 34-3). Such a **negative feedback** system is highly effective in preventing extreme oscillations in hormonal output. However, hormonal feedback systems are more complex than a rigid "closed-loop" system such as a thermostat that controls a house's central heating system, because hormonal feedback may be altered by input from the nervous system or by metabolites or other hormones.

Extreme oscillations in hormone output do sometimes occur under natural conditions. However, because they have the potential to disrupt finely tuned homeostatic mechanisms, such extreme oscillations, as a result of **positive feedback,** are highly regulated and possess an obvious shutoff mechanism. During positive feedback, the signal (or output of the system) feeds back to the control system and causes an increase in the initial signal. In this way the initial signal becomes progressively amplified to produce an explosive event. For example, hormones controlling parturition (childbirth) become elevated from their normal setpoint and are shut off by birth of offspring from the uterus (Chapter 7).

INVERTEBRATE HORMONES

In many metazoan phyla, the principal source of hormones is **neurosecretory cells,** specialized nerve cells capable of synthesizing and secreting hormones. Their products, called neurosecretions or neurosecretory hormones, are discharged directly into the circulation, and serve as a crucial link between the nervous and endocrine systems.

Neurosecretory hormones occur in all metazoan groups. An extensively studied neurosecretory process in invertebrates is control of development and metamorphosis of insects. In insects, as in other arthropods, growth is a series of steps in which the rigid, nonexpansible exoskeleton is periodically discarded and replaced with a new, larger one. Most insects undergo a process of metamorphosis (p. 410), in which a series of juvenile stages, each requiring formation of a new exoskeleton, end with a molt.

Insect physiologists discovered that molting and metamorphosis are primarily controlled by interaction of two hormones, one favoring growth and differentiation of adult structures and another favoring retention of juvenile structures. These two hormones are **molting hormone** or **ecdysone,** produced by the prothoracic gland, and **juvenile hormone,** produced by the corpora allata (Figure 34-4). The structures of both hormones have been determined. Extraction from 1000 kg (about 1 ton) of silkworm pupae was required to show that ecdysone is a steroid.

Ecdysone is controlled by **prothoracicotropic hormone** or **PTTH.** This hormone is a polypeptide (molecular weight about 5000) produced by neurosecretory cells of the brain, and transported by axons to the corpora cardiacum where it is stored. At intervals during juvenile growth, release of PTTH into the blood stimulates the prothoracic gland to secrete ecdysone. Ecdysone appears to act directly on the chromosomes to set in motion changes resulting in a molt, by favoring development of adult structures. It is held in check, however, by juvenile hormone, which favors development of juvenile characteristics. During juvenile life, juvenile hormone predominates and each molt yields another larger juvenile (Figure 34-4). Finally output of juvenile hormone decreases, allowing final metamorphosis to the adult stage.

The precise location of brain hormone in the brain of pupal tobacco hornworms was revealed by N. Agui by delicate microdissection. Using a human eyebrow hair, he was able to isolate the single cell in each brain hemisphere that contained brain hormone activity. Thus only two cells, each about 20 μm in diameter, produce this insect's total supply of PTTH. In an age when sophisticated instrumentation has removed much of the tedium (and some creativity) from research, it is refreshing to learn that certain biological mysteries succumb only to skillful use of the human hand.

Chemists have synthesized several potent analogs of juvenile hormone, which hold great promise as insecticides. Minute quantities of these synthetic analogs induce abnormal final molts or prolong or block development. Unlike chemical insecticides, they are highly specific and ecologically benign.

VERTEBRATE ENDOCRINE GLANDS AND HORMONES

In the remainder of this chapter we describe some of the best understood and most important of vertebrate hormones. While this discussion is limited principally to a brief overview of mam-

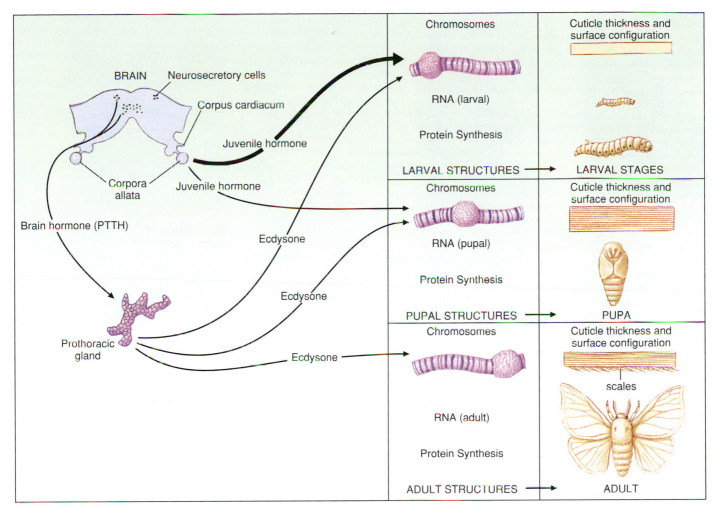

Figure 34-4

Endocrine control of molting in a moth, typical of insects having complete metamorphosis. Many moths mate in spring or summer, and eggs soon hatch into the first of several larval stages, called instars. After the final larval molt, the last and largest larva (caterpillar) spins a cocoon in which it pupates. The pupa overwinters, and an adult emerges in the spring to start a new generation. Juvenile hormone and ecdysone interact to control molting and pupation. Many genes are activated during metamorphosis, as seen by puffing of chromosomes (center column). Puffs form in sequence during successive molts. Changes in cuticle thickness and surface characteristics are shown at right.

malian hormonal mechanisms (since laboratory mammals and humans have always been the objects of the most intensive research), we will note some important differences in functional roles of hormones among different vertebrate groups.

Hormones of the Hypothalamus and Pituitary Gland

The pituitary gland, or **hypophysis,** is a small gland (0.5 g in humans) lying in a well-protected position between the roof of the mouth and floor of the brain (Figure 34-5). It is a two-part gland having a double embryological origin. The **anterior pituitary** (adenohypophysis) is derived embryologically from the roof of the mouth. The **posterior pituitary** (neurohypophysis) arises from a ventral portion of the brain, the **hypothalamus,** and is connected to it by a stalk, the **infundibulum.** Although the anterior pituitary lacks any anatomical connection to the

brain, it is functionally connected to it by a special portal circulatory system. A portal circulation is one that delivers blood from one capillary bed to another (Figures 34-5 and 34-6). In this case, the portal circulation provides a link between neurosecretory cells of the hypothalamus and anterior pituitary gland.

Hypothalamus and Neurosecretion

Because of the strategic importance of the pituitary in influencing most hormonal activities in the body, the pituitary was once called the "master gland." This description is not appropriate, however, because anterior pituitary hormones are regulated by a higher council, the neurosecretory centers of the hypothalamus. The hypothalamus is itself under ultimate control by the brain. The hypothalamus contains groups of neurosecretory cells, which are specialized nerve cells (Figure 34-5 and 34-6), that manufacture neurohormones, called **releasing hormones** or **release-inhibiting hormones** (or "factors").

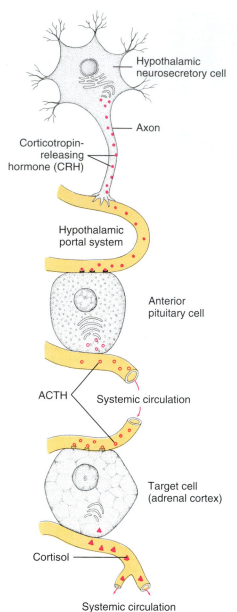

Figure 34-5

Human hypothalamus and pituitary gland. The posterior lobe is connected directly to the hypothalamus by neurosecretory fibers. The anterior lobe is indirectly connected to the hypothalamus by a portal circulation (shown in red) beginning in the base of the hypothalamus and ending in the anterior pituitary.

These neurohormones travel down nerve fibers to their endings in the median eminence. Here they enter a capillary network to complete their journey to the anterior pituitary by way of the pituitary portal system. The hypothalamic hormones then stimulate or inhibit release of various anterior pituitary hormones. Several hypothalamic releasing and release-inhibiting hormones have been discovered, characterized chemically, and isolated in pure state (Table 34-1), although the identification and action of some of the hypothalamic hormones listed in Table 34-1 is still tentative.

Figure 34-6

Relationship of hypothalamic, pituitary, and target-gland hormones. The hormone sequence controlling the release of cortisol from the adrenal cortex is used as an example.

Anterior Pituitary

The anterior pituitary consists of an **anterior lobe** (pars distalis) as shown in Figure 34-5, and an **intermediate lobe** (pars intermedia), which is absent in some animals (including humans). The anterior pituitary produces seven hormones, and all but one are released by the anterior lobe.

Four hormones of the anterior pituitary are **tropic hormones** (from the Greek *tropē,* to turn toward) that regulate other endocrine glands (Table 34-1). **Thyroid-stimulating hormone (TSH)** or **thyrotropin** stimulates production of thyroid hormones by the thyroid gland. Two tropic hormones are commonly called **gonadotropins** because they act on the gonads (ovaries of females, testes of males). These are **follicle-stimulating hor-**

mone (FSH) and **luteinizing hormone (LH).** FSH promotes egg production and secretion of estrogen in females, and supports sperm production in males. LH induces ovulation, corpus luteum production, and secretion of female sex steroids, progesterone and estrogen. In males, LH promotes production of male sex steroids (primarily testosterone). It once was called interstitial cell stimulating hormone (ICSH) in males, before it was discovered to be identical to LH in females. The fourth tropic hormone, **adrenocorticotropic hormone (ACTH),** increases production and secretion of steroid hormones from the adrenal cortex.

Prolactin and the structurally related **growth hormone (GH)** are proteins. Prolactin is essential for preparing mammary glands for lactation; after birth it is required for production of

TABLE 34.1

Hormones of the Vertebrate Pituitary

	Hormone	Chemical Nature	Principal Action	Hypothalamic Controls
Adenohypophysis				
Anterior lobe	Thyroid-stimulating hormone (TSH)	Glycoprotein	Stimulates thyroid hormone synthesis and secretion	TSH-releasing hormone (TRH)
	Follicle-stimulating hormone (FSH)	Glycoprotein	Female: follicle maturation and estrogen synthesis Male: stimulates sperm production	Gonadotropin-releasing hormone (GnRH)[1]
	Luteinizing hormone (LH)	Glycoprotein	Female: stimulates ovulation, corpus luteum formation, estrogen and progesterone synthesis Male: testosterone secretion	Gonadotropin-releasing hormone (GnRH)[1]
	Prolactin (PRL)	Protein	Mammary gland growth, milk synthesis, immune response and angiogenesis in mammals, parental behavior, electrolyte and water balance in lower vertebrates	Dopamine (prolactin release-inhibiting hormone or PIH) Prolactin-releasing factor (PRF)?
	Growth hormone (GH) (somatotropin)	Protein	Stimulates growth, protein synthesis, mobilization of glycogen and fat stores	Growth hormone-releasing hormone (GHRH) Growth hormone release-inhibiting hormone (GHIH) or somatostatin
	Adrenocorticotropic hormone (ACTH)	Polypeptide	Stimulates glucocorticoid synthesis by adrenal cortex	Corticotropin-releasing hormone (CRH)
Intermediate lobe[2]	Melanocyte-stimulating hormone (MSH)	Polypeptide	Increased melanin synthesis by melanocytes in epidermis of ectotherms; function not clear in endotherms	Melanocyte-stimulating hormone-inhibiting hormone (MSHIH)
Neurohypophysis				
(Posterior lobe)	Oxytocin	Octapeptide	Milk ejection and uterine contractions, sexual behavior and pair bonding in monogamous species	
	Vasopressin[3] (antidiuretic hormone or ADH)	Octapeptide	Water reabsorption in mammalian kidneys	
	Vasotocin[4]	Octapeptide	Increases water reabsorption	

[1] One GnRH hormone regulates both FSH and LH, but some recent studies suggest a separate FSH-releasing hormone (FSH-RH).
[2] Birds and some mammals lack an intermediate lobe. In these forms, MSH is produced by the anterior lobe.
[3] In mammals.
[4] In all vertebrate classes except mammals, although additional, related hormones have been identified.

milk. Prolactin also is implicated in parental behavior in a wide variety of vertebrates. Beyond its more traditional role in reproductive processes, prolactin regulates water and electrolyte balance in many species. More recently, prolactin has been shown to be a chemical mediator of the immune system and is important in formation of new blood vessels (angiogenesis). Unlike tropic hormones, prolactin acts directly on its target tissues rather than through other hormones.

GH (also called somatotropin) performs a vital role in governing body growth through its stimulatory effect on cellular mitosis, on synthesis of messenger RNA and protein, and on metabolism, especially in new tissue of young vertebrates. Growth hormone acts directly on growth and metabolism, as well as indirectly through a polypeptide hormone, **insulin-like growth factor (IGF)** or somatomedin, produced by the liver.

The only anterior pituitary hormone produced by the intermediate lobe is **melanocyte-stimulating hormone (MSH).** In cartilaginous and bony fishes, amphibians, and reptiles, MSH is a direct-acting hormone that promotes dispersion of the pigment melanin within melanocytes, causing darkening of the skin. In birds and mammals, MSH is produced by cells in the anterior pituitary rather than the intermediate lobe, but its physiological function remains unclear. MSH appears unrelated to pigmentation in endotherms, although it will cause darkening of the skin in humans if injected into the circulation. Until recently, many endocrinologists thought MSH was a vestigial hormone in mammals, but interest has been rekindled by studies showing that it enhances memory and growth of the fetus. In addition, MSH has been isolated from specific regions of the hypothalamus, where it has been linked to regulation of ingestive behaviors and metabolism in adult mammals. Future studies will determine if a similar role also exists for MSH during development. MSH and ACTH are derived from a precursor molecule (pro-opiomelanocortin or POMC) that is transcribed and translated from a single gene.

Posterior Pituitary

The hypothalamus is the source of two hormones of the posterior lobe of the pituitary (Table 34-1). They are formed in neurosecretory cells in the hypothalamus, whose axons extend down the infundibular stalk and into the posterior lobe. The hormones are secreted from axon terminals ending in close proximity to blood capillaries, which the hormones enter when released (see Figure 34-5). In a sense the posterior lobe is not a true endocrine gland, but a storage and release center for hormones manufactured entirely in the hypothalamus. The two posterior-lobe hormones of mammals, oxytocin and vasopressin, are chemically very much alike. Both are polypeptides consisting of eight amino acids (octapeptides, Figure 34-7). These hormones are among the fastest-acting hormones, since they are capable of producing a response within seconds of their release from the posterior lobe.

Oxytocin has two important specialized reproductive functions in adult female mammals. It stimulates contraction of uterine smooth muscles during parturition (birth of young). In clinical practice, oxytocin is used to induce delivery during a difficult labor and to prevent uterine hemorrhage after birth. A second action of oxytocin is that of milk ejection by the mammary glands in response to suckling. Recent work also has established a role for oxytocin in pair-bonding behavior in both sexes in monogamous voles.

Vasopressin, the second posterior-lobe hormone, acts on collecting ducts of the kidney to increase water reabsorption and thus restrict urine flow, as already described on p. 645. It is therefore often called **antidiuretic hormone** or **ADH.** Vasopressin also increases blood pressure through its generalized constrictor effect on smooth muscles of arterioles. Finally, vasopressin acts centrally to increase thirst, and therefore, drinking behavior.

All jawed vertebrates secrete two posterior-lobe hormones that are quite similar to those of mammals. All are octapeptides, but their structures vary because of amino-acid substitutions in three of eight amino-acid positions in the molecule.

Of all posterior-lobe hormones, **vasotocin** (Table 34-1) has the widest phylogenetic distribution and is believed to be the parent hormone from which other octapeptides evolved. It is found in all vertebrate classes except mammals. It is a water-balance hormone in amphibians, especially toads, in which it acts to conserve water by (1) increasing permeability of skin (to promote water absorption from the environment), (2) stimulating water reabsorption from the urinary bladder, and (3) decreasing urine flow. Action of vasotocin is best understood in amphibians, but it appears to play some water-conserving role in birds and reptiles as well.

Pineal Gland

In all vertebrates the dorsal part of the brain, the diencephalon Figure 33-13, p. 705 gives rise to a saclike evagination called the pineal complex, which lies just beneath the skull in a midline position. In ectothermic vertebrates the pineal complex contains glandular tissue and a photoreceptive sensory organ involved in pigmentation responses and in light-dark biological rhythms. In lampreys, many amphibians, lizards, and tuataras (*Sphenodon,* p. 550), the median photoreceptive organ is so well developed, containing structures analogous to the lens and cornea of lateral eyes, that it is often called a third eye. In birds and mammals, the pineal complex is an entirely glandular structure called the **pineal gland.** The pineal gland produces the hormone **melatonin.** Melatonin secretion is strongly affected by exposure to light. Its production is lowest during daylight hours and highest at night. In nonmammalian vertebrates, the pineal gland is responsible for maintaining **circadian rhythms**—self-generated (endogenous) rhythms that are about 24 hours in length. A circadian rhythm serves as a biological clock for many physiological processes that follow a regular pattern.

In mammals, an area of the hypothalamus called the **suprachiasmatic nucleus** has become the primary circadian pacemaker, although the pineal gland still produces melatonin nightly and serves to reinforce the circadian rhythm of the suprachiasmatic nucleus. In birds and mammals in which seasonal rhythms in reproduction are regulated by **photoperiod,** melatonin plays a critical role in regulating gonadal activity. In long-day breeders, such as ferrets, hamsters, and deer mice, reduced light stimulation with shortening day length in autumn increases melatonin secretion, and in these species reproductive activity is suppressed during winter months. Lengthening days in the spring have the opposite effect and reproductive activities are resumed. Short-day breeders, such as white-tailed deer, silver fox, spotted skunk, and sheep, are stimulated by reduced

Figure 34-7

Posterior-lobe hormones of mammals. Both oxytocin and vasopressin consist of eight amino acids (the two sulfur-linked cysteine molecules are considered a single amino acid, cystine). Oxytocin and vasopressin are identical except for amino-acid substitutions in the blue positions. Abbreviations represent amino acids.

day length in the fall; increasing melatonin levels in the fall are associated with increased reproductive activity. The role of melatonin is an indirect one in both cases since melatonin itself does not stimulate or inhibit the reproductive axis.

Only recently the pineal gland has been shown to produce subtle and incompletely understood effects on circadian and annual rhythms in nonphotoperiodic mammals (such as humans). For example, melatonin secretion has been linked to a sleeping and eating disorder in humans known as seasonal affective disorder (SAD). Some people living in northern latitudes, where day lengths are short in winter and when melatonin production is elevated, become depressed in winter, sleep long periods, and may go on eating binges. Often this wintertime disorder can be treated by exposure to sunlamps with full-spectrum light; such exposure depresses melatonin secretion by the pineal gland. Disturbed physiological rhythms associated with jet-lag, shift-work, and aging also have been linked to inappropriate melatonin rhythms.

Brain Neuropeptides

The blurred distinction between endocrine and nervous systems is nowhere more evident than in the nervous system, where a growing list of hormonelike neuropeptides have been discovered in central and peripheral nervous systems of vertebrates and invertebrates. In mammals, approximately 40 neuropeptides (short chains of amino acids) have been located using immunological labeling with antibodies that can be visualized in histological sections under a microscope, and the list is still growing. Many are known to lead double lives—to be capable of behaving both as hormones, carrying signals from gland cells to their targets, and as neurotransmitters, relaying signals between nerve cells. For example, both oxytocin and vasopressin have been discovered at widespread sites in the brain by immunochemical methods. Related to this discovery is the fascinating observation that people and experimental animals injected with minute quantities of vasopressin experience enhanced learning and improved memory. This effect of vasopressin in brain tissue is unrelated to its well-known antidiuretic function in the kidney (p. 645). Several hormones, such as gastrin and cholecystokinin (p. 689) (which long had been supposed to function only in the gastrointestinal tract), have been discovered in the cerebral cortex, hippocampus, and hypothalamus. In addition to its gastrointestinal actions, cholecystokinin is known to function in control of feeding and satiety and may serve other functions as a brain neuroregulator.

Among the dramatic developments in this field was the discovery in 1975 of endorphins and enkephalins, neuropeptides that bind with opiate receptors and influence perception of pleasure and pain (see note on p. 711). Endorphins and enkephalins are found also in brain circuits that modulate several other functions unrelated to pleasure and pain, such as control of blood pressure, body temperature, body movement, feeding and reproduction. Even more intriguing, endorphins are derived from the same prohormone (POMC) that gives rise to the anterior pituitary hormones ACTH and MSH.

The radioimmunoassay technique developed by Solomon Berson and Rosalyn Yalow about 1960 has revolutionized endocrinology and neurochemistry. First, antibodies to the hormone of interest (insulin, for example) are prepared by injecting a mammal, such as guinea pigs or rabbits, with the hormone. Then, a fixed amount of radioactively labeled insulin and unlabeled insulin antibodies is mixed with the sample of blood plasma to be measured. The native insulin in the blood plasma and the radioactive insulin compete for antibodies. The more insulin present in the sample, the less radioactive insulin will bind to the antibodies. Bound and unbound insulin are then separated, and their radioactivities are measured together with those of appropriate standards to determine the amount of insulin present in the blood sample. The method is so incredibly sensitive that it can measure the equivalent of a cube of sugar dissolved in one of the Great Lakes.

Prostaglandins and Cytokines

Prostaglandins

Prostaglandins are derivatives of long-chain unsaturated fatty acids that were discovered in seminal fluid in the 1930s. At first they were thought to be produced only by the prostate gland (hence the name) but now have been found in virtually all mammalian tissues. Prostaglandins act as local hormones and have diverse actions on many different tissues, making generalizations about their effects difficult. Many of their effects, however, involve smooth muscle. In some tissues prostaglandins regulate vasodilation or vasoconstriction by their action on smooth muscle in walls of blood vessels. They are known to stimulate contraction of uterine smooth muscle during childbirth. There also is evidence that overproduction of uterine prostaglandins is responsible for painful symptoms of menstruation (dysmenorrhea) experienced by many women. Several inhibitors of prostaglandins that provide relief from these symptoms have now been approved as medicines. Among other actions of prostaglandins is their intensification of pain in damaged tissues, mediation of the inflammatory response, and involvement in fever.

Cytokines

For some years we have known that cells of the immune system communicate with each other and that this communication was crucial to the immune response. Now we understand that a large group of polypeptide hormones called **cytokines** (p. 743) mediate communication between cells participating in an immune response. Cytokines can affect the cells that secrete them, affect nearby cells, and like other hormones, they can affect cells in distant locations. Their target cells bear specific receptors for the cytokine bound to the surface membrane. Cytokines coordinate a complex network, with some target cells being activated, stimulated to divide and often to secrete their own cytokines. The same cytokine that activates some cells may suppress division of other target cells. Cytokines also are involved in formation of blood, and more recently, their role in regulation of energy balance by the central nervous system is being explored.

Hormones of Metabolism

An important group of hormones adjusts the delicate balance of metabolic activities. Rates of chemical reactions within cells are often regulated by long sequences of enzymes. Although such sequences are complex, each step in a pathway is mostly self-regulating as long as the equilibrium between substrate, enzyme, and product remains stable. However, hormones may alter activity of crucial enzymes in a metabolic process, thus accelerating or inhibiting the entire process. The most important hormones of metabolism are those of the thyroid, parathyroid, adrenal glands, and pancreas, as well as the previously mentioned growth hormone of the anterior pituitary. Finally, a new hormone, leptin, has been added to the list of metabolic hormones.

Thyroid Hormones

Two hormones, **triiodothyronine** and **thyroxine** (T_3 and T_4, respectively) are secreted by the thyroid gland. This large endocrine gland is located in the neck of all vertebrates. The thyroid is composed of thousands of tiny spherelike units, called follicles, where thyroid hormones are synthesized, stored, and released into the bloodstream as needed. The size of the follicles, and amount of stored T_3 and T_4 they contain, depends on the activity of the gland (Figure 34-8).

A unique characteristic of the thyroid is its high concentration of **iodine;** in most animals this single gland contains well over half the body store of iodine. Epithelial cells of thyroid follicles actively trap iodine from the blood and combine it with the amino acid tyrosine, creating the two thyroid hormones. T_3 contains three iodine atoms, whereas T_4 contains four iodine atoms. T_4 is formed in much greater amounts than T_3, but in many animals T_3 is the more physiologically active hormone. T_4 is now considered a precursor to T_3. The most important actions of T_3 and T_4 are to (1) promote normal growth and development of the nervous system of growing animals, and (2) stimulate metabolic rate.

Undersecretion of thyroid hormones in fish, birds, and mammals dramatically impairs growth, especially of the nervous system. A human **cretin,** a mentally retarded dwarf, is a result of thyroid malfunction from a very early age. Conversely, oversecretion of thyroid hormones causes precocious development in all vertebrates, although its effect is particularly prominent in fish and amphibians. In frogs and toads, transformation from aquatic herbivorous tadpole without lungs or legs to semiterrestrial or terrestrial carnivorous adult with lungs and four legs, occurs when the thyroid gland becomes active at the end of larval development. Stimulated by rising thyroid hormone levels of the blood, metamorphosis and climax occur (Figure 34-9). Growth of frogs after metamorphosis is directed by growth hormone.

In birds and mammals, control of oxygen consumption and heat production is the best known action of thyroid hormones. The thyroid maintains metabolic activity of homeotherms (birds and mammals) at a normal level. Oversecretion of thyroid hor-

Inactive follicles

Active follicles

Figure 34-8

Appearance of thyroid gland follicles viewed through the microscope (approximately ×350). When inactive, follicles are distended with colloid, the storage form of thyroid hormones, and epithelial cells are flattened. When active, the colloid disappears as thyroid hormones are secreted into the circulation, and epithelial cells become greatly enlarged.

mones will speed up body processes by as much as 50%, resulting in irritability, nervousness, fast heart rate, intolerance of warm environments, and loss of body weight despite increased appetite. Undersecretion of thyroid hormones slows metabolic activities, leading to loss of mental alertness, slowing of heart rate, muscular weakness, increased sensitivity to cold, and weight gain. An important function of the thyroid gland is to promote adaptation to cold environments by increasing heat production. Thyroid hormones stimulate cells to produce more heat and store less chemical energy (ATP); in other words, thyroid hormones *reduce* efficiency of cellular oxidative phosphorylation (p. 64). Consequently many cold-adapted mammals have heartier appetites and eat more food in winter than in summer although their activity is about the same in both seasons. In winter, a larger portion of the food is being converted directly into body-warming heat.

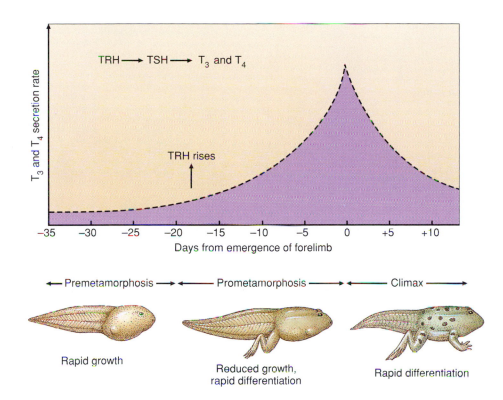

Figure 34-9

Effect of thyroid hormones (T_3 and T_4) on growth and metamorphosis of a frog. The release of TRH from the hypothalamus at the end of premetamorphosis sets in motion hormonal changes (increased TSH, T_3 and T_4) leading to metamorphosis. Thyroid hormone levels are maximal at the time forelimbs emerge.

Synthesis and release of thyroid hormones are governed by thyrotropic hormone (TSH) from the anterior pituitary gland (Table 34-1). TSH is regulated in turn by thyrotropin-releasing hormone (TRH) of the hypothalamus. As noted on p. 725, TRH is part of a higher regulatory council that controls the tropic hormones of the anterior pituitary. TRH and TSH control thyroid activity in an excellent example of negative feedback (p. 724). It can be overridden, however, by neural stimuli, such as exposure to cold, which directly stimulates increased release of TRH and thus TSH.

Some years ago, a condition called goiter was common among people living in the Great Lakes region of the United States and Canada, as well as some other parts of the world, such as the Swiss Alps. This type of goiter is an enlargement of the thyroid gland caused by deficiency of iodine in food and water. By striving to produce thyroid hormone with not enough iodine available, the gland hypertrophies, sometimes so much that the entire neck region becomes swollen (Figure 34-10). Goiter caused by iodine deficiency is seldom seen in North America because of the widespread use of iodized salt. However, it is estimated that even today 200 million people worldwide experience varying degrees of goiter, mostly in high mountains of South America, Europe, and Asia.

Hormonal Regulation of Calcium Metabolism

Closely associated with the thyroid gland and in some animals buried within it are **parathyroid glands.** These tiny glands occur as two pairs in humans but vary in number and position in other vertebrates. They were discovered at the end of the nineteenth century when the fatal effects of "thyroidectomy" were traced to unknowing removal of parathyroid glands together with the thyroid gland. In birds and mammals, including humans, removal of the parathyroid glands causes the level of calcium in the blood to decrease rapidly. This decrease in calcium leads to a serious increase in nervous system excitability, severe muscular spasms and tetany, and finally death. Subsequently, it was discovered that parathyroid glands secrete a hormone, **parathyroid hormone (PTH),** which is essential to

Figure 34-10

A large goiter caused by iodine deficiency. By enlarging enormously, the thyroid gland can extract enough iodine from the blood to synthesize the body's requirement for thyroid hormones.

maintenance of calcium homeostasis. Calcium ions are extremely important for formation of healthy bones. In addition, they are needed for neurotransmitter and hormone release, muscle contraction, intracellular signaling, and blood clotting.

Before considering how hormones maintain calcium homeostasis, it is helpful to summarize mineral metabolism in bone, a densely packed storehouse of both calcium and phosphorus. Bone contains approximately 98% of the calcium and 80% of the phosphorus in humans. Although bone is second only to teeth as the most durable material in the body (as evidenced by survival of fossil bones for millions of years) it is in a state of constant turnover in living vertebrates. Bone-building cells **(osteoblasts)** synthesize organic fibers of bone matrix which later become mineralized with a form of calcium phosphate called hydroxyapatite. Bone-resorbing cells **(osteoclasts)** are giant cells that dissolve the bony matrix, releasing calcium and phosphate into the blood. These opposing activities allow bone constantly to remodel itself, especially in a growing animal, for structural improvements to counter new mechanical stresses on the body. They additionally provide a vast and accessible reservoir of minerals that can be withdrawn as needed for general cellular requirements.

The level of calcium in the blood is maintained by three hormones that coordinate absorption, storage, and excretion of calcium ions. If blood calcium should decrease slightly, the parathyroid gland increases its secretion of **PTH.** This increase stimulates osteoclasts to dissolve bone adjacent to these cells, thus releasing calcium and phosphate into the bloodstream and returning blood calcium level to normal. PTH also decreases the rate of calcium excretion by the kidney and increases production of the hormone 1,25-dihydroxyvitamin D_3. PTH levels vary inversely with blood calcium level, as shown in Figure 34-11.

A second hormone involved in calcium metabolism in all tetrapods is derived from vitamin D_3. Vitamin D_3, like all vitamins, is a dietary requirement. But unlike other vitamins, vitamin D_3 may also be synthesized in the skin from a precursor by irradiation with ultraviolet light from the sun. Vitamin D_3 is then converted in a two-step oxidation to a hormonal form, **1,25-dihydroxyvitamin D_3.** This steroid hormone is essential for active calcium absorption by the gut (Figure 34-12). Production of 1,25-dihydroxyvitamin D_3 is stimulated by low plasma phosphate as well as by an increase in PTH secretion.

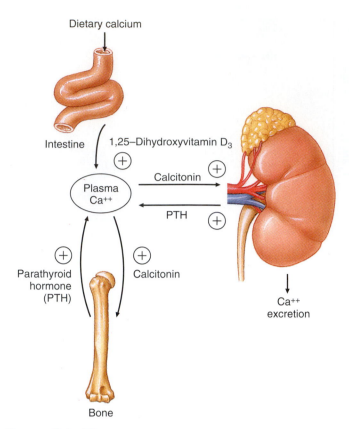

Figure 34-12

Regulation of blood calcium in birds and mammals.

In humans, a deficiency of vitamin D_3 causes rickets, a disease characterized by low blood calcium and weak, poorly calcified bones that tend to bend under postural and gravitational stresses. Rickets has been called a disease of northern winters, when sunlight is minimal. It once was common in the smoke-darkened cities of England and Europe.

A third calcium-regulating hormone, **calcitonin,** is secreted by specialized cells (C cells) in the thyroid gland of mammals and in the ultimobranchial gland of other vertebrates. Calcitonin is released in response to elevated levels of calcium in the blood. It rapidly suppresses calcium withdrawal from bone, decreases intestinal absorption of calcium, and increases excretion of calcium by the kidneys. Calcitonin therefore protects the body against an increase in level of calcium in the blood, just as parathyroid hormone protects it from a decrease in blood calcium (Figure 34-12). Calcitonin has been identified in all vertebrate groups, but its importance is uncertain because replacement of calcitonin is not required for maintenance of calcium homeostasis, at least in humans, if the thyroid gland is surgically removed (also removing the C cells).

Hormones of the Adrenal Cortex

The mammalian adrenal gland is a double gland composed of two unrelated types of glandular tissue: an outer region of adrenocortical cells, or **cortex,** and an inner region of specialized cells, the **medulla** (Figure 34-13). In nonmammalian verte-

Figure 34-11

How rate of secretion of parathyroid hormone (PTH) and calcitonin respond to changes in blood calcium level in a mammal.

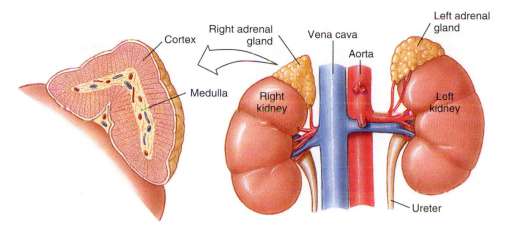

Figure 34-13

Paired adrenal glands of humans, showing gross structure and position on the upper poles of the kidneys. Steroid hormones are produced by the cortex. The sympathetic hormones epinephrine and norepinephrine are produced by the medulla.

brates homologs of adrenocortical and medullary cells are organized quite differently; they may be intermixed or distinct, but never arranged in a cortex-medulla relationship as in mammals.

At least 30 different compounds have been isolated from adrenocortical tissue, all of them closely related lipoidal compounds known as steroids. Only a few of these compounds are true steroid hormones; most are various intermediates in synthesis of steroid hormones from **cholesterol** (Figure 34-14). Corticosteroid hormones are commonly classified into two groups, according to their function: glucocorticoids and mineralocorticoids.

Glucocorticoids, such as **cortisol** (see Figure 34-14) and **corticosterone,** are concerned with food metabolism, inflammation, and stress. They promote synthesis of glucose from compounds other than carbohydrates, particularly amino acids and fats. The overall effect of this process, called **gluconeogenesis,** is to increase the level of glucose in the blood, thus providing a quick energy source for muscle and nervous tissue. Glucocorticoids are also important in diminishing the immune response to various inflammatory conditions. Because several diseases of humans are inflammatory diseases (for example, allergies, hypersensitivity, and rheumatoid arthritis), these corticosteroids have important medical applications.

Synthesis and secretion of glucocorticoids are controlled principally by ACTH of the anterior pituitary (see Figure 34-6), while ACTH is controlled in turn by corticotropin-releasing hormone (CRH) of the hypothalamus (Table 34-1). As with pituitary control of the thyroid, a negative feedback relationship exists between CRH, ACTH, and the adrenal cortex (Figure 34-3). An increase in release of glucocorticoids suppresses output of CRH

and ACTH; the resulting decline in blood level of CRH and ACTH then feeds back to the adrenal cortex to inhibit further release of glucocorticoids. An opposite sequence of events happens should the blood level of glucocorticoids drop: CRH and ACTH output increases, which in turn stimulates secretion of glucocorticoids. CRH is known to mediate stressful stimuli through the adrenal axis.

Adrenal steroid hormones, especially the glucocorticoids, are remarkably effective in relieving symptoms of rheumatoid arthritis, allergies, and various disorders of connective tissue, skin, and blood. Following the report in 1948 by P. S. Hench and his colleagues at the Mayo Clinic that cortisone dramatically relieved pain and crippling effects of advanced arthritis, steroid hormones were hailed by the media as "wonder drugs." Optimism was soon dimmed, however, when it became apparent that severe side effects always attended long-term administration of antiinflammatory steroids. Steroid therapy lulls the adrenal cortex into inactivity and may permanently impair the body's capacity to produce its own steroids. Today steroid therapy is applied with caution, because it is realized that the inflammatory response is a necessary part of the body's defenses.

Mineralocorticoids, the second group of corticosteroids, are those that regulate salt balance. **Aldosterone** (see Figure 34-14) is by far the most important steroid of this group. Aldosterone promotes tubular reabsorption of sodium and tubular

Cholesterol

Aldosterone

Cortisol

Figure 34-14

Hormones of the adrenal cortex. Cortisol (a glucocorticoid) and aldosterone (a mineralocorticoid) are two of several steroid hormones synthesized from cholesterol in the adrenal cortex.

secretion of potassium by the kidneys. Since sodium usually is in short supply in diets of many animals and potassium is in excess, mineralocorticoids play vital roles in preserving the correct balance of blood electrolytes. The salt-regulating action of aldosterone is controlled by the renin-angiotensin system, described on p. 644 and by blood levels of potassium ions. High levels of potassium ions in the blood are a potent direct stimulator of aldosterone release from the adrenal cortex, thus promoting reabsorption of sodium into the blood and secretion of excess potassium ions into the urine.

Adrenocortical tissue also produces **androgens** (Gr. *andros,* man, + *genesis,* origin), which, as the name implies, are similar in effect to the male sex hormone, testosterone. Adrenal androgens promote some developmental changes, such as the growth spurt, that occur just before puberty in human males and females. Recent development of so-called **anabolic steroids,** synthetic hormones related to testosterone, has led to widespread abuse of steroids among athletes (see following note).

Use of anabolic steroids by athletes became major news following Ben Johnson's drug-fueled win of the 100-meter race at the 1988 Olympics. Despite almost universal condemnation by Olympic, medical, and college sports authorities, an unscientific and clandestine program of experimentation with anabolic steroids has become popular with many amateur and professional athletes in many countries. These synthetics (and testosterone and its precursors) cause hypertrophy of skeletal muscle and may improve performance that depends on strength. Unfortunately, they also have serious side effects, including testicular atrophy (and infertility), periods of irritability, abnormal liver function, and cardiovascular disease. Recent data suggest that steroid abuse among adolescents is increasing. In 1999, self-reporting estimates of use were 2.7% in eighth- and tenth-graders and 2.9% in twelfth-graders. This represents a significant increase since 1991, when use was reported as 1.9% in eighth-graders, 1.8% in tenth-graders, and 2.1% in twelfth-graders (NIH-funded survey, http://www.nih.gov/researchreports/steroids/anabolicsteroid.html). Use among adults and professional athletes is not well documented, although anecdotal evidence suggests that such drugs are popular among athletes in a variety of sports (the most famous recent athlete being Mark McGwire of the St. Louis Cardinals).

Hormones of the Adrenal Medulla

Adrenal medullary cells secrete two structurally similar hormones: **epinephrine (adrenaline)** and **norepinephrine (noradrenaline).** The adrenal medulla is derived embryologically from the same tissue that gives rise to postganglionic sympathetic neurons of the autonomic nervous system (p. 707). Norepinephrine serves as a neurotransmitter at the endings of sympathetic nerve fibers. Thus functionally, as well as embryologically, the adrenal medulla can be considered a very large sympathetic ganglion.

It is not surprising then that adrenal medullary hormones and the sympathetic nervous system have the same general effects on the body. These effects center on responses to emergencies, such as fear and strong emotional states, flight from danger, fighting, lack of oxygen, blood loss, and exposure to pain. Walter B. Cannon, of homeostasis fame (p. 634), termed these "fight or flight" responses that are appropriate for survival. We are familiar with the increased heart rate, tightening of the stomach, dry mouth, trembling muscles, general feeling of anxiety, and increased awareness that attends sudden fright or other strong emotional states. These effects are attributable to increased activity of the sympathetic nervous system and to rapid release into the blood of epinephrine from the adrenal medulla.

Epinephrine and norepinephrine have many other effects of which we are not as aware, including constriction of arterioles (which, together with increased heart rate, increases blood pressure), mobilization of liver glycogen and fat stores to release glucose and fatty acids for energy, increased oxygen consumption and heat production, hastening of blood coagulation, and inhibition of the gastrointestinal tract. These changes prepare the body for emergencies and are activated in stressful conditions.

Insulin and Glucagon from Islet Cells of the Pancreas

The pancreas is both an exocrine and an endocrine organ (Figure 34-15). The exocrine portion produces pancreatic juice, a mixture of digestive enzymes and bicarbonate ions conveyed by

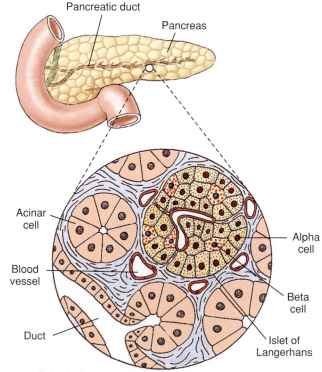

Figure 34-15

The pancreas is composed of two kinds of glandular tissue: exocrine acinar cells that secrete digestive juices that enter the intestine through the pancreatic duct, and endocrine islets of Langerhans. The islets of Langerhans secrete the hormones insulin and glucagon directly into the blood circulation.

a duct to the digestive tract (see Chapter 32). Scattered within the extensive exocrine portion of the pancreas are numerous small islets of tissue, called **islets of Langerhans** (see Figure 34-15 and photograph on p. 721). This endocrine portion of the pancreas is only 1% to 2% of the total weight of the organ. The islets are without ducts and secrete their hormones directly into blood vessels that extend throughout the pancreas.

Two polypeptide hormones are secreted by different cell types within the islets: **insulin,** produced by **beta cells,** and **glucagon,** produced by **alpha cells.** Insulin and glucagon have antagonistic actions of great importance in metabolism of carbohydrates and fats. Insulin is essential for uptake of blood-borne glucose by cells, especially skeletal muscle cells. Insulin promotes entry of glucose into body cells through its action on a glucose transporter molecule found in cell membranes. Although insulin-dependent glucose transporter molecules have been demonstrated on neurons of the central nervous system, neurons do not require insulin for glucose uptake. This independence from insulin is very important because unlike other cells of the body, neurons almost exclusively use glucose as an energy source. The exact role of insulin-dependent glucose transporters in the brain is not clear, but insulin is important in central regulation of food intake and body weight. Cells of the rest of the body, however, require insulin to use glucose; without insulin the level of glucose in the blood rises to abnormally high levels, a condition called hyperglycemia. When this level exceeds the transport maximum of the kidney (see note on p. 644) sugar (glucose) appears in urine. Insulin deficiency also inhibits uptake of amino acids by skeletal muscle, and fats and muscle are metabolized to provide energy. Body cells starve while the urine abounds in the very substance the body craves. The disease, called **diabetes mellitus,** afflicts nearly 5% of humans in varying degrees of severity. If left untreated, it can lead to severe damage to kidneys, eyes, and blood vessels, and it can greatly shorten life expectancy.

In 1982, insulin became the first hormone produced by genetic engineering (recombinant DNA technology, p. 95) to be marketed for human use. Recombinant insulin has the exact structure of human insulin and therefore will not stimulate an immune response, which has often been a problem for diabetics receiving insulin purified from pig or cow pancreas.

The first extraction of insulin in 1921 by two Canadians, Frederick Banting and Charles Best, was one of the most dramatic and important events in the history of medicine. Many years earlier two German scientists, J. Von Mering and O. Minkowski, discovered that surgical removal of the pancreas of dogs invariably caused severe symptoms of diabetes, resulting in the animal's death within a few weeks. Many attempts were made to isolate the diabetes preventive factor, but all failed because powerful protein-splitting digestive enzymes in the exocrine portion of the pancreas destroyed the hormone during extraction procedures. Following a hunch, Banting, in collabora-

Figure 34-16
Charles H. Best and Sir Frederick Banting in 1921 with the first dog to be kept alive by insulin.

tion with Best and his physiology professor J. J. R. Macleod, tied off the pancreatic ducts of several dogs. This caused the exocrine portion of the gland with its hormone-destroying enzyme to degenerate, but left the islets' tissues healthy long enough for Banting and Best to extract insulin successfully from these glands. Injected into another dog, the insulin immediately lowered the level of sugar in the blood (Figure 34-16). Their experiment paved the way for commercial extraction of insulin from slaughterhouse animals. It meant that millions of people with diabetes, previously doomed to disability or death, could look forward to more normal lives.

Glucagon, the second hormone of the pancreas, has several effects on carbohydrate and fat metabolism that are opposite to the effects of insulin. For example, glucagon raises the blood-glucose level (by converting liver glycogen to glucose), whereas insulin lowers blood glucose. Glucagon and insulin do not have the same effects in all vertebrates, and in some, glucagon is lacking altogether.

Growth Hormone and Metabolism

Growth hormone (GH) is a particularly important metabolic hormone in young animals during growth and development. It acts directly on long bones to promote cartilaginous growth and bone formation by cell division and protein synthesis, thus producing an increase in length and density of bone. GH increases release of fat from adipose tissue stores and glycogen from liver stores for energy metabolism. Thus, GH is considered a **diabetogenic hormone,** since oversecretion leads to an increase in blood glucose and can result in insulin insensitivity or diabetes. If produced in excess, GH causes giantism. A deficiency of this hormone in a human child leads to dwarfism. GH also acts indirectly on growth via stimulation of **insulin-like growth factor (IGF)** or somatomedin release from the liver. This polypeptide hormone promotes mobilization of glycogen and fat stores necessary for growth processes.

The Newest Hormone—Leptin

Following discovery of the *ob* gene in 1994, which codes for this newest hormone, J. Friedman and coworkers soon coined the term, **leptin** (Gr. *leptos,* thin), for the circulating hormone, produced by white fat cells (adipose tissue). Subsequently, receptors for leptin have been found in many tissues, but the primary site of action of leptin appears to be the brain, particularly the hypothalamus and brain stem. Leptin is an important hormone that regulates eating behavior and energy balance as part of a feedback system that informs the brain of the energy status of the periphery. Recent evidence suggests that leptin is more important during times of lower food, and consequently, energy availability, since decreased fat stores secrete less leptin. At such times, the brain responds by directing available energy away from nonessential processes, such as reproduction, and stimulates increased foraging and feeding behaviors. Leptin has become immensely significant in the study of satiety signals and energy expenditure, since these studies relate to the overall problem of human obesity (p. 689). It is of interest that blood-plasma levels of leptin mirror those of insulin, which also provides an important feedback signal to the brain.

SUMMARY

Hormones are chemical messengers synthesized by special endocrine and other cells and transported by blood or other body fluid to target cells where they affect cell function by altering specific biochemical processes. Specificity of response is ensured by the presence on or in target cells of protein receptors that bind only selected hormones. Hormone effects are vastly amplified in target cells by acting through one of two basic mechanisms. Many hormones, including epinephrine, glucagon, vasopressin, and some hormones of the anterior pituitary, cause production of a "second messenger," such as cyclic AMP, that relays the hormone's message from a surface receptor to a cell's biochemical machinery. Steroid hormones and thyroid hormones operate through mainly cytoplasmic or nuclear receptors. A hormone-receptor complex is formed that alters protein synthesis by stimulating or inhibiting gene transcription.

Most invertebrate hormones are products of neurosecretory cells. The best understood invertebrate endocrine system is that controlling molting and metamorphosis in insects. A juvenile insect grows by passing through a series of molts under control of two hormones, one (ecdysone) favoring molting to an adult and the other (juvenile hormone) favoring retention of juvenile characteristics. Ecdysone is under the control of a neurosecretory hormone (PTTH) from the brain.

The vertebrate endocrine system is orchestrated by the hypothalamus. Release of all anterior pituitary hormones is primarily regulated by hypothalamic neurosecretory products called releasing hormones (or release-inhibiting hormones). The hypothalamus also produces two neurosecretory hormones, which are stored in and released from the posterior lobe of the pituitary. In mammals these two hormones are oxytocin, which stimulates milk production and uterine contractions during parturition; and vasopressin (antidiuretic hormone), which acts on the kidney to restrict urine production, causes vasoconstriction of blood vessels, and increases thirst. In amphibians, reptiles, and birds, vasotocin replaces vasopressin as the water-balance hormone.

The anterior lobe of the pituitary produces seven well-characterized hormones. Four of these are tropic hormones that regulate subservient endocrine glands: thyrotropic hormone (TSH), which controls secretion of thyroid hormones; adrenocorticotropic hormone (ACTH), which stimulates release of steroid hormones by the adrenal cortex; and follicle-stimulating hormone (FSH) and luteinizing hormone (LH), which act on ovaries and testes. Three direct-acting hormones are (1) prolactin, which plays several diverse roles, including stimulation of milk production; (2) growth hormone, which governs body growth and metabolism; and (3) melanocyte-stimulating hormone (MSH), which controls melanocyte dispersion in ectothermic vertebrates.

The pineal gland, a derivative of the pineal complex of the diencephalon of the brain, produces the hormone melatonin. In many vertebrates, melatonin, which is released in response to darkness, maintains circadian rhythms. In birds and mammals that are seasonal breeders, melatonin level provides information regarding daylength, and thereby indirectly regulates seasonal reproductive activity.

Recent application of ultrasensitive radioimmunochemical techniques has revealed many neuropeptides in the brain, several of which behave as neurotransmitters in the brain but as hormones elsewhere in the body. The classical definition of a hormone has been modified to include other chemical messengers, such as prostaglandins and cytokines, which originate in sources other than clearly defined endocrine glands.

Several hormones play important roles in regulating cellular metabolic activities. Two thyroid hormones, triiodothyronine (T_3) and thyroxine (T_4), control growth, development of the nervous system, and cellular metabolism. Calcium metabolism is regulated principally by three hormones: parathyroid hormone from the parathyroid glands, a hormonal derivative of vitamin D_3, 1,25-dihydroxyvitamin D_3, and calcitonin from the thyroid gland. Parathyroid hormone and 1,25-dihydroxyvitamin D_3 increase plasma calcium levels; calcitonin decreases plasma calcium levels.

The principal steroid hormones of the adrenal cortex are glucocorticoids, which stimulate formation of glucose from non-glucose sources (gluconeogenesis), and mineralocorticoids, which regulate blood electrolyte balance. The adrenal medulla is the source of epinephrine and norepinephrine, which have many effects, including assisting the sympathetic nervous system in emergency responses. They also increase energy substrates in the blood for use in emergency situations.

Glucose metabolism is regulated by the antagonistic action of two pancreatic hormones. Insulin is needed for cellular use of blood glucose and uptake of amino acids in muscle. Glucagon opposes the action of insulin.

The most recently discovered hormone, leptin, is secreted by adipose tissue and feeds back to the hypothalamus to modulate food intake and energy balance.

REVIEW QUESTIONS

1. Outline the famous experiment of Berthold that marks the first endocrine experiment. What might the hypothesis have been?
2. Provide definitions for the following: hormone, endocrine gland, exocrine gland, hormone receptor molecule.
3. Hormone receptor molecules are the key to understanding specificity of hormone action on target cells. Describe and distinguish between receptors located on the cell surface and those located in the nucleus of target cells. Name two hormones whose action is mediated through each type of receptor.
4. What is the importance of feedback systems in the control of hormonal output? Offer an example of a hormonal feedback pattern.
5. Explain how the three hormones involved in insect growth—ecdysone, juvenile hormone, and PTTH—interact in molting and metamorphosis.
6. Name seven hormones produced by the anterior pituitary gland. Why are four of these seven hormones called "tropic hormones"? Explain how secretion of the anterior pituitary hormones is controlled by neurosecretory cells in the hypothalamus.
7. Describe the chemical nature and function of two posterior lobe hormones, oxytocin and vasopressin. What is distinctive about the way these neurosecretory hormones are secreted compared with the neurosecretory release and release-inhibiting hormones that control the anterior pituitary hormones?
8. What is the evolutionary origin of the pineal gland of birds and mammals? Explain the role of the pineal hormone, melatonin, in regulating seasonal reproductive rhythms in birds and mammals. Does melatonin have any function in humans?
9. What are endorphins and enkephalins? What are prostaglandins?
10. What are some functions of hormones called cytokines?
11. What are the two most important functions of the thyroid hormones?
12. Explain how you would interpret the graph in Figure 34-10 to show that PTH and calcitonin act in a complementary way to control blood-calcium level.
13. Describe the principal functions of the two major groups of adrenal cortical steroids, the glucocorticoids and the mineralocorticoids. To what extent do these names provide clues to their function?
14. Where are the hormones epinephrine and norepinephrine produced and what is their relationship to the sympathetic nervous system and its response to emergencies?
15. Explain the actions of hormones of the islets of Langerhans on the level of glucose in the blood. What is the consequence of insulin insufficiency as in the disease diabetes mellitus?
16. What is the function of the most recently characterized hormone, leptin? Why has its discovery proved important in the area of regulation of feeding?

SELECTED REFERENCES

Bentley, P. J. 1998. Comparative vertebrate endocrinology, ed. 3. Cambridge, Cambridge University Press. *Undergraduate text with good evolutionary perspective.*

Bolander, F. F. 1994. Molecular endocrinology, ed. 2. San Diego, Academic Press. *Excellent synthesis of a fast-moving field.*

Chester-Jones, I., P. M. Ingleton, and J. G. Phillips (eds.). 1987. Fundamentals of comparative vertebrate endocrinology. New York, Plenum Press. *Contributed chapters.*

Clapp, C., and G. M. de la Escalera. 1997. Prolactin. novel regulator of angiogenesis. News Physiol. Sci. **13**:231–237 (Oct.). *Review of a new function for prolactin.*

Gard, P. 1998 Human endocrinology. Bristol, Pennsylvania, Taylor & Francis Ltd. *Good treatment of human endocrinology.*

Gibbs, W. W. 1996. Gaining on fat. Sci. Am. **275**:88–94. *Research offers hope of ways to treat and prevent obesity, a costly epidemic spreading through the industrial world.*

Griffin, J. E., and S. R. Ojeda. 2000. Textbook of endocrine physiology, e. 4. Oxford, Oxford University Press. *Excellent upper-undergraduate and graduate text of human endocrinology.*

Hadley, M. E. 2000. Endocrinology, ed. 5. Englewood Cliffs, New Jersey, Prentice-Hall, Inc. *Undergraduate-level textbook in vertebrate endocrinology.*

Laufer, H., and G. H. Downer (eds.). 1988. Endocrinology of selected invertebrate types. New York, Alan R. Liss. *Eighteen contributed chapters and an introductory chapter on the comparative endocrinology of invertebrates, radiates through echinoderms.*

Lienhard, G. E., J. W. Slot, D. E. James, and M. M. Mueckler. 1992. How cells absorb glucose. Sci. Am. **266**:86–91 (Jan.). *How insulin regulates the function of a special transporter molecule that moves glucose across cell membranes.*

Norman, A. W., and G. Litwack. 1997. Hormones, ed. 2. San Diego, Academic Press. *Thorough treatment of vertebrate endocrinology.*

Woods, S. C., P. A. Rushing, and R. J. Seeley. 2001. Neuropeptides and the control of energy homeostasis. Nutrition and Brain **5**:93–115. *A readable discussion of food intake regulation and body weight control, incorporating our most recent hypotheses regarding leptin.*

ZOOLOGY LINKS TO THE INTERNET

Visit the textbook's Online Learning Center at www.mhhe.com/zoology to find live Internet links for each of the topics listed here.

The Endocrine System
Vertebrates: Macroscopic Anatomy of the Endocrine System
Vertebrates: Microscopic Anatomy of the Endocrine System
Human Endocrine Topics

Section through abcess in heart muscle.

Immunity

The Language of Cells in Immunity

For over 100 years scientists have known that certain cells in an animal could secrete substances that affected various processes in other, target cells, for example, metabolism, physiology, or differentiation; however, the means of this communication between cells remained a mystery. Much of the shroud was lifted by more recent discoveries. Specific signal molecules, often proteins or peptides, are secreted by certain cells. Target cells have receptors protruding through their outer membranes that specifically bind the signal molecules and only those molecules. Binding of a signal causes changes in the part of the receptor molecule (or in an associated membrane protein) that extends into the cytoplasm, and this initiates a cascade of activations involving protein kinases and phosphorylases (enzymes that transfer phosphate groups). Transcription factors are mobilized. In the nucleus transcription factors initiate transcription of formerly inactive genes, leading to synthesis of products they encode.

We now know that hormones affect target cells by this mechanism (Chapter 34), and it is also the scheme by which cells of the immune system communicate with each other and with other cells. Signal molecules of the immune system are called *cytokines*. Cytokines and their receptors are the language of communication in the immune system. They perform an intricate and elaborate ballet of activation and regulation, causing some cells to proliferate, suppressing proliferation of others, and stimulating secretion of additional cytokines or defense molecules. Precise signaling among cells and exact performance of their duties are essential to maintainence of human health and defense against invading viruses, bacteria, and parasites and for prevention of unrestrained cell division, as in cancer. Successful establishment of invaders in our bodies depends on evasion or subversion of our immune system, and inappropriate response of immune cells may itself produce disease. We have learned to manipulate the immune response so that we can transplant organs between individuals, but complete failure in its cell communication results in profound disease, such as AIDS.

The immune system is located throughout the body of an animal, and it is as crucial to survival as the respiratory, circulatory, nervous, skeletal, or any other system. Every animal's environment is filled with an incredible number of parasites and potential parasites: flatworms, nematodes, arthropods, single-celled eukaryotes, bacteria, and viruses. Whether any parasite can survive in that animal (the **host**), and severity of disease the parasite may cause, depends largely on the response of the host defense system.

SUSCEPTIBILITY AND RESISTANCE

A host is **susceptible** to a parasite if the host cannot eliminate a parasite before the parasite can become established. The host is **resistant** if its physiological status prevents establishment and survival of the parasite. Corresponding terms from the viewpoint of a parasite would be **infective** and **noninfective.**

These terms deal only with the success or failure of infection, not with mechanisms producing the result. Mechanisms that increase resistance (and correspondingly reduce susceptibility and infectivity) may involve either attributes of a host not related to active defense mechanisms or specific defense mechanisms mounted by a host in response to a foreign invader. It is important to remember that these terms are relative, not absolute; for example, one individual organism may be more or less resistant than another, and a single individual may be more or less resistant at different times of its life, depending on age, health, and environmental exposure.

The term **immunity** is often used as synonymous with resistance, but it is also associated with the sensitive and specific immune response exhibited by vertebrates. However, because many invertebrates can be immune to infection by various agents, a more concise statement is that *an animal demonstrates immunity if it possesses tissues capable of recognizing and protecting the animal against nonself invaders*. Most animals show some degree of **innate (nonspecific) immunity,** a mechanism of defense that does not depend on prior exposure to the invader. In addition to having innate immunity, vertebrates—and invertebrates to a lesser extent—develop **acquired (specific) immunity,** which is specific to a particular nonself material, requires time for its development, and occurs more quickly and vigorously on secondary response. Many innate mechanisms discussed in the next section *are dramatically influenced and strengthened in vertebrates as a consequence of acquired immune responses.*

Frequently resistance conferred by immune mechanisms is not complete. In some instances a host may recover clinically and be resistant to a specific challenge, but some parasites may remain and reproduce slowly, as in toxoplasmosis (p. 220), Chagas' disease (p. 218), and malaria (p. 220). This condition is called **premunition.**

INNATE DEFENSE MECHANISMS

Physical and Chemical Barriers

The unbroken surface of most animals provides a barrier to invading organisms. It may be tough and cornified, as in many terrestrial vertebrates, or sclerotized, as in arthropods. Soft outer surfaces are usually protected by a layer of mucus, which lubricates the surface and helps dislodge particles from it.

A variety of antimicrobial substances are present in body secretions of vertebrates. Chemical defenses present in many vertebrates include a low pH in the stomach and vagina and hydrolytic enzymes in secretions of the alimentary tract. Mucus is produced by mucous membranes lining the digestive and respiratory tract of vertebrates and contains parasiticidal substances such as **IgA** and **Lysozyme.** IgA is a class of antibody (p. 741) that can cross cellular barriers easily and is an important protective agent in mucus of the intestinal epithelium. IgA is present also in saliva and sweat. Lysozyme is an enzyme that attacks the cell wall of many bacteria.

Various cells, including those involved in the acquired immune response, liberate protective compounds. A family of low-molecular-weight glycoproteins, the **interferons,** are released by a variety of eukaryotic cells in response to invasion by intracellular parasites (including viruses) and other stimuli. **Tumor necrosis factor (TNF)** is produced mainly by cells called **macrophages** (see p. 740). TNF is a major mediator of inflammation (p. 746) and in sufficient concentration causes **fever.** Fever in mammals is one of the most common symptoms of infection. The protective role of fever, if any, remains unclear, but high body temperature may destabilize certain viruses and bacteria.

The intestine of most animals harbors a population of bacteria that seem not to be harmed by host defenses, nor do they elicit any protective defense response. In fact, the normal intestinal microflora in vertebrates tends to inhibit establishment of pathogenic microbes.

Substances in normal human milk can kill intestinal protozoa such as *Giardia lamblia* and *Entamoeba histolytica* (Chapter 11), and these substances may be important in protection of infants against these and other infections. Antimicrobial elements in human breast milk include lysozyme, IgA, interferons, and leukocytes (white blood cells, see Chapter 31).

Some species of mammals are susceptible to infection with parasites such as *Schistosoma mansoni* (Chapter 14), and others are partially or completely resistant. Without mediation by antibody, macrophages of more resistant species (rats, guinea pigs, rabbits) kill schistosome juveniles, but macrophages of susceptible species do not.

Complement is a series of enzymes that are activated in sequence as a host response to invading organisms. Activation of complement by the **classical pathway** (so called only because it was discovered first) depends on fixed antibody and so is an effector mechanism in the acquired immune response (p. 741). Complement activated by the **alternative pathway** is an important innate defense against invasion by bacteria and some fungi. Classical and alternative pathways share some, but not all, components. In the alternative pathway, the first component is activated spontaneously in the blood and binds to cell surfaces. This event initiates a cascade of activations, ultimately resulting in cell lysis. The host's own cells are not lysed by the alternative pathway because regulatory proteins rapidly inactivate the first active component of complement when it binds to host cells but not to foreign cells.

Cellular Defenses: Phagocytosis

For defense against an invader an animal's cells must recognize when a substance does not belong; they must recognize *nonself*. **Phagocytosis** illustrates nonself recognition, and it also serves as a process for removing senescent cells and cellular debris from the host. Phagocytosis occurs in almost all metazoa and is a feeding mechanism in many single-celled organisms (p. 214). A cell that has this ability is a **phagocyte.** Phagocytes engulf a particle within an invagination of the phagocyte's cell membrane. The invagination becomes pinched off, and the particle becomes enclosed within an intracellular vacuole. Other cytoplasmic vacuoles called **lysosomes** join with the particle-containing vacuole and pour in digestive enzymes to destroy the particle. Lysosomes of many phagocytes also contain enzymes that catalyze production of cytotoxic **reactive oxygen intermediates (ROIs)** and **reactive nitrogen intermediates (RNIs).** Examples of ROIs are superoxide radical (O_2^-), hydrogen peroxide (H_2O_2), singlet oxygen ($^1O_2$), and hydroxyl radical ($OH\bullet$). RNIs include nitric oxide (NO) and its oxidized forms, nitrite (NO_2^-) and nitrate (NO_3^-). All such intermediates are potentially toxic to invasive microorganisms or parasites.

Phagocytes and Other Defense Cells

Many invertebrates have specialized cells that function as itinerant troubleshooters within the body, acting to engulf or encapsulate foreign material (see Table 35-3) and repair wounds. Such cells are variously known as amebocytes, hemocytes, or coelomocytes in different animals. If a foreign particle is small, it is engulfed by phagocytosis; but if it is larger than about 10 µm, it is usually encapsulated. Arthropods can encapsulate a foreign object by deposition of melanin around it, either from cells of the capsule or by precipitation from the hemolymph (blood).

In vertebrates several categories of cells are capable of phagocytosis. **Monocytes** arise from stem cells in the bone marrow (Figure 35-1) and give rise to the **mononuclear phagocyte system (reticuloendothelial [RE] system),** which are phagocytic cells stationed around the body. The RE system includes **macrophages** in lymph nodes, spleen, and lung; **Kupffer cells** in sinusoids of the liver; and **microglial cells** in the central nervous system. Macrophages also have important roles in the specific immune response of vertebrates (see p. 741).

Some **polymorphonuclear leukocytes (PMNs),** a name that refers to the highly variable shape of their nucleus, are circulating phagocytes in blood. Another name for these leukocytes is **granulocytes,** which alludes to the many small granules that can be seen in their cytoplasm after treatment with appropriate stains. According to the staining properties of their granules, granulocytes are further subdivided into **neutrophils, eosinophils,** and **basophils.** Neutrophils are the most abundant, and they provide the first line of phagocytic defense in an infection. Eosinophils in normal blood account for about 2% to 5% of the total leukocytes, and basophils are the least numerous at about 0.5%. A high **eosinophilia** (eosinophil count in the blood) is often associated with allergic diseases and parasitic infections.

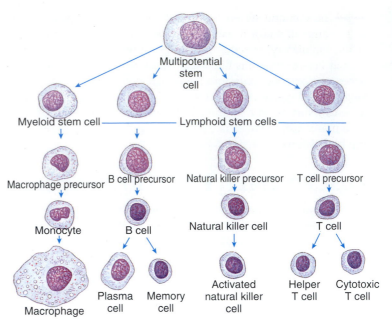

Figure 35-1

Lineage of some cells active in immune response. These cells, as well as red blood cells and other white blood cells are derived from multipotential stem cells in the bone marrow. B cells mature in bone marrow and are released into blood or lymph. Precursors of T cells go through a period in the thymus gland. Precursors of macrophages circulate in blood as monocytes.

Several other kinds of cells, including basophils, are not important as phagocytes but are important cellular components of the defense system. **Mast cells** are basophil-like cells found in the dermis and other tissues. When they are stimulated to do so (in inflammation, p. 746), basophils and mast cells release a number of pharmacologically active substances that affect surrounding cells. **Lymphocytes,** including **T lymphocytes (T cells)** and **B lymphocytes (B cells),** are crucial in the acquired immune response of vertebrates. **Natural killer (NK) cells** are lymphocytelike cells that can kill virus-infected and tumor cells in absence of antibody. They release substances onto the target-cell surface to lyse it.

Antimicrobial Peptides

Insects tend to be resistant to infection with many microbial pathogens. In the 1980s it was discovered that inoculation of moth larvae with bacteria caused a release of a barrage of antimicrobial agents that killed the bacteria, even without prior exposure to the invader. Since that time, hundreds of antimicrobial peptides have been described from a broad spectrum of animals, invertebrates and vertebrates. They are especially important at surfaces where organism meets environment, such as skin or mucous membranes. They do not have such high specificity as does the induced immune response of vertebrates, but rather each peptide is effective against a different category of microbe, for example gram-positive bacteria (bacteria that stain with "Gram stain"), gram-negative bacteria, and fungi. Release of the peptides is immediate in presence of the foreign organism and is not subject to prior immunizing experience with the microbe. Conventional antibiotics usually work by blocking a critical protein in an invading microbe, but these peptides interfere with internal signaling of a microbe or perforate its surface with holes.

Antimicrobial peptides in mammals have been called **defensins,** a name also given to some peptides from insects and plants that differ in structure from those in mammals. Defensins do not harm cells of the organism from which they originate. Macrophages, neutrophils, and cells around linings of intestinal, respiratory, and urogenital tracts pour out defensins in response to stimulation by molecules on the surface of microbes or, in some cases, to their metabolic products. Such molecules are conserved over a range of microbes but do not occur on host cells.

Release of peptides begins when receptors on a cell's surface recognize a microbial molecule. Many of these receptors are known as **Toll** proteins or **Toll-like receptors (TLRs),** so called because they are located in a cell membrane where they receive signals from the outside. At least nine TLRs have been described in humans, each of which recognizes a specific pattern of molecules from a class of microbes. Activation of a particular TLR signals the nucleus to synthesize a peptide against that particular microbe.

ACQUIRED IMMUNE RESPONSE IN VERTEBRATES

The specialized system of nonself recognition possessed by vertebrates results in increased resistance to *specific* foreign substances or invaders on repeated exposures. Investigations on mechanisms involved are currently intense, and our knowledge of them is increasing rapidly.

The immune response is stimulated by a specific foreign substance called an **antigen,** and, circularly, an antigen is any substance that will stimulate an immune response. Antigens may be any of a variety of substances with a molecular weight over 3000. They are most commonly proteins and are usually (but not always) foreign to the host. The immune response has two arms, known as **humoral** and **cellular.** Humoral immunity is based on **antibodies,** which are both on cell surfaces and dissolved in blood and lymph, whereas cellular immunity is entirely associated with cell surfaces. There is extensive communication and interaction among cells of the two arms.

Basis of Self and Nonself Recognition

Major Histocompatibility Complex

We have known for many years that nonself recognition is very specific. If tissue from one individual is transplanted into another individual of the same species, the graft will grow for a time and then die as immunity against it rises. In the absence of drugs that modify the immune system, tissue grafts will grow successfully only if they are between identical twins or between individuals of highly inbred strains of animals. The molecular basis for this nonself recognition involves certain proteins embedded in the cell surface. These proteins are coded by certain genes, now known as the **major histocompatibility complex (MHC).** MHC proteins are among the most variable known, and unrelated individuals almost always have different genes. There are two types of MHC proteins: class I and class II. Class I proteins are found on the surface of virtually all cells, whereas class II MHC proteins are found only on certain cells participating in immune responses, such as lymphocytes and macrophages.

The capability of an immune response develops over a period of time in the early development of an organism. All substances present at the time the capacity develops are recognized as self in later life. Unfortunately, the system of self and nonself recognition sometimes breaks down, and an animal may begin to produce antibodies against some part of its own body. This condition leads to one of several known *autoimmune diseases,* such as rheumatoid arthritis, multiple sclerosis, lupus, and insulin-dependent diabetes mellitus.

Recognition Molecules

We discuss the role of MHC proteins in nonself recognition on p. 743, but MHC proteins are not themselves the molecules that recognize foreign substances. This task falls to two basic types of molecules, the genes for which probably evolved from a common ancestor: **antibodies** and **T-cell receptors.** Each vertebrate individual has an enormous variety of antibodies, *each of which binds specifically* to one particular antigen (or part of an antigen), even though that antigen may have never been present in the body previously. There are probably an equal number of different T-cell receptors, each also specific for a particular antigen.

A major problem of immunology is understanding how the mammalian genome could contain information needed to produce at least a million different antibodies. The answer seems to be that antibody genes occur in pieces, rather than as continuous stretches of DNA, and that the antigen-recognizing sites (variable regions) of heavy and light chains of antibody molecules are pieced together from information supplied by separate DNA sequences, which can be shuffled to increase diversity of the gene products. The immense repertoire of antibodies is achieved in part by complex gene rearrangements and in part by frequent somatic mutations that produce additional variation in protein structure of the variable regions of the heavy and light antibody chains. Analogous processes occur in the production of genes for T-cell receptors.

Antibodies

Antibodies are proteins called **immunoglobulins.** They are borne in the surface of B lymphocytes or secreted by cells **(plasma cells)** derived from B cells. The basic antibody molecule consists of four polypeptide strands: two identical light chains and two identical heavy chains held together in a Y-shape

Figure 35-2

A, Antibody molecule is composed of two shorter polypeptide chains (light chains) and two longer chains (heavy chains) held together by covalent disulfide bonds. These are further subdivided into variable and constant regions that have independent folding units, or domains, of about 110 amino acids. The folding pattern is more complex than that shown here. Interchain disulfide bonds at the hinge region give the molecule flexibility at that point. The variable domains of both the light and heavy chains have hypervariable ends, which serve as the antigen-binding sites. **B,** Molecular model of antibody molecule.

by disulfide bonds and hydrogen bonds (Figure 35-2). The amino acid sequence toward the ends of the Y varies in both the heavy and light chains, according to the specific antibody molecule (the **variable region**), and this variation determines with which antigen the antibody can bind. Each of the ends of the Y forms a cleft that acts as the antigen-binding site (Figure 35-2), and specificity of the molecule depends on the shape of the cleft and properties of the chemical groups that line its walls. The remainder of an antibody is known as the **constant region.** The variable end of the antibody molecule is often called **Fab,** for **a**ntigen-**b**inding **f**ragment, and the constant end is known as **Fc,** for **c**rystallizable **f**ragment (Figure 35-2). The so-called constant region is not really constant: the light chains can be either of two types, and the heavy chains may be any of five types. The type of heavy chain determines the **class** of the antibodies, known as **IgM, IgG** (now familiar to many people as *gamma globulin*), **IgA, IgD,** and **IgE.** The class of the antibody determines the specific role of the antibody in the immune response (for example, whether the antibody is secreted or held on a cell surface) but not the antigen it recognizes.

Functions of Antibody in Host Defense   Antibodies can mediate destruction of an invader (antigen) in a number of ways. A foreign particle, for example, becomes coated with antibody molecules as their Fab regions become bound to it. Macrophages recognize the projecting Fc regions and are stimulated to engulf the particle. This is the process of **opsonization.** Antibodies also may be able to neutralize toxins that are secreted by an invader.

Another important process, particularly in destruction of bacterial cells, is interaction with complement activated by the classical pathway. As noted on p. 739, the first component in the classical pathway is activated by bound antibody. The end result in both pathways can be the same, that is, lysis of a foreign cell. Both pathways may also lead to opsonization or enhancement of inflammation. Binding of complement to antigen-antibody complexes can facilitate clearance of these potentially harmful masses by phagocytic cells.

Antibody bound to the surface of an invader may trigger contact killing of the invader by host cells in what is known as **antibody-dependent, cell-mediated cytotoxicity (ADCC).**

Receptors for Fc of bound antibody on a microorganism or tumor cell cause natural killer cells to adhere to them and pour forth the cytotoxic contents of their vacuoles.

T-Cell Receptors

T-cell receptors are transmembrane proteins on the surfaces of T cells. Like antibodies, T-cell receptors have a constant region and a variable region. The constant region extends slightly into the cytoplasm and the variable region, which binds with specific antigens, extends outward. Most T-cells also bear other transmembrane proteins closely linked to the T-cell receptors, which serve as **accessory** or **coreceptor** molecules. These are of one of two types: **CD4** or **CD8**.

Subsets of T Cells

Lymphocytes are **activated** when they are stimulated to move from their recognition phase, in which they simply bind with particular antigens, to a phase in which they proliferate and differentiate into cells that function to eliminate the antigens. We also speak of activation of effector cells, such as macrophages, when they are stimulated to carry out their protective function.

Communication between cells in the immune response, regulation of the response, and certain effector functions are performed by different kinds of T cells. Although morphologically similar, subsets of T cells can be distinguished by characteristic proteins in their surface membranes. For example, cells with the coreceptor protein CD4 (for **c**luster of **d**ifferentiation) are CD4$^+$ and those with CD8 are described as CD8$^+$. Until recently immunologists believed that certain CD4$^+$ cells (T helper or T$_H$) activated immune responses, and certain CD8$^+$ cells (T suppressors) downregulated such responses. Present evidence now suggests a more complicated web of interactions (Figure 35-3). Some T$_H$ cells (designated T$_H$1) activate cell-mediated immunity while suppressing the humoral response, and others (called T$_H$2) activate humoral and suppress cell-mediated immunity.

Cytotoxic T lymphocytes (CTLs) are CD8$^+$ cells that kill target cells expressing certain antigens. A CTL binds tightly to the target cell and secretes a protein that causes pores to form in the cell membrane, resulting in lysis.

Cytokines

The 1980s saw rapid advances in our knowledge of how cells of immunity communicate with each other. They do this by means of protein hormones called **cytokines** (Table 35-1). Cytokines can produce their effects on the same cells that produce them, on cells nearby, or on cells distant in the body from those that produced the cytokine.

> Interleukins were originally so called because they are synthesized by leukocytes and have their effect on leukocytes. We now know that some other kinds of cells can produce interleukins, and interleukins produced by leukocytes can affect other kinds of cells.

Figure 35-3

Major pathways involved in cell-mediated (T$_H$1) and humoral (T$_H$2) immune responses as mediated by cytokines. Solid arrows indicate positive signals and broken arrows indicate inhibitory signals. Broken lines without arrows indicate path of cellular activation. *IFN-γ*, interferon-γ, *Ig*, immunoglobulin; *IL* interleukin; *T$_H$*, T helper cells; TNF, tumor necrosis factor; *RNI* and *ROI*, toxic substances released onto invader.

Generation of a Humoral Response: T$_H$2 Arm

When an antigen is introduced into a body, it binds to a specific antibody on the surface of an appropriate B cell, but this binding is usually not sufficient to activate the B cell to multiply. Some of the antigen is taken up by **antigen-presenting cells (APCs)**, such as macrophages, that partially digest the antigen. The APCs then incorporate portions of the antigen into their own cell surface, bound in the cleft of MHC II protein (Figures 35-4 and 35-5). That portion of the antigen presented on the surface of the macrophage or other APC is called the **epitope** (or **determinant**). The macrophages also secrete IL-1, which stimulates T$_H$2 cells. The specific T-cell receptor for that particular epitope recognizes the epitope bound to the MHC II protein. Binding of the T-cell receptor to the epitope–MHC II complex is enhanced by the coreceptor CD4, which itself binds to the constant portion of the MHC II protein (Figure 35-5). Bound CD4 molecule also transmits a stimulation signal to the interior of the T cell. Activation of the T cell further requires interaction of additional costimulatory and adhesion signals from other proteins on the surface of the macrophage and T cell. CD8 coreceptors function in a similar way on CD8$^+$ cells; they enhance binding of the T-cell receptor and transmit a stimulatory signal into the T cell.

TABLE 35.1
Some Important Cytokines

Cytokine	Principal Source	Function
Interleukin-1 (IL-1)	Activated macrophages	Mediates inflammation, activates T and B cells
Interleukin-2 (IL-2)	CD4+ cells, some from CD8+ cells	Major growth factor for T and B cells, enhances cytolytic activity of natural killer cells, causing them to become **lymphocyte-activated (LAK) cells**
Interleukin-3 (IL-3)	CD4+ cells	Multilineage colony-stimulating factor; promotes growth and differentiation of all cell types in bone marrow
Interleukin-4 (IL-4)	Mostly by T_H2 CD4+ cells	Growth factor for B cells, some CD4+ T cells, and mast cells; suppresses T_H1 differentiation
Interleukin-5 (IL-5)	Certain CD4+ cells	Activates eosinophils; acts with IL-2 and IL-4 to stimulate growth and differentiation of B cells
Interleukin-6 (IL-6)	Macrophages, endothelial cells, fibroblasts, and T_H2 cells	Important growth factor for B cells late in their differentiation
Interleukin-8 (IL-8)	Antigen-activated T cells, macrophages, endothelial cells, fibroblasts, and platelets	Activating and chemotactic factor for neutrophils, and to a lesser extent, other PMNs
Interleukin-10 (IL-10)	T_H2 CD4+ cells	Inhibits T_H1, CD8+, NK, and macrophage cytokine synthesis
Interleukin-12 (IL-12)	Monocytes, macrophages, neutrophils, dendritic cells, and B cells	Activates NK cells and T cells; potently induces production of IFN-γ, shifts immune response to T_H1
Transforming growth factor-β (TGF-β)	Macrophages, lymphocytes, and other cells	Inhibits lymphocyte proliferation, CTL and LAK cell generation, and macrophage cytokine production
Interferon-γ (IFN-γ)	Some CD4+ and almost all CD8+ cells	Strong macrophage-activating factor; causes a variety of cells to express class II MHC molecules; promotes T- and B-cell differentiation; activates neutrophils and NK cells; activates endothelial cells to allow lymphocytes to pass through walls of vessels
Tumor necrosis factor (TNF)	Activated macrophages	Major mediator of inflammation; low concentrations activate endothelial cells, activate PMNs, stimulate macrophages and cytokine production (including IL-1, IL-6, and TNF itself); higher concentrations cause increased synthesis of prostaglandins, resulting in fever

Figure 35-4

Humoral immune response. (1) Macrophage consumes antigen, partially digests it, and displays portions on its surface, along with class II MHC protein, and secretes interleukin-1 (IL-1). (2) T_H2 cell, stimulated by IL-1, recognizes antigen and class II protein on macrophage, is activated, and secretes interleukins 4, 5, and 6 (IL-4, IL-5, IL-6). (3) T_H2 then activates B cell which carries antigen and class II protein on its surface. IL-4, IL-5, and IL-6 stimulate proliferation of B cell line. (4) Activated B cells finally produced many plasma cells that secrete antibody. (5) Some B-cell progeny become memory cells. (6) Antibody produced by plasma cells binds to antigen and stimulates macrophages to consume antigen (opsonization).

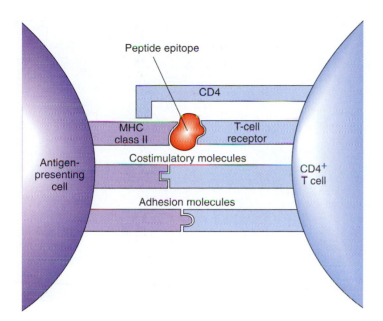

Figure 35-5
Interacting molecules during activation of a T helper cell.

Figure 35-6
Typical immunoglobulin response after primary and challenge immunizations. The secondary response is a result of the large numbers of memory cells produced after the primary B-cell activation.

Many aspects of immunology have been greatly assisted by the discovery of a method for producing stable clones of cells that will produce only one kind of antibody. Such monoclonal antibodies will bind only to *one kind* of antigenic determinant (most proteins bear many different antigenic determinants and thus stimulate the body to produce complex mixtures of antibodies). *Monoclonal* antibodies are made by fusing normal antibody-producing plasma cells with a continuously growing plasma cell line, producing a hybrid of the normal cell with one that can divide indefinitely in culture. This cell line is called a *hybridoma*. Clones are selected from among the hybrids and are grown to become "factories" that produce almost unlimited quantities of one specific antibody. Hybridoma techniques discovered in 1975 have become one of the most important research tools for immunologists.

Activated T_H2 cells secrete IL-4, IL-5 and IL-6, which activate the B cell that has the same epitope and class II MHC protein on its surface. The B cell multiplies rapidly and produces many plasma cells, which secrete large quantities of antibody for a period of time, then die. Thus if we measure the concentration of antibody **(titer)** soon after the antigen is injected, we can detect little or none. Titer then rises rapidly as plasma cells secrete the antibody, and it may decrease somewhat as they die and antibody is degraded (Figure 35-6). However, if we give another dose of antigen **(challenge),** there is little or no lag, and antibody titer rises quicker to a higher level than after the first dose. This is the **secondary** or **anamnestic response,** and it occurs because some of the activated B cells gave rise to long-lived **memory cells.** There are many more memory cells present in the body than the original B lymphocyte with the appropriate antibody on its surface, and they rapidly multiply to produce additional plasma cells. Existence of an anamnestic response has great practical value because it is the basis for protective vaccines.

Cell-Mediated Response: T_H1 Arm

Many immune responses involve little, if any, antibody and depend on the action of cells only. In cell-mediated immunity (CMI) the epitope of an antigen is also presented by ACPs, but the T_H1 arm of the immune response is activated and the T_H2 arm suppressed. Effector cells are macrophages, PMNs, cytotoxic T cells, and activated natural killer cells. The specific interaction of lymphocyte and antigen that generates a CMI greatly influences subsequent events in the nonspecific response we call **inflammation.**

Only a few years ago transplantation of organs from one person to another seemed impossible. Then physicians began to transplant kidneys and depress the immune response of the recipient. It was very difficult to immunosuppress the recipient enough that the new organ would not be rejected and at the same time not leave a patient defenseless against infection. Since discovery of a fungus-derived drug called cyclosporine, not only kidneys, but also hearts, lungs, and livers can be transplanted. Cyclosporine inhibits IL-2 and affects CTLs more than T_HS. It has no effect on other white cells or on healing mechanisms, so that a patient can still mount an immune response but not reject the transplant. However, the patient must continue to take cyclosporine because if the drug is stopped, the body will recognize the transplanted organ as foreign and reject it.

Like humoral immunity, CMI shows a secondary response due to large numbers of memory T cells produced from the original activation. For example, a second tissue graft (challenge) between the same donor and host will be rejected much more quickly than the first.

Inflammation

Inflammation is a vital process in mobilization of body defenses against an invading organism or other tissue damage and in repair of damage thereafter. The course of events in the inflammatory process is greatly influenced by the prior immunizing experience of the body with the invader and by the duration of the invader's presence or its persistence in the body. The processes by which an invader is actually destroyed, however, are themselves nonspecific. Manifestations of inflammation are **delayed type hypersensitivity (DTH)** and **immediate hypersensitivity,** depending on whether the response is cell mediated or antibody mediated.

The DTH reaction is a type of CMI in which the ultimate effectors are activated macrophages. The term *delayed type hypersensitivity* is derived from the fact that a period of 24 hours or more elapses from the time of antigen introduction until a response is observed in an immunized subject. This is because T_H1 cells with specific receptors in their surface for that particular antigen require some time to arrive at the antigen site, recognize epitopes displayed by antigen-presenting cells (APCs), and become activated and secrete IL-2, tumor necrosis factor (TNF), and interferon-γ (IFN-γ). TNF causes endothelial cells of the blood vessels to express on their surface certain molecules to which leukocytes adhere: first neutrophils and then lymphocytes and monocytes. TNF also causes endothelium to secrete inflammatory cytokines such as IL-8, which increase the mobility of leukocytes and facilitate their passage through the endothelium. TNF and IFN-γ also cause endothelial cells to change shape, favoring leakage of macromolecules and passage of cells. Escape of fibrinogen from blood vessels leads to conversion of fibrinogen to fibrin, and the area becomes swollen and firm.

As monocytes pass out of blood vessels, they become activated macrophages, which are the main effector cells of a DTH. They phagocytize particulate antigen, secrete mediators that promote local inflammation, and secrete cytokines and growth factors that promote healing. If the antigen is not destroyed and removed, its chronic presence leads to deposition of fibrous connective tissue, or **fibrosis.** Nodules of inflammatory tissue called **granulomas** may accumulate around persistent antigen and are found in numerous parasitic infections (Figure 35-7).

Immediate hypersensitivity is quite important in some parasitic infections. This reaction involves degranulation of mast cells in the area. Their surfaces bear receptors for Fc portions of antibody, especially IgE. Occupation of these sites by antigen-specific antibodies enhances degranulation of mast cells when the Fab portions bind the particular antigen. There is a rapid release of several mediators, such as histamine, that cause dilation of local blood vessels and increased vascular permeability. Escape of blood plasma into the surrounding tissue causes swelling **(wheal),** and engorgement of vessels with blood produces redness, the characteristic **flare** (Figure 35-8). Systemic immediate hypersensitivity is **anaphylaxis,** which may be fatal if not treated

Figure 35-7
Granulomatous reaction around eggs (*arrows*) of *Schistosoma mansoni* in mesenteries.

Figure 35.8
Wheal and flare reactions around sites of antigen injection for allergy testing.

rapidly. The swelling and change in permeability of capillaries allow antibodies and leukocytes to move from capillaries and easily reach the invader. The first phagocytic line of defense is neutrophils, which may last a few days, then macrophages (either fixed or differentiated from monocytes) become predominant.

Some degree of cell death (**necrosis**) always occurs in inflammation, but necrosis may not be prominent if the inflammation is minor. When necrotic debris is confined within a localized area, pus (spent leukocytes and tissue fluid) may increase in hydrostatic pressure, forming an **abscess.** An area of inflammation that opens out to a skin or mucous surface is an **ulcer.**

Immediate hypersensitivity in humans is the basis for allergies and asthma, which are quite undesirable conditions, leading one to wonder why they evolved. Some scientists believe that the allergic response originally evolved to help the body ward off parasites because only allergens and parasite antigens stimulate production of large quantities of IgE. Avoidance of or reduction in effects of parasites would have conferred a selective advantage in human evolution. The hypothesis is that in absence of heavy parasitic challenge, the immune system is free to react against other substances, such as ragweed pollen. People now living where parasites remain abundant are less troubled with allergies than are those living in relatively parasite-free areas.

Acquired Immune Deficiency Syndrome (AIDS)

AIDS is an extremely serious disease in which the ability to mount an immune response is disabled severely. It is caused by **human immunodeficiency virus (HIV).** The first case of AIDS was recognized in 1981, and by 1998, over 30 million people worldwide were infected with HIV/AIDS, of whom 90% were in developing countries of Asia and sub-Saharan Africa. HIV infection virtually always progresses to AIDS after a latent period of some years. To the best of our current knowledge, AIDS is a terminal disease. AIDS patients are continuously plagued by infections with microbes and parasites that cause insignificant problems in persons with normal immune responses. HIV preferentially invades and destroys CD4$^+$ lymphocytes. CD4 protein is the major surface receptor for the virus. Normally, CD4$^+$ cells make up 60% to 80% of the T-cell population; in AIDS they can become too rare to be detected. T_H1 cells are relatively more depleted than T_H2 cells, which upsets the balance of immunoregulation and results in persistent, nonspecific B-cell activation.

BLOOD GROUP ANTIGENS

ABO Blood Types

Blood cells differ chemically from person to person, and when two different (incompatible) blood types are mixed, **agglutination** (clumping together) of erythrocytes results. The basis of these chemical differences is naturally occurring antigens on the membranes of red blood cells. The best known of these inherited immune systems is the ABO blood group (p. 120). Antigens A and B are inherited as codominant alleles of a single gene. Homozygotes for a recessive allele at the same gene have type O blood, which lacks A and B antigens. Thus, as shown in Table 35-2, an individual with, for example, genes I^A/I^A or I^A/i develops A antigen (blood type A). The presence of an I^B gene produces B antigens (blood type B), and for the genotype I^A/I^B both A and B antigens develop on the erythrocytes (blood type AB). Epitopes of A and B also are present on the surfaces of many epithelial and most endothelial cells.

There is an odd feature about the ABO system. Normally we would expect that a type A individual would develop antibodies against type B cells only if cells bearing B epitopes were first introduced into the body. In fact, type A persons acquire anti-B antibodies soon after birth, even without exposure to type B cells. Similarly, type B individuals come to carry anti-A antibodies at a very early age. Type AB blood has neither anti-A nor anti-B antibodies (since if it did, it would destroy its own blood cells), and type O blood has both anti-A and anti-B antibodies. There is evidence that the antibodies develop as a response to A and B epitopes on intestinal microorganisms when the intestine becomes colonized with bacteria after birth. Presumably, small and unnoticed infections with the bacteria occur. The antibodies thus produced cross-react with the A and B epitopes on erythrocytes.

We see then that the blood-group names identify their *antigen* content. Persons with type O blood are called universal donors because, lacking antigens, their blood can be infused into a person with any blood type. Even though it contains antiA and anti-B antibodies, these are so diluted during transfusion that they do not react with A or B antigens in a recipient's blood. Persons with AB blood are universal recipients because they lack antibodies to A and B antigens. In practice, however, clinicians insist on matching blood types to prevent any possibility of incompatibility.

TABLE 35.2
Major Blood Groups

Blood Type	Genotype	Antigens on Red Blood Cells	Antibodies in Serum	Can Give Blood To	Can Receive Blood From	Whites	Blacks	Asians
O	i/i	None	Anti-A and anti-B	All	O	45	48	31
A	I^A/I^A, I^A/i	A	Anti-B	A, AB	O, A	41	27	25
B	I^B/I^B, I^B/i	B	Anti-A	B, AB	O, B	10	21	34
AB	I^A/I^B	AB	None	AB	All	4	4	10

Rh Factor

Karl Landsteiner, an Austrian—later American—physician discovered ABO blood groups in 1900. In 1940, 10 years after receiving a Nobel Prize, he made still another famous discovery. This was a blood group called the Rh factor, named after rhesus monkeys, in which it was first found. Approximately 85% of white individuals in the United States have the factor (positive) and the other 15% do not (negative). The Rh factor is encoded by a dominant allele at a single gene. Rh-positive and Rh-negative bloods are incompatible; shock and even death may follow their mixing when Rh-positive blood is introduced into an Rh-negative person who has been sensitized by an earlier transfusion of Rh-positive blood. Rh incompatibility accounts for a peculiar and often fatal **hemolytic disease of the newborn (erythroblastosis fetalis).** If an Rh-negative mother has an Rh-positive baby (father is Rh-positive) she can become immunized by fetal blood during the birth process. Anti-Rh antibodies are predominately IgG and can cross the placenta during a subsequent pregnancy and agglutinate fetal blood. Erythroblastosis fetalis normally is not a problem in cases of ABO incompatibility between mother and fetus because antibodies to ABO antigens are primarily IgM and cannot cross the placenta.

The genetics of the Rh factor are very much more complicated than it was believed when the factor was first discovered. Some authorities think that three genes located close together on the same chromosome are involved, whereas others adhere to a system of one gene with many alleles. In 1968 a revision of the single gene concept listed 37 alleles necessary to account for phenotypes then known. Furthermore, the frequency of the various alleles varies greatly between whites, Asians, and blacks.

Erythroblastosis fetalis can now be prevented by giving an Rh-negative mother anti-Rh antibodies just after the birth of her first child. These antibodies remain long enough to neutralize any Rh-positive fetal blood cells that may have entered her circulation, thus preventing her own antibody machinery from being stimulated to produce Rh-positive antibodies. Active, permanent immunity is blocked. The mother must be treated after every subsequent pregnancy (assuming the father is Rh-positive). If the mother has already developed an immunity, however, the baby may be saved by an immediate, massive transfusion of blood free of antibodies.

IMMUNITY IN INVERTEBRATES

One principal test of the ability of invertebrate tissues to recognize nonself is by grafting a piece of tissue from another individual of the same species **(allograft)** or a different species **(xenograft)** onto the host. If the graft grows in place with no host response, the host tissue is treating it as self, but if cell response and rejection of the graft occur, the host exhibits immune recognition. Most invertebrates tested in this way reject xenografts; and almost all can reject allografts to some degree (Table 35-3). Interestingly, nemertines and molluscs apparently do not reject allografts. It is curious that some animals with quite simple body organization, such as Porifera and Cnidaria, can reject allografts; this response may be an adaptation to avoid loss of integrity of an individual sponge or colony under conditions of crowding, with attendant danger of overgrowth or fusion with other individuals. American cockroaches (*Periplaneta americana*) reject allografts from the same source more quickly upon

TABLE 35.3

Some Invertebrate Leukocytes and Their Functions

Group	Cell Types and Functions	Phagocytosis	Encapsulation	Allograft Rejection	Xenograft Rejection
Sponges	Archaeocytes (wandering cells that differentiate into other cell types and can act as phagocytes)	+	+	+*	+*
Cnidarians	Amebocytes: "lymphocytes"	+		+	+
Nemertines	Agranular leukocytes; granular macrophagelike cells	+		−	±
Annelids	Basophilic amebocytes (accumulate as "brown bodies"), acidophilic granulocytes	+	+	+	+
Sipunculids	Several types	+	+	±	+
Insects	Several types, depending on family; e.g., plasmatocytes, granulocytes, spherule cells, coagulocytes (blood clotting)	+	+	−	±
Crustaceans	Granular phagocytes; refractile cells that lyse and release contents	+	+		+
Molluscs	Amebocytes	+	+	−	+
Echinoderms	Amebocytes, spherule cells, pigment cells, vibratile cells (blood clotting)	+	+	+	+
Tunicates	Many types, including phagocytes; "lymphocytes"	+	+	+	+

Data from Lackie, A. M. 1980. Parasitology **80**:393–412. (See Lackie's article for references.)
*Transplantation reactions occur, but the extent to which the leukocytes are involved is unknown.

second exposure; thus they show at least short-term immuno-logical memory.

Hemocytes of molluscs release degradative enzymes during phagocytosis and encapsulation, and bactericidal substances occur in body fluids of a variety of invertebrates. Substances functioning as opsonins have been reported in annelids, insects, crustaceans, echinoderms, and molluscs.

Bacterial infection in some insects stimulates production of antibacterial peptides (p. 740), but these peptides show broad-spectrum activity and are not specific for a single infective agent. Specific, induced responses that demonstrate memory upon challenge, resembling acquired immunity of vertebrates, have been found in American cockroaches.

Contact with infectious organisms can bring the defense systems of snails into enhanced levels of readiness that last for up to two months or more. Susceptibility of snail hosts of the trematode *Schistosoma mansoni* depends heavily on genotype of the snail. Excretory/secretory products of the trematode stimulate motility of hemocytes from resistant snails but inhibit motility of hemocytes from susceptible hosts. Hemocytes from resistant snails encapsulate the trematode larva and apparently kill it with superoxide and H_2O_2 and then destroy it by phago-cytosis (Figure 35-9). It appears that the cytokine interleukin-1 is present in resistant snails and is responsible for activating hemocytes.

Figure 35-9

Electron micrograph of a hemocyte (H) from a schistosome-resistant strain of a snail attacking a *Schistosoma mansoni* larva (S) under in vitro conditions. Note the hemocyte processes apparently engaged in phago-cytosis of portions of the larval tegument (*arrows*). (Scale bar = 1 μm.)

SUMMARY

A plethora of viral, prokaryotic, and eukaryotic parasites exists in every animal's environment, and a defense (immune) system is crucial to survival. Immunity can be defined concisely as possession of tissues capable of recognizing and protecting the animal against nonself invaders. Most animals have some amount of innate (non-specific) immunity, and vertebrates develop acquired (specific) immunity. The surface of most animals provides a physical barrier to invasion, and vertebrates have a variety of antimicrobial substances in their body secretions.

Phagocytes engulf particles and usually digest or kill them with enzymes and cytotoxic secretions. Many invertebrates have specialized cells that can perform defensive phagocytosis. Several kinds of vertebrate cells, especially macrophages and neutrophils, are important phagocytes, and cells of the mononuclear phagocyte system (RE system) reside in various sites in the body. Eosinophils are important in allergies and many parasitic infections. Basophils, mast cells, T and B lymphocytes, and natural killer cells are not phago-cytic but play vital roles in defense.

Exposure of many animals, vertebrate and invertebrate, to many microorganisms stimulates their innate immune response. This response is based on release of antimicrobial peptides, is immediate, not requiring prior immunizing exposure, and is nonspecific but related to the category of invading microbe.

An induced immune response is elicited by an antigen. Vertebrates demonstrate increased resistance to *specific* foreign substances (antigens) on repeated exposure, and the resistance is based on a vast number of specific recognition molecules: antibod-ies and T-cell receptors. Nonself recognition depends on markers in cell surfaces known as major histocompatibility (MHC) proteins. Antibodies are borne on the surfaces of B lymphocytes (B cells) and in solution in the blood after secretion by the progeny of B cells, plasma cells. T-cell receptors occur only on the surfaces of T lymphocytes (T cells).

The cells of immunity communicate with each other and with other cells in the body by means of protein hormones called cytokines such as interleukins, tumor necrosis factor, and interferon-γ. The two arms of the vertebrate immune response are the humoral response (T_H2), involving antibodies, and the cell-mediated response (T_H1), involving cell surfaces only. When one arm is activated or stimulated, its cells produce cytokines that tend to suppress activity in the other arm. Activation of either arm requires that the antigen be consumed by an APC (antigen-presenting cell, usually a macrophage), which partially digests the antigen and presents its determinant (epitope) on the surface of the APC along with an MHC class II protein. Extensive communication by cytokines and activation (and suppression) of various cells in the response leads to production of specific antibody or proliferation of T cells with specific receptors that recognize an antigenic epitope. After the initial response, memory cells remain in the body and are responsible for enhanced response on next exposure to the antigen.

Damage to the immune response done by HIV (human immun-odeficiency virus) in production of AIDS (acquired immune deficiency syndrome) is due primarily to destruction of a crucial set of T cells: those bearing the CD4 coreceptor on their surface.

Inflammation is an important part of the body's defense; it is greatly influenced by prior immunizing experience with an antigen.

People have genetically determined antigens in the surfaces of their red blood cells (ABO blood groups and others); blood types must be compatible in transfusions, or transfused blood will be agglutinated by antibodies in the recipient.

Many invertebrates show nonself recognition by rejection of xenografts or allografts or both. In some cases they may show enhanced response on repeated exposure.

REVIEW QUESTIONS

1. Distinguish susceptibility from resistance, and innate (nonspecific) from acquired (specific) immunity.
2. What are some examples of innate defense mechanisms that are chemical in nature? What is complement?
3. After a phagocyte has engulfed a particle, what usually happens to the particle?
4. What are some important phagocytes in vertebrates?
5. What is the molecular basis of self and nonself recognition in vertebrates?
6. What is the difference between T cells and B cells?
7. What is a cytokine? What are some functions of cytokines?
8. Outline the sequence of events in a humoral immune response from the introduction of antigen to the production of antibody.
9. Define the following: plasma cell, secondary response, memory cell, complement, opsonization, titer, challenge, cytokine, natural killer cell, interleukin-2.
10. What are the functions of CD4 and CD8 proteins on the surface of T cells?
11. In general, what are consequences of activation of the T_H1 arm of the immune response? Activation of the T_H2 arm?
12. Distinguish between class I and class II MHC proteins.
13. Describe a typical inflammatory response.
14. What is a major mechanism by which HIV damages the immune system in AIDS?
15. Give the genotypes of each of the following blood types: A, B, O, AB. What happens when a person with type A gives blood to a person with type B? With type AB? With type O?
16. What causes hemolytic disease of the newborn (erythroblastosis fetalis)? Why does the condition not arise in cases of ABO incompatibility?
17. Give some evidence that cells of many invertebrates bear molecules on their surface that are specific to the species and even to a particular individual animal.
18. Give an example of immunological memory in invertebrates.

SELECTED REFERENCES

Aderem, A., and R. J. Ulevitch. 2000. Toll-like receptors in the induction of the innate immune response. Nature **406:**782–787. *A good review of the role of Toll-like receptors in the innate immune response.*

Beck, G., and G. S. Habicht. 1996. Immunity and the invertebrates. Sci. Am. **275:**60–66 (Nov.). *Invertebrates lack lymphocytes, but they have cells like macrophages, such as coelomocytes, and they lack antibody-based humoral immunity. They have molecules similar to immunoglobulin, and several cytokines have been demonstrated, suggesting precursors for a vertebrate-type immune response.*

Benjamini, E., G. Sunshine, and S. Leskowitz. 1996. Immunology. A short course, ed. 3. New York, Wiley-Liss, Inc. *An excellent presentation of the essentials without excessive details.*

Cox, F. E. G., and E. Y. Liew. 1992. T-cell subsets and cytokines in parasitic infections. Parasit. Today **8:**371–374. *Describes interactions of T cells and cytokines in the T_H1 and T_H2 arms of the immune response.*

Dunn, P. E. 1990. Humoral immunity in insects. BioScience **40:**738–744. *There is evidence that cockroaches are capable of a vertebratelike, specific, adaptive humoral response.*

Engelhard, V. H. 1994. How cells process antigens. Sci. Am. **271:**54–61. *This article focuses on the roles of the MHC proteins.*

Garrett, L. 1995. The coming plague: newly emerging diseases in a world out of balance. New York, Penguin Books (orig. publ. 1994, Farrar, Straus, & Giroux, New York). *As Rachel Carson did with the environment, Garrett sounds a clarion call on infectious diseases. New pathogens are emerging, and familiar ones are developing multidrug resistance at a time when transmission favors the microbes: poor sanitation and increasing crowding in the world's cities and impairment of the immune response in millions of people by malnutrition and AIDS.*

Glausiusz, J. 1999. The chasm in care. Discover **20:**40,42. *More than 30 million people have HIV, and nearly 90% live in developing countries of Asia and sub-Saharan Africa. There are 16,000 people newly infected every day. In Zimbabwe, 25% of adult population has HIV.*

Golde, D. W. 1991. The stem cell. Sci. Am. **265:**86–93. *Undifferentiated cells in bone marrow give rise to white and red blood cells, macrophages, and platelets.*

Greene, W. C. 1993. AIDS and the immune system. Sci. Am. **269:**98–105 (Sept.). *There is some evidence that HIV triggers widespread apoptosis (programmed cell death) in CD4⁺ cells.*

Gura, T. 2001. Innate immunity: ancient system gets new respect. Science **291:**2068–2071. *Innate immunity evolved early in animal evolution.*

Karp, R. D. 1990. Cell-mediated immunity in invertebrates. BioScience **40:**732–737. *Experiments show that allograft rejection in insects has at least a short-term memory component; challenge allografts were rejected more quickly than third-party allograft controls.*

Lichtenstein, L. M. 1993. Allergy and the immune system. Sci. Am. **269:**116–124 (Sept.). *Describes what happens in an allergic response, including the role of mast cells, basophils, cytokines, and chemical mediators.*

Mann, J. M., and D. J. M. Tarantola. 1998. HIV 1998: The global picture. Sci. Am. **279:**82–83 (July). *This is the first of a series of articles on HIV in this issue. The authors estimate that more than 40 million people have contracted HIV since early 1980s. Two-thirds of all HIV-infected people and 90% of all infected children are in sub-Saharan Africa.*

Marrack, P., and J. W. Kappler. 1993. How the immune system recognizes the body. Sci. Am. **269:**80–89 (Sept.). *Part of the mechanism by which the immune system tolerates "self" antigens is by a mechanism known as clonal deletion.*

Medzhitov, R., and C. A. Janeway Jr. 2002. Decoding the patterns of self and nonself by the innate immune system. Science **296:**298–300. *The several strategies employed by innate immune systems for discrimination of self and nonself.*

Paul, W. E. 1993. Infectious diseases and the immune system. Sci. Am. **269:**90–97 (Sept.). *Describes the immune response in certain viral, microbial, and parasitic infections.*

Steinman, L. 1993. Autoimmune disease. Sci. Am. **269:**106–114 (Sept.). *In five percent of adults in Europe and North America, the immune system discrimination between "self" and "nonself" breaks down, usually with very serious results.*

Strange, C. 1995. Rethinking immunity. BioScience **45:**663–668. *Some immunologists believe that the self/nonself dogma of immunology is inadequate.*

Weiss, R. 1994. Of myths and mischief. Discover **15**(12):36–42. *Debunks the astonishing myths that have grown up around AIDS, including the most pernicious of all, that HIV is not the cause.*

ZOOLOGY LINKS TO THE INTERNET

Visit the textbook's Online Learning Center at www.mhhe.com/zoology to find live Internet links for each of the topics listed here.

The Immune System

Human Immune System Topics

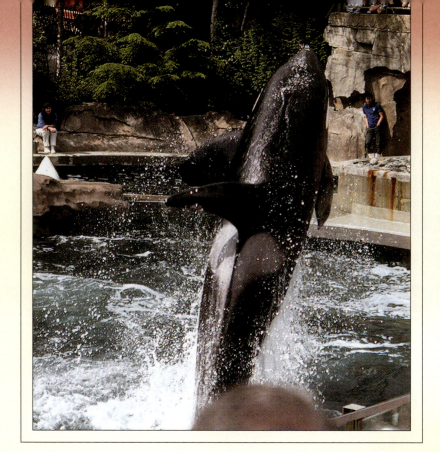

Killer whale in performance at the Vancouver Public Aquarium.

Animal Behavior

The Lengthening Shadow of One Person

Ralph Waldo Emerson said that an institution is the lengthening shadow of one person. For Charles Darwin the shadow is long indeed, for he brought into being entire fields of knowledge, such as evolution, ecology, and finally, after a long gestation, animal behavior. Above all, he altered the way we think about ourselves, the earth we inhabit, and the animals that share it with us.

Charles Darwin, with the uncanny insight of genius, prepared for the reception of animal behavior by showing how natural selection would favor specialized behavioral patterns for survival. Darwin's pioneering book, *The Expression of the Emotions in Man and Animals,* published in 1872, mapped a strategy for behavioral research still in use today. However, science in 1872 was unpre-

pared for Darwin's central insight that behavioral patterns, no less than bodily structures, are selected and have evolutionary histories. Another 60 years would pass before such concepts would begin to flourish within behavioral science.

In 1973, the Nobel Prize in Physiology or Medicine was awarded to three pioneering zoologists, Karl von Frisch, Konrad Lorenz, and Niko Tinbergen (Figure 36-1). The citation stated that these three were the principal architects of the new science of ethology, the scientific study of animal behavior, particularly under natural conditions. It was the first time any contributor to the behavioral sciences was so honored, and it meant that the discipline of animal behavior had arrived.

A

C

Figure 36-1

Pioneers of the science of ethology. **A,** Konrad Lorenz (1903 to 1989). **B,** Karl von Frisch (1886 to 1982). **C,** Niko Tinbergen (1907 to 1988).

Behavioral biologists have traditionally asked two kinds of questions about behavior: *how* animals behave and *why* they behave as they do. "How" questions are concerned with immediate or **proximate causation,** and are studied experimentally (p. 14). For example, a biologist might wish to explain the singing of a male white-throated sparrow in the spring in terms of hormonal

or neural mechanisms. Such physiological or mechanistic causes of behavior are proximate factors. Alternatively, a biologist might ask what function singing serves the sparrow, and then seek to understand those events in the ancestry of birds that led to springtime singing. These are "why" questions that focus on **ultimate causation,** the evolutionary origin and purpose of a behavior. Questions of ultimate causation are answered using comparative methodology (p. 14) applying phylogenetic analysis (p. 195) to understand evolutionary changes in behavior and their associated morphological and environmental contexts.

The study of animal behavior has arisen from several different historical roots, and there is no universally accepted term for the whole subject. **Comparative psychology** emerged from efforts to find general laws of behavior that would apply to many species, including humans. Early research that depended heavily on inference was later replaced by replicable experimental approaches that concentrated on a few species, particularly white rats, pigeons, dogs, and occasionally primates. Following criticisms that the discipline lacked an evolutionary perspective and focused too narrowly on the white rat as a model for other organisms, many comparative psychologists developed more truly comparative investigations, some of these conducted under natural conditions.

The aim of a second approach, **ethology,** has been to describe the behavior of an animal in its *natural habitat.* Most ethologists have been zoologists. Ethologists gather data by field observation and experiments, often with nature providing the variables, which are manipulated using animal models, playing recordings of animal vocalizations and altering habitats. Modern ethologists also conduct many experiments in laboratories where they can test their hypotheses under closely controlled conditions. Laboratory results often direct further testing of hypotheses using observations of free-ranging animals in undisturbed natural environments.

Ethology emphasizes the importance of ultimate factors affecting behavior. One of the great contributions of von Frisch, Lorenz, and Tinbergen was to demonstrate that behavioral traits are measurable entities like anatomical or physiological traits. This was to become the central theme of ethology: behavioral traits can be identified and measured, and their evolutionary histories investigated.

Much of the work by comparative psychologists and ethologists can be found under the discipline of **behavioral ecology.** Behavioral ecologists often focus on how individuals are expected to behave to maximize their reproductive success. They then concentrate on a particular aspect of behavior, such as mate choice, foraging, or parental investment. **Sociobiology,** the ethological study of social behavior, was formalized with the 1975 publication of E. O. Wilson's *Sociobiology: The New Synthesis.* Wilson describes social behavior as reciprocal communication of a cooperative nature (transcending mere sexual activity) that permits a group of organisms of the same species to become organized in a cooperative manner. In a complex system of social interactions, individuals are highly dependent on others for their daily living. While social behavior appears in many groups of animals, Wilson identified four "pinnacles" of complex social behavior. These are (1) colonial invertebrates, such as the

Portuguese man-of-war (p. 257), which is a tightlyknit composite of individual organisms; (2) social insects, such as ants, bees, and termites, which have developed sophisticated systems of communication; (3) nonhuman mammals, such as dolphins, elephants, and some primates, which have highly developed social systems; and (4) humans.

The inclusion by Wilson of human behavior in sociobiology, and his references to the genetic foundation of many human social behaviors, have been strongly criticized. Complex systems of human social interactions, including religion, economic systems, and such objectionable characteristics as racism, sexism, and war, are emergent properties (p. 5) of human culture and its history. Is it meaningful to search for a specific genetic basis or justification for such phenomena? Many answer "no," and look instead to the field of sociology, rather than sociobiology, to help us understand complex, emergent properties of human societies.

DESCRIBING BEHAVIOR: PRINCIPLES OF CLASSICAL ETHOLOGY

Early ethologists, through step-by-step analyses of the behavior of animals in nature, focused on relatively invariant components of behavior. From such studies emerged several concepts that were first popularized in Tinbergen's influential book. *The Study of Instinct* (1951).

Some basic concepts of animal behavior can be illustrated by considering the egg-retrieval response of greylag geese (Figure 36-2), described by Lorenz and Tinbergen in a famous paper published in 1938. If Lorenz and Tinbergen presented a female greylag goose with an egg a short distance from her nest, she would rise, extend her neck until the bill was just over the egg and then bend her neck, pulling the egg carefully into the nest.

Although this behavior appeared to be intelligent, Tinbergen and Lorenz noticed that if they removed the egg once the goose had begun her retrieval, or if the egg being retrieved slipped away and rolled down the outer slope of the nest, the goose would continue the retrieval movement without the egg until she was again settled comfortably on her nest. Then, seeing that the egg had not been retrieved, she would begin the egg-rolling pattern again.

Thus, the bird performed egg-rolling behavior as if it were a program that, once initiated, had to run to completion. Lorenz and Tinbergen viewed egg-retrieval as a "fixed" pattern of behavior: a motor pattern that is mostly invariable in its performance. A behavior of this type, performed in an orderly, predictable sequence is often called **stereotypical behavior.** Of course, stereotyped behavior may not be performed identically on all occasions, but it should be recognizable, even when performed inappropriately. Further experiments by Tinbergen disclosed that the greylag goose was not particularly discriminating about what she retrieved. Almost any smooth and rounded object placed outside the nest would trigger the egg-rolling behavior; even a small toy dog and a large yellow balloon were dutifully retrieved. But once the goose settled down on such objects, they obviously did not feel right and she discarded them.

Lorenz and Tinbergen realized that presence of an egg outside the nest must act as a stimulus, or trigger, that released egg-retrieval behavior. Lorenz termed the triggering stimulus a **releaser,** a simple signal in the environment that would trigger a certain innate behavior. Or, because the animal usually responded to some specific aspect of the releaser (sound, shape, or color, for example) the effective stimulus was called a **sign stimulus.** Ethologists have described hundreds of examples of sign stimuli. In every case the response is highly predictable. For example, the alarm call of adult herring gulls always releases a crouching freeze-response in their chicks. Certain nocturnal moths take evasive maneuvers or drop to the ground when they hear ultrasonic cries of bats that feed on them (p. 713); most other sounds do not release this response.

These examples illustrate the predictable and programmed nature of much animal behavior. This is even more evident when stereotyped behavior is released inappropriately. In spring a male three-spined stickleback, a small fish, selects a territory that it defends vigorously against other males. The underside of the male becomes bright red, and an approach of another red-bellied male will release a threat posture or even an aggressive attack. Tinbergen's suspicion that the red belly of the male served as a releaser for aggression was reinforced when a passing red postal truck evoked attack behavior from males in his aquarium. Tinbergen then performed experiments using a series of models, which he presented to the males. He found that they vigorously attacked any model bearing a red stripe, even a plump lump of wax with a red underside. Yet a carefully made model that closely resembled a male stickleback but lacked the red belly was attacked less frequently (Figure 36-3). Tinbergen discovered other examples of stereotyped behavior released by simple sign stimuli. Male English robins furiously attacked a bundle of red feathers placed in their territory but ignored a stuffed juvenile robin without the red feathers (Figure 36-4).

We have seen in these examples that there are costs to programmed behavior because it may lead to improper responses.

Figure 36-2
Egg-rolling behavior of the greylag goose *(Anser anser).*

Figure 36-3

Stickleback models used to study territorial behavior. The carefully made model of a stickleback (*left*), without a red belly, is attacked much less frequently by a territorial male stickleback than the four simple red-bellied models.

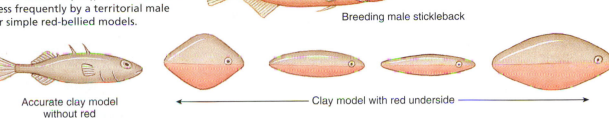

Breeding male stickleback

Accurate clay model without red

Clay model with red underside

Figure 36-4

Two models of an English robin. The bundle of red feathers is attacked by male robins, whereas the stuffed juvenile bird (*right*) without a red breast is ignored.

From N. Tinbergen, The study of instinct, Oxford University Press, Oxford, England, 1951; modified from D. Lack, The life of the robin, H. F. & G. Witherby Ltd., London, England, 1943. Reprinted by permission of Oxford University Press.

Fortunately for red-bellied sticklebacks and red-breasted English robins, their aggressive response toward red works appropriately most of the time because red objects are uncommon in the worlds of these animals. But why do these and other animals not use reasoning to choose the correct response rather than relying mostly on automatic responses? Under conditions that are relatively consistent and predictable, automatic preprogrammed responses may be most efficient. Even if they can or could, thinking about or learning the correct response may take too much time. Releasers have the advantage of focusing an animal's attention on the relevant signal, and release of a preprogrammed stereotyped behavior will enable an animal to respond rapidly when speed may be essential for survival.

CONTROL OF BEHAVIOR

From the beginning, the mostly invariable and predictable nature of stereotyped behavior suggested to ethologists that they were observing inherited, or **innate,** behavior. Many kinds of preprogrammed behavior appear suddenly in animals and are indistinguishable from similar behavior performed by older, experienced individuals. Orb-weaving spiders build their webs without practice, and male crickets court females without lessons from more experienced crickets or by learning from trial and error. To such behaviors the term innate, or instinctive, has been applied. Such words suggest that these behavioral patterns are absolutely committed and will develop in the same way regardless of environment. This idea, called instinct theory, has fallen from favor with behavioral scientists because it cannot be shown that a behavior develops independently of experience. Critics of instinct theory argue that all forms of behavior depend on an interaction of the organism and environment, beginning with the fertilized egg. Genes code for proteins and not directly for behavior. Even with web-building spiders and courting crickets, the environment is bound to have some influence. Given a different environment in which to develop, the resulting behavior may be different.

Nevertheless, it seems incontestable that many complex sequences of behavior in invertebrate animals are largely invariate in their execution, are not learned, and appear to be programmed by rules. It is easy to understand why programmed behavior is important for survival, especially for animals that never know their parents. They must be equipped to respond to the world immediately and correctly as soon as they emerge into it. It is also evident that more complex animals with longer lives and with parental care or other opportunities for social interactions may improve or change their behavior by learning.

The Genetics of Behavior

The hereditary transmission of most innate behavior is complex, with many interacting genes and environmental factors influencing each behavioral trait. However, there are a few examples of behavioral differences within species that show simple Mendelian transmission from parents to offspring. Perhaps the most convincing example is the inheritance of hygienic behavior in bees. Honey bees are susceptible to a bacterial disease, American foulbrood (*Bacillus larvae*). A bee larva that catches foulbrood dies. If the bees remove dead larvae from the hive they reduce the chance of the infection spreading.

Figure 36-5

The genetics of hygienic behavior in honey bees, as demonstrated by W. C. Rothenbuhler. The results are explained by assuming that there are two independently assorting genes, one associated with uncapping cells containing diseased larvae, and the other associated with removing the larvae from the cells. See text for further explanation.

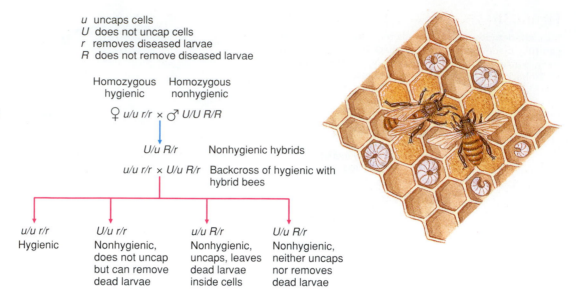

u uncaps cells
U does not uncap cells
r removes diseased larvae
R does not remove diseased larvae

Homozygous   Homozygous
hygienic   nonhygienic

♀ *u/u r/r* × ♂ *U/U R/R*

U/u R/r   Nonhygienic hybrids

u/u r/r × *U/u R/r*   Backcross of hygienic with hybrid bees

u/u r/r
Hygienic

U/u r/r
Nonhygienic, does not uncap but can remove dead larvae

u/u R/r
Nonhygienic, uncaps, leaves dead larvae inside cells

U/u R/r
Nonhygienic, neither uncaps nor removes dead larvae

Some strains of bees, called "hygienic," uncap hive cells containing rotting larvae and remove them from the hive. W. C. Rothenbuhler found that there are two components to this behavior: first removal of cell caps, and second removal of larvae. Hygienic bees have homozygous recessive genotypes for two different genes. Uncapping behavior is performed by individuals homozygous for a recessive allele, *u*, at one gene, and removal behavior is performed by individuals homozygous for a recessive allele, *r*, at a second gene. When Rothenbuhler crossed hygienic bees (*u/u r/r*) with a nonhygienic strain (*U/U R/R*), he found that all hybrids (*U/u R/r*) were nonhygienic. Thus only workers having both genes in the homozygous recessive condition show the complete behavior. Next, Rothenbuhler performed a "backcross" between the hybrids and the hygienic parental strain. As we should expect if hygienic behavior is transmitted by allelic variation of two genes, four different kinds of bees resulted (Figure 36-5). Approximately one-quarter of the bees were homozygous recessive for both *u* and *r* and showed the complete behavior: they both uncapped the cells and removed the infected larvae. Another quarter of the offspring (*u/u R/r*) uncapped but did not remove dead bees. Another quarter (*U/u r/r*) did not uncap, but would remove the larvae if another worker uncapped the cells. Workers that were heterozygous for the dominant allele at both genes (*U/u R/r*) would not perform either part of the cleaning behavior (Figure 36-5). The results showed clearly that each component of the cleaning behavior was associated with one, independently segregating gene.

Most inherited behaviors do not show simple segregation and independence; instead, hybrids of subspecies or species commonly show intermediate or confused behavior. A classic study of the effect of cross-breeding on behavior was performed by W. C. Dilger on nest-building behavior in different species of lovebirds. Lovebirds are small parrots of the genus *Agapornis* (Figure 36-6). Each species has its own method of courtship and technique for carrying nesting material. Fischer's lovebirds (*A. personata fischeri*) cut long strips of nesting material from vegetation, then carry this to the nest, one strip at a time. Peach-faced lovebirds (*A. roseicollis*) carry several strips of torn nesting material at one time by tucking them into feathers of the lower back and rump. Dilger, who was able to cross the two species successfully, found that hybrids displayed a confused conflict between a tendency to carry material in the feathers (inherited from the peach-faced lovebirds) and a tendency to carry material in the bill (inherited from Fischer's lovebird) (see Figure 36-6). The hybrids attempted both feather-tucking and bill-carrying but performed neither behavior correctly. The hybrids had inherited a behavior that was intermediate between that of the parents. With experience hybrids improved their carrying ability by tending to carry the material in their bills, like Fischer's lovebird.

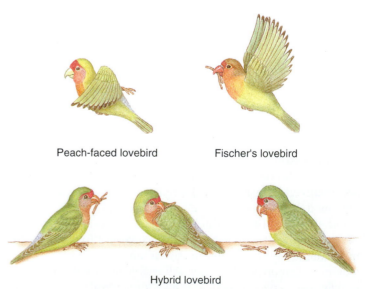

Peach-faced lovebird   Fischer's lovebird

Hybrid lovebird

Figure 36-6

Confused behavior in hybrid lovebirds (*Agapornis* sp.). The peach-faced lovebird carries nest-building material tucked into its feathers; Fischer's lovebird carries nest-building material in its beak. The hybrids attempted both carrying methods, neither method accomplished successfully.

Learning and the Diversity of Behavior

Another aspect of behavior is learning, which we define as the modification of behavior through experience. An excellent model system for studying learning processes has been the marine opisthobranch snail *Aplysia* (Figure 36-7), a subject of intense experimentation by E. R. Kandel and his associates. The gills of *Aplysia* are partly covered by the mantle cavity and open to the outside by a siphon (Figure 36-8). If one prods the siphon, *Aplysia* withdraws its siphon and gills and folds them in the mantle cavity. This simple protective response, called gill withdrawal reflex, will be repeated when *Aplysia* extends its siphon again. But if the siphon is touched repeatedly, *Aplysia* decreases the gill-withdrawal response and finally comes to ignore the

stimulus. This behavioral modification illustrates a widespread form of learning called **habituation.** If now *Aplysia* is given a noxious stimulus (for example, an electric shock) to the head at the same time the siphon is touched, it becomes **sensitized** to the stimulus and withdraws its gills as completely as it did before habituation occurred. Sensitization, then, can reverse any previous habituation.

The mechanisms of habituation and sensitization in *Aplysia* are known because these behaviors constitute a rare instance in which the nervous pathways involved have been completely revealed. Receptors in the siphon are connected through sensory neurons (black pathways in Figure 36-8) to motor neurons (blue pathway in Figure 36-8) that control the gill-withdrawal muscles and muscles of the mantle cavity. Kandel found that repeated stimulation of the siphon diminished the release of synaptic transmitter from the sensory neurons. Sensory neurons continue to fire when the siphon is probed but, with less neurotransmitter being released into the synapse, the system becomes less responsive.

Sensitization requires the action of a different kind of neuron called a facilitating interneuron. These interneurons make connections between sensory neurons in the animal's head and motor neurons that control muscles of the gill and mantle (see Figure 36-8). When sensory neurons in the head are stimulated by an electric shock, they fire the facilitating interneurons, which end on the synaptic terminals of the sensory neurons (red pathways in Figure 36-8). These endings in turn cause an *increase* in the amount of transmitter released by the siphon sensory neurons. This release increases the state of excitation in the excitatory interneurons and motor neurons leading to the gill and mantle muscles. The motor neurons now fire more readily than before. The system is now sensitized because any stimulus to the siphon will produce a strong gill-withdrawal response.

Kandel's studies indicate that strengthening or weakening of the gill-withdrawal reflex involves changes in levels of transmitter

Figure 36-7

The sea hare *Aplysia* sp., an opisthobranch gastropod used in many neurophysiological and behavioral studies.

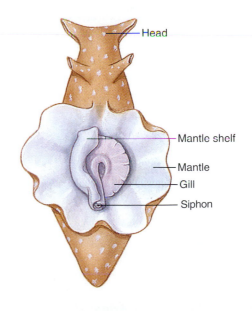

Figure 36-8

Neural circuitry concerned with habituation and sensitization of the gill-withdrawal reflex in the marine snail *Aplysia*. See text for explanation.

in existing synapses. However, we know that more complex kinds of learning may involve formation of new neural pathways and connections, as well as changes in existing circuits.

Imprinting

Another kind of learned behavior is **imprinting,** the imposition of a stable behavior in a young animal by exposure to particular stimuli during a critical period in the animal's development. As soon as a newly hatched gosling or duckling is strong enough to walk, it follows its mother away from the nest. After it has followed the mother for some time it follows no other animal. However, if the eggs are hatched in an incubator or if the mother is separated from the eggs as they hatch, the goslings follow the first large object they see. As they grow, the young geese prefer the artificial "mother" to anything else, including their true mother. The goslings are said to be imprinted on the artificial mother.

Imprinting was observed at least as early as the first century A.D. when the Roman naturalist Pliny the Elder wrote of "a goose

Figure 36-9
Ducklings imprinted on Konrad Lorenz follow him as faithfully as they would a natural mother.

which followed Lacydes as faithfully as a dog." Konrad Lorenz was the first to study imprinting systematically. When Lorenz hand-reared goslings, they formed an immediate and permanent attachment to him and waddled or swam after him wherever he went (see Figure 36-9). They could no longer be induced to follow their own mother or another human being. Lorenz found that the imprinting period is confined to a brief sensitive period in the individual's early life and that once established the imprinted bond usually is retained for life.

What imprinting shows is that the brain of a goose (or the brain of numerous other birds and mammals that show imprinting-like behavior) accommodates the imprinting experience. Natural selection favors evolution of a brain that imprints in this way, in which following the mother and obeying her commands are important for survival. The fact that a gosling can be made to imprint to a mechanical toy duck or a person under artificial conditions is a cost to the system that can be tolerated because goslings seldom encounter these stimuli in their natural environment. The disadvantages of the system's simplicity are outweighed by the advantages of its reliability.

One final example completes our consideration of learning. Males of many species of birds have characteristic territorial songs that identify the singers to other birds and announce territorial rights to other males of that species. Like many other songbirds, a male white-crowned sparrow must learn the song of its species by hearing the song of its father. If a sparrow is hand-reared in acoustic isolation in a laboratory, it develops an abnormal song (Figure 36-10). But if an isolated bird is allowed to hear recordings of normal white-crowned sparrow songs during a critical period of 10 to 50 days after hatching, it learns to sing normally. It even imitates the local dialect it hears.

It might appear from this result that characteristics of the song are determined by learning alone. However, if during the critical learning period, an isolated male white-crowned sparrow is played a recording of another species of sparrow, even a closely related one, it does not learn the song. It learns only the song appropriate to its own species. Thus although the song must be learned, the brain is constrained to recognize and to learn vocalizations produced by males of its species alone. Learning the wrong song would result in behavioral chaos, and natural selection favors a system that eliminates such errors.

Figure 36-10
Sound spectrograms of songs of white-crowned sparrows, *Zonotrichia leucophrys.* Top, natural songs of wild bird; *bottom,* abnormal song of isolated bird.

SOCIAL BEHAVIOR

When we think of "social" animals we are apt to think of highly structured honey bee colonies, herds of antelope grazing on the African plains (Figure 36-11), schools of herring, or flocks of starlings. But social behavior of animals *of the same species* living together is by no means limited to such obvious examples in which individuals influence one another.

In the broad sense, any interaction resulting from the response of one animal to another of the same species represents social behavior. Even a pair of rival males fighting for possession of a female displays a social interaction, despite our perceptual bias as people that might encourage us to label it antisocial. Social aggregations are only one kind of social behavior, and indeed not all aggregations of animals are social.

Clouds of moths attracted to a light at night, barnacles attracted to a common float, or trout gathering in the coolest pool of a stream are groupings of animals responding to environmental signals. Social aggregations, on the other hand, depend on signals from the animals themselves. They remain together and do things together by influencing one another.

Not all animals showing sociality are social to the same degree. While all sexually reproducing species must at least cooperate enough to achieve fertilization, among some animals breeding is about the only adult sociality to occur. Alternatively, swans, geese, albatrosses, and beavers, to name just a few, form strong monogamous bonds that last a lifetime. The most persistent social bonds usually form between mothers and their young and, for birds and mammals, these bonds usually terminate at fledging or weaning.

Advantages of Sociality

Living together may be beneficial in many ways. One obvious benefit for social aggregations is defense, both passive and active, from predators. Musk-oxen that form a passive defensive circle when threatened by a wolf pack are much less vulnerable than an individual facing the wolves alone.

As an example of active defense, a breeding colony of gulls, alerted by the alarm calls of a few, attack predators *en masse;* this collective attack is certain to discourage a predator more effectively than individual attacks. Members of a town of prairie dogs, although divided into social units called coteries, cooperate by warning each other with a special bark when danger threatens. Thus every individual in a social organization benefits from the eyes, ears, and noses of all other members of the group. Experimental tests using a wide variety of predators and prey support the notion that the larger the group, the less likely an individual within the group is to be eaten.

Sociality offers several potential benefits to reproduction. It facilitates encounters between males and females, which, for solitary animals, may consume much time and energy. Sociality also helps synchronize reproductive behavior through the mutual stimulation that individuals give one another. Among colonial birds the sounds and displays of courting individuals set in motion reproductive endocrine changes in other individuals. Because there is more social stimulation, large colonies of gulls produce more young per nest than do small colonies. Furthermore, parental care that social animals provide their offspring increases survival of the brood (Figure 36-12). Social living provides opportunities for individuals to give aid and to share food

Figure 36-11
Mixed herd of topi and common zebra grazing on a savanna of tropical Africa.

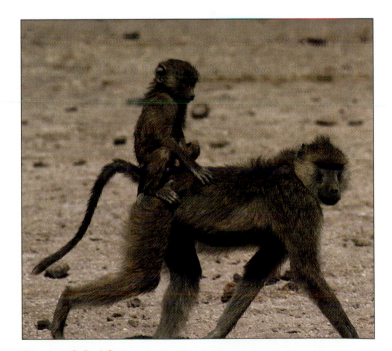

Figure 36-12
An infant yellow baboon (*Papio cyanocephalus*) "jockey rides" its mother. Later, as the infant is weaned, the mother-infant bond weakens and the infant will be refused rides.

with young other than their own. Such interactions within a social network have produced some intricate cooperative behavior among parents, their young, and their kin.

Of the many other advantages of social organization noted by ethologists, we mention only a few in this brief treatment: cooperation in hunting for food; huddling for mutual protection from severe weather; opportunities for division of labor, which is especially well developed in social insects with their caste systems; and the potential for learning and transmitting useful information through the society.

Observers of a seminatural colony of macaque monkeys in Japan recount an interesting example of acquiring and passing tradition in a society. The macaques were provisioned with sweet potatoes and wheat at a feeding station on the beach of an island colony. One day a young female named Imo was observed washing the sand off a sweet potato in seawater. The behavior was quickly imitated by Imo's playmates and later by Imo's mother. Still later when the young members of the troop became mothers they waded into the sea to wash their potatoes; their offspring imitated them without hesitation. The tradition was firmly established in the troop (Figure 36-13).

Some years later, Imo, an adult, discovered that she could separate wheat from sand by tossing a handful of sandy wheat in the water; allowing the sand to sink, she could scoop up the floating wheat to eat. Again, within a few years, wheat-sifting became a tradition in the troop.

Imo's peers and social inferiors copied her innovations most readily. The adult males, her superiors in the social hierarchy, would not adopt the practice but continued laboriously to pick wet sand grains off their sweet potatoes and scout the beach for single grains of wheat.

Social living also may have disadvantages as compared with a solitary existence for some animals. Species that survive by camouflage from potential predators profit by being dispersed. Large predators benefit from a solitary existence for a

Figure 36-13
Japanese macaque washing sweet potatoes. The tradition began when a young female named Imo began washing sand from the potatoes before eating them. Younger members of the troop quickly imitated the behavior.

different reason, their requirement for a large supply of prey. Thus there is no overriding adaptive advantage to sociality that inevitably disfavors a solitary way of life. It depends on the ecological situation.

Aggression and Dominance

Sociality requires cooperation. At the same time animals, like governments, tend to protect their own interests. They are in competition with each other because of limitations in common resources that all require for life. Animals may compete for food, water, sexual mates, or shelter when such requirements are limited in quantity and are therefore worth a fight.

Much of what animals do to resolve competition is called **aggression,** which we define as an offensive physical action, or threat, to force others to abandon something they own or might attain. Many ethologists consider aggression part of a somewhat more inclusive category called **agonistic** (Gr. contest) behavior, referring to any activity related to fighting, whether it be aggression, defense, submission, or retreat.

Contrary to the widely held notion that aggressive behavior aims to destroy or to defeat an opponent, most aggressive encounters lack the violence that we usually associate with fighting. Many species possess specialized weapons such as sharpened teeth, beaks, claws, or horns that are used for protection from, or predation on, other species. Although potentially dangerous, such weapons are seldom used in any severely damaging way against members of *their own species.*

Animal aggression within the species seldom results in injury or death because animals have evolved many symbolic **ritualized displays** that carry mutually understood meanings. A ritualized display is a behavior that has been modified through evolution to make it increasingly effective in serving a communicative function. Through **ritualization,** simple movements or traits become more intensive, conspicuous, or precise, and acquire the function of a signal. The result of such intensification is to reduce the possibility of misunderstanding. Fights over mates, food, or territory become ritualized jousts rather than bloody, no-holds-barred battles. When fiddler crabs spar for territory, their large claws usually are only slightly opened. Even in intense fighting when the claws are used, the crabs grasp each other in a way that prevents reciprocal injury. Rival male poisonous snakes engage in stylized bouts by winding themselves together; each attempts to butt the other's head with its own until one becomes so fatigued that it retreats. The rivals never bite each other. Many species of fish contest territorial boundaries with lateral displays, the males puffing themselves to look as threatening as possible. The encounter is usually settled when either animal perceives itself obviously inferior, withdraws, and swims away. Rival giraffes engage in largely symbolic "necking" matches in which two males standing side by side wrap and unwrap their necks around each other (Figure 36-14). Neither uses its potentially lethal hooves on the other, and neither is injured.

Thus animals fight as though programmed by rules that prevent serious injury. Fights between rival bighorn rams are spectacular to watch, and the sound of clashing horns may be heard for hundreds of meters (Figure 36-15), but the skull is so well protected by massive horns that injury occurs only by accident. Nevertheless, despite these constraints, aggressive encounters on occasion can be true fights to the death of a rival. If African male elephants are unable to resolve dominance conflicts painlessly with ritual postures, they may resort to incredibly violent battles, with each trying to plunge its tusks into the most vulnerable parts of the opponent's body.

More commonly, however, the loser of a ritualized encounter may simply run away, or signal defeat by a specialized subordination ritual. If it becomes evident to him that he is going to lose, he profits by communicating his submission as quickly as possible, thereby avoiding the cost of a thrashing. Such submissive displays that signal the end of a fight may be almost the opposite of threat displays (Figure 36-16). In his book, *The Expression of the Emotions in Man and Animals* (1872), Charles Darwin described the seemingly opposite nature of threat and appeasement displays as the "principle of antithesis." The principle remains accepted by ethologists today.

The winner of an aggressive competition is dominant to the loser, the subordinate. For the victor, dominance means enhanced access to all contested resources that contribute to reproductive success: food, mates, and territory. In a social species, dominance interactions often form a dominance hierarchy. One animal at the top wins encounters with all other members in the social group; the second in rank wins all but those with the top-ranking individual.

Such a simple, ordered hierarchy was first observed in chickens by Schjelderup-Ebbe, who called the hierarchy a "peck-order." Once social ranking is established, actual pecking diminishes and is replaced by threats, bluffs, and bows. Top hens and roosters get unquestioned access to feed and water, dusting areas, and the roost. The system works because it reduces social tensions that would constantly surface if animals had to fight constantly over social position.

However, not all dominance hierarchies have clear-cut dominant and subordinate individuals. In some hierarchies, dominant animals are frequently challenged by subordinates.

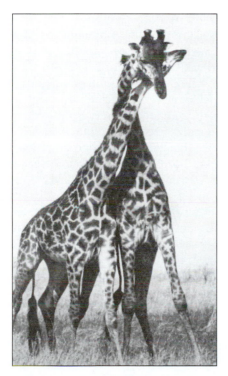

Figure 36-14
Male Masai giraffes. *Giraffa camelopardalis,* fight for social dominance. Such fights are largely symbolic, seldom resulting in injury.

Figure 36-16
Darwin's principle of antithesis as exemplified by the postures of dogs. **A,** A dog approaches another dog with hostile, aggressive intentions. **B,** The same dog is in a humble and conciliatory state of mind. The signals of aggressive display have been reversed.

Figure 36-15
Male bighorn sheep, *Ovis canadensis,* fight for social dominance during the breeding season.

Subordinates in any social order may be expendable. In many systems they never get a chance to reproduce, and are often the first to die. During times of food scarcity, death of weaker members protects resources for stronger members. Rather than sharing food, the excess population is sacrificed. This sacrifice is not viewed by contemporary ethologists as resulting from some purposeful process ensuring "good for the species"; it is a consequence of the individual advantage that the stronger, dominant individuals possess during such circumstances.

Territoriality

Territorial ownership is another facet of sociality in animal populations. A territory is a fixed area from which intruders *of the same species* are excluded. This exclusion involves defending the area from intruders and spending long periods of time being conspicuous on the site. Territorial defense has been observed in numerous animals: insects, crustaceans, fishes, amphibians, lizards, birds, and mammals, including humans.

> Sometimes the space defended moves with the individual. Individual distance, as it is called, can be observed in the spacing between swallows or pigeons on a wire, gulls on a beach, or people waiting for a bus.

Territoriality is generally an alternative to dominance behavior, although both systems may be observed operating in the same species. A territorial system may work well when the population is low, but it may fail with increasing population density and be replaced with dominance hierarchies in which all animals occupy the same space.

Like every other competitive endeavor, territoriality carries both costs and advantages. It is beneficial when it ensures access to limited resources, *unless* the territorial boundaries cannot be maintained with little effort. The presumed benefits of a territory are, in fact, numerous: uncontested access to a foraging area; enhanced attractiveness to females thus reducing problems of pair-bonding, mating, and rearing the young; reduced disease transmission; reduced vulnerability to predators. Advantages of holding a territory wane if an individual must spend most of its time in boundary disputes with neighbors.

Most of the time and energy required for territoriality are expended when the territory is first established. Once boundaries are located they tend to be respected, and aggressive behavior diminishes as territorial neighbors come to recognize each other. Indeed, neighbors may look so peaceful that an observer not present when the territories were established may conclude incorrectly that the animals are not territorial. A "beachmaster" sea lion (a dominant male with many females) seldom quarrels with neighbors, who have their own territories to defend. However, he must be constantly vigilant against bachelor bulls who challenge his mating privileges.

Birds are conspicuously territorial. Most male songbirds establish territories in early spring and defend these vigorously against males of the same species during spring and summer when mating and nesting are at their height. A male song sparrow, for example, has a territory of approximately three-fourths of an acre. In any given area, the number of song sparrows remains approximately the same each year. The population remains stable because the young occupy territories of adults that die. Any surplus in the song sparrow population is excluded from territories and thus not able to mate or to nest.

Sea birds such as gulls, gannets, boobies, and albatrosses occupy colonies that are divided into very small territories just large enough for nesting (Figure 36-17). Territories of these birds cannot include their fishing grounds, since they all forage in the sea where their food is always shifting in location and shared by all.

Territorial behavior is not as prominent with mammals as it is with birds. Mammals are less mobile than birds, making it more difficult for them to patrol a territory for trespassers. Instead, many mammals have **home ranges** (p. 599). A home range is the total area an individual traverses in its activities. It is not an exclusive, defended preserve but overlaps with home ranges of other individuals of the same species.

For example, home ranges of baboon troops overlap extensively, although a small part of each range becomes the recognized territory of each troop for its exclusive use. Home ranges may shift considerably with the seasons. A baboon troop may have to shift to a new range during the dry season to obtain water and better grass. Elephants, before their movements were restricted by humans, made long seasonal migrations across the African savanna to new feeding ranges. However, home ranges established for each season are remarkably consistent in size.

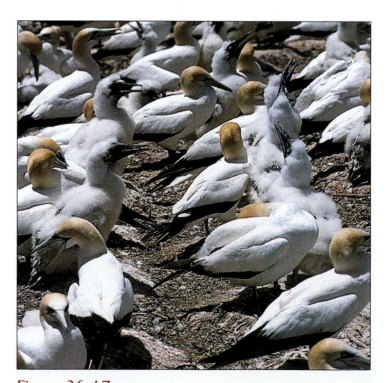

Figure 36-17

Gannet nesting colony. Note precise spacing of nests with each occupant just beyond pecking distance of its neighbors.

Mating Systems

Animals display diverse mating systems. Behavioral ecologists generally classify mating systems by the degree to which males and females associate during mating. **Monogamy** is an association between one male and one female at a time. **Polygamy** is a general term that incorporates all multiple mating systems where females and males have more than one mate. **Polygyny** refers to a male that mates with more than one female. **Polyandry** is a system in which a female mates with more than one male. There are specific types of polygyny. **Resource-defense polygyny** occurs when males gain access to females indirectly by holding critical resources. For example, female bullfrogs prefer to mate with males who are larger and older. These males defend territories of higher quality than smaller males because their territories have better temperature regimes for tadpoles to grow or because they are free of predatory leeches. **Female-defense polygyny** occurs when females aggregate and, consequently, are defendable. Thus, when female elephant seals occupy a small island, dominant males can defend and gain access to them for mating relatively easily (Figure 36-18). This situation was previously known as a "harem." **Male-dominance polygyny** occurs when females select mates from aggregations of males. For example, some animals form **leks**. A lek is a communal display ground where males congregate to attract and court females. Females choose and mate with the male having the most attractive qualities (Figure 36-19). Leks characterize some birds, including prairie chickens and sage grouse. In these systems, sexual selection (p. 123) is often intense, resulting in evolution of bizarre courtship rituals and exaggerated morphological traits.

Altruistic Behavior and Kin Selection

If, as Darwin suggested, animals should behave selfishly and strive to produce as many offspring as possible, why do some animals help others at some risk to themselves? Why do some

Figure 36-19
Male sage grouse, *Centrocercus urophasianus,* displaying at its lek.

individuals show strong cooperation with members of their social group and even forego breeding themselves? Why do some individuals appear to sacrifice themselves so that other members of their group can survive? Until the mid 1960s, scientists had trouble explaining in Darwinian terms how such **altruistic behavior** might persist in a population.

Most instances of altruistic behaviors were explained using a **group-selection** argument. Group selectionists suggested that animals that helped others or that failed to mate did so for the benefit of the group as a whole. Therefore, such behaviors produce increased survivorship of groups whose members behaved altruistically. According to proponents of this argument, selection occurs at the level of the group, not at the level of the individual as Darwin suggested. However, the group-selection argument as originally proposed by V. C. Wynne-Edwards in 1962 has been rejected by the vast majority of behavioral ecologists for several reasons.

For example, if alleles associated with a risky altruistic behavior, such as giving calls to warn others of predators, were distributed randomly in a social group, those lacking such genes would flourish. They would be warned with no risk to themselves; their chances of reproduction would be greater and, in time, the "selfish" alleles would eliminate the altruistic ones from the group's gene pool.

In 1964, W. D. Hamilton, based largely on his studies of insects, proposed a new way to explain altruistic behavior by modifying Darwin's original concept of fitness. He reasoned that fitness is measured not just by the number of offspring produced but by the increase or decrease in particular alleles in the gene pool. Thus, an individual may act altruistically, even at great risk, if it helps increase representation of its alleles in the gene pool. Alleles are shared by all relatives, including parents and offspring, brothers and sisters, cousins and other relations. Alleles

Figure 36-18

Two elephant seals, *Mirounga angustirostris,* fight to establish dominance. Males are much larger than females in this highly polygynous society.

that influence altruistic behavior among relatives would persist in future generations. Because the most closely related animals share the most genes by common descent, we expect that altruistic behavior would be most common among closely related individuals. Thus, if everything else were equal, brothers who on average share half their alleles would be more likely to aid one another than they would a cousin who shares on average only 25% of their alleles. Hamilton's hypothesis based on this genetic explanation for altruism and cooperation is called **kin selection.** Essentially, kin selection is the selection of genes by individuals assisting the survival and reproduction of individuals who possess the same genes by common descent.

Hamilton's hypothesis revolutionized evolutionary and behavioral biology. The main criterion of Darwinian fitness is the relative number of an individual's alleles that are passed to future generations. Hamilton, however, developed the concept of **inclusive fitness,** which is the relative number of an individual's alleles that are passed to future generations either as a result of an individual's own reproductive success *or that of related individuals*. Thus, kin selection and inclusive fitness may be able to explain many altruistic behaviors that have perplexed biologists for many years.

A good example of altruism and kin selection in nature is the remarkable cooperation and coordination among euscocial insects such as ants, bees, and wasps. Through haplodiploidy (p. 135), where males are haploid and females are diploid, sisters are genetically related on average by 75% rather than 50% (Figure 36-20). Sisters are more closely related to each other than to their own daughters! Therefore, they cooperate with other members of their social group, forego breeding themselves and aid their queen to produce more sisters who are more closely related (75% related) than potential offspring (50% related).

Female Belding's ground squirrels, found in the High Sierra of California, give alarm calls when a predator approaches (Figure 36-21). Alarm calling warns other members of the social group and is risky to the alarm caller. However, benefits to alarm calling outweigh the risks because alarm callers are warning *related* individuals. Thus, alarm-calling behavior, even if it is risky, may be favored by selection if it increases inclusive fitness of the caller.

Kinship theory suggests that animals may evolve an ability to recognize categories of relatives so that cooperation or aid-giving behavior is directed more efficiently toward relatives. Although kin recognition behavior was discussed by Hamilton, little was known about it until almost 20 years after he wrote his influential papers. Through a number of experimental studies we now know that animals of many species can discriminate kin and non-kin. Such species are found among invertebrates, including isopods and insects, and vertebrates, including fishes, frog and toad tadpoles, birds, squirrels, and monkeys. Individuals of some species can even discriminate between full siblings and half-siblings and between cousins and unrelated individuals. Thus, some species have a finely tuned ability to identify relatives of various degrees of relatedness. Cues used in kin recognition vary among species. Birds often use vocalizations, whereas chemical cues are commonly used in other groups.

Because altruistic behavior occurs among unrelated individuals in many natural populations, kin-selection theory does not

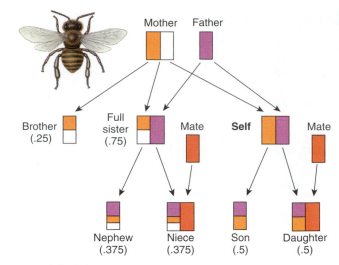

Figure 36-20

Haplodiploidy in honey bees, showing degrees of relatedness of a female worker bee (labeled SELF) to individuals she might raise. In honey bees, as in other haplodiploid animals, diploid females develop from fertilized eggs, and males develop from unfertilized eggs. Each daughter of a male gets all his genes (purple bar) and full sisters receive an identical one half of their genome from the same father. Open bars represent other, unrelated alleles. Because full sisters also share half the genes they receive from their common mother (yellow bar), the relatedness of SELF to a full sister is 0.75, the average of 0.5 and 1.0. (In a diploid-diploid system as in humans, the relatedness of siblings is 0.5 because both paternally and maternally inherited genes have a 50% chance of being present in a sibling.) Note that relatedness of female workers to a brother is only 0.25, because brothers are fatherless.

Figure 36-21

Belding's ground squirrel, *Spermophilus beldingi*, gives an alarm call to warn of the approach of a predator. This risky behavior endangers the callers more than noncallers.

explain all altruistic behavior. The theory of **reciprocal altruism,** formulated initially by Robert Trivers, provides an additional Darwinian foundation for explaining altruistic behaviors among individuals, including ones who are not close relatives. According to this theory, an individual may be selected to perform altruistic acts if doing so increases that individual's chances of receiving equally or more valuable favors from others. Reciprocal altruism is most likely to evolve in species that have stable social groupings whose members are mutually interdependent for defense, nutrition or reproduction, with frequent opportunities for altruistic interaction.

Gerard Wilkinson's studies of food sharing by vampire bats show that individuals do reciprocate altruistic behavior. Vampire bats aggregate in resting areas and leave those areas at night to obtain blood meals from large mammals, which are often difficult to find. A vampire bat that obtains a blood meal may perform the altruistic behavior of regurgitating blood to other hungry members of its group. Wilkinson used laboratory experiments to show that hungry bats were not fed at random, but that individual bats were more likely to be fed by those to whom they had previously provided the same service. These results confirm that vampire bats recognize each other as individuals, remember which individuals have performed altruistic behaviors, and then return favors. Despite such encouraging results, reciprocal altruism is difficult to study in nature because it generally requires long-term observations of marked individuals and may occur simultaneously with kin selection.

Darwinian selective explanations for social behavior have been investigated extensively using the mathematical theory of games to determine which behaviors may qualify as an **evolutionary stable strategy** or **ESS.** An ESS is expected to persist over long periods of evolutionary time because it prevails in competition with alternative strategies that might arise. An altruistic behavior would not be an ESS if it were subject to cheaters who could preferentially obtain altruistic behaviors from unrelated others without giving in return. Because altruistic behavior is hard to study in nature, results of ESS theory help to focus investigations on populations and behaviors that are most likely to show evolutionary stability. Ritualized displays of animal aggression (p. 760) are considered strong examples of evolutionary stable strategies because they prevent individuals from escalating conflicts to the point of serious injury and thereby enhance survivorship. Selection should favor ritualized displays over alternative behaviors that would subject an individual to violence. An important research topic is to determine how truthful are the displays used by animals to avoid violent conflict or to procure mates. Would it be an evolutionary stable strategy for organisms to evolve displays that deceptively overestimate their actual strength or desirability for mating?

Animal Communication

Only through communication can one animal influence the behavior of another. Compared with the enormous communicative potential of human speech, however, nonhuman communication is severely restricted. Animals may communicate by sounds, scents, touch, and movement. Indeed any sensory channel may be used, and in this sense animal communication has richness and variety.

Unlike our language, which is composed of words with definite meanings that may be rearranged to generate an almost infinite array of new meanings and images, communication of other animals consists of a limited repertoire of signals. Typically, each signal conveys one and only one message. These messages cannot be divided or rearranged to construct *new kinds* of information. A single message from a sender may, however, contain several bits of relevant information for a receiver.

A cricket's song announces to an unfertilized female the species of the sender (males of different species have different songs), his sex (only males sing), his location (source of the song), and social status (only a male able to defend the area around his burrow sings from one location). This information is crucial to a female and accomplishes a biological function. However, there is no way for a male to alter his song to provide additional information concerning food, predators, or habitat, which might improve his mate's chances of survival and thus enhance his own fitness.

Phylogenetic studies have been important for testing hypotheses of the evolution of mating-behavioral and morphological characters by sexual selection. Phylogenetic and behavioral studies of swordfish species by Alexandra Basolo show that female preference for the male sword evolved prior to evolution of the sword, consistent with the hypothesis that initial formation and elongation of the sword in males occurred through sexual selection by female preference. Another important phylogenetic study of mating behavior examines evolution of male displays in leks of neotropical bowerbirds. Richard Prum identified 44 behavioral characters that have been used to evolve species-specific behavioral displays in these birds, and he reconstructed the historical sequence by which these displays evolved. His results show a general evolutionary trend toward increased complexity of displays and a tendency for behavioral changes to precede changes in plumage. A new behavioral display that highlights a particular area of plumage subjects that plumage to sexual selection for morphological elaboration. These studies suggest that behavioral evolution may be a major factor in determining the action of selection on morphological characters. Some evolutionists have proposed that behavioral evolution generally accelerates morphological evolution, and that a change in behavior is often a critical factor permitting evolution of new adaptive zones.

Chemical Sex Attraction in Moths

Mate attraction in silkworm moths illustrates an extreme case of stereotyped, single-message communication that has evolved to serve a single biological function: mating. Virgin female silkworm moths have special glands that produce a chemical sex attractant to which males are sensitive. Adult males smell with their large bushy antennae, covered with thousands of sensory

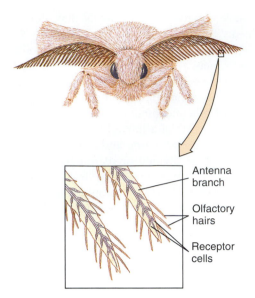

Figure 36-22

Large antennae of a male silkworm moth *Bombyx mori;* these are especially sensitive to the sex attractant (pheromone) released by the female moth.

hairs that function as receptors (Figure 36-22). Most of these receptors are sensitive to the chemical attractant **bombykol** (a complex alcohol named after the silkworm *Bombyx mori*) and to nothing else.

To attract males, females merely sit quietly and emit a minute amount of bombykol, which is carried downwind. When a few molecules reach a male's antennae, he is stimulated to fly upwind in search of the female. His search is at first random, but, when by chance he approaches within a few hundred yards of the female, he encounters a concentration gradient of the attractant. Guided by the gradient, he flies toward the female, finds her, and copulates with her.

In this example of chemical communication the attractant bombykol, a *pheromone* (p. 709), serves as a signal to bring the sexes together. Its effectiveness is ensured because natural selection favors evolution of males with antennal receptors sensitive enough to detect the attractant at great distances (several miles). Males with a genotype that produces a less sensitive sensory system fail to locate a female and thus are reproductively eliminated from the population.

Language of Honey Bees

One of the most sophisticated and complex of all nonhuman communication systems is the symbolic language of bees. Honey bees are able to communicate the location of food resources when these sources are too distant to be located easily by individual bees. They communicate by dances, which are mainly of two forms. The form having the most informational richness is the **waggle dance** (Figure 36-23). Bees most commonly execute these dances when a forager has returned from a rich source, carrying either nectar in her stomach or pollen grains packed in basketlike spaces formed by hairs on her legs. The waggle dance is roughly in the pattern of a figure-eight made against the vertical surface on the comb inside the hive. One cycle of the dance consists of three components: (1) a circle with a diameter about

Figure 36-23

Waggle dance of honey bees used to communicate both the direction and distance of a food source. The straight run of the waggle dance indicates direction according to the position of the sun (angles X and Y).

three times the length of the bee, (2) a straight run while waggling the abdomen from side to side and emitting a pulsed, low-frequency sound, and (3) another circle, turning in the opposite direction from the first. This dance is repeated many times with the circling alternating clockwise and counterclockwise.

The significance of bee dances was discovered in 1943 by German zoologist Karl von Frisch, one of the recipients of the 1973 Nobel Prize. Despite detailed and extensive experiments by von Frisch and others that supported his original interpretations of the honey-bee dances, the experiments have been criticized, especially by American biologist Adrian Wenner, who suggests that the correlation between dance symbolism and food location is accidental. He argued that foraging bees bring back odors characteristic of the food source, and that recruits are stimulated by dance to search for flowers bearing those odors. Wenner, with P. H. Wells, has reviewed his studies as a scientific autobiography and polemic (*Anatomy of a Controversy: The Question of a Language Among Bees,* 1990, Columbia University Press). Wenner and Wells' assertions have generated strong controversy and have stimulated more rigorously controlled research on the bee dances. Recently, researchers have constructed a robot bee that can be moved through the waggle dance while producing the dance song with vibrating metal "wings." When operated in a hive, the computer-directed robot successfully recruited attending bees to visit preselected food dishes outside the hive that had never been visited previously. These experiments provide convincing evidence that bee dances do communicate both direction and distance information to foraging bees.

The straight, waggle run is the important informational component of the dance. Waggle dances are performed almost always in clear weather, and the direction of the straight run is related to the position of the sun. If the forager has located food directly toward the sun, she will make her waggle run straight upward over the vertical surface of the comb. If food was located 60 degrees to the right of the sun, her waggle run is 60 degrees to the right of vertical. We see then that the waggle run points at the same angle relative to the vertical as the food is located relative to the sun.

Distance of the food source is also coded into bee dances. If the food is close to the hive (less than 50 m), the forager employs a simpler dance called the **round dance.** The forager simply turns a complete clockwise circle, then turns, and completes a counterclockwise circle, a performance that is repeated many times. Other workers cluster around the scout and become stimulated by the dance as well as by the odor of nectar and pollen grains from flowers she has visited. The recruits then fly out and search in all directions but do not stray far. The round dance carries the message that food is to be found in the vicinity of the hive.

If the food source is farther away, round dances become waggle dances, which provide both distance and directional information. The tempo of the waggle dance is related inversely to the food's distance. If the food is about 100 m away, each figure-eight cycle lasts about 1.25 seconds; if 1000 m away, it lasts about 3 seconds; and if about 8 km (5 miles) away, it lasts 8 seconds. When food is plentiful, the bees may not dance at all. But when food is scarce, the dancing becomes intense, and other workers cluster around returning scouts and follow them through the dance patterns.

Communication by Displays

A display is a kind of behavior or series of behaviors that serves a communicative purpose. The release of sex attractant by a female moth and the dances of bees just described are examples of displays; so are alarm calls of herring gulls, songs of the white-crowned sparrow, courtship dances of the sage grouse, and "eyespots" on the hind wings of certain moths that are exposed quickly to startle potential predators.

The elaborate pair-bonding displays of blue-footed boobies (Figure 36-24) are performed with maximum intensity when the birds come together after a period of separation. The male at right in the illustration is sky pointing: the head and tail are

Figure 36-24

A pair of Galápagos blue-footed boobies, *Sula nebouxii,* display to each other. A male (*right*) is sky pointing; a female (*left*) is parading. Such vivid, stereotyped, communicative displays serve to maintain reciprocal stimulation and cooperative behavior during courtship, mating, nesting, and care of the young.

pointed skyward and the wings are swiveled forward in a seemingly impossible position to display their glossy upper surfaces to the female. This display is accompanied by a high-piping whistle. The female at left, for her part, is parading. She steps with exaggerated slow deliberation, lifting each brilliant blue foot in turn, as though holding it aloft momentarily for the male to admire. Such highly personalized displays, performed with droll solemnity, appear comical, even inane to an observer. Indeed the boobies, whose name is derived from the Spanish word "bobo" meaning clown, presumably were so designated for their amusing antics.

The exaggerated nature of the displays ensures that the message is not missed or misunderstood. Such displays are essential to establish and maintain a strong pair bond between male and female. This requirement also explains the repetitious nature of displays that follow one another throughout courtship and until laying of eggs. Repetitious displays maintain a state of mutual stimulation between male and female, ensuring the degree of cooperation necessary for copulation and subsequent incubation and care of young. A sexually aroused male has little success with an indifferent female.

Communication between Humans and Other Animals

One uncertainty in studies of animal communication is understanding what sensory channel an animal is using. Signals may be visual displays, odors, vocalizations, tactile vibrations, or electrical currents (as, for example, among certain fishes). Even more difficult is establishing two-way communication between humans and other animals since an investigator must translate meanings into symbols that the animal can understand. Furthermore, people are poor social partners for most other animals.

Animal Cognition

One of the most fascinating subjects in animal behavior deals with animal intelligence and awareness. Animal cognition is a general term for mental function, including perception, thinking, and memory. Many biologists believe that mental processes of animals may be similar to those of humans. Recent studies on animal cognition with nonhuman primates and African Grey parrots have yielded fascinating results.

In the late 1960s Beatrix and Allen Gardner of the University of Nevada in Reno began using American Sign Language (ASL) to train a chimpanzee named Washoe to communicate with her hands the same way that deaf people do. By age five Washoe could sign 132 words, which she could put into strings forming sentences and phrases. She could answer questions, make suggestions, and convey moods. In one session, when asked what a swan was, Washoe answered "water bird." Washoe also taught signs to other chimpanzees. At first, signs were used as play but soon the chimpanzees used them to make spontaneous requests to trainers such as "drink," "tickle," and "hug." Similar work has been done with other primates including gorillas, orangutans, and pygmy chimpanzees.

Irene Pepperberg of the University of Arizona has worked for years with an African Grey parrot named Alex. Because parrots can vocalize like humans, Pepperberg was able to communicate with Alex using human vocal language. Over the years Alex learned a number of attributes including colors, shapes, and materials for more than 100 objects. Alex not only can identify objects by colors and shape; he can also distinguish the difference between two objects. Thus, if Alex is given two objects of the same color but one larger than the other, he could state that the difference between them was "size." Alex can also count and relate to the trainer how many objects of each particular category are present.

Conscious awareness is also part of cognition. Donald Griffin wrote two books suggesting that many animals are capable of self-awareness and can think and reason. The ability of apes, parrots, and other animals to use language-related skills is significant because it tells us about their cognitive abilities and we can begin to communicate with them. The possibility that animals may have thinking processes similar to humans and that they have a conscious awareness has shed new light on animal behavior studies and added new significance to our studies of animals in general. Studies of animal cognition remain highly controversial.

The animal behaviorist Irven DeVore reported how choosing the proper channel for dialogue can have more than academic interest:*

One day on the savanna I was away from my truck watching a baboon troop when a young juvenile came and picked up my binoculars. I knew if the glasses disappeared into the troop they'd be lost, so I grabbed them back. The juvenile screamed. Immediately every adult male in the troop rushed at me—I realized what a cornered leopard must feel like. The truck was 30 or 40 feet away. I had to face the males. I started smacking my lips very loudly, a gesture that says as strongly as a baboon can, "I mean you no harm." The males came charging up, growling, snarling, showing their teeth. Right in front of me they halted, cocked their heads to one side—and started lip-smacking back to me. They lip-smacked. I lip-smacked, "I mean you no harm." "I mean *you* no harm." It was, in retrospect, a marvelous conversation. But while my lips talked baboon, my feet edged toward the truck until I could leap inside and close the door.

The study of animal communication has made great strides in recent years, buoyed by the assimilation of a wealth of facts and information about communication in many species. The animal world is filled with communication. In recognizing that reasoning and insight are not required for effective, highly organized behavior, we should not conclude that other animals are, as Descartes proclaimed in the seventeenth century, nothing more than machines.

*DeVore, Irven. The marvels of animal behavior. 1972. Washington, D.C., National Geographic Society.

SUMMARY

Animal behavior has emerged as a scientific discipline from four different approaches. Comparative psychology emphasizes the identification of mechanisms controlling behavior, using relatively few species, with the intent that these mechanisms might have wide applicability among animals. Ethology is the study of behavior, both innate and learned, of animals in their natural habitats. Behavioral ecologists have shown that behavioral traits have evolutionary histories and may evolve by natural selection. Sociobiology aims to understand how and why social behavior in animals has evolved. Both ethology and sociobiology distinguish between studies that focus on the mechanisms of behavior (proximate causation) and those that focus on function or evolution of behavior (ultimate causation).

Students of animal behavior have observed and cataloged many behavioral patterns of animals that are highly predictable and almost invariable in performance. Often these patterns are triggered, or "released," by specific, and usually simple, environmental stimuli, called sign stimuli. Although such formalized behaviors may be released inappropriately at times, they are efficient and enable the animal to respond rapidly. Development of behavioral patterns depends on an interaction between an organism and the environment in which it lives. For this reason, behavioral scientists prefer not to describe behaviors—those that are largely invariable in their performance—as "instinctive" or "innate."

Behavior may be modified by learning through experience. Two simple kinds of learning behavior are habituation, which is reduction or elimination of a behavioral response in the absence of any reward or punishment; and sensitization, in which a repeated stimulus increases the strength of a behavioral response. The gill-withdrawal reflex of the marine mollusc *Aplysia* is described as a protective response that can be modified experimentally to show either habituation or sensitization. The modification of the alarm response of herring gull chicks is another example of habituation. Another form of learning is imprinting, the lasting recognition bond that forms early in life between the young of many social animals and their mothers.

Social behavior is behavior arising from interactions of members of a species with one another. In social organizations, animals tend to remain together, communicate with each other, and usually resist intrusions by "outsiders." Advantages of sociality include cooperative defense from predators, cooperative searching for food, improved reproductive performance and parental care of young, and transmission of useful information through the society. Because social animals compete with one another for resources (such as food, sexual mates, and shelter), conflicts are often resolved by a form of overt hostility called aggression. Most aggressive encounters between conspecifics are stylized bouts involving more bluff than intent to injure or kill. Dominance hierarchies, in which a priority of access to common resources is established by aggression, are common in social organizations. Territoriality is an alternative to dominance. A territory is a defended area from which intruders of the same species are excluded.

Mating systems include monogamy, the mating of an individual with only one partner of the opposite sex each breeding season, and polygamy, the mating of an individual with two or more partners in a breeding season. Two forms of polygamy are polygyny, the mating of a male with more than one female, and polyandry, the mating of a female with more than one male. Several types of polygyny are recognized.

A behavior in which one animal may reduce its own fitness to increase fitness of others is called altruistic behavior. Examples are risky behaviors such as one member of a social group warning others of a predator, and cooperative behavior among social insects in which an individual may sacrifice itself to benefit the colony. The favored explanation of altruism is kin selection, in which the recipient of an altruistic act is sufficiently closely related to the altruist that the recipient's survival would benefit the genes shared with the altruist.

Communication, often considered the essence of social organization, is the means by which animals influence behavior of other animals, using sounds, scents, visual displays, touch, or other sensory signals. As compared with the richness of human language, animals communicate with a very limited repertoire of signals. One of the most famous examples of animal communication is that of the symbolic dances of honey bees. Birds communicate by calls and songs and, especially, by visual displays. By ritualization, simple movements have evolved to become conspicuous signals having definite meanings.

REVIEW QUESTIONS

1. How do experimental approaches of comparative psychology and ethology differ? Comment on the aims and methods employed by each.
2. Egg-retrieval behavior of greylag geese is an excellent example of a highly predictable behavior. Interpret this behavior within the framework of classical ethology, using these terms: releaser, sign stimulus, and stereotyped behavior. Interpret the territorial defense behavior of male three-spined sticklebacks in the same context.
3. The idea that behavior must be *either* innate or learned has been called a "nature versus nurture" controversy. What reasons are there for believing that such a strict dichotomy does not exist?
4. Two kinds of simple learning are habituation and imprinting. Distinguish between these two types of learning, and offer an example of each.
5. Some strains of bees show hygienic behavior by uncapping cells containing larvae infected with a bacterial disease called foulbrood and removing the dead larvae from the hive. What evidence shows that this behavior is transmitted by two independently segregating genes?
6. Discuss some advantages of sociality for animals. If social living has so many advantages, why do many animals successfully live alone?
7. Suggest why aggression, which might seem counterproductive, exists among social animals.

8. What is the selective advantage to winners, as well as to losers, that aggressive encounters within species for social dominance are usually ritualized displays or symbolic fights rather than unrestrained fights to death?

9. Of what use is a territory to an animal, and how is a territory established and kept? What is the difference between territory and home range?

10. Polygyny is a form of polygamy in which a male mates with more than one female. Explain how three forms of polygyny differ from each other: resource-defense polygyny, female-defense polygyny, and male-dominance polygyny.

11. Give an example of an altruistic behavior and explain how such behavior conflicts with Darwin's expectation that animals will act selfishly to produce as many offspring as possible.

12. Earlier explanations of altruistic behavior as a form of group selection have been supplanted by Hamilton's hypothesis of kin selection. What distinguishes kin selection and how does it accord with the notion of inclusive fitness, the relative number of an individual's alleles that pass to the next generation?

13. Comment on limitations of animal communication as compared to those of human communication.

14. Dance language used by returning forager honey bees to specify location of food is an example of remarkably complex communication among "simple" animals. How is direction and distance information coded into the waggle dance of the bees? What is the purpose of the round dance?

15. What is meant by "ritualization" in display communication? What is the adaptive significance of ritualization?

16. Early efforts by humans to communicate vocally with chimpanzees were almost total failures. Recently, however, researchers have learned how to communicate successfully with apes. How was this task accomplished?

SELECTED REFERENCES

Alcock, J. 2001. Animal behavior: an evolutionary approach, ed. 7. Sunderland, Massachusetts, Sinauer Associates, Inc. *Clearly written and well-illustrated discussion of genetics, physiology, ecology, and history of behavior in an evolutionary perspective.*

Attenborough, D. 1990. The trials of life: a natural history of animal behavior. Boston, Little, Brown and Company. *Superb photographs and flowing text describe the life cycles of organisms, often focusing on unusual and exciting patterns of behavior.*

Basolo, A. L. 1996. The phylogenetic distribution of a female preference. Syst. Biol. **45**:290–307. *A phylogentic analysis of behavioral evolution.*

Bekoff, M., and D. Jamieson (eds.). 1996. Readings in animal cognition. MIT Press, Cambridge, Massachusetts. *Selected readings of papers in animal cognition by authors in the field.*

Bradbury, J. W., and S. L. Vehrencamp. 1998. Principles of animal communication. Sinauer Associates, Sunderland, Massachusetts. *A comprehensive text on animal communication.*

Drickamer, L. C., S. H. Vessey, and D. Meikle. 1996. Animal behavior: mechanisms, ecology and evolution, ed. 3. Dubuque, William C. Brown Publishers. *Comprehensive, with helpful discussions on the methods and experimental approaches used to answer behavioral questions.*

Gould, J. L., and C. G. Gould. 1994. The animal mind. New York, Scientific American Library. *Attractively illustrated, engagingly written exploration of animal behavior and efforts of researchers to determine what happens inside minds of animals.*

Greenspan, R. J. 1995. Understanding the genetic construction of behavior. Sci. Am. **272**:72–78 (Apr.). *Studies of courtship and mating in fruit flies indicate that behavior is regulated by many multipurpose genes, each of which handles diverse responsibilities in the body.*

Houck, L. D., and L. C. Drickamer (eds.). 1996. Foundations of animal behavior. The University of Chicago Press, Chicago. *Classic papers in animal behavior with commentaries.*

Kirchner, W. H., and W. F. Towne. 1994. The sensory basis of the honeybee's dance language. Sci. Am. **270**:74–80 (June). *Experiments with a robotic honey bee that can dance and emit sounds similar to living bees show conclusively that the dance language successfully recruits foragers to food outside the hive.*

Lorenz, K. Z. 1952. King Solomon's ring. New York, Thomas Y. Crowell Company, Inc. *One of the most delightful books ever written about behavior of animals.*

Manning, A., and M. S. Dawkins. 1998. An introduction to animal behaviour, ed. 5. Cambridge, England, Cambridge University Press. *Survey of animal behavior, drawing from ethology, physiology, and comparative psychology.*

Preston-Mafham, R., and K. Preston-Mafham. 1993. The encyclopedia of land invertebrate behavior. Cambridge, Massachusetts, The MIT Press. *Numerous examples of fascinating invertebrate behavior in a series of informative and beautifully illustrated essays. Highly recommended.*

Prum, R. O. 1990. Phylogenetic analysis of the evolution of display behavior in the neotropical manakins (Aves: Pipridae). *Ethology* **84**:202–231. *A phylogenetic analysis of behavioral evolution.*

Queller, D. C., and J. E. Strassmann. 1998. Kin selection and social insects. BioScience **48**(3):165–175. *How kin selection operates in social insects, and why most cases of altruism are found in social insects.*

Ridley, M. 1995. Animal behavior: an introduction to behavioral mechanisms, development, and ecology, ed. 2. Oxford, Blackwell Scientific Publications. *The principles of animal behavior presented with well-chosen examples and clear illustrations.*

Savage-Rumbaugh, S., S. G. Shanker, and T. J. Taylor. 1998. Ape language: from conditioned response to symbol. New York, Oxford University Press. *Details the general area of ape language.*

ZOOLOGY LINKS TO THE INTERNET

Visit the textbook's Online Learning Center at www.mhhe.com/zoology to find live Internet links for each of the topics listed here.

Animal Behavior
Sociobiology

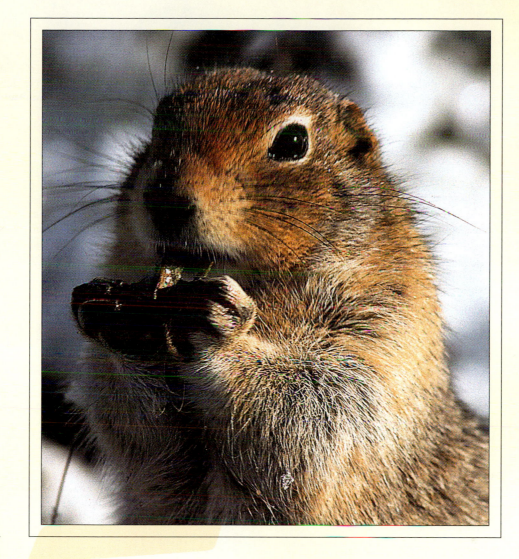

Arctic ground squirrel.

Animals and
Their Environment

37 The Biosphere and Animal Distribution
38 Animal Ecology

Spaceship earth.

The Biosphere and Animal Distribution

Spaceship Earth

All life is confined to a thin veneer of the earth called the **biosphere.** From the first remarkable photographs of earth taken from the *Apollo* spacecraft, revealing a beautiful blue and white globe lying against the limitless backdrop of space, viewers were struck and perhaps humbled by our isolation and insignificance in the enormity of the universe. The phrase "spaceship earth" became a part of our vocabulary as we realized that all the resources we will ever have for sustaining life are restricted to a thin layer of land and sea and a narrow veil of atmosphere above it. We could better appreciate just how thin the biosphere is if we could shrink the earth and all of its dimensions to a 1 m sphere. We would no longer perceive vertical dimensions on the earth's surface. The highest

mountains would fail to penetrate a thin coat of paint applied to our shrunken earth; a fingernail's scratch on the surface would exceed the depth of the ocean's deepest trenches.

Earth's biosphere and the organisms in it have evolved together. In the continuous interchange between organism and environment, both have been altered, and a favorable relationship preserved. Earth's biosphere, with its living and nonliving components, is not static but has undergone an evolution in every way as dramatic as the evolution of the animal kingdom. Today human activities are producing one of the greatest biotic disturbances the earth has ever known. Only the historical bombardment of the earth by asteroids has produced a greater disturbance of the earth's biota.

In a universe of billions of stars, our earth is a small planet circling an ordinary star. Thousands of other stars are like our sun with planetary systems that conceivably could support life. Yet, of all these, our planet is the only one that we *know* supports life. Until proven otherwise, the earth is unique, a true wonder of an infinite universe.

What makes earth an especially favorable environment for life? Most biologists would agree that foremost is the presence of liquid water on the earth's surface. Water, with its many extraordinary physical properties (p. 28) provided a medium for the origin of life and bestowed on earth a moderate climate suitable for life's continued evolution. Many other properties of earth make it optimal for life. Among these are a steady supply of light and heat from an unfailing sun; a suitable range of temperature for life, neither too hot nor too cold; a supply of the major and minor elements required by living matter; and a gravity force strong enough to hold an extensive gaseous atmosphere.

The many properties that make the earth wonderfully suitable for life were first recognized and examined in detail by Lawrence J. Henderson (1878 to 1942) in his book *The Fitness of the Environment*, published in 1913. The profound insights of this distinguished Harvard biochemist and physiologist were remarkable, appearing as they did before ecology had become a science. His insightful understanding of reciprocity between organism and environment has become a principle that underlies all ecological science. Henderson's book deserves a broader appreciation than it has received; it is, for example, seldom mentioned in ecology textbooks.

An organism and its environment share a reciprocal relationship. The environment is modified by organisms, and populations of organisms are modified by their evolutionary process to adapt them to the environment and its changes. As an open system, an animal is forever receiving and releasing materials and energy. Building materials for life are obtained from the physical environment, either directly by producers such as green plants or indirectly by consumers.

A living form is a transient link built of environmental materials, which are then returned to the environment to be used again in the recreation of new life. Life, death, decay, and re-creation have been the cycle of existence since life began.

The primitive earth of 4.5 billion years ago, barren, stormy, and volcanic with a reducing atmosphere of ammonia, methane, and water (Figure 37-1), was wonderfully fit for the prebiotic syntheses that led to life's beginnings. Yet, it was totally unsuited, indeed lethal, for the living organisms that inhabit the earth today, just as early forms of life could not survive in our present environment. The appearance of free oxygen in the atmosphere, produced largely if not almost entirely by life, is an example of the reciprocity between organism and environment. Although oxygen was poisonous to early forms of life, its gradual accumulation from photosynthesis forced protective biochemical alterations to appear that led eventually to complete dependence on oxygen by most organisms. As living organisms

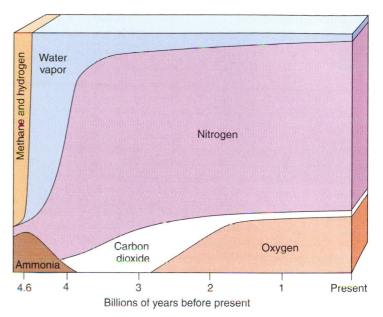

Figure 37-1

Changing composition of earth's atmosphere over time. The percentage of each gas in the atmosphere is proportional to its band width on the vertical axis. The primitive atmosphere formed as hydrogen, methane, and ammonia. Hydrogen, too light to be held by the earth's gravitational field, was lost to space. Nitrogen, carbon dioxide, sulfur dioxide, and water vapor emitted from volcanoes replaced the remaining primitive gases. The first free oxygen was formed by solar radiation acting on water molecules (photochemical dissociation) in the atmosphere. When oxygen-producing plants appeared 3 to 3.5 billion years ago, atmospheric oxygen gradually rose to its present level approximately 400 million years ago.

adapt and evolve, they change their environments. In so doing they must themselves change.

DISTRIBUTION OF LIFE ON EARTH

Biosphere and Its Subdivisions

The biosphere as usually defined is the thin outer layer of the earth capable of supporting life. It is probably best viewed as a global system that includes all life on earth and the physical environments in which living organisms exist and interact. The nonliving subdivisions of the biosphere include the lithosphere, hydrosphere, and atmosphere.

The **lithosphere** is the rocky material of the earth's outer shell and is the ultimate source of all mineral elements required by living organisms. The **hydrosphere** is the water on or near the earth's surface, and it extends into the lithosphere and the atmosphere. Water is distributed over the earth by a global hydrological cycle of evaporation, precipitation, and runoff. Five-sixths of the evaporation is from oceans, and more water is evaporated from oceans than is returned to them by precipitation. Oceanic evaporation therefore provides much of the rainfall that supports life on land. The gaseous component of the biosphere, the **atmosphere,** extends to some 3500 km above the earth's surface, but all life is confined to the lowest 8 to 15 km (troposphere). The screening layer in the atmosphere of oxygen-ozone is concentrated mostly between 20 and 25 km. The main gases present in the troposphere are (by volume) nitrogen, 78%; oxygen, 21%; argon, 0.93%; carbon dioxide, 0.03%; and variable amounts of water vapor.

Atmospheric oxygen has originated almost entirely from photosynthesis. As discussed in Chapter 2, the primitive earth contained a reducing atmosphere devoid of oxygen. When oxygen-producing photosynthesis appeared about 3 billion years ago (Figure 37-1), oxygen gradually accumulated in the atmosphere. The oxygen concentration probably reached its current level (21%) by the mid-Paleozoic era, some 400 million years BP. Since then, oxygen consumption by animals and plants has approximately equaled oxygen production. The current surplus of free oxygen in the atmosphere resulted from fossilization of plants before they could decay or be consumed by animals. As these vast stores of fossil fuels are burned by our industrialized civilization, the oxygen surplus that accumulated over the ages conceivably could be depleted. Fortunately, depletion is unlikely

for two reasons: (1) most of the total fossilized carbon exists as noncombustible shales and rocks, and (2) oxygen reserves in the atmosphere and in the oceans are so enormous that the supply could last thousands of years even if all photosynthetic replenishment suddenly were to cease.

The rapid input of carbon dioxide into the atmosphere from burning fossil fuels may significantly affect the earth's heat budget. Much of the sun's short-wave light energy absorbed by earth's surface reradiates as longer-wave infrared heat energy (Figure 37-2). Materials in the atmosphere, especially carbon dioxide and water vapor, impede this heat loss and raise the atmospheric temperature. This heating of the atmosphere is called the "greenhouse effect," since the atmosphere traps reradiated heat from the earth much as the glass of a greenhouse traps heat reradiated by plants and soil inside. While the greenhouse effect provides conditions essential for all life on earth, the gradual accumulation of carbon dioxide could increase the temperature of the biosphere as a whole (Figure 37-3).

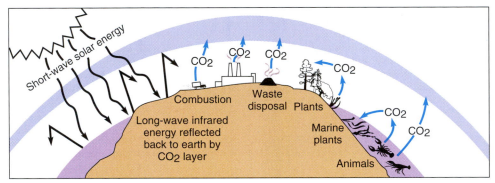

Figure 37-2
"Greenhouse effect." Carbon dioxide and water vapor in the atmosphere are transparent to sunlight but absorb heat energy reradiated from the earth, leading to warming of atmospheric air.

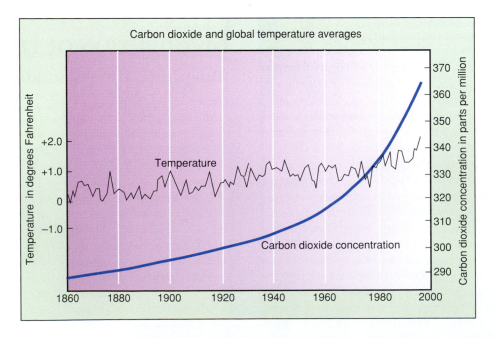

Figure 37-3
Rise in global atmospheric carbon dioxide and global temperature averages for the past 140 years. Data points before 1958 come from analysis of air trapped in bubbles in glacial ice from sites around the world. Atmospheric carbon dioxide has climbed steadily for more than a century while earth's temperature has followed a more erratic upward trend.

Concern over long-range effects of increasing atmospheric carbon dioxide, primarily from burning fossil fuel, stems not from mere conjecture. Atmospheric carbon dioxide increased from about 280 parts per million (ppm) before the Industrial Revolution to an average 370 ppm today. It may exceed 600 ppm in the next century. In the past century global temperature has increased 0.4° C and most experts agree that it will have increased 2° to 6° C when carbon dioxide and other heat-trapping greenhouse gases have doubled in the next century. If computer models are correct, rising global temperature will raise the sea level with gradual melting of the earth's ice caps (sea level has already risen 20 cm over the past century), alter crop yields, forests, and water supplies, and expand the world's desert—all of which would profoundly impact the earth's habitability.

Terrestrial Environments: Biomes

A biome is a major biotic unit bearing a characteristic and easily recognized array of plant life. Botanists long ago recognized that the terrestrial environment of the earth could be divided into large units having a distinctive vegetation, such as forests, prairies, and deserts. Animal distribution has always been more difficult to map, because plant and animal distributions do not exactly coincide. Over time zoogeographers came to accept plant distributions as the basic biotic units and recognized biomes as distinctive combinations of plants and animals. A biome is therefore identified by its dominant plant formation (Figure 37-4), but, because animals depend on plants, each biome supports a characteristic fauna.

Each biome is distinctive, but its borders are not. Anyone who has traveled across North America knows that plant communities grade into one another over broad areas. Moist deciduous forests of the Appalachians change gradually to drier oak forests of the upper Mississippi Valley, and then to oak woodlands with grassy understory, which yield to tall and mixed prairies (now corn and wheatlands), then to desert grasslands, and finally to desert shrublands. The indistinct boundaries where the dominant plants of adjacent biomes are mixed together form an almost continuous gradient called an **ecocline.** Thus biomes are in some sense abstractions, a convenient way for us to organize our concepts about different communities. Nevertheless, anyone can distinguish a grassland, deciduous forest, coniferous forest, or shrub desert by the dominant plants in each. And we can make reasonable assumptions about the animals that live in each biome.

Distinctiveness of a biome is determined mainly by climate, characteristic pattern of rainfall and temperature of each region, and amount of solar radiation it receives. Global variation in climate arises from uneven heating of the atmosphere by the sun. Because of the lower angle of the sun's rays striking higher latitudes, atmospheric heating is less there than at the equator (Figure 37-5). Air warmed at the equator rises and moves toward the poles. It is replaced by cold air moving away from the poles at lower levels. This pattern is complicated by earth's rotation, which produces a Coriolis effect that deflects moving air to the right in the Northern Hemisphere and to the left in the Southern Hemisphere. Air circulation in each hemisphere is broken into three latitudinal zones, called cells (Figure 37-6). In the Northern Hemisphere, for example, hot moist air at the equator cools and condenses as it rises, providing rainfall for the lush vegetation of

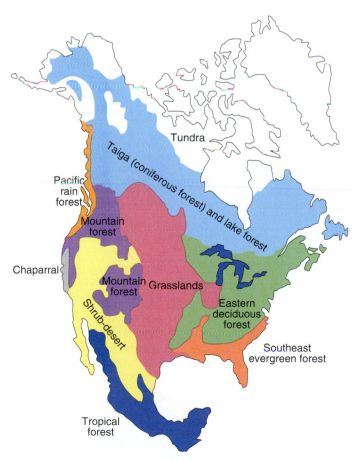

Figure 37-4

Major biomes of North America. Boundaries between biomes are not distinct as shown but grade into one another over broad areas.

Figure 37-5

Earth's climate is determined by differential solar radiation between the higher latitudes and equator. Solar energy is spread across a much larger, slanting surface area at high latitudes than is an equivalent amount of energy at the equator.

Figure 37-6

Earth as a heat engine. As the result of unequal heating across earth's surface, together with other factors such as earth's rotation, circulation of the oceans, and presence of landmasses, the earth acts like a giant heat engine that imposes a complex patchwork of climates on earth. See text for explanation.

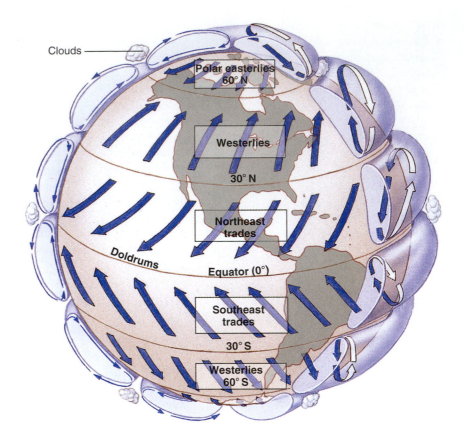

equatorial rain forests. Warm air then flows northward at high levels, cools, and sinks at 20° to 30° latitude. This air is very dry, having lost its moisture at the equator. As it heats it absorbs even more moisture, causing intense evaporation at earth's surface and producing a subtropical belt of deserts centered between 15° and 30° north (deserts of the American southwest, Saharan Africa, Arabian Peninsula, and India). The air then flows southward toward the equator, gathering moisture as it moves across the ocean, and being deflected to the right as the northeast trade winds. The cycle in this cell is completed when the air, now laden with moisture, reaches the equator.

A second circulation cell between 30° and 60° north arises when cool air sinking at around 30° moves northward at the surface. At 50° to 60° north it encounters cold air moving south from the North Pole, producing an unstable stormy area with abundant precipitation. Warmer air from the south is deflected upward and turns south at high altitude to complete the second cell. A third polar cell forms when cold, southward-moving Arctic air returns to the pole at high altitude.

The principal terrestrial biomes are temperate deciduous forest, temperate coniferous forest, tropical forest, grassland, tundra, and desert. In this brief survey, we will refer especially to biomes of North America and will consider predominant features of each.

Temperate Deciduous Forest

The temperate deciduous forest, best developed in eastern North America, encompasses several forest types that change gradually from the northeast to south. Deciduous, broad-leaved trees such as oak, maple, and beech that shed their leaves in winter predominate. Seasonal aspects are better defined in this biome than in any other. The deciduous habit is an adaptation of dormancy for low-energy levels from the sun in winter and freezing winter temperatures. In summer, the relatively dense forests form a closed canopy that creates deep shade below. Consequently there has been a selection for understory plants that grow rapidly in the spring and flower early before the canopy develops. Mean annual precipitation is relatively high (75 to 125 cm, or 30 to 50 inches), and rain falls periodically throughout the year. Mean annual temperatures range between 5° and 18° C (41° to 65° F).

Animal communities in deciduous forests respond to seasonal change in various ways. Some, such as insect-eating warblers, migrate. Others, such as woodchucks, hibernate during winter months. Others that are unable to escape survive by using available food (for example, deer) or stored food supplies (for example, squirrels). Hunting and habitat loss have eliminated virtually all large carnivores (mountain lions, bobcats, and wolves) that once roamed eastern forests. Deer, on the other hand, thrive in second-growth forests under protection of strict hunting management. Insect and invertebrate communities are abundant in deciduous forests because decaying logs and forest-floor debris provide excellent shelter.

Heavy exploitation of deciduous forests of North America began in the seventeenth century and reached a peak in the nineteenth century. Logging removed nearly all the once-magnificent stands of temperate hardwoods. With the opening of the prairie for agriculture, many eastern farms were abandoned and allowed to return gradually to deciduous forests.

Coniferous Forest

In North America coniferous forests form a broad, continuous, continent-wide belt stretching across Canada and Alaska, and south through the Rocky Mountains into Mexico. This biome continues across northern Eurasia, making it one of the largest plant formations on earth. It is dominated by evergreens—pine, fir, spruce, and cedar—which are adapted to withstand freezing and take full advantage of short summer growing seasons. Conical trees with their flexible branches shed snow easily. The northern area is the **boreal** (northern) **forest,** often called **taiga** (a Russian word, pronounced "tie-ga"). Taiga is dominated by white and black spruce, balsam, subalpine fir, larch, and birch. Mean annual precipitation is less than 100 cm (40 inches) and average temperatures range from –5° to +3° C (23° to 37° F).

In the central region of North America, the taiga merges into **lake forest,** dominated by white pine, red pine, and eastern hemlock. However, most of this forest was destroyed by exploitive logging and was replaced by shrubby second growth, which still characterizes much of Michigan, Wisconsin, southern Ontario, and Minnesota today. The large **southern evergreen forests** occupy much of the southeastern United States. The last old-growth coniferous forests of the Pacific northwest are rapidly falling to commercial logging.

Mammals of boreal and lake coniferous forests include deer, moose (Figure 37-7), elk, snowshoe hares, a variety of rodents, carnivores such as wolves, foxes, wolverines, lynxes, weasels, martins, and omnivorous bears. They are adapted physiologically or behaviorally for long, cold, snowy winters. Common birds are chickadees, nuthatches, warblers, and jays. One bird, the red crossbill, has a beak specialized for picking seeds from cones. Mosquitoes and flies are pests to both animals and humans in this biome. Southern coniferous forests lack many mammals found in the north, but they have more snakes, lizards, and amphibians.

Tropical Forest

The worldwide equatorial belt of tropical forests is an area of high rainfall (more than 200 cm [80 inches] per year), high humidity, relatively high and constant temperatures averaging more than 17° C (63° F), and little seasonal variation in day length. These conditions have nurtured luxurious, uninterrupted growth that reaches its greatest intensity in rain forests. In sharp contrast to temperate deciduous forests, dominated by relatively few tree species, tropical forests contain thousands of species, none of which is dominant. A single hectare typically contains 50 to 70 tree species as compared with 10 to 20 tree species in an equivalent area of hardwood forest in the eastern United States. Climbing plants and epiphytes are common among the trunks and limbs. A distinctive feature of tropical forests is stratification of life into six, and occasionally as many as eight, feeding strata (Figure 37-8).

Insectivorous birds and bats occupy the air above the canopy; below it birds, fruit bats, and mammals feed on leaves and fruit. In the middle zones are arboreal mammals (such as monkeys and sloths), numerous birds, insectivorous bats, insects, and amphibians. Climbing animals, such as squirrels and civets, move up and

Figure 37-7

A bull moose browses on dwarf birch in the coniferous forest biome. Note shedding of antler skin ("velvet"), signifying that antler growth is complete and that breeding season is approaching.

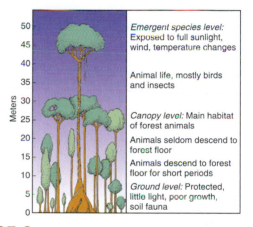

Figure 37-8

Profile of tropical forest, showing stratification of animal and plant life into six strata. The animal biomass is small compared with the biomass of plants.

down the trunks, feeding from all strata. On the ground are large mammals lacking climbing ability, such as the large rodents of South America (for example, capybara, paca, and agouti) and members of the pig family. Finally, a mixed group of small insectivorous, carnivorous, and herbivorous animals searches the litter and lower tree trunks for food. No other biome can match tropical forests in incredible variety of animal species. Food webs are intricate and notoriously difficult for ecologists to unravel.

Tropical forests, especially the enormous expanse centered in the Amazon Basin, are the most seriously threatened forest ecosystems. Large areas are being cleared for agriculture by "slash-and-burn" methods, but, because of low soil fertility, farms are soon abandoned. It may seem paradoxical that a biome as luxuriant as a tropical forest should have poor soil. This occurs because nutrients released by decomposition are rapidly recycled by plants, leaving no reservoir of humus. In many areas, once the plants are removed, the soil rapidly becomes a hard, bricklike crust called **laterite.** Tropical plants cannot recolonize such areas.

Each month an area of undisturbed tropical forest the size of Massachusetts is converted to other uses. Unlike forest clearing in temperate zones, which made possible sustained, productive agriculture, tropical soils quickly become depleted, forcing farmers to abandon them and to clear more forest. Other pressures on tropical forests include logging by multinational timber companies, which cut large tracts of timber to make furniture for developed countries, and clearing of land for cattle ranching.

Grassland

The North American prairie biome is one of the most extensive grasslands in the world, extending from the Rocky Mountain edge on the west to eastern deciduous forest on the east, and from northern Mexico in the south to the Canadian provinces of Alberta, Saskatchewan, and Manitoba in the north. Original grassland associations of plants and animals have been almost completely destroyed by humans. Prairies today have been transformed into the most productive agricultural region in the world, dominated by monocultures of cereal grains. In grazing lands virtually all major native grasses have been replaced by alien species. Vast areas of Arizona and New Mexico have been converted from lush grasslands to parched desert by more than a century of livestock overgrazing. Of the once dominant herbivore, bison (Figure 37-9), very few survive, but jackrabbits, prairie dogs, ground squirrels, and antelope remain. Mammalian predators include coyotes, ferrets, and badgers, although, of these, only coyotes are common. Rainfall on the North American prairie ranges from about 80 cm (31 inches) in the east to 40 cm (16 inches) in the west. Average annual temperatures range between 10° and 20° C (50° to 68° F).

Tundra

Tundra is characteristic of severe, cold climatic regions, especially treeless Arctic regions and high mountaintops. Plant life must adapt itself to a short growing season of about 60 days and to a soil that remains frozen for most of the year. Average annual

Figure 37-9
Bison grazing on a short-grass prairie.

Figure 37-10
A large male caribou on the Alaskan tundra. The gregarious caribou travel in large herds, feeding in summer on grasses, dwarf willow, and birch, but in winter almost exclusively on lichen.

precipitation is usually less than 25 cm (10 inches) and annual temperature averages about −10° C (14° F).

Most tundra regions are covered with bogs, marshes, ponds, and a spongy mat of decayed vegetation, although high tundras may be covered only with lichens and grasses. Despite thin soil and short growing season, vegetation of dwarf woody plants, grasses, sedges, and lichens may be quite profuse. Plants of the alpine tundra of high mountains, such as the Rockies and Sierra Nevadas, differ from the Arctic tundra in some respects. Characteristic animals of the Arctic tundra are lemmings, caribou (Figure 37-10), musk-oxen, arctic foxes, arctic hares, ptarmigans, and (during the summer) many migratory birds.

Desert

Deserts are arid regions where rainfall is low (less than 25 cm [10 inches] a year), and water evaporation is high. The North American desert is of two parts, the hot deserts of the southwest (Mojave, Sonoran, and Chihuahuan) and the cool, high desert in the rain shadow of the High Sierras and Cascade Mountains. Desert plants, such as thorny shrubs and cacti, have reduced foliage, drought-resistant seeds, and other adaptations for conserving water. Many large desert animals have developed remarkable anatomical and physiological adaptations for keeping cool and conserving water (p. 649). Most smaller animals avoid the most severe conditions by living in burrows or developing nocturnal habits. Mammals found there include mule deer, peccaries, cottontails, jackrabbits, kangaroo rats, and ground squirrels. Typical birds are roadrunners, cactus wrens, turkey vultures, and burrowing owls. Reptiles are numerous, and a few species of toads are common. Arthropods include a great variety of insects and arachnids.

Deserts are expanding rapidly. Between 1882 and 1952 the area of the earth's land surface occupied by desert increased from an estimated 9.4 to 23.3%. Since 1965, 650,000 km² of grazing land was added to the Sahara Desert of Africa, the largest desert on earth, because of an extended drought combined with overgrazing by livestock.

Aquatic Environments

Inland Waters

Of all the water in the world, a mere 2.5% is fresh water. Most fresh water is contained in polar ice caps, or stored underground in aquifers and soil moisture, leaving only 0.01% of the world's inland waters available as habitat for aquatic life. Yet a quarter of the world's vertebrates and nearly half of its fishes live in these fragile "islands" of water—water that must also supply human needs for irrigation, drinking water, hydroelectric power, and waste disposal.

Inland waters are divided broadly into running-water, or **lotic** (L. *lotus,* action of washing) habitats, and standing-water, or **lentic** (L. *lentus,* slow) habitats. Lotic habitats follow a gradient from mountain brooks to streams and rivers. Brooks and streams with high-velocity water flow are high in dissolved oxygen because of their turbulence. Energy input is chiefly in the form of organic detritus washed from adjacent terrestrial areas. More slowly moving rivers have less dissolved oxygen and more floating algae and plants. Their fauna is tolerant of lower oxygen concentration.

Lentic habitats, such as ponds and lakes, tend to have still lower concentrations of oxygen, particularly in the deeper areas. Animals living on the bottom or on submerged vegetation **(benthos)** include snails and mussels, crustaceans, and a wide variety of insects. Many swimming forms, called **nekton,** are found in lakes and larger ponds. Depending on the nutrients available, a large contingent of small floating or weakly swimming plants and animals **(plankton)** may occur. Ponds and lakes have short life spans—a few hundred to many thousands of years depending on size and rate of sedimentation—and undergo great physical change as they age. The Great Lakes of North America, which occupy depressions gouged by glacial advances of the Pleistocene epoch, became ice free about 5000 years ago.

> A striking exception to the short lifetimes of most lakes is Lake Baikal in southern Siberia. This enormous lake, 1741 m deep (more than 1 mile), is by far the oldest lake in the world, dating from at least the Paleocene—more than 60 million years BP. The speciation of sculpins in Lake Baikal is illustrated in Figure 6-20, p. 115.

Many freshwater habitats have been severely damaged by human pollution such as dumping of toxic industrial wastes and enormous quantities of sewage. Of the Great Lakes, Lake Erie has been the most seriously affected by the inflow of large amounts of nitrates and phosphates. These nutrients fertilize the lake, creating huge blooms of algae that die and sink to the bottom to decompose and rob the lake of oxygen. As a result, all levels of aquatic life are adversely affected.

Oceans

By almost any measure, oceans represent by far the largest portion of earth's biosphere. They cover 71% of earth's surface to an average depth of 3.75 km (2.3 miles), with their greatest depths reaching to more than 11.5 km (7.2 miles) below sea level. The marine world is relatively uniform as compared with land, and in many respects it is less demanding on life forms. However, the evident monotony of the ocean's surface belies the variety of life below. Oceans are the cradle of life, and this is reflected by the variety of organisms living there—more than 200,000 species of unicellular forms, plants, and animals. The vast majority of these forms, about 98%, live on the seabed **(benthic);** only 2% live freely in the open ocean **(pelagic).** Of the benthic forms, most occur in the intertidal zone or shallow depths of oceans. Less than 1% live below 2000 m.

The most productive areas are concentrated along continental margins and a few areas where waters are enriched by organic nutrients and debris lifted by upwelling currents into the sunlit, or **photic,** zone, where photosynthetic activity occurs. With certain notable exceptions (see box in Chapter 38, p. 801), all life below the photic zone must be supported by the light "rain" of organic particles from above.

Life in the ocean is divided into regions, or provinces, each with its own distinctive life forms (Figures 37-11 and 37-12). The **littoral,** or **intertidal,** zone, where sea and land meet, is paradoxically both the harshest and the richest of all marine environments. It, as well as the animals living there, is subjected to pounding surf, sun, wind, rain, extreme temperature fluctuations, erosion, and sedimentation. Yet because of the diversity of available habitats and bounteous supply of nutrients, animals such as barnacles, snails, chitons, limpets, mussels, sea urchins, sea stars, and many others flourish there. Below the littoral zone is the **sublittoral,** or **subtidal** zone, which is always submerged. It also supports a rich variety of animal life, as well as forests of brown algae.

An **estuary** is a semienclosed transition zone where fresh water flows into the sea. Despite unstable salinity caused by the variable entry of fresh water, an estuary is a nutrient-rich habitat that supports a diverse fauna.

The **neritic,** or shallow water, zone surrounds the continents and extends to the edge of the continental shelf—approximately to a depth of 200 m (Figure 37-11). This zone is more productive than open ocean because it benefits from nutrients delivered by rivers and by upwelling at the edge of the continental shelf. Algal growth is prolific, which in turn supports diverse animal life, including most of the world's fisheries.

The vast open ocean is known as the **pelagic** realm (Figure 37-12). Despite its size (comprising 90% of the total oceanic area), the pelagic realm is relatively impoverished biologically because, as organisms die, they sink out of the photic zone, carrying nutrients into the bathypelagic zone where they are immobilized.

Areas of **upwelling** and convergence of ocean currents are vital sources of nutrient renewal for the surface photic zone. The enormously productive polar seas are an example. Before their populations were overexploited by humans, baleen whales probably consumed around 77 million tons of krill (a shrimplike animal, Figure 19-25, p. 393) per year, far more than the entire catch of all fishes, crustaceans, and molluscs taken by all the world's fishing fleet in any single year. The enormous krill population was sustained by phytoplankton, the base of the food web (p. 800), which in turn flourished because of the abundance of nutrients in the Antarctic sea.

Figure 37-11

Major marine zones.

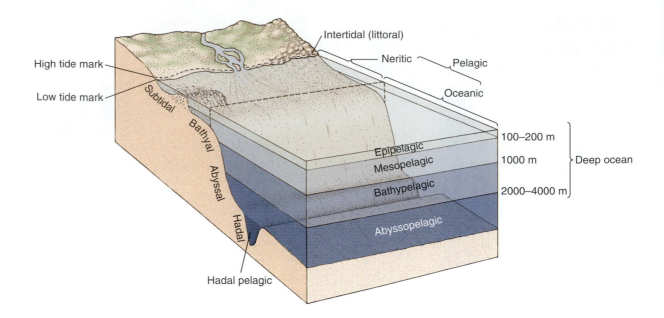

Some of the world's most productive fisheries are—or were—centered on upwelling regions. Before its collapse in 1972, the Peruvian anchovy fishery, which depended on the Peru Current, provided 22% of all fish caught in the world! Earlier, the California sardine fishery and Japanese herring fishery, both fisheries of upwelling regions, were intensively harvested to the point of collapse and have never recovered. The world's fisheries today are seriously imperiled by overfishing, degradation of fish habitats by trawling, wasteful fishing methods, and marine pollution. Some of the world's greatest fishing grounds, such as the Grand Banks and Georges Banks of eastern North America, have been fished to total collapse.

Below the surface, or **epipelagic,** layers of the pelagic realm are the great ocean depths, characterized by enormous pressure, perpetual darkness, and a constant temperature near 0° C. It remained a world unknown to humans until recently, when baited cameras, bathyscaphs, and deep-water trawls have been lowered to view and sample the ocean floor. There are several distinct habitats in the ocean depths (Figure 37-12). The **mesopelagic** is the "twilight zone," which receives dim light and supports a varied community of animals. Below the mesopelagic is a world of perpetual darkness, divided into three depth zones as shown in Figure 37-12: bathypelagic, abyssopelagic, and hadopelagic. Deep-sea forms depend on that meager portion of the gentle rain of organic debris from above that escapes consumption by organisms in the water column. On the sea floor exists the benthos, represented by sea anemones, sea urchins, crustaceans, polychaete worms, and fishes—indeed nearly all major metazoan groups. Most are deposit feeders characterized by very slow growth (because of scarcity of food) and long lives.

Recently self-contained benthic communities of animals that are completely independent of solar energy and the rain of organic debris from above were discovered adjacent to vents of hot water issuing from rifts in the ocean floor (see box in Chapter 38, p. 801).

ANIMAL DISTRIBUTION (ZOOGEOGRAPHY)

The study of zoogeography tries to explain why animals are distributed as they are, their patterns of dispersal, and the factors responsible for their dispersal. Most animals typically occupy limited geographic areas. Humans, however, and creatures such as house mice and cockroaches that share human habitations, are able to live almost anywhere on earth. It is not always easy to explain why animals are distributed as they are, since similar habitats on separate continents may be occupied by quite different kinds of animals. A particular species may be absent from a region that supports similar animals because of barriers that prevent it from getting there or because established populations of other animals prevent it from colonizing.

Thus we would like to discover why animals are found where they are or are not found where one thinks they ought to be. Usually this means studying the past. The fossil record plainly shows that animals once flourished in regions from which they are now absent. Extinction has played a major role, but many groups left descendants that migrated to other regions and survived. For example, ancestors of camels originated in North America, where their fossils are found. They spread during the Pleistocene epoch by way of Alaska to Eurasia and Africa, where they are represented today by true camels, and to South America, where their descendants survive as llamas, alpacas, guanacos and vicuñas. (The Pleistocene began about 1.7 million years BP and ended about 11 thousand years BP; see the geological time table on the back inside-cover.) Then camels became extinct in North America about 10,000 years BP at the close of the Ice Age. Thus the history of an animal species or its ancestor must be known before one can understand why it lives where it does. Earth's surface is undergoing constant change. Many areas that are now land were once covered with seas; fertile plains may be claimed by advancing desert; impassable

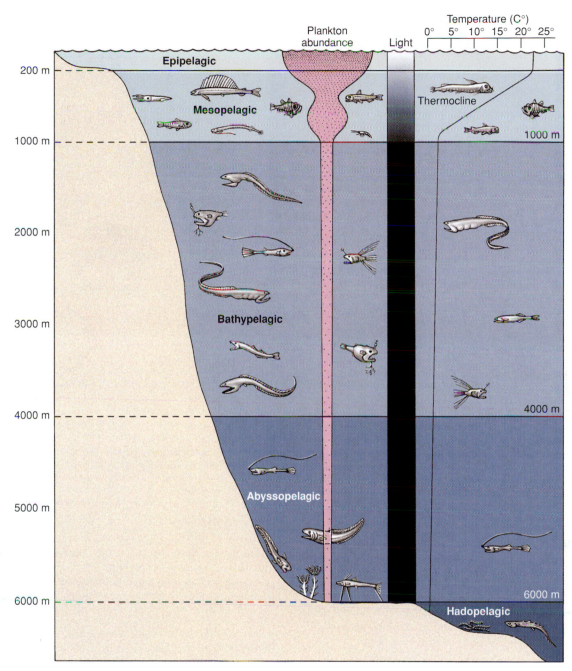

Figure 37-12

Life in pelagic zones. Each zone supports a distinct community of organisms. Animals in zones below the mesopelagic depend on the meager rain of food that sinks out of the epipelagic and mesopelagic zones.

mountain barriers may arise where none existed before; or inhospitable ice fields may retreat before a warmer climate to be replaced by forests. Geological change has been responsible for much of the alteration in animal (and plant) distribution and has been a powerful influence in shaping organic evolution.

Disjunct Distributions

A major problem for zoogeographers is to explain the numerous instances of discontinuous or **disjunct distributions:** closely related species living in widely separated areas of a continent, or even the world (Figure 37-13). How could a group of animals become so dispersed geographically? There are two possible ways for a disjunct distribution to arise. Either a population moves from its place of origin to a new location **(dispersal),** traversing intervening territory that is unsuited for long-term colonization, or the environment changes, breaking a once continuously-distributed species into geographically separated populations **(vicariance).** Vicariance may involve climatic changes that contract and fragment areas of habitat favorable for a species, or it may involve physical movement of landmasses or waterways that carry different populations of a species away from each other.

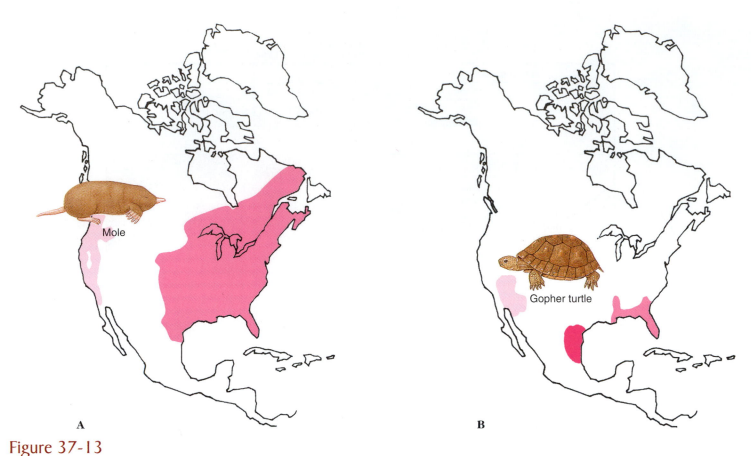

Figure 37-13

Disjunct distributions in North America. **A,** Moles of the family Talpidae probably entered North America across the Bering land bridge that once joined North America and Asia during the Tertiary period. Eastern and western populations are now separated by the Rocky Mountains. **B,** Gopher turtles of the genus *Gopherus* are now separated into three fully isolated populations.

Distribution by Dispersal

By dispersal, animals spread into new localities from their places of origin. Dispersal involves *emigration* from one region and *immigration* into another. Dispersal is a *one-way,* outward movement that must be distinguished from *periodic* movement back and forth between two localities, such as seasonal migration of many birds. Dispersing animals may move actively under their own power, or they may be passively dispersed by wind, by floating or rafting on rivers, lakes, or the sea, or by hitching rides on other animals. Animals are expected to expand their geographic distributions in this manner across all favorable habitat accessible to them. For example, as the last Pleistocene glaciers retreated northward, habitats favorable for many temperate species became available on formerly glaciated territory in North America, Europe, and Asia. Species that originated immediately south of the glaciated territory prior to glacial retreat then expanded northward as new habitats appeared. Because the reproductive rate of animal populations is great, there is a continuous pressure on populations to expand across all favorable habitats.

Dispersal easily explains movement of animal populations into favorable habitats that are geographically adjacent to their places of origin. This movement produces an expanded but geographically continuous distribution. Can dispersal also explain the origins of geographically disjunct distributions? For example,

flightless ratite birds (Figure 37-14) inhabit disjunct landmasses primarily of the Southern Hemisphere including Africa, Australia, Madagascar, New Guinea, New Zealand, and South America. These landmasses are separated from each other by ocean, a very strong barrier to ratite dispersal. To explain this distribution by dispersal, one must postulate a **center of origin** from which the group dispersed to reach all of the widely separated landmasses on which it is now found. Because ratites do not fly, a dispersalist hypothesis requires intermittent, passive rafting of individuals across the ocean. Is this hypothesis reasonable? We know from studies of the Galápagos Islands and Hawaii (Chapter 6) that occasional, long-distance dispersal of terrestrial animals and plants across oceans does occur. This is the only way terrestrial animals could colonize islands produced by oceanic volcanoes. For flightless birds and many other discontinuously distributed animals, however, there is an alternative to the hypothesis that disjunct distribution was produced by dispersal over unfavorable habitat. This is the hypothesis of vicariance (L. *vicarius,* a substitute).

Distribution by Vicariance

Disjunct distributions of animals may be created by physical changes in the environment that cause formerly continuous habitats to become disjunct. Areas once joined may become sep-

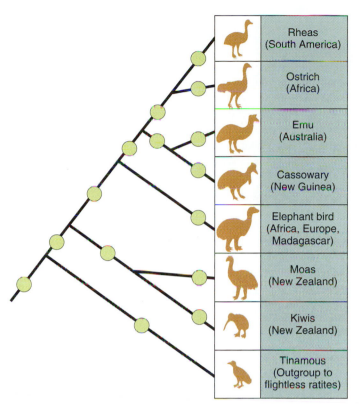

Figure 37-14

Phylogenetic relationships inferred for flightless birds (Chapter 6, p. 112). Vicariance biogeography proposes that these flightless species descended from an ancestral species that was widespread in the Southern Hemisphere when Africa, Australia, Madagascar, New Guinea, New Zealand, and South America were connected. By moving apart, these landmasses fragmented both themselves and the flightless populations they contained. If the vicariance hypothesis is correct, the phylogenetic branching sequence inferred for allopatric flightless species reflects the sequence by which their landmasses broke apart from each other. This hypothesis is tested by looking for similar phylogenetic patterns in other groups of animals and plants whose ancestral populations would have been fragmented by the same geological events. Widespread geographic distribution of the elephant bird suggests that it has dispersed following the fragmentation of landmasses.

arated by barriers that are effectively impenetrable for many animals that inhabit them. The study of fragmentation of biotas in this manner is called **vicariance biogeography.** At the species level, "vicariance" is often used as a synonym of "allopatry," which is simply a distribution of populations in geographically separated areas (p. 113). Lava flows from a volcano may cause a formerly continuous forest to become separated into geographically discontinuous patches, thereby breaking many species of plants and animals into geographically isolated populations.

Perhaps the most dramatic vicariant phenomenon in the earth's history is continental drift, through which a once continuous landmass was sequentially broken into continents and islands separated by ocean. All terrestrial and freshwater animal species that had spread across the initially continuous landmass became sequentially fragmented into many populations on different continents and islands separated by ocean. Vicariance by continental drift gives us another hypothesis for explaining the disjunct distribution of flightless birds; they may descend from an ancestral species that was widespread in the Southern Hemisphere when Africa, Australia, Madagascar, New Guinea, New Zealand, and South America were in closer contact than they are today. When these landmasses moved apart across the ocean, ancestral species would have fragmented into disjunct populations that evolved independently, producing the diversity of forms we observe today.

How do we test hypotheses of vicariance biogeography? Reconstructing past histories of environmental changes might seem impossible, but we do have a very powerful method for testing such hypotheses, and it is based on the systematic methods presented in Chapter 10.

Suppose that different species of flightless birds evolved allopatrically as continental drifting sequentially broke their terrestrial environment into isolated pieces. If we construct a cladogram or phylogenetic tree of these birds as shown in Figure 37-14, the earliest divergence should correspond to the first vicariant event that fragmented their common ancestral species. All subsequent branching events on the tree should correspond sequentially to subsequent vicariant events that fragmented major lineages further. Our tree hypothetically reconstructs the history of vicariant events for the group. If we erase the names of species from terminal branches of the tree and replace them with geographic areas in which each species is found, we have a hypothesis for the sequential separation of different geographic areas. We can test this vicariant hypothesis further by identifying other groups of terrestrial organisms that have different species in each of the same geographic areas as flightless birds. If our hypothesis is correct, these groups were fragmented geographically by the same vicariant events that fragmented flightless birds. We therefore predict that the cladogram or phylogenetic trees constructed for species in the other groups will show the same branching pattern as the flightless-bird tree when we replace the species names with those of the areas they inhabit. If this hypothesis is confirmed, we have a **general area cladogram** that depicts the history of fragmentation of the different geographic areas studied. This general area cladogram can be investigated further using geological and climatic studies.

In many groups of organisms, it is likely that both vicariance and dispersal have contributed to evolution of disjunct distributional patterns. Methods of vicariance biogeography will be very useful for finding such cases. Indeed, the cladogram of flightless birds is not just a simple grouping of birds that inhabit nearby areas. We can ask whether any branches on a cladogram representing a particular group of species are inconsistent with the general area cladogram for geographic areas that the species inhabit. Suppose that the cladogram for a particular taxon is consistent with the area cladogram except for placement of a single branch. We explain most geographic disjunctions within the taxon by vicariance but look for dispersal to explain the single branch that is not compatible with the general area cladogram. In this way, we can focus our study of dispersal on specific cases in which it is most likely to have occurred.

Continental Drift Theory

It is no accident that current enthusiasm for vicariance biogeography coincides with the recent acceptance of the continental drift theory by geologists. The continental drift theory is not new (it was proposed in 1912 by German meteorologist Alfred Wegener), but it remained controversial and largely neglected until the newly proposed theory of **plate tectonics** provided a mechanism to account for drifting continents (unfortunately, Wegener did not live to see his hypothesis accepted). According to the theory of plate tectonics (tectonics means "deforming movement"), earth's surface is composed of 6 to 10 rocky plates, about 100 km thick, that shift position on a more malleable underlying layer. Wegener proposed that earth's continents had been drifting like rafts following breakup of a single great landmass called Pangaea ("all land"). According to recent workers who have considerably revised Wegener's dating, the original breakup of Pangaea occurred approximately 200 million years ago. Two great supercontinents were formed: a northern Laurasia and a southern Gondwana, separated from each other by the Tethys Sea (Figure 37-15). At the end of the Jurassic period, some 135 million years ago, the supercontinents began to fragment and drift apart. Laurasia split into North America, most of Eurasia, and Greenland. Gondwana split into South America, Africa, Madagascar, Arabia, India, Australia, New Guinea, Antarctica, and numerous smaller fragments that now form Southeast Asia. The Arabian, Indian, and Southeast Asian fragments gradually moved across the Tethys Sea and eventually accreted to Laurasia, to which they are now attached. This theory is supported by the appearance of fit between the continents, by airborne paleomagnetic surveys, by seismographic studies, by the presence of mid-ocean ridges where the tectonic plates arise, and by a wealth of biological data.

Continental drift explains several otherwise puzzling distributions of animals, such as the similarity of invertebrate fossils in Africa and South America, as well as certain similarities in present-day faunas at the same latitudes on the two continents. However, the continents have been separated for all of the Cenozoic era and probably for much of the Mesozoic era as well, much too long to explain the distributions of some modern organisms such as placental mammals. Continental drift theory is, nevertheless, enormously useful in explaining interconnections between flora and fauna of the past.

The present distribution of marsupial mammals is an excellent example of the influence of continental breakup. Marsupials appeared in the Middle Cretaceous period, about 100 million years BP, probably in South America. Because South America was at that time connected to Australia through Antarctica (which was then much warmer than it is today), marsupials spread through all three continents. They also moved into North America, but there they encountered placental mammals, which had dispersed to that continent from Asia. Marsupials evidently could not coexist with placentals, and so became extinct in North America. (North American marsupials today, opossums, are relatively recent arrivals from South America.) Placentals followed marsupials into South America, but by that time marsupials had expanded and were too firmly established to be driven into

Many people consider Alfred Russel Wallace (Figure 6-1B) the founder of modern historical biogeography. Wallace conducted extensive field studies in the Malay Archipelago, where he discovered an abrupt faunal change between Asian faunal elements and Australian/New Guinean ones. Pheasants, parrots, monkeys, and numerous lizard groups are among the faunal elements whose geographic distributions show abrupt boundaries at this line. This biogeographic boundary is called "Wallace's Line," and bisects the current country of Indonesia. Wallace's Line has been a biogeographic mystery since he described it because there are no obvious environmental changes or barriers that would explain the abrupt displacement of faunas across the line. Plate tectonics currently offers the best explanation of Wallace's Line. Although currently close, the Southeast Asian plates separated from the Australian/New Guinean plate during the breakup of Gondwana, and these different plates spent many millions of years crossing the Tethys Sea in isolation from each other before finding their current locations. Wallace's Line marks the approximate boundary between the Southeast Asian and Australian/New Guinean plates, whose faunas diverged greatly during their long evolutionary separation. Recent molecular phylogenetic studies comparing lizard groups distributed on opposite sides of Wallace's Line support the interpretation that these groups were isolated at the time of Gondwanan fragmentation and only recently became geographic neighbors in the Malay Archipelago.

extinction. In the meantime, about 50 million years BP, Australia drifted apart from Antarctica, barring entrance to placentals. Australia remained in isolation, allowing marsupials to diversify into the present rich and varied fauna.

Temporary land bridges also have been important pathways of dispersal. An important and well-established land bridge that no longer exists connected Asia and North America across the Bering Strait. It was across this corridor that the placentals moved from Asia into North America.

Today a land bridge connects North and South America at the Isthmus of Panama. But from the mid-Eocene epoch (50 million years BP) to the end of the Pliocene epoch (3 million years BP), the two continents were completely separated by water. During this long period, the major groups of mammals evolved in distinctive directions on each continent. When the land bridge was reestablished at the end of the Pliocene epoch, a tide of mammals began to flow in both directions (Figure 37-16). This dispersal has been called the "Great American Interchange," one of the most important minglings of distinct continental faunas in the earth's history. For a period both continents gained in mammalian diversity, but extinction of large numbers of mammals on both continents soon followed. North American carnivores such as raccoons, weasels, foxes, dogs, cats (including sabercats), and bears began preying on South American mammals, which previously had evolved in an environment free of carnivores. Other North American invaders included hoofed mammals (horses, tapirs, peccaries, llamas, deer, antelopes, and mastodonts), rabbits, and several families of rodents. These mammals displaced many South American residents occupying

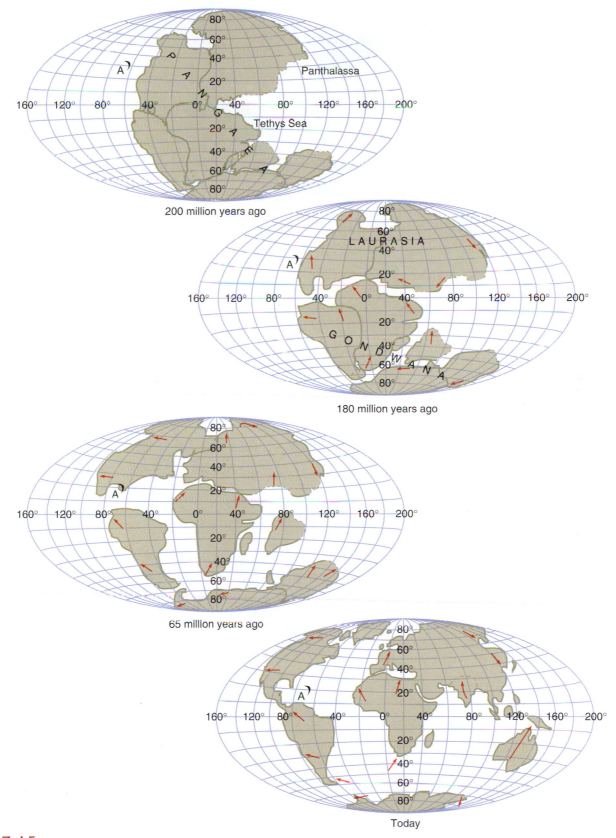

Figure 37-15

Hypothesized drift of continents over the past 200 million years from an original single landmass to their present positions. The single landmass Pangaea first separated into two smaller continents (Laurasia and Gondwana). These later separated into smaller continents. Arrows indicate vector movements of the continents. The black crescent labeled A is a modern geographical reference point representing the Antilles arc in the West Indies.

Figure 37-16

The Great American Interchange. The Isthmus of Panama emerged approximately 3 million years ago, permitting the extensive interchange of many families of mammals. At top are representatives of 38 South American genera that walked north across the isthmus. At bottom are representatives of 47 North American genera that migrated to South America. The North American immigrants diversified rapidly after entering South America. South American immigrants to North America diversified little and most became extinct.

similar habitats. Today nearly half of South American mammals are descendants of recent North American invaders. Only a few South American invaders survived in North America: porcupines, armadillos, and opossums. Several other South American groups, including giant ground sloths, glyptodonts, anteaters, giant aquatic capybaras, toxodonts (rhino-sized plant eaters), and giant armadillos, entered North America but subsequently became extinct there.

SUMMARY

The biosphere is a thin life-containing blanket surrounding the earth. The presence of life on earth is possible because numerous conditions for life are fulfilled on this planet. These include a steady supply of energy from the sun, presence of water, a suitable range of temperatures, the correct proportion of major and minor elements, and the screening of lethal ultraviolet radiation by atmospheric ozone. Earth's environment and living organisms have evolved together, each deeply marking the other.

The biosphere comprises lithosphere, the earth's rocky shell; hydrosphere, the global distribution of water; and atmosphere, the blanket of gas surrounding the earth.

Earth's terrestrial environment is composed of biomes that bear a distinctive array of plant life and associated animal life. Eastern deciduous forest is characterized by distinct seasons and autumn leaf fall. North of the deciduous forest is coniferous forest, which in its northern range is called taiga, an area dominated by needle-leafed trees adapted for heavy snowfall. Animals of the taiga are adapted for long, snowy winters.

The tropical forest is the richest biome, characterized in part by a great diversity of plant species and the vertical stratification of animal habitats. Most tropical forest soils rapidly deteriorate when the forest is removed.

The most modified biome is grassland, or prairie, which has been converted largely to agriculture and grazing. The tundra biome of the far north and the desert biome are both severe environments for animal life, but they are populated nevertheless with organisms that have evolved appropriate adaptations.

Freshwater habitats include rivers and streams (lotic habitats) and ponds and lakes (lentic habitats). All are geologically ephemeral habitats that are strongly influenced by nutrient input.

Oceans occupy 71% of earth's surface. The photic, or sunlit, zone supports photosynthetic activity by phytoplankton. A rain of nutrients from the photic zone supports the great diversity of life below on the seabed (benthos). The littoral, or tidal, zone is biologically rich but physically harsh. The neritic, or shallow-water, zone overlying the continental shelf is the locus of the world's great fisheries, which are especially productive in areas of upwelling where nutrients are constantly renewed. The open ocean, or pelagic zone, occupies most of the ocean's area but has low biological productivity.

Zoogeography is the study of animal distribution on earth. Animals have become distributed by dispersal, the spread of populations from the centers of origin, and by vicariance, the separation of populations by barriers. Continental drift, now strongly supported by plate tectonic theory, helps explain how animal groups become geographically separated so that evolutionary diversification could occur. It also explains how certain groups, such as marsupial mammals, can become isolated from others. Temporary land bridges have served as important pathways for animal dispersal.

REVIEW QUESTIONS

1. What are special conditions on earth that make this planet especially favorable for life?
2. What is the justification for saying that the earth and life on it have evolved together and that each has deeply influenced the other?
3. What is the biosphere? How would you distinguish between the following subdivisions of the biosphere: lithosphere, hydrosphere, atmosphere?
4. What is the origin of oxygen on earth? What would happen to earth's supply of oxygen if photosynthesis were suddenly to cease?
5. What is the evidence that increasing carbon dioxide levels in the atmosphere are responsible for the increase in the "greenhouse effect"?
6. What is a biome? Briefly describe six examples of biomes.
7. What are some very productive marine environments, and why are they so productive?
8. What is the source of nutrients for animals living in the deep-sea habitat?
9. What are some reasons why a species may be absent from a habitat or region to which it should adapt well?
10. Define and distinguish the alternative explanations for disjunct distributions among animals: dispersal and vicariance.
11. Who first proposed the continental drift theory and what finally convinced geologists that the theory was correct?
12. In what way does continental drift help to explain the current distribution of marsupial mammals on earth?
13. What was the Great American Interchange, when did it occur, and what were the results?

SELECTED REFERENCES

Berner, E. K., and R. A. Berner. 1996. Global environment: water, air, and geochemical cycles. Upper Saddle River, New Jersey, Prentice-Hall, Inc. *A geochemistry textbook with good coverage of global water and air circulation, greenhouse effect, acid rain, and geochemistry of rivers, lakes, and oceans.*

Cox, C. B., and P. D. Moore. 2000. Biogeography: an ecological and evolutionary approach, ed. 6. Boston, Blackwell Science Ltd. *Highly readable account with a strong ecological emphasis.*

Dietz, R. S., and J. C. Holden. 1970. The breakup of Pangaea. Sci. Am. **223:**30–41 (Oct.). *The sequence of continental drift since the early Mesozoic era is mapped.*

Henderson, L. J. 1913. The fitness of the environment. New York, Macmillan, Inc. *This short but influential book, one of the great classics of biological literature, explains how conditions on our planet made life possible.*

Lieberman, B. S. 2000. Paleobiogeography. New York, Kluwer Academic/Plenum Publishers. *A current coverage of historical biogeography with emphasis on use of fossils to study global change, plate tectonics, and evolution.*

Marshall, L. G. 1988. Land mammals and the Great American Interchange. Am. Sci. **76**:380–388. *Mammalian faunas of North and South America, having developed in isolation for millions of years, were suddenly allowed to intermingle when the Panamanian land bridge emerged 3 million years ago.*

MacDonald, G. 2003. Biogeography: introduction to space, time and life. New York, John Wiley and Sons, Inc. *A new introductory textbook of biogeography.*

Safina, C. 1995. The world's imperiled fish. Sci. Am. **273**:46–53. *The onslaught of modern industrial fishing methods has collapsed several world fisheries and imperiled those remaining.*

Schneider, D. 1997. The rising seas. Sci. Am. **276**:112–117 (Mar.). *The sea level is slowly rising with gradual melting of mountain glaciers. But whether or not global warming will lead to melting of polar ice sheets, and to calamitous coastline flooding, has been difficult to gauge.*

Van Oosterzee, P. 1997. Where worlds collide: the Wallace Line. Ithaca, New York, Cornell University Press. *An entertaining account of the founding of biogeography by Alfred Russel Wallace with emphasis on the paradox of an abrupt faunal change in the Malay Archipelago.*

ZOOLOGY LINKS TO THE INTERNET

Visit the textbook's Online Learning Center at www.mhhe.com/zoology to find live Internet links for each of the topics listed here.

Biomes and Environmental Habitats
Rocky Shore Communities
The Coast and the Continental Shelf
The Marine Deep Sea Zone
Tropical Rainforests
Temperate Forests
Tundra
Taiga
Grasslands
Deserts
Plate Tectonics

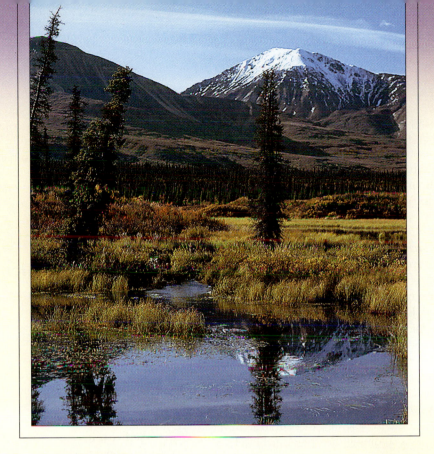

Talkeetna Mountain
Range, Alaska.

Animal Ecology

Every Species Has Its Niche

The lavish richness of the earth's biomass is organized into a hierarchy of interacting units: an individual organism, a population, a species, a community, and finally an ecosystem, that most bewilderingly complex of all natural systems. Central to ecological study is the habitat, the spatial location where an animal lives. An animal's use of resources and the conditions it tolerates in its habitat constitute its niche: how it gets its food, how it arranges for its reproductive perpetuity—in short, how it survives and propagates. The concept of niche applies also to populations and species and is usually studied by ecologists at these levels. The niche of a species, for example, comprises the resources collectively used and conditions tolerated by its members to sustain the species in its ecosystem.

The niche of a species is a product of its evolution and once it is established, no other species in the community can evolve to exploit exactly the same resources. The observation that no two species occupy the same niche in the same geographic area is called the "competitive exclusion principle": no two species may occupy the same niche. Different species are therefore able to form an ecological community in which each one has a different role in their shared environment.

In the mid-nineteenth century, the German zoologist Ernst Haeckel introduced the term **ecology,** defined as the "relation of the animal to its organic as well as inorganic environment." Environment here includes everything external to the animal but most importantly its immediate surroundings. Although we no longer restrict ecology to animals alone, Haeckel's definition is still basically sound. Animal ecology is now a highly synthetic science that incorporates everything we know about behavior, physiology, genetics, and evolution of animals to study interactions between populations of animals and their environments. The major goal of ecological studies is to understand how these diverse interactions determine geographical distributions and abundance of animal populations. Such knowledge is crucial for ensuring continued survival of many populations when their natural environments are altered by human activity.

THE HIERARCHY OF ECOLOGY

Ecology is studied as a hierarchy of biological systems in interaction with their environments. At the base of the ecological hierarchy is an **organism.** To understand why animals are distributed as they are, ecologists must examine the varied physiological and behavioral mechanisms that animals use to survive, to grow, and to reproduce. A near-perfect physiological balance between production and loss of heat is required for success of certain endothermic species (such as birds and mammals) under extreme temperatures as found in the Arctic or a desert. Other species succeed in these situations by escaping the most extreme conditions by migration, hibernation, or torpidity. Insects, fishes, and other ectotherms (animals whose body temperature depends on environmental heat) compensate for fluctuating temperatures by altering behavior and biochemical and cellular processes. Thus an animal's physiological capacities permit it to live under changing and often adverse environmental conditions. Behavioral responses are important also for obtaining food, finding shelter, escaping enemies and unfavorable environments, finding a mate, and caring for offspring. Physiological mechanisms and behaviors that improve adaptability to environments assist survival of species. Ecologists who focus their studies at the organismal level are called physiological ecologists or behavioral ecologists.

Animals in nature coexist with others of the same species; these groups are known as **populations.** Populations have properties that cannot be discovered by studying individual animals alone. These properties include genetic variability among individuals (polymorphism), growth in numbers over time, and factors that limit the density of individuals in a given area. Ecological studies at the population level help us to predict the future success of endangered species and to discover controls for pest species.

Just as individuals do not exist alone in nature, populations of different species co-occur in more complex associations known as **communities.** The complexity of a community is measured as **species diversity,** the number of different species that coexist to form the community. The populations of species in a community interact with each other in many ways, the most prevalent of which are **predation, parasitism,** and **competition. Predators** obtain energy and nutrients by killing and eating prey. **Parasites** derive similar benefits from their hosts, but usually do not kill the hosts. Competition occurs when food or space are in limited supply and members of the same or different species interfere with each other's use of shared resources. Communities are complex because all of these interactions occur simultaneously, and their individual effects on the community often cannot be isolated.

Most people are aware that lions, tigers, and wolves are predators, but the world of invertebrates also includes numerous predaceous animals. These predators include unicellular organisms, jellyfish and their relatives, various worms, predaceous insects, sea stars, and many others.

Ecological communities are biological components of even larger, more complex entities called **ecosystems.** An ecosystem consists of all populations in a community together with their physical environments. The study of ecosystems helps us to understand two key processes in nature, the flow of energy and the cycling of materials through biological channels. The largest ecosystem is the **biosphere,** the thin veneer of land, water, and atmosphere that envelops the great mass of the planet and supports all life on earth (see Chapter 37).

Environment and Niche

An animal's environment is composed of all conditions that directly affect its survival and reproduction. These factors include space, forms of energy such as sunlight, heat, wind and water currents, and also materials such as soil, air, water, and chemicals. The environment also includes other organisms, which can be an animal's food, or its predators, competitors, hosts, or parasites. The environment thus includes both abiotic (nonliving) and biotic (living) factors. Some environmental factors, such as space and food, are utilized directly by an animal and are called **resources.**

A resource may be expendable or nonexpendable, depending on how an animal uses it. Food is expendable, because once eaten it is no longer available. Food therefore must be continuously replenished in the environment. Space, whether total living area or a subset such as the number of suitable nesting sites, is not exhausted by being used, and thus is nonexpendable.

The physical space where an animal lives, which contains its environment, is its **habitat.** Size of a habitat is variable and depends upon the spatial scale under consideration. A rotten log is a normal habitat for carpenter ants. Such logs occur in larger habitats called forests where deer also are found. However, deer forage in open meadows, so their habitat is larger than the forest. On a larger scale, some migratory birds occupy forests of the north temperate region during summer and move to the tropics during winter. Thus, habitat is defined by the normal activity exhibited by an animal, rather than by arbitrary physical boundaries.

Animals of any species have certain environmental limits of temperature, moisture, and food within which they can grow, reproduce, and survive. A suitable environment therefore must simultaneously meet all requirements for life. A freshwater clam living in a tropical lake could tolerate the temperature of a tropical ocean, but would be killed by the ocean's salinity. A brittle star living in the Arctic Ocean could tolerate the salinity of the tropical ocean but not its temperature. Thus temperature and salinity are two separate dimensions of an animal's environmental limits. If we add another variable, such as pH, we increase our description to three dimensions (Figure 38-1). If we consider all environmental conditions that permit members of a species to survive and multiply, we define the role of that species in nature as distinguished from all others. This unique, multidimensional relationship of a species with its environment is called its **niche** (see the opening essay, p. 789). Dimensions of the niche vary among members of a species, making the niche subject to

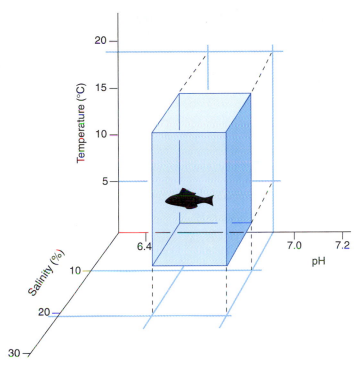

Figure 38-1
Three-dimensional niche volume of a hypothetical animal showing three tolerance ranges. This graphic representation is one way to show the multidimensional nature of environmental relations. This representation is incomplete, however, because additional environmental factors also influence growth, reproduction, and survival.

evolution by natural selection. The niche of a species undergoes evolutionary changes over successive generations.

Animals may be generalists or specialists with respect to tolerance of environmental conditions. For example, most fish are adapted to live in either fresh water or seawater, but not both. However, those that live in salt marshes, such as the minnow *Fundulus heteroclitus,* easily tolerate changes in salinity that occur over tidal cycles in these estuarine habitats as fresh water from land mixes with seawater. Similarly, whereas most snakes are capable of eating a wide variety of animal prey, others have narrow dietary requirements; for example, the African snake *Dasypeltis scaber* is specialized to eat bird eggs (Figure 32-3, p. 680).

However broad may be the tolerance limits of an animal, it experiences only a single set of conditions at a time. In fact, an animal probably will not experience in the course of its lifetime all environmental conditions that it potentially can tolerate. Thus, we must distinguish an animal's **fundamental niche,** which describes its potential role, and its **realized niche,** the subset of potentially suitable environments that an animal actually experiences. Likewise, we must distinguish fundamental niche from realized niche at the population and species levels, at which this phenomenon is mainly studied. For example, competition within a community may limit the realized niche of a species to a much smaller range of conditions than would be predicted by its fundamental niche.

Populations

Animals exist in nature as members of populations. As we saw in Chapter 6, a population is a reproductively interactive group of animals of a single species (p. 113). A species of animal may comprise a single, cohesive population or may contain many geographically disjunct populations, often called **demes.** Because members of a deme interbreed, they share a common gene pool. Migration of individuals among demes within a species can impart some evolutionary cohesion to the species as a whole.

Each population or deme has a characteristic **age structure, sex ratio,** and **growth rate.** The study of these properties and the factors that influence them is called demography. Demographic characteristics vary according to lifestyles of the species under study. For example, some animals (and most plants) are **modular.** Modular animals, such as sponges, corals, and bryozoans, consist of colonies of genetically identical organisms. Reproduction is by asexual **cloning,** as described for hydrozoans in Chapter 13 (p. 254). Most colonies also have distinct periods of gamete formation and sexual reproduction. Colonies propagate also by fragmentation, as seen on coral reefs during severe storms. Pieces of coral may be scattered by wave action on a reef, becoming seeds for formation of new reefs. For these modular animals, age structure and sex ratio are difficult to determine. Changes in colony size can be used to measure growth rate, but counting individuals is more difficult and less meaningful than in **unitary** animals, which are independently living organisms.

Most animals are unitary. However, even some unitary species reproduce by **parthenogenesis** (p. 134). Parthenogenetic species are found in many animal groups, including insects, reptiles, and fish. Such groups contain only females, which lay unfertilized eggs that hatch into daughters whose genotype comes entirely from their mothers. The praying mantid *Bruneria borealis,* common in the southeastern United States, is a parthenogenetic unitary animal.

Most metazoans are biparental (p. 133), and reproduction follows a period of organismal growth and maturation. Each new generation begins with a **cohort** of individuals born at the same time. Of course, individuals of any cohort do not all survive to reproduce. For a population to retain constant size from generation to generation, each adult female must replace herself on average with one daughter that survives to reproduce. If she produces more than one viable female offspring the population will grow; if fewer than one, the population will decline.

Animal species have different characteristic patterns of **survivorship** from birth until death of the last member of a cohort. The three principal types of survivorship are illustrated in Figure 38-2. Curve a, in which all individuals die at the same time, probably occurs only rarely in nature. Curve b, in which rate of mortality as a proportion of survivors is constant over all ages, is characteristic of some animals that care for their young, such as birds. Human populations generally fall somewhere between curves a and b, depending on nutrition and medical care.

The survivorship of most invertebrates, and of vertebrates like fish that produce great numbers of offspring, resembles curve c. For example, a mature female marine prosobranch snail,

Ilyanassa obsoleta, produces thousands of eggs each reproductive period. Zygotes develop into free-swimming planktonic veliger larvae, which can be scattered far from the mother's habitat by oceanic currents. They form part of the plankton and experience high mortality from numerous animals that feed on plankton. Furthermore, the larvae require a specific, sandy substrate on which to settle and metamorphose into adult snails. The probability of a larva surviving long enough to find a suitable habitat is very low, and most of the cohort dies during the veliger stage. We therefore see a rapid drop in survivorship in the first part of the curve. The few larvae that do survive to become snails have improved odds of surviving further, as reflected by the more gentle slope of the curve for older snails. Thus, high reproductive output balances high juvenile mortality in such animals.

Many animals survive to reproduce only once before they die, as seen in many insect species of the temperate zone. Here, adults reproduce before the onset of winter and die, leaving only their eggs to overwinter and repopulate their habitat the following spring. Similarly, Pacific salmon after several years return from the ocean to fresh water to spawn only once, after which all adults of a cohort die. However, other animals survive long enough to produce multiple cohorts of offspring that may mature and reproduce while their parents are still alive and reproductively active.

Populations of animals containing multiple cohorts, such as robins, box turtles, and humans, exhibit **age structure.** Analysis of age structure reveals whether a population is actively growing, stable, or declining. Figure 38-3 shows age profiles of two idealized populations. On a global scale, humans exhibit an age structure similar to curve a in Figure 38-2, although age structures vary among regions.

Population Growth and Intrinsic Regulation

Population growth is the difference between rates of birth and death. As Darwin recognized from an essay of Thomas Malthus (p. 106), all populations have an inherent ability to grow exponentially. This ability is called the **intrinsic rate of increase,** denoted by the symbol **r.** The steeply rising curve in Figure 38-4 shows this kind of growth. If species actually grew in this fashion unchecked, earth's resources soon would be exhausted and mass extinction would follow. A bacterium dividing three times per hour could produce a colony a foot deep over the entire

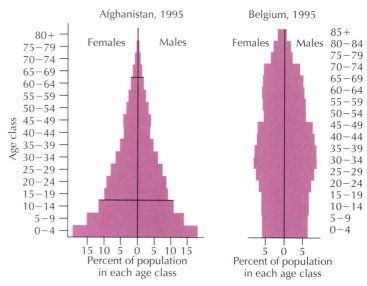

Figure 38-3

Age structure profiles of the human populations of Afghanistan and Belgium in 1995 contrast the rapidly growing, youthful population of Afghanistan with the stable population of Belgium, where the fertility rate is below replacement. Countries such as Afghanistan with a large fraction of the population as children are strained to provide adequate child services. With so many children soon to enter their reproductive years, the population will continue to grow rapidly for many years to come.

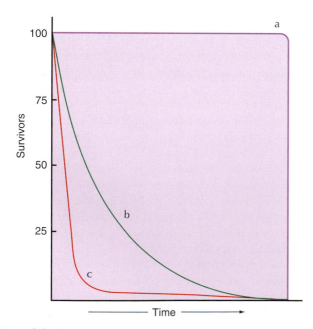

Figure 38-2

Three principal types of theoretical survivorship curves. See text for explanation.

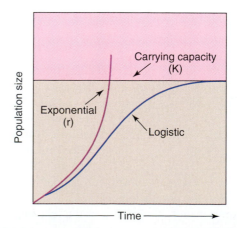

Figure 38-4

Population growth, showing exponential growth of a species in an unlimited environment, and logistic growth in a limited environment.

earth after a day and a half, and this mass would be over our heads only one hour later. Animals have much lower potential growth rates than bacteria, but could achieve the same kind of result over a longer period of time, given unlimited resources. Many insects lay thousands of eggs each year. A single codfish may spawn 6 million eggs in a season, and a field mouse can produce 17 litters of five to seven young each year. Obviously, unrestricted growth is not the rule in nature.

Even in the most benign environment, a growing population eventually exhausts food or space. Exponential increases such as locust outbreaks or planktonic blooms in lakes must end when food or space is expended. Actually, among all resources that could limit a population, the one in shortest supply relative to the needs of the population will be depleted before others. This one is termed the **limiting resource.** The largest population that can be supported by the limiting resource in a habitat is called the **carrying capacity** of that environment, symbolized **K.** Ideally, a population will slow its growth rate in response to diminishing resources until it just reaches K, as represented by the sigmoid curve in Figure 38-4. The mathematical expressions of exponential and sigmoid (or logistic) growth curves are compared in the box on page 794. Sigmoid growth occurs when there is negative feedback between growth rate and population density. This phenomenon is called density dependence, and is the mechanism for intrinsic regulation of populations. We can compare density dependence by negative feedback to the way endothermic animals regulate their body temperatures when the environmental temperature exceeds an optimum. If the resource is expendable, as with food, carrying capacity is reached when rate of resource replenishment equals rate of depletion by the population; the population is then at K for that limiting resource. According to the logistic model, when population density reaches K, rates of birth and death are equal and growth of the population ceases. Thus, a population of grasshoppers in a green meadow may be at carrying capacity even though we see plenty of unconsumed food.

Although experimental populations of protozoa may fit the logistic growth curve closely, most populations in nature tend to fluctuate above and below carrying capacity. For example, after sheep were introduced to Tasmania around 1800, their numbers changed logistically with small oscillations around an average population size of about 1.7 million; we thereby infer the carrying capacity of the environment to be 1.7 million sheep (Figure 38-5A). Ring-necked pheasants introduced on an island in Ontario, Canada exhibited wider oscillations (Figure 38-5B).

Why do intrinsically regulated populations oscillate this way? First, the carrying capacity of an environment can change over time, requiring that a population change its density to track a limiting resource. Second, animals always experience a lag between the time that a resource becomes limiting and the time that the population responds by reducing its rate of growth. Third, **extrinsic** factors occasionally may limit a population's growth below carrying capacity. We consider extrinsic factors below.

On a global scale, humans have the longest record of exponential population growth (Figure 38-5C). Although famine and war have restrained growth of populations locally, the only dip in global human growth resulted from bubonic plague ("black

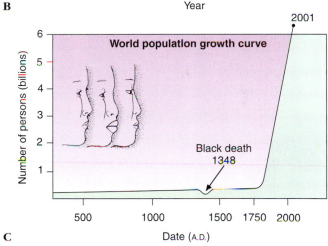

Figure 38-5

Growth curves for sheep **(A)**, ring-necked pheasant **(B)**, and world human populations **(C)** throughout history. Note that the sheep population on an island is stable because of human control of the population, but the ring-necked pheasant population oscillates greatly, probably because of large changes in carrying capacity. Where would you place the carrying capacity for the human population?

death"), which decimated much of Europe during the fourteenth century. What then is the carrying capacity for the human population? The answer is far from simple, and several important factors must be considered when estimating the human K.

With development of agriculture, the carrying capacity of the environment increased, and the human population grew steadily from 5 million around 8000 B.C., when agriculture was introduced, to 16 million around 4000 B.C. Despite the toll taken by terrible famines, disease, and war, the population reached 500 million by 1650. With the coming of the Industrial Revolution in Europe and England in the eighteenth century, followed by a medical revolution, discovery of new lands for colonization, and better agricultural practices, the human carrying capacity increased dramatically. The population doubled to 1 billion around 1850. It doubled again to 2 billion by 1927, to 4 billion in 1974, passed 6 billion in October 1999, and is expected to reach 8 billion by the year 2030. Thus, the growth has been exponential and remains high (Figure 38-5C).

Recent surveys provide hope that the growth of the human population is slackening. Between 1970 and 2000 the annual growth rate decreased from 1.9% to 1.33%. At 1.33%, it will take nearly 53 years for the world population to double rather than 36.5 years at the higher annual growth rate. The decrease is credited to better family planning. Nevertheless, half the global population is under 25 years old and most live in developing countries where access to reliable contraception is limited or nonexistent. Thus, despite a drop in growth rate, the greatest surge in population lies ahead, with a projected 3 billion people added within the next three decades, the most rapid increase ever in human numbers.

Although rapid advancements in agricultural, industrial, and medical technology have undoubtedly increased the earth's carrying capacity for humans, it also has widened the difference between birth and death rates to increase our rate of exponential growth. Each day we add 208,000 people (net) to the approximately 6.3 billion people currently alive. Assuming that growth remains constant (certainly not a dependable assumption, based on the history of human population growth), by the year 2030 more than half a million people will be added each day. In other words, less than ten days will be required to replace all people who inhabited the world in 8000 B.C.

In trying to arrive at an estimate of carrying capacity for the human species, we must consider not only quantity of resources, but quality of life. Approximately 2 billion of the 6.3 billion people alive today are malnourished. At present 99% of our food comes from the land, and the tiny fraction that we derive from the sea is decreasing due to overexploitation of fish stocks (p. 780). Although what would constitute the maximum sustainable agricultural output is uncertain, scientists do not expect food production to sustain exponential population growth.

Extrinsic Limits to Growth

We have seen that the intrinsic carrying capacity of a population for an environment prevents unlimited exponential growth of the population. Population growth also can be limited by extrinsic biotic factors, including predation, parasitism (including disease-causing pathogens), and interspecific competition, or by abiotic influences such as floods, fires, and storms. Although abiotic factors certainly can reduce populations in nature, they cannot truly regulate population growth because their effect is wholly independent of population size; abiotic limiting factors are **density-independent.** A single hailstorm can kill most

Exponential and Logistic Growth

We can describe the sigmoid growth curve (see Figure 38-4) by a simple model called the logistic equation. The slope at any point on the growth curve is the growth rate, how rapidly the population size is changing with time. If **N** represents the number of organisms and **t** the time, we can, in the language of calculus, express growth as an instantaneous rate:

$d\text{N}/d\text{t}$ = the rate of change in the number of organisms per time at a particular instant in time.

When populations are growing in an environment of unlimited resources (unlimited food and space, and no competition from other organisms), growth is limited only by the inherent capacity of the population to reproduce itself. Under these ideal conditions growth is expressed by the symbol **r**, which is defined as the intrinsic rate of population growth per capita. The index **r** is actually the difference between the birth rate and death rate per individual in the population per unit time. The growth rate of the population as a whole is then:

$$d\text{N}/d\text{t} = r\text{N}$$

This expression describes the rapid, **exponential growth** illustrated by the early upward-curving portion of the sigmoid growth curve (see Figure 38-4).

Growth rate for populations in the real world slows as the upper limit is approached, and eventually stops. At this point **N** has reached its maximum density because the space being studied has become "saturated" with animals. This limit is called the carrying capacity of the environment and is expressed by the symbol **K.** The sigmoid population growth curve now can be described by the logistic equation, which is written as:

$$d\text{N}/d\text{t} = r\text{N}([\text{K} - \text{N}]/\text{K})$$

This equation states that the rate of increase per unit of time $(d\text{N}/d\text{t})$ = rate of growth per capita **(r)** × population size **(N)** × unutilized freedom for growth ([**K** − **N**]/**K**). We can see from the equation that when the population approaches carrying capacity, **K** − **N** approaches 0, $d\text{N}/d\text{t}$ also approaches 0, and the curve will flatten.

Populations occasionally overshoot the carrying capacity of the environment so that **N** exceeds **K.** The population then exhausts some resource (usually food or shelter). The rate of growth, $d\text{N}/d\text{t}$, then becomes negative and the population declines.

young of wading bird populations, and a forest fire can eliminate entire populations of many animals, regardless of how many individuals exist.

In contrast, biotic factors can and do act in a **density-dependent** manner. Predators and parasites respond to changes in density of their prey and host populations, respectively, to maintain populations at fairly constant sizes. These sizes are below carrying capacity, because populations regulated by extrinsic factors are not limited by their resources. Competition between species for a common limiting resource lowers the effective carrying capacity for each species below that of either one alone.

Community Ecology

Interactions among Populations in Communities

Populations of animals that form a **community** interact in a variety of ways that can be detrimental (−), beneficial (+), or neutral (0) to each species, depending on the interaction. For instance, we can consider a predator's effect on its prey as (−), because survival of the prey animal is reduced. However, the same interaction benefits the predator (+) because food obtained from prey increases a predator's ability to survive and to reproduce. Thus, the predator-prey interaction is + −. Ecologists use this shorthand notation to characterize interspecific interactions because it shows the direction in which the interaction affects each species.

We see other kinds of + − interactions. One is **parasitism,** in which the parasite benefits by using the host as a home and source of nutrition, and the host is harmed. **Herbivory,** in which an animal eats a plant, is another + − relationship. **Commensalism** is an interaction that benefits one species and neither harms nor benefits the other (0+). Most bacteria that normally inhabit our intestinal tracts do not affect us (0), but the bacteria benefit (+) by having food and a place to live. Some evidence suggests that the harmless bacteria in our intestinal tracts may prevent entry of more harmful bacterial species, in which case this commensalism grades into **mutualism.** A classic example of commensalism is the association of pilot fishes and remoras with sharks (Figure 38-6). These fishes get the "crumbs"

Figure 38-6
Four remoras, *Remora* sp., attached to a shark. Remoras feed on fragments of food left by their shark host, as well as on pelagic invertebrates and small fishes. Although they actually are good swimmers, remoras prefer to be pulled through the water by marine creatures or boats. The shark host may benefit by having embedded copepod skin parasites removed by remoras.

remaining when the host shark makes its kill, but we now know that some remoras also feed on ectoparasites of sharks. Therefore, this commensalism also grades into mutualism.

Organisms engaged in mutalism have a friendlier arrangement than commensalistic species, because the fitness of both is enhanced (++). Biologists are finding mutualistic relationships far more common in nature than previously believed (Figure 38-7). Some mutualistic relationships are not only beneficial, but necessary for survival of one or both species. An example is the relationship between a termite and protozoa inhabiting its gut. The protozoa can digest wood eaten by the termite because the protozoa produce an enzyme, lacking in the termite, that digests cellulose; the termite lives on waste products of protozoan metabolism. In return, the protozoa gain a habitat and food supply. Such absolute interdependence among species can be a liability if one of the participants is lost. *Calvaria* trees native to the island of Mauritius have not reproduced successfully for over 300 years, because their seeds germinate only after being eaten and passed through the gut of a dodo bird, now extinct.

The Biodiversity Crisis

We are experiencing the most serious environmental crisis in our history. The single most direct measure of the planet's health is biodiversity. At unprecedented rates, we are losing its three components: species diversity, genetic diversity, and habitat diversity. These losses arise primarily from habitat destruction, especially rapid destruction of tropical rain forests. Excessive fishing and hunting, illegal pet trade, pollution, and spread of invasive exotic species also intensify the biodiversity crisis.

Earth's biodiversity provides us with food and fiber; it underpins our health services (some 25% of drugs prescribed in the United States include chemical compounds from wild organisms); it provides genes for maintaining vigor of crops and livestock; it provides natural pest control and creates our soils. Loss of keystone species (p. 799) such as African elephants, sea otters, or sea stars means significant changes and lowering of diversity within their ecosystems. With extinction of animal species we often lose beauty and wonder. We will miss the breaching of whales, the howling of wolves, and the grace of cheetahs. Only through significant conservation efforts and through laws that limit environmental degradation will we preserve biodiversity.

Figure 38-7

Among the many examples of mutualism that abound in nature is the whistling thorn acacia of the African savanna and the ants that make their homes in the acacia's swollen galls. The acacia provides both protection for the ants' larvae *(lower photograph of opened gall)* and honeylike secretions used by the ants as food. In turn, the ants protect the tree from herbivores by swarming out as soon as the tree is touched. Giraffes, however, which love the tender acacia leaves, seem immune to the ants' fiery stings.

Competition between species reduces fitnesses of both (− −). Many biologists, including Darwin, considered competition the most common and important interaction in nature. Ecologists have constructed most of their theories of community structure from the premise that competition is the chief organizing factor in species assemblages. Sometimes the effect on one species in a competitive relationship is negligible. This condition is called **amensalism,** or **asymmetric competition** (0−). For example, two species of barnacles that commonly occur in rocky intertidal habitats, *Chthamalus stellatus* and *Balanus balanoides,* compete for space. A famous experiment by Joseph Connell* demonstrated that *B. balanoides* excluded *C. stellatus* from a portion of the habitat, while *C. stellatus* had no effect on *B. balanoides*.

We have treated interactions as occurring between pairs of species. However, in natural communities containing populations of many species, a predator may have more than one prey and several animals may compete for the same resource. Thus, eco-

---

*Connell, J. H. 1961. The influence of interspecific competition and other factors on the distribution of the barnacle *Chthamalus stellatus*. Ecology **42:**710–723.

logical communities are quite complex and dynamic, a challenge to ecologists who wish to study this level of natural organization.

Competition and Character Displacement

Competition occurs when two or more species share a **limiting resource.** Simply sharing food or space with another species does not produce competition unless the resource is in short supply relative to the needs of the species that share it. Thus, we cannot prove that competition occurs in nature based solely on sharing of resources. However, we find evidence of competition by investigating the different ways that species exploit a resource.

Competing species may reduce conflict by reducing overlap of their niches. **Niche overlap** is the portion of a niche's resources shared by two or more species. For example, if two species of birds eat seeds of exactly the same size, competition eventually will exclude one species from the habitat. This example illustrates the principle of **competitive exclusion:** strongly competing species cannot coexist indefinitely. To coexist in the same habitat, species must specialize by partitioning a shared resource and using different portions of it. Specialization of this kind is called **character displacement.**

Character displacement usually appears as differences in organismal morphology or behavior related to exploitation of a resource. For example, in his classic study of Galápagos finches (p. 116), English ornithologist David Lack noticed that bill sizes of these birds depended on whether they occurred together on the same island (Figure 38-8). On the islands Daphne and Los Hermanos, where *Geospiza fuliginosa* and *G. fortis* occur separately and therefore do not compete with each other, bill sizes are nearly identical; on the island Santa Cruz, where both *G. fuliginosa* and *G. fortis* coexist, their bill sizes do not overlap. These results suggest resource partitioning, because bill size determines the size of seeds selected for food. Recent work by

American ornithologist Peter Grant confirms what Lack suspected: *G. fuliginosa* with its smaller bill selects smaller seeds than does *G. fortis* with its larger bill. Where these two species coexisted, competition between them has led to evolutionary displacement of bill sizes to diminish competition. An absence of competition today has been called appropriately "the ghost of competition past."

Character displacement promotes coexistence by reducing niche overlap. When several species share the same general resources by such partitioning, they form a **guild.** Just as a guild in medieval times constituted a brotherhood of men sharing a common trade, species in an ecological guild share a common livelihood. The term guild was introduced to ecology by Richard Root in his 1967 paper on niche patterns of the blue-gray gnatcatcher.[†] A classic example of a bird guild is Robert MacArthur's study of a feeding guild consisting of five species of warblers in spruce woods of the northeastern United States.[‡] At first glance, we might ask how five birds, very similar in size and appearance, could co-exist by feeding on insects in the same tree. However, on close inspection MacArthur found subtle differences among these birds in sites of foraging (Figure 38-9). One species searched only on outer branches of spruce crowns; another species used the top 60% of the tree's outer and inner branches, although not next to the trunk; another species concentrated on inner branches closer to the trunk; another species used the midsection from the periphery to the trunk; and still another species foraged in the bottom 20% of the tree. These observations suggest that each warbler's niche within this guild is defined by structural differences in habitat.

Guilds are not limited to birds. For example, a study done in England on insects associated with Scotch broom plants revealed nine different guilds of insects, including three species of stem miners, two gall-forming species, two that fed on seeds and five that fed on leaves. Another insect guild consists of three species of praying mantids that avoid both competition and predation by differing in sizes of their prey, timing of hatching, and height of vegetation in which they forage.

Predators and Parasites

Ecological warfare waged by predators against their prey causes coevolution: predators get better at catching prey, and prey get better at escaping predators. This is an evolutionary race that predators cannot afford to win. If a predator became so efficient that it exterminated its prey, the predator species would become extinct. Because most predators feed on more than a single species, specialization on a single prey species to the point of extermination is uncommon.

However, when a predator does rely primarily on a single prey species, both populations tend to fluctuate cyclically. First prey density increases, then that of the predator until prey become scarce. At that point, predators must adjust their population size

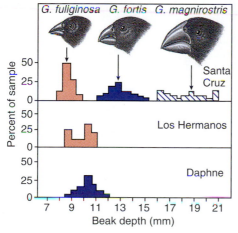

Figure 38-8

Displacement of beak sizes in Darwin's finches from the Galápagos Islands. Beak depths are given for the ground finches *Geospiza fuliginosa* and *G. fortis* where they occur together (sympatric) on Santa Cruz Island and where they occur alone on the islands Daphne and Los Hermanos. *Geospiza magnirostris* is another large ground finch that lives on Santa Cruz.

[†]Root, R. B. 1967. The niche exploitation pattern of the blue-gray gnatcatcher. Ecological Monographs **37**:317–350.
[‡]MacArthur, R. H. 1958. Population ecology of some warblers of northeastern coniferous forests. Ecology **39**:599–619.

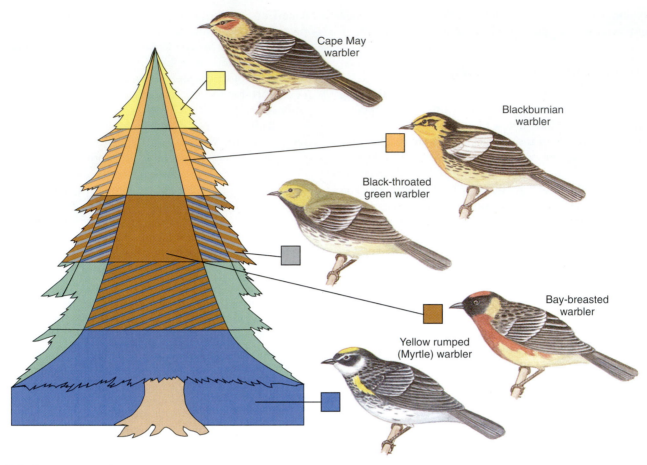

Figure 38-9

Distribution of foraging effort among five species of wood warblers in a northeastern spruce forest. The warblers form a feeding guild.

downward by leaving the area, lowering reproduction, or dying. When density of the predator population falls enough to allow reproduction by prey to outpace mortality from predation, the cycle begins again. Thus, populations of both predators and prey show cycles of abundance, but increases and decreases in predator abundance are slightly delayed relative to those of prey because of the time lag in a predator's response to changing prey density. We can illustrate this process in the laboratory with protozoa (Figure 38-10). Perhaps the longest documented natural example of a predator-prey cycle is between Canadian populations of snowshoe hares and lynxes (see Figure 28-27, p. 600).

The war between predators and prey reaches high art in the evolution of defenses by potential prey. Many animals that are palatable escape detection by matching their background, or by resembling some inedible feature of the environment (such as a stick). Such defenses are called cryptic. In contrast to cryptic defenses, animals that are toxic or distasteful to predators actually advertise their strategy with bright colors and conspicuous behavior. Such defenses are termed aposematic. These species are protected because predators learn to recognize and to avoid them after one or more distasteful encounters.

When distasteful prey adopt warning coloration, advantages of deceit arise for palatable prey. Palatable prey can deceive potential predators by mimicking distasteful prey, a phenome-

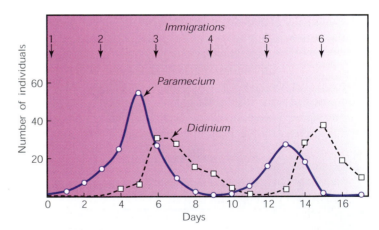

Figure 38-10

Classic predator-prey experiment by Russian biologist G. F. Gause in 1934 shows the cyclic interaction between predator (*Didinium*) and prey (*Paramecium*) in laboratory culture. When the *Didinium* find and eat all the *Paramecium*, the *Didinium* themselves starve. Gause could keep the two species coexisting only by occasionally introducing one *Didinium* and one *Paramecium* to the culture (*arrows*); these introductions simulated migration from an outside source.

Figure 38-11

Artful guises abound in the tropics. **A,** A palatable butterfly *(top photograph)* mimics a poisonous butterfly of a different family *(lower photograph).* **B,** A harmless clearwing moth *(top photograph)* mimics a yellowjacket wasp, which is armed with a stinger *(lower photograph).* Both **A** and **B** are examples of Batesian mimicry. **C,** Two unpalatable tropical butterflies of different taxonomic families resemble one another, an example of Müllerian mimicry.

non called **Batesian mimicry.** Coral snakes and monarch butterflies are both brightly colored, noxious prey. Coral snakes have a venomous bite, and monarch butterflies are poisonous because caterpillars store poison (cardiac glycoside) from milkweed they eat. Both species serve as **models** for other species, called **mimics,** that do not possess toxins of their own but look like the model species that do (Figure 38-11A and B).

In another form of mimicry termed **Müllerian mimicry,** two or more toxic species resemble each other (Figure 38-11C). We can ask why an animal that has its own poison should gain by evolving resemblance to another poisonous animal. The answer is that a predator needs only to experience the toxicity of one species to avoid all similar prey. A predator can learn one warning signal more easily than many! The benefits that two distasteful species derive from mutual mimicry are not always equal, for example, when a mildly distasteful species mimics a highly poisonous one. Such cases illustrate a continuum between Batesian and Müllerian mimicry.

Sometimes the influence of one population on others is so pervasive that its absence drastically changes the character of an entire community. We call such a population a **keystone species.**[*] For example, in 1983, a mysterious epidemic swept through Caribbean populations of the sea urchin *Diadema antillarum,* destroying more than 95% of the animals. The immediate effect was on the algal community, no longer grazed by urchins. In some reefs the algae grew from a thin mat to a thick canopy of altered composition. Both productivity and diversity on the coral reefs declined. *Diadema antillarum* clearly was a keystone

predator species for those communities. On the West Coast, the sea star *Pisaster ochraceous* is a keystone species. Sea stars are a major predator of the mussel *Mytilus californianus.* When sea stars were removed experimentally from a patch of Washington State coastline, mussels expanded in numbers, occupying all space previously used by 25 other invertebrate and algal species (Figure 38-12). Keystone predators act by reducing prey populations below the level at which resources such as space are limiting. The original notion that all keystone species were predators has been broadened to include any species whose removal causes extinction of others.

By reducing competition, keystone species may allow more species to coexist on the same resource. Consequently they contribute to maintaining diversity in a community. Keystone species can be considered a special case of a more general phenomenon, disturbance. Periodic natural disturbances such as fires and hurricanes also can prevent monopolization of resources and competitive exclusion by a few broadly adapted competitors. Ecologists now believe that disturbances permit more species to coexist in such highly diverse communities as coral reefs and rain forests.

Parasites are often considered freeloaders because they appear to get something from their hosts for nothing. Coevolution between parasite and host may be expected to generate an increasingly benign, less virulent relationship. Selection would favor a benign relationship, because a parasite's fitness is diminished if its host dies. This traditional view has been challenged in recent years. Virulence is correlated, at least in part, with availability of new hosts. When alternative hosts are common and transmission rates high, a host's life is of less value to a parasite, which may become more virulent.

[*]Paine, R. T. 1969. A note on trophic complexity and community stability. *American Naturalist* **103:**91–93.

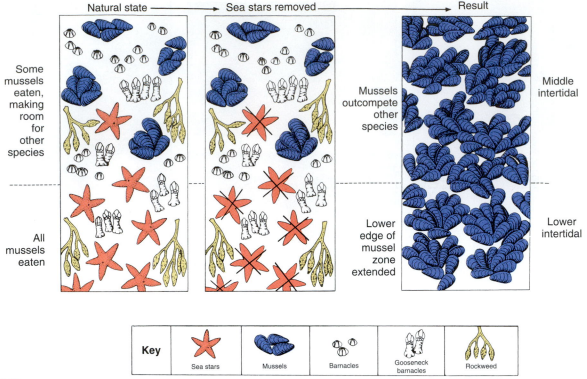

Natural state ──────→ Sea stars removed ──────────→ Result

Some mussels eaten, making room for other species

All mussels eaten

Mussels outcompete other species

Lower edge of mussel zone extended

Middle intertidal

Lower intertidal

| Key | Sea stars | Mussels | Barnacles | Gooseneck barnacles | Rockweed |

Figure 38-12

The experimental removal of a keystone species, the predatory sea star *Pisaster ochraceus,* from an intertidal community completely changes the structure of the community. With their principal predator missing, mussels form dense beds by outcompeting and replacing other intertidal species.

Ecosystems

Transfer of energy and materials among organisms within ecosystems is the ultimate level of organization in nature. Energy and materials are required to construct and to maintain life, and their incorporation into biological systems is called **productivity.** Productivity is divided into component **trophic levels** based on how organisms obtain energy and materials. Trophic levels are linked together into **food webs** (Figure 38-13), which are pathways for transfer of energy and materials among organisms within an ecosystem.

Primary producers are organisms that begin productivity by fixing and storing energy from outside the ecosystem. Primary producers usually are green plants that capture solar energy through **photosynthesis** (but see an exception in the box titled "Life without the Sun"). Powered by solar energy, plants assimilate and organize minerals, water, and carbon dioxide into living tissue. All other organisms survive by consuming this tissue, or by consuming organisms that consumed this tissue. **Consumers** include **herbivores,** which eat plants directly, and **carnivores,** which eat other animals. The most important consumers are **decomposers,** mainly bacteria and fungi that break dead organic matter into its mineral components, returning it to a soluble form that can be used by plants to re-enter the cycle. Although important chemicals such as nitrogen and carbon are reused endlessly through biological cycling, all energy ultimately is lost from the ecosystem as heat and cannot be recycled. Thus, no ecosystems, including the earth's biosphere, are truly closed.

Energy Flow

Every organism in nature has an **energy budget.** Just as we each must partition our income for housing, food, utilities, and taxes, each organism must obtain enough energy to meet its metabolic costs, to grow, and to reproduce.

Ecologists divide the budget into three main components: **gross productivity, net productivity,** and **respiration.** Gross productivity is like gross income; it is the total energy assimilated, analogous to your paycheck before deductions. When an animal eats, food passes through its gut and nutrients are absorbed. Most energy assimilated from these nutrients serves the animal's metabolic demands, which include cellular metabolism and regulation of body heat in endotherms. Energy required for metabolic maintenance is respiration, which is deducted from gross productivity to arrive at net productivity, an animal's take-home pay. Net productivity is energy stored by an animal in its tissues as **biomass.** This energy is available for growth, and also for reproduction, which is population growth.

The energy budget of an animal is expressed by a simple equation, in which gross and net productivity are represented by P_g and P_n, respectively, and respiration is **R:**

$$P_n = P_g - R$$

This equation is a way of stating the first law of thermodynamics (p. 57) in the context of ecology. Its important messages are that the energy budget of every animal is finite and may be limiting, and that energy is available for growth of individuals and populations only after maintenance is satisfied.

Figure 38-13
Midwinter food web in *Salicornia* salt
marsh of San Francisco Bay area.

Third-level
carnivores

Hawks,
owls

Second-level
carnivores

Rat

Sparrow

Migrant shore birds

Mice, voles

Rails, ducks

First-level
carnivores

Shrew

Egrets,
herons

Fishes

Herbivores

Terrestrial invertebrates

Fishes

Intertidal marine
invertebrates

Producers

Terrestrial
and salt marsh
plants

Marine
plants and
algae

Life without the Sun

For many years, ecologists believed that all animals depended directly or indirectly on primary production from solar energy. However, in 1977 and 1979, dense communities of animals were discovered living on the sea floor adjacent to vents of hot water issuing from rifts (Galápagos Rift and East Pacific Rise) where tectonic plates on the sea floor slowly spread apart. These communities (see photo) include several species of molluscs, some crabs, polychaete worms, enteropneusts (acorn worms), and giant pogonophoran worms. The temperature of seawater above and immediately around vents is 7° to 23° C where it is heated by basaltic intrusions, whereas the surrounding normal seawater is 2° C.

The producers in these vent communities are chemoautotrophic bacteria that derive energy from oxidation of hydrogen sulfide abundant in the vent water and fix carbon dioxide into organic carbon. Some animals in the vent communities—for example, bivalve molluscs—are filter feeders that ingest bacteria. Others, such as giant pogonophoran tubeworms (see p. 429), which lack mouths and digestive tracts, harbor colonies of symbiotic bacteria in their tissues and use the organic carbon that these bacteria synthesize.

A population of giant pogonophoran tubeworms grows in dense profusion near a Galápagos Rift thermal vent, photographed at 2800 m (about 9000 feet) from the deep submersible *Alvin.* Also visible in the photograph are mussels and crabs.

The second law of thermodynamics, which states that total disorder or randomness of a system always increases, is important when we study energy transfers between trophic levels in food webs. Energy for maintenance, R, usually constitutes more than 90% of the assimilated energy (P_g) for animal consumers. More than 90% of the energy in an animal's food is lost as heat, and less than 10% is stored as biomass. Each succeeding trophic level therefore contains only 10% of the energy in the next lower trophic level. Most ecosystems are thereby limited to five or fewer trophic levels.

Our ability to feed a growing human population is influenced profoundly by the second law of thermodynamics (p. 57). Humans, who occupy a position at the end of the food chain, may eat the grain that fixes the sun's energy; this very short chain represents an efficient use of potential energy. Humans also may eat beef from animals that eat grass that fixes the sun's energy; the addition of a trophic level decreases available energy by a factor of 10. Ten times as much plant biomass is needed to feed humans as meat eaters as to feed humans as grain eaters. Consider a person who eats a bass that eats a sunfish that eats zooplankton that eats phytoplankton that fixes the sun's energy. The tenfold loss of energy occurring at each trophic level in this five-step chain requires that the pond must produce 5 tons of phytoplankton for a person to gain a pound by eating bass. If the human population depended on bass for survival, we would quickly exhaust this resource.

These figures must be considered as we look to the sea for food. Productivity of oceans is very low and limited largely to estuaries, marshes, reefs, and upwellings where nutrients are available to phytoplankton producers. Such areas constitute a small part of the ocean. The rest is a watery void.

Marine fisheries supply about one-fifth of the world's dietary protein, but much of this protein is used to feed livestock and poultry. If we remember the rule of 10-to-1 loss in energy with each transfer of material between trophic levels, then use of fish as food for livestock rather than humans is poor use of a valuable resource in a protein-deficient world. Fishes that we prefer to eat include flounder, tuna, and halibut, which are three or four levels up the food chain. Every 125 g of tuna requires one metric ton of phytoplankton to produce. If humans are to derive greater benefit from oceans as a food source, we must eat more of the less desirable fishes from lower trophic levels.

When we examine the food chain in terms of biomass at each level, we can construct **ecological pyramids** either of numbers, energy, or biomass. A pyramid of numbers (Figure 38-14A), also called an **Eltonian pyramid,** depicts numbers of organisms transferred between each trophic level. This pyramid provides a vivid impression of the great difference in numbers of organisms involved in each step of the chain, and supports the observation that large predatory animals are rarer than the small animals on which they feed. However, a pyramid of numbers does not indicate actual mass of organisms at each level.

Figure 38-14

Ecological pyramids of numbers, biomass, and energy. Pyramids are generalized, since the area within each trophic level is not scaled proportionally to quantitative differences in units given.

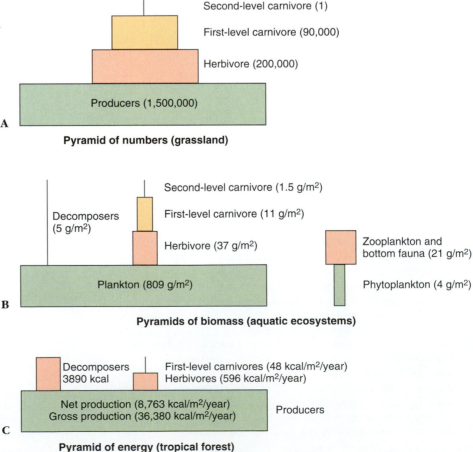

The concepts of food chains and ecological pyramids were invented and first explained in 1923 by Charles Elton, a young ecologist at Oxford University. Working for a summer on a treeless arctic island, Elton watched arctic foxes as they roamed, noting what they ate and, in turn, what their prey had eaten, until he was able to trace the complex cycling of nitrogen in food throughout the animal community. Elton realized that life in a food chain comes in discrete sizes, because each form evolved to be much bigger than the thing it eats. He thus explained the common observation that large animals are rare while small animals are common.

More instructive are pyramids of biomass (Figure 38-14B), which depict the total bulk, or "standing crop," of organisms at each trophic level. Such pyramids usually slope upward because mass and energy are lost at each transfer. However, in some aquatic ecosystems in which the producers are algae, which have short life spans and rapid turnover, the pyramid is inverted. Algae can tolerate heavy exploitation by zooplankton consumers. Therefore, the base of the pyramid (biomass of phytoplankton) is smaller than the biomass of zooplankton it supports. We could liken this inverted pyramid to a person who weighs far more than the food in a refrigerator, but who is sustained from the refrigerator because food is constantly replenished.

A third type of pyramid is a pyramid of energy, which shows rate of energy flow between levels (Figure 38-14C). An energy pyramid is never inverted because energy transferred from each level is less than what went into it. A pyramid of energy gives the best overall picture of community structure because it is based on production. In the example above, productivity of phytoplankton exceeds that of zooplankton, even though biomass of phytoplankton is less than biomass of zooplankton (because of heavy grazing by the zooplankton consumers).

Nutrient Cycles

All elements essential for life are derived from environmental air, soil, rocks, and water. When plants and animals die and their bodies decay, or when organic substances are burned or oxidized, elements and inorganic compounds essential for life processes (nutrients) are released and returned to the environment. Decomposers fulfill an essential role in this process by feeding on remains of plants and animals and on fecal material. The result is that nutrients flow in a perpetual cycle between biotic and abiotic components of an ecosystem. Nutrient cycles are often called **biogeochemical cycles** because they involve exchanges between living organisms (bio-) and rocks, air, and water of the earths' crust (geo-). Continuous input of energy from the sun keeps nutrients flowing and the ecosystem functioning (Figure 38-15).

We tend to think of biogeochemical cycles in terms of naturally occurring elements, such as water, carbon, and nitrogen. In recent times, however, humans have added to the biosphere synthetic materials that have entered food webs, sometimes with disastrous consequences. Probably most harmful of these materials,

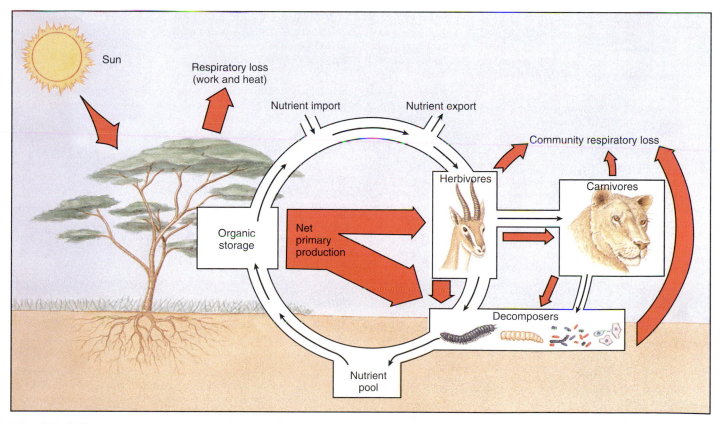

Figure 38-15

Nutrient cycles and energy flow in a terrestrial ecosystem. Note that nutrients are recycled, whereas energy flow (red) is one way.

"R.F.D.2."

SORRY. I KNOW IT SEEMS UNFAIR.

YOU SEE, THIS EGG IS A LINK IN THE FOOD CHAIN

AND I, A HUMAN, AM AT THE END OF THE CHAIN.

I HOPE.

Steve Stinson and Roanoke Times and World-News

in terms of ecosystemic processes, are pesticides. We currently produce about 2.5 million *tons* of pesticides worldwide, mainly to protect crops from insects.

Despite such extensive use of poison, more than half our crops are lost either before or after harvest to pests. The role of pesticides in natural food webs can be insidious for three reasons. First, many pesticides become concentrated as they travel up succeeding trophic levels. The highest concentrations will occur in the biomass of top carnivores such as hawks and owls, diminishing their ability to reproduce. Second, many species that are killed by pesticides are not pests, but merely innocent bystanders, called nontarget species. Nontarget effects happen when pesticides move outside the agricultural field to which they were applied, through rainwater runoff, leaching through the soil, or dispersal by wind. The third problem is persistence; some chemicals used as pesticides have great longevity in the environment, so that nontarget effects persist long after pesticide application. Scientists are working to create new pesticides that are more specific in their effects and decompose faster in the environment.

SUMMARY

Ecology is the study of relationships between organisms and their environments to explain the distribution and abundance of life on earth. The physical space that contains an animal's environment is its habitat. Within the habitat are physical and biological conditions suitable to survival and reproduction, which constitute the niche of an animal, population, or species.

Animal populations consist of demes of interbreeding members sharing a common gene pool. Cohorts of animals have characteristic patterns of survivorship that represent adaptive trade-offs between parental care and numbers of offspring. Animal populations consisting of overlapping cohorts have age structure that indicates whether they are growing, declining, or at equilibrium.

Every species in nature has an intrinsic rate of increase that gives it the potential for exponential growth. The human population is growing exponentially at about 1.33% each year, and is expected to increase from 6.3 billion to 8 billion by the year 2030. Population growth may be regulated intrinsically by the carrying capacity of the environment, extrinsically by competition between species for a limiting resource, or by predators or parasites. Density independent abiotic factors can limit, but not truly regulate, population growth.

Communities consist of populations that interact with one another in any of several ways, including competition, predation, parasitism, commensalism, and mutualism. These relationships are the results of coevolution among populations within communities. Guilds of species avoid competitive exclusion by character displacement, the partitioning of limited resources by morphological specialization. Keystone predators are those that control community structure and reduce competition among prey, which increases species diversity. Parasites and their hosts evolve a benign relationship that ensures their coexistence.

Ecosystems consist of communities and their abiotic environments. Animals occupy the trophic levels of herbivorous and carnivorous consumers within ecosystems. All organisms have an energy budget consisting of gross and net productivity, and respiration. For animals, respiration usually is at least 90% of this budget. Thus, the transfer of energy from one trophic level to another is limited to about 10%, which in turn limits the number of trophic levels in an ecosystem. Ecological pyramids of energy depict how productivity decreases in successively higher trophic levels of food webs.

Ecosystem productivity is a result of energy flow and material cycles within ecosystems. All energy is lost as heat, but nutrients and other materials including pesticides are recycled. No ecosystem, including the global biosphere, is closed because all depend upon exchanges of energy and materials from outside sources.

REVIEW QUESTIONS

1. The term ecology is derived from the Greek meaning "house" or "place to live." However, as used by scientists, the term "ecology" is not the same as "environment." What is the distinction between these terms?
2. How would you distinguish between ecosystem, community, and population?
3. What is the distinction between habitat and environment?
4. Define the niche concept. How does the "realized niche" of a population differ from its "fundamental niche"? How does the concept of niche differ from the concept of guild?
5. Populations of independently living (unitary) animals have a characteristic age structure, sex ratio, and growth rate. However, these properties are difficult to determine for modular animals. Why?

6. Explain which of the three survivorship curves in Figure 38-2 best fits the following: (a) a population in which mortality as a proportion of survivors is constant; (b) a population in which there is little early death and most individuals live to old age; (c) a population that experiences heavy mortality of the very young but with the survivors living to old age. Offer an example from the real world of each survivorship pattern.

7. Contrast exponential and logistic growth of a population. Under what conditions might you expect a population to exhibit exponential growth? Why cannot exponential growth be perpetuated indefinitely?

8. Growth of a population may be hindered by either density-dependent or density-independent mechanisms. Define and contrast these two mechanisms. Offer examples of how growth of the human population might be curbed by either agent.

9. Herbivory is an example of an interspecific interaction that is beneficial for the animal (+) but harmful to the plant it eats (−). What are some + − interactions among animal populations? What is the difference between commensalism and mutualism?

10. Explain how character displacement can ease competition between coexisting species.

11. Define predation. How does the predator-prey relationship differ from the parasite-host relationship? Why is the evolutionary race between predator and prey one that the predator cannot afford to win?

12. Mimicry of monarch butterflies by viceroys is an example of a palatable species resembling a toxic one. What is the advantage to the viceroy of this form of mimicry? What is the advantage to a toxic species of mimicking another toxic species?

13. A keystone species has been defined as one whose removal from a community causes the extinction of other species. How does this extinction happen?

14. What is a trophic level, and how does it relate to a food chain?

15. Define *productivity* as the word is used in ecology. What is a primary producer? What is the distinction between gross productivity, net productivity, and respiration? What is the relation of net productivity to biomass (or standing crop)?

16. What is a food chain? How does a food chain differ from a food web?

17. How is it possible to have an inverted pyramid of biomass in which the consumers have a greater biomass than the producers? Can you think of an example of an inverted pyramid of *numbers* in which there are, for example, more herbivores than plants on which they feed?

18. The pyramid of energy has been offered as an example of the second law of thermodynamics (p. 57). Why?

19. Animal communities surrounding deep-sea thermal vents apparently exist in total independence of solar energy. How can this existence be possible?

SELECTED REFERENCES

Brooks, D. R., and D. A. McLennan. 1991. Phylogeny, ecology, and behavior. Chicago, University of Chicago Press. *A discussion of how systematic methods can be used to enhance the study of ecology and animal behavior.*

Dodson, S. I., T. F. H. Allen, S. R. Carpenter, A. R. Ives, R. L. Jeanne, J. F. Kitchell, N. E. Langston, and M. G. Turner. 1998. Ecology. New York, Oxford University Press. *Eight authors to this ecology textbook contribute their expertise in specific fields.*

Kates, R. W. 1994. Sustaining life on the earth. Sci. Am. **271:**114–122 (Oct.). *Major technological revolutions—toolmaking, agriculture, and manufacturing—have triggered geometric growth in the human population. The author asks whether we can learn enough about biological, physical, and social reality to fashion a future that our planet can sustain.*

Krebs, C. J. 2001. Ecology: the experimental analysis of distribution and abundance, ed. 5. San Francisco, Benjamin-Cummings. *Important treatment of population ecology.*

Molles, M. C. Jr. 2002. Ecology: concepts and applications, ed. 2. New York, McGraw-Hill. *A concise and well-illustrated survey of ecology.*

Pianka, E. R. 2000. Evolutionary ecology. San Francisco, Benjamin-Cummings. *An introduction to ecology written from an evolutionary perspective.*

Ricklefs, R. E. 1997. The economy of nature, ed. 4. New York, W. H. Freeman. *Clearly written, well-illustrated general ecology text.*

Smil, V. 1997. Global population and the nitrogen cycle. Sci. Am. **277:**78–81 (July). *The sudden growth of the human population worldwide in the 20th century parallels global consumption of synthetically-produced nitrogen-rich fertilizers upon which humans now heavily depend for food production. But there are adverse consequences for the environment.*

Smith, R. L. and T. M. Smith. 2001. Ecology and field biology. San Francisco, Benjamin-Cummings. *Clearly written, well-illustrated general ecology text.*

Tunnicliffe, V. 1992. Hydrothermal-vent communities of the deep sea. Am. Sci. **80:**336–349. *At hot vents along mid-ocean ridges, nuclear and chemical energy make possible exotic ecosystems that have evolved in near-total isolation.*

ZOOLOGY LINKS TO THE INTERNET

Visit the textbook's Online Learning Center at www.mhhe.com/zoology to find live Internet links for each of the topics listed here.

General Ecology Sites
Animal Population Ecology
Life Histories
Population Growth

Community Ecology
Competition
Parasitism, Predation, Herbivory
Mutualism
Food Webs
Global Ecology

This glossary lists definitions, pronunciations, and derivations of the most important recurrent technical terms, units, and names (excluding taxa) used in the text.

A

abiotic (ā′bī-äd′ik) (Gr. *a*, without, + *biōtos*, life, livable). Characterized by the absence of life.

abomasum (ab′ō-mā′səm) (L. *ab*, from, + *omasum*, paunch). Fourth and last chamber of the stomach of ruminant mammals.

aboral (ab-o′rəl) (L. *ab*, from, + *os*, mouth). A region of an animal opposite the mouth.

abscess (ab′ses) (L. *abscessus*, a going away). Dead cells and tissue fluid confined in a localized area, causing swelling.

acanthodians (a′kan-thō′dē-əns) (Gr. *akantha*, prickly, thorny). Along with placoderms, these groups were among the earliest known true jawed fishes from Lower Silurian to Lower Permian.

acanthor (ə-kan′thor) (Gr. *akantha*, spine or thorn, + *or*). First larval form of acanthocephalans in the intermediate host.

acclimatization (ə-klī′mə-də-zā-shən) (L. *ad*, to, + Gr. *klima*, climate). Gradual physiological adaptation in response to relatively long-lasting environmental changes.

acetabulum (as′ə-tab′ū-ləm) (L. a little saucer for vinegar). True sucker, especially in flukes and leeches; the socket in the hip bone that receives the thigh bone.

acicula (ə-sik′ū-lə) (L. *acicula*, a small needle). Needlelike supporting bristle in parapodia of some polychaetes.

acid A molecule that dissociates in solution to produce a hydrogen ion (H$^+$).

acinus (as′ə-nəs), pl. **acini** (as′ə-nī) (L. grape). A small lobe of a compound gland or a saclike cavity at the termination of a passage.

acoelomate (ā-sēl′ə-māt′) (Gr. *a*, not, + *koilōma*, cavity). Without a coelom, as in flatworms and proboscis worms.

acontium (ə-kän′chē-əm), pl. **acontia** (Gr. *akontion*, dart). Threadlike structure bearing nematocysts located on mesentery of sea anemone.

acquired immune deficiency syndrome An eventual consequence of infection with the human immunodeficiency virus in which the immune response is severely disabled. The disease is ultimately fatal, and no cure has been discovered.

acrocentric (ak′rō-sen′trək) (Gr. *akros*, tip, + *kentron*, center). Chromosome with centromere near the end.

acron (a′krän) (Gr. *akron*, mountaintop, fr. *akros*, tip). Preoral region of an insect.

actin (Gr. *aktis*, ray). A protein in the contractile tissue that forms the thin filaments of striated muscle.

actinotroch (ək-tin′ə-trōk) (Gr. *aktis*, ray, beam, + *trochos*, wheel). Larval form found in Phoronida.

active transport Mediated transport in which a transmembrane protein transports a molecule across a cell membrane against a concentration gradient; requires expenditure of energy; contrast with **facilitated diffusion.**

adaptation (L. *adaptatus*, fitted). An anatomical structure, physiological process, or behavioral trait that evolved by natural selection and improves an organism's ability to survive and leave descendants.

adaptive radiation Evolutionary diversification that produces numerous ecologically disparate lineages from a single ancestral one, especially when this diversification occurs within a short interval of geological time.

adaptive value Degree to which a characteristic helps an organism to survive and reproduce or lends greater fitness in its environment; selective advantage.

adaptive zone A characteristic reaction and mutual relationship between environment and organism ("way of life") demonstrated by a group of evolutionarily related organisms.

adductor (ə-duk′tər) (L. *ad*, to, + *ducere*, to lead). A muscle that draws a part toward a median axis, or a muscle that draws the two valves of a mollusc shell together.

adenine (ad′nēn, ad′ə-nēn) (Gr. *adēn*, gland, + *ine*, suffix). A purine base; component of nucleotides and nucleic acids.

adenosine (ə-den′ə-sen) **(di-, tri) phosphate** (ADP and ATP). A nucleotide composed of adenine, ribose sugar, and two (ADP) or three (ATP) phosphate units; ATP is an energy-rich compound that, with ADP, serves as a phosphate bond-energy transfer system in cells.

adipose (ad′ə-pōs) (L. *adeps*, fat). Fatty tissue; fatty.

adrenaline (ə-dren′ə-lən) (L. *ad*, to, + *renalis*, pertaining to kidneys). A hormone produced by the adrenal, or suprarenal, gland; epinephrine.

adsorption (ad-sorp′shən) (L. *ad*, to, + *sorbeo*, to absorb). The adhesion of molecules to solid bodies.

aerobic (a-rō′bik) (Gr. *aēr*, air, + *bios*, life). Oxygen-dependent form of respiration.

afferent (af′ə-rənt) (L. *ad*, to, + *ferre*, to bear). Adjective meaning leading or bearing toward some organ, for example, nerves conducting impulses toward the brain or blood vessels carrying blood toward an organ; opposed to **efferent.**

aggression (ə-gres′hən) (L. *aggressus*, attack). An offensive action or procedure.

agonistic behavior (Gr. *agōnistēs*, combatant). An offensive action or threat directed toward another organism.

AIDS See **acquired immune deficiency syndrome.**

alate (ā′lāt) (L. *alatus*, wing). Winged.

albumin (al-byū′mən) (L. *albumen*, white of egg). Any of a large class of simple proteins that are important constituents of

---

bat / ā̆pe / ärmadillo / herring / fēmale / finch / līce / crocodile / crōw / duck / ūnicorn / ə indicates unaccented vowel sound "uh" as in mammal, fishes, cardinal, heron, vulture / stress as in bi-ol′o-gy, bi′o-log′i-cal vertebrate blood plasma and tissue fluids and also present in milk, whites of eggs, and other animal substances.

alimentary (al′ə-men′tə-rē) (L. *alimentum,* food, nourishment). Having to do with nutrition or nourishment.

allantois (ə-lan′tois) (Gr. *allas,* sausage, + *eidos,* form). One of the extraembryonic membranes of the amniotes that functions in respiration and excretion in birds and reptiles and plays an important role in the development of the placenta in most mammals.

allele (ə-lēl′) (Gr. *allēlōn,* of one another). Alternative forms of genes coding for the same trait; situated at the same locus in homologous chromosomes.

allograft (a′lō-graft) (Gr. *allos,* other, + graft). A piece of tissue or an organ transferred from one individual to another individual of the same species, not identical twins; homograft.

allometry (ə-lom′ə-trē) (Gr. *allos,* other, + *metry,* measure). Relative growth of a part in relation to the whole organism.

allopatric (Gr. *allos,* other, + *patra,* native land). In separate and mutually exclusive geographical regions.

alpha-helix (Gr. *alpha,* first, + L. *helix,* spiral). Literally the first spiral arrangement of the genetic DNA molecule; regular coiled arrangement of polypeptide chain in proteins; secondary structure of proteins.

altricial (al-tri′shəl) (L. *altrices,* nourishers). Referring to young animals (especially birds) having the young hatched or born in an immature, dependent condition.

alula (al′yə-lə) (L. dim. of *ala,* wing). The first digit or thumb of a bird's wing, much reduced in size.

alveolus (al-vē′ə-ləs) (L. dim. of *alveus,* cavity, hollow). A small cavity or pit, such as a microscopic air sac of the lungs, terminal part of an alveolar gland, or bony socket of a tooth.

ambulacra (am′byə-lak′rə) (L. *ambulare,* to walk). In echinoderms, radiating grooves where podia of water-vascular system characteristically project to outside.

amebocyte (ə-mē′bə-sīt) (Gr. *amoibē,* change, + *kytos,* hollow vessel). Cell in metazoan invertebrate, often functioning in defense against invading particles.

ameboid (ə-mē′boid) (Gr. *amoibē,* change, + *oid,* like). Ameba-like in putting forth pseudopodia.

amictic (ə-mik′tic) (Gr. *a,* without, + *miktos,* mixed or blended). Pertaining to female rotifers, which produce only diploid eggs that cannot be fertilized, or to the eggs produced by such females. Compare with **mictic.**

amino acid (ə-mē′nō) (amine, an organic compound). An organic acid with an amino group (—NH$_2$). Makes up the structure of proteins and peptides.

amitosis (ā′mī-tō′səs) (Gr. *a,* not, + *mitos,* thread). A form of cell division in which mitotic nuclear changes do not occur; cleavage without separation of daughter chromosomes.

ammocoetes (am-mō-cēt′ēz) (Gr. *ammo,* sand + *coet,* bed). The filter-feeding, larval stage of lampreys.

amniocentesis (am′nē-ō-sin-tē′səs) (Gr. *amnion,* membrane around the fetus, + *centes,* puncture). Procedure for withdrawing a sample of fluid around the developing embryo for examination of chromosomes in the embryonic cells and other tests.

amnion (am′nē-än) (Gr. *amnion,* membrane around the fetus). The innermost of the extraembryonic membranes forming a fluid-filled sac around the embryo in amniotes.

amniote (am′nē-ōt). Having an amnion; as a noun, an animal that develops an amnion in embryonic life, that is, reptiles, birds, and mammals.

amphiblastula (am′fə-blas′chə-lə) (Gr. *amphi,* on both sides, + *blastos,* germ, + L. *ula,* small). Free-swimming larval stage of certain marine sponges; blastula-like but with only the cells of the animal pole flagellated; those of the vegetal pole unflagellated.

amphid (am′fəd) (Gr. *amphidea,* anything that is bound around). One of a pair of anterior sense organs in certain nematodes.

amphipathic (am-fi-pa′thək) (Gr. *amphi,* on both sides, + *pathos,* suffering, passion). Adjective to describe a molecule with one part soluble in water (polar) and another part insoluble in water (nonpolar).

amplexus (am-plek′səs) (L. embrace). The copulatory embrace of frogs or toads.

ampulla (am-pūl′ə) (L. flask). Membranous vesicle; dilation at one end of each semicircular canal containing sensory epithelium; muscular vesicle above tube foot in water-vascular system of echinoderms.

amylase (am′ə-lās′) (L. *amylum,* starch, + *ase,* suffix meaning enzyme). An enzyme that breaks down starch into smaller units.

anabolism (ə-na′bə-li′zəm) (Gr. *ana,* up, + *bol,* to throw, + *ism,* suffix meaning state of condition). Constructive metabolism.

anadromous (an-ad′rə-məs) (Gr. *anadromos,* running upward). Refers to fishes that migrate up streams from the sea to spawn.

anaerobic (an′ə-rō′bik) (Gr. *an,* not, + *aēr,* air, + *bios,* life). Not dependent on oxygen for respiration.

analogy (L. *analogus,* ratio). Similarity of function but not of origin.

anaphylaxis (an′ə-fə-lax′əs) (Gr. *ana-,* up, + *phylax,* guard). A systemic (whole body) immediate hypersensitivity reaction.

anapsid (ə-nap′səd) (Gr. *an,* without, + *apsis,* arch). Amniotes in which the skull lacks temporal openings, with turtles the only living representatives.

anastomosis (ə-nas′tə-mō′səs) (Gr. *ana,* again, + *stoma,* mouth). A union of two or more blood vessels, fibers, or other structures to form a branching network.

androgen (an′drə-jən) (Gr. *anēr, andros,* man, + *genēs,* born). Any of a group of vertebrate male sex hormones.

androgenic gland (an′drō-jen′ək) (Gr. *anēr,* male, + *gennaein,* to produce). Gland in Crustacea that causes development of male characteristics.

aneuploidy (an′ū-ploid′ē) (Gr. *an,* without, not, + *eu,* good, well, + *ploid,* multiple of). Loss or gain of a chromosome, cells of the organism have one fewer than normal chromosome number, or one extra chromosome, for example, trisomy 21 (Down syndrome).

angiotensin (an′jē-o-ten′sən) (Gr. *angeion,* vessel, + L. *tensio,* to stretch). Blood protein formed from the interaction of renin and a liver protein, causing increased blood pressure and stimulating release of aldosterone and ADH.

Angstrom (after Ångström, Swedish physicist). A unit of one ten-millionth of a millimeter (one ten-thousandth of a micrometer); it is represented by the symbol Å.

anhydrase (an-hī′drās) (Gr. *an,* not, + *hydōr,* water, + *ase,* enzyme suffix). An enzyme involved in the removal of water from a compound. Carbonic anhydrase promotes the conversion of carbonic acid into water and carbon dioxide.

anisogametes (an′īs-ō-gam′ēts) (Gr. *anisos,* unequal, + *gametēs,* spouse). Gametes of a species that differ in form or size.

anlage (än′lä-gə) (Ger. laying out, foundation). Rudimentary form; primordium.

annulus (an′yəl-əs) (L. ring). Any ringlike structure, such as superficial rings on leeches.

antenna (L. sail yard). A sensory appendage on the head of arthropods, or the second pair of the two such pairs of structures in crustaceans.

antennal gland Excretory gland of Crustacea located in the antennal metamere.

anterior (L. comparative of *ante,* before). The head end of an organism, or (as an adjective) toward that end.

anthracosaurs (an-thrak′ə-sors) (Gr. *anthrax,* coal, carbon, + *sauros,* lizard). A group of Paleozoic labyrinthodont amphibians.

antibodies (an′tē-bod′ēz). Proteins (immunoglobulins) in cell surfaces and dissolved in blood, capable of combining with the antigens that stimulated their production.

anticodon (an′tī-kō′don). A sequence of three nucleotides in transfer RNA that is complementary to a codon in messenger RNA.

antigen (an′ti-jən). Any substance capable of stimulating an immune response, most often a protein.

antigenic determinant See **epitope.**

aperture (ap′ər-chər) (L. *apertura* from *aperire,* to uncover). An opening; the opening into the first whorl of a gastropod shell.

apex (ā′peks) (L. summit). Highest or uppermost point; the lower pointed end of the heart.

apical (ā′pə-kl) (L. *apex,* tip). Pertaining to the tip or apex.

apical complex A certain combination of organelles found in the protozoan phylum Apicomplexa.

apocrine (ap′ə-krən) (Gr. *apo,* away, + *krinein,* to separate). Applies to a type of mammalian sweat gland that produces a viscous secretion by breaking off a part of the cytoplasm of secreting cells.

apoptosis (a′pə-tō′səs) (Gr. *apo-,* prefix meaning away from, + *ptōsis,* a falling). Genetically determined cell death, "programmed" cell death.

apopyle (ap′ə-pīl) (Gr. *apo,* away from, + *pylē,* gate). In sponges, opening of the radial canal into the spongocoel.

appendicular (L. *ad,* to, + *pendere,* to hang). Pertaining to appendages; pertaining to vermiform appendix.

arboreal (är-bōr′ē-al) (L. *arbor,* tree). Living in trees.

archaeocytes (ärk′ē-ō-sites) (Gr. *archaios,* beginning, + *kytos,* hollow vessel). Ameboid cells of varied function in sponges.

archenteron (ärk-en′tə-rän) (Gr. *archē,* beginning, + *enteron,* gut). The main cavity of an embryo in the gastrula stage; it is lined with endoderm and represents the future digestive cavity.

archinephros (ärk′ē-nəf′rōs) (Gr. *archaois,* ancient, + *nephros,* kidney). Ancestral vertebrate kidney, existing today only in the embryo of hagfishes.

archosaur (är′kə-sor) (Gr. *archōn,* ruling, + *sauros,* lizard). Advanced diapsid vertebrates, a group that includes the living crocodiles and the extinct pterosaurs and dinosaurs.

areolar (a-rē′ə-ler) (L. *areola,* small space). A small area, such as spaces between fibers of connective tissue.

arginine phosphate Phosphate storage compound (phosphagen) found in many invertebrates and used to regenerate stores of ATP.

Aristotle's lantern Masticating apparatus of some sea urchins.

arteriole (är-tir′ē-ōl) (L. *arteria,* artery). A small arterial branch that delivers blood to a capillary network.

artery (ärt′ə-rē) (L. *arteria,* artery). A blood vessel that carries blood away from the heart and toward a peripheral cavity.

artiodactyl (är′ti-o-dak′təl) (Gr. *artios,* even, + *daktylos,* toe). One of an order of ungulate mammals with two or four digits on each foot.

asconoid (Gr. *askos,* bladder). Simplest form of sponges, with canals leading directly from the outside to the interior.

asexual Without distinct sexual organs; not involving formation of gametes.

assimilation (L. *assimilatio,* bringing into conformity). Absorption and building up of digested nutriments into complex organic protoplasmic materials.

athecate (ā′thē-kāt′) (Gr. *a-,* an absence of something, + *thēkē,* a case, box). An organism without a theca.

atherosclerosis (a′thə-rō-sklə-rō′səs) (Gr. *athērōma,* tumor full of gruel-like material, + *sklērōs,* hard). Disease characterized by fatty plaques forming in the inner lining of arteries.

atoke (ā′tōk) (Gr. *a,* without, + *tokos,* offspring). Anterior, nonreproductive part of a marine polychaete, as distinct from the posterior, reproductive part (epitoke) during the breeding season.

atoll (ə-tol′) (Maldivian, *atolu*). A coral reef or island surrounding a lagoon.

atom The smallest unit of an element, composed of a dense nucleus of protons and (usually) neutrons surrounded by a system of electrons.

ATP Adenosine triphosphate. In biochemistry, an ester of adenosine and triphosphoric acid.

atrium (ā′trē-əm) (L. *atrium,* vestibule). One of the chambers of the heart; also, the tympanic cavity of the ear; also, the large cavity containing the pharynx in tunicates and cephalochordates.

auricle (aw′ri-kl) (L. *auricula,* dim. of *auris,* ear). One of the less muscular chambers of the heart; atrium; the external ear, or pinna; any earlike lobe or process.

auricularia (ə-rik′u-lar′e-ə) (L. *auricula,* a small ear). A type of larva found in Holothuroidea.

autogamy (aw-täg′ə-me) (Gr. *autos,* self, + *gamos,* marriage). Condition in which the gametic nuclei produced by meiosis fuse within the same organism that produced them to restore the diploid number.

autosome (aw′tə-sōm) (Gr. *autos,* self, + *sōma,* body). Any chromosome that is not a sex chromosome.

autotomy (aw-täd′ə-me) (Gr. *autos,* self, + *tomos,* a cutting). The breaking off of a part of the body by the organism itself.

autotroph (aw′tə-trōf) (Gr. *autos*, self, + *trophos*, feeder). An organism that makes its organic nutrients from inorganic raw materials.

autotrophic nutrition (Gr. *autos*, self, + *trophia*, denoting nutrition). Nutrition characterized by the ability to use simple inorganic substances for the synthesis of more complex organic compounds, as in green plants and some bacteria.

avicularium (L. *avicula*, small bird, + *aria*, like or connected with). Modified zooid that is attached to the surface of the major zooid in Ectoprocta and resembles a bird's beak.

axial (L. *axis*, axle). Relating to the axis, or stem; on or along the axis.

axocoel (ak′sə-cēl) (Gr. *axon*, an axle, + *koilos*, hollow). The most anterior of three coelomic spaces that appear during larval echinoderm development.

axolotl (ak′sə-lot′l) (Nahuatl, *atl*, water, + *xolotl*, doll, servant, spirit). Larval stage of any of several species of the genus *Ambystoma* (such as *Ambystoma tigrinum*) exhibiting neotenic reproduction.

axon (ak′sän) (Gr. *axōn*). Elongate extension of a neuron that conducts impulses away from the cell body and toward the synaptic terminals.

axoneme (aks′ə-nēm) (L. *axis*, axle, + Gr. *nēma*, thread). The microtubules in a cilium or flagellum, usually arranged as a circlet of nine pairs enclosing one central pair; also, the microtubules of an axopodium.

axopodium (ak′sə-pō′di-um) (Gr. *axon*, an axis, + *podion*, small foot). Long, slender, more or less permanent pseudopodium found in certain amebas. (Also **axopod.**)

B

B cell A type of lymphocyte that is most important in the humoral immune response.

barrier reef A coral reef that runs approximately parallel to the shore and is separated from the shore by a lagoon.

basal body Also known as kinetosome and blepharoplast, a cylinder of nine triplets of microtubules found basal to a flagellum or cilium; same structure as a centriole.

base A molecule that dissociates in solution to produce a hydroxide ion.

basis, basipodite (bā′səs, bā-si′pə-dīt) (Gr. *basis*, base, + *pous*, *podos*, foot). The distal or second joint of the protopod of a crustacean appendage.

bathypelagic (bath′ə-pe-laj′ik) (Gr. *bathys*, deep, + *pelagos*, open sea). Relating to or inhabiting the deep sea.

benthos (ben′thäs) (Gr. depth of the sea). Organisms that live along the bottom of the seas and lakes; adj., **benthic.** Also, the bottom itself.

bilirubin (bil′ə-ru-bən) (L. *bilis*, bile, + *rubeo*, to be red). A breakdown product of the heme group of hemoglobin, excreted in the bile.

binary fission A mode of asexual reproduction in which the animal splits into two approximately equal offspring.

biogenesis (bī′ō-jen′ə-səs) (Gr. *bios*, life, + *genesis*, birth). The doctrine that life originates only from preexisting life.

biological species concept A reproductive community of populations (reproductively isolated from others) that occupies a specific niche in nature.

bioluminescence Method of light production by living organisms in which usually certain proteins (luciferins), in the presence of oxygen and an enzyme (luciferase), are converted to oxyluciferins with the liberation of light.

biomass (Gr. *bios*, life, + *maza*, lump or mass). The weight of total living organisms or of a species population per unit of area.

biome (bī′ōm) (Gr. *bios*, life, + *ōma*, abstract group suffix). Complex of plant and animal communities characterized by climatic and soil conditions; the largest ecological unit.

biosphere (Gr. *bios*, life, + *sphaira*, globe). That part of earth containing living organisms.

biotic (bī-äd′ik) (Gr. *biōtos*, life, livable). Of or relating to life.

bipinnaria (L. *bi*, double, + *pinna*, wing, + *aria*, like or connected with). Free-swimming, ciliated, bilateral larva of the asteroid echinoderms; develops into the brachiolaria larva.

biramous (bī-rām′əs) (L. *bi*, double, + *ramus*, a branch). Adjective describing appendages with two distinct branches, contrasted with uniramous, unbranched.

bivalent (bī-vāl′ənt) (L. *bi*, double, + *valen*, strength, worth). The pairs of homologous chromosomes at synapsis in the first meiotic division, a tetrad.

blastocoel (blas′tə-sēl) (Gr. *blastos*, germ, + *koilos*, hollow). Cavity of the blastula.

blastocyst (blast′ō-sist) (Gr. *blastos*, germ, + *kystis*, bladder). Mammalian embryo in the blastula stage.

blastomere (Gr. *blastos*, germ, + *meros*, part). An early cleavage cell.

blastopore (Gr. *blastos*, germ, + *poros*, passage, pore). External opening of the archenteron in the gastrula.

blastula (Gr. *blastos*, germ, + L. *ula*, dim.). Early embryological stage of many animals; consists of a hollow mass of cells.

blending See **polygenic inheritance.**

blepharoplast (blə-fä′rə-plast) (Gr. *blepharon*, eyelid, + *plastos*, formed). See **basal body.**

blood plasma The liquid, noncellular fraction of blood, including dissolved substances.

blood type Characteristic of human blood given by the particular antigens on the membranes of the erythrocytes, genetically determined, causing agglutination when incompatible groups are mixed; the blood types are designated A, B, O, AB, Rh negative, Rh positive, and others.

Bohr effect A characteristic of hemoglobin that causes it to dissociate from oxygen in greater degree at higher concentrations of carbon dioxide.

boreal (bōr′ē-əl) (L. *boreas*, north wind). Relating to a northern biotic area characterized by a predominance of coniferous forests and tundra.

B.P. Before the present.

brachial (brak′ē-əl) (L. *brachium*, forearm). Referring to the arm.

brachiation (brak′ē-ā′shən) (L. *brachium*, arm). Locomotion by swinging by the arms from one hold to another.

brachiolaria (brak′ē-ō-lār′ē-ə) (L. *brachiola*, little arm, + *aria*, pertaining to). This asteroid larva develops from the bipinnaria larva and has three preoral holdfast processes.

---

bat / āpe / ärmadillo / herring / fēmale / finch / līce / crocodile / crōw / duck / ūnicorn / ə indicates unaccented vowel sound "uh" as in mammal, fishes, cardinal, heron, vulture / stress as in bi-ol′o-gy, bi′o-log′i-cal

bradyzoite An individual coccidian (a single-celled parasite) such as *Toxoplasma gondii* that is encased in a tissue cyst and divides slowly.

brain hormone See **ecdysiotropin.**

branchial (brank′ē-əl) (Gr. *branchia,* gills). Referring to gills.

bronchiole (brän′kē-ōl) (Gr. *bronchion,* dim. of *bronchos,* windpipe). Small, thin-walled branch of the bronchus.

bronchus (brän′kəs) pl. **bronchi** (Gr. *bronchos,* windpipe). Either of two primary divisions of the trachea that lead to the right and left lung.

brown fat Mitochondria-rich, heat-generating adipose tissue of endothermic vertebrates.

buccal (buk′əl) (L. *bucca,* cheek). Referring to the mouth cavity.

budding Reproduction in which the offspring arises as an outgrowth from the parent and is initially smaller than the parent. Failure of the offspring to separate from the parent leads to colony formation.

buffer Any substance or chemical compound that tends to keep pH levels constant when acids or bases are added.

bursa pl. **bursae** (M.L. *bursa,* pouch, purse made of skin). A saclike cavity. In ophiuroid echinoderms, pouches opening at bases of arms and functioning in respiration and reproduction (genitorespiratory bursae).

C

caecum See **cecum.**

calciferous glands (kal-si′fə-rəs). Glands in an earthworm that secrete calcium ions into the gut.

calorie (kal′ə-rē) (L. *calere,* to be warm). Unit of heat defined as the amount of heat required to heat 1 g of water from 14.5 to 15.5° C; 1 cal = 4.184 joules in the International System of Units.

calyx (kā-′liks) (L. bud cup of a flower). Any of various cup-shaped zoological structures.

cancellous (kan′səl-əs) (L. *cancelli,* latticework, + *osus,* full of). Having a spongy or porous structure.

capitulum (ka-pi′tə-ləm) (L. small head). Term applied to small, headlike structures of various organisms, including projection from body of ticks and mites carrying mouthparts.

captacula (kap-tak′ū-lə) (L. *captare,* to lie in wait for). Tentacles extending from head of scaphopod molluscs, used in feeding.

carapace (kar′ə-pās) (F. from Sp. *carapacho,* shell). Shieldlike plate covering the cephalothorax of certain crustaceans; dorsal part of the shell of a turtle.

carbohydrate (L. *carbo,* charcoal, + Gr. *hydōr,* water). Compounds of carbon, hydrogen, and oxygen having the generalized formula $(CH_2O)_n$; aldehyde or ketone derivatives of polyhydric alcohols, with hydrogen and oxygen atoms attached in a 2:1 ratio.

carboxyl (kär-bäk′səl) (carbon + oxygen + yl, chemical radical suffix). The acid group of organic molecules (—COOH.)

cardiac (kär′dē-ak) (Gr. *kardia,* heart). Belonging or relating to the heart.

carinate (kar′ə-nāt) (L. *carina,* keel). Having a keel, in particular the flying birds with a keeled sternum for the insertion of flight muscles.

carnivore (kar′nə-vōr′) (L. *carnivorous,* flesh eating). One of the flesh-eating mammals of the order Carnivora. Also, any organism that eats animals. Adj., **carnivorous.**

carotene (kär′ə-tēn) (L. *carota,* carrot, + *ene,* unsaturated straight-chain hydrocarbons). A red, orange, or yellow pigment belonging to the group of carotenoids; precursor of vitamin A.

carrying capacity The maximum number of individuals that can persist under specified environmental conditions.

cartilage (L. *cartilago;* akin to L. *cratis,* wickerwork). A translucent, specialized connective tissue that makes up most of the skeleton of embryos, very young vertebrates, and adult cartilaginous fishes, such as sharks and rays; in higher forms much of it is converted into bone.

caste (kast) (L. *castus,* pure, separated). One of the polymorphic forms within an insect society, each caste having its specific duties, as queen, worker, soldier, and so on.

catabolism (Gr. *kata,* downward, + *bol,* to throw, + *ism,* suffix meaning state of condition). Destructive metabolism; process in which complex molecules are reduced to simpler ones.

catadromous (kə-tad′rə-məs) (Gr. *kata,* down, + *dromos,* a running). Refers to fishes that migrate from fresh water to the ocean to spawn.

catalyst (kad′ə-ləst) (Gr. *kata,* down, + *lysis,* a loosening). A substance that accelerates a chemical reaction but does not become a part of the end product.

caudal (käd′l) (L: *cauda,* tail). Constituting, belonging to, or relating to a tail.

caveolae (ka-vē′ə-lē) (L. *cavea,* a cave, + dim. suffix). The invaginated vesicles and pits in potocytosis.

cDNA See **complementary DNA.**

cecum, caecum (sē′kəm) (L. *caecus,* blind). A blind pouch at the beginning of the large intestine; any similar pouch.

cell-mediated immune response Immune response involving cell surfaces only, not antibody production, specifically the T_H1 arm of the immune response. Contrast **humoral immune response.**

cellulose (sel′ū-lōs) (L. *cella,* small room). Chief polysaccharide constituent of the cell wall of green plants and some fungi; an insoluble carbohydrate $(C_6H_{10}O_5)_n$ that is converted into glucose by hydrolysis.

centriole (sen′trē-ol) (Gr. *kentron,* center of a circle, + L. *ola,* small). A minute cytoplasmic organelle usually found in the centrosome and considered to be the active division center of the animal cell; organizes spindle fibers during mitosis and meiosis. Same structure as basal body or kinetosome.

centrolecithal (sen′tro-les′ə-thəl) (Gr. *kentron,* center, + *lekithos,* yolk, + Eng. *al,* adjective). Pertaining to an insect egg with the yolk concentrated in the center.

centromere (sen′trə-mir) (Gr. *kentron,* center, + *meros,* part). A localized constriction in a characteristic position on a given chromosome, bearing the kinetochore.

centrosome (sen′trə-sōm) (Gr. *kentron,* center, + *sōma,* body). Microtubule organizing center in nuclear division in most eukaryotic cells; in animals and many unicellular organisms it surrounds the centrioles.

cephalization (sef′ə-li-zā-shən) (Gr. *kephale,* head). The process by which specialization, particularly of the sensory organs and appendages, become localized in the head end of animals.

cephalothorax (sef′ə-lä-thō′raks) (Gr. *kephale,* head, + thorax). A body division found in many Arachnida and higher Crustacea, in which the head is fused with some or all of the thoracic segments.

cercaria (ser-kar′ē-ə) (Gr. *kerkos,* tail, + L. *aria,* like or connected with). Tadpolelike larva of trematodes (flukes).

cervical (sər′və-kəl) (L. *cervix,* neck). Relating to a neck.

character (kar′ik-tər). A component of phenotype (including specific molecular, morphological, behavioral or other features) used by systematists to diagnose species or higher taxa, or to evaluate phylogenetic relationships among different species or higher taxa, or relationships among populations within a species.

charging In protein synthesis, a reaction catalyzed by tRNA synthetase, in which an amino acid is attached to its particular tRNA molecule.

chelicera (kə-lis′ə-rə) pl. **chelicerae** (Gr. *chēlē,* claw, + *keras,* horn). One of a pair of the most anterior head appendages on the members of the subphylum Chelicerata.

chelipeds (kēl′ə-peds) (Gr. *chēlē,* claw, + L. *pes,* foot). Pincerlike first pair of legs in most decapod crustaceans; specialized for seizing and crushing.

chemoautotroph (ke′mō-aw′tō-trōf) (Gr. *chemeia,* transmutation, + *autos,* self, + *trophos,* feeder). An organism utilizing inorganic compounds as a source of energy.

chemotaxis (kē′mō-tak′səs) (Gr. *chēmeia,* an infusion, + *taxō* > *tassō,* to put in order). Orientation movement of cells or organisms in response to a chemical stimulus.

chemotroph (kem′ə-trōf) (Gr. *chēmeia,* an infusion, + *tropē,* to turn). An organism that derives nourishment from inorganic substances without using chlorophyll.

chiasma (kī-az′mə), pl. **chiasmata** (Gr. cross). An intersection or crossing, as of nerves; a connection point between homologous chromatids where crossing over has occurred at synapsis.

chitin (kī′tən) (Fr. *chitine,* from Gr. *chitōn,* tunic). A horny substance that forms part of the cuticle of arthropods and is found sparingly in certain other invertebrates; a nitrogenous polysaccharide insoluble in water, alcohol, dilute acids, and digestive juices of most animals.

chlorocruorin (klō′rō-kroo′ə-rən) (Gr. *chlōros,* light green, + L. *cruor,* blood). A greenish iron-containing respiratory pigment dissolved in the blood plasma of certain marine polychaetes.

chlorogogen cells (klōr′ə-gog-ən) (Gr. *chlōros,* light green, + *agōgos,* a leading, a guide). Modified peritoneal cells, greenish or brownish, clustered around the digestive tract of certain annelids; apparently they aid in elimination of nitrogenous wastes and in food transport.

chlorophyll (klō′rə-fil) (Gr. *chlōros,* light green, + *phyllōn,* leaf). Green pigment found in plants and in some animals; necessary for photosynthesis.

chloroplast (klō′rə-plast) (Gr. *chlōros,* light green, + *plastos,* molded). A plastid containing chlorophyll and usually other pigments, found in cytoplasm of plant cells.

choanocyte (kō-an′ə-sīt) (Gr. *choanē,* funnel, + *kytos,* hollow vessel). One of the flagellate collar cells that line cavities and canals of sponges.

cholinergic (kōl′i-nər′jik) (Gr. *chōle,* bile, + *ergon,* work). Type of nerve fiber that releases acetylcholine from axon terminal.

chorion (kō′rē-on) (Gr. *chorion,* skin). The outer of the double membrane that surrounds the embryo of reptiles, birds, and mammals; in mammals it contributes to the placenta.

choroid (kōr′oid) (Gr. *chorion,* skin, + *eidos,* form). Delicate, highly vascular membrane; in vertebrate eye; the layer between the retina and sclera.

chromatid (krō′mə-tid) (Gr. *chromato,* from *chrōma,* color, + L. *id,* feminine stem for particle of specified kind). A replicated chromosome joined to its sister chromatid by the centromere; separates and becomes daughter chromosome at anaphase of mitosis or anaphase of the second meiotic division.

chromatin (krō′mə-tin) (Gr. *chrōma,* color). The nucleoprotein material of a chromosome; the hereditary material containing DNA.

chromatophore (krō-mat′ə-fōr) (Gr. *chrōma,* color, + *pherein,* to bear). Pigment cell, usually in the dermis, in which usually the pigment can be dispersed or concentrated.

chromomere (krō′mō-mir) (Gr. *chrōma,* color, + *meros,* part). One of the chromatin granules of characteristic size on the chromosome; may be identical with a gene or a cluster of genes.

chromonema (krō-mə-nē′mə) (Gr. *chrōma,* color, + *nēma,* thread). A convoluted thread in prophase of mitosis or the central thread in a chromosome.

chromoplast (krō′mə-plast) (Gr. *chrōma,* color, + *plastos,* molded). A plastid that contains pigment.

chromosome (krō′mə-sōm) (Gr. *chrōma,* color, + *sōma,* body). A complex body, spherical or rod shaped, that arises from the nuclear network during mitosis, splits longitudinally, and carries a part of the organism's genetic information as genes composed of DNA.

chrysalis (kris′ə-lis) (L. from Gr. *chrysos,* gold). The pupal stage of a butterfly.

chyme (kīm) (Gr. *chymos,* juice). Semifluid mass of partly digested food in stomach and small intestine as digestion proceeds.

cilium (sil′i-əm), pl. **cilia** (L. eyelid). A hairlike, vibratile organelle process found on many animal cells. Cilia may be used in moving particles along the cell surface or, in ciliate protozoans, for locomotion.

cinclides (sing′klid-əs), sing. **cinclis** (sing′kləs) (Gr. *kinklis,* latticed gate or partition). Small pores in the external body wall of sea anemones for extrusion of acontia.

circadian (sər-kād′ē-ən) (L. *circa,* around, + *dies,* day). Occurring at a period of approximately 24 hours.

cirrus (sir′əs) (L. curl). A hairlike tuft on an insect appendage; locomotor organelle of fused cilia; male copulatory organ of some invertebrates.

cisternae (sis-ter′nē) (L. *cista,* box). Space between membranes of the endoplasmic reticulum within cells.

cistron (sis′trən) (L. *cista,* box). A series of codons in DNA that code for an entire polypeptide chain.

clade (klād) (Gr. *klados,* branch). A taxon or other group consisting of an ancestral species and all of its descendants, forming a distinct branch on a phylogenetic tree.

cladistics (klad-is'-təks) (Gr. *klados,* branch, sprout). A system of arranging taxa by analysis of evolutionarily derived characteristics so that the arrangement will reflect phylogenetic relationships.

cladogram (klād'ə-gram) (Gr. *klados,* branch, + *gramma,* letter). A branching diagram showing the pattern of sharing of evolutionarily derived characters among species or higher taxa.

clathrin (kla'thrən) (L. *chathri,* latticework). A protein forming a lattice structure lining the invaginated pits during receptor-mediated endocytosis.

cleavage (O.E. *cleofan,* to cut). Process of nuclear and cell division in animal zygote.

climax (klī'maks) (Gr. *klimax,* ladder). Stage of relative stability attained by a community of organisms, often the culminating development of a natural succession. Also, orgasm.

climax community (Gr. *klimax,* ladder, staircase, climax). A self-perpetuating, more-or-less stable community of organisms that continues as long as environmental conditions under which it developed prevail.

clitellum (klī-tel'əm) (L. *clitellae,* packsaddle). Thickened saddlelike portion of certain midbody segments of many oligochaetes and leeches.

cloaca (klō-a'kə) (L. sewer). Posterior chamber of digestive tract in many vertebrates, receiving feces and urogenital products. In certain invertebrates, a terminal portion of digestive tract that serves also as respiratory, excretory, or reproductive tract.

clone (klōn) (Gr. *klōn,* twig). All descendants derived by asexual reproduction from a single individual.

cnidoblast (nī'də-blast) (Gr. *knidē,* nettle, + *blastos,* germ). See **cnidocyte.**

cnidocil (nī'də-sil) (Gr. *knidē,* nettle, + L. *cilium,* hair). Modified cilium on nematocyst-bearing cnidocytes in cnidarians; triggers nematocyst.

cnidocyte (nī'də-sīt) (Gr. *knidē,* nettle, + *kytos,* hollow vessel). Modified interstitial cell that holds the nematocyst; during development of the nematocyst, the cnidocyte is a cnidoblast.

coacervate (kō'ə-sər'vət) (L. *coacervatus,* to heap up). An aggregate of colloidal droplets held together by electrostatic forces.

coagulation (kō-ag'ū-lā-shən). Process in which a series of enzymes are activated, resulting in clotting of blood.

cochlea (kōk'lē-ə) (L. snail, from Gr. *kochlos,* a shellfish). A tubular cavity of the inner ear containing the essential organs of hearing; occurs in crocodiles, birds, and mammals; spirally coiled in mammals.

cocoon (kə-kun') (Fr. *cocon,* shell). Protective covering of a resting or developmental stage, sometimes used to refer to both the covering and its contents; for example, the cocoon of a moth or the protective covering for the developing embryos in some annelids.

codominance A condition in which each allele maintains its distinctive homozygous expression in the heterozygous condition, not a blending of the separate homozygous phenotypes (contrast intermediate inheritance). Genes for A and B blood types show codominance (p. 120).

codon (kō'dän) (L. code, + on). In messenger RNA a sequence of three adjacent nucleotides that codes for one amino acid.

coelenteron (sē-len'tər-on) (Gr. *koilos,* hollow, + *enteron,* intestine). Internal cavity of a cnidarian; gastrovascular cavity; archenteron.

coelom (sē'lōm) (Gr. *koilōma,* cavity). The body cavity in triploblastic animals, lined with mesodermal peritoneum.

coelomocyte (sē'lō'mə-sīt) (Gr. *koilōma,* cavity, + *kytos,* hollow vessel). Another name for amebocyte; primitive or undifferentiated cell of the coelom and the water-vascular system.

coelomoduct (sē-lō'mə-dukt) (Gr. *koilos,* hollow, + L. *ductus,* a leading). A duct that carries gametes or excretory products (or both) from the coelom to the exterior.

coenecium, coenoecium (sə-nēs[h]'ē-əm) (Gr. *koinos,* common, + *oikion,* house). The common secreted investment of an ectoproct colony; may be chitinous, gelatinous, or calcareous.

coenenchyme (sēn'ən-kīm) (Gr. *koinos,* shared in common, + *enchyma,* something poured in). Extensive mesogleal tissue between the polyps of an octocorallian (phylum Cnidaria) colony.

coenocytic (sē-nə-sit'ik) (Gr. *koinos,* common, + *kytos,* hollow vessel). A tissue in which the nuclei are not separated by cell membranes; syncytial.

coenosarc (sē'nə-särk) (Gr. *koinos,* shared in common, + *sarkos,* flesh). The inner, living part of hydrocauli in hydroids.

coenzyme (kō-en'zīm) (L. prefix, *co,* with, + Gr. *enzymos,* leavened, from *en,* in, + *zymē,* leaven). A required substance in the activation of an enzyme; a prosthetic or nonprotein constituent of an enzyme.

collagen (käl'ə-jən) (Gr. *kolla,* glue, + *genos,* descent). A structural protein, the most abundant protein in the animal kingdom, characterized by high content of the amino acids glycine, alanine, proline, and hydroxyproline.

collenchyme (käl'ən-kīm) (Gr. *kolla,* glue, + *enchyma,* infusion). A gelatinous mesenchyme containing undifferentiated cells; found in cnidarians and ctenophores.

collencyte (käl'ən-sīt) (Gr. *kolla,* glue, + *en,* in, + *kytos,* hollow vessel). A type of cell in sponges that is star shaped and apparently contractile.

colloblast (käl'ə-blast) (Gr. *kolla,* glue, + *blastos,* germ). A glue-secreting cell on the tentacles of ctenophores.

colloid (kä'loid) (Gr. *kolla,* glue, + *eidos,* form). A two-phase system in which particles of one phase are suspended in the second phase.

columella (kä'lə-mel'ə) (L. small column). Central pillar in gastropod shells.

comb plate One of the plates of fused cilia that are arranged in rows for ctenophore locomotion.

commensalism (kə-men'səl-iz'əm) (L. *cum,* together with, + *mensa,* table). A relationship in which one individual lives close to or on another and benefits, and the host is unaffected; often symbiotic.

community (L. *communitas,* community, fellowship). An assemblage of organisms that are associated in a common environment and interact with each other in a self-sustaining and self-regulating relation.

competition Some degree of overlap in ecological niches of two populations in the same community, such that both depend on the same food source, shelter, or other resources, and negatively affect each other's survival.

complement Collective name for a series of enzymes and activators in the blood, some of which may bind to antibody and may lead to rupture of a foreign cell.

complementary DNA (cDNA) DNA prepared by transcribing the base sequence from mRNA into DNA by reverse transcriptase; also called **copy DNA.**

compound A substance whose molecules are composed of atoms of two or more elements.

condensation reaction A chemical reaction in which reactant molecules are combined by the removal of a water molecule (a hydrogen from one and a hydroxyl from the other reactant).

condyle (kän'dīl) (Gr. *kondylos,* bump). A process on a bone used for articulation.

conjugation (kon'ju-ga'shən) (L. *conjugare,* to yoke together). Temporary union of two ciliate protozoa while they are exchanging chromatin material and undergoing nuclear phenomena resulting in binary fission. Also, formation of cytoplasmic bridges between bacteria for transfer of plasmids.

conodonts (kon'-ə-donts) (Gr. *con,* cone + *odont,* tooth). Microscopic, toothlike fossils belonging to an extinct vertebrate-like animal known from the Cambrian to Triassic periods.

conspecific (L. *com,* together, + *species*). A member of the same species.

contractile vacuole A clear fluid-filled cell vacuole in protozoa and a few lower metazoa; takes up water and releases it to the outside in a cyclical manner, for osmoregulation and some excretion.

control That part of a scientific experiment to which the experimental variable is not applied but which is similar to the experimental group in all other respects.

coprophagy (kə-prä'fə-jē) (Gr. *kopros,* dung, + *phagein,* to eat). Feeding on dung or excrement as a normal behavior among animals; reinjestion of feces.

copulation (Fr. from L. *copulare,* to couple). Sexual union to facilitate the reception of sperm by the female.

copy DNA See **complementary DNA.**

coralline algae Algae that precipitate calcium carbonate in their tissues; important contributors to coral reef mass.

corium (kō're-um) (L. *corium,* leather). The deep layer of the skin; dermis.

cornea (kor'nē-ə) (L. *corneus,* horny). The outer transparent coat of the eye.

corneum (kor'nē-əm) (L. *corneus,* horny). Epithelial layer of dead, keratinized cells. Stratum corneum.

cornified (kor'nə-fīd) (L. *corneus,* horny). Adjective for conversion of epithelial cells into nonliving, keratinized cells.

corona (kə-rō'nə) (L. crown). Head or upper portion of a structure; ciliated disc on anterior end of rotifers.

corpora allata (kor'pə-rə əl-la'tə) (L. *corpus,* body, + *allatum,* aided). Endocrine glands in insects that produce juvenile hormone.

corpora cardiaca (kor'pə-rə kar-dī'ə-cə) (L. *corpus,* body, + Gr. *kardiakos,* belonging to the heart). Paired organs behind the brain of insects, serve as storage and release organs for brain hormone.

cortex (kor'teks) (L. bark). The outer layer of a structure.

covalent bond A chemical bond in which electrons are shared between atoms.

coxa, coxopodite (kox'ə, kəx-ä'pə-dīt) (L. *coxa,* hip, + Gr. *pous, podos,* foot). The proximal joint of an insect or arachnid leg; in crustaceans, the proximal joint of the protopod.

creatine phosphate High-energy phosphate compound found in the muscle of vertebrates and some invertebrates, used to regenerate stores of ATP.

cretin (krēt'n) (Fr. *crétin,* [dialect], fr. L. *christianus,* Christian, to indicate idiots so afflicted were also human). A human with severe mental, somatic, and sexual retardation resulting from hypothyroidism during early stages of development.

crista (kris'ta), pl. **cristae** (L. *crista,* crest). A crest or ridge on a body organ or organelle; a platelike projection formed by the inner membrane of mitochondrion.

crossing over Exchange of parts of nonsister chromatids at synapsis in the first meiotic division.

cryptobiotic (Gr. *kryptos,* hidden, + *bioticus,* pertaining to life). Living in concealment; refers to insects and other animals that live in secluded situations, such as underground or in wood; also tardigrades and some nematodes, rotifers, and others that survive harsh environmental conditions by assuming for a time a state of very low metabolism.

ctenidia (te-ni'dē-ə) (Gr. *kteis,* comb). Comb-like structures, especially gills of molluscs; also applied to comb plates of Ctenophora.

ctenoid scales (ten'oid) (Gr. *kteis, ktenos,* comb). Thin, overlapping dermal scales of the more advanced fishes; exposed posterior margins have fine, toothlike spines.

cupula (kū'pū-lə) (L. little tub). Small inverted cuplike structure housing another structure; gelatinous matrix covering hair cells in lateral line and equilibrium organs.

cuticle (kū'ti-kəl) (L. *cutis,* skin). A protective, noncellular, organic layer secreted by the external epithelium (hypodermis) of many invertebrates. In higher animals the term refers to the epidermis or outer skin.

cyanobacteria (sī-an-ō-bak-ter'ē-ə) (Gr. *kyanos,* a dark-blue substance, + *bakterion,* dim. of *baktron,* a staff). Photosynthetic prokaryotes, also called blue-green algae, cyanophytes.

cyanophyte (sī-an'ō-fit) (Gr. *kyanos,* a dark-blue substance, + *phyton,* plant). A cyanobacterium, blue-green alga.

cyclin A protein important in the control of the cell division cycle and mitosis.

cycloid scales (sī'-kloid) (Gr. *kyklos,* circle). Thin, overlapping dermal scales of the more primitive fishes; posterior margins are smooth.

cydippid larva (sī-dip'pid) (Gr. *kydippe,* mythological Athenian maiden). Free-swimming larva of most ctenophores; superficially similar to the adult.

cynodonts (sin'ə-dänts) (Gr. *kynodōn,* canine tooth). A group of mammal-like carnivorous synapsids of the Upper Permian and Triassic.

cyrtocyte (ser′tō-sīt) (Gr. *kyrtē*, a fish basket, cage, + *kytos*, hollow vessel). A protonephridial cell with a single flagellum enclosed in a cylinder of cytoplasmic rods.

cystacanth (sis′tə-kanth) (Gr. *kystis*, bladder, pouch, + *akantha*, thorn). Juvenile stage of an acanthocephalan that is infective to the definitive host.

cysticercoid (sis′tə-ser′koyd) (Gr. *kystis*, bladder, + *kerkos*, tail, + *eidos*, form). A type of juvenile tapeworm composed of a solid-bodied cyst containing an invaginated scolex; contrast with **cysticercus.**

cysticercus (sis′tə-ser′kəs) (Gr. *kystis*, bladder, + *kerkos*, tail). A type of juvenile tapeworm in which an invaginated and introverted scolex is contained in a fluid-filled bladder; contrast with **cysticercoid.**

cystid (sis′tid) (Gr. *kystis*, bladder). In an ectoproct, the dead secreted outer parts plus the adherent underlying living layers.

cytochrome (sī′tə-krōm) (Gr. *kytos*, hollow vessel, + *chrōma*, color). Several iron-containing pigments that serve as electron carriers in aerobic respiration.

cytokine (sī′tə-kīn) (Gr. *kytos*, hollow vessel, + *kinein*, to move). A molecule secreted by an activated or stimulated cell, for example, macrophages, that causes physiological changes in certain other cells.

cytokinesis (sī′tə-kin-ē′sis) (Gr. *kytos*, hollow vessel, + *kinesis*, movement). Division of the cytoplasm of a cell.

cytopharynx (Gr. *kytos*, hollow vessel, + *pharynx*, throat). Short tubular gullet in ciliate protozoa.

cytoplasm (sī′tə-plasm) (Gr. *kytos*, hollow vessel, + *plasma*, mold). The living matter of the cell, excluding the nucleus.

cytoproct (sī′tə-prokt) (Gr. *kytos*, hollow vessel, + *prōktos*, anus). Site on a protozoan where undigestible matter is expelled.

cytopyge (sī′tə-pīj) (Gr. *kytos*, hollow vessel, + *pyge*, rump or buttocks). In some protozoa, localized site for expulsion of wastes.

cytosol (sī′tə-sol) (Gr. *kytos*, hollow vessel, + L. *sol*, from *solutus*, to loosen). Unstructured portion of the cytoplasm in which the organelles are bathed.

cytosome (sī′tə-sōm) (Gr. *kytos*, hollow vessel, + *sōma*, body). The cell body inside the plasma membrane.

cytostome (sī′tə-stōm) (Gr. *kytos*, hollow vessel, + *stoma*, mouth). The cell mouth in many protozoa.

cytotoxic T cells (Gr. *kytos*, hollow vessel, + toxin). A special T cell activated during cell-mediated immune responses that recognizes and destroys virus-infected cells.

D

dactylozooid (dak-til′ə-zō-id) (Gr. *dakos*, bite, sting, + *tylos*, knob, + *zōon*, animal). A polyp of a colonial hydroid specialized for defense or killing food.

Darwinism Theory of evolution emphasizing common descent of all living organisms, gradual change, multiplication of species and natural selection.

data sing. **datum** (Gr. *dateomai*, to divide, cut in pieces). The results in a scientific experiment, or descriptive observations, upon which a conclusion is based.

deciduous (də-sij′ə-wəs) (L. *decidere*, to fall off). Shed or falling off at end of a growing period.

deduction (L. *deductus*, led apart, split, separated). Reasoning from the general to the particular, that is, from given premises to their necessary conclusion.

defensins Antimicrobial peptides produced abundantly by cells in and around the linings of mammalian intestine, urogenital tract, and respiratory tract and by neutrophils. Some antimicrobial peptides produced by insects and plants are also called defensins, but they differ in structure.

definitive host The host in which sexual reproduction of a symbiont takes place; if no sexual reproduction, then the host in which the symbiont becomes mature and reproduces; contrast **intermediate host.**

delayed type hypersensitivity Inflammatory reaction based primarily on cell-mediated immunity.

deme (dēm) (Gr. populace). A local population of closely related animals.

demography (də-mäg′rə-fē) (Gr. *demos*, people, + *graphy*). The properties of the rate of growth and the age structure of populations.

dendrite (den′drīt) (Gr. *dendron*, tree). Any of nerve cell processes that conduct impulses toward the cell body.

deoxyribonucleic acid (DNA) The genetic material of all organisms, characteristically organized into linear sequences of genes.

deoxyribose (dē-ok′sē-rī′bōs) (L. *deoxy*, loss of oxygen, + *ribose*, a pentose sugar). A 5-carbon sugar having 1 oxygen atom less than ribose; a component of deoxyribonucleic acid (DNA).

dermal (Gr. *derma*, skin). Pertaining to the skin; cutaneous.

dermis The inner, sensitive mesodermal layer of skin; corium.

desmosome (dez′mə-sōm) (Gr. *desmos*, bond, + *sōma*, body). Buttonlike plaque serving as an intercellular connection.

determinate cleavage The type of cleavage, usually spiral, in which the fate of the blastomeres is determined very early in development; mosaic cleavage.

detritus (də-trī′tus) (L. that which is rubbed or worn away). Any fine particulate debris of organic or inorganic origin.

Deuterostomia (dū′də-rō-stō′mē-ə) (Gr. *deuteros*, second, secondary, + *stoma*, mouth). A group of higher phyla in which cleavage is indeterminate (regulative) and primitively radial. The endomesoderm is enterocoelous, and the mouth is derived away from the blastopore. Includes Echinodermata, Chordata, and Hemichordata. Compare with Protostomia.

dextral (dex′trəl) (L. *dexter*, right-handed). Pertaining to the right; in gastropods, shell is dextral if opening is to right of columella when held with spire up and facing observer.

diapause (dī′ə-pawz) (Gr. *diapausis*, pause). A period of arrested development in the life cycle of insects and certain other animals in which physiological activity is very low and the animal is highly resistant to unfavorable external conditions.

diapsids (dī-ap′səds) (Gr. *di*, two, + *apsis*, arch). Amniotes in which the skull bears two pairs of temporal openings; includes reptiles (except turtles) and birds as living representatives.

diastole (dī-as′tə-lē) (Gr. *diastolē*, dilation). Relaxation and expansion of the heart during which the chambers are filled with blood.

diffusion (L. *diffusus,* dispersion). The movement of particles or molecules from area of high concentration of the particles or molecules to area of lower concentration.

digitigrade (dij′ə-də-grād) (L. *digitus,* finger, toe, + *gradus,* step, degree). Walking on the digits with the posterior part of the foot raised; compare plantigrade.

dihybrid (dī-hī′brəd) (Gr. *dis,* twice, + L. *hibrida,* mixed offspring). A hybrid whose parents differ in two distinct characters; an offspring having two different alleles at two different loci, for example, *A/a B/b.*

dimorphism (dī-mor′fizm) (Gr. *di,* two, + *morphē,* form). Existence within a species of two distinct forms according to color, sex, size, organ structure, and so on. Occurrence of two kinds of zooids in a colonial organism.

dioecious (dī-ē′shəs) (Gr. *di,* two, + *oikos,* house). Having male and female organs in separate individuals.

diphycercal (dif′i-ser′kəl) (Gr. *diphyēs,* twofold, + *kerkos,* tail). A tail that tapers to a point, as in lungfishes; vertebral column extends to tip without upturning.

diphyodont (di′fi-ə-dänt) (Gr. *diphyēs,* twofold, + *odous,* tooth). Having deciduous and permanent sets of teeth successively.

diploblastic (di′plə-blas′tək) (Gr. *diploos,* double, + *blastos,* bud). Organism with two germ layers, endoderm and ectoderm.

diploid (dip′loid) (Gr. *diploos,* double, + *eidos,* form). Having the somatic (double, or *2n*) number of chromosomes or twice the number characteristic of a gamete of a given species.

disaccharides (dī-sak′ə-rīds) (Gr. *dis,* twice, + L. *saccharum,* sugar). A class of sugars (such as lactose, maltose, and sucrose) that yield two monosaccharides on hydrolysis.

distal (dis′təl). Farther from the center of the body than a reference point.

DNA See **deoxyribonucleic acid.**

dominance hierarchy A social ranking, formed through agonistic behavior, in which individuals are associated with each other so that some have greater access to resources than do others.

dominant An allele that is expressed regardless of the nature of the corresponding allele on the homologous chromosome.

dorsal (dor′səl) (L. *dorsum,* back). Toward the back, or upper surface, of an animal.

Down syndrome A congenital syndrome including mental retardation, caused by the cells in a person's body having an extra chromosome 21; also called trisomy 21.

dual-gland adhesive organ Organs in the epidermis of most turbellarians, with three cell types; viscid and releasing gland cells and anchor cells.

duodenum (dū-ə-dēn′əm) (L. *duodeni,* twelve each, fr. its length, about 12 fingers' width). The first and shortest portion of the small intestine lying between the pyloric end of the stomach and the jejunum.

dyad (dī′əd) (Gr. *dyas,* two). One of the groups of two chromosomes formed by the division of a tetrad during the first meiotic division.

E

eccrine (ek′rən) (Gr. *ek,* out of, + *krinein,* to separate). Applies to a type of mammalian sweat gland that produces a watery secretion.

ecdysiotropin (ek-dē-zē-o-tro′pən) (Gr. *ekdysis,* to strip off, escape, + *tropos,* a turn, change). Hormone secreted in brain of insects that stimulates prothoracic gland to secrete molting hormone. Prothoracicotropic hormone; brain hormone.

ecdysis (ek′də-sis) (Gr. *ekdysis,* to strip off, escape). Shedding of outer cuticular layer; molting, as in insects or crustaceans.

ecdysone (ek-dī′sōn) (Gr. *ekdysis,* to strip off). Molting hormone of arthropods, stimulates growth and ecdysis, produced by prothoracic glands in insects and Y organs in crustaceans.

ecocline (ek′ō-klīn) (Gr. *oikos,* home, + *klino,* to slope, recline). The gradient between adjacent biomes; a gradient of environmental conditions.

ecology (Gr. *oikos,* house, + *logos,* discourse). Part of biology that deals with the relationship between organisms and their environment.

ecosystem (ek′ō-sis-təm) (eco[logy] from Gr. *oikos,* house, + system). An ecological unit consisting of both the biotic communities and the nonliving (abiotic) environment, which interact to produce a stable system.

ecotone (ek′ō-tōn) (eco[logy] from Gr. *oikos,* home, + *tonos,* stress). The transition zone between two adjacent communities.

ectoderm (ek′tō-derm) (Gr. *ektos,* outside, + *derma,* skin). Outer layer of cells of an early embryo (gastrula stage); one of the germ layers, also sometimes used to include tissues derived from ectoderm.

ectognathous (ek′tə-nā′thəs) (Gr. *ektos,* outside, without, + *gnathos,* jaw). Derived character of most insects; mandibles and maxillae not in pouches.

ectolecithal (ek′tō-les′ə-thəl) (Gr. *ektos,* outside, + *lekithos,* yolk). Yolk for nutrition of the embryo contributed by cells that are separate from the egg cell and are combined with the zygote by envelopment within the eggshell.

ectoneural (ek′tə-nu-rəl) (Gr. *ektos,* outside, without, + *neuron,* nerve). Oral (chief) nervous system in echinoderms.

ectoplasm (ek′tō-plazm) (Gr. *ektos,* outside, + *plasma,* form). The cortex of a cell or that part of cytoplasm just under the cell surface; contrasts with **endoplasm.**

ectothermic (ek′tō-therm′ic) (Gr. *ektos,* outside, + *thermē,* heat). Having a variable body temperature derived from heat acquired from the environment; contrasts with **endothermic.**

edema (ē-dē′mə) (Gr. *oidēma,* swelling). Buildup of fluid in interstitial spaces, causing swelling.

effector (L. *efficere,* bring to pass). An organ, tissue, or cell that becomes active in response to stimulation.

efferent (ef′ə-rənt) (L. *ex,* out, + *ferre,* to bear). Leading or conveying away from some organ, for example, nerve impulses conducted away from the brain, or blood conveyed away from an organ; contrasts with **afferent.**

egestion (ē-jes′chən) (L. *egestus,* to discharge). Act of casting out indigestible or waste matter from the body by any normal route.

---

bat / āpe / ärmadillo / herring / fēmale / finch / līce / crocodile / crōw / duck / ūnicorn / ə indicates unaccented vowel sound "uh" as in mammal, fishes, cardinal, heron, vulture / stress as in bi-ol′o-gy, bi′o-log′i-cal

electron A subatomic particle with a negative charge and a mass of 9.1066 × 10⁻²⁸ gram.

eleocyte (el′ē-ə-sīt) (Gr. *elaion,* oil, + *kytos,* hollow vessel). Fat-containing cells in annelids that originate from the chlorogogen tissue.

elephantiasis (el-ə-fən-tī′ə-səs). Disfiguring condition caused by chronic infection with filarial worms *Wuchereria bancrofti* and *Brugia malayi.*

embryogenesis (em′brē-ō-jen′ə-səs) (Gr. *embryon,* embryo, + *genesis,* origin). The origin and development of the embryo; embryogeny.

emergence (L. *e,* out, + *mergere,* to plunge). The appearance of properties in a biological system (at the molecular, cellular, organismal, or species levels) that cannot be deduced from knowledge of the component parts taken separately or in partial combinations; such properties are termed **emergent properties.**

emigrate (L. *emigrare,* to move out). To move *from* one area to another to take up residence.

emulsion (ə-məl′shən) (L. *emulsus,* milked out). A colloidal system in which both phases are liquids.

endemic (en-dem′ik) (Gr. *en,* in, + *demos,* populace). Peculiar to a certain region or country; native to a restricted area; not introduced.

endergonic (en-dər-gän′ik) (Gr. *endon,* within, + *ergon,* work). Used in reference to a chemical reaction that requires energy; energy absorbing.

endite (en′dīt) (Gr. *endon,* within). Medial process on an arthropod limb.

endochondral (en′dō-kän′drōl) (Gr. *endon,* within, + *chondros,* cartilage). Occurring with the substance of cartilage, especially bone formation.

endocrine (en′də-krən) (Gr. *endon,* within, + *krinein,* to separate). Refers to a gland that is without a duct and that releases its product directly into the blood or lymph.

endocytosis (en′dō-sī-tō-səs) (Gr. *endon,* within, + *kytos,* hollow vessel). The engulfment of matter by phagocytosis, potocytosis, receptor-mediated endocytosis, and by bulk-phase (nonspecific) endocytosis.

endoderm (en′də-dərm) (Gr. *endon,* within, + *derma,* skin). Innermost germ layer of an embryo, forming the primitive gut; also may refer to tissues derived from endoderm.

endognathous (en′də-nā-thəs) (Gr. *endon,* within, + *gnathous,* jaw). Ancestral character in insects, found in orders Diplura, Collembola, and Protura, in which the mandibles and maxillae are located in pouches.

endolecithal (en′də-les′ə-thəl) (Gr. *endon,* within, + *lekithos,* yolk). Yolk for nutrition of the embryo incorporated into the egg cell itself.

endolymph (en′də-limf) (Gr. *endon,* within, + *lympha,* water). Fluid that fills most of the membranous labyrinth of the vertebrate ear.

endometrium (en′də-mē′trē-əm) (Gr. *endon,* within, + *mētra,* womb). The mucous membrane lining the uterus.

endoplasm (en′də-pla-zm) (Gr. *endon,* within, + *plasma,* mold or form). The portion of cytoplasm that immediately surrounds the nucleus.

endoplasmic reticulum A complex of membranes within a cell; may bear ribosomes (rough) or not (smooth).

endopod, endopodite (en′də-päd, en-dop′ə-dīt) (Gr. *endon,* within, + *pous, podos,* foot). Medial branch of a biramous crustacean appendage.

endopterygote (en′dəp-ter′i-gōt) (Gr. *endon,* within, + *pteron,* feather, wing). Insect in which the wing buds develop internally; has holometabolous metamorphosis.

endorphin (en-dor′fin) (contraction of endogenous morphine). Group of opiate-like brain neuropeptides that modulate pain perception and are implicated in many other functions.

endoskeleton (Gr. *endon,* within, + *skeletos,* hard). A skeleton or supporting framework within the living tissues of an organism; contrasts with **exoskeleton.**

endostyle (en′də-stīl) (Gr. *endon,* within, + *stylos,* a pillar). Mucus-secreting, ciliated groove(s) in the floor of the pharynx of tunicates, cephalochordates, and larval jawless fishes useful for accumulating and moving food particles to the stomach.

endothelium (en-də-thē′lē-əm) (Gr. *endon,* within, + *thēlē,* nipple). Squamous epithelium lining internal body cavities such as heart and blood vessels. Adj., **endothelial.**

endothermic (en′də-therm′ic) (Gr. *endon,* within, + *thermē,* heat). Having a body temperature determined by heat derived from the animal's own oxidative metabolism; contrasts with **ectothermic.**

enkephalin (en-kef′ə-lin) (Gr. *endon,* within, + *kephale,* head). Group of small brain neuropeptides with opiate-like qualities.

enterocoel (en′tər-ō-sēl′) (Gr. *enteron,* gut, + *koilos,* hollow). A type of coelom formed by the outpouching of a mesodermal sac from the endoderm of the primitive gut.

enterocoelic mesoderm formation Embryonic formation of mesoderm by a pouchlike outfolding from the archenteron, which then expands and obliterates the blastocoel, thus forming a large cavity, the coelom, lined with mesoderm.

enterocoelomate (en′ter-ō-sēl′ō-māte) (Gr. *enteron,* gut, + *koilōma,* cavity, + Eng. *ate,* state of). An animal having an enterocoel, such as an echinoderm or a vertebrate.

enteron (en′tə-rän) (Gr. intestine). The digestive cavity.

entomology (en′tə-mol′ə-jē) (Gr. *entoma,* an insect, + *logos,* discourse). Study of insects.

entozoic (en-tə-zō′ic) (Gr. *entos,* within, + *zōon,* animal). Living within another animal; internally parasitic (chiefly parasitic worms).

entropy (en′trə-pē) (Gr. *en,* in, on, + *tropos,* turn, change in manner). A quantity that is the measure of energy in a system not available for doing work.

enzyme (en′zīm) (Gr. *enzymos,* leavened, from *en,* in, + *zyme,* leaven). A substance, produced by living cells, that is capable of speeding up specific chemical transformations, such as hydrolysis, oxidation, or reduction, but is unaltered itself in the process; a biological catalyst.

eocytes (ē′ə-sīts) (Gr. *eōs,* the dawn, + *kytos,* hollow vessel). A group of prokaryotes currently classified among the Archaebacteria but possibly a sister group of eukaryotes.

ephyra (ef′ə-rə) (Gr. *Ephyra,* Greek city). Refers to castlelike appearance. Medusa bud from a scyphozoan polyp.

---

bat / āpe / ärmadillo / herring / fēmale / fĭnch / līce / crocodile / crōw / duck / ūnicorn / ə indicates unaccented vowel sound "uh" as in mammal, fishes, cardinal, heron, vulture / stress as in bi-ol′o-gy, bi′o-log′i-cal

epidermis (ep′ə-dər′məs) (Gr. *epi,* on, upon, + *derma,* skin). The outer, nonvascular layer of skin of ectodermal origin; in invertebrates, a single layer of ectodermal epithelium.

epididymis (ep′ə-did′ə-məs) (Gr. *epi,* on, upon, + *didymos,* testicle). Part of the sperm duct that is coiled and lying near the testis.

epigenesis (ep′ə-jen′ə-sis) (Gr. *epi,* on, upon, + *genesis,* birth). The embryological (and generally accepted) view that an embryo is a new creation that develops and differentiates step by step from an initial stage; the progressive production of new parts that were nonexistent as such in the original zygote.

epigenetics (ep′ə-je-net′iks) (Gr. *epi,* on, upon, + *genesis,* birth). Study of the relationship between genotype and phenotype as mediated by developmental processes.

epipod, epipodite (ep′ē-päd, e-pip′ə-dīt) (Gr. *epi,* on, upon, + *pous, podos,* foot). A lateral process on the protopod of a crustacean appendage, often modified as a gill.

epistasis (e-pis′tə-səs) (Gr. *epi,* on, upon, + *stasis,* standing). Prevention of expression of an allele at one locus by an allele at another locus.

epistome (ep′i-stōm) (Gr. *epi,* on, upon, + *stoma,* mouth). Flap over the mouth in some lophophorates bearing the protocoel.

epithelium (ep′ə-thē′lē-əm) (Gr. *epi,* on, upon, + *thēlē,* nipple). A cellular tissue covering a free surface or lining a tube or cavity.

epitoke (ep′ə-tōk) (Gr. *epitokos,* fruitful). Posterior part of a marine polychaete when swollen with developing gonads during the breeding season; contrast with **atoke.**

epitope That portion of an antigen to which an antibody or T-cell receptor binds. Also called **antigenic determinant.**

erythroblastosis fetalis (ə-rith′rə-blas-tō′səs fə-tal′əs) (Gr. *erythros,* red, + *blastos,* germ, + *osis,* a disease; L. *fetalis,* relating to a fetus). A disease of newborn infants caused when Rh-negative mothers develop antibodies against the Rh-positive blood of the fetus. See **blood type.**

erythrocyte (ə-rith′rə-sīt) (Gr. *erythros,* red, + *kytos,* hollow vessel). Red blood cell; has hemoglobin to carry oxygen from lungs or gills to tissues; during formation in mammals, erythrocytes lose their nuclei, those of other vertebrates retain the nuclei.

esthete (es-thēt′) (Gr. *esthēs,* a garment). Light sensory receptor on a shell of a chiton (phylum Mollusca).

estrus (es′trəs) (L. *oestrus,* gadfly, frenzy). The period of heat, or rut, especially of the female during ovulation of the egg. Associated with maximum sexual receptivity.

estuary (es′chə-we′rē) (L. *aestuarium,* estuary). An arm of the sea where the tide meets the current of a freshwater drainage.

ethology (e-thäl′-ə-jē) (Gr. *ethos,* character, + *logos,* discourse). The study of animal behavior in natural environments.

euchromatin (ū′krō-mə-tən) (Gr. *eu,* good, well, + *chrōma,* color). Part of the chromatin that takes up stain less than heterochromatin, contains active genes.

eukaryotic, eucaryotic (ū′ka-rē-ot′ik) (Gr. *eu,* good, true, + *karyon,* nut, kernel). Organisms whose cells characteristically contain a membrane-bound nucleus or nuclei; contrasts with **prokaryotic.**

euploidy (ū′ploid′ē) (Gr. *eu,* good, well, + *ploid,* multiple of). Change in chromosome number from one generation to the next in which there is an addition or deletion of a complete set of chromosomes in the progeny; the most common type is polyploidy.

euryhaline (ū′-rə-hā′līn) (Gr. *eurys,* broad, + *hals,* salt). Able to tolerate wide ranges of saltwater concentrations.

euryphagous (yə-rif′ə-gəs) (Gr. *eurys,* broad, + *phagein,* to eat). Eating a large variety of foods.

eurytopic (ū-rə-täp′ik) (Gr. *eurys,* broad, + *topos,* place). Refers to an organism with a wide environmental range.

eutely (u′te-lē) (Gr. *euteia,* thrift). Condition of a body composed of a constant number of cells or nuclei in all adult members of a species, as in rotifers, acanthocephalans, and nematodes.

evagination (ē-vaj′ə-nā′shən) (L. *e,* out, + *vagina,* sheath). An outpocketing from a hollow structure.

evolution (L. *evolvere,* to unfold). Organic evolution encompasses all changes in the characteristics and diversity of life on earth throughout its history.

evolutionary duration The length of time that a species or higher taxon exists in geological time.

evolutionary species concept A single lineage of ancestral-descendant populations that maintains its identity from other such lineages and has its own evolutionary tendencies and historical fate; differs from the biological species concept by explicitly including a time dimension and including asexual lineages.

evolutionary taxonomy A system of classification, formalized by George Gaylord Simpson, that groups species into Linnean higher taxa representing a hierarchy of distinct adaptive zones; such taxa may be monophyletic or paraphyletic but not polyphyletic.

excision repair Means by which cells are able to repair certain kinds of damage (dimerized pyrimidines) in their DNA.

exergonic (ek′sər-gon′ik) (Gr. *exō,* outside of, + *ergon,* work). An energy-yielding reaction.

exite (ex′īt) (Gr. *exō,* outside). Process from lateral side of an arthropod limb.

exocrine (ek′sə-krən) (Gr. *exō,* outside, + *krinein,* to separate). A type of gland that releases its secretion through a duct; contrasts with **endocrine.**

exocytosis (eks′ə-sī-tō′səs) (Gr. *exo,* outside, + *kytos,* hollow vessel). Transport of a substance from inside a cell to the outside.

exon (ex′ən) (Gr. *exō,* outside). Part of the mRNA as transcribed from the DNA that contains a portion of the information necessary for final gene product.

exopod, exopodite (ex′ə-päd, ex-äp′ə-dīt) (Gr. *exō,* outside, + *pous, podos,* foot). Lateral branch of a biramous crustacean appendage.

exopterygote (ek′səp-ter′i-gōt) (Gr. *exō,* without, + *pteron,* feather, wing). Insect in which the wing buds develop externally during nymphal instars; has hemimetabolous metamorphosis.

exoskeleton (ek′sō-skel′ə-tən) (Gr. *exō,* outside, + *skeletos,* hard). A supporting structure secreted by ectoderm or epidermis; external, not enveloped by living tissue, as opposed to **endoskeleton.**

experiment (L. *experiri*, to try). A trial made to support or disprove a hypothesis.

exteroceptor (ek′stər-ō-sep′tər) (L. *exter*, outward, + *capere*, to take). A sense organ excited by stimuli from the external world.

F

facilitated diffusion Mediated transport in which a transmembrane protein makes possible diffusion of a molecule across a cell membrane in the direction of a concentration gradient; contrast with **active transport.**

FAD Abbreviation for flavin adenine dinucleotide, an electron acceptor in the respiratory chain.

fascicle (fas′ə-kəl) (L. *fasciculus*, small bundle). A small bundle, usually referring to a collection of muscle fibers or nerve axons.

fatty acid Any of a series of saturated organic acids having the general formula $C_nH_{2n}O_2$; occurs in natural fats of animals and plants.

fermentation (L. *fermentum*, ferment). Enzymatic transformation, without oxygen, or organic substrates, especially carbohydrates, yielding products such as alcohols, acids, and carbon dioxide.

fiber (L. *fibra*, thread). A fiberlike cell or strand of protoplasmic material produced or secreted by a cell and lying outside the cell.

fibril (L. *fibra*, thread). A strand of protoplasm produced by a cell and lying within the cell.

fibrillar (fī′brə-lər) (L. *fibrilla*, small fiber). Composed of or pertaining to fibrils or fibers.

fibrin Protein that forms a meshwork, trapping erythrocytes, to become blood clot. Precursor is fibrinogen.

fibrosis (fī-brō′səs). Deposition of fibrous connective tissue in a localized site, during process of tissue repair or to wall off a source of antigen.

filipodium (fī′li-pō′de-əm) (L. *filum*, thread, + Gr. *pous, podos*, a foot). A type of pseudopodium that is very slender and may branch but does not rejoin to form a mesh.

filter feeding Any feeding process by which particulate food is filtered from water in which it is suspended.

fission (L. *fissio*, a splitting). Asexual reproduction by a division of the body into two or more parts.

fitness Degree of adjustment and suitability for a particular environment. Genetic fitness is relative contribution to one genetically distinct organism to the next generation; organisms with high genetic fitness are naturally selected and become prevalent in a population.

flagellum (flə-jel′em) pl. **flagella** (L. a whip). Whiplike organelle of locomotion.

flame cell Specialized hollow excretory or osmoregulatory structure of one or several small cells containing a tuft of flagella (the "flame") and situated at the end of a minute tubule; connected tubules ultimately open to the outside. See **solenocyte, protonephridium.**

fluke (O.E. *flōc*, flatfish). A member of class Trematoda or class Monogenea. Also, certain of the flatfishes (order Pleuronectiformes).

FMN Abbreviation for flavin mononucleotide, the prosthetic group of a protein (flavoprotein) and a carrier in the electron transport chain in respiration.

food vacuole A digestive organelle in the cell.

foraminiferan (for′əm-i-nif′-ər-ən) (L. *foramin*, hole, performation, + *fero*, to bear). Granuloreticulosean amebas bearing a test with many openings.

fossil (fos′əl). Any remains or impression of an organism from a past geological age that has been preserved by natural processes, usually by mineralization in the earth's crust.

fossorial (fä-sōr′ē-əl) (L. *fossor*, digger). Characterized by digging or burrowing.

fouling Contamination of feeding or respiratory areas of an organism by excrement, sediment, or other matter. Also, accumulation of sessile marine organisms on the hull of a boat or ship so as to impede its progress through the water.

founder event Establishment of a new population by a small number of individuals (sometimes a single female carrying fertile eggs) that disperse from their parental population to a new location geographically isolated from the parental population.

fovea (fō′vē-ə) (L. small pit). A small pit or depression; especially the fovea centralis, a small pit containing only cones in the retina of some vertebrates, a point of acute vision.

free energy The energy available for doing work in a chemical system.

frontal plane A plane parallel to the main axis of the body and at right angles to the sagittal plane.

fusiform (fū′zə-form) (L. *fusus*, spindle, + *forma*, shape). Spindle shaped; tapering toward each end.

G

gamete (ga′mēt, gə-mēt′) (Gr. *gamos*, marriage). A mature haploid sex cell; usually, male and female gametes can be distinguished. An egg or a sperm.

gametic meiosis Meiosis that occurs during formation of the gametes, as in humans and other metazoa.

gametocyte (gə-mēt′ə-sīt) (Gr. *gametēs*, spouse, + *kytos*, hollow vessel). The mother cell of a gamete, that is, immature gamete.

ganglion (gang′lē-ən) pl. **ganglia** (Gr. little tumor). An aggregation of nerve tissue containing nerve-cell bodies.

ganoid scales (ga′noid) (Gr. *ganos*, brightness). Thick, bony, rhombic scales of some primitive bony fishes; not overlapping.

gap junction An area of tiny pores communicating the cytoplasm between two cells.

gastrodermis (gas′tro-dər′mis) (Gr. *gastēr*, stomach, + *derma*, skin). Lining of the digestive cavity of cnidarians.

gastrolith (gas′trə-lith) (Gr. *gastēr*, stomach, + *lithos*, stone). Calcareous body in the wall of the cardiac stomach of crayfish and other Malacostraca, preceding the molt.

gastrovascular cavity (Gr. *gastēr*, stomach, + L. *vasculum*, small vessel). Body cavity in certain lower invertebrates that functions in both digestion and circulation and has a single opening serving as both mouth and anus.

---

bat / āpe / ärmadillo / herring / fēmale / finch / līce / crocodile / crōw / duck / ūnicorn / ə indicates unaccented vowel sound "uh" as in mammal, fishes, cardinal, heron, vulture / stress as in bi-ol′o-gy, bi′o-log′i-cal

gastrozooid (gas′trə-zō-id) (Gr. *gastēr*, stomach, + *zōon*, animal). The feeding polyp of a hydroid, a hydranth.

gastrula (gas′trə-lə) (Gr. *gastēr*, stomach, + L. *ula*, dim.). Embryonic stage, usually cap or sac shaped, with walls of two layers of cells surrounding a cavity (archenteron) with one opening (blastopore).

gastrulation (gas′trə-lā′shən) (Gr. *gastēr*, stomach). Process by which an early metazoan embryo becomes a gastrula, acquiring first two and then three layers of cells.

gel (jel) (from gelatin, from L. *gelare*, to freeze). That state of a colloidal system in which the solid particles form the continuous phase and the fluid medium the discontinuous phase.

gemmule (je′mūl) (L. *gemma*, bud, + *ula*, dim.). Asexual, cystlike reproductive unit in freshwater sponges; formed in summer or autumn and capable of overwintering.

gene (Gr. *genos*, descent). A nucleic acid sequence (usually DNA) that encodes a functional polypeptide or RNA sequence.

gene pool A collection of all of the alleles of all of the genes in a population.

genetic drift Random change in allelic frequencies in a population occurring by chance. In small populations, genetic variation at a locus may be lost by chance fixation of a single allelic variant.

genome (jē′nōm) (Gr. *genos*, offspring, + *ōma*, abstract group). All the DNA in a haploid set of chromosomes (nuclear genome), organelle (mitochondrial genome, chloroplast genome) or virus (viral genome, which in some viruses consists of RNA rather than DNA).

genomics (jē-nō′miks). Mapping and sequencing of genomes (= structural genomics). Functional genomics is development and application of genome or systemwide experimental approaches to assess gene function. Functional genomics uses information derived from structural genomics.

genotype (jēn′ō-tīp) (Gr. *genos*, offspring, + *typos*, form). The genetic constitution, expressed and latent, of an organism; the total set of genes present in the cells of an organism; contrasts with **phenotype.**

genus (jē-nus), pl. **genera** (L. race). A group of related species with taxonomic rank between family and species.

germ layer In the animal embryo, one of three basic layers (ectoderm, endoderm, mesoderm) from which the various organs and tissues arise in the multicellular animal.

germ plasm Cell lineages giving rise to the germ cells of a multicellular organism, as distinct from the somatoplasm.

germovitellarium (jer′mə-vit-ə-lar′ē-əm) (L. *germen*, a bud, offshoot, + *vitellus*, yolk). Closely associated ovary (germarium) and yolk-producing structure (vitellarium) in rotifers.

gestation (jes-tā′shən) (L. *gestare*, to bear). The period in which offspring are carried in the uterus.

globulins (glo′bū-lənz) (L. *globus*, a globe, ball, + *-ulus*, ending denoting tendency). A large group of compact proteins with high molecular weight; includes immunoglobulins (antibodies).

glochidium (glō-kid′e-əm) (Gr. *glochis*, point, + *idion*, dim.). Bivalved larval stage of freshwater mussels.

glomerulus (glä-mer′u-ləs) (L. *glomus*, ball). A tuft of capillaries projecting into a renal corpuscle in a kidney. Also, a small spongy mass of tissue in the proboscis of hemichordates, presumed to have an excretory function. Also, a concentration of nerve fibers situated in the olfactory bulb.

gluconeogenesis (glü-cō-nē-ō-gən′ə-səs) (Gr. *glykys*, sweet, + *neos*, new, + *genesis*, origin). Synthesis of glucose from protein or lipid precursors.

glycogen (glī′kə-jən) (Gr. *glykys*, sweet, + *genēs*, produced). A polysaccharide constituting the principal form in which carbohydrate is stored in animals; animal starch.

glycolysis (glī-kol′ə-səs) (Gr. *glykys*, sweet, + *lysis*, a loosening). Enzymatic breakdown of glucose (especially) or glycogen into phosphate derivatives with release of energy.

gnathobase (nāth′ə-bās′) (Gr. *gnathos*, jaw, + base). A median basic process on certain appendages in some arthropods, usually for biting or crushing food.

gnathostomes (nath′ə-stōms) (Gr. *gnathos*, jaw, + *stoma*, mouth). Vertebrates with jaws.

Golgi complex (gōl′jē) (after Golgi, Italian histologist). An organelle in cells that serves as a collecting and packaging center for secretory products.

gonad (gō′nad) (N.L. *gonas*, primary sex organ). An organ that produces gametes (ovary in the female and testis in the male).

gonangium (gō-nan′jē-əm) (N.L. *gonas*, primary sex organ, + *angeion*, dim. of vessel). Reproductive zooid of hydroid colony (Cnidaria).

gonoduct (Gr. *gonos*, seed, progeny, + duct). Duct leading from a gonad to the exterior.

gonopore (gän′ə-pōr) (Gr. *gonos*, seed, progeny, + *poros*, an opening). A genital pore found in many invertebrates.

grade (L. *gradus*, step). A level of organismal complexity or adaptive zone characteristic of a group of evolutionarily related organisms.

gradualism (graj′ə-wal-iz′əm). A component of Darwin's evolutionary theory postulating that evolution occurs by the temporal accumulation of small, incremental changes, usually across very long periods of geological time; it opposes claims that evolution can occur by large, discontinuous or macromutational changes.

granulocytes (gran′ū-lə-sīts) (L. *granulus*, small grain, + Gr. *kytos*, hollow vessel). White blood cells (neutrophils, eosinophils, and basophils) bearing "granules" (vacuoles) in their cytoplasm that stain deeply.

green gland Excretory gland of certain Crustacea; the antennal gland.

gregarious (L. *grex*, herd). Living in groups or flocks.

guanine (gwä′nēn) (Sp. from Quechua, *huanu*, dung). A white crystalline purine base, $C_5H_5N_5O$, occurring in various animal tissues and in guano and other animal excrements.

guild (gild) (M.E. *gilde*, payment, tribute). In ecology, a group of species that exploit the same class of environment in a similar way.

gynandromorph (ji-nan′drə-mawrf) (Gr. *gyn*, female, + *andr*, male, + *morphē*, form). An abnormal individual exhibiting characteristics of both sexes in different parts of the body; for example the left side of a bilateral organism may show characteristics of one sex and the right side those of the other sex.

---

bat / āpe / ärmadillo / herring / fēmale / finch / līce / crocodile / crōw / duck / ūnicorn / ə indicates unaccented vowel sound "uh" as in mammal, fishes, cardinal, heron, vulture / stress as in bi-ol′o-gy, bi′o-log′i-cal

gynocophoric canal (gī′nə-kə-fōr′ik) (Gr. *gynē*, woman, + *pherein*, to carry). Groove in male schistosomes (certain trematodes) that carries the female.

H

habitat (L. *habitare*, to dwell). The place where an organism normally lives or where individuals of a population live.

habituation A kind of learning in which continued exposure to the same stimulus produces diminishing responses.

hair cell An important sensory component of several kinds of mechano- and auditory receptors found in both invertebrate (statocyst) and vertebrate (vestibular organ, organ of Corti) organs of equilibrium and hearing. The "hairs" are cilia, or sensory endings that project from the cell surface, and when bent by mechanical stimuli they generate nerve impulses or action potentials, communicating a signal to the central nervous system.

halter (hal′tər), pl. **halteres** (hal-ti′rēz) (Gr. leap). In Diptera, small club-shaped structure on each side of the metathorax representing the hindwings; believed to be sense organs for balancing; also called balancer.

haplodiploidy (Gr. *haploos*, single, + *diploos*, double, + *eidos*, form). Reproduction in which haploid males are produced parthenogenetically, and diploid females are from fertilized eggs.

haploid (Gr. *haploos*, single). The reduced, or *n*, number of chromosomes, typical of gametes, as opposed to the diploid, or *2n*, number found in somatic cells. In certain groups, mature organisms may have a haploid number of chromosomes.

Hardy-Weinberg equilibrium Mathematical demonstration that the Mendelian hereditary process does not change the populational frequencies of alleles or genotypes across generations, and that change in allelic or genotypic frequencies requires factors such as natural selection, genetic drift in finite populations, recurring mutation, migration of individuals among populations, and nonrandom mating.

hectocotylus (hek-tə-kät′ə-ləs) (Gr. *hekaton*, hundred, + *kotylē*, cup). Specialized, and sometimes autonomous, arm that serves as a male copulatory organ in cephalopods.

hemal system (hē′məl) (Gr. *haima*, blood). System of small vessels in echinoderms; function is probably distribution of nutrients to specific body regions.

hemerythrin (hē′mə-rith′rin) (Gr. *haima*, blood, + *erythros*, red). A red, iron-containing respiratory pigment found in the blood of some polychaetes, sipunculids, priapulids, and brachiopods.

hemimetabolous (he′mi-mə-ta′bə-ləs) (Gr. *hēmi*, half, + *metabolē*, change). Refers to gradual metamorphosis during development of insects, without a pupal stage.

hemocoel (hēm′ə-sēl) (Gr. *haima*, blood, + *koiloma*, cavity). Major body space in arthropods replacing the coelom, contains the blood (hemolymph).

hemoglobin (Gr. *haima*, blood, + L. *globulus*, globule). An iron-containing respiratory pigment occurring in vertebrate red blood cells and in blood plasma of many invertebrates; a compound of an iron porphyrin heme and globin proteins.

hemolymph (hē′mə-limf) (Gr. *haima*, blood, + L. *lympha*, water). Fluid in the coelom or hemocoel of some invertebrates that represents the blood and lymph of vertebrates.

hemozoin (hē-mə-zo′ən) (Gr. *haima*, blood, + *zōon*, an animal). Insoluble digestion product of malaria parasites produced from hemoglobin.

hepatic (hə-pat′ik) (Gr. *hēpatikos*, of the liver). Pertaining to the liver.

herbivore ([h]ərb′ə-vōr′) (L. *herba*, green crop, + *vorare*, to devour). Any organism subsisting on plants. Adj., **herbivorous.**

heredity (L. *heres*, heir). The faithful transmission of biological traits from parents to their offspring.

hermaphrodite (hə[r]-maf′rə-dīt) (Gr. *hermaphroditos*, containing both sexes; from Greek mythology, Hermaphroditos, son of Hermes and Aphrodite). An organism with both male and female functional reproductive organs. **Hermaphroditism** may refer to an aberration in unisexual animals; **monoecy** implies that this is the normal condition for the species.

hermatypic (hər-mə-ti′pik) (Gr. *herma*, reef, + *typos*, pattern). Relating to reef-forming corals.

heterocercal (het′ər-o-sər′kəl) (Gr. *heteros*, different, + *kerkos*, tail). In some fishes, a tail with the upper lobe larger than the lower, and the end of the vertebral column somewhat upturned in the upper lobe, as in sharks.

heterochromatin (het′ə-rō-krōm′ə-tən) (Gr. *heteros*, different, + *chrōma*, color). Chromatin that stains intensely and appears to represent inactive genetic areas.

heterochrony (het′ə-rō-krōn-y) (Gr. *heteros*, different, + *chronos*, time). Evolutionary change in the relative time of appearance or rate of development of characteristics from ancestor to descendant.

heterodont (het′ə-ro-dänt) (Gr. *heteros*, different, + *odous*, tooth). Having teeth differentiated into incisors, canines, and molars for different purposes.

heterostracans (Gr. *heteros*, different, + *ostrakon*, shell). A group of extinct fishes with dermal armor and no jaws or paired fins known from the Ordovician to Devonian periods.

heterotroph (het′ə-rō-träf) (Gr. *heteros*, different, + *trophos*, feeder). An organism that obtains both organic and inorganic raw materials from the environment in order to live; includes most animals and those plants that do not carry on photosynthesis.

heterozygote (het′ə-rō-zī′gōt) (Gr. *heteros*, different, + *zygōtos*, yoked). An organism in which homologous chromosomes contain different allelic forms (often dominant and recessive) of a locus; derived from a zygote formed by union of gametes of dissimilar allelic constitution.

hexamerous (hek-sam′ər-əs) (Gr. *hex*, six, + *meros*, part). Six parts, specifically, symmetry based on six or multiples thereof.

hibernation (L. *hibernus*, wintry). Condition, especially of mammals, of passing the winter in a torpid state in which the body temperature drops nearly to freezing and the metabolism drops close to zero.

---

bat / āpe / ärmadillo / herring / fēmale / finch / līce / crocodile / crōw / duck / ūnicorn / ə indicates unaccented vowel sound "uh" as in mammal, fishes, cardinal, heron, vulture / stress as in bi-ol′o-gy, bi′o-log′i-cal

hierarchical system A scheme arranging organisms into a series of taxa of increasing inclusiveness, as illustrated by Linnean classification.

histogenesis (his-tō-jen'ə-sis) (Gr. *histos,* tissue, + *genesis,* descent). Formation and development of tissue.

histone (hi'stōn) (Gr. *histos,* tissue). Any of several simple proteins found in cell nuclei and complexed at one time or another with DNA. Histones yield a high proportion of basic amino acids on hydrolysis; characteristic of eukaryotes.

holoblastic cleavage (Gr. *holo,* whole, + *blastos,* germ). Complete and approximately equal division of cells in early embryo. Found in mammals, amphioxus, and many aquatic invertebrates that have eggs with a small amount of yolk.

holometabolous (hō'lō-mə-ta'bə-ləs) (Gr. *holo,* complete, + *metabolē,* change). Complete metamorphosis during development.

holophytic nutrition (hōl'ō-fit'ik) (Gr. *holo,* whole, + *phyt,* plant). Occurs in green plants and certain protozoa and involves synthesis of carbohydrates from carbon dioxide and water in the presence of light, chlorophyll, and certain enzymes.

holozoic nutrition (hōl'ō-zō'ik) (Gr. *holo,* whole, + *zoikos,* of animals). Type of nutrition involving ingestion of liquid or solid organic food particles.

home range The area over which an animal ranges in its activities. Unlike territories, home ranges are not defended.

homeobox (hō'mē-ō-box) (Gr. *homoios,* like, resembling, + L. *buxus,* boxtree [used in the sense of enclosed, contained]). A highly conserved 180-base-pair sequence found in homeotic genes, regulatory sequences of protein-coding genes that regulate development.

homeostasis (hō'mē-ō-stā'sis) (Gr. *homeo,* alike, + *stasis,* state or standing). Maintenance of an internal steady state by means of self-regulation.

homeothermic (hō'mē-ō-thər'mik) (Gr. *homeo,* alike, + *thermē,* heat). Having a nearly uniform body temperature, regulated independent of the environmental temperature; "warm blooded."

homeotic genes (hō-mē-ät'ik) (Gr. *homoios,* like, resembling). Genes, identified through mutations, that give developmental identity to specific body segments.

hominid (häm'ə-nid) (L. *homo, hominis,* man). A member of the family Hominidae, now represented by one living species, *Homo sapiens.*

hominoid (häm'ə-noid). Relating to the Hominoidea, a superfamily of primates to which the great apes and humans are assigned.

homocercal (hō'mə-ser'kəl) (Gr. *homos,* same, common, + *kerkos,* tail). A tail with the upper and lower lobes symmetrical and the vertebral column ending near the middle of the base, as in most telost fishes.

homodont (hō'mō-dänt) (Gr. *homos,* same, + *odous,* tooth). Having all teeth similar in form.

homograft See **allograft.**

homolog, homologue (hōm'ə-log) One member of a set of homologous structures or one of a pair of homologous chromosomes.

homology (hō-mäl'ə-jē) (Gr. *homologos,* agreeing). Similarity of parts or organs of different organisms caused by evolutionary derivation from a corresponding part or organ in a remote ancestor, and usually having a similar embryonic origin. May also refer to a matching pair of chromosomes. Serial homology is the correspondence in the same individual of repeated structures having the same origin and development, such as the appendages of arthropods. Adj., **homologous.**

homoplasy (hō'mō'plā'sē). Phenotypic similarity among characteristics of different species or populations (including molecular, morphological, behavioral or other features) that does not accurately represent patterns of common evolutionary descent (= nonhomologous similarity); it is produced by evolutionary parallelism, convergence and/or reversal, and is revealed by incongruence among different characters on a cladogram or phylogenetic tree.

homozygote (hō-mə-zī'gōt) (Gr. *homos,* same, + *zygotos,* yoked). An organism having identical alleles at one or more genetic loci. Adj., **homozygous.**

humoral (hū'mər-əl) (L. *humor,* a fluid). Pertaining to an endocrine secretion.

humoral immune response Immune response involving production of antibodies, specifically the T_H2 arm of the immune response. Contrast **cell-mediated immune response.**

hyaline (hī'ə-lən) (Gr. *hyalos,* glass). Adj., glassy, translucent. Noun, a clear, glassy, structureless material occurring, for example, in cartilage, vitreous body, mucin, and glycogen.

hybridoma (hī-brid-ō'mah) (contraction of hybrid + myeloma). Fused product of a normal and a myeloma (cancer) cell, which has some of the characteristics of the normal cell.

hydatid cyst (hī-da'təd) (Gr. *hydatis,* watery vesicle). A type of cyst formed by juveniles of certain tapeworms (*Echinococcus*) in their vertebrate hosts.

hydranth (hī'dranth) (Gr. *hydor,* water, + *anthos,* flower). Nutritive zooid of hydroid colony.

hydrocaulus (hī'drə-kä'ləs) (Gr. *hydor,* water, + *kaulos,* stem of a plant). Stalks or "stems" of a hydroid colony, the parts between the hydrorhiza and the hydranths.

hydrocoel (hī'-drə-sēl) (Gr. *hydor,* water, + *koilos,* hollow). Second or middle coelomic compartment in echinoderms; left hydrocoel gives rise to water vascular system.

hydrocorals Members of phylum Cnidaria, class Hydrozoa, with massive calcareous skeletons.

hydrogen bond A relatively weak chemical bond resulting from unequal charge distribution within molecules, in which a hydrogen atom covalently bonded to another atom is attracted to the electronegative portion of another molecule.

hydroid The polyp form of a cnidarian as distinguished from the medusa form. Any cnidarian of the class Hydrozoa, order Hydroida.

hydrolysis (Gr. *hydor,* water, + *lysis,* a loosening). The decomposition of a chemical compound by the addition of water; the splitting of a molecule into its groupings so that the split products acquire hydrogen and hydroxyl groups.

hydrorhiza (hī'drə-rī'zə) (Gr. *hydor,* water, + *rhiza,* a root). Rootlike stolon that attaches a hydroid to its substrate.

hydrosphere (Gr. *hydōr*, water, + *sphaira*, ball, sphere). Aqueous envelope of the earth.

hydrostatic pressure The pressure exerted by a fluid (gas or liquid), defined as force per unit area. For example, the hydrostatic pressure of one atmosphere (1 atm) is 14.7 lb/in^2.

hydrostatic skeleton A mass of fluid or plastic parenchyma enclosed within a muscular wall to provide the support necessary for antagonistic muscle action; for example, parenchyma in acoelomates and perivisceral fluids in pseudocoelomates serve as hydrostatic skeletons.

hydrothermal vent A submarine hot spring; seawater seeping through the sea bottom is heated by magma and expelled back into the sea through the hydrothermal vent.

hydroxyl (hydrogen + oxygen, + yl). Containing an OH$^-$ group, a negatively charged ion formed by alkalies in water.

hyomandibular (hī-ō-mən-dib'yə-lər) (Gr. *hyoeides* [shaped like the Gr. letter upsilon Y, + *eidos*, form], + L. *mandere*, to chew). Bone derived from the hyoid gill arch, forming part of articulation of the lower jaw of fishes, and forming the stapes of the ear of amniotic vertebrates.

hyperosmotic (Gr. *hyper*, over, + *ōsmos*, impulse). Refers to a solution whose osmotic pressure is greater than that of another solution to which it is compared; contains a greater concentration of dissolved particles and gains water through a selectively permeable membrane from a solution containing fewer particles; contrasts with **hypoosmotic.**

hyperparasitism (hī'pər-par'ə-sid-iz-əm) (Gr. *hyper*, over, + *para*, beside, + *sitos*, food). Parasitism of a parasite by another parasite.

hypertrophy (hī-pər'trə-fē) (Gr. *hyper*, over, + *trophē*, nourishment). Abnormal increase in size of a part or organ.

hypodermis (hī'pə-dər'mis) (Gr. *hypo*, under, + L. *dermis*, skin). The cellular layer lying beneath and secreting the cuticle of annelids, arthropods, and certain other invertebrates.

hypoosmotic (Gr. *hypo*, under, + *ōsmos*, impulse). Refers to a solution whose osmotic pressure is less than that of another solution with which it is compared or taken as a standard, contains a lesser concentration of dissolved particles and loses water during osmosis; contrasts with **hyperosmotic.**

hypophysis (hī-pof'ə-sis) (Gr. *hypo*, under, + *physis*, growth). Pituitary body.

hypostome (hī'pə-stōm) (Gr. *hypo*, under, + *stoma*, mouth). Name applied to structure in various invertebrates (such as mites and ticks), located at posterior or ventral area of mouth.

hypothalamus (hī-pō-thal'ə-mis) (Gr. *hypo*, under, + *thalamos*, inner chamber). A ventral part of the forebrain beneath the thalamus; one of the centers of the autonomic nervous system.

hypothesis (Gr. *hypothesis*, foundation, supposition). A statement or proposition that can be tested by experiment.

hypothetico-deductive (Gr. *hypotithenai*, to suppose, + L. *deducere*, to lead). Scientific process of making a conjecture and then seeking empirical tests that potentially lead to its rejection.

I

imago (ə-mā'gō). The adult and sexually mature insect.

immediate hypersensitivity Inflammatory reaction based primarily on humoral immunity.

immunity Ability by tissues in an organism to recognize and defend against nonself invaders. **Innate immunity** is a mechanism of defense that does not depend on prior exposure to the invader; **acquired immunity** is specific to a nonself material, requires time for development, and occurs more quickly and vigorously on secondary response.

immunoglobulin (im'yə-nə-glä'byə-lən) (L. *immunis*, free, + *globus*, globe). Any of a group of plasma proteins, produced by plasma cells, that participates in the immune response by combining with the antigen that stimulated its production. Antibody.

imprinting (im'print-ing) (L. *imprimere*, to impress, imprint). Rapid and usually stable learning pattern appearing early in the life of a member of a social species and involving recognition of its own species; may involve attraction to the first moving object seen.

inbreeding The tendency among members of a population to mate preferentially with close relatives.

incomplete dominance See **intermediate inheritance.**

incus (in'kəs) (L. *incus*, anvil). The middle of a chain of three bones of the mammalian middle ear.

indeterminate cleavage A type of embryonic development in which the fate of the blastomeres is not determined very early as to tissues or organs, for example, in echinoderms and vertebrates; regulative cleavage.

indigenous (ən-dij'ə-nəs) (L. *indigena*, native). Pertains to organisms that are native to a particular region; not introduced.

induction (L. *inducere, inductum,* to lead). Reasoning from the particular to the general, that is, deriving a general statement (hypothesis) based on individual observations. In embryology, the alteration of cell fates as the result of interaction with neighboring cells.

inductor (in-duk'ter) (L. *inducere*, to introduce, lead in). In embryology, a tissue or organ that causes the differentiation of another tissue or organ.

inflammation (in'fləm-mā'shən) (L. *inflammare*, from *flamma*, flame). The complicated physiological process in mobilization of body defenses against foreign substances and infectious agents and repair of damage from such agents.

infraciliature (in-frə-sil'e-ə-tər) (L. *infra*, below, + *cilia*, eyelashes). The organelles just below the cilia in ciliate protozoa.

infundibulum (in'fun-dib'u-ləm) (L. funnel). Stalk of the neurohypophysis linking the pituitary to the diencephalon.

innate (i-nāt') (L. *innatus*, inborn). A characteristic based partly or wholly on genetic or epigenetic constitution.

instar (inz'tär) (L. form). Stage in the life of an insect or other arthropod between molts.

instinct (L. *instinctus*, impelled). Stereotyped, predictable, genetically programmed behavior. Learning may or may not be involved.

integument (ən-teg'ū-mənt) (L. *integumentum*, covering). An external covering or enveloping layer.

intercellular (in-tər-sel'yə-lər) (L. *inter*, among, + *cellula*, chamber). Occurring between body cells.

bat / āpe / ärmadillo / herring / fēmale / finch / līce / crocodile / crōw / duck / ūnicorn / ə indicates unaccented vowel sound "uh" as in mammal, fishes, cardinal, heron, vulture / stress as in bi-ol'o-gy, bi'o-log'i-cal

interferons Several cytokines encoded by different genes, important in mediation of natural immunity and inflammation.

interleukin-1 A cytokine produced by macrophages that stimulates T helper lymphocytes.

interleukin-2 A lymphokine produced by T helper lymphocytes that leads to proliferation of T helper cells and other T lymphocytes.

interleukins A series of cytokines produced primarily by various leukocytes, such as macrophages and T cells, whose target cells are various leukocytes and other cells. Given the name "interleukins" when it was believed that they were produced only by leukocytes and their target cells were limited to leukocytes.

intermediary meiosis Meiosis that occurs neither during gamete formation nor immediately after zygote formation, resulting in both haploid and diploid generations, such as in foraminiferan protozoa.

intermediate host A host in which some development of a symbiont occurs, but in which maturation and sexual reproduction do not take place.

intermediate inheritance Neither of alternate alleles of a gene are completely dominant, and heterozygote shows a condition intermediate between or different from homozygotes for each allele.

interstitial (in-tər-sti′shəl) (L. *inter,* among, + *sistere,* to stand). Situated in the interstices or spaces between structures such as cells, organs, or grains of sand.

intracellular (in-trə-sel′yə-lər) (L. *intra,* inside, + *cellula,* chamber). Occurring within a body cell or within body cells.

intrinsic growth rate Exponential growth rate of a population, that is, the difference between the density-independent components of the birth and death rates of a natural population with stable age distribution.

intron (in′trän) (L. *intra,* within). Portion of mRNA as transcribed from DNA that will not form part of mature mRNA, and therefore does not encode an amino-acid sequence in the protein product.

introvert (L. *intro,* inward, + *vertere,* to turn). The anterior narrow portion that can be withdrawn (introverted) into the trunk of a sipunculid worm.

invagination (in-vaj′ə-nā′shən) (L. *in,* in, + *vagina,* sheath). An infolding of a layer of tissue to form a saclike structure.

inversion (L. *invertere,* to turn upside down). A turning inward or inside out, as in embryogenesis of sponges; also, reversal in order of genes or reversal of a chromosome segment.

ion An atom or group of atoms with a net positive or negative electrical charge because of the loss or gain of electrons.

ionic bond A chemical bond formed by transfer of one or more electrons from one atom to another; characteristic of salts.

iridophore (ī-rid′ə-fōr) (Gr. *iris,* rainbow, or iris of eye). Iridescent or silvery chromatophores containing crystals or plates of guanine or other purine.

irritability (L. *irritare,* to provoke). A general property of all organisms involving the ability to respond to stimuli or changes in the environment.

isogametes (īs′o-gam′ēts) (Gr. *isos,* equal, + *gametēs,* spouse). Gametes of a species in which gametes of both sexes are alike in size and appearance.

isolecithal (ī′sə-les′ə-thəl) (Gr. *isos,* equal, + *lekithos,* yolk, + *al*). Pertaining to a zygote (or ovum) with yolk evenly distributed. Homolecithal.

isosmotic A liquid having the same osmotic pressure as another, reference liquid.

isotonic (Gr. *isos,* equal, + *tonikos,* tension). Pertaining to solutions having the same or equal osmotic pressure; isosmotic.

isotope (Gr. *isos,* equal, + *topos,* place). One of several different forms (species) of a chemical element, differing from each other in atomic mass but not in atomic number.

J

juvenile hormone Hormone produced by the corpora allata of insects; among its effects are maintenance of larval or nymphal characteristics during development.

juxtaglomerular apparatus (jək′stə-glä-mer′yə-lər) (L. *juxta,* close to, + *glomus,* ball). Complex of sensory cells located in the afferent arteriole adjacent to the glomerulus and a loop of the distal tubule, which produces the enzyme renin.

K

kentrogon (ken′trə-gən) (Gr. *kentron,* a point, spine, + *gonos,* progeny, generation). A larva of the cirripede order Rhizocephala (subphylum Crustacea) that functions to inject the parasite cells into the host hemocoel.

keratin (ker′ə-tən) (Gr. *kera,* horn, + *in,* suffix of proteins). A scleroprotein found in epidermal tissues and modified into hard structures such as horns, hair, and nails.

keystone species A species (typically a predator) whose removal leads to reduced species diversity within the community.

kinesis (kə-nē′səs) (Gr. *kinēsis,* movement). Movements by an organism in random directions in response to stimulus.

kinetochore (kī-nēt′ə-kōr) (Gr. *kinein,* to move, + *choris,* asunder, apart). A disc of proteins located on the centromere, specialized to interact with the spindle fibers during mitosis.

kinetodesma (kə-nē′tə-dez′mə). pl. **kinetodesmata** (Gr. *kinein,* to move, + *desma,* bond). Fibril arising from the kinetosome of a cilium in a ciliate protozoan, and passing along the kinetosomes of cilia in that same row.

kinetosome (kən-ēt′ə-sōm) (Gr. *kinētos,* moving, + *sōma,* body). The self-duplicating granule at the base of the flagellum or cilium; similar to centriole, also called basal body or blepharoplast.

kinety (kə-nē′tē) (Gr. *kinein,* to move). All the kinetosomes and kinetodesmata of a row of cilia.

kinin (kī′nin) (Gr. *kinein,* to move, + *in,* suffix of hormones). A type of local hormone that is released near its site of origin; also called parahormone or tissue hormone.

K-selection (from the K term in the logistic equation). Natural selection under conditions that favor survival when populations are controlled primarily by density-dependent factors.

Kupffer cells Phagocytic cells in the liver, part of the reticuloendothelial system.

kwashiorkor (kwash-ē-or′kər) (from Ghana). Malnutrition caused by diet high in carbohydrate and extremely low in protein.

L

labium (lā′bē-əm) (L. a lip). The lower lip of the insect formed by fusion of the second pair of maxillae.

labrum (lā′brəm) (L. a lip). The upper lip of insects and crustaceans situated above or in front of the mandibles; also refers to the outer lip of a gastropod shell.

labyrinth (L. *labyrinthus,* labyrinth). Vertebrate internal ear, composed of a series of fluid-filled sacs and tubules (membranous labyrinth) suspended within bone cavities (osseous labyrinth).

lachrymal (lak′rə-məl) (L. *lacrimia,* tear). Secreting or relating to tears.

lacteal (lak′te-əl) (L. *lacteus,* of milk). Noun, one of the lymph vessels in the villus of the intestine. Adj., relating to milk.

lacuna (lə-kū′nə), pl. **lacunae** (L. pit, cavity). A sinus; a space between cells; a cavity in cartilage or bone.

lagena (lə-jē′nə) (L. large flask). Portion of the primitive ear in which sound is translated into nerve impulses; evolutionary beginning of cochlea.

Lamarckism Hypothesis, as expounded by Jean Baptiste de Lamarck, of evolution by the acquisition during an organism's lifetime of characteristics that are transmitted to offspring.

lamella (lə-mel′ə) (L. dim. of *lamina,* plate). One of the two plates forming a gill in a bivalve mollusc. One of the thin layers of bone laid concentrically around an osteon (Haversian canal). Any thin, platelike structure.

lappets Lobes around the margin of scyphozoan medusae (phylum Cnidaria).

larva (lar′və), pl. **larvae** (L. a ghost). An immature stage that is quite different from the adult.

larynx (lar′inks) (Gr., the larynx, gullet). Modified upper portion of respiratory tract of air-breathing vertebrates, bounded by the glottis above and the trachea below; voice box; adj., **laryngeal** (lə-rin′j(ē)əl), relating to the larynx.

lateral (L. *latus,* the side, flank). Of or pertaining to the side of an animal; a *bilateral* animal has two sides.

laterite (lad′ə-rīt) (L. *later,* brick). Group of hard, red soils from tropical areas that show intense weathering and leaching of bases and silica, leaving aluminum hydroxides and iron oxides; adj. **lateritic.**

lecithotrophy (le′sə-thə-trō′fē) (Gr. *lekithos,* yolk of egg, + *trophos,* one who feeds). Nutrition of an embryo directly from the yolk of an ovum.

lek (lek) (Sw. play, game). An area where animals assemble for communal courtship display and mating.

lemniscus (lem-nis′kəs) (L. ribbon). One of a pair of internal projections of the epidermis from the neck region of Acanthocephala, which functions in fluid control in the protrusion and invagination of the proboscis.

lentic (len′tik) (L. *lentus,* slow). Of or relating to standing water such as swamp, pond, or lake.

lepidosaurs (lep′ə-dō-sors) (L. *lepidos,* scale, + *sauros,* lizard). A lineage of diapsid reptiles that appeared in the Permian and that includes the modern snakes, lizards, amphisbaenids, and tuataras, and the extinct ichthyosaurs.

lepospondyls (lep′ə-spänd′ls) (Gr. *lepos,* scale, + *spondylos,* vertebra). A group of Paleozoic amphibians distinguished by the possession of spool-shaped vertebral centra.

leptocephalus (lep′tə-sef′ə-ləs) pl. **leptocephali** (Gr. *leptos,* thin, + *kephalē,* head). Transparent, ribbonlike migratory larva of eels and related teleosts.

leukemism (lū′kə-mi-zəm) (Gr. *leukos,* white, + *ismos,* condition of). Presence of white pelage or plumage in animals with normally pigmented eyes and skin.

leukocyte (lū′kə-sīt) (Gr. *leukos,* white, + *kytos,* hollow vessel). Any of several kinds of white blood cells (for example, granulocytes, lymphocytes, monocytes), so called because they bear no hemoglobin, as do red blood cells.

library In molecular biology, a set of clones containing recombinant DNA. Obtained from and representing the genome of the organism.

ligament (lig′ə-mənt) (L. *ligamentum,* bandage). A tough, dense band of connective tissue connecting one bone to another.

ligand (lī′gənd) (L. *ligo,* to bind). A molecule that specifically binds to a receptor; for example, a hormone (ligand) binds specifically to its receptor on the cell surface.

limax form (lī′məx) (L. *limax,* slug). Form of pseudopodial movement in which entire organism moves without extending a discrete pseudopodium.

lipase (lī′pās) (Gr. *lipos,* fat, + *ase,* enzyme suffix). An enzyme that accelerates the hydrolysis or synthesis of fats.

lipid, lipoid (li′pid) (Gr. *lipos,* fat). Certain fatlike substances, often containing other groups such as phosphoric acid; lipids combine with proteins and carbohydrates to form principal structural components of cells.

lithosphere (lith′ə-sfir) (Gr. *lithos,* rock, + *sphaira,* ball). The rocky component of the earth's surface layers.

littoral (lit′ə-rəl) (L. *litoralis,* seashore). Adj., pertaining to the shore. Noun, that portion of the sea floor between the extent of high and low tides, intertidal; in lakes, the shallow part from the shore to the lakeward limit of aquatic plants.

lobopodium (lō′bə-pō′de-əm) (Gr. *lobos,* lobe, + *pous, podos,* foot). Blunt, lobelike pseudopodium.

locus (lō′kəs), pl. **loci** (lō′sī) (L. place). Position of a gene in a chromosome.

logistic equation A mathematical expression describing an idealized sigmoid curve of population growth.

lophocyte (lō′fə-sīt) (Gr. *lophos,* crest, + *kytos,* hollow vessel). Type of sponge amebocyte that secretes bundles of fibrils.

lophophore (lōf′ə-fōr) (Gr. *lophos,* crest, + *phoros,* bearing). Tentacle-bearing ridge or arm within which is an extension of the coelomic cavity in lophophorate animals (ectoprocts, brachiopods, and phoronids).

lorica (lo′rə-kə) (L. corselet). Protective external case found in some protozoa, rotifers, and others.

lotic (lō′tik) (L. *lotus,* action of washing or bathing). Of or pertaining to running water, such as a brook or river.

lumbar (lum′bär) (L. *lumbus,* loin). Relating to or near the loins or lower back.

lumen (lū′mən) (L. light). The cavity of a tube or organ.

lymph (limf) (L. *lympha,* water). The fluid in the lymphatic system.

lymphocyte (lim′fō-sīt) (L. *lympha,* water, goddess of water, + Gr. *kytos,* hollow vessel). Cell in blood and lymph that has central role in immune responses. See **T cell** and **B cell.**

lymphokine (limf′ə-kīn) (L. *lympha,* water, + Gr. *kinein,* to move). A molecule secreted by an activated or stimulated lymphocyte that causes physiological changes in certain other cells.

lysosome (lī′sə-sōm) (Gr. *lysis,* loosing, + *sōma,* body). Intracellular organelle consisting of a membrane enclosing several digestive enzymes that are released when the lysosome ruptures.

M

macroevolution (L. *makros,* long, large, + *evolvere,* to unfold). Evolutionary change on a grand scale, encompassing the origin of novel designs, evolutionary trends, adaptive radiation, and mass extinction.

macrogamete (mak′rə-gam′ēt) (Gr. *makros,* long, large, + *gamos,* marriage). The larger of the two gamete types in a heterogametic organism, considered the female gamete.

macromere (mak′rə-mer′) (Gr. *makros,* long, large, + *meros,* part). The largest size class of blastomeres in a cleaving embryo when the blastomeres differ in size from one another.

macromolecule A very large molecule, such as a protein, polysaccharide, or nucleic acid.

macronucleus (ma′krō-nū′klē-əs) (Gr. *makros,* long, large, + *nucleus,* kernel). The larger of the two kinds of nuclei in ciliate protozoa; controls all cell functions except reproduction.

macrophage (mak′rə-fāj) (Gr. *makros,* long, large, + *phagō,* to eat). A phagocytic cell type in vertebrates that performs crucial functions in the immune response and inflammation, such as presenting antigenic epitopes to T cells and producing several cytokines.

madreporite (ma′drə-pōr′īt) (Fr. *madrépore,* reef-building coral, + *ite,* suffix for some body parts). Sievelike structure that is the intake for the water-vascular system of echinoderms.

major histocompatibility complex (MHC) Complex of genes coding for proteins inserted in the cell membrane; the proteins are the basis of self-nonself recognition by the immune system.

malacostracan (mal′ə-käs′trə-kən) (Gr. *malako,* soft, + *ostracon,* shell). Any member of the crustacean subclass Malacostraca, which includes both aquatic and terrestrial forms of crabs, lobsters, shrimps, pillbugs, sand fleas, and others.

malleus (mal′ē-əs) (L. hammer). The ossicle attached to the tympanum in middle ears of mammals.

Malpighian tubules (mal-pig′ē-ən) (Marcello Malpighi, Italian anatomist, 1628–1694). Blind tubules opening into the hindgut of nearly all insects and some myriapods and arachnids, and functioning primarily as excretory organs.

mantle Soft extension of the body wall in certain invertebrates, for example, brachiopods and molluscs, which usually secretes a shell; thin body wall of tunicates.

manubrium (man-ū′bri-əm) (L. handle). The portion projecting from the oral side of a jellyfish medusa, bearing the mouth; oral cone; presternum or anterior part of sternum; handle-like part of malleus of ear.

marasmus (mə-raz′məs) (Gr. *marasmos,* to waste away). Malnutrition, especially of infants, caused by a diet deficient in both calories and protein.

marsupial (mär-sū′pē-əl) (Gr. *marsypion,* little pouch). One of the pouched mammals of the subclass Metatheria.

mast cells Inflammatory cells in a variety of locations. Upon activation by an antigen they release pharmacologically active compounds leading to redness and swelling.

mastax (mas′təx) (Gr. jaws). Pharyngeal mill of rotifers.

matrix (mā′triks) (L. *mater,* mother). The intercellular substance of a tissue, or that part of a tissue into which an organ or process is set.

maturation (L. *maturus,* ripe). The process of ripening; the final stages in the preparation of gametes for fertilization.

maxilla (mak-sil′ə) (L. dim. of *mala,* jaw). One of the upper jawbones in vertebrates; one of the head appendages in arthropods.

maxilliped (mak-sil′ə-ped) (L. *maxilla,* jaw, + *pes,* foot). One of the pairs of head appendages located just posterior to the maxilla in crustaceans, a thoracic appendage that has become incorporated into the feeding mouthparts.

medial (mē′dē-əl). Situated, or occurring, in the middle.

mediated transport Transport of a substance across a cell membrane mediated by a carrier molecule in the membrane.

medulla (mə-dul′ə) (L. marrow). The inner portion of an organ in contrast to the cortex or outer portion. Also, hindbrain.

medusa (mə-dū′sə) (Gr. mythology, female monster with snake-entwined hair). A jellyfish, or the free-swimming stage in the life cycle of cnidarians.

Mehlis′ gland (me′ləs). Glands of uncertain function surrounding the ootype of trematodes and cestodes.

meiofauna (mī′ō-faw-nə) (Gr. *meion,* smaller, + L. *faunus,* god of the woods). Small invertebrates found in the interstices between sand grains.

meiosis (mī-ō′səs) (Gr. from *mieoun,* to make small). The nuclear changes by means of which the chromosomes are reduced from the diploid to the haploid number; in animals, usually occurs in the last two divisions in the formation of the mature egg or sperm.

melanin (mel′ə-nin) (Gr. *melas,* black). Black or dark-brown pigment found in plant or animal structures.

melanophore (mel′ə-nə-fōr, mə-lan′ə-fōr) (Gr. *melania,* blackness, + *pherein,* to bear). Black or brown chromatophore containing melanin.

memory cells Population of long-lived B lymphocytes remaining after initial immune response that provides for the secondary response.

meninges (mə-nin′jez), sing. **meninx** (Gr. *mēninx,* membrane). Any of three membranes (arachnoid, dura mater, pia mater) that envelop the vertebrate brain and spinal cord. Also, solid connective tissue sheath enclosing the central nervous system of some vertebrates.

menopause (men′ō-pawz) (Gr. *men,* month, + *pauein,* to cease). In the human female, that time of life when ovulation ceases; cessation of the menstrual cycle.

menstruation (men'stroo-ā'shən) (L. *menstrua,* the menses, from *mensis,* month). The discharge of blood and uterine tissue from the vagina at the beginning of a menstrual cycle. It occurs during the first few days of the ovarian cycle.

meroblastic (mer-ə-blas'tik) (Gr. *meros,* part, + *blastos,* germ). Partial cleavage occurring in zygotes having a large amount of yolk at the vegetal pole; cleavage restricted to a small area on the surface of the egg.

merozoite (me'rə-zō'īt) (Gr. *meros,* part, + *zōon,* animal). A very small trophozoite at the stage just after cytokinesis has been completed in multiple fission of a protozoan.

mesenchyme (me'zən-kim) (Gr. *mesos,* middle, + *enchyma,* infusion). Embryonic connective tissue; irregular or amebocytic cells often embedded in gelatinous matrix.

mesentery (mes'ən-ter'ē) (L. *mesenterium,* mesentery). Peritoneal fold serving to hold the viscera in position.

mesocoel (mez'ō-sēl) (Gr. *mesos,* middle, + *koilos,* hollow). Middle body coelomic compartment in some deuterostomes, anterior in lophophorates, corresponds to hydrocoel in echinoderms.

mesoderm (me'zə-dərm) (Gr. *mesos,* middle, + *derma,* skin). The third germ layer, formed in the gastrula between the ectoderm and endoderm; gives rise to connective tissues, muscle, urogenital and vascular systems, and the peritoneum.

mesoglea (mez'ō-glē'ə) (Gr. *mesos,* middle, + *glia,* glue). The layer of jellylike or cement material between the epidermis and gastrodermis in cnidarians and ctenophores; also may refer to jellylike matrix between epithelial layers in sponges.

mesohyl (me'sə-hil) (Gr. *mesos,* middle, + *hylē,* a wood). Gelatinous matrix surrounding sponge cells; mesoglea, mesenchyme.

mesolecithal (me'zō-ləs'ə-thəl) (Gr. *mesos,* middle, + *lekithos,* yolk). Pertaining to a zygote (or ovum) having a moderate amount of yolk concentrated in the vegetal pole.

mesonephros (me-zō-nef'rōs) (Gr. *mesos,* middle, + *nephros,* kidney). The middle of three pairs of embryonic renal organs in vertebrates. Functional kidney of embryonic anamniotes; its collecting duct is a Wolffian duct. Adj., **mesonephric.**

mesosome (mez'ə-sōm) (Gr. *mesos,* middle, + *sōma,* body). The portion of the body in lophophorates and some deuterostomes that contains the mesocoel.

messenger RNA (mRNA) A form of ribonucleic acid that carries genetic information from the gene to the ribosome, where it determines the order of amino acids as a polypeptide is formed.

metabolism (Gr. *metabolē,* change). A group of processes that includes digestion, production of energy (respiration), and synthesis of molecules and structures by organisms; the sum of the constructive (anabolic) and destructive (catabolic) processes.

metacentric (me'tə-sen'trək) (Gr. *meta,* between, among, after, + *kentron,* center). Chromosome with centromere at or near the middle.

metacercaria (me'tə-sər-ka'rē-ə) (Gr. *meta,* between, among, after, + *kerkos,* tail, + L. *aria,* connected with). Fluke juvenile (cercaria) that has lost its tail and has become encysted.

metacoel (met'ə-sēl) (Gr. *meta,* between, among, after, + *koilos,* hollow). Posterior coelomic compartment in some deuterostomes and lophophorates; corresponds to somatocoel in echinoderms.

metamere (met'ə-mēr) (Gr. *meta,* after, + *meros,* part). A repeated body unit along the longitudinal axis of an animal; a somite, or segment.

metamerism (mə-ta'mə-ri'zəm) (Gr. *meta,* between, among, after, + *meros,* part). Condition of being made up of serially repeated parts (metameres); serial segmentation.

metamorphosis (Gr. *meta,* between, among, after, + *morphē,* form, + *osis,* state of). Sharp change in form during postembryonic development, for example, tadpole to frog or larval insect to adult.

metanephridium (me'tə-nə-fri'di-əm) (Gr. *meta,* between, among, after, + *nephros,* kidney). A type of tubular nephridium with the inner open end draining the coelom and the outer open end discharging to the exterior.

metanephros (me'tə-ne'frōs) (Gr. *meta,* between, among, after, + *nephros,* kidney). Embryonic renal organs of vertebrates arising behind the mesonephros; the functional kidney of reptiles, birds, and mammals. It is drained from a ureter.

metasome (met'ə-som) (Gr. *meta,* after, behind, + *sōma,* body). The portion of the body in lophophorates and some deuterostomes that contains the metacoel.

metazoa (met-ə-zō'ə) Gr. *meta,* after, + *zōon,* animal). Multicellular animals.

MHC See **major histocompatibility complex.**

microevolution (mī-krō-ev-ə-lü'shən). (L. *mikros,* small, + *evolvere,* to unfold). A change in the gene pool of a population across generations.

microfilament (mī'krō-fil'ə-mənt) (Gr. *mikros,* small, + L. *filum,* a thread). A thin, linear structure in cells; of actin in muscle cells and others.

microfilariae (mik'rə-fil-ar'ē-ē) (Gr. *mikros,* small, + L. *filum,* a thread). Partially developed juveniles borne alive by filarial worms (phylum Nematoda).

microgamete (mīk'rə-gam'et) (Gr. *mikros,* small, + *gamos,* marriage). The smaller of the two gamete types in a heterogametic organism, considered the male gamete.

microglial cells Phagocytic cells in the central nervous system, part of the reticuloendothelial system.

micromere (mīk'rə-mer') (Gr. *mikros,* small, + *meros,* part). The smallest size class of blastomeres in a cleaving embryo when the blastomeres differ in size from one another.

micron (μ) (mī'krän) (Gr. neuter of *mikros,* small). One one-thousandth of a millimeter; about 1/25,000 of an inch. Now largely replaced by micrometer (μm).

microneme (mī'krə-nēm) (Gr. *mikros,* small, + *nēma,* thread). One of the types of structures composing the apical complex in the phylum Apicomplexa, slender and elongate, leading to the anterior and thought to function in host cell penetration.

micronucleus A small nucleus found in ciliate protozoa; controls the reproductive functions of these organisms.

micropyle (mīk'rə-pīl) (Gr. *mikros,* small, + *pileos,* a cap). The small opening through which the cells emerge from a gemmule (phylum Porifera).

bat / āpe / ärmadillo / herring / fēmale / finch / līce / crocodile / crōw / duck / ūnicorn / ə indicates unaccented vowel sound "uh" as in mammal, fishes, cardinal, heron, vulture / stress as in bi-ol'o-gy, bi'o-log'i-cal

microthrix See **microvillus.**

microtubule (Gr. *mikros,* small, + L. *tubule,* pipe). A long, tubular cytoskeletal element with an outside diameter of 20 to 27 nm. Microtubules influence cell shape and play important roles during cell division.

microvillus (Gr. *mikros,* small, + L. *villus,* shaggy hair). Narrow, cylindrical cytoplasmic projection from epithelial cells; microvilli form the brush border of several types of epithelial cells. Also, microvilli with unusual structure cover the surface of cestode tegument (also called **microthrix** [pl. **microtriches**]).

mictic (mik′tik) (Gr. *miktos,* mixed or blended). Pertaining to haploid egg of rotifers or the females that lay such eggs.

mineralocorticoids (min(ə)rəl-ō-kord′ə-koids) (M. E. *minerale,* ore, + L. *cortex,* bark, + *oid,* suffix denoting likeness of form). Hormones of the adrenal cortex, especially aldosterone, that regulate salt balance.

miracidium (mīr′ə-sid′ē-əm) (Gr. *meirakidion,* youthful person). A minute ciliated larval stage in the life of flukes.

mitochondrion (mīd′ə-kän′drē-ən) (Gr. *mitos,* a thread, + *chondrion,* dim. of *chondros,* corn, grain). An organelle in the cell in which aerobic metabolism takes place.

mitosis (mī-tō′səs) (Gr. *mitos,* thread, + *osis,* state of). Nuclear division in which there is an equal qualitative and quantitative division of the chromosomal material between the two resulting nuclei; ordinary cell division.

molecule A configuration of atomic nuclei and electrons bound together by chemical bonds.

monocyte (mon′ə-sīt) (Gr. *monos,* single, + *kytos,* hollow vessel). A type of leukocyte that becomes a phagocytic cell (macrophage) after moving into tissues.

monoecious (mə-nē′shəs) (Gr. *monos,* single, + *oikos,* house). Having both male and female gonads in the same organism; hermaphroditic.

monogamy (mə-näg′ə-mē) adj. **monogamous** (Gr. *monos,* single, + *gamos,* marriage). The condition of having a single mate at any one time.

monohybrid (Gr. *monos,* single, + L. *hybrida,* mongrel). A hybrid offspring of parents different in one specified character.

monomer (mä′nə-mər) (Gr. *monos,* single, + *meros,* part). A molecule of simple structure, but capable of linking with others to form polymers.

monophyly (män′ə-fī-lē) (Gr. *monos,* single, + *phyle,* tribe). The condition that a taxon or other group of organisms contains the most recent common ancestor of the group and all of its descendants; contrasts with **polyphyly** and **paraphyly.**

monosaccharide (män′nə-sa′kə-rīd) (Gr. *monos,* one, + *sakcharon,* sugar, from Sanskrit *sarkarā,* gravel, sugar). A simple sugar that cannot be decomposed into smaller sugar molecules; the most common are pentoses (such as ribose) and hexoses (such as glucose).

monozoic (mo′nə-zō′ik) (Gr. *monos,* single, + *zōon,* animal). Tapeworms with a single proglottid, do not undergo strobilation to form chain of proglottids.

morphogenesis (mor′fə-je′nə-səs) (Gr. *morphē,* form, + *genesis,* origin). Development of the architectural features of organisms; formation and differentiation of tissues and organs.

morphology (Gr. *morphē,* form, + L. *logia,* study, from Gr. *logos,* work). The science of structure. Includes cytology, the study of cell structure; histology, the study of tissue structure; and anatomy, the study of gross structure.

morula (mär′u-lə) (L. *morum,* mulberry, + *ula,* dim.). Solid ball of cells in early stage of embryonic development.

mosaic cleavage Embryonic development characterized by independent differentiation of each part of the embryo; determinate cleavage.

mucin (mū′sən) (L. *mucus,* nasal mucus). Any of a group of glycoproteins secreted by certain cells, especially those of salivary glands.

mucus (mū′kəs) (L. *mucus,* nasal mucus). Viscid, slippery secretion rich in mucins produced by secretory cells such as those in mucous membranes. Adj., **mucous.**

Müller's larva Free-swimming ciliated larva that resembles a modified ctenophore, characteristic of certain marine polyclad turbellarians.

multiple fission A mode of asexual reproduction in some single-celled eukaryotes in which the nuclei divide more than once before cytokinesis occurs.

mutation (mū-tā′shən) (L. *mutare,* to change). A stable and abrupt change of a gene; the heritable modification of a characteristic.

mutualism (mū′chə-wə-li′zəm) (L. *mutuus,* lent, borrowed, reciprocal). A type of interaction in which two different species derive benefit from their association and in which the association is necessary to both; often symbiotic.

myelin (mī′ə-lən) (Gr. *myelos,* marrow). A fatty material forming the medullary sheath of nerve fibers.

myocyte (mī′ə-sīt) (Gr. *mys,* muscle, + *kytos,* hollow vessel). Contractile cell (pinacocyte) in sponges.

myofibril (Gr. *mys,* muscle, + L. dim. of *fibra,* fiber). A contractile filament within muscle or muscle fiber.

myogenic (mī′o-jen′ik) (Gr. *mys,* muscle, + N.L., *genic,* giving rise to). Originating in muscle, such as heartbeat arising in vertebrate cardiac muscle because of inherent rhythmical properties of muscle rather than because of neural stimuli.

myomere (mī′ə-mer) (Gr. *mys,* muscle, + *meros,* part). A muscle segment of successive segmental trunk musculature.

myosin (mī′ə-sin) (Gr. *mys,* muscle, + *in,* suffix, belonging to). A large, actin-binding protein of contractile tissue that forms the thick filaments of striated muscle. During contraction it combines with actin to form actomyosin.

myotome (mī′ə-tōm) (Gr. *mys,* muscle, + *tomos,* cutting). That part of a somite destined to form muscles; the muscle group innervated by a single spinal nerve.

N

nacre (nā′kər) (F. mother-of-pearl). Innermost lustrous layer of mollusc shell, secreted by mantle epithelium. Adj., **nacreous.**

NAD Abbreviation of nicotinamide adenine dinucleotide, an electron acceptor or donor in many metabolic reactions.

nares (na′rēz), sing. **naris** (L. nostrils). Openings into the nasal cavity, both internally and externally, in the head of a vertebrate.

natural killer cells Lymphocyte-like cells that can kill virus-infected cells and tumor cells in the absence of antibody.

natural selection A nonrandom reproduction of varying organisms in a population that results in the survival of those best adapted to their environment and elimination of those less well adapted; leads to evolutionary change if the variation is heritable.

nauplius (naw′plē-əs) (L. a kind of shellfish). A free-swimming microscopic larval stage of certain crustaceans, with three pairs of appendages (antennules, antennae, and mandibles) and median eye. Characteristic of ostracods, copepods, barnacles, and some others.

nekton (nek′tən) (Gr. neuter of *nēktos,* swimming). Term for actively swimming organisms, essentially independent of wave and current action. Compare with **plankton.**

nematocyst (ne-mad′ə-sist′) (Gr. *nēma,* thread, + *kystis,* bladder). Stinging organelle of cnidarians.

neo-Darwinism (nē′ō′ där′wə-niz′əm). A modified version of Darwin's evolutionary theory that eliminates elements of the Lamarckian inheritance of acquired characteristics and pangenesis that were present in Darwin's formulation; this theory originated with August Weismann in the late nineteenth century and, after incorporating Mendelian genetic principles, has become the currently favored version of Darwinian evolutionary theory.

neopterygian (nē-äp′tə-rij′ē-ən) (Gr. *neos,* new, + *pteryx,* fin). Any of a large group of bony fishes that includes most modern species.

neotenine See **juvenile hormone.**

neoteny (nē′ə-tē′nē, nē-ot′ə-nē) (Gr. *neos,* new, + *teinein,* to extend). An evolutionary process by which organismal development is retarded relative to sexual maturation; produces a descendant that reaches sexual maturity while retaining a morphology characteristic of the preadult or larval stage of an ancestor.

nephridiopore (nə-frid′ē-ə-pōr) (Gr. *nephros,* kidneys, + *porus,* pore). An external excretory opening in invertebrates.

nephridium (nə-frid′ē-əm) (Gr. *nephridios,* of the kidney). One of the segmentally arranged, paired excretory tubules of many invertebrates, notably the annelids. In a broad sense, any tubule specialized for excretion and/or osmoregulation; with an external opening and with or without an internal opening.

nephron (ne′frän) (Gr. *nephros,* kidney). Functional unit of kidney structure of vertebrates, consisting of a Bowman's capsule, an enclosed glomerulus, and the attached uriniferous tubule.

nephrostome (nef′rə-stōm) (Gr. *nephros,* kidney, + *stoma,* mouth). Ciliated, funnel-shaped opening of a nephridium.

neritic (nə-rid′ik) (Gr. *nērites,* a mussel). Portion of the sea overlying the continental shelf, specifically from the subtidal zone to a depth of 200 m.

nested hierarchy A pattern in which species are ordered into a series of increasingly more inclusive clades according to the taxonomic distribution of synapomorphies.

neural crest Populations of ectodermally derived embryonic cells that differentiate into many skeletal, neural, and sensory structures unique to vertebrates.

neurogenic (nū-rä-jen′ik) (Gr. *neuron,* nerve, + N.L. *genic,* give rise to). Originating in nervous tissue, as does the rhythmical beat of some arthropod hearts.

neuroglia (nū-räg′lē-ə) (Gr. *neuron,* nerve, + *glia,* glue). Tissue supporting and filling the spaces between the nerve cells of the central nervous system.

neurolemma (nū-rə-lem′ə) (Gr. *neuron,* nerve, + *lemma,* skin). Delicate nucleated outer sheath of a nerve cell; sheath of Schwann.

neuromast (Gr. *neuron,* sinew, nerve, + *mastos,* knoll). Cluster of sense cells on or near the surface of a fish or amphibian that is sensitive to vibratory stimuli and water.

neuron (Gr. nerve). A nerve cell.

neuropodium (nū′rə-pō′de-əm) (Gr. *neuron,* nerve, + *pous, podos,* foot). Lobe of parapodium nearer the ventral side in polychaete annelids.

neurosecretory cell (nu′rō-sə-krēd′ə-rē). Any cell (neuron) of the nervous system that produces a hormone.

neutron A subatomic particle lacking an electrical charge and having a mass 1839 times that of an electron and found in the nucleus of atoms.

niche The role of an organism in an ecological community; its unique way of life and its relationship to other biotic and abiotic factors.

nictitating membrane (nik′tə-tā-ting) (L. *nicto,* to wink). Third eyelid, a transparent membrane of birds and many reptiles and mammals, that can be pulled across the eye.

nitrogen fixation (Gr. *nitron,* soda, + *gen,* producing). Reduction of molecular nitrogen to ammonia by some bacteria and cyanobacteria, often followed by **nitrification,** the oxidation of ammonia to nitrites and nitrates by other bacteria.

nondisjunction Failure of a pair of homologous chromosomes to separate during meiosis, leading to one gamete with $n + 1$ chromosomes (see **trisomy**) and another gamete with $n - 1$ chromosomes.

notochord (nōd′ə-kord′) (Gr. *nōtos,* back, + *chorda,* cord). An elongated cellular cord, enclosed in a sheath, which forms the axial skeleton of chordate embryos, jawless fishes, and adult cephalochordates.

notopodium (nō′tə-pō′de-əm) (Gr. *nōtos,* back, + *pous, podos,* foot). Lobe of parapodium nearer the dorsal side in polychaete annelids.

nucleic acid (nu′klē′ik) (L. *nucleus,* kernel). One of a class of molecules composed of joined nucleotides; chief types are deoxyribonucleic acid (DNA), found in cell nuclei (chromosomes) and mitochondria, and ribonucleic acid (RNA), found both in cell nuclei (chromosomes and nucleoli) and in cytoplasmic ribosomes.

nucleoid (nu′klē-oid) (L. *nucleus,* kernel, + *oid,* like). The region in a prokaryotic cell where the chromosome is found.

nucleolus (nu-klē′ə-ləs) (dim. of L. *nucleus,* kernel). A deeply staining body within the nucleus of a cell and containing RNA; nucleoli are specialized portions of certain chromosomes that carry multiple copies of the information to synthesize ribosomal RNA.

nucleoplasm (nu′klē-ə-plazm′) (L. *nucleus,* kernel, + Gr. *plasma,* mold). Protoplasm of nucleus, as distinguished from cytoplasm.

nucleoprotein A molecule composed of nucleic acid and protein; occurs in the nucleus and cytoplasm of all cells.

bat / āpe / ärmadillo / herring / fēmale / finch / līce / crocodile / crōw / duck / ūnicorn / ə indicates unaccented vowel sound "uh" as in mammal, fishes, cardinal, heron, vulture / stress as in bi-ol′o-gy, bi′o-log′i-cal

nucleosome (nu′klē-ə-som) (L. *nucleus,* kernel, + *sōma,* body). A repeating subunit of chromatin in which one and three-quarter turns of the double-helical DNA are wound around eight molecules of histones.

nucleotide (nu′klē-ə-tīd). A molecule consisting of phosphate, 5-carbon sugar (ribose or deoxyribose), and a purine or a pyrimidine; the purines are adenine and guanine, and the pyrimidines are cytosine, thymine, and uracil.

nucleus (nū′klē-əs) (L. *nucleus,* a little nut, the kernel). The organelle in eukaryotes that contains the chromatin and which is bounded by a double membrane (nuclear envelope).

nuptial flight (nup′shəl). The mating flight of insects, especially that of the queen with male or males.

nurse cells Single cells or layers of cells surrounding or adjacent to other cells or structures for which the nurse cells provide nutrient or other molecules (for example, for insect oocytes or *Trichinella* spp. juveniles).

nymph (L. *nympha,* nymph, bride). An immature stage (following hatching) of a hemimetabolous insect that lacks a pupal stage.

O

ocellus (ō-sel′əs) (L. dim. of *oculus,* eye). A simple eye or eyespot in many types of invertebrates.

octomerous (ok-tom′ər-əs) (Gr. *oct,* eight, + *meros,* part). Eight parts, specifically, symmetry based on eight.

odontophore (ō-don′tə-fōr′) (Gr. *odous,* tooth, + *pherein,* to carry). Tooth-bearing organ in molluscs, including the radula, radular sac, muscles, and cartilages.

olfactory (äl-fakt′(ə)-rē) (L. *olor,* smell, + *factus,* to bring about). Pertaining to the sense of smell.

omasum (ō-mā′səm) (L. paunch). The third compartment of the stomach of a ruminant mammal.

ommatidium (ä′mə-tid′ē-əm) (Gr. *omma,* eye, + *idium,* small). One of the optical units of the compound eye of arthropods.

omnivore (äm′nə-vōr) (L. *omnis,* all, + *vorare,* to devour). An animal that uses a variety of animal and plant material in its diet.

oncogene (än′kə-jen) (Gr. *onkos,* protuberance, tumor, + *genos,* descent). Any of a number of genes that are associated with neoplastic growth (cancer). The gene in its benign state, either inactivated or carrying on its normal role, is a **proto-oncogene.**

oncomiracidium (än′kō-mīr′ə-sid′ē-əm) (Gr. *onkos,* barb, hook, + *meirakidion,* youthful person). A ciliated larva of a monogenetic trematode.

oncosphere (än′kəs-fər) (Gr. *onkinos,* a hook, + *sphaira,* ball). Rounded larva common to all cestodes, bears hooks.

ontogeny (än-tä′jə-nē) (Gr. *ontos,* being, + *geneia,* act of being born, from *genēs,* born). The course of development of an individual from egg to senescence.

oocyst (ō′ə-sist) (Gr. *ōion,* egg, + *kystis,* bladder). Cyst formed around zygote of malaria and related organisms.

oocyte (ō′ə-sīt) (Gr. *ōion,* egg, + *kytos,* hollow). Stage in formation of ovum, just preceding first meiotic division (primary oocyte) or just following first meiotic division (secondary oocyte).

ooecium (ō-ēs′ē-əm) (Gr. *ōion,* egg, + *oikos,* house, + L. *ium,* from). Brood pouch; compartment for developing embryos in ectoprocts.

oogenesis (ō-ə-jen′ə-səs) (Gr. *ōion,* egg, + *genesis,* descent). Formation, development, and maturation of a female gamete or ovum.

oogonium (ō′ə-gōn′ē-əm) (Gr. *ōion,* egg, + *gonos,* offspring). A cell that, by continued division, gives rise to oocytes; an ovum in a primary follicle immediately before the beginning of maturation.

ookinete (ō-ə-kī′nēt) (Gr. *ōion,* egg, + *kinein,* to move). The motile zygote of malarial parasites.

ootid (ō-ə-tid′) (Gr. *ōion,* egg, + *idion,* dim.). Stage of formation of ovum after second meiotic division following expulsion of second polar body.

ootype (ō′ə-tīp) (Gr. *ōion,* egg, + *typos,* mold). Part of oviduct in flatworms that receives ducts from vitelline glands and Mehlis' gland.

operculum (ō-per′kū-ləm) (L. cover). The gill cover in bony fishes; horny plate in some snails.

operon (äp′ə-rän). A genetic unit consisting of a cluster of genes under the control of other genes, found in prokaryotes.

ophthalmic (äf-thal′mik) (Gr. *ophthalamos,* an eye). Pertaining to the eye.

opisthaptor (ä′pəs-thap′tər) (Gr. *opisthen,* behind, + *haptein,* to fasten). Posterior attachment organ of a monogenetic trematode.

opisthonephros (ō-pisth′ō-nef-rōs) (Gr. *opisth,* back + *nephros,* kidney). A kidney that develops from the middle and posterior portions of the nephrogenic region of vertebrates and is drained by the Wolffian duct or accessory ducts. Functional kidney of most adult anamniotes.

opisthosoma (ō-pis′thə-sō′mə) (Gr. *opisthe,* behind, + *sōma,* body). Posterior body region in arachnids and pogonophorans.

opsonization (op′sən-i-zā′shən) (Gr. *opsonein,* to buy victuals, to cater). The facilitation of phagocytosis of foreign particles by phagocytes in the blood or tissues, mediated by antibody bound to the particles.

organelle (Gr. *organon,* tool, organ, + L. *ella,* dim.). Specialized part of a cell; literally, a small organ that performs functions analogous to organs of multicellular animals.

organizer (or′gan-ī-zer) (Gr. *organos,* fashioning). Area of an embryo that directs subsequent development of other parts.

orthogenesis (ōr′thō-jen′ə-səs). A unidirectional trend in the evolutionary history of a lineage as revealed by the fossil record; also, a now discredited, anti-Darwinian evolutionary theory, popular around 1900, postulating that genetic momentum forced lineages to evolve in a predestined linear direction that was independent of external factors and often led to decline and extinction.

osculum (os′kū-ləm) (L. *osculum,* a little mouth). Excurrent opening in a sponge.

osmole Molecular weight of a solute, in grams, divided by the number of ions or particles into which it dissociates in solution. Adj., **osmolar.**

osmoregulation Maintenance of proper internal salt and water concentrations in a cell or in the body of a living organism, active regulation of internal osmotic pressure.

---

osmosis (oz-mō′sis) (Gr. *ōsmos,* act of pushing, impulse). The flow of solvent (usually water) through a semipermeable membrane.

osmotic potential Osmotic pressure.

osmotroph (oz′mə-trōf) (Gr. *ōsmos,* a thrusting, impulse, + *trophē,* to eat). A heterotrophic organism that absorbs dissolved nutrients.

osphradium (äs-frā′dē-əm) (Gr. *osphradion,* small bouquet, dim. of *osphra,* smell). A sense organ in aquatic snails and bivalves that tests incoming water.

ossicles (L. *ossiculum,* small bone). Small separate pieces of echinoderm endoskeleton. Also, tiny bones of the middle ear of vertebrates.

osteoblast (os′tē-ō-blast) (Gr. *osteon,* bone, + *blastos,* bud). A bone-forming cell.

osteoclast (os′tē-ō-clast) (Gr. *osteon,* bone, + *klan,* to break). A large, multinucleate cell that functions in bone dissolution.

osteocyte (os′tē-ə-sīt) (Gr. *osteon,* bone, + *kytos,* hollow). A bone cell that is characteristic of adult bone, has developed from an osteoblast, and is isolated in a lacuna of the bone substance.

osteoderm (äs′tē-ə-dərm) (Gr. *osteon,* bone, + *derma,* skin). A bony, dermal plate located under and supporting an epidermal scale.

osteon (os′tē-on) (Gr. bone). Unit of bone structure; Haversian system.

osteostracans (os-tē-os′trə-kəns) (Gr. *osteon,* bone, + *ostrakon,* shell). A group of jawless, extinct fishes with dermal armor and pectoral fins known from the Silurian and Devonian periods.

ostium (L. door). Opening.

otolith (ōd′əl-ith′) (Gr. *ous, otos,* ear, + *lithos,* stone). Calcareous concretions in the membranous labyrinth of the inner ear of lower vertebrates, or in the auditory organ of certain invertebrates.

outgroup In phylogenetic systematic studies, a species or group of species closely related to but not included within a taxon whose phylogeny is being studied, and used to polarize variation of characters and to root the phylogenetic tree.

oviger (ō′vi-jər) (L. *ovum,* egg, + *gerere,* to bear). Leg that carries eggs in pycnogonids.

oviparity (ō′və-pa′rəd-ē) (L. *ovum,* egg, + *parere,* to bring forth). Reproduction in which eggs are released by the female; development of offspring occurs outside the maternal body. Adj., **oviparous** (ō-vip′ə-rəs).

ovipositor (ō′və-päz′əd-ər) (L. *ovum,* egg, + *positor,* builder, placer, + *or,* suffix denoting agent or doer). In many female insects a structure at the posterior end of the abdomen for laying eggs.

ovoviviparity (ō′vo-vī-və-par′ə-dē) (L. *ovum,* egg, + *vivere,* to live, + *parere,* to bring forth). Reproduction in which eggs develop within the maternal body without additional nourishment from the parent and hatch within the parent, or immediately after laying. Adj., **ovoviviparous** (ō′vo-vī-vip′ə-rəs).

ovum (L. *ovum,* egg). Mature female germ cell (egg).

oxidation (äk′sə-dā-shən) (Fr. *oxider,* to oxidize, from Gr. *oxys,* sharp, + *ation*). The loss of an electron by an atom or molecule; sometimes addition of oxygen chemically to a substance. Opposite of reduction, in which an electron is accepted by an atom or molecule.

oxidative phosphorylation (äk′sə-dād′iv fäs′fər-i-lā′shən). The conversion of inorganic phosphate to energy-rich phosphate of ATP, involving electron transport through a respiratory chain to molecular oxygen.

P

p53 protein A tumor suppressor protein with critical functions in normal cells. A mutation in the gene that encodes it, *p53,* can result in loss of control over cell division and thus cancer.

paedogenesis (pē-dō-jen′ə-sis) (Gr. *pais,* child, + *genēs,* born). Reproduction by immature or larval animals caused by acceleration of maturation. Progenesis.

paedomorphosis (pē-dō-mor′fə-səs) (Gr. *pais,* child, + *morphē,* form). Retention of ancestral juvenile features in later stages of the ontogeny of descendants.

pair bond An affiliation between an adult male and an adult female for reproduction. Characteristic of monogamous species.

pangenesis (pan-jen′ə-sis) (Gr. *pan,* all, + *genesis,* descent). Darwin's hypothesis that hereditary characteristics are carried by individual body cells that produce particles that collect in the germ cells.

papilla (pə-pil′ə) (L. nipple). A small nipplelike projection. A vascular process that nourishes the root of a hair, feather, or developing tooth.

papula (pa′pū-lə) (L. pimple). Respiratory processes on skin of sea stars; also, pustules on skin.

parabiosis (pa′rə-bī-ō′sis) (Gr. *para,* beside, + *biosis,* mode of life). The fusion of two individuals, resulting in mutual physiological intimacy.

paramylon granules (par′ə-mī-lən) (Gr. *para,* beside, + *mylos,* mill, grinder). Organelles containing the starchlike substance paramylon; in some algae and flagellates.

paraphyly (par′ə-fī-lē) (Gr. *para,* beside, + *phyle,* tribe). The condition that a taxon or other group of organisms contains the most recent common ancestor of all members of the group but excludes some descendants of that ancestor; contrasts with **monophyly** and **polyphyly.**

parapodium (pa′rə-pō′dē-əm) (Gr. *para,* beside, + *pous, podos,* foot). One of the paired lateral processes on each side of most segments in polychaete annelids; variously modified for locomotion, respiration, or feeding.

parasitism (par′ə-sīd′izəm) (Gr. *parasitos,* from *para,* beside, + *sitos,* food). The condition of an organism living in or on another organism (host) at whose expense the parasite is maintained; destructive symbiosis.

parasympathetic (par′ə-sim-pə-thed′ik) (Gr. *para,* beside, + *sympathes,* sympathetic, from *syn,* with, + *pathos,* feeling). One of the subdivisions of the autonomic nervous system, whose fibers originate in the brain and enter the periphery through the anterior and posterior parts of the spinal cord.

parenchyma (pə-ren′kə-mə) (Gr. anything poured in beside). In lower animals, a spongy mass of vacuolated mesenchyme cells filling spaces between viscera, muscles, or epithelia; in some, cell bodies of muscle cells. Also, the specialized tissue of an organ as distinguished from the supporting connective tissue.

parenchymula (pa′rən-kī′mū-lə) (Gr. *para,* beside, + *enchyma,* infusion). Flagellated, solid-bodied larva of some sponges.

parietal (pä-rī-ə-təl) (L. *paries,* wall). Something next to, or forming part of, a wall of a structure.

parthenogenesis (pär′thə-nō-gen′ə-sis) (Gr. *parthenos,* virgin, + L. from Gr. *genesis,* origin). Unisexual reproduction involving the production of young by females not fertilized by males; common in rotifers, cladocerans, aphids, bees, ants, and wasps. A parthenogenetic egg may be diploid or haploid.

pathogenic (path′ə-jen′ik) (Gr. *pathos,* disease, + N.L. *genic,* giving rise to). Producing or capable of producing disease.

PCR See **polymerase chain reaction.**

peck order A hierarchy of social privilege in a flock of birds.

pecten (L. comb). Any of several types of comblike structures on various organisms, for example, a pigmented, vascular, and comblike process that projects into the vitreous humor from the retina at a point of entrance of the optic nerve in the eyes of all birds and many reptiles.

pectines (pek′tīnz) (L. comb, pl. of **pecten**). Sensory appendage on abdomens of scorpions.

pectoral (pek′tə-rəl) (L. *pectoralis,* from *pectus,* the breast). Of or pertaining to the breast or chest; to the pectoral girdle; or to a pair of horny shields of the plastron of certain turtles.

pedal laceration Asexual reproduction found in sea anemones, a form of fission.

pedalium (pə-dal′ē-əm) (L. *pedalis,* of or belonging to the foot). Flattened blade at the base of the tentacles in cubozoan medusae (Cnidaria).

pedicel (ped′ə-sel) (L. *pediculus,* little foot). A small or short stalk or stem. In insects, the second segment of an antenna or the waist of an ant.

pedicellaria (ped′ə-sə-lar′ē-ə) (L. *pediculus,* little foot, + *aria,* like or connected with). One of many minute pincerlike organs on the surface of certain echinoderms.

pedipalps (ped′ə-palps′) (L. *pes, pedis,* foot, + *palpus,* stroking, caress). Second pair of appendages of arachnids.

pedogenesis See **paedogenesis.**

peduncle (pē′dun-kəl) (L. *pedunculus,* dim. of *pes,* foot). A stalk. Also, a band of white matter joining different parts of the brain.

pelage (pel′ij) (Fr. fur). Hairy covering of mammals.

pelagic (pə-laj′ik) (Gr. *pelagos,* the open sea). Pertaining to the open ocean.

pellicle (pel′ə-kəl) (L. *pellicula,* dim. of *pellis,* skin). Thin, translucent, secreted envelope covering many protozoa.

pelvic (pel′vik) (L. *pelvis,* a basin). Situated at or near the pelvis, as applied to girdle, cavity, fins, and limbs.

pelycosaur (pel′ə-kō-sor) (Gr. *pelyx,* basin, + *sauros,* lizard). Any of a group of Permian synapsids characterized by homodont dentition and sprawling limbs.

pentadactyl (pen-tə-dak′təl) (Gr. *pente,* five, + *daktylos,* finger). With five digits, or five fingerlike parts, to the hand or foot.

pentamerous symmetry (pen-tam′ər-əs) (Gr. *pente,* five, + *meros,* part). A radial symmetry based on five or multiples thereof.

peptidase (pep′tə-dās) (Gr. *peptein,* to digest, + *ase,* enzyme suffix). An enzyme that breaks down simple peptides, releasing amino acids.

peptide bond A bond that binds amino acids together into a polypeptide chain, formed by removing an OH from the carboxyl group of one amino acid and an H from the amino group of another to form an amide group—CO—NH—.

perennibranchiate (pə-ren′ə-brank′ē-āt) (L. *perennis,* throughout the year, + Gr. *branchia,* gills). Having permanent gills, relating especially to certain paedomorphic salamanders.

pericardium (pə-ri-kär′dē-əm) (Gr. *peri,* around, + *kardia,* heart). Area around heart; membrane around heart.

periostracum (pe-rē-äs′trə-kəm) (Gr. *peri,* around, + *ostrakon,* shell). Outer horny layer of a mollusc shell.

peripheral (pə-ri′fər-əl) (Gr. *peripherein,* to move around). Structure or location distant from center, near outer boundaries.

periproct (per′ə-präkt) (Gr. *peri,* around, + *prōktos,* anus). Region of aboral plates around the anus of echinoids.

perisarc (per′ə-särk) (Gr. *peri,* around, + *sarx,* flesh). Sheath covering the stalk and branches of a hydroid.

perissodactyl (pə-ris′ə-dak′təl) (Gr. *perissos,* odd, + *daktylos,* finger, toe). Pertaining to an order of ungulate mammals with an odd number of digits.

peristalsis (per′ə-stal′səs) (Gr. *peristaltikos,* compressing around). The series of alternate relaxations and contractions that serve to force food through the alimentary canal.

peristomium (per′ə-stō′mē-əm) (Gr. *peri,* around, + *stoma,* mouth). Foremost true segment of an annelid; it bears the mouth.

peritoneum (per′ə-tə-nē′əm) (Gr. *peritonaios,* stretched around). The membrane that lines the coelom and covers the coelomic viscera.

petaloids (pe′tə-loids) (Gr. *petalon,* leaf, + *eidos,* form). Describes flowerlike arrangement of respiratory podia in irregular sea urchins.

pH (*potential of hydrogen*). A symbol referring to the relative concentration of hydrogen ions in a solution; pH values are from 0 to 14, and the lower the value, the more acid or hydrogen ions in the solution. Equal to the negative logarithm of the hydrogen ion concentration.

phagocyte (fag′ə-sīt) (Gr. *phagein,* to eat, + *kytos,* hollow vessel). Any cell that engulfs and devours microorganisms or other particles.

phagocytosis (fag′ə-sī-tō-səs) (Gr. *phagein,* to eat, + *kytos,* hollow vessel). The engulfment of a particle by a phagocyte or a protozoan.

phagosome (fa′gə-sōm) (Gr. *phagein,* to eat, + *sōma,* body). Membrane-bound vesicle in cytoplasm containing food material engulfed by phagocytosis.

phagotroph (fag′ə-trōf) (Gr. *phagein,* to eat, + *trophē,* food). A heterotrophic organism that ingests solid particles for food.

pharynx (far′inks), pl. **pharynges** (Gr. *pharynx,* gullet). The part of the digestive tract between the mouth cavity and the esophagus that, in vertebrates, is common to both digestive and respiratory tracts. In cephalochordates the gill slits open from it.

---

bat / āpe / ärmadillo / herring / fēmale / finch / līce / crocodile / crōw / duck / ūnicorn / ə indicates unaccented vowel sound "uh" as in mammal, fishes, cardinal, heron, vulture / stress as in bi-ol′o-gy, bi′o-log′i-cal

phasmid (faz′mid) (Gr. *phasma,* apparition, phantom, + *id*). One of a pair of glands or sensory structures found in the posterior end of certain nematodes.

phenetic (fə-ne′tik) (Gr. *phaneros,* visible, evident). Refers to the use of a criterion of overall similarity to classify organisms into taxa; contrasts with classifications based explicitly on a reconstruction of phylogeny.

phenotype (fē′nə-tīp) (Gr. *phainein,* to show). The visible or expressed characteristics of an organism, controlled by the genotype, but not all genes in the genotype are expressed.

phenotypic gradualism The hypothesis that new traits, even those that are strikingly different from ancestral ones, evolve by a long series of small, incremental steps.

pheromone (fer′ə-mōn) (Gr. *pherein,* to carry, + *hormōn,* exciting, stirring up). Chemical substance released by one organism that influences the behavior or physiological processes of another organism.

phosphagen (fäs′fə-jən) (phosphate + gen). A term for creatine phosphate and arginine phosphate, which store and may be sources of high-energy phosphate bonds.

phosphatide (fäs′fə-tīd′) (phosphate + ide). A lipid with phosphorus, such as lecithin. A complex phosphoric ester lipid, such as lecithin, found in all cells. Phospholipid.

phosphorylation (fäs′fə-rə-lā′shən). The addition of a phosphate group, that is, —PO₃, to a compound.

photoautotroph (fōt-ō-aw′-tō-trōf) (Gr. *photōs,* light, + *autos,* self, + *trophos,* feeder). An organism requiring light as a source of energy for making organic nutrients from inorganic raw materials.

photosynthesis (fōt-ō-sin′thə-sis) (Gr. *phōs,* light, + *synthesis,* action or putting together). The synthesis of carbohydrates from carbon dioxide and water in chlorophyll-containing cells exposed to light.

phototaxis (fōt′ō-tak′sis) (Gr. *phōs,* light, + *taxis,* arranging, order). A taxis in which light is the orienting stimulus. An involuntary tendency for an organism to turn toward (positive) or away from (negative) light.

phototrophs (fōt′-ō-trōfs) (Gr. *phōs, phōtos,* light, + *trophē,* nourishment). Organisms capable of using CO₂ in the presence of light as a source of metabolic energy.

phyletic gradualism A model of evolution in which morphological evolutionary change is continuous and incremental and occurs mainly within unbranched species or lineages over long periods of geological time; contrasts with **punctuated equilibrium.**

phyllopodium (fī′lə-pō′dē-əm) (Gr. *phyllon,* leaf, + *pous, podos,* foot). Leaflike swimming appendage of branchiopod crustaceans.

phylogenetic species concept An irreducible (basal) cluster of organisms, diagnosably distinct from other such clusters, and within which there is a parental pattern of ancestry and descent.

phylogenetic systematics See **cladistics.**

phylogeny (fī-loj′ə-nē) (Gr. *phylon,* tribe, race, + *geneia,* origin). The origin and diversification of any taxon, or the evolutionary history of its origin and diversification, usually presented in the form of a dendrogram.

phylum (fī′ləm), pl. **phyla** (N.L. from Gr. *phylon,* race, tribe). A chief category, between kingdom and class, of taxonomic classifications into which are grouped organisms of common descent that share a fundamental pattern of organization.

physiology (L. *physiologia,* natural science). A branch of biology dealing with the organic processes and phenomena of an organism or any of its parts or of a particular bodily process.

phytophagous (fī-täf′ə-gəs) (Gr. *phyton,* plant, + *phagein,* to eat). Organisms that feed on plants.

pinacocyte (pin′ə-kō-sīt′) (Gr. *pinax,* tablet, + *kytos,* hollow vessel). Flattened cells composing dermal epithelium in sponges.

pinacoderm (pə-nak′ə-dərm) (Gr. *pinax,* plank, tablet, + *derma,* skin). The layer of pinacocytes in sponges.

pinna (pin′ə) (L. feather, sharp point). The external ear. Also a feather, wing, or fin or similar part.

pinocytosis (pin′o-sī-tō′sis, pīn′o-sī-to′sis) (Gr. *pinein,* to drink, + *kytos,* hollow vessel, + *osis,* condition). Taking up of fluid by endocytosis; cell drinking.

placenta (plə-sen′tə) (L. flat cake). The vascular structure, embryonic and maternal, through which the embryo and fetus are nourished while in the uterus.

placentotrophy (plə-sent′ə-trō′fē) (L. *placenta,* flat cake, + *trophos,* one who feeds). Nutrition of an embryo from a placenta.

placode (pla′kōd) (Gr. *plakos,* flat round plate). Localized, plate-like thickening of vertebrate head ectoderm from which a specialized structure develops; such structures include eye lens, special sense organs, and certain neurons.

placoderms (plak′ə-dərm) (Gr. *plax,* plate, + *derma,* skin). A group of heavily armored jawed fishes of the Silurian and Devonian periods.

placoid scale (pla′koid) (Gr. *plax, plakos,* tablet, plate). Type of scale found in cartilaginous fishes, with basal plate of dentin embedded in the skin and a backward-pointing spine tipped with enamel.

plankton (plank′tən) (Gr. neuter of *planktos,* wandering). The passively floating animal and plant life of a body of water; compares with **nekton.**

plantigrade (plan′tə-grād′) (L. *planta,* sole, + *gradus,* step, degree). Pertaining to animals that walk on the whole surface of the foot (for example, humans and bears); compares with **digitigrade.**

planula (plan′yə-lə) (N.L. dim. from L. *planus,* flat). Free-swimming, ciliated larval type of cnidarians; usually flattened and ovoid, with an outer layer of ectodermal cells and an inner mass of endodermal cells.

planuloid ancestor (plan′yə-loid) (L. *planus,* flat, + Gr. *eidos,* form). Hypothetical form representing ancestor of Cnidaria and Platyhelminthes.

plasma cell (plaz′mə) (Gr. *plasma,* a form, mold). A descendant cell of a B cell, functions to secrete antibodies.

plasma membrane (plaz′mə) (Gr. *plasma,* a form, mold). A living, external, limiting, protoplasmic structure that functions to regulate exchange of nutrients across the cell surface.

plasmalemma (plaz′mə-lem-ə) (Gr. *plasma,* a form, mold, + *lemma,* rind, sheath). The cell membrane or plasma membrane.

plasmid (plaz′məd) (Gr. *plasma,* a form, mold). A small circle of DNA that may be carried by a bacterium in addition to its genomic DNA.

plasmodium (plaz-mō′dē-əm) (Gr. *plasma,* a form, mold, + *eidos,* form). Multinucleate ameboid mass, syncytial.

plastid (plas′təd) (Gr. *plast,* formed, molded, + L. *id,* feminine stem for particle of specified kind). A membranous organelle in plant cells functioning in photosynthesis and/or nutrient storage, for example, chloroplast.

plastron (plast′trən) (Fr. *plastron,* breast plate). Ventral body shield of turtles; structure in corresponding position in certain arthropods; thin film of gas retained by epicuticle hairs of aquatic insects.

platelet (plāt′lət) (Gr. dim. of *plattus,* flat). A tiny, incomplete cell in the blood that releases substances initiating blood clotting.

pleiotropic (plī-ə-trō′pic) (Gr. *pleiōn,* more, + *tropos,* to turn). Pertaining to a gene producing more than one effect; affecting multiple phenotypic characteristics.

pleopod (plē′ə-päd) (Gr. *plein,* to sail, + *pous, podos,* foot). One of the swimming appendages on the abdomen of a crustacean.

plesiomorphic (plē′sē-ə-mōr′fik). An ancestral condition of a variable character.

pleura (plu′rə) (Gr. side, rib). The membrane that lines each half of the thorax and covers the lungs.

plexus (plek′səs) (L. network, braid). A network, especially of nerves or blood vessels.

pluteus (plū′dē-əs), pl. **plutei** (L. *pluteus,* movable shed, reading desk). Echinoid or ophiuroid larva with elongated processes like the supports of a desk; originally called "painter's easel larva."

pneumostome (nū′mə-stōm) (Gr. *pneuma,* breathing, + *stoma,* mouth). The opening of the mantle cavity (lung) of pulmonate gastropods to the outside.

podium (pō′de-əm) (Gr. *pous, podos,* foot). A footlike structure, for example, the tube foot of echinoderms.

poikilothermic (poi-ki′lə-thər′mik) (Gr. *poikilos,* variable, + thermal). Pertaining to animals whose body temperature is variable and fluctuates with that of the environment; cold blooded; compares with **ectothermic.**

polarity (Gr. *polos,* axis). In systematics, the ordering of alternative states of a taxonomic character from evolutionarily ancestral to derived conditions. In developmental biology, the tendency for the axis of an ovum to orient corresponding to the axis of the mother. Also, condition of having opposite poles; differential distribution of gradation along an axis.

polarization (L. *polaris,* polar, + Gr. *iz,* make). The arrangement of positive electrical charges on one side of a surface membrane and negative electrical charges on the other side (in nerves and muscles).

Polian vesicles (pō′le-ən) (from G. S. Poli, Italian naturalist). Vesicles opening into ring canal in most asteroids and holothuroids.

polyandry (pol′ē-an′drē) (Gr. *polys,* many, + *anēr,* man). Condition of having more than one male mate at one time.

polygamy (pə-lig′ə-mē) (Gr. *polys,* many, + *gamos,* marriage). Condition of having more than one mate at a time.

polygenic inheritance Inheritance of traits influenced by multiple alleles; traits show continuous variation between extremes; offspring are usually intermediate between the two parents; also known as **blending** and **quantitative inheritance.**

polygyny (pə-lij′ə-nē) (Gr. *polys,* many, + *gynē,* woman). Condition of having more than one female mate at one time.

polymer (pä′lə-mər) (Gr. *polys,* many, + *meros,* part). A chemical compound composed of repeated structural units called monomers.

polymerase chain reaction (PCR) A technique for preparing large quantities of DNA from tiny samples, making it easy to clone a specific gene as long as part of the sequence of the gene is known.

polymerization (pə-lim′ər-ə-zā′shən). The process of forming a polymer or polymeric compound.

polymorphism (pä′lē-mor′fi-zəm) (Gr. *polys,* many, + *morphē,* form). The presence in a species of more than one structural type of individual.

polynucleotide (poly + nucleotide): A nucleotide of many mononucleotides combined.

polyp (pä′lip) (Gr. *polypous,* many-footed). Individual of the phylum Cnidaria, generally adapted for attachment to the substratum at the aboral end, often form colonies.

polypeptide (pä-lē-pep′tīd) (Gr. *polys,* many, + *peptein,* to digest). A molecule consisting of many joined amino acids, not as complex as a protein.

polyphyly (päl′ē-fī′lē) (Gr. *polys,* many, + *phylon,* tribe). The condition that a taxon or other group of organisms does not contain the most recent common ancestor of all members of the group, implying that it has multiple evolutionary origins; such groups are not valid as formal taxa and are recognized as such only through error. Contrasts with **monophyly** and **paraphyly.**

polyphyodont (pä′lē-fī′ə-dänt) (Gr. *polyphyes,* manifold, + *odous,* tooth). Having several sets of teeth in succession.

polypide (pä′li-pīd) (L. *polypus,* polyp). An individual or zooid in a colony, specifically in ectoprocts, which has a lophophore, digestive tract, muscles, and nerve centers.

polyploid (pä′lə-ploid′) (Gr. *polys,* many, + *ploidy,* number of chromosomes). An organism possessing more than two full homologous sets of chromosomes.

polysaccharide (pä′lē-sak′ə-rid, -rīd). (Gr. *polys,* many, + *sakcharon,* sugar, from Sanskrit *sarkarā,* gravel, sugar). A carbohydrate composed of many monosaccharide units, for example, glycogen, starch, and cellulose.

polysome (polyribosome) (Gr. *polys,* many, + *sōma,* body). Two or more ribosomes connected by a molecule of messenger RNA.

polytene chromosomes (pä′li-tēn) (Gr. *polys,* many, + *tainia,* band). Chromosomes in the somatic cells of some insects in which the chromatin replicates repeatedly without undergoing mitosis.

polyzoic (pä′lē-zō′ik) (Gr. *polys,* many, + *zōon,* animal). A tapeworm forming a strobila of several to many proglottids; also, a colony of many zooids.

pongid (pän′jəd) (L. *Pongo,* type genus of orangutan). Of or relating to the primate family Pongidae, comprising the anthropoid apes (gorillas, chimpanzees, gibbons, orangutans).

population (L. *populus,* people). A group of organisms of the same species inhabiting a specific geographical locality.

populational gradualism The observation that new genetic variants become established in a population by increasing their frequencies across generations incrementally, initially from one or a few individuals and eventually characterizing a majority of the population.

porocyte (pō′rə-sīt) (Gr. *porus;* passage, pore, + *kytos,* hollow vessel). Type of cell found in asconoid sponges through which water enters the spongocoel.

portal system (L. *porta,* gate). System of large veins beginning and ending with a bed of capillaries; for example, hepatic portal and renal portal system in vertebrates.

posterior (L. latter). Situated at or toward the rear of the body; situated toward the back; in human anatomy the upright posture makes posterior and dorsal identical.

potocytosis (pä′tə-sī-tō′səs) (Gr. *potos,* a drinking, + *kytos,* hollow vessel). Endocytosis of certain small molecules and ions bound to specific receptors limited to small areas on the cell surface. The areas of the receptors are invaginated and pinch off to form tiny vesicles. See **caveolae.**

preadaptation The possession of a trait that coincidentally predisposes an organism for survival in an environment different from those encountered in its evolutionary history.

prebiotic synthesis The chemical synthesis that occurred before the emergence of life.

precocial (prē-kō′shəl) (L. *praecoquere,* to ripen beforehand). Referring (especially) to birds whose young are covered with down and are able to run about when newly hatched.

predaceous, predacious (prē-dā′shəs) (L. *praedator,* a plunderer, *praeda,* prey). Living by killing and consuming other animals; predatory.

predator (pred′ə-tər) (L. *praedator,* a plunderer, *praeda,* prey). An organism that preys on other organisms for its food.

prehensile (prē-hen′səl) (L. *prehendere,* to seize). Adapted for grasping.

premunition A resistance to reinfection by an animal (host) when some infective organisms remain in the host's body.

primary bilateral symmetry Usually applied to a radially symmetrical organism descended from a bilateral ancestor and developing from a bilaterally symmetrical larva.

primary radial symmetry Usually applied to a radially symmetrical organism that did not have a bilateral ancestor or larva, in contrast to a secondarily radial organism.

primate (prī-māt) (L. *primus,* first). Any mammal of the order Primates, which includes the tarsiers, lemurs, marmosets, monkeys, apes, and humans.

primitive (L. *primus,* first). Primordial; ancient; little evolved; said of characteristics closely approximating those possessed by early ancestral types.

proboscis (prō-bäs′əs) (Gr. *pro,* before, + *boskein,* feed). A snout or trunk. Also, tubular sucking or feeding organ with the mouth at the end as in planarians, leeches, and insects. Also, the sensory and defensive organ at the anterior end of certain invertebrates.

producers (L. *producere,* to bring forth). Organisms, such as plants, able to produce their own food from inorganic substances.

production In ecology, the energy accumulated by an organism that becomes incorporated into new biomass.

progesterone (prō-jes′tə-rōn′) (L. *pro,* before, + *gestare,* to carry). Hormone secreted by the corpus luteum and the placenta; prepares the uterus for the fertilized egg and maintains the capacity of the uterus to hold the embryo and fetus.

proglottid (prō-gläd′əd) (Gr. *proglōttis,* tongue tip, from *pro,* before, + *glōtta,* tongue, + *id,* suffix). Portion of a tapeworm containing a set of reproductive organs; usually corresponds to a segment.

prohormone (prō′hor-mōn) (Gr. *pro,* before, + *hormaein,* to excite). A precursor of a hormone, especially a peptide hormone.

prokaryotic, procaryotic (pro-kar′ē-ät′ik) (Gr. *pro,* before, + *karyon,* kernel, nut). Not having a membrane-bound nucleus or nuclei. Prokaryotic cells characterize the bacteria and cyanobacteria.

promoter A region of DNA to which the RNA polymerase must have access for transcription of a structural gene to begin.

pronephros (prō-nef′rəs) (Gr. *pro,* before, + *nephros,* kidney). Most anterior of three pairs of embryonic renal organs of vertebrates, functional only in adult hagfishes and larval fishes and amphibians, and vestigial in mammalian embryos. Adj., **pronephric.**

proprioceptor (prō′prē-ə-sep′tər) (L. *proprius,* own, particular, + receptor). Sensory receptor located deep within the tissues, especially muscles, tendons, and joints, that is responsive to changes in muscle stretch, body position, and movement.

prosimian (prō-sim′ē-ən) (Gr. *pro,* before, + L. *simia,* ape). Any member of a group of arboreal primates including lemurs, tarsiers, and lorises, but excluding monkeys, apes, and humans.

prosoma (prō-sōm′ə) (Gr. *pro,* before, + *sōma,* body). Anterior part of an invertebrate in which primitive segmentation is not visible; fused head and thorax of arthropod; cephalothorax.

prosopyle (präs′ə-pīl) (Gr. *prosō,* forward, + *pyle,* gate). Connections between the incurrent and radial canals in some sponges.

prostaglandins (präs′tə-glan′dəns). A family of fatty-acid hormones, originally discovered in semen, known to have powerful effects on smooth muscle, nerves, circulation, and reproductive organs.

prostomium (prō-stōm′ē-əm) (Gr. *protos,* first, + *stoma,* mouth, + *-idion,* dim. ending). Anterior closure of a metameric animal, anterior to the mouth.

protandrous (prō-tan′drəs) (Gr. *prōtos,* first, + *anēr,* male). Condition of hermaphroditic animals and plants in which male organs and their products appear before the corresponding female organs and products, thus preventing self-fertilization.

protease (prō′tē-ās) (Gr. *protein,* + *ase,* enzyme). An enzyme that digests proteins; includes proteinases and peptidases.

protein (prō′tēn, prō′tē-ən) (Gr. *protein,* from *proteios,* primary). A macromolecule of carbon, hydrogen, oxygen, and nitrogen and sometimes sulfur and phosphorus; composed of chains of amino acids joined by peptide bonds; present in all cells.

prothoracic glands Glands in the prothorax of insects that secrete the hormone ecdysone.

prothoracicotropic hormone See **ecdysiotropin.**

prothrombin (prō-thräm′bən) (Gr. *pro,* before, + *thrombos,* clot). A constituent of blood plasma that is changed to thrombin by a catalytic sequence that includes thromboplastin, calcium, and plasma globulins; involved in blood clotting.

protist (prō′tist) (Gr. *protos,* first). A member of the paraphyletic kingdom Protista, generally considered to include the protozoan groups and eukaryotic algae.

protocoel (prō′tə-sēl) (Gr. *protos,* first, + *koilos,* hollow). The anterior coelomic compartment in some deuterostomes, corresponds to the axocoel in echinoderms.

protocooperation A mutually beneficial interaction between organisms in which the interaction is not physiologically necessary to the survival of either.

proton A subatomic particle with a positive electrical charge and having a mass of 1836 times that of an electron; found in the nucleus of atoms.

protonephridium (prō′tə-nə-frid′ē-əm) (Gr. *protos,* first, + *nephros,* kidney). Primitive osmoregulatory or excretory organ consisting of a tubule terminating internally with flame bulb or solenocyte; the unit of a flame bulb system.

proto-oncogene See **oncogene.**

protoplasm (prō′tə-plazm) (Gr. *protos,* first, + *plasma,* form). Organized living substance; cytoplasm and nucleoplasm of the cell.

protopod, protopodite (prō′tə-päd, prō-top′ə-dīt) (Gr. *protos,* first, + *pous, podos,* foot). Basal portion of crustacean appendage, containing coxa and basis.

Protostomia (prō′tə-stō′mē-ə) (Gr. *protos,* first, + *stoma,* mouth). A group of phyla in which cleavage is determinate, the coelom (in coelomate forms) is formed by proliferation of mesodermal bands (schizocoelic formation), the mesoderm is formed from a particular blastomere (called 4d), and the mouth is derived from or near the blastopore. Includes the Annelida, Arthropoda, Mollusca, and a number of minor phyla. Compares with **Deuterostomia.**

proventriculus (prō′ven-trik′ū-ləs) (L. *pro,* before, + *ventriculum,* ventricle). In birds the glandular stomach between the crop and gizzard. In insects, a muscular dilation of foregut armed internally with chitinous teeth.

proximal (L. *proximus,* nearest). Situated toward or near the point of attachment; opposite of distal, distant.

proximate cause (L. *proximus,* nearest, + *causa*). The factors that underlie the functioning of a biological system at a particular place and time, including those responsible for metabolic, physiological, and behavioral functions at the molecular, cellular, organismal, and population levels.

pseudocoel (sū′do-sēl) (Gr. *pseudēs,* false, + *koilōma,* cavity). A body cavity not lined with peritoneum and not a part of the blood or digestive systems, embryonically derived from the blastocoel.

pseudopodium (sū′də-pō′dē-əm) (Gr. *pseudēs,* false, + *podion,* small foot, + *eidos,* form). A temporary cytoplasmic protrusion extended out from a protozoan or ameboid cell, and serving for locomotion or for taking up food.

puff Strands of DNA spread apart at certain locations on giant chromosomes of some flies where that DNA is being transcribed.

pulmonary (pul′mən-ner-ē) (L. *pulmo,* lung, + *aria,* suffix denoting connected to). Relating to or associated with lungs.

punctuated equilibrium A model of evolution in which morphological evolutionary change is discontinuous, being associated primarily with discrete, geologically instantaneous events of speciation leading to phylogenetic branching; morphological evolutionary stasis characterizes species between episodes of speciation; contrasts with **phyletic gradualism.**

pupa (pū′pə) (L. girl, doll, puppet). Inactive quiescent stage of the holometabolous insects. It follows the larval stages and precedes the adult stage.

purine (pū′rēn) (L. *purus,* pure, + *urina,* urine). Organic base with carbon and nitrogen atoms in two interlocking rings. The parent substance of adenine, guanine, and other naturally occurring bases.

pygidium (pī-jid′e-əm) (Gr. *pygē,* rump, buttocks, + *-idion,* dim. ending). Posterior closure of a metameric animal, bearing the anus.

pyrimidine (pī-rim′ə-dēn) (alter. of pyridine, from Gr. *pyr,* fire, + *id,* adj. suffix, + *ine*). An organic base composed of a single ring of carbon and nitrogen atoms; parent substance of several bases found in nucleic acids.

Q

quantitative inheritance See **polygenic inheritance.**

queen In entomology, the single fully developed female in a colony of social insects such as bees, ants, and termites, distinguished from workers, nonreproductive females, and soldiers.

R

radial canals Canals along the ambulacra radiating from the ring canal of echinoderms; also choanocyte-lined canals in syconoid sponges.

radial cleavage Embryonic development in which early cleavage planes are symmetrical to the polar axis, each blastomere of one tier lying directly above the corresponding blastomere of the next layer; indeterminate cleavage.

radial symmetry A morphological condition in which the parts of an animal are arranged concentrically around an oral-aboral axis, and more than one imaginary plane through this axis yields halves that are mirror images of each other.

radiolarian (rā′dē-ə-la′rē-ən) (L. *radius,* ray, spoke of a wheel, + *Lar,* tutelary god of house and field). Amebas with actinopodia and beautiful tests.

radioles (rā′dē-ōlz) (L. *radius,* ray, spoke of a wheel). Featherlike processes from the head of many tubicolous polychaete worms (phylum Annelida), used primarily for feeding.

radula (re′jə-lə) (L. scraper). Rasping tongue found in most molluscs.

bat / āpe / ärmadillo / herring / fēmale / finch / līce / crocodile / crōw / duck / ūnicorn / ə indicates unaccented vowel sound "uh" as in mammal, fishes, cardinal, heron, vulture / stress as in bi-ol′o-gy, bi′o-log′i-cal

Ras protein A protein that initiates a cascade of reactions leading to cell division when a growth factor is bound to the cell surface. The gene encoding Ras becomes an oncogene when a mutation produces a form of Ras protein that initiates the cascade even in the absence of the growth factor.

ratite (ra′tīt) (L. *ratis,* raft). Referring to birds having an unkeeled sternum; compares with **carinate.**

recapitulation Summing up or repeating; hypothesis that an individual repeats its phylogenetic history in its development.

receptor-mediated endocytosis Endocytosis of large molecules, which are bound to surface receptors in clathrin-coated pits.

recessive An allele that must be homozygous for the allele to be expressed.

recombinant DNA DNA from two different species, such as a virus and a mammal, combined into a single molecule.

redia (rē′dē-ə), pl. **rediae** (rē′dē-ē) (from Redi, Italian biologist). A larval stage in the life cycle of flukes; it is produced by a sporocyst larva, and in turn gives rise to many cercariae.

reduction In chemistry, the gain of an electron by an atom or molecule of a substance; also the addition of hydrogen to, or the removal of oxygen from, a substance.

regulative development Progressive determination and restriction of initially totipotent embryonic material.

releaser (L. *relaxare,* to unloose). Simple stimulus that elicits an innate behavior pattern.

renin (rē′nin) (L. *ren,* kidney). An enzyme produced by the kidney juxtaglomerular apparatus that initiates changes leading to increased blood pressure and increased sodium reabsorption.

rennin (re′nən) (M.E. *renne,* to run). A milk-clotting endopeptidase secreted by the stomach of some young mammals, including bovine calves and human infants.

replication (L. *replicatio,* a folding back). In genetics, the duplication of one or more DNA molecules from the preexisting molecule.

reproductive barrier (L. *re + producere,* to lead forward; M.F. *barriere,* bar). The factors that prevent one sexually propagating population from interbreeding and exchanging genes with another population.

repugnatorial glands (L. *repugnare,* to resist). Glands secreting a noxious substance for defense or offense, for example, as in the millipedes.

respiration (L. *respiratio,* breathing). Gaseous interchange between an organism and its surrounding medium. In the cell, the release of energy by the oxidation of food molecules.

restriction endonuclease An enzyme that cleaves a DNA molecule at a particular base sequence.

rete mirabile (rē′tē mə-rab′ə-lē) (L. wonderful net). A network of small blood vessels so arranged that the incoming blood runs countercurrent to the outgoing blood and thus makes possible efficient exchange between the two bloodstreams. Such a mechanism serves to maintain the high concentration of gases in the fish swim bladder.

reticular (rə-tīk′ū-lər) (L. *reticulum,* small net). Resembling a net in appearance or structure.

reticuloendothelial system (rə-tic′u-lō-en-dō-thēl′i-əl) (L. *reticulum,* dim. of net, + Gr. *endon,* within, + *thele,* nipple). The fixed phagocytic cells in the tissues, especially the liver, lymph nodes, spleen, and others; also called RE system.

reticulopodia (rə-tik′ū-lə-pō′dē-ə) (L. *retiulum,* dim. of *rete,* net, + *podos, pous,* foot). Pseudopodia that branch and rejoin extensively.

retina (ret′nə, ret′ən-ə) (L. *rete,* net). The posterior sensory membrane of the eye that receives images.

rhabdite (rab′dit) (Gr. *rhabdos,* rod). Rodlike structures in the cells of the epidermis or underlying parenchyma in certain turbellarians. They are discharged in mucous secretions.

rheoreceptor (rē′ə-rē-cep′tər) (Gr. *rheos,* a flowing, + receptor). A sensory organ of aquatic animals that responds to water current.

rhinarium (rī-na′rē-əm) (Gr. *rhis,* nose). Hairless area surrounding the nose of a mammal.

rhinophore (rī′nə-fōr) (Gr. *rhis,* nose, + *pherein,* to carry). Chemoreceptive tentacles in some molluscs (opisthobranch gastropods).

rhopalium (rō-pā′lē-əm) (N.L. from Gr. *rhopalon,* a club). One of the marginal, club-shaped sense organs of certain jellyfishes; tentaculocyst.

rhoptries (rōp′trēz) (Gr. *rhopalon,* club, + *tryō,* to rub, wear out). Club-shaped bodies in Apicomplexa composing one of the structures of the apical complex; open at anterior and apparently functioning in penetration of host cell.

rhynchocoel (ring′kō-sēl) (Gr. *rhynchos,* snout, + *koilos,* hollow). In nemerteans, the dorsal tubular cavity that contains the inverted proboscis. It has no opening to the outside.

ribosome (rī′bə-sōm). Subcellular structure composed of protein and ribonucleic acid. May be free in the cytoplasm or attached to the membranes of the endoplasmic reticulum; functions in protein synthesis.

ritualization In ethology, the evolutionary modification, usually intensification, of a behavior pattern to serve communication.

RNA Ribonucleic acid, of which there are several different kinds, such as messenger RNA, ribosomal RNA, and transfer RNA (mRNA, rRNA, tRNA), as well as many structural and regulatory RNAs.

RNA world Hypothetical stage in the evolution of life on earth in which both catalysis and replication were performed by RNA, not protein enzymes and DNA.

rostellum (räs′tel′ləm) (L. small beak). Projecting structure on scolex of tapeworm, often with hooks.

rostrum (räs′trəm) (L. ship's beak). A snoutlike projection on the head.

rumen (rū′mən) (L. cud). The large first compartment of the stomach of ruminant mammals. Serves as a fermentation chamber in which bacteria break down cellulose.

ruminant (rūm′ə-nənt) (L. *ruminare,* to chew the cud). Cud-chewing artiodactyl mammals with a complex four-chambered stomach.

S

saccule (sa′kūl) (L. *sacculus,* small bag). Small chamber of the membranous labyrinth of the inner ear.

sacrum Adj. **sacral** (sā′krəm, sā′krəl) (L. *sacer,* sacred). Bone formed by fused vertebrae to which pelvic girdle is attached; pertaining to the sacrum.

sagittal (saj′ə-dəl) (L. *sagitta,* arrow). Pertaining to the median anteroposterior plane that divides a bilaterally symmetrical organism into right and left halves.

salt (L. *sal,* salt). The reaction product of an acid and a base; dissociates in water solution to negative and positive ions, but not H⁺ or OH⁻.

saprophagous (sə-präf′ə-gəs) (Gr. *sapros,* rotten, + *phagos,* from *phagein,* to eat). Feeding on decaying matter; saprobic; saprozoic.

saprophyte (sap′rə-fīt) (Gr. *sapros,* rotten, + *phyton,* plant). A plant living on dead or decaying organic matter.

saprozoic nutrition (sap-rə-zō′ik) (Gr. *sapros,* rotten, + *zōon,* animal). Animal nutrition by absorption of dissolved salts and simple organic nutrients from surrounding medium; also refers to feeding on decaying matter.

sarcolemma (sär′kə-lem′ə) (Gr. *sarx,* flesh, + *lemma,* rind). The thin, noncellular sheath that encloses a striated muscle fiber.

sarcomere (sär′kə-mir) (Gr. *sarx,* flesh, + *meros,* part). Transverse segment of striated muscle forming the fundamental contractile unit.

sarcoplasm (sär′kə-plaz′əm) (Gr. *sarx,* flesh, + *plasma,* mold). The clear, semifluid cytoplasm between the fibrils of muscle fibers.

sauropterygians (so-räp′tə-rij′ē-əns) (Gr. *sauros,* lizard, + *pteryginos,* winged). Mesozoic marine reptiles.

schizocoel (skiz′ō-sēl) (Gr. *schizo,* from *schizein,* to split, + *koilōma,* cavity). A coelom formed by the splitting of embryonic mesoderm. Noun, **schizocoelomate,** an animal with a schizocoel, such as an arthropod or mollusc. Adj., **schizocoelous.**

schizocoelous mesoderm formation (skiz′ō-sēl-ləs). Embryonic formation of the mesoderm as cords of cells between ectoderm and endoderm; splitting of these cords results in the coelomic space.

schizogony (skə-zä′gə-nē) (Gr. *schizein,* to split, + *gonos,* seed). Multiple asexual fission.

sclerite (skler′it) (Gr. *sklēros,* hard). A hard chitinous or calcareous plate or spicule; one of the plates making up the exoskeleton of arthropods, especially insects.

scleroblast (skler′ə-blast) (Gr. *sklēros,* hard, + *blastos,* germ). An amebocyte specialized to secrete a spicule, found in sponges.

sclerocyte (skler′ə-sīt) (Gr. *sklēros,* hard, + *kytos,* hollow vessel). An amebocyte in sponges that secretes spicules.

sclerotic (skler-äd′ik) (Gr. *sklēros,* hard). Pertaining to the tough outer coat of the eyeball.

sclerotization (sklər′ə-tə-zā′shən). Process of hardening of the cuticle of arthropods by the formation of stabilizing cross linkages between peptide chains of adjacent protein molecules.

scolex (skō′leks) (Gr. *skōlēx,* worm, grub). The holdfast, or so-called head, of a tapeworm; bears suckers and, in some, hooks, and posterior to it new proglottids are differentiated.

scrotum (skrō′təm) (L. bag). The pouch that contains the testes in most mammals.

scyphistoma (sī-fis′tə-mə) (Gr. *skyphos,* cup, + *stoma,* mouth). A stage in the development of scyphozoan jellyfish just after the larva becomes attached, the polyp form of a scyphozoan.

sebaceous (sə-bāsh′əs) (L. *sebaceus,* made of tallow). A type of mammalian epidermal gland that produces a fatty substance.

sedentary (sed′ən-ter-ē). Stationary, sitting, inactive; staying in one place.

selectively permeable Permeable to small particles, such as water and certain inorganic ions, but not to larger molecules.

seminiferous (sem-ə-nif′rəs) (L. *semen,* semen, + *ferre,* to bear). Pertains to the tubules that produce or carry semen in the testes.

semipermeable (L. *semi,* half, + *permeabilis,* capable of being passed through). Permeable to small particles, such as water and certain inorganic ions, but not to larger molecules.

sensillum, pl. **sensilla** (sin-si′ləm) (L. *sensus,* sense). A small sense organ, especially in the arthropods.

septum, pl. **septa** (L. fence). A wall between two cavities.

serial homology See **homology.**

serosa (sə-rō′sə) (N.L. from L. *serum,* serum). The outer embryonic membrane of birds and reptiles; chorion. Also, the peritoneal lining of the body cavity.

serotonin (sir′ə-tōn′ən) (L. *serum,* serum). A phenolic amine, found in the serum of clotted blood and in many other tissues, that possesses several poorly understood metabolic, vascular, and neural functions; 5-hydroxytryptamine.

serous (sir′əs) (L. *serum,* serum). Watery, resembling serum; applied to glands, tissue, cells, fluid.

serum (sir′əm) (L. whey, serum). The liquid that separates from the blood after coagulation; blood plasma from which fibrinogen has been removed. Also, the clear portion of a biological fluid separated from its particulate elements.

sessile (ses′əl) (L. *sessilis,* low, dwarf). Attached at the base; fixed to one spot, not able to move about.

seta (sīd′ə), pl. **setae** (sē′tē) (L. bristle). A needlelike chitinous structure of the integument of annelids, arthropods, and others.

sex chromosomes Chromosomes that determine gender of an animal. They may bear a few or many other genes.

sibling species Reproductively isolated species that are so similar morphologically that they are difficult or impossible to distinguish using morphological characters.

sickle cell anemia A condition that causes the red blood cells to collapse (sickle) under oxygen stress. The condition becomes manifest when an individual is homozygous for the gene for hemoglobin-S (HbS).

siliceous (sə-li′shəs) (L. *silex,* flint). Containing silica.

simian (sim′ē-ən) (L. *simia,* ape). Pertaining to monkeys or apes.

sinistral (si′nə-strəl, sə-ni′stral) (L. *sinister,* left). Pertaining to the left; in gastropods, shell is sinistral if opening is to left of columella when held with spire up and facing observer.

sinus (sī′nəs) (L. curve). A cavity or space in tissues or in bone.

siphonoglyph (sī′fän′ə-glif′) (Gr. *siphōn,* reed, tube, siphon, + *glyphē,* carving). Ciliated furrow in the gullet of sea anemones.

siphuncle (sī′fun-kəl) (L. *siphunculus,* small tube). Cord of tissue running through the shell of a nautiloid, connecting all chambers with body of animal.

sister group The relationship between a pair of species or higher taxa that are each other's closest phylogenetic relatives.

sociobiology Ethological study of social behavior in humans or other animals.

---

solenia (sō-len′ē-ə) (Gr. *sōlēn,* pipe). Channels through the coenenchyme connecting the polyps in an octocorallian colony (phylum Cnidaria).

solenocyte (sō-len′ə-sīt) (Gr. *sōlēn,* pipe, + *kytos,* hollow vessel). Special type of flame bulb in which the bulb bears a flagellum instead of a tuft of flagella. See **flame cell, protonephridium.**

soma (sō′mə) (Gr. body). The whole of an organism except the germ cells (germ plasm).

somatic (sō-mat′ik) (Gr. *sōma,* body). Refers to the body, for example, somatic cells in contrast to germ cells.

somatocoel (sə-mat′ə-sēl) (Gr. *sōma,* the body, + *koilos,* hollow). Posterior coelomic compartment of echinoderms; left somatocoel gives rise to oral coelom, and right somatocoel becomes aboral coelom.

somatoplasm (sō′mə-də-pla′zm) (Gr. *sōma,* body, + *plasma,* anything formed). The living matter that makes up the mass of the body as distinguished from germ plasm, which makes up the reproductive cells. The protoplasm of body cells.

somite (sō′mīt) (Gr. *soma,* body). One of the blocklike masses of mesoderm arranged segmentally (metamerically) in a longitudinal series beside the neural tube of the embryo; metamere.

sorting Differential survival and reproduction among varying individuals; often confused with natural selection which is one possible cause of sorting.

speciation (spē′sē-ā′shən) (L. *species,* kind). The evolutionary process or event by which new species arise.

species (spē′shez, spē′sēz) sing. and pl. (L. particular kind). A group of interbreeding individuals of common ancestry that are reproductively isolated from all other such groups; a taxonomic unit ranking below a genus and designated by a binomen consisting of its genus and the species name.

spermatheca (spər′mə-thē′kə) (Gr. *sperma,* seed, + *thēkē,* case). A sac in the female reproductive organs for the reception and storage of sperm.

spermatid (spər′mə-təd) (Gr. *sperma,* seed, + *eidos,* form). A growth stage of a male reproductive cell arising by division of a secondary spermatocyte; gives rise to a spermatozoon.

spermatocyte (spər-mad′ə-sīt) (Gr. *sperma,* seed, + *kytos,* hollow vessel). A growth stage of a male reproductive cell; gives rise to a spermatid.

spermatogenesis (spər-mad′ə-jen′-ə-səs) (Gr. *sperma,* seed, + *genesis,* origin). Formation and maturation of spermatozoa.

spermatogonium (spər′mad-ə-gō′nē-əm) (Gr. *sperma,* seed, + *gonē,* offspring). Precursor of mature male reproductive cell; gives rise directly to a spermatocyte.

spermatophore (spər-mad′ə-fōr′) (Gr. *sperma, spermatos,* seed, + *pherein,* to bear). Capsule or packet enclosing sperm, produced by males of several invertebrate groups and a few vertebrates.

sphincter (sfingk′tər) (Gr. *sphinkter,* band, sphincter, from *sphingein,* to bind tight). A ring-shaped muscle capable of closing a tubular opening by constriction.

spicule (spi′kyul) (L. dim. *spica,* point). One of the minute calcareous or siliceous skeletal bodies found in sponges, radiolarians, soft corals, and sea cucumbers.

spiracle (spi′rə-kəl) (L. *spiraculum,* from *spirare,* to breathe). External opening of a trachea in arthropods. One of a pair of openings on the head of elasmobranchs for passage of water. Exhalent aperture of tadpole gill chamber.

spiral cleavage A type of embryonic cleavage in which cleavage planes are diagonal to the polar axis and unequal cells are produced by the alternate clockwise and counterclockwise cleavage around the axis of polarity; determinate cleavage.

spongin (spun′jin) (L. *spongia,* sponge). Fibrous, collagenous material making up the skeletal network of horny sponges.

spongioblast (spun′je-o-blast) (Gr. *spongos,* sponge, + *blastos,* bud). Cell in a sponge that secretes spongin, a protein.

spongocoel (spun′jō-sēl) (Gr. *spongos,* sponge, + *koilos,* hollow). Central cavity in sponges.

spongocyte (spun′jō-sīt) (Gr. *spongos,* sponge, + *kytos,* hollow vessel). A cell in sponges that secretes spongin.

sporocyst (spō′rə-sist) (Gr. *sporos,* seed, + *kystis,* pouch). A larval stage in the life cycle of flukes; it originates from a miracidium.

sporogony (spor-äg′ə-nē) (Gr. *sporos,* seed, + *gonos,* birth). Multiple fission to produce sporozoites after zygote formation.

sporozoite (spō′rə-zō′it) (Gr. *sporos,* seed, + *zōon,* animal, + *ite,* suffix for body part). A stage in the life history of many sporozoan protozoa; released from oocysts.

squalene (skwā′lēn) (L. *squalus,* a kind of fish). A liquid acyclic triterpene hydrocarbon found especially in the liver oil of sharks.

squamous epithelium (skwā′məs) (L. *squama,* scale, + *osus,* full of). Simple epithelium of flat, nucleated cells.

stapes (stā′pēz) (L. stirrup). Stirrup-shaped innermost bone of the middle ear.

statoblast (stad′ə-blast) (Gr. *statos,* standing, fixed, + *blastos,* germ). Biconvex capsule containing germinative cells and produced by most freshwater ectoprocts by asexual budding. Under favorable conditions it germinates to give rise to new zooid.

statocyst (Gr. *statos,* standing, + *kystis,* bladder). Sense organ of equilibrium; a fluid-filled cellular cyst containing one or more granules (statoliths) used to sense direction of gravity.

statolith (Gr. *statos,* standing, + *lithos,* stone). Small calcareous body resting on tufts of cilia in the statocyst.

stenohaline (sten-ə-hā′līn, -lən) (Gr. *stenos,* narrow, + *hals,* salt). Pertaining to aquatic organisms that have restricted tolerance to changes in environmental saltwater concentration.

stenophagous (stə-näf′ə-gəs) (Gr. *stenos,* narrow, + *phagein,* to eat). Eating few kinds of foods.

stenotopic (sten-ə-tä′pik) (Gr. *stenos,* narrow, + *topos,* place). Refers to an organism with a narrow range of adaptability to environmental change; having a restricted environmental distribution.

stereogastrula (ste′rē-ə-gas′trə-lə) (Gr. *stereos,* solid, + *gastēr,* stomach, + L. *ula,* dim.). A solid type of gastrula, such as the planula of cnidarians.

stereom (ster′ē-ōm) (Gr. *stereos,* solid, hard, firm). Meshwork structure of endoskeletal ossicles of echinoderms.

stereotyped behavior A pattern of behavior repeated with little variation in performance.

---

bat / āpe / ärmadillo / herring / fēmale / finch / līce / crocodile / crōw / duck / ūnicorn / ə indicates unaccented vowel sound "uh" as in mammal, fishes, cardinal, heron, vulture / stress as in bi-ol′o-gy, bi′o-log′i-cal

sternum (ster′nəm) (L. breastbone). Ventral plate of an arthropod body segment; breastbone of vertebrates.

sterol (ste′rōl), steroid (ste′roid) (Gr. *stereos,* solid, + L. *ol,* from *oleum,* oil). One of a class of organic compounds containing a molecular skeleton of four fused carbon rings; it includes cholesterol, sex hormones, adrenocortical hormones, and vitamin D.

stigma (Gr. *stigma,* mark, tatoo mark). Eyespot in certain protozoa. Spiracle of certain terrestrial arthropods.

stolon (stō′lən) (L. *stolō, stolonis,* a shoot, or sucker of a plant). A rootlike extension of the body wall giving rise to buds that may develop into new zooids, thus forming a compound animal in which the zooids remain united by the stolon. Found in some colonial anthozoans, hydrozoans, ectoprocts, and ascidians.

stoma (stō′mə) (Gr. mouth). A mouthlike opening.

stomochord (stō′mə-kord) (Gr. *stoma,* mouth, + *chordē,* cord). Anterior evagination of the dorsal wall of the buccal cavity into the proboscis of hemichordates; the buccal diverticulum.

strobila (strō′bə-lə) (Gr. *strobilē,* lint plug like a pine cone [*strobilos*]). A stage in the development of the scyphozoan jellyfish. Also, the chain of proglottids of a tapeworm.

strobilation (strō′bə-lā′shən) (Gr. *strobilos,* a pine cone). Repeated, linear budding of individuals, as in scyphozoans (phylum Cnidaria), or sets of reproductive organs, as in tapeworms (phylum Platyhelminthes).

stroma (strō′mə) (Gr. *strōma,* bedding). Supporting connective tissue framework of an animal organ; filmy framework of red blood corpuscles and certain cells.

structural gene A gene carrying the information to construct a protein.

subnivean (səb-ni′vē-ən) (L. *sub,* under, below, + *nivis,* snow). Applied to environments beneath snow, in which snow insulates against a colder atmospheric temperature.

substrate The substance upon which an enzyme acts; also, a base or foundation (substratum); and the substance or base on which an organism grows.

sycon (sī′kon) (Gr. *sykon,* fig). A type of canal system in certain sponges. Sometimes called syconoid.

symbiosis (sim′bī-ōs′əs, sim′bē-ōs′əs) (Gr. *syn,* with, + *bios,* life). The living together of two different species in an intimate relationship. Symbiont always benefits; host may benefit, may be unaffected, or may be harmed (mutualism, commensalism, and parasitism).

sympatric (sim′pa′-trik) (Gr. *syn,* with, + *patra,* native land). Having the same or overlapping regions of geographical distribution. Noun, **sympatry.**

symplesiomorphy (sim-plē′sē-ə-mōr′fē). Sharing among species of ancestral characteristics, not indicative that the species comprise a monophyletic group.

synapomorphy (sin-ap′o-mor′fē) (Gr. *syn,* together with, + *apo,* of, + *morphe,* form). Shared, evolutionarily derived character states that are used to recover patterns of common descent among two or more species.

synapse (si′naps, si-naps′) (Gr. *synapsis,* contact, union). The place at which a nerve impulse passes between neuron processes, typically from an axon of one nerve cell to a dendrite of another nerve cell.

synapsids (si-nap′sədz) (Gr. *synapsis,* contact, union). An amniote lineage comprising the mammals and the ancestral mammal-like reptiles, having a skull with a single pair of temporal openings.

synapsis (si-nap′səs) (Gr. *synapsis,* contact, union). The time when the pairs of homologous chromosomes lie alongside each other in the first meiotic division.

synaptonemal complex (sin-ap′tə-nē′məl) (Gr. *synapsis,* a joining together, + *nēma,* thread). The structure that holds homologous chromosomes together during synapsis in prophase of meiosis I.

syncytium (sən-sish′e-əm) adj. **syncytial** (Gr. *syn,* with, + *kytos,* hollow). A multinucleated cell.

syndrome (sin′drōm) (Gr. *syn,* with, + *dramein,* to run). A group of symptoms characteristic of a particular disease or abnormality.

syngamy (sin′gə-mē) (Gr. *syn,* with, + *gamos,* marriage). Fertilization of one gamete with another individual gamete to form a zygote, found in most animals with sexual reproduction.

synkaryon (sin-ker′e-on) (Gr. *syn,* with, + *karyon,* nucleus). Zygote nucleus resulting from fusion of pronuclei.

syrinx (sir′inks) (Gr. shepherd's pipe). The vocal organ of birds located at the base of the trachea.

systematics (sis-tə-mat′iks). Science of classification and reconstruction of phylogeny.

systole (sis′tə-lē) (Gr. *systolē,* drawing together). Contraction of heart.

T

T cell A type of lymphocyte important in cellular immune response and in regulation of most immune responses.

T-cell receptors Receptors borne on surfaces of T cells. The variable region of a T-cell receptor binds with a specific antigen.

tactile (tak′til) (L. *tactilis,* able to be touched, from *tangere,* to touch). Pertaining to touch.

tagma, pl. **tagmata** (Gr. *tagma,* arrangement, order, row). A compound body section of an arthropod resulting from embryonic fusion of two or more segments; for example, head, thorax, abdomen.

tagmatization, tagmosis Organization of the arthropod body into tagmata.

taiga (tī′gä) (Russ.). Habitat zone characterized by large tracts of coniferous forests, long, cold winters, and short summers; most typical in Canada and Siberia.

tantulus (tan′tə-ləs) (Gr. *tantulus,* so small). Larva of a tantulocaridan (subphylum Crustacea).

taxis (tak′sis), pl. **taxes** (Gr. *taxis,* arrangement). An orientation movement by a (usually) simple organism in response to an environmental stimulus.

taxon (tak′son), pl. **taxa** (Gr. *taxis,* arrangement). Any taxonomic group or entity.

taxonomy (tak-sän′ə-mi) (Gr. *taxis*, arrangement, + *nomos*, law). Study of the principles of scientific classification; systematic ordering and naming of organisms.

tectum (tek′təm) (L. roof). A rooflike structure, for example, dorsal part of capitulum in ticks and mites.

tegmen (teg′mən) (L. *tegmen*, a cover). External epithelium of crinoids (phylum Echinodermata).

tegument (teg′ū-ment) (L. *tegumentum,* from *tegere,* to cover). An integument: specifically external covering in cestodes and trematodes, formerly believed to be a cuticle.

telencephalon (tel′en-sef′ə-lon) (Gr. *telos*, end, + *encephalon,* brain). The most anterior vesicle of the brain; the anterior-most subdivision of the prosencephalon that becomes the cerebrum and associated structures.

teleology (tel′ē-äl′ə-jē) (Gr. *telos*, end, + L. *logia*, study of, from Gr. *logos*, word). The philosophical view that natural events are goal directed and are preordained, as opposed to the scientific view of mechanical determinism.

telocentric (tē′lō-sen′trək) (Gr. *telos*, end, + *kentron*, center). Chromosome with centromere at the end.

telolecithal (te-lō-les′ə-thəl) (Gr. *telos*, end, + *lekithos*, yolk, + *al*). Having the yolk concentrated at one end of an egg.

telson (tel′sən) (Gr. *telson*, extremity). Posterior projection of the last body segment in many crustaceans.

temnospondyls (tem-nō-span′dəls) (Gr. *temnō*, to cut, + *spondylos*, vertebra). A large lineage of amphibians that extended from the Carboniferous to the Triassic.

template (tem′plət). A pattern or mold guiding the formation of a duplicate; often used with reference to gene duplication.

tendon (ten′dən) (L. *tendo*, tendon). Fibrous band connecting muscle to bone or other movable structure.

tentaculocyst (ten-tak′u-lō-sist) (L. *tentaculum*, feeler, + *kystis*, pouch). One of the sense organs along the margin of medusae; a rhopalium.

tergum (ter′gəm) (L. back). Dorsal part of an arthropod body segment.

territory (L. *territorium*, from *terra*, earth). A restricted area preempted by an animal or pair of animals, usually for breeding purposes, and guarded from other individuals of the same species.

test (L. *testa*, shell). A shell or hardened outer covering.

tetrad (te′trad) (Gr. *tetras*, four). Group of two pairs of chromatids at synapsis and resulting from the replication of paired homologous chromosomes; the bivalent.

tetrapods (te′trə-päds) (Gr. *tetras*, four, + *pous, podos,* foot). Four-footed vertebrates; the group includes amphibians, reptiles, birds, and mammals.

theca (thē′kə) (Gr. *thēkē*, a case for something, a box). A protective covering for an organism or an organ.

thecate (thē′kāt) (Gr. *thēkē*, a case, box). An organism bearing a theca.

thecodonts (thēk′ə-dänts) (Gr. *thēkē*, box, + *odontos*, tooth). A large assemblage of Triassic archosaurian diapsids of the order Thecodontia and characterized by having teeth set in sockets.

therapsids (thə-rap′sidz) (Gr. *theraps*, an attendant). Extinct amniotes, from the Permian to Triassic, from which true mammals evolved.

thermocline (thər′mō-klīn) (Gr. *thermē,* heat, + *klinein,* to swerve). Layer of water separating upper warmer and lighter water from lower colder and heavier water in a lake or sea; a stratum of abrupt change in water temperature.

thoracic (thō-ra′sək) (L. *thōrax,* chest). Pertaining to the thorax or chest.

thrombin Enzyme catalyzing fibrinogen transformation into fibrin. Percursor is **prothrombin.**

Tiedemann's bodies (tēd′ə-mənz) (from F. Tiedemann, German anatomist). Four or five pairs of pouchlike bodies attached to the ring canal of sea stars, apparently functioning in production of coelomocytes.

tight junction Region of actual fusion of cell membranes between two adjacent cells.

tissue (ti′shu) (M.E. *tissu*, tissue). An aggregation of cells, usually of the same kind, organized to perform a common function.

titer (tī′tər) (Fr. *titrer*, to titrate). Concentration of a substance in a solution as determined by titration.

Toll-like receptors (TLRs) Named for the Toll protein family discovered in *Drosophila*, Toll-like receptors are located on cell membranes of vertebrates. When activated by binding a microbe, they signal the cell to synthesize an appropriate antimicrobial peptide. Recognizing patterns rather than specific molecular configurations, they are a vital part of innate immune defenses.

tornaria (tor-na′rē-ə) (L. *tornare*, to turn). A free-swimming larva of enteropneusts that rotates as it swims; resembles somewhat the bipinnaria larva of echinoderms.

torsion (L. *torquere*, to twist). A twisting phenomenon in gastropod development that alters the position of the visceral and pallial organs by 180 degrees.

toxicyst (tox′i-sist) (Gr. *toxikon*, poison, + *kystis*, bladder). Structures possessed by predatory ciliate protozoa, which on stimulation expel a poison to subdue the prey.

trabecular reticulum (trə-bek′ū-lər rə-tik′ū-ləm) (L. *trabecula*, a small beam; *reticulum*, a net). A bilayered, syncytial tissue forming the main body structure of hexactinellid sponges (phylum Porifera).

trachea (trā′kē-ə) (M.L. windpipe). The windpipe. Also, any of the air tubes of insects.

transcription Formation of messenger RNA from the coded DNA.

transduction Condition in which bacterial DNA (and the genetic characteristics it bears) is transferred from one bacterium to another by the agent of viral infection.

transfer RNA (tRNA) A form of RNA of about 70 or 80 nucleotides, which are adapter molecules in the synthesis of proteins. A specific amino acid molecule is carried by transfer RNA to a ribosome-messenger RNA complex for incorporation into a polypeptide.

transformation Condition in which DNA in the environment of bacteria somehow penetrates them and is incorporated into their genetic complement, so that their progeny inherit the genetic characters so acquired.

translation (L. a transferring). The process in which the genetic information present in messenger RNA is used to direct the order of specific amino acids during protein synthesis.

transporter See **permease.**

bat / āpe / ärmadillo / herring / fēmale / finch / līce / crocodile / crōw / duck / ūnicorn / ə indicates unaccented vowel sound "uh" as in mammal, fishes, cardinal, heron, vulture / stress as in bi-ol′o-gy, bi′o-log′i-cal

transverse plane (L. *transversus,* across). A plane or section that lies or passes across a body or structure.

trichinosis (trik-ən-o′səs). Disease caused by infection with the nematode *Trichinella spiralis.*

trichocyst (trik′ə-sist) (Gr. *thrix,* hair, + *kystis,* bladder). Sac-like protrusible organelle in the ectoplasm of ciliates, which discharges as a threadlike weapon of defense.

triglyceride (trī-glis′ə-rīd) (Gr. *tria,* three, + *glykys,* sweet, + *ide,* suffix denoting compound). A triester of glycerol with one, two, or three acids.

trimerous (trī′mə-rəs) (Gr. *treis,* three, + *meros,* a part). Body in three main divisions, as in lophophorates and some deuterostomes.

tripartite (trī-par′tīt). See **trimerous.**

triploblastic (trip′lō-blas′tik) (Gr. *triploos,* triple, + *blastos,* germ). Pertaining to metazoa in which the embryo has three primary germ layers—ectoderm, mesoderm, and endoderm.

trisomy 21 See **Down syndrome.**

trochophore (trōk′ə-fōr) (Gr. *trochos,* wheel, + *pherein,* to bear). A free-swimming ciliated marine larva characteristic of most molluscs and certain ectoprocts, brachiopods, and marine worms; an ovoid or pyriform body with preoral circlet of cilia and sometimes a secondary circlet behind the mouth.

trophallaxis (trōf′ə-lak′səs) (Gr. *trophē,* food, + *allaxis,* barter, exchange). Exchange of food between young and adults, especially certain social insects.

trophi (trō′fī) (Gr. *trophos,* one who feeds). Jaw-like structures in the mastax of rotifers.

trophic (trō′fək) (Gr. *trophē,* food). Pertaining to feeding and nutrition.

trophoblast (trōf′ə-blast) (Gr. *trephein,* to nourish, + *blastos,* germ). Outer ectodermal nutritive layer of blastodermic vesicle; in mammals it is part of the chorion and attaches to the uterine wall.

trophosome (trof′ə-sōm) (Gr. *trophē,* food, + *sōma,* body). Organ in poganophorans bearing mutualistic bacteria, derived from midgut.

trophozoite (trōf′ə-zō′īt) (Gr. *trophē,* food, + *zōon,* animal). Adult stage in the life cycle of a protozoan in which it is actively absorbing nourishment.

tropic (trä′pic) (Gr. *tropē,* to turn toward). Related to the tropics (tropical); in endocrinology, a hormone that influences the action of another hormone or endocrine gland (usually pronounced trō′pic).

tropomyosin (trōp′ə-mī′ə-sən) (Gr. *tropos,* turn, + *mys,* muscle). Low-molecular weight, actin-binding protein surrounding the actin filaments of striated muscle. It works with troponin to regulate muscle contraction.

troponin (trə-pōn′in). Complex of globular, actin-binding proteins positioned at intervals along the actin filament of skeletal muscle; serves as a calcium-dependent switch in muscle contraction.

tube feet (podia) Numerous small, muscular, fluid-filled tubes projecting from body of echinoderms; part of water-vascular system; used in locomotion, clinging, food handling, and respiration.

tubercle (tū′bər-kəl) (L. *tuberculum,* small hump). Small protuberance, knob, or swelling.

tubulin (tū′bū-lən) (L. *tubulus,* small tube, + *in,* belonging to). Globular protein forming the hollow cylinder of microtubules.

tumor necrosis factor A cytokine, the most important source of which is macrophages, that is a major mediator of inflammation.

tumor suppressor gene A gene whose products act as restraints on cell division by triggering apoptosis, controlling transcription of other genes, restraining progression in phases of the cell cycle, or by other means.

tundra (tun′drə) (Russ. from Lapp, *tundar,* hill). Terrestrial habitat zone, located between taiga and polar regions; characterized by absence of trees, short growing season, and mostly frozen soil during much of the year.

tunic (L. *tunica,* tunic, coat). In tunicates, a cuticular, cellulose-containing covering of the body secreted by the underlying body wall.

tympanic (tim-pan′ik) (Gr. *tympanon,* drum). Relating to the tympanum that separates the outer and middle ear (eardrum).

type specimen A specimen deposited in a museum that formally defines the name of the species that it represents.

typhlosole (tif′lə-sōl′) (Gr. *typhlos,* blind, + *sōlēn,* channel, pipe). A longitudinal fold projecting into the intestine in certain invertebrates such as the earthworm.

typology (tī-päl′ə-jē) (L. *typus,* image). A classification of organisms in which members of a taxon are perceived to share intrinsic, essential properties, and variation among organisms is regarded as uninteresting and unimportant.

U

ulcer (ul-sər) (L. *ulcus,* ulcer). An abscess that opens through the skin or a mucous surface.

ultimate cause (L. *ultimatus,* last, + *causa*). The evolutionary factors responsible for the origin, state of being, or purpose of a biological system.

umbilical (L. *umbilicus,* navel). Refers to the navel, or umbilical cord.

umbo (um′bō), pl. **umbones** (əm-bō′nēz) (L. boss of a shield). One of the prominences on either side of the hinge region in a bivalve mollusc shell. Also, the "beak" of a brachiopod shell.

ungulate (un′gū-lət) (L. *ungula,* hoof). Hooved. Noun, any hooved mammal.

uniformitarianism (ū′nə-fōr′mə-ter′ē-ə-niz′əm). Methodological assumptions that the laws of chemistry and physics have remained constant throughout the history of the earth, and that past geological events occurred by processes that can be observed today.

ureter (ūr′ə-tər) (Gr. *ouētēr,* ureter). Duct carrying urine from kidney to bladder.

urethra (ū-rē′thrə) (Gr. *ourethra,* urethra). The tube from the urinary bladder to the exterior in both sexes.

uropod (ū′rə-pod) (Gr. *oura,* tail, + *pous, podos,* foot). Posteriormost appendage of many crustaceans.

utricle (ū′trə-kəl) (L. *utriculus,* little bag). That part of the inner ear containing the receptors for dynamic body balance; the semicircular canals lead from and to the utricle.

V

vacuole (vak′yə-wōl) (L. *vacuus,* empty, + Fr. *ole,* dim.). A membrane-bound, fluid-filled space in a cell.

valence (vā′ləns) (L. *valere,* to have power). Degree of combining power of an element as expressed by the number of atoms of hydrogen (or its equivalent) that the element can hold (if negative) or displace in a reaction (if positive). The oxidation state of an element in a compound. The number of electrons gained, shared, or lost by an atom when forming a bond with one or more other atoms.

valve (L. *valva,* leaf of a double door). One of the two shells of a typical bivalve mollusc or brachiopod. In cardiovascular and lymphatic systems, valves allow one-way flow of blood or lymph.

variation (L. *varius,* various). Differences among individuals of a group or species that cannot be ascribed to age, sex, or position in the life cycle.

vector (L. a bearer, carrier, from *vehere, vectum,* to carry). Any agent that carries and transmits pathogenic microorganisms from one host to another host. Also, in molecular biology, an agent such as bacteriophage or plasmid that carries recombinant DNA.

veins (vānz) (L. *vena,* a vein). Blood vessels that carry blood toward the heart; in insects, fine extensions of the tracheal system that support the wings.

velarium (və-la′rē-əm) (L. *velum,* veil, covering). Shelf-like extension of the subumbrella edge in cubozoans (phylum Cnidaria).

veliger (vēl′ə-jər) (L. *velum,* veil, covering). Larval form of certain molluscs; develops from the trochophore and has the beginning of a foot, mantle, shell, and so on.

velum (vē′ləm) (L. veil, covering). A membrane on the subumbrella surface of jellyfish of class Hydrozoa. Also, a ciliated swimming organ of the veliger larva.

ventral (ven′trəl) (L. *venter,* belly). Situated on the lower or abdominal surface.

venule (ven′ūl) (L. *venula,* dim. of *vena,* vein). Small vessel conducting blood from capillaries to vein; small vein of insect wing.

vermiform (ver′mə-form) (L. *vermis,* worm, + *forma,* shape). Adjective to describe any wormlike organism; an adult (nematogen) rhombozoan (phylum Mesozoa).

vestige (ves′tij) (L. *vestigium,* footprint). A rudimentary organ that may have been well developed in some ancestor or in the embryo.

vibrissa (vī-bris′ə), pl. **vibrissae** (L. nostril-hair). Stiff hairs that grow from the nostrils or other parts of the face of many mammals and that serve as tactile organs; "whiskers."

vicariance (vī-kar′ē-ənts) (L. *vicarius,* a substitute). Geographical separation of populations, especially as imposed by discontinuities in the physical environment that fragmented populations that were formerly geographically continuous.

villus (vil′əs), pl. **villi** (L. tuft of hair). A small fingerlike process on the wall of the small intestine that increases the surface area for absorption of digested nutrients. Also one of the branching, vascular processes on the embryonic portion of the placenta.

virus (vī′rəs) (L. slimy liquid, poison). A submicroscopic noncellular particle composed of a nucleoprotein core and a protein shell; parasitic; will grow and reproduce in a host cell.

viscera (vis′ər-ə) (L. pl. of *viscus,* internal organ). Internal organs in the body cavity.

visceral (vis′ər-əl). Pertaining to viscera.

vitalism (L. *vita,* life). The discredited viewpoint that natural processes are controlled by supernatural forces and cannot be explained through the laws of physics and chemistry alone, as opposed to mechanism.

vitamin (L. *vita,* life, + *amine,* from former supposed chemical origin). An organic substance required in small amounts for normal metabolic function; must be supplied in the diet or by intestinal flora because the organism cannot synthesize it. An exception is vitamin D_3, which is manufactured in the skin in the presence of sunlight.

vitellaria (vi′təl-lar′e-ə) (L. *vitellus,* yolk of an egg). Structures in many flatworms that produce vitelline cells, that is, cells that provide eggshell material and nutrient for the embryo.

vitelline gland See **vitellaria.**

vitelline membrane (və-tel′ən, vī′təl-ən) (L. *vitellus,* yolk of an egg). The noncellular membrane that encloses the egg cell.

viviparity (vī′və-par′ə-dē) (L. *vivus,* alive, + *parere,* to bring forth). Reproduction in which eggs develop within the female body, with nutritional aid of maternal parent as in therian mammals, many reptiles, and some fishes; offspring are born as juveniles. Adj., **viviparous** (vī-vip′ə-rəs).

W

water-vascular system System of fluid-filled closed tubes and ducts peculiar to echinoderms; used to move tentacles and tube feet that serve variously for clinging, food handling, locomotion, and respiration.

weir (wer) (Old English *wer,* a fence placed in a stream to catch fish). Interlocking extensions of a flame cell and a collecting tubule cell in some protonephridia.

X

xanthophore (zan′thə-for) (Gr. *xanthos,* yellow, + *pherein,* to bear). A chromatophore containing yellow pigment.

xenograft (zē′nə-graft). Graft of tissue from a species different from the recipient.

X-organ Neurosecretory organ in eyestalk of crustaceans that secretes molt-inhibiting hormone.

Y

Y-organ Gland in the antennal or maxillary segment of some crustaceans that secretes molting hormone.

Z

zoecium, zooecium (zō-ē′shē-əm) (Gr. *zōon,* animal, + *oikos,* house). Cuticular sheath or shell of Ectoprocta.

zoochlorella (zō′ə-klōr-el′ə) (Gr. *zōon,* life, + *Chlorella*). Any of various minute green algae (usually *Chlorella*) that live symbiotically within the cytoplasm of some protozoa and other invertebrates.

zooid (zō-id) (Gr. *zōon,* life). An individual member of a colony of animals, such as colonial cnidarians and ectoprocts.

zooxanthella (zo′ə-zan-thəl′ə) (Gr. *zōon,* animal, + *xanthos,* yellow). A minute dinoflagellate alga living in the tissues of many types of marine invertebrates.

zygote (Gr. *zygōtos,* yoked). The fertilized egg.

zygotic meiosis Meiosis that takes place within the first few divisions after zygote formation; thus all stages in the life cycle other than the zygote are haploid.

CREDITS

PHOTOS

Part Openers

1: Cleveland P. Hickman, Jr.; 2: © Tom Tietz/Stone Images/Getty; 3: Larry S. Roberts; 4: © Corbis; 5: Cleveland P. Hickman, Jr.

Chapter 1

Opener: Cleveland P. Hickman, Jr.; 1.1a: © Dave Fleetham/Visuals Unlimited; 1.1b: © Steve McCutcheon/Visuals Unlimited; 1.1c: © Peter Ziminski/Visuals Unlimited; 1.1d: © Link/Visuals Unlimited; 1.1d(inset): © T.E. Adams/Visuals Unlimited; 1.2a: Courtesy of IBM U.K. Scientific Centre; 1.3: © John D. Cunningham/Visuals Unlimited; 1.4: © David M. Phillips/Visuals Unlimited; 1.5a: © N.P. Salzman; 1.5b: © Ed Reschke; 1.5c: © Ken Highfill/Photo Researchers, Inc.; 1.5d bottom: Larry S. Roberts; 1.5d: © William C. Ober; 1.6a: © A. C. Barrington Brown/Photo Researchers, Inc.; 1.7a: © M. Abbey/Visuals Unlimited; 1.7b: © S. Dalton/National Audubon Society/Photo Researchers, Inc.; 1.8a, b: © D. Kline/Visuals Unlimited; 1.11a, b: © Michael Tweedie/Photo Researchers, Inc.; TA01: Foundations For Biomedical Research; 1.12: Courtesy American Museum of Natural History, Neg. #326669; 1.16a, b: Courtesy Gregor Mendel Museum, Bmo. Czechoslavakia; 1.18: Prèvost and Dumas; 1.19: © Carolina Biological Supply/Phototake

Chapter 2

Opener: Larry S. Roberts; 2.12: Courtesy Kevin Walsh, U.S.C.D.; TA 2.4: © G. I. Bernard/Animals Animals/Earth Scenes; 2.13: Courtesy R.M. Syren and S.W. Fox, Institute of Molecular Evolution/University of Miami, Coral Gables, Florida; 2.14: Cleveland P. Hickman, Jr.

Chapter 3

Opener: © Bill Ober; 3.1a: © John D. Cunningham/Visuals Unlimited; 3.1b: From C. R. Morgan and R. A. Jersild, Jr., 1970.Anat. Rec. 166:575-586; 3.5: Courtesy A. Wayne Vogl; 3.7: Courtesy of G. E. Palade, University of California School of Medicine; 3.8b: Courtesy Richard Rodewald; 3.9b, 3.11b: Courtesy of Charles Flickinger; 3.12: Courtesy A. Wayne Vogl; 3.13: © K.G. Murti/Visuals Unlimited; 3.14b: Courtesy Kent McDonald; 3.16: Courtesy Susumo Ito; 3.24: © Times Mirror Higher Education Group, Inc./Kingsley Stern, photographer

Chapter 4

Opener: © Gary W. Carter/Visuals Unlimited

Chapter 5

Opener: Larry S. Roberts; 5.1: Courtesy Gregor Mendel Museum, Brno, Czechoslavakia; 5.8a: © Peter J. Bryant/Biological Photo Service

Chapter 6

Opener: © John N. A. Lott/Biological Photo Service; 6.1a: Courtesy American Museum of Natural History, New York, Neg. # 32662; 6.1b, 6.2, 6.3: The Natural History Museum, London; 6.5a: © Bridgeman/Art Resource; 6.5b: © Stock Montage; 6.6, 6.7: Cleveland P. Hickman, Jr.; 6.8a: © Ken Lucas/Biological Photo Service; 6.8b: © A. J. Copley/Visuals Unlimited; 6.8c: © Roberta Hess Poinar; 6.8d: Courtesy G. O. Poinar, University of California at Berkeley; 6.9a: Courtesy W. Boehm; 6.10: Cleveland P. Hickman, Jr.; 6.14: Courtesy Library of Congress; 6.18: Courtesy M. K. Kelley, Courtesy of Harvard University Press; 6.22b: Cleveland P. Hickman, Jr.; 6.23: Courtesy of Storrs Agricultural Experiment Station, University of Connecticut at Storrs; 6.26: Fritz Goro; 6.28: © Timothy W. Ranson/Biological Photo Service; 6.29: © Krasemann/Photo Researchers, Inc.; 6.30b: Courtesy Dr. Robert K. Selander; 6.34: Courtesy of the Canada Center for Remote Sensing, Energy, Mines, and Resources, Canada

Chapter 7

Opener: © Francis Leroy, Biocosmos/SPL/Photo Researchers, Inc.; 7.3: © Robert Humbert/Biological Photo Service; 7.7: From R. G. Kessel and R. H. Kardon, Tissues and Organs: A Text-Atlas of Scanning Electron Microscopy, 1979, W. H. Freeman and Co.

Chapter 8

Opener: Marie A. Vodicka & John C. Gerhart/University of California, Berkeley; 8.5: Courtesy G. Schatten; 8.15: © F. R. Turner/Biological Photo Service

Chapter 9

Opener: Larry S. Roberts; 9.4 top, center, bottom, 9.5 top left, bottom, 9.6 top left, bottom left: © E. Reschke; 9.6 top right: Cleveland P. Hickman, Jr.; 9.6 bottom right: © E. Reschke; 9.7 top, center, bottom: © E. Reschke

Chapter 10

Opener: Cleveland P. Hickman, Jr.; 10.1: Courtesy Library of Congress; 10.3: © Kjell Sandved; 10.7: Courtesy American Museum of Natural History, Neg. #334101; 10.8a: © M. Coe/OSF/Animals Animals/Earth Scenes; 10.8b: © D. Allen/OSF/Animals Animals/Earth Scenes; 10.10: Courtesy of Dr. George W. Byers, University of Kansas

Chapter 11

Opener: © M. Abbey/Visuals Unlimited; 11.2: Courtesy L. Tetley; 11.3b: Courtesy Dr. Ian R. Gibbons; 11.5: © M. Abbey/Visuals Unlimited; 11.6a: Courtesy L. Evans Roth; 11.27a: © Manfred Kage/Peter Arnold; 11.27b: © A. M. Siegelman/Visuals Unlimited; 11.28: Courtesy J. and M. Cachon. From Lee, J.J., S. H. Hutner, and E. C. Bovee (editors). 1985. An Illustrated Guide to the Protozoa, Society of Protozoologists, Allen Press, Lawrence, KS

Chapter 12

Opener: Larry S. Roberts; 12.6, 12.8, 12.15a, b, c: Larry S. Roberts

Chapter 13

Opener: Larry S. Roberts; 13.1a: © William Ober; 13.5: © R. Harbo; 13.6: © Carolina Biological Supply/Phototake; 13.8: © Cabisco/Visuals Unlimited; 13.10a: © Larry S. Roberts; 13.11: © D. W. Gotshall; 13.14: © Peter Parks/OSF/Animals Animals/Earth Scenes; 13.15a, b: Larry S. Roberts; 13.16, 17: © R. Harbo; 13.19: Larry S. Roberts; 13.21: © D. W. Gotshall; 13.22a: © Jeff Rotman Photography; 13.22b: Larry S. Roberts; 13.24a, b: © R. Harbo; 13.25, 13.26a, b, c, 13.28: Larry S. Roberts; 13.29: © Mary Wicksten; 13.30a, 13.32, 13.33a, b, c: Larry S. Roberts; 13.34b: Cleveland P. Hickman, Jr.; 13.35a: © William Ober; 13.35b: © Kjell Sandved

Chapter 14

Opener: Larry S. Roberts; 14.2: ©John D. Cunningham/Visuals Unlimited, Inc.; 14.10, 14.11: Larry S. Roberts; 14.13a: H. Zaiman, M.D./From H. Zaiman A Pictorial Presentation of Parasites; 14.14: Courtesy A. W. Cheever./From H. Zaiman A Pictorial Presentation of Parasites; 14.15: © Arthur M. Seigelman/Visuals Unlimited; 14.20: © Cabisco/Visuals Unlimited; 14.21: Courtesy Ana Flisser; 14.23: © Stan Elems/Visuals Unlimited; 14.25: Cleveland P. Hickman, Jr.

Chapter 15

Opener: Courtesy D. Despommier/From H. Zaiman A Pictorial Presentation of Parasites; 15.5a: Frances M. Hickman; 15.5b, 15.6: H. Zaiman, M.D./From H. Zaiman A Pictorial Presentation of Parasites; 15.8: © Larry S. Roberts; 15.9a: © R. Calentine/Visuals Unlimited; 15.9b: H. Zaiman, M.D./From H. Zaiman A Pictorial Presentation of Parasites; 15.11: Contributed by E. L. Schiller, AFIP; 15.12: Courtesy Sharon Patton

Chapter 16

Opener: Larry S. Roberts; 16.1a: © Kjell B Sandved/Visuals Unlimited, Inc.; 16.1b, c: © R. Harbo; 16.1d: © D. W. Gothshall; 16.1e, 16.3b: Larry S. Roberts; 16.7: © Kjell Sandved/Visuals Unlimited, Inc.; 16.10: ©Daniel Gotshall/Visuals Unlimited, Inc.; 16.15a: © Gerald and Buff Corsi/Visuals Unlimited, Inc.; 16.15b: © David Wrobel /Visuals Unlimited, Inc.; 16.16a, b: © A. Kerstitch; 16.19a: © R. Harbo; 16.19b: © Tom Phillipp; 16.20a: © R. Harbo; 16.20b: Larry S. Roberts; 16.21a, b: Cleveland P. Hickman, Jr.; 16.22, 16.23a: Larry S. Roberts; 16.23b: Cleveland P. Hickman, Jr.; 16.24a: © R. Harbo; 16.24b: © D. P. Wilson/Frank Lane Picture Agency Ltd.; 16.25, 16.27a, b: Larry S. Roberts; 16.28a: © R. Harbo; 16.32: Larry S. Roberts; 16.35b: Richard J. Neves; 16.36a: Courtesy of M. Butschler, Vancouver Public Aquarium; 16.37: Larry S. Roberts; 16.38b: © Dave Fleetham/Tom Stack & Associates

Chapter 17

Opener: Photo gear, #CRAB 02.TIF; 17.2a, b: Larry S. Roberts; 17.3e: General Biological Supply; 17.8: © S. Elems/Visuals Unlimited; 17.9: Larry S. Roberts; 17.17g: © G. L. Twiest/Visuals Unlimited; 17.19: Photograph by T. Branning; 17.21: Cleveland P. Hickman, Jr.

Chapter 18

Opener: © A. J. Copley/Visuals Unlimited; 18.1a, b: © A. J. Copley/Visuals Unlimited; 18.3b: Cleveland Hickman; 18.6: © J. H. Gerard/Nature Press; 18.7: © J. H. Gerard/Nature Press; 18.8: © Todd Zimmerman/Natural History Museum of Los Angeles County; 18.9a, b: © J. H. Gerard/Nature Press; 18.10a: © Todd Zimmerman/Natural History Museum of Los Angeles County; 18.10b: Cleveland P. Hickman, Jr.; 18.11a, b: © J. H. Gerard/Nature Press; 18.12: From G. W. Wharton, "Mites and commercial extracts of house dust, " in Science 167:1382-1383. Copyright © 1970 by AAAS; 18.13: Larry S. Roberts; 18.14: © D. S. Snyder/Visuals Unlimited; 18.15: © A. Siegelman/Visuals Unlimited; 18.16: © John D. Cunningham/Visuals Unlimited

Chapter 19

Opener: © T. E. Adams/Visuals Unlimited; 19.19a: © William C. Ober; 19.19b: © R. Harbo; 19.21a: Cleveland P. Hickman, Jr.; 19.22: Larry S. Roberts; 19.23a: © R. Harbo; 19.23b, c: © Kjell Sandved; 19.24a, b: © Larry S. Roberts; 19.26a: © Cleveland P. Hickman, Jr; 19.26b: © R. Harbo; 19.26c: © Cleveland P. Hickman, Jr; 19.26d, e, 19.27: Larry S. Roberts

Chapter 20

20.1a, 20.2a: © James L. Castner; 20.3b: © Dan Kline/Visuals Unlimited; 20.4b: © Larry S. Roberts; 20.7a, b: © Ron West/Nature Photography; 20.9a, b: © Kjell Sandved; 20.10: Cleveland P. Hickman, Jr.; 20.11: © J. H. Gerard/Nature Press; 20.12: © James Castner; 20.15: © John D. Cunningham/Visuals Unlimited; 20.16: Courtesy Jay Georgi; 20.17: © James L. Castner; 20.18a: Cleveland P. Hickman, Jr.; 20.18b: © J. H. Gerard/Nature Press; 20.23a: Cleveland P. Hickman, Jr.; 20.23b: Cleveland P. Hickman, Jr.; 20.25a, b: © Robert Brons/Biological Photo Service; 20.26a, b: © James L. Castner 20.28a: Cleveland P. Hickman, Jr.; 20.28b: © J. H. Gerard/Nature Press; 20.28c: © Carolina Biological Supply/Phototake; 20.29a, b: © J. H. Gerard/Nature Press; 20.30a, b, c: © Kjell Sandved; 20.31: © J. H. Gerard/Nature Press; 20.32: Courtesy J. E. Lloyd; 20.33: © K. Lorenzen /Andromeda /Educational Images; 20.34a: © J. H. Gerard/Nature Press; 20.34b, 20.35a: © James L. Castner; 20.35b: Larry S. Roberts; 20.36a, b, c: © James Castner; 20.37a: © L. L. Rue, III; 20.37b: © James L. Castner; 20.37c: © J. H. Gerard/Nature Press; 20.38, 20.39, 20.41, 20.42: © James Castner; 20.43a, b: © Kjell Sandved; 20.43c: Cleveland P. Hickman, Jr.; 20.43d: © Kjell Sandved

Chapter 21

Opener: Larry S. Roberts; 21.6: Courtesy J.F. Grassle/Woods Hole Oceanographic Institution; 21.11a, b: Larry S. Roberts; 21.12: © Ken Lucas/Biological Photo Service; 21.13a, b: © Robert Brons/Biological Photo Service; 21.14: Larry S. Roberts; 21.18: Courtesy J. Ubelaker; 21.19b: © James L. Castner; 21.21, 21.23: Courtesy D. R. Nelson; 21.24: From R. M. Sayre, Trans. Am. Microsc.88:266-274, 1969. 21.25b: Thuesen, E. V., and R. Bieri, 1987

Chapter 22

Opener: © Ken Lucas/Visuals Unlimited; 22.1a, b, c, d: Larry S. Roberts; 22.4f, 22.5a: © R. Harbo; 22.5b: © D. W. Gotshall; 22.7: Larry S. Roberts; 22.10a: © Rick Harbo; 22.10b: Larry S. Roberts; 22.13a, b, 22.14: © R. Harbo; 22.15a: © A. Kerstitch; 22.15b, c: © R. Harbo; 22.15d: © W. C. Ober; 22.15e: © Kjell Sandved; 22.16a, b: © A. Kerstitch; 22.17a, b: Larry S. Roberts; 22.20a, b, c, 22.23a, b: © R. Harbo; 22.23c, 22.25: Larry S. Roberts

Chapter 23

Opener: © Heather Angel; 23.4: Courtesy of R.P.S. Jeffries, The Natural History Museum, London; 23.6a, b, 23.8: Larry S. Roberts; 23.10a: Cleveland P. Hickman Jr.

Chapter 24

Opener: © Jonathan Green; 24.4: © Berthoule-Scott/Jacana/Photo Researchers; 24.9: © Jeff Rotman Photography; 24.12a: © William Ober; 24.12b: © Jeff Rotman Photography; 24.20a, b: Courtesy John G. Shedd Aquarium/ Patrice Ceisel; 24.21a: © James D. Watt/Animals, Animals; 24.21b: © Biological Photo Service; 24.21c: © Jeff Rotman Photography; 24.21d: © Fred McConnaughey/Photo Researchers; 24.30: © D. W. Gotshall; 24.31: © Mary Beth Angelo/Photo Researchers, Inc.; 24.33: © Will Troyer/Visuals Unlimited; 24.34: © D. W. Gotshall; 24.35: Courtesy of F. McConnaughey

Chapter 25

Opener: Cleveland P. Hickman, Jr.; 25.6a, b: Courtesy L. Houck; 25.9: Cleveland P. Hickman, Jr.; 25.11: Allan Larson; 25.12a: © Ken Lucas/Biological Photo Service; 25.12b, 25.13: Cleveland P. Hickman, Jr.; 25.14: Courtesy American Museum of Natural History, Neg. #125617; 25.15, 25.18, 25.25: Cleveland P. Hickman, Jr.

Chapter 26

Opener: Courtesy of Ron Magill/Miami Metrozoo; 26.7, 26.8: Cleveland P. Hickman, Jr.; 26.9: Jonathan Green; 26.10: © OSF LTD.; 26.12: © John Mitchell/Photo Researchers, Inc.; 26.13: Cleveland P. Hickman, Jr.; 26.14: © Stephen Dalton/Photo Researchers, Inc; 26.15, 26.16: © L. L. Rue, III; 26.18b: © Austin J. Stevens/ Animals, Animals/ Earth Scenes; 26.19: Cleveland P. Hickman, Jr.; 26.20: © Joe McDonald/Visuals Unlimited; 26.22: Cleveland P. Hickman, Jr.; 26.23: © Renee Lynn/Stone Images/Getty; 26.26: © Zig Leszczynski/Animals, Animals; 26.27a, b: Cleveland P. Hickman, Jr.

Chapter 27

Opener: William J. Weber/Visuals Unlimited; 27.1a: Courtesy American Museum of Natural History, Neg. #125065; 27.4: Cleveland P. Hickman, Jr.; 27.6: © CORBIS; 27.22: © D. Poe/Visuals Unlimited; 27.23a, b: © L.L. Rue, III; 27.26: © John Gerland/Visuals Unlimited; 27.27: © Richard R. Hansen/ Photo Researchers, Inc.; 27.29a: © L. L. Rue, III; 27.30: Courtesy of Culver Pictures; 27.31, 27.32, 27.33, 27.34: Cleveland P. Hickman, Jr.

Chapter 28

Opener: © L. L. Rue, III; 28.6: © L. L. Rue, III; 28.7a: © PhotoDisc; 28.7b: © CORBIS; 28.8: R. E. Treat; 28.12: © L. L. Rue, III; 28.13, 28.15: © Gerlach/Visuals Unlimited; 28.16a: Cleveland P. Hickman, Jr.; 28.18: S. Malowski/Visuals Unlimited; 28.20: © Kjell Sandved/Visuals Unlimited; 28.23: © L. L. Rue, III; 28.25: © M. H. Tierney, Jr./Visuals Unlimited; 28.26: © G. Herben/Visuals Unlimited; 28.28: Cleveland P. Hickman, Jr.; 28.29: © L. L. Rue, III; 28.30: Cleveland P. Hickman, Jr.; 28.31a: Courtesy of Zoological Society of San Diego; 28.31b: © Timothy Ransom/Biological Photo Services; 28.32: © Milton H. Tierney, Jr./Visuals Unlimited; 28.33: © John Reader; 28.36, 28.37, 28.38, 28.39: Cleveland P. Hickman, Jr.; 28.41: © William Ober

Chapter 29

Opener: © Stephen Dalton/Photo Researchers, Inc.; 29.11: Courtesy Dr. Ian R. Gibbons; 29.13a: © G. W. Willis, M. D./Biological Photo Service; 29.13b: © E. Reschke; 29.13c: © G. W. Willis M. D./Biological Photo Service

Chapter 30

Opener: Cleveland P. Hickman, Jr.; 30.1: From J. F. Fulton and L. G. Wilson, Selected Readings in the History of Physiology, 1966. Courtesy of Charles C. Thomas, Publisher, Springfield, Illinois; 30.11: From R. G. Kessel and R. H. Kardon, Tissues and Organs: A Text- Atlas of Scanning Electron Microscopy, 1979 W. H. Freeman and Co.; 30.21: © L. L. Rue, III

Chapter 31

Opener: © David M. Phillips/Visuals Unlimited; 31.2: From J. F. Fulton and L. G. Wilson, Selected Readings in the History of Physiology, 1966. Courtesy of Charles C. Thomas, Publisher, Springfield, Illinois; 31.4a, b: Courtesy P. P. C. Graziadei; 31.5: © David M. Phillips/Visuals Unlimited

Chapter 32

Opener: Cleveland P. Hickman, Jr.; 32.3a, b: Courtesy Carl Gans; 32.6: Cleveland P. Hickman, Jr.; 32.11: Courtesy of Wyeth-Ayerst Laboratories; 32.12: From R. G. Kessel and R. H. Kardon, Tissues and Organs: A Text-Atlas of Scanning Electron Microscopy, 1979 W. H. Freeman and Co.; 32.13b: Courtesy J. D. Berlin; 32.17: Hospital Tribune 8:1, 1974

Chapter 33

Opener: © D. H. Ellis/Visuals Unlimited

Chapter 34

Opener: © Ed Reschke; 34.1a, b: From J. F. Fulton and L. G. Wilson, Selected Readings in the History of Physiology, 1966. Courtesy of Charles C. Thomas, Publisher, Springfield, IL.; 34.10, 34.16: From J. A. Prior, et al., Physical Diagnosis, 1981 Mosby-Year Book, Inc.

Chapter 35

Opener: © Dr. F.G. Skvara/Peter Arnold, Inc; 35.7: Courtesy of H. Zaiman, M.D.; 35.8: © SUI/Visuals Unlimited; 35.9: From Van der Knapp, W. P. W., and E. S. Loker, "Immune mechanisms in trematode-snail interactions, " Parasit. Today, 6:175-182, 1990.

Chapter 36

Opener: Cleveland P. Hickman, Jr.; 36.1a: Thomas McAvoy/ Life Magazine © 1995. Time Inc.; 36.1b, c: Courtesy of W. S. Hoar; 36.7: Cleveland P. Hickman, Jr.; 36.9: Nina Leen/Life Magazine © Time Inc.; 36.11, 36.12: Cleveland P. Hickman, Jr.; 36.14: © L. L. Rue, III; 36.15: © CORBIS; 36.17: Cleveland P. Hickman, Jr.; 36.18: © Tom McHugh/Photo Researchers, Inc.; 36.19: © Ray Richardson/Animals Animals; 36.21: © Richard R. Hansen/ Photo Researchers, Inc.

Chapter 37

Opener: NASA; 37.7: Cleveland P. Hickman, Jr.; 37.9: © F. Gohier/Photo Researchers, Inc.; 37.10: Cleveland P. Hickman, Jr.

Chapter 38

Opener: Cleveland P. Hickman, Jr.; 38.6: © Noble Proctor/Photo Researchers, Inc.; 38.7a, b: Cleveland P. Hickman, Jr.; 38.11a, b, c: © James L. Castner; TA-38-1: © D. Foster/WHOI/Visuals Unlimited

LINE ART AND TEXT

Chapter 1

Figure 1-11: Source: After P. M. Brakefield, Industrial melanism: Do we have the answers?, *Trends in Ecology and Evolution* 2:117-122, 1987. Figure 1-13: Source: From S. Gould, *Ontogeny and Phylogeny.* Harvard University Press, 1977. Figure 1-14: Source: After W. Bock, *Evolution.* 24:704-122, 1970.

Chapter 2

p. 31: "When water freezes at 0 C . . ." From Peter H. Raven and George B. Johnson, *Biology,* 4e. Copyright © 1996 Times Mirror Higher Education Group, Inc., Dubuque, Iowa. Reprinted by permission. All rights reserved. p. 31: "When a crystal of sodium chloride dissolves . . ." From Peter H. Raven and George B. Johnson, *Understanding Biology,* 3e. Copyright © 1995 Times Mirror Higher Education Group, Inc., Dubuque, Iowa. Reprinted by permission. All rights reserved. Figure 2-11: From Peter H. Raven and George B. Johnson, *Understanding Biology,* 3e. Copyright © 1995 Times Mirror Higher Education Group, Inc., Dubuque, Iowa.

Chapter 14

Figure 14-5: After L. T. Threadgold in L. S. Roberts and J. J. Janovy, Jr., *Foundations of Parasitology,* 6th ed. McGraw-Hill Higher Education, Dubuque, Iowa, 2000. Figure 14-16: Source: From J. F. Mueller and H. J. Van Cleave, *Roosevelt Wildlife Annals,* 1932. Figure 14-17: Source: After D. J. Morseth in *Journal of Parasitology* 53:492-500, 1967. Figure 14-27: From W. E. Sterrer, Systematics and evolution within the Gnathostomulida, *System. Zool.* 21: 151, 1971. Reprinted by permission.

Chapter 15

Text, p. 295: "If all the matter in the nematode parasites." Source: From N. A. Cobb, *Yearbook of the United States Department of Agriculture,* 1914, p. 472. Figure 15-15: Source: After R. M. Kristensen, Loricifera, a new phylum with Aschelminthes characters from the meiobenthos, *Zeitsch. Zool. Syst. Evol.* 21:163, 1983. Figure 15-22: From *Synopsis and Classification of Living Organisms,* edited by S. P. Parker. Copyright © 1982 McGraw-Hill, Inc. Reprinted by permission. Figure 15-23A: Source: After C. Cori, Kamptozoa in *Klassen und Ordnungen des Tier-Reichs,* vol. 4, part 2, edited by H. G. Bronn. Akademische Verlagsgesselschaft, Leipzig, 1936.

Chapter 18

Figure 18-3a: From *Synopsis and Classification of Living Organisms,* edited by S. P. Parker. Copyright © 1982 McGraw-Hill, Inc. Reprinted by permission.

Chapter 19

Figure 19-8: Source: After G. B. Moment, *General Zoology.* Houghton Mifflin, Boston, 1967. Figure 19-17: Source: After G. A. Boxshall and R. J. Lincoln, *J. Crust. Biol.* 3:1-16, 1983.

Chapter 20

Text, p. 397: ("Tunis, Algeria—Treating is as an invading army . . .") Source: From the *New York Times,* 20 April 1988. Figure 20-19b, d, e: From Peter H. Raven and George B. Johnson, *Biology,* 4e. Copyright © 1996 Times Mirror Higher Education Group, Inc., Dubuque, Iowa. Reprinted by permission. All rights reserved. Figure 20-22: From Peter H. Raven and George B. Johnson, *Biology,* 4e. Copyright © 1996 Times Mirror Higher Education Group, Inc., Dubuque, Iowa. Reprinted by permission. All rights reserved. Figure 20-45: Redrawn from J. Kukalová-Peck in R. A. Raff, *The Shape of Life,* University of Chicago Press, Chicago, 1996.

Chapter 22

Figure 22-4a–e: Source: Courtesy of Tim Doyle. Figure 22-8: Source: After W. D. Russell-Hunter, *A Biology of Higher Invertebrates.* Macmillan, New York, 1969. Figure 22-26: Source: After A. N. Baker, F. W. E. Row, and H. E. S. Clark. A new class of Echinodermata from New Zealand, *Nature* 321:862-864, 1986.

Chapter 23

Text, p. 467: "IT'S A LONG WAY FROM AMPHIOXUS" © Alpha Music Inc. All Rights Reserved. Used by permission. Figure 23-11: Source: After S. J. Gould, *Wonderful Life.* W. W. Norton, New York, 1989. Figure 23-17: Source: After R. Zangerl and M. E. Williams, *Paleontology* 18:333-341, 1975.

Chapter 24

Figure 24.3b: Source: After R. Conniff, *Audubon,* March 1991. Figure 24-3c: Source: After F. H. Pough et al., *Vertebrate Life,* Macmillan, 1989; and D. Jensen, *Sci. Am.* 214(2):82-90, 1966. Figure 24-3d: Source: After D. Jensen, *Sci. Am.* 214(2):82-90, 1966. Figure 24.23: From Peter Castro and Michael E. Huber, *Marine Biology.* Copyright © 1992 Mosby-Year Book. Reprinted by permission of Times Mirror Higher Education Group, Inc., Dubuque, Iowa. All rights reserved.

Chapter 25

Figure 25-5: Source: After W. E. Duellman and L. Trueb, *Biology of Amphibians.* McGraw-Hill, New York, 1986. Figure 25-20: Source: After M. S. Gordon et al., *Animal Function: Principles and Adaptations.* Macmillan, New York, 1968.

Chapter 26

Figure 26-11: Source: After R. M. Alexander, *The Chordates.* Cambridge University Press, England, 1975.

Chapter 27

Text, p.564: "When an eagle grips in earnest . . . talons in such a clutch." Source: From L. Brown, *Eagles.* Arco Publishing, New York, 1970. Figure 27-1: From Peter H. Raven and George B. Johnson, *Understanding Biology,* 3d ed. Copyright © 1995 Times Mirror Higher Education Group, Inc., Dubuque, Iowa. Reprinted by permission. All rights reserved. Figure 27-7b: Source: After P. Wellenhofer, Archaeopteryx, *Sci. Am.* 262:70-77, May 1990. Figure 27-12b: Source: After K. Schmidt-Nielsen, *Animal Physiology,* 4th ed. Cambridge University Press. Figure 27-29b: Source: After S. R. Johnson and I. T. McCowan, Thermal adaptation as a factor affecting colonizing success of introduced Sturnidae (Aves) in North America, *Canadian Journal of Zoology* 52:1559–1576, 1974.

Chapter 28

Figure 28-4: Source: After J. Z. Young, *The Life of Mammals.* Oxford University Press, Oxford, 1975. Figure 28-11: Source: After E. Rogers, *Looking at Vertebrates.* Longman Group, Essex, England, 1986. Figure 28-19: Source: After N. Suga, Biosonar and neural computation in bats, *Sci. Am.* 262:60-68, June 1990. Figure 28-21: Source: After J. A. Lillegraven et al., The origin of eutherian mammals, *Biol. Jour. Linn. Soc.* 32:281-336, 1987. Figure 28-22: Source: After C. R. Austin and R. V. Short, editors, *Reproduction in mammals: Volume 4, Reproductive patterns.* Cambridge University Press, New York, 1972.

Chapter 29

Figure 29-7: From Kent M. Van De Graaff and Stuart Ira Fox, *Concepts of Human Anatomy & Physiology,* 4e. Copyright © 1995 Times Mirror Higher Education Group, Inc., Dubuque, Iowa. Reprinted by permission. All rights reserved. Figure 29-9: From Kent M. Van De Graaff and Stuart Ira Fox, *Concepts of Human Anatomy & Physiology,* 4e. Copyright © 1995 Times Mirror Higher Education Group, Inc., Dubuque, Iowa. Reprinted by permission. All rights reserved. Figure 29-10: Source: After A. A. Biewener, Mammalian terrestrial locomotion and size, *BioScience* 39(11): 776-783, 1989; and R. M. Alexander, How dinosaurs ran, *Sci. Am.* 264: 130-136, April 1991. Figure 29-12a: Source: After M. A. Sleigh, *The Biology of Cilia and Flagella.* Pergamon Press, Oxford, 1962. Figure 29-12b: Source: After M. A. Sleigh and D. I. Barlow, Metachronism and control of locomotion in animals with many propulsive structures, *Aspects of Animal Movement* ed. by H.Y. Elder and E. R. Trueman, Cambridge University Press, 1980.

Chapter 30

Figure 30-3: Source: After D. Webster and M. Webster, *Comparative Vertebrate Morphology.* Academic Press, New York, 1974. Figure 30-4: Source: After D. Webster and M. Webster, *Comparative Vertebrate Morphology.* Academic Press, New York, 1974. Figure 30-13: Source: After R. F. Pitts, *Physiology of the Kidney and Body Fluids,* 3e. Mosby-Year Book, St. Louis, 1974. Figure 30-15, *left:* Source: After H. Wirz, B. Hargitay, and W. Kuhn, *Helv. Physiol. Acta* 9:196-207, 1951. Figure 30-15, *right:* Source: After K. J. Ullrich and K. H. Jarausch, *Pflugers Archiv.* 262:537-550, 1956. Figure 30-20: Source: After R. C. Lasiewski, *Physiol. Zool.* 36:122-140, 1963.

Chapter 31

Figure 31-23: Source: After M. S. Gordon et al., *Animal Function: Principles and Adaptations*. Macmillan, New York, 1968.

Chapter 32

Text, p. 678: "A humpback whale...for a few hours at most." Source: From N. J. Berrill, *You and the Universe*. Dodd, Mead & Co., New York, 1958. Figure 32-10: From Peter H. Raven and George B. Johnson, *Biology*, 4e. Copyright © 1996 Times Mirror Higher Education Group, Inc., Dubuque, Iowa. Reprinted by permission. All rights reserved. Figure 32-16: Source: After M. Winick, *Malnutrition and Brain Development*. Oxford University Press, New York, 1976.

Chapter 33

Figure 33-11: From Peter H. Raven and George B. Johnson, *Biology*, 3e. Copyright © 1992 Mosby-Year Book. Reprinted by permission of Times Mirror Higher Education Group, Inc., Dubuque, Iowa. All rights reserved. Figure 33-14: From Kent M. Van De Graaff and Stuart Ira Fox, *Concepts of Human Anatomy & Physiology*, 4e. Copyright © 1995 Times Mirror Higher Education Group, Inc., Dubuque, Iowa. Reprinted by permission. All rights reserved. Figure 33-15: From Peter H. Raven and George B. Johnson, *Understanding Biology*, 3e. Copyright © 1995 Times Mirror Higher Education Group, Inc., Dubuque, Iowa. Reprinted by permission. All rights reserved. Figure 33-17: From Peter H. Raven and George B. Johnson, *Biology*, 4e. Copyright © 1996 The McGraw-Hill Companies, New York. Reprinted by permission. All rights reserved. Figure 33-30: Source: After F. Lenci and G. Colombetti, *Photoreception and Sensory Transduction in Aneural Organisms*. Plenum Press, New York, 1980. Figure 33-32: From Peter H. Raven and George B. Johnson, *Biology*, 4e. Copyright © 1996 Times Mirror Higher Education Group, Inc., Dubuque, Iowa. Reprinted by permission. All rights reserved. Figure 33-33: From Peter H. Raven and George B. Johnson, *Understanding Biology*, 2e. Copyright © 1991 Mosby-Year Book. Reprinted by permission of Times Mirror Higher Education Group, Inc., Dubuque, Iowa. All rights reserved. Figure 33-34: From Peter H. Raven and George B. Johnson, *Understanding Biology*, 2e. Copyright © 1991 Mosby-Year Book. Reprinted by permission of Times Mirror Higher Education Group, Inc., Dubuque, Iowa. All rights reserved.

Chapter 34

Figure 34-9: Source: After P. J. Bentley, *Comparative Vertebrate Endocrinology*, 2d ed. Cambridge University Press, 1982. Figure 34-11: Source: After D. J. Copp, *J. Endocrinol.* 43:137-161, 1969.

Chapter 35

Figure 35-2a: From Peter H. Raven and George B. Johnson, *Understanding Biology*, 3e. Copyright © 1995 Times Mirror Higher Education Group, Inc., Dubuque, Iowa. Reprinted by permission. All rights reserved. Figure 35-2b: From Peter H. Raven and George B. Johnson, *Biology*, 4e. Copyright © 1996 Times Mirror Higher Education Group, Inc., Dubuque, Iowa. Reprinted by permission. All rights reserved.

Chapter 36

Figure 36-2: Source: After K. Lorenz and N. Tinbergen, *Zeit. Tierpsychol.* 2:1-29, 1938. Figure 36-4: Two models of the English robin . . . From N. Tinbergen, *The study of instinct*, Oxford University Press, Oxford, England, 1951; modified from D. Lack, *The life of the robin,* H. F. & G. Witherby Ltd., London, England, 1943. Reprinted by permission of Oxford University Press. Figure 36-5: Source: After N. Rothenbuhler, Behavior genetics of next cleaning honey bees. IV. Responses of F1 and backcross generations to disease-killed brood, *Am. Zool.* 4:111-123, 1964. Figure 36-6: Source: After W. C. Dilger, The behavior of lovebirds, *Sci. Am.* 206:89-98, January 1962. Figure 36-10: Source: After J. Alcock, *Animal Behavior: An Evolutionary Approach,* 3e., Sinauer Associates, Sunderland, MA, 1984, from a photography by Masakasu Konishi. Figure 36-16: Source: From C. Darwin, *Expression of the Emotions in Man and Animals*. Appleton and Co., New York, 1872. Text, p. 768: "One day on the savanna . . . leap inside and close the door." Source: From Irven DeVore, *The Marvels of Animal Behavior*. National Geographic Society, Washington, DC, 1972.

Chapter 37

Figure 37-2: From Peter Castro and Michael E. Huber, *Marine Biology,* Copyright © 1992 Mosby-Year Book. Reprinted by permission of Times Mirror Higher Education Group, Inc., Dubuque, Iowa. All rights reserved. Figure 37-3: With permission from *Natural History,* March 1990, Copyright the American Museum of Natural History. Figure 37-6: From Peter Castro and Michael E. Huber, *Marine Biology*. 3e. Copyright © 1999 The McGraw-Hill Companies. Reprinted by permission. Figure 37-11: From Peter Castro and Michael E. Huber, *Marine Biology*. Copyright © 1992 Mosby-Year Book. Reprinted by permission of Times Mirror Higher Education Group, Inc., Dubuque, Iowa. All rights reserved. Figure 37-12: From Peter Castro and Michael E. Huber, *Marine Biology*. Copyright © 1992 Mosby-Year Book. Reprinted by permission of Times Mirror Higher Education Group, Inc., Dubuque, Iowa. All rights reserved. Figure 37-13a: Source: After W. H. Burt, *Zoogeography* ed. by C. L. Hubbs. AAAS Pub. No. 51, Washington, D.C., 1958. Figure 37-13b: Source: W. F. Blair, *Zoogeography* ed. by C. L. Hubbs, AAAS Pub. No. 51, Washington, D.C., 1958. Figure 37-14: Source: After J. Cracraft, *Ibis* 116:294-521, 1974. Figure 37-15: Adapted from "The breakup of Pangaea" by Robert S. Dietz and John C. Holden. Copyright © October 1970 by Scientific American, Inc. All rights reserved. Adapted by permission. Figure 37-16: Drawings by Marlene Hill Werner, reprinted courtesy of Larry G. Marshall.

Chapter 38

Figure 38-3: Source: Data from E. Bos, et al. 1994. *World Population Projections 1994-95*. Baltimore, Johns Hopkins University Press for the World Bank. Figure 38-8: Source: Data from D. Lack, *Darwin's Finches*. Cambridge University Press, 1947. Figure 38-10: Source: Data from G. F. Gause, 1934. *The Struggle for Existence*. New York, Williams and Wilkins. Figure 38-12: From Peter Castro and Michael E. Huber, *Marine Biology*. Copyright © 1992 Mosby-Year Book. Reprinted by permission of Times Mirror Higher Education Group, Inc., Dubuque, Iowa. All rights reserved. Figure 38-14: Source: Data from R. L. Smith, *Biology and Field Biology,* 3e. Harper and Row. New York, 1980; and E. P. Odum, *Fundamentals of Ecology,* 3e., W. B. Saunders, Philadelphia, 1971.

INDEX

Note: **Boldface** page numbers refer to figures, illustrations, or tables. *Italic* page numbers refer to chapter opener illustrations.

Aardvark, 608
Abalones, 320, 325, **326,** 328
Abdomen
   of insects, 401
   of spiders, 369
ABO blood types, 747, **747**
Abomasum, 592
Aboral end, 250
Aboral sense organ, 268
Aboral system, 448
Abortion, spontaneous, 146
Abscess, 738, 747
Absorption, in digestion, 678, 688
Absorption spectrum, for color vision, 718, **718**
Acanthamoeba palestinensis, **209**
Acanthaster planci, **444,** 447
Acanthobdella, 360
Acanthocephala, 309–10, **310,** 312
Acanthodians, 483, **484,** 489
Acanthor, 310
Acanthostega, 516, **517**
Acari, **372,** 372–74
Accessory chromosomes, 78
Accessory heart, 337
Accessory molecules, 743
Accessory sex characteristics, 144
Accessory sex organs, 134, 141
Acetyl coenzyme A, 62
   fatty acids and, 68
   formation of, **64**
   oxidation of, 64
Acetylcholine, 628, 699
Acetylcholinesterase, 701
Acicula, 350
Acid stomach, 685
Acinar cells, **734**
Acinetopsis, 226
Acipenser, 510
Acipenser oxyrhynchus, **500**
Acoela, 280, **280**
Acoelomates, **187**
   Bilateral. *See* Bilateria
   body plan of, **186, 296**
Acontia threads, 261
Acorn barnacles, **387**
Acorn worms, **461,** 461–62, **462**
Acquired characteristics, inheritance of, 103
Acquired immune deficiency syndrome (AIDS), 747
   animal research and, 12–13
   apoptosis and, 53
   coccidiosis in, 220
   toxoplasmosis in, 220
Acquired immunity, 739, 741–47
Acraniata, 469
Acron, **395**
Acropora, 267
Acrosome, 139
Actin, 42, 622, 626
Actinophrys, 228, **229,** 231
Actinopoda, 211, 228, 231
Actinopterygii, 482, 489, 497–500, **500,** 510
Actinosphaerium, 228
Actinosphaerium nucleofilum, **212**
Action potential, 697, **698**
Actitis macularia, 575
Activated lymphocytes, 743
Activation energy, 58
Activation, fertilization and, 153–55

Active transport, 47, 636, 643, 688
Actomyosin system, 622
Adaptation, 15, **16,** 118–20
Adaptive radiation, 115, **116**
   of Annelida, 362
   of Arthropoda, 375
   of Bilateria, 292
   of Cnidaria, 271
   of Crustacea, 394
   of Ctenophora, 271
   of Echinodermata, 460
   of Hemichordata, 464
   of Insecta, 420–24
   of Mollusca, 340–42
   of Porifera, 244
   of protozoa, 230
   of pseudocoelomates, 314
   of Reptilia, 535–38
Adaptive zone, 197
ADCC. *See* Antibody-dependent cell-mediated cytotoxicity
Adductor muscles, 625
Adenine, 89, **89,** 90
   hydrogen bond of, 90, **90**
Adenohypophysis, 725, **727**
Adenosine triphosphate (ATP)
   anaerobic glycolysis and, 65–66
   chemical energy transfer by, 60–61
   formation of, 60–61, **61**
   locomotion and, 622
   muscle contraction and, 630
   oxidative phosphorylation and, 64–65, **66**
   protozoa and, 211
   space-filling model of, **61**
   structure of, 60, **61**
   total generation of, **68**
Adhesive cells, 268
Adhesive sac, 433
Adipose tissue, 181
   brown, 689
   white, 689
Adrenal glands, 732–34, **733**
Adrenal steroid hormones, 733–34
Adrenaline, 734
Adrenocorticotropic hormone, 726, **727**
Aeolidae, 328
Aeolosoma, 357, **358,** 360
Aequipecten, 334
Aerial respiration, 667
Aerobes, 62
Aerobic metabolism, 31–32, 62, 630
Afferent arteriole, 642
Afferent division, of peripheral nervous system, 707
Afferent neurons, 696, 702–3
African clawed frogs, 525, **526**
African house snake, **549**
African lungfish, 500, **501**
African sleeping sickness, 218, 417
Afterbirth, 146
Aftershaft, **561**
Agapornis, 756, **756**
Age structure, 791, **792**
Agglutination, 747
Aggression, 119, 760–62, **761**
Agnatha, 468–69, 482, 487, 489–93, 510
Agonistic behavior, 760
Agranulocytes, 658
Agriculture, molecular genetics and, 96
Agriolimax, 329
Agui, N., 724
AIDS. *See* Acquired immune deficiency syndrome
Air, buoyant density of, 515
Air capillaries, 668

Air sacs, 407, 566, 668
Alarm-calling behavior, 764
Alaska, **778,** 789
Albatross, 579
Albinism, 121–22, 588
Albumin, 574, 658
Alciopidae, 350
Alcoholic fermentation, 65
Alcyonaria. *See* Octocorallia
Alcyonium, 267
Aldosterone, 644, **733,** 733–34
Algae, 230
   coralline, 264
   green, 216, 229
   oxygen produced by, 31–32
   suctorians in, 226
Alimentary canal, 170, 683, **683**
   absorption region of, 685–88
   concentration of solids and, 688
   conduction and storage region of, 684
   derivatives, 170, **171**
   grinding region of, 684–85
   of insects, **683**
   mammalian, **687**
   receiving region of, 683–84
   terminal digestion and, 685–88
   of vertebrates, **683,** 686, **687**
   water absorption and, 688
Alimentary system, 682
Alkaline phosphatase, 686
Allantois, 166
Allantosoma, 226
Alleles, 76
   multiple, 83
   rare, 121–22
Allelic frequency, 120–21, **121**
Alligator, 550–52, **552**
   brain of, **704**
   sex determination in, 137
Alligator mississipiensis, 552, **552**
Alligator snapping turtle, **542**
Allografts, 167, 748
Allopatric speciation, 113–14
Allopauropus, 399
Allosaurus, 546–47
Alouatta, 606
Alouatta seniculus, **602**
Alpha cells, 735
Alpha-helix structure, of proteins, 25, **26**
Alpheus, 386
Alternative pathway, 739
Altricial young, 575, **576**
Altruistic behavior, 763–65
Alula, 569
Alvarez, Walter, 127–28
Alveolar ducts, 669
Alveolata, 219, 230
Alveoli, 219, 668–69
Amber, 106
Amblyrhynchus cristatus, **543**
Amboseli Reserve, Kenya, **2**
Ambulacra, 443
Ambulacral area, 445
Ambulacral groove, 445, 456
Ambulacrum, 445
Ambystoma, 522
Ambystoma mexicanum, **523**
Amebas, **209,** 226–29, **227,** 231
   feeding processes of, **8,** 226–27
   fossil records of, 228–29
   locomotion of, 43, **211,** 226, 622–23
   nutrition of, **214**
Ameiotic parthenogenesis, 135
Amensalism, 796
American Civil Liberties Union, 11

Amia, 498, 510
Amia calva, **500**
Amictic eggs, 308
Amino acid, pool of, 69
Amino acids, 24–26, **25**
   bonds between, 4, **5**
   degradation of, 69–70
   requirements for, in human nutrition, **691**
   sequence of, 7
   synthesis of, 27, **27**
Aminopeptidase, 686
Amiskwia, **108**
Amitosis, 215
Ammocoete larva, 480, **480,** 491–92
Ammonia, 27, 70
Ammonoidea, 340
Amniocentesis, 87–88
Amnion, 165
Amniota, 165, 468–69
   cladogram of, **537**
   classification of, 551
   evolution of, **536**
   skull, 583–84, **584**
Amniotic egg, 165, **167,** 539, **539**
Amoeba proteus, 227
Amoeba radiosa, 227
Amphibians, 482, 514–33. *See also* Frogs; Salamanders; Tetrapoda
   biological contributions of, 515
   circulatory system of, **661**
   classification of, 531
   comparison with reptiles, 538–40
   eggs of, 155
   evolution of, **519**
   eyes of, 530, **530**
   gastrulation in, 158
   heart of, 661
   integument of, **614**
   lateral line system of, 712
   lungs of, 668–69
   modern, 519–32
      characteristics of, 522
      and movement onto land, 515
   osmoregulation in, 636, **637**
   populations of, 525
   position in animal kingdom, 515
   red blood cells of, **658**
   thyroid gland of, 730
Amphiblastula, 242
Amphids, 298
Amphineura, 321
Amphioxus, 467, *467,* 475–77, **476,** 480, 482
Amphipholis, 452
Amphipoda, 391, **391,** 392
Amphiporus, 290, **290,** 291
Amphiporus bimaculatus, **290**
Amphiprion chrysopterus, **262**
Amphisbaena alba, **545**
Amphisbaenians, 544, **545,** 551
Amphiscolops, **280**
Amphitrite, 348, **349,** 351, 360, **680**
Amphiumas, 521
Amplexus, 530
Ampulla, 145, 446, 715
Ampullae of Lorenzini, 495
Amylase, 684
Anabas, 504
Anabolic processes, 7–8
Anabolic reaction, 60
Anabolic steroids, 734
Anadromous fishes, 508–9
Anadromous lampreys, 491
Anaerobes, 62
Anaerobic glycolysis, 65–66, **68,** 630
Anaerobic metabolism, 62

Anal canals, 269
Anal fin, 494
Analogous structures, 473
Anamnestic response, 745
Anamniota, 469
Anaphase
  in meiosis, **77,** 78
  in mitosis, **50,** 51, **52**
Anaphase A, 51
Anaphase B, 51
Anaphylaxis, 746
Anapsids, 535, 540–52, 583–84, **584**
Anarma, 226
Anarrhichthys ocellatus, **506**
Anaspids, 481, **481**
Ancestral character states, 195
Ancon sheep, 117, **117**
Ancylostoma duodenale, **300–301**
Androctonus, 371
Androgenic glands, 385
Androgens, 734
Anemia, sickle cell, 97, 672
Anemone fish, 262, **262**
Anemone-like animal, fossil of, **108**
Anemones, *247,* 248–49, 260, **260–61,**
  261–63, **262**
Aneuploidy, 87
Angiotensin, 644
Anglerfishes, 500
Anguilla, 507, 510
Anguilla anguilla, **508**
Anguilla rostrata, 508, **508**
Anguilla vulgaris, 508
Anguispira, 329
Anilocra, **391**
Animal kingdom, subdivisions of, 203–4, **204**
Animal pole, 155
Animal rights, 12–13
Anisogametes, 215
Anisogammarus, **391**
Annelida, 345–63
  adaptive radiation of, 362
  biological contributions of, 346
  body plan of, **186,** 346–47, **347**
  characteristics of, 349
  circulatory system of, **660**
  cladogram of, **361**
  classification of, 360
  cleavage in, 156
  comparison with arthropods, 360–61, 366
  feeding habits of, **680**
  immunity in, **748**
  nervous system of, 346, **702**
  phylogeny of, 360–61
  position in animal kingdom, 346
  reproduction of, 133, 141
Annuli, 346, 358
Anodonta, 343
Anolis carolinensis, 192
Anomalocaris canadensis, **108**
Anopheles, 221, 417
Anopheles quadrimaculatus, **418**
Anopla, 291
Anoplura, **405,** 420, 424
Anostraca, 387, **387,** 392
Anser anser, 754, **754**
Anseriformes, 579
Ant lions, 420, **420**
Antagonistic muscles, 617
Anteaters, 607
Antedon, **457,** 460
Antelopes, 127, **127,** 608
Antennae
  of crustaceans, 378
  of insects, 400–401, **401**
  of Onychophora, 435
Antennal glands, 382, **383,** 640, **640**
Antennapedia (gene), 163, **163**
Antennules, 379, **380–81**
Anterior, 185, **187**
Anterior ganglia, 276
Anterior lobe, of anterior pituitary gland, 726
Anterior pituitary gland, 725–28, **727**
Anthopleura, 267
Anthozoa, 249, 260–67
Anthracosaurs, 516
Antibodies, 95, 741–43, **742**
  in host defense, 742–43
  monoclonal, 745

Antibody-dependent cell-mediated
    cytotoxicity (ADCC), 742–43
Anticodon, 93
Antidiuretic hormone, 644–45, **727,** 728
Antigen-presenting cells (APC), 743
Antigens, 95, 741
  blood groups and, **747,** 747–48
Antimicrobial peptides, 740–41
Antiparallel strands, 90
Antipatharia, **265**
Antipathes, **265,** 267
Antithesis, principle of, 761, **761**
Antlers, 589, **589**
Antrostomus vociferus, 580
Anura, 523–32
  circulatory system of, 528, **529**
  coloration of, 526, **527**
  digestive system of, 529
  distribution of, 524–26
  feeding habits of, 529
  habitat of, 524–26
  integument of, 526, **526**
  muscular system of, 527
  nervous system of, 529
  reproduction of, 519–20, **520,** 530–32, **531**
  respiratory system of, 527–28, **528**
  sensory organs of, 529–30
  skeletal system of, 527, **527**
  vocalization in, 528
Anus, **142–43**
Anvil, 713
Aorta, 476, 662
Aortic arches, 354
Apanteles, **405**
Apatosaurus, 551, 622
APC. *See* Antigen–presenting cells
Aperture, 324
Apes, 109, **112,** 606
Apex, of shell, 324, **324**
Aphids, **416,** 420
Aphrodita, 360
Apical complex, 219, **219**
Apical ectodermal ridge, 164
Apicomplexa, 208, 219–22, 230–31
Aplacophora, 321
Aplidium cratiferum, **280**
Aplysia, 324, 326, 328, 343, 757, **757**
Aplysia dactylomela, **329**
Aplysina fistularis, 233
Apocrine glands, 590
Apoda, 520, 531
Apodemes, 366
Apodiformes, 580
Apoptosis, 53
Apopyles, 238
Aposematic coloration, 413, **413**
Appendages
  of Arthropoda, 366
  biramous, 378–79, **379**
  of Crustacea, 378, **379,** 379–80, **380–81**
  development of, 170
  diversity of, **111**
  of insects, 400, **401**
  jointed, 365–66
  morphogenesis of, 164, **165**
  paired, 468
  of tetrapods, 516, **517**
  unjointed, 435
  of Vertebrata, variation in, **16**
Appendicular skeleton, 619–20
Apterygiformes, 578
Apterygota, 421–22
Aquatic chelicerates, 374
Aquatic environments, 779–80, **780**
Aquatic mandibulates, 377–94
Aquatic respiration, 667–68, **668**
Aqueous humor, 717
Arachnida, 368–74
Araneae, 369–71
Arbacia, 452, 460
Arcella, **215,** 227, **227,** 231
Archaea, 32, 203
Archaebacteria, 32
Archaeocytes, 239–40, **240**
Archaeopteryx, 556, **557,** 562, **563,** 569, 578
Archaeornithes, 578
Archenteron, 158, 170
Archinephric duct, 641

Archinephros, 641, **641**
Archipelagoes, 115
Architectural patterns, 176–89, **186**
Architeuthis, 317, 335–36
Archosauria, 535, 538, 551
Arctic, The (Bruemmer), 600
Arctic ground squirrel, *771*
Ardea herodias, 579
Ardipithecus ramidus, 603
Arenicola, 348, **349,** 351, 360
Argiope, 374
Argiope aurantia, **370**
Argonauta, 339
Argulus, 392
Ariolimax columbianus, **330**
Aristotle, 103, 191, 353
Aristotle's lantern, 453, **455**
Armadillium, 390, 392
Armadillium vulgare, **391**
Armadillos, 607
Armillifer, 434–35, **435**
Arms, 258
Arrowworms. *See* Chaetognatha
Artemia, 392
Arteries, 661–62, **664,** 664–65
Arterioles, 661, 664
Arteriosclerosis, 664
Arthropoda, 364–75
  adaptive radiation of, 375
  biological contributions of, 365
  body plan of, **186, 188**
  characteristics of, 366
  circulatory system of, 366, 660, **660**
  cladogram of, **439**
  classification of, 374
  comparison with annelids, 360–61, 366
  diversity of, 365–66
  eggs of, 155
  exoskeleton of, 617–18
  eyes of, 716, **716**
  fossils of, **108, 367**
  integument of, 613, **614**
  kidneys of, 640–41
  nervous system of, 702, **702**
  phylogeny of, 374–75
  population of, 365–66
  position in animal kingdom, 365
Arthropodization, 364
Articulata, 434
Artificial selection, 119
Artiodactyla, 608–9
Ascaphus, 523
Ascaris, **298,** 302
Ascaris lumbricoides, 299–300, **300**
Ascaris megalocephala, 299
Ascaris suum, 299, **300**
Ascidiacea, 473–75, **475**
Ascidians, cleavage in, 156, **158**
Asconoid, 237–38, **238**
Asexual reproduction, 6, 48, 132, **132,** 133
  of Asteroidea, 449
  of hydra, 254
  lack of, 135–36
  of Metazoa, 235, **235**
  of Nemertea, 291
  of paramecium, 224, **225**
  of platyhelminthes, 278
  of protozoa, 208, 214–15
  of sponges, 240–42
  of *Volvox,* **217,** 217–18
Asiatic scrub typhus, 373
Aspidosiphon, 428
Asplanchna, 309
Asplanchna priodonta, **307**
Asses, 608
Assimilation, 678
Association areas, 706
Association for Assessment and Accreditation
    of Laboratory Animal
    International, 13
Aster, 42
Asterias, 445–46, **447,** 448, 460
Asterias rubens, **330**
Asteroid bombardment, 127–28, **128**
Asteroidea. *See* Sea stars
Asters, 50
Astrangia, 267
Astrocytes, 697
Astrophyton, 452

Astrophyton muricatum, **450**
Astropyga magnifica, **453**
Asymmetric competition, 796
Asymmetry, bilateral, 325
Asynchronous muscles, 403
Ateles, 606
Athecate hydroids, 255, **255**
Atherosclerosis, 664, 691
Atlantic salmon, 508
Atlantic sturgeon, **500**
Atlas vertebrae, 620
Atmosphere, 773–75
  of earth, 773, **773**
  primitive, 26–27
Atolls, 265
Atoms, 4
ATP. *See* Adenosine triphosphate
Atractaspididae, 548
Atrax, 371
Atretic, 144
Atrial cavity, 474
Atriolum robustum, **474**
Atriopore, 476
Atrioventricular valves, 662
Atrium, 476, 661
Atrophy, 627
Attach-pull-release cycle, 628, **629**
Auditory canal, 713
Auditory reception, 408
Augmented muscular activity, 650
Auks, 579
Aurelia, 258, **259,** 267
Aurelia aurita, 258, **260**
Auricularia, **450,** 456
Australian lungfish, 500, **501**
Australopithecus afarensis, **603,** 603–4, **605**
Australopithecus africanus, **605**
Australopithecus anamensis, 603–4, **605**
Australopithecus garhi, 604, **605**
Australopithecus ramidus, **605**
Autogamy, 215, 225
Autoimmunity, 741
Autolytus prolifer, **351**
Autonomic nervous system, **707,** 707–9, **708**
Autosomal linkage, 85–86
Autosomes, 78
Autotomy
  of Asteroidea, 448–49
  of Ophiuroidea, 452
Autotrophs, 31, **31,** 214, 678
Avocets, 579
Avoidance reaction, 224, **224**
Axial skeleton, **171,** 619–20
Axis vertebrae, 620
Axocoels, 449
Axoneme, 210, **210**
Axons, 169, **171, 185,** 251, 355, 696, **696**
Axopodia, 211, **212**
Axostylata, 216, 230
Axostyle, 216
Aye-aye, 606
Aysheaia, **108,** 435

B lymphocytes, 740, **740**
Babesia, 231
Baboons, 602, **759**
Bacillus thuringiensis, 418
Bacteria, 203
  chemoautotrophic, 801
  reproduction of, 133
Bacteriophages, 96
Baculum, 142
Balanced Treatment for Creation-Science and
  Evolution-Science Act, 11
Balanglossus, 461
Balantidium, 225, **226,** 231
Balanus, 392
Balanus balanoides, **389,** 796
Baleen whales, 609
Banana slug, **330**
Bandicoots, 606
Bank reefs, 266
Bankia, 334
Banting, Frederick, 735, **735**
Barbs, 560
Barbules, 560
Bar-headed geese, 567

Bark beetles, 414, 417
Bark lice, 419
Barnacles, 389, **389**, 392, **393**
    competition among, 796
    feeding habits of, **679**
    reproduction of, **132**
Barrier reef, 265, **267**
Basal body, 43, 210, **210**
Basal disc, 252
Base sequences, in DNA, 91, **92**
Baseodiscus, 291
Basilar membrane, 714
Basipodella, 392
Basipodite, 379
Basis, 379
Basket stars, **450**, 460
Basolo, Alexandra, 765
Basophils, **657**, 658, 740
Batesian mimicry, **413**, 799, **799**
Bats, 111, 594–96, **596**, 606
Bayliss, W. H., 721, **722**
Bdelloidea, 308–9
Bdellostoma, 510
Beach hopper, 386, 391
Beadle, 88
Beagle, **104**, 104–6, **105**
Beak of birds, **16**, 562, 565, **566**, 614–15, **615**,
    797, **797**
Beardworms, **429**, 429–30
Bears, 582, *582,* 607, **607**
Beaumont, William, 685, **685**
Beavers, **588**, 599, **599**
Bed bugs, 418
Beef tapeworm, 286–88, **287**
Bees. *See* Honey bees
Beetles, 401, 403–4, 413–14, 420
Behavior, 752–70
    altruistic, 763–65
    of Arthropoda, 366
    communication and, 765–68
    control of, 755–58
    diversity and, 757–58
    genetics of, 755–56, **756**
    inherited, 755
    innate, 755
    of Insecta, 413–16
    learning and, 757–58
    science of, 753–54
        principles of, 754–55
    social, 759–68
    stereotypical, 754
    territorial, 754–55, **755**, 762, **762**
Behavioral ecology, 753
Békésy, Georg von, 714
Belding's ground squirrels, 764, **764**
Belted kingfisher, 580
Benthic zone, 779
Benthos, 779
Bernard, Claude, 634, 656, **657**
Beroe, 270, **270**
Berrill, N. J., 678
Berson, Solomon, 729
Berthold, Arnold Adolph, 722
Best, Charles, 735, **735**
Beta cells, 735
Bicarbonate, 657
Bichirs, 498, **500**, 510
Bicoid (gene), 162
Bicuspid, 662
Big bang model, 22
Big brown bats, 606
Bighorn rams, 760–61
Bilateral asymmetry, 325
Bilateral cleavage, 156, **158**
Bilateral nervous system, 701–2
Bilateral symmetry, 164, 182, 184–85, **186–87,**
    234, 274
Bilateria, 185, 274–94
    acoelomate, 185, **187**
    adaptive radiation of, 292
    biological contributions of, 275
    classification of, 204, **205**
    eucoelomate, 185–87, **187**
    phylogeny of, 292
    position in animal kingdom, 275
    pseudocoelomates, 185, **187**
Bile, 686, 688
Bile duct, 688
Bile pigments, 688

Bile salts, 688
Bilharzia, 283
Bilirubin, 658
Bill of birds, **16**, 562, 565, **566**, 614–15, **615**,
    797, **797**
Binary fission, **132**, 133, 214, **215**, 224, **225**
Binomial nomenclature, 192
Biochemistry, comparative, 196
Biodiversity, 795
Biogenetic law, 111–12
Biogeochemical cycles, 803
Biogeography, vicariance, 783
Biological control of insects, 418
Biological species concept, 194
Biological time, **33**
Biology
    evolutionary developmental, 164–65
    subdivisions of, 5
    subfields of, 14
    zoology as part of, 10
Bioluminescence, 226, 414
Biomass, 800
    pyramid of, **802**, 802–3
Biomes, **775**, 775–78
Biomphalaria, 283
Biosphere, 772, 790
    subdivisions of, 773–75
Biparental reproduction, 133–34, **134**
Bipinnaria, 449, **450**, **462**
Biradial symmetry, 184, 268
Biramous appendages, 378–79, **379**
Birds, 482, 555–81
    behavior of, 754–55, **754**, **755,**
        756, **756**
    bills of, **16**, 562, 565, **566**, 614–15, **615**,
        797, **797**
    biological contribution of, 556
    characteristics of, 560
    circulatory system of, 565–66
    classification of, 578–80
    cognition and, 768
    communication by, **767**, 767–68
    development of, 575–76
    digestive system of, 565
    dominance in, 761
    eggs of, 140, 155
    embryos of, **113**
    evolution of, **558**
    excretory system of, 567
    feathers of, 560–62, **561**
    feeding habits of, 565
    flight of, 569–70, **570**
    flightless, 557, 560
    fossil record of, 556, **557**
    gastrulation in, 159
    guilds of, 797, **798**
    hunting of, 576, **577**
    hypothermia in, 651
    imprinting and, 758, **758**
    lungs of, 668, **669**
    mating systems of, 574–75, **575**, 763, **763**
    migration of, 555, *555,* 570–73, **572**
    muscular system of, 564, **565**
    nervous system of, 567–68, **568**
    nesting of, 575–76, **576**
    origins and relationships of, 556–60
    perching mechanism of, 564, **565**
    phylogenetic tree of, 110, **112**
    populations of, 576–77, **793**
    position in animal kingdom, 556
    reproduction of, 573–76, **574–75**
    reptiles and, 556–57
    respiratory system of, 566–67, **567**
    sensory organs of, **568**, 568–69
    sexual selection in, 123, **124**
    skeleton of, 562–64, **563**
    social behavior of, **574**
    songs of, 758, **758**
    specimens of, **193**
    submarine flight of, 197–98, **200**
    territoriality and, 762, **762**
    thyroid gland of, 730
    young, 575–76, **576**
Birth control, 145
Birth, multiple, 148, **149**
Bisexual reproduction, 133–34, **134**
Bison, **778**
Bispira brunnea, **347**
Biston betularia, **11,** 11–12

Bitterns, 579
Bivalent chromosomes, 78
Bivalvia, 317, **318**, 329–35, **330–32**, 343
    body plan of, 331
    boring by, 334–35
    development of, 334
    feeding habits of, 332, **334–35**, **679**
    foot of, 319
    gills of, 331
    internal structure and function of, 332–34
    locomotion of, 331
    mantle cavity of, 342
    pearls in, 317, 320, **320**, 330–31
    reproduction of, 321, 334
    shell of, **320**, 330–31
Black coral, **265**
Black widow spider, 370–71, **371**
Bladder, 382
Bladder worms, 288
Blarina brevicauda, **594**
Blastocoel, 158
Blastocyst, 145, 158, **159**, 167
Blastomeres, 155–56
Blastula, **159**
Blastulation, 158
Blending, 16–17, **18**, 79, 84
Blepharisma, 213, 231
Blood, 181. *See also* Circulatory system
    coagulation of, 658–59, **659**
    composition of, **657**, 657–59
    types of, 120–21, **121**, 747, **747**
Blood cells, 655
Blood flukes, **282**, 283–84
Blood groups, **747**, 747–48
Blood plasma, 178, 656–57
Blood pressure, 664–65
Blood tissue, **180**
Bloodsucking bugs, 417
Blow flies, 405
Blue marlin, **501**
Bluefin tuna, 502, **503**
Blue-footed boobies, **767**, 767–68
Blue-green algae. *See* Cyanobacteria
Boa constrictors, 549
Boaedon fuluginosus, **549**
Bobolink, migration of, **572**
Bobwhite quail, 579
Body cavities, 185–87
Body cells, 48
Body plans, 182–88, **186**
    of acoelomates, **186**
    of Annelida, **186**, 346–47, **347**
    of Arthropoda, **186**, **188**
    of Bivalvia, 331
    of Chordata, **188**
    of Echinodermata, **186**
    of Enterocoelomate, **186**
    of eucoelomates, **186**, **296**
    of eumetazoans, **186**
    of flatworms, **186**
    hexamerous, 260
    of Hydra, 252
    of Mollusca, **186**
    of multicellular organisms, **186**
    of Nemertea, **186**
    octomerous, 260
    of pseudocoelomates, **186**, **296**
    of Reptilia, 540
    of Schizocoelomate, **186**
    tube-within-a-tube, 185, **186**
    of unicellular organisms, **186**
    of Vertebrata, **186**, 480–81
Body size, 177–78, **179**
Body weight, metabolic rate and, **594**
Bohadschia argus, **457**
Bohr effect, 673
Bolitoglossa, 523, **524**
Bombardier beetles, 413
Bombykol, 766
Bombyx mori, 766, **766**
Bonasa umbellus, 579
Bond energy, electron transport and, 61–62
Bone morphogenetic protein-4 (gene), 164
Bone tissue, **180**
Bonellia viridis, 428
Bones, 181, **183**, 477, 618–19
    body size and, 622, **622**
    pneumatized, 562
    of wings, 564, **564**

Bony fishes, 489, 497–502
    buoyancy of, 503
    fossils of, 497–98
    gills of, 504
    integument of, 614, **615**
    origin and evolution of, 497–98
    osmotic regulation in, 46, 505,
        636–37, **637**
Boobies, 579, **767**, 767–68
Book gills, 367
Book lice, 419
Book lungs, 369
Boophilus annulatus, 374, **374**
Boreal forest, 777
Boring, by Bivalvia, 334–35
Bowfins, 498, **500**, 510
Bowman's capsule, 642
Box tortoise, 541–42
Brachial canals, 258
Brachiolaria, 449, **450**
Brachiopoda, 156, 427, 430, 433–34, **434**, 439
Brachiosaurus, 546–47, 551
Brachypelma vagans, **371**
Braconid wasp, 410
Bradyzoites, 220
Brain
    of birds, 567–68, **568**
    of fishes, 487
    of frog, 529, **529**
    of humans, **706**
    in metamorphosis, 411–12
    neuropeptides and, 729
    of Rotifera, 308
    tripartite, 478–79
    of turtles, 541
    of Vertebrata, 478–79, 702, **704,**
        704–7, **705**
Branchial heart, 337
Branchial system, of Enteropneusta, 462
Branchial tufts, 668
Branchiobdellida, 360
Branchiomeres, 478
Branchionecta, 392
Branchiopoda, 387, **387**, 392
Branchiostoma, 475
Branchiura, 389, **389**, 392, 439
Bread mold, 88
Breathing. *See also* Respiratory system
    coordination of, 670–71
Breeding
    controlled, 194
    mutations and, 117, **117**
    true, 17
Brenner, Sydney, 297
Brine shrimp, 392
Bristletails, 419
Bristowe, W. S., 371
Brittle stars. *See* Ophiuroidea
Brno, Czech Republic, 74
Broad-billed hummingbird, *611*
Bronchi, 669
Bronchioles, 669
Brown adipose tissue, 689
Brown bats, 606
Brown, L., 564
Brown recluse spider, 370–71, **371**
Browsers, 591
Bruemmer, Fred, 600
Brugia malayi, 302
Bruneria borealis, 791
Bryozoa. *See* Ectoprocta
Buccal cavity, 683
Buccal diverticulum, 461
Buccinum, 325
Budding, **132**, 133, 214, 240
Buds, 252
Buffon, Georges Louis, 103
Bufo americanus, **525**
Bufo marinus, 525
Bufonidae, 523
Bugs, 411
Bulbourethral glands, 142, **142**
Bulinus, 283
Bulk-phase endocytosis, 48
Bull sharks, 493–94
Bullfrog, 523–24, **524**, **527**
Bullia, 325
Buoyancy, of fishes, 503–4, **504**
Burgess Shale of British Columbia, 107, **108**

Burgess, Thornton, 599
Burns, John, 286
Burrowing shipworms, 317
Bursa of Fabricius, 565
Bursae, 451, **451**
Bush babies, 606
Busycon, **324,** 328, 343
Busycon carica, **324**
Busycon contrarium, **324**
Butterflies, 421, **421**
  antennae of, 401
  chemoreception in, 408
  feeding habits of, 404
  metamorphosis in, 410, **410**
  mimicry in, 413, **413, 799**
  mouthparts of, 405, **406**
  muscles of, 403
  sperm of, **139**
Buzzards, 579

Caddisflies, 421
Caecidotea, 390, **391,** 392
Caecilians, 516, 520, **520,** 531
Caenorhabditis, 302
Caenorhabditis elegans, 53, 297, **299**
Calanus, 377, 388, 392
Calcarea, 237, **239,** 242, 245
Calcareous spicules, **241**
Calcichordata, 458, 473
Calciferous glands, 353
Calcification, 613
Calcispongiae, 242, 245
Calcitonin, 619, 732, **732**
Calcium carbonate, 264
Calcium, metabolism of, 731–32, **732**
Calcium salts, 366
California grey whale, **393**
Caligus, 392
Calliostoma annulata, **318**
Callitrichidae, 606
Callorhinus ursinus, **595**
Callyspongia, **241,** 245
Calmodulin, 723
Calorie, 28
Calypte anna, **576**
Calyx, 312, 456
Cambered wings, 569
Cambrian explosion, 32, 34
Cambrian period, 32, 107, **108**
Camels, 608
Camouflage, **413**
Campodeids, 419
Canadia, **108**
Canal systems, of sponges, 237–39, **239, 241**
Canaliculi, 618
Cancellous bone, 618
Cancer, 392
Cancer, molecular genetics of, 98
Canidae, 608
Canine teeth, 590, 680
Cannon, Walter B., 634, **635,** 734
Cantbon pilularis, 414, **414**
Capillaries, 661, **665,** 665–66
Capillary exchange, 665–66
Capitulum, 372
Caprella, **391**
Caprimulgiformes, 580
Captacula, 323
Captorhinida, 551
Carapace, 367, 378, 540
Carapus, 456
Carbaminohemoglobin, 674
Carbohydrates, 4, 22–24, 691
Carbon, 22
Carbon dioxide
  in atmosphere, **774,** 774–75, **775**
  transport of, in blood, 674, **674**
Carbon monoxide, 673
Carboxypeptidase, 686
Carcharhiniformes, 493
Carcharhinus levcas, 494
Carcharias, **495**
Carcharodon, 494
Carchesium, 231
Cardiac center, 663
Cardiac muscle, **180,** 182, **184,** 624–25, **625,** 663
Cardiac output, 662

Cardiac sphincter, 685
Cardinalis cardinalis, 73
Cardinalis sinuatus, 73
Caribou, 594, **595, 778**
Carncinoscorpius, 367
Carnivores, 34, 506–7, **591–92,** 592–93, 607–8, 678, 800
Carolina chickadee, 192
Carotenoid, 616
Carrier, 84, 121
Carrying capacity, 793
Cartilage, 181, **183,** 477, 618
Cartilaginous fishes, 487, 493–97, 510
Carybdea, **260,** 267
Casauriiformes, 578
Cassiopeia, 258, 267
Cassowaries, 578
Caste system, 415
Castor canadensis, **588**
Castoridae, 607
Catabolic processes, 7–8
Catabolic reaction, 60
Catadromous eels, 507
Catalysts, enzymes as, 58, **58**
Catalytic RNA, 30
Catastrophic species selection, 128
Caterpillars, 404, 410, 413
Catfishes, 499
Cats, 607
  parasites of, 220, 300
  sperm of, **18**
Cattle, 608
Cattle ticks, 374
Caudal fins, 494, **498**
Caudal vertebrae, 620
Caudata, 520–23, 531
Caudipteryx, 557
Caudofoveata, 321, 343
Causality
  immediate, 13
  proximate, 13
  ultimate, 14
Causation, 753
Caveolae, 47
Cavolina, 328
CCK, 689, **690**
CD4, 743
CD8, 743
Cebidae, 606
Cebus, 606
Cecum, 565, 591
Cell(s)
  acinar, **734**
  adhesive, 268
  alpha, 735
  beta, 735
  chief, 685
  ciliated epithelial, **5**
  division of labor in, 208
  emergent properties of, **6**
  endothelial, 665
  epitheliomuscular, 252, **253**
  eukaryotic, 39–42, **39**
  flame, 276–77, 290, 308, 639, **639**
  flux of, 53
  germ. *See* Germ cell(s)
  gland, 252–53
  glial, 696–97
  hair, 712
  interstitial, 142, 144, 252–53
  islet, 734–35
  Kupffer, 740
  of liver, **37**
  in living systems, 4, **5**
  mast, 740
  memory, 745
  microglial, 697, 740
  mucus-secreting, **5**
  multinucleate, 50
  natural killer, 740, **740**
  nerve, 253, **253**
  neural crest, 169, 479
  neurosecretory, 355, 385, 724
  nurse, 301
  nutritive-muscular cells, 253
  organization of, 39–48, **40**
  pacemaker, 663
  parietal, 685

plasma, 741
prokaryotic, 39, **39**
red blood, 657, **657,** 658, **658**
salt-absorbing, 505
salt-secretory, 505
Schwann, **185,** 697, **697**
sensory, 253
sex, 49, 75, 132
somatic, 48, 136
Spemann organizer, *151*
of sponges, 239–40, **240–41**
study of, **6,** 37–39
surfaces of, 44–45
sustentacular, 138, **138,** 142, 144
target, 722
timescale for reproduction of, **6**
white blood, 657, **657,** 658
Cell aggregate, **186**
Cell biology, heredity and, 19
Cell cycle, 51–52, **53**
Cell division, 48–53
Cell membranes, 39
  fluid-mosaic model of, 39–41, **40**
Cell metabolism, 56–72
Cell organelles, 37
  separation of, 38, **38**
Cell theory, 37
Cell walls, 10
Cell-tissue grade of organization, **178**
Cellular bone, 487
Cellular grade of organization, **178**
Cellular immunity, 740, **740,** 741, **743,** 745–46
Cellular level, reproduction at, 6, **7**
Cellular respiration, 61–66, **63,** 666
Cellulase, 684
Cellulose, 22, 24, 591
Center of origin, 782
Centipedes, 375, 398, **398,** 418
Central nervous system, 702–3
Centrioles, 42, 210
Centrolecithal egg, 155
Centrocercus urophasianus, **575, 763**
Centrolecithal egg, 155
Centromeres, 49, **49,** 78
Centrosome, 42
Centuroides, 371, 374
Cephalaspidomorphi, 482, 489, 491–93, 510
Cephalaspis, 481
Cephalization, 185, 188
Cephalocarida, 387, **387,** 392
Cephalochordata, 475–77, 482
Cephalodiscus, 463, **463**
Cephalopholis fulvus, **391**
Cephalopoda, 317, 335–40, **337,** 343
  circulatory system of, 337–38
  communication by, 339
  egg of, 155
  fossil record of, 335
  locomotion of, 320, 336–37
  major groups of, 340
  nervous system of, 338
  reproduction of, 321, 339, **340**
  respiratory system of, 337
  sense organs of, 338–39
  shell of, 336, 341
Cephalothorax, 369, 371, 378
Ceranthiapathas, **265**
Cerata, 328
Ceratium, 226, **227,** 231
Cercariae, 281
Cerci, 401
Cercopithecidae, 607
Cerebellum, 568, 705, **705**
Cerebral cortex, 567, 706
Cerebral ganglia, 355
Cerebral malaria, 221
Cerebratulus, 290–91
Cerebrum, **705,** 706
Cerianthus, 267
Ceriantipatharia, 260, **264,** 267
Cermatia, 418
Cervical bones, 562
Cervical vertebrae, 620
Cervix, 143, **143**
Cestoda, 276–77, 285–89, **287**
Cestum, 268, 270, **270**
Cetacea, 609
Chaetoderma, 321, 343
Chaetogaster, 357

Chaetognatha, 427, 438, 440
Chaetonotus, **311**
Chaetopoda, 360
Chaetopterus, 351–53, **352,** 360
Chaetura pelagia, 580
Chagas' disease, 218–19, 417
Chalk deposits, 229
Challenge, 745
Chamaedorea, 373
Chamaleo chamaeleon, **544**
Chameleon, **8,** 543, **544**
Change, in evolution, 14–15
Chaos carolinense, 227
Character displacement, 797
Character variation, 195–96
Charadriiformes, 579, **579**
Charcarodon, 510
Charging, 93
Cheetahs, genetic variation in, 122, **123**
Chelicerae, 369
Chelicerata, 367–75, 420
  aquatic, 374
  cladogram of, **375**
  classification of, 374
Cheliceriformes, 375
Cheliped, **380–81**
Chelonia, 540–42, 551
Chelonia mydas, **542**
Chelydra serpentina, **541,** 542
Chemical barriers, in immunity, 739
Chemical bond energy, 36
  electron transport and, 61–62
Chemical energy transfer, by ATP, 60–61
Chemical evolution, 26–30
Chemical mutagens, 98
Chemical reaction, direction of, 60
Chemical signals, 414
Chemical synapses, 699
Chemical uniqueness, of living systems, 4
Chemoautotrophic bacteria, 801
Chemoreception, 408–9, 709–11
Chemotaxis, 224, 709
Chemotrophs, 678
Chestnut weevil, **421**
Chewing lice, 420
Chewing mouthparts, 406
Chickens
  blastula and gastrula of, **159**
  dominance in, 761
  egg of, **156**
  embryo of, **166**
  gastrulation in, **159**
  inheritance in, 81, **81**
Chief cells, 685
Chiggers, **372,** 373
Chilopoda, 398, **398,** 418
Chimaeras, 482, 487, 497, **497,** 510
Chimney swift, 580
Chimpanzees, 606
  cognition and, 768
  taxonomy of, 199
Chinese liver flukes, **282**
Chipmunk, **593**
Chironex, 267
Chironex fleckeri, 260
Chiropsalmus, 267
Chiroptera, 606
Chitin, 365
Chitons, 317, **318,** 321–23, **323,** 343
Chlamydomonas, **217,** 230
Chlamydophrys, **227,** 231
Chlamys, 334
Chlamys opercularis, **330**
Chloeia, 345
Chloragogen tissue, 353
Chloride, 657
Chloride shift, 674
Chlorocruorin, 673
Chlorohydra, 253
Chlorophyta, 216–18, **217,** 230
Chloroplasts, 216, 218
Choanoblasts, 243, **244**
Choanocytes, 237, 239, **240–41**
Cholesterol, 24, **25,** 733, **733**
Chondrichthyes, 482, 487, 493–97, 510
Chondrocytes, 618
Chondrophora, 256
Chondrosteans, **498, 500**

Chonopeltis, 392
Chordata, 467–83, 510
  biological contributions of, 468
  body plan of, **188**
  characteristics of, 469
  cladogram of, **471**
  classification of, 468–69, 482
  cleavage in, 156
  divisions of, **472**
  evolution of, 472–73
  fossil of, **108**
  hallmarks of, 469–72
  larval evolution of, 479, **479**
  phylogeny of, **470**
  position in animal kingdom, 468
Chordeiles minor, 580
Chordin (gene), 162, 164
Choriaster granulatus, **444**
Chorioallantoic membrane, 166
Chorion, 146, 166
Chorionic villi, 167
Choriovitelline placenta, 597
Choroid coat, 717
Christmas tree worm, **239**, **347**
Chromatids, 50, 78
Chromatin, 32, 41
Chromatography, 38–39
Chromatophores, 339, 526, **526**, 538, 616, **616**
Chromodoris, 318
Chromosomal theory of inheritance, 17–19
Chromosomes, **18**, 19, 32
  aberrations in, 87–88
  accessory, 78
  bivalent, 78
  diploid number, 76, 133
  haploid number, 76, 133
  homologous, 76
  inheritance and, 75–79
  structure of, 49, **49**
Chrondrostei, 510
Chthamalus stellatus, 796
Chyme, 686
Chymotrypsin, 686
Cicada, **411**, 414, 420
Cichlid species, 114–15
Ciconiiformes, 579, **579**
Ciguatera, 226
Cilia, 43
  in alimentary canal, 683
  of polychaetes, 351, **352**
  of protozoa, **210**, 210–11
Ciliary filter feeders, 312
Ciliary movement, **623**, 623–24, **624**
Ciliary muscles, 717
Ciliary tracts, 318
Ciliated epithelial cells, **5**
Ciliated pits, 290
Ciliates, 208, **210**, **222**, 222–26, 231
  structure of, 222, **223**
  symbiotic, 225, **226**
Ciliophora. *See* Ciliates
Cimex, 418
Ciona, 474
Circadian rhythms, 728
Circulatory system, 655, 659–66
  of Amphibia, **661**
  of Anura, 528, **529**
  of Arthropoda, 366, 660, **660**
  of birds, 565–66
  of Bivalvia, 333, **333**
  of Brachiopoda, 434
  of cephalopods, 337–38
  closed, 660–61
  coronary, 664
  of Crustacea, 381–82, **382**
  of earthworms, 353–54
  of Echiura, 429
  of Enteropneusta, 462
  of fish, **661**
  of Hirudinea, 359–60
  of Insecta, 406, **661**
  of Mammalia, **661**
  of Mollusca, 317–18, 320
  of Nemertea, 290–91, **291**
  of Onychophora, 436
  open, 660–61
  of polychaetes, 349

of Reptilia, 539
  of Tetrapoda, 515
  of Vertebrata, 661–64
Cirri, **222**, 443, 456
Cirripedia, **389**, 389–90, **390**, 392, **393**
Cirrus, 279
Cisternae, 41
Citric acid cycle. *See* Krebs cycle
Clades, 195–96
Cladistics, 201–3
Cladocera, 387, **387**
Cladogram, 196, **196–97**
  of Amniota, **537**
  of Annelida, **361**
  of Arthropoda, **439**
  of Chelicerata, **375**
  of Chordata, **471**
  of Cnidaria, **271**
  of Crustacea, **395**
  of Deuterostomia, **454**
  of Echinodermata, **459**
  of fishes, **489**
  general area, 783, **783**
  of Insecta, **423**
  of Mollusca, **342**
  of Onychophora, **439**
  of Platyhelminthes, **293**
  of Saurischia, **559**
  of synapsids, **586**
  of Tardigrada, **439**
  of Tetrapoda, **518**
  of Uniramians, **422**
Clam shrimps, 392
Clam worms, 348, 350
Clams, *316*, 317, *318*, 330, **331–36**
Clark, R. B., 360
Clasper, 493–94
Classical pathway, 739
Classification, 190–206, **204**
  development of, 191–92
  family-level, 197, **200**
Clathrin, 48
Clathrina, 238, 242, 245
Clathrina canariensis, **239**
Clathrin-coated pits, 48
Clathrulina, 228, **229**, 231
Clavelina puertosecensis, **474**
Claws, 614–15, **615**
Clay, 29
Clearwater Lakes, **128**
Cleavage, 155–58
  bilateral, 156, **158**
  discoidal, 157
  holoblastic, 155
  inherited patterns of, 156–58
  meroblastic, 155, 157
  mosaic, 161, **161**
  patterns of, 155–58, **156**
  radial, 156
  regulative, 161, **161**
  rotational, 156–57
  spiral, 156
  superficial, 157, **157**
  yolk and, 155–56
Cleavage furrow, 51
Climate, 775, **775–76**
Cliona, 245
Clione, 328
Clitellata, 360
Clitellum, 356, **357**
Clitoris, 143, **143**
Cloaca, 142, 308, 456, 565, 567
Clones, 96, 135, 161, 791
Clonorchis, 288
Clonorchis sinensis, **279**, **281**, 282, **282**
Closed circulatory system, 660–61
Clostridium, 31
Clotting, 658–59, **659**
Cnidaria, *176*, 184, 247–67, **249**
  adaptive radiation of, 271
  characteristics of, 250
  cladogram of, **271**
  classification of, 267
  comparison with Ctenophora, 268
  form and function of, 249–51
  gastrulation in, 160
  immunity in, **748**

nervous system of, 250–51
  phylogeny of, 270–71
  reproduction of, 133
Cnidoblast, 250
Cnidocil, 250
Cnidocytes, 248, 250, **250**, 253
Coagulation, 658–59, **659**
Cobb, N. A., 295
Cobra, 548, **549**
Coccidea, 219–22, 231
Coccidiosis, 220
Coccyx, 620
Cochlea, 568, 713, **714**
Cochlear duct, 713
Cockroaches, 401, 419
Cocoon, earthworm, 356, **357**
Cod, 509, **704**
Coding, in DNA, 91, **92**
Codonosiga, **214**, **229**, 234
Codons, 91, **92**
Coelacanth, 489, 497, 500, 502, **502**, 510
Coelenterata. *See* Cnidaria
Coelom
  of Asteroidea, 445–46
  of Bilateria, 185
  formation of, 160, **188**
  metamerism and, 360
  of Mollusca, 317
  of pseudocoelomates, 296
  reduced, 366
  tripartite, 439
  of Vertebrata, 479
Coelomic vesicles, 158
Coelophysis, 546–47
Coeloplana, **270**
Coenenchyme, 264
Coenosarc, 254
Coenzyme A, 64
Coenzymes, 59
Cofactors, 59
Cognition, 768
Cohorts, 791
Coiling, of shells, 325, **325**
Cold-blooded animals, 647–48
Coleoidea, 340
Coleoptera, **402**, 403, **414**, **417**, 420, **421**
Coliiformes, 580
Colinus virginianus, 579
Collagen, 180–81, 297
Collar bodies, 243, **244**
Collecting duct, 642
Collembola, 419
Collenchyme, 268
Collenocyte, 240, **240**
Colloblasts, 268
Colloid osmotic pressure, 665–66
Colloidal systems, 209
Colobocentrotus atratus, **453**
Colonial flagellate hypothesis, 234
Colonial gorgonian coral, **266**
Colonies, hydroid, 254–55
Color blindness, 84–85, **85**
Color patterns, geographic variation in, 192, **192**
Color vision, 718, **718**
Coloration, 615–16
  of Anura, 526, **527**
  of feathers, 562
  protective, 413
  warning, 413
Colpoda, 215, 231
Colubridae, 548
Columbiformes, 579
Columella, 324, **324**, 530, 568
Columnar epithelium, **180–81**
Comantheria, 460
Comantheria briareus, **457**
Comb jellies, 267, **268–69**
Comb plates, 268
Comb rows, 268
Comet Shoemaker–Levy 9, 127
Commensal sponges, 237
Commensalism, 795
Common descent, 109–12, 197–201
  in evolution, 15
  public ridicule of, 110, **112**
  species recognition and, 193

Common robin, 192
Communication, 765–68
  by birds, **767**, 767–68
  by cephalopods, 339
  by display, 767–68
  by honey bees, **766**, 766–67
  between humans and other animals, 768
  by insects, 413–16
Community, 790
  ecology of, 795–99
  interactions within, 795–97
Compact bone, 618, **619**
Compact nuclei, 209
Comparative biochemistry, 196
Comparative causation, 753
Comparative cytology, 196
Comparative method, in evolutionary sciences, 14
Comparative morphology, 196
Competition, 790, 796–97
Competitive exclusion, 797
Complement, 739
Complementary strands, 90
Complete septa, 261
Complexity, 177–78, **179**
  hierarchical organization of, 177–78, **178**
  of living systems, 4–6, **5–6**
  of reproduction, 6, **7**
Compound eyes, 383, **383**, 409, 716, **716**
Computer modeling, in research, 12
Concentration gradient, 46
Concentration, prebiotic synthesis and, 29–30
Concentricycloidea, 457, **458**, 460
Concertina movement, 545, **548**
Conch, 324, 328
Condensation, 29
  thermal, 30
Condensed nuclei, 209
Condors, 579
Conductance, decreased, 649
Conduction, of nerve impulses, 698–99, **699**
Condylactis gigantea, *247*
Cones, of eye, 530, 717
Coney, **391**, 600
Coniferous forest, 777, **777**
Conispiral shells, 325
Conjugation, 134, 215, 225, **225**
Connecting transverse tubules, **628**
Connective tissue, **180**, 180–81, **183**
Connective tissue proper, 180
Connectives, 355
Connell, Joseph, 796
Conodonts, 481, **483**
Conotoxins, 326
Conraua goliath, 524, **525**
Constant region, of antibodies, 742
Consumers, 800
Contact chemical receptors, 709
Continental drift theory, 784–86, **785**
Contour feathers, 560–61, **561**
Contraception, 145
Contractile heart, 366
Contractile proteins, 622
Contractile vacuole, 41, 213, **213**, 218, 223, **224**, 227, 639
Control center, 663
Controls, in experiments, 14
Conus, 326, **327**
Conus arteriosus, 528
Convergence, 339
Convergent evolution, 339
Coots, 579
Cope, Edward Drinker, 177
Copepoda, *377*, 388, **388**, 392, 394
Cope's law of phyletic increase, 177
Coprophagy, 591
Copulatory spicules, 298–99
Coral, **239**, 248–49, 260, **263–64**
  hermatypic, 264–65
  hexacorallian, 263, **264**
  horny, 266
  octocorallian, 264, **265–66**
  soft, **266**
  thorny, 264, **265**
Coral reef, *175*, 249, 264–67
Coralline algae, 264

Corbicula fluminea, 333
Coreceptor molecules, 743
Cormorants, **560,** 579
Corn ear worms, **417**
Cornea, 383, 717
Cornification, 613
Corona, 306
Coronary circulation, 664
Corpora allata, 411–12
Corpora cardiaca, 411–12
Corpus luteum, 144, **147**
Corrodentia, 419
Cortex, 642
    of adrenal gland, 732, **733**
Cortical reaction, 154
Corticosterone, 733
Cortisol, 733, **733**
Corymorpha, **255**
Cosmine, 500
Countercurrent flow, 504, 668
Countercurrent heat exchange, 649, **650**
Countercurrent multiplication, 645
Coupled reaction, 60–61, **61**
Coupled septa, 260
Coxa, 379
Coxal glands, 369
Coxopodite, 379
Crabs, 392–93, **394**
    parasites of, 390, **390**
    sperm of, **139**
Cranes, 579
Cranial nerves, 479
Craniata. *See* Vertebrata
Craspedacusta sowerbii, 256, **256**
Crayfish, 393, **640**
    appendages of, 379, **380–81**
    development of, 384
    excretory system of, 382
    external structure of, **379**
    feeding habits of, 386
    internal structure of, **382**
    muscular system of, 380
    respiratory system of, 380
    sense organs of, 383
Creatine phosphate, 630
Creation myths, 103
Creationism, 111
Creation-science, 11
Cretaceous extinction, 127, **128**
Cretin, 730
Cricetidae, 607
Crick, Francis, **8,** 74, 90
Crickets, **141,** 400, 414, 419
Crinoidea, 443, 456–57, **457,** 458, 460
Cristae, 41–42
Cristatella, **433**
Crocodiles, **540,** 550–52, **552,** 614
Crocodylus acutus, 552
Crocodylus niloticus, 552, **552**
Crocodylus porosus, 552
Crop, 353, 565, 684
Cross-breeding, 17
Cross-fertilization of crops, insects and,
    416–17
Crossing over, 86–88, **87**
Crotalinae, 548–49
Crotalus molossus, **548**
Crow, skeleton of, **563**
Crown, 456
Crown-of-thorns star, **444,** 447
Crustacea, 374, 377–94
    adaptive radiation of, 394
    circulatory system of, 381–82, **382**
    cladogram of, **395**
    classification of, 392
    development of, 384
    endocrine function of, 384–86
    excretory system of, 382, **383**
    external structure of, 378–80, **379**
    feeding habits of, 386, **386**
    form and function of, 378–86
    immunity in, **748**
    internal structure of, 380–83
    muscular system of, 380
    nature of, 378
    nervous system of, 382–83
    phylogeny of, 393–94
    reproduction of, 383–86

respiratory system of, 380
sensory organs of, 383
subdivisions of, 386–93
Crypsis, 413, **413**
Cryptic defenses, 798
Cryptobiosis, 437
Cryptochiton, 322
Cryptosporidium, 231
Cryptosporidium parvum, 220
Crystalline style, 333
Crystallography, 38
Ctenidium, 318, 320, **320,** 326, **327, 333**
Ctenoid scales, 499, **499**
Ctenophora, 184, 267–70
    adaptive radiation of, 271
    characteristics of, 268
    classification of, 270
    comparison with Cnidaria, 268
    diversity of, **270**
    gastrulation in, 160
    phylogeny of, 270–71
Ctenoplana, 270
Cuboidal epithelium, **181**
Cubozoa, 249, 259–60, 267
Cuckoos, 579
Cuculiformes, 579
Cuculus canorus, 579
Cucumaria, 455, 460
Cucumaria frondosa, 455
Cucumaria miniata, **455**
Culcita navaeguineae, **444**
Culex, 222, **411**
Cupula, 712, 715
Curculio proboscideus, **421**
Curculionidae, **421**
Cutaneous leishmaniasis, 219
Cutaneous respiration, 667
Cuticle, 297, 365, 384, **384,** 613
Cuticular exoskeleton, 364–66
Cuttlefish, 336, **339–40**
Cyanea, 257
Cyanea capillata, **258**
Cyanobacteria, 31–32
Cybister fimbriolatus, 404
Cyclic AMP, 723
Cyclic GMP, 723
Cyclin, 52, **53**
Cyclin-dependent kinases, 52
Cycliophora, 312
Cycloid scales, 499, **499**
Cyclops, 388, 392
Cyclospora, 231
Cyclospora cayetanensis, 220
Cyclosporine, 745
Cydippid larva, 269
Cynodonts, 584, **587**
Cynomys ludovicianus, **599**
Cyphoma gibbosum, 326, **329**
Cypridina, 392
Cypris, 389, 392
Cystacanth, 310
Cysteine, **25**
Cysticerci, 288
Cysticercosis, 288, **289**
Cystid, 432
Cystoidea, 460
Cysts, formation of, 215–16
Cytokines, 729–30, 738, 743, **743–44**
Cytokinesis, 50–51
Cytology, 37–39
    comparative, 196
Cytopharynx, 222–23
Cytoplasm
    gray-crescent, 160
    of protozoa, 209, **209**
Cytoplasmic division, 51
Cytoplasmic localization, 153
Cytoplasmic specification, 161
Cytoplasmic streaming, 43–44
Cytoproct, 214, 223
Cytopyge, 214
Cytosine, 89, **89**
    hydrogen bond of, 90, **90**
    methylation of, 95, **95**
Cytoskeleton, 42
    of cell, **44**
Cytostome, 214, 223
Cytotoxic T lymphocytes, 743

Dactylogyrus, 284, 288
Dactylozooids, 257
Daddy longlegs, 372
Daily torpor, 651, **651**
Damselfishes, 262
Damselflies, 407, 410, 419
Danaus plexippus, 404, **410**
Daphnia, 253, **387,** 392
Daphnia pulex, **3**
Darwin, Charles, 14, **14,** 103, **103, 105,** 601
    Beagle and, **104,** 104–6, **105**
    on behavior, 752, 761
    on earthworms, 353
    house of, **106**
    theory of evolution by. *See* Evolutionary
        theory, Darwinian
Darwin's frog, 520
Dasyatis americana, **496**
Dasypeltis, **680**
Dasypeltis scaber, 791
Dasypus novemcinctus, 607
DDS. *See* Dosage-sensitive sex reversal
Decapoda, 392–93, **394**
Deciduous forest, temperate, 776
Deciduous teeth, 590
Decomposers, 800
Deer, **589,** 608, **609**
Defecation, 688. *See also* Excretory system
Defenses, 798–99, **799**
    of cephalopods, 339
    of insects, 413
    social behavior and, 759
Defensins, 741
Definitive host, 281
Dehydration, 638, 645–46
Dehydration reactions, 30
Deinonychus, 551
Delayed implantation, 597
Delayed type hypersensitivity (DTH), 746
Deletion, 88
Demes, 791
Demodex, 373
Demodex folliculorum, **373**
Demographics, 791
Demography, 791
Demospongiae, 237
Dendrites, **185,** 696
Dendrobatid frogs, 526
Dendrocystites, **458**
Dendronephthya, 176, 264, **266**
Dendrosoma, 226
Dense connective tissue, 180, **183**
Density gradient, 38, **38**
Density, of water, 28
Density-dependent growth factors, 795
Density-dependent population factors, 600
Density-independent growth factors, 794–95
Density-independent population factors, 600
Densraster, 452
Dentalium, 323, **323,** 343
Deoterthron, 392
Deoxyhemoglobin, 674
Deoxyribose, 88, **89**
Deoxyribose nucleic acid (DNA), 6–7, 26, 74
    base sequences in, 91, **92**
    chemical components of, **89**
    damage to, 91, 98
    mitochondrial, 33
    modifications to, in eukaryotes, 95
    nuclear, 33
    parasitic, 97
    plastid, 33
    recombinant, **95,** 95–96
    replication of, 6, **92**
    structure of, **8, 89,** 89–90, **91**
Depolarization, 698
Deposit feeders, 678, **680**
Derived character states, 195
Dermacentor, 373–74
Dermacentor variabilis, **372**
Dermal bone, 614
Dermal branchiae, 445
Dermal endoskeleton, 443, 445
Dermal ostia, 237
Dermal papulae, 668
Dermaptera, 419, 424
Dermasterias, 262, 445
Dermatophagiodes farinae, 373, **373**
Dermis, 587, **588,** 613–14

Dermoptera, 606
Dero, 357, **358**
Derocheilocaris, 392
Descent of Man, The (Darwin), 109–10, 601
Desert, 778
Desmognathus, **521**
Desmognathus wrighti, **521**
Desmosomes, 44, **45**
Desmospongiae, **233, 239, 241,** 242, **243,**
    243–44, **244,** 245
Determination, 152, 743
Detorsion, 324
Deuterostomia, 427, 443, 468, 473
    cladogram of, **454**
    classification of, 204, **205**
    cleavage in, 156
    development of, **157**
    gastrulation in, 158
Development, 8–9, **9**
    of Asteroidea, 449–50
    of birds, 575–76
    of Bivalvia, 334
    of Crustacea, 384
    direct, 155
        of insects, 411
    early, 155–58
    of earthworms, 356, **357**
    gene expression during, 162–65
    indirect, 155–56
    of Insecta, 410–12
    key events in, 152, **152**
    of Mammalia, 166–68
    mechanisms of, 160–65
    mosaic, 161, **161**
    of organs and systems, 168–71
    of Polychaeta, 350
    principles of, 151–73
    regulative, 161, **161**
    of Reptilia, 539
    of Vertebrata, 165–68
Developmental biology, evolutionary, 164–65
Developmental decisions, hierarchy of, 152
Developmental mode, yolk and, 155–56
Devonian period, tetrapods in, 515–16
DeVore, Irven, 768
Dextral shell, 324, **324**
Dextrose. *See* Glucose
DHT. *See* Dihydrotestosterone
Diabetes mellitus, 644, 735
Diabetogenic hormone, 735
Diacylglycerol, 723
Diadema, 452–53
Diadema antillarum, 453, **453,** 799
Diapause, 412–13
Diapheromera femorata, **402**
Diaphragm, 670
Diapsida, 535, 542–52, 583–84, **584**
Diaptomus, 388, 392
Diarrhea, 220
Diastole, 662, **663**
Dicrostonyx, **600**
Didelphis marsupialis, 194, **598**
Didelphis virginiana, 606
Didinium, 214, 222, **798**
Diencephalon, **705**
Dientamoeba, 230
Difflugia, 227, **227,** 231
Diffusion, 45–46
    facilitated, 47
    lungs and, **671,** 671–72, **672**
Digenia, 281–82
Digestive glands, 447
Digestive system, 682–83. *See also* Alimentary
        canal; Feeding habits; Nutrition
    of Anura, 529
    of arthropods, 366
    of Asteroidea, 446–48
    of birds, 565
    of Bivalvia, 332–33, **333**
    of Ctenophora, 269
    of Enteropneusta, 462
    extracellular, 277, 682
    of fishes, 507
    of hydra, 253–54, **254**
    of Insecta, 404
    intracellular, 240, 277, 682, **682**
    of Mammalia, **592**
    of Nemertea, 290–91

of platyhelminthes, 277
regulation of, 689–90, **690**
of sponges, 240
Digestive tube, 170
Dihydrotestosterone (DHT), 136, 144
1,25-dihydroxyvitamin D, 732
Dileptus, **222**
Dilger, W. C., **756**
Diloboderus abderus, **402**
Dimer, **51**
Dimorphism
in Cnidaria, 249–50
sexual, 428
Dinoflagellates, 226, **227,** 231
Dinomischus, **108**
Dinosaur Provincial Park, Alberta, **109**
Dinosaurs, 546, **551**
birds and, 556–57
extinction of, **127**
skeleton of, 107, **109**
Dioctophyme, 302
Diodora, 327–28
Diodora aspera, **329**
Dioecious animals, 133, 509
Diphycercal fin, **498,** 500, **502**
Diphyllobothrium, 288, 388
Diphyllobothrium latum, **287,** 288
Diphyodont, 584
Diploblastic animals, 160, 250
Diplodocus, 551
Diploid amictic eggs, 308
Diploid chromosomes, 76, 133
Diplomonadea, 230
Diplomonadida, 229–30
Diplopoda, 393, 398–99, **399,** 418, 420
Diplostraca, 392
Diplura, 419
Diptera, 402–3, 405, 421
Dipylidium caninum, **287**
Direct development, 155, 521
of insects, 411
Direct diffusion, 667
Direct flight muscles, 403
Direction, of chemical reaction, 60
Directional selection, 125, **125**
Dirofilaria immitis, 303, **304**
Disaccharidases, 686
Disaccharides, 22–23, **23**
Discoidal cleavage, 157
Disease, definition of, 221
Disjunct distribution, 781, **782**
Dispersal, distribution by, 781–82
Display, communication by, 767–68
Disruptive selection, 125, **125**
Distal, 185
Distal convoluted tubule, 642
Distance chemical receptors, 709
Distribution
disjunct, 781, **782**
by dispersal, 781–82
of life on earth, 773–86
by vicariance, 782–83
Disulfide bonds, 25
Diversity
behavior and, 757–58
of species, 790
Diving beetles, 404, 407
Dizygotic twins, 148, **149**
DNA. *See* Deoxyribose nucleic acid
DNA ligase, 91, 95, **95**
Dobsonflies, 420
Dobzhansky, T., 97, 194
Dog tapeworm, **287–89**
Dogfish shark, 493–94, **495–96**
Dogs, 607
parasites of, 300, 303, **304**
sperm of, **18**
Doliolaria, **450,** 457
Dolomedes triton, **370**
Dolphins, 609
Domestic fowl, 579
Domesticated mammals, 601
Dominance, 760–62, **761, 763**
incomplete, 81, **81**
Dominant traits, 79
Dormancy, 412–13
Dorsal, 185, **187**
Dorsal aorta, 476
Dorsal fin, 494, 500

Dorsal tubular nerve chord, 469–72
Dorsal ventricular ridge, 567
Dorsal vessel, 354
Dosage-sensitive sex reversal (DDS), 137
Double circulation, 515, 661
Double helix structure, **8,** 90, **91**
Doves, 579
Down feathers, 561, **561**
Down, John Langdon, 87
Down syndrome, 87–88
Dracunuculus, 388
Dragonflies, 401, 403–4, 407, **412,** 419
Dreissena polymorpha, 333
Driesch, Hans, 160
Dromeosaurs, 557
Dromidia, 393
Dromidia antillensis, **394**
Drone, 415
Drosophila
embryo of, **157**
flight of, 403
homeotic genes in, 163, **163**
inheritance of eye color in, 85–86, **86**
melanogaster, 17
mutations in, 98
pattern formation in, 162–63
ventral nerve cord development in, 164
Drosophila melanogaster, 17, 75, **86,** 87
DTH. *See* Delayed type hypersensitivity
Dual-gland, 276, **277**
Dubois, Eugene, 601
Duck-billed platypus, 597, 606
Ducks, 579
Dugesia, 276, 288
Dung beetles, 414, **414**
Duodenum, 685–86
Duplication, 88
Dusky salamander, **521**
Dwarf tapeworm, **287**
Dwarfing, 117
Dwarfism, 735
Dyads, 78
Dynein, 623
Dysentery, 225, 227
Dysonia, **413**
Dytiscus, 407

Eagles, 579
Eardrum, 568
Ears, 712–15. *See also* Hearing
of birds, 568
of frogs, 530, **530**
human, 713, **714**
of moths, 712–13, **713**
Earth, *772*
atmosphere of, 773, **773**
distribution of life on, 773–86
Earthworms, 353–56, **354**
behavior of, 356
circulatory system of, 353–54, **660**
excretory system of, 354–55,
639–40, **640**
form and function of, 353–56
locomotion of, 616–17, **617**
nervous system of, 355, **355–56**
nutrition of, 353
reproduction of, 356, **357**
respiratory system of, 354
sensory organs of, 356
Earwigs, 419
Eccrine glands, 589–90
Ecdysial glands, 411–12
Ecdysis, 366, 384–85, **385,** 411, **411,** 412
Ecdysone, 412, 724
Ecdysozoa, 296, 312, 340, 427, 434–38
Echeneis naucrates, **501**
Echidnas, 589, 606
Echinarachnius, 452
Echinaster luzonicus, **449**
Echiniscus, 436–37
Echinococcus granulosus, **286–87,** 289, **289**
Echinococcus multilocularis, **287**
Echinodera, 305
Echinoderes, 305, **305**
Echinodermata, 443–60
adaptive radiation of, 460
biological contributions of, 443
body plan of, **186**
characteristics of, 445

chordates and, 473
cladogram of, **459**
classification of, 460
cleavage in, 156
fossils of, **108, 473**
immunity in, **748**
larvae of, **450**
phylogeny of, 458–60
position in animal kingdom, 443
reproduction of, 133
Echinoidea. *See* Sea urchins
Echinopluteus larva, **450,** 454
Echiura, 156, 427, **428,** 428–29, **429,** 439
Echiurida, 439
Echiurus, 428, **428**
Echolocation, 595–96, **596**
Ecocline, 775
Ecological pyramids, **802,** 802–3
Ecology, 9, 789–805
of communities, 795–99
definition of, 789
hierarchy of, 790–804
Ecosystems, 790, 800–804
Ectapracta, 439
Ectoderm, 158
derivatives of, 169–70
Ectognathous, 422
Ectolecithal development, 279
Ectoneural system, 448
Ectoplasm, 209, 223, 227, 623
Ectoprocta, 133, 296, 312, **426,** 427, 430–33,
432–33
Ectotherms, 647–49, **648**
Ectyoplasia ferox, **244**
Eels
feeding habits of, 506, **506**
lack of scales in, 499
locomotion of, 503, **503**
migration of, 507–8, **508**
respiratory system of, 504
Effect macroevolution, 127
Effectors, 696, 703
Efferent arteriole, 642
Efferent division, of peripheral nervous
system, 707
Efferent neurons, 696, **696,** 702–3
Egestion, 678
Egg(s), 48, 75, 133, 139
activation of, 154–55
amictic, 308
centrolecithal, 155
contact with sperm, 153, **153**
of gastropoda, 327, **328**
human, *131,* 140
isolecithal, 155, **156**
mesolecithal, 155, **156**
number of chromosomes in, 78
telolecithal, 155, **156**
Egg-recognition protein, 153
Eimeria, **219,** 220, 231
Eimeria tenella, 220
Ejaculatory duct, *142,* 282
Elapidae, 548
Elasmobranchii, 493–97, 510, 638
Elassochirus gilli, **394**
Elastic arteries, 664
Eldonia, **108**
Eldredge, Niles, 117
Electra pilosa, **433**
Electric eel, 504
Electric rays, 496–97, **497**
Electrical energy, 27
Electrical synapses, 699
Electron microscope, 37–38, **38**
Electron transport chain, 64
Electron transport, chemical bond energy
and, 61–62
Electrophoresis, **124–25**
Electrophorus, 504
Electroreceptors, 495
Eleocytes, 353
Elephant seals, **763**
Elephant shrews, 606
Elephantiasis, 303, **304**
Elephants, 608, **681,** 761
trunk of, 617, **617**
Elephas maximus, 608
Eleutherodactylus iberia, 524
Eleutherozoa, 460

Elliptical wings, 570, **571**
Ellisella, **266**
Elton, Charles, 803
Eltonian pyramid, 802–3
Embioptera, 419, 424
Embolus, 664
Embryo(s), **168**
similarity of, **113**
of vertebrates, **166**
Embryonic diapause, 597
Embryonic induction, 161–62
Embryonic period, 168
Emergence, 5
Emergent properties, 5, **6**
Emigration, 782
Emlen, Stephen, 572–73
Empedocles, 103
Emperor scorpion, **372**
Emus, 578
Encephalization, 702
Enchytraeus, 357
Encope grandis, **453**
Encope micropora, **453**
Encystment, protozoa, 215–16
End sac, 382
Endamoeba, 227
Endamoeba blattae, 227
Endangered species, 583
amphibians, 525
coral reefs, 267
humans and, 601
songbirds, 576–77
Endergonic reaction, 58
Endites, 379
Endochondral bone, 497, 618
Endocrine events, in reproduction, 143–48
Endocrine glands, 722
of vertebrates, 724–36
Endocrine system, 721–37
of Crustacea, 384–86
of Vertebrata, 479
Endocuticle, 365
Endocytosis, 45, 47–48
bulk–phase, 48
receptor–mediated, 48, **49**
Endoderm, 158
derivatives of, 170
Endognathy, 422
Endolecithal development, 279
Endometrium, 143
Endonucleases, restriction, 95
Endoplasm, 209, 223, 227, 623
Endoplasmic reticulum, 41, **41–42**
rough, **37,** 41, **42**
smooth, 41
Endopod, 379, **380–81**
Endopodite, 379
Endorphins, 729
Endoskeleton, **617**
of Anura, 527
of Chordata, 468
dermal, 443, 445
of Vertebrata, 477, 479
Endostyle, 472, 474
Endothelial cells, 665
Endotherms, 647–48, **649,** 649–50, **650**
Energy
activation, 58
budget for, 800–803
chemical bond, 36
electrical, 27
flow of, 800–803, **803**
free, 26, 58
kinetic, 57
laws governing, 9–10, 36, 57, 802
for muscle contraction, 630, **631**
potential, 57
pyramid of, **802,** 802–3
solar, 56, **57**
storage in tendons, 630–31, **631**
transfer of, by ATP, 60–61
Energy-coupling agent, 61
Enkephalins, 729
Enopla, 291
Enoplea, 302
Ensatina, geographic variation in, 192, **192**
Entamoeba, 227, 231
Entamoeba coli, 227
Entamoeba dispar, 227

Entamoeba gingivalis, 227
Entamoeba histolytica, 215, 225, 227
Enterobius, 302
Enterobius vermicularis, **300,** 302, **302**
Enterocoelomate, body plan of, **186**
Enterocoelous development, 185–87, **188**
Enterocoely, 160
Enteropneusta, **461,** 461–62, **462**
Entodesma saxicola, **332**
Entodinium, 225, **226**
Entomology, 400
Entoprocta, 296, **311,** 311–12
Entropy, 10, 56, **57**
Environmental influence
    moths and, **11,** 11–12
    in sex determination, 78–79
Environmental interaction, of living systems, 9
Enzyme-catalyzed reactions, 60
Enzymes, 58–60
    action of, **59,** 59–60
    activation energy and, 58
    as catalysts, 58, **58**
    digestive, 682–**83**
    main-line, 59–60
    nature of, 59
    proteins as, 25–26
    regulation of, 70, **70**
    specificity of, 60, **60**
Enzyme-substrate complex, 59
Eocene epoch, horses in, 109, **110**
Eon, 107
Eosinophilia, 740
Eosinophils, **657,** 658, 740
Ephelota, 226, 231
Ephemeroptera, 419, **419**
Ephyrae, 259
Epicuticle, 365, 384, 613
Epidemiology, definition of, 221
Epidermal nerve plexus, 448
Epidermal placodes, 478–79
Epidermis, 252, 587, **588,** 613
Epididymis, 142, **142**
Epidinium, 231
Epigenesis, versus preformation, 152–53
Epinephrine, 734
Epipelagic zone, 780
Epiphanes, 309
Epipod, 379
Epipodite, 379
Epistasis, 83
Epistome, 431
Epithelial cells, ciliated, **5**
Epithelial tissue, 180, **180**–**82**
Epitheliomuscular cells, 252, **253**
Epithelium, 180, **181–82**
Epitokes, 350, **350–51**
Epitope, 743
Epochs, 107
Eptatretus stouti, 490
Eptesicus, 606
Equilibrium, 715
    genetic, 121–22
        upset of, 122–24
    Hardy-Weinberg, 121
    punctuated, 117–18, **118**
Equus, 109
Eras, 107
Erethizontidae, 607
Ergasilus, 392
Erythroblastosis fetalis, 748
Erythroblasts, 658
Erythrocytes, 479, 566, 657–58
Eschrichtius robustus, **393**
Esophagus, 684
ESS. *See* Evolutionary stable strategy
Essential nutrients, 690
Esthetes, 322
Estigena pardalis, **413**
Estivation, 412
Estradiol, **144**
Estrogen, 136, 144, **144, 146**
Estrous cycle, 143, 596
Estrus, 143, 596
Estuarine crocodile, 552
Estuary, 779
Ethology, 753
    principles of, 754–55
Eubacteria, 32
Eucalyptus tree, **31**

Eucarya, 203
Euchromatin, 97
Eucidaris thouarsi, **7**
Eucidaris tribuloides, 7, **453**
Eucoelomates, **187**
    Bilateria, 185–87, **187**
    body plan of, **186, 296**
Eudorina, **217**
Euglena, 10, **10,** 218, **218,** 230
    fission in, **215**
    nutrition of, 214
    reproduction of, **132**
Euglena oxyuris, 218
Euglena viridis, 218
Euglenicha, 218
Euglenida, 214, 230
Euglenoidea, 230
Euglenozoa, 218–19, 230–31
Euglypha, 211, **215**
Eukaryotes, **33**
    appearance of, 32–34
    definition of, 10
    gene regulation in, 94–95
    oxygen production and, 32
Eukaryotic cells, components of, 39–42
Eumenes, 414
Eumetazoans, body plan of, **186**
Eunice viridis, **350**
Eupentacta quinquisemita, **457**
Euphausiacea, 391–92, **393**
Euplectella, **241,** 242–43, 245
Euploidy, 87
Euplotes, **222,** 223, 231
European corn ear worms, 414
European gypsy moth, 414
Eurycea longicauda, **523**
Euryhaline animals, 635
Euryhaline fishes, 505
Euryphagous animals, 565
Eurypterida, 367, **367,** 374
Euspongia, **238**
Eustachian tube, 713
Eusthenopteron, 516, **517**
Eutely, 296, 309
Eutheria, **591,** 597–98, 606
Evaginations, 667
Evaporative cooling, 649
Even-toed hoofed animals, 608–9
Evergreen forest, 777
Eversible proboscis, 290
Evo-devo, 164–65
Evolution, convergent, 339
Evolutionary developmental biology, 164–65
Evolutionary sciences, 13–14
Evolutionary species concept, 194
Evolutionary stable strategy (ESS), 765
Evolutionary taxonomy, 197–201
Evolutionary theory, 3, 12–13
    chemical, 26–30
    versus creation–science, 11
    Darwinian, 14–17
        common descent in, 15, 109–12
        evidence for, 106–20
        gradualism in, 15, **16,** 115–18, **117**
        modern, 120
        multiplication of species in, 15,
            112–15
        natural selection in, 15–16, **16,** 118–20
        origins of, 103–6
        perpetual change in, 14–15, 106–9
        revisions of, 120
    founders of, 103, **103**
    genetic code and, 7
    of kidneys, 641
    of Mammalia, 583–87
    pre-Darwinian, 103–4
    transformational, 103
    variational, 103
Evolutionary tree, 15
Evolutionary trends, 109
    diversity in, 109, **111**
Excision repair, of DNA, 91
Excitation–contraction coupling, 628–30, **629**
Excitatory synapses, 701
Exclusion, competitive, 797
Excretory canals, 286
Excretory system, 678
    of arthropods, 366
    of Asteroidea, 445–46

of birds, 567
of Crustacea, 382, **383**
of earthworms, 354–55
of Enteropneusta, 462
of Hirudinea, 359
of Insecta, 408, **408**
of invertebrates, 639–40
of Nemertea, **291**
of platyhelminthes, 277–78
of polychaeta, 349
of protozoa, 213
of spiders, 369
Excretory tubule, 382
Excurrent siphon, 474
Excystment, protozoa, 215–16
Exergonic reaction, 58
Exites, 379
Exocrine acinar cells, **734**
Exocrine glands, 722
Exocuticle, 365, 384
Exocytosis, 48
Exons, 92
Exopod, 379, **380–81**
Exopodite, 379
Exoskeleton, 617
    cuticular, 364–66
    of protozoa, 208
    of Vertebrata, 477
Experimental sciences, 13–14
Expiration, 670
Exponential population growth, 794
*Expression of the Emotions of Man and
    Mammals, The* (Darwin), 752, 761
Extensors, 380
External buds, 240
External ear, 568
External nares, 669
External respiration, 666
Exteroceptors, 709
Extinction
    mass, 127–28
    through geological time, 126–27
Extracellular digestion, 277, 682
Extracellular fluids, 656, **656**
Extracellular space, 178
Extraembryonic membranes, 165–67, **167**
Extrinsic limits on population growth, 793–95
Eye(s), 715–18. *See also* Vision
    of amphibians, 530, **530**
    of birds, 568–69
    of Bivalvia, 334
    of cephalopods, 338–39, **339**
    color of, 83–85, **86**
    of crustaceans, 383, **383**
    of Gastropoda, 326–27
    of insects, 409
    of maxillopods, 388
    of Mollusca, 317
    parietal, **550**
    of Polychaeta, 350
    of spiders, 369
Eyespots, 218, 308
Eyestalks, crustacean, 385

F_1 generation, in Mendelian heredity, 17
F_2 generation, in Mendelian heredity, 17
Fab, 742
Facilitated diffusion, 47
Facilitated transport, **47**
Fairy shrimp, **387,** 392
Falconiformes, 579
Falcons, 579
Fangs
    of snakes, 548–49
    of spiders, 369
Fanworms, 351, **679**
Fascicle, 405, 624
Fasciola, 288
Fasciola hepatica, **277,** 282, **282**
Fasciolaria, 326
Fasciolopsis buski, **282,** 284
Fast block, 154
Fast fibers, 630
Fast flight muscles, of insects, 625
Fats
    neutral, 24, **24**
    stored, 68–69
Fat-soluble vitamins, 690, **691**
Fatty acids, 24, **24,** 68–69, **691**

Fc, 742
Featherduster worms, **347,** 351
Feathers, 556, 560–62, **561**
    colors of, 562
    molting of, 562, **562**
    origin and development of, 561–62
    types of, 561
Feather-stars, 456–57, 460
Feces, 688
Federal Animal Welfare Act, 12
Feedback, 145
Feeding habits. *See also* Digestive system;
        Nutrition
    of Annelida, **680**
    of Anura, 529
    of Asteroidea, 446–48
    of birds, 565
    of Bivalvia, 332, **334–35, 679**
    of Crustacea, 386, **386**
    of Ctenophora, 269
    of Enteropneusta, 462
    of fishes, 505–7, **679**
    fluids and, 681–82
    food masses and, 680–81
    of Gastropoda, 325–26
    of hydra, 253, **254**
    of lampreys, 492, **492**
    of Mammalia, 590–94, **591**
    of Nemertea, 291
    particulate matter and, 678
    of polychaeta, 351–53
    of snakes, 548, **679**
    of spiders, 369–70
Feeding mechanisms, 678–82
Feeding processes, 7–8, **8**
Felidae, 608
Female-defense polygyny, 763
Fermentation, 21
    alcoholic, 65
Fertilization, 132–33, 153–55
    and activation, 153–55
    timing of events in, **155**
Fertilization cone, 153, **154**
Fertilization membrane, 154
Fetal period, 168
Fetus, 168
Fever, 739
FGF. *See* Fibroblast growth factor
Fibrillar muscles, 625
Fibrillar portion of muscle cell, 297
Fibrin, 658, **659**
Fibrinogen, 658
Fibroblast growth factor (FGF), 164, **165**
Fibrosis, 746
Fiddler crab, **394,** 760
Fight or flight responses, 734
Filarial worms, 302–3, **303–4**
Filoplumes, 561, **561**
Filopodia, 169, 211
Filter feeding, 330, 678
Filtrate, 642
Filtration, 642
Fimbriae, 143, **143**
Final electron acceptor, 62
Final host, 281
Finches, **580**
    Galápagos, 115, **116**
Fins, 494, **498**
Fire corals, 249
Fireflies, 410, 414, **415,** 420
Firework, 351, **351**
First law of thermodynamics, 9–10, 57
First messenger, 723
First polar body, 139
Fish lice, 392
Fish louse, **387**
Fish tapeworm, **287,** 288
Fisher spiders, 370
Fishes, 486–513
    ancestry of, 487–89
    behavior of, 760
    biological contributions of, 487
    bony. *See* Bony fishes
    buoyancy of, 503–4, **504**
    cartilaginous, 487, 493–97, 510
    circulatory system of, **661**
    cladogram of, **489**
    classification of, 510
    definition of, 486

digestive system of, 507
eggs of, 155
embryos of, **113,** 165–66, **166**
family tree of, **488**
feeding habits of, 505–7, **679**
fossils of, 107
freshwater, 505, **506,** 509–10, 636
heart of, 661
hermaphroditism in, 134
integument of, 614, **615**
jawless, 468–69, 482, 487, 489–93, 510
larvae of, **167**
lateral line system of, 712, **712**
lobe-finned, 482, 489, 497, 500–502, 510, 515–16
locomotion of, **502,** 502–3, **503**
migration, 507–9
osmotic regulation in, 46, 505, **506,** 636–38, **637**
parasitic, 507
parthenogenesis in, 135
position in animal kingdom, 487
ray-finned, 482, 489, 497–500, **500,** 506–7, 510
reproduction of, 509–12
respiratory system of, 504
saltwater, 46, 505, **506,** 509–10, 636–38
scales of, 493–94, 499, **499,** 512, **512**
sex determination in, 78
skeleton of, 619–20, **620**
teleost. *See* Teleost fishes
vestibular apparatus of, 713, **713**
Fission, 214–15, **215**
Fitness of the Environment, The
(Henderson), 773
FitzRoy, Robert, 104
Flaccisagitta hexaptera, **438**
Flagella, 43, 218
of protozoa, **210,** 210–11
Flagellar movement, **623,** 623–24, **624**
Flagellated canals, 238
Flagellated chambers, 238–39
Flagellated spongocoels, 237–38
Flagellates, fission in, **215**
Flame cell, 276–77, 290, 308, 639, **639**
Flamingo tongues, 326
Flamingoes, 579, **579**
Flapping flight, 569, **571**
Flares, 746, **746**
Flatworms. *See* Platyhelminthes
Fleas, 404, **405,** 409, 417, 421
Flectonotus pygmaeus, **520**
Flexors, 380
Flies, **107,** 403, 405–6, 408, 410, 414, 417
Flight
of birds, 569–70, **570**
of insects, 402–4
of mammals, 595–96
Flight feathers, 561
Flight muscles, **404**
Flightless birds, 557, 560, 782, **783**
Flower animals. *See* Anthozoa
Fluid compartments, **656**
Fluid-mosaic model, of cell membranes, 39–41, **40**
Fluids
composition of, 656–57
feeding habits and, 681–82
internal environment of, 656–57
Flukes. *See* Trematoda
"Flying foxes," 606
Flying lemurs, 606
Flying squirrel, 595, **596**
Follicle, 143
Follicle-stimulating hormone (FSH), 144, **145,** 726, **727**
Follicular phase, of menstrual cycle, 144
Food chain, 802–3
Food intake, regulation of, 689–90
Food production, molecular genetics and, 96
Food vacuole, 41, 214, 223, 227
Food webs, 800, **801**
Foot
of Mollusca, 317–19
of Rotifera, 307
Foraminiferans, 228, **228**
Fore reef slope, 266
Forebrain, **705,** 706–7

Forelimbs
diversity of, **111**
in vertebrates, variation in, **16**
Forepart, 429
Forest(s)
boreal, 777
coniferous, 777, **777**
lake, 777
southern evergreen, 777
temperate deciduous, 776
tropical, **777,** 777–78
Formation of Vegetable Mould Through the Action of Worms, The
(Darwin), 353
Fossils, 30, *102,* 106–7, **107–8,** 117–18, 340
of amebas, 228–29
of antelopes, 130, **130**
of Arthropoda, **367**
of birds, 556, **557**
of bony fishes, 497–98
of cephalopods, 335
of chimaeras, 497
of Echinodermata, **473**
evolutionary trends and, 109
of insects, 421
interpretation of, 107
of sponges, 237
of Vertebrata, 478
Fossorial animals, 649
Fouling, 324
Founder event, 113–14
Four-chambered stomach, 592
Four-o'clock flower, 81
Fovea, 568
Fovea centralis, 717
Fox, Sidney, 30
Fragmentation, 133
Franklin, Rosalind, 90
Fraternal twins, 148, **149**
Free energy, 26, 58
Free-tailed bats, 606
Fresh water, 636
Freshwater fishes, 505, **506,** 509–10, 636
Freshwater medusae, 256
Freshwater oligochaetes, 357, **358**
Friedman, J., 736
Fringing coral reefs, 265
Frog-legs, 523
Frogs, 516, 523–32, **524**
blastula and gastrula of, **159**
brain of, 529, **529,** **704**
ears of, 530, **530**
eggs of, 140, **156**
feeding and digestion in, 529
forelimbs of, **111**
gastrulation in, 158
habitats and distribution of, 524–26
heart of, 528, **529, 662**
integument and coloration of, 526
life cycle of, **531**
lungs of, **669**
metamorphosis of, 514, *514,* 523
osmoregulation in, 636, **637**
reproduction of, **132,** 519–20, **520**
respiratory system of, 669, **670**
thyroid gland of, 730, **731**
vision of, 530
Frontal plane, 185, **187**
Fructose, 23, **23**
Frugivores, **591**
Fruit flies
allopatric speciation and, 114
in chromosomal genetics investigation, 75
flight of, 403
homeotic genes in, 163, **163**
inheritance of eye color in, 85–86, **86**
mouthparts of, 405
mutations in, 98
pattern formation in, 162–63
ventral nerve cord development in, 164
FSH. *See* Follicle-stimulating hormone
Fulmars, 579
Fulton, P., 312
Funch, P., 312
Functional genomics, 97
Fundamental niche, 791
Fundulus heteroclitus, 791
Fungi, 10, 234

Funnel, 336
Funnelweb spiders, 371
Fur seals, migration of, 594, **595**
Furcula, 562

Galactose, 23, **23**
Galápagos finches, 115, **116**
Galápagos Islands, 105, **105,** 115
Galápagos tortoises, 105, **105,** 541, **541**
Galen, 655
Galeocerdo, 494
Galeopithecus, 606
Gallbladder, 688
Galliformes, 579
Gallinules, 579
Galvanotaxis, 224
Gametes, 17, 48, 75, 132. *See also* Egg(s); Sperm
reduction division of, 75–78, **77**
reproduction with, 133–35
reproduction without, 133
Gametogenesis, 136, 138–40
Gamma globulins, 742
Gammarus, 391–92
Ganglia, 696
Gannets, **574,** 579
Ganoid scales, **499**
Ganoin, 498
Gap genes, 163
Gap junctions, 44, **45**
Garden centipedes, 418
Garden peas, Mendel's experiments on, 75, **76**
Gardner, Allen, 768
Gardner, Beatrix, 768
Gars, 498–99, **500,** 510
Garstang, Walter, 479
Gas chambers, 336
Gas exchange. *See also* Respiratory system
by direct diffusion, 667
of insects, 406–7, **407**
lungs and, **671,** 671–72, **672**
through tubes, 667–68
Gas gland, 504
Gastric filaments, 258
Gastric glands, 308
Gastric juice, 685
Gastric mill, 386, **386**
Gastric pouches, 258
Gastrin, 689, **690**
Gastrodermis, 252–53
Gastroliths, 384
Gastropoda, 317, **318,** 321, 324–29, 341, 343
coiling and, 325, **325**
feeding habits of, 325–26
head of, **319**
internal form of, 326–27, **327–28**
major groups of, 327–29
reproduction of, 327
shell of, 324, **324,** 325
torsion and, 324–25, **325**
Gastrotricha, 310–11, **311**
Gastrovascular cavity, 250, 252, 261, 277
Gastrozooids, 255
Gastrula, **159**
Gastrulation, 158–60, **159**
Gause, G. F., **798**
Gavia immer, 578
Gaviiformes, 578
Gecko, 542, **543**
Geese, 579
Gekko gecko, **543**
Gel electrophoresis, genetic variation and, **124**
Gemmulation, 133
Gemmules, 240, **242**
Gene expression
during development, 162–65
regulation of, 94–95
Gene mapping, 87, 97
Gene pool, 120
Gene regulatory protein, 723
Gene theory, 88
Gene therapy, 96
General area cladogram, 783, **783**
Genes, 75
concept of, 88
definition of, 88
in eukaryotes, 94–95
homeotic, 163

interaction of, 83–84
mutations of, 97–98
rearrangement of, 95
regulation of, 94–95
Genetic approach, to heredity, 17, **18**
Genetic code, 7
Genetic drift, 122–23, **123,** 124
Genetic equilibrium, 121–22
upset of, 122–24
Genetic information
storage of, 88–97
transportation of, 88–97
Genetic program, 6–7
Genetic variation, measurement of, **124,** 124–25
Genetics
of behavior, 755–56, **756**
molecular, 95–97
of cancer, 98
principles of, 74–101
Genital ligaments, 310
Genital ridge, 136
Genomics, 97
Genotypes, 79
Genus, 192
Geographical isolation, speciation and, 113
Geological time, 107–9, 126–27
Geomydiae, 607
Geophilus, 418
Geospiza fortis, 797, **797**
Geospiza fuliginosa, **580,** 797, **797**
Geospiza magnirostris, 797
Geotaxis, 224
Germ cell(s), 132, **138,** 168
maturation of, 136–40
migration of, 136
origin of, 136–40
primordial, 136, **137**
Germ cell line, 136
Germ layers, 158, 168
derivatives of, **169**
formation of, 158–60
Germ plasm, 136
Germinal period, 168
Germinal vesicle, 153
Germinative zone, 285
Germiovitellaria, 308
Gerris, 402, **403**
GH. *See* Growth hormone
Ghostfish, 497
Giant axons, 355
Giant horned beetle, **402**
Giantism, 735
Giardia, 216, **216,** 230
Giardia lamblia, 216
Gibbons, 606
Gigantocypris, 392
Gigantorana goliath, **525**
Gila monster, **544**
Gilbert, William, 97
Gill arches, survival of, 170
Gill pores, 462
Gill slits, 462, 504
Gills, 667–68, **668**
of Bivalvia, 331
of cephalopods, 337
of Chordata, 472
of Crustacea, 378, 380
of fishes, 504, **505**
of Gastropoda, 326
of insects, 407
of Mollusca, 317–18, 320, **320**
of rays, 496
of salamanders, 521
Giraffa camelopardalis, 760, **761**
Giraffes, 760, **761**
Girdles, 429
Gizzard, 353, 565, 684
Gland cells, in hydra, 252–53
Glans penis, **142**
Glass lizard, **544**
Glass sponges, 242–43
Glial cells, 696–97
Gliricola porcelli, **405**
Global temperature, **774,** 774–75
Globigerina, **227,** 229, 231
Globin, 672
Globulins, 658

Glochidium larva, 334, **336**
Glomerular filtration, 643
Glomerulus, 462, 642
Glossina, 218
Glottidia, 434
Glottis, 669
Glucagon, 735
Glucocorticoids, 733
Gluconeogenesis, 733
Glucose, 22–23, **23**
    chair representation of, **23**
    oxidation of, 64–65, **67**
    reabsorption of, 644, **644**
    structure of, **24**
    transport of, **24**
Glutamic acid, **25**
Glutathione, 253–54
Glycera, 349, 360
Glycerol, **24**
Glycine, **25**
Glycogen, 23–24, 630
Glycogen granules, **37**
Glycogen particles, **41**
Glycolysis, 62–64, **63**
    anaerobic, 65–66, **68**, 630
Glycoproteins, 41
Gnathostomata, 469, 482–83, 510
Gnathostomula jenneri, **292**
Gnathostomulida, 291–92
Gnathostome fishes, 487
Gnawers, 591, 607
Goats, 608
Goatsuckers, 580
Goiter, 731, **731**
Golden garden spider, **370**
Golden plover, migration of, **572**
Goldschmidt, Richard, **16**
Golfingia, 428
Golgi complex, 41, **42**
Gonadal steroids, 144, **144**
Gonadotropin-releasing hormone, 144
Gonadotropins, 726
    pituitary, 144
Gonads, 134
Gonangium, 255
Gondwana, 784, **785**
Goniobasis, 328
Gonionemus, **256**
Gonium, **217**
Gonophores, **252, 255**, 257
Gonopores, 379
Goose
    behavior of, 754, **754**
    brain of, **704**
Goosefish, **507**
Gooseneck barnacles, **387**
Gordius, 304
Gorgonia, 267
Gorgonian coral, **261, 266**
Gorgonocephalus, 452, 460
Gorgonocephalus eucnemis, **452**
Gorgonorhynchus, 289
Gorilla, **191,** 199, **201, 603,** 606–7
Gould, Stephen Jay, 117, 126
Graafian follicle, 144
Grade, 197
Gradualism, 115–18, **117**
    in evolution, 15, **16**
    phenotypic, 117, **117**
    phyletic, 117
    populational, 117
    punctuated equilibrium and, 117–18
Grant, Peter, 797
Grantia, 242
Granulocytes, 658, 740
Granulomas, 746, **746**
Granuloreticulosea, 228, 231
Grapsus, 392
Grapsus grapsus, **394**
Grasshoppers, 419
    external features of, 400, **401**
    internal features of, **404**
    mouthparts of, 406, **406**
    muscles of, 612
    nervous system of, **702**
    nutrition of, 404
    sound production in, 414
    in spider web, **370**
Grassland, 778, **778**

Gray tree frog, **527**
Gray whales, 594
Gray-crescent cytoplasm, 160
Grazers, 591
Great American interchange, 784, **786**
Great blue heron, 579
Great Lakes, lampreys in, 492–93
Great white shark, 493–94, **494**
Grebes, 579
Greek philosophy, 103
Green algae, 216, 229
Green anole, 192
Green frog, *1,* 524, **531**
Green glands, 382, **383**
Green sea turtle, **542**
Green tree frog, **524**
Green turtles, 541
Greenhouse effect, **774,** 774–75
Gregarina, 231
Gregarinea, 231
Grell, K. G., 236
Griffin, Donald, 768
Grizzly bear, *582, 607*
Gross productivity, 800
Ground finch, **580**
Ground substance, 180
Group selection, 763
Grouse, **575–76,** 579, **763**
Growth cone, 169, **171**
Growth hormone (GH), 726–27, **727,** 735
Growth rate, 791
Gruiformes, 579
Guanine, 89, **89**
    hydrogen bond of, 90, **90**
Guard hair, 587, **588**
Guide fossils, 107
Guilds, 797, **798**
Guinea pig, sperm of, **18**
Gulf shrimp, **384**
Gullet, 222–23
Gulls, 579, **579**
Guppies, 509
Gymnodinium, **227**
Gymnophiona, 520, 531
Gynecophoric canal, 283
Gypsy moths, 417
Gyrodactylus, 284, **285,** 288

Habitat, 790
Habituation, 757
Haeckel, Ernst, **15,** 111, 267, 789
Haeckelia rubra, 267
Haementeria, 358
Haementeria ghilianii, **358**
Hagfishes, 482, 487, 489, **490,** 490–91, 510
Haikouella, 478, **479**
Haikouichthys, 481
Hair, 582, 587–89, **588,** 614
Hair cells, 712
Haldane, J. B. S., 22, 26
Hales, Stephen, 664
Halichondrites, **108**
Haliclona loosanoffi, 242
Halicryptus higginsi, 306
Haliotis, 320, 325, 327–28
Haliotis refuscens, **326**
Hallucigenia, **108**
Halobates, 400, 402
Halteres, 402
Hamilton, W. D., 763
Hammer, 713
Hammerhead shark, *486,* 493–94, **494**
Haplodiploidy, 135, 302, 764, **764**
Haploid chromosomes, 76, 133
Haploid mictic eggs, 309
Haplopharynx, **277**
Haplorhini, 606–7
Hardy-Weinberg equilibrium, 121–22
Harem system of mating, 126
Hares, **588,** 600, **600,** 607
Harrison, Ross G., 169–70
Hartsoeker, Niklass, **152**
Harvestmen, 372, **372,** 374
Harvey, William, 131, 655
Hasler, A. D., 508
Haversian system, 618
Hawaii, allopatric speciation and, 114
Hawaiian honeycreepers, bill shapes in, **16**
Hawks, 579

Head
    of gastropods, **319**
    of insects, 400
    of molluscs, 318–19
    of vertebrates, 478
Head slits, 290
Head-foot, of molluscs, 318–19
Hearing, 712–15. *See also* Ears
    in birds, 568
    in frogs, 530, **530**
    human, 713, **714**
    of Insecta, 712–13
Heart, 170–71, 661. *See also* Circulatory system
    accessory, 337
    of amphibians, 661
    of Bivalvia, 333, **333**
    branchial, 337
    of cephalopods, 337
    contractile, 366
    control of, **663,** 663–64
    of Crustacea, 381, **382**
    excitation of, 663–64
    of fishes, 661
    of frogs, **662**
    human, **662–63**
    mammalian, 662
    myogenic, 664
    neurogenic, 664
    of Mollusca, 317
    of sharks, 495
    of Vertebrata, 478–79
Heart rate, 662
Heart urchins, 452–53
Heartworms, 303, **304**
Heat production, increased, 649
Hectocotylus, 339
Hedgehogs, 606
    sperm of, **18**
    spiny armor of, 589
Helicobacter pylori, 685
Helicoplacus, **458**
Heliosoma, 329
Helix, 329, 343
Helix aspersa, **124**
Heloderma suspectum, **544**
Hemal system, 443, 445, 448, **449**
Heme, 672
Hemerythrin, 673
Hemichordata, 460–64, 473
    adaptive radiation of, 464
    biological contributions of, 461
    characteristics of, 464
    cleavage in, 156
    phylogeny of, 464
    position in animal kingdom, 461
    reproduction of, 133
Hemimetabolous metamorphosis, 411, **411–12**
Hemiptera, 78, **403,** 403, **417,** 420, 424
Hemizygous organisms, 85, **86**
Hemocoel, 380, 435, 660
Hemocyanin, 382, 673
Hemocytes, 749, **749**
Hemoglobin, 25, 382, 479, 672, **673**
Hemoglobin saturation curves, 673, **673**
Hemolymph, 381–82, 406, 660
Hemolytic disease of the newborn, 748
Hemophilia, 84–85, 659
Hemorrhagin venom, 549
Hemostasis, 658–59
Hemozoin, 220
Hench, P. S., 733
Henderson, Lawrence J., 773
Hennig, Willi, 201, **202**
Hennigian systematics, 201
Hepatic cecum, 476
Herbivores, 34, 404, 507, 591, **591–92,** 678, 795, 800
Heredity, 6, 75. *See also* Inheritance
    cell biology and, 19
    genetic approach to, 17, **18**
    Mendelian, 17–19
    natural selection and, 16
Hermaphroditism, **132,** 133–34, **135**
Hermatypic coral, 264–65
Hermissenda, 328
Hermit crab, **249,** 393, **394**
Hermodice carunculata, 351, **351**

Herons, 579
Herrerasaurus, 546–47
Herring, 504, **679**
Hesperonoe adventor, **351**
Heterocentrotus mammilatus, **453**
Heterocercal fin, 493–94, **498,** 500
Heterochromatin, 97
Heterochrony, 112
Heterodonts, 590
Heteroptera, 420
Heterostracans, 481, **481**
Heterotrophs, 31, **31,** 214, 678
Heterozygotes, 79
Heterozygous organisms, **86**
Hexacorallia, 260, 262, **264,** 267
Hexacorallian corals, 263, **264**
Hexactinellida, 237, **241,** 242–43, **244,** 245
Hexamerous body plan, 260
Hexoses, 23, **23**
Hibernation, 412, 525, 651–52, **652**
Hierarchy
    of classification, 191–92
    of developmental decisions, 152
    of ecology, 790–804
    of living systems, 4–6, **5–6**
    nested, 110–11
High-energy bonds, 60
High-lift wings, 570, **571**
High-speed wings, 570, **571**
Hindbrain, 705, **705**
Hippocampus, 706
Hippopotamus, 608, **609**
Hirudinea, 346, 357–60, **358–59,** 360
    circulatory system of, 359–60
    excretory system of, 359
    form and function of, 358–60
    nervous system of, 359
    nutrition of, 358–59
    reproduction of, 359
    respiratory system of, 359
    sensory system of, 359
Hirudo, 360
Hirudo medicinalis, 359, **359**
Histology, 179
Histones, 32, 41
HIV. *See* Human immunodeficiency syndrome
Hodgkin, A., 338
Holoblastic cleavage, 155
Holocephali, 497, **497,** 510
Holometabolous metamorphosis, **410,** 411
Holothuria difficilis, **455**
Holothuroidea, 443, 454–56, **455–56,** 460
Holozoic feeding, 214, 222–23, 226
Homalozoa, 458
Homarus, 392
Homarus americanus, **385,** 394
Home ranges, 600, 762
Homeobox genes, 163, **163,** 164, 478
Homeostasis, 634–53
Homeothermic animals, 647
Homeotic genes, 163–64
Homeotic mutations, 163
Homing pigeons, 572
Homing salmon, 508–9, **509**
Hominidae, 201, **201,** 607
Hominids, 603–4, **605**
Homo, 199–200, **201,** 604, 607
Homo erectus, 601, 604, **605**
Homo habilis, 604, **605**
Homo sapiens, **603,** 604, **605**
Homocercal fin, **498,** 499–500
Homodonts, 590
Homologous chromosomes, 76
Homologous structures, 473
Homology, 109–11, 195
    nester hierarchy and, 110–11
    serial, 380
    skeletal, **111**
Homoptera, 408, 410, **411, 413, 416–17,** 420, **420,** 424
Homozygotes, 79
Homozygous organisms, **86**
Honey bees, 421
    behavior of, 414, 755–56, **756**
    communication by, **766,** 766–67
    insecticides and, 418
    kin selection and, **764**
    legs of, 400, **402**

mouthpart of, 405
parthenogenesis in, 135
reproduction of, 410
social behavior of, 415, **415**, 416
Honeydew, 408, **416**
Honeyguides, 580
Hoofed animals, 608–9
Hooke, Robert, 37
Hookworms, 300, **300–301**
Hormones, 721–22
and calcium metabolism, 731–32, **732**
of human pregnancy, 145–48, **147**
of invertebrates, 724
mechanisms of, 696, 722–24, **723**
of metabolism, 730–36
of pituitary gland, **727**
and reproductive cycles, 143–44
secretion of, 723–24
thyroid, **730**, 730–31, **731**
tropic, 726
of vertebrates, 724–36
Hornbills, 580
Horns, 589, 615, **615**
Hornworm, **405**
Horny coral, **266**
Horse, 608, **609**
evolution of, 109, **110**
forelimb of, **111**
sperm of, **18**
Horse flies, 404–6
Horsehair worms. *See* Nematomorpha
Horseshoe crabs, 367–68, **368**, 374
Host, 739
Host defense, antibodies in, 742–43
House flies, 405, **406**, 417
Hox genes, 163, **163**, 164, 478
Human(s), 606–7
age structure of, **792**
appendages of, **111**
birth of, **147**
brain of, 705, **706**
cestodes and, 286–88, **287**
communicating with other animals, 768
digestive system of, 684, **684**
ears of, 713, **714**
eggs of, *131*, 140
embryo of, **113**, **166**, **168**
evolution of, 601–4
eyes of, **717**
flukes and, **282**, 282–84
hearing in, 713, **714**
heart of, 662–63
insects and, 416–18
integument of, **614**
lungs of, **669**
lymphatic system of, **666**
muscles of, 626
nervous system of, **708**
nutrient requirements of, **691**
osmoregulation in, 638, **638**
parasites of, **300**
population of, **793**, 793–94
position on food chain, 802
relationships with other mammals, 601
respiratory system of, **671**
skeleton of, **621**
sperm of, *131*, 139, **139**
spinal cord of, **703**
taxonomy of, **191**, 198–200
urinary system of, **642**
Human chorionic gonadotropin, 146
Human chorionic somatomammotropin, 146
Human Genome Project, 97
Human Genome Sequencing Consortium, 97
Human immunodeficiency syndrome
(HIV), 747
Human placental lactogen, 146
Hummingbirds, 580, *611*
flight of, 569, **571**
nesting, 575, **576**
torpor in, 651, **651**
UV-sensitivity in, 569
Humoral immunity, 741, **743**, 743–45, **744–45**
Humpback whale, *36*, 678
Hunger center, 689
Hunting, of birds, 576, **577**
Hutchinsoniella, 392
Huxley, A., 338

Huxley, Julian, 103
Huxley, Thomas Henry, 193, 556
Hyalella, 391–92
Hyaline cap, 211
Hyaline cartilage, 618
Hyalonema, 245
Hyalophora cecropia, **421**
Hyalospongiae, 242–43, 245
Hyaluronidase, 139
Hybridization, 114, **114**
Hybridoma, 745
Hybrids, 75, 114, **114**
in Mendelian heredity, 17
Hydatic cyst, 289
Hydra, **252**, 252–54, **253–54**, 267
body plan of, 252
body wall of, 252
digestion in, 253–54, **254**
feeding habits of, 253, **254**
locomotion of, 253
reproduction of, **132**, 254
Hydractinia milleri, **249**
Hydranths, 255, **255**
Hydrocauli, 254
Hydrocephalus, 705
Hydrochloric acid, 685
Hydrocoels, 449
Hydrocorals, 257
Hydrogen bonds, 25
in nucleic acids, 90, **90**
Hydrogenosomes, 216
Hydroid, 248, **252**, **255**
Hydroid colonies, 254–55
Hydrolagus, 510
Hydrolagus collei, **497**
Hydrolases, 682
Hydrolysis, 29, 60, 682
triglycerides and, 68, **68**
Hydrolytic enzymes, 682
Hydrophiidae, 548
Hydropore, 445
Hydrorhiza, 254
Hydrosphere, 773
Hydrostatic pressure, 46, 251, 297, 672
Hydrostatic skeleton, 616–17
of Annelida, 347
of hydra, 253
metamerism and, 360
Hydrothermal vents, 27
Hydrozoa, 249, 252–57, 267
Hyla cinerea, **524**, **531**
Hyla versicolor, **527**
Hylidae, 523
Hylobates, **201**
Hylobatidae, **201**, 607
Hyman, Libbie, 442
Hymen, 143
Hymenolepis, 288
Hymenolepis nana, **287**
Hymenoptera, 403, 410, 415, 421, **421**
Hyperosmotic animals, 46
Hyperosmotic regulators, 505, 636
Hyperparasitism, 404
Hyperpolarization, 224
Hypersensitivity, 746
Hyperventilation, 671
Hyphae, 10
Hypodermal cords, 297
Hypodermis, 297, 613
Hyponeural system, 448
Hypoosmotic animals, 46
Hypoosmotic regulators, 505, 636
Hypophysis, 725–28
Hypostome, 252, 373
Hypothalamus, 144, **705**, 706, 725–28, **726**
Hypothermia, adaptive, 651–52
Hypotheses, 11–12
Hypothetical ancestral mollusc, 340
Hypothetico-deductive method, 11
Hypsibius, 436
Hypsurus caryi, **509**
Hyracoidea, 608
Hyraxes, 608

Ibises, 579
Ichneumon wasp, **403**, 410
Ichthyomyzon, 491, 510
Ichthyophthirius, 225, **226**

Ichthyosauria, 551
Ichthyostega, 516, **517**
"Ick," 225
Identical twins, 148, **149**
IgA, 739
IGF. *See* Insulin-like growth factor
Iguanas, 542–43, **543**
Ilyanassa obsoleta, 792
Immediate causes, 13
Immediate hypersensitivity, 746
Immigration, 782
Immunity, 738–51
acquired, 739, 741–47
cellular, 740, **740**, 741, **743**, 745–46
definition of, 739
humoral, 741, **743**, 743–45, **744–45**
inflammation and, 746–47
innate, 739–41
in invertebrates, **748**, 748–49
Immunizations, 745, **745**
Immunoglobulins, 741, **745**
Implantation, 145
delayed, 597
Imprinting, 758, **758**
Inarticulata, 434
Inbreeding, 123
Incipient tissues, 234
Incisors, 590, 680, **681**
Inclusive fitness, 764
Incomplete dominance, 81, **81**
Incomplete septa, 261
Incubation period, 220
Incurrent canals, 238
Incurrent siphon, 474
Incus, 713
Independent assortment, law of, 81–83, **82**, **84**
Index fossils, 107
Indian climbing perch, 504
Indirect development, 155–56
Indirect flight muscles, 403
Inducer, 214
Induction, 153
embryonic, 161–62
Infective parasites, 739
Inflammation, 745–47
Infraciliature, 222
Infundibulum, 574, **725**
Inheritance. *See also* Heredity
of acquired characteristics, 103
of behavior, 755
chromosomal basis of, 75–79
chromosomal theory of, 17–19
fidelity of, 6
gene interaction and, 83–84
intermediate, 81, **81**
Mendelian laws of, 79–88
law of independent assortment,
81–83, **82**, **84**
law of segregation, 79–81
molecular basis of, 88–91
particulate, 17, **18**
polygenic, 83
quantitative. *See* Blending
sex-linked, 84–85, **86**
variation and, 118
Inhibin, 144
Inhibitory synapses, 701
Ink gland, 339
Inland waters, 779
Innate behavior, 755
Innate immunity, 739–41
Inner cell mass, 157
Inner chamber, of eye, 717
Inner ear, 568
Inositoltrisphosphate, 723
Insecta, 374, 400–416, 418
adaptability of, 400
adaptive radiation of, 420–24
behavior of, 413–16
beneficial, 416–17, **417**
circulatory system of, 406, **661**
cladogram of, **423**
classification of, 419–21
communication by, 413–16
control of, 418
defenses of, 413
development of, 410–12

digestive system of, 404, **683**
direct development of, 411
distribution of, 400
eggs of, 155
endocrine system of, 724
excretory system of, 408, **408**
external form and function of,
400–404
eyes of, **716**
fast flight muscles of, 625
fossils of, 106, **107**, 421
gas exchange of, 406–7, **407**
harmful, **417**, 417–18
hearing in, 712–13
homeotic genes in, **163**
human welfare and, 416–18
immunity in, 740, **748**, 749
internal form and function of, 404–10
life cycle of, **9**
locomotion of, 401–4
malpighian tubules of, **640**
metamorphosis of, 410–12
nervous system of, 408
neuromuscular coordination of, 409
nutrition of, 404–6
parasitic, 404
phylogeny of, 420–24
populations of, 397
reproduction of, 141–42, 410, **410**
respiratory system of, **667**
sensory organs of, 408–9
water balance in, 408
Insecticides, 418
Insectivores, 590, **591–92**, 596, 606
Instar, 410
Insulin, 734–35
Insulin-like growth factor (IGF), 727, 735
Integrated pest management, 418
Integument, 479, 613–16, **614**
of Anura, 526, **526**
derivatives of, 614–15, **615**
effects of sunlight on, 616
of invertebrates, 613
of Mammalia, 587–90, **588**
pigmentation of, 84
of Reptilia, **538**, 538–39
of Vertebrata, 613–16
Integumentary glands, 589–90
Interactions, within communities, 795–97
Interbreeding, 15, 113, 194
Intercellular fluid, 656
Interferons, 739, **744**, 746
Interleukins, 743, **744**, 746
Intermediary meiosis, 215
Intermediate filaments, 42
Intermediate host, 281
Intermediate inheritance, 81, **81**
Intermediate lobe, of anterior pituitary
gland, 726
Intermediate neurons, 717
Internal buds, 240
Internal fluids, 656–57
Internal gills, 472, 668
Internal nares, 669
International Union for the Conservation of
Nature and Natural Resources
(IUCN), 583
Interneurons, 696
Interoceptors, 709
Interphase, **50**, 51, **52–53**
Interstitial cells, 142, 144, 252–53
Interstitial fluid, 178, 656
Intertidal zone, 779
Intestinal ceca, 447
Intestinal flukes, **282**
Intestinal mucosa, 686
Intestinal roundworm, 299–300, **300**
Intestine, 353, 495, **686**, 686–88
Intracellular digestion, 240, 277, 682, **682**
Intracellular fluid, 656, **656**
Intracellular space, 178
Intracellular specialization, 208
Intramembranous bone, 618
Intrinsic rate of increase, 792–93
Introns, 88, 92
Introvert, 427
Invagination, 158, 667
Inversion, 88, 242

Invertebrates
  excretory system of, 639–40
  exoskeleton of, 617–18
  hormones of, 724
  immunity in, **748**, 748–49
  integument of, 613
  marine, salt and water balance in, **635**, 635–36
  muscles of, 625
  nervous system of, 701–2, **702**
  reproduction of, **141**, 141–42
Iodine, 730, **731**
Ionizing radiation, 98
Iridophores, 526, 616
Iris, 717
Irish elk, 119, **120**
Iron pyrite, 29–30
Irritability, 9, 696
Islands, evolutionary diversity on, 115
Islet cells, of pancreas, 734–35
Islets of Langerhans, *721*, **734**, 735
Isogametes, 215
Isolecithal egg, 155, **156**
Isopoda, 390–91, **391**, 392
Isoptera, 415, **416**, 419, 424
Isosmotic animals, 46
Isospora, 220
Isurus, 494
IUCN. *See* International Union for the Conservation of Nature and Natural Resources
Ixodes, 373–74

Jacobson's organs, 545
Japanese beetles, **417**
Japanese crab, 365
Japygids, 419
Jarvik, Erik, 516
Java man, 601
Jawfish, **509**
Jawless fishes, 468–69, 482, 487, 489–93, 510
Jawless ostracoderms, 481–82
Jaws
  of birds, 562
  of chimaeras, 497
  of fishes, 482, **484**, 506
  origin of, 482–83, **483**
  pharyngeal, 500
  of reptiles, 539
  of sharks, 495
  of snakes, 544, **545**
Jellyfishes, 248–49, 257–59, **258**, **260**
Johannsen, W., 88
Johanson, Donald, 602–3
Johnson, Ben, 734
Jointed appendages, 365–66
Julus, 399, 418
Jumping bristletails, 419
Jumping spiders, **369**, 370
Juvenile hormone, 412, 724
Juxtaglomerular apparatus, 644

K (carrying capacity), 793
Kangaroo rat, osmoregulation in, 638, **638**
Kangaroos, 597, **598**, 606
Katydid, **191**, 401, **413**
Keeton, W. T., 572
Kentrogon, 390
Kenya, Amboseli Reserve in, **2**
Keratin, 526, 587, 613–15
Keratinization, 613
Keystone species, 799
Kidney
  of Arthropoda, 640–41
  of Bivalvia, 333
  of fishes, 505
  of Gastropoda, 326
  opisthonephric, 505
  osmotic concentration in, **647**
  of vertebrates, 641–46
    embryology of, 641, **641**
    evolution of, 641
    function of, 642
    glomerular filtration and, 643
    tubular reabsorption and, 643–45, **644**
    tubular secretion and, 645
    water excretion and, 645–46
Killer whale, *752*
Kin selection, 763–65, **764**

Kinase, 723
  cyclin-dependent, 52
Kinetic energy, 57
Kinetic skull, 542, **543**, 544, **545**
Kinetochore, 49, **49**
Kinetochore fiber, 50
Kinetoplast, 218
Kinetoplasta, 218–19, 230–31
Kinetosome, 43, 210, **210**, 218, **223**
Kinety, 222, **223**
King, 415
King snake, **7**
Kingfishers, 580
Kinorhyncha, 305, **305**
Kinorhynchus, 305
Kissing bugs, 218
Kiwis, 578
Klinefelter syndrome, 88
"Knee-jerk" reflex, 703, **703**
Koala, **31**, 606
Komodo lizard, *534*
Kramer, Gustav, 572, **573**
Krebs cycle, 64, **65**
Krebs, Hans, 62
Krill, 391–92, **393**
Kristensen, R. M., 312
Kupffer cells, 740
Kwashiorkor, 692

Labia majora, 143, **143**
Labia minora, 143, **143**
Labor (pregnancy), 146, **147**
Labyrinth, 382, 713, 715
Lac insects, 416
Lacerta nilotica, 194
Lacertilia, 551
Lacewings, 420
Lack, David, 115, 797
Lactase, 686
Lacteals, 688
Lactose, 23
Lactose intolerance, 688
Lacunae, 358, 618
Lacunar system, 309
Ladybird beetles, 414, **417**
Lagena, 530, 713
Lagomorpha, 607
LAK. *See* Lymphocyte-activated cells
Lake Baikal, 114, **115**, 779
Lake forest, 777
Lake Turkana, 118
Lakes, 779
  lampreys in, 492–93
  speciation in, 114, **115**
de Lamarck, Jean Baptiste, 103, **103**, *190*
Lamarckism, 103
Lamellae, 331, 504
Lamellisabella, **430**
Lamniformes, 493
Lamp shells. *See* Brachiopoda
Lampetra, 491
Lampetra tridentatus, 491
Lampreys, 482, 487, 489, **491**, 491–93, 510
  brain of, **704**
  characteristics of, 491
  feeding habits of, 492, **492**
  in Great Lakes, 492–93
  larva of, 480, **480**, 491–92
  life cycle of, **492**
  parasitic, 492
Lampsilis ovata, **336**
Lancelets. *See* Amphioxus
Land, movement onto, 515
Landsteiner, Karl, 748
Lankester, E. R., 428
Lappets, 258
Lapping mouthparts, 405–6
Lapwings, 579
Larra bicolor, **417**
Larus atricilla, **579**
Larva
  ammocoete, 480, **480**, 491–92
  of Chordata, 479, **479**
  cydippid, 269
  of Echinodermata, **450**
  echinopluteus, **450**, 454
  of fishes, **167**
  glochidium, 334
  of insects, 404

of lampreys, 480, **480**, 491–92
  pilidium, 290
  planula, 250
  trochophore, 317–18, 321, **321**
  veliger, 318, 321
Larvacea, 473, 475, **476**
Larynx, 528, 669
Lateral, 185
Lateral canals, 446
Lateral force, 503
Lateral line system, 495, 712, **712**
Lateral undulation, 545, **548**
Laterite, 777
Latimeria, 502, **502**, 510
Latrodectus, 370
Latrodectus mactans, **371**
Laurasia, 784, **785**
Law of independent assortment, 81–83, **82**, **84**
Law of segregation, 79–81
Law of specific nerve energies, 709
Law of stratigraphy, 107
Laws of thermodynamics, 9–10, 36, 57, 802
LD_{50}, 550
Leafhoppers, 420
Learning, behavior and, 757–58
Leatherbacks, 541
Lecithin, 24, **25**
Lecithotrophy, 542
Leeches, 346, 357–60, **358–59**. *See also* Hirudinea
Left atrioventricular valves, 662
Left atrium, 662
Left ventricle, 662
Legs. *See* Appendages
Leidyopsis, **214**
Leishmania, 219, **219**, 231
Leishmaniasis, 219
Lek, 575, 763, **763**
Lemmings, 600, **600**
Lemnisci, 309
Lemurs, 602, 606
Lens, of eye, 717
Lentic habitat, 779
Leopard frog, 524, **531**
Lepas anatifera, **389**
Lepidoptera, **405**, 408, **413**, **417**, 421, **421**
Lepidosauria, 535, 551
Lepidosiren, 500, **501**, 510
Lepidurus, 392
Lepisma, **419**
Lepisosteus, 498, 510
Lepisosteus osseus, **500**
Lepospondyls, 516
Leptin, 689, 736
Leptocephali, 507
Leptodora, 392
Leptophis ahaetulla, **548**
Leptosynapta, 455
Lepus americanus, **588**
Lernaea, 392
Leucochloridium, 284
Leuconia, 240
Leuconoids, **238**, 238–39
Leucosolenia, 237, **238**, **241**, 242, 245
Leukemism, 588
Leukocytes, 657–58, **748**
LH. *See* Luteinizing hormone
Libellula pulcella, **412**
Libinia, 393
Lice, 404, **405**, 417
Life
  conditions necessary for, **22**
  definitions of, 3–4
  fundamental properties of, 3–10
  major divisions of, 202–3
  origins of, 4, 27
  spontaneous generation of, 21–22
  water and, 28–29
Life cycle, sexual, **134**
Lift-to-drag ratio, 569
Ligament sacs, 310
Ligands, 48
Light microscope, 37, **38**
Lightningbugs, 414
Lightning, 27
Ligia, 390–92
Lima, 334
Limax, 211
Limax, 211, 329

Limbic system, 706
Limbs. *See* Appendages
Limenitis archippus, **413**
Limestone, 229
Limifossor, 343
Limiting resources, 793, 797
Limnoscelis, **517**
Limpets, 324, 328
Limulus, 367–68, **368**, 374
Lineus, 289, 291
Lineus longissimus, 289
Linguatula, **435**
Linguatula serrata, 434–35
Lingula, 433–34, **434**
Linkage, autosomal, 85–86
Linnaeus, Carolus, **191**, 191–92
Linoleic acids, 24, **24**
Lioness, **593**
Lionfish, **501**
Lions, **597**
Lipids, 24, 691
  bonds of, 4
  metabolism of, 68–69
Lissamphibians, 516
Lissomyema, 428
Lithobius, 418
Lithosphere, 773
Little brown bats, 606
Littoral zone, 779
Littorina, 328
Liver, **37**, 170, 495
Liver fluke, **279**, **281**, 282, **282**
Living systems
  chemical uniqueness of, 4
  complexity of, 4–6, **5–6**
  development of, 8–9, **9**
  environmental interaction and, 9
  general properties of, 4–9
  genetic programs and, 6–7
  hierarchical organization of, 4–6, **5–6**
  metabolism and, 7–8, **8**
  organic molecular structure of, 22–26
  origin of, 30–32
  physical laws and, 9–10
  reproduction and, 6
Lizards, 542–44, **543–44**, 551
  integument of, 614, **615**
  lungs of, **669**
  skull of, **543**
  temperature regulation in, **9**
Lobe-finned fishes, 482, 489, 497, 500–502, 510, 515–16
Lobopodia, 211, 227
Lobsters, 392–93, **394**
Locomotion, 170–71, 622–31
  ameboid, 43, **211**, 226, 622–23
  Limbs of Arthropoda, 366
  of Bivalvia, 331
  of cephalopods, 336–37
  cilia and, **623**, 623–24, **624**
  of earthworms, 616–17, **617**
  of fishes, **502**, 502–3, **503**
  flagella and, **623**, 623–24, **624**
  of hydra, 253
  of Insecta, 401–4
  muscular system and, 624–30
  of Nemertea, 291
  of Ophiuroidea, 451
  of paramecium, 223–24
  of protozoa, 208, 210–13, **211–12**
  of snakes, 545, **548**
Locus, of genes, 78
Locusts, 397, 400, 407, 419
Loggerhead sponges, 236, 243
Logistic population growth, 794
Loligo, 335, 338–39
Longitudinal fibers, 297
Longitudinal muscles, 304
Longitudinal nerve cords, 276, 278, 286
Longtailed mealybug, **417**
Loons, 578
Loop of Henle, 642
Loose connective tissue, 180, **183**
Lophius piscatorius, **507**
Lophocytes, 240
Lophophorates, 430–31
Lophophore, 431–33, **433**, **434**
Lophotrochozoa, 156, 235, 296, 312, 340, 427–34, 439

Lorenz, Konrad, 752, **753**, 754, 758, **758**
Lorica, 307
Loricifera, 305, **306**, 312
Lorises, 602, 606
Lotic habitat, 779
Loudness, 715
Loxodonta africana, 608
Loxosceles, 370
Loxosceles reclusa, **371**
Loxosoma, 312
Loxosomella, **311**, 312
Lucy, 603, **603**
Lugworm, 348, **349**, 351
Lumbar vertebrae, 620
Lumbricus, 355, 360
Lumbricus terrestris, 353
Lumen, 682
Luminescent beetles, 414
Lung flukes, **282**, 284
Lungfish, 489, 497, **498**, 500, **501**, 504, 510, 515
Lungs, 668–69, **669**, **671**–72
  diffusion and partial pressure and, 671–72
  as endoderm derivative, 170
  of fishes, 487, 504, 515
  of Gastropoda, 326
  of Mollusca, 317–18
  and pulmonary circuit, 661
  of reptiles, 539
  of salamanders, 521
Luteal phase, 144
Luteinizing hormone (LH), 144, **145**, 726, **727**
Lyell, Charles, 104, **104**
Lyme disease, 373–74
Lymnaea, 329
Lymph, 181, 666
Lymph capillaries, 666
Lymph nodes, 666
Lymphatic system, 661, 666, **666**
Lymphocyte-activated (LAK) cells, **744**
Lymphocytes, 95, **657**, 740, **740**
Lynceus, 392
Lynx, **600**
Lysosomes, 41, 214, 740
Lysozymes, **5**, 739
Lytechinus, 452, 460

Macaca, 607
Macaque, 760, **760**
MacArthur, Robert, 797
Maccabeus, 306
McClung, C., 78
Machrocheira, 365
Mackerel, 504
Macleod, J. J. R., 735
Macracanthorhynchus hirudinaceus, 309–10
Macrobdella, 360
Macrobiotus, 436
Macrobiotus bufelandii, **437**
Macrochaetus longipes, **307**
Macroclemys temmincki, **542**
Macrodasys, **311**
Macroevolution, 120, **126**, 126–28
Macrogametes, 218
Macrogametocytes, 221
Macromeres, 242
Macromolecules, 4
Macronucleus, 209, 222–23
Macrophages, 658, 739–40, **740**
Macroscelidea, 606
Madreporite, 445
Maggots, 417
Magnesium, 657
Maiasaura, 546
Main-line enzymes, 59–60
Major histocompatibility complex (MHC), 741
Major histocompatibility proteins, 167
Makaira nigricans, **501**
Mako sharks, 493–94
Malacosoma, 415
Malacostraca, 378, **378**, 386, 390–93, **391**, **393**
Malaria, 220–22, 417, **418**
Male-dominance polygyny, 763
Mallard ducks, **3**
Malleus, 713
Mallophaga, **405**, 420, 424
Malnutrition, 692, **692**

Malpighi, Marcello, 665
Malpighian tubules, 366, 369, 408, **408, 640,** 640–41
Maltase, 686
Malthus, T. R., 106
Maltose, 23, **23**, 684
Mammalia, 482, 582–610
  alimentary canal of, **687**
  biological contributions of, 583
  body weight of, 593–94, **594**
  characteristics of, 584
  circulatory system of, **661**
  classification of, 606–9
  cleavage, 156–57
  development of, 166–68
  digestive system of, **592**
  echolocation in, 595–96, **596**
  eggs of, 155
  extraembryonic membranes of, 167, **167**
  feeding habits of, 590–94, **591**
  flight of, 595–96
  gastrulation in, 159–60
  glands of, 589–90
  heart of, 662
  homeotic genes in, **163**
  hypothermia in, 651–52
  immunity in, 741
  integument of, 587–90, **588**
  lungs of, 668, **669**
  migration of, 594, **595**
  origins and evolution of, 583–87
  placenta of, 167–68
  populations of, 600
  position in animal kingdom, 583
  red blood cells of, **658**
  reproduction of, 596–99
  respiratory system of, 669–74
  teeth of, 590, 680, **681**
  territoriality of, **599**, 599–600
  thyroid gland of, 730
Mammary glands, 146–48, 590
Manatees, 608
Mandibles, 378, 435
Mandibulates
  aquatic, 377–94
  terrestrial, 397–424
Mandrillus, 607
Mangold, Hilde, 151, 161
Mantle, 317–20, 331, 474
Mantle cavity, 318–20
Manubrium, 255
Marasmus, 692
Margulis, Lynn, 33–34
Marine animals
  families in, **126**
  hypoosmotic, 46
  invertebrate, salt and water balance in, **635**, 635–36
Marine turtles, 541
Marlin, **501**, 502
Marmosets, 606
Marmota monax, **652**
Marrela splendens, **108**
Marsupialia, 166, 597, **598**, 606
Marsupium, 606
Martens, Conrad, **105**
Martini, E., 309
Mary, Ernst, 194
Mass extinctions, 127–28
Mast cells, 740
Mastax, 308
Mastication, 680
Mastigophora, 229
Mating, 763. *See also* Sexual reproduction
  of birds, 574–75, **575**
  communication and, 765–66, **766**
  harem system of, 126
  nonrandom, 123
  positive assortative, 123
Matrix, 180
Maxam, Allan, 97
Maxillae, 378
Maxillary glands, 382
Maxilliped, **380–81**
Maxillopoda, 387–90, **388–90**, 392
Maxillopodan eye, 388
Maxilloposa, **393**
Maxillules, 379, **380–81**
Mayflies, 407, 419, **419**, **422**

Mayr, Ernst, 14, 113, **113**, 117, 197
Meadowlark, **576**
Mechanoreception, 408, 711–15
Mecoptera, 420–21
Medial, 185
Mediated transport, 45, 47
Medical research, animal use in, 12–13
Medicinal leeches, 359, **359**
Medulla, 642
  of adrenal gland, 732, 734
Medulla oblongata, **705**
Medusa, **249**, 249–50, **254**, 255, **255**–56
  freshwater, 256
Megaceryle alcyon, 580
Meganyctiphanes, 392, **393**
Megaptera novaeangliae, 36, **609**
Mehlis' gland, 282
Meiosis, 49, 75–78, **77**, 133
  anaphase, **77**, 78
  intermediary, 215
  prophase, **77**, 78
  zygotic, 215
Meiotic parthenogenesis, 135
Melanin, 616
Melanocyte-stimulating hormone (MSH), 727, **727**
Melanophores, 526
Melatonin, 728
Meleagrina, 317, 331
Melithaea, **266**
Mellita, 460
Melongena, 326
Membrane enzymes, 686–88
Membrane-bound receptors, 723
Membranelles, **222**
Membranipora, **433**
Membranous layer, 384
Memory cells, 745
Mendel, Gregor, 17, **17**, 75, **76**
  garden of, 74
  heredity theory of, 17–19
  laws of inheritance, 79–88
Meninges, 703
Menopause, 618
Menstrual cycle, 143–45, **145**, 597
Meoma, **454**
"Mermaid's purse," 496
Meroblastic cleavage, 155, 157
Merostomata, 367–68, 374
Merozoites, 219–20
Mesencephalon, 704, **705**
Mesenteries, 261, 346
Mesocoel, 431
Mesoderm, 158, **188**
  derivatives of, 170–71
Mesoglea, 252–53
Mesolecithal eggs, 155, **156**
Mesonephric duct, 142
Mesonephros, 641, **641**
Mesopelagic zone, 780
Mesosoma, 371
Mesosome, 431
Mesothorax, 400
Mesozoa, 234–35
  parasitic, 234–35, **236**
  phylogeny, 235
  position in animal kingdom, 234
  reproduction of, 235, **235**
Messenger RNA, 91–93, **92**, **94**
Metabolic rate, body weight and, **594**
Metabolic water, 638
Metabolism, 7–8, **8**
  aerobic, 62, 630
  anaerobic, 62
  of calcium, 731–32, **732**
  cellular, 56–72
  growth hormone and, 735
  hormones of, 730–36
  of lipids, 68–69
  management of, 70, **70**
  origin of, 31
  oxidative, 31–32
  of proteins, **69**, 69–70
Metacercariae, 282
Metacoel, 431
Metameres, 188, 346
Metamerism, 163, 188, **188**, 345–46, 360
Metamorphosis, 9, **9**, 156
  of Arthropoda, 366–67

  of ascidians, **475**
  of Asteroidea, 449–50, **450**
  of Crustacea, 384
  of Echinodermata, 443
  of frogs, 514, *514*, 523
  hemimetabolous, 411, **411–12**
  holometabolous, **410**, 411
  of Insecta, 410–12
  physiology of, 411–12
Metanephridia, 318, 320–21, 349, 354, 639
Metanephros, 641, **641**
Metaphase
  in meiosis, 77
  in mitosis, **50**, 51, **52**
Metaphasic plate, 51
Metasoma, 371
Metasome, 431
Metatheria, 606
Metathorax, 400
Metazoa, 233–46
  biological contributions of, 234
  extracellular components of, 178–79
  nervous system of, 701
  organization of, 177
  origins of, 234
  position in animal kingdom, 234
Metencephalon, **705**
Methane, origin of life and, 27
Methylation, of cytosine, 95, **95**
Metridium, 261, 267
MHC. *See* Major histocompatibility complex
Micelles, 688
Michaelis, Leonor, 59
Microevolution, 120–25
Microfilaments, 42
Microfilariae, 302
Microgametes, 218
Microgametocytes, 221
Microglial cells, 697, 740
Microhydra ryderi, 256
Micromeres, 242
Micronemes, 219
Micronucleus, 222–23
Micropyle, 240
Microscopes, in study of cells, 37–39
Microspheres, proteinoid, 30, **30**
Microstomum, **279**, 288
Microtriches, 285
Microtubules, 42, **44**, 50, **51**
Microvilli, 45, **45**, 686
Mictic eggs, 309
Midbrain, **705**, 706
Middle ear, 568
Migration, 123–24
  of birds, 555, *555*, 570–73, **572**
    navigation in, 572–73, **573**
    routes of, 570–72
    stimulus for, 572
  of fishes, 507–9
  of mammals, 594, **595**
Milk, secretion of, 146–48
Milk teeth, 590
Millepora, **257**, 264
Miller, Stanley, 27, **27**
Millipedes, 375, 398–99, **399**, 418
Mimicry, 413, **413**, 799, **799**
Mindanao tarsier, **602**
Mineralocorticoids, 733–34
Minerals, **691**
Minkowski, O., 735
Mirabilis, 81
Miracidium, 281
Mirounga angustirostris, **763**
Miscarriage, 146
Mites, 372, 372–74, **373**
Mitochondria, 33, **37**, 41, 41–42, **43**
Mitochondrial DNA, 33
Mitopus, **372**
Mitosis, 48
  phases of, **50**, 50–51, **52**
  stages of, **52**
Mnemiopsis, **268**
Mnemiopsis leidyi, 270
Models
  of behavior, 754–55, **755**
  defenses and, 799
Modular animals, 791
Molars, 590, 680, **681**
Mole crickets, 400

Mole vipers, 548
Molecular genetics, 95–97
 of cancer, 98
Molecular level, reproduction at, 6, **7**
Molecular sequencing, 32
Molecular structure, of living systems, 22–26
Molecules, 4
Moles, 606
Mollies, 509
Mollusca, 316–43, **318–19**
 adaptive radiation of, 340–42
 biological contributions of, 317
 body plan of, **186**
 cladogram of, **342**
 classification of, **341,** 343
 cleavage in, 156
 destructive, 317
 eggs of, 155
 exoskeleton of, 617
 feeding habits of, **679**
 form and function of, 318–21
 immunity in, **748**
 integument of, 613
 internal features of, 320–21
 muscles of, 625
 nervous system of, 702
 phylogeny of, 340–42
 position in animal kingdom, 317
 reproduction of, 321
 shell of, *190,* 318, 320
Molting, 366, 384–85, **385**
 of feathers, 562, **562**
 of hair, 588
Molting hormone, 385, 412, 724, **725**
Molt-inhibiting hormone, 384
Monanchora unguifera, **244**
Monarch butterfly, **9,** 404, 410, 413, **413**
Monastrea cavernosa, **263**
Monera, 32, 203
Mongolism, 87–88
Monitor lizard, skull of, **543**
Monkeys, **602,** 606
Monoclonal antibodies, 745
Monocystis, 231
Monocytes, **657,** 740, **740**
Monoecious animals, 133–34, 242
Monoestrous animals, 596
Monogamy, 574, 763
Monogenea, 276–77, 284, **285,** 288
Monogononta, 308, **308,** 309
Monohybrid cross, 79
Mononuclear phagocyte system, 740
Monophyletic groups, 468
Monophyly, 197–98, **200**
Monoplacophora, 321–22, **322,** 340–41, 343
Monosaccharides, 22–23, **23**
Monosomic, 87
Monotremata, 606
Monotremes, 155, 166, 597
Monozygotic twins, 148, **149**
Montastrea annularis, **239, 264**
Moon snails, **326,** 328
Mopalia, 343
Mopalia muscosa, **323**
Moray eels, **3**
Morgan, Thomas Hunt, 75, 85
Morphogenesis, 253
 of appendages, 164, **165**
 of organs, 164, **164**
Morphogenetic determinants, 153
Morphogens, 162, 253
Morphological species concept, 193–94
Morphology
 comparative, 196
 organismal, 193
Morus bassanus, **574**
Mosaic cleavage, 161, **161**
Mosaic development, 161, **161**
Mosquitos
 eggs of, **411**
 larvae of, 407, **411**
 and malaria, 220–22, 417, **418**
 mouthpart of, 405, **406**
Moss animals. *See* Ectoprocta
Moths, 421, **421**
 antennae of, 401
 chemical sex attraction in, 765–66, **766**
 chemoreception in, 408
 ear of, 712–13, **713**

endocrine system of, **725**
 environmental influence and, **11,** 11–12
 flight of, 404
 mimicry in, 413, **799**
 mouthpart of, 405
 nutrition of, 404
 reproduction of, 410
Motor cortex, 706, **707**
Motor division, of peripheral nervous
 system, **707**
Motor neurons, 696, **696,** 702–3
Motor unit, 628
Motor unit recruitment, 628
Mountain pine beetle, 397
Mouse
 blastula and gastrula of, **159**
 egg of, **156**
 homeotic genes in, **163**
 sperm of, **18**
Mousebirds, 580
Mouth, 307
Mouthparts, 683
 of insects, 404–6, **406**
Movement. *See also* Locomotion
 of cells, 43
 onto land, 515
MSH. *See* Melanocyte-stimulating hormone
Mucous glands, 526
Mucus-secreting cells, **5**
Mud shrimp, 386
Mudskippers, **501**
Müller, Johannes, 709
Müllerian mimicry, 799, **799**
Multicellular organisms, 233
 body plans of, **186**
 life cycle of, 8–9
Multilocular hydatid, **287**
Multinucleate cell, 50
Multiparous animals, 148
Multiple births, 148, **149**
Multiple fission, 133, 214–15
Multiplication of species, 112–15
Mummification, 106
Murex, 328
Muridae, 607
Muscle(s). *See also specific muscle*
 of Anura, 527
 of arthropods, 366
 atrophy of, 627
 of birds, 564, **565**
 contraction of
  control of, 627–28
  energy for, 630, **631**
  sliding filament model of, 626–27, **627**
 of Crustacea, 380
 development of, 170, **171**
 excitation-contraction coupling,
  628–30, **629**
 of Insecta, 403, **404**
 of invertebrates, 625
 locomotion and, 624–30
 myoneural junction and, 628
 performance of, 630–31
 of platyhelminthes, 276–77
 of vertebrates, 624–25, **625**
Muscle arm, 297
Muscle fibers, 182, 276, 624
Muscle tissue, **180,** 182, **184**
Muscular arteries, 664
Muscular pharynx, 479
Musophagiformes, 579
Mussels, **330,** 333
Mustache shrimps, 392
Mustelidae, 608
Mustelus, 494
Mutation, 87–88, 117
 frequency of, 98
 genetic, 97–98
 homeotic, 163
Mutualism, 795, **796**
Mycale laevis, **239**
Myelencephalon, **705**
Myelin, 251, 696–97
Myelin sheath, **697**
Myenia, 243, 245
Myllokunmingia, 481
Myocytes, 239
Myofibrils, 182, 626

Myogenic hearts, 664
Myomeres, 477, 479, **502,** 502–3
Myoneural junction, 628, **628**
Myosin, 42, 622, 626
Myosin filaments, 626, **627**
Myotis, 606
Myotis lucifugus, **596**
Myriapods, 374–75, 398, 420
Myrmica, 413
Mystacocarida, 388, **388,** 392
Mytilus californicanus, **799**
Mytilus edulis, **330**
Myxine, 510
Myxine glutinosa, 490, **490**
Myxini, 469, 482, 489–91, 510
Myxobolus cerebralis, 357

Nacreous layer, 320
Nais, 357
Naja naja, **549**
Nannopterum harrisi, **560**
Nanoloricus mysticus, **306**
Nares, 669
Nasal chamber, 669
National Research Council, 12
Natural killer (NK) cells, 740, **740**
Natural resources, limits on, 118
Natural selection, 15–16, **16,** 106, 118–20
 allelic frequency and, 123
 challenges to, 119–20
 hypothesis of, **11,** 11–12
 origin of life and, 30
Nauplius, 384
Nautiloidea, 340
Nautilus, 320, 335–36, **337,** 338, 340
Neanderthals, 604
Necator, 302
Necator americanus, 300, **300**
Necrosis, 747
Nectalia, 257
Nectonema, 304
Necturus, 522, **523**
Negative feedback, 145, 724, **724**
Negative pressure, 669
Nekton, 779
Nematocysts, 247–48, 250–51, **251, 254, 261**
Nematoda, 295, *295,* 296–303, **298–300**
 adaptive radiation of, 314
 classification of, 302
 form and function of, 297–303
 parasitic, 298–303, **300**
Nematodinium, 716, **716**
Nematogens, 235
Nematomorpha, 303–5, **304**
Nemertea, 289–91, **290–91**
 blastula and gastrula of, **159**
 body plan of, **186**
 characteristics of, 290
 circulatory system of, 290–91, **291**
 classification of, 291
 cleavage in, 156
 digestive system of, 290–91
 egg of, **156**
 excretory system of, **291**
 feeding habits of, 291
 form and function of, 290–91
 gastrulation in, 158
 immunity in, **748**
 locomotion of, 291
 reproduction of, 291
Neoceratodus, 500, 510
Neoceratodus forsteri, **501**
Neocortex, 706
Neo-Darwinism, 17, 120
Neodermata, 277–78
Neognathae, 557, 578–79
Neomenia, 343
Neopilina, 321–22, **322,** 340, 343
Neoplastic growth, 98
Neopterygians, 498
Neopterygii, **500,** 510
Neornithes, 557, 578
Neoteny, 480
Nephridia, 346, 354, **355,** 436, 639–40.
  See also Kidney
Nephridiopore, 354
Nephrons, 642
Nephrops norvegicus, 312
Nephrostome, 321, 349, 354, 640

Nereis, 348–50, 360, 680
Nereis diversicolor, **348**
Nereis virens, **348**
Neritic zone, 779
Nerve(s), 696–97
 growth of, 169–70
 structure of, **696**
 synapses and, 699–701, **700–701**
Nerve cells, in hydra, 253, **253**
Nerve cords, 298
 longitudinal, 276, 278, 286
Nerve energies, law of specific, 709
Nerve fibers, **185**
Nerve ganglion, 312, 474
Nerve impulses, 697–99
 conduction of, 698–99, **699**
Nerve net, **178,** 250–51
Nerve ring, 286, 448
Nerve-muscle synapse, **628**
Nervous lamella, 304
Nervous system, 696–709
 of Annelida, 346
 of Anura, 529
 of arthropods, 366
 of Asteroidea, 448
 autonomic, **707,** 707–9, **708**
 of birds, 567–68, **568**
 of Bivalvia, 333
 central, 702–3
 of Cephalopoda, 338
 of Cnidaria, 250–51
 of Crustacea, 382–83
 of Ctenophora, 269
 development of, 169–70
 of earthworms, 355, **355–56**
 of Enteropneusta, 462
 evolution of, 701–9
 of Gastropoda, 326
 of Hirudinea, 359
 of Insecta, 408
 of invertebrates, 701–2, **702**
 of jellyfishes, 258
 of Mollusca, 321, 702
 parasympathetic, 707–9
 peripheral, 702, 707–9
 of platyhelminthes, 278
 of Polychaeta, 349
 of Reptilia, 540
 somatic, 707
 sympathetic, 707–9
 of Vertebrata, 478, 702–9
Nervous tissue, **180,** 182
Nested hierarchy, 110–11, 196, **196**
Nesting, 575–76, **576,** 756, **756**
Nesting colonies, 762, **762**
Net productivity, 800
Neural crest, **170,** 478
Neural crest cells, 169, 479
Neural mechanisms, 696
Neural plate, 169
Neural tube, 169, **170**
Neurogenic hearts, 664
Neuroglia, 182, 696–97
Neurohypophysis, 725, **727,** 728, **728**
Neuromasts, 495, 712, **712**
Neuromuscular coordination, 409
Neuromuscular junction, 628, **628**
Neuromuscular system, 251
Neurons, 182, **185,** 696–99
 intermediate, 717
 postsynaptic, 699
 presynaptic, 699
Neuropeptides, 729
Neuropodium, 350
Neuroptera, 420, **420**
Neurosecretion, hypothalamus and, 725–26
Neurosecretory cells, 355, 385, 724
Neurospora, 88
Neurotoxic venom, 549
Neurotransmitters, 699, 722
Neutral fats, 24, **24**
Neutrophils, **657,** 658, 740
New World monkeys, 606
Newt, **522**
Niche, 790–91, **791**
 overlap of, 797
Nictitating membrane, 530
Night crawlers. *See* Earthworms
Nighthawks, 580

Nile crocodile, 552, **552**
Nile monitor lizard, 194
Nitrogenous base, 88–89
Nitrogenous waste, 70
NK cells. *See* Natural killer cells
NMR. *See* Nuclear magnetic resonance
    spectroscopy
Noctiluca, 226, **227**, 231
Nocturnal animals, 649
Nodes of Ranvier, **185**, 699
Nomenclature, 194
    binomial, 192
Nomeus, 257
Nonallopatric speciation, 114–15
Nondisjunction, 87
Nonidentical twins, 148, **149**
Noninfective parasites, 739
Nonrandom mating, 123
Nonself, recognition of, 741
Nonshivering thermogenesis, 650, 689
Nonspecific immunity, 739–41
Noradrenaline, 734
Norepinephrine, 734
Nostrils, 494
Notochord, 468, **468**, 469, 618
Notophthalmus viridescens, **522**
Notopodium, 350
Notostraca, 387, **387**, 392
Nucella lamellosa, **328**
Nuclear DNA, 33
Nuclear envelope, 39, 41, **41**
Nuclear equivalence, 160–61
Nuclear fusion, 154–55, **155**
Nuclear magnetic resonance (NMR)
    spectroscopy, 38
Nuclear receptors, 723
Nucleases, 686
Nucleic acid, 6, 26, 88–91
    bonds of, 4
    sugars of, **89**
Nucleoid, of prokaryotes, 32
Nucleoli, 41, **41**, 209
Nucleotide bases, 8
Nucleotides, 7, 26, 88–89
Nucleus, 39, 41, **41**, 227
    constancy of, 309
    DNA and, 33
    of protozoa, 209, **209**
Nucula, 332
Nuda, 270
Nudibranch, **318**, 328, **328–29**
Numbers, pyramid of, **802**, 802–3
Numbats, 606
Nurse cell, 301
Nüsslein-Volhard, Christiane, 162
Nutrient cycles, **803**, 803–4
Nutrients, essential, 690
Nutrition, 690–92, **691**. *See also* Digestive
        system; Feeding habits
    of earthworms, 353
    of Hirudinea, 358–59
    of Insecta, 404–6
    of platyhelminthes, *277*
    of Polychaeta, 348–49
    of protozoa, 208, 214, **214**
    of sponges, 240, **241**
Nutritive-muscular cells, in hydra, 253
Nyctotherus, 225
Nymphon, 368, **369**
Nymphs, 411, **412**

Oak treehoppers, **420**
Obelia, 252, **254**, 254–55, 267
Obesity, 69, 689
Oceans, 779–80
Ocelli, 255, 278
Ochotona princeps, **607**
Octocorallia, 260, 267
Octocorallian coral, 264, **265–66**
Octomerous body plan, 260
Octopods, 338–40
Octopus, **318**, 320, 335, 337–38, 343
Octopus briareus, **318**
Odd-toed hoofed animals, 608
Odocoileus virginianus, 56
Odonata, **412**, 419
Odontogriphus, **108**
Odontophore, 319
Old World monkeys, 606

Oleic acids, 24, **24**
Olfaction, 709–10, **710**
Oligochaeta, 346, 353–57, 360
    freshwater, 357, **358**
Oligodendrocytes, 697
Omasum, 592
Ommatidia, 383, 716
Omnivores, 507, **591**, 593, 678
*On the Origin of Species by Means of
        Natural Selection* (Darwin),
        14, 105–6, 193
Onchocerciasis, 303
Oncogenes, 98
Oncomelania, 283
Oncomiracidium, 284
Oncorhynchus, 508, 510
Oncorhynchus nerka, **509**
Oncospheres, 288
One gene-one enzyme hypothesis, 88
Ontogeny, 111–12
Onychophora, **108**, 375, 420, 427, 435–36,
        436, 439, 439–40
Oocyst, 219, **219**, 221
Oocytes, 143
    maturation of, 153
    primary, 139, **140**
    secondary, 139, **140**
Oogenesis, 139–40, **140**
Oogonia, 139, **140**
Ookinete, 221
Ootid, 139, **140**
Ootype, 282
Opabinia, **108**
Oparin, Alexander I., 22, 26
Open circulatory system, 660–61
    of arthropods, 366
    of Brachiopoda, 434
    of Crustacea, 381
    of Mollusca, 317–18, 320
    of Onychophora, 436
Operculum, 250, 324, 432, 497, 504
Ophioderma, 452
Ophiopholis aculeata, **452**
Ophiothrix, **451**, 452
Ophiothrix suensoni, **244**
Ophisaurus, **544**
Ophiopluteus, **450**
Ophiura, 452, 460
Ophiura lutkeni, **450**
Ophiuroidea, 443, **450**, 450–52, 460
Opiliones, 372, **372**
Opisthaptor, 284
Opisthobranchs, 328
Opisthonephric kidney, 495, 505
Opisthosoma, 369, 429
Opistognathus macrognathus, **509**
Opistonephros, 641
Opossum, 194, **598**, 606
Opsin, 718
Opsonization, 742
Optic lobes, 568, **705**
Oral arms, 258
Oral contraceptives, 145
Oral disc, 261
Oral end, 250
Oral groove, 223
Oral papillae, 435
Oral suckers, 282
Orangutan, 199, 606
Orchestia, 386, 391–92
Organ of Corti, 713
Organ systems, 4, 177
Organ transplantation, 745
Organelles, 4, 37
Organic compounds, 22
Organic molecular structure, of living
        systems, 22–26
Organismal morphology, 193
Organisms, 790
    emergent properties of, **6**
    first living, 30
    levels of organization in, 177–78, **178**
    reproduction and, 6, **7**
    study of, **6**
    timescale for reproduction of, **6**
Organization
    cell-tissue grade of, **178**
    cellular grade of, **178**
    organ-system grade of, **178**

protoplasmic grade of, **178**
    tissue-organ grade of, **178**
Organs, 4, 177
    development of, 168–71
    morphogenesis of, 164, **164**
Organ-system grade of organization, **178**
Origin of Species, The (Darwin). *See On the
        Origin of Species by Means of
        Natural Selection* (Darwin)
Ornithischia, 546–47, 551
Ornithodelphia, 606
Ornithorhynchus anatinus, 606
Orthasterias koehleri, **449**
Orthogenesis, 119
Orthonectida, 234–35, **236**
Orthoptera, **401–2**, 408, **413**, 419, 424
Oscula, 237
Osmometer, **46**
Osmoregulation, 635–39, **637**
    in earthworms, 355
    in fishes, 46, 505, **506**
    in marine invertebrates, **635**, 635–36
    in platyhelminthes, 277–78
    in protozoa, 213
    in terrestrial animals, 638–39
Osmosis, 46, **46**
Osmotic concentration, kidneys and, **647**
Osmotic conformers, 635
Osmotic potential, 46
Osmotic pressure, 46, 251
    colloid, 665–66
Osmotrophs, 214
Osphradia, 323, 327
Osprey, **562**
Ossicles, 445, 455, **455**
Osteichthyes. *See* Bony fishes
Osteoblasts, 619, **619**, 732
Osteoclasts, 619, 732
Osteocytes, 618, **619**
Osteoderms, 539
Osteons, 618, **619**
Osteoporosis, 618
Osteostracans, 481, **481**
Ostia, 237
Ostracoda, 387, **388**, 392
Ostracoderms, **481**, 481–82
Ostrich, 557, 578, **578**
Otariidae, 608
Ottoia, **108**
Outer chamber, of eye, 717
Outgroup comparison, 195
Oval window, 713
Ovale, 504
Ovarian balls, 310
Ovary, 134, 143, **143**, 282
Overhydration, 646
Overpopulation, 106
Overton, William R., 11
Oviduct, 142–43, 279, 282
Ovigers, 368
Oviparous animals, 140, 495, 509–10,
        550, 606
Oviraptor, 546
Ovis canadensis, **761**
Ovisacs, **388**
Ovoviviparous animals, 140–41, 496, 509
Ovulation, 144
Ovum. *See* Egg
Owen, Richard, 109, 546
Owls, 579
Oxidation-reduction reactions, 62, **62**
Oxidative metabolism, 31–32, 62
Oxidative phosphorylation, 64–65, **66**
Oxidization, 678
Oxygen, photosynthesis and, 31–32
Oxyhemoglobin, 673
Oxytocin, 146, **727**, 728, **728**
Oyster borers, 326, 328
Oyster catchers, 579
Oysters, 317, **336**
Ozone, 27

p53, 98
Pacemaker, 171
Pacemaker cells, 663
Pacific salmon, 508, 510, **511**
Pacific sea star, **449**
Pacinian corpuscles, 711, **711**

Paddlefish, 498, **500**, 510
Paedomorphosis, 112, 479–80
    in salamanders, 522–23, **523**
Pagurus, 392
Pain, 711–12
Paired appendages, 468
Paired septa, 260
Pair-rule genes, 163
Palaeoniscids, 498
Paleocortex, 706
Paleognathae, 557, 578
Paleozoic era, 107
Paleozoic rock, *102*
Palolo worm, 350, **350**
Pan, 199, **201**, 607
Pancreas, 170, 495, **734**
    islet cells of, 734–35
Pancreatic amylase, 686
Pancreatic enzymes, 686
Pancreatic juice, 686
Pancreatic lipase, 686
Pandion haliaetus, **562**
Pandorina, **211**, **217**
Pangaea, 784, **785**
Pangolins, 607
Paninus imperator, **372**
Panope abrupta, **318**
Panthera leo, **593**
Panulirus, 392
Panulirus argus, **394**
Paper wasp, **421**
Papilio krishna, **421**
Papilionidae, **421**
Papio, 607
Papio cynocephalus, **2**
Papio hamadryas, **602**
Papulae, 445
Parabasal body, 216
Parabasalea, 216, 229–30
Parabronchi, 566, 668
Paradigms, 13
Parafossarulus, 282
Paragonimus, **282**, 284
Paragonimus kellicotti, 284
Paragonimus westermani, **282**, 284, **285**
Paragordius, **304**
Parahormones, 722
Parakeets, 579
Paramecium, **207**, 223–25, 231, **798**
    contractile vacuole of, 213, 223, **224**
    form and function of, 223–24
    locomotion of, 223–24, **224**
    reproduction of, 224–25, **225**
Paramecium caudatum, 223
Paramecium multimicronucleatum, 223
Paramylon granules, 218
Paranthropus robustus, 604, **605**
Paraphyletic group, 468, 538
Paraphyly, 197–98, **200**, 201–2
Parapodia, 346–47, 350
Pararthropoda, 427
Parasaurolophus, 546–47, 551
Parasites, 790, 795, **795**, 797–99
    fishes, 507
    infective, 739
    insects, 404
    lampreys, 492
    Mesozoa, 234–35, **236**
    noninfective, 739
    phylum Platyhelminthes, 276, 280–89
    protozoa, 218–22, 226–27
    Pseudocoelomates, 297–303, **300**
Parasitic DNA, 97
Parasitoids, 404
Parastichopus, 455, 460
Parastichopus californicus, **455, 457**
Parasympathetic nervous system, 707–9
Parathyroid glands, 731
Parathyroid hormone (PTH), 619,
        731–32, **732**
Parazoa, 234, 236
Parchment worm, 351–53
Parenchyma, 177, 185, 275–76
Parenchymula, 242
Parietal cells, 685
Parietal eye, **550**
Parietal pleura, 670
Parrot snake, **548**
Parrots, 579

Parthenogenesis, 133–35, 791
  ameiotic, 135
  meiotic, 135
Partial pressure, lungs and, **671**, 671–72, **672**
Particulate inheritance, 17, **18**
Parturition, 146, **147**
Parus carolinensis, 192
Passenger pigeons, 576, **577**
Passeriformes, 580
Pasteur, Louis, 21
Patch reefs, 265–66
Patella, 328
Pattern formation, 162–63
Pauropoda, 398–99, **399**, 418, 420
Pauropus, 399, 418
PCR. *See* Polymerase chain reaction
Pearls, 317, 320, **320**, 330–31
Pecten, 568
Pectines, 371
Pectoral, definition of, 185
Pectoral fins, 487, 494
Pectoralis muscle, 564
Pedal glands, 307
Pedal laceration, 263
Pedalium, 259
Pedicel, 433–34
Pedicellariae, 445, **448**, 452
Pedicularia, **321**
Pediculus humanus, **405**
Pedipalps, 369
Pelage, 587, **588**
Pelagic fishes, 507
Pelagic zone, 779, **781**
Pelecaniformes, 579
Pelecanus onocrotalus, **574**
Pelecypoda. *See* Bivalvia
Pelicans, **574**, 579
Pellicle, 218, 223, **223**, 227
Pelmatozoa, 460
Pelomyxa carolinensis, 227
Pelvic, definition of, 185
Pelvic fins, 487, 494
Pelycosauria, 551, 584
Pen, 336
Penaeus, **384**
Penaeus, 392
Penguins, 197–98, **200**, 557, 578
Penis, 142, **142**
Pentadactyl, 620
Pentastomida, 427, 434–38, **435,** 439
Pentatrichomonas hominis, 216
Pepperberg, Irene, 768
Peppered moths, colored variants of, **11,** 11–12
Pepsin, 685
Pepsinogen, 685
Peptide bonds, 24
Peranema, 218, **219**
Perca, 510
Perch, **498,** 504, **509,** 510
Perching mechanism, of birds, 564, **565**
Perching songbirds, 580
Perennibranchiate, 522, **523**
Perforated pharynx, 468, 472
Pericardial sinus, 381
Pericardium, 662
Perihemal channels, 448
Perihemal sinuses, 445
Periods, 107
Periophthalmus, **501**
Periostracum, 320
Peripatus, **436**
Peripheral nervous system, 702, 707–9
Periproct, 452
Perisarc, 255
Perissodactyla, 608
Peristalsis, 683, **683**
Peristomium, 348
Peritoneum, **142,** 296
Periwinkles, 324, 328
Perla, 412
Permeability, 46
Permeases, 47
Permian extinction, 127
Perpetual change, 14–15, 106–9
Perviata, 429
Pesticides, 804
Pesudobiotus, **437**
Petaloids, 454
Petrels, 579

Petromyzon, 510
Petromyzon marinus, 491, **492,** 492–93
Petromyzontes, 491–93
Phacus, **219**
Phagocata, **279**
Phagocytes, 566, 740
Phagocytosis, 47, **49,** 226, 740, **740**
Phagosome, 41, 214
Phagotrophs, 214
Phalaropes, 579
Pharyngeal grooves, 472
Pharyngeal jaws, 500
Pharyngeal pouches, 469, 472
Pharyngeal sheath, 277
Pharyngeal slits, 472
Pharynx, 261, 474, 477–79, 495, 669, 683
Phascolosoma, 428
Phasmids, 298
Pheasants, **561,** 579
Phenetic taxonomy, 200
Phenotypes, 79
  variation in, 97–98
Phenotypic effect, 121
Phenotypic gradualism, 117, **117**
Pheromones, 414–15, 709, **709,** 766
Philippine Trench, 443
Philodina, 306–7, **307,** 308–9
Phoenicopterus ruber, **579**
Pholas, 335
Pholidota, 607
Phonoproduction, of insects, 414
Phonoreception, of insects, 414
Phoronida, 427, 430–31, **431,** 439
Phoronis, **431**
Phoronis architecta, 431
Phoronopsis californica, 431
Phosphate group, 88–89
Phosphate ions, 657
Phosphatidyl choline. *See* Lecithin
Phosphoanhydride bonds, 60
Phospholipids, 24, 39–41
Phosphorylation, oxidative, 64–65, **66**
Photic zone, 779
Photinus tanytoxus, **415**
Photoperiod, 728
Photophore, 392
Photoreception, 715–18
Photosynthesis, 10, 22, 31–32, 800
Phototaxis, 224
Phototrophs, 678
Photuris, 414
Photuris versicolor, **415**
Phronmia, **391**
Phyletic gradualism, 117
Phyletic increase, Cope's law of, 177
Phyllidia ocellata, **329**
Phyllobates, 526
Phyllobates bicolor, **520**
Phyllophaga, 310
Phyllopodia, 387
Phylogenetic information, sources of, 196–97
Phylogenetic reconstruction, 109–11, 195–97
Phylogenetic species concept, 194–95
Phylogenetic systematics, 201–2
Phylogenetic tree, 196, **197**
Phylogeny, 15, 109, 111–12, 190–206
  of Annelida, 360–61
  of Arthropoda, 374–75
  of Bilateria, 292
  character variation in, 195–96
  of Chordata, **470**
  of Cnidaria, 270–71
  of Crustacea, 393–94
  of Ctenophora, 270–71
  of Echinodermata, 458–60
  of flightless birds, 782, **783**
  of Hemichordata, 464
  of Insecta, 420–24
  of Mesozoa, 235
  of Mollusca, 340–42
  of Porifera, 244
  of protozoa, 229–30
  of pseudocoelomates, 312–14, **313**
  of smaller protostomes, 438–40
Physa, 329, 343
Physalia, 256–57, **257,** 267
Physical barriers, in immunity, 739
Physical laws, 9–10

Physiology, 8
Physoclistous fishes, 504
Physopsis, 283
Physostomous fishes, 504
Phytomastigophorea, 216, 229
Phytophagous animals, 404
Piciformes, 580
Pickerel frog, *514,* 524
Piddock, 335
Pigeons, 579
Pigmentation, 615–16
  inheritance and, 84
Pika, 607, **607**
Pikaia, **108,** 478
Pilidium larva, 290
Pill bugs, 390, **391**
Pinacocytes, 239, **240**
Pinacoderm, 239
Pincers, 392
Pincushion star, **444**
Pineal gland, 728–29
Pinnules, 456
Pinocytosis, 48, 214
Pinworms, **300,** 302, **302**
Pisaster, 445, 460
Pisaster ochraceus, 442, 799, **800**
Pisces, 469
Piscivores, **591**
Pistol shrimp, 386
Pit viper, 548–49, **549**
Pitch discrimination, place hypothesis of, 714–15
Pituitary gland, 725–28, **726–27**
Pituitary gonadotropins, 144
Pitx2 (gene), 162
Place hypothesis, of pitch discrimination, 714–15
Placenta, 143, 146, **147,** 496
Placental mammals, 167–68, 597–98, **598**
Placentotrophy, 542
Placobdella, **359,** 360
Placoderms, 483, **484,** 487
Placoid scales, 493–94, **499**
Placozoa, 236, **236**
Plague, 417
Planarians, **276–78,** 288
Plankton, 226, 678, 779
Plankton feeders, 507
Planocera, 288
Planospiral shells, 325
Plants
  distinctions from animals, 10
  oxygen production and, 31–32
Planula larva, 250
Planuloid ancestors, 292
Plasma, blood, 178
Plasma cells, 741
Plasma membrane, **40**
  function of, 45–48
  transport across, 46
Plasma proteins, 658
Plasmids, 95
Plasmodium, 235
Plasmodium, 220–22, 231
Plasmodium falciparum, 220–21
Plasmodium malariae, 220
Plasmodium ovale, 220
Plasmodium vivax, 220
Plastid DNA, 33
Plastids, 39
  DNA and, 33
Plastron, 540
Plate tectonics, 784, **785**
Platelets, 657, **657**
Platycotis vittata, **420**
Platyhelminthes, 275–89
  body plan of, **186**
  characteristics of, 276
  cladogram of, **293**
  classification of, 288
  digestive system of, 277
  excretory system of, 277–78
  flame cell system of, 639, **639**
  muscles of, 276–77
  nervous system of, 278, **702**
  nutrition of, 277
  osmoregulation in, 277–78
  parasitic, 276, 280–89
  regeneration in, 278–79

  reproduction of, 278–79
  sense organs of, 278
  tegument of, 277
Platyrrhine monkeys, 606
Plecoptera, **412,** 419
Pleiotropy, 83
Plesiomorphy, 196
Plesiosauria, 551
Plethodon jordani, **114**
Plethodon teyabalee, **114**
Plethodontids, 521–22, **523–24**
Pleural cavity, 670
Pleurobrachia, 268–69
Pleurobrachia, **268,** 269, **269**
Pleuroploca, 328
Pleuroploca gigantea, 324
Plexaura, 267
Pliny the Elder, 758
Plovers, 579
Plumatella, **432**
Plumatella repens, **433**
PMN. *See* Polymorphonuclear leukocytes
Pneumatized bones, 562
Pneumatophore, 257
Pneumostome, 326, **330**
Podia, 445–46
  of Holothuroidea, 455–56
  of Ophiuroidea, 451
Podicipediformes, 579
Podilymbus podiceps, 579
Podo, 215
Podophrya, **214,** 226, 231
Pogonophora, 427, **429,** 429–30, **430,** 439
Poikilothermic animals, 647
Poison arrow frog, **520**
Polar bears, **3**
Polar body, first, 139
Polarity, 155, 188
Polian vesicles, 446
Policipes, 392
Polinices, 326, 328, 343
Polinices lewisii, **326**
Pollination, insects and, 416–17
Polyandry, 575, 763
Polychaeta, *345,* 346–53, 360
  circulatory system of, 349
  excretory system of, 349
  feeding habits of, 351–53
  form and function of, 348–50
  fossils of, **108**
  nervous system of, 349
  nutrition of, 348–49
  reproduction of, 141, 350
  respiratory system of, 349
  sensory organs of, 350
Polycladida, *274,* 280, **280**
Polycystinea, **228**
Polyestrous animals, 596
Polygamy, 574, 763
Polygenic inheritance, 83
Polygyny, 575, 763
Polygyra, 329
Polymerase chain reaction (PCR), 96, **96**
Polymers
  concentration and, 29–30
  formation of, 28–30
  thermal condensation and, 30
Polymorphism, 120
  in cnidaria, 249–50
  protein, **124,** 124–25, **125**
Polymorphonuclear leukocytes (PMN), 740
Polynoidae, 351
Polyodon, 510
Polyodon spathula, **500**
Polyorchis penicillatus, **255**
Polyp, 255, **255**
Polypeptide chains, **94**
Polyphyletic origin, of metazoans, 234
Polyphyly, 197, **200**
Polypide, 432
Polyplacophora, 317, **318,** 321–23, **323,** 343
Polyploidy, 87
Polyps, **249,** 249–50, **254,** 265
Polypterus, 498, 510
Polypterus bichir, **500**
Polyribosome, 93, **93**
Polysaccharides, 22–24
Polysome, 93, **93**
Polyspermy, prevention of, 153–54

Polystoma, 284, 288
Polyunsaturated fatty acids, **691**
Polyzoic, 285
Pomacentridae, 262
Pongidae, 201, **201**
Pongo, 199, **201,** 607
Pons, 705, **705**
Popillia japonica, **417**
Population, 106, 790–95
    of birds, 576–77
    emergent properties of, **6**
    growth of, 692, **692, 792,** 792–94, **793**
        exponential, 794
        extrinsic limits on, 794–95
        logistic, 794
    of mammals, 600
    regulation of, 792–94
    size of, 118
    study of, **6**
    timescale for reproduction of, **6**
Population genetics, 120
Populational gradualism, 117
Porcellio, 390, 392
Porcupines, 589, **589**
Porifera, 233, 236–45
    adaptive radiation of, 244
    canal systems of, 237–39, **239, 241**
    cells of, 239–40, **240–41**
    characteristics of, 237
    classification of, 245
    digestion in, 240
    embryogenesis in, 242
    form and function of, 237
    fossil of, **108**
    immunity in, **748**
    nutrition of, 240, **241**
    phylogeny of, 244
    physiology of, 240
    regeneration in, 242
    reproduction of, 240–42
    skeleton of, 240, **241**
    structure of, **238**
    types of, 236, **237**
Pork tapeworm, **287,** 288
Porpoises, **111,** 609
Portrait of a Meadow Mouse (Burgess), 599
Portuguese man-of-war, 248–49, 256–57, **257**
Positive assortative mating, 123
Positive feedback, 145, 724
Positive pressure, 669
Postabdomen, 371
Postanal tail, 469, 472
Postdisplacement, 480
Posterior, 185, **187**
Posterior pituitary gland, 725, **727,** 728, **728**
Postganglionic fibers, 708
Postmating barriers, 114
Postsynaptic neurons, 699
Posture, 622, **622**
Potassium, 657
Potassium-argon dating, 107
Potential energy, 57
Poterion, **241,** 245
Potocytosis, 47–48, **49**
Pottos, 606
Powder-down feathers, 561
Prairie, 778, **778**
Prairie dogs, **599**
Praying mantis, 400, **402,** 419
Preabdomen, 371
Prebiotic synthesis
    concentration and, 29–30
    of small organic molecules, **27,** 27–28
    thermal condensation and, 30
Precambrian era, 107–9
Precambrian life, 32–34
Precocial young, 575, **576**
Predators, 386, 404, 468, 790, 797–99, **798**
Preformation, versus epigenesis, **152,** 152–53
Preganglionic fibers, 708
Pregnancy
    human, hormones of, 145–48, **146–47**
    Toxoplasma and, 220
Premating barriers, 114
Premolars, 590, 680
Premunition, 739
Prepuce, **142**
Presbytis, 607
Presynaptic neurons, 699

Prey, 797–99, **798**
Priapulida, **108,** 305–6, **306**
Priapulus, 306, **306**
Primary heterotrophs, 31
Primary induction, 162
Primary oocytes, 139, **140**
Primary organizer, 151, 161–62, **162**
Primary producers, 800
Primary reticulum, 243, **244**
Primary septa, 261
Primary sex organs, 134, 141
Primary spermatocytes, 138, **138**
Primary structure, of proteins, 25, **26**
Primates, 606
    evolutionary radiation of, 602
Primitive atmosphere, 26–27
Primitive streak, 159
Primordial germ cells, 136, **137**
Principal layer, 384
Principle of antithesis, 761, **761**
Principles of Geology (Lyell), 104
Prismatic layer, 320
Probability, 83
Proboscidea, 608
Proboscis, 290, 368, 427, 461
Proboscis pore, 461–62
Proboscis receptacle, 309
Procellariiformes, 579
Procuticle, 365, 384, 613
Procyonidae, 608
Product rule, 83, **84**
Productivity, 800
Progenesis, 480
Progesterone, 94, 144, **144, 146**
Proglottid, 285
Prokaryotes, 32, **33**
Prolactin, 146, 726–27, **727**
Proliferative phase, of menstrual cycle, 144
Proline, **25**
Prometaphase, **50**
Pronephros, 641, **641**
Pronuclei, fusion of, 154–55
Prophase
    in meiosis, **77,** 78
    in mitosis, 50, **50, 52**
Proprioceptors, 709
Pröscholdt, Hilde, 151
Prosencephalon, 704, **705**
Prosimians, 602, **602,** 606
Prosobranchs, 327–28
Prosoma, 369
Prosopyles, 238
Prostaglandins, 146, 729
Prostate, 142, **142**
Prostoma, 291
Prostoma rubrum, 290
Prostomium, 346, 348
Protandrous, 263
Protarchaeopteryx, 557
Protease, 685
Protective coloration, 413
Protein(s), 24–26, **26,** 657, 691–92
    contractile, 622
    as enzymes, 25–26
    functions of, 25–26
    metabolism of, **69,** 69–70
    structure of, 4, **5,** 25, **26**
Protein electrophoresis, 125
Protein polymorphism, **124,** 124–25, **125**
Proteinoid microspheres, 30, **30**
Proteome, 97
Proteomics, 97
Proterospongia, **229,** 234
Proterozoic eon, 107
Prothoracic glands, 411–12
Prothoracicotropic hormone (PTTH), 412, 724
Prothorax, 400
Prothrombin, 658
Protista, 10, 203, 208
Protocells, 30–31
Protochordata, 469, 472–73
Protocoel, 431
Proton pumps, 213, **213**
Protonephridia, 276–77, 305, 309, 639
Protonephridial tubules, 308
Proto-oncogenes, 98
Protoplasm, 37
Protoplasmic grade of organization, **178**
Protopod, 379, **380–81**

Protopodite, 379
Protopterus, 500, **501,** 510
Protostome, 317
Protostomia, 473
    classification of, 204, **205**
    cleavage in, 156
    development of, **157**
    gastrulation in, 158
    smaller, 426–40
Prototheria, 606
Prototroch, 321
Protozoa, 203, 207–32
    adaptive radiation of, 230
    characteristics of, 209
    classification of, 230–31
    encystment in, 215–16
    excretion in, 213
    excystment in, 215–16
    form and function of, 209–16
    integument of, 613
    locomotion of, 210–13
    nutrition of, 214, **214**
    osmoregulation in, 213
    phylogeny of, 229–30
    position in animal kingdom, 208
    reproduction of, 214–15
    types of, 216–29
    unicellular, organization of, 177
Protura, 419
Proventriculus, 565
Proximal, 185
Proximal convoluted tubule, 642
Proximate causation, 753
Proximate causes, 13
Pseudoceratina crassa, **244**
Pseudoceros hancockanum, **280**
Pseudococcus longuspinus, **417**
Pseudocoel, 185, 296–97, 308
Pseudocoelomates, **187,** 295–314
    adaptive radiation of, 314
    Bilateria, 185, **187**
    biological contributions of, 296
    body plan of, **186, 296**
    characteristics of, 297
    parasitic, 297–303, **300**
    phylogeny of, 312–14, **313**
    position in animal kingdom, 296
Pseudopodia, 43, 226, 229, 623
    of protozoa, **211,** 211–13, **212**
Psittaciformes, 579
Psocids, 419
Psocoptera, 419, 424
Psolus chitonoides, **455**
Psyllophryne didactyla, 524
Ptarmigan, 579
Pterobranchia, **463,** 463–64
Pterois, **501**
Pteromyzon marinus, **491**
Pteropods, 328
Pteropus, 606
Pterosauria, 551
PTH. *See* Parathyroid hormone
Ptilosarcus, 264
Ptilosarcus gurneyi, **261**
PTTH. *See* Prothoracicotropic hormone
Ptychodiscus, 231
Pubic bone, **142–43**
Puffbirds, 580
Puffins, 579
Pulex irritans, **405**
Pulmonary artery, 662
Pulmonary circuit, 661
Pulmonates, 328–29, **330**
Punctuated equilibrium, 117–18, **118**
Punnett square, 79–80, **82,** 83
Pupa, 412
Pupil, 717
Purines, 88–89, **89**
Purkinje fibers, 663
Purkinje, J., 37
Pycnogonida, 368, **369,** 374–75
Pycnogonum, 368, **369,** 374
Pycnophyes, 305
Pycnopodia, 446
Pycnopodia helianthoides, **448–49**
Pygidium, 346
Pygmy marsupial frog, **520**
Pygmy salamanders, **521**
Pygostyle, 562

Pyloric ceca, 447, 507
Pyloric sphincter, 685–86
Pyramids, ecological, **802,** 802–3
Pyrimidines, 88–89, **89**
Pyruvic acid, 63
    acetyl coenzyme A and, **64**
Python, **545,** 549

Q_{10}, 647
Quail, 579
Quantitative inheritance. *See* Blending
Quantitative variation, 125, **125**
Quarternary structure, of proteins, 25, **26**
Queen, 415, **415–16**
Queen substance, 415
Quills, 560, 589, **589**

Rabbitfish, 497
Rabbits, **587–88,** 607
    humans and, 601
Radial canals, 238, 258, 446
Radial cleavage, 156
Radial nerve, 445, 448
Radial symmetry, 183, **186–87,** 234, 250, 443, 445
Radiate animals, 184, 247–73
Radiating canals, 223
Radiation, 98
    adaptive. *See* Adaptive radiation
Radioactive clocks, 107
Radioimmunoassay, 729
Radiolarian ooze, 229
Radiolarians, **228,** 228–29
Radioles, 351
Radiometric dating methods, 107
Radula, 317–19, **319**
Rails, 579
Raja, 510
Raja eglonteria, **496**
Rajidae, 496
Rajiformes, 493
Raleigh, Sir Walter, 677
Ralpharia gorgoniae, **255**
Ram, **18,** 760–61
Ram ventilation, 504
Rana catesbeiana, 524, **524, 527**
Rana clamitans, 1, 524
Rana palustris, 514, 524
Rana pipiens, 524, 528
Rana sylvatica, 524
Rangifer tarandus, **595, 601**
Ranidae, 523, **525**
Ras protein, 98
Ratfish, 497, **497,** 510
Ratite, 578
Rats, **601,** 607
Rattlesnake, **548–49**
Rattus norvegicus, **601**
Ray-finned fishes, 482, 489, 497–500, **500,** 506–7, 510
Rays, 482, 487, 493–97, **496,** 510, 638
Razor clam, 326, **331**
Reabsorption, 642
Reactive force, 503
Reactive nitrogen intermediates (RNIs), 740
Reactive oxygen intermediates (ROIs), 740
Realized niche, 791
Recapitulation, 111–12
Receptor molecules, 722
Receptor potential, 711
Receptor-mediated endocytosis, 48, **49**
Receptors, 696, 703, 709
Recessive traits, 79, 121–22
Reciprocal altruism, 765
Recognition molecules, 741–43
Recombinant DNA, **95,** 95–96
Recruitment, of motor unit, 628
Rectal glands, 495, 688
Rectilinear movement, 545, **548**
Rectum, **142–43**
Red blood cells, 657, **657,** 658, **658**
Red howler monkeys, **602**
Red kangaroos, 597
Red muscles, 630
Red tides, 226
Redbugs, 373
Rediae, 281
Redox reactions, 62, **62**
Red-spotted newt, **522**

Reducing mixture, 27
Reduction, in animal research, 12
Red-water fever, 374
Reef crest, 266
Reef fishes, hermaphroditism in, 134
Reef flat, 266
Reef front, 266
Refinement, in animal research, 12
Reflex act, 704
Reflex arc, **703,** 703–4
Regeneration
    of Asteroidea, 448–49
    of hydra, 253
    of Ophiuroidea, 452
    of platyhelminthes, 278–79
    of sponges, 242
Regulative cleavage, 161, **161**
Regulative development, 161, **161**
Reighardia sternae, 434
Reindeer, **601**
Relative fitness, 123
Relaxin, 146
Release-inhibiting hormones, 725–26
Releaser, 754
Releasing hormones, 725–26
Religion, 11
    evolutionary theory and, 103, 111
Remipedia, 379, 386, **387,** 392–93
Remora, **795**
Renal artery, 642
Renal corpuscle, 642, **643**
Renal pelvis, 642
Renal tubule, 642
Renal vein, 642
Renilla, 264, 267
Renin, 644
Rennin, 685
Replacement bone, 618
Replacement, in animal research, 12
Replication, 90
    of DNA, **92**
    origins of life and, 4
Repolarization, 698
Reproduction, 131–50. *See also* Asexual
            reproduction; Sexual
            reproduction
    complexity of, 6, **7**
    endocrine events in, 143–48
    in living systems, 6
    nature of, 132–36
    patterns of, 140–41
    plan of, 141–43
Reproductive barriers, 113–14
Reproductive community, species recognition
        and, 193
Reproductive cycles, hormonal control of,
        143–44
Reproductive system
    of crickets, **141**
    female, 143, **143**
    human, **142–43**
    male, 142, **142**
Reproductive tissue, **180**
Reptilia, 468, 482, 534–54
    adaptive radiation of, 535–38
    biological contributions of, 535
    birds and, 556–57
    body plan of, 540
    characteristics of, 538
    circulatory system of, 539
    classification of, 535–38, 551
    comparison with amphibians, 538–40
    development of, 539
    eggs of, 155
    embryos of, **113**
    gastrulation in, 159
    integument of, **538,** 538–39, 614
    jaws of, 539
    nervous system of, 540
    origins of, 535–38
    position in animal kingdom, 535
    respiratory system of, 539
    water conservation in, 539
Repugnatorial glands, 399
RER. *See* Rough endoplasmic reticulum
Resemblance, 75
Reservoir, 218
Resilin, 409

Resistance, 739
Resource-defense polygyny, 763
Resources, 790
    limiting, 793, 797
Respiratory pigments, 672
Respiratory system, 666–74
    of amphibians, 668–69
    of Anura, 527–28, **528**
    aquatic, 668, **668**
    aquatic versus aerial, 667
    of Arthropoda, 366
    of Asteroidea, 445–46
    of birds, 566–67, **567**
    cellular, 61–66, 666
    of Cephalopoda, 337
    coordination of, 670–71
    of Crustacea, 380
    cutaneous, 667
    of earthworms, 354
    energy budget and, 800
    external, 666
    of fishes, 504
    of Gastropoda, 326, **327**
    of Hirudinea, 359
    of Mammalia, 668–74
    organs of, 667–69
    of Polychaeta, 348–49
    of Reptilia, 539
    of salamanders, 521–22
    of spiders, 369
    transport of gases in, 672–74, **674**
    of turtles, 540–41
    of Vertebrata, 477–78
Respiratory tree, 456
Resting membrane potential, 697–98, **698**
Restriction endonucleases, 95
Rete mirabile, 504
Reteporella graffei, **432**
Reticulitermes flavipes, **416**
Reticuloendothelial system, 740
Reticulopodia, 211, 228
Reticulum, 592
    primary, 243, **244**
    secondary, 243, **244**
    trabecular, 242–43, **244**
Retina, 530, 568, 717, **717**
Retinal, 718
Retortamonada, 216, 230
Rh factor, 748
Rhabdites, 276, 302
Rhabdopleura, 463, **463**
Rheas, 578
Rheiformes, 578
Rheotaxis, 224
Rhineura florida, 544
Rhinoceros, 608, **609**
Rhinoceros horn, 589
Rhinoderma darwinii, **520**
Rhinophores, 328
Rhipidistians, 501
Rhizocephala, 390, **390**
Rhizopoda, 227, 231
Rhizostoma, 267
Rhodnius, 417
Rhodopsins, 718
Rhombencephalon, 704, **705**
Rhombozoa, 234–35
Rhopalium, 258
Rhoptries, 219
Rhynchocephalia, 551
Rhynchocinetes rigens, **394**
Rhynchocoel, 291
Rhynchocoela, 289–91
Ribbon worms. *See* Nemertea
Ribose, 88, **89**
Ribosomes, 41, 93
Ribozymes, 30
Ribs, 620
Riftia pachyptila, 429
Right atrioventricular valves, 662
Right atrium, 662
Right ventricle, 662
Rigid skeletons, 617–22
Ring canal, 258, 446
"Ring nerves," 251
Ritualized displays, 760
River blindness, 303
RNA, 26

catalytic, 30
    chemical components of, **89**
    messenger, 91–93, **92, 94**
    ribosomal, 93
    structure of, 90
    transfer, 93, **94**
RNA polymerase, 91
RNIs. *See* Reactive nitrogen intermediates
Roadrunners, 579
Robin, **139,** 192
Rock shells, 328
Rocky Mountain spotted fever, 373
Rodents, 601, 607
Rods, of eye, 530, 717
ROIs. *See* Reactive oxygen intermediates
Romalea guttata, **401**
Root, Richard, 797
Rostellum, 288
Rostrum, 373, 378, 494
Rotaria, 309
Rotational acceleration, 715
Rotational cleavage, 156–57
Rothenbuhler, W. C., 756, **756**
Rothschildia jacobaea, **421**
Rotifera, 306–9, **307**
    classification of, 309
    external features of, 307
    internal features of, 307–8
    reproduction of, **308,** 308–9
Rough endoplasmic reticulum (RER), **37,**
        41, **42**
Round dance, of honey bees, 767
Round window, 713
Roundworms. *See* Nematoda
Roux-Weismann hypothesis, 160
Royal jelly, 415
Royal Society of London, 37
Ruffed grouse, 579
Rumen, 592
Ruminants, 592, **592,** 609

Sabella, 352
Saccoglossus, 461, **461,** 463
Saccule, 382, 530, 713, 715
Sacculina, 390, **390,** 392
Sacral vertebrae, 620
Sacrosomatida, 428
Sacrum, 620
Sagitta, 438, **438**
Sagittal plane, 185, **187**
Sahelanthropus tchadensis, 603
St. Hilaire, Etienne Geoffroy, 164
St. Martin, Alexis, 685, **685**
Salamanders, 516, 520–23, **523,** 531
    embryo of, **166**
    foot structure of, **524**
    geographic variation in, 192, **192**
    lungs of, **669**
    paedomorphosis in, 522–23, **523**
    pattern formation in, 162
    reproduction of, 521, **521**
    respiratory system of, 521–22
    speciation in, **114**
Salicornia, **801**
Salientia, 523–32
Salivary glands, 308, 683
Salminocola, 392
Salmo salar, 508–9
Salmon
    migration of, 508–9, **509**
    spawning by, 510, **511**
Salps, **391,** 475
Salt
    kidneys and, 644
    reabsorption of, 644
Salt balance
    in marine invertebrates, **635,** 635–36
    in terrestrial animals, 638–39
Salt glands, 567, **568,** 639
Salt lick, **681**
Salt-absorbing cells, 505
Saltatory conduction, 699
Salt-secretory cells, 505
Saltwater fishes, 46, 505, **506,** 509–10, 636–38
Sand dollars, 452–53, **453,** 460
Sand flies, 219
Sand worms. *See* Clam worms
Sandpipers, 579

Sanger, Frederick, 97
Saprophagous animals, 404, 678
Saprozoic feeding, 214
Sarcodina, **215,** 229
Sarcolemma, 626
Sarcomastigophora, 208, 216, 229
Sarcomere, 626
Sarcoplasm, 182
Sarcoplasmic portion of muscle cell, 297
Sarcoplasmic reticulum, 628, **628**
Sarcopterygii, 482, 489, 497, 500–502,
        510, 516
Sarcoptes scabiei, 373, **373**
Saturated fatty acids, 24
Saturation effect, 59
Saturniid moth, **421**
Sauer, E., 572
Sauria, 542–44, 551
Saurischia, 546–47, 551, **559**
Sauropodomorphs, 546, 551
Sauropterygia, 535, 551
Säve-Söderberg, Gunnar, 516
Sawyer, Roy K., **358**
Scale insects, 420
Scale worms, 351, **351**
Scales, 493–94, 499, **499,** 512, **512,** 614
Scallops, **330,** 331, 334
Scandentia, 606
Scanning electron microscope, 38
Scaphopoda, 323, **323,** 342–43
Scavengers, 386, 507
Scent glands, 590
Schistocerca gregaria, **401**
Schistosoma, **282–83,** 283–84, 288
Schistosoma haematobium, **282,** 283
Schistosoma japonicum, **282,** 283, **283,** 284
Schistosoma mansoni, **282,** 283, **283,** 746,
        749, **749**
Schistosome dermatitis, 284
Schistosomiasis, 283
Schizocielomate, body plan of, **186**
Schizocoel, 347
Schizocoelous development, 185–87, **188**
Schizocoely, 160
Schizogony, 215
Schleiden, Matthias, 37
Schmidt, Johann, 507–8
Schwann cells, **185,** 697, **697**
Schwann, Theodor, 37
Science
    evolutionary, 13–14
    experimental, 13–14
    nature of, 10–11
    principles of, 10–14
Scientific method, 11–13
Scientific revolution, 13
Sciuridae, 607
Sciurus carolinensis, **607**
Sclera, 717
Scleractinia, 263, **263**
Sclerites, 400
Sclerocytes, 240
Sclerodactyla, **456,** 460
Sclerodactyla briareus, 455
Sclerospongiae, 237
Sclerotic ring, 568
Sclerotin, 613
Sclerotization, 613
Scolex, 285, **287**
Scolopendra, **398**
Scorpiones, 371–72
Scorpionfishes, 500
Scorpionflies, 420–21
Scorpions, *364,* 371–72, **372,** 374
Screwworm flies, 418
Scrotum, **142**
Sculpins, speciation in, 114, **115**
Scutigera, **398**
Scutigerella, 399, **399,** 418
Scypha, 242
Scyphistoma, 258–59
Scyphozoa, 249, 251, 257–59, 267
Sea anemones, *247,* 248–49, 260, **260–61,**
        261–63, **262**
Sea biscuits, 460
Sea butterflies, 324, 328
Sea cows, 608
Sea cucumbers. *See* Holothuroidea

Sea daisies, 457, 460
Sea fan, **266**
Sea hare, 324, 328, **329**, 757, **757**
Sea lamprey, **492**, 492–93
Sea lilies, 456–57, **457**, 460
Sea lions, 607, **608**
Sea pen, **261**
Sea slugs, 324, 328
Sea snakes, 548
Sea spiders, 368, **369**, 374
Sea squirt, 474–75
Sea stars, **262**, **442**, 443, **444**, 444–50, **448–49**, 460, 799, **800**
autotomy of, 448–49
blastula and gastrula of, **159**
cleavage of, 156
development of, 449–50
digestive system of, 446–48
egg of, **156**
endoskeleton of, 443, 445
excretory system of, 445–46
external features of, 445, **446**
feeding habits of, 446–48
gastrulation in, 158
hemal system of, 448, **449**
internal features of, **447**
metamorphosis of, 449–50, **450**
nervous system of, 448
regeneration of, 448–49
reproduction of, 448–49
respiratory system of, 445–46
water–vascular system of, 446
Sea urchins, 443, **448**, **452–54**, 460, 799
development of, **155**
eggs of, 153, **153–54**
form and function of, 452–54
formation of new species in, **7**
Sea walnuts, 184, 267
Sea wasp, 260
Seals, 594, **595**, 607
Sebaceous glands, 590
Sebum, 590
Second law of thermodynamics, 10, 36, 57, 802
Second messenger, 723
Secondary induction, 162
Secondary oocyte, 139, **140**
Secondary response, 745
Secondary reticulum, 243, **244**
Secondary sex characteristics, 144
Secondary spermatocytes, 138, **138**
Secondary structure, of proteins, 25, **26**
Secretin, 689, **690**, 721
Secretion, 642
Secretory phase, of menstrual cycle, 144
Segmentation, 162–63, 188, 469
of arthropods, 366
digestion and, 683, **683**
Segmentation genes, 163
Segmented trunk musculature, 477
Segmented worms. *See* Annelida
Segment-polarity genes, 163
Segments, 346
Segregation, law of, 79–81
Seisonidea, 308–9
Selection
catastrophic species, 128
directional, 125, **125**
disruptive, 125, **125**
interactions of, 124
species, 126–27
stabilizing, 125, **125**
Selective permeability, 46
Self, recognition of, 741
Self-awareness, 768
Self-fertilization, 225
Self-propagation, of nerve impulses, 698
Semicircular canals, 713, **713**, 715, **716**
Semilunar valves, 662
Seminal receptacle, 282
Seminal vesicle, 142, **142**, 279, 282
Seminiferous tubules, 138, **138**, 142
Sense organs, 709–18
of Bivalvia, 333–34
of cephalopods, 338–39
of Gastropoda, 326–27
of platyhelminthes, 278
Sensillae, 359, 408
Sensitization, 757
Sensory cells, in hydra, 253

Sensory cortex, 706, **707**
Sensory division, of peripheral nervous system, 707
Sensory neurons, 696, 702–3
Sensory organs
of Anura, 529–30
of Arthropoda, 366
of birds, **568**, 568–69
of Crustacea, 383
of Ctenophora, 269
of earthworms, 356
of Enteropneusta, 462
of Hirudinea, 359
of Insecta, 408–9
of Polychaeta, 350
of sharks, 494–95, **495**
of spiders, 369
of Vertebrata, 478
Sensory papillae, 298
Sensory setae, 369
Sepia, 339
Sepia, 339, **339**, 343
Sepioidea, 340
Sepioteuthis, 343
Sepioteuthis, 339
Sepioteuthis lessoniana, **338**
Septa, 260–61, 347
Septal filament, 261
Septibranch, 332, **332**
Sequential hermaphrodites, 134
Serial homology, 380
Serous glands, 526
Serpent stars. *See* Ophiuroidea
Serpentes, 544–51
Sertoli cells. *See* Sustentacular cells
Serum, 658
Setae, 346, **355**, 369
Sex cells, 49, 75, 132
Sex characteristics
accessory, 144
secondary, 144
Sex determination, **78**, 78–79, **79**, 136–37
temperature-dependent, 137, **137**
Sex hormones, 144, **144**
Sex organs
accessory, 134, 141
primary, 134, 141
Sex ratio, 791
Sex-determining region (SRY), 136
Sexes, 133
Sex-linked inheritance, 84–85, **86**
Sex-reversing X chromosome (SRX), 137
Sexual dimorphism, 428
Sexual life cycle, **134**
Sexual reproduction, 6, 48–49, 132, **132**, 133–35
abundance of, 135–36
of Anura, 519–20, **520**, 530–32, **531**
of aquatic vertebrates, 142
of arthropods, 366
of Asteroidea, 448–49
of birds, 573–76, **574–75**
of Bivalvia, 334
of Cephalopoda, 339, **340**
of Crustacea, 383–86
of Ctenophora, 269
of earthworms, 356, **357**
of Enteropneusta, 462
of fishes, 509–12
of Gastropoda, 327
of hagfishes, 491
of Hirudinea, 359
of hydra, 254
of insects, 410, **410**
of invertebrates, **141**
of lampreys, 491
of Mammalia, 596–99
of Mesozoa, 235, **235**
of Mollusca, 321
of Nemertea, 291
of paramecium, 225
phenotypic variation and, 97
of platyhelminthes, 278–79
of Polychaeta, 350
of protozoa, 208, 215
of Rotifera, **308**, 308–9
of salamanders, 521, **521**
of Scyphozoa, 258–59
of spiders, 370

of sponges, 242
of turtles, 541
of vertebrates, 142–43
in *Volvox*, **217**, 218
Sexual selection, 123, **124**
Shaft, 560
Sharks, 482, **486**, 487, 493–97, 510
brain of, **704**
buoyancy of, 503
diversity of, **494**
fins of, 493, **498**
form and function of, 494–96, **495–96**
osmoregulation in, 638
reproduction of, 509
violent, 493–94
Sharksucker, 500, **501**
Shaw, George Bernard, 97
Shearwaters, 579
Sheep, 608, 760–61
mutations in, 117, **117**
population growth of, **793**
Sheep liver fluke, **282**
Shells
of bivalves, 330–31
of cephalopods, 324
of gastropods, 324, **324**, 325
of molluscs, **190**, 318, 320
Shh. See Sonic hedgehog
Shipworms, 317, **332**, **332**, 334
Short gastrulation (gene), 162, 164
Shrews, **594**, 606
Shrimp, **384**, 392–93, **394**
Sibling species, 114, 194
Sickle cell anemia, 97, 672
Side-winding, 545, **548**
Sidneyia, **108**
Sign stimulus, 754
Siliceous spicules, **241**
Silk glands, 370
Silkworms, 416–17
Silverfish, 419, **419**
Simians, 602
Simple epithelium, 180, **181**
Simple eyes, 369
Simpson, George Gaylord, 109, 194, 197, **200**
Sinasauropteryx, 557
Sinistral shell, 324, **324**
Sinoatrial node, 171, 663
Sinus gland, 385
Sinus venosus, 661
Siphon, 336, 453, 474
Siphonaptera, 421
Siphonoglyph, 261
Siphonophora, 256, **257**
Siphuncle, 336
Sipuncula, **427**, 427–28, 439
Sipunculids, immunity in, **748**
Sipunculus, **427**, 428
Sirenia, 608
Sister groups, 202
Sitta carolinensis, 192
Size, 177–78, **179**
bones and, 622, **622**
Skates, **139**, 482, 493–97, **496**, 510
Skeletal muscle, **180**, 182, **184**, 624, **625**, 627–28
Skeletal system, 616–22
of Anura, 527, **527**
appendicular, 619–20
axial, 619–20
of birds, **563**, 563–64
of fishes, 619–20, **620**
of humans, **621**
hydrostatic, 347, 616–17
metamerism and, 360
rigid, 617–22
of sponges, 240, **241**
of vertebrates, 619–22
Skeleton shrimp, **391**
Skimmers, 579
Skin. *See* Integument
Skin gills, 445
Skinks, 543
Skuas, 579
Skull
of amniotes, 583–84, **584**
of birds, 562
of hominids, **605**
kinetic, 542, **543**, 544, **545**

Sleeping sickness, 218
Sliding filament model, of muscle contraction, 626–27, **627**
Sliding microtubule hypothesis, 211
Slime glands, 435
Slime, of hagfishes, 490
Slit sense organs, 369
Sloths, 607
Slow fibers, 630
Slugs, 317, 324, **330**, 343
Small intestine, 686–88
Smaller protostomes, 426–40
phylogeny of, 438–40
position in animal kingdom, 427
Smallest distinct groupings, 193
Smell, sense of, 709–10, **710**
Smooth endoplasmic reticulum, 41
Smooth muscle, **180**, 182, **184**, 624, **625**
Snails, 317, **318**, 324–26, **326**, 328, **329–30**, 343
anatomy of, **328**
foot of, 319
fossil record of, 118
genetic variation in, **124**
immunity in, 749, **749**
radula of, 319
reproduction of, **135**, 321
veliger of, **321**
Snakes, 544–50, **548–50**, 551
behavior of, 760
feeding habits of, **680**
integument of, 614
locomotion of, 545, **548**
skull of, 544, **545**
Snapping turtle, **541**, 542, **542**
Snipe, 579
Snow fleas, 419
Snowshoe hares, **588**, 600
Soaring wings, 570, **571**
Social behavior, 759–68
advantages of, 759–60
aggression and dominance, 760–62
of birds, **574**
of insects, 414–16
mating systems and, 763
territoriality, 762
Social organization, of arthropods, 365
Society of Protozoologists, 208
Sociobiology, 753–54
Sociobiology: The New Synthesis (Wilson), 753
Sockeye salmon, 508, **509**
Sodium, 657
Sodium pump, **697**, 698
Sodium-potassium pump, 47, **48**, **697**, 698
Soft coral, **266**
Solar energy, 56, **57**
Solar system, **22**
Soldier substance, 416
Soldiers, 415
Sole, 455
Solenia, 264
Solenocytes, 305, 310
Solenogastres, 321–22, 340, 343
Solute, diffusion of, **57**
Solution, solutes and, **57**
Solvent, water as, 29
Soma, **185**
Somatic cells, 48, 136
Somatic nervous system, 707
Somatocoels, 449
Somatotropin, 726–27, **727**
Somites, **171**, 188, 346, 365, 469
Songbirds, 576–77, 580
Songs, of birds, 758, **758**
Sonic hedgehog (Shh) (gene), 162
Sorting, 119
Sound production, of insects, 414
Sound reception, of insects, 414
South American lungfish, 500, **501**
Southern evergreen forest, 777
Southern leopard frog, taxonomic categories of, **191**
Sow bugs, 390, **391**
Spadella, 438
Speciation, 113, **115**
allopatric, 113–14
nonallopatric, 114–15
through geological time, 126–27
vicariant, 113–14

Species, 15, 193–95
  biological concept of, 194
  changes within, 120–25
  criteria for, 193
  definition of, 113
  diversity of, 790
  dynamism of, 195
  emergent properties of, **6**
  evolutionary concept, 194
  multiplication of, 15, 112–15
  phylogenetic concept of, 194–95
  recognition of, 113
  sibling, 114
  study of, **6**
  timescale for reproduction of, **6**
  typological concept of, 193–94
Species epithet, 192
Species level, reproduction at, 6, **7**
Species selection, 126–27
  catastrophic, 128
Specific heat, of water, 28
Specific immunity, 739
Specific nerve energies, law of, 709
Spectroscopy, nuclear magnetic resonance, 38
Speculation, 12
Speleonectes, 392
Spemann, Hans, 151, 160–61
Spemann organizer cells, *151*
Spencer, Herbert, 119
Sperm, 48, 75, 78, 133, 139
  contact with egg, 153, **153**
  diversity of, **18**
  human, *131*
  hypothesis about, 19
  size of, 139, **139**
  types of, **139**
Spermatids, **138**, 139
Spermatocytes
  primary, 138, **138**
  secondary, 138, **138**
Spermatogenesis, **138**, 138–39
Spermatogonia, 138, **138**
Spermatophore, 521
Spermatozoa. *See* Sperm
Spermophilus beldingi, 764, **764**
Sphaerophrya, 226
Sphenisciformes, 578
Sphenodon, 535, 550, **550**, 551
Sphenodonta, 550, **550**, 551
Spherical symmetry, 183, **187**
Sphinx moths, 404, **405**, 410
Sphongophorus, **413**
Sphygmomanometer, 665
Sphyrna, 494, 510
Spicules, 236, 240, **241**, 242–43, 445
Spider mites, 373
Spiders, **369**, 369–71, **370**, 374
  dangerous, 370–71
  form and function of, 369–70
  reproduction of, 370
  web-spinning by, 370
Spinal cord, 487, 702–3, **703**
Spinal nerve roots, 477
Spindle, 50, 297
Spines, 445
Spiny anteater, 606
Spiny-headed worms. *See* Acanthocephala
Spiracles, 407, **407**, 436, 667
Spiral cleavage, 156
Spiral valve, 495, 528
Spirobolus, 399, 418
Spirobranchus giganteus, **239, 347**
Spirotrichonympha, **216**
Sponges. *See* Porifera
Spongilla, 243, 245
Spongillidae, **242**
Spongin, 237, **241**
Sponging mouthparts, 405–6
Spongocoels, flagellated, 237–38
Spongocytes, 240
Spongy bone, 618
Spontaneous abortion, 146
Spontaneous generation, 21–22
Spookfish, 497
Spoonbills, 579
Spoonworms, 428, **428**
Spores, 133, 219, **219**
Sporocyst, 281
Sporogony, 133, 215

Sporozoites, 219–21
Sports, 117
Spotted ratfish, **497**
Spotted sandpipers, 575
Springtails, 414, 419
Spruce budworms, 397, 417
Squalene, 503
Squaliformes, 493
Squalorophrya, 226
Squalus, 494, 510
Squalus acanthias, **496**
Squamata, 542–51
Squamous epithelium, **181–82**
Squatinella rostrum, **307**
Squids, 320, 335–36, 338, **338**, 343
Squirrels, 595, **596**, 607, **607**, 764, **764**, *771*
SRX. *See* Sex-reversing X chromosome
SRY. *See* Sex-determining region
Stabilizing selection, 125, **125**
Stalk, 456
Stapes, 713
Star coral, **264**
Starches, 22–23
Starfishes. *See* Sea stars
Starling, **577**
Starling, E. H., 721, **722**
Statoblasts, 433
Statocysts, 255, 268, 383, 715
Statoliths, 383, 715
Stearic acid, **24**, 66
Stearin, **24**
Stegosaurus, 546–47, 551
Stenohaline, 635
Stenophagous animals, 565
Stenostomum, **279**
Stensiö, Erik, 482
Stentor, **222**, 223, 226, 231
Stephalia, 257
Stephanoceros fimbriatus, **307**
Stereom, 445
Stereo-olfaction, 494
Stereotypical behavior, 754
Sterile-male approach to insect control, 418
Sternal sinus, 382
Sternum, 378
Steroid hormones, 94
Steroids, 24, **25**, 723, 733–34
  gonadal, 144, **144**
  therapy, 733
Stichopathes, 267
Stickleback, behavior of, 754–55, **755**
Stigma, 218
Stimulus, 709
Stingrays, 496, **496**
Stink bugs, 413, **417**
Stirrup, 713
Stomach, 684
  four-chambered, 592
Stomatopoda, 386
Stomochord, 460
Stomphia, 262
Stomphia didemon, **262**
Stone canal, 446
Stoneflies, 407, **412**, 419, 424
Storage, of food energy, 678
Storks, *555*, 579
Straminopiles, 230
Stratified epithelium, 180, **180, 182**
Stratigraphy, law of, 107
Stratum corneum, 613, 616
Strength, 612
Strepsiptera, 420
Strepsirhini, 606
Stretch reflex, 703, **703**
Striated muscle, 182, 366, 380, 624, 626
Strigiformes, 579
Strobila, 259, 285
Strobilation, 259
Stroke volume, 662
Stroma, 177
Strombus, 324
Strongylocentrotus, **448**, 452, 460
Strongylocentrotus franciscanus, **449**
Strongylocentrotus purpuratus, **452**
Structural color, 615
Structural genomics, 97
Struggle for existence, 118
Struthio camelus, **578**
Struthioniformes, 578, **578**

Sturgeon, 498, **500**, 510
Sturnus vulgaris, **577**
Stylaria, 357, **358**, 360
Stylaster roseus, **257**
Style sac, 332–33
Stylet, 291
Stylops, 420
Subepidermal muscles, 307
Subergorgia mollis, **266**
Subesophageal ganglia, 382
Sublittoral zone, 779
Subneural gland, 474
Subnivean environment, 650
Subspecies, 192
Substrate, 59, **59**
Subtidal zone, 779
Succinea, 284, 329
Suckers, 282, 446
Sucking mouthparts, 404–5
Suckling lice, 420
Sucrase, 686
Sucrose, 23
Suctorians, 226
Sugars, 22–24, 88–89, **89**
Suina, 609
Sula nebouxii, **767**
Sun-azimuth orientation, 572, **573**
Sunlight, 56, **57**, 801
  injurious effects of, 616
Superficial cleavage, 157, **157**
Supersaurus, 546
Support, 170–71
Suprabranchial chamber, 331
Suprachiasmatic nucleus, 728
Supracoracoideus muscle, 564
Supraesophageal ganglia, 382
Surface tension, of water, 28–29
Surinam frog, **520**
Survival of the fittest, 119
Survivorship, 791, **792**
Susceptibility, 739
Suspension feeders, 386, 492, 507, 678, **679**
Sustentacular cells, 138, **138**, 142, 144
Swallowing, 684, **684**
Swallowtail butterfly, **421**
Swans, 579
Sweat glands, 589
Swifts, 580
Swim bladder, 487, 500, 503–4, **504**
Swimmeret, **379, 381**
Swimmer's itch, 284
Swimming, **502**, 502–3, **503**
Swine, 608
Swordfish, 502
Sycon, **238–39**, 242, **243**, 245
Syconoids, 238, **238**
Symbiosis, eukaryotes and, 33
Symbiotic ciliates, 225, **226**
Symmetry, 183–85, **187**
  bilateral, 164, 182, 184–85, **186–87**, 234, 274
  biradial, 184, 268
  radial, 183, **186–87**, 234, 250, 443, 445
  spherical, 183, **187**
Sympathetic nervous system, 707–9
Sympatric speciation, 114–15
Symphyla, 398–99, 418
Symplesiomorphy, 196
Synapomorphy, 196
Synapses, 699–701, **700–701**
Synapsids, 535, 551
  cladogram of, **586**
  evolution of, **585**
  skull of, 583–84, **584**
Synapsis, 78
Synaptic cleft, 628, 699
Synaptic vesicles, 628, 700
Synaptonemal complex, 86
Synchronous muscles, 403
Syncoid sponge, **243**
Syncytial epidermis, 276, 297, 307
Syncytical ciliate hypothesis, 234
Syncytium, 50
Syndrome, 87
Syngamy, 215
Synsacrum, 562
Systema Naturae (Linnaeus), 191
Systemic circuit, 661
Systole, 662, **663**

T lymphocytes, 740, **740**
  subsets of, 743
Tachyglossus, 606
Tachypleus, 367
Tactile communication, of insects, 414
Tactile hairs, 383
Tadarida, 606
Tadpole, morphogenesis in, **164**
Tadpole shrimp, **387**
Taenia, 288
Taenia pisiformis, **288**
Taenia saginata, 286–88, **287**
Taenia solium, **287**, 288, **289**
Taenidia, 407, **407**
Tagelus, 326, 343
Tagelus plebius, **331**
Tagmata, 365–66
Tagmosis, 365
Taiga, 777
Tail, postanal, 472
Talkeetna Mountain Range, 789
Tamias striatus, **593**
Tantulocarida, 388, **388**, 392
Tantulus, 388
Tapeworms. *See* Cestoda
Tapirs, 608
Tarantulas, 370, **371**
Tardigrada, 375, 427, 436–37, **437, 439,** 440
Target cells, 722, 738
Target-gland hormones, **726**
Tarsiers, 602, 606
Tarsiidae, 606
Tarsius syrichta carbonarius, **602**
Tasmanian wolves, 606
Taste, 709–10
Taste buds, **709**, 710
Tatum, 88
Taxis, 224
Taxon, 191–92
Taxonomy
  categories, **191**
  characters, 195–97
  current state of, 202
  phenetic, 200
  ranks, 191–92
  theories of, 197–202
  traditional evolutionary, 197–201
TCA cycle. *See* Krebs cycle
T-cell receptors, 741, 743
Tealia, 267
Tealia piscivora, **260**
Tectorial membrane, 530, 714
Tectum, 373, **705**, 706
Teeth, 590, 680, **681**
Tegmen, 456
Tegument, of platyhelminthes, 277
Telencephalon, **705**
Teleost fishes, 499–500, **501–2**, 507, 510
  buoyancy of, **504**
  reproduction of, 509
  vestibular apparatus of, 713, **713**
Telolecithal egg, 155, **156**
Telophase
  in meiosis, **77**
  in mitosis, **50**, 51, **52**
Telson, 367, 378
Temnospondyls, 516
Temperate deciduous forest, 776
Temperature compensation, 648
Temperature regulation, 646–52
  in cold environments, 649–50, **650**
  in hot environments, 649, **650**
  hypothermia and, 651–52
  in lizards, **9**
Temperature sensing, insects, 409
Temperature-dependent sex determination, 137, **137**
Template, for replication, 90
Tendons, energy storage in, 630–31, **631**
Tenrecs, 606
Tent caterpillars, 415
Tentacle sheath, 268
Tentacles, 247, **254**, 258, 268, **330**, 429, **430**
Tentaculata, **268**, 268–70
Terebratella, 434, **434**
Teredo, 334, 343
Tergum, 378
Termites, 404, 414–16, **416**, 419
Terns, 579

Terrapene carolina, 541–42
Terrapin, 540
Terrestrial mandibulates, 397–424
Territoriality, 754–55, **755**, 762, **762**
  of Mammalia, **599**, 599–600
Tertiary structure, of proteins, 25, **26**
Test, 226–27
Testcross, 80
Testes, 134, 142, 279, 282
Testicle, **142**
Testosterone, 136, 142, 144, **144**
Testudines, 540–42, 551
Tethys Sea, 784, **785**
Tetrads, 78
Tetrahymena, **222**, 231
Tetrahymena thermophila, **210**
Tetranychidae, 373, **373**
Tetrapoda, 469, 482
  carboniferous radiation of, 516
  cladogram of, **518**
  in Devonian period, 515–16
  early, 514–33
  evolution of, **519**
  legs of, 516, **517**
Teuthoidea, 340
Texas cattle fever, 374
Thalamus, **705**, 706
Thalicea, 473, 475, **475**
Thaumatoscyphus hexaradiatus, **258**
Thecodonts, 551–52
Themiste, **427**
Thenea, 245
Theory, 12–13
Therapsids, 551, 584
Theria, 606
Thermal condensations, 30
Thermodynamics, laws of, 9–10, 36, 57, 802
Thermogenesis, nonshivering, 650, 689
Thermogenin, 689
Thermotaxis, 224
Theropods, 551
  birds and, 556–57
Thigmotaxis, 224
Thoracic vertebrae, 620
Thoracica, 389, **389**, 393
Thorax, of insects, 400
Thorny corals, 264, **265**
Thresher shark, **494**
Threshold current, 711
Thrips, 420
Thrombocytes, 657
Thromboplastin, 659
Thrombus, 664
Thrust, 503
Thymine, 89, **89**
  hydrogen bond of, 90, **90**
Thyroid gland, 469, 472, **730**, 730–31, **731**
Thyroid hormones, **730**, 730–31, **731**
Thyroid-stimulating hormone (TSH), 726, **727**, 731
Thyrotropin, 726
Thyrotropin-releasing hormone (TRH), 731
Thyroxine, 730
Thysanoptera, 420, 424
Thysanozoon nigropapillosum, 274
Thysanura, 419, **419**, 422
Tibicen davisi, **411**
Tickle, Cheryll, 164
Ticks, **372**, 372–74, 695
Tiedemann's bodies, 446
Tiger moth, 135
Tiger sharks, 493–94, **495**
Tight junctions, 44, **45**
Timbre, 715
Time, biological, **33**
Tinamiformes, 578
Tinamous, 578
Tinbergen, Niko, 752, **753**, 754
Tissue(s), **177**, **178**
  connective, **180**, 180–81, **183**
  epithelial, 180, **180–82**
  incipient, 234
  muscle, **184**
  muscular, 182
  nervous, **180**, 182
  types of, 179–82, **180**
Tissue fluid, 178, 181
Tissue-organ grade of organization, **178**
Tissues, 4

Titanosaurus, 547
Titer, 745
TLRs. *See* Toll-like receptors
TMAO. *See* Trimethylamine oxide
TNF. *See* Tumor necrosis factor
Toads, **139**, 523–32, **525**, 730
Tobacco hornworms, 724
Toes, 307
Tokay, **543**
Toll, 741
Toll-like receptors (TLRs), 741
Tongue worms, *See* Pentastomida
Tonicella, 343
Tonicella lineata, **318**
Tooth shells, 323
Toothed whales, 596, 609
Topi, 677, **759**
Tornaria, 462–63, **462**
Torpedo, **497**
Torpor, 651, **651**
Torsion, 324–25, **325**
Tortoises, 540, **541**
  embryo of, **166**
  Galápagos, 105, **105**, 541, **541**
Tosia queenslandensis, **444**
Toucans, 580
Touch, 711, **711**
Tough tissue cysts, 220
Toxicysts, 222
Toxocara, 300
Toxoplasma, 231
Toxoplasma gondii, 220
Trabecular reticulum, 242–43, **244**
Tracheae, 365, 407, **407**, 667, 669
Tracheal gills, 407
Tracheal system, 406–7, 436, **667**, 667–68
Tracheoles, 407, **407**, 667
Transcription, 91–93, **92–93**, 94
Transcription factors, 94
Transfer RNA (tRNA), 93, **94**
Transformational evolutionary theory, 103
Transforming growth factor, **744**
Transgenic animals, 96
Transitional epithelium, **182**
Translation, 93, 95
Translocation, 88
Transplantation of organs, 745
Transport maximum, 644
Transporters, 47
Transverse plane, 185, **187**
Tree frog, **524**, 525
Tree of life, **15**
Tree shrews, 606
Treehoppers, **413**, 420, **420**
Trematoda, 276–77, **277**, 280–84, **281–82**, **285**, 288
Trembley, Abraham, 253
Trepang, 444
TRH. *See* Thyrotropin-releasing hormone
Triatominae, 218
Tricarboxylic acid cycle. *See* Krebs cycle
Triceratops, 546–47, 551
Trichina worm, **300**, 300–301, **302**
Trichinella, **300**, 302
Trichinella spiralis, 295, 300–301, **302**
Trichinosis, 300
Trichobothria, 369
Trichocysts, 222–23, **223**
Trichodina, 231
Trichomonadida, 216, 230
Trichomonas, **216**, 230
Trichomonas tenax, 216
Trichomonas vaginalis, 216
Trichonympha, **216**
Trichophyra, 226
Trichoplax adhaerens, 236, **236**
Trichoptera, 421
Trichuris, 302
Trichuris trichura, **300**
Tricladida, 276, 280, **280**
Tricuspid, 662
Tridacna, 329
Tridacna gigas, 317, **335**
Tridacna squamosa, 316
Triglycerides, 24
  hydrolysis and, 68, **68**
Triiodothyronine, 730
Trilobites, *102*, 367, **367**, 374
Trimbicula, 373

Trimerous coelom, 439
Trimethylamine oxide (TMAO), 496, 638
Triops, 392
Tripartite brain, 478–79
Tripartite coelom, 439
Tripedalia, 267
Tripedalia cystophora, 259
Triphyllozoon, **432**
Triploblastic animals, 160, 250, 268, 275
Trisomy, 87
Trisomy 21, 87–88
Trivers, Robert, 765
Trochophore larva, 317–18, 321, **321**
Trochozoa, 321, 439
Trogoniformes, 580
Trogons, 580
Trombicula, **372**
Trombidium, **372**
Trophallaxis, 416
Trophi, 308
Trophic levels, 800
Trophoblast, 145, **146**, 157
Trophosome, 430
Trophozoites, 220
Tropic hormones, 726
Tropical forest, **777**, 777–78
Tropomyosin, 626
Troponin, 626
Trout, 493, 502, **503**
True breeding, 17
True bugs, 420
True flies, 421
True horns, 589
Trunk, 427, 429
Trypanosoma, **215**, 218–19, **219**, 231
Trypanosoma brucei gambiense, 218
Trypanosoma brucei rhodesiense, 218
Trypanosoma cruzi, 218–19
Trypanosomatidae, 230–31
Trypsin, 686
  specificity of, **60**
Tryptophan, **25**
Tsetse flies, 218, 417
TSH. *See* Thyroid–stimulating hormone
T-tubules, 628
Tuatara, 535, 550, **550**
Tubastrea, **263**
Tubastrea aurea, **280**
Tube anemones, 260, 264, **264**
Tube dwellers, 351
Tube feet, 445–46
  of Holothuroidea, 455–56
  of Ophiuroidea, 451
Tube-within-a-tube body plan, 185, **186**
Tubifex, 357, **358**, 360
Tubiluchus, 305
Tubipora, 267
Tubulanus, 291
Tubular reabsorption, 643–45, **644**
Tubular secretion, 645
Tubularia, 267
Tubularia crocea, **252**
Tubulidentata, 608
Tubulin, 42, 50, **52**, 623
Tularemia, 373–74
Tumor necrosis factor (TNF), 739, **744**
Tumor suppressor genes, 98
Tuna, 502, **503**
Tundra, 778, **778**
Tunic, 473
Tunicates, 473–75, **474**, 482
  cleavage in, 156
  immunity in, **748**
Turacos, 579
Turbanella, **311**
Turbellaria, *274*, 276, **277**, **279**, 279–80, **280**, 288
Turbellarian flatworms, cleavage in, 156
Turbinatebone, 584
Turdus migratorius, 192
Turkeys, 135, 579
Turner syndrome, 88
Turnstones, 579
Turtles, 535, 540–42, **541–42**, 551, 614
Tusk shells, 323, 343
Twins, 148, **149**
Twisted wing parasites, 420
Tylopoda, 609

Tympanic canal, 713
Tympanic membrane, 530, 713
Type specimen, 193, **193**
Typhlosole, 353, 686
Typhoid, 417
Typhus fever, 417
Typological species concept, 193–94
Tyrannosaurus, 546, 551, 622

Uca, 393, **394**
Ulcer, 747
Ultimate causation, 14, 753
Ultraviolet radiation, 98
Umbilical cord, 168
Umbo, 330
Underhair, 587
Undulating membrane, 222
Undulipodia, 43, 210
Ungulates, 608, **609**, 649
Unicellular eukaryotes, reproduction in, 133
Unicellular organisms. *See also* Protozoa
  body plans of, **186**
  classification of, 203
Uniformitarianism, 104
Unilocular hydatid, **287**, 289
Uniparous animals, 148
Uniramia, 374, 397–424
  cladogram of, **422**
  classification of, 418
Unitary animals, 791
Univalves, 324
Unjointed legs, 435
Unsaturated fatty acids, 24, **24**
Upogebia, 386
Upwelling, 779
Uracil, 89, **89**
Urea, 70, 496, 638
Urechis, **351**, 428
Ureters, 142, **142**, 567, 641–42
Urethra, 142, **142–43**
Urey, Harold, 27
Uric acid, 70, 567
Urinary bladder, **142–43**, 642
Urinary meatus, **142**
Urine concentration, 645–46, **646**
Urnatella, 311, **311**
Urochordata, 473–75, **474**, 482
Urodela, 520–23, 531
Urogenital systems, male, 142
Uropods, 378, **380–81**
Urosalpinx, 317, 328
Urosalpinx cinerea, 326
Urostyle, 527
Ursidae, 608
Ursus horribilis, **607**
Usher, James, 103
Uterine bell, 310
Uterine duct, 142
"Uterine milk," 496
Uterine tubes, 143, **143**
Uterus, 143, **143**, 167, 279, 282
Utricle, 530, 713, 715
UV-sensitivity, 569

Vaccination, animal research and, 12
Vacuoles, 41
Vagina, 143, **143**
Vaginitis, 216
Vagus nerves, 663
Valves, 330
Vampire bats, 765
Vampyromorpha, 340
van Helmont, Jean Baptiste, 21
van Leeuwenhoek, A., 37
Vane, 560
Vaporization, of water, 28
Varanus, **543**
Varanus niloticus, 194
Variable region, of antibodies, 742
Variation, 6, 75, 118
  natural selection and, 15
  quantitative, **125**
  sexual reproduction and, 135
  sources of, 97–98
Variational evolutionary theory, 103
Vas deferens, 142, **142**, 279, 282
Vasa efferentia, 142, 279, 282
Vasa recta, 645
Vascular tissue, 181

Vasopressin, **727,** 728, **728**
Vasotocin, **727,** 728
Vectors, 96
Vedalia beetles, 418
Vegetal plate, 158
Vegetal pole, 155
Veins, 661, **664,** 666
Velarium, 259
Velella, 248
Veliger larva, 318, 321
Velociraptor, 546–47, 551
Velum, 255, 457
Velvet, 589
Velvet worms. *See* Onychophora
Venom
    of scorpions, 371
    of snakes, 548–50
    of spiders, 370
Ventral, 185, **187**
Ventral aorta, 476
Ventral nerve cord, 355
Ventral suckers, 282
Ventral vessel, 354
Ventricle, 528, 661
Venules, 661
Venus, 343
Venus' flower basket, 243, 245
Vermiforms, 235
Vertebrae, 451, 620
Vertebral column, 487
Vertebralima, 231
Vertebralima striata, **228**
Vertebrata, 469, 477–83, 510
    acquired immunity in, 741–47
    ancestry of, 478–80
    aquatic, reproductive systems of, 142
    body plan of, **186,** 480–81
    brain of, **704–5**
    characteristics of, 479
    circulatory system of, 661–64
    development of, 165–68
    digestive system of, **683,** 686, **687**
    embryos of, **166**
    endocrine glands of, 724–36
    endoskeleton of, 617–18
    evolution of, 477–78
    eyes of, 718
    fossils of, 478
    hearing in, 713
    hormones of, 724–36
    integument of, 613–16
    jawed, 482–83, **483–84**
    kidneys of, 641–46
    morphogenesis of limbs in, 164, **165**
    muscles of, 624–25, **625**
    nervous system of, 702–9
    reproductive systems of, 142–43
    skeletal system of, 619–22
    terrestrial, evolution of, 515–16
Vesicles, 227
Vesicular nucleus, 209
Vespertilionidae, 595–96
Vestibular apparatus, 713, **713**
Vestibular canal, 713
Vestimentifera, 429
Veterinary medicine, 12
Vibrissae, 589
Vicariance, 781
    distribution by, 782–83

Vicariance biogeography, 783
Vicariant speciation, 113–14
Villi, 686
Viperidae, 548–49
Virginia opossum, 606
Visceral larva migrans, 300
Visceral leishmaniasis, 219
Visceral mass, 318–20, 331
Visceral muscle, 624
Visceral pleura, 670
Viscosity, of water, 29
Vision, 715–18. *See also* Eye(s)
    of birds, 568–69
    of cephalopods, 338–39, **339**
    chemistry of, 718
    color, 718, **718**
    of frogs, 530
    of insects, 409
    of Polychaeta, 350
    of sharks, 495
    of spiders, 369
    UV-sensitivity and, 569
Visual signals, 414
Vitalism, 9
Vitamins, 690, **691**
Vitellaria, 279, 282
Vitelline duct, 279
Vitelline gland, 286
Vitreous humor, 717
Viviparous animals, 141, 496, 509, 606
Viviparus, 328
Vocal cords, 528
Volcanic activity, 27
Voluntary muscle, 624
Volvox, 10, 230
Volvox globator, **5,** 216–18, **217**
von Baer, K. E., 112
von Frisch, Karl, 752, **753,** 767
Von Mering, J., 735
von Uexküll, Jakob, 695
Voronus komodoensis, 534
Vorticella, **222,** 223, 231
Voyage of the Beagle, The (Darwin), 105
Vultures, 579
Vulva, 143

Waggle dance, of honey bees, **766,** 766–67
Walking legs, 369
Walking sticks, **402,** 419
Walking worms. *See* Onychophora
Wallace, Alfred Russel, 14, 103, **103,** 106
Walruses, 608
Warblers, guilds of, 797, **798**
Warm-blooded animals, 647–48
Warning coloration, 413
Wasps, **403,** 404, 408, 410, 414, **417,** 421, **421**
Water
    buoyant density of, 515
    condensation of, 29
    density of, 28
    excretion of, 645–46
    fresh, 636
    hydrolysis and, 29
    inland, 779
    life and, 28–29
    locomotion in, **502,** 502–3, **503**
    movement from, to land, 515
    osmotic regulation and, 635–39
    as solvent, 29

specific heat of, 28
supply of, *21*
surface tension of, 28–29
vaporization of, 28
viscosity of, 29
Water balance, in insects, 408
Water bugs, 400
Water conservation, in reptiles, 539
Water currents, 234
Water fleas, 392
Water striders, 400, 402, **403**
Waterfowl, poisoning of, 576
Water-soluble vitamins, 690, **691**
Water-vascular system, 443, 445–46
Watson, James, **8,** 74, 90
Way of an Investigator, The (Cannon), 634
Weasels, 588, 607
Weaver ant, **416**
Webspinners, 419
Web-spinning, 370, **370**
Weevils, 420, **421**
Wegener, Alfred, 784
Weir, 277–78
Weismann, August, 120
Wenner, Adrian, 767
Whale shark, **494**
Whalebone whale, **679**
Whales, 594, 609, **609, 679,** *752*
Whaling, 583
Wheals, 746, **746**
Whelks, 324, **324,** 328, **328**
Whippoorwills, 580
Whiptail lizards, parthenogenesis in, 135, **136**
Whipworm, **300**
Whiskers, 589
White adipose tissue, 689
White blood cells, 657, **657,** 658
White-breasted nuthatch, 192
Whitefish, **52**
White-tailed deer, *56*
Whorl, 324
Wilkins, Maurice, 90
Wilkinson, Gerard, 765
Williamson, Peter, 118
Wilson, E. O., 753–54
Wing slot, 569
Wings, 569–70, **570–71**
    bones of, 564, **564**
    elliptical, 570, **571**
    high-lift, 570, **571**
    high-speed, 570, **571**
    of insects, 400, 402–4
    soaring, 570, **571**
    types of, 569–70, **571**
Wiwaxia, **108**
Woese, Carl, 31–32
Wolf eel, 506, **506**
Wolf spiders, 370
Wolff, Kaspar Friedrich, 152
Wolffian duct, 142
Wolves, 607
Wombats, 606
Wood ducks, **124**
Wood frogs, 524
Wood tick, 695
Wood wasp, 410
Wood-boring beetle, 410
Woodchucks, 607, **652**
Woodcocks, 579

Woodpecker, 565, 580
Woodpecker finch, **116**
Worker jelly, 415
Worker substance, 416
Workers, 415
Worm lizards, 544, **545**
Worms. *See* Platyhelminthes
    segmented. *See* Annelida
Wrasses, hermaphroditism in, 134
Wuchereria, 302
Wuchereria bancrofti, 302, **303–4**
Wynne-Edwards, V. C., 763

X chromosome, 78
    sex determination and, 137
Xanthophores, 526, 616
Xenarthra, 607
Xenografts, 748
Xenophanes, 103
Xenopus laevis, 525, **526**
X1Hbox 1 protein, **164**
Xianguangia, **108**
Xiphosurida, 367–68
X-organ, 384–85
X-ray crystallography, 38
XX-XO sex determination, 78, **78**
XX-XY sex determination, 78, **78**
Xyloplax, **458,** 460

Y chromosome, 78
    sex determination and, 136
Yalow, Rosalyn, 729
Yellow baboons, **2**
Yellow perch, **498**
Yohoia, **108**
Yoldia, 332
Yolk, 140
    cleavage and, 155–56
    developmental mode and, 155–56
Yolk sac, 165–66, **167**
Y-organs, 385
Young, Thomas, 718
Yunnanozoon, 478

Z line, 626
Zalophus californianus, **608**
Zebra, 608, 677, **759**
Zebra mussels, 333
Zoantharia. *See* Hexacorallia
Zoecium, 432
Zonitoides, 329
Zonotrichia leucophrys, **758**
Zoogeography, 780–86
Zooids, 217, 432, **432**
Zoology
    as part of biology, 10
    principles of, 2
    variety in, 3, **3**
Zooplankton, 388
Zoothamnium, **222**
Zooxanthella, 231
Zooxanthellae, 226, 258, 265
Zoraptera, 419, 424
Zygote nucleus, 154–55, **155**
Zygotes, 48, 76, 132–33, 145, 153
Zygotic meiosis, 215

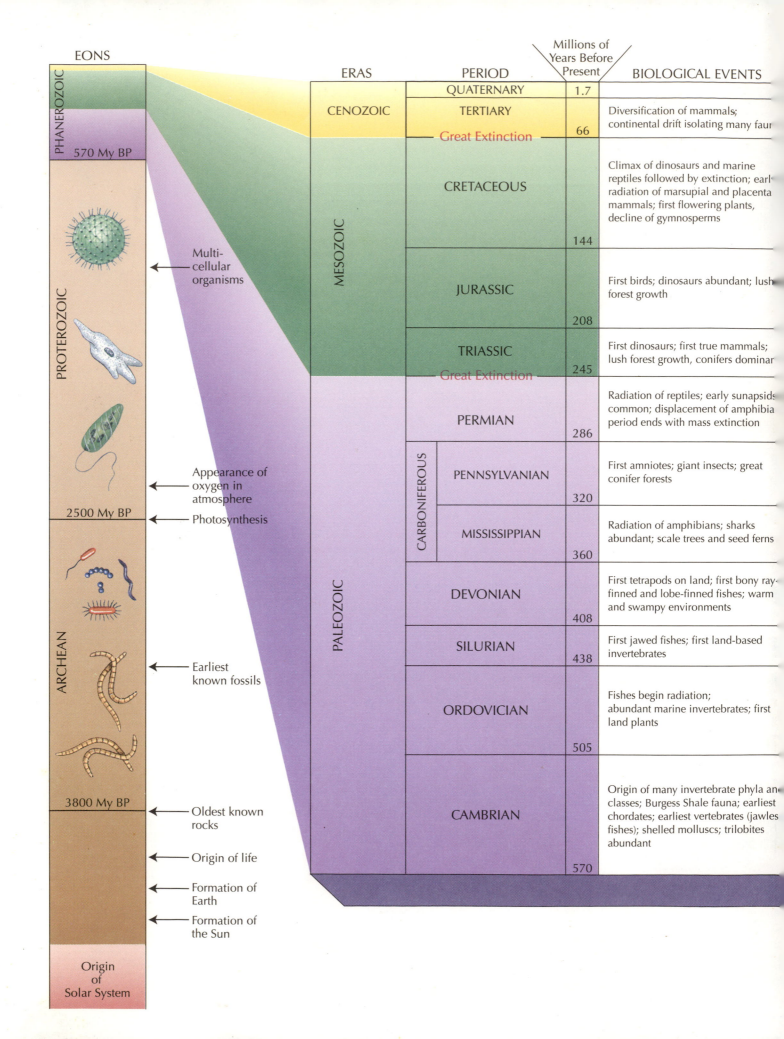

EONS		ERAS	PERIOD		Millions of Years Before Present	BIOLOGICAL EVENTS
PHANEROZOIC		CENOZOIC	QUATERNARY		1.7	
			TERTIARY			Diversification of mammals; continental drift isolating many faun
			Great Extinction		66	
		MESOZOIC	CRETACEOUS			Climax of dinosaurs and marine reptiles followed by extinction; earl radiation of marsupial and placenta mammals; first flowering plants, decline of gymnosperms
					144	
			JURASSIC			First birds; dinosaurs abundant; lush forest growth
					208	
			TRIASSIC			First dinosaurs; first true mammals; lush forest growth, conifers dominan
			Great Extinction		245	
		PALEOZOIC	PERMIAN			Radiation of reptiles; early sunapsids common; displacement of amphibia period ends with mass extinction
					286	
			CARBONIFEROUS	PENNSYLVANIAN		First amniotes; giant insects; great conifer forests
					320	
				MISSISSIPPIAN		Radiation of amphibians; sharks abundant; scale trees and seed ferns
					360	
			DEVONIAN			First tetrapods on land; first bony ray-finned and lobe-finned fishes; warm and swampy environments
					408	
			SILURIAN			First jawed fishes; first land-based invertebrates
					438	
			ORDOVICIAN			Fishes begin radiation; abundant marine invertebrates; first land plants
					505	
			CAMBRIAN			Origin of many invertebrate phyla an classes; Burgess Shale fauna; earliest chordates; earliest vertebrates (jawles fishes); shelled molluscs; trilobites abundant
					570	

PHANEROZOIC

570 My BP

PROTEROZOIC

← Multi-cellular organisms

← Appearance of oxygen in atmosphere

2500 My BP ← Photosynthesis

ARCHEAN

← Earliest known fossils

3800 My BP ← Oldest known rocks

← Origin of life

← Formation of Earth

← Formation of the Sun

Origin of Solar System